NUMERICAL CHEMISTRY

How to Solve Problems in General Chemistry

With Solutions

NUMERICAL CHEMISTRY

How to Solve Problems in General Chemistry

WITH SOLUTIONS

A.S. BHATIA
B. Tech (IITD)

DEEP & DEEP PUBLICATIONS PVT. LTD.
F-159, Rajouri Garden, New Delhi - 110 027

NUMERICAL CHEMISTRY

ISBN 978-81-8450-201-5

Typeset by RAHUL COMPOSERS
358, Pocket-B, Phase-II, Sector-16B, Dwarka, New Delhi - 110 078

Printed in India at MAYUR ENTERPRISES
WZ Plot No. 3, Gujjar Market, Tihar Village, New Delhi - 110 018

Published by DEEP & DEEP PUBLICATIONS PVT. LTD.
F-159, Rajouri Garden, New Delhi - 110 027 • Phone : 25435369, 25440916
E-mail : ddpubs@gmail.com • ddpbooks@yahoo.co.in
Showroom :
2/13, Ansari Road, Daryaganj, New Delhi - 110 002 • Telefax : 23245122

Dedicated to
My Parents
Sardar Sarup Singh Bhatia
and
Sardarni Parkash Kaur Bhatia

CONTENTS

Part IV
SOLUTIONS

PREFACE

The basic principles of chemistry are few in number yet powerful and wide ranging in their application. However, these principles are not usually mastered simply by reading about chemistry. They must be reinforced through laboratory practice and through problem solving. The purpose of this book is to make available a self-teaching and to master problem solving of a very broad range of problems encountered in chemistry.

This book is to assist students in developing insight into solving numerical problems in General Chemistry and Organic Chemistry by offering *solutions* to various problems of every type, even difficult ones. These problems should be approached only after the relevant text of the chapters has been studied carefully. A serious individual effort should be made to solve the problems. The student may find it tempting to turn directly to solutions offered but be may succumb to this temptation and will lose out on two counts. Firstly, he will reduce his chance of the problem solving aspect in chemistry to just additional reading about chemistry and secondly, he will miss opportunities to discover by himself more original solutions.

Key concepts before problems in every chapter are given, which will be useful to students for quick reference. Various charts for reactions of organic chemistry will also help to revise the reactions studied in text books.

The solution of many different types of problems in thermo-chemistry, thermodynamics, kinetics and colligative properties of solutions is given. Problem solving in Chemistry requires an ability to defuse terms clearly to explain chemical phenomenon to represent chemical entities—atoms, irons, molecules, formulas and geometrical sketches; to graph data and interpret these graphs and so on.

HOW TO SOLVE A PROBLEM ?

To the Student

When you go about solving any problem in this or from any other book or in any test or examination every problem you encounter should be solved in the fashion as :

(1) *Read the Problem Carefully* : Note exactly what is given and what is sought, what you are supposed to do and what you have to do it with. Note that most chemistry problems

contain more information than is explicitly given. For example if a problem specifies a given mass of water, it is also specifying a given number of moles, molecules or atoms. Note any and all special conditions. Every problem in this book is designed to illustrate some principle, some relationship, some law, some definition or some fact. If you understand these, you should have no difficulty to solving the problem. The only reason that students face difficulty with chemistry problems is their failure to understand, exactly and well, the chemical principles involved and the meaning and value of all terms and units that are used in the problem.

(2) *Plan* in detail, how the problem is to be solved. Make a habit of visualizing the entire solution. Insist on knowing what you are going to do and why you are going to do it. Aim to learn to solve every problem in the most efficient manner i.e. the shortest way, with fewest steps.

(3) *Checking Units* : Specify definitely what each number represents and the units in which it is expressed, when actually carrying out mathematical operation of solving the problem. Do not just write 192/32 = 6.
Write, 192g of sulphur/32g of sulphur per mole of sulphur = 6 moles of sulphur or whatever the case may be. Always divide and multiply the units as well as the numbers. This is a good way to help avoid errors.

(4) *Examine the Answer* : Having solved the problem, examine the answer to see if it is reasonable and sensible.

(5) *Decide Earliest* : If you do not understand how to solve the problem, have it explained to you at the very earliest possible time. To be able to solve the later problems you must understand the earlier ones. After a problem is understood by you, fix the explanation in you mind by working other similar problems at once. Test yourself to be sure that you can apply your understanding of the solution.

New Delhi

A.S. BHATIA

PART I

Chapter 1

ATOMIC STRUCTURE

Important Formulae

$$c = v\lambda, = \frac{c}{v}$$

Wave number $(\bar{v}) = \frac{1}{\lambda}$

$$\lambda = \frac{h}{p},\ p = \text{mv}$$

For hydrogen atom

$$E_n = \frac{-2\pi^2 e^4 m}{h^2}\left(\frac{1}{n^2}\right)$$

$$= \frac{-21.79 \times 10^{-12}}{n^2} \text{ erg per atom}$$

$$= \frac{-13.6}{n^2} \text{ eV per atom}$$

$$= \frac{-313.6}{n^2} \text{ k cal mol}^{-1} = \frac{1310.85}{\text{n}^2} \text{KJ mol}^{-1}$$

$$r_n = \frac{h^2}{4\pi^2 me^2} \times n^2 = 0.4529 \times n^2 \text{ Å}$$

Angular momentum, $mvr = \dfrac{nh}{2\pi}$

Velocity, $v_n = \dfrac{2\pi e^2}{nh}$

$$\Delta E = E_t - E_i = \frac{2\pi^2 e^4 m}{h^2}\left(\frac{1}{n_i^2} - \frac{1}{n_f^2}\right)$$

$$= 2.179 \times 10^{-11}\left(\frac{1}{n_i^2} - \frac{1}{n_f^2}\right)$$

erg per atom
where $n_i < n_f$

Wave rtumber $(\overline{v}) = R_H\left(\frac{1}{n_i^2} - \frac{1}{n_f^2}\right)$ cm^{-1}

where R_H = Rydberg constant = 109677 cm^{-1}

Time for one revolution $= \dfrac{2\pi r}{v} = \dfrac{2\pi}{w}$ and number of revolutions

For unit time $= \dfrac{w}{2\pi}$

STRUCTURE OF ATOMS AND MOLECULES

Fundamental particles of atom An atom consists of three fundamental particles namely electrons protons and neutrons. The charge and mass of these particles are as follows:

Particle	*Symbol*	*Charge in Coulombs*	*Mass*	
			kg	*a.m.u.*
Electron	$-1e^0$	-1.602×10^{-19}	9.1091×10^{-31}	0.000548
Proton	$+1p^1$	$+1.602 \times 10^{-19}$	1.6725×10^{-27}	1.00757
Neuton	0^{n^1}	Zero (0)	1.6748×10^{-27}	1.00893

Atomic number (Z): The atomic no. of an element
Z represent the no. of protons in the nucleus of an element.

$\because$ Z = No. of protons in the nucleus of an atom

= No. of electrons in the extra nuclear part of neutral atom.

Mass number (*m*): It is equal to sum of numbers of protons and no. of neutrons in an atom.

$m = p + n$

Dual nature of matter (de Broglie equation) de Broglie proposed (on the basis of quantum theory) a relationship between wavelength (λ) of a moving particle with its velocity (v) as

$$\lambda = \frac{h}{mv} = \frac{h}{p} = \frac{h}{\sqrt{2mK_E}} \qquad \left(\because K_E = \frac{1}{2}mv^2\right)$$

where h is plan c k's constant. Its value is 6.6262×10^{-34} J s or 6.662×10^{-34} kg m^2 s^{-1}; m is mass of moving particle, v is its velocity

and P is momentum of particle equal to mu or $\sqrt{2m(KE)}$

Note: The de Broglie equation is insignificant for bigger particles and is useful only for smaller particles like electrons because:

(i) For a large mass
Consider a stone of mass 100 g moving with a velocity of 1000 cm/sec. The de Broglie's wavelength is given as follows

$$\lambda = \frac{h}{mv} = \frac{6.625 \times 10^{-27}}{100 \times 1000} = 6.625 \times 10^{-32} \text{ cm}$$

This is too small to be measured by any instrument. Hence it has no significance.

(ii) For a small mass
Consider an electron in hydrogen atom it has a mass 9.11×10^{-28} g and moves with a velocity 2.18×10^8 cm/sec. The de Broglie's wave length is given as follows

$$\lambda = \frac{h}{mv} = \frac{6.625 \times 10^{-27}}{9.11 \times 10^{-28} \times 2.18 \times 10^8}$$

$$= 3.34 \times 10^{-8} \text{ cm.}$$

This value is almost equal to the wave length of X-rays

HEISENBERGS'S UNCERTANITY PRINCIPLE

It states that "it is imposible to determine the exact position and the exact momentum or velocity simultaneously of a particle like electron in an atom".
Mathematically

$$\Delta x \cdot \Delta p \approx h$$

$$\Delta x \cdot \Delta p \geq \frac{h}{2\pi}$$

But $$p = \text{mv}$$

$$\Delta x \cdot \Delta \text{mv} \geq \frac{h}{2\pi} \quad \text{or} \quad \Delta x \cdot \Delta v \geq \frac{h}{2\pi m}$$

where Δp is uncertainty in momentum
Δx is uncertainty in position
Δv is uncertainty in velocity.

If position of electron is detemined more accurately, *i.e.*, Δx is small (= 0 say) then, uncertainty with regards to momentum, i.e. Δp would be large as, $\Delta p = h/\Delta x = h/0 = \infty$ (infinity)

On the other hand, if momentum is determined more accurately, *i.e.*, Δp small (= 0 say) then, uncertainty with regard to momentum, *i.e.*, Δx would be large as, $\Delta p = 0$, then $\Delta x = \infty$ (infinity)

Hence, more certainty in one leads to less certainty in another and *vice versa*.

QUANTISATION OF ANGULAR MOMENTUM

Consider an electron moving in a circular orbit around a nucleus. Bohr has assumed that the angular momentum of an electron is an integral multiple of $h/2\pi$.

i.e., $$\text{mvr} = n\frac{h}{2\pi}$$

On rearranging, $$2\pi r = n\frac{h}{mv}$$

But according to de Broglie, $$\frac{h}{mv} = \lambda$$

Hence $$2\pi r = n\lambda$$

The circumference ($2\pi r$) of the orbit must be equal to integral multiple of its wave length (λ).

SCHRODINGER WAVE EQUATION

Schrodinger, proposed an equation to describe the wave associated with an electron. The differential equation is

$$\frac{\partial^2 \Psi}{\partial x^2} + \frac{\partial^2 \Psi}{\partial y^2} + \frac{\partial^2 \Psi}{\partial z^2} + \frac{8\pi^2 m}{h^2} (\text{E - V}) \Psi = 0$$

where, x, y and z are cartesian coordinates of the electron.
m = mass of the electron
E = total energy of the electron
V = potential energy of the electron
h = Planck's constant
Ψ = wave function of the electron

This equation gives both the wave character and probability of the particle (electron) The wave function Ψ is regarded as amplitude function and may have +ve or –ve values. So we consider Ψ^2 +ve as –ve values are not to be considered.

Thus Ψ^2 is the probability function. It describies the probability of finding an electron within a small space. That is the probability of finding the electron at a point can be indicated by Ψ^2.

Quantum Numbers

There are four quantum numbers as a result of Schrodinger wave equation

(1) Principle quantum number It gives the avg. distance of an electron from the nucleus & so the size of electroncloud is known.

(a) Denoted by 'n'

(b) The values of 'n' are from 1 to n

$n = 1$	K shell
$n = 2$	L = shell
$n = 3$	M shell
$n = 4$	N shell

(c) 'n' signify for the size and energy level of major energy shell, Any energy level can accommodate at the most $2n^2$ electrons. Thus first second, third, fourth energy levels can accommodate $2 \times 1^2 = 2$, $2 \times 2^2 = 8$, $2 \times 3^2 = 18$, $2 \times 4^2 = 32$ electrons respectively.

(2) Azimuthal or angular quantum number;

(a) Denoted by 'l'

(b) The values of 'l' are from 0 to $n - 1$. The total values of l is equal to the number of principle quantum numbers.

For		
$n = 1$	$l = 0$	s subshell
$n = 2$	$l = 1, k = 0$	p subshell
$n = 3$	$l = 2, l = 1$ & $l = 2$	d subshell
$n = 4$	$l = 3, l = 2, l = 1$ $l = 0$	f subshell

(c) 'l' signify for shape and energy level of subshells (sherical, dumb-bell, double dumb bell etc). The numbr of subshells present in an energy level in due to its angular momentum $= \sqrt{l(l+1)}\,.\,h/2\pi$.

In a given energy level, the order of increasing energy of various sub-shells is $s < p < d < f$.

3. Magnetic quanum number

(a) Denoted by 'm'

(b) The values of 'm' are from $\pm l$ to $\mp l$ $\therefore$ Total number of possible values of m is $(2l + 1)$. Each value represents a particular orbital.

$l = 0$	$m = 0$	s subshell		p subshell	d subshell
Let $l = 1$	$m = -1$	0		+1	
	p_x or p_y	p_z		p_y or p_x	
Let $l = 2$	$m = -2$	-1	0	+1	+2
	d_{xy} or $d_{x^2} - y^2$	d_{xz} or d_{z^2} d_{yz}		d_{yz} or $d_{x^2} - y^2$ d_{xz} or d_{xy}	

and for $l = 3$, we have $m = -3, -2, -1, 0, +1, +2, +3$ ... (f subshell)

(c) 'm' signify for the possible no. of orientations of subshells space.

Since each orientation of electron in space corresponds to an orbital, the magnetic quantum number determines, the total number of orbitals presents in any subshell.

(4) Spin quantum no:

(a) Denoted by s Since and electron can spin either in the cockwise direction or anticlockwise direction, therfore, for a given value of magnetic quantum number, spin quantum number can have two values.

(b) The values of 's' are $+\frac{1}{2}$ (also represented by upward arrow mark for clockwise spin) and $-\frac{1}{2}$ (also represented by downward arrow mark for antic clockwise spin).

(c) 's' signify for the direction of spin of electron in a sub subshell or orbital.

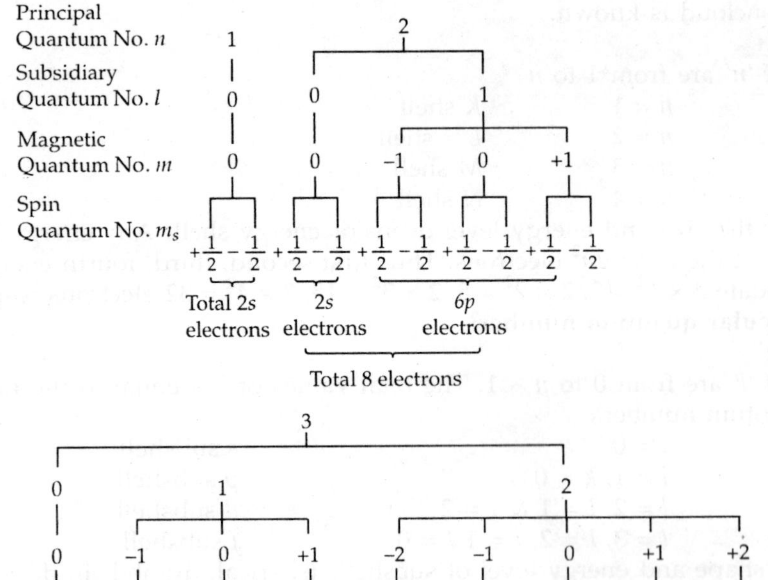

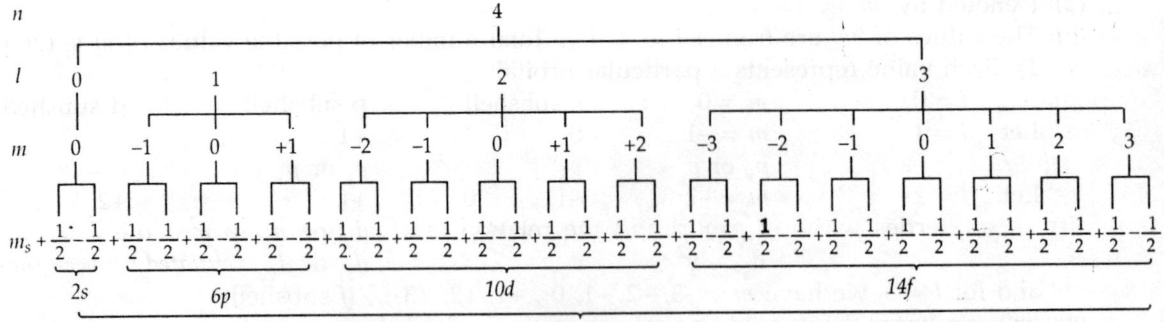

Quantum numbers, the permissible number of electrons and the shape of the periodic table,

The Periodic Table

	s-block												*p-block*					
Group	*1*	*2*											*13*	*14*	*15*	*16*	*17*	*18*
1	^{1}H																^{1}H	^{2}He
2	^{3}Li	^{4}Be					*d*-block						^{5}B	^{6}C	^{7}N	^{8}O	^{9}F	^{11}Ne
3	^{11}Na	^{12}Mg	3	4	5	6	7	8	9	10	11	12	^{13}A1	^{14}Si	^{15}P	^{16}S	^{17}Cl	^{18}Ar
4	^{19}K	^{20}Ca	^{21}Se	^{22}Ti	^{23}V	^{24}Cr	^{25}Mn	^{25}Fe	^{27}Co	^{29}Ni	^{29}Cu	^{30}Zn	^{31}Ga	^{32}Ge	^{33}As	^{34}Se	^{35}Br	^{36}Kr
5	^{37}Rb	^{38}Sr	^{39}Y	^{40}Zr	^{41}Nb	^{42}Mo	^{43}Tc	^{44}Ru	^{45}Rh	^{46}Pd	^{47}Ag	^{48}Cd	^{49}In	^{50}Sn	^{51}Sb	^{52}Te	^{33}I	^{54}Xe
6	^{55}Cs	^{54}Ba	^{57}La•	^{72}Hf	^{73}Ta	^{74}W	^{75}Re	^{76}Os	^{77}Ir	^{78}Au	^{80}Hg	^{81}Tl	^{82}Pb	^{83}Bi	^{84}Po	^{85}At	^{86}Rn	
7	^{87}Fr	^{88}Ra	^{89}Ac•															
							f-Block											
Lanthanides			↓	^{88}Ce	^{59}Pr	^{60}Nd	^{61}Pm	^{61}Sm	^{64}Eu	^{64}Gd	^{65}Tb	^{66}Dy	^{67}Ho	^{68}Er	^{69}Tm	^{70}Yb	^{71}Lu	
Actinides			↓	^{90}Th	^{91}Pa	^{92}U	^{93}Np	^{94}Am	^{95}Am	^{96}Cm	^{97}Cm	^{98}Cf	^{99}Es	^{100}Fm	^{101}Md	^{102}No	^{103}Lr	

ARRANGEMENT OF THE ELEMENTS IN GROUPS IN THE PERODIC TABLE

In the periodic table, the elements are arranged in order of increasing atomic number, that is in order of increased nuclear charge, or increased number of orbital electrons. Thus each elements contains one more orbital electron than the preceding element. Instead of listing the 103 elements as one long list, the periodic table arranges them into several horizontal rows or periods, in such a way that each row begins with an alkali metal and ends with a noble gas. The sequance in which the various energy levels are filled determines the number of elements in each period.

Thus the periodic table individed into four main block according to whether the s.p.d or f levels are being filled.

1st period	1*s*				elements in this period 2
2nd period	2*s*			2*p*	elements in this period 8
3rd period	3*s*			3*p*	elements in this period 8
4th period	4*s*		3*d*	4*p*	elements in this period 18
5th period	5*s*		4*d*	5*p*	elements in this period 18
6th period	6*s*	4*f*	5*d*	6*p*	elements in this period 32

The alkali metals appear in a vertical column labelled Group 1, in which all elements have one s electron in their outer shell, and hence have similar properties. Thus when one elements in a group racts with a reagent, the other elemnts in the goup will probably react similarly, forming compounds which have similar formulae. Thus reactions of new compounds and their formulae may be predicted by analogy with known compounds, *the noble gases all appear in a vertical column labelled Group 18, and all have a complete outer shell of electrons.* This is called the long form of the periodic table. It has many advantages, the most important being that it eimphasizes the similarity of properties within a group and the relation between the goup and the electron structure. *The d-block elements are referred to as the transition elements as they are situated between the s- and p-blocks.*

Hydrogen and helium differ from the rest of the elements because there are no p orbitals in the first shell. Helium obviously belongs to Group 18, the noble gases, which are chemcally inactive because their outer shell of electrons is full. Hydrogen is more difficult to place in a group. It could be included in Group 1 because it has one *s* electron in tis outer shell, is univalent and commonly forms unvalent positive ions. Howerver, hydrogen is not a metal and is a gas whist Li, Na, K, Rb and Cs are metals and are solids. Similarly, hydrogen could be included in Group 17 because it is one electron sort of a complete shell, or in Group 14 because its outer shell is half full. Hydrogen does not resemble the alkali metals, the halogens or Group 14 very closely. Hydrogen atoms are extermely small, and have may unique properties. Thus there is a case for placing Hydrogen in a group on its own.

Rules for filling of orbitals (in an atom or ion)

1. **Pauli Exclusion Principle:** is as follows:
   (1) *"No* two electrons in an atom may have all the four quantum numbers alike".
   Thus, three quantum numbers for two electrons in an atom can be identical at the most but the fourth quantum number must be different, e.g., if quantum number *'n'*, *'l'* and *'m'* are alike, the fourth quanum number *'s'* must be different.
   (2) *e.g.*, [↑↓] is correct for $1s^2$
   [↓↓] or [↑↑] is wrong for $1s^2$
   (3) Following results have been obtained by Pauli exclusion principle
   (a) Maximum no. of electrons ina shell can be $2n^2$
   (b) Maximum no. of electrons in a subshell can be 2, 6, 10, 14, in *s, p, d, f* respectively
   (c) Maximum no. of electrons in a sub-subshell is 2 only.
2. **Aufbau principles**
   According to this principle elctrons occupy various orbitals of an atom in the order of their increasing energy.
   The order of increasing energy of various orbital in an atom or ion (other than hydrogen or hydrogen like species) is as follows;
   1s, 2s, 2p, 3s, 3p, 4s, 3d, 3p, 4p, 5s, 4d, 5p, 6s, 4f, 5d, 6p, 7s
3. **(*n* + 1) Rule:** According to this rule a subshell having lowest value of ($n + 1$) has lower energy and thus is filled up first e.g.,
   (a) *2s and 2p orbitals.*
   For 2*s* orbital, $n + 1 = 2 + 0 = 2$, For 2*p* orbital $n + l = 2 + l = 3$
   Thus, 2*s* orbital having lower energy is filled up first than 2*p* orbital
   (b) 4*s*, 3*d* orbital.
   For 4*s* orbital, $n + 1 = 4 + 0 = 4$, For 3*d* orbital $n + l = 3 + 2 = 5$
   Thus, 4*s* orbital having lower energy is filled up first than 3*d* orbital,
   e.g., $_{19}$K; $1s^2$, $2s^2\ 2p^6$, $3s^2\ 3p^6\ 3d^1$ is wrong
   $1s^2$, $2s^2\ 2p^6$, $3s^2 3p^6$, $4s^1$ is correct
   $n + l$ of $4s = 4 + 0 = 4$
   $n + l$ of $3d = 3 + 2 = 5$
   Thus, 4*s* shoud be filled first.
   (c) If ($n + l$) value for two orbitals is same, an orbital with lower value of *'n'* filled up first, *e.g.*, 2*p* orbital is fillled up first than 3*s* orbital although ($n + 1$) value for both the orbitals is 3,
   e.g, $_{21}$Sc: $1s^2$, $2s^2\ 2p^6$, $3s^2\ 3p^6$, $4s^2\ 4p^1$ is wrong
   $1s^2$, $2s^2\ 2p^6$, $3s^2\ 3p^6\ 3d^1$, $4s^2$ is correct
   $n + l$ of $4p = 4 + 1 = 5$
   $n + l$ of $3d = 3 + 2 = 5$
   Thus, 3*d* should be filled first $\because$ *n* of 3*d* < *n* of 4*s*

4. **Hund's rules:**

(a) If filling a group of orbitals of equal energy (or subshells) it is preferred to assign electrons to empty orbitals rather than pair them in a particular subshell, because the former arrangement leads to lower energy level.

(b) Same spin of unpaired electrons in sub-subshell also gives rise to lwore energy level.
e.g., $_7N$: $1s^2$, $2s^2$, $2p^2$

For $2p^3$ [↑|↑|↑] is coorect

[↑↓|↑|] is wrong (statements *a*)

[↑|↓|↑] is wrong (statements *b*)

Stability of Completely Filled and Exactly Half Filled Orbitals

A subshell having nearly full filled or nearly half filled configuration tend to acquire exactly full fileld or exactly half filled nature in order to attain stability, *i.e.*, lower energy level.

e.g., in Cr and Cu atoms one electrons promotes from 4*s* to 3*d* orbitals. Thus, $4s^1\ 3d^5$ configuration is more stable than $4s^2\ 3d^4$ configuration in Cr and $4s^1\ 3d^{10}$ configuration is more stable than $4s^2\ 3d^9$ configuration in Cu.

Thus $_{24}Cr$ $1s^2, 2s^2, 2p^6, 3s^2\ 3p^6\ 3d^4, 4s^2$ is wrong
$1s^2, 2s^2, 2p^6, 3s^2\ 3p^6\ 3d^5, 4s^1$ is correct

$_{29}Cu$: $1s^2, 2s^2, 2p^6, 3s^2\ 3p^6\ 3d^9, 4s^2$ is wrong
$1s^2, 2s^2\ 2p^6, 3s^2\ 3p^6\ 3d^{10}, 4s^1$ is corret

$_{46}Pd$: $1s^1, 2s^2\ 2p^6, 3s^2\ 3p^6\ 3d^{10}, 4s^2\ 4p^6\ 4d^8, 5s^2$ is wrong
$1s^2, 2s^2\ 2p^6, 3s^2\ 3p^6\ 3d^{10}, 4s^1\ 4p^6\ 4d^{10}$ is correct.

SEQUENCE OF ENERGY LEVELS

The sequence in which the enrgy levels are filled is: 1*s*, 2*s*, 2*p*, 3*s*, 3*p*, 4*s*, 3*d*, 4*p*, 5*s*, 4*d*, 5*p*, 6*s*, 4*f*, 5*d*, 6*p*, 7*s* etc.

After the 1*s*, 2*s*, 2*p*, 3*s* and 3*p* levels have been filled at argon, the new two electron go into the 4*s* level. This gives the elements potassium and calcium. Once the 4*s* level is full the 3*d* level is the next lowest in energy. not the 3*p* level. Thus the 3*d* starts to fill at scandium. The elements from scandium to copper have two electrons in the 4*s* level and an incomplete 3*d* level, and all behave in a similar manner chemically. Such a series of atoms is known as transition series.

A second transition series starts after the 5*s* orbital has been filled, at strontium, because in the next element, ythrium, the 4*d* level begins to fill up. A third transition series starts at lanthanum where the electrons start to fill the 5*d* level after the 6*d* level has been filled with two electrons.

A further complication arises here because after lanthanum, which has one electron in the 5*d* level, the 4*f* level begins to fill, giving the elements from cerium to luetium with from one to 14*f* electrons. These are sometimes called the innertransition elements last are usually known as the lanthanides or Rare earthmetals.

Electronic Configurations of Elements in the Ground State

Atomic Number	*Element*	*Electronic configuration*
1.	H	$1s^1$
2.	He	$1s^2$
3.	Li	$[He]^2\ 2s^1$
4.	Be	$[He]^2\ 2s^2$
5.	B	$[He]^2\ 2s^2\ 2p^1$
6.	C	$[He]^2\ 2s^2\ 2p^2$
7.	N	$[He]^2\ 2s^2\ 2p^3$
8.	O	$[He]^2\ 2s^2\ 2p^4$
9.	F	$[He]^2\ 2s^2\ 2p^5$
10.	Ne	$[He]^2\ 2s^2\ 2p^6$
		or $1s^2$, $2s^2$, $2p^6$
11.	Na	$[Ne]^{10}\ 3s^2$
12.	Mg	$[Ne]^{10}\ 3s^2$
13.	Al	$[Ne]^{10}\ 3s^2\ 3p^1$
14.	Si	$[Ne]^{10}\ 3s^2\ 3p^2$
15.	P	$[Ne]^{10}\ 3s^2\ 3p^3$
16.	S	$[Ne]^{10}\ 3s^2\ 3p^4$
17.	Cl	$[Ne]^{10}\ 3s^2\ 3p^5$
18.	Ar	$[Ne]^{10}\ 3s^2\ 3p^6$
		or $1s^2$, $2s^2$, $2p^6$
		$3s^2$, $3p^6$
19.	K	$[Ar]^{18}\ 4s^1$
20.	Ca	$[Ar]^{18}\ 4s^2$
21.	Sc	$[Ar]^{18}\ 3d^1\ 4s^2$
22.	Ti	$[Ar]^{18}\ 3d^2\ 4s^2$
23.	V	$[Ar]^{18}\ 3d^3\ 4s^2$
24.	Cr	$[Ar]^{18}\ 3d^5\ 4s^1$
25.	Mn	$[Ar]^{18}\ 3d^5\ 4s^2$
26.	Fe	$[Ar]^{18}\ 3d^6\ 4s^2$
27.	Co	$[Ar]^{18}\ 3d^7\ 4s^2$
28.	Ni	$[Ar]^{18}\ 3d^8\ 4s^2$
29.	Cu	$[Ar]^{18}\ 3d^{10}\ 4s^1$
30.	Zn	$[Ar]^{18}\ 3d^{10}\ 4s^2$
31.	Ga	$[Ar]^{18}\ 3d^{10}\ 4s^2\ 4p^1$
32.	Ge	$[Ar]^{18}\ 3d^{10}\ 4s^2\ 4p^2$
33.	As	$[Ar]^{18}\ 3d^{10}\ 4s^2\ 4p^3$
34.	Se	$[Ar]^{18}\ 3d^{10}\ 4s^2\ 4p^4$
35.	Br	$[Ar]^{18}\ 3d^{10}\ 4s^2\ 4p^5$
36.	Kr	$[Ar]^{18}\ 3d^{10}\ 4s^2\ 4p^6$
		or $1s^2$, $2s^2\ 2p^2\ 3s^2$
		$3p^6\ 3d^{10}\ 4s^2\ 4p^6$
37.	Rb	$[Kr]^{36}\ 5s^1$
38.	Sr	$[Kr]^{36}\ 5s^2$
39.	Y	$[Kr]^{36}\ 4d^1\ 5s^2$
40.	Zr	$[Kr]^{36}\ 4d^2\ 5s^2$
41.	Nb	$[Kr]^{36}\ 4d^4\ 5s^1$
42.	Mo	$[Kr]^{36}\ 4d^5\ 5s^1$
43.	Tc	$[Kr]^{36}\ 4d^5\ 5s^2$
44.	Ru	$[Kr]^{36}\ 4d^7\ 5s^1$
45.	Rh	$[Kr]^{36}\ 4d^8\ 5s^1$
46.	pd	$[Kr]^{36}\ 4d^{10}\ 5s^0$
47.	Ag	$[Kr]^{36}\ 4d^{10}\ 5s^1$
48.	Cd	$[Kr]^{36}\ 4d^{10}\ 5s^2$
49.	In	$[Kr]^{36}\ 4d^{10}\ 5s^2\ 5p^1$
50.	Sn	$[Kr]^{36}\ 4d^{10}\ 5s^2\ 5p^2$
51.	Sb	$[Kr]^{36}\ 4d^{10}\ 5s^2\ 5p^3$
52.	Te	$[Kr]^{36}\ 4d^{10}\ 5s^2\ 5p^4$
53.	I	$[Kr]^{36}\ 4d^{10}\ 5s^2\ 5p^5$
54.	Xe	$[Kr]^{36}\ 4d^{10}\ 5s^2\ 5p^6$
		or $1s^2$, $2s^2$, $2p^6$, $3s^2$,
		$3p^6$, $3d^{10}$, $4s^2\ 4p^6$, $4d^{10,}\ 5s^2\ 5p^6$
55.	Cs	$[Xe]^{54}\ 6s^1$
56.	Ba	$[Xe]^{54}\ 6s^2$
57.	La	$[Xe]^{54}\ 5d^1\ 6s^2$
58.	Ce	$[Xe]^{54}\ 4f^1\ 5d^1\ 6s^2$
59.	Pr	$[Xe]^{54}\ 4f^3\ 5d^0\ 6s^2$
60.	Nd	$[Xe]^{54}\ 4f^4\ 6s^2$
61.	Pm	$[Xe]^{54}\ 4f^5\ 6s^2$
62.	Sm	$[Xe]^{54}\ 4f^6\ 6s^2$
63.	Eu	$[Xe]^{54}\ 4f^7\ 6s^2$
64.	Gd	$[Xe]^{54}\ 4f^7\ 5d^1\ 6s^2$
65.	Th	$[Xe]^{54}\ 4f^9\ 6s^2$
66.	Dy	$[Xe]^{54}\ 4f^{10}\ 6s^2$
67.	Ho	$[Xe]^{54}\ 4f^{11}\ 6s^2$
68.	Er	$[Xe]^{54}\ 4f^{12}\ 6s^2$
69.	Tm	$[Xe]^{54}\ 4f^{13}\ 6s^2$
70.	Yb	$[Xe]^{54}\ 4f^{14}\ 6s^2$
71.	Lu	$[Xe]^{54}\ 4f^{14}\ 5d^1\ 6s^2$
72.	Hf	$[Xe]^{54}\ 4f^{14}\ 5d^2\ 6s^2$
73.	Ta	$[Xe]^{54}\ 4f^{14}\ 5d^3\ 6s^2$
74.	W	$[Xe]^{54}\ 4f^{14}\ 5d^4\ 6s^2$
75.	Re	$[Xe]^{54}\ 4f^{14}\ 5d^5\ 6s^2$
76.	Os	$[Xe]^{54}\ 4f^{14}\ 5d^6\ 6s^2$
77.	Ir	$[Xe]^{54}\ 4f^{14}\ 5d^7\ 6s^2$
78.	Pr	$[Xe]^{54}\ 4f^{14}\ 5d^8\ 6s^2$
79.	Au	$[Xe]^{54}\ 4f^{14}\ 5d^{10}\ 6s^2$
80.	Hg	$[Xe]^{54}\ 4f^{14}\ 5d^{10}\ 6s^2$
81.	Tl	$[Xe]^{54}\ 4f^{14}\ 5d^{10}\ 6s^2\ 6p^1$
82.	Pb	$[Xe]^{54}\ 4f^{14}\ 5d^{10}\ 6s^2\ 6p^2$
83.	Bi	$[Xe]^{54}\ 4f^{14}\ 5d^{10}\ 6s^2\ 6p^3$

(Contd.)

(Contd.)

84.	Po	$[Xe]^{54}\ 4f^{14}\ 5d^{10}\ 6s^2\ 6p^4$	93.	Np	$[Rn]^{86}\ 5f^4\ 6d^1\ 7s^2$
85.	At	$[Xe]^{54}\ 4f^{14}\ 5d^{10}\ 6s^2\ 6p^5$	94.	Pu	$[Rn]^{86}\ 5f^6\ 7s^2$
86.	Rn	$[Xe]^{54}\ 4f^{14}\ 5d^{10}\ 6s^2\ 6p^6$ or $1s^2, 2s^2, 2p^6, 3s^2, 3p^6, 3d^{10}$ $4s^2, 4p^6, 4d^{10}, 4f^{14}, 5s^2, 5p^6,$ $5d^{10}, 6s^2, 6p^6$	95.	Am	$[Rn]^{86}\ 5f^7\ 7s^2$
			96.	Cm	$[Rn]^{86}\ 5f^7\ 6d^1\ 7s^2$
			97.	Bk	$[Rn]^{86}\ 5f^9\ 7s^2$
			98.	Cf	$[Rn]^{86}\ 5f^{10}\ 7s^2$
			99.	Es	$[Rn]^{86}\ 5f^{11}\ 7s^2$
87.	Fr	$[Rn]^{86}\ 7s^1$	100.	Fm	$[Rn]^{86}\ 5f^{12}\ 7s^2$
88.	Ra	$[Rn]^{86}\ 7s^2$	101.	Md	$[Rn]^{86}\ 5f^{13}\ 7s^2$
89.	Ac	$[Rn]^{86}\ 6d^1\ 7s^2$	102.	No	$[Rn]^{86}\ 5f^{14}\ 7s^2$
90.	Th	$[Rn]^{86}\ 6d^2\ 7s^2$	103.	Lr	$[Rn]^{86}\ 5f^{14}\ 6d^1\ 7s^2$
91.	Pa	$[Rn]^{86}\ 5f^2\ 6d^1\ 7s^2$	104.	Ku	$[Rn]^{86}\ 5f^2\ 6d^2\ 7s^2$
92.	U	$[Rn]^{86}\ 5f^3\ 6d^1\ 7s^2$	105.	Ha	$[Rn]^{86}\ 5f^2\ 6d^3\ 7s^2$

Theory of relativity and velocity of particle: According to the theory of relativity, the mass (m_1) of a particle (electron) at high speed is given by

$$m_1 = \frac{m}{\sqrt{\left[1-\left(\frac{u}{c}\right)^2\right]}} \qquad \ldots\ldots (1)$$

where m is mass in rest; u is velocity and c is velocity of light

If $\quad u = c$

Then $\quad m_1 = \infty$

Planck's quantum theory: Radiant energy is emitted or absorbed only in discrete units or packets of energy called photon (quantum). The energy 'E' associated with a quantum is given by $E = h\nu$ where h is Planck's constant and ν is frequency of radiations.

$$E = h\nu = \frac{hc}{\lambda} = hc\bar{\nu} \quad \left(\because \frac{1}{\lambda} = \bar{\nu}\right) \qquad \ldots\ldots (2)$$

The value of h, Planck's constant = 6.625×10^{-34} J sec,
and of c is velocity of light = 3.0×10^8 m sec^{-1}

THE BOHRMODEL

Bohr's modle for H or H like atoms, i.e., one elctron systems

(1) The electrons are in continuous motions round the nucleus in closed orbits of definite energy level known as shells. Shells are named as $K, L,$,M N ... or numbered as 1, 2, 3, 4, ... from the nucleus. As the distance of shell increases from the nucleus, energy level of shell increases.

(2) As long as an electron occupy a definite energy level, it does not radiate out energy. The emission or absorption of energy occurs only when electron jumps from one level to other

$$\boxed{\Delta E = E_{n_2} - E_{n_1} = h\nu} \qquad (1)$$

If $n_2 > n_1$ then an emission spectra
but If $n_2 < n_1$ then an absorption spectra

(3) The angular momentum of electron in closed shell is always an integer multiple of $h/(2\pi)$

$$\text{Angular momentum} = n\ \frac{h}{2\pi}$$

or

$$mVr = n \cdot \frac{h}{2\pi}$$

where m is mass of electron, v is its tangential velocity, r is radius of orbit and n is an integer

Some important results of Bohr's model

For H atom and H like atoms, *i.e.*, He^+, Li^{2+} ...

Radius: $r_n = n^2 \times r_1$

and

$$r_n = \frac{n^2 h^2}{4\pi^2 me^2 Z} \tag{2}$$

Thus, $r_1 = 0.529$ Å,for H atom, Z is atomic number, e is charge on electron, m is mass of electron and n is number of the shell

Energy: $E_n = \text{PE} + \text{KE}$

where E_n is total energy of an electron in a shell.

$$\text{PE is potential energy} = \frac{Ze^2}{r_n}$$

$$\text{KE is kinetic energy} = \frac{1}{2}\frac{Ze^2}{r_n}$$

$$\therefore \quad E_n = \frac{Ze^2}{r_n} + \frac{1}{2}\frac{Ze^2}{r_n} = -\frac{Ze^2}{2r_n} \tag{3}$$

$$E_n = \frac{1}{2}\ \text{PE and KE} = -\frac{PE}{2}$$

Aslo by Eqs. (2) and (3), we get

$$E_n = -\frac{2\pi^2 me^4 Z^2}{n^2 h^2} \tag{4}$$

where Z is the atomic number, m is the mass and e is the charge on an electron n is the number of energy level to which the electron belongs.

In case of hydrogen atom Z = 1, On substituting all the values.

$$E_1 = -\frac{21.72 \times 10^{-19}}{n^2}\text{joule} \qquad \text{for H}$$

$$= -\frac{13.6}{n^2}\text{eV} \qquad \text{for H}$$

These equations also reveal that

$$E_n \propto -\frac{1}{n^2} \text{ or, } E_n = -\frac{E_1}{n^2} \quad \ldots\ldots (5)$$

where E_n and E_1 are energy levels in nth shell and 1st shell.

Also $r_{\text{H like atom}} = \dfrac{r_{for}\ H}{Z}$ ...... (6)

and $E_{n\ \text{H like atom}} = E_{n\ \text{for H}} \times Z^2$

The velocity of electron in an orbit is given by

$$V_n = \frac{2\pi Ze^2}{nh} \quad \ldots\ldots (8)$$

For **H atom:** $\boxed{V_n = \dfrac{2\pi e^2}{nh}}$ ...... (9)

$$\therefore \quad V_n = \frac{V_1}{n} \quad \ldots\ldots (10)$$

where V_1 is velocity of electron in 1st orbit.

Now as: $\Delta E = E_{n2} - E_{n1} = \dfrac{hc}{\lambda}$

$$\therefore \quad \boxed{\Delta E = \frac{hc}{\lambda} = \frac{2\pi me^4 Z^2}{h^2}\left[\frac{1}{n_1^2} - \frac{1}{n_2^2}\right]}$$

or

$$\frac{1}{\lambda} = \bar{v} = \frac{2n^2 mZ^2 e^4}{ch^3}\left[\frac{1}{n_1^2} - \frac{1}{n_2^2}\right]$$

$$\boxed{\frac{l}{\lambda} = Z^2 R_H\left[\frac{1}{n_1^2} - \frac{1}{n_2^2}\right]} \quad \ldots\ldots (11)$$

where λ is the wavelength of radiations during electronic transition from n_2 to $n_1 \cdot R_H$ *is Rydberg constant and is equal to 109678 cm.*

(1) When $n_1 = 1$ $n_2 = 2, 3, 4$ ... Lyman series in UV region
(2) When $n_1 = 2$ $n_2 = 3, 4, 5,$ ... Balmer series in visible region
(3) When $n_1 = 3$ $n_2 = 4, 5, 6$ ... Paschen series in IR region
(4) When $n_1 = 4$ $n_2 = 5, 6, 7$ ... Brackett serie in IR region
(5) When $n_1 = 5$ $n_2 = 6, 7, 8$ ... Pfund series in IR region

Total spin and magnetic moment: The total spin (s) of an atom is given by $s = \dfrac{1}{2} \times n$, where n is number of unpaired electrons.

The spin magnetic moment (μ) of electron (excluding orbital magnetic moment) in Bohr Magneton (B.M.) is given by:

$$\mu_{\text{effective}} = \sqrt{4s(s+1)}$$

If
$$s = \frac{1}{2} \times n$$

$$\mu_{\text{effective}} = \sqrt{n(n+2)} \text{ B.M.}$$

Photo Electric Effect

When a photon strikes the metallic surface, it gives up its energy to the electron. Part of this energy (say W) is used by the electrons to escape from the metal, the remaining imparts the kinetic enrgy $\left(\frac{1}{2}mu^2\right)$ to the photeelectrons. If the incident radiation has frequency ν, then its photons have enrgy $h\nu$, it follows from the conservation of enrgy principle that

$$h\nu = W + \left(\frac{1}{2}\right)mu^2 \quad \text{or} \quad \left(\frac{1}{2}\right)mu^2 = h\nu - W$$

The equation shws that if KE is plotted against frequency of incident radiations, a straight line is obtained with a slope equal to Planck's constant. The equation expresses the fact that if a photon strikes a metal then it can release an electron from the metal provided the photon energy (*i.e.*, $h\nu$) is greater than the binding energy or work function (W) of the electron in the metal. Further the released electron will escape out with kinetic energy equalt to ($h\nu - W$).

Instead of iradiating a metal, one can irradiate atoms with photons of known frequency, the above equation may be written as: $h\nu$ = IE + KE.

This suggests that the photon energy is partly used to knock out an electron from the atom (*i.e.*, IE and the remainder shows up as the kinetic energy of the released photoelectron.

The potential applied on the surface to reduce the velocity of photoelectron to zero is known as stopping potential V_0, thus kinitic energy = eV_0

Thus
$$h\nu = W + (\text{stopping potential} \times \text{charge})$$
$$= W + eV_0$$

where e is electronic charge and V_0 is stopping potential.

Number of nodes

Total number of nodes in a shell = (n – 1)

Angular nodes = l and Spherical nodes = $n - l - 1$

Problems

Q. 1. Calculate the frequency and wave number of yellow light having wave length 580 nm.

Q. 2. Calculate the wave number and energy of the spectral line (emitted by an excited hydrogen atom) coresponding to $n_1 = 3$ and $n_2 = 5$ Rydberg constant $R = 10.97 \times 10^4\ cm^{-1}$

Q. 3. Calculate the radius of tenth orbit of Hydrogen atom.

Q. 4. An electron has a kinetic energy of 2.8×10^{-25}. Find its velocity.

Q. 5. Calculate the momentum of a particle having deBroglie wavelength of 10^{-17} m.

Q. 6. Calculate the uncertainty in velocity of a ball of mass 0.10 Kg when its uncertainity in position is given by 10^{-10m}

Q. 7. Calculate the velocity of the electron in the intermost orbit of hydrogen atom (The Bohr radius = 0.0529 nm).

Q. 8. A communication satelite weighing 1.0 Kg is travelling as the rate of 100 m sec^{-1}. If the uncertainty in is velocity is 0.1 m sec^{-1}, calculate the uncertainty in the location of its position.

Q. 9. The electron energy in hydrogen atom is given by $E = \dfrac{-21.7 \times 10^{-12}}{n^2}$ erg

(a) Calculate the energy required to remove an electron completely from the $n = 2$ orbit in Joules.

(b) What is the wave length of the light that can be used to cause this transition?

Q. 10. A metal of Atomic weight 54.94 has a density of 7.42 g/cm^3. Calcualte (i) the apparent volume occupied by one atom of the metal and (ii) radius of the atom assuming it to be a sphere.

Q. 11. A typical atomic nucleus is about 5×10^{-15} m in radius. Use the uncertainty principle to place a lower limit on the energy an electron must have if it is to be part of a nucleus.

Q. 12. A hydrogen atom is 5.3×10^{-11} m in radius. Use the uncertainty principle to estimate the minimum energy an electron can have in this atom.

Q. 13. Experiments indicate that 13.6 eV is required to separate a hydrogen atom into a proton and an electron; that is, its binding energy is $E = -13.6$ eV. Find the orbital radius and velocity of the electron in a hydrogen atom.

Q. 14. Find the de Broglie wavelengths of (a) a 46-golf ball with a velocity of 30 m/s, and (b) an electron with a velocity of 10^7 m/s.

Q. 15. Electromagnetic radiation with a wavelength of 700 nm is visible red light. (a) What is the frequency and wave number of this radiation? (b) What is the wavelength in angstroms?

Q. 16. Calculate the wavelength, frequency and wave number of the fifth line in the Balmer series . In what region of the specirum does this line occur?

Q. 17. What is the wavelength and frequency of the limiting line of the Lyman series in the spectrum of atomic hydrogen? In what region of the spectrum does this limiting line occur?

Q. 18. How much energy is required to move one electron from the ground sate of the H atom to the first excited state? How much energy is required to cause 1 mol of H atoms to undergo the transition from the ground state to the first excited state? What is the frequency of the radiation that must be absorbed to cause this transition to occur?

Q. 19. A beam of electrons is accelerated through a potential difference of 8000 V. Calculate the energy, the speed, the linear momentum, and the wavelength of these electrons.

Q. 20. What is the wavelength associated with a (142 g) baseball moving with a speed of 30.0 $m \cdot s^{-1}$?

Q. 21. **(a)** Calculate the energy of the photon emitted when the electron in a Be^{3+} ion drops from the $n = 2$ level to the ground state in electron volts and in joules. How much energy is emitted if a mole of Be^{3+} ions undergoes this trnasition? In what region of the spectrum will the radiation emitted during this transition be observed?

(b) Calculate the energy required to ionize the electron in a B^{3+} ion.

Q. 22. In the photoelectic effect, electrons are ejected from a surface when light strikes it. A certain minimum energy, E_{min} is required to eject an electron. Any energy absorbed because that minimum gives kinetic energy to the electron. It is found that when light at a wavelength of 540 nm falls on a cesium surface an electron in ejected with a kinetic energy of 2.60×10^{-20} J. When the wavelength is 400 nm, the kinetic energy is 1.54×10^{-19} J.

(a) Calculate E_{min} for cesium in joules.

(b) Calculate the longest wavelength, in nanometers, that will eject electrons from cesium.

Q. 23. The energy required to dissociate the Cl_2 molecule to Cl atoms is 239 kJ/mol Cl_2 . If the dissociation of a Cl_2 molecule were accomplished by the absorption of a single photon whose energy was exactly the quantity required, what would be its wavelength (in meters)?

Q. 24. A microwave oven heats by radiating food with microwave radiation, which is absorbed by the food and converted to heat. Suppose an oven's radiation wavelength is 12.5 cm. A container with 0.250 L of water was placed in the oven and the temperature of the water rose from 293 K to 373 K. How many photons of this microwave radiation were required? Assume that all the energy from the radiation was used to raise the temperature of the water.

Q. 25. Light with a wavelength of 425 nm fell on a potesium surface, and electrons were ejected as a speed of 4.88×10^5 m/s. What energy was expended in removing the electron from the metal? Express the answer in joules (per electron) and in kilojoules per mole (of electrons).

Q. 26. When an electron is accelerated by a voltage difference, the kinetic energy acquired by the electron equals the voltage times the charge on the electron. Thus one volt imparts a kinetic energy of 1.602×10^{-19} volt-coulombs, which equals 1.602×10^{-19} J. What is the wavelength associated with electrons accelerated by 4.00×10^3 volts?

Q. 27. The energy of any one-electron species in its nth state (n = principal quantum number) is given by $E = -BZ^2/n^2$, where-Z is the charge on the nucleus and B is 2.180×10^{-18} J. Find the ionization energy of the Li^{2+} ion in its first excited state in kilojoules per mole.

Q. 28. In 1885, Johann Balmer, a mathematician, derived the following relation for the wavelength of lines in the visible spectrum of hydrogen

$$\lambda = \frac{364.5\,\mathbf{n}^2}{(\mathbf{n}^2 - 4)}$$

whre λ is nanometers and n is an integer that can be 3, 4, 5, ... Show that this relation follows from the Bohr equation and the equation using the Rydberg constant. Note taht in the Balmer series, the electron is returning to the **n** = 2 level.

Q. 29. Suppose the rules for assigning quantum numbers were as follows

$$\mathbf{n} = 1, 2, 3, \ldots$$
$$\ell = 0, 1, 2, \ldots \mathbf{n}$$
$$\mathbf{m}_\ell = 0, 1, 2, \ldots \ell + 1$$
$$\mathbf{m}_s = \frac{1}{2} \text{ or } -\frac{1}{2}$$

Prepare a table based on these rules for **n** = 1 and **n** = 2. Give the electron configuration for an atom with eight electrons.

Q. 30. Suppose that the spin quantum number could have the values $\frac{1}{2}$, 0, and $-\frac{1}{2}$. Assuming that the rules governing the values of the other quantum numbers and the order of filling sublevels were unchanged,

(a) what would be the electron capacity of an *s* sublevel? A *p* sublevel? a *d* sublevel?
(b) how many electrons could fit in the **n** = 3 level?
(c) what would be the electron configuration of the element with atomic number 8? 17?

Q. 31. Consider the following transitions

1. **n** = 3 to **n** = 1 **2.** **n** = 2 to **n** = 3
3. **n** = 4 to **n** = 3 **4.** **n** = 3 to **n** = 5

(a) For which of the trnaitions is energy absorbed?
(b) For which of the transitions is energy emitted?
(c) Which transition involve the ground state?
(d) Which transition absorbs the most energy?
(e) Which transition emits the most energy?

Q. 32. State the relationship, if any, between the total number of orbitals and

(a) n **(b)** ℓ **(c)** m_s

Q. 33. Explain in your own words what is meant by

(a) the pauli exclusion principle.
(b) Hand's rule.
(c) a line in a atomic spectrum.
(d) the principal quanntum number.

Q. 34. Criticize or comment on the following statements:

(a) The energy of a photon is inversely proportional to its wavelength.

(b) The energy of the hydrogen electron is inversely proportional to the quantum number ℓ

(c) Electrons start to enter the fifth principal level as soon as the fourth is full.

Q. 35. Explain why

(a) negative ions are larger than corresponding atoms.

(b) scandium, a transition metal, forms an ion with a noble-gas structure.

(c) electronegativity decreases down a group in the periodic table.

Q. 36. When d is the diameter of the molecules of a gas in a confined space, the coliston frequency will be proportional to

(A) d (B) $\frac{1}{d}$ (C) d^2 (D) $\frac{1}{d^2}$

Q. 37. Gas is being pumped into a spherical balloon at the rate of 100 cubic metres per minute. When the radius r = 10 metres, at what rate in m^2/min is the surface, area changing

(A) 5π (B) 10 (C) 10π (D) 20

Q. 38. A base ball having mass 100g moves with velocity 100 m/sec. Find out the value of wave length of base ball.

Q. 39. The schrodinger wave equation for hydrogen atom is

$$\Psi_{2s} = \frac{1}{4(2\pi)^{1/2}}\left(\frac{1}{a_0}\right)^{3/2}\left(2-\frac{r}{a_0}\right)e^{-r/a}$$

Where a_0 is Bohr's radius. Let the radial node in 2s be at r_0. then find r in terms of a_0

Q. 40. Ther energy of the electron in the second and the third Bohr's orbits of the hydrogen atoms is -5.42×10^{-12} erg and -2.41×10^{-12} erg respectively. calculate the wavelength of the emitted radiation when the electron drops from the third to the second orbit.

Q. 41. Calculate the wave number for the shortest wave length transition in the balmer series of atomic hydrogen. (ii) According to Bohr's theory

Q. 42. An electron beam can undergo diffraction by crystals. Through what potential should a bean of electrons be accelerated so that its wavelength becomes equal to 1.54 A.

Q. 43. According to Bohr's theory, the electronic energy of hydrogen atom to the nth Bohr's orbit is given by

$$E_s = \frac{-21.76 \times 10^{-19}}{n^2} \text{ J.}$$

Calculate the longest wavelength of light that will be needed to remove an electron from the third Bohr orbit of the He^+ ion.

Q. 44. Find out the number of waves made by Bohr's electron in one complete revolution in its 3rd orbit.

Q. 45. What transition in the hydrogen spectrum would have the same wave length as the balmer transition, n = 4 to n = 2 of He^+ spectrum?

Q. 46. The electron energy in hydrogen atom is given by $E = (-21.7 \times 10^{-18})/n^2$ ergs. Calculate the energy in Joules required to $n = 2$ orbit. What is the longest wavelength (in cm) of light that can be used to cause this transition?

Q. 47. Calculate the wavelength in Angstroms of the photon that is emitted when an electron in the bohr orbit, $n = 2$ returns to the orbit, $n = 1$ in the hydrogen atom. The ionization potential of the ground state hydrogen atom is 2.17×10^{-11} erg per atom.

Q. 48. Give reason why the ground state outermost electronic configuration of silicon is—

	3s	3p
	[↑↓]	[↑][↑][ ]
and not	[↑↓]	[↑][↓][ ]

Q. 49. Wavelength of high energy transition of H- atoms is 91.2 nm. Calculate the corresponding wavelength of He atoms.

Q. 50. Estimate the difference in energy between 1st and 2nd Bohr orbit of a hydrogen atom. At what minimum atomic number, a transition from $n = 2$ to $n = 1$ energy level would result in the emission of X-rays with $\lambda = 3.0 \times 10^{-8}$ m? Which hydrogen atom-like species does this atomic number correspond to?

Q. 51. **(a)** What is the maximum number electrons that may be present in all the atomic orbitals with principal quantum number 3 and azimuthal quantum number 2?

(b) A compound of vanadium has a magnetic moment of 1.73 BM. Work out the electronic configuration of the vanadium ion in the compound.

Q. 52. Consider the hydrogen atom to be proton embedded in a cavity of radius a_0 (Bohr radius). whose structure is neutralised by the addition of an electron to the cavity in vaccum, infinitely slowly. Estimate the average total energy of an electron in its ground state in a hydrogen atom as the work done in the above neutralisation proces. Also, if the magnitude of the average kinetic energy is half the magnitude of the average potential energy, find the average potential energy.

Q. 53. Iodine molecule dissociates into atoms after absorbing light of 4500. Å. If one quantum of radiation is absorbed by each molecule, calculate the kinetic energy of iodine atoms.

(Bond energy of I_2 = 240 KJ mol^{-1})

Q. 54. The dipole moment of KCl is 3.336×10^{-29} coulomb metre which indicate that it is highly polar molecule. the inter atomic distance between K^+ and Cl^- in thi molecule is 2.6×10^{-10} m. Calculate the dipole moment of KCl molecule if there were opposite charges of one fundamental unit located at each nucleus. Calculate the percentage ionic character of KCl.

Q. 55. How may sigma bonds and how many pi bonds are present in a benzene molecule?

Q. 56. Arrange the following compounds in the increasing order of bond length of O—O bond

$$O_2, O_2[AsF_6], KO_2$$

Explain on the basis of ground state electronic configuration of dioxygen in these molecultes.

Q. 57. Using the VSEPR theory identify the type of hybridization and draw the structure of OF_2. What are the oxidation states of O and F?

Q. 58. Interpret the non-linear shape of H_2S molecule and non-planar shape of PCl_3 using valence shell electron pair repulsion (VSEPR) theory.

(Atomic numbers: h = 1; P = 15; S = 16; Cl = 17)

Q. 59. Using VSEPR Theory draw the shape of PCl_5 and BrF_5.

Q. 60. Write the Lewis dot structure formula for each of the following. Give, also, the formula of a neutral molecule, which has the same geometry and the same arrangement of the bonding electrons as in each of the following. An example is given below in the case of H_3O^+:

$$\left[\begin{array}{c} \text{H} \\ \text{H} : \ddot{\text{O}} : \text{H} \end{array}\right] \qquad \left[\begin{array}{c} \text{H} \\ \text{H} : \ddot{\text{N}} : \text{H} \end{array}\right]$$

Lewis dot structure Neutral molecule

(i) O_2^{2-} (ii) CO_3^{2-} (iii) CN^- (iv) NCS

Chapter **2**

CHEMICAL BONDING AND MOLECULAR STRUCTURE

The forces that hold togather atoms in a molecule are called chemical forces and the association between atoms in a molecule by these forces is called a Chemical Bond

Three major types of bonds formed are: Ionic, Covalent and Co-ordinate bonds which is due to attaining a stable configuration by atoms in three different ways (a) by losing electrons (b) by sharing electrons (c) by gaining electrons.

(A) Ionic Bond

This is due to complete transfer of one or more electrons from one atom to another., thus making +ve or –ve ions. In other words ionic bonds are formed when electro positive elements react with electro negative elements e.g. is NaCl

No atom has electronic configuration $1s^2\ 2s^2\ 2p^6\ 3s^1$ while

Chlorine atom has electronic configuration $1s^2\ 2s^2\ 2p^6\ 3s^2\ 3p^5$

In NaCl, the electron $3s^1$ of Na, is transfered to the $3p^5$ so that the Na^+ & Cl^- (ions) are produced respectively. Electrostatic attraction of Na & Cl ions (+ve & –ve) hold these ions togather in the crystal lattice.

Note: The number of electrons that an atom can give up or gain, when it forms an ion is called its ionic valence. It is equal to the net change on the ion.

Conditions for Ionic Bonding

(a) One of the atoms must have a low ionisation energy (i.e. metals of group IA, II A & transition elements)

(b) The other atom should have a high electron affinity (so that it can take up the additional electron easily).

Compounds formed by ionic bond are mostly solids. The +ve & –ve ions (in such compounds) are held together in the form of a geometic array called Lattice. Energy released on formation of such compounds is called lattice energy; the higher the lattice energy the more stable is ionic bond. Thus the cation (compounds have high M.P. & B.P.) should be smaller & anion should be larger and if magnitude of charge is more on cation or anion, the lattice energy will be more.

Lewis Symbols

The outer electrons of elements are shown as dots. Thus a Lewis structure is a combination of Lewis symbols in a chemical bond.

The Convalent Bond

It can be achieved in two ways.

The Lewis Theory (The octet rule)

(a) By sharing of electrons between atoms of same kind, formation of, H_2, O_2, Cl_2 etc.
(b) By sharing of electrons between atoms of different kind, formation of HCl, CO_2, CH_4 etc

Sharing of a single pair of electrons between bonded atoms produce a single covalent bond (–) A pair of electrons which make a covalent bond is termed *Bond pair* and the pair which do not make covalent bond are called lone pairs. e.g.

$$:\ddot{Cl} + \ddot{Cl}: \rightarrow :\ddot{Cl}:\ddot{Cl}: \text{ OR } :\ddot{Cl}—\ddot{Cl}:$$

lone pairs

Bond pair

In oxygen molecule, the (O atom) has six electrons in valence shell is so requires 2 more to achieve stability, it shares two electrons with two more electrons of the other O atom, thus forming a *Double Covalent bond* (=)
In nitrogen; (which has five electrons), shares 3 electrons with (N) and forms a *triple covalent bond* (≡)

Exceptions to the octet rule

The octet rule is broken in a significant number of cases:

1. For example, for atoms such as Be and B which have less than four outer electrons. Even if all the outer electrons are used to form bonds an octet cannot be attained.

$$\dot{Be} + 2\left[\cdot\ddot{F}:\right] \longrightarrow :\ddot{F}:\ Be\ :\ddot{F}:$$

$$\cdot\dot{B}\cdot + 3\left[\cdot\ddot{F}:\right] \longrightarrow :\ddot{F}:\ \underset{}{\overset{:\ddot{F}:}{Be}}\ :\ddot{F}:$$

2. The octet rule is also broken where atoms have an extra energy level which is close in energy to the *p* level and may accept electrons and be used for bonding. PF_3 obeys the octet rule, but PF_5 does not. PF_5 has ten outer electrons and uses one 3*s*, three 3*p* and one 3*d* orbitals.
3. The octet rule does not work in molecules which have an odd number of electrons such as NO and ClO_2, nor does it explain why O_2 is paramagnetic and has two unpaired electrons.

Polar Covalent Bond

Covalent bond when formed between two alike atoms, bonded pair of electrons is equally shared. But when it is formed between two atoms of different kinds bonded pair is not equally shared. An atom which has more tendency to attract the shared pair of electrons towards it self (high electronegativity) will pull the bond pair towards itself. Consider a single bond between A –B. Let B be more electronegative than A, so that bond pair is shifted towards B. Thus B acquires a partial –ve charge ($-\delta$), and hence A will acquire a ($+\delta$) charge. Such a state where +ve & –ve charges are seperated by a certain distance is known as polarised state and process is called polarisation, hence the covalent bond is known as polar covalent bond. (Here, $A^{\delta+} - B^{\delta-}$ are called dipoles)

Now the molecule with +ve & –ve charge centres in equilibrium is called Dipole and is chacterised by possessing a quantity **Dipole MOMENT** μ.

Covalent molecules having bond dipoles are of two types:

(a) Polar molecules (b) non-polar molecules.

A molecule is polar if dipole moment is not equal to zero & (eq. NH_3, H_2O, HF etc. the molecule is nonpolar if dipolemoment is zero. eg H_2, Cl_2, ...

This also depends upon the structure or shape of the molecule.

Definition : *Dipole moment* μ is a measure of the extent to which a sepration of charges exists within a molecule. It is the product of the magnitude of the charge and the distance seperating the charge centres. The unit of μ is debye, = 3.34×10^{-30} c.m. (coloumbs.meter)

MOLECULAR SHAPES

I Sidgwick-Powell Theory

According to the theory in the molecules and ions that only contain single bonds, the approximate shape can be predicted from the number of electron pairs in the outer or valence shell of the central atom. The outer shell contains one or more bond pairs of electrons, but it may also contain unshared pairs of electrons (lone pairs). Because call electron pairs repel each other, so repulsion is minimized if the electron pairs are oriented in space as far apart as possible.

Molecular Shapes Predicted by Sidgwick-Powell Theory

Number of electron pairs in outer shell	*Shape of molecule*		*Bond angles*
2	linear		180°
3	plane triangle	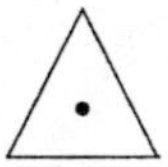	120°
4	tetrahedron	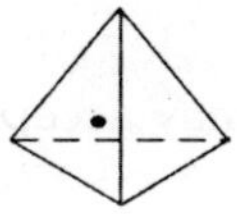	109°28′
5	trigonal bipyramid	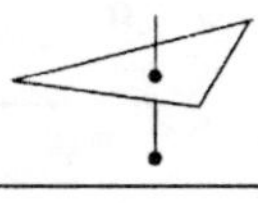	120° and 90°

(Contd.)

(Contd.)

Number of electron pairs in outer shell	Shape of molecule		Bond angles
6	octahedrom		90°
7	pentagonal bipyramid		72° and 90°

II Valence-Shell Electron-Pair Repulsion (VSEPR) Theory

This theory is summarized as:

1. The shape of the molecule is determined by repulsions between all of the electron pairs present in the valence shell (Some as I above)
2. A lone pair of electrons takes up more space round the central atom than a bond pair, since the lone pair is attracted to one nucleus whilst the bond pair is shared by two nuclei. It follows that repulsion between two lone pairs is greater than repulsion between a lone pair and a bond pair, which in turn is greater than the repulsion between two bond pairs. Thus the presence of lone pairs on the central atom causes slight distortion of the bond angles from the ideal shape. If the angle between a lone pair, the central atom and a bond pair is increased it follows that the actual bond angles between the atoms must be decreased. The force of repulsion between two electron pairs is much stronger at 90° than at 120° or 180°. The repulsion of one Ione pair of electron per another lone pair is greater than between two bond pairs. Thus, order of repulsive forces (Stronger to weakers is: Ione piar – lone pair > lone pair – bond pair > bond pair – bond pair.
3. The magnitude of repulsions between bonding pairs of electrons depends on the electronegativity difference between the central atom and the other atoms.
4. Double bonds cause more repulsion than single bonds, and triple bonds cause more repulsion than a double bond.

Thus, to have a complete description of the shape of a molecule we need to know two quantities.

- **bond lengths,** the distances between the nuclei of bonded atoms. In Figure 10-15 the bond lengths are d_1 and d_2.

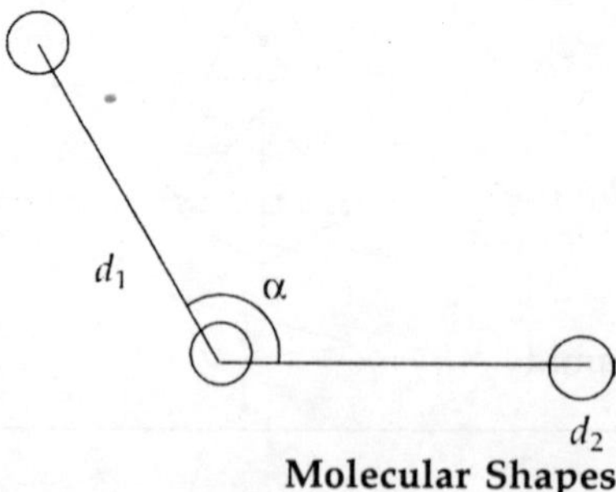

Molecular Shapes

- **bond angles,** the angle between adjacent lines representing chemical bonds. The triatomic molecule in Figure has just one bond angle, α.

MOLECULAR SHAPES

To establish the shape of a molecule we must determine the distances between the centers of the bonded atoms (d_1 and d_2) and the angle between the adjacent bonds α (see figure).

Molecular Geometry as a Function of Geometrial Distribution of Valence-Shell Electron Pairs and Number of Lone Pairs. The geometrical shapes of CH_4, NH_3, NH_3 and H_2O are shown in Figure below, together with space-filling molecular models. A special notation is used in the figure that has this meaning. A represents the central atoms, X presents a terminal atom bonded to the central atom, and E represents a lone pair of electrons. Thus, the symbolism AX_2E_2 signifies that *two* atoms (X) are bonded to the central atom (A). The central atom also has *two lone pairs of electrons*. H_2O is an example of a molecule of the AX_2E_2 type.

The methane molecule, CH_4, in which the central C atom has acquired the Ne electron configuration by forming covalent bonds with four H atoms.

$$\begin{array}{c} H \\ H:\ddot{\underset{..}{C}}:H \\ H \end{array}$$

The method prodicts that CH_4 should be a *tetrahedral* molecule, with a C atom at the center of the tetrahedron and H atoms at the corners.

In NH_3, the central atom is surrounded by four pairs of electrons $\begin{array}{c} H \\ H:\ddot{\underset{..}{N}}:H \end{array}$

∴ In the NH_3 molecule only *three* of the electron pairs are bond pairs; the fourth pair is a *lone pair*. When the nucleus of the N atom is joined to those of the H atoms, the geometrical figure obtained is a *pyramid* (called a trigonal pyramid), not a tetrahedron. The pyramid has H atoms at the base and an N atom at the apex.

In H_2O the central atom is also surrounded by four pairs of electrons $H:\ddot{\underset{..}{O}}:H$

In the H_2O molecule two of the four electron pairs about the O atom are *bonding pairs and two are* lone pairs. The figure obtained by joining the nuclei of the two H atoms with that of the O atom is a *Planar* figure. The molecule is *angular* or *bent*.

Number of electron pairs	*Geometrical distribution of electron pairs*	*Number of lone pairs*	*VSEPR notation*	*Molecular geometry*	*Ideal bond angles*	*Example*
2	Linear	0	AX_2	X—A—X (linear)	180°	$BeCl_2$
3	trigonal planar	0	AX_3	X—A(—X)—X (trigonal planar)	120°	BF_3
	trigonal planar	1	AX_2E	X—A(:)—X (angular)	120°	SO_2

(*Contd.*)

(Contd.)

Number of electron pairs	Geometrical distribution of electron pairs	Number of lone pairs	VSEPR notation	Molecular geometry	Ideal bond angles	Example
4	tetrahedral	0	AX_4	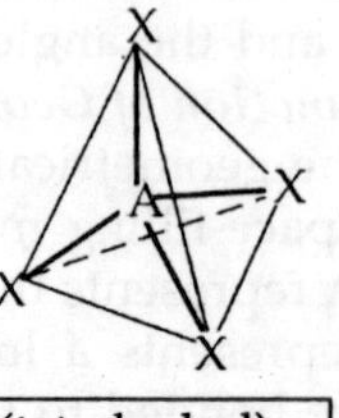 (tetrahedral)	109.5°	CH_4
	tetrahedral	1	AX_3E	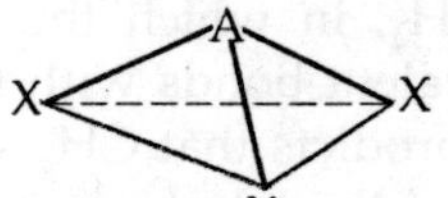 (trigonal pyramidal)	109.5°	NH_3
	tetrahedral	2	AX_2E_2	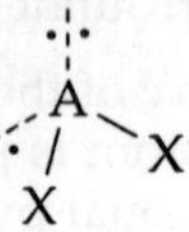 (angular)	109.5°	(H_2O) OH_2
5	trigonal bipyramidal	0	AX_5	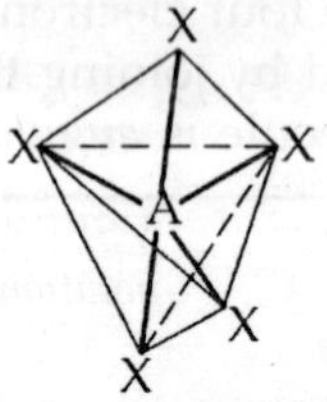 (trigonal bipyramidal)	90°, 120°	PCl_5
	trigonal bipyramidal	1	AX_4E	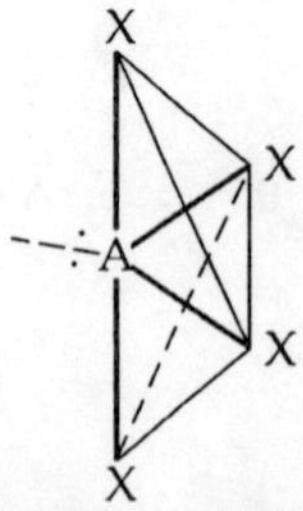 (irregular tetrahedral)	90°, 120°	SF_4

(Contd.)

(Contd.)

Number of electron pairs	*Geometrical distribution of electron pairs*	*Number of lone pairs*	*VSEPR notation*	*Molecular geometry*	*Ideal bond angles*	*Example*
	trigonal bipyramidal	2	AX_3E_2	(T-shaped)	90	ClF_3
	trigonal bipyramidal	3	AX_2E_3	(linear)	180°	XeF_2
6	octahedral	0	AX_6	(octahedral)	90°	SF_6
	octahefral	1	AX_5E	(square pyramidal)	90°	BrF_5
	octahedral	2	AX_4E_2	(square planar)	90°	XeF_4

Applying VSEPR Theory

In applying VSEPR theory let us use this four-step procedure for assesing molecular shape

1. Draw a plausible Lewis structure of the species (molecule or polyatomic ion).
2. Determine the number of electron pairs around the central atom and identify each pair as being either a *bond* pair or a *lone* pair.
3. Establish the geometrical orientation of the electron pairs around the central atom i.e. linear, trigonal planar, tetrahedral, trigonal bi pyramidal, or octahedral.
4. Use the position around the central atom that are occupied by other atoms (notlonepair electrons) to specify the geometrical shape.

HYBRIDIZATION OF ATOMIC ORBITALS

Hybridization and the mixing of orbitals is a most useful concept. Mixing of *s* and *p* orbitals is well accepted, but the involvement of *d* orbitals is controversial. For effective mixing, the energy of the orbitals must be nearly the same, since the 3*d* orbitals is much larger it is much higher in energy than the 3*s* and 3*p* orbitals. It seems unlikely that hybridization involving *s*, *p* and *d* orbitals could possibly occur.

Several factors effect the size of orbitals. The most important is the charge on the atom. If the atom carries a formal positive charge then all the electrons will be pulled in towards the nucleus. The effect is greatest for the outer electrons. If the central P atom is bonded to a highly electronegative element such as F, O or Cl, then the electronegative element attracts more than its share of the bonding electrons and the F or Cl atom attains a $\delta-$ charge. This leaves a $\delta+$ charge on P, which makes the orbitals contract,. Since the 3*d* orbital contracts in size very much more thatn the 3*s* and 3*p* orbitals, the energies of the 3*s* , 3*p* and 3*d* orbitals may become close enough to allow hybrdization to occur in PCl_5. Hydrogen does not cause this large contraction, so PH_5 does not exist.

III Molecular Shapes by Hybridization

Number of orbitals and type of hybridization and shape of molecules

Number of outer orbitals	*Type of hybridization*	*Bond Angle*	*Distribution in space*	*Examples*
2	*sp*	180°	Linear	$BeCl_2$, $ZnCl_2$
3	sp^2	120°	Plane triangle	BF_3, NO_3^-
4	sp^3	109.5	Tetrahedron	NH_3, CH_4, H_2O
5	sp^3d	120; 90	Trigonal bipyramid	PCl_5, PF_5
6	sp^3d^2	90	Octahedred	SF_6
7	sp^3d^3		Pentagonal bipyramid	
8	dsp^2		Square planar	

Approximate strengths of bonds formed by various orbitals

Orbital	*Relative bond strength*
s	1.0
p	1.73
sp	1.93
sp^2	1.99
sp^3	2.00

sp Hybridization

It shoud in principle be possible to calculate the relative strength of bonds formed using *s* , *p* or various hybrid orbitals.

Because of their shape these *sp* orbitals overlap more effectively and result in stronger bonds than the original atomic orbitals. Repulsion is minimized if these two hybrid orbitals are oriented at 180° to each other. If these orbitals overlap with *p* orbitals of F atoms, a linear BeF_2 molecule is obtained.

The structure of a gaseous molecule of beryllium fluoride BeF_2 is linear F—Be—F. Be is the central atom in this molecule and determines the shape of the molecule formed. The ground state electronic configuration of Be is $1s^2 2s^2$. This has no unpaired electrons and so can form no bonds. If energy is supplied, an exited state will be formed by unpairing and promoting a 2s electron to an empty 2*p* level, giving is $2s^1\ 2P_x^1$.

There are now two unpaired electrons (i.e. two Half filled orbitals) so the atom can form the required two bonds which are oriented at angle of 180°.

	1*s*	2*s*	2*p*
Benyllium atom – ground state	↑↓	↑↓	[ ][ ][ ]
Beryllium atom – excited state	↑↓	↑	[↓][ ][ ]
BeF_2 molecule having gained a share in two electrons by bonding to two F atoms	↑↓	↑↓	[↑↓][ ][ ]

(*sp* hybridization of the two orbitals in the outher shell, hence structure is linear)

Hybridizing the 2*s* and $2p_x$ atomic orbitals gives two equivalent *sp* hybrid orbitals.

These *s*-p hybridized orbital of Be overlap with half filled orbitals of two fluorine atoms to give a linear shape.

*sp*² Hybridization

Consider BF_3 molecule [sp^2 hybridized orbital of B overlaps with three *p*-orbitals of fluorine.)

The B atom is the central atom in the molecule, and it must be excited to give three unpaired electrons so that it can form three covalent bonds. With three fluorine atoms, one 2*s* an 2*p* orbitals of B atom undergo sp^2 hybridization to form three half-filled sp^2-hybrid orbitals. these sp^2 hybridized orbitals are oriented at an angle of 120°, three σ-bonds are formed. The shape of BF_3 molecule is thus *Trigonal Planar,*

	1s	2*s*	2*p*
Boron atom – ground state	↑↓	↑↓	[↑][ ][ ]
BF_3 molecule having gained a share in three electrons by bonding to three F atoms	↑↓	↑↓	[↑↓][↑↓][ ]

(sp^2 hybridization of the three orbitals in outer shell; hence structure is a planar triangle).

sp³ Hybridization

	1s	2s	2p
Now consider CH_4. Here C in ground state is	↑↓	↑↓	↑ ↑ _
$1s^2\ 2s^2,\ 2p_x^1\ 2p_y^1$ and in Excited state	↑↓	↑	↑ ↑ ↑

it is $1s^2\ 2s^1\ 2p_x^1\ 2p_y^1\ 2p_z^1$.

Thus 1s and 3*p* orbitals in this way gives four sp^3 hybrid orbitals. These four sp^3 hybridized orbitals of carbon over laps with four 1s orbitals of hydrogen giving a tetrahedral shape of CH_4 molecule. Sp^3 hybrid orbitals from stronger bonds than the original atomic orbitals. These orbitals are oriented at an angle of 109° 28′. These sp^3 hybridized orbitals are directed along the four corners of a regular tetrahedron

sp³d Hybridization

Consider PCl_5 molecule: The phosphorus atom has electronic configuration in ground state as $1s^2\ 2s^2\ 2p^6\ 3s^2\ 3p_x^1\ 3p_y^1\ 3p_z^1$ and in excited $1s^2\ 2s^2\ 2p^6\ 3s^1\ 3p_x^1\ 3p_y^1\ 3p_z^1\ 3d_z^1$

So, the bonding in PCl_5 may be described using hybrids of the 3*s*, 3*p* and 3*d* atomic orbitals for P (see below). The five half-filled orbitals undergo hybridization to give five sp^3d hybrid orbitals. These five sp^3d hybridized orbitals then overlap with half-filled orbitals of five chlorine atoms to form five P—Cl sigma bonds

The five sp^3d hybridized orbitals point towards the corners of a *trigonal bipyramid*. Three chlorine atoms lie at an angle of 120° and other two chlorine atoms at an angle 90°

		3s	3p	3d
Electronic structure of phosphorus atom —ground state	full inner shell	↑↓	↑ ↑ ↑	_ _ _ _ _
Phosphorus atom – excited state		↑	↑ ↑ ↑	↑ _ _ _ _
Phosphorus having gained five electrons from chlorine atoms in PCl_5 molecule		↑↓	↑↓ ↑↓ ↑↓	↑↓ ↑↓ _ _ _

(sp^3d hybridization, trigonal bipyramid)

sp³d² Hybridization

In a similar way the structure of SF_6 can be described by mixing the 3*s* three 3*p* and two 3*d* orbitals, that is sp^3d^2 hybridization.

		3s	3p	3d
Electronic structure of sulphur atom-ground state $1s^2\ 2s^2\ 2p^6\ 3s^2\ 3p_x^2\ 3p_y^1\ 3p_z^1$	full inner shell	↑↓	↑↓ ↑ ↑	_ _ _ _ _

Electronic structure of sulphur atom-excited state

$1s^2\, 2s^2\, 2p^6\, 3s^1\, 3p_x^1\, 3p_y^1\;\; 3d_{x^2-y^2}\,.3d_{z^2}^1$

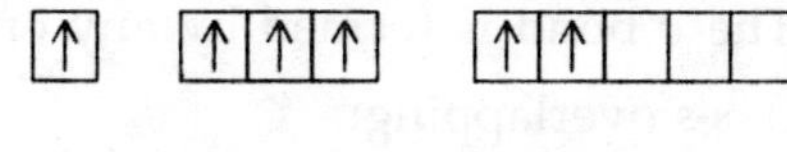

Sulphur atom having gained six electrons from fluorine atoms in SF_6 molecule

(sp^3d^2 hybridization, octahedral structure)

These six half-filled orbitals *viz.* one 3*s*, three 3*p* and two 3*d*-orbitals undergo sp^3d^2 hybridization to generate six hybrid orbitals, which then overlap with $3p_z$ orbitals of six F-atoms to form SF_6 molecule. The six sp^3d^2 hybridized orbitals point towards the six corners of a *regular octahedron*. The bond angles are all equal to 90°.

PREDICTION OF HYBRIDIZATION (IN A MOLECULE WHICH CONTAINS HYDROGEN ATOM)

$$\text{The number of hybridization} = \frac{\text{Total number of valence electrons}}{2}$$

Example	*Total no. of Valance electrons*	*No. of hybridization hybridization*	*Hybridization*
NH_3	$5 + 3 \times 1 = 8$	$\frac{8}{2} = 4$	sp^3
H_2O	$2 \times 1 + 6 = 8$	$\frac{8}{2} = 4$	sp^4
CH_4	$4 + 4 \times 1 = 8$	$\frac{8}{2} = 4$	sp^3

TYPES OF OVERLAPPING AND NATURE OF COVALENT BONDS

Sigma and Pi Bonds

(I) In bonds formed from end to end overlap of orbitals are called (sigma) σ bonds. They have a cylindrical symmetry about the bond axis.

In σ bonds the electron density is concentrated in between the two atoms, and on a line joining the two atoms.

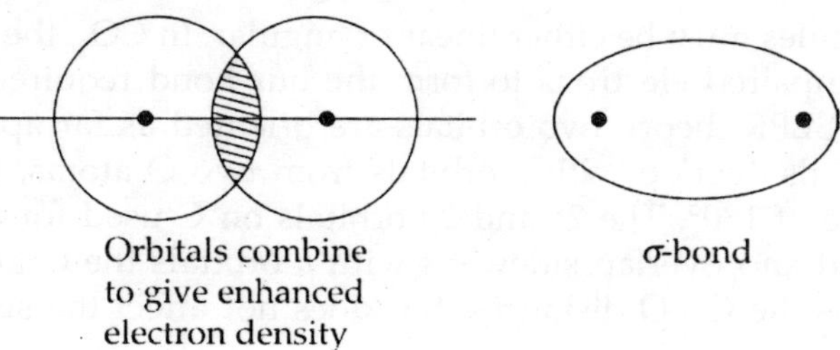

The σ bond is formed by any one of the following types of combination of atomic orbitals:

(i) *s-s* overlapping:

(ii) *s-p* overlapping:

(iii) *p-p* overlapping:

(II) The bond produced by the side by side overlap of half filled *p–orbitals* is called a pi bond (π bond)

The atomic orbitals overlap in such a way that their axes remain parallel to each other and perpendicular to the internuclear axis.

The orbitals obtained as a result of sidewise overlap consists of two saucer type charged clouds above and below the plane of the participating atoms.

(a) + → or

Double or triple bonds occur by the sideways overlap of orbitals giving pi π bonds. In π bonds the electron density also concentrates between the atoms, but on either side of the line joining the atoms. The shape of the molecule is determined by the σ bonds (and lone pairs) but not by the π bonds. Pi bonds merelv shorten the bond lengths.

The π bonds do not alter the shape, but merely shorten the bond lengths. The strength of the π-bond is significantly less than that of the σ-bond as the overlapping in π bonds occurs to a smaller extent.

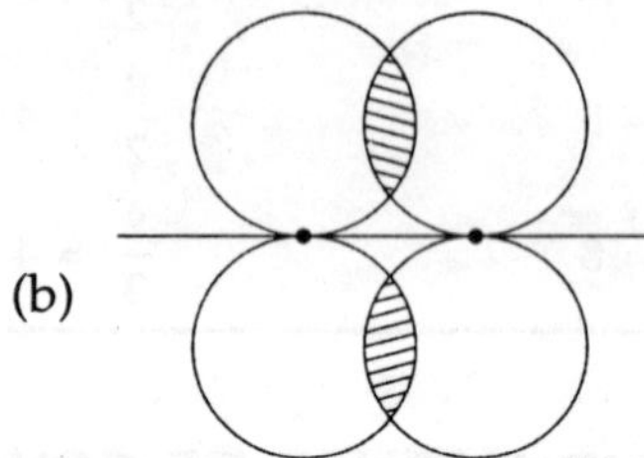

Sigma and pi overlap: *(a) sigma overlap (lobes point along the nuclei)*
(b) pi overlap (lobes are at right angles to the line joining the nuclei)

Examples

(a) Consider the structure of the carbon dioxide molecule. Since C is typically four-valent and O is typically two-valent, the bonding can be simply represented

$$O{=}C{=}O$$

Triatomic molecules must be either linear or angular. In CO_2, the C atom must be exacted to provide four unpaired electrons to form the our bond required.

According to VSEPR theory two orbitals are oriented as far apart as possible

These two orbitals overlap with *p* orbitals from two O atoms, forming linear molecule with a bond angle of 180°. The 2*s* and 2*p* orbitals on C used for a π bonding are at right angles to the bond and overlap sideways with *p* orbitals the O atoms at either side. This π overlap shortens the C—O distances, but does not affect the shape.

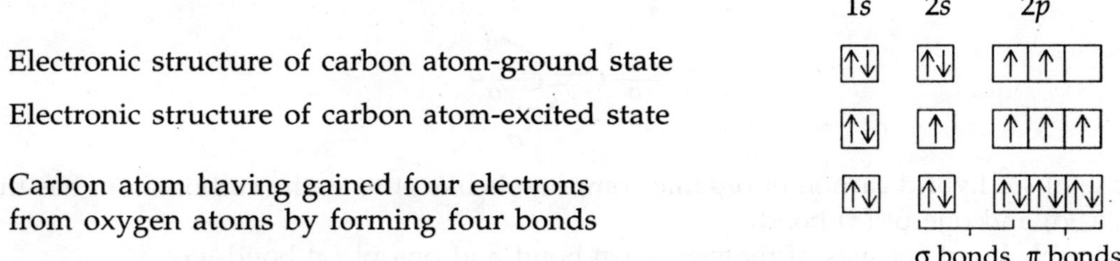

There are two σ bond and two π bonds in the molecule.

(b) Consider sulphur dioxide molecule SO_2. It shows oxidation states of (+II), (+IV) and (+VI), whilst O is two-valent.

The structure may be represented as: O═S═O

Triatomic molecules are either linear or bent. The S atom must be excited to provide four unpaired electrons.

The two electron pairs which form the π bonds do not affect the shape of the molecule. The remaining three orbitals point to the corners of a triangle, and result in a planar triangular structure for the molecule with two corners occupied by O atoms and one corner occupied by a lone pair. The SO_2 molecule is this angular or V shaped

Electronic structure of sulphur atom-ground state

Electronic structure of sulphur atom-excited state

Sulphur atom having gained four electrons Iron four bonds to oxygen atoms in SO_2 molecule

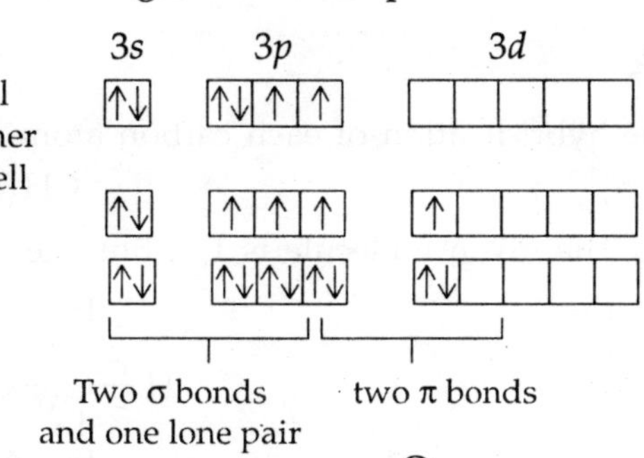

(c) Similarly for SO_3 molecule. The structure may be represented as O=S(=O)=O

The central S atom must be excited to provide six unpaired electrons to form six bonds.

The three π bonds are ignored in determining the shape of the molecule. The three σ orbitals are directed towards the corners of an equilateral triangle, and the SO_3 molecule is a completely regular plane triangle. The π bonds shorten the bond lengths, but do not affect the shape.

Electronic structure of sulphur atom-excited state

Sulphur atom having gained six electrons from six bonds to oxygen along in SO_3 molecule

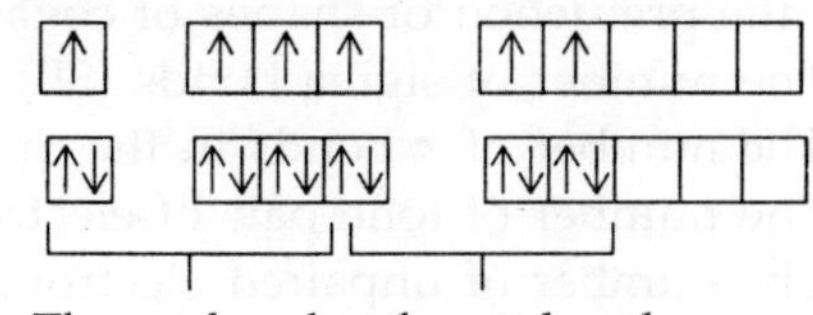

Now,

(i) In case of *sp* hybrid carbon of organic compounds like acetylene (C_2H_2), the carbon atom has two sigma (σ) and two pi (π) bonds. The triple bond consists of one sigma and two pi bonds i.e.,

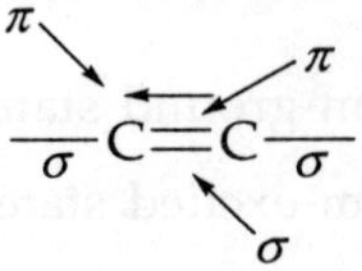

π ↘ ↙ π
—C=C—
σ σ
σ

(ii) In case of sp^2 hybrid carbon of organic compounds like ethene, the carbon atom has three sigm a (σ) and one pi (π) bond.
The double bond consists of the sigma (σ) bond and one pi (π) bond, i.e.,

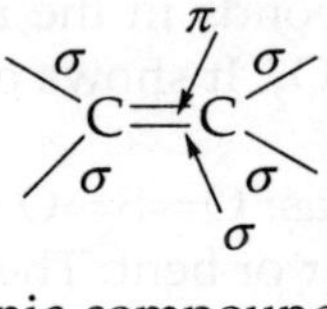

π
σ σ
C=C
σ σ
σ

(iii) In case of sp^3 hybrid carbon of organic compounds like methane, the carbon has only four sigma (σ) bonds, i.e.,

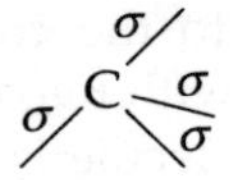

σ
C σ
σ σ

Example

Write the hybridization of each carbon atom of the following molecule

$$CH_3—CH=C=CH_2$$

Solution The given molecule is 1, 2-Butadiene

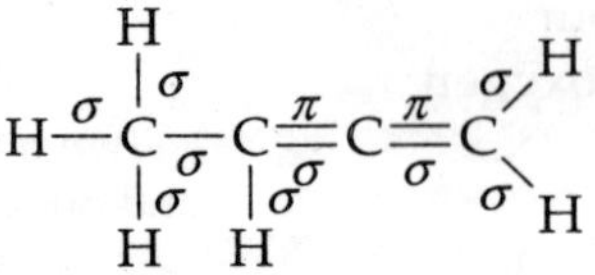

H
| σ
H—σ—C—σ—C=π σ=C=π σ=C(σ H)(σ H)
| σ | σ
H H

4σ	3σ	2σ	3σ
0π	1π	2π	1π
sp^3	sp^2	sp	sp^2

PREDICTION OF SHAPES OF COVALENT MOLECULES

For the prediction of shapes of covalent molecules, we need to determine:

(a) The number of sigma bonds (SB) made by the central atom,
(b) The number of π bonds in the molecule,
(c) The number of lone pair of electrons (LP) on central atom,
(d) The number of unpaired electrons (Ue)
(e) Hybridisation and
(f) Approximate bond angles.

Calculation of the Number of p-electrons in the Molecule

(a) *For molecules having hydrogen atoms like* H_2O, C_2H_2, C_2H_4, NH_3 etc.
The number of π electrons = 6A – 6H + 2 – VE . . . . (1)
where A = Total number of atoms in the molecule,
H = Number of hydrogen atoms of the molecule,
VE = Sum of valence electrons of the molecule.

Examples

S. No.	Molecule	Number of π electrons (πe) = 6A – 6H + 2 – VE	Number of π - bonds = $\frac{\pi e}{2}$
1.	H_2O	(6 × 3) – (6 × 2) + 2 – 8 = 20 – 20 = 0	0
2.	C_2H_2	(6 × 4) – (6 × 2) + 2 – (2 × 4 + 2 × 1) = 26 – 22 = 4	2
3.	C_2H_4	(6 × 6) – (6 × 4) + 2 – (2 × 4 + 4 × 1) = 38 – 36 = 2	1
4.	NH_3	(6 × 4) – (6 × 3) + 2 – (1 × 5 + 3 × 1) = 26 – 26 = 0	0

(b) *For molecules that do not contain hydrogen atoms, like* F_2O, CO_2, C_3^{2-} etc.
The number of π electrons = 6 A + 2 – VE . . . . (2)
where A = Total number of atoms is the molecule,
VE = Sum of valence electrons of the molecule.

Examples

S. No.	Molecule	Number of π electrons (πe) = 6 A + 2 – VE	Number of π bonds = $\pi e/2$
5.	F_2O	(6 × 3) + 2 – (2 × 7 + 6) = 20 – 20 = 0	0
6.	CO_2	(6 × 3) + 2 – (1 × 4 + 2 × 6) = 20 – 16 = 4	2
7.	CO_3^{2-}	(6 × 4) + 2 – (1 × 4 + 3 × 6 + 2) = 26 – 24 = 2	1

Calculation of SB, LP, Ue, Hybridization and Approximate Bond Angle

Let VE = Sum of valence electrons of the molecule,
SB = Number of sigma bonds made by central atom,
LP = Number of Ione pair of electrons on central atom,
Ue = Unpaired electrons.

Then **Hybridisation = SB + LP + Ue** (3)

and, **Approximate bond angle = standard bond angle for Hybridization – 2.5 × Lp** (4)

SB + LP + Ue	2	3	4
Hybridization	*sp*	sp^2	sp^3
Standard bond angle	180°	120°	109.5°

From VE we can calculate SB, LP and Ue by the following method
Let VE = 21

Now proceed as follows: Always Divide VE by 8, then divide the remainder by the quotient

$$\begin{array}{r} 2\,(SB) \\ 8\overline{)\,21} \\ 16 \\ 2\overline{)\,5}\;(2\,(LP) \\ 4 \\ \hline 1\,(Ue) \\ \hline \end{array}$$

Thus, SB = 2, LP = 2 Ue = 1

Example Consider F_2O molecule, Here we have

$$VE = (2 \times 7 + 6) = 20$$

So,

$$\begin{array}{r} 2 \\ 8\overline{)\,20} \\ 16 \\ 2\overline{)\,4}\;(2 \\ 4 \\ \hline 0 \\ \hline \end{array}$$

∴ SB = 2, LP = 2, Ue = 0

Thus Hybridization = SB + LP + Ue = 2 + 2 + 0 = 4, i.e., sp^3; and approximate bond angle = 109.5° – 2.5 × 2 = 104.5° But π in F_2O = 0 (See calculations of π bonds)

Thus, the structure of F_2O is $\ddot{O}$ (F 104.5° F)

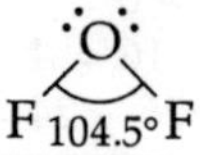

Example :
Let us consider CO_2 molecule,
Here we have,

$$VE = 4 + 2 \times 6 = 16.$$

So,

$$\begin{array}{r} 2 \\ 8\overline{)\,16} \\ 16 \\ \hline 0 \\ \hline \end{array}$$

∴ SB = 2, LP = 0, Ue = 0

Thus, Hybridisation = SB + LP + Ue = 2 + 0 + 0 = 2, *i.e.*, *sp* and Bond angle = 180° – 2.5 × 0 = 180°
But number of π bonds in the CO_2 molecule are two (se calculation of π bonds)
Thus, the structure of CO_2 molecule is.

O=C=O
180°

Example :
Now, consider C_3^{2-} ion,

Here we have,

$$VE = 4 + 3 \times 6 + 2 = 24.$$

So,

$$8\overline{)24}\;3$$
$$\underline{24}$$
$$0$$

$\therefore$ SB = 3, LP = 0, Ue = 0

Thus Hybridisation = SB + LP + Ue = 3 + 0 + 0 = 3, *i.e.*, sp^2

Thus, CO_3^{2-} is sp^2 hybridised, and approximate bond angle = 120° – 2.5 × 0 = 120° we know that number of π bonds in CO_3^{2-} is 1 (as given in π bond calculating)

Thus, the structure of & CO_3^{2-} is

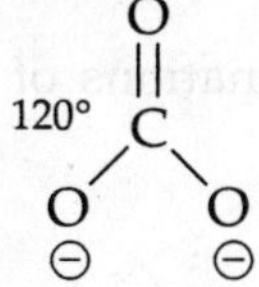

Molecular Orbital Method (Bonding & Antibonding, Orbitals)

Molecular Orbital Theory describes the covalent bonds in a molecule by considering that the atomic orbitals of the component atoms are replaced by electron orbitlas (interms of charge densities or probabilities) extending over the entire molecule. A set of rules is used to assign electrons to these M.O., thereby yielding the electronic structure of the molecule.

By wave mechanics, a combination of two 1s orbitals produces a *bonding molecular orbital* (σ_{1s}^{b}). Another combination produces an antibonding orbital (σ_{1s}^{*})

The Electron charge density concentrated in the internuclear region reduces the repulsion between the positively charged nuclei. This permits bonding between the atoms—hence the term bonding molecular orbital.

The electron charge density in an antibonding molecular orbitals is concentrated away from the internuclear region, where it is ineffective in reducing internuclear repulsion—hence the term *antibonding*.

LCAO Method Linear combination of atomic orbitals

s–s combination of orbitals

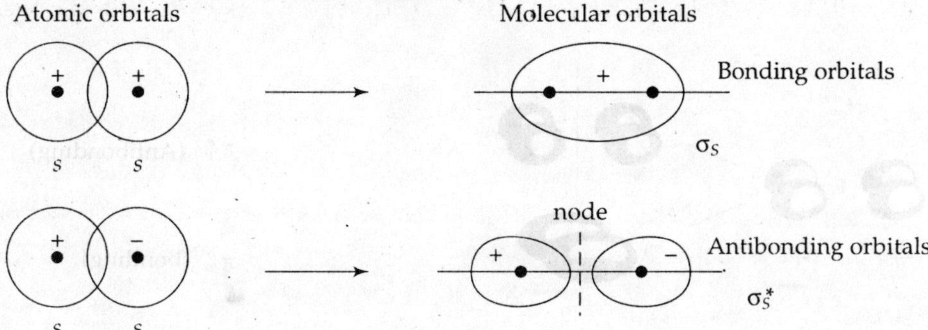

s–p combinations of orbitals

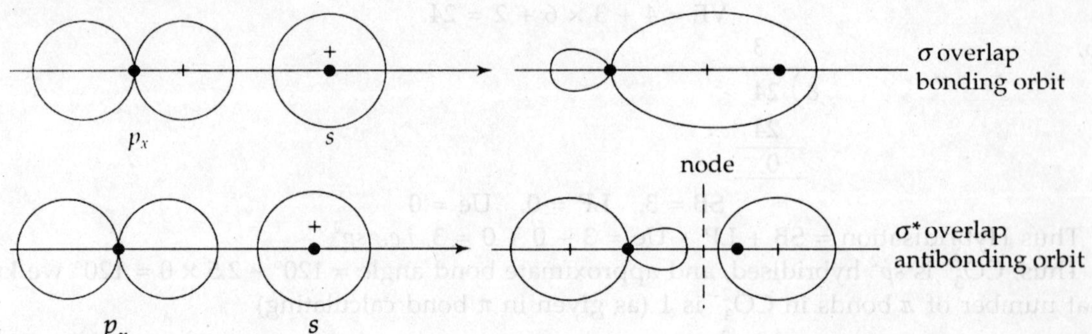

p–p combinations of orbitals

Representation of the molecular orbitals formed by a combinations of $2p$ atomic orbitals.

(i) axially

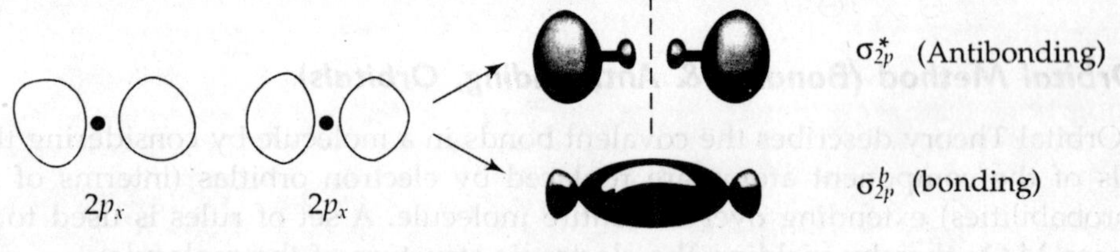

(ii) Sideways

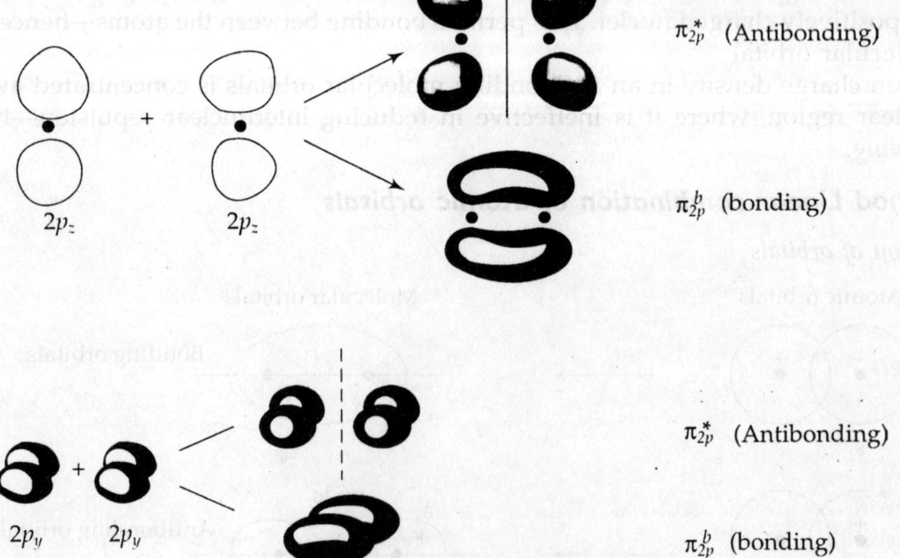

p–d combinations of orbitals

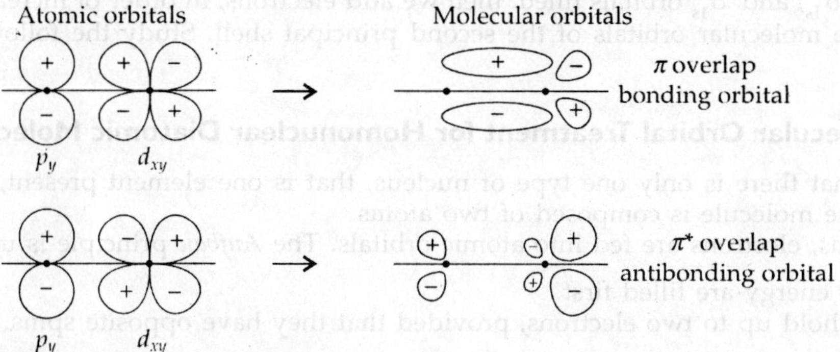

ENERGY LEVEL DIAGRAMS

The energy levels of diatomic molecules are conventionally represented in the form of a **molecular orbital energy level diagram**. The orbital is represented by a horizontal line whose vertical position indicates the relative energy of that orbital. The atomic orbital energy level of the two constituent atoms are arranged either side of the molecular orbital energy levels, usually, with lines linking related molecular and atomic orbitals.

The energies of the 1*s* atomic orbitals are represented by the segments on either side of the diagram. The energy of the bonding molecular orbital corresponds to the line segment at the lower level, and that of the antibonding molecular orbital by the line segment at the higher level.

The energy of the bonding molecular orbital is lower than that of the atomic orbital by an amount Δ.

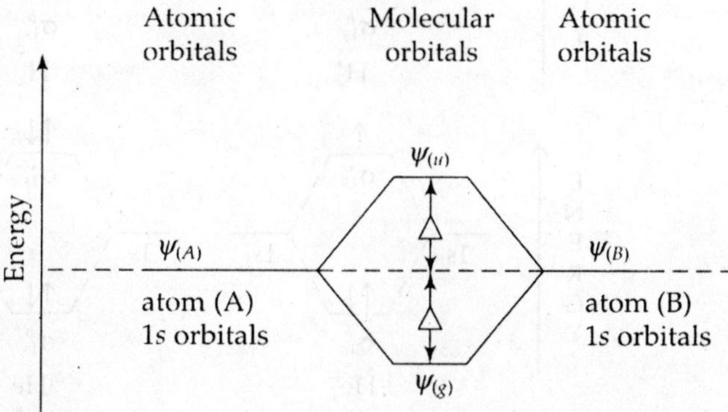

How to Assign M.O

- The energies of bonding molecular orbitals are *lower* than those of antibonding molecular orbitals.
- The energies of molecular orbitals formed from 2*s* atomic orbitals are *lower* than those formed from 2*p* atomic orbitals. (This is the same relationship as between the 2*s* and 2*p* atomic orbitals.)
- The energies of σ-type bonding orbitals are lower than those of π-type orbitals [End-to-end overlap of 2*p* orbitals (σ-type) is more extensive than sidewise overlap of 2*p* orbitals (πtype), resulting in a lower energy.]

To describe the assignment of electrons to the molecular orbitals in X_2, we can proceed in this way. We start with the σ^b_{1s}, and σ^*_{1s} orbitals filled. then we add electrons, in order of increasing energy, to the available molecular orbitals of the second principal shell. Study the following examples:

(A) Examples of Molecular Orbital Treatment for Homonuclear Diatomic Molecules

Homonuclear means that there is only one type of nucleus, that is one element present, and diatomic means that the molecule is composed of two atoms.
In the build-up of atoms, electrons are fed into atomic orbitals. The *Aufbau* principle is used:

1. Orbitals of lowest energy are filled first.
2. Each orbital may hold up to two electrons, provided that they have opposite spins.

	MOs	
H_2^+ molecule ion	$\sigma 1s^1$	The electron occupies the σ1s bonding MO. The energy of this ion is thus lower that that of the constituent atoms and ion, by an amount Δ, so there is some stabilization. It produces a bond between two H-atoms.
H_2 molecule	$\sigma 1s^2$	There is one electron from each atom, and hence there are two electrons in the molecular both electrons one in $\sigma 1s$ orbital. A single covalent bond is formed. These occupy the lowest energy MO.

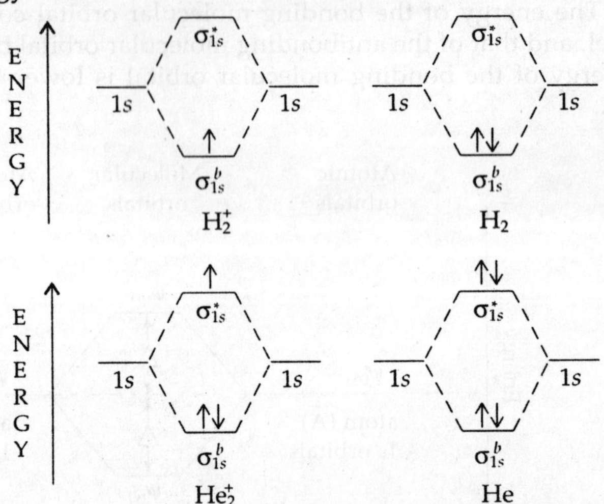

He_2^+ molecule ion	$\sigma 1s^2$, $\sigma^* 1s^1$	This ion has three electrons. Two are in the σ^b_{1s} orbital, and one in the σ^*_{1s}, orbital. The net number of bondings is 2 – 1 = 1 This may be considered as a combination of a He atom and a He^+ ion. There are three electrons in the molecular ion, which are arranged in MO_s. This is a stable ion.
He_2 molecule	$\sigma 1s^2$, $\sigma^* 1s^2$	There are two electrons from each atom, and the four electrons are arranged in MOs: The 2Δ stabilization energy from filling the $\sigma 2s$ MO is cancelled by the 2Δ destabilization energy from filling the σ^*1s MO. Thus a bond is not formed and the molecule does not exist.

(B) Examples of Molecular Orbital Treatment for Diatomic Molecules in IInd Period

We can think of the first shell (K shell) electrons as not involved in the bonding (that is as nonbonding electrons). This allows us simply to consider an assignment of the valence-shell electrons of the two atoms. Figure below presents a generalized energy-level diagram for the actual and hypothetical homonuclear diatomic molecules of the second period elements and the assignment of electrons to their molecular orbitals.

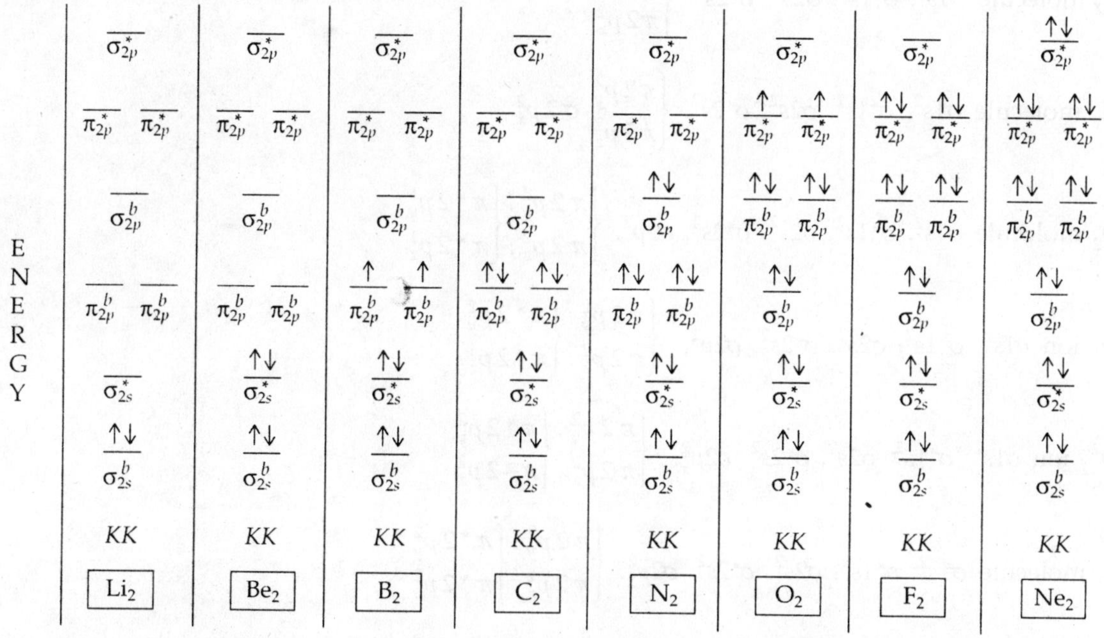

ENERGY	Li_2	Be_2	B_2	C_2	N_2	O_2	F_2	Ne_2
	σ^*_{2p}	σ^*_{2p}	σ^*_{2p}	σ^*_{2p}	σ^*_{2p}	σ^*_{2p}	σ^*_{2p}	↑↓ σ^*_{2p}
	π^*_{2p} π^*_{2p}	π^*_{2p} π^*_{2p}	π^*_{2p} π^*_{2p}	π^*_{2p} π^*_{2p}	π^*_{2p} π^*_{2p}	↑ π^*_{2p} ↑ π^*_{2p}	↑↓ π^*_{2p} ↑↓ π^*_{2p}	↑↓ π^*_{2p} ↑↓ π^*_{2p}
	σ^b_{2p}	σ^b_{2p}	σ^b_{2p}	σ^b_{2p}	↑↓ σ^b_{2p}	↑↓ π^b_{2p} ↑↓ π^b_{2p}	↑↓ π^b_{2p} ↑↓ π^b_{2p}	↑↓ π^b_{2p} ↑↓ π^b_{2p}
	π^b_{2p} π^b_{2p}	π^b_{2p} π^b_{2p}	↑ π^b_{2p} ↑ π^b_{2p}	↑↓ π^b_{2p} ↑↓ π^b_{2p}	↑↓ π^b_{2p} ↑↓ π^b_{2p}	↑↓ σ^b_{2p}	↑↓ σ^b_{2p}	↑↓ σ^b_{2p}
	σ^*_{2s}	↑↓ σ^*_{2s}	↑↓ σ^*_{2s}	↑↓ σ^*_{2s}	↑↓ σ^*_{2s}	↑↓ σ^*_{2s}	↑↓ σ^*_{2s}	↑↓ σ^*_{2s}
	↑↓ σ^b_{2s}	↑↓ σ^b_{2s}	↑↓ σ^b_{2s}	↑↓ σ^b_{2s}	↑↓ σ^b_{2s}	↑↓ σ^b_{2s}	↑↓ σ^b_{2s}	↑↓ σ^b_{2s}
	KK	*KK*	*KK*	*KK*	*KK*	*KK*	*KK*	*KK*

Molecular orbital diagrams for actual (Arrow) and hypothetical diatomic molecules of the second period elements.

Concerning the relative placement of the σ^b_{2p} and π^b_{2p} orbitals more extensive overlap occurs when *p* atomic orbitals are combined end-to-end rather than side-by-side suggesting that σ^b_{2p}, lies at a lower energy than π^b_{2p}. This is the situation, confirmed by experiment when the energy difference between 2*s* and 2*p* orbitals is large (as in O, F, and Ne). When this difference is smaller (as in Li through N), there is a mixing of 2*s* and 2*p* atomic orbitals in the formation of molecular orbitals that results in a reversal of the σ^b_{2p}, and π^b_{2p} levels. This variation in the energy level diagrams of the diatomic molecules of the second period elements has little effect.

The symbol KK means that electrons in the first electronic shells (KK) are not involved in the bonding.

Li molecule $\sigma 1s^2 . \sigma^* 1s^2 . \sigma 2s^2$
Be_2 molecule $\sigma 1s^2 . \sigma^* 1s^2 . \sigma 2s^2 . \sigma^* 2s^2$

B_2 molecule $\sigma 1s^2 . \sigma^* 1s^2 . \sigma 2s^2 . \sigma^* 2s^2 . \begin{cases} \pi 2p_y^1 \\ \pi 2p_z^1 \end{cases}$

C_2 molecule $\sigma 1s^2 . \sigma^* 1s^2 . \sigma 2s^2 . \sigma^* 2s^2 . \begin{cases} \pi 2p_y^2 \\ \pi 2p_z^2 \end{cases}$

N_2 molecule $\sigma 1s^2 . \sigma^* 1s^2 . \sigma 2s^2 . \sigma^* 2s^2 . \begin{cases} \pi 2p_y^2 \\ \pi 2p_z^2 \end{cases} \sigma 2p_x^2$

O_2 molecule $\sigma 1s^2 . \sigma^* 1s^2 . \sigma 2s^2 . \sigma^* 2s^2 . \sigma 2p^2 . \begin{cases} \pi 2p_y^2 . \\ \pi 2p_z^2 . \end{cases} \begin{cases} \pi^* 2p_y^1 \\ \pi^* 2p_z^1 \end{cases}$

O_2^- ion $\sigma 1s^2 . \sigma^* 1s^2 . \sigma 2s^2 . \sigma^* 2s^2 . \sigma 2p^2 . \begin{cases} \pi 2p_y^2 . \\ \pi 2p_z^2 . \end{cases} \begin{cases} \pi^* 2p_y^2 \\ \pi^* 2p_z^1 \end{cases}$

O_2^{2-} ion $\sigma 1s^2 . \sigma^* 1s^2 . \sigma 2s^2 . \sigma^* 2s^2 . \sigma 2p_x^2 . \begin{cases} \pi 2p_y^2 . \\ \pi 2p_z^2 . \end{cases} \begin{cases} \pi^* 2p_y^2 \\ \pi^* 2p_z^2 \end{cases}$

F_2 molecule $\sigma 1s^2 . \sigma^* 1s^2 . \sigma 2s^2 . \sigma^* 2s^2 . \sigma 2p_x^2 . \begin{cases} \pi 2p_y^2 . \\ \pi 2p_z^2 . \end{cases} \begin{cases} \pi^* 2p_y^2 \\ \pi^* 2p_z^2 \end{cases}$

CHARACTERISTICS OF COVALENT BOND

(I) Bond Length

It is defined on the average distance between the nuclei of two bonded atoms in a molecule. Bond length can be determined by x-ray diffraction (in crystals) and by spectroscopy in gases.

Bond length decreases with multiplicity of the bond formed between the 2 atoms, e.g., Bond length: C—C >C=C >C≡C

(II) Bond Angle

It may be defined as the internal angle between the orbitals containing electron pairs in the valence shell of the central atom in a molecule.

e.g.,	Molecules	CH_4	NH_3	H_2O
	Bond Angles	109.5	107°	104.5°

$H \overset{\ddot{O}}{\frown} H$ (104.5°)

The bond angle gives an idea of the distribution of the orbitals in three-dimensional space around the central atom in the molecule and thus gives an idea of the shape of the molecule.

(III) Bond strength or Bond Energy

Bond strength or Bond Energy of a particular type of bond is energy required to break one mole of bonds of that type in a substance in gaseous state.

It indicates stability of the bond e.g., N≡N is more stable then O=O.

Bond energy (i) C≡C >C=C >C—C

(IV) Bond Order

Bond order is defined as one-half of the difference between the number of electrons present in the bonding and the antibonding molecular orbitals, i.e.,

$$\text{Bond order} = \frac{1}{2}(N_b - N_a)$$

N_b and N_a are the number of electrons present in bonding and antibonding molecular orbitals

(a) If $N_b > N_a$, the molecule is stable, and
(b) If $N_b \le N_a$, the molecule is ustable.

Bond order is inversely proportional to bond length but is directly proportional to bond strength.

(V) (i) Characteristics of H_2^+, H_2 and H_2^-

Species	Electronic	Bond order configuration	Stability	Magnetic character
H_2	$[\sigma(1s]^2$	1	Quite stable	Diamagnetic
H_2^+	$[\sigma(1s)]^1$	1/2	Somewhat stable	Paramagnetic
H_2^-	$[\sigma(1s)]^2\ [\sigma^*(1s)]^1$	1/2	Somewhat stable	Paramagnetic

The decreasing order of stability is $H_2 > H_2^+ > H_2^-$

(VI) Characteristics of N_2, N_2^+, N_2^-, N_2^{2-}

Species	Bond order	Stability	Magnetic character
N_2	3	Most stable	Diamagnetic
N_2^+	$2\frac{1}{2}$	Less stable	Paramagnetic
N_2^-	$2\frac{1}{2}$	Less stable	Paramagnetic
N_2^{2-}	2	Less stable	Paramagnetic

The decreasing order of stability is $N_2 > N_2^+ > N_2^- > N_2^{2-}$

Electronic configurations:
N_2: $kk[\sigma 2s)]^2$ $[\sigma^*(2s)]^2$ $[\pi(2p_x)]^2$ $[\pi(2p_y)]^2$ $[\sigma(2p_z)]^2$
N_2^+ : kk $[\sigma(2s)]^2$ $[\sigma^*(2s)]^2$ $[\pi(2p_x)]^2$ $[\pi(2p_y)]^2$ $[\sigma(2p_z)]^1$
N_2^- : kk $[\sigma(2s)]^2$ $[\sigma^*(2s)]^2$ $[\pi(2p_z)]^2$ $[\pi(2p_x)]^2$ $[\sigma(2p_y)]^2$ $[\pi^*(2p_x)]^1$

(VII) Characteristics of O_2^-, O_2^{2-}

Speciles	*Bond order*	*Stability*	*Magnetic character*
O_2^+	$2\frac{1}{2}$	Most stable	paramagnetic
O_2	2	Less stable	Paramagnetic
O_2^-	$1\frac{1}{2}$	Still less stable	paramagnetic
O_2^{2-}	1	Least stable	Diamagnetic

The decreasing order of stability is $O_2^+ > O_2 > O_2^- > O_2^{2-}$

Problems

Q. 1. The percentage ionic character of a single bond may be estimated from the ratio of the observed dipole moment to the calculated moment, assuming oppositely charged ions located at a distance equal to the bond length. The observed dipole moments of HCl, HBr, and HI are 1.03, 0.79 and 0.38 D, respectively. Calculate the percentage ionic character of the bond in each of these compounds. Given the radii of H = 0.28 Å; Cl = 0.99 Å; Br = 1.4 Å and I = 1.33 Å

Q. 2. Deduce the geometry and diagram the electronic structure of (a) I_3^- (b) ClO_3^- (c) ClO_3^+ (d) F_2SeO (e) $IO_2F_2^-$

Q. 3. (a) What are the bond ordrs for CN^-, CN, and CN^+? (b) Which of these species should have the shortest bond length?

Q. 4. Represent the following equation pictorially (see pictorial drawing), using squares to represent A and circles to B,

$$A_3 + B_2 \rightarrow AB_4$$

After have "down" the equation, use your drawing as a guide to balance it.

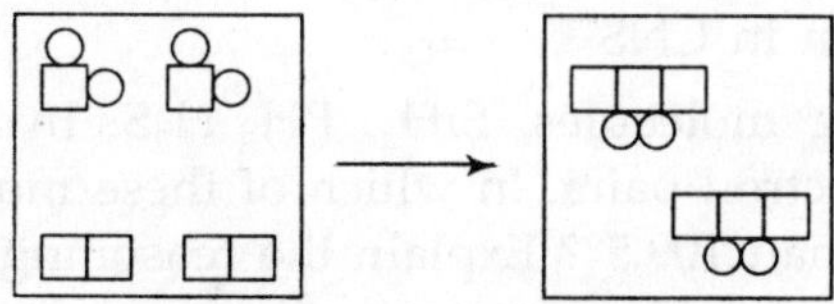

Q. 5. Consider the diagram where atom *X* is represented by a square and atom *Y* is represented by a circle.

(a) Write the equation for the reaction represented by the diagram.

(b) If each circle stands for a mole of *Y* and each square a mole of *X*, how may moles of *X* did one start with? How many moles of *Y*?

(c) Using the same representation described to part (b), how many moles of product are formed? How many modes of *X* and *Y* are left unreacted?

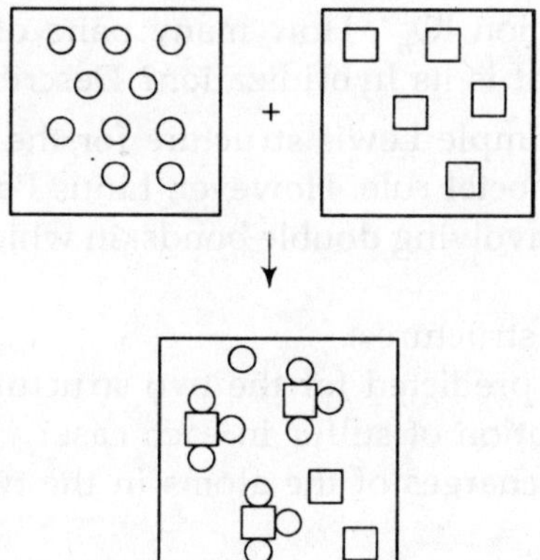

Q. 6. When 4.0 mol of CCl_4 reacts with an excess of HF, 3.0 mol of CCl_2F_2 (Freon) is obtained. The equation for the reaction is

$$CCl_4(l) + 2HF(g) \rightarrow CCl_2F_2(l) + 2HCl(g)$$

State which of the statements are true about the reaction and make the false statements true.

(a) The theoretical yield for CCl_2F_2 is 3.0 mol.
(b) The theoretical yield for HCl is 71 g.
(c) The precent yield for the reaction is 75%
(d) Half a mole of HF is consumed for every mole of CCl_4 used.

Q. 7. (a) Suppose that Si-28 ($^{14}_{28}Si$) is taken as the standard for expressing atomic masses and assigned an atomic mass of 10.00 amu. Estimate the molar mass of lithium nitride.
(b) Chlorophyll, the substance responsible for the green color of leaves, has one magnesium atom per chlorophyll molecule and contains 2.72% magnesium by mass. What is the molar mass of chlorophyll?

Q. 8. Given the following electronegativities

$$C = 2.5 \quad N = 3.0 \quad S = 2.6$$

what is the central atom in CNS^-?

Q. 9. Consider the following molecules, SiH_4, PH_3 H_2S. In each case, a central atom is surrounded by four electron pairs. In which of these molecules would you expect the bond angle to be less than 109.5°? Explain the reasoning.

Q. 10. Give the formula of an ion or molecule in which an atom of
(a) N forms three bonds using sp^3 hybrid orbitals.
(b) N forms two pi bonds and one sigma bond.
(c) O forms one sigma and one pi bond.
(d) C forms four bonds in three of which it uses sp^2 hybrid orbitals.
(e) *Xe* forms two bonds using sp^3d^2 hybrid orbitals.

Q. 11. Draw the Lewis structure and describe the geometry of the hydrazine molecule, N_2H_4. Would you expect this molecule to be polar?

Q. 12. Consider the polyatomic ion IO_6^{5-}. How many pairs of electrons are there around the central iodine atom? What is its hybridization? Describe the geometry of the ion.

Q. 13. It is possible to write a simple Lewis structure for the SO_4^{2-} ion, involving only single bonds, which follows the octet rule. However, Linus Pauling and other have suggested an alternative structure, involving double bonds, in which the sulfur atom is surrounded by six electron pairs.
(a) Draw the two Lewis structures.
(b) What geometries are predicted for the two structures?
(c) What is the hybridization of sulfur in each case?
(d) What are the formal charges of the atoms in the two structures?

Q. 14. Explain why
(a) oxalic acid removes stains.
(b) there are no geometric isomers of tetrahedral complexes.
(c) cations such as Co^{2+} act as Lewis acid.

(d) $C_2O_4^{2-}$ is a chelating agent.
(e) NH_3 can be a ligand but NH_4^+ is not.

Q. 15. A certain coordination compounds has the simplest formula $PtN_2H_6Cl_2$. It has a molar mass of about 600 g/mol and contains both a complex cation and a complex anion. What is its structure?

Q. 16. Two coordination compounds decompose at different temperatures but have the same mass percent analysis data: 20.25% Cu, 15.29 C, 7.07% H, 26.86% N, 10.23% S, and 20.39% O. Each contains Cu^{2+}.
(a) Determine the simplest formula of the compounds.
(b) Draw the structural formulas of the complex ion in each case.

Q. 17. A molecular compounds is composed of 60.4% Xe, 22.1% O, and 17.5% F, by mass. If the molecular weight is 217.3 amu, what is the molecular formula? What is the Lewis formula? Predict the molecular geometry using the VSEPR model. Describe the bonding using valence bond theory.

Q. 18. A compound of chlorine and fluorine, ClF_n, reacts at about 348 K with uranium metal to produce uranium hexafluoride. UF_6 and chlorine monofluoride ClF(*g*). A quantity of uranium produced 3.53 g UF_6 and 343 ml ClF at 348 K and 2.50 atm. What is the formula (*n*) of the compound? Describe the bonding in the molecule, using valence bond theory.

Q. 19. Each of the following compounds has nitrogen-nitrogen bond N_2, N_2H_4, N_2F_2. Match each compound with one of the following bond lengths: 110 pm, 122 pm, 145 pm. Describe the geometry about one of the N atoms in each compound. What hybrid orbitals are needed to describe the bonding in valence bond theory?

Q. 20. Draw resonance formula of the nitric acid molecule, HNO_3. What is the geometry about the N atom? What is the hybridization on N? Use bond energies and one Lewis formula for HNO_3 to estimate ΔH_f° for $HNO_3(g)$. The actual value of ΔH_f° for $HNO_3(g)$ is—135 kJ/mol, which is lower than the estimated values because of stabilization of HNO_3 by resonance. The *resonance energy* is defined as ΔH_f° (estimated ΔH_f° (actual). What is the resonance energy of HNO_3?

Q. 21. Phosphoryl chloride, $POCl_3$, has the skeleton structure

```
      O
      |
 Cl—P—Cl
      |
     Cl
```

Write
(a) a Lewis structure of $POCl_3$ following the octet rule. Calculate the formal charges in this structure.
(b) a Lewis structure where all the formal charges are zero. (The octet rule need not be followed.)

Q. 22. Phosphorus acid, H_3PO_3, has the structure $(HO)_2PHO$, in which one H atom is bonded to P atom, and two H atom are bonded to O atoms. For each bond to an H atom, decide whether it is polar or nonpolar. Assume that only polar-bonded H atoms are acidic.

Write the balanced equation for the complete neutralization of phosphorous acid with sodium hydroxide. A 200-ml sample of H_3PO_3 requires 22.50 ml, of 0.1250 M NaOh for complete neutralization. What is the molarity of the H_3PO_3 solution?

Q. 22. An ionic compound has the following composition (by mass): Mg, 109% Cl, 31.8%; O, 57.3%. what are the formula and name of the compound? Write the Lewis formula for the ions.

Q. 24. A gaseous compound has the following composition by mass: C, 25.0%; H, 2.1%; F, 39.6%, O, 33.3%. Its molecular weight is 48.0 amu. Write the Lewis formula for the molecule.

Q. 25. A compound of tin and chlorine is colorless liquid. The vapor has a density of 7.49 g/L at 424 K and 1.00 atm. What is the molecular weight of the compound? Who do you think the compound is molecular and not ionic? Wirte the Lewis formula for the molecule.

Q. 26. Using Mulliken's formula, calculate a value for the electronegativity of chlorine. Use values of the ionization energy and values of the electron affinity from tables of your text book. Divide this result (in kJ/mol) by 230 to get a value comparable to Pauling's scale.

Q. 27. Draw Lewis structures for the following substances: (a) methanol, CH_3OH, (b) chlorate ion, ClO_3^-, (c) N_2O (which has a nitrogen atom as the central atom). (d) phosgene, $COCl_2$ (both O and Cl are bonded to C), and (e) CO.

Q. 28. Calculate the formal charges on the atoms in the Lewis structure of (a) $COCl_2$ (b) CO_2 (c) N_2O and (d) CrO_4^{2-}.

Q. 29. Draw appropriate resonance structures for SO_3. Spectroscopic data indicate that all three S—O bonds are identical, and that the bond length is shorter than a typical S—O single bond, but longer than a typical S—O double bond.

Q. 30. Experimental evidence indicates that the bond between Cr and O in the CrO_4^{2-} ion is in between a single and a double bond. Draw resonance structures that represent the actual structure of CrO_4^{2-}.

Q. 31. Draw an electron dot structure for XeF_2.

Q. 32. Predict the geometry of (a) SO_4^{2-} ion, (b) PCl_3 and (c) SCl_2.

Q. 33. Predict the geometry of (a) ethylene, C_2H_4 (b) I_3^- and (c) ICl_4^-.

Q. 34. Give the electronic configuration of the C_2 molecule and predict whether or not it is paramagnetic. What is the bond order? Do you expect the bond in C_2 to be stronger or weaker than that in B_2?

Q. 35. State whether or not each of the following symbols is an acceptable designation for an atomic orbital. Explain what is wrong with the unacceptable symbols. (a) 2*d*, (b) 6*g*, (c) 7*s* and (d) 5*h*.

Q. 36. Given in this example are several electronic configuration that may for correct for the nitrogen atom. Indicate whether each of these configurations represents the ground state, or an excited stae, or is an impossible (forbidden) configuration for nitrogen.

	1s	2s	2p	2p	2p	3s	3p	3p	3p
(a)	$\uparrow\downarrow$	$\uparrow\downarrow$	$\uparrow$	$\downarrow$		$\downarrow$			
(b)	$\uparrow\downarrow$	$\uparrow\downarrow$	$\uparrow$		$\uparrow$	$\uparrow$			
(x)	$\uparrow\downarrow$	$\uparrow\downarrow$	$\uparrow$	$\uparrow$	$\uparrow$				
(d)	$\uparrow\downarrow$	$\uparrow\downarrow$	$\uparrow$	$\downarrow$	$\uparrow$				
(e)	$\uparrow\downarrow$	$\uparrow\downarrow$	$\downarrow$	$\downarrow$	$\downarrow$				
(f)	$\uparrow\downarrow$	$\uparrow\downarrow$		$\downarrow$		$\downarrow$	$\downarrow$		
(g)	$\uparrow\downarrow$	$\uparrow\downarrow$	$\uparrow\downarrow$	$\uparrow$					

Chapter 3

THE GASEOUS STATE

FORMULAE (Note : The symbols below have their usual meanings)

$$PV = \text{Constant} \qquad \text{(at constant temp.)}$$

or $$P_1V_1 = P_2V_2$$

$$\frac{V}{T} = \text{Constant} \qquad \text{(at constant press.)}$$

or $$\frac{V_1}{T_1} = \frac{V_2}{T_2}$$

$$PV = nRT = \frac{w}{m} RT$$

$$\frac{P_1V_1}{T_1} = \frac{P_2V_2}{T_2}$$

$$P = P_1 + P_2 + P_3 + P_4 + \dots$$

$$P_1 = P_{\text{Total}} \cdot X_1 \qquad (X \text{ is mole fraction})$$

$$R \propto \sqrt{\frac{1}{d}} \text{ or } \sqrt{\frac{1}{M}}$$

or $$\frac{r_1}{r_2} = \sqrt{\frac{d_2}{d_1}} = \sqrt{\frac{M_2}{M_1}}$$

$$P = P^\circ e^{-\text{Mgh}/\text{RT}}$$

or $$\log \frac{P}{P^\circ} = \log \frac{d}{d^\circ} = \log \frac{n}{n^\circ} = \frac{1}{2.303} \frac{\text{Mgh}}{\text{RT}}$$

$$PV = \frac{1}{3}\ \text{mnu}^2 \qquad (u = \text{R.M.S. velocity})$$

or
$$u = \sqrt{\frac{3PV}{M}} = \sqrt{\frac{3RT}{M}} = \sqrt{\frac{3P}{d}}$$

$$\text{K.E.} = \frac{1}{2}\ \text{mnu}^2 = \frac{3}{2}\ RT = 12\ \text{T J}$$

Average velocity, $V = \sqrt{\frac{8RT}{\pi M}} = 0.9213\ \text{u}$

Most probable velocity,

$$\alpha = \sqrt{\frac{2RT}{\pi M}}$$

The ratio $\alpha : v : u = 1 : 1.128 : 1.234$

$$\left(P + \frac{a}{V^2}\right)(V - b) = RT$$

$$\left(P + \frac{an^2}{V^2}\right)(V - nb) = nRT$$

$$\left(\alpha + \frac{3}{\beta^2}\right)(3\beta - 1) = 8\gamma$$

$$T_c = \frac{8a}{27bR}$$

$$P_c = \frac{a}{27b^2}$$

$$V_c = 3b$$

$$a = 3P_cV_c^2 = \frac{9}{8}RT_cV_c = \frac{27}{64}\frac{R^2T_c^2}{P_e}$$

$$b = \frac{1}{3}V_c = \frac{1}{8}\frac{RT_c}{P_e}$$

$$b = 4\text{N}\left[\frac{4}{3}\pi r^3\right]$$

$$T_i = \frac{2a}{bR}$$

$$\gamma = \frac{C_p}{C_V} = \frac{2.5R + \text{x}}{1.5R + \text{x}}$$

x = 0, R, 1.5R for mono-; di- and tri or poly atomic gases respectively.

Boyle's Law

At constant temperature the volume of a given mass of a gas is inversely, proportional to its pressure.
Mathematically,

$$V \propto \frac{1}{P} \quad \text{or } PV = k \text{ (Constant)}.$$

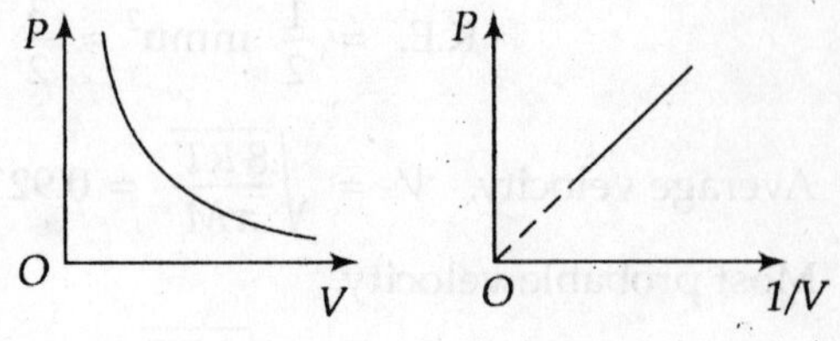

Graphically, for two gases 1 and 2.

$\therefore$ $P_1V_1 = P_2V_2$ or $\frac{P_1}{d_1} = \frac{P_2}{d_2}$, where d_1 and d_2 are densities of the gas at P_1 & P_2.

Charle's Law

At constant pressure the volume of a given mass of gas is directly proportional to its absolute temperature (expressed in kelvin 'K').
Mathematically,

$$V \propto T \quad \text{or} \quad V/T = \text{Constant.}$$

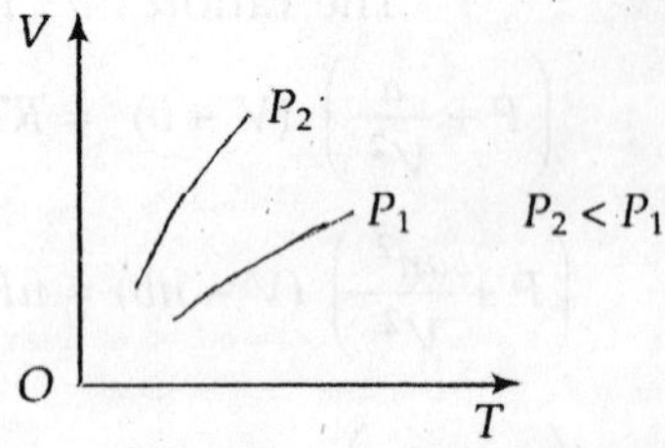

Graphically, for two gases 1 and 2

$\therefore$ $V_1/T_1 = V_2/T_2$ or $V_1/V_2 = T_1/T_2$

Absolute Zero

The lowest attainable temperature, –273°C, equals O on the Kelvin temperature scale.

i.e. Temperature in K = Temperature in °C = 273.

At –273°C, the volume of any gas is zero.

Gay Lussac's Law

At constant volume the pressure of a given mass of gas is directly proportional to its temperature.
Mathematically,

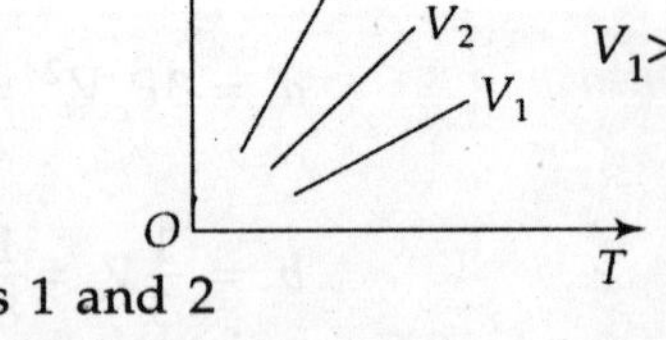

$P \propto T$ or P/T = Constant. Graphically, for two gases 1 and 2

$\therefore$ $P_1/T_1 = P_2/T_2$ or $P_1/P_2 = T_1/T_2$

S.T.P.

Gas volumes are usually reported at standard temperature (273 K) and and pressure (1 atm). Also, 1 ml. of gas = 22.4 litres at S.T.P.

Ideal Gas Equation

$PV = nRT$ is the ideal gas equation, where R is universal gas constant. The value of R in different units is = 0.081 lit. atm. K^{-1} mol^{-1} (SI Units)

= 8.341 Joules. K^{-1} mol^{-1} (SI Units)

$= 8.31 \times 10^3$ g.m^2/s^2. mol.K (SI Units)
$= 8.341 \times 10^7$ erg. K^{-1} · mol^{-1} (old units) [1 J = 4.18 Calories]
$= 1.987$ Calorie · K^{-1} · mol^{-1} (old units)
$= 62.63$ L . mmHg · K^{-1}· mol^{-1} (old units)

we have, $$PV = \frac{g}{M} \cdot RT$$

or, Molecular mass $$M = \frac{\rho RT}{P}$$ ($\because \rho = g/V$, where ρ is gas density and $n = g/M$)

AVAGADRO'S Law

Equal volumes of gases at same temperature and pressure contain equal number of molecules.

$V \propto n$ (P and T Constants)

One mole contains 6.023×10^{23} molecules. (Avagadro's number) thus if two gases contain equal number of molecules, they also contain same number of moles.

Dalton's Law of Partial Pressures

The total pressure exerted by a gaseous mixture (P) is equal to the sum of partial pressure of each gas present in the mixture ($P = P_A + P_B + \cdots$). The pressure exerted by each component in a gaseous mixture is known as the partial pressure. For a perfect gas the partial pressure P for n moles of each component x, is given by $P = \frac{nRT}{V}$

$\therefore$ Total Pressure $$P = \frac{nRT}{V} = P_A + P_B + P_C + \cdots$$

$$= (n_A + n_B + n_C + \ldots)\frac{RT}{V}$$

$\therefore$ $$\frac{Pa}{P} = \left(\frac{n_A}{n_A + n_B + \ldots}\right) = \frac{n_A}{n_{total}} \Rightarrow P_A = P \times \text{mole fraction of } A \text{ in mixture} = P \times x_A.$$

i.e. $$P_A = x_A \cdot P_{\text{total}}.$$

Diffusion

The inherent property of gases to mix with each other to form homogeneous mixture *irrespective* of a gravity is called diffusion. Lighter gases diffuses more rapidly than heavier gases.

Rate of diffusion : The rate or velocity of diffusion of a gas is the volume of the gas diffused in unit time, $r = \frac{V}{t}$, where, r = rate of diffusion in time t and V = volume of gas

Graham's Law of Diffusion:

At similar conditions of temperature and pressure, the relative rates of diffusion of gases are inversely proportional to the square root of their densities or vapour densities or molecular weights.

i.e. $$r \propto \sqrt{\frac{P}{M}}$$

For two gases 1 and 2, we have,

$$\therefore \quad \frac{r_1}{r_2} = \sqrt{\frac{d_2}{d_1}};\ \frac{r_1}{r_2} = \sqrt{\frac{VD_2}{VD_1}};\ \frac{r_1}{r_2} = \sqrt{\frac{M_2}{M_1}} \qquad \text{(at Constant } P \text{ \& } T\text{)}$$

Also:,

(i) $$\frac{V_1}{V_2} = \sqrt{\frac{M_2}{M_1}}$$

where V_1 and V_2 are the volumes of the gases that diffuse in the same time and M_1 and M_2 are the molecular weights of the two gases.
and

(ii) $$\frac{t_2}{t_1} = \sqrt{\frac{M^2}{M_1}}$$

where t_1 and t_2 are the times of diffusion of the same volume of the two gases respectively.

EFFUSION

Effusion is the escape of a gas through an orfice. The rate of escape is related to the root mean square speed of the molecules. By Graham's Law of effusion for two gases 1 and 2

$$\frac{u_1}{u_1} = \sqrt{\frac{M_2}{M_1}} = \sqrt{\frac{\rho_2}{\rho_1}}$$

The Kinetic Gas Equation

The Kinetic equation is:

$$PV = \frac{1}{3}\text{mnu}^2 = \frac{2}{3}\text{K}_\text{E} = RT$$

where,

m = mass of one molecule.
n = number of molecules
u = root mean square velocity of the gas molecules.

$$\Rightarrow \quad K_E = \frac{3}{2}\text{RT or } K_E \propto T.$$

where, K_E = Transactional kinetic energy/mole.
K = Boltzman constant.

Thus K.E of a gas depends on its mass but does not depend upon its volume or pressure.

Transational K_E/molecule $= \frac{3}{2}\frac{RT}{n} = \frac{3}{2}KT$

R.M.S. Velocity, U

It is the square root of the mean of the square of the velocities of all the molecules present is given gas.

Mathematically, $U = \sqrt{\frac{U_1^2 + U_2^2 + \ldots + U_m^2}{n}}$

Where, U_1, U_2 are velocities of different molecules and n is the total number of molecules. Also,

$$U = \sqrt{\frac{3RT}{M}} = \sqrt{\frac{3PV}{M}} = \sqrt{\frac{3P}{D}}.$$

Average Velocity $\bar{U}$

It is the average of the velocities of all the molecules (1, 2, . . . m) present in a given gas

Mathematically, $\bar{U} = \frac{U_1 + U_2 + \ldots + U_m}{n} = \sqrt{\frac{8RT}{\pi M}}$

Most Probable Velocity U_{MP}

It is the velocity possessed by the maximum number of molecules present in a given gas.

Mathematically, $U_{MP} = \sqrt{\frac{2RT}{M}}$

RATIO: $U : \bar{U} : U_{MP} = 1.224 : 1.128 : 1$

Ideal Gases

Ideal gases are gases which obey the gas laws or the general gas equation PV = n RT at all conditions of temperature and pressure.

Real Gases

Real gases are gases which do not obey the gas laws or gas equation rigidly.
Real gases approach ideal behaviour at low temperature and high pressure.

The Vanderwalls Equation

The behaviour of real gases differ from behaviour of ideal gases because gases molecules occupy space in the container & also there are weak attractive forces between the molecules.

Vanderwall's equation is:

$$\left(P + \frac{a}{V^2}\right)(V - b) = RT \text{ for 1 mole of gas.}$$

or, $\left(P + \frac{an^2}{V^2}\right)(V - nb) = n\,RT$ for n moles of gas, where a & b are called Vanderwalls constant and have units as: a . . . liter^2 atm. mol^{-2}; b . . . liter. mol^{-1}.

Critical Constants

These are mathematically as:

(i) Critical Temperature $T_C = 8a/27Rb$. The temperature at which it is not possible to liquify a gas with any amount of pressure is called its critical temperature. (T_C). The higher the critical temperature of a gas, greater is the attractive force between its molecules.

(ii) Critical Pressure $P_C = a/27b^2$. The pressure required to liquify a gas at its critical temperature is called its critical pressure (P_C).

(iii) Critical Volume $V_C = 3b$. The volume occupied by one mole of a gas under critical temperature and critical pressure is called its critical volume (V_C).

Molar Heat Capacity of Gases

The quantity of heat required to raise temperature of one mole of the gas by 1°C is known as Molar heat capacity of the gas. It is of two types.

1. Molar heat capacity at constant pressure. (C_P)
2. Molar heat capacity at constant volume. (C_V)

Atomicity

Number of atoms present in the molecule is known as "atomicity" of the gas molecule. Atomicity can be known from values of C_P/C_V.

For monatomic gases

$$C_V = 3,\ C_P = 5 \text{ Cal. } K^{-1} \text{ mol}^{-1}$$

Poissons Ratio $\quad \gamma = \frac{C_P}{C_V} = \frac{5}{3} = 1.67$

Also, $\quad C_P - C_V = R.$

Diatomic gases

$$C_V = 5,\ C_P = 7$$

$$\gamma = \frac{C_P}{C_V} = \frac{7}{5} = 1.4$$

Polyatomic gases

$$C_V = 6,\ C_P = 8$$

$$\gamma = \frac{C_P}{C_V} = 1.33$$

Inversion Temperature

For every gas there is a certain temperature below which only the gas cools when allowed to expand. This temperature is known as "inversion temperature". It is denoted by T_i

Where, $$T_i = \frac{2a}{Rb}$$

a and b are Vander Waals' constants; R = gas constant.

Joule-Thomson Effect

The phenomena of decrease in temperature when a gas is made to pass through a porus plug from a region of high pressure to a region of low pressure is known as Joule-Thomson Effect. Mathematically,

$$T_1 - T_2 = 0.276 \left(\frac{273}{T_1}\right)^2 (P_1 - P_2)$$

where

T_1 = Initial temperature
T_2 = Final temperature
P_1 = Pressure of the gas before expansion
P_2 = Pressure of the gas after expansion

From the equation it can be seen that decrease in temperature is more when (1) the initial temperature is less (2) the pressure different is more.

Problems

Q. 1. A gas has a volume of 500 cc at a pressure of 700 mm of mercury. What volume will it occupy if the pressure if increased to 800 mm, the temperature remaining constant?

Q. 2. A gas occupies a volume of 200 cc at 273 K and 760 mm. What volume will it occupy at 373 K and 760 mm?

Q. 3. The volume of a gas is 200 litres at 285 K and 750 mm. What volume will it occupy at 313 K and 720 mm?

Q. 4. What volume in liters will 2.71 moles of He gas occupy at STP?

Q. 5. What volume in liters will 100 g of CH_4 gas occupy at STP?

Q. 6. Calculate the mass in grams of 40.0 liters of NO gas at standard conditions.

Q. 7. Flask A contains 20.0 of CH_4 gas. Flask B contains 30.0 litres of CO gas. Each volume is measured at the same temperature and pressure. If flask *A* contains 6.18 moles of CH_4, how many moles of CO are there in flask *B*?

Q. 8. A sample of carbon monoxide occupies a volume of 2.2 litre and exerts a pressure of 0.95 atm. The pressure of CO is increased to 1.1 atm. What is the new volume of CO. Assume that the gas is maintained at constant temperature.

Q. 9. A sample of gas at room temperature is placed in an evacuated bulb of volume 0.51 litre and is found to exert a pressure of 0.24 atm. This bulb is connected to another evacuated bulb whose volume is 0.63 litre and the gas is allowed to fill in the both bulbs. What is the new pressure of the gas at room temperature?

Q. 10. A glass bulb is filled with nitrogen gas at 300 K and 1 atmosphere pressure. To what temperature must the bulb be heated if the pressure of the gas inside the bulb is to become 3 atmospheres. The thermal expansion of bulb is ignored.

Q. 11. The volume of a certain mass of gas at 290 K is 400 ml. What temperature must be attained in order to (a) double the volume and (b) reduce the volume to one half.

Q. 12. 500 ml. of oxygen at 300 K was cooled to 258 K without change in pressure. Calculate the contraction in volume.

Q. 13. A balloon whose volume is 5 litre contained 6.1 g. of N_2. What mass of nitrogen must be added to the balloon to expand its volume to 11 litre at the same temperature and pressure.

Q. 14. A certain mass of a gas occupies 38 ml. at 300 K and 750 mm. of Hg. pressure. What volume will it occupy at 290 K and 760 mm. of Hg. pressure.

Q. 15. In an experiment 190 ml. of dry-hydrogen gas is collected at 289 K and 740 mm. of Hg pressure. Calculate the volume of the gas at STP.

Q. 16. What is the volume occupied by 11 g. of carbon dioxide at 300 K temperature and 780 mm. of Hg. pressure?

Q. 17. 0.3 mole of a gas is enclosed in a cubical vessel at STP. Calculate the length of the cube.

Q. 18. Calculate the RMS velocity of the molecules of carbon dioxide at 300 K.

Q. 19. The density of a gas at 1.5 atmospheres pressure is 1.52 g/litre. Calculate RMS velocity of the molecules of the gas.

Q. 20. Calculate the RMS velocity of the molecules of ethane gas of 1.5 litre at 750 mm of Hg. pressure?

Q. 21. At what temperature will the RMS velocity of hydrogen be the same as that of oxygen at 298 K?

Q. 22. Calculate the root mean square velocity of ozone kept in a closed vessel at 293 K and 82 cm Hg. pressure.

Q. 23. Calculate the average velocity, RMS velocity and most probable velocity for molecules of sulphur dioxide gas at 300 K.

Q. 24. When 2 g. of a gas. *A* is introduced into an evacuated flask kept at 298 K the pressure is found to be one atmosphere. If 3 g of another gas *B* is then added to the same flask, the total pressure becomes 1.5 atm. Assuming ideal gas behaviour, calculate the ratio of the molecular weights.

Q. 25. Acalculate the average kinetic energy in Joules of the molecules in 8 g of methane at 300 K.

Q. 26. Oxygen is present in 1 litre flask at a pressure of 7.6×10^{-10} mm. of Hg. Calculate the number of oxygen molecules in the flask at 273 K?

Q. 27. At room temperature, ammonia gas at 1 atm. Pressure and hydrogen chloride gas at *P* atm. Pressure are allowed to effuse through identical pin holes from opposite ends of a glass tube of one metre length and of uniform cross section. Ammonium chloride is first formed at a distance of 60 cm. from the end through which HCI gas is sent in. What is the value of *P*.

Q. 28. An open vessel at 300 K is heated until 3/5th of the air in it has been expelled. Assuming the volume of the vessel remains constant, find the temperature to which the vessel has to be heated?

Q. 29. The pressure in bulb dropped from 2000 to 1500 mm. of Hg in 47 minutes when the contained oxygen leaked through a small hole. The bulb was then completely evacuated. A mixture of oxygen and another gas of molecular weight 79 in the molar ratio 1 : 1 at a total pressure of 4000 mm. of Hg was introduced. Find the molar ratio of the two gases remaining in the bulb after a period of 74 minutes.

Q. 30. From 200 mg. of CO_2, 10^{21} molecules are removed. How many moles of CO_2 are left?

Q. 31. Calculate the density of ammonia 303 K and 5 atm. pressure?

Q. 32. 1 litre of a mixture of CO and CO_2 is taken. This mixture is passed through a tube containing red hot charcoal. The volume now becomes 1.6 litre. The volumes are measured under the same conditions. Find the composition of the mixture by volume?

Q. 33. A straight glass tube has two inlets *x* and *y* at the two ends. The length of the tube is 200 cm. HCl gas through inlet '*x*' and NH_3 gas through inlet '*y*' are allowed to enter the tube at the same time. White fumes first appear at a point '*P*' inside the tube. Find the distance of *P* from *x*?

Q. 34. A gas occupies 50 ml at STP. At what temperature will it occupy 111 ml if the pressure is kept constant?

Q. 35. Oxygen gas is filled in 10 lit. vessel at 350 K until pressure becomes 1087 mm of Hg. Calculate the weight of oxygen present in the vessel.

Q. 36. 4 gm of hydrogen were filled in 10 litre flask provided with a stopcock at 300 K.

(a) Calculate the pressure of the gas

(b) What is the density of the gas?

(c) How much gas should be removed from the vessel so that the pressure is reduced to 1 atmosphere?

Q. 37. Assume the centre of sun to consist of gas whose average molecular weight is 2.0. The density and pressure of gases are 1.3 gm/c.c. and 1.12×10^9 atm. respectively. Calculate the temp. of the sun.

Q. 38. Air contains 21% of Oxygen and 79% of nitrogen by volume. Calculate the density of the air in gms/litre at 300 K and 760 mm pressure.

Q. 39. An open vessel at 300 K is heated until there fifth of the air of it has been expelled. Assuming that the volume of the vessel remains constant, find the temperatures to which, vessel has been heated.

Q. 40. A sample of gaseous hydrocarbon occupying 1.12 lit at NTP when completely burnt in air produced 2.2 gm of CO_2 and 1.8 gm of H_2O. Calculate the weight of the compound taken and the volume of O_2 at NTP required for its burning. Find the molecular formula of the hydrocarbon.

Q. 41. Spherical balloons, each of 21 cm diameter are to be filled with hydrogen gas at NTP from a cylinder containing the gas at 100 atmosphere and 300 K. If the cylinder can hold 2.82 litres of water calculate the number of balloons that can be filled up.

Q. 42. 100 lit. capacity vessel contains 3.2 gm of O_2 and 2.2 g of CO_2. Calculate the total pressure of O_2 and CO_2, if each alone occupies, the total volume of the mixture is 100 lit. at 300 K.

Q. 43. A vessel contains 1.6 gm of O_2, 4.4 gm of CO_2 and 2.3 gm of NO_2 at 298 K. If the pressure of the gaseous mixture is 1 atm., calculate the partial pressures of the gases.

Q. 44. When 2 gm of a gaseous substance *A* is introduced into an initially evacuated flask at 298 K the pressure is found to be 1 atm. Three grams of another gaseous substance *B* is then added to it at the same temperature and pressure. The final pressure is found to be 1.5 atm. Assuming ideal gas behaviour calculate the ratio of the molecular weights of *A* and *B*.

Q. 45. Gaseous benzene reacts with hydrogen gas in the pressure of nickel catalyst to form gaseous cyclohexane according to the reaction

$$C_6H_6\ (g) + 3H_2\ (g) \longrightarrow C_6H_{12}\ (g)$$

A mixture of benzene and excess hydrogen had a pressure of 60 mm Hg in an unknown volume. After the gas had been passed over a nickel catalyst and all the benzene converted to cyclohexane, the pressure of the gas was 30 mm Hg in the same volume and same temperature. What fraction (by volume) of the original mixture was benzene?

Q. 46. Equal masses of gases H_2 and Cl_2 ar taken in two different flasks of same capacity at constant temperature.

(a) Which flask contains more molecules? How many times more?

(b) In which flask the pressure is greater? How many times greater?

Q. 47. 0.17 gm of a volatile organic compound in Victor Meyer's experiment displaced 26.6 ml of air at 288 K and 758 mm. The saturated vapours pressure of water is 12 mm at 288 K. Calculate the molecular weight of the compound.

Q. 48. In Victor Meyer's apparatus 0.2901 gm of anhydrous aluminium chloride was heated at 293 K and 770 mm pressure. The vapour collected over mercury measured 25.8 ml. Determine the molecular formula of anhydrous aluminium chloride.

Q. 49. The relative density of gaseous mixture containing N_2 and O_2 is 15.2 at 298 C and 1 atm. Calculate the composition of the mixture **(a)** by volume and **(b)** by weight.

Q. 50. 80 ml of a hydrocarbon diffuses through a porous membrane is 15 minutes while 120 ml of sulphur dioxide, under identical conditions diffuse in 20 minutes. What is the molecular weight of the hydrocarbon?

Q. 51. A mixture containing 1.12 litres of H_2 and 1.12 litres of D_2 at NTP is taken inside a bulb connected to another bulb by a stopcock with a small opening. The second bulb is full evacuated, the stopcock opened for a certain time and then closed. The first bulb us now found to contain 0.05 gm of H_2. Determine the percentage composition by weight of the gases in the second bulb.

Q. 52. A certain volume of a gaseous mixture containing 22% by volume of O_3 & remaining oxygen diffused in 180 sec. An equal volume of oxygen diffused in 170 sec. under similar conditions of temperature and pressure. Find the density of ozone.

Q. 53. A mixture in which the mole ratio of hydrogen to oxygen 2 : 1 is used to prepare water by the reaction.

$$2H_2(g) + O_2(g) \rightarrow 2H_2O(g)$$

The total pressure in the container is 0.950 atm at 298 K before the reaction. What is the final pressure in the container at 398K, after the reaction, assuming an 88.0% yield and no volume change?

Q. 54. The volume of fraction of a gas A in mixture is defined by the equation

$$\text{Volume fraction } A = \frac{V_A}{V}$$

Where V is the total volume and V_A is the volume that gas would occupy alone at the same temperature and pressure. Assuming ideal gas behaviour, show that the volume fraction is the same as the mole fraction. Explain why the volume fraction differs from the mass fraction.

Q. 55. What is the minimum amount of sodium azide, NaN_3, that can be added to an automobile airbag to give a volume of 20.0 L of $N_2(g)$ on inflation? Make any reasonable assumptions required to obtain an answer, but state what these assumptions are.

Q. 56. A 2.50-g sample of barium reacted completely with water. What is the equation for the reaction? How many milliliters of dry H_2 evolved at 294 K and 748 mmHg?

Q. 57. A mixture of N_2 and Ne contain equal moles of each gas and has a total mass of 10.0 g. What is the density of this gas mixture at 500 K and 10.00 atm? Assume ideal gas behaviour.

Q. 58. An ideal gas with a density of 30.0 g/L has a pressure of 675 mmHg at 298 K. What is the root-mean-square-speed of the molecules of this gas.

Q. 59. A rigit sealed cylinder has seven molecules of neon (Ne).

(a) Make a sketch of the cylinder with the neon molecules at 298 K. Make a similar sketch of the same seven molecules in the same cylinder at 153 K.

(b) Make the same sketches asked for in part (a), but this time attach a pressure gauge to the cylnider.

Q. 60. Two tanks have the same volume and are kept at the same temperature. Compare the pressure in both tanks if

(a) tank *A* has 2.00 mol of carbon dioxide and tank *B* has 2.00 mol of helium.

(b) tank *A* has 2.00 g of carbon dioxide and tank *B* has 2.00 mol of helium. (Try to do this without a calculator)

Q. 61. A sample of natural as a 85.02% methane, CH_4, and 14.8% ethane, C_2H_6, by ass. What is the density of this mixture at 291 K and 748 mmHg?

Q. 62. A sample of sodium peroxide, Na_2O_2, was reacted with an excess of water.

$$2Na_2O_2(s) + 2H_2O(l) \rightarrow 4NaOH(aq) + O_2(g)$$

Al the sodium peroxide reacted, and the oxygen was collected over water at 294 K. The barometric pressure was 771 mmHg. The level of water inside the tube was 25.0 cm above the level of water outside the tube. If the volume of the tube is 31.0 mL, how many grams of sodium peroxide were in the sample?

Q. 63. A mixture contained calcium carbonate, $CaCo_3$, and magnesium carbonate, $MgCO_3$. A smple of this mixture weighing 7.85 g was reacted with excess hydrochloric acid. The reactions are

$$CaCO_3(g) + 2HCl(aq) \rightarrow CaCl_2(aq) + H_2O(l) + CO_2(g)$$
$$MgCO_3(s) + 2HCl(aq) \rightarrow MgCl_2(aq) + H_2O(l) + CO_2(g)$$

If the sample reacted completely and produced 1.94 L of carbon dioxide, CO_2, at 298 K and 785 mmHg, what were the percentages of $CaCO_3$ and $MgCO_3$ in the mixture?

Q. 64. The tube 1.5 m. long is evacuated. Sample of NH_3 and HCl, at the same temperature and pressure, are introduced simultaneously through tiny openings at opposite ends of the tube. When the two gases meet, a white ring of $NH_4Cl(s)$ forms. How far from the end at which ammonia was introduced will the ring form?

Q. 65. The Rankine temperature scale resembles the Kelvin scale in that 0° is taken to be the lowest attainable temperature (0°R = 0 K). However, the Rankine degree is the same size as the farenheit degree, whereas the Kelvin degree is the same size as the Celcius degree. What is the value of the gas constant L-atm/(mol-°R)?

Q. 66. A 0.2500-g sample of an Al-Zn alloy reacts with HCl to form hydrogen gas.

$$Al(s) + 3H^+(aq) \rightarrow Al^{3+}(aq) + \frac{3}{2}H_2(g)$$

$$Zn(s) + 2H^+(aq) \rightarrow Zn^{2+}(aq) + H_2(g)$$

The hydrogen produced has a volume of 0.147 L at 298 K ad 755 mm Hg. What is the percentage of zinc in the alloy?

Q. 67. The buoyant force on a balloon is equal to the mass it displaces. The gravitational force in the balloon is equal to the sum of the masses of the balloon, the gas it contains, and weight of balloonist. If the balloon and the balloonist together weigh 168 kg, what would the diameter of a spherical hydrogen-filled balloon have to be in meters if the rig is to get off the ground at 295 K and 758 mm Hg? (Take $\mathcal{M}_{air}$ = 29.0 g/mol.)

Q. 68. The following figure shows three 1.00-L bulbs connected by valves. Each bulbs contains neon gas with amounts proportional to the number of atoms pictorially represented in each chamber. All three bulbs are maintained at the same temperature. Unless states otherwise, assume that the valves connecting chambers are closed and seal the gases in their respective chambers. Assume also that the volume between chambers is negligible.

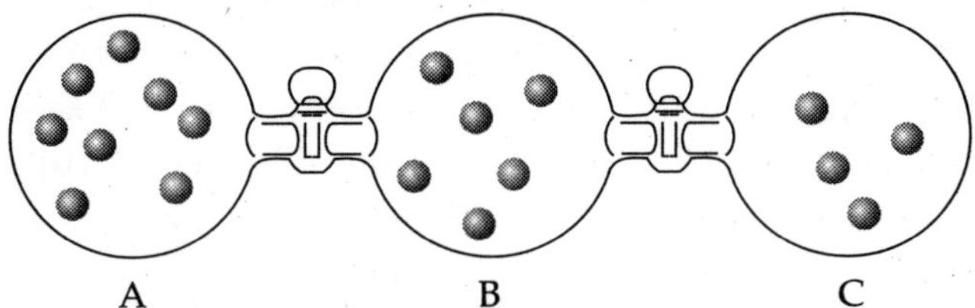

(a) Which bulb has the lowest pressure?
(b) If the pressure in bulb A is 2.00 atm, what is the pressure in bulb C?
(c) If the pressure in bulb A is 2.00 atm, what is the sum of $P_A + P_B + P_C$?
(d) If the pressure in bulb A is 2.00 atm and the valve between bulbs A and B is opened, redraw the above figure to accurately represent the gas atoms in all the bulbs. What is the $P_A + P_B$? What is $P_A + P_B + P_C$? Compare your answer with your answer in part (c).
(e) Follow the same instructions as in (d) if all the valves are open.

Q. 69. Consider an ideal gas that exerts a pressure 23.76 mm Hg at 298 K. Assuming n and V are held constant, what would its pressure be at 313 K? 343 K? 373 K? Compare the numbers you have just calculated with vapor pressures of water at these temperatures. Can you suggest a reason why the two sets of numbers are so different?

Q. 70. The escape velocity required for gas molecules to overcome the Earth's gravity and go off to outer space is 1.12×10^3 m/s at 288 K. Calculated the molar mass of species with that velocity. Would you expect to find H_e and H_2 molecules in the Earth's atmosphere? How would argon atoms?

Q. 71. A flask with a volume of 10.0 L contains 0.400 g of hydrogen gas and 3.20 g of oxygen gas. The mixture is ignited and the reaction

$$2H_2(g) + O_2(g) \rightarrow 2H_2O$$

goes to completion. The mixture is cooled to 300 K Assuming 100% yield,
(a) What physical state(s) of water is (are) present in the flask?
(b) What is the final pressure in the flask?
(c) What is the pressure in the flask if 3.2 g of each gas is used?

Q. 72. When the temperature drops from 293 K and 283 K, the pressure of a cylinder of compressed N_2 drops by 3.4%. The same temperature change decreases the pressure of a propane (C_3H_8) cylinder by 42%. Explain the difference in behaviour.

Q. 73. It has been suggested that the pressure exerted on a skate blade is sufficient to melt the ice beneath it and form a thin film of water, which makes it easier for the blade to slide over the ice. Assume that a skater weighs 120 kg and the blade has an area of 0.10 in^2.

Calculate the pressure exerted on the blade (1 atm 15 kg/m^2). From information in the text, calculate the decrease in melting point at this pressure. Comment on the plausibility of this explanation and suggest another mechanism by which the water film might be formed.

Q. 74. A compound of chlorine and fluorine, ClF_x, reacts at about 248K with uranium to produce uranium hexafluoride and chlorine fluoride, Clf. A certain amount of uranium produced 5.63 g of uranium hexafluoride and 457 ml of chlorine fluoride at 248K and 3.00 atm. What is x? Describe the geometry, polarity, and bond angles of compound and the hybridization of chlorine. How many sigma and pi bonds are there?

Q. 75. **(a)** Calculate the volume of 3.50 mol of an ideal gas at 16.0 atm and 320 K.

(b) At 320 K and 16.0 atm, 3.50 mol of ammonia gas, NH_3, occupy a volume of 5.20 mL. Calculate the percentage error made when the ideal gas law is used to determine the volume of 3.50 mol of ammonia at 320 K and 16.0 atm.

(c) Is the percentage error made in using the ideal gas law to compute the volume of 3.50 mol of NH_3 at 298 K and 1.00 atm greater than, less than, or the same as the error made at 320 K and 16.0 atm? Explain your answer.

Q. 76 A glass bulb of volume 2.0 L contains 0.625 mole of helium gas at 298 K. Helium is a monatomic gas; the only kind of kinetic energy monatomic gases possess is translational kinetic energy.

(a) What is the total kinetic energy of this sample of He gas, expressed in joules.

(b) What is the average kinetic energy of an atom of He in this sample? Express the answer in joules.

(c) How much energy, expressed in joules, is required to heat this sample of gas from 298 K to 398 K?

Q. 77. At a given instant of time the speeds of four flies kept in a jar are 3.00, 4.00, 7.00, and 10.00 m · s^{-1}. Calculate u, u^2, and u_{rms} for the flies at that instant.

Q. 78. The gas with the lowest molecular weight is hydrogen H_2. A substance with a high molecular weight that is gaseous at firely low temperatures is uranium hexafluoride, UF_6. Uranium is a heavy metal, and most compounds of uranium are solids at room temperature. Uranium hexafluoride, however, is a liquid at room temperature, and boils at 329.6 K. Calculate the root-mean square speeds of H_2 and UF_6 at 353K.

Q. 79. A known volume of nitrogen gas at a fixed temperature and pressure effuses through a small orifice 1 min and 24 s. It takes 1 min and 43 s for the same volume of an unknown gas to effuse at the same temperature and pressure. What is the approximate molecular weight of the unknown gas? If the empirical formula of this gas is determined from its elemental analysis to be $(CH_2)_n$, determine its molecular formula and exact molecular weight.

Q. 80. **I.** The root-mean-square speed of O_2 at 273K is 4.61×10^2 m · s^{-1}. What is the toot-mean-square speed of SO_2, (a) at 273 K and (b) at 373 K?

II. (a) Calculate the pressure of 10.00 mol of ethane, C_2H_6, in a 4.86-L flask at 300 K using (1) the ideal gas law and (2) van der Waals' equation.

(b) The observed pressure for 10.00 mol of C_2H_6 at 300.0 K in a volume of 4.86 L is 34.0 atm. Calculate the percentage error using (1) the ideal gas law and (20 van der Waals' equation.

Q. 81. A 2.640-g sample of an unknown gas occupies a volume of 705.6 mL at 273 K and 1 atm (STP). What is the molecular weight of this gas?

Q. 82. An evacuated gas bulb of volume 952 mL and mass 73.2684 g was filled with the gaseous borane at 273 K and a pressure of exactly 1 atm. The mass of the gas-filled bulb was 75.5398 g. This sample of the compound was then burned in excss oxygen, which completely converted it to water, H_2O, and the solid white boron oxide, B_2O_3. He mass of B_2O_3 was found to be 5.9316 g. What is the correct molecular formula of the borane?

Q. 83. A sample of gas contained in a bulb supports a column of mercury 680.2 mm high. What is the pressure of this gas in (a) atmospheres, (b) newtons per square meter or pascals.

Q. 84. What is the height of a column of water than can be supported in a barometer tube by a pressure of exactly 1 atm at 298 K? The density of water is 0.997 $g \cdot cm^{-3}$ at 298 K.

Q. 85. A sample of gas occupies a volume of 400.0 mL at 298 K and 1.00 atm. What volume will the gas occupy at 473 K at the same pressure?

Q. 86. A gas volume "thermometer" is filled with a sample of N_2 gas at 1-atm pressure, and immersed in an ice-water bath at 273.15 K. The volume of the gas is found to be 58.7 mL. The bulb containing the gas is then immersed in boiling liquid isopropyl alcohol, and the volume of the same sample of gas is found to be 76.4 mL at a pressure of 1 atm. What is the boiling point of isopropyl alcohol?

Q. 87. A gas is found to occupy a volume of 346 mL at 298 K when the pressure is 751 mmHg. What volume will this gas occupy if the pressure is increased to 1.40 atm and the temperature is increased to 308 K?

Q. 88. A bulb is filled with 20.0 g of CO_2 gas at 296 K and 1 atm. The bulb is heated to 336 K, and the stopcock is opened. How much CO2 escapes, if the external pressure is 1 atm when the stopcock is opened, and the bulb is kept at 336 K? When the pressure inside the bulb reached 1 atm, the stopcock is closed again.

Q. 89. An evacuated bulb of unknown volume is filled with a sample of H_2 gas at a temperature *T*. The pressure of the gas in the bulb is 756 mmHg. A portion of the H_2 gas is transferred to a different flask and found to occupy a volume of 40.0 mL at 1.00 atm and the same temperature, *T*. The pressure of the H_2 remaining in the original bulb drops to 625 mmHg at the same temperature, *T*. Assuming H_2 is an ideal gas, what is the volume of the bulb?

Q. 90. A sample of butane of mass 3.728 g is placed in an evacuated bulb of volume 489 mL at 298 K. What is the pressure of the gas in the bulb?

Q. 91. At room temperature, pyridine is a colorless liquid with an extremely unpleasant odor. Pyridihe ocurs naturally in tobacco and hemlock. It contains only the elements C, H and N, and is 75.92% C and 6.37% H by weight. At 383 K and a pressure of 630 mmHg, the density of gaseous pyridine has been measured as 2.12 $g \cdot L^{-1}$.

(a) Calculate the molecular formula of pyridine.

(b) How great an error is made by using the ideal gas equation to calculate the molecular weight of pyridine from the gas density at this temperature and pressure?

Q. 92. Bdenzoic acis is a white, crystalline solid, important as the starting point in the manufacture of a large group of chemicals used in medicine, food preservatives, dyes, and insect repellents. A 1.5388-g sample of benzoic acid, C_6H_5COOH, is completely burned with excess oxygen in an 6.00-L container. The equation for the combustion reaction is

$$C_6H_5COOH(s) + \frac{15}{2} O_2(g) \rightarrow 7\,CO_2(g) + 3H_2O(l)$$

When the reaction is complete, the container and its contents are maintained at a constant temp 298 K.

(a) What is the partial pressure of CO_2 in the container after the reaction is complete? Assume that the volume of liquid water formed is negligible compared to the volume of the container.

(b) The total pressure in the container is 814 mmHg after the combustion. Calculate the partial pressure of the excess (unused) oxygen in the container. How many moles of O_2 remain in the container after the combustion is complete?

Q. 93. A closed bulb contains 0.0100 mol of inert helium gas and a sample of solid white ammonium chloride, NH_4Cl. Assume that the volume of the solid NH_4Cl is negligible compared to the volume of the bulb. The pressure of the He is measured at 300 K and is found to be 114 mmHg. The bulb is then heated to 600 K. All the NH_4Cl decomposes according to the equation $NH_4\,Cl(s) \longrightarrow NH_3(g) + HCl(g)$

(a) What is the partial pressure of HCL(g) in the bulb at 600 K when reaction is complete?

(b) How many grams of $NH_4Cl(s)$ were in the bulb at 300 K?

Q. 94. A mixture of propane, C_3H_8 and methane, CH_4 gas is contained in a vessel of unknown volume, *V*, at a temperature *T*, and exerts a pressure of 320 mmHg. The gas is burned in excess oxygen and all the carbon is recovered as CO_2. The CO_2 is found to have a pressure of 448 mmHg in a volume, *V*, at the same temperature, *T*. Calculate the mole fraction of propane in the original mixture. Assume all gases are ideal.

Q. 95. How many grams of nitrogen are there in 100 liters of NO measured at STP?

Q. 96. A volume of 2.0 liters of a gas, measured at STP, has a mass of 5.71 g. Calculate the approximate molecular weight of the gas.

Q. 97. What volume will 45.0 g of CH_4 gas occupy at 300 K and 800 mm?

Q. 98. The hydrogen gas in a 2.0-liter steel cylinder at 298 K was under a pressure of 4.00 atm. How many moles of H_2 were in the cylinder?

Q. 99. What pressure, in atmospheres, will 26.0 g of He gas exert when placed in a 3.24-lter steel cylinder at 473 K?

Q. 100. Calculate the mass in grams of the pure H_2S gas contained in a 60.0-liter cylinder at 293 K under a pressure of 2.0 atm.

Q. 101. A 32.4-g sample of an ideal gas in a 12.0-liter container at 298 K exerted a pressure of 1.50 atm. Calculate its molecular weight.

Q. 102. Calculate the volume occupied by 5.0 g. of acetylene gas at 323 K and 740 mm pressure.

Q. 103. What is density of C_2H_2 gas at STP?

Q. 104. A reaction vessel contained 5 liters of a mixture of N_2 and O_2 gases at 298 K and 2 atm pressure. The oxygen in the mixture was completely removed by causing it to oxidize and excess of electrically heated zinc wire contained in the vessel to solid ZnO (nonvolatile.) The pressure of the nitrogen gas that remained in the vessel, measured at 298 K, was 1.5 atm. What was the mole percent of oxygen in the original mixture; the mass percent?

Q. 105. The volume of a dry gas is 600 cc at 298 K and 750 mm. What volume would this gas occupy if collected over water at 746 mm and 305 K? Vapour pressure of water at 305 K = 35.4 mm.

Q. 106. In a gaseous mixture of equal grams of CH_4 and CO what is the ratio of moles of CH_4 to CO? What is the mole fraction of CH_4?

Q. 107. A mixture of 50.0 g of oxygen gas and 50.0 g of CH_4 gas is placed in a contained under a pressure of 600 mm. What is the partial pressure of the oxygen gas in the mixture.

Q. 108. Gas *A* is nine times as dense as gas *B*. In a given diffusion apparatus and at a certain temperature and pressure, gas *B* diffuses 15 cm in 10 sec. In the same apparatus and at the same temperature and pressure, how fast will *A* difuse?

Q. 109. How do the rates of diffusion of HBr and SO_2 compare?

Q. 110. A student cut a 4.60 cm piece of pure magnesium ribbon. The student then prepared a gas collecting tube with HCl at the bottom and water at the top and placed the magnesium in a copper cage into the water end. The tube was then inverted (see Figure below. The HCl being more dense flowed down the tube and then reacted with magnesium, producing bubbles of gas. After all of the magnesium has reacted, the student then transferred the gas collecting tube carefully to a large cylinder filled with water and adjusted the gas collecting tube so that the water level in the tube was even with the water

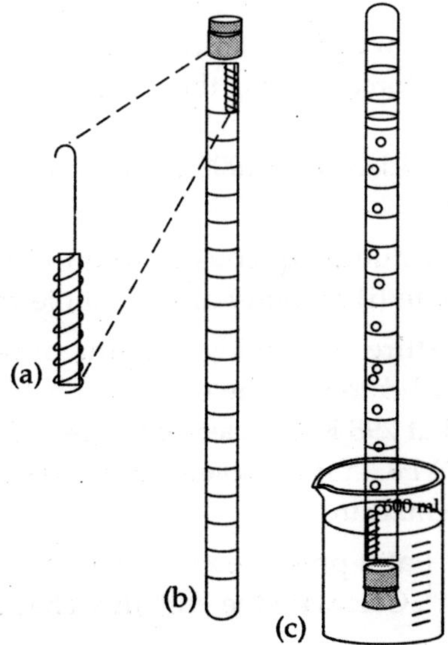

level in the cylinder. The student then read the volume of the gas in the collecting tube as 40.44 cm^3. The temperature of the water was 299.3 K and air pressure of the room was 743 mm of Hg. Answer the following 1 to 9.

1. Given that 1.00 mm of the pure Mg ribbon weighs 0.816 g, determine the mass of Mg used in the experiment.
2. Calculate the number of moles of Mg used in the experiment.
3. Calculate the partial pressure of the hydrogen gas in the mixture. The vapour pressure of the water at 299.3 K is 25.7 mm Hg.
4. Calculate the volume of the hydrogen gas at STP assuming that the gas temperature equals the temperature of the aqueous HCl.
5. Write the balanced equation for the reaction.
6. Determine the number of moles of $H_2(g)$ produced, based on the moles of Mg used.
7. Calculate the molar volume of the $H_2(g)$ at STP.
8. What was the percent error in this experiment?
9. Determine the number of moles of hydrogen gas by using the Ideal Gas Law.

Q. 111. The pressure is a bulb dropped from 2000 to 1500 mm of mercury in 4 minutes when the contained oxygen leaked through a small hole. The bulb was then evacuated. A mixture of oxygen and another gas of molecular weight 79 in the molar ratio 1 : 1 at a total pressure of 4000 m of mercury was introduced. Find the molar ratio of the two gases remaining in the bulb after a period of 74 minutes.

Q. 112. Calculate the energy required to excite one litre of hydrogen gas at 1 atm and 298 K to the first excited state of atomic hydrogen. The energy for the dissociation of H-H bond is 436 KJ mol^{-1}.

Q. 113. The density of the vapour of a substance at 1 atm pressure and 500 K is 0.36 kgm^{-3}. The vapour effuses through a small hole at a rate of 1.33 times faster then oxygen under the same condition.

(a) determine (i) molecular weight, (ii) molar volume, (iii) compression factor (Z) of the vapour and (iv) which forces among the gas molecules are dominating, the attractive or the repulsive?

(b) If the vapour behaves ideally at 1000 K, determine the average translation kinetic energy of a molecule.

Q. 114. The vapour pressure of a dilute aqueous solution of glucose ($C_6H_{12}O_6$) is 750 mm of mercury at 373 K. Calculate (i) molality and (ii) mole fraction of the solute.

Q. 115. A 2.0 g sample of a mixture containing sodium carbonate, sodium bicarbonate and sodium sulphate is gently heated till the evolution of CO_2 ceases. The volume of CO_2 at 750 mm Hg pressure and at 298 K is measured to be 123.9 ml. A 1.5g of the same sample requires 150 ml of (M/10) HCl for complete neutralisation. Calculate the % composition of the components of the mixture.

Q. 116. A sample of argon gas at 1 atm pressure and 300 K expands reversibly and adiabatically from 1.25 dm^3 to 2.50 dm^3. Calculate the enthalpy change in this process. $C_{v.m}$ for argon is 12.48 JK^{-1} mol^{-1}.

Q. 117. The degree of dissociation of $Ca(NO_3)_2$ in a dilute aqueous solution, containing 7.0 g of the salt per 100 gm of water at 373 K is 70%. If the vapour pressure of water at 373 K is 760 mm., calculate the vapour pressure of the solution.

Q. 118. The vapour pressure of two miscible liquids A and B are 300 and 500 mm of Hg respectively. In a flask 10 moles of A is mixed with 12 moles of B. However, as soon as B is added, A starts polymerising into a completely insoluble solid. The polymerisation follows first-order kinetics. After 100 minutes, 0.525 mole of a soluble is dissolved which arrest the polymerisation completely. The final vapour pressure of the solution is 400 mm of Hg. Estimate the rate constant of the polymerisation reaction. Assume negligible volume change of mixing and polymerisation, and ideal behaviour for the final solution.

Q. 119. For the reaction, $N_2O_5(g) = 2NO_2(g) + 0.5\ O_2(g)$,
calculate the mole fraction of $N_2O_5(g)$ decomposed at a constant volume and temperature, if the initial pressure is 600 mm Hg and the pressure at any time is 950 mm Hg. Assume ideal gas behaviour.

Q. 120. A 20.0 cm^3 mixture of CO, CH_4. and He gases is exploded by an electric discharge at room temperature with excess of oxygen. The volume contraction is found to be 13.0 cm^3. A further contarction of 14.0 cm^3 ocurs when the residual ga sis treated with KOH solution. Find out the composition of the gaseous mixture in terms of volume percentage.

Q. 121. 1 litre of a mixture of CO and CO_2 is taken. This mixture is passed through a tube containing red hot charcoal. The volume now becomes 1.6 litre. The volumes are measured under the same conditions.

Find the composition of the mixture by volume.

Q. 122. A 5.0 cm^3 solution H_2O_2 liberates 0.508 g of iodine from an acidified KI solution. Calculate the strength of H_2O_2 solution in terms of volume strength at STP.

Q. 123. A mixture of ethane (C_2H_6) and ethene (C_2H_6) occupies 40 litres at 1.00 atm and at 400 K. The mixture reacts completely with 130 g of O_2 to produce CO_2 and H_2O. Assuming ideal as behaviour, calculate the mole fraction of C_2H_4 and C_2H_6 in the mixture.

Q. 124. The pressure exerted by a 12 g of an ideal gas at temperature (t+273) K in a vessel of volume *V* litre is one atm. When the temperature is increased by 10 degrees at the same volume, the pressure increases by 10%. Calculate the temperature *t* and volume *V*. (Molecular weight of the gas = 120)

Q. 125. An evacuated gas vessel weighs 50.0 g when empty, 148.0 g when filled with a liquid of density 0.98 g ml^{-1} and 50.5 g when filled with an ideal gas at 760 mm Hg at 300 K. Determine the molecular weight of the gas.

Q. 126. Oxygen is present in 1 litre flask at a pressure of 76×10^{-10} mm of Hg. Calculate the number of oxygen molecules in the flask at 273 K.

Q. 127. Calculate the average kinetic energy, of the molecules in 8.0 g of methane at 300 K.

Q. 128. When 2 g of a gas *A* is introduced into an evacuated flask kept at 298 K the pressure is found to be one atmosphere. If 3 g of another gas *B* is then added to the same flask, the total pressure becomes 1.5 atm. Assuming ideal gas behaviour, calculate the ratio of the molecular weights.

Q. 129. At room temperature the following reactions proceed nearly to completion:

$$2NO + O_2 \rightarrow 2NO_2 \rightarrow N_2O_4.$$

The dimer, N_2O_4, solidifies at 262 K. A 250 ml flask and a 100 ml flask are separated by a stop-clock. At 300 K, the nitric oxide in the larger flask exerts a pressure of 1.053 atm and he smaller one contains oxygen at 0.789 atm. The gases are mixed by opening the stopcock and after the end of the reaction the flasks are cooled to 220 K. Neglecting the vapour pressure of the dimer, find out the pressure and composition of he gas remaining at 220 K. (Assume the gases to behave ideally).

Q. 130. A spherical balloons of 21 cm diameter is to be filled up with hydrogen at N.T.P. from a cylinder containing the gas at 20 atmosphere at 300 K. If cylinder can hold 2..82 litres of water calculate the number of balloons that can be filled up.

Q. 131. Calculate the pressure exerted by one mole of CO_2 gas at 273 K if the van der Waals constant a = 3.592 dm^6 atm mol^{-2}.

$$a = 3.592 \text{ dm}^6 \text{ atm mol}^{-2}$$

Assume that the volume occupied by CO_2 molecules is negligible.

Q. 132. An LPG (liquified petroleum gas) cylinder weights 14.8 kg when empty. When full, it weights 29.0 kg and shows a pressure of 2.5 atm. In the course of use at 300 K, the weight of the full cylinder reduced to 23.2 kg. Find out the volume of the gas in cubic meters used up at the normal usage conditions, and the final pressure inside the cylinder. Assume LPG to be n-butane with normal boiling point of 273 K.

Q. 133. The compression factor (compressibility factor) for one mole of a van der Waals gas at 0°C and 100 atmosphere pressure is found to be 0.5. Assuming that the volume of a gas molecule is negligible, calculate the van der Waals constant(a).

Q. 134. One way of writing the equation of state for real gases is

$$pV = RT\left[1 + \frac{B}{V} + \dots\right]$$

where B is a constant. Derive on approximate expression for B in terms of the Van der Waals constant a and b.

Q. 135. Using van der Waals equation, calculate the constant, 'a' when two moles of gas confined in a four litre flask exerts a pressure of 11.0 atmospheres at a temperature of 300 K. The value of 'b' is 0.05 lit mol^{-1}.

Q. 136. Calculate the root mean square velocity of ozone kept in a closed vessel at 293 K and 82 cm mercury pressure.

Q. 137. The average velocity at T_1K, and the most probable at T_2K of CO_2 gas is 9.0×10^4 cm sec^{-1}. Calculate the value of T_1 and T_2.

Q. 138. A gas bulb of 1 litre capacity contains 2.0×10^{21} molecules of nitrogen exerting a pressure of 7.57×10^3 NM^{-2}. Calculate the root mean square (r.m.s) speed and the temperature of the gas molecules. If the ratio of the most probable speed to the root mean square speed is 0.82, calculate the most probable speed for these molecules at this temperature.

Q. 139. The ratio between the root mean square velocity of H_2 at 50 K and that of O_2 at 800 K is,

(a) 4 (b) 2 (c) 1 (d) 1/4

Q. 140. The average velocity of gas molecules is 400 m/sec. Calculate its rms velocity at the same temperature.

Q. 141. The vapour pressures of ethanol and methanol are44.5 and 88.7 mm respectively. A ideal solution is formed at the same temperature by mixing 60 g of ethanol with 40 g of methanol. Calculate the total vapour pressure of the solution and the mole fraction of methanol in the vapour.

Q. 142. While studying the decomposition of gaseous N_2O_5 it is observed that a plot of logarithm of its partial pressure versus time is linear. What kinetic pressure versus time is linear. What kinetic parameters can be obtained from this observation.

Q. 143. A 4 : 1 molar mixture of He and CH_4 is contained in a vessel at 20 atm pressure. Due to a hole in the vessel the gas mixture leaks out. What is the composition of he mixture effusing out initially?

Q. 144. A hydrocarbon contains 10.5 gm of carbon per gram of hydrogen 1 litre of the vapour of the hydrocarbon at 400 K and 1 atmosphere pressure weighs 2.8 gm. Find the molecular formula.

Q. 145. At room temperature, ammonia gas at 1 atm pressure and hydrogen chloride gas at *P* atm pressure are allowed to effuse through identical pin holes from opposite ends of a glass tube of one metre length and of uniform cross-section. Ammonium chloride is first formed at a distance of 60 cm from the end through which HCl gas is sent in. What is the value of P?

Q. 146. One mole of nitrogen gas at 0.8 at takes 38 s to diffuse through a pinhole, whereas one mole of an unknown compound of xenon with fluorine at 1.6 atm takes 57 s to diffuse through the same hole Calculate the molecular formula of the compound.

Q. 147. At 300 K hydrogen is leaked through a tiny hole into a vessel for 20 minutes. Another unknown gas at the same temperature and pressure as that of H_2 is leaked through the same hole for 20 minutes. After the effusion to the gases the mixture exerts a pressure of 6 atmosphere. The hydrogen content of the mixture is 0.7 mole. If the volume of the container is 3 liters, what is the molecular weight of the unknown gas?

Chapter **4**

STOICHIOMETRY

FORMULAE

1 mole $= 6.023 \times 10^{23}$

Mass of 1 atom $= \dfrac{\text{Atomic mass (g mol}^{-1})}{6.03 \times 10^{23}\ (\text{mol}^{-1})}$

Mass of 1 molecule $= \dfrac{\text{Molecular mass (g mol}^{-1})}{6.03 \times 10^{23}\ (\text{mol}^{-1})}$

No. of moles $= \dfrac{\text{Mass of the substance (g)}}{\text{Molecular mass of substance (g mol}^{-1})}$

No. of moles of a gas $= \dfrac{\text{Volume of gas at S.T.P. (mL)}}{22400\ \text{mL}}$

Loss of electrons = Oxidation
Increase in Ox. No. = oxidation
Gain of electrons = Reduction
Decreases in Ox. No. = Reduction
Reduction + Oxidation = Redox
Oxidants→ Oxidise other → Reduced themselves → Gain electrons
Reductants → Reduce other → Oxidised themselves → Loss electrons

Equiv. Wt. of redox reagents $= \dfrac{\text{Molecular weight}}{\text{Change in Ox. No.}}$

DEFINITIONS, TERMS AND FUNDAMENTAL LAWS

Law of Conservation of Mass

Matter can neither be created nor destroyed. In a chemical change the total mass of the reactants is equal to the total mass of the products.

Law of Constant Proportions

A chemical compound has a fixed composition by weight.

Law of Multiple of Proportions

When two elements combine to form two or more compounds, the different weights of one more elements combining with the constant weight of the other bear a simple numerical ratio to one another.

Law of Reciprocal Proportions

The ratio between the weights of two elements *A* and *B* which combine with the fixed weight of the third element *C*, will be the same or simple multiplies of the ratio in which the elements *A* and *B* combine together.

Gay-Lussac's Law of Combining Volumes

When gases combine they do so in volumes which bear a simple ratio to one another and to the volume of the product, if gaseous, provided all the gases are measured under the same conditions of temperature and pressure.

Dalton's Atomic Theory

(a) Every element is composed of atoms.
(b) Atoms are indivisible. They can neither be created not destroyed.
(c) Atoms of the same element are identical.
(d) Atoms combine in simple ratio to form compound atoms.
(e) Compound atoms of same compound are identical.

Atom

A atom is the smallest particle of an element that takes part in a chemical reaction.

Molecule

A molecule is the smallest particle of an element or compound that can exist in free state.

Atomic Mass

The atomic mass of an element is the ratio of the mass of one atom of an element to 1/12 part of the mass of a carbon-12 atom.

Atomic Mass Unit (AMU)

The atomic mass unit is defined as one twelfth of the mass of nuclide ${}^{12}_{6}C$. [1 amu = 1.66×10^{-24} g]

Avogadro's Number, N

Avogadro's number is the number of molecules contained in one gram molecular weight of the substance or the number of atoms contained in one gram atom of an element. Its value of 6.023×10^{23}.

Mole

A mole is the amount of substance. Which contains one Avogadro number of particles (atoms, molecules, ions or electrons).

Gram Atom (Atomic Weight in Grams)

The gram atom of an element is the weight of Avogadro number (6.02×10^{23}) of atoms of it in grams.

The term gram atom is an abbreviation for gram-atomic weight, since one gram-atomic weight of an element is 6.023×10^{23} atoms. As one mole of atoms means 6.023×10^{23} atoms, the number, 6.023×10^{23} is called the Avogadro number, has symbol *N*. Hence, when applied to a particular element, the terms, *one mole of atoms, one gram atom, one gram-atomic weight, 6.023×10^{23} atoms, N atoms;* all mean the same thing.

Specifically, 32.064 g of sulfur is:

one mole of sulfur atoms, one gram atom of sulfur, one gram-atomic weight of sulfur, 6.023×10^{23} atoms of sulfur, *N* atoms of sulfur.

Molecular Weight

The molecular weight of a substance (element or compound) is the ratio of the substance to 1/12 part of the mass of a carbon-12 atom. It is the sum of atomic weight of all the atoms which form the molecule.

Gram Mole (Gram Molecular Weight)

The gram molecular weight of a substance in the weight of Avogadro number of molecules or its weight in grams. For oxygen Gram molecular weight = $16 \times 2 = 32$ amu = $32 \times 1.66 \times 10^{-24} = 53.2 \times 10^{-24}$ gms.

Gram Molar Volume (GMV)

The volume occupied by one gram molecular weight of any gas or vapour at STP is called gram molar volume. Its value is 22.4 litres or 22400 ml. It is also defined as the volume occupied by Avogadro's number (6.02×10^{23}) of molecules.

Avogadro's Hypothesis

Equal volumes of all gases under the same conditions of temperature and pressure contain equal number of molecules.

Relative Density (R.D)

$$\text{R.D} = \frac{\text{density of gas}}{\text{density of hydrogen}}$$

Vapour Density (V.D)

The ratio between the density of vapour and density of hydrogen.

$$\text{Vapour density} = \frac{\text{weight of 1 lit. of vapour at STP}}{\text{weight of 1 lit. of } H_2 \text{ at STP}}$$

Also, 2 × V.D= Molecular weight
Density × 11.2= Vapour density
Density × 22.4= Molecular weight

Emperical Formula

Shows the ratio of atoms of different elements in one molecule of a compound, eg. emprical formula of benzene is CH.

Molecular Formula

Shows the actual number of atoms of the different elements in one molecule of a compound, eg. molecular formula of benzene is C_6H_6.

Structural Formula

Shows the arrangement of atoms in a molecule; eg. structural formula benzene is 

Oxidation & Reduction

Oxidation is a process which liberates electrons i.e. it is the loss of electrons operations by an atom or an ion.
Reduction is a process which gains electrons. i.e. it is the gain of electrons by an atom or an ion.

Oxidation Number & Reduction

Oxidation number indicates the charge which an atom appears to have when electrons are counted in its compound. It may be +ve, –ve or 0.

Oxidation is increase in oxidation number and reduction a decrease in oxidation number. Oxidation number is an arbitrary number assigned to element according to some rules and conventions.

Oxidation numbers are assigned as follows:

(i) free element = 0
(ii) oxygen in a compound = –2 (–1 for peroxides and +2 in OF_2)
(iii) hydrogen in a compound = +1 (–1 for metallic hydrides)
(iv) group IA metal in a compound = +1
(v) group IIA metals in a compound = +2
(vi) sum of the oxidation number of a neutral molecule = 0
(vii) sum of the oxidation number of an ion = charge on the ion.

Oxidation number per atom is called *oxidation state.*
Oxidation number equations are balanced as follows:

(a) assign oxidation numbers to the elements
(b) determine change in oxidation number per atom and per molecule
(c) make increase in oxidation number equal decrease in oxidation number by use of coefficients
(d) balance the respect of equation by inspection.

eg: in $KMnO_4$, oxidation number of Mn = –7 while oxidation state of Mn = Mn^{7+}

Balancing Oxidation—Reduction Equations

The steps in balancing redox equations by the $\frac{\text{half reaction}}{\text{equation}}$ method are illustrated by the equation

$$MnO_4^- + H^+ + C_2O_4^{2-} \rightarrow Mn^{2+} + CO_2 + H_2O$$

(a) separate the equation into half-reactions
(b) balance all atoms except oxygen and hydrogen
(c) balance oxygen by adding water; balance hydrogen by adding H^+
(d) balance charges by adding e^-
(e) add balanced half-reactions to e^- cancel
(f) cancel terms and neutralise H^+ and OH^- if necessary.

$$MnO_4^- + 8H^+ + 5e^- \rightarrow Mn^{2+} + 4H_2O] \times 2$$
$$C_2O_4^{2-} \rightarrow 2CO_2 + 2e^-] \times 5$$
$$\overline{2MnO_4^- + 16H+ + 5\,C_2O_4^{2-} \rightarrow 2Mn^{2+} + 10CO_2 + H_2O}$$

Equivalent Weight

Equivalent Weight of a substance is that weight of the substance which will combine with or displace directly or indirectly 1.008 parts by weight of hydrogen, 8 parts by weight of oxygen, 35.5 parts by weight of chlorine and 108 parts by weight of silver.

It is also defined as a quantity of acid that produces 1 mol of H^+ or a quantity of base that reacts with 1 mol. of H^+, *OR.* A quantity of substance associated with transfer of 1 mol. of electrons in an oxidation-reduction reaction.

1g-equivalent of a substance means its equivalent weight in grams.

i.e. equivalent weight $= \dfrac{\text{weight in gms.}}{\text{no. of equation}}$

Equivalent Weight of an Acid

It is the ratio of M.wt. to its basicity.

$E_{Acid} = \dfrac{\text{M. wt.}}{\text{Basicity}}$, Basicity is the number of replaceable hydrogen atom represent in an acid.

eg.

$E_{HCl} = 36.5/1 = 36.5$ (monobasic),

$E_{H_2SO_4} = 98/2 = 49.0$ (dibasic),

$E_{Fe(CH)_3} = 108/3 = 36.0$ (tribasic) $\left[HO-\overset{\overset{O}{\uparrow}}{\underset{\underset{OH}{|}}{P}}-OH \right]$

Equivalent Weight of Base

It is the ratio of M.wt. to its acidity.

$E_{Base} = \dfrac{\text{M. wt.}}{\text{Acidity}}$, Acidity is the number of replaceable-OH groups *OR* the number of monobasic acid molecules with which one molecule of the base reacts. eg.

$E_{NaOH} = 40/1 = 40$ (monobasic),

$E_{Ca(OH)_2} = 74/2 = 49.0$ (dibasic),

$E_{Fe(OH)_3} = 107/3 = 35.7$ (tri acidic)

In $H_3PO_3 + 2NaOH \longrightarrow Na_2HPO_3 + 2H_2O$ $\left(E_{H_3PO_3} = \dfrac{M}{2} \ \& \ E_{NaOH} = \dfrac{M}{1}\right)$

(Basicity = 2) (Acidity = 1)

Equivalent Weight of a Salt

It is the ratio of M.wt. to the total charge of the Cation or Anion.

$E_{salt} = \dfrac{\text{M. wt.}}{\text{Total change of cotion or Anion}}$

e.g. $E_{NaCl} = 58.1/1 = 58.5$ (Monovalent),

$E_{BaCl_2 \cdot 2H_2O} = 244/2 = 122.15$ (Divalent),

$E_{AlCl_3} = \dfrac{133.5}{3} = 44.5$ (Trivalent)

Equivalent weight by Oxidation-Reduction Method

Equivalent weight by *Oxidising Agent or reducing Agent*: It is the ratio of M.wt. to the change ion oxidation number. (i.e. of the number of electrons gained or lost in reaction)

$$E_{\text{Red-ox.agent}} = \frac{\text{M.wt.}}{\text{Total change in oxidation number}} = \frac{\text{M. wt.}}{\text{No. of electrons gained or lost}}$$

eg: $E_{KMnO_4} = \frac{158}{5} = 31.6$ (in acidic medium)

$= \frac{158}{1} = 158$ (in alkali)

$= \frac{2.\text{M.wt.}}{6} = \frac{2 \times 158}{6} = 52.6$

Equivalent weight of an ion

$$\text{Eq. Wt. of ion} = \frac{\text{Formula weight of ion}}{\text{charge on ion}}$$

PROBLEMS

Q. 1. How many grams of sulfur are there in 2.2 moles of H_2S?

Q. 2. How many gram atoms are there in 144 g. of sulphur?

Q. 3. How many moles of O are there in 182 g of $KClO_3$?

Q. 4. How many grams of phosphorus are there in 160 g of P_4O_{10}?

Q. 5. Calculate the percent of carbon in CO_2.

Q. 6. How many grams and gram atoms of nitrogen are contained in 0.25 mole of HNO_3?

Q. 7. How many gram atoms are there in 100 g. of magnesium (at. Wt = 25.3)?

Q. 8. Calculate the mass of a single atom of sulphur and a single molecule of carbon monoxide?

Q. 9. How many atoms of hydrogen and sulphur are contained in 1.6 moles of hydrogen sulphide?

Q. 10. What is the volume occupied by 2.5 moles of carbon dioxide at STP?

Q. 11. Which of the following weigh maximum? (a) 50 g. iron (b) 5 g. atoms of nitrogen (c) 0.1 g. atom of silver (d) 1×10^{23} atoms of carbon.

Q. 12. What is the number of molecules present in 1.8 g. of water?

Q. 13. Calculate the volume at STP occupied by each of the following?

(a) 0.5 mole nitrogen

(b) 6.023×10^{22} molecules of H_2S

(c) 0.25 mole of ammonia

(d) 8 g of oxygen

Q. 14. A tetra-atomic gas occupies 1.4 litres at 273 K and 1 atm pressure. Find the number of atoms in the gas?

Q. 15. How many atoms are present in 21 g. of Ozone?

Q. 16. How many electrons are to be added to convert one mole of permanganate ions into manganous ions?

Q. 17. How many atoms are present in 5.6 lit. of carbon dioxide at STP.

Q. 18. Calculate the difference in the no. of Carbon atoms in 1.00 g of C-14 isotope (atomic mass = 14.01 g. mol^{-1}) and 1 g. of C-12 isotope.

Q. 19. Calculate the total no, of electrons present in 1.6 g. of methane (At. Wt.s H = 1 C, 12, Avogadro no. = 6.023×10^{23})

Q. 20. One atom of an element 'X' weighs 6.644×10^{-23} g. Calculate the no. of gram atoms in 40 Kg of it (N = 6.02×10^{23})

Q. 21. How many molecules are there in 64 g. of sulphur.

Q. 22. What is the weight of 3.015×10^{20} molecules of phosphorus?

Q. 23. Calculate the number of moles present in 8 g of oxygen and 48 g. of sulphur dioxide in gaseous mixture.

Q. 24. Weight of 350 ml. of diatomic gas at 273 K and 2 atmosphere pressure. is 1 g. Calculate the weight in gram of one atom.

Q. 25. Density of water at room temperature is 1.0 g/ml. How many molecules are there in 1 ml. of water?

Q. 26. The volume of a gas in an electron tube is 1.12×10^{-7} cc. N.T.P. Calculate the number of molecules of the gas in the tube.

Q. 27. From 200 mg. of CO_2; 10^{21} molecules are removed. How many molecules of CO_2 are left?

Q. 28. The mass of one sodium atom is 3.82×10^{-23} g. Calculate the number of sodium atoms present in 2 moles of sodium.

Q. 29. A compound containing one gram atom of oxygen per mole of the compound has 57.14% of oxygen. Calculate the molecular weight of the compound.

Q. 30. A metal M of atomic weight 54.94 has a density of 7.42 g cm^{-3}. Calculate (i) the apparent volume occupied by one atom of the metal and (ii) the radius of the atom assuming it to be a sphere.

Q. 31. If a mole were to contain 1.00×10^{24} particles. Which would be the mass (i) of a single oxygen molecule and (ii) of one mole of oxygen?

Q. 32. Calculate the weight of Mg in grams which contain same number of atoms as are present in 4.0 g of Ca.

Q. 33. How many atoms are present in a mixture of 2g. of hydrogen and 16g. of oxygen.

Q. 34. A tomato weighing 60.0 g is analyzed for sulfur content by digestion in concentrated HNO_3 followed by precipitation with Ba to form $BaSO_4$. 0.466 g of $BaSO_4$ is obtained. If all the sulfur is converted to $BaSO_4$, what is the percentage of sulfur in the tomato?

Q. 35. How many moles are there in one atom?

Q. 36. How many moles of Mg are present in a sample containing 1.00×10^{22} Mg atoms?

Q. 37. A sample of oxygen contains 1.87×10^{27} atoms of oxygen. What would be the weight of this collection of atoms?

Q. 38. In a chemical reaction requiring three atoms of magnesium (Mg) for two atoms of nitrogen (N), how many grams of N are required by 4.86 of Mg? (The atomic weight of Mg is 24.3 amu, and the atomic weight of N is 14.0 amu.)

Q. 39. What are the minimum amounts of $Fe(NH_4)_2(SO_4)_2 \cdot 6H_2O$ and $K_2C_2O_4$ needed for the synthesis of 10.0 g of $K_3Fe(C_2O_4)_3$?

Q. 40. During the course of world war I, 1.1×10^8 kg of poison gas was fired on Allied soldiers by German troops. If the gas is assumed to be phosgene ($COCl_2$), how many molecules of gas does this correspond to?

Q. 41. A flask containing H_2 at 273 K was sealed off at a pressure of 1 atm and the gas was found to weigh, 4512 g. Calculate the number of moles and the number of molecules of H_2 present

Q. 42. For a single year, the motor vehicles in a large city produced a total of 9.1×10^6 kg of the poisonous gas carbon monoxide (CO). How many moles of CO does this correspond to?

Q. 43. When a weighed piece of metal M (atomic weight 121.75 amu) is heated with excess oxygen (atomic weight 32.06 amu), a chemical reaction occurs between the metal and sulfur. The excess sulfur is then driven off leaving only the compound, consisting of combined metal and sulfur. It is found that :

Weight of metal	2.435 g
Weight of compound	3,397 g
Weight of sulfur	0.962 g

What is the simplest formula of the compound?

Q. 44. On being heated, 4.90 g of $KClO_3$ shows a weight loss of 0.384 g. What percentage of the original $KClO_3$ has decomposed:

Q. 45. Caffeine, the main stimulant in coffee, can be analyzed be burning it in a stream of oxygen and then collecting and weighing the oxides formed. Analysis such a method shows that caffeine consists of 49.48% by weight C, 5.19% H, 28.85% N, and 16.48% O. What is its simplest formula?

Q. 46. A 2.574-g of sample of hydrated copper (H) sulfate, $CuSO_4 \cdot xH_2O$, was heated to 413 K, cooled to room temperature and reweighed. Te resulting solid was reheated to 673 K, cooled to room temperature and reweighed. Finally, this solid was heated to 1273 K; the resulting black residue was cooled to room temperature and weighed for a final time. The measured masses were

original sample:	2.576 g
after heating to 413 K:	1.883 g
after reheating to 673 K:	1.647 g
after reheating to 1273:	0.812 g

1. Assuming that all the water of hydration is driven off after heating to 100°C, what must be the formula of the hydrate $CuSO_4 \cdot xH_2O$?
2. What is the formula of the hydrate obtained after the original solid is heated to 413 K?
3. The black residue obtained 1273 K is an oxide of copper. What is its percentage composition and its empirical formula?

Q. 47. Potassium-40 is one of the new naturally occuring radioactive isotopes of elements of elements of low atomic. Its percent natural abundance is 0.012%. How many K-40 atoms do you ingest by drinking one cup of whole milk containing 370. mg K?

Q. 48. The compound calcium cyanamide, $CaCN_2$, is prepared in considerable amounts for use as a solid fertilizer. The solid mixed with the water in the soil slowly adds ammonia and $CaCO_3$, to the soil.

$$CaCN_2 + 3H_2O + C2CO_3 + 2NH_3$$

What weight of ammonia if produced as 200 kg of $CaCN_2$ reacts?

Q. 49. $NaNO_2$ is prepared by passing nitrogen and monoxide and oxygen gases into an aqueous solution of sodium carbonate. Carbon dioxide gas is another produce of the reaction.

What mass of $NaNO_2$ should result from the reaction of 748 g Na_2CO_3 with an excess of NO(g) and O_2(g)?

In another reaction the reactants are 225 mL of 1.5 M Na_2CO_3(aq), 22.1 g NO, and a large excess of O_2.

(a) What mass of $NaNO_2$ should be produced if the reaction has a95% yield?
(b) What is the theoretical yield of $NaNO_2$? Simply convert the amount of $NaNO_2$ determined in (a) to a mass in grams.
(c) What is the actual yield of $NaNO_2$?

Q. 50. How many grams of (a) O_2 (b) NO_2 are required to burn 1.00 g of caffeine?

Q. 51. Suppose in burning of caffeine, according to the equation given above. 1.00 g of CO_2 is produced. How many grams of H_2O must be produced simultaneously?

Q. 52. How much magnesium sulfide can be made from 1.00 g of magnesium and 1.00 g of sulfur by reaction $Mg + S \rightarrow MgS$?

Q. 53. Wen you see tip of a match flare, the chemical reaction is likely to be $P_4S_3 + 8O_2 \rightarrow P_4O_{10} + 3SO_2$. What is the minimum amount o P_4S_3 that would have to be burned by this reaction to produce at least 1.0 g of P_4O_{10} and at least 1.0 g of SO_2?

Q. 54. When caffeine $C_8H_{10}N_4O_2$ is burned in a limited supply of O_2, the products are CO, H_2O and NO. How many grams of H_2O would be produced in such a reaction if the starting amount of $C_8H_{10}N_4O_2$ is limited to 1.0 g and the supply of O_2 is limited to 2.0 g?

Q. 55. In the reaction of gaseous carbon monoxide (CO) with solid iron oxide (Fe_2O_3), the products are gaseous CO_2 and solid Fe. How many grams of Fe can be formed from 1.00 g of Co and 5.0 g of Fe_2O_3?

Q. 56. **1.** Calculate the oxidation number of manganese in potassium permanganate, $KMnO_4$.
2. Find the oxidation number of hydrogen in the following.
(a) H_2 (b) HF (c) H_2O_2 (d) KH

Q. 57. Indicate in the following reactions which element is oxidised and which element is reduced.
(a) $H_2S + Cl_2 \longrightarrow 2HCl + S$
(b) $3Br_2 + 6NaOH \longrightarrow NaBrO + 5NaBr + dH_2O$
(c) $3Cu + 8HNO_3 \longrightarrow 3Cu(NO_3)_2 + 4H_2O + 2NO$
(d) $2H_2O_2 \longrightarrow 2H_2O + O_2$

Q. 58. On being heated, the white solid $KClO_3$ decomposes to form the white solid KCl and oxygen gas O_2. How many grams of $KClO_3$ must be decomposed to give 0.96 g of O_2?

Q. 59. The reaction

$$4HF(aq) + SiO_2(aq) \rightarrow SiF_4(aq) + 2H_2O$$

can be used to release gold that is distributed in certain quartz (SiO_2) veins of hydrothermal origin. If the quartz contains 1.0×10^{-3}% Au by weight and the gold has a market value of Rs. 12750 per gm, would the process be economically feasible if commercial HF (50% by weight, d = 1.17 gcm^{-3}) costs Rs. 75 a liter?

Q. 60. The amount of sodium hypochlorite in a bleach solution can be determined by using a given volume of bleach to oxidize excess iodide ion to iodine; ClO^- is reduced to Cl^-. The amount of iodine produced by the redox reaction is determined by tetration with sodium thiosulfate, $Na_2S_2O_3$; I_2 is reduced to I^-. The sodium thosulfate is oxidized to sodium tetrahionate, $Na_2S_4O_6$. In this analysis, potassium iodide was added in excess to 5.0 mL of bleach (d = 1.0 g cm^{-3}). If 25 mL of 0.07 M Na_2SO_3 was required to reduce all the iodine produced by the bleach back to iodide, what is the mass percent of NaClO in the bleach?

Q. 61. A child eats 10.0 g of paint containing 5.0% Pb. How many grams of the sodium salt of EDTA, Na_4(EDTA), should be receive to bring the lead into solution Pb · EDTA?

Q. 62. Rust, which you can take to be $Fe(OH)_3$, can be dissolved by treating it with oxalic acid. An acid-base reaction occurs, and a complex ion is formed.

(a) Write a balance equation for the reation.

(b) What volume of 0.10 *M* $H_2C_2O_4$ would be required to remove a rust stain weighing 1.0 ?

Q. 63. By x-ray diffraction it is possible to determine the geometric pattern in which atoms are arranged in a crystal and the distances between atoms. In a crystal of silver, four atoms effectively occupy the volume of a cube 0.409 nm on an edge. Taking the density of silver to be 10.5 g cm^{-3}, calculate the number of atoms in one mole of silver.

Q. 64. A 5.025-g sample of calcium is burned in air to produce a mixture of two ionic compounds calcium oxide and calcium nitride. Water is added to this mixture. It reacts with calcium oxide to form 4.832 g of calcium hydroxide. How many grams of calcium oxide are formed. How many grams of calcium nitride?

Q. 65. A mixture of potassium chloride and potassium bromide weighing 3.595 g is heated with chlorine, which converts the mixture completely to potassium chloride. The total mass of potassium chloride after the reaction is 3.129 g. What percentage of the original mixture was potassium bromide?

Q. 66. A sample of an oxide of vanadium weighing 4.589 g was heated with hydrogen gas to form water and another oxide of vanadium weighing 3.782 g. The second oxide was treated further with hydrogen until only 2.573 g of vanadium metal remained.

(a) What are the simplest formulas of the two oxides?

(b) What is the total mass of water formed in the successive reactions?

Q. 67. A sample of cocaine, $C_{17}H_{21}O_4N$, is diluted with sugar $C_{12}H_{22}O_{11}$. When a 1.0-mg sample of this mixture is burned, 1.0 mL of carbon dioxide (d = 1.8 g/L) is formed. What si te percentage of cocaine in this mixture?

Q. 68. Elemental bromine is the source of bromine compounds. The element is produced from certain brine solutions that occur naturally. These brines are essentially solutions of calcium bromide that, when treated with chlorine gas, yield bromine in a displacement reaction. What are the molecular equation and net ionic equation for the reaction? A solution containing 40.0 g of calcium bromide requires 14.2 g of chlorine to react completely with it, and 22.2 g of calcium chloride is produced in addition to whatever bromine is obtained. How many grams of calcium bromide are required to produce 10.0 grams of bromine?

Q. 69. Mercury (II) nitrate is treated with hydrogen sulfide, H_2S, forming a precipitate and a solution. Write the molecular equation and the ionic equation for the reaction. An acid is formed; is it strong or weak? Name each of the products. If 81.15 g of mercury (II) nitrate and 8.52 g of hydrogen sulfide are mixed in 550 g of water to form 58.16 g of precipitate, what is the mass of the solution after the reaction?

Q. 70. A 0.5-g mixture of Cu_2O and CuO contains 0.425 g Cu. What is the mass of CuO in the mixture?

Q. 71. A 3.41-g sample of a metallic element, *M*, reacts completely with 0.0158 mol of a gas, X_2, to form 4.52 g *MX*. What are the identities of *M* and X?

Q. 72. An alloy of iron (54.7%), nickel (45.0%), and manganese (0.3%) has a density of 8.17 g cm^{-3}. How many iron atoms are there in a block of alloy measuring 10.0 cm × 10.0 cm × 15.0 cm?

Q. 73. What volume of a solution of ethanol, C_2H_6O, that is 94.0% ethanol by mass contain 0.200 mol C_2H_6O? The density of the solution 0.807 g mL^{-1}.

Q. 74. A metal, M, was converted to the sulfate, $M_2(SO_4)_3$. Then a solution of the sulfate was treated with barium chloride to give barium sulfate crystals, which were filtered off.

$$M_2(SO_4)_3(aq) + 3BCl_2(aq) \rightarrow 2MCl_3(aq) + 3BaSO_4(s)$$

If 1.2 g of the metal gave 6.02 g of barium sulfate, what is the atomic weight of the metal? What is the metal?

Q. 75. Phosphoric acid is prepared by dissolving phosphorus(V) oxide, P_4O_{10}, in water. What is the balanced equation for this reaction? How many grams of P_4O_{10} are required to make 1.50 L of aqueous solution containing 5.00% phosphoric acid by mass? The density of the solution is 1.025 g mL^{-1}.

Q. 76. An alloy of aluminum and magnesium was treated with sodium hydroxide solution, in which only aluminum reacts.

$$2Al(s) + 2NaOH(aq) + 6H_2O(l) \rightarrow 2NaAl(OH)_4(aq) + 3H_2(g)$$

If a sample of alloy weighing 1.118 g gate 0.1068 g of hydrogen, what is the percentage aluminum in the alloy?

Q. 77. Determine the volume of sulfuric acid solution needed to prepare 18.7 g of aluminum sulfate, $Al_2(SO_4)_3$, by the reaction.

$$2Al(s) + 3_2SO_4(aq) \rightarrow Al_2(SO_4)_3(aq) + 3H_2(g)$$

The sulfuric acid solution, whose density is 1.104 g mL^{-1}, contains 15.0% H_2SO_4 by mass.

Q. 78. The active ingredients of an antacid tablet contained only magnesium hydroxide and aluminum hydroxide. Complete neutralization of a sample of the active ingredients required 48.5 mL of 0.187 *M* -hydrochloric acid. The chloride salts from this neutralization were obtained by evaporation of the filtrate from the titration; the weighed 0.42 g. What was the percentage by mass of magnesium hydroxide in active ingredients of the antacid tablet.

Q. 79. Using circles to represent cations and squares to represent anions, represent the reactions that occur between aqueous solutions of

(a) Na^+ and Cl^- **(b)** Ag^+ and Cl^-

Q. 80. Using squares to represent atoms of one element (or cations) and circles to represent the atoms of the other element (or anions), represent the principal species in the following pictorially. (You may represent the hydroxide anion as a single circle.)

(a) a solution of HCl **(b)** a solution of HF

(c) a solution of KOH **(d)** a solution of HNO_2

Q. 81. A student is asked to identity the metal nitrate present in an aqueous solution. The cation in the solution can either be Na^+, Ba^{2+}, Ag^+, or Ni^{2+}. Results of solubility experiments are as follows:

unknown + chloride ions–no precipitate
unknown + carbonate ions—precipitate
unknown + sulfate ions—precipitate

What is he cation in the solution?

Q. 82. Calcium in blood or urine can be determined by precipitation as calcium oxalate, CaC_2O_4. The precipitate is dissolved in strong acid and titrated with potassium permanganate. The products of the reaction are carbon dioxide and manganese(II) ion. A 24-hour urine sample is collected from an adult patient, reduced to a small volume, and titrated with 26.2 mL of 0.0946 *M* $KMnO_4$. How many grams of calcium oxalate are in the sample? Normal range for Ca^{2+} output for an adult is 100 to 300 mg per 24 hours. Is the sample within the normal range?

Q. 83. Stomach acid is approximately 0.02 *M* HCl. What volume of this acid is neutralized by an antacid tablet that weighs 330 mg and contains 41.0% $Mg(OH)_2$, 36.2% $NaHCO_3$, and 22.8% NaCl? The reactions involved are

$$Mg(OH)_2(s) + 2H^+ (aq) \rightarrow Mg^{2+} (aq) + 2H_2O$$
$$HCO_3^{\ -}(aq) + H^+(aq) \rightarrow CO_2(g) + H_2O$$

Q. 84. Copper metal can reduce silver ions to metallic silver. The copper is oxidized to copper ions according to the reaction

$$2Ag^+ (aq) + Cu(s) \rightarrow Cu^{2+} (aq) + 2Ag(s)$$

A copper strip with a mass of 2 g is dipped into a solution of $AgNO_3$. After some time has elapsed, the copper strip is coated with silver. The strip is removed from the solution, dried, and weighed. The coated strip has a mass of 4.18 g. What are the masses of copper and silver metals in the strip? (*Hint:* Remember that the copper metal is being used up as silver metal forms.)

Q. 85. One mole of Na_2S is represented as ⊟O where □ represents Na and O represents S. Complete the picture showing only the sodium and sulfide ions. The water molecules need not be shown.

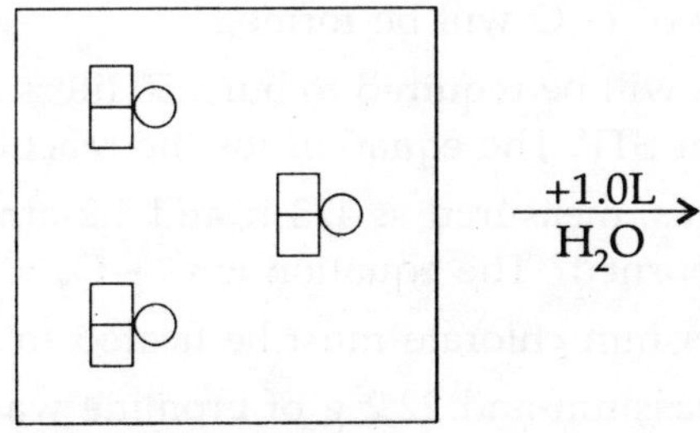

What is the molarity of Na^+? of S^{2-}?

Q. 86. Use Figure to answer the following questions.

(a) How many grams of source can be dissolved in 80 g of water at 313 K to make a saturated solution?

(b) A supersaturated solution is prepared by dissolving 750 g of source in125 g of boiling water (373 K). The solution is then slowly cooled to 298 K. One milligram of sucrose is added. How many grams of sucrose crystallize out?

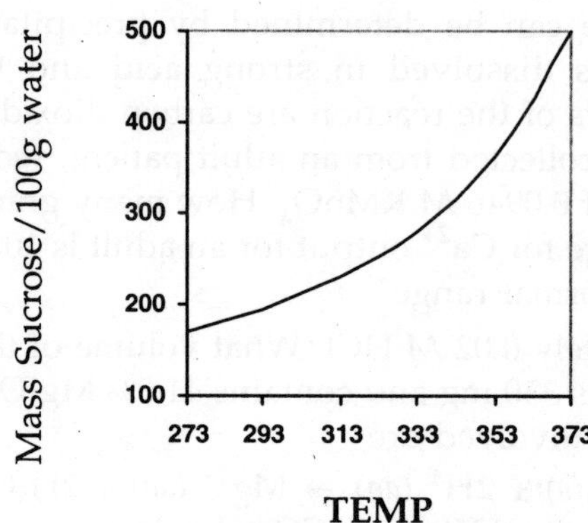

Solubility of Sucrose

Q. 87. How many moles of O_2 will be obtained by heating 3.5 moles of $KClO_3$?

Q. 88. When antimony is burned in oxygen the following reaction occurs:

$$4\ Sb + 3\ O_2 = 2\ Sb_2O_3$$

How many *moles* of oxygen will be needed to burn 18 moles of antimony? How many grams of Sb_2O_3 will be formed?

Q. 89. When C_3H_8 is burned in O_2 gas, CO_2 and H_2O are formed as products. If 2.40 moles of C_3H_8 are burned in a plentiful supply of oxygen, how many grams of H_2O and how many liters of CO_2, measured at STP, will be formed?

Q. 90. When C_4H_{10} is burned in excess oxygen the following reaction occurs.

$$2C_4H_{10} + 13O_2 = 8CO_2 + 10\ H_2O$$

How many *grams* of O_2 will be needed to burn 36 *grams* of C_4H_{10}? How many *grams* of CO_2 and how many *grams* of H_2O will be formed?

Q. 91. How many liters of O_2 gas will be required to burn 50 liters of H_2 gas? The volumes of both gases are measured at STP. The equation for the reaction s $2\ H_2 + O_2 = 2\ H_2O$.

Q. 92. How many liters of CO_2 gas, measured at 473 K and 1.2 atm pressure, will be formed when 40 g of carbon are burned? The equation is $C + O_2 = CO_2$.

Q. 93. How many grams of potassium chlorate must be heated to give 60 g oxygen?

Q. 94. A mixture of 12.2 g of potassium and 22.2 g of bromine was heated until the reaction was completed. How many grams of KBr were formed?

Q. 95. (a) When 2.451 g of pure, dry MXO_3 are heated, 0.96 g of oxygen gas is liberated. The other product is 1.491 g of solid MX. When this MX is treated with excess $AgNO_3$, all of its reacts with the $AgNO_3$ to form solid AgX; 2.869 g of AgX are formed. Knowing that the atomic weights of O and Ag are respectively, 16.00 and 1089.0, calculate the atomic weights of M and X.

(b) How many grams of copper will be formed when the hydrogen gas liberated when 41.6 g of aluminium are treated with excess HCl is passed over excess CuO?

Q. 96. How many grams of oxygen will be required to prepare 200 g of P_4O_{10} from elemental phosphorus?

Q. 97. Ammonia gas is oxidized by oxygen gas in the presence of a catalyst as follows:

$$4NH_3 + 5O_2 \longrightarrow 6H_2O + 4NO$$

How many liters of oxygen will be necessary to oxidise 500 liters of NH_3 gas? How many liters of NO and how many liters of steam will be formed?

All gases are measured under same conditions of temperature and pressure.

Q. 98. Exactly 2 g of NH_3 were neutralized by HCl, NH_4Cl being formed as a product. In a separate experiment exactly 20 g of $AgNO_3$ were formed by the action of HNO_3 on 12.70 g of silver. All of the NH_4Cl formed in the first experiment was exactly sufficient to react with all of the $AgNO_3$ formed in the second experiment. Knowing that the atomic weight of H is 1.008 and of O is 15.999, calculate the atomic weights of N and Ag.

Q. 99. Equals volumes of hydrogen and oxygen, both at room temperature and atmospheric pressure, were introduced into a completely evacuated reaction bomb. The bomb was sealed and was heated to 393 K; the pressure of the mixture gases in the bomb was found to be 100 mm at 393 K. An electric arc inside the container was turned on which caused the reaction, $2H_2 + O_2 = 2H_2O$, to take place. When the reaction was over the bomb was cooled until the temperature was again 393 K. An examination revealed that there was no liquid water in the bomb. Water was the new pressure inside the bomb.

Q. 100. A compound X contains C and H. When 12 mole of X is burned to CO_2, and H_2O, the total mass of the CO_2 and H_2O is 160 g greater than the mass of the 1 mole of X. The mass of the CO_2 is 60 g greater than the mass of the H_2O. Calculate the chemical formula of the compound.

Q. 101. A mixture of O_2 and N_2 contains 30.0 mole percent oxygen. To this mixture of gases is added just enough carbon so that, on heating, all the oxygen and al of the carbon are consumed ad, in the reaction, half of the carbon is converted to CO and half to CO_2. What is the average molecular weight of the resulting mixture of CO, CO_2, and N_2?

Q. 102. A 10-g mixture of H_2S and CS_2 was burned in oxygen to form a mixture of H_2O, SO_2, and CO_2. The dried mixture on being separated into its pure components yielded 0.275 mole of SO_2 and 0.0774 mole of CO_2. How many grams of H_2S were there in the original mixture?

Q. 103. A 12-g mixture of carbon and sulfur, when burned in air, yielded mixture of CO_2 and SO_2 in which the moles of CO_2 was one half the moles of the SO_2. How many grams of carbon were there in the mixture?

Q. 104. A mixture of CS_2 and CH_4 was burned to a mixture of CO_2, SO_2, and H_2O at 723 K. The partial pressure of the CO_2, SO_2, and H_2O in the mixture were 175 mm, 200 mm, and 150 mm, respectively. What was the mass percent CH_4 in the original mixture.

Q. 105. A mixture of sulfur and carbon weighing 2.0 g, when burned, gave a mixture of SO_2 and CO_2 weighing 6.0 g. How many grams of carbon were there in the original mixture?

Q. 106. A mixture of Mg and Zn having a mass of 1.0 g, when burned in oxygen, gave a mixture of MgO and ZnO which had a mass of 1.409 g. How much Zn was there in the original mixture?

Q. 107. A mixture of NaBr and NaI has a mass of 1.62 g. When treated with excess $AgNO_3$, unit yields a mixture of AgBr and AgI which has a mass 2.822 g. How many grams of NaI were there in the original mixture.

Q. 108. A mixture CO_2 and SO_2 has a mass of 2.952 g and contains a total of 5.3×10^{-2} moles. How many moles of CO_2 are there in the mixture?

Q. 109. A mixture of pure AgCl and pure AgBr contains 66.35% silver. What is the mass percent of bromine in the mixture?

Q. 110. What fraction of a mole of zinc atom is 12 g of Zn?

Q. 111. How many atoms are there in 20 g of boron?

Q. 112. An amount of 4.63×10^{21} atoms is how many moles of Sn?

Q. 113. It was found that 56 g of iron combined with 32 g of sulfur. Calculate the empirical formula of the compound that was formed.

Q. 114. When heated, 433.22 g of a pure compound yielded 401.22 g of mercury and 32 g of oxygen. Calculated the empirical formula of the compound.

Q. 115. When burned, 4.04 g of magnesium combined with 2.66 g of oxygen to form 6.7 g of magnesium oxide. Calculate the empirical formula of the oxide.

Q. 116. A compound was found to have the empirical formula CH_2 and a molecular weight of 71. Calculate the molecular formula.

Q. 117. A pure compound was found on analysis to contain 31.9% potassium, 28.9% chlorine, and 39.2% oxygen. Calculate its empirical formula.

Q. 118. A student was given 2.982 grams of a sample of a pure anhydrous lead chloride, compound *A*. The student added a small amount of water to it in a test tubve and heated it thus liberating chlorine gas and creating compound *B*. The chlorine gas is traped in a flask filled with 0.5 M NaOH.

The student then heated compound *B* driving the off the remaining water, and determined the mass of compound *B* as 2.364 g. He then reduce the Compound *B* with hydrogen gas to form elemental lead. The chlorine Compound *B* was driven off as $HCl_{(g)}$. This lead was massed and found to weigh 1.77 g. Now,

1. Determine the mass of chlorine in Compounds *A* and *B*.
2. Using the masses of lead and chlorine from the experiment, what relationship exists for compounds *A* and *B*?
3. Using information from Question 2, calculate the number of grams of chlorine that would combine with 1 mole of lead in Compounds *A* and *B*.
4. Calculate the number of moles of chlorine that combined with one mole of lead in Compounds *A* and *B*.
5. What is the empirical formula of Compound *A*?
6. What is the empirical formula of Compound *B*?

7. Write the equations for the experiment and then identify what is being oxidized and what is being reduced.

Q. 119. A student placed some blue $CoCl_2$, on a watch glass and observed that over time it changed to violet and then to red. Upon heating the red compound gently, she observed that it changed back to violet and then to blue. She then took a sample of sodium sulfate decahydrate whose large crystals appeared colorless and transparent and observed over time that they changed to a fine white powder. She then added water to the fine white to the fine white powder, dissolving it. Upon gently heating this mixture the large, colorless transparent crystals reappeared. Then, she took 1.7 g of a green nickel sulfate hydrate and heated the hydrate completely in a crucible, allowed it to cool in a desiccator and then reweighed it and discovered it weighed 1.006 grams. The anhydride produced was yellow. Now,

1. What was the formula of the green nickel sulfate hydrate?
2. What was the percentage, by mass, of the water in the hydrate based upon this experiment.
3. What would be the theoretical percent of water in the hydrate and what was the percent error?
4. What was the formula for the yellow anhydride produced?

Q. 120. The active ingredient in commercial bleach is NaClO. A student was given a sample of commercial bleach which was labeled 5.20% NaClO by mass. The student was then directed to prepare 300. ML of a 0.5 M NaClO solution. The student was also directed to prepare 300. ML of a 0.5 M solution of sodium thiosulfate, $Na_2S_2O_3$ in 0.2 M sodium hydroxy. Both solutions were then allowed to reach a room temperature of 298 K. Keeping a constant final volume of 50 mL of solution, the student mixed various amounts of sodium hypochlorite and sodium thiosulfate together, stirring well, and recorded the maximum temperature (±0.2°K) of the solution with a calibrated temperature probe. The change in temperature (ΔT) is equal to T_{final} – 290 K. A chart and graph of the data is given below. Answer the following 1-6.

Time #	DT°C	mL NaClO	mL $Na_2S_2O_3$	vol ratio
1	0	0	50	x
2	3.9	5	45	1:9
3	8.1	10	40	1:4
4	11.3	15	35	1:2.3
5.	14.7	20	30	1:1.5
6.	18.8	25	25	1:1
7.	22.6	25	25	1.5:1
8.	27.6	35	15	2.3:1
9.	31.9	40	10	4:1
10	12	45	5	9:1
11	0.1	50	0	x

Temperature change vs. volume ratios

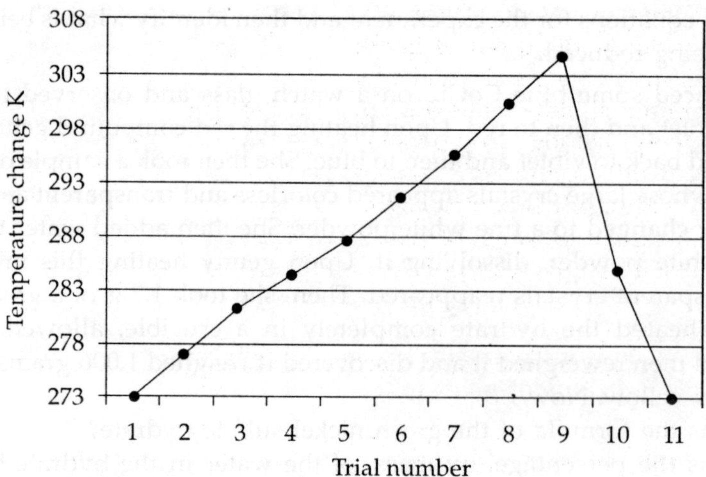

1. Write the net ionic equations involved in this experiment.
2. Show the steps necessary to produce the 0.5 M sodium hypochlorite solution.
3. Show the steps necessary to produce the 300. ML of 0.5 M sodium thiosulfite in 0.2 M NaOH.
4. If 20. ML of the sodium hypochlorite solution was at 298 K and 30 mL of the sodium thiosulfate solution was initially at 308 K before they were mixed, and after they were mixed together the final temperature was 307 K. What would be the actual final temperature of the mixture?
5. What conclusions can be drawn from Trial 9?

Q. 121. A student obtained a Büchner funnel, a piece of Whatman No. 42 filter paper, and a filter flask. She then weighed out 1.059 grams of alum and dissolved it in 50 mL of distilled water. She then calculated how much 0.2 M $Ba(NO_3)_2$ she would need to totally precipitate all of the sulfate ion present in solution. She carefully added twice this amount to the alum solution, stirring constantly. She then heated the solution to just under the boiling point for 15 minutes and then allowed the solution to stand overnight. The next day the discovered that fine crystals had appeared in the solution. She weighed the filter paper and determined its mass on an analytical balance as 1.675 g. She then filtered the solution through the Büchner funnel containing the filter paper and collected the crystals. The filter paper was removed and allowed to dry in a drying oven set at 323 K to avoid charring of the paper. She returned later and weighed the paper and crystals and determined their total mass to be 2.715 grams. Answer 1 to 4.

1. Calculate how much 0.200 M $Ba(NO_3)_2$ would be needed to totally precipitate all of the sulfate ion present in the alum solution.
2. What was the percent sulfate ion in the alum based upon the experiment?
   2.715 g paper + barium sulfate – 1.675 g paper = 1.040 $BaSO_4$
3. What would the theoretical % sulfate be in the alum?
4. What was the % error in this investigation?

Q. 122. A compound contains 28 per cent of nitrogen and 72 per cent of a metal by weight. 3 atoms of the metal combine with 2 atoms of N. Find the atomic weight of the metal.

Q. 123. A plant virus is found to consist of uniform cylindrical particles of 150 Å in diameter and 5000 Å long. The specific volume of the virus is 0.75 $cm^3\ g^{-1}$. If the virus is considered to be a single particle, find its molecular weight.

Q. 124. In reaction $H_2(g) + I_2(g) \rightarrow 2HI(g)$, 0.1, mole of H_2 and 0.2 mole of I_2 react so that only 20% of H_2 is transformed into HI. Calculate the weight of HI formed (I = 127).

Q. 125. An organic X, analysis gives 24.24 per cent carbon and 4.04 per cent hydrogen. Further, sodium extract of 1.0 g of X gives 2.9 g of silver chloride with acidified silver nitrate solution. The compound X may be represented by two isomeric structures, Y and Z. Y on treatment with aqueous potassium hydroxide solution gives a dihydroxy compound while Z on similar treatment gives ethanal. Find out the molecular formula of X and give the structures of Y and Z.

Q. 126. A mixture contains NaCl and an unknown chloride MCl.

(i) 1 gm. of tis is dissolved in water. Excess of acid fied $AgNO_3$ solution is added to it. 2.567 g of a while precipitate is formed.

(ii) 1 gm. of original mixture is heated to 573 k. Some vapours come out which are absorbed in acidified $AgNO_3$ solution. 1.341 gm. of a while precipitate is obtained. Find the molecular weight of the unknown chloride.

Q. 127. An aqueous solution containing 0.10 g KIO_3 (formula wt 214.0) was trated with an excess of KI solution. The solution was acidified with HCl. The liberated I_2 consumed 45.0 ml of thiosulphate solution to decolourise the blue starch-iodine complex. Calculate the molarity of the sodium thiosulphate solution.

Q. 128. Upon mixing 45 ml 0.25 M lead nitrare solution with 25 ml of 0.1 M chromic sulphate solution. Precipitation of lead sulphate takes place. How many moles of lead sulphate are formed? Also calculate the molar concentrations of the species left behind in the final solution. Assume that lead sulphate is completely insoluble.

Q. 129. (a) A sugar syrup of weight 214.2 g contains 34.2 g of sugar ($C_{123}H_{22}O_{11}$). Calculate (i) molar concentration and (ii) mole fraction of sugar in the syrup.

(b) Calculate the molarity of water if its density is 1000 kg m^{-3}.

Q. 130. The sodium salt of a carboxylic acid, *A*, was produced by passing a gas. *B*, into an aqueous solution of caustic alkali at an elevated temperature and pressure. On heating in presence of sodium hydroxide followed by treatment with suphuric acid gave a dibasic acid, *C*. *A* sample of 0.4 g of acid *C*, on combustion gave 0.08 g of water and 0.39 g of carbon dioxide. The silver salt of the acid weighing 1.0 g on ignition yielded 0.71 g of silver as residue. Identify *A*, *B* and *C*.

Q. 131. One litre of a sample of hard water contains 1 mg of $CaCl_2$ and 1 mg of $MgCl_2$. Find the total hardness in terms of parts of $CaCO_3$ per 10^6 parts of water by weight.

Q. 132. 19 gm of molten $SnCl_2$ is electrolysed for some time. Inert electrodes are used. 0.119 gm of Sn in deposited at the cathode. No substance is lost during the electrolysis. Find the ratio of the weights of $SnCl_2$: $SnCl_4$ after electrolysis.

Q. 133. 1 gm of Mg is burnt in a closed vessel which contains 0.5 gm of O_2.

(i) Which reactant is left in excess?
(ii) Find the weight of the excess reactant.
(iii) How many millilitres of 0.5 N H_2SO_4 will dissolve the residue in the vessel.

Q. 134. 2.68×10^{-3} moles of a solution containing an ion A^{+n} require 1.61×10^{-3} moles of MnO_4^- for the oxidation of A^{+n} to AO_4^- in acid medium. What is the value of n?

Q. 135. A mixture of FeO and Fe_3O_4 when heated in air to a constant weight gains 5% in its weight. Find the composition of the initial mixture. (Fe = 55.8; O = 16).

Q. 136. One gram of an alloy of aluminium and magnesium when heated with excess of dil. HCl forms magnesium chloride, aluminuim chloride and hydrogen. The evolved hydrogen collected over mercury at 273 K has a volume of 1.20 litres at 0.92 atm. pressure. Calculate the composition of the alloy. (Al = 27; Mg = 24.3)

Q. 137. 1,000 g. of a mixture, consisting of equal number of moles of carbonates of two alkali metals, required 44.4 ml. of 0.5 N-hydrochloric acid for complete reaction. If the at. Wt. of one of the metals is 7, find the at. wt. of the other metal. What will be the total amount of sulphate formed on quantitative conversion of 1.0 g. of mixture into sulphates. (Ca = 12, S = 32, O = 16)

Q. 138. One litre of an acidified solution of $KMnO_4$ containing 15.8 of $KMnO_4$ is decolorised by passing sufficient amount of SO_2. If the SO_2 is produced by roasting of iron pyrites (FeS_2) what will be the amount of pyrites required to produce the necessary amount of SO_2? (K = 39, Mn = 55, S = 32, Fe = 56, O = 16)

Q. 139. A sample of magnesium metal containing some magnesium oxide as impurity was dissolved in 125 ml. of 0.1 N-H_2SO_4. The volume of hydrogen evolved at 300.3 K and 1 atm. was 120.1 ml. The resulting solution was found to be 0.02 N with respect to sulphuric acid.

Calculate

(i) the wt. of the sample dissolved

(ii) the percentage by weight of magnesium in the sample. Neglect any change in the volume of the solution.

Q. 140. 408 g of mixture of BaO and unknown carbonate MCO_3 was heated strongly. The residue weighed 3.64 g. This was dissolved in 100 ml of 1 N HCl. The excess acid required 16 ml of 2.5 N NaOH solution for complete neutralization. Identity the metal M.

Q. 141. Hydroxylamine reduces iron (III) according to the equation

$2\ NH_2OH + 4\ Fe^{3+} \rightarrow N_2O(g) + H_2O + 4\ Fe^{2+} + 4H^+$.

Iron (II) thus produced is estimated by titration with a standard permanganate solution. The reaction is:

$$MnO_4^- + 5\ Fe^{2+} + 8\ H^+ \rightarrow Mn^{2+} + 5\ Fe^{3+} + 4\ H_2O.$$

A 10 ml sample of hydroxylamine solution was diluted to 1 litre. 50 ml of this diluted solution was boiled with an excess of iron (III) solution. The resulting solution required 12 ml of 0.02 M $KMnO_4$ solution for complete oxidation of iron (II). Calculate the weight of hydroxylamine in one litre of the original solution.

Q. 142. Hydrogen peroxide solution (20 ml) reacts quantitatively with a solution of $KMnO_4$ (20 ml) acidified with dilute H_2SO_4. The same volume of the $KMnO_4$ solution is just decolourised by 10 ml of $MnSO_4$ in neutral medium simultaneously forming a dark brown precipitate by hydrated MnO_2. The brown precipitate is dissolved in 10 ml of 0.2 M sodium oxalate under boiling condition in the presence of dilute H_2SO_4. Write the balanced equations involved in the reactions and calculate the molarity of H_2O_2.

Q. 143. An organic compound $C_xH_{2y}O_y$ was burnt with twice the amount of oxygen needed for complete combustion to CO_2 and H_2O. The hot gases when cooled to 273 K and 1 atom pressure, measured 2.24 litres. The water collected during cooling weighed 0.9 g. The vapour pressure of pure water at 293 K is 17.5 mm Hg and is lowered by 0.104 mm when 50 g of the organic compound are dissolved in 1000 g of water. Give the molecular formula of the organic compound.

Q. 144. 1.0 g of a moist sample of a mixture of potassium chloride and potassium chlorate was dissolved in water and made up to 250 ml. 25 ml. of this solution was treated with SO_2 to reduce the chlorate to chloride and excess SO_2 was removed by boiling. The total chloride was pptd. as silver chloride. The weight of the ppt. was 0.1435 g. In another experiment, 25 ml. of the original solution was heated ferrous sulphate required 37.5 ml. of 0.08 N solution of an oxidising agent for complete oxidation. Calculate the molar ratio of the chlorate to the chloride in the given mixture. Fe^{+2} reacts with ClO_3^- according to the equation—

$$ClO_3^- + 6F^{+2} + 6H^+ \rightarrow Cl^- + 6Fe^{+3} + 3H_2O$$

Q. 145. Upon having a litre of a N/2-HCl solution, 2.675 g. of hydrogen chloride is lost and the volume of the solution shrinks to 750 ml. calculate (i) normality of the resultant solution, and (ii) the number of milli equivalent of HCl in 100 ml. of the original solution.

Q. 146. An equal volume of a reducing agent is titrated separately with 1 M $KMnO_4$ in acid neutral and alkaline media. The volumes of $KMnO_4$ required are 20 ml in acid, 33.4 ml out the oxidation state of manganese in each reduction product. Give the balanced equations for all the three half reactions. Find out the volume of 1 M $K_2Cr_2O_7$ consumed, if the same volume of the reducing agent agent is titrated in acid medium.

Q. 147. Find the equivalent weight of H_3PO_4 in the reaction:

$$Ca(OH)_2 + H_3PO_4 = CaHPO_4 + 2H_2O$$

Q. 148. The density of a 3 M solution thiosulphate solution ($Na_2S_2O_3$) is 1.25 g per ml. Calculate (i) the percentage by weight of sodium thiosulphate, (ii) the mole fraction of sodium thiosulphate and (iii) the molalities of Na^+ and $S_2O_3^{2-}$ ions.

Q. 149. A sample of hard water contains 20 mg of Ca^{++} ions per litre. How many milli-equivalent of Na_2CO_3 would be required to soften 1 litre of the sample?

Q. 150. (a) A mixture contains NaCl and an unknown chloride MCl. (i) 1 gm of this is dissolved in water. Excess of acidified $AgNO_3$ solution if added to it. 2.567 gm. of a white ppt. is formed. (ii) 1 gm. Of the original mixture is heated to 573 K. Some vapours come out which are absorbed in acidified $AgNO_3$ solution. 1.341 gm. of a white ppt. is obtained. Find the mol. wt. of the unknown chloride.

Q. 151. Consider the reaction

$$2\,{}^{2}_{1}\mathrm{H} \rightarrow {}^{4}_{2}\mathrm{He}$$

(a) Calculate ΔE in kilojoules per gram of deuterium fused.

(b) How much energy is potentially available from the fusion of all the deuterium in sea water? The percentage of deuterium in water is about 0.0017%. the total mass of water in the oceans is 1.3×10^{24} g.

(c) What fraction of the deuterium in the oceans would have to be consumed to supply the annual energy requirements of the world (2.3×10^{17} kJ)?

Q. 152. The vapour pressure of a volatile liquid can be determined by slowly bubbling a known volume of gas through the liquid at a given temperature and pressure. In an experiment, a 5.4-L sample of nitrogen gas, N_2, at 293 K and 745 mmHg is bubbled through liquid isopropyl alcohol, C_3H_8O, at 293 K. Nitrogen containing the vapor of C_3H_8O at its vapour pressure leaves the vessel at 293 K and 745 mmHg. It is found that 0.6149 g C_3H_8O has evaporated. How many moles of N_2 are in the gas leaving the liquid? How many moles of alcohal are in this gaseous mixture? What is the mole fraction of alcohol vapour in the gaseous mixture? What is the partial pressure of the alcohol in the gaseous mixture? What is the vapor pressure of C_3H_8O at 293 K?

Q. 153. Acedic acid, CH_3COOH, forms stable pairs of molecules held together by two hydrogen bonds.

```
     H                              H
     |        O---H—O               |
  H—C—C//              \C—C—H
     |      \O---H—O/               |
     H                              H
```

Such molecules—themselves formed by the association of two simpler molecules—are called dimers. The vapor over liquid acetic acid consists of a mixture of monomers (single acetic acid molecules) and dimers. At 373 K the total pressure of vapor over liquid acetic acid is 436 mmHg. If the vapor consists of 0.630 mole fraction of the dimer, what are the mass of monomer and dimer in 1.000 L of the vapor? What is the density of the vapor?

Chapter **5**

SOLUTIONS

I. Formulae

The homogeneous mixture of two or more substances is known as solution. The concentration of solutions may be expressed by the following different ways:

1. *Percentage of solute* $= \dfrac{\text{Mass of solute} \times 100}{\text{Mass of solution}}$
2. *Grams per litre* Amount of solute in grams present in 100 mL or 1 litre of solution
3. *Normality (N)* $= \dfrac{\text{Mass of solute (g)}}{\text{Eq. mass of solute} \times \text{volume of solution (L)}}$
4. *Molarity (M)* $= \dfrac{\text{Mass of solute (g)}}{\text{Mol. mass of solute } \times \text{ Volume of solution (L)}}$
5. *Molality (m)* $= \dfrac{\text{Mass of solute (g)}}{\text{Mol. mass of solute } \times \text{ Mass of solvent (kg)}}$
6. *Formality* $= \dfrac{\text{Masss of solute (g)}}{\text{Formula mass of solute } \times \text{ mass of solution (L)}}$
7. *Mole fraction* $= \dfrac{\text{No. of moles of one component (i.e. solute or solvent)}}{\text{Total no. of moles of the solution}}$

   or $x_1 = \dfrac{n_1}{n_1 + n_2}, \; x_2 = \dfrac{n_2}{n_1 + n_2}$

   where x_1 and x_2 are the mole fractions of solute and solvent respectively, n_1 and n_2 are the number of moles of solute and solvent respectively.

II. Colligative Properties

Formulae

1. ***Osmotic Pressure***

$$PV = nRT$$
$$PV = inRT$$
$$C_1 = C_2 \text{ (For Isotonic Solutions)}$$

2. ***Raoult's law***

$$p_A = p_A^0 X_A; \; p_B = p_B^0 X_B$$
$$\frac{p - p_s}{p} = \frac{n}{n+N} = \frac{w/m}{w/m + W/M}$$
$$\frac{p - p_s}{p} = \frac{PM}{RdT}$$

3. ***Elevation of Boiling Point:***

$$\Delta T_b = K_b \times \text{molality}$$
$$\Delta T_b = \frac{1000\, K_b\, w}{\text{mW}}$$
$$\Delta T_b = I \times K_b \times \text{molality}$$
$$K_b = \frac{0.002\, T^2}{L_v \text{ in cal.}}$$

4. ***Depression of Freezing Point***

$$\Delta T_f = K_f \times \text{molality}$$
$$\Delta T_f = \frac{1000\, K_f\, w}{\text{mW}}$$
$$\Delta T_f = iK_f \text{ molality}$$
$$K_f = \frac{0.002\, T^2}{L_f \text{ in cal.}}$$

5. ***Van't Hoff Factor***

$$i = \frac{\text{Observed Colligative Properties}}{\text{Calculated Colligative Properties}}$$
$$= \frac{\text{Cal. Mol. Wt.}}{\text{Obs. Mol. Wt.}}$$

6. ***Henry's Law*** $m = kp$
7. ***Distribution Law***

(i) $K_D = \dfrac{C_1}{C_2}$ (ii) $K_D = \dfrac{C_1}{\sqrt[n]{C_2}}$ (For Association of solute)

(iii) $K_D = \dfrac{C_1}{C_2(1-\alpha)}$ (For Dissociation of solute) (iv) $w = W\left[\dfrac{K_D V}{K_D V + v_1}\right]^n$

III. Solution Stoichiometry (Summary)

Solution Solute & Solvent

A solution is a homogeneous mixture of two or more components. The smaller proportion of the solution is called solute & larger proportion is called solvent. The solvent of a liquid solution is the liquid component. All other components of the solution are solute.

Types of Solution

There are two type of solution.

(a) *Homogeneous Solutions eg:*

(i) Gas into liquid
(ii) Solid into liquid
(iii) Solid into liquid
(iv) Solid into solid

(b) *Hetrogeneous Solutions (colloidal solutions) eg:*

(i) Gases into Gases—Air

Concentration of Solution

The relative amount of substance in a solution of a mixture is known as its concentration. The concentration can be expressed in different units or measures.

The are: (i) Percentage by weight , (ii) Percentage by volume, (iii) Molarity, (iv) Molality, (v) Mole fraction, (vi) Normality, (vii) Formality (viii) P.P.M. (Parts per million)

Percent by Weight

It is defined as the amount of a substance in grams present in 100 gms. of the solution.

$$\text{Weight Percent} = \frac{\text{weight of solute}}{\text{weight of solution}} \times 100. \text{ or weight fraction} = \frac{\text{weight of solute}}{\text{weight of solution}}$$

Percent by Volume

Here, molecule of reacting gases & the products formed are compared to their volumes and the ratio amongst the molecules is taken as the ratio amongst their volumes. Convertion of moles to volume is considered ($\because$ by gram-molecular volume law, a gram molecule of every gas at STP occupies 22.4 litres). Mathematically.

$$\% \text{ by volume} = \frac{\text{weight of solute}}{\text{volume of solution}} \times 100;$$

$$\% \text{ by strength} = \frac{\text{Volume of solute}}{\text{volume of solution}} \times 100$$

Molarity (M)

The molarity of a solution is the number of moles of solute per litre of solution or the number of gram molecular weight per litre of solution. The units of M are mole litre^{-1}. Mathematically,

$$\text{Molarity (M)} = \frac{\text{No. of moles of solute}}{\text{Volume of solution in litres}} = \frac{\text{milli moles of solute}}{\text{milli litres of solution}}$$

$$= \frac{\text{weight of solute in grams}}{\text{weight of solute} \times \text{volume of solution in litres}}$$

$$= \frac{\text{strength in g.litre}^{-1}\text{ of soution}}{\text{molecular weight of solute}} = \frac{\text{weight of solute} \times 1000}{\text{mol. wt. of solute} \times \text{vol. in ml.}}$$

Or, Molarity (M) × Volume of Solution in litres (V) = moles of solute = $\frac{\text{weight of solute}}{\text{M. wt. of solute}}$

Molality (m)

It is the number of moles of the solute per 1000 gms (1 kg) of the solvent. Mathematically

$$\text{Molality (m)} = \frac{\text{No. of moles of solute} \times 1000}{\text{weight of solvent in grams}} = \frac{\text{weight of solute in grams} \times 1000}{\text{mol. wt. of solute} \times \text{weight of solvent in grams}}$$

$$= \frac{\text{strength in g litre}^{-1}\text{ of solvent}}{\text{gm. mol. wt. of solvent}} = \frac{\text{weight of solute}}{\text{M. wt. of solute}} \times \frac{1000}{\text{wt. of solvent in grams}}$$

Note: Weight of solvent = weight of solution – weight of solute

Mole fraction (x)

Mole fraction is the ratio (or fraction) of number of moles of one component to the total number of moles of the solution.

eg. Suppose in a solution there are present n_1 moles of A & n_2 moles of B. i.e. total $n_1 + n_2$ moles, then:

$$\text{mole fraction of } A;\ x_A = \frac{\text{moles of } A}{\text{moles of } A + \text{moles of } B} = \frac{n_1}{n_1 + n_2} = \frac{w/m}{w/m + \frac{W}{M}}$$

$$\text{mole fraction of } B;\ x_B = \frac{\text{moles of } B}{\text{moles of } A + \text{moles of } B} = \frac{n_2}{n_1 + n_2} = \frac{W/M}{W/M + \frac{w}{n}}$$

where, w = weight of solute, W = weight of solvent

m = M. wt. of solute, M = M. wt. of solvent

∴ mole fraction of A + mole fraction of $B = x_A + x_B = 1$

Normality (N)

It is defined as the number of gram equivalents of a solute per litre of solution. This unit is used in volumetric analysis.

Mathematically,

$$\text{Normality (N)} = \frac{\text{No. of grams equivalent of solute}}{\text{Volume of solution in litres}}$$

$$= \frac{\text{weight of solute in grams}}{\text{gram equivalent weight of solute} \times \text{vol of solution in litres}}$$

$$= \frac{\text{Strength in grams litre}^{-1}}{\text{Gram equivalent weight of solute}}$$

$$\therefore \quad \text{Equivalent} = \text{N} \times \text{V in litres} = \frac{\text{wt of solute}}{\text{Eq. wt. of solute}}$$

$$\text{and milli equivalent} = \text{N} \times \text{V in ml} = \frac{\text{wt. of solute}}{\text{Eq. wt of solute}} \times 1000$$

Law of Normality

It states that the product of volume and normality of one solution is equal to the product volume and normality of another solution.

$N_1V_1 = N_2V_2$ i.e. milli equivalent (1) = milli equivalent (2)

or milli equivalent (acid) = milli equivalent (base)

Note: From the normality we can deduce that

(a) 1 litre of 1N solution contains 1 gram equivalent of solute
(b) x litres of yN solution contains $x \times y$ gram equivalents of solute
(c) x litres of yN solution contains $x \times y \times E$ grams of solute (E = Equivalent wt.)

eg: 0.5 litres of 0.3 N H_2SO_4 contains
$0.3 \times 0.5 = 0.15$ g eq. wts. of H_2SO_4
$0.3 \times 0.5 \times 49 = 7.35$ grams of H_2SO_4

Relationship between M & N

$$\text{Molarity} = \text{Normality} \times \frac{\text{Equivalent weight}}{\text{Molecular weight}}$$

In this preparing the standard solution the concept of molarity (M) can be used e.q.

(a) 1 N in 1000 ml. requires 1 gm eq. wt $\therefore$ xN in y mol require $\frac{xy}{1000}$ × gm eq. wt.

(b) 1 M in 1000 ml. require 1 gm M. wt. $\therefore$ xM in y ml. require $\frac{xy}{1000}$ × gm M. wt.

Hence,
When 1 mol. of H_2SO_4 = 98 g and 1 equivalent = 49 g
which means 98 g of H_2SO_4 dissolved in 1 litre of solution is 1 M and 49 g dissolved in 1 litres is 1N.

∴ 98 g H_2SO_4 in 1 litre solution 1 M is 2 N
So we conclude

$$\text{Eq. wt} = \frac{\text{M. wt.}}{2} \text{ in case of } H_2SO_4 \text{ or N} = 2\text{ M} \qquad (\because \text{ mol contains 2 equivalents})$$

$$\text{and Eq. wt} = \frac{\text{M. wt.}}{5} \text{ in case of kMnO}_4 \text{ or N} = 5\text{ M} \qquad (\because \text{ 1 mol contains 5 equivalents})$$

$$\text{and Eq. wt} = \frac{\text{M. wt.}}{6} \text{ in case of } K_2Cr_2O_7 \text{ or N} = 6\text{ M} \qquad (\because \text{ 1 mol contains 6 equivalents})$$

Formality (F)

It is the number of gram formula weights per litre of solution

$$\text{Formality} = \frac{\text{wt. of solute}}{\text{Formula weight} \times \text{V in litres}}$$

$$\text{Formality of A} = \frac{\text{number of moles added to the solution}}{\text{litres of solution}}$$

Use of Double Indicator

In the titration of alkali mixtures e.g., ($NaOH + Na_2CO_3$) or ($Na_2CO_3 + NaHCO_3$) two indicators phenolphthalein and methyl orange are used. The indicator phenolphthalein is a weak organic acid and gives end point between pH 8 to 10, while methyl orange, a weak base indicates end point sharply between pH 3.1 to 4.4. Following points are to be remembered therefore.

1. Phenolphthalein is not a good indicator for weak alkali titrations.
2. Methyl orange is not a good indicator for weak acid titrations.
3. Titration in between strong acid and strong base using phenolphthalein or methyl orange as indicator gives reading for complete neutralization i.e. Meq. of acid = Meq. of base
4. Titration between strong acid and strong base such as Na_2CO_3 using phenolphthalein as indicator, the Meq. of acid re utilized only for the end point upto $NaHCO_3$ state.

$$\underset{\text{weak}}{Na_2CO_3} + H^+ \rightarrow NaHCO_3 + Na^+$$

i.e. Meq. of acid= $\frac{1}{2}$ Meq. of Na_2CO_3 (if E. wt. of Na_2CO_3 is taken as $M/2$)

or Meq. of Acid= Meq. of Na_2CO_3 (if E. wt. of Na_2CO_3 is taken as $M/1$ since one H is replaced)

However in titration with methyl orange as indicator, the Meq. of acid corresponds to total Meq. of alkali present at that time in mixture.

5. It is to be clearly noted that methyl orange is used as indicator in titration for fresh mixture or in continuation of phenolphthalein *e.g.*,
Na_2CO_3 NaOH mixture Vs HCl

Case I: I end point is determined using phenolphthalein as indicator and then methyl orange is used to get II end point-in continuation

I end point: Meq. of Acid = Meq. of NaOH + $\frac{1}{2}$ Meq. of Na_2CO_3

II end point: Meq. of Acid = $\frac{1}{2}$ Meq. of Na_2CO_3

Case II: End point is determined using phenolphthalein as indicator. Next time end point is determined by taking another (fresh) same volume of mixture using methyl orange as indicator.

For phenolphthalein: Meq. of Acid = Meq. of NaOH + $\frac{1}{2}$ Meq. of Na_2C_3

For methyl orange: Meq. of Acid = Meq. of NaOH + Meq. of Na_2CO_3

COLLIGATIVE PROPERTIES OF SOLUTION

Vapour Pressure Lowering

(a) The pressure exerted by the vapours of a solvent when they are in equilibrium with its liquid at a temperature is known as **vapour pressure.**

(b) Vapour pressure is influenced with:
1. Temperature—Increases with increase in temperature.
2. Nature of solvent—High boiling point liquid has low vapour pressure and vice-versa.
3. Addition of a non volatile solute: to a solvent always lowers the vapour pressure, *i.e.*,

V.P. of solvent > V.P. of solution
(P^n) (P_s)

(c) The decrease in V.P. of a solvent on addition of a non volatile solute (*i.e.*, $P^0 - P_s$) is known as lowering in vapour pressure ($\Delta P = P^0 - P_s$)

Rault's Law

It states that the fractional lowering of the vapour pressure of a solvent is equal to the mole fraction of solute present.

$$\frac{P_A^0 - P_A}{P_A^0} = x_B$$ where, p_A^0 is vapour pressure of pure solvent

P_A is partial pressure of solution

x_B is mole fraction of solute, x_A = mole fraction of solvent

As, $x_A + x_B = 1 \quad \therefore \quad \frac{P_A^0 - P_A}{P_A^0} = 1 - x_A \Rightarrow x_A = \frac{P_A}{P_A^0} \Rightarrow P_A = x_A p_A^0$

$\therefore P_A \propto x_A$. (*i.e.* vapour pressure of solution is proportional to the mole fraction of solvent). Thus, as the concentration of solute is increase the mole fraction of the solvent is decreased & its vapour pressure decreases proportionately.

As, $P_A = x_A p_A^0$; $P_B = x_B p_B^0$ $\therefore$ By Dalton's law of partial pressure $P_s = P_A + P_B$

$$\text{or } P_s = P_A^0 x_A + P_B^0 x_B = \frac{n_A}{n_A + n_B} \times P_A + \frac{n_B}{n_A + n_B} \times P_B$$

n_A = moles of A, n_B = moles of B

Similarly, Roults law for a mixture of volatile liquids is

$$P_M = P_A^0 \cdot \frac{n_A}{n_A + n_B + \ldots} + P_B^0 \cdot \frac{n_B}{n_A + n_B + \ldots} + \ldots$$

Elevation in Boiling Point

(a) B. pt. of liquid is characteristic temperature at which its vapour pressure becomes equal to 1 atm pressure or 76 cm or 760 mm pressure.

(b) Addition of a non-volatile solute in a solvent lowers its vapour pressure and thus more heat is required to increase the V.P. of solution upto 1 atm. That is why b. pt. of solution increases. This increase in b.pt. is known is elevation in b.pt

Elevation in b.pt. = $\Delta T_b = T_1 - T_0$

where T_1 = b.pt. of solution

T_0 = b.pt. of solvent

(c) The elevation in b.pt. (ΔT_b) of solution is given by

$$\Delta T_b = \frac{1000\, K_b' w}{m' W} = \frac{1000\, K_b w}{m' W}$$

$$= k_b \times m$$

where ΔT_b= elevation in b. pt.; w = weight of solute; W = weight of solvent

K_b' = molal elevation constant in K mol^{-1} kg

k_h= molecular elevation const. in K mol^{-1} 100 g

and m' = mol. weight of solute; m = molality of solution

i.e. $\Delta T_b = K_b \times m$

or Boiling point elevation is proportional to solute molality.

Note: k_b of a solvent is defined as the elevation in boiling point of that solvent caused by dissolving 1 mole of non-volatile, non electrolyte solute in 1000 g (1 kg) of solvent.

Depression in Freezing Point (K_f)

(a) F. pt. of a liquid is a characteristic temperature at which vapour pressure of solvent in its liquid and solid phase becomes same.

(b) Addition of a non-volatile solute to a solvent lowers the freezing point. The decrease in freezing point is known as depression in freezing point.

Depression in f. pt. = $\Delta T_f = T_0 - T_1$

where, T_1= f. pt. of solution and T_0 = *f*. pt. of solvent

(c) The depression in f.pt. of solution is given by $\Delta T_f = \dfrac{1000\,K'_f w}{nW} = \dfrac{100\,kK_f w}{mW}$

where K'_f = molal depression constant in K mol^{-1} kg

K_f = molecular depression constant in K mol^{-1} 100 g

Rest all terms have usual meanings.

(d) k_f *of a solvent* is defined as the depression in freezing point caused by dissolving 1 mole of nonvolatile, non electrolyte solute in 1000 g (1 kg) of that solvent.

$$\because \quad \left(\text{Molality (m)}\ \frac{\text{moles of solute}}{\text{kg of solvent}} = \frac{\text{wt. of solute} \times 100}{\text{M. wt. of soute} \times \text{wt of solvent}}\right)$$

So, Mol. wt.= $\dfrac{1000 \cdot k_b w}{\Delta T_b W}$ where W = wt. of solvent; w = wt. of solute

Volt Hoff's Factor

In a solution of an electrolyte, more particles are present per mole of dissolved solute than is indicated by the molality of the solution hence abnormal colligative properties such as freezing point depressions are observed.

Since a colligative property like ΔT_f or ΔT_b etc. is dependent on the number of dissolved particles, the extent to which a substance is ionized may be estimated from the observed values (experimental) of ΔT_f and ΔT_b.

$$\text{Von't Hoff's factor } (i) = \frac{\text{Observed value of a colligative property}}{\text{Calculated value of a colligative property}} \quad \text{(For an electrolyte).}$$

Von't Hoff (Theory of dilute solutions):

A substance in dilute solution behave as a gas and all gas laws are applicable to dilute solutions.

i.e. $\pi V = nRT = \dfrac{\text{w}}{\text{M}} RT$ w = weight of solute, m = M. wt. of solute,

V = volume of solution, n = number of moles of solute.

or $\pi = \dfrac{\text{n}}{\text{V}} RT \Rightarrow \pi = \text{CRT}$ $\left(\text{c} = \dfrac{\text{n}}{\text{V}} = \text{molarity of solution or conc. of solution}\right)$

∴ Osmotic pressure depends only on conc. of solute in the solution and the temperature of the solution. i.e. π is directly proportional to molar concentration of solute and the absolute temp.

The constant of proportionality turns out to be same as the gas constant R.

Osmosis and Osmotic Pressure (TT)

When two solutions of unequal vapour pressure are separated by a semi-permeable membrane, solvent molecules will pass from the solution of higher vapour pressure into a solution of lower vapour (Osmosis).

The flow of solvent form a dilute to a concentrated solution (separated by a semi-permeable membrane like that of copper ferrocyanide, gall bladder etc.) is called osmosis, and the pressure applied to stop osmosis is called osmotic pressure.

The osmotic pressure is also defined as the hydrostatic pressure developed in a **vertical** column as a result of osmosis, when a solution is separated with its solvent by SPM. Or the minimum external pressure which must be applied on solution side in order to prevent osmosis against pure solvent separated by SPM. *i.e.* Osmoticpress ure (π) = hdg, d = density in kg m^{-3}

Isotonic or iso-osmotic Solution

Two solutions of different substances having same osmotic pressure at same temperature are known as isotonic solution.

i.e. $\pi_1 = \pi_2$

and $C_1 = C_2$ or $\frac{n_1}{V_1} = \frac{n_2}{V_2}$ or $\frac{w_1}{m_1 v_1} = \frac{w_2}{m_2 v_2}$

If volume is same

then $n_1 = n_2$ or $\frac{\text{wt of Ist}}{\text{M. wt. of Ist}} = \frac{\text{wt. of 2nd}}{\text{M. wt. of 2nd}}$

PROBLEMS

Q. 1. Calculate the molarity of water in pure water.

Q. 2. Calculate the molarity oxalic acid when 1.575 g. of it is dissolved in 250 ml of solution

Q. 3. Calculate the weight of anhydrous sodium carbonate (Mol. wt. 106) required to prepare decimolar solution in 500 ml. of solution.

Q. 4. To prepare 100 ml of 0.03 M cane sugar solution, how many grams of the cane sugar is required? (Molecular formula of cane sugar is $C_{12}H_{22}O_{11}$)?

Q. 5. 5.3 g. of Na_2CO_3 is present in 200 ml of the solution. What is the molarity?

Q. 6. What is the Molarity of 20% NaOH solution?

Q. 7. How much water is to be added to 100 ml of 0.2 M NaOH to be changed to 0.1 M solution?

Q. 8. Calcualte the normality of a solution containing 0.5 g. of NaOH in 500 ml. (Mol. wt of NaOH = 4)

Q. 9. Calculate the weight of oxalic acid required to prepare 0.05 normal solution in 400 ml (Mol. wt. of oxalic acid = 126)

Q. 10. The maximum safe concentration of fluoride ion in drinking water is 2.0 ppm (parts per million). Express this concentration in milligram %

Q. 11. Calculate the molality of aqueous solution containing 3.g of urea per 250 g. of water (Mol wt. of Urea = 60)

Q. 12. Calculate the weight of glucose, $C_6H_{12}O_6$ required to prepare 0.025 molal soln. in 750 g. of water (Mol. wt. 180)

Q. 13. Calculate the mole fraction of glucose and water when 18 g. of glucose is dissolved in 180 g of water.

Q. 14. Calculate the mole fraction of N_2, Ne and He in a gaseous solution of 7g N_2, 10g of Ne, and 3.g. of Helium (Mol. wt. N_2 = 28; Ne = 20; He = 4).

Q. 15. Calculate the volume of decinormal sulphuric acid required to neutralize 25 ml 0.125 N sodium hydroxide.

Q. 16. Calculate the normality of sodium carbonate when 20 ml of it neutralizes 24 ml. of 0.05 N hydrochoric acid.

Q. 17. Calculate the molarity of sodium corbonate when 25 ml of it neutralizes 30 ml of decimolar hydrochloric acid.

Q. 18. What is the molarity of a solution made by dissolving 15.0 g of HNO_3 in enough water in make 0.20 liter of solution?

Q. 19. Suppose you need 0.837 mol of HNO_3 for a particular reaction. You have a solution that is labeled 6.0 M HNO_3. How much of the solution should you take?

Q. 20. Calculate the volume of mohr's salt solution required to reduce 22 ml. of 0.125 molar. potassium permanganate solution. The molarity of ferrous ammounium sulphate solution is 0.105 M.

Q. 21. Calculate the volume of water to be added to dilute 250 ml of a normal solution to decinormal solution.

Q. 22. 400 ml of 0.25 molar HCl is diluted to 2,600 ml. What is the molarity of the solution?

Q. 23. Calculate the vapour pressure of a solution at 373 K containing 3 g. of cane sugar in 33 g. of water. (Mol. wt of sugar = 342)

Q. 24. 5 g. of a substance with molecular weight 200 have been dissolved in 50 g. of a solvent (Mol. wt 60 and vapour pressure 40 cms.). Find the vapour pressure of the solution.

Q. 25. An aqueous solution containing 28% by weight of a liquid A (mol. wt = 140) has a vapour pressure of 160 mm, at 310 K. Find the vapour pressure of the pure liquid *A* ("The vapour pressure of water at 310 K is 150 mm)

Q. 26. Two liquids *A* and *B* from ideal solutions. At 300 K, the vapour pressure of a solution containing 1 mole of *A* an 3 moles of *B* is 550 mm of Hg. At the same temperature if one more mole of *b* is added to this solution, the vapour pressure of the solution increases by 10 mm Hg. Determine the vapour pressure of *A* and *B* in their pure states.

Q. 27. If in a given solution 49 g. of sulphuric acid (H_2SO_4) is dissolved per litre (sp. gr is 1.049 g/ml.). Express the strength of the solution in terms of normality, molarity, molality and mole fraction.

Q. 28. A solution of NaOH is prepared by dissolving 2 g. of NaOH in 100 ml. If the specific gravity of the solution is 1.96, calculate normality, molarity molality and percentage of sodium hydroxide in solution. Also calculate the mole fractions of NaOH and water.

Q. 29. If 50 ml of 0.5 *M* H_2SO_4 is added to 75.0 ml of 0.25 *M* H_2SO_4, what will be the concentration of the final solution assuming its volume is 125 ml.

Q. 30. 6.90 M solution of KOH in water contains 30% by weight of KOH. Calculate the density of the solution.

Q. 31. A solution is prepared by mixing 10 g of CH_3OH and 100 g of H_2O. Calculate the mole fraction of each component.

Q. 32. A solution is prepared at 298 K by mixing 15 g of Na_2SO_4 with 125 cc of H_2O. The density of H_2O at this temperature is 1.0 g/cc. What is the molality of Na_2SO_4 in the solution?

Q. 33. How many grams of NaOH will be required to prepare 1.0 liter of 1.0 *M* NaOH?

Q. 34. If 12 g of NaOH are dissolved in enough water to give 500 ml of solution, calculate the molarity of the solution.

Q. 35. A solution of $Cu(NO_3)_2$ contains 100 mg of the salt per milli- liter. Calculate the molarity of the solution.

Q. 36. How many grams of K_2SO_4 will be required to prepare 1.0 liter of 0.5 *M* K_2SO_4?

Q. 37. A 1.0 *M* solution of CH_3OH in water has a density of 0.91 g/ml. Calculate the mole fraction, mass percent and molality of the CH_3OH.

Q. 38. How many moles of hydrogen will be liberated from 320 ml of 0.5 *M* H_2SO_4 by an excess of magnesium?

Q. 39. How many moles of hydrogen will be liberated fro 400 ml of 0.4 *M* HCl by an excess of zinc?

Q. 40. A 447-ml solution of Na_2CO_3 was warmed with an excess of sulfuric acid until all action ceased. Five liters of dry CO_2 gas, measured at standard conditions, were given off. Calculate the molarity of the sodium carbonate solution.

Q. 41. How many milliliters of 0.25 *M* HCl will be required to neutralize 120 ml of 0.8 *M* KOH?

Q. 42. A quantity of 60.0 g of a nonelectrolyte dissolved in 1000 g of H_2O lowered the freezing point 1.02°K. Calculate the approximate molecular weight of the nonelectrolyte.

Q. 43. When 4.2 g of a nonelectrolyte were dissolved in 40.0 g of water, a solution which froze at –1.52°K was obtained. Calculate the approximate molecular weight of the nonelectrolyte.

Q. 44. A 20-g quantity of $C_6H_{10}O_5$, a nonelectrolyte, is dissolved in 250 g of H_2O. Calculate the boiling point of the solution at 760 mm.

Q. 45. A solution is prepared by dissolving 2.37 g of NaCl in 100 g of water. What is the boiling point of the solution?

Q. 46. A solution is prepared by dissolving 1.27 g of an electrolyte whose molecular weight is 310 in 50 g of water, and the freezing point of the solution is found to be –0.762°K. How many moles of solute particles are formed in solution from each mole of solute?

Q. 47. When H_2SO_3, a weak acid, is dissolved in water it undergoes partial dissociation according to the equation $H_2SO_3 \rightleftharpoons H^+ + H_2SO_3^-$. A solution is prepared which contains 1.64 g of H_2SO_3 in 100 g of water. The freezing point of this solution is found to be –0.649°K. Calculate the fraction of H_2SO_3 which dissociates.

Q. 48. A solution of 42.2 g of glucose, $C_6H_{12}O_6$, a nonelectrolyte, in 200 g of water is prepared at 298 K. What is the vapor pressure of water above the solution. (Given, $P^oH_2O = 23.6$)

Q. 49. A solution is prepared from 27.1 g of K_2SO_4 and 1000 g of water. Calculate the vapor pressure of water above the solution at 373 K.

Q. 50. A solution is prepared at 343 K from 80 g of C_2H_6O, ethanol, and 100 g of water. Calculate the vapor pressure of each component above the solution and the mole fraction of ethanol i the vapor. The vapor pressure of pure ethanol at this temperature is 543 mm and of pure water is 234 mm.

Q. 51. A 1.0-liter volume of a solution of 10.0 g of hemoglobin, the oxygen-carrying substance in red blood corpuscles, in water is found to have an osmotic pressure of 2.8 mm of Hg of 300 K. Calculate the molecular weight of hemoglobin.

Q. 52. (a) What is the stable phase of CO_2 (i) at 4 atm and 198 K and (ii) at 2 atm and 263 K in the phase diagram for CO_2.

(b) Describe the phase change that occurs if CO_2 at 1.00=atm pressure is heated from 173 K to room temperature in the phase diagram.

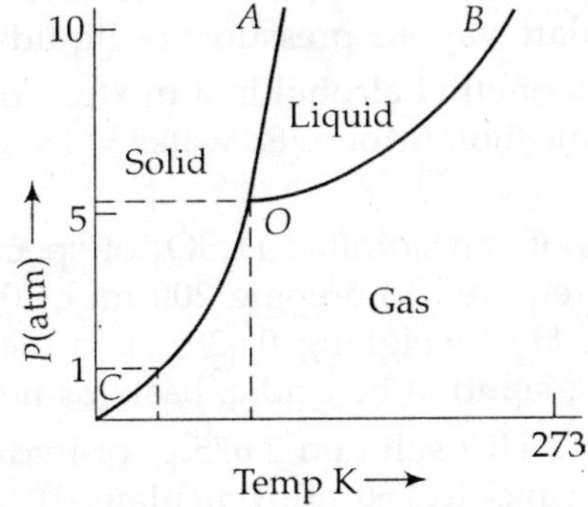

Q. 53. A solution is prepared by dissolving 16.6624 g of solid naphthalene, $C_{10}H_8$, in 200 mL of liquid benzene, C_6H_6, at 293 K. The density of benzene at 293 K is 0.87865 g.mL^{-1}. Calculate the mole fraction and mole percentage of benzene and naphthalene in this solution.

Q. 54. If 1.275 g of glucose are dissolved in 63.59 g of water, what is the molality of the resulting solution? What are the mole fractions of glucose and of water in this solution?

Q. 55. How do you prepare 50.0 mL of a 0.040 *M* Na_2SO_4 solution from a stock solution of 0.1 *M* Na_2SO_4?

Q. 56. If 40 mL of 0.2 *F* $NaNO_3$ is mixed with 60 mL of 0.1 *F* $Cu(NO_3)_2$, what is the concentration of each solute in the resulting solution?

Q. 57. The lable on a commercially available concentrated hydrochloric acid solution reads "37.4% HCl by weight, density 1.18 g.mL^{-1}." Calculate the molality, formality and mole fraction of HCl in this solution.

Q. 58. An aqueous solution of sucrose is prepared in which the molality of sucrose is 0.475 *m*.

(a) What is the vapor pressure of H_2O above this solution at 299.4 K? The vapor pressure of pure H_2O at 299.4 K is 25.812 mm Hg.

(b) What is the magnitude of the vapor pressure lowering?

Q. 59. A bottle of commercial sulphuric acid (density 1.787 g/c.c) is labeled as 86% by weight. What is the molarity of the acid? What volume of the acid has to be used to make it 0.2 M H_2SO_4 in one litre?

Q. 60. A 100 ml solution Na_2CO_3 is prepared by dissolving 8.653 g. of the salt in water. The density of the solution is 1.0816 g/c.c. What is the molarity and molality fo the solution?

Q. 61. A solution of ethanol in water is 1.54 molal. How many grams of it are there in 500 g. of the solvent?

Q. 62. What is the molarity of a solution of sulphuric acid which has a density of 1.84 g./c.c. at 298 K and contains 98% by weight?

Q. 63. What is the molarity and molality of a 13% solution (by weight) of sulphuric acid with a density of 1.09 g/c.c.? To what volume should 1000 c.c. of this acid be diluted to prepare 1.5 N solution.

Q. 64. How much ethyl alcohol must be added to 1.00 litre of water, so that the solution will freeze at 263 K? (K_f for water = 1.86°C/mole).

Q. 65. 40 g of liquid *A* (molar mass 120 g.mol^{-1}) was dissolved in 500 g of liquid *B* (molar mass 160 g mol^{-1}). The vapour pressure of liquid *B* in this solution is 400 mm. If total vapour pressure is 475 mm, calculate vapour pressures of liquids *A* and *B*.

Q. 66. Calculate the mole fraction of ethyl alcohol in a mixture of 27 g. of water 15 g. of acetic acid, and 11.5 g. of ethyl alcohol. (Mol. wts: water = 18, ethyl alcohol = 46, acetic acid = 60).

Q. 67. (a) How many milli litres of concentrated H_2SO_4 of specific gravity 1.84 contains 98% H_2SO_4 by weight are required to prepare 200 ml of 0.5 N solution?

(b) A sample of Na_2CO_3, H_2O weighing 0.62 g. is added to 100 ml of 0.1 N H_2SO_4 solution. Will resulting solution be acidic, basic or neutral?

Q. 68. Upon heating a litre of N/2 HCl solution 2.675 g. of hydrogen chloride is lost and the volume of the solution shrinks to 750 ml. Calculate (i) the normality of the resultant solution and (ii) the number of milli equivalents of HCl in 100 ml of the original solution.

Q. 69. 20 ml of 10 N, HCl are mixed with 10 ml of 36 N. H_2SO_4 and the mixture is made up to one litre with water. What is the normality of the mixture?

Q. 70. A mixture of $H_2C_2O_4$ (oxalic acid) and $NaHC_2O_4$ weighting 2.02 g was dissolved in water and the solution made upto one litre. Ten milliliters of the solution required 3.0 ml of 0.1 N sodium hydroxide solution for complete neutralization. In another experiment 10.0 ml of the same solution, in hot dilute sulphuric acid medium, required 4.0 m of 0.1 N potassium permanganate solution for complete reaction. Calcualte the amount of $H_2C_2O_4$ and $NaHC_2O_4$ in the mixture.

Q. 71. (a) A solid mixture (5.0 g) consisting of lead nitrate and sodium nitrate was heated 873 K C until the weight of the residue was constant. If the loss in weight is 28.0 percent find the amount of lead nitrate and sodium nitrate the mixture.

(b) Calculate the molality of 1 litre solution of 93% H_2SO_4 (weight/volume). The density of the solution is 1.84 g/ml.

Q. 72. Calculate the boiling point of a solution containing 0.45 g of camphor (mol. mass = 152) dissolved in 34 g acetone (B.P 329.3 K) if the molar elevation constant per 100 g acetone is 290.2 K.

Q. 73. Calculate the boiling point of a one-molar aqueous solution (density 1.04 g mL^{-1}) of KCl (K_b for water = 0.52 K kg m^{-1}]

Q. 74. The vapour pressure of pure benzene at a contain temperature is 640 mm Hg. A non-volatile non-electrolyte solid weighing 2.175 g. is added to 39.0 g. of benzene. The vapour pressure of the solution is 600 mm Hg. What is the molecular weight of the solid substance?

Q. 75. The vapour pressure of a dilute aquesous solution of glucose ($C_6H_{12}O_8$) is 750 mm of mercury at 373 K. Calculate (i) molality and (ii) mole fraction of the solute.

Q. 76. A sample of hydrazine sulphate ($N_2H_6SO_4$) was dissolved in 100 ml of water. 10 ml this solution was made to react with excess of ferric chloride solution and warmed to complete the reaction. Ferrous ion formed was estimated and it required 20 ml of 1/50 potassium permanganate solution. Estimate the amount of hydrogen sulphate in one litre of the solution.

Q. 77. **(a)** What is the weight of sodium bromate and molarity of solution necessary to prepare 85.5 mol 0.672 N solution when the half cell reaction is

$$BrO_3 + 6H^+ + 6e^- \rightarrow Br + 3H_2O$$

(b) What would be th weight as well as molarity if the half cell reaction is

$$2BrO_3 + 12H^+ + 10e^- \rightarrow Br_2 + 6H_2O$$

Q. 78. The vapour pressure of ethanol and methanol are 44.5 mm and 88.7 mm. Hg respectively. An ideal solution is formed at the same temperature by mixing 60 g. of ethanol with 40 g. of methhanol. Calculate the total vapour pressure of the solution and the mole fraction of methanol in the vapour.

Q. 79. The solubility of $Mg(OH)_2$ in pure water is 9.57×10^{-3} $glit^{-1}$. Calculate its solubility in 0.02 M $Mg(NO_3)_2$ solution.

Q. 80. Five ml of 8 N Nitric acid, .8 ml of 5 N HCl and certain volume of 17 M sulphuric acid are mixed together and made upto 2 litre 30 ml of this acid mixture exactly neutralizes 42.9 ml of sodium carbonate solution containing one gram of Na_2CO_3 $10H_2O$ in 100 ml of water. Calculate the amount in grams of the sulphate ions in solution.

Q. 81. One litre solution of N/2 HCl is heated. The volume of the solution reduced to 750 mL and 2.675 g HCl is lost. Calculate (i) the normality of the resultant solution and (ii) the number of milli equivalents of HCl in 100 mL of the original solution.

Q. 82. Two liquids *X* and *Y* on mixing form an ideal solution. At 303 K, the vapour pressure of the solution containing 3 mol of *X* and 1 mole of *Y* is 550 mm Hg but when 4 mole of *X* and 1 mol of *Y* are mixed, the vapour pressure of the solution thus formed is 560 mm Hg. What would be vapour pressure of pure X and pure *Y* at this temperature?

Q. 83. The density of a 3 M sodium thiosulphate solution ($Na_2S_2O_3$) is 1.25 g per ml. Calculate

(a) The % by weight of $Na_2S_2O_3$.

(b) Mole fraction of $Na_2S_2O_3$ and

(c) The molalities of Na^+ and $S_2O_3^{2-}$ ions

Q. 84. The vapour pressure of pure benzene at 298K is 639.7 mm, of mercury and the vapour pressure of a solution in benzene at the same temperature is 631.9 mm, of Hg. Calculate the molality of the solution.

Q. 85. 0.5 g. of fuming H_2SO_4 (Oleum) is diluted with water. This solution is completely neutralized by 26.7 ml of 0.4 N NaOH. Find the % of free SO_3 in the sample of oleum.

Q. 86. A solution contains Na_2CO_3 and $NaHCO_3$. 10 ml of he solution requires 2.5 ml of 0.1 MH_2SO_4 for neutralization using phenolphthale in as an indicator. Methyl orange is then used when a further 2.5 ml of 0.2 M H_2SO_4 was required. Calculate the amount of Na_2CO_3 and $NaHCO_3$ in one litre of the solution.

Q. 87. Two litres of ammonia of 303K and at 0.20 atm. pressure neutralized 134 ml of a solution of H_2SO_4. Calculate the normality of the acid?

Q. 88. How many milli litres of conc. H_2SO_4 of specific gravity 1.84 containing 98% by wt. are required to prepare 200 ml of 0.5 N H_2SO_4?

Q. 89. The vapour pressure of water is 23.55 mm. What would be the vapour pressure of a solution containing 6 g. of urea (mol. wt. 60) in 100 g. of water at the same temperature?

Q. 90. 10 ml of pure ethanol of density 0.785 $g.ml^{-1}$ was diluted with water to a final volume of 100 ml. The density of the resultant solution was 0.9866 $g.ml^{-1}$. Calculate (i) the percentage by weight of ethanol and (i) the molarity of ethanol in the solution.
[At. wts. H = 1, C = 12; O = 16]

Q. 91. 8.0575×10^{-2} kg of Glouber's slat is dissolved in water to obtain 1 dm^3 of a solution of density 1077.2 kg m^{-3}. Calculate the molarity, molality and mole fraction of Na_2SO_4 in the solution.

Q. 92. A volume of 22.5 ml of 2.5 N NaOH was required to neutralize 10.5 ml of a solution H_2SO_4 of unknown strength. The density of the H_2SO_4 solution was 1.16 g ml^{-1}. Calculate the % by weight of H_2SO_4 in the solution.

Q. 93. An excess of $BaCl_2$ solution was added to 25.0 ml of a solution of Na_2SO_4. The precipitated $BaSO_4$ weighed 0.864 g. What was the molarity of the Na_2SO_4 solution?

Q. 94. A solution containing 28 g of 'p' molecules in 315 g. CS_2) (B.P. 219.3 K) boils at (320.98 K). The molal elevation constant for CS_2 is 2.34°K. Calculate the mol. wt. of 'p' and deduce its molecular formula in CS_2.

Q. 95. The vapour pressure of benzene and toluene at 293 K are 75 mm and 22 mm respectively. 23.4 g. of benzene and 64.4 g of toluene are mixed. If these two form an ideal solution, calculate the mole fraction of benzene in the vapour phase (if the vapours are in equilibrium with liquid mixture at the same temperature).

Q. 96. The freezing point of water was lowered by 0.19°K by dissolving 0.152 g of a substance in 25 g of water. Calculate the molecular weight of the substance. Molecular depression constant for water (100 g) is 18.5.

Q. 97. The freezing point of solution containing 11 g. of sugar (mol. wt. 342) per 100 g. of water is –0.597 K. A solution of a substance '*x*' containing 15 gr. per 100 gr of water freezes at –2.657°K. What is the mol. wt. of *x*?

Q. 98. Find the weight of 95% pure Na_2CO_3 to prepare 2 liters of 1.5 N solution.

Q. 99. 10 g of a substance dissolved in 100 g of water lowered its freezing point by 1.2 k. Find the molecular mass of the substance.
The molar depression constant for water is 18.5 K (for 100 g).

Q. 100. What volume of water should be added to make 50 ml HCl having sp. gr. 1.14 and 37% strength to give 1 N solution.

Q. 101. If it takes 25.0 ml of 0.198 M HNO_2 neutralize 37.5 ml of solution of NaOH. What is the molarity of the NaOH solution?

Q. 102. When 10.0 ml of ethyl alcohol (C_2H_5OH, density 0.7893 g mL^{-1}) is mixed with 20.0 ml of water (density 0.9971 g mL^{-1}) at 298K, the final solution has a density of 0.9571 g mL^{-1}. Calculate the percentage change in total volume on mixing, and calculate also the molarity of the final solution.

Q. 103. What is the molarity of ethanol in 90-proof whiskey? Density of alcohol = 0.8 g mL^{-1}.

Q. 104. Calculate the osmotic pressure of decinormal sodium chloride, (NaCl) solution which is ionized to the extent of 80% at 300 K.

Q. 105. A tube of uniform cross sectional area 1 cm^2 is closed at one end with semipermeable membrane. 5% solution of glucose is taken in the tube and dipped in pure water at 300 K. Calculate Osmotic pressure of solution and height of solution in vertical column of equilibrium. Given that the density of final glucose solution is 1 g mL^{-1}.

Q. 106. Suppose a rain cloud has a water volume of 10^5 liters. If the cloud forms over a chemical factory which puts SO_3 into the air and each SO_3 molecule bonds with one out of 10^4 H_2O molecules (to form H_2SO_4), what is the molarity of the H_2SO_4, in the cloud?

Q. 107. A lab technician prepared a calcium hydroxide solution by dissolving 1.48 g of $Ca(OH)_2$ in water. How many milliliters of 0.125 N HCl solution would be required to neutralize this calcium hydroxide solution?

Q. 108. How many ml of 0.015 M $Ca(OH)_2$ would be required to neutralize 35.0 ml of 0.36 M H_3PO_4

Q. 109. Assuming complete neutralization, calculate the number of milliliters of 0.025 M H_3PO_4, required to neutralize 25 ml of 0.03 M Ca(OH).

Q. 110. Determine the molarity of an H_2SO_4 solution, 15.0 ml of which is completely neutralized by 26.5 ml of 0.1 M NaOH

Q. 111. Suppose you spill 40 ml of 0.1 M H_2SO_4. How much solid $NaHCO_3$ would you have to dump on it to neutralize the acid by the following reaction?

$$2NaHCO_3(s) + H_2SO_4 \rightarrow Na_2SO_4 + 2H_2O + 2CO_2(g)$$

Q. 112. How many mL of 0.0252 M HNO_3(aq) are required to neutralize 25.0 mL (0.025L) of 0.0148 M $Sr(OH)_2$(aq)?

Q. 113. A common method for commercially peeling potatoes is to soak them for 1-5 minutes in a 10-20 g solution for NaOH (molecular weight = 40.0 g mol^{-1} at 333–361K, and to spray off the peel once the potatoes are removed from solution. A manufacturer titrates the NaOH solution with standardized H_2SO_4 (molecular weight = 98.0 g mol^{-1}) at the end of each day to determine whether the solution is still capable fo peeling potatoes. If, at the end of the day, he finds that it takes 64.0 ml of a 0.2 M solution of H_2SO_4 to titrate a 10.0 ml sample of NaOH solution to neutrality, what concentration of NaOH did he find?

Q. 114. 50 ml of rhubarb juice is titrated against 0.25 N NaOH 20 ml of NaOH solution is required for neutralization. Assuming the acidity of the juice is due to oxalic acid ($H_2C_2O_4$) determine (a) the weight of oxalic acid per liter of juice, (b) normality of the juice.

Q. 115. A chemist performs a neutralization reaction. She finds that 1 g of $C_6H_{10}O_4$, an acid, requires 0.768 g of KOH for complete neutralization. Determine the number of neutralizable protons in this molecule.

Q. 116. How much water should be added to 1.38 liters of 3.0 *M* HNO_3 to make a solution that is 1.40 *M* HNO_3? Assume volume are additive.

Q. 117. A sample consisting of 25.0 ml of 3.0 *M* HNO_3 is mixed with 75.0 ml of 4.0 *M* HNO_3. Assuming volumes are additive, what would be the concentration of the final mixture?

Q. 118. A solution is made by mixing 10.0 ml of 0.02 *M* $Ca(NO_3)_2$ and 15.0 ml of 0.03 M $NaNO_3$. The final volume is 25.0 ml and that dissociation is complete, calculate the concentrations of Na^+ and NO_3^- in the final solution.

Q. 119. An unknown white molecular solid has been sent out for elemental analysis. From the weight percentage the empirical formula is determined to be $(C_3H_4)_n$. A 0.527 g sample of the compound is dissolved in 40.19 g of $CHCl_3$, and it is found that the solution boils 0.392 K higher than does the pure $CHCl_3$ used as solvent. The unknown neither associates not dissociates in $CHCl_3$ solution. Calculate the molecular formula and exact molecular weight of this compound.

Q. 120. (a) Aqueous solutions of ethylene glycol, CH_2OHCH_2OH, are commonly used as antifreeze for automobile radiators. Pure ethylene glycol is a liquid with density 1.11 g · mL^{-1}. What should be the approximate molality of ethylene glycol in an aquous solution, if the solution is to freeze at a temperature no higher than 258 K?

(b) What volume of ethylene glycol should be added to 20 L of water (density 1.0 g · mL^{-1}) to produce a solution with the molality calculated in part (a)?

Q. 121. Arrange the following solutions in order of decreasing freezing point.

(a) 0.05 *F* $NaNO_3$ **(b)** 0.075 *F* $CuSO_4$

(c) 0.06 *F* $(NH_4)_2SO_4$ **(d)** 0.14 *m* sucrose, $C_{12}H_{22}O_{11}$ **(e)** 0.04 F $BaCl_2$

Q. 122. Insulin is a hormone that regulates carbohydrate metabolism by controlling the amount of glucose (a simple sugar) in the blood. A 2.0-g sample of insulin is dissolved in enough water to make 250.0 mL of solution, and the osmotic pressure of the solution is measured and found to be 26.1 mm Hg at 303 K. What is the molecular weight of this insulin?

Q. 123. Two liquids *A* and *B* from ideal solutions. A solution of *A* and *B* is prepared at 303 K in which the mole fraction of *A* is 0.256. In the vapor phase in equilibrium with this solution, the mole fraction of *A* is 0.318 and the total pressure is 673 mmHg.

(a) Calculate the partial pressure of *A* and *B* in the vapor phase.

(b) Calculate the vapor pressure of pure liquid *A* and pure liquid *B* at 303 K.

Q. 124. Calculate the concentration of dissolved O_2 in equilibrium with the air at 1-atm pressure at 298 K. The mole fraction of O_2 in air is 0.21 (21.0 mol %). Also calculate molality from mole fraction obtained.

Q. 125. A sample of an unknown volatile liquid is placed into a 0.264 flask. The flask is covered with a piece of aluminium foil which has a very small pin hole. The flask is now immersed into a boiling water bath at 373.5 K; the liquid is allowed to vaporize thus forcing out the air in the flask. When there is no more liquid visible in the flask, the flask is placed into cold water so that the vapour could condense. The flask is then dried thoroughly and reweighed. The mass of the condensed vapor is 0.43 g at the room pressure of 750 mm of Hg. Calculate the molecular mass of the unknown.

Q. 126. A student set-up an apparatus to determine the molecular mass through freezing point depression, of naphthalene.

Naphthalene, $C_{10}H_8$

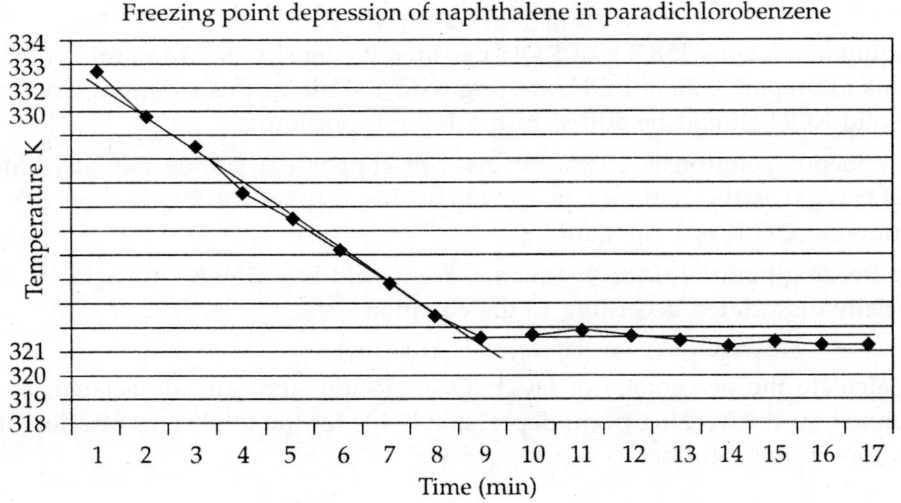

The student measured out 52 grams of paradichlorobenzene and 4 grams of naphthalene, placed them in a test tube in a hot-water bath and allowed the mixture to completely melt. The mixture was stirred well. The tube was then removed from the hot-water bath, dried and allowed to cool while it was gently stirred to minimize supercooling. Temperature readings were taken every 60 seconds until the mixture solidified and the temperature stabilized. A graph of the results is given below. Now, Answer 1 to 5

1. What is the freezing point of the solution?
2. What was the freezing point depression of the solution? The standard freezing point for paradichlorobenzene is 326 K.
3. What was the molality of the naphthalene? The freezing point depression constant for paradichlorobenzene is 7.1 $K \cdot m^{-1}$.
4. What was the molecular mass of naphthalene?
5. What was the percent error of the results compared to the actual molecular mass of 128 $g \cdot mol^{-1}$?

Q. 127. A student diluted 50 mL of a commercial bleach to 250 mL in a volumetric flask and then titrated a 20 mL aliquot. The titration required 35.5 mL of 0.1 M $Na_2S_2O_3$ solution. The density of the bleach was .15 g . mL^{-1}. Answer the following 1-7

1. Calculate the number of moles of $S_2O_3^{2-}$ ion required for titration.
2. Calculate the number of moles of I_2 produced in the titration mixture.
3. Calculate the number of moles of OCl^- ion present in the diluted bleaching solution that was titrated.
4. Calculate the mass of NaOCl present in the diluted bleaching solution titrated
5. Determine the volume of commercial bleach present in the diluted bleaching solution titrated (aliquot).
6. Calculate the mass of commercial bleach titrated.
7. Determine the percent NaOCl in the commercial bleach.

Q. 128. What is the density of an aqueous solution of potassium nitrate that has a normal boiling point of 376 K and an osmotic pressure of 122 atm at 298 K?

Q. 129. A solution contains 158.2 g of KOH per liter, its density is 1.13 g/mL. A lab technician wants to prepare 0.25 *m* KOH, starting with 100 mL of this solution. How much water or solid KOH should be added to the 100 mL portion.

Q. 130. An aqueous solution is 15.0% by mass of copper (II) sulfate pentahydrate. $CuSO_4$ $5H_2O$. What is the molarity of $CuSO_4$ in this solution at 293 K. The density of this solution at 293 K is 1.167 $g.mL^{-1}$.

Q. 131. The freezing point of 0.083 *m* aqueous acetic acid is –0.159 K. Acetic acid, $HC_2H_3O_2$, is partially dissociated according to the equation

$$HC_2H_3O_2(aq) \rightleftharpoons H^+(aq) + C_2H_3O^-(aq)$$

Calculate the percentage of $HC_2H_3O_2$ molecules that are dissociated, assuming the equation for the freezing-point depression holds for the total concentration of molecules and loss in the solution.

Q. 132. A compound of carbon, hydrogen, and oxygen was burned in oxygen, and 1.000 g of the compound produced 1.434 g CO_2 and 0.783 g H_2O. In another experiment, 0.1107 g of the compound was dissolved in 25 g of water. This solution had is freezing point of 0.0894°K. What is the molecular formula of the compound?

Q. 133. A certain gaseous solute dissolves in water, evolving 12 kJ of heat, its solubility at 298 K and 4 atm is 0.02 *M*. Would you expect the solubility to be greater or less than 0.02 *M* at

(a) 278 K and 6 atm? **(b)** 323 K and 2 atm?
(c) 293 K and 4 atm? **(d)** 298 K and 1 atm?

Q. 134. The freezing point of 0.1 *M* $KHSO_3$ is –0.38°K. Which of the following equations best represent what happens when $KHSO_3$ dissolves in water?
(a) $KHSO_3(s) \rightarrow KHSO_3(aq)$
(b) $KHSO_3(s) \rightarrow K^+(aq)$ + HSO**(aq)
(c) $KHSO_3s) \rightarrow K^+\ (aq) + SO_3^{2-}\ (aq) + H^+\ (aq)$

Q. 135. A solution is made up by dissolving 15 g $MgSO_4 . 7H_2O$ in 100 g of water. What is the molality of $MgSO_4$ in this solution?

Q. 136. The water soluble nonelectrolyte *X* has a molar mass of 410 g mol^{-1} 0.1 g mixture containing this substance and sugar (M = 342 g mol^{-1}) is added to 1 g of water to give a solution whose freezing point is –0.5 K. Estimate the mass percent of X in the mixture.

Q. 137. A martini, weighing about 5.0 oz (142 g), contains 30% by mass of alcohol. About 15% of the alcohol in the martini passes directly into the bloodstream (7.0 for an adult. Estimate the concentration of alcohol in the blood (g/cm^3) of a person who drinks two martinis before dinner. (A concentration of 0.001 g cm^{-3} or more is frequently considered indicative of intoxication in a "normal" adult.)

Q. 138. When water is added to a mixture of aluminium metal and sodium hydroxide, hydrogen gas is produced. This is the reaction used in commercial drain clearners:

$$2Al(s) + 5H_2O(l) + 2OH^-(aq) \rightarrow 2Al(OH)_4\ (aq) + 3H_2(g)$$

A sufficient amount of water is added to 49.92 g of NaOH to make 0.6 L of solution; 41.28 g of Al is added to this solution and hydrogen gas is formed.
(a) Calculate the molarity of the initial NaOH solution
(b) How many moles of hydrogen were formed?
(c) The hydrogen was collected over water at 298 K and 758.6 mm Hg. The vapor pressure of water at this temperature is 23.8 mm Hg. What volume of hydrogen was generated?

Q. 139. Phosphorous acid, H_3PO_3, and phosphoric acid, H_3PO_4, have approximately the same acid strengths. From this information and noting the possibility that one or more hydrogen atoms may be directly bonded to the phosphorus atom, draw the structural formula of phosphorous acid. How many grams of sodium hydroxide would be required to completely neutralize 1.0 g of this acid?

Q. 140. Boron trifluoride, BF_3, acid ammonia, NH_3 react to produce $BF_3.NH_3$. A coordinate covalent bond is formed between the boron atom on BF_3 and the nitrogen atom on NH_3. Write the equation for this reaction, using Lewis electron-dot formulas. Label the Lewis acid and the Lewis base. Determine how many grams. of BF_3 <.> NH_3 are formed when 10 g BF_3 and 10 g NH_3 are placed in a reaction vessel, assuming that the reaction goes to completion.

Q. 141. (a) Each figure below represent an acid solution in equilibrium. Squares represent H^+ ions, and circles represent the anion. Water molecules are not shown. Which figure represents a strong acid? Which a weak acid?

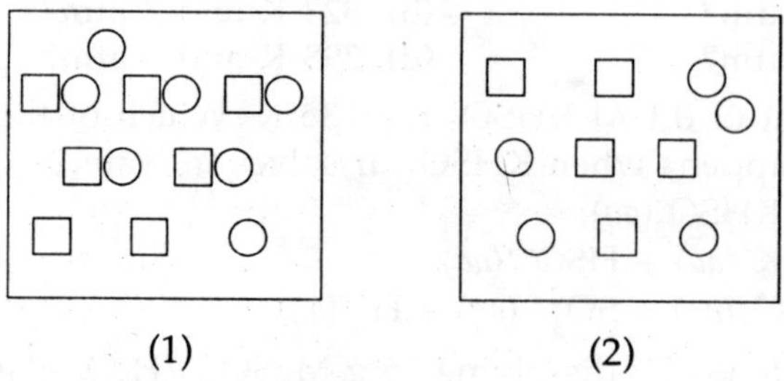

(1) (2)

(b) If □○ represent weak acid HA(□ = H^+, ○ = A^-) in Figure a below, fill in Figure *b* with □○, □, and/or ○ to represent HA as being 10% ionized. (Water molecules are omitted).

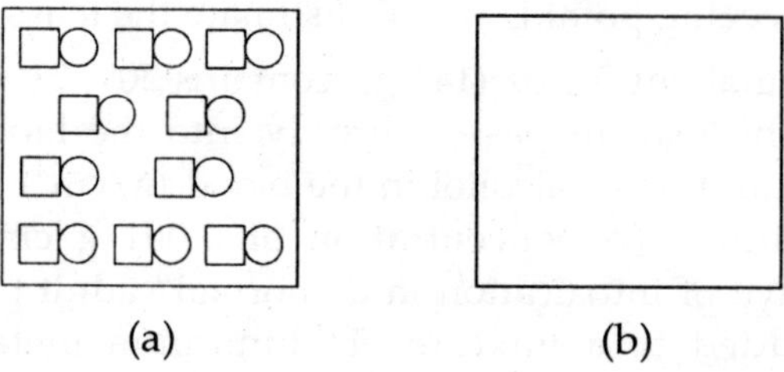

(a) (b)

Q. 142. To 500 cm^3 of water, 3.0×10^{-3} kg of acetic acid is added. If 23% of acetic acid is dissociated, what will be the depression in freezing point? K_f and density of water are 1.86 K kg^{-1} mol^{-1} and 0.997 g cm^{-3} respectively.

Q. 143. A very small amount of a non volatile solute (hat does not dissociate) is dissolved in 56.8 cm^3 of benzene (density 0.889 g cm^{-3}). At room temperature vapor pressure at this solution 98.88 mm Hg while that of benzene is 100 mm Hg. Find the molality of this solution, if the freezing temperature of this solution is 0.73 degree lower than that of benzene, what is the value of molal freezing point depression constant of benzene?

Q. 144. A solution of a non volatile solute in water freezes at 243 K. The vapour pressure of pure water at 298 K is 23.51 mm Hg and K_f for water is 1.85 degree/molal. Calculate the vapour pressure of this solution at 298 K.

Q. 145. When 5.5 g of the minor product is dissolved in 45 g of benzene, the boiling point of the solution is 1.84°K higher than that of pure benzene. Calculate the molecular weight of the minor product and determine its molecular and structural formula.

(Molal boiling point elevation constant of benzene is 2.53 K kg mol^{-1})

Q. 146. The molar volume of liquid benzene (density –0.887 g ml^{-1}) increases by a factor of 2750 as it vaporizes at 293 K and that of liquid toluene (density = 0.867 g ml^{-1}) increases by a factor of 7720 at 293 K. A solution of benzene and toluene at 293 K has a vapour pressure of 46 mm Hg. Find the mole fraction of benzene in the vapour above the solution.

Q. 147. Addition of 0.643 g of a compound to 50 ml of benzene (density: 0.879 g mL^{-1}) lowers the freezing point from 278.51 K to 278.03 K. If K_f for benzene is 5.12, calculate the molecular weight of the compound.

Q. 148. 28 Kg of nitrogen is mixed with 17.6 Kg of Carbondioxide at pressure of 250 KPa and 300 K. Assuming the gases to be ideal calculate:

(i) partial pressures of gases in the mixture.

(ii) density of gas mixture.

Q. 149. Sodium tetraborate decahydrate, $Na_2B_4O_7 \cdot 10H_2O$, commonly called borax, is a white crystalline solid that reacts with strong acids as follows:

$$Na_2B_4O_7 \cdot 10H_2O(s) + 2H^+(aq) \rightarrow 4H_3BO_3 + 5H_2O + 2Na^+(aq)$$

The boric acid formed, H_3BO_3, is a weak acid, soluble in water. How many milliliters of 0.1082 *F* HCl are required to reach the equivalence point in a titration of a 0.6941-g sample of borax?

Q. 150. Nitrous acid, HONO or HNO_2, is a weak acid. The freezing point of an 0.0750 *m* aqueous solution of HONO is 0.150 K. Calculate the percentage of ionization of 0.0750 *m* nitrous acid.

Q. 151. Calculate the concentrations of all species in the solution that results when 20 mL of 0.12 *F* NaCl, 40 mL of 0.15 *F* KNO_3, and 60 mL of 0.08 *F* $CaCl_2$ are mixed.

Q. 152. A 25-mL portion of 0.112 *F* $Pb(NO_3)_2$ is mixed with a 45-mL portion of 0.084 *F* K_2CrO_4, and a bright yellow precipitate of lead chromate is formed. Assuming the reaction has gone to completion, calculate the molarity of the principal ions present in solution after the reaction.

Q. 153. A 10 mL sample of 0.093 *F* $AgNO_3$ is mixed with a 15-mL sample of 0.12 *F* Na_2CO_3, and pale yellow precipitate of silver carbonate is formed. Assuming the reaction has gone to completion, calculate the mass of the silver carbonate precipitate and the concentrations of the principal ions remaining in solution.

Q. 154. A 1.00 g. sample of H_2O_2 solution containing *x* per cent H_2O_2 by weight required *x* ml. of a $KMnO_4$ solution for complete oxidation under acidic condition. Calculate the normality of the $KMnO_4$ solution.

Q. 155. Find out the volume in ml. of 0.1 N hydrochloric acid required to react completely with 1 g. of mixture of sodium carbonate and sodium bicarbonate containing equi-molar amounts of the two compounds.

Q. 156. (i) A solution of a mixture of potassium chloride and potassium hydroxide was neutralised with 120 ml. of 0.12 N-hydrochloric acid. Calculate the amoung of potassium hydroxide in the mixture.

(ii) After titration the resultant solution was made acidic with nitric acid. Then excess of silver nitrate solution was added to precipitate the silver chloride which weighed 3.7 g. after drying. Calculate the percentage of potassium hydroxide in the original mixture.

Q. 157. **(a)** $CuSO_4$ reacts quantitaively with KI in an acidic medium to liberate I_2. Write a balanced equation for this reaction.

(b) Mercuric iodate $[Hg_5(IO_6)_2]$ reacts with a mixture of KI and HCl according to the following equation:

$$Hg_5(IO_6)_2 + 34KI + 24HCl \rightarrow 5K_2HgI_4 + 8I_2 + 24KCl + 12H_2O$$

The liberated iodine is treated with $Na_2S_2O_3$ solution, one ml. of which is equivalent to 0.499 g. of $CuSO_4 \cdot 5H_2O$. What volume (in ml.) of the $Na_2S_2O_3$ solution will be required to react with iodine liberated from 0.7245 g. of $Hg_5(IO_6)_2$? Hg = 200.5, Cu = 63.5; I = 127)

Q. 158. 0.108 g of finely-divided copper was treated with an excess of ferric sulphate solution until copper was completely dissolved. The solution of er the addition of excess dilute sulphuric acid required 33.7 ml. of 0.1 N-$KMnO_4$ for complete oxidation. Find the equation which represents the reaction between the metallic copper and ferric sulphate solution. (Cu = 63.5, Fe = 56)

Q. 159. One gram of commercial $AgNO_3$ is distilled in 50 ml of water. It is treated with 50 ml of a KI solution. The silver iodide thus precipitated is filtered off. Excess of KI in the filterate is titrated with (M/10) KIO_3 solution in presence of 6m HCl till all I^- ions are converted into ICl. It requires 50 ml of (M/10) KIO_3 solution. 20 mL of the same stock solutions of KI requires 30 ml of (M/10)KIO_3 under similar conditions. Calculate the percentage of $AgNO_3$ in the sample.

Q. 160. 0.5 gm of fuming H_2SO_4 (oelum) is diluted with water. The solution is completely neutralized by 26.7 ml of 0.4 *N* NaOH. Find the percentage of free SO_3 in the sample of oleum.

Q. 161. A 0.5 g. sample of an iron-containing mineral mainly in the form of $CuFeS_2$ was reduced suitably to convert all the ferric-ion into the ferrcus form and was obtained as a solution. In the absence of only interfering matter the solution required 42 ml. of 0.01 M-$K_2Cr_2O_7$ solution of titration. Calculate the percentage of $CuFeS_2$ in the mineral. (Cu = 63.5; Fe = 55.8; S = 32; O = 16)

Q. 162. A commercial sample (2.013 g.) of NaOH containing Na_2CO_3 as an impurity was dissolved to give 250 ml. of solution. A 10 ml. portion of this solution required 20 ml. of 0.1 N-H_2SO_4 for complete neutralization. Calculate the percentage by weight of Na_2CO_3 in the sample.

Q. 163. **(i)** A sample of $MnSO_4 \cdot 4H_2O$ is strongly heated in air. The residue is Mn_3O_4.

(ii) The residue is dissolved in 100 ml of 0.1 N $FeSO_4$ containing dilute H_2SO_4.

(iii) The solution reacts completely with 50 ml of $KMnO_4$ solution.

(iv) 25 ml of the $KMnO_4$ solution used in step (iii) requires 30 ml of 0.1 N F2SO4 solution for complete reaction.

Find the amount $MnSO_4 \cdot 4H_2O$ present in the sample.

Q. 164. To a 25 mL H_2O_2 solution, excess of acidified solution of potassium iodide was added. The iodine liberated required 20 mL of 0.3 N sodium thiosulphate solution. Calculate the volumé strength of H_2O_2 solution.

Q. 165. A mixture containing only Na_2CO_3 and K_2CO_3 weighing 1.22 g. was dissolved in water to form 100 ml. of a solution. 20 ml of this solution required 40 ml of 0.1 N-HCl for neutralization. Calculate the weight of Na_2CO_3 in the mixture. If another 20 ml. portion of solution is heated with excess $BaCl_2$, what will be the weight of the precipitate obtained? (Na = 23; K = 39.1; C = 12; O = 16; Ba = 137.3; Cl = 35.5)

Q. 166. 3.0 g of sample containing Fe_3O_4, Fe_2O_3 and an inert impure substance, is treated with excess of Kl solution in presence of dilute H_2SO_4. The entire iron is converted Fe^{2+} along with the liberation of iodine. The resulting Slution is diluted to 100 ml. A 20 ml of diluted solution requires 11 ml of 0.5 M $Na_2S_2O_3$ solution to reduce the iodine present. A 50 ml of the diluted solution, after complete extraction of the iodine requires 12.8 ml of 0.25M $KMnO_4$ solution in dilute H_2SO_4 medium for the oxidation of Fe^{2+}. Calculate the percentages of Fe_2O_3 and Fe_3O_4 in the original sample.

Q. 167. A polyvalent metal weighing 0.1 g and having atomic weight 51, reacted with dil. H_2SO_4 to give 43.9 ml of hydrogen at S.T.P. This solution containing the metal in the lower oxidation state, was found to require 58.8 ml. of 0.1 N permanganate for complete oxidation. What are the valencies of the metal?

Q. 168. 5.5 g of a mixture of $FeSO_4 \cdot 7H_2O$ and $Fe_2(SO_4)_3 \cdot 9H_2O$ required 5.4 ml. of 0.1 N-$KMnO_4$ solution for complete oxidation. Calculate the number of g · moles of hydrated ferric sulphate in the mixture.

Q. 169. A 10.0-mL sample of potassium iodide solution was analyzed by adding an excess of silver nitrate solution to produce silver iodide crystals, which filtered from the solution.

$$KI(aq) + AgNO_3(aq) \rightarrow KNO_3(aq) + AgI(s)$$

If 2.29 g of silver iodide was obtained, what was the molarity of the original KI solution?

Q. 170. 8.0575×10^{-2} kg of Glauber's salt is dissolved in water to obtain 1 dm^3 of a solution of density 1077.2 kg m^{-3}. Calculate the molarity, molality and mole fraction of Na_2SO_4 in the solution.

Q. 171. In a quantitative determination of iron in an ore, an analyst converted 0.42 g of the ore into its ferrous form. This required 42 ml. of 0.1 N solution $KMnO_4$ for titration.

(i) How many equivalents of $KMnO_4$ does 42 ml. of 0.1 N-solution represent?

(ii) How many equivalents of iron were present in the sample of the or taken for analysis.

(iii) How many grams of iron were present in the sample?

(iv) What is the percentage of iron in the ore?

(v) What is the molarity of the $KMnO_4$ solution used?

(vi) How many moles of $KMnO_4$ were used for titration? (Fe = 56)

Q. 172. A sample of hard water contains 96 ppm of SO_4^- and 183 ppm of HCO_3^- with Ca^{2+} as the only cation. (a) How many moles of CaO will be required to remove HCO_3^- from 1000 kg of this water? (b) If 1000 kg of this water is treated with the amount of CaO calculated above, what will be the concentration (in ppm) of residual Ca^{2+} ions? (Assume $CaCO_3$ to be completely insoluble in water). (c) If the Ca^{2+} ions in one litre of the treated water are completely exchanged with hydrogen ions, what will be its *p*H? (One ppm means one part of the substance in one million part of water by weight.

Q. 173. A 100 ml. solution of Na_2CO_3 is prepared by solving 8.653 g of the salt in water. The density of the solution is 1.816 gm. per millilitre. What are the molality and molarity of the solution? (At. of Na = 23, C = 12 and O = 16).

Q. 174. (a) A zinc rod weighing 25 g. was kept in 100 ml. of 1 M-$CuSO_4$ soln. After a certain time the molarlity of Cu^{++} in the soln. was 0.8. What was the molarity of the sulphate (SO_4^{-2})? What was the weight of zinc rod after cleaning?

(b) If the above experiment is done with coper rod of weight 25 g. and 50 ml. of 2M-ZnSO4 solution, what would be the molarity of Zn^{++} at the end of the same in interval? (Zn = 56.4; Cu = 63.5)

Q. 175. A piece of aluminium weighing 2.7 g. is heated with 75 ml. of sulphuric acid (sp. gr 1.18, containing 24.7 per cet H_2SO_4 by weight). After the metal is carefully dissolved, the solution is diluted to 400 ml. calculate the molarity of the free H_2SO_4 in the resulting solution.

Q. 176. The reaction: $Zn + CuSO_4 = Cu + ZnSO_4$ goes completely to the right. In one experiment 10 g of metallic zinc was added to 200 ml. of copper sulphate solution. After al the copper was precipitated it was found that not all the zic had dissolved. After filtration the total solid at the end of the reaction was 9.81 g.

(i) the weight of copper deposited.

(ii) the molarity of copper sulphate in the original solution.

Chapter **6**

THERMODYNAMICS

FORMULAE

$\Delta H = H_P - H_R$
ΔH= (–)ve For exothermic process
= (+)ve For endothermic process
$\Delta E = q - w$

$$w = 2.303 \text{ nRT} \log \frac{V_2}{V_1} = 2.303 \text{ nRT} \log \frac{P_1}{P_2}$$

For isothermal process $\Delta E = 0$
For adiabatic process $q = 0$
For isobaric process $\Delta P = 0$
For isobaric process $\Delta V = 0$
$\Delta H = \Delta E + P\Delta V = \Delta E + \Delta nR$

$$\Delta C_P = \frac{\Delta H_2 - \Delta H_1}{T_2 - T_1}$$

$$\Delta C_V = \frac{\Delta E_2 - \Delta E_1}{T_2 - T_1}$$

$$\eta = \frac{w}{q_2} = \frac{T_2 - T_1}{T_2} = \frac{q_2 - q_1}{q_2}$$

$$\Delta S = \frac{q_{rev}}{T}$$

$$= 2.303 \text{ nR} \log \frac{V_2}{V_1} = 2.303 \text{ nR} \log \frac{P_1}{P_2} = 2.303 \text{ nC}_P \log \frac{T_2}{T_1} = 2.303 \text{ mC}_v \log \frac{T_2}{T_1}$$

$$= 2.303 \text{ nC}_P \log \frac{T_2}{T_1} + 2.303 \text{ nR} \log \frac{P_1}{P_2} = 2.303 \text{ nC}_v \log \frac{T_2}{T_1} + 2.303 \text{ nR} \log \frac{V_2}{V_1}$$

$$\Delta G^\circ = \Delta H - T\Delta S$$

$$= -2.303 \text{ R} \log \text{K} = -2.303 \text{ nRT} \log \frac{P_1}{P_2} = -2.303 \text{ nRT} \log \frac{V_2}{V_1} = -nFE^\circ_{\text{cell}}$$

$$\Delta G = \Delta H + \left[\frac{\partial(\Delta G)}{\partial T}\right]_P$$

$$\frac{d \ln P}{dT} = \frac{\Delta H}{RT^2}$$

$$\log \frac{P_2}{P_1} = \frac{\Delta H}{2.303\, R}\left[\frac{T_2 - T_1}{T_1 T_2}\right]$$

SUMMARY

Energy

Chemical reactions are accompanied by energy changes for example combustion of coal and natural gas involves liberation of heat. These reactions can be known with the help of certain in thermodynamic functions such as internal energy, enthalpy, entropy and free energy.
Energy is the capacity for doing work. There are different forms of energy such as potential energy, kinetic energy, heat energy, light energy, electrical energy, chemical energy, nuclear energy, etc. All forms of energy are governed by the law of conservation of energy. According to this 'energy can neither be created nor destroyed' but it can be changed from one form to another.
A process that occurs without heat exchange is ADIABATIC. An isothermal process occurs at constant temperature.

Internal Energy, E

Every substance is associated with definite amount of energy known as internal energy or intrinsic energy. It is denoted by the symbol E

The internal energy is made up of

1. the energy associated with the interactions of the electrons within the molecule (electronic energy).
2. the energy of internal motions within the molecule (transactional energy).
3. the energy due to molecular rotations (rotational energy) and
4. the energy due to vibratory options (vibrational energy).

The change in internal energy, ΔE is dependent only on the initial and final states of the system and on the path followed between two states. The symbol Δ stands for the change.

For example, if E_1 is the internal energy in the initial state and E_2 that of the final state, then

$$\Delta E = E_2 - E_1 \qquad (1)$$

In a chemical reaction ΔE is the difference between internal energies of the products and reactants.

$$\Delta E = E_P - E_R \qquad (2)$$

Where ΔE = change in internal, energy, E_R = internal energy of the reactants and E_P = internal energy the products.

Work

Work under different conditions:
Note that—Work is expressed as the product of two factors, i.e.,

$$W = \text{intensity factor} \times \text{capacity factor}$$

Intensity factor is measure of force against which work is done. Capacity factor is the extent for which work is done.

eg:

(i) Mechanical work W = force × displacement = $F \times d$ Joules.

(ii) Gravitational work W = gravitational force × height = $mg \times h$ If we raise mass 'm' to height 'h' against gravitational force 'g', we apply force mg over a distance h.)

(iii) Electrical work = pot. difference × quantity of current = $E \times Q$

(iv) Work of expansion W = pressure × change in volume = $P \times \Delta V = P(V_2 - V_1)$ litre atm.

i.e. (a) A gas is expanded from initial volume V_1 litres to final V_2 litres against a constant pressure of "P" atmosphere.

(b) Instead of volume of initial pressure P_1, temperature T_1 and final pressure P_2 and temperature T_2 are given then

$$V_1 = \frac{nRT_1}{P_2} \text{ liter; } V_2 = \frac{nRT}{P_2} \text{ litres, where, } R = 0.0821 \text{ litre atm. mol}^{-1}\text{ K}^{-1}.$$

(c) When the reaction takes place or when phase changes from solid gas (sublimation) or liquid to vapour (vaporisation), then

$$w = P\Delta V = \Delta nRT$$

where Δn = moles of products (gaseous) minus moles of reactants (gaseous)

The general expression for all types of PV work can be written as $W = -P\Delta V$

For expansion $\Delta V = V_2 - V_1 = +\text{ve}$ ($\because V_2 > V_1$)

$\therefore$ $W = -\text{ve}$

Thus, for of expansion or work done by the system is –ve.

For compression: $\Delta V = V_2 - V_1 = -\text{ve}$ ($\because V_2 < V_1$)

$\therefore$ $W = +\text{ve}$

Thus, work of compression or work done on the system is +ve.

A –ve sign is given for work done *by* the system.

A +ve sign is given for work done *on* the system.

Units of work are W = newton or joule

For an expanding system involving pressure volume work.

$$w = \int_{V_1}^{V_2} P\Delta V \text{ and } \Delta E = q - \int_{V_1}^{V_2} P\Delta V$$

At constant pressure,

$$w = P(V_2 - V_1), \text{ and } \Delta E = q - P(V_2 - V_1)$$

At constant volume, $\Delta V = 0,\ w = 0$

$\therefore$ $\Delta E = q - w$ becomes

$$\Delta E = q_v$$

i.e., all heat absorbed at constant volume goes to increase the internal energy of the system. A process which is carried out a constant volume is called an *isochoric process*.

The variation of ΔH or ΔE with temperature is express in terms of Kirchoff's equations

$$\Delta H - \Delta H_1 = \Delta C_P(T_2 - T_1) \text{ or } \Delta H = C_P\,\Delta T \text{ and } \Delta E_2 - \Delta E_1 = \Delta C_v(T_2 - T_1)$$

Where, ΔH_2 and ΔE_2 are heats of reaction at temperature T_2,

ΔH_1 and ΔE_1 are heats of reaction at temperature T_1,

$$C_P = C_P \text{ of products} - C_P \text{ of reactants,}$$
$$C_V = C_V \text{ of products} - C_V \text{ of reactants}$$

C_P and C_V are molar specific heats at constant P & V respectively.

If pressure is not constant, then $\Delta H = \int_{T_1}^{T_2} C_p\, dT$

Also,

(a) $C_P = \dfrac{\Delta E}{\Delta T} = \dfrac{q_P}{\Delta T}$ (at constant Pressure)

(b) $C_V = \dfrac{\Delta E}{\Delta H} = \dfrac{q_v}{\Delta T}$ (at constant Pressure)

Thermodynamic Process

The operations by which the changes from one state to other are made, are known as thermodynamic process. The operations are made in either of the way

1. **Isothermal process:**
   (a) A process in which temperature of the system remains constant throughout the studies.
   (b) $\Delta T = 0$ and also $\Delta E = 0$
   (c) Achieved by thermostatic bath.
2. **Adiabatic process:**
   (a) A process during which no exchange of heat takes place in between system and surroundings.
   (b) $q = 0$.
   (c) Achieved by insulating the system boundaries.
3. **Cyclic process:**
   (a) A process in which initial state of system is regained after a series of operations.
   (b) $\Delta E = 0$; $\Delta H = 0$.
4. **Isobaric process:**
   (a) A process in which pressure of the system remains constant throughout the investigations.
   (b) $\Delta P = 0$
5. **Isochoric process:**
   (a) A process in which volume of the system remains constant through the investigations.
   (b) $\Delta V = 0$.

Work Done in Irreversible Isothermal Process

Suppose an ideal gas expands against external pressure P and its volume changes by an amount dV then work done W can be given by

$$W = -PdV$$

For a finite change V_1 to V_2,

Total work done the system W is derived by

$$W = -P(V_2 - V_1)$$

or $$W_{irr} = -P(V_2 - V_1)$$

Thus, at constant V

$$W_{irr} = 0$$

Work Done in Isothermal Reversible Process

Consider a system under isothermal condition, showing reversible expansion of an ideal gas by a volume dV, then

$\because \; \Delta E = 0$

$\therefore \; dq = -dW = +PdW \qquad (\because dW = -PdV)$

The total work done during expansion of gas from V_1 to V_2

$$\int dW = \int_{V_1}^{V_2} -PdV = \int_{V_1}^{V2} -\frac{nRT}{V}\cdot dV$$

$$W_{rev} = -nRT \log_e \frac{V_2}{V_1} = -2.303\ nRT \log_{10} \frac{V_2}{V_1}$$

Work Done in Adiabatic Reversible Process

Consider a system under adiabatic conditions, showing reversible expansion of an ideal gas by a volume dV, then I law of thermodynamics,

$(\because dq = 0)$

$$dE = dW \quad dE = nC_v\,dT$$

or $$C_v n \times dT = dW$$

This equation reveals on further treatment within temperatrue limits of T_1 and T_2

$$dW = C_v \times n \times dT$$

Now $$C_p - C_v = R \text{ or } \frac{C_p}{C_v} - \frac{C_v}{C_v} = \frac{R}{C_v} \Rightarrow \gamma - 1 = \frac{R}{C_v}$$

$\therefore$ $$C_v = \frac{R}{\gamma - 1}$$

$\therefore$ $$dW = \frac{R}{(\gamma - 1)} \times n \times dT = \frac{nR}{\gamma - 1}\cdot dT$$

on integration $$\int dW = \frac{nR}{\gamma - 1}\int_{T_1}^{T_2} dT$$

$$w_{rev} = \frac{nR}{\gamma - 1}[T_2 - T_1]$$

if $T_2 > T_1$ then W_{rev} = +ve, i.e., work done on the system
if $T_2 < T_1$ then w_{rev} = –ve, i.e., work done on the system

where γ is Poisson's ratio $= \frac{C_p}{C_v}$

Some important results of adiabatic expansions

$$PV^{\gamma} = \text{constant}$$
$$T^{\gamma}P^{1-\gamma} = \text{constant}$$
$$V^{\gamma-1}T = \text{constant}$$

EXOTHERMIC AND ENDOTHERMIC REACTIONS

Exothermic Reactions

Chemical reactions which take place with the evolution of heat are known as exothermic reactions. In exothermic reactions, the total energy of the products is less than the total energy of the reactants. The difference in the energy is given out to the surroundings in the form of heat. eg.

1 : $C(s) + O_2(g) \rightarrow CO_2(g)$ 94.1 K · Cal

(or)

$C(s) + O_2(g) \rightarrow CO_2(g)$; $\Delta H = -94.1$ K · Cal

2 : $N_2(g) + 3H_2(g) \rightleftharpoons 2NH_3(g)$ + 22 K · Cal

(or)

$N_2(g) + 3H_2(g) \rightleftharpoons 2NH_3(g)$; $\Delta H = -22$ K · Cal

Endothermic Reactions

Chemical reactions which take place with the absorption of heat are called endothermic reactions. In endothermic reactions, the total energy of the products is more than the total energy of the reactants. The difference in the energy is absorbed by the system from the surroundings. Eg.

1 : $N_2(g) + O_2(g) \rightarrow 2NO(g)$ + –43.23 K · Cal

(or)

$N_2(g) + O_2(g) \rightarrow 2NO(g)$; $\Delta H = +\ 43.23$ K · Cal

2 : $C(s) + H_2O(g) \rightarrow CO(g) + H_2(g)$ –31.4 K · Cal

(or)

$C(s) + H_2O(g) \rightarrow CO(g) + H_2(g)$; $\Delta H = +31.4$ K · Cal

In a reversible reaction, if the forward reaction is exothermic, the reverse reaction will be endothermic and vice versa.

Also, If ΔE = +ve; the reaction is endothermic

If ΔE = –ve; the reaction is exothermic.

Ist law of Theromodynmamic

When a system exchanges heat or work with its surroundings, the total energy of the system and its surroundings remain constant.

$\Delta E = q + w$ where ΔE is change in internal energy, q = heat & w = work done

(i) For heat absorbed by system $q > 0$, work is done on system, $w > 0$, energy is entering system $E > 0$

(ii) For heat given off by system $q < 0$, work is done by system $w < 0$, energy is leaving system $E < 0$.

Enthalpy or Heat Content, H

Like the internal energy, enthalpy is also an extensive property ΔE is the heat evolved or absorbed in a reaction carried out at constant volume and at a given temperature. Thus ΔE represents the change in heat content of the reaction taking place at constant volume.

The heat content of a system at constant pressure is called is its enthalpy.

For example, if H_P is the total heat content of the products and H_R that of the reactants, then $\Delta H = H_P - H_R$ where ΔH is known as the change in heat content or change in enthalpy of the reaction.

If H_P is greater than H_R, ΔH is positive. In such a reaction heat is absorbed by the system from the surroudings.

If H_P is less than H_R, ΔH is negative. In such a raction heat isgiven out by the system to the surroundings.

If the reaction is carried out at constant pressure at a temperature then the change in heat content or enthalpy of the reaction is equal to the sum of the change in internal energy and the work done during expansion against atmospheric pressure.

Work done $P\,\Delta V$

Where P is the pressure and ΔV is the change in volume.

$\therefore$ $\mathbf{\Delta H = \Delta E + P\Delta V}$ for a reaction at constant pressure. ΔH represents the difference between the enthalpies of the products and those of the reactants.

$$\Delta H = H_P - H_R$$

In reactions involving solids and liquids and the volume changes are very small and may be neglected.

For ideal gases: $\Delta H = \Delta E + \Delta n(RT)$ where, Δn is change in number of moles of gaseous substances.

Heat of Reaction

The heat of reaction is the amount of heat evolved or absorbed when the reactants have completely reacted as indicated by the chemical equation.

$$\Delta H = H_P - H_R$$

i.e. Heat of reaction = Total heat content of products – Total heat content of reactants.

Factors affecting the heat of reaction. These are:

1. As $\Delta H = \Delta E + P\Delta V$

   (a) $\Delta H = q_V$ (at constant pressure) i.e. Heat of reaction at constant pressure & temperature is equal to enthalpy change.

(b) $\Delta E = q_v$ (at constant pressure) i.e. Heat of reaction at constant volume and temperature is equal to change in internal energy of the system.

2. Physical state (of the reactants and products.)

eg. For Reactants $C(diamond) + O_2(g) \rightarrow CO_2(g); \Delta H = -395.4$ KJ

$C(graphite) + O_2(g) \rightarrow CO_2(g); \Delta H = -393.5$ KJ

For Products $H_2(g) + \frac{1}{2}O_2(g) \rightarrow H_2O(l)\ \Delta H = -285.5$ KJ

For Products $H_2(g) + \frac{1}{2}O_2(g) \rightarrow H_2(g)\ \Delta H = -238$ KJ

Therefore it is necessary to write physical state of reactants & products while writing thermochemical equations

3. The heat of reaction depends upon stoichiometry. This is illustrated by the following equation.

$1/2\ H_2(g) + 1/2\ Cl_2(g) \rightarrow HCl(g);\quad \Delta H = -92.05$ KJ

$H_2(g) + Cl_2(g) \rightarrow 2\ HCl(g);\quad \Delta H = -184.1$ KJ.

4. The heat of reaction also depends upon the *temperature*. This is evident from the following thermochemical equation.

$N_2(g) + 3H_2(g) \rightleftharpoons 2NH_3(g)$

ΔH at 300 K is –91.94 KJ while at 353 K, ΔH = –92.85 KJ.

5. The heat reaction also depends upon the concentration of solution or pressures of gases if any one stated.

$NaOH(1M) + HCl(1M) \rightarrow NaCl + H_2O; \Delta H = -57.3$ KJ

DIFFERENT TYPES OF HEATS OF REACTION OR ENTHALPY CHANGES

I. Heat of Formation

It is defined as the change in heat enthalpy or change in internal energy when **one mole** of a substance is formed from its standard or **initial component.**

From the standard enthalpies, the standard heat of formation of compound can be determined.

eg. (a)

Thermochemical equation for the formation of CO_2 (g) is

$C(graphite) + O_2(g) \rightarrow CO_2(g)\ \Delta H_f^\circ = -94.1$ K.Cal

The standard heat of formation of $O_2(g)$ is –94.1

$$\text{But } \Delta H_f^\circ(CO_2) = H^\circ_{CO_2} - (H^\circ_C + H^\circ_{O_2})$$

By convention H°_C and $H^\circ_{O_2}$ are zero.

$$\therefore \quad \Delta H_f^\circ(CO_2) = H^\circ_{CO_2} = -94.1 \text{ K.Cal}$$

$$\text{Similarly} \quad \Delta H_f^\circ(H_2O) = (H^\circ_{H_2O}) = -68.3 \text{ K.Cal}$$

$$\Delta H_f^\circ(HCl) = H^\circ_{HCl} = -22.06 \text{ K.Cal}$$

Therefore the standard heat of formation of compounds is equal to its enthalpy.

(b) $Mg(s) + \frac{1}{2}O_2(g) \rightarrow MgO(s)$ $\Delta H = -602$ kJ mole^{-1}.

The term "standard" implies that one is considering the substance at 298 k and 1 atm. pressure. *The heat of formation of an element is taken as zero.*

(c) (i) $H_2(g) + \frac{1}{2}O_2(g) \rightarrow H_2O(l)$; $\Delta H = -68.3$ K.Cal

The heat of formation of liquid water from hydrogen and oxygen gases is –68.3 KCal. The statement implies that when one mole of liquid water is formed from its elements, the quantity of heat liberated is 68.3 K.Cal.

(ii) $N_2(g) + O_2(g) \rightarrow 2\,NO(g)$; $\Delta H + 43.2$ K.Cal.

43.2 K.Cal of heat of absorbed when 2 moles of nitric is formed. Hence when one mole is formed 43.2/2 = 21.6 K.Cal of heat is absorbed. Therefore the heat of formation of nitric oxide is +21.6 K.Cal.

Note: H_2SO_4 has H, S and O atoms therefore, heat of formation H_2SO_4 is

$$H_2(g) + S(s) + 2O_{2(g)} + H_2SO_{4(l)};\ \Delta H_{(f)} = q$$

and not
$$H_2O + SO_3 \rightarrow H_2SO_4 \rightarrow H_2SO_4;\ \Delta H = q$$

II. Enthalpy of Heat of Combustion

Enthalpy of heat of combustion is heat energy change when 1 mole substance is burned in oxygen e.g.

$$C + O_2 \rightarrow CO_2,\ \Delta H = -394 \text{ kJ mol}^{-1}$$

The enthalpy of combustion of reactant and product materials may be used to calculate the enthalpy change associated with of reaction in a similar manner to that of the enthalpy of formation.

$$\Sigma H_{reactant} = \Sigma \Delta H_R \text{ (reactant)} - \Sigma \Delta H_P \text{ (product)}$$

III. Heat of Neutralisation

The amount of heat evolved when one gram equivalent weight of an acid is neutralised by one gram equivalent weight of a base in solution is called the heat of neutralisation.
Eg:

1. $HCl(aq) + NaOH(aq) \rightarrow NaCl(aq) + H_2O(l)$; $\Delta H = -13.7$ K.Cal mole^{-1}
2. $HNO_3(aq) + KOH(aq) \rightarrow KNO_3(aq) + H_2O(l)$; $\Delta H = -13.7$ K.Cal mole^{-1}
3. $CH_3COOH(aq) + NaOH(aq) \rightarrow CH_3COONa(aq) + H_2O(l)$; $\Delta H = -13.2$ K.Cal mole^{-1}
4. $HCN(aq) + NaOH(aq) \rightarrow NaCN(aq) + H_2O(l)$; $\Delta H = -12.3$ K.Cal mole^{-1}

IV. Heat of Solution

Heat of solution is the amount of heat evolved or absorbed when 1 mole of a substance dissolves in large amount of solvent, water.

OR

The heat of solution is the changes in the heat content of the system when one mole of the solute is dissolved in large excess of water, so that further dilution of the solution produce no heat exchange.

The heat of solution may be expressed as

$NaCl(s) + aqueous \rightarrow Na^{+}(aq) + Cl^{-}(aq), \quad \Delta H = +5kJ\ mol^{-1}$

$KCl(s) + aq \rightarrow KCl\ (aq); \quad \Delta H = 4.44\ K.Cal$

$MgSO_4(s) + aq \rightarrow MgSO_4(aq); \quad \Delta H = -20.08\ K.Cal$

The symbol 'aq' stands for aqueous solution.

V. Heat of Hydration

The heat of hydration is the change in enthalpy when one mole of anhydrous substance combines with requisite number of moles of water to form the hydrate.

VI. Heat of Vaporisation

The heat of vaporisation is the enthalpy change when one mole of a liquid changes into vapour state without change of temperature.

Eg: $H_2O(l) \rightarrow H_2O(g); \Delta H = +10.5\ K.Cal$

VII. Heat of Fusion

The heat of fusion is the enthalpy change when one mole of a solid melts or fuses to a liquid without change of temperature.

VIII. Heat of Sublimation

The heat of sublimation is the enthalpy change when one mole of a solid is converted into a vapour phase at given temperature below its melting.

Most solids that sublime readily are molecular solids. Eg: Iodine and napthalene

From Hess's law it follows that the heat of sublimation is equal to the sum of heat of fusion and heat of vaporisation.

$$\Delta H_{(sub)} = \Delta H_{(fus)} + \Delta H_{(vap)}$$

This equation can be used to calculate the heat of sublimation provided heat of fusion and heat of vaporisation are known.

HESS'S LAW OF CONSTANT HEAT SUMMATION

It states that the heat of reaction is the same whether the reaction takes place in one or several steps.

Suppose a chemical system changes from the state *A* to the state *Z* in following multiple stages.

$A \rightarrow B; \quad \Delta H = -q_1$

$B \rightarrow C; \quad \Delta H = -q_2$

$C \rightarrow Z; \quad \Delta H = -q_3$

Net change $A \rightarrow Z \quad \Delta H = (-q_1) + (-q_2) + (-q_3)$

A to *B*, *B* to *C* and *C* to *Z* liberating q_1, q_2 and q_3. K.Cal of heat respectively. Then according to Hess's Law

$$-Q = (-q_1) + (-q_2) + (-q_3)$$

$$A \xrightarrow{Q} Z$$

$$A \xrightarrow{q_1} B \xrightarrow{q_2} C \xrightarrow{q_3} Z$$

$$(Q = q_1 + q_2 + q_3)$$

example carbon can be burnt to carbon dioxide directly or it may first be changed to carbon monoxide and then oxidised to carbon dioxide.

Ist way (One step)

$$C(s) + O_2(g) \rightarrow CO_2(g); \ \Delta H = -94.05 \text{ K.al}$$

IInd way (Two steps)

$$C(s) + 1/2\ O_2(g) \rightarrow CO(g); \ \Delta H = -26.42 \text{ K.Cal}$$

$$CO(g) + 1/2\ O_2(g) \rightarrow CO_2(g); \ \Delta H = -67.71 \text{ K.Cal}$$

$$C(s) + O_2(g) \rightarrow CO_2(g); \ \Delta H = -94.13 \text{ K.Cal}$$

Bond Energy

Bond energy is the average energy required to break all the bonds of a particular type in a mole of gas to form gaseous atoms. Bond energies are useful in estimation of heats of reactants for which experimental data can be determined. Bond energies are average values only.

For example, the bond energy of C-H bond in methane is one fourth of the energy required to dissociate one mole of methane into carbon and hydrogen atoms.

$$CH_4(g) \rightarrow C(g) + 4H(g) - 1665 \text{ KJ or } \Delta H = 1665 \text{ KJ or } E_{(C\text{-}H)} = 1665/4 = 416 \text{ KJ}$$

Bond energy values remain nearly constant and independent of different bonds present in a molecule. The C-H bond energy has nearly the same value in all the hydrocarbons as in methane. There are a few exceptions. For example the C=O bond energy in carbon dioxide is 799 KJ but the average value of C=O in carbonyl compounds is 724 KJ.

Carnot Cycle

Carnot cycle having efficiency η

$$\eta = \frac{W}{Q_2} = \frac{T_2 - T_1}{T_2} = \frac{Q_2 - Q_1}{Q_2} = 1 - \frac{Q_1}{Q_2}$$

where Q_2 is the heat absorbed from the source at T_2 and Q1 is the heat lost to the sink at T_1. Thus "It is impossible to convert heat into equivalent amount of work without compensation.

(a) $\eta = 1$ (100% efficiency) only when T_1 = O K (i.e., –273°C) which is not practically possible.

(b) Greater the difference between T_2 and T_1 greater is the value of η. Hence superheated steam (under high pressure) is used in an engine with a boiler.

Spontaneous Process

Spontaneous Process is a process that occurs in a system by itself (i.e. without the need for input of work into the system) eq. expansion of gas into volume; flow of heat from a hot body to a cold body.

Non-spontaneous

Non-spontaneous is a process where some external actions has to be continuously applied i.e. energy on the form, of work must be put into the system eg: compression of gas into smaller volume; raising of weight against gravity.

Entropy

It is a thermodynamic measure of disorder or randomness of a system, and is defined in terms of entropy changes (ΔS), rather then its absolute value. More the disorder or randomness is a system, high is is its entropy. For a reversible process at constant temperature, the change in entropy $\Delta S = \frac{\Delta_{rev}}{T}$

For a reaction entropy change (ΔS) is given by $\Delta S = \Sigma S_{\text{products}} - \Sigma S_{\text{Reactants}}$

Entropy of one mole of a substance in pure state at one atmosphere and 298 K is termed as standard entropy, S^0.

For a reaction in standard state, $A \rightarrow B$ $\qquad \Delta S^0 = S_B^0 - S_A^0$

Also,

$$\Delta S = \frac{q_{rev}}{T} = \frac{W_{rev}}{T} = \frac{nRT \log_e \frac{V_2}{V_1}}{T} = nR \log_e \frac{V_2}{V_1}$$

$$= 2.303\, nR \log_{10} \frac{V_2}{V_1} = 2.3030\, nR \log_{10} \frac{P_1}{P_2}$$

where, q_{rev} is the heat supplied to a system at temperature $T(K)$ under reversible conditions. We have,

At constant P, $\qquad q_{rev} = \Delta H_{rev}$

$$\therefore \qquad \Delta S = \frac{\Delta H_{rev}}{T}$$

Entropy change of fusion: $\Delta S_f = \frac{\Delta H_f}{T}$ where T is freezing point

entropy change of vaporizations: $\Delta S_v = \frac{\Delta H_v}{T}$ where T is boiling point

SECOND LAW OF THERMODYNAMICS

It is stated in several forms as given below:

1. All the spontaneous process are irreversible in nature.

2. Heat amount of itself pass from a colder body to hotter body without the intervention of external energy. i.e. the spontaneous process cannot reversed.
3. It is impossible to obtain work by cooling a body below its lowest temperature.
4. It is impossible to construct a machine functioning in cycle which can convert heat completely into equivalent amount of work without producing changes elsewhere, i.e., perpetual motions are not allowed.
5. Entropy of the universe is always increasing. It increases in the course of every spontaneous change.
6. It is impossible to take heat from a hot reservoir and convert it completely into work by a cyclic process without transferring a part of its to a cold reservoir.

Free Energy (G)

A thermodynamic quantity involving both entropy and enthalpy is called Gibbs' energy.

(G) is also a state function and is a measure of maximum work done. It is expressed as

$$G = H - TS = E + PV - TS$$

or $\Delta G = \Delta H - T\Delta S$ at constant T (Gibb's Helmholtz equation)

$$= \Delta E + \Delta(PV) - \Delta TS$$

Changes in free energy for a reaction are.

$$\Delta G^\circ = \Sigma(\Delta G_f^\circ)_{\text{products}} - \Sigma(\Delta G_f^\circ)_{\text{reactants}}$$

where G_f° is formation of species form its elements.

The standard free energy (ΔG°) is given the relation $\Delta G = \Delta G^\circ + RT \ln K$, so when $\Delta G = 0$ at equilibrium

Then $G^\circ = -RT \ln K \Rightarrow \Delta G^\circ = -2.303\ RT \log_{10} K\ e^{(-\Delta G^\circ/RT)}$

and., The decrease in free energy ($-\Delta G$) is given as:

$$-\Delta G = W_{\text{net}} = 2.303\ nRT \log_{10} \frac{V_2}{V_1}$$

Also, we have, $\Delta G^\circ = \Delta H^\circ - T\Delta S^\circ = \Delta E^\circ + \Delta(PV) - T\Delta S^\circ$

Conditions for Spontaniety of a Reaction

For a spontaneous change ΔG = –ve. Thus following results can also be evaluated for spontaneous process.

ΔH	ΔS	ΔG = ΔH – TΔS	*Nature*
–ve	+ve	∴ ΔG = –ve	Spontaneous
+ve	–ve	∴ ΔG = +ve	Non spontaneous
+ve	+ve	if $\Delta H < T\Delta S$; ΔG = –ve	Spontaneous
–ve	–ve	ΔG = –ve if $\Delta H > T\Delta S$ i.e., T is low	Spontaneous
		ΔG = +ve if $\Delta H + < T\ \Delta S$ i.e., T is high	Non spontaneous

Thus is can be concluded that the sign of ΔG gives the real idea about the spontaniety of the process. It can be said that

(i) ΔG is negative i.e., less than zero, the prices is spontaneous.
(ii) ΔG is zero, the process does not occur. Actually the system is in equilibrium
(iii) ΔG is positive the process does not occur or may process only in the reverse direction.

Eg 1: The formation of hydrogen chloride

$$H_2(g) + Cl_2(g) \rightarrow 2HCl(g);\ \Delta H \text{ is } -ve$$

Here ΔS factor is zero. The value of ΔG will be negative and the process is spontaneous.

Eg 2: Formation of water

$$H_2(g) + 1/2O_2(g) \rightarrow H_2O\ (l)\ ;\ \Delta H = -68.3 \text{ K.Cal}$$

Here ΔS in negative as liquid phase is less random than gaseous phase. But ΔH is stronger than ΔS and ΔS will be negative. Hence this reaction will be spontaneous.

Eg 3: Dissolution of ammonium chloride in water

$$NH_4Cl(s) + \text{water} \rightarrow NH_4^+(aq) + Cl\ (aq);\ \Delta H = 3.6 \text{ K.Cal}$$

In this case, the ΔS is positive., and is stronger than ΔH. Therefore, the value of ΔG will be negative and the reaction will be spontaneous

THIRD LAW OF THERMODYNAMICS

It states that the entropy of a perfectly crystalline solid at absolute zero of temperature is zero.

PROBLEMS

A student constructed a 'coffee cup calorimeter' (see Figure). To determine the heat capacity of the calorimeter, the student places 50 mL of room temperature distilled water in the calorimeter. A calibrated temperature probe recorded the temperature as 296 K. The student then added 50 mL of warm distilled water 334 K) to the calorimeter and recorded the temperature every 30 seconds for the next three minutes. The calorimeter was then emptied and dried. Next, the student measured temperature change when 50 mL of 2.0 M HCl was added to 50 mL of 2.0 M NaOH. Again, temperature change was recorded every 30 seconds for three minutes. The same procedure was followed for 2.0 M NH_4Cl with 2.0 M NaOH and finally, 2.0 M NH_3 with 2.0 M HCl. **Answer the following. (1 to 4)**

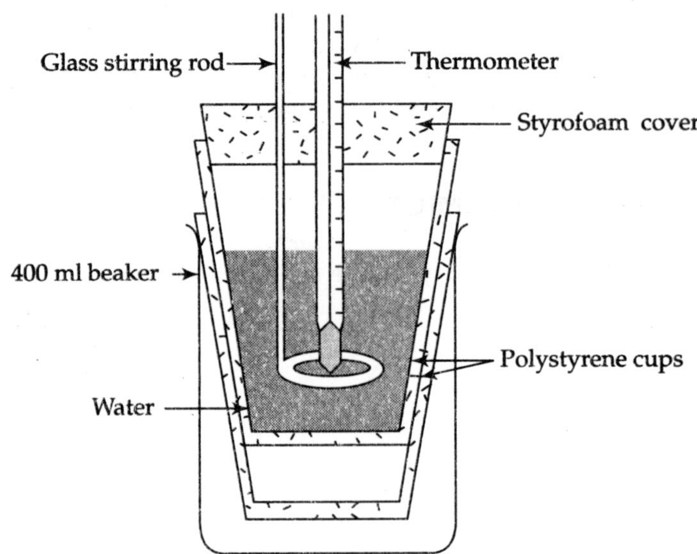

Q. 1. A graph (see on the next page) was constructed to determine the heat capacity of the calorimeter (see Figure). From the graph, determine the heat capacity of the calorimeter. The specific gravity of water at 296 K is 0.998 and at 334 K it is 0.983.

Q. 2. Graph are constructed for the temperature changes for each of the three reactions given below, in the table. Calculate the heat evolved in each reaction (kJ mol^{-1} of product). Assume the density of each solution = 1.0 g · mL^{-1}.

	$T_{initial}$ (K)	$T_{at\ mixing}$ (K)
HCl + NaOH	296	308.6 K
NH_4Cl + NaOH	295.9 K	297.1 K
NH_3 + HCl	296	306.1 K

Q. 3. Write the net ionic equation, including ΔH's for the first two reactions studied and rearrange the equation(s) where necessary to produce the third reaction and its ΔH.

Q. 4. Calculate the percent error between the measured ΔH and calculated ΔH.

	A	B	C	D	E	F	G	H	I	J	K
1	30	314.5									
2	60	313.9									
3	90	313.1									
4	120	313.1									
5	150	312.9									
6	180	312.4									
7	Time in Sec.	Temperature °C									
8											
9											
10											
11											
12											
13											
14											
15											
16											
17											
18											
19											

Heat capacity of calorimeter

Tempertaure: 314.5, 314.0, 313.5, 313.0, 312.5, 312.0

Time (sec): 30, 60, 90, 120, 150, 180

Q. 5 **(a)** The heat of combustion of ethylene at constant volume at 290 K is 1388.5 KJ. Calculate the heat of combustion at constant pressure considering water to be in liquid state.

(b) The heat of formation of methane at 298 K and at constant pressure is –17890 Cal. Calculate its heat of formation at constant volume.

($R = 8.314\ JK^{-1}\ mol^{-1}$)

Q. 6. Calculate the heat of formation of benzene, given that the heats of combustion of benzene, carbon and hydrogen are 3256 KJ, 393.5 KJ and 285.8 KJ respectively.

Q. 7. Calculate the heat of formation of ethyl alcohol from the following data:

$$C(s) + O_2(g) \rightarrow CO_2(g);\ \Delta H = -397\ KJ \quad (1)$$

$$H_2(g) + 1/2O_2(g) \rightarrow H_2O(l);\ \Delta H = -292.6\ KJ \quad (2)$$

$$C_2H_5OH(l) + 3O_2(g) \rightarrow 2\ CO_2(g) + 3H_2O\ (1)\ \Delta H = -1380\ KJ \quad (3)$$

Q. 8. From the following data find the heat of formation of potassium hydroxide.

$$K(s) + H_2O(l) + aq \rightarrow KOH(aq) + 1/2H_2(g)\ \Delta H = -201\ KJ \quad (1)$$

$$H_2(g) + 1/2O_2(g) \rightarrow H_2O(l);\ \Delta H = -286\ KJ \quad (2)$$

$$KOH(s) + aq \rightarrow KOH(aq);\ \Delta H = -55.6\ KJ \quad (3)$$

Q. 9. 0.5 gm. of benzoic acid wad subjected to combustion in a bomb calorimeter at 288 K when the temperature of the calorimeter system (including water) was found to rise by 0.55 KL. Calculate the heat of combustion of benzoic acid (i) at constant volume and (ii) at constant pressure. The thermal capacity of the calorimeter including water was found to be 23.85 KJ.

Q. 10. 150 ml of 0.5 N hydrochloric acid solution at 298.35 K were mixed with 150 ml of sodium hydroxide solution at the same temperature taken in a Dewar flask. The final temperature was recorded to be 301.2 K. calculate the heat of neutralisation of hydrochloric acid with sodium hydroxide. The water equivalent of the calorimeter was found to be 60g (Specific heat of water = 4.184 Joules.)

Q. 11. Calculate the enthalpy of formation of carbon disulphide given the enthalpies of combustion of carbon sulphur and carbon disulphide to be –395 KJ, –297KJ and –1076 KJ respectively?

Q. 12. Both methane and ethane can be used as fuels. Which of them is more efficient as a fuel? The standard heats of formation CH_4, C_2H_6, CO_2 and H_2O are –74.8, –84.4, –393, –241.6 KJ respectively?

Q. 13. Calculate the lattice energy of calcium oxide given that the heat of atomization of calcium is 177 KJ mol^{-1}. The bond dissociation energy of oxygen is 500 KJ mol^{-1}; the I and II ionization energies of calcium are 587 KJ and 1156 KJ mol^{-1} respectively and the heat of formation CaO is –637 KJ mol^{-1}. The energy required to form 1 mole O^{-2} ion gaseous oxygen atoms (electron affinity) is 704 KJ.

Q. 14. Calculate the enthalpy of ionisation of hydrocyanic acid, given the enthalpies of formation of HCN, NaOH, NaCN and water to be +105.5, –469.5 KJ, –88.7KJ and –286 KJ respectively. The enthalpy of neutalisation of a strong acid by a strong base is –57.3 KJ mol^{-1}.

Q. 15. Benzene burns in oxygen according to the following reaction.

$$C_6H_6 + 15/2\ O_2 \rightarrow 3H_2O + 6CO_2$$

If the standard heats of formation (ΔH_1) of C_6H_6, H_2O and CO_2 are +48.9, –285.5 and –393 KJ mol^{-1} respectively, calculate the amount of heat that will be liberated by burning 1 Kg of benzene.

Q. 16. If the heats of formation Al_2O_3 and Cr_2O_3 are 1596 KJ and 1134 KJ (both exothermic) respectively calculate ΔH in the thermite reaction.

$$2\ Al + Cr_2O_3 \rightarrow 2\ Cr + Al_2O_3;\ \Delta H = ?$$

Q. 17. Calculate the enthalpy of formation of anhydrous Al_2Cl_6 form the following data.

$$2Al + 6HCl(aq) \rightarrow Al_2Cl_6(aq\_ + 3H_2 + 1005\ KJ$$

$$H_2 + Cl_2 \rightarrow 2\ HCl + 184.2\ KJ$$

$$HCl + aq \rightarrow HCl(aq) + 73.3\ KJ$$

$$Al_2Cl_6 + aq \rightarrow Al_2Cl_6(aq) + 643.5\ KJ$$

Q. 18. On an average, a person requires 11996.6 KJ of energy for leading normal life. How many kilograms of cane sugar [$C_{12}\ H_{22}O_{11}$], must be consumed, if all the required energy has to be derived from cane sugar? The heats of formation of $C_{12}H_{22}O_{11}$, CO_2 and H_2O are –2216.6, –393 and –285.5 KJ respectively?

Q. 19. The heat of combustion of napthalene $C_{10}H_6$ is 5145.6 KJ mol^{-1}. The heat of formation of CO_2 and H_2O are 394.6 and 286 KJ mol^{-1}. Calculate heat of formation of napthalene.

Q. 20. Calculate the heat evolved by burning 1 gm, of lead in O_2 according to the following reaction? $PbS + O_2(g) \rightarrow PbO_2(s)$ $\Delta H = –276.4$ KJ

Q. 21. The heat evolved by burning 1g. of graphite in oxygen is 29.34 KJ. How much heat is evolved by burning graphite in 1 mole of oxygen?

Q. 22. Heat of combustion of sucrose ($C_{12}H_{22}O_{11}$) is 5643 KJ mol^{-1}. Calculate the heat evolved by burning 12 g. of sucrose?

Q. 23. The heat if combustion of sucrose ($C_{12}H_{22}O_{11}$) is 5643 KJ mol^{-1}. Calculate (a) amount of CO_2 evolved and (b) volume of oxygen consumed by burning 12 g. of sucrose.

Q. 24. Using the following thermochemical equation, calculate the heat evolved by burning 2.8 litres of methane at STP and calculate the volume of CO_2 evolved in litres in STP.

$CH_4 + 2O_2(g) \rightarrow CO_2(g) + 2H_2O(l)$, $\Delta H = -890$ KJ

Q. 25. Calculate the heat of reaction for the reaction $CaO + CO_2 \rightarrow CaCO_3$ using the following data.

$CaO + H_2CO_3 \rightarrow CaCO_3 + H_2O + 127.52$ KJ

$CO_2 + H_2O \rightarrow H_2CO_3 + 48$ KJ

Q. 26. Heat of combustion of 1 gram of diamond and graphite are 394, 396 KJ respectively. Calculate the heat of transformation of C (graphite) → C (diamond).

Q. 27. The heat of combustion of acetylene, carbon (graphite) and H_2 are –1296, 393.3, 285.5 KJ respectively. Calculate the heat of formation of acetylene.

Q. 28. The heats of formation of methane, carbon dioxide, water are –74.8, –393.3 and –285.5 KJ respectively. Calculate the heat of combustion of methane?

Q. 29. $NaOH(aq) + HCl(aq) \rightarrow NaCl(aq) + H_2O \; \Delta H = -57.26$ KJ

$NaOH(aq) + CH_3COOH(aq) \rightarrow CH_3COONa\ (aq) + H_2O \; \Delta H = -56$ KJ.

By using the above two equations calculate the heat required for the dissociation of CH_3COOH.

Q. 30. When 1.89 g. of benzoic acid (C_6H_5COOH) is burned in a bomb calorimeter in presence of excess of oxygen, the liberated heat increases the temperature of 18.49 kilograms of water by 0.632°K. At 298 K the specific heat of water 4.17 Jg^{-1} K^{-1}. Calculate heat of combustion of benzoic acid.

Q. 31. Heat of combustion of methanol CH_3OH is 714.8 KJ 3.2 g. of methanol is burnt in a calorimeter containing 18.94 Kg of water at 298 K. Calculate the rise in the temperature of water. At 298 K the specific heat of water is 4.17 Jg^{-1} K^{-1}.

Q. 32. Using the data (all values are in kilo Joules per mole at 298 K) given below. Calculate the bond energy of C—C and C—H bonds.

$\Delta H^{\circ}_{combustion}$ (ethane) = –1555 KJ   Bond energy of H—H = 434.7 KJ

$H^{\circ}_{combustion}$ (propane) = –2215.4 ΔH°_f of H_2O (l) = –284.12 KJ

H°_f for (graphite) → C(g) = 719 ΔH°_f of CO_2(g) = –393 KJ

Q. 33. The standard enthalpy of combustion at 298 K of hydrogen, cyclohexane C_6H_{10} and cyclohexane C_6H_{12} are –241, –3800 and –3920 KJ mol^{-1} respectively. Calculate the heat of hydrogenation of cyclohexane.

Q. 34. An intimate mixture of ferric oxide, Fe_2O_3 and aluminium, Al, is used in solid fuel sockets. Calculate the fuel values per gram. of the mixture. Heats of formation and densities are as follows:

$H^{\circ}_f(Al_2O_3) = 1667.8$ KJ mol^{-1}. H°_f $(Fe_2O_3) = 831.8$ KJ mol^{-1}

Density of Fe_2O_3 = 5.2 g/c.c; Density of AL = 2.7 g/c.c

Q. 35. The standard molar heat of formation of ethane, carbon dioxide and liquid water are –88, –393.3 and –285.5 KJ respectively. Calculate standard molar heat of Combustion of ethane?

Q. 36. The bond dissociation energies of gaseous H_2, Cl_2 and HCl are 434.7, 242.5 and 430.5 KJ mol^{-1} respectively. Calculate the enthalpy of formation of HCl gas.

Q. 37. Given the following standard heats of reaction.

(a) Heat of formation of water = –285.5

(b) Heat of combustion of acetylene = –1298.3

(c) Heat of combustion of ethylene = –1409.5

Calculate the heat of reaction for the hydrogenation of acetylene at constant volume at 298 K.

Q. 38. The molar heats of combustion of C_2H_2, C and H_2 are –1296, 393 and 285 KJ respectively. Calculate the standard heat of formation of C_2H_2.

Q. 39. The standard heats of formation at 298 K for CCl_4; H_2O, CO_2 and HCl are –106.6, –241.6, 393 and 92.4 KJ mol^{-1} respectively. Calculate ΔH for the following reaction at 298 K.

$$CCl_4 + 2H_2O \rightarrow CO_2 + 4HCl; \Delta H = ?$$

Q. 40. The enthalpy for the following reacitons ($\Delta H°$) at 298 K are given

$$1/2\ H_2 + 1/2\ O_2 \rightarrow OH - \Delta H^o = -42\ KJ$$

$$H_2 \rightarrow 2H - \Delta H^o = -435.5\ KJ$$

$$O_2 \rightarrow 2O - \Delta H^o = -494.6\ KJ$$

Calculate the O—H bond energy in the hydroxyl radical.

Q. 41. Calculate q, w, Δ E and ΔH when one mole of an ideal gas expended isothermally and reversibly at 300 K from a volume of 2.28 litres to 4.56 litres.

Q. 42. Calculate the entropy change when one mole of liquid water evaporates at 373 K. Latent heat of vaporisation of water is 40.63 KJ mol^{-1}.

Q. 43. Calculate the entropy change when 1 mole of an ideal gas expands reversibly from an initial volume of 2 litre to a final volume of 20 litres at 298 K.

Q. 44. Six moles of an ideal gas are compressed isothermally at 30 K form 2 atmospheres to 4 atm pressure. Calculate the free energy change for the process?

Q. 45. Calculate the entropy change for the following reversible processes.

(a) 1 mole of liquid water at 1 atm. 313 K evaporates to 1 mole of water vapour at 1 atm, 373 K (ΔH Vap = 40850 J mol^{-1}).

(b) 1 mole of tin at 1 atm. 404 K changes to 1 mole of tin at 1 atm and 286 K (ΔH = 2090 J mol^{-1} transition)

Q. 46. What is the entropy change for the conversion of a mole of ice to water at 273 K and 1 atm?

Q. 47. A sample of biphenyl $(C_6H_5)_2$ weighing 0.526 g was ignited in a bomb calorimeter initially at 298 K producing temperature rise of 1.91 K. In a separate calibration experiment, a sample of benzaoic acid C_6H_5COOH weighing 0.825 g was ignited under identical conditions produced a temperature rise in 1.94 K. For benzoic acid, the heat of combustion a constant pressure is known to be 3226 kJ mol^{-1}. Calculate the amount of heat lost by the system.

Q. 48. Assuming that the heat of combustion of ethyl alcohal is –330 kJ. If the heat of formation of CO_2(g) and H_2O(l) be –94.3 kJ and –68.5 respectively, the heat of formation of ethyl alcohal will be?

Q. 49. Enthalpy of neutralisation of CH_3COOH by NaOH is –50.7 kJ mol^{-1}. ΔH for the ionisation of CH_3OOH is?

Q. 50. Heats of combustion of rhombic sulphur and monoclinic sulphur are –291.4 kJ and –295.4 kJ. So the heat of transition from monoclinic sulphur to rhombic sulphur will be?

Q. 51. Calculate $\Delta H°$ for the reaction $C_2H_4(g) + 3O_2(g) \rightarrow 2CO_2(g) + 2H_2O(g)$
The average energies of various bonds are

Bond Energy	414	499	724	460	619
Bond	C—H	O=O	C=O	O—H	C=C

Q. 52. Compare the value of $\Delta H°$ obtained by using the enthalpy of formation data: $\Delta_f H°$ $(CO_2(g))] = -393.5$ kJ mol^{-1}, $\Delta_f H°$ $[H_2O(g)] = -241.8$ kJ mol^{-1} and $\Delta H°$ $(C_2H_4(g)] = 52.3$ kJ mol^{-1}

Q. 53. Using the following bond energy data, compute $\Delta_f H°$ for ethyl alcohol (C_2H_5OH). ΔH (C—H bond) = 413 kJ, ΔH (C—C) = 348 kJ. ΔH(C—O) = 351 kJ, ΔH (O—H) = 463 kJ, ΔH_{atom} = 2990.45 kJ

Q. 54. Calculate the heat formation HCl at 384 K from the following data

$$\frac{1}{2} H_2(g) + \frac{1}{2} Cl_2(g) \rightarrow HCl(g);\ \Delta H^{\circ}_{298k} = 92300 \text{ J}$$

The mean heat capacities over this temperature range are

$$H_2(g);\ C_p = 28.53 \text{ JK}^{-1} \text{ mol}^{-1}$$
$$Cl_2(g);\ C_p = 32.26 \text{ JK}^{-1} \text{ mol}^{-1}$$
$$HCl(g);\ C_p = 28.49 \text{ JK}^{-1} \text{ mol}^{-1}$$

Q. 55. The standard heats of formation of $C_2H_5OH(l)$, $CO_2(g)$ and $H_2O(l)$ are –227.0, –393.5 and –285.5 kJ mol^{-1} respectively. Calculate the standard heat change for the reaction.

$$C_2H_5OH(g) + 3O_2(g) \rightarrow 2CO_2(g) + 3H_2O(l)$$

Q. 56. The enthalpy of combustion of benzoic acid (C_6H_5COOH) at 298 K and 1 atm pressure is –2546.0 kJ mol^{-1}. Find the ΔE for the reaction.

Q. 57. Given the enthalpy of combustion of ethylene, hydrogen and ethane as 337.2, 68.3 and 372.8 kJ respectively. Calculate the rate of reaction at 298 K.

$$C_2H_4(g) + H_2(g) \rightarrow C_2H_6(g)$$

Q. 58. The enthalpies of formation $H_2O(l)$, $CO_2(g)$ and $C_3H_8(g)$ are 285, –393 and 104 kJ mol^{-1} respectively. Calculate the enthalpy of combustion of propane.

Q. 59. ΔH values for the combustion reaction of ethyl alcohol, carbon and hydrogen are –330, –94.3 and 68.5 kJ respectivgely. Find, the heat of formation of ethyl alcohol.

Q. 60. Enthalpies of formation of FeO(s) and $Fe_2O_3(s)$ are –267.9 and –822.2 kJ respectively, calculate the enthalpy of the reaction:

$$2FeO(s) + \frac{1}{2} O_2(g) \rightarrow Fe_2O_3(s)$$

Q. 61. Calculate the heat of formation of benzene at 298 *K* if the heats of combustion of benzene, carbon and hydrogen are –780.98, –94.05 and –68.32 kJ respectively (at 298 K).

Q. 62. Calculate the heat of the reaction, ΔH, of the following reaction

$$CH_4(g) + 4F_2(g) \rightarrow CF_4(g) + 4HF(g)$$

Bond energies of C—H, F—F, C—F and H—F bonds are 99.3, 38,116 and 135 kJ mol^{-1} respectively.

Q. 63. Bond energies of F_2 and Cl_2 are 36.6 and 580 kJ mol^{-1} respectively. Heat liberated in the reaction is 26.6 kJ, Find the bond energy of F—Cl bond will be:

Q. 64. Given at 101.3 k Pa(1 atm) and 298 K, enthalpy of formation per mole of methane = 74.75; ΔH_a° atomization per mole of C(s) $\rightarrow$ C(g) = 716.68 kJ, ΔH° atomization per mole of hydrogen atom = 217.97 kJ. Calculate the bond dissociation energy of C—H bond.

Q. 65. From the following data at 298 K, Calculate the bond energy of O—H bond.
From the given data

(i) $H_2(g) \rightarrow 2H(g);\ \Delta H_1 = 436.08$ kJ

(ii) $O_2(g) \rightarrow 2O(g);\ \Delta H_2 = 495.17$ kJ

(iii) $H_2(g) + \frac{1}{2}O_2(g);\ \rightarrow H_2O(g);\ \Delta H_3 = -241.84$ kJ.

Q. 66. Hydrogen bromide dissociation into its gaseous elements, H_2 and Br_2, at elevated temperature. Calculate the percent dissociation at 648 K and 1.00 atm. What would be the percent dissociation at 648 K and 10.0 atm?

Q. 67. A 20.-L vessel is filled with 1.0 mol of ammonia, NH_3. What percent of ammonia dissociates to the elements if equilibrium is reached at 618 K?

Q. 68. K_a for acetic acid at 298 K is 1.754×10^{-5}. At 323 K, K_a is 1.633×10^{-5}. What are ΔH° and ΔS° for the ionization of acetic acid?

Q. 69. Write thermochemical equations for the following.

(a) mercury thawing

(b) bromine vaporizing

(c) benzene freezing

(d) mercury condensing

(e) napthalene subliming

Q. 70. On a hot day, you take a six-pack of soda on a picnic, cooliing it with ice. Each empty (aluminium) can weighs 12.5 g and contains 340 g of soda. The speficic heat of aluminium is 0.902 J/g · K; take that of soda to be 4.10 J/g . K.

(a) How much heat must be absorbed from the six-pack to lower the temperature from 298 K to 278 K?

(b) How much ice must be melted to absorb this amount of heat? (Take ΔH_{fus} of ice from Table in text book).

Q. 71. A cafeteria sets out glasses of tea at room temperature. The customer adds ice. Assuming that the customer wants to have some ice left when the tea cools to 273 K what fraction of total volume of the glass should be left empty for adding ice? Make any reasonable assumptions needed to work this problem.

Q. 72. A sample containing 2.0 mol of graphite reacts completely with a limited quantity of oxygen at 298 K and 1.0 atm pressure, producing 481 kJ of heat and a mixture of CO and CO_2. Calculate the masses of CO and CO_2 produced.

Q. 73. How much heat is released when a mixture containing 10 g NH_3 and 20.0 g O_2 reacts by the equation

$$4NH_3(g) + 5O_2(g) \rightarrow 4NO(g) + 6H_2(O)g; \Delta H° = -906 \text{ kJ}$$

Q. 74. Consider the Haber process:

$$N_2(g) + 3H_2(g) \rightarrow 2NH_3(g); \Delta H° = -91.8 \text{ kJ}$$

The density of ammonia at 298 K and 1.0 atm is 0.696 g L^{-1}. The density of nitrogen, N_2 is 1.145 g L^{-1}, and the molar heat capacity is 29.12 J K^{-1} mol^{-1} · °C). **(a)** How much heat is evolved in the production of 1.0 L of ammonia at 298 K and 1.0 atm? (b) What percentage of this heat is required to heat the nitrogen required for this reaction (0.500 L) from 298 K and 673 K, the temperature at which the Haber process is run?

Q. 75. The carbon dioxide exhaled in the breath of astronauts is often removed from the spacecraft by reaction with lithium hydroxide.

$$2LiOH(s) + CO_2(g) \rightarrow Li_2CO_3(s) + H_2O(l)$$

Estimate the grams of lithium hydroxide required per astronaut per day. Assume that each antronaut requires 10.45×10^3 kJ of energy per day. Further assume that this energy can be equated to the heat of combustion of a quantity of glucose, $C_6H_{12}O_6$, to $CO_2(g)$ and $H_2O(l)$. From the amount of glucose required to give 10.45×10^3 kJ of heat, calculate the amount of CO_2 produced and hence the amount of LiOH required. The ΔH_f° for glucose(s) is –1273 kJ mol^{-1}.

Q. 76. The relationship between $\Delta G°$ and T is linear. Draw the graph of a reaction with properties described below. You needed only label the point at which $\Delta G° = 0$

- the reaction is exothermic.
- $\Delta n_g < 0$ (Δn_g = moles of gas products – moles of gas reactants).
- at 300 K, the system is at equilibrium and $K = 1$.

Q. 77. The normal boiling point for ethyl alcohol is 351.4 K; S° for $C_2H_5OH(g)$ is 282.7 J mol^{-1} K^{-1}. At what temperature is the vapour pressure of ethyl alcohol 357 mm Hg?

Q. 78. ΔH_f° iodine gas is 62.4 kJ mol^{-1}, and $S°$ is 260.7 J mol^{-1} · K. Calculate the equilibrium partial pressure of $I_2(g)$, $H_2(g)$, and HI(g) for the system.

$$2HI(g) \rightleftharpoons H_2(g) + I_2(g)$$

at 773 K if the initial partial pressures are all 0.2 atm.

Q. 79. The heat of fusion of ice is 333 J/g. For the process

$$H_2O(s) \rightarrow H_2O(l)$$

determine

(a) $\Delta H°$ (b) $\Delta G°$ at 273 K (c) $\Delta S°$

(d) $\Delta G°$ at 253 K (e) $\Delta G°$ at 293 K

Q. 80. The overall reaction that occurs when sugar is metabolized is

$$C_{12}H_{22}O_{11}(s) + 12O_2(g) \rightarrow 12CO_2(g) + 11H_2O(l)$$

For this reaction, $\Delta H°$ is –5650 kJ and $\Delta G°$ is –5790 kJ at 298 K.

(a) If 25% of the free energy change is actually converted to useful work, how many kilojoules of work are obtained when one gram of sugar is metabolized at body temperature, 310 K?

(b) How many grams of sugar would a 54.4 kg woman have to eat to get the energy to climb a mountain which is 4158 m high? (w = 9.,79 × 10^{-3} mh, where w = work in kilojoules,. m is body mass in kilogram, and h is height in meters.)

Q. 81. A sample of sucrose, $C_{12}H_{22}O_{11}$, is contaminated by sodium chloride. When the contaminated sample is burned in a bomb calorimeter, sodium chloride does not burn. What is the percentage of sucrose in the sample if a temperature increase of 1.67° is observed when 3.0 g of the sample is burned in the calorimeter? Sucrose gives off 5.64 × 1203 kJ mol^{-1} when burned. The heat capacity of the calorimeter is 22.51 kJ K^{-1}.

Q. 82. Calculated the temperature at which the equilibrium constant (K) for the following reaction is 1.0.

$$MnO_2(s) \rightarrow Mn(s) + O_2(g)$$

Q. 83. What will be the final temperature of a mixture made from 25.0 g of water 288 K from 45.0 g of water 323 K and from 15.0 g of water at 310 K?

Q. 84. Graphite is burned in oxygen to give carbon monoxide and carbon dioxide. If the product mixture is 33% CO and 67% CO_2 by mass, what is the heat from the combustion of 1.0 g of graphite?

Q. 85. Trichloroethene, $C_2H_3Cl_2$, is the active ingredient in aerosols that claim to stain proof men's ties. Trichloroethane has a vapor pressure of 100 mm Hg at 293 K and boils at 347.3 K. An uncovered cup (236.5 ml.) of trichloroethane (d = 1.325 g ml^{-1}). is kept in an 510 L refrigerator at 277 K. What percentage (by mass) of the trichloroethane is left as a solid when equilibrium is established?

Q. 86. Using the Tables in text book, calculate $\Delta \dot{H}$ for the reaction of the following.

(a) 1.0 L of 0.1 M NaoH with 1.0 L of 0.1 M HCl

(b) 1.0 L of 0.1 M NaoH with 1.0 L of 0.1 M HF, taking the heat of formation of HF(aq) to be –320.1 kJ mol^{-1}.

Q. 87. Hydrogen has been suggested as the fuel of the future. One way to store it is to convert it to a compound that can be heated to release the hydrogen. One such compound is calcium hydride, CaH_2. This compound has a heat of formation of –186.2 kJ mol^{-1} and a standard entropy of 42 J $mol^{-1} \cdot K^{-1}$. What is the minimum temperature to which calcium hydride would have to be heated to produce hydrogen at one atmosphere pressure?

Q. 88. When a copper wire is exposed to air at room temperature, it becomes coated with a black oxide, CuO. If the wire is heated above a certain temperature, the black oxide is converted to a red oxide, Cu_2O. At a still high er temperature, the oxide coating disappears. Explain these observations in terms of the thermodynamics of the reactions

$$2CuO(s) \rightarrow Cu_2O(s) + \frac{1}{2}O_2(g)$$

$$Cu_2O(s) \rightarrow 2Cu(s) + \frac{1}{2}O_2(g)$$

and estimate the temperature at which the changes occur.

Q. 89. The lattice enthalpy of sodium chloride, $\Delta H°$ for

$$NaCl(s) \rightarrow Na^+(g) + Cl^-(g)$$

Is 787 kJ mol^{-1}; the heat of solution in making up 1 *M* NaCl(*aq*) is +42.9 kJ mol^{-1}. From these data, obtain the sum of the heats of hydration of Na^+ andCl^-. That is, obtain the sum of $\Delta H°$ values for

$$Na^+(g) \rightarrow Na^+(aq)$$
$$Cl^-(g) \rightarrow Cl^-(aq)$$

If the heat of hydration of Cl^- is 338 kJ mol^{-1}, what is the heat of hydration of Na^+?

Q. 90. The normal boiling point of acetone, CH_3COCH_3, is 329.2 K is and that of nitrobenzene, $C_6H_5NO_2$, is 483.8; Estimate ΔH_{vap} for these two liquids.

For problems 91 to 95, use required values from tables in text books.

Q. 91. (a) Calculate $\Delta S°$ for $Zn(s) + \frac{1}{2}O_2(g) \rightleftharpoons ZnO(s)$.

(b) Calculate $\Delta S°$ for $2NO_2(g) \rightleftharpoons N_2O_4(g)$.

Q. 92. (a) Calculate $\Delta G°$ for $2NO_2(g) \rightleftharpoons N_2O_4(g)$.

(b) Calculate $\Delta G°$ for $CaCO_3(s) \rightleftharpoons CaO(s) + CO_2(g)$.

Q. 93. Calculate $\Delta G_f^°\cdot(N_2O_4, g)$.

Q. 94. Calculate $\Delta G°$ for the following reactions:

(a) $2NO_2(g) \rightleftharpoons N_2O_4(g)$

(b) $CaCO_3(s) \rightleftharpoons CaO(s) + CO_2(g)$

Q. 95. Calculate the equilibrium constant at 238 K for the following reactions:

(a) $2NO_2(g) \rightleftharpoons N_2O_4(g)$

(b) $CaCO_3(s) \rightleftharpoons CaO(s) + CO_2(g)$

Q. 96. If NO_2 and N_2O_4 are mixed in a flasks so that P_{NO_2} = 0.1 atm and $P_{N_2O_2}$ = 0.5 atm at 298 K, will the reaction $2NO_2(g) \rightleftharpoons N_2O_4(g)$ proceed to the right to the left?

Q. 97. A mixture of the three gases NO, Br_2, and NOBr is placed in a contained at 298 K so that at the instant of mixing.

$$P_{Br_2} = 0.16 \text{ atm} \quad P_{NO} = 0.1 \text{ atm} \quad \text{and} \quad P_{NOBr} = 0.8 \text{ atm}$$

will the reaction

$$NOBr(g) \rightleftharpoons NO(g) + \frac{1}{2}Br_2(g)$$

Proceed spontaneously to the right or to the left? The standard free energies of formation at 298 K in kilojoules per mole for these three gases are

NO(g) 86.57 $\quad Br_2$(g)3.142 $\quad$ NOBr(g) 82.42

Q. 98. Using the data of Tables from text book calculate $\Delta G°$ at both 298 and 600 K for the reaction $CaCO_3(s) \rightleftharpoons CaO(s) + CO_2(g)$.

Q. 99. (a) Calculate $\Delta G°_{1000}$ and $\Delta H°$ for this reaction between 900 and 1000 K.

(b) Assuming $\Delta H°$ remains constant between 900 and 1200 K, calculate K_{1200}, the equilibrium constant at 1200 K. Also calculate $\Delta G°_{1200}$.

(c) Using Tables calculate $\Delta H°_{298} - \Delta G°_{298}$.

Q. 100. A sample of an ideal gas, maintained at a constant temperature of 298 K is contained in a cylinder sealed by a piston. The gas is initially at a pressure of 6 atm and occupies a volume of 400 mL. If the external pressure on the piston is constant at 1.5 atm, and the gas expands until is own pressure is equal to the external pressure, how much work does the gas do on expanding?

Q. 101. An experiment uses two bulbs connected by a stopcock. The larger bulb, with a volume of 1.2 L, is evacuated, and an ideal gas is placed in the smaller bulb, which has a volume of 0.4 L. The gas has a pressure of 6 atm, and is kept at 298 K. What is the final volume of the gas when the stopcock is opened, and how much work does the gas do on expanding?

Q. 102. Calculate $\Delta H - \Delta E$ for the following reactions at 298 K

(a) $C_4H_{10}(g) + \frac{13}{2}O_2(g) \rightarrow 4CO_2(g) + 5H_2O(l)$

(b) $C_6H_{12}O_6(s) + 6O_2(g) \rightarrow 6CO_2(g) + 6H_2O(l)$

(c) $H_2O(l) \rightarrow H_2O(g)$

Q. 103. Calculate $\Delta H°$ for the reaction, use ΔH_f values from tables in text book.

$$CH_4(g) + 3Cl_2(g) \rightarrow CHCl_3(l) + 3HCl(g)$$

Q. 104. The reaction of a mixture of powered aluminium with iron oxide, Fe_2O_3,

$$2Al(s) + Fe_2O_3(s) \rightarrow 2Fe(s) + Al_2O_3(s)$$

is so exothermic that it is used for welding iron. The amount of heat released by the reaction melts the iron formed, and the molten iron can be directed into the space between the pieces that are to be welded. The use of this reaction is called the thermite process.

Calculate the amount of heat released when 10 g of powered Al reacts with excess Fe_2O_3 at 298 K in the thermite process.

Q. 105. Calculate $\Delta H°$ for the reaction

$$2C(gr) + 2H_2O(l) \rightarrow CH_3COOH\ (l)$$

using both heats of combustion and heats of formation.

Q. 106. The heat of combustion of liquid ethanol, CH_3CH_2OH, is -1366.8 kJ · mol^{-1}. Calculate the standard heat of formation of ethanol.

Q. 107. Calculate the O—H enthalpy as one half of the enthalpy change for the dissociation reaction

$H_2O(g) \rightarrow 2H(g) + O(g)$, $\Delta(H_2) = 435.9$; $\Delta\ (O_2) = 495$ and $\Delta H_f°\ (H_2O)(g) = -241.8$ KJ mol^{-1} respectively.

Q. 108. Calculate C—O bond enthalpy from the enthalpy change for the dissociation reaction $CH_3CH_2OH(g) \rightarrow 2C(g) + O(g)$, and the C—H, C—C, and O—H bond enthalpies. Given

$H_2(g) \rightarrow 2(g)$; $\Delta(H_2) = 435.9$ kJ · mol^{-1}

$O_2(g) \rightarrow 2O(g)$; $\Delta(O_2) = 495.0$ kJ · mol^{-1}

$C(gr) \rightarrow C(g)$; $\Delta H_{subl}(C) = 716.7$ kJ · mol^{-1}

$2C(s) + 3H_2(g) + \frac{1}{2}O_2(g) \rightarrow CH_3CH_2OH(g)$; $\Delta H_f°(C_2H_5OH, g) = -235.1$ kJ · mol^{-1}

Q. 109. (a) Estimate ΔH_{298} for the isomerization reaction

$$CH_3—O—CH_3(g) \rightarrow CH_3CH_2OH(g)$$

From bond enthalpy values From Tables in the text book.

(b) Calculate ΔH_{298} for the isomerization reaction of part(a) using standard heats of formation from Tables and compare it with the estimated value. What is the percentage error of the estimated value?

Q. 110. The C_p values at 298 K in joules per mole per kelvin are given for the following substances: Ar(g) 20.8, N_2(g) 29.3, $H_2O(l)$ 75.4, and Pb(s) 26.4. What is C_v for each of these substances?

Q. 111. (a) Write a balanced equation for the combustion of sucrose.

(b) The heat of combustion of sucrose is –5647 kJ · mol^{-1} at 298 K. A 2.0026-g sample of sucrose was burned in a bomb calorimeter at 298 K and the temperature was observed to rise 2.966°. Calculate the heat capacity of the calorimeter.

(c) In the same calorimeter, 0.8021 g of benzene, C_6H_6, was burned at 298 K, and the temperature was observed to rise 3.009°. Write a balanced equation for the combustion of benzene and calculate ΔE and ΔH for this combustion.

Q. 112. An athlete is given 100 of glucose ($C_6H_{12}O_6$) of energy equivalent to 1560 kJ. He utilizes 50 per cent of this gained energy in the event. In order to avoid storage of energy in the body, calculate the weight of water he would need to perspire. The enthalpy of evaporation of water is 44 kJ mol^{-1}.

Q. 113. The bond dissociation energies of gaseous H_2, Cl_2 and HCl are 434.7, 242.5, and 430.5 KJ mol^{-1} respectively. Calculate the enthalpy of formation of HCl gas.

Q. 114. Determine the enthalpy of the reaction,

$C_3H_8(g) + H_2(g) \rightarrow C_2H_6(g) + CH_4(g)$, at 298 K using the given heat of combustion values under standard conditions.

Compound H_2(g) CH_4(g) and C_2H_6(g) C(graphite)

$\Delta H°$(KJ mol^{-1}) – 285.8 – 890.0 –1560.0 –393.5

The standard heat of formation of C_3H_8(g) is 103.8 kJ mol^{-1}

An aqueous solution NaCl on electrolysis gives H_2(g), Cl_2(g) and NaOH according to the reaction:

$$2Cl^-_{(aq)} + 2H_2O = 2OH^-_{(aq)} + H_{2(g)} + Cl_{2(g)}.$$

Q. 115. A gas mixture of 3.67 litres of ethylene and methane on complete combustion at 298 K produces 6.11 litres of CO_2. Find out the amount of heat evolved on burning one litre of the gas mixtured. The heats of combustion of ethylene and methane are –1243 and –891 kJ mol^{-1} at 298 K.

Q. 116. The polymeration of ethylene to linear polyethylene is represented by the reaction

$$nCH_2 = CH_2 \rightarrow (—CH_2—CH_2—)_n.$$

where n has a large integral value. Given that the average enthalpies of bond dissociation for C=C and C—C at 298 K are +590 and +331 KJ mol^{-1} respectively, calculate the enthalpy of polymerisation per mole of ethylene at 298 K.

Q. 117. In order to get maximum calorific output, a burner should have an optimum fuel to oxygen ratio which corresponds to 3 times as much oxygen as is required theoretically for complete combustion of methane as fuel. A burner which has been adjusted for methane as fuel (with x litre/ hour of CH_4 and $6x$ litre/hour of O_2) is to be readjusted for butane. C_4H_{10}. In order to get the rate of supply of butane and oxygen? Assume that losses due to incomplete combustion etc. are the same for both fuels and that the gases behave ideally. Heats of combustion. CH_4 = 804 KJ mol^{-1}; C_4H_{10} = 2878 KJ mol^{-1}.

Q. 118. Diborane is a potential rocket fuel which undergoes combustion according to the reaction.

$$B_2H_6(g) + 3O_2(g) \rightarrow B_2O_4(s) + 3H_2O(g)$$

From the following data, calculate the enthalpy change for the combustion of diborane

$$2B(s) + (3/2)O_2(g) \rightarrow B_2O_3(s) \qquad \Delta H - 1273 \text{ kJ mol}^{-1}$$
$$H_2(g) + (1/2)O_2(g) \rightarrow H_2O(l) \qquad \Delta H - 286 \text{ kJ mol}^{-1}$$
$$H_2O(l) \rightarrow H_2O(g) \qquad \Delta H = 44 \text{ kJ mol}^{-1}$$
$$2B(s) + 3H_2(g) \rightarrow B_2H_6(g) \qquad \Delta H = 36 \text{ kJ mol}^{1}$$

Q. 119. From the following data, calculate the enthlpy change for the combustion of cyclopropane at 298 K. The enthalpy of formation of $CO_2(g)$, $H_2O(l)$ and propane (g) are -393.5, –285.8 and 20.42 kJ mol^{-1} respectively. The enthalpy of is isommeriation of cyclopropane to propene is –33 kJ mol^{-1}.

Q. 120. The enthalpy change involved in the oxidation of glucose is –2880 kJ mol^{-1}. Twenty five percent of this energy is available for muscular work. If 100 kJ of muscular work is needed to walk one kilometer, what is the maximum distance that a person will be able to walk after eating 120 g of glucose?

Q. 121. Estimate the average S-F bond energy in SF_6. The standard heat of formation values of $SF_6(g)$, S(g) and F(g) are: –1100, 275 and 80 kJ mol^{-1} respectively.

Q. 122. The standard molar enthalpies of formation of cyclohexane (l) and benzene (l) at 298 K are –156 and +49 kJ mol^{-1} respectively. The standard enthalpy of hydrogenation of cyclohexane (l) at 298 K is –119 kJ mol^{-1}. Use these data to estimate the magnitude of the resonance energy of benzene.

Q. 123. Show that the reaction

$CO(g) + (1/2)O_2(g) \rightarrow CO_2(g)$ at 300 K, is spontaneous and exothermic, when the standard entropy changes is –0.094 kJ mol^{-1} K^{-1}. The standard Gibbs free energies of formation for CO_2& CO are –394.4 & –137.2 kJ mol–1, respectively.

Q. 124. (a) In the following equilibrium

$$N_2O_4 \rightleftharpoons 2NO_2(g)$$

When 5 moles of each are taken, the temperature is kept at 298 K the total pressure was found to be 20 mm Hg. Given that

$$\Delta G_f^\circ (N_2O_4) = 100 \text{ kJ}$$
$$\Delta G_f^\circ (NO_2) = 50 \text{ kJ}$$

(i) Find ΔG of the reaction

(ii) The direction in which the equilibrium shifts

(b) A graph is plotted for a real which follows vander. Wall's equation with PV_m taken on Y-axis & P on X-axis. Find the intercept of the line where V_m is molar volume.

Q. 125. Two moles of a perfect gas undergo the following processes:

(a) a reversible isobaric expansion from (1 atm, 20 L) to (1 atm, 40 L);

(b) a reversible isochoric change of state from (1 atm, 40 L) to (0.5 atm, 40 L);

(c) a reversible isothermal compression from (0.5 atm, 40 L) to (1 atm, 20 L);

(i) Sketch with labels each of the processes on the same P-V diagram.

(ii) Calculate the total work (w) and the total heat energy (q) involved in the above processes.

(iii) What will be the values of ΔU, ΔH and ΔS for the overll process?

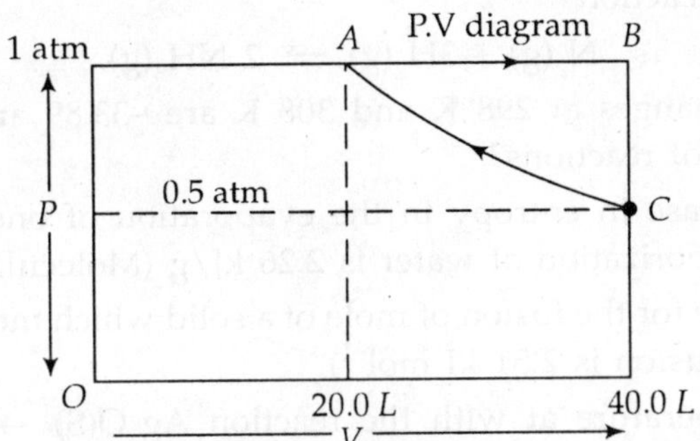

Q. 126. (a) A Carnot engine operates between a hot reservoir a 320 K and a cold reservoir at 260 K. If it absorbs 500 J of heat per cycle at the hot reservoir, how much work poor cycle does it deliver? (b) If the same engine, working in reverse, functions as a refrigerator between the same two reservoirs, how much work per cycle must be supplied to remove 1000 J of heat from the cold reservoir?

Q. 127. In a Carnot cycle, the isothermal expansion of an ideal gas takes place at 400 K and the isothermal compression at 300 K. During the expansion 2090 Joules of heat energy are transferred to the gas. Determine (a) the work performed by the gas during the isothermal expansion, (b) the heat rejected from the gas during the isothermal compression, and (c) the work done on the gas during the isothermal compression.

Q. 128. A brass rod is in thermal contact with a heat a reservoir at 400 K at one end and a heat reservoir at 300 K at the other end. (a) Compare the total change in the entropy arising from the process of conduction of 5 KJ of heat through the rod. (b) Does the entropy of the rod change in the process.

Q. 129. An ideal heat engine operates according to the Carnot cycle and receives 2.5 kJ from a hot source each cycle. The temperature of the hot source is 400°K and that of the cold sink 300°K. Find the work performed by the engine per cycle and the amount of heat rejected to the cold sink per cycle.

Q. 130. Hydrogen amounting to 6.6 grammes expands isobarically until its volume doubles. Find the change in entropy during expansions.

Q. 131. Helium of mass m = 1.7 g of expanded adibatically n = 3.0 times and then compressed isobarically down to the initial volume. Find the entropy increment of the gas in this process.

Q. 132. Find the change in entropy upon the isothermal-expansion of 6 grammes of hydrogen from 10^5 to 0.5×10^6 N/m^2.

Q. 133. Find the change in entropy during the transition of 8 grammes ofoxygen from a volume of 10 litres at a temperature of 353 K toa volume of 40 litres at a temperature of 573 K.

Q. 134. An ice-making machine operated in a Carnot cycle; it takes heat from water 273 K and discards heat to a room at 295 K. Suppose that 40.0 kg of water at 273 K are converted to ice at 273 K. (a) How much heat is discarded to the room? (b) How much energy must be supplied to the device?

Q. 135. For the following reaction

$$N_2(g) + 3H_2(g) \rightleftharpoons 2\,NH_3(g)$$

The free energy changes at 298 K and 308 K are –33.89 and –28.018 kJ respectively. Calculate the heat of reactions?

Q. 136. Calculate the increase in entropy in the evaporation of one mole of water at 373 K. Latent heat of vapporization of water is 2.26 kJ/g (Molecular mass of water = 18)

Q. 137. The entropy change for the fusion of mole of a solid which melts at 300 K will be, (Given the latent heat of fusion is 2.51 kJ mol^{-1}).

Q. 138. Calculate the temperature at with the reaction $Ag_2O(S) \rightarrow 2Ag(s) + \frac{1}{2}O_2$ at 1 atm pressure will be in equilibrium. The values of ΔH and Δs for the reaction are 30.58 kJ and 66.11 JK^{-1} respectively).

Q. 139. Calculate the entropy change involved for the isothermal reversible expansion of 5 moles of an ideal gas from a volume of 10 litres to a volume of 100 litres at 300 K.

Q. 140. Calculate the free energy change accompanying the compression of 1 mole of CO_2 at 330 K from 5 atm to 50 atm. (assume the CO_2 behaves like an ideal gas.)

Q. 141. Calculate the free energy change which occurs when one mole of an ideal gas expands reversibly and isothermally at 300 K from the initial volume of 5 litres of 50 litres.

Q. 142. When 1-pentyne (A) is treated with 4N alcohlic KOH at 448 K, it is converted slowly into an equilibrium mixture of 1.3% 1-pentyne (A), 95.2% 2-pentyne (B) and 3.5% of 1, 2-pentadiene (C). The equilibrium was maintained at 448 K. Calculate $\Delta G°$ for the following equilibria.

$$B \rightleftarrows A \quad \Delta G_1^0 = ?$$

$$B \rightleftarrows C \quad \Delta G_2^0 = ?$$

From the calculated value of ΔG_1^0 and ΔG_2^0 indicate the order of stability of *A*, *B* and C. Write a reaction mechanism showing all intermediates leading to *A*, *B* and *C*.

Q. 143. Calculate the enthalpy of reaction for

$$HCN(g) \rightarrow H(g) + C(g) + N(g)$$

from enthalpies of formation (see your text book tables). Given that C—H bond energy is 411 kJ/mol, obtain a value for the C≡N bond energy. Compare your result with the value given in table of your text book.

Q. 144. According to Pauiling the A—B bond energy is equal to the average of the A—A and B—Bond energies plus an energy contribution from the polar character of the bond.

$$BE(A\text{—}B) = \frac{1}{2}\,[BE\ (A\text{—}A) + BE(B\text{—}B)] + k(X_A - X_B)^2$$

Here X_A and X_B are the electronegativities of atoms A and B and k is a constant equal to 98.6 kJ. Assume that the electronegativity of H is 2.1. Use the formula to calculate the electronegativity of oxygen.

Q. 145. How much heat is needed to vaporize 10.0 mL of liquid hydrogen cyanide, HCN, at 298 K? The density of the liquid is 0.687 g mL^{-1}. Use standard heats of formation, which are given in tables.

Q. 146. How much heat must be added to 25.0 g of solid white phosphorus, P4, at 298 K to give the liquid at its melting point, 317.1K? The heat capacity of solid white phosphorus is 95.4 J K^{-1} mol^{-1}; its heat of fusion is 2.63 kJ mol^{-1}.

Chapter 7

CHEMICAL KINETICS

I. FORMULAE

$$\text{Rate of reaction} = \frac{\text{Amount of substance transformed}}{\text{Time taken in transformation}}$$

For a reaction

$aA + bB \rightarrow cC + dD$

$$\text{Rate of reaction} = \frac{1}{a}\frac{d[A]}{dt} = \frac{1}{b}\frac{d[B]}{dt} = \frac{1}{c}\frac{d[C]}{dt} = \frac{1}{d}\frac{d[D]}{dt}$$

or, $r = k[A]^x \times [B]^y$

where, order of reaction = $x + y$; unit of rate constant = $[A]^x$

and k= Rate of constant; = $\text{mol}^{(1-n)}\ L^{n-1}\ \text{time}^{-1}$

Equations for rate constant,

For 1st order, $k = \dfrac{2.303}{t(a-b)}\log\dfrac{a}{a-x}$

For IInd order, $k = \dfrac{1}{t}\cdot\dfrac{x}{a(a-x)}$ (when initial conc. of reactants are same)

$k = \dfrac{2.303}{t(a-b)}\log\dfrac{b(a-x)}{a(b-x)}$ (when initial conc. of reactants are different)

For IIIrd order, $k = \dfrac{1}{2t}\left[\dfrac{1}{(a-x)^2} - \dfrac{1}{a^2}\right] = \dfrac{1}{2t}\left[\dfrac{x(2a-x)}{a^2(a-x)^2}\right]$

order of reaction	**Unit of k**
0	mol L^{-1} time^{-1}
1	time^{-1}
2	L mol^{-1} time^{-1}
3	L^2 mol^{-2} time^{-1}

$$t_{1/2} \alpha \frac{1}{a^{n-1}}$$ (n = order of reaction)

$$N_t = N_0\left(\frac{1}{2}\right)^n$$ (n = No. of half-life periods)

$$k = A \cdot e^{-E_a/RT}$$

or $$\log k = \log A - \frac{E_a}{2.303\,RT}$$

or $$\log \frac{k_2}{k_1} = \frac{E_a}{2.303\,R}\left[\frac{T_2 - T_1}{T_1 T_2}\right]$$

II.

Chemical kinetics is the study of rates of a chemical reaction. The reaction speed tells us how fast the concentration of reactants change with time, eg. To know, how long it takes to convert 2 moles of $H_2(g)$ and 1 mole of $O_2(g)$ into 2 moles of $H_2O(l)$ ot room temp & pressure i.e. the rate at which the reaction occurs, is the study chemical kinetics.

Role of Chemical Reaction

Role of chemical reaction is defined as instantaneous rate of change in concentration of any one or more of the reactants or products at a particular instant of time.

Let $A + 3B \rightarrow 2C + 2D$. Here, in the reaction, the rate of disappearance of B is three times as fast as A and formation of C & D. So these rates are not same as the rate of disappearance of A, as 3 moles of B are consumed for every mole of A.

Therefore, the rate of disappearance of B = 3 × rate of disappearance of A.

Similarly, rate of formation of C = rate of formation of D = 2 × rate of disappearance of A.

Note: Role of Disappearance are negative (conc. decrease with time) and rate of formation are positive (conc. increase with time)

rate of reaction (reaction rate) = – (rate of disappearance A) $(= -\Delta A/dt)$

$$= \frac{1}{3}\text{(rate of disappearance of } B) = \frac{1}{3}(-\Delta B/dt) = \frac{1}{2}\text{(rate of formation of C)} = \frac{1}{2}(+\Delta C/dt)$$

$$= \frac{1}{2}\text{(rate of formation of } D) = \frac{1}{2}(+\Delta D/dt)$$

The –ve sign indicates that the concentration decreases with time while +ve sign indicates that the concentration increases with time.

Units

Usually, the *concentration is expressed in moles per litre and time in seconds* ∴ unit of rate is mol litre^{-1} sec^{-1} (mol $L^{-1}S^{-'}$)

Rate law or Rate Equation

An equation which relates the rate of a reaction to the concentration of the reactants is called the rate law or rate equation. Rate of reaction = $\dfrac{\text{Amount of substance transformed}}{\text{Time taken in transformation}} = \dfrac{-dx}{dt}$

For the reaction $mA + nB \rightarrow$ products, experiment shows that the rate of reaction is proportional to the concentration of A to the power m. i.e. rate $\propto [A]^m$. Similarly it may also be proportional to the concentration of B to the power n. Hence over all rate is proportional $[A]^m$ $[B]^n$ i.e. rate $\alpha\ [A]^m\ [B]^n$ or rate = $k\ [A]^m\ [B]^n$ . . . . (i)

Where k is the specific reaction rate or rate constant. This equation is termed rate law or rate equation.

The reaction is said to have an order of 'm' with respect to the reactant A and 'n' with respect to the reactant B. The sum, $m + n$ denotes the total order of the reaction. The values m and n may have integral values i.e. 1, 2 or 3 for simple reactions, but they can take any value which is not necessarily integral. The order can only be determined experimentally.

Rate Constant or Velocity Constant:

The rate constant is the rate of the reaction (at given temperature) when the concentrations of all the reactants are equal to unity. (i.e. Rate = k, when $[A] = [B] = 1$)

Factors Influencing rate of Reaction

(1) Conc. of reactant, (2) Temperature, (3) Pressure, (4) Surface area of reactant, (5) Exposure to light (Radiation) (6) analyst, (7) Order of reaction.

I. Effect of Concentration

The effect of concentration or pressure on the rate of a reaction is given by the *Law of Mass Action according to which*

The rate of a reaction is proportional to the active mass or molar concentration of each of the reacting substances. Molar concentration is represented by putting the formula of the reactant within square brackets.

For example in the formation of NH_3 from N_2 and H_2 according to the equation:

$3H_2 + N_2 \rightleftharpoons 2NH_3$ or $3/2\ H_2 + 1/2\ N_2 \rightleftharpoons NH_3$

It can be experimentally proved that the rate of the formation of NH_3 increases 3 times if the concentration of H_2 is doubled and its are of formation increases two times if the concentration of N_2 is doubled. Thus the rate of the forward reaction may be represented as under:

Rate of formation of NH_3 (r)

$$\alpha\ [H_2]^{3/2}\ [N_2] = k[H_2]^{3/2}[N_2]$$

where k is called the proportionally constant or the rate constant of velocity constant Graphically,

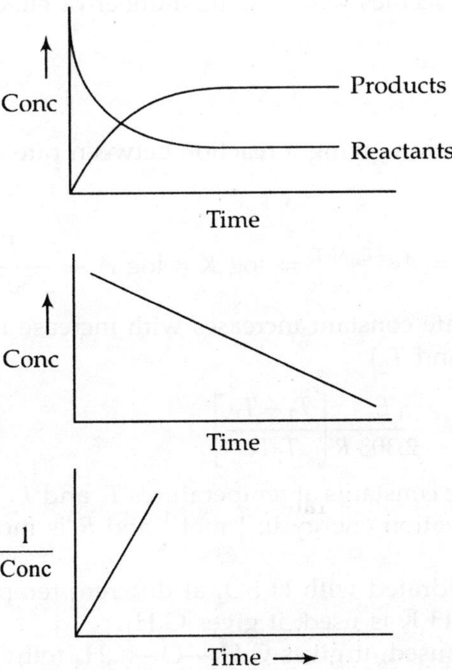

Rate constant relationship with concentration (changes over finite time periods).

Let initial concentration of the reactant A be a moles/liter when $t = 0$. After t seconds, let z moles per liter of A be already converted to product. Then, the remaining concentration of A that is $(a - x)$ moles/liter will determine the rate at the instant time (l), and is given by the expression

$$\frac{dx}{dt} = k(a - x)$$

Integrating the above relationship between the time limit $t = 0$ and $t = t$, we will have a relationship between k, a, x and t.

$$\int_{x=0}^{x=x} \frac{dx}{n - x} = \int_{i=0}^{t=t} kdt$$

$$-\ln (a - x) + \ln a = k (t - 0)$$

or $$\ln \frac{a}{a - x} = kt.$$

or $$2.303 \log_{10} \frac{a}{a - x} = kt.$$

It may be observed here that the velocity constant of a first order reaction is independent of the units of concentration and given by sec^{-1} or $time^{-1}$.

II. Effect of Temperature on Reaction Rate

The reaction rate increases with the increase in temperature. This is attributed to the increase in kinetic energy of colliding molecules whereby the number of effective collisions per second increases.

Arrhenius Equation

Arrhenius equation is an equation giving a reaction between rate constant and temperature. Mathematically,

(a) For one temperature (T) $K = Ae^{-E_a/RT} \Rightarrow \log K = \log A - \frac{E_a}{2.303\ RT}$

The equation shows that rate constant increases with increase in temperature.

(b) For two temperatures (T_1 and T_2)

$$\log \frac{K_2}{K_1} = \frac{E_a}{2.303\ R}\left[\frac{T_2 - T_1}{T_1 T_2}\right]$$

where K_1 and K_2 are the rate constants at temperatures T_1 and T_2 (in Kelvin), A is frequency (constant factor), E_a is activation energy in J mol^{-1} and R is molar constant = 8.314 J K^{-1} mol^{-1}.

eg : when C_2H_5OH is dehydrated with H_2SO_4 at different temperature & concentrations;

(a) If concentrated H_2SO_4 at 443 K is used, it gives C_2H_4.

(b) It dilute H_2SO_4 at 413 K is used, it gives C_2H_5—O—C_2H_5 (ethyl ether)

III. Order of a Reaction

The order of a reaction is the number of molecules whose concentrations determine the rate of the reaction at a given temperature [e.g. $NH_4NO_2 \rightarrow N_2 + 2H_2O$; rate = $k[NH_4NO_2]$ ∴ O.R = 1]

OR

The *order of a reaction may be defined as the sum of the powers of concentrations of reactants in the rate law equation*

Order of a reaction may be zero, fractional or integer (1, 2, 3, 4 . . . . . . n)

The order of a reaction depends upon pressure and temperature.

TYPES OF ORDER OF REACTIONS

EXAMPLES	RATE	CHARACTERISTICS
Zero Order Reaction		
Zero order reactions are those in which the rate of a reaction is independent of the concentration of the reactants.	Rate = $k[\ ]^0$ for $A \rightarrow$ Product $\frac{dx}{dt} = k$ a at $t = 0$ $(a - x)$ at $t = t$ On integrating, $\frac{x}{t} = K$	1. Rate remains constant through out the course of reaction. 2. Unit of rate constant = mol $litre^{-1}$ $time^{-1}$ 3. Half life period $t_{1/2} \propto (a)$, where a is initial concentration
The decomposition of HI on the surface of gold $2\,HI \rightleftharpoons H_2 + I_2$	rate = $k[H]^2$	
$CH_3COOC_2H_5 + H_2O \rightarrow CH_3COOH + C_2H_5OH$	Rate = $k[CH_3COOC_2H_5]^1 [H_2O]^0$	
$NH_4NO_2 \rightarrow N_2 + 2H_2O$	rate = $k[NH_4NO_2]$	
1 order Reaction		
The reactions are said to be of the first order if the rate of reaction is directly proportional to one power of the concentration of the reacting substance	Rate = $k[\ ]^1$ for $A \rightarrow$ Product $\frac{dx}{dt} = k[\ ]^1$ a at $t = 0$ $(a - x)$ at $t = t$ On integrating $k \times t = 2.303 \log_{10} \frac{a}{a - x}$	1. Rate continuously decreases with time 2. Unit of the constant $Time^{-1}$ 3. Half life period: $t_{1/2} \propto (a)^0$ half life for 1 order is independent of initial concentration
1. $N_2O_5(g) \rightarrow N_2O_4(g) + 1/2O_2(g)$	rate = $k[N_2O_5]$	
2. $SO_2Cl_2(g) \rightarrow SO_2(g) + Cl_2(g)$:	rate = $k[SO_2Cl_2]$	

EXAMPLES	RATE	CHARACTERISTICS
3. $H_2O_2(aq) \rightarrow H_2O(l) + 1/2\ O_2(g)$	Rate = $k[H_2O_2]$	
II Order Reaction **Case 1:** If one reactant or two reactants with same initial concentration;	Rate = $k[\ ]^2$ | $2A \rightarrow$ Products $\frac{dx}{dt} = k(a-x)^2$ | a at $t = 0$ $(a - x)$ at $t = t$ On integrating $k \times t = \frac{1}{a}\frac{x}{(a-x)}$	1. Rate continuously decreases with time 2. Unit of rate constant mol^{-1} $litre^1$ time 3. Half life period: $t_{1/2} \propto (a)^{-1} \propto 1/a$
2 A $\rightarrow$ products . . . . $2O_3(g) \rightarrow Cl_2(g) + O_2(g)$ $2Cl_2O(g) \rightarrow 3O_2(g)$ $2N_2O(g) \rightarrow 2N_2(g) + O_2(g)$	Rate = $k[A]^2$	
Case II: If reactants are two and initial concentration are different;	Rate $k[\ ]^1[\ ]^1$ | $A + B \rightarrow$ Products $\frac{dx}{dt} = k(a - x)(b - x)$ | $(a - x)(b - x)$ at $t = t$ On integrating $k \times t = \frac{2.303}{(a-b)} \log_{10}\frac{b(a-x)}{a(b-x)}$	
$A + A$ or $A + B \rightarrow$ products $CH_3COOCH_3 + NaOH \rightarrow CH_3COONa + CH_3OH$	rate = $k[A][B]$ rate = $k[CH_3COOCH_3][NaOH]$	
Third Order Reactions $2A + B \rightarrow$ products The order with respect to A is 2 and with respect to B is 1 The overall order is 2 + 1 = 3	rate = $k[A]^2[B]$	
e.g. $2NO\ (g) + 2H_4 \rightarrow N_2(g) + 2H_2O$	rate = $k[NO]^2\ [H_2]^1$	
$2\ NO(g) + O_2(g) \rightarrow 2NO_2(g)$	rate = $k[NO]^2\ [O_2]^1$	
$2NO(g) + Cl_2(g) \rightarrow 2NOCl(g)$ $2FeCl_3 + SnCl_2 \rightarrow 2\ FeCl_2 + SnCl_4$		

EXAMPLES	RATE	CHARACTERISTICS

Fraction Order Reactions

The order of a reaction may not always be a whole number. It can be fractional quality also

$H_2(g) + Br(g) \rightarrow 2\ HBr(g)$ rate = $k[H_2]\ [B_2]^{1/2}$

order $= 1 + \frac{1}{2} = \frac{3}{2}$ or 1.5

Methods of Determination of Order of Reactions

(A) Salient features for determination of order of reaction

1. The rate expression is directly given i.e.,

$2NO + O_2 \rightarrow 2NO_2$

and Rate $= K[NO]^2O_2]^1$ $\therefore$ OR = 2 + 1 = 3

i.e., O.R. is also defined as sum of powers raised on concentration terms in order to write rate expression.

2. The problem is provided up in form of language, i.e,

$2A + B \rightarrow$ Product and given

A shows first order kinetics and B shows zero order kinetics.

$\therefore$ O.R. = 1 + 0 = 1 and rate $= K[A]^1\ [B]^0$

3. Either of reactant is mentioned as excess, i.e.,

(a) $2A + \underset{\text{excess}}{B} \rightarrow$ Products

O.R. = 2 + 0 = 2 and rate $= K[A]^2[B]^0$

(b) $\underset{\text{excess}}{2A} + B \rightarrow$ Products

$\therefore$ O.R. = 0 + 1 = 1 and rate $= K[A]^0[B]^1$

4. Experimental data is given:

Rate is given using different conc. of A and B at same temperature

Given that,

$[A]$	$[B]$	Rate
a_1	b_1	r_1
a_2	b_2	r_2
a_3	b_3	r_3

$\therefore$
$$r_1 = K[a_1]^m[b_1]^n$$
$$r_2 = K[a_2]^m[b_2]^n$$
$$r_3 = K[a_3]^m[b_3]^n$$

where m and n are order for A and B respectively

Solve these to get m and m

$\therefore$ O.R. = $[m\ \ n]$

Graphical Method

In case of the reactions of first order straight lines are obtained when log $(a - x)$ or log $a/(a\text{-}x)$ is plottaed against time. In the case of second order reactions linear graphs are obtained. When $1/(a\text{-}x)$ or. $x/(a\text{-}x)$ is plotted against time.

Order	Rate	Rate law in the Integrated From	Integrated Rate Equation	Rate law in the Differential From	Dimensions of K	Half life Proportional	Graphs
O	K	$\frac{dx}{dt} = K$	$[A_t] = -kt + [A]_0$	$kt = x$	mol. lit^{-1} sec^{-1}	a^1	$[\bar{A}]$ / $[A]$ vs Time, Slope = –ve
1	$K[A]$	$\frac{dx}{dt} = K(a-x)$	$\ln[A]_t - kt + \ln[A_0]$	$kt = 2.303 \log \frac{a}{a-x}$	sec^{-1}	a^0	$\ln[A]$ vs Time, Slope = –ve
2	$K[A]^2$	$\frac{dx}{dt} = K[a-x]^2$	$\frac{1}{[A]_t} = kt + \frac{1}{[A]_0}$	$kt = \frac{1}{a-x} - \frac{1}{a}$	lit. mol^{-1} sec^{-1}	a^{-1}	$\frac{1}{[\bar{A}]}$ vs Time, Slope = –ve
3		$\frac{dx}{dt} = K(a-x)^3$		$kt = \frac{1}{2(a-x)^2} - \frac{1}{2a^2}$	lit^2 · mol^{-2}/sec^{-1}	a^{-2}	

IV Half Life of a Reaction

Half life of reaction is *time required to reduce the concentration of A(a) to one half (x = a/2) its initial value at any instant is called its half-life* period or the time in which half of a reaction is completed is called half life period or simply half life of the reaction.

For the first order reaction the rate equation is, $K = (2{,}303/\text{K})\log_{10}\dfrac{a}{(a-x)}$

At the end of half life period, the initial concentration 'a' will be reduced to a/2. The x and $(a\text{-}x)$ will both be equal to $a/2$. Let the half change time be denoted by $t_{1/2}$.

Then the above equation is reduced to

$$t_{1/2} = \frac{2.303}{K}\log\frac{a}{\left(a-\frac{a}{2}\right)} = (2.303/\text{K})\log 2 = (2.303/\text{K})0.3010 = 0.693/\text{K}$$

Since initial concentration of the reactant does not occur on the right hand side of this expression, follows that for a first order reaction, time for half change is independent of the initial concentration of the reactant.

V Catalyst

Catalyst is a substance that alters the rate of a chemical reaction, itself remaining chemically unchanged at the end of the reaction. The process is termed as catalysis. Generally a small amount of catalyst is only needed.

A catalyst may increases or decreases the rate of a reaction and are the refer termed +ve or –ve catalyst respectively.

e.g. +ve catalyst MnO_2 is a positive catalyst in the decomposition of $KClO_3$

Finely divided iron is a positive catalyst in the synthesis of ammonia.

Glycerine decreases the decomposition of hydrogen peroxide –ve catalyst

Ethyl alcohol acts as a negative catalyst in the oxidation of chloroform into phosgene.

Rate Reaction Kinetics to Solve Problem (To the Student)

1. If a question deals with the Rate of a Reaction and values of K, m, n, are given then use the rate law: rate of reaction = $K[A]^m [B]^n$.
2. If the value of K, m, n, are not given but you have to find the rate of reaction, then establish the rate of reaction from:
   (a) The slope of an instantaneous tangent line to the graph of $[A]$ vs t. This is instantaneous rate of reaction.
   (b) The expression rate of reaction = $\frac{-\Delta[A]}{\Delta t}$, if the time interval is short. When applied to initial stages of a reaction, the rate of reaction is the Initial rate of reaction.
3. If order of reaction is to be found depending on data given, the (a) find method of initial rates (b) a graph of rate data to yield a st lime.
4. To relate reactant conc. & times (for 0, 1 or 2) use appropriated integrated rate equation for determining K.

PROBLEMS

A student prepared a data table (see below) for the results that he got in doing an experiment on the kinetics or reaction mechanisms, and their rates, when iodine is added to acetone:

$$CH_3-\overset{O}{\overset{||}{C}}-CH_3(aq) + I_2(aq) \rightarrow CH_3-\overset{O}{\overset{||}{C}}-CH_2I(aq) + H^+(aq) + 1\ (aq)$$

Mixture	*Acetone*	H^+	$(I_2)_0$	*Time (sec)*	*Temp K*	*rate* = $(I_2)_0 \cdot sec^{-1}$
I	0.8 M	0.2 M	0.01 M	240	298 K	4.2×10^{-4}
II	1.6 M	0.2 M	0.001 M	120	298 K	8.3×10^{-4}
III	0.8 M	0.4 M	0.001 M	120	298 K	8.3×10^{-6}
IV	0.8 M	0.2 M	0.0005 M	120	298 K	4.2×10^{-6}

Answer the following 1-4

Q. 1. Determine the rate order for each reactant.

Q. 2. Determine the rate specific constant, *k*, for the reaction.

Q. 3. Given the following information, predict the time required for the raction to reach completion:

Acetone	*HCl*	t_2	H_2O
25 mL	25 mL	25 mL	25 mL
3.2 M	4.0 M	0.02 M	—

Q. 4. The student used the ratios in Mixture I and ran the experiment at two different temperatures. Calculate the rate, the rate constant, log *k* and 1/T for each temperature studied. From the data, plot *k* versus 1/T and determine the activation energy. Given that the activation energy for the reaction is 8.6×10^4 Joules, calculate the % error.

Q. 5. At equilibrium at a given temperature and in a liter reaction vessel HI is 20 mole percent dissociated into H_2 and I_2 according to the equation, $2HI \rightleftharpoons H_2 + I_2$. If 1 mole of pure HI is introduced into a liter reaction vessel at the given temperature, how many moles of each component will be present when equilibrium is established?

Q. 6. A reaction vessel in which the following reaction had reached a state of equilibrium, $CO + Cl_2 \rightleftharpoons COCl_2$, was found, on analysis, to contain 0.3 mole of CO, 0.2 mole of 0.8 mole of $COCl_2$, in a liter of mixture. Calculate the equilibrium constant for the reaction.

Q. 7. A reaction vessel with a capacity of 1 liter in which the following reaction had reached a state of equilibrium $2SO_2 + O_2 \rightleftharpoons 2SO_3$, was found to contain 0.6 mole of SO_3, 0.2 mole of SO_2, and 0.3 mole of O_2. Calculate the equilibrium constant.

Q. 8. The equilibrium mixture, $SO_2 + NO_2 \rightleftharpoons SO_3 + NO$, in a liter vessel, was found to contain 0.6 mole of SO_3, 0.4 mole of NO, 0.1 mole of NO_2, and 0.8 mole of SO_2. How many moles of NO would have to be forced into the reaction vessel, volume and temperature being kept constant, in order to increase he amount of NO_2 to 0.3 mole?

Q. 9. A mixture of 2 moles of CH_4 gas and 1 mole of H_2S gas was placed in an evacuated container which was then heated to and maintained at a temperature of 1000 K. When equilibrium was established in the gaseous reaction, $CH_4 + 2H_2S \rightleftharpoons CS_2 + 4H_2$, the total pressure in the container was 0.92 atm and the partial pressure of the hydrogen gas was 0.2 atm. What was the volume of the container?

Q. 10. When N_2O_5 gas is heated it dissociate into N_2O_3 gas and O_2 gas according to the reaction, $N_2O_5 \rightleftharpoons N_2O_3 + O_2$. K_1 for this reaction at a specific temperature, t°C, is 7.75 moles liter^{-1}. The N_2O_3 dissociates to give N_2O gas and O_2 gas according to the reaction $N_2O_3 \rightleftharpoons N_2O + O_2$. K_2 for this reaction at the same specific temperature, 1°K is 4 moles liter^{-1}.

When 4 moles of N_2O_5 are heated in a 1.0 liter reaction vessel at t°C the concentration of O_2 at equilibrium is 4.5 moles liter^{-1}. Calculate the concentrations in moles per liter of all other species in the equilibrium system.

Q. 11. In an equilibrium mixture, $CO_2 + H_2 \rightleftharpoons CO + H_2O$, contained in a 6 liter reaction vessel at 1280 K the partial pressure of the reactants are: CO_2 = 63.1 atm, H_2 = 21.1 atm, CO = 84.2 atm, H_2O = 31.6 atm. Enough CO_2 was then removed from the vessel to reduce the partial pressure of the CO to 63 atm, temperature being kept constant.

(a) Calculate the partial pressure of the CO_2 in the new equilibrium system.

(b) For the above reaction how does the numerical value of k in which concentration is expressed in moles per liter, compare with the numerical value of K_c, in which concentrations are expressed in atmospheres?

(c) Suppose the volume of the new equilibrium system was reduced to 3 liters by depressing a piston, what would the partial pressure of the CO_2 be?

Q. 12. To the system $LaCl_3$ (solid) + H_2O(gas) + heat $\rightleftharpoons$ LaClO (solid) + 2HCl (gas) already at equilibrium we add more water vapor without changing either the temperature or the volume of the system. When equilibrium is re-established the pressure of water vapor is found to have been doubled. Hence, the pressure of HCl present in the system has been multiplied by what factor?

Q. 13. The equilibrium mixture $SO_2 + NO_2 \rightleftharpoons SO_3 + NO$, was found to contain 0.6 mole of SO_3, 0.4 mole of NO, 0.8 mole of SO_2 and 0.1 mole of NO_2 per liter. One mole of NO was then forced into the reaction vessel, temperature and volume being kept constant. Calculate the number of moles of each gas in the new equilibrium mixture.

Q. 14. The standard free energy of formation (ΔG_f°) of NO_2 is 51.83 KJ mol^{-1}. Calculate the value of K at 298 K for the reaction.

$$2O_{2(g)} + N_{2(g)} \rightleftharpoons 2NO_{2(g)}$$

Q. 15. A large sample of $CaCO_3$(g) is placed in an evacuated container and heated to 813 K, to bring about the reaction

$$CaCO_{3(s)} \rightarrow CaO_{(s)} + CO_{2(g)}$$

and the pressure of CO_2 is monitored. The pressure is found to be 100 mm after 5 minutes, and 200 mm after 10 minutes. Calculate the average rate of appearance of CO_2 during each 5-minute interval and during the 10-minute interval.

Q. 16. It is found that $N_2O_3(g)$ decomposes according to $N_2O_3(g) \rightarrow NO_2(g) + NO(g)$. A bulb is filled with N_2O_3 to a pressure of 100 mm and the decomposition is monitored by measuring the total pressure. after 75 seconds the pressure is measured to be 150 mm and after 150 seconds the pressure is measured to be 175 mm. Calculate the average rate of disappearance of $N_2O_3(g)$ during these time intervals.

Q. 17. The reaction $C_4H_9I + H_2O \rightarrow C_4H_{10}O + H^+ + I^-$ is first order in C_4H_9I. Calculate the value of k, the specific rate constant, from the following dta obtained by starting with $[C_4H_9I]$ – 0.2 *M*.

$[H^+]$ moles/liter	~0	0.0587	0.1	0.13	0.15
t min	0	30	60	90	120

Q. 18. Calculate the specific rate constant from the data of Problem 17 using the integrated form of the first order rate equation.

Q. 19. Use the data of Problem 17 to find how much time is required for the $[C_4H_9I]$ to be reduced from 0.2 *M* to 0.125 *M*.

Q. 20. The following rate data are obtained for the reaction $2A(g) \rightarrow B(l)$, which is second order.

P_A mm	300	200	151	99
t sec	0	250	500	1000

Calculate the specific rate constant.

Q. 21. Dinitrogen pentoxide, N_2O_5, undergoes first-order decomposition in chloroform solvent to yield NO_2 and O_2. The rate constant 348 K is 6.2×10^{-4}/min. Calculate the volume of O_2 obtained from the reaction of 1.0 ol N_2O_5 at 318 K and 780 mmHg after 20.0 hr.

Q. 22. Hydrogen peroxide in aqueous solution decomposes by a first-order reaction to water and oxygen. The rate constant for this decomposition is 7.4×10^{-4}/s. What quantity of heat energy is initially liberated per second from 2 L of solution that is 1.5 M H_2O_2? Given, ΔH_f°; $H_2O_2(aq) = -191.$kJ mol^{-1} and ΔH_f°; $H_2O\ (l) = -285.5$ kJ mol^{-1}

Q. 23. Nitric oxide reacts with oxygen to give nitrogen dioxide.

$$2NO(g) + O_2(g) \rightarrow 2NO_2(g)$$

The rate law is $-\Delta(NO)/\Delta t = k[NO]^2[O_2]$, where the rate constant is 1.16×10^{-5} L^2 mol^{-2} s^{-1} at 612 K. A vessel contains NO and O_2 at 612 K. The initial partial pressure of NO and O_2 are 155 mmHg and 345 mmHg., respectively. What is the rate of decreases of partial pressure of NO (in mmHg per second)? (*Hint:* From the ideal gas law, obtain an expression for the molar concentration of a particular gas in terms of its partial pressure.)

Q. 24. Consider the decomposition of *A* represented by circles where each circle represents a molecule of *A*.

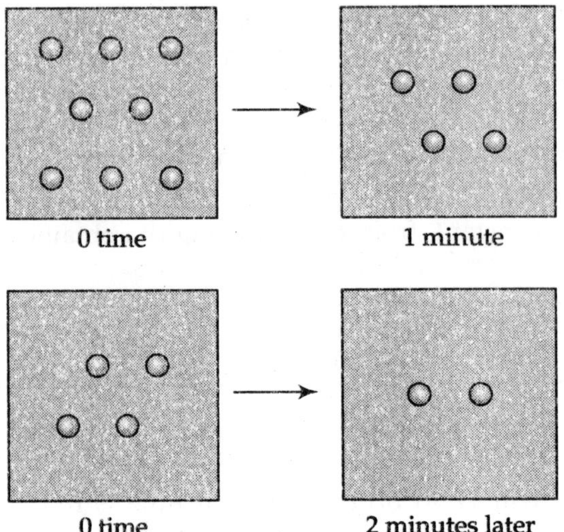

Is the reaction zero-order, first-order, or second-order?

Q. 25. The following reaction is second-order in A and first-order in B.

$A + B \rightarrow$ products

(a) Write the rate expression.

(b) Consider the following one-liter vessel in which each square represents a mole of A and each circle represents a mole of B Fig. (i).

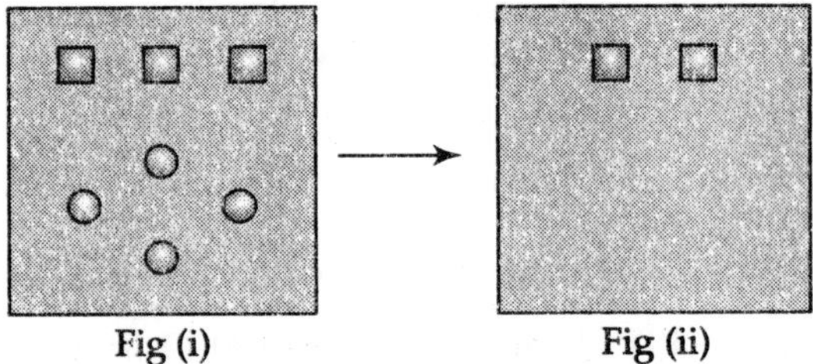

Fig (i) Fig (ii)

What is the rate of he reaction in terms of k?

(c) Assuming the same rate and k as (b), fill the similar one-liter vessel shown in the figure (ii) with an appropriate number of circles (representing B).

Q. 26. For the reaction

$$A + B \rightarrow C$$

the rate expression is rate = $k[A][B]$

(a) Given three test tubes, with different concentrations of A and B which test tube has the smallest rate?

(1) 0.1 M A; 0.1 M B

(2) 0.15 M A; 0.15 M B

(3) 0.06 M A; 1.0 M B

(b) If the temperature is increased, describe (using the words *increases, decreases, or remains the same)* what happens to the rate, the value of k, and E_a.

Q. 27. The gas-phase reaction between hydrogen and iodine

$$H_2O(g) + I_2(g) \rightleftharpoons 2HI(g)$$

proceeds with a rate constant for the forward reaction at 973 k of 138 L mol^{-1} s^{-1} and an activation energy of 165 kJ mol^{-1}.

(a) Calculate the activation energy of the reverse reaction given that ΔH_f° for HI is 26.48 kJ/mol and ΔH_f° for $I_2(g)$ is 62.44 kJ mol^{-1}.

(b) Calculate the rate constant for reverse reaction at 973 k (Assume A in the equation $k = Ae^{-E_a/RT}$ is the same for both forward and reverse reactions.).

(c) Calculate the rate of the reverse reaction if the concentration of HI is 0.2 M. The reverse reaction is second order in HI.

Q. 28. For a first-order reaction $aA \rightarrow$ products, where $a \neq 1$, the rate is $-\Delta(A)/a\,\Delta t$, or in derivative notation, $\frac{1}{a}\frac{d[A]}{dt}$. Derive the integrated rate law for the first-order decomposition of a moles of reactant.

Q. 29. Using calculus derive the equation for

(a) the concentration-time relation for a second-order reaction

(b) the concentration-time relation for a third-order reaction, $A \rightarrow$ product.

Q. 30. The following data apply to the reaction

$$A(g) + 3B(g) + 2C(g) \rightarrow \text{products}$$

[A]	[B]	[C]	Rate
0.2	0.4	0.1	X
0.4	0.4	0.2	8X
0.2	0.2	0.2	X
0.4	0.4	0.1	4X

Determine the rate law for the reaction.

Q. 31. In a first-order reaction, suppose that a quantity X of a reactant is added at regular intervals of time, Δt. At first the amount of reactant in the system builds up; eventually, however it levels off at a saturation value given by the expression

$$\text{saturation value} = \frac{X}{1-10^{-a}}, \text{ where } a = 0.3\frac{\Delta t}{t_{1/2}}$$

This analysis applies to prescription drugs, where you take a certain amount each day. Suppose you take 0.1 g of a drug three times a day and that the half-life for elimination is 2 days. Using this equation calculate the mass of the drug in the body at saturation. Suppose further that side effect show up when 0.5 g of the drug accumulates in the body. As a pharmacist, what is the maximum dosage you could assign to a patient for an 8-h period without casing side effects?

Q. 32. Write expression for both K_p and K_c for the reaction

$$NO_2(g) \rightleftharpoons NO(g) + \frac{1}{2}O_2(g)$$

Give the units of each of these equilibrium constants and the relation between them.

Q. 33. Ammonia is produced from the elements hydrogen and nitrogen by the reaction $3H_2(g)N_2(g) \rightleftharpoons 2NH_3(g)$. It is carried out at elevated temperatures because it is a very slow reaction at room temperature. For this reaction K_p is 6.5×10^{-3} atm^{-2} at 723 K. Each of the following mixtures may or may not be at equilibrium at 723 K at the instant of measurement. For each mixture, state whether or not it is in a sate of equilibrium, and if it is not, indicate the direction in which reaction must proceed to achieve equilibrium.

(a) A mixture in which the partial pressure of the three gases are.

$$P_{NH_3} = 60 \text{ atm} \qquad P_{H_2} = 150 \text{ atm} \qquad P_{N_2} = 80 \text{ atm}$$

(b) A mixture in which the partial pressures of the three gases are

$$P_{NH_3} = 60 \text{ atm} \qquad P_{H_2} = 2.0 \text{ atm} \qquad P_{N_2} = 5.0 \text{ atm}$$

Q. 34. At 548 K the equilibrium constant for the reaction

$$NH_4Cl(s) \rightleftharpoons NH_3(g) + HCl(g)$$

is 1.04×10^{-2} atm^2

A 0.98-g sample of solid NH_4Cl is placed in a 1.0-L closed container and heated to 548 K.

(a) Will there be any decomposition of the NH_4Cl when the system comes to equilibrium at 548 K?

(b) At equilibrium at 548 K what will the partial pressures of NH_3 and HCl in the container be?

(c) What will be the mass of solid NH_4Cl in the container at equilibrium at 598 K?

Q. 35. At 548 K the equilibrium constant for the reaction

$$NH_4Cl(s) \rightleftharpoons NH_2(g) + HCl(g)$$

is 1.04×10^{-2} atm^2

The gases HCl and NH_3 both at 548 K, are introduced ino an evacuated 1.0-L flask. at the instant of mixing, the partial pressure of HCl in the flask is 0.8 atm and the partial pressure of NH_3 is 0.5 atm. The temperature of the flask is maintained constant at 548 K.

(a) Show that the reaction between NH_3 and HCl to form solid NH_4Cl will take place when these gases are mixed.

(b) Calculate the partial pressure of NH_3 and Cl at equilibrium at 548 K.

Q. 36. Consider the exothermic formation of sulfur trioxide form sulfur dioxide and oxygen in the gas phase:

$$2SO_2(g) + O_2(g) \rightleftharpoons 2SO_3(g)$$

At 900 K, K_p for this reaction is 40.5 atm^{-1} and $\Delta H = -198$ kJ.

(a) Write the expression for the equilibrium constant for this reaction.

(b) Will the equilibrium constant for this reaction at room temperature (–300 K) be greater than, less than, or equal to the equilibrium constant at 900 K? Explain your answer.

Q. 37. Consider the dissolution of the insoluble compound manganese(II) sulfide in water

$$MnS(s) \rightleftharpoons Mn^{2+}(aq) + S^{2-}(aq)$$

(a) Write the expression for the equilibrium constant for this reaction and specify its units.

(b) will adding excess $Mn(NO_3)_2$ solution to a saturated solution of MnS in equilibrium with excess solid MnS increase, decrease, or leave unchanged the solubility of MnS?

(c) Will adding excess HCL solution to a saturated solution of MnS in equilibrium with excess solid MnS increase, decrease, or leave unchanged the solubility of MnS? Explain.

Q. 38. Calculate k_p for the reaction

$$2SO_3(g) \rightleftharpoons 2SO_2(g) + O_2(g)$$

both at 298 and at 900 K, using data given in Q 40.

Q. 39. For the reaction

$$NO(g) + \frac{1}{2}Cl_2(g) \rightleftharpoons NOCl(g) \qquad \text{.....(I)}$$

the equilibrium constant $k_{p(1)}$, is 3.7×10^3 atm$^{-1/2}$ at 298 K.

Calculate $k_{p(2)}$ for the reaction

$$2NOCl(g) \rightleftharpoons 2NO(g) + Cl_2(g) \qquad \text{.....(II)}$$

Give the expression for $k_{p(2)}$ its units and its value at 298 K

Q. 40. The equilibrium constant for the reaction

$$2HCl(g) \rightleftharpoons H_2(g) + Cl_2(g) \qquad \text{.....(I)}$$

is $k_1 = 4.17 \times 10^{-34}$ at 298 K

The equilibrium constant for the reaction

$$I_2(g) + Cl_2(g) \rightleftharpoons 2ICl(g) \qquad \text{.....(II)}$$

is $k_2 = 2.1 \times 10^5$ at 298 K

Calculate the equilibrium constant for the reaction

$$2HCl(g) + I_2(g) \rightleftharpoons 2ICl(g) + H_2(g) \qquad \text{.....(III)}$$

Q. 41. The equilibrium constant for the reaction

$$N_2(g) + \frac{1}{2}O_2(g) \rightleftharpoons N_2O(g) \qquad \text{.....(1)}$$

is $k_1 = 7.1 \times 10^{-19}$ atm$^{-1/2}$ at 298 K

The equilibrium constant for the reaction

$$N_2(g) + O_2(g) \rightleftharpoons 2NO(g) \qquad \text{.....(2)}$$

is $k_2 = 4.23 \times 10^{-31}$ at 298 K

Calculate the equilibrium constant at 298 K for the reaction

$$N_2O(g) + \frac{1}{2}O_2(g) \rightleftharpoons 2NO(g) \qquad \text{.....(3)}$$

Q. 42. Some $SO_2(g)$ and $O_2(g)$ are mixed together in a flask at 1100 K in such a way that at the instant of mixing their partial pressure are, respectively, 1.0 and 0.5 atm. When the system comes to equilibrium at 1100 K, the total pressure in the flask is found to be 1.35 atm. Calculate K_p at 1100 K for the reaction

$$2SO_2(g) + O_2(g) \rightleftharpoons 2SO_3(g)$$

Q. 43. At a given temperature the equilibrium constant of the reaction

$$H_2(g) + I_2(g) \rightleftharpoons 2HI(g)$$

is 6.0.

If 1.0 mol of H_2 and 1.0 mol of I_2 are introduced into a 1.0-L flask at this temperature, what will be the concentrations of all three species when the system comes to equilibrium?

Q. 44. Consider the equilibrium mixture of H_2, I_2, and HI. If this equilibrium is disturbed by adding more HI so that the partial pressure of HI is suddenly increased to 1.0 atm, what will the partial pressure of each of the gases be when the system returns to equilibrium? The equilibrium constant for the reaction $H_2 + I_2 \rightleftharpoons 2HI$ is 48.7 at 730.8 K.

Q. 45. When 2.94 moles of iodine and 8.1 mole of hydrogen are heated at 717 K while equilibrium is established, 5.64 moles of hydroiodic acid are formed. Find out the equilibrium constant.

Q. 46. The value of K_c is 60 for the reaction $H_2 + I_2 \rightleftharpoons 2HI$ at a certain temperature. What will be the amount of HI formed at equilibrium if the experiment is started with 5 moles of I_2 and 8 moles liter^{-1} of H_2?

Q. 47. Calculate the equilibrium constant for the reaction $PCl_5 \rightleftharpoons PCl_3 + Cl_2$, If PCl_5 dissociates to an extent of 70% at 2 atmospheres pressure.

Q. 48. A mixture of hydrogen and iodine vapours in molecular proportion of 2 : 3 was heated at 718 K till the change represented by the equation $H_2 + I_2 \rightleftharpoons 2HI$ reached the equilibrium state. the equilibrium constant at this temperature is 0.02. Calculate the percentage of iodine converted into hydrogen iodide.

Q. 49. At 713 K the equilibrium constant (K) for the following reaction is 49.5.

$$H_2(g) + I_2(g) \rightleftharpoons 2hi(g)$$

If 0.2 mole of H_2 and 0.2 mole of I_2 are placed in a 10 L vessel and permitted to react at this temperature. What will be the concentration of each substance at equilibrium?

Q. 50. Calculate the rate constant at 590 K for the following first order reaction. $SO_2Cl_2 \rightarrow SO_2 + Cl_2$. The time for half change is 60 minutes.

Q. 51. The velocity constant for the decomposition of N_2O_5 is 6.2×10^{-4} sec^{-1}. What is the half life period for the reaction.

Q. 52. A first order reaction is 50% completed in 20 min. Calculate time taken for 75% completion of the reaction.

Q. 53. The concentration of a reactant a changed from 0.044 M to 0.032 M in 25 minutes. What was the average rate of reaction during this interval?

Q. 54. A first order reaction takes 69.3 minutes for 50% completion. How much time it will be needed for 80% completion?

Q. 55. A combination of hydrogen and iodine was carried out by heating 2.94 moles of iodine and 8.1 moles of hydrogen at 730.6 K. At equilibrium 5.64 moles of hydroiodic acid were formed. Calculate the equilibrium constant.

Q. 56. The following data were reported for the decomposition of nitrous oxide, N_2O on gold at 1263 K

Time in min.	30	53	65	80	100	120
Percent decomposed	32	50	57	65	73	78

Show that the reaction is of the first order and calculate the velocity constant.

Q. 57. Calculate the composition of equilibrium mixture with 240 g: of acetic acid when warmed with 46 g. of ethyl alcohol at 298 K, the equilibrium constant at this temperature being 4.

Q. 58. Solid ammonium carbamate NH_4COONH_2 dissociates into NH_3 and CO_2 according to the reaction. $NH_4COONH_2(s) \rightarrow 2NH_3(g) + CO_2(g)$ at 300 k. The total pressure of the gases in equilibrium with the solid is 0.225 atm. What is the value of equilibrium constant change if the pressure of the system is increased to 1 atm. by adding an inert gas?

Q. 59. N_2 and H_2 are added to 5 liter flask under pressure. the flask was sealed and heated at 620 K. the equilibrium mixture contained 19.0 g. of NH_3, 0.16 g of H_2 and 3.4 g of N_2. Calculate the equilibrium constant K_p for the reaction. $N_2 + 3H_2 \rightleftharpoons 2NH_3$.

Q. 60. Ammonia under a pressure of 15 atm at 32 k is heated to 620 K in a closed vessel in the presence of a catalyst. Under these condition NH_3 is partially decomposed according to the equation

$$2NH_3 \rightleftharpoons N_2 + 3H_2$$

The vessel is such that the volume remains effectively constant whereas pressure increases to 50 atm. Calculate the percentage of NH_3 actually decomposed.

Q. 61. 0.5 mole of hydrogen and 0.5 mole of iodine react in a 10 litre evacuated vessel at 723 K hydrogen iodide is formed. The equilibrium constant (K_c) for the reaction is 50:

(i) Calculate the value of K_p.

(ii) Calculate the moles of iodine which are in excess.

Q. 62. NH_3 under a pressure of 15 am. at 300 K is heated to 620 K in a closed vessel in the presence of a catalyst. Under these conditions NH_3 is a partially decomposed according to the equation $2NH_3 \rightleftharpoons N_2 + 3H_2$. The vessel is such that the volume remains effectively constant. Whereas pressure increases to 50 atm. Calculate the percentage of NH_3 actually decomposed.

Q. 63. Two gram moles of PCl_5 are heated in a closed two litre vessel. When equilibrium is attained the pentachloride is 40% dissociated into PCl_3 and Cl_2. Calculate the equilibrium constant.

Q. 64. The combination of hydrogen and iodine is carried out by heating 60 ml of hydrogen with 42 ml of iodine in a closed vessel. At equilibrium it s found that the vessel contains 28 ml. of HI. Calculate the degree of dissociation of HI.

Q. 65. The degree of dissociation of 5 moles of HI at 717 K was found to be 22%. Calculate the no of moles of HI, H_2 and I_2 formed at equilibrium.

Q. 66. Consider three complexes of Ag^+ and their formation contain.

Complex	k_f
$Ag(NH_3)_2^+$	1.6×10^7
$Ag(CN)_2^-$	5.6×10^{18}
$AgBr_2^-$	1.3×10^2

Which statement are true?

(a) $Ag(NH_3)_2^+$ is more stable than $Ag(CN)_2^-$.

(b) Adding a strong acid (HNO_3) to a solution that is 0.01 M in $Ag(NH_3)_2^+$ will tend to dissociate the complex ion into Ag^+ an NH_4^+.

(c) Adding a strong acid (HNO_3) to a solution that is 0.01 *M* in $AgBr_2^-$ will tend to dissociate the complex ion into Ag^+ and Br^-.

(d) To dissolve AgI, one can add either NaCN or HCN as a source of the cyanide-complexing ligand. Fewer moles of NaCN would be required.

(e) Solution A is 0.1 *M* in Br^- and contains the complex ion $AgBr_2^-$. Solution *B* will have more particles of complex ion per particle of Ag^+ than solution A.

Q. 67. There aqueous acidic solutions of acetone and iodine at 303 K are prepared. These solution are described in the following table:

Solution	*(CH_3COCH_3)(M)*	*[I_2](M)*	*[H_3O^+](M)*
I	0.1	0.08	0.02
II	0.3	0.03	0.01
III	0.2	0.01	0.02

Acetone and I_2 react, as shown in Eq. $CH_3COCH_3 + I_2 + H_2O \rightarrow CH_3COC_2 + H_3O + I^-$, and the rate of reaction is given by Eq.

$$\text{rak} = -\frac{-d[CH_3COCH_3]}{dt} = \frac{-d[I_2]}{dt} = k_2' \, CH_3COCH_3][H_3O^+]$$

(a) In which of these solutions is the rate of reaction fastest?

(b) If more water is added to each of these solutions, keeping the temperature constant at 303 K, will the rate of reaction increase, decrease, or remain the same? Will the specific rate constant, k_2', increase, decrease, or remain the same?

(c) If each of these three solution is cooled to 283 K will the rate of reaction increase, decrease, or remain the same? Will the specific rate constant k_2' increase, decrease, or remain the same?

Q. 68. The initial rate of a reaction

$$A + 2B \rightarrow \text{products}$$

is measured for a series of different initial concentrations of *A* and *B*. The following table lists the data obtained.

Experiment Number	*Initial Concentrations* $[A]_0$ *(M)*	$[B]_0$ *(M)*	*Intial Rate (mol · L⁻¹ min⁻¹*
I	0.1	0.1	3.20×10^{-3}
II	0.1	0.2	3.20×10^{-1}
III	0.2	0.1	1.28×10^{-2}
IV	0.3	0.2	2.88×10^{-2}

Determine the rate expression for this reaction

Q. 69. A study of the rate of formation of molecular iodine by the reaction

$$2I(g) \rightarrow I_2(g)$$

in the presence of argon produced the following data:

Experiment Number	Initial Concentrations $[I]_o$ (M)	$[Ar]_o$ (M)	Initial Rate (mol · L^{-1} s^{-1})
1	1.0×10^{-5}	1.0×10^{-3}	8.70×10^{-4}
2	2.0×10^{-5}	1.0×10^{-3}	3.48×10^{-3}
3	4.0×10^{-5}	1.0×10^{-3}	1.39×10^{-2}
4	1.0×10^{-5}	5.0×10^{-3}	4.35×10^{-3}
5	2.0×10^{-5}	5.0×10^{-3}	1.74×10^{-2}

Determine the order and the specific rate constant of this reaction.

Q. 70. The rate of dimerization of gaseous butadiene, $CH_2{=}CH{-}CH{=}CH_3$, has been studied at 599 K. The following data giving the pressure of butadiene as a function of time were obtained.

t (min)	0.0	17.3	42.5	55.08	68.05	90.05	119.0	176.67
P(atm)	0.832	0.661	0.509	0.458	0.417	0.361	0.307	0.235

What is the order of this reaction, and what is the rate constant?

Q. 71. From the values of specific rate constant for the reaction between methyl iodine and sodium ethoxide in ethanole, at 285 K and at 303 K are given as $k_1 = 2.45 \times 10^{-4}$ and $k_2 = 20.8 \times 10^{-4}$ respectively. Calculate the activation energy for this reaction.

Q. 72. The activation energy for the reaction between methyl iodine and sodium ethoxide calculated from the slope of a plot of ln *k* versus $1/T$ using all the data of Q. 71 is 82.6 kJ. What is the specific rate constant for reaction at 293 K?

Q. 73. For the given reaction

$$A + B \rightarrow \text{Products}$$

following data were given

Initial conc. [A] (m/L)	Initial conc. [B] (m/L)	Initial rate (ml^{-1} s^{-1})
0.1	0.1	0.05
0.2	0.1	0.1
0.1	0.2	0.05

(a) Write the rate equation

(b) Calculate the rate constant.

Q. 74. Rate of a reaction $A + B \rightarrow$ products is given below as a function of different initial concentrations of *A* and *B*:

[A] (mol/1)	[B] (mol/1)	Initial rate (ml^{-1} s^{-1})
0.01	0.01	0.005
0.02	0.01	0.010
0.01	0.02	0.005

Determine the order of the reaction with respect to A an with respect to B. What is the half life of A in the reaction?

Q. 75. A first order reaction is 20% complete in 10 minutes. Calculate (i) the specific rate constant of the reaction , and (i) the time taken for the reaction to go to 75% completion.

Q. 76. Two reactions (i) A $\rightarrow$ products (ii) B $\rightarrow$ products, follows first order kinetics. The rate of the reaction (i) is doubled when the temperature is raised from 300 K to 310 K. The half life for this reaction at 310 K is 30 minutes. At the same temperature B decomposes twice as fast as A. If the energy of activation for the reaction (ii) is half that of reaction (i) calculate the rate constant of the reaction (ii) at 300 K.

Q. 77. A first order reaction is 50% complete in 30 minutes at 300 K, and in 10 minutes 320 K. Calculate the reaction rate constant at 300 K and the energy of activation of the reaction in kJ/mole.

Q. 78. A first order reaction $A \rightarrow B$, requires activation energy of 70 kJ mol^{-1}. When a 20% solution of A was kept at 298 K for 20 minutes, 25% decomposition took place. What will be the per cent decomposition in the same time in a 30% solution maintained at 313 K? Assume that activation energy remains constant in this range of temperature.

Q. 79. The rate constant for a isomerisation reaction, $A \rightarrow B$ is 4.5×10^{-3} min min^{-1}. If the initial concentration A is 1 M, calculate the rate of the reaction after 1 h.

Q. 80. The ionization constant of NH_4^+ in water is 5.6×10^{-10} at 298 K. The rate constant for the reaction of NH_4^+ and OH^- to form NH_3 and H_2O at 298 K is 3.4×10^{10} L mol^{-1} s^{-1}. Calculate the rate constant for proton transfer from water to NH_3.

Q. 81. In the Arrhenius equation for a certain reaction, the value of A and E_a (activation energy) are 4×10^{12} sec^{-1} and 98.6 kJ mol^{-1} respectively. If the reaction is of first order, at what temperature will its half-life period be ten minutes?

Q. 82. The rate of first-order reaction is 0.04 mole $litre^{-1}$ s^{-1} at 20 minutes after initiation. Find the half life of the reaction.

Q. 83. The rate constant of a reaction is 1.5×10^7 s^{-1} at 323 K and 4.5×10^7 s^{-1} at 373 K. Evaluate the Arrhenius parameters A and E.

Q. 84. From the following data for the reaction between A and B

	[A] *mol lit^{-1}*	*[B]* *mol lit^{-1}*	*initial rate mole lit^{-1} s^{-1} at*	
			300 K	320 K
I	2.5×10^{-4}	3.0×10^{-5}	5.0×10^{-4}	2.0×10^{-3}
II	5.0×10^{-4}	6.0×10^{-5}	4.0×10^{-3}	
III	1.0×10^{-3}	6.0×10^{-5}	1.6×10^{-2}	

Calculate

(i) the order of the reaction with respect to A and with respect to B,

(ii) the rate constant at 300 K.

(iii) the energy of activation, and

(iv) the pre-exponential factor.

Q. 85. The rate constant for the first order decomposition of a certain reaction is described by the equation

$$\log k(s^{-1}) = 14.34 - \frac{1.25 \times 10^4 \text{ K}}{T}$$

(i) what is the energy of activation for this reaction?

(ii) At what temperature will its half life period be 256 minutes?

Q. 86. The time required for 10% completion of a first order reaction at 298 K is equal to that required for its 25% completion at 308 K. I the prexponential factor for the reaction is $3.56 \times 10^9 \text{ s}^{-1}$, calculate its rate constant at 318 K and also the energy of activation.

Q. 87. The gas phase decomposition of dimethyl ether follows first order kinetics

$$CH_3—O—CH_{3(g)} \rightarrow CH_{4(g)} + H_{2(g)} + CO_{(g)}$$

The reaction is carried out in a constant volume container at 773 K and has a half life of 14.5 minutes. Initially only dimethyl ether is present at a pressure of 0.4 atmosphere. What is the total pressure of the system after 12 minuttes? Assume ideal gas behavior.

Chapter **8**

CHEMICAL EQUILIBRIUM

FORMULAE

Active mass of x,

$$[x] = \frac{\dfrac{\text{Mass of } x \text{ g}}{\text{Mol. mass } x \text{ g mol}^{-1}}}{\text{Volume (L)}}$$

For the reaction, $aA + bB \rightleftharpoons cC + dD$

$$K_c = \frac{[C]^c\,[D]^d}{[A]^a\,[B]^b} \text{ and } K_P = \frac{(P_C)^c\,(P_D)^d}{(P_A)^a\,(P_B)^b}$$

K_P and K_c are the equilibrium constants when the concentrations are expressed in terms of partial pressure and molarity respectively.

Also, $K_P = K_c(RT)^{\Delta n}$

Where, $\Delta n = (c + d) - (a + b)$

Unit of $K_c = (\text{mol } L^{-1})^{\Delta n}$

Unit of $K_P = (\text{atm})^{\Delta n}$

$$\log \frac{K_{P_2}}{K_{P_1}} = \frac{\Delta H}{2.303\,R}\left[\frac{T_2 - T_1}{T_1 T_2}\right]$$

$$\log \frac{K_{c_2}}{K_{c_1}} = \frac{\Delta E}{2.303\,R}\left[\frac{T_2 - T_1}{T_1 T_2}\right]$$

Equilibrium constant depends upon

(i) Temperature
(ii) Stoichiometry of reaction
(iii) Mode of representing the chemical equation

$$\Delta G = -2.30 \text{ RT} \log K_p$$

$$\alpha = \frac{D-d}{(n-1)d}$$

where α = Degree of dissociation = $\dfrac{\text{Molesdissociated}}{\text{Total moles present before dissociation}}$

D = Density of reactant initially
d = Density of the product at equilibrium
n = No. of particles obtained on dissociation

For heterogeneous system the active mass of a pure solid is constant irrespective of amount of solid. It is taken as unity.

Reverse Reactions

All chemical reactions can be divided into two types namely 1. reversible and 2. irreversible reactions.

Reactions in which the product can recombine under a similar set of conditions to give back the reactants are called reversible reactions. Reactions which take place simultaneously in both forward and backward directions are known as reversible reactions. example: $H_2(g) + I_2(g) \rightleftharpoons 2HI(g)$

Here two reactions take place at same time.

$$H_2(g) + I_2(g) \rightarrow 2HI\ (g)$$

is a forward reaction and

$$2\ HI(g) \rightarrow H_2(g) + I_2(g)$$

is a backward reaction. If the reaction is allowed to take place in a closed vessel both the reactants and the products will always be present in the vessel.

The following are some more examples of reversible reactions.

$$3\ Fe\ (s) + 4\ H_2O(g) \rightleftharpoons Fe_3O_4(s) + 4\ H_2(g)$$

$$N_2(g) + 3H_2(g) \rightleftharpoons 2NH_3(g)$$

$$2SO_2(g) + O_2(g) \rightleftharpoons 2SO_3(g)$$

$$PCl_5(g) \rightleftharpoons PCl_3(g)\ Cl_2(g)$$

$$N_2(g) + O_2(g) \rightleftharpoons 2\ NO(g)$$

$$CH_3COOH(l) + C_2H_5OH(l) \rightleftharpoons CH_3COOC_2H_5(l) + H_2O(l)$$

Note: $\Delta G > 0$ for reversible process.

Irreversible Reactions

Reactions which proceed only in one direction till one of the reactants is used up are called irreversible or unidirectional reactions example; $2KClO_3(s) \rightarrow 2KCl(s) + 3O_2(g)$

Here, Potassium chloride and oxygen cannot react together to form potassium chlorate.

The following are some more examples of irreversible reactions.

$$2Mg(s) + O_2(g) \rightarrow 2\ MgO(s)$$

$$C_2H_5OH(l) + 3O_2(g) \rightarrow 2CO_2(g) + 3H_2O(g)$$

$$NH_4NO_2(s) \rightarrow N_2(g) + 2H_2O(g)$$

X-Law of Mass Action:

According to the **Law of Mass Action** the rate at which a substance reacts is proportional to its active mass, and hence the rate of a chemical reaction is proportional to the product of the active masses of different substances involved in the reaction. Eg. For $A + B \rightleftharpoons C + D$

$$V_f \propto [A][B]$$

or $V_f = K_f[A][B]$ where K_f is velocity constant of forward reaction.

Similarly, $V_b = [C][D]$

or $V_b = K_b[C][D]$ where K_b is velocity constant of backward reaction.

$\therefore$ $\dfrac{K_f}{K_b} = K_c = \dfrac{[C][D]}{[A][B]}$ where K_c is called Equilibrium Constant.

Active mass, denoted by square brackets, means molar concentration (molarity) (i.e. moles/litre)

Equilibrium

Most physical and chemical phenomenas like solubility, reactions, ionization, dissociation etc. proceed to some equilibrium state where concentrations of reactants and products do not change in time.

It is attained when rate of forward reactions becomes equal to rate of backward reaction.

Mathematically, for a reaction $aA + bB + \ldots \rightleftharpoons cC + dD + \ldots\ldots$ it is found experimentally, that,

$$\frac{[C]^c\,[D]^d}{[A]^a\,[B]^b} = K_c$$

K_c called the Equilibrium Constant (is constant at a given temperature). This experimental result in known as the **Equilibrium Law:**

Equilibrium of Gasesous Reactions

There are two types of gaseous reactions:

(a) Those in which there is no change in volume. eg. : $H_2 + I_2 \rightleftharpoons 2HI$
(b) Those in which there is a change in volume. eg. : $PCl_5 \rightleftharpoons PCl_3 + Cl_2$

In the case of gases, active mass is taken proportional to partial pressure.

Since the partial pressure of a gas is proportional to its concentration, we can express the equilibrium constant of gaseous reaction in terms of partial pressures. For the above reaction, K_p (equilibrium constant in terms of partial pressures) can be written as:

$$\frac{[P_C]^c\,[P_D]^d}{[P_A]^a\,[P_B]^b} = K_p,$$ where P_A, P_B, P_C & P_D are partial pressures of A, B, C, D respectively.

As an example, for the reaction $H_2(g) + I_2(g) \rightleftharpoons 2HI(g)$, we can write either

$$K_c = \frac{[HI]^2}{[H_2][I_2]} \text{ or } K_p = \frac{[pHI]^2}{[pH_2][pI_2]}$$

(a) Now, In the formation of hydriodic acid, let a and b be the initial molar concentrations of H_2 and I_2 and x be the no. of moles of each that combine to produce $2x$ moles of HI and let, V be the total volume of the vessel in litres in which the reaction takes place. At equilibrium:

$$\begin{array}{cccc} H_2 & + & I_2 \rightleftharpoons & 2HI \\ (a-x) & & (b-x) & 2x \end{array}$$

$$\therefore [H_2] = \frac{a-x}{V} \text{ moles litres}^{-1},\ [I_2] = \frac{b-x}{V} \text{ moles litres}^{-1} \ \&\ [HI] = \frac{2x}{V} \text{ moles litres}^{-1}$$

$$\text{Hence, } K_c = \frac{[HI]^2}{[H_2][I_2]} = \frac{(2x)^2 \times V \times V}{V^2 \times (a-x) \times (b-x)} = \frac{4x^2}{(a-x)(b-x)}$$

Thus K_c is independent of the volume of the vessel in which the reaction takes place.

(b) For the dissociation of PCl_5 where 1 mole of pentachloride is vaporised in a vessel of the total volume of V litres. If x is its degree of dissociation, then at equilibrium

$$\underset{(1-x)}{PCl_5} \rightleftharpoons \underset{x}{PCl_3} + \underset{x}{Cl_2}$$

$\therefore$ $[PCl_5] = \dfrac{1-x}{V}$ moles litre^{-1}, and $[PCl_3] = [Cl_2] = x$ moles litre^{-1}

Hence, $K_c = \dfrac{[PCl_3]\times[Cl_2]}{[PCl_5]} = \dfrac{x^2}{(1-x)V}$

Thus K_c involves the volume of the vessel in which the reaction takes place.

Characteristics of Chemical Equilibrium

1. Chemical equilibrium can be approached from either direction.
2. An equilibrium may be shifted one way or the other by changing the conditions like pressure, temperature concentration, etc.
3. Chemical equilibrium is dynamic in nature.
4. Chemical equilibrium at a given temperature is characterised by contancy of certain observable properties such as pressure, concentration, density or colour.

Kinds of Chemical Equilibria

A chemical equilibrium may be homogeneous or heterogeneous.

A homogeneous equilibrium is one in which the reactants and the products are all in one phase (gas or liquid)

Examples of homogeneous equilibrium are:

$$H_2(g) + I_2(g) \rightleftharpoons 2\ HI(g)$$

$$PCl_5(g) \rightleftharpoons PCl_3(g) + Cl_2(g)$$

$$N_2(g) + 3H_2(g) \rightleftharpoons 2NH_3(g)$$

$$CH_3COOH(l) + C_2H_5OH(l) \rightleftharpoons CH_3COOC_2H_5(l) + H_2O(l)$$

A heterogeneous equilibrium is one in which the components are in different phases. (solid, liquidor gas) Examples of heterogeneous equilibrium are:

$$CaCO_3(s) \rightleftharpoons CaO(s) + CO_2(g)$$

$$3Fe(s) + 4H_2O(g) \rightleftharpoons Fe_3O_4(s) + 4H_2(g)$$

$$Zn(s) + CuSO_4(aq) \rightleftharpoons Cu(s) + ZnSO_4(aq)$$

Factors Influencing the Equilibrium State

The factors influencing the equilibrium state are

1. Concentration
2. Pressure
3. Temperature and
4. Catalyst

Characteristics of Equilibrium Constant

1. The equilibrium constant, K has a constant value at the given temperature.
2. Its values also remains constant under the given conditions of temperature and pressure irrespective of the concentration of the reactants and the products
3. Its value depends upon the nature and temperature of the reaction and is independent of the presence of a catalyst or inert materials.
4. The value of K tells us the extent to which the reaction can proceed in a particular direction at a given temperature.
   If K is large, the equilibrium mixture will contain a high proportion of products and the reaction will be nearly complete at equilibrium.

Relation between K_c & K_p

$$K_c = \frac{K_P}{[RT]^{\Delta n}} \quad \text{or} \quad K_p = K_c[RT]^{\Delta n},\ R = \text{gas constant} = 0.0821 \text{ litre-atm mol}^{-1}\text{ K}^{-1}.$$

Where K_c and K_p are the equilibrium constants concentration (molarity) and partial pressures respectively) and Δn is the difference between the number of product molecules (moles) and reactant molecules (mole) in the balanced equation. The value of Δn is calculated for say the hypothetical reaction.

$aA + bB \rightleftharpoons cC + dD$ as $\quad \Delta n = (c + d) - (a + b)$

Also if $\quad \Delta n = 0,\ K_P = K_c$, and if $\quad \Delta n < 0,\ K_P < K_c$

and, $\Delta n > 0,\ K_P > K_c$

Unit of K_P = (Unit of pressure)$^{\Delta n}$ = (atm)$^{\Delta n}$

Unit of K_c = (Unit of concentration)$^{\Delta n}$ = (mol. litre^{-1})$^{\Delta n}$

Equilibrium Constant for Homogeneous liquid Systems

Consider esterfication of Ethyl alcohal.

$$\overset{a}{CH_3COOH} + \overset{b}{C_2H_5OH} \rightleftharpoons \overset{0}{CH_3COOC_2H_5} + \overset{0}{H_2O}$$

At equilibrium:— $\quad (a\text{-}x) \quad (b\text{-}x) \quad (x) \quad (x)$

If the total volume is 'v' litres, then the molar concentration per unit volume of all the constituents at the equilibrium will be—

$$[CH_3COOH] = \frac{a-x}{v};\ [C_2H_5OH] = \frac{b-x}{v}$$

$$[CH_3COO_2H_5] = x/v : [H_2O] = x/v,$$

$$\text{Whence, } K_c = \frac{x/v \times x/v}{(a-x)\,v \times (b-x)v} = \frac{x^2}{(a-x)(b-x)} \quad \because K_c = \frac{[CH_3COOC_2H_5][H_2O]}{[CH_3COOH][C_2H_5OH]}$$

Equilibrium Constant for Heterogeneous Equilibria

Since the concentration of a solid or a pure liquid is constant, we can write modified equilibrium constants for heterogeneous equilibria excluding these constant concentration. Thus, for the equilibrium,

$$\underset{(s)}{3Fe} + \underset{steam}{4H_2O} \rightleftharpoons \underset{(s)}{Fe_3O_4} + 4H_2(g)$$

$$K_p = \frac{[H_2]^4}{[H_2O]^4}$$

Equilibrium Constant (Thermodynamics)

(i) The standard Gibbs energy ($\Delta G°$) of a reaction is the free energy change when each reactant and produce has an activity (or molar concentration) of one. The relationship between the standard free energy and the equilibrium constant is

$$\log k = \frac{\Delta G°}{2.303\,RT} \quad \text{or} \quad \Delta G° = -RT \ln K$$

(ii) The standard enthalpy change for the forward reaction can be obtain form the variation of ln K with temperature.

$$\left(\frac{d \ln K}{dT}\right) = \frac{\Delta H°}{RT^2}$$

(iii) The following equation describes the effect of temperature of equilibrium constant.

$$\log \frac{K_2}{K_1} = \frac{\Delta H}{2.303\,E}\left[\frac{1}{T_1} - \frac{1}{T_2}\right]$$

where, ΔH = heat of reaction
K_2 = equilibrium constant at temperature T_2
K_1 = equilibrium constant at temperature T_1
R = molar gas constant.

Relation between degree of dissociation and moles at equilibrium

For the reaction, $2HI \rightleftharpoons H_2 + I_2$

$$\text{Degree of dissociation } (\alpha) = \frac{\text{Mole dissociated}}{\text{Tota moles present before dissociation}}$$

We should made use of 'α' very carefully. The reaction ratio if HI : H_2 : I_2 :: 2 : 1 : 1.

Case I.

$2HI$	$\rightleftharpoons$	H_2	+	I_2	
1		0		0	(Moles initially)
$1 - x$		$\frac{x}{2}$		$\frac{x}{2}$	(Moles at equil.)

$$\therefore \quad \alpha = \frac{x}{1} = x$$

Case II.

$2HI$	$\rightleftharpoons$	H_2	+	I_2	
1		0		0	(Moles initially)
$1 - 2x$		x		x	(Moles at equil.)

$$\therefore \quad \alpha = \frac{2x}{1} = 2x$$

Case III. $2HI \rightleftharpoons H_2 + I_2$

	$2HI$	H_2	I_2	
	a	0	0	(Moles initially)
	$a - x$	$\frac{x}{2}$	$\frac{x}{2}$	(Moles at equil.)

$\therefore \quad \alpha = \frac{x}{a}$

Case IV. $2HI \rightleftharpoons H_2 + I_2$

	$2HI$	H_2	I_2	
	a	0	0	**(Moles initially)**
	$a - 2x$	x	x	**(Moles at equil.)**

$\therefore \quad \alpha = \frac{2x}{a}$

Case V. $2HI \rightleftharpoons H_2 + I_2$

	$2HI$	H_2	I_2	
	a	0	0	**(Moles initially)**
	$a - ax$	$\frac{ax}{2}$	$\frac{ax}{2}$	**(Moles at equil.)**

$\therefore \quad \alpha = \frac{ax}{a} = x$

Calculation of degree of dissociation from vapour density: Let us consider the dissociation of AB which occurs as follows:

AB	$\rightleftharpoons$	A	+	B	
1		0		0	(Initially)
$1 - x$		x		x	(At equil.)

Now since density is inversely proportional to volume, therefore,

$$D \propto \frac{1}{V}$$

and $$d \propto \frac{1}{(1+x)V}$$

where D and d are densities of the reactants before and after equilibrium. Thus

$$\frac{D}{d} = 1 + x$$

or $$x = \frac{D-d}{d}$$

If 1 mole of a substance dissociates into 'n' particles, then the above equation may be written as:

$$x = \frac{D-d}{(n-1)d}$$

PROBLEMS

Q. 1. At a certain temperature the vapour density of N_2O_4 is 24.8. Calculate the percentage dissociation at this temperature.

Q. 2. The equilibrium constant in the reaction between ethyl alcohol and acetic acid to form ethyl acetate and water is 4 at 298 K. What weight of ethyl acetate will be obtained from 92 g. of alcohol and 120 g. of acetic acid, the reaction being carried out at 298 K in a closed vessel?

Q. 3. If one mole of acetic acid and one mole of ethyl alcohal are mixed and the reaction proceeds to equilibrium, the concentration of the acid and water are found to be 1/3 and 2/3 moles. If one mole of ester and three moles of water are mixed. How much ester is present when equilibrium is reached?

Q. 4. A mixture of N_2 and H_2 in the ratio 1 : 3 by volume was subjected to a pressure of 30 atm and a temperature of 723 K. After equilibrium was established, cooling and analysis indicated that the mixture contained 6% by volume of NH_3. Calculate K_p assuming ideal behaviour if the gases.

Q. 5. The value of K_p for the reaction, $CO(g) + H_2O(g) \rightleftharpoons CO_2(g) + H_2(g)$ is 0.63. A mixture of 1 mole of water vapour and 3 moles of CO is allowed to come to equilibrium at a total pressure of 2 atm. Calculate the moles of hydrogen that are present at equilibrium and partial pressure of each gas at equilibrium.

Q. 6. Ammonium hydrogen sulphide dissociates as $NH_4HS(s) \rightleftharpoons NH_3(g) + H_2S(g)$. The equilibrium pressure is 0.66 atm. At 298 K. Find the value of K_p.

Q. 7. Calculate the equilibrium constant at 523°K and 1 atm. pressure when PCl_5 is decomposed into PCl_3 and Cl_2?

Q. 8. A first reaction is 40% completed at the end of 50 minutes. What is the value of rate constant. In how many minutes will the reaction be 80% completed.

Q. 9. Rs. 75% of a reaction of the first order was completed in 72 minutes. When was it half completed?

Q. 10. A 1st order reaction is 25% completed in 30 minutes. Calculate (a) specific reaction rate and (b) time required for 75% of reaction to be completed?

Q. 11. The 1st order rate constant for the decomposition of H_2O_2 is 4.29×10^{-2} min^{-1}. What is the half life period for this decomposition?

Q. 12. 20% of N_2O_4 molecules air dissociated in a sample of gas at 300 K and 1 atm. Calculate the density of the equilibrium mixture.

Q. 13. At 273 K and 1 atm pressure '*a*'; litre of N_2O_4 decomposes of NO_2 as:

$$N_2O_4(g) \rightleftharpoons 2NO_2(g)$$

To what extent has the decomposition proceeded when the original volume is 25% less than that of existing volume.

Q. 14. At 1085 K, the specific rate constant of the reaction is 3.76×10^3 $hour^{-1}$. At what temperature will it be 1.16×10^{-4}, if energy of activation is 68400 Cal. $Mole^{-1}$?

Q. 15. For the reaction, $CO(g) + 2H_2(g) \rightleftharpoons CH_3OH(g)$, hydrogen gas is introduced in to a five-litre flask at 600 K, containing 0.2 mole of CO(g) and a catalyst until the pressure is 4.92

atm. At this point 0.1 mole of $CH_3OH(g)$ is formed. Calculate the equilibrium constants, K_p and K_c.

Q. 16. In the Arrhenius equation for a certain reaction, the value of A and E_a (activation energy) are $4 \times 10^{13}\ s^{-1}$ and 98.6 kj mol^{-1} respectively. If the reaction is of first order, at what temperature will its half-life period be ten minutes?

Q. 17. A first order reaction is 50% complete in 30 minutes at 300 K and in 10 minutes at 320 K. Calculate the reaction rate constant at 300 K and the energy of activation of the reaction in K mol^{-1}.

Q. 18. The equilibrium constant K_p of the reaction, $2SO_2(g) + O_2(g) \rightleftharpoons 2SO_3(g)$, is 900 atm at 800 mixture containing SO_3 and O_2 having initial partial pressures of 1 and 2 atm respectively is heated at constant volume of equilibrate. Calculate the partial pressure of each gas at 800 K.

Q. 19. Freshly precipitated aluminium and magnesium hydroxides are stirred vigorously in a buffer solution containing 0.25 mol $litre^{-1}$ of ammonium chloride and 0.05 mol $litre^{-1}$ of ammonium hydroxide. Calculate the concentration of aluminium and magnesium ions in solution. $K_b[NH_4OH] = 1.8 \times 10^{-5}$, $K_{sp}[Mg(OH)_2] = 6 \times 10^{-10}$, $K_{sp}[Al(OH)_3] = 6 \times 10^{-32}$

Q. 20. The equilibrium constant of the reaction $H_2 + I_2 = 2HI$ at 693 K is $K_c = 50.25$. Calculate the yield of hydrogen iodide if 0.846×10^{-3} kg I_2 and 0.0212×10^{-3} kg H_2 have been placed in a vessel having a capacity of $10^{-3}\ m^3$.

Q. 21. N_2O_4 is 25% dissociated at 310 K and one atmosphere pressure. Calculate: (i) K_p and (ii) the percentage dissociation at 0.1 atmosphere and 310 K.

Q. 22. The equilibrium constant of the reaction $A_2 + B_2 \rightleftharpoons 2AB(g)$ at 373 K is 50. If a one litre flask containing one mole of A is connected to a two litre flask containing two moles of B, how many moles of AB will be formed at 310 K.

Q. 23. While studying the decomposition of gaseous N_2O_5, it is observed that a plot of logarithm of its partial pressure versus time is linear. What kinetic parameters can be obtained from the observation.

Q. 24. A first order reaction is 20% complete in 10 minutes. Calculate (i) the specific rate constant of the reaction and (ii) the time taken for the reaction to go to 75% completion.

Q. 25. One mole of N_2 and 3 moles of PCl_5 are placed in 100 litres vessel heated to 500 K. The equilibrium pressure is 2.05 atm. Assuming ideal behaviour, calculate the degree of dissociation for PCl_5 and K_p for the reaction.

Q. 26. Rate of a reaction $A + B \rightarrow$ products is given below as a function of different initial concentration A and B.

A *mol lit.*$^{-1}$	B *mol lit.*$^{-1}$	*initial rate mol lit.*$^{-1}$ *min*$^{-1}$
0.01	0.01	0.005
0.02	0.01	0.010
0.01	0.02	0.005

Determine the order of the reaction with respect to A and w.r.t B. What is the half life of 'A' in the reaction.

Q. 27. One mole of nitrogen is mixed with three moles of hydrogen in a 4 litre container. If 0.25% of nitrogen is converted to ammonia by the following reaction

$$N_2 + 3H_2 \rightleftharpoons 2NH_3$$

Calculate the equilibrium constant (K_C) in concentration units. What will be the value of K_C for following equilibrium

$$\frac{1}{2} N_2 + \frac{3}{2} H_2 \rightleftharpoons NH_3$$

Q. 28. At 900 K and 1 atm pressure SO_3 is partially dissociated into SO_2 and O_2. The density of the equilibrium mixture is 0.925 g L^{-1}. Calculate the degree of dissociation of SO_3.

Q. 29. The K_p for the reaction: $N_2O_4 \rightleftharpoons 2NO_2$, is 640 mm at 775 K. Calculate the percentage dissociation of N_2O_4 at equilibrium pressure of 160 mm. At what pressure, the dissociation will be 50%?

Q. 30. The value of the rate constant of a certain reaction is 0.1 lit mol^{-1} sec^{-1}. What is the value in ml mol^{-1} min^{-1}?

Q. 31. **(a)** A closed vessel having a capacity of 0.05 m^3 contains one mole at substance A and 12 moles of substance B. The reaction $A + 4B = D$ results in equilibirium. The total pressure in the equilibrium state of the system, is 4.51×10^5 Pa. Calculate the equilibrium concentration of all reactants at 298 K. The system obeys the laws of ideal gaseous state.

(b) Calculate the equilibrium constant K_c of the reaction $A + 4B = D$ if the volume of the reaction vessel is $V = 0.05$ m^3.

Q. 32. Calculate the equilibrium yield of NO_2 in the reaction $N_2O_4 \rightleftharpoons 2NO_2$ at 1.0133×10^5 Pa and 298 K. Given,

Substance	NO_2	N_2O_4
$\Delta H^\circ_{f,\ 298}$, kJ mol^{-1}	33.89	9.37
S°_{298}, J $mole^{-1}$ K^{-1}	240.45	304.3
$\Delta H^\circ_{298} = 58.41$ kJ,	$\Delta S^\circ_{298} = 176.6$ J	

Q. 33. Express the equilibrium constant K_P of the reaction $H_2 + 0.5O_2 = H_2O(g)$ in terms of total pressure P and degree of dissociation α of the water vapour.

Q. 34. Determine the equilibrium constant K_P of the reaction of ammonia oxidation.

$$NH_2 + \frac{3}{4}O_2 = \frac{1}{2}N_2 + \frac{3}{2}H_2O$$

at 1000 K. Given, the following equilibrium constants of the reactions of ammonia and water formation from simple substances at 1000 K:

$$\frac{1}{2}N_2 + \frac{3}{2}H_2\text{—}NH_3 \qquad K_{P_1} = 5.69 \times 10^{-4}$$

$$\frac{1}{2}O_2 + H_2 == H_2O_{(g)} \qquad K_{P_2} = 1.18 \times 10^{10}$$

(pressure is given in atmosphere).

Q. 35. For the system

$$SO_3(g) \rightleftharpoons SO_2(g)\ \frac{1}{2}O_2(g)$$

at 100 K, $K = 0.45$. Sulfur trioxide, originally at 1.00-atm pressure, partial dissociation to SO_2 and O_2 at 100 K. What is its partial pressure at equilibrium.

Q. 36. At a certain temperature, the reaction

$$Xe(g) + 2F_2(g) \rightleftharpoons XeF_4(g)$$

Gives a 50.0% yield of XeF_4, starting with Xe ($P_{Xe} = 0.20$ atm) and F_2 ($P_{F_2} = 0.40$ atm). Calculate K at this temperature. What must the initial pressure of F_2 be to convert 75.0% of the xenon to FeF_4?

Q. 37. A student studies the equilibrium

$$I_2(g) \rightleftharpoons 2I(g)$$

at a high temperature. She finds the total pressure at equilibrium is 40% greater than is was originally, when only I_2 was present. What is K for this reaction at temperature?

Q. 38. Benzaldehyde, a flavoring agent, is obtained by the dehy-drogenation of benzyl alcohol.

$$C_6H_5H_2OH(g) \rightleftharpoons C_6H_5CHO(g) + H_2(g)$$

K for the reaction at 523 K is 0.56. If 1.50 g of benzyl alcohol is placed in a 2.0-L flask and heated to 523 K.

(a) what is the partial pressure of the benzaldehyde when equilibrium is established?

(b) how many grams of benzyl and alcohol remain at equilibrium?

Q. 39. Ammonia can decompose into its constituent element according to the reaction

$$2NH_3(g) \rightleftharpoons N_2(g) + 3H_2(g)$$

The equilibrium constant for the decomposition at a certain temperature is 2.5. Calculate the partial pressures all the gases at equilibrium if ammonia with a pressure of 1.00 atm is sealed in a 3.0 L-flask.

Q. 40. Hydrogen iodide gas decomposes to hydrogen gas and iodine gas:

$$2HI(g) \rightleftharpoons H_2(g) + I_2(g)$$

To determine the equilibrium constant of the system, identical one-liter glass bulbs are filled with 3.20 g of HI and maintained at a certain temperature. Each bulb is periodically opened and analyzed for iodine formation by titration with sodium thiosulpate, $Na_2S_2O_3$.

$$I_2(aq) + 2S_2O_3^{2-}(aq) \rightarrow S_4O_6^{2-}(aq) + 2I^-(aq)$$

It is determined that when equilibrium is reached, 37 ml of 0.2 M Na_2SO_3 is required to titrate the iodine. What is K at the temperature of the experiment?

Q. 41. The following equilibrium was studied by analyzing the equilibrium mixture for the amount of H_2S produced.

$$Sb_2S_3(s) + 3H_2(g) \rightleftharpoons 2Sb(s) + 3H_2S(g)$$

A vessel whose volume was 2.50 L was filled with 0.01 mol of antimony(III) sulfide, Sb_2S_3, and 0.01 mol H_2. After the mixture came to equilibrium the closed vessel at 713 K, the gaseous mixture was removed, and the hydrogen sulfide was completely with the H_2S to precipitate lead(II) sulfide, PbS. If 1.029 g PbS was obtained, what is the value of K_c at 713 K.

Q. 42. Phosphorous (V) chloride, PCl_5, dissociated on heating to give phosphorous (III) chloride, PCl_3, and chlorine.

$$PCl_5(g) \rightleftharpoons PCl_3(g) + Cl_2(g)$$

A closed 2-L vessel initially contains 0.01 mol PCl_5. What is the total pressure 523 K when equilibrium is achieved? The value of K_c at 523 K is 4.15×10^{-2}.

Q. 43. Which of the following is/are true about a 0.10 *M* solution of a strong axis, HY?

(a) $[Y^-]$ **(b)** $[HY] = 0.1$ M

(c) $[H^+] = 0.1$ M **(d)** pH = 1.0

(e) $[H^+] + [Y^-] = 0.2$ M

Q. 44. The following data are for the system

$$A(g) \rightleftharpoons 2B(g)$$

Time (s)	0	20	40	60	80	100
P_A (atm)	1.0	0.83	0.72	0.65	0.62	0.62
P_B (atm)	0.0	0.34	0.56	0.70	0.76	0.76

Prepare a graph of P_A and P_B versus time and use it to answer the following questions:

(a) Estimate P_A and P_B after 30 s.

(b) Estimate P_A after 150 s.

(c) Estimate P_B when P_A = 0.7 atm.

Q. 45. For the reaction

$$C(s) + CO_2(g) \rightleftharpoons 2CO(g)$$

K = 1.68 at 1273 K. If one starts with 0.3 atm of CO_2 and 12 g of C at 1273 K, will the equilibrium mixture contain

(a) mostly CO_2? **(b)** mostly CO?

(c) roughly equal amounts of CO_2 and CO? **(d)** only C?

Q. 46. Consider the graph below:

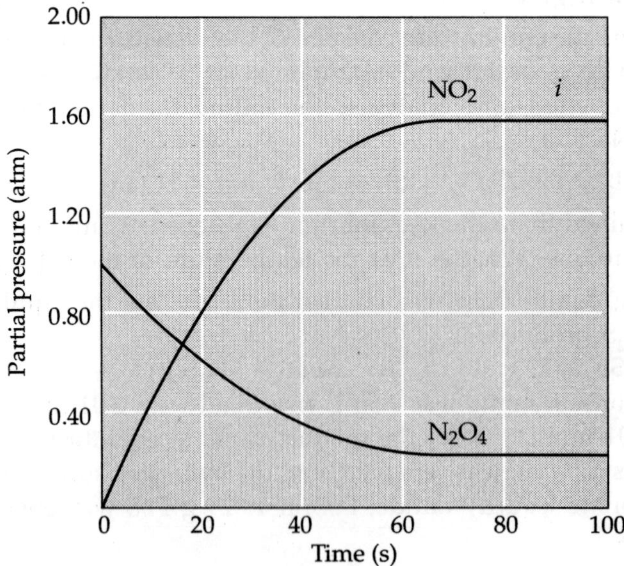

If after 100 s have elapsed the partial pressure of N_2O_4 is increased to 1.0 atm, what will the graph for N_2O_4 look like beyond 100 s?

Q. 47. The degree of dissociation is 0.4 at 400 K and 1.0 atm for the gaseous reaction

$$PCl_5 \rightleftarrows PCl_3 + Cl_{2+}$$

Assuming ideal behaviour of all the gases, calculate the density of equilibrium mixture at 400 K and 1.0 atmosphere. (Atomic mass of P = 31.0 and Cl = 35.5)

Q. 48. When 3.06 g of solid NH_4HS is introduced into a two-evacuated flask at 300 K, 30% of the solid decomposes into gaseous ammonia and hydrogen sulphide. (i) Calculate K_c and K_p for the reaction at 300 K. (ii) What would happen to the equilibrium when more solid NH_4HS is introduced into the flask?

Q. 49. The optical rotation sucrose in presence of dil HCl at various intervals is given in following table

T sec.	0	600	1200	2400	6000	∞
rotation (deg)	32.4	28.8	25.5	19.6	6.7	–14

show that the reaction is of 1st order and find when the solution will become optically inactive?

Q. 50. 5 ml. of ethyl acetate was added to a flask containing 100 ml. of 0.1 HCl placed in a thermostat, maintained at 303 K. 5 ml. of the reaction mixture was withdrawn at different intervals of time and after chillling, titrated against a standard alkali the following data were obtained.

Time [min]	0	75	119	183	∞
ml (alkali)	9.62	12.10	13.10	14.75	21.05

From the above data show that the hydrolysis of ethyl acetate is a first order reaction.

Q. 51. A sample of air consisting of N_2 and O_2 was heated 2500 K until the equilibrium

$$N_{2(g)} + O_{2(g)} \rightleftharpoons 2NO_{(g)}$$

was established with an equilibrium constant $K_c = 2.1 \times 10^{-3}$. At equilibrium, the mole % of NO was 1.8. Estimate the initial composition of air in mole fraction of N_2 and O_2.

Q. 52. For the reaction

$$[Ag(CN)_2] \leftrightarrow Ag^+ + 2CN^-$$

the equilibrium constant at 298 K is 4.0×10^{-19}.

Calculate the silver ion concentration in a solution which was originally 0.10 molar in KCN and 0.03 molar in $AgNO_3$.

Q. 53. The progress of the reaction $A \leftrightarrow nB$, with time is presented in the figure Determine

(i) the value of n,

(ii) the equilibrium constant, K, and

(iii) the initial rate of conversion of A.

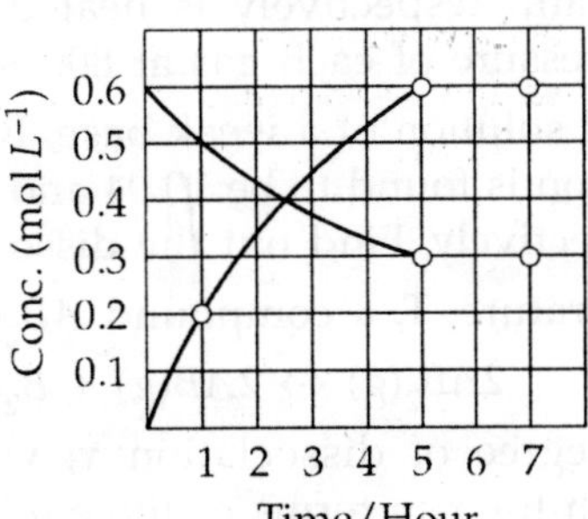

Q. 54. For the reaction,

$$CO(g) + 2H_2(g) \leftrightarrow CH_3OH(g),$$

Hydrogen gas is introduce into a five-litre flask at 600 K, containing 0.2 mole of CO(g) and a catalyst, until the pressure is 4.92 atm. At this point 0.1 mole of $CH_3OH(g)$ is formed. Calculate the equilibrium constant, K_p and k_c.

Q. 55. N_2O_4 is 25% dissociated at 310 K and one atmosphere pressure. Calculate (i) K_p and (ii) the percentage dissociation at 0.1 atmosphere and 310 K.

Q. 56. At a certain temperature equilibrium constant (K_c) is 16 for the reaction

$$SO_2(g) + NO_2(g) \rightleftharpoons SO_3(g) + NO(g)$$

If we take one mole of each of all the four gases in a one litre container, what would be the equilibrium concentrations of NO and NO_2?

Q. 57. The equilibrium constant of the reaction $A_2(g) + B_2(g) \rightleftharpoons 2AB(g)$ at 373 K is 50. If a one litre flask containing once mole of A_2 is connected to a two litre flask containing of two mole of B_2, how many mole of *AB* will be formed at 373 K?

Q. 58. One mole of nitrogen is mixed with three moles of hydrogen moles of hydrogen in a 4 litre container. If 0.25 per cent of nitrogen is converted to ammonia by the following reaction

$$N_2(g) + 3H_2(g) \rightleftharpoons 2NH_3(g)$$

Calculate the equilibrium constant (K_c) in concentration units. What will be the value of K_c for the following equilibrium?

$$\frac{1}{2}N_2(g) + \frac{3}{2}H_2(g) \rightleftharpoons NH_3(g)$$

Q. 59. 0.15 mole of CO taken in a 2.51 flask is maintained at 750 K along g with a catalyst so that the following reaction can take place:

$$CO_{(g)} + 2H_{2(g)} \Leftrightarrow CH_3OH_{(g)}$$

Hydrogen is introduced until the total pressure of the system is 8.5 atmosphere at equilibrium and 0.08 mole of methanol is formed. Calculate (i) K_p and K_c and (ii) the final pressure if the same amount of CO and H_2 as before are used, but with no catalyst so that the reaction does not take place.

Q. 60. The equilibrium constant K_p of the reaction.

$$2SO_2(g) + O_2(g) \rightleftharpoons 2SO_3(g),$$

is 900 atm. at 800 K. A mixture containing SO_3 and O_2 having initial partial pressure of 1 and 2 atm respectively is heated at constant volume to equilibrate. Calculate the partial pressure of each gas at 800 K.

Q. 61. A 40.0 ml solution of a weak base, BOH is titrated with 0.1 N HCl solution. The pH of the solution is found to be 10.04 and 9.14 after the addition of 5.0 ml and 20.0 ml of the acid respectively. Find out the dissociation constant of the base.

Q. 62. At temperature, T, a compound $AB_2(g)$ dissociates according to the reaction

$$2AB_2(g) \leftrightarrow 2AB(g) + B_2(g)$$

with a degree of dissociation *x*, which is small compared with unity. Deduce the expression for *x* in terms of the equilibrium constant. K_p and the total pressure, *P*.

Q. 63. Match the following if the molecular weight of *X*, *Y* and *Z* are same

Boiling Point	K_b
x 100	0.68
x 27	0.53
z 253	0.98

Q. 64. How many gram-mole of HCl will be required to prepare one litre of buffer solutioning (containing NaCN and HCl) of pH 8.5 using 0.01 gram formula weigfht of NaCN?
K dissociation (HCN) = 4.1×10^{-10}

Q. 65. The concentration of hydrogen ions is a 0.2 molar solution of formic acid 6.4×10^{-3} mol litre^{-1}. To this solution sodium formate is added so as to adjust the concentration of sodium formate to one mole per litre. What will be the pH of this solution? The dissociation constant of formic acid is 2.4×10^{-4} and the degree of dissociation of sodium formate is 0.75.

Q. 66. Calculate the pH of an aqueous solution of 1.0 M ammonium formate assuming complete dissociation (pk_a of formic acid = 3.8 and pk_b of ammonia = 4.8).

Q. 67. What is the pH of a M solution of acetic acid? To what volume must one litre of this solution be diluted so that the pH of the resulting solution will be twice the original value? Given: $K_a = 1.8 \times 10^{-5}$.

Q. 68. What is the pH of the solution when 0.20 mole of hydrochloric acid is added to one litre of a solution containing.

(i) 1 M each of acetic acid and acetate ion?

(ii) 0.1 M each of aceti cacid and acetate ion?

Assume the total volume is one litre. K_a for acetic acid = 1.8×10^{-5}

Q. 69. The pH of blood stream is maintained by a proper balance of H_2CO_3 and $NaHCO_3$ concentrations. What volume of 5 M $NaHCO_3$ solution should be mixed with a 10 ml sample of blood which is 2 M in H_2CO_3, in order to maintain a pH of 7.4? K_a for H_2CO_3 in blood is 7.8×10^{-7}.

Q. 70. How many moles of sodium propionate should be added to one litre of an aqueous solution containing 0.020 mole of propionic acid to obtain a buffer solution of pH 4.75? What will be the pH if 0.010 mole of hydrogen chloride is dissolved in the above buffer solution. Compare the last pH value with the pH of 0.010 molar HCl solution. Dissociation constant of Propionic acid Ka at 298 K.

Q. 71. 500 mL of 0.2 M aqueous solution of acetic acid is mixed with 500 mL of 0.2 M HCl at 289 K.

(i) Calculate the degree of dissociation of acetic acid in the resulting solution and pH of the solution.

(ii) If 6 g NaOH is added to the above solution, determine the final pH.

[Assume there is no change in volume on mixing; K_a of acetic acid is 1.75×10^{-6} mol L^{-1}].

Q. 72. Calculate the percent ionization of 0.60 M hydrofluoric acid $K_a = 7.1 \times 10^{-4}$

Chapter **9**

IONIC EQUILIBRIA

FORMULAE

For the reaction, $AB \rightleftharpoons A^+ + B^-$

$$K = \frac{[A^+][B^-]}{[AB]} = \frac{C\alpha^2}{(1-\alpha)}; \ \alpha = \frac{\lambda_v}{\lambda_\infty}$$

$$\frac{\alpha_1}{\alpha_2} = \sqrt{\frac{K_{a_1}}{K_{a_2}}} = \sqrt{\frac{K_{b_1}}{K_{b_2}}}$$

For the electrolyte, A_aB_b

$$K_{sp} = a^a b^b S^{a+b} = [A^{+b}]^a [B^{-a}]^b$$

$$K_w = [H^+][OH^-]$$

At 298 K $\quad [H^+] = [OH^-] = 10^{-7}$

$$[H^+] = 10^{-pH} \text{ and } [OH^-] = 10^{-pOH}$$

$$p^H = -\log [H^+] = \log \frac{[\text{Salt}]}{[\text{Acid}]}$$

$$\text{and } p^{OH} = -\log [OH^-] = \log \frac{[\text{Salt}]}{[\text{Acid}]}$$

$$p^H + p^{OH} = 14$$

For *acidic buffers,*

$$P^{H} = pK_a + \log\frac{[\text{Salt}]}{[\text{Acid}]}$$

For *basic buffers,*

$$P^{OH} = pK_b + \log\frac{[\text{Salt}]}{[\text{Acid}]}$$

$$\text{Buffer capacity } \varphi = \frac{d(b)}{dP^{H}}$$

where, b = No. of moles of acid or base added to one litre solution

dP^{H} = change in pH value

For the *salt of weak acid and strong base,* and vice versa

$$K_H = \frac{K_w}{K_a} = \frac{K_w}{K_b}\frac{h^2C}{(1-h)}; \text{ where, } h = \sqrt{\frac{K_w}{K_aC}} = \sqrt{\frac{K_w}{K_bC}}$$

$$P^{H} = \frac{1}{2}[pK_w + pK_a + \log C] = \frac{1}{2}[pK_w + pK_b - \log C]$$

For the *salt of weak acid and weak base*

$$K_H = \frac{K_w}{K_aK_b} = \frac{h^2}{(1-h)^2}$$

$$P^{H} = \frac{1}{2}[pK_w + pK_a + pK_b]$$

Strong base + strong acid	*No hydrolysis – neutral*
NaOH + HCl	Salt formed – Nacl $H^+ + OH^- \Leftrightarrow H_2O$ $K = \frac{1}{[H^+][OH^-]} = \frac{1}{K_w}$ Methyl Red (end point = pH 5) Bromthymol Blue (end point = pH 7) Phenolphthalein (end point = pH 9)
Strong base + weak acid	*Basic – only anion hydrolyzes*
NaOH + HAc	Salt formed – NaAc $Ac^- + H_2O \Leftrightarrow Hac + OH^-$ $K_b = \frac{[HAc][OH^-]}{[Ac^-]} \times \frac{[H^+]}{[H^+]} = \frac{K_w}{K_{HAc}}$ Phenolphthalein (end point = pH 9)

(*Contd.*)

Weak base + strong acid	*Acidic – only cation hydrolyzes*
$NH_3 + HCl$	Salt formed – NH_4Cl $NH_4^+ + H_2O \Leftrightarrow NH_3 + H_3O^+$ $K_a = \frac{[NH_3][H_3O^+]}{[NH_4^+]} \times \frac{[OH^-]}{[OH^-]} = \frac{K_w}{K_{NH_3}}$ Methyl Red (end point = P^H 5) or Methyl Orange (end point = P^H 4)
Weak base + weak acid	*Variable P^H – both ions hydrolyze.* P^H depends upon extent of hydrolysis of each ion involved.
$NH_3 + HCN$	$NH_3 + HCN$ Salt formed – NH_4CN $NH_4^+ + H_2O \Leftrightarrow NH_3 + H_3O^+$ $CN^- + H_2O \Leftrightarrow HCN + OH^-$ $K_a = \frac{[NH_3][HCN]}{[NH_4^+][CN^-]} \times \frac{[H^+][OH^-]}{[H^+][OH^-]} = \frac{K_w}{K_{NH_3} \times K_{HCN}}$ P^H of NH_4CN (aq) is greater than 7 because CN^- ($K_b = 2.0 \times 10^{-5}$) is more basic than NH_4^+ ($K_a = 5.6 \times 10^{-10}$) is acidic.

ACID AND BASE

Bronsted Lowry Theory

An acid-base reaction is a proton transfer from an acid to a base. An acid is a proton donor and a base is a proton acceptor, example HCl acts as an acid as it gives a proton to water molecule.

$HCl + H_2O \rightarrow H_3O^+ + Cl^-$ and $NH_4^+ + OH^- \rightarrow NH_3 + H_2O$; NH_3 acts as a base as it accepts a proton from water.

These reactions are reversible. Therefore the products combine to form the reactants. In the first equation H_3O^+ lose a proton to act as an acid and Cl^- accepts a proton to act as a base. Similarly NH_4^+ and OH^-. So each acid-base reaction involves two acids and two bases. Moreover the acid base pair differ only by a proton. Such acid-base pairs which differ by a proton are called ***Conjugate Acid-base Pairs***

The examples of conjugate acid base pairs are:

$$\text{Acid} \xrightarrow[\text{(Or} - H^+)]{\text{takeout-oneproton}} \text{Base (Conjugate)} \qquad \text{Acid} \xrightarrow[\text{(Or} + H^+)]{\text{add oneproton}} \text{Base}$$

$$HCl \rightleftharpoons Cl^- \qquad H_2O \rightleftharpoons OH^-$$

$$H_3O^+ \rightleftharpoons H_2O \qquad NH_3 \rightleftharpoons NH_2^-$$

$$NH_4^+ \rightleftharpoons NH_3 \qquad CH_3COOH \rightleftharpoons CH_3COO^-$$

The overall reactions involving two conjugate pairs can be shown as follows

$$HC_2H_3O_2 + H_2O \rightleftharpoons H_3O^+ + C_2H_3O_2^-$$

$$\underset{(\text{Acid}_1)}{HCl} + \underset{(\text{Base}_2)}{H_2O} \rightleftharpoons \underset{(\text{Acid}_2)}{H_3O^+} + \underset{(\text{Base}_1)}{Cl^-}; \qquad \underset{(\text{Base}_1)}{NH_3} + \underset{(\text{Acid}_2)}{H_2O} \rightleftharpoons \underset{(\text{Acid}_1)}{NH_4^+} + \underset{(\text{Base}_2)}{OH^-}$$

In Bronsted-Lowry concept, H_2O and NH_3 can act both as an acid as well as a base. Therefore they are described as amphoteric substances

$$\underset{\text{Acid}}{H_2O} \xrightleftharpoons{-H^+} \underset{\text{Base}}{OH^-} \qquad \underset{\text{Acid}}{NH_3} \xrightleftharpoons{-H^+} \underset{\text{Base}}{NH_2^-}$$

$$\underset{\text{Base}}{H_2O} \xrightleftharpoons{+H^+} \underset{\text{Acid}}{H_3O^+} \qquad \underset{\text{Base}}{NH_3} \xrightleftharpoons{+H^+} \underset{\text{Acid}}{NH_4^+}$$

Strength of Acids and Bases

The strength of an acid depends upon its tendency to loose the proton and that of a base depends upon its tendency to attract the proton and because HCl is a strong acid whereas CH_3COOH is a weak acid so HCl easily donates a proton but it is difficult with CH_3COOH.

Hence the conjugate base of a strong acid is weak and the conjugate base of a weak acid is strong.

Lewis Concept

According to Lewis acids are electron pair acceptor, e.g., BF_3, $AlCl_3$. Whereas, Bases are electron pair donor, e.g., NH_3, PCl_3.

Lewis Acids are of the following types:

1. Compounds whose central atom has an incomplete octet. In other words. Lewis acids are electron defficient molecules such as BF_3, $AlCl_3$, $GaCl_3$

$$\begin{array}{c}F\\F\,B\\F\end{array} + :NH_3 \rightarrow F_3B{:}NH_3$$

electron acceptor — electron donar — salt

2. Compounds in which the central atom has available d-orbitals and may acquire more than an octet of valence of electrons. Silicon tetrafluoride and tin (IV) chloride.

$$SiF_4 + 2F^- \rightarrow [SiF_6]^{2-}$$

3. All simple cations, e.g. Na^+, Ag^+, Cu^{2+}, Al^{3+}, Fe^{3+}, they combine with electron pair.

$$Ag^+ + 2{:}NH_3 \rightarrow [H_3N : Ag : NH_3]^+$$

$$Fe^{2+} + 6CN^- \rightarrow [Fe\,(CN)_6]^{-4}$$

4. Molecules with a multiple bond between atoms of dissimilar electronegativities.

Eg :O = C = O, O = S = O etc

$$\ddot{O} = C = \ddot{O} + [:\ddot{O} - H]^- \rightleftharpoons C(:O:)(:O:^-)-\ddot{O}-H$$

5. Elements with an electron sextet: Oxygen and sulphur atoms contain six electrons in their valency shells.

Lews bases may be of following types

1. All the simple negative ions, such as $:\ddot{C}l^-:$ $:\ddot{F}^-:$
2. Molecules with one or two unshared pairs of electrons.
   Eg: H_2O, NH_3, ROH, R_2O, R_2S, pyridine.
3. Multiple bonded compounds which form coordination compounds with transition metals.
   Eg: CO, NO, ethylene, acetylene etc.

$$CH_2 = CH_2 + Ag^+ \rightarrow \left[\begin{array}{c}CH_2-CH_3\\ \diagdown\ \diagup\\ Ag\end{array}\right]^+$$

Relative Strength of Lewis Acids and Bases:

For weak acids:

$$\text{Relative strength} = \frac{\text{Strength of I acid}}{\text{Strength of II acid}}$$

For acids HA_1 if concentration is C_1 and degree of dissociation α_1

$$HA_1 \rightleftharpoons H^+ + A_1^-$$

$$[H^+] = C_1\alpha_1$$

For HA_2 if concentration is C_2 and degree of dissociation is α_2

$$HA_2 \rightleftharpoons H^+ + A_2^-$$

$$[H^+] = C_2\alpha_2$$

$$\text{Then, relative strength} = \frac{[H^+] \text{ furnished by I acid}}{[H^+] \text{ furnished by II acid}} = \frac{C_1\alpha_1}{C_2\alpha_2}$$

$$\text{R.S.} = \frac{C_1}{C_2} \times \sqrt{\left(\frac{K_{a_1} C_2}{K_{a_2} C_1}\right)} = \sqrt{\left(\frac{K_{a_1} C_1}{K_{a_2} C_2}\right)}$$

If C_1 and C_2 are same, then

$$\text{R.S.} = \sqrt{\left(\frac{K_{a_1}}{K_{a_2}}\right)}$$

For strong acids:

$$\text{Relative strength} = \frac{K_1}{K_2} = \frac{\text{rate constant for a reaction catalysed by I acid}}{\text{rate constant for a reaction catalysed by II acid}}$$

Arrhenius Concept

An acid is an electrolyte that reacts with water to produce a hydrated H^+ ions. A base dissociates in water to form OH ion. According to this theory, acids are represented by the general formula HX and bases by MOH where 'X' is an anion and 'M' is a metal. The neutralization reaction between a strong acid and a strong base is

$$H^+(aq) + OH^-(aq) \rightleftharpoons H_2O$$

Acid: Acid is a substance that gives hydrogen ions (H^+) in aqueous solution. In aqueous solution acid dissociates giving hydrogen ion and an anion.

The undissociated acid is always in equilibrium with the ions.

$$HX \rightleftharpoons H^+ + X^-$$

This ions are responsible for the electrolytic conductivity. The strength of an acid is directly proportional to the rate or extent of ionisation. Examples HCl, HNO_3, H_2SO_4 CH_3 COOH

Base

Base is a substance that yields hydroxyl ions (OH^-) in aqueous solution. In aq solution base dissociated giving hydroxylion and a cation. The undissociated base is always in equilibrium with the ions.

$$MOH \rightleftharpoons M^+ + OH^-$$

The strength of a base depends upon the extent of ionisation. These ions are responsible for the electrolytic conductivity. The examples NaOH, NH_4OH, $Ca(OH)_2$, $Ba(OH)_2$.

Salt

Salt is a substance that can not yield hydrogen or hydroxylions in solution during primary ionisation examples NaCl, KNO_3, $NaHCO_3$, Na_2SO_4.

Strength of Acids and Bases

The strength of an acid or base in Arrhenisus concept is based on their rates of ionisation or the equilibrium constants of acids and bases.

The equilibrium constant, K_a for the acid HX = $\dfrac{[H^+]}{[X^-][HX]}$

The equuilibrium constant, K_b for the base MOH = $\dfrac{[M^+][OH^-]}{[MOH]}$

The value of K_a (1 to 0.1) is very high for strong acids such as $HClO_4$, HNO_2, H_2SO_4, HCl etc. It is low for weak acids like CH_3COOH, H_3PO_4, H_2CO_3, H_3BO_3 etc.

The value of K_b (1 to 0.1) is very high for strong bases like NaOH, KOH, RbOH etc. It is low for weak bases like NH_4OH, $Al(OH)_3$, $Fe(OH)_3$ etc.

Ionic Product of Water

Pure water is a very weak electrolyte. It ionises to a small extent to form hydronium or hydrogen ions and hydroxide ions.

$$H_2O \rightleftharpoons H^+ + OH^-$$

But actually the hydrogen ion is hydrated to form the hydronium ion. Then the ionisation follows the Bronsted-Lowry concept of amphi-protic reaction.

$$\underset{\text{acid}}{H_2O} + \underset{\text{base}}{H_2O} \rightleftharpoons \underset{\text{acid}}{H_3O^+} + \underset{\text{base}}{OH^-}$$

By applying Law of Mass Action to the above equilibria

$$K = \frac{[H^+][OH^-]}{[H_2O]} \text{ or } K = \frac{[H_3O^+][OH^-]}{[H_2O]^2}$$

The terms in square brackets are the corresponding molar concentrations at the equilibrium

$$\therefore \quad K[H_2O] = [H^+][OH^-] = K_w \text{ or } K[H_2O]^2 = [H_3O^+][OH^-] = K_w$$

As water is present in large excess and the dissociation is very small, its concentration remains almost constant.

$$K[H_2O]^2 = K_w = [H_3O^+][OH^-] \text{ or } [H^+][OH^-] = 1.0 \times 10^{-14} \text{ at 298 K}$$

Where K_w is called ionic product of water. The unit of K_w is $mole^2\ litre^{-2}$

The extent of dissociation of any weak electrolyte depends upon the temperature. The value of K_w depends upon temperature.

In acidic solutions, the concentration of hydrogen ions far exceeds the concentration of hydroxyl ions. Nevertheless, the value of K_w (ionic product) remains constant at a particular temperature by the equilibrium relationship. That is, with the increase of $[H_3O^+]$, the value of $[OH^-]$ falls.

Because $[H_3O^+] = \dfrac{K_w}{[OH^-]}$ and *vice versa.*

In such solutions, the value of $[H_3O^+]$ varies from 10^{-7} M (neutral) to 10^0 M varying through the expressions 10^{-6} M, 10^{-5} M etc. Thus its value increases as the solution becomes more and more strong with respect to the acid.

Similarly, in alkaline solutions, when the concentration of $[OH^-]$ ions increases, the concentration of $[H_3O^+]$ ions falls. However, when the concentration of $[OH^-]$ increases, the concentration of $[H_3O^+]$ falls from 10^{-7} M to 10^{-14} M through the expressions 10^{-8} M, 10^{-9} M etc.

Also, in pure water as well as in dilute solutions, the product of $[H^+]$ and [OH] must be equal to 1×10^{-14}. Moreover, the H^+ ions are equal to OH^- ions in pure water.

$$[H^+] = [OH^-] = \sqrt{K_w} = \sqrt{1\times 10^{-14}} = 1 \times 10^{-7} \text{ mole litre}^{-1}$$

pH value (scale):

is to express very meaqre (small) concentrations of hydrogen ion in solution.

Definition

The P^H of an aqueous solution is defined as the negative value of logarithm (to the base 10) of the concentration of hydrogen ions $[H^+]$ or $[H_3O^+]$ in solution.

$$P^H = -\log_{10}[H^+] \text{ or } \log\frac{1}{[H^+]} \text{ or } [H^+] = 10^{-pH}$$

Similarly $$P^{OH} = -\log_{10}[OH] \text{ or } \log\frac{1}{[OH]} \text{ or } [OH] = 10^{-pOH}$$

and $$P^{OH} = -\log K_w;\quad pK_a = -\log K_a;\quad pK_b = -\log K_b$$

Also, In pure water $[H^+] = [OH^-] = 10^{-7}$ mole-litre^{-1}

$$P^H = -\log[1 \times 10^{-7}] = 7$$

$$P^{OH} = -\log[1 \times 10^{-7}] = 7$$

If hydrogen ion concentration is one molar, the P^H becomes zero. Similarly if the concentration of hydroxyl ion is one molar, the P^H is 14.

The *p*H of a neutral solution is 7.0. & solutions with P^H 0 to 7 are acidic (i.e. P^H lower than 7) and 7 to 14 are basic i.e. P^H higher than 7). The hydrogen ion concentration above one molar connot be indicated in P^H ion has always positive values. There are no negative values for p^H.

The hydrogen-ion concentration of a solution of a weak acid is

$$[H^+] = \sqrt{K_b C} \text{ If } C \gg [H^+]$$

where K_a is the acid dissociation constant and C the molar concentration of the acid

$$p^H = \frac{1}{2} pK_a - \frac{1}{2} \log C$$

For a solution of a weak base,

$$[OH^-] = \sqrt{K_b C} \quad \text{If } C \gg [OH^-]$$

The K_b value of a base is related to the acid dissociation constant of its conjugate acid,

$$K_b = \frac{K_w}{K_a}\,;\; [H^+] = \frac{K_a}{[OH^-]} = \frac{K_w}{\sqrt{K_b}}$$

or $$p^H = 14 - \frac{1}{2} pK_b + \frac{1}{2} \log C$$

pK_a and pK_b for a conjugate acid-base pair

For an acid HX: $HX \rightleftharpoons H^+ + X^-$

$$K_a = \frac{[H^+][X^-]}{[HX]} \qquad \text{........ (i)}$$

For conjugate base: X^- of acid HX

$$X^- + H_2O \rightleftharpoons HX + OH^-$$

$$K_b = \frac{[HX][OH^-]}{[X^-]} \qquad \text{....... (ii)}$$

By Eqs. (i) and (ii)

$$K_a \times K_b = [H^+][OH^-] = K_w$$

or $$pK_a + pK_b = pK_w = 14$$

Note: 1. Stronger is acid, weaker is its conjugate base.

2. Higher is the value of pK_a of an acid, lower is acid strength and higher is basic strength of its conjugate base.

Buffer Solutions

Solutions which possess reserve acidic nature or alkaline nature or solutions which resist change in pH due to dilution or addition of small quantities of acid or alkali are known as buffer solutions.

Types of Buffer Solutions

There are two types of buffer solutions

(1) Simple buffer: Salts of weak acid and weak base, e.g., CH_3COONH_4, NH_4CN, etc.

(2) Mixed buffers: These are of two types.

(a) Acidic buffer mixtures: A weak acid + its salts
e.g., $CH_3COOH + CH_3COONa$

(b) Basic buffer mixtures: A weak base + its salt
e.g., $NH_4OH + NH_4Cl$

The process by which added H^+ and OH^- ions are removed so that P^H remains constant. This is known as buffer action for example Boric acid + sodium tetraborate (borax)

Handerson equation for pH and pOH of mixed buffers

A buffer solution contains a weak acid and its salt with a strong base ($CH_3COOH + CH_3COONa$) or a weak base and its salt with a strong acid (NH_4OH and NH_4Cl)

$$K_a \text{ of } CH_3COOH = 1.80 \times 10^{-5},\ p^{K_a} = 4.76$$

$$K_b \text{ of } NH_3OH = 1.80 \times 10^{-5},\ p^{K_b} = 4.76$$

$$[H^+] = K_a \frac{[HA]}{[A^-]} \text{ or } P^H = p^{K_a} + \log\frac{[salt]}{[acid]} \quad \text{(i)}$$

$$[OH^-] = K_b \frac{[BOH]}{[B^-]} \text{ or } P^{OH} = p^{K_b} + \log\frac{[salt]}{[acid]}$$

(i) is called Henderson's Equation. A buffer resists changes in pH due to dilution, addition of an acid, or addition of a base.

Buffer Capacity

The property of a buffer solution to which it resist the change in pH on addition of acid or alkali. It is expressed as the ratio of number of moles of acid for base added per litre to the change in p^H

Mathematically, $\phi = \frac{d(b)}{dpH}$, where b is no. of mole of acid or base added to one litre solution and dpH is change in pH.

Thus higher the amount of acid or base used to produce a definite change of P^H in a buffer solution, higher will be its buffer capacity. In general *buffer capacity is maximum when both the components are equimolar* (i.e. when the molar ratio is 1). Its value decreases as the ratio of salt to acid increases or decreases. A given acid may be used for preparing buffer solutions having P^H values in the range of $p^{K_a} \pm 1$. In other words any buffer can be used as a buffer up to two P^H units only depending upon the value of p^{K_a} or p^{K_b}. Since the limit of P^H values is given by $P^H = p^{K_a} \pm 1$. The best buffer will have $P^H = p^{K_a}$.

For example, the best buffer of acetic acid (p^{K_a} = approximately 4.8) with sodium acetate will be when CH_3COOH and CH_3 COONa are present in equivalent amounts.

in such case $P^H = p^{K_a} = 4.8$ and P^H limits $= p^{K_a} \pm 1 = 3.8$ to 5.8

Common ion effect

is the phenomenon in which degree of dissociation of a **weak electrolyte** is suppressed due to the presence of **common ion** present in it. OR. **The decrease in the ionisation of a weak electrolyte by the addition of a strong electrolyte** having an ion common with the weak electrolyte is known as common ion effect examples:

(a) The solubility of AgCl is suppressed by adding KCl to its saturated solution at equilibrium.
$AgCl \rightleftharpoons Ag^+ + Cl^-$ and $KCl \rightleftharpoons K^+ + Cl^-$
The increase in concentration of Cl^- ions (common ion) shifts the equilibrium towards left. This results in lowering of solubility of AgCl.

(b) The ionisation of acetic acid CH_3COOH is suppressed by addition of sodium acetate CH_3COONa.

$$CH_3COOH \rightleftharpoons CH_3COO^- + H^+$$
$$CH_3COONa \rightleftharpoons CH_3COO^- + Na^+.$$

For acetic acid $K_a = \dfrac{[CH_3COO^-][H^+]}{[CH_3COOH]}$

Thus, $[CH_3COO^-]$ in solution becomes more in presence of CH_3COONa. However K_a is constant. Thus in order to have K_a constant, $[H^+]$ in solution must decrease or $[CH_3COOH]$, i.e., undissociated acid must increase or in other words degree of disociation of CH_3COOH decreases.

(c) When NH_4Cl is added to NH_4OH, the dissociation of NH_4OH is suppressed.
as $NH_4OH \rightleftharpoons NH_4^+ + OH^-$ and $NH_4Cl \rightleftharpoons NH_4^+ + Cl^-$
So the concentration of NH_4^+ ions increase (common ions) and equilibrium shifts towards left.

Ostwald Dilution Law

For weak electrolytes:

	BA	$\rightleftharpoons$	B^+	+	A^-
	1 mole in V litres		—		—
	$1 - x$		x		x

$$\therefore \quad K = \frac{x^3/V^3}{\dfrac{(1-x)}{V}} = \frac{x^2}{(1-x)V} = \frac{x^2}{V} \quad (x \ll 1)$$

$$x \text{ (degree of dissociation)} = \sqrt{KV} = K'\sqrt{V}$$

Also,
$$K = \frac{x^2}{V} = x^2C \left[\because C = \frac{1}{V}\right]$$

$$x = \sqrt{\frac{K}{C}} = K'\sqrt{V}$$

Hydrolysis of Salts

A reaction in which anion or cation of a salt or both, with water to produce acidity or alkalinity is called SALT HYDROLYSIS. It usually affects the P^H of a solution. Examples: Sodium Acetate, KCN, Borax, Sodium Sulphide etc.

There are four types of salts reaction:

(1) ***Salts of strong acids and strong bases*** (e.g. NaCl) do not undergo hydrolysis. Aqueous solutions of these salts are neutral. Example, NaCl, KNO_3, Na_2S, KI.

So, (a) their P^H is exactly equal to 7.

(b) they have no action on litmus

(c) they neither react with acids or bases.

(2) ***Salts of weak acids and strong bases*** (e.g. CH_3COONa contain an anion which undergoes hydrolysis whereas the cation is unaffected. Representing such as anion as A^-

$$A^- + H_2O \rightleftharpoons HA + OH^-$$

Thus the solution is basic as a result of the excess OH^-

Therefore,

(a) the P^H of aqueous solution is greater than 7

(b) they turn red litmus to blue

(c) they can be titrated against acids.

The hydrolysis constant K_h is

$$K_h = \frac{[HA][OH^-]}{[A^-]}$$

If h is the degree of hydrolysis and c is the concentration of the salt solution, then

$$[A^-] = (1 - h)c;\ [HA] - hc;\ [OH^-] = hc$$

$$K_h = \frac{h^2c^2}{(1-h)c} = \frac{h^2c}{1-h} = h^2c(h \ll 1)$$

$$\text{or } h(\text{degree of hydrolysis}) = \sqrt{\frac{K_h}{c}}$$

Since HA ionises as $HA \rightleftharpoons H^+ + A^-$

$$K_h \times K_a = \frac{[HA][OH^-]}{[A^-]} \times \frac{[H^+][A^-]}{[HA]} = K_w$$

or

$$\boxed{K_h = \frac{K_w}{K_a}}$$

(3) ***salt of a strong acid and weak base*** e.g.: NH_4Cl contain a cation which undergoes hydrolysis while the anion is unaffected. Hydrolysis of such salts form weak base and strong acid example: NH_4Cl, $BaCl_2$, $Ca(NO_3)_2$, Ferric chloride etc. Therefore aqueous solutions of such salts are acidic in nature. So

(a) their P^H is less than 7
(b) they turn blue litmus to red
(c) they are titrated against bases

Representing such a cation as B^+, we have $B^+ + H_2O \rightleftharpoons BOH + H^+$

Solution is acidic due to excess H^+

$$K_h = \frac{[BOH][H^+]}{[B^-]}$$

It can be shown that

$$K_h = \frac{K_w}{K_a}$$

If h is the degree of hydrolysis,

$$K_h = \frac{h^2 c^2}{(1-h)c} = h^2 c$$

But $$K_h = \frac{K_w}{K_a}$$

$$h = \sqrt{\frac{K_h}{c}} = \sqrt{\frac{K_w}{K_{bC}}}$$

And $$[H^+] = h \times c$$

$$pH = -\log[H^+]$$

(4) ***Salts of weak acids and weak bases:*** The aqueous solutions of these salts are almost neutral. Examples: Ammonium Acetate, $Ba(CO_3)_2$, Ammonium Phosphate, Calcium Acetate. Therefore

(a) their p^H is almost equal to 7.
(b) they have no action on litmus.
(c) they neither react with acid or bases.

If CH_3COONH_4 contain a cation B^+ and an anion A^-, both of which undergo hydrolysis.

$$B^+ + A^- + H_2O \rightleftharpoons HA + BOH$$

Solution is nearly neutral if $K_a = K_b$ as in CH_3COONH_4

But if K_a are not equal, we can proceed as:

$$K_b = \frac{[HA][BOH]}{[H^+][A^-]}$$

It can be shown that

$$K_h = \frac{K_w}{K_a \cdot K_b}$$

If h is degree of hydrolysis, then

$$\begin{array}{ccccccccc} B^+ & + & A^- & + & H_2O & \rightleftharpoons & BOH & + & HA \\ 1 \text{ mole} & & 1 \text{ mole} & & & & — & & — \\ 1-h & & 1-h & & & & & & \end{array}$$

$$K_h = \frac{K_w}{K_a K_b} = \frac{h^2}{(1-h)^2}$$

Suppose K_a is greater than K_b then $[H^+]$ and pH can be calculated as

$$HA \rightleftharpoons H^+ + A^-$$

$$Ka = \frac{[H^+][A^-]}{[HA]}$$

$$[H^+] = \frac{[HA]}{[A^-]}$$

But
$$\frac{[HA]}{[A^-]} = \frac{h}{1-h} = \sqrt{K_h} = \sqrt{\frac{K_w}{K_a K_b}}$$

∴
$$[H^+] = K_a \cdot \sqrt{\frac{K_w}{K_a K_b}}$$

Hence
$$pH = \log_{10}[H^+]$$

Solubility Product

When a sparingly soluble salt like AgCl is dissolved in water, its saturated solution is formed equilibrium exists between undissolved solid, undissociated molecules and ions in solution.

$$AgCl\ (s) \rightleftharpoons AgCl\ (aq) \rightleftharpoons Ag^+(aq) + Cl(aq)$$

We have,
$$AgCl(s) \rightleftharpoons Ag^+(aq) + Cl^-(aq)$$

∴ Applying the law of mass action $K_C = \dfrac{[Ag^+][Cl^-]}{[AgCl]\,[solid]}$

The concentration of any solid is constant and it does not change. It is taken as unity

$$K_C[AgCl]\ [solid] = [Ag^+][Cl^-]$$

Hence
$$K_C\ [constant] = [Ag^+][Cl^-]$$

$$\text{or,}\quad K_{sp} = [Ag^+][Cl^-]$$

Where K_{sp} is another constant known as solubility product of AgCl.

Examples: For magnesium hydroxide, $Mg(OH)_2$ the equilibrium equation is

$$Mg(OH)_2(s) \rightleftharpoons Mg^{2+}\ (aq) + 2\ OH^-\ (aq),\ K_{sp} \text{ is written as}$$

$$K_{sp} = [Mg^{2+}][OH^-]^2$$

For calcium phosphate, $Ca_3(PO_4)_2$ equilibrium equation is

$$Ca_3(PO_4)_2(s) \rightleftharpoons 3\ Ca^{2+}\ (aq) + 2\ PO_4^3\ (aq)$$

The K_{sp} is written as

$$K_{sp} = [Ca^{2+}]^3[PO_4^{3}]$$

It may be noted that the solubility product is only a kind of equilibrium constant. It is therefore constant at a given temperature. With the rise in temperature K_{sp} usually increases owing to increased ionisation.

Thus,

1. When the ionic product is equal to the solubility product, the solution is saturated.
2. When the ionic product is less than the solubility product, the solution is super saturated and more of the salt can be dissolved in it.
3. When the ionic product exceeds the solubility product is super saturated. To keep the ionic product equal to the solubility product, the excess of the ions will recombine to form solid and thus the precipitation takes place. In other words, precipitation occurs when the product of the ionic concentration exceeds its solubility product.

Relation between solubility and solubility product

The term solubility is applicable to all solutes but solubility product is only applicable to electrolytes.

Consider a sparingly soluble binary electrolyte AB (ionises to two ions in solution). It ionises in solution as $AB = A^+ (aq) + B^- (aq)$

If the solubility of the electrolyte AB be 's' moles per litre, the concentration of A^+ and B^- ions will be 's' g ions per litre each. The electrolyte AB is assumed to be completely ionised.

$$[A^+][B^-] = K_{sp} \text{ or } s \times s = K_{sp} \text{ or } s^2 = K_{sp} \text{ or } s = \sqrt{K_{sp}}$$

Thus the solubility of the sparingly soluble binary electrolyte is equal to the square root of its solubility product.

Consider a sparingly soluble electrolyte of the type AB_2 it ionises in solution as

$$AB_2 \rightarrow A^{2+} + 2B^-$$

If the solubility of the electrolyte AB_2 be 's' moles per litre, the solubility of A^{2+} and B^- ions will be 's' and 2s gram ions per litre respectively.

The electrolyte AB_2 is assumed to be completely ionised

$$[A^{2+}][B^-]^2 = K_{sp} \text{ or } s \times (2s)^2 = K_{sp} \text{ or } s \times 4s^2 = K_{sp}$$

or

$$4s^3 = K_{sp} \text{ or } s = \sqrt[3]{\frac{K_{sp}}{4}}$$

POINTS TO REMEMBER

The ionization constant, dissociation constant, and hydrolysis constant are all equilibrium constants. The problems can all be solved using the basic principles for solving for equilibrium compositions. Do not conclude from the different names assigned these equilibria that there are different principles involved. In each case the problem can be solved by the following steps:

1. Write the reaction.
2. Write the equation relating the equilibrium constant to compositions.
3. Write each composition in terms of a single variable from the stoichiometry.
4. Solve the equation for the single variable and calculate the equilibrium compositions.

For doing mixing problems, keep the following in mind:

(a) Calculate millimoles of each, whichever is less will react with same number of millimoles of other forming salt. Find out which is in excess.

(b) If strong acid or strong base remains unused, calculate the molar concentration of H^+ or OH^- directly from millimoles left. Note the volume of two solution is taken into account while computing molarity.

(c) If weak acid or weak base is left unused, find millimoles of salt formed and use Henderson's equation for buffer.

(d) If acid and base are completely neutralised, calculate the molar concentration of salt and use the hydrolysis equation to compute H^+, OH^- and hence P^{OH}.

PROBLEMS

Q. 1. Calculate the P^H of the following solution?

(a) 0.1 NHCl (b) 0.0032 M H_2SO_4 (c) 0.05 N HNO_3

Q. 2. Calculate the hydrogen on concentration of the following solutions?

(a) P^H = 5.81 (b) P^H = 6.3 (c) P^H = 11.8

Q. 3. Calculate the P^H of the following solutions?

(a) 0.001 M NaOH (b) 0.02 KOH

Q. 4. Calculate the P^H of 0.98% sulphuric acid when it is 98% ionized?

Q. 5. Calculate the P^H of 0.12 M ammonia solution while its degree of ionization is 0.014.

Q. 6. 100 ml of 0.1 M H_2SO_4 is diluted to 250 ml. Calculate the change in P^H?

Q. 7. Calculate the P^H of a 0.50 M nitrous acid solution $K_4 = 4.5 \times 10^{-4}$

Q. 8. The P^H of a 0.100 M monoprotic acid HA is 2.85. What is the K_a the acid?

Q. 9. (a) The concentration of H^+ ions in a bottle of table wine was 3.2×10^{-4} night after the cork was removed. Only half of the wine was consumed. The other half after it has been standing open to the air for a month, was found to have a hydrogen ion concentration equal to 1.0×10^{-3}. Calculate the P^H of the wine on these two occasions.

(b) Calculate the P^H in the titration of a 25.0 ml of 0.1 M acetic acid by sodium hydroxide after the addition to the acid solution of 10 ml. of 0.1 M NaoH.

Q. 10. Describe how you would prepare a "phosphate buffer" at P^H of about 7.40.

Data given

$H_3PO_4(aq) \rightleftharpoons H^+(aq) + H_2PO_4^-$ $K_1 = 7.5 \times 10^{-3}$; $P_1^K = 2.12$

$H_2PO_4^-(aq) \rightleftharpoons H^+(aq) + HPO_4^{2-}$ $K_2 = 6.2 \times 10^{-8}$; $P_2^K = 7.21$

$HPO_4^{2-}(aq) \rightleftharpoons H^+(aq) + PO_4^{3-}$ $K_3 = 4.8 \times 10^{-13}$; $P_3^K = 12.32$

Q. 11. The P^H of blood plama is 7.40. Assuming the principal buffer system is HCO_3^- and H_2CO_3. Calculate the ratio HCO_3^-/H_2CO_3. Is this buffer more effective against added acid or an added base? $Ka_1 = 4.2 \times 10^7 = 6.38$.

Q. 12. What is the P^H of 0.002N HCl solution?

Q. 13. Calculate the P^H of 0.005M NaOH solution.

Q. 14. The P^H of an aqueous solution of an acid is 4.62. Calculate the hydrogen ion concentration P^H = 4.62.

Q. 15. The P^H of an aqueous solution of NaOH is 8.62. What is the concentration of NaOH?

Q. 16. If the dissociation constant of HCN is 7.2×10^{-10}, what is the P^{H-} of 0.1 M aqueous solution?

Q. 17. How much amount of water has to be added to 50 ml of 0.1 M sulphuric acid to change its P^H to 2.5.

Q. 18. Find the [H^+ ion concentration for the following pH values
(i) 5.34 (ii) 2.3 (iii) 8.9

Q. 19. A buffer solution is prepared by mixing acetic acid 0.01 mole and sodium acetate 0.1 mole^{-1} liter^{-1}. What is the P^H for the buffer if the dissociation constant of acetic acid is 1.75×10^{-5}.

Q. 20. What is the ratio of acetic acid and sodium acetate to the maintained to prepare a P^H = 4 buffer solution?

Q. 21. What is the P^H of a 1.0 M solution of acetic acid? To what volume must one litre of this solution be diluted so that the P^H of the resulting solution will be twice the original value.

Q. 22. How any gram moles of HCl will be required to prepare 1 litre of a buffer solution (containing NaCN and HCl) of P^H 8.5 using 0.01 gram formula weight of NaCN.

$$K_{(dissociation)\ (HCN)} = 4.1 \times 10^{-10}$$

Q. 23. What is the P^H of the solution when 0.2 mole of hydrochloric acid is added to one litre of a solution containing (i) 1 M each of acetic acid and acetate ion (ii) 0.1 M each of acetic acid and acetate ion.

Q. 24. The dissociation constant of a weak acid HA is 4.9×10^{-8}. After making the necessary approximations calculate (i) the percentage ionization, (ii) P^H and OH^- concentration in decimolar solution of the acid water has P^H of 7.

Q. 25. 20 ml of 0.2 M NaOH is added to 50 ml of 0.2 M acetic acid to give 70 ml of the solution. What is the P^H of this solution? Calculate the addition volume of 0.2 M NaOH required to make the P^H of the solution 4.74. The ionization constant of acetic acid is 1.8×10^{-5}.

Q. 26. How many moles of Sodium propionate should be added to one litre of an aqueous solution containing 0.02 mole of propionic acid to obtain a buffer solution P^H = 4.75? What will be the P^H if 0.01 mole of HCl is dissolved in the above buffer solution? Compare the last P^H of 0.01 molar HCl solution. Dissociation constant of proponic acid K_a at 298 K = 1.34×10^{-5}.

Q. 27. Calculate the P^H of 0.4833% solution of hydrochloric acid.

Q. 28. Calculate the P^H of the resulting solution obtained by mixing 20 ml of 0.01 M acetic acid with 12.2 ml of 0.01 M caustic soda solutions. (neglect the hydrolysis of the salt)

Q. 29. 0.002 molar solution of carbonic acid has P^H equal to 4.54 (he dissociation constant of the acid is 4.3×10^{-7} at 298 K.) To what extent the acid is dissociated in this solution?

Q. 30. Calculate the P^H of a solution containing 50 ml. of 0.12 N acetic acid and 100 ml. of 0.12 N potassium acetate. Dissociation constant of acetic acid is 1.8×10^5.

Q. 31. A buffer solution was obtained by adding 8.3 gm o acetic acid and 16 gm of sodium acetate. The buffer is diluted to one litre. Calculate P^H of the solution if 0.7 ml of 1.0 M HCl is added to it. The dissociation constant of acetic acid is 1.8×10^{-5} (Mol. wt of acetic acid is 60 and solution acetate is 82).

Q. 32. A buffer solution contain 2.0 gm of acetic acid in 250 ml of the solution. find the amount of acetate salt required to get the hydrogen ion concentration same as the dissociation constant of the acid, 1.8×10^{-5} (mol wts. of CH_3COOH and $NaCH_3COO$ are 60 and 82 respectively.)

Q. 33. Potassium acetate salt is 0.014 percent hydrolysed in 0.013 molar solution. Calculate the P^H of the solution.

Q. 34. Calculate the P^H of N/100 solution of a weak acid which is dissociated to the extent of 1.1 percent in N/10 solution at 298 K.

Q. 35. Calculate the amount of acetic acid that must be dissolved in 100 ml of the solution to get P^H equal to 2.4. The dissociation constant of the acid is 1.85×10^{-5}.

Q. 36. 500 ml N/5 solution of a monobasic acid HA, is mixed with 500 ml of N/5 solution of another monobasic acid HA_2. What is the P^H of the resulting solution? ($K_{HA} = 1.6 \times 10^{-5}$ acid $K_{HA_2} = 1.8\ 10^{-5}$)

Q. 37. the second dissociation constant the reaction $HSO_4^- \rightleftharpoons H^+ + SO_4^{2-}$ is 1.26×10^{-12}. Find the $[H^+]$ of 0.15 M solution of H_2SO_4. Assume that the first dissociation,

$$H_2SO_4 \rightarrow HSO_4^- + H^+ \text{ is 100 per cent.}$$

Q. 38. Calculate the P^H of the solution obtained by mixing 25 ml of 0.1 M HCl and 30 ml of 0.1 M NH_3

$$(K_b \text{ for } NH_3 + H_2O \rightleftharpoons NH_4^+ + OH^- \text{ is } 1.81 \times 10^{-5})$$

Q. 39. The indicator constant of phenolphihalein is 10^{-9}. what will be its useful P^H range? The indicator is colourless in acid solution and red in alkaline medium. Explain its colour change with P^H?

Q. 40. The P^H values of two solution are 8.5 and 11.0 will they give different colour with litmus and with phenolphthalein? (P^H Range is 8 to 10)

Q. 41. Phenolphthalein is colorless in acidic form and red (or pink) in alkaline form. If a very small amount of phenolphthalein ($K_{In} = 3.16 \times 10^{-10}$) were added to a decinormal solution of sodium acetate at 298 K. What fraction of the indicator would exist in the coloured form? Dissociation constant of CH_3COOH is 1.8×10^{-5}.

Q. 42. Indicator is red in acid solution and yellow in alkaline solution. Calculate what fraction of it will be in the red in solution of P^H 4.5 to 6. The indicator constant is 1×10^{-5}.

Q. 43. An indicator, HIn has an ionization constant of 50×10^{-8}. The acid colour of the indicator is yellow and the alkaline colour is blue. The yellow colour is visible when the ratio of the yellow form to the blue form is 20 to 1. The blue colour is predominant when the ratio of blue form to yellow form is 2 to 1. What is the P^H range of colour change for indicator?

Q. 44. **(a)** The concentration of hydrogen ions in a 0.2 molar solution of formic acid is 6.4×10^{-3} moles litre^{-1}. To this solution sodium formate is added so as a adjust the concentration of sodium formate to one mole per litre. What will be the P^H of this solution? The dissociation constant of formic acid is 2.4×10^{-4} and the degree of dissociation of sodium formate is 0.75.

(b) Find K_b and p^{K_b} for the acetate ion CH_3COO^-..The ionization constant of CH_3COOH is $K_a = 1.75 \times 10^{-5}$; $K_w = 1.0 \times 10^{-14}$.

Q. 45. When 1 litre of a saturated solution of $CaCO_3$ is evaporated to dryness the residue is found to weigh 7.0 mg. what is the K_{sp} of $CaCO_3$?

Q. 46. At 293 K a saturated solution of $PbCl_2$ is found to contain 4.5 g of $PbCl_2$ per litre dissolved in lt. What is K_{sp} $(PbCl_2)$?

Q. 47. The K_{sp} of (AgCl) is known to be 1.6×10^{-10} at 293 K. Calculate the solubility of (AgCl) in water at 293 K?

Q. 48. A solution contains a mixture of Ag (0.1 M) and Hg^{2++} (0.1 M) which are to be separated by selective precipitation. Calculate the maximum concentration of iodine ion at which one of them gets precipitated almost completely. What percentage of that metal ion is precipitated?

$[K_{s.p.}: AgI = 8.5 \times 10^{-17}; Hg_2I_2 = 2.5 \times 10^{-26}]$.

Q. 49. A sample of AgCl was treated with 5 mL of 1.5 M Na_2CO_3 solution to give Ag_2CO_3. The remaining solution contained 0.0026 g of Cl^- per litre. Calculate the solubility product of AgCl (K_{sp} $Ag_2CO_3 = 8.2 \times 10^{-12}$).

Q. 50. The solubility of $Mg(OH)_2$ in pure water is 9.57×10^{-3} g $litre^{-1}$. Calculate its solubility (in g $litre^{-1}$) in 0.02 M Mg $(NO_3)_2$ solution.

Q. 51. Given: $Ag(NH_3)_2^+ \rightleftarrows Ag^+ + 2NH_3$,

$K_c = 6.2 \times 10^{-8}$ and

K_{sp} of $AgCl = 1.8 \times 10^{-10}$

At 298 K. Calculate the concentration of the complex in 1.0 M aqueous ammonia.

Q. 52. The solubility product (K_{sp}) of $Ca(OH)_2$ at 298 K is 4.52×10^{-5}. A 500 ml of saturated solution of $Ca(OH)_2$ is mixed with equal volume of 0.4 M NaOH. How much $Ca(OH)_2$ in milligrams is precipitated?

Q. 53. Freshly precipitated aluminium and magnesium hydroxides are stired vigorously in a buffer solution containig 0.25 mole L^{-1} of ammonium chloride and 0.05 mole L^{-1} of ammonium hydroxide. Calculate the concentration of aluminium and magnesium ions in solution.

$K_b\,[NH_4OH] = 8 \times 1^{-6}$

$K_{sp}[Mg(OH)_2] = 6 \times 10^{-10}$

$K_{sp}[Al(OH)_2] = 6 \times 10^{-32}$

Q. 54. The solubility product of $Ag_2C_2O_4$ at 298 K is 1.29×10^{-11} $mol^3\ l^{-3}$. A solution of $K_2C_2O_4$ containing 0.152 mole in 500 ml water is shaken at 298 K with excess of Ag_2CO_3 till the folowing equilibrium is reached:

$$Ag_2CO_3 + K_2C_2O_4 \leftrightarrow Ag_2C_2O_4 + K_2CO_3$$

At equilibrium the solution contains 0.0358 mole of K_2CO_3. Assuming the degree of dissociation of $K_2C_2O_4$ and K_2CO_3 to be equal, calcualte the solubility product of Ag_2CO_3

Q. 55. An aqueous solution of a metal bromine MBr_2 (0.05 M) is saturate with H_2S. What is the minimum P^H which MS will precipitate.

Q. 56. The solubility of $Pb(OH)_2$ in water in 6.7×10^{-6} M.

Calculate the solubility of $Pb(OH)_2$ in a buffer solution of $P^H = 8$.

Q. 57. What will be the resultant P^H when 200 ml of an aqueous solution of HCl (P^H =) is mixed with 30 ml of an aqueous solution of NaOH ($P^H = 12.0$)?

Q. 58. An acid type indicator, HIm differs in colour from its conjugate base (In^-). The human eye is sensitive to colour differences only when the ratio $[In^-]/[HIn]$ is grater than 10 or smaller than 0.1. What should be the minimum change in the P^H of the solution to observe a complete colour change ($K_a = 1.0 \times 10^{-5}$)?

Q. 59. Twenty ml of 0.2 M sodium hydroxide is added to 50 ml of 0.2 M acetic acid to give 70 ml of the solution. What is the P^H of this solution? Calculate the additional volume of 0.2 M NaOH required to make the P^H of the solution 4.74. The ionization constant of axis is 1.8×10^{-5}.

Q. 60. The dissociation constant of a weak acid HA is 4.9×10^{-8}. After making the necessary approximations, calculate (i) percentage ionization, (ii) pH and (iii) OH concentration in a decimolar solution of the acid. Water has a P^H of 7.

Q. 61. What is the P^H of a gas 0.50 M aqueous NaCN Solution?

PK_b of CN^- is 4.70.

Q. 62. 1.22 g C_6H_5 COOH is added into solvent and data of ΔT_b and K_b are given as:—

1. In 100 g CH_3 COH_3 $\Delta T_b = 0.17$

   $K_b = 1.7$ kg Kelvin mol^{-1}

2. In 100 g benzene

   $\Delta T_b = 0.13$ and $K_b = 2.6$ kg Kelvin mol^{-1}

(a) Find out the molecular weight of C_6H_5 COOH in both the cases

(b) What do you deduce out of it in terms of structure of benzoic acid?

(ii) 0.1 M of HA is titrated with 0.1 M NaOH, calculate the P^H at end point. Given Ka (HA) = 5×10^{-6} and $\alpha \ll 1$.s

Q. 63. What is the solubility of $PbBr_2$ at 293 K?

Q. 64. A few drops of a solution of $AgNO_3$ is added o a large amount of a 0.01 M solution of NaCl and a ppt. of AgCl is formed. On analysis it is found that the solution contains 0.01 of moles $litre^{-1}$ of Cl^- ions. If K_{sp} (AgCl) = 1.6×10^{-10}. Calculate the $[Ag^+]$ left in the solution.

Q. 65. What concentration of Ag^+ ions will be in equilibrium with a saturated solution containing a ppt. of Ag_2CrO_4 and CrO_4^{2-} ion concentration of 0.40 mole $litre^{-1}$? (K_{sp} (Ag_2CrO_4) = $1.1. \times 10^{-11}$)

Q. 66. A solution contains a concentration of Pb^{2+} ions of 2×10^{-3} mole $litre^{-1}$. To this is added enough solid NaCl to bring the Cl^- in concentration to 3×10^{-2} mole $litre^{-1}$. Will a ppt. of $PbCl_2$ form, given that $K_{sp}(PbCl_2) = 1.7 \times 10^{-5}$ (ignore volume change)

Q. 67. A solution is contaminated with Pb^{2+} ions. By adding Na_2SO_4; $PbSO_4$ may be made to ppt. What concentration of SO_4^{2-} ions is needed to reduce the $[Pb^{2+}]$ ions to 2×10^{-6} moles $litre^{-1}$. $K_{sp}(PbSO_4) = 1.8 \times 10^{-8}$

Q. 68. Calculate the solubilities in mole lit^{-1} of each of the following

(a) $BaSO_4$; $K_{sp} = 1.5 \times 10^{-9}$ (b) CaF_2, $K_{sp} = 1.08 \cdot 10^{-10}$

(c) $Al(OH)_3$; $K_{sp} = 1 \times 10^{-39}$

Q. 69. Given their solubilities calculate the solubility product of the following sparingly soluble salts.

salt solubility

(a) **AgI** 1.23×10^{-8} mole lit^{-1}
(b) Ag_2CO_3 1.27×10^{-4} mole lit^{-1}
(c) SrF_2 1.22×10^{-2} g. lit^{-1}

Q. 70. Calculate the P^H of a saturated solution of magnesium hydroxide if K_{sp} for $Mg(OH)_2$ 8.9×10^{-12}.

Q. 71. Calculate the $[H_3O^+]$, $[CH_3COO^-]$, $[CH_3COOH]$, $[OH^-]$, P^{OH}, and the degree or dissociation, α, of a 0.1 F CH_3COOH solution at 298 K.

Q. 72. Calculate K_b at 298 K for (a) sulfide ion, S^{2-}, and (b) monohydrogen phosphate ion, HPO_4^{2-}.

Q. 73. Calculate the $[OH^-]$, $[NH_4^+]$, $[H_3O^+]$, and the P^H of a 0.25 F NH_3 solution at 298 K. For ammonia K_b= 1.8×10^{-5} at 298 K.

Q. 74. Methylamine is a colorless gas at room temperature with a strong unpleasant odor. It dissolves readily in water, with which it undergoes the proton-transfer reaction

$$CH_3NH_2(aq) + H_2O \rightleftharpoons CH_3NH_3^+(aq) + OH^-(aq)$$

At 298 K, K_b =-4.2×10^{-4} for methylamine. Calculate the degree of dissociation of a 0.2 F solution of methylamine at 298 K.

Q. 75. A solution is prepared by dissolving 0.12 mol of acetic acid and 0.1 mol of HCl in enough water to make exactly 1 L solution. Calculate the P^H of this solution and the acetate ion concentration of equilibrium.

Q. 76. A solution a acidified with HCl so that the $[H_3O^+]$ = 0.3 M, and is kept continuously saturated with H_2S gas at 298 K so that $[H_2S]$ = 0.1. Calculate the $[S^{2-}]$ in this solution.

Q. 77. Calculate the $[NH_4^+]$, $[Br^-]$, $[NH_3]$, $[H_3O^+]$, $[OH^-]$, and P^H of a 0.3 F NH_4Br solution.

Q. 78. The student prepares a buffer solution by adding 4 g of sodium acetate of 200 mL of 0.25 M acetic acid. The solution is then diluted to 100 mL with distilled water. The student then obtains 4 beakers labeled 1, 2, 3 and 4. To beakers 1 and 2 she adds 50 mL of the buffer solution. to breakers 3 an 4 she adds 50 mL of distilled water. She then uses a calibrated pH meter and obtains the P^H of the solution in the beakers. To beakers 1 and 3, she pipets 25 mL of 0.25 M HCl, mixes well and determines the P^H using the calibrated P^H meter. The P^H of each beaker was follows (1) 4.32, (2) 1.01, (3) 5.25, (4) 12.88. She also selects an appropriate in dicator and records the color produced. To breakers 2 and 4 she adds 25 mL f 0.25 M. NaOH and repeats the process as the did with the HCl. Answer the following 1–9

1. A solution from Part I was determined to have a P^H of 1.0. Determine the P^{OH}, the $[H_3O^+]$ and the $[OH^-]$.
2. A solution for Part I had an $(OH^-]$ of 2×10^{-12} M. Determine the P^H.
3. The p^H of a 0.1 M acetic acid solution in Part II was found to be 2.87. Calculate the K_a, of acetic acid.
4. calculate the hydrolysis constant K_b for the ammonium chloride solution from Part III. Assume that $[NH_4^+]$ is the same as the initial concentration of ammonium chloride.
5. Calculate the theoretical P^H of the original buffer solution assuming the volume change due to the addition of the solid is negligible.
6. Calculate the theoretical P^H of the buffer solution with the added HCl (breaker 1).

7. Calculate the theoretical P^H of the buffer solution with the added NaOH (breaker 2).
8. Calculate the theoretical P^H of the distilled water and added HCl (breaker 3).
9. Calculate the theoretical P^H of the distilled water and added NaOH (breaker 4).

Q. 79. Calculate both the $[H_3O^+]$ and $[OH^-]$ in a 0.4 *F* HCl solution at 298 K.

Q. 80. Calculate the P^H of (a) 2.00 *F* HCl, (b) 0.200 *F* $HClO_4$ and (c) 2.00×10^{-4} *F* HNO_3.

Q. 81. The P^H of a sample of skim milk is found to be 6.6 at 298 K. what are the concentration of hydronium ion and hydroxide ion in this skim milk?

Q. 82. Calculate the P^H of a neutral solution at (a) 273 K and (b) 333 K

Q. 83. Calculate the $[OH^-]$, $[H_3O^+]$, P^H , and P^{OH} in 0.06 *F* $Ba(OH)_2$ at 298 K.

Q. 84. Calculate the $[Na^+]$, $[OAc^-]$, $[H_2O^+]$, $[OH^-]$, and P^H of a solution of 0.045 *F* NaOAc at 298 K

Q. 85. Arrange the following solutions in order of increasing P^H ; **(a)** 0.2 *F* NH_4NO_3, **(b)** 0.2 *F* KNO_2, **(c)** 0.2 *F* KNO_3, **(d)** 0.2 *F* $HClO_4$, **(e)** 0.2 *F* Na_2CO_3 and **(f)** 0.2 *F* HNO_2.

Q. 86. Classify the following salts as acidic, basic, or neutral. Explain your reasoning **(a** $(NH_4)_2CO_3$, **(b)** $KClO_4$, **(c)** $KHSO_4$, **(d)** BaO, **(e)** $NaHSO_3$, and **(f)** $CH_3NH_3NO_3$.

Q. 87. The box contains ten moles of a week acid, □○ in a liter of solution. Using the symbols (○ = atom, △ = OH^-), show what happens on

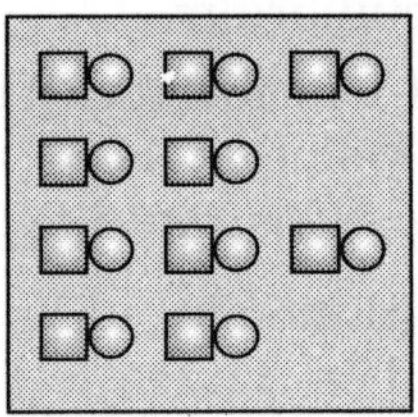

(a) the addition of two moles of OH^- (2Δ)
(b) the addition of five moles OH^- (SΔ)
(c) the addition of ten moles of OH^- (10Δ)
(d) the addition of 12 moles of OH^- (12Δ)

Which addition, (a)–(d), represent neutralization halfway to the equivalence point?

Q. 88. Carbonic acid, H_2CO_3, is a solution of CO_2 in water, Use L of Chateller's principle to explain the build-up of CO_2 in the blood when H^+ ions are added.

Q. 89. Indicate whether each of the following statements is true or false. If the statement is false, restate it t make it true.

(a) The formate ion (CHO_2^-) concentration in 0.10 *M* $HCHO_2$ is the same as in 0.10 *M* $NaCHO_2$.

(b) Because K_a for HCO_3^- is 4.7×10^{-11}, K_b for HCO_3^- is 2.1×10^{-4}.

Q. 90. The P^H of a white vinegar solution is 2.45. This vinegar is an aqueous solution of acetic acid with a density of 1.09 g per mL. What is the mass percentage of acetic acid in the solution?

Q. 91. What is the freezing point of 0.92 *M* aqueous acetic acid? The density of this solution is 1.008 gcm^{-3}.

Q. 92. A chemist needs a buffer with P^H 4.35. How many milliliters of pure acetic acid (density = 1.049 g per mL) must be added to 0.465 mL of 0.0941 *M* NaOH solution to obtain such a buffer?

Q. 93. The species called *glacial acetic acid* is 98% acetic acid by mass (*d* = 1.0542 g per mL). What volume of glacial acetic acid must be added to 100 mL of 1.25 M NaOH to give a buffer with a p^H of 4.2?

Q. 94. Explain why it is not possible to prepare a buffer with a P^H of 6.5 by mixing NH_3 and NH_4Cl.

Q. 95. Four grams of a monoprotic weak acid are dissolved in water to make 250 mL of solution with a P^H of 2.56. The solution is divided into two equal parts. *A* and *B*. Solution *A* is titrated with strong base to its equivalence point. Solution *B* is added to solution *A* after solution *A* is neutralized. The P^H of the resulting solution is 4.26. What is the molar mass of the acid?

Q. 96. Fifty cm^3 of 1.0 *M* nitrous acid is titrated with 0.85 NaOH. What is the P^H of the solution

(a) before any NaOH is added?

(b) at half-neutralization?

(c) at the equivalence point?

(d) when 0.1 mL less than the volume of NaOH to reach the equivalence point is added?

(e) when 0.1 mL more than the volume of NaOH to reach the equivalence point is added?

Q. 97. In a titration of 50 mL of 1.0 *M* $HC_2H_3O_2$ with 1.0 *M* NaOH, a student used bromcresol given as an indicator ($K_a = 1.0 \times 10^{-5}$). About how many milliliters of NaOH would it take to reach the end point with this indicator?

Q. 98. What is the P^H of a 0.15 *M* H_2SO_4 solution if

(a) the ionization of HSO_4^- is ignored?

(b) the ionization of HSO_4^- is taken into account? ($_a$ for HSO_4^- is 1.1×10^{-2}).

Q. 99. A solution is 1.5×10^{-4} *M* Zn^{2+} and 0.2 *M* HSO_4^-. The solution also contains Na_2SO_4. What should be the minimum molarity of Na_2SO_4 to prevent the precipitation of zinc suitable when the solution is saturated with hydrogen sulfide (0.1 *M* H_2S)?

Q. 100. what is the solubility of calcium fluoride in a buffer solution containing 0.45 *M* $HCHO_2$ (formic acid) and 0.2 *M* $NaCHO_2$? (*Hint:* Consider the equation $CaF_2(s) + 2H^+(aq) \rightleftharpoons Ca^{2+}(aq) + 2HF(aq)$, and solve the equilibrium problem.)

Q. 101. A 67 mL sample of 0.35 *M* $MgSO_4$ is added to mL of 0.25 *M* $Ba(OH)_2$. What is the net ionic equation for the reaction that occurs? What are the concentration of ions in the mixture at equilibrium?

Q. 102. The box below represent one liter of a saturated solution of the species □, where squares represent the cation and circles represent the atom. Water molecules, through present are not shown.

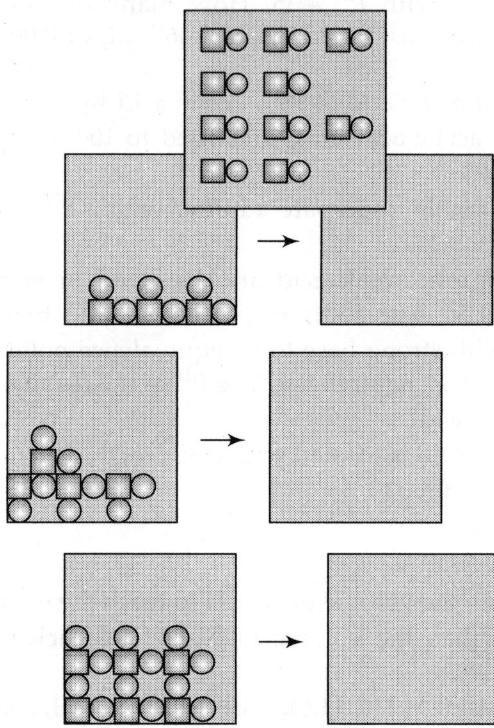

Complete the next three figures by filling one-liter boxes to the right of the arrow, showing the state of the ions after water is added to form saturated solutions. The species represent to the left of the arrow is the solid form of the ions represented above. Do not show the water molecules.

Q. 103. Marble used for statuary is almost pure $CaCO_3$. When marble is exposed to H^+ ions, the following reactions occurs.

$$CaCO_3(s) + H^+(aq) \rightleftharpoons HCO_3^-(aq) + Ca^{2+}(aq)$$

Why is acid rain the primary culprit in the disintegration of marble statuary exposed to the elements? Assume a P^H o f 3.0 for acid rain and a P^H of 5.4 for that of normal rainwater.

Q. 104. Ammonium chloride solutions are slightly acidic, so they are better solvents than water for insoluble hydroxides such as $Mg(OH)_2$. Find the solubility of $Mg(OH)_2$ in moles per liter in 0.2 *M* NH_4Cl and compare with the solubility in water. *Hint:* Find *K* for the reaction

$$Mg(OH)_2 + 2NH_4^+(aq) \rightarrow Mg^{2+}(aq) = 2NH_3(aq) + 2H_2O$$

Q. 105. What is the solubility of CaF_2 in a buffer solution containing 0.3 *M* $HCHO_2$ and 0.2 *M* $NaCHO_2$? *Hint:* Consider the equation

$$CaF_2(s) + 2H^+(aq) \rightarrow Ca^{2+}(aq) + 2HF(aq)$$

and solve the equilibrium problem.

Q. 106. What is the I^- concentration just a AgCl begins to precipitate 1.0 *M* $AgNO_3$ is slowly added to a solution containing 0.2 *M* Cl^- and 0.2 *M* I^-?

Q. 107. The concentrations of various cations in sea water, in moles per liter, are

cation	Na^+	Mg^{2+}	Ca^{2+}	Al^{3+}	Fe^{3+}
Molarity (*M*)	0.46	0.056	0.01	4×10^{-7}	2×10^{-7}

(a) At what $[OH^-]$ does $Mg(OH)_2$ start to precipitate?
(b) At this concentration, will any of the other ions precipitate?
(c) If enough OH^- is added to precipitate 50% of the Mg^{2+}, what percentage of each of the other ions will precipitate?
(d) Under the condition in (c), what mass of precipitate will be obtained from one liter of sea water?

Q. 108. Consider the equilibrium

$$Zn(NH_3)_4^{2+}(aq) + 4OH^+(aq) \rightleftharpoons Zn(OH)_4^{2-}(aq) + 4NH_3(aq)$$

(a) Calculate K for this reaction.
(b) What is the ratio $[Zn(NH_3)_4^{2+}]/[Zn(OH)_4^{2-}]$ in a solution 1.0 M in NH_3?

Q. 109. Use the equilibrium constant from tables to calculate K for the reaction

$$Ag(NH_3)_2^+(aq) + 2H^+(aq) + Cl^-(aq) \rightleftharpoons AgCl(s) + 2NH_4^+(aq)$$

Q. 110. An experimenter makes up a solution of 0.375 mol $NaCO_3$, 0.125 mol $Ca(NO_3)_2$ and 0.2 mol $AgNO_3$ in 2 L of water solution. Note any precipitations that occur, writing a balanced equation for each. Then calculate the molarities of each ion the solution.

Q. 111. What is the freezing point of vinegar, which is an aqueous solution of 5.0% acetic acid, $HC_2H_3O_2$ mass (*d* –1.006 g cm^{-3})?

Q. 112. Show by calculation that when the concentration of a weak acid decreases by factor of 10, its percent ionization increases by a factor of $10^{1/2}$.

Q. 113. A student prepared three solutions by adding 12, 18.5, and 35 mL of a 7.50×10^{-2} M NaOH solution to 50 mL of a 0.1 M solution of a weak monoprotic acid, HA. The solutions were labeled X, Y and Z, respectively. Each of the solution were diluted to a total volume of 100 mL with distilled water. The p^H readings of these solutions, obtained through the use of a calibrated pH meter, were (X) 6.5, (Y) 6.7 and (Z) 7.1.
Answer the following 1–9

1. Convert the p^H of each solution to an equivalent H_2O^+ concentration.
2. For each solution, calculate the number of moles of acid added.
3. For each solution, calculate the number of moles of OH^- added.
4. For each solution determine the concentration of HA and of A^- ion present at equilibrium.
5. Determine the reciprocal of the A^- ion concentration for each solution.
6. Plot the reciprocal of the A^- ion concentration on the ordinate against the H_3O^+ ion concentration on the abscissa. Draw a straight "best fit" line (see Figure 1).
7. Determine the slope of the line.
8. Determine the initial concentration of the acid, HA.
9. Determine the K_a for the weak acid.

Q. 114. A student mixed 5 mL of 0.015 M $Pb(NO_3)_2$ with 3 mL of 0.03 M KI in 2 mL of 0.2 M KNO_2. The test tube was shaken for 15 minutes and the solid precipitate of lead (II) iodide was allowed to settle. The tube was centrifuged to remove any excess lead (II) iodide from the supernatant. To the supernatant, she added KNO_2 to oxidize the I^- to I_2. The supernatant was then analyzed for I^- by using a spectrophotometer. She then made known dilutions of a 0.1 M potassium iodide solution in acidified KNO_2 to create a calibration curve to be able to determine the I^- in the supernatant.

1. How many moles of Pb^{2+} were initially present?
2. How many moles of I^- were initially present?
3. From known dilutions of KI in acidified KNO_2, a calibration curve was created. From the calibration curve, it was determined that the concentration of I^- at equilibrium in the supernatant was 1.4×10^{-3} mole litre^{-1}. How many moles of I^- were present in 10 mL. of the final solution?
4. How many moles of I^- precipitated?
5. How many moles of Pb^{2+}precipitated?
6. How many moles of Pb^{2+} were in the solutions?
7. What was the molar concentration of Pb^2 in the solution at equilibrium?
8. Determine K_{sp}.
9. The published value of $_{sp}$ at 298 K for PbI_2 is 7.1×10^{-9}. Determine the percent error.

Q. 115. A student determined that the optimal wavelength for the absorbance of $FeSCN^{2+}$ experiment was 445 mm. then the student prepared samples of known concentration of $FeSCN^{2+}$ ranging from 4.0×10^{-5} M to 1.4×10^{-4} M. The samples were then examined by means of a spectrophotometer and their transmittances recorded. From the transmittance, the 5 mL of 2×10^{-3} M KSCN. This solution was then analyzed in the spectrophotometer and through extrapolation, he was able to determine that the concentration of $FeSCN^{2+}$at equilibrium was 1.3×10^{-4} M.

1. How many moles of Fe^{3+} and SCN^- were initially present?
2. How many moles of $FeSCN^{2+}$ were in the mixture at equilibrium?
3. How many moles of Fe^{3+} and SCN^- were used up in making the $FeSCN^{2+}$?
4. How many moles of Fe^{3+} and SCN^- remain in the solution at equilibrium?
5. What are the concentration of Fe^{3+}, SCN^-, and $FeSCN^{2+}$ at equilibrium?
6. Determine K_c for the reaction,

Q. 116. A student measured out approximately 10 mL of 6 M NaOH and diluted the base to approximately 600 ml_2. The student then performed an acid-base titration (see Figure and determined that 48.7 mL of NaOH solution were needed to neutralize 50 mL of a 0.1 M HCl solution.

Once the student had determined the exact concentration of the base, the student than proceeded to determine the equivalent mass of an unknown acid. To do this, the student measured out 0.5 grams of an unknown slid acid and titrated it with the standardized base, recording pH with a calibrated pH meter as the base was added. the student added 43.2 mL of the base but went too far past the end point and needed to back-titrate with 5.2 mL of the 0.1 M HCl to exactly reach the end point. Answer the following 1–4

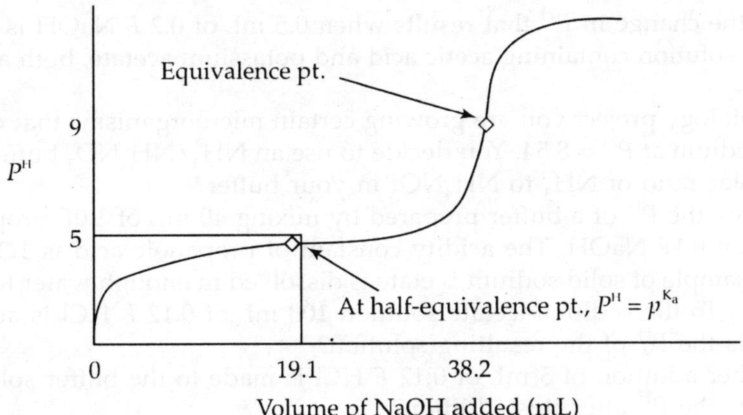

1. Calculate the molarity of the NaOH solution from Part 1.
2. The graph of the titration from this experiment is presented here. Determine the K_a of K_a's.
3. What is the P^H at the equivalence point?
4. Determine the equivalent mass of the solid acid.

Q. 117. Buffer solution is made by mixing 0.1 mol of acetic acid and 0.4 mol of sodium acetate per liter of solution. Calculate the P^H of this buffer.

Q. 118. If 100 mL of water added to 100 mL of a buffer that is 0.2 *F* NH_4NO_3 and 0.2 *F* NH_3, what effect does this dilution have on the P^H of the buffer?

Q. 119. A 25 mL portion of an aqueous solution of HBr of unknown concentration is titrated with 0.15 *F* NaOH. The equivalence point is reached when 32.8 mL of base has been added. Calculate the concentration of the hydrobromic acid.

Q. 120. In order to determine the strength of a commercial antacid, one tablet was dissolved in 50 mL of 1.0 *F* HCl. The resulting solution, which was still acidic, was titrated with standardizd 0.5 *F* NaOH. the volume of the NaOH solution required to reach the equivalence point of the titration was 43.72 mL. Calculate the number of moles of acid that had been neutralized by the antacied tablet.

Q. 121. In titrating 25 mL of 0.01 *F*$HClO_4$ with 0.01 *F* NaOH, a beginning student used two drops of bromcresol green as an indicator. When 24.85 mL of base have been added, the solution turns from yellow to green, and the titration is stopped. Calculate the pH and the titration error when 24.85 mL of base have been added.

Q. 122. A 25 mL portion of a solution of the weak base methylamine, CH_3NH_2, is titrated with 0.1009 *F* HCl, and the equivalence point is reached when 25.57 mL of the acid have been added:

(a) What is the concentration of the methylamine?

(b) What is the P^H at the equivalence point in this titration? what indicator should be used?

Q. 123. The average pressure of CO_2 in the air is 3×10^{-4} atm. Calculate the P^H of water in equilibrium with the air at 298 K and the concentration of all species present in this solution.

Q. 124. Calculate the concentration of all species present in a 0.1 *F* $NaHCO_3$ solution at 298 K.

Q. 125. Calculate the concentrations of all species present in a 0.05 *F* Na_2Co_3 solution.

Q. 126. Calculate the change in P^H that results when 0.5 mL of 0.2 *F* NaOH is added to 10 mL of a buffer solution containing acetic acid and potassium acetate, both are concentration of 1.00 *F*.

Q. 127. (a) For a biology project you are growing certain microorganisms that can only survive in a medium at $P^H = 8.54$. You decide to use an NH_3/NH_4NO_3 buffer. What must be the molar ratio of NH_3 to NH_4NO_3 in your buffer?

(b) Calculate the P^H of a buffer prepared by mixing 40 mL of 1.0F propanoic acid with 60 mL of 0.1F NaOH. The acidity constant of propanoic acid is 1.3×10^{-5}.

Q. 128. A 2.461-g sample of solid sodium acetate is dissolved in enough water to make 50.00 mL of solution. To the sodium acetate solution 100 mL of 0.12 *F* HCl is added.

(a) What is the P^H of the resulting solution?

(b) A further addition of 5 mL of 0.12 *F* HCl is made to the buffer solution of part (a) What is the P^H after this addition?

Q. 129. Direction for preparing a buffer by mixing a weak acid and a strong base. What volume of 2.0 *F* NaOH must be added to 100 mL of 2.5 *F* HCOOH to prepare a buffer with $P^H = 4.0$?

Q. 130. The solubility of CaF_2 in water at 298 K is 1.7×10^{-3} g per 100 mL. Calculate the solubility product of CaF_2 at 298 K.

Q. 131. The value of K_{ap} for aSO_4 at 298 K is 1.1×10^{-10}. Calculate the molar solubility of $BaSO_4$ in water, and its solubility in grams per 100 mL, at 298 K.

Q. 132. T The solubility product of lead iodate, $Pb(IO_3)_2$, is 2.6×10^{-13} at 298 K. Calculate the molar solubility of $Pb(O_3)_2$ in H_2O at 298 K

Q. 133. The solubility product of white, insoluble $La(OH)_3$ at 298 K is 1×10^{-19}. Calculate the molar solubility of $La(OH)_3$ at 298 K.

Q. 134. The solubility of AgCl at 298 K is 1.8×10^{-10}. Calculate the molar solubility of AgCl in 0.04 *F* NaCl at 298 K.

Q. 135. The solubility product of mercurous bromide is 1.3×10^{-21} at 298 K. Calculate the molar solubility of mercurous bromide, Hg_2Br_2, in 0.034 *F* KBr at 298 K.

Q. 136. The solubility product of calcium sulfate is 2.4×10^{-5} at 298 K. Calculate the molar solubility of $CaSO_4$ in 0.008 $FCaCl_2$ at 298 K.

Q. 137. If we mix 70.0 mL of 0.05 *F* $a(NO_3)_2$ with 30 mL of 0.02 *F* NaF, will any BaF_2 precipitate? K_{ap} of BaF_2 is 1.7×10^{-6} at 298 K.

Q. 138. Calculate the $[Zn^{2+}]$ and $[C_2O_4^{2-}]$ remaining in solution after 15 mL of 0.12 *F* $Zn(NO_3)_2$ are mixed with 10 mL of 0.1 *F* $Na_2C_2O_4$. The K_{sp} of ZnC_2O_4 is 2.5×10^{-9}.

Q. 139. It is possible to dissolve MnS in excess HCl?

Q. 140. How many grams of $BaSO_4$ dissolve in 1 L of 2 *F* HCl?

Q. 141. Will any MnS precipitate from a solution that is 0.2 *M* in Mn^{2+} ion if it is acidified with HCl so that the $[H_3O^+] = 0.3$ *M*, and is saturated with H_2S so that $[H_2S] = 0.1$ *M*?

Q. 142. The solubility product of MnS is 5×10^{-15} at 298 K. Calculate the molar solubility of MnS in water at 298 K, including the effect of the proton-transfer reaction between S^{2-} and H_2O.

Q. 143. A given solution is 0.08 *M* in Ca^{2+} and 0.06 *M* in Mg^{2+} ions.

(a) Is it possible to precipitate 99.9% of the Ca^{2+} as $CaCO_3$ without precipitating a $MgCO_3$?

(b) What is the upper limit we must set on the $[CO_3^{2-}]$ in this solution in order to prevent any MgO_3 from precipitating?

(c) If we keep the $[CO_3^{2-}]$ at this upper limit, what is the $[Ca^{2+}]$ in solution at the end of the separation?

Chapter 10

ELECTRO CHEMISTRY

FORMULAE

$$W = z.i.t.$$

$$\text{Eq. mass} = \text{E.C.E.} \times 96500$$

$$\frac{m_1}{E_1} = \frac{m_2}{E_2} = \frac{m_3}{E_3}$$

$$E^\circ_{cell} = E^\circ_{cathode} - E^\circ_{anode} = E^\circ_{right} - E^\circ_{left}$$

$$E = E^\circ - \frac{2.303\,RT}{nF}\log\frac{[\text{Product}]}{[\text{Reactant}]} = E^\circ - \frac{0.0591}{n}\log\frac{[\text{Product}]}{[\text{Reactant}]}$$

$$E^\circ_{cell} = \frac{0.0591}{n}\log K$$

$$\Delta H = nF\left[T\cdot\frac{dE}{dT} - E\right]$$

$$\kappa = x\cdot\frac{1}{\rho}$$

$$\lambda_v = \frac{100\,\kappa}{N}$$

$$\mu_v = \frac{100\,\kappa}{M}$$

$$\lambda = \frac{\lambda_v}{\lambda_\infty}$$

$$\lambda_\infty = \lambda_a + \lambda_c$$

the conversion of electrical energy into chemical energy or vice versa is known as ***electrochemistry.***

DEFINITIONS

Electro Chemical Cell : (a) The device in which chemical energy changes into electrical energy (b) The change in free energy = electrical work done i.e. $-\Delta G$ = n.F.E.

Electrolytic Cell : The device in which electrical energy changes into chemical energy

Ohm's law. The potential difference across the ends of a conductor is directly proportional to the current flowing through it.

i.e. $$V \propto I$$

or $$\boxed{V = IR}$$

where V is the potential difference in volts, *I is the current in amperes and* R is the resistance in ohms (designated as Ω).

Conductance (C)

The ease with which current flows through a solution (or conductor) is known as conductance It is the reciprocal of the resistance, i.e.,

$$\boxed{C = \frac{1}{R}}$$

The unit of conductance is ohm^{-1} or mho or S (siemens) or Ω^{-1}.

Specific Conductance

Specific conductance of a solution of an electrolyte may be defined as the conductance of one centimeter cube of the solution.

The resistance offered to the passage of current in directly proportional to long and inversely proportional to the area of cross-section.

$\therefore$ $$R \propto \frac{l}{a}$$

$R = \rho \times \frac{l}{a}$, where ρ (rho) is specific resistance or resistively. R is the resistance, l is the length of a conductor and a is the area of the cross-section of the conductor.

$\therefore$ Specific resistance $\rho = R\ \alpha\ a/l$

Also specific conductor = k (k- kappa) = $\frac{1}{\rho} = \frac{1}{R} \times \frac{l}{a} = C \times \frac{l}{a}$

(A good conductor of electricity has a large value of K)

$\therefore$ Specific conductance (k) = observed conductance $\times$ cell constant

C is the conductance and $\frac{l}{a}$ is the cell constant. Cell constant is expressed in cm^{-1}. The unit of specific conductance (k) is ohm^{-1} cm^{-1}.

Equivalent conductivity of an electrolyte is defined as *the product of the specific conductivity of the solution and the volume in ml, which contains 1g, equivalent of the electrolyte.*

It may also be defined as the conductivity of a solution containing 1 g. eq. of the electrolyte, when placed between two electrodes one centimeter apart. It is usually denoted by λ (Lambda).

Equivalent Conductance (λ_{eq})

It may be defined as conducting power of all the ions produced by dissolving one gram equivalent (i.e., gram equivalent mass) of an electrolyte in a known volume of the solution.

It is representyed by λ_{eq} (lambda).

Mathematically

$$\boxed{\lambda_{eq} = k \times V}$$

where λ_{eq} is equivalent conductance, k is specific conductance and $V$is the volume of a solution containing one gram equivalent of a solute.

If C g equivalent of an electrolyte are dissolved in V cm^3 solution,

$\therefore$ 1 g equivalent of the electrolyte would be present in $\frac{V}{C}$ cm^3 solution.

$$\therefore \quad \boxed{\lambda_{eq} = k \times \frac{V}{C}}$$

But $V = \frac{1000}{C}$, where C is normality of the solution

$$\boxed{\lambda_{eq} = k \times \frac{1000}{C}} \text{ mhos cm}^2\text{/equivalent.}$$

Remember, here one g equivalent of an electrolyte is dissolved in 1000 cm^3 solution.

The unit of equivalent conductance in ohm^{-1} cm^2 (g eq)$^{-1}$ or S cm^2 (g eq)$^{-1}$

Molar Conductance (λ_m)

It may be defined as the conducting power of all the ions produced by dissolving one mole of an electrolyte in a known volume of the solution.

It is represented by λ_m

mathematically, $\lambda_m = k \times V$

where A_m is molar conductance, k is specific conductance and V is the volume of a solution containing one mole of an electrolyte.

If M mole of an electrolyte are dissolved in V cm^3 solution,

$\therefore$ 1 mole of the electrolyte would be present in $\frac{V}{M}$ cm^3 solution.

$$\therefore \quad \lambda_m = k \times \frac{V}{M}$$

If M = 1 mol and V = 1000 cm^3,

$$\boxed{\lambda_m = k \times 1000}$$

Remember, have one mole of an electrolyte is dissolving in 1000 cm^3 solution.

The unit of molar conductance in ohm^{-1} cm^2 mol^{-1} or S cm^2 mol^{-1}.

On dilution, specific conductance decreases due to decrease in the number of ions in one cm^3 solution but both equivalent conductance and molar conductance increase with dilution due to increase in ionization with dilution.

Molecular Conductance

The conducting power of all the ions sent in a solution having 1 mole of electrolyte (μ).

$$\mu = K \times V_{\text{in mL}} = K \times \frac{1000}{C}$$

Where V is volume having 1 mole of electrolyte; C is concentration in mol lit^{-1}, μ is expressed in ohm^{-1} cm^2 mol^{-1}.

Infinite Dilution: It means that the electrolyte is 100% ionized. The μ value increase with dilution and become constant at infinite dilution for weak electrolytes.

$$\text{Degree of dissociation } '\alpha' = \frac{\lambda_v}{\lambda_\infty}$$

where λ_1, λ_∞ are equivalent conductances of electrolyte at dilution v and respectively.

The equivalent conductivity at infinite dilution λ_∞ can be indirectly determined with the help of Kohlrausch's law (see below)

$$\lambda_\infty = \lambda_a + \lambda_0$$

where λ_0 corresponds to the conductance due to cation and λ_a corresponds to the conductance due to anions. Values of λ_c and λ_a can be determined separately.

Kohlrausch's Law

This law states that at infinite dilution where dissociation of all electrolytes is complete and where all interonic effects disappear, each ion migrates independently of its co-ion and contributes to the total equivalent conductance of an electrolyte a definite share which depends only on its own nature and not at all on the ions with which it is associated. Thus equivalent conductance at infinite dilution is equal to the sum of ionic conductances of its component ions a infinite dilution

i.e., $$\lambda^\infty_{\text{electrolyte}} = \lambda^\infty_{\text{cation}} + \lambda^\infty_{\text{anion}}$$

where λ^∞_c and λ^∞_a are ionic equivalent conductance of cation and anion respectively at infinite dilution expressed in ohm^{-1} cm^2 eq^{-1}

Thus,

$$\lambda^\infty_{\text{NaCl}} = \lambda^\infty_{\text{Na}} + \lambda^\infty_{\text{Cl}} \quad (1)$$

$$\lambda^\infty_{\text{BaCl}_2} = \frac{1}{2}\lambda^\infty_{\text{Ba}^{+2}} + \frac{1}{1}\lambda^\infty_{\text{Cl}} \quad (2)$$

$$\lambda^\infty_{\text{AlCl}_3} = \frac{1}{3}\lambda^\infty_{\text{Al}^{+3}} + \frac{1}{1}\lambda^\infty_{\text{Cl}} \quad (3)$$

$$\lambda^{\infty}_{Al_2\ (SO_4)_3} = \frac{1}{3}\lambda^{\infty}_{Al^{+1}} + \frac{1}{2}\lambda^{\infty}_{SO_4^{2}} \qquad (4)$$

In General

$$\boxed{\lambda^{\infty}_{m} = v_c\,\lambda^{\infty}_{c} + v_a\,\lambda^{\infty}_{a}}$$

where λ^{∞}_{m} is the molar conductance of the electrolyte at infinite dilution (zero concentration), λ^{∞}_{c} and λ^{∞}_{a} are the molar conductance of cations and anions present in one formula unit of the electrolyte.

e.g.

$$\boxed{\lambda^{\infty}_{m}(BaCl_2) = \lambda^{\infty}_{m}(Ba^{2+}) + 2\,\lambda^{\infty}_{m}(Cl^{-})}$$

Similarly,

$$\boxed{\lambda^{\infty}_{eq} = v_c\lambda^{\infty}_{c} + v_a\lambda^{\infty}_{a}}$$

where λ^{∞}_{eq} is the equivalent conductance of the electrolyte at infinite dilution, λ^{∞}_{c} and λ^{∞}_{a} are the electrolyte at infinite dilution λ^{∞}_{c} and λ^{∞}_{a} are the equivalent conductance of cation and anion respectively at infinite dilution.

For the molar conductance (μ) at infinite dilution

$$\mu^{\infty}_{BaCl_2} = \mu^{\infty}_{Ba^{+2}} + 2\,\lambda^{\infty}_{Cl^-}$$

$$\mu^{\infty}_{AlCl_3} = \lambda_{Al^{+3}} + 3\,\lambda^{\infty}_{Cl}$$

$$\mu^{\infty}_{Al_2(SO_4)} = 2\,\lambda^{\infty}_{Al^{+3}} + 3\,\lambda^{\infty}_{SO_4^{-2}}$$

Thus in general

$$\mu^{\infty} = \gamma_+\,\lambda^{\infty}_{+} + \gamma_-\,\lambda^{\infty}_{-}$$

where γ_+ and γ_- are the number of +ve and –ve ions furnished by each formula unit of electrolyte.

Electrolysis

The process of decomposition of an electrolyte (in the fused state or in the form of their solution) by the passage of electric current is known as electrolysis.

Faraday's Laws of Electroloysis

Faraday's First law of Electrolysis

The amount of substance liberated or decomposited at the respective electrode during electrolysis is directly proportional to the quantity of electricity that passes through the electrolyte.

$$W \propto Q \text{ or } W = ZQ$$

and $$W \propto It \text{ or } \boxed{W = ZIt}$$

where W is the amount of the substance liberated or deposited; Z is electrochemical equivalent: I is the strength of the current in amperes; t is the time in seconds and Q is the quantity of electricity in coulombs.

when $I = 1$ ampere, $t = 1$ sec, $W = Z$

Thus, Electrochemical equivalent may be defined as the amount of the substance liberated or deposited during electrolysis, when one coulomb of electricity is passed through the electrolytic cell.

Thus unit of $Z = \text{kg C}^{-1}$

Faraday's Second law of Electrolysis

When same quantity of electricity is passed through the solution of different electrolysis connected in series the masses of different substances deposited or liberated at the electrodes are directly proportional to their equivalent weights.

i.e., A given quantity of electricity produces the same number of equivalents of any substances during electrolysis.

$$W \propto E \text{ or } \frac{W}{E} = \text{constant}$$

where W is the mass of the substance liberated or deposited and E is the equivalent weight.

It W_A and W_B be the masses of the substances A and B deposited at the electrodes and E_A and E_B be their equivalent weights,

$$\boxed{\frac{W_A}{W_B} = \frac{E_A}{E_B}}$$

Combining the two laws, we get

$$W \propto EQ$$

or $$\boxed{W = \frac{EQ}{F}}$$

or $$\boxed{\frac{W}{E} = \frac{Q}{F} = \frac{1 \times t}{F}}$$

where F is a constant know as **FARADAY**. It may be defined as the quantity of electricity required to deposit one g equivalent of an element during electrolysis.

As 1 Faraday (F) is the quantity of change carried by 1 mole of electrons

So $1F = N \times e(1 \text{ mole})\ 6.023 \times 10^{23} \times 1.602 \times 10^{-19} = 96514 \neq 96500$ coulombs, where N is Avagdro number & e is charge of 1 electron.

Thus, 96500 C discharge E g of ion

$$1 \text{ C discharge } \frac{E}{96500}, \text{ g ion} = Z$$

$$\therefore \quad w = \frac{E.i.t}{96500}$$

Note: 1. If E is given in g: w is in g
(If E is given in kg, w is in kg)

2. and Equivalent of an ion discharged, $\left(\frac{w}{E}\right) = \frac{i.t}{96500}$

Relation between FARADAY and Number of Electrons

Consider the electrode equations: $Cu^{2+} + 2e \rightarrow Cu(s)$ or $Na^{+} + e^{-} \rightarrow Na$.

2F	1 mole	IF	1 mole
1 F	$\frac{1}{2}$ mole or		lg equivalent

In case of Na only 1 electron is taken while in case of copper two electrons are taken. In general, to deposit 1 mole of atoms of any metal of valency Z; Z moles of electrons are needed i.e. if 1F electricity is passed through molten NaCl, 23 g or 1 mole or 1 gram equivalent of Na is deposited at cathode OR 1 g equivalent of chlorine is formed at anode.

ELECTRODE POTENTIAL

The difference of potentials between the electrodes of a galvanic call can be measured by using a voltameter. The potential difference is known as electromotive force (emf of EMF) of the cell.

Each electrode makes a contribution to the cell potential which is known as electrode potential. This is caused by the electrode reaction taking place at the electrodes when the cell is working.

Oxidation takes place at an electrode, the electrode potential is called reduction electrode potential.

For reduction potentials, electrode reaction is a reduction reaction.

The general reduction electrode reaction is $A^{n+} + ne^{-} \rightarrow A$

Electrode Potential Depends Upon

(i) The nature of the electrode (ii) the concentration of the ions or the electrode material in solution, and (iii) the temperature.

Electrode potential is given by the symbol E. The relationship between the electrode potential and the concentration of the ions is solution is given by Nernst equation. (See below)

The list of standard reduction electrode potentials arranged in order of increasing value is often called activity, series or electro chemical series because it is also the order of activity of the respective elements from given pages.

The most active metals are at the top of the list.

Therefore, the more negative the reduction electrode potential the more reactive the metal will be in displacing other metals from their salts. If the possible to demonstrate in the laboratory that any metal given in the table disposes the metal given below it from the salt solution. For example, copper lies just above silver. If copper foils are added to silver nitrate solution silver is displaced from the solution.

If Lithium is added to a solution containing H^+ ions, H^+ ions are reduced to hydrogen gas and lithium is oxidised to Li.

GALVANIC CELL OR VOLTAIC CELL

It is a device for converting chemical energy into electrical energy. Daniell cell is an example of Galvanic cell.

In galvanic cell anode has –ve sign and cathode has +ve sign. Galvanic cell is a source of EMF.

The galvanic cell consists of two half-cell—a cathode and an anode. A reduction half-reaction occurs in the cathode and an oxidation reaction occurs in the anode when an external circuit connects the two half-cells. A porous barrier prevents contact between the oxidizing agent and reducing agent in the cell while allowing the migration of ions between the half-cells.

Cell notation: A single vertical line sands for an interface between two phases. A dashed line represents a liquid-liquid juncture. The anode appears on the left and the cathode on the right.

A Daniell cell is made up of two half cells: One containing zinc electrode in contact with 1 M solution of $ZnsO_4$ and the other containing copper electrode in contact with 1 M solution of $CuSO_4$. The two half cells are separated by porous pot or salt bridge.

$$Zn \rightarrow Zn^{2+} + 2e^- \qquad \text{(oxi. half cell reaction at left electrode)}$$

$$Cu^{2+} + 2e^- \rightarrow Cu \qquad \text{(red. half cell reaction at right electrode)}$$

Thus ***Daniell cell is represented as***

$$Zn|Zn^{2+}(1M) \,||\, Cu^{2+}(1M) \,|\, Cu$$

In Daniell cell, if the external EMF is greater than cell EMF, cell reaction is reversed. It is reversible cell

$$Zn^{2+} + 2e^- \rightarrow Zn: Cu \rightarrow Cu^{2+} + 2e^-$$

The L.H.S. is anode and is negative electrode (electrons are produced).

The R.H.S. the cathode positive electrode (electrons are consumed).

The **potential** of the above cell, E_{cell}, is given by:

1st Method:

$$\begin{aligned} E_{cell} &= E_{left} + E_{right} \\ &= E_{oxidation} + E_{reduction} \\ &= E_{anode} + E_{cathode} \end{aligned}$$

This is a method used in most standard text-books

2nd Method. The E_{cell} is the algebraic difference between the oxidation potentials of the two half-cells making up the cell.

$$E_{cell} = E_{left} + E_{right}$$

(Use only oxidation potentials for the two half-reactions)

3rd Methods $E_{cell} = E_{right} + E_{left}$

(use only reduction potentials for the two half-reactions)

A positive value of E_{cell} indicates that the reaction as written shall occur spontaneously.

Electromotive Force (EMF)

1. The difference in the potential of the two half cells of a cell.
2. It is directly measured with a voltmeter.
3. It is exposed in volts.
4. It can be calculated from standard potentials of the half cells.

Cell potential of EMF of the cell can be calculated in three ways.

(a) Cell potential = reduction potential of cathode – reduction potential of anode.

$$E_{cell} = E_{cathode} - E_{anode} \text{ or } E_{cell} = E_{right} - E_{left}$$

(b) E_{cell} = oxidation potential of anode – oxidation potential of cathode i.e.,

$$E_{cell} = E_{anode} - E_{cathode}$$

(c) E_{cell} = oxidation potential of anode + reduction potential of cathode.

Nernst Equation

Nernst equation for electrode potential.

The effect of the concentration of the ions on the voltage of an electrode may be determined by means of Nernst equation:

I. (a) For the reduction reaction.

$$M^{n+} + ne^- \rightarrow M$$

The Nernst equation for the electrode potentials at 298 K is

$$E = E^\circ - \frac{0.0591}{n} \log \frac{1}{[M^{n+}]}$$

where E = Eelctrode potential under given concentration of M^{n+} ions

E° = Standard electrode potential

n = Number of moles of electrons involved in the electrode reaction

(b) For the overall cell reaction

$$aA + bB = xX + yY$$

The Nernst equation at 298 K is:

$$E_{cell} = E^\circ_{cell} - \frac{0.0591}{n} \log \frac{[X]^x\,[Y]^y}{[A]^a\,[B]^b}$$

where n is the number of moles of electrons exchanged in the reaction.

II. If electrode potential is given as:

$$A \rightleftharpoons A^{+n} + ne$$

Thus, $$E_{OP} = E^\circ_{OP} - \frac{RT}{nF} \log_e \frac{a_{\text{oxidized state}}}{a_{\text{reduced state}}} \qquad \ldots\ldots (1)$$

and $$E_{RP} = E^{\circ}_{RP} + \frac{RT}{nF}\log_e \frac{a_{OS}}{a_{RS}} \qquad \ldots\ldots (2)$$

Also $$E_{OP} = -E_{RP} \qquad \ldots\ldots (3)$$

and $$E^{\circ}_{OP} = E^{\circ}_{RP} \qquad \ldots\ldots (4)$$

where, E_{OP} and E_{RP} are oxidation potential and reduction potential respectively.

E°_{OP} and E°_{RP} are standard O.P. and standard R.P. defined as equal to E_{OP} and E_{RP} respectively when, $\frac{a_{OS}}{a_{gS}} = 1$

where

R is molar gas constant = 8.314 J K^{-1} mol^{-1} (M.K.S. system since E in volt)
T is temperature in Kelvin
n is no. of electrons lost or gained during oxidation or reduction in redox change.
F is one Faraday i.e. 96500 C
a_{OS} = active mass of oxidized state in solution
a_{RS} = active mass of reduced state in solution.

Since $a = fc$

where, f is activity coefficient

For dilute solutions $f = 1$

$\therefore$ a = concentration in molarity

Note, Eq. (1) may be written as

$$E_{OP} = \frac{RT}{nF}\log_e \frac{[O.S.]}{[R.S.]} \qquad \ldots\ldots (5)$$

$$E_{OP} = E^{\circ}_{OP} - \frac{2.303\,RT}{nF}\log_{10} \frac{[O.S.]}{[R.S.]} \qquad \ldots\ldots (6)$$

Numerical value of $\frac{2.303\,RT}{F}$ = 0.058 at 288 K

= 0.059 at 298 K

= 0.060 at 308 K

i.e. there is no significant change with temperature

Thus, By Eq. (6)

$$\therefore \quad E_{OP} = E^{\circ}_{OP} - \frac{0.059}{n}\log_{10} \frac{[O.S.]}{[R.S.]} \qquad \ldots\ldots (7)$$

Similarly,

and $$E_{RP} = E^{\circ}_{RP} + \frac{0.059}{n}\log_{10} \frac{[O.S.]}{[R.S.]} \qquad \ldots\ldots (8)$$

Formation of Equation for Different Electrodes

Case I. $\boxed{M|M^{+n}_{eq}}$ i.e., $M \rightleftharpoons M^{+n} + ne$

$$E_{OP} = E^{\circ}_{OP} - \frac{0.059}{n}\log_{10}\frac{[M^{+n}]}{[M]}$$

or
$$E_{OP} = \frac{0.059}{n} - \frac{0.059}{n}\log_{10}(M^{+n}) \qquad [\therefore M \text{ in solid state = unity}]$$

and
$$E_{RP} = \frac{0.059}{n}\log_{10}[M^{+n}]$$

II. $\boxed{P_{H_2}|H^+_{aq}\ (P)}$ i.e., $H_2 \rightleftharpoons 2H^+ + 2e$

$$\frac{1}{2}H_2 \rightleftharpoons H^+ + e$$

gaseous phase concentration is reported as pressure, i.e., $[H_2]$ as P_{H_2}

$$E_{OP} = E^{\circ}_{OP} - \frac{0.059}{2}\log_{10}\frac{[H^+]^2}{P_{H_2}} \qquad \ldots\ldots (9)$$

$$E_{OP} = E^{\circ}_{OP} - \frac{0.059}{1}\log_{10}\frac{[H^+]}{(P_{H_2})^{1/2}} \qquad \ldots\ldots (10)$$

Eqs (9) and (10) are same and thus it is evident the stoichiometry of range in half cell e.m.f. has no effect on Nernst expression.

Similarly,
$$E_{RP} = E^{\circ}_{RP} + \frac{0.059}{2}\log_{10}\frac{[H^+]^2}{P_{H_2}}$$

III. $\boxed{Pt_{Cl_2}|Cl^-_{aq}}$ i.e., $2Cl^- \rightleftharpoons Cl_2 + 2e$

$$E_{OP} = E^{\circ}_{OP} - \frac{0.059}{2}\log_{10}\frac{P_{Cl_2}}{[Cl^-]^2}$$

Similarly,
$$E_{RP} = E^{\circ}_{RP} + \frac{0.059}{2}\log_{10}\frac{P_{Cl_2}}{[Cl^-]^2}$$

Some Applications of Nernst Equation

(I) ***Equilibrium Constant form Nernst Equation:*** When the cell reaction at equilibrium, the system does no no work and the cell e.m.f. is zero

$$-\Delta G = nF.\ E$$
$$-\Delta G = 0 \qquad (\because E = 0)$$

Consider the following reaction in equilibrium

$$Zn + Cu^{+2} \rightleftharpoons Zn^{+2} + Cu$$

At equilibrium

$$K_C = \frac{[Zn^{+2}]}{[Cu^{+2}]} \quad \text{.....(1)}$$

$$E_{cell} = E^{\circ}_{cell} + \frac{RT}{nF}\log_e \frac{[Cu^{+2}]}{[Zn^{+2}]} \quad \text{.....(2)}$$

Zn is oxidized and Cu^{+2} is reduced

At equilibrium

$$E_{cell} = 0$$

$$\therefore \quad -E^{\circ}_{cell} = \frac{RT}{nF}\log_e \frac{1}{K_C}$$

$$E^{\circ}_{cell} = \frac{RT}{nF}\log_e K_C$$

or $$n.F.\ E^{\circ}_{cell} = RT \log_e K_C$$

or $$-\Delta G^{\circ} = RT \log_e K_C$$

$$-\Delta G^{\circ} = 2.303\ RT \log_{10} K_C \quad \text{.....(3)}$$

where, ΔG° is change in standard free energy

II. ***Heat of reaction for cell reaction:*** The heat of reaction for cell reaction (ΔH) at a temperature is calculated by Gibb's Helmholtz equation.

$$\Delta G = \Delta H + T\left(\frac{\delta}{\delta T}\Delta G\right)_P$$

$$-\Delta G = nEF$$

$$-nEF = \Delta H + T\left[\frac{\delta}{\delta T}(-nEF)\right]_P$$

or $$E = \frac{\Delta H}{nF} + T\left(\frac{\delta E}{\delta T}\right)_P$$

$$\text{or } \Delta H = nF\left[T\left(\frac{\delta E}{\delta T}\right)_P - E\right]$$

here, $\left(\frac{\delta E}{\delta T}\right)_P$ is called temperature coefficient of e.m.f. i.e., rate of change e.m.f. with temperature.

III. ***To decide spontaniety of a cell reaction:*** Compute E°_{cell} for the given reaction e.g.,

$$A + B^{+2} \rightarrow A^{+2} + B$$

$$E^{\circ}_{cell} = E^{\circ}_{OP_A} + E^{\circ}_{EP_B}$$

If E°_{cell} comes to be +ve, cell reaction is spontaneous and if E°_{cell} comes be –ve, cell reaction is not spontaneous.

IV. ***The maximum work that can be obtained from galvanic cell,***

$$\Delta G^\circ = -nF\,E^\circ_{cell} = -2.303\ RT \log K_c$$

where ΔG° is standard free energy change; n is the number of moles of electrons involved in the cell reaction; F is Faraday (1 F = 96,500 columbs); E°_{cell} is standard cell potential; R is gas constant and K_c is the equilibrium constant.

Concentration Cell

A voltaic cell in which two electrodes are same is known as concentration cell.

For such cell, $E^\circ_{cell} = E^\circ_1 - E^\circ_2 = 0$

The potential of a concentration cell depends on the difference in concentration between two half-cells, both of which utilize the same half-reaction. The measured potential of a concentration cell can be used to estimate concentrations.

PROBLEMS

Q. 1. Calculate the number of coulombs required to deposit 50 g. of silver at cathode from silver nitrate solution.

Q. 2. An electric current of 100 amperes is passed through a molten liquid of sodium chloride for 5 hours. Calculate the volume of chlorine gas liberated at the electrode at STP.

Q. 3. A certain quantity of electricity deposits 0.54 of Ag from silver nitrate solution. What volume of hydrogen will be liberated by the same quantity of electricity at 300 K and 750 mm of Hg pressure (H = 1, Ag = 108)

Q. 4. If 3.0 faradays of electricity is passed through an iron (II) bromide solution, how many grams of iron metal will be deposited? (At. wt. of Fe = 56)

Q. 5. An electric current is passed through a solution of 10 g. of $CuSO_4 \cdot 5H_2O$ crystals in 500 ml. of water, platinum electrodes being used in each case. After 30 minutes it was found that 1.307 g. of silver has been deposited. What was the concentration of copper expressed as grams of copper per litre in the copper sulphate solution after electrolysis?

Q. 6. The same quantity of electricity that caused the deposition 10 g of silver from a solution of $AgNO_3$ was passed through a solution of gold salt containing cations of unknown charge. An amount of 6.08 g of gold was deposited. What is the charge on the gold cations?

Q. 7. How many grams of silver could be plated out on a serving tray be electrolysis of a solution containing silver in +1 oxidation state for a period of 8.0 hours at a current of 8.46 Amperes? What is the area of the tray if the thickness of the silver plating is 0.00254 cm? Density of silver is 10.5 g cm^{-3}.

Q. 8. Molten aluminium chloride is electrolysed with a current of 0.5 ampere to produce 27.0 g. aluminium.

(i) How many gram equivalents of aluminum were produced?
(ii) How many gram atoms of aluminium were produced?
(iii) How many atoms of aliuminium were produced?
(iv) How many electrons were produced?
(v) What is the number of faradays of electricity consumed?
(vi) How long did the electrolysis take?
(vii) How many litres of chlorine at S.T.P. were produced?
(A. wt. of At = 27; Avogadro No. = 6.02×10^{23})

Q. 9. 0.2964 g of copper was deposited on passage of current 0.5 amperes for 30 minutes through a solution of copper sulphate. Calculate the atomic weight of copper.

Q. 10. In the electrolysis of an aqueous cupric bromide solution how many grams of bromine are formed on passing a current of 1 ampere for 16 minutes and 5 seconds? Write the anode and the cathode half-reactions.

Q. 11. A current of 3.7 amperes is passed for 6 hours between nickel electrodes in 0.5 litres of a 2M solution of nickel nitrate. what will be the molarity of the solution at the end of electrolysis? [F = 96500; Ni 58.7].

Q. 12. Write the chemical reaction of following cells.
(a) $Cd \mid Cd^{2+}$ (1 M) || Cl^- (1 M), AgCl (s) | Ag

(b) $Zn \mid Zn^{2+}$ (1 M) $\parallel$ Cl $\mid$ Pt, Cl_2
(c) Pt, H_2 (1 atm) $\mid$ HCl (0.1 M) $\mid$ Cl_2, Pt

Q. 13. Construct electric cells for the following reactions.
(a) $Fe + Cu^{2+} = Cu + Fe^{2+}$
(b) $Cu + 2\ Ag^{+} = 2\ Ag + Cu^{2+}$
(c) $6\ Fe^{2+} + Cr_2O_7^{2-} + 14\ H^{+} = 6\ Fe^{3+} + 2\ Cr^{3+} + 7H_2O$

Q. 14. For the following cells, calculate EMF
(a) $Cu \mid Cu^{2+} \parallel Ag^{+} \mid Ag$
Electrode reduction potentials $Cu^{2+}/Cu = 0.34$ V, $Ag^{+}/Ag = 0.8$ V
(b) $Zn \mid Zn^{2+} \parallel Fe^{2+} \mid Fe$
Electrode reduction potentials $Fe^{2+}/Fe = -0.44$ Zn^{2+}/Zn 0.76V

Q. 15. Explain wheather follwong cell can produce electricity or not.
$Ag \mid Ag^{+} \parallel Zn^{2+} \mid Zn$ $\qquad E^{\circ}_{Ag^{+}/Ag} = 0.80$ V, $E_{Zn^{2+}/Zn} = -0.76$ V

Q. 16. An aqueous solution of NaCl on electrolysis given H_2 (g), Cl_2 (g) and NaOH according to the reaction:

$$2Cl^{-}(aq) + 2H_2O(l) \rightarrow 2OH^{-}(aq) + H_2(g) + Cl_2\ (g)$$

A direct current of 25 amperes with a current efficiency of 62% is passed through 20 litres of NaCl solution (20% by mass). Write down the reactions taking place at the anode and the cathode. How long will it take to produce 1 kg of Cl_2? What will be the molarity of the solution with respect to hydroxide ion?
(Assume no loss due to evaporation)

Q. 17. The Edison storage cell is represented as:

Fe (s) $\mid$ FeO (s) $\mid$ KOH (aq) $\mid$ Ni_2O_3 s) $\mid$ Ni (s)

the half –cell reaction are

$$Ni_2O_3(s) + H_2O\ (l) + 2e^{-} \rightleftharpoons 2\ NiO(s) + 2OH; \quad E^{\circ} = +0.4\ V$$
$$FeO(s) + H_2O(l) + 2e^{-} \rightleftharpoons Fe(s) + 2OH; \quad E^{+} = -0.87\ V$$

(i) What is the cell reaction?
(ii) What is the cell e.m.f. How does it depend on concentration of KOH?
(iii) What is the maximum amount of electrical energy that can be obtained from one mole of Ni_2O_3?

Q. 18. Indicate wheather the reaction occurs or not by determining cell EMF

$$Cd + Cu^{2+} = Cd^{2+} + Cu$$

$$E^{\circ}_{cd^{2+}/Cd} = -0.403V;\ E^{2+}_{Cu^{2+}/Cu} = 0.3$$

Q. 19. A 200 watt, 110 volt in-candescent lamp is connected in series with an electrolytic cell of negligible resistance containing a solution of zinc chloride. What weight of zinc will be deposited from the solution on passing the current for 30 minutes?

Q. 20. Calculate the emf of half cells given below.
(a) Pt, H_2 $\mid$ HCl $\mid$ $E^{\circ} = 0$ V;
2 atm $\quad$ a = 0.02
(b) Fe $\mid$ $FeSO_4$ $\quad$ $E^{\circ} = 0.33$ V
a = 0.1

(c) Pt, Cl_2 | HCl E° = −1.35 V
10 atm a = 0.1

Q. 21. (a) Calculate E for a Zn electrode in which Zn^{2+} = 0.025 M, E° for Zn → Zn^{2+} + $2e^-$ is 0.763 volts

(b) Calculate E_{cell} for the Deniell cell given below at 298 K

E° for Cu^{2+} | Cu I − 0.34 Volts.

Zn | zn^{2+} 0.5 M || Cu^{2+} (0.20 M) | Cu

Q. 22. A current of 5 amperes is employed to plate nickel in a $NiSO_4$ bath for 2 hours. The current efficiency with respect to formation of Ni is 70%

(a) How many grams of Nickel are plated on cathode

(b) What is the thickness of the plating of the Ni sheet of an side of 4.0 cm. The density of Nickel is 8.9 g/c.c.

Q. 23. A current of 3 amperes was passed for 2 hours through a solution of copper sulphate when 3 gm. of copper was deposited. Calculate the current efficiency. (Cu = 63.5)

Q. 24. The same quantity of electricity is passed between Pt-electrodes dipped in the following solutions kept in separate beakers. (a) 0.03 molar $AlCl_3$, (b) 0.01 molar $AlCl_3$, (c) 0.1 molar $AgNO_3$ an (d) 0.1 molar NaOH. If 0.36 g of Ag is deposited in beaker C, what is the mass of Al deposited in beaker *A* and *B* and volume of O_2, gas evolved in cell *D* at S.T.P.?

Q. 25. Calculate the e.m.f. of the following cell

Fe | Fe^{2+} || Sn^{2+} | Sn E°_{Fe} = −0.44 volt

a = 0.6 *a* = 0.2 E°_{Sn} = +0.14 volt

Q. 26. Calculate the quantity of current necessary to produce oxygen gas at the rate of 1 ml per second?

Q. 27. (a) 100 ml of 0.4M $CuSO_4$ solution is electrolysed with a current of 1.5 amperes for 50 min. The electrode reactions are: Cathode Cu^{2+} + $2e^-$ → Cu. Anode $2H_2O \rightarrow O_2 + 4H^+ + 4e^-$. Assuming that volume of solution does not change during electrolysis. calculate molar conc. of Cu^{2+}, H^+ and SO_4^{2+} after electrolysis.

(b) A current of 60 micro amperes is passed through a solution of copper sulphate for 20 minutes using Pt. electrodes. Cu is uniformly deposited and covers 53% of the cathode surface. The thickness layer deposited is a single atom copper. What is the total surface area of cathode if each copper atom covers an area of 8.7×10^{-6} cm^2?

Q. 28. A current of 40 micro ampere is passed through a solution of silver nitrate for 32 minutes using platinum electrodes. a uniform single atom thick layer of silver is deposited covering 43% of the cathode surface. What is the total surface area of cathode if each silver atom curves 5.4×10^{-16} cm^2?

Q. 29. Cadmium amalgam is prepared by electrolysis of a solution of calcium chloride, using a mercury cathode. for how long should electrolysis be carried out in order to prepare a 12 percent (by wt.) calcium amalgam using 22 g. of Hg and a current of 5 ampere?

Q. 30. The standard reduction potential of the Ag^+/Ag electrode at 298 K is 0.799 V. Given that for AgI, $K_{sp} = 8.7 \times 10^{-17}$, evaluate the potential of the Ag^+/Ag electrode in a standard solution of AgI. Also calculate the standard reduction potential of the $I^-/AgI/Ag$ electrode.

Q. 31. Calculate the standard electrode potentials of the following half cells:
(i) Zn | $ZnSO_4$ (0.1 M), E_x° = +0.76 V
(ii) Pt, Cl_2 (10 atm.) | HCl (0.1 M); E_x° = 1.36 V
(iii) Pt, H_2 (4 atm.) HCl (0.02 M); E_x° = 0.0 V

Q. 32. Calculate the pH of the following solutions
(a) Pt, H_2 | HCl E = 0.25 V
(b) Pt, H_2| H_2SO_4 E = 0.3 V

Q. 33. Calculate the equilibrium constant for the reaction.

Fe + $CuSO_4$ = $FeSO_4$ + Cu at 298 K E°_{FE} = 0.44, E°_{Cu} = –0.33 V

Q. 34. The standard reduction potential of Cu^{2+}/Cu and Ag^+/Ag electrodes are 0.337 and 0.799 volt respectively. Construct a galvanic cell using these electrodes so that it standard emf is positive. For what concentration of Ag^+ will the emf of the cell, at 298 K, be zero if the concentration of Cu^{2+} is 0.01 M.

Q. 35. Calculate the quantity of electricity that would be required to reduce 12.3g of Nitrobenzene and the current of the process is 50 percent. If the potential drop across the cell is 3.0 volts how much energy will be consumed.

Q. 36. An acidic solution of Cu^{2+} salt containing 0.4 g of Cu^{2+} is electrolysed until all the copper is electrolysed. The electrolysis is continued for seven more minutes with the volume of solution kept at 100 ml and the current at 1.2 amp. Calculate the volume of gases evolved at NTP during the entire electrolysis.

Q. 37. The standard reduction potential at 298 K of the reaction $2H_2O + 2e^- \rightarrow H_2 + 2OH$ is 0.8277 V. Calculate the equilibrium constant for the reaction. $2H_2O \rightarrow H_3O^+ + OH^-$.

Q. 38. A cell contains two hydrogen electrodes. The negative electrode is in contact with a solution of 10^{-4} M hydrogen ions. The emf of the cell is 0.118 v at 298 K. Calculate the concentration of hydrogen ions at the positive electrodes.

Q. 39. In a fuel cell hydrogen and oxygen react to produce electricity. In the process hydrogen gas is oxidized at the anode and oxygen at the cathode. If 67.2 litres of H_2 at STP react in 15 minutes, what is the average current produced? If the entire current is used for electro deposition of copper from copper (II) solution, how many grams of copper will be deposited.

Anode reaction: $H_2 + 2OH^- \rightarrow 2H_2O + 2e^-$

Cathode reaction: $O_2 + 2H_2O + 4e^- \rightarrow 4OH^-$

Q. 40. A 100 W, 110 V incandescent lamp is connected in series with an electrolysis cell containing cadmium sulphate solution. What weight of cadmium will be deposited by the current flowing for 10 hrs.

Q. 41. The molar volume of liquid benzene (density = 0.877 g mL^{-1}) increases by a factor of 2750 as it vaporizes at 293K can that of liquid toluene (density = 0.867 g ml^{-1}) increases by a factor of 7720 at 293K. A solution of benzene and toluene at 293K has a vapour pressure of 605×10^{-4} atm. Find the mole fraction of benzene in the vapour above the solution.

Q. 42. During the discharge of a lead storage battery the density of sulphuric acid fell from 1.294 to 1.139 g ml. sulphuric acid of density 1.294 g mL^{-1} is 39% H_2SO_4 by weight and that of density 1.139 g mL^{-1} is 293K H_2SO_4 by weight. The battery holds 3.5 litres of the acid and the volume remained practically constant during the discharge.

Calculate the number of ampere – hours for which the batter must have been used. The changing and discharging reactions are

$$Pb + SO_4^{2-} = PbSO_4 + 2e^- \text{ (charging)}$$

$$PbO_2 + 4H^+ + SO_4^{2-} + 2e^- = PbSO_4 + 2H_2O \text{ (discharging)}$$

Q. 43. How long a current of 3 ampere has to be passed through a solution of $AgNO_3$ to coat a metal surface of 80 cm with 0.005 mm thick layer (Density of silver is 10.5 g mL^{-1})

Q. 44. In an electrolysis experiment current was passed for 5 hours through two cells connected in series. The first cell contains a solution of gold and the second contains copper sulphate solution. 9.85 g. of gold was deposited in the first cell. If the oxidation number of gold is +3 find the amount of copper deposited on the cathode of this second cell. Also calculate the magnitude of current in amperes.

Q. 45. 19 g, of molten $SnCl_2$ electrolysed for some time. Inert electrodes are used 0.119 g. of Sn is deposited at the cathode. No. substance is lost during the electrolysis. find the ratio of the weights of $SnCl_2$ and $SnCl_4$ after electrolysis.

Q. 46. A hot solution of NaCl in water is electrolysed. Iron electrodes are used. Diaphragm cell is not used. Give equations for all the chemical reactions that take place during electrolysis.

Q. 47. Find the charge in coulombs of 1 g. ion of N^{3-}.

Q. 48. **(a)** The standard oxidation potential of the half-cell $Zn \mid Zn^{2+}$ is 0.763 volt. It is connected with another half cell and the total e.m.f. of the cell is also 0.763 volts. What is other half cell? Write down the other half cell if Zn is being oxidized to Zn^{2+}

(b) Calculate the E° value for the half-cell $2Hg^{2+} + 2e^- \rightarrow Hg_2^{2+}$ from the given data:

$2Hg \rightarrow 2Hg_2^{2+} + 2e^-; -0.796$ V $Hg \rightarrow Hg^{2+} + 2e^-$; E° = –0.851 V.

Q. 49. The density of copper is 8.94 g mL^{-1}. Find out the number of coulombs needed to plate an area 10 mm × 10 mm to a thickness of 10 mm using $CuSO_4$ solution as an electrolyte? [Cu = 63.5]

Q. 50. Silver is electro-deposited on a metallic vessel of surface area 800 cm by passing a current of 0.20 amp for 30 hours. Calculate the thickness of Silver deposited, given its density as 10.47 g cm^{-3}.

Q. 51. Calculate the quantity of electricity that will be required for liberating 710 gm of chlorine gas by the electrolysis of a concentrate solution of NaCl. What (i) weight of NaOH and (ii) volume of hydrogen measured at 300 K and 1 atm. pressure will be obtain during the process.

Q. 52. After electrolysis of a NaCl solution with inert electrodes for a certain period of time, 600 ml of the solution was left which was found to be 1 N in NaOH. During the same time 31.8 g of copper was deposited in copper voltmeter in series with the electrolytic cell. Calculate the percentage of the theoretical yield of the sodium hydroxide obtained (Cu = 63.5)

Q. 53. 100 ml of 0.8 M cusO_4 is electrolysed for 30 minute by passing a current of 5 amperes. Calculate the amount of $CuSO_4$ in grams in the solution. [Cu = 63.5 5 = 32]

Q. 54. For the galvanic cell Ag|AgCl(s) | KCl (0.02 M) || KBr (0.01 M) | AgBr (s) | Ag calculate the EMF generated and assign correct polarity to each electrode for spontaneous process after taking into account the cell reaction at 298 K.
$[K_{sp}(AgCl) = 2.8 \times 10^{-10}, K_{sp}(AgBr) = 3.3 \times 10^{-13}]$

Q. 55. 10 grams of a fairly concentrated solution of cupric sulphate is electrolysed using 0.01. Faraday of electricity. Calculate.
(i) the weight of the resulting solution and
(ii) the number of equivalents of acid or alkali in the solution
[Cu = 63.5; F = 96,500]

Q. 56. (a) A from the values of a faraday and of Avogadro number calculate the coulombic charge on an electron.
(b) Electric current is passed through two cells *A* and *B* in series. Cell *A* contains an aqueous solution of Ag_2SO_4 and platinum electrodes. Cell *B* contains aqueous solution of $CuSO_4$ and copper electrodes. The current is passed till 1 g of oxygen is liberated at the anode of cell *A*.
(i) Give equations for the reaction taking place at each electrode.
(ii) Calculate the quantities of substances at the cathode of the two cells.

Q. 57. A for the electrolytic production of $NaClO_4$ from $NaClO_3$ as per the reaction

$$ClO_3^- + H_2O \rightarrow ClO_4^- + 2H^+ + 2e^-$$

(i) How many faradays of electricity would be required to produce 1 gram molecule of $NaClO_4$?
(ii) What volume of H_2 at S.T.P. will be berated at the carthode in the time that maxes to form 12.25 grams of $NaClO_4$?

Q. 58. 40 ml of 0.125 M $NiSO_4$ solution is electrolysed by a current of 0.05 ampere for 40 minute
(i) Write equations for the reactions occurring at each electrode.
(ii) How many coulombs of electricity passed through the solution?
(iii) How many grams of the product deposited on cathode?
(iv) How long will the same current have to pass through the solution?
(v) At the end of the electrolysis how many grams of the product would appear at the anods.

Q. 59. The electrode reactions for charging of a lead storage battery are

$$PbSO_4 + 2e^- = Pb + SO_4^{2-}$$
$$PbSO_4 + 2H_2O = PbO_2 + SO_22 + 4H^+ + 2e^-$$

the electrolyte in the battery is an aqueous solution of sulphuric acid. Before charging, the Sp.gr. of the liquid was found to be 1.11 (15.7% H_2SO_4 by weight]. After charging for 100 hours the Sp.gr. of the liquid was found to be 1.28 (36.9% H_2SO_4 by weight). If the battery contained two litres of the liquid, calculate the average current used in charging the battery.

N.B. [e^- denotes electrons. Assume that the volume of the battery liquid remained constant during charging. Faraday = 96,500 coulombs]

Q. 60. An electric current is passed through a solution of (i) Silver nitrate and (ii) a solution of 10 gm of $CuSO_4 \cdot 5H_2O$ crystals in 500 ml. of water, platinum electrodes being used in each case. After 30 minutes it was found that 1.307 gm silver has been deposited. What was the concentration of copper, expressed as grams of copper per litre in the copper sulphate solution after the electrolysis.

Q. 61. The normal potential of Zn referred to the standard Hydrogen electrodes is 0.76 V and that of copper is –0.34 V at 298K. When excess of Zn is added to a solution of copper sulphate. the zinc displaces copper till equilibrium is reached. What is the ratio of concentration of Zn^{2+} to Cu^{2+} ions at equilibrium?

Q. 62. 0.5 N solution of a salt surrounding two platinum electrodes 2.1 cm apart and 4.2 square cms. in area was found to other a resistance of 25 ohms. Calculate the equivalent conductance of the solution?

Q. 63. Find the solubility product of saturated solution of Ag_2CrO_4 in water at 298 K if the emf of the cell:

$$Ag \,||\, Ag^+ \text{ (sat. } Ag_2CrO_4 \text{ soln.)} \,||\, Ag^+ \text{ (0.1 M)} \,|\, Ag \text{ is 0.164 v at 298 K.}$$

Q. 64. A Daniell cell is prepared using a Zn electrode that weighs 23.72 g immersed in a 1.0 F $ZnSO_4$ solution, and a copper electrode immersed in 600 mL of a 1.0 F $CuSO_4$ solution. Assuming that a steady current of 0.1 A is drawn from the cell, how long will the cell continue to supply electricity?

Q. 65. For the cell $Zn(s)|Zn^{2+}(1\ M)||Cu^{2+}(1\ M)|Cu(s)$ the measured EMF is 1.103 V. Write the equation for the anode reaction, the cathode reaction and the net cell reaction, and calculate $\varepsilon^o_{Cu^{2+}/Cu^+}$.

Q. 66. Will permanganate ion oxidize bromide ion when all concentrations are 1 *M*?

Q. 67. Will Pb reduce Zn^{2+} ions to ZN when all concentrations are 1 *M*?

Q. 68. The ε^o values of the following two half-reactions, both of which occur in basic solution, have been obtained from tables.

$$Ni(OH)_2(s) + 2e^- \rightarrow Ni(s) + 2\ OH^- \qquad \varepsilon^o = -0.72\ V$$
$$CrO_4^{2-} + 4H_2O + 3e^- \rightarrow cr(OH)_3(s) + 5\ OH^- \qquad \varepsilon^o = -0.12\ V$$

Write the balanced equation for the oxidation-reduction reaction that occurs spontaneously and involves these half-reactions.

Q. 69. For each of the following galvanic cells write the anode, cathode, and net cell reactions. Using tables calculate ΔE°_{cell}. Does the net cell reaction you have written proceed spontaneously to the right when all substances are in their standard states.

(a) $Pt(s) \mid Cl_2(g) \mid Cl^- \parallel Fe=, Fe^{2+} \mid Pt(s)$

(b) $Cu(s) \mid CuCl_2(aq) \parallel MnCl_2(aq), HCl(aq) \mid MnO_2(s) \mid Pt(s)$

Q. 70. From the value of ΔG° caluclate ΔE°_{cell} for the cell

$Pt(s) \mid I_2(s) \mid I^- \parallel Fe^{3+}, Fe^{2+} \mid Pt(s)$

Q. 71. **(a)** Calculate the EMF of the galvanic cell

$$Zn(s) \mid Zn^{2+}(M_1) \parallel Ag^+(M_2) \mid Ag(s) \text{ at } 298 \text{ K}$$

when

(a) $M_1 = 1.0\ M$ and $M_2 = 1.0\ M$

(b) $M_1 = 0.1\ M$ and $M_2 = 0.01\ M$

(c) $M_1 = 0.012\ M$ and $M_2 = 0.2\ M$

Q. 72. If we want the galvanic cell

$$Cu(s) \mid Cu^{2+}(M_1) \parallel Cl^-(0.1\ M) \mid Cl_2(g, 1 \text{ atm}) \mid Pt(s)$$

To have an EMF of 1.122 V at 298 K, what value should be used for M_1, the concentration of Cu^{2+} ions?

Q. 73. **(a)** What is the EMF of the following concentration cell at 298 K?

$$Zn(s) \mid Zn^{2+}(0.024\ M) \parallel Zn^{2+}(0.480\ M) \mid Zn(s)$$

(b) If water is added to the more dilute solution.so that the $[Zn^{2+}]$ is decreased to 0.012 *M*, will the cell voltage increase, decrease or remain the same? Explain you answer.

Q. 74. Calculate the equilibrium constants for the following reactions at 298 K

(a) $Cu(s) + Br_2(aq) \rightleftharpoons Cu^{2+}(aq) + 2Br^-(aq)$

(b) $2Ag(s) + 2H^+(aq) \rightleftharpoons 2Ag^+(aq) + H_2(g)$

(c) $Ag(s) + H^+(aq) \rightleftharpoons Ag^+(aq) + \frac{1}{2}H_2(g)$

Q. 75. Is the oxidation of ferrous sulfate in sulfuric acid by O_2 in the air a spontaneous reaction? Calculate ΔG° and the equilibrium constant for the reaction between Fe^{2+} ions and O_2 in acid solution.

Q. 76. **(a)** Consider the cell $Co(s) \mid Co^{2+}(c_1) \parallel Ni^{2+}(c_2) \parallel Ni(s)$ at 298 K. Write the net cell reaction.

(b) Calculate $\Delta \mathcal{E}^\circ_{cell}$, ΔG°, and K_{eq}.

(c) Calculate Q, $\Delta \mathcal{E}_{cell}$, and ΔG when $c_1 = 1.0\ M$ and $c_2 = 0.001\ M$.

(d) Will the reaction proceed spontaneously from left to right when all substances are in their standard states?

(e) Will the reaction proceed spontaneously from left to right when $c_1 = 1.0\ M$ and $c_2 = 1.0 \times 10^{-3}\ M$?

Q. 77. Calculate the solubility product for $Cu(OH)_2$ using data from tables.

Q. 78. **(a)** How many faradays and how many coulombs must be passed through a molten mixture of Al_2O_3 and Na_3AlF_6 to produce 1 kg of aluminium metal?

(b) If 100 A of current are passed through the cell, how long will it take to produce 1 kg of aluminium?

(c) How much current is required if it is desired to product 1 kg of aluminium in 5 min?

Q. 79. Write the anode and cathode reactions for the electrolysis of aqueous solutions of (a) potassium sulfate, K_2SO_4, (b) calcium iodide, CaI_2, (c) mercuric nitrate, $Hg(NO_3)_2$, and (d) cupric bromide, $CuBr_2$.

Q. 80. In biological systems, acetate ion is converted to ethyl alcohol in a two-step process:

$$CH_3COO^-(aq) + 3H^+(aq) + 2e^- \rightarrow CH_3CHO(aq) + H_2O \quad E^{\circ\prime} = -0.581 \text{ V}$$

$$CH_3CHO(aq) + 2H^+(aq) + 2e^- \rightarrow C_2H_5OH(aq) \quad E^{\circ\prime} = -0.197 \text{ V}$$

($E^{\circ\prime}$ is the standard reduction voltage at 298 K and pH of 7.0.)

(a) Calculate $\Delta G^{\circ\prime}$ for each step and for the overall conversion.

(b) Calculate $E^{\circ\prime}$ for the overall conversion.

Q. 81. Consider the cell

$$Pt \mid H_2 \mid H^+ \parallel H^+ \mid H_2 \mid Pt$$

In the anode half-cell, hydrogen gas at 1.0 atm is bubbled over a platinum electrode dipping into a solution that has a pH of 7.0. The other half-cell is identical to the first except that the solution around the platinum electrode has a pH of 0.0;. What is the cell voltage?

Q. 82. A student constructed an electrochemical cell as shown in Figure 1:

The student began by constructing a half-cell using Zn as a reference electrode in 1.0 M $Zn^{2+}_{(aq)}$. The voltage between the Zn electrode and other electrodes in 1.0 M solutions of their metallic ion at 298K are presented in Table 1 below. A salt bridge of 1.0 M KNO_3 connected the two half-cells.

Table 1

	Voltage	*Anode*	*Cathode*
Zn–Ag	1.40	Zn	Ag
Zn–Cu	0.99	Zn	Cu
Zn–Fe	0.55	Zn	Fe
Zn–Mg	0.60	Mg	Zn
Zn–Pb	0.48	Zn	Pb

Next, the student used three different combinations of metals for both anodes and cathodes and measured the potential difference. The results are presented in Table 2.

Table 2

Anode	*Cathode*	*Measured Potential Difference*
Fe	Cu	0.5
Mg	Pb	1.0
Pb	Cu	0.51

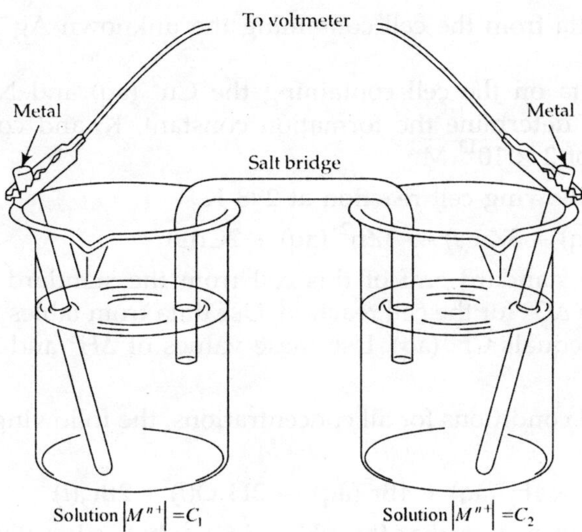

Fig. 10.1

A 0.01 M $Cu(NO_3)_2$ solution was then made up and used in the Cu^{2+} |Cu half cell. The $[Zn^{2+}(aq)]$ was kept at 1.0 M in the Zn | Zn^{2+} half-cell. The voltage was then determined and measured as 1.01 volts.

The student then added 10 mL of 1.0 NaCl solution to an empty beaker. He then added one drop of 1.0 M $AgNO_3$ to the beaker and stirred well. A silver electrode was immersed in this solution and connected through the salt bridge to the Zn|Zn^{2+} half-cell. The potential difference was measured as 0.91 volts.

And finally, the same procedure was followed as above, but instead of obtaining an unknown concentration of silver ion, the student prepared a solution containing 2 drops of 1.0 M $Cu(NO_3)_2$ in mL of 6.0 M NH_3. The student determined that it took 20 drops to equal 1 mL. This solution was then added to the cell containing the copper electrode. The voltage was read as 0.56 V. The cell can be represented as:

$$Zn_{(s)} | Zn^{2+}_{(aq)}(1.00) \text{ M}) || Cu^{2+}_{(aq)}(? \text{ M}) | Cu_{(s)}$$

Answer the following: 1–5

1. In reference to Table 1 above and using a chart of E°_{red} potentials, determine the potentials had hydrogen been used as a reference electrode instead of the Zn electrode and calculate the differences.
2. For each reaction

$$Cu^{2+}_{(aq)} + Fe_{(s)} \rightarrow Cu_{(s)} + Fe^{2+}_{(aq)}$$
$$Pb^{2+}_{(aq)} + Mg_{(s)} \rightarrow Cu_{(s)} + Mg^{2+}_{(aq)}$$
$$Cu^{2+}_{(aq)} + Pb_{(s)} \rightarrow Cu_{(s)} + Pb^{2+}_{(aq)}$$

   Predict the potentials from the data obtained and found in Table 1 and compare it to the measured values found in Table 2.
3. Using the Nernst equation, predict the expected voltage of the cell which contained the 0.01 M Cu^{2+}(aq) solution and compare it to the actual voltage obtained.

4. Using the data from the cell containing the unknown $Ag^+_{(aq)}$ solution, calculate the K_{sp} of AgCl.
5. From the data on the cell containing the Cu^{2+}(aq) and $NH_{3(aq)}$ mixture, write the reaction and determine the formation constant, K_f and compare it to the accepted value of K_f of 2×10^{13} M.

Q. 83. Consider the following cell reaction at 298 K

$$2Cr^{3+}(aq) + 3Zn(s) \rightarrow 3Zn^{2+}(aq) + 2Cr(s)$$

Calculate the standard emf of this cell from the standard electrode potentials, and from this obtain $\Delta G°$ for the cell reaction. Use data from tables to calculate $\Delta H°$; note that $Cr(H_2O)_6^{3+}$ (aq) equals Cr^{3+}(aq). Use these values of $\Delta H°$ and $\Delta G°$ to obtain $\Delta S°$ for the cell reaction.

Q. 84. Under standard conditions for all concentrations, the following reactions is spontaneous at 298 K.

$$O_2(g) + 4H^+(aq) + 4Br^-(aq) \rightarrow 2H_2O(l) + 2Br_2(l)$$

If $[H^+]$ is decreased so that the pH = 3.60, what value will E_{cell} have, and will the reaction be spontaneous at this $[H^+]$?

Q. 85. Under standard conditions for all concentrations, the following reactions is spontaneous at 298 K.

$$O_2(g) + 4H^+(aq) + 4Br^-(aq) \rightarrow 2H_2O(l) + 2Br_2(l)$$

If $[H^+]$ is adjusted by adding a buffer of 0.1 M NaOCN and 0.1 *M* HOCN ($K_a = 3.5 \times 10^{-4}$), what value will E_{cell} have, and will the reaction be spontaneous at this $[H^+]$?

Q. 86. An electrode is prepared by dipping a silver strip into a solution saturated with silver thiocyanate, AgSCN, and containing 0.1 *M* SCN^-. The emf of the voltaic cell constructed by connecting this electrode as the cathode to the standard hydrogen half-cell as the anode is 0.45 V. What is the solubility product of silver thiocyanate?

Q. 87. Which of the changes below will increase the voltage of the following cell?
$Co|Co^{2+}$ (0.01 M) $||H^+$ (0.01 M) | H_2 (0.5 atm) |Pt

(a) Increase the volume of $CoCl_2$ solution from 100 mL to 300 mL.
(b) Increase $[H^+]$ from 0.01 *M* to 0.5 *M*.
(c) Increase the pressure of H_2 from 0.5 atm to 1 atm.
(d) Increase the mass of the Co electrode from 15 g to 25 g.
(e) Increase $[Co^{2+}]$ from 0.01 *M* to 0.5 *M*.

Q. 88. The standard potential for the reaction of AgSCN is 0.0895 V.

$$AgSCN(s) + e^- \rightarrow Ag(s) + SCN^-(aq)$$

Find another electrode potential to use together with the above value and calculate K_{sp} for AgSCN.

Q. 89. In a fully charged lead storage battery, the electrolyte consists of 38% sulfuric acid by mass. The solution has a density of 1.286 g cm^{-3}. Calculate *E* for the cell. (Assume all the H^+ ions come from the first dissociation of H_2SO_4, which is complete, K_a $HSO_4^- = 1.0 \times 10^{-2}$.)

Q. 90. Consider a voltaic cell in which the following reaction occurs.

$$Zn(s) + Sn^{2+}(aq) \rightarrow Zn^{2+}(aq) + Sn(s)$$

(a) Calculate $E°$ for the cell.

(b) When the cell operates, what happens to the concentration of Zn^{2+}? The concentration of Sn^{2+}?

(c) When the cell voltage drops to zero, what is the ratio of the concentration of Zn^{2+} to that Sn^{2+}?

(d) If the concentration of both cations is 1.0 *M* originally, what are the concentrations when the voltage drops to zero?

Q. 91. Electrolysis of a solution of $MnSO_4$ in aqueous sulphuric acid is a method for the preparation of MnO_2 as per the reaction

$$Mn^{2+}_{aq} + 2H_2O \rightarrow MnO_{2(s)} + 2H^+_{aq} + H_{2(g)}$$

Passing a current of 27 A for 24 hours gives one Kg of MnO_2. What is the value of current efficiency? Write the reactions taking place at the cathode and the anode

Q. 92. Find the charge in coulombs of one gram ion N^{3-}.

Q. 93. The standard reduction potential at 298 of reaction, $2H_2O + 2e^- \rightleftharpoons H_2 + 2OH^-$ is –0.8277 V. Calculate the equilibrium constant for the reaction $2H_2O \rightleftharpoons H_3O^+ + OH^-$ at 298 K.

Q. 94. Calculate the equilibrium constant for the reaction, $2Fe^{3+} + 3I^- \rightleftarrows 2Fe^{2+} + I_3^-$. The standard reduction potentials in acidic conditions are 0.77 and 0.54 V respectively for Fe^{3+}/Fe^{2+} and I_3^-/I^- couples.

Q. 95. A current of 1.70 A is passed through 300 ml of 0.16 M solution of a $ZnSO_4$ for 230 sec. with a current efficiency of 90%. Find out the molarity of Zn^{2+} after the deposition of Zn. Assume the volume of the solution to retain constant during the electrolysis.

Q. 96. Copper sulphate solution (250 mL) was electrolyzed using a platinum anode and a copper cathode. A constant current of 2 mA was passed for 16 minutes. It was found that after electrolysis the absorbance of the solution was reduced to 50% of its original value. Calculate the concentration of copper sulphate in the solution to begin with.

Q. 97. A cell, $Ag | Ag^+ || Cu^{2+} | Cu$, initially contains 1 M Ag^+ and 1 m Cu^{2+} ions.

Calculate the change in the cell potential after the passage of 9.65 A of current for 1 h.

Q. 98. Calculate the equilibrium constant for the reaction

$$Fe^{2+} + Ce^{4+} \rightleftarrows Fe^{3+} + Ce^{3+}$$

(given $E^{\circ}_{Ce^{4+}/Ce^{3+}} = 1.44$ V; $E_{Fe^{3+}/Fe^{2+}} = 0.68$ V).

Q. 99. The standard potential of the following cell is 0.23 eV at 288 K and 0.21 V at 308 K $Pt/H_2(g)/HCl(aq)|AgCl/Ag(s)$

(i) Write the cell reaction

(ii) Calculate $\Delta H°$ and $\Delta S°$ for the cell reaction by assuming that these quantities remains unchanged in the range of 288 K to 308 K.

(iii) Calculate their solubility of AgCl in water at 298 K Given: The standard reduction potential of the $Ag^+(aq)/Ag(s)$ couple is 0.8 V at 298 K.

Q. 100. Find the equilibrium constant for the reaction

$$Cu^{+2} + In^{+2} \rightleftharpoons Cu^{+} + In^{+3}$$

Given that $E^{\circ}_{Cu^{2+}/Cu^{+}} = 0.15$ V

$E^{\circ}_{In^{+2}/In^{+}} = -0.4$ V and $E^{\circ}_{In^{3+}/In^{+}} = -0.42$ V

Q. 101. Two students make Daniel cells in laboratory. They take $ZnSO_4$ from common stock with Cu as positive electrode. The emf of one cell is 0.03 V more than the other. The concentration of $CuSO_4$ in cell of higher emf is 0.5 M. Find the concentration of $CuSO_4$ in second cell $\left(\text{Given, } \frac{2.303\,RT}{F} = 0.06\text{ V}\right)$

Q. 102. The standard reduction potential for the half-cell.

$$NO^{-}_{3(aq)} + 2H^{+}_{(aq)} + e^{-} \rightarrow NO_{2(g)} + H_2O \text{ is } 0.78 \text{ V}.$$

(i) Calculate the reduction potential in 8M H^{+}.

(ii) What will be the reduction potential of the half-cell in a natural solution? Assume all the other species to be at unit concentration.

Q. 103. Zinc granules are added in excess to a 500 ml of 1.0 M nickel nitrate solution at 298 K until the equilibrium is reached. If the standard reduction potential of Zn^{2+} |Zn and Ni^{2+}| Ni are –0.75 V and –0.24 V respectively, find out the concentration of Ni^{2+} in solution at equilibrium.

Q. 104. Consider the cell

$$Zn \,||\, Zn^{2+}_{(aq)} \,(1.0\text{ M}) \,||\, Cu^{2+}_{(aq)} \,(1.0\text{ M})/Cu$$

The standard reduction potentials are:

$+0.350$ volts for $2e^{-} + Cu^{2+}_{(aq)} \rightarrow Cu$ and

-0.763 volts for $2e^{-} + Zn^{2+}_{(aq)} \rightarrow Zn$.

(i) Write down the cell reaction.

(ii) Calculate the emf of the cell.

Q. 105. An acidic solution Cu^{2+} salt containing 0.4 g of Cu^{2+} salt containing 0.4 g of Cu^{2+} electrolysed until all the copper is deposited. The electrolysis is continued for seven more minutes with the volume of solution kept at 100 ml and the current at 1.2 amp. Calculate the volume of gases evolved at NTP during the entire electrolysis.

Q. 106. The following electrochemical cell has been set up

$Pt(1)|Fe^{3+}, Fe^{2+}$ $(a = 1)\,|\,Ce^{4+}$

Ce^{3+} $(a = 1)$ | Pt (2)

$E°(Fe^{3-}/Fe^{2+}) = 0.77$ V

$E°(Ce^{4+}/Ce^{3+}) = 1.61$ V

If an ammeter if connected between the two platinum electrodes, predict the direction of flow of current. Will the current increase or decrease with time?

Q. 107. The EMF of a cell corresponding to the reaction

$$Zn(s) + 2H+(aq) \rightarrow [Zn^{2+} + (0.1\text{ M}) + H_2\text{ (g, 1 atm) is 0.28 volt is 298 K.}$$

Write the half-cell reactions and calculate the pH of the solution at the hydrogen electrode.

$$E^{\circ}_{Zn^{2+}/Zn} = -0.76\text{ volt}; \; E^{\circ}_{H^{+}/H_2} = 0$$

Q. 108. An excess of liquid mercury is added to an acidified solution 1.0×10^{-3} M Fe^{3+}. It is found that 5% of Fe^{3+} remains at equilibrium at 298 K. calculate $E^{\circ}_{(Hg^{2+}/Hg)}$, assuming that the only reaction that occurs is

$$2\text{ Hg} + 2Fe^{3+} \rightarrow Hg_2^{2+} + 2\text{ }Fe^{2+} \text{ [Given } E^{\circ}_{(Fe^{3+}/Fe^{2+})} = 0.77\text{ V]}$$

Chapter 11

NUCLEAR CHEMISTRY

IMPORTANT FORMULAE

$$\text{Activity} = -\frac{dN}{dt} = \lambda N; \ \lambda = \frac{2.303}{t}\log\frac{N_o}{N}$$

$$\textbf{Half life period } (t_{1/2}) = \frac{0.693}{\lambda}$$

Average life (T) = 1.443 × Half life period

Binding energy = mass defeat × c^2

1 curie (Ci) = 3.7×10^{10} distintegrations per second

1 rd = 10^6 disintegrations per second

Ci = $3.7 \times 10^4 \times$ rd.

1 MeV = 10^6 eV

1 amu = 9.3148×10^2 MeV

RADIOACTIVITY

The spontaneous emission of certain radiations from a nuclide is known as radioactivity. Unstable radioactive elements like Ra, Th, U, etc., emit α, β and γ rays on spontaneous disintegration to form new elements (called daughter elements). α-rays consist of nuclei of Helium atoms, that is, they posses positive charge of +2 and a mass of 4 units and with the loss of each α-particle the resultant nucleus is deficient of atomic number by two units and atomic mass of by four units as compared with the disintegrating atom. The loss of a β-particle, which is nothing but a unit negative charge, results in the gain of atomic number by one unit in the resultant nucleus. A β-particle from the nucleus results from a neutron changing into a proton.

(Neutron = Proton + Electron). γ-rays are only ethereal vibrations produced due to the loss of α and β-particles from radioactive nuclei. Their loss produces no effect with respect to atomic number or atomic mass. This

$$\underset{\text{Uranium}}{{}_{92}U^{238}} \xrightarrow[({}_2He^4)]{-\alpha} {}_{90}Th^{234} \text{ and } \underset{\text{Neptunium}}{{}_{93}Np^{239}} \xrightarrow{-\beta} \underset{\text{Plutonium}}{{}_{94}PU^{1289}}$$

Nuclear Equations

To solving problems involving nuclear reactions; balance the sum of the atomic numbers and the sum of atomic weights, in amu (atomic mass units), on each side of the equation only (but do not balance equations to obtain the same number of elements.

We write a nuclear equation as ${}^{A}_{Z}X$, where A = mass numbers (or number of nuclear) and Z is atomic number eg. ${}^{238}_{92}U \rightarrow {}^{234}_{90}Th + {}^{4}_{2}He$

Here

1. the sum of mass numbers must be same on both side of the equation and
2. the sum of atomic numbers must be same on both sides of the equation.

Consider the following:

$${}^{14}_{7}N + {}^{4}_{2}He \rightarrow {}^{17}_{8}O + {}^{1}_{4}H$$

Here, Nitrogen always has an atomic of 7; H always has an atomic number of 1; etc. However, it is convenient to carry the atomic number specifically in nuclear reactions since they must balance. The superscripted number preceeding the chemical symbol is the atomic weight in amu and differs for different isotopes of the same element. The atomic weight may be thought of as the sum of neutrons and protons in the nucleus.

Identification of Emissions

S.No.	*particle name*	*(Notations)*	*n/P ratio*
1.	alpha particle (^{4}He nucleus)	${}^{4}_{2}\alpha$ (also ${}^{4}_{2}$He)	increased during α-emission
2.	beta particle (a stream of electrons)	${}^{0}_{-1}\beta$ (an electron)	decreases during β-emission
3.	positron	${}^{0}_{1}\beta$ (or ${}^{0}_{1}e$)	
4.	gamma ray (electromagnetic rays of short wave length x-ray)	${}^{0}_{0}\gamma$ (a gamma ray has no mass only energy)	
5.	proton (H^1 nucleus)	${}^{1}_{1}p$ or (${}^{1}_{1}H$)	
6.	neutron	${}^{0}_{1}n$	
7.	electron	${}^{0}_{-1}e$	

For balancing reactions, it is often convenient, though not precisely correct, to consider a neutron as a proton plus an electron and a proton as a neutron plus a positron.

Penetrating power: Gamma > Beta > Alpha.

Ionising power: Alpha > Beta> Gamma.

Radioactive disintegration series

Radioactive disintegration series is a series involving transformations of the parent radioactive element into the non-radioactive end product.

(i) During this integration series, either α or β particle is emitted at each step. There are four known disintegration series. (Three natural occurring series and one artificial series).

Natural — $4n + 0$ series: Thorium series from Th-232 to Pb-208 (value of n for Th is 58 and for Pb is 52 Parent element is Th^{232}.

Artificial— $4n + 1$ series: Nepturium series from Np-217 to Bi-209 (value of n for Np is 59 and for Bi is 52). Parent element Np^{217}

Natural — $4n + 2$ series: Uranium series from U-238 to Pb-206 (value of n for U is 59 and for Pb is 51). Parent element Li^{238}

Natural — $4n + 3$ series: Actinium series from U-235 to Pb-207 (value of n for U is 59 and for Pb is 51). Parent element U^{235}

The significance of 'n' lies in the fact that mass of each member of given series is an integral multiple of n with residue 0, 1, 2, 3 respectively for $4n$, $4n + 1$, $4n + 2$ and $4n + 3$ series.

Characteristics of Disintegration Series

Series	*Long lived isotope*	*Stable end product*	*Particle lost*		*Half life of parent (million years)*
			Alpha	*Beta*	
$n + 0$ or Th	$^{232}_{90}Th$	$^{208}_{82}Ph$	6	4	13,000
$n + 1$ or Np	$^{237}_{93}Np$	$^{209}_{83}Bi$	7	5	22
$n + 2$ or U	$^{238}_{92}U$	$^{206}_{82}Pb$	8	6	4,500
$n + 3$ or Ac	$^{235}_{92}U$	$^{207}_{82}Pt$	7	4	707

Group Displacement Law (soddy and Fajans)

(i) *In alpha ray change,* the atomic number is decreased by 2 units and mass number by 4 units, and the daughter element occupies 2 places behind the parent element in the periodic Table. This is due to the fact that atomic number of new element formed decreases by two.

$$^{235}_{92}U \rightarrow {}^{232}_{90}Th + {}^{4}_{2}He \quad {}^{226}_{88}Ra \rightarrow {}^{229}_{86}Rn + {}^{4}_{2}He \quad {}^{215}_{84}Po \rightarrow {}^{211}_{90}Pb + {}^{4}_{2}He$$

(ii) *In Beta ray change,* the atomic number is increased by one unit but mass number remains the same with the result the daughter element occupies one place ahead of the parent element in the periodic table. This is due to the fact that atomic number increases by 1.

$$^{234}_{90}Th \rightarrow {}^{234}_{91}Pa \quad {}^{0}_{-1}e \qquad {}^{212}_{83}Bi \rightarrow {}^{212}_{74}Po + {}^{0}_{-1}e \qquad {}^{12}_{6}C \rightarrow {}^{14}_{7}N + {}^{0}_{-1}e$$

The total effect is that one neutron becomes a proton by losing the negative charge.

$$^{1}_{0}n \rightarrow {}^{1}_{1}p + {}^{0}_{-1}e$$

$$\underset{\substack{144n \\ \textit{Thorium} \\ \textit{nuleaus}}}{90p} \longrightarrow \underset{\substack{143n \\ \textit{Protactium} \\ \textit{nucleus}}}{91p}$$

NUCLEAR TRANSMUTATION

Nuclear Transformation or transmutations is conversion of one element into another. There are two types of nuclear transmutation:

(a) Natural Transmutation: elements possessing natural radioactivity undergo a series of transmutation until a non-radioactive element is formed. For example, U-238 undergoes natural transmutation until non-radioactive element, Pb-206 is formed.

Also, ${}^{209}_{83}Bi$ is the nuclide of highest atomic and mass number that is stable. All known nuclides beyond it in atomic & mass nos are radioactive. Naturally occurring ${}^{238}_{92}U$ is radioactive and disintegrates by the loss of α-particles.

${}^{238}_{92}U \rightarrow {}^{234}_{90}Th + {}^{4}_{2}He$; ${}^{234}_{90}Th$ is also radioactive given ${}^{-}\beta$ emission.

${}^{234}_{90}Th \rightarrow {}^{234}_{91}Pa + {}^{0}_{-1}e$; ${}^{234}_{91}Pa$ also decays giving ${}^{-}\beta$ emission

${}^{234}_{91}Pa \rightarrow {}^{238}_{92}U + {}^{0}_{-1}e$ ${}^{234}_{92}U$ is also radioactive

(b) Artificial transmutation: The conversion of one element into another by artificial means is called artificial transmutation.

In artificial transmutation, stable nuclei of atoms of an element are bombarded with highly energetic particles such as α-particles, deuterons $({}^{2}_{1}D)$, neutrons $({}^{1}_{0}n)$ example;

Bombarding N-14 nuclei with α-particles. ${}^{14}_{7}N + {}^{4}_{2}He \rightarrow {}^{17}_{8}O + {}^{1}_{1}H$

This method of nuclear bombardment is written as $^{14}N(\alpha, p)^{17}O$. The target and product nuclei are written on left and right of a parenthetical expression. Within that the bombarding particle is written first followed by the ejected particle.

Similarly,

$${}^{27}_{13}Al + {}^{4}_{2}He \rightarrow {}^{30}_{15}P + {}^{1}_{0}n \quad \text{or} \quad {}^{27}Al(\alpha, n)^{30}P.$$

$${}^{30}_{15}P \rightarrow {}^{30}_{14}Si + {}^{0}_{1}e$$

${}^{238}_{92}U + {}^{1}_{0}n \rightarrow {}^{239}_{92}U$, the nucleas formed here ${}^{239}_{92}U$ is unstable and it undergoes

${}^{239}_{92}U \rightarrow {}^{239}_{93}Np + {}^{0}_{-1}e$ β decay yielding element neptunium with Z = 93.

${}^{40}_{19}K \rightarrow {}^{40}_{20}Ca + {}^{0}_{-1}e$

Note: Radioactivety and Isotopes of high atomic number is common, but are rare in naturally occuring lighter isotopes eq. ^{40}K, ^{50}V, ^{138}La.

Whereas, radioactive isotope formed by artificial transmutation having usually very short half-life period changes to element with stable nucleus.

KINETICS OF RADIOACTIVE DECAY

Activity Law: The activity of a radioactive sample is defined as $R = \lambda N$, where $R = -\frac{dN}{dt}$.

The rate of radioactive disintegration (or Rate of Decay) depends upon the amount of element present and is given by.

$$\boxed{N = N_0 e^{-\lambda t}}$$

or $\log N_0 - \log N = 0.4343 \times t$; where λ is decay constant or disintegration constant, N_0 is initial number of atoms (at time $t = t_0$), N is final number of atoms (at time $t = t$).

Characteristics of Rate of Decay

(a) ***Half Life:*** The time required to complete half of the decay.

i.e. If $t = t_{1/2}$, $N = \frac{N_0}{2}$. Here, $t_{\frac{1}{2}}$ = Half life.

or substituting these in $N = N_0 e^{-\lambda \tau}$

$$\lambda = \frac{0.693}{t_{1/2}} \quad \text{or} \quad t_{\frac{1}{2}} = \frac{0.693}{\lambda}$$

e.g. $t_{1/2}$ for U_{206} is 4.6×10^9 yr and $t_{1/2}$ for $^{206}_{82}Pb$ is infinite.

Important Note :

If Amount left after 1 half $= \frac{N_0}{2}$

Then, Amount left after 2 halves $= \frac{N_0}{2^2}$

and, Amount left after n halves $= \frac{N_0}{2^n}$

$\therefore$ Amount used in n halves $= N_0 - \frac{N_0}{2^n} = \frac{N_0\,[2^n - 1]}{2^n}$

Hence, No. of halves $(n) = \frac{\text{total time}}{\text{half time period}} = \frac{T}{t_{1/2}}$

(b) ***Average life:*** Average life $(t)_{avg} = \frac{1}{\lambda} = \frac{1}{\text{disintegration constant}}$

(c) The time required to disintegrate a definite fraction is independent initial concentration, i.e., $t_{1/n} \propto (N_0)^0$, where $t_{1/n}$ is time required to complete $1/n$ decay. Therefore, half decay is also written as,

$$t_{1/2} \propto (N_0)^0$$

For a decaying nuclide

$$R = \lambda N,$$

$$\therefore \quad \frac{R}{R_0} = \frac{\lambda N}{\lambda N_0} = \frac{N}{N_0} = \frac{m}{m_0}$$

Where, R and R_0 are activities; N and N_0 are numbers of atoms and M and M_0 are masses; and are all of the same decaying nuclide.

(d) Radioactive equilibrium: A state ultimately reached when a radioactive substance of slow decay yield a radioactive product on disintegration. In Radio-active equilibrium:

$$\frac{N_1}{N_2} = \frac{\lambda_2}{\lambda_1} = \frac{t_1(\text{ave})}{t_2(\text{ave})}$$

(e) Maximum yield of daughter element: A radioactive element A decays to give a daughter element B further decays to another daughter element C and so n till a stable element is formed ($A \rightarrow B \rightarrow$ C). Also if number of daughter atoms at $t = 0$ is zero and parent atom is much more lived than daughter (i.e., $\lambda_A < \lambda_B$), where λ_A and λ_B are decay constant of A and B respectively, then number of atoms of daughter element B after time t is

$$N_B = \frac{N_0 \lambda_A}{\lambda_B - \lambda_A}[e^{-\lambda_A t} - e^{-\lambda_B t}] \quad (1)$$

Maximum activity of daughter element can be expressed at t_{max}:

$$t_{max} = \frac{2.303}{\lambda_B - \lambda_A} \log_{10}\left[\frac{\lambda_B}{\lambda_A}\right] \quad (2)$$

Unit of Radioactivity

The unit of radioactivity of an element is measured by the rate at which it changes into daughter element. It has been derived on the scale of disintegration of Ra.

Consider 1 g Ra ($t_{1/2}$ = 1600 years) undergoes decay, then

Rate of decay = $K \times$ no. of atoms of Ra in 1 g

$$= \frac{0.693}{1600 \times 365 \times 24 \times 60 \times 60} \times \frac{6.023 \times 10^{23}}{226}$$

$= 3.7 \times 10^{10}$ dps (disintegrations per second) $= 3.7 \times 10^{10}$ Becquerrel

= 1 curie ($\because$ 3.7×10^{10} dps = 1*ci*)

$= 3.7 \times 10^4$ Rutherford ($\because$ 10^6 dps = 1 *rd*)

The S.I. unit of radioactivity is dps or Becquerrel. The other units to express rate of decay are,

Micro curie = 10^{-6} curie = 3.7×10^4 dps

Milli curie = 10^{-3} curie = 3.7×10^7 dps

Radio Carbon Dating

It is the process of determining the age of a piece of wood from the amount of its radioactivity. Atmospheric O_2 contains a small proportion of radioactive isotopes of carbon, namely $_6Cl^{14}$ which is assimilated by living plants which have a constant level of this isotope as in the atmosphere at that time. In dead plants, the level of this isotope decreases constinusouly die to disintegration. Measure of the radioactivity of the dead plant indicates the length of the time which elapsed after the death of the living plant and thus serves to find out the time when it was felled. Hence it helps to find out the age of wood.

$$\text{Half-time period of } {}_6C^{14} = \frac{0.693}{k} = 5{,}580 \text{ years}$$

or
$$k = \frac{0.693}{5{,}580}$$

Let the time taken for $_6C^{14}$ to change to $_6C^{12}$ to become 13% of the atmosphere be t years. It is given by the expression.

$$t = \frac{2.303}{k}\log_{10}\frac{100}{13}$$

Substituting the value of k as found above, we have

$$t = \frac{2.303 \times 5{,}580}{0.693} \times (2 - 1.1139)$$

ENERGIES OF A NUCLEAR REACTION

(A) Binding Energy

1. The total energy given out during binding up of a nucleons in nucleus is known as binding energy.
2. $\text{B.E. per nucleons} = \dfrac{\text{Total B.E.}}{\text{No. of nucleons}}$

(B) Mass Defeat

1. A stable nucleus hass less mass than its constituent particles. This difference is known as mass defect, i.e.,
$$\Delta m = \text{sum of the masses of constituents–mass of stable nucleus}$$
2. The difference in mass is converted into energy (known as B.E.) according to Einstein mass-energy relationship i.e.,
$$E = mc^2$$
3. Thus, Binding energy = $\Delta m \times c^2$

(C) Stability of Nucleus

1. Greater is the mass defect, more is B.E., lesser is the energy level of nucleus, more is its stability.
2. If *neutron-proton ratio*, i.e., $n/p > 1.5$ nucleus is unstable.

3. The no. of stable nuclide is maximum when both at. no. and no. of neutrons are even numbers.

Stable nuclei occurs whenever the no. of neutrons or the no. of protons equals one of the magic nos. 2, 8, 20, 28, 50, 82 or 126, 184, 196 etc. Nuclei which contain a magic no. of protons and a magic no. of neutrons are most stable eq. ${}^{4}_{2}He$, ${}^{14}_{8}O$, ${}^{40}_{20}Ca$ and ${}^{208}_{82}Pb$.

NUCLEAR FISSION

It is the splitting up of a large nucleus into smaller fragments. OR, It is a radioactive decay in which a heavy nucleus breaks up into lighter nuclei and several neutrons, with liberation of a large amount of energy eg.: When the neutron bombardment of U-235 nuclei under goes fission approx. 220 MeV of energy is produced by one atom of uranium. (1 Mev = 1.602×10^{-13} Joules). The break up can be in several ways.

Fragents number Energy

$${}^{235}_{92}U + {}^{1}_{0}n \rightarrow {}^{236}_{92}U \rightarrow \begin{bmatrix} {}^{139}_{56}Ba + {}^{94}_{36}Kr + 3\,{}^{1}_{0}n \\ {}^{144}_{54}Xe + {}^{90}_{38}Sr + 2\,{}^{1}_{0}n \end{bmatrix} + 3.2 \times 10^{-11}\ \text{J (approx 220 MeV)}$$

∴ From 1 gm of Uranium, energy released = 8.2×10^{7} kJ i.e. energy released by burning 3 tons of coal. As ${}^{238}_{92}U$ is abundant and it undergoes the following:

$${}^{238}_{92}U + {}^{1}_{0}n \rightarrow {}^{239}_{92}U \rightarrow {}^{239}_{9}N_p \quad {}^{0}_{-1}e \rightarrow {}^{239}_{94}Pu + {}^{0}_{-1}e$$

The energy released is due to loss in mass.

Critical Mass

The energy released is due to loss in mass. The minimum quantity of an element needed to support a self-sustaining chain reaction is called the **critical mass.**

NUCLEAR FUSION

Nuclear fusion; the merger of lighter nuclei to form heavier ones,

examples : ${}^{2}_{1}H + {}^{1}_{1}H \rightarrow {}^{4}_{2}He + {}^{1}_{0}n$ + energy

${}^{1}_{1}H + {}^{7}_{3}Li \rightarrow {}^{4}_{2}He + {}^{4}_{2}He$ + energy

Since fusion occurs at entremely high temperatures, these are called thermo nuclear reaction. Fusion reaction are the basis of H-bomb.

The mass defect (ΔM) is the difference between the expected mass and the actual mass of the substance. The mass defeat is related with energy as 1 a.m.u. = 931 MeV. Therefore

Binding energy = 931 × Mass defect MeV

$$\text{Packing fraction} = \frac{\text{Isotopic atomic mass} - \text{Mass No.}}{\text{Mass number}} \times 10^{4}$$

PROBLEMS

Q. 1. Find the mass defect, total binding energy and average binding energy per nucleon in $_{14}Si^{28}$.

Mass of silicon atom = 27.9769 a.m.u
Mass of hydrogen atom = 1.007825 a.m.u
Mass of neutron = 1.008655 a.m.u

Q. 2. Find the mass defect of $_{20}Ca^{40}$. Calculate the average binding energy of a nucleon.

Mass of calcium atom = 40.08 a.m.u
Mass of hydrogen atom = 1.007826 a.m.u
Mass of neutron = 1.008665 a.m.u

Q. 3. Balance the following nuclear reactions.

$$_{11}Na^{23} + \rightarrow {}_{13}Al^{26} + {}_{0}n^{1}$$

$$_{3}Li^{7} + {}_{2}He^{4} \rightarrow {}_{5}B^{10} + \text{—}$$

Q. 4. Calculate alpha and beta particles emitted in thorium series in reaching lead –208 from thorium-232.

Q. 5. The half life of a radioactive element is 28 days. Calculate the time taken by a given sample of the element to reduce to 1/10 of its initial activity.

Q. 6. The observed mass of $_{26}Fe^{56}$ is 55.9375 a.m.u. Using masses of the atom of hydrogen and neutron as 1.007825 and 1.008665 respectively, calculate the binding energy per nucleon in MeV.

Q. 7. The half-life period of Pb^{210} is 22 yrs. If 2 g of Pb^{210} is taken, after 11 yrs. How much of Pb^{210} will present.

Q. 8. Find the radius of a nuclide whose mass number is 27?

Q. 9. Find the mass number of a nuclide with a nuclear radius is 5.6 Fm?

Q. 10. $^{M}_{Z}X$ emits an α-particle. What is its mass number and atomic number?

Q. 11. What is the value of disintegration constant if the half life of a nuclide is 11 mi. 33 sec.

Q. 12. Calculate the mass number of nuclide when its nuclear radius is 8.4×10^{-13} cm.

Q. 13. The nuclear radius of an atom is 2.52×10^{-13} cm. Find its mass number.

Q. 14. The mass of a nuclide is 64. Find its nuclear radius in Fermis?

Q. 15. The life of a radioactive element is 5.8 years. Calculate its disintegration constant.

Q. 16. The half life of a nuclide is 50 months. Find its disintegration constant.

Q. 17. What is the value of disintegration constant of a nuclide when its half-life is 15 hours?

Q. 18. The disintegration constant of a nuclide is 0.0581 disint. $Year^{-1}$. Find its half life.

Q. 19. The disintegration constant of a radioactive element is 0.0078 disint. $Hour^{-1}$. Calculate its half life.

Q. 20. What is the energy equivalent to 0.05 g. of a substance?

Q. 21. How much energy is obtained by destructing 0.08 g. of a substance?

Q. 22. Calculate the mass number of a nuclide when its nuclear radius is 3.5 Fm.

Q. 23. What is the mass equivalent to 20.9 × 10^{10} Joules?

Q. 24. Calculate the mass defect, binding energy and average binding energy of $^{28}_{14}Si$ whose atomic weight is 27.9769 a.m.u.

Q. 25. Calculate the energy associated with the following nuclear reactions. Masses are given in a.m.u.

(a) $^{27}_{13}Al + ^{2}_{+1}H \rightarrow ^{25}_{12}Mg + ^{4}_{2}He$

26.9815 2.0141 24.9858 4.0026

(b) $^{10}_{5}B + ^{1}_{0}n \rightarrow ^{7}_{3}Li + ^{4}_{2}He$

10.01294 1.008665 7.0160 4.0026

(c) $^{12}_{6}C + ^{2}_{1}H \rightarrow ^{13}_{7}N + ^{1}_{+0}n$

12 2.0141 13.00574 1.008665

Q. 26. Calculate the nuclear radius of an atom whose mass number is 125.

Q. 27. The half life period of Radon is 3.825 days. Compute the weight of 1 curie of Radon (At. wt of Radon = 222).

Q. 28. The half life period of radio-active isotope of calcium is 1.1 × 10^5 yrs. What weight of this isotope would give 3.3 × 10^{10} disintegration per day?

Q. 29. What weight of Radium *A* would be in equilibrium with 4 gm of Radium *B*? The half life periods of Radium *A* and Radium *B* are respectively 3.05 minutes and 26.8 minutes.

Q. 30. Radioactive Th has a half life of 54.5 seconds. What fraction of Thorium (Th) would be left behind at the end of 1 minute?

Q. 31. How many curies of Neptunium would be required to produce a sample of 0.1 milligram of plutonium (At. wt. 239)? ($T_{1/2}$ for Neptunium = 2.33 days.) (Atomic weight of Neptunium = 239)

Q. 32. A freshly cut sample of a plant gives 20.4 counts per minute per gm. If it is assumed that these counts are due to the presence of radio-active carbon (At. wt. 14) only. How many counts would it give after 100 yrs., if the half life period of carbon 14 is 5760 yrs?

Q. 33. A piece of wood recovered after 10,000 yrs. is found to have lost 71.28 percent of radioactive carbon (At. wt. = 14). What is the half life period of C^{14} atom?

Q. 34. The half life period of Uranium (238) is 4.5 × 109 yrs. and that of uranium (235) is 7.1 × 10^8 yrs. The relative abundance of these two isotopes at present is found to be 99.2% and 0.80% respectively. What is the age of the earth?

Q. 35. From the following particle masses calculate the binding energy of $^{18}_{8}O$ per nucleon in MeV.

O^{18} = 18.00369; H^+ = 1.00759

e^- = 0.000555; 1 a.m.u = 931 MeV

Q. 36. Calculate the ratio of charge (q) to mass (m) of a particle from the following data.

Electronic charge = 4.8 × 10^{-10} esu

Avogadro's Number = 6.02 × 10^{23}

Atomic weight of Helium = 4.0

Q. 37. $^{234}_{90}Th$ disintegrates to give $^{206}_{82}Pb$ as the final product. How many alpha and beta particles are emitted during this process?

Q. 38. Radioactive decay is a first order process. Radioactive carbon in wood sample decays with a half life of 5770 yrs. What is the rate constant (in yrs^{-1}.) for the decay. What fraction would remain after 11540 yrs?

Q. 39. The half life period of uranium 237 is 2.5×10^5 years. After how many years will the mass of uranium becomes 25% of its original mass?

Q. 40. Mass number and atomic number of an atom are 232 and 90 respectively. After emitting two β-particles how many alpha particles it should emit so as to get an atom having mass number and atomic number 212 and 82 respectively.

Q. 41. Calculate the Q value of the reaction
N^{14} (n, p) C^{14}. Mass of N^{14} = 13.9824 and C^{14} = 13.9817, n^1 = 1.008665 and hydrogen = 1.007825

Q. 42. The amount of C-14 isotope in a piece of wood is found to be one sixth of its amount present in a fresh of wood. Calculate the age of wood. (half life of C–14 = 5600 yrs.)

Q. 43. **(a)** The half life of a radio active element is 4.8 minutes. Starting with 1.0 gm. of the element how much would remain after 24 minutes?

(b) Half life period of a radioactive element is 1.39 × 109 yrs., eah atom of which disintegrates with the emission of a single alpha particle. Alculate the number of alpha particles emitte per sec by one gram of the substanec if the atomic weight is 232.

Q. 44. Calculate nuclear radii of the following atoms:
(i) Thorium (mass number = 232)
(ii) Astatine (mass number = 210)
(R_0 = 1.5f)

Q. 45. Uranium, ^{238}U (Z = 92) during a series of disintegration steps loses 8 alpha and 6 beta particles. What will be the atomic mass and atomic number of the new product?

Q. 46. Uranium-235, ($^{235}_{92}U$) decays with emission of alpha and beta particles to from ultimately Lead-207, ($^{207}_{82}Pb$). How many alpha and beta particles are emitted per atom of Lead-207 produced?

Q. 47. (a) Calculate the weight of $^{66}_{31}Ga$ after 5 hours when its initial weight is 30 gms. The disintegration constant is 0.0737 disintegration $hour^{-1}$.

(b) How many grams is disintegrated in 200 days when 15 grams of $^{65}_{30}Zn$ is taken. The disintegration constant is 0.00283 disintegration day^{-1}.

Q. 48. If 6 grams of bromine is taken, 4.226 grams remain after 20 hours what will be its disintegration rate?

Q. 49. 7 g. of a nuclide has disintegrated from 15 g. in 5 years. Find its disintegration constant?

Q. 50. Find the activity of 1 mg (10^{-6} kg) of radon.

Q. 51. How long does it take for 60% of a sample of radon to decay?

Q. 52. A piece of wood from the ruins of an ancient dwelling was found to have a ^{14}C activity of 13 disintegrations per minute per gram of its carbon content. The ^{14}C activity of living wood is 16 disintegrations per minute per gram. How long ago did the tree die from which the wood sample came?

Q. 53. A sample of uraninite, a uranium-containing mineral, was found on analysis to contain 0.214 g of lead for every of uranium. Assuming that the lead all resulted from the radioactive disintegrations of the uranium since the geological formation of the uraninite and that all isotopes of uranium other than ^{238}U can be neglected, estimate the date when the mineral was formed in the earth's crust. The half-life of 238U is 4.5×10^9 *y*.

Q. 54. Upon irradiating californium with neutrons, a scientist discovered a new nuclide having mass number of 250 and a half-life of 0.50 h. Three hours after the irradiation, the observed radioactivity due to the nuclide was 10 dis min^{-1}. How many atoms of the nuclide were prepared initially?

Q. 55. A radionuclide has an initial activity of 2.0×10^6 dis min^{-1}, and after 4 days its activity is 9 × 105 dis min^{-1}. Calculate the activity in the sample after 40 days.

Q. 56. The activity of 30 μg of ^{247}Cm is 2.8 nCi. Calculate the disintegration rate constant and the half-life of ^{247}Cm.

Q. 57. A sample of river water was found to contain 8×10^{-18} tritium atoms, $^{3}_{1}H$, per atom of ordinary hydrogen. Tritium decomposes radioactivity with a half-life of 12.3*y*. (a) What will be the ratio of tritium to normal hydrogen atoms 49 *y* after the original sample was taken if the sample is stored in a place where additional tritium atoms cannot be formed? (b) How many tritium atoms would 10 g of such a sample contains 40 *y* after the initial sampling?

Q. 58. A solution contains a mixture of two isotopes $A(t_{1/2} = 14$ days) and $B(t_{1/2} = 25$ days). Initially the total activity is 1 curie. The activity reduces to 50% in 20 days. Calculate (i) the initial activities of *A* and *B*.

Q. 59. Calculate the energy released in joules and MeV in the following nuclear reaction:

$$^{2}_{1}H + ^{2}_{1}He \rightarrow ^{3}_{2}He + ^{1}_{0}n$$

Assume that the masses of $^{2}_{1}H$, $^{3}_{2}He$ and neutron (n) respectively are 2.0141, 3.0160 and 1.0087 in amu.

Q. 60. Calculate the energy equivalent of 1 amu.

Q. 61. Calculate the binding energy per nucleon for $^{4}_{2}He$.

Q. 62. Natural nitrogen is 99.63% ^{14}N, with mass 14.00307 amu, and 0.37% ^{15}N, with mass 15.00011 amu. Which isotope is more stable, ^{14}N or ^{15}N? Offer an explanation for their relative stabilities.

Q. 63. The isotope $^{173}_{71}Lu$ is neutron deficient for stability. Can it decay by β^+ emission or electron capture of both? The mass of $^{173}_{71}Lu$ is 172.9390 amu. It decays to a stable isotope of ytterbium, $^{173}_{70}Yb$, with mass 172.9383 amu.

Q. 64. List the most likely modes of decay for the following unstable nuclei (whose half-lives are given for information only) and write a balanced nuclear equation for each decay process.

(a) ${}^{13}_{7}N\,(t_{\frac{1}{2}} = 10\,\text{min})$ **(b)** ${}^{25}_{11}Na\,(t_{\frac{1}{2}} = 60s)$

(c) ${}^{132}_{57}La\,(t_{\frac{1}{2}} = 4.5\,h)$ **(d)** ${}^{210}_{85}At\,(t_{\frac{1}{2}} = 8.3\,h)$

Q. 65. What is the wavelength of the 0.1427 MeV γ ray emitted by ${}^{99m}Tc$?

Q. 66. By definition, one curie (Ci) is an activity of 3.7×10^{10} dps (disintegrations per second). The unstable nucleus 14C is a β^- emitter with a half-life of 5730 yr. What mass of 14C is required in order to have an activity of 1 Ci?

Q. 67. A radioactive isotope commonly used in biotracer studies of phosphorus metabolism is ${}^{32}P$, which has a half-life of 14.3 days. Suppose we have been doing experiments with ${}^{32}P$ and accumulate waste material that had an activity of 1 millicurie (1 mCi = 1×10^{-3} Ci). The radiation safety officer in our laboratories says it is unsafe to dispose of this waste material until the activity has fallen to 0.01 μCi [1 microcurie (μCi) = 1×10^{-6} Ci]. How long must we store the waste in a lead container before it is safe to dispose of it?

Q. 68. In order to determine the volume of blood in an animal without killing it, a 1.0-mL sample of an aqueous solution containing tritium is injected into the animal's bloodstream. The sample injected has an activity of 1.8×10^6 cps. After sufficient time for the sample to be completely mixed with the animal's blood due to normal blood circulation, 2.0 mL of blood are withdrawn from the animal, and the activity of the blood sample withdrawn is found to be 1.2×10^4 cps. Calculate the volume of the animal's blood.

Q. 69. A specimen of a tree is observed to emit β^- rays with a specific activity of 10.8 dpm · g^{-1} of carbon. The radioactivity is due to ${}^{14}C$, which has a half-life of 5730 yr. How long ago did the tree die?

Q. 70. A sample of a rock is analyzed and found to contain 0.232 g of ${}^{206}Pb$ and 1.605 g of ${}^{238}U$. Assuming all the ${}^{206}Pb$ now present came from the decay of ${}^{238}U$, calculate the time since the solidification of this rock. (Assume half life of ${}^{238}U = 4.5 \times 10^9$ yrs.)

Q. 71. At 328 K the gas-phase decomposition N_2O_5 is first order in N_2O_5 with specific rate constant $1.42 \times 10^{-3}\ s^{-1}$.

(a) Determine the half-life of this decomposition.

(b) If the initial partial pressure N_2O_5 is 2.84 atm at 328 K, how long will it take for the partial pressure to fall to 0.355 atm?

Q. 72. The radioactivity of a sample of a transuranium nuclide is counted on a radiation detection device. The following data are obtained:

Time (h)	0.0	2.0	4.0	6.0	8.0	12.0
Activity (cpm)	8780	7527	6452	5531	4741	3484

Determine the half-time of the nuclide.

Q. 73. A sample of sodium phosphate, Na_3PO_4, weighing 54.5 mg containing radioactive phosphoris-32 (with mass 32.0 amu). If 15.6% of the phosphorus atoms in the compound is phosphorus-32 (the remainder is naturally occurring phosphorus), how many disintegrations of this nu cleus occur per second in this sample? Phosphorus-32 has a half-time of 14.3 d.

Q. 74. 1 Polonium-210 has a half-life of 138.4 days, decaying by alpha emission. Suppose the helium gas origating from the alpha particles in this decay were collected. What volume of helium at 298K and 735 mmHg could be obtained from 1.0 g of polonium dioxide, PoO_2, in a period of 48 h?

Q. 75. What is the energy (in joules) evolved when 1 mol of helium-4 nuclei is produced from protons and neutrons? How many liters of ethane, $C_2H_6(g)$, at 298K and 725 mmHg are needed to evolve the same quantity of energy when the ethane is burned in oxygen to $CO_2(g)$ and $H_2O(g)$? See Tables for data.

Q. 76. Americium-241 is present in the homes smoke detectors. It has decay constant of $1.51 \times 10^{-3}\ y^{-1}$. You are urged to check the battery in the detector at least once a year. You are, however, ever encouraged to check how much Am-241 remains under analysed. Explain why.

Q. 77. The following data are obtained for a radioactive isotope. Plot the data and determine the half-life of the isotope.

Time (h)	0.0	0.5	1.0	1.5	2.0	2.5
Activity (disintegrations h^{-1})	14,472	13,095	11,731	10,615	9,605	8,504

Q. 78. An activity of 20 picocuries (20×10^{-12} Ci) of radon-222 per liter of air in a house constitutes a health hazard to anyone living there. The half-life of radon-222 is 3.82 days. Calculate the concentration of radon in air (moles per liter) that corresponds to a 20-picocurie activity level.

Q. 79. Plutonoium-239 decays by the reaction

$$^{239}_{94}Pu \rightarrow ^{235}_{92}U + ^{4}_{2}He$$

Its rate constant is $5.5 \times 10^{-11}\ min^{-1}$. In a one-gram sample of Pu-239,

(a) how many grams decompose in 45 minutes?

(b) how much energy in kilojoules is given off in 45 minutes?

(c) what radiation dosage in rems is received by a 75-kg man exposed to a gram of Pu–239 for 45 minutes?

Q. 80. It is possible to estimate the activation energy for fusion by calculating the energy required to bring two deutrons close enough to one another to form an alpha particle. This energy can be obtained by using Coulomb's law in the form $E = 8.99 \times 10^9\ q_1q_2/r$, where q_1 and q_2 are the charges of the deutrons (1.60×10^{-19} C), r is the radius of the He nucleus, about 2×10^{-15} m, and E is the energy in joules.

(a) Estimate E in joules per alpha particle.

(b) Using the equation $E = mv^2/2$, estimate the velocity (metres per second) each deuteron must have if a collision between the two of them is to supply the activation energy for fusion (m is the mass of the deuteron in kilograms).

Q. 81. Find:

(i) the total number of neutrons and

(ii) the total mass of neutrons in 7 mg of ^{14}C.

(Assume that mass of neutron = mass of a hydrogen atoms)

Q. 82. $^{234}Th_{90}$ disintegrates to give $^{206}Pb_{82}$ as the final product. How many alpha and beta particles are emitted during this process?

Q. 83 One of the hazards of nuclear explosion is the generation of ^{90}Sr and its subsequent incorporation in bones. The nuclide's has a half-life of 28.1 years. Suppose one microgram was absorbed by a new-born child, how much ^{90}Sr will remain in his bones after 20 years?

Q. 84. A first order gas reaction has $K = 1.5 \times 10^{-6}$ per second at 473 K. If the reaction is allowed to run for 10 hours, what percentage of the initial concentration would have changed in the product? What is the half life of this reaction?

Q. 85. ^{227}Ac has a half-life of 22 yrs with respect to radioactive decay. The decay follows two parallel paths, one leading to ^{227}Th and the other to ^{223}Fr. The percentage yields of these two daughter nuclides are 2.0 and 98.0 respectively. What are the decay constants (λ) for each of the separate paths?

Q. 86. At 653 K, the half-life period for the first order decomposition of H_2O_2 is 360 min. The energy of activation of the reaction is 200 KJ mol^{-1}. Calculate the time required for 75% decomposition at 723 K.

Q. 87. An experiment require minimum beta activity produced at the rate of 346 beta particles per minute. The half-life period of $^{99}_{42}Mo$, which is a beta emitter if 66.6 hours. Find the minimum amount $^{99}_{42}Mo$ required to carry out the experiment in 6.909 hours.

Q. 88. ^{64}Cu (half life = 12.8 h) decays by β^- emission (38%), β^+ emission (19%) amu electron capture (43%). Write the decay products and calculate partial half-lives for each of the decay processes.

Q. 89. The half life period of Bi–212 is one hour. If you started work with 1 g of this isotope at 8 A.M, how much would remain 4 PM on the same day.

Chapter **12**

CRYSTALLOGRAPHY

CRYSTAL SYSTEMS

1. Cubic
2. Tetragonal
3. Orthorhomic
4. Monoclinic
5. Hexagonal
6. Rhombohedral
7. Triclinic

Unit Cell

$$\text{Density } d = \frac{ZM}{a^3 N_A}$$

$$\text{Radii of ion} = \frac{a}{4}$$

Radii of Atoms for,

(i) Simple cubic lattice = $a/2$

(ii) B.C.C. Lattice = $\sqrt{3}a/4$

(iii) F.C. Lattice = $\dfrac{a}{2\sqrt{2}}$

Limiting Radius Ratio's for Coordination Number, 3 = 0.155
4 = 0.255
6 = 0.414.

CRYSTALLOGRAPHY

Crystallography deals with the geometry (the external form and internal structure) of crystals and crystalline substances, their development and growth as well as related physical properties.

THE SOLID STATE

The physical state of matter in which the constituent molecules, atoms or ions have no translatory motion although they vibrate about the fixed position, that they occupy in a crystal lattice. Thus solids have definite volume, shape, slow diffusion, low vapour pressure and possesses the unique property of being rigid.

A solid is said to posses cohension forces, remaining in the same shape unless changed by certain forces.

CLASSIFICATION OF SOLIDS

Solids are of two types:

1. **Crystalline solids:** "A crystal is a solid composed of atoms (ions or molecules) arranged in an orderly repetitive array". A crystalline solid has a definite melting point at which it becomes liquid. Crystals of a substance are bounded by well defined plane facts, having same gerometry irrespective of the source form which they have been obtained. The most important and essential feature of a crystal is the periodicity of the arrangement along with regularity. **eg:** Iron, copper, silver, NaCl, KNO_3 etc.
2. **Amorphous solids:** Amorphous solids have no precise melting point but when heated, become increasingly pliable until they assume the properties usually associated with liquids; they may, therefore, be through of as super-cooled liquids in which the force of attraction holding the molecules together is so great that the material is rigid but there is no regularity of structure.
   eg.: glass, amorphous silica etc.
3. Polymorphs: are solids which adopt different structural arrangements under different conditions eg. Graphite, diamond etc.

Different Types of Solids

Crsytalline solids based on the nature of bonding are classified into four main catergories viz. covalent ionic, metallic and molecular solids. Table below provides example of different types of solids. The physical properties like standard molar enthalpy of fusion, electrical and thermal conductivity provide information to which of these four categories the solids belongs.

Types of solid	*Constituent particles*	*Bonding/ Attractive forces*	*Approximate binding energy /kJ/mol⁻¹*	*Examples*	*Physical nature*	*mp/K*	*Electrical conductivity*
1. Ionic	Ions	Columbic or electrostatic	400–4000	NaCl, MgO, Kl. $BaCl_2$	Hard but brittle	High –500	Insulator
2. Covalent	Atoms	Covalent	150–500	SIO_2, SIC, C(diamond)	Hard	~4000	Insulator
3. Molecular	Molecules	(i) Dispersion	0.05–40	Argon	Soft	84	Insulator
		(ii) Dipole interaction	5–25	HCl	Soft	158	Insulator
		(iii) Hydrogen bonding	10–40	H_2O(ice)	Hard	273	Insulator
4. Metallic	Atoms	Positive ions and electrons (delocalized electrons)	70–1000	Ag, Cu, Mg	Hard	~800 to 1000	Conductor

Determination of Structure of Solids

Determination of Structure of Solids a diffraction pattern in produced by passing X-rays through a crystal of copper sulphate. W.W. Bragg and his father W.H Bragg determined the cubic structure of sodium chloride using X-rays.

THE BRAGG'S EQUATION

The diffraction pattern obtained during X-ray diffraction resulting from the scattering of X-rays by a regular arrangement of atoms or ions has been interpreted in terms of Bragg's equation:

$$N\lambda = 2d \sin \theta \quad \text{....... (1)} \quad \text{(Bragg's Equation)}$$

where n is an integer (1, 2, 3, ...); λ is the wavelength of X-rays used, d is the distance of separation between the planes of atoms, θ is the angle at which constructive interference occurs.

It is evident from Bragg's equation, if $2d < n\lambda$ then $\sin \theta > 1$ which is not possible.

These days, the structure of extremely complex substances such as proteins and nucleic acids are determined using this technique.

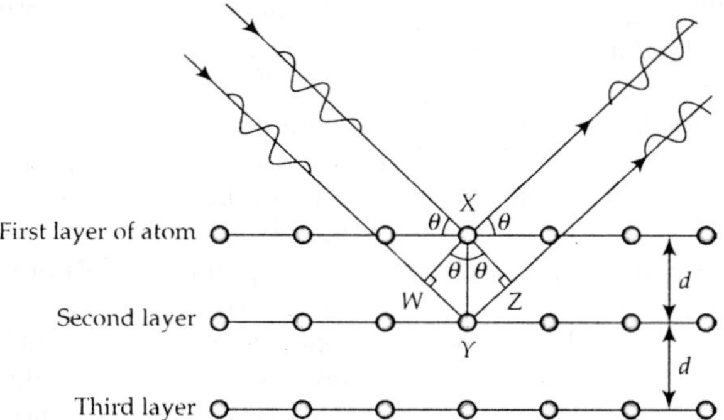

The diffraction of two waves of X-rays by the regularly spaced atoms in a crystal.

LATTICE AND UNITS CELLS

Crystal structure are best described in terms of lattices & unit cells. A lattice is a regular arrangement of points two dimensional or three dimension (called space lattice). Choose four points in two dimensional lattice and connect them. The square, rectangle or parallelogram formed is called ***Unit cell.***

However, for a hexagonal lattice, a rhombus with an angle of 60° is chosen as the unit cell.

A unit cell with an interior point is called a **Centroid Unit Cell.** In the case of rhombic lattices a centred rectangular unit cell is chosen.

A regular three dimensional arrangement of points in space is called a space lattice.

OR

The array of points showing how molecules, atoms or ions are arranged in different sites in three dimensional space is referred as space lattice.

For every crystal lattice, it is possible to select a group of lattice points which sets the pattern for a whole lattice. This three dimensional group of points is called Unit Cell of a crystal and is characterized by the distances a, b and c and angled α, β and γ. The whole crystal can be developed by a step wise shifting of the unit cells in all the three dimensions. In simple cubic there is only one atom, in body-centred cubic (b.c.c) there are two atoms and in face-centred cubic (f.c.c.) there are four atoms.

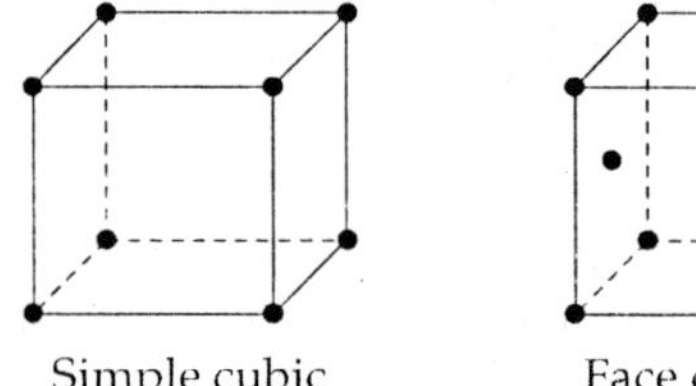

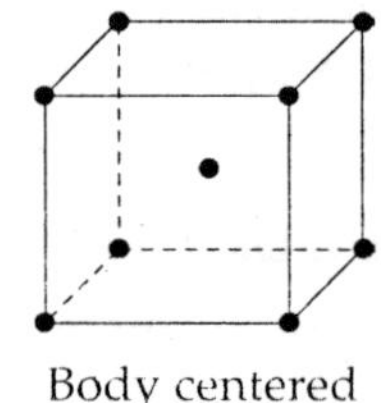

Simple cubic Face centered Body centered

Three different types of cubic unit cells.

Crystal System

The different possible geometrical shape of crystal lattice (or crystals) are called crystal systems. There are mainly seven such systems:

System	*Cell Parameters*		*Maximum $2H_2O$, elements*	*Examples*
	Axial distance	*Axial angles*		
Cubic	$a = b = c$	$\alpha = \beta = \gamma = 90°$	Nine planes, thirteen axis	NaCl, KCl, CaF_2 ZnS, diamond, alums
Tetragonal	$a = b \neq c$	$\alpha = \beta = \gamma = 90°$	Five planes, five axis,	TiO_2, ZnO_2, SO_2
Orthorhombic	$a \neq b \neq c$	$\alpha = \beta = \gamma = 90°$, $\beta \neq 90°$	Three planes three axis	Rhombic, Sulphur, $BaSO_4$, KNO_3,
Monoclinic	$a \neq b \neq c$	$\alpha = \gamma = 90°$, $\beta = 90°$	One plane, one axis	Na_2SO_4. 10 H_2O $CaSO_4$, $2H_2O$, $Na_2B_4O_7$. $10H_2O$ monoclinic sulphur

(Contd.)

(Contd.)

Hexagonal	$a = b \neq c$	$\alpha = \beta = 90°, \gamma = 120°$	Seven planes, seven axis	Mg, PbI_2, CdS, ZnO, HgS, graphite
Rhombohedral	$a = b = c$	$\alpha = \beta = \gamma \neq 90°$	—do—	Calcite, $NaNO_3$, ICl
Triclinic	$a \neq b \neq c$	$\alpha \neq \beta \neq \gamma \neq 90°$	No plane, no axis	$CuSO_4$ $5H_2O$ $K_2Cr_2O_7$, H_3BO_3

Cubic crystals have three patterns of *lattice in which crystal are formed:*

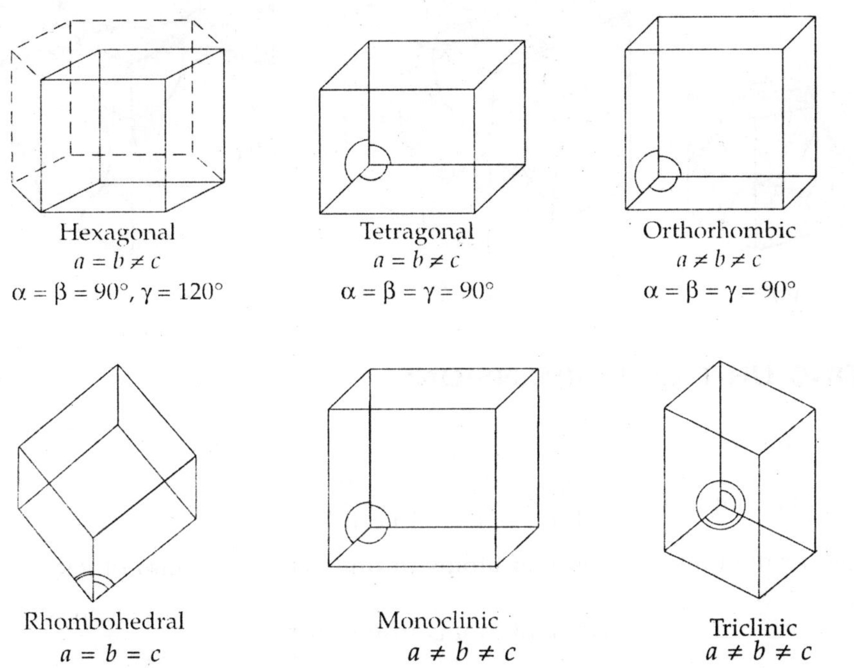

Number of Atoms in a Unit Cell

In various unit cells, three are kinds of **lattice points:** Points located at the corners; points in the face centers and points that lie entirely with in the unit cell. In a crystal, atoms located at the corners and face center of a unit cell are shared by other cells and only a portion of such an atom actually lies within a given unit cell:

(i) Simple Cubic a points that lies at the corner of unit cell is shared among eight unit cells and thus, it contributes for 1/8 of each such points to unit cell. i.e. there is only one atom per unit cell in a simple cubic structure.

(ii) A point along an edge is shared by four unit cells and thus contribute for 1/4 of each such point to unit cell.

(iii) A face-centred point is shared by two unit cells and thus contributed for ½ of each such point to cell. so in F.C. C. there are 1 + 3 = 4 atoms in 1 unit cell.

(iv) A body-centred point lies entirely within the unit cell and thus contributes for one each such point to unit cell. Thus in b.c.c. there are 2 atoms in one unit cell.

Types of cell	*Corners*	*Number of atoms at faces*	*Center*	*Total*	*Volume occupied by particles (%)*
(a) Simple	8 × 1/8 = 1	0	0	1	52.4
(b) Body-centred cubic (bcc)	8 × 1/8 = 1	0	1	2	68
(c) Face-centred cubic (fcc)	8 × 1/8 = 1	6 × 1/2 = 3	0	4	94

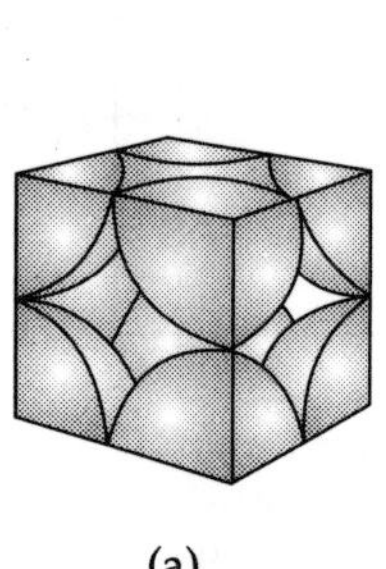
(a)

(b)

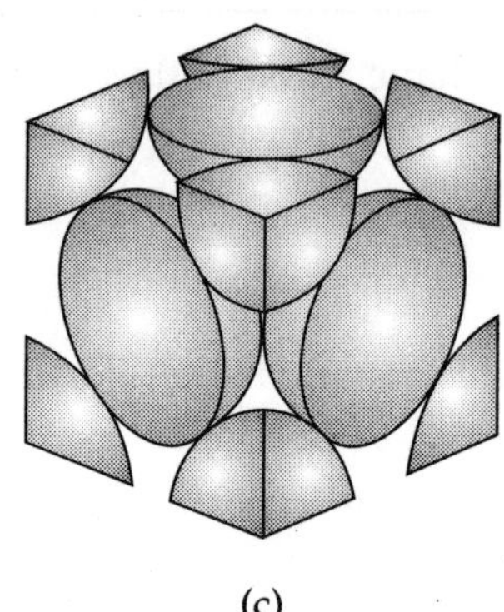
(c)

CALCULATING UNIT CELL DIMENSIONS

(i) Density of unit cell

$$\text{Density of the unit cell} = \frac{\text{Mass of the unit cell}}{\text{Volume of the unit cell } (V)} \qquad \ldots\ldots (1)$$

Also Mass of the unit cell = No. of atoms in unit cell (Z) × Mass of each atom (m)

$$= \text{No. of atoms in unit cell (Z)} \times \frac{\text{Atomic mass}}{\text{Av. No.}} \frac{(M)}{N_A} \qquad \ldots\ldots (2)$$

∴ By Eqs. (1) and (2),

$$\text{Density of unit cell 'd'} = \frac{Z \times \text{at. wt}}{\text{Av. no.} \times V} = \frac{Z \times \text{at. wt}}{\text{Av. no.} \times a^3} = \frac{ZM}{a^3 N_A}$$

Where volume of cubic unit cell = a^3 and a is edge length of unit cell, N_A is Avagadro number & M in molar mass.

(ii) Radii of ion:

If the edge length of the cubic cell is known ionic radii for the ion in the crystal lattice can be calculated.

Where a is edge length of cubic cell.

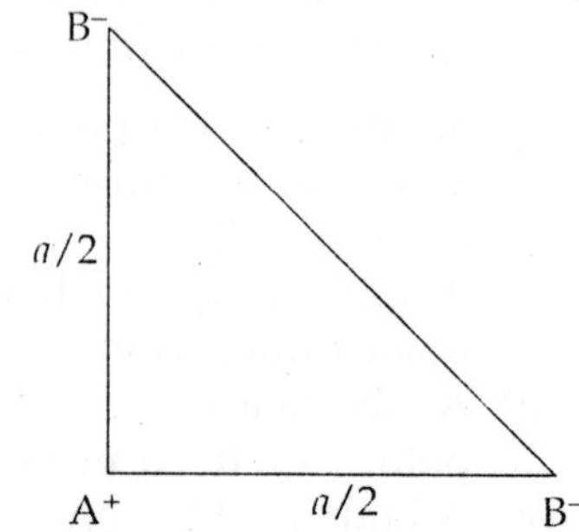

$$\text{Thus, Radius of } B^- - B^- = \frac{\sqrt{\left(\frac{a}{2}\right)^2 + \left(\frac{a}{2}\right)^2}}{2} = \frac{a}{4}$$

(iii) Atomic Radius or Radii of metal atoms: If a is the length of unit cell, then

(a) For simple cubic lattice: Radius of metal atom = $\frac{a}{2}$

(b) For b.c.c. lattice: Radius of metal atom = $\frac{\sqrt{3}a}{4}$

(c) For f.c. lattice Radius of metal atom = $\frac{a}{2\sqrt{2}}$

CLASSIFICATION OF IONIC STRUCTURES

It is convenient to divide ionic compounds into groups AX, AX_2, AX, depending on the relative numbers of positive and negative ions.

Ionic compounds of the type AX (ZnS, NaCl, C_sCl)

Ionic Compounds of the type AX_2(CaF_2, SiO_2, TiO_2)

Some structure based on close packing

Formula		*Type of p*	*tetrahedral*	*Octahedral*	*Coordination No.*		
					A		*X*
AX	NaCl	ccp	none	all	6	:	6
	NiAs	hcp	none	all	6	:	6
	ZnS zinc blende	ccp	$\frac{1}{2}$	none	4	:	4
	ZnS wurtzite	hcp	$\frac{1}{2}$	none	4	:	4
AX_2	F_2Ca* fluoride	ccp*	all	none	8	:	4
	CdI_2	hap	none	$\frac{1}{2}$	6	:	3
	$CdCl_2$	ccp	none	$\frac{1}{2}$	6	:	3
	b-$ZnCl_2$	hcp	$\frac{1}{4}$	none	4	:	2
	HgI_2	cp	$\frac{1}{4}$	none	4	:	2
MX_3	BiI_3	hcp	none	$\frac{1}{3}$	6	:	2
	$CrCl_3$	ccp	none	$\frac{1}{3}$	6	:	2
MX_4	SnI_4	hcp	$\frac{1}{8}$	none	4	:	1
MX_6	α-WCl_6 and UCl_6	ccp	none	$\frac{1}{6}$	6	:	1
M_2X_3	α-Al_2O_3 corundum	hcp	none	$\frac{2}{3}$	6	:	4

Structures of Ionic Solids

The structures of many ionic solids can be accounted for by considering the relative sizes of the positive and negative ions, and their relative numbers. Simple geometric calculations allow us to work out how many ions of a given size can be in contact with a smaller ion. Thus we can predict the coordination number form the relative sizes of the ions.

The Coordination Number

The number of nearest neighbours for any particular atom in a space lattice is called the co-ordination number. In diamond the co-ordination number is 4. In simple cubic crystal like NaCl it is 6. In b.c.c. crystal like CsCl it is 8. Here Cs^+ is at the body-centre of a cube of Cl^- and Cl^- at the body-centred of cube of C_s^+. In cubic close packed and hexagonal close packed, the co-ordination number is 12.

Radius Ratio Rules

The ionic radius ratios of cations and anions can determine the nature of crystal structure of the ionic substance. The coordination number of 8 is expected in AB type structure when radius ratio exceeds 0.732.

The ratio of the radius of cation and anions (r^+r^-)

The co-ordination number of an ion in closed packed structures depends on the ratio of the radius of ion with respect to the radius of the ions surrounding it. In case of ionic solids, the radius ratio (r^+/r^-) is the ratio of the cation to anion raidus. In a octahedral closed pack the ratio of the small cations to big anion comes out be 0.414 to 0.732 and in a tetrahedral close pack it is 0.225.

Thus, it has been concluded that co-ordination number can be predicted as:

Limiting radius ratios and structures

Limiting radius ratio (r^+/r^-)	*Coordination number*	*Shape*
< 0.155	2	Linear
0.155 → 0.225	3	Planar triangle
0.225 → 0.414	4	Tetrahedral
0.414 → 0.732	4	Square planar
0.414 → 0.732	6	Octahedral
0.732 → 0.999	8	Body-centred cubic

PACKING IN METALLIC CRYSTALS

Close Pack Structure

Here the atoms are stacked together occupying maximum space leaving minimum vacant space. These are of two types.

1. Cubic close packes (ccp). The resulting structure has an ABCABC...... patterns of layers.
2. Hexagonal close packed (hcp). The resulting structure has ABAB ...... patterns of layers.

Voids for (Holes) Created in Packing of Atoms in a Crystal)

In close packed structure, voids or holes are there in many crystals. These are of two types. (a) Octahedral void (or hole) (b) Tetrahedral voids (or hole)
When six atoms are in contact on octahedral hole (or void) is formed but when four atoms are in contact an tetrahedral hole (or void) is formed.

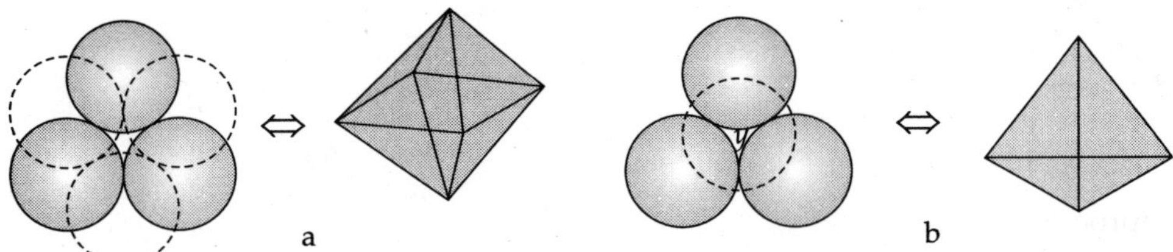

Two types of voids in crystals
(a) Octahedral void, *(b) Tetrahedral void.*

In both these arrangement i.e., in hcp and ccp structures, 74% of the total space is occupied by the atoms while 26% space remains vacant. In contrast to this, packing in body centred cubic (bcc) structure is not that efficient and only 68% space is occupied by the atoms.

The voids or holes in crystals are also called **interstices**. Their size acquires importance in relation to the formation of transition metal hydrodes, bordies, carbides and nitrides in which the respective non-metals (H, B, C and N) are accomodated in the interstices.

AB_2 Type structure eg. C_2F_2, BaF_2, $BaCl_2$ etc.
A_2B Type structure eg. Na_2O, Li_2O, K_2O etc.

CALCULATION OF SOME LIMITING RADIUS RATIO VALUES

(I) Coordination Number 3 (Planar Triangle)

Figure (a) shows the smaller positive ion of radius r^+ in contact with three negative ions of radius r^-. Painly $AB = BC = AC = 2r^-$, $BE = r^-$, $BD = r^+ + r^-$. Further, the angle $A - B - C$ is 60°, and the angle $D - B - E$ is 30°. By trigonometry

$$\cos 30° = BE/BD$$

$$BD = BE/\cos 30°$$

$$r^+ + r^- = r^-/\cos 30° = r^-/0.866 = r^- \times 1.155$$

$$r^+ = (1.55\ r^-) - r^- = 0.155r^-$$

hence $$r^+/r^- = 0.155$$

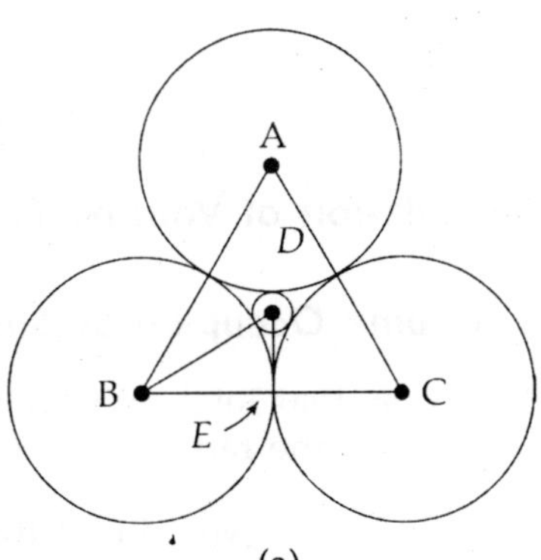

(a)

(2) Coordination Number of 4 (Tetrahedral)

Figure (b) shows a tetrahedral arrangement inscribed in a cube. Part of this tetrahedral arrangement is drawn in Figure (c). It can be seen that the angle *ABC* is the trtrahedral angle of 109°28 and hence the angle *ABD* is half of this, that is 54°44. In the triangle *ABD*.

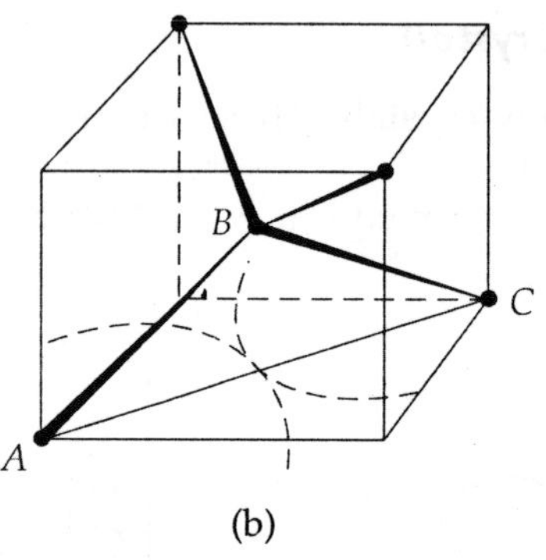

(b)

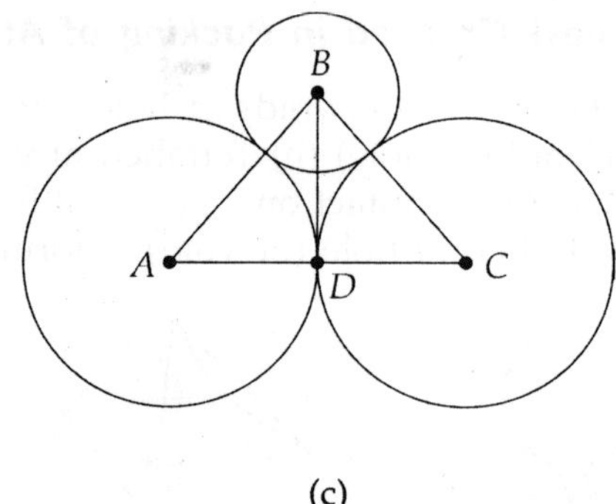

(c)

Taking reciprocals

$$\frac{r^+ + r^-}{r^-} = \frac{1}{0.8164} = 1.225 \text{ or } \frac{r^+}{r^-} + 1 = 1.225$$

hence

$$r^+/r^- = 0.225$$

(3) Coordination Number 6 (Octahedral)

A cross-section through an octahedral site is shown in Figure (d), and the smaller positive ion (of radius r^+) touches six larger negative ions (of radius r^-). (Note that only four negative ions are shown in this section, and one is above and another below the plane of the paper.) It is obvious that $AB = r^+ + r^-$ and $BD = r^-$. The angle AB is 45°. In the triangle ABD

Taking reciprocals

$$\frac{r^+}{r^-} = \frac{1}{07071} = 1.414 \text{ or } \frac{r^+}{r^-} + 1 = 1.414$$

hence

$$r^+/r^- = 0.414$$

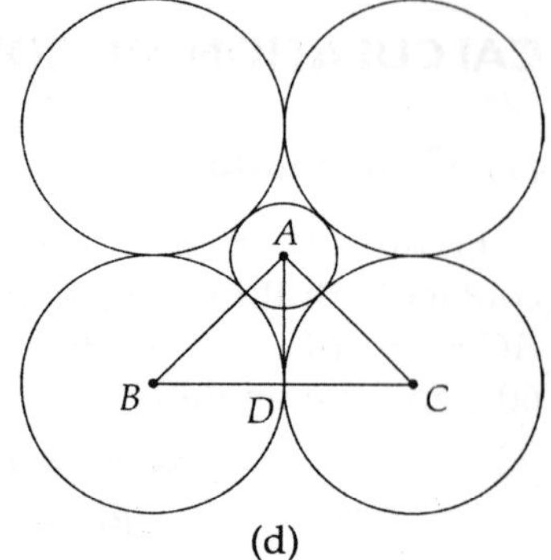

(d)

Calculation of Volume Occupied

I. Volume Occupied by Simple Cube

We know that total number of atoms per unit cell for simple cube is 1.

Let r be the radius of the atom (or sphere).

$\therefore$ Volume of the sphere $= \frac{4}{3}\pi r^3$

or Volume occupied by atoms per unit cell $= \frac{4}{3}\pi r^3$

In simple cube atoms at the corner of the cube are touching with each other hence length of the edge of the simple cube is 2r.

$\therefore$ Volume of the cube $= (2r)^3 = 8r^3$

Now Percentage of occupied space = $\dfrac{\text{volume of spheres}}{\text{volume of cube}} \times 100$

$$= \frac{\frac{4}{3}\pi r^3}{8r^3} \times 100 = 52.4\%$$

∴ Volume occupied by simple cube = 52.4%

and Volume of free space = 100 – 523.4 = 47.6%

II. IN F.C.C. Structures

Let us calculate the efficiency of packing in ccp structure. In (Fig. Below let the unit cell edge length be 'a' and face diagonal AC = b.

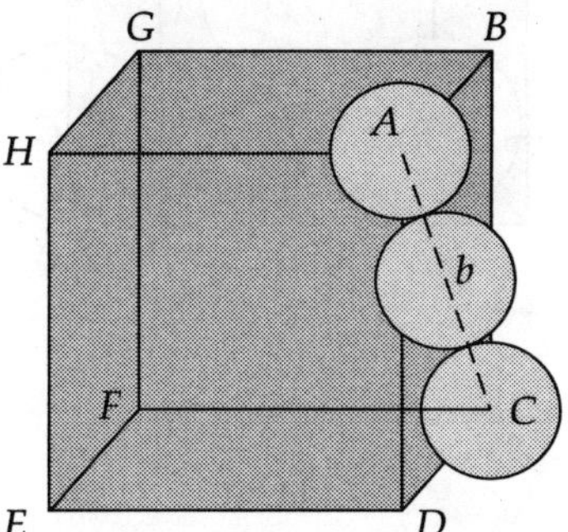

In ΔABC

$$AC^2 = b^2 = BC^2 = AB^2 = a^2 + a^2 = 2a^2$$

or $$b = \sqrt{2}a$$

If r is the radius of sphere, we find

$$b = 4r = \sqrt{2}a$$

or $$a = \frac{4r}{\sqrt{2}} = 2\sqrt{2}r \Rightarrow \left(r = \frac{a}{2\sqrt{2}}\right) \qquad \ldots\ldots (1)$$

We know, a unit cell in ccp structure, has 4 spheres and the total volume of four spheres is $4 \times (4/3)\pi r^3$ and volume of the cube is a^3 or $(2\sqrt{2}r)^3$

Therefore,

$$\therefore \text{Percentage efficiency} = \frac{\text{Volume occupied by four spheres in the unit cell} \times 100}{\text{Total volume of the unit cell}}\%$$

$$= \frac{4 \times (4/3)\pi r^3 \times 100}{(2\sqrt{2}\,r)^3}\% = \frac{(16/3)\pi r^3 \times 100}{16\sqrt{2}\,r^3}\% = 74\%$$

Hence in close packing (hcp and ccp) 74% of the available volume is occupied by atoms, & Volume of free space = 100 – 74 = 26%

III. Volume Occupied in Body Centred Cubic Structures (B.C.C)

Let unit fell length be 'a' and face diagonal $AC = b$. From the diagram below. It is clear that the atom at the center will be in touch with other two atoms diagonally arranged.

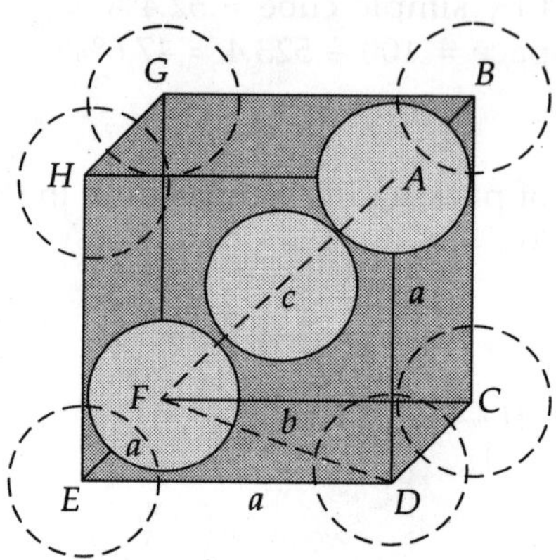

In ΔEFD,

$$b^2 = a^2 + a^2 = 2a^2 \Rightarrow b = \sqrt{2}a$$

Now in ΔAFD

$$c^2 = a^2 + b^2 = a^2 + 2a^2 = 3a^2$$

$$\therefore c = \sqrt{3}a$$

The length of the body diagonal c is equal to $4r$. where r is the radius of the sphere (atom). But $c = 4r$, (as all the three spheres along the diagonal touch each other).

Therefore, $\sqrt{3}a = 4r \Rightarrow a = \dfrac{4r}{\sqrt{3}}$ or $r = \dfrac{\sqrt{3}}{4}a$ (2)

In this structure total number of atoms is two and their volume is $2 \times (4/3)\pi r^3$.

Volume of the cube, a^3 will be $\left(\dfrac{4}{\sqrt{3}}r\right)^3$ or $a^3 = \left(\dfrac{4}{\sqrt{3}}r\right)^3$.

$$\therefore \text{Percentage efficiency} = \frac{\text{Volume occupied by two spheres in the unit cell} \times 100}{\text{Total volume of the unit cell}}$$

$$= \frac{2 \times (4/3)\pi r^3 \times 100}{[(4/\sqrt{3})r]^3}\% = \frac{(8/3)\pi r^3 \times 100}{[64/(3\sqrt{3})]r^3}\% = 68\%$$

So, volume of free space = 100 – 68 = 32%

DENSITY OF THE CRYSTAL

If the length of edge of the unit cell is known then the density of the crystal is calculated as follows:

Let length of edge of the unit cell be l.

$\therefore$ Volume of the unit cell = l^3 = V cm^3

$$\text{Density of the unit cell} = \frac{\text{Mass of unit cell}}{\text{Volume of unit cell}}$$

Mass of the unit cell = Number of atoms in a unit cell × Mass of one atom

$= n \times m\ g$

$$\text{But mass of one atom} = \frac{\text{Atomic mass}}{\text{Avogadro Number}} = \frac{M}{N_0}$$

$$\therefore \quad \text{Density of the unit cell } \rho = \frac{n \times M}{V \times N_0}\ \text{g cm}^{-3}$$

PROBLEM

Q. 1. Calculate the value of Avogadro's number from the internuclear distance of adjacent ions in NaCl, 0.282 nm, and the density of solid NaCl, 2.17×10^3 kg m^{-3}.

Q. 2. Calculate the ionic radius of fluoride ion using only the following interionic distances and the radius of the iodide ion, 2.19Å. RbI, 3.67 Å; RbBr, 3.44 Å; KB3, 3.29 Å; KCl 3.14 Å; NaCl, 2.79 Å; NaF, 2.31 Å.

Q. 3. Potassium crystallizes in a body-centred cubic lattice (see Figure below), with a unit cell length a = 5.20 Å.

(a) What is the distance between nearest neighour?

(b) What is the distance between next-nearest neighbour?

(c) How many nearest neighbours does each K atom have?

(d) How many next-nearest neighbours does each K have?

(e) What is the calculated density of crystalline K?

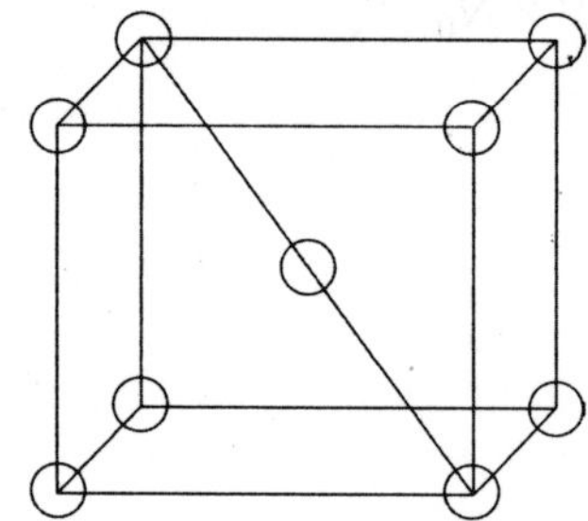

Q. 4. The NaCl lattice has the cubic unit cell shown in Figure below KBr crystallizes in this lattice.

(a) How many K^+ ions and how many Br^- ions are in each unit cell?

(b) Assuming the additivity of ionic radii, what is a?

(c) Calculate the density of a perfect KBr crystal.

(d) What minimum value of r_+/r_- is needed to prevent anion-anion contact in this structure?

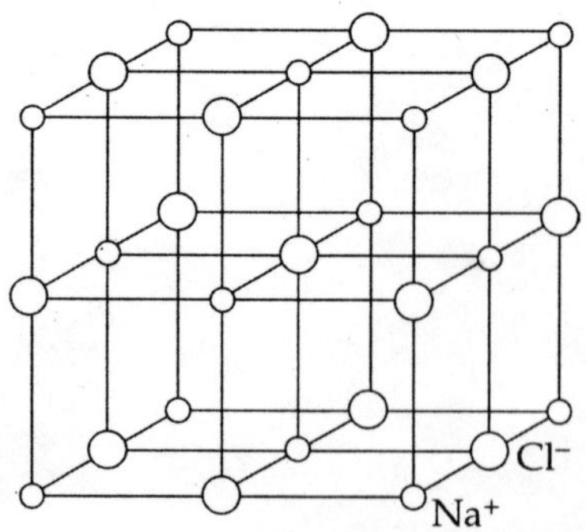

Q. 5. The hexagonal close-packed lattice can be represented by Figure (a) below. If $c = a\sqrt{8/3} = 1.633a$. There is an atom at each corner of the unit cell and another atom which can be located by moving one-third the distance along the diagonal of the rhombus base, starting at the lower left-hand corner and moving perpendicularly upward by $c/2$ (see Figure below.) Mg crystallizes in this lattice and has a density of 1.74 g cm^{-3}.

(a) What is the volume of the unit cell?

(b) What is a?

(c) What is the distance between nearest neighbours?

(d) How many nearest neighbours does each atom have?

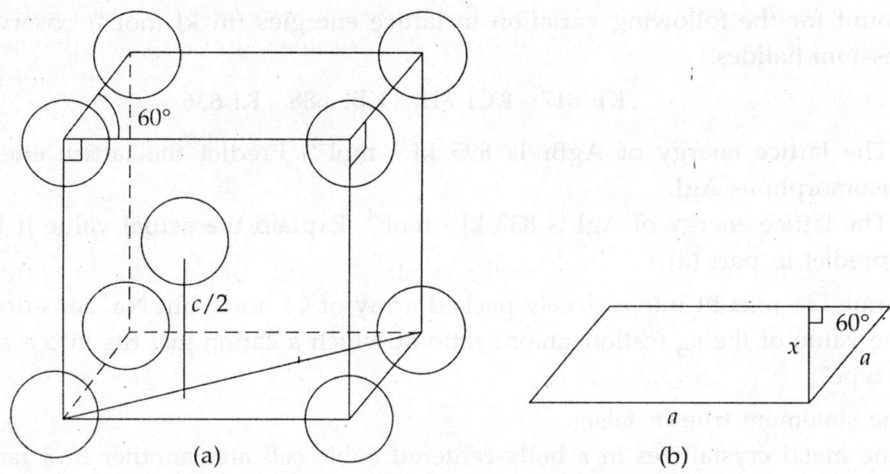

Q. 6. The length of a unit cell in the nickel crystal; is 0.352 nm. Diffraction of x-rays of 0.154 nm wavelength from a nickel crystal occurs at 22.2°, 25.9°, and 38.2°. Show that the these data are consist6ent with a face-centred cubic crystal structure.

Q. 7. The density of KBr is 2.75 g cm^{-3}. The length of the edge of the unit cell is 654 pm. Show that KBr has a face centred cubic structure.

Q. 8. An element crystallizes in a structure having f.c.c. unit cell of an edge 200 pm. Calculate its density, if 200 g of this element contains 24×10^{23} atoms.

Q. 9. A solid AB has a NaCl structure. If the radius of cation A^+ is 110 pm, calculate the maximum possible value of radius of the anion B^-.

Q. 10. Ionic compound *AB* has a ZnS type of structure. If the radius of A^+ is 22.5 pm. What is the ideal radius of B^-?

If the radius of A^+ is enlarged by some special technique and becomes 50 pm, in what type of structure would *AB* crystallize and that would AB be its coordination number?

Q. 11. A metallic element cryatallize into a lattice containing a sequence of layers of *ABABAB*. Any packing of spheres leaves out voids in the lattice. What percentage of volume of this lattice is empty space?

Q. 12. If molecules occupy the lattice points of a face-centred unit cell, how many molecules are there per unit cell?

Q. 13. Calculate the fraction of a body-centered unit cell that is occupied by atoms.

Q. 14. Platinum crystallizes in a face-centered cubic crystal with a unit cell length of 3.923, Å. Calculate the density and the atomic radius of platinum.

Q. 15. Lithium chloride crystallizes in a face-centred cubic structure. The unit cell length is 5.14 Å. Assuming Li^+ ions are so much smaller than Cl^- ions that there is anion–anion contact, and the Li^+ ions fit into the interstices between the Cl^- ions, calculate values for the radii of lithium and chloride ions.

Q. 16. Rubidium chloride cryatallizes in a face-centred cubic lattice with unit it cell length = 6.54 Å. If r_{cl} is 1.81 Å, calculate r_{Rb^+}.

Q. 17. Account for the following variation in lattice energies (in kJ mol^{-1}) observed for the potassium halides:

KF 817 KCl 718 KBr 688 Ki 636

Q. 18. **(a)** The lattice energy of AgBr is 895 kJ · mol^{-1}. Predict the lattice energy of the isomorphous AgI.

(b) The lattice energy of AgI is 833 kJ · mol^{-1}. Explain the actual value if larger than predict in part (a).

Q. 19. Assume Li+ ions fit into a closely packed array of Cl^- ions, but Na^+ ions do not. What is the value of the r_{ca} (cation–anion) ratio at which a cation just fits into a structure of this type?

Q. 20. Is the statement true or false:

One metal crystallizes in a body-centered cubic cell and another in a face-centered cubic cell of the same volume. The two atomic radii are related by the factor $\sqrt{1.5}$.

Q. 21. Iron crystallizes in a body-centered unit cell. Its atomic radius is 0.124 nm. Its density is 7.86 g cm^{-3}. Using this information, estimate Avogadro's number.

Q. 22. AB crystallizes in a rock salt structure with $A : B = 1 : 1$. The shortest distance between A and B is $y^{1/3}$ nm. The formula mass of AB is 6.023 Y amu where Y is any arbitrary constant. Find the density in kg m^{-3}

Q. 23. Chromium metal crystallizes with a body centered cubic lattice. The length of the unit cell edge is found to be 287 pm. Calculate the atomic radius. What would be the density of chromium in g cm^{-3}?

Q. 24. A metal crystallizes into two cubic phases, face centred cubic (FCC) and body centred cubic (BCC), whose unit cell lengths are 3.5 and 3.0 A, respectively. Calculate the ratio of densities of FCC and BCC.

Q. 25. The figures given below show the location of atoms in three crystallographic planes in a FCC lattice. Draw the unit cell of the corresponding structure and identify these planes in your diagram.

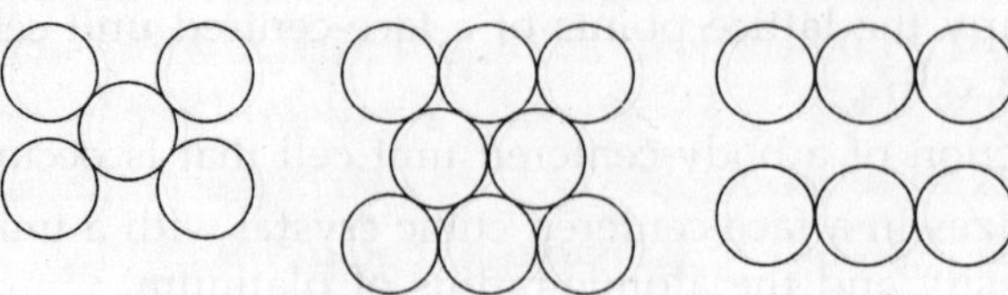

Q. 26. You are given marbles of diameter 10 mm. They are to be placed such that their centers are lying in a square bound by four lines each of length 40 mm. What will be the arrangements of marbles in a plane so that maximum number of marbles can be places inside the area?

Q. 27. 1 gm of charcoal absorbs 100 ml 0.5 M acetic acid to form a monolayer, and thereby the molarity of CH_3COOH reduces to 0.49. Calculate the surface area of the charcoal absorbed by each molecule of acetic acid. Surface area of charcoal = $3.01 \times 10^2 m^2\ gm^{-1}$.

Q. 28. The density of mercury is 13.6 g ml^{-1}. Calculate approximately the diameter of an atom of mercury assuming that each atom is occupying a cube of edfe length equal to the diameter of the mercury atom.

Q. 29. A metallic element crystallises into a containing a sequence of layers of *ABABAB* ...... Any packing of spheres leaves out voids in the lattice. What percentage by volume of this lattice is empty space?

Q. 30. The ionic radii of the alkali metal ions are:
Li^+ + 0.68 Å, Na^+ = 0.95Å, K^+ = 1.33 Å, Rb^+ = 1.48 Å and Cs = 1.69 Å. These four halide ions are: F^- = 1.36 Å, Cl^- = 1.81 Å, Br^- = 1.95 Å and I^- = 2.16 Å. If the radius ratio rules predict the structures exactly, give the expected cation coordination number in the following cases: (i) NaCl, (ii) RbBr, (iii) RbI, (iv) sCl and (v) LiF.

Q. 31. A solid *AB* has the NaCl structure. If the radius of the cation is 100 mp, what is the radius of the anion *B*?

Q. 32. Two ions A^+ and B^- have radii 88 pm and 200 pm respectively . In the close packed crystal of compound *AB*, predict the coordination number of A^+.

Q. 33. If the radius of an octahedral void is *r* and the radius of the atoms in close-packing is *R*, device the relation between *r* and *R*.

Q. 34. Magnetite, a magnetic oxide of iron use don recording tapes, crystallizes with iron atoms occuring both $\frac{1}{8}$ of the tetrahedral holes and $\frac{1}{2}$ of the octahedral holes in a closest packed array of oxide ions. What is the formula of magnetite?

Q. 35. The edge length of the unit cell of KCl (*NaCl structure. face centred cubic*) is 6.28 Å. Assuming anion–cation contact, calculate the ionic radius for the potassium ion, if that of the chloride ion is 1.81 Å.

Q. 36. The edge length of the unit of Lil (NaCl—like structure, face centered cubic) is 5.14 Å. Assuming anion-anion contact, calculate the ionic radius for the chloride ion.

Q. 37. Calculate the value of the Avogadro's number from the following data:

Density of Nal = 2.165 g cm^{-3}
Distance between Na^+ and Cl^- in NaCl = 281 pm

Q. 38. The face-centered unit cell of nickel has an edge length of 3.5239 Å as determined by X-rays. The atomic mass of Ni is 58.71 amu and the density is 8.9 g cm^{-3}. From these data, compute Avogadro's number.

Q. 39. KF has NaCl structure. What is the distance between K^+ and F^- in KF, if the density is 2.48 g^{-3}?

Q. 40. The compound CuCl has the ZnS (cubic) structure. Its density is 3.4 g cm^{-3}. What is the length of the edge of the unit cell?

Q. 41. The effective radius of an iron atom is 1.24 Å. Iron occurs both in a BCC structure and a FC structure. Calculate the density of each in Kg m^{-3}.

Q. 42. Silver crystallizes in a face-centered cubic unit cell with a silver atom on each lattice point. (a) If the edge length of the unit cell is 4.0862 Å, what is the atomic radius of silver? (b) Calculate the density of silver.

Q. 43. What is the density of silver, if the nearest neighbour atoms in its crystal are 2.88 Å apart?

Q. 44. An element crystallizes into a structure which may be described by a cubic type of unit cell having one atom on each corner of the cube and two atoms on one of its diagonals. If the volume, of the unit cell is 24×10^{-24} cm^3 and density of the element is 7.2 g cm^{-3}, calculate the number of atoms present in 200 g of the element.

Q. 45. An element 'A' (atomic mass 100) having B.C.C. structure has cell edge 400 pm. Calculate the density of A and the number of unit cells for 10 g of A.

Q. 46. A crystal of lead (II) sulphide has NaCl structure. In this crystal the storest distance between Pb^{2+} and S^{2-} ions is 297 pm. What is the length of the edge of the unit cell in lead sulphide? Also calculate the unit cell volume. (N = 6.023×10^{23} mol^{-1}; atomic mass : Pb = 207.3, S = 32.

Q. 47. An element (atomic mass = 60) having face-centered cubic structure has a density of 6.23 g cm^{3-}. What is the edge length of the unit cell?

Chapter **13**

SURFACE CHEMISTRY

ADSORPTION

Frundlich isotherms

$$\frac{x}{m} = kp^{\frac{1}{n}} \quad (n > 1)$$

Langmuir isotherms

$$\frac{x}{m} = \frac{ap}{1+bp}$$

ADSORPTION

I. The adsorption of a gas at T = const is given by Langmuir's adsorption isotherm equation.

$$y = \theta_{-} = \frac{ap}{1+bP} \qquad y_{\infty} = \frac{kP}{1+kP} \qquad \ldots\ldots (1)$$

in which y is the amount of gas adsorbed by 1 g of the adsorbent of 1 cm^2 of its surface, y_{∞} is the maximum amount of gas that can be adsorbed by 1 g of the adsorbent (or 1 cm^2 of its surface) at a given temperature, k is the adsorption equilibrium constant, and P is the equilibrium gas pressure. Equilibrium (1) expressed in terms of degree of filling θ takes the form

$$\theta = kP/(1 + kP)$$

The degree of filling is determined from the following ratio:

$$\theta = y/y_{\infty}$$

II. ***Langmuir adsorption isotherm*** can be expressed as the adsorbed ***gas volume***:

$$V = V_m kP/(1 + kP)$$

The constants the Langmuir's equation are calculated using the relation

$$\frac{P}{y} = \frac{1}{ky_\infty} + \frac{1}{y_\infty}P$$

or

$$\frac{1}{V} = \frac{1}{V_m} + \frac{k}{V_m P}$$

where V is the volume of the adsorbed gas, and V_m is that of the adsorbed gas covering completely the surface of 1 g of the adsorbent, both volume being reduced to normal conditions.

III. The area S of the surface of 1 g of the adsorbent is

$$S = \frac{V_m N_A}{22.414} \text{s}$$

where s is the area occupied by one molecule of the adsorbate on the surface of the adsorbent.

IV. Adsorption from solution is given by Langmuir's equation of the following type:

$$Y = y_\infty kc/(1 + kc)$$

where c is the equilibrium concentration of the adsorbate in the solution.

V. Feeundlich's Equation

Is to determine the mean degree of filling of the adsorbent surface and is called ***Freundlich isotherm equation which is***

$$X/m = kc^n$$

or

$$\text{Log}\ (x/m) = \log k + n \log c$$

In which x is the amount of the absorbate, m is the weight of the absorbent, and k and n are constants.

SURFACE CHEMISTRY

Surface chemistry is the chemistry at theboundary separating two bulk phases. This boundary which is also known as surface or interface between a solid and liquid may be represented by solid-liquid or solid/liquid. Important features features of surface chemistry are: adsorption, catalysis, colloids, emulsions and micelles.

Adsorption

The phenomenon of attracting and retaining the molecules of a substance son the surface of a liquid or a solid resulting into a higher concentration of the molecules on the surface is called adsorption. *As a result of adsorption there is a decrease of surface energy.*

Solids, particularly when finely divided, have a large surface area and, therefore, show this property of adsorption to a much larger extent. Charcoal, silica gel, alumina gel, clay etc. are very good adsorbents because they have highly porous structures and hence large surface areas. The substitute on the surface of which adsorption takes place is called **adsorbent** and the substance adsorbed is known **adsorbate.**

Adsorption generally occurs with release of energy i.e., it is exothermic in nature. A spontaneous process requires a negative ΔG. Since the translational freedom of the adsorbate is reduced when it is adsorbed, ΔS is generally negative. Therefore, in order for $\Delta G = \Delta H - T\Delta S$ to be negative, ΔH must be negative and hence the process is exothermic. In an adsorption process, the enthalpy change for the adsorption of one mole of an adsorbate on an adsorbent surface is called the **Enthalpy of adsorption.**

Adsorption can be classified into Physisorption (Physical adsorption) and Chemisorption (chemical adsorption). In physisorption the adsorbate is held on the surface by weak van der Waals attraction while in chemisorption, the binding is of chemical nature i.e., the molecules/ atoms of adsorbate form chemical bonds with the adsorbent surface. Example behaviour of N_2 on Iron. At 83 K, N_2 is physisorbed on iron surface as N_2 molecules while at 773 K or more, it is chemisorbed on iron surface as N_2 atoms.

Adsorption of Gases on Solids

eg: Characteristics chlorine and polluting gases silica gel absorbs moisture (hence used for drying)

Factors—Adsoroption of Gases on Solid

Charcoal absorbs chlorine and polluting gases. Silicagel absorbs moisture (hence used for drying).

The extent of adsorption of a gas on a solid adsorbent depends upon the following factors: **(a) Nature of the adsorbate (b) Nature of the adsorbent (c) Specific area of the adsorbent (d) Pressure of the gas (e) Temperature (f) Activation of adsorbent.**

(a) Nature of the absorbate: Every gas gets absorbed on the surface of any solid to a lesser or greater extent. At given conditions of temperature and pressure, the easily liquefiable gases like NH_3, HCl, CO_2, etc. are absorbed to a greater extent than the permanent gases such as H_2, O_2, N_2, etc. The ease with which a gas can be liquefied is primarily determined by its critical temperature. T_c. The higher the T_c of a gas, the move easily it will be liquefied and more readily it will be absorbed on the solid. e.g. 16 mL of methane (T_c = 430 K) is absorbed by 1g of activated charcoal.

(b) Nature of the absorbent: Most common absorbents are activated carbon, metal oxides (silica gel and aluminium oxide) and clay. Each of these have their characteristic adsorption properties.

(c) Specific area of the adsorbent: Specific area of an adsorbent is the surface area available for adsorption per gram of the adsorbent. The greater the specific area of the solid, the greater would be its adsorbing capacity.

(d) Pressure of the Gas

Adsorption isotherms : A relation or a graph between x/m and the pressure p of the gas at a constant temperature is called adsorption isotherm where m is the mass of the adsorbent and x is the number of moles of the adsorbate when dynamic equilibrium has been attained between the free gas and the adsorbed gas.

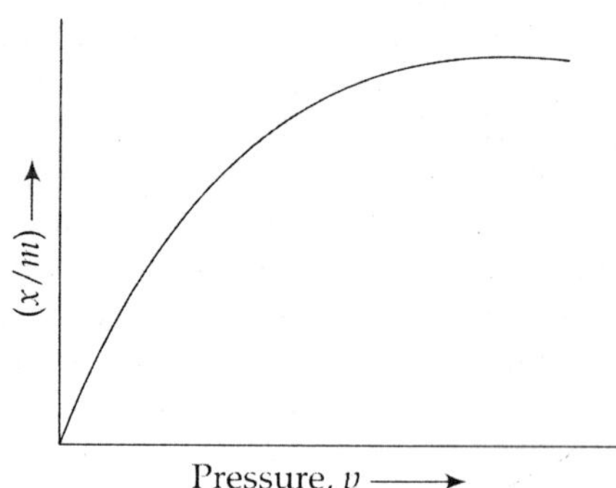

Freundlich Adsorption isotherms: In the case of adsorption of gases on solids, the relation between x/m and the pressure p of the gas at constant temperature is given by the equation.

$$\frac{x}{m} = kp^{1/n} \ (n > 1) \qquad \ldots\ldots \text{(i)}$$

where k and n are parameters of the equation depending upon the nature of the gas and the solid. The graph of the isotherm for $\frac{x}{m}$ & P is shown according to this, x/m increases with increase of p but since $n > 1$, therefore, x/m does not increase as rapidly, as p as evident from the isotherm.

Taking logarithms on both sides of Eq. (i) We get

$$\log\frac{x}{m} = \log k + \frac{1}{n}\log p$$

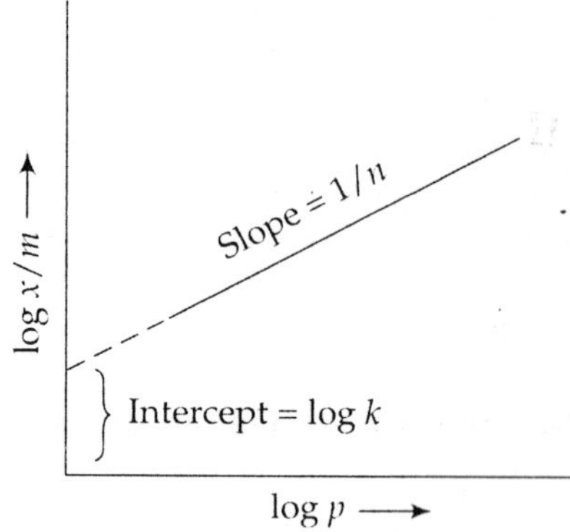

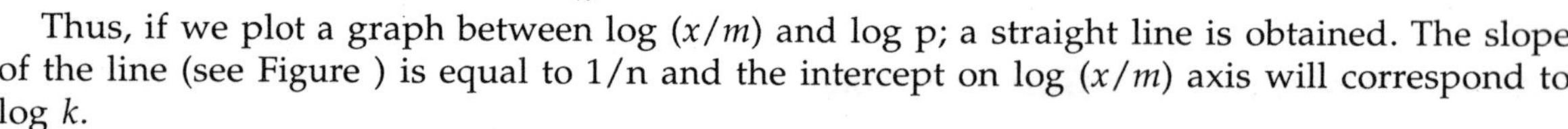

Thus, if we plot a graph between log (x/m) and log p; a straight line is obtained. The slope of the line (see Figure) is equal to 1/n and the intercept on log (x/m) axis will correspond to log k.

Langmuir Adsorption Isotherm

The Langmuir adsorption isotherm is represented by the relation

$$\frac{x}{m} = \frac{ap}{1+bp} \qquad (1)$$

where a and b are two Langmuir parameters. At very high pressure, the above isotherm acquires the limiting form.

$$\frac{x}{m} = \frac{a}{b} \text{ (at very high pressure)} \qquad (2)$$

At very low pressure, eq. (1) is reduced to

$$x/m = a\,p \text{ (at very low pressure)} \quad (3)$$

In order to determine the parameters a and b, eq. (1) may be written in its inverse form:

$$\frac{m}{x} = \frac{1+bp}{ap} = \frac{b}{a} + \frac{1}{ap} \quad (4)$$

A plot of m/x against $1/p$ gives a straight line with slope and intercept equal to $1/a$ and b/a, respectively. Thus, both parameters can be determined.

The graph of x/m versus p is shown. At low pressures, according to eq. (3) pressure x/m increases linearly with p. At high pressure according to eq. (2), x/m becomes constant i.e. the surface is fully covered and change in pressure has no effect and no further adsorption takes place.

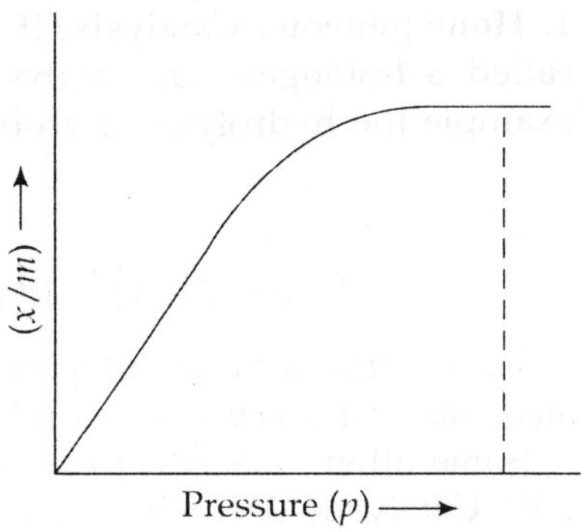

(e) Effect of temperature: Adsorption isobar Adsorption is generally temperature dependent. Mostly adsorption processes are exothermic and, therefore, adsorption decreases with increasing temperature. However, as expected for endothermic adsorption processes, adsorption increases with increase in temperature.

ADSORPTION FROM SOLUTIONS

Solids can adsorb dissolved substances from solutions also. Common examples are the adsorption of colour impurities by activated carbon in decolourising of solutions.

Freundlich adsorption and Langmuir adsorption isotherms have been found to be applicable in the adsorptions from solutions.

Bu there, we use equilibrium concentrations of the adsorbates in the solution and the isotherms take the form.

$$\frac{x}{m} = Kc^{1/n} \ (n > 1) \quad \ldots\ldots\ (5)$$

(Freundlich adsorption isotherms)

$$\frac{x}{m} = ac/(1 + bc) \quad \ldots\ldots\ (6)$$

(Langmuir adsorption isotherms)

CATALYSIS

A catalyst is a substance that increase the rate if reaction without being consumed in the reaction.

In general a catalyst functions by lowering the energy of activation which in turn makes the rate constant larger and hence the rate of the reaction higher.

Two important points

(a) A catalyst speeds up the forward and reverse reactions to the same extent and therefore, does into affect the equilibrium constant.
(b) A catalyst lowers the energy of activation by providing a different mechanism for the reaction.

Types of Catalysis

There are two homogenbeous and heterogeneous

1. Homogeneous Catalysis: If the catalyst is present in the same phase as the reactants, it is called a homogeneous catalyst and this type of catalysis is called homogeneous catalysis. example the hydrolysis of an organic ester (RCOOR′).

$$R-\overset{\overset{\ddot{O}:}{||}}{C}-\ddot{\underset{..}{O}}-R'(l) + H_2O(L) \xrightleftharpoons{H^+(aq)} R-\overset{\overset{\ddot{O}:}{||}}{C}-\ddot{\underset{..}{O}}H(l) + R'\ddot{\underset{..}{O}}H(l)$$

The reaction rate is low at room temperature but can be increases by adding a small amount of a strong inorganic acid which provides H^+ ion, that acts as a catalyst in the reaction.

Some other *examples of homogeneous catalysts are:*

(i) Catalystic decomposition of ozone by chlorine atoms in the gas phase.

$$O_3 + O \xrightarrow{Cl} 2O_2$$

(ii) Oxidation of CO by O_2 in the presence of NO as catalyst

$$2CO(g) + O_2(g) \xrightarrow{NO} 2CO_2(g)$$

(iii) Hydrolysis of sugarcane solution in presence of dilute sulphuric acid.

$$\underset{\text{Sucrose}}{C_{12}H_{22}O_{11}(aq)} + H_2O(l) \xrightarrow{H_2SO_4} \underset{\text{Glucose}}{C_6H_{12}O_6(aq)} + \underset{\text{Fructose}}{C_6H_{12}O_6(aq)}$$

2. Heterogeneous Catalysis: In this type of catalysis the catalyst is present in a different phase than that of the reactants.

In heterogeneous catalyst, catalyst is generally a solid and the reactants are generally gases but sometimes liquid reactants are also used. It is also known as surface catalysis as the *reaction starts at the surface of the solid catalyst.*

example : the addition of H_2 to the C=C bonds of organic compounds to form C—C bonds i.e. ethylene to ethane:

$$CH_2{=}CH_2(g) + H_2(g) \rightarrow CH_3{-}CH_3(g)$$

Some other examples of heterogeneous catalysis are:

(i) Manufacturer of ammonia from N_2 and H_2 by Harber's process using iron as catalyst.

$$N_2(g) + 3H_2(g) \rightleftharpoons 2NH_3(g)$$

(ii) Synthesis of methyl alcohal (CH_3OH) from CO and H_2 using a mixture of copper, ZnO and Cr_2O_3 are catalyst.

$$CO(g) + 2H_2(g) \xrightarrow{Cu,\ ZnO \cdot Cr_2O_3} CH_3OH(l)$$

COLLOIDS

In solutions we have homogeneous heterogeneous mixtures. There are systems where neither homogeneous not heterogeneous mixtures are formed. Such intermediate systems are called *colloids* and such a solution is termed colloidal solution. eg: fog, emulsions, foam, egg albuim etc.

Imagine a process in which a sample of solid is placed in a liquid and progressively subdivided. After some time, the visible particles of the solid will separate out (or can be separated by filtration). But at further progressive subdivision continues, a state is reached where the particles breakdown to individual molecules or atoms. After certain limit, the two phases (liquid & solid) can no longer be distinguished; a solution is produced and the particles in it do not separate on filteration. These particles are so small that they do not form a separate phase. This intermidiate state of particles is called the The Colloidal State. Such a dispersion is called a Colloid; a colloidal suspension or a colloidal solution.

The diameter of particles in a colloidal state should be between 10^{-7} and 10^{-4} cms. Or 1 to 100 nm. If the dimensions are less than 1 nm, the particles are in Molecules Size, range but if they are greater than 1 nm, they are of macroscopic or ordinary size. The particles are spherical, rod, disc or randomly coiled shaped, eg. Silica in water; tobacco mosic; gamma globium is human blood; oil silk on water.

CLASSIFICATION OF COLLOIDAL

They are classified as: sols, emulsions, gels, aerosols etc.

Depending upon whether the dispersed phase and the dispersion medium are solids, liquids or gases, **eight types of collidal systems are possible.** A gas mixed with another gas forms a homogeneous mixture and hence is not colloidal system. The examples of the various types of colloids along with their typical names are:

Types of Colloidal Systems

	Dispersed phase	*Dispersion medium*	*Type of colloid*	*Examples*
1.	Solid	Solid	Solid sol	Some coloured glasses, Blue rock salt, black diamond and Ruby glass gem stone
2.	Solid	Liquid	Sol	Paints, cell fluids, colloidal gold,
3.	Solid	Gas	Aerosol	Smoke, dust, laden air
4.	Liquid	Solid	Gel	Cheese, butter, jellies
5.	Liquid	Liquid	emusion	Milk, hair cream, mayonnaise
6.	Liquid	gas	Aerosol	Fog, mist, cloud, insecticide sprays
7.	gas	Solid	Solid sol	Pumica stone, lava, valconicash
8.	gas	Liquid	Foam	Froth, whipped, cream, detergent suds, soap lather.

PREPARATION OF COLLOIDAL SOLS

Sols can be prepared by two types of methods:
1. Condensation methods: (a) Chemical methods (b) Physical methods.
2. Dispersion methods

1. Condensation methods

(a) Chemical methods. Colloidal solutions can be prepared by chemical reactions leading to formation of molecules by double decomposition, oxidation, reduction or hydrolysis. These molecules then aggregate leading to formation of sols.

$$As_2O_3 + 3H_2S \xrightarrow{\text{Double decomposition}} As_2S_3(sol) + 3H_2O$$

$$SO_2 + 2H_2S \xrightarrow{\text{Oxidation}} 3S(sol) + 2H_2O$$

$$2\ AuCl_3 + 3\ HCHO + 3H_2O \xrightarrow{\text{Reduction}} 2Au(sol) + 3HCOOH + 6HCl$$

$$FeCl_3 + 3H_2O \xrightarrow{\text{Hydrolysis}} Fe(OH)_3\ (sol) + 3HCl$$

(b) Physical methods

(i) Exchange of solvent: When a solution is mixed with an excess of the other solvent in which the solute is insoluble but solvent is miscible, a colloidal sol is obtained. For example, when a solution of sulphur in alcohol is poured in excess of water, a colloidal sol of sulphur is obtained.

(ii) Excessive cooling: The colloidal sol of ice in an organic solvent such as $CHCl_3$ or ether can be obtained by freezing a solution of water in the solvent. The molecules of water which can no longer be held in solution separately combine to form particles of colloidal size.

2. Dispersion Methods

In this, large particles are broken down by grinding etc. until particles are sufficiently small to remain suspended. There are three methods:

(a) Mechanincal disperion
(b) Bredig' Arc method or Electrical disintegration
(c) Peptization

Properties of Colloids:

1. The freezing point of colloidal mixture of silica in water are slightly below O°
2. These mixture must contain small no. of particles compared to true solutions with comparable quantities of solute.
3. Though the number of particles in colloidal dispersions are small. This physical dimensions and masses are huge compared to solute particles.
4. Osmotic pressures of colloids are very small but give very high figures of molecular weight.
5. The colloidal systems show optical (Tyndall effect) mechanical (Brownian movement) & electrical properties (electrophoresis)

Stability of Colloids: They have high surface area to volume ratio. This due to surface tension). The atoms, ions or molecules behave differently from those in the interior; the pressure of these at the surface being very small, than in the interior.

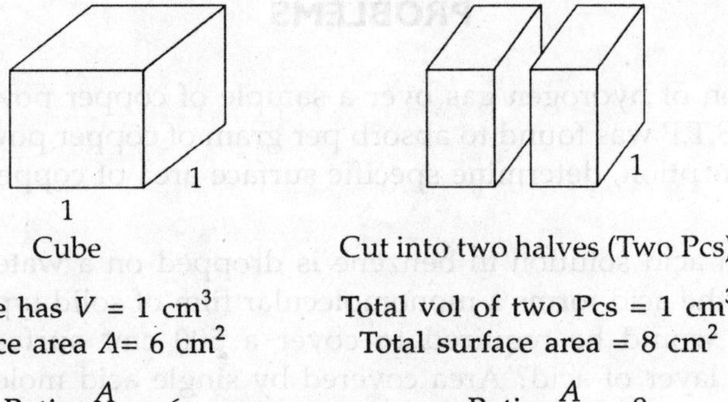

Cube | Cut into two halves (Two Pcs)

Cube has $V = 1\ cm^3$ | Total vol of two Pcs = $1\ cm^3$

Surface area $A = 6\ cm^2$ | Total surface area = $8\ cm^2$

$\therefore$ Ratio $\frac{A}{V} = 6$ | $\therefore$ Ratio $\frac{A}{V} = 8$

Repeated cuts will increase the surface area, though the total volume will remain the same. Due to the properties of surfaces being able to attach spices to themselves, a phenomenon called ADSOPTION, same colloidal particles adsorb large no. of ions from solution and become electrically charged eg.silica particles in colloidal silica mixture adsorb OH of ions, thus acquiring a recharge. Having like charges they repel each other. These mutual repulsions overcome forçe of gravity and keep particles suspended; thus a stabely colloid is formed.

Also, a high concentration to dissolved ions can bring coagulation or precipitation of a colloid. (specially those carrying the opposite change to the colloidal particles).

PURIFICATION OF COLLOIDS SOLS

The colloidal cols obtained by various methods are impure and contain impurities of electrolytes and other soluble substances. These impurities may destabilize the sol. Hence, they have to be removed. A very important method of removal of soluble impurities from sols by a semipremeable membrane is known as **dislysis**.

Another method is by ultra filtration and ultra centrifugation.

PROBLEMS

Q. 1. In the adsorption of hydrogen gas over a sample of copper powder, 1.36 cm^3 of $H_2(g)$ measured over S.T.P was found to absorb per gram of copper powder. Considering only mono-layer adsorption, determine specific surface area of copper. Density of liquid H_2 is 0.07 g cm^{-3}.

Q. 2. A 0.016 M of an acid solution in benzene is dropped on a water surface, the benzene evaporates and the acid forms a monomolecular film of solid type. What volume of the above solution would be required to cover a 500 cm^2 surface area of water with monomolecular layer of acid? Area covered by single acid molecule is 0.2 nm^2.

Q. 3. One gram of activated charcoal has a surface area of 1000 m^2. Considering complete coverage as well as monomolecular adsorption, how much ammonia in mL at S.T.P. would be adsorbed on the surface of 25 g charcoal? Diameter of a ammonia molecules is 0.3 nm.

Q. 4. Calculate surface area of a catalyst of which adsorb 100 cm^3 of $N_2(g)$ at S.T.P. per gram in order to form a monomolecular layer of $N_2(g)$. The effective surface area occupied by nitrogen molecule on the surface of catalyst is 0.16 nm^2.

Q. 5. In an experimental with adsorption of argon by coke at 194.7 K, the following results have been obtained:

P, Pa	31.9×10^3	130.5×10^3	290×10^3
P, mm Hg	24	98.4	218
Y, mg/g	5	15.4	24

Calculate the constants in Langmuir's equation.

Q. 6. The volume V_m of gaseous nitrogen at 1.04×10^5 Pa (750 mm Hg) and 273 K, necessary to cover a silica gel matrix with a monomolecular layer, is 129 ml g^{-1}. Calculate the surface of 16.2×10^{-20} m^2.

Q. 7. Write down the heterogeneous catalyst involved in the polymerization of ethylene.

PART II

Chapter 14

ORGANIC CHEMISTRY

TO THE STUDENT

Organic reactions form the foundation of organic chemistry. In order to do any synthesis or analysis of organic compounds. You should memorize general type-reactions.

For example, to work out a synthesis of acetic acid from methyl alcohol, reactions as given below as should be familiar to you. Also you should be able to picture or draw the structures of isobutane, propylene, acetic acid, etc., and check back the valence rule. Without special efforts to memorize these structures you should be able to write them almost subconsciously.

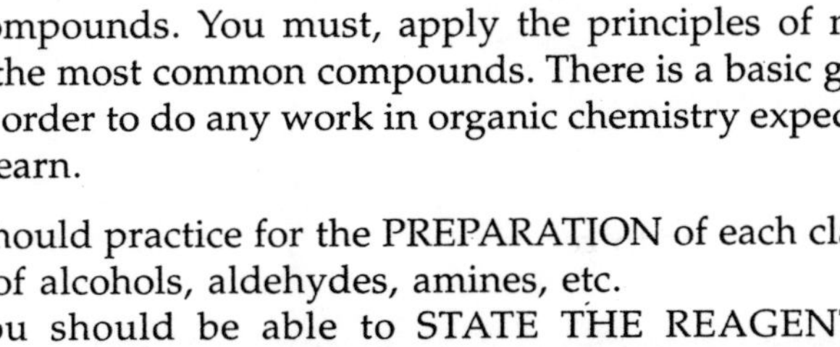

CH_3CN

KCN ↗ ↘ H_2O^+ Δ

$CH_3OH \xrightarrow{PBr_2} CH_3Br$ CH_3CN

Mg ↘ ↗ CO_2 than H_2O

CH_3MgBr

For complex compounds. You must, apply the principles of nomenclature to arrive at the structure, but for, the most common compounds. There is a basic group of organic reactions that must be known in order to do any work in organic chemistry expeditiously, and this is the group which you must learn.

(i) Firstly you should practice for the PREPARATION of each class of compound studied, i.e., preparation of alcohols, aldehydes, amines, etc.

(ii) Secondly you should be able to STATE THE REAGENTS AND CONDITIONS for transforming a given class of compound directly into another given class, that is, to indicate, for example, how an alkyl halide would be changed into a nitroalkane, or how a carboxylic acid would be reduced to an alcohol.

(iii) Thirdly you should be able to use these reactions for the qualitative identification of a given class or for the qualitative distinction between two or more classes.

(iv) Fourthly you should be able to figure out a sequence of reactions for multistep syntheses.: for example, to work out the preparation of *n*-butyl alcohol from *n*-propyl alcohol or the preparation of adipic acid from phenol.

Thus this requires greatest recollection of the type reactions. Study the given examples in how to solve problems in organic chemistry.

For Weak Students

Students who have been away from organic chemistry for an interval of time or those doing poorly in the chemistry course should review as follows:

1. Write down the general equations for the preparation and chemical properties of each class of compound.
2. On a separate sheet, make up a list of questions asking for a reproduction of these general equations; for example, "give three general equations for the preparation of primary alcohols." Then write examples of each of these equation using real substances, i.e., not just general equations in which R groups are used, but with specific structural formulas.
3. Familiarize yourself with the physical properties of each class of compound taken up, attention being given to water solubility, ether or sulfuric acid solubility acid or base character and whether associated or not. Such properties from the basis for carrying out *separations* of different classes of compounds.
4. Note the differences in chemical properties of each class of compound

Always remember qualitative test results.

Know the reagents and conditions.

Practice by writing a series of reactions that could be used to prepare alkanes, alkanes. alkynes, alkylhalides, alcohols, aldelydes, nitrites, amides, esters, carboxylic acid, acid chlorides, amines and nitroalkanes. Also try to specify catalyst for such reactions.

ORGANIC CHEMISTRY

How to Solve Problems in Organic Chemistry

To solve problems for conversions, transformations or predictions in organic chemistry the following steps in followed will be useful.

Step I: (i) Look for the functional group of the reactants (for Predict problems)

(ii) Look for the connectivity of the two compounds (for conversions/transformations). How many carbon atoms does each contain. Also, rings and position of branches.

Step II: Look for the change in functions groups in going from starting material to product. Does any of the reactant (or starting material) has a leaving group.

Step III (i) Are the reactants acids, bases, nucleopliles or electrophiles (∵ most common reactions can be classified as reactions of a nucleophile with an electrophile).

(ii) What is the most likely first step for the reactions: protonation or deprotonatios, ionization, attack by a nucleophile or an electrophile.

(iii) What are the poperties of the spices present in the reaction mixture after the first step. What is likely to happen next.

Step IV: From the structure of the starting material and products. Find out which bonds are broken and which bonds are formed. Check, if any part of the product molecule is coming from a good nucleophile or an electrophile.

Step V: What type of compound will be a good processor to the product.

Step VI After step V, we find that we have converted starting material to product. If not, we have to analyze the structure obtained in step V by applying steps IV and V to it again.

The examples given below should be efficient to give you an idea of solving the conversion, transformation or predict problems.

(a) Example Predict the product(s) the following reaction.

$$CH_3CH_2CH_2Br + C_6H_5{-}S^-Na^+ \longrightarrow ?$$

t-Bromopropane

Solution *Step I*: One reactant is an alkyl halide, the other is a sulfur anion (sodium ions are usually spectator ions i.e. thy do not participate is a reaction)

Step II: The bromine atom in I-bromopropane is leaving group.

Step III: ∵ Most common reactions can be classified has reactions of a nucleo phile with an electrophile.

Here, the sulfur anion is a nucleophile. The alkyl halide contains an electrophilic carbon atom.

Step IV: the reaction of nucleophile sulphur atom with electrophile carbonatom is:

$$CH_3CH_2\overset{\delta+}{CH_2}{-}\overset{\delta-}{Br} + :\ddot{S}:{-}C_6H_5 \longrightarrow CH_3CH_2CH_2{-}\ddot{\underset{..}{S}}{-}C_6H_5 + :\ddot{\underset{..}{Br}}:^-$$

Step V: As all the species formed are stable, no further reaction will occur. So the complete equation is:

$$CH_3—CH_2CH_2Br + \text{C}_6\text{H}_5—S^-NA^+ \longrightarrow CH_3CH_2CH_2—S—\text{C}_6\text{H}_5 + Na^+Br^-$$

(b) Example How would you synthesize $CH_3CH_2C{\equiv}CH$ from CH_3CH_2I?

Solution We examine this problem as:

Step I Here, The starting material has a two carbon chain, an ethyl group, with an iodine atom bonded to one carbon. The product has a four-carbon chain in which the two carbons that have been added o the ethyl group are bounded to each other by a triple bond.

Step II: The iodine atom in the starting material is a leaving group and has been replaced in the product by an alkyne functional group.

Step III: From the structures of the starting material and product which bonds must be broken and which are formed;

we have

$CH_3CH_2 + I$ (bond to be broken) $CH_3CH_2 + C{\equiv}CH$ (new bond to be formed)

$CH_3CH_2—$ $—C{\equiv}CH$

(the pieces that have to be brought together)

Step IV: The ethyl group is attached to a leaving group in the starting material and so must serve as the electrophilic center.

—CH group is related to the nucleophile $HC{\equiv}C\!:^-$, the conjugate base of an alkyne.)

Step V: a displacement of the iodine atom by the nucleophile will give us the product we want.

$$\overset{\delta+}{CH_3CH_2}—\overset{\delta-}{\ddot{\underset{..}{I}}}: \longrightarrow CH_3CH_2C{\equiv}CH$$

$$^-:C{\equiv}CH \qquad\qquad :\ddot{\underset{..}{I}}:^-$$

Step VI After this last step we have converted starting material to product?

If we do not get this than, we need to analyze the structure, obtained in step V.

(c) Example *mso*-2,3-Butanediol is prepared from (Z)-2-butene. Suggest a reagent that can be used for this transformation.

Solution

Step I: Writing in structural from we have

(Z) 2-butene

meso-2, 3-butanediol drawn in the eclipsed conformation to show the plane of symmetry

more stable anti conformation of meso-2, 3-butanediol

Here, the carbon skeletons of the two compounds are identical. The two carbon atoms that are joined by a double bond in the starting material have hydroxyl groups on them in the product. The methyl groups in the starting alkene and in the eclipsed conformation of the resulting diol are on the same side of the two carbon atoms to which they are bonded.
Step II: The double bond disappears, and the two carbon atoms of the double bond end up with hydroxyl groups on them. There is no leaving group present in the starting material.
Step III: From the structures of the starting material and product we find that bonds broken and formed are:

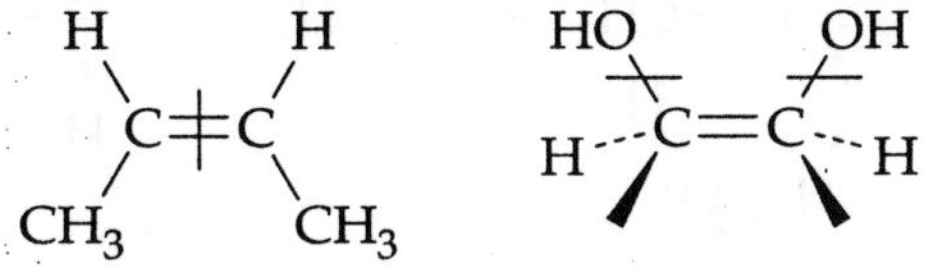

(bond broken) *(bonds formed)*

Step IV: A diol results from reaction of an alkene with an oxidizing agent (an electrophile) such as permanganate or osmium tetroxide. The reaction of permanganate or osmium tetroxide is a syn addition and therefore will give the desired stereochemistry.
Step V: What type of compound would be a good *precursor* to the product?
No other precursor is needed. The starting material is converted directly to the product.

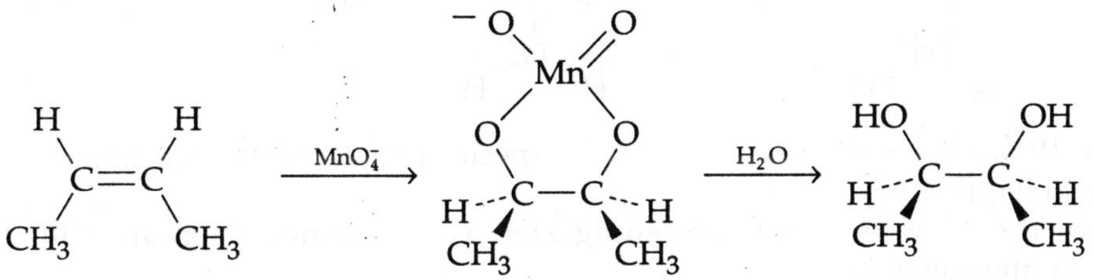

$$\text{CH}_3\text{CH=CHCH}_3 \xrightarrow{\text{MnO}_4^-} \text{cyclic manganate ester} \xrightarrow{\text{H}_2\text{O}} \text{CH}_3\text{CH(OH)CH(OH)CH}_3$$

(d) Example Write a detailed mechanism for the addition of water to 2-butyne.

$$\text{CH}_3\text{C}\equiv\text{CCH}_3 \xrightarrow[\text{H}_2\text{SO}_4]{\text{H}_2\text{O}} \text{CH}_3\overset{\text{O}}{\overset{\|}{\text{C}}}\text{CH}_2\text{CH}_3$$

Solution We proceed similarly as in above examples:
Step I: How have connectivities changed in going from reactant to product? How many carbon atoms does each contain? What bonds must be broken and formed to transform reagent into product?

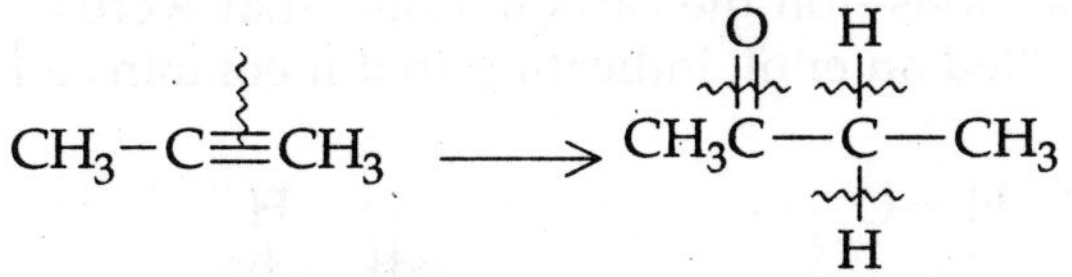

$$\text{CH}_3-\text{C}\equiv\text{CH}_3 \longrightarrow \text{CH}_3\overset{\text{O}}{\text{C}}-\overset{\text{H}}{\underset{\text{H}}{\text{C}}}-\text{CH}_3$$

Here, the reactant and the product have the same number of carbon atom arranged is the same way. The π bonds in the alkyne must be broken. A double bond between carbon and oxygen and two carbon-hydrogen bonds must be formed.

Step II: What reagents are present? Are they good acids, bases, nucleophilies, or electrophiles?

(Note, H_2SO_4 is a strong acid, so in the presence of a strong acid, H_2O behaves as a weak base. It is also a nucleophile. The mixture of water and sulfuric acid contains the hydronium ion, H_3O^+. The organic regent is an alkyne. The triple bond has a cloud of π electrons, which are nucleophilic)

Step III: What is the most likely first step for the reaction: protonation or deprotonation, ionization, attack by a nucleophile, or attack by an electrophile?

(The π bond will be protonated.)

$$CH_3-C\equiv C-CH_3 + H_3O^+ \longrightarrow CH_3-\overset{+}{C}=C(H)-CH_3 + H_2O$$

Step IV: What are the properties of the species present in the reaction mixture after the first step? What is likely to happen next?

(The vinyl cation is an electrophile. It will react with the nucleophile, water.)

$$CH_3-\overset{+}{C}=C(H)-CH_3 + H_2\ddot{O}: \longrightarrow CH_3-C(\overset{+}{O}H_2)=C(H)-CH_3$$

Step V: Repeat IV what are the properties of the species present in the reaction mixture. Now what is likely to happen next?

(The species resulting from the preceding step is an oxonium ion-an acid. It will be deprotonated by the solvent.)

$$CH_3-C(\overset{+}{O}H_2)=C(H)-CH_3 + H_2O \rightleftharpoons CH_3-C(\ddot{O}H)=C(H)-CH_3 + H_3O^+$$

Comparing this with the starting material and product, we see a carbon-oxygen bond and a carbon-hydrogen bond at the right locations—on the carbon atoms that were originally part of the triple bond. This intermediate is called an **enol**, indicating that it contains a hydroxyl group (ol) on a double bond (ene).)

$$CH_3-C\equiv C-CH_3 \quad CH_3-C(OH)=C(H)-CH_3 \quad CH_3C(=O)-C(H)(H)-CH_3$$

an enol

To get from the enol to the product, a hydrogen must be added at the double bond, a hydrogen must be removed from oxygen, and the double bond must be shifted to the oxygen.)
Step VI : Repeat. What are the properties of the species present in the reaction mixture now? What is likely to happen next?

(The enol will be protonated in the acidic solution. *Two sides of protonation are available*)

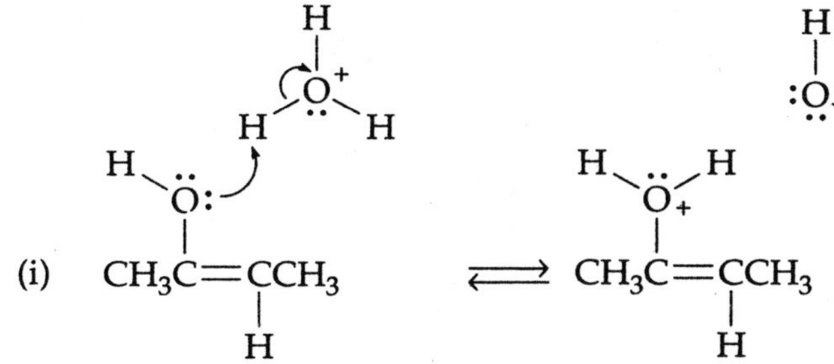

(i) $CH_3C(OH){=}C(H)CH_3 + H_3O^+ \rightleftharpoons CH_3C(OH_2^+){=}C(H)CH_3 + H_2O$

(gain of proton at oxygen)

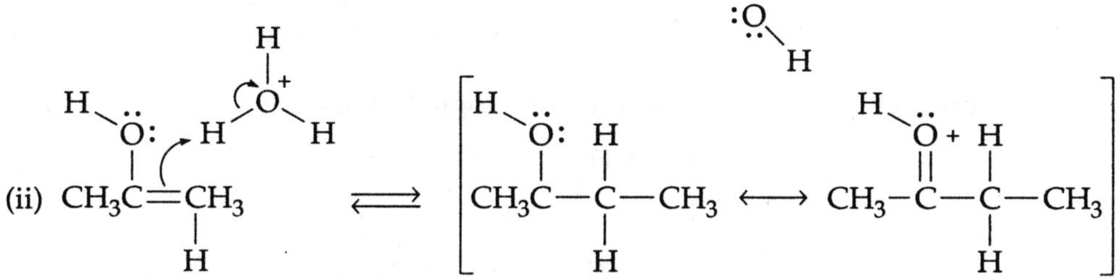

(ii) $CH_3C(OH){=}CHCH_3 + H_3O^+ \rightleftharpoons [CH_3C^+(OH){-}CH_2{-}CH_3 \longleftrightarrow CH_3{-}C({=}OH^+){-}CH_2{-}CH_3] + H_2O$

(gain of a proton at carbon)

Protonation at the oxygen is the reverse of the deprotonation reaction of the preceding step. It occurs, but does not further the overall reaction that gives he desired product. Protonation of the double bond in an enol is particularly easy because the positive charge in the resulting carbocation is adjacent to the oxygen atom and stabilized by resonance. In one of the resonance contributors for this cation, carbon and oxygen each have an octet of electrons, and the positive charge is delocalized to the oxygen atom. *A close inspection of this resonance-stabilized cation reveals that it is a protonated ketone. loss of the proton to a base gives 2-butanone.*)

The complete, correct answer is as follows:

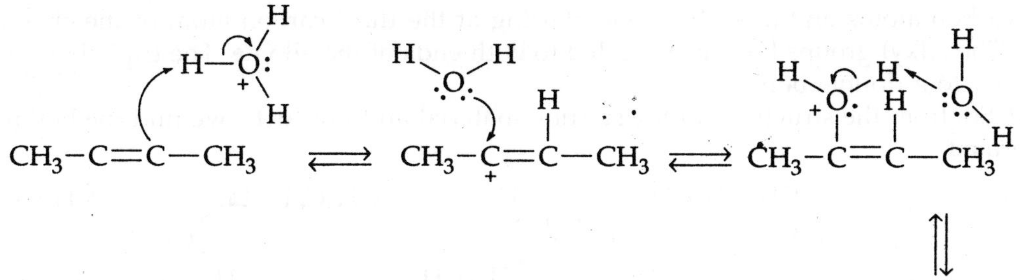

$CH_3{-}C{=}C{-}CH_3 + H_3O^+ \rightleftharpoons CH_3{-}\overset{+}{C}{=}C(H){-}CH_3 + H_2O \rightleftharpoons CH_3{-}C(\overset{+}{O}H_2){=}C(H){-}CH_3 + H_2O$

$\Updownarrow$

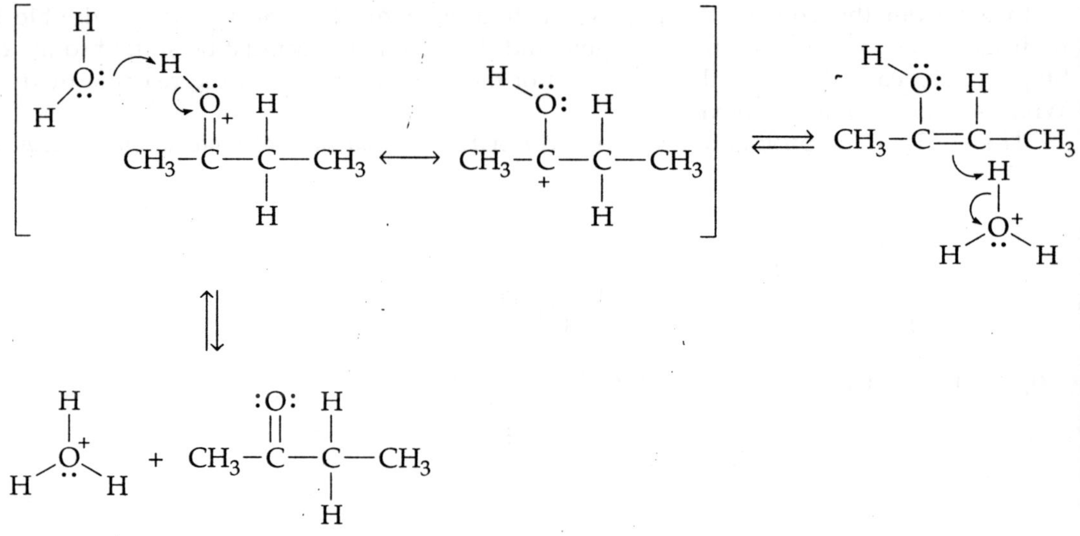

$$\left[H_2\ddot{O}: \curvearrowright H-\overset{+}{\ddot{O}}=\underset{CH_3}{C}-\underset{H}{\overset{H}{C}}-CH_3 \longleftrightarrow H-\ddot{O}-\underset{+}{C}(CH_3)-\underset{H}{\overset{H}{C}}-CH_3 \right] \rightleftharpoons CH_3-\overset{\ddot{O}H}{C}=\overset{H}{C}-CH_3 \;\; H-\overset{+}{O}H_2$$

$$\rightleftharpoons$$

$$H_3\overset{+}{O} \; + \; CH_3-\overset{:O:}{\overset{\|}{C}}-\underset{H}{\overset{H}{C}}-CH_3$$

(e) **Example** How would you synthesize (*E*)- and (*Z*)-3-heptene from acetylene and any other chemicals?

Solution: Again going through the same procedure we adopted above:

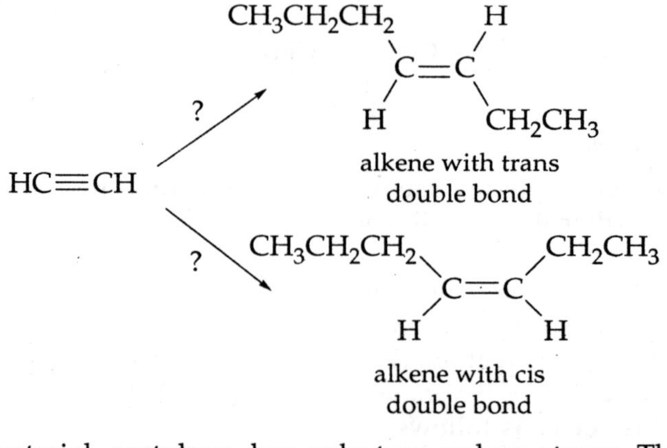

$$HC\equiv CH \xrightarrow{?} \underset{H}{\overset{CH_3CH_2CH_2}{}}C=C\underset{CH_2CH_3}{\overset{H}{}}$$

alkene with trans double bond

$$HC\equiv CH \xrightarrow{?} \underset{H}{\overset{CH_3CH_2CH_2}{}}C=C\underset{H}{\overset{CH_2CH_3}{}}$$

alkene with cis double bond

Here the starting material, acetylene, has only two carbon atoms. The products each have seven carbon atoms and a double bond starting at the third carbon atom of the chain.

Step II: The Alkyl groups have been added to both ends of the alkyne. The triple bond has been converted to a double bond.

Step III: So, from the structures of the starting material and products, we find the bonds broken and formed :

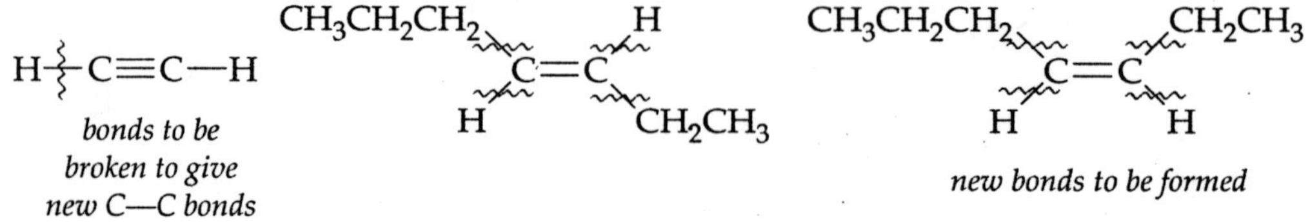

$$H\text{-}\!\!\wr\!\!\text{-}C\equiv C-H \qquad \underset{H}{\overset{CH_3CH_2CH_2}{}}C=C\underset{CH_2CH_3}{\overset{H}{}} \qquad \underset{H}{\overset{CH_3CH_2CH_2}{}}C=C\underset{H}{\overset{CH_2CH_3}{}}$$

bonds to be broken to give new C—C bonds

new bonds to be formed

The pieces we need are:

$$CH_3CH_2CH_2— \quad —C{\equiv}C— \quad —CH_2CH_3 \quad —H \quad —H$$

Step IV: As the alkyne $CH_3CH_2CH_2C{\equiv}CCH_2CH_3$ can be converted into either the cis or the trans alkene, the carbon-carbon bonds in the alkyne must be the result of nucleophilic substitution reactions. Acetylene can be converted into a good nucleophile. The alkyl groups, attached to leaving groups. must supply the electrophiles. The alkyalkyne can be put together from these reagents.

$$CH_3CH_2H_2Br \quad HC{\equiv}CH \quad BrCH_2CH_3$$

Step V: Thus,
The alkyne $CH_3CH_2CH_2C{\equiv}CCH_2CH_3$ is the precursor.
Step VI After the above step, do we se how to get from starting material to product? If not, we need to analyze the structure obtained in step V. Hence, the complete synthesis is as follows:

$$HC \equiv HC \xrightarrow[NH_3\ (liq)]{NaNH_2} Na^+ \ddot{C} \equiv CH + CH_3\,CH_2\,CH_2\,Br$$

$$\downarrow$$

$$CH_3CH_2CH_2C \equiv CH$$

$$\downarrow$$

$$CH_3CH_2CH_2C \equiv C^{-}\!: Na^+ + Br\,CH_2CH_3$$

$$\downarrow NH_3\ (liq)$$

$$\underset{\text{(cis: } CH_3CH_2CH_2,\,H \text{ on one C; } CH_2CH_3,\,H \text{ on the other)}}{CH_3CH_2CH_2(H)C{=}C(H)CH_2CH_3} \xleftarrow[\text{Pd/CaCO}_3\text{ quinodine}]{H_2} CH_3CH_2CH_2C{\equiv}CCH_2CH_3 \xrightarrow{Na} \underset{\text{(trans)}}{CH_3CH_2CH_2(H)C{=}C(H)CH_2CH_3}$$

(f) Example SYNTHESIZE 5-methyl – 1-hexanol from acetylene, using any other reagents you need.

Solution

Step I: Writing 5-methyl-1-hexanol in structural from.

$$\underset{\text{alkyne}}{HC{\equiv}CH} \xrightarrow{?} \underset{\text{primary alcohol}}{CH_3\overset{\displaystyle CH_3}{\overset{|}{C}}HCH_2CHCH_2CH_2OH}$$

Here acetylene has only two carbon atoms. The product has a chain of six carbon atoms with a hydroxyl group on the first carbon and a methyl group on the fifth carbon.

Step II: The triple bond disappears. The rest of the chain must have been added to the alkyne, probably on one side.

$$CH_3\overset{\displaystyle CH_3}{\overset{|}{C}}HCH_2CH_2C{\equiv}CH$$

Step III: From the structures of the starting material and product we find that bonds broken and formed are:

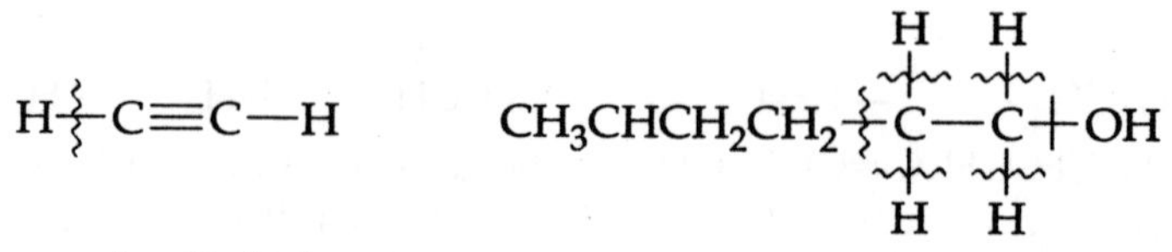

$$H\text{-}C\equiv C\text{—}H \qquad CH_3CHCH_2CH_2\text{—}\underset{H}{\overset{H}{C}}\text{—}\underset{H}{\overset{H}{C}}\text{—}OH$$

(bond broken) (new bonds formed)

Thus the pieces we need are:

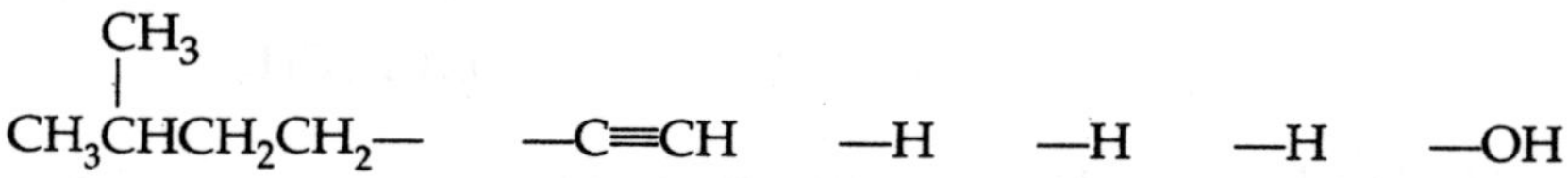

$$CH_3CH(CH_3)CH_2CH_2\text{—} \quad \text{—}C\equiv CH \quad \text{—}H \quad \text{—}H \quad \text{—}H \quad \text{—}OH$$

Step IV: Now, we know that when an electrophile reacts with a nucleophile new bonds are formed.

Te carbon-carbon bond must have resulted from a nucleophile substitution reaction. acetylene can be converted into a good nucleophile. An alkyl group attached to a leaving group be the electrophile. The other pieces that have been added look like hydrogen, H—H, and water, H—OH. This molecule can be put together from the following regents.

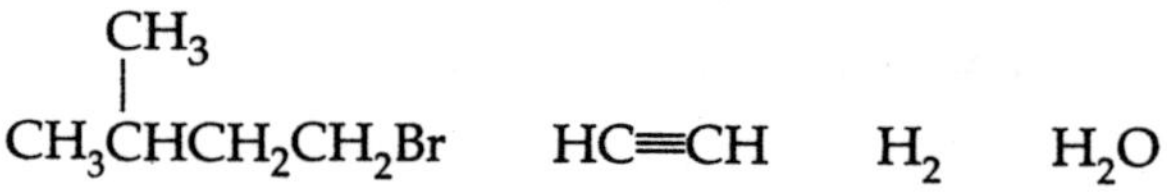

$$CH_3CH(CH_3)CH_2CH_2Br \quad HC\equiv CH \quad H_2 \quad H_2O$$

Step V: What type of compound would be a good precursor to the product?

Anti Markovnikov addition of water to a double bond will give a primary alcohol. If we had a terminal alkene having the carbon connectivity of the product, we could make the alcohol.

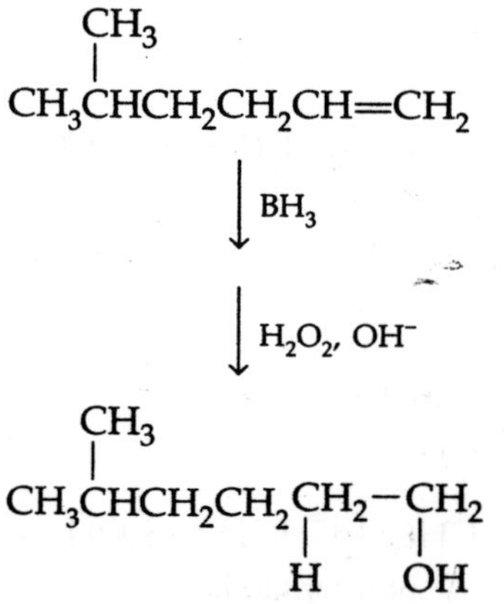

$$CH_3CH(CH_3)CH_2CH_2CH{=}CH_2$$

$$\downarrow BH_3$$

$$\downarrow H_2O_2,\ OH^-$$

$$CH_3CH(CH_3)CH_2CH_2\underset{H}{CH_2}\text{—}\underset{OH}{CH_2}$$

Step VI After this step, we see how to get from starting material to product? If not, we need to analyze the structure obtained in step V

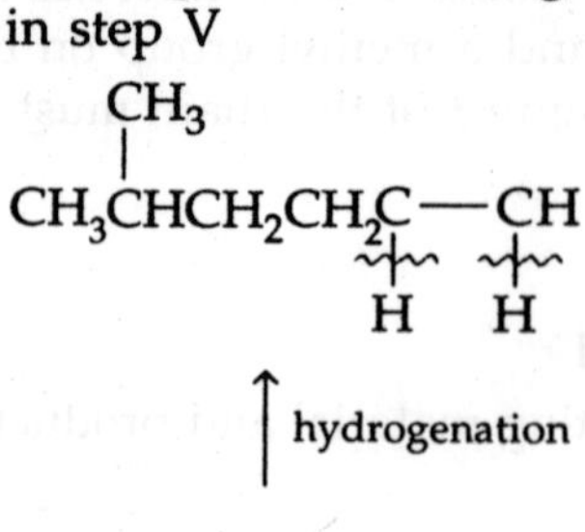

$$CH_3CH(CH_3)CH_2CH_2\underset{H}{C}\text{—}\underset{H}{CH}$$

$$\uparrow \text{hydrogenation}$$

$$\begin{array}{c} CH_3 \\ | \\ CH_3CHCH_2CH_2{-}C{\equiv}CH \end{array}$$

$$\uparrow$$

$$\begin{array}{c} CH_3 \\ | \\ CH_3CHCH_2CH_2Br \end{array} + {}^{-}{:}\,C{\equiv}CH$$

The complete synthesis is as follows:

$$HC{\equiv}CH \xrightarrow[NH_2\ (liq)]{NaNH_2} \begin{array}{c} CH_3 \\ | \\ CH_3{-}CHCH_2CH_2Br \end{array} + Na^+\ {:}C{\equiv}CH$$

$$\downarrow NH_3(liq)$$

$$\begin{array}{c} CH_3 \\ | \\ CH_3CHCH_2CH_2C{\equiv}CH \end{array}$$

$$H_2 \downarrow \begin{array}{c} pd/CaCO_3 \\ quinoline \end{array}$$

$$\begin{array}{c} CH_3 \\ | \\ CH_3CHCH_2CH_2CH{=}CH_2 \end{array}$$

$$\downarrow \begin{array}{l} BH_3 \\ tetrahydrofuran \end{array}$$

$$\downarrow H_2O_2,\ OH^-$$

$$\begin{array}{c} CH_3 \\ | \\ CH_3CHCH_2CH_2CH_2CH_2CH_2OH \end{array}$$

(g) Example How could we carry out the following transformation?

$$C_6H_5{-}CH_2CH_2CH{=}CH_2 \longrightarrow C_6H_5{-}CH_2CH_2CH_2\overset{\overset{\displaystyle O}{\|}}{C}H$$

Step I, Here,

Both compounds have a straight chain of four carbon atoms with an arylgroup at one end. The starting material has a double bond at the end of the chain; the product is an aldehyde.
Step II: An aldehyde functional group is formed at the first carbon atom of the double bond. There is no good leaving group in the starting material.
Step III: From the structures of the starting material and product we find bonds broken and formed :

$C_6H_5-CH_2CH_2CH{=}CH-H$ (bonds broken) $C_6H_5-CH_2CH_2CH(H)-C(=O)-H$ (bonds formed)

A π bond and a carbon-hydrogen bond must be broken. Carbon-hydrogen and carbon-oxygen bonds are formed.
Step V: We know that when an electrophile reacts with a nucleophile new bonds are formed.

The formation of bonds to a hydrogen and an oxygen an adjacent carbon atoms looks like an addition of H—OH to the double bond, in an anti-Markovnikov fashion.

$$C_6H_5-CH_2CH_2CH{=}CH_2 \xrightarrow{?} C_6H_5-CH_2CH_2\underset{H}{CH}-\underset{OH}{CH_2}$$

Step V: What type of compound would be a good precursor to the product?
(An aldehyde is formed by the oxidation of a primary alcohol)

$$C_6H_5-CH_2CH_2CH_2\underset{OH}{CH_2} \xrightarrow{\text{oxidation}} C_6H_5-CH_2CH_2CH_2\overset{O}{\overset{\|}{C}}H$$

Step VI After this step, do we how to get from starting material to product? If not, we ned analyze the structure obtained in step V.

$$C_6H_5-CH_2CH_2CH{=}CH_2 \xrightarrow[\text{2. } H_2O_2,\ NaOH,\ H_2O]{\text{1. } BH_3,\ \text{tetrahydrofuran}} C_6H_5-CH_2CH_2CH_2CH_2OH$$

$$C_6H_5-CH_2CH_2CH_2CH_2OH \xrightarrow[\text{or } CH_3\overset{O}{\overset{\|}{S}}CH_3;\ Cl\overset{O}{\overset{\|}{C}}-\overset{O}{\overset{\|}{C}}Cl;\ (CH_3CH_2)_3N]{C_5H_5N^+H\ CrO_3Cl^-\ \text{(dichloromethane)}} C_6H_5-CH_2CH_2CH_2\overset{O}{\overset{\|}{C}}H$$

(h) Example: We would you carry out the following transformation?

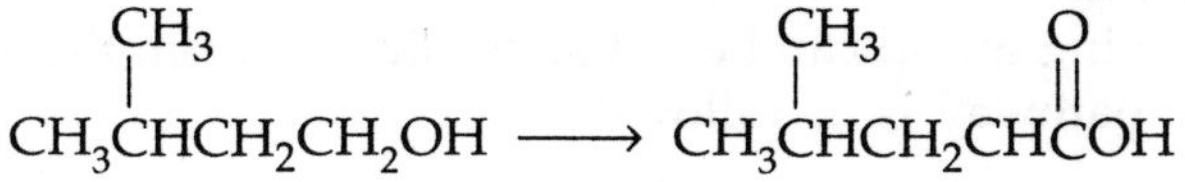

$$CH_3\underset{\displaystyle |}{\overset{CH_3}{C}}HCH_2CH_2OH \longrightarrow CH_3\overset{CH_3}{\overset{|}{C}}HCH_2CH\overset{O}{\overset{||}{C}}OH$$

***Solution** Step I* The main chain of the carbon skeleton of the starting material is four carbon long, with a methyl group on the third carbon and a hydroxyl group on the first carbon. The main chain of the product is five carbons long, with a branch at the fourth carbon atom. The chain in the reagent has been lengthened by one carbon atom, which appears as a carboxylic acid group.

Step II: A hydroxyl group has been converted into a carboxylic acid group.

$$—OH \longrightarrow \overset{O}{\overset{||}{C}}OH$$

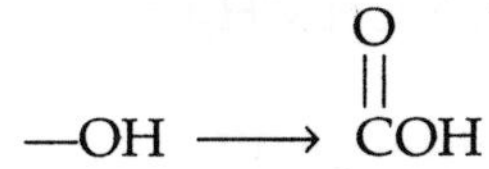

The starting material does not have a good leaving group, but the hydroxyl group can be converted into one

Step III: From the structure of the starting material and product we find bond broken and bonds formed are:

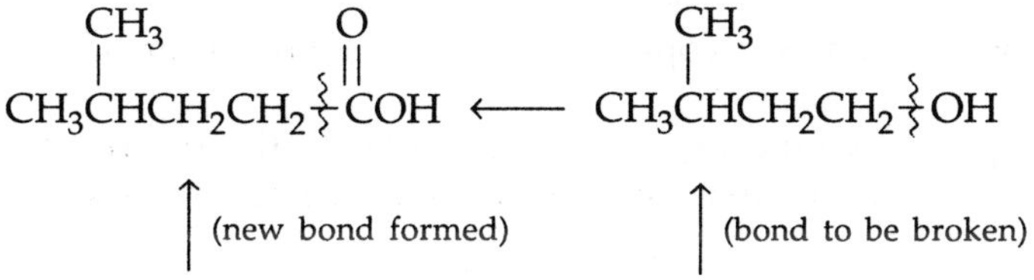

$$CH_3\overset{CH_3}{\overset{|}{C}}HCH_2CH_2 \wr \overset{O}{\overset{||}{C}}OH \longleftarrow CH_3\overset{CH_3}{\overset{|}{C}}HCH_2CH_2 \wr OH$$

↑ (new bond formed) ↑ (bond to be broken)

Step IV: We know that when an electrophile reacts with a nucleophile new bonds are formed.

A carboxylic acid can be prepared from a nitrile.

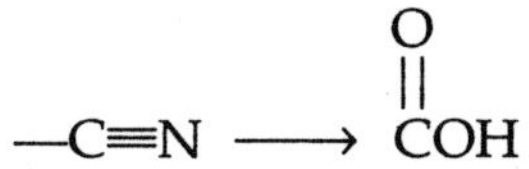

$$—C{\equiv}N \longrightarrow \overset{O}{\overset{||}{C}}OH$$

The cyano group comes from cyanide ion, a good nucleophile

Step V: What type of compound would be a good precursor to the product? The corresponding cyano compound would be a good precursor.

$$CH_3\overset{CH_3}{\overset{|}{C}}HCH_2CH_2—\overset{O}{\overset{||}{C}}OH \longleftarrow CH_3\overset{CH_3}{\overset{|}{C}}HCH_2CH_2—C{\equiv}N$$

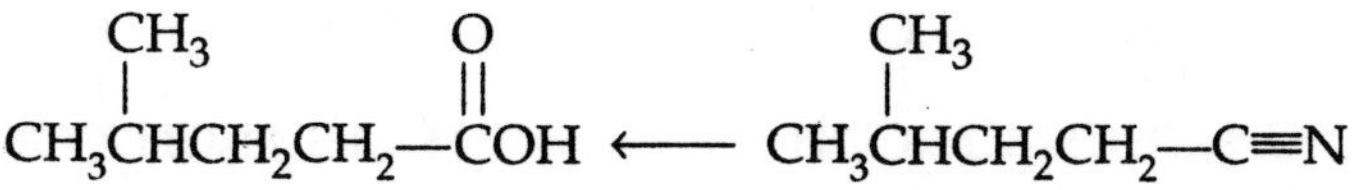

The best precursor for the cyano compound would have a good leaving group where the —C≡N group is

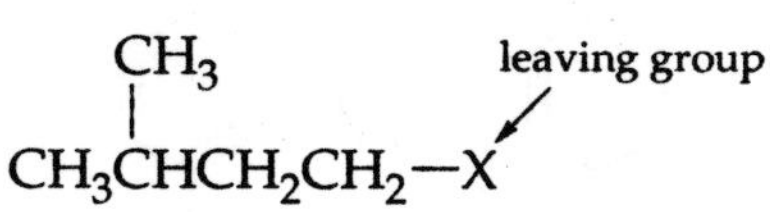

$$CH_3\overset{CH_3}{\overset{|}{C}}HCH_2CH_2—X$$

leaving group (→ X)

Step V: After this step, do we see how to get from starting material to product? If not, we need to analyze the structure obtained in step V.

The precursor with a good leaving group can be obtained directly from the starting primary alcohol, so a complete synthetic pathway is as follows.

$$CH_3CH(CH_3)CH_2CH_2OH \xrightarrow[\text{pyridine}]{\text{TsCl}} CH_3CH(CH_3)CH_2CH_2OTs$$

$$\downarrow NaC{\equiv}N$$

$$CH_3CH(CH_3)CH_2CH_2COOH \xleftarrow[\Delta]{H_3O^+} CH_3CH(CH_3)CH_2CH_2C{\equiv}N$$

OR

There is at least one other set of answer to Step IV through VI

Step IV: A carboxylic acid having one more carbon atom than the starting material can also be prepared by the reaction of an organometallic reagent with carbon dioxide.

$$RMgX + CO_2 \longrightarrow RC(=O)OMgX \longrightarrow RC(=O)OH$$

The Grignard regent is a good nucleophile, and the carbon atom of carbon dioxide is electrophilic.

Step V: What type of compound would be a good precursor to the product?

A good precursor would be an alkylhalide, from which a Grignad reagent can be prepared.

$$CH_3CH(CH_3)CH_2CH_2Br$$

Step VI: After this step, do we see how to get from starting material to product? If not e need to analyze the structure obtained to step V.

Hence, another complete synthesis is as follows:

$$CH_3CH(CH_3)CH_2CH_2OH \xrightarrow{PBr_3} CH_3CH(CH_3)CH_2CH_2Br \xrightarrow[\text{diethyl ether}]{\text{Mg}} CH_3CH(CH_3)CH_2CH_2MgBr$$

$$\downarrow CO_2$$

$$CH_3CH(CH_3)CH_2CH_2COOH \xleftarrow{H_3O^+} CH_3CH(CH_3)CH_2CH_2C(=O)OMgBr$$

(i) Example: How would you carry out the following conversion?

Benzene $\longrightarrow$ 4-chlorobenzonitrile (benzene ring with Cl and CN in para positions)

Solution:

Step I: The starting material is benzene, an aromatic hydrocarbon. The product has a chlorine atom and a cyano group as substituents para to each other on the aromatic ring. The product has one more carbon atom than the starting material; this carbon is part of the cyano group. A chlorine atom has also been substituted on the ring.

Step II: Here, the aromatic ring has undergone substitution at two positions.

Step III: e find bond broken and formed are:

H H H H H H — H H H H Cl CN

(bonds to be broken) *(bonds to be formed)*

Step IV: We know that when nucleophiles react with electrophiles new bonds are formed.

Here chlorine atom can be introduced by direct electrophile aromatic substitution. The cyano group must be introduced indirectly, by first producing a diazonium ion.

Step V: What type of compound would be a good precursor to the product?

(The amine corresponding to the cyano compound is needed)

Cl, NH_2 $\xrightarrow[\text{298 K}]{NaNO_2,\ HCl,\ H_2O}$ Cl, $\overset{+}{N}\equiv NCl^-$ $\xrightarrow{Na_2CO_3}$ $\xrightarrow{CuCN}$ Cl, CN

Step VI We have converted the starting material to product? If not, we have to analyze the structure obtained in step 5 by repeating step IV to V

The amine is prepared by reducing of a nitro compound which is the product of an electrophile substitution. Because the chloro group is ortho, para-directing, it is introduced first

Benzene $\xrightarrow[Fe]{Cl_2}$ Cl $\xrightarrow[H_2SO_4]{HNO_3}$ Cl, NO_2 $\xrightarrow[\text{catalyst}]{H_2}$ Cl, NH_2

To do this sequence the other way, that is, to make the amine first and then chlorinate it, would mean that the amino group would have to be protected as an amide and then deprotected later.

NO_2 $\xrightarrow[\text{catalyst}]{H_2}$ NH_2 $\xrightarrow{(CH_3\overset{O}{\overset{\|}{C}})_2O}$ $NH\overset{O}{\overset{\|}{C}}CH_3$ $\xrightarrow[Fe]{Cl_2}$ $NH\overset{O}{\overset{\|}{C}}CH_3$, Cl $\xrightarrow[\Delta]{H_3O^+}$ $\xrightarrow{NaOH}$ NH_2, Cl

Note: Direct chlorination of aniline to get the monochloro product is difficult. The amino group activities the ring strong and forms polychloro products.

(j) Example: How would you carry out the following conversion?

$$C_6H_6 \longrightarrow 3,5\text{-}(O_2N)_2C_6H_3COOH$$

Step I: Here, The starting material is benzene, an aromatic hydrocarbons. The product is an aromatic carboxylic acid with two nitro groups meta to the carboxyl group. One carbon atom, that of the arboxyl group has been added to benzene. Two nitro groups have been substituted meta to the carboxylic acid function.

Step II: The aromatic hydrocarbon has undergone substitution at three position/

Step III: Bonds broken and bonds formed are:

(bonds to be broken) (bonds to be formed)

Step IV: New bonds are created when an electrophile reacts with a nucleophile.

The nitro groups can be introduced by a an electrophilic aromatic substitution reaction on the carboxylic acid, which is meta-directing. The carbon-carbon bond to the carboxylic acid group would be most formed by the reaction o a nucleophile carbon atom with the electrophile carbon atom of carbon dioxide. An organometalic reagent such as a Grignard reagent has a nucleophile carbon atom.

Step V: What type of compound would be a good precursor to the product?

$$C_6H_5MgBr \xrightarrow{CO_2} \xrightarrow{H_3O^+} C_6H_5COOH \xrightarrow[H_2SO_4,\ \Delta]{HNO_3} 3,5\text{-}(O_2N)_2C_6H_3COOH$$

Step VI After this last step, do we see how to get from starting material to product? If not we need to analyze the structure obtained in step V.

An electrophilic substitution reaction on benzene will give bromobenzene, which can be used to make the organometallic regent necessary for the synthesis.

$$C_6H_6 \xrightarrow[Fe]{Br_2} C_6H_5Br \xrightarrow[\text{diethyl ether}]{Mg} C_6H_5MgBr$$

A potential alternate pathway that might seem feasible is preparation of 3,5-dinitrobromobenzene and then formation of theacid.

$$C_6H_6 \xrightarrow[H_2SO_4,\ \Delta]{HNO_3} \text{m-}C_6H_4(NO_2)_2 \xrightarrow[\text{FeBr}_3\ \text{(difficult reaction)}]{Br_2} \text{3,5-dinitrobromobenzene} \;\not\rightarrow\; \text{3,5-}(NO_2)_2C_6H_3COOH$$

(k) Example: The following transformation was used in the synthesis of compounds needed in a study of reactin mechanisms. How would you carry out this transformation?

$$C_6H_5CH(CH_3)CH_2CH_2COH(=O) \longrightarrow C_6H_5CH(CH_3)CH_2CH_2CH_2CCH_3(=O)$$

Here, *Step I*: The starting·material has a four-carbon chain with a phenyl on the third carbon atom and a carboxyl group for the first one. The product has a six-carbon chain with a phenyl group on the fifth carbon atom. The chain has been lengthened by two carbon atoms and contains a carbonyl group as the second carbon atom.

Step II: The carbon atom of the carboxylic acid has been charged to a methylene group in the product.

$$-\overset{O}{\overset{\|}{C}}OH \longrightarrow CH_2\overset{O}{\overset{\|}{C}}CH_3$$

↑ derived from carbon atom of carbonyl group

(The hydroxyl group of the acid can be converted into a leaving group).

Step III: From the structure of the starting material and product, we find bonds broken and formed are:

$$C_6H_5CH(CH_3)CH_2-\overset{O}{\overset{\|}{C}}\text{-}\!\!\!/\!\!-OH \longrightarrow C_6H_5CH(CH_3)CH_2\overset{H}{\underset{H}{C}}\text{-}\!\!\!/\!\!-\overset{O}{\overset{\|}{C}}CH_3$$

(bonds to be broken) *(bonds to be formed)*

Step IV: We know that when an electrophile reacts with a nucleophile new bonds are formed: but here, the replacement of carbon-oxygen bonds in the carboxylic acid with carbon-hydrogen bonds in the product requires a reduction.

$$C_6H_5-CH(CH_3)CH_2CH_2\overset{O}{\overset{\|}{C}}OH \xrightarrow[\text{diethyl ether}]{LiAlH_4} \xrightarrow{H_2O} C_6H_5-\overset{CH_3}{\overset{|}{C}}HCH_2CH_2{+}OH$$

(*bond to be broken*)

Step V: What type of compound would be a good precursor the product?
(a ketone is obtained by oxidizing a secondary alcohol.)

$$C_6H_5-\overset{CH_3}{\overset{|}{C}}HCH_2CH_2-\overset{O}{\overset{|}{C}}HCH_3 \xrightarrow{\text{oxidation}} C_6H_5-\overset{CH_3}{\overset{|}{C}}HCH_2CH_2-\overset{O}{\overset{\|}{C}}CH_3$$

Step VI After this step, do we see how to get from starting material to product? If not, we need to analyze the structure obtained in step V.

The main bond to the formed in the secondary alcohol will come from the reaction of an organometallic reagent with an aldehyde.

$$C_6H_5-\overset{CH_3}{\overset{|}{C}}HCH_2CH_2{+}\overset{O}{\overset{|}{C}}HCH_3$$

(*bonds to be formed*)

This allows us to complete the synthesis of the alcohol.

$$C_6H_5-\overset{CH_3}{\overset{|}{C}}HCH_2COH \xrightarrow[\text{diethyl ether}]{LiAlH_4} \xrightarrow{H_2O^+} C_6H_5-\overset{CH_3}{\overset{|}{C}}HCH_2CH_2OH \xrightarrow{PBr_3} C_6H_5-\overset{CH_3}{\overset{|}{C}}HCH_2CH_2Br$$

$$\downarrow \text{Mg diethyl ether}$$

$$C_6H_5-\overset{CH_3}{\overset{|}{C}}HCH_2CH_2\overset{O^-Mg^{2+}Br^-}{\overset{|}{C}}HCH_3 \xleftarrow{CH_3\overset{O}{\overset{\|}{C}}H} C_6H_5-\overset{CH_3}{\overset{|}{C}}HCH_2CH_2MgBr$$

$$\downarrow H_3O^+$$

$$C_6H_5-\overset{CH_3}{\overset{|}{C}}HCH_2CH_2\overset{OH}{\overset{|}{C}}HCH_3 \xrightarrow[\text{dichloromethane}]{C_5H_5NH^+\ CrO_3Cl^-} C_6H_5-\overset{CH_3}{\overset{|}{C}}HCH_2CH_2\overset{O}{\overset{\|}{C}}CH_3$$

REACTION OF ORGANIC COMPOUNDS

I. Alkenes (Addition Reaction)

Alkene	*Reagent*	*Intiation Conditions*	*Redical Intermediates*	*Products*
$R—C(R')=CH_2$	HBr	peroxides	$:\ddot{Br}\cdot$ and $R—\dot{C}(R')—CH_2Br$	$R—C(R')(H)—CH_2Br$
	CH_3Br or CCl_4	products, $h\nu$	$Cl_3\dot{C}$ and $R—\dot{C}(R')—CH_2CCl_3$	$R—C(R')(X)—CH_2CCl_3$ X = Br or Cl
	$R''SH$	Δ	$R''\ddot{S}\cdot$ and $R—\dot{C}(R')—CH_2SR''$	$R—C(R')(H)—CH_2SR''$
	R''_3SiH	peroxides, Δ	R''_3Si and $R—\dot{C}(R')—CH_2SiR''_3$	$R—C(R')(H)—CH_2SiR''_3$
$R—C(R')=CH_2$		$CH_3C(CN)(CH_3)N{=}NC(CN)(CH_3)CH_3$	$R''CH_2C(R')(R)\cdot$ and $R''CH_2C(R')(R)CH_2C(R')(R)\cdot$	$R''[—CH_2C(R')(R)—]R'''$
(Br) (intramolecular reaction)		$CH_3C(CN)(CH_3)N{=}NC(CN)(CH_3)CH_3$ and $(CH_3CH_2CH_2CH_2)_3SnH$	$CH_3C(CN)(CH_3)\cdot$ and $(CH_3CH_2CH_2)_3\dot{S}n$ (cyclopentyl) and (alkyl)	and —H
			and (alkyl) and (cyclopentyl)	

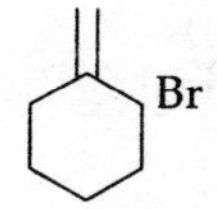

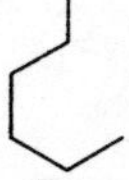

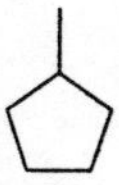

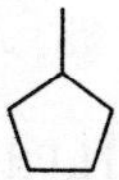

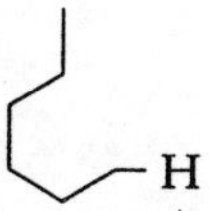

II (Addition; Oxidation & Reduction Reactions)

Electriophilic Addition Reactions

Alkyne	*Electrophile*	*Intermediates*		*Product*
RC°CR'				
R may be H	X_2	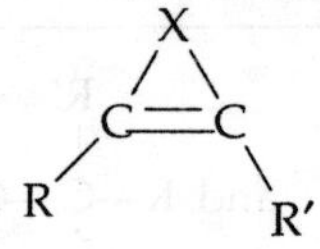 $RC{=}CR'$ (bridged by X)	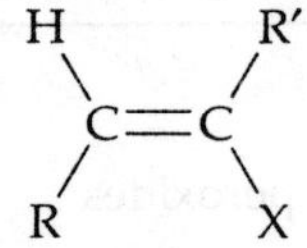 $(H)(R)C{=}C(R')(X)$	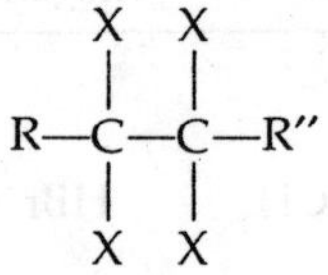 $R{-}CX_2{-}CX_2{-}R''$
	HX	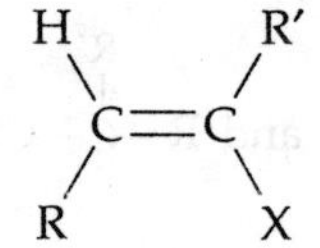 $(H)(R)C{=}C(R')(X)$	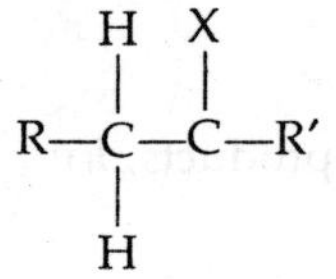 $R{-}CH_2{-}CX{-}R'$	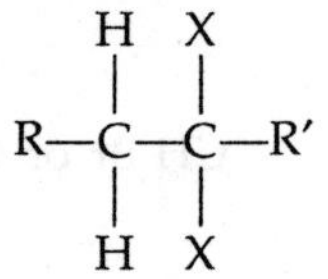 $R{-}CH_2{-}CX_2{-}R'$
	H_3O^+	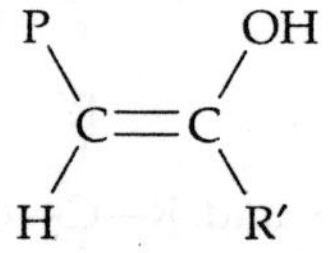 $(P)(H)C{=}C(OH)(R')$		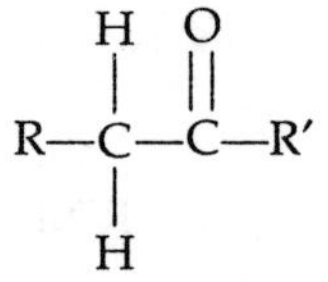 $R{-}CH_2{-}C({=}O){-}R'$

Oxidation Reaction

	O_3			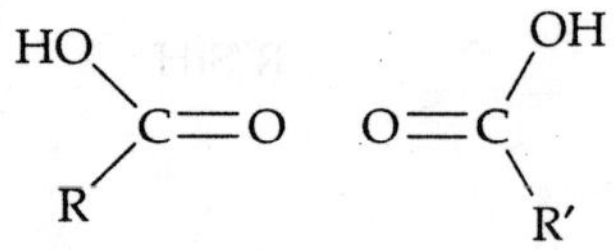 $R{-}C({=}O){-}OH$ $\quad$ $HO{-}C({=}O){-}R'$

Reduction reactions

	H_2, PiO_2 or $RhCl[P(C_6H_5)_3]_3$			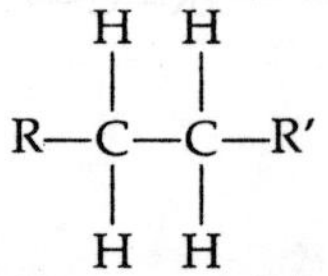 $R{-}CH_2{-}CH_2{-}R'$
	H_2Pd, $BaSO_4$ quinoline			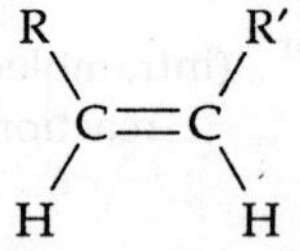 $(R)(H)C{=}C(R')(H)$ (syn. addition)
	Na, NH_3	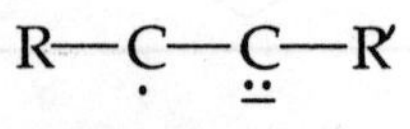 $R{-}\dot{C}{-}\ddot{C}^{-}{-}R'$	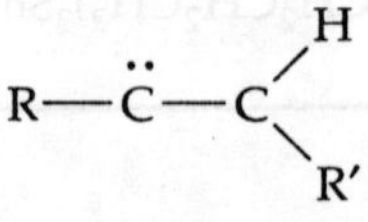 $R{-}\ddot{C}{-}C(H)(R')$	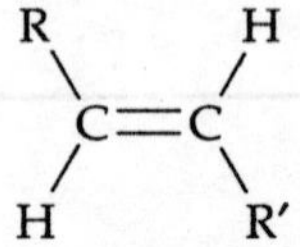 $(R)(H)C{=}C(H)(R')$ anti addition

Alkyne	*Reagent*	*Intermediate*	*Reagent*	*Product*
$RC{\equiv}CH$	$NaNH_2$, NH_3(liq) strong base	$C{\equiv}C{:}^-$ nucleophile	$R'CH_2X$ electrophile	$RC{\equiv}CCH_2R'$

III Halogenstion Reactions

Compound Reacting	*Reagent*	*Initiation Conditions*	*Radical Intermediate (s)*	*Product*
—C—H tetrary > secondary > primary > methane	Cl_2 or Br_2	*hv* or Δ	:Cl· or :Br: —C·	—C—Cl or —C—Br
H—C (allylic) C=C	N—Br (NBS)	*hv*, Δ peroxides	C, C=C	Br—C, C=C
Ph—C—H	N—Br (NBS)	*hv*, Δ peroxides	:Br: Ph—C	Ph—C—Br
	Cl_2 or Br_2	*hv* or Δ	:Cl· or :Br: Ph—C·	Ph—C—Cl or Ph—C—Br

IV Reactions of Alcohols

Alcohol	*Reagent*	*Intermediates*	*Product*
		Conversion of Alkyl Halds	
ROR	HX	$ROH_2^+ X^-$	RX
	$SOCl_3$	$R-O-\overset{O}{\underset{Cl}{S^+}}\ \ Cl^-$	RCl
	PBr_3	$R-O-\overset{Br}{\underset{Br}{P^+}}\ \ Br^-$	RBr
		Oxidation Reactions	
RCH_2OH	$Cr_2O_7^{2-}$, H_2O, H_3O^+	$R\underset{H}{C}H-O-\overset{O}{\underset{O}{Cr}}-OH$	$R\overset{O}{\overset{\|}{C}}H$, $R\overset{O}{\overset{\|}{C}}OH$
	$CH_3\overset{O}{\overset{\|}{S}}CH_3$, $Cl\overset{O}{\overset{\|}{C}}-\overset{O}{\overset{\|}{C}}Cl$ $(CH_3CH_2)_3N$ or $C_5H_5\underset{H}{N^+}\ CrO_3Cl^-$	$R\overset{H}{C}H-O-\overset{CH_3}{\overset{\|}{S^+}}-CH_3$ $R\underset{H}{C}H-O-\overset{O}{\underset{O}{Cr}}-OH$	 $R\overset{O}{\overset{\|}{C}}H$
$R\overset{R}{\underset{H}{C}}-OH$	(a) $CH_3\overset{O}{\overset{\|}{S}}CH_3$, $Cl\overset{O}{\overset{\|}{C}}-\overset{O}{\overset{\|}{C}}Cl$ $(CH_3CH_2)_3N$	$R\overset{R}{\underset{H}{C}}-O-\overset{CH_3}{\overset{\|}{S^+}}-CH_3$	$R\overset{O}{\overset{\|}{C}}R$
	(b) $Cr_2O_7^{2-}$, H_2O, H_3O^+		
	(c) $C_5H_5\underset{H}{N^+}\ CrO_3Cl^-$	$R\overset{R}{\underset{H}{C}}-O-\overset{O}{\underset{O}{Cr}}-OH$	

(*Contd.*)

(*Contd.*) TABLE IV

Alcohol	*Reagent*	*Intermediates*	*Product*
$R_3C{-}OH$	(a) $CH_3\overset{\overset{O}{\|}}{S}CH_3$, $Cl\overset{\overset{O}{\|}}{C}{-}\overset{\overset{O}{\|}}{C}Cl$ $(CH_3CH_2)_3N$ (b) $Cr_2O_3^{2-}$, H_2O, HO^+ (c) $C_5H_5\overset{+}{N}H$ CrO_3Cl^-	not oxidized	
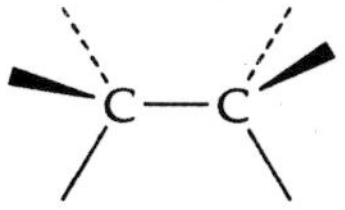 $>C{-}C<$ (with wedged substituents)	HIO_4	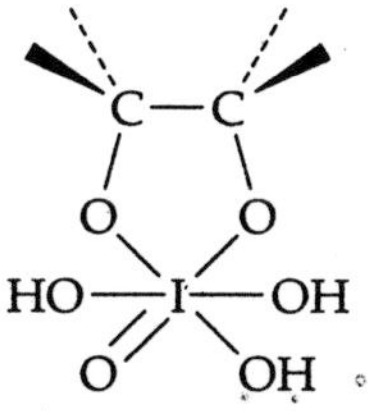 cyclic ester: C—C, O, O bonded to I; $HO{-}I{-}OH$, $=O$, OH	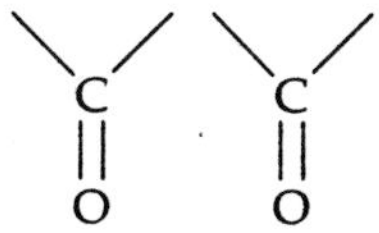 $>C{=}O$ $\quad$ $>C{=}O$
		Conversion to Ethers	
ROH	Na or NaH	$RO^- + R'X$ (primary or secondary	ROR′
 ring with OH and X	NaOH	 ring with O^- and X (five or six membered ring)	 cyclic ether with O (five or six membered ring)
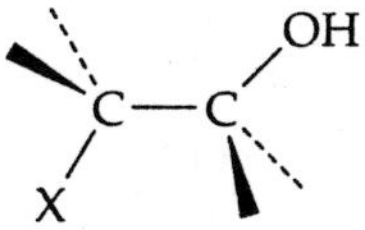 $X{-}C{-}C{-}OH$ (with wedged substituents)	NaOH	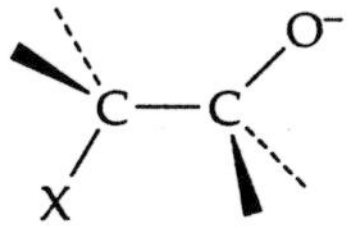 $X{-}C{-}C{-}O^-$ (with wedged substituents)	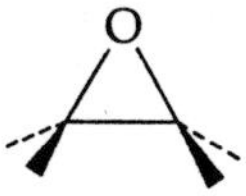 epoxide (three-membered ring with O)

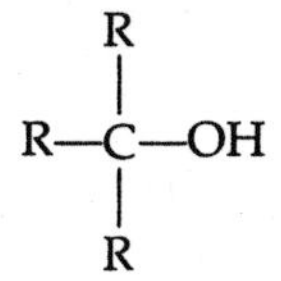

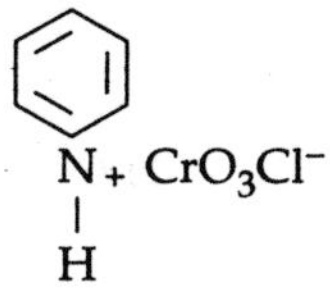

V Reaction of Aldehydes and Ketones

Starting Material	*Reagent*	*Intermediate*	*Reagent in Second Step*	*Product*
Nucleophilic Addition Reactions				
$R-C(=O)-R''$ R″ may be H	$NaBH_4$ $LiAlH_4$	$R-C(O^-)(H)-R''$	H_2O	$R-C(OH)(H)-R''$
	NaCN	$R-C(O^-)(R'')-C\equiv N$	H_3O^+	$R-C(OH)(R'')-C\equiv N$
	R′MgX, R′Li	$R-C(O^-)(R'')-R'$	H_3O^+	$R-C(OH)(R'')-R'$
	$CH_2=CHCH_2X$, Zn, NH_4Cl	$R-C(O^-)(R'')-CH_2CH=CH_2$	H_3O^+	$R-C(OH)(R'')-CH_2CH=CH_2$
ethylene oxide (epoxide)	R′MgX, R′Li	$R'CH_2CH_2O^-$	H_3O^+	$R'CH_2CH_2OH$
$R-C(=O)-R''$ R″ may be H	R′OH	$R-C(O^-)(\overset{+}{H}OR')-R''$		$R-C(OH)(OR')-R''$
	R′OH, HB^+	$R-C(OH)(OR')-R''$	R′OH, HB^+	$R-C(OR)(OR')-R''$
Nucleophilic Addition allowed by Elimentation of Water				
$R-C(=O)-R''$ R″ may be H	H_2NR'	$R-C(OH)(HNR')-R''$		$RR''C=NR'$

(Contd.)

(*Contd.*) TABLE IV

Starting Material	*Reagent*	*Intermediate*	*Reagent in Second Step*	*Product*
	H_2NOH	R—C(OH)—R″ with HN—OH		R(R″)C=NR′
	H_2NNH–C_6H_4–R′ HB^+	R–C(OH)–R″ with H_2NNH–C_6H_4–R′		R(R″)C=NNH–C_6H_4–R′
	Reduction of Merthylene Groups			
R–C(=O)–R″ R″ may be H	Zn(Hg), HCl, Δ			R—CH₂—R″ (H—C—H)
	H_2NNH_2, OH^-	R(R″)C=NNH_2	OH^-	R—C(H)(H)—R″
	$HSCH_2CH_2SH$, BF_3	1,3-dithiolane: S–CH_2CH_2–S ring with C(R)(R″)	Reney Ni (H_2)	R—C(H)(H)—R″

VI Reactions of Ethers

Ether	*Reagent*	*Intermediate*	*Product*
ROR	HI	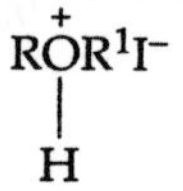 $R\overset{+}{O}R^1I^-$ (H on O)	ROH + R′I RI + R′OH
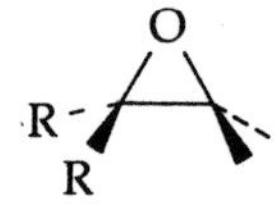 epoxide (R, R; O)	H_2O, H_3O^+	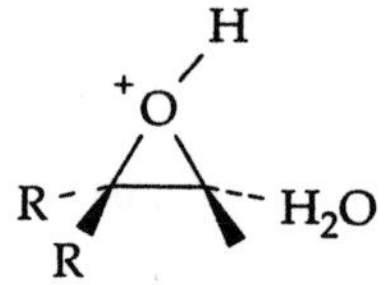 protonated epoxide (^+O–H), R, R, H_2O	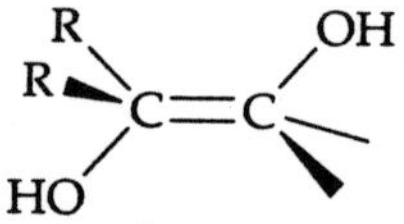 R, R, OH, HO, C=C
	HX	protonated epoxide (^+O–H), R, R, X^-	R, R, OH, HO, C=C major product OH, R, R, X, C=C major product

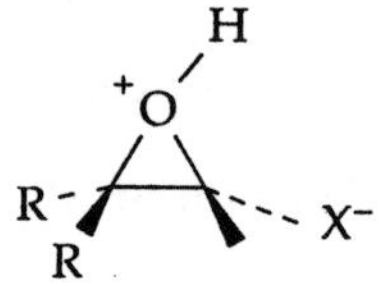

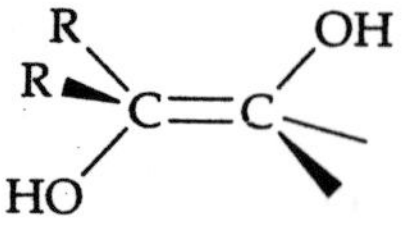

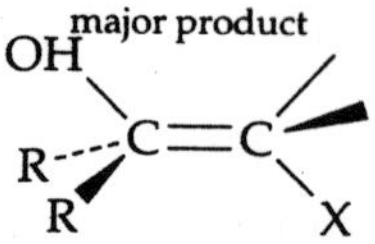

Nucleophiles

	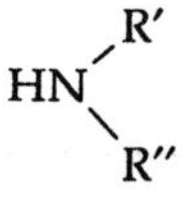	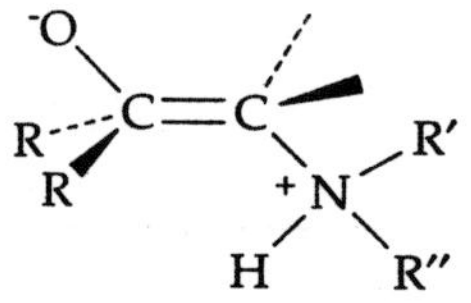	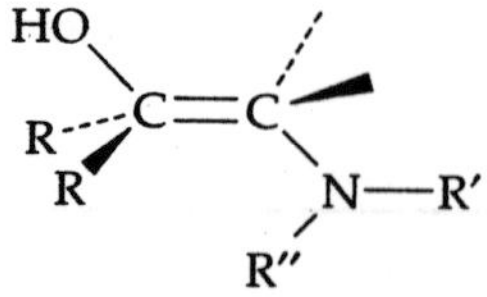
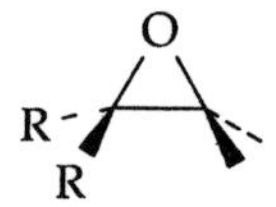 epoxide (R, R; O)	R′O⁻, R′OH	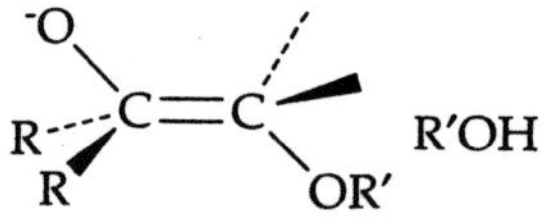 ⁻O, R, R, C=C, OR′ R′OH	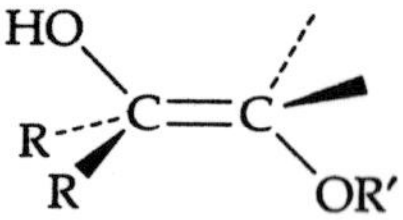 HO, R, R, C=C, OR′
	HN(R′)(R″)	⁻O, R, R, C=C, ^+N(H)(R′)(R″)	HO, R, R, C=C, N—R′, R″
	HS—R′	⁻O, R, R, C=C, ^+S—R′, H	HO, R, R, C=C, S—R′

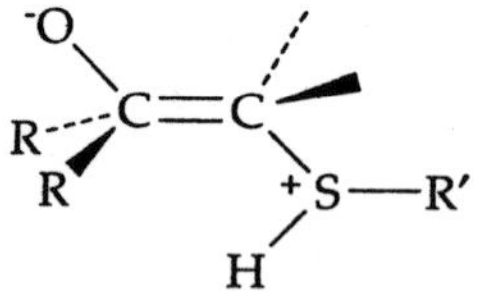

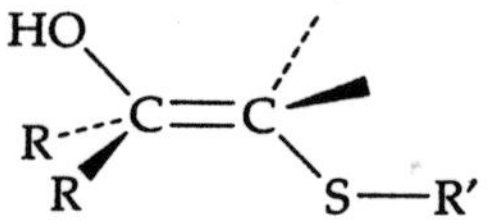

VII Reactions of Acids and Acid Derivatives

Starting Material	*Reagent*	*Intermediate*	*Reagent in Second Step*	*Product*
		Organometallic Reagents		
		$R-C(O^-Mg^{2+}X^-)(OR')-R''$		
$RC(=O)OR'$	$R''MgX$	$R-C(O^-Mg^{2+}X^-)(OR')-R''$	H_3O^+	$R-C(OH)(R'')-R''$
$RC(=O)OH$	$R'Li$	$RC(=O)O^-Li^+$		
		$R-C(O^-Li^+)(O^-Li^+)-R'$	H_2O	$R-C(OH)(OH)-R' \rightleftarrows RC(=O)R'$
$RC(=O)N(OCH_3)CH_3$	$R'M$	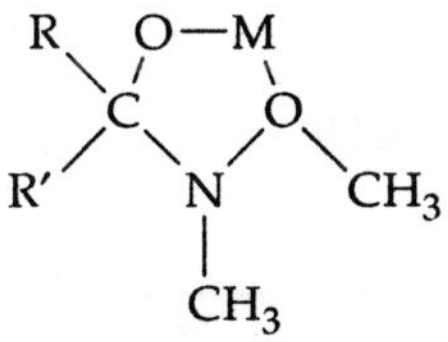 cyclic chelate: $R(R')C$ bonded to $O-M$ and $N(CH_3)$, M coordinated to $O-CH_3$ on N	H_3O^+	$RC(=O)R'$
		Metal Hydrides		
$RC(=O)OH$	$LiAlH_4$	$R-C(O-M)(H)-H$	H_3O^+	RCH_2OH
$RC(=O)NR'_2$	$LiAlH_4$	$RC(H)=\overset{+}{N}R'_2$	$LiAlH_4$	$CH_2NR'_2$
$RC(=O)OR'$	$LiAlH_4$	$R-C(O-M)(H)-H + R'O^-$	H_3O^+	$RCH_2OH + R'OH$

(Contd.)

(*Contd.*) TABLE VII

Starting Material	*Reagent*	*Intermediate*	*Reagent in Second Step*	*Product*
$R\overset{O}{\overset{\|\|}{C}}OR'$	$[(CH_3)_2CHCH_2]_2AlH$ excess	$R-\underset{H}{\overset{O-M}{C}}-H + R'O^-$	H_3O^+	$RCH_2OH + R'OH$
	$[(CH_3)_2CHCH_2]_2AlH$ 1 equivalent	$R-\underset{H}{\overset{O-M}{C}}-OR'$	H_2O	$R\overset{O}{\overset{\|\|}{C}}H + R'OH$
$RC\equiv N$	$LiAlH_4$	$R-\underset{H}{\overset{H}{C}}-\underset{M}{N}-M$	H_2O	$RCH_2\ NH_2$
	$[(CH_3)_2CHCH_3)_2AlH$ 1 equivalent	$R\underset{H}{C}=N-M$	H_3O^+	$R\underset{H}{C}=O$
$R\overset{O}{\overset{\|\|}{C}}\underset{CH_3}{N}OCH_3$	$(CH_3)_2CHCH_2]_2AlH$	cyclic: C(R)(R′) bonded to O—M and to N(CH₃); O—M coordinated to O, which bears CH₃ and is bonded to N	H_3O^+	$R\underset{H}{C}=O$

VIII Reagents for Interconverting Acids and Acid Derivatives

Form ↓ \ To Make →	$\mathrm{R\overset{O}{\overset{\|}{C}}Cl}$	$\mathrm{R\overset{O}{\overset{\|}{C}}O\overset{O}{\overset{\|}{C}}R}$	$\mathrm{R\overset{O}{\overset{\|}{C}}OR'}$	$\mathrm{R\overset{O}{\overset{\|}{C}}OH}$	$\mathrm{R\overset{O}{\overset{\|}{C}}NR_2}$	$\mathrm{R\overset{O}{\overset{\|}{C}}O^-}$
$\mathrm{R\overset{O}{\overset{\|}{C}}Cl}$	—	$\mathrm{R\overset{O}{\overset{\|}{C}}CO^-}$	R'OH	H_2O	R_2NH	H_2O, OH^-
$\mathrm{R\overset{O}{\overset{\|}{C}}O\overset{O}{\overset{\|}{C}}R}$	—	—	R'OH	H_2O	R_2NH	H_2O, OH^-
$\mathrm{R\overset{O}{\overset{\|}{C}}OR''}$	—	—	R'OH, Δ HB^+ or B:	H_2O, H_3O^+ Δ	R_2NH	H_2O, OH^- Δ
$\mathrm{R\overset{O}{\overset{\|}{C}}OH}$	$SOCl_2$	Δ – H_2O Cyclic anhydride	R'OH HB^+, Δ	—	R_2NH, Δ	OH^-
$\mathrm{R\overset{O}{\overset{\|}{C}}NR_2}$	—	—	—	H_2O, H_3O^+ Δ	—	H_2O, OH^- Δ
$\mathrm{R\overset{O}{\overset{\|}{C}}O^-}$	—	$\mathrm{R\overset{O}{\overset{\|}{C}}Cl}$	—	H_3O^+	—	—

IX Aromatic Substitution Reactions

Reaction	*Reagents*	*Electrophile*	*Substitution Product of Benzene*
Halogenation	Cl_2Br_2 FeX_3	$R—Fe(X)_2^{\delta-}—X\cdots X^{\delta+}$	C_6H_5X
Nitration	HNO_3 H_2SO_4	NO_2	$C_6H_5NO_2$
Sulfonation	SO_3 H_2SO_4	SO_3	$C_6H_5SO_3H$
Friedel-Crafts alkylation	ROH HB^+ or RX $AlCl_3$	R^+	C_6H_5R
	$(CH_3)_2C{=}C(CH_3)_2$ HB^+	$(CH_3)_2C^+—C(CH_3)_2H$	$C_6H_5C(CH_3)_2C(CH_3)_2H$
Friedel-Crafts acylation	RCCl (C=O) $AlCl_3$	$RC\equiv O^+$	$C_6H_5C(=O)R$
	RCOCR (O=C, C=O) $AlCl_3$	RCO^+	$C_6H_5C(=O)R$
	cyclic anhydride $AlCl_3$	$C\equiv O^+$ … $C(=O)OAlCl_2$ (cyclic)	$C_6H_5C(=O)$ … $C(=O)OH$ (HOC)

X Reaction of Aromatic Compounds other than Substitution

Nucleophilic Substitution				
Aromatic Compound	Nucleophile	Intermediate	Reagents for Second Step	Product
1-X-2,4-dinitrobenzene (X, NO_2, NO_2)	RNH_2	X, NHR, NO_2; N^+, ^-O, O^-	—	NHR, NO_2, NO_2
	RO^-	X, OR, NO_2; N^+, ^-O, O^-	—	OR, NO_2, NO_2
benzene–$Cr(CO)_3$	$R_3C{:}^-\ Li^+$ from R_3CH $pK > 20$	CR_3, H, $Cr(CO)_3$	I_2	CR_2
X–benzene–$Cr(CO)_3$	$R'{:}^-\ M^+$ (may be $R_2C{:}^-\ Li^+$ or enolates)	R′–benzene–$Cr(CO)_3$	I_2	R′
CHR_2, X	$NaCr_2O_7$ H_2SO_4 H_2O Δ	—	—	COH (C=O), X
	$KMnO_4$ H_2O Δ	$CO^-\ K^+$ (C=O), X	H_3O^+	COH (C=O), X

XI Reactions of Amines

Amine	*Electroghile*	*Intermediate Second Step*	*Reagents for*	*Product*
RNH_2	R′X	$RN^{+}H_2R'X^-$	RNH_2	$RNHR'$
R_2NH	R′X	$R_2N^{+}HR'X^-$	R_2NH	R_2NR'
R_3N	R′X	—	—	$R_3N^{+}R'X^-$
RNH_2	HONO (NO^+)	$R—N_2^+$	Cl^-, H_2O	RCl, ROH
R-substituted aniline ($RC_6H_4NH_2$)	NO^+	$RC_6H_4N_2^+$	(see Section 23.1)	
R_2NH	NO^+			$R'_2N—NO$
$RCH_2NR'_2$	NO^+	—	—	$RCH{=}O + R'_2N—NO$
$C_6H_5NR_2$	NO^+	—	—	$p\text{-}ON{-}C_6H_4{-}NR_2$

XII Reactions of Diazonium Ions

Ion	*Reagent*	*Product*
$R'-C_6H_4-\overset{+}{N}\equiv N$	CuCl	$R-C_6H_4-Cl$
	CuBr	$R-C_6H_4-Br$
	Na_2CO_3, CuCN	$R-C_6H_4-CN$
	KI	$R-C_6H_4-I$
	BF_4^-	$R-C_6H_4-\overset{+}{N}\equiv N\ BF_4^-$
$R-C_6H_4-\overset{+}{N}\equiv N\ BF_4^-$	Δ	$R-C_6H_4-F$
$R-C_6H_4-\overset{+}{N}\equiv N$	H_2O, Δ	$R-C_6H_4-OH$
	H_2PO_2	$R-C_6H_4-H$
	R′, O	R, N=N, R′, OH
	R′, N, R′	R, N=N, R, N – R′, R′

XIII Reaction of Quetemary Ammonium Compounds

Ammonium Salt	Reagent	Product of First Reaction	Reaction Conditions	Proucts
$(CH_3)_4N^+I^-$	Ag_2O, H_2O	$(CH_3)_4N^+OH$	Δ	$(CH_3)_2N$ + I CH_3OH
RCH_2CH-CH_3 with $^-I\ \overset{+}{N}(CH_3)_3$	Ag_2O, H_2O	RCH_2CH-CH_3 with $HO^-\ \overset{+}{N}(CH_3)_3$	Δ	$RCH_2CH=CH_2$
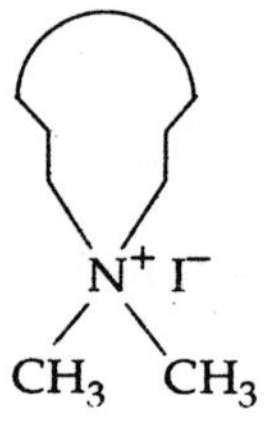 Cyclic $N^+\ I^-$ (CH_3, CH_3)	Ag_2O, H_2O	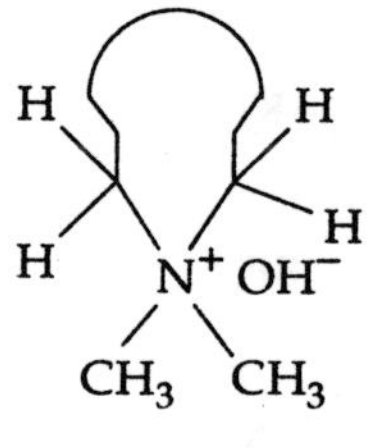 Cyclic $N^+\ OH^-$ (CH_3, CH_3), H, H, H, H	Δ	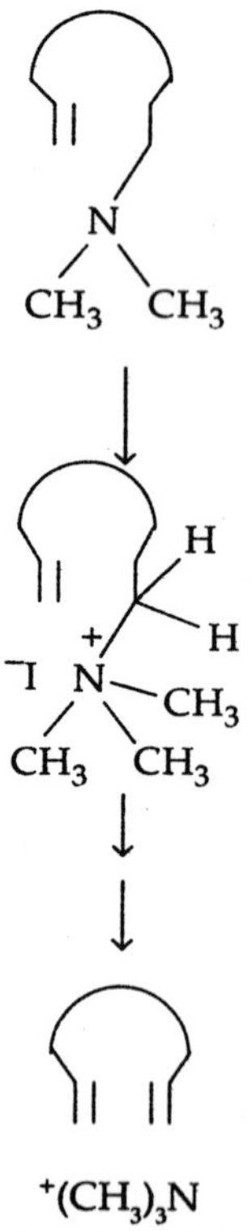 Ring-opened N (CH_3, CH_3) → $^-I\ \overset{+}{N}-CH_3$ (CH_3, CH_3), H, H → → diene + $(CH_3)_3N$

XIV Oxidation Reactions

Compound Reacting	*Reagent(s)*	*Radical Intermediates*	*Product*
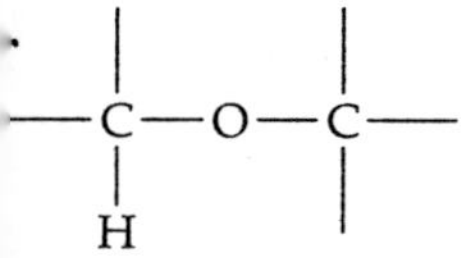	O_2, other R·, metal ion	ROO · and 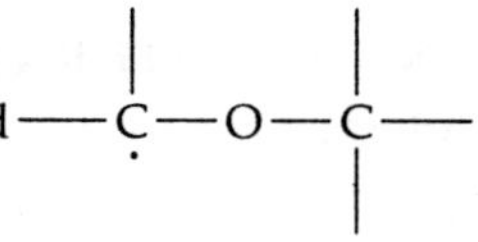	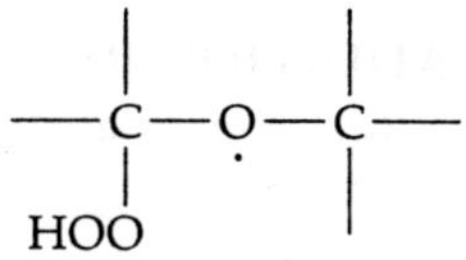
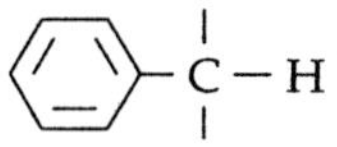	O_2, other R·, metal ion	ROO · and 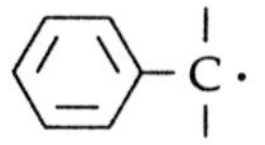	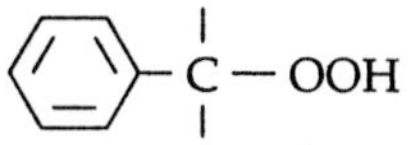
	O_2	HOO · and 	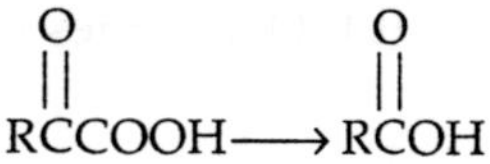
	O_2 other R·	ROO · and 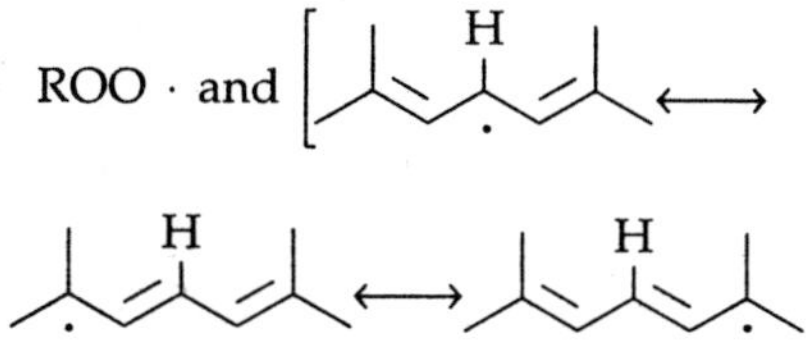	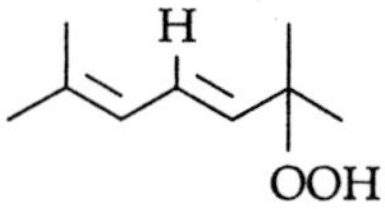
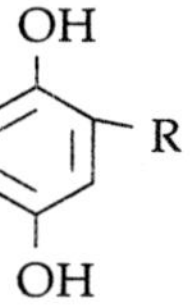	Ag^+, OH^-	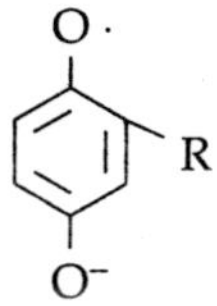	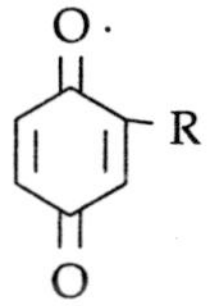
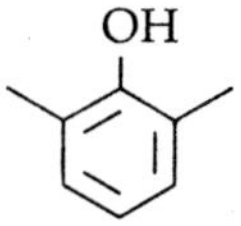	$FeCl_3$	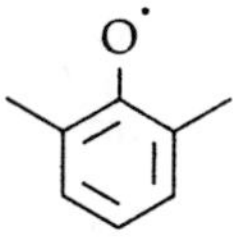	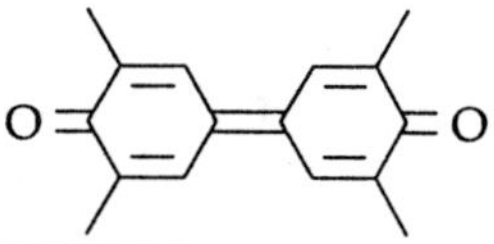

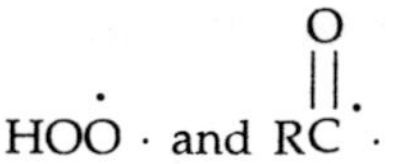

KEY POINTS

Solving Structure and reaction and based problems

A. ADDITION REACTIONS (UNSATURATION IN A MOLECULE) *i.e. (double or triple bonds).*

(a) Shows addition of H_2

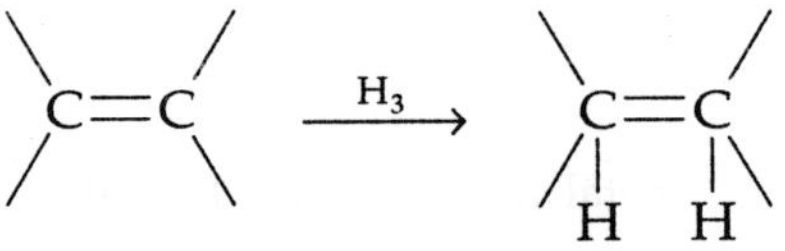

$$>C=C< \xrightarrow{H_3} >C(H)-C(H)<$$

(b) Discharges or decolourizes Br_2, in CCl_4 or Br_2 water

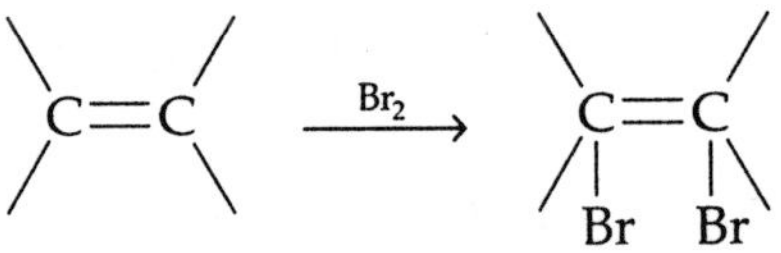

$$>C=C< \xrightarrow{Br_2} >C(Br)-C(Br)<$$

(c) Decolourizes Baeyer's reagent, i.e., 1% cold alk. $KMnO_4$

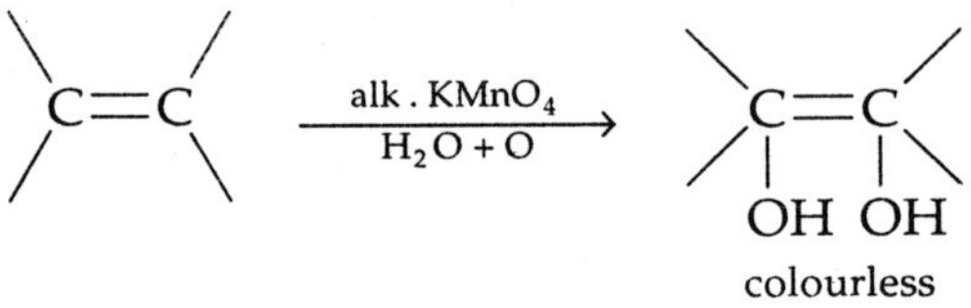

$$>C=C< \xrightarrow[H_2O + O]{alk.\, KMnO_4} >C(OH)-C(OH)<$$

colourless

(d) Undergoes ozonolysis

(i) 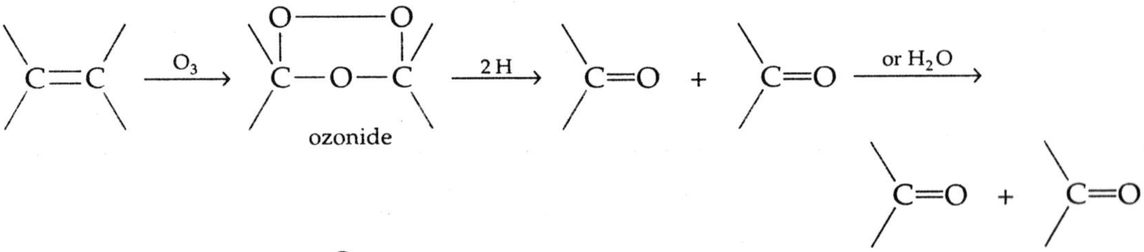 $$>C=C< \xrightarrow{O_3} \text{ozonide} \xrightarrow{2H} >C=O + >C=O \xrightarrow{\text{or } H_2O} >C=O + >C=O$$

(ii) 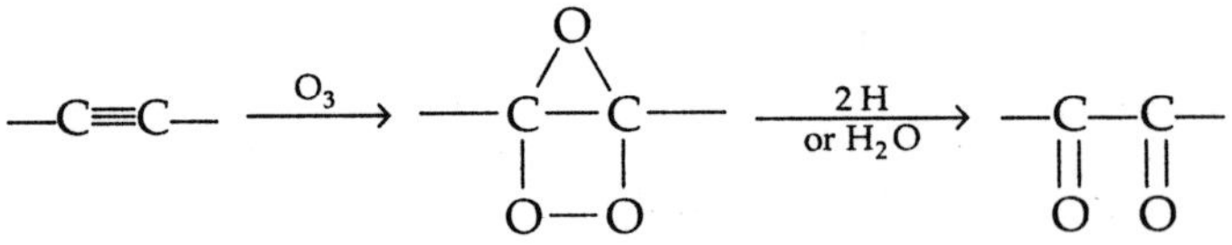 $$-C\equiv C- \xrightarrow{O_3} \text{ozonide} \xrightarrow[\text{or } H_2O]{2H} -\underset{\|}{\underset{O}{C}}-\underset{\|}{\underset{O}{C}}-$$

Remember

(i) The products formed during ozonolysis also help in deciding the nature and position of unsaturation in molecule.

(ii) If organic compound is terminal alkyne, it

(a) gives H_2 with Na

$$-C\equiv CH \xrightarrow[\text{in } NH_3]{Na} -C\equiv CNa + \frac{1}{2}H_2$$

(b) reduces amm. $AgNO_3$ to give white ppt.

$$-C{\equiv}CH + \underset{\substack{\text{or}\\ \text{amm. AgNO}_3}}{Ag_2O} \longrightarrow \underset{\substack{\text{Silver alkylide}\\ \text{while ppt. white ppt.}}}{-C{\equiv}C{\cdot}Ag} + H_2O$$

(c) gives red ppt. with ammonical Cu_2Cl_2.

$$2[-C{\equiv}CH] + Cu_2O \longrightarrow \underset{\text{Red ppt.}}{2[-C{\equiv}C{\cdot}Cu]} + H_2O$$

B. HALIDES

Monohalides with aqueous KOH give alcohal.

$$R{-}X \xrightarrow{\text{KOH (aq.)}} ROH$$

With alcohalic KOH give alkene.

$$R{\cdot}CH_2{\cdot}CH_2X \xrightarrow{\text{(alc.) KOH}} R{\cdot}CH{=}CH_2$$

Dihalides: with aqueous KOH give—

(a) **Diol**—It is dihalide with halogens on different carbon atoms.

$$CH_2X{-}CH_2X \xrightarrow{\text{KOH (aq.)}} CH_2OH{\cdot}CH_2OH$$

(b) **Aldehyde**—with a gem dihalide having position of halogens.

$$CH_3{\cdot}CH_2{\cdot}CHX_2 \xrightarrow{\text{KOH (aq.)}} CH_3{\cdot}CH_2{\cdot}CHO$$

(c) **Ketone**—It is a dihalide having non terminal position of halogens.

$$CH_3{\cdot}CX_2{\cdot}CH_3 \xrightarrow{\text{KOH (aq.)}} CH_3{\cdot}CO{\cdot}CH_3$$

Trihalides with aqueous KOH give acid, then all the three halogens are no terminal carbon atoms.

$$CH_3CX_3 \xrightarrow{\text{KOH (aq.)}} CH_3COOH$$

C. ALCOHALS AND PHENOLS

(i) Alcohals give effervescences of H_2 gas with Na or K.

$$R{-}OH + Na \longrightarrow RONa + \frac{1}{2}H_2$$

(ii) Alcohals give red colour with ceric ammonium nitrate.

(iii) *Dehydration:*

Alcohols on dehydration give:

(a) fat lower temperature ether.

$$2R{-}OH \xrightarrow[\text{agent}]{\text{Dehydrating}} ROR$$

(b) at higher temperature alcohol give alkene. The order of ease of dehydration of alcohol is

tertiary alcohol > secondary alcohol > primary alcohol

$$CH_3{\cdot}CH_2OH \xrightarrow[\text{agent}]{\text{Dehydrating}} CH_2{=}CH_2.$$

(iv) *Esterification*

Alcohols from esters with acids, acid chlorides and acid anhydrides.

$$\underset{\text{Alcohol}}{R{-}OH} + \underset{\text{acid}}{HOOCR'} \longrightarrow \underset{\text{ester}}{R'COOR} + H_2O$$

(v) *Distinction*

Distinction in *primary, secondary and tertiary alcohols.*

(a) Primary alcohols on oxidation give aldehydes and then acid. (Both contain the same number carbon atoms as the parent alcohol)

$$R\text{—}CH_2OH \xrightarrow{[O]} RCHO \xrightarrow{[O]} RCOOH$$

(b) Secondary alcohols on oxidation give ketone with same number of carbon atoms and then acids on prolong oxidation containing lesser number of carboin atoms than the parent alcohol.

$$R \cdot CHOH \cdot R' \xrightarrow{[O]} R \cdot COR'$$

(c) Tertiary alcohols on vigorous oxidation form ketones and then acids. (Each with lesser number of carbon atoms than the parent alcohol.)

$$(CH_3)_3 \cdot COH \longrightarrow CH_3 \cdot CO_3 \cdot CH_3$$

(d) Lucas reagent test: Primary alcohols react with Lucas reagent ($ZnCl_{2anhy.}$ + $HCl_{conc.}$) only on heating. Secondary alcoholc give white turbidity only after 5–10 minutes of reaction. Tertiary alcohols give white turbidity immediately.

(vi) *Phenols*:

(i) Phenols are soluble in alkali.

(ii) They turn blue litmus to red.

(iii) They give no effervescences with $NaHCO_3$, although they are acidic in nature.

(iv) On distillation with Zn dust phenols give benzene.

$$C_6H_5OH \xrightarrow[\Delta]{Zn/dust} C_6H_6$$

(v) Phenols also show Some characteristics colours with $FeCl_{3(aq.)}$.

D. ETHERS

(a) (i) An ether with conc. HI—

(a) gives an alcohol and alkyl iodine (when cold)

$$R\text{—}O\text{—}R + HI \longrightarrow RI + ROH$$

(ii) In case of mixed ethers, the iodine attaches to the smaller alkyl group

$$CH_3 \cdot OC_2H_5 + HI \longrightarrow CH_3I + C_2H_5OH$$

(b) An ether gives two mole of alkyl iodine (when hot)

$$ROR + 2HI \longrightarrow 2RI + H_2O$$

$$R\text{—}O\text{—}R' \xrightarrow{2\,HI} RI + R'I + H_2O$$

E. ALDEHYDES AND KETONES ***or carbonyl compounds***

(a) *Addition Tests*

(i) A carbonyl compound shows addition reactions with sodium bisulphite forming white ppt.

$$>C{=}O + NaHSO_3 \longrightarrow >C<^{OH}_{SO_3Na}$$

Important: Only methyl ketones and all aldehydes show this test:

(ii) Carbonyl compounds form oximes and hdrazones.

$$>C{=}O + H_2NOH \longrightarrow >C{=}NOH$$

$$>C{=}O + H_2N{\cdot}NHC_6H_5 \longrightarrow >C{=}N{\cdot}NHC_6H_5$$

(b) *Reduction Tests*

A carbonyl compound is aldehyde when it;

(i) Reduces Fehling solution (except bendaldehyde).

$$R\text{—}CHO + 2CuO \longrightarrow RCOOH + Cu_2O \text{ (red ppt.)}$$

(ii) Reduces Tollen's reagent.

$$R\text{—}CHO + Ag_2O \longrightarrow RCOOH + 2Ag \text{ (silver mirror)}$$

(iii) On reduction gives primary alcohol.

(iv) Reduces Schiffs reagent and gives pink colour.

(v) Give red ppt. with Benedict's reagent.

$$R\text{—}CHO \xrightarrow{2H} RCH_2OH$$

Important: A carbonyl compound, if does not show above tests for aldehyde, is ketone.

(c) (i) Iodoform test: Organic compounds having (Ketone) $CH_3\text{—}\underset{\underset{O}{||}}{C}\text{—}$ unit attached to cabron or hydrogen give yellow ppt. (iodoform) with I_2 + NaOH

$$CH_3{\cdot}CO{\cdot}CH_3 + 3I_2 + 4NaOH \longrightarrow CHI_3 + CH_3COONa + 3NaI + 3H_2O$$

Important: acetaldehyde and all methyl ketones respond to this test.

(ii) With 2, 4-dinitrophenyldrazine: Both aldehydes and ketones give an orange-yellow solid with 2,4-dinitrophenydrazine.

F. ACIDS

(i) An acid *turns blue litmus red.*

(ii) An acid *gives effervescences with* $NaHCO_3$.

$$RCOOH + NaHCO_3 \longrightarrow RCOONa + H_2O + CO_2\uparrow$$

(iii) An acid *forms ester with alcohols* in presence of concentrated H_2SO_4.

$$RCOOH + R'OH \xrightarrow{\text{Conc. } H_2SO_4} RCOOR' + H_2O$$

(iv) An acid on decarboxylation (on heating with sodalime) loses CO_2 and forms hydrocarbon.

$$RCOOH \xrightarrow{NaOH + CaO} R{-}H + CO_2\uparrow$$

(v) Dr distillation of Ca salt of an acid gives *aldehydes and ketones*.

$$2(HCOO)_2Ca \longrightarrow 2HCHO + 2CaCO_3$$

$$2(CH_3COO)_2Ca \longrightarrow 2CH_3COCH_3 + 2CaCO_3$$

$$(HCOO)_2Ca + (CH_3COO)_2Ca \longrightarrow 2CH_3CHO + 2CaCO_3$$

(vi) Acids *form amides on heating with* NH_3 which on further heating with P_2O_5 give cyanides.

$$RCOOH + NH_3 \longrightarrow \underset{\text{amm. salt}}{RCOONH_4} \xrightarrow{\Delta} \underset{\text{amide}}{RCONH_2} \xrightarrow{P_2O_5} RCN$$

G. ESTERS

(i) All the esters posses fruity smell.

(ii) Esters on reduction produce two molecules of alcohol.

$$RCOOR' \xrightarrow{\text{Reduction}} RCH_2OH + R'OH$$

(iii) Esters on hydrolysis give acid and alcohol.

$$RCOOR' \xrightarrow{HOH} RCOOH + R'OH$$

H. ACID CHLORIDES

(i) On hydrolysis give HCl and carboxylic acid.

$$RCOCl \xrightarrow{HOH} RCOOH + HCl$$

(ii) On reduction give alcohols.

$$RCOCl \xrightarrow{\text{Reduction}} RCH_2OH$$

I. ACID AMIDES AND ACID ANHYDRIDES

(i) On heating with HNO_2 give acids and N_2 gas.

$$RCONH_2 + HONO \longrightarrow RCOOH + N_2 + H_2O$$

(ii) On boiling with alkali give ammonia.

$$RCONH_2 + NaOH \longrightarrow RCOOH + N_2 + H_2O$$

(iii) On heating with P_2O_5, yield nitriles.

$$RCONH_2 \xrightarrow[-H_2O]{P_2O_5,\ \Delta} RCN$$

(iv) Amides, on heating with Br_2 and $KOH_{(aq)}$, undergo Hoffmann's bromamide reaction to give primary amines.

$$RCONH_2 + Br_2 + 4KOH \longrightarrow RNH_2 + 2KBr + K_2CO_3 + 2H_2O$$

(v) On hydrolysis an acid anhydride gives an acid.

$$(RCO)_2O \xrightarrow{HOH} 2RCOOH$$

(vi) On reacting with alcohols an acid anhydride gives an acid and an ester.

$$(RCO)_2O + R'OH \longrightarrow RCOOR' + RCOOH$$

J. AMINES

(i) Primary amines ungoes carbylamine reaction.

(ii) Action of HNO_2 on amines

(a) **Primary amine:**

$$R{\cdot}CH_2NH_2 \xrightarrow{HNO_2} RCH_2OH + N_2 + H_2O$$

(b) **Secondary amine:**

$$R_2NH \xrightarrow{HNO_2} R_2N{-}N{=}O$$

Yellow oil, nitroso compound

(c) **Tertiary amine:**

$$(R)_3N \xrightarrow{HNO_2} R_3N{\cdot}HNO_2 \xrightarrow{\Delta} R_2N{\cdot}NO + ROH$$

(iii) Action of $C_6H_5SO_2Cl$

(a) **Primary amines.** Give sulphonamide, soluble in alkali

$$C_6H_5SO_2Cl + RNH_2 \longrightarrow RNHSO_2C_6H_5 \xrightarrow{NaOH} RNNaSO_2C_6H_5$$

(b) **Secondary amines.** Gives sulphonamide in soluble in alkali.

$$C_6H_5SO_2Cl + R_2NH \longrightarrow R_2NSO_2C_6H_5$$

(c) **Tertiary amines.** Do not react

(iv) Aromatic amines having $—NH_2$ group in the nucleus:

(a) With $NaNO_2$ + HCl or $NaNO_2 + H_2SO_4$, they undergodiazoization at 273 K.

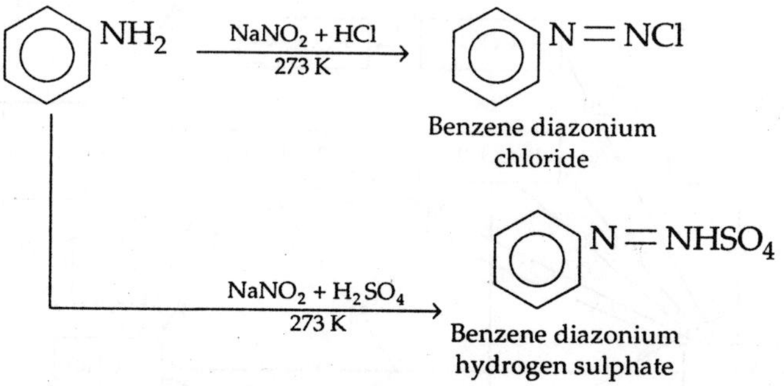

$$C_6H_5NH_2 \xrightarrow[273\ K]{NaNO_2 + HCl} C_6H_5N{=}NCl$$

Benzene diazonium chloride

$$C_6H_5NH_2 \xrightarrow[273\ K]{NaNO_2 + H_2SO_4} C_6H_5N{=}NHSO_4$$

Benzene diazonium hydrogen sulphate

Aromaticaminasane easily oxidised by air

K. CYANIDES AND ISOCYANIDES

(a) *Reduction*

On reduction they give primary amines and secondary amines respectively.

$$\underset{\text{Cyanides}}{RCN} \xrightarrow{\text{Reduction}} RCH_2NH_2$$

$$RNC \xrightarrow{\text{Reduction}} RNHCH_3$$

(b) *Hyrdysis*
On hydrolysis cyanides & isocyanide give acid and primary amine respectively.

$$RCN \xrightarrow{\text{HOH}} RCOOH$$

$$RNC \xrightarrow{\text{HOH}} RNH_2 + HCOOH$$

L. NITRO COMPOUNDS

(i) with NaOH solution they give yellow colour
(ii) On reduction in acid medium (Sn + HCl), nitr ocompounds give primary amines.

$$R\text{—}NO_2 \xrightarrow{\text{Sn/HCl}} RNH_2$$

TEST YOUR SELF BEFORE SOLVING PROBLEMS

(Fill in numbers 1 to 44 by the reagents you will use for conversions)

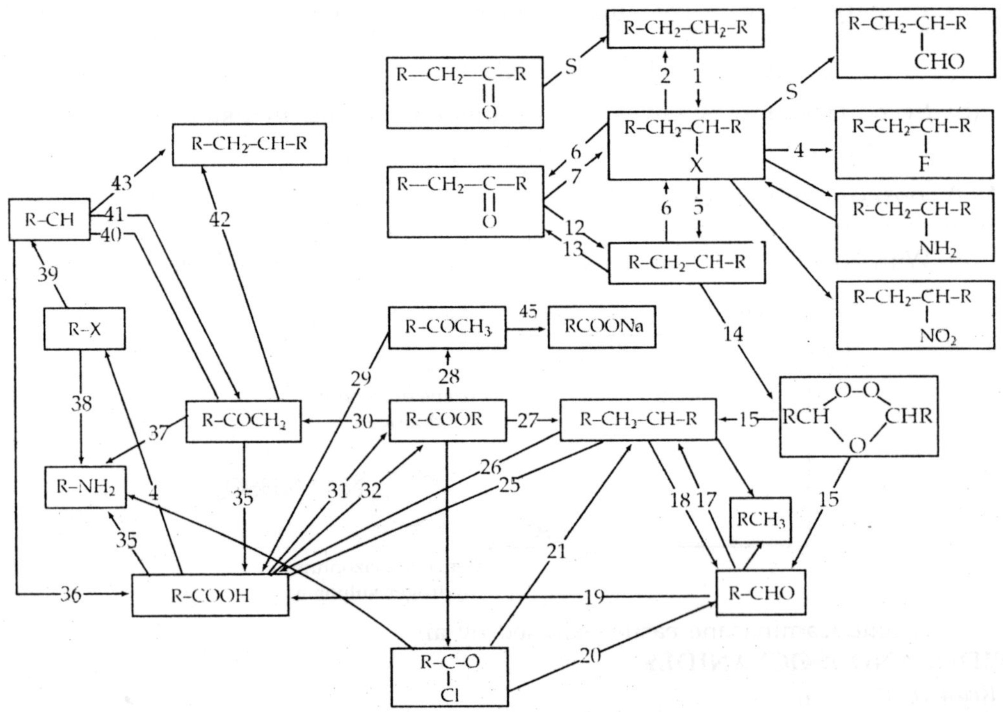

(*Note*: Solution is not given; you have to attempt yourself See Annexure A)

PROBLEMS (Section I)

Q. 1. Should compound A be optically active? Would you predict its oxime to be optically active?

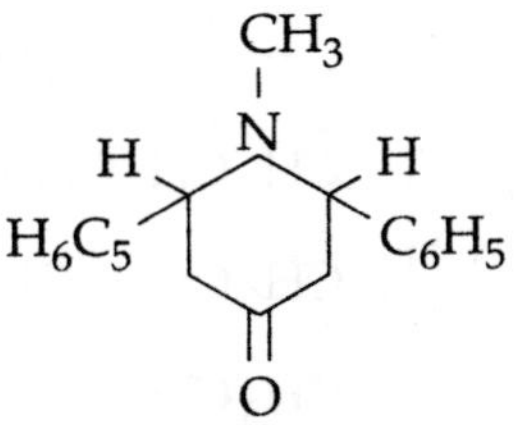

CH_3 N; H, H_6C_5 / H, C_6H_5; O

A

Q. 2. Give the R-S nomenclature for

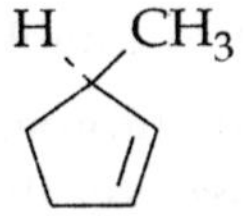

H CH_3

Q. 3. Predict the order of increasing solubility in water for chloroform, bromoform, and carbon tetrachloride. Explain the basis for your prediction and compare it with the observed facts. (Solubilities in g/100 g H_2O at 293 K. $CHCl_3$, 0.82; $CHBr_3$, 0.1; CCl_4, 0.08.)

Q. 4. Name one substance in list B which would be chosen for each use listed in column A.

A. Uses	*B. Substances*
(a) Aerosol bomb propellant	*p*-Dichlorobenzene
	Phenol
(b) Insecticide	Iodoform
	Methanol
(c) Anesthetic	Methyl bromide
	Buna S
(d) Antiseptic	Vinyl chloride
	Butyl rubber
(e) Solvent for grease	Carbon tetrachloride
(f) Inner tube synthetic rubber	Freon
	Napthalene
	Cycloropropane

Q. 5. Five unbalanced liquids are known to be ethyl bromide, ethyl ether, n-hexane, butyraldehyde, and acetone. How could one proceed to label each?

Q. 6. Give the structural formula for 1, 1-dibromo-4-isopropyl-4-*n*-propyl-5-tert-butyl-mo-6-chloromethylnoane.

Q. 7. Give a name for the following compound:

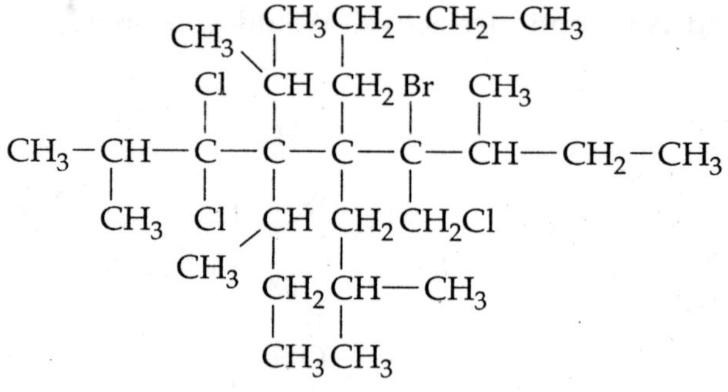

$$
\begin{array}{ccccccccc}
 & & CH_3 & & CH_3 & CH_2-CH_2-CH_3 & & & \\
 & & & \searrow & | & | & & & \\
 & & Cl & & CH & CH_2 & Br & CH_3 & \\
 & & | & & | & | & | & | & \\
CH_3- & CH- & C- & & C- & C- & C- & CH- & CH_2-CH_3 \\
 & | & | & & | & | & | & & \\
 & CH_3 & Cl & & CH & CH_2 & CH_2Cl & & \\
 & & & \swarrow & | & | & & & \\
 & & CH_3 & & CH_2 & CH-CH_3 & & & \\
 & & & & | & | & & & \\
 & & & & CH_3 & CH_3 & & &
\end{array}
$$

Q. 8. A pure hydrocarbon is known to be either 2, 5-dimethylhexane or 3, 4-dimethylhexane. What synthetic method could be used to establish its structure? Give equations.

Q. 9. How could one explain the fact that trichloroacetic acid is a much stronger acid than acetic acid, and trimethylacetic acid is weaker than acetic acid? On this basis, predict the stronger acid in the following pairs:

$Br\ CH_2\ COOH$ and $Cl\ CH_2\ COOH$

$CH_3\ CHCl\ COOH$ and $Cl\ CH_2\ CH_2\ COOH$

Q. 10. Give a sequence of reactions you could carry out for the following transformations

(a) *n*-propyl bromide to 1,2-dichloropropane

(b) isopropyl alcohol to *n*-propyl bromide

(c) ethyl bromide to ethylene chlorohydrin

(d) 1-bromo-2-methylbutane to $CH_3\ CH_2\ CO\ CH_3$

(e) isopropyl bromide to *n*-hexane

(f) 3-methyl-1-butene to 2,3-dimethylbutane

Q. 11. What reactions would you use to prepare 2,2,7,7-tetra-methyloctane starting with 2,2-dimethyl-3-bromobutane?

Q. 12. Starting with 1,6-dibromo-5-methyl-3-hexene, what actions would you use to prepare 1-methyl-1, 3-cyclohexadiene?

Q. 13. What reactions could one use to prepare: (a) β-bromo-propionic acid from acetic acid; (b) γ-bromovaleric acid from valeric acid?

Q. 14. What reactions would you use to prepare 1,2,3-trinitro-benzene from aniline?

Q. 15. Indicate the reactions one could use to transform

(a) $(CH_3)_2CH\ CH_2CH_2CH_2OH$ to $(CH_3)_2CH\ CH_2CH_2CH_2CH_2NO_2$

(b) $CH_3CH_2CH_2CH_2CH_2CO_2C_2H_5$ to $CH_3CH_2CH_2CH_2CN$

Q. 16. Give formulas for compounds I-VII formed in the following reactions sequence:

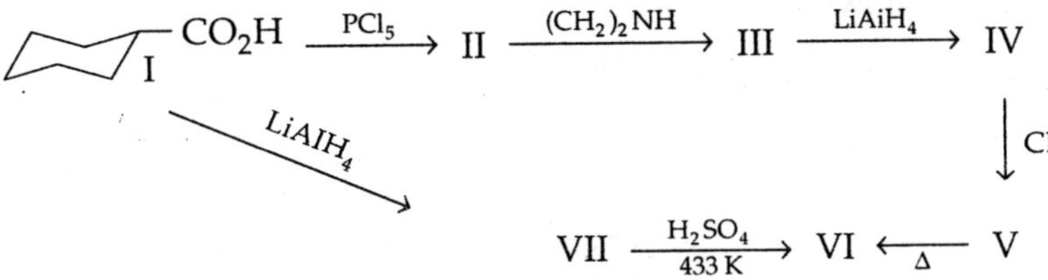

$$\text{(cyclohexane)}-CO_2H\ (I) \xrightarrow{PCl_5} II \xrightarrow{(CH_3)_2NH} III \xrightarrow{LiAlH_4} IV \xrightarrow{CH_3I} V \xrightarrow{\Delta} VI \xleftarrow[433\,K]{H_2SO_4} VII;\quad I \xrightarrow{LiAlH_4} VII$$

Q. 17. How many stereoisomers are possible with the formula CH_3 CH=CH $CH(CH_3)$—COOH?

Q. 18. Indicate the series of reactions you would use in the laboratory to prepare (a) *m*-flurobenzoic acid starting with benzene and (b) compound (A) starting with any open-chain compound.

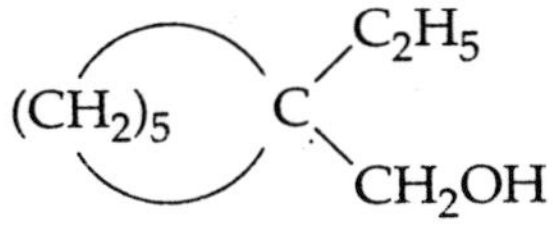

$$(CH_2)_5\ C(C_2H_5)(CH_2OH)\quad (A)$$

Q. 19. What series of laboratory reactions could be used to convert (a) isopropyl alcohol to propyne, (b) 1-butene to 2-butyne, (c) *n*-pentyl and bromide to 1-pentyne, (d) 4-methyl-2-hexyne to 2,3-dimethylhexane?

Q. 20. Write the structural formulas for the products that would be present after the hydrolysis of each of the following compounds in a 15% aqueous solution of sulfuric acid:

(a) C_6H_5 CONH CH_2 $CONH_2$.

(b) CH_3 $(CH_2)_3$ NH CONH CO $(CH_2)_3$ CH_3.

(c) $(C_2H_5)_2C{=}N$ NHC_6H_5

(d) 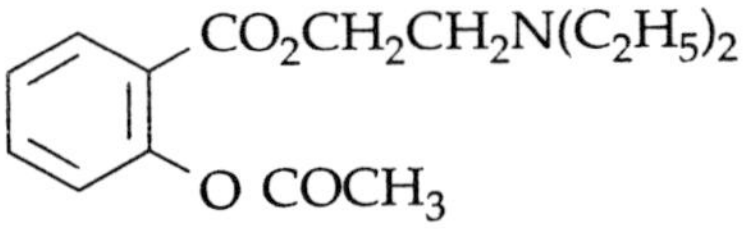 $C_6H_4(CO_2CH_2CH_2N(C_2H_5)_2)(O\,COCH_3)$ (ortho)

(e) 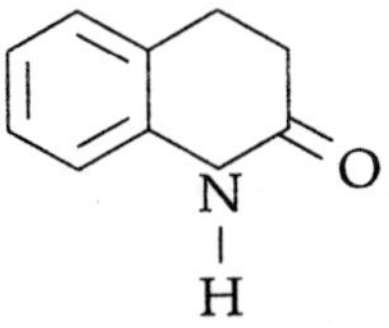 3,4-dihydro-2(1H)-quinolinone (benzene ring fused to ring with C=O and N–H)

Q. 21. Predict the products of each of the following reactions. Analysis of each problem to identify acids and bases or electrophiles and nucleophiles will be helpful in most cases.

(a) $CH_3CH_2NH_2 + HCl \rightarrow$

(b) $CH_3CH_2C{\equiv}CH + NaNH_2 \rightarrow$

(c) $(CH_3CH_2CH_2)_2NH + CH_3CH_2Br \rightarrow$

(d) $CH_3CH_2CH{=}CHCH_2CH_3 + HBr \rightarrow$

(e) $CH_3CH_2CH_2SH + NaOH \rightarrow$

(f) $CH_3CH_2CH{=}CHCH_3 + Br_2 \rightarrow$

(g) $CH_3CH_2CH_2C{\equiv}CCH_3 + H_2$ (2 equivalents) $\xrightarrow{PtO_2}$

(h) $CH_3CH_3 + Cl_2 \xrightarrow{hv}$

1 volume 1 volume

Q. 22. Predict the products of he following reactions

(a) $CH_3CH_2CH_2CH_2CH{=}CH_2 \xrightarrow[H_2SO_4]{H_2O}$

(b) $CH_3C(CH_3)(OH)CH_3 \xrightarrow[\Delta]{H_3PO_4}$

(c) $CH_3CH(OH)CH_3 \xrightarrow[\Delta]{H_3PO_4}$

Q. 23. Complete the following equations.

(a) $CH_3CH_2CH{=}CHCH_2CH_3 \xrightarrow[Pt]{H_2}$

(b) $CH_3CH_2CH_2\overset{\overset{\displaystyle CH_3}{|}}{C}{=}CH_2 \xrightarrow[\text{benzene}]{H_2 \atop RbCl[P(C_6H_5)_3}$

(c) $CH_3CH_2CH_2CH_2CH_2CH{=}CH_2 \xrightarrow[RbCl(P(C_6H_5)_3)_3]{H_2}$

Q. 24. Complete the following equations.

(a) $(E)\text{-}CH_3CH_2CH{=}CHCH_2CH_3 \xrightarrow[\text{carbo tetrachloride}]{Br_2}$

(b) $(Z)\text{-}CH_3CH_2CH{=}CHCH_2CH_3 \xrightarrow[\text{carbon tetrachloride}]{Br_2}$

Q. 25. Complete the following equations.

(a) $CH_3\overset{\overset{\displaystyle CH_3}{|}}{C}{=}CHCH_3 \xrightarrow{O_3} \xrightarrow[\text{water}]{Zn}$

(b) $CH_3\overset{\overset{\displaystyle CH_3}{|}}{C}HCH{=}\overset{\overset{\displaystyle CH_2CH_3}{|}}{C}CH_2CH_3 \xrightarrow{O_3, CH_3OH} \xrightarrow{(CH_3)_2S}$

(c) $CH_3\overset{\overset{\displaystyle CH_3}{|}}{C}HCH{=}\overset{\overset{\displaystyle CH_2CH_3}{|}}{C}CH_2CH_3 \xrightarrow[\text{water}]{O_3} \xrightarrow{H_2O,\ NaOH} \xrightarrow{H_3O^+}$

(d) $CH_3CH_2\overset{\overset{\displaystyle CH_3}{|}}{C}{=}CH_2 \xrightarrow{O_3} \xrightarrow[\text{water}]{Zn}$

(e) $CH_3(CH_2)_3CH{=}CH_2 \xrightarrow[\text{water}]{O_3} \xrightarrow{H_2O_2,\ NaOH} \xrightarrow{H_3O^+}$

(f) $(CH_3CH_2)(CH_3)C{=}C(CH_3)(H) \xrightarrow{HBr}$

(g) $(HO\overset{O}{\overset{\|}{C}})(H)C{=}C(\overset{O}{\overset{\|}{C}}OH)(H) \xrightarrow[H_2O]{Br_2}$

(h) $CH_3CH_2\underset{\underset{\displaystyle OH}{|}}{C}HHCH_2CH_3 \xrightarrow[\Delta]{H_3PO_4}$

(i) $CH_3\overset{\overset{\displaystyle CH_3}{|}}{\underset{\underset{\displaystyle OH}{|}}{C}}CH_2CH_3 \xrightarrow[\Delta]{H_2SO_4}$

Q. 26. Complete the following equations.

(a) $CH_3CH_2CH_2C{\equiv}CH + NaNH_2 \xrightarrow[NH_3(liq)]{}$

(b) $CH_3CH_2CH_2C{\equiv}CH + CH_3CH_2MgBr \xrightarrow[\text{diethyl ether}]{}$

(c) $CH_3\overset{\overset{CH_3}{|}}{C}HCH_2CH_2\overset{\overset{O}{||}}{C}H + CH_3OH \xrightarrow[\text{benzene, }\Delta]{TsOH}$

(d) $CH_3\overset{\overset{O}{||}}{C}CH_2\overset{\overset{CH_3}{|}}{C}HCH_3 + CH_3\overset{\overset{CH_3}{|}}{C}HCH_2NH_2 \xrightarrow[HCl]{} \xrightarrow{NaOH}$

(e) $CH_3CH_2C{\equiv}C\overset{\overset{OTHP}{|}}{C}HCH_2CH_2CH{=}CH\overset{\overset{O}{||}}{C}CH_3 + CH_3ONH_3^+Cl^- \xrightarrow{base}$

(f) $HOCH_2CH_2\underset{\underset{CH_3}{|}}{\overset{\overset{CH_3}{|}}{C}}CH_2CH{=}CH\overset{\overset{O}{||}}{C}OCH_3 \xrightarrow[\text{dichloromethene}]{C_5H_5N^+H\ CrO_3Cl^-}$

(g) $CH_3\overset{\overset{O}{||}}{C}(CH_2)_8CH_3 \xrightarrow[\Delta]{Zn(Hg),\ HCl}$

(h) $CH_3\overset{\overset{CH_3}{|}}{C}HCH_2\overset{\overset{O}{||}}{C}CH_2\overset{\overset{CH_3}{|}}{C}HCH_3 \xrightarrow[\text{diethylene glycol, }\Delta]{H_2NNH_2,\ NaOH}$

Q. 27. Complete the following equations by supplying structural formulas for the products and reagents symbolized by the letters A–M.

(a) $CH_3CH_2CH_2C{\equiv}CH \xrightarrow[NH_3(liq)]{NaNH_2} A \xrightarrow{CH_3CH_2CH_2Br} B$

(b) $C \xrightarrow[NH_3(liq)]{D} CH_3\overset{\overset{CH_3}{|}}{C}HCH_2CH_2C{\equiv}CH$

(c) $CH_3CH_2CH_2C{\equiv}CH \xrightarrow[NH_3(liq)]{E} F \xrightarrow{(CH_3)_2SO_4} G$

[Hint: $(CH_3)_2SO_4$ is dimethyl sulfate with connectivity $CH_3OSO_2OCH_3$. What is the leaving group in $(CH_3)_2SO_4$?]

(d) $CH_3C{\equiv}CCH_3 \xrightarrow{HBr(excess)} H$

(e) $CH_3CH_2CH_2CH{=}CH_2 \xrightarrow{HCl} I$

(f) $HC{\equiv}CH \xrightarrow{HBr(excess)} J$

(g) $CH_3\overset{CH_3}{\overset{|}{C}}=\overset{CH_3}{\overset{|}{C}}CH_3 \xrightarrow[\text{diethyl ether}]{OsO_4} \xrightarrow[\text{water}]{KOH} K$

(h) $CH_3\overset{CH_3}{\overset{|}{C}}=CHCH_2CH_2\overset{O}{\overset{||}{C}}OCH_3 \xrightarrow[\substack{\text{dichloromethane}\\ -78\,°C}]{O_3} \xrightarrow{(CH_3)_2S} L + M$

Q. 28. Complete the following equations. Be sure to show the stereochemistry of the product where it is pertinent.

(a) $CH_3CH_2CH_2CH_2C\equiv CH \xrightarrow[\substack{Pd/BaSO_4\\ \text{quinoline}}]{H_2}$

(b) $CH_3CH_2CH_2CH_2C\equiv CCH_3 \xrightarrow[\substack{Pd/BaSO_4\\ \text{quinoline}}]{H_2}$

(c) $CH_3CH_2CH_2CH_2C\equiv CCH_3 \xrightarrow[NH_3(liq)]{Na}$

(d) $CH_3CH_2CH_2C\equiv CCH_3 \xrightarrow[RhC(P(C_6H_5)_3)_3]{H_2}$

(e) $CH_3CH_2CH_2CH_2C\equiv CCH_3 \xrightarrow[\text{Pt (acetic acid)}]{H_2\text{ (excess)}}$

(f) $CH_3\overset{CH_3}{\overset{|}{C}}HCH_3\equiv CCH_2CH_3 \xrightarrow[\text{carbon tetrachloride}]{O_3} \xrightarrow{H_2O}$

(g) $CH_3C\equiv C\overset{CH_3}{\overset{|}{C}}HCH_2CH_3 \xrightarrow[\text{carbon tetrachloride}]{O_3} \xrightarrow{H_2O}$

Q. 29. Provide products of the following reactions:

(a) 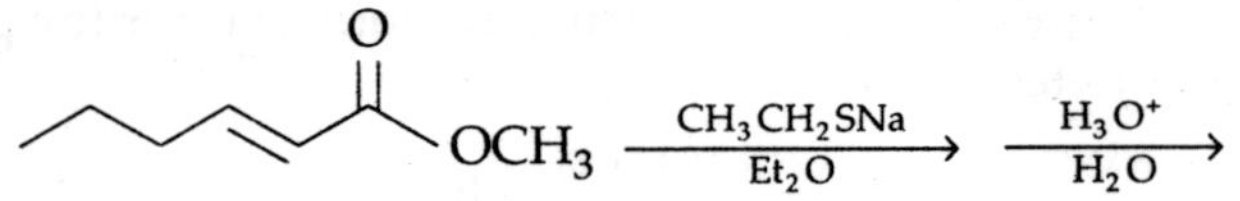 [structure: α,β-unsaturated methyl ester, OCH$_3$] $\xrightarrow[Et_2O]{CH_3CH_2SNa} \xrightarrow[H_2O]{H_3O^+}$

(b) 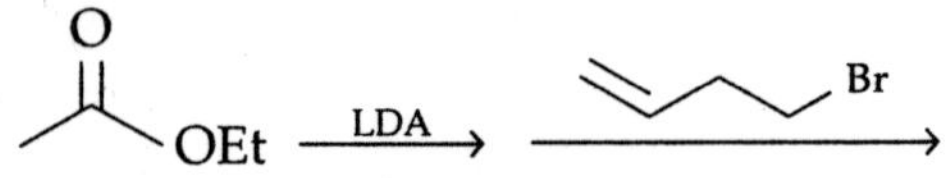 [structure: ethyl ester, OEt] $\xrightarrow{LDA} \xrightarrow{\text{allyl-type bromide, Br}}$

(c) 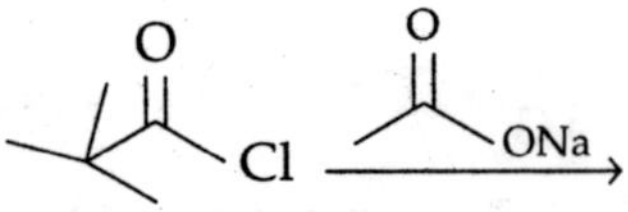 [structure: acid chloride, Cl] $\xrightarrow{\text{sodium acetate, ONa}}$

Q. 30. Predict products in the following reactions:

(a) 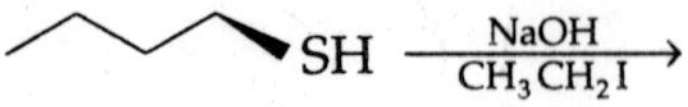 [structure: thiol, SH] $\xrightarrow[CH_3CH_2I]{NaOH}$

(b) 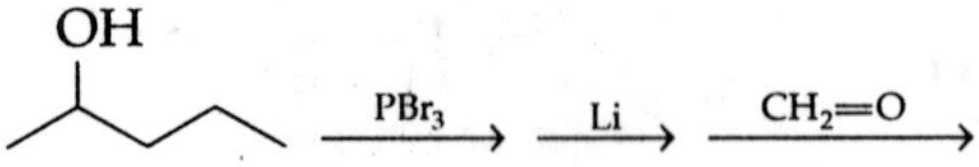 [structure: secondary alcohol, OH] $\xrightarrow{PBr_3} \xrightarrow{Li} \xrightarrow{CH_2=O}$

(c) 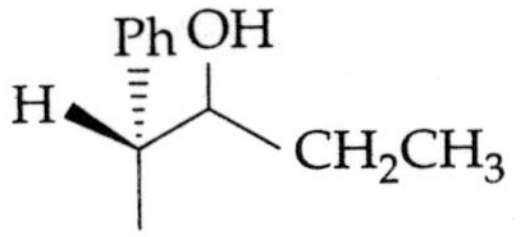 H, Ph, OH (stereo structure with CH_2CH_3) $\xrightarrow{PCC}$ $\xrightarrow[\text{(ii) } H_3O^+]{\text{(i) } CH_3MgBr}$

(d) $CH_3\overset{O}{\overset{||}{C}}CHO + OH \longrightarrow$

(e) $CH_3\overset{O}{\overset{||}{C}}CH_2CH_2CHO + \underset{\text{Excess}}{CH_3OH} \xrightarrow{H^+} A \xrightarrow{LiAlH_4} B \xrightarrow[\text{Reflux}]{H^+} C$

(f) $C_6H_5\overset{O}{\overset{||}{C}}CH_3 \xrightarrow{NH_2OH} y + D \xrightarrow{H^+} G \xrightarrow{^-OH} H + I$

$\downarrow H^+$

$E \xrightarrow{OH^-} C_6H_5NH_2 + F$

Q. 31. Complete the following reactions:

(i) $CH_2{=}CH\ CN + PhCH_2OH \xrightarrow{H_2SO_4} ?$

(ii) $PhCH_2COCHN_2 + HCl(g) \xrightarrow{Et_2O} ?$

(iii) $MeNH_2 + ? \longrightarrow MeNHCOOEt$

(iv) $HOCH_2CH_2CN \xrightarrow[\text{Heat}]{HBr} ?$

(v) $C_2H_5CH{=}CH_2 \xrightarrow{B_2H_6} A \xrightarrow{ClNH_2} B$

(vi) $C_2H_5NH_2 + KCN + Br_2 \xrightarrow{KOH} C \xrightarrow{LiAlH_4} D$

(vii) $CH_2C{=}O + HCN \longrightarrow E \xrightarrow{H_2C\,C{=}O} F$

Q. 32. Complete the following reactions

(a) $CH_3-\underset{OH}{\overset{H}{C}}-\underset{CH_3}{\overset{CH_3}{C}}-CH_3 \xrightarrow{TsCl} \xrightarrow{Ph/NH_2}$

(b) $CH_3-\underset{CH_3}{\overset{CH_3}{C}}-\underset{OH}{CH}\,CH_2CH_3 \xrightarrow{HBr}$

(c) 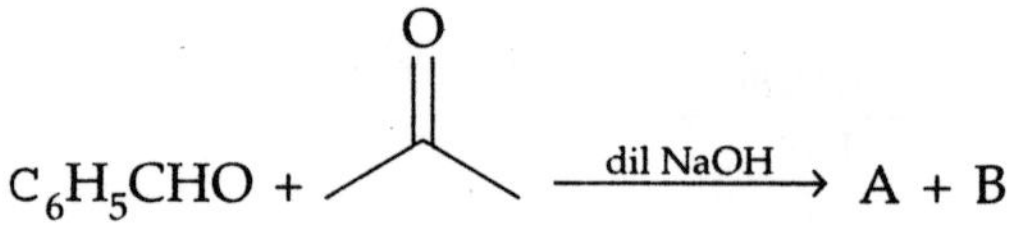 C_6H_5CHO + (acetone, CH_3COCH_3) $\xrightarrow{\text{dil NaOH}}$ A + B

(d) $C_6H_5CHO + C_6H_5COCH_3 \xrightarrow{\text{dil. NaOH}}$ C + D

(e) $C_6H_5CHO + CH_3NO_2 \xrightarrow{\text{dil NaOH}}$

(f) 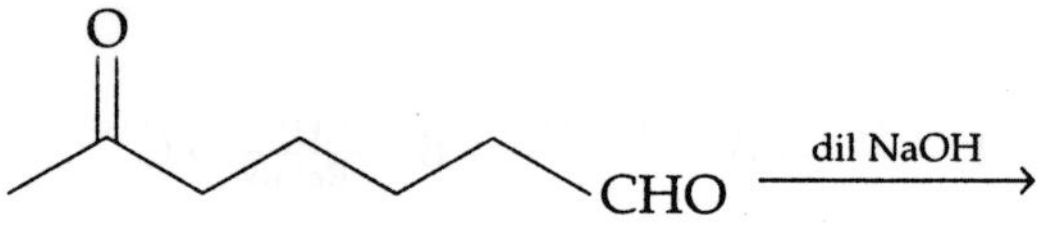 $CH_3CO(CH_2)_4CHO \xrightarrow{\text{dil NaOH}}$

(g) 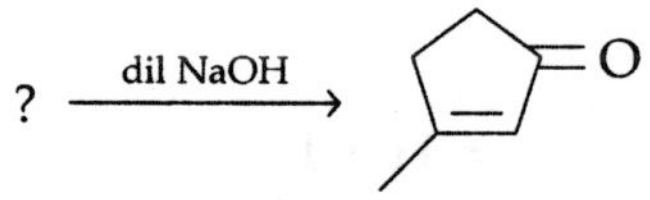 ? $\xrightarrow{\text{dil NaOH}}$ 3-methylcyclopent-2-en-1-one

Q. 33. Provide products in the following reactions:

(a) 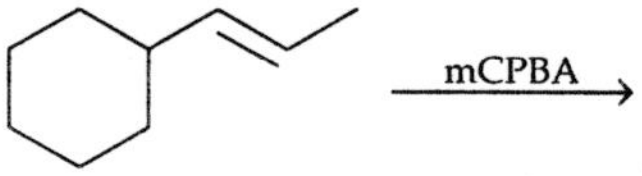 cyclohexyl–CH=CH–CH_3 $\xrightarrow{\text{mCPBA}}$

(b) 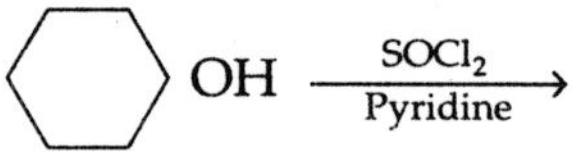 cyclohexyl–OH $\xrightarrow[\text{Pyridine}]{SOCl_2}$

(c) 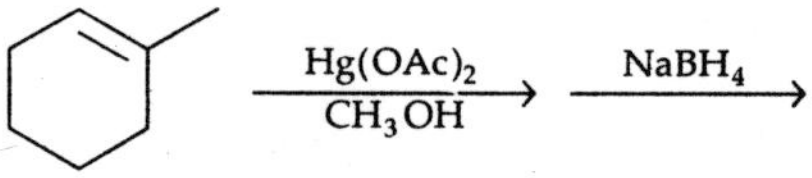 1-methylcyclohexene $\xrightarrow[CH_3OH]{Hg(OAc)_2}$ $\xrightarrow{NaBH_4}$

(d) 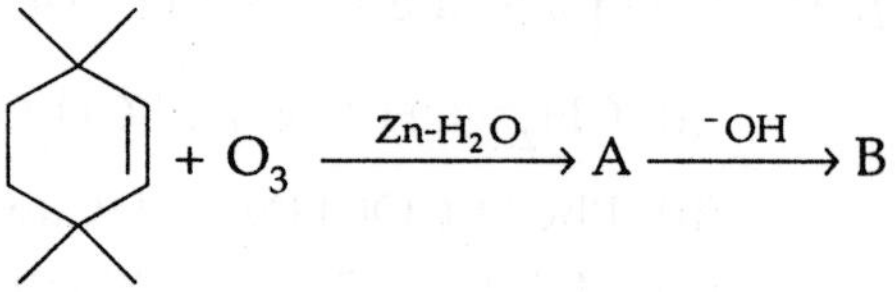 3,3,6,6-tetramethylcyclohexene + O_3 $\xrightarrow{\text{Zn-}H_2O}$ A $\xrightarrow{^-OH}$ B

(e)  H_2N–C_6H_4–CHO + HOC–C_6H_4–NO_2 $\xrightarrow{^-OH}$ Only cross product

(f) 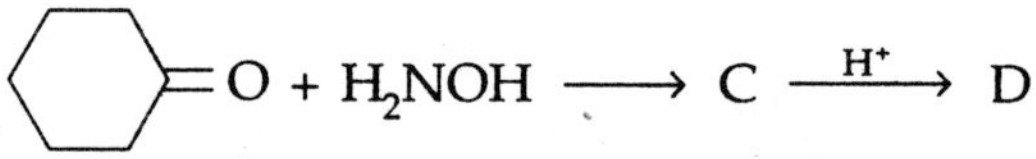 cyclohexanone (=O) + $H_2NOH \longrightarrow$ C $\xrightarrow{H^+}$ D

Q. 34. Predict the major product in the following reactions:

(a) 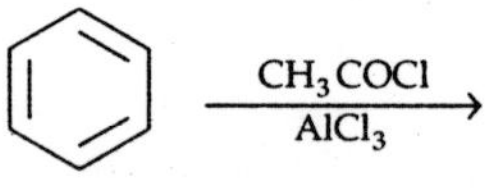 benzene $\xrightarrow[AlCl_3]{CH_3COCl}$

(b) 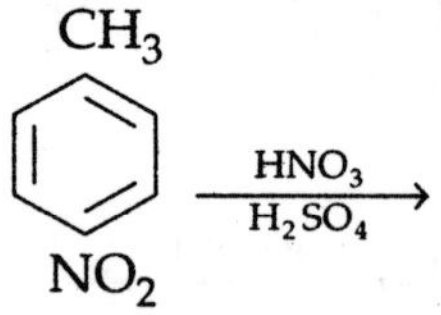 CH_3–C_6H_4–NO_2 (p-nitrotoluene) $\xrightarrow[H_2SO_4]{HNO_3}$

(c) OMe–C_6H_5 (anisole) $\xrightarrow[H_2SO_4]{SO_3}$

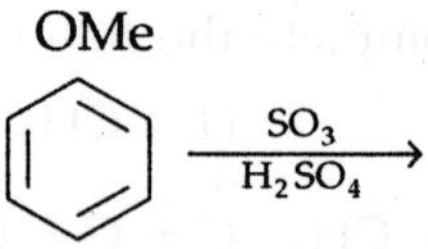

(d) 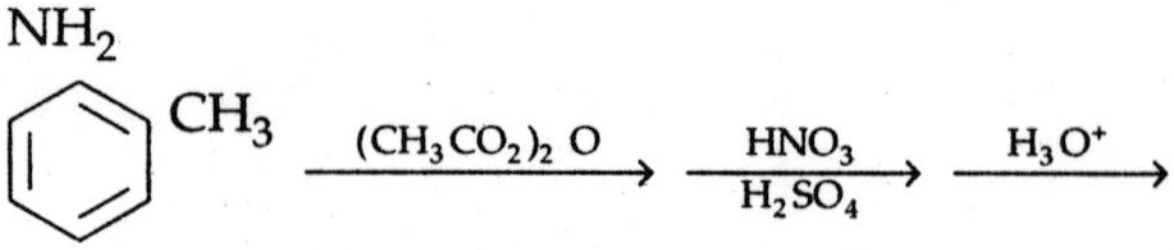 NH_2, CH_3 (o-toluidine) $\xrightarrow{(CH_3CO_2)_2O}$ $\xrightarrow[H_2SO_4]{HNO_3}$ $\xrightarrow{H_3O^+}$

Q. 35. (I) Provide reagents for accomplishing the following transformations:

$PhC(O)CH_3 \xrightarrow{A} Ph(CH_3)C(OH)OCH_3$

$PhC(O)CH_3 \xrightarrow{B} Ph(C{=}CH_2)N(CH_3)_2$

$PhC(O)CH_3 \xrightarrow{C} Ph(CH_3)C(OH)CH_2CH{=}CH_2$

$PhC(O)CH_3 \xrightarrow{D} Ph(CH_3)C(OCH_3)_2$

Q. 35. (II) (a) $H_3C(H)C{=}C(H)CH_2Cl \longrightarrow H_3C(H)C{=}C(H)CH_2CH_2Br$

(b) $H_3C(H)C{=}C(H)COOC_2H_5 \longrightarrow H_3C(H)C{=}C(H)CHO$

(c) $CH_3CH_2CH(CH_3)COOCH_3 \longrightarrow CH_3CH_2CH(CH_3)CHO$

Q. 36. Predict products.

(a) Furan-2-CHO + CH_3COCH_3 $\xrightarrow[\text{Heat}]{\text{NaOH}}$

(b) 3-Chlorophthalic acid (COOH, Cl, COOH) $\xrightarrow[H_2SO_4]{HNO_3}$

(c) 4-Phenylphenol (biphenyl–OH) $\xrightarrow{Br_2/CHCl_3}$

(d) 2,6-Dimethylacetophenone (CH_3, $COCH_3$, CH_3) $\xrightarrow[H_2SO_4]{HNO_3}$

(e) 4-Methylanisole (OMe, CH_3) + isobutylene $\xrightarrow{H_2SO_4}$

Q. 37. Predict the products of each of the following reactions. Identify acids and bases or electrophiles & neuclophiles.

(a) 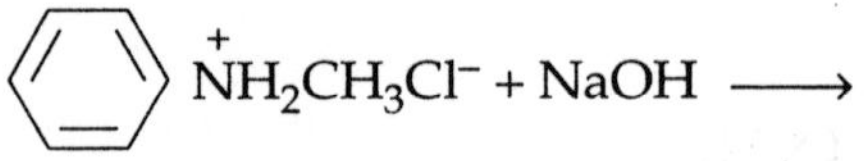 $C_6H_5\overset{+}{N}H_2CH_3Cl^- + NaOH \longrightarrow$

(b)  $C_6H_5SNa + CH_3CH_2CH_2Br \longrightarrow$

(c) $Br\text{-}C_6H_4\text{-}\overset{O}{\overset{||}{C}}OH + NaHCO_3 \longrightarrow$

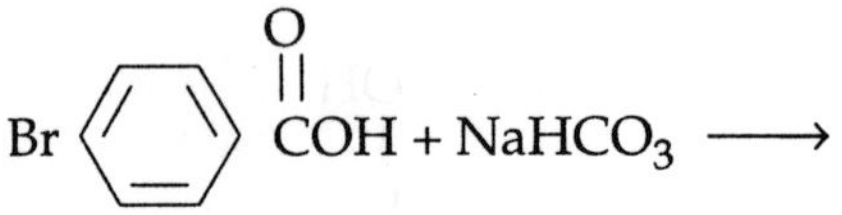

Q. 38. Predict the products of the following reactions.

(a) 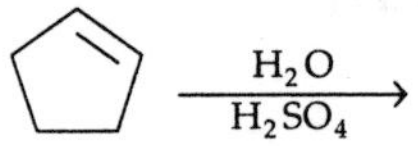 cyclopentene $\xrightarrow[H_2SO_4]{H_2O}$ (b) 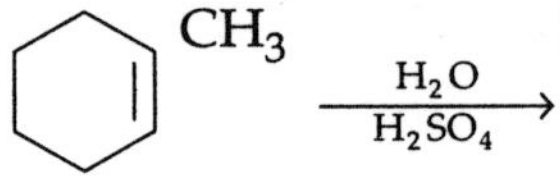 1-methylcyclohexene (CH_3) $\xrightarrow[H_2SO_4]{H_2O}$

Q. 39. Predict the major product of each of the following reactions. Write a structure showing the stereochemistry when it is pertinent.

(a) 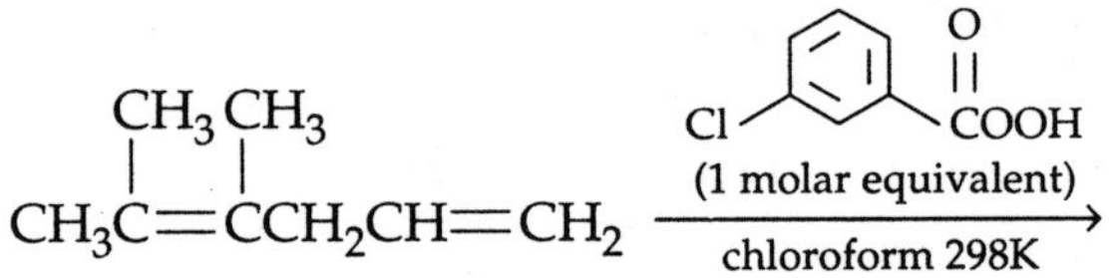 $CH_3\overset{CH_3}{\overset{|}{C}}=\overset{CH_3}{\overset{|}{C}}CH_2CH=CH_2 \xrightarrow[\text{chloroform 298K}]{m\text{-}ClC_6H_4CO_3H \text{ (1 molar equivalent)}}$

(b) 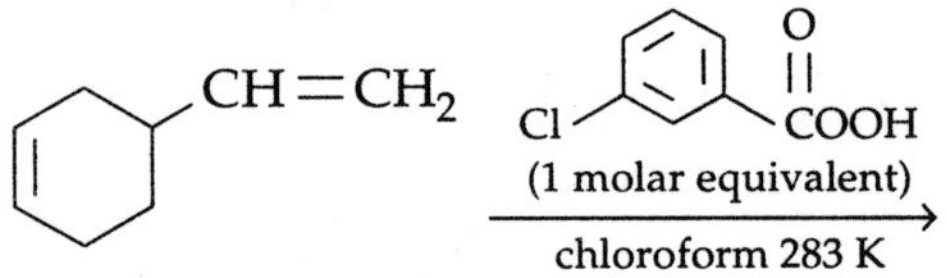 cyclohexenyl$-CH=CH_2 \xrightarrow[\text{chloroform 283 K}]{m\text{-}ClC_6H_4CO_3H \text{ (1 molar equivalent)}}$

(c) $CH_3CH_2CH_2CH=CH_2 \xrightarrow[\text{dichloromethane 298 K}]{Cl\text{-}C_6H_4\text{-}COOH}$

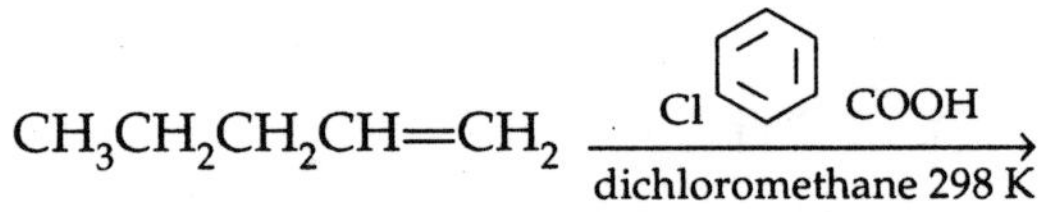

(d) 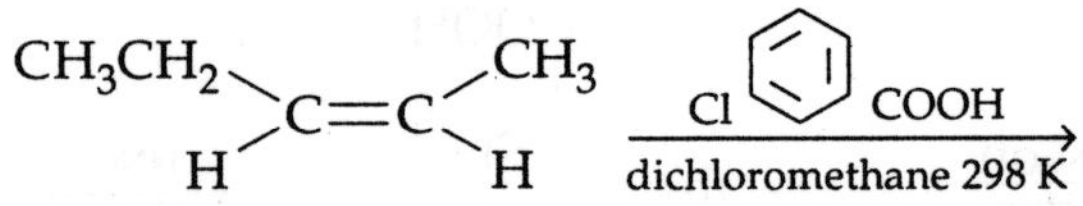 $\underset{H}{\overset{CH_3CH_2}{}}C=C\underset{H}{\overset{CH_3}{}} \xrightarrow[\text{dichloromethane 298 K}]{Cl\text{-}C_6H_4\text{-}COOH}$

(e) 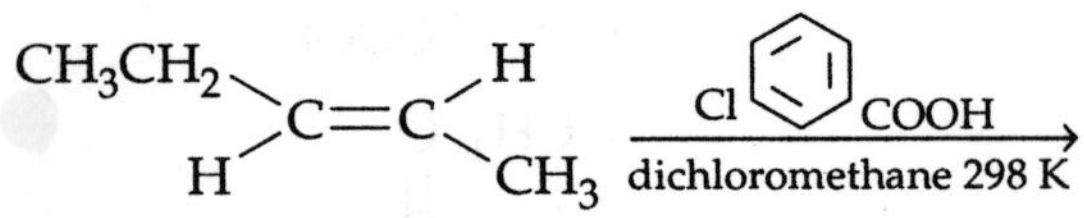 $\underset{H}{\overset{CH_3CH_2}{}}C=C\underset{CH_3}{\overset{H}{}} \xrightarrow[\text{dichloromethane 298 K}]{Cl\text{-}C_6H_4\text{-}COOH}$

Q. 40. **(I)** Complete the following equations, showing the stereochemistry wherever it is known.

(a) 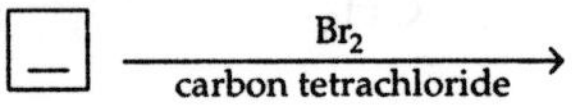 cyclobutene $\xrightarrow[\text{carbon tetrachloride}]{Br_2}$

(b) 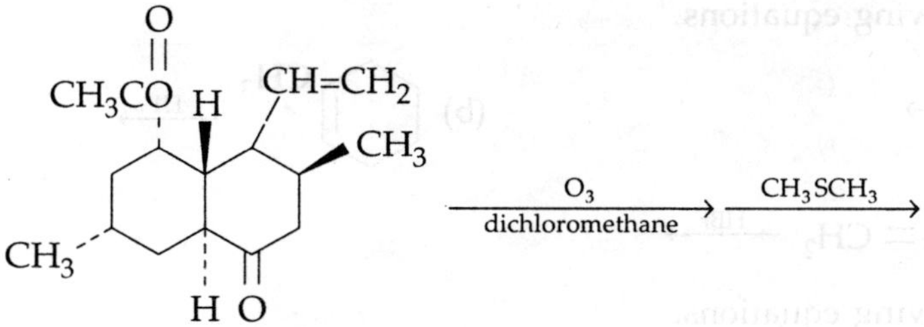 [structure: $CH_3\overset{O}{\overset{\|}{C}}O$, H, $CH{=}CH_2$, CH_3, CH_3, H, O] $\xrightarrow[\text{dichloromethane}]{O_3}$ $\xrightarrow{CH_3SCH_3}$

(c) 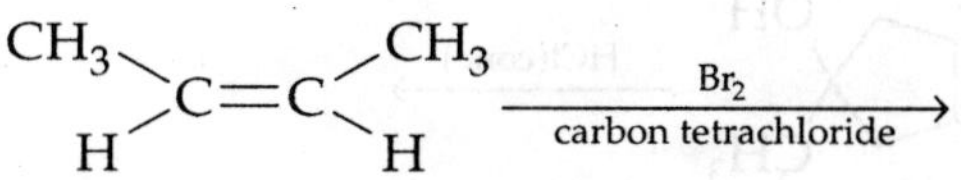 $\overset{CH_3}{\underset{H}{}}\!\!>C{=}C<\!\!\overset{CH_3}{\underset{H}{}}$ $\xrightarrow[\text{carbon tetrachloride}]{Br_2}$ (d) $\overset{CH_3}{\underset{H}{}}\!\!>C{=}C<\!\!\overset{H}{\underset{CH_3}{}}$ $\xrightarrow[\text{carbon tetrachloride}]{Br_2}$

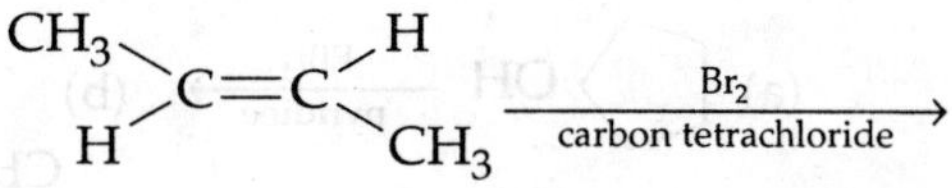

(e) 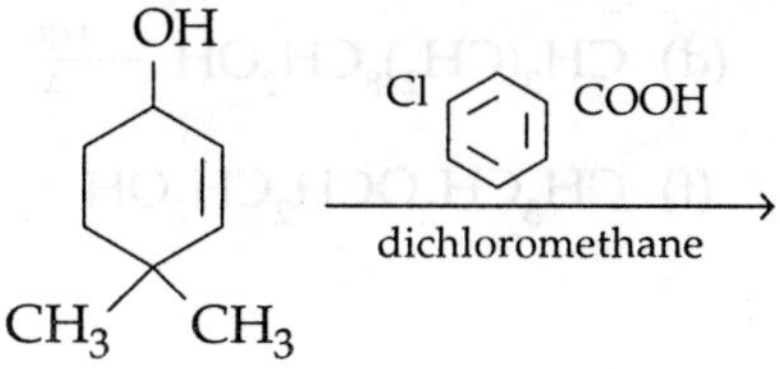 [structure: OH, CH_3, CH_3] $\xrightarrow[\text{dichloromethane}]{\text{Cl-C}_6\text{H}_4\text{-COOH}}$ (f) [cyclohexene] $\xrightarrow[H_2SO_4]{H_2O}$

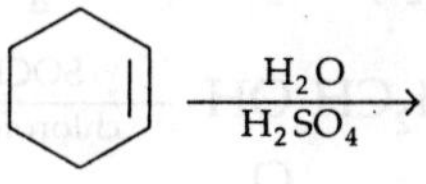

(g) 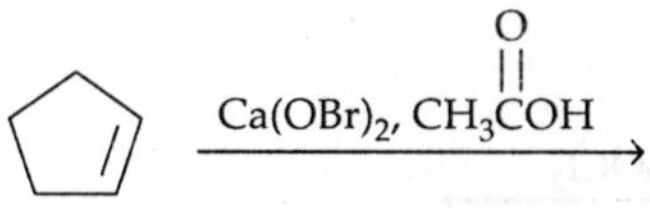 [cyclopentene] $\xrightarrow{Ca(OBr)_2,\ CH_3\overset{O}{\overset{\|}{C}}OH}$ (h) $C_6H_5CH{=}CH_2$ $\xrightarrow[\text{carbon tetrachloride}]{Br_2}$

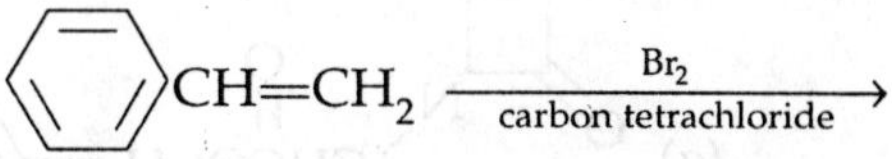

(i) 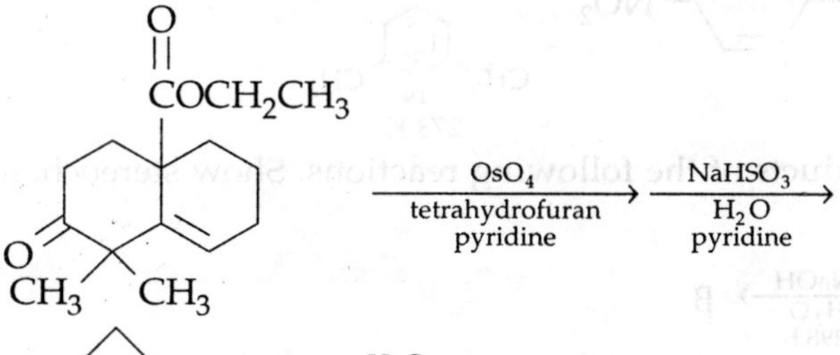 [structure: $\overset{O}{\overset{\|}{C}}OCH_2CH_3$, O, CH_3, CH_3] $\xrightarrow[\text{tetrahydrofuran, pyridine}]{OsO_4}$ $\xrightarrow[H_2O,\ \text{pyridine}]{NaHSO_3}$

Q. 40. (II) (a) 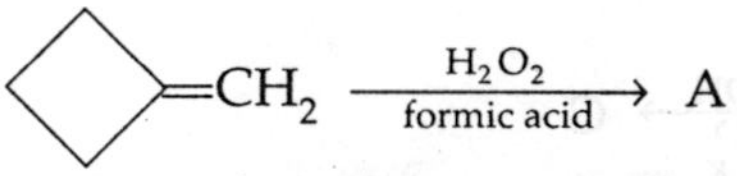 [cyclobutane]$=CH_2$ $\xrightarrow[\text{formic acid}]{H_2O_2}$ A

(b) [structure: CH_3, CH_3, CH_3, CH_3] $\xrightarrow[\text{chloroform 253 K}]{O_3}$ $\xrightarrow{H_2O}$ B

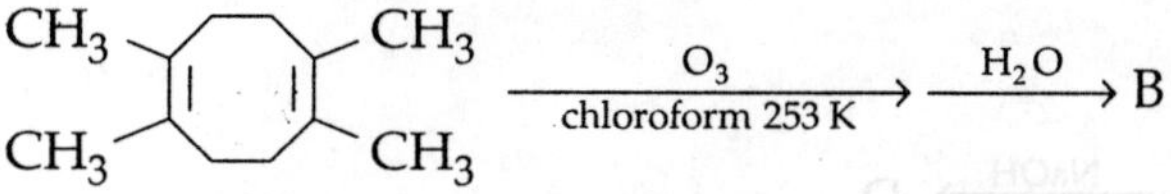

(c) 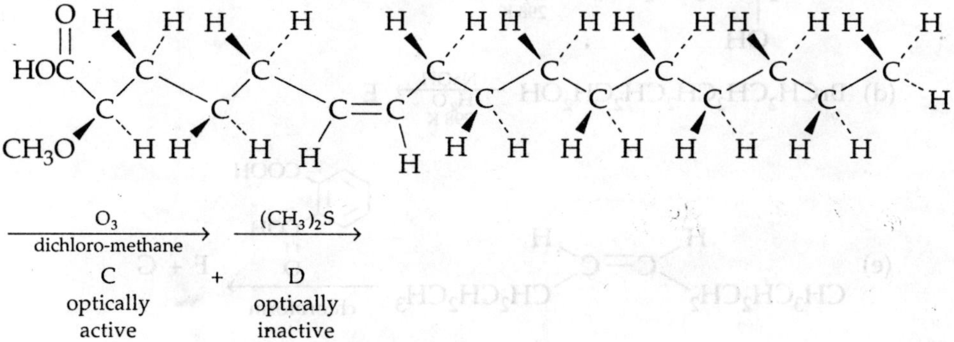 [structure: $HO\overset{O}{\overset{\|}{C}}$, CH_3O, H, C, C=C, H] $\xrightarrow[\text{dichloro-methane}]{O_3}$ $\xrightarrow{(CH_3)_2S}$

C optically active + D optically inactive

Q. 41. Complete the following equations.

(a) 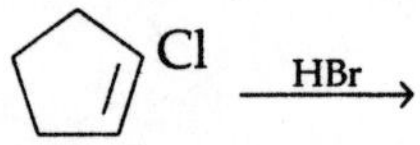 1-chlorocyclopentene $\xrightarrow{HBr}$

(b) 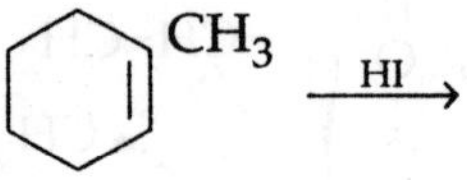 1-methylcyclohexene $\xrightarrow{HI}$

(c) 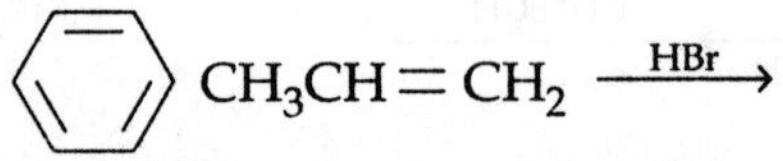 $C_6H_5CH_3CH=CH_2 \xrightarrow{HBr}$

Q. 42. Complete the following equations.

(a) 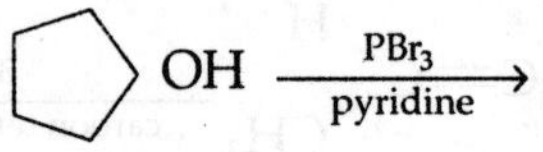 cyclopentyl–OH $\xrightarrow[\text{pyridine}]{PBr_3}$ (b) 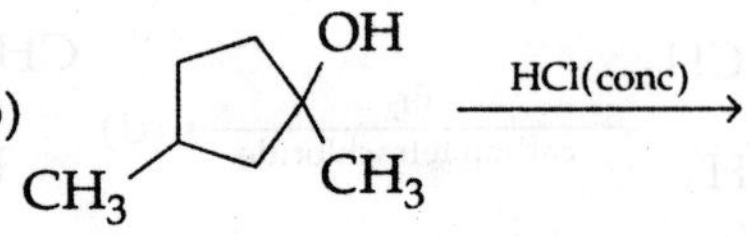 1,3-dimethylcyclopentan-1-ol $\xrightarrow{HCl(conc)}$

(c) $CH_3(CH_2)_9CH_2OH \xrightarrow[\Delta]{SOCl_2}$ (d) $CH_3(CH_2)_8CH_2OH \xrightarrow[\Delta]{HBr}$

(e) $CH_3SCH_2CH_2OH \xrightarrow[\text{chloroform}]{SOCl_2}$ (f) $CH_3CH_2OCH_2CH_2OH \xrightarrow[\text{pyridine}]{PBr_3}$

(g) 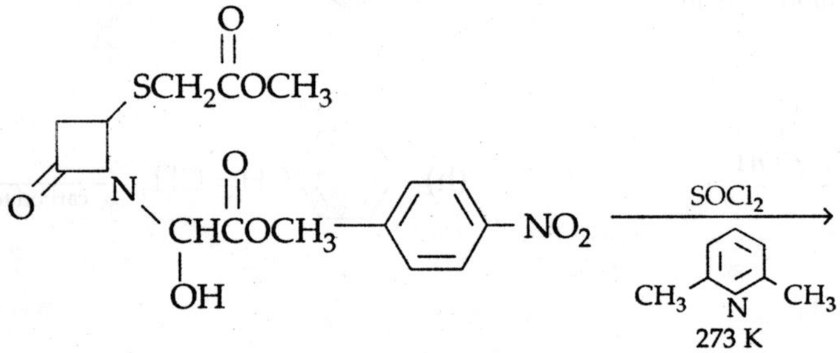 β-lactam (azetidin-2-one) bearing SCH_2COCH_3 (with C=O) at C-4 and N-substituted by $CHCOCH_2$–C₆H₄–NO_2 (ester, CH bearing OH) $\xrightarrow[\text{2,6-dimethylpyridine, 273 K}]{SOCl_2}$

Q. 43. Give structures for the products of the following reactions. Show stereochemistry if it is known.

(a) 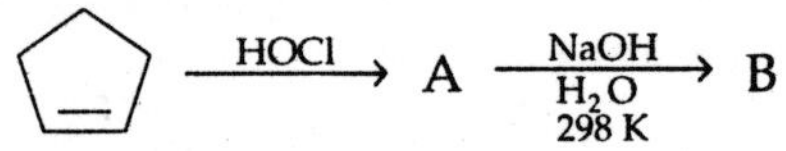 cyclopentene $\xrightarrow{HOCl}$ A $\xrightarrow[\substack{H_2O \\ 298\ K}]{NaOH}$ B

(b) $CH_3CH_2CH_2CH(Cl)CH(OH)CH_2CH_2CH_3 \xrightarrow[\substack{H_2O \\ 298\ K}]{NaOH}$ C

(c) $ClCH_2C(CH_3)(OH)CH_2CH_2CH_3 \xrightarrow[\substack{H_2O \\ 298\ K}]{NaOH}$ D

(d) $BrCH_2CH_2CH_2CH_2CH_2OH \xrightarrow[\substack{H_2O \\ 298\ K}]{NaOH}$ E

(e) 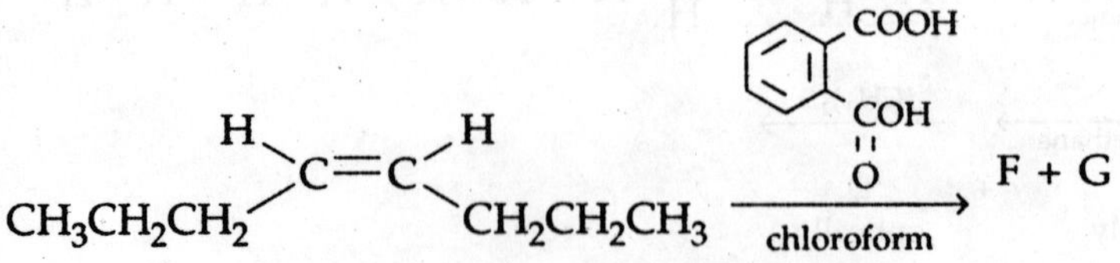 cis-$CH_3CH_2CH_2(H)C=C(H)CH_2CH_2CH_3$ $\xrightarrow[\text{chloroform}]{\text{C}_6\text{H}_4(\text{COOH})(\text{COOH with C=O, peroxy acid})}$ F + G

Q. 44. Use the curved-arrow notations to show how each of the following substitution reactions will take place. Also, show that the products will be.

(a) $CH_3CH(CH_2)CH_2CH_2I$ + $C_6H_5O^-Na^+$ $\xrightarrow{\text{sectone}}$

(b) CH_3I + piperidine (N–H) (excess) $\longrightarrow$

(c) $CH_3CH_2CH_2Br + CH_3S^-Na^+ \longrightarrow$

Q. 45. Complete the following equations.

(a) 1-methylcyclobutene $\xrightarrow[\text{Pb/C}]{H_2}$

(b) benzene $\xrightarrow[\text{Pb/C}]{H_2\text{(excess)}}$

(c) $CH_3C(CH_3)_2CH{=}CH_2 \xrightarrow[\text{benzene}]{H_2,\ RhCl[P(C_6H_5)_3]_3}$

(d) cyclopentene $\xrightarrow[\text{benzene}]{H_2,\ RhCl[P(C_6H_5)_3]_3}$

(e) 4-methylcyclohexene (CH_3 on wedge) $\xrightarrow[\text{carbon tetrachloride}]{Br_2}$

(f) 1,2-dimethylcyclopentene (CH_3, CH_3) $\xrightarrow[\text{carbon tetrachloride}]{Br_2}$

(g) 1-methylcyclohexene (CH_3) $\xrightarrow{O_3,\ CH_3OH} \xrightarrow{(CH_3)_2S}$

(h) cyclic diene $\xrightarrow[\text{dichloromethane, 298 K}]{O_3} \xrightarrow{H_2O_2,\ NaOH}$

Q. 46. (I) Complete the following equations, predicting the products and showing stereochemistry wherever possible.

(a) $CH_3CH(CH_3)CH{=}CH_2 \xrightarrow[\text{water, 273 K}]{KMnO_4,\ KOH} A$

(b) cyclobutene $\xrightarrow[\text{diethyl ether, pyridine}]{OsO_4} B \xrightarrow[\text{mannitol water}]{KOH} C$

(c) 1,2-dimethylcyclohexene (CH_3, CH_3) $\xrightarrow[\text{diethyl ether, pyridine}]{OsO_4} D \xrightarrow[\text{mannitol water}]{KOH} E$

(d) $\underset{H}{\overset{CH_3}{}}\!\!>C{=}C<\!\!\underset{CH_3}{\overset{H}{}}$ $\xrightarrow[\text{diethyl ether, pyridine}]{OsO_4}$ F $\xrightarrow[\text{manmitol water}]{KOH}$ G

(e) $\underset{H}{\overset{CH_3CH_2CH_2}{}}\!\!>C{=}C<\!\!\underset{H}{\overset{CH_3}{}}$ $\xrightarrow[\text{water 298 K}]{KMnO_4,\ KOH}$ H

Q. 46.(II) Complete the following equations

(a) $CH_3\underset{OH}{CH}\overset{CH_3}{CH}\overset{CH_3}{CH}CH_2CH_3 \xrightarrow[H_2SO_4,\ H_2O,\ 313\ K]{NaC_2O_7}$

(b) 2-phenylcyclohexanol $\xrightarrow[CH_3COH\ (O),\ (318\text{-}323)K]{Na_2C_2O_7}$

(c) cyclooctanol $\xrightarrow[H_2SO_4,\ H_2O,\ \text{acetone}]{CrO_3}$

(d) bicyclic compound bearing CH_3, CH_3, OH and $\underset{CH_3}{\overset{CH_3}{C}}HCH_2\overset{O}{\overset{||}{C}}CH_3$ $\xrightarrow[\text{dichloromethane, sodium acetate}]{C_5H_5N^+H\ CrO_3Cl^-}$

(e) 4-methyl-1-isopropyl (CH_3CHCH_3) cyclohexanol (OH) $\xrightarrow[H_2SO_4,\ H_2O]{NaCr_3O_7}$

(f) fused bicyclic compound with CH_3, OH, H, O, H $\xrightarrow[\text{dichloromethane}]{C_5H_5N^+H\ CrO_3Cl^-}$

Q. 46. (III) Predict the product of the following equations

(a) C_6H_5—CH=CHCH$_2$OH $\xrightarrow[\text{dichloromethane, 213 K}]{CH_3\overset{O}{\overset{||}{S}}CH_3,\ Cl\overset{O}{\overset{||}{C}}-\overset{O}{\overset{||}{C}}Cl}$ $\xrightarrow[\text{dichloromethane, 213K}]{(CH_3CH_3)_3N}$

(b) 1,2-dimethylcyclohexane (CH_3, CH_3) $\xrightarrow[\text{dichloromethane, 263 K}]{CH_3\overset{O}{\overset{||}{S}}CH_3,\ Cl\overset{O}{\overset{||}{C}}-\overset{O}{\overset{||}{C}}Cl}$ $\xrightarrow[\text{dichloromethane, 263 K}]{(CH_3CH_3)_3N}$

(c) 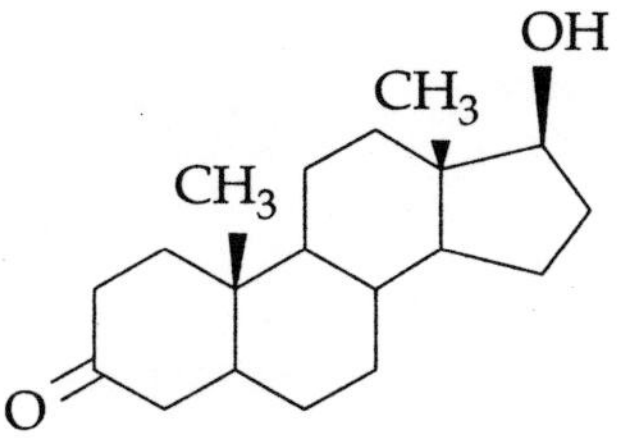 [steroid structure: 17β-hydroxy-3-keto steroid with angular CH_3 groups] $\xrightarrow[\text{dichloromethane, } -10°C]{\overset{O}{\parallel}\ CH_3SCH_3,\ \overset{O}{\parallel}\ \ \overset{O}{\parallel}\ ClC-CCl}$ $\xrightarrow[\text{dichloromethane, } -10°C]{(CH_3CH_3)_3N}$

Q. 46. (IV) Predict the products of the following reactions.

(a) $HO\overset{O}{\overset{\parallel}{C}}\underset{HO}{\underset{|}{C}H}\underset{OH}{\underset{|}{C}H}\overset{O}{\overset{\parallel}{C}}OH \xrightarrow[H_2O]{HIO_4}$

(b) 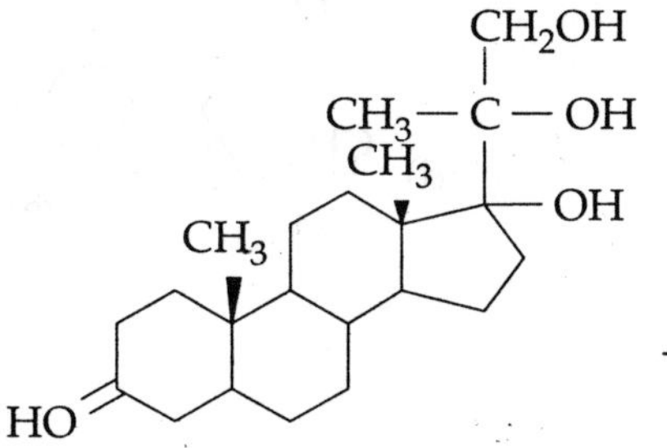 [steroid structure with CH_3 groups, HO at C-3, and at C-17: OH and $CH_3-C(OH)-CH_2OH$] $\xrightarrow[\text{methanol}]{HIO_4}$

(c) 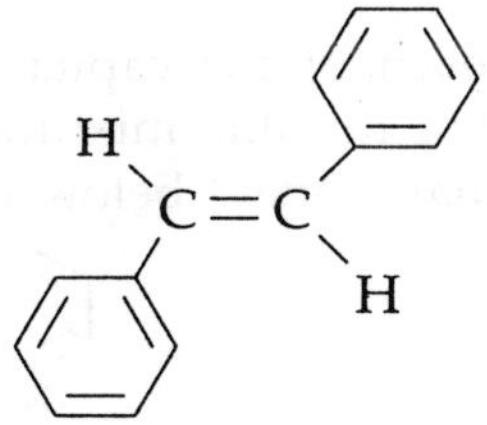 [(E)-stilbene: $C_6H_5(H)C=C(H)C_6H_5$] $\xrightarrow[\substack{H_2O \\ \text{dioxane} \\ 298\text{ K}}]{OsO_4,\ NaO_4}$

Q. 47. (I) Complete the following equations.

(a) $CH_3\overset{CH_3}{\overset{|}{C}}HCH_2OH \xrightarrow[\text{pyridine}]{SOCl_2}$

(b) $CH_3\underset{OH}{\underset{|}{\overset{CH_3}{\overset{|}{C}}}}CH_2CH_3 \xrightarrow{HCl\text{ (conc)}}$

(c) $CH_3CH_2\underset{OH}{\underset{|}{C}}HCH_3 \xrightarrow{PBr_2}$

(d) 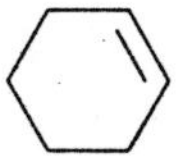 [cyclohexene]$CH_2CH_3 \xrightarrow[\text{diglyme}]{BH_3} \xrightarrow[H_2O]{H_2O_2,\ NaOH}$

(e) $HOCH_2(CH_2)_4CH_2OH \xrightarrow{HBr(g),\text{ excess}}$

(f) 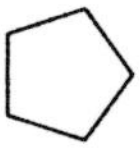 [cyclopentane]$OH \xrightarrow{HI}$

(g) $CH_3CH(OH)C(CH_3)_2CH(CH_3)CH_2OCH_3$ $\xrightarrow[\text{dichloromethane}]{C_5H_5N^+H\ CrO_3Cl^-}$

(h) $C_6H_5CH(OH)CH_3$ $\xrightarrow{CH_3SCH_3,\ ClC(=O)-C(=O)Cl}$ $\xrightarrow{(CH_3CH_2)_3N}$

Q. 47. (II) Complete the following equations.

(a) $C_6H_{11}CH_2CH_2CH_2CH_2OH$ $\xrightarrow[\text{pyridine}]{PBr_3}$

(b) (tetrahydrofuran-2-yl)CH_2OH $\xrightarrow[\text{pyridine, }\Delta]{SOCl_2}$

(c) cyclohexene $\xrightarrow[\text{H}_2\text{O, diethyl ether}]{OsO_4,\ NaIO_4}$

(d) CH_3, H-substituted epoxide $\xrightarrow[\text{ethanol}]{C_6H_5S^-Na^+}$

(h) 1,2,3-trimethoxybenzene (OCH_3, OCH_3, OCH_3) $\xrightarrow[\Delta]{\text{HI (excess)}}$

Q. 48. (I) Find the products or reagents corresponding to each of the capital letters. Show stereochemistry by appropriate conventions. When a racemic mixture forms, show the structure for one enantiomer and write "and enantiomer" below it.

(a) $C_6H_5CH_2CH_2CH{=}CH_2$ $\xrightarrow[\text{tetrahydrofuran}]{BH_3}$ A $\xrightarrow[H_2O]{H_2O_2;\ NaOH}$ B $\xrightarrow[\text{dichloromethane}]{C_5H_5N^+H\ CrO_3Cl^-}$ C

(b) (cyclopentenyl)CH_3 $\xrightarrow{\text{9-BBN}}$ D + E; D $\xrightarrow[H_2O]{H_2O_2,\ NaOH}$ F; E $\xrightarrow[H_2O]{H_2O_2,\ NaOH}$ G

major *minor*

(c) 1-phenylcyclohexene $\xrightarrow{\text{9-BBN}}$ H $\xrightarrow[H_2O]{H_2O_2,\ NaOH}$ I $\xrightarrow[H_2O]{CrO_3,\ H_2SO_4}$ K

1-phenylcyclohexene $\xrightarrow{H_2O,\ H_2SO_4}$ L

I $\xrightarrow{PBr_3,\ \text{pyridine}}$ J

(d) 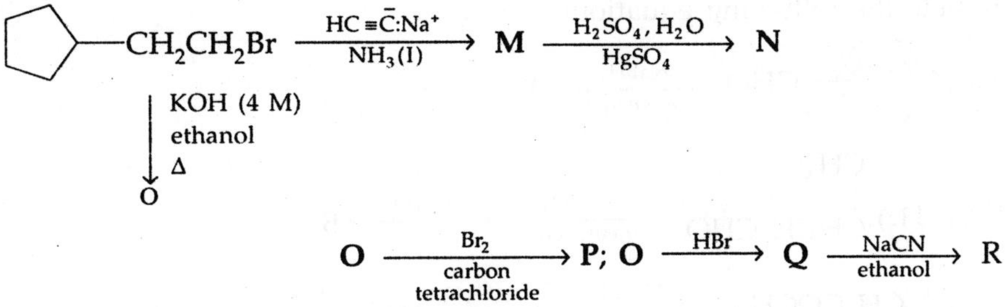 Cyclopentyl—CH_2CH_2Br $\xrightarrow[NH_3(l)]{HC\equiv\bar{C}:Na^+}$ **M** $\xrightarrow[HgSO_4]{H_2SO_4,\ H_2O}$ **N**

Cyclopentyl—CH_2CH_2Br $\xrightarrow{KOH\ (4\ M),\ ethanol,\ \Delta}$ O

O $\xrightarrow[carbon\ tetrachloride]{Br_2}$ **P**; **O** $\xrightarrow{HBr}$ **Q** $\xrightarrow[ethanol]{NaCN}$ R

Q. 48. (II)

(a) 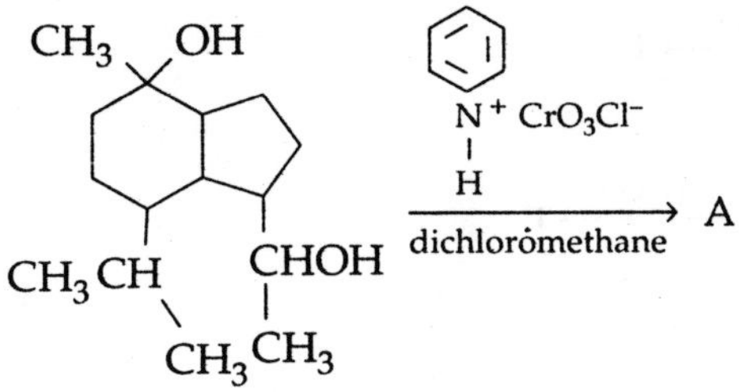 [bicyclic compound with CH_3, OH; CH_3CH, $CHOH$, CH_3, CH_3] $\xrightarrow[dichloromethane]{C_5H_5N^+H\ CrO_3Cl^-}$ A

(b) C_6H_5—CH=CHCOCH$_3$ $\xrightarrow[ethanol,\ water,\ 233\ K]{KMnO_4}$ B

(c) Indene $\xrightarrow[diethyl\ ether]{OsO_4}$ C $\xrightarrow[water]{KOH}$ D

(d) $PhCH_2O$—CH_2—CH_2—CH=CH_2 $\xrightarrow{E}$ $PhCH_2O$—CH_2—CH_2—CH(OH)—CH_2Cl

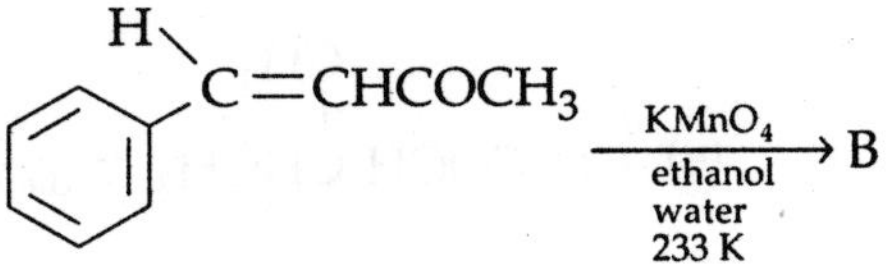

$\xrightarrow{F}$ $PhCH_2O$—CH_2—CH_2—CH(epoxide)—CH_2

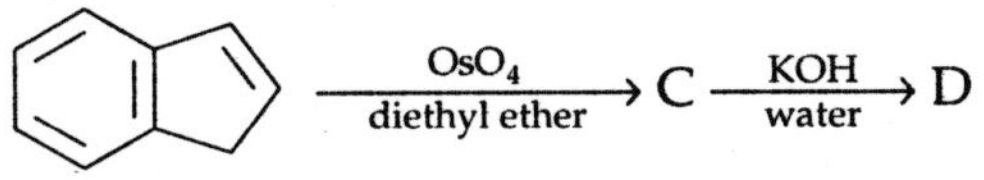

(e) HO—$CHCH_2CH_2CH_3$ (on bicyclic N ring) $\xrightarrow{G}$ O=$CCH_2CH_2CH_3$ (on bicyclic N ring)

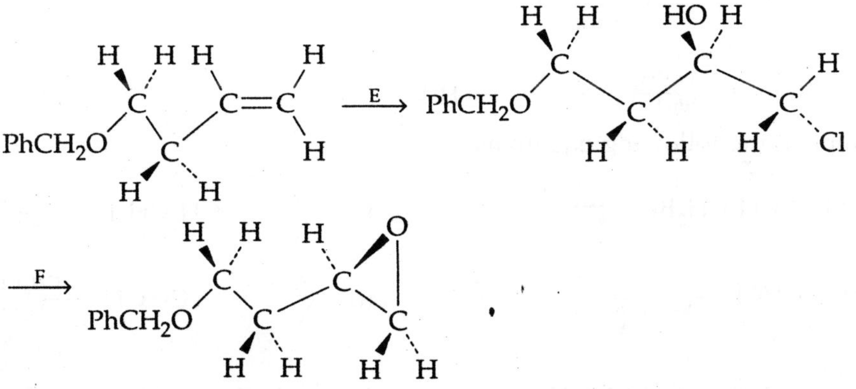

(f) $(C_6H_5)_2$CHOH $\xrightarrow{CH_3\overset{O}{\overset{\|}{S}}CH_3,\ Cl\overset{O}{\overset{\|}{C}}-\overset{O}{\overset{\|}{C}}Cl}$ $\xrightarrow{(CH_3CH_2)_3N}$ H

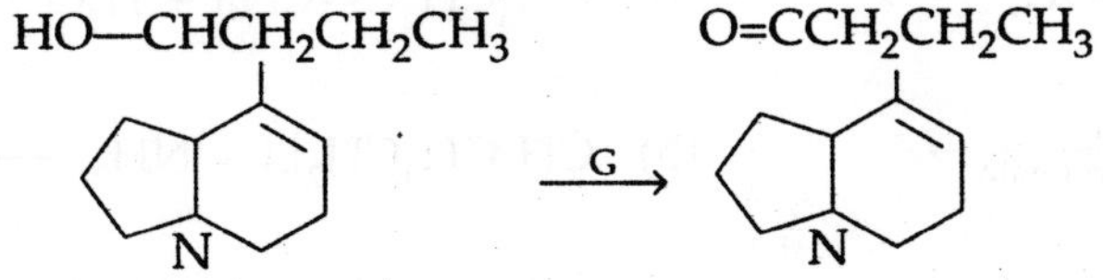

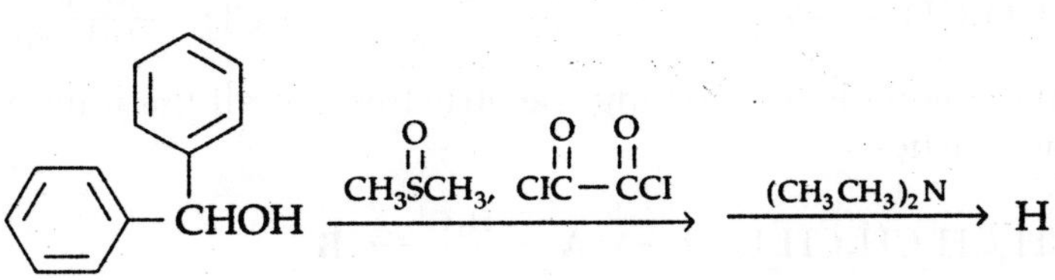

Q. 49. (I) Complete the following equations.

(a) $CH_3O-C_6H_4-CHO \xrightarrow[\text{methanol}]{NaBH_4} A$

(b) $CH_3(CH_2)_5CH(CH_3)CH_2CHO \xrightarrow[\text{diethyl ether}]{LiAlH_4} \xrightarrow{H_3O^+} B$

(c) $C_6H_5CH_2COCH_3 \xrightarrow[\text{diethyl ether}]{LiAlH_4} \xrightarrow{H_3O^+} C$

(d) $C_6H_5CHO \xrightarrow[\text{dimethyl ether}]{LiAlH_4} \xrightarrow{H_3O^+} D$

(e) $CH_3COCH_2CH(CH_3)CH_3 \xrightarrow[\text{diethyl ether}]{LiAlH_4} \xrightarrow{H_3O^+} E$

(f) $CH_3CH{=}CHCOCH_3 \xrightarrow[\text{CeCl}_3,\,7H_2O,\ \text{methanol}]{NaBH_4} F$

(g) (bicyclic enone-ketone with angular methyl) $\xrightarrow[\text{CeCl}_3,\,7H_2O,\ \text{methanol}]{NaBH_4\,(\text{excess})} G$

(h) cyclobutanone ($\square{=}O$) $\xrightarrow[\text{diethyl ether}]{LiAlH_4} \xrightarrow{H_3O^+} H$

Q. 49. (II) Complete the following equations.

(a) $CH_3CH_2CH_2CH_2Br \xrightarrow[\text{tetrahydrofuran}]{Li}$

(b) $C_6H_5CH_2CH_2Br \xrightarrow[\text{diethyl ether}]{Mg}$

(c) $CH_2{=}CHCl \xrightarrow[\text{tetrahydronium},\ \Delta]{Mg}$

(d) $C_6H_5-C{\equiv}CH \xrightarrow[\text{diethyl ether}]{CH_3CH_2MgBr}$

(e) $C_6H_5MgBr + CH_3COOH \longrightarrow$

(f) $CH_3CH_2C{\equiv}CNa + H_2O \longrightarrow$

(g) $C_6H_5CH_2Br \xrightarrow[\text{tetrahydrofuram}]{Li}$

(h) $CH_3CH_2CH_2Li + NH_3 \longrightarrow$

(i) $CH_3CH_2MgI + CH_3OH \longrightarrow$

(j) $C_6H_5C{\equiv}CH \xrightarrow[NH_3(\text{liq})]{NaNH_2}$

Q. 50. Complete the following equations, showing the structures of all the intermediates and final products in the reactions.

(a) cyclohexanone ($C_6H_{10}{=}O$) $+ CH_3CH_2CH_2CH_2Li \longrightarrow A \xrightarrow{H_2O^+} B$

(b) 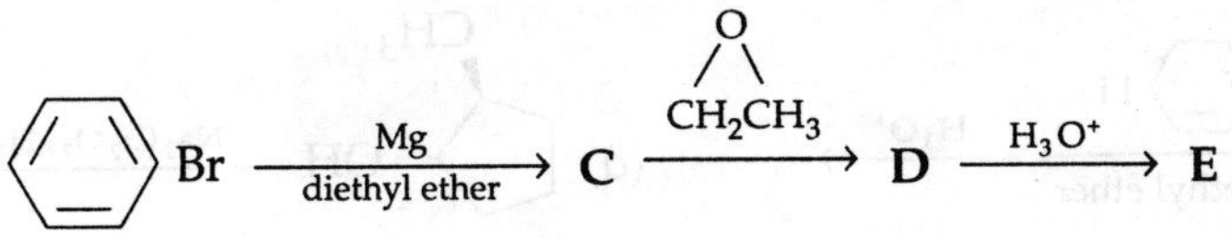 Bromobenzene $\xrightarrow[\text{diethyl ether}]{Mg}$ C $\xrightarrow{\text{epoxide } CH_2CH_3}$ D $\xrightarrow{H_3O^+}$ E

(c) 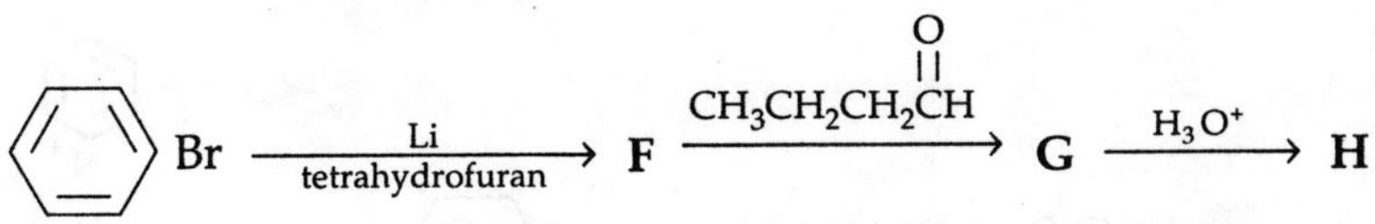 Bromobenzene $\xrightarrow[\text{tetrahydrofuran}]{Li}$ F $\xrightarrow{CH_3CH_2CH_2CHO}$ G $\xrightarrow{H_3O^+}$ H

(d) 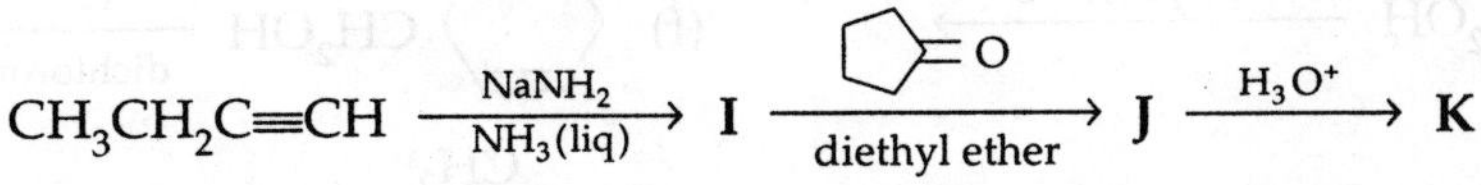 $CH_3CH_2C{\equiv}CH \xrightarrow[NH_3(liq)]{NaNH_2}$ I $\xrightarrow[\text{diethyl ether}]{\text{cyclopentanone}}$ J $\xrightarrow{H_3O^+}$ K

(e) 5,5-dimethylcyclohex-2-en-1-one (CH_3, CH_3) $\xrightarrow[\text{diethyl ether}]{CH_3Li}$ L $\xrightarrow{H_3O^+}$ M

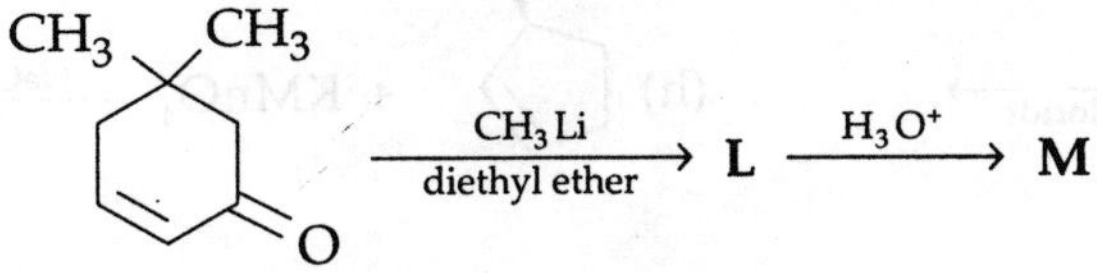

Q. 51. Complete the following reactions:

(a) 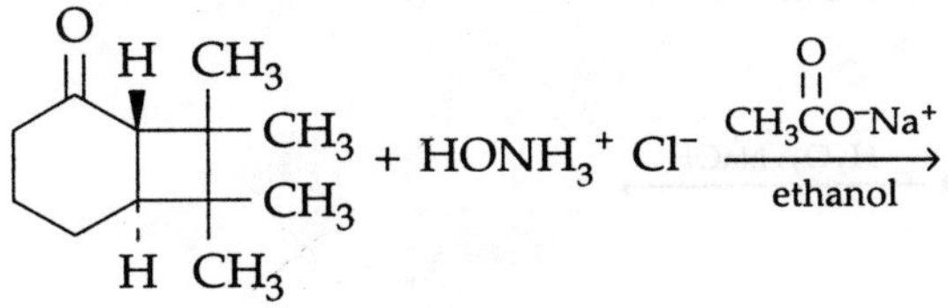 Bicyclic ketone (H, CH_3, CH_3, CH_3, H, CH_3) + $HONH_3^+\ Cl^-$ $\xrightarrow[\text{ethanol}]{CH_3CO^-Na^+ \text{ (O)}}$

(b) 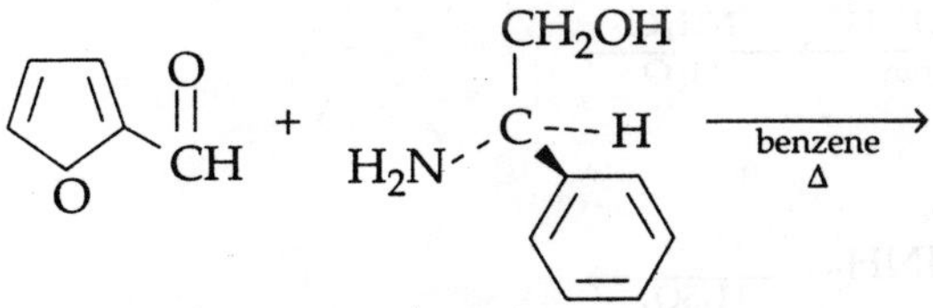 Furan-2-carbaldehyde ($\overset{O}{\|}$CH) + H_2N–C(–H)(CH_2OH)(phenyl) $\xrightarrow[\Delta]{\text{benzene}}$

(c) Cyclohexanone $\xrightarrow[\text{diethyl ether}]{CH_3CH_2MgBr}$ $\xrightarrow[H_2O]{NH_4Cl}$

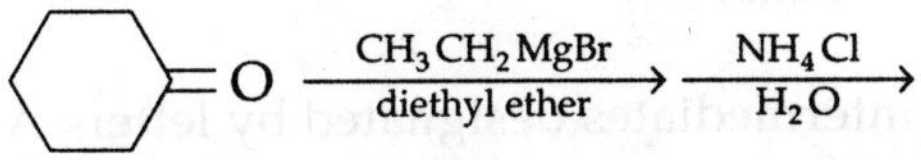

(d) 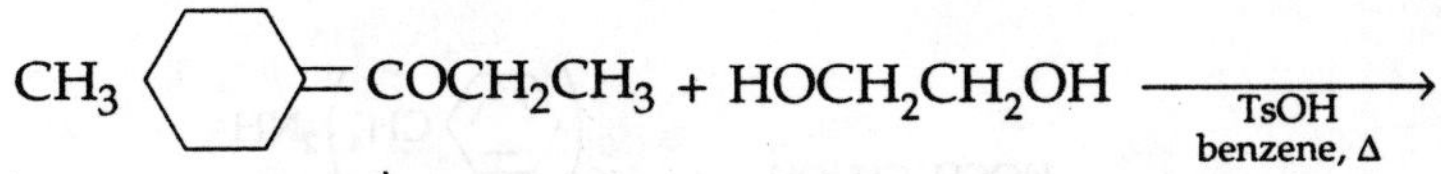 CH_3–cyclohexane=$COCH_2CH_3$ + $HOCH_2CH_2OH$ $\xrightarrow[\text{benzene, } \Delta]{TsOH}$

(e) 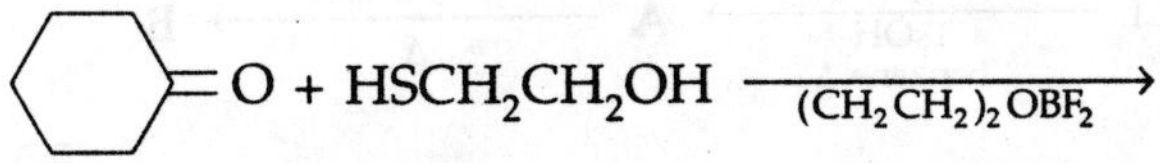 Cyclohexanone + $HSCH_2CH_2OH$ $\xrightarrow{(CH_3CH_2)_2OBF_3}$

Q. 52. (I) Complete each of the following equations, showing the structures of the products that will be formed.

(a) 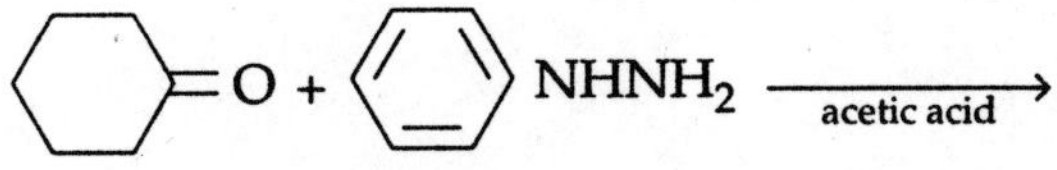 Cyclohexanone + $C_6H_5NHNH_2$ $\xrightarrow{\text{acetic acid}}$

(b) 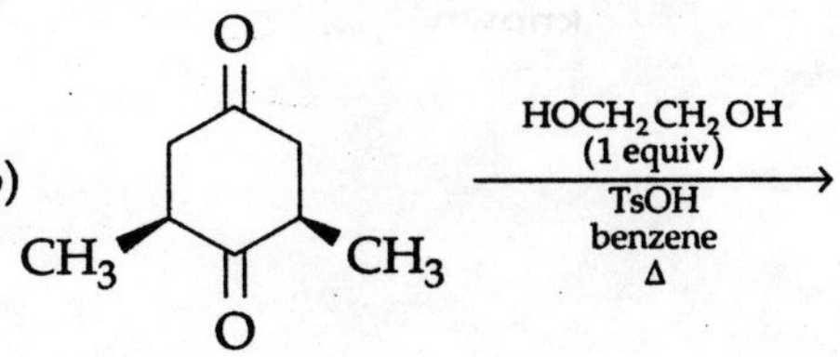 2,6-dimethylcyclohexane-1,4-dione (CH_3, CH_3) $\xrightarrow[\substack{\text{benzene} \\ \Delta}]{\substack{HOCH_2CH_2OH \\ (1\text{ equiv}) \\ TsOH}}$

(c) $CH_2{=}CH\overset{O}{\overset{||}{C}}H \xrightarrow[\text{diethyl ether}]{C_6H_5Li} \xrightarrow{H_3O^+}$

(d) (cyclopentanol with CH_3) $\xrightarrow{Na_2Cr_2O_7,\ H_3O^+}$

(e) (cyclohexyl)$CH_2OH \xrightarrow{Na_2Cr_2O_7,\ H_3O^+}$

(f) (cyclohexenyl)$CH_2OH \xrightarrow[\text{dichloromethane}]{C_5H_5N^+H\ CrO_3Cl^-}$

(g) (3-methylcyclopentene, CH_3) $+ Br_2 \xrightarrow{\text{carbon tetrachloride}}$

(h) (3-methylcyclopentene, CH_3) $+ KMnO_4 \xrightarrow[H_2O]{NaOH}$

(i) (3-methylcyclopentene, CH_3) + (3-chlorophenyl)$COOH \xrightarrow{\text{dichloromethane}}$

(j) (cyclopentyl)$CH_3 \xrightarrow[\text{tetrahydrofuran}]{BH_3} \xrightarrow{H_2O_2,\ NaOH}$

(k) $C_6H_5\overset{O}{\overset{||}{C}}CH_3 \xrightarrow[\text{tetrahydrofuran}]{CH_3CH_2CH_2CH_2Li} \xrightarrow[H_2O]{NH_4Cl}$

(l) (cyclopentyl)$\overset{O}{\overset{||}{C}}H + O_2N$(C$_6H_3$, NO_2)$NHNH_2 \xrightarrow[\text{ethanol}]{H_2SO_4}$

Q. 52. (II) Predict the structuresof products or intermediates designated by letters A to R, in the following :

(a) $C_6H_5COCH_2CH_2CH_2Cl \xrightarrow[\substack{TsOH \\ \text{benzene } \Delta}]{HOCH_2CH_2OH} A \xrightarrow[\Delta]{(C_6H_5CH_4)_2NH} B$

(b) $CH_3CH_2CH_2COCH_3 \xrightarrow[H_2O]{NaBH_4} C \xrightarrow[\text{pyridine}]{TsCl} D \xrightarrow[\text{ethanol, } \Delta]{KOH\ (4M)} E \xrightarrow[\text{dichloromethane}]{C_6H_5CHOOH} F$

(c) $CH_2{=}CHCH_2CH_2$–C($HOCH_2$)(OH)–CH_2CH_3 $\xrightarrow[\text{TsOH}]{CH_3COCH_3,\ CH_3C(OCH_3)_2CH_3}$ G $\xrightarrow[\text{dichloromethane}]{O_3}$ $\xrightarrow{CH_3SCH_3}$ H + I

(d) $\overset{O}{\overset{\|}{CH}}$ $\xrightarrow[\text{tetrahydrofuran}]{\text{LiH}}$ J $\xrightarrow{H_3O^+}$ K $\xrightarrow[\text{acetone}]{CrO_3,\ H_3O^+}$ L $\xrightarrow[\text{acetic acid}]{NH_2NH_2}$ M

(e) $CH_3\overset{CH_3}{\overset{|}{C}}{=}CHCH_2$ $\xrightarrow[\text{tetrahydrofuran}]{BH_3}$ N $\xrightarrow{H_2O_2,\ NaOH}$ O $\xrightarrow[H_2O]{Na_2Cr_2O,\ H_2SO_4}$ P $\xrightarrow{NaCN}$ Q $\xrightarrow{H_2O^+}$ R

(f) $HC{\equiv}CH$ $\xrightarrow[NH_3\,(liq)]{NaNH_2}$ S $\xrightarrow[\text{diethyl ether}]{CH_2Br}$ T $\xrightarrow[\text{diethyl ether}]{CH_3CH_2MgBr}$ U $\xrightarrow{O}$ V $\xrightarrow[H_2O]{NH_3Cl}$ W

(g) O $\xrightarrow{H_2NNH_2}$ X $\xrightarrow[\text{diethylene glycol, } \Delta]{KOH}$ Y

(h) OH $\xrightarrow{CH_3\overset{O}{\overset{\|}{S}}CH_3,\ Cl\overset{O}{\overset{\|}{C}}-\overset{O}{\overset{\|}{C}}Cl}$ $\xrightarrow{(CH_3CH_2)_3N}$ Z

Q. 53. For each products or intermediate designated by a letter or letters in the following equations give the structure.

(a) $HC{\equiv}CCH_2CH_2OH$ $\xrightarrow[\text{TsOH, } \Delta]{O}$ A $\xrightarrow{CH_3CH_2MgBr}$ B $\xrightarrow{CH_3(CH_2)_5\overset{O}{\overset{\|}{C}}H}$ C $\xrightarrow{H_3O^+}$ D $\xrightarrow[\text{Pd/CaCO}_3\text{, quinoline, ethyl acetate}]{H_2}$ E

(b) Br $\xrightarrow[\text{diethyl ether}]{Mg}$ F $\xrightarrow{O}$ G $\xrightarrow{H_3O^+}$ H $\xrightarrow[\text{dichloromethane}]{N^+CrO_3Cl,\ H}$ I

(c) $CH_3\overset{CH_3}{\overset{|}{C}}CHCH_2CH_2COCH_3 \xrightarrow[\text{TsOH, benzene, }\Delta]{HOCH_2CH_2OH} J \xrightarrow[\text{tetrahydofura}]{BH_3} K \xrightarrow{H_2O_2,\ NaOH} L \xrightarrow{H_3O^+} M$

(d) cyclopentanone $\xrightarrow[\text{tetrahydrofuran}]{C_6H_5Li} N \xrightarrow[H_2O]{NH_4Cl} O \xrightarrow[\text{cold}]{HCl\ (conc)} P \xrightarrow[\text{ethanol, }\Delta]{KOH(4\ M)} Q$

(e) (bicyclic enone bearing CH_3 and OTBDMS) $\xrightarrow[\text{methanol}]{NaBH_4,\ CeCl_3 - 7H_2O} R \xrightarrow[RhCl[P(C_6H_5)_3]_3]{H_2} S$

(f) $CH_3CH_2\overset{OTHP}{\overset{|}{C}}{\equiv}CCH_2CH_3CH_2CH_2OTBDMS \xrightarrow{(CH_3CH_2CH_2CH_2)_4N^+F^-} T \xrightarrow{CH_3\overset{O}{\overset{||}{S}}CH_3,\ Cl\overset{O}{\overset{||}{C}}-\overset{O}{\overset{||}{C}}Cl} \xrightarrow{(CH_3CH_2)_3N} U$

(g) cyclohex-2-enone $\xrightarrow[\text{diethyl ether}]{C_6H_5Li} V \xrightarrow{H_3O^+} W$

(h) $CH_2{=}CHCH_2CH_2$–(cyclohexanone) $\xrightarrow[\text{chloroform}]{Br_2} Y \xrightarrow[\text{TsOH, benzene, }\Delta]{HOCH_2CH_2OH} Z$

(i) $CH_3\overset{CH_3}{\overset{|}{C}}HCH_2CH_2COCH_2CHCOOH \xrightarrow[\text{diethylene glycol, }\Delta\Delta]{H_2,NNH_2,\ NaOH} AA \xrightarrow{H_3O^+} BB$

(j) $CH_3CH_2CH_2CH_2\overset{O}{\overset{||}{C}}C_6H_5 \xrightarrow[\Delta]{HCl,\ Zn(Hg)} CC$

(k) (hexahydroindane bearing $=CH_2$, CH_3CHCH_3 and CHO) $\xrightarrow[\text{diethyl ether}]{CH_3Li} DD \xrightarrow{H_3O^+} EE$

Q. 54. Complete the following equations

(a) $CH_3\overset{CH_3}{\underset{CH_3}{C}}OH + CH_3\overset{O}{\overset{||}{C}}\overset{O}{\overset{||}{C}}OCH_3 \longrightarrow$

(b) $CH_3CH_2CH_2CH_2OH + CH_3CH_2CH_2\overset{O}{\overset{||}{C}}Cl \longrightarrow$

(c) $C_6H_5\overset{O}{\overset{||}{C}}Cl$ + cyclohexyl$-OH \longrightarrow$

(d) maleic anhydride (cyclic, C=C ring with two C=O and ring O) $+ CH_3CH_2OH \longrightarrow$

1 equivalent   1 equivalent

Q. 55. Complete the following equations

(a) $C_6H_5CH_2CH_2COOH \xrightarrow[\Delta]{SOCl_2}$

(b) $CH_3CH_2CH_2COOH \xrightarrow{PCl_3}$

(c) $CH_3CH_2CH_2COCl + Na^+ \ {}^-OCOCH_2CH_2CH_3 \longrightarrow$

(d) $CH_3(CH_2)_{10}COOH \xrightarrow[\Delta]{SOCl_3}$

(e) $HOOCH_2CH_2COOH \xrightarrow{CH_3\overset{O}{\overset{||}{C}}O\overset{O}{\overset{||}{C}}CH_3}$

(f) $CH_3CH_2COOH \xrightarrow[\substack{\Delta \\ 923\ K}]{clay}$

(g) $HO\overset{O}{\overset{||}{C}}(CH_2)_3\overset{O}{\overset{||}{C}}OH \xrightarrow[\Delta]{\text{excess } SOCl_2}$

Q. 56. Complete the following equations by writing structures for the expected products.

(a) $CH_3\underset{}{\overset{CH_3}{\overset{|}{C}}}HCH_2CH_2\overset{O}{\overset{||}{C}}H + HONH_3Cl^- \xrightarrow{CH_3\overset{O}{\overset{||}{C}}O^-Na^+}$

(b) $CH_3\overset{O}{\overset{||}{C}}CH_2CH_2CH_3 + C_6H_5NHNH_2 \xrightarrow[\text{acetic acid}]{}$

(c) cyclopentanone $(C_5H_8{=}O) + C_6H_5NH_2 \xrightarrow[\Delta]{}$

(d) $C_6H_5CHO + O_2N-C_6H_3(NO_2)-NHNH_2 \xrightarrow[H_2SO_4]{}$

(e) 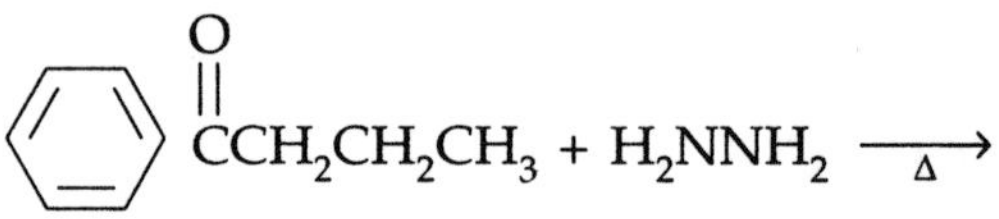 $C_6H_5-\overset{O}{\overset{||}{C}}CH_2CH_2CH_3 + H_2NNH_2 \xrightarrow{\Delta}$

(f) $CH_3\overset{O}{\overset{||}{C}}CH_2CH_3 + CH_3\overset{CH_3}{\overset{|}{C}}HCH_2NH_2 \xrightarrow{HCl} \xrightarrow{NaOH}$

(g) 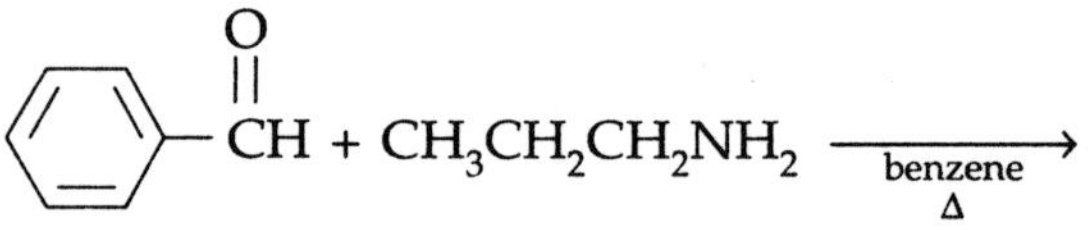 $C_6H_5-\overset{O}{\overset{||}{C}}H + CH_3CH_2CH_2NH_2 \xrightarrow[\Delta]{benzene}$

Q. 57. Predict what the products of the following reactions will be.

(a) 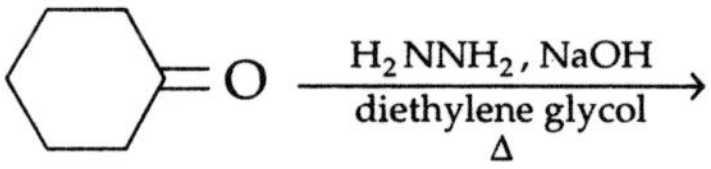 Cyclohexanone $=O \xrightarrow[\text{diethylene glycol, } \Delta]{H_2NNH_2,\ NaOH}$

(b) $C_6H_5-\overset{O}{\overset{||}{C}}CH_3 \xrightarrow[\Delta]{HCl,\ Zn(Hg)}$

(c) $HO-C_6H_4-\overset{O}{\overset{||}{C}}H \xrightarrow[\Delta]{HCl,\ Zn(Hg)}$

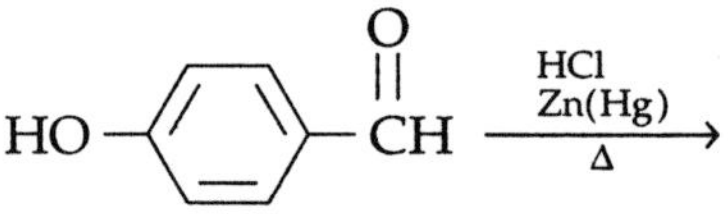

(d) 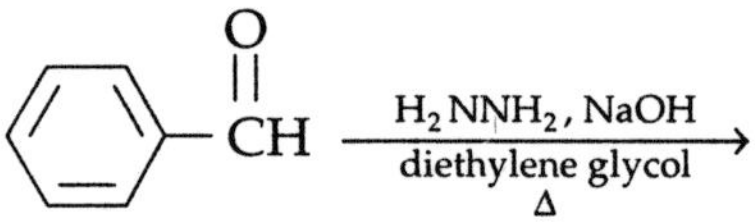 $C_6H_5-\overset{O}{\overset{||}{C}}H \xrightarrow[\text{diethylene glycol, } \Delta]{H_2NNH_2,\ NaOH}$

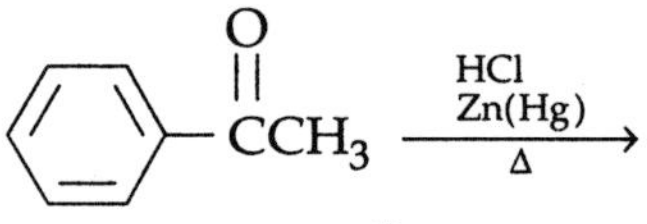

(e) 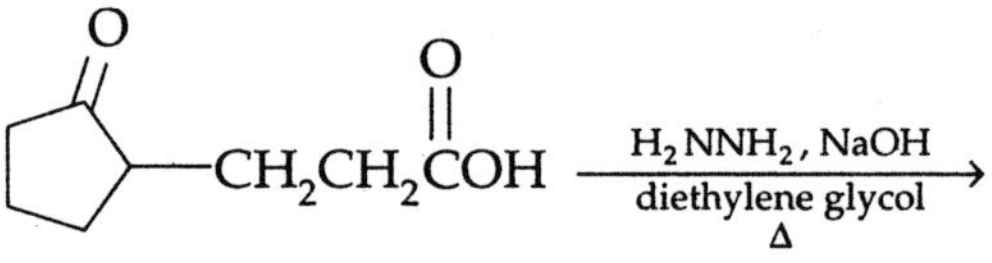 2-oxocyclopentyl$-CH_2CH_2\overset{O}{\overset{||}{C}}OH \xrightarrow[\text{diethylene glycol, } \Delta]{H_2NNH_2,\ NaOH}$

Q. 58. Provide reagent for the following reactions :

(a) 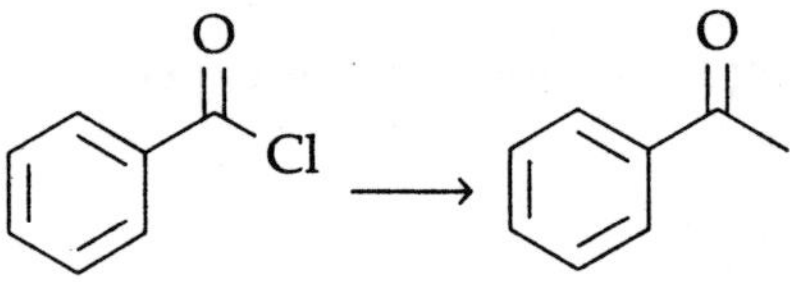 $C_6H_5-\overset{O}{\overset{||}{C}}-Cl \longrightarrow C_6H_5-\overset{O}{\overset{||}{C}}-CH_3$

(b) $C_6H_5NH(CH_3) \longrightarrow C_6H_5-\underset{CH_3}{\underset{|}{N}}-CH_2CH_3$

(c) $C_6H_5NH_2 \longrightarrow C_6H_5CH_2NHCH_2NH_3$

(d) 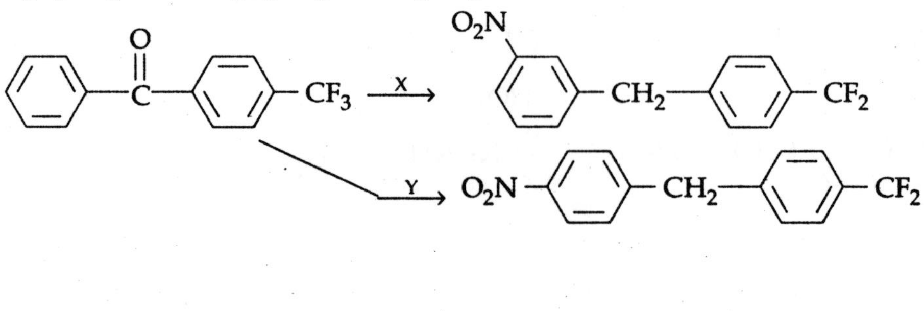 $C_6H_5-\overset{O}{\overset{||}{C}}-C_6H_4-CF_3 \xrightarrow{X}$ O_2N-C_6H_4(meta)$-CH_2-C_6H_4-CF_2$

$\xrightarrow{Y} O_2N-C_6H_4-CH_2-C_6H_4-CF_2$

Q. 59. Bring about the following transformations:

(a) 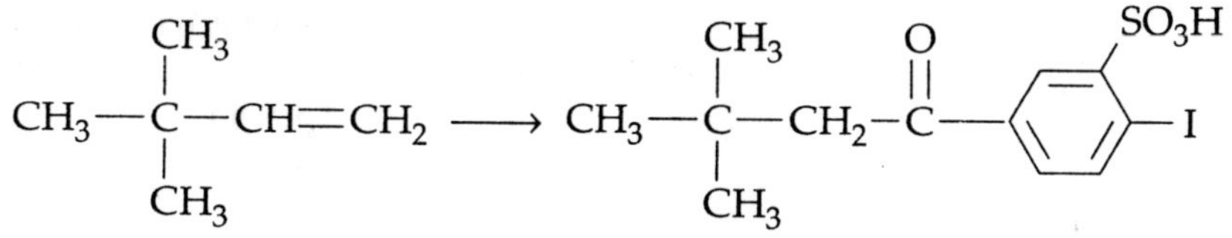 $CH_3-C(CH_3)_2-CH=CH_2 \longrightarrow CH_3-C(CH_3)_2-CH_2-C(=O)-C_6H_3(SO_3H)(I)$

(b) $PhCH_2CH_3 \longrightarrow Ph-CH(Br)-CH_2OH$

(c) $CH_3CH_2CH_2CH_2Br \longrightarrow CH_3CH_2-C(OH)(CH_3)-CH_2CH_3$

(d) 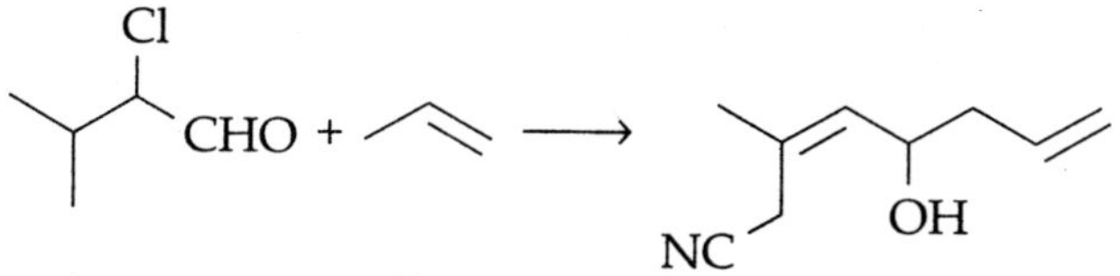 (2-chloro-3-methylbutanal) CHO + (propene) $\longrightarrow$ NC, OH (product structure)

(e) 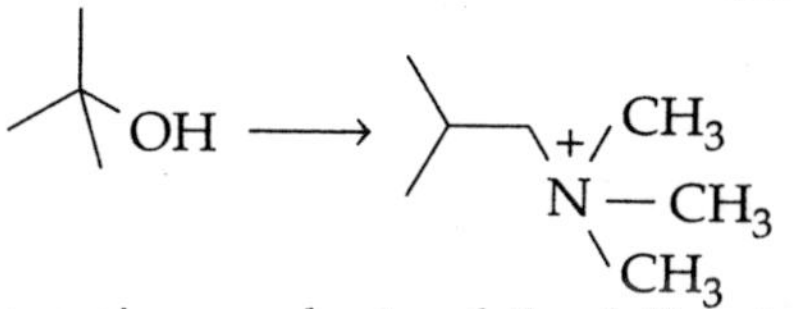 (tert-butyl alcohol) OH $\longrightarrow$ (isobutyl) $\overset{+}{N}(CH_3)_3$

Q. 60. Give major products of the following reactions :

(a) 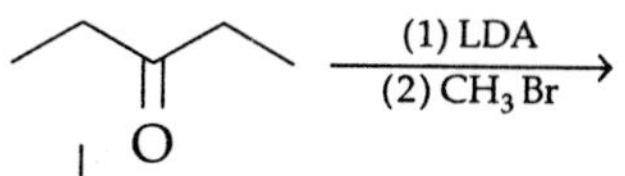 (3-pentanone) O $\xrightarrow[(2)\,CH_3Br]{(1)\,LDA}$

(b) 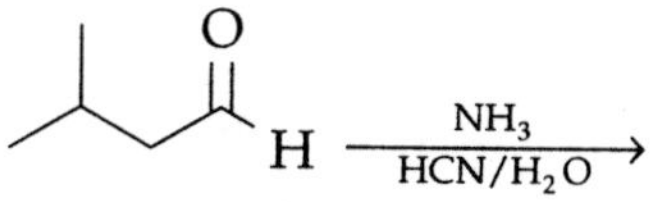 (3-methylbutanal) O, H $\xrightarrow[HCN/H_2O]{NH_3}$

(c) 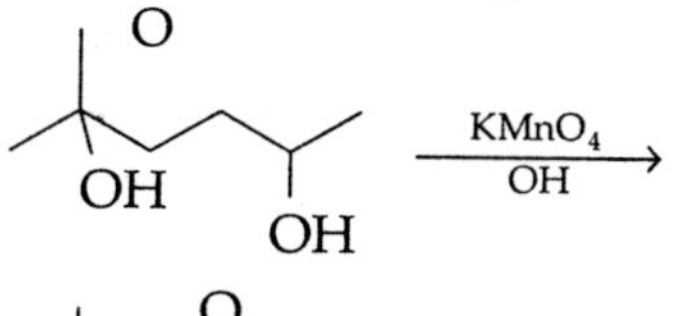 (2-methylhexane-2,5-diol) OH, OH $\xrightarrow[OH]{KMnO_4}$

(d) $2PhCH_2CHO \xrightarrow[warm]{HO^-}$

(e) 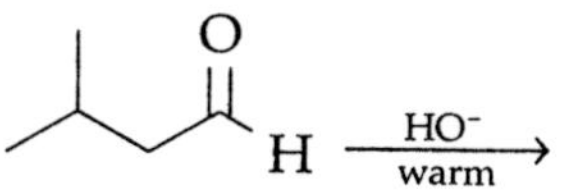 (3-methylbutanal) O, H $\xrightarrow[warm]{HO^-}$

(f) (phenylacetone) O $\xrightarrow[H_2O]{KCN}$

(g) (benzoic acid) COOH $\xrightarrow[excess]{CH_3Li} \xrightarrow[H^+]{H_2O}$

(h) $C_6H_5-COOH \xrightarrow{SOCl_2} \xrightarrow{CH_3NH_2}$

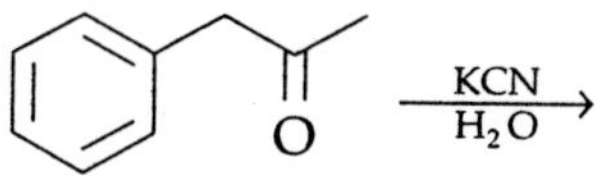

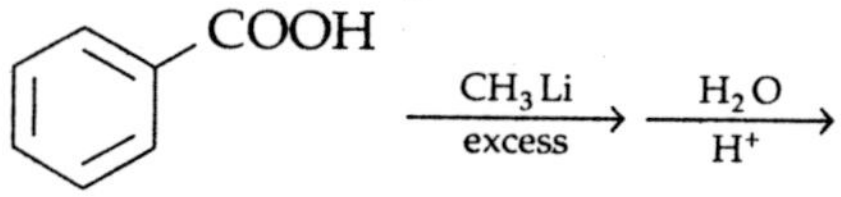

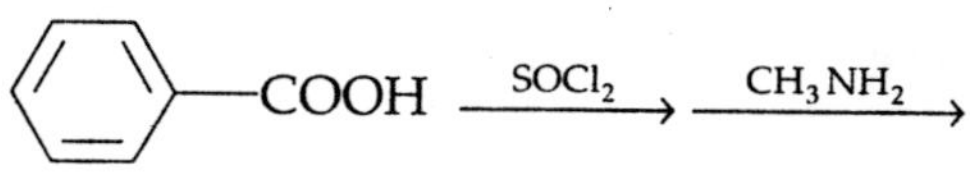

(i) 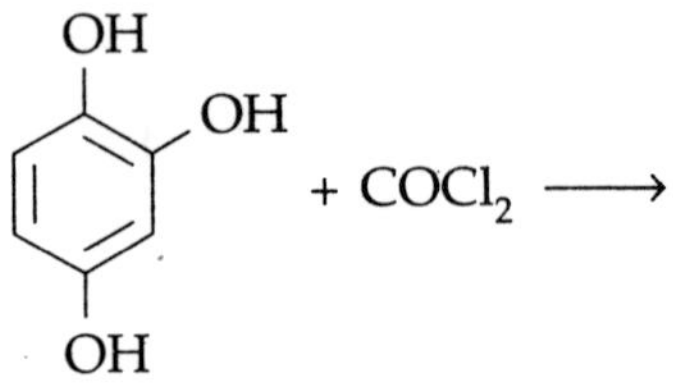 (benzene ring with OH, OH, OH) + $COCl_2$ $\longrightarrow$

(j) 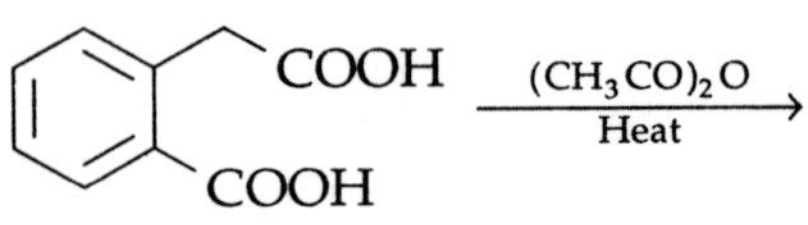 (benzene ring with CH_2COOH and COOH) $\xrightarrow[\text{Heat}]{(CH_3CO)_2O}$

(k) 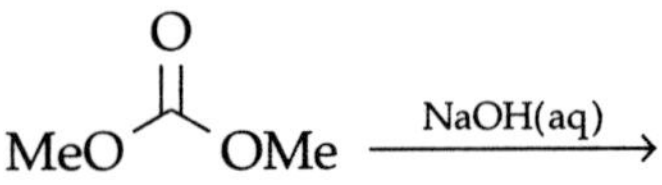 MeO–C(=O)–OMe $\xrightarrow{\text{NaOH(aq)}}$

(l) 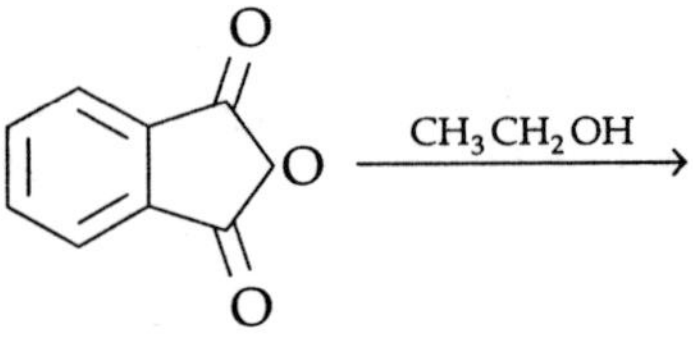 (phthalic anhydride: O, O, O) $\xrightarrow{CH_3CH_2OH}$

(m) $PhCH_2COOEt \xrightarrow[\text{PhCOOEt}]{\text{EtONa}}$

(n) $EtO_2CCH_2COOEt \xrightarrow[CH_3CH_2I]{\text{EtONa}}$

(o) 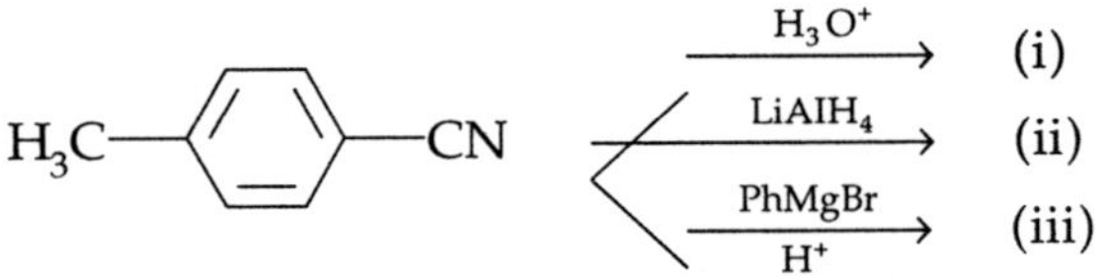 H_3C–C_6H_4–CN
- $\xrightarrow{H_3O^+}$ (i)
- $\xrightarrow{LiAlH_4}$ (ii)
- $\xrightarrow[H^+]{PhMgBr}$ (iii)

Q. 61. Bring about the following transformations:

(a) $CH_3(CH_2)_4CH_2CH_2Br \longrightarrow CH_3(CH_2)_4CH(OH)–CH_2CN$

(b) $PhCH_2CH_2CH_2Br \longrightarrow$ (H_3C)(H)C=C(H)–C(OH)(CH_3)–CH_2Ph

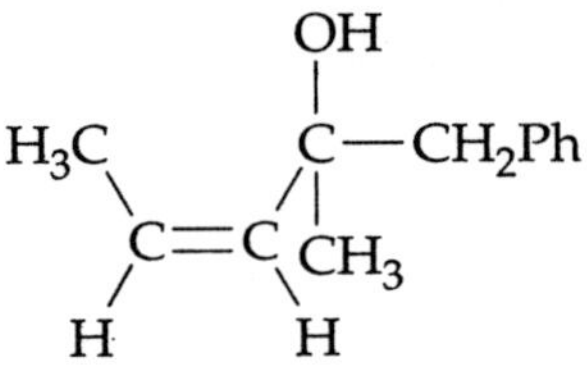

(c) (butan-1-ol) OH $\longrightarrow CH_3CH_2–C(=O)–CH_2CN$

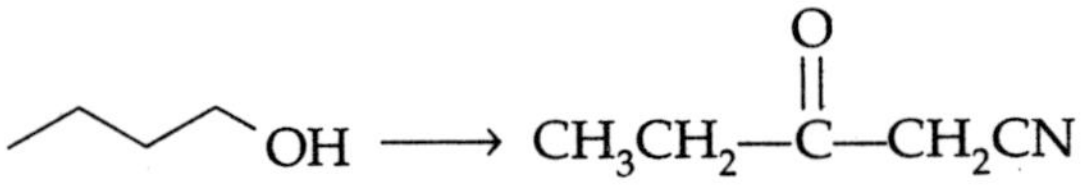

(d) 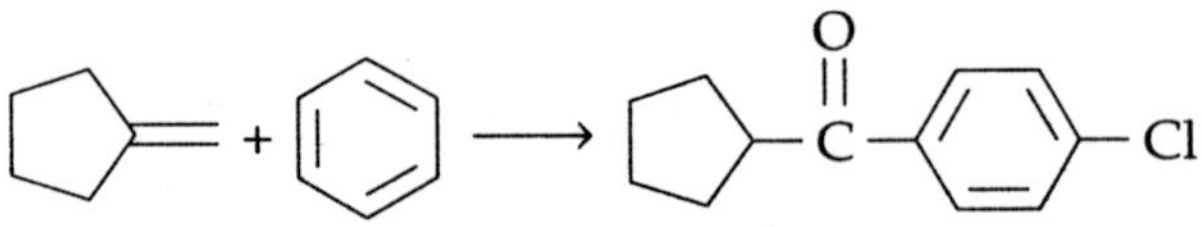 (methylenecyclopentane) + (benzene) $\longrightarrow$ (cyclopentyl)–C(=O)–C_6H_4–Cl

(e) 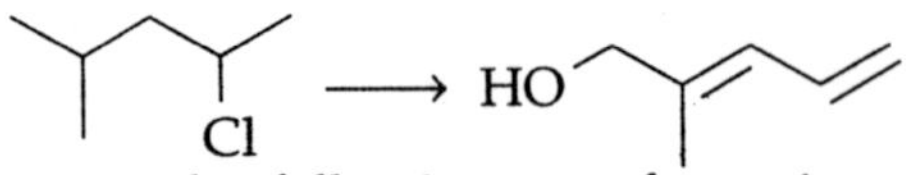 (chloro compound) Cl $\longrightarrow$ HO (dienol)

Q. 62. Carry out the following transformations:

(a) 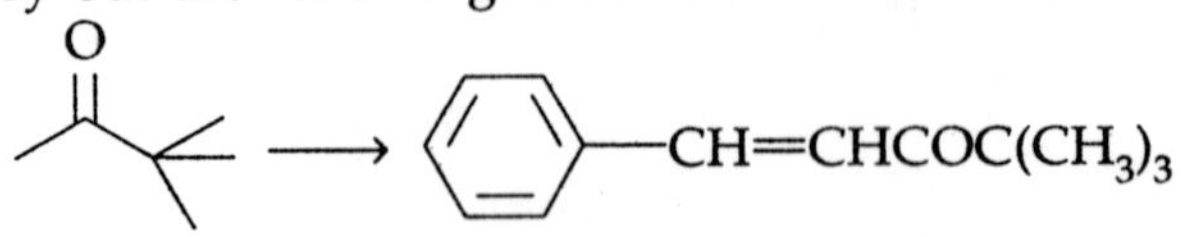 (pinacolone) O $\longrightarrow$ C_6H_5–CH=CHCOC$(CH_3)_3$

(b) $C_6H_5CHO \longrightarrow C_6H_5CH=$(2-cyclopentanone, ylidene)

(c) $C_6H_5CHO \longrightarrow C_6H_5CH_2CH(NH_2)CH_3$

(d) 2,6-octanedione $\longrightarrow$ 1-(2-ethylcyclopent-1-enyl)ethanol (CH(OH)CH₃ substituent)

Q. 63. Provide the reagents that will effect the following transformations:

(a) 1-bromo-cyclohexanol (HO, Br) $\longrightarrow$ (CH_3O, CH_2N_3) cyclohexane

(b) $C_6H_5CH_2Br \longrightarrow C_6H_5COOCH_3$

(c) $C_6H_5C(CH_3)_2CH_2OH \longrightarrow C_6H_5C(CH_3)_2CN$

(d) 3-methylcyclohexanol (OH, CH_3) $\xrightarrow{\text{(i)}}$ 3-methylcyclohexyl chloride (Cl dashed, CH_3 wedged)

$\xrightarrow{\text{(ii)}}$ 3-methylcyclohexyl chloride (Cl wedged, CH_3 wedged)

(e) 2,6-bis(hydroxymethyl)cyclohexanone ($HOCH_2$, CH_2OH) $\xrightarrow{H^+}$

Q. 64. Bring about the following transformations:

(a) nitrobenzene ⟶ 4-nitrobenzoic acid (HOOC, NO_2)

(b) benzaldehyde (CHO) ⟶ 4-chlorobenzaldehyde (OHC, Cl)

(c) 1,3-dinitrobenzene (NO_2, NO_2) ⟶ 3-chloronitrobenzene (Cl, NO_2)

(d) 2-chloronitrobenzene (NO_2, Cl) ⟶ 2-cyanobenzoic acid (CN, COOH)

(e) 4-methylaniline (NH_2, CH_3) ⟶ 2-bromo-4-methylaniline (NH_2, Br, CH_3)

(f) 4-amino-3-bromotoluene (CH_3, Br, NH_2) ⟶ 3-bromotoluene (CH_3, Br)

(g) 2-nitrobenzaldehyde (NO_2, CHO) ⟶ 2-(aminomethyl)benzaldehyde (CH_2NH_2, CHO)

(h) toluene (CH_3) ⟶ 3,4-dinitrotoluene (CH_3, NO_2, NO_2)

(i) aniline (NH_2) ⟶ 1,3,5-tribromobenzene (Br, Br, Br)

Q. 65. Bring about the following transformations:

(a) 4-chlorotoluene (CH_3, Cl) ⟶ 4-methylbenzyl alcohol (CH_3, CH_2OH)

(b) 2-bromoanisole (OMe, Br) ⟶ 3-methoxyaniline (OMe, NH_2)

(c) phenol (OH) ⟶ 3-methylphenol (CH_3, OH)

(d) benzene ⟶ 2-nitroaniline (NH_2, NO_2)

(e) 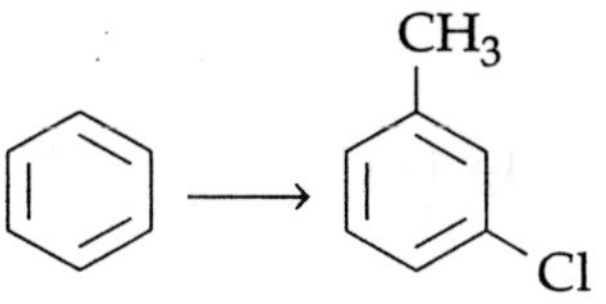 ⟶ CH_3 Cl

(f) 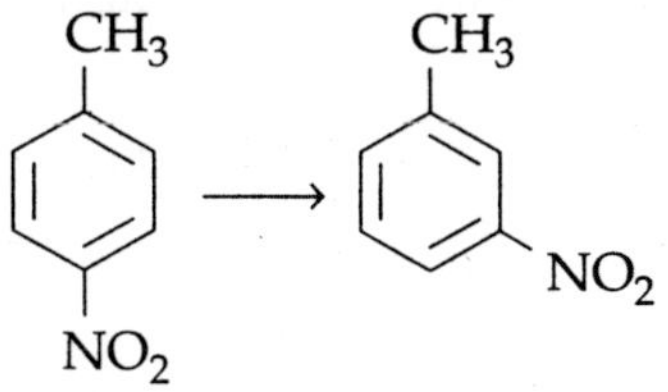 CH_3 NO_2 ⟶ CH_3 NO_2

Q. 66. Bring about the following transformations:

(a) 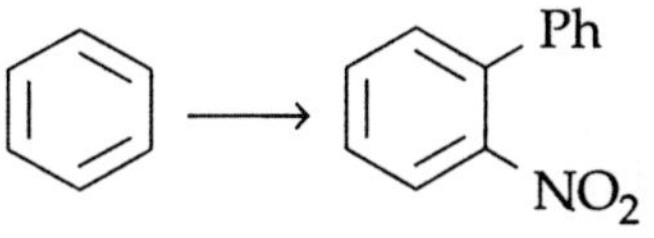 ⟶ Ph NO_2

(b) 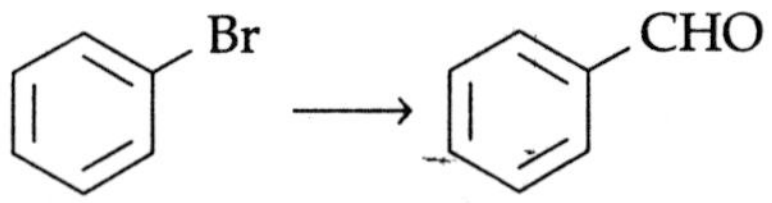 Br ⟶ CHO

(c) 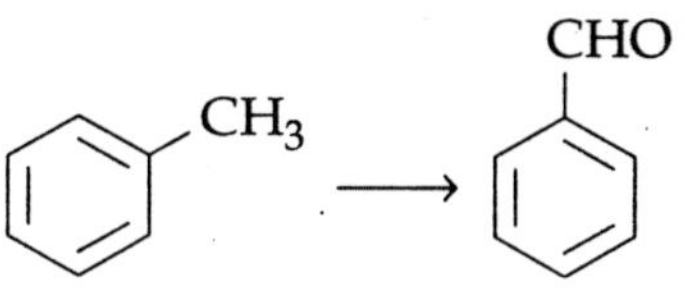 CH_3 ⟶ CHO

(d) 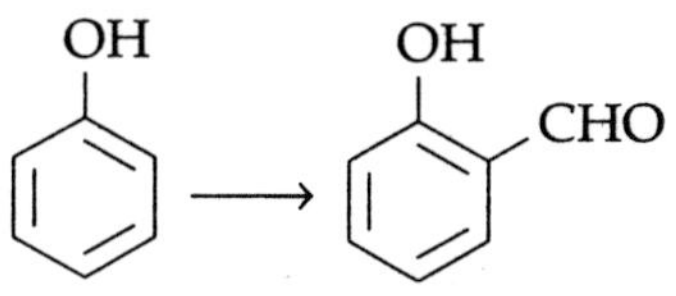 OH ⟶ OH CHO

(e) 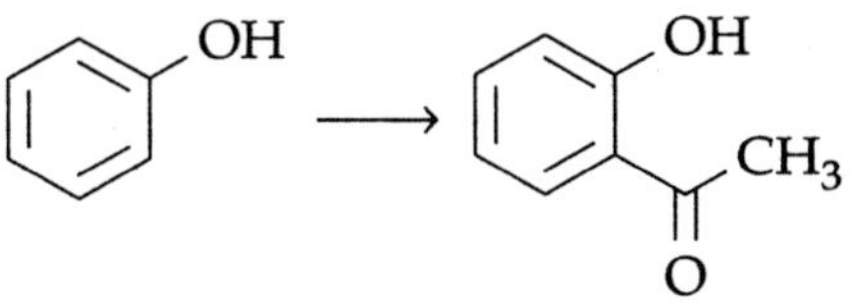 OH ⟶ OH CH_3 O

(f) 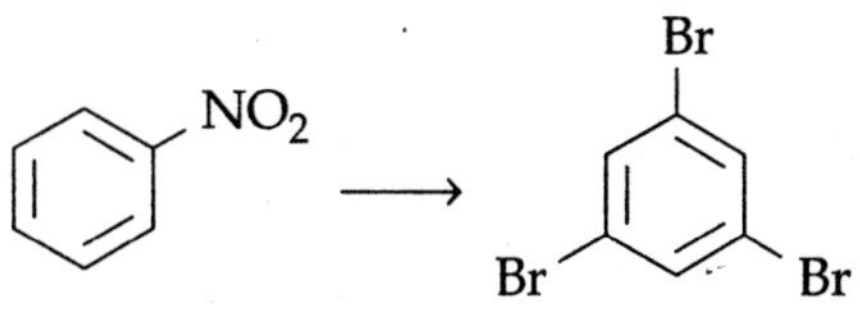 NO_2 ⟶ Br Br Br

Q. 67. Bring about the following transformations:

(a) 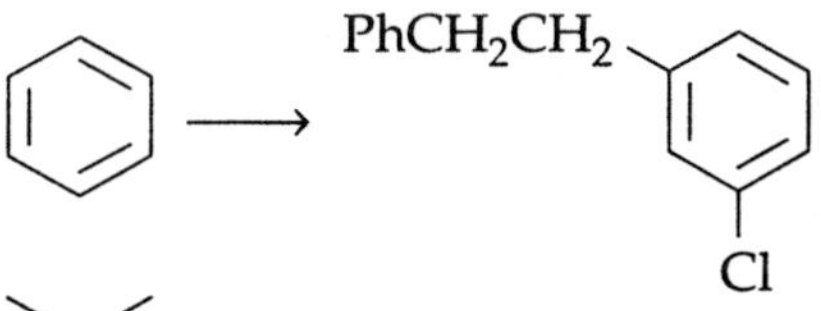 ⟶ $PhCH_2CH_2$ Cl

(b) 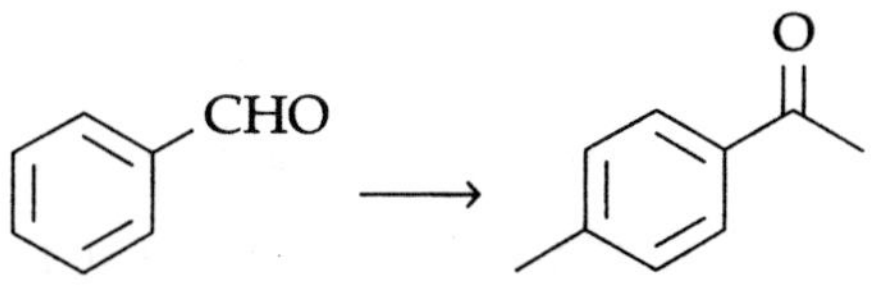 CHO ⟶ O

(c) 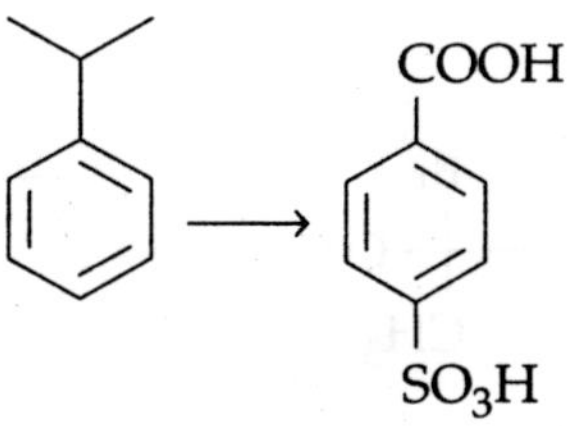 ⟶ COOH SO_3H

(d) 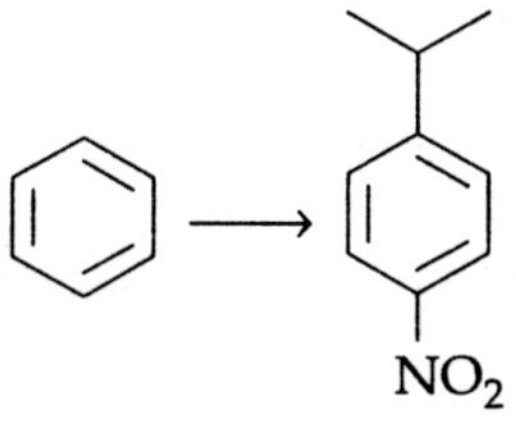 ⟶ NO_2

(e) 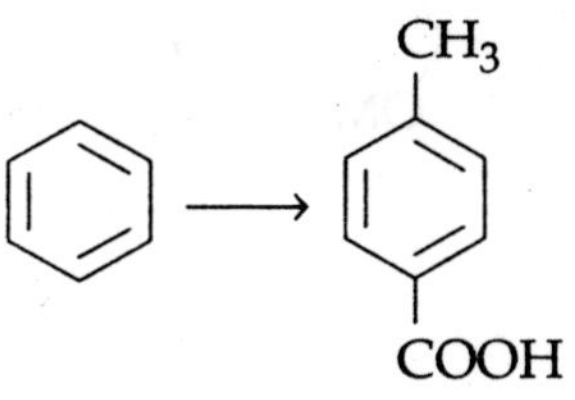 ⟶ CH_3 COOH

(f) 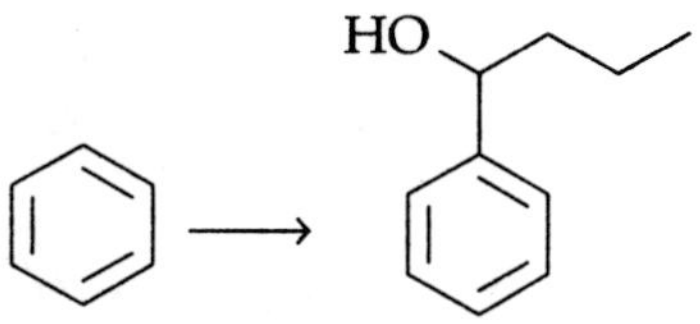 ⟶ HO

Q. 68. Bring about the following transformations:

(a) 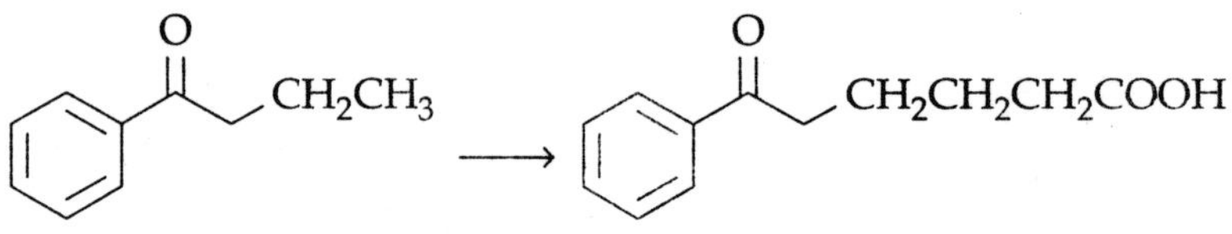 O, CH_2CH_3 ⟶ O, $CH_2CH_2CH_2COOH$

(b) 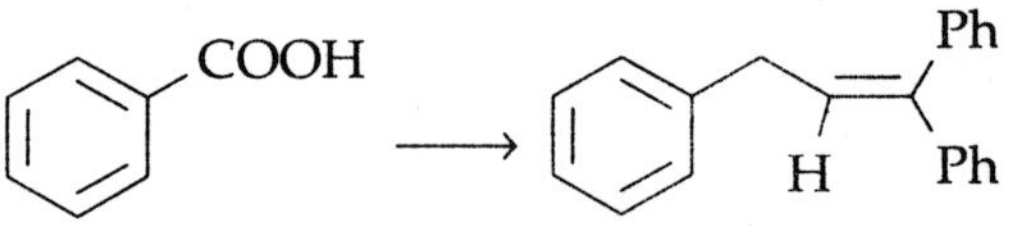 COOH ⟶ Ph, H, Ph

(c) $H_2C(COOCH_3)_2$ ⟶ Ph, $COOCH_3$, $COOCH_3$

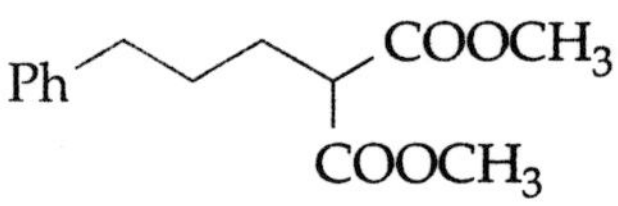

Q. 69. Convert:

(a) $CH_3CH_2C{\equiv}CH$ ⟶ H, H, CH_2CH_2OH

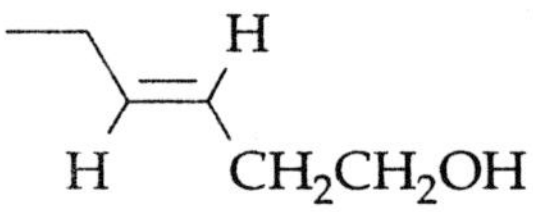

(b) 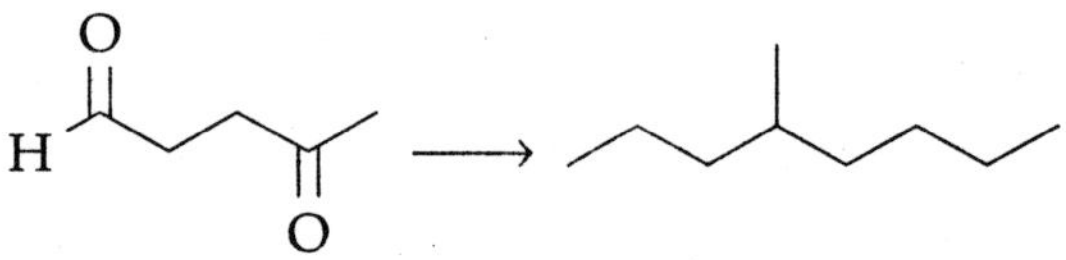 O, H, O ⟶

(c) CH_3COOH ⟶ EtOOC COOEt

(d) 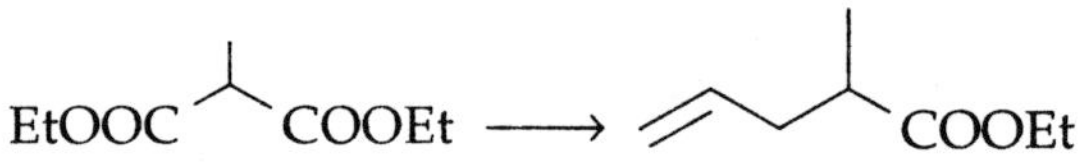 EtOOC COOEt ⟶ COOEt

(e) PhCOOH ⟶ $PhC(CH_3)_2OH$

(f) $PhCH_2OH$ ⟶ $PhCH_2CN$

Q. 70. Convert:

(a) 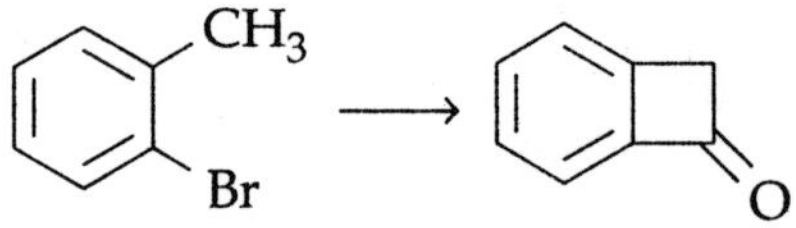 CH_3, Br ⟶ O

(b) 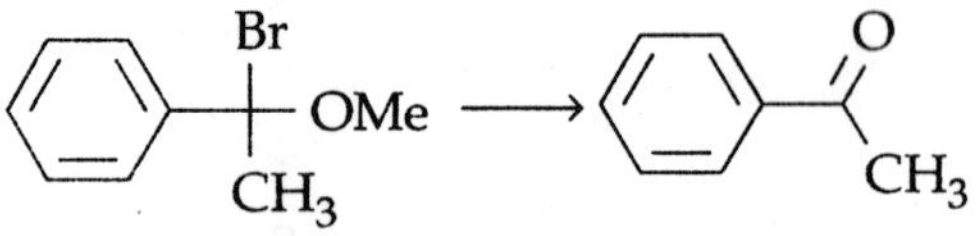 Br, OMe, CH_3 ⟶ O, CH_3

(c) 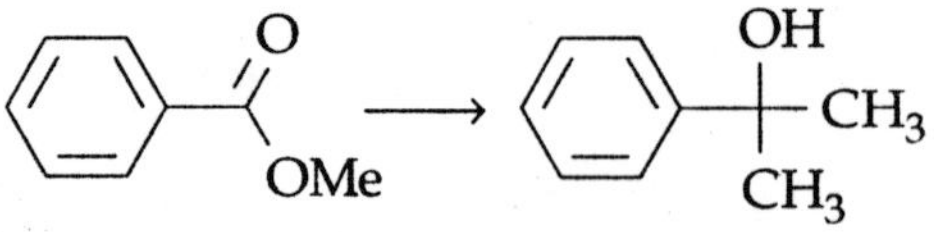 O, OMe ⟶ OH, CH_3, CH_3

(d) 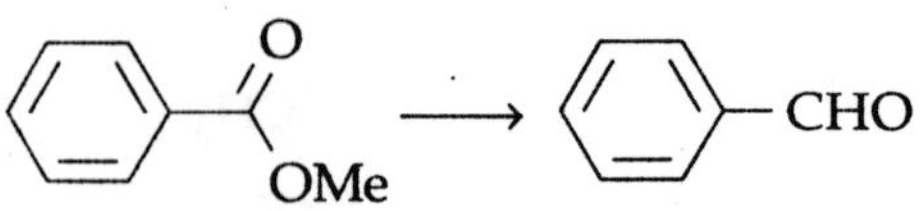 O, OMe ⟶ CHO

(e) 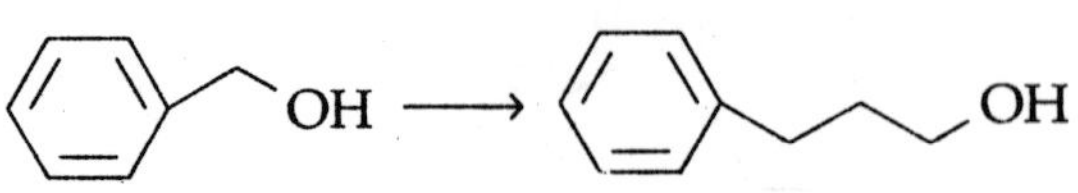 OH ⟶ OH

Q. 71. (a) Rank the following compounds in increasing order of the electrophilicity:

$F_3C-CH=O$ (I), $CH_3-C(=CH_2)-H$ (II), $CH_3-C(=NH)-CH_3$ (III)

(b) Arrange the following hydrates in order of increasing K_{eq}:

$$R-C(=O)-A \underset{}{\overset{H_2O}{\rightleftharpoons}} R-C(OH)_2-A$$

(c) Arrange the following in increasing order of reactivity towards nucleophile:

(i) $H_3C-C(=O)-NH_2$(I), $CH_3-C(=O)-Cl$(II), $CH_3-C(=O)-OAc$(III), $H_3C-C(=O)-OCH_3$(IV)

(ii) $CH_3-C(=O)-Cl$(I), $CH_3CH_2-C(=O)-Cl$(II), $(CH_3)_2CH-C(=O)-Cl$(III),

$(CH_3)_3C-C(=O)-Cl$(IV)

(iii) $CH_3-C(=O)-OCH_3$(I), $CH_3-C(=O)-OCH_2CH_3$(II), $H_3C-C(=O)-OCH(CF_3)_2$(III)

Q. 72. (A) Which of the following pairs has larger equilibrium constant for addition of water?

(a) Cl_3CCHO or $Cl_3C-C(=O)-CH_3$

(b) $Cl_3CCH_2C(=O)-CH_3$ or $CH_3-CCl_2-C-CH_3$

(c) $C_6H_5-C(=O)-H$ or $MeO-C_6H_4-C(=O)-H$

(d) $C_6H_5-C(=O)-C_6H_5$ or $O_2N-C_6H_4-C(=O)-C_6H_4-NO_2$

(e) (o-$NO_2C_6H_4$)$-C(=O)-$(o-$NO_2C_6H_4$) or (m-$NO_2C_6H_4$)$-C(=O)-$(m-$NO_2C_6H_4$)

(B) Within each set, select he compound which is more reactive in nucleophoilic addition at carbonyl carbon:

(a) $CH_3COCH_2CH_2Br$ or CH_3COCH_2Br

(b) CH_3CHO or CH_3COCH_3

(c) CH_3COCH_3 or $CH_3—CO—CF_3$

(d) PhCHOC or CH_3CHO

Q. 73. (a) Starting from propane, propanal methylbromide as the only organic reagents, and any reagent, synthesize the following compound:

$$CH_2{=}CH—CH_2—\underset{}{\overset{OCH_3}{\overset{|}{C}H}}—CH_2CH_3$$

(b) Beginning with alcohols containing not more than three carbons, synthesize:

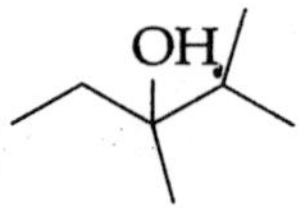

OH

(c) Starting from propanone, suggest synthesis of the following compounds:

(i) 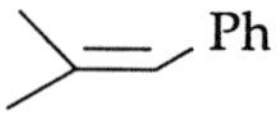 Ph **(ii)** 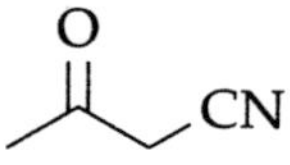 O, CN **(iii)** 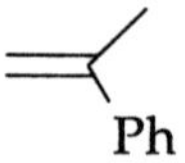 Ph **(iv)** 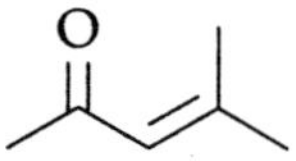 O

Q. 74. Identify the compounds labeled *A* to *F*:

$$A(C_8H_{10}) \xrightarrow[hv/CCl_4]{NBS} B(C_8H_9Br) \xrightarrow[(CH_3)_3\,COH]{(CH_3)_2\,COK} C \xrightarrow[Zn—H_2O]{O_3} D(C_7H_6O) \xrightarrow{MeMgBr}$$

$$\xrightarrow[H_2O]{H_3O^+} E \xrightarrow[CH_2Cl_2]{PCC} F \xrightarrow[OH^-/\Delta]{N_2N_4} A$$

Q. 75. Syhthesize products from indicated reactants and other inorganic reagents needed:

(a) $CH_3CH_2—\underset{Br}{\underset{|}{C}H}{=}CH_2 + CH_3—\underset{OH}{\underset{|}{C}H}—CH_3 \longrightarrow CH_3CH_2—C{\equiv}C—\overset{CH_3}{\overset{|}{C}}{=}CH_2$

(b) $CH_3—\underset{CH_3}{\underset{|}{\overset{CH_3}{\overset{|}{C}}}}—OH \longrightarrow CH_3—\underset{Br}{\underset{|}{\overset{CH_3}{\overset{|}{C}}}}—COOH$

(c) $CH_3CH_2—\overset{CH_3}{\overset{|}{C}}—CH{=}CH_2 + CH_2{=}CH_2 \longrightarrow CH_3—CH_2—\overset{CH_3}{\overset{|}{C}H}—\underset{OH}{\underset{|}{\overset{CH_3}{\overset{|}{C}}}}—C{\equiv}CH$

(d) Isobutyl amine $\longrightarrow$ N-isobutyl formamide

(e) Ethylacetate $\longrightarrow$ 3-methyl-3-pentanol.

Q. 76. (a) Starting with alcohal containing not more than six carbons syhthesize:

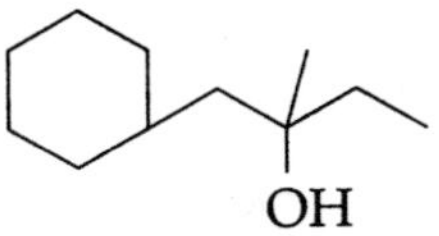

OH

(b) Synthesize the indicated product from the supplied starting materials:

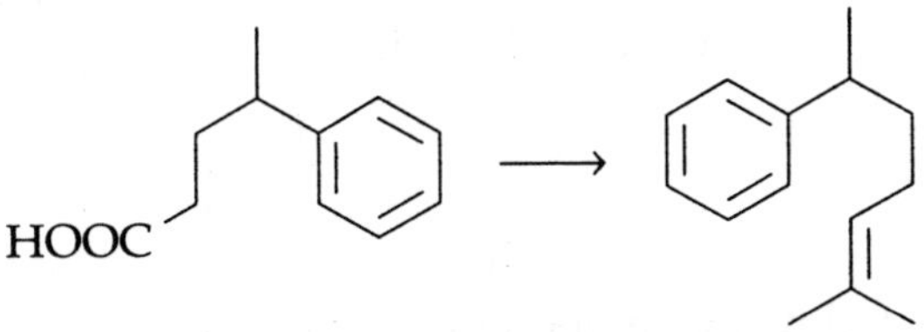

HOOC $\longrightarrow$

(c) Syhtnesize 3-pentanone using alkyl bromide containing not more than two carbons.

(d) Synthesize

(i) 2-butanone from ethyl bromide as only starting organic compounds.

(ii) 5-nonenone from 1-butanol

Q. 77. Bring about the following synthesis starting from benzene:

(a) 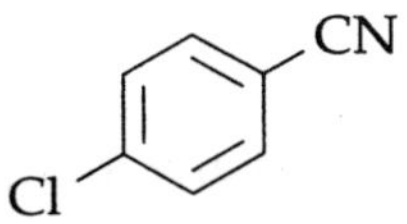 CN, Cl

(b) 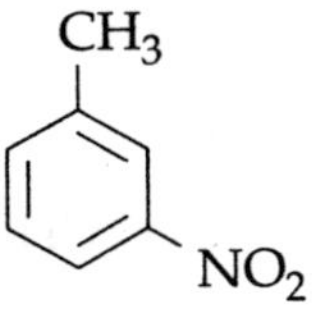 CH_3, NO_2

(c) 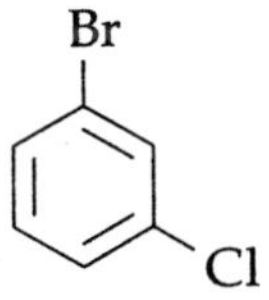 Br, Cl

(d) 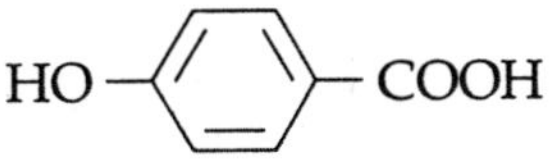 HO—C$_6$H$_4$—COOH

(e) 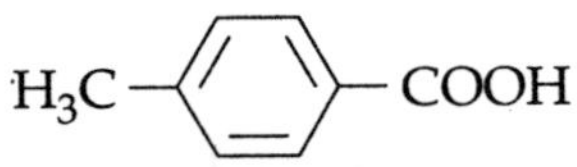 H_3C—C$_6$H$_4$—COOH

(f) 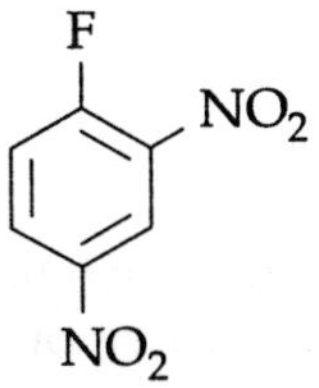 F, NO_2, NO_2

(g) 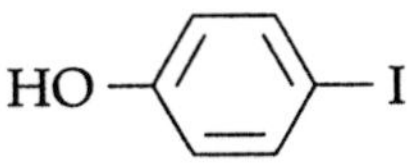 HO—C$_6$H$_4$—I

Q. 78. Synthesize the following compounds starting from benzene or toluene:

(a) 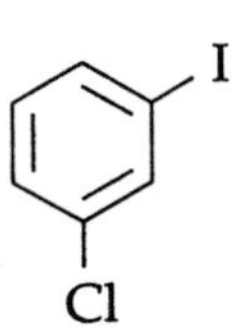 I, Cl

(b) 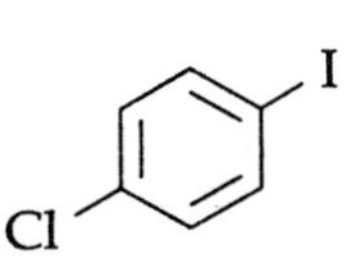 I, Cl

(c) 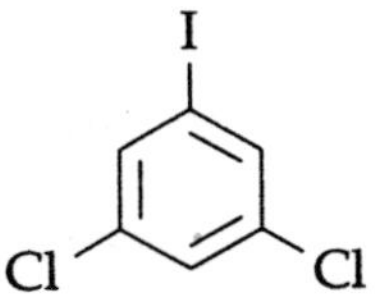 I, Cl, Cl

(d) 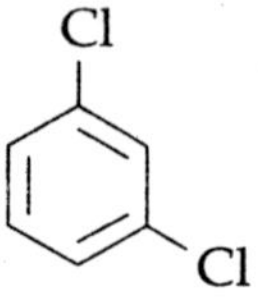 Cl, Cl (e) 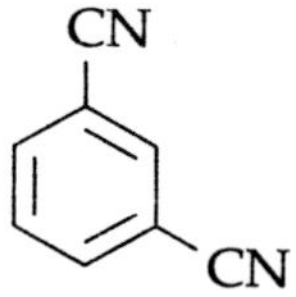 CN, CN (f) 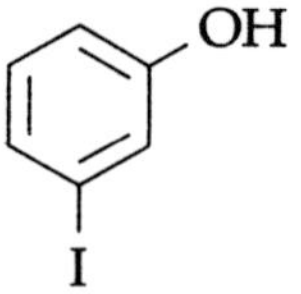 OH, I

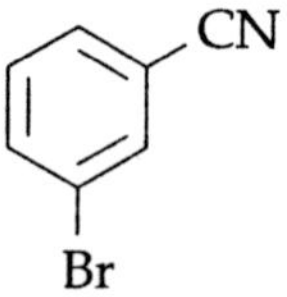

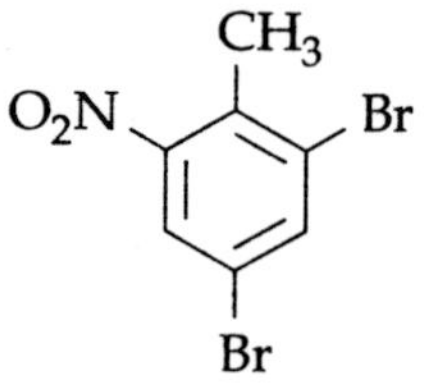

(g) CN, Br (h) CH_3, O_2N, Br, Br (i) NH_2, Br, Br

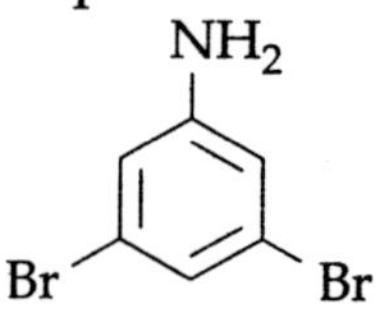

Q. 79. (a) Synthesize the following compounds starting from benzene:

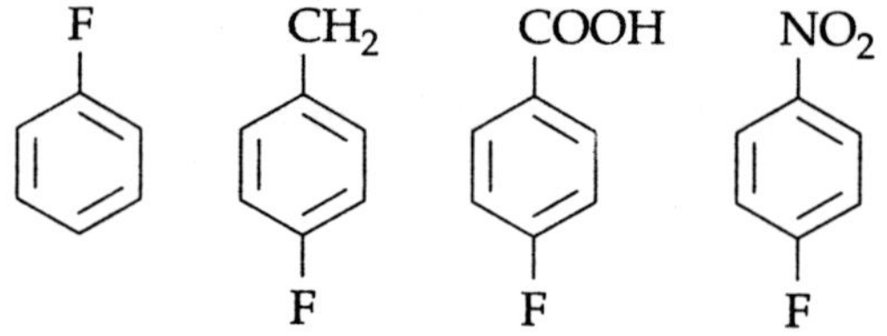

F CH_2 / F   COOH / F NO_2 / F

(b) 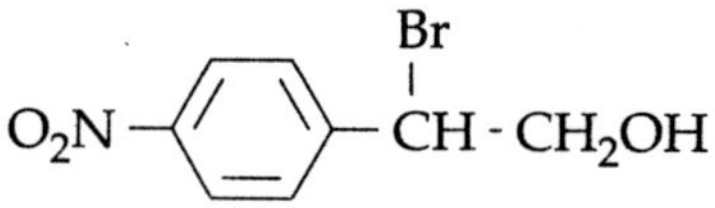 $O_2N-C_6H_4-\underset{}{\overset{Br}{CH}}-CH_2OH$ from benzene, chloroethane and any inorganic reagent

(c) $C_6H_5CH_2COC_6H_5$ from benzene, oxirane and inorganic reagent

(d) $Ph-CH_2-\underset{CH_3}{\overset{OH}{C}}-\underset{H}{C}=\underset{H}{C}-CH_3$ from $Ph-CH_2CH_2CH_2Br$

(e) 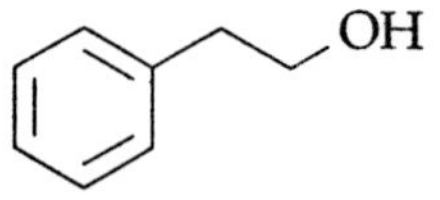 OH

Q. 80. Starting with bromo benzene, synthesize the following compounds using any organic reagents containing three or fewer carbon atoms and any organic reagents you need.

(a) 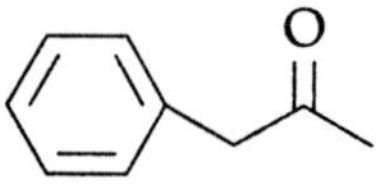 O (b) 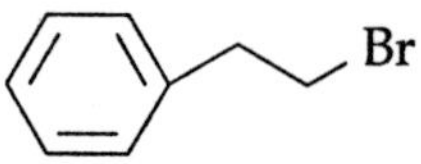 Br

(c) 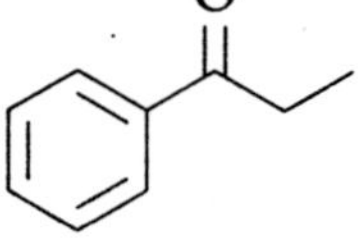 O (d)

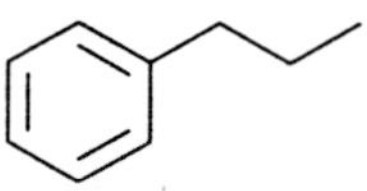

Q. 81. **I.** Show how would you accomplish the following multisteps synthesis?

(a) 6-hepten-1-ol ⟶ caprolactum

(b) 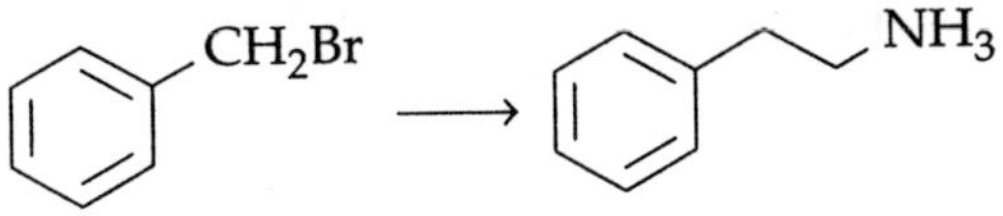 $C_6H_5CH_2Br$ ⟶ $C_6H_5CH_2CH_2NH_3$

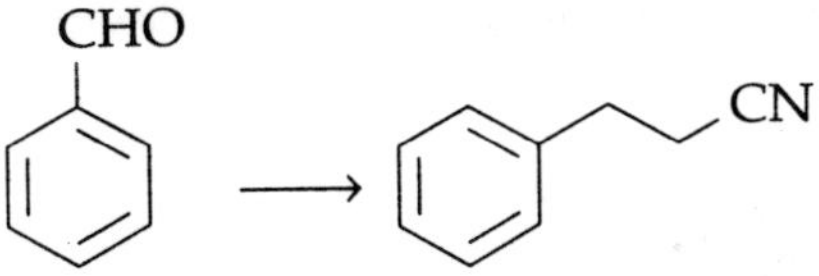

(c) C_6H_5CHO ⟶ $C_6H_5CH_2CH_2CN$

II. Give structural formula of all possible products in the following reactions:

(a) 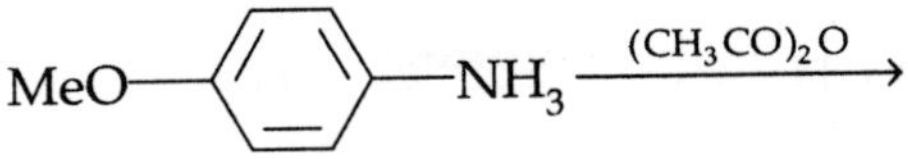 $MeO{-}C_6H_4{-}NH_3 \xrightarrow{(CH_3CO)_2O}$

(b)  $C_6H_5{-}COOEt \xrightarrow{NH_2OH}$

(c) $o\text{-}HOOC{-}C_6H_4{-}CH_2COOH \xrightarrow[\text{heat}]{(CH_3CO)_2O}$

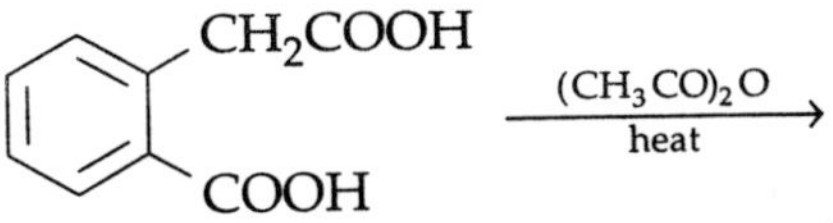

Q. 82. Assign a structure to each compound indicated by a letter in the following equations.

(a) $HO\overset{O}{\overset{||}{C}}(CH_2)_9\overset{O}{\overset{||}{C}}OH \xrightarrow[H_2SO_4,\ \Delta]{CH_3OH\,(\text{excess})} A$

(b) 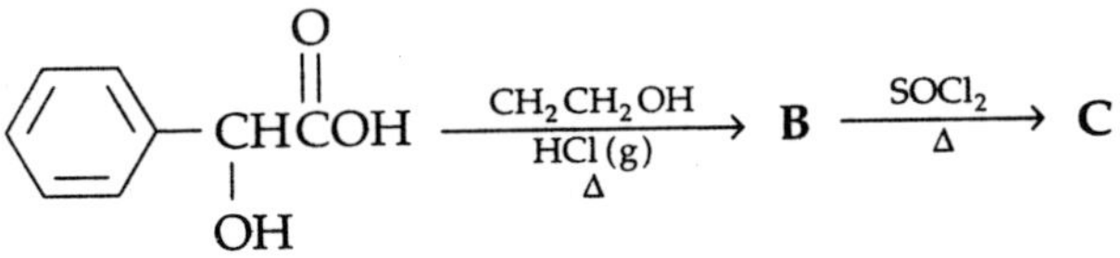 $C_6H_5{-}\underset{OH}{\underset{|}{C}H}\overset{O}{\overset{||}{C}}OH \xrightarrow[HCl\,(g),\ \Delta]{CH_2CH_2OH} B \xrightarrow[\Delta]{SOCl_2} C$

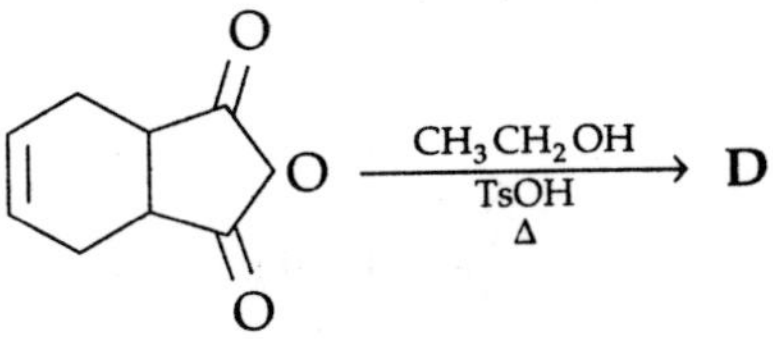

(c) (cyclohexene-fused cyclic anhydride, tetrahydrophthalic anhydride) $\xrightarrow[TsOH,\ \Delta]{CH_3CH_2OH} D$

(d) $HO\overset{O}{\overset{||}{C}}C{\equiv}C\overset{O}{\overset{||}{C}}OH \xrightarrow[H_2SO_4,\ \Delta]{CH_3OH\,(\text{excess})} E$

(e) $BrCH_2\overset{O}{\overset{||}{C}}OH \xrightarrow[H_2SO_4,\ \Delta]{CH_3CH_2OH} F$

(f) $CH_3CH{=}CHCOH$ (with C=O) $\xrightarrow[\text{H}_2\text{SO}_4,\ \text{benzene},\ \Delta]{CH_3CH_2CHCH_3 (\text{OH})}$ G

Q. 83. Assign structures to the compounds designated by letters in the following equations.

(a) $CH_3CH_2OCCH{=}CHCOCH_2CH_3$ (two C=O) $\xrightarrow[\text{NH}_4\text{Cl},\ \text{H}_2\text{O}]{\text{NH}_3\ (\text{excess})}$ A

(b) $CH_3CH_2OCCH_2CH_2CH_2CH_2COCH_2CH_3$ (two C=O) $\xrightarrow{\text{H}_2\text{NNH}_2\ (\text{excess})}$ D

(c) $CH_3(CH_2)_4CHCOH$ (C=O; CH_2CH_3 on CH) $\xrightarrow[\Delta]{\text{SOCl}_2}$ E $\xrightarrow[\text{H}_2\text{O}]{\text{NH}_3}$ F

(d) $N{\equiv}CCH_2COCH_2CH_3$ (C=O) $\xrightarrow[\text{H}_2\text{O},\ 273\ \text{K}]{\text{NH}_3}$ H

(e) C_6H_5CCl (C=O) + HN(cyclohexane ring) $\xrightarrow[\text{H}_2\text{O}]{\text{NaOH}}$ G

(f) Cyclopentyl–COH (C=O) $\xrightarrow[\Delta]{\text{SOCl}_2}$ B $\xrightarrow{(\text{CH}_3)_2\text{NH}\ (\text{excess})}$ C

Q. 84. Give structural formulas for all of the organic intermediates and products indicated by letters in the following equations.

(a) CH_3O–C_6H_4–NH_2 $\xrightarrow[\text{acetic acid},\ \text{H}_2\text{O},\ 273\text{-}278\text{K}]{\text{CHCOCCH}_3\ (\text{O O})}$ A

(b) C_6H_5COH (C=O) $\xrightarrow[\Delta]{\text{PBr}_3}$ B

(c) C_6H_5CCl (C=O) $\xrightarrow[\text{H}_2\text{O}]{\text{NaOH}}$ C

(d) $CH_2(CH_2)_{25}CCl$ (C=O) $\xrightarrow{\text{CH}_3\text{CH}_2)_3\text{CO}^-\text{Na}^+\ (\text{C=O})}$ D

(e) C_6H_5–$CHCNH_2$ (C=O; CH_2CH_3 on CH) $\xrightarrow[\text{H}_2\text{SO}_4,\ \Delta]{\text{H}_2\text{O}}$ E

(f) 2-CH_3–C_6H_4–CN $\xrightarrow[\text{H}_2\text{SO}_4,\ \Delta,\ 5\ \text{h}]{\text{H}_2\text{O}}$ F

(g) $CH_3CH_2\overset{O}{\overset{||}{C}}Cl \xrightarrow{NH_3\text{(excess)}} \mathbf{G}$ (h) $C_6H_5-\overset{O}{\overset{||}{C}}OCH_2CH_3 \xrightarrow{H_2NOH} \mathbf{H}$

Q. 85. (a) $CH_3-C(\text{cyclopentyl})-CH_2CH_2\overset{O}{\overset{||}{C}}OH + CH_3Li \xrightarrow[\text{ether}]{\text{diethyl}} \mathbf{A} \xrightarrow[\text{ether}]{CH_3Li,\ \text{dethyl}} \mathbf{B} \xrightarrow{H_3O^+} \mathbf{C}$

(b) $C_6H_{11}-\overset{O}{\overset{||}{C}}N(CH_3)OCH_3 + CH_3CH_2CH_2CH_2MgBr \xrightarrow{\text{tetrahydrofuran}} \mathbf{D} \xrightarrow{H_3O^+} \mathbf{E + F}$

(c) $C_6H_{11}-\overset{O}{\overset{||}{C}}N(CH_3)OCH_3 + C_6H_5-Li \xrightarrow{\text{tetrahydrofuran}} \mathbf{G} \xrightarrow{H_3O^+} \mathbf{H + I}$

(d) $C_6H_5-Br \xrightarrow[\text{diethyl ether}]{Mg} \mathbf{I} \xrightarrow{CH_3\overset{O}{\overset{||}{C}}OCH_2CH_3} \mathbf{J} \xrightarrow{NH_4Cl\ H_2O} \mathbf{K}$

(e) $C_6H_{11}-\overset{O}{\overset{||}{C}}OH \xrightarrow[\text{dim ethoxyethane}]{LiH} \mathbf{L} \xrightarrow[\text{diethyl ether}]{CH_3Li} \mathbf{M} \xrightarrow{H_3O^+} \mathbf{N}$

Q. 86. Complete the following equations.

(a) $o\text{-}H_2N-C_6H_4-\overset{O}{\overset{||}{C}}OH \xrightarrow[\text{diethyl ether}]{LiAlH} \xrightarrow{H_3O^+}$

(b) N-phenylglutarimide (2,6-dioxopiperidin-1-yl benzene) $\xrightarrow[\text{diethyl ether}]{LiAlH_4\text{ (excess)}} \xrightarrow{H_2O}$

(c) $CH_3CH{=}CHCH_2CH_2\overset{O}{\overset{||}{C}}OCH_3 \xrightarrow[\text{diethyl ether}]{LiAlH_4} \xrightarrow{H_3O^+}$

(d) $CH_3CH_2O\overset{O}{\overset{||}{C}}(CH_2)_4\overset{O}{\overset{||}{C}}OCH_2CH_3 \xrightarrow[\text{diethyl ether}]{LiAlH_4} \xrightarrow{H_3O^+}$

(e) $CH_3CH_2\overset{O}{\overset{\|}{C}}NHCH_2CH_3 \xrightarrow[\text{diethyl ether}]{LiAlH_4} \xrightarrow{H_3O^+}$

Q. 87. Complete the following equations.

(a) [cyclic diester (1,4-dioxane-2,5-dione)] $\xrightarrow[H_2O,\ \Delta]{NaOH}$

(b) $CH_3\overset{O}{\overset{\|}{C}}CH_2CH_2CH_2\overset{O}{\overset{\|}{C}}OH \xrightarrow[H_2O]{NaBH_4} \xrightarrow{H_3O^+}$

(c) [5,5-dimethyl lactone ring with CH_3, CH_3] $\xrightarrow[H_2O,\ \Delta]{NaOH}$

(d) [cyclopentane with O, OH, O] $\xrightarrow[H_2SO_4]{CrO_3}$

(e) [cyclobutanone] $\xrightarrow[\text{chloroform}]{C_6H_5CO_3H \text{ (peroxybenzoic acid, COOH)}}$

Q. 88. Give structural formulas for all the organic intermediates and products indicated by letters in the following equations.

(a) $C_6H_5-CH_2-C(H)(NH_2)-\overset{O}{\overset{\|}{C}}OCH_2CH_3 \xrightarrow[\text{diethyl ether}]{LiAlH_4}$ **A + B**

(b) [o-aminobenzoic acid: $C_6H_4(\overset{O}{\overset{\|}{C}}OH)(NH_2)$] $\xrightarrow[Na_2CO_3,\ H_2O,\ \Delta]{TsCl}$ **C** $\xrightarrow{H_3O^+}$ **D**

(c) $C_6H_5-\overset{O}{\overset{\|}{C}}-\underset{OH}{\underset{|}{C}H}-C_6H_5 \xrightarrow{CH_3\overset{O}{\overset{\|}{C}}Cl}$ **E**

(d) [indane-5-carbaldehyde, with CHO drawn as C(=O)H] $\xrightarrow[H_2O, \Delta]{KMnO_4}$ **F** $\xrightarrow{H_3O^+}$ **G**

(e) $CH_3(CH_2)_{14}CH_2OH \xrightarrow[\text{pyridine}]{TsCl}$ **H**

(f) [2-bromo-1,3,5-trimethylbenzene: benzene ring with Br and CH_3 groups] $\xrightarrow[\text{diethyl ether}]{Mg}$ **I** $\xrightarrow{CO_2}$ **J** $\xrightarrow{H_3O^+}$ **K**

Q. 89. This problem gives more practice in recognizing reactions.

(a) [methyl 3-nitrobenzoate: benzene ring with O_2N and $\overset{O}{\overset{\|}{C}}OCH_3$] $\xrightarrow[H_2O, \Delta]{NaOH}$ **A + B**

(b) [ring lactone with C=O and CH_3 substituent] $\xrightarrow[\text{dichloromethane}]{\text{Cl-}C_6H_4\text{-}\overset{O}{\overset{\|}{C}}OOH}$ **C**

(c) [$CH_2{=}C(CH_3){-}CH_2{-}CH_2{-}CH(COOH){-}CH(CH_3)_2$, drawn as a structure with CH_3, $C{=}CH_2$, CH_2, CH_2, $\overset{O}{\overset{\|}{C}}OH$, CH, CH, CH_3, CH_3] $\xrightarrow[\text{diethyl ether}]{LiAlH_4}$ $\xrightarrow{H_2O}$ **D** $\xrightarrow[\substack{CH_3\overset{O}{\overset{\|}{C}}ONa \\ \text{dichloromethane}}]{C_5H_5\overset{+}{N}{-}H\ CrO_3Cl^-}$ **E**

(d) [$CH_3CH_2CH_2O$ and CH_3O substituted benzene ring bearing $CH(OH)CH_2CH_2\overset{O}{\overset{\|}{C}}O^-K^+$] $\xrightarrow[\text{toluene}, \Delta]{C_5H_5N{-}H\ +\ ^-OTs}$ **F** + $H_2O\uparrow$

(e) 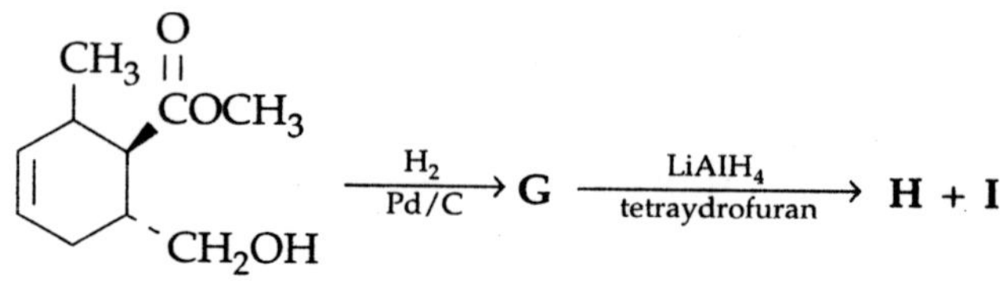 [cyclohexene with CH_3, $\overset{O}{\|}{C}OCH_3$ and CH_2OH substituents] $\xrightarrow[\text{Pd/C}]{H_2}$ G $\xrightarrow[\text{tetraydrofuran}]{LiAlH_4}$ H + I

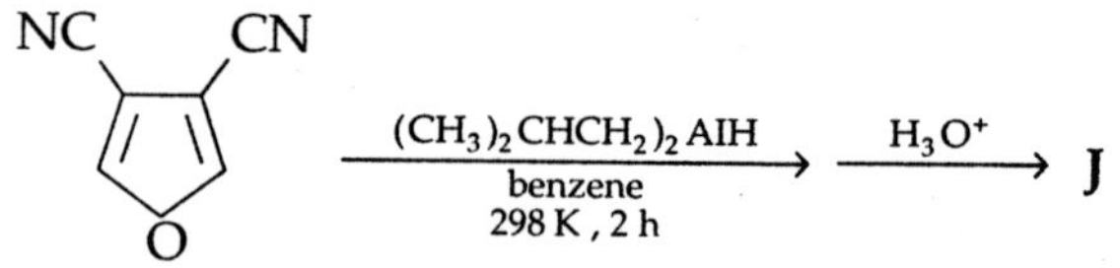

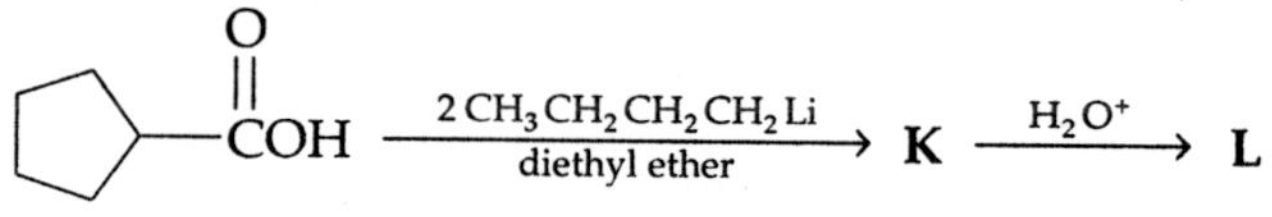

(f) [furan ring with NC and CN substituents] $\xrightarrow[\substack{\text{benzene} \\ 298\text{ K}, 2\text{ h}}]{((CH_3)_2CHCH_2)_2AlH}$ $\xrightarrow{H_3O^+}$ J

(g) [cyclopentyl]$-\overset{O}{\|}{C}OH$ $\xrightarrow[\text{diethyl ether}]{2\,CH_3CH_2CH_2CH_2Li}$ K $\xrightarrow{H_2O^+}$ L

(h) [bicyclic compound with H, H and $\overset{O}{\|}{C}CH$ substituents] $\xrightarrow[\text{chloroform}]{\text{m-ClC}_6\text{H}_4\overset{O}{\|}{C}OOH}$ M

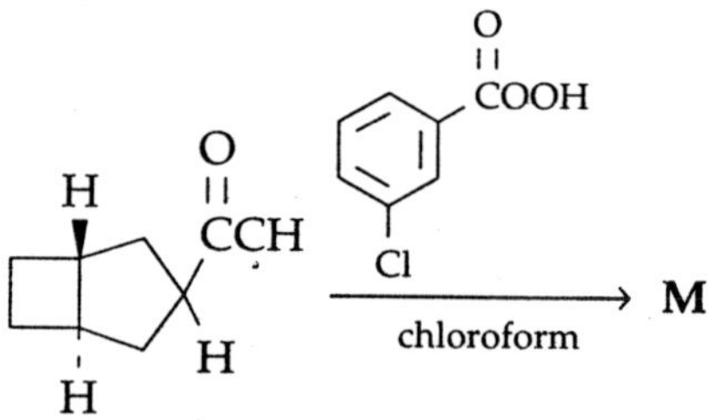

Q. 90. Give structural formulas for the intermediates and products indicated by letters in the following equations.

(a) $CH_2{=}CHCH{=}CH\overset{O}{\overset{\|}{C}}OH \xrightarrow[\text{dethyl ether}]{LiAlH_4} \xrightarrow{H_2O}$ A $\xrightarrow[\substack{\text{toluene} \\ \Delta}]{C_5H_5\overset{+}{N}H\ CrO_3Cl^-}$ B

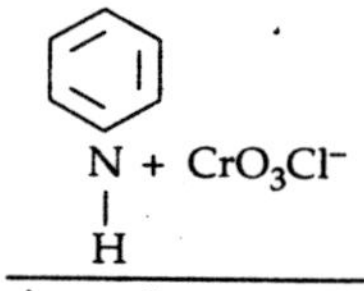

(b) $CH_3CH_2CH_2CH_2\overset{O}{\overset{\|}{C}}OCH_2CH_3 \xrightarrow[\text{diethyl ether}]{2\,CH_3CH_2MgBr}$ C $\xrightarrow[H_2O]{NH_4Cl}$ D

(c) $CH_3CH_2\overset{O}{\overset{\|}{C}}NHCH_2CH_3 \xrightarrow[\text{diethyl ether}]{LiAlH_4} \xrightarrow{H_2O}$ E

(d) $CH_3CH{=}CHCH_2C{\equiv}CCH_2CH_2\overset{O}{\overset{\|}{C}}OH \xrightarrow[\substack{NH_2(\text{liq}) \\ \text{ethanol}}]{Li} \xrightarrow{H_2O^+}$ F

(e) F $\xrightarrow[\substack{NH_2(\text{liq}) \\ \text{ethanol}}]{Li} \xrightarrow{H_2O^+}$ G $\xrightarrow{C_5H_5\overset{+}{N}H\ CrO_3Cl^-}$ H

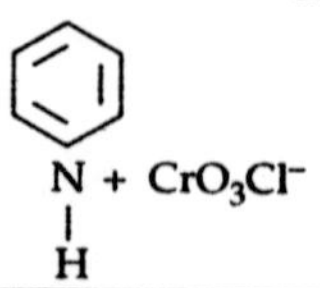

(f) $CH_3CH_2NHC(=O)CH_2-C_6H_5 \xrightarrow[H_2O,\ \Delta]{NaOH}$ **I + J**

(g) $CH_3OC(=O)(CH_2)_3C(=O)N(CH_3)OCH_3 \xrightarrow[\text{terahydrofuran}]{C_6H_5Br\ \text{(6 equivalents)}}$ **K + L + M**

(h) $CH_3C(=O)CH_2OC(=O)H + CH_3OH \xrightarrow[\Delta]{KOH}$ **A + B**

(i) $CH_3CH(CH_3)CH_2CO^-Na^+ + CH_3CH_2CH_2C(=O)Cl \longrightarrow$ **C**

(j) $CH_3C(CH_3)_2-C(=O)O^-Li^+ + C_6H_5-Li \xrightarrow{\text{diethyl ether}}$ **D** $\xrightarrow{H_3O^+}$ **E**

(k) $C_6H_5-C(=O)Cl + H_2NCH_2CH_2CH_2CH_2CH_2C(=O)OH \xrightarrow[H_2O]{NaOH}$ **F** $\xrightarrow{H_3O^+}$ **G**

(l) [N-methyl amide of 4-methylpent-3-enoic acid, N-substituted with 4-methylcyclohex-3-enyl] $\xrightarrow[\text{diethyl ether}]{LiAlH_4} \xrightarrow{H_2O}$ **H**

(m) [bicyclic diketone with CH₃ groups, N bearing C(CH₃)(H)Ph] $\xrightarrow[\substack{(CH_2CH_2)O^-BF_3 \\ \text{chloroform}}]{HSCH_2CH_2CH_2SH}$ **I** $\xrightarrow[\text{tetrahydrofuran}]{LiAlH_4} \xrightarrow{H_2O}$ **J**

$\xrightarrow[\text{ethanol}]{\text{Raney Ni}}$ **K**

(n) $C_6H_5-\overset{O}{\overset{||}{C}}N(CH_3)OCH_3 \xrightarrow[\text{tetrahydrofuran, 293 K}]{C_6H_5-C\equiv CLi} L \xrightarrow{H_3O^+} M + N$

Q. 91. Supply structural formulas for the regents, intermediates and products indicated by letters.

(a) (cyclic lactone: 2-oxo-tetrahydropyran ring, O and C=O) $\xrightarrow[\text{tetrahydrofuran, 195K}]{A} B \xrightarrow[\text{hexamethylplosphoric triamide, 195 K}]{HC\equiv CCH_2Br}$ (lactone ring with $CH_2C\equiv CH$ on the carbon adjacent to C=O)

(b) $C \xrightarrow[\text{hexane, diethyl ether, 213 K}]{(CH_3\overset{O}{\overset{||}{C}}H)_2N^-Li^+}$ 2,2-dimethylcyclohexanone (cyclohexanone with CH_3, CH_3)

(c) $CH_3CH_2CH_2\overset{O}{\overset{||}{C}}OCH_3 \xrightarrow[\text{tetraydofuran, 195 K}]{D} E \xrightarrow[\text{hexamethylphophoric triamide, 195 K}]{F} CH_3CH_2\underset{CH_2OCH_3}{\underset{|}{C}H}\overset{O}{\overset{||}{C}}OCH_3$

Q. 92. Give structural formulas for all species symbolized by letters in the following equations.

(a) $CH_2(\overset{O}{\overset{||}{C}}OCH_2CH_3)_3 \xrightarrow[\text{ethanol}]{CH_3CH_2ONa} A \xrightarrow{CH_3\overset{Br}{\overset{|}{C}}HCH_3} B$

(b) $CH_3CH_2CH(\overset{O}{\overset{||}{C}}OCH_2CH_3)_2 \xrightarrow[\text{tert-butyl alcohol}]{CH_3\overset{CH_3}{\overset{|}{\underset{CH_3}{\underset{|}{C}}}}ONa} C \xrightarrow{BrCH_2CH{=}CH_3} D$

(c) $CH_2(\overset{O}{\overset{||}{C}}OCH_2CH_3)_2 \xrightarrow[\text{ethanol}]{CH_2CH_2ONa} E \xrightarrow{C_6H_5-CH_2Cl} F$

(d) $CH_3\overset{O}{\overset{||}{C}}CH_2\overset{O}{\overset{||}{C}}OCH_2CH_3 \xrightarrow[\text{ethanol}]{CH_3CH_2ONa} G \xrightarrow{BrCH_2CH_2\overset{O}{\overset{||}{C}}OCH_2CH_3} H$

(e) $CH_3\overset{O}{\overset{||}{C}}CH_2\overset{O}{\overset{||}{C}}OCH_2CH_3 \xrightarrow[\text{ethanol}]{CH_3CH_2ONa\text{ (1 molar equivalent)}} I \xrightarrow{ClCH_2CH_2CH_2Br} J$

(f) $CH_3\overset{O}{\overset{||}{C}}CH_2\overset{O}{\overset{||}{C}}OCH_2CH_3 \xrightarrow[\text{ethanol}]{CH_2CH_2ONa} K \xrightarrow{CH_2{=}CHCH_2CH_2Br} L \xrightarrow[\text{toluene}]{M} N$

$\xrightarrow{CH_3\overset{|}{C}HCH_3} O$

Q. 93. Give structural formulas for all compounds symbolized by letters in the following equations.

(a) $CH_3\overset{O}{\overset{\|}{C}}CH_2\overset{O}{\overset{\|}{C}}OCH_2CH_3 \xrightarrow[\text{ethanol}]{CH_2CH_3ONa} A \xrightarrow[\Delta]{ClCH_3\overset{O}{\overset{\|}{C}}OCH_2CH_3} B \xrightarrow[\Delta]{H_3O^+} C$

(b) $C_6H_5-\overset{O}{\overset{\|}{C}}CH_2\overset{O}{\overset{\|}{C}}OCH_2CH_3 \xrightarrow[\text{ethanol}]{CH_3CH_2ONa} D \xrightarrow[\Delta]{BrCH_2C{\equiv}CH} E \xrightarrow[\Delta]{H_3O^+} F$

(c) $CH_2(\overset{O}{\overset{\|}{C}}OCH_2CH_3)_2 \xrightarrow[\text{ethanol}]{CH_3CH_2ONa\text{ (molar equivalent)}} G \xrightarrow{BrCH_2CH_2CH_2Cl} H$

$\xrightarrow[\text{ethanol}]{CH_2CH_2ONa\text{ (1 molar equivalent)}} I \xrightarrow[\Delta]{H_3O^+} J$

Q. 94. Complete the following equations

(a) (furan-2-yl)$-\overset{O}{\overset{\|}{C}H}$ $+ CH_3\overset{O}{\overset{\|}{C}}CH_3 \xrightarrow[H_2O]{NaOH} \xrightarrow{H_2O^+}$

(b) $C_6H_5-\overset{O}{\overset{\|}{C}}H + CH_3\overset{O}{\overset{\|}{C}}CH_3 \xrightarrow[H_2O]{NaOH} \xrightarrow{H_3O^+}$

2 equivalents   1 equivalents

(c) $C_6H_5-\overset{O}{\overset{\|}{C}}H + CH_3\overset{O}{\overset{\|}{C}}CH_2CH_3 \xrightarrow{HCl}$

(d) $2\ C_6H_5-CH_2\overset{O}{\overset{\|}{C}}H \xrightarrow[H_2O]{NaOH}$

Q. 95. Give the structures of the starting materials that would be required to synthesize the following compounds.

(a) $(p\text{-}CH_3O\text{-}C_6H_4)(H)C{=}C(CH_3)(\overset{}{C}H{=}O)$

(b) $CH_3CH_2CH_2CH_2\overset{OH}{\overset{|}{C}H}\underset{CH_2CH_2CH_3}{\underset{|}{C}H}\overset{O}{\overset{\|}{C}}H$

(c) [cyclohexene ring bearing a phenyl group and $\overset{O}{\overset{||}{C}}-C(CH_3)_3$]

(d) [$C_6H_5-CH=CH-\overset{O}{\overset{||}{C}}-C_6H_5$, with H and H trans about C=C]

Q. 96. Give structures for the intermediates and products designed by the letters.

(a) $C_6H_5\overset{O}{\overset{||}{C}}C_6H_5 + \begin{matrix} CH_2\overset{O}{\overset{||}{C}}OCH_2CH_3 \\ | \\ CH_2\underset{O}{\underset{||}{C}}OCH_2CH_3 \end{matrix} \xrightarrow[\text{tert-butyl alcohol}]{(CH_3)_3COK} A$

(b) [cyclooctanone] $\xrightarrow[\text{benzene, } \Delta]{NaH} B \xrightarrow{CH_3CH_2O\overset{O}{\overset{||}{C}}OCH_2CH_3} C \xrightarrow{CH_3\overset{O}{\overset{||}{C}}OH} D$

(c) $CH_3CH\begin{matrix} CH_2\overset{O}{\overset{||}{C}}OCH_2CH_3 \\ CH_2\underset{O}{\underset{||}{C}}OCH_2CH_3 \end{matrix} + \begin{matrix} \overset{O}{\overset{||}{C}}OCH_2CH_3 \\ | \\ \underset{O}{\underset{||}{C}}OCH_2CH_3 \end{matrix} \xrightarrow[\text{ethanol}]{CH_3CH_2ONa} E \xrightarrow{H_3O^+} F$

Q. 97. Give structural formulas for all intermediates and products designed by letters in the following equations.

(a) [cyclohexanone] $=O + N\equiv CCH_2\overset{O}{\overset{||}{C}}OCHCH_3 \xrightarrow[\text{acetic acid, benzene, } \Delta]{CH_3\overset{O}{\overset{||}{C}}O^-NH_4^+} A$

(b) $CH_2(\overset{O}{\overset{||}{C}}OCH_2CH_3)_2 \xrightarrow[\text{ethanol}]{CH_3CH_2ONa} B \xrightarrow{CH_3\overset{CH_3}{\overset{|}{C}}HCH_2Br} C \xrightarrow[\Delta]{H_3O^+} D$

(c) [2,2-dimethylcyclohexanone (CH_3, CH_3, $=O$)] $\xrightarrow[\text{acetic acid}]{Br_2} E$

(d) [Structure: bicyclic lactone with $CH_2{=}CH$ substituent and $OCH_2OCH_2C_6H_5$ group] $\xrightarrow[\text{tetrahydrofuran, 195 K}]{(CH_3\overset{CH_3}{|}CH)_2N^-Li^+}$ F $\xrightarrow{ClCH_2OCH_2C_6H_5}$ G

(e) $O_2N-C_6H_4-\overset{O}{\overset{||}{C}}CH_3 + CH_3\overset{O}{\overset{||}{C}}OCH_3 \xrightarrow{CH_3ONa\ (excess)} \xrightarrow{H_3O^+}$ H + I

(f) [Structure: cyclohexanone ring bearing $-\overset{O}{\overset{||}{C}}-OCH_2CH_3$ side chain] $\xrightarrow[(2)\ H_3O^+]{(1)\ CH_3CH_2ONa}$ J + K

(g) $CH_2(\overset{O}{\overset{||}{C}}OCH_2CH_3)_2 \xrightarrow[\text{ethanol}]{CH_3CH_2Na}$ L $\xrightarrow{\text{3-chlorocyclopentene (Cl)}}$ M

Q. 98. Give the structural formulas of the designated letters.

(a) [1-indanone] $\xrightarrow[\text{diethyl ether}]{NaNH_2}$ A $\xrightarrow{C_6H_5\overset{O}{\overset{||}{C}}C_6H_5}$ B

(b) $CH_3\overset{O}{\overset{||}{C}}CH_2\overset{O}{\overset{||}{C}}OCH_2CH_3 \xrightarrow[H_2O]{NaOH}$ C $\xrightarrow[NaOH]{C_6H_5\overset{O}{\overset{||}{C}}Cl}$ D

(c) $CH_3\overset{O}{\overset{||}{C}}CH_2\overset{O}{\overset{||}{C}}OCH_2CH_3$ + [benzaldehyde with $\overset{O}{\overset{||}{C}}H$, OCH_3 and CH_3O substituents] $\xrightarrow[\text{acetic acid}]{\text{piperidine (N–H)}}$ E

(d) $CH_3\overset{O}{\overset{||}{C}}CH_2\overset{O}{\overset{||}{C}}CH_2CH_2CH_2CH{=}CH_2 \xrightarrow[\text{tetrahydrofuran}]{(CH_2CH)_2N^-LI^+} \xrightarrow[H_2O]{NH_4Cl} \xrightarrow[\text{tetrahydrofuran}]{CH_3CHO}$ G

(e) $C_6H_5-CH_2C{\equiv}N \xrightarrow[\text{ethanol}]{CH_3CH_2ONa}$ H $\xrightarrow[\text{toluene, }\Delta]{CH_3CH_2O\overset{CH_2}{\overset{||}{C}}OCH_2CH_3}$ I

(f) 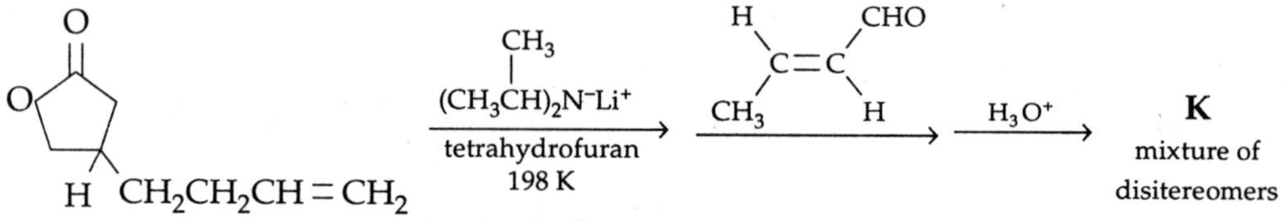 [structure: γ-lactone with H and $CH_2CH_2CH=CH_2$ substituents] $\xrightarrow[\text{tetrahydrofuran, 198 K}]{(CH_3CH)_2N^-Li^+ \text{ (}CH_3\text{)}}$ $\xrightarrow{\text{H(CH}_3\text{)C=C(H)CHO}}$ $\xrightarrow{H_3O^+}$ **K** mixture of disitereomers

(g) 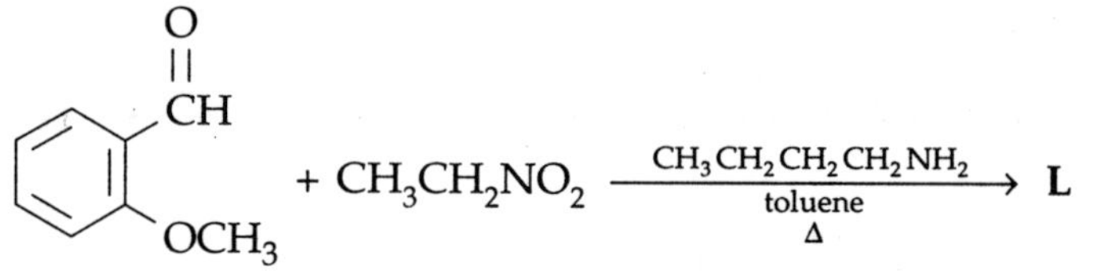 [2-methoxybenzaldehyde: $C_6H_4(CHO)(OCH_3)$] + $CH_3CH_2NO_2$ $\xrightarrow[\text{toluene, }\Delta]{CH_3CH_2CH_2CH_2NH_2}$ **L**

(h) $CH_3(CH_2)_4CH_2\overset{O}{\overset{||}{C}}CH_2CH_2CH_2OH$ $\xrightarrow[\text{dichloromethane}]{C_5H_5\overset{+}{N}H\ CrO_3Cl^-}$ **M** $\xrightarrow[\text{H}_2\text{O, tetrahydrofuran}]{KOH}$ **N**

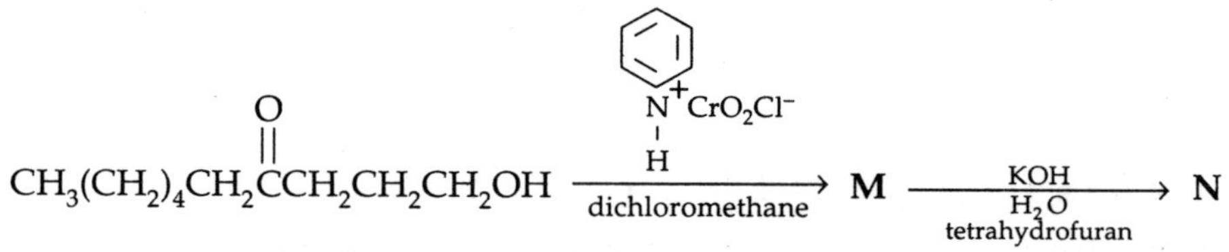

Q. 99. The same compound is formed when either of the two alcohols shown below is treated with boron trifluoride etherate. Propose a structure for the product.

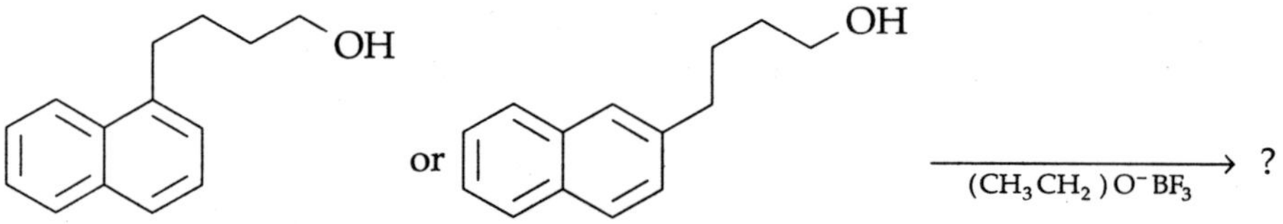

[structure: 4-(1-naphthyl)butan-1-ol] or [structure: 4-(2-naphthyl)butan-1-ol] $\xrightarrow[(CH_3CH_2)O^-BF_3]{}$?

Q. 100. Predict the products, designated by letters, that will be formed when 1-oxo-1,2,3,4-tetrahydro-naphthalene reacts with the following reagents.

(a) $NaBH_4$, CH_3CH_2OH $\longrightarrow$ **A**

(b) CH_3CH_2MgBr, diethyl ether $\longrightarrow$ **B** $\xrightarrow[H_2O]{NH_4Cl}$ **C**

(c) $NaC{\equiv}CCH_3$, NH_3(liq) $\longrightarrow$ **D** $\xrightarrow[H_2O]{NH_4Cl}$ **E**

(d) 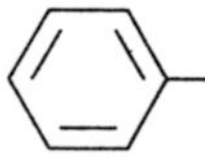 C_6H_5–$NHNH_2$, acetic acid $\longrightarrow$ **F**

(e) $HOCH_2CH_2OH$, TsOH, toluene $\xrightarrow[\Delta]{}$ **G**

(f) **H** $\xrightarrow[H_3PO_4,\ \Delta]{}$ **I** $\xrightarrow[\text{tetrahydrofuran}]{BH_3}$ $\xrightarrow{H_3O_2,\ OH^-}$ **J**

(g) **K** $\xrightarrow[H_3PO_4,\ \Delta]{}$ **L**

(h) **M** $\xrightarrow[\text{Pd/BeSO}_4\text{ quinatine}]{H_3}$ **N** $\xrightarrow{\text{9-BBN}}$ **O** $\xrightarrow{H_2O_2,\ OH^-}$ **P**

Q. 101. Complete the following equations

(a) Benzene + glutaric anhydride $\xrightarrow{AlCl_3}$

(b) Benzene + $CH_2(CH_2COCl)_2$ $\xrightarrow{AlCl_3\,(excess)}$

2 equivalents 1 equivalents

(c) $C_6H_5CH_2COCl_2$ + Benzene $\xrightarrow{AlCl_3}$

(d) $C_6H_5OCH_3 + CH_3COOCCH_3$ $\xrightarrow{AlCl_3}$

Q. 102. Complete the following equations.

(a) $C_6H_5COCH_3$ $\xrightarrow[\text{diethylene glycol}, \Delta]{H_2NNH_2,\ KOH}$

(b) $C_6H_5CH_2COCH_2CH_3$ $\xrightarrow[\Delta]{Zn(Hg)\,HCl}$

(c) 2-Naphthyl–CCH_2CH_2COH (with both C=O) $\xrightarrow[\Delta]{Zn(Hg)\,HCl}$

(d) 3,4,5-Trimethoxybenzaldehyde (CH_3O, OCH_3, OCH_3 substituted benzene bearing CHO) $\xrightarrow[\text{diethylene glycol}, \Delta]{H_2NNH_2,\ KOH}$

(e) 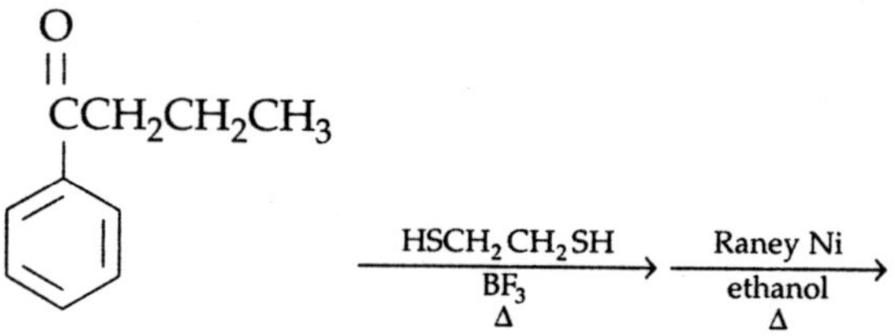 $C_6H_5COCH_2CH_2CH_3 \xrightarrow[BF_3,\ \Delta]{HSCH_2CH_2SH} \xrightarrow[\text{ethanol},\ \Delta]{\text{Raney Ni}}$

Q. 103. Predict the major product(s) of the following reactions.

(a) 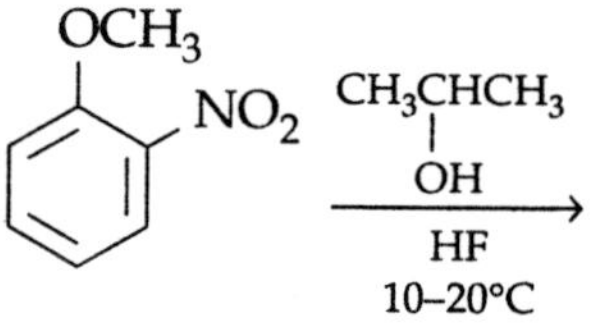 2-nitroanisole (OCH_3, NO_2 on benzene) $\xrightarrow[HF,\ 10\text{–}20°C]{CH_3CH(OH)CH_3}$

(b) 4-nitrophenol (OH, NO_2 para on benzene) $\xrightarrow[\text{acetic acid},\ 25°C]{Br_2}$

(c) o-xylene (CH_3, CH_3 ortho on benzene) $\xrightarrow[Fe]{Br_2}$

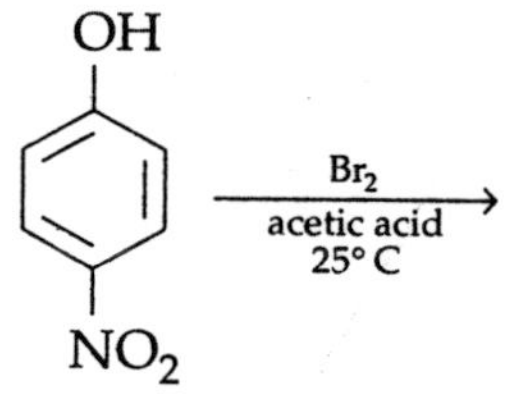

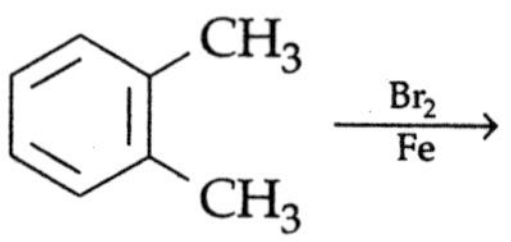

Q. 104. Complete the following equations.

(a) 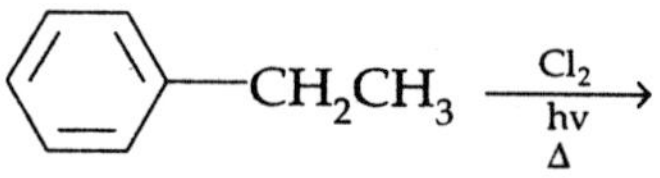 $C_6H_5CH_2CH_3 \xrightarrow[hv,\ \Delta]{Cl_2}$

(b) indene $\xrightarrow[\text{benzeyl peroxide, carbon tetrachloride},\ \Delta]{NBS}$

(c) 2-methylnaphthalene (CH_3) $\xrightarrow[\text{crbon tetrachloride},\ \Delta]{NBS}$

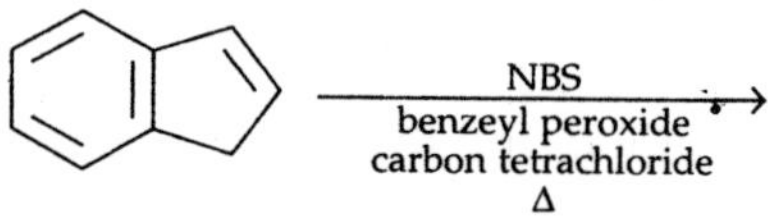

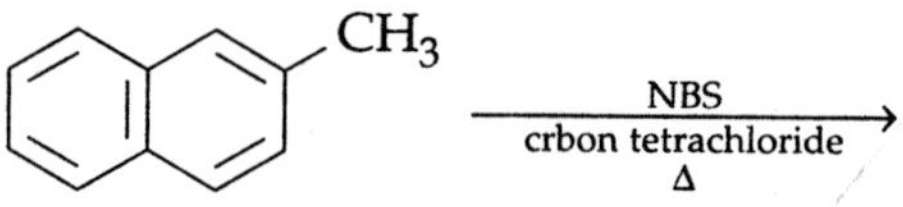

(d) 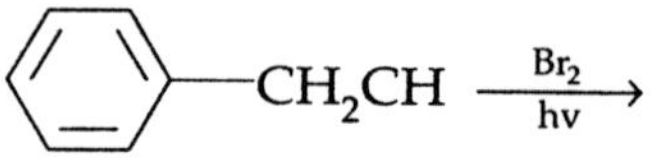 $C_6H_5CH_2CH \xrightarrow[hv]{Br_2}$

Q. 105. Complete the following equations.

(a) 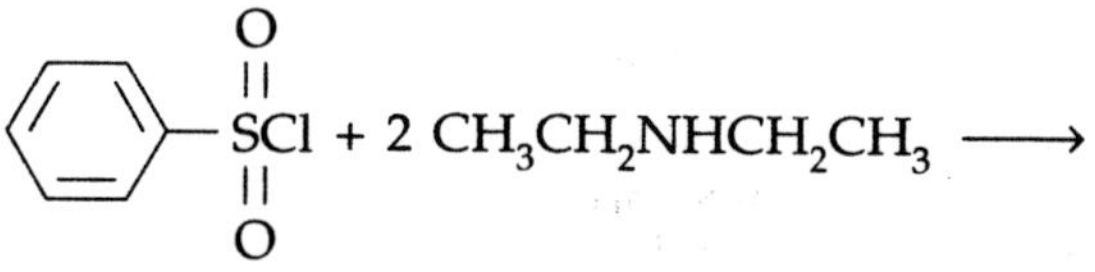 $C_6H_5SO_2Cl + 2\ CH_3CH_2NHCH_2CH_3 \longrightarrow$

(b) 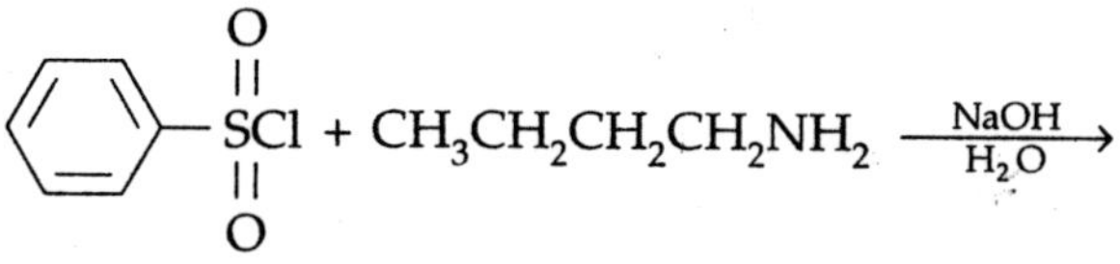 $C_6H_5SO_2Cl + CH_3CH_2CH_2CH_2NH_2 \xrightarrow[H_2O]{NaOH}$

(c) product of part (b) $\xrightarrow{H_3O^+}$

(d) 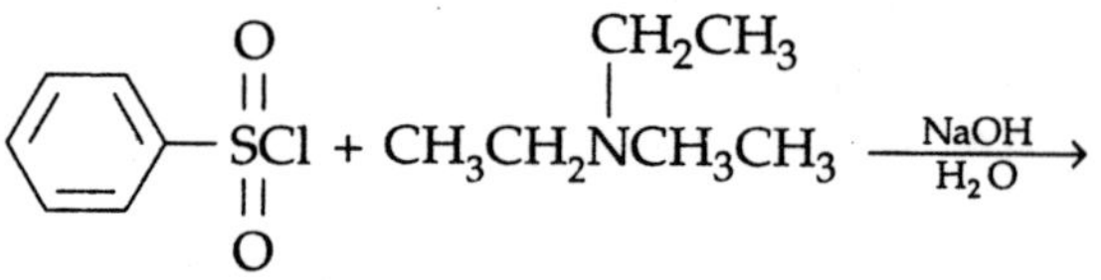 $C_6H_5SO_2Cl + CH_3CH_2N(CH_2CH_3)CH_3CH_3 \xrightarrow[H_2O]{NaOH}$

Q. 106. Give structures for compounds designated by letters in the following equations.

(a) 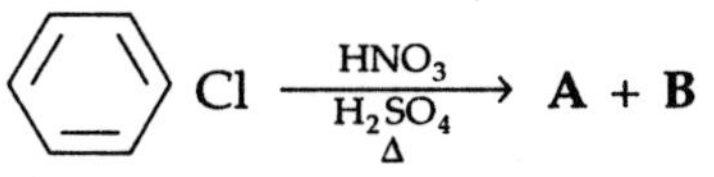 $C_6H_5Cl \xrightarrow[\Delta]{HNO_3, H_2SO_4}$ **A + B**

(b) $p\text{-}CH_3C_6H_4CH(CH_3)CH_3 \xrightarrow[AlCl_3]{CH_3COCl}$ **C + D**

(c) $CH_3C_6H_4COCH_2CH_2CH_3 \xrightarrow[\text{ethylene glycol}, \Delta]{H_2NH_2, KOH}$ **E**

(d) $CH_3C_6H_4COCH_2CH_2CH_3 \xrightarrow[FrBr_3, \Delta]{Br_2}$ **F**

(e) $C_6H_5COCH_3$ (with CH_3) $\xrightarrow[\Delta]{Zn(Hg), HCl}$ **G**

(f) $C_6H_5CH(CH_3)CH_3 \xrightarrow[AlCl_3]{CH_3CHClCH_3}$ **H + I**

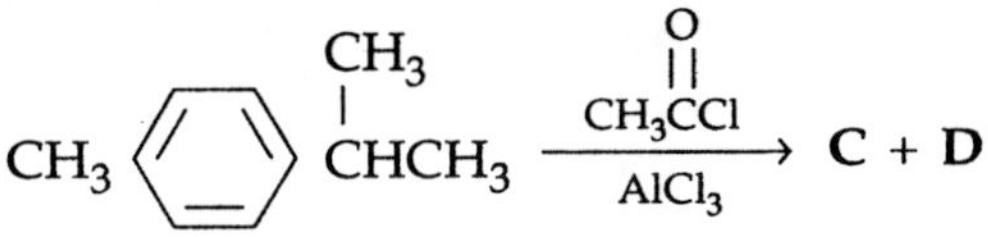

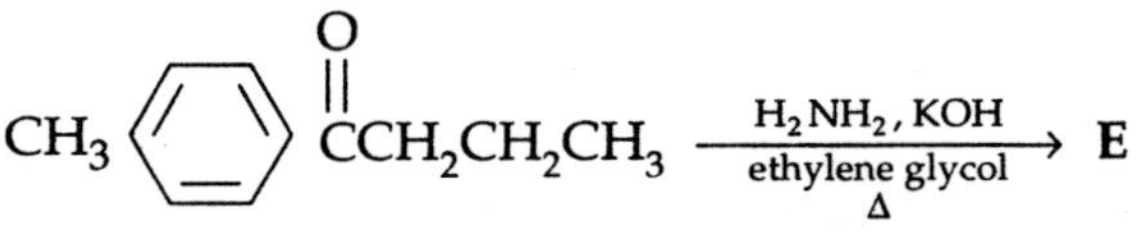

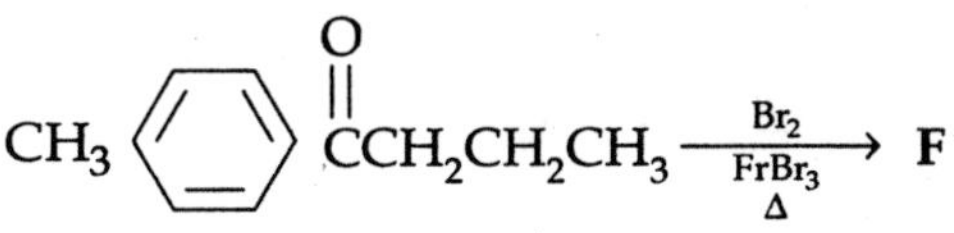

(g) $C_6H_5CH_3$ + cyclopentene $\xrightarrow{HF}$ **J + K**

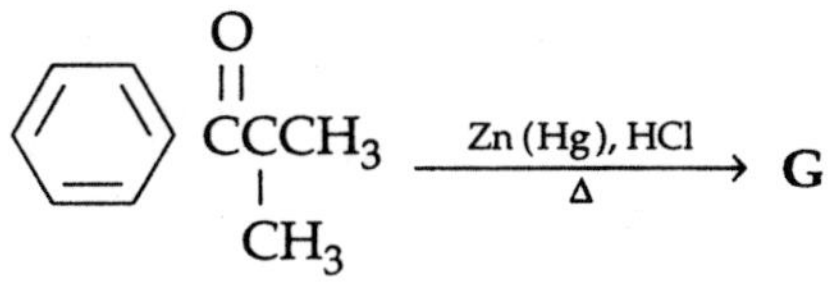

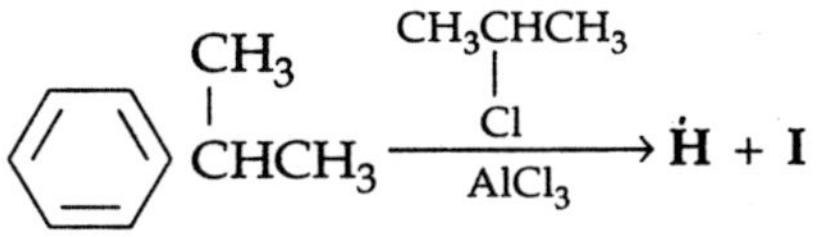

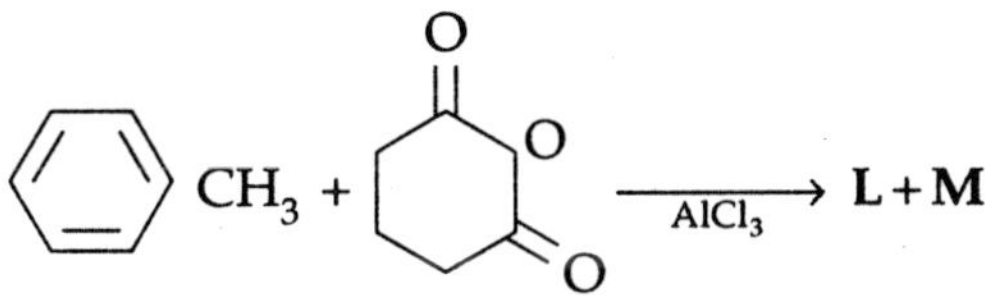

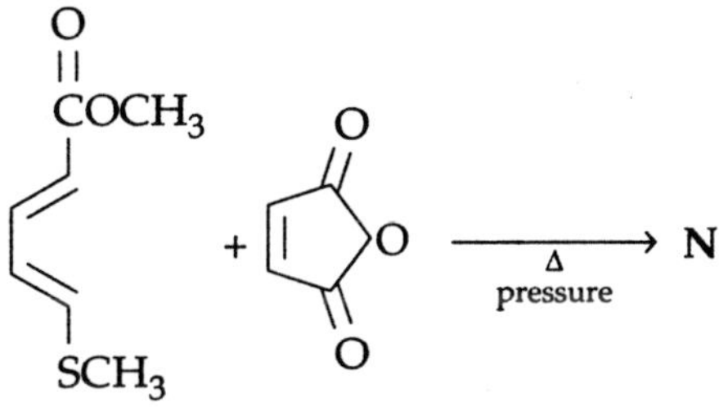

(h) $C_6H_5CH_3$ + (cyclic anhydride) $\xrightarrow{AlCl_3}$ **L + M**

(i) $CH_3OOC\text{-}CH{=}CH\text{-}CH{=}CH\text{-}SCH_3$ + maleic anhydride $\xrightarrow[\text{pressure}]{\Delta}$ **N**

(j) $\xrightarrow{O}$ $\xrightarrow[\text{(no } H_2)]{Pd/C, \Delta}$ **P** $C_{26}H_{16}$

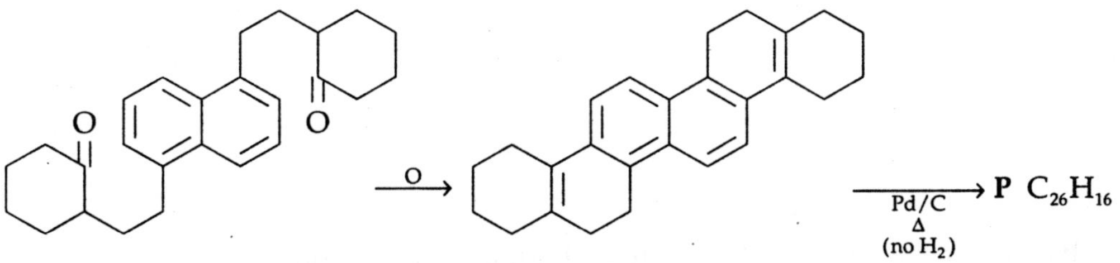

Q. 107. Give structures for the compounds designated by letters in the following equations

(a) 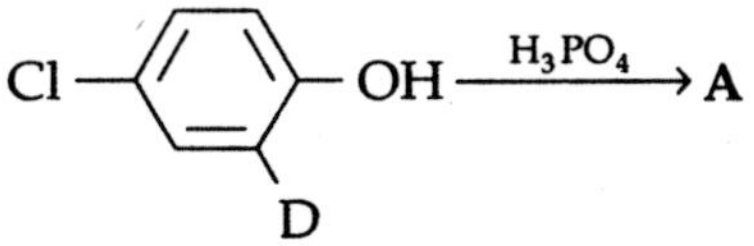 $Cl\text{-}C_6H_3(D)\text{-}OH \xrightarrow{H_3PO_4}$ **A**

(b) $C_6H_5NO_2 \xrightarrow[FeBr_2, \Delta]{Br_2}$ **B**

(c) $C_6H_5OCH_3 \xrightarrow[H_2SO_4]{SO_3}$ **C + D**

(d) $C_6H_5CH_3$ + phthalic anhydride $\xrightarrow{AlCl_3}$ **E + F**

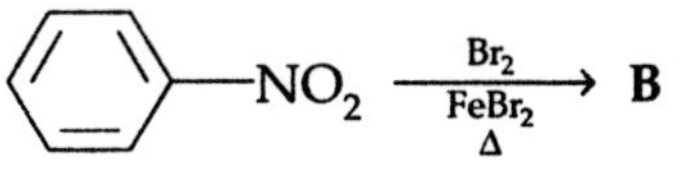

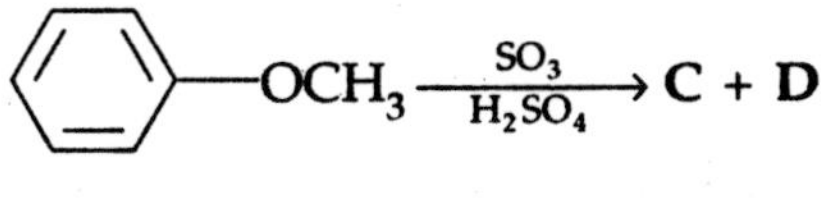

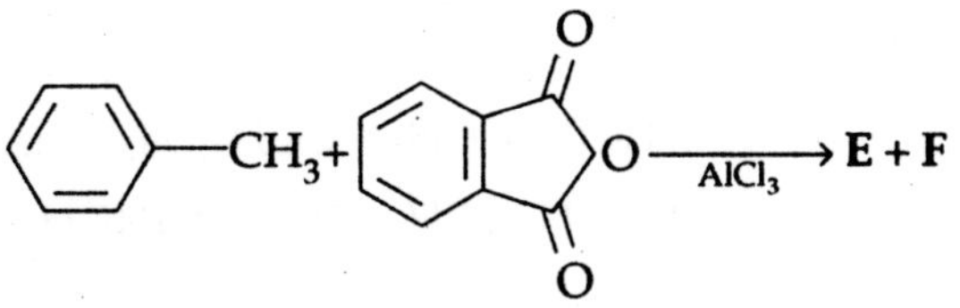

(e) 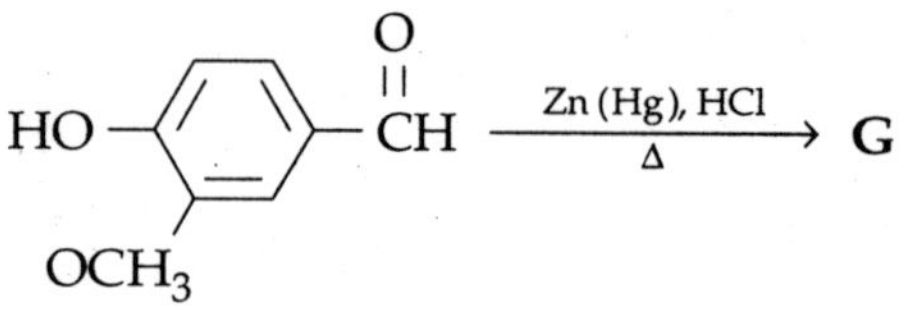 HO–(benzene ring, OCH_3)–$\overset{O}{\overset{||}{C}}H$ $\xrightarrow[\Delta]{Zn(Hg),\ HCl}$ **G**

(f) 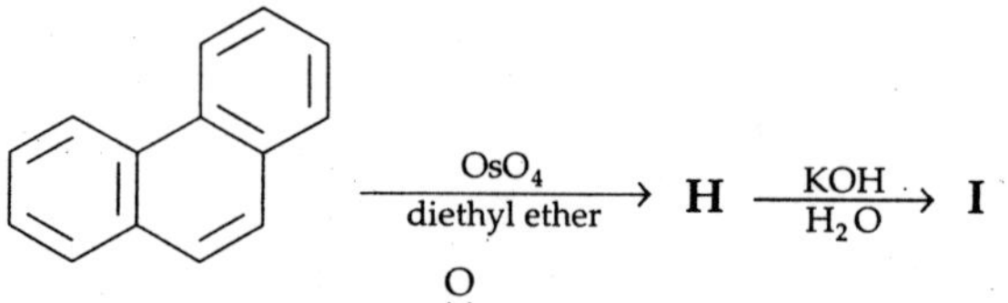 (phenanthrene) $\xrightarrow[\text{diethyl ether}]{OsO_4}$ **H** $\xrightarrow[H_2O]{KOH}$ **I**

(g) 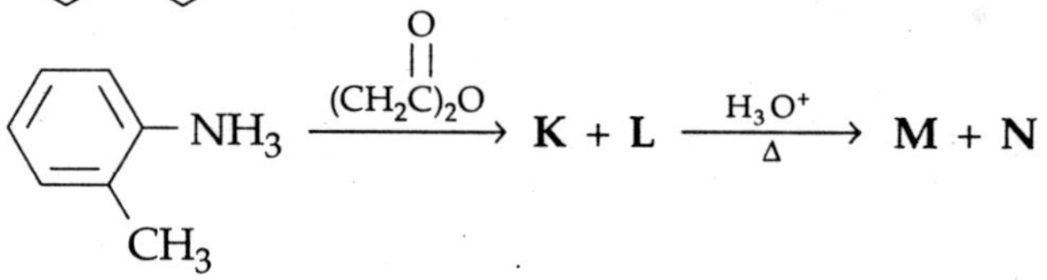 (benzene ring, CH_3)–NH_3 $\xrightarrow{(CH_3\overset{O}{\overset{||}{C}})_2O}$ **K** + **L** $\xrightarrow[\Delta]{H_3O^+}$ **M** + **N**

(h) 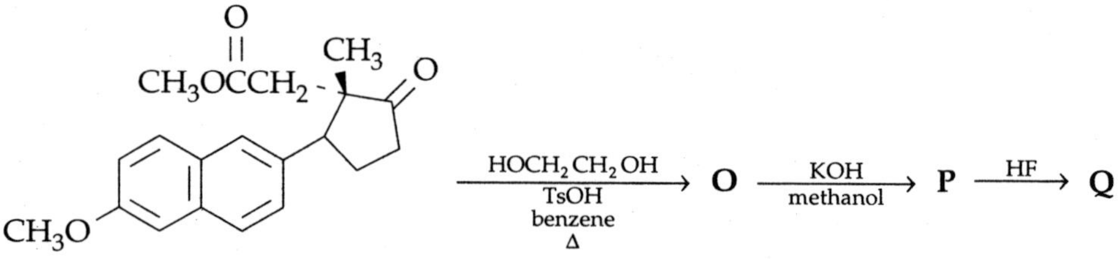 CH_3O–(naphthalene)–(cyclopentanone bearing CH_3 and $CH_3O\overset{O}{\overset{||}{C}}CH_2$–) $\xrightarrow[\text{TsOH, benzene, }\Delta]{HOCH_2CH_2OH}$ **O** $\xrightarrow[\text{methanol}]{KOH}$ **P** $\xrightarrow{HF}$ **Q**

(i) (benzene ring)–$\overset{O}{\overset{||}{C}}OH$ $\xrightarrow[H_2SO_4,\ \Delta]{SO_3\text{(excess)}}$ **R**

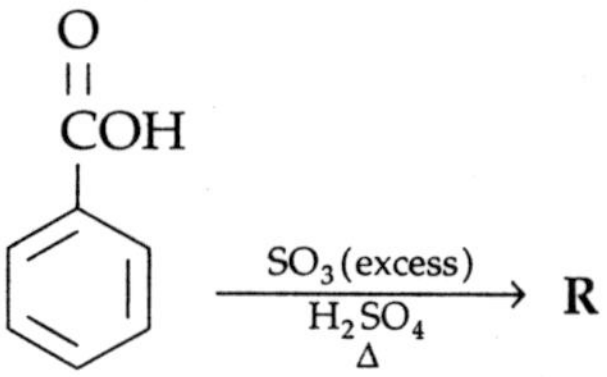

(j) 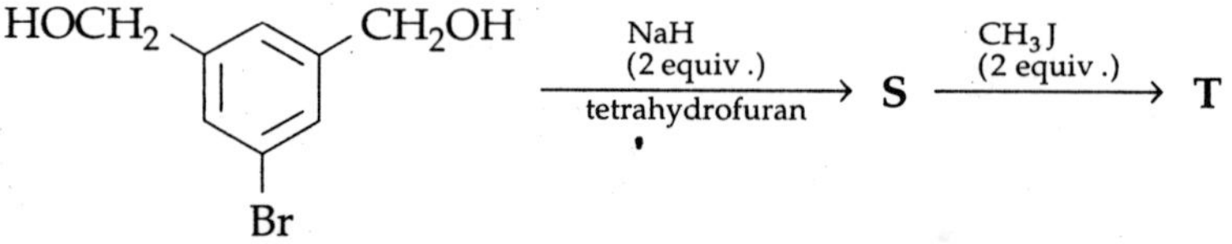 $HOCH_2$–(benzene ring, Br)–CH_2OH $\xrightarrow[\text{tetrahydrofuran}]{\text{NaH (2 equiv.)}}$ **S** $\xrightarrow{CH_3J\text{ (2 equiv.)}}$ **T**

Q. 108. Predict the product(s) of each of the following reactions.

(a) 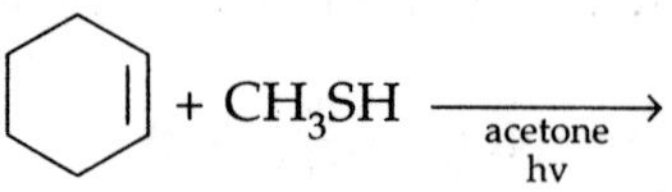 (cyclohexene) + CH_3SH $\xrightarrow[h\nu]{\text{acetone}}$

(b) $CH_3(CH_3)_5CH{=}CH_2$ $\xrightarrow[\text{peroxides}]{HBr}$

(c) CH_3CBr + $CH_2{=}\underset{}{\overset{CH_2CH_3}{\overset{|}{C}}}CH_2CH_3$ $\xrightarrow[\Delta]{}$

(d) $CH_3(CH_2)_4SiH_3$ + $CH_2{=}CH(CH_2)_5CH_3$ $\xrightarrow[\text{peroxides}]{}$

Q. 109. Predict the products of the following reactions.

(a) 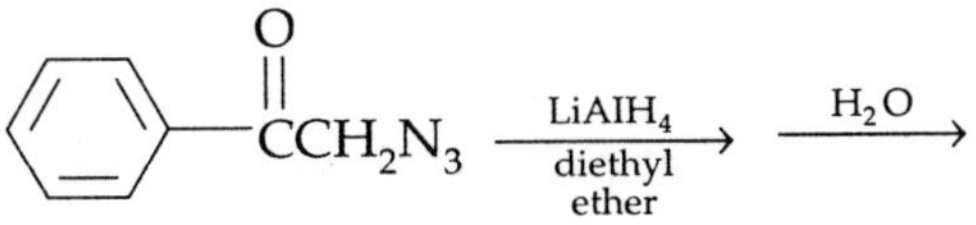 $C_6H_5\overset{O}{\overset{\|}{C}}CH_2N_3 \xrightarrow[\text{diethyl ether}]{LiAlH_4} \xrightarrow{H_2O}$

(b) $CH_3(CH_2)_4C{\equiv}CH_2)_8N_3 \xrightarrow[\text{Pd/CaCO}_3\text{ quinoline}]{H_2}$

Q. 110. The following reactions were carried out in the synthesis of compounds used to investigate intramolecular hydrogen bonding between amines and carboxylic acids. Provide structural formulas for the compounds designated by letters.

$H_3CO\overset{O}{\overset{\|}{C}}$–(decalin, H_3C, ethylene ketal) $\xrightarrow[\substack{CH_3OH \\ H_2O}]{HCl}$ A $\xrightarrow[\text{pyridine methanol}]{HONH_2 \cdot HCl}$ B $\xrightarrow[\substack{PrO_2 \\ \text{acetic acid}}]{H_2}$ $\xrightarrow{HCl\,(g)}$ C

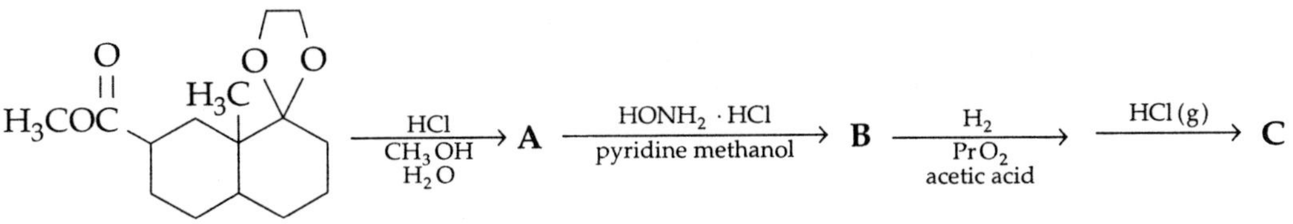

Q. 111. Amino acid often serve as chiral starting materials in syntheses. The methyl ester of the amino acid senine is the starting point for this synthesis. Provide structural formulas for the regions and products designated by letters.

$HOCH_2CH(NH_3^+Cl^-)\overset{O}{\overset{\|}{C}}OCH_2$ (H) $\xrightarrow{pH\,8}$ A $\xrightarrow[\substack{\text{methanol} \\ \text{30 min}}]{C_6H_5\overset{O}{\overset{\|}{C}}H}$ B $\xrightarrow{NaBH_4}$ $\xrightarrow{HCl\,(g)}$ C

$\xrightarrow{D}$ $HOCH_2CH(NHCH_2C_6H_5)\overset{O}{\overset{\|}{C}}NH_2$ (H)

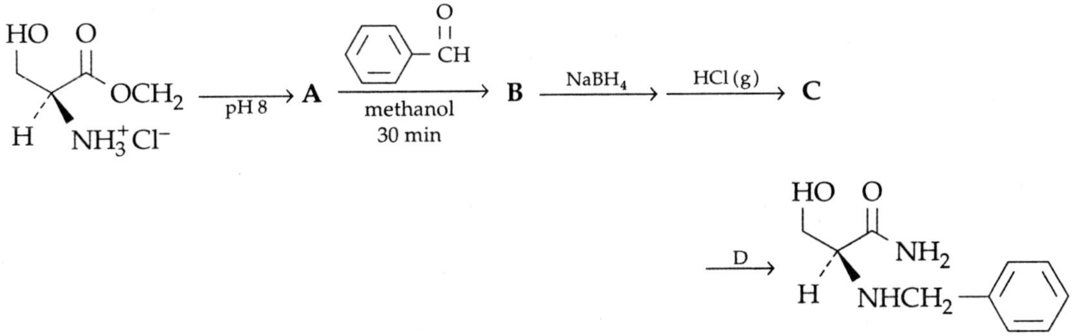

Q. 112. Complete the following equations.

(a) 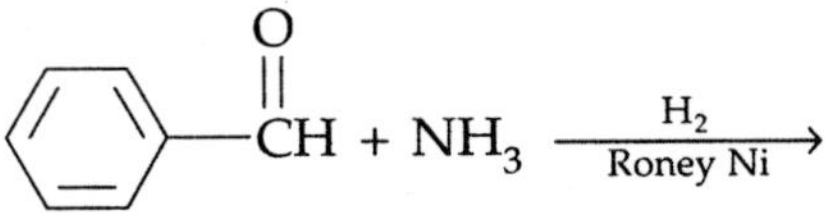 $C_6H_5\overset{O}{\overset{\|}{C}}H + NH_3 \xrightarrow[\text{Roney Ni}]{H_2}$

(b) $C_6H_5CH_2CH_2CH_3 \xrightarrow[\text{diethyl ether}]{LiAlH_4} \xrightarrow{H_2O}$

(c) $O_2N{-}C_6H_4{-}CH_2\overset{O}{\overset{\|}{C}}OH \xrightarrow[Ni]{H_2}$

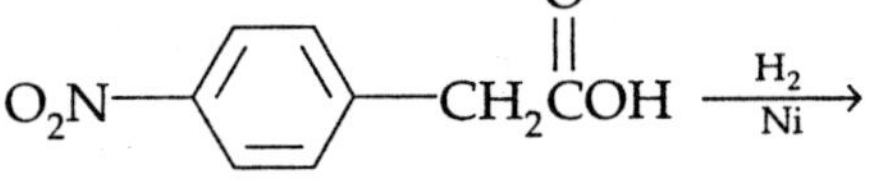

(d) 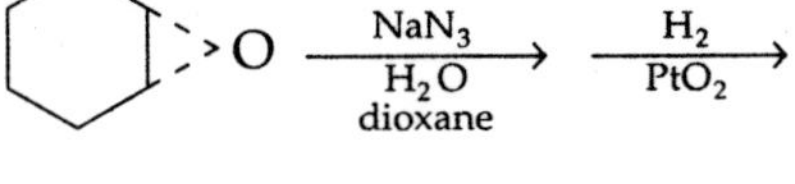 (cyclohexene oxide) $\xrightarrow[\substack{H_2O \\ \text{dioxane}}]{NaN_3} \xrightarrow[PtO_2]{H_2}$

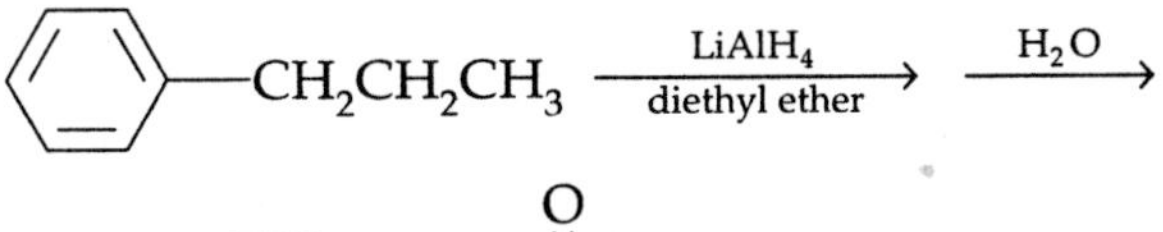

(e) Cyclohexenyl–CH_2CN $\xrightarrow[\text{diethyl ether}]{LiAlH_4}$ $\xrightarrow{H_2O}$

Q. 113. Complete the following equations, showing what product(S) you expect in each case.

(a) $CH_3CH_2CH_2CH(NH_2)CH_3 \xrightarrow[\substack{H_2O\\273K}]{NaNO_2,\ HCl}$

(b) Piperidine (N–H) $\xrightarrow[\substack{H_2O\\273\ K}]{NaNO_2,\ HCl}$

(c) $CH_3CH_2CH_2N(CH_3)CH_3 \xrightarrow[\substack{CH_3CO^-Na^+ \ (C{=}O)\\ \Delta}]{NaNO_2,\ CH_3COOH}$

(d) $CH_3NHC(=O)NHCH_3 \xrightarrow[\substack{H_2O\\273K}]{NaNO_2,\ HCl}$

(e) $CH_3C(CH_3)_2NH_2 \xrightarrow[\substack{H_2O\\273K}]{NaNO_2,\ HCl}$

Q. 114. Complete the following equations.

(a) $ClC(=O)Cl + NH_3\ (excess) \longrightarrow$

(b) $ClC(=O)Cl + C_6H_5CH_2OH$ (1 molar equiv) $\longrightarrow$

(c) $C_6H_5N{=}C{=}O + CH_3CH_2CH_2NH_2 \longrightarrow$

(d) $CH_3CH_2OCONH_2 + H_2O \xrightarrow[\Delta]{H_3O^+}$

(e) $C_6H_5N{=}C{=}S + C_6H_5NH_2 \longrightarrow$

phethyl isothiocyanate

Q. 115. More practice in recognizing reactions follows

(a) Br–cyclohexene–Br $\xrightarrow{CH_3NHCH_3\ (2\ molar\ equiv)}$ **A** $\xrightarrow{BaO}$ **B** $\xrightarrow{CH_3\ (2\ molar\ equiv)}$

C $\xrightarrow[\substack{H_2O\\ \Delta}]{Ag_2O}$ **D + E**

(b) 2,2-diphenylcyclohexanone (ring C=O with two C_6H_5 groups on the adjacent carbon) $\xrightarrow[\text{pyridine, }\Delta]{HO\overset{+}{N}H_3Cl^-}$ **F** $\xrightarrow[\text{diethyl ether}]{LiAlH_4}$ $\xrightarrow{H_3O^+}$ **G** $\xrightarrow[H_2O]{NaOH}$ **H**

(c) ^-O_2S–C_6H_4–$\overset{+}{N}H_3$ $\xrightarrow[H_2O]{NaCO_3}$ **I** $\xrightarrow[\substack{H_2O \\ 273\text{–}278\ K}]{NaNO_2,\ HCl}$ **J**

(d) 1-azidonaphthalene (N_3) $\xrightarrow[\text{diethyl ether}]{LiAlH_4}$ $\xrightarrow{H_2O}$ **K**

(e) 3,3,5-trimethylcyclohexanone (CH_3, CH_3, CH_3, =O) $\xrightarrow[\substack{Ni \\ \Delta,\ \text{pressure}}]{NH_3,\ H_2}$ **L**

(f) $CH_3\overset{O}{\overset{||}{C}}CH_3$ $\xrightarrow[\text{HCl(catalyst)}]{\text{cyclohexyl}-NH_2}$ **M** $\xrightarrow{NaOH}$ **N**

(g) 3-bromobenzamide (Br–C_6H_4–$\overset{O}{\overset{||}{C}}NH_2$) $\xrightarrow[H_2O]{Br_2,\ NaOH}$ **O**

(h) F–$CH_2CH_2CH_2$–OH $\xrightarrow{P}$ F–$CH_2CH_2CH_2$–OTs $\xrightarrow{O}$ product with N (pyrrolidinyl)–CH_2CH_2–N($CH_2CH_2CH_2$F)–CH_2CH_2–(3,4-dichlorophenyl) (Cl, Cl)

Q. 116. Write structural formulas for all compounds symbolized by letters in the following equations.

(a) $CH_3\underset{Cl}{\underset{|}{C}H}\overset{O}{\overset{||}{C}}OCH_2CH_3$ $\xrightarrow[\text{dimethyl sulfoxide}]{NaNO_2}$ **A**

(b) 2-methylbenzonitrile ($C\equiv N$, CH_3) $\xrightarrow[\text{diethyl ether}]{LiAlH_4}$ $\xrightarrow{H_2O}$ **B**

(c) $CH_3\overset{O}{\overset{||}{C}}NH_2$ $\xrightarrow[H_2O]{Br_2,\ KOH}$ **C**

(d) 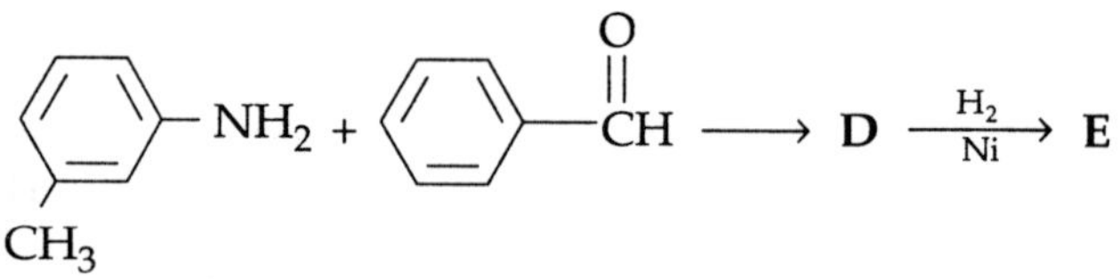 $\text{(3-}CH_3\text{)}C_6H_4\text{–}NH_2 + C_6H_5\text{–}\overset{O}{\overset{\|}{C}}H \longrightarrow D \xrightarrow[Ni]{H_2} E$

(e) $CH_3CH_2\underset{\underset{OH}{|}}{C}HCH_2CH_2CH_3 \xrightarrow[\text{pyridine}]{TsCl} F \xrightarrow{CH_3NHCH_3} \xrightarrow{NaOH} G \xrightarrow[\text{acetontrile}]{CH_3I} H$

(f) $Br(CH_2)_{10}Br + CH_3\overset{\overset{CH_3}{|}}{N}CH_3 \xrightarrow[\substack{\text{methanol} \\ 298\text{ K} \\ 2\text{ weeks}}]{} I \xrightarrow[H_2O]{Ag_2O} J \xrightarrow{373\text{–}473\text{ K}} K$

1 equivalent 2 equivalent

(g) 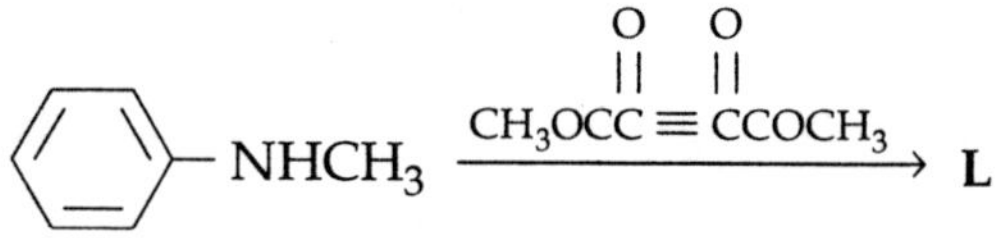 $C_6H_5\text{–}NHCH_3 \xrightarrow{CH_3O\overset{O}{\overset{\|}{C}}C\equiv C\overset{O}{\overset{\|}{C}}OCH_3} L$

(h) 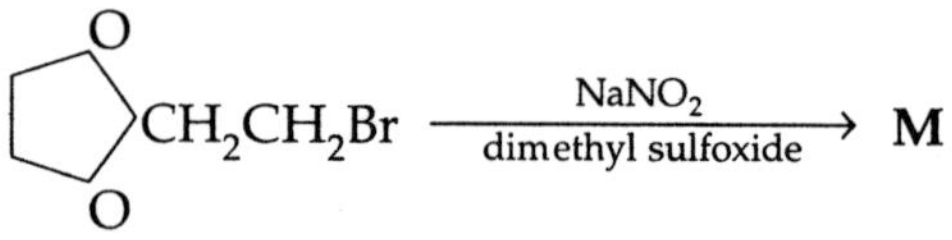 (1,3-dioxolan-2-yl)$CH_2CH_2Br \xrightarrow[\text{dimethyl sulfoxide}]{NaNO_2} M$

(i) 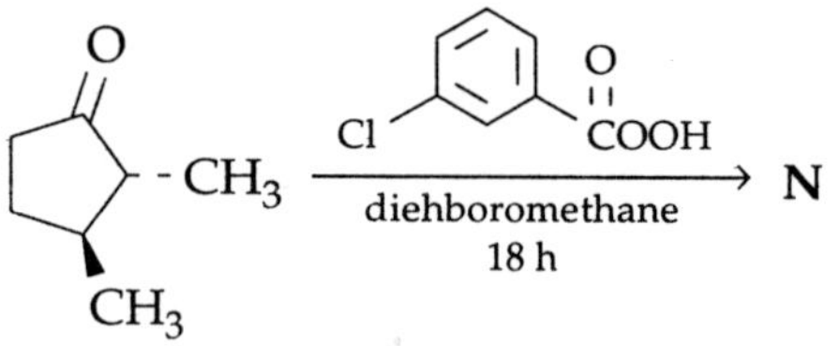 2,3-dimethylcyclopentanone $\xrightarrow[\substack{\text{diehboromethane} \\ 18\text{ h}}]{\text{Cl–}C_6H_4\text{–COOH (}m\text{-chloroperbenzoic acid)}} N$

(j) 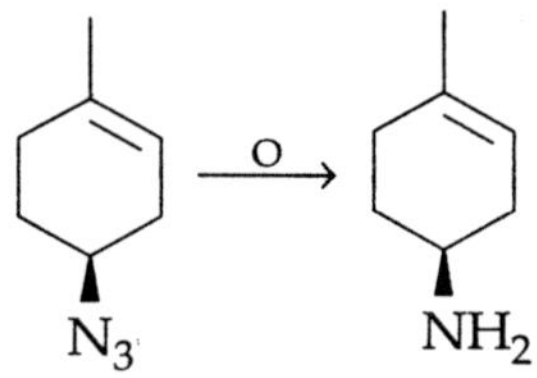 4-methylcyclohex-3-enyl–N_3 $\xrightarrow{O}$ 4-methylcyclohex-3-enyl–NH_2

Q. 117. Show how you would carry out each of the following transformations. Some of them may require more than one step.

(a) 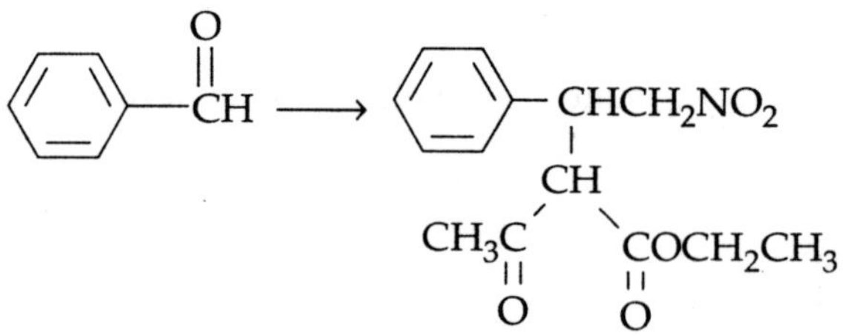 $C_6H_5\text{–}\overset{O}{\overset{\|}{C}}H \longrightarrow C_6H_5\text{–}\underset{\underset{\displaystyle CH(\overset{O}{\overset{\|}{C}}CH_3)(\overset{O}{\overset{\|}{C}}OCH_2CH_3)}{|}}{C}HCH_2NO_2$

(b) $HO\overset{O}{\overset{\|}{C}}CH_2\overset{\overset{CH_3}{|}}{C}HCH_2CH_2CH_2\overset{O}{\overset{\|}{C}}OH \longrightarrow H_2NCH_2\overset{\overset{CH_3}{|}}{C}HCH_2CH_2NH_2$

(c) $CH_3CH_2\overset{O}{\overset{\|}{C}}CH_2CH_3 \longrightarrow CH_3CH_2\overset{O}{\overset{\|}{C}}NHCH_2CH$

(d) 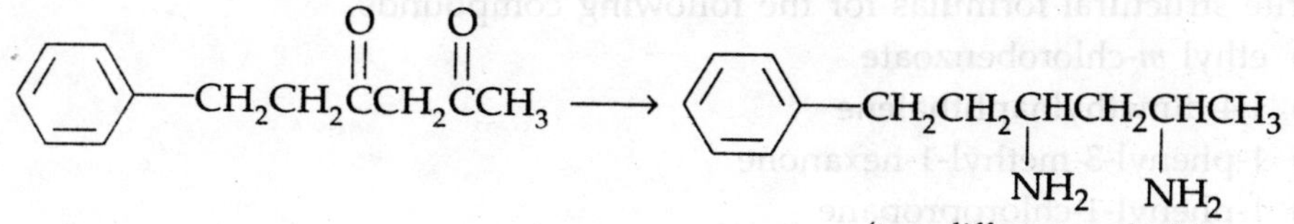 $C_6H_5CH_2CH_2\overset{O}{\overset{\|}{C}}CH_2\overset{O}{\overset{\|}{C}}CH_3 \longrightarrow C_6H_5CH_2CH_2\underset{NH_2}{\underset{|}{C}H}CH_2\underset{NH_2}{\underset{|}{C}H}CH_3$

(two different methods)

(e) 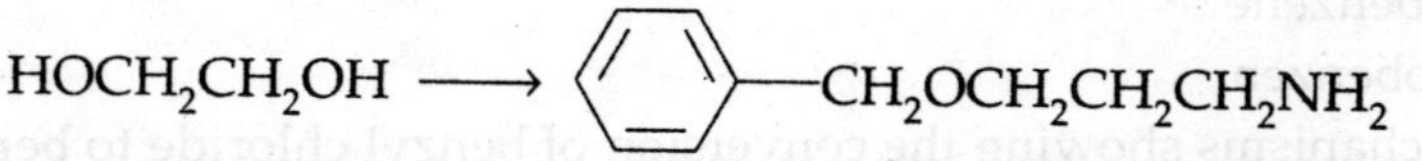 $HOCH_2CH_2OH \longrightarrow C_6H_5CH_2OCH_2CH_2CH_2NH_2$

(two different methods)

Q. 118. The following sequence of reactions was used to synthesize an amino acid. Write structural formulas for intermediates or products represented by letters.

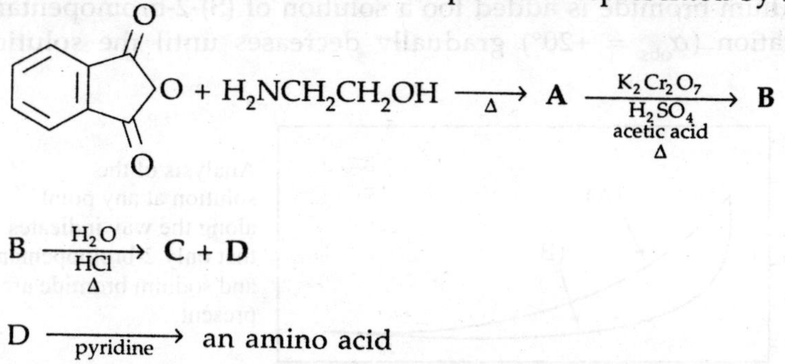

$$\text{(phthalic anhydride)} + H_2NCH_2CH_2OH \xrightarrow{\Delta} A \xrightarrow[\text{acetic acid, }\Delta]{K_2Cr_2O_7,\ H_2SO_4} B$$

$$B \xrightarrow[HCl,\ \Delta]{H_2O} C + D$$

$$D \xrightarrow{\text{pyridine}} \text{an amino acid}$$

Q. 119. A bicyclic amine has been synthesized by the following reactions. Provide structure formulas for the compound designated by letters.

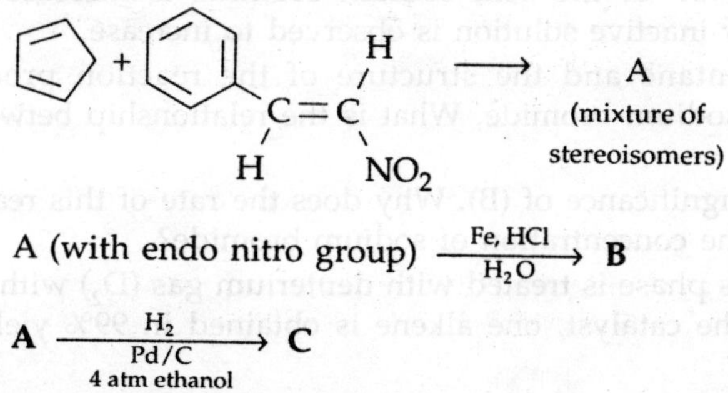

$$\text{(cyclopentadiene)} + C_6H_5(H)C{=}C(H)NO_2 \longrightarrow A \text{ (mixture of stereoisomers)}$$

$$A \text{ (with endo nitro group)} \xrightarrow[H_2O]{Fe,\ HCl} B$$

$$A \xrightarrow[\text{4 atm ethanol}]{H_2,\ Pd/C} C$$

The structure of compound C was conformed by the reactions that follow. Give structural formulas for compounds D to H.

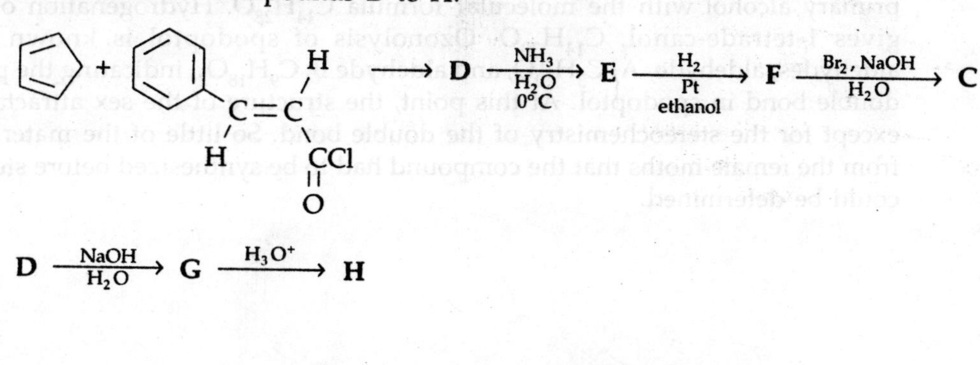

$$\text{(cyclopentadiene)} + C_6H_5(H)C{=}C(H)\overset{O}{\overset{\|}{C}}Cl \longrightarrow D \xrightarrow[H_2O,\ 0^\circ C]{NH_3} E \xrightarrow[\text{Pt, ethanol}]{H_2} F \xrightarrow[H_2O]{Br_2,\ NaOH} C$$

$$D \xrightarrow[H_2O]{NaOH} G \xrightarrow{H_3O^+} H$$

Q. 120. Write structural formulas for the following compounds.

(a) ethyl *m*-chlorobenzoate
(b) 1,4-dimethylnaphthalene
(c) 1-phenyl-3-methyl-1-hexanone
(d) 1-phenyl-1-chloropropane
(e) *m*-bromonitrobenzene
(f) *p*-chlorobromobenzen

Q. 121. Write detailed mechanisms showing the conversion of benzyl chloride to benzyl alcohol by both S_N1 and S_N2 pathways.

Q. 122. The specific rotation of (S)-2-bromopentane in ethanol is $[\alpha] = +31$

(A) When sodium bromide is added too a solution of (S)-2-bromopentane, the initially observed rotation ($\alpha_{obs} = +20°$) gradually decreases until the solution is optically inactive.

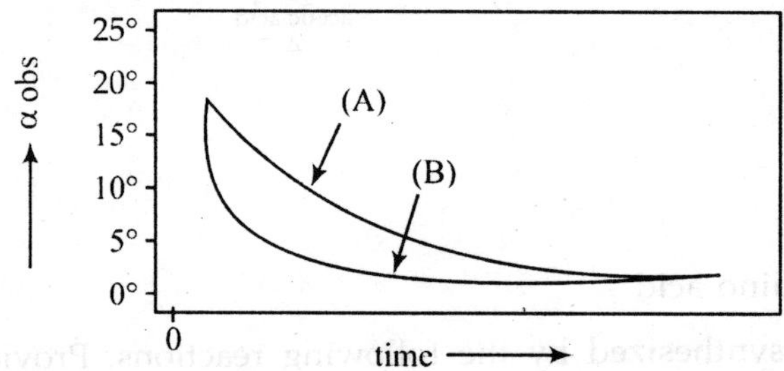

Analysis of the solution at any point along the way indicates that only 2-bromopentane and sodium bromide are present.

(B) When the concentration of the acid sodium bromide is increased, the rate of formation of the optically inactive solution is observed to increase.

(a) Draw (S)-2-bromopentane and the structure of the reaction product of (S)-2-bromopentane with sodium bromide. What is the relationship between these two compounds?

(b) Briefly describe the significance of (B). Why does the rate of this reaction increase with an increase in the concentration of sodium bromide?

Q. 123. When 2-butyne in the gas phase is treated with deuterium gas (D_2) with palladium on an alumina support as the catalyst, one alkene is obtained in 99% yield. Predict the structure of the alkene.

Q. 124. The sex attractant of the fall armyworm moth is produced by the female. This material which attracts the male moth at very low concentrations, is called spodoptol and is a primary alcohol with the molecular formula $C_{14}H_{28}O$. Hydrogenation of this alcohol gives 1-tetrade-canol, $C_{14}H_{30}O$. Ozonolysis of spodoptol is known to give two aldehydes, aldehyde, A, $C_3H_{10}O$, and aldehyde B, $C_9H_{18}O_2$, indicating the position of the double bond in spodoptol. At this point, the structure of the sex attractant is known, except for the stereochemistry of the double bond. So little of the material is isolated from the female moths that the compound had to be synthesized before stereochemistry could be determined.

(a) Draw the structural formula for compounds (or intermediates) D, E, and F formed in the synthesis of spodoptol.

$$HOCH_2CH_2CH_2CH_2CH_2CH_2CH_2CH_2C{\equiv}CH \xrightarrow[NH_3(liq)]{Na^+NH_2^-\ (excess)}$$

$$\underset{\text{dianion}}{D} \xrightarrow[(2)\ H_3O^+]{(1)\ CH_2CH_2CH_2CH_2Br} E \xrightarrow[\text{poisoned catayst}]{H_2} \underset{\text{spodopal}}{F}$$

(b) As a check on the stereochemistry of the double bond, compound E, prepared above, was also reduced with sodium metal in liquid ammonia. The alkene that resulted attracted no male moths, indicating that this was not the stereochemistry found in the natural product. What is the structure of the alkene formed in the sodium metal reduction of E?

(c) Assign structures to aldehyde A and aldehyde B.

$$\text{spodoptol} \xrightarrow[(2)\ CH_3SCH_3]{(1)\ O_3} \underset{C_5H_{10}O}{\text{aldehyde A}} + \underset{C_9H_{18}O_2}{\text{Aldehyde B}}$$

Q. 125. Supply structural formulas for the intermediates and products designated by capital letters. Show the stereochemistry of the products (s) when known.

(a) (Ph)HC=CH(Ph) $\xrightarrow[\text{dimethyl sulfoxide}]{\text{N-bromosuccinimide (N—Br), } H_2O}$ A

(b) 4-methylcyclohexene (cyclohexene ring —CH_3) $\xrightarrow[Pt]{H_2}$ B

(c) cyclopentene ring with wedged CH_3 $\xrightarrow{CH_3(CH_2)_{10}COOOH\ (O,\ ||)}$ C + D

(d) (Ph)(H)C=C(CH_3)($CH_2CH_2CH_3$) $\xrightarrow{O_3,\ CH_3OH}$ $\xrightarrow{(CH_3)_2S}$ E + F

(e) H(CH$_2$CH$_2$CH$_2$OCCH$_3$, with C=O)C=C(CH$_3$)(H) $\xrightarrow{BD_3}$ G + H $\xrightarrow{H_2O_2,\ NaOH}$ I + J

(f)  Methylenecyclohexane ($=CH_2$) $\xrightarrow[\text{benzene}]{H_2,\ RhCl[P(C_6H_5)_3]_3}$ K

(g)

$$\begin{matrix} CH_3 & & CH_3 \\ & \diagdown\ \diagup & \\ & C & \\ & \| & \\ & CH & \\ & \diagup & \\ CH_2 & & CH_2OH \\ | & & | \\ CH_2 & & CH_2 \\ & \diagdown\ \diagup & \\ & CH & \\ & | & \\ & CH_3 & \end{matrix} \xrightarrow{NaH} L \xrightarrow{C_6H_5CH_2Br} M$$

(Hint: L is a good nucleophile)

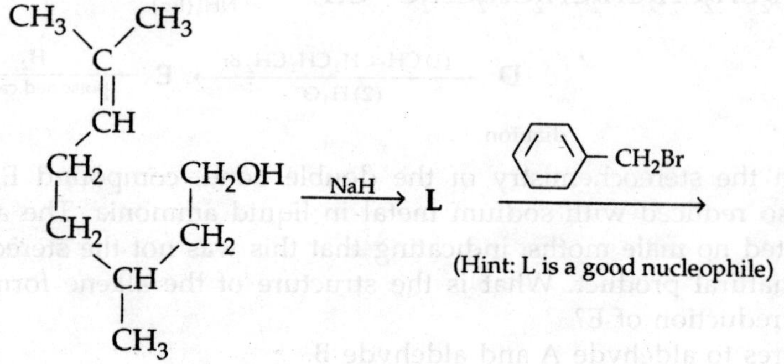

Q. 126. Fill in structural formulas for the starting material or regents that would be necessary the following transformations.

(a) 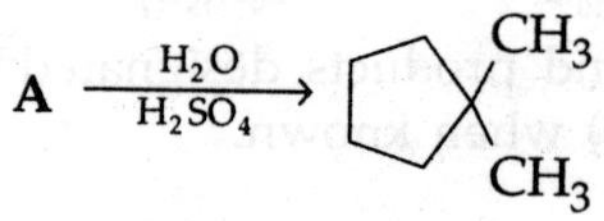 A $\xrightarrow[H_2SO_4]{H_2O}$ 1,1-dimethylcyclopentane (CH_3, CH_3)

(b) B $\xrightarrow{C}$ 2-bromocyclopentanol (Br, OH) + 2-bromocyclopentanol (HO, Br)

(c) D $\xrightarrow[\text{carbon tetrachloride}]{Br_2}$ $CH_3\overset{CH_3}{\overset{|}{C}}HCH_2-C(CH_2Br)(Br)-H$ + $H-C(CH_2Br)(Br)-CH_2\overset{CH_3}{\overset{|}{C}}HCH_3$

(d) E $\xrightarrow[\text{tetrahydofuran}]{9\text{-BBN}}$ $\xrightarrow[H_2O]{NaOH,\ H_2O_2}$ $CH_3\overset{CH_3}{\overset{|}{C}}-\underset{OH}{\underset{|}{C}}HCH_3$ (with H on the central C)

(e) F $\xrightarrow[\text{dichloromethane}]{m\text{-}ClC_6H_4COOH}$ $CH_3CH_2O\underset{CH_3CH_2O}{\underset{|}{C}}HCH\overset{O}{-}CH_2$

(f) G $\xrightarrow[\text{carbon tetrachloride}]{Br_2}$ $H-C(C_6H_5)(Br)-C(Br)(H)(CH_3)$ + $CH_3-C(H)(Br)-C(Br)(CH_3)(H)$

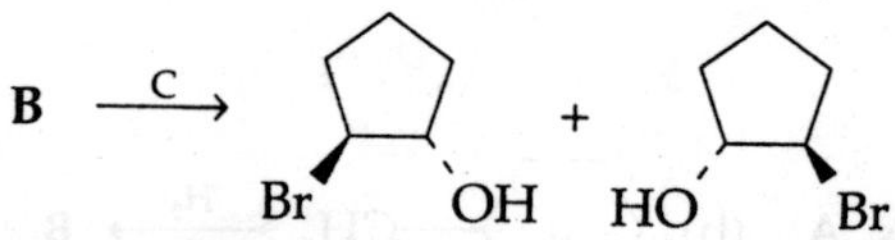

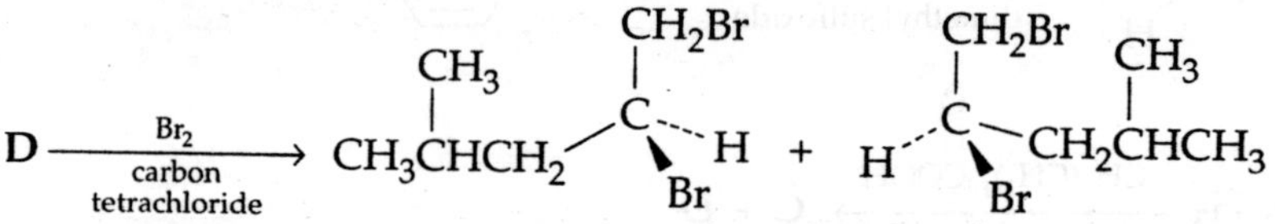

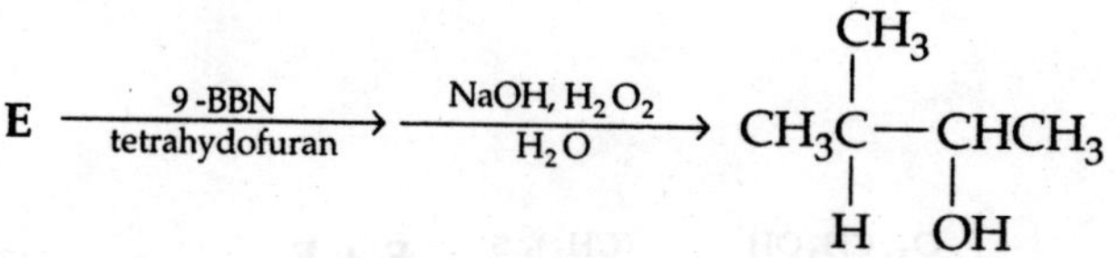

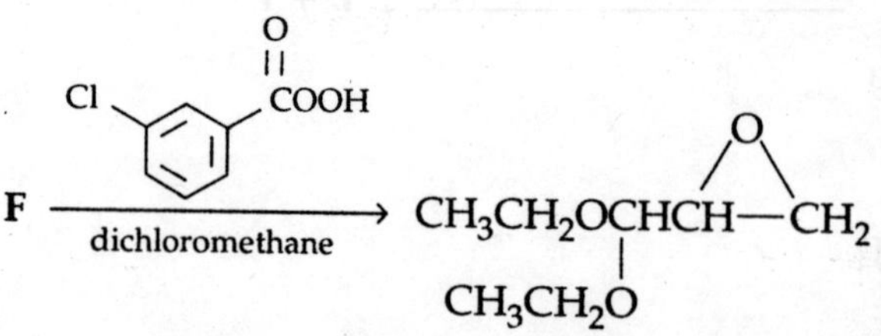

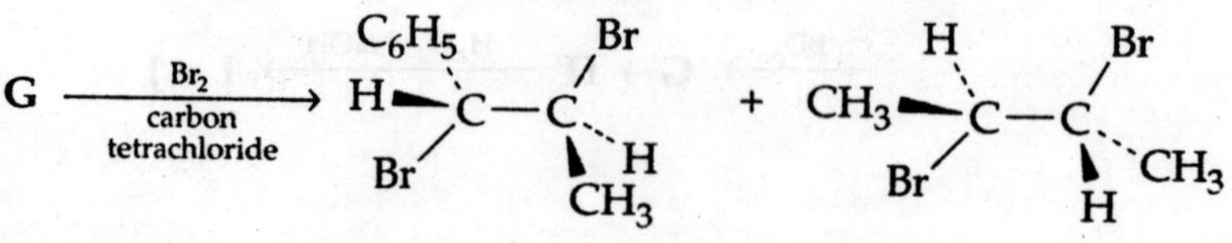

(g) $CH_3(CH_2)_6CH{=}CH_2 \xrightarrow{H} CH_3(CH_2)_6CH(Br)CH_2CH_3$

(h) I $\xrightarrow[\text{dichloromethane}]{}$ [3-(trimethylstannyl)cyclohexane epoxide, two stereoisomers] + ; Sn(CH₃)₃ products

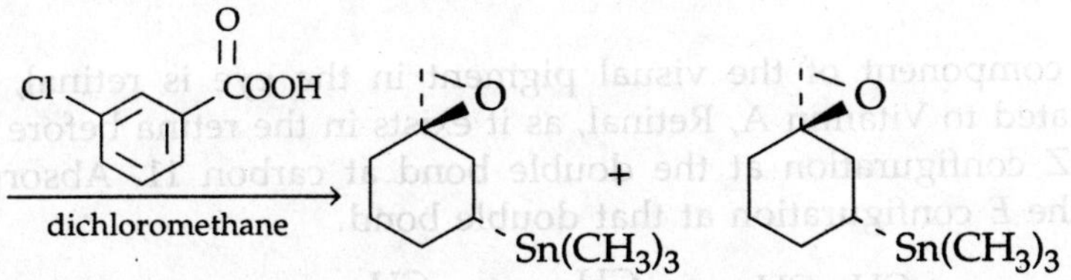

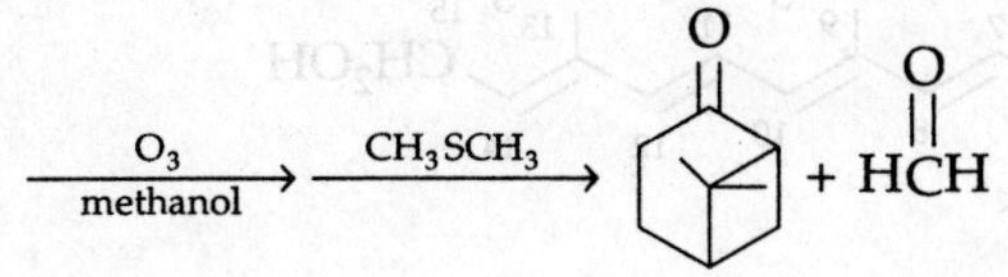

(i) J $\xrightarrow[\text{methanol}]{O_3}$ $\xrightarrow{CH_3SCH_3}$ [bicyclic ketone] + HCH (=O)

(j) [H₂C=CH– on N-containing ring with PhCH₂O and OCH₂Ph substituents] $\xrightarrow{K}$ [HO–CH(H)–C(H)(OH)– on ring with PhCH₂O and OCH₂Ph]

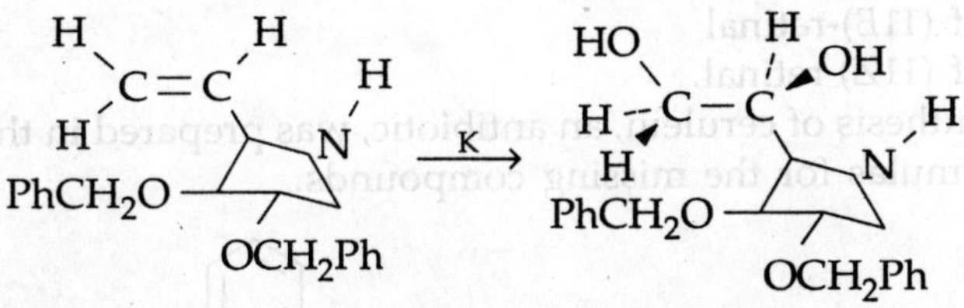

Q. 127. The two columns on the left contain structural formulas for some organic compounds. The right-hand column is a list of regents that may react with some of them. For each compound on the left, list all the regents on the right that would when mixed with the compound in a test tube produce a reaction detectable by the human senses. (The human senses, for example, can detect a change in color, the formation of a solution, the evolution of a gas, the evolution of heat, or a change in odor.)

(a) 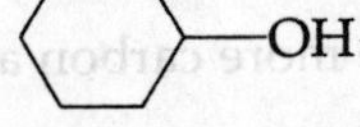 cyclohexyl–OH

(b) 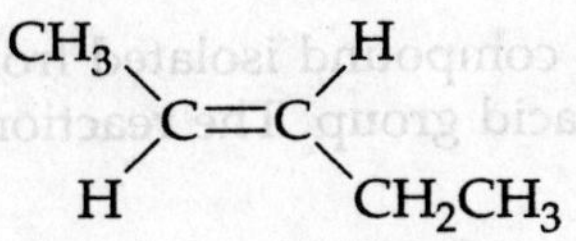 $CH_3(H)C{=}C(H)CH_2CH_3$

(c) $CH_3(H)C{=}C(H)CH_2Cl$

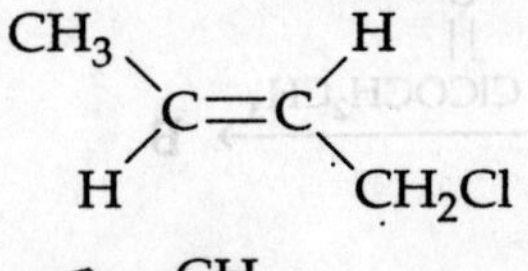

(d) 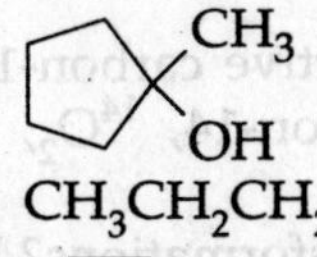 1-methylcyclopentanol (CH_3, OH)

(e) 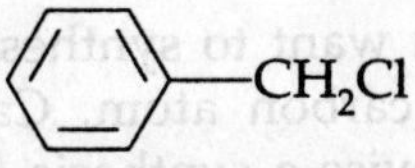 C_6H_5–CH_2Cl

(f) $CH_3CH_2CH_2CH_2CH_2OH$

(g) $CH_3C{\equiv}CCH_2CH_3$

(h) cyclohexane

(i) $HOCH_2CH_2OH$

1. H_2O, cold (solubility)
2. H_2SO_4, cld, conc (protonation)
3. Br_2, carbon tetrachloride
4. CrO_3, H_2O, H_2SO_4
5. $AgNO_3$, CH_3CH_2OH
6. NaI, acetone

(j) $CH_3CH_2C(CH_3)(Br)CH_3$ (k) cyclopentyl–OCH_2CH_3 (l) $CH_3CH_2CH_2CH_2CH_2CH_2Cl$

Q. 128. An important component of the visual pigment in the eye is retinal, the aldehyde structurally related to Vitamin A, Retinal, as it exists in the retina before absorption of light, has the Z configuration at the double bond at carbon 11. Absorption of light converts it to the *E* configuration at that double bond.

Vitamin A

(a) Draw the structure of (11*E*)-retinal

(b) Draw the structure of (11Z)-retinal.

Q. 129. An intermediate in the synthesis of cerulein, an antibiotic, was prepared in the following way. Fill in structural formulas for the missing compounds.

$$CH_3C{\equiv}CCH_2C{\equiv}CCH_2CH_2CH_2OH \xrightarrow[\text{tert-butyl alcohol, ammonium sulfate}]{\text{Li, } NH_3(\text{liq})} X \xrightarrow[\text{dichloromethane}]{C_5H_5\overset{+}{N}H\ CrO_3Cl^-} Y$$

Q. 130. Acetone reacts with glycerol (1,2,3-propanetriol) in the presence of *p*-toluenesulfonic acid to give a product that has the molecular formula $C_6H_{12}O_3$. Propose a structure of this compound.

Q. 131. Pick an alcohol and convert it into an aldehyde containing one more carbon atom, using the sequence of reactions.

Q. 132. In this synthesis of an alkaloid a nitrogen-containing compound isolated from plants, a method given below is used to activate a carboxylic acid group. The reaction is shown. Product the structures of compounds A and B.

$$\text{(2-oxocyclopentyl)}CH_2COOH \xrightarrow[\text{tetralydofuran}]{(CH_3CH_3)_3N} A \xrightarrow{ClCOCH_2CH_3} B$$

Q. 133. Suppose you want to synthesis 3-ethyl-3-pentanol with radioactive carbon-14 atom as the tertiary carbon atom. Carbon dioxide labeled with carbon-14, $^{14}O_2$, is readily available. Devise a synthesis for 3-ethyl-3-pentanol-3-^{14}C.

Q. 134. Which regents could be used to make each of the following transformations? More than one step may be necessary in some cases.

(a) $\xrightarrow{CH_3NH_2} CH_3CH_2\overset{+}{N}H_2CH_3Br^-$

(b) $CH_2{=}CH_2 \xrightarrow{?} CH_3CH_2OCH_3$

(c) $CH_3CH_2CH_2CH_2NH_2 \xrightarrow{?} CH_3CH_2CH_2CH_2\overset{+}{N}H_3Cl^-$

(d) $? \xrightarrow[\text{(excess)}]{Br_2} CH_3CH_2C(Br)_2{-}CH(Br)_2$

(e) $CH_3CH_2CH_2CH_2OH \xrightarrow{?} CH_3CH_2CH_2CH_2OCH_3$

(f) $CH_3CH{=}CHCH_3 \xrightarrow{?} CH_3CH(CN)CH_2CH_3$

(g) $CH_3CH_2CH(OH){=}CH_2 \xrightarrow{?} CH_3CH(OH)CH_2CH_2CH_3$

Q. 135. The following incomplete reaction sequences give a starting material and a product that can be prepare from it. Slow the reagents that would be necessary to carry out each transformation and any major products that would be formed in the intermediate stage. There may be more than one correct way to complete a sequence.

(a) $CH_3CH{=}CH_2 \longrightarrow CH_3CH_3CH(SCH_2CH_3)CH_3$

(b) $CH_3CH_2CH_2CH_2CH_2Br \longrightarrow CH_3CH_2CH_2CH_2CH(\overset{O}{\overset{\|}{C}}OCH_2CH_3)_2$

(c) $HC(=O){-}C_6H_2(I)(OCH_3){-}OH \longrightarrow HC(=O){-}C_6H_2(I)(OCH_3){-}OCH_2CH_2CH_3$

Q. 136. The following series of reactions was carried out during research.

$$HOCH_2CH_2CH_2CH_2CH_2\overset{O}{\overset{\|}{C}}OCH_2CH_3 \xrightarrow[\text{dichloromethane, sodium acetane}]{C_5H_5\overset{+}{N}H\ CrO_3Cl^-} A \xrightarrow[\text{tetrahydrofuran}]{CH_2{=}CHMgBr\ \text{(1 molar equivalent)}} \xrightarrow[H_2O]{NH_4Cl}$$

$$B \xrightarrow[\text{ethanol}]{\text{KOH}/H_2O} C \xrightarrow{H_3O^+} D \xrightarrow[\text{pyridine}]{(CH_3\overset{O}{\overset{\|}{C}})_2O} E$$

Match the structure of compounds A, B, C, D, and E

Q. 137. Chemists wanted to synthesize some α-chloroketo acids with the general structure

$$ClCH_2\overset{O}{\overset{\|}{C}}(CH_2)_n\overset{O}{\overset{\|}{C}}OH$$

They decide a route, which is outlined below.

$$CH_2{=}CH(CH_2)_9CH_2OH \xrightarrow[\text{acetone}]{m\text{-}ClC_6H_4COOH} A \xrightarrow[\text{tetrahydroufuran}]{CH_3\overset{O}{\overset{\|}{C}}OH,\ LiCl} B \xrightarrow[H_2SO_4,\ H_2O,\ \text{acetone}]{CrO_3\ (\text{excess})} \underset{\alpha\text{-chloroketo acid}}{C}$$

Give the structural formulas for compounds A, B, and C

Q. 138. The following sequence of reactions was used to synthesize to optically active cyclic either, compound E. Give the structural formula, for each product designed by a letter in these transformations.

$$\text{H}\cdots\underset{\text{HO}}{\overset{CH_3}{C}}-CH_2\overset{O}{\overset{\|}{C}}OCH_2CH_3 \xrightarrow[CF_3\overset{O}{\overset{\|}{C}}OH]{CH_3CH_2OCH{=}CH_2} A \xrightarrow[\text{diethyl ether}]{LiAlH_4} \xrightarrow{H_2O} B$$

$$\xrightarrow[\text{pyridine, dichloromethane}]{TaCl} C \xrightarrow[\text{tetrahydrofuran, cold}]{H_3O^+} D \xrightarrow{KOH} E\ (C_4H_8O)$$

Q. 139. The following incomplete reaction sequence give only a starting material and product that can be prepared from it. Show the regents that would be necessary to carry out each transformation and any major products that would be formed in the intermediate stages. There may be more than one correct way to complete a sequence.

(a) $C_6H_5{-}CH_2OH \longrightarrow C_6H_5{-}CH_2SCH_2CH_3$

(b) $CH_3CH_2CH_2CH_2CH_2OH \longrightarrow CH_3CH_2CH_2CH_2CH_2I$

(c) $CH_3CH_2CH_3CH_2Br \longrightarrow CH_3CH_2CH_2CH_2C{\equiv}CH$

(d) $SH \longrightarrow C_6H_5{-}SCH_2CH_2{-}C_6H_5$

(e) $HO{-}\overset{CH_3}{C}(H)(CH_2CH_3) \longrightarrow H\cdots\overset{CH_3}{C}(NH_2)(CH_3CH_2)$

(f) $C_6H_5{-}CH_2CH_2CH{=}CH_2 \longrightarrow C_6H_5{-}CH_2CH_2\overset{N_3}{\overset{|}{C}}HCH_3$

(g) $CH_3\overset{O}{\overset{\|}{C}}OCH_2\overset{CH_3}{\overset{|}{C}}=CHCH_2Cl \longrightarrow CH_3\overset{O}{\overset{\|}{C}}OCH_2\overset{CH_3}{\overset{|}{C}}=CHCH_2NH\overset{CH=CH_2}{\overset{|}{C}H}CH_2OH$

140. Dissect the following alcohols and propose regents that would lead to their formation.

(a) $CH_3\overset{CH_3}{\overset{|}{C}}=CH\overset{CH_3}{\overset{|}{\underset{OH_3}{\underset{|}{C}}}}CH_2CH_2CH_2CH_3$

(b) $CH_3\overset{CH_3}{\overset{|}{C}}=CHCH_2CH_2\overset{CH_3}{\overset{|}{C}}=CHCH_2CH_2\underset{OH}{\underset{|}{C}H}C\equiv CH$

(c) 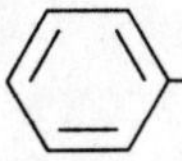 C_6H_5—CH_2CH_2OH

Q. 141. Answer the questions below for this S_N2 reaction:

$$CH_3CH_2CH_2CH_2Br + NaOH \xrightarrow[\text{ethanol}]{} CH_3CH_2CH_2CH_2OH + NaBr$$

(a) What is the rate expression for the reaction?
(b) What will be the effect on the rate of the reaction of doubling the concentration of π-butyl bromide?
(c) What will be the effect on the rate of the reaction of having the concentration of sodium hydroxide?

Q. 142. Answer the questions below for this S_N1 reaction:

$$C_6H_5\overset{C_6H_5}{\overset{|}{\underset{Br}{\underset{|}{C}}}}CH_3 \xrightarrow{CH_3CH_2OH} C_6H_5\overset{C_6H_5}{\overset{|}{\underset{OCH_2CH_3}{\underset{|}{C}}}}CH_3 + HBr$$

1-bromo-1,1-dipenylethane

What is the rate expression for the reaction?

Q. 143. Show how you would carry out each of the following transformations.

(a) $HC\equiv CH \xrightarrow{??} CH_3CH_2CH_2\underset{Br}{\underset{|}{C}H}CH_3$

(b) $CH_3CH_2C\equiv CH \xrightarrow{??} CH_3CH_2CH_2\overset{O}{\overset{\|}{C}}CH_2CH_3$

(c) $CH_3CH_2CH_2C\equiv CH \xrightarrow{??}$ (H)(HO)(CH₃CH₂CH₂)C—C(OH)(H)(CH₂CH₃) + enantiomer

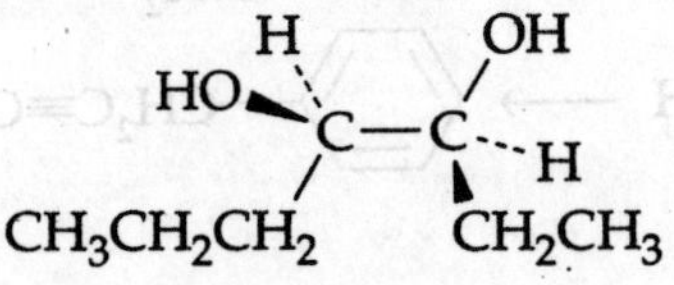

Q. 144. How would you carry out each of the following transformations? For some, more than one step may be necessary.

(a) Cyclohexane ring bearing H, $COCH_3$ (C=O), OH, H, $CH_2CH{=}CH_2$, H $\longrightarrow$ cyclohexane ring bearing H, $COCH_3$ (C=O), H, N_3, $CH_2CH_2CH_2OH$, H

(b) $CH_3CH_2C{\equiv}CCH_2CH_3 \longrightarrow$ H---C—C---H epoxide (O bridging), with CH_2CH_3 and CH_2CH_3

(c) $CH_3CH_2C{\equiv}CCH_2CH_3 \longrightarrow$ CH_2CH_3, $CH_3CH_2CH_2$—C---H, OH + enantiomer

(d) 1-methylcyclopentene (CH_3) $\longrightarrow$ $HCCH_2CH_2CH{-}CH$ (O, CH_3, O)

(e) 1-methylcyclohexene (CH_3) $\longrightarrow$ 1-bromo-1-methylcyclohexane (CH_3, Br)

(f) Indene $\longrightarrow$ indane diol (OH, OH) + enantiomer

Q. 145. How would you carry out each of the following transformation?

(a) 1-methylcyclohexene (CH_3) $\longrightarrow$ 2-methylcyclohexanone (CH_3, O)

(b) $CH_3CH_2CH{=}CH_2 \longrightarrow CH_3CH_2CHCHCH_3$ (NH_2)

(c) C_6H_5—$CH_2OH \longrightarrow C_6H_5$—$CH_2C{\equiv}CCH_3$

Q. 146. Supply the reagents that are necessary and the intermediate compounds that will form in the following transformation. There may be more than one good way to carry out each synthesis.

(a) $CH_3CH(CH_3)CH_2CH{=}CH_2 \longrightarrow CH_3CH(CH_3)CH_2C(=O)CH_3$

(b) $CH_3(CH_2)_4CH{=}CH_2 \longrightarrow CH_3(CH_2)_4CH_2CH(=O)$

(c) $C_6H_5{-}CH_2CH_2CH{=}CH_2 \longrightarrow C_6H_5{-}CH_2CH_2CH_2CH_2O{-}C_6H_5$

(d) N-acetyl spiro compound (cyclohexene ring with C=C) $\longrightarrow$ N-acetyl spiro compound (saturated ring)

(e) $CH_3CH_2C(H)(D)OH \longrightarrow CH_3CH_2C(D)(H)N_3$

(f) $CH_3(CH_2)_3CH{=}CH_2 \longrightarrow CH_3(CH_2)_4CH_2OCH_3$

(g) $C_6H_5{-}CH{=}CH_2$ C $\longrightarrow C_6H_5{-}CH(OH)CH_2N(CH_3)_2$

Q. 147. The pheromone of the ambosia beetle, 6-methyl-5-hepten-2-ol, can be prepared by the sodium borohydride reduction of 6-methyl-5-hepten-2-one but has also been synthesised another route:

$$CH_3C(CH_3) = CHCH_2CH{=}CH_2 \longrightarrow A \longrightarrow B \longrightarrow CH_3C(CH_3){=}CHCH_2CH_2CH(OH)CH_3$$

Supply the structures of compounds A and B and the regents necessary to transform 5-methyl 1,4-hexadiene into 6-methyl-5-hepten-2-ol

Q. 148. Suggest synthetic routes for the following transformations

(a) $(CH_3)(H)C{=}C(H)(CH_2Cl) \longrightarrow (CH_3)(H)C{=}C(H)(CH_2CH_2Br)$

(b) $(CH_3(CH_2)_3)(H)C{=}C(H)(COCH_2CH_3)$, with C=O $\longrightarrow (CH_3(CH_2)_3)(H)C{=}C(H)(CH)$

(c) $\underset{\displaystyle OH}{CH_3\overset{|}{C}H}CH_2CH_2CH_2\underset{\displaystyle OCH_3}{\overset{|}{C}H}OCH_3 \longrightarrow CH_3\underset{\underset{\displaystyle \overset{||}{O}}{\displaystyle OCH_3}}{\overset{|}{C}H}CH_2CH_2CH_2\overset{\displaystyle O}{\overset{||}{C}H}$

(d) $HC{\equiv}C\overset{\displaystyle O}{\overset{||}{C}}OCH_2CH_3 \longrightarrow CH_3(CH_2)_7\overset{\displaystyle OH}{\overset{|}{C}H}HC{\equiv}C\overset{\displaystyle O}{\overset{||}{C}}OCH_2CH_3$

(e) $CH_3(CH_2)_7\overset{\displaystyle OH}{\overset{|}{C}H}C{\equiv}C\overset{\displaystyle O}{\overset{||}{C}}OCH_2CH_3 \longrightarrow CH_3(CH_2)_7$–(γ-lactone ring: five-membered ring containing O, with C=O)

Q. 149. How would you carry out the following transformation?

(a) $CH_3\overset{\displaystyle CH_3}{\overset{|}{C}}{=}\overset{\displaystyle \overset{\displaystyle O}{\overset{||}{C}}OCH_2CH_3}{\overset{|}{C}}CH_2\underset{\displaystyle O}{\underset{||}{C}}OCH_2CH_3 \longrightarrow CH_3\overset{\displaystyle CH_3}{\overset{|}{C}H}\underset{\displaystyle CH_2O\underset{\displaystyle O}{\underset{||}{C}}CH_3}{\underset{|}{C}H}CH_2CH_2O\overset{\displaystyle O}{\overset{||}{C}}CH_3$

(b) HO–CH(CH)–CH₂–C(=O)OCH₃ (CH on wedge bond) $\longrightarrow$ OH (cyclic ketal: five-membered ring with O, gem-dimethyl carbon, wedge bond, side chain CH₂CH₂I)

Q. 150. Show how each of the following transformations could be carried out. Several steps may be required.

(a) (cyclohexanone) $\longrightarrow$ (cyclohexyl)–$\overset{\displaystyle CH_3}{\overset{|}{N}}CH_3$

(b) CH_3O–(benzene) $\longrightarrow$ CH_3O–(benzene)–NH_2

(c) $CH_3CH_2CH_2\overset{\displaystyle O}{\overset{||}{C}}H \longrightarrow$ (benzene)–$\overset{\displaystyle CH_2CH_2CH_2CH_3}{\overset{|}{N}}CH_2CH_2CH_2CH_3$

(d) $CH_3CH_2CH_2CH_2CH_2CH_2Br \longrightarrow CH_3(CH_2)_6CH_2NH_2$

Q. 151. How would you convert *tert*-butyl acetate into the following compound? Show all reactants that you would need and reactive intermediates that would form.

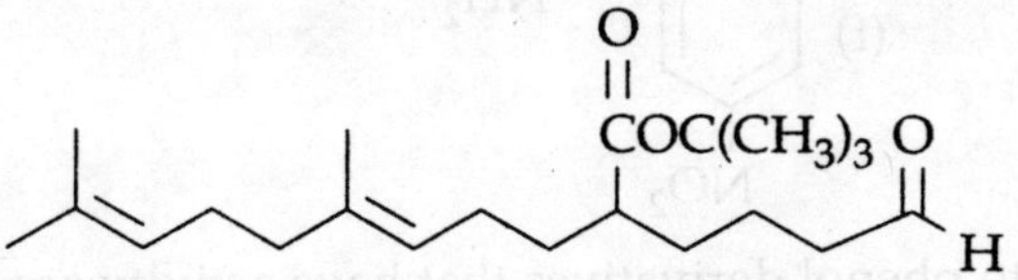

$COC(CH_3)_3$ (C=O), O, H

Q. 152. Plan a synthesis for each of the following compounds. You have bromobenzene, any organic reagents containing there of fewer carbon atoms, and any inorganic, reagents you need. There may be more than one acceptable route for a given product. Try to find the shortest route.

(a) 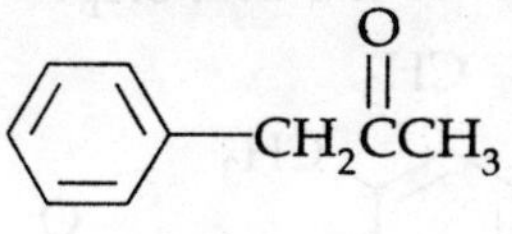 $C_6H_5CH_2CCH_3$ (C=O)

(b) 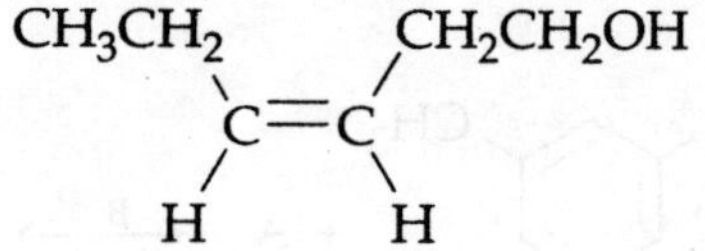 $CH_3CH_2(H)C{=}C(H)CH_2CH_2OH$

(c) $CH_3CH_2\overset{O}{\overset{\|}{C}}CH_2CH_2CH_3$

(d) $CH_3\overset{CH_3}{\overset{|}{C}}{=}CHCH_2CH_2CH_3$

(e) 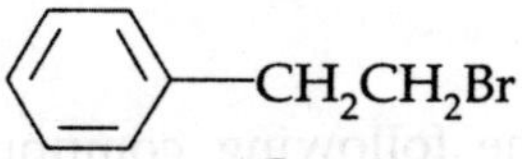 $C_6H_5CH_2CH_2Br$

(f) 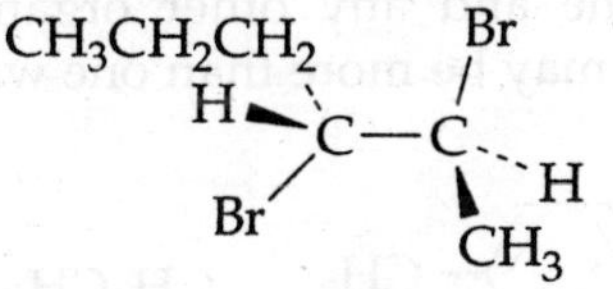 $CH_3CH_2CH_2$, H, Br on C—C with Br, H, CH_3

(g) $C_6H_5\overset{O}{\overset{\|}{C}}CH_2CH_3$

(h) H, HO, CH_3 on C—C with OH, H, CH_3

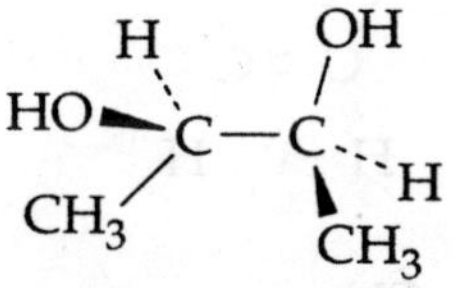

(i) $CH_3\underset{OH}{\underset{|}{C}}HCH_2CH_2\overset{CH_3}{\overset{|}{C}}HCH_3$

(j) 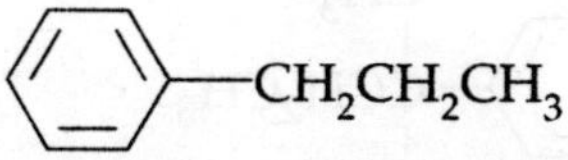 $C_6H_5CH_2CH_2CH_3$

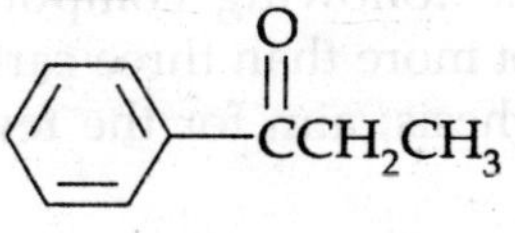

153. How would you synthesize the following compounds from benzene or toluene and any other organic reagents containing not more than there carbon atoms? You may assume that ortho and para isomers are easily separable in the laboratory.

(a) 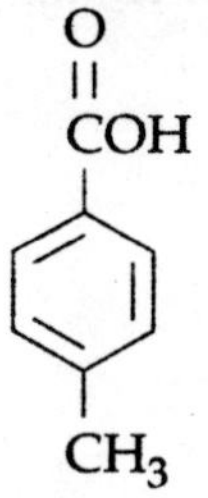 *p*-$CH_3C_6H_4COOH$ (COH with C=O, para CH_3)

(b) 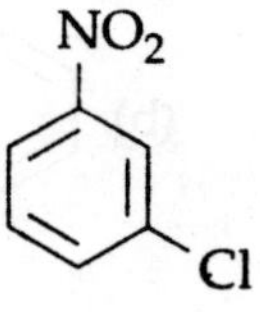 NO_2, Cl (meta) on benzene

(c) Br, Cl (para) on benzene

(d) 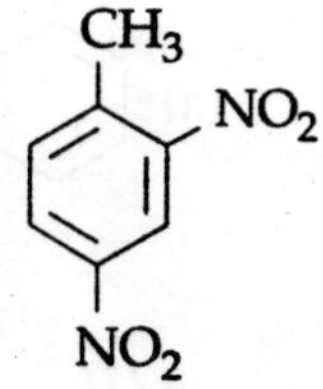 CH_3, NO_2 (ortho), NO_2 (para) on benzene

(e) 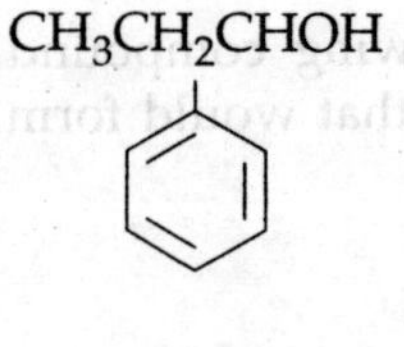 CH_3CH_2CHOH attached to a benzene ring

(f) 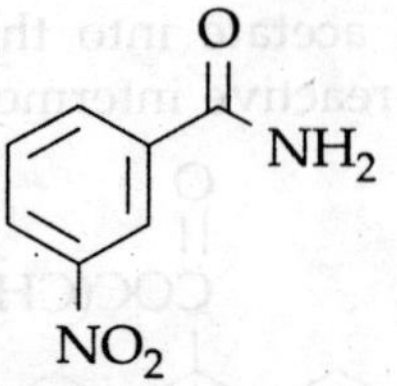 3-nitrobenzamide (C_6H_4 ring bearing $C(=O)NH_2$ and NO_2)

Q. 154. During research into phenol derivatives that have activity against hormones involved is blood clotting, the following synthesis was carried out in two steps. The first is a Friedel-Crafts alkylation reaction starting with an alcohol. Provide structural formulas for the regents that would be necessary for the transformation of the starting material into the final product, as well as for the structure of the product of the first step.

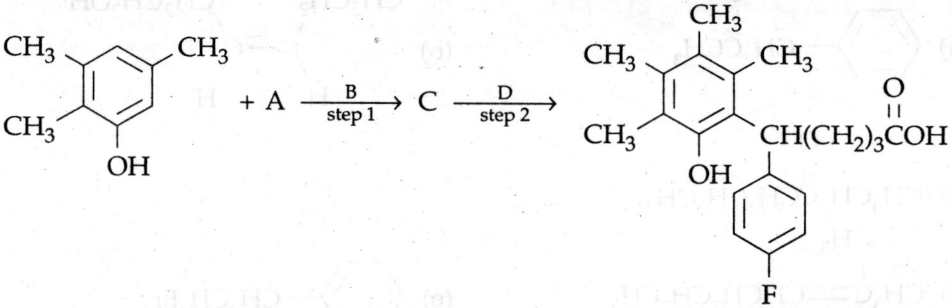

Starting material: benzene ring with CH_3, CH_3, CH_3 and OH substituents

$$\text{Starting material} + A \xrightarrow[\text{step 1}]{B} C \xrightarrow[\text{step 2}]{D}$$

Product: benzene ring with CH_3 (four of them), OH, and $CH(CH_2)_3\overset{O}{\overset{||}{C}}OH$, where the CH bears a 4-fluorophenyl group (F)

Q. 155. Synthesize the following compounds from benzene and any other organic regions containing not more than three carbon atoms. There may be more than one way to carry out each synthesis; aim for the fewest steps.

(a) 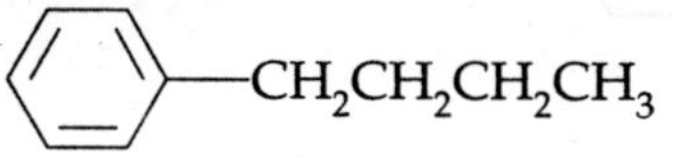 C_6H_5—$CH_2CH_2CH_2CH_3$

(b) C_6H_5—CH_2(H)C=C(H)CH_2CH_3

(c) CH_3-substituted tetralone (ring with C=O)

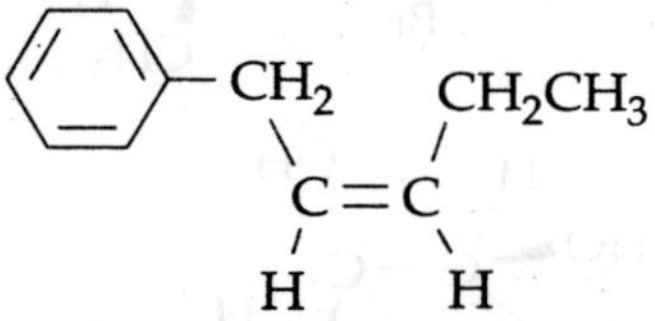

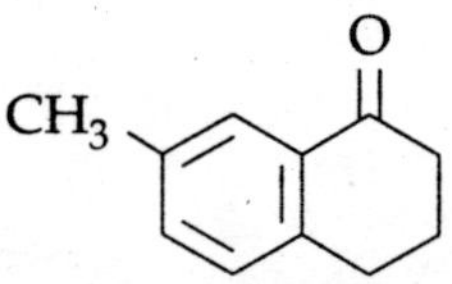

(d) 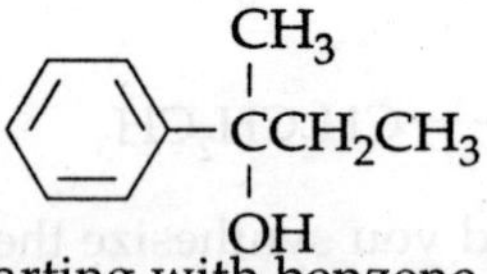 C_6H_5—$C(CH_3)(OH)CH_2CH_3$

Q. 156. Propose syntheses for the following compounds starting with benzene, toluene, and any other organic compounds having no more that three carbon atoms. There may be more than one way to synthesize wach compound; try to find the shortest route.

(a) 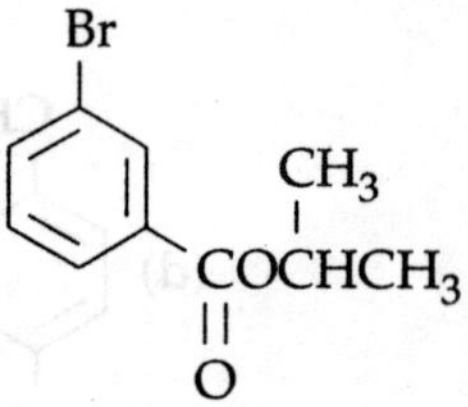 Br-substituted benzene ring with $\underset{\underset{O}{||}}{C}OCH(CH_3)CH_3$ (COCHCH$_3$ with CH_3)

(b) 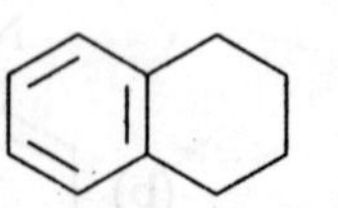 tetralin (fused benzene and cyclohexane rings)

(c) 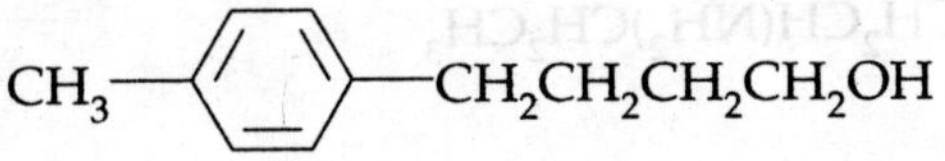 $CH_3-C_6H_4-CH_2CH_2CH_2CH_2OH$ (d) 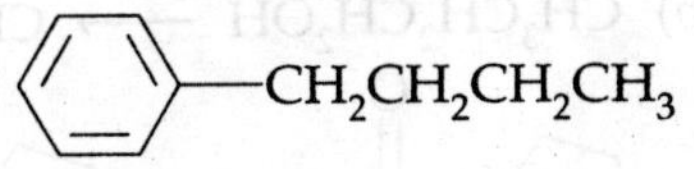 $C_6H_5-CH_2CH_2CH_2CH_3$

(e) 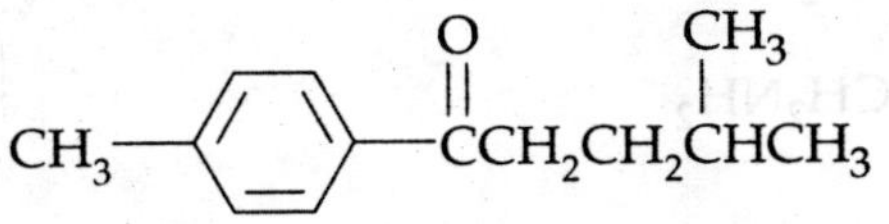 $CH_3-C_6H_4-\overset{O}{\overset{||}{C}}CH_2CH_2\overset{CH_3}{\overset{|}{C}}HCH_3$ (f) 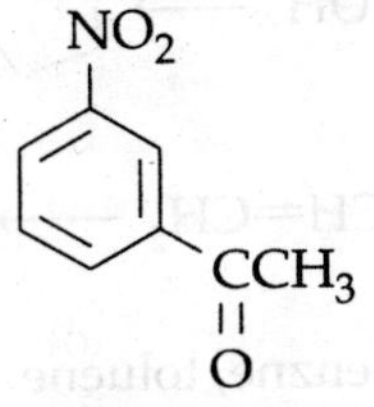 NO_2, $\overset{}{C}CH_3$ (=O)

(g) 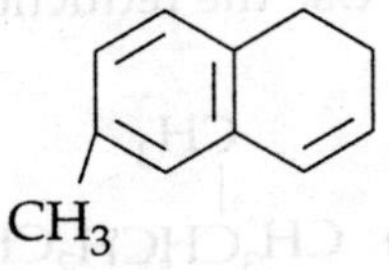 CH_3 (h) 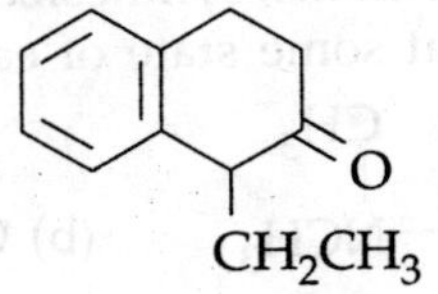 O, CH_2CH_3

(i) 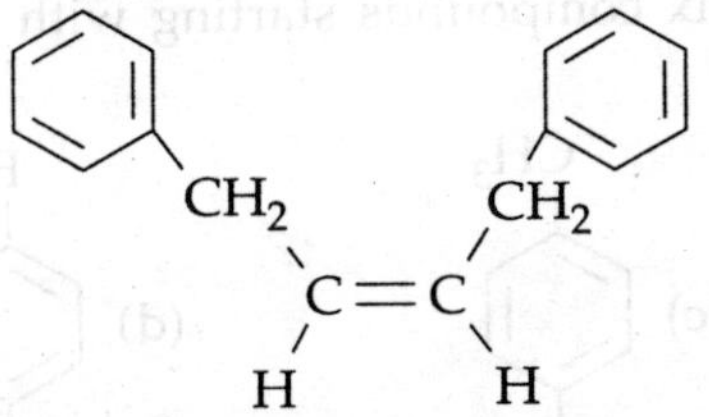 CH_2, CH_2, C=C, H, H

Q. 157. 1,3,5-Trimethylbenzene undergoes electrophilic aromatic substitution with iodine monochloride, ICl. Write an equation for this reaction showing the product you expect.

Q. 158. The following reactions were carried out in studies into the synthesis of alkaloid, Provide structural formulas for the products designated by letters.

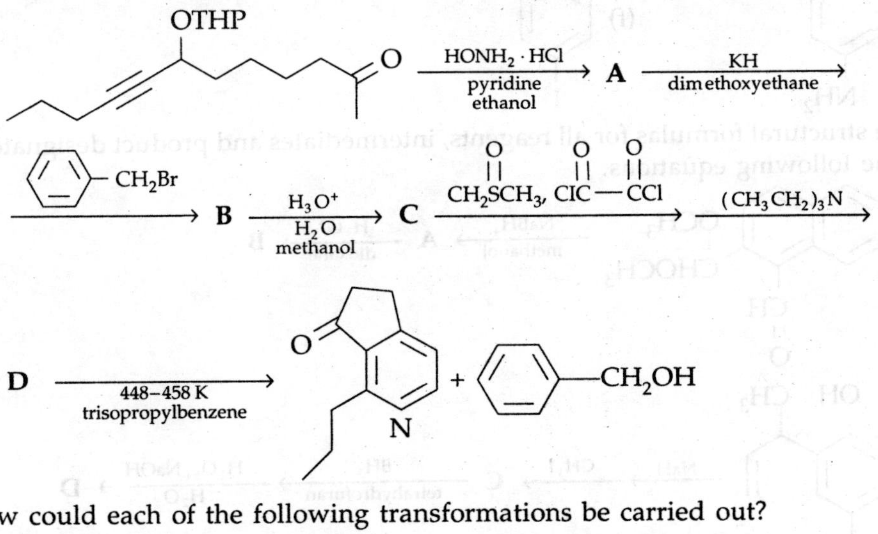

OTHP ... O $\xrightarrow[\text{pyridine ethanol}]{HONH_2 \cdot HCl}$ A $\xrightarrow[\text{dimethoxyethane}]{KH}$

$\xrightarrow{C_6H_5-CH_2Br}$ B $\xrightarrow[\text{H}_2\text{O, methanol}]{H_3O^+}$ C $\xrightarrow{CH_3SCH_3,\ Cl\overset{O}{\overset{||}{C}}-\overset{O}{\overset{||}{C}}Cl}$ $\xrightarrow{(CH_3CH_2)_3N}$

D $\xrightarrow[\text{trisopropylbenzene}]{448-458\ K}$ O, N + $C_6H_5-CH_2OH$

Q. 159. How could each of the following transformations be carried out?

(a) 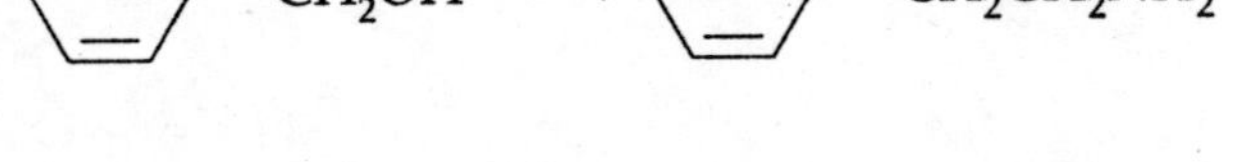 $C_6H_5-CH_2OH \longrightarrow C_6H_5-CH_2CH_2NH_2$

(b) $CH_3CH_2CH_2OH \longrightarrow CH_3CH_2CH_2CH(NH_2)CH_2CH_3$

(c) 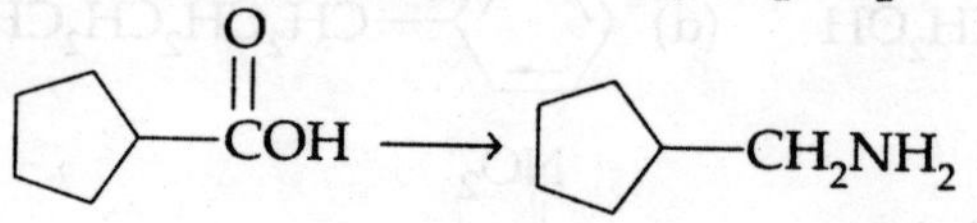 cyclopentyl–$\overset{O}{\overset{||}{C}}OH \longrightarrow$ cyclopentyl–CH_2NH_2

(d) $CH_2{=}CHCH{=}CH_2 \longrightarrow$ cyclohexane-1,2-di(CH_2NH_2)

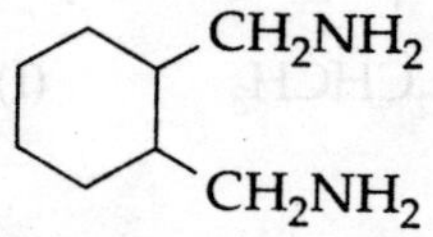

Q. 160. Starting with benzne, toluene, and any other organic compound containing to more than three carbon atoms, synthesize each of the following amines. Use the reduction of a nitro compound at some state of each synthesis.

(a) 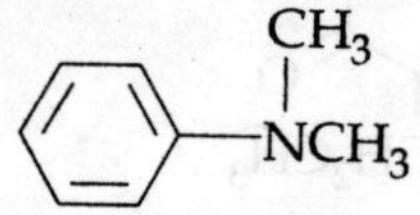 C_6H_5–$N(CH_3)CH_3$ (b) O_2N–C_6H_4–NH_2 (c) $CH_3\overset{CH_3}{\overset{|}{C}}HCH_2CH_2NH_2$

Q. 161. Devise a synthesis for each of the following six compounds starting with benzene or toluene.

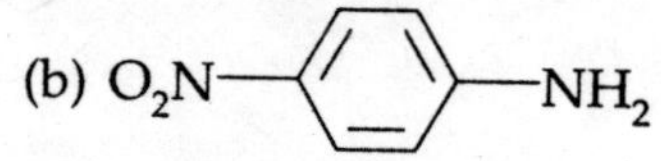

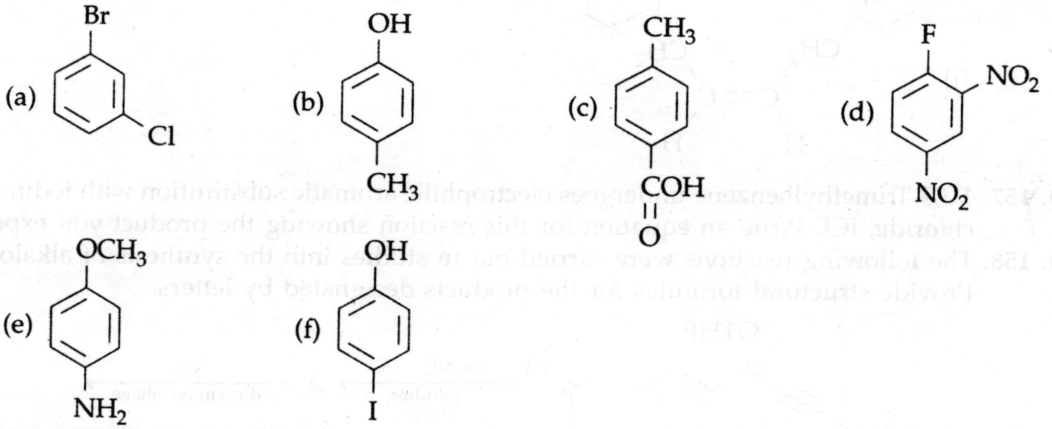

(a) 1-Br-3-Cl-benzene (b) 4-CH_3-phenol (OH, CH_3) (c) CH_3–C_6H_4–$\overset{O}{\overset{||}{C}}OH$ (d) F, NO_2, NO_2 substituted benzene (e) OCH_3–C_6H_4–NH_2 (f) OH–C_6H_4–I

Q. 162. Give structural formulas for all reagents, intermediates and product designated by letter in the following equations.

(a) 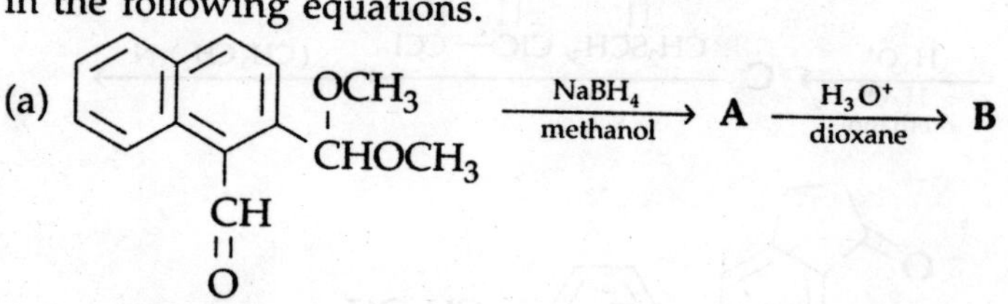 naphthalene bearing $\overset{OCH_3}{\overset{|}{C}}HOCH_3$ and $\underset{O}{\underset{||}{CH}}$ $\xrightarrow[\text{methanol}]{NaBH_4}$ A $\xrightarrow[\text{dioxane}]{H_3O^+}$ B

(b) benzene ring bearing OH, $CH(CH_3)CH{=}CH_2$ and isopropyl $\xrightarrow{NaH}\xrightarrow{CH_3I}$ C $\xrightarrow[\text{tetrahydrofuran}]{BH_3}\xrightarrow[H_2O]{H_2O_2,\ NaOH}$ D

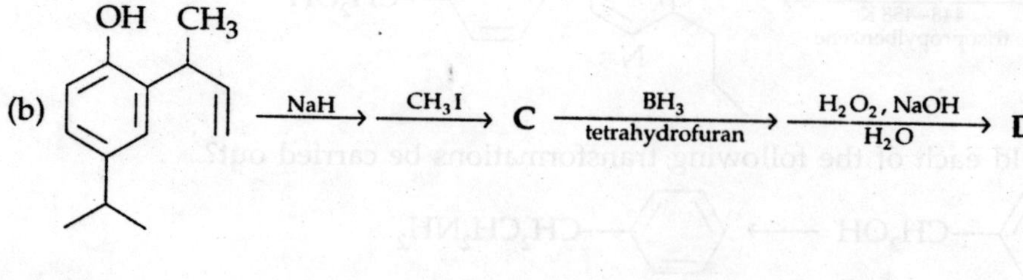

(c) D $\xrightarrow[H_2SO_4,\ H_2O,\ \Delta]{CrO_3}$ **E** $\xrightarrow[\Delta]{H_3PO_4}$ **F** $\xrightarrow{LiAlH_4}$ $\xrightarrow{H_2O}$ **G**

(d) $HOCCH_2CH_3$ (C=O) –(1,3-phenylene)– CH_2CH_2COH (C=O) $\xrightarrow[\Delta]{H_3PO_4}$ **H** + **I**

(e) O_2N–(benzene ring with NO_2)–F + CH_3ONa $\xrightarrow[\text{methanol}]{}$ **J**

(f) 2-methylpyridine (N, CH_3) $\xrightarrow[H_2O,\ \Delta]{KMnO_4}$ **K**

(g) (benzene)$Cr(CO)_3$ $\xrightarrow[\text{tetrahydrofuran}]{LiCH_2CN}$ $\xrightarrow[\text{tetrahydrofuran}]{I_2}$ **L**

(h) benzene ring with CH(=O), OH, OCH_3 $\xrightarrow[\text{ethylene glycol}]{H_2NNH_2,\ KOH}$ **M**

(i) benzene ring with CH(=O), OH, OCH_3 $\xrightarrow{NaBH_4}$ **N** $\xrightarrow[\text{tetrahydrofuran}]{NaH\ (2\ equiv)}$ $\xrightarrow{CH_3I\ (1\ equiv)}$ **O**

(j) $CH_2(COCH_2CH_3)_2$ (C=O) $\xrightarrow[\text{diethyl ether}]{Na}$ **P** $\xrightarrow{\text{Br–(benzene ring with } NO_2,\ Cl,\ NO_2)}$ **Q**

(k) HO, OH, COH (C=O) benzene ring $\xrightarrow{R}$ CH_3O, OCH_3, $COCH_3$ (C=O) benzene ring $\xrightarrow{S}$ CH_3O, OCH_3, CH_2OH benzene ring

$\xrightarrow{T}$ (2-bromo-3,5-dimethoxybenzyl alcohol: Br, CH_3O, CH_2OH, OCH_3)

Q. 163. Give structures for the intermediates and products represented by letters in the following equations.

(a) $CH_3\overset{CH_3}{\overset{|}{C}}HC{\equiv}N \xrightarrow[\text{tetrahydrofuran}]{(CH_3CH_2)\overset{CH_3}{\overset{|}{N}}{}^- Li^+} A \xrightarrow[\text{tetrahydrofuran}]{C_6H_6Cr(CO)_3} B \xrightarrow[\text{tetrahydrofuran}]{I_2} C$

(b) $C_6H_6Cr(CO)_3 \xrightarrow[\text{tetrahydrofuran}]{CH_3-C_6H_4-Li} D \xrightarrow{I_2} E$

(c) $C_6H_6Cr(CO)_3 \xrightarrow[\text{tetrahydrofuran, hexanehylphophoric triamide}]{LiCH_2\overset{O}{\overset{||}{C}}O\overset{CH_3}{\underset{CH_3}{C}}CH_3} F \xrightarrow[\text{tetrahydrofuran}]{I_2} G$

(d) $C_6H_5Cl\,Cr(CO)_3 \xrightarrow[\text{tetrahydrofuran}]{Li\overset{CN}{\overset{|}{C}}(CH_3)_2} \xrightarrow[\text{tetrahydrofuran}]{I_2} H$

Q. 164. Complete the following equations.

(a) $O_2N-C_6H_3(NO_2)-Br + HN(CH_2)_5 \xrightarrow{\text{methanol}}$

(b) $O_2N-C_6H_3(NO_2)-F + C_6H_5-S^-Na^+ \xrightarrow{\text{methanol}}$

(c) 2,4,6-trinitrophenyl tosylate: O_2N–$C_6H_2(NO_2)_2$–OTs + H_2N–C_6H_5 $\xrightarrow[\Delta]{\text{benzene}}$

(d) O_2N–$C_6H_2(NO_2)_2$–Cl + H_2N—N$(C_6H_5)_2$ $\xrightarrow{\text{chloroform}}$

(e) O_2N–$C_6H_3(NO_2)$–F + $H_3\overset{+}{N}CH(CH_3)CO^-$ (C=O) $\xrightarrow{H_2O}$

Q. 165. What reagents would be needed to carry out each of the following transformations? More than one step may be required in some cases.

(a) C_6H_5–CH_3 ⟶ F–C_6H_4–$\overset{O}{\overset{||}{C}}OCH_2CH_3$

(b) C_6H_5–CH_3 ⟶ HO–C_6H_3(Br)–CH_3

(c) C_6H_5–CH_2Br ⟶ C_6H_5–$CH_2CH(CH_2CH_3)CH_2NH_2$

(d) C_6H_5–Br ⟶ I–C_6H_4–Br

(e) C_6H_5–Cl ⟶ H_2N–C_6H_4–Cl

(f) 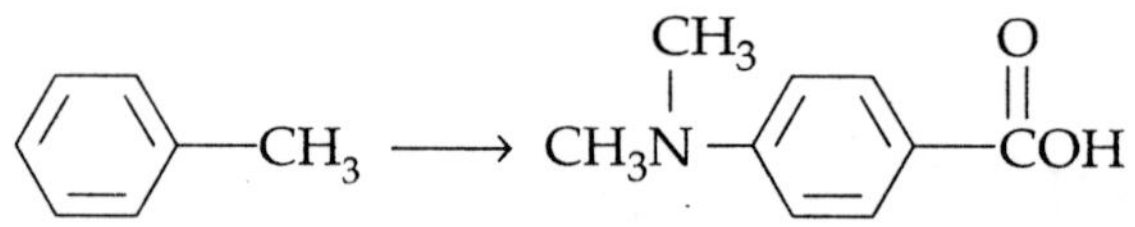 $C_6H_5-CH_3 \longrightarrow (CH_3)_2N-C_6H_4-COOH$

Q. 166. The painkiller ibuprofen can be synthesized in a number of steps from benzene. Some key intermediates are shown below. A number of steps may be necessary to go from one to the other. Fill in the steps necessary for the synthesis.

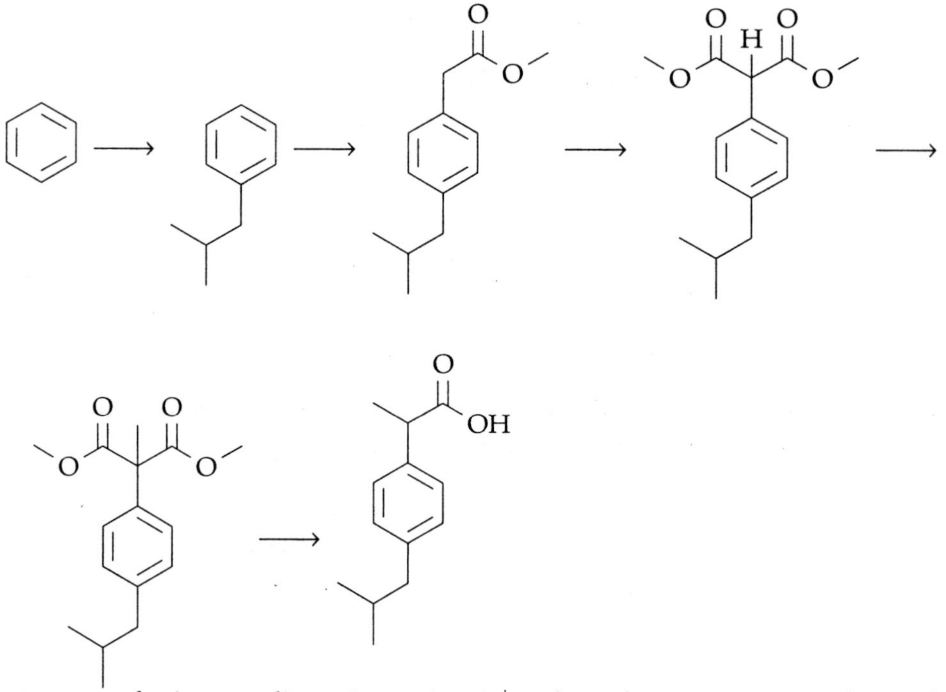

O H O OH

Q. 167. Give structures for intermediates & starting & products for representation letters below.

(a) 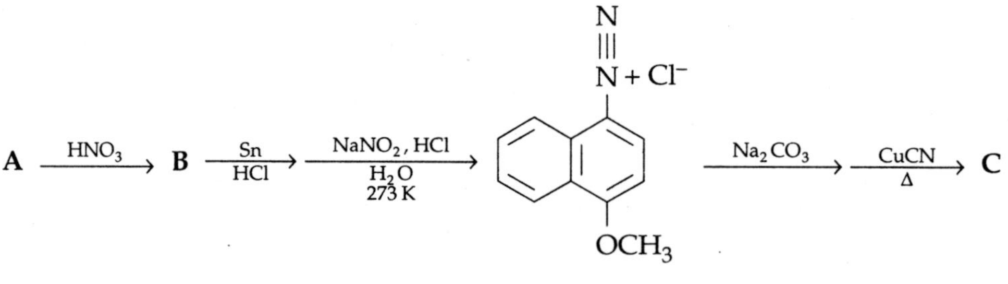 A $\xrightarrow{HNO_3}$ B $\xrightarrow[HCl]{Sn}$ $\xrightarrow[\substack{H_2O \\ 273\,K}]{NaNO_2,\,HCl}$ (1-methoxynaphthalene-4-diazonium, $N\equiv N^+ Cl^-$, OCH_3) $\xrightarrow{Na_2CO_3}$ $\xrightarrow[\Delta]{CuCN}$ C

(b) $O_2N-C_6H_3(NO_2)-O-C_6H_4-NO_2$ + HN(cyclohexane ring) $\xrightarrow[\text{dioxane}]{\text{water}}$ D + E

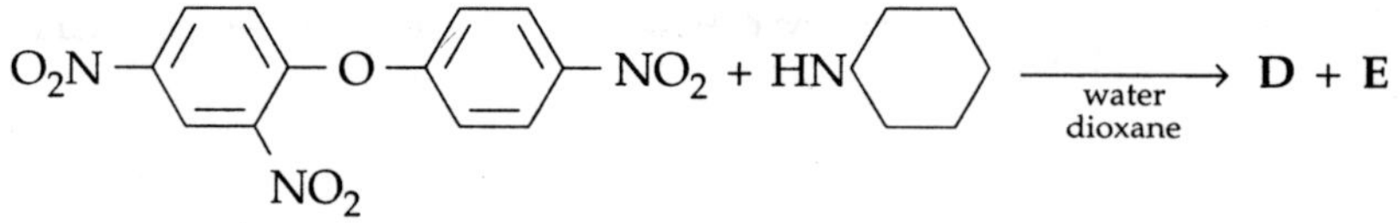

(c) 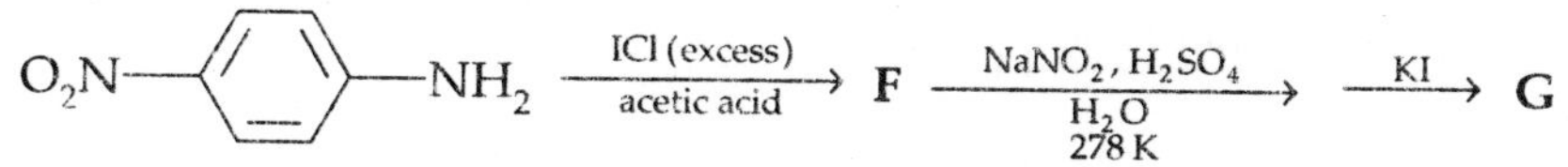 $O_2N-C_6H_4-NH_2 \xrightarrow[\text{acetic acid}]{\text{ICl (excess)}} F \xrightarrow[\substack{H_2O \\ 278\text{ K}}]{NaNO_2,\, H_2SO_4} \xrightarrow{KI} G$

(d) 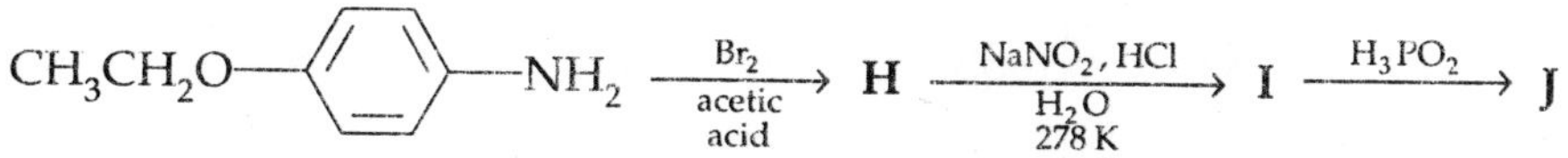 $CH_3CH_2O-C_6H_4-NH_2 \xrightarrow[\text{acetic acid}]{Br_2} H \xrightarrow[\substack{H_2O \\ 278\text{ K}}]{NaNO_2,\, HCl} I \xrightarrow{H_3PO_2} J$

(e)  $HO-C_6H_4-NH_2 \xrightarrow[\substack{H_2O \\ 278\text{ K}}]{NaNO_2,\, HCl} K \xrightarrow[\substack{Cu \\ \Delta}]{KI} L$

(f) 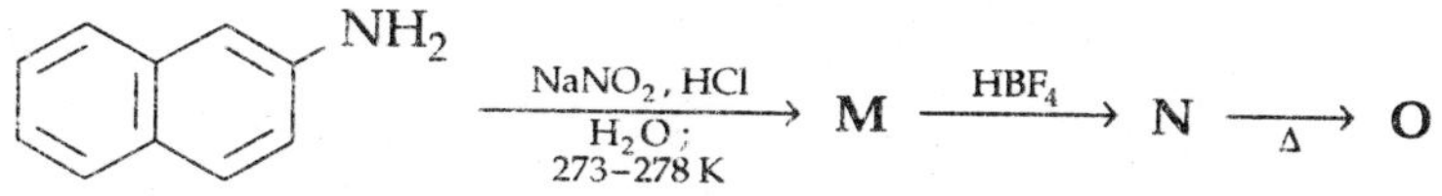 2-Naphthylamine ($C_{10}H_7NH_2$) $\xrightarrow[\substack{H_2O; \\ 273-278\text{ K}}]{NaNO_2,\, HCl} M \xrightarrow{HBF_4} N \xrightarrow[\Delta]{} O$

(g) 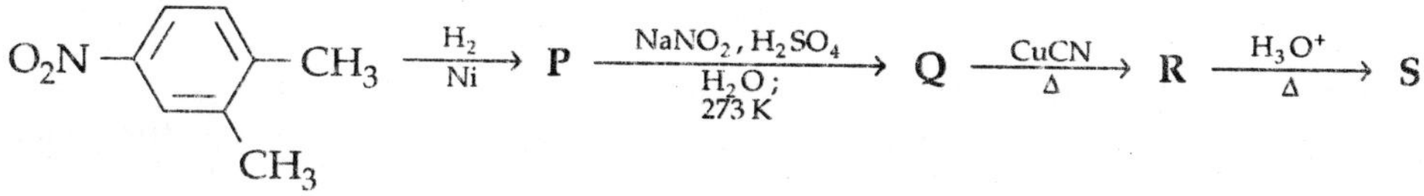 $O_2N-C_6H_3(CH_3)-CH_3 \xrightarrow[Ni]{H_2} P \xrightarrow[\substack{H_2O; \\ 273\text{ K}}]{NaNO_2,\, H_2SO_4} Q \xrightarrow[\Delta]{CuCN} R \xrightarrow[\Delta]{H_3O^+} S$

(h) $H_2N-C_6H_3(OCH_3)-C_6H_3(OCH_3)-NH_2 \xrightarrow[\substack{H_2O; \\ 278-283\text{ K}}]{NaNO_2\text{ (excess) } HCl} T \xrightarrow{H_3PO_2\text{ (excess)}} U$

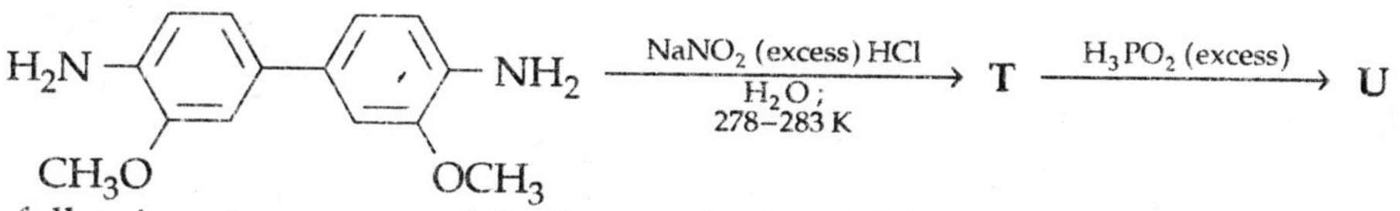

Q. 168. The following steps are used in the synthesis of the controversial sweetener sachacim. Give structures for the intermediate products, compounds A and B

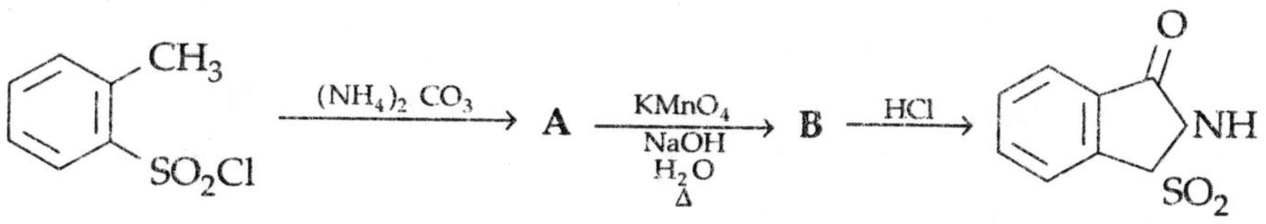

$o\text{-}CH_3C_6H_4SO_2Cl \xrightarrow{(NH_4)_2CO_3} A \xrightarrow[\substack{NaOH \\ H_2O \\ \Delta}]{KMnO_4} B \xrightarrow{HCl}$ saccharin

Q. 169. Identify *A* to (*K*) as reactant, reagent, product as name of the following:

(i) $3(CH_3)_2C{=}O \xrightarrow{(A)} (B)$ *Aldol condensation*

(ii) $CH_3COCl + \xrightarrow{(C)} D$ *Rosenmund's reaction*

(iii) $(E) \xrightarrow[C_2H_5ONa]{NH_2NH_2} CH_3CH_2CH_3\ F$

(iv) $C_6H_5OH \xrightarrow{(G)} (H_1) + (H_2)$ *Elb's oxidation*

(v) $C_6H_5N_2Cl \xrightarrow[HCl]{(J)} (K)$ *Gattermann reaction*

Q. 170. Predict the major product in the following reactions.

(i) $C_6H_6 + (CH_3)_2CHCH_2OH \xrightarrow{H_2SO_4} A$

(ii) $C_6H_5CH_2COOCH_3 \xrightarrow[2\,H^+]{1.\ CH_3MgBr\text{ (excess)}} B$

(iii) $C_6H_2C_2H_5 \xrightarrow[\text{2 NaCN}]{\text{1 Br}_2\text{, Heat Light}} C$

(iv) $CH_3CH_2(H)C{=}O \xrightarrow[\text{2 LiAlH}_4]{\text{1. KCN/H}_2\text{SO}_4} D$

(v) Predict the structure of the product in the following reaction.

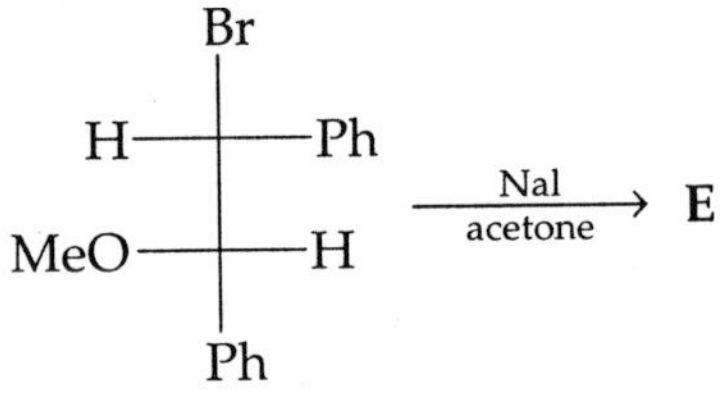

Fischer projection: top Br; H—C—Ph; MeO—C—H; bottom Ph $\xrightarrow[\text{acetone}]{\text{NaI}}$ **E**

(vi) Suggest appropriate structures of the missing compounds.
(The number of carbon atoms remains the same throughout the reactions.)

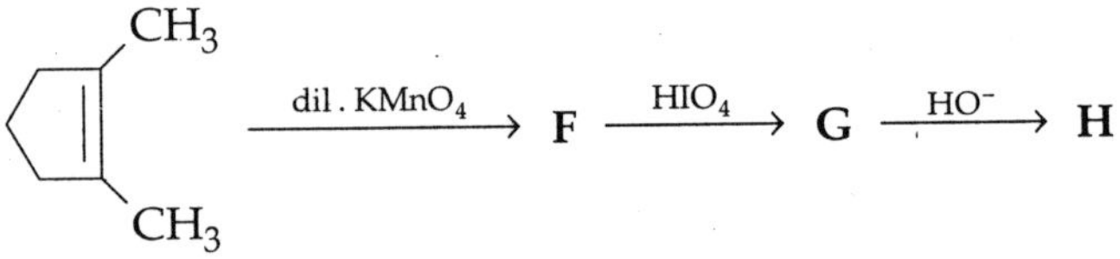

1,2-dimethylcyclopentene $\xrightarrow{\text{dil. KMnO}_4}$ **F** $\xrightarrow{\text{HIO}_4}$ **G** $\xrightarrow{\text{HO}^-}$ **H**

Q. 171. (a) Carry out the following transformation in not more than three steps.

$$CH_3—CH_2—C{\equiv}C—H \rightarrow CH_3—CH_3—CH_2—\overset{O}{\overset{||}{C}}—CH_3$$

(b) Discuss the hybridiation of carbon atoms in the allene (C_3H_4) and show the π-orbital overlaps.

(c) Explain why o-hydroxybenzaldehyde is a liquid at room temperature while p-hydroxbenzaldehyde is a high melting solid.

Q. 172. Explain briefly the formation of the products giving the structures of the intermediates.

(a) $H_2C{=}CH—CH_2—CH_2—OH \xrightarrow{\text{HCl}} H_2C{=}CH—CH_2—CH_2—Cl$ + (cyclopropyl: CH_2, CH_2 ring)$CH—CH_2—Cl$ + etc.

$H_2C{=}CH—CH_2—CH(CH_3)—OH \xrightarrow{\text{HCl}} H_2C{=}CH—CH_2—CH(CH_3)—Cl$ (only)

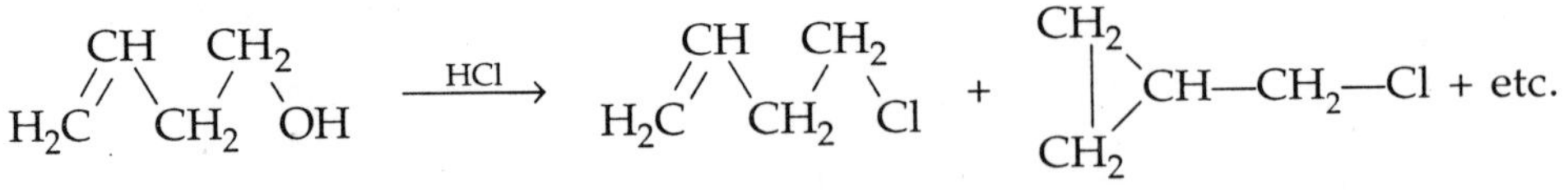

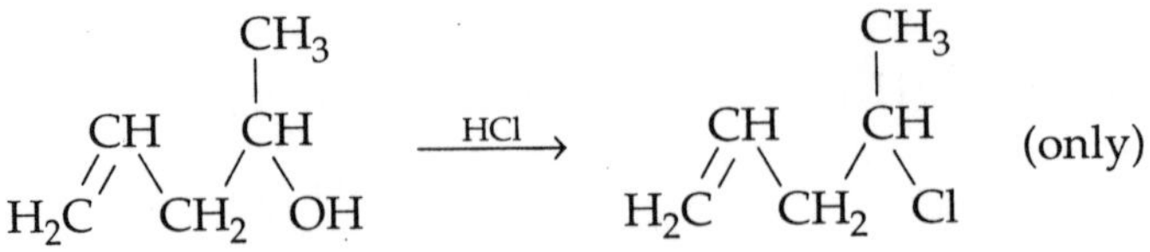

(b) (o-bromoanisole: benzene ring with OCH_3 and Br) $\xrightarrow[\text{NH}_3]{\text{NaNH}_2}$ (m-anisidine: benzene ring with OCH_3 and NH_2)

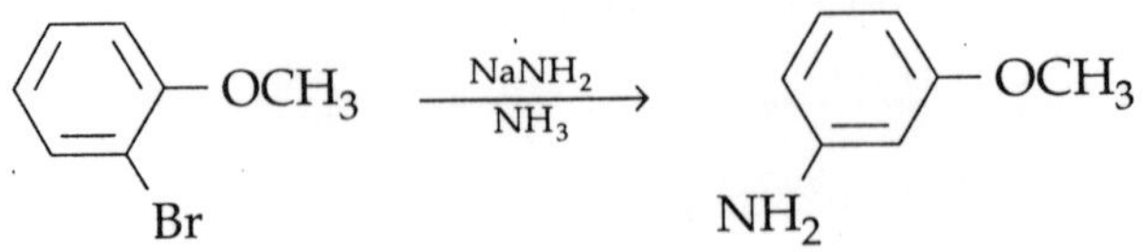

(c) 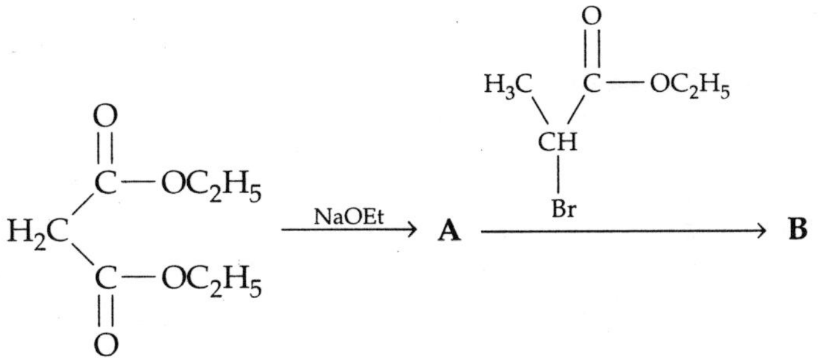 $H_2C(COOC_2H_5)_2 \xrightarrow{NaOEt} A \xrightarrow{CH_3CH(Br)COOC_2H_5} B$

(B) $\xrightarrow[(ii)\ H^+]{(i)\ OH^-} H_3C-CH(COOH)-CH_2-COOH$

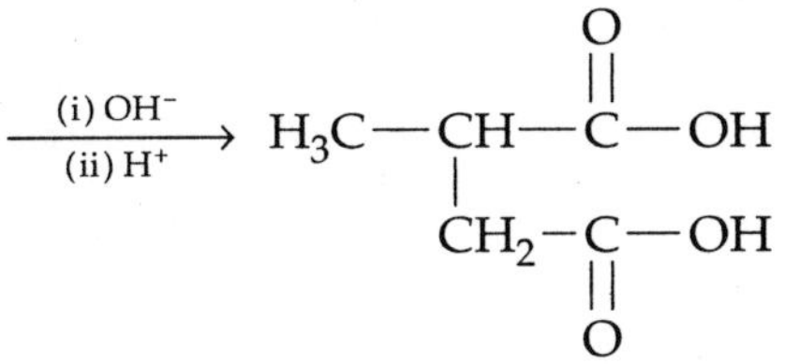

Q. 173. (a) Write the intermediate steps for each of the following reaction

(i) $C_6H_5C(OH)C{\equiv}C \longrightarrow C_6H_5CH{=}CHCHO$

(ii) 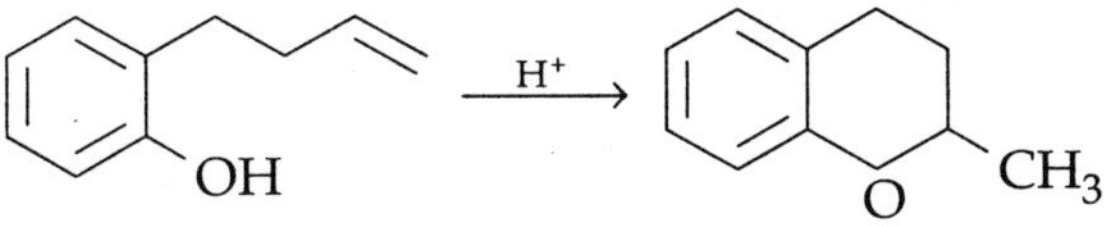 2-(but-3-enyl)phenol $\xrightarrow{H^+}$ 2-methylchromane (ring O, CH_3)

(b) Sow the steps to carry out the following transformations.

(i) Ethylbenzene $\longrightarrow$ benzene

(ii) Ethylbenzene $\longrightarrow$ 2-phenylpropionic acid

Q. 174. Complete the following:

(a) 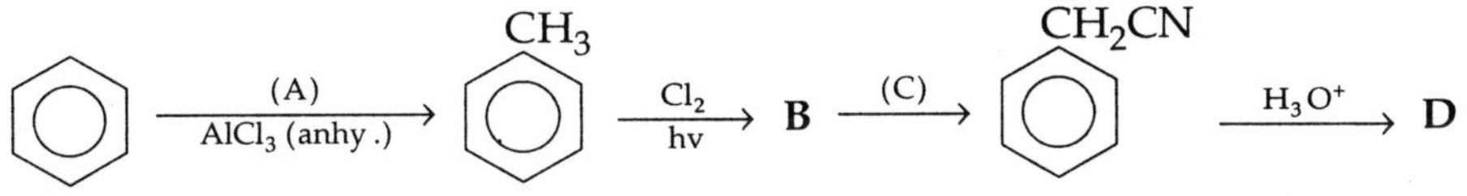 $C_6H_6 \xrightarrow[AlCl_3\ (anhy.)]{(A)} C_6H_5CH_3 \xrightarrow[hv]{Cl_2} B \xrightarrow{(C)} C_6H_5CH_2CN \xrightarrow{H_3O^+} D$

(b) $C_6H_5NO_2 \xrightarrow{Br_2/Fe} A \xrightarrow{Sn/HCl} B \xrightarrow[HCl,\ 273-278\ K]{NaNO_2} C \xrightarrow{Cu_2Cl_2} D$

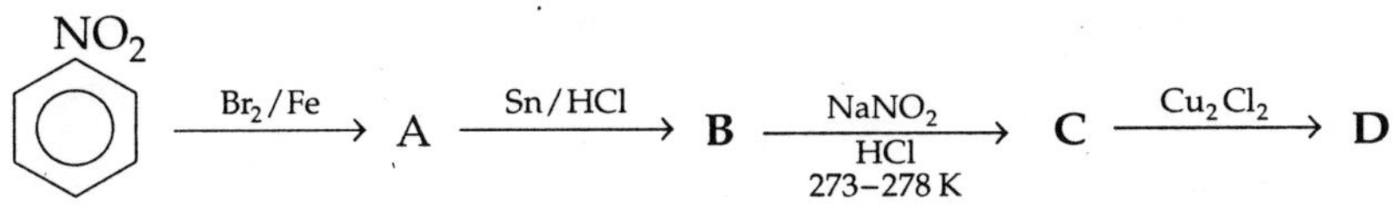

(c) $CH_3CH_2CH_2OH \xrightarrow{P + Br_2} A \xrightarrow{KOH_{alc.}} B \xrightarrow{HBr} C \xrightarrow[\Delta]{NH_{3(g)}} D$

(d) $C_6H_5COOC_6H_5 \xrightarrow{Mononitration} A$

(e) $C_6H_5CONH_2 \xrightarrow{P_2O_5} A \xrightarrow[H_2O,\ \Delta]{H^+} B$

(f) 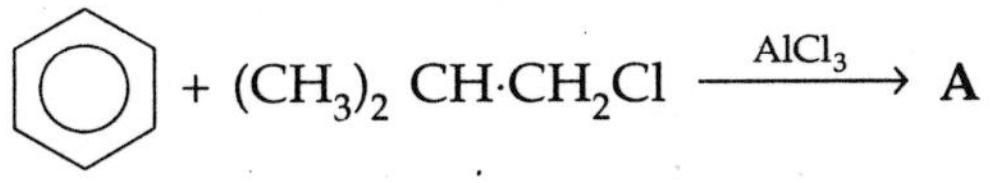 $C_6H_6 + (CH_3)_2\ CH{\cdot}CH_2Cl \xrightarrow{AlCl_3} A$

(g) $C_6H_4(CHO)(OCH_3) + HCHO \xrightarrow{KOH} A + B$

(h) $CH_3—CH_2—COOH \xrightarrow[\text{bromine}]{\text{P and}} A \xrightarrow[\text{2.H+}]{\text{1. alcoholic KOH (excess)}} B$

(i) 2,4-Dinitroanilne $\xrightarrow[\text{anisole}]{NaNO_2 \text{ and HCl at 278 K}} C$

(j) $C_6H_5—CHO + CH_3—COOC_2H_5 \xrightarrow[\text{C } H_5OH \text{ and heat}]{NaOC_2H_5 \text{ inobsolute}} D$

(k) $o—HOOC—C_6H_4—CH_2—C_6H_5 \xrightarrow{SOCl_2} G \xrightarrow[AlCl_3]{\text{anhydrous}} H \xrightarrow[HCl]{Zn-Hg} I$

(l) $A \xrightarrow{PCl_5} B \xrightarrow{LiAlH_4} C \xrightarrow[H_2SO_4]{HBr} D \xrightarrow[\text{ether}]{Na} CH_3{\cdot}CH_2{\cdot}CH_2{\cdot}CH_3$

(m) $R—C(=NOH)—R' \xrightarrow{\text{Acidic reagent}} \ldots$

(n) $R—C(=O)—Cl + H_2 \xrightarrow{Pd/BaSO_4} \ldots$

(o) $BrCH_2{\cdot}CH_2Br \xrightarrow{?} C_2H_2$

(p) $C_6H_{11}—C\equiv CH \xrightarrow[H_2SO_4]{Hg^{+2}} \ldots$ (cyclohexyl)

(q) benzene + 1,3-butadiene $\longrightarrow \ldots$

(r) $C_6H_5CH_2CClC_6H_5 \xrightarrow{\text{alcoholic KOH, heat}} A + B$

(s) $CH_3CH_2NH_2 \xrightarrow{(CH_3CO)_2O, \text{ heat}} A + B$

(t) $(CH_3)_2CHOCH_3 \xrightarrow{\text{HI (excess), heat}} A + B$

(u) $CH_3CONHC_6H_5 \xrightarrow{Br_2, Fe} A + B$

(v) $C_6H_5CH=CH_2 \xrightarrow{Br_2} A \xrightarrow[\text{(ii) } CH_3I]{\text{(i) } NaNH_2 \text{ (3.0 equiv.)}} B$

(w) cyclohexanone $\xrightarrow{[C]}$ 2-(CHC_6H_5)cyclohexanone (=CHC_6H_5 at C-2, C=O at C-1) $\xrightarrow[\text{(ii) } H^+, \text{ heat}]{\text{(i) } LiAlH_4} D$

(x) methylenecyclopentane $\xrightarrow{1} \xrightarrow{2} \xrightarrow{3}$ cyclohexene

(y) $(D)(H_3C)C=C(D)(H)$ with H_3C, CH_3 $\xrightarrow{4} \xrightarrow{5} \xrightarrow{6}$ $(HO)(D)(H_3C)C=C(CH_3)(H)(D)$ with CH_3, CH_3

(z) $C_6H_5-NH-\overset{O}{\overset{||}{C}}-CH_3 \xrightarrow{7} \xrightarrow{8} \xrightarrow{9} \xrightarrow{10}$ 3,5-diiodo-nitrobenzene (I, I, NO_2)

Section II

Q. 175. Two unknown compounds, A and B, with the same molecular formula, $C_5H_{11}Br$, give isopentane when reduced with HI plus phosphorus. When refluxed over sodium metal, A gives 3,3,4,4-tetramethylhexane and *B* gives 2,7dimethloctane. What are the structure formulas of *A* and *B*?

Q. 176. An alkyl dihalide R ($C_6H_{12}Br_2$) was treated with excess alcoholic KOH to produces S (C_6H_{10}). When *S* is hydrated in the presence of sulfuric acid, compound *T* is obtained:

$$CH_3-\overset{O}{\overset{||}{C}}-CH_2-CH(CH_3)_2$$

T

Give structural formulas for *R* and *S* that are consistent with the facts above.

Q. 177. An unsaturated compound *W*($C_9H_{17}Br$) was treated with alcoholic KOH to produce an alkadiene *X*(C_9H_{16}). When *X* was treated with one mole of bromine, compound *Y* was obtained. Then the structure of compound *Y is,*

Q. 178. An alkene A was prepare by the dehalogenation of an alkyl dihalide with zinc dust. The alkene was then treated with ozone to produce

$$CH_3-CH_2-\underset{CH_2}{\underset{|}{C}}=O \quad \text{and} \quad (CH_3)_2CH-\underset{H}{\underset{|}{C}}=O$$

What would be the structure of the glycol obtained if the alkene *A* is oixidized with hydrogen peroxide in glacial acetic acid?

Q. 179. An alkene W as oxidised with $KMnO_4$ in aqueous Na_2CO_3 to yield

$$CH_3-CH_2-CH_2-\underset{H_3C}{\underset{|}{\overset{OH}{\overset{|}{C}}}}-\underset{CH_3}{\underset{|}{\overset{OH}{\overset{|}{C}}}}-CH_2-CH_3$$

If the alkene W is treated with ozone and the resulting ozonide is hydrolyzed to yield two carbonyl compounds, what would be the structures of the two carbonyl compounds?

Q. 180. Given three unlabeled bottled *A*, *B*, and *C*, each containing a colorless liquid. The liquid in *A* is inert to aqueous $KMnO_4$, and it is identical to the compound produced by the reduction of isopentyl bromide with HI + phosphorus. The liquid in *B* is found to have a molecular formula C_6H_{10}, adds one mole of Br_2 per mole of C_6H_{10}, and upon treatment of *B* with ozone and hydrolsis of the ozonide, one gets $O{=}CH{-}(CH_2)_4{-}CHO$. Compound *C*, also with molecular formula C_6H_{10}, adds two moles of Br_2 and gives a precipitate with ammonical silver nitrate. Deduce the structures of *A*, *B* and *C*.

Q. 181. Two geometric isomers (*A* and *B*) of 3,4-dimethyl-1,-cyclopentanedicarboxylic acid **(I)** are possible Draw their structures., Upon decarboxylation, isomer A yields two isomeric acids of 3,4-dimethylcyclopentanecarboxylic acid and *B* yield only one isomer. Assign structures to *A* and *B*.

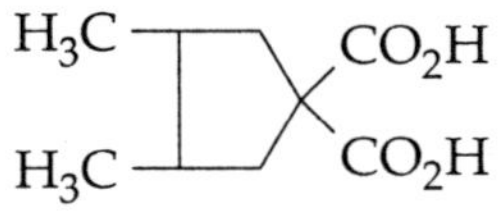

H_3C CO_2H

H_3C CO_2H

(I)

Q. 182. A compound X, C_7H_7Br, prepared from 1,3,5-cycloheptatriene, is soluble in water but insoluble in nonpolar solvents, and aqueous solutions give an immediate precipitate of AgBr when aqueous silver nitrate is added. Reduction of *X* yields cycloheptane. Propose a structure for *X* and hive the basis of your answer.

Q. 183. A hydrocarbon R, C_9H_8, rapidly decolorizes cold aqueous permangate and readily absorbs one mole of hydrogen. Vigorous hydrogenation of C_9H_8 gives a compound of formula of C_9H_{16}. Strong oxidation of *R* yields phthalic acid. What is the structure for *R* and give the basis for your answer.

Q. 184. A neutral compound *X*, $C_4H_8O_2$, reduces Tollen's reagent, liberates H_2 when treated with sodium metal, and slowly gives a positive iodoform test. What is a possible structural formula for *X*?

Q. 185. Two isomeric compounds *A* and *B* were prepared and characterized. What n.m.r. spectral characteristics would you seek ion order to differentiate the two? Should there by any significant different between the ratios of K_1 and K_2 for *A* and *B*, where K_1 and K_2 refer to the first and second acid dissociation constants?

Q. 186. From the following facts, determine a possible structure formula of A, $C_4H_8O_2$, and show you reached your conclusion.

(a) A reacts with 2,4-dinitrophenylhydrazine to yield a red precipitate, and A gives a red coloration with aq. Ceric ammonium nitrate.

(b) A reacts with acetyl chloride to produce a compound of molecular formula $C_6H_{10}O_3$.

(c) A can be oxidized to a monocarboxylic acid *B* without the loss of carbon atoms.

(d) B is optically active and loses waste when heated alone to give an unsaturated acid.

Q. 187. An optically active compound *A*, $C_7H_{14}O_2$, is hydrolyzed to give two compounds *B*, $C_5H_{10}O_2$, and C, C_2H_6O. *B* is optically active, water insoluble, but alkali soluble. *C* is water soluble. What is the structural formula of *A*.

Q. 188. An unknown substance, $C_7H_{15}NO$, hydrolyzes to give an amine and an acid. The amine reacts with nitrous acid to give a yellow-green, water-insoluble liquid. The acid is optically active. What is the structural formula of the unknown compound?

Q. 189. A compound A $C_{10}H_{12}O_2$, has infrared absorption bands at 1685 and 1250 cm^{-1}. It is insoluble in dilute acid and base. It does not reduce Tollens' reagent but forms a 2,4-dinitrophenylhydazone. Upon vigorus oxidation it yields an acidic product, $C_9H_8O_5$, which forms an acid anhydride upon heating. Treatment of A with sodium hypobromite gives a different acid which, when heated with soda-lime, yields *p*-$CH_3O—C_6H_4—CH_3$. Propose a structural formula for A.

Q. 190. A compound A, $C_5H_{10}O_4$, reacts with sodium metal to liberate hydrogen gas, gives a negative test for a 1,2-glycol, gives a positive iodoform test, does not give a 2, 4-dinitrophenylhydrazone, and when reacted with excess acetyl chloride, A gives B, $C_9H_{14}O_6$. When A is refluxe in dilute alkali for one hour and the reaction mixture acidified, two compounds C and D are recovered. C is neutral to litmus and when gently warmed with periodic acid, the mixture gives a white precipitate upon the addition of aqueous silver nitrate. D is acidic to litmus, turns ceric ammonium nitrate solution red, gives a positive iodoform test, but gives no 2,4-dinitrophenylhydrazone. What are the structuers of A, C, and D?

Q. 191. An unknown compound *A*($C_{15}H_{11}O_5N$) gives a precipitate with 2,4-dinitrophenylhydrazine, does not reduce Tollens' reagent, and give a bright red color when a solution in $CHCl_3$ is poured on sublimed $AlCl_3$. When *A* is heated with hydroxylamine hydrochloride in alcoholic KOH and then neutralized, the reaction mixture gives a blood-red color with aq. $FeCl_3$. Upon saponification and then acidification, two compounds *B*($C_9H_8O_3$) and *C*($C_6H_5O_3N$) are recovered. *C* gives a purple color with aq. $FeCl_3$, decolorizes a CCl_4 solution of Br_2 rapidly with the evolution of HBr, gives a brick-red color with treated with ferrous ammonium sulfate and alcoholic KOH in an oxygen-free vessel, and it steam distillable. *B* gives a precipitate with 2,4-dinitro-phenylhydrazine, does not redsuce Tollen's reagent, gives positive iodoform test, and dissolves in alkali to be precipitated unchanged upon acidification. Upon oxidation of B with permanagenate, the same acid, $C_3H_6O_4$, is obtained as results form the vigorous oxidation of napthalene with permanagenate. What is the structure of *A*.

Q. 192. From the following facts, give the structural formula of the unknown compound A.

(a) A is an optically active compound, molecular formula $C_5H_{11}O_2N$.
(b) Upon hydrolysis of *A*, an optically active acid, $C_3H_6O_3$, and an amine are isolated.
(c) The amine reacts with benzenesulfonyl chloride to give an alkali-insoluble amide.

Q. 193. An optically active compound A, $C_3H_5O_2Cl$, dissolves in dilute alkali and is recovered unchanged upon acidification. What is its structural formula?

Q. 194. A ketone having five carbon atoms yields an oxime whose reduction produces an amine with one asymmetric carbon atom. Write the structural formulas for possible original ketones, and for one whose structure is ruled out by the conditions imposed.

Q. 195. An optically active amide $A(C_5H_{11}ON)$ yields an acid and ammonia when hydrolyzed. When A undergoes the Hofmann degradation and the product is treated with nitrous acid, N_2 and an optically active alcohol are produced. The alcohol gives a positive iodoform test. What is the structure of A?

Q. 196. A compound $A(C_7H_6O_3)$, reduces Tollen's reagent and gives deep coloration with aq. $FeCl_3$. When 1.38 g of A are treated with excess acetic anhydride, 40 ml. of 0.5 N NaOH are required to neutralize the liberated acid. What is a possible structural formula for A? What additional information would you need to determine the identity of A?

Q. 197. An unknown compound A, C_8H_8ClBr, is insoluble in conc. H_2SO_4 but quickly gives a precipitate when treated with alcoholic silver nitrate. When heated with an aqueous solution of sodium hydroxide, A is converted into B, C_8H_9OBr, which is soluble in conc. H_2SO_4. Treatment of B with benzoyl chloride yields C, $C_{15}H_{13}O_2Br$, and vigorous oxidation of B produces D, $C_8H_5O_4Br$, which in soluble in dil. aq. NaOH and has a neutral equivalent of 121. D remains unchanged when heated. Give a possible structure for A.

Q. 198. A compound A, $C_{10}H_{14}O_2$, does not react with Tollen's solution, hot aq. NaOH, acetyl chloride, or sodium, metal. Dilute hydrochloric acid changes A into B, C_8H_8O, which does react with Tollen's reagent. Vigorous oxidation converts A and B into C_8H_6O, which loses water upon heating. What is a probable structures for A?

Q. 199. A compound A, which contains nitrogen, is insoluble in water, but soluble in dilute acid and dilute alkali. It reacts with benzensulfonyl chloride to give a product that is soluble in dilute alkali but insoluble in dilute acid. This latter compound has a neutral equivalent of 290. Compound A also reacts with benzoyl chloride to give a derivative with a neutral equivalent of 255. A does not envolve nitrogen when treated with nitrous acid, but reacts with bromine in carbon tetrachloride with the evolution of hydrogen bromide. What is a possible structure for A?

Q. 200. A compound A, $C_{11}H_8O_3NCl$, is insoluble in dil. Acid and base, and does not react with 2,4-dinitrophenyldrazine or acetyl chloride. Upon reduction with tin and hydrochloric acid, A is converted into B, $C_{11}H_{10}ONCl$, which is soluble in dilute hydrochloric acid and liberates nitrogen upon treatment with nitrous acid. Vigorous oxidation of A yields C, $C_9H_7O_7N$, which dissolves in dilute alkali, precipitates unchanged upon acidification, and has an equivalent weight of 121. Vigorous oxidation of A produces D, $C_8H_5O_4Cl$, which is soluble in aq. Alkali and has a neutral equivalent of about 101. C and D each lose a molecule of water when heated. Give a possible structure for A.

Q. 201. Analysis, quantitative hydrogenation, and molecular weight measurements on a hydrocarbon A indicates it to be a cyclononatriene. It has the following spectral properties: IR: v = 1615, 1647, 670, and 717 cm^{-1}; UV: λ_m = 223 mμ, ε 3890; n.m.r.: τ = 4.23, 7.4 (C=C—CH_2—C=C), and 7.96 (C=C—CH_2) in the ratio 3 : 1 : 2. Propose a structure for A and give the basis for your reply.

Q. 202. In the oligomerization of butadiene, a compound was obtained by b.p. 117°. It analysis and molecular weight in benzene indicated a dimer of butadinene. It absorbed 3 moles of hydrogen. The infrared spectrum of the hydrogen product is identical to that of 3-methylheptane. The UV spectrum of the dimer has a λ_m at 227 mμ, ε 25,100. Its CH_3 group gave a doublet (r = 8.9) in the n.m.r. Propose a structure for the dimer.

Q. 203. In a reaction by which it was expected to obtain a trifluoromethyl ketone, CF_3COR, there were recovered two isomeric products *A* and *B* ($C_5H_5OF_3$). *A* and *B* each give a positive Baeyer unsatruration test, form 2,4-dinitrophenylhydrazones, and give a positive haloform test. A has strong infrared absorption bands at 1780, 1630 and 3510 cm^{-1}, and *B* has strong bands at 1730 and 1630 cm^{-1}. Assign structures to *A* and *B* and account for the infrared absorption bands.

Q. 204. A compound *S* when treated with ozone gives compound *T*. When *T* is treated with water it gives the same two compounds as does the oxidation of CH_3—$CH(CH_3)$—CHOH—CHOH—CH_2CH_3 with periodic acid. What are the structures of *S*, *T* and the two oxidation products?

Q. 205. A compound X, $C_7H_{14}O$, has an infrared band in dilute CCl_4 at 3628 cm^{-1}, now in the 1680–1725 cm^{-1} range, and is inert to cold aq. $KMnO_4$. Upon oxidation with hot aq. potassium dichromate in sulfuric acid, a compound. $C_7H_{12}O$ is recovered which neutralizes two equivalents of sodium hydroxide. Propose two possible structures for X and give the basis for your answer.

Q. 206. A compound *A*, $C_8H_{16}O_3$, liberates hydrogen when treated with sodium metal, is insoluble in alkali, and decolorizes cold dilute permanganate water. When *A* is reacted with $Pb(OAc)_4$, two compounds are obtained: *B*, $C_5H_{10}O_2$ and *C*, C_3H_4O. *C* reduces Tollen's reagent and decolorizes cold permanaganate water. *B* does not reduce Tollen's reagent, forms a bisulfate addition compound, gives a positive iodoform test, gives a turbidity with Lucas reagent within 1 to 2 min, and liberates H_2 when treated with sodium metal. When B is heated with copper metal, compound *D*, $C_5H_8O_2$, is isolated. *D* is soluble in aq. NaOH, gives a color with aq. $FeCl_3$, adds Br_2, and forms a semicarbazone. What is the structure of *A*?

Q. 207. (a) When *A* was treated with Bi_2O_3 in hot acetic acid, two isomeric products, *B* and *C* were obtained. *B* is yellow, m.p. = 383.5–385.5 K, gives no immediate color with $FeCl_3$ in alcohol, has no IR band in the 3330–4000 cm^{-1} range but has a band at 1715 cm^{-1}. *C* is colorless, m.p. 371.5–373.5 K, gives an immediate dark green colour with $FeCl_3$, and has strong absorption bands in the IT at 3420, 1630, 1230 cm^{-1} and a shoulder at 1655 cm^{-1}. Give probable structures for *B* and *C*.

Q. 208. A compound A, C_5H_8, reacts with ammonical silver nitrate to form a precipitate. Oxidation of A with excess $KMnO_4$ yields an acid of the structure, CH_3—$CH(CH_3)$—COOH. What is the structural formula of A?

Q. 209. Given three unlabeled bottled *A*, *B* and *C*, each containing a color less liquid. The liquid in *A* is inert to aqueous $KMnO_4$, and it is identical to the compound produced by the reduction of neopentyl chloride with HI and phosphorous. The liquid in *B* is found to have a molecular formula of C_6H_{10}, adds two moles of bromine per mole of C_6H_{10}, and produces CH_3—CHBr—CH=CH—CHOH—CH_3 upon the addition of HOBr. The liquid in C also has a molecular formula of C_6H_{10}, is inert to ammonical silver nitrate, and upon hydration in sulfuric acid produces CH_3—CH_2—CO—CH_2—CH_2—CH_3.

Q. 210. An unknown, optically active compound A, $C_9H_{12}O_5$, turns aq. ceric ammonium nitrate red, gives a red coloration with alc. $FeCl_3$, and givesa negative Lucas test. When 3.36 g of A are reacted with excess acetic anhydride, it requires 40 ml of 1 N NaOH to neutralize the liberated acetic acid. Reaction of A with dimethylsulfate in alkali yields

B, $C_{11}H_{16}O_3$, which when oxidized with potassium permanaganate yields C, *p*-anisic acid, p-$CH_3O—C_6H_4—COOH$.

Q. 211. A methanolc solution of a halide of an alkane *A* on treatment with zinc dust gave an alkene *B* which on ozonolysis and subsequent hydrolysis gave formaldehyde and propionaldehyde. The compound 'B' on hydration and further oxidation gave a ketone C. Identify A , B and C. Give the reactions involved.

Q. 212. Compound A is treated with methyl magnesium iodide and the product is hydrolysed to get a compound B. When sodium metal is added to B, hydrogen gas is liberated along with compound C. C on treatment with methyl iodide gives methoxyethane. What are compounds A, B and C?

Q. 213. Formaldehyde on reduction gives a compound A. This on treating with Pl_3 gives a compound B which on treating with KCN gives another compound C. C on hydrolysis gives D. What are A, B, C and D.

Q. 214. Ethyl alcohol is oxidised to get a compound. A on oxidation gives another compound B which on treatment with ethyl alcohol in presence of mineral acids gives a compound C. What are A, B and C?

Q. 215. Benzoic acid reacts with ammonia to give a compound A which on heating gives B. B on dehydration with P_2O_5 gives C. What are A, B and C?

Q. 216. An organic compound A ($C_7H_6Cl_2$) yields another compound B (C_7H_6O) when treated with caustic soda solution. On oxidation 'B' gives an acid C ($C_7H_6O_2$). Give the possible structure of A B and C. Explain the reactions.

Q. 217. The three pentane isomers C_5H_{12} are each chlorinated at 573 K. Compound 'A' gives three different monochloropentanes. B gives one monochloropentane and 'C' gives four different monochloropentanes. Assign structure to A, B and C.

Q. 218. What is the end product in the following reaction?

$$CO \xrightarrow[Ni]{2H_2} A \xrightarrow{Pl_3} B \xrightarrow{CH_3ONa} C$$

Q. 219. 0.369 g of a bromo derivative of a hydrocarbon, when vaporized occupied 67.2 mL at NTP. *A* on reaction with aqueous NaOH gives *A*. *B* when passed over alumina at 523 K gives a neutral compound *C* while at 623 K it gives a hydrocarbon *D*. *D* when heated with HBr gives an isomer of *A*. When *D* is treated with conc. H_2SO_4 and the product is diluted with water and distilled, *E* is obtained. Identify *A* to *E* and explain the reactions.

Q. 220. When 0.088 g of compound *A* was dissolved in 0.50 g camphor, the melting point of camphor was lowered by 8°K. Analysis of *A* gave 68.18% C and 13.16% H. Compound *A* showed the following reactions:

(i) It reacted with acetyl chloride and evolved hydrogen with sodium.

(ii) When reacted with HCl + $ZnCl_2$, a dense oily layer separated out immediately.

Compound *A* was passed over Al_2O_3 at 623 K to give compound *B*. *B* on ozonolysis followed by hydrolysis gave two neutral compounds *C* and *D*, which gave positive test with carbonyl reagents but only *C* gave a positive test with Fehling solution and resinous substance with NaOH. Identify *A*, *B*, *C* and *D* with proper reasoning, K_f for camphor = 40 K mol^{-1} Kg.

Q. 221. An ester has molecular weight 102. On aqueous hydrolysis it produces a monobasic acid and an alcohal. If 0.185 g of the acid produced completely neutralizes 25 mL of 0.1N NaOH, find out the structure formulae of the produced alcohal, acid and the ester with proper reasoning.

Q. 222. A compound *A* $C_4H_{10}O_4$ yields on acetylation *B* of formula $C_{12}H_{18}O_8$. How many hydroxyl gps are present in compound? Also write structures of *A* and *B*.

Q. 223. An organic acid *A* $C_5H_{10}O_2$ reacts with Br_2 in the presence of phosphorus to give *B*. Compound *B* contains an asymmetric carbon atom and yields *C* on dehydration. Compound *C* does not show geometric isomerism and an decarboxylation gives an alkene *D* which on ozonolysis *E* and *F*. Compound *E* gives a positive Schiff's test but *F* does not. Give structures of *A* to *F* with reasons.

Q. 224. An organic compound *A* fumes in moist air and reacts with cold water give an acid *B*. Acid *B* reacts with NH_3 to give an amide *C*. *C* on heating with P_2O_5 gives propannitrile. What are *A*, *B* and *C*? Also report three isomers of *A*.

Q. 225. An unsaturated hydrocarbon *A*, C_6H_{10} readily gives *B* on treatment with $NaNH_2$ in liquid NH_3. When *B* is allowed to react with 1-chloropropane, a compound *C* is obtained. On partial hydrogenation in the presence of Lindlar's catalyst, *C* gives *D*, C_9H_{18}. O ozonolysis *D* gives 2, 2-dimethylpropanal and 1-butanal. Give structures of *A*, *B*, *C* and *D* with proper reasoning.

Q. 226. An acidic compound *A*, $C_4H_8O_3$ loses its optical activity on strong heating yielding *B*, $C_4H_6O_2$ which reacts readily with $KMnO_4$. *B* forms a derivative *C* with $SOCl_2$, which on reaction with $(CH_3)_2NH$ gives *D*. The compound *A* on oxidation with dilute chromic acid gives an unstable compound *E* which decarboxilates readily to give *F*, C_3H_6O. The compound *F* give s hydrocarbon *G* on treatment with amalgamated Zn and HCl. Give structures of *A* to *G* with proper reasoning.

Q. 227. The freezing point constant of C_6H_6 is 4.90 and its melting point 278.51 K. A solution of 0.816 g of compound *A* when dissolved in 7.5 g of benzene freezes at 274.59 K. The compound *A* has C = 70.58%, *H* = 5.88%. Determine the molecular weight and molecular formula of *A*. Compound *A* on heating with sodaline gives another compound *B* which on oxidation and subsequent acidificatiopn gives and acid *C* of equivalent weight 122. *C* on heating with sodaline gives benzene. Identify *A*, *B* and *C* and explain the reactions involved.

Q. 228. A ketone *A* which undergoes haloform reaction gives compound *B* on reduction. *B* on heating with H_2SO_4 gives *C* which forms forms monoozonide *D*. *D* on hydrolysis in presence of zinc dust gives only acetaldehyde. Identify *A*, *B* and *C*. Write reactions involved.

Q. 229. A liquid (*X*) having molecular formula $C_6H_{12}O_2$ is hydrolysed with water in presence of an acid to give a carboxylic acid (*Y*) and an alcohol (Z). Oxidation of (Z) with chromic acid gives (*Y*). What are (X), (*Y*) and (Z)?

Q. 230. The sodium salt of a carboxylic acid, *A* was produced by passing a gas *B* into aqueous solution of caustic alkali at an elevated temperature and pressure. *A* on heating in presence of sodium hydroxide followed by treatment with sulphuric acid gave a dibasic acid C. A sample of 0.4 g of *C* on combustion gave 0.08 g of H_2O and 0.39 g of CO_2. The silver salt of the acid, weighing 1.0 g, on ignition yielded 0.71 g of Ag as residue. Identify *A*, *B* and *C*.

Q. 231. Compound '*X*' containing chlorine, on treatment with NH_3 gives a solid '*Y*' which is free form chlorine. (*Y*) analysed as *C* = 49.31%, *H* = 9.59% and *N* = 19.18% and reacts with Br_2 and caustic soda to give a basic compound (Z), (Z) reacts with HNO_2 to give ethanol. Suggests structures for (*X*), (*Y*) and (Z)

Q. 232. An organic compound *A* on treatment with acetic acid in presence of sulphuric acid produces an ester *B*. *A* on mild oxidation gives *C*. *C* with 50% KOH followed by acidification with dilute HCl generates *A* and *D*. *D* with PCl_5 followed by reaction with ammonia gives *E*. *E* on dehydration produces hydrocyanic acid. Identify *A* to *E*.

Q. 233. An organic compound *A* on treatment with ethyl alcohal gives a carboxylic acid *B* and compound *C*. Hydrolysis of *C* under acidic conditions gives *B* and *D*. Oxidation of *D* with $KMnO_4$ also gives *B*. *B* on heating with $Ca(OH)_2$ gives *E* (Molecular formula C_3H_6O). *E* does not give Tollen's test and does not reduce Fehling solution but forms 2, 4-dinitrophenylydrazone *A* to *E*.

Q. 234. A hydrocarbon *A* (molecular formula C_5H_{10}) yields 2-methylbutane on catalytic hydrogenation. *A* adds HBr in accordance with Markownikoff's rule to form a compound *B* which on reaction with silver hydroxide forms an alcohol *C*, $C_5H_{12}O$. Alcohol *C* on oxidation gives a ketone *D*. Deduce *A* to *D* and give the reactions involved.

Q. 235. An organic compound (*X*) on analysis gives 24.24% *C*, 4.04% *H*. Further sodium extract of 1.0 g of (*X*) gives 2.9 g of AgCl with acidified $AgNO_3$ solution. The compound (*X*) may be represented by two isomeric structures (*X*) and (Z). (*Y*) on treatment with aqueous KOH solution gives a dihydroxy compound, while (*X*) on similar treatment gives ethanal. Find out (*X*), (*Y*) and (Z).

Q. 236. The percentage of carbon, hydrogen and nitrogen in a disubstituted aromatic compound *A* is 71.1, 6.67 and 10.37 respectively. Its molecule contains a single atom of nitrogen. Compound *A* evolves NH_3 when heated with NaOH to form salt *B*, from which an acid *C* is obtained on acidification. If an electrophilic substitution is carried on *C*, an entering electrophilic can occupy either of the two identical positions available in the ring. Acid *C* on heating with PCl_5, gives *D* which reacts with *B* to give an organic product *E*. Give structures of *A* to *E* with proper reasoning.

Q. 237. Write the structures of but-2-ene and 2-methylpropene and show addition of HBr on both of them in presence and absence of peroxide.

Q. 238. A hydrocarbon *A* of molecular weight 54 reacts with an excess of Br_2 in CCl_4 to give a compound *B* whose molecular weight is 593% more than that of *A*. However, on catalytic hydrogenation with excess of hydrogen, *A* forms *C* whose molecular weight is only 7.4% more than that of *A*. *A* reacts with CH_3CH_2Br in the presence of $NaNH_2$ to give another hydrocarbon *D* which on ozonolysis yields diketone *E*. *E* on oxidation gives propanoic acid. Give structures of *A* to *E* with reason.

Q. 239. 0.90 g of an organic compound $C_4H_{10}O_2(A)$ when treated with sodium, 224 mL of hydrogen were evolved at NTP. Compound *A* could be separated into fractions *B* and *C* by crystallization of which, fraction *B* could only be resolved into optical isomers *D* and *E*. Write down the structure formulae of *A* to *E* with proper reasoning.

Q. 240. An alkane *A*, C_5H_{12} on chlorination at 573 K gives a mixture of four different monochlorinated derivatives, *B*, *C*, *D* and *E*. Two of these derivatives give the same stable alkene *F* on dehydrohalogenation. On oxidation with hot alkaline $KmnO_4$, followed by acidification, *F* gives two products *G* and *H*. Give structures of *A* to *H*.

Q. 241. An organic compound *A*, $C_9H_{12}O$ was subjected to a series of tests in the laboratory. It was found that this compound

(i) Rotates the plane of polarized light.
(ii) Evolutes hydrogen with sodium.
(iii) Reacts with I_2 and NaOH to produce a pale yellow solid compound.
(iv) Does not react with Br_2/CCl_4.
(v) Reacts with hot $KMnO_4$ to form compound *B*, $C_7H_6O_2$ which can also be synthesized by the reaction of benzene and carbonyl chloride followed by hydrolysis.
(vi) Loses optical activity as a result of formation of compound *C* on being heated with *HI* and *P*.
(vii) Reacts with Lucas reagent in about 5 min.

Give structures of *A* to *C* with proper reasoning and draw Fischer projections for *A*. Give reactions for the steps wherever possible.

Q. 242. An organic *A* on heating with $AlPO_4$ at 973 K forms *B*. Compound *B* also reacts with *A* to give *C*. Compound *C* on reaction with 1,3,5-trimethyl benzene in presence of $AlCl_3$ gives a ketone *D* and CH_3COOH. *D* on treatment with Na(Hg)/HCl gives an aromatic hydrocarbon *E*. Give structures of *A* to *E* with proper reasoning.

Q. 243. Compound *A* ($C_6H_{12}O_2$) on reduction with $LiAlH_4$ yielded two compounds *B* and *C*. The compound *B* on oxidation gave *D* which on treatment with alkali (aqueous) and subsequent heating furnished *E*. The later in catalytic hydrogenation gave *C*. The compound *D* was oxidized further to give a *F* which was found to be monobasic acid (m., wt. 60). Deduce structures of *A* to *E*.

Q. 244. An organic compound *A* of molecular weight 135 on boiling with NaOH evolves a gas which gives white dense fumes on bringing a rod dipped in HCl near it. The alkaline solution thus, obtained on acidification gives the precipitate of a compound *B* having molecular weight 136. Treatment of *A* with HNO_2 also yields *B*, whereas its treatment with Br_2/KOH gives *C*. Compound *C* reacts with cold HNO_2 to give *D*, which gives red colour with ceric ammonium nitrate. On the other hand *E* an isomer of *A* on boiling with dil. HCl gives an acid *F*, having molecular weight 136. On oxidation followed by heating, *F* gives an anhydride *G* which condenses with benzene in presence of $AlCl_3$ to give anthraquione. Give structures of *A* to *G* with proper reasoning.

Q. 245. Hydrocarbon *A*, C_6H_{10} on treatment with H_2/Ni, H_2/Lindlar catalyst and Na/Liquid ammonia forms three different reduction products *B*, *C* and *D* respectively. *A* does not form any salt with ammonical $AgNO_3$ solution, but forms a salt *E* on heating with $NaNH_2$ in an inert solvent. Compound *E* reacts with CH_3I to give *F*. Compound *D* on oxidative ozonolysis gives butanoic acid along with other product.
Give structure ozonolysis *A* to *F* with proper reasoning.

Q. 246. An organic compound *A*, $C_{18}H_{20}O$ on ozonolysis gives *B*, $C_{10}H_{12}O$ and *C*, $C_8H_8O_2$. Compound *B* gives iodoform reaction and produces and oxime *D*, $C_{10}H_{13}ON$ on treatment with NH_2OH. Compound *D* reacts with PCl_5 in dry ether to give *E* which on hydrolysis gives *F*, $C_{87}H_{11}N$ and acetic acid. *F* on treatment with HNO_2 followed by oxidation gives phthalic acid. Compound *C* on mild oxidation gives *G* which gives effervescence with $NaHCO_3$. *G* on treatment with HI produces p-hydroxybenzoic acid and CH_3I. Give structures of *A* to *G* with proper reasoning.

Q. 247. A 10 g mixture of isobutane and isobutene requires 20 g of Br_2 (in CCl_4) for complete addition. If 10 g of the mixture is catalytically hydrogenated and the entire alkane is monobrominated in the presence of light at 400 K, which exclusive product and how much of it would be formed? (Atomic weight of bromine = 80).

Q. 248. Two isomeric compounds *A* and *B* have the same formula $C_{11}H_{13}OCl$. Both are unsaturated, yield the same compound *C* on catalytic hydrogenation and produce 4-chloro-3-ethoxybenzoic acid on vigorous oxidation. *A* exists in geometrical isomers, *D* and *E* but not *B*. Give structures of *A* to *E* but not *B*. Give structure of *A* to *E* with proper reasoning.

Q. 249. An organic compound *A*, $C_6H_{12}O_3$ on treatment with concentrated H_2SO_4 gives CO_2, H_2O and *B*. Compound *B* can be prepare by passing vapours of pentanol-1 over heated copper at 570 K. Compound *A* on heating gives *C*, $C_{12}H_{20}O_4$. Give structures of *A* to *C* with proper reasoning.

Q. 250. Compound *A* with empirical formula C_7H_9N on diazotization gives a product which undergoes Sand Meyer's reaction with Cu_2Cl_2 and HCl to give a compound *B*. *B* on oxidation gives a compound *C* which when heated with sodalime gives chlorobenzene. Give the structure formula of *A*, *B* and *C* and explain the reactions involved.

Q. 251. Compound *A*, $C_{10}H_{12}O$ gives off hydrogen on treatment with sodium and also decolourises Br_2 in CCl_4 to give *B*, $C_{10}H_{12}OBr_2$. *A* on treatment with I_2 in NaOH gives iodoform and an acid *C* after acidification. Give structures of *A* to *C* and also of all the geometrical and optical isomers of *A*.

Q. 252. An organic compound *A* has 76.6% as C and 6.38% *H*. Its vapour density is 47. It gives characteristics colour with aq. $FeCl_3$ solution. *A* when treated with CO_2 and NaOH at 413 K under pressure gives *B* which on acidification gives *C*. *C* reacts with acetyl chloride to give *D* which is a well known pain killer. Identity *A*, *B*, *C* and *D* and explain the reactions involved.

Q. 253. Two isomeric forms of an organic compound *A* $C_{11}H_{13}OCl$ readily decolourise bromine water and give some compound *B* on catalytic hydrogenation. Both the isomeric forms on vigorous oxidation gives *C* which on treatment with Ni/Al alloy in alkaline medium gives 3-ethoxy benzoic acid. Only one of the isomers of *A* give geometrical isomers *D* and *B*. Identify *A* to *E* with proper reasoning.

Q. 254. Two isomeric mononitro derivatives *B* and *C* are obtained by the nitration of an organic compound *A*, C_7H_8O. Treatment of *A* with acetyl chloride produces *D* which on reaction with CrO_2Cl_2 gives *E* whose oxidation with neutral $KMnO_4$ followed by acidification gives *F*. The compound *F* on heating gives phenol. *A* on treatment with alkaline $C_6H_5SO_2Cl$ produces *G* which on oxidation with $KMnO_4$ gives *H*. Hydrolysis of *H* also gives *F*. Give structures of *A* to *H* with proper reasoning.

Q. 255. An organic compound *A* of molecular weight 140.5 has 68.32%, C and 6.4% and 25.26% Cl. Hydrolysis of *A* with dilute acid gives compound *B* $C_8H_{10}O$. Compound *B* can be oxidized under mile conditions to compound *C*, C_8H_8O. Compound *C* forms a phenyl hydrozone *D* with $PhNHNH_2$ and gives positive iodoform test. Deduce the structures of *A* to *D* with proper reasoning.

Q. 256. Compound *A* and *B* on reaction in ether medium and subsequent acidification and oxidation give 2,5-dimethyl-3-hexanone. What are *A* and *B*?

Q. 257. An organic compound *A* gives positive Liebermann reaction and on treatment with $CHCl_3/KOH$ followed by hydrolysis gives *B* and *C*. Compound *B* gives colour with Schiff's reagent but not *C*, which is steam volatile. *C* on treatment with $LiAlH_4$ gives *D*, $C_7H_8O_2$ which on oxidation gives *E*. Compound *E* reacts with $(CH_3CO)_2O/CH_3COOH$ to give a pain reliever *F*. Give structures of *A* to *F* with proper reasoning.

Q. 258. An aromatic compound *A*, C_8H_9Br reacts with $H_2C(COOC_2H_5)$ in the presence of C_2H_5ONa to give *B*. Compound *B* on refluxing with dilute H_2SO_4 gives *C* which on vigorous oxidation gives *D*. The compound *D* is a dibasic acid but on heating does not give an anhydride. It forms a mononitro derivative *E*, in which all the substituents are at equidistant from one another. Give structures of *A* to *E* with proper reasoning.

Q. 259. When bromobenzene is monochlorinated, two isomeric compound *A* and *B* are obtained. Mono bromination of *A* gives several isomeric products of formula $C_6H_3ClBr_2$ while mono bromination of *B* yields only two isomers *C* and *D*. Compound *C* is identical with one of the compounds obtained from the bromination of *A*, however, *D* is totally different from any of the isomeric compounds obtained from bromination of *A*. Give structures of *A*, *B* and *C* and also suggest structures of four isomeric mono brominated products of *A*. Support answer with reasoning.

Q. 260. Compound *A* having empirical formula C_7H_8 is chlorinated in sunlight to give a product *B* which on hydrolysis gives a compounds *C*. *C* gives a positive test with Tollen's reagent and with sodium acetate and acetic anhydride gives an acid *D* with an equivalent weight 148.15. Identify *A*, *B*, *C* and *D* and explain the reactions involved.

Q. 261. A hydrocarbon *A* [C = 90.56%; vapour density 53] was subjected to vigorous oxidation to give a dibasic acid *B*. 0.10 g of *B* required 24.1 mL of 0.05 N NaOH for complete neutralization. When *B* was heated strongly with sodalime, it gave benzene. Nitration of *B* gave a single mono nitro derivative. Identify *A* and *B* with proper reasoning and also give their structures.

Q. 262. A phenolic compound *A*, $C_7H_8O_2$ on mild oxidation gives a highly volatile oil *B*. *A* forms *C* on reaction with dimethyl sulphate in alkali. Oxidation of *C* with hot $KMnO_4$ gives *D* which then reacts with bromine water to give *E* containing about 72% bromine. Give structures of *A* to *E* with proper reasoning.

Q. 263. (a) An organic compound C_8H_{18} on monochlorination gives a single monochloride. Write the structure of hydrocarbon.

(b) Write the structure of possible major monosubstituted products formed when Bromine attacks the following molecules. Justify your answer.

$C_6H_5-COO-C_6H_5$; $C_6H_5-C_6H_4-OCH_3$

$C_6H_5-C_6H_4-C_6H_5$; $HO_3S-C_6H_4-NHCOCH_3$

Q. 264. An aromatic compound *A*, $C_7H_5NO_3Cl_2$on reduction with Sn/HCl gives *B* which on reaction with $NaNO_2$/HCl gives *C*. Compound *B* does not form a dye with β-naphthol. However, *C* gives red colour with ceric ammonium nitrate and on oxidation gives on acid *D* of equivalent weight 191. Decarboxylation of *D* gives *E* which forms a signle mononitro derivative *F* on nitration. Gives structure of *A* to *F* with proper reasoning.

Q. 265. Compound *A* contains, Hydrogen, nitrogen and oxygen. When treated with caustic soda and bromine, it gives a compound (*B*). *B* on reaction with sodium nitrite and hydrochloric acid gives ethanol. What are the structures of *A* and *B*. Support your answers with proper reasoning and give the reactions involved.

Q. 266. An organic *A*, $C_5H_8O_2$ on heating with sodalime gives *B*, which reacts with HCl to give *C*. The compound *C* reacts with thionyl chloride to produce *D* which on reaction with KCN gives compound *E*. Alkaline hydrolysis of *E* gives a salt *F* which on heating with sodalime produces n-butane. Careful oxidation of *A* with dichromate gives acetic acid and malonic acid. Give structures of *A* to *F* with proper reasoning.

Q. 267. Compound *A* $C_5H_8O_2$ liberated CO_2 in reaction with sodium bicarbonate. It exists in two forms neither of which is optically active. It yielded compound *B*, $C_5H_{10}O_2$ on hydrogenation. Compound *B* can be separated into enantiomorphs. Write structures of *A* and *B*.

Q. 268. A white precipitate was formed slowly when $AgNO_3$ was added to a compound *A* with molecular formula $C_6H_{13}Cl$. Compound *A* on treatment with hot alcoholic KOH gave a mixture of two isomeric alkenes *B* and *C* having formula C_6H_{12}. The mixture of *B* and *C* on ozonolysis furnished four compounds.

(i) CH_3CHO (ii) C_2H_5CHO

(iii) CH_3COCH_3 (iv) $(CH_3)_2CH \cdot CHO$

What are *A*, *B* and *C*?

Q. 269. A hydrocarbon *A*, of the formula C_8H_{10}, on ozonolysis gives compound *B*, ($C_4H_6O_2$) only. The compound *B* can also be obtained from the cyclo-alkyl bromide *C* (C_3H_5Br) upon treatment with magnesium in dry ether, followed by carbon dioxide and acidification. Identify *A*, *B* and *C* and also give equations for the reactions.

Q. 270. Acetophenone on reaction with hydroxylamine-hydro-chloride can produce two isomeric oximes. Write structures of the oximes.

Q. 271. One mole of each bromoderivative *A* and NH_3 react to give one mole of an organic compound *B*. *B* reacts with CH_3I to give *C*. Both *B* and *C* reacts with HNO_2 to give compounds *D* and *E*, respectively. *D* on oxidation and subsequent decarboxylation gives 2-methyox-2-methylpropane. Give structures of *A* to *E*.

Q. 272. An organic compound *A* of molecular formula C_5H_8 when treated with sodium in liquid ammonia followed by reaction with *n*-propyl iodide yields *B*, C_8H_{14}. *A* gives a ketone *C*, $C_5H_{10}O$, when treated with dilute H_2SO_4 and $HgSO_4$. *B* on oxidation with alkaline $KMnO_4$ gives two isomeric acids *D* and *E*, $C_4H_8O_2$. Give structures of compounds *A* to *E* with proper reasoning.

Q. 273. Alkenes *A* and *B* yield the same alcohol *C* on hydration. On vigorous oxidation with $KMnO_4$ *A* gives a carbonyl compound *D* and an acid *E*, each containing four carbon atoms. On the other hand *B* gives an acid *F* and a carbonyl compound *G*. In *G* no two identical groups are attached to the same carbon atom. Give structure of *A* to *G*.

Q. 274. An organic compound *A*. C_6H_{10} on reduction first gives *B*, C_6H_{12} and finally *C* C_6H_{14}. *A* on ozonolysis followed by hydrolysis gives two aldehydes *D* C_2H_4O and *E* $C_2H_2O_2$. Oxidation of *B* with acidified $KMnO_4$ gives acid *F*, $C_4H_8O_2$. Determine the structures of *A* to *F*.

Q. 275. Compound *A* contains only carbon and hydrogen. It decolourizes bromine in CCl_4 solution and reacts with concentrate H_2SO_4. Compound *A* reacts with HBr to form *B*. *B* reacts with NaOH to form *C*. On oxidation *C* gives hexanone-3. Write the structures of *A*, *B* and *C* and give reactions.

Q. 276. Write the stereoisomers formed when Cis-2-butene reacts with (i) bromine (ii) Baeyer's reagent.

Q. 277. (a) Catalytic dehydrogenation of methylcyclohexane, obtained from petroleum gives a liquid which on treatment with chlorosulophonic acid at 370 K yields a mixture of two isomers *A* and *B*. $C_7H_7SO_2Cl$. The major isomer *A* reacts with ammonia to form *C*, which on oxidation with permanganate gives compound *D*. On heating compound *D* gives a well known sweetening agent *E*. The minor isomer *B* also reacts with ammonia to give a compound *F* which on treatment with NaCl/NaOH gives an antiseptic *G*. Identify *A* to *G* with proper reasoning.

Q. 278. A liquid *A* is reacted with hot aqueous sodium carbonate solution. A mixture of two salts *B* and *C* are produced in the solution. The mixture on acidification with sulphuric acid and distillation produces the liquid *A* again. Identify *A*, *B* and *C* and write the equations involved.

Q. 279. 1, 4-Pentadiene reacts with excess of HCl in the presence of benzoyl peroxide to give compound *X* which upon reaction with excess of Mg in dry ether forms *Y*. Compound *Y* on treatment with ethyl acetate followed by dilute acid yields *Z*. Identify the structures of compounds *X*, *Y* and *Z*.

Q. 280. The alkyl halide $C_4H_9Br(A)$ reacts with alcoholic KOH and gives an alkene *B*, which reacts with bromine to give dibromide *C*. *C* is transformed with sodamide to a gas *D* which forms a precipitate when passed through an ammonical silver nitrate solution. Give the structure formulae of the compounds *A*, *B*, *C* and *D* and explain reactions involved.

Q. 281. One mole of the compound *A* (molecular formula C_8H_{12}), incapable of showing stereoisomerism, reacts with only one mole of H_2 on hydrogenation over Pd. A undergoes ozonolysis to give a symmetrical diketone $B(C_8H_{12}O_2)$. What are the structure of *A* and *B*?

Q. 282. Write down the structure of stereoisomers of the following:
(a) 2, 3-dihydroxybutane
(b) 3-phenyl-2-propenoic acid.

Q. 283. An organic compound *A* composed of *C*, *H* and *O* gives characteristic colour with ceric ammonium nitrate. Treatment of *A* with PCl_5 gives *B* which reacts with KCN to form *C*. The reduction of *C* with warm Na/C_2H_5OH produces *D* which on heating gives *E* with evolution of ammonia. Pyridine is obtained on treatment of *E* with nitrobenzene. Give structure of *A* to *E* with proper reasoning.

Q. 284. Write the structure formulae and IUPAC names for the different alkanes formed when a mixture of 1-bromopropane and 2-bromo-propane is reacted with sodium in presence of ether. What is the name of the reaction?

Q. 285. Write the structure and give IUPAC name of the alkene which gives only pentan-2-one on ozonolysis.

Q. 286. (a) An alkyne A, with 5 carbon atoms per molecule when passed through dilute sulphuric acid containing mercuric sulphate gives a compound which forms an oxime, but has no effect on Fehling's solution.; The compound on oxidation gives dimethyl acetic acid. It reacts with sodamide to form a hydrocarbon. What is the structure of the alkyne?

(b) An organic compound *A*, C_5H_9Br which readily decolourises bromine water and $KMnO_4$ solution, gives *B*, $C_5H_{11}Br$ on treatment with Sn/HCl. The reaction of *A* with $NaNH_2$ produce C with evolution of ammonia. C neither reacts with sodium nor forms any metal acetylide but reacts with Lindlar's catalyst to give *D* and on reaction with Na/liq. NH_3 produces *E*. Both the compounds *D* and *E* are isomeric. Give structures of *A* to *E* with proper reasoning.

Q. 287. Dehydrobromination of compounds *A* and *B* yield the same alkene *C*. Alkene *C* can regenerate *A* and *B* by the addition of HBr in the presence and absence of peroxide respectively. Hydrolysis of *A* and *B* give isomeric products, *D* and *E* respectively, 1, 1-diphenylethane is obtained on reaction of *C* with benzene in presence of H^+. Give structures of *A* to *E* with reasons.

Q. 288. An organic compound *A*, C_4H_9Cl on reacting with aqueous KOH gives *B* and on reaction with alcoholic KOH gives *C* which is also formed on passing the vapours of *B* over heated copper. The compound *C* readily decolourises bromine water. Ozonolysis of *C* gives two compounds *D* and *E*. Compound *D* reacts with NH_2OH to give *F* and the compound *E* reacts with NaOH to give an alcohol *G* and sodium salt *H* of an acid. *D* can also be prepared from propyne on treatment with water in presence of Hg^{2+} and H_2SO_4. Identify *A* to *H* with proper reasoning.

Q. 289. Write down the structure of stereoisomers formed when cis 2-butene is reacted with Br_2.

Q. 290. Compound A (C_8H_8O) on treatment with $NH_2OH{\cdot}HCl$ gives *B* and *C*. *B* and *C* rearrange to give *D* and *E*, respectively, on treatment with acid. *B*, *C*, *D* and *E* are all isomers of molecular formula (C_8H_9NO). When *D* is boiled with alcoholic KOH on oil *F* (C_6H_7N) separates out. *F* reacts rapidly with CH_3COCl to give back *D*. One their other hand, *E* on boiling with alkali followed by acidification gives a white solid *G* ($C_7H_6O_2$). Identify *A* to *G*.

Q. 291. As ester A ($C_4H_8O_2$), on treatment with excess methyl magnesium chloride followed by acidification, gives on alcohol *B* as the sole organic product, Alcohol *B*, on oxidation with NaOCl followed by acidification gives acetic acid. Deduce the structures of *A* and *B*. Shows the reactions involved.

Q. 292. An organic compound *A*, C_8H_6, on treatment with dilute sulphuric acid containing mercuric sulphate gives a compound *B*, which can also be obtained from a reaction of benzene with an acid chloride in the presence of anhydrous aluminium chloride. The compound *B*, when treated with iodine in aqueous KOH, yields *C* and a yellow compound *D*. Identify *A*, *B*, *C* and *D* with justification. Show how *B* formed from *A*?

Q. 293. A organic compound containing *C*. *H* and *O* exists two isomeric forms *A* and *B*. 0.108 of an isomer gives 0.308 g of CO_2 and 0.072 g. of H_2O. A, is insoluble in NaOH and $NaHCO_3$ while *B* is soluble in NaOH. *A* reacts with conc. HI to give *C* and *D*. *C*, can be seperated from *D* by ethanolic $AgNO_3$ solution and *D* is soluble in NaOH. *B* reacts readily with bromine water to give compound *E* of molecular formula $C_7H_5OBr_3$. Identify *A* to *E*.

Q. 294. An aldehyde *A* ($C_{11}H_8O$), which does not undergo self aldol condensation gives benzaldehyde and two mole of *B* on ozonolysis Compound *B*, on oxidation with silver ion, gives oxidic acid. Identify the compounds *A* and *B*.

Q. 295. Compound *A* C_7H_8O is insoluble in aqueous sodium bicarbonate but dissolves in aqueous sodium hydroxide and gives characteristic colour with aqueous ferric chloride. When treated with bromine, *A* forms a compound *B* $C_7H_5OBr_3$.

(i) Give structure formulae of *A* and *B*.

(ii) What would be structure of *A* if it neither dissolves in aquous sodium hydroxide not gives a characteristic colour with $FeCl_3$?

Q. 296. A mixture of two oromatic compound *A* and *B* was separated by dissolving in chloroform followed by extraction with aqueous, KOH solution. The organic layer containing *A*, when heated with alcoholic solution of KOH produce *C* C_7H_5N associated with unpleasant odour. The alkanine aqueous layer on the other hand, when heated with chloroform and then acidified gave a mixture of two isomeric compounds *D* and *E* of molecular formula $C_7H_6O_2$. Identify *A* to *E*.

Q. 297. An organic compound *A* containing *C*, *H*, *N* and *O*, on analysis gives 49.32% *C*, 9.59% *H* and 19.18% *N*. *A* on boiling with NaOH gives off NH_3 and a salt which on acidification gives a monobasic nitrogen free acid *B*. The silver salt of *B* contains 59.67% silver. Deduce structures of *A* and *B*.

Q. 298. A basic volatile, nitrogen compound gave a foul smelling gas when treated with $CHCl_3$ and alcoholic KOH. A 0.295 g sample of the substance dissolved in aqueous HCl and treated with $NaNO_2$ solution at 273 K liberated a colourless; odourless gas whose volume correspond to 112 mL at STP. After the evolution o the gas was complete, the aqueous solution was distilled to give an organic liquid which did not contain nitrogen and which on warming with alkali and iodine gave yellow precipitate. Identify the original substance. Assume it contains one *N* atom per molecule.

Q. 299. A compound has two isomers *A* and *B* of formula $C_5H_{10}O$. Isomer *A* on treating with $NaOH_{(aq)}$ gives 2,2-dimethylpropan-1-*ol* and 2,2-dimethylpropanoic acid salt. The isomer *B* on treating with $NaOH_{(aq)}$ gives 3-hydroxy, 2-propylheptanal. What are *A* and *B*.

Q. 300. A compound *X* of molecular formula $C_4H_8Cl_2$ yields a compound *Y* on hydrolysis. *Y* gives red ppt. with Fehling solution. Oxidation of *Y* gives *Z* which when reacts with $LiAlH_4$ forms 2-methyl, 1-propanol. What are *X*, *Y*, and *Z*?

Q. 301. An organic compound *A* contains 40% carbon and 6.7% hydrogen. Its vapor density is 15. On reacting with a concentrated solution of potassium hydroxide, it gives two compound *B* and *C*. When *B* is oxidized original compound *A* is obtained. When *C* is treated with concentrated HCl, it gives a compound *D* which reduces Fehling and ammonical $AgNO_3$ solution and also gives effervescences with sodium bicarbonate solution. Write structures of *A*, *B*, *C* and *D* and explain reactions.

Q. 302. An organic compound *A* C_5H_6O adds Br_2 to give $C_5H_8Br_2O$. It does not react with Tollen's reagent but enters into reaction with phenyl hydrazine. Ozonolysis of *A* gives acetaldehyde and $C_3H_4O_2$ which readily loses CO to form acetaldehyde. What is *A*?

Q. 303. Write the products in the following reactions

(i) $CH_3CH{=}CH_2 + BrCCl_3 \xrightarrow{\text{Peroxide}}$

(ii) $CH_3CH{=}C(CH_3) + HCl \xrightarrow{\text{Peroxide}}$

Q. 304. Two mole of an ester *A* are condensed in presence of sodium ethoxide to give a β-keto ester *B* and ethanol. On heating in an acidic solution *B* gives ethanol and a β-keto acid *C*. On decarboxylation *C* gives 3-pentanone. Identify *A*, *B* and *C* with proper reasoning and give reactions.

Q. 305. An organic compound *A* has molecular formula $C_8H_{16}O_2$ and it can be resolved into enantiomers. *A* on treatment with acidified dichromate solution yields $B(C_7H_{14}O_4)$ which is achiral. A on treatment with PCC/CH_2Cl_2 yields another achiral compound $C(C_8H_{14}O_2)$. *C* on refluxing with dilute Na_2CO_3 solution yields $D(C_8H_{14}O_2)$ which is distereomeric. *D* on heating in presence of H_2SO_4 yields $E(C_8H_{12}O)$ which is still chiral. *E* in heating strongly in stream of H_2/Pt yields 1,3-dimethylcyclohexane. Identify *A* to *E*.

Q. 306. A hydrocarbon $A(C_7H_{14})$ exist in two stereomeric forms. A on reductive ozonolysis yields $B(C_3H_6O)$ and $C(C_4H_8O)$. *B* on catalytic hydrogenation yields $D(C_3H_8O)$ that can't be resolved into enantiomers whereas *C* on catalytic hydrogenation yields $E(C_4H_{10}O)$, which can be resolved into enantiomers. *B* on treatment with HCN followed by hydrolysis of product yields $F(C_4H_8O_3)$ which can be resolved into enantiomers. *C* on treatment with HCN followed by hydrolysis of product yields $G(C_5H_{10}O_3)$, which is also resolvable. Identify *A* to *G* and show the streometric structures of *A*.

Q. 307. An organic compound $A(C_6H_{10})$ on catalytic hydrogenation with H_2/Pt yields C_6H_{12}. *A* on ozonolysis followed by work-up with Zn-H_2O yields $B(C_6H_{12}O_2)$, *B* on refluxing with dilute NaOH solution gives $C(C_6H_8O)$. *C* on catalytic reduction yields $D(C_6H_{12}O)$. *D* on treatment with chromic acid yields $E(C_6H_{10}O_2)$ which on heating with soda lime yields $F(C_5H_{10})$. *F* on monochlorination yields a single isomer. Identify *A* to *F*.

Q. 308. An organic compound *A* has molecular formula $C_{11}H_{14}O$. A treatment with hydroxyl amine yields two stereomers *B* and *C* with their molecular formula $C_{11}H_{15}NO$. *B* on treatment with concentrated sulphuric acid solution yields *D*, an isomer of *B* where *C* on treatment with concentrated sulphuric acid solution yields *E*, another isomer of *C*. *D* on alkaline hydrolysis yields a white crystalline solid and an optically active amine $F(C_4H_{11}N)$. E on alkaline hydrolysis produces aniline as one product. Identify *A* to *F*.

Q. 309. An organic compound $A(C_{12}H_{20})$ decolourise bromine water, and on reductive ozonolysis produces two molecules of $B(C_6H_{10}O)$. *B* on aldol condensation followed by dehydration of product gives $C(C_{12}H_{18}O)$ which on reduction with N_2H_2/OH^- regenerate *A*. *B* on reduction with Zn(Hg)/CHCl produces $C_6H_{12}(D)$ which on monochlorination gives $C_6H_{11}Cl(E)$ as a sole product. Identify *A* to *E*.

Q. 310. An alkene *A* on reductive ozonolysis yields acetone and an aldehyde. The aldehyde is easily oxidised to an acid *B*. *B* when treated with Br_2/P yields a compound *C* which on hydrolysis gives *D*. *D* can also be obtained acetone by reaction with HCN/H_2O. Identify *A* to *D*.

Q. 311. A compound having molecular formula C_6H_{12} is treated with Cl_2 to form a compound $B(C_6H_{12}C_{12})$. *B* is treated with alcoholic KOH followed by $NaNH_2$ resulting in the formation of $C(C_6H_{10})$. *C* when treated with Pt/H_2 gave 2-methyl pentane. Compound *C* does not react with $NaNH_2$ and with $NH_4OH/AgNO_3$. On ozonolysis *A* gave two aldehydes *D* and ethanal. Deduce the structure of *A* to *D*.

Q. 312. A hydrocarbon *A* containing *C* = 88.24% (MW = 68) when reacted with Na followed by *n*-propylbromide, produces a compound $B(C_8H_{14})$. It gave a ketone $C_5H_{16}O(C)$ with Hg^{2+}/H_2SO_4. B on refluxing with $H^+/KMnO_4$ gave two isomeric acid *D* and *E*. Identify *A* to *E*.

Q. 313. A neutral, resolvable organic compound *A* has molecular formula $(C_8H_{16}O_2)$. A on treatment with $LiAlH_4$ gives isomeric *B* and $C(C_4H_{10}O)$ of which only *B* is optically active. *B* on treatment with acidified dichromate solution gives $D(C_4H_8O)$ which on refluxing with dilute NaOH followed by acidification of product gave $E(C_8H_{14}O)$. *E* on heating with N_2H_4 in alkaline medium affords $F(C_8H_{16})$. *F* on treatment with $B_2H_6/H_2O_2/OH^-$ produced a resolvable $G(C_{18}H_8O)$. *G* on treatment with acidified dichromate solution produced $H(C_8H_{16}O)$ which on treatment with "*m*-chloroperbenzoic acid" produces *A*. Deduce structures of *A* to *H*.

Q. 314. An organic compound *A* has molecular formula $(C_7H_{12}O)$ and does not decolourise Bayer's reagent and evolves no gas with Na-metal. *A* also does not react with either ammonical silver nitrate or phenyl hydrazine solution. *A* on refluxing with methanol in presence of trace amount of H_2SO_4 produces a non-resolvable $B(C_8H_{16}O_2)$. *B* on treatment with CrO_3/pyridine produced $C(C_8H_{14}O_2)$ which is non-resolvable. *C* on refluxing in HI solution produced another non-resolvable compound $D(C_7H_{11}OI)$. *D* on treatment with alcoholic KOH followed by oxidative ozonolysis of product produced $E(C_7H_{10}O_4)$ which does not show optical activity. *E* on simple heating produced $C_6H_{10}O_2$ which on refluxing with dilute NaOH gave: O= . Deduce structures *A* to *E*.

Q. 315. An optically active compound *A* has molecular formula $C_{10}H_{12}O$. A decolourises Bayer's reagent as well as it forms salt with NaOH. *A* on treatment with HBr forms $B(C_{10}H_{13}OBr)$ as major product which is still enantiomeric. *B* on treatment with alcoholic solution of KOH gives a steriomeric *C* that is also an isomer *A*. *A* on treatment with acidic solution of $KMnO_4$ gives ortho hydroxy benzoic ascid as one of the product. A on treatment with $B_2H_6/H_2O_2/NaOH$ produces $D(C_{10}H_{14}O_2)$ which on refluxing with dilute sulphuric acid produced $E(C_{10}H_{12}O)$, *E* neither decolourises Br_2-H_2O solution nor forms any salt with NaOH. Deduce structures of *A* to *E*.

Q. 316. An organic compound A has molecular formula "$C_9H_{16}O$" and it gives an yellow precipitate on treatment with alkaline solution of iodine. An on treatment with hydroxyl amine (H_2NOH) yields two stero-isomeric compound *B* and *C* with their molecular formula "$C_9H_{17}NO$". B on treatment with concentrated sulophuric acid undergo rearrangement to yield a stable product *D* which is an isomer of *B*. *C* on similar treatment undergo similar type of rearrangement to yield a stable product *E* which is an isomer of *C*. Hydrolysis of *E* in acid medium yield $F(C_7H_{15}N)$ as one of the product., *F* on treatment with nitrous acid followed by oxidation with acidified dichromate solution yields $G(C_7H_{12}O_2)$. *G* on heating with sodalime yields a product C_6H_{12} which on heating with nitric acid affords a single mono-nitroderivative. Identify *A* to *G*.

Q. 317. An optically active organic compound $A(C_4H_8O_2)$ gives an orange precipitate with 2,4-dinitrophenylhydrazine and a silver mirror with ammonical silver nitrate solution. A is converted into $B(C_4H_6O)$ on treatment with concentrated sulphuric acid. *B* may exist as *cis* or *trans* isomers. *B* reacts with HCl to give $C(C_4H_7OCl)$ as major product and with chlorine to give $D(C_4H_8OCl_2)$. Reduction of *C* with Sn/HCl gives $E(C_4H_9OCl)$ which on hydrolysing with aqueous alkali yields $F(C_4H_{10}O_2)$. *F* gives a yellow precipitate with I_2 and NaOH solution but *A* does not. Identify *A* to *F*.

Q. 318. An organic compound A has molecular formula $C_7H_{14}O_2$ and it can be resolved into enantiomers. A on treatment with $H_2Cr_2O_4$ yields $B(C_7H_{12}O_2)$ which can't be resolved into enantiomers. Refluxing *B* with dilute alkali solution yields "*C*" which is an isomers of *B*. *C* evolves a gas with NaH but does not change colour of acidified dichromate solution. C on heating with concentrated sulphuric acid yields $D(C_7H_{10}O)$ which on heating with Pt at high pressure yields methyl cyclohexane. Identify *A* to *D*.

Q. 319. An organic compound a has molecular formula $C_{11}H_{22}$. A on treatment with concentrated solution of $KMnO_4$ in acidic medium yields $B(C_3H_6O)$ and $C(C_8H_{16}O)$, both form orange colour precipitate with 2,4-dinitrophenyl hydrazine but neither gives positive Tollen's test. C on treatment with $LiAlH_4$ yields $D(C_8H_{18}O)$ which on heating with concenbtrated sulphuric acid solution gives $E(C_8H_{16})$. *E* on further treatment with acidified permanganate solution yielded *B* and 2,2-dimethyl propanoic acid. Identify *A* to *E*.

Q. 320. A carboxylic acid *A* of unknown structure was found to contain only *C*, *H* and *O*. 150 mg of A required 11.9 mL of 0.22 N NaOH to reach the equivalent point. Gentle heating of A evolves CO_2 and a new carboxylic acid *B* was formed whose equivalent weight was found toe be 74. Identify *A* and *B*.

Q. 321. An organic acid $A(C_5H_{10}O_2)$ reacts with Br_2 in presence of phosphorus to produce a resolvable *B*. *B* on dehydrobromination yields *C*. *C* does not show geometrical isomerism and on decarboxylation gives an alkene *D*. *D* on ozonolysis gives *E* and *F*. Compound *E* gives positive Schiff's test but *F* does not. Give structures of *A* to *F*.

Q. 322. A monbobasic acid A has neutralization equivalent 116 and does not decolourise Bayer's reagent. Also A is enantiomeric. *A* on treatment with Br_2/red phosphorus produces *B* which is still resolvable. *B* on dehydrobromination produced *C* which show stereo isomerism. *C* on decarboxylation produced *D* which doe not show stereoisomerism. Deduce structures of *A* to *D*.

Q. 323. A neutral organic compound *A* has formula $C_9H_{16}O_3$ and rotates plane polarized light. A on acid hydrolysis produces B and $C(C_3H_8O)$. *C* on partial oxidation with PCC gives $D(C_3H_6O)$ which does not gives iodoform test. *D* on treatment with dilute solution of NaOH gives $E(C_6H_{12}O_2)$ which is diastereomeric. *E* on treatment with acidic solution of $KMnO_4$ produces *B*. Deduce structures of *A* to *E*.

Q. 324. An organic compound A($C_9H_6O_3$) does not reacts with aqueous solution of $NaHCO_3$ and does not change the colour of litmus paper. A on acid hydrolysis gives *B*($C_9H_8O_4$) whereas *A* on treatment with methanol in acid medium gives C($C_{10}H_{14}O$) as the major alcoholysis product; *B* on heating with excess of soda-line produces toulene. Deduce structures of *A* to *C*.

Q. 325. An organic compound contain *C*, *H* and oxygen and its aqueous solution is neutral. 1.44 g of A on refluxing with dilute H_2SO_4 produces two organic compound B and C of which only *B* is resolvable and gives effervescence with $NaHCO_3$. Also neutralization of entire *B* require 0.4 g of NaOH. *B* on reduction with HI/red phosphorous produces a hydrocarbon *D* which is nonresolvable. Also *D* on monochlorination yields three alkyl halide in which only one is resolvable. Deduce structures of *A* to *D*.

Q. 326. An organic compound exist in three isomeric forms *A*, *B* and *C* with their molecular formula $C_{10}H_{12}O_2$ and all are resolvable Decarboxylation of either *B* or *C* gives the same *D*(C_9H_{12}) whereas decarboxylation of *A* gives *E* an isomer of *D*. Both *D* and *E* are non-resolvable. Treatment of either *D* or *E* with $KMnO_4$ in acidic medium produces benzoic acid as one product. Monobromination of *E* in presence of sunlight yields only two isomeric products whereas *D* on mobobromination yields three isomers. Also *B* is more reactive than C towards decarboxylation reaction. Deduce structures of *A* to *E*.

Q. 327. AN optically active organic compound A has molecular formula ($C_7H_{12}O_3$) and found to produce an yellow precipitate with alkaline solution of Iodine. Also A neither decolourised brown colour of bromine water nor evolved any gas with aqueous solution $NaHCO_3$ but evolved a gas on heating with sodium metal. On hydrolysing in acidic medium, A produced another optically active compound B($C_7H_{14}O_4$) which also produced yellow ppt with NaOH/I_2. Also *B* evolved a colourless, acidic gas on treatment with aqueous $NaHCO_3$ solution. A on treating with CrO_3/HCl/Pyridine in CH_2Cl_2 produced another optically active compound C($C_7H_{10}O_4$). C on refluxing with aqueous Ag_2O produced an optically inactive compound *D*($C_7H_{10}O_5$). *D* does not decarboxylate (does not evolve CO_2) on simple heating. However, heating *D* with sodalime gave E($C_5H_{10}O$). In a separate analysis, 3-oxoethylbutanoate was heated with excess of sodium and then with CH_3I to produce *X*($C_8H_{14}O_3$). *X* on hydrolysis followed by simple heating of product yielded *E*. Identify *A* to *E* and X.

Q. 328. An oily liquid A is insoluble in water but on heating with aqueous NaOH for half an hour it dissolves. From the reaction mixture a liquid *B* can be distilled, which gaves an yellow precipitate with iodine and NaOH. On careful oxidation, *B* gives an aldehyde, *C*, which also gives an yellow precipitate with iodine and NaOH solution. If sulphuric acid is added to the solution obtained from heating A with NaOH solution, a white precipitate *D* is obtained. *D* liberates a gas on treatment with aqueous $NaHCO_3$. Heating *D* with sodalime produced benzene. Identify *A*–*D*.

Q. 329. 2.81 g of an optically active diester *A*, containing only *C*, *H* and oxygen was saponified using 30 mL 1.0 M NaOH solution. Following sponification, the solution required 6 mL 1.0 M HCl solution to titrate unreacted base. The saponification products were an optically inactive diacid *B*, methanol and an optically active alcohol *C*. Alcohol C reacted with alkaline solution producing an yellow precipitate and C_6H_5COONa. Also, the diacid B reacted with Br_2 in CCl_4 to give a single, optically inactive product *D*. Ozonolysis of *B* gave only one product. Also, the diester A reacts with Br_2/CCl_4 to give a mixture of *E* and *F*, both optically active. Identify *A* to *F*.

Q. 330. An organic compound A reacts with I_2 in basic medium to give a yellow precipitate. *A* on treatment with Br_2 in acetic acid medium gives *B* which readily undergoes reaction with cyanide ion to give *C*. *C* on reduction with $NaBH_4$ forms *D*—an optically active substance. Acid hydrolysis of *D* followed by treatment with acidic ethanol gives *E*. Gentle oxidation of *E* with potassium chlorochromate gives *F*, which on treatment with potassium ethoxide in ethanol followed by reaction with ethyl iodine and then heating the final product yields 1phenyl-1-butanone. Identify *A* to *F*.

Q. 331. An organic compound *A* has molecular C_5H_{13} *N* and it reacts with benzene sulphonyl chloride in aqueous potassium hydroxide to give a clear solution, acidificatrion of solution gives a precipitate. Also A can be resolved into enantiomers. *A* on refluxing with nitrous acid gives $B(C_5H_{12}O)$ which is still resolvable. *B* on treatment with CrO_3 in pyridinine solution of HCl yields another resolvable compound $C(C_5H_{10}O)$ which on further treatment with Zn(Hg)) in concentrated HCl solution yields a non-resolvable $D(C_5H_{12})$. Identify *A* to *D*.

Q. 332. An organic compound *A* has molecular C_7H_9N and it forms a clear solution when dissolved in aqueous KOH solution of benzene sulphonyl chloride. *A* on treatment with $NaNO_2$ and HCl at 273 K, and then with 2-napthol forms an intensity coloured compound. Also A on treatment with CH_3COCl followed by electrophilic substitution yield a single substitution product. Identify *A*.

Q. 333. An optically active organic compound *A* has molecular formula $C_{11}H_{16}O_2$ and it does not forms salt with NaOH as well as it does not decolourise Br_2—H_2O solution but evolve a gas on treatment with MeMgBr. *A* on treatment with HI gives another optically active compound $B(C_9H_{10}O_2)$. *B* on treatment with aqueous KOH yields another optically active compound $C(C_6H_{12}O_2)$. *C* on treatment with PCC yields $D(C_9H_8O_2)$ which gives positive iodoforms test as well as positive Tollen's test, *D* is optically inactive and on treatment with $Cl_2/AlCl_3$ gives one and only one possible monochloro derivative $E(C_6H_7O_2Cl)$. Deduce structures of *A* to *E*

Q. 334. An organic compound *A* (C_8H_{10}) on treatment with fuming sulphuric acid yields two isomeric products *B* and *C*. *B* on fusion with KCN gives $D(C_9H_9N)$ which on hydrolysis yields $E(C_9H_{10}O_2)$. *E* on treatment with hot alkaline $KMnO_4$ yields *F* $(C_8H_6O_4)$. *F* on heating with P_2O_5 undergo intamolecular dehydration to yield $G(C_8H_4O_3)$. Identify *A* to *G*.

PART III

Chapter **15**

PRACTICE PAPER 1

Directions: Clearly show the methods used and steps involved in ariving at your answer. Alternation should be paid to significant figures.

Solve the following problem

1. Ethylamine reacts with water as follows:
   $C_2H_3NH_2(aq) + H_2O(l) \rightarrow C_2H_5NH_3^+(aq) + OH^-(aq)$
   The base-dissociation constant K_b, for the ethylamine ion is 5.6×10^{-4}
   (a) In 65.987 mL of a 0.250 M solution of ethylamine, calculate the OH^- ion concentration.
   (b) Calculate the pOH of the solution
   (c) Calculate the % ionization of the ethylamine in the solution in part (a).
   (d) What would be the pH of a solution made by addig 15 grams of ethylammoniuim bromide ($C_2H_5NH_3Br$) to 250 ml of a 0.1-molar of ethylamine?
   (e) If 0.125 grams of solid silver nitrate to the solution in part (a) is added will silver hydroxide form as a precipitate? The value of K_{sp} for silver hydroxide is 1.52×10^{-8}.
2. Water is introduced into a test tube that contains 2.51 grams of $SbCl_3$. The products of the reaction are collected, analyzed, and found to be
   (i) 1.906 gram of a solid containing only antimony, oxygen, and chlorine.
   (ii) 0.802 grams of a single gas that is found to be 97.20% by weight chlorine and 2.75% by weight hydrogen
   (a) Determine the simplest formula for the gas.
   (b) What fraction of the chorine atoms are found in the solid compound, and what fraction are found in the gas phase, after the reaction?
   (c) What is the formula of the solid product?
   (d) Write a balanced equation for the reaction. Assume that the empirical formula of the gas is the true formula.
3. Methyl alcohol oxidizes to produce methanoic (formic) acid and water according to the following reaction and structural diagram.

$CH_3OH(aq) + O_2(g) \rightarrow HCOOH(aq) + H_2O(l)$

$$\begin{array}{c} \text{H} \\ | \\ \text{H—C—O—H} \\ | \\ \text{H} \end{array} + O_2 \rightarrow \begin{array}{c} \text{O} \\ \| \\ \text{H—C—O—H} \end{array} + H_2O$$

Given the following data:

Substance	ΔH_f° *(kJ mole*$^{-1}$*)*	S° *(J K*$^{-1}$ $\cdot$ *mole*$^{-1}$*)*
$CH_3OH(aq)$	–238.6	129
$O_2(g)$	0	205.0
$HCOOH(aq)$	–409	127.0
$H_2O(l)$	–285.84	69.94

(a) Calculate ΔH° for the oxidation of methyl alcohol.
(b) Calculate ΔS° for the oxidation of methyl alcohol.
(c) (1) Is the reaction spontaneous at 298 K.
(2) If the temperature were increased to 373 would the reaction be spontaneous?
(d) The heat of fusion of methanoic acid is 12.7 kJ mole^{-1}, and its freezing point is 273 K. Calculate ΔS° for the reaction

$$HCOOH(l) \rightarrow HCOOH(s)$$

(e) What is the standard molar entropy of HCOOH(*s*)?

$$S^\circ\ HCOOH(l) = 109.1 \text{ J mole}^{-1} \cdot \text{K}^{-1}$$

(f) Calculate ΔG° for the ionization of methanoic acid at 298 K $K_4 = 1.9 \times 10^{-4}$.

4. **Directions** : Assume that solutions are aqueous unless otherwise indicated. You need not balance the equations.
Example: A strip of magnesium is added to a solution of silver nitrate.

$$Mg + Ag^+ \rightarrow Mg^{2+} + Ag$$

(a) a piece of solid tin is heated in the presence of chlorine gas.
(b) Ethane is burned completely in air.
(c) Solid copper shavings are added to a hot, dilute nitric acid solution.
(d) Dilute sulfuric acid is added to a solution of mercuric nitrate.
(e) Sulfur trioxide gas is heated in the presence of solid calcium oxide.
(f) Copper sulfate pentahydrate is strongly heated.
(g) A strong ammonia solution is added to suspension of zinc hydroxide.
(h) Ethane gas is heated in the presence of bromine gas to yield a monobrominated product.

5. Give a brief explain for each of the following:
(a) Water can act either as an acid or as a base.
(b) HF is a weaker acid then HCl.
(c) For the tripotic acid H_3PO_4, K_{a1} is 7.5×10^{-3} whereas K_{a2} is 6.2×10^{-8}
(d) Pure HC is not an acid.
(e) $HClO_4$ is a stronger acid than $HClO_3$, HSO_3^-, or HSO_3.

6. Interpret each of the following four examples using modern bonding principles.
   (a) C_2H_2 and C_2H_6 both contain two carbon atoms. However the bond between the two carbons in C_2H_2 is significantly shorter than that between the two carbons in C_2H_6.
   (b) The band angle in the hydronium ion H_2O^-, is less than 109.5°, the angle of a tetrahedrom.
   (c) The lengths of the bonds between the carbon and the oxygens in the carbonate ion, CO_3^{2-}, are all equal and are longer than one might expert to find in the carbon monoxide molecule, CO.
   (d) The CNO^- ion is linear.
7. If one completely vaporizes a measured amount of a volatile liquid, the molecular weight of the liquid can be determined by measuring the volume, temperature, and pressure of the resulting gas. When using this procedure, one most use the ideal gas equation and assume that the gas behaves ideally. However, if the temperature of the gas is only slightly above the boiling point of the liquid, the gas deviates from ideal behavior. Explain the postulates of the ideal gas equation and explain why, if measured just above the boiling point the molecular weight deviates from the true value.
8. The boiling points of the following compounds increase in the order in which they are listed bellow.

$$F_2 < PH_3 < H_2O$$

Discuss the theoretical considerations involved, and use then to account for this order.

PRACTICE PAPER 2

1. (a) Write the ground-state electron configuration for a phosphrous atom.
   (b) Write the four quantum numbers the describe all the valence electrons in the phosphorus atom.
   (c) Explain whether a phosphorus atom, in its ground state, is paramagnetic or diamagnetic.
   (d) Phosphorus can be found in such diverse compounds as PCl_3, PCl_5, PCl_4^+, PCl_6^-, and P_4. How can phosphorus, in its ground state, bond is so many different arrangements? Be specific in terms of hybridization, type of bonding, and geometry.
2. Given ClO_2^-, ClO_4^-, Cl_2O, ClO_3^- and ClO_2.
   (a) Draw Lewis structures for all species.
   (b) Predict the bond angle of all species.
   (c) Predict the geometry of the chlorine in all species.
   (d) Predict the hybridization of the chlorine in all species.
   (e) Identify any of the species that might dimerize.
   (f) Identify any species that might be polar.
3. The first three ionization energies (I_1, I_2, I_3) for beryllium and neon are given in the following table:

(KJ mol)	I_1	I_2	I_3
Be	900	1757	14,840
Ne	2080	3963	6376

   (a) Write the complete electron configuration for beryllium and for neon.
   (b) Explain any tends or significant discrepancies found in the ionization energies for beryllium and neon.
   (c) If chlorine gas is passed into separate containers of heated beryllium and heated neon, explain what compounds, if any, might be formed, and explain your answer in terms of the electron configurations of these two elements.
   (d) An unknown element, X, has the following three ionization energies:

(KJ mole^{-1})	I_1	I_2	I_3
X	419	3069	4600

On the basis of the ionization energies given, what is most likely to be the compound produced when chlorine reacts with element X?

4. Assume that 185 grams of fluorine gas and 4 moles of xenon gas are contained in a flask at 273 K and 2.5 atm of pressure.

(a) Calculate (1) the volume of the flask and (2) the partial pressure of each gas.

(b) 23 grams of lithium metal is introduced into the flask, and a violent reaction occurs in which one of the reactants is entirely consumed. What weight of lithium fluoride is formed?

(c) Calculate the partial pressures of any gas(es) present after the reaction in part (b) is complete and the temperature has been brought back to 273 K (The volumes of solid reactants and products may be ignored.)

5. A student performed the following experiment in the laboratory; She suspended a clean piece of silver metal in an evacuated test tube. The empty test tube weighed 42.8973 grams. The silver weighed 1.7838 grams. Next, she introduced a stream of chlorine gas into test tube and allowed it to react with the silver. After a few minutes, a white compound was found to have formed on the silver strip, coating it uniformly. She then opened the apparatus, weighed the coated strip, and found it to weigh 1.9342 grams. Finally, she washed the silver strip, and then dried the compound and the strip and reweighed. She discovered that the silver strip weighed 1.3258 grams.

(a) Show how she would determine:

(1) the number of moles of chlorine gas that reacted.

(2) the number of moles of silver that racted.

(b) Show how she could determine the simplest formula for the silver chloride.

(c) Show how her results would have been affected if:

(1) some of the white compound had been washed down the sink before it was dried and reweighed.

(2) the silver strip was not thoroughly dried when it was reweighed.

6. An unknown hydrocarbon is burned in the presence of oxygen in order to determine its empirical formula. An other sample of the hydrocarbon is subjected to colligative property tests in order to determine its molecular mass.

(a) Calculate the empirical formula of the hydrocarbon, if upon combustion at STP, 9.01 grams of liquid H_2O and 11.2 liters of CO_2 gas are produced.

(b) Determine the mass of the oxygen gas that is used.

(c) The hydrocarbon dissolves easily in CCl_4. A solution prepared by mixing 135 grams of CCl_4 and 4.36 grams of the hydrocarbon has a boiling point of 351.7 K. The molal boiling-point-elevation constant of CCl_4 is 5.02°K/m and its normal boiling point is 349.8 K. Calculate the molecular weights of the hydrocarbon.

(d) Determine the molecular formula of the hydrocarbon.

7. To produce molten iron abroad ships, the Navy uses the thermite reaction, which consists of mixing iron (III) oxide with powered aluminium, igniting it, and producing aluminium oxide as a by-product. ΔH_f° for aluminium oxide is –1669.8 kJ mole^{-1}, and that for iron (III) oxid is –822.2 kJ mole^{-1}.

(a) Write a balance equation for the reaction.
(b) Calculate ΔH for the reaction.
(c) The specific heat of alumium oxide is 0.79 J $g^{-1} \cdot K^{-1}$, and that or iron is 0.48 J g^{-1} K^{-1}. If one were to start the reaction at room temperature, how hot would the aluminium oxide *and* the iron become?
(d) If the temperature needed to melt iron is 1808 K and the heat of fusion for iron is 270 J g^{-1}, confirm through calculations that the reaction will indeed produce molten iron.

8. Given the equation $N_2O_4(g) \rightarrow 2NO_2(g)$ and the following data:

Species	ΔH_f° *(KJ · mole⁻¹)*	G_f° *(KJ · mole⁻¹)*
$N_2O_4(g)$	9.16	97.82
$NO_2(g)$	33.2	51.30

(a) Calculate ΔG°.
(b) Calculate ΔH°.
(c) Calculate the equilibrium constant K_p at 298 K 1 atm.
(d) Calculate K at 773 K and 1 atom.
(e) Calculate ΔS° at 298 K and 1 atm.
(f) Calculate the temperature at which ΔG° is equal to zero at 1 atom, assuming that ΔH° and ΔS° do not change significantly as the temperature increases.

9. The reaction

$$2NO_2(g) + Cl_2(g) \rightarrow 2NO_2Cl(g)$$

was studied at 293 K and the following data were obtained:

Experiment	*Initial* $[NO_2]$ *(mole · liter⁻¹)*	*Initial* $[Cl_2]$ *(mole · liter⁻¹)*	*Initial Rate of increase of* $[NO_2Cl]$ *(mole · liter⁻¹ · sec⁻¹)*
1	0.1	0.005	1.35×10^{-7}
2	0.1	0.01	2.70×10^{-7}
3	0.2	0.01	5.40×10^{-7}

(a) Write the rate law for the reaction.
(b) What is the overall order for the reaction? Explain.
(c) Calculate the rate-specific constant, including units.
(d) In Experiment 3, what is the initial rate of decrease of $[Cl_2]$?

10. 250 grams of solid copper (II) nitrate is placed in an empty 4-liter flask. Upon heating the flask to 535 K some of the solid decomposes into solid copper (II) oxide, gaseous nitrogen (IV) oxide, and oxygen gas. At equilibrium, the pressure is measured and found to be 5.5 atmospheres.
(a) Write the balanced equation for the reaction.
(b) Calculate the number of moles of oxygen gas present in the flask at equilibrium.
(c) Calculate the number of grams of solid copper (II) nitrate that remained in the flask at equilibrium.
(d) Write the equilibrium expression for K_p and calculate the value of the equilibrium constant.
(e) If 420 grams of copper (II) nitrate had been placed into the empty flask at 523 K, what would the total pressure have been at equilibrium?

11. In an experiment to determine the equivalent mass of an unknown acid, a student measured out a 0.25-gram sample of an unknown solid acid and then used 45.77 mL of 0.15 M NaOH solution for neutralization to a phenolphthalein end point. Phenolphthalein is colorless in acid solutions but becomes pink when the pH of the solution reaches 9 or higher. During the course of the experiment, a back-titration was further required using 1.5 mL of 0.01 M HCl.
    (a) How many moles of OH^- were used in the titration?
    (b) How many moles of H^+ were used in the back-titration?
    (c) How many moles of H^+ are there in the solid acid?
    (d) What is the equivalent mass of the unknown acid?
12. A student places a copper electrode in a 1 M solution of $CuSO_4$ and in another beaker places a silver electrode in a 1 M solution $AgNO_3$. A salt bridge composed of Na_2SO_4 connects the two beakers. The voltage measured across the electrodes is found to be +0.42 volt.
    (a) Draw a diagram of this cell.
    (b) Describe what is happening at the cathode. (Include any equations that may be useful.)
    (c) Describe what is happening at the anode. (Include any equations that may be useful.)
    (d) Write the balanced overall cell equation.
    (e) Write the balanced cell notation.
    (f) The student adds 4 M ammonia to the copper sulfate solution producing the complex ion $Cu(NH_3)_4^{2+}$ (aq). The student remeasures the cell potential and discovers the voltage to be 0.88 volt at 298 K. What is the Cu^{2+}(aq) concentration in the cell after the ammonia has been added?
13. $^{208}_{81}Tl$ undergoes β-decay. The half-life is 3.1 min.
    (a) Write the nuclear equation.
    (b) How long will it take for 1.0 gram of $^{208}_{81}Tl$ to be reduced to 0.2 gram by decay?
    (c) Given 11 grams of pure $^{208}_{81}Tl$ initially, how many grams of pure $^{208}_{81}Tl$ will remain after 23 min?

PRACTICE PAPER 3

1. Explain each of the following in terms of (1) inter-and intra-atomic or molecular forces and (2) structure.
   (a) ICl has a boiling point of 1250 K whereas NaCl has a boiling point of 1673 K.
   (b) KI(s) is very soluble in water, whereas I_2(s) has a solubility of only 0.03 gram per 1200 grams of water.
   (c) Solid Ag conducts as electric current, whereas solid $AgNO_3$ does not.
   (d) PCl_3 has a measurable dipole moment, whereas PCl_5 does not.
2. Three students, measured an empty flask and a tight-fitting stopper and found the mass to be 62.371 g. The students then carefully measured out 4.4 mL of a volatile liquid, poured the liquid into the flask, and heated the flask containing the liquid in a 374.1 K boiling water bath. As soon as all the liquid had vaporized, the students covered the flask with the stopper and set it aside to cool. Then, after a few minutes, they very quickly removed and replaced the stopper and reweighted the flask. The mass of the flask, stopper, and condensed vapor was 63.088 g. At the conclusion of the experiment, the students rinsed out the flask, filled it with water to the top, replaced the tight-fitting stopper (being sure there were no air bubbles), and obtained a volume of 261.9mL. The barometric pressure that day was 733 mm Hg.
   (a) Calculate the pressure of the vapor inside the flask (in atm) after the vapor had cooled and the students opened the flask momentarily.
   (b) What was the mass of the vapor in the flask?
   (c) Calculate the mass of 1 mole of the vapor.
      (1) There was a hole in the stopper.
      (2) There were a few drops of water left on the flask from the water bath when the final mass was taken.
3. Three compounds, *D*, *E* and *F*, all contain element *G*. The percent (by weight of element *G* in each of the compounds was determined by analysis. The experimental data are presented in the following chart.

Compound	*% by Weight of Element G*	*Molecular Weight*
D	53.9	131.7
E	64.2	165.9
F	47.7	74.5

(a) Determine the mass of element *G* contained in 1.0 mole of each of compounds *D*, *E*, and *F*.

(b) What is the most likely value for the atomic weight of element *G*?

(c) Compound *F* contains carbon, hydrogen, and element G. When 2.19 g of compound F is completely burned in oxygen gas 3.88 g of carbon dioxide gas and 0.8 g of water are produced. What is the most likely formula for compound F?

4. Your pack six aluminum cans of cola in a cooler filled with ice. Each aluminium can (empty) weighs 50 grams. When filled, each can contains 355 mL of cola. The density of the cola is 1.23 g mL^{-1}. The specific heat of aluminium is 0.902 J g^{-1}. · K and that of the cola is 4 J g^{-1} · K.

(a) If the cola is initially at 303 K and you wish to cool it to 283 K, how much heat must be absorbed by the ice?

(b) What is the minimum amount of ice (at 273 K) needed to cool the cola? ΔH_{fus} for ice is 6.00 kJ $mole^{-1}$.

5. From each of the pairs of substances listed, and assuming 1 mole of each substance, choose the once that would be expected to have the lower absolute entropy. Explain your choice in each case.

(1) $H_2O(s)$ or $SiC(s)$ at the temperature and pressure

(2) $O_2(g)$ at 3.0 atm or $O_2(g)$ at 1.0 atm, both at the same temperature.

(3) $NH_3(\ell)$ or $C_6H_6(\ell)$ at the same temperature and pressure

(4) $Na(s)$ or $SiO_2(s)$

6. Hydrogen peroxide, H_2O_2, decomposes by first-order decomposition and has a rate constant of 0.015 min^{-1} at 473 K. Starting with a 0.5 M solution of H_2O_2, calculate:

(a) The molarity of H_2O_2 after 10 min.

(b) The time it will take for the concentration of H_2O_2 to go from 0.5 M to 0.15 M

(c) The half-life

7. Magnesium hydroxide has a solubility of 9.24×10^{-4} grams per 100 mL H_2O when measured at 298 K

(a) Write a balanced equation representing magnesium hydroxide in equilibrium in a water solution.

(b) Write an equilibrium expression for magnesium hydroxide in water.

(c) Calculate the value of K_p at 298 K for magnesium hydroxide.

(d) Calculate the value of pH and pOH for a saturated solution of magnesium hydroxide at 298 K.

(e) Show by the use of calculations whether a precipitate would form if one were to add 75 mL of a 4×10^{-4} M aqueous solution of magnesium chloride to 75 mL of a 4×10^{-4} M aqueous solution of potassium hydroxide.

8. Acetic acid, $HC_2H_3O_2$, which is represented as HA, has an acid ionization constant K_a of 1.74×10^{-5}.

(a) Calculate the hydrogen ion concentration, $[H^+]$, in a 0.5-molar solution of acetic acid.

(b) Calculate the pH and pOH of the 0.5-molar solution.
(c) What percent of the acetic acid molecules do not ionize?
(d) A buffer solution is designed to have a pH of 6.5. What is the [HA]; [A^-] ratio in this system?
(e) 0.5 liter of a new buffer is made using sodium acetate. The concentration of sodium acetate in this new buffer is 0.356 M. The acetic acid concentration is 0.5 M. Finally, 1.5 grams of LiOH is added to the solution. calculate the pH of this new buffer.

9. A student wanted to determine the molecular weight of a monoprotic, solid acid, symbolized as HA. The student carefully measured out 25,000 grams of HA and dissolved it in distilled H_2O to bring the volume of the solution exactly 500 mL. The student next measured out several fifty-mL aliquots of the acid solution and then titrated it against standardized 0.1 M NaOH solution. The results of the three titrations are given in the table.

Trial	*Milliliters of HA Solution*	*Milliliters if NaOH Solution*
1	49.12	87.45
2	49.00	84,68
3	48.84	91.23

(a) Calculate the number of moles of HA in the fifty-mL aliquots.
(b) Calculate the molecular weight of the acid, HA.
(c) Calculate the pH of the fifty-mL aliquots solution (assume complete ionization).
(d) Calculate the pOH of the fifty-mL aliquot solution (assume complete ionization.
(e) Discuss how each of the following errors would affect the determination of the molecular weight of the acid, HA.
   (1) The balance that the student used in measuring out the 25 grams of HA was reading 0.01 gram too low.
   (2) There was an impurity in the acid, HA.
   (3) The NaOH solution used in titration was actually 0.15 M instead of 0.1 M.

10. The ferrous ion, $Fe^{2+}(aq)$, reacts with the permanganate ion, $MnO_4^-(aq)$, in an acidic solution to produce to the ferric ion, $Fe^{3+}(aq)$. A 6.893-gram sample of orw was mechanically crushed and then treated with concentrated hydrochloric acid, which oxidized all of the iron in the ore to the ferrous ion, Fe^{2+} (aq). Next, the acid solution containing all of the ferrous ions was titrated with 0.1 M $KMnO_4$ solution. The end point was reached when 13.889 mL of the potassium permanganate solution was used.
(a) Write the oxidation half-reaction.
(b) Write the reduction half-reaction.
(c) Write the balanced final redox reaction.
(d) Identify the oxidizing agent, the reducing agent, the species oxidized, and the species reduced.
(e) Calculate the number of moles of iron in the sample of ore.
(f) Calculate the mass percent of iron in the ore.

11. Radon-222 can be produced from the α-decay of radium-226.
(a) Write the nuclear reaction.

(b) Calculate ΔE (in KJ) when 7 g of ${}^{226}_{88}Ra$ decays.

$${}^{4}_{2}He = 4.0015 \text{ g mole}^{-1}$$

$${}^{222}_{86}Rn = 221.9703 \text{ g mole}^{-1}$$

$${}^{226}_{88}Ra = 225.9771 \text{ g mole}^{-1}$$

(c) Calculate the mass defect of ${}^{226}_{88}Ra$

$$1 \text{ mole protons} = 1.00728 \text{ g}$$

$$1 \text{ mole neutrons} = 1.00867 \text{ g}$$

$$\text{atomic } {}^{226}_{88}Ra = -225.9771 \text{ g mole}^{-1}$$

(d) Calculate the binding energy (in kJ mole^{-1}) of ${}^{226}_{88}Ra$

(e) ${}^{226}_{88}Ra$ has a half-life of 1.62×10^3 year. Calculate the first-order rate constant.

(f) Calculate the fraction of ${}^{226}_{88}Ra$ that will remain after 100 year.

PART IV
SOLUTIONS

Chapter 1 : ATOMIC STRUCTURE

1. 580 nm = 580 × 10^7 cm

$$C = 3 \times 10^{10} \text{ cm/sec} = \frac{3\times10^{10}}{580\times10^{-7}} = 5.17 \times 10^{14} \text{ Hz}$$

$$\therefore \text{ Wave number} = \frac{1}{580\times10^{-7}} = 1.72 \times 10^4 \text{ cm}^{-1}$$

2. We know that $\bar{v} = R\left(\frac{1}{n_1^2} - \frac{1}{n_2^2}\right)$

Here, $R = 10.97 \times 10^4 \text{ cm}^{-1}$; $n_1 = 3$, $n_2 = 5$

$$\therefore \quad \bar{v} = 10.97 \times 10^4 \left(\frac{1}{3^2} - \frac{1}{5^2}\right) = 10.97 \times 10^4 \times \frac{16}{225} = 7.8 \times 10^3 \text{ cm}^{-1}$$

3. We know that the radius is calculated by the expression

$$r = \frac{n^2 h^2}{4\pi^2 mze^2}$$

Here, $n = 10$ $h = 6.625 \cdot 10^{-27}$ erg-sec

$m = 9.1 \times 10^{28}$ z $= 1$ $e = 1.602 \times 10^{-19}$ Coulombs

∴ Radius of tenth orbit of hydrogen atom

$$= \frac{10^2 \times (6.625\times10^{-27})\times7^2}{4\times22^2\times9.1\times10^{-28}\times1\times(1.602\times10^{-19})^2} = 5.29 \times 10^{-7} \text{ cm.}$$

4. We have energy, $E = \frac{1}{2} mv^2$

$$\Rightarrow v^2 = \frac{2E}{m};$$

$$\text{or, } v = \sqrt{\frac{2E}{m}} = \sqrt{\frac{2\times2.8\times10^{-25}}{9.1\times10^{-31}}} = 786.46 \text{ m sec}^{-1}$$

5. According deBroglie wave equation

$$\text{mv} = \frac{h}{\lambda} = \frac{6.625\times10^{-34}}{10^{-17}} = 6.625 \times 10^{-17} \text{ Kg m sec}^{-1}$$

6. According to uncertainty principle

$$\Delta P \cdot \Delta x = \frac{h}{4\pi}; \ \Delta V = \frac{h}{4\pi\Delta x \cdot m} \quad (\because \Delta p = \Delta vxm)$$

$$= \frac{6.625\times10^{-34}}{4(3.14)\times10^{-10}\times0.1} = 5.27 \times 10^{-24} \text{ m sec}^{-1}$$

7. According to Bohr's theory the angular momentum of an electron if mass m moving with a velocity v along a path with radius of curvature r, is quantized and can have only values multiple of $\frac{h}{2\pi}$ i.e. $mvr = \frac{nh}{2\pi}$

for inner most orbit $n = 1$ substituting the values in $v = \frac{nh}{2\pi.m.r}$

$$= \frac{6.625 \times 10^{-34}}{2 \times 3.14 \times 9.1 \times 10^{-31} \times 0.0529 \times 10^{-9}} = 2.19 \times 10^6 \text{ m/sec}$$

8. We know that $\Delta x = \frac{h}{4\pi \cdot p} = \frac{h}{4\pi \cdot \Delta V \times m}$

Values given are m = 1.0 Kg v = 100 m sec^{-1}; Δv = 0.1 m sec^{-1}
Substituting the values

$$\Delta x = \frac{6.625 \times 10^{-34}}{4 \times 3.14 \times 0.1 \times 1000} = 5.27 \times 10^{-37} \text{ m}$$

9. (a) Given $E = \frac{-2.17 \times 10^{-12}}{n^2}$ erg

∴ Energy of the electron in the second orbit of hydrogen atom

$$= \frac{-2.17 \times 10^{-12}}{4} = -5.425 \times 10^{-12} \text{ erg}$$

∴ The energy required to remove the electron in the 2nd orbit of hydrogen atom
$= +5.425 \times 10^{12}$ erg $= 5.425 \times 10^3$ Joules (∵ 1 Joule = 10^7 ergs)
(Note: The sign is changed to indicate endothermic nature)

(b) $E = hv = h\frac{c}{v}$

$$\therefore v = \frac{hc}{E} = \frac{6.625 \times 10^{-27} \times 3 \times 10^{10}}{5.425 \times 10^{-12}} = 3.668 \times 10^{-8} \text{ cm} = 3.668 \text{ Å}$$

10. (i) Volume of 1 gm. atom or 6.023×10^{23} atoms

$$= \frac{\text{atomic weight}}{\text{density}} = \frac{54.94}{7.2} = 7.40 \text{ c.c.}$$

Volume of one atom of the metal

$$= \frac{\text{Volume of 1g atom}}{\text{Avogadro number}} = \frac{7.40}{6.02 \times 10^{23}} = 1.23 \times 10^{-23}$$

(ii) Radius of the atom

Volume of a sphere $= \frac{4}{3}\pi r^3 = 1.23 \times 10^{-23}$ c.c

$$\therefore\ r^3 = \frac{1.23\times10^{-23}\times3}{4\pi} \qquad \therefore\ r = \left(\frac{1.23\times10^{-23}\times3}{4(3.14)}\right)^{\frac{1}{3}} = 1.432\times10^{-8}\ \text{cm}.$$

11. Letting $\Delta x = 5\times10^{-15}$ m we have

$$\Delta p = \frac{h}{2\,\Delta x} \geq \frac{1.054\times10^{-34}\ \text{J}\cdot\text{s}}{(2)(5\times10^{-15}\ \text{m})}$$

$\geq 1.1\times10^{-20}$ kg · m/s

If this is the uncertainty in a nuclear electron's momentum, the momentum p itself must be at least comparable in magnitude. An electron with such a momentum has a kinetic energy K many times greater than its rest energy m_0c^2. Let $K = pc$. Therefore

$K = pc \geq (1.1\times10^{-20}$ kg · m/s) $(3\times10^8$ m/s) $\geq 3.3\times10^{-12}$ J

Since 1 eV = 1.6×10^{-19} J, the kinetic energy of an electron must exceed 20 MeV if it is to be a nuclear constituent. Expriments indicate that even the electrons associated with unstable atoms never have more than a fraction of this energy, and we conclude that electrons are not present within nuclei.

12. Here we find that with $\Delta x = 5.3\times10^{-11}$ m,

$$\Delta p \geq \frac{h}{2\,\Delta x} \geq 9.9\times10^{-25}\ \text{kg}\cdot\text{m/s}$$

An electron whose momentum is of this order of magnitude behaves like a classical particle, and its kinetic energy is

$$K = \frac{p^2}{2m} \geq \frac{(9.9\times10^{-25}\ \text{kg}\cdot\text{m/s})^2}{(2)(9.1\times10^{-31}\ \text{kg})}$$

$$\geq 5.4\times10^{-19}\ \text{J}$$

which is 3.4 eV. The kinetic energy of an electron in the lowest energy level of a hydrogen atom is actually 13.6 eV.

13. Since 13.6 eV = 2.2×10^{-8} J,

$$\text{therefore, } r = \frac{e^2}{8\pi\varepsilon_0 E} = \frac{(1.6\times10^{-19}\ \text{C})^2}{(8\pi)(8.85\times10^{-12}\ \text{F/m})(-2.2\times10^{-18}\ \text{J})} = 5.3\times10^{-11}\ \text{m}$$

An atomic radius of this magnitude agrees with estimates made in other ways. The electron's velocity is found from

$v = \dfrac{e}{\sqrt{4\pi\varepsilon_0 mr}}$. Substituting the values,

$$v = \frac{1.6\times10^{-19}\ \text{C}}{\sqrt{(4\pi)(8.85\times10^{-12}\ \text{F/m})(9.1\times10^{-31}\ \text{kg})(5.3\times10^{-11}\ \text{m})}} = 2.2\times10^6\ \text{m/s}$$

Since $v \gg c$, we can ignore special relativity when cosidering the hydrogen atom.

14. (a) Since $v \ll c$, we can let $m = m_0$. Hence

$$\lambda = \frac{h}{mv} = \frac{6.63 \times 10^{-34}\ \text{J} \cdot \text{s}}{(0.046\ \text{kg})\ (30\ \text{m/s})} = 4.8 \times 10^{-34}\ \text{m}$$

The wavelength of the golf ball is so small compared with its dimensions that we would not expect to find any wave aspects in its behavior.

(b) Again $v \ll c$, so with $m = m_0 = 9.1 \times 10^{-31}$ kg, we have

$$\lambda = \frac{h}{mv} = \frac{6.63 \times 10^{-34}\ \text{J} \cdot \text{s}}{(9.1 \times 10^{-31}\ \text{kg})\ (10^7\ \text{m/s})} = 7.3 \times 10^{-11}\ \text{m}$$

The dimensions of atoms are comparable with this figure—the radius of the hydrogen atom for instance, is 5.3×10^{-11} m. It is therefore not surprising that the wave character of moving electrons is the key to understanding atomic structure and behavior.

15. (a) Since the speed of light is given in meters per second, we must convert the wavelength to meters

$\lambda = 700\ \text{nm} = 700 \times 10^{-9}\ \text{m} = 7.00 \times 10^{-7}\ \text{m}$

$$v = \frac{c}{\lambda} = \frac{2.998 \times 10^8\ \text{m} \cdot \text{s}^{-1}}{7.0 \times 10^{-7}\ \text{m}} = 0.428 \times 10^{15} = 4.28 \times 10^{14}\ \text{s}^{-1}$$

$\bar{v} = 1/\lambda = 1/(7.0 \times 10^{-7}\ \text{m}) = 1.43 \times 10^6\ \text{m}^{-1}$

(b) We convert the wavelength to angstroms,

$$700\ \text{nm} = (700\ \text{nm}) \left(\frac{1\ \text{Å}}{1 \times 10^{-1}\ \text{nm}} \right) = 7000\ \text{Å}$$

16. The fifth line in the Balmer series is the line for which $n = 7$. The wave number, $\bar{v}$, is given by

$$\bar{v} = \frac{1}{\lambda} = 1.08678 \times 10^{-7} \left(\frac{1}{2^2} - \frac{1}{7^2} \right) = 1.09678 \times 10^7 \left(\frac{1}{4} - \frac{1}{49} \right) = 2.51812 \times 10^6\ \text{m}^{-1}$$

The wavelength, λ, is therefore $1/(2.51812 \times 10^6)$ m, or

$\lambda = 3.97122 \times 10^{-7}\ \text{m} = 397.122\ \text{nm}$

The frequency, v, is given by $v = c/\lambda$, so that

$$v = \frac{2.997925 \times 10^8\ \text{m} \cdot \text{s}^{-1}}{3.97122 \times 10^{-7}\ \text{m}} = 7.54913 \times 10^{14}\ \text{s}^{-1}$$

17. For the Lyman series $n_L = 1$. As n_H gets larger and larger and approaches infinity ($n_H \rightarrow \infty$), $1/n_H^2$ gets smaller and smaller and approaches zero ($1/n_H^2 \rightarrow 0$), so that $1/\lambda \rightarrow \mathcal{R}/1^2 = \mathcal{R} = 1.09678 \times 10^7\ \text{m}^{-1}$. Therefore, for the limiting line

$\lambda = 1/(1.09678 \times 10^7)\ \text{m} = 9.1176 \times 10^{-8}\ \text{m} = 91.176\ \text{nm}$

$$\upsilon = \frac{c}{\lambda} = \frac{2.997925 \times 10^8 \text{ m} \cdot \text{s}^{-1}}{9.1176 \times 10^{-8} \text{ m}} = 3.2881 \times 10^{15} \text{ s}^{-1}$$

A line with a wavelength of 91.176 nm falls in the middle of the far UV region of the spectrum.

18. To move the electron of a hydrogen atom from the ground state, where it has energy E_1, to the first excited state where it has energy E_2, we must supply the difference in energy, that is $E_2 - E_1$.

∴ energy required for transition from ground state to first excited state = $E_2 - E_1$ = –3.40 eV – (–13.60 eV) = 10.2 ev.atom^{-1}

To express this in joules,

$$10.20 \text{ eV} = (10.20 \text{ eV})\ (1.6022 \times 10^{-19} \text{ J} \cdot \text{eV}^{-1}) = 1.634 \times 10^{-18} \text{ J}$$

To obtain the energy required to cause a mole of H atoms to undergo this transition multiply the energy needed per atom by Avogadro's number.

$$\left(1.634 \times 10^{-18} \frac{\text{J}}{\text{atom}}\right)\left(6.022 \times 10^{23} \frac{\text{atom}}{\text{mol}}\right) = 9.840 \times 10^5 \frac{\text{J}}{\text{mol}} = 984. \frac{\text{kJ}}{\text{mol}}$$

We could also have calculated this directly in kilojoules per mole as:

$$E_2 - E_1 = \frac{-K}{2^2} - \left(\frac{-K}{1^2}\right) = \frac{-K}{4} + K = \frac{3}{4}K = \frac{3}{4}\left(1312 \frac{\text{J}}{\text{mol}}\right) = 984.0 \frac{\text{kJ}}{\text{mol}}$$

In using the Bohr frequency condition, to calculate the frequency of the radiation that causes this transition, it is most convenient to give the energy difference in joules.

$$\upsilon = \frac{\Delta E}{h} = \frac{1.634 \times 10^{-18} \text{ J}}{6.6261 \times 10^{-34} \text{ J} \cdot \text{s}} = 2.466 \times 10^{15} \text{ s}^{-1}$$

This is the frequency of the first and most intense line of the Lyman series.

19. An electron volt is, by definition, the energy an electron possesses when it is accelerated through a potential difference of 1 V. An electron accelerated through a potential difference of 8000 V therefore has 8000 eV of energy. This value can be converted to joules.

$$\text{Energy} = (8000 \text{ eV})\ (1.6022 \times 10^{-19} \text{ J} \cdot \text{eV}^{-1}) = 1.2818 \times 10^{-15} \text{ J}$$

The definition of the kinetic energy of a particle of mass m travelling with speed v is

$$\text{kinetic energy} = E = \frac{1}{2} mv^2$$

This can be rearranged to solve for the speed. We obtain

$$v^2 = 2E/m$$

Using the known mass of the electron we can calculate the speed of an electron with a kinetic energy of 1.2818 × 10^{-15} J. We most be carefull to use only SI units for all quantities. Mass must be expressed in kilograms, and the speed will be given in meters per second. We obtain

$$v^2 = \frac{2(1.2818 \times 10^{-23})J}{9.1094 \times 10^{-31} \text{ kg}} = 2.814 \times 10^{15} \text{ m}^2\text{·s}^{-2}$$

Taking the square root we find that $v = 5.305 \times 10^7$ m · s^{-1}, the speed of the electrons. The linear momentum, p, is given by

$$p = mv = (9.1094 \times 10^{-31} \text{ kg})(5.305 \times 10^7 \text{ m} \cdot \text{s}^{-1}) = 4.832 \times 10^{-23} = \text{kg}\cdot\text{m}\cdot\text{s}^{-1}$$

The de Broglie relation can now be utilized to obtain the wavelength:

$$\lambda = \frac{h}{mv} = \frac{h}{p} = \frac{6.6261 \times 10^{-34} \text{ J} \cdot \text{s}}{4.832 \times 10^{-23} \text{ kg} \cdot \text{m} \cdot \text{s}^{-1}}$$

$$\lambda = 1.371 \times 10^{-11} \text{ m} = 0.01371 \text{ nm} = 0.371 \text{ Å}$$

20. Using the de Broglie relation, we obtain:

$$\lambda = \frac{h}{mv} = \frac{6.6261 \times 10^{-34} \text{ J} \cdot \text{s}}{(0.142 \text{ kg})(300 \text{ m} \cdot \text{s}^{-1})} = 1.56 \times 10^{-34} \frac{\text{J} \cdot \text{s}^2}{\text{kg} \cdot \text{m}} = 1.56 \times 10^{-34} \text{ m}$$

$$= 1.56 \times 10^{-25} \text{ nm} = 1.56 \times 10^{-24} \text{ Å}$$

21. (a) The atomic number of Be is 4. The Be^{3+} ion has only a single electron and is therefoe an H-like ion. Since $hc\mathcal{R}$ is 13.60 eV,

$$\Delta E = (13.60)4^2 \left(\frac{1}{1^2} - \frac{1}{2^2}\right) = \frac{3}{4} (13.6)(16) = (12)(13.6) = 163.2 \text{ eV}$$

The photon emitted during this transition therefore has 163.2 eV of energy. In joules this is

$$(163.2 \text{ eV})(1.6022 \times 10^{-19} \text{ J} \cdot \text{eV}^{-1}) = 2.614 \times 10^{-17} \text{ J}$$

This is energy of a single photon.

If a mole of ions undergoes this transition the energy emitted is

$$(2.614 \times 10^{-17} \text{ J} \cdot \text{ion}^{-1})(6.022 \times 10^{23} \text{ ions} \cdot \text{mol}^{-1}) = 1.574 \times 10^7 \text{ J} \cdot \text{mol}^{-1}$$

or 1.574×10^4 kJ · mol^{-1}, which is a very large amount of energy!

The wavelength of the radiation can be calculated as

$$1/\lambda = (1.09678 \times 10^7 \text{ m}^{-1})(4^2)(3/4) = 12(1.09678 \times 10^7) \text{ m}^{-1} = 1.316 \times 10^8 \text{ m}^{-1}$$

Therefore

$$\lambda = (1/1.316) \times 10^{-8} \text{ m} = 7.598 \times 10^{-9} \text{ m} = 7.598 \text{ nm}$$

(b) The amount of energy required to remove the electron from a single Be^{3+} ion is the difference in energy between the ground state ($n_L = 1$) and the state with $n_H = \infty$. Thus,

$$\text{IE} = Hc\mathcal{R}Z^2 = (13.60 \text{ eV})(4^2) = 217.6 \text{ eV}$$

In other units the ionization energy is

$$\left(217.6 \frac{\text{eV}}{\text{ion}}\right)\left(1.6022 \times 10^{-19} \frac{\text{J}}{\text{eV}}\right)\left(6.022 \times 10^{23} \frac{\text{ions}}{\text{mol}}\right) = 2.099 \times 10^7 \frac{\text{J}}{\text{mol}}$$

Note that the requires 2.099×10^7 J to ionize a mole of Be^{3+} ions, compared with only 1.312×10^6 J to ionize a mole of H atoms. The reason so much more energy is needed to ionize Be^{3+} ions is that the nucleus of the Be^{3+} ion has four times as much positive charge as the nucleus of the H atom, and therefore attracts the single electron much more strongly.

22. (a) To calculate the energy of the light, the difference between the energy and the kinetic energy is the energy needed to eject the electron (E_{min}) is.

$$E = \frac{hc}{\lambda} = \frac{(6.626 \times 10^{-34}\ \text{J} \cdot \text{s})(2.998 \times 10^{8}\ \text{m/s})}{5.4 \times 10^{-7}\ \text{m}} = 3.68 \times 10^{-19}\ \text{J}$$

$E_{min} = 3.68 \times 10^{-19}$ J – 2.60×10^{-20} J = 3.42×10^{-19} J

While the calculations above used the longer wavelength, we also could have used the shorter wavelength and gotten the same answer (within significance).

$$\therefore E_{min} = \frac{hc}{\lambda} = \frac{(6.626 \times 10^{-34}\ \text{J} \cdot \text{s})(2.998 \times 10^{8}\ \text{m/s})}{4.0 \times 10^{-7}\ \text{m}} = 4.97 \times 10^{-19}\ \text{J}$$

$E_{min} = 4.97 \times 10^{-19}$ J – 1.54×10^{-19} J = 3.43×10^{-19} J

(b) Since longer wavelength correspond to lower energies, E_{min} would correspond to the longest wavelength

$$\lambda = \frac{hc}{E_{min}} = \frac{(6.626 \times 10^{-34}\ \text{J} \cdot \text{s})(2.998 \times 10^{8}\ \text{m/s})}{3.42 \times 10^{-19}\ \text{J}} = 5.81 \times 10^{-7}\ \text{m} = 581\ \text{nm}$$

23. First use Avogadro's number to calculate the energy for one Cl_2 molecult

$$\frac{239\ \text{kJ}}{1\ \text{mol}} \times \frac{1000\ \text{J}}{1\ \text{kJ}} \times \frac{1\ \text{mol}}{6.022 \times 10^{23}\ \text{mole}} = 3.9687 \times 10^{-19}\ \text{J/mole}$$

Then convert energy to frequency and finally to wavelength.

$$v = \frac{E}{h} = \frac{3.9687 \times 10^{-19}\ \text{J}}{6.626 \times 10^{-34}\ \text{J} \cdot \text{s}} = 5.9897 \times 10^{14}/\text{s}$$

$$\lambda = \frac{c}{v} = \frac{2.998 \times 10^{8}\ \text{m/s}}{5.9897 \times 10^{14}\ /\text{s}} = 5.0052 \times 10^{-7}\ \text{m (501 nm; visible region)}$$

24. First calculate the energy needed to heat the 0.250 L of water from 293 K to 373 K.

$$0.250\ \text{L} \times \frac{1000\ \text{g}}{1\text{L}} \times \frac{4.184\ \text{J}}{(\text{g} \cdot \text{k})} \times (373\ \text{K} - 293\ \text{K}) = 8.368 \times 10^{4}\ \text{J}$$

Then calculate the frequency, the energy of one photon, and the number of photons.

$$v = \frac{c}{\lambda} = \frac{2.998 \times 10^{8}\ \text{m/s}}{0.125\ \text{m}} = 2.398 \times 10^{9}/\text{s}$$

E of 1 photon = hv = (6.626×10^{-34} J · s) × (2.398×10^{9}/s) = 1.589×10^{-24} J

$$\text{No. photons} = hv = 8.368 \times 10^{4}\ \text{J} \times \frac{1\,\text{photon}}{1.589 \times 10^{-24}\ \text{J}} = 5.2656 \times 10^{28}$$

$$= 5.27 \times 10^{28}\ \text{photons}$$

25. First write the following equality for the energy to remove one electron, $E_{removal}$

$$E_{removal} = E_{425\ nm} - E_k \text{ of ejected photon}$$

Use $E = h\nu$ to calculate the energy of the photon.
Now, kinetic energy $E_k = 1/2\ mv^2$. Use this to calculate E_k.

$$E_{425\ nm} = h\nu = \frac{hc}{\lambda} = \frac{(6.626\times10^{-34}\ J\cdot s)(2.998\times10^{8}\ m/s)}{4.25\times10^{-7}\ m} = 4.6740 \times 10^{-19}\ J$$

and $E_k = 1/2\ mv^2 = 1/2 \times (9.1095 \times 10^{-31}\ kg) \times (4.88 \times 10^5\ m/s)^2$
$= 1.0846 \times 10^{-19}\ J$

Subtract to find $E_{removal}$ and convert it to kJ/mol:

$E_{removal} = 4.6740 \times 10^{-19}\ J - (1.0847 \times 10^{-19}\ J) = 3.\ 5893 \times 10^{-19}$
$= 3.59 \times 10^{-19}\ J/electron$

$$\text{or } E_{removal} = \frac{3.5893\times10^{-19}\ J}{1e^-} \times \frac{6.022\times10^{23}\ e^-}{1\ mol} \times \frac{1KJ}{1000\ J}$$

$= 2.161 \times 10^2\ kJ/mol$

26. First calcualtion the energy, E, in joules using the product of voltage and charge
$E = (4.00 \times 10^3\ V) \times (1.602 \times 10^{-19}\ C) = 6.408 \times 10^{-16}\ J$
Now use the kinietic energy equation $E_k = 1/2\ mv^2$ and solve for velocity

$$v = \sqrt{\frac{2E_k}{m}} = \sqrt{\frac{2\times6.408\times10^{-16}\ J}{9.1095\times10^{-31}\ kg}} = 3.7508 \times 10^7\ m/s$$

$$\lambda = \frac{h}{mv} = \frac{6.626\times10^{-34}\ J\cdot s}{(9.1095\times10^{-31}\ kg)\times(3.7508\times10^{7}\ m/s)} = 1.939 \times 10^{-11}$$

$= 1.94 \times 10^{-11}\ m\ (19.4\ pm)$

27. For one electron species, the ground stae is $\mathbf{n} = 1$, so the first excited state is $\mathbf{n} = 2$

$$E = \frac{-Bz^2}{n^2} = \frac{-(2.180\times10^{-18}\ J)(3)^2}{2^2} = -4.905 \times 10^{-18}\ J$$

$$E = -\ 4.905 \times 10^{-18}\ J \times \frac{1\ kJ}{1000\ J} \times \frac{6.022\times10^{23}}{1\ mol} = -2.954 \times 10^2\ kJ/mol$$

E is the energyof the electron. The energy needed to ionize (or remove) an electron is $+2.954 \times 10^3$ kJ/mol.

28. We have

$$\lambda = \frac{hc}{\Delta E} \text{ and } \Delta E = -\ R_H\left[\frac{1}{\mathbf{n}_{hi}^2} - \frac{1}{\mathbf{n}_{lo}^2}\right]$$

Substituting for ΔE and $\mathbf{n}_{lo} = 2$, we get $\lambda = \dfrac{hc}{-R_H\left[\dfrac{1}{\mathbf{n}_{hi}^2} - \dfrac{1}{2^2}\right]} = -\dfrac{hc}{R_H}\dfrac{1}{\left[\dfrac{1}{n_{hi}^2} - \dfrac{1}{4}\right]}$

$$\lambda = -\frac{(6.626\times10^{-34}\ \text{J}\cdot\text{s})(2.998\times10^{8}\ \text{m/s})}{2.180\times10^{-18}\ \text{J}}\frac{1}{\left[\frac{1}{\mathbf{n}_{hi}^2}-\frac{1}{4}\right]} = -9.112\times10^{-8}\ \text{m}\frac{1}{\left[\frac{1}{\mathbf{n}_{hi}^2}-\frac{1}{4}\right]}$$

$$\lambda = 9.112\times10^{-8}\ \text{m}\frac{1}{\left[\frac{1}{4}-\frac{1}{\mathbf{n}_{hi}^2}\right]} = 9.112\times10^{-8}\ \text{m}\frac{4\mathbf{n}_{hi}^2}{\frac{4\,\mathbf{n}_{hi}^2}{4}-\frac{4\,\mathbf{n}_{hi}^2}{\mathbf{n}_{hi}^2}}$$

$$\lambda = 9.112\times10^{-8}\ \text{m}\ \frac{4\,\mathbf{n}_{hi}^2}{\mathbf{n}_{hi}^2-4} = \frac{(3.645\times10^{-7}\ \text{m})\,\mathbf{n}_{hi}^2}{\mathbf{n}_{hi}^2-4}$$

$$\text{Converting to nanometers } \lambda = \frac{(3.645\times10^{2}\ \text{nm})\,\mathbf{n}_{hi}^2}{\mathbf{n}_{hi}^2-4}$$

29.

n	1		2		
l	0	1	0	1	2
sublevel	1*s*	1*p*	2*s*	2*p*	2*d*
$\mathbf{m}_l$	0 1	0 1 2	0 1	0 1 2	0 1 2 3

electron configuration with eight electrons $1s^4 1p^4$

30. (a) If there were 3 values for $\mathbf{m}_s$, then each orbital could hold 3 electrons.
s-sublevel one orbital ($m_l = 0$), 3 electrons
p-sublevel: 3 orbital ($\mathbf{m}_l = -1, 0, 1$) 9 electrons
d-sublevel: 5 orbitals ($\mathbf{m}_l = -2, -1, 0, 1, 2$) 15 electrons

(b) The $\mathbf{n} = 3$ level could hold 27 electrons, 3 in the s-sublevel, 9 in the *p*-sublevel and 15 the *d*-sublevel

(c) AN = 8; $1s^3 2s^3 2p^2$
AN = 17; $1s^3 2s^3 2p^9 3s^2$

31. (a) Energy is absorbed when the electrons are excited from lower levels to a higher one thus energy is absorbed for **transitions 2 and 4**

(b) Energy is emitted when the electron relaxes from higher level to a lower one, thus energy emitted for **transitions 1 and 3.**

(c) We cannot answer this question without knowing what the element is since one must know the number of electrons to determine the ground state configuration. If we assume the atom to be hydrogen, then **transition 1** involves the ground state.

(d) The transition with largest energy difference will absorb the most energy, thus **transition 2.**

(e) The transition with largest energy difference will emtted the most energy, thus **transition 1.**

32. (a) The number of orbitals is equal to $\mathbf{n}^2$.

(b) The number of orbitals is equal to $\mathbf{2}l + \mathbf{1}$

(c) The number of spin states (m_s) in **unrelated** to the number of orbitals, it is always 2.

33. (a) No two electrons in an atom can have the same four quantum numbers. Also, no two electrons in the same orbital can have the same spin.

(b) It takes more energy to put two electrons in the same orbital than it does to put each electron in a separate orbital (within a given sublevel). Also electrons enter orbitals of a sublevel singly (untill each orbital has one electron) before electrons are paired.

(c) A line an atomic spectrum represents the transition of an electron from a higher energy level to a lower energy level.

(d) The principal quantum number gives the relative position and energy of electrons, with smaller **n** representing electrons closer to the nucleus and with lower energy.

34. (a) The statement in **true**. Photons with very short wavelength are high energy.

(b) This statement is **false**. The energy of an electron is inversely proportional to $\mathbf{n}^2$, not l; $E_n = -\frac{R_H}{\mathbf{n}^2}$

(c) This statement is **false**. Electrons start entering the 5th principal level before the fourth is filled. The order of filling is [Kr] $5s$ $4d$ $5p$.

35. (a) Ionic size depends on the ratio of electrons to protons. The protons pull the electrons toward the nucleus, but the electrons also repel each other. In neutral atoms, these two forces balance one another. In cations, the excess of protons relative to electrons causes the attraction to dominate and the cation is smaller than the atom. In anions, the excess of electrons relative to protons causes the repulsion to dominate and the anion is larger than the atom.

(b) Se has an electron configuration of [Ar] $4s^2$ $3d^1$. thus only 3 electron are lost to form the Se^{+3} ion with a noble gas configuration. By contrast, a later transition metal such as Fe must loss 8 electrons to form an ion with a noble gas configuratio, a more formidable tsk.

(c) Electrongativity is a measure of the ability of an atom in a molecule to attract electrons to itself. The attraction is due to the protons of the nucleus. Thus for larger atoms (those down a group) in a molecule, the electron will be further from the nucleus and will experience less attraction.

36. (C) The colision frequency is the number of collision experienced by the molecules of a gas per cc per second and is given by

$$Z = \sqrt{2}\pi v d^2 N$$

where v = average molecular velocity
d = diameter of molecules
N = numebr of molecules per cc.

37. (D) Volume $V = \frac{3}{4}\pi r^3$

Differentiating w.r.t. 't' we get, $\frac{dV}{dt} = 4\pi r^2 \frac{dr}{dt}$

or $100 = 4\pi \times 10^2 \cdot \frac{dr}{dt}$

or $\frac{dr}{dt} = \frac{1}{4\pi}$

Also surface $s = 4\pi r^2$

hence $\frac{ds}{dt} = 8\pi r\frac{dr}{dt} = 80\pi\frac{dr}{dt} = 80\pi \times \frac{1}{dt} = 20$ m/min

38. $\lambda = \frac{h}{mv} = \frac{6.626 \times 10^{-34}}{100 \times 10^{-3} \times 100}$

$\lambda = 6.626 \times 10^{-35}$ m $= 6.626 \times 10^{-25}$ Å

39. Here Ψ^2_{2s} = probability of finding electrons at any place

$\Psi^2 = 0$, at node

$$\therefore \quad \Psi^2 = 0 = \frac{1}{4}\frac{1}{\sqrt{2\pi}}\left(\frac{1}{a^0}\right)^3\left(2 - \frac{r}{a_0}\right)^2 \times e^{-r/a_0}$$

or $\left(2 - \frac{1}{a_0}\right) = 0 \Rightarrow 2 = \frac{r}{a_0} \Rightarrow 2a_0 = r$

40. The energy (E_2) of the electron in 2nd Bohr's orbit = -5.42×10^{-12} erg
(Note 10^7 erg = 1 Joule)
The energy (E_3) of the electron in 3rd Bohr's orbit = -2.41×10^{-12} erg
Since, $E_3 - E_2 = h\nu$

$$= \frac{hc}{\lambda}$$

$$\therefore \lambda = \frac{hc}{E_3 - E_2} = \frac{6.626 \times 10^{-27} \times 3 \times 10^{10}}{-2.41 \times 10^{-12} + 5.42 \times 10^{-12}}$$

$$= \frac{6.626 \times 3 \times 10^{-17}}{3.02 \times 10^{-12}} = 6.5820 \times 10^{-5} \text{ cm} = 6582 \text{ Å}$$

41. Wave number $(\overline{v}) = \frac{1}{\lambda} = RH\left[\frac{1}{n_1^2} - \frac{1}{n_2^2}\right]$

For Balmer series $n_1 = 2, n_2 = \infty$

$\therefore \overline{v} = RH\left[\frac{1}{2^2} - \frac{1}{\infty^2}\right] = 27419.25 \text{ cm}^{-1}$

42. For the beam of desired wavelength accelerated through a potential 'V' following condition has to be satisfied.

$eV = h\nu$

or $eV = h\frac{c}{\lambda}$

where $e \rightarrow$ charge of electron (in beam)
$c \rightarrow$ velocity of light.
$\lambda \rightarrow$ Plank's constant
$V \rightarrow$ Desired potential

Thus $V = \frac{hc}{\lambda e}$ Substituting the values we get;

$$V = \frac{6.62 \times 10^{-34}\ J \cdot S^{-1} \times 3 \times 10^{8}\ m/s}{1.54 \times 10^{-10}\ m \times 1.6 \times 10^{-19}\ C} = 8060 \text{ volts} = 8.06 \text{ KV}$$

43. The expression of electrons energy is

$$E = Z^2 E_H$$

For He^{2+} ion $Z = 2$.

Hence $$E = \frac{4 \times 21.76 \times 10^{-19}}{n^2} \text{ J}$$

To have longest wavelength , the transiition would be from $n = 3$ to $n = 4$,

$$\therefore \Delta E = -4 \times 21.76 \times 10^{-19} \left(\frac{1}{16} - \frac{1}{9}\right) \text{ J} = 4.231 \times 10^{-19} \text{ J}$$

Now, $$\lambda = \frac{hc}{\Delta E} = \frac{(6.626 \times 10^{-34})(3 \times 10^{18})}{4 \times 231 \times 10^{-19}} = 4.7 \times 10^{-7} \text{ m.}$$

44. For the wave to be in phase the circumference of the circle must be on integral multiple of wavelength i.e. $2\pi r = n\lambda$

Where r = radius of the orbit; v = velocity of the electron
m = mass of the electron; n = Number of waves

By de Broglie equation

$$\lambda = \frac{h}{mv} \text{ i.e. } 2\pi r = \frac{nh}{mv} \quad \ldots\ldots \text{(i)}$$

r_n = radius of the nth circular orbit = $(0.53 \times 10^{-10}\ m) \times n^2$

$\therefore$ r_3 = radius of the 3rd orbit = $9 \times 0.53 \times 10^{-10}$ m

According to the Bohr's postulate.

Velocity of electron in any orbit $V_n = \frac{1}{\varepsilon_0} \cdot \frac{e^2}{2nh}$

$\therefore$ By equation (i), $$n = \frac{2\pi r \times mv}{h} = \frac{2\pi r_3 \times m}{h} \times v_3$$

$$\therefore n = \frac{2\pi r_3 \times m}{h} \times \frac{1}{\varepsilon_0} \times \frac{e^2}{2 \times 3 \times h}$$

$$= \frac{2 \times 3.14 \times 9 \times 0.53 \times 10^{-10} \times 9.1 \times 10^{-31} \times (1.6 \times 10^{-19})^2}{8.85 \times 10^{-12} \times 2 \times 3 \times (6.63 \times 10^{-34})^2} = 3$$

Hence, Number of waves = 3

45. We have, $\frac{1}{\lambda} = RZ^2\left(\frac{1}{n_1^2} - \frac{1}{n_2^2}\right)$

$$\therefore\ RZ_1^2\left(\frac{1}{n_1^2} - \frac{1}{n_2^2}\right) = RZ_2^2\left(\frac{1}{n_1'^2} - \frac{1}{n_2'^2}\right)$$

(For He^+ spieces) (For Hydrogen atom)

Now R is common on both the side and here

$Z_1 = 2(H_e^+)$, $Z_2 = 1$ (for H atom)

$n_1 = 2$ $n_2 = 4$

Substituting the values in the above equation

$$4 \times \frac{3}{16} = \frac{1}{n_1'^2} - \frac{1}{n_2'^2} = \frac{3}{4}$$

So if $n_1' = 1$ and $n_2' = 2$

then $\frac{1}{n_1'^2} - \frac{1}{n_2'^2} = \frac{3}{4}$

$\therefore$ transition $n = 2$ to $n = 1$ (Lyman series) in hydrogen spectrum has the same wavelength as the Balmer transition $n = 4$ to $n = 2$ of He^+ spectrum.

46. We have, $\Delta E = E_\infty - E_2$

$$= [0] - \left[-\frac{2.17 \times 10^{-12}}{4}\right]$$

Since $E_n = \frac{2.17 \times 10^{-12}}{n^2}$ ergs. [Note 10^7 ergs = 1 Joule]

$$= \frac{2.17 \times 10^{-12}}{4}$$

$$\Delta E = h\nu = h\frac{c}{\lambda}$$

Substituting the values we have;

$$\frac{2.17 \times 10^{-12}}{4} = 6.627 \times 10^{-27} \times \frac{3 \times 10^{10}}{\lambda}$$

$\therefore\ \lambda = 3.67 \times 10^{-5} = 3670$ Å

47. $E_1 = 2.17 \times 10^{-12}$ ergs

$$E_2 = \frac{2.17\times10^{-5}}{2^2} \text{ ergs [Note } 10^{-7} \text{ ergs} = 1 \text{ Joule]}$$

$$\text{Difference in energy} = E_1 - E_2 = 2.17 \times 10^{-11} - \frac{2.17\times10^{-2}}{2^2} \text{ ergs}$$

$$= 2.17 \times 10^{-11}\left(1 - \frac{1}{4}\right) \text{ erg} = 6.13 \times 10^{-11} \text{ erg}$$

$$E_1 - E_2 = h\nu = \frac{hc}{\lambda}$$

$$\therefore \lambda = \frac{hc}{E_1 - E_2} \text{ Substituting the value we get,}$$

$$\lambda = \frac{6.62\times10^{-27}\times3.0\times10^{10}}{1.63\times10^{-11}} \text{ cm} = 1.12 \times 10^{-5} \text{ cm} = 1120 \text{ Å}$$

The wavelength of the light emitted is 1120 Å

48. Ground state electronic configurations of Si:

↑↓	↑	↑	
$3s$	$3p_x$	$3p_y$	$3p_z$

is in accord with Hund's rule which state that electrons are given singly to degenerate orbitals before pairing occurs.

49. For high energy transition $n_1 = 1$ and $n_2 = \infty$

$$\text{we know that, } \bar{v} = \frac{1}{\lambda} = RZ^2\left(\frac{1}{n_1^2} - \frac{1}{n_2^2}\right)$$

(where R is a constant, Z is atomic number

For H atom, $\frac{1}{\lambda_H} = R_H Z_H^2\left(\frac{1}{1^2} - \frac{1}{\infty^2}\right)$

$$\frac{1}{\lambda_H} = R_H Z_H^2 \qquad \text{......(1)}$$

For He^+ ion, $\frac{1}{\lambda_{He^+}} = R_{He}Z_{He}^2$ ......(2)

Dividing equation (1) by (2), we get

$$\frac{l_{He^+}}{l_H} = \frac{R_H}{R_{He^+}}\frac{Z_H^2}{Z_{He}^2}$$

(R_H and R_{He^+} are equal and constt.)

$$\frac{\lambda_{He^+}}{\lambda_H} = \frac{(1)^2}{(2)^2} = \frac{1}{4}$$

$$\lambda_{He^+} = \frac{\lambda_H}{4} = \frac{91.2}{4} = 22.8 \text{ n,m.}$$

50. $\Delta E = RhCZ^2 \left(\frac{1}{n_1^2} - \frac{1}{n_2^2}\right)$

$R = 1.0867 \times 10^7 \text{ m}^{-1}$; $C = 3 \times 10^8$ m/sec

$h = 6.626 \times 10^{-34}$ J sec; $n_1 = 1, n_2 = 2$ and for H-atom, $Z = 1$

$$\therefore E_2 - E_1 = 1.09677 \times 10^7 \times 6.626 \times 10^{-34} \ldots \times 3 \times 10^8 \left(\frac{1}{1} - \frac{1}{4}\right)$$

$$\Delta E = 1.0967 \times 6.626 \times 3 \times \frac{3}{4} \times 10^{-19} \text{ J} = 16.3512 \times 10^{-19} \text{ J}$$

$$= \frac{16.3512 \times 10^{-19}}{1.6 \times 10^{-19}} \text{ eV} = 10.22 \text{ eV}$$

$$\Delta E = \frac{hc}{\lambda} = RhCZ^2 \left(\frac{1}{n_1^2} - \frac{1}{n_2^2}\right)$$

$$\frac{1}{\lambda} = RZ^2\left(\frac{1}{1} - \frac{1}{4}\right) = RZ^2 \times \frac{3}{4}$$

Given $\lambda = 3 \times 10^{-8}$

$$\therefore \frac{1}{3 \times 10^{-8}} - 1.09677 \times z^2 \times \frac{3}{4} \times 10^7$$

$$\therefore Z^2 = \frac{10^8 \times 4}{3 \times 3 \times 1.09677 \times 10^7} = \frac{40}{9 \times 1.09677} \simeq 4$$

$\therefore Z = 2$ (since atom with atomic number 2 is He). So it correspond to He.

51. (a) We have, $n = 3$

$l = 2$

$m = -2, \quad -1, \quad 0, \quad +1, \quad +2$

$s = +\frac{1}{2} - \frac{1}{2} \quad +\frac{1}{2} - \frac{1}{2} \quad +\frac{1}{2} - \frac{1}{2} \quad +\frac{1}{2} - \frac{1}{2} \quad +\frac{1}{2} - \frac{1}{2}$

or $S = \pm\frac{1}{2}, \pm\frac{1}{2}, \pm\frac{1}{2}, \pm\frac{1}{2}, \pm\frac{1}{2}$

Maximum no. of electrons in all the five *d*-orbitals is 10.

(b) Magnetic moment is given by $\mu = \sqrt{n(n+2)}$ B.M. where n is the number of unpaired electrons.

when $n = 1$, $\mu = \sqrt{1(1+2)}$ BM = 1.73 BM

Hence the vanadium ion has **d** electronic configuration

52. E (according to Bohr) $= \dfrac{e^2}{4\pi\varepsilon_0 a_0}$

E= Potential energy + Kinetic energy. But according to question, K.E = $\frac{1}{2}$ P.E.

Thus, -2 K.E. + K.E = E

So, K.E. $= \dfrac{e^2}{4\pi\varepsilon_0 a_0}$

$\therefore$ P.E. $= -\dfrac{2e^2}{4\pi\varepsilon_0 a_0}$

53. Bond energy of I_2 molecule

$$= \frac{240 \times 10^3}{6.023 \times 10^{23}} \text{ J/molecule} = 3.9 \times 10^{-19} \text{ J/molecule}$$

$$\text{Energy absorbed} = h\nu = \frac{hc}{\lambda} = \frac{6.626 \times 10^{-34 \times 3 \times 10^8}}{4500 \times 10^{10}} \text{ J} = 4.41 \times 10^{-19} \text{ J}$$

Kinetic energy of Iodine atom

= Energy absorbed – Bond energy = $4.41 \times 10^{-19} - 3.98 \times 10^{-19} = 4.37 \times 10^{-20}$ J

$$\therefore \text{Kinetic energy per atom} = \frac{4.37 \times 10^{-20}}{2} = 2.186 \times 10^{-2} \text{ J}$$

54. Dipoie moment $\mu = e \times d$ coulombs metre

For KCl $d = 2.6 \times 10^{-10}$ m

For complete separation of unit charge

Electronic charge (e) = 1.602×10^{-19} C

Hence $\mu = 1.602 \times 10^{-19} \times 2.6 \times 10^{-8}$

$= 4.1652 \times 10^{-29}$ Cm

$\mu_{KCl} = 3.336 \times 10^{-29}$ m

55. Benzene has 12 sigma and 6 pi bonds

56. For O_2; M.O. configuration:

$\sigma 1s^2, \sigma^* 1s^2, \sigma 2s^2, \sigma^* 2s^2, \sigma 2p_z^2$

$$\begin{bmatrix} \pi 2p_x^2 \\ \pi 2p_y^2 \end{bmatrix} \begin{bmatrix} p^* 2p_x^1 \\ p^* 2p_y^1 \end{bmatrix}$$

Bond order = $\frac{1}{2}$ $[N_b - N_a]$ = $\frac{1}{2}$ (10 –6) = 2

For O_+^2; M.O. configuration:

$\sigma 1s^2, \sigma^* 1s^2, \sigma 2s^2\ \sigma^* 2s^2, \sigma 2p_2^2$

$$\begin{bmatrix} \pi 2p_x^2 \\ \pi 2p_y^2 \end{bmatrix} \begin{bmatrix} p^* 2p_x^1 \\ p^* 2p_y^0 \end{bmatrix}$$

Bond order = $\frac{1}{2}$ (10 –5) = 2.5

For O_2^-; M.O. configuration:

$\sigma 1s^2, \sigma^* 1s^2, \sigma 2s^2\ \sigma^* 2s^2, \sigma 2P_z^2$

$$\begin{bmatrix} \pi 2p_x^2 \\ \pi 2p_y^2 \end{bmatrix} \begin{bmatrix} p^* 2p_x^{1+1} \\ p^* 2p_y^1 \end{bmatrix}$$

Bond order = $\frac{1}{2}$ (10 –7) = 1.5

Higher the bond order smaller the bond length because maximum attraction,

Hence, $O_2[AsF_6] < O_2 < KO_2$

57. (a) For the application of the VSEPR theory we first write the dot Lewis diagram of the OF_2.

$$:\overset{**}{\underset{**}{F}}-\overset{**}{\underset{**}{O}}-\overset{**}{\underset{**}{F}}:$$

In the above structure the central atom O has the electronic configuration $1s^2 2s^2 2p^6$. Two electrons of O are shared to one electron of each F atom; and this completing the octate. Two pair of electrons are present as lone pairs on oxygen. This the FOF bond angle is not equal to the ideal tetrahedral angle i.e. 109° 28′ in spite of its sp^3 hybridization and deviated due to the presence of two lone pair electrons. The bounding electrons are more near to F atom because of its higher electronegativity. Therefore repulsion between the *lp's* exceeds that between the *bp's* and consequently the bond angle in OF_2 is less than 109° 28′ and it is are arranged in "V" shaped like structure. In this compound the oxidation number of O is +2 and oxidation number of F is –1

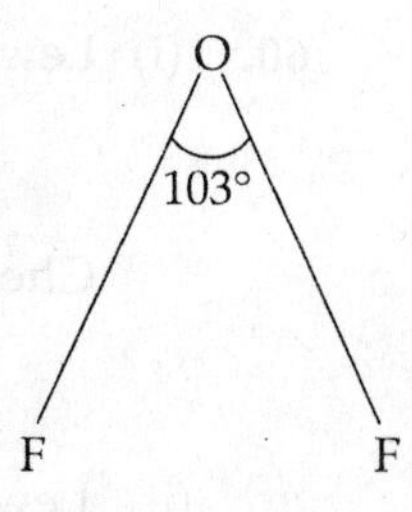

58. **H_2S**

hybridization –sp^3

Bond angle –90°

Shapes – V– shaped

No. of Ione pairs-two

No. of bond pairs - two

Due to repulsion in between two Ione pairs shape

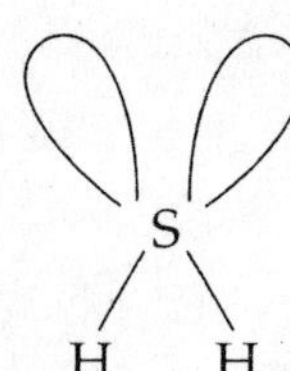

is V - shaped OR bent.

PCl_3

Hybridisation – sp^3

Bond angle – 106

Shape - pyramidal

No. of Ione pairs – one

No. of Bond pairs – 3

Due to the presence of one Ione pair at a corner of tetrahedron shape is pyrammidal

59. PCl_5 : $sp^3d \rightarrow$ Trigonal bipyramidal

BrF_5 : $sp^3d^2 \rightarrow$ Square pyramidal

Shape of **PCl_5**

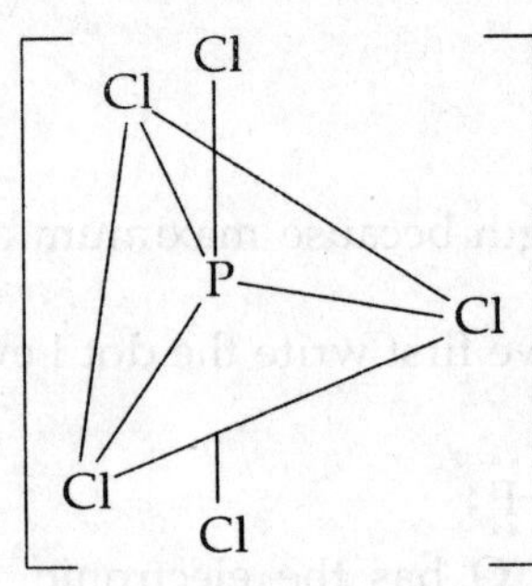

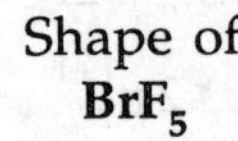

Shape of **BrF_5**

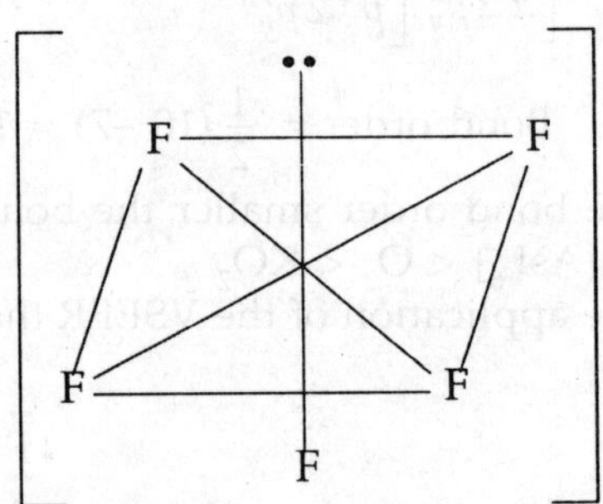

60. (i) Lewis dot structure of O_2^{2-}

$[:\ddot{O}:\ddot{O}:]^{2-}$

Chemical formula of a netural molecule (F_2)

$[:\ddot{F}:\ddot{F}:]$

(ii) Lewis dot structure of CO_3^{2-}

$[\ddot{O}{-}C{=}\ddot{O}\,,\ \ddot{O}]^{2-}$ $\quad$ $[\ddot{O}{-}C{=}\ddot{O}\,,\ \ddot{O}]$

Chemical formula of a neutral molecule (SO_2)

$[\ddot{O}\,{+}\,S{=}\ddot{O}\,,\ \ddot{O}]$

(iii) Lewis dot structure of CN^-

$$\left[\,{}^{\circ}_{\circ}C{}^{\circ}_{\circ}\,\underset{\bullet\times}{{}^{\times}_{\times}}\,N{}^{\times}_{\times}\right]^-$$

Chemical formula of a neutral molecule (N_2)

$$\left[\,{}^{\circ}_{\circ}N{}^{\circ}_{\circ}\,\underset{\bullet\times}{{}^{\times}_{\times}}\,N{}^{\times}_{\times}\right]$$

(iv) Lewis dot structure of NCS^-

$$\left[\,{}^{\circ}_{\circ}\overset{\circ\circ}{N}{}^{\circ}_{\bullet}{}^{\times}_{\times}C{}^{\times}_{\times}\;\;\overset{\circ\circ}{\underset{\circ\circ}{S}}{}^{\circ}_{\circ}\right]^-$$

Chemical formula of a neutral molecule (CO_2)

$$\left[\,{}^{\circ}_{\circ}\overset{\circ\circ}{O}{}^{\circ}_{\circ}{}^{\times}_{\times}C{}^{\times}_{\times}{}^{\circ}_{\circ}\overset{\circ\circ}{\underset{\circ\circ}{O}}\right]$$

Chapter 2 : CHEMICAL BONDING

1. The bond distances are calculated as follows:

H—Cl	0.28 Å + 0.99 Å = 1.27 Å
H—Br	0.28Å + 1.14 Å = 1.42 Å
H—I	0.28 Å + 1.33 Å = 1.61 Å

Now,

For HCl: $$d = \frac{\mu}{d} = \frac{1.03 \times 10^{-18} \text{ esu} \cdot \text{cm}}{1.27 \times 10^{-8} \text{ cm}} = 8.11 \times 10^{-11} \text{ esu}$$

$$[8.11 \times 10^{-11} \text{ esu})/(4.80 \times 10^{-10} \text{ esu})] \times 100\% = 16.9\% \text{ ionic}$$

For HBr $$\delta = \frac{0.79 \times 10^{-18} \text{ esu} \cdot \text{cm}}{1.42 \times 10^{-8} \text{ cm}} = 5.6 \times 10^{-11} \text{ esu or } 12.0\% \text{ ionic}$$

For HI: $$\delta = \frac{0.38 \times 10^{-18} \text{ esu} \cdot \text{cm}}{1.61 \times 10^{-8} \text{ cm}} = 2.4 \times 10^{-11} \text{ esu or } 5.0\% \text{ ionic}$$

Thus the % ionic character of the bonds are 16.9%, 12% and 5%

2. (a) Linear sp^3d The electrons are directed toward the corners of a trigonal bipyramid [Fig. (a)].

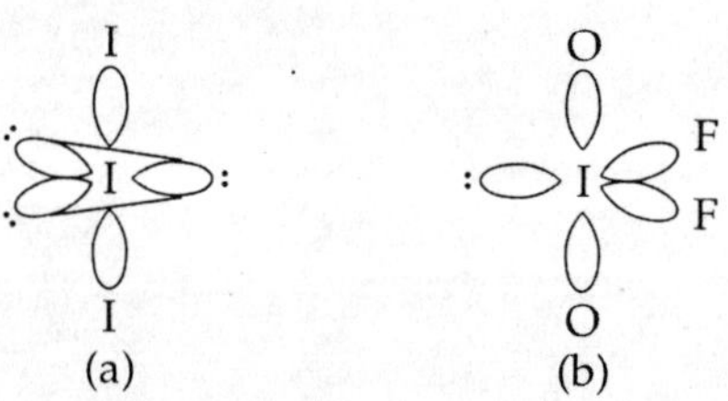

(a) (b)

(b) pyramidal sp^3 The electros are arranged in a tetrahedron.

(c) trigonal sp^2 In this unstable ion, Cl—O double bonding is expected

$$O{-}Cl{=}O^{+}$$
(with O bonded below Cl)

(d) pyramidal sp^3 The structure is analogous to SO_3^{2-}

:Fe: Se :Fe:
:O:

(e) seesaw sp^3d^2 The electrons are located at the corners of a trigonal bipramid bipyramid, but one of the equatorial pairs is unshared [Fig. (b)]

3. 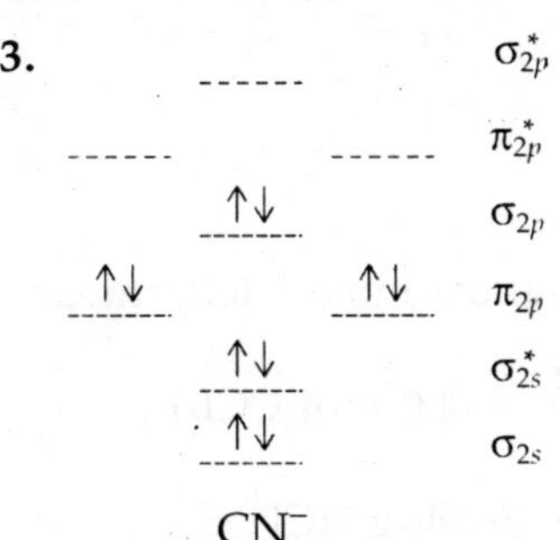

(a) CN^- has 4 bonding pairs and 1 antibonding pair, for a net bond order of 3. CN has one fewer bonding electron, for a bond order of $2\frac{1}{2}$, and CN^+ has still one fewer, for a bond order of 2.

(b) CN^- has the shortest bond because it has the highest bond order.

4. $A_3 \quad + \quad B_2 \quad \rightarrow \quad AB_4$

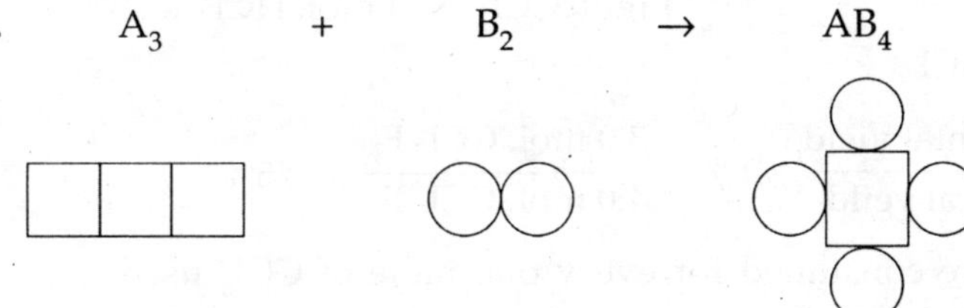

From the product, it is evident that we need 4 ◯ s for each □ Tus we need 6 ◯◯ s for one □□□

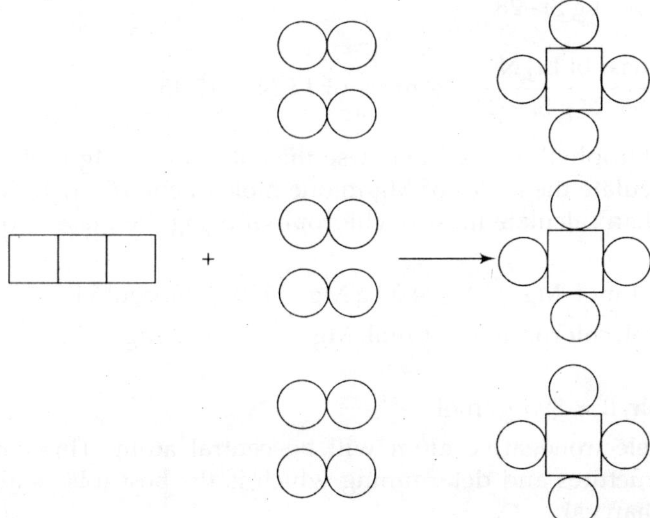

Thus; $A_3 + 6B_2 \rightarrow 3AB_4$.

5. Subtract unreacted product from the reactants to get the balanced equation, then simplify the conficients to their least common denominator.

(a) 5□ + 10◯ → 3□◯$_3$ + 2□ + ◯

That the balanced reaction is.

3□ + 9◯ → 3□◯$_3$

which simplifies to:

$$\square + 3\bigcirc \rightarrow \square\bigcirc_3$$

thus: $X + 3Y \rightarrow XY_3$

(b) Started with 5 moles of X and 10 moles Y.

(c) The reaction yielded 3 moles product (XY_3) with 2 moles X and 1 mole Y left unreacted.

6. (a) **False**. theoretical yield $= 4.0 \text{ mol } CCl_4 \times \dfrac{1 \text{ mol. } CCl_2F_2}{1 \text{ mol. } CCl_4} = 4.0 \text{ mol } CCl_2F_2$

(b) **False**. theoretical yield $= 4.0 \text{ mol } CCl_4 \times \dfrac{2 \text{ mol. HCl}}{1 \text{ mol. } CCl_4} \times \dfrac{36 \cdot 46 \text{ g HCl}}{1 \text{ mol. HCl}} = 2.9 \times 10^2$ g HCl

(c) **True**. % yield $= \dfrac{\text{experimental yield}}{\text{theoretical yeild}} \times 1\,\% = \dfrac{3.0 \text{ mol. } CCl_2F_2}{4.0 \text{ mol. } CCl_2F_2} = 75\%$

(d) **False**. Two moles of HF are consumed for every one mole of CCl_4 used.

7. (a) First we calculate the mass ratio of Si to Li_3N using standard masses. Then use this ratio to calculate the mass of Li_3N assuming a molar mass of 10 for Si.

$$\text{ratio} = \frac{34.83 \text{ } Li_3N}{27.9694 \text{ g Si - 28}} = 1.245.$$

$$1.245 = \frac{\text{mass of } Li_3N}{27.96924 \text{ g Si} - 28} \Rightarrow \text{mass of } Li_3N = 12.45 \text{ g}$$

(b) We calculate the mass of chlorophyll in one mole. Use the chlorophyll Mg mole ratio as a conversion factor to calculate the moles of Mg in one mole of chlorophyll. Covert moles of Mg to grams and than calculate mass of chlorophyll using the mass % of Mg in chlorophyll.

$1 \text{ mol. chlorophyll} \times \dfrac{1 \text{ mol. Mg}}{1 \text{ mol. chlorophyll}} \times \dfrac{24.30 \text{ g Mg}}{1 \text{ mol. Mg}} \times \dfrac{100 \text{ g chlorophyll}}{2.72 \text{ Mg}} = 895$ chlorophyll

Thus the molar mass of chlorophyll = 893 g/mol

8. As a rule of thumb, the least electronegative atom will be central atom. This can be confirmed by drawing Lewis structures and determining which is the best (obeys all the rules and has the least formal charges)

$[\ddot{N}{=}C{=}\ddot{S}]^-$ The only formal charge is on the N, the most electronegative atom.

$[\ddot{N}{=}S{\equiv}C]^-$ The N still bears a negative formal charge, but now the S and C also bear formal charges, so that is a step in the wrong direction.

$[:C{\equiv}N{-}\ddot{S}:]^-$ The nitrogen now bears a positive charge, and the S and C still bear formal charges, so this structure is even worse than the one above.

9. Draw Lewis structure of each molecule. Electron pairs in unshared pairs exert greater repulsion than electron pairs in bonds. Thus the structures with unshared pairs will have the bond angles smaller than 109.5°.

H—Si(H)(H)—H   H—P̈(H)—H   H—S̈:(H)

PH_3 and H_2S have at least one unshared pair and will have bond angles less than 109.5°.

10. (a) H—N̈(H)—H — The sp^3 hybridization indicates that 4 orbitals (1*s* and 3*p*) are involved. If there are 3 bonds, then there must also be one unshared pair of electrons.

(b) :N≡N: — Two pi and one sigma bond indicates the pressure of a triple bond.

(c) Ö≡Ö — One pi and one sigma bond indicates a double bond is present.

(d) $H_2C{=}CH_2$ — The three sp^3 hybrid orbitals indicate that the carbon is bonded to 3 other atoms. The remaining *p*-orbital must be involved in a pi bond. Thus the carbon atom will have 2 single bonds, and 1 double bond.

(e) :F̤̈—Xe—F̤̈: — The sp^3d^2 hybrid indicates that *Xe* has 6 electron pairs around it. Since 2 are bonding pairs, the other 4 must be unshared pairs.

The examples given in each of the problems above are but one of many possibilities.

11. Lewis Structure H—N̈(H)—N̈(H)—H

Each N is AX_3E, and thus would be **triabgular pyramid**. Due to the Ione pairs of electrons on the nitrogens, the bond angles will be slightly less than 109.5°.

Thus molecules is not pair, since the dipoles cancel one another, as can be seen from the perspective drawings at right.

12. There are 6 group around the central iodine atom.
The hybridization sp^3d^2.
The geometry is octahedron.

$[IO_6]^{5-}$

13.

	SO_4 (all single bonds)	SO_4 (two S=O double bonds)
Geometry:	AX_4:tetrahedron	AX_4: tetrahedron
Hybridization	sp^3	sp^3

C_f(S)	$6 - (0 + ½(8)) = 2$	$6 - (0 + ½ (12)) = 0$
C_f(O) (with single bond)	$6 - (6 + ½(2)) = -1$	$6 - (6 + ½(2)) = -1$
C_f(O) (with double bond)		$6 - (4 + ½(4)) = 0$

Since the atom in the second structure have charges closer to zero, it is the better structure.

14. (a) Rust contains iron in the from of Fe^{3+}, which will from a complex with the oxalate ion, $[Fe(C_2O_4)_3]^{3-}$. this complex is water soluble, while rust is not.

(b) There are no geometric isomers of tetrahedral complexes because any two ligands are 109.5° apart. Thus all spatial orientations are identical. Contrast this with square planar or octahedral, in which two ligands could be either 90° or 180° apart.

(c) CO^{2+} has empty orbitals (such as the 4*s*), which can accept a pair of electros from a Lewis.

(d) 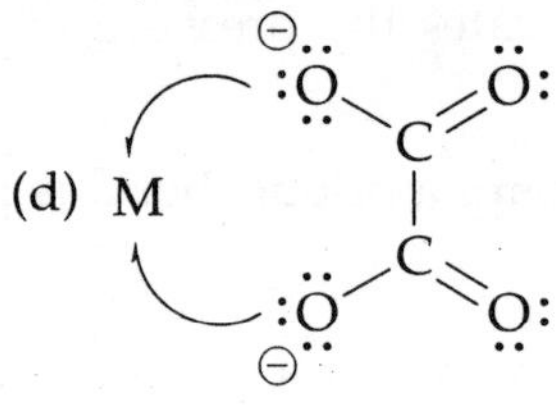

It should be clear from the Lewis structure the $C_2O_4^{2-}$ has two atoms with unshared electron pairs, which can act as Lewis bases. Also those atom are held in the correct position for both of them to bond to the metal simultaneously.

(e) H—N̈—H (with H below N)   H—N$^{\oplus}$—H (with H above and below N)   Ammonia has an unshared pair of electrons, which allows it to bond to metals. By contrast the ammonium ion does not.

15. Here we first calculate the molar mass of the simplest formula, then determine the molecular formula. From that, determine the legand(s) present and cation and anion.

$$M(PtN_2H_6Cl_2) = 300.2 \text{ g/mol}$$

600 is 2 × 300, thus the molecular formula is $(Pt_{1\times2}N_{2\times2}H_{6\times2}C_{2\times2}) = Pt_2N_4H_{12}Cl_4$

Since we have a complex cation and a complex anion, it is likely one Pt is with the cation and the other is with the anion. Aslo N_4H_{12} is likely to be $4NH_3$. Pt has a +2 charge and a coordination number of 4. Thus a likely candidate is

$$[Pt(NH_3)_4]^{2+}\ [PtCl_4]^{2-} \Rightarrow [Pt(NH_3)_4]\ [PtCl_4]$$

A second candidate is :

$$[Pt(NH_3)_4]^{+}\ [Pt(NH_3)Cl_3]^{-} \Rightarrow [Pt(NH_3)_3Cl)][Pt(NH_3)Cl_3]$$

16. (a) Assume a 100 g sample.

$$20.25 \text{ g Cu} \times \frac{1 \text{ mol.}}{63.55 \text{ g}} = 0.3186 \text{ mol Cu.}$$

$$15.29 \text{ g C} \times \frac{1 \text{ mol.}}{12.01 \text{ g}} = 1.273 \text{ mol. C}$$

$$7.07 \text{ g H} \times \frac{1 \text{ mol.}}{1.008 \text{ g}} = 7.01 \text{ mol. H}$$

$$26.86 \text{ g N} \times \frac{1 \text{ mol.}}{14.01 \text{ g}} = 1.917 \text{ mol. N}$$

$$20.39 \text{ g O} \times \frac{1 \text{ mol.}}{16.00 \text{ g}} = 1.274 \text{ mol. O}$$

$$10.23 \text{ g S} \times \frac{1 \text{ mol.}}{32.07} = 0.3190 \text{ mol. S}$$

0.3186/0.3186 = 1 mol. Cu
1.273/0.3186 = 4 mol. C
7.07/0.3186 = 22 mo. H
1.917/0.3186 = 6 mol. N
1.274/0.3186 = 4 mol. O
0.3190/0.3186 = 1 mol. S
Thus the simplest formula is $CuC_4H_{22}N_6O_4S$

(b) To deduce the structural formula, one must determine the possible ligands that are present. The O_4S could be the sulfate atom. the –2 charge of the sulfate atom would balance the +2 charge that accompanies most copper cations, which would indicate that the remaining ligands are neutral. A ligand containing C, H and N is ethylenediamine (*en*). If we subtract two on units from the simplest formula, we are left with H_6N_2, which is likely $2NH_3$. This gives $[Cu(C_2H_8N_2)_2(NH_3)_2]^{2+}SO_4^{2-}$.

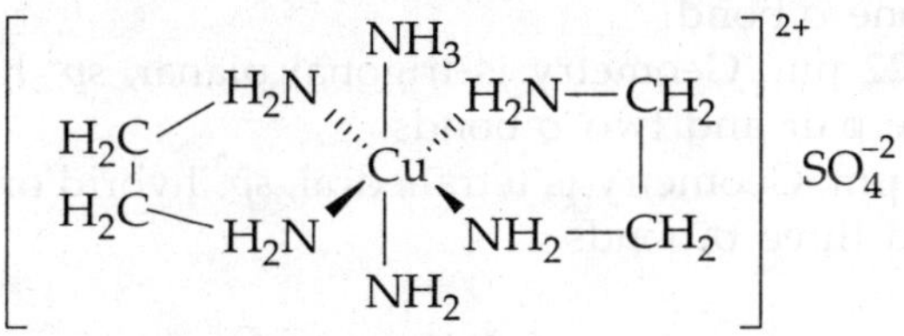

(NH_3 trans) 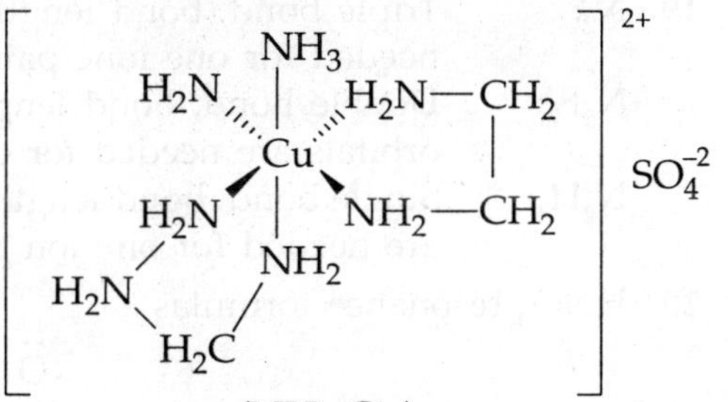 (NH_3 Cis)

17. After assuming a 100 g sample, convert to moles:

$$60.4 \text{ g Xe} \times \frac{1 \text{ mol Xe}}{131.29 \text{ g Xe}} = 0.46005 \text{ mol Xe}$$

$$22.1 \text{ g O} \times \frac{1 \text{ mol O}}{16.00 \text{ g O}} = 1.381 \text{ mol O}$$

$$17.5 \text{ g F} \times \frac{1 \text{ mol F}}{18.99 \text{ g F}} = 0.9215 \text{ mol F}$$

Divide by 0.46

$$\text{Xe: } \frac{0.46}{0.46} = 1 \qquad \text{O: } \frac{1.381}{0.46} = 300; \qquad \text{F: } \frac{0.9215}{0.46} = 2$$

the simplest formula is XeO_3F_2. this is also the molecular formula. The Lewis formula is

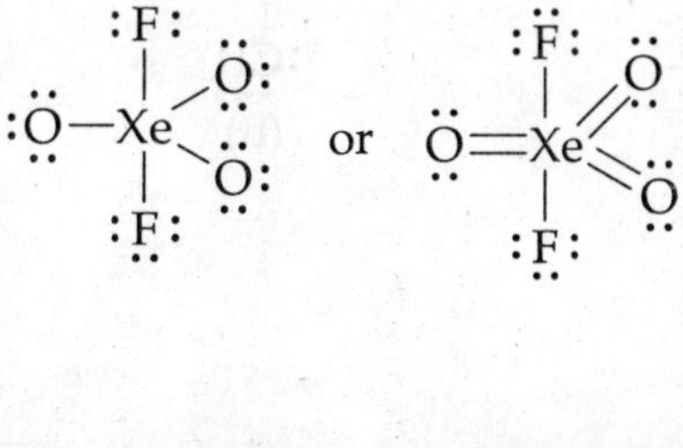

Number of electrons pairs = 5, number of Ione pairs = 0, hence the geometry is trigonal bipyramidal. Because xenon has five single bonds. It will require five orbitals to describe the bonding. This suggests sp^3d hybridization

18. $U(s) = ClF_n \rightarrow UF_6 + ClF(g)$

mol $UF_6 = 3.53 \text{ g } UF_6 \times \dfrac{1 \text{ mol } UF_6}{352.07 \text{ g } UF_6} = 0.01003 \text{ mol } UF_6$

mol $ClF = n = \dfrac{PV}{RT} = \dfrac{2.5 \text{ atm} \times 0.343 \text{ L}}{0.082057 \text{ L} \bullet \text{atm/(K} \bullet \text{mol)} \times 348 \text{ K}} = 0.3003 \text{ mol ClF}$

0.01 mol UF_6 = 0.06 mol F and 0.03 mol ClF = 0.03 mol F; therefore, the total moles of F from ClF_n = 0.09 mol F. Because mol ClF_n must equal mol ClF, mol ClF_n = 0.03 mol and n = 0.09 mol F + 0.03 mol ClF_n = 3. The Lewis formula is

```
:F̤—C̤l—F̤:
     |
    :F:
```

Number of electron pairs = 5, number of ione pairs = 2, hence, the geometry is T-shaped. Because chlorine has five electron pairs, it will require five orbitals to describe the bonding. This suggest sp^3d hybridization.

19. N_2: Triple bond, bond length = 110 pm. Geometry is linear, *sp* hybrid orbitals are needed for one ione pair and one σ bond.

N_2F_2: Double bond, bond length = 122 pm. Geometry is trigonal planar, sp^2 hybrid orbitals are needed for one ione pair and two σ bonds.

N_2H_4: Single bond, bond length = 145 pm. Geometry is tetrahedral, sp^3 hybrid orbitals are needed for one ion pair and three σ bonds.

20. HNO_3 resonance formulas

```
          :Ö:                          Ö:
          /                           //
  H—Ö—H          ←———→         H—Ö—N
          \\                           \
          Ö:                           :Ö:
```

The geometry around the nitrogen is trigonal planar, therefore, the hybridization is sp^2

Formation reaction: $H_2(g) + 3O_2(g) + N_2(g) \rightarrow 2HNO_3(g)$

$2 \times \Delta H_f^\circ = [BE(H—H) + 3BE(O_2) + BE(N_2)] - [2BE(H—O) + 4BE(N—O) + 2BE(N{=}O)]$

$= [(432 + 3 \times 494 + 942) - (2 \times 459 + 4 \times 201 + 2 \times 607) \text{ kJ/2 mol}]$

$= -80 \text{ kJ/2 mol} = -40 \text{ kJ/mol}$

Resonance energy = –4.0 kJ – (–135 kJ) = 95 kJ

21. The lewis structure is:

```
       :Ö:                  :Ö:
        |                    ||
  :C̤l—P—C̤l:          :C̤l—P—C̤l:
        |                    |
       :C̤l:                 :C̤l:
       (a)                  (b)
```

$C_t(P)$ $5 - (0 + 4) = +1$ $5 - (0 - 5) = 0$

$C_t(O)$: $6 - (6 + 1) = -1$ $6 - (4 + 2) = 0$

$C_t(Cl)$: $7 - (6 + 1) = 0$ $7 - (6 + 1) = 0$

Since the second structure (b) has no formal charges, it is the better structure.

22. The electronegativity differences and bond polarities are:

P—H	0.0	nonpolar
O—H	1.5	polar (acidic)

$H_3PO_3(aq) + 2NaOH(aq) \rightarrow Na_2HPO_3(aq) + 2H_2O(1)$

$$\frac{\text{mol}}{\text{L}} = \frac{0.125 \text{ mol NaOH}}{\text{L NaOH}} \times 0.0225 \text{ LNaOH} \times \frac{1}{0.2 \text{ LH}_3\text{PO}_3}$$

$$\times \frac{1 \text{ mol H}_3\text{PO}_3}{2 \text{ mol NaOH}} - 0.0070312 = 0.007031 \text{ M H}_3\text{PO}_3$$

23. After assuming a 100-g sample, convert to moles:

$$10.9 \text{ g Mg} \times \frac{1 \text{ mol Mg}}{24.3 \text{ g Mg}} = 0.4485 \text{ mol Mg}$$

$$31.8 \text{ g Cl} \times \frac{1 \text{ mol Cl}}{35.453 \text{ g Cl}} = 0.89696 \text{ mol Cl}$$

$$57.3 \text{ g O} \times \frac{1 \text{ mol O}}{16.00 \text{ g } O} = 3.581 \text{ mol O}$$

Divide by 0.4485:

$$\text{Mg: } \frac{0.4485}{0.4485} = 1; \text{ Cl: } \frac{0.89696}{0.4485} = 2; \text{ O: } \frac{3.581}{0.4485} = 7.98$$

The simplest formula is $MgCl_2O_8$. However since $Cl_2O_8^{2-}$ is not a well known ion, write the simplest formula as $Mg(ClO_4)_2$, magnesium perchlorate. The Lewis formulas are Mg^{2+} and ClO_4^-

$$\left[\begin{array}{c} :\ddot{O}: \\ :\ddot{O}-Cl-\ddot{O}: \\ :\ddot{O}: \end{array}\right]^- \quad \text{or} \quad \left[\begin{array}{c} :\ddot{O}: \\ :\ddot{O}=C=\ddot{O}: \\ :\ddot{O}: \end{array}\right]^-$$

24. After assuming a 100.0-g sample, convert to moles:

$$25 \text{ g C} \times \frac{1 \text{ mol C}}{12.01 \text{ g C}} = 2.081 \text{ mol C}$$

$$2.1 \text{ g H} \times \frac{1 \text{ mol H}}{1.008 \text{ g H}} = 2.08 \text{ mol H}$$

$$39.6 \text{ g F} \times \frac{1 \text{ mol F}}{18.99 \text{ g F}} = 2.085 \text{ mol F}$$

$$33.3 \text{ g O} \times \frac{1 \text{ mol O}}{16 \text{ g O}} = 2.081 \text{ mol O}$$

The simplest formula is CHOF. Because the molecular mass of 48 divide by the formula mass of 48 = 1 molecular formula is also CHOF. the Lewis formula is

```
    H
     \    ..
      C=O:
     /
  :F:
   ..
```

25. First calculate the number of moles in one liter.

$$n = \frac{PV}{RT} = \frac{1 \text{ atm} \times 1 \text{ L}}{0.0821 \text{ L} \bullet \text{atm}/(\text{K} \bullet \text{mol}) \times 424 \text{ K}} = 0.287727 \text{ mol}$$

$$\text{MW} = \frac{\text{g}}{\text{mol}} = \frac{7.49 \text{ g}}{0.028727} = 260.7 \text{ g/mol}$$

$$\text{MW} = 260.7 \text{ amu} = 118.69 \text{ amu Sn} + n(35.453 \text{ amu Cl})$$

$$n = \frac{260.7 - 118.69}{35.453} = 4.0055$$

The formula is $SnCl_4$ which is molecular because the electronegativity difference is 1.2. The Lewis formula is

```
         ..
        :Cl:
   ..    |    ..
  :Cl—Sn—Cl:
   ..    |    ..
        :Cl:
         ..
```

26. Using values from Tables we get,

$$X = \frac{\text{I.E - E.A}}{2} = \frac{(1250 - (-349))\,\text{kJ mol}^{-1}}{2} = 788.5 \text{ kJ mol}^{-1}$$

$$\frac{799.5 \text{ kJ mol}^{-1}}{230 \text{ kJ mol}^{-1}} = 3.47 \text{ (Pauling's X = 3.0)}$$

27. (a) The total number of valence electrons for CH_3OH is 14 (4 from C, 6 from O, 1 from each of the 4 H atoms). the structure is

```
     H
     |    ..
  H—C—O—H
     |    ..
     H
```

(b) There are 26 valence electrons for the ClO_3^- ion (7 from Cl, 6 from each of 3 oxygen atoms, and 1 from the single negative charge on the ion). The Lewis structure is

$$\left(\begin{array}{c} :\ddot{O}: \\ | \\ :\ddot{\underset{..}{Cl}}—C—\ddot{\underset{..}{Cl}}: \end{array}\right)$$

(c) There are 16 valence electrons for N_2O. The Lewis structure is $:\ddot{N}{=}N{=}\ddot{O}:$. Note that this is isoelectronic with CO_2. A second possible Lewis structure for N_2O is $:N::N::\ddot{\underset{..}{O}}:$ or $:N{\equiv}N—\ddot{\underset{..}{O}}:$

(d) In $COCl_2$ there are 24 valence electrons. The central atom is C, the least electronegative of these three atom. the Lewis structure is

$$\begin{array}{c} :O: \\ \| \\ :\ddot{\underset{..}{Cl}}—C—\ddot{\underset{..}{Cl}}: \end{array}$$

(e) There are 10 valence electrons in CO, and the structure is isoelectronic with that of N_2, $:C{\equiv}O:$.

28. (a) The structure of $COCl_2$ is

$$\begin{array}{c} :O: \\ \| \\ :\ddot{\underset{..}{Cl}}—C—\ddot{\underset{..}{Cl}}: \end{array}$$

There are three Ione pairs on each Cl atom, and one of the electrons in the pair shared with C is assigned to each Cl, for a total of seven valence electrons, just as in an isolated Cl atom. Hence the formal charge on Cl is 0. The C atom has four shared pairs, and is therefore assigned four electrons, just as in an isolated C atom, so its formal charge is also 0. Similarly, oxygen has a zero formal charge with two Ione pairs and two shared pairs.

(b) The formal charge on each atom in CO_2, $:\ddot{O}{=}C{=}\ddot{O}:$, is 0.

(c) The formal charges in the two possible structures for N_2O are indicated by

$$:\overset{-}{\underset{..}{N}}{=}\overset{+}{N}{=}\underset{..}{O}: \quad \text{and} \quad :\underset{..}{N}{\equiv}\overset{+}{N}—\overset{..\,-}{\underset{..}{O}}:$$

(d) The atomic number of Cr is 24. Eighteen of its electrons ae in closed shells (the argon configuration), so Cr has 6 valence electrons. Each of the four O atoms has 6 valence electrons. There are 2 additional electrons because the charge on the ion is –2, making a total of 32 valence electrons for the CrO_4^{2-} ion. The Lewis structure showing the format charges on each atom is

$$\begin{array}{c} :\ddot{O}:^- \\ |\;{}^{2+} \\ {}^-:\ddot{\underset{..}{O}}—Cr—\ddot{\underset{..}{O}}:^- \\ | \\ :\underset{..}{O}:^- \end{array}$$

Note that the sum of the formal charges is –2, the charge on the CrO_4^{2-} ion.

29. There are 24 valence electrons in SO_3, and we can draw thre structure similar to those used for CO_3^{2-} or NO_3^-.

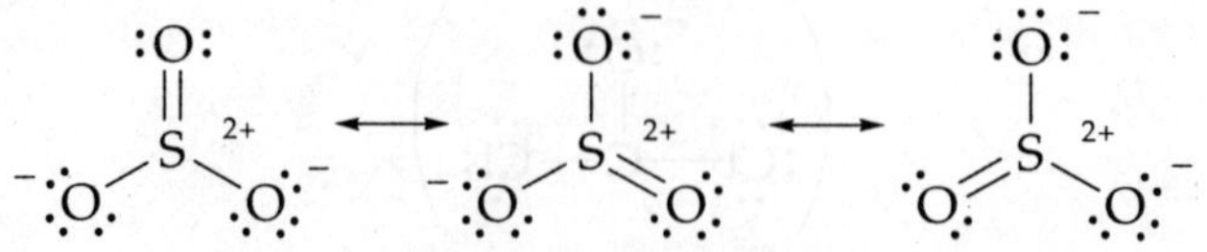

$$:O: \; \| \; S^{2+} (^-:O:)(:O:^-) \longleftrightarrow :O:^- \; | \; S^{2+} (^-:O:)(=O:) \longleftrightarrow :O:^- \; | \; S^{2+} (O=)(:O:^-)$$

These three structures place a formal charge of +2 on s. Sulfur is certainly less electronegative than oxygen, but we can avoid a formal charge as large as +2 by including structures with more than 8 electrons about S, such as

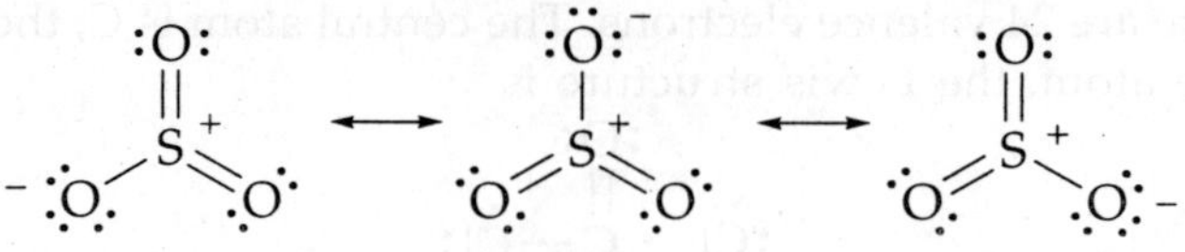

$$:O: \; \| \; S^{+} (^-:O:)(=O) \longleftrightarrow :O:^- \; | \; S^{+} (O=)(=O) \longleftrightarrow :O: \; \| \; S^{+} (O=)(:O:^-)$$

30. The Lewis structure with four Cr—O single bonds is shown below has resonance structures with double bonds between Cr and O necessitate having 10 electrons around Cr. Since *d* orbitals can be used by Cr for bonding an expanded valence shell is possible. The required resonance structure are

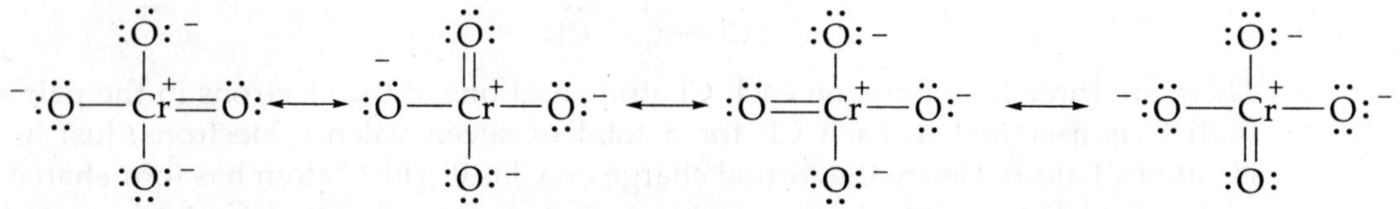

$$^-:O{-}\overset{+}{Cr}{=}O: \text{ (with } :O:^- \text{ above and below)} \longleftrightarrow :O{-}\overset{+}{Cr}{-}O:^- \text{ (} =O: \text{ above}, :O:^- \text{ below)} \longleftrightarrow :O{=}\overset{+}{Cr}{-}O:^- \text{ (} :O:^- \text{ above and below)} \longleftrightarrow {}^-:O{-}\overset{+}{Cr}{-}O:^- \text{ (} :O:^- \text{ above}, =O: \text{ below)}$$

31. Xenon has 8 electrons in its valence shell and each F atom has 7, so there are a total of 22 valence electrons for XeF_2. There can only be octet of electrons about each F atom which uses 16 electrons, so there must be 6 more electrons about Xe. These must be ione pairs, giving Xe an expanded valence shell with 10 electrons. The structure is

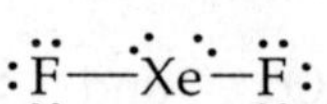

$$:F{-}Xe{-}F:$$

32. (a) There are 32 valence electrons for the sulfate ion, 6 from S and each of the four O atoms, plus 2 from the charge on the anion. the Lewis structure that satisfies the octet rule,

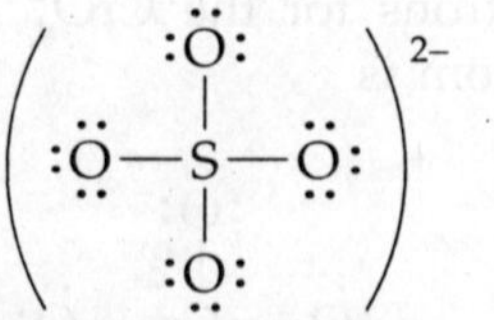

$$\left(:O{-}S{-}O: \text{ with } :O: \text{ above and below S} \right)^{2-}$$

has four bonded pairs and no lone pairs about S. It is therefore predicted to be tetrahedral, in agreement with observation.

Note that the CrO_4^{2-} ion, has 32 valence electrons, and the same Lewis structure as SO_4^{2-}. The chromate ion is also tetrahedral, as predicated by VSEPR theory. Thus structure of CrO_4^{2-}, ion is shown in Fig. below:

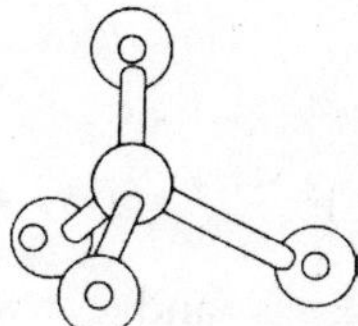

Fig. The tetrahedral geometry of the CrO_4^{2-} ion. there are four bonded pairs and no Ione pairs around the Cr atom.

(b) The Lewis structure for PCl_3 is

```
        ..
       :Cl:
        |
  ..    |    ..
 :Cl—P—Cl:
  ..   ..    ..
```

so that there are three bonded pairs and one Ione pair about P. the ideal geometry for four bonded electron pairs is tetrahedral; VSEPR theory predicts that the bond angle in PCl_3 should be less than 109.5°, and the geometry should be **trigonal pyramidal**. like that the ammonia. The observed bond angle is 100°, and the geometry is indeed trigonal pyramidal.

(c) The Lewis structure for SCl_2 is :Cl—S—Cl:, with two bonded pairs and two lone pairs about S. We therefore predict that the molecule is bent, and the bond angle should be less than 109.5°. The observed bond angle is 102°.

33. (a) There are 12 valence electrons for ethylene, and the Lewis structure is $H_2C{=}CH_2$. Since the double bond is counted as one bonded pair, there are three bonded pairs about C and the ideal geometry is trigonal planar, with 120° bond angles. Because the double bond takes up more room than a single bond, we expect the H—C—H bond angles to be slightly less tan 120°, and this is indeed what is observed. The actual geometry of ethylene is shown below.

```
   ↙ 121.5° ↘
  H         H
   \       /
    C == C       117°
   /       \
  H         H
```

(b) There are 22 valence electrons for the triiodide ion I_3^-. The Lewis structure is

:I—I—I: (with lone pairs: three on each terminal I, three on the central I)

so that the central I atom has two bonded pairs and three lone pairs. We expect the three lone pairs to occupy the equatorial positions of the trigonal bipyramidal geometry. The three iodine atoms should therefore lie on a straight line, and the ion is indeed observed to be linear.

(c) Each halogen atoms has 7 valence electrons, and 1 electrons has been added to make the anion, so that ICl_4^- has 36 valence electrons. Putting complete octets about each atom uses only 32 electrons. The remaining 4 are in two lone pairs on the I atom, which therefore has a total of 6 electron pairs around it. The geometry will be derived from distortion of an octahedral structure. There are two possibilities to consider.

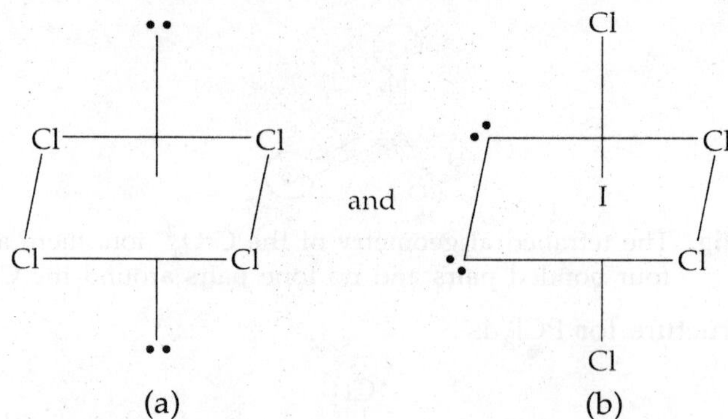

Since lone pairs repel one another more than bonded pairs do, structure (*a*) which has the lone pairs 180° apart is preferred over structure (*b*), which has the lone pairs only 90° apart. The observed square-planar geometry of ICl_4^- is indeed that of structure (*a*).

34. The electronic configuration of C is $1s^2 2s^2 2p^2$ so that each C has four valence electrons and there are eight electrons to occupy molecular orbitals in C_2. The energy level diagram for C_2 is shown below. The $\pi 2p$ molecular orbitals are filled and there are no unpaired electrons so C_2 should be diamagnetic, which is in agreement with observation. There are six electrons in bonding molecular orbitals, and two in antibonding molecular orbitals, so that the bond order is $\frac{1}{2}(6 - 2) = 2$. Note that the net bonding electrons are in π-molecular orbitals. A bond order of 2 corresponds to a double bond, and we therefore expect the bond in C_2 to be considerably stronger than the single bond in B_2. Again this agrees with observation. the bond energy of C_2 is 630 kJ. mol^{-1}, somewhat more than twice the bond energy of B_2. The bond length in C_2, 124 pm, is shorter than the bond length in B_2, 159 pm. A double bond is shorter than a single bond. the electronic configuration of C_2 is $KK(\sigma 2s)^2 (\sigma^* 2s)^2 (\pi 2p)^4$.

σ^*2p

π^*2p

$\sigma 2p$

↑↓ ↑↓ $\pi 2p$

↑↓ σ^*2p

↑↓ $\sigma 2s$

35. (a) The symbol 2*d* is unacceptable. For a *d* electron, $l = 2$. The value of l cannot exceed $n - 1$. thus if $n = 2$, the largest value of l allowed is 1. the only possible orbitals for $n = 2$ are the 2*s* and 2*p*.

(b) The symbol 6*g* is acceptable. It stands for the nine degenerate orbials with $n = 6$ and $l = 4$.

(c) The symbol 7*s* is acceptable. It stands for the atomic orbital with $n = 7$ and $l = 0$.

(d) The symbol 5*h* is unacceptable. For an *h* electron $l = 5$. If $n = 5$, the largest possible value of l is 4. The only possible orbitals for $n = 5$ are the 5*s*, 5*p*, 5*d*, 5*f*, and 5*g*.

36. (a), (b), (d), (f), and (g) are all configurations of different excited states of N. The ground state electronic configuration of N is (e). configuration (c) is forbidden. It violates the Pauli Principle as the two electrons in the 2*s* orbitals have parallel spin.

Chapter 3 : THE GASEOUS STATE

1. Using the Boyle's law equation

$$\frac{V_1}{V_2} = \frac{P_2}{P_1}$$

Here, volume, V_1 = 500 sec.; pressure, P_1 = 700 mm.
volume, V_2 = ?; pressure, P_2 = 800 mm.
Substituting these values in the formula above we have

$$\frac{500 \text{ cc}}{V_2} = \frac{800 \text{ mm}}{700 \text{ mm}}$$

Solving, $$V_2 = \frac{500 \text{ cc} \times 700 \text{ mm}}{800 \text{ mm}} = 437 \text{ cc}$$

2. Since the pressure is constant, this is a problem involving temperature change only. We can solve by substituting in the Charles' law formula,

$$\frac{V_1}{V_2} = \frac{T_1}{T_2}$$

Here, V_1 = 200 cc; T_1 = 273K; T_2 = 373K

$$\frac{200 \text{ cc}}{V_2} = \frac{273 \text{K}}{373 \text{K}}$$

$$V_2 = \frac{200 \text{ cc} \times 373 \text{ K}}{273 \text{K}} = 273 \text{ cc}$$

3. By substituting in the general formula

$$\frac{V_1 P_1}{T_1} = \frac{V_2 P_2}{T_2}$$

Here, V_1 = 200 liter; P_1 = 750 mm; T_1 = 285 K;
V_2 = ?; P_2 = 720 mm; T_2 = 313 K
Substituting the values we get,

$$\therefore \quad \frac{200 \text{ liter} \times 750 \text{ mm}}{285 \text{K}} = \frac{V_2 \times 720 \text{ mm}}{313 \text{ K}}$$

or $$V_2 = 200 \text{ liters} \times \frac{313 \text{K} \times 750 \text{ mm}}{285 \text{K} \times 720 \text{ mm}} = 229 \text{ liters}$$

4. The molar volume of gas at STP is 22.4 liters. That means that 1 mole of He will occupy a volume of 22.4 liters at STP.

$$2.71 \text{ moles} \times \frac{22.4 \text{ liters}}{1 \text{ mole}} = 60.7 \text{ liters}$$

Note that "moles" cancel, leaving the answer in "liters."

5. In solving this problem our first question is "How many moles of CH_4 are there?" We know that one mole of a gas occupies a volume of 22.4 liters at STP. Therefore, if we know how many moles of CH_4 gas there are in 100 g of CH_4, we can multiply this number of moles by 22.4 liters/mole.
   The molecular weight of CH_4 is 16. Therefore, 100 g of CH_4 is 100/16 moles

$$\frac{100}{16} \text{ mole} \times \frac{22.4 \text{ liters}}{1 \text{ mole}} = 140 \text{ liters}$$

   The complete calculation, in one operation is:

$$\frac{100 \text{ g of } CH_4}{16.0 \text{ g of } CH_4 \text{ /mole of } CH_4} \times \frac{22.4 \text{ liters of } CH_4}{1 \text{ mole of } CH_4} = 140 \text{ liters of } CH_4$$

   Note that "g of CH_4" and "mole of CH_4" cancel.

6. In solving this problem our first question is "How many *moles* of NO do we have?" If we know the number of moles, we can multiply this number of moles by the number of grams of NO in 1 mole.
   22.4 liters of NO at STP is 1 mole. Therefore, 40.0 liters at STP is 40.0/22.4 moles of NO. The molecular weight of NO is 30.0.

7. At the same temperature and pressure, the number of moles of gas is directly proportional to the volume of the gas. Since the volume of *B* is 1.50 times the volume of *A* the number of moles of CO in *B* will be 1.50 times the moles of CH_4 in *A*.

8. Here, P_1 = 0.95 atm; P_2 = 1.1. atm.; V_1 = 2.2 litre; V_2 = ?
   Applying Boyle's law, $P_1V_1 = P_2V_2$

$$0.95 \times 2.2 = 1.1 \times V_2$$

$$V_2 = \frac{0.95 \times 2.2}{1.1} = 1.90$$

   Or The volume of CO = 1.90 litre.

9. Here, P_1 = 0.24 atm. P_2 = ?
   V_1= 0.51 litre V_2 = 0.51 + 0.63 = 1.14 litre
   Applying Boyle's law, $P_1V_1 = P_2 V_2$
   Substituting the values we get,

$$0.24 \times 0.51 = P2 \times 1.14$$

$$\therefore P_2 = \frac{0.24 \times 0.51}{1.14} = 0.1074 \text{ atm.}$$

   New pressure of the gas = 0.1074 atm.

10. Here, P_1 = 1 atm.; P_3 = 3 atm.; T_1 = 300 K; T_2 = ?

   Applying Charles' law $\frac{P_1}{T_1} = \frac{P_2}{T_2}$;

$$\frac{1}{300}=\frac{3}{T_2} \Rightarrow T_2=\frac{300\times 3}{1}=900\text{ K}$$

The temperature to which the bulb be heated is 900 K.

11. (a) Here, V_1 = 400 ml; V_2 = 400 × 2 = 800 ml.

T_1= 290 K; T_2 = ?

Applying Charles' law

$\frac{V_1}{T_1}=\frac{V_2}{T_2}$; we have on substituting the values,

$$\frac{400}{290}=\frac{800}{T_2}$$

$$\therefore\ T_2=\frac{800\times 290}{400}=850\text{ K}$$

(b) Note the temperature to be attained to double the volume of the gas is 580 K

Here, V_1 = 400 ml.; V_2 = 400 × 1/2 = 200 ml

T_1 = 290 K; T_2 = ?

$\frac{V_1}{T_1}=\frac{V_2}{T_2}$; By substituting the above values we get

$$\frac{400}{290}=\frac{800}{T_2}$$

or $$T_2=\frac{200\times 290}{400}=145\text{ K}$$

∴ The temperature to be attained to reduce the volume of the gas to half is 145 K

12. Here, V_1 = 500 ml. V_2 = ?

T_1 = 300 K; T_2 = 258 K

Applying Charles' law and substituting the above values

$$\frac{V_1}{T_1}=\frac{V_2}{T_2} \text{ or } \frac{500}{300}=\frac{V_2}{258}$$

$\therefore\ V_2=\frac{500\times 258}{300}$ = 430 ml. Hence, contraction in volume = 500 – 430 = 70 ml.

A similar relation is found to exist between pressure and temperature, when the volume of the gas is maintained constant.

This also goes by the name of Charles' law. Thus, $\frac{P_1}{T_1}=\frac{P_2}{T_2}$

Where P_1 is the pressure of a sample of gas at temperature T_1 and T_2 is the pressure at temperature T_2.

13. Here, V_1 = 5 litre, V_2 = 1 litre; for N_2, m = 28g. So the number of moles,

$$n_1=\frac{6.1}{28}=0.22\text{ mole, } n_2=?$$

Using the relation, $\frac{V_1}{n_1} = \frac{V_2}{n_2}; \frac{5}{0.22} = \frac{11}{n_2}$

or $n_2 = \frac{11 \times 0.22}{5} = 0.484$

The total number of moles of a gas required for the volume to be 11 litre is 0.484. Moles of N_2 to be added = 0.484 – 0.22 = 0.264

Mass of nitrogen to be added = $\frac{0.264 \times 28}{1 \text{ mole}} = 7.39$ g.

14. Initial conditions

P_1 = 750 mm. of Hg.; V_1 = 38 ml.; T_1 = 300 K; T_2 = 290 K

Final conditions

P_2 = 760 mm of Hg; V_2 = ?

Applying the gas equation, $\frac{P_1V_1}{T_1} = \frac{P_2V_2}{T_2}$

$$\therefore \quad \frac{750 \times 38}{300} = \frac{760 \times V_2}{290}$$

$$\text{or} \quad V_2 = \frac{750 \times 38 \times 290}{300 \times 760} = 36.3 \text{ ml.}$$

Volume of gas at 290 K and 760 mm. pressure = 36.6 ml.

15. Experimental conditions

P_1 = 740 mm.; V_1 = 190 ml.; T_1 = 289 K

STP conditions

P_2 = 760 mm.; V_2 = ?; T_2 = 273 K

$$\frac{P_1V_1}{T_1} = \frac{P_2V_2}{T_2}$$

$$\frac{740 \times 190}{289} = \frac{760 \times V_2}{273}$$

Substituting the values we get,

$$\text{or} \quad V_2 = \frac{740 \times 190 \times 273}{289 \times 760} = 174.8 \text{ ml.}$$

Volume of gas at STP = 174.8 ml.

16. Here, P = 780 mm. = $\frac{780}{760}$ atm.; V = ?; T = 300 K; R = 0.0821 litre. atm. deg^{-1}. mole^{-1}

$$n = \frac{W}{m} = \frac{11}{44} = 0.25 \text{ mole}$$

Applying gas equation, $PV = nRT$

$$\Rightarrow V = \frac{nRT}{P} = \frac{0.25 \times 0.0821 \times 300 \times 760}{780} = 5.998 \text{ litre.}$$

Volume occupied by CO_2 = 6 lit.

17. We know that

1 mole of gas occupied 22400 ml. at STP.

0.3 mole would occupy = 22400 × 0.3 = 6720 ml.

∴ Volume of the cube = 6720 cm^3.

or Length of the cube = $\sqrt[3]{6720}$ = 18.9

Thus, Length of the cube = 18.9 cm.

18. Here, T = 300 K; C = ?

M = molecular weight of CO_2 = 44

R = 8.314 × 10^7 erg. degree^{-1}. mole^{-1}

Substituting the values in the equaiton.

$$C = \sqrt{\frac{3RT}{M}} = \sqrt{\frac{3 \times 8.314 \times 10^7 \times 300}{44}} = 4.123 \times 10^4$$

RMS velocity of CO_2 molecules = 4.123 × 10^4 cm/sec.

Note : For calculation of RMS velocity when pressure and volume are given

The kinetic gas equation is $PV = \frac{1}{3}$ mnC^2 or $PV = \frac{1}{3}$ MC^2

Where M = the molecular weight of the gas

$$MC^2 = 3\,PV;\; C^2 = \frac{3PV}{M};\; C = \sqrt{\frac{3PV}{M}}$$

Where P = pressure in dynes/cm^2 = hdg = h × 13.6 × 980 dynes/cm^2.

(h is the height of mercury column in cm.)

19. Here, P = 1.5 atms = 1.5 × 76 × 13.6 × 980 dyne/cm^2

d = 1.52 g/litre. = 0.00152 g/ml; C = ?

$$C = \sqrt{\frac{3P}{d}} = \sqrt{\frac{3 \times 1.5 \times 76\, x\, 13.6 \times 980}{0.00152}} = 5.476 \times 10^4$$

RMS velocity= 5.476 × 10^4 cm/sec.

20. M = molecular weight of ethane, = 2 × 12 + 6 × 1 = 30

P = 750 mm. = 75 × 13.6 × 980 dyne/cm^2

V = 1.5 litre = 1500 ml.; C = ?

$$C = \sqrt{\frac{3\,PV}{M}} = \sqrt{\frac{3 \times 75 \times 13.6 \times 980 \times 1500}{30}} = 1.225 \times 10^4$$

RMS velocity of ethane molecules

= 1.225 × 10^4 cm/sec.

21. $C = \sqrt{\frac{3RT}{M}}$

For two gases (1) and (2) i.e. H_2 and O_2 respectively,

$$C_1 = \sqrt{\frac{3RT_1}{M_1}}\,;\; C_2 = \sqrt{\frac{3RT_2}{M_2}}$$

When $C_1 = C_2$

$$\sqrt{\frac{3RT_1}{M_1}} = \sqrt{\frac{3RT_2}{M_2}} \quad \text{or} \quad \frac{T_1}{M_1} = \frac{T_2}{M_2}$$

Here, T_1 = temperature of hydrogen = ?; T_2 = temperature of oxygen = 298 K
M_1 = mol. wt. of hydrogen = 2; M_2 = mol. of wt. of oxygen = 32

$$\frac{T_1}{2} = \frac{298}{32}\,;\; T_1 = \frac{298 \times 2}{32} = 18.62 \text{ K.}$$

The temperature at which the RMS velocity of hydrogen is equal to that of oxygen at 298 K is 18.62 K

22. We know, RMS Velocity = $\sqrt{\frac{3RT}{M}}$

R = gas constant = 8.134×10^7 erg deg^{-1} mole^{-1}
T = 293 K; M = 48

$$C = \sqrt{\frac{3 \times 8.314 \times 10^7 \times 293}{48}} = 3.902 \times 10^4 \text{ cm/sec.}$$

23. Here, T = 300 K

M = mol. wtl of SO_2 = 32 + 2 × 16) = 64
$R = 8.314 \times 10^7$ erg degree^{-1} mole^{-1}

Average velocity, $C = \sqrt{\frac{8RT}{\pi M}}$

$$C = \sqrt{\frac{8 \times 8.314 \times 10^7 \times 300}{3.14 \times 64}} = 3.15 \times 10^4 \text{ cm/sec.}$$

Average velocity of SO_2 molecules = 3.15×10^4 cm/sec.

RMS velocity, $C = \sqrt{\frac{3RT}{M}}$

$$C = \sqrt{\frac{3 \times 8.314 \times 10^7 \times 300}{64}} = 3.419 \times 104 \text{ cm/sec.}$$

RMS velocity of SO_2 molecules = 3.419×10^4 cm/sec.
Most probable velocity,

$$C_p = \sqrt{\frac{2RT}{M}} = \sqrt{\frac{2 \times 8.314 \times 10^7 \times 300}{64}} = 2.792 \times 10^4 \text{ cm/sec.}$$

Most probable velocity of SO_2 molecules = 2.792×10^4 cm/sec.

24. Wt. of A = 2 g;   Wt. of B = 3 g.
Pressure = 1 atm.; Pressure of the mixture = 1.5 atm.
Increase in pressure = 1.5 – 1 = 0.5
Now, $PV = nRT$
At constant temperàtrure and volume $P \propto n$
Wt. of gas A which exerts 1 atm pressure = 2 gm
Wt. of gas B which exerts 1 atm pressure = 6 gm
(3 g exerts 0.5 atm. pressure)

$$\frac{\text{Mol. wt. of } A}{\text{Mol. wt. of } B} = \frac{2}{6} = \frac{1}{3}$$

$$M_A : M_B = 1 : 3$$

25. We have, KE = $\frac{3}{2}$ RT (for one mole)
Here, T = 300 K; R = 8.314 Joule
8 g methane = 0.5 moles

$$\text{KE} = \frac{3}{2} \times 0.5 \times 300 \times 8.314 = 1871 \text{ Joule}$$

26. Here we have, $P = 7.6 \times 10^{-10}$ mm = 10^{-12} atm.; n = ?
V = 1 litre ; T = 273 K, R = 0.082 liter.atm.k^{-1}.mol^{-1}.

no. of moles. $$n = \frac{PV}{RT} = \frac{10^{-12} \times 1}{0.082 \times 273}$$

1 mole of gas contains 6.02×10^{23} molecules.

So, no. of molecules $$= \frac{10^{-12} \times 1 \times 6.023 \times 10^{23}}{0.082 \times 273}\ 2.69 \times 10^{10} \text{ molecules}$$

27. Given, Glass tube length = 100 cm
HCl travels a distance of 60 cm. and NH_3 travels a distance 40 cm.

The rate of diffusion of HCl = $\frac{\text{distance}}{\text{time}} = \frac{60}{t}$

He rate of diffusion of NH_3, $r_2 = \frac{40}{t}$

The rate of diffusion is proportional to RMS

i.e., $r \ \alpha \ \sqrt{C^2}$ or $r \ \alpha \ \sqrt{\frac{3P}{d}}$

$$\frac{HCl}{NH_3} = \frac{60t}{40t} = \sqrt{\frac{3P_{(HCl)}/d_{(HCl)}}{3P_{(NH_3)}/d_{(NH_3)}}}$$

or, $$\left(\frac{3}{2}\right)^2 = \frac{3P_{HCl}}{d_{HCl}} \times \frac{d_{NH_3}}{3P_{NH_3}}$$

where, $P_{NH_3} = 1$ atm

$$d_{NH_3} = \frac{\text{Mol. wt.}}{2} = \frac{17}{2} = 8.5$$

$$d_{HCl} = \frac{36.5}{2} = 18.25 ; \quad P_{HCl} = ?$$

$$\frac{9}{4} = \frac{P_{HCl} \times 18.25}{18.25 \times 1}$$

$$\therefore \quad P_{HCl} = \frac{9 \times 18.25}{4 \times 8.5} = 4.84 \text{ atm.}$$

28. Let the no. of moles of air at 300 K be 1. The no. of moles of air at T_2

$$T_2 = (1 - 3/5) = 2/5$$

Since the vessel is an open open and the volume of the vessel is also constant, the pressure will be constant.

$$PV = n_1RT_1 = n_2RT_2 \quad \text{or} \quad \frac{T_1}{T_2} = \frac{n_2}{n_1}$$

i.e.,
$$T_2 = \frac{T_1 n_1}{n_2} = \frac{300 \times 1}{2/5} = 750 \text{ K}$$

29. The molar ratio of oxygen and another gas = 1 : 1

The total pressure of the mixture = 4000 mm.

The partial pressure of oxygen = 2000 mm. and the partial pressure of another gas = 2000 mm.

The drop in the pressure of oxygen after 47 minutes = 500 mm.

The drop in the pressure of oxygen after 74 minutes. = $\frac{500}{47} \times 74 = 787.2$ mm. of Hg. Let the rate of diffusion of oxygen and another gas be r_1 and r_2.

Applying Graham's law of diffusion.

$$\frac{r_2}{r_1} = \sqrt{\frac{M_2}{M_1}} = \sqrt{\frac{32}{79}}; \quad r \propto p \quad \therefore \quad \frac{r_2}{r_1} = \frac{P_2}{P_1}$$

$$P_2 = P_1\sqrt{\frac{32}{79}} = \frac{787.2 \times \sqrt{32}}{\sqrt{79}} = 501.8 \text{ mm. of Hg}$$

The pressure of the other gas after 74 minutes = 2000 – 501.8 = 1498.2 mm. of g.

$$\text{The mole ratio} = \frac{\text{Pressure of unknown gas}}{\text{Pressure of oxygen}} = \frac{1498.2}{1212.8} = 1.24 : 1$$

30. The molecular weight of CO_2 = 2 + 32 = 44.

44 g CO_2 at NTP contain 6.02×10^{23} molecules

200 mg. of CO2 will contain

$$\frac{6.02 \times 10^{23} \times 200 \times 10^{-3}}{44} = \frac{6.02 \times 2 \times 10^{21}}{44} = 2.636 \times 1022 \text{ molecules}$$

No. of molecules remaining = $2.736 \times 10^{21} - 1.0 \times 10^{21} = 1.736 \times 10^{21}$

No. of moles left = $\dfrac{1.736 \times 10^{21}}{6.023 \times 10^{23}} = 2.884 \times 10^{-3}$

31. Use, $\dfrac{P_1 V_1}{T_1} = \dfrac{P_2 V_2}{T_2}$ for a mole of gas

Here,
P_1 is the pressure at NTP = 1 atm.
V_1 is the volume of the gas at NTP = 22.4 lit.
T_1 is the temperature at NTP = 273 K
P_2 = 5 at m,.; T_2 = 303 K
V_2 is the volume of 1 mole of NH_3 at 303 K and 5 atm.

$$= \frac{P_1 V_1 T_2}{P_2 T_2} = \frac{1 \times 22.4 \times 303}{273 \times 5} \text{ 4.97 lit.}$$

$$\text{Density of Ammonia} = \frac{\text{Molecular weight of } NH_3}{\text{Volume of 1 mole}} = \frac{17}{4.97} = 3.42 \text{ g/litre.}$$

32. Let the volume of CO_2 be x ml.

The volume of CO = $(1000 - x)$ ml.

On passing this mixture through red hot charcoal CO_2 is reduced to C:

$$CO_2 + C \rightarrow 2\,CO$$
$$\text{1 vol.} \qquad \text{2 vol.}$$

From this equation,

1 volume of CO_2 gives 2 volumes of CO.

x volumes of CO_2 gives $2x$ volumes of CO

The total volume = $2x + (1000 - x)$

or, $2x + (1000 - x) = 1600$ ml.

$$x = 600 \text{ ml.}$$

600 ml. CO_2 and 400 ml. CO are present in the mixture.

33. Use, $\dfrac{R_1}{R_2} = \sqrt{\dfrac{M_2}{M_1}}$

Where,

R_1 is the rate of diffusion of HCl
R_2 is the rate of diffusion of NH_3
M_1 is the molecular weight of HCl = 36.5

M_2 is the molecular weight of NH_3 = 17

$$\frac{R_1}{R_2} = \sqrt{\frac{17}{36.5}}$$

Let the distance between P and x be l cm.

$$\frac{1}{200-l} = \sqrt{\frac{17}{36.5}} = 0.6824$$

$$1 = (200 - l)\ (0.6824)$$

$$1.6824 \times l = 136.48 \text{ cm.}$$

$$l = \frac{136.48}{1.6824} = 81.12 \text{ cm.}$$

34. Here, $V_1 = 50$ ml; $V_2 = 111$ ml; $T_1 = 273$ K; $T_2 = ?$

By Charles' law, at constant pressure, $\frac{V_2}{T_2} = \frac{V_1}{T_1}$

Substituting the values, $\therefore T_2 = T_1 \times \frac{V_2}{V_1} = 273 \times \frac{111}{50} = 608$ K

35. Here, $P = 1087$ mm of Hg $= \frac{1087}{760} = 1.43$ atm.

$V = 10$ lit.; $T = 350$ K

$\therefore PV = \frac{W}{M} \times RT$

Substituting the values we get,

$$W_{O_2} = \frac{P \times V \times M}{R \times T} = \frac{1.43 \times 10 \times 32}{0.082 \times 350} = 16.95 \text{ gm}$$

36. (a) Given, $W_{H_2} = 4.0$ gm; $M = 2$; $V = 10$ lit; $T = 300$ K $P = ?$

$$PV = \frac{W}{M} \times R \times T$$

Substituting the values we get,

$$P = \frac{W \times R \times T}{M \times V} = \frac{4 \times 0.082 \times 300}{2 \times 10} = 4.92 \text{ atm.}$$

(b) Density of the gas can be represented as, $\frac{W}{V} = \frac{P \times M}{R \times T} = \frac{4.92 \times 2}{0.082 \times 300} = \frac{4}{V}$

Calculating V and then d we get

$d = 0.4$ gm/litre

(c) Let W_1 be the weight of H_2 left when the pressure is 1 atm. then

$$PV = \frac{W_1}{M} \times RT$$

$$W_1 = \frac{PV \times M}{RT} = \frac{1 \times 10 \times 2}{0.082 \times 300} = 0.813 \text{ gm}$$

$\therefore$ Wt. of H_2 removed = 4.0 – 0.813 = 3.187 gm

37. From, $PV = \frac{W}{M} \times RT \Rightarrow \frac{W}{V} = \frac{P \times M}{R \times T}$

Here

P = 1.12×10^9 atm; M = 2 gm mole

R = 0.082 lit.atm $K^{-1}.mol^{-1}$

T = Absolute temperature

Then, $\frac{W}{V}$ will be in gm/lit.

Since the given density d' is in gm/cc.

$$\therefore \quad \frac{W}{V} = \frac{P \times M}{R \times T} \times \frac{1}{1000} \text{ gm/cc}$$

$$\text{or} \quad d' = \frac{P \times M}{R \times T} \times \frac{1}{1000}$$

Substituting the values,

$$\therefore \quad 1.3 = \frac{1.12 \times 10^9 \times 2}{0.082 \times T \times 1000}$$

$$\Rightarrow \quad T = 2.1 \times 10^7 \text{ K}$$

38. Consider 1 litre of air at 300 K and 1 atm. This will consist of 0.21 litres of O_2 and 0.79 lit. of N_2. The weights of O_2 and N_2 can be calculated as follows:

P = 760 mm = 1 atm; T = 300 K

The volume of O_2 is 0.21 lit.

Weight of O_2, $\quad {}^{W}O_2 = \frac{P \times V \times M}{R \times T} = \frac{1 \times 0.21 \times 32}{0.082 \times 300} = 0.2735$ gm

Weight of N_2, $\quad {}^{W}N_2 = \frac{P \times V \times M}{R \times T} = \frac{1 \times 0.79 \times 28}{0.082 \times 300} = 0.8993$ gm.

$\therefore$ Total weight, $\quad W = W_{O_2} + W_{N_2} = 0.2735 + 0.8993 = 1.1728$ gm

$\therefore \quad$ Density of air $= \frac{W}{V} = \frac{1.1728}{1}$

$$d = 1.1728 \text{ gm/litre}$$

39. For an open vessel volume and pressure are constant.

$\therefore \quad P \times V = nRT$ = Constant.

Where n are number of moles of air pressure in the vessel.

As, $n \propto \frac{1}{T}$

$\therefore \quad \frac{n_1}{n_2} = \frac{T_2}{T_1}$

If $\quad n_1 = 1,\ n_2 = \left(1 \cdot \frac{3}{5}\right) = \frac{2}{5}$

$T_1 = 300$ K

$\therefore \quad T_2 = T_1 \times \frac{n_1}{n_2} = 300 \times \frac{1}{2/5} = 750$ K

40. Let the hydrocarbon be represented as C_xH_y

$$C_xH_y + O_2 = CO_2 + H_2O$$
1.12 lit 2.2 gm 18 gm

$\because$ 44 gm of CO_2 contains 32 gm O_2

$\therefore$ 2.2 gm of CO_2 contains $\frac{32 \times 2.2}{44} = 1.6$ gm. of O_2

$\because$ 18 gm of H_2O contains 16 gm O_2

$\therefore$ 1.8 gm of H_2O contains $\frac{1.8 \times 16}{18} = 1.6$ gm. of O_2

$\therefore$ Total weight of $O_2 = 1.6 + 1.6 = 3.2$ gm

Now, $\because$ 32 gm of O_2 at NTP occupy 22.5 lit.

$\therefore$ 3.2 gm of O_2 at NTP will occupy

$\frac{22.4 \times 3.2}{32} = 2.24$ lit. Hence volume O_2 required is 2.24 lit.

$\therefore$ Weight of Carbon in 2.2 gm $CO_2 = 22 - 1.6 = 0.6$ gm

and Weight of Hydrogen in 1.8 gm of $H_2O = 1.8 - 1.6 = 0.8$ gm

$\therefore$ Weight of Hydrocarbon $= 0.6 + 0.2 = 0.8$ gm

$\because$ 1.12 lit of hydrocarbon at NTP weight 0.8gm

$\therefore$ 22.4 lit of hydrocarbon will weigh $\frac{0.8 \times 22.4}{1.12} = 16$ gm

$\therefore$ The molecular weight of hydrocarbon is 16. If molecule contains one carbon atom and four hydrogen atoms, the molecular weight will be 16.

Hence molecular formula is CH_4.

41. Volume of cylinder = 2.82 litres

Hence the initial volume of H_2 is 2820 ml at 300 K and 100 atm. Converting this volume to NTP conditions.

$$V_2 = \frac{P_1 \times V_1}{T_1} \times \frac{T_2}{P_2} = \frac{100 \times 2.82 \times 273}{300 \times 1} = 256.62 \text{ lit.}$$

$\therefore$ Volume of H_2 available for filling $= 256.62 - 2.82 = 253.80$ lit.

Volume of the balloon $= \frac{4}{3}\pi r^3 = \frac{4}{3} \times 3.14 \times \left(\frac{21}{2}\right)^3 = 4846.6$ cc $= 4.8466$ lit.

$\therefore$ Number of balloons that can be filled $= \dfrac{253.8}{4.8466} = 52.36 = 52$

Hence 52 balloons can be filled.

42. Partial pressure O_2, $\quad {}^{P}O_2 = \dfrac{W}{M} \times \dfrac{R \times T}{V}$

$$= \frac{3.2}{32} \times \frac{0.082 \times 300}{100} = 0.0246 \text{ atm.}$$

Similarly $\quad {}^{P}CO_2 = \dfrac{2.2}{44} \times \dfrac{0.082 \times 300}{100} = 0.0123$ atm.

$\therefore$ Total pressure $\quad P = P_{O_2} + P_{CO_2} + 0.0246 + 0.0123 = 0.0369$ atm.

43. Number of moles of each gas is calculated as follows:

$${}^{n}O_2 = \frac{W}{M} = \frac{1.6}{32} = 0.05 \text{ moles of oxygen}$$

$${}^{n}CO_2 = \frac{4.4}{44} = 0.1 \text{ moles of } CO_2$$

$${}^{n}NO_2 = \frac{2.3}{46} = 0.05 \text{ moles of } NO_2$$

$\therefore$ Total number of moles = 0.05 + 0.1 + 0.05

Partial pressure of O_2, ${}^{P}O_2 = \dfrac{\text{No. of moles of } O_2}{\text{Total No. of moles}} \times$ Total pressure

$$= \frac{0.05}{0.2} \times 1 = 0.25 \text{ atm.}$$

Similarly, $\quad {}^{P}CO_2 = \dfrac{0.1}{0.2} \times 1 = 0.5$ atm.

$${}^{P}NO_2 = \frac{0.05}{0.2} \times 1 = 0.25 \text{ atm.}$$

44. Under similar conditions of volume and temperature, the pressure of the gases in the flask will be proportional to the total number of moles. ($PV = nRT$)

Let M_A and M_B be the molecular weights of A and B respectively, then the number of moles,

$$N_A = \frac{2}{M_A} \text{ and } n_B = \frac{3}{M_B}$$

Total number of moles initially present $= n_A = \dfrac{2}{M_A}$

Total number of moles finally present,

$$n_A + n_B = \frac{2}{M_A} + \frac{3}{M_B}$$

$$\therefore \qquad \frac{n_A}{n_A + n_B} = \frac{P_1}{P_2}$$

$$\text{Or} \qquad \frac{\frac{2}{M_A}}{\frac{2}{M_A} + \frac{3}{M_B}} = \frac{1}{1.5}$$

$$\Rightarrow \quad 2\, M_B + 3\, M_A = 3\, M_B$$

$$\therefore \qquad \frac{M_A}{M_B} = \frac{1}{3}$$

45. Let us consider 10 moles of a mixture of C_6H_6 and H_2. Let x moles of C_6H_6 be present in the mixture. Then number of moles of H_2 will be $(10 - x)$

$$\underset{1\text{ mole}}{C_6H_6} + \underset{3\text{ mole}}{3H_2} = \underset{1\text{ mole}}{C_6H_{12}}$$

Now, x moles of C_6H_6 will combine with $3x$ moles of H_2 and x moles of C_6H_{12} and there will be $(10 - x - 3x)$ moles of unused H_2.

We have Number of moles before reaction, $n_1 = 10$

Number of moles after reaction,

$$n_2 = (10 - x - 3x) + x = 10 - 3x$$

$$\text{Now,} \qquad \frac{P_1 \times V}{P_2 \times V} = \frac{n_1 \times R \times T}{n_2 \times R \times T}$$

$$\therefore \qquad \frac{P_1}{P_2} = \frac{n_1}{n_2}; \quad \frac{60}{30} - \frac{10}{10 - 3x}$$

$$\therefore \qquad x = 1.66 \text{ moles}$$

$\therefore$ Number of moles of C_6H_6 in 10 moles = 1.66

Number of moles of H_2 in 10 moles = 10 × 1.66 = 8.34

$$\therefore \text{ mole fraction of } C_6H_6 = \frac{1.66}{10} = 0.166$$

$$\text{and mole fraction of } H_2 = \frac{8.34}{10} = 0.834$$

Consider 100 volume of a mixture.

By Amagat's Law, volume of C_6H_6 = mole fraction of C_6H_6 × Total volume

$$= 0.166 \times 100 = 16.6$$

$$\therefore \qquad \text{Volume of } H_2 = 100 - 16.6 = 83.4$$

Hence original mixture contains 16.6% C_6H_6 and 83.4% of H_2 by volume.

46. (a) Let the weight of each gas be x gm. Then the number of moles of H_2 and C_2 will be,

$$^{n}H_2 = \frac{x}{2} \quad \text{and} \quad ^{n}Cl_2 = \frac{x}{71}.$$

$$\frac{{}^nH_2}{{}^nCl_2} = \frac{\text{Number of molecules of } H_2}{\text{Number of molecules of } Cl_2}$$

$$= \frac{\frac{x}{2} \times 6.02 \times 10^{23}}{\frac{x}{71} \times 6.02 \times 10^{23}} = \frac{71}{2} = 35.5$$

$\therefore$ $\quad {}^nH_2 = 35.5 \times {}^nCl_2$

Hence the number of H_2 molecules is 35.5 times more than Cl_2 molecules.

(b) Let P_{H_2} and P_{Cl_2} be the pressure of two gases, then

$$\frac{P_{H_2} \times V}{P_{Cl_2} \times V} = \frac{{}^nH_2 \times RT}{{}^nCl_2 \times RT}$$

$$\therefore \quad \frac{P_{H_2}}{P_{Cl_2}} = \frac{{}^nH_2}{{}^nCl_2} = 35.5$$

$$\therefore \quad P_{H_2} = 35.5 \times {}^PCl_2$$

Thus the pressure of H_2 is 35.5 items greater than that of Cl_2.

(Since pressure is directly proportional to number of moles it can be obtained directly.)

47. Since the displaced air is collected over water, it also contains water vapour. Hence the total pressure is the pressure of air and water vapour which is equal to 758 mm.

$\therefore$ Pressure of air= (758 – 12 =) 746 mm

$$P = \frac{746}{760} = 0.981 \text{ atm.} \qquad V = 0.0266 \text{ lit}$$

$$T = 288\text{K} \qquad W = 0.17 \text{ gm}$$

$$\therefore \quad M = \frac{W \times RT}{P \times V} = \frac{0.17 \times 0.082 \times 288}{0.981 \times 0.0266} = 153.8$$

$\therefore$ $\quad$ Molecular weight = 153.8

48. We have, $\quad M = \frac{W \times R \times T}{P \times V}$,

Here, $\quad W = 0.2901$; $T = 293$ K; $P = \frac{770}{760}$ atm ; $V = \frac{25.8}{1000} = 0.02588$

Substituting the values, $M = \frac{0.2901 \times 0.082 \times (293)}{\frac{770}{760} \times 0.0258} = 262.1$

Molecular weights of $AlCl_3 = 27 + (3 \times 35.5) = 133.5$

If this is considered as empirical formula then, $\frac{262.1}{133.5} = 1.96 = 2$

Hence anhydrous aluminium chloride must be dimer having formula $Al_2\,Cl_6$.

49. (a) Consider 100 cc of the gaseous mixture. Let the volume of N_2 be x cc. Molecular weights of N_2 and O_2 are 28 and 32 respectively and therefore the relative vapour densities of N_2 and O_2 will be 14 and 16 respectively.

$\therefore$ weight of x cc of $N_2 = 14 \times x$

$\therefore$ weight of $(100 - x)$ cc of $O_2 = 16 \times (100 - x)$

Total weight of the mixture = 15.2×100

$\therefore (14 \times x) + 16 \times (100 - x) = 15.2 \times 100$

$\therefore \quad x = 40 \quad$ and $\quad (100 - x) = 60$

Hence N_2 = 40% by volume and O_2 = 60% by volume.

(b) Consider 100 gm of a gaseous mixture. Let weight of N_2 be x gm then the weight of O_2 will be $(100 - x)$ gm

$$\therefore V_{N_2} = \frac{x}{M} \text{ and } V_{O_2} = \frac{(100-x)}{16}; \quad \left(\because \text{Volume} = \frac{\text{Weight}}{\text{Density}}\right)$$

$$\text{Volume of the mixture} = \frac{100}{15.2}$$

$$\therefore \quad \frac{x}{14} + \frac{(100-x)}{16} = \frac{100}{15.2}$$

$\therefore 16 \times 15.2x + 14 \times 15.2\,(100 - x) = 100 \times 16 \times 4$

$\therefore \quad x = 36.72$ and $(100 - x) = 63.28$

Hence $\quad N_2$ = 36.72% by weight and O_2 = 63.28% by weight

50. For SO_2, $V_1 = 120$ ml, $t_1 = 20$ min. $M_1 = 64$

For hydrocarbon, $V_2 = 180$ ml,

$t_2 = 15$ min. $M_2 = ?$

$$\frac{V_1 \times t_2}{V_2 \times t_1} = \sqrt{\frac{M_2}{M_1}}$$

$\therefore \quad M_2 = 16$

Molecular weight of hydrocarbon is 16.

51. Before diffusion the first bulb contained 1.12 lit of H_2 and 1.12 lit of D_2 at NTP.

After diffusion 0.05 gm of H_2 remained in the first bulb. From this the volume of H_2 diffused can be obtained.

Volume of 2 gm of H_2 at NTP is 22.4 lit.

$\therefore$ Volume of 0.05 gm of H_2 at NTP is

$$\frac{22.4 \times 0.05}{2} = 0.56 \text{ lit.}$$

$\therefore$ The volume of H_2 diffused in the second bulb is $(1.12 - 0.56) = 0.56$ lit.

Hence the weight of the H_2 present in the second bulb is 0.05 gm. Now, let V_1 and V_2 be the volumes of H_2 and D_2 diffused and let M_1 and M_2 be their molecular weights then,

$$\frac{V_2}{V_1} = \sqrt{\frac{M_1}{M_2}}$$

$\therefore \quad \frac{V_2}{0.056} = \sqrt{\frac{2}{4}}$

$\therefore \quad V_2 = 0.396$ litres

Hence volume of D_2 diffused in the second bulb is 0.396 lit.

$\because$ 22.4 lit of D_2 at NTP weight 4 gm.

$\therefore$ 0.396 lit of D_2 at NTP will weigh $\frac{4 \times 0.396}{22.4} = 0.07$ gm

Total weight of the mixture = 0.05 + 0.07 = 0.12 gm

$\therefore$ Percentage by weight of $H_2 = \frac{0.05 \times 100}{0.12} = 41.66$

and % of D_2 = 100 – 41.66 = 58.34

52. Let t_1 and t_2 be the times of diffusion of mixture O_3 and O_2 and D_1 and D_2 be their densities.

$$\frac{t_1}{t_2} = \sqrt{\frac{D_1}{D_2}}$$

$$\therefore \quad \frac{180}{170} = \sqrt{\frac{D_1}{16}}$$

$\therefore \quad D_1 = 17.92$

Let the density of O_3 be x. Thus, 100 cc of the given mixture will contain 22 cc of O_3 and 78 cc of O_2. Then,

Wt. of O_2 + Wt. of O_3 = Wt. of the mixture

$78 \times 16 + 22 \times x = 100 \times 17.92$

or $\quad x = 24.7$

$\therefore$ density of O_3 is 24.7

53. We can solve this problem using the ideal gas law for two gases.

$\frac{P_i V_i}{n_i T_i} = \frac{P_f V_f}{n_f T_f}$. Since V is constant, we get: $\frac{P_i}{n_i T_i} = \frac{P_f}{n_f T_f}$.

$$\therefore \quad P_f = \frac{p_i n_f T_f}{n_i T_i}$$

Here, $P_i = 0.95$ atm, $T_i = 298$ K, $T_f = 398$ K

n_f = unreacted moles of H_2 and O_2 + moles of product.

(Remember that 3 moles of reactants yields only 2 moles of product in the given reaction.)

Here, unreacted moles of H_2 and O_2 = (1–0.88) ni = 0.12ni

Thus, $\quad n_f = (0.12)(n_i) + (0.88)\left(\frac{2}{3} n_i\right) = 0.707\ n_I$

Hence, $$P_f = \frac{(0.950\text{ atm})(0.707n_i)(398\text{ K})}{(n_i)(298\text{ K})} = 0.897\text{ atm}$$

54. Use the ideal gas law, solving for V_A and V, and substitute into the equation for volume fraction. Since R, T and P are constant, they cancel out.

$$\text{Volume fraction } A = \frac{V_A}{V} = \frac{\frac{n_A RT}{P}}{\frac{nRT}{p}} = \frac{n_A}{n} = \text{mole fraction } A$$

Volume fraction (and mole fraction) differ from mass fraction because different gases have different molar masses.

55. Assume that none of the N_2 escaps from the bag. Then use the ideal gas law to calculate the moles of N_2 needed to fill 20.0 L, and the amount NaN_3 to product that N_2.

$$n = \frac{PV}{RT} = \frac{(1\text{ atm})(20\text{ L})}{(0.0821\text{ L}\cdot\text{atm/mol}\cdot\text{K})(298\text{ K})} = 0.817\text{ mol. } N_2$$

$$2NaN_3(s) \rightarrow 2Na(s) + 3N_2(g)$$

$$0.817\text{ mol. } N_2 \times \frac{2\text{ mol } NaN_3}{3\text{ mol. } N_2} \times \frac{65.01\text{ g } NaN_3}{1\text{ mol. } NaN_3} = 35.4\text{ g } NaN_3$$

56. The equation is:

$$Ba(s) + 2H_2O(I) \rightarrow Ba(OH)_2(aq) + H_2(g)$$

Using the equation, calculate the moles of H_2; then use the ideal gas law to convert to volume.

$$\text{Mol } H_2 = 2.50\text{ g Ba} \times \frac{1\text{ mol Ba}}{137.33\text{ g Ba}} \times \frac{1\text{ mol } H_2}{1\text{ mol Ba}} = 0.018204\text{ mol}$$

$$V = \frac{nRT}{P} = \frac{[0.018204\text{ mol}][0.082057\text{ L}\bullet\text{atm/(K}\bullet\text{mol)}][294\text{ K}]}{(748/760)\text{ atm}} = 0.44541\text{ L (447 mL)}$$

57. Write a mass-balance equation to solve for the moles of each gas, using 28.01 g/mol for the molar mass N_2 and 20.18 g/mol for the molar mass of Ne. Let y = mol of each gas:

$$28.01\,y + 20.18\,y = 10.0\text{ g}$$

$$\text{Total mass}(n) = 0.41502\text{ mol}$$

Use the ideal gas law to calculate the volume, which can then be used to calculate density.

$$\text{Vol} = \frac{(0.41502\text{ mol})(0.082\text{ L}\bullet\text{atm/K}\bullet\text{mol})(500\text{ K})}{10\text{ atm}} = 1.7028\text{ L}$$

$$d = 10.0\text{ g} \div 1.7028\text{ L} = 5.872 = 5.87\text{ g/L}$$

58. Here since density is given we rearrange the equation $PM_m = dRT$ to find the quantity RT/M_m. Then put into the equation for the root-mean-square speed

$$\frac{RT}{M_m} = \frac{P}{d}; \text{ Here d = 3.0g/L, P = 675 mm} = \frac{675}{760}\text{atm}; T = 298K$$

Substituting the values we have,

$$u = \sqrt{\frac{3RT}{M_m}} = \sqrt{\frac{3P}{d}} = \sqrt{\frac{3\left(\frac{675}{760}\right)\text{atm} \times (8.31\,\text{kg} \bullet \text{m}^2/s^2 \bullet \text{K} \bullet \text{mol})}{(3.0 \times 10^{-3}\,\text{kg/L})\,(0.08206\,\text{L} \bullet \text{atm/K} \bullet \text{mol})}}$$

$$= 2.99 \times 102 = 3.0 \times 102 \text{ m/s}$$

59. Since this a rigid, scaled container, neither the volume nor number of moles can change. Thus, as the temperature is lowered, only the pressure can change, which also drops. (see figures below)

(a)

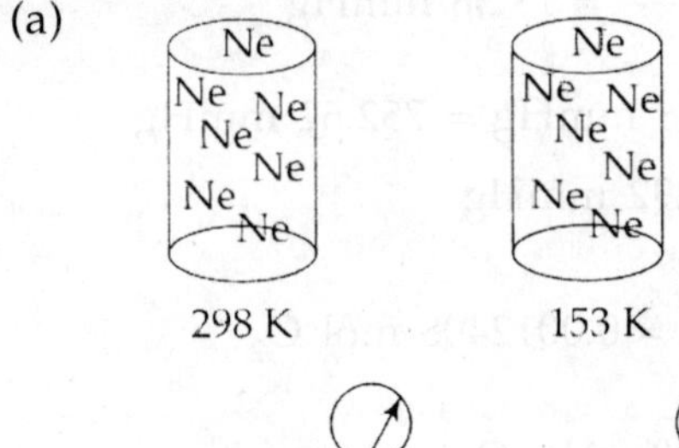

(b)

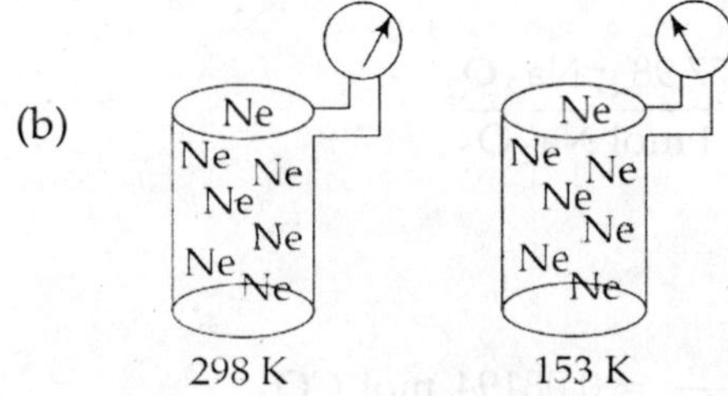

60. We can solve these problems using the ideal gas law, so solving for R, and setting the equations equal to one another we have,

$$\frac{PV}{nT} = R, \text{ thus } \frac{P_A V_A}{n_A T_A} = R = \frac{P_B V_B}{n_B T_B}. \text{ Since } V_A = V_B \text{ and } T_A = T_B, \text{ then: } \frac{P_A}{n_A} = \frac{P_B}{n_B}.$$

(a) If $n_A = n_B$, then: $P_A = P_B$

(b) CO_2 has a much larger mass than He, thus if these are equal grams of each, these will be much more He than CO_2, and $n_A < n_B$. Thus $P_A < P_B$.

61. Assume a 100-g sample, giving 85.2 g CH_4 and 14.8 g C_2H_6. Convert each to moles:

$$85.2 \text{ g } CH_2 \times \frac{1 \text{ mol } CH_4}{16.04 \text{ g } CH_4} = 5.311 \text{ mol } CH_4$$

$$14.8 \text{ g } C_2H_6 \times \frac{1 \text{ mol } C_2H_6}{30.07 \text{ g } C_2H_6} = 0.4922 \text{ mol } C_2H_6$$

$$V_{CH_4} = \frac{(5.311\ \text{mol})(0.082\ \text{L} \bullet \text{atm/K} \bullet \text{mol})(291\ \text{K})}{(748/760)\ \text{atm}} = 128.9\ \text{L}$$

$$V_{CH_2H_6} = \frac{(0.4922\ \text{mol})(0.082\ \text{L} \bullet \text{atm/K} \bullet \text{mol})(291\ \text{K})}{(748/760)\ \text{atm}} = 11.94\ \text{L}$$

The density is calculated as follows:

$$d = \frac{85.2\ \text{g}\ CH_4 + 14.8\ \text{g}\ C_2H_6}{(128.9 + 11.94)\ \text{L}} = 0.71002 = 0.71\ \text{g/L}$$

62. First subtract the height of mercury equivalent to the 25.00 cm (250 mm) of water inside the tube from 771 mmHg to get P_{gas}. Then subtract the vapor pressure of water 18.7 mmHg, from P_{gas} to get P_{O_2}.

$$h_{H_g} = \frac{(h_W)(d_W)}{d_{Hg}} = \frac{250\ \text{mm} \times 0.99987\ \text{g/cm}^3}{13.596\ \text{g/cm}^3} = 18.38\ \text{mmHg}$$

$$P_{gas} = P - P_{25\ \text{cm water}} = 771\ \text{mmHg} - 18.38\ \text{mmHg} = 752.62\ \text{mmHg}$$

$$P_{O_2} = 752.62\ \text{mmHg} - 18.7\ \text{mmHg} = 733.92\ \text{mmHg}$$

$$n = \frac{PV}{RT} = \frac{(733.92/760\ \text{atm})\ (1.94\ \text{L})}{(0.082\ \text{L} \bullet \text{atm/K} \bullet \text{mol})\ (298\ \text{K})} = 0.0012408\ \text{mol}\ O_2$$

$$\text{Mass} = (2 \times 0.0012408)\ \text{mol}\ Na_2O_2 \times \frac{77.98\ \text{g}\ Na_2O_2}{1\ \text{mol}\ Na_2O_2}$$

$$= 0.1935 = 0.194\ \text{g}\ Na_2O_2$$

63. First find the moles of CO_2:

$$n = \frac{PV}{RT} = \frac{(785/760\ \text{atm})\ (1.94\ \text{L})}{(0.082\ \text{L} \bullet \text{atm/K} \bullet \text{mol})\ (198\ \text{K})} = 0.08194\ \text{mol CO}$$

Set up one equation is one unknown: x = mol $CaCO_3$; $(0.08194 - x)$ = mol = $MgCO_3$

$$7.58\ \text{g} = (100.1\ \text{g/mol})x + (84.32\ \text{g/mol})\ (0.08194 - x)$$

$$x = \frac{(7.85 - 6.9092)}{(100.1 - 84.32)} = 0.05962\ \text{mol}\ CaCO_3$$

$$(0.08194 - x) = 0.02232\ \text{mol}\ MgCO_3$$

$$\%\ CaCO_3 = \frac{0.05962\ \text{mol}\ CaCO_3 \times 100.1\ \text{g/mol}}{7.85\ \text{g}} \times 100\% = 76.02 = 76\%$$

$$\%\ MgCO_3 = 100.00\% - 76.02\% = 23.98 = 24\%$$

64. Set up algebraic equation, using x for the distance ammonia travels and $(1.5 - x)$ for the distance HCl travels, correlating distance traveled with molar mass.

$$\frac{\text{distance } NH_3}{\text{distance HCl}} = \left(\frac{\mathcal{M}_{HCl}}{\mathcal{M}_{NH_3}}\right)^{\frac{1}{2}} \Rightarrow \frac{x}{1.5-x} = \left(\frac{32.5}{17}\right)^{\frac{1}{2}}$$

$$\frac{x}{1.5-x} = 1.45 \Rightarrow x = 2.175 - 1.45x$$

$$2.45x = 2.175$$

$$x = 0.88 \text{ m}$$

65. Kelvin and °C are the same size and °R and °F are also the same size. Thus if 1.8°C = 1°F, then 1.8 K = 1°R.

$$273.15 \text{ K} = 273.15 \times 18 = 491.67°\text{R}$$

$$R = \frac{PV}{nT} = \frac{(1.0 \text{ atm})(22.4 \text{ L})}{(1.0 \text{ mol.})(491.67°\text{R})} = 0.0456 \text{ L} \cdot \text{atm/mol. °R}$$

66. Calculate the total amount of H_2 produced, then calculate the moles of H_2 prodeuced by each metal using x grams Zn and (0.2500 – x) grams Al. Solve for x to get the grams of Zn, then calculate the mass percent.

$$P = 755 \text{ mm Hg} \times \frac{1 \text{ atm}}{760 \text{ mm Hg}} = 0.993 \text{ atm}$$

$$n_{H_2} = \frac{PV}{RT} = \frac{(0.93 \text{ atm})(0.147 \text{ L})}{(0.0821 \text{ L} \cdot \text{atm/mol.} \cdot \text{K})(298 \text{ K})} = 0.00597 \text{ moles } H_2$$

$$x \text{ g Zn} \times \frac{1 \text{mol Zn}}{65.39 \text{g Zn}} \times \frac{1 \text{mol } H_2}{1 \text{mol Zn}} = 0.0153 \text{ mol. } H_2$$

$$(0.25 - x) \text{ g Al} \times \frac{1 \text{ mol Al}}{26.98 \text{ g Al}} \times \frac{\frac{3}{2} \text{ mol. } H_2}{1 \text{ mol. Al}} = (0.0139 - 0.0556\,x) \text{ mol. } H_2$$

$$0.00597 = 0.0153x + 0.0139 - 0.0556x$$

$$0.0403x = 0.00793$$

$$x = 0.197 \text{ g Zn}$$

$$\text{mass \% Zn} = \frac{0.197 \text{ g Zn}}{0.2500 \text{ g total}} \times 100\% = 78.7\%$$

67. To get off the ground, the buoyant force of the balloon must exceed the gravitational force holding it down. We can find the volume of the balloon by equating the two forces, converting mass to ($d \times V$) and solving got V. Although we are not given densities, we can solve the ideal gas law for d and substitute. Now we can calculate a numerical value for V and calculate the radius and diameter of the balloon.

$$\text{mass}_{(\text{air})} = \text{mass}_{(H_2)} + \text{mass}_{(\text{man + balloon})} = \text{mass}_{(H_2)} + 1.68 \times 10^5 \text{ g}$$

$$d_{(\text{air})} \times V = d_{(H_2)} \times V + 1.68 \times 10^5 \text{ g}$$

$$V = \frac{1.68 \times 10^5 \text{ g}}{d_{air} - d_{H_2}} = \frac{1.68 \times 10^5 \text{ g}}{\dfrac{PM_{air}}{RT} - \dfrac{PM_{H_2}}{RT}} = \frac{(1.68 \times 10^5 \text{ g})\, RT}{P(M_{air} - M_{H_2})}$$

$$V = \frac{(1.68 \times 10^5 \text{ g})(0.0821 \text{ L} \cdot \text{atm/mol.} \cdot \text{K})(295 \text{ K})}{\left(758 \text{ mm Hg} \times \dfrac{1 \text{ atm}}{760 \text{ mm Hg}}\right)(29 \text{ g/mol.} - 2.01 \text{ g/mol.})} = 1.51 \times 10^5 \text{ L}$$

$$V = 1.51 \times 10^5 \ L \times \frac{1 \text{ m}^3}{1000 \text{ L}} = 1.51 \times 10^2 \text{ m}^3$$

$$V = 1.51 \times 10^2 \text{ m}^3 = \frac{4}{3}\pi r^3 \Rightarrow r = 3.30 \text{ m} \qquad \therefore \ \text{diameter} = 2r = 6.60 \text{ m}$$

68. (a) Pressure is directly proportional to the number of moles, so bulb C (the one with the fewest molecules) would have the lowest pressure.

(b) The relative pressures are directly proportional to the relative moles. Bulb C contains half as many molecules (or moles) as bulb A, so the pressure would be half that A : 1.00 atm.

(c) As mentioned in part (b) above, the pressure in bulb C is 1.0 atm. From similar reasoning, pressure in bulb B is [(6/8)(2.0)] 1.50 atm.
Total pressure = 2.0 + 1.50 + 1.00 = 4.50 atm.

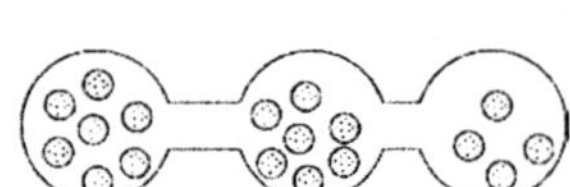

(d) Bulbs A and B now contain 7 molecules each, so $P_A = P_B = (7/8)(2.0 \text{ atm}) = 1.75$ atm.

$P_A + P_B + 1.75 + 1.75 = 2.5$ atm

$P_A + P_B + P_C = 1.75 + 1.75 + 1.0 = 4.5$ atm

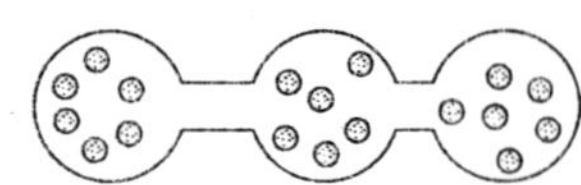

Thus the total pressure is unchanged

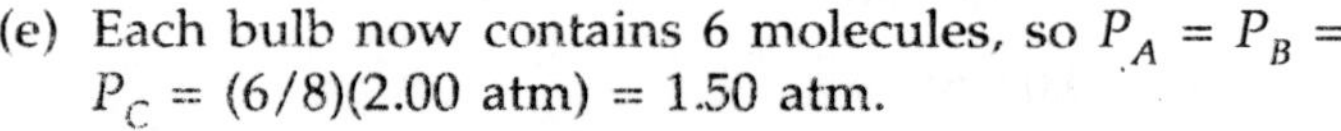

(e) Each bulb now contains 6 molecules, so $P_A = P_B = P_C = (6/8)(2.00 \text{ atm}) = 1.50$ atm.

$P_A + P_B = 1.50 + 1.50 = 3.00$ atm

$P_A + P_B + P_C = 1.50 + 1.50 + 1.50 = 4.50$ atm

Thus the total pressure is unchanged

69. We have, pressure are $P_1 = 23.76 \text{ mm Hg} \times \dfrac{1 \text{ atm}}{760 \text{ mm Hg}} = 0.0313$ atm;

$T_1 = 298$; $P_2 = ?$; $T_2 = 313$

$$\frac{P_1}{T_1} = \frac{P_2}{T_2} \Rightarrow \frac{0.0313}{298} = \frac{P_2}{313} \Rightarrow P_2 = 0.0329 \text{ atm}$$

$$0.0329 \text{ atm} \times \frac{760 \text{ mm Hg}}{1 \text{ atm}} = 25 \text{ mm Hg}$$

Similarly,

$$\frac{P_1}{T_1} = \frac{P_3}{T_3} \Rightarrow \frac{0.0313}{298} = \frac{P_3}{343} \Rightarrow P_3 = 0.036 \text{ atm}$$

$$0.036 \text{ atm} \times \frac{760 \text{ mm Hg}}{1 \text{ atm}} = 27.4 \text{ mm Hg}$$

and,

$$\frac{P_1}{T_1} = \frac{P_4}{T_4} \Rightarrow \frac{0.0313}{298} = \frac{P_4}{373} \Rightarrow P_4 = 0.0392 \text{ atm}$$

$$0.0392 \text{ atm} \times \frac{760 \text{ mm Hg}}{1 \text{ atm}} = 29.8 \text{ atm Hg}$$

Temperature	*P(ideal gas)*	*P(water vapor)*
T_1 = 298 K	23.76 mm Hg	23.76 mm Hg
T_2 = 313 K	25.0 mm Hg	55.3 mm Hg
T_3 = 343 K	27.4 mm Hg	233.7 mm Hg
T_4 = 373 K	29.8 mm Hg	760.0 mm Hg

These numbers are so different because the number of moles (n) is not constant for the water vapor. As the temperature increases, the amount of water that is vaporised increases. Since pressure is proportional to the moles of gas, the pressure also increases with increasing temperature.

70. Use the equation for average speed of a gas to calculate the molar mass. Compare that value to the molar masses of H_2, He and Ar.

$$u^2 = \frac{3RT}{M} \Rightarrow M = \frac{3RT}{u^2}$$

$$M = \frac{3RT}{u^2} = \frac{(3)(8.31 \times 10^3 \text{ g} \cdot \text{m}^2/\text{s}^2 \cdot \text{mol} \cdot \text{K})(288 \text{ K})}{(1.12 \times 10^3 \text{ m/s})^2} = 5.72 \text{ g/mol.}$$

Note that molar mass and speed are inversely proportional, thus heavier atoms and molecules will move more slowly. Those atoms and molecules that will have sufficient speed to escape gravity are those with a molar mass of less than 5.72 g/mol., such as H_2 and He. Ar's molar mass is greater than 5.72 g/mol., thus it's good speed will be insufficient.

71. First, determine the limiting reagent in the reaction. Then calculate the moles of water that would be produced, and the pressure that would result if all the rate existed as vapor. Compare that to the vapor pressure of water at 300 K.

(a) $$0.5 \text{ g } H_2 \times \frac{1 \text{ mol. } H_2}{2.016 \text{ g}} \times \frac{2 \text{ mol. } H_2O}{2 \text{ mol. } H_2} = 0.198 \text{ mol. } H_2O$$

$$3.2 \text{ g } O_2 \times \frac{1 \text{ mol. } O_2}{32 \text{ g}} \times \frac{2 \text{ mol } H_2O}{1 \text{ mol } O_2} = 0.200 \text{ mol. } H_2O$$

Thus O_2 and H_2 are present in nearly stoichiometric quantities. Consequently, the only gas in the flask will be H_2O. Out calculations will be based on H_2.

$$P = \frac{nRT}{V} = \frac{(0.198\text{ mol.})(0.0821\text{ L}\cdot\text{atm/mol}\cdot\text{K})(300\text{ K})}{10\text{ L}} = 0.488\text{ atm}$$

$$P_{H_2O(300\,K)} = 26.74\text{ mm Hg} \times \frac{1\text{ atm}}{760\text{ mm Hg}} = 0.0352\text{ atm}$$

$P > P_{H_2O}$ Since total pressure cannot exceed the vapor pressure of the water at 300 K, much of water will exist as liquid, and both the liquid and vapor phases will be present.

(b) The final pressure in the flask will be 0.0352 atm (26.7 mm Hg), the vapor pressure of the H_2O at 300 K

(c) $3.2\text{ g } H_2 \times \dfrac{1\text{ mol. }H_2}{2.016\text{ g}} \times \dfrac{2\text{ mol }H_2O}{2\text{ mol. }H_2} = 1.6\text{ mol. }H_2O$

$$3.2\text{ g } O_2 \times \frac{1\text{ mol. }O_2}{32\text{ g}} \times \frac{2\text{ mol }H_2O}{1\text{ mol. }O_2} = 0.20\text{ mol. }H_2O$$

Thus O_2 is the limiting reactant and 1.4 mol. H_2 will be left over. The total pressure will be P_{H_2O} (0.0352 atm, see part a) plus P_{H_2}.

$$P = \frac{nRT}{V} = \frac{(1.4\text{ mol.})(0.082)\text{ L}\cdot\text{atm/mol}\cdot\text{K})(300\text{ K})}{10\text{ L}} = 3.4\text{ atm}$$

$$P_{total} = 3.4\text{ atm} + 0.0352\text{ atm} = 3.4\text{ atm}$$

72. The critical temperature of N_2 is 126 K thus N_2 exists entirely as a gas at both 293 K and 283 K. Propane, on the other hand, has a critical temperature of 370 K and thus is present as an equilibrium between the vapor and liquid phases. He change in pressure observed with the N_2 is due to a "contraction" of the gas, or a reduction in the thermal motion of the N_2 molecules, while the pressure change with the propane is due to a change in vapor pressure. Changes in vapor pressure are much more sensitive to temperature changes are changes in gas pressure.

73. The pressure exerted by a skater is:

$$\text{Pressure} = \frac{120\text{ kg}}{0.1\text{ m}^2} \times \frac{1\text{ atm}}{15\text{ kg/m}^2} = 80\text{ atm}$$

Statement #2 under the heading melting point states "An increase in pressure of 134 atm is required to lower the melting point of ice by 1°K." Thus, unless the ice is only a half degree below freezing, the concept that pressure melts the ice is implausible.

Another possible scenario is heat conduction, given that metals are good conductors of heat. The explanation also falls short, however, since skating is frequently enjoyed when ambient temperatures are well below the freezing point of water.

Friction also fails as an explanation since skating is relatively frictionless.

The most likely explanation for the ease at which a skater glides over ice is a low coefficient of friction between ice and steel.

74. By calculating the moles of Cl and the total moles of *F* in the products, one can determine the mole ratio of Cl to F and thus the formula of ClF_x. Then one can proceed to address the questions of geometry and polarity.

$$n=\frac{PV}{RT}=\frac{(3\text{ atm})(0.457\text{ L})}{(0.082\text{ L}\cdot\text{atm/mol}\cdot\text{K})(348\text{ K})}=0.048\text{ mol. Cl}$$

$$5.6\text{ g UF6}\times\frac{1\text{ mol UF}_6}{352\text{ g UF}_6}\times\frac{6\text{ mol. F}}{1\text{ mol. UF}_6}=0.096\text{ mol. F}$$

moles Cl = 0.048 moles.
Moles F = 0.048 + 0.096 = 0.144 moles.

$$\frac{0.144\text{ mol. F}}{0.048\text{ mol. Cl}}=3\text{ mol. F/1 mol. Cl}$$

Thus, $x = 3$ ClF_3 F—$\ddot{\text{Cl}}$—F (with F bonded below Cl)

Geometry: AX_3E_2 T-shaped.
This molecule is polar.
Bond angles are about 90° and 180°.
Cl: sp^3d hybridized.
Hence 3 σ bonds, 0 π bonds.

75. (a) For an ideal gas PV = nRT. Substituting the given values we have,

$$\therefore\ V_{ideal}=\frac{nRT}{P}=\frac{(3.5\text{ mol})}{(16\text{ atm})}\left(0.082\frac{\text{L}\cdot\text{atm}}{\text{mol}\cdot\text{K}}\right)(320\text{ K})=5.75\text{ L}$$

(b) $$\text{percentage error}=\left(\frac{V_{ideal}-V_{real}}{V_{real}}\right)\times 100=\left(\frac{5.75-5.2}{5.2}\right)\times 100=10.5\%$$

(c) The error made using the ideal gas law depends of the average distance between molecules, which depends on the number of molecules per cubic centimeter. The larger the gas density, the more the molecules are crowded together and the large the error we will make, if we use the ideal gas law. For 3.50 mol of an ideal gas occupying a volume of 5.75 L or 5.75×10^3 cm^3,

$$\frac{\text{No. molecules}}{\text{cubic centimeter}}=\frac{(3.5)(6.022\times 10^{23})}{5.75\times 10^3}=3.67\times 10^{20}\text{ molecules}\cdot\text{cm}^{-3}$$

For 3.5 mol of an ideal gas at 298 K and 1.0 atm, the volume is

$$V=\frac{(3.5\text{ mol})}{(1.0\text{ atm})}\left(0.08206\text{ L.atm mol}^{-1}\text{.K}^{-1}\right)(298.2\text{ K})=85.6\text{ L}$$

The number of molecules per cubic centimeter is, therefore,

$$\frac{(3.5)(6.022 \times 10^{23})\text{ molecules}}{85.6 \times 10^3 \text{ cm}^3} = 2.46 \times 10^{19} \text{ molecules} \cdot \text{cm}^{-3}$$

Because the number of molecules per cubic centimeter is significantly less at 1.0 atm and 298 K the average distance between molecules is greater, and the percentage error made using the ideal gas law is smaller at 298 K and 1.0 atm than it is at 320 K and 16 atm.

76. (a) The total translational kinetic energy for a sample of n moles of gas is $\frac{3}{2}nRT$. In this case, $n = 0.625$ mol and $T = 298$ K. Since monatomic gases possess only translational kinetic energy,

$$\text{total kinetic energy} = \frac{3}{2}(0.625 \text{ mol})\ (R)\ (298 \text{ K})$$

By choosing the proper value of R, the total kinetic energy can be given in any desired unit. Thus

$$\text{Total kinetic energy} = \frac{3}{2}(0.625 \text{ mol})\left(8.3145 \text{ mol}^{-1} \cdot \text{K}\right)(298 \text{ K}) = 2.23 \times 10^3 \text{ J} = 2.32 \text{ kJ}$$

(b) The average translational kinetic energy per He atom is independent of the size of the sample of gas, and is

$$E_K = \frac{3}{2}kT = \frac{3}{2}(1.38066 \times 10^{-23} \text{ J} \cdot \text{atom}^{-1} \text{ K}^{-1})(298 \text{ K}) = 6.175 \times 10^{-21} \text{ J} \cdot \text{atm}^{-1}$$

(c) The kinetic energy of the sample at 298 K is $\frac{3}{2}nR$ (298 K), whereas after heating it to 398 K the kinetic energy is $\frac{3}{2}nR$ (298 K). It is the difference between these two values that must be supplied in order to raise the temperature to 398 K. Thus the amount of energy required is

$$\frac{3}{2}nR(398 - 298) = \frac{3}{2}nR(100)$$

$$\begin{pmatrix}\text{energy to heat sample}\\ \text{from 298 to 398 K}\end{pmatrix} = \frac{3}{2}(0.625 \text{ mol})\left(8.3145 \text{ mol}^{-1} \cdot \text{K}^{-1}\right)(100 \text{ K}) = 779 \text{ J}$$

Note that this method of calculation takes less time and less numerical calculation than if you calculate the kinetic energy at 298 K and also at 298 K, and then subtract.

77. The mean speed, u, is obtained simply by adding the four speed s and dividing by 4:

$$u = \frac{3+4+7+10}{4} \text{ m} \cdot \text{s}^{-1} = 6 \text{ m} \cdot \text{s}^{-1}$$

The mean-square speed is obtained by squaring each speed first, and then averaging:

$$u^2 = \frac{3^2+4^2+7^2+10^2}{4} = \frac{9+16+49+100}{4} = \frac{174}{4} = 43.5\ \text{m}^2 \cdot \text{s}^{-2}$$

Hence,

$$u_{rms} = u^{½} = (43.5)^{1/2}\ \text{m} \cdot \text{s}^{-1} = 6.6\ \text{m} \cdot \text{s}^{-1}$$

Note that while u_{rms} and u are not identical, they ar fairly close to one another. For any sample of a gas, u_{rms} is always greater than u.

78. We calculate u_{rms}. Here, $u_{rms} = (3RT/M)^{1/2}$. In SI units, the molecular weight, M must be in kilograms per mole, and $R = 8.31\ \text{J} \cdot \text{mol}^{-1}\ \text{K}^{-1}$. For H_2, M = 2.016×10^{-3} kg· mol^{-1}. For UF_6, M = 0.35202 kg · mol^{-1}. Thus, for H_2,

$$u_{rms} = \left[\frac{(3)(8.31)(353)}{2.016 \times 10^{-3}}\right]^{1/2} = 2.09 \times 10^3\ \text{m} \cdot \text{s}^{-1} = 2.09\ \text{km} \cdot \text{s}^{-1}$$

For UF_6,

$$u_{rms} = \left[\frac{(3)(8.3145)(353)}{0.35202}\right] = 158.2\ \text{m} \cdot \text{s}^{-1} = 0.1582\ \text{km} \cdot \text{s}^{-1}$$

79. We have,

$$\frac{\text{time for gas 2}}{\text{time for gas 1}} = \frac{\text{rate for gas 1}}{\text{rate for gas 2}} = \left(\frac{M_2}{M_1}\right)^{1/2}$$

The molecular weight of nitrogen is 28.013 g · mol^{-1}. Expressing the times in minutes we obtain $t_1 = \frac{84}{60} = 1.4$ and $t_2 = \frac{103}{60} \approx 1.72$

$$\frac{1.72}{1.4} = \left(\frac{M_2}{28.013}\right)^{1/2}$$

Squaring both sides yields

$$1.509 = \frac{M_2}{28.013}$$

and M_2 = (28.013 g · mol^{-1}) (1.509) = 42.3 g · mol^{-1}. This is only approximately correct due to experimental uncertainties in measuring the times of gas flow. Since the formula weight of CH_2 is 14, the molecular weight of $(CH_2)_n$ is 14n, and 14n ~ 42.3. Therefore n is exactly 3, and the molecular formula is C_3H_6. The exact molecular weight is 42.08 g · mol^{-1}.

80. I (a) It is much easier to make use of the ratio. The 3R factor cancels when you take the ratio. Since both gases are at the same temperature, so we have,

$$\frac{u_{rms}\ \text{for}\ SO_2}{u_{rms}\ \text{for}\ O_2} = \left(\frac{M_{O_2}}{M_{SO_2}}\right)^{1/2} = \left(\frac{32.00}{64.06}\right)^{1/2} = 0.7068$$

Thus,

$$u_{rms} \text{ for } SO_2 \text{ at } 273 \text{ K} = (0.7068)\ (u_{rms} \text{ for } O_2 \text{ at } 273 \text{ K})$$
$$= (0.7068)\ (4.61 \times 10^2) = 3.26 \times 10^2 \text{ m} \cdot \text{s}^{-1}$$

(b) $$\frac{u_{rms} \text{ for } SO_2 \text{ at } 373 \text{ K}}{u_{rms} \text{ for } SO_2 \text{ at } 272\ K} = \left(\frac{373}{273}\right)^{1/2} = 1.169$$

II (a) $$P_{ideal} = \frac{nRT}{V} = \frac{(10)\,(0.08206)\,(300)}{4.86} = 50.7 \text{ atm}$$

To evaluate P using vander Waals' equation, we obtain a and b for ethane from Table. Substituting values in the vanderwaal's equation

We have

$$P_{(v \text{ der } W)} = \frac{(10)(0.08206)\,(300)}{4.86-(10)\,(0.0638)} - \frac{(100)\,(5.489)}{(4.86)^2} = 58.3 - 23.2$$

$$= 35.1 \text{ atm}$$

(b) (1) Percentage error $= \left(\frac{P_{ideal} - P_{real}}{P_{real}}\right) \times 100 = \left(\frac{50.7-34.0}{34.0}\right) \times 100 = 49.1\%$

(2) percentage error $= \left(\frac{P_{V \text{ der } W} - P_{real}}{P_{real}}\right) \times 100 = \left(\frac{35.1-34.0}{34.0}\right) \times 100 = 3.24\%$

At this relatively high pressure, the vander Waals' equation closer to reality than is the ideal gas law.

81. Using Avogadro's law, we can find the number of moles of gas in the sample, since 1 mol of any gas occupies 22.4 L at STP. The volume of the gas should be expressed in liters.

$$\text{No. mol gas} = (0.7056 \text{ L}) \left(\frac{1 \text{ mol}}{22.4 \text{ L}}\right) = 0.0315 \text{mol}$$

$$\therefore \text{ molecular weight} = \frac{\text{mass in g}}{\text{No. mol}} = \frac{2.640 \text{ g}}{0.0315 \text{ mol}} = 83.8 \text{ g} \cdot \text{mol}^{-1}$$

(*Note:* Do *not* use 22.4 L as the molar volume at temperatures pressures other than 273 K and 1 atm.)

82. To find the molecular formula we need the molar ratio of boron atoms to hydrogen atoms. Let us first calculate the number of moles of B atoms in the sample of the unknown barone. This must be the same as the number of moles of B in the boron oxide, by the law of conservation of mass. This formula weight of B_2O_3 is 69.62 g · mol^{-1}.

$$\text{No. of mol } B_2O_3 = \frac{5.9316 \text{ g}}{69.62 \text{ g} \cdot \text{mol}^{-1}} = 8.520 \times 10^{-2} \text{ mol}$$

$$\therefore \quad \text{No. mol } B = 2(\text{No. of mol } B_2O_3) = 1.704 \times 10^{-1} \text{ mol}$$

$$\therefore \quad \text{mass of } B = (1.704 \times 10^{-1} \text{ mol})\ (10.81 \text{ g} \cdot \text{mol}^{-1}) = 1.842 \text{ g of } B$$

The number of moles of hydrogen in the barone sample, can be calculated if we know the mass of hydrogen in the sample.

Now, mass of barone sample = mass of B + mass of H

The mass of the sample is obtained by subtracting the mass of the evacuated bulb from the mass of the gas-filled bulb:

Mass of borane sample = 75.5398 g – 73.2684 g = 2.2714 g

Hence, Mass of H= 2.2714 g – 1.842 g = 0.429 g of H atoms

The number of moles of H atoms is:

$$\text{No. mol H} = \frac{0.429 \text{ g}}{1.008 \text{ g} \cdot \text{mol}^{-1}} = 0.426 \text{ mol}$$

The molar ratio of B : H in the unknown borane is therefore

$$\frac{\text{mol B}}{\text{mol H}} = \frac{0.1704}{0.426} = \frac{1}{2.5} = \frac{2}{5}$$

and the simplest formula is B_2H_5. The empirical formula is $(B_2H_5)_n$.

We can determine the value of n in the empirical formula by utilizing Avogadro's law at 273 K and 1-atm pressure The molar volume of any gas at STP is 22.4 L. This 2.2414-g sample of the gaseous borane occupies a volume of 952 mL at STP. Therefore the number of moles of gas in the sample is

$$\text{No. of mol borane} = \frac{0.952 \text{ L}}{22.4 \text{ L} \cdot \text{mol}^{-1}} = 4.25 \times 10^{-2} \text{ mol}$$

$$\text{Thus, molecular weight of borane} = \frac{2.2714 \text{ g}}{4.25 \times 10^{-2} \text{ mol}} = 53.4 \text{ g} \cdot \text{mol}^{-1}$$

The formula weight of B_2H_5 is 26.66, and therefore the molecular weight of a gas with empirical formula $(B_2H_5)_n$ is 26.66n, where n must be an integer. The only integral value for which 26.66n is close to 53.4 is n = 2. The correct molecular formula of this compound is B_4H_{10}.

Note particularly that the exact molecula: weight of B_4H_{10} is 53.32 g · mol^{-1}. Using Avogardo's law we obtained 53.4 g · mol^{-1} as the molecular weight of the gas, which is very close to the true value, but is not exactly correct. But it is not necessary to know the molecular weight exactly in order to determine the correct molecular formula from the empirical formula.

83. (a) Since 1 atm supports a column of mercury exactly 760 mm high,

$$(680.2 \text{ mmHg}) \left(\frac{1 \text{ atm}}{760 \text{ mm Hg}}\right) = 0.8950 \text{ atm}$$

(b) To convert in to pascals, we multiply pressure in atm. by 1.01325 × 10^5 N.m^{-2}.

$$(0.895 \text{ atm}) \left(\frac{1.01325 \times 10^5 \text{ N} \cdot \text{m}^{-2}}{1 \text{ atm}}\right) = 9.069 \times 10^4 \text{ N} \cdot \text{m}^{-2} = 9.069 \times 10^4 \text{ Pa}$$

84. One atmosphere supports a column of mercury 76.0 cm high; the density of mercury is 13.6 $g \cdot cm^{-3}$. If we let h = height of he column of water supported by 1 atm we obtain

$$(13.6\ g \cdot cm^{-3})\ (76\ cm) \begin{pmatrix} \text{acceleration} \\ \text{of gravity} \end{pmatrix} = (0.997\ g \cdot cm^{-3})\ (h) \begin{pmatrix} \text{acceleration} \\ \text{of gravity} \end{pmatrix}$$

Each side of this equation is equal to atm. The acceleration of gravity cancels out of the calculation, and we obtain.

$$h = \frac{(13.6)(76)}{(0.997)} = 1.037 \times 10^3\ cm = 10.4\ m$$

85. Since both the pressure and the number of moles of gas are constant, $V_1/T_1 = V_2/T_2$,

$\Rightarrow\ V_2 = T_2(V_1/T_1)$

so that by substituting the given values we have,

$$V_2 = \left(\frac{473.15\ K}{298.15\ K}\right)\ (4.0\ mL) = 634.8\ mL$$

86. In this problem, both the number of moles of N_2 gas and the pressure remain constant. If we rearrange the ideal gas law so that all the constant terms are on side of the equation, we have $nR/P = V/T$. In the initial state (state 1) the gas has a volume $V_1 = 58.7$ mL and a temperature $T_1 = 273.15$ K. When the gas is immersed in boiling isopropyl alcohal (state 2), its volume $V_2 = 76.4$ mL, and its temperature, T_2, is to be determined. We therefore have

$$\frac{nR}{P} = \text{constant} = \frac{V_1}{T_1} = \frac{V_2}{T_2} \qquad \text{(Charles' Law)}$$

Rearranging this to solve for T_2, we obtain

$$T_2 = \left(\frac{V_2}{V_1}\right) T_1 = \frac{76.4}{58.7}\ (273.15\ K) = 355.5\ K$$

The boiling point of isopropyl alcohal is 355.5 K at 1 atm.

87. In this problem

$$\frac{P_1 V_1}{T_1} = \frac{P_2 V_2}{T_2}$$

Here, $P_1 = 751$ mmHg, $P_2 = (1.4\ atm)\ (760\ mmHg \cdot atm^{-1}) = 1064$ mmHg
$T_1 = 298$ K $\quad T_2 = 308$ K

Solving the equation given above for V_2 we obtain

$$V_2 = \left(\frac{P_1}{P_2}\right)\left(\frac{T_2}{T_1}\right) \quad \therefore\ V_2 = \frac{(751\ mmHg)(308.2\ K)}{(1064\ mmHg)\ (298.2\ K)}\ (364\ mL) = 252\ mL$$

88. In this problem we must focus on the gas inside the bulb. The quantities that are the same at the beginning and the end of the experiment are the volume of the bulb and the pressure inside the bulb. Heating the gas to 336 K increases the pressure in the bulb above 1 atm, so that when the stopcock is opened, gas escapes from the bulb until the pressure in the bulb is again equal to the external pressure of 1 atm.

$$\frac{PV}{R} = \text{constant} = n_1 T_1 = n_2 T_2$$

where n_2 is the number of moles of gas remaining inside the bulb after the pressure has returned to 1 atm at 340 K.

$$\therefore \quad n_1 = \frac{20\text{ g}}{44.01\text{ g}\cdot\text{mol}^{-1}} = 0.454\text{ mol};\; n_2 = ?;\; T_1 = 296\text{ K};\quad T_2 = 336\text{ K}$$

Therefore,

$$(0.454\text{ mol})\,(296\text{ K}) = n_2(336\text{ K})$$

$$\text{or,}\quad n_2 = \frac{(0.454)\,(296\text{ K})}{(336)} = 0.4\text{ mol}$$

Since 0.4 mol remains in the bulb after the stopcock has been opened, 0.454 mol of CO_2, must have escaped. Therefore the mass of CO_2 that escaped is (0.054 mol) (44.01 g · mol^{-1}) = 2.4 g.

89. In this problem, the temperature and the total number of moles of H_2 both remain constant throughout, but the sample is divided into two portions. If you do not know the numerical value of a quantity, assign it a symbol. Then write the ideal gas equations using both symbols and numbers. After you look at the equations you have written, a method of solving them algebraically will be more apparent. Let V = volume of the bulb (mL) n = total number of moles of H_2

Then, for the initial condition, 756 $V = nRT$. There is no reason to convert the 756-mmHg pressure into atmospheres, because no numerical value of R has been selected. The only restriction on the unit of pressure used is that it must be consistent with the pressure unit used in the value of R.

We don't know how big a sample was transferred to the 40.0-mL flask. Let n_1 = the number of moles transferred to the 40.0-mL flask. Then, for the gas in that flask (760) (40.0) = n_1RT. In this equation, we have written the 1-atm pressure as 760 mmHg, so that we use the same pressure unit in all the equations we write.

Let n_2 = the number of moles of H_2 remaining in the original bulb. Then, for the n_2 mol of H_2 after the transfer, $635V = n_2RT$.

Since the total number of moles of H_2, $n = n_1 + n_2$, then $nRT = n_1RT + n_2RT$. We have an expression for each term in this equation:

$$nRT = 756V = (760)\,(40.0) + 625V$$

Solving for V we obtain

$$(756V - 625V) = 131V = (760)\,(40.0)\text{ mmHG}\cdot\text{mL}$$

Hence,

$$V = \frac{3.04\times10^4\text{ mL}}{131} = 232\text{ mL}$$

Note that to calculate V we did not need to use any numerical value of R, or know the value of T.

90. We will need the number of moles of butane in order to use the ideal gas law. The molecular weight of butane, C_4H_{10}, is 58.12 g · mol^{-1}. Hence,

$$\text{No. mol butane} = \frac{3.728\text{ g}}{58.12\text{ g}\cdot\text{mol}^{-1}} = 6.414\times10^{-2}\text{ mol}$$

We must decide what value of R to use. We will do the calculation with two different choices for R, to show that as long as the units of all quantities are consistent with the units of R, it makes no difference which value of R is used.

Choice 1. Use $R = 62.36\ \text{L} \cdot \text{mmHg} \cdot \text{mol}^{-1}\ \text{K}^{-1}$. Then the volume must be expressed in liters, and the pressure calculated will be in millimeters of mercury.

$$P = \frac{nRT}{V} = \frac{(6.414 \times 10^{-2}\ \text{mol})}{(0.489\ \text{L})}\left(62.36\ \frac{\text{L} \cdot \text{mmHg}}{\text{mol} \cdot \text{K}}\right)(298\ \text{K}) = 2439\ \text{mmHg}$$

$$= (2.44 \times 10^3\ \text{mmHg})\left(\frac{1\ \text{atm}}{760\ \text{mmHg}}\right) = 3.21\ \text{atm}$$

Choice 2. Use $R = 0.082\ \text{L} \cdot \text{atm} \cdot \text{mol}^{-1}\ \text{K}^{-1}$. Then the volume must be in liters and the pressure calculated will be in atmospheres.

$$P = \frac{nRT}{V} = \frac{(6.414 \times 10^{-2}\ \text{mol})}{(0.489\ \text{L})}\left(0.082\ \frac{\text{L} \cdot \text{atm}}{\text{mol} \cdot \text{K}}\right)(298\ \text{K}) = 3.21\ \text{atm}$$

91. (a) The percentage of N in pyridine is obtained by difference:

$$100 - 75.92 - 6.37 = 17.71\%\ \text{N}.$$

Consider exactly 100 g of pyridine. This sample contains 75.92 g of C, 6.37 g of H, and 17.71 g of N.

$$\text{no. mol C} = \frac{75.92\ \text{g}}{12.011\ \text{g} \cdot \text{mol}^{-1}} = 6.321\ \text{mol of C}$$

$$\text{no. mol H} = \frac{6.37\ \text{g}}{1.008\ \text{g} \cdot \text{mol}^{-1}} = 6.32\ \text{mol of H}$$

$$\text{no. mol N} = \frac{17.71\ \text{g}}{14.007\ \text{g} \cdot \text{mol}^{-1}} = 1.264\ \text{mol of N}$$

The molar ratio C : N : N = 6.321 : 6.32 : 1.264 = 5 : 5 : 1. Thus the simplest formula of pyridine is C_5H_5N, and the empirical formula is $(C_5H_5N)_n$, where n is an integer. An approximate value of the molecular weight is obtained

$$M = \left(2.12\ \frac{\text{g}}{\text{L}}\right)\left(8.2 \times 10^{-2}\ \frac{\text{L} \cdot \text{atm}}{\text{mol} \cdot \text{K}}\right)\frac{(383\ \text{K})}{(630/760\ \text{atm})} = 8.04\ \text{g} \cdot \text{mol}^{-1}$$

The pressure of 630 mmHg was converted to atmospheres in order to be consistent with the unit of pressure in the value of R chosen. The formula weight of C_5H_5N is 79.10 g · mol^{-1}. The only value n for which $79.10n \sim 80.4$ is $n = 1$. Thus the molecular formula of pyridine is C_5H_5N, and the correct molecular weight is 79.10 g · mol^{-1}.

(b) The error made is given by

$$\text{pecentage error} = \frac{(\text{calculated value} - \text{true value})}{(\text{true value})} \times 100$$

The calculated value of the molecular weight using the ideal gas equation was 80.4 g · mol^{-1}. The percentage error made is therefore

$$\text{Pecentage error} = \frac{(80.4 - 79.1)}{79.1} \times 100 = 1.6\%$$

The error of 1.6% is due to the fact that pyridine is not an ideal gas, particularly close to the temperature and pressure where it condenses to a liquid. The normal boiling point of pyridine is 388 K at 1-atm pressure.

92. (a) The partial pressure of CO_2 in the container is given by $P_{CO_2} = n_{CO_2}(RT/V)$. We must therefore calculate the number of moles of CO_2 formed in the combustion reaction. The equation for the combustion reaction tells us that the number of moles of CO_2 formed is seven times the number of moles of benzoic acid burned. The molecular weight of C_6H_5COOH is 122.12 g · mol^{-1}. The number of moles of benzoic acid used in the combustion reaction is

$$\frac{1.5388 \text{ g}}{122.12 \text{ g} \cdot \text{mol}^{-1}} = 1.26 \times 10^{-2} \text{ mol of } C_6H_5COOH$$

Hence,
No. of mol CO_2 formed = $7(1.26 \times 10^{-2}) = 8.820 \times 10^{-2}$ mol of CO_2
Therefore,

$$P_{CO_2} = \left(2.12 \frac{\text{g}}{\text{L}}\right)\left(8.2 \times 10^{-2} \frac{\text{L} \cdot \text{atm}}{\text{mol} \cdot \text{K}}\right) \frac{(383 \text{ K})}{(630/760 \text{ atm})} \quad (298 \text{ K})$$

$$= 0.36 \text{ atm} = 273 \text{ mmHg}$$

(b) The contents of the container at the end of the reaction are CO_2, O_2, water vapor, and liquid water. There will always be water vapor in the gas phase in a closed container that has any liquid water in it. The vapor pressure of water depends only on the temperature. From Table P_{H_2O} = 23.8 mmHg at 298 K. Dalton's law, applied to mixture of three gases is

$$P_{\text{total}} = P_{CO_2} + P_{H_2O} + P_{O_2}$$

$$814 \text{ mmHg} = 273.3 \text{ mmHg} + 23.8 \text{ mmHg} + P_{O_2}$$

Therefore P_{O_2} = 517 mmHg. We can calculate the number of moles of excess oxygen in the container from the partial pressure of O_2 as follows:

$$n_{O_2} = \frac{P_{O_2} \cdot V}{RT} = \frac{(517 \text{ mmHg})(6 \text{ L})}{(62.36 \text{ L} \cdot \text{mmHg} \cdot \text{mol}^{-1} \text{ K}^{-1})(298 \text{ K})} = 0.167 \text{ mol of } O_2$$

93. (a) The contents of bulb at 600 K are the threee gases He, NH_3, and HCl. Using Dalton's law, the total pressure is

$$P_{\text{total}} = 908 \text{ mmHg} = P_{He} + P_{NH_3} + P_{HCl}$$

We can calculate the pressure of the He at 600 K, because we know its pressure at 300 K. The number of moles of He and the volume of the bulb remain constant, so that

$$P_2 = P_1(T_2/T_1) = \frac{(114 \text{ mmHg})(600 \text{ K})}{(300 \text{ K})} = 228 \text{ mmHg} = P_{He} \text{ at } 600 \text{ K}$$

The equation for the reaction tells us that NH_3 and HCl are formed in a 1 : 1 molar ratio. Since the number of moles NH_3 and HCl are equal, and they are at the same temperature and occupy the same volume, both gases must be at the same pressure. Let

$$x = P_{NH_3} = P_{HCl}$$

Then Dalton's law for the total pressure becomes

$$908 \text{ mmHg} = 228 \text{ mmHg} + 2x \quad \text{so that} \quad x = P_{HCl} = 340 \text{ mmHg}$$

(b) Since all the NH_4Cl(s) was converted to HCl(g) and NH_3(g), if we calculate the number of moles of HCl(g) formed, we will know the number of moles of solid NH_4Cl originally present. We do not know the volume of the bulb, so we cannot use the ideal gas law to calculate the number of moles of HCl(g). We do know the number of moles of He present, and it temperature and pressure. We can calculate the volume of the bulb from the data we have about the helium, but it is not necessary to do so. Of the terms in the ideal gas law, V and T are the same for all the gases, so write the term V/RT, which is constant, for both He and HCl
$V/RT = n_{He}/P_{He} = n_{HCl}/P_{HCl}$ Solve this for the number of moles of *HCl* $n_{HCl} = n_{He}$ $(P_{HCl}/P_{He}$
Thus

$$n_{HCl} = (0.01 \text{ mol})(340 \text{ mmHg})/(228 \text{ mmHg}) = 0.0149 \text{ mol}$$

There was, therefore, 0.0149 mol of NH_4Cl(s) in the bulb originally. The formula weight of NH_4Cl is 53.49 g · mol^{-1}. Hence the mass of solid NH_4Cl in the bulb was (0.0149 mol) (53.49 g · mol^{-1}) = 0.798 g of NH_4Cl.

94. Let n_1= No. mol C_3H_8 in the mixture
n_2= No. mol CH_4 in the mixture

Then the mole fraction of C_3H_8, X_1, is given by $X_1 = \frac{n_1}{n_1 + n_2}$. Writing the ideal gas equation for the original mixture of the two gases we have,

$$P_{total} = 320 \text{ mmHg} = n_{total}(RT/V) = (n_1 + n_2)(RT/V)$$

The pressure of the collected CO_2 can be expressed as:

$$P_{CO_2} = n_{CO_2}(RT/V)$$

Now, what is the relation between n_1, n_2, and the number of moles of CO_2 formed? The number of moles of CO_2 produced by the combustion is determined by the law of conservation of C atoms. Each mole of C_3H_8, after reaction with O_2, yields 3 mol of CO_2. Since we started with n_1 mol of C_3H_8, the combustion produces $3n_1$ mol of CO_2. Similarly, each mole of CH_4, when burned in O_2, produces 1 mol of CO_2, so that burning n_2 mol of CH_4 yields n_2 mol of CO_2. Hence the total number of moles of CO_2 produced by burning n_1 mol of C_3H_8 and n_2 mol of CH_4 is

$$n_{CO_2} = 3n_1 + n_2$$

The equation for the pressure of CO_2 is,

$$P_{CO_2} = 448 \text{ mmHg} = n_{CO_2}(RT/V) = (3n_1 + n_2)\ (RT/V)$$

We now have two equations involving the unknowns n_1, n_2, T and V:

$$320 \text{ mmHg} = (n_1 + n_2)\ (RT/V)$$

and

$$448 \text{ mmHg} = (3n_1 + n_2)\ (RT/V)$$

By dividing one of these equations by the other, the factor (RT/V) cancels out, and we obtain

$$\frac{320}{448} = \frac{n_1 + n_2}{3n_1 + n_2} = \frac{5}{7}$$

Our equation has therefore been simplified to $\frac{5}{7} = (n_1 + n_2)/(3n_1 + n_2)$. Multiplying this fraction out yields $15n_1 + 5n_2 = 7n_1 + 7n_2$,

$\therefore\ 8n_1 = 2n_2$, or $4n_1 = n_2$. Substituting $4n_1$ for n_2 in the definition of the mole fraction, X_1:

$$X_1 = \frac{n_1}{(n_1 + n_2)} = \frac{n_1}{(n_1 + 4n_1)} = \frac{n_1}{5n_1} = \frac{1}{5}$$

The mole fraction of propane in the original mixture is $\frac{1}{5}$ or 0.2 (20.0 mol%).

95. Since 1 mole of NO contains 14.0 g of N. There is 1 mole (1 gram-atom) of N in 1 mole of NO. One mole of N has a mass of 14 g. 100 liters of NO is 100/22.4 moles of NO and 100/22.4 moles of NO contain 100/22.4 moles of N.

$$\frac{100}{22.4} \text{ moles of N} \times \frac{14 \text{ g of N}}{1 \text{ mole of N}} = 62.5 \text{ g of N}$$

The solution, in one operation, is

$$\frac{100 \text{ liters of NO}}{22.4 \text{ liters of NO/mole of NO}} \times \frac{1 \text{ mole of N}}{1 \text{ mole of NO}} \times \frac{14 \text{ g of N}}{1 \text{ mole of N}} = 62.5 \text{ g of N}$$

Note that all units except "g of N" cancel.

96. The molecular weight of a gas is the mass in grams of 1 mole of the gas. One mole of a gas occupies a volume of 22.4 liters at STP. Therefore, the molecular weight of a gas is the mass in grams of 22.4 liters at STP.

Since 2 liters have a mass of 5.72 g, the mass of 1 liter will be 5.71/2 g and the mass of 22.4 liters will be 22.4 × 5.71/2 g, Thus, the calculation, in one operation, is:

$$\frac{5.71 \text{ g}}{2 \text{ liters}} \times \frac{22.4 \text{ liters}}{1 \text{ mole}} = 64 \text{ g/mole}$$

Since the gas law are not exact, the molecular weight calculated in this manner is not exact.

97. We will solve by substituting in the equation, $PV = nRT$.

$$n = \text{moles of } CH_4 = \frac{45\text{ g}}{16\text{ g/moles}} = \frac{45}{16}\text{ mole}$$

$P = 800$ mm; $T = 300$K

$$V = \frac{nRT}{P} = \frac{45}{16}\text{ mole} \times \frac{62.3\dfrac{\text{liters} \times \text{mm}}{\text{mole} \times \text{K}} \times 300\text{ K}}{800\text{ mm}}$$

$= 65.6$ liters

Note that moles, mm, and K cancel.

98. $PV = nRT$

$$n = \frac{PV}{RT} = \frac{4\text{ atm} \times 2\text{ liters}}{0.082\dfrac{\text{liters} \times \text{atm}}{\text{mole} \times \text{K}} \times 298\text{ K}} = 0.33\text{ mole}$$

99. $PV = \dfrac{\text{g}}{\text{MW}}\,RT$

$$P = \frac{gRT}{MWV} = \frac{26\text{ g}}{\dfrac{4\text{ g}}{1\text{ mole}}} \times 0.082\frac{\text{liters} \times \text{atm}}{\text{mole} \times \text{K}} \times \frac{473\text{ K}}{3.24\text{ liters}} = 78\text{ atm}$$

Note that g, liters, K, and moles cancel.

100. $PV = \dfrac{gRT}{\text{MW}}$

$$g = \frac{PV\text{MW}}{RT} = \frac{2\text{ atm} \times 60\text{ liter} \times 34.1\text{ g/mole}}{0.082\dfrac{\text{liters} \times \text{atm}}{\text{mole} \times \text{K}} \times 293\text{ K}} = 170\text{ g}$$

101. $\text{MW} = \dfrac{gRT}{PV} = \dfrac{32.4\text{ g} \times 0.082\dfrac{\text{liters} \times \text{atm}}{\text{mole} \times \text{K}} \times 298\text{ K}}{1.5\text{ atm} \times 12\text{ liters}}$

$= 44.0$ g/mole

102. Here pressure $= \dfrac{740}{760}$ atm; Temperature = 323k.

Number of moles of acetylene (n) $= \dfrac{5}{26}$

And from the expression $PV = nRT$

We have $V = \dfrac{nRT}{P}$

So putting he above values in this expression we get

$$V = \frac{\left(\frac{5}{26}\right)(0.082)(323)}{\frac{740}{760}} = 5.231$$

103. The density of a gas is the mass of a unit volume of the gas. The volume normally selected is either the molar volume, 22.4 liters, or 1 liter. One mole of C_2H_2 at STP weighs 26.0 g and occupies 22.4 liters. Therefore, the density is 26.0 g per 22.4 liters or 1.16 g/liter.

104. We have, $P_T = P_{O_2} + P_{N_2}$, or $2 = P_{O_2} + 1.5$, $\therefore$ $P_{O_2} = 0.5$ atm. When V and T are constant, the number of moles of a gas is directly proportional to its partial pressure. Therefore,

$$\frac{\text{moles of } O_2}{\text{moles of } N_2} = \frac{0.50 \text{ atm}}{1.5 \text{ atm}} \quad \text{Mole\% of } O_2 = \frac{0.50}{2.0} \times 100\% = 25\%$$

To convert mole percent to mass percent convert moles of O_2 and N_2 to grams. Mass % of $O_2 = \frac{0.50 \times 32}{(0.50 \times 32) + (1.5 \times 28)} \times 100\% = 28\%$.

105. Since the gas is to be collected over water, the vapor pressure of water at 305 K must be subtracted from the barometric pressure of 746 mm.
But, vapour pressure of water at 305 K = 35.4 mm
$\therefore$ actual new pressure = 746 – 35.4 = 710.6 mm
Using this pressure of 710.6 mm as the new pressure in place of the 746 mm, we can solve as.

$$\text{volume} = 600 \text{ cc} \times \frac{750 \text{ mm}}{710.6 \text{ mm}} \times \frac{305° \text{ K}}{298° \text{ K}} = 648 \text{ cc}$$

106. Let $X = g$ of CH_4 = g of CO
Moles of CH_4 = X/16. Moles of CO = X/28

$$\text{ratio} = \frac{\text{moles of } CH_4}{\text{moles of CO}} = \frac{X/16}{X/28} = \frac{28}{16}$$

and ,

$$\text{Mole fraction of } CH_4 = \frac{\text{moles of } CH_4}{\text{moles of } CH_4 + \text{moles of CO}} = \frac{28}{44} = 0.64$$

Note that, in a mixture of *equal grams* of two gases, the number of *moles* of the gases are in the inverse ratio of their molecular weights. Mole fractions and mole percents represent *mole ratios*.

107. The partial pressure of a gas is proportional to its mole fraction. Therefore, we must first and the number of moles of O_2 and CH_4 in the mixture.

$$\text{moles of } O_2 = \frac{50 \text{ g}}{32 \text{ g/mole}} = 1.56 \text{ moles of } O_2$$

$$\text{moles of } CH_4 = \frac{50 \text{ g}}{16 \text{ g/mole}} = 3.12 \text{ moles of } CH_4$$

$$\text{total moles} = 1.56 + 3.12 = 4.68 \text{ moles}$$

$$\text{mole fraction of } O_2 = \frac{1.56}{4.68}$$

$$\text{partial pressure of } O_2 = \text{mole fraction of } O_2 \times \text{total pressure}$$

$$= \frac{1.56}{4.68} \times 600 \text{ mm} = 200 \text{ mm}$$

108. $$\frac{\text{rate of } A}{\text{rate of } B} = \frac{\sqrt{\text{density of } B}}{\sqrt{\text{density of } A}}$$

Substituting in this formula, $$\frac{\text{rate of } A}{15 \text{ cm/10 sec}} = \frac{\sqrt{1}}{\sqrt{9}} = \frac{1}{3}$$

$$\text{rate of } A = \frac{15 \text{ cm/10 sec}}{3} = 5 \text{ cm in 10 sec}$$

109. $$\frac{\text{rate of } A}{\text{rate of } B} = \frac{\sqrt{\text{density of } B}}{\sqrt{\text{density of } A}}$$ (Let A = HBr and B = SO_2)

Since the density of a gas is directly proportional to its molecular weight we can write:

$$\frac{\text{rate of HBr}}{\text{rate of } SO_2} = \frac{\sqrt{\text{molecular weight of } SO_2}}{\sqrt{\text{molecular weight of HBr}}}$$

110. 1. $$\frac{4.6 \text{ cm Mg}}{1} \times \frac{1.0 \text{ m Mg}}{100 \text{ cm Mg}} \times \frac{0.816 \text{ g Mg}}{1.0 \text{ m Mg}} \doteq 0.0375 \text{ g Mg}$$

2. $$\frac{0.0375 \text{ g.Mg}}{1} \times \frac{1 \text{ mol Mg}}{24.3 \text{ g.Mg}} = 0.00154 \text{ mol Hg}$$

3. $P_{H_2} = P_{tot} - P_{H_2O} = 743 \text{ mm Hg} - 26 \text{ mm Hg} = 717 \text{ mm Hg}$

4. $$V_2 = V_1 \frac{P_1}{P_2} \times \frac{T_2}{T_1} = 40.44 \text{ cm}^3 \times \frac{717 \text{ mm Hg}}{760 \text{ mm Hg}} \times \frac{273 \text{ K}}{299.3} = 34.8 \text{ cm}^3$$

5. $Mg_{(s)} + 2H^+_{(aq)} \rightarrow H_{2(g)} + Mg^{2+}_{(aq)}$

6. $$\frac{0.00154 \text{ mol.Mg}}{1} \times \frac{1.0 \text{ mol } H_2}{1.0 \text{ mol.Mg}} = 0.00154 \text{ mol } H_{2(g)}$$

7. $V_{STP} = \dfrac{34.8\,\text{cm}^3\,H_2}{0.00154\,\text{mol}\,H_2} \times \dfrac{1.0\,\text{L}}{1000\,\text{cm}^3} = 22.6\ \text{L} \cdot \text{mol}^{-1}$

8. $\%\ \text{error} = \dfrac{\text{obs} - \text{exp}}{\text{exp}} \times 100\% = \dfrac{22.6\,\text{L.mol}^{-1} - 22.4\,\text{L.mol}^{-1}}{22.4\,\text{L.mol}^{-1}} \times 100\% = 0.9\%$

9. Here, PV = nRT

$\therefore \quad n = \dfrac{P \cdot V}{R \cdot T} = \dfrac{717/760\ \text{atm} \times 0.04044\ \text{L mol}^{-1}\text{K}}{0.082\ \text{L.atm} \times 299\,\text{K}} = 1.55 \times 10^{-3}\ \text{mol}\ H_2$

111. From the Gas Law

$$PV = nRT$$

If volume and temperature are constant then

$$P \propto n$$

or $$\frac{P_1}{P_2} = \frac{n_1}{n_2}$$

Let us suppose that bulb originally contained one mole of oxygen,

i.e., $$n = 1$$

Then $$n_2 = \frac{P_2}{P_1} = \frac{1500}{2000} = \frac{3}{4}\ \text{moles}$$

$\therefore$ gas leaked $= 1 - \dfrac{3}{4}$ mole $= \dfrac{1}{4}$ mole

Rate of leakage$= \dfrac{1}{4 \times 47}$ moles/minute (1)

From Graham's Law of diffusion.

$$\frac{r_{O_2}}{r_x} = \sqrt{\frac{M_x}{M_{O_2}}} = \frac{79}{32}$$

where r_x and M_x are rate of diffusion and molecular weight of the other gas

Substituting the value of r_{O_2} from (1)

$$r_x = \frac{1}{4 \times 47} \times \sqrt{\frac{32}{79}}$$

[Assuming that one mole of each gas is present at start]

Number of moles of gas (x) leaked in 74 minutes $= \dfrac{74}{4 \times 47}\sqrt{\dfrac{32}{79}}$

Number of moles of (X) left

$$= 1 - \frac{74}{4 \times 47}\sqrt{\frac{32}{79}}\ \text{mole} = 1 - 0.25 = 0.75\ \text{moles}$$

Number of moles O_2 leaked in 74 minutes

$$= \frac{74}{4 \times 47} = 0.40 \text{ mole}$$

Number of moles of O_2 left = 0.6 mole
Molar ratio of the two gases

$$\frac{n_{O_2}}{n_z} = \frac{0.6}{0.75} = \frac{4}{5}$$

Thus molar ratio of oxygen and other gas is 4 : 5.

112. From ideal gas equation

$$PV = nRT$$

$$n = 0.0409$$

Thus Energy required to dissociate = 436 × 0.0409 = 17.83 kJ
Energy required to exite electron from G.S to E.S

$$= \left(1 - \frac{1}{4}\right) 13.6 \text{ eV} = 1.6 \times 10^{-21} \text{ kJ}$$

Thus total E required = $17.83 + 1.6 \times 10^{-21} \times 0.0409 \times 2 \times 6.02 \times 10^{23} = 98.1$ kJ

113. (a) Substituting the values we get, (i) We have, $\frac{r_1}{r_{oxygen}} = \sqrt{M_{oxygen}/M_1}$

$$\frac{1.33}{1} = \sqrt{32/M_1}$$

or $$1.33 \times 1.33 = \frac{32}{M_1}$$

∴ $$M_1 = \frac{32}{1.33 \times 1.33} = 18.09$$

(ii) Molar volume $\nabla = \frac{\text{melecular weight}}{\text{Molar density}} = \frac{18.09 \times 10^{-3}}{0.36} = 50.25 \times 10^{-3}\ m^3$

(iii) Compression factor $Z = \frac{PV}{RT} = \frac{1.01325 \times 10^5 \times 50.25 \times 10^{-3}}{8.314 \times 500} = 1.224$

(iv) Repulsive forces dominate since the actual density is less than the given density. If it was ideal.

(b) Average translational kinetic energy

= 3/2kT. (Where k = Boltzmann constant = 1.381×10^{-23} JK^{-1})

$$= \frac{3}{2} \times 1.381 \times 10^{-23} \times 1000 \text{ J} = 2.07 \times 10^{-20} \text{ J}$$

114. From Rault's law,

$$p_1 = x_1/p_1^*$$

Hence, $$x_1 = \frac{p_1}{p_1^*} = \frac{750}{760} = 0.9868 = 0.0132$$

$$m = \frac{n_2}{n_1 M_1} = \frac{0.0132 \text{ mol}}{\left(0.9868 \times \frac{18}{1000}\right) \text{kg}} = 0.7407 \text{ mol kg}^{-1}.$$

115. The liberation of CO_2 is due to the reaction

$$2NaHCO_3 \rightarrow Na_2CO_3 + H_2O + CO_2$$

Number of moles of CO_2 liberated

$$= \frac{PV}{RT} = \frac{\left(\frac{750}{760}\right)(0.1239)}{(0.0821)(298)} = 0.005$$

Therefore number of moles of $NaHCO_3$ in the given mixture = 0.01 mol.

% composition of $NaHCO_3$ = $\frac{0.01 \times 84}{2} \times 100 = 42$

The neutralization reactions

$$NaHCO_3 + HCl \rightarrow NaCl + H_2O + CO_2$$
$$Na_2HCO_3 + 2HCl \rightarrow 2NaCl + CO_2 + H_2O$$

In a and b are the moles of $NaHCO_3$ and Na_2CO_3 present in 2.0 g sample of mixture, we have

$a + 2b$= number of moles of HCI in 200 ml (= 150 × 2.0/15) of M/10 solution

$$= 0.02 \text{ mol}$$

Now $$a = 0.01 \text{ mol.}$$
$$b = 0.005 \text{ mol.}$$

% composition of Na_2CO_3 = $\frac{0.005 \times 106}{2} \times 100 = 26.5$

∴ of Composition of Na_2SO_4

$$100 - (42 + 26.5) = 31.5$$

116. $P.V = nRT$, Substituting the values we get,

$$n = \frac{1 \times 1.25}{0.082 \times 300} = 0.05$$

For adiabatic expression

$$\left(\frac{V_1}{V_2}\right)^{r-1} = \frac{T_2}{T_1}$$

OR

$$\frac{C_V}{R} \ln\frac{T_2}{T_1} = \ln\frac{V_1}{V_2} \quad \text{.....(i)}$$

Here, $T_1 = 300$ K; $C_V = 12.48$

$$\frac{V_1}{V_2} = \frac{1.25\,dm^2}{2.5\,dm^3} = 0.5$$

Hence, on substituting and solving equation (i) we get,

$$T_2 = 189 \text{ K.}$$

Now $\Delta T = T_2 - T_2 = -111\text{K}$

and $C_P - C_V = R$

$\therefore$ $C_P - C_V + R = 20.7 \text{ JK}^{-1}$ or $C_p = 20.7 + 12.48 - 0.082 = 33.098$

ΔH = Change in enthalpy = $\Delta n\, C_{P} \cdot \Delta T$

Solving we get, $\Delta H = -115.4$ J

117. The reaction is

$$Ca(NO_3)_2 \leftrightarrow Ca^{2+} + 2NO_3^-$$

No. of moles $(1 - 0.7)\frac{7}{164}$ $\quad 0.7 \times \frac{7}{164}$ $\quad (2 \times 0.7)\frac{7}{164}$

The above is the number of mole when 7.0 g of salt is ionized. Now the number of moles due to ionization of $Ca(NO_3)_2$

$$Ca(NO_3)_2 = \frac{7}{164}(0.3 + 0.7 + 1.4) = 0.102$$

Number of moles of water $= \frac{100}{18} = 5.556$

Now let χ_A to be the mole fraction of solvent and χ_B that of solute so that

$$\chi_A = \frac{n_A}{n_A + n_B} = \frac{5.556}{5.556 + 0.102} = \frac{5.556}{5.658}$$

And according to Rasult's law $\{P_A = \chi_A P_A^*\}$

Where P_A is the pressure of solvent and P_A^* is partial pressure of solvent. $P_A^* = 760$ mmHg. Putting the values of mole fraction and partial pressure in Rault's law expression we get

$$P_A^* = \frac{5.556}{5.658} \times 760 = 746.3 \text{ mmHg}$$

Alternate method:-

We have $$\chi_B = \frac{n_B}{n_A + n_B}$$

So $$\chi_B = \frac{0.102}{5.556 + 1.102} = \frac{0.102}{5.658} = 0.0180$$

And by the Rault's law we have $\frac{P^* - P_A}{P_A^*} = \chi_B$

$$\Rightarrow \quad \frac{760 - P_A}{760} = 0.0180$$

$$\Rightarrow \quad P_A = 760 - (760 \times 0.0180) = 746.3 \text{ mmHg}$$

118. Let the number of moles of A left after 100 min be Z

$\therefore$ Total number of moles after 100 min = Z + 12 + 0.525

$$P_{mix} = P_A + P_B = P_A{}^0 Z_A + P_B{}^0 Z_B$$

Where, Z_A = mole fraction of $A = \frac{Z}{12+Z}$; Z_B = mole fraction of $B = \frac{12}{12+Z}$

P_A^0 = Pure vapour pressure of A = 300; P_B^0 = pure vapour pressure of B = 500

$$\therefore \; P_{mix} = \left(\frac{Z}{12+Z} \times 300\right) + \left(\frac{12}{12+Z} \times 500\right) = \frac{6000 + 300Z}{12+Z}$$

According to the Raoult's law, the relative lowering of vapour pressure is equal to the mole fraction of the solute

Thus,

$$\frac{\left(\frac{6000+300Z}{12+Z}\right) - 400}{\frac{6000+300Z}{12+Z}} = \frac{0.525}{Z + 12 + 0.525} \Rightarrow Z = 9.9$$

According to the first order kinetics $K = \frac{2.303}{t} \log \frac{[A]_0}{[A]}$

Here, $[A]_0 = 10$, $[A] = 9.9$ & $t = 100$ min

$$\therefore \quad K = \frac{2.303}{100} \log \frac{10}{9.9} = 1.005 \times 10^{-4} \text{ min}^{-1}$$

The final solution is ideal in behaviour because

(a) It obeys Raoult's law

(b) ΔV_{mixing} is neglibible

(c) (b) ΔH_{mixing} is negligile

119. We know that, $\quad N_2O_5 \rightleftharpoons 2NO_2 + \frac{1}{2}O_2$

		N_2O_5	$2NO_2$	$\frac{1}{2}O_2$
initial pressure	$\longrightarrow$	600	0	0
at time t	$\longrightarrow$	600–P	$2P$	$\frac{1}{2}P$

$$\text{Total } P \text{ at time } t = 600 - P + 2P + \frac{1}{2}P = 600 + \frac{3}{2}P = 960$$

So $\quad P = 240$

Mole fraction N_2O_5 decomposed

$$x = \frac{P}{600} = \frac{240}{600} = 0.4$$

120. The two equation are

$$CO(g) + \frac{1}{2}O_2(g) \rightarrow CO_2(g)$$

$$CO_4(g) + 2O_2(g) \rightarrow CO_2(g) + H_2O(l)$$

Let V_1 be the volume CO and V_2 be the volume of CH_4
So the volume of He is $20 - (V_1 + V_2)$
The Ist contraction is 13. The total volume is

$$\left[V_1 + \frac{1}{2}V_1 + V_2 + 2V_2 + 2O - (V_1 + V_2)\right] - 13$$

$$= V_1 + V_2[20 - (V_1 + V_2)$$

$$\frac{1}{2}V_1 + 2V_2 = 13, \ V_1 + V_2 = 14$$

$$V_1 = 10 \text{ cm}^2, \ V_2 = 4 \text{ cm}^3$$

Thus the percentages are $CH_4 = \frac{4}{20} \times 100 = 20\%$

& percentage $CO = \frac{10}{20} \times 100 = 50\%$

$$V_{He} = 20 - (V_1 + V_2) = 20 - 14 = 16$$

$$\% \text{ He} = \frac{6}{20} \times 100 = 3\%$$

121. The reaction is

$$C + CO_2 \rightarrow 2\ CO$$

Volume of the mixture of CO and CO_2 = 1 litre
Let x be the volume of CO_2 in the mixture.
∴ Volume of CO produced from x of $CO_2 = 2x$.
Volume of CO originally present in the mixture = $1 - x$.
Total volume of CO after the reaction

$$= (1 - x) + 2x = 1 + x$$

Total volume after the reaction = 1.6 litre (given)

∴ $(1 + x) = 1.6$

or $x = 0.6$

Thus volume of CO_2 = 0.6 litre.
And volume of CO = 0.4 litre
Composition is CO_2 and CO are in the ratio of 3 : 2

122. At STP volume strength of H_2O_2

$$= \frac{0.508}{127} \times \frac{1}{5} \times \frac{1}{2} \times \frac{1}{2} \times 22400 = 4.48 \text{ volumes}$$

123. Let the volume of ethane by 'v' litres

∴ the vol. of ethene is $(40 - v)$L

Reaction

$$C_2H_6 + C_2H_{4(g)} + \frac{13}{2} \rightarrow 4CO_{2(g)} + 5H_2O_{(l)}$$

Here the moles of O_2 present

$$= \left(\frac{7v}{2}\right) + \left[\frac{(40 - v)}{RT}\right] \times 3$$

∴ mass of O_2

$$= \left(\frac{7v}{2}\right) + \left[\frac{(40 - v)}{RT} \times 3\right] \times 32 = 130 \quad \text{(given)}$$

∴ Solving we get $\quad v = 26.5$

∴ $$\%C_2H_2 = \frac{26.5}{40} \times 100 = 66.25\% \ \&$$

$$\%C_2H_4 = 33.75\%$$

124. We have $$\frac{P_1}{T_1} = \frac{P_2}{T_2}$$

or $$\frac{1}{t + 273} = \frac{1.1}{t + 283}$$

∴ $$t = -173°C = 100 \text{ K}$$

On applying ideal gas equation

$$P.V = nRT$$

$$V = \frac{nRT}{P} = \frac{0.1 \times 0.082 \times 100}{1} = 0.821 \text{ lit}$$

125. Here, $P = 760$ mm $= 1$ atm, $T = 300$ K

$$n = \frac{w}{m} \quad \therefore \quad w = 0.5 \text{ g}, m = ?$$

$$V = \text{vol of vessel} = \frac{m}{d} = \frac{(148 - 50)}{0.98} = 100 \text{ ml} = 0.1 \text{ L}$$

$$P.V = nRT = \frac{\text{w}}{\text{m}} RT$$

∴ $$m = \frac{W.R.T}{P.V} = \frac{0.5 \times 0.082 \times 300}{1 \times 0.1} \quad \text{(substituting the values)}$$

$\Rightarrow \quad m = 123$ g/mol

$M = 123$ a.m.u

126. Volume of oxygen in the flask $V = 1$ litre

Pressure in the flask $P = 7\ 7.6 \times 10^{-10}$ mm of Hg

$$= \frac{7.6 \times 10^{-10}}{760} \text{ atm} = 10^{-12} \text{ atm}$$

Universal gas constant $R = 0.082$ atm litre K^{-1} $mole^{-1}$

Temperature of oxygen in the flask $T = 273$ K

The gas equation is $PV = nRT$,

Where n is the number of moles.

$\therefore\ 10^{-12} \times 1 = n \times 0.082 \times 273$ (Substituting the values)

or $$n = \frac{10^{-12} \times 1}{0.082 \times 273} \text{ moles}$$

or $$n = \frac{10^{-21} \times 1 \times 6.023 \times 10^{23}}{0.082 \times 273} \text{ molecules}$$

[$\because$ 1 mole of a gas at S.T.P. contains 6.023×10^{23} molecules]

$= 2.69 \times 10^{10}$ molecules.

127. Average kinetic energy $= \frac{3}{2}\ n\ RT$

Here, $T = 300$ K; $R = 8.31$ Joules K^{-1} mol^{-1}

Molecular weight of methane (CH_4) $= 12 + (4 \times 1) = 16$

moles of methane in 8 g $= 0.5$ mol

Average kinetic energy $= \frac{3}{2} \times 0.5 \times 8.31 \times 300$ Joules $= 1869.75$ Joules.

128. We know from the gas equation

$$PV = n\ RT$$

that temperature and volume remaining constant, the pressure of a gas is directly proportional to its number of moles

Wt of gas A introduced into the flask = 2 g

Wt of gas B introduced into the flask = 3 g

Pressure when gas A is introduced = 1 atm.

Pressure when gas B is also introduced = 1.5 atm.

$\therefore$ Increase in pressure when gas B is introduced into the flask

$$= 1.5 - 1 = 0.5 \text{ atm} = \frac{1}{2} \text{ atm.}$$

Clearly, 3 g of gas B will have half as many moles at 0.5 atm. pressure as 2 g of gas A has at 1 atm. pressure.

$\therefore$ 6 g of gas B at 1 atm pressure will have moles equal its number to those which 2 g of gas A has at 1 atm.

$$\therefore \quad \frac{\text{Molecular weight of } A}{\text{Molecular weight of } B} = \frac{2\text{ g}}{6\text{ g}}$$

or $$\frac{M_A}{M_B} = \frac{1}{3}$$

$$\therefore \quad M_A : M_B = 1 : 3.$$

129. Number of moles of NO $= \dfrac{PV}{RT} = \dfrac{(1.053 \text{ atm}) \times (0.250 \text{ L})}{(0.082 \text{ L} - \text{atm mol}^{-1} \text{ K}^{-1}) \times 300 \text{ K}} = 0.0107$

Number of moles of O_2 $= \dfrac{(0.789 \text{ atm}) \times (0.10 \text{ L})}{(0.082 \text{ L} - \text{atm mol}^{-1} \text{ K}^{-1}) \times 300 \text{ K}} = 0.0032$

For 1 mol of O_2, 2 moles of NO are consumed
For 0.0032 mole of O_2, 0.0064 mol of NO are consumed..
Amount of NO left unreacted = 0.0107 – 0.0064
= 0.0043 mol.

Since, N_2O_4 solidifies at 220 K, the only gas left in the flask will be NO. Therefore, pressure exerted by NO is given by

$$P = \frac{nRT}{V} = \frac{(0.0043 \text{ mol}) \times (0.082 \text{ L . atm mol}^{-1} \text{ K}^{-1}) \times (220 \text{ K})}{0.35 \text{ L}} = 0.2216 \text{ atm.}$$

130. Let V the volume hydrogen in the cylinder at N.T.P. It is given:

$P = 1$ atm	$P_1 = 20$ atm
$T = 273$	$T_1 = 300$
$V = ?$	$V = 2.82\ l$

Now, $$\frac{PV}{T} = \frac{P_1 V_1}{T_1}$$

Substituting the values, we get

$$V = \frac{20 \times 2.82 \times 273}{300 \times 1} = 51.324\ l = 51324 \text{ ml}$$

The volume which will be actually transferred = 51324 – 2820 = 48504 ml

Volume of each bolloon $= \dfrac{4}{3} r^3 = \dfrac{4}{3} \times 3.14 \times (10.5)^3 = 4847$ ml

$\therefore$ No. of balloon which can be filled up $= \dfrac{48504}{4847} = 10$

131. $\left(P + \dfrac{a}{v^2}\right)(V - b) = RT$

for 1 mole of real gas, if b negligible.

$$\left(P+\frac{n}{v^2}\right) V = RT$$

or $$\left(P+\frac{3.592}{22.4\times 22.4}\right) 22.4 = 0.0821 \times 273$$

$$P = \frac{0.0821\times 273}{22.4}-\frac{3.592}{22.4\times 22.4}; \; P = 0.9935 \text{ atm}$$

132. Molecular wt. of LPG gas = 12 × 4 + 10 × 1.01 = 58.1 gm
Weight of empty cylinder = 14.98 kg.
Weight of full cylinder = 29.0 kg.
Weight of gas initially in the cylinder = 29.0 – 14.8 = 14.2 kg
Pressure in the cylinder = 2.5 atm
Weight of the cylinder after use = 23.2 kg
Weight of the gas after use = 23.2 – 14.8 = 8.4 kg
Let V be the volume of gas initially present in the cylinder.
By the equation of state,

$$V = \frac{nRT}{P} = \frac{14.2\times 10^3 \times 0.0821\times 300}{58.1\times 2.5} = 2.408 \times 10^3 \text{ litre}$$

wt of gas used up = 14.2 – 8.4 = 5.8 kg
Volume of gas used up at normal usage conditions.

$$V' = \frac{5.8\times 10^3 \times 0.0821\times 300}{58.1\times 2.5} = 0.988 \text{ m}^3$$

Final pressure inside the cylinder,

$$P' = \frac{8.4\times 10^3 \times 0.0821\times 300}{58.1\times 2.408\times 10^3} = 1.478 \text{ atm.}$$

133. We know that c.f or Compressibility factor Z, = $\frac{PV}{nRT}$

Here n = 1 mole, c.f = 0.5, P = 100 atm; T = 273°k
Substituting the values

$$0.5 = \frac{100\times V}{0.082\times 273}$$

∴ $V = 0.1119$ L

Now, at moderate pressure, van der Waal's equation is

$$PV = RT - \frac{a}{V} \text{ or } a = RTV - PV^2$$

Substituting the values

$a = (0.082 \times 0.1119 \times 273) - (100 \times 0.1119 \times 0.1119) = 1.253$ atm L^2 mol^{-2}

134. Real gas equation is

$$\left(P + \frac{q}{V^2}\right)(V - b) = RT, \text{ for one mole}$$

or $$\left(P + \frac{q}{V^2}\right) = \frac{RT}{V - b}$$

$$\Rightarrow \quad P = \frac{RT}{V-b} - \frac{a}{V^2} \Rightarrow \frac{RT}{V\left(1 - \frac{b}{V}\right)} - \frac{a}{V^2}$$

$$\Rightarrow \quad PV = RT\left(1 - \frac{b}{V}\right)^{-1} - \frac{a}{V^2}$$

$$\Rightarrow \quad PV = RT\left(1 + \frac{b}{V} + \frac{b^2}{V^2} - - -\right) - \frac{a}{v} \text{ (By binomial theorem)}$$

$$\Rightarrow \quad PV = RT\left(1 + \frac{b - a/RT}{V} + \frac{b^2}{V^2} + \frac{b^3}{V^3} - - -\right)$$

$$\Rightarrow \quad PV = RT\left(1 + \frac{B}{V} + - - -\right)$$

Here $$B = \frac{b - a/RT}{v}$$

135. According to Vander Waals equation

$$\left(P + \frac{an^2}{V^2}\right)(V - nb) = nRT$$

Here, $P = 11$ atm; $n = 2$. $V = 4$ L, $T = 300$ K; $b = 0.05$ lit mol^{-1}; $R = 0.0832$

Substituting values we get,

$$\left(11 + \frac{a \times 4}{4 \times 4}\right)(4 - 2 \times 0.05) = 2 \times 0.0832 \times 300 = 49.92$$

or $$\left(\frac{44 + a}{4}\right)(3.9) = 49.92$$

$\therefore$ $a = 6.46$ atm l^2 mol^{-2}

136. We have

$$U_{ms} = \sqrt{\frac{3PV}{M}}$$

where $P = 82 \text{ cm} = 82 \times 13.6 \times 981$ dynes

V = Volume of gas under 82 cm and at 293 K

$$= \frac{76.0 \times 22400}{273} \times \frac{293}{82} = 26664 \text{ c.c}$$

M = molecular weight = 48

$$\therefore \quad u_{ms} = \sqrt{\frac{3 \times (82 \times 13.6 \times 981) \times 26664}{48}}$$

$$= 3.9 \times 10^1 \text{ cm sec}^{-1}$$

137. The average velocity, $\bar{U}$, and the most probable velocity. U_{mp}, are related to R (Gas const.), T (temperature), and M (molecular weight) by the following equation:

$$\bar{U} = \sqrt{\frac{8RT}{\pi M}} \text{ and } U_{mp} = \sqrt{\frac{2RT}{M}}$$

By substitution in $\bar{U} = \sqrt{\frac{8RT}{\pi M}}$

$$9 \times 10^4 \text{ cm sec}^{-1} = \sqrt{\frac{8 \times 8.314 \times 10^7 \times T_1}{3.14 \times 44}} \quad (\text{Mol. wt. of } CO_3 = 44)$$

Squaring both side,

$$81 \times 10^8 = \frac{8 \times 8.314 \times 10^2 \times T_1}{3.14 \times 44}$$

Whereby,

$$T_1 = \frac{81 \times 10^8 \times 3.14 \times 44}{8 \times 8.314 \times 10^7} = 1682 \text{ K}$$

By substitution in $C_{mp} = \sqrt{\frac{2RT}{M}}$

$$9 \times 10^4 \text{ cm sec}^{-1} = \sqrt{\frac{2 \times 8 \times 8.314 \times T_2}{44}}$$

Squaring both side;

$$81 \times 10^8 = \frac{2 \times 8 \times 8.314 \times 10^7 \times T_2}{44}$$

Whereby,

$$T_2 = \frac{81 \times 10^8 \times 44}{2 \times 8 \times 8.314 \times 10^7} = 2679 \text{ K}$$

138. $U_{rms} = \sqrt{(3RT/m} = \sqrt{(3p/d}$

Here $\quad d$ = density of gas molecules = $\dfrac{\text{Mass}}{\text{Volume}}$

No. of molecules = 2×10^{21}

Mass of 6.023×10^{23} molecules = $14 \times 2 = 28$ gm = 28×10^{-3} kg

$$\text{Mass of } 2 \times 10^{21} \text{ molecules} = \frac{28 \times 2 \times 10^{21} \times 10^{-3}}{6.023 \times 10^{23}} = \frac{56 \times 10^{-5}}{6.023} = 0.093 \times 10^{-3} \text{ kg.}$$

$$\therefore d = \frac{0.093}{1} = 0.093 \times 10^{-3} \text{ kg/l} = \frac{0.093 \times 10^{-3} \text{ kg}}{10^{-3} \text{ m}^3} = 0.093 \text{ kg/m}^3$$

$$\text{Root mean square speed } U_{rms} = \sqrt{\left(\frac{3 \times 7.57 \times 10^3}{0.093}\right)} = 494.16 \text{ m/sec}$$

$$\text{Now,} \quad U_{rms}^2 = \frac{3RT}{M} = \frac{3P}{d} = \frac{3 \times 7.57 \times 10^3}{0.093}.$$

$$\text{or} \quad = \frac{RT}{M} = \frac{7.57 \times 10^3}{0.093}$$

$$\therefore \quad T = \frac{7.57 \times 10^3}{0.093} \times \frac{28 \times 10^{-3}}{8.314}$$

$$\Rightarrow \quad T = 274.13 \text{ K}$$

$$\frac{\text{Most Probable vel. (U)}}{\text{Root mean sq. vel. } (U_{rms})} = 0\ 82 = \frac{u}{494.16\, m/\text{sec}}$$

$$\therefore \quad V = 0.82 \times 494.16 = 405.2 \text{ m/sec.}$$

139. (c) Root means square velocity (v_r)

$v_r = \sqrt{\dfrac{3RT}{M}}$; T = Temperature; M = Molecular weight; R = Gas constant

Substituting the values for H_2 and O_2 and finding their ratios we get the required result.

140. We know that,

$$U_{rms} = \sqrt{\frac{3RT}{M}} \text{ and, } U_{av} = \sqrt{\frac{8RT}{\pi M}}$$

$$\text{Now,} \quad \therefore \frac{U_{rms}}{U_{av}} = \sqrt{\frac{3RT}{M}} \times \sqrt{\frac{\pi M}{8RT}} = \sqrt{\frac{3\pi}{8}} = 1.085$$

$$\text{or} \quad U_{rms} = 1.085 \times U_{av} = 1.085 \times 400 = 434 \text{ m/sec}$$

141. Vapour pressure of ethanol, $P°_{C_2H_5OH}$ = 44.5 mm

Vapour pressure of methanol, $P°_{CH_3OH}$ = 88.7 mm

Moles of C_2H_5OH, $= \frac{60}{46} = 1.304$

Moles of CH_3OH, $= \frac{40}{32} = 1.250$

Moles fraction of C_2H_5OH, $= \frac{1.304}{1.304 + 1.250} = 0.5107$

Mole fraction of CH_3OH = 1 – 0.5107 = 0.4893

Vapour pressure due to C_2H_5OH = 44.5 × 0.5107 = 22.73 mm Hg

Vapour pressure due to CH_3OH = 88.7 × .4893 = 43.40 mm Hg

Total pressure = 22.73 + 43.40 = 6.13 mm Hg

Mole fraction of CH_3OH in the vapour

$$= \frac{\text{Vapour pressure due to } CH_3OH}{\text{Total pressure}} = \frac{43.4}{66.13} = 0.6563$$

142. Ist order kinetic equation is

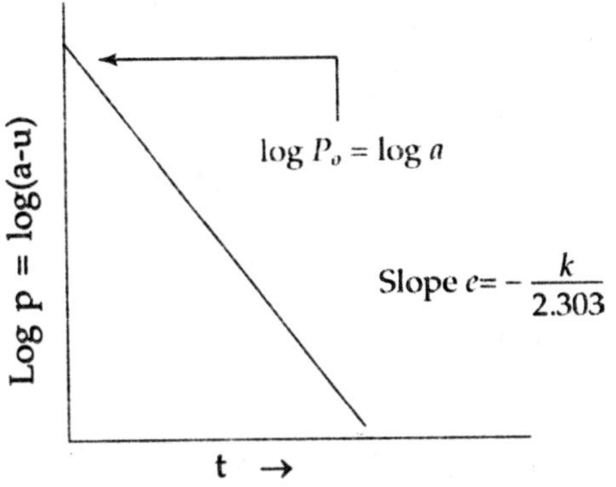

$$2.303 \log \frac{P}{P_0} = -kt \Rightarrow \log \frac{P}{P_0} = -\frac{kt}{2.303}$$

or $$\log \frac{P}{P_0} = -\frac{kt}{2.303}$$

From, the slope, value of K (Specific reaction rate) can be evaluated.

The linear graph indicates that decomposition of N_2O_5 obeys Ist order kinetic.

143. Ratio of moles of the He and CH_4 = 4 : 1

$\therefore$ pressure $\propto$ number of moles at constant V and T

$\therefore$ Partial pressure of He $= \frac{4}{5} \times$ Total pressure $= \frac{4}{5} \times 20 = 16$ atm.

Partial pressure of $CH_4 = \frac{1}{5} \times 20 = 4$ atm.

Molecules wt. of He = 4
Molecules wt. of CH_4 = 16

Rate of diffusion is given by, $r = \frac{KP}{\sqrt{M}}$, where K is constant at constant T

$$\therefore \quad r_{He} = \text{Rate of diffusion of He} = \frac{K \times 16}{\sqrt{4}} = 8\text{ K.}$$

$$r_{CH_4} = \text{Rate of diffusion of } CH_4 = \frac{K \times 4}{\sqrt{16}} = \text{K}$$

$\therefore$ Molar ratio of the two gases effusing out initially is He : CH_4 = 8 : 1.

144. Since

$$PV = nRT$$

Here T = 400 K; P = 1 atm; n = 1; V = 1 lit.; R = 0.082 L.atm.mol^{-1} K^{-1}

$$\therefore \quad n = \frac{1 \times 1}{0.082 \times 400}$$

Now, molecular weight= $\frac{\text{weight}}{n}$

$\therefore$ molecular weight of hydrocarbon $= \frac{2.8 \times 400 \times 0.082}{1} = 91.84$

Element	*Weight of element*	*Relative number of atoms*	*Ratio of atoms*	*Whole number of atoms*
Carbon.	10.5	10.5 ÷ 12 = 0.875	1	7
Hydrogen	1.0	1.0 ÷ 1 = 1.0	1.1428	8

Thus emperical formula is C_7H_8.
Emperical formula weight = (7 × 12) + (1 × 8) = 84 + 8 = 92

$$n = \frac{\text{molecular weight}}{\text{emperical formula weight}} = \frac{91.84}{92} = 1 \text{ (nearest whole number)}$$

$\therefore$ Emperical formula is the molecular formula

145. Rate of diffusion (R) of gas at constant temperature i.e. C_7H_8

$$= \frac{KP}{\sqrt{M}}$$

where P is the pressure of the gas, K is a constant and M is the molecular weight of the gas.
Rate of diffusion (R) is directly proportional to the distance travelled by a gas.
Ammonium chloride is formed at a distance = 60 cm
Molecular weight (M) of HCl

$$= 1 + 35.5 = 36.5$$

Therefore, rate of diffusion of HCl gas at pressure

$$P = 60 = \frac{KP}{\sqrt{36.5}}$$

Distance travelled by ammonia = 40 cm.
Molecular weight of NH_3 = 14 + (3 × 1) = 17

Therefore, rate of diffusion of ammonia at atm pressure $= 40 = \dfrac{K \times 1}{\sqrt{17}}$

or $\qquad K = 14 \times \sqrt{17}$

$$\therefore \qquad P = \frac{60 \times \sqrt{36.5}}{40 \times \sqrt{14}} = 2.2 \text{ atm.}$$

146. According to Graham's law of diffusion

$$\text{Rate of dissuson} \propto \sqrt{\frac{1}{\text{density}}}$$

$$\text{Rate of diffusion} = \frac{\text{Volume diffusion}}{\text{time}}$$

$$\frac{r_A}{r_B} = \frac{\sqrt{m_B}}{m_A} = \sqrt{\frac{d_B}{d_A}} = \sqrt{\frac{t_B}{t_A}} = \sqrt{\frac{v_A}{v_B}}$$

where, r_A = rate of diffusion of A; r_B = rate of diffusion of B
m_A = m.wt of A; m_B = m.wt of B
d_A = diffusion of A; d_B = diffusion of B
t_A = time taken by A; t_B = time taken by B
v_A = vol. of A diffused; v_B = vol. of B diffused

$$\text{Thus } \frac{r_{N_2}}{r_{XeF_n}} = \frac{P_{N_2}}{P_{XeF_n}} \sqrt{\frac{M_{XeF_n}}{M_{N_2}}} = \frac{t_{XeF_n}}{t_{N_2}}$$

$$\frac{57}{38} = \frac{0.8}{1.6} \sqrt{\frac{M_{XeF_n}}{28}}$$

m.wt. of XeF_n = 252

Thus molecular formula for XeF_n is

$$252 = Xe + n \times F, \Rightarrow n = 6 \; XeF_6$$

147. We have

$$\text{Number of moles of gases} = \frac{pV}{RT} = \frac{6 \times 3}{0.0821 \times 300} = 0.731$$

Number of moles of unknown gas = 0.731 – 0.7 = 0.031

Now

$$\frac{r_1}{r_2} = \frac{0.7}{0.031} = \sqrt{\frac{M_2}{M_1}}$$

or $$M_1 = \left(\frac{0.7}{0.031}\right)^2 (2) = 1033\ g.$$

Chapter 4 : STOICHIOMETRY

1. One mole of H_2S contains 1 mole of S. One mole of S has a mass 32.1 g.

$$2.2 \text{ moles of } H_2S \times \frac{1 \text{ mole of S}}{1 \text{ mole of } H_2S} \times \frac{32.1 \text{ g of S}}{1 \text{ mole of S}} = 706 \text{ g of S}$$

Since it is obvious that one mole of H_2S contains 32.1 g of S, the second factor in the above equation can be omitted to give

$$2.2 \text{ mole of } H_2S \times \frac{32.1 \text{ g of S}}{1 \text{ mole of } H_2S} = 70.6 \text{ g of S}$$

2. 1 gram atom of sulphur = 32 g. (or)
32 g. of sulphur = 1 gram atom of sulphur

$$144 \text{ g. of sulphur} = \frac{1}{32} \times 144 = 4.5 \text{ gram atoms.}$$

3. $$\frac{182 \text{ g of } KClO_3}{122.6 \text{ g of } KClO_3/\text{mole of } KClO_3} \times \frac{3 \text{ moles of O}}{1 \text{ mole of } KClO_3} = 4.46 \text{ moles of O}$$

4. One mole of P_4O_{10} contains 4 moles of P. Moles of P = 4 × moles of P_4O_{10}.

$$\frac{160 \text{ g of } P_4O_{10}}{284 \text{ g of } P_4O_{10}/\text{mole of } P_4O_{10}} = \frac{160}{284} \text{ moles of } P_4O_{10}$$

$$\text{moles of P} = \frac{4 \times 160}{284}$$

$$\therefore \quad \frac{4 \times 160}{284} \text{ mole of P} \times \frac{31.0 \text{ g of P}}{1 \text{ mole of P}} = 69.8 \text{ g of P}$$

5. By definition, percent means parts per 100 parts by mass. So all we need do in this problem is find how many grams of C there are in 100 g of CO_2; the answer wil then be the percent of C in pure CO_2. A more general definition of percent, called fraction, is that it is the ratio of the number of parts by the mass of the particular thing you want to find to the parts by mass by the whole thing. That is,

$$\text{fraction of C in } CO_2 = \frac{\text{mas of the C in } CO_2}{\text{mass in the } CO_2}$$

A mole of CO_2 has a mass of 44 g and contains 12 g of C. Therefore,

$$\text{fraction of C in } CO_2 = \frac{12 \text{ g of C}}{44.0 \text{ g of } CO_2} = 0.273$$

This gives a decimal (fractional) percent. To change this to the standard notation, we simply multiply by 100%. Doing the whole calculation in one operation.

$$\text{Percent of C in } CO_2 = \frac{12}{44} \times 100\% = 27.3\%$$

6. Molecular weight of HNO_3 = 1 + 14 + (3 × 16) = 63.

(a) 1 mole of HNO_3 = 63 g.

0.25 mole of HNO_3 = 63 × 0.25 = 15.75 g.

1 mole of HNO_3 contains 14 g. of N

0.25 mole HNO_3 = 14 × 0.25 = 3.5 nitrogen

(b) 1 mole of HNO_3 = 1g. atom of nitrogen

0.25 mole of HNO_3= 1 × 0.25 = 0.25 gram atom of nitrogen

7. Weight of magnesium = 100 g.

Atomic weight of magnesium = 24.3

$$\text{No. of gram atoms of magnesium} = \frac{\text{mass of the element in grams}}{\text{Atomic weight}} = \frac{100}{24.3} = 4.115$$

8. 1. Atomic weight of sulphur = 32

gram atom of sulphur = 32 g.

$$\text{Mass of 1 atom of sulphur} = \frac{32}{6.02 \times 10^{23}} = 5.33 \times 10^{-23} \text{ g.}$$

2. Formula of carbon monoxide = CO

Molecular weight of CO = 12 + 26 = 28

1 mole of CO = 28 g.

$$\text{Mass of 1 molecule of CO} = \frac{28}{6.02 \times 10^{23}} = 4.67 \times 10^{-23} \text{ g.}$$

9. 1 mole of H_2S contains

$2 \times 6.02 \times 10^{23}$ atoms of hydrogen.

∴ 1.6 moles of H_2S

$= 1.6 \times 2 \times 6.02 \times 10^{23}$ atoms of hydrogen.

$= 19.624 \times 10^{23}$ atoms of hydrogen

Similarly 1 mole of H_2S contains 6.02×10^{23} atoms of sulphur.

∴ 1.6 moles of H_2S

$= 1.6 \times 6.02 \times 10^{23} = 9.632 \times 10^{23}$ atoms of sulphur

10. 1 mole of CO_2 occupies 22.4 litres at STP, ∴ 2.5 moles of CO_2 = 22.4 × 2.5 = 56.00 lit.

11. (a) weight of iron = 50 g.

(b) atomic weight of nitrogen = 14

or one gram atom of nitrogen = 14 g.

∴5 gram atoms of nitrogen weighs = 14 × 5 = 70 g.

(c) atomic weight of silver = 108

one gram atom of silver= 108 g.

0.1 gram atom of silver= 108 × 0.1 = 10.8 g.

(d) 6.02×10^{23} atoms of carbon = 12 g

or 1×10^{23} atoms of carbon = $\frac{12}{6.02 \times 10^{23}} \times 1 \times 10^{23} = 1.992$ g.

Hence we conclude that 5 gram atoms of nitrogen weigh maximum.

12. 18 g of water contains 6.023×10^{23} molecules.

∴ The number of molecules persent in 1.8 g = $\frac{1.8}{18} \times 6.023 \times 1023 = 6.023 \times 10^{22}$

13. (a) 1 mole of nitrogen at STP = 22.4 lit.

∴ 0.5 mole of $N_2 = 22.4 \times 0.5 = 11.2$ lit.

(b) 1 mole of H_2S at STP = 22.4 lit

But 1 mole of H_2S contains 6.023×10^{23} molecules

i.e., 6.023×10^{23} molecules of H_2S = 22.4 lit.

∴ 6.023×10^{22} molecules of $H_2S = \frac{22.4 \times 6.023 \times 10^{22}}{6.023 \times 10^{23}} = 2.24$ lit. at STP

(c) 1 mole of ammonia at STP = 22.4 lit.

Volume at STP of 0.25 mole of ammonia = $22.4 \times 0.25 = 5.6$ lit. at STP

(d) 1 mole of oxygen = 32 g.

32 g. of oxygen = 22.4 lit. at STP

8 g. of oxygen = $\frac{22.4 \times 8}{32} = 5.6$ lit. at STP

14. At 273 K and 1 atm pressure,

Number of molecules present in 22.4 litre of a gas = 6.023×10^{23}

Since the gas is tetra atomic, no. of atoms present in 22.5 litre = $6.023 \times 10^{23} \times 4$

∴ No. of atoms present in 1.4 litres of the gas at 273 K and 1 atm.

$$= \frac{6.023 \times 10^{23} \times 4 \times 1.4}{22.4} = 1.505 \times 10^{23}$$

15. Ozone is triatomic and its molecular wt. 48

No. of molecules present in 48 g. of ozone = 6.023×10^{23}

No. of atoms present in 48 g. of ozone = $6.023 \times 10^{23} \times 3$

No. of atoms present in 48 g. of ozone = $\frac{6.023 \times 10^{23} \times 12}{48} = 4.515 \times 10^{23}$

16. No. of electrons that are to be added to convert permanganate ion into manganous ion = 5

No. of electrons that are to be added to convert one mole of MnO_4^- ions into Mn^{2+} ions.
= $5 \times 6.023 \times 10^{23} = 3.0115 \times 10^{24}$

17. No. of CO_2 molecules in 22.4 lit at STP = 6.023×10^{23}

No. of CO_2 molecules in 5.6 lit. at STP = $\frac{6.023 \times 10^{23}}{22.4} \times 5.6 = 1.505 \times 10^{23}$

No. of atoms in CO_2 molecule = 3
No. of atoms in 1.505×10^{23} molecules = $3 \times 1.505 \times 10^{23} = 4.515 \times 10^{23}$

18. No. of moles of C-14 of in 1.0 g. = $\frac{1.0}{14.01}$ = 0.0714 moles

No. of moles of C-12 of in 1.0 g. = $\frac{1.0}{12}$ = 0.0833 moles

The difference in no. of moles between the two samples 0.0833 – 0.0714 = 0.0119 moles
Difference in no. of Carbon atoms = $0.0119 \times 6.023 \times 10^{23} = 7.0 \times 10^{21}$ atoms

19. Molecular weight of methane CH_4 = 12 + 4 = 16
Weight of methane = 1.6 g.
No. of moles of methane = 1.6/16 = 0.1
No. of molecules of methane = $0.1 \times 6.023 \times 10^{23} = 6.023 \times 10^{22}$
One molecules of methane contains = 6 + 4 = 10 electrons
1.6 g. of methane contains
= no. of molecules × 10
= $6.023 \times 10^{22} \times 10 = 6.023 \times 10^{23}$

20. We know that the weight of Avogadro number of atoms is equal to atomic weight in grams.
Wt. of one atom = 6.644×10^{23} g.
Wt. of Avogadro No. of atoms = $6.644 \times 10^{-23} \times 6.02 \times 10^{23}$ = 40 g.
∴ Atomic wt. of the element = 40
No. of gram atoms present in 40 Kg

$$= \frac{\text{Wt. of the element in grams}}{\text{Atomic weight}}$$

$$= \frac{40 \times 1000}{40} = 1000.$$

21. Molecular weight of sulphur = 256
No. of molecules present in 256 g. of sulphur = 6.023×10^{23}

No. of molecules present ion 64 g. of sulphur = $\frac{6.023 \times 10^{23} \times 64}{256} = 1.505 \times 10^{23}$

22. Phosphorus is tetra atomic molecule.
Weight of 6.023×10^{23} molecules of phosphorus = 124 g.

Weight of 3.0115×10^{20} molecules of phosphorous = $\frac{124 \times 3.0115 \times 10^{20}}{6.023 \times 10^{23}} = 6.2 \times 10^{2}$ g.

23. No. of moles of oxygen = $\frac{8}{32}$ = 0.25

and no. of moles of sulphur dioxide = $\frac{48}{64}$ = 0.75

∴ No. of moles present in the mixture = 0.25 + 0.75 = 1.0

24. Let V_1 be the volume of the gas at N.T.P

Here, $P_1 = 1$ atm. $P_2 = 2$ atm.

$V_1 = ?$ $V_2 = 350$ ml.

$T_1 = 273$ K $T_2 = 273$ K

Substituting these values in the gas equation:

$$\frac{P_1\ V_1}{T_1} = \frac{P_2\ V_2}{T_2}$$

we get $$V_1 = \frac{2 \times 350 \times 273}{273 \times 1} = 700 \text{ ml}$$

Now 700 ml. of the gas at N.T.P weights 1 g. 22400 ml. of the gas at N.T.P weighs

$$\frac{1}{700} = 22400 = 32 \text{ g.}$$

Hence molecular weight of the diatomic gas = 32

32 g. of the contains $2 \times 6.02 \times 10^{23}$ atoms (since the gas is diatomic)

$\therefore$ Weight of one atom of the gas = $\dfrac{32}{2 \times 6.02 \times 10^{23}} = 2.66 \times 10^{-23}$ g.

25. Weight of 1 ml. of water = 1 g.

Molecular weight of $H_2O = 18$

18 g. of water contains 6.02×10^{23} molecules

1 g. of water contains $\dfrac{6.02}{18} \times 10^{23} = 3.344 \times 10^{22}$ molecules.

26. 22400 c.c. of the gas at N.T.P contains 6.02×10^{23} molecules.

$\therefore$ 1.12×10^{-7} c.c. of the gas at N.T.P. contains

$$\frac{6.02}{22400} \times 1.12 \times 10^{-7} \times 10^{23} = 3.01 \times 10^{12} \text{ molecules.}$$

Hence the number of molecules of the gas at N.T.P. in the tube = 3.01×10^{12}.

27. Molecular weight of $CO_2 = 44$

$\therefore$ 44 g. of CO_2 contains 6.02×10^{23} molecules.

0.2 g. (or 200 mg) of CO_2 contains $\dfrac{6.02}{44} \times 0.2 \times 10^{23} = 2.73 \times 10^{21}$ molecules.

$\therefore$ Number of molecules left after the removal of 10^{21} molecules.

$= 2.73 \times 10^{21} - 10^{21} = 10^{21}\ (2.73 - 1) = 1.73 \times 10^{21}$

28. Atomic weight of sodium = 23

1 mole of sodium atom = 23 g.

2 mole of sodium atom = 2 × 23 = 46 g.

3.82×10^{-23} g. of sodium represents 1 atom therefore, 46 g. of sodium represents

$$\frac{1 \times 46}{3.82 \times 10^{-23}} = 12.04 \times 10^{23} \text{ atoms}$$

29. Atomic weight of oxygen = 16

100 g of the compound contains 57.14 g. of oxygen

57.14 g. of oxygen corresponds to 100 – 57.14 = 42.86 g. of other atoms.

16 g. of oxygen corresponds to $\frac{42.86}{57.14} \times 16 = 12$ g. of other atoms.

Since 1 mole of the compound contains 1 gram atom of oxygen, therefore, the molecular weight of the compound is 12 + 16 = 28.

30. (i) Atomic weight of M = 54.94

1 gram atom of M = 54.94 g.

$$\text{Volume of 1 gram atom of } M = \frac{\text{Weight}}{\text{Density}} = \frac{54.95}{7.42} = 7.4 \text{ c.c.}$$

Since, 6.02×10^{23} atoms of M occupies a volume of 7.4 c.c. therefore, 1 atom of M occupies a volume of $\frac{42.86}{57.14} = 1.23 \times 10^{-23}$ c.c.

(ii) Volume of the sphere= $\frac{4\pi r^3}{3}$ where, r = radius of the sphere.

$$\therefore \frac{4\pi r^3}{3} = 1.23 \times 10^{-23} \text{ or } r^3 = \frac{1.23 \times 10^{-23} \times 3 \times 7}{4 \times 22}$$

or $$r = \left(\frac{1.23 \times 10^{-23} \times 3 \times 7}{4 \times 22}\right) \quad r = 1.432 \times 10^{-8} \text{ cm.}$$

31. (i) 6.02×10^{23} molecules of oxygen weighs 23 g.

1 molecules of oxygen weighs $\frac{32}{6.02 \times 10^{23}} = 5.23 \times 10^{-23}$ g.

(ii) As per the question, 1 mole of oxygen contains 1.0×10^{24} particles.

Hence 1.0×10^{24} particles weight

$= 5.32 \times 10^{-23} \times 1.00 \times 10^{24} = 5.32 \times 10 = 53.2$ g.

32. Weight of Ca = 4 g

No. of atoms in 40 g of ca = 6.023×10^{23}

No. of atoms in 4 g of Ca = 6.023×10^{22}

No. of atoms of Mg = 6.023×10^{22}

Wt. of 6.023×10^{23} atoms of Mg = 24 g.

Wt. of 6.023×10^{22} atoms of Mg = 2.4 g.

33. The number of atoms in 2 g. of hydrogen = $\frac{2}{1} \times 6.023 \times 10^{23} = 12.046 \times 10^{23}$

The number of atoms in 16 g of oxygen = $\frac{16}{18} \times 6.023 \times 10^{23} = 6.023 \times 10^{23}$

$\therefore$ Total number of atoms in the mixture = $(12.046 + 6.023) \times 10^{23} = 1.8069 \times 10^{24}$

34. We have, 32 g (1 mole) of sulfur are consumed to form 233 g (1 mole) of $BaSO_4$. The fraction of S in $BaSO_4$ is

$$\frac{\text{MW of S}}{\text{MW of BaSO}_4} = \frac{32\ \text{g/mole}}{233\ \text{g/mole}} = 0.14.$$

The weight of sulfur in 0.466 g of $BaSO_4$ is thus,

$$(0.14)(0.466\ \text{g}) = 0.065\ \text{g}.$$

Now, that the weight of sulfur is known, the percentage of sulfur in the tomato is,

$$\frac{0.065\ \text{g of S}}{60.0\ \text{g of tomato}} \times 100 = 0.11.$$

35. A mole of atom is defined as containing Avogadro's number of atoms. Avogadro's number is 6.02×10^{23}. Therefore, the number of moles in one atom is equal to 1 atom divided by 6.02×10^{23} atoms/mole.

$$\text{No. of mole} = \frac{1\ \text{atom}}{6.2 \times 10^{23}\ \text{atoms/mole}} = 1.66 \times 10^{-24}\ \text{moles.}$$

36. $$\text{No. mol Mg} = 1.00 \times 10^{22}\ \text{Mg atoms} \times \frac{1\ \text{mol Mg}}{6.022 \times 10^{23}\ \text{Mg atoms}}$$

$$= 0.0166\ \text{mol mg}$$

37. We know that 1 mol of O atoms weighs 16 g. All we have to do is to figure out how many moles of O atoms we have.

$$\frac{1.87 \times 10^{27}\ \text{atoms}}{6.02 \times 10^{23}\ \text{atoms/mol}} = 3.11 \times 10^{3}\ \text{mol of O atoms}$$

$$(3.11 \times 10^{3}\ \text{mol of O})\left(\frac{16\ \text{g}}{\text{mol of O}}\right) = 4.97 \times 10^{4}\ \text{g}$$

38. Figure out how many moles of Mg atoms we have, calculate from this how many moles of N atoms are required, then convert this to grams.

$$\frac{4.86\ \text{g of Mg}}{24.3\ \text{g/mol of Mg}} = 0.2\ \text{mol of Mg}$$

$$(0.2\ \text{mol of Mg})\left(\frac{2\ \text{mol of N}}{3\ \text{mol of Mg}}\right) = 0.133\ \text{mol of N}$$

$$(0.133\ \text{mol of N})\left(\frac{14\ \text{g of N}}{1\ \text{mol of N}}\right) = 1.87\ \text{g of N}$$

39. $$10\ \text{g of } K_3Fe(C_2O_4)_3 = \frac{100\ \text{g}}{437.2\ \text{g/mol}} = 0.0229\ \text{mole of } K_3Fe(C_2O_4)_3$$

1 mole of $K_3Fe(C_2O_4)_3$ requires 3 mole atoms, or 1 mole $K_2C_2O_4$, and 1 mole of Fe atoms, or 1 mol of $Fe(NH_4)_2(SO_4)_2$ $6H_2O$.

$$\text{Amount of } K_2C_2O_4 \text{ required} = 0.0229 \text{ mol of } K_3Fe(C_2O_4)_3 \times \frac{\frac{3}{2} \text{ mol of } K_2C_2O_4}{1 \text{ mol of } K_3Fe(C_2O_4)_3}$$

$$= 0.0343 \text{ mol of } K_2C_2O_4 = (0.0343 \text{ mol}) \left(166.2 \frac{R}{\text{mol}}\right) = 5.70 \text{ g}$$

Amount of $Fe(NH_4)_2(SO_4)_2 \cdot 6H_2O$ required

$$= 0.0229 \text{ mol of } K_3Fe(C_2O_4)_3 \times \frac{1 \text{ mol of } Fe((NH_4)_2(SO_4)_2 \cdot 6H_2O}{1 \text{ mol } K_3Fe(C_2O)_3}$$

$$= 0.0229 \text{ mole of } Fe(NH_4)_2 \text{ mol of } Fe(NH_4)_2(SO_4)_2 \cdot 6H_2O$$

$$= (0.0229 \text{ mol}) \left(392.1 \frac{g}{\text{mol}}\right) = 9.98 \text{ g}$$

40. The number of moles of gas is given by

$$\text{moles} = \frac{\text{mass (grams)}}{\text{molecular weight of } COCl_2 \text{ (g.mol}^{-1})}$$

The mass of gas is 1.1×10^8 kg = 1.1×10^8 kg $\times 10^3$ g/kg = 1.1×10^{11} g. the molecule weight of $COCl_2$ is = 12 g.mol^{-1} + 16 g.mol^{-1} + 2 × 35.5 g.mol^{-1} = 99 g.mol^{-1}.
Hence, the number of moles of gas is

$$\text{moles} = \frac{\text{mass}}{\text{molecular weight}} = \frac{1.1 \times 10^{11} \text{ g}}{99 \text{ g.mol}^{-1}} \cong 1.1 \times 10^9 \text{ moles.}$$

Multiplying the number of moles by Avogadro's number, we obtain the number of molecules of gas:

Number of molecules = moles × Avogadro's number
= 1.1×10^9 moles × 6×10^{23} molecules/mole = 6.6×10^{32} molecules.

41. ∴ 1 mole of H_2 weighs 2.016 grams. So, number of moles present is

$$= \frac{\text{grams}}{\text{M.wt. } H_2} = \frac{.4512 \text{ g}}{2.016 \text{ g.mol}^{-1}} = .2238 \text{ mole of } H_2$$

To calculate the number of molecules, recall that 1 mole of any gas at STP has 6.023×10^{23} molecule/mole.
Therefore,

$$\text{No. of molecules} = 6.023 \times 10^{23} \frac{\text{moleucles}}{\text{mole}} \times \text{no. of moles}$$

$$= 6.023 \times 10^{23} \frac{\text{moleucles}}{\text{mole}} \times 0.2238 \text{ moles} = 1.348 \times 10^{23} \text{ molecules of } H_2.$$

42. The mass of CO is 9.1×10^6 kg = 9.1×10^6 kg × 1000 g/kg = 9.1×10^9 g. The molecular weight of CO is the sum of the atomic weight of C and the atromic weight O, or molecular weight of (CO) = 12 $g.mol^{-1}$ + 16 $g.mol^{-1}$ = 28 $g.mol^{-1}$.

$$\text{Hence, moles} = \frac{\text{mass (g)}}{\text{molecular weight } (g.mol^{-1})} = \frac{9.1 \times 10^9 \text{ g}}{28 \text{ g.mol}^{-1}} = 3.3 \times 10^8 \text{ moles.}$$

43. Since the simplest formula gives the relative number of moles of atoms of the different kinds in the compound, we must calculate how many moles of *M* atoms and how many moles of *S* atoms have combined.

$$\frac{2.435 \text{ g of M}}{121.75 \text{ of M/mol of M}} = 0.02 \text{ mol of M and } \frac{0.952 \text{ g of S}}{32.06 \text{ g of S/mol of S}} = 0.03 \text{ mol of S}$$

The simplest formula shows this ratio as subscripts, and so we write $M_{0.0200}S_{0.0300}$. We can simplify the subscripts by dividing all of them by the smallest subscript to get $M_1S_{1.5}$ and then doubling to get M_2S_3.

44. The weight loss of 0.384 g is due to the fact that a gas is driven off. The only gas formed in the reaction is oxygen.

$2KClO_3(s) \rightarrow 2KCl(s) + 3O_2(g)$

$$\text{Moles of } O_2 \text{ driven off} = \frac{0.384 \text{ g of } O_2}{32.0 \text{ g.mol}^{-1}} = 0.012 \text{ mol}$$

$$\text{Moles of } KClO_3 \text{ decomposed} = (0.012 \text{ mol of } O_2) \frac{2 \text{ mol of } KClO_3}{3 \text{ mol of } O_2} = 0.008 \text{ mol}$$

$$\text{Moles of } KClO_3 \text{ originally available} = \frac{4.9 \text{ g of } KClO_3}{122.5 \text{ g.mol}^{-1}} = 0.04 \text{ mol}$$

$$\text{Percent decomposed} = \frac{\text{moles decomposd}}{\text{moles available}} \times 100 = \frac{0.008}{0.04} \times 100 = 20.0\%$$

45. To get the simplest formula, we need to know the relative number of moles C, N, and O atoms in some fixed weight of compound. We can take any convenient weight of compound, 1 g, 12.011 g or any other weight. In practice it is most convenient to solve this problem by taking 100 g of compound since then the numbers giving percentage in weights correspond directly to the grams of each element in the sample. All we need to do then is to convert each of the weights into moles of atoms. We do this by dividing each weight of element in the number of grams needed to make 1 mol of atoms. Thus, in 100.0 g of compound there would be 49.48 g of C 5.19 g of H, 28.85 g of N_2 and 16.48 g O_2 or

$$\frac{49.48 \text{ g of C}}{12.011 \text{ g of C/mol of C atoms}} = 1420 \text{ mol of C atom}$$

$$\frac{5.19 \text{ g of H}}{1.0079 \text{ g H/mol of H atoms}} = 5.15 \text{ mol of H atom}$$

$$\frac{28.86 \text{ g of } N_2}{14.0067 \text{ g of N/mol of N atoms}} = 2.060 \text{ mol of N atom}$$

$$\frac{16.48 \text{ g of } O_2}{15.999 \text{ g O/mol of O atoms}} = 1.03 \text{ mol of O atom}$$

The simplest formula can be written as $C_{1.42}\ H_{5.15}N_{2.06}O_{1.03}$, Now dividing all the subscripts by 1.03 we get $C_4H_5N_2O$.

46. 1. The 1.647-g mass corresponds to the anhydrous compound $CuSO_4$. The loss of mass (2.754 – 1.647) g = 0.927 g is the mass of water of hydration in the original sample. Convert the masses of the anhydrous compound and of the water to number of moles.

$$\text{No. mol } CuSO_4 = 1.647 \text{ g } CuSO_4 \times \frac{1 \text{ mol } CaSO_4}{159.6 \text{ g } CuSO_4} = 0.01032 \text{ mol } CuSO_4$$

$$\text{No. mol } H_2O = 0.927 \text{ g } H_2O \times \frac{1 \text{ mol } H_2O}{18.02 \text{ g } H_2O} = 0.05144 \text{ mol } H_2O$$

Now, write a formula based on these molar amounts

$$(CuSO_4)_{0.01032} \cdot (H_2O)_{0.05144}$$

and reduce to the simplest formula.

$$(CusO_4)_{\frac{0.01032}{0.01032}} \cdot (H_2O)_{\frac{0.05144}{0.01032}} = (CuSO_4) \cdot (H_2O)_{4.98} = CuSO_4 \cdot 5H_2O$$

2. The loss of water occurring at 413 K is (2.574 – 1.833) = 0.741 g. The mass of water *remaining* at 413 K is the *total* loss of water (0.927 g) minus the loss that occurs at 413 K (0.741 g): 0.927 – 0.741 = 0.186 g H_2O. Now express this amount of water on a mole basis.

$$\text{no. mol } H_2O = 0.186 \text{ g } H_2O \times \frac{1 \text{ mol } H_2O}{18.02 \text{ g } H_2O} = 0.01032 \text{ mol } H_2O$$

The number of moles of $CuSO_4$ was determined in part 1. It is 0.01032 mol. The $CuSO_4$ and the H_2O are in a 1:1 mol ratio. The formula of the hydrate obtained at 413 K is $CuSO_4\ H_2O$.

3. Recall that the 1.647-g anhydrous $CuSO_4$ sample corresponds to 0.01032 mol. The 0.01032 mol Cu present in this sample appears in the 0.812 g of copper oxide residue. The mass of Cu, percentage, and the percent O in copper oxide are

$$\text{no. of g Cu} = 0.01032 \text{ Cu} \times \frac{63.546 \text{ g Cu}}{1 \text{ mol Cu}} = 0.6558 \text{ g Cu}$$

$$\% \text{ Cu} = \frac{0.6558 \text{ g Cu}}{0.812 \text{ g}} \times 100 = 80.8\%$$

$$\% \text{ O} = 100.0 - 80.8 = 19.2\%$$

We can readily determine the empirical formula from the percentage composition.

$$\text{no. mol Cu} = 80.8 \text{ g Cu} \times \frac{1 \text{ mol Cu}}{63.55 \text{ g Cu}} = 1.27 \text{ mol Cu}$$

$$\text{no. mol O} = 19.2 \text{ g O} \times \frac{1 \text{ mol O}}{16.00 \text{ g O}} = 1.20 \text{ mol O}$$

The empirical formula is $Cu_{1.27}O_{1.20} = Cu_{1.06}O = CuO$.

47. First, convert the mass of K to number of moles of K.

$$\text{no. mol } K = 370 \text{ mg K} \times \frac{1.00 \text{ g K}}{100 \text{ mg K}} \times \frac{1 \text{ mol K}}{39.10 \text{ g K}} = 9.46 \times 10^{-3} \text{ mol K}$$

Then convert the number of moles of K to number of K atoms.

$$\text{No. K atoms} = 9.46 \times 10^{-3} \text{ mol K} \times \frac{6.022 \times 10^{23} \text{ K atoms}}{1 \text{ mol K}}$$

$$= 5.7 \times 10^{21} \text{ K atoms.}$$

Finally, use the percent natural abundance of K-40 to formulate a factor to convert from number of K atoms to number of K-40 atoms.

$$\text{No. of K-40 atoms} = 5.7 \times 10^{21} \text{ K atoms} \times \frac{0.012 \text{ K-40 atoms}}{100 \text{ K atoms}} = 6.84 \times 10 \text{ atoms.}$$

48. From the stoichiometry of this equation, 1 mole of $CaCN_2$ produces 2 moles of ammonia. The number of moles of $CaCN_2$ (MW = 80 g mol^{-1}) in 200 kg is

$$\frac{200{,}000 \text{ g } CaCN_2}{80 \text{ g mole}^{-1}} = 2{,}500 \text{ moles } CaCN_2$$

Thus, (2,500 × 2=) 5,000 moles of ammonia is produced and the weight of ammonia is its molecular weight (MW = 17 g.mol^{-1}) times the number of moles produced: (17 g.mol^{-1}) (5000 moles) = 85,000 g = 85 kg of ammonia produced.

49. 1. The series of conversion required here is

$$\text{g } Na_2CO_3 \rightarrow \text{mol } Na_2CO_3 \rightarrow \text{mol } NaNO_2 \rightarrow \text{g } NaNO_2$$

These conversion can be done on a step-by-step basis or in a single setup.

$$\text{No. of } NaNO_2 = 748 \text{ g } Na_2CO_3 \times \frac{1 \text{ mol } Na_2CO_3}{106.0 \text{ g } Na_2CO_3} \times \frac{4 \text{ mol } NaNO_2}{2 \text{ mol } Na_2CO_3} \times \frac{69 \text{ g } NaNO_2}{1 \text{ mol } NaNO_2}$$

(g Na_2CO_3 → mol Na_2CO_3 → mol $NaNO_2$

$$= 974 \text{ g } NaNO_2$$

2. (a) This problem is best broken is best down into three smaller problems, Which is the limiting reagent? O_2(g) is stated to be in excess. We must choose between Na_2CO_3 and NO.

Assuming Na_2CO_3 is the limiting reagent:

$$\text{No. mol } NaNO_2 = 0.225 \text{ L} \times \frac{1.5 \text{ mol } Na_2CO_3}{1 \text{ L}} \times \frac{4 \text{ mol } NaNO_2}{2 \text{ mol } Na_2CO_3} = 0.675 \text{ mol } NaNO_2$$

Assuming NO is the limiting reagent:

$$\text{No. mol } NaNO_2 = 22.1 \text{ g NO} \times \frac{1 \text{ mol NO}}{30.01 \text{ g NO}} \times \frac{4 \text{ mol } NaNO_2}{4 \text{ mol NO}} = 0.736 \text{ mol } NaNO_2$$

Since 0.675 mol $NaNO_2$ is the smaller of the two results, Na_2CO_3 is the limiting reagent.

(b) no. g $NaNO_2$ = 0.675 mol $NaNO_2 \times \frac{69 \text{ g NaNO}_2}{1 \text{ mol NaNO}_2}$ = 16.6 g $NaNO_2$

(c) actual yield = $\frac{\% \text{ yield} \times \text{theoretical yield}}{100} = \frac{95 \times 46.6 \text{ g}}{100}$ = 44.3 g $NaNO_2$

50. (a) We have:

$$2C_8H_{10}N_4O_2 + 27O_2 \rightarrow 16CO_2 + 10H_2O + 8NO_2$$

or in words, that 2 mol of $C_8H_{10}N_4O_2$ required 27 mol of O_2 for the reaction.

$\therefore$ No. of moles of $C_8H_{10}N_4O_2 = \frac{1.0 \text{ g of } C_8H_{10}N_4O_2}{194.1 \text{ g of } C_8H_{10}N_4O_2/\text{mol of } C_8H_{10}N_4O_2}$

= 0.00515 mol of $C_8H_{10}N_4O_2$

The equation tells us we need 27 mol of O_2 per 2 mol of $C_8H_{10}N_4O_2$, and so we have

$$(0.00515 \text{ mol of } C_8H_{10}N_4O_2)\left(\frac{27 \text{ mol } O_2}{2 \text{ mol } C_8H_{10}N_4O_2}\right) = 0.0695 \text{ mol of } O_2$$

To convert this number of moles O_2 to mass, we need to multiply by the number of grams per mole of O_2, 32 g.

$$(0.0695 \text{ mol of } O_2)\left(\frac{32 \text{ g of } O_2}{1 \text{ mol of } O_2}\right) = 2.22 \text{ g}$$

(b) $\frac{1.0 \text{ g of } C_8N_{10}N_4O_2}{194.1 \text{ g of } C_8H_{10}N_4O_2/\text{mol of } C_8H_{10}N_4O_2}$ = 0.00515 mol of $C_8H_{10}N_4O_2$

The equation tells us that there are 8 mol of NO_2 produced when 2 mol of $C_8H_{10}N_4O_2$ are consumed. Therefore, we can write

$$(0.00515 \text{ mol of } C_8H_{10}N_4O_2)\left(\frac{8 \text{ mol of } NO_2}{2 \text{ mol of } C_8H_{10}N_4O_2}\right) = 0.0206 \text{ mol of } NO_2$$

As noted above, 1 mol of NO_2 weighs 46.0 g. The final answer is given by

$$(0.0206 \text{ mol of } NO_2)\left(\frac{46.0 \text{ g of } NO_2}{1 \text{ mol of } NO_2}\right) = 0.948 \text{ g of } NO_2$$

51. The equation tells us that there are 10 mol of H_2O produced per 16 mol of CO_2 and so all we have to do is figure out how many moles of CO_2 we have, multiply that by $\frac{10}{16}$, and then convert to grams of H_2O

$$\frac{1.0 \text{ g of } CO_2}{44.0 \text{ g of } CO_2/\text{mol of } CO_2} = .00228 \text{ moles of } CO_2$$

$$(0.0227 \text{ mol of } CO_2)\left(\frac{10 \text{ mol of } H_2O}{16 \text{ mol of } CO_2}\right) = 0.0142 \text{ mol of } H_2O$$

$$(0.0142 \text{ mol of } H_2O)\left(\frac{18.0 \text{ g of } H_2O}{1 \text{ mol of } H_2O}\right) = 0.256 \text{ mol of } H_2O$$

52. The first step is to convert the weight into moles:

$$\frac{1.0 \text{ g of Mg}}{\text{g.mol}^{-1}} = 0.0412 \text{ mol of Mg}$$

$$\frac{1.0 \text{ g of S}}{32.1 \text{ g.mol}^{-1}} = 0.0312 \text{ mol of S}$$

The equation Mg + S $\rightarrow$ MgS tells us that we need 1 mol of Mg per 1 mol of S. We are given more moles of Mg than of S. Therefore, the Mg is in excess, and not all of its will be used up. The given moles of S control the amount of produce. We get 1 mol of MgS per 1 mol of S. and so we can write

$$(0.0132 \text{ mol of S})\left(\frac{1 \text{ mol of MgS}}{1 \text{ mol of S}}\right) = 0.0312 \text{ mol of MgS}$$

To get the weight of MgS produced, we multiply by the weight of 1 mol of MgS:

$$(0.0132 \text{ mol of MgS})\left(\frac{56.4 \text{ g of MgS}}{1 \text{ mol of MgS}}\right) = 1.76 \text{ g of MgS}$$

53. The trick about this equation lies in the implification that more than 1.0 g of P_4O_{10} or of SO_2 may be produced. However, we do not know at first which product will be produced in excess over that required. The most direct way to solve the problem is to calculate how much P_4S_3 would be required to produce 1.0 g of P_4O_{10} and how much would be needed to produce 1.0 g of SO_2. The larger of the two answers would satisfy both requirements.

To solve, we convert the designed weights into moles and use the chemical equation to determine how many moles of P_4S_3 are needed.

(a) To make the 1.0 g of P_4O_{10}:

$$\frac{1.0 \text{ g of } P_4O_{10}}{283.9 \text{ g of } P_4O_{10}/\text{mol of } P_4O_{10}} = 0.00352 \text{ mol of } P_4O_{10}$$

The equation tells us 1 mol of P_4S_3 is required to produce 1 mol of P_4O_{10}:

$$(0.00352 \text{ mol of } P_4O_{10})\left(\frac{1 \text{ mol of } P_4S_2}{1 \text{ mol of } P_4O_{10}}\right) = 0.00352 \text{ mol of } P_4S_3 \text{ required}$$

(b) To make the 1.0 g of SO_2:

$$\frac{1.0 \text{ g of } SO_2}{64.06 \text{ g of } SO_2/\text{mol of } SO_2} = 0.0156 \text{ mol of } SO_2$$

The equation tells us 1 mol of P_4S_3 is required to produce 3 mol of SO_2:

$$(0.0156 \text{ mol of } SO_2)\left(\frac{1 \text{ mol of } P_4S_2}{3 \text{ mol of } SO_2}\right) = 0.0052 \text{ mol of } P_4S_3 \text{ required}$$

As we can see it takes more P_4S_3 to make 1.0 g of SO_2 than it does to make 1.0 g of P_4O_{10}. To satisfy both required products, we need to take the **larger** of the two amounts (the amount of P_4S_3) required:

$$(0.0052 \text{ mol } P_4S_3)\left(\frac{220.1 \text{ g of } P_4S_3}{1 \text{ mol of } P_4S_2}\right) = 1.14 \text{ g of } P_4S_3$$

54. The products here are different, and so we have a new equation to balance. Given

$$C_8H_{10}N_4O_2 + ?O_2 \rightarrow ?CO + ?H_2O + ?NO$$

The balanced equation will be: $C_8H_{10}N_4O_2 + \frac{15}{2} \rightarrow 8CO + 5H_2O + 4NO$

Now, we are given a quantity of $C_8H_{10}N_4O$ and a quantity of O_2. Only one of these will control the reaction; the other reagent will be excess. The problem is to decide which reagent will control the amount of reaction, i.e., which is the limiting reagent. The way to tell is to convert each of the given weight to moles and compare with the desired ratio of $1 : \frac{15}{2}$.

$$\frac{1.0 \text{ g of } C_8H_{10}N_4O_2}{194.1 \text{ g of } C_8H_{10}N_4O_2/\text{mol of } C_8H_{10}N_4O_2} = 0.00515 \text{ mol of } C_8H_{10}N_4O_2$$

$$\frac{2 \text{ g of } O_2}{32.0 \text{ g of } O_2/\text{mol of } O_2} = 0.0625 \text{ mol of } O_2$$

According to the equation, we need $\frac{15}{2}$ mol of O_2 per mole of $C_8H_{10}N_4O_2$, and so 0.00515 mol of $C_8H_{10}N_4O_2$ would require

$$(0.00515 \text{ mol of } C_8H_{10}N_4O_2)\left(\frac{\frac{15}{2} \text{ mol of } O_2}{1 \text{ mol of } C_8H_{10}N_4O_2}\right) = 0.0386 \text{ mol of } O_2$$

Clearly, 0.0386 mol of O_2 is less than the 0.0675 mol of O_2 we are being provided with, and so the given quantity of O_2 is an excess over that required for reaction. Some of the given O_2 will be left over, and calculations based on the quantity of O_2 given will not tell us how much product actually forms. To find how much products is generated, we need to base out calculation one the limiting reagent, the $C_8H_{10}N_4O_2$. Actually, once we decide which is the limiting reagent, we have practically solved the problem. We know how many moles of $C_8H_{10}N_4O_2$ are provided, 0.00515. We can also see from the balanced equation that 5 mol of H_2O are produced per mole of $C_8H_{10}N_4O_2$.

$$\therefore \quad (0.00515 \text{ mol of } C_8H_{10}N_4O_2)\left(\frac{5 \text{ mol of } H_2O}{1 \text{ mol of } C_8H_{10}N_4O_2}\right) = 0.0258 \text{ mol of } H_2O$$

To convert this answer to grams, we multiply by the weight of 1 mol of H_2O:

$$(0.0258 \text{ mol of } H_2O)\left(\frac{18 \text{ g of } H_2O}{1 \text{ mol of } H_2O}\right) = 0.464 \text{ g of } H_2O$$

55. In solving this problem, the first thing to do is to write the balanced equation:

$$3CO(g) + Fe_2O_3(s) \rightarrow 3CO_2(g) + 2Fe(s)$$

Next we decide which reactant limits the amount of product and which reactant is present in excess. To do this, we convert the given data into moles. The formula weight of CO is 28 amu; the formula weight of Fe_2O_3 is 159.7 amu.

$$\frac{1.0 \text{ g of CO}}{28.0 \text{ g.mol}^{-1}} = 0.0357 \text{ mol of CO}$$

$$\frac{5.0 \text{ g of } Fe_2O_3}{159.7 \text{ g.mol}^{-1}} = 0.0313 \text{ mol of } Fe_2O_3$$

The equation tell us there should be 1 mol of Fe_2O_3 per 3 mole of CO, so that 0.0357 mol CO would require

$$(0.00357 \text{ mol of CO}) \frac{1 \text{ mol of } Fe_2O_3}{3 \text{ mol of CO}} = 0.0119 \text{ mol of } Fe_2O_3$$

Clearly, there is more than enough Fe_2O_3 to supply its needed amount, so that the Fe_2O_3 is excess. CO is the limiting reagent. The calculation of Fe product should be based on CO.

$$(0.00357 \text{ mol of CO}) \frac{2 \text{ mol of Fe}}{3 \text{ mol of CO}} = 0.0238 \text{ mol of Fe}$$

$$(0.0238 \text{ mol of Fe})\left(55.85 \text{ g.mol}^{-1}\right) = 1.33 \text{ g}$$

56. 1. Let the oxidation number of Mn be x. The sum of the oxidation numbers of atoms of a molecules is zero.

K MnO_4

$$+1 + x + (-2 \times 4) = 0$$

$$1 + x - 8 = 0$$

$$\Rightarrow \quad x = 8 - 1 = +7$$

2. Find the oxidation number of hydrogen in the following.

	H_2	HF	H_2O_2	KH	CaH_2
	0	$x - 1 = 0$	$(2x - 1) \times 2 = 0$	$+1 + x = 0$	$2 + 2x = 0$
$\Rightarrow$		$x = +1$	$2x = +2$	$x = -1$	$x = -1$

58. (a) Oxidation number of sulphur increases from –2 to 0. H_2S is oxidised to sulphur, oxidation number of chorine decreases from 0 to –1. Chlorine is reduced HCl.

(b) $3Br_2 + 6NaOH \rightarrow NaBrO_3 + 5\,NaBr + 3H_2O$

Bromine is both oxidised and reduced. It is oxidised to sodium bromate (oxidation number of bromine $0 \rightarrow 5$) and it is reduced to sodium bromide (oxidation number of bromine $0 \rightarrow -1$).

(c) $3Cu + 8HNO_3 \rightarrow 3Cu(NO_3)_2 + 4H_2O + 2NO$

The oxidation number of Cu has increased from 0 to +2. Therefore, Cu is oxidised. The oxidation number of nitrogen in HNO_3 has decreased from +5 to +2 in NO. Therefore 2 nitrogen atoms are reduced. The oxidation number nitrogen in $(CuNO_3)_2$ is also +5. There is no change in the oxidation number. Hence there is neither reduction nor oxidation in the case of six nitrogen atoms.

(d) $2H_2O_2 \rightarrow 2H_2O + O_2$

The oxygen in H2O2 has an oxidation number of –1. Therefore 2 oxygen atoms are reduced (oxidation number –1 to –2) and 2 oxygen atoms are oxidised (oxidation number –1 to 0). Since no other substance is involved in the reaction, it is called auto oxidation and auto reduction.

58. First we need to balance the equation

$$?KClO_3(s) \rightarrow ?KCl(s) + O_2(g)$$

+5 – 2 (in $KClO_3$); –1 (in KCl); 0 (in O_2)

$6e^-$ (Cl: +5 → –1); $2e^- \times 3$ (O: –2 → 0)

The Cl changes from +5 to –1; ait appears to gain 6 electrons. The O changes from –2 to 0; it appears to lose 2 electrons. There are three oxygen atoms for every chlorine atom in $KClO_3$, and so the compound itself has taken care of the electron gain and electron loss. To fix the right side of the equation, we note that one $KClO_3$ on the left produces one KCl on the right. Three oxygen atoms on the left require three oxygen atoms on the right. We can get these three oxygen atoms on the right by placing the coefficient $\frac{3}{2}$ before the formula O_2, giving

$$KClO_3(s) \rightarrow KCl(s) + \frac{3}{2}O_2(g)$$

Multiplying through by 2 to get rid of the fraction gives

$$2KClO_3(s) \rightarrow 2KCl(s) + 3O_2(g)$$

We now have the balanced equation and can proceed to solve the problem by converting the given oxygen weight to moles and calculating how much $KClO_3$ is required to produce this:

$$0.96 \text{ g } O_2 = \frac{0.96 \text{ g}}{32 \text{ g.mol}^{-1}} = 0.03 \text{ mol of } O_2$$

$$(0.03 \text{ mol of } O_2)\ \frac{2 \text{ mol of } KClO_2}{3 \text{ mol of } O_2} = 0.02 \text{ mol of } KClO_3$$

$$(0.02 \text{ mol of } KClO_3)\ (122.5 \text{ g.mol}^{-1}) = 2.5 \text{ g}$$

59. Use a given number of quartz (i.e. 1000 g) to calculate the amount of gold present and the amount of HF solution needed. Then calculate the cost of each.

$$1000 \text{ g } SiO_2 \times \frac{1.0 \times 10^3 \text{ g Au}}{100 \text{ g } SiO_2} = 0.01 \text{ g Au}$$

$$1000 \text{ g } SiO_2 \times \frac{1 \text{ mol } SiO_2}{0.08 \text{ g } SiO_2} \times \frac{4 \text{ mol HF}}{1 \text{mol } SiO_2} \times \frac{20.01 \text{ g}}{1 \text{ mol HF}} = 1330 \text{ g HF}$$

$$1330 \text{ g HF} \times \frac{100 \text{ g sol'n}}{50 \text{ g HF}} \times \frac{1 \text{ mL}}{1.17 \text{ g sol'n}} = 2270 \text{ L solution}$$

$$2270 \text{ mL} \times \frac{1 \text{ L}}{1000 \text{ mL}} \times \frac{\text{Rs } 75}{1 \text{ L}} = \text{Rs } 170.25$$

$$0.01 \text{ g Au} \times \frac{12750}{1.0 \text{ g}} = \text{Rs } 127.5$$

Thus Rs. 170 of HF solution is needed to recover Rs. 127 of gold. Definitely not economical

60. Determine the moles ratios of the redundant and oxidant for each redox couple. Then use mass relationships of thiosulfate to iodine and iodine to NaClO to determine the mass of NaClO and from that, the mass percent of NaClO in the bleach solution.

$S^{2+} \rightarrow S^{2.5+} + 1/2e^-$

$2S_2O_3^{2-} \rightarrow S_4O_6^{2-} + 2e^-$

$I^0 + e^- \rightarrow I^-$

$I_2 + 2e^- \rightarrow 2I^-$

$2S_2O_3^{2-} + I_2 \rightarrow 2I^- + S_4O_6^{2-}$

This gives the mole ratio $S_2O_3^{2-}$ to I_2 for the calculation of moles of I_2.

$$25 \text{ mL} \times \frac{1 \text{ L}}{1000 \text{ mL}} \times \frac{0.07 \text{ mol } S_2O_3^{2-}}{1 \text{ L}} \times \frac{1 \text{ mol } I_2}{2 \text{ mol } S_2O_3^{2-}} = 8.75 \times 10^{-4} \text{ mol } I_2$$

$Cl^+(aq) + 2e^- \rightarrow Cl^-(aq)$

$2H^+(aq) + ClO^-(aq) + 2e^- \rightarrow Cl^-(aq) + H_2O(l)$

$2I^- \rightarrow I_2 + 2e^-$

$2H^+(aq) + ClO^-(aq) + 2I^-(aq) \rightarrow I_2(aq) + Cl^-(aq) + H_2O(l)$

This gives the mole ratio of $ClO^-(aq)$ (or NaClO) to I_2 for the calculation of mass of NaClO.

$$8.75 \times 10^{-4} \text{ mol. } I_2 \times \frac{1 \text{ mol NaClO}}{1 \text{ mol } I_2} \times \frac{74.44 \text{ g NaClO}}{1 \text{ mol. NaClO}} = 0.0651 \text{ g NaClO}$$

$$\text{mass \%} = \frac{0.0651 \text{ g NaClO}^-}{5 \text{ g sol'n}} \times 100\% = 1.30\%$$

61. Calculate the amount Pb that was consumed, and then the amount of EDTA need bind the lead. Recall the EDTA forms a 1:1 complex with Pb.

$$10 \text{ g paint} \times \frac{5 \text{ g Pb}}{100 \text{ g paint}} \times \frac{1 \text{ mol. Pb}}{207.2 \text{ g}} = 2.41 \times 10^{-3} \text{ mol. Pb}$$

$Na_4(EDTA) = Na_4(C_{10}H_{12}N_2O_8)$

$M_{EDTA} = 380.2 \text{ g.mol}^{-1}$

$$2.41 \times 10^{-3} \text{ mol. Pb} \times \frac{1 \text{ mol EDTA}}{1 \text{ mol Pb}} \times \frac{1 \text{ mol } Na_4 \text{ EDTA}}{1 \text{ mol EDTA}} \times \frac{380.2 \text{ g}}{1 \text{ mol } Na_4 \text{ EDTA}} = 0.92 \text{ g EDTA}$$

62. (a) The oxalic acid is a chelating ligand, and Fe^{3+} has 6 coordination sites, suggesting that a complex with 3 oxalates forms. Recall the oxalic acid is a diprotic acid (has 2 H^+)

$$Fe(OH)_3(s) + 3H_2C_2O_4(aq) \rightarrow [Fe(C_2O_4)_3]^{3-}(aq) + 3H_2O(l) + 3H^+(aq)$$

(b) $1.0 \text{ g Fe(OH)} \times \dfrac{1 \text{ mol Fe(OH)}_3}{106.9 \text{ g}} \times \dfrac{3 \text{ mol }_2C_2O_4}{1 \text{ mol. Fe(OH)}_3} \times \dfrac{1 \text{ L solution}}{0.10 \text{ mol. } H_2C_2O_4} = 0.28 \text{ L:}$

63. volume of a cube = s^3

$$s = 0.409 \text{ nm} \times \frac{1 \text{ m}}{1 \times 10^9 \text{ mm}} \times \frac{100 \text{ cm}}{1 \text{ m}} = 4.09 \times 10^{-8} \text{ cm}$$

$$V = (4.09 \times 10^{-8} \text{ cm})^3 = 6.84 \times 10^{-23} \text{ cm}^3$$

$$\frac{4 \text{ atoms}}{684 \times 10^{-23} \text{ cm}^3} \times \frac{1 \text{ cm}^3}{10.5 \text{ g}} \times \frac{107.87 \text{ g Ag}}{1 \text{ mole Ag}} = 6.01 \times 10^{23} \text{ atoms mol}^{-1}$$

64. First write balance chemical equation. Then calculate the amount CaO formed and the amount of Ca needed to form the CaO. The remaining Ca will have formed the Ca_3N_2.

Reactions:

$$2Ca + O_2 \rightarrow 2Ca$$

$$3Ca + N_2 \rightarrow Ca_3N_2$$

$$CaO + H_2O \rightarrow Ca(OH)_2$$

$$4.832 \text{ g Ca(OH)}_2 \times \frac{1 \text{ mol Ca(OH)}_2}{74.10 \text{ g Ca(OH)}_2} \times \frac{1 \text{ mol CaO}}{1 \text{ mol. Ca(OH)}_2} \times \frac{56.08 \text{ g CaO}}{1 \text{ mol CaO}} = 3.657 \text{ g CaO}$$

$$3.657 \text{ g CaO} \times \frac{1 \text{ mol. CaO}}{56.08 \text{ g CaO}} \times \frac{2 \text{ mol Ca}}{2 \text{ mol. CaO}} \times \frac{40.08 \text{ g Ca}}{1 \text{ mol Ca}} = 2.614 \text{ Ca}$$

5.025 g Ca(original –2.614 g Ca (to form CaO) = 2.411 g Ca (to form Ca_3N_2)

$$2.411 \text{ g Ca} \times \frac{1 \text{ mol. Ca}}{40.08 \text{ Ca}} \times \frac{1 \text{ mole } Ca_3N_2}{3 \text{ mol Ca}} \times \frac{148.26 \text{ g } Ca_3N_2}{1 \text{ mol. } Ca_3N_2} = 2.973 \text{ g } Ca_3N_2$$

65. First, calculate the mass lost on converting KBr to KCl. This corresponding to the difference in masses between Br and Cl. Then calculate the mass the of KBr and the percentage.
79.90 $g.mol^{-1}$. Br –35.45 $g.mol^{-1}$. Cl = 44.45 $g.mol^{-1}$ (of Br replaced with Cl)
3.595 g (original mixture) –3.129 g (final KCl) = 0.466 g

$$0.466 \times \frac{1 \text{ mol}}{44.45 \text{ g}} = 0.0105 \text{ mol. KBr converted to KCl.}$$

$$0.0105 \text{ mol. KBr} \times \frac{119.0 \text{ g KBr}}{1 \text{ mol. KBr}} = 1.25 \text{ g KBr}$$

$$\text{mass \%} = \frac{1.25 \text{ g}}{3.595 \text{ g}} \times 100\% \; 34.8\%$$

66. (a) Both oxides contain 2.573 g V. The first also contains (4.589 g VO_X – 2.573 g V) 2.016 g O, while the second contains (3.782 g VO_Y – 2.573 g V) 1.209 g O.

$$2.573 \text{ g V} \times \frac{1 \text{ mol. V}}{50.94 \text{ g V}} = 0.005051 \text{ mol. V}$$

$$2.016 \text{ g O} \times \frac{1 \text{ mol O}}{16 \text{ g O}} = 0.1260 \text{ mol. O} \qquad \text{(1st oxide)}$$

$$1.209 \text{ g O} \times \frac{1 \text{ mol. O}}{16.00 \text{ g O}} = 0.07556 \text{ mol. O} \qquad \text{(2nd oxide)}$$

0.05051 mol./0.05051 mol. = 1.0 mol. V

0.07556 mol./0.05051 mol. = 1.496 mol. O

0.1260 mol./0.05051 mol. = 2.496 mol. O

Ist oxide: V_2O_3

2nd oxide: V_2O_5

(b) Calculate the mass of H_2O formed using the mass of O lost. The second heating converting all the oxygen to water.

$$2.016 \text{ g O} \times \frac{1 \text{ mol. O}}{16 \text{ g O}} \times \frac{1 \text{ mol } H_2O}{1 \text{ mol. O}} \times \frac{18.02 \text{ g } H_2O}{1 \text{ mol. } H_2O} = 2.271 \text{ g } H_2O$$

67. First, calculate the amount of CO_2 expected from burning cocaine and sucrose using variables for the masses of those compounds. Remember that cocaine has 17 C atoms and will thus produce 17 CO_2 atoms; similarly, sucrose will produce 12 CO_2's. Then set up a algebraic equation and solve for the mass of cocaine in the sample. Then calculate the ass percent.

Molar mass of $C_{17}H_{21}O_4N$ = 303.39 g/mol.
Molar mass of $C_{12}H_{22}O_{11}$ = 342.34 g/mol.

$$\text{mass } CO_2 = 1.0 \text{ mL} \times \frac{1 \text{ L}}{100 \text{ mL}} \times \frac{1.8 \text{ g}}{1 \text{ L}} = 1.8 \times 10^{-3} \text{ g}$$

$$\text{mass of sugar + cocaine} = 1.0 \text{ mg} \times \frac{1 \text{ g}}{1000 \text{ mg}} = 1.0 \times 10^{-3} \text{ g}$$

Let X = mass of cocaine, then

$(1.0 \times 10^{-3} - X)$ = mass of sucrose

$$X \text{ g cocaine} \times \frac{1 \text{ mol cocaine}}{303.39 \text{ g cocaine}} \times \frac{17 \text{ mol. } CO_2}{1 \text{ mol cocaine}} \times \frac{44.01 \text{ g } CO_2}{1 \text{ mol } CO_2} = 2.466X \text{ g } CO_2$$

$$(1.0 \times 10^{-3} - X) \text{ g sucrose} \times \frac{1 \text{ mole sucrose}}{342.34 \text{ g sucrose}} \times \frac{12 \text{ mol. } CO_2}{1 \text{ mol. sucrose}} \times \frac{44.01 \text{ g } CO_2}{1 \text{ mol. } CO_2}$$

$= (1.54 \times 10^{-3} - 1.54X)$ g CO_2

total CO_2 = CO_2 from cocaine + CO_2 from sucrose

or $1.8 \times 10^{-3} = 2.466X + 1.54 \times 10^{-3} - 1.54X$

or $2.6 \times 10^{-4} = 0.93X$

$\therefore$ X = 2.8×10^{-4} g cocaine

$$\text{mass \% cocaine} = \frac{2.8 \times 10^{4}\ \text{g}}{1.0 \times 10^{3}\ \text{g}} \times 100\% = 28\%$$

68. Net ionic equation: $2Br^-(aq) + Cl_2(g) \rightarrow 2Cl^-(aq) + Br_2(l)$

Molecular equation: $CaBr_2(aq) + Cl_2(g) \rightarrow CaCl_2(aq) + Br_2(l)$

Masses: 40 g 14.2 g 22.2 g (40 + 14.2 – 22.2) g

Combining the three known masses gives the unknown mass of Br_2.

$$10\text{g } Br_2 \times \frac{40\text{ g } CaBr_2}{(40 + 14.2 - 22.2)\text{ g } Br_2}\ 12.5\text{g } CaBr_2$$

69. Molecular equation: $Hg(NO_3)_2 + H_2S + HgS(s) + 2\ HNO_3(aq)$

Net ionic equation: $Hg^{2+} + H_2S \rightarrow HgS(s) + 2H^+(aq)$

The acid formed is nitric acid, a strong acid. The other product is mercury (II) sulfide

Mass HNO_3 = (81.15 g + 8.52 g) – 58.16 g = 31.51 g

Mass of solution = 550 g H_2O + 31.51 g HNO_3 = 581.51 = 581.5 g

70. Let y = the mass of CuO in the mixture. Then 0.5 g – y = the mass of Cu_2O in the mixture. Multiplying the appropriate conversion factors for Cu times the mass of each oxide will hive one equation in one unknown for the mass of 0.425 g Cu:

$$0.425 = \left[\frac{63.55\text{ g Cu}}{79.55\text{ g CuO}}\right] + (0.5 - y)\left[\frac{127.1\text{ g Cu}}{143.1\text{ g } Cu_2O}\right]$$

Simplifying the equation by dividing the conversion factors and combining terms gives:

$$0.425 = 0.79887\ y + 0.888190\ (0.500 - y)$$

$$0.08932\ y = 0.019095$$

$$y = 0.2139 = 0.21\text{ g} = \text{mass of CuO}$$

71. Use the data to find the molar mass of the metal and anion. Start with X_2.

Mass X_2 in MX= 4.52 – 3.41 = 1.11 g

Molar mass X_2 = 1.11 g ÷ 0.0158 mol = 70.25 g.mol^{-1}.

Molar mass X= 70.25 ÷ 2 = 35.14 = 35.1 g.mol^{-1}.

Thus *X* is Cl, chlorine

Moles of M in 4.52g MX = 0.0158 × 2 = 0.0316 mol.

Molar mass of M= 3.41 g ÷ 0.0316 mol = 107.9 = 108 g.mol^{-1}.

Thus M is AG, silver.

72. After finding the volume of the alloy, convert it to mass Fe using density and % Fe. Then use Avogadro's number and the atomic weight for the number of atoms.

Vol. = 10 cm × 20 cm × 15 cm = 3×10^3 cm^3

$$\text{Mass Fe} = 3 \times 10^3\ \text{cm}^3 \times \frac{8.17\text{ g alloy}}{1\text{ cm}^3} \times \frac{54.7\text{ g Fe}}{100\text{ g alloy}} = 1.3407 \times 10^4\text{ g}$$

$$\text{No. of Fe atoms} = 1.3407 \times 104 \text{ g Fe} \times \frac{1 \text{ mol Fe}}{55.85 \text{ gFe}} \times \frac{6.022 \times 10^{23} \text{ Fe atoms}}{1 \text{ mol Fe}}$$

$$= 1.4456 \times 10^{26} = 1.45 \times 10^{26} \text{ Fe atoms}$$

73. Use the density, formula weight, and percentage to convert molarity. Then combine the 0.2 mol with mol L^{-1} to obtain the volume in liters.

$$\frac{0.807 \text{ g soln}}{1 \text{ mL}} \times \frac{0.94 \text{ g ethanol}}{1.0 \text{ g soln}} \times \frac{1 \text{ soln ethanol}}{46.07 \text{ g ethanol}} \times \frac{1000 \text{ mL}}{1\text{L}} = \frac{16.465 \text{ mol ethanol}}{1 \text{ L ethanol}}$$

$$\text{L ethanol} = 0.2 \text{ mol ethanol} \times \frac{1 \text{ L ethanol}}{16.465 \text{ mol ethanol}} = 0.012146 = 0.0121 \text{ L}$$

74. Convert the 6.026 g of $BaSO_4$ to mol $BaSO_4$; then from the equation deduce that 3 mol $BsSO_4$ is equivalent to 1 mol $M_2(SO_4)_3$ and is equivalent to 2 mol of M. use that with 1.2g of the metal M to calculate the atomic weight of M.

$$6.026 \text{ g BaSO}_4 \times \frac{1 \text{ mol BaSO}_4}{233.4 \text{ g BaSO}_4} \times \frac{2 \text{ mol M}}{3 \text{ mol BaSO}_4} = 0.017212 \text{ mol M}$$

$$\text{Atomic Wt of M in g.mol}^{-1} = \frac{1.2\text{gM}}{0.017212 \text{ molM}} = 69.719 \text{ g.mol}^{-1} = \text{(gallium)}$$

75. Use the density, formula weight, percentage, and volume to convert to mol H_3PO_4. Then, from the equation $P_4O_{10} + 6H_2O \rightarrow 4H_3PO_4$, deduce that 4 mol H_3PO_4 is equivalent to 1 mol of P_4O_{10}, and use that to convert to mol P_4O_{10}.

$$15 \text{ mL} \times \frac{1.025 \text{ g soln}}{1 \text{ mL}} \times \frac{0.05 \text{ gH}_3\text{PO}_4}{1 \text{ g soln}} \times \frac{1 \text{ mol H}_3\text{PO}_4}{98 \text{ g H}_3\text{PO}_4} = 0.7884 \text{ mol H}_3\text{PO}_4$$

$$0.7844 \text{ mol H}_3\text{PO}_4 \times \frac{1 \text{ mol P}_4\text{O}_{10}}{4 \text{ mol H}_3\text{PO}_4} = 0.1961 \text{ mol P}_4\text{O}_{10}$$

$$\text{Mass P}_4\text{O}_{10} = 0.1961 \text{ mol P}_4\text{O}_{10} \times \frac{284.92 \text{ g P}_4\text{O}_{10}}{1 \text{ mol P}_4\text{O}_{10}} = 55.67 = 55.7 \text{ g P}_4\text{O}_{10}$$

76. Convert the 0.1068 g of hydrogen to mol H_2; then deduce from the equation that 3 mol H_2 is equivalent to 2 mol Al. Use the moles of Al to calculate mass Al and the percentage Al.

$$0.1068 \text{ g H}_2 \times \frac{1 \text{ mol H}_2}{2.016 \text{ g H}_2} \times \frac{2 \text{ mol Al}}{3 \text{ mol H}_2} = 0.0353174 \text{ mol Al}$$

$$\% \text{ Al} = \frac{0.0353174 \times \dfrac{26.98 \text{ g Al}}{\text{mol Al}}}{1.118 \text{ g alloy}} \times 100\% = 85.229 = 85.23\%$$

77. Use the formula weight of $Al_2(SO_4)_3$ to convert to mol $Al_2(SO_4)_2$. Then deduce from the equation that 1 mol $Al_2(SO_4)_3$ is equivalent to 3 mol H_2SO_4, and calculate moles of H_2SO_4 needed. Combine density, percentage and formula weight to obtain molarity of H_2SO_4. Then combine molarity and moles to obtain volume

$$18.7 \text{ g } Al_2(SO_4)_3 \times \frac{1 \text{ mol } Al_2(SO_4)_3}{342.17 \text{ g } Al_2(SO_4)_3} \times \frac{3 \text{ mol}_2\ H_2SO_4}{1 \text{ mol } Al_2(SO_4)_3} = 0.1639 \text{ mol } H_2SO_4$$

$$\frac{1.104 \text{ g soln}}{1 \text{ mL}} \times \frac{15.0 \text{ g } H_2SO_4}{100 \text{ g soln.}} \times \frac{1 \text{ mol} H_2SO_4}{98.09 \text{ g} H_2SO_4} \times \frac{1000 \text{ mL}}{1 \text{ L}} = 1.688 \text{ mol } H_2SO_4 \text{ L}^{-1}$$

$$0.1693 \text{ mol } H_2SO_4 \times \frac{1 \text{ L } H_2SO_4}{1.688 \text{ mol } H_2SO_4} = 0.0971 \text{ L (97.1 mL)}$$

78. The equations for neutralization are $2HCl + Mg(OH)_2 \rightarrow MgCl_2 + 2H_2O$ and $3HCl + Al(OH)_3 \rightarrow AlCl_3 + 3H_2O$. Calculate the moles of HCl and set equal to the total moles of hydroxide ion, OH^-:

$$0.0485 \text{ L HCl} \times \frac{0.187 \text{ mol HCl}}{\text{L HCl}} = 0.0090695 \text{ mol HCl}$$

$0.0090695 \text{ mol HCl} = 2\,[\text{mol } Mg(OH)_2] + 3[\text{mol } Al(OH)_3]$

Rearrange the last equation and solve the moles of $Al(OH)_3$.

(Eq 1) mol $Al(OH)_3$ = 0.0030231 mol HCl – 2/3 [mol $Mg(OH)_2$]

The total mass of chloride salts is equal to the sum of the masses of $MgCl_2$ (molar mass = 95.21 $g.mol^{-1}$) and $AlCl_3$ (molar mass = 133.33 $g.mol^{-1}$)

$[95.21 \text{ g.mol}^{-1} \times \text{mol } MgCl_2] + 133.33 \text{ g.mol}^{-1} \times \text{mol } AlCl_3] = 0.4200 \text{ g}$

Since the mol $Mg(OH)_2$ = mol $MgCl_2$, and mol $Al(OH)_3$ = mol $AlCl_3$, you can substitute these quantities into the last equation and get

$[95.21 \text{ g.mol}^{-1} \times \text{mol } Mg(OH)_2] + [133.33 \text{ g.mol}^{-1} \times \text{mol } Al(OH)_3] = 0.4200 \text{ g}$

Substitute Eq 1 into this equation for the mol of $Al(OH)_3$:

$[95.21 \text{ g.mol}^{-1} \times \text{mol } Mg(OH)_2] \times [133.33 \text{ g.mol}^{-1} \times (0.0030231 \text{ mol} - 2/3 \text{ mol } Mg(OH)_2)] = 0.42 \text{ g}$

Rearrange the equation and solve for the moles of $Mg(OH)_2$.

$6.323 \text{ mol } Mg(OH)_2 + 0.40307 = 0.42$

$\text{mol } Mg(OH)_2 = 0.01693 \div 6.323 = 0.0026728 \text{ mol}$

Calculate the mass of $Mg(OH)_2$ in the antacid tablet (molar mas = 58.33 $g.mol^{-1}$).

$$0.0026728 \text{ mol } Mg(OH)_2 \times \frac{58.33 \text{ g } Mg(OH)_2}{1 \text{ mol } Mg(OH)_2} = 0.15590 \text{ g } Mg(OH)_2$$

Use Eq 1 to find the moles and mass of $Al(OH)_3$ (molar mass 78.00 $g.mol^{-1}$) in a similar fashion

$$\text{mol } Al(OH)_3 = 0.0030231 - \frac{2}{3}\left[0.0026728 \text{ mol } Mg(OH)_2\right] = 0.0012412 \text{ mol}$$

$$0.0012412 \text{ mol Al(OH)}_3 \times \frac{78.00 \text{ g Al(OH)}_3}{1 \text{ mol Al(OH)}_3} = 0.096681 \text{ g Al(OH)}_3$$

The percent $Mg(OH)_2$ in the antacid is

$\% \ Mg(OH)_2 = [0.1559 \text{ g} \div (.1559 + 0.096681) \text{ g}] \times 100\% = 61.72 = 61.7\%$

79. Recall that NaCl is soluble and AgCl is insoluble

(a) $Na^+ + Cl^-$:

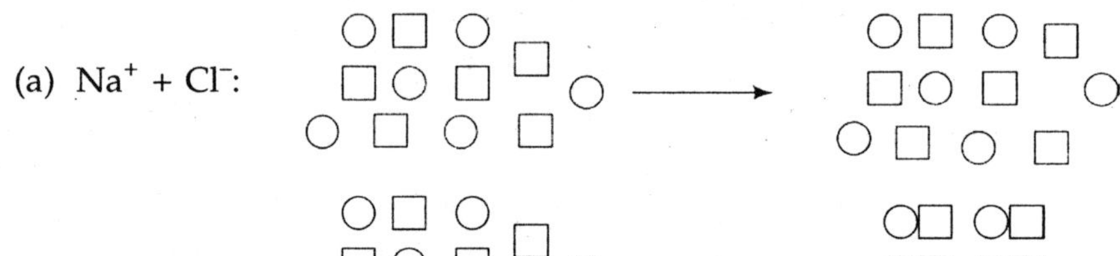

(b) $Ag^+ + Cl$:

80. (a) HCl

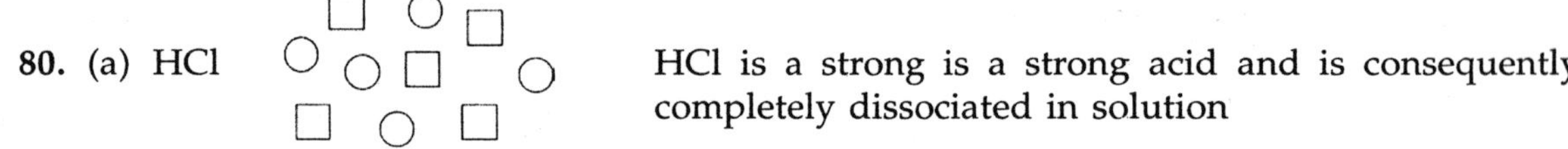

HCl is a strong is a strong acid and is consequently completely dissociated in solution

(b) HF

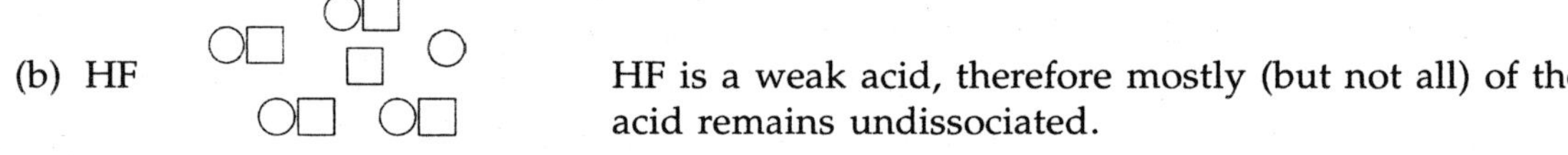

HF is a weak acid, therefore mostly (but not all) of the acid remains undissociated.

(c) KOH

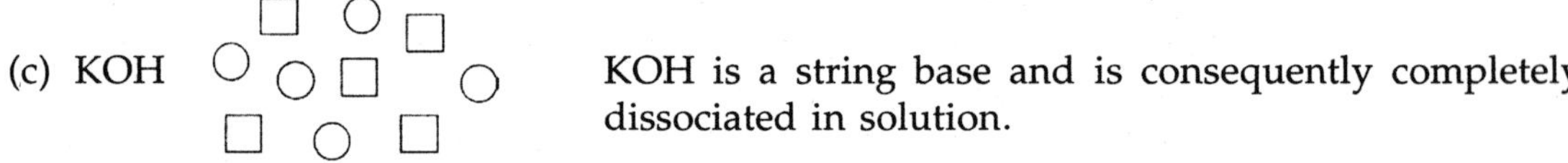

KOH is a string base and is consequently completely dissociated in solution.

(d) HNO_2

HNO_2 is a weak acid, therefore most (but not all) of the acid remains undissociated. Note that NO_2 is represented with a single circle for the anion.

81. No precipitate with Cl– indicates the metal ion is not Ag^+.
A precipitate with CO_3^{2-} indicates the metal ions is not Ni^+.
A precipitate with SO_4^{2-} indicates the metal ions is not Ni^{+2}.
Thus, by elimination, the ion must be Ba^{+2}.

82. First write a balanced redox equation for the oxidation of oxalate to CO_2. Then use the amount of $KMnO_4^{2-}$ to calculate the amount of CaC_2O_4 present in the sample.

$CaC_2O_4(s) + 2H^+(aq) \rightarrow H_2C_2O_4^{2-}(aq) + Ca^{2+}(aq)$

$H_2C_2O_4(aq) + MnO_4^-(aq) \rightarrow CO_2(g) + Mn^{2+}(aq)$

$C^{3+} \rightarrow C^{4+} + e^-$ Oxidation half-reaction

$H_2C_2O_4(aq) \rightarrow 2CO_2(g) + 2e^-$

$H_2C_2O_4(aq) \rightarrow 2CO_2(g) + 2e^- + 2H^+(aq)$ . . . . . eq. (1)

$Mn^{7+} + 5^{e-} \rightarrow Mn^{2+}$ Reduction half reaction

$MnO_4^-(aq) + 5e^- \rightarrow Mn^{2+}(aq)$

$8H^+(aq) + MnO_4^-(aq) + 5e^- \rightarrow Mn^{2+}(aq)$

$8H^+(aq) + MnO_4^-(aq) + 5e^- \rightarrow Mn^{2+}(aq) + 4H_2O(l)$ . . . . . . (2)

To balance the electrons, multiply eq (1) by 5 and eq. (2) by 2 and add the results.

$$16H^+(aq) + 2MnO_4^-(aq) + 10e^- + 5H_2C_2O_4(aq)$$
$$\rightarrow 10CO_2(g) + 10e^- + 10H^+(aq) + 2M_n^{2+}(aq) + 8H_2O(l)$$
$$6H^+(aq) + 2MnO_4^-(aq) + 5H_2C_2O_4(aq) \rightarrow 10CO_2(g) + 2Mn^{2+}(aq) + 8H_2O(l)$$

$$26.2\text{ mL} \times \frac{1\text{ L}}{1000\text{ mL}} \times \frac{0.0946\text{ mol KMnO}_4}{1\text{ L}} \times \frac{1\text{ mol MnO}_4^-}{1\text{ mol KMnO}_4} \times \frac{5\text{ mol C}_2\text{O}_4^{2-}}{2\text{ mol MnO}_4^-}$$

$$\times \frac{1\text{ mol CaC}_2\text{O}_4}{1\text{ mol C}_2\text{O}_4^{2-}} \times \frac{128.10\text{ g CaC}_2\text{O}_4}{1\text{ mol CaC}_2\text{O}_4} = 0.794\text{ g CaC}_2\text{O}_4$$

or $$0.794\text{ g} \times \frac{1\text{ mol CaC}_2\text{O}_4}{128.1\text{ g}} \times \frac{1\text{ mol Ca}^{2+}}{1\text{ mol. CaC}_2\text{O}_4} \times \frac{40.08\text{ g Ca}^{2+}}{1\text{ mol Ca}^{2+}} = 0.248\text{ g Ca}^{2+}$$

This is within the normal range.

83. Calculate the amount of $Mg(OH)_2$ and $NaHCO_3$ present and the moles of acid each could neutralize. Then calculate the volume of acid this corresponds to.

33 mg = 0.33 g

mass $Mg(OH)_2$ = 0.41 × 0.33 g = 0.135 g

mass $NaHCO_3$ = 0.362 × 0.33 g = 0.119 g

$$0.135\text{ g Mg(OH)}_2 \times \frac{1\text{ mol Mg(OH)}_2}{58.3\text{ g Mg(OH)}_2} \times \frac{2\text{ mol H}^+}{1\text{ mol. Mg(OH)}_2} = 0.00463\text{ mol. H}^+$$

$$0.119\text{ g NaHCO}_3 \times \frac{1\text{ mol NaHCO}_3}{84.0\text{ g NaHCO}_3} \times \frac{1\text{ mol H}^+}{1\text{ mol NaHCO}_3} = 0.00142\text{ mol. H}^+$$

Total moles H^+ neutralized = 0.00643 + 0.00142 = 0.00605 mol. H^+

$$0.0065\text{ mol. H}^+ \times \frac{1\text{ L acid}}{0.020\text{ mol H}^+} = 0.3\text{ L acid}$$

84. The difference in mass between the original copper strip and the silver coated strip corresponds to the mass of copper lost and the mass of silver gained. Bear in mind that two moles of silver are gained for each mole of copper lost. Set up an algebraic equation.

Let x = mass of Cu lost,

then $2.00 - x$ = mass of Cu remaining.

$$\text{mass of Ag} = x \text{ g Cu} \times \frac{1 \text{ mol Cu}}{63.55 \text{ g Cu}} \times \frac{2 \text{ mol. Ag}}{1 \text{ mol Cu}} \times \frac{107.9 \text{ g Ag}}{1 \text{ mol. Ag}} = 3.40x$$

$4.18 = (2 - x) + 3.4x$
$4.18 \text{ g} = 2 + 2.4x$
$x = 0.9078 \text{ g}$
Mass of Cu remaining = 2 – 0.908 = 1.09 g
Mass of Ag remaining = 3.40 × 0.908 = 34.09 g

85. 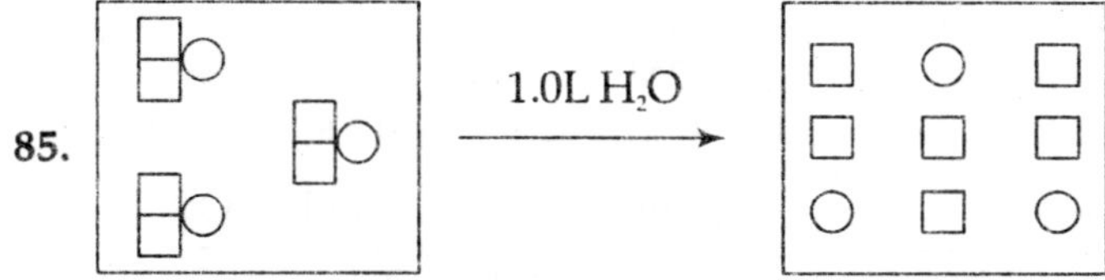

There are 3 moles Na_2S in 1.00 L, thus the molarity is 3.

$$3 \text{ mol.} \times \frac{1 \text{ L}}{3 \text{ mol}} = 1 \text{ L}$$

$Na_2S \rightarrow 2Na^+ + S^{2-}$

$$\frac{3.0 \text{ mol } Na_2S}{1 \text{ L}} \times \frac{2 \text{ mol } Na^+}{1 \text{ mol } Na_2S} = 6 \; M \; Na^+$$

$$\frac{3.0 \text{ mol } Na_2S}{1 \text{ L}} \times \frac{1 \text{ mol } S^{2-}}{1 \text{ mol } Na_2S} = 3 \; M \; S^{2-}$$

86. (a) At 313 K, approximately 240 g sucrose will dissolve in 100 g water.

$$80 \text{ g } H_2O \times \frac{240 \text{ g sucrose}}{100 \text{ g } H_2O} = 190 \text{ g sucrose}$$

(b) At 298 K approximately 210 g sucrose dissolves in 100 g water.

$$125 \text{ g } H_2O \times \frac{210 \text{ g sucrose}}{100 \text{ g } H_2O} = 260 \text{ g sucrose (remains in solution)}$$

750 g – 260 g = 490 g sucrose crystallizes out of sodium.

87. The equations for the liberation of O_2 from $KClO_3$ is

$$\underset{\text{2 moles}}{2\,KClO_3} = \underset{\text{2 moles}}{2\,KCl} + \underset{\text{3 moles}}{3O_2}$$

In order to be able to solve this problem at all we must know that heating liberates all of the oxygen from $KClO_3$. That is, we must *know the reaction* that occurs.
The equation tells us that 3 *moles* of O_2 are liberated from 2 moles of $KClO_3$.
The means that 1 *mole* $KClO_3$ liberates 1.5 *moles* of oxygen.
It should be pointed out that the equation can be written in the form

$$KClO_3 = KCl + 1\frac{1}{2}\, O_2$$

This tells us, at a glance, that the *moles* of O_2 liberated is 1.5 times the moles of $KClO_3$ heated, and 1.5 times 3.5 moles is 5.25 moles.

88. A glance at the equation tells us that 3 moles of O_2 are needed for every 4 *moles* of Sb burned. That means that it takes $\frac{3}{4}$ *mole* of O_2 to burn 1 *mole* of Sb. Therefore, $18 \times \frac{3}{4}$ *mole* or 13.5 *moles* of O_2 will be needed to burn 18 *moles* of Sb.

To find the number of *grams* of Sb_2O_3 we first find the number of *moles* of Sb_2O_3. The equation tells us that the number of *moles* of Sb_2O_3 formed is $\frac{1}{2}$ the number of *moles* of Sb burned. Therefore, 9 moles of Sb_2O_3 will be formed.

$$9 \text{ moles of } Sb_2O_3 \times \frac{291.6 \text{ g of } Sb_2O_3}{1 \text{ mole of } Sb_2O_3} = 2628 \text{ g of } Sb_2O_3$$

89. The formula, C_3H_8 tells us that, since 1 molecule of C_3H_8 contains 3 atoms of C and 8 atoms of H, while 1 molecule of CO_2 contains 1 atom of C and 1 molecule of H_2O contains 2 atoms of H, *1 mole of* C_3H_8 *will yield* 3 *moles of* CO_2 *and* 4 *moles of* H_2O. Therefore, 2.40 moles of C_3H_8 will yield 2.40 × 3 or 7.2 moles of CO_2 and 2.4 × 4 or 9.6 moles of H_2O.

At STP 1 mole of CO_2 occupies a volume of 22.4 liters and 7.2 moles will occupy a volume of 7.2 × 22.4 liters or 161 liters.

Since the molecular weight of H_2O is 18, 1 mole of H_2O weighs 18 g and 9.60 moles will weigh 173 g.

Each calculation can be carrier out in one operation.

$$2.4 \text{ moles of } C_3H_8 \times \frac{3 \text{ moles of } CO_2}{1 \text{ mole of } C_3H_8} \times \frac{22.4 \text{ liters of } CO_2}{1 \text{ mole of } CO_2} = 161 \text{ liters of } CO_2$$

$$2.4 \text{ moles of } C_3H_8 \times \frac{4 \text{ moles of } H_2O}{1 \text{ mole of } C_3H_8} \times \frac{18 \text{ of } H_2O}{1 \text{ mole of } H_2O} = 173 \text{ g of } H_2O$$

90. The equation tells us that 6.5 *moles* of O_2 are consumed for every *mole* of C_4H_{10} burned.

The equation, or the formula, C_4H_{10}, tells us that 1 *mole* of C_4H_{10} yields 4 *moles* of CO_2 and 5 *moles* of H_2O. The molecular weights of C_4H_{10}, O_2, CO_2, and H_2O are 58.0, 32.0, 44.0, and 18.0 respectively. Therefore, we have:

$$\frac{36 \text{ g of } C_4H_{10}}{58 \text{ of } C_6H_{10}/\text{mole of } C_4H_{10}} \times \frac{6.5 \text{ mole of } O_2}{1 \text{ mole of } C_4H_{10}} \times \frac{32 \text{ g of } O_2}{1 \text{ mle of } O_2} = 129 \text{ g of } O_2$$

$$\frac{36}{58} \text{ moles of } C_4H_{10} \times \frac{4 \text{ mole of } CO_2}{1 \text{ mole of } C_4H_{10}} \times \frac{44 \text{ g of } CO_2}{1 \text{ mole of } CO_2} = 109 \text{ g of } CO_2$$

$$\frac{36}{38} \text{ moles of } C_4H_{10} \times \frac{5 \text{ mole of } H_2O}{1 \text{ mole of } C_4H_{10}} \times \frac{18.0 \text{ g of } H_2O}{1 \text{ mole of } H_2O} = 55.9 \text{ g of } H_2O$$

91. The equation tells us that 1 *mole* of O_2 reacts with 2 *moles* of H_2. At STP 1 mole of H_2 is 22.4 liters. Therefore, 50 liters is 50/22.4 moles of H_2.

$$\frac{1}{2} \text{ of } \frac{50}{22.4} \text{ moles, or } \frac{50}{2 \times 22.4} \text{ moles of } O_2 \text{ will be required}$$

$$\text{or } \frac{50}{2 \times 22.4} \text{ moles of } O_2 \times \frac{22.4 \text{ liters of } O_2}{1 \text{ mole of } O_2} = 25 \text{ liters of } O_2$$

We see that the volume of O_2 required is $\frac{1}{2}$ the volume of H_2 just as the number of moles of O_2 is $\frac{1}{2}$ the number of moles of H_2. In other words, at STP *the volumes of the two gases are to each other as the number of moles*. Since both gases will respond equally to changes of temperature and pressure it follows that, *when measured at the same temperature and pressure, the volumes of gases involved in a reaction are to each other as the number of moles of the gases in the equation for the reaction.*

92. *Moles* of CO_2 formed = moles of C burned.

$$\frac{40 \text{ g of C}}{120 \text{ g.mol}^{-1}} = \frac{40}{12} \text{ moles of C burned}$$

Therefore, 40/12 moles of CO_2 will be formed.

To find the volume occupied by 40/12 moles of CO_2 at 473 K and 1.2 atm we will use the ideal gas equation, $PV = nRT$.

$$V = \frac{nRT}{P} = \frac{40}{12} \text{ moles} \times 0.082 \frac{\text{liters} \times \text{atm}}{\text{mole} \times \text{deg}} \times \frac{473 \text{ deg}}{1.20 \text{ atm}} = 108 \text{ liters}$$

93. $2\ KClO_3 = 2\ KCl + 3\ O_2$

or

$KClO_3 = KCl + 1.5\ O_2$

1 mole of $KClO_3$ will yield 1.5 atoms of O_2. That means that 1/1.5 or 0.667 mole of $KClO_3$ will yield 1 mole of O_2. In other words, moles of $KClO_3$ needed = 0.667 × moles of O_2 produced.

The formula weight of O_2 is 32, of $KClO_3$ is 122.6.

$$\text{Moles of } O_2 = \frac{60 \text{ g } O_2}{32 \text{ g of } O_2 / 1 \text{ mole of } O_2}$$

$$\text{Moles of } KClO_3 \text{ needed} = 0.667 \times \frac{60}{32}$$

$$\text{grams of } KClO_3 \text{ needed} = 0.667 \times \frac{60}{32} \times 122.6 = 153 \text{ g}$$

The entire calculation, in one operation, is:

$$\frac{60 \text{ g of } O_2}{32 \text{ g } O_2 / 1 \text{ mole of } O_2} \times \frac{1 \text{ mole of } KClO_2}{1.5 \text{ moles of } O_2} \times \frac{122.6 \text{ g of } KClO_3}{1 \text{ mole of } KClO_3} = 153 \text{ g of } KClO_3$$

94. From the formula KBr and the equation,

$$2\ K + Br_2 = 2\ KBr$$

we see that potassium and Br_2 combine in the ratio of 2 moles of K to 1 mole of Br_2. The weight we start with are not in this molar ratio; since 12.2 g of K is 12.2/39.1 or 0.312 mole of K and 22.2 g of Br_2 is 22.2/159.8 or 0.139 mole of Br_2. Since 0.139 moles of Br_2 require only 0.278 mole of K for complete reaction, the K is present in excess and Br_2 is said to be the limiting reagent because it limits the quantity of KBr which will form. We now ignore the K and note that the 0.139 mole of Br_2 will form 0.278 mole of KBr.

Therefore, 0.278 mole × 119 g of KBr per mole = 33.1 g of KBr

95. (a) $MXO_3 = MX + 1.5\ O_2$

$MX + AgNO_3 = AgX + MNO_3$

Moles of O_2 = 0.96 g/32 g per mole = 0.3 mole.

Moles of AgX = mole of X = 1/1.5 × moles of O_2 = 1/1.5 × 0.03 = 0.02.

Molecular weight of AgX = 2.869 g/0.02 mole = 143.45 $g.mol^{-1}$

Atomic weight of X = 143.45 – 108.0 = 35.45.

Molecular weight of MX = 1.491 g/0.02 mole = 74.55 $g.mol^{-1}$.

Atomic weight of M = 74.55 – 35.45 = 39.10.

The entire calculation, in one operation, is

$$\frac{40 \text{ g of C}}{12 \text{ of C/mole of C}} \times \frac{1 \text{ mole of } CO_2}{1 \text{ mole of C}} \times 0.082 \frac{\text{liters} \times \text{atm}}{\text{mole} \times \text{deg}} \times \frac{473 \text{ deg}}{1.2 \text{ atm}} = 108 \text{ liters}$$

(b) The reaction that occur are:

$2\ Al + 6\ HCl = 3\ H_2 + 2\ AlCl_3$

$H_2 + CuO = Cu + H_2O$

We note that 2 *moles* of Al liberates 3 *moles* of H_2; that means that 1 *mole* of Al liberates 1.5 *moles* of H_2.

One *mole* of H_2, when it react with CuO, will produce 1 *mole* of Cu. That means that the 1,5 moles of H_2 that are liberated by 1 mole of Al will produce 1.5 moles of Cu.

In short, 1 mole of Al will liberate enough hydrogen to produce 1.5 moles of Cu. The Al that reacts and the Cu that is produced are in the *ratio of* 1 *mole* of Al to 1.5 *moles* of Cu. The 41.6 g of Al is 41.6/27 moles.

Therefore, 1.5 × 41.6/27 moles of Cu will be produced.

$$\frac{1.5 \times 41.6}{27} \text{ moles of Cu} \times \frac{63.5 \text{ g of Cu}}{1 \text{ mole of Cu}} = 147 \text{ g of Cu}$$

96. The molecular weight of P_4O_{10} is 284.

$$200 \text{ g of } P_4O_{10} \text{ is } \frac{200}{284} \text{ mole of } P_4O_{10}.$$

Inspection of the formulas P_4O_{10} and O_2 tells us that 5 moles of O_2 will be needed to produce 1 *mole* of P_4O_{10}.

Therefore $5 \times \frac{200}{284}$ moles of O_2 will be needed.

The molecular weight of O_2 is 32.

Therefore $5 \times \frac{200}{284}$ $32 = 113$ g of O_2

The entire calculation, in one operation, is:

$$\frac{200 \text{ g of } P_4O_{10}}{284 \text{ g of } P_4O_{10} / 1 \text{ mole of } P_4O_{10}} \times \frac{3 \text{ moles of } O_2}{1 \text{ mole of } P_4O_{10}} \times \frac{32 \text{ g } O_2}{1 \text{ mole of } O_2} = 113 \text{ g of } O_2$$

97. At constant temperature and pressure the *volumes* of gases are directly proportional; to the number of *moles* of each gas in the equation for the reaction. Therefore, in this problem the volumes of NH_3, O_2, H_2O (steam), and NO will be in the ratio, respectively, of 4 to 5 to 6 to 4. Accordingly, $\frac{5}{4} \times 500$ or 625 liters of O_2 will be needed, and $\frac{4}{4} \times 500$ or 750 liters of H_2O and $\frac{4}{4} \times 500$ or 500 liters of NO will be formed.

98. $NH_3 + HCl \rightarrow NH_4Cl$

$Ag \rightarrow AgNO_3$ $\quad NH_4Cl + AgNO_3 \rightarrow AgCl + NH_4NO_3$

Moles of NH_3 = moles of NH_4Cl = moles of $AgNO_3$ = moles of NO_3 = moles of Ag

Let N represents the atomic weight of nitrogen.

$$\text{Moles of } NH_3 = \text{g of } NH_3 \div \text{mol wt. of } NH_3 = \frac{2 \text{ g}}{(N + 3.024) \text{ g.mol}^{-1}}$$

$$\text{Moles of } NO_3 = \text{g of } NH_3 \div \text{mol wt. of } NO_3 = \frac{7.30 \text{ g}}{(N + 48) \text{ g.mol}^{-1}}$$

Since moles of NH_3 = moles of NO_3 we can solve for N.

Having calculated N we can then solve for the atomic weight of Ag, knowing that moles of Ag = moles of NH_3.

99. At the same temperature and pressure equal volumes of O_2 and H_2 contains numbers of moles of O_2 and H_2. Since, from the equation $2 H_2 + O_2 = 2 H_2O$, we see that H_2 and O_2 combine in the ratio of 2 moles of H_2 with 1 mole of O_2 to form 2 moles of H_2O and since, at constant volume and temperature, the pressure is directly proportional to the number of moles, 50 mm worth of H_2 will combine with 25 mm worth of O_2 to form 50 mm worth of H_2O. There will be an excess of 25 mm worth of O_2, so the final pressure will be 75 mm.

100. Let y = atoms of C and z = atoms of H in 1 molecule of X. We can then write the following balanced equation for the reaction the occurs:

$1\ C_yH_z + (y + z/4)O_2 = y\ CO_2 + z/2H_2O$

Knowing the atomic weights of C, H, and O we can then write the following expressions:

Mass of y moles of $CO_2 = 44y$

Mass of $z/2$ moles of $H_2O = 18z/2 = 9z$

Mass of $(y + z/4)$ moles of $O_2 = 32(y + z/4) = 32y + 8z$

On the basis of the facts given in the problem we can then write the following equations:

$$44y = 9z + 60$$
$$32y + 8z = 160$$

Solving for y and $z \Rightarrow y = 3$ and $z = 8$.

101. The average molecular weight of a mixture of gases is calculated by dividing the total mass of the mixture by the total number of moles of gases in the mixture. This mean that, in this particular problem, we must know the *relative number* of moles of CO, CO_2 and N_2 in the final mixture; we need not know the actual number of moles of each species.

In the original mixture of N_2 and O_2 the ratio of moles of N_2 to moles of O_2 = 70/30 or $\frac{7}{3}$. From the equations

$$C + O_2 = CO_2$$
$$C + 0.5O_2 = CO$$

and the fact that $\frac{1}{2}$ the C is converted to CO_2 and $\frac{1}{2}$ to CO, we see that 1.5 moles of O_2 will yield 1 mole of CO_2 and 1 mole of CO. That means that the 3 moles of O_2 in the above $\frac{7}{3}$ ratio will be replaced by 2 moles of CO_2 and 2 moles of CO to give a final *ratio* of 7 moles of N_2 to 2 moles of CO_2 to 2 moles of CO.

$$\text{Av mol wt.} = \frac{(7 \times 28) + (2 \times 44) + (2 \times 28)}{11} = 30.9$$

102. From the equations

$$H_2S + 1.5O_2 = H_2O + SO_2$$
$$CS_2 + 3O_2 = CO_2 + 2SO_2$$

we note that CO_2 and SO_2 are formed from CS_2 in the ratio of 1 mole of CO_2 to moles of SO_2. Since 0.0774 mole of CO_2 are present, and this CO_2 was derived from the CS_2, 2(0.0774) = 0.155 mole of SO_2 must also have come from the CS_2. Therefore, (0.275 – 0.155) = 0.120 mole of SO_2 must have come from the H_2S.

103. Let X = g of C and Y = g of S. then $X + Y = 2$

$$C + O_2 = CO_2$$
$$S + O_2 = SO_2$$

We note that:

1. moles of CO_2 = moles of C burned
2. moles of SO_2 = moles of S burned
3. $X/12$ = moles of C = moles of CO_2
4. $Y/32$ = moles of S = moles of CO_2

Since moles of $CO_2 = \frac{1}{2}$ moles of SO_2, that is,

$$\frac{X}{12} = \frac{1}{2}\left(\frac{Y}{32}\right)$$

We know have two equations in two unknowns. The first is derived from the mass relationship; the total mass is the sum of the mass of the components. The second is derived from the mole relationship and the fact that moles = g/MW. Many mixture peoblems can be solved by setting up and solving two such equations. In this case solving gives X = 1.9 g of carbon.

104. Since the reaction takes place in one reaction vessel the three gases, CO_2, H_2O, and SO_2, are at the same temperature and occupy the same volume. Therefore, the number of moles of the three gases are in the ratios of their partial pressures.

$$CS_2 + 3O_2 = CO_2 + 2SO_2$$
$$CH_4 + 2O_2 = CO_2 + 2HO_2$$

Since the partial pressure of the SO_2 and H_2O are, respectively, 200 mm 150 mm they must be present in the *ratio* of 2 moles of SO_2 to 1.5 moles of H_2O. Therefore, as the equations testify, the CS_2 and CH_4 must have been present in the ratio of 2 moles of CS_2 to 1.5 moles of CH_4. Knowing the mole ratio, the mass percent of CH_4 can then be calculated.

105. In solving this particular problem we should first note that each of the following is true:

X = g of C Y = g of S ∴ X + Y = 2 ...... (i)

Now, $C + O_2 = CO_2$ and

$S + O_2 = SO_2$

∴ moles of CO_2 = moles of C and, moles of SO_2 = moles of S

or moles of C + moles of S = moles of O_2 ...... (ii)

or $\frac{x}{12} + \frac{y}{32} = \frac{4}{32} g$

Also, moles of C = $\frac{X}{12}$ ...... (iii)

moles of S = $\frac{Y}{32}$ ...... (iv)

moles of CO_2 = $\frac{X}{12}$ and, moles of SO_2 = $\frac{Y}{32}$

∴ grams of CO_2 + gram of SO_2 = 6 g ...... (v)

$\frac{x}{12} \times 44 + \frac{y}{32} \times 64 = 6g$

or g of (CO_2 + SO_2) – g of (C + S) = g of O_2 = 4 g [∴ g og C + g of S = 2]

moles of O_2 = $\frac{40}{32}$ ...... (vi)

g of CO_2 = moles of CO_2 × $\frac{\text{g of } CO_2}{\text{mole of } CO_2} = \frac{X}{12} \times 44$ ...... (vii)

g of SO_2 = moles of SO_2 × $\frac{\text{g of } SO_2}{\text{mole of } SO_2} = \frac{Y}{32} \times 64$ ...... (viii)

g of O_2 in CO_2 + g of O_2 in SO_2 = 4 g ...... (ix) or $\frac{32x}{12}+\frac{32}{32}=4g$

g of O_2 in CO_2 = $\frac{32}{12}$ × g of C = $\frac{32}{12}X$ ...... (x)

Since 1 mole CO_2 contains 1 mole of C and 1 of O_2 it follows that, in CO_2

$$\frac{\text{mass of the } O_2}{\text{mass of the C}}=\frac{\text{mol weight of } O_2}{\text{mol weight of C}}=\frac{32}{12}$$

∴ Mass of the O_2 = $\frac{32}{12}$ × mass of the C

Now, g of O_2 in SO_2 = $\frac{32}{12}$ × g of S = $\frac{32}{32}$ × Y ...... (xi)

and g of CO_2 = $\frac{44}{12}$ × g of C = $\frac{44}{12}X$ ...... (xii)

Since 1 mole of CO_2 contains 1 mole of C it follows that

$$\frac{\text{mass of the } CO_2}{\text{mass of the C}}=\frac{\text{mol weight of } CO_2}{\text{mol weight of C}}=\frac{44}{12}$$

∴ Mass of the CO_2 = $\frac{44}{12}$ × mass of the C

To solve for the value of X we need two equations in two unknowns. One is given by equation (i). This is the mass equation. A second equation can be constructed by various combinations of the 12 relationships. In one way or another way any second equation will incorporate a mole relationship.

106. Let, X = g of Zn. Y = g of Mg, X + Y = 1.0. By examining the formulas, MgO and ZnO, we see that 1 mole of Mg will yield 1 mole of MgO and 1 mole of Zn will yield 1 mole of ZnO. Since X g of Zn = X/65.4 mole of Zn and Y g of Mg = Y/24.3 mole of Mg; X/65.4 mole of ZnO and Y/24.3 mole of MgO have formed. The total mass of ZnO and MgO are known, so we convert these molar quantities to masses by multiplying by the appropriate molecular weight: g of ZnO = 81.4 (X/65.4) and g of MgO = 40.3 (Y/24.3). The sum of these masses is given to be 1.409 g. This gives the second equation,

$$\frac{81.4}{65.4}X+\frac{40.3}{24.3}=1.409 \text{ g} \Rightarrow X=0.569 \text{ g of Zn}$$

107. Let, X = g of NaI; Y = g of NaBr;

∴ X + Y = 1.62 (the mass equation)

NaBr = 102.9; NaI = 149.9; AgBr = 187.8; AgI = 234.8

Since 1 mole of NaBr and 1 mole of AgBr each contain 1 mole of Br and since 1 mole of NaI and 1 mole of AgI each contains 1 mole of I, 1 mole of NaBr will yield 1 mole of AgBr and 1 mole of NaI will yield 1 mole of AgI.

It follows, therefore, that

$$\frac{\text{mass of the AgBr}}{\text{mass of the NaBr}}=\frac{\text{mol weight of Agbr}}{\text{mol weight of NaBr}}=\frac{187.8}{102.9}$$

$$\text{mass of AgBr}=\frac{187.8}{102.9}\times\text{mass of NaBr}$$

Likewise,

$$\text{mass of AgI} = \frac{234.8}{149.9} \times \text{mass of NaI}$$

$$\text{g of AgI} = \frac{234.8}{149.9} \times \text{of NaI} = \frac{234.8}{149.9} X$$

$$\text{g of AgBr} = \frac{187.8}{102.9} \times \text{g of NaBr} = \frac{187.8}{102.9} Y$$

g of AgI + g of AgBr = 2.822 g

$$\frac{234.8}{149.9} X + \frac{187.8}{102.9} Y = 2.822 \text{ (the molar relationship equation)}$$

$X = 0.520$ g of NaI

108. Let, X = g of CO_2; Y = g of SO_2

$\therefore$ $X + Y = 2.952$ (the mass equation)

$$\frac{X}{44.01} + \frac{Y}{64.06} = 5.3 \times 10^{-2} \text{ (the mass equation)}$$

$X = 0.9682$ g; moles of CO_2 $44.01/0.9682 = 2.2 \times 10^{-2}$ moles

$44.01X + 64.06(5.3 \times 10^{-2} - X) = 2.952$ g

Solving, $X = 2.2 \times 10^{-2}$ moles of CO_2

109. We will assume that we have 100 g of mixture. The answer, in grams, will then equal, numerically, the mass percent. Let

X = g of Br $\quad$ Y = g of Cl $\quad$ g of Ag = 66.35

$X + Y = 100 - 66.35$ (the mass equation)

$$\text{Moles of AgBR} = \text{moles of Br} = \frac{X}{79.91}$$

$$\text{Moles of AgCl} = \text{moles of Cl} = \frac{Y}{35.45}$$

$$\text{Moles of (AgCl + AgBr)} = \text{moles of Ag} = \frac{66.35}{107.9}$$

So that we can write the mole relationship equation:

$$\frac{X}{79.91} + \frac{Y}{35.45} = \frac{66.35}{107.9} \Rightarrow X = 21.3 \text{ g} = 21.3\%$$

110. The atomic weight of zinc is 65.37. Therefore, 65.37 g is 1 mole of Zn, and 12 g is 12/65.37 mole of Zn.

$$12 \text{ g of Zn} \div 65.37 \text{ g of Zn/mole of Zn} = \frac{12}{65.37} \text{ mole of Zn}$$

Since any fraction represents a process of division, the calculation can take the form

$$\frac{12 \text{ g of Zn}}{65.37 \text{ g Zn/1 mole of Zn}} = \frac{12}{65.37} \text{ mole of Zn}$$

Note that moles in the numerator and denominator cancel. The answer is in grams.

111. We know that there are 6.023×10^{23} atoms in 1 mole of B. If we know how many *moles* of B there are in 20 g of B we can then multiply moles of B by 6.023×10^{23} atoms per mole. The atomic weight of boron is 10.811. That means that there are 10.811 g in 1 mole of B. Therefore

The entire calculation can be combined in one operation:

$$\frac{20\,\text{g}}{10.811\,\text{g.mol}^{-1}} \times 6.023 \times 10^{23} \text{ atoms mol}^{-1} = 1.1 \times 10^{24} \text{ atoms}$$

Note that grams cancel grams and moles cancel moles.

112. 6.023×10^{23} atoms is 1 *mole* of Sn. Therefore, 4.63×10^{21} atoms is

$$\frac{4.63 \times 10^{21}}{6.023 \times 10^{23}} \text{ mole of Sn} = \frac{4.63 \times 10^{23}}{6.023 \times 10^{23}} \text{ mole} = 0.772 \times 10^{-2} \text{ mole} = 7.72 \times 10^{-3} \text{ moles}$$

113. The formula of a compound gives the number of moles of atoms of each element in one mole of the compound. The atomic weight of iron is 56 and of sulfur is 32. Therefore, 56 g is 1 mole of Fe and 32 g is 1 mole of S. therefore, Fe and S are combined in the ratio of 1 mole of Fe to S. Since 1 mole of iron atoms is represented by the symbol, Fe, and 1 mole of sulfur atoms by the symbol, S, the empirial formula for compound, iron sulfide is FeS.

114. To find the number of moles of atoms of each element present, we will divide the quantity in grams of each element by the mass of 1 mole of atoms of the element, that is, by the atomic weight of that element.

$$\frac{401.22 \text{ g of Hg}}{200.61 \text{ g of Hg per mole of Hg}} = 2 \text{ moles of Hg}$$

$$\frac{32 \text{ g of O}}{16 \text{ g per mole of O}} = 2 \text{ moles of O}$$

The formula would thus appear to be Hg_2O_2. However, since we are interested in getting the simplest formula (the empirical formula), we will take the simplest ratio. Since 2 is to 2 as 1 is to 1, the empirical formula is HgO.

115. To find the number of moles of Mg in 6.7 g of magnesium oxide we will divide the mass of the magnesium by the gram-atomic weight of Mg, and to find the number of moles of O we will divide the mass in grams of the oxygen by the gram-atomic weight of O.

$$\text{moles of Mg} = \frac{4.04 \text{ g of Mg}}{24.3 \text{ g per mole of Mg}} = 0.166 \text{ mole of Mg}$$

$$\text{moles of O} = \frac{2.66 \text{ g of O}}{16.0 \text{ g per mole of O}} = 0.166 \text{ mole of O}$$

Therefore, the magnesium and oxygen are combined in the ratio of 0.166 mole of Mg to 0.166 mole of O. But 0.166 is to 0.166 as 1 is to 1. Therefore, the simplest formula is MgO.

116. The formula weight of $CH_2 = 1 \times 12 + 2 \times 1 = 14$. Now, calculate how many times the empirical formula goes into the molecular formula by dividing the empirical formula weight into the molecular weight i.e. $\frac{71}{14} = 5.07$ and round the answer off to the nearest integer, which is 5. Multiply each subscript in the empirical formula by this integer to get the molecular formula; $CH_2 \times 5 = C_5H_{10}$.

117. To say a compound contains 31.9% potassium, 28.9% chlorine, and 39.2% oxygen is equivalent to saying that 100 g of that compound contain 31.9 g of potassium, 2.9 g of chlorine, and 39.2 g of oxygen. Therefore, to find the number of moles of each element in a mole of the compound, we will divide the percent of each element by its atomic weight.

$$K = \frac{31.9 \text{ g of K}}{39.1 \text{ g/mole of K}} = 0.815 \text{ mole}$$

$$Cl = \frac{28.9 \text{ g of Cl}}{35.5 \text{ g/mole of Cl}} = 0.815 \text{ mole}$$

$$O = \frac{39.2 \text{ of O}}{16 \text{ g/mole of Cl}} = 2.45 \text{ mole}$$

The mole ratio of K to Cl to O is 0.815 to 0.815 to 2.45. To simplify this, we divide all three of these numbers by the smallest.

$$K = \frac{0.815}{0.815} = 1 \quad C = \frac{0.815}{0.815} = 1 \quad O = \frac{2.45}{0.815} = 3$$

The simplest formula of the compound is, therefore, $KClO_3$.

118. 1. 2.982 g A – 1.77 g Pb = 1.212 g Cl in A

2.982 g B – 1.77 g Pb = 0.594 g Cl in B

2. A: 1.770 g Pb : 1.212 g Cl

B: 1.770 g Pb : 0.594 g Cl

3. A: $\frac{1.212 \text{ g Cl}}{1.770 \text{ g Pb}} kx \frac{207.2 \text{ g Pb}}{1 \text{ mol Pb}} = 14.19 \text{ g Cl. (mole Pb)}^{-1}$

B: $\frac{0.594 \text{ g Cl}}{1.770 \text{ g Pb}} \times \frac{207.2 \text{ g Pb}}{1 \text{ mol Pb}} = 69.5 \text{ g Cl (mole Pb)}^{-1}$

4. A: $\frac{141.9 \text{ g Cl}}{35.451 \text{.mol}^{-1}} = 4.003 \text{ mol Cl: 1 mol Pb}$

B: $= \frac{69.5 \text{ g Cl}}{35.451 \text{.mol}^{-1}} = 1.96 \text{ mol Cl: 1 Pb}$

5. $PbCl_4$
6. $PbCl_2$
7. $PbCl_4 \rightarrow PbCl_2 + Cl_2$
   Oxidation: Chlorine (from –1 to 0)
   Reduction: lead (from +4 + +2)

   $PbCl_2 + H_2 \rightarrow Pb + 2HCl$
   Oxidation: Hydrogen (from 0 to +1)
   Reduction: Lead (from +2 to 0)

119. 1. 1.7 g $NiSO_4 \cdot ?H_2O$

$$-1.006\text{g } NiSO_4 = \frac{1.006 \text{ g } NiSO_4}{1} \times \frac{1 \text{ mol } NiSO_4}{154.76 \text{ g } NiSO_4} = 0.0065 \text{ mol } NiSO_4$$

$$0.694 \text{ g } H_2O = \frac{0.694 \text{ g } H_2O}{1} \times \frac{1 \text{ mol } H_2O}{18.02 \text{ g } H_2O} = 0.0385 \text{ mol } H_2O$$

$$\frac{0.0385 \text{ mol } H_2O}{0.0065 \text{ mol } NiSAO_4} = \frac{5.93}{1.0} = NiSO_4.6H_2O$$

$NiSO_4$ exists in two forms. The β-form is stable at 313 K and changes to the blue α-form at room temperature. The α-form is an aquamarine to bluish-green color.

2. $$\% \text{ water} = \frac{\text{mass of } H_2O \text{ in sample}}{\text{mass of hydrate}} \times 100\%$$

$$\frac{0.6904 \text{ g } H_2O}{1.7 \text{ g } NiSO_4 \cdot 6H_2O} \times 100\% = 40.8\% \; H_2O$$

3. $$\% \; H_2O = \frac{\text{part water}}{\text{whole mass}} \times 100\% = \frac{6 \times 18.02}{262.87} \times 100\% = 41.1\%$$

$$\% \text{ error} = \frac{\text{obs} - \text{exp}}{\text{exp}} \times 100\% = \frac{40.8 - 41.1}{41.1} \times 100\% = -0.7\%$$

4. $NiSO_4$

120. 1. $4OCl^-_{(aq)} + S_2O_3^{2-}{}_{(aq)} + 2OH^-{}_{(aq)} \rightarrow 2SO_4^{2-}{}_{(aq)} + 4Cl^-{}_{(aq)} + H_2O_{(1)}$

2. The commercial bleach was 5.20% NaClO. The directions required making a 0.5 M solution. MM of NaClO = 74.44 g · mol^{-1}. 300. mL of solution should contain 11.2 g NaClO:

$$\frac{300. \text{ mL sol'n}}{1} \cdot \times \frac{1 \text{ L sol'n}}{1000 \text{ mL sol'n}} \times \frac{0.500 \text{ mole NaClO}}{1 \text{ mole NaClO}} \times \frac{74.44 \text{ g NClO}}{1 \text{ mole NaClO}} = 11.2 \text{ g NaClO}$$

To obtain 11.2 g of NaClO, using a 5.20% solution, would require 215 g of bleach solution, diluted to 300. ML with distilled water.

$$\frac{11.2 \text{ g NaClO}}{1} \times \frac{100 \text{ g sol'n}}{5.20 \text{ g NaClO}} = 215 \text{ g bleach sol'n diluted to 300.mL}$$

3. (a) 0.2 M NaOH

$$\frac{300.\ \text{mL sol'n}}{1} \times \frac{1\ \text{L sol'n}}{1000\ \text{mL sol'n}} \times \frac{0.2\ \text{mole NaOH}}{1\ \text{L sol'n}} \times \frac{40\ \text{g NaOH}}{1\ \text{mol NaOH}} = 2.4\ \text{g NaOH}$$

(b) 0.5 M $Na_2S_2O_3$

$$\frac{300\ \text{mL sol'n}}{1} \times \frac{1\ \text{L sol'n}}{1000\ \text{mL sol'n}} \times \frac{0.5\ \text{mol}\ Na_2S_2O_3}{1\ \text{L sol'n}} \times \frac{158.12\ \text{g}\ Na_2S_2O_3}{1\ \text{mol}\ Na_2SO_3} = 23.7\ \text{g}\ Na_2SO_3$$

To make the solution, add 2.4 g of NaOH and 23.7 g $Na_2S_2O_3$ to a graduated cylinder. Dilute to 300.mL with distilled water.

4. A temperature correction factor would be needed:

$$\frac{(20\ \text{mL} \cdot 298\ \text{K}) + (30\ \text{mL} \cdot 308\ \text{K})}{50\ \text{mL}} = 304\ \text{K}$$

If the measured temperature is 307 K. ΔT should be reported as 307 K – 304 K = 3.0

5. According to the balanced net ionic equation

$$4OCl^-_{(aq)} + S_2O_3^{2-}{}_{(aq)} + 2OH^-_{(aq)} \rightarrow 2SO_3^{2-}{}_{(aq)} + 4Cl^-\ (aq) + H_2O_{(1)}$$

4cmoles of OCl^- are required for every for every 1 mole of $S_2O_3^{2-}$. In trial 9, the trial which showed the greatest temperature change, 40. ML of NaClO was used for 10 mL of $Na_2S_2O_3$.

$$\frac{40\ \text{mL NaClO}}{1} \times \frac{1\ \text{L}}{100\ \text{mL}}\ kx\ \frac{0.5\ \text{mol NaClO}}{1\ \text{L sol'n}}\ 0.02\ \text{ml}\ ClO^-$$

$$\frac{10\ \text{mL}\ Na_2SO_2O_3}{1} \times \frac{1\ \text{L}}{1000\ \text{mL}} \times \frac{0.5\ \text{mol}\ Na_2S_2O_3}{1\ \text{L sol'n}} = 0.005\ \text{mol}\ S_2O_3^{2-}$$

$$\frac{0.02\ \text{mol}\ ClO^-}{0.05\ \text{mol}\ S_2O_3^{2-}} = 4{:}1\ \text{molar ratio}$$

121. (1) $$\frac{1.059\ \text{g}\ KAl(SO_4)_2 \cdot 12H_2O}{1} \times \frac{1\ \text{mol}\ KAl(SO_4)_2 \cdot 12H_2O}{474.46\ \text{g}\ KAl(SO_4)_2 \cdot 12H_2O}$$

$$\times \frac{2\ \text{mol}\ Ba(NO_3)_2}{1\ \text{g}\ KAl(SO_4)_2 \cdot 12H_2O} \times \frac{1\ \text{L}\ Ba(NO_3)_2}{0.2\ \text{mol}\ Ba(NO_3)_2} \times \frac{1000\ \text{mL}}{1\ \text{L}} = 22.3\ \text{mL}$$

(2) $$\frac{1.04\ \text{g}\ BaSO_4}{1} \times \frac{1\ \text{mol}\ BaSO_4}{233.37\ \text{g}\ BaSO_4} \times \frac{1\ \text{mol}\ SO_4^{2-}}{1\ \text{mol}\ BaSO_4} \times \frac{96.04\ \text{g}\ SO_4^{2-}}{1\ \text{mol}\ SO_4^{2-}} = 0.4281\ \text{g}\ SO_3^{2-}$$

$$\% \text{ sulfate} = \frac{\text{part}}{\text{whole}} \times 100\% = \frac{0.4281 \text{ g sO}_4^{2-}}{1.059 \text{ g KAl(SO}_4)_2 \cdot 12\,\text{H}_2\text{O}} \times 100\% = 40.43\%$$

$$(3) = \frac{192.08 \text{ g SO}_4^{2-}}{474.46 \text{ g KAl(SO}_4)_2 \cdot 12\,\text{H}_2\text{O}} \times 100\% = 40.4\%$$

$$(4)\quad \% \text{ error} = \frac{\text{obs} - \text{exp}}{\text{exp}} \times 100\% = \frac{40.43 - 40.48}{40.48} \times 100\% = -0.15\%$$

122. Formula of the compound M_3N_2

Valency of nitrogen in the compound $M_3N_2 = 3$

$\therefore$ Equivalent weight of nitrogen $= \frac{14}{3}$

28 g Nitrogen combines with metal = 72 g

1 g Nitrogen combines with metal $= \frac{72}{28}$ g

$\frac{14}{3}$ g nitrogen combines with metal $= \frac{72}{28} \times \frac{14}{3} = 12$

$\therefore$ equivalent weight of metal = 12

Valency of metal in $M_3N_2 = 2$

$\therefore$ Atomic weight of metal = 12 × 2 = 24

123. Volume of virus $= \pi r^2 h$

(Considering its cylindrical shape) $= 3.14 \times \left(\frac{150}{2} \times 10^{-8}\right)^2 \times 5000 \times 10^{-8}$

Mass of one virus= $\frac{\text{Volume of virus}}{\text{specific volume}}$

$$= \frac{3.14 \times (75)^2 \times 5000 \times (10^{-8})^3}{0.75} = 1.178 \times 10^{-16} \text{ g}$$

m.wt. of virus= 6.02×10^{23} × Mass of one virus

$= 6.02 \times 10^{23} \times 1.178 \times 10^{-16}$ g $= 7.07 \times 10^7$ g mol^{-1}

124. Out of the two reacting constituents, the lesser one can react completely leaving the excess of the other in the reaction mixture.

Accordingly, the number of moles of H_2 consumed when 20% of it is transformed into HI = 0.1 × 20/100 = 0.02

According to the equation, the number of moles of HI formed = 0.02 × 2 = 0.04

Thus, Mol. wt of HI being 128, the wt. of HI formed = 0.04 × 128 = 5.12 g

125. Mass % C = 24.24; Mass % H = 4.04

$$\therefore \text{Mass \% Cl} = \frac{35.5}{143.5} \times \frac{2.9}{1.0} \times 100 = 71.72$$

Empirical Formula can, thus be calculated

Element	Mass	Amount of element	Simple ratio
C	24.24 g	$\frac{24.24}{12} = 2.02$ mol	1
H	4.04 g	$\frac{4.04}{1} = 4.04$ mol	2
Cl	71.72 g	$\frac{71.72}{35.5} = 2.02$ mol	1

∴ Empirical formula of *X* = CH_2Cl

Isomers of *X* are *Y* and *Z*

$$Y \xrightarrow[\text{KOH}]{\text{aqueous}} \text{Dihydroxy compound}$$

∴ *Y* should contain two chlorine atoms. Molecular formula of *X* should be $C_2H_4Cl_2$ Structure of *Y* should be $ClCH_2 \cdot CH_2Cl$

$$Z \xrightarrow[\text{KOH}]{\text{aqueous}} H_3C \cdot CHO$$

∴ Structure of *Z* should be $H_3C \cdot CHCl_2$

126. Total weight of AgCl obtained = 2.567 g

NaCl does not decompose on heating to 573 K. Therefore, amount of AgCl formed due to MCl = 1.341 g

∴ Weight of AgCl formed due to NaCl = 2.567 – 1.341 g = 1.226 g

NaCl ≡ AgCl ≡ MCl

23 + 35.5 = 58.5   108 + 35.5 = 143.5

143.5 g of AgCl is obtained from NaCl = 58.5

∴ 1.226 g of AgCl is obtained from NaCl = $\frac{58.5}{143.5} \times 1.226$ g = 0.4948 g

∴ Weight of MCl in 1 g of mixture = 1.0 – 0.4948 g = 0.5052 g

1.341 g AgCl is obtained from MCl = 0.5052 g

143.5 g AgCl is obtained from MCl = $\frac{0.5052}{1.341} \times 143.5$ g = 42.029 g

Molecular weight of MCl = 42.029.

127. $2IO_3^- + 12H^+ + 10I^- \rightarrow 6I_2 + 6H_2O$

$2S_2O_3^{-2} + I_2 \rightarrow S_4O_6^{-2} + 2I^-$

Thus $n(S_2O_3^{-2}) = eq\ (S_2O_3^{-2}) = eq\ (I_2) = 2 \times n(I_2) = 2 \times 3 \times n\ (KIO_3)$

$$\Rightarrow \quad M \times \frac{45\ \text{ml}}{1000\ (\text{ml/L})} = \frac{2 \times 3 \times 0.1\ (\text{g})}{1000\ (\text{mL/L})} = \frac{2 \times 3 \times 0.1\ (\text{g})}{214\ (\text{g.mol}^{-1})}$$

$$M = 0.0623\ (\text{mol L}^{-1})$$

128. Total volume of solution = 45 ml Lead nitrate + 25 ml of chronic sulphate = 70 ml.
Now, No of moles of 0.25 M $Pb(NO_3)_2$

$$= 0.25 \times \frac{45}{100} = 0.01125 \text{ moles.}$$

No. of moles of Pb^{2+} in $Pb(NO_3)_2$ = 0.01125 I

and No. of moles of NO_3^- in $Pb(NO_3)_2$ = 0.01125 × 2 = 0.0225

No. of moles 0.1 M $[Cr^2(SO_4)_3] = 0.1 \times \frac{25}{1000} = 0.0025$ moles

No. of moles of SO_4^{2-} in $[Cr_2\ (SO_4)_3]$ = 0.025 × 3 = 0.075 moles

No. of moles of Cr^{3+} in $[Cr_2\ (SO_4)_3]$ = 0.025 × 2 = 0.005

Now since SO_4^{2-} ions are present in less number of moles (i.e. 0.0075) that Pb^{2+} (0.01125 moles) and they combine in a ratio of 1:1 ($PbSO_4$). So the no. of moles of $PbSO_4$ formed = 0.0075 and the no. of moles Pb^{2+} left behind = 0.01125 – 0.0075 = 0.0035.

No. of moles of NO^{3-} and Cr^{3+} remain unchanged ∴ the molar concentration of species left are

(i) $Pb^{2+} = \frac{0.00375}{70} \times 1000 = 0.05357$ M

(ii) $NO_3^- = \frac{0.0225}{70} \times 1000 = 0.3214$ M

(iii) $Cr^{3+} = \frac{0.005}{70} \times 1000 = 0.714$ M

129. (a) Weight of sugar syrup = 214.2 g

Weight of sugar in the syrup = 34.2 g

∴ Weight of water in the syrup = 214.2 – 34.2 = 180 g

Mol. wt. of sugar, $C_{12}H_{22}O_{11}$ = 342

(i) ∴ Molar concentration $= \frac{34.2 \times 1000}{342 \times 180} = 0.56$

Mol. wt of water, H_2O = 180

(ii) ∴ Mole fraction $= \frac{34.2/342}{180/18 + 34.2/342} = 0.0099$

(b) 1 litre of water = 1000 g water

If its density = 1000 kg/m^3

Then, Number of moles of water $= \frac{1000}{18} = 55.55$

So, molarity of water = 55.55 M.

130. First we calculation of the empirical formula, molecular weight, and molecular formula of C.

$$0.08 \text{ g of water} \equiv 0.08 \times \frac{2}{18} \text{ g H} = 0.009 \text{ g H}$$

0.39 g of $CO_2 \equiv 0.39 \times \frac{12}{44}$ g C = 0.1064 g C

$\therefore \quad \%C = \frac{.1064}{.4} \times 100 = 26.6$

$\%H = \frac{0.08}{.4} \times 100 = 2.22$

and $\quad \% O = 100 - (26.6 + 2.22) = 71.2$

$\therefore \quad C : H : O = \frac{26.6}{12} : \frac{2.2}{1} : \frac{71.2}{16}$

$= 2.2 : 2.2 : 4.4 = 1 : 1 : 2$

$\therefore$ Empirical formula of C is CHO_2:

The equivalent formula of C is CHO_2:

$$\frac{E + 107}{108} = \frac{1.0\,g}{0.71\,g} \Rightarrow E = \frac{32.03}{0.71} = 45$$

Since C is a dibasic acid, its molecular weight = 2 × Ėq. 2 × 45 = 90

Empirical formula weight of (CHO_2) is 12 + 1 + 32 = 45.

C, being a dibasic acid must contain two carboxyl groups.

So. C is $(COOH)_2$.

Hence, A is H · COONa (sodium formate)

B is CO (carbon monoxide)

And C is COOH
|
COOH (oxalic acid)

131. $CaCl_2 \equiv CaCO_3 \equiv MgCl_2$

$40 + (2 \times 35.5) = 111 \qquad 40 + 12 + (3 \times 16) = 100 \qquad 24 + (2 \times 35.5) = 95$

111 mg $CaCl_2$ will gives $CaCO_3$ = 100 mg

1 g $CaCl_2$ will gives $CaCO_3 = \frac{100}{111}$ mg

Similarly

1 mg $MgCl_2$ will give $CaCO_3 = \frac{100}{95}$

Total $CaCO_3$ formed due to 1 mg $CaCl_2$ and 1 mg of $MgCl_2$

$= \frac{100}{111} + \frac{100}{95}$ mg = 2.03 mg

$CaCO_3$ per liter of water = 2.03 mg

Weight of 1 ml of water = 1 g = 10^3 mg

Weight of 1000 ml of water = $10^3 \times 10^3$ mg = 10^6 mg

2.03 mg of $CaCO_3$ is present in 10^6 mg of water

or 2.03 parts of $CaCO_3$ in 10^6 part of water.

132. The reaction is:

$$\begin{array}{ccc} 2SnCl_2 & \rightarrow Sn & + SnCl_4 \\ 2\{119 + (2 \times 35.5)\} = 380 & 119 & 119 + (4 \times 35.5) = 261 \end{array}$$

119 g Sn deposits on decomposition of $SnCl_2$ = 380 g

$\therefore$ 0.119 of Sn formed due to the decomposition of $SnCl_2$

$$= \frac{380}{119} \times 0.119 \text{ g} = 0.38 \text{ g}$$

380 g $SnCl_2$ decmopose to give $SnCl_4$ = 261 g

$\therefore$ 0.38 g $SnCl_4$ decompose to give SnCl $= \frac{261}{380} \times 0.38 \text{ g} = 0.261 \text{ g}$

Weight of $SnCl_2$ left after deocomopsition = 19.0 – 0.38 g = 18.62 g

Weight of $SnCl_4$ formed = 0.261 g.

Ratio of $SnCl_2$: $SnCl_4$ is 18.62 : 0.261

or **71.34 : 1**

133. The reaction is

$$\begin{array}{cccc} 2Mg & + \; O_2 & \rightarrow & 2MgO \\ 2 \times 24 = 48 & 32 & & 2(24 + 16) = 80 \end{array}$$

32 g oxygen reacts with Mg = 48g

$\therefore$ 0.5 g oxygen reacts with Mg $= \frac{48}{32} \times 0.5 = 0.75$ g

Weight of unreacted Mg = 1.0 – 0.75 g = 0.25 g

Weight of MgO formed $= \frac{80}{48} \times 0.75 = 1.25$ g

$MgO + H_2SO_4 \rightarrow MgSO_4 + H_2O$

40 g MgO will be dissolved in 100 ml of 1 N H_2SO_4

40 g MgO will be dissolved in 2000 ml of 0.5 N H_2SO_4

$\therefore$ 1.25 g MgO will be dissolved in 0.5 N H_2SO_4

$$= \frac{2000 \times 1.25}{40} = 62.5 \text{ ml.}$$

134. Since in acidic solution,

$MnO_4^- + 8H^+ + 5e^- \rightarrow Mn^{2+} + 4H_2O$

MnO_4^- gains $5e^-$, its eq. wt. $= \frac{\text{Mole. wt.}}{5}$

It means 1.61×10^{-3} moles of $MnO_4^- = 5 \times 1.61 \times 10^{-3}$ equivalent

$A^{n+} \rightarrow AO_3^-$

[ox. No. of A in AO_3^- is 5]

Loss of electrons = $(5 - n)$

$\therefore$ 2.68×10^{-3} moles of the solution containing A^{n+} ion

$= (5 - n) \times 2.68 \times 10^{-3}$ equivalents

Equivalents of O.A. $\equiv$ equivalents of R.A.

$(1.61 \times 10^{-3}) \times 5 = (5 - n) \times 2.68 \times 10^{-3}$

$$5 - n = \frac{1.61 \times 10^{-3} \times 5}{2.68 \times 10^{-3}} = 3$$

whence $n = 2$

135. Magnetic oxide of iron (Fe_3O_4) is a compound oxide containing FeO and Fe_2O_3, the former part (FeO) of which is affected on heating. Reactions involved being

(i) $2FeO + \frac{1}{2}O_2 \rightarrow Fe_2O_3$

143.6g 159.6 g

(ii) $2Fe_3O_4 + \frac{1}{2}O_2 \rightarrow 3Fe_2O_3$

462.8 g 478.8 g

Let the weight of FeO in the mixture be x g. and that of Fe_3O_2 would be $(100 - x)$ g., if the total wt. of mixture taken is 100 g.

Wt. of Fe_2O_3 produced from x g. of FeO according to equation (i)

$$= \frac{x \times 159.6}{143.6}$$

Also, the wt. of Fe_2O_3 produced from $(100 - x)$ g. of Fe_3O_4 according to equation (ii)

$$= \frac{(100 - x) \times 478.8}{462.8} \text{ g}$$

Total weight of Fe_2O_3 produced $= \frac{159.6x}{143.6} + \frac{(100 - x) \times 478.8}{462.8} = 105$ (given)

(Increase in weight being due to conversion of the contents of the mixture to Fe_2O_3.)

whence $x = 19.92$

Hence, the percentage of FeO is 19.92 and that of $Fe_3O_4 = 80.08$

136. Reactions $2Al + 6HCl \rightarrow 2AlCl_3 + 3H_2$
$Mg + 2HCl \rightarrow MgCl_2 + H_2$

Let the wt. of Al be x g. and that of Mg be $(1 - x)$ g.

Moles of Al $= \frac{x}{27}$ and those of Mg $= \frac{(1 - x)}{24.3}$

According to equations

moles of H_2 that will be produced by $\frac{x}{27}$ moles of Al $= \frac{3}{2} \times \frac{x}{27}$

also moles of H_2 that be produced by $\frac{(1-x)}{24.3}$ moles of Mg $= \frac{(1-x)}{24.3}$

Vol. of H_2 actually produced at N.T.P. $= 1.2 \times \frac{273}{273} \times \frac{0.92 \times 760}{760} = 1.104$ litres

moles of H_2 actually produced $= \frac{1.104}{22.4} = 0.0493$

Thus, $\frac{3}{2} \times \frac{x}{27} + \frac{1-x}{24.3} = 0.0493$

whence, $x = 0.566$ g.

Hence, Al = 0.566 gm. and Mg = 0.434 gm.

137. 44.4 ml. of 0.5 N—HCl

$$= \frac{0.5 \times 44.4}{1000} = 0.0222 \text{ g. equiv. of HCl}$$

1 mol. of Alkali metal carbonate (M_2CO_3) would require 2 moles of 2 g. equiv. of HCl for complete reaction according to the equation

$$M_2CO_3 + 2HCl \rightarrow 2MCl + H_2O + CO_2$$

Hence, 0.0222 g. eq. of HCl will correspond to 0.0222 × 1/2 or 0.0111 moles of the carbonates of two alkali metals of 1.000 g. of the mixture. (as given)

Mol. wt. of one carbonate, (M_2CO_3) when the at. wt. of the alkali metal is 7 = 14 + 12 + 48 = 74.

Mol. wt. of other carbonate (M_2CO_3) when the at. wt. of the alkali metal is $x = 2x + 60$.

Since the mixture contains equal no. of moles of each carbonate, the wt. of one of the carbonates would be

$\frac{74 \times 0.0111}{2}$ and that of the other carbonate would be $(2x + 60) \times \frac{0.0111}{2}$ g

Thus, $\frac{74 \times 0.0111}{2} + \frac{(2x+60) \times 0.0111}{2} = 1.0$ (given)

$\Rightarrow$ $x = 23.09$

Hence, the at. wt. of the other metal = 23.09.

For the formation of sulphate the following reaction will take place, where *M* stands or the metal *A* or *B*.

M_2CO_3	$+ H_2SO_4 \rightarrow M_2SO_4 + H_2O + CO_2$
1st carbonate 74 g	110 g
2nd carbonate 106.18 g	142.18 g

Amount of sulphate formed from 1 g. of mixture or $\frac{0.0111}{2}$ moles of each carbonate will thus be equal to—

$$\frac{0.0111 \times 110}{2} + \frac{0.0111 \times 142.18}{2} = 1.4 \text{ g.}$$

138. Reactions are:

$$2KMnO_4 + 5SO_2 + 2H_2O \rightarrow K_2SO_4 + 2MnSO_4 + 2H_2SO_4 \quad \ldots\ldots \text{(i)}$$

$$4FeS_2 + 11O_2 \rightarrow 2Fe_2O_3 + 8SO_2 \quad \ldots\ldots \text{(ii)}$$

According to equation (i) 2 moles of KMnO≡5 moles of SO_2

According to equation (ii) 4 moles of FeS_2≡8 moles of SO_2

From (i) and (ii)

$$2 \text{ moles of } KMnO_4 \equiv 5 \text{ moles of } SO_2 \equiv \frac{4}{8} \times 5 \text{ moles of } FeSO_2$$

$$\text{Moles of } KMnO_4 \text{ taken} = \frac{15.8}{158} = 0.1$$

$$\text{Hence, the moles of } FeS_2 \text{ required} = \frac{5}{2} \times \frac{0.1}{2} = 0.125$$

$$\text{Wt. of } FeS_2 \text{ required} = 0.125 \times 120 = 15 \text{ g}$$

139. Volume of H_2 at S.T.P $= 120.1 \times \dfrac{273}{300.3} = 109.2$ ml.

Reactions involved:

$Mg + H_2SO_4 \rightarrow MgSO_4 + \uparrow H_2$

$MgO + H_2SO_4 \rightarrow MgSO_4 + H_2O$

Wt. of Mg. corresponding to 109.2 ml. of H_2 at S.T.P. $= \dfrac{24.3 \times 109.2}{22,400} = 0.1185$ g

G. eq. of Mg corresponding to 0.1185 g. $= \dfrac{0.1185}{12.15} = 0.009753$

No. of g. eq. of H_2SO_4 originally present $= \dfrac{125 \times 0.1}{1000} = 0.0125$

No. of eq. of H_2SO_4 after reaction $\dfrac{125 \times 0.02}{1000} = 0.0025$

No. of eq. of H_2SO_4 used by Mg + MgO = 0.0125 – 0.0025 = 0.01
No. fo g. eq. of MgO present = 0.01 – 0.009753 = 0.000247

Wt. of MgO present $= 0.000247 \times \dfrac{40.3}{2} = 0.005$ g.

Total wt. of the sample dissolved = 0.0050 + 0.1185 = 0.1235 g.

Percentage of weight of magnesium in the sample $= \dfrac{0.185 \times 100}{0.1235} = 95.95$.

140. The wt. of the mixture of BaO and MCO_3 = 4.08 g
The wt. of the residue = 3.64 g
∴ CO_2 emitted = 4.08 – 3.64 = 0.44 g
Molecular wt. of CO_2

∴ CO_2 emitted $= \dfrac{0.44}{44} = \dfrac{1}{100} = 0.01$ mole

Number of moels of $MCO_3 \equiv 0.01$
Vol. of 1 N HCl which dissolved the residue = 0.01 ml
The volume of 1N HCl which was required for dissolution of the residue

$= (100 - 2.5 \times 16)$ ml $= (100 - 40)$ ml $= 60$ ml $= \dfrac{60}{1000} = 0.06$ equivalent

The dissolution is given by the equation

$BaO + MO + 4HCl + BaCl_2 + MCl_2 + 2H_2O$

Where 2 moles of BaO and MO are dissolved.

Number of moles of BaO and MO $= \dfrac{0.06}{2} = 0.03$

Number of moles of MO = 0.01

∴ Number of moles of BaO = 0.03 – 0-01 = 0.02

Molecular wt. of BaO = 138 + 16 = 154

∴ Wt of 1 mole of BaO = 154

∴ Wt. of 0.02 mole of BaO = 154 × 0.02 g = 3.08 g

∴ Wt. of 0.01 mole of MCO_3 = Wt. of the mixture – Wt. of BaO

= 4.08 – 3.08 = 1.00 g

∴ Molecular wt. of $MCO_3 = \frac{1.0 \times 1}{0.01} = 100.$

Molecular wt. of CO_3 = 60

∴ At wt. of metal M

= Mol. wt. of MCO_3 – Mol. wt. of CO_3

= 100 – 60 = 40

Since the atomic wt. of calcium is 40, therefore, the metal M is calcium.

141. The equations of the reactions are:

$$2NH_2OH + 4Fe^{3+} \rightarrow N_2O + H_2O + 4Fe^{2+} + 4H^+ \quad (1)$$

and $$MnO_4^- + 5Fe^{2+} + 8H^+ \rightarrow Mn^{2+} + 5Fe^{3+} + 4H_2O \quad (2)$$

Multiplying the equation (1) by 5 and the equation (20 by 4 and adding the two equations gives

$$10NH_2OH + 4MnO_4^- + 12H^+ \rightarrow 5N_2O + 21H_2O + 4Mn^{2+}$$

That is,

$$10\ NH_2OH \equiv 4MnO_4^-$$

$$10 \text{ moles} \equiv 4 \text{ moles}$$

or $$33 \times 10 \text{ g} \equiv 4 \times 1000 \text{ ml } 1 \text{ M } MnO_4^-$$

(because molecular weight of NHOH is 33)

$$330 \text{ g} \equiv 4000 \text{ ml of } 1M/MnO_4^-$$

This 4000 ml of 1 M MnO_4^- will react with 330 g of NH_2OH.

Therefore, 12 ml of 0.2 M $KMnO_4$ will react with

$$\frac{330 \times 0.02 \times 12}{4000} \text{ g of } NH_2OH.$$

Thus amount of NH_2OH in 50 ml of diluted solution

$$= \frac{330 \times 0.02 \times 12}{4000} \text{ g.}$$

Therefore, amount of NH_2OH in 1000 ml of the diluted solution

$$= \frac{330 \times 0.02 \times 12 \times 1000}{4000 \times 50} \text{ g} = 0.396 \text{ g.}$$

Thus, the amount of NH_2OH in 10 ml of the original solution (see problem) = 0.396 g

Therefore, the amount of NH_2OH in 1 litre of the original solution

$$= \frac{0.396 \times 1000}{10} = 39.6 \text{ g.}$$

142. According to the given problem
m. eq of sodium oxalate

$$= x \times \text{molarity} \times \text{Vol. in ml.}$$
$$= 2 \times 0.2 \times 10 = 4$$

$\therefore$ m. eq. of $H_2O_2 = 4$

$$\text{m. moles of } H_2O_2 = \frac{4}{2} = 2 = \text{Molarity} \times \text{Vol. inml.}$$

$$\therefore \quad \text{Molarity} = \frac{2}{20} = 0.1M$$

Balanced equations involved in chemical reactions

$$2MnO_4^- + 5H_2O_2 + 6H^+ \rightarrow 2Mn^{2+} + 5O_2 + 8H_2O$$
$$2MnO_4^- + 3Mn^{2+} + 2H_2O \rightarrow 5MnO_2 + 4H^+$$
$$MnO_2 + C_2O_4^{2-} + 4_5H^+ \rightarrow Mn^{2+} + 2H_2O + 2CO_2$$

143. The chemical equation for the combustion of organic compound $C_xH_{2y}O_y$ is

$$C_xH_{2y}O_y + 2xO_2 = xCO_2 + Y\ H_2O + xO_2$$

The gases obtained after cooling $= x + x = 2x$

$\therefore \quad 2x = 2.24$

or $x = \dfrac{2.24}{2} = 1.12$ litres

$$\text{Number of moles of } CO_2 = \frac{1.12 \text{ litres}}{22/4 \text{ litres mole}^{-1}} = \frac{1}{20} \text{ mole} = 0.05 \text{ mole}$$

The empirical formula of the organic compound is $C(H_2O)$ (1)
The mole fraction of the solute = relative decrease in vapour pressure of the solvent

Let the relative decrease in vapour prssure of the solvent be $\dfrac{p-p'}{p}$.

(where W_A and W_B are respective wts of A and B M_A and M_B are respective molecular weights of A and B and the term on R.H.S. is mole fraction of soulte A).

$$\therefore \quad \frac{p-p'}{p} = \frac{\frac{W_A}{M_A}}{\frac{W_A}{M_A} + \frac{W_B}{M_B}};$$

$$\text{So,} \quad \frac{0.104}{17.5} = \frac{\frac{50}{M_A}}{\frac{50}{M_A} + \frac{1000}{18}}$$

$$\text{or} \quad \frac{0.104}{17.5} = \frac{50}{M_A\left(\frac{50 \times 18 + 1000\,M_A}{18\,M_A}\right)}$$

or $\frac{104}{17500} = \frac{50 \times 18}{900 + 1000\, M_A}$

$\Rightarrow M_A = 105.6$

Molecular wt. of the organic compound $(CH_2O)_n = 150$

Molecular wt. of $CH_2O = 12 + 2 + 16 = 30$

$\therefore\ 30 \times n = 150$

or $n = \frac{150}{30} = 5$

$\therefore$ Molecular formula of the given organic compound is $(CH_2O)_5$ or $C_5H_{10}O_5$.

144. $ClO_3^- + 6F^{+2} + 6H^+ \rightarrow Cl^- + 6Fe^{+3} + 3H_2O$

25 mls of the mixture solution

$\equiv 0.1435$ g. or $\frac{0.1435}{143.5}$ moles AgCl

$= 1 \times 10^{-3}$ moles of AgCl $\equiv 1 \times 10^{-3}$ moles of Cl^- ion

(10^{-3} moles of Cl^- ions are obtained from Cl^- and ClO_2^-

Again, 37.5 ml. of 0.08 N oxidixing agent

$\equiv$ V ml. (say) of 0.2 N-$FeSO_4$ (unused)

whence, $V = \frac{37.5 \times 0.08}{0.2} = 15$ ml.

Thus, the volume of 0.2 N $FeSO_4$ used up by $KClO_3 = 30 - 15 = 15$ ml.

Now,15 ml. of 0.2 N-$FeSO_4$

$= \frac{0.2 \times 15}{1000}$ g. equiv. $FeSO_4 = 3.0 \times 10^{-3}$ g. equiv. or moles $FeSO_4$

Since 1 mol. of ClO_3^- uses 6 moles of $FeSO_4$ for reduction.

$\therefore$ 3.0×10^{-3} moles of $FeSO_4$

$\equiv \frac{3.0}{6} \times 10^{-3} = 0.5 \times 10^{-3}$ moles ClO_3^-.

Total no. of moles of Cl^- and $ClO_3^- = 1.0 \times 10^{-3}$ (as above)

Hence, the no. of moles of Cl^- present $= (1.0 - 0.5) \times 10^{-3}$

$=$ **0.5×10^{-3} moles**

Hence the chloride and the chlorate and present in equi-molar proportions.

145. (i) HCl contained in original solution $= 36.5/2 = 18.25$ g.litre^{-1}

HCl contained in the resultant solution (750 ml.) after evaporation

$= 18.25 - 2.675$ g. $= 15.575$ g $= 15.575/36.5$ g. eq.

Normality of the resultant solution

$= \frac{15.575}{36.5} \times \frac{1000}{750} = 0.569$ N

(ii) No. *of g equivalents* in 1 litre of original solution $\left(\frac{N}{2}\right) = 0.5$

and no. of *milli equivalents* in 1 litre

$= 0.5 \times 1000 = 500$

Hence, 100 ml. of original solution contains = 50 milli eq.

Since 1 ml. of 0.5 N acid contains 0.5 milli-equiv.

∴ 100 ml. of 0.5 N acid will contain 0.5 × 100 = 50 milli-eq.

146. Let N_1, N_2 and N_3 be the normalities of the 1 M $KMnO_4$ solution in acidic, neutral and alkaline, media, respectively.

Then,

$$(20 \text{ ml}) N_1 \equiv (33.4 \text{ ml}) N_2 \equiv (100 \text{ ml}) N_3 \quad \text{(i)}$$

In acidic medium, we have:

$$KMnO_4^- + 8H^+ + 5e^- \rightarrow Mn^{2+} + 4H_2O$$

i.e. 1 M $KMnO_4 \equiv$ 5N $KMnO_4$

Thus, from relation (i),

$$N_2 = \frac{20}{33.4} \times N_1 = \frac{20}{33.4} \times 5N = 3N$$

$$N_2 = \frac{20}{100} \times N_1 = \frac{20}{100} \times 5N = 1N$$

In other words,

$$MnO_4^- + 5e^- \xrightarrow{\text{acidic}} Mn^{2+}$$

$$MnO_4^- + 3e^- \xrightarrow{\text{neutral}} Mn^{4+}$$

$$MnO_4^- + e^- \xrightarrow{\text{alkaline}} Mn^{6+}$$

Falanced equations are:

$$MnO_4^- + 8H^+ + 5e^- \rightarrow Mn^{2+} + 4H_2O \text{ (in acidic medium)}$$

$$MnO_4^- + 2H_2O + 3e^- \rightarrow MnO_2 + 4OH^- \text{ (in neutral medium)}$$

$$MnO_4^- + e^- \rightarrow MnO_4^{2-} \text{ (in alkaline medium)}$$

The reaction involving $K_2Cr_2O_7$ is acidic medium is as under:

$$Cr_2O_7^{2-} + 14H^+ + 6e^- \rightarrow 2Cr^{3+} + 7H_2O$$

i.e. 1 M $Cr_2O_7^2 \equiv 6NCr_2O_7^{2-}$

∴ Volume of 1 M $Cr_2O_7^{2-}$ consumed, if the same volume of the reducing agent is titrated, will be given by

$$V = \frac{(20 \text{ m})}{6N} = \frac{20 \text{ ml} \times 5N}{6N}$$

$$= 16.7 \text{ ml.}$$

147. In the reaction

$Ca(OH)_2 + H_3PO_4 + CaHPO_4 + 2H_2O$

two hydrogen atoms of H_3PO_4 are replaced, therefore, the equivalent weight of H_3PO_4 be half of its molecular weight.

Molecular weight of H_3PO_4 = (1 × 3) + 31 + (4 × 16) = 98

∴ **Equivalent weight** = $\frac{\mathbf{98}}{\mathbf{2}}$ **= 49**

148. Molecular wt. of sodium thiosulpate solution (Na_2SO_3)

= 23 × 2 + 32 × 2 + 16 × 3

= 46 + 64 + 48 = 158

(i) The percentage by weight of Na_2SO_3

$$= \frac{3 \times 158 \times 100}{1000 \times 1.25} = \frac{3 \times 158 \times 10}{125} = 8 = \frac{3 \times 158 \times 2}{25} = 37.92$$

(ii) Mass of litre solution

$= 1.25 \times 1000$ g $= 1250$ g

Mole fraction of moles

$$= 3 + \frac{\text{Number of moles of } Na_2S_2O_3}{\text{Total number of moles}}$$

$$\text{Total number of moles} = 3 + \left(\frac{1250 - 158 \times 3}{18}\right)$$

$$= 3 + \left(\frac{1250 - 474}{18}\right) = \frac{776}{18} = \frac{830}{18} = \frac{415}{9}$$

$\therefore$ Mole fraction of $Na_2S_2O_3$

$$= \frac{3 \times 9}{415} = \frac{27}{415} = 0.65$$

(iii) 1 mole of sodium thiosuphate ($Na_2S_2O_3$) yields 2 moles of Na^+ and 1 mole of $S_2O_3^{2}$.

Molality of $Na_2S_2O_3$

$$= \frac{3 \times 1000}{776} = \frac{3 \times 125}{97} = \frac{375}{97} = 3.87$$

Molality of $Na^+ = 3.87 \times 2 = 7.74$

Molality of $S_2O_3^{2-} = 3.87$.

149. $Ca^{2+} + Na_2CO_3 \rightarrow CaCO_3 + 2Na^+$

$$\text{Equivalent weight of } Ca^{2+} = \frac{40}{2} = 20$$

1 milliequivalents of Ca^{2+} = 20 mg

1 milli eq. of $Ca^{2+} \equiv$ 1 milli eq. of Na_2CO_3

$\therefore$ 1 mlli equivalents of Na_2CO_3 is required to soften 1 litre of hard water.

150. Let the mol. wt. of unknown chloride be x and its weight in the mixture be y gm.
Wt. of NaCl in the mixture = $(1 - y)$ gm.

$$\text{Moles of MCl present} = y/x \text{ and those of NaCl} = \frac{1-y}{58.5}$$

Since NaCl and MCl will precipitate corresponding equimolar quantities of AgCl, therefore, the moles of AgCl privated from the mixture

$$= \frac{y}{z} + \frac{1-y}{58.5} = \frac{2.567}{143.5} \text{ (given)} \qquad \text{(I)}$$

Again, in the 2nd experiment, MCl is supposed to vaporize and form corresponding amount of AgCl after reaction with acidified $AgNO_3$. Thus

$$\frac{y}{z} = \frac{1.341}{143.5} \text{ (given)} \qquad \text{(II)}$$

Subtracting II from I, we have

$$\frac{1-y}{58.5} = \frac{2.567 - 1.341}{143.5} = \frac{1.226}{143.5}$$

hence, $y = 0.5$ g.

151. Using the data from tables we get

(a) $\Delta m = 4.0015 - 2(2.01335) = -0.0256$ g (per 2 mol. H-2)

$$\Delta m_{(1\ g)} = 1.0 \times \frac{1\ \text{mol.}}{2.01355\ \text{g}} \times \frac{-0.0256\ \text{g}}{2\ \text{mol}} = -0.006357\ \text{g}$$

$\Delta E = 9.00 \times 10^{10}$ kJ $g^{-1} \times (-0.006357$ g$) = -5.72 \times 10^{8}$ kJ (per gram of deuterium fused)

(b) $$1.3 \times 10^{4}\ \text{g} \times \frac{0.0017\ \text{g}}{100\ \text{g}} = 2.2 \times 10^{19}\ \text{g H-2}$$

$2.2. \times 10^{19}$ g $\times -5.72 \times 10^{8}$ kJ $g^{-1} = -1.3 \times 10^{28}$ kJ

(c) $$\frac{2.3 \times 10^{17}\ \text{kJ}}{1.3 \times 10^{28}\ \text{kJ}} = 1.8 \times 10^{-11}$$

152. Use the ideal gas law to calculate n, the number of moles of N_2

$$N_2\text{: } n = \frac{PV}{RT} = \frac{(745/760\ \text{atm})(5.4\ \text{L})}{(0.082\ \text{L} \bullet \text{atm}\,\text{K}^{-1} \bullet \text{mol}^{-1})(293\ \text{K})} = 0.22\ \text{mol}$$

$$C_3H_8O\ n = 0.6149\ \text{g}\ C_3H_6O \times \frac{1\ \text{mol}/C_3H_8O}{60.09\ \text{g}\ C_3H_8O} = 0.01023\ \text{mol}$$

$$X_{C_3H_8O} = \frac{0.01023\ \text{mol}\ C_3H_8O}{(0.01023\ \text{mol} + 0.22\ \text{mol})} = \frac{0.01023}{0.23023} = 0.044\ \text{mol fraction}$$

153. Use the ideal gas law to calculate the total number of moles of monomer and dimer

$$n = \frac{PV}{RT} = \frac{(436/760)\ \text{atm} \times 1.0\ \text{L}}{(0.08206\ \text{L} \bullet \text{atm}\ (\text{K} \bullet \text{mol}^{-1}) \times 373.6\ \text{K}} = 0.0187\ \text{mol monomer + dimer}$$

$$(0.0187\ \text{monomer + dimer}) \times \frac{0.63\ \text{mol dimer}}{1\ \text{mol dimer + monomer}} = 0.01178\ \text{mol dimmer}$$

0.0187 mol both – 0.01178 mol dimer = 0.00692 mol monomer

$$\text{Mass dimer} = 0.01178\ \text{mol dimer} \times \frac{120\ \text{g dimer}}{\text{mol dimer}} = 1.414\ \text{g dimer}$$

$$\text{Mass monomer} = 0.00692\ \text{mol monomer} \times \frac{60\ \text{monomer}}{\text{mol monomer}} = 0.415\ \text{g monomer}$$

$$\text{Density} = \frac{1.414\ \text{g} + 0.415\ \text{g}}{1.0\ \text{L}} = 1.829 = 1.83\ \text{g L}^{-1}\ \text{vapor}$$

Chapter 5 : SOLUTIONS

1. According to the definition molarity is the number of moles of water per liter of pure water.

 1 mole of water = 18 g.

 Weight of 1 litre of pure water = 1000 g.

 $$\text{Molarity of } H_2O \text{ in pure water} = \frac{1000}{18} = 55.55$$

2. $$\text{Molarity} = \frac{\text{Wt of the solute}}{\text{Mol. Weight of solute}} \times \frac{1000}{\text{Volume (litres)}} = \frac{1.575}{126} \times \frac{1000}{250} = 0.05 \text{ M}$$

3. Amount of solute = Molarity × Mol. Wt. × Vol. in lit. = 0.1 × 106 × 0.5 = 5.3 g.

4. Volume of solution = 100 ml; Molarity = 0.03 M, Molecular weight = 342

 $$\text{Weight of cane sugar in 100 ml} = \frac{\text{Molarity} \times \text{Mol. Wt.} \times 100}{1000} = \frac{0.03 \times 342 \times 100}{1000} = 1.026 \text{ g.}$$

5. Weight of Na_2CO_3; (W) = 5.3 g.; Volume of solution; (V) = 200 ml; Molecular weight = 106

 $$\text{Molarity} = \frac{W}{\text{Mol. wt.}} \times \frac{1000}{V} = \frac{5.3}{106} \times \frac{1000}{200} = 0.25 \text{ M}$$

6. Weight of NaOH; (W) = 20 g.; Volume of solution; (V) 100 ml; Molecular weight = 40

 $$\text{Molarity} = \frac{W}{\text{Mol. wt.}} \times \frac{1000}{V} = \frac{20}{40} \times \frac{1000}{100} = 5$$

 ∴ Molarity of 20% NaOH solution = 5M

7. Before dilution / After dilution

 $V_1 = 100$ ml $V_2 = ?$

 $M_2 = 0.2$ M $M_2 = 0.1$

 $V_1M_1 = V_2M_2$

 $100 \times 0.2 = V_2 \times 0.1$ $\therefore V_2 = \frac{100 \times 0.2}{0.1} = 200$ ml

 Hence, volume of water to be added = 200 – 100 = 100 ml.

8. $$\text{Eq. wt. of NaOH} = \frac{\text{Mol. wt}}{\text{acidity}} = \frac{40}{1} = 40$$

 $$\text{Normality} = \frac{\text{Wt. of the solute}}{\text{Gm. et. wt}} \times \frac{1000}{V\text{(ml)}} = \frac{05}{40} \times \frac{1000}{500} = 0.025 \text{ N}$$

9. $$\text{Eq. wt. of oxalic acid} = \frac{\text{Mol. wt.}}{\text{basicily}} = \frac{126}{2} = 63$$

 Weight of oxalic acid = Normality × Eq. wt. × Vol. = (0.05 × 63 × 400)/1000 = 1.26 g.

10. Milligrams percent is the number of milligrams of solute in 100 ml. of liquid. It is given that there are 2 g. of fluoride in 1.0×10^6 g H_2O. Because 1 ml of H_2O weighs 1 g, there are 2 g of fluoride in 1×10^6 ml of H_2O.

$$\therefore \text{ milligram \%} = \frac{100 \text{ ml} \times 2\text{g} \times 1000 \text{ mg g}^{-1}}{1 \times 10^6 \text{ ml}} = 0.2 \text{ mg \%}$$

11. $\text{Molarity (m)} = \frac{\text{Wt. of solute}}{\text{Mol. wt. of solute}} \times \frac{1000}{\text{Wt. of solvent in g.}} = \frac{3}{60} \times \frac{1000}{250} = 0.2$

12. Weight of solute = Molarity × Mol. Wt. × Wt. of solvent in Kg.
= 0.025 × 180 × 750/1000 = 3.375 g.

13. Mole fraction of glucose $(x_1) = \frac{n_1}{n_1 + n_2} = \frac{w/m}{w/m + W/M}$

Substituting the values w = 18; m = 180; W = 180; M = 18

$$\frac{18/180}{18/180 + 180/80} = \frac{0.1}{0.1 + 10} = \frac{0.1}{10.1} = 0.0099$$

Mole fraction of glucose = 0.0099
Mole fraction of water $(x_2) = (1 - x_1)$ = 1–0.0099 = 0.9901

14. No. of moles of Nitrogen (n_1) = Weight/Mol. wt. = 7/28 = 0.25
No. of moles of Neon (n_2) = Weight/Mol. wt = 10/20 = 0.5
No. of moles of Helium (n_2) = Weight/Mol. wt. = ¾ = 0.75

Mole fraction of $N_2(x_1) = \frac{n_1}{n_1 + n_2 + n_3} = \frac{0.25}{0.25 + 0.50 + 0.75} = 0.1666$

Mole fraction of Ne $(x_2) = \frac{n_2}{n_1 + n_2 + n_3} = \frac{0.5}{1.0} = 0.3333$

Mole fraction of He $(x_3) = 1 - (x_1 + x_2)$ = 1 – (0.1666 + 0.3333) = 0.5

15. H_2SO_4 Vs NaOH
$V_1 = ?$ $V_2 = 25$ ml
$N_1 = 0.1$ $N_2 = 0.125$

Applying rule of normal solutions $V_1N_1 = V_2N_2$

$$V_1 = \frac{V_2 N_2}{N_1} = \frac{25 \times 0.125}{0.1} = 31.25 \text{ ml}$$

Volume of H_2SO_4 required = 31.25 ml.

16. Na_2CO_3 Vs HCl
$V_1 = 20$ ml $V_2 = 24$ ml
$N_1 = ?$ $N_2 = 0.05$

Applying rule of normal solutions

$V_1N_1 = V_2N_2$; $N_1 = \frac{V_2 N_2}{N_1} = \frac{24 \times 0.05}{20} = 0.06$

Normality of Na_2CO_3 = 0.06 N

17. Molar ratio of sodium carbonate and hydrochloric acid is known from the equation:

$$Na_2CO_3 + 2\,HCl \rightarrow 2NaCl + H_2O + CO_2$$

1 mole 2 moles

For, Na_2CO_3	For, HCl
$V_1 = 25$ ml	$V_2 = 30$ ml
$M_1 = ?$	$M_2 = 0.1$
$n_1 = 1$	$n_2 = 2$

Applying rule of molar solutions $\dfrac{V_1 M_1}{n_1} = \dfrac{V_2 M_2}{n_2}$

$$\therefore\ M_1 = \frac{V_2 M_2 n_1}{V_1 n_2} = \frac{30 \times 0.1 \times 1}{25 \times 2} = 0.06$$

∴ Molality of sodium carbonate = 0.06 M

18. Convert the mass of HNO_2 into moles and divide by the number of liters of solution:

$$\frac{15 \text{ g of } HNO_3}{63 \text{ g mol}} = 0.238 \text{ mol of } HNO_3$$

$$\frac{0.238 \text{ mol of } HNO_2}{0.2 \text{ liter of solution}} = 1.19\ M\ HNO_3$$

19. Divide the moles needed by the strength of the solution and that will tell you how much solution to take:

$$\frac{0.837 \text{ mol}}{6.0 \text{ mol liter}} = 0.14 \text{ liter}$$

20. The redox reaction between mohr's salt and potassium permanganate in acid medium is

$$2KMnO_4 + 10\,FeSO_4 + 8H_2SO_4 \rightarrow K_2SO_4 + 2MnSO_4 + Fe_2(SO_4)_3 + 8H_2O$$

2 moles 10 moles

For, $FeSO_4$	$V_1 = ?$	$M_1 = 0.105$	$n_1 = 10$
For $KMnO_4$	$V_2 = 22$ ml	$M_2 = 0.125$	$n_2 = 2$

Applying rule of molar solutions, $\dfrac{V_1 M_1}{n_1} = \dfrac{V_2 M_2}{n_2}$

$$M_1 = \frac{V_2 M_2 n_1}{V_1 n_2} = \frac{22 \times 0.125 \times 10}{0.105 \times 2} = 130.95 \text{ ml}$$

21. Applying rule of normal solutions, $V_1N_1 = V_2N_2$; $250 \times 1 = V_2 \times 0.1$

$$V_2 = \frac{250 \times 1}{0.1} = 2500 \text{ ml};$$ V_2 is the final volume of the solution.

22. For the same solution, even in molar concept, we can apply:

$V_1M_1 = V_2M_2$; i.e. $400 \times 0.25 = 2{,}600 \times M_2$

$$\text{or, } M_2 = \frac{400 \times 0.25}{2600} = 0.03846$$

The resultant molarity of the solution = 0.03846 M

23. Vapour pressure of pure water at 373 K; i.e. p = 760 mm 373 K being the B.P. of water)
Vapour pressure of solution (p) = ?

Weight of solvent (W) = 33 g.; Weight of solute (w) = 3 g.

Molecular weight of water (M) = 18; Molecular weight of sugar (m) = 342

Applying Raolt's law, $(p° - p)/p° = n/\ (n + N)$

or, (W/m)/(W/m + W/M) = (w/m)/(W/M)

(w/m is negligible in the denominator)

$$\therefore\ \ P = P° - \frac{w \times M}{m \times W} \times P° = 760 - \frac{3 \times 18}{342 \times 33} \times 760 = 760 - 3.10 = 756.9 \text{ mm.}$$

24. Weight of the solute (w) = 5 g.; Weight of the solvent (W) = 50 g.
Mol. wt of the solute (m) = 200; Molecular weight of the solvent (M) = 60
Vapour pressure of solvent ($P°$) = 40 cms.

Applying Roult's law, $\frac{p°\cdot p}{p°} = \frac{n}{N} = \frac{w/n}{W/M} = \frac{wM}{mW}$ (n is negligible in the denominator)

$$\frac{p° - p}{40} = \frac{5 \times 60}{200 \times 50}$$

$$p° - p = \frac{5 \times 60}{200 \times 50} \times 40 = 1.2 \text{ cm.}$$

Thus vapour pressure of solution, (P) = $p° - (p° - P)$ = 40 – 1.2 = 38.8 cm.

25. For two miscible liquids

$$P_{total} = P_A \times x_A + P_B \times x_B \text{(i)}$$

Where x_A and x_B are the mole fractions of liquids A and B.

Moles of (A) = 28/140 = 0.2

Moles of water (B) = 72/18 = 4.0

Total number of moles = 0.2 + 4.0 = 4.2

Mole fraction of (A) = 0.2/4.2 = 0.048

Mole fraction of (B) = (1 – 0.048) = 0.952

P_{Total} = 160 mm

Substituting the values in equation (i)

$$160 = P_A \times 0.048 + 150 \times 0.952$$

$$P_A = 160 - \frac{142.8}{0.048} = 358.3 \text{ mm}$$

26. At 300 K, the vapour pressure of the solution contains
1 mole of A + 3 moles of B = 550 mm of Hg and
1 mole of A + 4 moles of B = 560 mm of Hg

Let the vapour pressure of $A = P_1^o$ Let the vapour pressure of $B = P_2^o$

Let the mole fraction of $A = x_1$ Then the vapour pressure of the solution,

$$P_{total} = P_1^o \times x_1 + P_2^o \times x_2$$

Thus, in solution – 1; P_{total} = 550 mm; $x_1 = \frac{1}{1+3}; x_2 = \frac{3}{1+3}$

$\therefore\ 550 = P_1^\circ \times \frac{1}{4} + P_2^\circ \times \frac{3}{4}$ or $550 = \frac{P_1^0}{4} + \frac{3P_2^0}{4} \Rightarrow 2200 = P_1^\circ + 3P_2^\circ$ .... (i)

In solution – 2; $P_{total} = 560$ mm, $x_1 = \frac{1}{1+4}$; $x_2 = \frac{4}{1+4}$

$\therefore\ 560 = P_1^\circ \times 1/5 + P_2^\circ \times 4/5$ or $560 = \frac{P_1^\circ}{5.4} + \frac{P_2^\circ}{5}$; $2{,}800 = P_1^\circ + 4P_2^\circ$ .... (ii)

we get from (i) and (ii),

$P_1^\circ = 600$ mm of Hg $\qquad P_2^\circ = 400$ mm of Hg

i.e. vapour pressure of A = 400 mm of Hg. and vapour pressure of B = 600 mm of Hg.

27. Molecular weight of $H_2SO_4 = 98$; Equivalent weight $= \frac{\text{Mol. wt.}}{\text{Basicily}} = 49$

No. of moles of H_2SO_4 present in solution = 49/98 = 0.5

Weight of 1 litre of the solution = 1.049 × 1000 = 1049 g.

Weight of solute present = 49 g.

Weight of water present = (1049 – 49) = 1000 g. = 1 lit.

(i) Molarity of the solution (No. of g,. moles per litre) = 0.5 M

(ii) Normality of the solution (No. of gm. equivalent per litre) = 49/49 = 1 N

(iii) Molarity of the solution (No. of moles per 1000 gm of solvent) = 0.5 Molal (m)

(iv) Mole fraction of the acid $= \frac{n}{n+N} = \frac{49/98}{49/98 + 100/18} = \frac{0.5}{0.5+55.5} = 0.09$

28. We know that

Mol. wt of NaOH = Eg. wt of NaOH = 40

Normality = Molarity

$$\frac{2/40}{100/1000} = \frac{2}{40} \times \frac{1000}{100} = 0.5\text{ M} = 0.5\text{ N}$$

Weight of solution = volume × specific gravity = 100 × 1.96 = 196 g.

Weight of water = 196 – 2 = 194 g.

$$\text{Molality} = \frac{2}{10} \times \frac{1000}{100} = 0.26$$

$$\%\text{ of NaOH} = \frac{\text{Wt} \times 100}{\text{Wt. of solution}} = \frac{2 \times 100}{198} = 1.02$$

$$\text{Moles of NaOH} = \frac{\text{Wt}}{\text{Mol. wt.}} = \frac{2}{40}\ 0.05$$

Moles of H_2O = Wt/Mol. wt = 194/18 = 10.80

Total number of moles = 0.05 + 10.80 = 10.85

Mole fraction of NaOH = 0.05/10.85 = 0.0046

Mole fraction of H_2O = (1 – 0.0046) = 0.9985

29. 0.5 ml of 0.5 M H_2SO_4 contains (0.05 liter) (0.5 mol liter^{-1}).

25 ml of 0.25 M H_2SO_4 contains (0.75 liter) (0.2 mol liter^{-1})

Total moles = (0.05 liter) (0.5 mol/liter^{-1}) + (.07 liter) (0.25 mol/liter^{-1})

$$\therefore \quad \text{Concentration} = \frac{0.044 \text{ mole}}{0.125 \text{ liter}} = 0.35\ M\ H_2SO_4$$

30. 6.9 M solution contains 6.9 mole of KOH in 1000 ml. of solution. Weight of KOH = 6.9 × 56 g. (Mol. wt. KOH = 56) = 386.4 g.

Weight of KOH in 1000 ml. of solution = 386.4 g.

As the solution of KOH is 30%, 30 g. of KOH is present in 100 g. of solution. 386.4 g. of KOH is present in (100/30) × 386.4 g = 1288 g. of solution

$$\text{since, density} = \frac{\text{Weight}}{\text{Volume}} = \frac{1288}{1000} = 1.288 \text{ g mL}^{-1}.$$

31. To calculate the mole fraction of a component of a solution we must know the number of moles of that component and the total number of moles of all the component in the solution. The moles of CH_3OH, $n_{CH_3OH} = \frac{10}{32} = 0.31$; the moles of H_2O, $n_{H_2O} = \frac{100}{18} = 5.5$

Then

$$f_{CH_3OH} = \frac{n_{CH_3OH}}{n_{CH_3OH} + n_{H_2O}} = \frac{0.31}{0.31 + 5.5} = .053$$

Once the fraction of all the components of a solution but one are known, the remaining fraction is most easily calculated by remembering that the sum of the fractions is 1. In this case $f_{CH_3H} + f_{H_2O} = 1$ or $f_{H_2O} = 1 - 0.053 = 0.947$.

32. Molarity refers to the number of moles of solute which are dissolved in 1000 g of solvent. Thus, there are 15 g/142 = 0.106 moles of Na_2SO_4 and there are 125 cc × 1.0 g/cc = 125 g of H_2O. Substituting into the expression for molarity

$$m = \frac{0.16 \text{ moles}}{125 \text{ g}} \times 1000 = 0.84$$

33. 1.0 M NaOH will contain 1.0 mole of NaOH dissolved in enough water to make 1 liter of solution. One mole of NaOH is 40 g of NaOH. Therefore 40 g will be required.

How many grams of $Al_2(SO_4)_3$ will be required to prepare 300 ml of 0.2 M $Al_2(SO_{43}$?

One liter of 0.2 M $Al_2(SO_4)_3$ ill contain 0.2 mole of $Al_2(SO_3)_3$. 300 ml is 0.3 liter; 0.3 liter of 0.2 M $Al_2(SO_4)_3$ will contain 0.3 × 0.2 or 0.06 mole of $Al_2(SO_4)_3$. one mole of $Al_2(SO_4)_3$ is 342 g. Therefore, 0.06 × 342 g = 20.6 g.

What has been stated above can be summarized, briefly, as follows;

$$0.3 \text{ liter} \times \frac{0.2 \text{ mole}}{\text{liter}} \times \frac{342 \text{ g}}{\text{mole}} = 20.6 \text{ g}$$

Note that liters and moles cancel, leaving the answer in grams

34. To find the molarity means to find the number of moles of solute that are present in 1000 ml (1 liter) of solution.

1 mole of NaOH = 40 g of NaOH

Therefore,

$$12 \text{ g of NaOH} = \frac{12}{40} \text{ mole of NaOH} = 0.3 \text{ mole of NaOH.}$$

The 0.3 mole of NaOH is present in 500 ml of solution. Since the molarity is the number of moles per 1000 ml, then,

$$100 \text{ ml} \times \frac{0.3 \text{ mole}}{500 \text{ ml}} = 0.6 \text{ mole}$$

35. 100 mg – 0.1 g. 0.1 g in 1 ml would be the same concentration s 100 g in 1000 ml. Therefore, this is a solution containing 100 g of $Cu(NO_3)_2$ per liter. To find the molarity we must find the number of moles per liter. There are 187.5 g of $Cu(NO_3)_2$ in a mole therefore,

$$\frac{100 \text{ g } CuNO_3)_2}{187.5 \text{ g.mol}^{-1}} = 0.53 \text{ mole of } Cu(NO_3)_2$$

Therefore, the molarity is 0.53.

36. One liter of 0.500 *M* K_2SO_4 will contain half a mole of K_2SO_4 dissolved in enough water to make liter of solution. Half a mole of K_2SO_4 is 87.1 g. Therefore, 87.1 g of K_2SO_4 will be required.

37. From the density, 1 liter of solution weighs (0.91) g/ml × 1000 ml = 910 g. It contains 1 mole = 32 g of CH_3OH and therefore 910 – 32 = 878 g of H_2O

$$\text{Thus, } f_{CH_3OH} = \frac{n_{CH_3OH}}{n_{CH_3OH} + n_{H_2O}} = \frac{1}{1 + \frac{878}{18}} = 0.02$$

$$\text{mass percent} = \frac{32}{910} \times 100 = 3.52\%$$

$$\text{molality} = \frac{1 \text{ mole}}{(910 - 32) \text{ g}} \times 1000 = 1.13\ m$$

38. The reaction is : $Mg + H_2SO_4 = MgSO_4 + H_2$
0.5 *M* H_2SO_4 contains 0.5 mole of H_2SO_4 per liter

$$0.32 \text{ liter of } H_2SO_4 \times \frac{0.5 \text{ mole of } H_2SO_4}{1 \text{ liter of } H_2SO_4} = 0.16 \text{ mole of } H_2SO_4$$

1 mole of H_2SO_4 liberates 1 mole of H_2
Therefore, 0.16 mole of H_2 will be liberated.

39. The reaction is : $Zn + 2HCl = ZnCl_2 + H_2$
2 moles of HCl yield 1 mole of H_2
1 mole of HCl yields 0.5 mole of H_2
0.4 *M* HCl contains 0.4 moles of HCl per liter
∴ 0.4 liter of HCl will contain

$$= 0.4 \text{ liter} \times \frac{0.4 \text{ mole}}{1 \text{ liter}} = 0.16 \text{ mole of HCl}$$

Therefore, $0.08 \text{ mole } \left(\frac{1}{2} \text{ of } 0.16 \text{ mole}\right)$ of H_2 will be liberated.

40. The analysis of this problem could run something like this: To find the molarity we must find the number of moles of Na_2CO_3 per liter of solution. If we knew how many moles of Na_2CO_3 there were in 447 ml, we could calculate the number in 1000 ml (1 liter). From the equation.

$$Na_2CO_3 + H_2SO_4 = Na_2SO_4 + H_2O + CO_2$$

or, more simply, from the formulas, Na_2CO_3 and CO_2, we can see that 1 mole of Na_2CO_3 yields 1 mole of CO_2. Therefore, if we knew how many moles of CO_2 were formed we would know how many moles of Na_2CO_3 were present in the 447 ml of solution.
We know that 5 liter of CO_2 were evolved. Since 1 mole of CO_2 occupies a volume of 11.4 liters at STP, 5 liters of CO_2 is 5/22.4 mole. Therefore, 5/22.4 mole of Na_2CO_3 is present in the 447 ml (0.447 liter) of solution.

$$\frac{5}{22.4} \text{ mole} \div \text{liter} = 0.5 \text{ mole/liter}$$

Therefore, the solution is 0.5 *M*.

41. One mole of HCl will neutralize 1 mole of KOH. Therefore, moles of HCl required = moles of KOH present in the solution. 0.8 *M* KOh contains 0.8 mole of KOH per liter; 0.25 *M* HCl contains 0.25 mole of HCl per liter.

$$0.12 \text{ liter} \times \frac{0.8 \text{ mole of KOH}}{1 \text{ liter}} = \text{moles of KOH present}$$

$$X \text{ liter} \times \frac{0.25 \text{ mole of HCl}}{1 \text{ liter}} = \text{moles of HCl required}$$

Therefore, since moles of HCl required = moles of KOH present

$$X \text{ liters of HCl} \times \frac{0.25 \text{ mole}}{1 \text{ liter}} = 0.12 \text{ liter} \times \frac{0.8 \text{ mole}}{1 \text{ liter}}$$

$X = 0.384$ liter = 384 ml

42. One mole of nonelectrolyte in 1000 g of water would have depressed the freezing point 1.86 K. Since a depression of 1.02 K was observed, 1.02/1.84 mole of nonelectrolyte must have been dissolved.

$$\frac{1.02}{1.86} \text{ mole} = 60 \text{ g}$$

$$\Rightarrow 1 \text{ mole} = \frac{1.86}{1.02} \times 60 \text{ g} = 109 \text{ g}$$

The molecular weight is approximately 109.

43. The molecular weight is the mass of solute that will depress the freezing point 1.86 K when dissolved in 1000 g of solvent. Therefore, we will first find the concentration in grams of solute per 100 g of water

$$1000 \text{ g of water} \times \frac{4.2 \text{ g of solute}}{40 \text{ g of water}} = 105 \text{ g of solute}$$

A solution containing 105 g of solute in 1000 g of water is the same *concentration* as one containing 4.2 g of solute in 40 g of water. Therefore, 105 g of solute will depress the freezing point of 1000 g of solvent exactly the same number of degrees that 4.2 g of solute will depress the freezing point of 40 g of solvent, namely, 1.52 K.

$$1 \text{ mole} \times \frac{1.52}{1.86} = 105$$

$$\therefore \quad 1 \text{ mole} = \frac{105 \times 1.86}{1.52} = 124$$

$$\therefore \quad \text{M.wt} = 124$$

44. First we calculate the molarity of the solution. Since the molecular weight of $C_6H_{10}O_3$ is 162.

$$m = \frac{20/162}{2.5} \times 1000 = 0.494$$

ΔT is the boiling point of the solution in K minus 373 K, the boiling point of water. Thus

$$\Delta T = \text{B.P.}_{solution} - 373 \text{ K} = 0.52(m) = (0.52)(0.494) = 0.26$$

$$\text{B.P.}_{solution} = 0.26 + 373 = 373.26 \text{ K}$$

45. Here the fact is each mole of NaCl gives two moles of solute particles when it is dissolving in water. In order to calculate the boiling point we use

$$\therefore \quad m = 2 \times \frac{2.37 \times 1000}{58.4 \times 100} = 0.811$$

46. First calculate the freezing point depression assuming no dissociation

$$\Delta T = K_f m = (1.86) \frac{(1.27/310 \times 1000)}{50} = 0.152$$

Then divide this value into the observed value: 0.762.0.152 = 5.01 suggesting that each mole of solute has dissociated into 5 moles of solute particles.

47. First calculate the freezing point depression assuming no dissociation occurs.

$$\Delta T = (1.86) \left(\frac{1.64/82.1}{100} \times 1000 \right) = 0.372°\text{C}$$

Divide this calculated ΔT into the observed ΔT: 0.649/0.372 = 1.74. This verities that dissociation was not complete since if it were we would find that the observed ΔT is twice the calculate one.

To calculate the fraction dissociated let X = fraction dissociated. From the equation given for the reaction, for each mole of H_2SO_3 which dissociates, two moles of particles are formed. If X dissociates, $2X$ particles are formed and $1 - X$ particles of H_2SO_3 remain. Thus the total number of particles is $2X + (1 - X) = X + 1$. We have also calculated that this number of particles gives 1.74 times as much of a freezing point depression as if there were no dissociation. Thus $X + 1 = 1.74$ or, $X = 0.74$, is the fraction of H_2SO_3 which has dissociated.

48. Assuming ideal solution behavior we can substitute into the Raoult's law expression; using the given value of $P^{\circ}_{H_2O} = 23.6$.

Substituting the values :

$$P_{H_2O} = P^{\circ}_{H_2O} f_{H_2O} = (23.6)\left(\frac{\frac{200}{18}}{\frac{200}{18}+\frac{42.2}{180}}\right) = 23.1 \text{ mm}$$

49. Since pure water boils at 373 K its vapor pressure is 760 mm. The vapor pressure above the solution will depend on the mole fraction of water based on the number of moles of solute in the solution. Since one mole of K_2SO_4 gives 3 moles of solute particles, 27.1 g gives 3 × 27.1/174 = 0.467 moles of solute particles.

$$f_{H_2O} = \frac{\frac{1000}{18}}{\frac{1000}{18}+0.467} = 0.992$$

and $P = (760)(0.992) = 754$ mm.

Note that the solution does not boil at 373 K since its vapor pressure is less than 760 mm.

50. For a solution with more than one volatile component Roult's law can be applied to each separately. First we calculate the mole fraction of each component:

$$f_{H_2O} = \frac{\frac{100}{18}}{\frac{100}{18}+\frac{80}{46}} = 0.762$$

$$f_{C_2H_6O} = 1 - f_{H_2O} = 0.238$$

$$P_{H_2O} = P^{\circ}_{H_2O} f_{H_2O} = (238)(0.762) = 178 \text{ mm}$$

$$P_{C_2H_6O} = P^{\circ}_{C_2H_0O} f_{C_2H_6O} = (543)(0.238) = 129 \text{ mm}$$

Since for a given set of condition moles are directly proportional to pressure the $f_{C_2H_6O}$ in the vapor can be calculated directly from the pressure.

$$f_{C_2H_6O} = \frac{129}{129+178} = 0.42$$

51. The osmotic pressure depends on the number of moles of solute particles in a given volume of solution. Assuming hemoglobin does not dissociate, there are 10/MW moles in 1.0 liters of solution.

Substituting into the osmotic pressure relationship gives:

$$n = \frac{\pi V}{RT} = \frac{10}{MW} = \left(2.8\,\text{mm} \times \frac{1\text{ atm}}{760\text{ mm}}\right) \times 1.0 \text{ lit.} \times 0.082 \text{ lit. atm deg. mol.} \times \frac{1}{300}\text{deg}$$

∴ MW = 66800

52. (a) The point $P = 4.0$ atm and $t = 198$ K lies in the region to the left of curve OC; in that region the solid phase is the stable phase. (2) The point $P = 2.0$ atm and $t = 263$ K lies to the right of curve OC and below curve OB. In that region the gas phase is the stable phase.

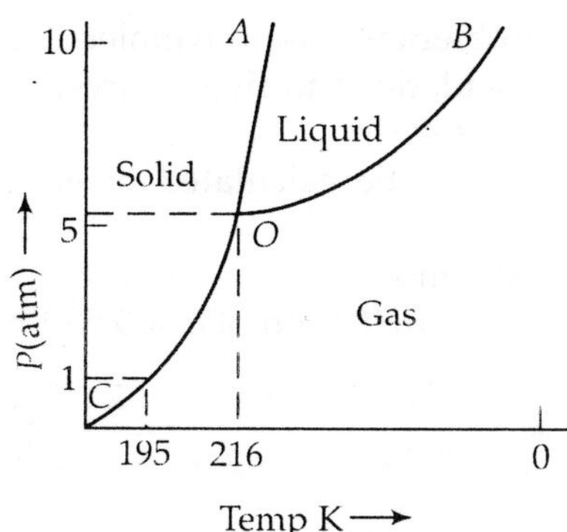

Fig. 5.23 Phase diagram for CO_2. Point *O* is the triple point, at which solid, liquid and gas are in equilibrium.

(b) At 173K and 1 atm, solid CO_2 is stable. If we keep the pressure constant at 1 atm and increase the temperature, we proceed along the horizontal dashed line at P = 1 atm. From 173 to 195 K, the solid phase is stable. At 195 K the vapor forms and solid CO_2 sublimes. Above 195 K the gas phase is stable. At a pressure of 1 atm, no liquid CO_2 will from at any temperature.

53. The molecular weight of $C_{10}H_8$ is 128.17 g · mol^{-1}. The number of moles of naphthalene in a sample of mass 16.6624 g is therefore

$$\frac{16.6624 \text{ g}}{128.17 \text{ g} \cdot \text{mol}^{-1}} = 0.13 \text{ mol of naphthalene}$$

The mass of benzene used is (200 mL) (0.87865 g · mL^{-1}) = 175.7 g. The molecular weight of C_{66} is 78.113 g · mol^{-1}, so that the number of moles of benzene in the 200 mL is

$$\frac{175.7 \text{ g}}{78.113 \text{ g} \cdot \text{mol}^{-1}} = 2.249 \text{ mol of benzene}$$

The mole fraction of naphthalene is

$$\frac{\text{No.molC}_{10}\text{H}_7}{\text{No. mol C}_{10}\text{H}_6 + \text{No. mol C}_6\text{H}_6} = \frac{0.13}{0.13 + 2.249} = 0.05464$$

The mole percentage of naphthalene is 5.464%

The mole fraction of benzene is calculated from the relation

$$X_{C_6H_6} = 1 - X_{C_{10}H_6} = 1 - 0.05464 = 0.94546$$

The mole percentage of benzene is therefore 94.546%

54. The molecular weight of glucose, $C_6H_{12}O_6$, is 180.16.· mol^{-1}. Therefore the sample of 1.275 g of glucose contains 7.077×10^{-3} mol of glucose. The ratio of moles of glucose to kilograms of H_2O in this solution is

$$\frac{7.077 \times 10^{-3} \text{ mol C}_6\text{H}_{12}\text{O}_6}{63.59 \times 10^{-3} \text{ kg H}_2\text{O}} = \frac{0.1113 \text{ mol C}_6\text{H}_{12}\text{O}_6}{1 \text{ kg H}_2\text{O}} = 0.113\ m$$

Note that in solving this problem we use the molecular weight of glucose to five figures, and carry all intermediate calculations to five figures, so that we round only once, at the end to the four significant figures.

The mole fraction of glucose can be calculated in either of two ways. We can use the original data as follows:

No. mol $C_6H_{12}O_6$= 7.077×10^{-3} mol

No. mol H_2O= $(63.59\ g \cdot mol^{-1})/18.015\ gmol^{-1} = 3.530$ mol

$$X_{C_6H_{12}O_6} = \frac{7.077 \times 10^{-3}}{7.077 \times 10^{-3} + 3.53} = \frac{7.077 \times 10^{-3}}{3.537} = 2.001 \times 10^{-3}$$

$$X_{H_2O} = 1 - 0.002 = 0.998$$

Alternatively, we can use the molality

$$X_{C_6H_{12}O_6} = \frac{0.1113}{0.1113 + 1000/18.015} = \frac{0.1113}{55.621} = 2.001 \times 10^{-3}$$

55. The question can be rephrased as follows: To what volume V, of the 0.1 M solution should water be added so that the resulting solution has a concentration of 0.04 M and a volume of 50 mL? The volume of the 0.1 M solution must contain the same number of millimoles of Na_2SO_4 as the final solution, because adding water cannot change the number of millimoles of Na_2SO_4.

Now How many millimoles of Na_2SO_4 are in 50 mL of a 0.04 M solution?

No. mmol Na_2SO_4 in the final solution $= (50\ mL)\left(0.04 \frac{mmol}{mL}\right) = 2$ mmol

What volume, V, of 0.1 M Na_2SO_4 contains 2 mmol of Na_2SO_4?

$$(V\ mL)\left(0.1 \frac{mmol}{mL}\right) = 2\ mmol$$

and therefore $V = 20$ mL.

Note that one can simply equate the number of millimoles of Na_2SO_4 in the final solution and in the volume, V, of the 0.1 M solution:

$$(50\ mL)\left(0.04 \frac{mmol}{mL}\right) = (V\ mL)\left(0.01 \frac{mmol}{mL}\right)$$

to obtain $V = 20$ mL.

Thus for preparing the required dilute solution. Use a 20 mL **volumetric pipet** transfer 20 mL of the stock 0.1 M Na_2SO_4 solution to a 50-mL volumetric flask. Add distilled water to just below the mark, invert and mix thoroughly. Add water drop wise to the mark, mixing thoroughly.

56. Because two dilute solutions are being mixed, the volume of the final solution is 100 mL. The $NaNO_3$ solution contains Na^+ ions and NO_3^- ions, while the $Cu(NO_3)_2$ solution contains Cu^{2+} ions and NO_3^- ions. The final solution therefore contains three solutes: Na^+, Cu^{2+}, and NO_3^- ions. We need to know how many millimoles of each ion are present in the final solution.

No. mmol Na^+ in 40 mL 0.2 F $NaNO_3$

$$= (40 \text{ mL}) \left(0.2 \frac{\text{mmol NaNO}_3}{\text{mL}}\right)\left(\frac{1 \text{ mmol Na}^+}{1 \text{ mmol NaNO}_3}\right) = 8 \text{ mmol Na}^+$$

No. mmol NO_3^- in 40 mL 0.2 *F* $NaNO_3$

$$= (40 \text{ mL}) \left(0.2 \frac{\text{mmol NaNO}_3}{\text{mL}}\right)\left(\frac{1 \text{ mmol NO}_3^-}{1 \text{ mmol NaNO}_3}\right) = 8 \text{ mmol NO}_3^-$$

No. mmol Cu^{2+} in 60 mL 0.1 *F* $Cu(NO_3)_2$

$$= (60 \text{ mL}) \left[0.1 \frac{\text{mmol Cu(NO}_3)_2}{\text{mL}}\right]\left[\frac{1 \text{ mmol Cu}^{2+}}{1 \text{ mmol Cu(NO}_3)_2}\right] = 6 \text{ mmol Cu}^{2+}$$

No. mmol NO_3^- is 60 mL 0.1 *F* Cu $(NO_3)_2$

$$= (60 \text{ mL}) \left[0.1 \frac{\text{mmol Cu(NO}_3)_2}{\text{mL}}\right]\left[\frac{2 \text{ mmol NO}_3^-}{1 \text{ mmol Cu(NO}_3)_2}\right] = 12 \text{ mmol NO}_3^-$$

The total number of millimoles of NO_3^- in the solutions is 20 mmol, 8 mmol from the $NaNO_3$ and 12 mmol from the $Cu(NO_3)_2$. Hence the concentration of each of these three ions in the final solution, which has a volume of 100 mL, is

$$[\text{Na}^+] = \frac{8 \text{ mmol}}{100 \text{ mL}} = 0.08\ M; \quad [\text{NO}_3^-] = \frac{20 \text{ mmol}}{100 \text{ mL}} = 0.2\ M$$

$$[\text{Cu}^{2+}] = \frac{6 \text{ mmol}}{100 \text{ mL}} = 0.06\ M$$

57. Since HCl is extensively dissociated into H^+(aq) and Cl^-(aq) ions, we use the term formality rather than molarity. To calculate the formality we need to know the number of moles of HCl in 1 L of this solution. The mass of 1 L of this solution is obtained from the density, as follows:

$$\text{mass of 1 L of solution} = \left(1.18 \frac{\text{g}}{\text{mL}}\right)(1000 \text{ mL}) = 1180 \text{ g}$$

Of this 1180 g, only 37.4% is HCl, the rest is water.

Thus the mass of HCl is (0.374) (1180 g) 441 g. The formula weight of HCl is 36.46 g · mol^{-1}, so that there are 441/36.46 = 12.1 mol of HCl in a liter of this solution. The solution is therefore 12.1 *F* HCl.

The mass of water in one liter of this solution is:

$$= \text{mass of solution} - \text{mass of HCl} = 1180 \text{ g} - 441 \text{ g} = 739 \text{ g}$$

To calculate the molality, we need the mass of water in kilograms, which is 0.739 kg. The molality of HCl is therefore 12.1-mol HCl per 0.739-kg H_2O.

$$\frac{12.1 \text{ mol HCl}}{0.739 \text{ kg H}_2\text{O}} = 16.4\ m$$

Since the molecular weight of water is 18.015 g · mol^{-1}, the number of moles of water in this liter of solution is

$$(739\text{ g})/(18.015\text{ g}\cdot\text{mol}^{-1}) = 41\text{ mol of }H_2O$$

The mole fraction of HCl is therefore

$$X_{HCl} = \frac{12.1}{12.1 + 41.0} = \frac{12.1}{53.1} = 0.228$$

58. (a) We must calculate the mole fraction of H_2O in the solution. If there are 0.475 mol of sucrose, there are exactly 1000 g of H_2O, or
(1000 g)/(18.015 g · mol^{-1}) = 55.509 mol of H_2O
Therefore

$$X_{H_2O} = \frac{55.509}{55.509 + 0.475} = \frac{55.509}{55.984} = 0.99152$$

$$P_{H_2O} = (25.812\text{ mmHg})\ (0.99152) = 25.593\text{ mmHg}$$

(b) ΔP = 25.812 mmHg – 25.593 mm Hg = 0.219 mmHg. Alternatively, we can calculate the vapor pressure lowering as follows:

$$X_{sucrose} = 1 - X_{H_2O} = 1 - 0.99152 = 0.00848$$

$$\Delta P = (0.00848)\ (25.812\text{ mm Hg}) = 0.219\text{ mmHg}$$

59. Molarity = No of moles of solute per litre of the solution.
= wt. of H_2SO_4 in one litre of he solution/Mol. wt. of H_2SO_4
Wt. of given H_2SO_4 in one litre = 1000 × density × strength = 1000 × 1.787 × 86/100 = 1537g.
Molarity of given H_2SO_4 = 1537/98 = 15.68 M
Calculation of the volume of acid for making 1 litre of 0.2 M. H_2SO_4. New applying:
$M_1V_1 = M_2V_2$ ∴ 15.68 × V_1 = 0.2 × 1000; or V_1 = 0.2 × 1000/15.68 = 12.75 ml

60. Wt. of Na_2CO_3 dissolved = 8.653 g; Volume of the solution = 100 ml
Density of the solution = 1.0816 g/c.c.

(i) Calculation fo molarity:
Weight of solution = Volume · density = 100 × 1.0816 = 108.16 g.
Weight of salt is 8.653 g. The wt. of H_2O = 108.16 – 8.653 = 99.507 g.
No. of moles of Na_2CO_3 = 8.653/106 = 0.0816 moles
99.507 g. of water contains 0.0816 moles of Na_2CO_3
1000 g. of water contains = 1000 × 0.0816/92.507 = 0.82 mole
Hence molality = 0.82 m

(ii) Calculation of molarity:
100 ml of the solution contains 0.0816 moles of Na_2CO_3
100 ml of the solution contains = 1000 × 0.0816/100 ≐ 0.816 moles
Hence molarity = 0.816

61. 1.54 molal solution means 1000 g. of the solvent contains 1.54 moles of the solute.

Wt. of 1.54 moles C_2H_5OH = 1.54 × 46 (Mol. wt of C_2H_5OH = 46)
= 70.84 g. of C_2H_5OH
1000 g. of water contains = 70.84 g. of C_2H_5OH
500 g. of water contains = 500 × 70.84/1000 = 35.42 g.

62. Weight of the solution in 1000 ml = 1000 × density = 1000 × 1.84 = 1840 g.
Weight of sulphuric acid in 1840 g. of acid = 1840 × 98/100 = 1803.2 g.
Molarity is equal to number of moles at the solute per litre = 1803.2/98 = 18.4

63. Weight of sulphuric acid in one litre = 1000 × density = 1000 × 1.090 = 1900 g.
Weight of sulphuric acid in 13% solution = 1090 × 13/100 = 141.7 g.
No. of g. moles of H_2SO_4 per litre = 141.7/98 = 1.445 moles
Hence molarity = 1.445 M
Calculation of molality:
Weight of sulphuric acid in 13% solution = 141.7 g.
Weight of water in 1050 g. of solution = (1090 – 141.7) = 948.3 g.

Molarity of H_2SO_4 = moles of H_2SO_4, in 1000 g. of solvent)

$$= \frac{141.7\text{ g}/98\text{ mole}}{948.3\text{ g}/1000\text{ g.}} = \frac{141.7 \times 1000}{98 \times 948.3} = 1.524 \text{ mole}$$

Hence molarity = 1.524 m
Normality of H_2SO_4 = 2 × molarity = 2 × 1.445 = 2.890 N
Calculation of dilution: $V_1N_1 = V_2N_2$
Substituting the values

100 ml × 2.890 N = V=2 × 1.5

or V_2 = 100 × 2.89/1.5 = 186.6 ml

64. Here $\Delta T = 273 - 263 = 10$ K

Since $\Delta T = K_f$, molality

$$10 = 1.86 \times \frac{w}{46}$$

or

$$w = \frac{46 \times 10}{1.86} = 247.3 \text{ g}$$

65. We know that

$$n_A = \frac{40\text{ g}}{120\text{ g mol}^{-1}} = 0.333 \text{ mole}; \; n_B = \frac{500\text{ g}}{160\text{ g mol}^{-1}} = 3.125 \text{ mole}$$

$$\therefore \quad X_A = \frac{n_A}{n_A + n_B} = \frac{0.333}{0.333 + 3.125} = 0.096$$

Thus, $X_B = 1 - X_A = 1 - 0.096 = 0.904$

Since, $p_B = X_B p_B^\circ$

$$\therefore \quad p_B^\circ = \frac{p_b}{X_b} = \frac{400\text{ mm}}{0.904} = \mathbf{442.5\ mm}$$

Now, $p_T = X_A p_A^\circ + X_B p_B^\circ$

$\therefore \qquad 475 = 0.096\ p_A^\circ + 400$

$\Rightarrow \qquad p_A^\circ = \mathbf{781\ mm}$

66. No. of moles of water = 27/18 = 1.5 moles

No. of moles of acetic acid = 15/60 = 0.25 moles

No. of moles of ethyl alcohol = 11.5/46 = 0.25 moles

Total number of moles = 1.5 + 0.25 + 0.25 = 2 moles

∴ Mole fraction of water = 1.5/1 = 0.75

Mole fraction of acetic acid = 0.25/2 = 0.125

Mole fraction of ethyl alcohol = 0.25/2 = 0.125

67. (a) Eq. wt. of H_2SO_4 = Mol. wt/Basicity = 98/2 = 49

0.5 N solution requires 49 × 0.5 = 245 g. of H_2SO_4 litre.

Hence for 200 ml of 0.5 N solution, the amount of acid required

= (24.5 × 200)/1000 = 4.9 ≃ 5g.

$$\text{Volume of 98\% } H_2SO_4 \text{ required} = \frac{\text{Mass}}{\text{Sp. gravity}} = \frac{5.0}{1.84} = 2.72 \text{ ml}$$

(b) Eq. wt. of $Na_2CO_3 . H_2O$ = Mol wt/Acidity = 124/2 = 62

Equivalents of $Na_2CO_3 . H_2O$ contained in 0.62 g. = 0.62/62 = 0.01

Equivalents of H_2SO_4 in 100 ml of 0.1 N solution = 0.1 × 100/1000 = 0.01

Since the resulting solution contains equivalent quantities of the acid and the base, it will be neutral.

68. (i) HCl contained in original solution = 36.5/2 = 18.25 g per litre.

HCl contained in the resultant solution after evaporation (in 750 ml).

No. of gm equivalent of HCl = 15.575/36.5

$$\text{Normality of the resultant solution} = \frac{15.575}{36.5} \times \frac{1000}{750} = 0.569 \text{ N}$$

(ii) No. of g,. equivalents in one litre of original solution = 0.5

No. of milli equivalents in 1 litre = 0.5 × 1000 = 500

Hence, 100 ml of original solution contains = 50 milli equivalents.

69. Normality of each acid should be calculated separately and then they will be added to get the combined normally of the mixture.

(i) Normality due to HCl:

20 ml of 10N HCl = 1000 ml of N/? HCl

$$\text{New Normality of HCl} = \frac{20 \times 10}{1000} = 0.2 \text{ N}$$

(ii) Normality due to H_2SO_4

$$\text{ml of 36 N } H_2SO_4 = \frac{10 \times 36}{1000} = 0.36 \text{ N}$$

$$\text{New Normality of } H_2SO_4 = \frac{10 \times 36}{1000} = 0.36\text{N}$$

Normality of the mixture = 0.2 + 0.36 = 0.56 N

70. Weight of $H_2C_2O_4$ + $NaHC_2O_4$ = 2.02 g.
Let the weight of $H_2C_2O_4 = x$g.
∴ Weight of $NaHC_2O_4$ = (2.03 – x) g.
I. Titration with NaOH

Mixed Solution	NaOH
$V_1 = 10$ ml	$V_2 = 3$ml
$N_1 = ?$	$N_2 = 0.1$ N

$\because V_1N_1 = V_2N_2$:

$$\therefore N_1 = \frac{V_2 N_2}{V_1} = \frac{3 \times 0.1}{10} = 0.03 \text{ N}$$

∴ No of equivalents of H^+ ions present in the liter of the solution = 0.03
II. Titration with $KMnO_4$

Mixed solution	$KMnO_4$
$V_1 = 10$ ml	$V_2 = 4$ ml
$N_1 = ?$	$N_2 = 0.1$ N

We know that $V_1N_1 = V_2N_2$

$$\therefore \quad N_1 = \frac{V_2 N_2}{V_1} = \frac{4 \times 0.1}{10} = 0.04 \text{ N}$$

No of equivalents of $C_2O_4^{-2}$ in 1 litre of solution = 0.04
∴ No. of equivalents of $NaHC_2O_4$ = 0.04 – 0.03 = 0.01
i.e. Normality of $NaHC_2O_4$ = 0.01 N
Now, Equivalent weight of $NaHC_2O_4$ = 112
Amount of $NaHC_2O_4$ = 0.01 × 112 = 1.12 g.
But amount of NaC_2O_4 = (2.02 – x) g. = 2.02 – x = 1.12 or x = 0.9 g.
∴ Amount of $H_2C_2O_4$ present in the mixture = 0.90 g.
Amount of $NaHC_2O_4$ present in the mixture = 1.12 g.

71. **(a)** $Pb(NO_3)_2$ and $NaNO_3$ both decompose below 873 K
$Pb(NO_3)_2$ decomposes according to the equation

$$2\,Pb(NO_3)_2 \rightarrow 2PbO + 4NO_2 + O_2$$
$$664 \text{ g.} \qquad\qquad 448 \text{ g.}$$

$NaNO_3$ decomposes according to the equation.

$$2NaNO_3 \rightarrow 2NaNO_2 + O_2$$
$$170 \text{ g.} \qquad\qquad 138 \text{ g.}$$

Total weight of $Pb(NO_3)_2$ and $NaNO_3$ = 5 g.
Let the weight of $Pb(NO_3)_2$ be x g.
∴ Weight of $NaNO_3$ = (5 – x) g.
Loss in weight of the mixture = (28/100) × 5 = 1.4
Weight of residue = 5 – 14 = 3.6 g.
($PbO + NaNO_2$)
664 g. of Pb $(NO_3)_2$ on heating gives 448 g. of PbO
x g. of $Pb(NO_3)_2$ on heating gives 448 × x/664 g. of PbO.

170 g. of $NaNO_3$ on heating gives 138 g. of $NaNO_3$.

(5-x) g. of $NaNO_3$ on heating gives 138 × (5-x)/170 g. of $NaNO_2$.

$$\text{Total weight of residue} = \frac{448}{664} + \frac{138(5-x)}{170} = \frac{448\,x}{664} + \frac{138(5-x)}{170} = 3.6$$

or, $0.6746x + 4.058 - 0.817x = 3.6$

or, $0.1317x = 4.058 - 3.6 = 0.458$

∴ $x = 0.458/0.1371 = 3.340$ g.

∴ Weight of $Pb(NO_3)_2$ = 3.340 g.

Weight of $NaNO_3$ = 5 – 3.340 = 1.66 g.

(b) Weight of H_4SO_4 in 100 ml = 93 g.

∴ Weight of H_2SO_4 in 1000 ml = 93 × 10 = 930 g.

Weight of 1 ml of solution = 1.84 g

∴ Weight of 1000 ml of solution = 1.84 × 1000 = 1840 g.; weight of H_2O = 1840 – 930 = 910 g.

Molecular weight of H_2SO_4 = 98

$$\text{Molality} = \frac{W}{M} \times \frac{1000}{\text{weight of solvent in g.}} = \frac{930}{98} \times \frac{1000}{910} = 10.428 \text{ m}$$

∴ Molality of 98% H_2SO_4 = 10.428 m

72. Given

$m = 152 \text{ g mol}^{-1}$, $\Delta T = ?$ $K_b = 17.2$

$w = 0.450$ g $W = 34.0$ g

Substituting these values in following equation:

$$\Delta T = \frac{100 \times K_b \times w}{m \times W} \quad \because \; m = \frac{100 \times K_b \times w}{\Delta T \times W}$$

$$= \frac{100 \text{ g} \times 290.2 \text{ C m}^{-1} \times 0.450 \text{ g}}{152 \text{ g mol}^{-1} \; 34.0 \text{ g}} = 2.59 \text{ K}$$

Hence B.P. of the solution = B.P. of acetone + ΔT = 329.3 + 2.58 = 331.88 K

73. Molecular mass of KCl= 39 + 35.5 = 74.5 g mol^{-1}

Mass of 1 L KCl = 100 mL × 1.04 g mL^{-1} = 1040 g

∴ Mass of water = 1040 g – 74.5 g = 965.5 g

Now, $\Delta T_b = 1\, K_b m = (2)(0.52 \text{ K kg mol}^{-1}) \dfrac{1}{0.9655 \text{ kg}} = 1.077$ K

∴ Boiling point of solution = 373 + 1.077 = 374.077 K

74. Vapour pressure of pure benzene P_0 = 640 mm

Vapour pressure of solution = P_s = 640

Lowering of vapour pressure P_0–P_s = 640–600 = 40 mm

Weight of benzene W_1 = 39 g given

Molecular weight of benzene M_1 = 78

weight of solute W_2 = 2.175 g.
Molecular weight of solute, M_2 = ?

According to Raoult's law, $\frac{P_0 - P_s}{P_0} = \frac{W_2}{M_1} \times \frac{M_2}{W_1}$; or $\frac{40}{640} = \frac{2.175}{M_2} \times \frac{78}{39}$

$$\Rightarrow \quad M_2 = \frac{2.175 \times 78 \times 640}{39 \times 40} = 69.6$$

∴ Molecular weight of solute = 69.6

75. At 373 K, the vapour pressure of water can be assumed as 1 atm = 760 m of Hg. Due to the presence of solute the vapour pressure is lowered to 750 mm of Hg. Relative lowering of vapour pressure = mole fraction of solute

or, (760 – 750)/760 = 10/760 = 1/76 is the mole fraction of solute

i.e. 1 mole of glucose in 75 moles of water (total moles = 76)

molality = moles/1000 g. of water

18 g. of water = 1 mole

∴ 1000 g of water = 55.55 mole

75 moles of water = 1 molal glucose

55.55 moles of water = 1000 g. of water = 55.55 × 1/75 = 0.741 m

76. Reaction

$$4Fe^{3+} + N_2H_4 \rightarrow N_2 + 4Fe^{2+} + 4H^+$$
$$MnO_4^- + 5Fe^{2+} + 8H^+ \rightarrow Mn^{2+} + 5Fe^{3+} + 4H_2O$$
$$[4Fe^{3+} + N_2H_4 \rightarrow N_2 + 4Fe^{2+} + 4H^+] \times 5$$
$$[MnO_4^- + 5Fe^{2+} + 8H^+ \rightarrow Mn^{2+} + 5Fe^{3+} + 4H_2O]\ 4$$
$$\overline{4MnO_4^- + 5N_2H_4 + 32H^+ \rightarrow 4\ Mn^{2+} + 2OH^+ + 16H_2O}$$

4 moles MnO_4^-= 5 moles N_2H_4 = 5 moles $N_2H_6SO_4$ = 5 × 130 g $N_2H_6SO_4$
(Since Mol. wt. of $N_2H_6SO_4$ is 130 g) = 650 g. $N_2H_6SO_4$

4000 ml 1 MnO_4^- = 650 g $N_2H_6SO_4$

$$20 \text{ ml } \frac{M}{50}\ MnO_4^- = \frac{20 \times (1/50) \times 650}{4000} = 0.065 \text{ g.}$$

Since 10 ml contain = 0.065 g. $N_2H_6SO_4$

one litre contain = 0.065 × 100/10 = 6.5 g.

77. (a) From the half cell reaction Eq. wt of $NaBrO_3$ = Mol. weight/6

$$= \frac{23 + 80 + 48}{6} = 25.17$$

To prepare 1000 ml of 1 N Na BrO_3 solution, amount required = 25.17 g.
Thus, to prepare 85.5 ml of 0.672 N $NaBrO_3$ solution amount required

= (85.5 × 0.672 × 25.17)/1000 = 1.446 g.

Molarity = 0.672/6 = 0.112 M

(b) From the half cell reaction Eq. wt. of $NaBrO_3$ = $\frac{\text{Mo. weight}}{5} = \frac{23 + 80 + 48}{5}$ = 30.2

To prepare 1000 ml of 1 $NNaBrO_3$ solution, amount required = 30.2
Thus, to prepare 85.5 ml 0.672 N $NaBrO_3$ solution

$$\text{amount required} = \frac{30.2 \times 0.672 \times 85.5}{1000} = 1.7352$$

$$\text{Molarity} = 0.672/5 = 0.1344 \text{ M}$$

78. Vapour pressure of ethanol, $p° = 44.5$ mm,
Vapour pressure of methanol $p° = 88.7$ mm.
Moles of $C_2H_5 OH(n_1) = 60/46 = 1.304$
Moles of $CH_3OH(n_3) = 40/32 = 1.25$

$$\text{moles fraction of } CH_3OH = \frac{n_2}{n_1 + n_2} = \frac{1.25}{1.304 \ 1.25} = 0.4693$$

Moles fraction of $C_2H_5OH = 1 - 0.4893 = 0.5107$
Vapour pressure due to $C_2H_5OH = 44.5 \times 0.5107 = 22.73$ mm
Vapour pressure due to $CH_3OH = 88.7 \times 0.4893 = 43.40$ mm
Total pressure = 22.73 + 43.40 = 66.13 mm.

$$\text{Mole fraction of } CH_3 \text{ OH in the vapour} = \frac{43.40}{22.73 + 43.40} = 0.663.$$

79. Solubility of Mg $(OH)_2$ 'S' = 9.57×10^{-3} gm $L^{-1} = \dfrac{9.57 \times 10^{-3}}{58} = 1.65 \times 10^{-4}$ mol^{-1}

$$Mg\,(OH) \rightleftharpoons Mg^{2+} + 2OH^-$$

$$K_{sp} = 4S^3 = 4\,(1.65 \times 10^{-4})^3 = 1.8 \times 10^{-11} \text{ (approx)}$$

$$K_{sp} = [Mg^{2+}][OH^-]^2;\ [OH^-]^2 = \frac{1.8 \times 10^{-11}}{0.02} = 9 \times 10^{-10} \text{ (in 0.02 Mg } (NO_3)_2]$$

$$[OH^{-1}] = 3 \times 10^{-5};\ [Mg^{2+}] = \frac{1}{2}\ [OH^-] = \frac{3 \times 10^{-5}}{2}$$

$= 1.5 \times 10^{-5}$ mol $litres^{-1} = 1.5 \times 10^{-5} \times 58 = 8.7 \times 10^{-4}$ g L^{-1}

80. Total volume of the acid mixture = 2000 ml
New normality of HNO_4 dilution = $(5 \times 8)2000 = 40/2000$ N

$$\text{New normality of HCl on dilution} = \frac{4.8 \times 5}{2000} = \frac{24}{2000} \text{ N}$$

$$\text{New normality of } H_2SO_4 \text{ on dilution} = \frac{17 \times 2x}{2000} = \frac{34x}{2000} \text{ N; Let volume of } H_2SO_4 \text{ is } x)$$

$$\text{Combined normality of the mixture} = \frac{40}{2000} + \frac{24}{2000} + \frac{34x}{2000} \text{N} = \frac{64}{2000} + \frac{34x}{2000} \text{ N}$$

Equivalent weight of Na_2CO_3, $10H_2O = 286/2 = 143$
Strength of $Na_2CO_3 \cdot 10H_2O = 1/14.3$
By applying rule of normal solutions

$$\frac{64}{2000} + \frac{34x}{2000}30 = \frac{1}{14.3}\ (42.9)$$

After evaluation $x = 4$ ml

Weight of sulphate ions present in 4 ml of 17 M $H_2SO_4 = \frac{4 \times 48 \times 17 \times 2}{1000} = 6.528$ g.

81. (i) Since the eq. mass of HCl = 36.5 g equiv^{-1}

∴ Mass of HCl in 1 litre of N/2 HCl = 36.5/2 = 18.25 g

Mass of HCl lost during heating = 2.675 g

∴ Mass of HCl left after heating = 18.25 – 2.675 = 15.575 g

Now, Normality = $\frac{\text{Mas of solute (g)/ Eq. mass of solute}}{\text{Volume of solution (L)}} = \frac{15.575/36.5}{0.75} =$ **0.57 N**

(ii) Since, equivalents of HCl in 1000 mL of the original solution = ½

∴ Milli equivalent of HCl in 1000 mL

= (1/2) × 1000 = 500

or Milli equivalent of HCl in 100 mL = $\frac{500 \times 100}{1000} =$ **50**

82. Let the vapour pressure of pure X and pure Y be P_X° and P_Y° respectively

Since, $X_X = \frac{n_X}{n_X + n_Y} = \frac{3}{3+1} = 0.75$

and $X_Y = 1 - X_X = 1 - 0.75 = 0.25$

According to Raoult's law,

$$P = X_X P_X^\circ + X_Y P_Y^\circ$$

$$550 = 0.75\ P_X^\circ + 0.25\ P_Y^\circ \qquad \ldots\ldots \text{(i)}$$

For first solution,

Similarly for second solution,

$$X_X = \frac{n_X}{n_X + n_Y} = \frac{4}{4+1} = 0.8$$

and $X_Y = 1 - X_Y = 1 - 0.8 = 0.2$

∴ $560 = 0.80\ P_X^\circ + 0.2\ P_Y^\circ \qquad \ldots\ldots$ (ii)

On solving eq. (i) and (ii) we get

$$P_X^\circ = 600 \text{ mm Hg}$$

$$P_Y^\circ = 400 \text{ mm Hg}$$

83. **(a)** % by weight of $Na_2S_2O_3$

Density of solution = 1.25 g ML^{-1}.

Wt. of 1 ml. of solution is = 1.25 g.

Wt. of 1000 ml. of solution is = 1250 g.

Mol. w.1 of $Na_2S_2O_3$ = 158

Wt. of $Na_2S_2O_3$ present in 3 M solution = 3 × 158 = 474 g.

100 g. of type contains $= \frac{474 \times 100}{1250} = 37.92$ g.

(b) The mole fraction of solution thiosulphate:

No. of moles of $Na_2S_2O_3$ present in 1 lit. or 1250 g. of 3 M $Na_2S_2O_3$ = 3 moles
Wt. of water in 1 lit. of solution = 1250 – 474 = 776 g.
No. of moles of water in 1 lit. of solution = 776/18 = 43.1 moles
Total No. of moles = 3 + 46.1 = 0.065
Mole fraction = 3/46.1 = 0.065

(c) The molalities of Na^+ and $S_2O_3^{2-}$ ions.

776 g. of water contains 3 moles of $Na_2S_2O_3$
∴ 1000 g. of water contains = (3/776) × 1000 moles of $Na_2S_2O_3$ = 3.865 moles (or) 3.865 molal solution
1 mole of $Na_2S_2O_3$ will contain 2 moles of Na^+ ions and 1 mole of $S_2O_3^{2-}$ ions.
3.865 mole of $Na_2S_2O_3$ will contain 2 × 3.865 moles of Na^+ ions.
The molality of Na^+ ions is 7.73 and of $S_2O_3^{2-}$ ions is 3.865

84. Vapour pressure of pure benzene $P°$ = 639.7 mm

Vapour pressure of the solution P = 631.9 mm
Lowering of vapour pressure ($P° - P$) = 639.7 – 631.9 = 7.8 mm
Mol. wt of benzene = 78
Let the molality of solution be "m"
i.e., "m" moles of solute is present in 1000 g. of benzene
No. of moles benzene N = 1000/78

Using Raoult's law $\frac{p^0 - p}{p^0} = \frac{\text{m}}{\text{N}}; \frac{7.8}{639.7} = \frac{\text{m} \times 78}{1000}$

Molality of the solution, $= \frac{1000 \times 7.8}{78 \times 639.7} = \frac{100}{639.7} = 0.1563$

85. (a) Calculation of Wt. of H_2SO_4:

26.7 ml of 0.4 N NaOH neutralizes (26.7 ml of 0.4 N H_2SO_4)
Wt. of H_2SO_4 = (0.4 × 49 × 26.7)/1000 = 0.5233 g.
i.e. 0.5 g. of fuming $H_2SO_4 \rightleftharpoons$ 0.5233 g. (H_2SO_4)

(b) The dilution reaction is

$H_2S_2O_7 + H_2O \rightarrow 2H_2SO_4$
178 g. of oleum = 196 g. of H_2SO_4
0.5 g. of oleum = (196 × 0.5)/178 = 0.551 g. H_2SO_4
Actual amount produced = 0.5233 g.
% of free SO_3 in the sample = 0.5233/0.551 × 100 = 95%

86. Eq. Wt. of H_2SO_4 = 49

0.1 M H_2SO_4 = 0.2 N H_2SO_4
0.2 M H_2SO_4 = 0.4 N H_2SO_4

Using phenolphthalein indicator:
In this titration, $Na_2\ CO_2$ is only half neutralized. Hence it is converted to $NaHCO_3$

$$Na{=}2CO_3 + H^+ \rightarrow Na^+ + NaHCO_3$$

The volume of 0.2 NH_2SO_4 used for this = 2.5 ml
The volume of 0.2 N H_2SO_4 required for complete neutralization is 5 ml.

$N_1V_1 = N_2V_2$
V_1 = 5 ml; $V{=}2$ = 10 ml
N_1 = 0.2; N_2 = Normality of Na_2CO_3

$$N_2 = \frac{V_1N_1}{V_2} = \frac{5 \times 0.2}{10} = 0.1 \text{ N}$$

amount of 0.1 N Na_2CO_3 present in a lt. (- eq. wt. × normality = 53 × 0.1) = 5.3 g.
Using Methyl orange indicator.
In this titration, all the alkali is neutralised at the end point

The total acid used = 2.5 ml of 0.2 N H_2SO_4 + 2.5 of 0.4 N H_2SO_4
= 2.5 ml of 0.2 N H_2SO_4 + 5 ml of 00.2 N H_2SO_4
= 7.5 ml of 0.2 N H_2SO_4

The volume of H_2SO_4 used to neutralise $NaHCO_3$ = (7.5 – 5.0) = 2.5 ml of 0.2 N H_2SO_4

$N_1V_1 = N_2V_2$
V_1 = 2.5 ml; V_2 = 10 ml
N_1 = 0.2 ml;
N_1 = Normality of $NaHCO_3$

$$N_2 = \frac{V_1N_1}{V_2} = \frac{2.5 \times 0.2}{10} = 0.05 \text{ N}$$

The amount of $NaHCO_3$ present in 1 litre = 84 × 0.05 = 4.3 g.

87. No. of moles $n = \frac{PV}{RT} = \frac{0.2 \times 2}{0.082 \times 303} = 0.0161$ moles

0.0161 eq. of NH_3 = 0.0161 eq. of H_2SO_4;
∴ 134 ml of H_2SO_4 contain 0.0161 eq. of H_2SO_4

$$1000 \text{ ml of } H_2SO_4 \text{ contain} = \frac{0.0161 \times 1000}{134} = 0.12 \text{ eq.}$$

Normality of H_2SO_4 = 0.12 N

88. Molarity of $H_2SO_4 = \frac{\text{Sp. gravity} \times \% \times 10}{\text{Mol. wt.}} = \frac{1.84 \times 98 \times 10}{98} = 18.4$ M

Vol. of H_2SO_4 required to prepare 200 ml of 0.5 N H_2SO_4

N_1 = 2 × 18.4 N_2 = 0.5
V_1 = ? V_2 = 200

$$\therefore \quad V_1 = \frac{200 \times 0.5}{2 \times 18.4} = 2.717 \text{ ml}$$

89. Since, $p°\text{-}p/p° = wM/mW$

$p°$ = Vapour pressure of water = 23.55 m = Mol. wt. of Urea 60

p = Vapour pressure of solution = ? W = Wt. of water = 100 g

w = Weight of urea = 6 g M= Mol. wt. of water = 18

$$p° - p = \frac{p^0 \times w \times M}{mW} = \frac{23.55 \times 6 \times 18}{60 \times 1000} \; 0.0424 \text{ mm}$$

or, $p = p° - 0.0424 = 23.55 - 0.0424 = 23.5076$ mm.

90. (i) Calculation of the percentage by weight of ethanol;

Density of ethanol = 0.785 g. mL^{-1}

1 ml of ethanol weighs 0.785 g

10 ml of ethanol weight 0.785 × 10 = 7.85 g.

Density of ethanol water solution = 0.9866 g.mL^{-1}

The weight of 100 ml solution = 0.9866 × 100 = 98.66 g.

Now, 98.66 g. of the solution contains 7.85 g. of ethanol.

∴ 100 g. of the solution contains (7.85/98.66) × 100 = 7.957 g.

Percentage of ethanol in the solution = 7.957

(ii) Calculation of the molarity of ethanol in the solution.

100 ml of the solution contains 7.85 g. ethanol.

1000 ml of the solution contains = 7.85 × 1000/100 = 78.5 g.

The molecular weight of C_2H_5OH = [2 × 12 + 6 × 1 + 1 × 16] = 46

$$\text{Molarity of ethanol} = \frac{\text{Wt. of ethanol in 1000 ml}}{\text{gram molecular weight}} = \frac{78.5}{46} = 1.71$$

i.e. the molarity of the solution = 1.71 M

91. Density of solution = 1077.2 kg m^{-3}

Volume of solution = 1 dm^3 = 10^{-3} m^3 = 1077.2 × 10^{-3} kg ≐ 1077.2 g

$$\text{Mass of } Na_2SO_4 = 8.0575 \times 10^{-2} \text{ kg} \times \frac{142}{322} = 35.53 \text{ g}$$

(Mol. mass of Na_2SO_4 = 142 g mol^{-1}, and Na_2SO_4 10 H_2O = 322 g mol^{-1})

Mass of solvent = (1077.2 – 35.53) g = 1041.67 g

$$\text{No. of moles of } Na_2SO_4 = \frac{35.53 \text{ g}}{142 \text{ g mol}^{-1}} = 0.25$$

$$\text{No. of moles of solvent} = \frac{1041.67 \text{ g}}{18 \text{ g mol}^{-1}} = 57.87 \text{ moles}$$

$$\text{Molarity} = \frac{\text{No. of moles of solute}}{\text{Vol. of solution in litre}} = \frac{0.26 \text{ mole}}{1 \text{ L}} = 0.25 \text{ M}$$

$$\text{Molality} = \frac{\text{No. of moles of solute}}{\text{Mass of solvent in kg}} = \frac{0.25}{1.04167} = 0.24 \text{ m}$$

$$\text{Mole fraction of solute} = \frac{n}{n+N} = \frac{0.25}{0.25+57.87} = \mathbf{0.0043}$$

92. $N_1V_1 = N_2V_2$

or $22.5 \times 2.5 = 10.5 \times N_2$

$\therefore$ $$N_2 = \frac{22.5 \times 2.5}{10.5} = 5.35;$$

Now, $$\frac{\text{Density} \times \%\text{weight} \times 10}{\text{Equivalent weight}} = \text{Normality}$$

$$\therefore \ \% \text{ Weight of } H_2SO_4 \text{ in solution} = \frac{\text{N} \times \text{Eq. wt.}}{\text{density} \times 10} = \frac{5.35 \times 49}{1.16 \times 10} = 22.59\%$$

93. $BaCl_2 + Na_2SO_4 \rightarrow 2NaCl + BaSO_4$

1.0 mole 1 mole

142.1 g 233.4 g

From the above equation we see that equal mole quantities of Na_2SO_4 and $BASO_4$ are involved.

$\because$ 233.4 g of $BaSO_4 \rightarrow 1$ mole

$\therefore$ 0.864 g of $BaSO_4 \rightarrow 0.864/233.4 = 0.0037$ moles of $BaSO_4$

Molarity (moles per litre solution) of Na_2SO_4;

0.0037 moles/0.025 litre = 0.150 M Na_2SO_4

94. $$\text{Molarity (m)} = \frac{\text{No. of moles of P}}{\text{No. of kg of solvent}} = \frac{28/\text{M}}{315/1000} = \frac{28 \times 1000}{\text{M} \times 315}$$

$\Delta T_b = K_b \times m;$ or, $320.98 - 2193 = 2.34 \text{ m} \Rightarrow \text{m} = \dfrac{1.68}{2.34}$,

$\therefore$ $$M = \frac{2.34 \times 28 \times 1000}{315 \times 1.68} = 123.8$$

No. of phosphorous atoms in '*p*' (molecule = 123.8/31 = 4)

Molecular formula of '*p*' in $CS_2 = P_4$

95. According to Raoult's law, partial pressure of liquid benzene in a mixture.

$$P_A = P_A^\circ \times X_A = \frac{\frac{23.4}{78}}{\frac{64.4}{92} + \frac{23.4}{78}} P_A^\circ = \frac{0.3}{0.3+0.7} \times P_A^\circ = 0.3 \ P_A^\circ$$

Similarity partial pressure of toluene = $0.7 \times P_B^\circ$

In the vapour phase mole fraction of benzene

$$= \frac{\text{Partial pessure of benzene}}{\text{Total pressure}} = \frac{0.3 \times p_A^\circ}{0.3\ p_A^\circ + 0.7\ p_B^\circ} = \frac{0.3 \times 75}{(0.3 \times 75) + (0.7 \times 22)} = 0.594$$

96. Wt. of solute (w) = 0.152 g.

Wt. of water (w) = 25 g, ΔTf = 0.19°C; kf = 18.5; M = ?

Hence $\Delta Tf = K_f \times \text{Molarity} = \frac{K_f \times 0.152/M}{25} \times 100; M = \frac{18.5 \times 0.152 \times 100}{25 \times 0.19} = 59.2$

97. $m = \frac{1000 \times K_f \times W}{W \times \Delta t}$; or, $342 = \frac{1000 \times K_f \times 110}{0.597 \times 1000}$ $\therefore K_f = 1.855$

Using 'k_f' in second case: m (1000 × 1.855 × 150)/2.657 × 1000 = 104.8

98. Grams of pure Na_2CO_3 required = (106.2) × (1.5/1) × 2 = 159 g.

Therefore

95% Na_2CO_3 required = (100 × 159)/95 = 167.37 g.

99. Let m be the molecular mass of substance

Now, m = ? w = 10 g W = 100.0 g

K_f = 18.5 K ΔT = 1.2 K

Substituting all these values in the following equation

$$m = \frac{100 \times K_f \times w}{\Delta T \times W} = \frac{100 \times 18.5\ K \times 10\ g}{1.2\ K \times 100\ g} = \mathbf{154.16\ g\ mol^{-1}}$$

100. Sp. gr. of HCl = 1.14

1 ml of solution= 1.14 g. [Water + HCl]

Amount of HCl in 1.14 gm (1 ml solution] = (37/100) × 1.14 ($\because$ 37% strength)

$\therefore$ Normality= $\frac{37.0 \times 1.14 \times 1000}{100 \times 36.5} = 11.556$ N

Now 50 ml of 11.56 HCl= V ml of 1 N HCl.

50 × 11.556 N = V × N

V = 50 × 11.556/1 = 578 ml

101. Figure out how many moles o HNO_3 we have use the chemical equation to calculate how many moles of NaOH are needed. Since we know the volume over which the NaOH moles are distributed, we can calculate the concentration:

$$\text{Moles of } HNO_3 = \left(\frac{25\ \text{ml}}{1000\ \text{ml/liter}}\right)\left(\frac{0.198\ \text{mol of } HNO_3}{\text{liter}}\right) = 0.00495$$

The balanced equation

$$NaOH + HNO_3 \rightarrow NaNO_2 + H_2O$$

shows that 1 mol of NaOH is required per mole of HNO_3 and so the 0.0495 mol of HNO_3 requires 0.00495 mol of NaOH. This is distributed over 37.50 ml of solution so that we can write

$$\text{NaOH concentration} = \frac{0.00495 \text{ mol of NaOH}}{0.0375 \text{ liter}} = 0.132\ M \text{ NaOH}$$

102. We start with 10 ml of one component plus 20 ml of the other component, so that the initial total volume is 30 ml. To find the final total volume, we can use the given density but we need to know the final mass of solution. *Masses are additive*, and so we need to know the mass of each component.

$$(10 \text{ ml of ethyl alcohol}) \left(0.7893 \text{ g.ml}^{-1}\right) = 7.893 \text{ g of ethyl alcohol}$$

$$(20 \text{ ml of water}) \left(0.9971 \text{ g.ml}^{-1}\right) = 19.94 \text{ g of water}$$

$$\text{Total mass} = 7.893 + 19.94 \text{ g} = 27.83 \text{ g of solution}$$

$$\text{Final volume} = \frac{27.83 \text{ g}}{0.9571 \text{ g.ml}^{-1}} = 29.08 \text{ ml}$$

$$\% \text{ volume change} = \left(\frac{30 - 29.08}{30}\right) 100 = 3.1\% \text{ contraction}$$

To get the final molar concentration of ethyl alcohol, we divide the moles of ethyl alcohol by the final total volume.

$$\text{Moles of ethyl alcohol} = \frac{7.893 \text{ g}}{46.07 \text{ g mol}^{-1}} = 0.1713 \text{ mol}$$

$$\text{Molarity} = \frac{0.1713 \text{ mol}}{0.02908 \text{ liter}} = 5.891 \text{ M } C_2H_3OH$$

103. Molarity is defined as the number of moles of solute per liter of solution. By definition, 90-proof of whisky is 45% ethanol by volume. Thus in 1 liter = 1000 ml. of whiskey there is (45%) (1000 ml) = 450 ml of ethanol. To determine the number of moles of ethanol in 1 liter of whiskey first calculate the weight of ethanol in 1 liter by multiplying the density of ethanol by the volume of ethanol in 1 liter. Then, divide the weight of ethanol in 1 liter by its molecular weight (MW ethanol = 46 g.mol^{-1}). Therefore,
(450 ml ethanol) (0.8 g.mol^{-1}) = 360 g ethanol

$$\text{molarity} = \frac{\text{number moles ethanol}}{\text{1 liter whiskey}} = \frac{360 \text{ g}/46 \text{ g.mol}^{-1}}{1 \text{ liter}} = 7.83 \text{ M}$$

104. Since for NaCl, mol. weight = equivalent weight, hence decinormal solution may be taken as 0.1 molar solution.
According to equation $PV = nRT$

$$P = \frac{nRT}{V} = \frac{(0.1 \text{ mol})(0.0821 \text{ L atm K}^{-1} \text{ mol}^{-1}) \times 300 \text{ K}}{1 \text{ L}} = \mathbf{2.463 \text{ atm}}$$

Let x be the degree of dissociation of NaCl, then

$$NaCl \rightleftharpoons Na^+ + Cl^-$$

1	0	0	(Before dissociation)
$1 - x$	x	x	(After dissociation)

No. of particles after dissociation $= 1 - x + x + x = 1 + x = 1 + 0.80 = 1.8$

($\because$ NaCl is 80% dissociated, hence $x = 0.8$)

Now,

$$\frac{\text{Observed Osmotic Pressure}}{\text{Calculated Osmotic Pressure}} = \frac{\text{No. of particles after dissociation}}{\text{No. of particles before dissociation}}$$

i.e., $$\frac{\text{Observed O.P.}}{2.463} = \frac{1.8}{1}$$

or, Observed osmotic pressure $= 1.8 \times 2.463 =$ **4.43334 atm**

105. (i) We know that $PV = nRT = \frac{w}{m} RT$

or $$P \times 0.1 \text{ L} = \frac{5\text{ g}}{180 \text{ g mol}^{-1}} \times 0.0821 \text{ L atm K}^{-1} \text{ mol}^{-1} \times 300 \text{ K} = 6.842 \text{ atm}$$

(ii) Since, $P = h.d\ g$

$$6.842 \text{ atm} = h \times 1 \text{ g mL}^{-1} \times 981 \text{ cm s}^{-2}$$

or $$6.842 \times 1.01 \times 105 \text{ Nm}^{-2} = h \times 10^3 \text{ kg m}^{-3} \times 9.81 \text{ ms}^{-2}$$

or $$h = \frac{6.842 \times 1.01 \times 10^5 \text{ kg ms}^{-2} \text{ m}^{-2}}{10^3 \text{ kg m}^{-3} \times 9.81 \text{ m s}^{-2}} = \mathbf{70.43 \text{ m}}$$

106. The equation for the reaction of SO_3 with H_2O is

$$H_2O + SO_3 \rightarrow H_2SO_4$$

According to this equation for every mole of SO_3 consumed one mole of H_2SO_4 is produced or, in other words, if the number of moles of SO_3 consumed is know, then the number of moles of H_2SO_4 produced is also known. Now the no. of moles of SO_3 is equal to the number of moles of H_2O.

$\therefore$ $$\text{Number of moles of } H_2O = \frac{(10^5 \text{ liter})(1000 \text{ g liter}^{-1})}{18.02 \text{ g.mol}^{-1}} = 5.5 \times 10^{-5} \text{ moles}$$

(Note 1000 g liter^{-1} is the density of water (rain), and number of H_2O molecules

$= (5.5 \times 1^6 \text{ moles})(6.02 \times 10^{23} \text{ molecules/mol}^{-1}) = 3.3 \times 10^{30}$ molecules.

Now, the problem also states that SO_3 molecules bond with one out of 10^4 H_2O molecules. From this, the number of SO_3 molecules is determined,

$$\frac{3.3 \times 10^{30} \text{ H}_2\text{O molecules}}{10^4 \frac{\text{H}_2\text{O molecules}}{\text{SO}_3 \text{ mlecules}}} = 3.3 \times 10^{26} \text{ SO}_3 \text{ molecules}$$

The number of moles of SO_3 can now be obtained.

$$\text{number of moles of } SO_3 = \frac{(3.3\times10^{26}\ SO_3 \text{ molecules})}{6.02\times10^{23}\ \dfrac{\text{molecules}}{\text{mole}}} = 5.5 \times 10^2 \text{ moles}$$

or 5.5×10^2 moles of H_2SO_4 are formed.
The molarity of the H_2SO_4 produced is thus

$$\frac{5.5\times10^2 \text{ moles}}{10^5 \text{ liter}} = 5.5 \times 10^{-3} \text{ M}.$$

107. For a neutralization reaction to occur, the number of equivalents of acid must equal the number of equivalents of base. Therefore, one can solve this problem by computing how many equivalents are present in 1.48 g of $Ca(OH)_2$; the equivalents of acid (HCl) must equal this number.

An equivalent is defined as the weight of a substance in grams that releases one mole of protons (H^+) or hydroxyl ions. When one mole of Ca(OH); dissolves in water, it produces 2 moles of hydroxyl ions (OH^-) according to the equation $Ca(OH)_2 \rightarrow Ca^{2+} + 2OH^-$. The weight of 1 mole of $Ca(OH)_2$ is equal to its molecular weight. Therefore, $Ca(OH)_2$ has a weight of 74 grams.

Thus, 74 grams of $Ca(OH)_2$ produces 2 moles of OH^- ions. Recalling the definition of equivalency, 74/2 = 37 grams is the equivalent weight of $Ca(OH)_2$. The number of equivalents of $Ca(OH)_2$ is,

$$= \frac{\text{weight in grams}}{\text{equivalent weight}} = \frac{1.48\text{g}}{37\text{g}} = 0.04 \text{ equiv.}$$

The number of equivalents of HCl must also be .04 for neutralization to occur. The normality of HCl is given as 0.125. Normality is defined as the number of equivalents divided by liters of solution, i.e., N = equivalents/volume. This means that 0.125 N = 0.04 equivalents/volume for HCl. The volume in liters required for neutralization can be obtained by solving this equation.

$$\text{Liters} = \frac{0.04}{0.125} = 0.32 \text{ l}$$

To convert to milliliters multiply by the conversion factor 100 ml/l. One has
0.32 l × 1000 ml/l = 320 ml.

108. Moles of H_3PO_4 = (0.035 liter) $\left(0.36 \dfrac{\text{mol}}{\text{liter}}\right)$ = 0.00126

The balanced equation is

$3Ca(OH)_2 + 2H_3PO_4 \rightarrow Ca_3)PO_4)_2 + 6H_2O$

and tells us we need 3 mol of $Ca(OH)_2$ per 2 mol of H_3PO_4.

$$(0.00126 \text{ mol of } H_3PO_4)\left(\frac{3 \text{ mol of } Ca(OH)_2}{2 \text{ mol of } H_3PO_4}\right) = 0.00189 \text{ mol of } Ca(OH)_2 \text{ required}$$

The $Ca(OH)_2$ solution is given to be 0.0150 M, so that we can write

$$\text{Volume required} = \frac{0.00189\,\text{mol}}{0.015\,\text{mol L}^{-1}} = 0.126 \text{ liter, or } 126 \text{ ml}$$

109. For the neutralization reaction. The balanced equation is

$$3Ca(OH)_2 + 2H_3PO_4 \rightarrow Ca_3(PO_4)_2 + 6H_2O.$$

As given, the $Ca(OH)_2$ solution is 25 ml or 0.025 liters. Therefore, the number of moles of $Ca(OH)_2$ is (0.03) (0.025) = 0.00075 moles. As the balanced equation indicates, the number of moles of H_3PO_4 is 2/3 the moles of $Ca(OH)_2$ or 2/3 (0.00075) = 0.0005 moles of H_3PO_4.

From the definition of molarity for H_3PO_4 one has

$$0.025 \text{ M} = \frac{0.0005 \text{ moles}}{x \text{ liters (say)}},$$

$$\therefore x = \frac{0.0005}{0.025} = 0.02 \text{ liters or } 20 \text{ ml.}$$

110. The balanced reactions is: $H_2SO_4 + 2NaOH \rightarrow Na_2SO_4 + 2HOH$.

26.5 ml of 0.1 M NaOH = 0.0265 litres of 0.1 M NaOH

or $$0.1 = \frac{\text{no. of moles of NaOH}}{0.0265 \text{ liters}}. \text{ Solving}$$

moles = 0.00265 of NaOH.

This, then, must be twice the number of moles of acid. For the acid, therefore, molarity $= \frac{0.00265}{2}$ liters. The volume of the H_2SO_4 solution is 15 ml, or 0.015 liters.

$$\text{Molarity of the acid } \frac{0.001325}{0.015} = 0.088 \text{ M},$$

111. Moles of H_2SO_4 spilled = (0.04 liter) 0.1 mol liter^{-1} = 0.004 mol

$$\text{Moles of } NaHCO_3 \text{ required} = (0.004 \text{ mol of } H_2SO_4)\left(\frac{2 \text{ mol of } NaHCO_3}{1 \text{ mol of } H_2SO_4}\right)$$

$$= 0.008 \text{ mol of } NaHCO_3$$

$$\therefore \text{Grams of } NaHCO_3 = (0.008 \text{ mol})\left(84.0\,\text{g.mol}^{-1}\right) = 0.672 \text{ g}$$

112. Method 1: Using the molecular equation

$$2HNO_3(eq) + Sr(OH)_2\ Sr(NO_3)_2(aq) + 2H_2O(l)$$

$$\text{no. mol } Sr(OH)_2 = 0.025 \text{ L} \times \frac{0.0148 \text{ mol } Sr(OH)_2}{\text{L}} = 3.7 \times 10^{-4} \text{ mol } Sr(OH)_2$$

$$\text{no. mol } HNO_3 = 3.7 \times 10^{-4} \text{ mol sr}(OH)_2 \times \frac{2 \text{ mol } HNO_3}{1 \text{ mol } Sr(OH)_2} = 7.4 \times 10^{-4} \text{ mol } HNO_3$$

$$\text{no. ml } HNO_3 = 7.4 \times 10^{-4} \text{ and } HNO_3 \times \frac{1 \text{ L } HNO_3}{0.0252 \text{ mol } HNO_3} \times \frac{1000 \text{ mL } HNO_3}{1 \text{ L } HNO_3}$$

$$= 29.1 \text{ ml. } HNO_3$$

Method 2: Using the net ionic equation. Here the net ionic equation

$$H^+(aq) + OH^-(aq) \rightarrow H_2O(l)$$

states that one H^+ is required for every OH^- in the neutralization reaction. Again, this factor arises in the second step.

$$\text{no. mol } OH^- = 0.025 \text{ L} \times \frac{0.0148 \text{ mol } Sr(OH)_2}{\text{L}} \times \frac{2 \text{ mol OH}}{1 \text{ mol } Sr(OH)_2} = 7.4 \times 10^{-4} \text{ mol } (OH)^-$$

$$\text{no. mol } H^+ = 7.4 \times 10^{-4} \text{ mol } (OH)^- \times \frac{1 \text{ mol } N^+}{1 \text{ mol (OH)}} = 7.40 \times 10^{-4} \text{ mol } H^+$$

$$\text{no mL } HNO_3 = 7.4 \times 10^{-4} \text{ mol } H^+ \times \frac{1 \text{ mol } HNO_3}{1 \text{ mol } H^+} \times \frac{1 \text{ L } HNO_2}{0.0252 \text{ mol } HNO_2}$$

$$\times \frac{1000 \text{ mL } HNO_3}{1 \text{ L } HNO_3} = 29.1 \text{ mol } HNO_3$$

113. At the neutralization point, the number of equivalents of NaOH is equal to the number of equivalents of H_2SO_4. But since the number of equivalents is equal to the product of the normality, N, and the volume, V, this condition may be stated as

$$N_{NaOH} V_{NaOH} = N_{H_2SO_4} V_{H_2SO_4} \quad \ldots\ldots \text{(i)}$$

To solve this problem, we must solve for N_{NaOH},

Since H_2SO_4 has two ionizable protons, its normality is equal to twice its molarity, or $N_{H_2SO_4} = 2 \times 0.2 \text{ M} = 0.4 \text{ N}$, NaOH has only one hydroxyl group, so that its normality is equal to its molarity. Then from (i)

$$N_{NaOH} = \frac{N_{H_2SO_4} V_{H_2SO_4}}{NaOH} = \frac{0.4 \text{ N} \times 64.0 \text{ ml}}{10.0 \text{ ml}} = 2.56 \text{ N}.$$

The concentration of NaOH solution is therefore 2.56 N, or 2.56 M.

114. (a) For neutralization in the reaction which takes when NaOH is added,

$$NaOH \rightleftharpoons Na^+ + OH^-$$

Now, in a 0.25 N NaOH solution are 0.25 equivalents of OH^-, which is equal to the number of moles.

But, no. of moles = no. of liters × normality

$$= 0.02 \text{ liters} \times \frac{.25 \text{ moles}}{\text{liter}} = 0.005 \text{ moles.} \because 20 \text{ ml} = 0.02 \text{ liters.}$$

Thus, there must be 0.005 moles of H^+ ions from the oxalic acid to neutralize the NaOH.

$\because H_2C_2O_4 = 2H^+ + C_2O_4$

Therefore, each mole of oxalic acid ionizes 2 moles of NaOH. Thus, for neutralization to occur, only one half as much oxalic acid is needed as NaOH which will be .005/2 = 0.0025 moles. Hence 0.0025 moles of oxalic acid are these in the 50 ml of rhubarb juice.

$$\text{no. of moles liter}^{-1} \text{ of oxalic acid} = \frac{0.0025 \text{ moles}}{50 \text{ ml}} \frac{100 \text{ ml}}{1 \text{ liter}} = 0.05 \text{ moles liter}^{-1}.$$

Since there are 0.05 moles of $H_2C_2O_4$ in 1 liter of juice, one can find the weight of $H_2C_2O_4$ in this quantity of juice by multiplying 0.05 moles by the MW of $H_2C_2O_4$ = 90. Weight of $H_2C_2O_4$ in 1 liter = 0.05 moles × 90 g.moles^{-1} = 4.5 g

(b) For titrations, $N_aV_a = N_bV_b$

Here

$N_a = ?$ $\quad N_b = 0.25$ N $\quad \therefore N_a \times 50$ ml. = 0.25 N × 20 ml

$V_a = 50$ ml $\quad V_b = 20$ ml $\quad \Rightarrow N_a = \dfrac{0.25 \text{ N} \times 20 \text{ ml}}{50 \text{ ml}} = 0.10$ N.

115. The number of grams/equiv. for KOH is 56.1, since only 1 mole of OH^- can be furnished. The number of equivalents is

$$\frac{0.768 \text{ g}}{56.1 \text{ g/equiv}} = 0.0137 \text{ equiv of KOH.}$$

The number of equivalents of base must equal that of the acid. The number of equivalents of acid = 0.0137. The number of grams/equiv is

$$\frac{1.00 \text{ g of } C_5H_{10}O_4}{0.0137 \text{ equiv.}} = 73.0 \text{ g/equiv.}$$

One mole of $C_6H_{10}O_4$, weighs 146.1 g, since a mole = weight in grams/molecular weight. Therefore, the number of moles/equiv is

$$\frac{73.0 \text{ g/equiv}}{146.1 \text{ g.mole}^{-1}} = 0.5 \text{ moles/equiv.}$$

It follows that each mole of acid furnishes two H_3O^+. as such, there are two neutralizable protons in $C_6H_{10}O_4$.

116. Figure out how many moles of HNO_3 you have, then calculate over what volume you need to spread it out to get a final concentration of 1.40 *M*.

$$(1.38 \text{ liters}) \left(\frac{30.0 \text{ mol of } HNO_3}{\text{liter}}\right) = 4.14 \text{ mol of } HNO_3.$$

$$\frac{4.14 \text{ mol of } HNO_3}{x \text{ liter}} = 1.4 \text{ mol liter}^{-1}$$

$$\therefore \quad x = \frac{4.14}{1.4} = 2.96$$

But this is the total volume, of which 1.38 liters is contributed by the original 3.0 *m* HNO_3. To get the volume of water needed to be added, subtract 1.38 liters from 2.96 liters. Final answer: 1.58 liters of water.

117. Calculate the moles of solute contributed by each sample and divide the total number of moles by the total volume

First find the moles of HNO_3 in 25 ml of 3.0 M HNO_3.

$$(0.025 \text{ liter})\left(\frac{3.0 \text{ mol}}{\text{liter}}\right) = 0.075 \text{ mol}$$

Then find the moles of HNO_3 in 75 ml of 4.0 M HNO_3.

$$(0.075 \text{ liter})\left(\frac{4.0 \text{ mol}}{\text{liter}}\right) = 0.3 \text{ mol}$$

Total moles of HNO_3 is 0.075 plus 0.3 = 0.375 mol
Total final volume is 0.025 liter plus 0.075 liter, or 0.1 liter.

$$\text{Final concentration} = \frac{0.375 \text{ mol of } HNO_3}{0.1 \text{ liter of solution}} = 3.75\ M\ HNO_2$$

118. Ca^+ (0.01 liter) (0.02 mol/liter) = 0.0002 mol

$$\text{Concentration of } Ca^{2+} = [Ca^{2+}] = \frac{0.0002 \text{ mol}}{0.025 \text{ liter}} = 0.008 \text{ M}$$

Na^+ (0.015 liter) (0.03 mol liter^{-1}) = 0.00045 mol

$$\text{Concentration of } Na^+ = [Na^+] = \frac{0.00045 \text{ mol}}{0.025 \text{ liter}} = 0.018 \text{ M}$$

NO_3^- (0.01 liter) (2 × 0.02 mol/liter) + (0.015 liter) (0.03 mol/liter
= 0.0004 + 0.00045 = 0.00085 mol)

$$\text{Concentration of } NO_3 = [NO_3^-] = \frac{0.00085 \text{ mol}}{0.025 \text{ liter}} = 0.034 \text{ M}$$

119. We can easily calculate the molality of the solution m as:

$\Delta T_b = 0.392 \text{ K} = 3.63\ m$ so that $m = 0.392/3.63 = 0.108$ molal

Let, the molecular weight of this compound, be M. So, 0.527 g of compound is (0.527/M) moles of compound. Now, molality is the ratio of moles of solute to kilograms of solvent.

$$\text{i.e.} \quad \frac{(0.527/M) \text{ mol solute}}{40.19 \times 10^{-3} \text{ kg } CHCl_3} = 0.108\ m$$

Solving for the molecular weight we obtain

$$M = \frac{527}{(40.19)(0.108)} = 121 \text{ g} \cdot \text{mol}^{-1}$$

The formula weight of C_5H_4 is 64.087, and the molecular weight of the compound is 64.087n, where n is an integer. The only value of n that yields a molecular weight close to

121 is $n = 2$. The correct molecular formula is therefore $C_{10}H_8$, and the exact molecular weight is 128.17 g · mol^{-1}.

120. (a) Pure water freezes at 273 K and has a molal freezing point depression constant, K_f of 1.86 kg · K · mol^{-1}. If the freezing point is lowered to 258 K; $\Delta T_f = +15$ K.

$\therefore \quad +15 = 1.86\, m \quad$ so that $\quad m = 8.1$ molal

(b) Since the density of water is 1.0 g · mL^{-1}, 20 L of water have a mass of 20×10^3 g or 20 kg. The molality is the ratio of moles of solute to kilograms of solvent. Let x equal the number of moles of ethylene glycol that must be added to 20 kg of water to produce a solution that is 8.1 m. Then

$$\frac{x \text{ mol } CH_2OHCH_2OH}{20 \text{ kg } H_2O} = \frac{8.1 \text{ mol } CH_2OHCH_2OH}{1.0 \text{ kg } H_2O}$$

Solving for x, we obtain $x = (2.0)\ (9.1) = 1.6 \times 10^2$ mol of ethylene glycol. The molecular weight of CH_2OHCH_2OH is 62.07 g · mol^{-1}. Thus the mass of ethylene glycol needed to prepare this 8.1 m solution is (62.07 g · mol^{-1}) (1.6×10^2 mol) = 9.9×10^3 g. To find the volume required, we need the density of ethylene glycol.

$$\text{volume in mL} = \frac{\text{mass in g.}}{\text{density to g} \cdot \text{mL}^{-1}} = \frac{9.9 \times 10^3 \text{ g}}{1.11 \text{ g} \cdot \text{mL}^{-1}} = 8.9 \times 10^3 \text{ mL}$$

Hence we must mix 8.9 L of ethylene glycol with 20 L of water to prepare a solution that is 8.1 *mi* n ethylene glycol.

121. We will tabulate the calculations required to obtain the total concentration of all solute particles in each solution.

Substance	*v ions/formula)*	*Solute Particle Concentrations (M)*	*Total Concentration of All Solute Particles (M)*
$NaNO_3$	2	$[Na^+] = 0.050$	0.100
		$[NO_3^-] = 0.050$	
$CuSO_4$	2	$[Cu^{2+}] = 0.075$	0.150
		$[SO_4^{2-}] = 0.075$	
$NH_4)_2SO_4$	3	$[NH_4^+] = 0.120$	0.180
		$[NH_4^+] = 0.060$	
Sucrose	(nonelectroylete)	[sucrose] = 0.140	0.140
$BaCl_2$	3	$[Ba^{2+}] = 0.40$	0.120
		$[Cl^-] = 0.080$	0.120

The larger the total concentration of all solute particles, the lower the freezing point. Since we are only arranging the solutions in order of decreasing freezing point, the concentration unit is not important. The correct arrangement, starting with the solution with the highest freezing point, is as follows:

highest freezing point (a) 0.050 F $NaNO_3$

next (e) 0.040 F $BaCl_2$

next (d) 0.140 M sucrose

next (b) 0.075 F $CuSO_4$

lowest freezing pint (c) 0.060 F $(NH_4)_2SO_4$

122. Let M = molecular weight of insulin. The number of moles of insulin in this solution is therefore (2.0/M) mol, and the volume of the solution is 0.25 L. Therefore the concentration is

$$\frac{(2.0/M)\text{ mol of insulin}}{0.25\text{ L}} = (8.0/\text{M})\text{ mol} \cdot \text{L}^{-1}$$

If R is expressed in liter · atm · mol^{-1} K^{-1}, the osmotic pressure must be given in atmospheres

$$\pi = 26.1\text{ mmHg} = (26.1/760)\text{ atm}$$

Substituting in $\pi = cRT$ yields

$$\frac{26.1}{760}\text{ atm} = \left(\frac{8.0}{M}\frac{\text{mol}}{\text{L}}\right)\left(0.08206\frac{\text{L}\cdot\text{atm}}{\text{mol}\cdot\text{K}}\right)(303\text{ K})$$

Solving this for M, we obtain

$$M = (8.0)\ (.08206)\ (303.2)\ (760)/26.1 = 5.8 \times 10^3$$

The molecular weight of insulin is about 5800.

123. (a) $P_A = X_A^{gas} P_{total} = (0.318)\ (673\text{ mmHg}) = 214\text{ mmHg}$

$P_B = P_{total} - P_A = 673\text{ mmHg} - 214\text{ mmHg} = 459\text{ mmHg}$

(b) $P_A = P_A^{\circ} P_A^{liq} = 214\text{ mmHg} = P_A^{\circ}\ (0.256)$

Therefore,

$$P_A^{\circ} = \frac{214\text{ mmHg}}{0.256} = 836\text{ mmHg};$$

$$P_B^{\circ} = \frac{P_B}{1 - X_A^{liq}} = \frac{459\text{mm Hg}}{0.744} = 617\text{ mmHg}$$

124. We can calculate the partial pressure of O_2 in air by using Dalton's law:

$$P_{O_2} = P_{O_2}^{gas} \cdot P_{total} = (0.12)\ (760\text{ mmHg}) = 159.6\text{ mmHg}$$

The mole fraction of O_2 in aquous solution is given by Henry's law (see values in the table below).

$$159.6 = (3.30 \times 10^7)\, X_{O_2} \quad \text{so that } X_{O_2} = 4.84 \times 10^{-6}$$

Table Henry's Law Constants for Gases in *Aqueous Solution* at 298 K

Gas	k_H *(mmHg)*
Carbon dioxide, CO_2	1.25×10^6
Hydrogen, H_2	5.34×10^7
Nitrogen, N_2	6.51×10^7
Oxygen, O_2	3.30×10^7

We can calculate the molality, m, from the mole fraction equation:

$$X_{O_2} = \frac{m}{m + (1000/18.015)} = \frac{m}{55.5}$$

Now, neglecting the number of moles of dissolved O_2 compared to the number of moles of H_2O as $m \ll 55.5$. For this solution the molarity of dissolved O_2 is

$$m = (55.5)\ (4.84 \times 10^{-6}) = 2.69 \times 10^{-4} \text{ molal}$$

This solution is sufficiently dilute that its density is the same as that of water, 1.0 g · mL^{-1}, and it is a very good approximation that 1000 g of solution contains 1000 g of H_2O, so that the molarity and molality are identical, and the concentration of dissolved O_2 is

$$(O_2) = 2.69 \times 10^{-4}\ M$$

125. We have by substituting the given values

$$MM = \frac{g \cdot R \cdot T}{P \cdot V} = \frac{0.043\ g \cdot 0.821\ \cancel{L \cdot atm} \cdot 373.5\ \cancel{K}}{mol \cdot \cancel{K}\ (750/760\ \cancel{atm}) \cdot 0.264\ \cancel{L}} = 50.6\ g \cdot mol^{-1}$$

126. 1. The intersection of the two lines which is at approximately 321.4 K

2. $\Delta T_f = T_f^{\circ} - T_f = 326\ K - 321.4\ K = 4.6\ K$

3. $m = \dfrac{\Delta T_f}{k_f} = \dfrac{4.6\,K}{7.1\,K \cdot m^{-1}} = 0.65\ m$

4. $MM_{solute} = \dfrac{k_f \cdot g\ solute}{kg\ solvent \cdot \Delta T_f} = \dfrac{7.1\ \cancel{K}\ \cancel{kg} \cdot 4.0\ g}{mol \cdot 0.0520\ \cancel{kg} \cdot 4.6\ \cancel{K}} = 120\ g \cdot mol^{-1}$

5. $\%\ error = \dfrac{observed - expected}{expected} \times 100\% = \dfrac{120 - 128}{128} \times 100\% = -6\%$

The results were about 6% below expected.

127. 1. The number of moles of $S_2O_3^{2-}$ ion required for titration are :

$$= \frac{35.5\ mL\ \cancel{Na_2S_2O_3\ sol'n}}{1} \times \frac{1\ L\ \cancel{sol'n}}{1000\ \cancel{mL\ Na_2S_2O_3\ sol'n}} \times$$

$$\frac{0.1\ \cancel{mol\ Na_2S_2O_3}}{1\ \cancel{L\ sol'n}} \times \frac{1\ mol\ S_2O_3^{2-}}{1\ \cancel{mol\ Na_2S2O_3}} = 3.55 \times 10^{-3}\ mol\ S_2O_3^{2-}$$

2. The number of moles of I_2 produced in the titration mixture is

$$= \frac{3.55 \times 10^{-3}\ \cancel{mol\ S_2O_3^{2-}}}{1} \times \frac{1\ mol\ I_2}{2\ \cancel{mol\ S_2O_3^{2-}}} = 1.775 \times 10^{-3}\ mol\ I_2$$

3. The number of moles of OCl^- ion present in the diluted bleaching solution

$$= \frac{1.775 \times 10^{-3} \cancel{\text{mol } I_2}}{1} \times \frac{1 \text{ mol } OCl^-}{1 \cancel{\text{mol } I_2}} = 1.775 \times 10^{-3} \text{ mol } OCl^-$$

4. The mass of NaOCl present in the diluted bleaching·solution

$$= \frac{1.775 \times 10^{-3} \cancel{\text{mol } OCl^-}}{1} \times \frac{1 \cancel{\text{mol NaOCl}}}{1 \cancel{\text{mol. OCl}}} \times \frac{74.44 \text{ g NaOCl}}{1 \cancel{\text{mol NaOCl}}} = 0.1321 \text{ g NaOCl}$$

5. The volume of bleach present in diluted bleaching solution (aliquot) is

$$= \frac{20 \text{ mL } \cancel{\text{diluted bleach sol'n}}}{1} \times \frac{50 \text{ mL commercial bleach}}{250 \text{ mL } \cancel{\text{diluuted bleach sol'n}}} = 4 \text{ mL}$$

6. The mass of commercial bleach titrated is :

$$= \frac{4 \text{ mL } \cancel{\text{commercial bleach}}}{1} \times \frac{1.15 \text{ g commercial bleach}}{1.0 \cancel{\text{mL commercial bleach}}} = 4.6 \text{ g commercial bleach}$$

7. The % NaOCl is

$$= \frac{\text{part}}{\text{whole}} \times 100\% \times \frac{0.1321 \text{ g NaOCl}}{4.6 \text{ g commercial bleach}} \times 100\% = 2.87\%$$

128. From the change in boiling point, one can calculate the molality and from that the mass of solute in 1.0 kg of solvent. Thus also yields the total mass of the solution. The calculate density, one also needs to know the volume of solution. From osmotic pressure, one can calculate molarity. Using the moles of solute and mass of solution calculated from molarity, one can now calculate density.

$$KNO_2(s) \rightarrow K^+(aq) + NO_3(aq) \; (i = 2)$$
$$\Delta T_b = 376 \text{ K} - 373 \text{ K} = 3.0 \text{ K}$$
$$\Delta T_b = ik_b m$$
$$3.0 \text{ K} = (2)(0.52 \text{ K}/m)(m) \qquad \therefore [K_b = 0.52 \text{ k/m}]$$
$$m = 2.884 \; m$$

$$\frac{2.884 \text{ mol. } KNO_3}{1.0 \text{ kg solvent}} \times \frac{101 \text{ g}}{1 \text{ mol. } KNO_3} \simeq 290 \text{ g } KNO_3/1.0 \text{ kg solvent}$$

Thus the total mass of solution is 290 g solute + 1000 g solvent = 1290 g solution

$$\pi = MRTi$$
$$122 \text{ atm} = (M)(0.0821 \text{ L} \cdot \text{atm mol}^{-1} \cdot \text{K}^{-1})(2)\ 298$$
$$M = 2.49$$

$$\therefore \text{Density} = \frac{2.49 \text{ mol. } KNO_3}{1.0 \text{ L sol'n}} \times \frac{1290 \text{ g sol'n}}{2.9 \text{ mol } KNO_3} \times \frac{1 \text{ L}}{1000 \text{ mL}} = 1.1 \text{ g/mL}$$

129. Molality is moles solute per kilogram of solvent, so convert mass of solute to moles, and liters of solution to kg of solvent.

Now, 158.2 g KOH contains

$$= 158.2 \text{ g KOH} \times \frac{1 \text{ mol KOH}}{56.11 \text{ g KOH}} = 2.82 \text{ mol KOH}$$

$$\text{and } 1.0 \text{ L solution} \times \frac{1000 \text{ mL}}{1 \text{ L}} \times \frac{1.13 \text{ g}}{1 \text{ mL}} = 1130 \text{ g solution}$$

$$\therefore 1130 \text{ g solution} - 158.2 \text{ g solute} = 972 \text{ g solvent} = 0.972 \text{ kg solvent}$$

$$\text{Thus, molality} = \frac{2.829 \text{ mol.}}{0.972 \text{ kg}} = 2.90 \ m$$

This solution is about 2.9 m, approximately 10 times stronger that the solution technician needs. Thus he will need to increase the volume approximately 10 fold (to 1L).

If there are 2.820 moles KOH in 1.000 L, then there are 0.2820 moles in the 100 L sample. Likewise, the 100 mL sample also has only 0.0972 kg of solvent. The problem can then be solved by setting up a ratio

$$\frac{0.25 \text{ mol. KOH}}{1.0 \text{ kg solvent}} = \frac{0.282 \text{ moles KOH}}{(0.0972 \text{ kg} + x \text{ kg}) \text{ solvent}}$$

$$0.0972 \text{ kg } H_2O + x \text{ kg } H_2O + 1.128 \text{ kg } H_2O$$
$$x = 1.03 \text{ kg } H_2O$$

Thus the technician must add 1.903 kg (1.03 L) H_2O to the 100 mL of KOH solution.

130. $15 \text{ g } CuSO_4 \cdot 5H_2O \times \frac{1 \text{ mol}}{249.7\text{g}} = 0.06007 \text{ mol } CuSO_4 \cdot 5H_2O$

$$100 \text{ g soln} \times \frac{1 \text{ mL soln}}{1.167 \text{ g soln}} \times \frac{1 \text{ L}}{1000 \text{ mL}} = 0.085689 \text{ L}$$

$$M = \frac{0.06007 \text{ mol } CuSO_4}{0.085689 \text{ L}} = 0.70102 = 0.701 \text{ mol L}^{-1}$$

131. $0.159 \times \frac{\text{m}}{1.858\text{K}} = 0.08557 \text{ m} = \frac{0.08557 \text{ mol AA} + H^+}{1000 \text{ g (or 1000 mL) } H_2O}$

Note that 0.083 mol AA + mol H^+ = 0.08557 mol/L (AA + H^+)

$$\% \text{ dissoc.} = \frac{\text{mol } H^+}{\text{mol AA}} (100) = \frac{0.00257 \text{ mol}}{0.083 \text{ mol}} (100) = 3.09 = 3.1\%$$

132. Calculate the empirical formula first, using the mass of C, O, and H in 1.0 g

$$1.434 \text{ g } CO_2 \times \frac{12.01 \text{ g C}}{44.01 \text{ g } CO_2} = 0.39132 \text{ g C}$$

$$0.783 \text{ g } H_2O \times \frac{2.016 \text{ g H}}{18.01 \text{ g } H_2O} = 0.08761 \text{ g H}$$

g O = 1.0 g – 0.39132 g – 0.08761 g = 0.5211 g O

mol C = 0.39132 g C × 1mol/1 (n.01) 2.01 g = 0.03258 mol C (lowest integer = 1)

Mol H = 0.08761 g H × 1 mol/1.008 g = 0.08691 mol H (lowest integer = 8/3)

Mol O = 0.5211 g O × 1 mol/16 g = 0.03257 mol O (lowest integer = 1)

Therefore, the empirical formula is $C_3H_8O_3$. The formula weight from the freezing point is calculated by fist finding the molality.

0.0894 K × (m/1 858 K) = 0.04811 = (0.04811 mol/1000 g H_2O)

(0.04811 mol/1000 g H_2O) × 25.0 g H_2O = 0.001203 mol (in 25 g H_2O)

Molar mass = M_m = 1.1107 g/0.001203 mol = 92.02 g mol^{-1}

Because this is also the formula weight the molecular formula is $C_3H_8O_3$ also.

133. Since heat is evolved during dissolution, $\Delta H_{solution}$ is exothermic. Consequently, adding heat (increasing temperature) would decrease solubility, while removing heat (lowering temperature) would make it more soluble. According to Heny's law, increasing pressure will increase stability, while decreasing pressure will decrease solubility.

(a) $\dfrac{C_{g1}}{C_{g2}} = \dfrac{(k)(P_{g1})}{(k)(P_{g2})} \Rightarrow \dfrac{0.02}{C_{g2}} = \dfrac{(4\text{ atm})}{(6\text{ atm})}$

C_{g2} = 0.03 M, thus solubility is **greater** than 0.02 *M*

(b) Temperature is higher (thus decreased solubility) and pressure is less (also decreased solubility), thus the combined effect is less solubility.

(c) Temperature is lower (thus increased solubility) and pressure is the same (thus no change in solubility), thus the combined effect is greater solubility.

(d) Temperature is the same (thus no change in solubility) and pressure is lower (thus decreased solubility), thus the combined effect is less solubility.

134. We have, for dilute solution, $M \approx m; \Delta T_f = k_f mi$

∴ 0.38 K= (1.86 K/m) (0.1 *m*)(*i*)

Hence, i = 2

Thus two moles of ions are formed, indicating that the best equation is **b**.

135. 15 g $MgSO_4{\cdot}7H_2O \times \dfrac{1\text{ mol}}{246.5\text{ g}}$ = 0.60854 mol $MgSO_4{\cdot}7H_2O$

$$0.060854\text{ mol } MgSO_4{\cdot}7H_2O \times \frac{7\text{ mol } H_2O}{1\text{ mol hydrade}} \times \frac{18\text{ g}H_2O}{1\text{ mol } H_2O} = 7.667\text{ g } H_2O$$

$$kgH_2O = (100\text{ g } H_2O + 7.667\text{ g}H_2O) \times \frac{1\text{ kg } H_2O}{1000\ H_2O} = 0.10786\text{ kg } H_2O$$

$$\therefore \quad m = \frac{0..060854\text{ mol } MgSO_4}{0.10766\text{ kg } H_2O} = 0.5652\text{ mol kg}^{-1} = 0.565\text{ m}$$

136. From freezing point depression one can calculate the molality of the solution. Since the mass of the solvent is given, one can also calculate the modes of solute. From the molar masses of the two solutes and the mass of the mixture, one can them determine the mole ratio.

$$\Delta T_f = k_f m$$

$$0.5 \text{ K} = 1.85 \text{ K}/m)(m) \Rightarrow m = 0.269 \text{ m}$$

$$\therefore \; 0.269\ m = \frac{\text{mol. solute}}{1.0 \times 10^{-3} \text{ kg solvent}} \Rightarrow \text{moles of solute} = 2.69 \times 10^{-4} \text{ mol.}$$

Thus the total moles of solute (moles sugar + mass X) = 2.69×10^{-4} mol.
If mass of solute = x g, then mass of sugar = $(0.100 - x)$ g.
Combining to above we get

$$xg \times \frac{1 \text{ mol X}}{410 \text{ g}} + (0.1 \text{ g} - xg) \times \frac{1 \text{ mol sugar}}{342 \text{ g}} = 2.69 \times 10^{-4} \text{ mol}$$

$$\frac{x \text{ mol}}{410} + \frac{0.1}{342} - \frac{x \text{ mol}}{342} = 2.69 \times 10^{-4} \text{ mol} \Rightarrow \frac{x \text{ mol.}}{410} - \frac{x \text{ mol}}{342} = -2.34 \times 10^{-5} \text{ mol.}$$

$$\frac{(410)(x \text{ mol})}{410} - \frac{(410)(x \text{ mol})}{342} = (4.10)(-2.34 \times 10^{-5} \text{ mol.})$$

x mol. – $1.20x$ mol. = -9.59×10^{-3} mol.
$-0.20x$ mol. = -9.59×10^{-3} mol.
$x = 0.048$ and

$$\text{mass percent of X} = \frac{0.048 \text{ g}}{0.1 \text{ g}} \times 100\% = 48\%$$

137. Calculate the amount of alcohol that centres the blood (recall that to convert a percent to a decimal, one divides by 100). Then calculate the concentration of alcohol in the blood. Now, 142 g × 2 × 0.3 (30% alcohol) × 0.15 (15% enter bloodstream) = 13 g alcohol
7.0 L blood = 7000 mL

$$\text{Conc. of alcohol in blood} = \frac{13 \text{ g alcohol}}{7000 \text{ mL blood}} = 0.0019 \text{ g cm}^{-3}$$

Thus the person is legally intoxicated.

138. (a) $$49.92 \text{ g NaOH} \times \frac{1 \text{ mol}}{40 \text{ g NaOH}} = 1.248 \text{ mol. NaOH}$$

$$\frac{1.248 \text{ mol. NaOH}}{0.6 \text{ L}} = 2.08\ M$$

(b) $$1.248 \text{ mol. NaOH} \times \frac{1 \text{ mol OH}^-}{1 \text{ mol NaOH}} \times \frac{3 \text{ mol. H}_2}{2 \text{ mol OH}^-} = 1.872 \text{ mol. H}_2$$

(c) Determined he limiting reactant first.

$$41.28 \text{ g Al} \times \frac{1 \text{ mol OH}^-}{26.98 \text{ g Al}} \times \frac{3 \text{ mol H}_2}{2 \text{ mol. Al}} = 2.30 \text{ mol H}_2$$

The amount of H_2 that could be produced by the Al is less than the actual amount produced, thus the limiting reactant is NaOH

$$P = (758.6 \text{ mm Hg} - 23.8 \text{ mm Hg}) \times \frac{1 \text{ atm}}{760 \text{ mm Hg}} = 0.967 \text{ atm}$$

$$V = \frac{\pi RT}{P} = \frac{(1.872 \text{ mol.})(0.0821 \text{ L} \cdot \text{atm/mol.} \cdot \text{K}(298 \text{ K})}{0.967 \text{ atm}} = 47.4 \text{ L}$$

139. For $(HO)_mYO_n$ acids, acid strength increases with n, regardless of the number of OH's. The structure of H_3PO_3 is $(HO)_3$ PO; because H_3PO_3 and H_3PO_4 have about the same acidity, H_3PO_3 must also have $n = 1$, thus $m = 2$. This leaves one H, which must bond to phosphorus, giving a structure of $(HO)_2(O)PH$. Assuming that only 2H's react with NaOH, the mass of NaOH that reacts with 1.0 g of H_3PO_3(PA) is calculated as follows:

$$1.0 \text{ g PA} \times \frac{1 \text{ mol PA}}{81.994 \text{ g PA}} \times \frac{2 \text{ mol NaOH}}{1 \text{ mol PA}} \times \frac{40 \text{ g NaOH}}{1 \text{ mol NaOH}}$$

$$= 0.9756 \text{ g NaOH}$$

140. BF_3 acts as a Lewis acid, accepting an electron pair from NH_3

$$BF_3 + :NH_3 \rightarrow F_3B{:}NH_3$$

The NH_3 acts as a Lewis base in donating an electron pair to BF_3. When 10.0 g of each are mixed, the BF_3 is the limiting regent because it has the higher formula weight. The mass of F_3 NH_3 formed is

$$10 \text{ g } BF_3 \times \frac{1 \text{ mol } BF_3}{67.81 \text{ g } BF_3} \times \frac{1 \text{ mol } BF_3 \, NH_3}{1 \text{ mol } BF_3} \times \frac{84.84 \text{ g } BF_3 \, NH_3}{1 \text{ mol } BF_3 NH_3}$$

$$= 12.51 = 12.5 \text{ g } BF_3NH_3$$

141. (a) Box one shows $2H^+$, $2B^-$ and 5HB. This is an acid that is partially ionized in solution, and thus must be a weak acid.

Box two shows $5H^+$, $5B^-$ and no HB. This is an acid that is completely ionized in solution, and thus must be a strong acid.

(b) Figure (a) shows ten molecules in the box. If 10% dissociates, then one of those molecules will dissociate into H^+ and A, leaving the remaining nine as HA.

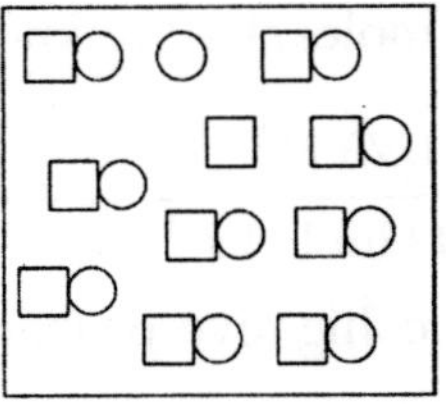

142. $CH_3COOH \rightarrow H_3COO^- + H^+$

Weight of water = mass of water × density of water

$$\text{molality} = \frac{w}{m} \times \frac{1000}{\text{wt of solvent (water)}} = \frac{3 \times 10^{-3} \times 1000 \times 1000}{60 \times 500 \times 0.9}$$

Vant Holf factor $i = 1 + x = 1 + \frac{23}{100} = 1.23$

$$\Delta T_f = i \cdot m \cdot K_f = \frac{1.23 \times 1.86 \times 3 \times 10^{-3} \times 1000 \times 1000}{60 \times 500 \times 0.9}$$

Hence $\Delta T_f = 0.0229$

143. Given

Volume of Benzene $v = 56.8$ cc

Density $\rho = 0.889$ gm/cc

Vapour pressure of solution = 98.88 mm Hg

Vapour Pressure of Benzene = 100 mm Hg

$\Delta T_f = 0.73$ k

To find $K_f = ?$ $M = ?$

From Rault's law

$$\frac{\Delta \rho}{p_0} = \chi_B \text{ (relative lowering of vapour pressure of solution = mole fraction of solute)} \quad \ldots\ldots \text{(i)}$$

By convention A of stands for solvent (Benzene)

$B \rightarrow$ solute, $n \rightarrow$ no. of moles

$$\therefore \quad \frac{100 - 98.88}{100} = 0.0112 = \chi_B \quad \ldots\ldots \text{(ii)}$$

Also, $\chi_B = \frac{n_B}{n_A + n_B} \cong \frac{n_B}{n_A}$ ($\therefore$ $n_B << n_A$ 2 hence ignored in denominator)

Now, n_A= no. of moles of Benzene

mass of Benzene in solution = $v \times \rho$ = 56.8 cc × 0.889 g/cc = 50.5 gm.

$\therefore$ no. of moles

$$n_A = \frac{50.5}{78} = 0.65 \text{ moles (78 = molar mass)} \quad \ldots\ldots \text{(iii)}$$

But, $n_B = \chi_B \times n_A = 0.65 \times 0.0112 = 7.25 \times 10^{-3}$ mole $\times 0.0112 = 7.25 \times 10^{-3}$ moles. . . . . (iv)

$\therefore$ Molality of solution (no of moles of solute per kg. of solvent)

$$= \frac{7.25 \times 10^{-3}}{50.5} \times 1000 = 0.143 \text{ moles/kg We know } \Delta T_f = k_f m_B \Rightarrow k_f = \frac{\Delta T_f}{m_n}$$

Substituting the values, $K_f = \frac{0.73 \text{ k}^{-1}}{0.143 \text{ mole kg}^{-1}} = 5.10 \text{ kg mol}^{-1} \text{ k}^{-1}$

144. Depression in freezing point is a colligative property

We have, $\Delta T_f = iK_f \cdot m$

Here, $\Delta T_f = 0.3$; $K_f = 1.86$ and $i = 1$

$\therefore$ $m = 0.3/1.86$ mol kg^{-1}

Now, $$\frac{P_A^\circ - P_A}{P_A^\circ} = x_B = \frac{n_B}{n_A}$$

Thus, $\Delta P = \frac{n_B}{n_A} \times P_A^\circ = \frac{0.3}{1.86} \times \frac{23.51}{55.5} = 0.068$

$\therefore \; P_A = 23.51 - 0.661 = 23.44$ mm Hg

145. Calculation for molecular weight

$\Delta T_B = \frac{w}{M} \times \frac{1000}{W} \times K_B$, where, ΔT_B = elevation in B. pt = 1.84 K;

w = weight of solute = 5.5g, W = weight of solvent = 45 g,

$\therefore$ m = m.wt. of solute = $\frac{5.3}{1.84} \times \frac{1000}{45} \times 2.53 = 1.64$ g

Equivalent weight of $C_3H_2NO_2 = 3 \times 12 + 2 \times 1 + 14 + 2 \times 16 =$ **84**

Calculation for Molecular formula

$$n = \frac{\text{m. wt.}}{\text{e. wt.}} = \frac{168}{84} = 2$$

Thus, MF = $(C_3H_2NO_2)_2 = C_6H_4N_2O_4$

Thus structure of minor product is

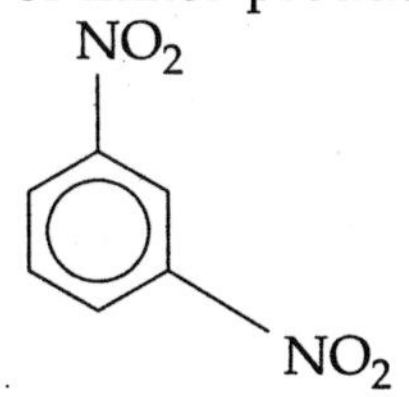

NO$_2$ / NO$_2$ m-Dinitrobenzene

Benzene $\xrightarrow[373\text{ K}]{HNO_3 + H_2SO_4}$ Nitrobenzene (NO$_2$) + m-Dinitrobenzene (NO$_2$, NO$_2$) minor

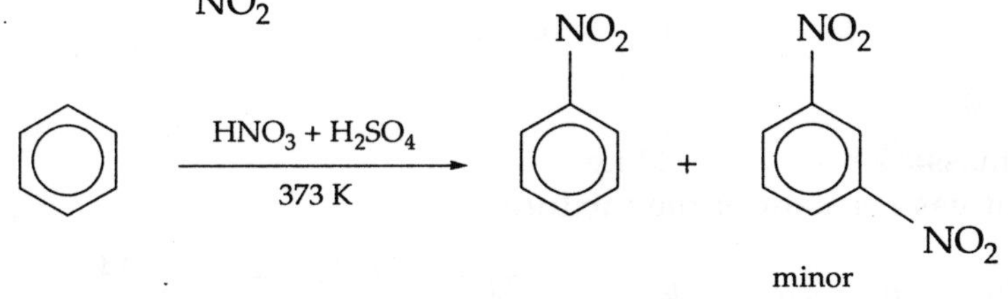

146. Density = Mass/Volume.

So, Volume of 1 mole of benzene in liquid form = $\frac{78}{0.817}$ = 88.94 ml

Because on vaporization it increases by a factor 2750

Thus Volume of 1 mole of benzene (g) = 88.94 × 2750 = 244.59 lit

and Volume of 1 mole of toluene (g) = $\frac{92 \times 7720}{0.867}$ = 819.19 lit

From P.V. = nRT

$$P^\circ_{(\text{Benzene})} = \frac{0.082 \times 293}{244.58} = 0.098 \text{ atm.}$$

$$P^\circ_{(\text{toluene})} = \frac{0.082 \times 293}{819.19} = 0.029 \text{ atm}$$

According to Dalton's Raoult's law

$$P_T = x_A P_A^\circ + x_B P_B^\circ$$
$$x_B = 1 - x_A$$

Thus, $46/760 = 0.06 \text{ atm} = X_{\text{benzene}} \cdot P^\circ_{\text{benzene}} + (1 - X_{\text{benzene}}) \cdot P^\circ_{\text{toluene}}$

$$\Rightarrow X_{\text{benzene}} = 0.45$$

$\therefore$ Mole fraction (X_{benzene}) in vapour $= \dfrac{0.45 \times 0.098}{0.06} = 0.74$

147. Mass of benzene $= 500 \text{ mL} \times 0.897 \text{ g mL}^{-1} = 43.95 \text{ g}$

$$\Delta T_f = 5.51 - 5.03 = 0.48$$

$$\Delta T_f = K_f m = K_f \frac{W_B}{W_A M_B} \times 1000$$

or $M_B = \dfrac{K_f W_B}{W_A} \times \dfrac{1000}{\Delta T_f} = \dfrac{5.12 \times 0.643}{43.95} \times \dfrac{1000}{0.48} = 156$ amu.

148. Mole fraction, $x_i = \dfrac{n_p}{\Sigma n_1}$

$$xN_2 = \frac{\frac{2.8}{28}}{\frac{2.8}{28} + \frac{1.76}{44}} = \frac{\frac{1}{10}}{\frac{1}{10} + \frac{4}{10}} = \frac{\frac{1}{10}}{\frac{5}{10}} = \frac{1}{5} = 0.2$$

$\therefore x_{CO_2} = 0.8$

Total mass= 2.8 + 17.6 = 20 kg

Equivalent gas constant of the mixture,

$$R = \frac{m_{N_2} \cdot R_{N_2} + m_{CO_2} \cdot R_{CO_2}}{m} = \frac{2.8 \times \frac{8.3143}{28} + 17.6 \times \frac{8.3143}{44}}{2.8 + 17.6}$$

$$= \frac{8.3143\left(\frac{2.8}{28} + \frac{17.6}{44}\right)}{20} = \frac{8.3143}{20} \times \left(\frac{1}{10} + \frac{4}{10}\right) = \frac{8.3143}{40} \text{ kJ/kg K}$$

Now partial pressure of gases are given by

$$p_{N_2} = z_{N_2} \cdot p = 0.2 \times 250 = 50 \text{ kPa}$$

$$\therefore \quad p_{CO_2} = 250 - 50 = 200 \text{ kPa}$$

Total volume of mixture,

$$V = \frac{nRT}{p} = \frac{(2.8 + 17.2)\left(\frac{8.3143}{40}\right) \times 300}{250} = \frac{20 + 8.3143 \times 300}{40 \times 250} = 0.6 \times 8.3143 \text{ m}^3$$

Density of the mixture $\rho = \rho_{N_2} + \rho_{CO_2} = \dfrac{20}{(0.6 \times 8.3143)} = 4.009 \text{ kg/m}^3$

149. The equivalence point is attained when the number of moles of acid added is exactly enough to rect completely with the borax, with no acid in excess. Therefore we need to know the number of moles of borax (formula weight = 381.37 g· mol^{-1}) in the sample.

$$\text{No. mol borax} = \frac{0.6941 \text{ g}}{381.37 \text{ g} \cdot \text{mol}^{-1}} = 1.820 \times 10^{-3} \text{ mol}$$

From the net ionic equation for the reaction, the following molar ratio is obtained.

$$\frac{\text{mol } H^+ \text{ (aq) used}}{\text{mol } Na_2B_4O_7 \cdot 10\, H_2O(s) \text{ used}} = \frac{2}{1}$$

We therefore need 2(1.82 × 10^{-3}) = 3.64 × 10^{-3} mol, or 3.64 mmol, of $H^+(aq)$ to reach the equivalence point. The volume of 0.1082 *F* HCl required is

$$\text{mL of } 0.0182\ F \text{ HCl} = \frac{3.64 \text{ mmol}}{0.1082 \text{ mmol} \cdot \text{mL}^{-1}} = 33.64 \text{ mL.}$$

150. The proton-transfer reaction between water and nitrous acid is

$$\underset{}{H_2O} + \underset{\text{Nitrous acid}}{HONO} \rightleftharpoons H_3O^+ + \underset{\text{Nitrite ion}}{ONO^-}$$

Let x = molality of nitrite ion in this solution

Then x is also the molality of hydronium ion in this solution, because the molar ratio of H_3O^++ ions; ONO^- ions is 1:1. The molality of nitrous acid molecules in the solution is less than 0.075 *m* because some of the nitrous acid has ionized. Since

$$\frac{\text{No. mol } ONO^- \text{ formed}}{\text{No. mol HONO ionized}} = \frac{1}{1}$$

the molality of the nitrous acid that has *not* ionized is (0.075 – x)

The total molality of all solute particles is solution is

$$m_{\text{total}} = \text{molality HONO} + \text{molality } H_3O^+ + \text{molality } ONO^-$$
$$= (0.075 - x) + x + \text{i}x = 0.075 + x$$

We can calculate m_{total} from freezing point depression. For H_2O, K_f is 1.86 (see Tables). Therefore

$$m_{\text{total}} = \frac{\Delta T_f}{K_f} = \frac{0.15}{1.86} = 0.0806$$

Hence,

$$0.075 + x = 0.0806 \quad \text{and} \quad x = 0.0056\ m \text{ or } 5.6 \times 10^{-3}\ m$$

$$\text{percentage nitrous acid ionized} = \frac{\text{molality of } H_2O^+}{\text{initial molality}} \times 100$$

$$\left(\frac{5.6\times10^{-1}}{7.5\times10^{-2}}\right) = 7.5\%\text{ ionized}$$

151. The ions in the mixture are Na^+, K^+, Ca^{2+}, Cl^-, and NO_3^-. No precipitate is formed because nitrates and chlorides are generally soluble, as are salts of the alkali metals. NO reaction occurs, when these three solutions are mixed.

First we must determine the number of millimoles of each of these five ions in the resulting mixture:

$$\text{No. mmol Na}^+ = (20\text{ mL})\left(0.12\frac{\text{mmol}}{\text{mL}}\right) = 2.4\text{ mmol}$$

$$\text{No. mmol Cl}^- = (20\text{ mL})\left(0.12\frac{\text{mmol}}{\text{mL}}\right) + (60\text{ mL})(2)\left(0.08\frac{\text{mmol}}{\text{mL}}\right)$$

$$= 2.4\text{ mmol} + 9.6\text{ mmol} = 12\text{ mmol}$$

There are two sources of chloride ions, both the NaCl and the $CaCl_2$. We must and the number of millimoles of Cl^- from each source. Because the molar ratio of Cl^- ions to $CaCl_2$ is 2 : 1, the $[Cl^-]$ in 0.08 *F* $CaCl_2$ is 2(0.08 *M*) = 0.16 *M*.

$$\text{No. mmol K}^+ = \text{No. mmol NO}_3^- = (40\text{ mL})\left(0.15\frac{\text{mmol}}{\text{mL}}\right) = 6\text{ mmol}$$

$$\text{No. mmol Ca}^{2+} = (60\text{ mL})\left(0.08\frac{\text{mmol}}{\text{mL}}\right) = 4.8\text{ mmol}$$

The total of solution is 120 mL. Hence the final concentrations are

$$[\text{Na}^+] = \frac{2.4\text{ mmol}}{120\text{ mL}}\ 0.02\ M$$

$$[\text{Cl}^-] = \frac{12\text{ mmol}}{120\text{ mL}} = 0.1\ M$$

$$[\text{K}^+] = [\text{NO}_3^-] = \frac{6\text{ mmol}}{120\text{ mL}}\ 0.05\ M$$

$$[\text{Ca}^{2+}] = \frac{4.8\text{ mmol}}{120\text{ mL}}\ 0.04\ M$$

152. The net ionic equation for the reaction is

$$Pb^{2+}(aq) + CrO_4^{2-}(aq) \rightarrow PbCrO_4\downarrow$$

Thus Pb^{2+} ions and CrO_4^{2-} ions rect in a 1 : 1 molar ratio.

$$\text{No. mmol Pb}^{2+}\text{ ions added} = (25\text{ mL})\left(0.112\frac{\text{mmol}}{\text{mL}}\right) = 2.8\text{ mmol}$$

$$\text{No. mmol CrO}_4^{2-}\text{ added} = (45\text{ mL})\left(0.084\frac{\text{mmol}}{\text{mL}}\right) = 3.78\text{ mmol}$$

Consequently, there is an excess of CrO_4^{2-} ions. If virtually 100% of the Pb^{2+} ions are used up to form the $PbCrO_4$ precipitate, there will be 3.78 – 2.8 = 0.98 mmol of CrO_4^{2-} in excess. It is not possible to precipitate very close to that, say 99.999%. The volume of the final solution is 70 mL. The chromate ion concentration in the solution after reaction is therefore

$$[CrO_4^{2-}] = \frac{0.98 \text{ mmol}}{70 \text{ mL}} = 0.014\ M$$

The NO_3^- and K^+ ions are spectator ions, and their concentrations are calculated as following

$$[NO_3^-] = \frac{(25 \text{ mL})(2)(0112 \text{ mmol} \cdot \text{mL}^{-1})}{70 \text{ mL}} = 0.08\ m$$

$$[K^+] = \frac{(45 \text{ mL})(2)(0.084 \text{ mmol} \cdot \text{mL}^{-1})}{70 \text{ mL}} = 0.108\ M$$

153. The net ionic equation for the reaction is

$$2Ag^+(aq) + CO_3^{2-}(aq) \rightarrow Ag_2CO_3\downarrow$$

Silver ions and carbonate ions react in a 2 : 1 molar ratio

$$\text{No. mmol } Ag^+ \text{ added} = (10 \text{ mL})\left(0.093 \frac{\text{mmol}}{\text{mL}}\right) = 0.93 \text{ mmol}$$

$$\text{No. mmol } CO_3^{2-} \text{ added} = (15 \text{ mL})\left(0.12 \frac{\text{mmol}}{\text{mL}}\right) = 1.8 \text{ mmol}$$

To precipitate the Ag^+ ions in solution completely, 0.465 mmol of CO_3^{2-} ions are needed, because

$$\text{mmol } CO_3^{2-} \text{ in precipitate} = \frac{1}{2}(\text{mmol } Ag^+ \text{ is precipitate})$$

Therefore there is excess CO_3^{2-} in the solution. The amount of the excess is 1.8 – 0.465 = 1.335 mmol. The final volume of solution is 25 mL, so that

$$[CO_3^{2-}] = \frac{1.335 \text{ mmol}}{25 \text{ mL}} = 0.534\ M$$

Since

$$\text{No. mmol } Ag_2CO_3 \text{ formed} = \frac{1}{2}(\text{No. mmol } Ag^+ \text{ used up})$$

there are 0.465 mmol, or 4.65×10^{-4} mol, of Ag_2CO_3 in the precipitate. The formula weight of Ag_2CO_3 is 275.75 g · mol^{-1}, so that the mass of the precipitate is

$$(4.65 \times 10^{-4} \text{ mol})(275.75 \text{ g} \cdot \text{mol}^{-1}) = 0.128 \text{ g}$$

The Na^+ and NO_3^- ions are spectator ions,

$$[Na^+] = \frac{(15 \text{ mL})(2)(0.12 \text{ mmol} \cdot \text{mL}^{-1})}{(25 \text{ mL})} = 0.14\ M$$

$$[NO_3^-] = \frac{(10 \text{ mL})(0.093 \text{ mmol} \cdot \text{mL}^{-1})}{(25 \text{ mL})} = 0..0372 \ M$$

You can check your arithmetic by verifying the electroneutrality of the final solution. The total amount of negative charge must equal the total amount of positive charge. Since each CO_3^{2-} ion carries two negative charges, we must multiply the amount of CO_3^{2-} ions by 2 to obtain the negative charge on those ions. As the number of moles is directly proportional to the concentration of each ion in solution, the electroneutrality condition is expressed as

$$[NO_3^-] + 2[CO_3^{2-}] = [Na^+]$$
$$0.0372 + (0.534) = 0.144$$

Since 0.0372 + 0.1068 does equal 0.144, we have verified that the final solution is electrically neutral, as it must be.

154. 1.0 gm of H_2O_2 solution contains = 0.01 x gm H_2O_2 = $\frac{0.01x}{17}$ gm eqv. H_2O_2

Since the reaction takes place in equivalent amounts therefore

$$\frac{0.01x}{17} \text{ gm eqv } H_2O_2 \equiv \frac{0.01x}{17} \text{ gm eqv. } KMnO_4$$

Thus X ml of $KMnO_4$ solution contains = $\frac{0.01x}{17}$ gm eqv.

Therefore normality of $KMnO_4$ solution = $\frac{0.01x}{17} \times \frac{1000}{x} = 0.588$ N

155. Mol. wt. of Na_2CO_3 = 106 and that of $NaHCO_3$ = 84.

Let the no. of moles present be x each.

Wt. of Na_2CO_3 present in the mixture = $x \times 106 = 106x$ g. and that of $NaHCO_3 = x \times 84 = 84x$ g.

Total wt. of the mixture = $106x + 84x = 1$ g. (given)

Whence, $x = 1/190$ moles each of Na_2CO_4 and $NaHCO_3$

Reactions for neutralisation

$$Na_2CO_3 + 2HCl \rightarrow NaCl + CO_2 + H_2O$$
$$NaHCO_3 + 2HCl \rightarrow NaCl + CO_2 + H_2O$$

Amount of HCl required by 1/190 moles of Na_2CO_3

$$\frac{1}{190} \times 2 = \frac{2}{190} \text{ moles.}$$

Also, the amount of HCl required by 1/190 moles of $NaHCO_3 = \frac{1}{190} \times 1 =' \frac{2}{190}$ moles.

Total amount of HCl required = 2/190 + 1/190 = 3/190 moles

Now, 1 mole of HCL≡1 litre of 1 N-HCl

≡ 10 litres of 0.1 N-HCl

Hence, 3/190 moles of HCl≡10 × 3/190 litres of 0.1 N-HCl

or 157.9 ml of 0.1 N-HCl.

156. In the first instance, only KOH will be neutralised by HCl Hence, the g. eqs. of KOH present in the mixture

$$= \frac{120 \times 0.12}{1000} = 0.0144$$

and the wt. of KOH present = 0.0144 × 56 = 0.8064 g.
AgCl formed in the second operation is due to both KOH and KCl because KOH originally present has since changed to KCl by the first operation.

Equivalents of AgCl formed = $\frac{3.7}{143.5}$ = 0.0258 (eq. wt. of AgCl being 143.5)

Equivalents of KCl originally present = 0.0114
Hence, the weight of KCl originally present

$$= 0.0114 \times 74.5 = 0.8493 \text{ g.}$$

$$\%KOH = \frac{0.8064}{(0.8064 + 0.8493)} \times 100 = 48.7\%$$

157. Equations of reactions involved:

(a) $2CuSO_4, 5H_2O + 4Kl \rightarrow \downarrow Cu_2I_2 + I_2 + 2K_2SO_4 + 10H_2O$
(2495 × 2 g.) (127 × 2 g.)

$$I_2 \times 2Na_2S_2O_3 \rightarrow Na_2S_4O_6 + 2NaI$$

(b) $Hg_5(IO_6)_2 + 34KI + 24KCl \rightarrow 5K_2HgI_4 + 8I_2 + 24KCl + 12H_2O$
1448.5 g

From the above equations—

$$1448.5 \text{ g. } Hg_5(IO_6)_2 \equiv 8I_2 \equiv 16CuSO_4 \cdot 5H_2O \equiv 16 \times 249.6 \text{ g. } CuSO_4 \cdot 5_2O$$

$$\therefore \quad 0.7245 \text{ g. } Hg_5(IO_6)_2 \equiv \frac{16 \times 249.5 \times 0.7245}{1448.5} \text{ g. } CuSO_4 \cdot 5H_2O$$

Again, 0.0499 g. $CuSO_4 \cdot 5H_2O$

$\equiv$ 1 ml. of the given $Na_2S_2O_3$ soln.

Hence the volume of the given $Na_2S_2O_3$ solution required

$$= \frac{16 \times 249.5 \times 8.7245}{1448.5 \times 0.0499} = 40 \text{ ml.}$$

158. We have,

$$33.7 \text{ ml. of } 0.1 \text{ N-KMnO}_4 \equiv 33.7 \text{ ml. of } 0.1 \text{ N-FeSO}_4$$
$$\equiv 33.7 \text{ ml. of } 0.1\text{N } Fe^{+2}$$

($FeSO_4$ or Fe^{+2} being formed on reduction of $Fe_2(SO_4)_3$ by Cu powder)

Thus, the amount of Fe^{++} = $\frac{33.7 \times 0.1}{1000}$ g. eq. = 0.00337g. atoms.

$$\text{No. of g. atoms Cu that have reacted} = \frac{0.108}{63.5} = 0.0017$$

Thus, 0.0017 g. atom Cu produce 0.00337 g. atom Fe^{+2}

$\therefore$ 1 g. atom Cu will produce $= \frac{0.00337}{00017}$

$\approx$ 2 g. atom Fe^{+2} or 2 g. mol. $FeSO_4$

Hence, the equation may be given as under—

$$Cu + Fe_2(SO_4)_3 \rightarrow CuSO_4 + 2FeO_4$$

159. Reaction:

$KIO_3 + 2KI + 6HCl \rightarrow 3ICl + 3KCl + 3H_2O$

30 ml (M/10) $KIO_3 \equiv$ 3m mol $KIO_3 \equiv$ 6 mmol KI

$\therefore$ 20 mL KI = 6 mmol KI

$\therefore$ 50 mL KI $= \frac{6}{20} \times 50 = 15$ mmol KI

50 mL (M/10) $KIO_3 \equiv$ 5 mmol $KIO_3 \equiv$ 10 mmol KI

Amount of KI reacted with $AgNO_3$ = 15 – 10 = 5 mmol

$AgNO_3 + KI \rightarrow AgI + KNO_3$

5 mmol KI $\equiv$ 5 mmol $AgNO_3 \equiv 5 \times 170 \times 10^{-3} = 0.85$ g $AgNO_3$

$\therefore$ % of $AgNO_3$ = 85%

160. Molecular weight of H_2SO_4 = 2 + 32 + 64 = 98

Equivalent weight of $H_2SO_4 = \frac{98}{2} = 49$

26.7 ml of 0.4 N H_2SO_4 is equal to V ml of 1 N H_2SO_4.

Using formula $N_1V_1 = N_2V_2$

$\therefore$ $1 \times V_1 = 26.7 \times 0.4 = 10.68$ ml of 1 N H_2SO_4

$\therefore$ 49 g of H_2SO_4 will be neutralise by 1 N NaOH = 1000 ml

$\therefore$ 0.5 g of H_2SO_4 will be neutralised by 1 N NaOH

$$= \frac{1000}{49} \times 0.5 \text{ ml} = 10.20 \text{ ml}$$

Volume of 1 N NaOH used by dissolved SO_3 = 10.68 – 10.20 = 0.48 ml.

$SO_3 + 2NaOH \rightarrow Na_2SO_4 + H_2O$

$\therefore$ Eq. wt. of $SO_3 = \frac{32 + (3 \times 16)}{2} = \frac{32 \times 48}{2} = \frac{80}{2} = 40$

Hence, Weight of $SO_3 = \frac{40}{1000} \times 0.8 = 0.0192$ g

and, Percentage of $SO_3 = \frac{0.0192}{0.5} \times 100 = 3.84\%$

161. Eq. wt. of $K_2Cr_2O_7$ being M/6 because 6 equivalents of oxygen are generated by it during oxidation or 1 mole of $Cr_2O_7^{2-}$ ion requires 6 electrons for reduction, according to the equation

$$Cr_2O_7^{2-} + 14H^+ + 6e^- \rightarrow 2Cr^{3+} + 7H_2O$$

or

$$K_2Cr_2O_7 + 4H_2SO_4 \rightarrow K_2SO_4 + Cr_2(SO_4)_3 + 4H_2O + 3\ O$$

Thus, 0.01 M-$K_2Cr_2O_7$ = 0.06 N-$K_2Cr_2O_7$

No. of equivalents of $K_2Cr_2O_7$ present in 42 ml. soln.

$$= \frac{0.06 \times 42}{1000} = 0.00252$$

According to the law of normalities,

0.00252 equiv. $K_2Cr_2O_7$ ≡ 0.00252 equiv. Fe^{2+} salt ≡ 0.00252 equiv. $CuFeS_2$

Eq. wt. of $CuFeS_2$ is the same as its mol. wt. according to the reactions:

$$CuFeS_2 \rightarrow \downarrow CuS + FeS$$

$$FeS + H_2SO_4 \rightarrow FeSO_4 + \uparrow H_2S$$

$$2\ FeSO_4 + H_2SO_4 + O \rightarrow Fe_2(SO_4)_3 + H_2O$$

Mol. wt. of $CuFeS_2$ = 63.5 + 55.8 + 64 = 183.3

Amount of $CuFeS_2$ present in the mineral = 183.3 × 0.00252 = 0.4619 g.

$$\text{Percentage of } CuFeS_2 = \frac{0.4619 \times 100}{0.5} = 92.38\%$$

162. Normality of the alkali solution $= \frac{20 \times 0.1\text{ n}}{10} = 0.20$ N

Let the wt. of Na_2CO_3 persent as impurity be x gm./250 ml. or $4x$./liter. The wt. of NaOH in the sample

= (2.013 – x) gm./250 ml. = 4(2.013 – x) gm.litre^{-1}

Eq. wt. of NaOH = 40, Eq. wt. of $Na_2CO_3 = \frac{106}{2} = 53$

$$\text{Normality due to } Na_2CO_3 = \frac{4x}{53}$$

$$\text{Normality due to NaOH} = \frac{4(2.013 - x)}{40}$$

Hence, *total normality* $= \frac{4x}{53} + \frac{4(2.013 - x)}{40} = 0.2$ (as above)

whence, $x = 0.052$ g.

$$\text{Percentage of } Na_2CO_3 \text{ in the sample} = \frac{0.052 \times 100}{2.013} = 2.58\%$$

163. Reactions involved

$$3MnSO_4 \cdot 4H_2O \rightarrow Mn_3O_4 + 4H_2O + 3SO_2 + O_2$$

$$Mn_3O_4 + 2FeSO_4 + 4H_2SO_4 \rightarrow Fe_4(SO_4)_3 + 3MnSO_4 + 4H_2O$$

Milliequivalent of $FeSO_4$ in 30 ml of 0.1 N = F_3SO_4 = 30 × 0.1 = 3

25 ml of $KMnO_4$ reacts with = 3 m eq of $FeSO_4$

∴ 50 ml of $KMnO_4$ reacts with $= \frac{3 \times 50}{25}$ m eq of $FeSO_4$ = 6 meq of $FeSO_4$

Milliequivalent of 100 ml of 0.1 N $FeSO_4$ = 100 × 0.1 = 10 m eq

Ferrous sulphate reacted with Mn_3O_4 = 10 – 6 m eq = 4 m eq

Since milli eq. of oxidising agent = mille eq reducing agent

∴ milli equivalent of Mn_3O_4 formed: 4 meq

Since

$Mn_3O_4 \equiv 3\ MnSO_4,\ 4H_2O$

1 m eq = 3 m eq

∴ 4 m eq = 12 m eq

$$\text{Equivalent weight of } MnSO_4 \cdot 4H_2O = \frac{55 + 32 + (4 \times 16) + (4 \times 18)}{2} = \frac{223}{2}$$

$$\therefore \quad \text{Weight of } MnSO_4 \cdot 4H_2O \text{ in the sample} = \frac{\mathbf{12 \times 223}}{\mathbf{2}} \text{ mmg} = 1.338 \text{ g}$$

164. Volume × Normality of H_2O_2 solution = Volume × Normality of hypo solution

$$\therefore \quad \text{Normality of } H_2O_2 \text{ sol.} = \frac{20 \times 0.3}{25} = 0.24 \text{ N}$$

And hence, the volume strength of H_2O_2 sol. = 0.24 × 5.6 = 1.344

165. Let the wt. of Na_2CO_3 be x g. and therefore, that of K_2CO_3 will be $(1.22 - x)$ g.

Equivalent of Na_2CO_3 and k_2CO_3 dissolved in 100 ml. of solution

$$= \frac{x}{53} + \frac{1.22 - x}{69} \qquad \left(\text{Eq. wt.} = \frac{\text{Mol. wt.}}{\text{Total + valency}}\right)$$

$$\text{Equivalents of the mixture in 20 ml} = \frac{x}{53 \times 5} + \frac{1.22 - x}{69 \times 5}$$

$$\text{Equivalents of HCl in 40 ml of 0.1 N soln.} = \frac{40 \times 0.1}{1000} = 0.004$$

According to the law of normalities

$$\frac{x}{265} + \frac{1.22 - x}{345} = 0.004$$

whence, $x = 0.53$ g.

wt. of Na_2CO_3 present in the mixture is 0.53 g.

Again, according the law of equivalents

0.04 eq. HCl≡ 0.005 eq. mixture of carbonates ≡ 0.004 eq. of $BaCO_3$ which is likely to be precipitated by the action of $BaCl_2$ on carbonate mixture.

$$\text{Eq. wt. of } BaCO_3 = \text{Mol. wt.}/2 = \frac{197.3}{2} = 98.65$$

Eq. wt. of $BaCO_3$ ppt. = 0.004 × 98.65 = 0.3946 g.

166. In the mixture iron in present in two forms—

Fe (II) i.e. FeO in Fe_3O_4

Fe (III) i.e. Fe_2O_3 in $Fe_3O_4 + Fe_2O_3$

The total amount Fe_2O_3 is calculated as

$$\text{Equivalents of } Na_2SO_3 = \frac{11 \times 0.5}{1000} = 0.0055$$

Equivalents of Fe_2O_3 in 20 ml = 0.0055
Equivalents of Fe_2O_3 in 20 ml = 0.0055 × 5 = 0.0275

$$\text{Moles of } Fe_2O_3 = \frac{0.0275}{2} = 0.01375$$

Total moles of FeO is calculated as—

$$\text{Equivalent (Moles) of } Fe^{+2} = \frac{0.25 \times 5 \times 12.8 \times 2}{1000} = 0.032$$

Moles of FeO = Moles of Fe_3O_4 – Moles of Fe_2O_3 = 0.032 – 0.0275 = 0.0045
Thus moles of Fe_2O_3 existing separately = 0.01375 – 0.0045 = 0.00925
Mass of Fe_3O_4= 0.0045 × 232 = 1.044 g
Mass of Fe_2O_3= 0.00925 × 232 = 1.48 g
∴ % of Fe_2O_4= 34.8% % % of Fe_2O_3 = 49.3%

167. 43.9 ml. of H_2 at S.T.P. being produced by 0.1 g. of the metal. 11,200 ml. of H_2 (1 g. eq) will be produced from g. eq. wt. of the metal which is

$$= \frac{0.1 \times 11,200}{43.9} = 25.5 \text{ g.}$$

$$\text{Valency of the metal in its lower oxidation state} = \frac{\text{At. wt.}}{\text{Eq. wt.}} = \frac{51}{25.5} = 2$$

Again, 0.1 g. of the metal in the lower oxidation state required 58.8 ml of 0.1 N-$KMnO_4$ soln. For further oxidation.

∴1 g. at. wt. or 51.0 g. of the metal will require for further oxidation. $KMNO_4$ soln.

$$= \frac{58.8 \times 51}{0.1} \text{ ml. of 0.1 N soln} \approx 30,000 \text{ ml. of 0.1 N soln.}$$

≈ 3,000 ml. of 1 N soln ≈ 3 equivalents of KnO_4

Hence, the increase in valency during further oxidation is three. Thus the higher oxidation state is 5.

The valencies of the metal are 2 and 5.

168. $KMnO_4$ oxides Fe^{++} salt only.

Amount of $KMnO_4$ used up in milli-equivalents = 5.4 × 0.1 = 0.54 milli-equivalents

$$\left(\because \text{ 1 ml. of 1 N solution contains 1 milll-equivalent of the solute as } \frac{1 \times 10^3}{10^3} = 1\right)$$

According to the law of normalities

0.54 milli-equiv. $KMnO_4$ ≡ 0.54 milli-equiv. $FeSO_4$ $7H_2O$

Eq. wt. of $FeSO_4 \cdot 7H_2O$ = Mol. wt. of $FeSO_4 \cdot 7H_2O$ = 278

$[2FeSO_4 + H_2SO_4 + O \rightarrow Fe_2(SO_4)_3 + H_2O]$.

Hence, the wt. of $FeSO_4 \cdot 7H_2O$ present in the mixture = 0.54 × 278 × 10^{-3} g. = 0.15 g.

Amount of $Fe_2(SO_4$ 3 · $9H_2O)$ = 5.5 – 0.15 = 4.35 g.

Mol. wt. of $Fe_2(SO_4)_3 \cdot 9H_2O$ = 56 × 2 + 96 × 3 = 18 × 9 = 562.

Hence, the number of moles of the Ferric salt present in the mixture

$$= \frac{4.05}{362} = 0.00774 \text{ or } 7.74 \text{ milli-moles.}$$

169. Convert the 2.29 g of Ag to mol Agl, which is chemically equivalents to moles of Kl. Use that to calculate the molarity of the Kl.

$$2.29 \text{ g Agl} \times \frac{1 \text{ mol Agl}}{234.77 \text{ g Agl}} = 9.7542 \times 10^{-3} \text{ mol Agl (eq. to mol Kl)}$$

$$\text{Molarity} = \frac{7542 \times 10^{-3} \text{ mol Kl}}{0.01 \text{ L}} = 0.97542 = 0.975 \text{ mol L}^{-1}$$

170. The formula of Glauber's salt is $Na_2SO_4 \cdot 10H_2O$

Molecular mass of $Na_2SO_4 = 2 \times 23 + 32.1 + 4 \times 16 = 46 + 32.1 + 64 = 142.1$ g.

Molecular mass of $10H_2O = 10(1.01 \times 2 + 16) = 1890.2$ gm.

Molecular mass of Glauber's salt = 142.1 + 180.2 = 322.3 g mol^{-1}

Weight of the Glauber's salt taken = 80.575 gm

Out of 80.575 g of salt weight of anhydrous $Na_2SO_4 = \frac{142.1}{322.3} \times 80.575 = 32.535$ g

Number of moles Na_2SO_4 per dm^3 of the solution $= \frac{35.525}{142.1} = 0.25$

Molality of the solution = 0.25 M

Density of solution $= 1077.2 \text{ k gm}^{-3} = \frac{1077.2 \times 10^3}{10^6} \text{ gm cm}^{-3} = 1.0772 \text{ g cm}^{-3}$

Total weight of sol $= V \times D = 1\text{dm}^3 \times D$

$= 1000 \text{ cm}^3 \times 1.0772 \text{ g cm}^{-3} = 10.77.2$ g

Weight of water = 1077.2 – 35.525 = 1041.67 g

$$\text{Molality of sol.} = \frac{0.25}{1041.675 \text{ g}} \times 1000 \text{ g} = 0.2399 = 0.24 \text{ mg}$$

$$\text{Number of moles of water in the solution} = \frac{1041.67}{18} = 57.87$$

$$\text{Mole fraction of } Na_2SO_4 = \frac{\text{No. of moles of } Na_2 sO_4}{\text{Total numbr of moles}}$$

$$= \frac{0.25}{0.25 + 57.87} = 0.0043 = 4.3 \times 10^{-3}$$

171. (i) 1 litre of 0.1 N-$KMnO_4$ contains 0.1 × 1000 = 100 milli-equivalents of $KMnO_4$. Hence, 42.00 ml. of that solution contains

$$\frac{100 \times 42}{1000} = 4.2 \text{ milli-equivalents.}$$

(ii) 4.2 milli-equivalents of $KMnO_4$ correspond to 4.2 milli-equivalents of 4.2/1000 = 0.0042 equivalents of iron in the sample of the ore taken.

(iii) Wt. of iron corresponding to 0.0042 equivalents = 0.0042 × 56 = 0.2352 g.

(iv) Percentage of iron in the ore— $= \dfrac{0.2452 \times 100}{0.42} = 56.0$

(v) $KMnO_4$ acts as an oxidising agent according to the equation—

$$MnO_4^- + 8H^+ + 5e^- \rightarrow Mn^{++} + 4H_2O$$

or $$2KMnO_4 + 3H_2SO_4 \rightarrow K_2SO_4 + 2MnSO_4 + 3H_2O + 5O$$

2 moles 10 equiv.

Thus, 1 mole of $KMnO_4$ corresponding to 5 equivalents of it, and 0.1 N or 0.1 equivalents will corresponding to

$$\frac{1}{5} \times 0.1 = 0.02 \text{ moles.}$$

Hence, the molarity of the $KMnO_4$ solution = 0.02 M.

(vi) Number of moles of $KMnO_4$ contained in 42 ml. of its 0.02 M solution

$$= \frac{0.02 \times 42}{1000} = 0.00084 \text{ moles} = 8.4 \times 10^{-4}$$

172. (a) Sample is $\rightarrow Ca_2 + SO_4^{2-} + Ca^{2+} (H + CO_3)_2^-$

96 ppm 183 ppm.

When CaO is added to the above aquous following reaction occur

$$CaO + H_2O \rightarrow Ca(OH)_2 \quad \text{(i)}$$

$$Ca(OH)_2 + Ca(HCO_3)_2 \rightarrow 2\ CaCO_3(\downarrow) + 2H_2O \quad \text{(ii)}$$

Ppt

From equation (i) and (ii) it is clear that 2 moles of HCO_3^- will be neutralised by 1 mole of CaO (iii)

Now, 1000, kg of hard water contains

$\therefore$ $1000 \times 1000 \times 183 \times 10^{-6} = 183$ m of Bicarbonate ion

($\because$ $x$ppm = $x \times 10^{-6}$ parts)

So, 183 gm of $HCO_3^- = \dfrac{183}{61} = 3$ moles (iv)

(61 gm B molar mass of HCO_3^- ion)

Thus 3 moles of HCO_3^- present n 100 kg of hard water will require $\dfrac{3}{2}$ moles of CaO

From equation (iii) and (iv)

(b) Since $Ca(1 + CO_3)_2$ has been completely neutralised only $CaSO_4$ remain in solution

$\Rightarrow Ca^{2+} = SO_4^{2-}$ conc.

Hence residual $[Ca^{2+}]$ = 96 ppm.

$\Rightarrow$ In 1000 kg of residual water 96. Gm of Ca^{2+} is present.

(c) 96 gm of Ca^{2+} ion $= \dfrac{96}{40} = 2.4$ moles

Present in 1000 kg of water

or 2.4 moles of Ca^{2+} present in 1000 L of water

($\because$ for water density $\rho \cong 1$)

$\Rightarrow$ Conc. of $Ca^{2+} = \frac{2.4}{100}$ moles/L = 2.4×10^{-3} (v)

If now Ca^{2+} replaced by H^+ ions

$$[H^+] = 2.4 \times 10^{-3} \text{ M}$$

$\therefore$ $pH = -\log[H^+] = -\log(2.4 \times 10^{-3}) = 2.62$

173. (i) Molarity—

Mol. wt. of Na_2CO_3 = 106

Strength of the solution = 8.653 g./100 ml. = 86.53 g.litre^{-1}

Molarity of the solution is the strength of the solution in terms of g. mol./litre = 86.53/106 = 0.816 M.

(ii) Molality—

Density of the solution = 1.0816 g./ml = 1081.6 g.litre^{-1}

Wt. of 1 litre of the solution = 10981.6 g.

Wt. of 1the solute per litre = 86.53 g.

Weight of water contained in one litre of the solution = 1081.6–86.53 = 995.07 gm.

Thus the weight of solute dissolved in 1000 g. of the solvent

$$= \frac{86.53 \times 1000}{995.07} = 86.96 \text{ g.}$$

Hence, the number of gm. mole dissolved per 1000 g. of the solvent, i.e., the molality of the solution

$$= \frac{86.96}{106} = 0.83 \text{ Molal.}$$

174. (a) Oxidation potential of Zn is higher than that of Cu, that is, zinc is more electro-positive than copper. Thus Zn will displace Cu from $CuSO_4$ according to the equation

$$Zn + CuSO_4 \rightarrow ZnSO_4 + \downarrow Cu$$

Loss of molarity of $CuSO_4$ by the above reaction in the given interval of time

= 1 M – 0.8 M = 2 M

Equivalents of Cu^{++} that have reacted = 0.2 ×2 = 0.4

Equivalents of Zn that have dissolved = 0.4

Hence, wt. of Zn dissolved = 0.4 × 65.4/2 = 13.08.

Wt. of the zinc rod after cleaning = 25.0 – 13.08 = **11.92 g.**

There is no change in the concentration of SO_4^{-2} ion, because it does not take part in the reaction. Hence the molarity of the solution with respect to sulphate ion is 1 M.

(b) Since the total concentration of $ZnSO_4$ in the same as that of $CuSO_4$ in the previous experiment and copper is less electropositive than zinc, hence Cu cannot displace Zn from $ZnSO_4$. Thus, the molarity of Zn^{++} will be same as before, i.e., 2 M.

175. Weight of H_2SO_4 contained in 75 ml. of the given acid

$$= \frac{75 \times 1.18 \times 24.7}{100} = 21.86 \text{ g.}$$

Reaction: $\underset{54 \text{ g.}}{2Al} + \underset{(3 \times 98 = 294 \text{ g})}{3H_2SO_4} \rightarrow Al_2(SO_4)_3 + \uparrow 3H_2$

Amount of H_2SO_4 used up by 2.7 g. Al = $\frac{294 \times 2.7}{9}$ = 14.7 g

(Amount of H_2SO_4 used by Al can also be calculated as under:

Equivalents of Al = $\frac{2.7}{9}$ = 0.3

H_2SO_4 consumed = 0.3 equiv. = 0.3 × 49 = 14.7 g.

Amount of unreacted H_2SO_4 diluted to 400 ml. = 21.86–14.7 = 7.16 g.

Strength in gms. litre^{-1} = $\frac{7.16 \times 100}{400}$ = 17.9

∴ Molarity of solution = $\frac{17.9}{9.8}$ = 0.183 M

176. Let morality of the solution of $CuSO_4$ be '*m*'.

Amount of Cu contained or liable to be precipitated from 1 litre of '*m*' molar solution = *m* × 63.5 g.

∴ Amount of Cu pptd. from 200 ml. of '*m*' molar solution

$$= \frac{m \times 63.5 \times 200}{1000} = \frac{63.5\text{ m}}{5}$$

∴ Amount of zinc dissolved for $\frac{63.5\text{ m}}{5}$ g. Cu precipitation

$= \frac{63.4\text{ m}}{5}$ g. (substances react in eq. proportions)

Amount of unreacted zinc = $\left(10 - \frac{63.4\text{ m}}{5}\right)$g

Total amount of deposit after reaction—

$= \frac{63.5\text{ m}}{5}$ Cu + $\left(10 - \frac{63.4\text{ m}}{5}\right)$g Zn = 9.810 g. (given),

whence, molarity of CuSO4 solution (m) = 0.5.

and the wt. of Cu deposited = $\frac{63.5 \times 0.5}{5}$ = 6.35 g.

Chapter 6 : THERMODYNAMICS

1. Extrapolating the regression line to the Y axis (0 seconds) gives a temperature of 314.4 K at the moment the room temperature and warm and water were mixed.

$$= \frac{298 \text{ K} + 334 \text{ K}}{2} = 315 \text{ K}$$

$$C_{calorimeter} = \frac{q_{calorimeter}}{(T_{mix} - T_{initial})}$$

$$q_{calotrimeter} = -q_{water}$$

$$q_{water} = (\text{mass water}) \cdot (\text{specific heat}) \cdot (T_{mix} - T_{avg})$$

At 296 K: $\frac{50 \text{ mL } H_2O}{1} \times 0.983 \text{ g} \cdot \text{mL}^{-1} = 49.9 \text{ g } H_2O$

At 334 K: $\frac{50.0 \text{ mL } H_2O}{1} \times 0.983 \text{ g} \cdot \text{mL}^{-1} = 49.1 \text{ g } H_2O$

Total mass = $49.9 \text{ g} + 49.1 \text{ g} = 99 \text{ g } H_2O$

$= (99 \text{ g}) \cdot (4.18 \text{ J/g} \cdot \text{K}) \cdot (314.4 \text{ K} - 315 \text{ K}) = -25 \times 10^2 \text{ J}$

$\therefore$ Heat gained by calorimeter = $-q_{water} = 2.5 \times 10^2$ J

$$C_{calorimeter} = \frac{q_{calorimeter}}{(T_{mix} - T_{initial})} \times \frac{2.5 \times 10^2 J}{(314.4 - 296) K} = 14 \text{ J} \cdot \text{K}^{-1}$$

2. (a) We have, for; HCl + NaOH

$$q_{rxn} = \frac{-[(\text{mass}_{sol'n}) \times (\text{specific heat}_{sol'n}) \times (\Delta T_{sol'n}) + (C_{calorimeter} \times \Delta T_{sol'n})}{\text{volume}_{sol'n} \times \text{molarity}}$$

$$= \frac{[100 \text{ g}_{sol'n} \times 4.18 \text{ J/g} \cdot \text{K} \times (308.6 \text{ K} - 296 \text{ K}) + (14 \text{ J/K} \times 12.5 \text{ K})}{0.500 \not{L} \times 2 \text{ mole}/\not{L}}$$

$$= \frac{-5.4 \times 10^3 \text{ J}}{0.05 \not{L} \times 2 \text{ mol}/\not{L}} = -54 \text{ kJ mol}^{-1}$$

(b) for, NH_4Cl + NaOH

$$q_{rxn} = \frac{-[100 \text{ g}_{sol'n} \times 4.18 \text{ J/g} \cdot \text{K} \times (297.1\text{K} - 295.9\text{K})] + (14 \text{ J/g} \cdot \text{K} \times 1.2 \text{ K})}{0.05 \text{ L} \times 2.0 \text{ mol}/\not{L}}$$

$$= \frac{-5.2 \times 10^2 \text{ J}}{0.05 \not{L} \times 2 \text{ mol}/\not{L}} = -5.2 \text{ kJ mol}^{-1}$$

(c) for, NH_3 + HCl

$$q_{rxn} = \frac{[100 \text{ g}_{sol'n} \times 4.18 \text{ J/g}^\circ \not{C} \times (306.1\text{K} - 296\text{K})] + (13.6 \text{ J/K} \times 10.1\text{K})}{0.05 \not{L} \times 20 \text{ mole}/\not{L}}$$

$$= \frac{-4.4 \times 10^3 \text{ J}}{.00500 \cancel{L} \times 2.0 \text{ mol}/\cancel{L}} = -44 \text{ kJ/mol}$$

3. The net ionic equation is given below :

$$H^+_{(aq)} + \cancel{OH^-_{(aq)} \rightarrow H_2O_{(1)}} \qquad \Delta H = -54 \text{ kJ} \cdot \text{mol}^{-1}$$

$$NH_{3(aq)} + \cancel{OH^-_{(aq)}} \rightarrow NH^+_{4(aq)} + \cancel{OH^-_{(aq)}} \qquad \Delta H = +5.2 \text{ kJ} \cdot \text{mol}^{-1}$$

$$NH_{3(aq)} + H^+_{(aq)} \rightarrow NH^+_{4(aq)} + OH^-_{(aq)} \qquad \Delta H = -49 \text{ kJ} \cdot \text{mol}^{-1}$$

4. $\% \text{ error} = \frac{\text{obs} - \text{exp}}{\text{exp}} \times 100\% = \frac{-44 - (-49)}{-49} \times 100\% = -10\%$

5. (a) The thermochemical equation for the reaction is

$$C_2H_4(g) + 3O_2(g) \rightarrow 2CO_2(g) + 2H_2O(1);$$

1 mole 3 moles 2 moles

$\Delta E = -1388.5$ KJ; $\Delta n = 2 - 4 = -2$; $R = 8.314 \times 10^{-3}$ KJ mol^{-1} K^{-1}; $T = 290$ K

Now, $\Delta H = \Delta nRT = -1388.5 + (-2)(8.314 \times 10^{-3})(290) = 1393.3$ KJ

∴ The heat of combustion of constant pressure = –1393.3

(b) The thermochemical equation for the reaction is

$$C(s) + 2H_2(g) \rightarrow CH_4(g); \qquad \Delta H = -74.78 \text{ KJ}$$

Negligible volume 2 moles 1 mole

$\Delta n = 1 - 2 = -1$; $\Delta H = -74780$ J; $R = 8.314$ J mol^{-1} K^{-1}; $T = 298$ K

∴ $\Delta H = \Delta E + \Delta nRT$

∴ $-74780 = \Delta E + (-1)(8.314)(298)$

or, $\Delta E = -74780 + 2477.6 = 72302 = 72.3$ KJ

∴ Heat of formation of methane at constant pressure = –72.3 KJ

6. According to the data given

$$C_6H_6(l) + 7\tfrac{1}{2} O_2(g) \rightarrow 6CO_2(g) + 3H_2O(l); \qquad \Delta H = -3156 \text{ KJ} \qquad (1)$$

$$C(s) + O_2(g) \rightarrow CO_2(g); \qquad \Delta H = -393.5 \text{ KJ} \qquad (2)$$

$$H_2(g) + 1/2\, O_2(g) \rightarrow H_2O(l) \qquad \Delta H = -285.8 \text{ KJ} \qquad (3)$$

The required equation is $6C(s) + 3H_2(g) \rightarrow C_6H_6(l)$, $\Delta H = ?$

Multiply equation (2) by 6 and (3) by 3

$$6C(s) + 6O_2(g) \rightarrow 6CO_2(g); \Delta H = -2361 \text{ KJ}$$

$$3H_2(g) + 3/2\, O_2(g) \rightarrow 3H_2O(l); \Delta H = -587.4 \text{ KJ}$$

$$6C(s) + 3H_2(g) + 15/2\, O_2(g) \rightarrow 6\, CO_2(g) + 3H_2O(l); \Delta H = -3218 \text{ KJ} \qquad (4)$$

$$C_6H_6(l) + 3H_2(g) \rightarrow C_6H_6(l); \Delta H = -62.4 \text{ KJ}$$

On subtracting (1) from (4)

$$6C(s) + 3H_2(g) \rightarrow C_6H_6(l); \Delta H = -62.4 \text{ KJ}$$

∴ The heat of formation of benzene is –62.4 KJ

7. ΔH of the following reaction is to be calculated

$$2C(s) + 3H_2(g) + 1/2\ O_2(g) \rightarrow C_2H_5OH \quad (1)$$

Multiplying equation (1) by 2 and equation (2) and 3 and adding them together.

$$2C(s) + 2O_2(g) \rightarrow 2CO_2(g); \Delta H = -794.2 \text{ KJ}$$
$$3H_2(g) + 3/2\ O_2(g) \rightarrow 3H_2O(l); \Delta H = -877.8 \text{ KJ}$$

$$2C(s) + 3H_2\ (g) + 7/2\ O_2(g) \rightarrow 2CO_2(g) + 3H_2O(l); \Delta H = -1672 \text{ KJ} \quad (4)$$

Subtracting equation (3) from equation (4) and rearranging we get

$$2C(s) + 3H_2(g) + 7/2\ O_2(g) \rightarrow 2CO_2(g) + 3H_2O(l); \Delta H = -1672 \text{ KJ}$$
$$C_2H_5OH(l) + O_2(g) \rightarrow 2CO_2(g) + 3H_2O(l); \Delta H = -1380 \text{ KJ}$$

$$2C(s) + 3H_2(g) + 1/2\ O_2(g) \rightarrow C_2H_5OH(l); \Delta H = -292 \text{ KJ}$$

The heat of formation of ethyl alcohal is –292 KJ

8. The required thermo chemical equation is

$$K(s) + 1/2\ O_2(g) + 1/2\ H_2(g) \rightarrow KOH(s)\ ; \Delta H = ?$$

Adding equations (1) and (2) we get

$$K(s) + 1/2\ O_2(g) + 1/2\ H_2(g) + aq \rightarrow KOH\ (aq)\ ; \Delta H = -487 \text{ KJ} \quad (4)$$

Subtracting equation (3) from equation (4) and on rearrangement we get –431.4 KJ

9. Heat of combustion at constant volume, ΔE = ?

Weight of benzoic acid = 0.5 gm

Rise in temperature = 0.55K

Mol. Wt. of benzoic acid = 122

Thermal capacity of the calorimeter system = 23.85 KJ.

Heat of combustion at a constant volume ΔE

$$= \text{Heat capacity of calorimeter} \times \frac{\text{molecular weight of substance}}{\text{weight of substance}} \times \text{rise in temp.}$$

= 23.05 × 0.55 × (122/0.5

i.e., $\Delta E = -3200.7$ KJ

Heat of combustion at constant pressure, ΔH = ?

The combustion of benzoic acid at 298 K involves the following reaction

$$C_6H_5COOH(s) + 7.5\ O_2(g) \rightarrow 7CO_2(g) + 3H_2O\ (l)\ \Delta H = ?$$

We know that, $\Delta H = \Delta E + \Delta nRT$

Here $\Delta E = -3200.7$ KJ ; $\Delta n = 7 - 7.5 = -0.5$;

$R = 8.314 \times 10^{-3}$ KJ ; $T = 298$ K

$$\Delta H = -3200.7 + 8.314 \times 10^{-3} \times 298 \times (-0.5) = -3201.9$$

Therefore heat of combustion at constant pressure, ΔH = –3201.9 KJ

10. Total weight of water = 150 + 150 = 300 gm

Water equivalent of calorimeter = 60 g

Rise in temperature = (301.2 – 298.35) = 2.85°K
Heat produced = 360 × 4.184 × 2.85 = 4255.1 Joules
Heat evolved for the neutralisation of 150 ml of 5 N HCl = 4255.1 Joules
Heat evolved for the neutralisation of 1000 ml N HCl = 4255.1 × (1000/150) × 0.05 Joules = 56735 Joules.
Heat of neutralised of hydrochloric acid = –56.735 KJ
(Since heat is evolved it carries a negative sign)

11. Required equation is $C + 2S \rightarrow CS_2$; $\Delta H = ?$

$$C + O_2 \rightarrow CO_2;\ \Delta H = -394 \text{ KJ} \quad (1)$$
$$S + O_2 \rightarrow SO_2;\ \Delta H = -297 \text{ KJ} \quad (2)$$
$$CS_2 + 3O_2 \rightarrow CO_2 + 2SO_2 : \Delta H = -1076 \text{ KJ} \quad (3)$$

Multiplying equation (2) with 2 and adding to equation (1)

$$C + 2S + 3O_2 \rightarrow CO_2 + 2SO_2 \qquad \Delta H = -988 \quad (4)$$

Subtracting equation (3) from equation (4) and on rearranging

$$C + 2S + 3O_2 \rightarrow CO_2 + 2SO_2; \qquad \Delta H = -988 \text{ KJ}$$
$$CS_2 + 3O_2 \rightarrow CO_2 + 2SO_2; \qquad \Delta H = -1076 \text{ KJ}$$
$$\overline{C + 2S \rightarrow XS_2; \qquad \Delta H = +88 \text{ KJ}}$$

12. Heats of combustion of methane and ethane.

$$CH_4 + 2O_2 \rightarrow CO_2 + 2H_2O;\ \Delta H_M = ?$$
$$C_2H_6 + 7/2\ O_2 \rightarrow 2CO_2 + 3H_2O;\ \Delta H_E = ?$$

Heat of combustion of methane

$$\Delta H_M = H_P - H_R = [H_{CO_2} + 2H_{H_2O}] - [H_{CH_2} + 2H_{O_2}]$$
$$= [-393 + (-2 \times 241.6)] - [(-74.8) + (-2 \times 0)] = -801.4 \text{ KJ}$$

Heat of combustion of ethane

$$\Delta H_E = H_P - H_R = [2H_{CO_2} + 3H_{H_2O}] \cdot [H_{C_2H_6} + 7/2\,H_{O_2}]$$
$$= [(-2 \times 393 - 3 \times 241.6)] - (-84.4) = -1426.4 \text{ KJ}$$

Heat evolved per gram of methane = 801.4/16 = 50.08 KJ
Heat evolved per gram of ethane = 1426.4/30 = 47.54 KJ
So methane is more efficient fuel than that of ethane.

13. Required equation is

$$Ca_2(g) + O^{2-}(g) \rightarrow CaO(s)\ ;\ \Delta H = ?$$

Given:

$$Ca(s) \rightarrow Ca(g);\ \Delta H_s = +177 \text{ KJ}$$
$$1/2\ O_2(g) \rightarrow O(g);\ \Delta H_D = 500/2 = 250 \text{ KJ}$$
$$Ca(g) \rightarrow Ca^{2+}(g) + 2e;\ \Delta H_1 = 587 + 1156 = 1743 \text{ KJ}$$
$$O(g) + 2e \rightarrow O^{2-}(g);\ \Delta H_E = -704 \text{ KJ}$$
$$Ca(s) + 1/2\ O_2(g) \rightarrow CaO(s);\ \Delta H_f = -637 \text{ KJ}$$
$$\Delta H_f = \Delta H_s + \Delta H_D + \Delta H_1 + \Delta H_E + \Delta H_U$$
$$\therefore \quad \Delta H_U = \Delta H_f - (\Delta H_s + \Delta H_D + \Delta H_1 + \Delta H_E)$$
$$\therefore \quad \Delta H = -637 - (177 + 250 + 1743 - 704) = -2113$$

14. $HCN + NaOH \rightarrow NaCN + H_2O$

The enthalpy change, $\Delta H = H_P - H_R = [-88.7 - 286] - [105.5 - 469.5] = -374.7 + 364 = -10.7$ KJ

Enthalpy of ionisation of HCN, ΔH

= [enthaly of neutralisation of HCN by NaOH]

– [enthalpy of neutralisation of strong acid by strong base]

$= -10.7 - (-57.3) = +46.6$ KJ

15. $C_6H_6 + 15/2\ O_2 \rightarrow H_2O + 6CO_2$

$$\Delta H = H_P - H_R$$

$$\Delta H = [3(-285.5) + 6(-393)]-(48.9) = -3163.4 \text{ KJ mol}^{-1}$$

No. of moles of benzene = 1000/78 = 12.82 moles

Energy liberated by burning 1 Kg of benzene = $3163.4 \times 12.82 = 40575$ KJ

16. $2\ Al + 3/2\ O_2 \rightarrow Al_2O_3$; $\Delta H_1 = -1596$ KJ

$2Cr + 3/2\ O_2 \rightarrow Cr_2\ O_3$; $\Delta H_2 = -1134$ KJ

$\Delta H = H_P - H_R$ = heat of combustion of Al – Heat of combustion of Cr

$= -1596 - (-1134) = -462$ KJ

17. Required equation is $2Al + 3Cl_2 \rightarrow Al_2Cl_6$ $\Delta H = ?$

Heat change = Enthalpy of $3Cl_2$ molecules + Enthalpy of 6 HCl molecules + Enthalpy of $Al_2Cl_6(aq)$ – Enthalpy Al_2Cl_6 solid.

$= (3 \times 184.2) + (6 \times 73.3) + 1005 - 143.5 = +1358.8$ KJ

$\Delta H = -1358.8$ KJ

18. The required equation of (Heat of combustion of cane sugar).

$$C_{12}H_{22}O_{11} + 12O_2(g) \rightarrow 12CO_2(g) + 11H_2O(l) \ ; \ \Delta H = ?$$

$$\Delta H = 12H_{CO_2} + 11H_{HO} - H_{\text{cane sugar}} = 12(-393) + 11(-285.5)-(-2216.6) = -5640 \text{ KJ}$$

–5640 KJ of energy is liberated from 1 mole (0.342 kg) of cane sugar.

11996.6 KJ of energy is liberated from $11996.6 \times 0.342/5640 = 0.727$ Kg. of cane sugar.

19. The equation for the formation of napthalene is

$$10C + 4H_2 \rightarrow C_{10}H_5; \ \Delta H = ?$$

From the given data

$$C_{10}H_5 + 12\ O_2 \rightarrow 10\ CO_2 + 4H_2O + 5145.6 \text{ KJ} \quad (1)$$

$$C + O_2 \rightarrow CO_2 + 394.6 \text{ KJ} \quad (2)$$

$$H_2 + 1/2\ O_2 \rightarrow H_2O + 286 \text{ KJ} \quad (3)$$

Multiply equation (2) by 10 and equation (3) by 4

$$10 + 10O_2 \rightarrow 10\ CO_2 + 3946 \text{ KJ} \quad (4)$$

$$4H_2 + 2O_2 \rightarrow 4H_2O + 1144 \text{ KJ} \quad (5)$$

Adding (4) and (5) and subtracting equation (1)

$$10C + 4H_2 - C_{10}H_4 \rightarrow -55.6 \text{ KJ}$$

$\therefore$ $10C + 4H_8 \rightarrow C_{10}H_5; \ \Delta H = -55.6$ KJ

$\therefore$ Heat of formation of napthalene, $\Delta H = + 55.6$ KJ

20. $Pb(s) + O_2(g) \rightarrow PbO_2(s)$ $\Delta H = -276.4$ KJ

Heat evolved when 208 g. (1 gram atom) of lead is burnt = $\frac{276.4}{208} = 1.33\,KJ$

∴ Heat evolved by burning 1 g. of lead = (66.12/208) 1000 = 319.3 Cal.

21. $C(graphite) + O_2(g) \rightarrow CO_2(g)$
1 mole   1 mole
Heat evolved when 1 g. of graphite is burnt = 29.34 KJ
1 mole of oxygen consumed = 1 mole of graphite oxidised = 12 g. of graphite oxidised
∴ Heat evolved when 1 mole of O_2 used = 29.34 × 12 = 351.6 KJ

22. 1 gram mole of sucrose = 342 g.
Heat evolved by burning 342 g. of sucrose = 5643 KJ
Heat evolved by burning 12 g. sucrose = (5643/342) × 12 = 198 KJ

23. $C_{12}H_{22}O_{11}(s) + 12O_2(g) \rightarrow 12CO_2(g) + 11H_2O(l)$, $\Delta H = -5643$ KJ
1 mole   12 moles   12 moles
342 g.   12 × 22400 ml   12 × 44 g. = 528

(a) Amount of CO_2 evolved by burning 342 g. of sucrose = 528 g.
∴ Amount of CO_2 equivalent by burning 12 g. of sucrose = (528/342) × 12 = 18.5 g.

(b) Volume of oxygen consumed at STP by burning 342 g. of sucrose = 12 × 22400 ml
∴ Volume of oxygen consumed at STP by burning 12 g. of sucrose = (12 × 22400 × 12)/342 = 943.1 ml

24. $CH_4(g) + 2O_2(g) \rightarrow CO_2(g) + 2H_2O(l)$, $\Delta H = -890$ KJ
1 mole   1 mole
22.4 litres   22.4 litres

(a) Heat evolved by burning 22.4 litres of CH_4 at STP = 890 KJ
∴ Heat evolved by burning 2.8 litres CH_4 at STP = (890 KJ × 2.8)/22.4 = 111.25 KJ

(b) 22.4 litres of methane produces 2.4 litres of CO_2 at STP
∴ 2.8 litres of methane produces = (22.4/22.4) × 2.8 = 2.8 litres of CO_2 at STP.

25.
$$CaO + H_2CO_3 \rightarrow CaCO_3 + H_2O + 127.5\ KJ$$
$$CO_2 + H_2O \rightarrow H_2CO_3 + 48\ KJ$$
Adding
$$CaO + CO_2 \rightarrow CaCO_3 + 175.5\ KJ$$

26.
1. $C(diamond) + O_2 \rightarrow CO_2$ $\Delta H = -394$ KJ
2. $C(graphite) + O_2 \rightarrow CO_2 \Delta H = -396$ KJ

Subtracting equation (2) from (1), we get
$$C(graphite) \rightarrow C(diamond)\ \Delta H = -2\ KJ$$

27. The heat of combustion of acetylene is given by the equation.
$$C_2H_2 + 3/2O_2 \rightarrow 2CO_2 + H_2O\ \Delta H = -1296\ KJ$$
The required equation $2C + H_2 \rightarrow C_2H_2$; $\Delta H = ?$

We know that $\Delta H = H_P - H_r = [2H_{CO_2} + H_{H_2O}] - (H_{C_2H_2} + \frac{3}{2}H_{O_2})$

or, $-1296 = [2 \times (-393.3) + (-285.5)] - [H_{C_2H_2} + 0]$

∴ $H_{C_2H_2} = -786 - 285.5 + 1296 = 224.5$ KJ

∴ Heat of formation C_2H_2 = 22.45 KJ

28. The required equation is

$$CH_4 + 2O_2 \rightarrow CO_2 + 2H_2O \ \Delta H = ?$$

We know that $\Delta H = H_P - H_r$

$= [H_{CO_2} + 2H_{H_2O}] \cdot [H_{CH_4} + 2H_{O_2}] = [-393.3 + 2(-285.5)]-[(-74.8) + 0] = -890$ KJ

∴ Heat of combustion of $CH_4 = -890$ KJ

29. $NaOH(aq) + CH_3COOH(aq) \rightarrow CH_3COONa(aq) + H_2O \ \Delta H = -56$ KJ

$NaOH(aq) + HCl(aq) \rightarrow NaCl(aq) + H_2O \ \Delta H = -57.26$ KJ

Subtracting and rearranging

$CH_3COOH(aq) + NaCl(aq) \rightarrow CH_3COONa(aq) + HCl(aq) \ \Delta H + 1.26$ KJ

(or)

$CH_3COOH(aq) + Na^+(aq) + Cl^-(aq) \rightarrow CH_3COO^-(aq) + Na^+(aq) + H+(aq) + Cl^-(aq)$

$\Delta H = 1.26$ KJ

Cancelling common ions on both sides,

$CH_3COOH(aq) \rightarrow CH_3COOH(aq) + H^+(aq) \ \Delta H = +1.26$ KJ

∴ Heat of dissociation of CH_3COOH = +1260 Joules

30. Rise in temperature (t) = 0.632K
Mass of water (m) 18.94 kg
Specific heat of water (s) = 4.17 $Jg^{-1} K^{-1}$
Heat energy gained by water = mst = (18.94 × 1000) × 4.17 × 0.632
Heat energy given by 1.89 g. of benzoic acid = 18940 × 4.17 × 0.632 J

Heat energy given by 122 g. benzoic acid $= \dfrac{18940 \times 4.17 \times 0.632}{1.89} \times 122 = 3230$ J

∴ Heat of combustion of benzoic acid = 3230 J

31. Rise in temperature of water = t°K

Mass of water = 18.94 Kg
Specific heat of water = 4.17 $Jg^{-1} K^{-1}$
Heat energy liberated = mst

When 3.2 g of methanol is burnt, heat liberated = (18.94 × 1000) × 4.17 × t.Joules
Energy liberated by burning 32 g. of of methanol on combustion = 714.8 KJ
Energy liberated by 3.2 g. of methanol of combustion = 71.48

∴ $18.94 \times 4.17 \times t = 71.4$

or $t = \dfrac{71.4}{18.94 \times 4.17} = 0.946$

∴ Rise in temperature = 0.90 K

32. Heat of combustion of ethane = –1555

i.e. $C_2H_6 + 1/2\ O_2 \rightarrow 2CO_2 + 3H_2O \ \Delta H = -1555$

$$\Delta H = H_P - H_R$$

$$\Delta H = 2\ \Delta H_f^o\ CO_2 + 3\Delta H_f^o\ H_2O - \Delta H_f^o\ C_2H_6$$

$$-1555 = -786 - 852.6 - H^o\ C_2H_6$$

∴ $\Delta H_f^o\ C_2H_6 = -83.6$

∴ Heat of formation of ethane = –83.6

Heat of combustion propane = –2215.4

i.e. $C_3H_8 + 5\ O_2 \rightarrow 3CO_2 + 4H_2O\ \Delta H = -530.0$

$$\Delta H = 3\Delta H_f^\circ\ CO_2 + 4\Delta H_f^\circ\ H_2O - \Delta\Delta H_f^\circ\ C_3H_8$$

$$\Delta H_f^\circ\ C_3H_8 = 3\Delta H_f^\circ\ CO_2 + 4\Delta H_f^\circ\ H_2O - \Delta H = -1179 - 1136.8 + 2215.4 = -100.4$$

∴ Heat of formation of propane = –100.4

$$2C(g) + 3H_2(g) \rightarrow C_2H_6 \qquad \Delta H_f^\circ = -83.6 \qquad (1)$$

$$2C(gr) \rightarrow 2C(g) \qquad \Delta H = +1438 \qquad (2)$$

$$3H_2(g) \rightarrow 6H(g) \qquad \Delta H = +1304 \qquad (3)$$

Adding eq (2) and eq (3) and subtracting eq (1) from sum and on rearranging.

$$C_2H_6 \rightarrow 2C(gr) + 6H(g)\ \Delta H = 282.5$$

Since C_2H_6 has 6 C—H bonds 1 C—C bond, the sum of these bond energies is equal to 282.5 KJ.

$$3C(g) + 4H_2(g) \rightarrow C_3H_8 \qquad \Delta H_f^\circ = -100.4\ KJ \qquad (1)$$

$$3C(gr) \rightarrow 3C(g) \qquad \Delta H = 2157\ KJ \qquad (2)$$

$$4H_2(g) \rightarrow 8H(g) \qquad \Delta H = 1738.8\ KJ \qquad (3)$$

Adding eq (2) and eq (3) and subtracting eq (1) from sum

$$C_3H_8 \rightarrow 4C(gr) + 8H(gr) \qquad \Delta H = 3996\ KJ$$

Since C_3H_8 has 8 C—H bonds 2 C—C bond, the sum of these bond energies is equal to 3996 KJ.

$$6(C—H) + (C—C) = 2825\ KJ$$

$$8(C—H) + 2(C—C) = 3160\ KJ$$

On simplification1C—C bond is equal to 342.7 KJ mol^{-1}

1H—H bond is equal to 413.8 KJ mol^{-1}

33. $C_6H_{10} + H_2 \rightarrow C_6H_{12};\ \Delta H = ?$

Given that

$$H_2 + 1/2O_2 \rightarrow H_2O;\ \Delta H = -241 \qquad (1)$$

$$C_6H_{10} + 17/2O_2 \rightarrow 6CO_2 + 5H_2O;\ OH = -3800 \qquad (2)$$

$$C_6H_{12} + 9O_2 \rightarrow 6O_2 + 6H_2O;\ \Delta H = -3920 \qquad (3)$$

Adding equation (1) and (2)

$$C_6H_{10} + H_2 + 9O_2 \rightarrow 6CO_2 + 6H_2O;\ \Delta H = -4041 \qquad (4)$$

Subtracting equation (3) from equation (4) and on rearrangement

$$C_6H_{10} + H_2 \rightarrow C_6H_{12} \qquad \Delta H = -121\ KJ$$

∴ Heat of hydrogenation of cyclohexane is –121 KJ/Mole

34. $Fe_2O_3 + 2Al \rightarrow Al_2O_3 + 2Fe$

Heat of reaction= $H_P - H_R = H_f^\circ(Al_2O_3) - H_f^\circ(Fe_2O_3) = 1667.8 - 831.8 = 836$ KJ mol^{-1}

214 g. intimate mixture of Al and Fe_2O_3 give → 836 KJ mol^{-1}

1 g. intimate mixture of Al and Fe_2O_3 gives → 836/214 = 3.9 KJ

35. Given data is,

$$2C(s) + 3H_2(g) \rightarrow C_2H_6(g) \; ; \quad \Delta H_f = -88 \text{ KJ} \quad \text{(i)}$$

$$C(s) + O_2(g) \rightarrow CO_2(g) \; ; \quad \Delta H_f = -393.3 \text{ KJ} \quad \text{(ii)}$$

$$H_2(g) + 1/2O_2(g) \rightarrow H_2O(l) \; ; \quad \Delta H_f = -256 \text{ KJ} \quad \text{(iii)}$$

Required equation

$$C_2H_6 + 7/2O_2 \rightarrow 2CO_2 + 3H_2O$$

$$\Delta H = H_p - H_r$$

Multiplying (ii) by 2, (iii) by 3 subtract (i)

$$\Delta H = -(2 \times 393.3) - (3 \times 285.5) - (-88) = 1555 \text{ KJ}$$

∴ Heat of combustion of ethane is –1555 KJ

36. Required reaction is: $1/2 \; H_2 + 1/2 \; Cl_2 \rightarrow HCl$

ΔH = Bond dissociation energy of reactants—Bond dissociation energy of products.

$$= \left(\frac{434.7}{2} + \frac{242.5}{2}\right) - 430.5 = 91.9 \text{ KJ mol}^{-1}$$

37. The required equation is $C_2H_2(g) + H_2(g) \rightarrow C_2H_4(g)$; $\Delta H = ?$

Given; $H_2 + 1/2 \; O_2 \rightarrow H_2O \qquad \Delta H = -285.5$ (1)

$$C_2H_2 + 5/2 \; O_2 \rightarrow CO_2 + H_2O \qquad \Delta H = -1298.3 \quad (2)$$

$$C_2H_4 + 3O_2 \rightarrow 2CO_2 + 2H_2O \qquad \Delta H = -1409.5 \quad (3)$$

To get required reaction add eq(1) and eq(2) and subtract (3)

$$H_2 + C_2H_2 - C_2H_4 \rightarrow O \qquad \Delta H = -174.3 \text{ KJ}$$

$$H_2 + C_2H_2 \rightarrow C_2H_4 \qquad \Delta H = -174.3 \text{ KJ}$$

Calculation of ΔE

$$\Delta H = \Delta E + \Delta n \; RT$$

$$\therefore \quad \Delta E = \Delta H - \Delta n \; RT = -174.3 - (-1)\,(8.314 \times 10^{-3}) \times 298] = 171.8 \text{ KJ}$$

$$(\text{when } \Delta n = n_p - n_r = 1 - 2 = -1)$$

38. Required equation is $2C + H_2 \rightarrow C_2H_2$; $\Delta H = ?$

Given, $C_2H_2 + \frac{5}{2} O_2 \rightarrow 2CO_2 + H_2O$; $\Delta H = -1296$ KJ ..... (1)

$$C + O_2 \rightarrow CO_2 \qquad \Delta H = -393 \text{ KJ} \quad \text{..... (2)}$$

$$H_2 + 1/2 \; O_2 \rightarrow H_2O \qquad \Delta H = -285.5 \text{ KJ} \quad \text{..... (3)}$$

To get required reaction; multiply eq (2) with 2 and add to eq. (3)

$$2C + H_2 + 5/2 \; O_2 \rightarrow 2 \; CO_2 + H_2O \; \Delta H = -1071 \text{ KJ} \quad \text{..... (4)}$$

Subtract eq (1) from eq (4) and on re-arranging + 225 KJ

$$2C + H_2 \rightarrow C_2H_2; \; \Delta H = +225 \text{ KJ}$$

∴ Heat of formation of acetylene = ΔH = +225 KJ

39. $\Delta H = H_P - H_R = [H_{CO_2} + 4_{HCl}] - [H_{CCl_4} + 2H_{H_2O}]$

$$= [-393 - 4 \times 92.4] - [-106.6 - 2 \times 241.6] = -762.6 + 589.8 = -172.8 \text{ KJ mol}^{-1}.$$

40. O + H → OH

Bond energy of O_2 = –494.6 KJ; for one 'O' atom = –272.3 KJ

Bond energy of H_2 = –435.5 KJ; for one 'H' atom = –217.8 KJ

$1/2\ H_2 + 1/2\ O_2 \rightarrow OH(g)\ \Delta H^o = -42$ KJ

Bond energy of OH radial

= Bond energy of H atom + Bond energy of O atom – enthalphy of OH

= –217.8–272.3–(–42)=–448.1 KJ

41. A. Calculation of W.

$$W = \int_{v_1}^{v_2} pdV = \int_{v_1}^{v_2} \frac{RT}{V}\,dV = RT\ln\frac{v_2}{v_1}$$

$$= 2.3030\ RT\ \frac{v_2}{v_1} = 2.303 \times 8.314 \times 300 \times \log 4.56/2.28 = 1729\ \text{J mole}^{-1}.$$

B. Calculation of ΔE

For isothermal reversible expansion of ideal gas $E_2 = E_1$, hence $\Delta E = 0$

C. Calculation of $q = \Delta E + W = 0 + 1729 = 1729$ J

D. Calculation of ΔH:

$$\Delta H = \Delta E + \Delta(PV) = \Delta E + \Delta(RT) = 0 + 0 = 0.\ \text{(as } R \text{ \& } T \text{ being constant)}$$

42. $\Delta S_{vap} = \dfrac{\Delta H_{vap}}{T_b} = \dfrac{40.63\ \text{KJ mole}^{-1}}{373\ \text{K}} = 0.1088\ \text{KJ K}^{-1}\ \text{mole}^{-1}$

43. $\Delta S = nR\ 2.303 \log \dfrac{V_2}{V_1} = nR\ 2.303 \log \dfrac{V_2}{V_1}$

$= 1 \times 8.314 \times 2.303 \times \log 20/2 = 19.15\ \text{J} \cdot \text{Mol}^{-1}\ \text{K}^{-1}$

44. $n = 6$ $\quad p_1 = 2$ atm;

$G = 300°K$ $\quad P_2 = 4$ atm; $\quad \Delta G = ?$

$\Delta G = nRT \log \dfrac{P_2}{P_1} = 21.303\ nRT \log \dfrac{P_2}{P_1}\ 2.303 \times 6 \times 8.314 \times 300 \log 4/2 = 10373.9$

45. The changes can be represent as 1 mole $H_2O \rightleftharpoons$ 1 mole water vapour

$T_b = 373$ K; $\Delta H_{vap} = 40850\ \text{J mol}^{-1}$

$\Delta S_{vap} = 40850/373 = 109.5\ \text{JK}^{-1}\ \text{mole}^{-1}$

For surroundings, ΔS surrounding = $-\Delta S$ system = –109.5 J.

1 mole Sn (a) =1 mole of Sn (b)

$T_{trans} = 286$K

$\Delta S_{trans} = 2090\ \text{J mol}^{-1}$

$\Delta S_{trans} = 2090/286 = 7.32\ \text{J K}^{-1}\ \text{mol}^{-1}$.

46. $\Delta S_{fus} = \dfrac{\Delta H_{fus}}{T_{fus}} = \dfrac{6.025}{273}$ K.J mole^{-1}

(ΔH of ice = 6.025 K.J mol^{-1}) = 0.022 K.J. K^{-1}. mole^{-1}

47. The calorimeter constant is given

$$\frac{(-326 \text{ kJ mol}^{-1})(.825 \text{ g})}{(1.94 \text{ K})(123 \text{ g mol}^{-1})} = -11.1 \text{ kJ}^{-1}$$

The heat released by the combustion of the biphenyl is then

$$\frac{(-11.1 \text{ kJ K}^{-1})(1.94 \text{ K})(154 \text{ g mol}^{-1})}{0.526 \text{ g}} = -6240 \text{ kJ mol}^{-1}$$

(The negative sign indicates that heat is released in this process.) From the reaction equation

$$(C_6H_5)_2(s) + 19/2\ O_2(g) \rightarrow 12\ CO_2(g)\ 5H_2O(l)$$

We have $\Delta n = 12 - (19/2) = -5/2$. Converting to ΔH, we substitute into

$\Delta H = q_v + \Delta n\ RT$

$\therefore\ \Delta H^\circ = \Delta E^\circ - (5/2)\ (8.314 \text{ J mol}^{-1} \text{ K}^{-1}) = -6246 \text{ J mol}^{-1}$

This is the amount of heat that is lost by the system if reaction takes place at constant pressure and the temperature is restored to its initial value.

48. The given thermochemical data is :

(i) $C_2H_5OH(l) + 3O_2(g) \rightarrow 2\ CO_2(g) + 3H_2O(l),\ \Delta H = -330.0$ kJ

(ii) $C(s) + O_2(g) \rightarrow CO_2(l),\ \Delta H = -94.3$ kJ

(iii) $H_2(g) + \frac{1}{2}O_2(g) \rightarrow H_2O(l),\ \Delta H = -68.5$ kJ

The required equation is

$$2C(s) + 3H_2(g) + \frac{1}{2}O_2(g) \rightarrow C_2H_5OH(l)\ \Delta H = ?$$

Multiplying (ii) by 2 and (iii) by 3 and adding up these, we get

$$2C(s) + 3H_2(g) + \frac{1}{2}O_2 \rightarrow 2CO_2(g) + 3H_2O(l)\ \Delta H = -394.1 \text{ kJ}$$

Subtracting equation (i) from (iv), we get

$$2C(s) + 3H_2(g) + \frac{1}{2}O_2(g) \rightarrow C_2H_5OH(l)$$

$$\Delta H = -394.1 - (330.0) = -64.1 \text{ kJ}$$

Thus the enthalpy of formation of ethyl alcohol is –64.1 kJ.

49. The neutralisation of a strong acid with a strong base is represented as

$$H^+(aq) + OH^-(aq) \rightarrow H_2O(l);\ \Delta H = -55.9 \text{ kJ mol}^{-1}$$

The neutralisation of acetic acid.

$CH_3COOH \rightleftharpoons CH_3COO^- + H^+;\ \Delta H_{ion}$

$H^+(aq) + OH^-(aq) \rightarrow H_2O(l);\qquad \Delta H = -55.9 \text{ kJ mol}^{-1}$

obviously,

$$\Delta H_{ionisation} = \Delta H_{ionisation} + (-55.9 \text{ kJ mol}^{-1})$$
$$-50.6 \text{ kJ mole}^{-1} = \Delta H_{ionisation} + (-55.9 \text{ kJ mol}^{-1})$$
$$\Delta H_{ionisation} = -50.6 + 55.9 = 5.3 \text{ kJ mol}^{-1}$$

50. The thermochemical date given to us is

1. $S_{rhombic} + O_2(g) \rightarrow SO_2(g); \Delta H = -291.4$
2. $S_{monoclinic} + O_2(g) \rightarrow SO_2(g); \Delta H = -295.4$

Subtracting eq (2) from eq (1) we get

$$S_{rhombic} + \not{O}_2(g) - \not{O}_2(g)\ S\not{O}_2(g) \rightarrow S_{monoclinic} \rightarrow S\not{O}_2(g) - S\not{O}_2(g)$$
$$\Delta H = -291.4 - (-295.4)$$
$$= 4.0 \text{ kJ}$$

Thus $S_{rhombic} \rightarrow S_{monoclonic}$; $\Delta H = 4$ kJ

51. In this reaction four C—H, one C=C and three O=O bonds are broken and four C=O and four O—H bonds are formed

```
H   H
|   |
C = C  + 3O2 → 2CO2 + 2H—O
|   |
H   H
```

$\therefore$ $\Delta H° = 4\Delta H(\text{C—H}) + \Delta H(\text{C=C}) + 3\,\Delta H(\text{O=O}) — 4\,\Delta H(\text{C=O}) \quad - 4\,\Delta H$
(OH)

$$= 4(414) + 619 + 3(499) - 4(724) - 4(460) = -964 \text{ kJ mol}^{-1}$$

52. Using the enthalpy of combustion data

$$\Delta H° = 2\Delta_f H°\,(CO_2) + 2\Delta_f H°(H_2O) - (\Delta_f H°(C_2H_4) + 3\Delta_f H°(O_2)]$$
$$= 2 \times (-393.5) + 2\,(-241.8) - [52.3 + 3 \times O] = -1322.9 \text{ kJ}$$

Hence, the enthalpy of reaction calculated from bond enthalpies is much lower than the actual value and therefore, the values so calculated are approximate.

53. The structure of ethyl alcohal can be represented by the following structure

```
    H   H
    |   |
H — C = C — O — H
    |   |
    H   H
```

Total bond energy is = $5 \times \Delta H$ (C—H) + ΔH(C—C) + ΔH(C—O) + ΔH(O—H)
= 5 × 413 kJ + 348 kJ + 351 kJ + 363 kJ = 3227 kJ

Thus when 3227 kJ of energy is supplied then 2C(g), 6H(g) and O(g) will be the product given by C_2H_5OH

Heat of atomisation = 6H(g) + 2C(g) + O(g) = 2990.5 (Given)

$\Delta H°$= Enthalpy of atomisation – Bond Enthalpy = 2990.4 – 3227 = –236.6 kJ.

54. The reaction is $\frac{1}{2}H_2(g) + \frac{1}{2}Cl_2(g) \rightarrow HCl(g)$

For this reaction

$$\Delta C_p = C_p(HCl) - \left(\frac{1}{2}C_p(H_2) + \frac{1}{2}C_p(Cl_2)\right) = 28.49 - \frac{1}{2}(28.53) - \frac{1}{2}(32.26) = 1.91\ JK^{-1}\ mol^{-1}$$

further, $\Delta H_1 = -92300$ J; $T_1 = 398$ K; $T_2 = 348$ K; $\Delta H_2 = ?$
using Kirchoff's equation

$$\frac{\Delta H_2 - \Delta H_1}{T_2 - T_1} = \Delta C_p$$

$$\Delta H_2 = \Delta H_1 + \Delta C_p(T_2 - T_1)$$
$$= -92300 + (-1.91)(348 - 298) = -92395.5J = -92.395\ kJ$$

55. As we know

$$\Delta H_r^\circ = \Delta_f^\circ H\ (\text{products}) - \Delta_f H^\circ\ (\text{reactants})$$
$$= [2 \times \Delta_f H^\circ [CO_2\ (g)] + 3 \times \Delta_f H^\circ (H_2{}^\circ O\ (l)] - \Delta_f H^\circ [C_2H_5OH)\ (l)] + 3\ \Delta_f H^\circ [O_2(g)]$$
$$= [2 \times (-393.5) + 3 \times 285.5] - [-277.0 - 0] = -1643.5 - (-277) = -1366.5\ kJ$$

56. $C_6H_5\ COOH(S) + \frac{15}{2}O_2(g) \rightarrow CO_2(g) + 3H_2O(l)$

$$\Delta H = \Delta E + \Delta nRT$$

$$-2546.0 = \Delta E + \left(7 - \frac{15}{2}\right) \times 8.314 \times 10^{-3} \times 298$$

$$-2546 + \left(\frac{1}{2} \times 8.314 \times 298 \times 10^{-3}\right) = \Delta E \times 10^{-3} = -2544.76\ kJ.$$

57. Enthalpy of combustion of ethylene

$$C_2H_4(g) + 3O_2(g) \rightarrow 2CO_2(g) + 2H_2O(l) \qquad \Delta H_1 = -337.2\ kJ$$

Enthalpy of combustion H_2.

$$H_2(g) + \frac{1}{2}O_2(g) \rightarrow H_2O(l)$$

$$C_2H_6(g) + \frac{7}{2}O_2 \rightarrow 2\ CO_2 + 3H_2O \qquad \Delta H_2 = -68.3\ kJ$$

Enthalpy of combustion of ethane.

$$C_2H_6(g) + \frac{1}{2}O_2 \rightarrow 2\ CO_2 + 3H_2O. \qquad \Delta H_3 = -372.8\ k$$

Applying Hess's law,
for the reaction

$$C_2H_4(g) + H_2(g) \rightarrow C_2H_6(g);\ \Delta H = ?$$

$$\therefore\ \Delta H = \Delta H_1 + \Delta H_2 - \Delta H_3 = -337.2 - 68.3 - (372.8) = -32.7\ kJ$$

58. $H_2(g) + \frac{1}{2}O_2(g) \rightarrow H_2O(l)\Delta H_1 = 285\ kJ\ mol^{-1}$

Enthalpy of formation of $CO_2(g)$

$$\text{(graphite)} + O_2(g) \rightarrow CO_2(g) \qquad \Delta H_2 = 393 \text{ kJ mol}^{-1}$$

Enthalpy of formation of propane $C_3H_8(g)$

$$3\text{ C(graphite)} + 4H_2(g) - C_3H_8(g) \qquad \Delta H_3 = -104 \text{ kJ mol}^{-1}$$

Enthalpy of formation of propane

$$C_3H_8 + 5O_2 \rightarrow 3CO_2 + 4H_2O \qquad \Delta H = ?$$

Applying Hess's law

$\Delta H = 3\Delta H_2 + 4\Delta H_1 - \Delta H_3 = (3 \times -393) + 4(-285) - (-104)$

$= -1179 - 1140 + 104 = -2215.0 \text{ kJ mol}^{-1}$

59. The combustion reaction for liquid ethanol can be written as

$$C_2H_5OH(l) + 3O_2(g) \rightarrow 2CO_2(g) + 3H_2O(l);\ \Delta H_1 = -3330 \text{ kJ} \qquad \ldots\ldots \text{(i)}$$

Enthalpy of Combustion of carbon,

$$\text{C (graphite)} + O_2(g) \rightarrow O_2(g);\ \Delta H_2 = -94.3 \text{ kJ} \qquad \ldots\ldots \text{(ii)}$$

Enthalpy of combustion of H_2

$$H_2(g) + \frac{1}{2}O_2(g) \rightarrow H_2O(l);\ \Delta H_3 = -68.5 \text{ kJ} \qquad \ldots\ldots \text{(iii)}$$

Enthalpy of formation of alcohal is given by equation (application of Hess's law) 2(ii) + 3(iii) – (i)

$2\text{C (graphite)} + 2O_2(g) + 3H_2(g) + \frac{3}{2}O_2(g) + 2CO_2(g) + 3H_2O(l) \rightarrow 2CO_2(g) + 3H_2O(l) + C_2H_5OH(l) + 3O_2(g)$

$$2\text{C (graphite)} + \frac{1}{2}O_2(g)\ \frac{1}{2}O_2(g) + 3H_2(g) \rightarrow C_2H_5OH(l) - (4) \qquad \Delta H = ?$$

$\Delta H = 2 \times (-94.3) + 3(-68.5) - (-33) = -188.6 - 205.5 + 330 = -64.1$

60. Enthalpy of formation FeO (S)

$$Fe(s) + \frac{1}{2}O_2(g) \rightarrow Fe\,O(S) \qquad \Delta H_1 = -267.9 \text{ kJ}$$

Enthalpy of formation Fe_2O_3 (S)

$$2\ Fe(S) + \frac{3}{2}O_2(g) \rightarrow Fe_2O_3\ (S) \qquad \Delta H_2 = -822 \text{ kJ}$$

For the reaction; $2\ FeO(s) + \frac{1}{2}O_2 \rightarrow FeO_3(g) \quad \Delta H = ?$

Applying Hess's law, we get

$\therefore\ \Delta H = \Delta H_2 - 2\Delta H_1 = -822.2 - 2\,(-267.9) = -822.2 + 535.8 = -286.4 \text{ kJ}$

61. Heat of Combustion of benzene

$$C_6H_6 + \frac{15}{2}O_2 \rightarrow 6CO_2 + 3H_2O \qquad \Delta H_1 = -780.98 \text{ kJ}$$

Enthalpy of Combustion of carbon

$$\text{C (graphite) } O_2(g) \rightarrow CO_2(g) \qquad \Delta H_2 = -94.05 \text{ kJ}$$

Enthalpy of Combustion of Hydrogen

$$H_2(g) + \frac{1}{2}O_2(g) \rightarrow H_2O(l) \qquad \Delta H_3 = -68.32 \text{ kJ}$$

Hence 6 C (graphite) + $3H_2(g)$ + → $C_6H_6(l)$

$\therefore\ \Delta H = 6\Delta H_2 - \Delta H_1 = (6 \times 94.05) + 3(-68.32) - (-780.98) = 11.72$ kJ

62. Heat of reaction for the reaction

$$CH_4(g) + 4F_2(g) \rightarrow CF_4(g) + 4HF(g)$$

$$\Delta H = H_{\text{reactants}} - H_{\text{products}}$$

4 C—H bonds = 4 × 99.3 = 397.2 kJ
4 F—F bonds = 4 × 38 = 152 kJ] H_R = 549.2 kJ

4 C—F bonds = 4 × 116 = 464 kJ
4 H—F bonds = 4 × 135 = 540 kJ] H_p = 1004 kJ

$\therefore\ \Delta H = 549.2 - 1004 = -454.8$ kJ

63. F—F Bond enthalpy = 36.6 kJ mol^{-1}

Cl — Cl Bond enthalpy = 580 kJ mol^{-1}

Heat of reaction

$$F_2 + Cl_2 \rightarrow 2\ (F\text{—}Cl) \qquad \Delta H = 26.6 \text{ kJ}$$

$$\Delta H = H_{\text{reactant}} - H_{\text{products}}$$

$26.6 = 36.6 + 580 - 2\Delta H_{F\text{—}Cl} \Rightarrow \Delta H_{F\text{-}Cl} = 295 \text{ kJ}$

64. Writing thermochemical equations from the given data on

$$C(s) + 2H_2(g) \rightarrow CH_4(g)\ \Delta H_f^\circ (H_4) = -74.75 \text{ kJ} \qquad \text{(i)}$$

$$C(s) \rightarrow C(g);\ \Delta H^\circ \text{ atomization} = 716.68 \text{ kJ} \qquad \text{(ii)}$$

$$\frac{1}{2}H_2(g) \rightarrow H(g);\ \Delta H^\circ \text{ atomization} = 217.97 \text{ kJ} \qquad \text{(iii)}$$

The bond dissociation energy of C—H bond is 1/4th of the enthalpy of reaction

$$\Delta H\text{—}H = \frac{1}{4}\Delta H^\circ$$

Now, $$CH_4(g) \rightarrow C(g) + 2H_2(g) \qquad \text{(iv)}$$

$$\Delta H^\circ = \Delta H_{C\text{—}H}$$

To obtain reaction (iv), we multiply (iii) by 4 and add the product to (ii) so,

$$2H_2(g) + C(S) \rightarrow 4H(g) + C(g) \qquad \text{(v)}$$

or $$\Delta H^\circ = 871.88 + 716.68 = 1588.56$$

subtracting (i) from (v)

$$CH_4(g) \rightarrow C(g) + 4H(g) \Rightarrow \Delta H^\circ = 1588.56 + 74.5 = 1663.31 \text{ kJ}$$

Hence, Bond dissociation energy of C—H = $\frac{1663.31}{4}$ = 145.83 kJ mol^{-1}

65. The given data is

(i) $H_2(g) \rightarrow 2H(g);\ \Delta H_1 = 436.08$ kJ

(ii) $O_2(g) \rightarrow 2O(g);\ \Delta H_2 = 495.17$ kJ

(iii) $H_2(g) + \frac{1}{2}O_2(g) \rightarrow H_2O(g); \Delta H_3 = -241.84$ kJ

Bond energy of the O—H bond can be obtained from the enthalpy change of the following reaction in which the two O—H bonds of a $H_2O(g)$ molecule are broken.

$H_2O(g) \rightarrow 2H(g) + O(g)$

Applying Hess's law; ΔH of this reaction is given by

$$\Delta H = \Delta H_1 + \frac{1}{2}\Delta H_2 - \Delta H_3 = 436.08 + \frac{495.17}{2} - (-241.84) = 925.51 \text{ kJ}$$

Hence $\qquad \Delta H_{O—H} = \frac{1}{2}(925.51) = 462.76$ kJ

66. For the dissociation of HBr, assume that ΔH and ΔS are constant over the temperature range from 298 K to 648 K, and calculate the value of each to use to calculate K at 648 K.

Start by calculating $\Delta H°$ and $\Delta S°$ at 298 K, using $\Delta H_f°$ and $S°$ values.

	$2HBr(g)$	$\rightarrow$	$H_2(g)$	+	$Br_2(g)$
$\Delta H_f°$:	$2 \times (-36)$		0		30.91
$\Delta S°$:	2×198.59		130.6		245.38

$\therefore \qquad \Delta H° = [(30.91 + 0) - 2(-36)]$ kJ = 102.91 KJ $\qquad (\cong \Delta H$ at 648 K)

$\Delta S° = [(245.38 + 130.6) - 2(198.59)]$ J/K = –21.20 J/K = –0.02120 KJ/K

Substitute $\Delta H°$ and $\Delta S°$ into the Gibbs equation at 648 K.

$$\Delta G° = \Delta H° - T\Delta S \cong 102.91 \text{ KJ} - (648 \text{ K})(-0.02120 \text{ KJ/K}) \cong 116.6 \text{ KJ}$$

Substitute $\Delta G°$ into the K equation at 648 K, using R = 0.008314 KJ/K:

$$\log K = \frac{\Delta G°}{-2.303\,RT} = \frac{116.6 \text{ kJ}}{(-2.303)(0.008314 \text{ kJ/K})(648 \text{ K})} = -9.3976$$

$\therefore \qquad K = 4.0 \times 10^{-10}$

Assuming $x = [H_2] = [Br_2]$, and assuming [HBr] = $(1.0 - 2x)$ = 1.0 atm, substitute into the equilibrium expression.

$$\frac{[H_2][Br_2]}{[HBr]^2} \cong \frac{(x)(x)}{(1.00)^2} \cong 4.0 \times 10^{-10}$$

$$\therefore \qquad x = \sqrt{(4.0 \times 10^{-10})(1.0)^2} = 2.0 \times 10^{-5} \text{ atm}$$

The percent dissociation at 1.0 atm is

$$\% \text{ dissociation} = \frac{2(2.0 \times 10^{-5} \text{ atm})}{1.0 \text{ atm}} \times 100\% = 0.004\% \text{ (at 1.0 atm)}$$

Based on Le Chatelier's principle, pressure has no effect on equilibrium, so the % dissociation is 4.0×10^{-3} = 0.004%.

67. For the dissociation NH_3, assume that ΔH and ΔS are constant over the temperature range from 298 K to 618 K and calculate values of each to calculate K at 618 K.

	$2NH_3(g)$	$\rightarrow$	$3H_2(g)$	+	$N_2(g)$
$\Delta H_f°$:	$2 \times (-45.9)$		0		0
S°:	2×193		3×130.6		191.5

First calculate $\Delta H°$ and $\Delta S°$, using $\Delta H_f°$ and $S°$ values

$$\Delta H° = [0 + 0 - 2(-45.9)] \text{ KJ} = 91.8 \text{ KJ } (\simeq \Delta H \text{ at } 618 \text{ K})$$

$$\Delta S° = [(3 \times 130.6 + 191.5) - 2(193)] \text{ KJ} = 197.3 \text{ J/K}$$

$$= 0.1973 \text{ KJ/K } (\cong \Delta S \text{ at } 618 \text{ K})$$

Substitute $\Delta H°$ and $\Delta S°$ into the Gibbs equation at 618 K

$$\Delta G° \cong \Delta H - T\Delta S \cong 91.8 \text{ KJ} - (618 \text{ K})(0.1973 \text{ KJ/K}) \cong -30.13 \text{ KJ}.$$

Substitute $\Delta G°$ into the equation at 618 K, using $R = 0.008314$ KJ.

$$\log K = \frac{\Delta G°}{-2.303\,RT} = \frac{-30.13 \text{ kJ}}{(-2.303)(0.008314 \text{ kJ K}^{-1})(618 \text{ K})} = 2.5462$$

$$\therefore \quad K = 3.517 \times 10^2 = K_p$$

Now obtain K_C

$$K_C = K_p(RT)^{-2} = (3.517 \times 10^2)(0.0821 \times 618)^{-2} = 0.1366$$

The starting concentration of NH_3 is 1.0 mol/20.L = 0.05 M.

$$2NH_3(g) \rightarrow 3H_2(g) + N_2(g) \qquad \ldots\ldots (i)$$

	$2NH_3(g)$	$3H_2(g)$	$N_2(g)$
Starting	0.05	0	0
Change	$-2x$	$+3x$	$+x$
Equilibrium	$0.05 - 2x$	$3x$	x

The equilibrium equation for (i) is

$$\frac{[H_2]^3[N_2]}{[NH_3]^2} = K_c$$

or

$$\frac{(3x)^2}{(0.05 - 2x^2)} = 0.1366 \text{ or } \frac{x^4}{(0.05 - 2x)^2} = \frac{0.1366}{27} = 5.06 \times 10^{-3}$$

Taking the square root of both sides of this equation gives $\dfrac{x^2}{0.05 - 2x} = 7.113 \times 10^{-2}$

which rearranges to

$$x^2 + (1.423 \times 10^{-1}) \times - (3.556 \times 10^{-3}) = 0$$

From the quadratic formula, you obtain

$$x = \frac{-0.1423 \pm \sqrt{(0.1423)^2 + 4x \times 3.556 \times 10^{-3}}}{2}$$

The positive root is $x = 0.02168$ M

Hence,

$$[NH_3] = 0.05 - 2(0.02168) = 6.64 \times 10^{-3} \text{ M}$$

$$\%NH_3 \text{ dissociated} = \left(1 - \frac{6.64 \times 10^{-3}\ M}{0.05\ M}\right) \times 100\% = 86.7\% = 87\%$$

68. First calculate $\Delta G°$ at each temperature, using $\Delta G° = -2.303\ RT \log K_a$:

298 K : $\Delta G° = -(2.303)\ (0.008314\ kJ\ K^{-1})\ (298\ K)(\log 1.754 \times 10^{-5}) = 27.1551\ kJ$

323 K : $\Delta G° = -(2.303)\ (0.008314\ kJ\ K^{-1})\ (323\ K)(\log 1.633 \times 10^{-5}) = 29.6237\ kJ$

Next solve two equations in two unknowns, assuming that $\Delta H°$ and $\Delta S°$ are constant over the range of 298 K to 323 K. Use 0.298 K and 0.323 K to convert $\Delta S°$ in J to $T\Delta S°$ in KJ. Applying $\Delta G = \Delta H° - T\Delta S°$, we get,

$27.1551\ kJ = \Delta H° - [0.298\ K\ \Delta S°]$ ..... (i)

$29.6237\ kJ = \Delta H° - [0.323\ K\ \Delta S°]$

Solve for $\Delta S°$: (Substrat (i) from (ii) ..... (ii)

$$\Delta S° = \frac{(29.6237 - 27.1551)\ kJ}{(0.2982 - 0.323)\ K)} = -98.74 = -98.7\ KJ\ K^{-1}$$

Substitute the value into equation (ii) and solve for $\Delta H°$:

$$\Delta H° = [(0.0323\ K \times (-98.74\ KJ\ K^{-1})] + 29.623\ kJ = -2.298 = -2.29\ kJ$$

69. Write the equations. Remember that changes which require the input of energy (such as melting and vaporization) will have a positive (+) ΔH, while those that release energy (such as freezing and condensing) will have a negative (–) ΔH.

(a) $Hg(s) \rightarrow Hg(l)$ $\quad \Delta H = +2.33\ kJ\ mol^{-1}$

(b) $Br_2(l) \rightarrow Br_2(g)$ $\quad \Delta H = +29.6\ kJ\ mol^{-1}$

(c) $C_6H_6(l) \rightarrow C_6H_6(s)$ $\quad \Delta H = -9.84\ kJ\ mol^{-1}$

(d) $Hg(g) \rightarrow Hg(l)$ $\quad \Delta H = -59.4\ kJ\ mol^{-1}$

(e) This phase change is not in the table and must be calculated with Hess's Law.

$C_{10}H_8(s) \rightarrow C_{10}H_8(l)$ $\quad \Delta H = +43.3\ kJ\ mol^{-1}$

$C_{10}H_8(l) \rightarrow C_{10}H_8(g)$ $\quad \Delta H = +19.3\ kJ\ mol^{-1}$

$C_{10}H_8(s) \rightarrow C_{10}H_8(g)$ $\quad \Delta H = +62.6\ kJ\ mol^{-1}$

70. (a) Mass of soda = 6 cans × 340 g × = 2.04×10^3 g

$$q = -mc\Delta t$$

$$q_{soda} = -(2.04 \times 10^{-3}\ g)(4.10\ J/g \cdot K)\ (278\ K - 298\ K) = 1.68 \times 10^5\ J$$

$$q_{Al} = -(6\ cans \times 12.5\ g/can)(0.902\ J/g \cdot °C)\ (278\ K - 298\ K) = 1.35 \times 10^3\ J$$

$$q_{total} = q_{soda} + q_{Al} = 1.68 \times 10^5\ J + 1.35 \times 10^3\ J = 1.69 \times 10^5\ J$$

(b) $\Delta H_{fus} = 6\ kJ\ mol^{-1} = 6 \times 10^3\ J\ mol^{-1}$

$$1.69 \times 10^5\ J \times \frac{1\ mol\ ice}{6 \times 10^3\ J} \times \frac{18.02\ g\ ice}{1\ mol.\ ice} = 508\ g\ ice$$

71. Some reasonable assumptions to make are

density of tea = density of H_2O = 1.0 g mL^{-1}

density of ice = 1.0 g mL^{-1} (although, since ice floats, the density is actually less than 1)

room temperature = 298 K

volume of the glass = 480 mL

Then, let x = mass of ice and $(480 - x)$ = mass of tea.

$$q_{ice} = \text{heat to melt ice + heat to warm ice to 298 K}$$

$$q_{melt} = x \text{ g } H_2O \times \frac{1 \text{ mol } H_2O}{18.02 \text{ g } H_2O} \times \frac{6.00 \text{ kJ}}{1 \text{ mol } H_2O} \times \frac{1000 \text{ J}}{1 \text{ kJ}} = 333x \text{ J}$$

$$q_{warning} = mc\Delta t$$

(note that the sign is positive since the water is warming, and thus gaining heat)

$$q_{warning} = (x \text{ J})(4.18 \text{ J/g K} \cdot (298 \text{ K} - 273 \text{ K})$$

$$q_{warning} = 105x \text{ J}$$

$$q_{ice} = 333x \text{ J} + 105x \text{ J} = 438x \text{ J}$$

$$q_{tea} = -(480 \text{ g} - x \text{ g})(4.18 \text{ J/g K}) \cdot (273 \text{ K} - 298 \text{ K})$$

$$q_{tea} = 50200 - 105x$$

$$q_{ice} = q_{tea}$$

$$438x = 50200 - 105\,x$$

$$543x = 50200$$

$$x = 92.4 \text{ g (of ice)}$$

$$(480 - x = 480 - 92.4 = 388 \text{ g tea})$$

Since the densities of the tea and water (from the ice) equal 1.00 g mL^{-1}, density = mass and

$$\text{fraction to be left empty} = \frac{92.4 \text{ mL}}{480 \text{ mL}} \times 100\% = 19.3\%$$

72. The $\Delta H°$ for the reactions to produce CO and CO_2 in each case is equal to the $\Delta H_f°$ of the respective gases. Thus $\Delta H°$ to produce CO is –110.5 KJ/mol and $\Delta H°$ to produce CO_2 is –393.5 KJ/mol. Since one mol of graphite is needed to produce one mol of either CO or CO_2, one equation is one unknown can be written of the heat produced, using m for the moles of CO and (2.00 – m) for moles of CO_2.

$$-481 \text{ KJ} = \text{m} \times (-110.5 \text{ KJ}) + (23.00 - \text{m}) \times (-393.5 \text{ KJ})$$

$$-481 = -110.5 \text{ m} - 787 + 393.5 \text{ m}$$

mol CO = m= 1.081; mass CO = 1.081 × 28 g/mol = 30.26 = 30.6 g

mol CO_2 = (2.0 – m) = 0.919; mass CO_2 = 0.919 mol × 44.0 g mol^{-1} = 40.4 = 40. g

73. The equation is

$$4NH_3(g) + 5O_2(g) \rightarrow 4NO(g) + 6H_2O(g);\ \Delta H° = -906 \text{ kJ}$$

First determine the limiting reactant by calculating the moles of NH_3 and of O_2; then, assuming one of the reactants is totally consumed, calculate the moles of the other reactant needed for the reaction.

$$10 \text{ g } NH_3 \times \frac{1 \text{ mol } NH_3}{17.03 \text{ g } NH_3} = 0.5872\ NH_3$$

$$20 \text{ g } O_2 \times \frac{1 \text{ mol } O_2}{32.0 \text{ g } O_2} = 0.265 \text{ mol } O_2$$

$$0.625 \text{ mol } O_2 \times \frac{4 \text{ mol } NH_3}{5 \text{ mol } O_2} = 0.5 \text{ mol } NH_3 \text{ needed if } O_2 \text{ is totally consumed}$$

Because NH_3 is present in excess of what is needed, O_2 must be the limiting reactant. Now calculate the heat released on the basis of the complete reaction of 0.62259 and O_2.

$$\Delta H = \frac{-906 \text{ kJ}}{5 \text{ mol } O_2} \times 0.625 \text{ mol } O_2 = -113.25 = -113 \text{ KJ}$$

The heat released by the complete reaction of the 20 g (0.2650) mol of $O_2(g)$ is 113 KJ.

74. The equation is

$$N_2(g) + 3H_2(g) \rightarrow 2NH_3(g);\ \Delta H° = -91.8 \text{ kJ}$$

(a) To find the heat evolved from the production of 1.0 L of NH_3, convert the 1.0 L to mol NH_3, using the density and molar mass (17.03 g mol^{-1}). Then convert the moles to heat (ΔH) using $\Delta H°$.

$$\text{moles } NH_3 = \frac{d \times V}{\text{Mwt}} = \frac{0.696 \text{ g L}^{-1} \times 1\text{L}}{17.03 \text{ g mol}^{-1}} = 0.04087 \text{ mol } NH_3$$

$$\Delta H = \frac{-91.8 \text{ kJ}}{2 \text{ mol } NH_3} \times 0.04087 \text{ mol } NH_3 = -1.876 \text{ (1.88 kJ heat evolved)}$$

(b) First find the moles of N_2 using the density and molar mass (28.02 g mol^{-1}.) Then convert to the heat needed to raise the N_2 from 298 K to 673 K:

$$\text{moles } N_2 = \frac{d \times V}{\text{Mwt}} = \frac{1.145 \text{ g/L} \times 0.5 \text{ L}}{28.02 \text{ g mol}^{-1}} = 0.02043 \text{ mol } N_2$$

$$0.02043 \text{ mol } N_2 \times \frac{2912.1}{(\text{mol K})} \times (673\text{–}298) \text{ K} = 223.2 \text{ J } (0.2232 \text{ kJ})$$

$$\% \text{ heat for } N_2 = \frac{0.2232 \text{ kJ}}{1.876 \text{ kJ}} \times 100\% = 11.89 = 11.9\%$$

75. The glucose equation is

$$C_6H_{12}O_6 + 6O_2 \rightarrow 6CO_2 + 6H_2O;\ \Delta H° = -2.802.8 \text{ kJ}$$

Convert the 10.45×10^3 KJ to mol of glucose, using the $\Delta H°$ of –2802.8 KJ for the reaction into moles of glucose.

$$10.45 \times 10^3 \text{ kcal} \times \frac{1 \text{ mol glucose}}{2802.8 \text{ kJ}} = 3.73 \text{ mol glucose}$$

Next convert mol glucose to mol LiOH using the above equation for glucose and the equation for LiOH: $2LiOH(s) + CO_2(g) \rightarrow Li_2CO_3(s) \rightarrow H_2O(l)$.

$$3.73 \text{ mol glucose} \times \frac{6 \text{ mol } CO_2}{1 \text{ mol glucose}} \times \frac{2 \text{ mol LiOH}}{1 \text{ mol } CO_2} = 44.64 \text{ mol LiOH}$$

Finally, use the molar mass of LiOH to convert moles to mass:

$$44.64 \text{ mol LiOH} \times \frac{23.95 \text{ g LiOH}}{1 \text{ mol LiOH}} = 1.07 \times 10^3 \text{ g (1.07 kg) LiOH}$$

76. The reaction is exothermic. This indicates that ΔH is negative.

$\Delta n_g < 0$. This indicates that ΔS is negative.

The system is at equilibrium, and K = 1 at 573 K. This indicates that $\Delta G° = 0$ at $T = 300$ K.

The three above points taken together indicate that $\Delta G°$ is negative below 300 K and positive above 300 K.

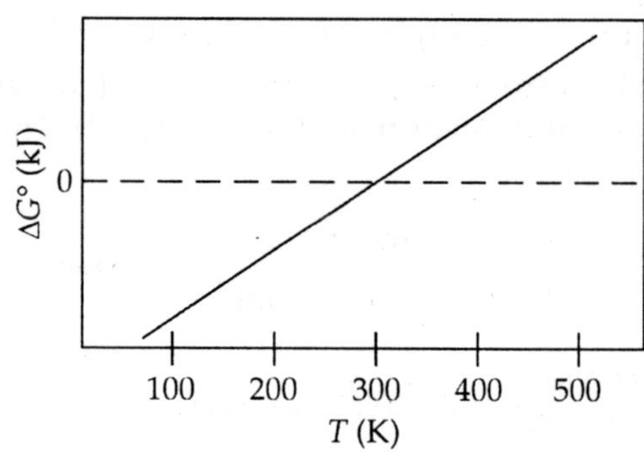

77. For the equation $C_2H_5OH(l) \rightleftharpoons C_2H_5OH(g)$, $\Delta G° = 0$ (since this is a phase change)

$$\Delta G° = \Delta H° - T\Delta S° \Rightarrow \Delta H° = T\Delta S°$$

Calculate ΔS, ΔH_{vap} and the use of the Claysius-Clapeyron equation is made.

$\Delta S° = \Sigma S°_{products} - \Sigma S°_{reactants}$

$\Delta S° = [(1 \text{ mol.})(0.2872 \text{ kJ mol}^{-1} \cdot \text{K})] - [(1 \text{ mol.})(0.1607 \text{ kJ mol}^{-1} \cdot \text{K})] = 0.122 \text{ kJ K}^{-1}$

$\Delta H° = T\,\Delta S° = (351.6 \text{ K})(0.122 \text{ KJ K}^{-1})$

$\Delta H° = \Delta H_{vap} = 42.89 \text{ kJ}$

$$\ln\left(\frac{P_2}{P_1}\right) = \frac{+\Delta H_{vap}}{R}\left(\frac{1}{T_1} - \frac{1}{T_2}\right)$$

$$\ln\left(\frac{760 \text{ mm Hg}}{357 \text{ mm Hg}}\right) = \frac{42890 \text{ J/mol}}{8.31 \text{ J/mol} \cdot \text{K}}\left(\frac{1}{T_1} - \frac{1}{351.6 \text{ K}}\right)$$

$$0.756 = (5161 \text{ K})\left(\frac{1}{T_1} - 0.00285\right) \Rightarrow \frac{1}{T_1} = 0.00300 \text{ K}^{-1}$$

or, $T_1 = 333$ K

78. Let us

Calculate $\Delta S°$ and $\Delta H°$, then $\Delta G°$. Use $\Delta G°$ to calculate K. Finally, calculate the final pressures of the gases using the equilibrium expression.

$\Delta S° = \Sigma S°_{products} - \Sigma S°_{reactants}$

$= [(1 \text{ mol.})(0.1306 \text{ kJ/mol.} \cdot \text{K})] + [(1 \text{ mol.})(0.2607 \text{ kJ/mol.} \cdot \text{K})]$

$- [(2 \text{ mol.})(0.2065 \text{ kJ/mol.} \cdot \text{K})]$

or, $\Delta S° = -0.0217 \text{ kJ K}^{-1}$

$$\Delta H^\circ = \Sigma \Delta H^\circ_{f\,\text{products}} - \Sigma \Delta H^\circ_{f\,\text{reactants}}$$

$$\Delta H^\circ = [(1 \text{ mol.})(0.0 \text{ kJ/mol}^{-1}.) + [(1 \text{ mol.})(62.4 \text{ kJ mol}^{-1}.)] - [(2 \text{ mol.})(26.5 \text{ kJ mol}^{-1}.)]$$

$$= 9.4 \text{ kJ}$$

$$\Delta G^\circ = \Delta H^\circ - T\,\Delta S^\circ = 9.4 \text{ kJ} - (773 \text{ K})(-0.0217 \text{ kJ K}^{-1}) = 26.2 \text{ kJ}$$

$$\Delta G^\circ = -RT \ln K$$

$$26.2 \text{ kJ} = -(8.31 \text{ J mol}^{-1} \cdot \text{K}^{-1})(0.001 \text{ kJ K}^{-1})(773 \text{ K}) \ln K$$

or $\ln K = -4.08$

$\therefore$ $K = 0.0169$

Since ΔG° is positive, the reaction will proceed in the reverse direction, giving

$$[H_2] = [I_2] = 0.2 - x \text{ and } [HI] = 0.2 + 2x$$

$$K = \frac{(P_{H_2})(P_{I_2})}{(P_{HI})^2} \Rightarrow 0.0169 = \frac{(0.2 - x)^2}{(0.2 + 2x)^2}$$

$$\sqrt{0.0169} = \frac{0.2 - x}{0.2 + 2x} \Rightarrow 0.026 + 0.26\,x = 0.2 - x$$

$\therefore$ $x = 0.138$ atm

$$[H_2] = [I_2] = 0.2 - x = 0.062 \text{ atm}$$

$$[HI] = 0.2 + 2x = 0.476 \text{ atm}$$

79. (a) Assume the phase change is at equilibrium

$$\Delta H^\circ = \Delta H_{\text{fusion}} = \frac{333 \text{ J}}{1 \text{ g}} \times \frac{18.02 \text{ g}}{1 \text{ mol.}} = 6 \times 10^3 \text{ J mol}^{-1}. = 600 \text{ kJ mol}^{-1}.$$

(b) $\Delta G^\circ = -RT \ln K$. Since neither solids not liquids appear in the expression for K (both are one), we get: $\Delta G^\circ = -RT \ln 1 = 0$.

(c) $\Delta G^\circ = \Delta H^\circ - T\,\Delta S^\circ$

$$0 = 6 \text{ kJ} - (273 \text{ K})(\Delta S^\circ)$$

$$\Delta S^\circ = 0.022 \text{ kJ/K}$$

(d) $\Delta G^\circ = \Delta H^\circ - T\,\Delta S^\circ$

$$\Delta G^\circ = 6 \text{ kJ} - (273 \text{ K})(0.022 \text{ kJ/K})$$

$\Delta G^\circ = 0.43$ kJ Thus, ice does not melt spontaneously at 253 K

(e) $\Delta G^\circ = \Delta H^\circ - T\,\Delta S^\circ$

$$\Delta G^\circ = 6 \text{ kJ} - (293 \text{ K})(0.022 \text{ kJ/K})$$

$\Delta G^\circ = -0.45$ kJ But ice does melt spontaneously at +293 K

80. (a) Calculate ΔS° and 298 K using the Gibbs-Helmoltz equation, then calculate ΔG° at 310 K and calculate the amount of energy from one gram sugar with 25% efficiency.

$$\Delta G^\circ \doteq \Delta H^\circ - T\,\Delta S^\circ$$

$$-5790 \text{ kJ} = -5650 \text{ kJ} - (298 \text{ K})(\Delta S^\circ)$$

$$\Delta S^\circ = 0.47 \text{ kJ/K}$$

$$\Delta G^\circ = \Delta H^\circ - T\,\Delta S^\circ$$

$$\Delta G^\circ = -5650 \text{ kJ} - (310 \text{ K})(0.47 \text{ kJ K}^{-1} = -5796 \text{ kJ}$$

$$\therefore \quad \frac{-5796 \text{ kJ}}{\text{mol}} \times \frac{1 \text{ mol}}{342.0 \text{ g}} \times 1.0 \text{ g} \times 0.25 = -4.2 \text{ kJ}$$

Thus one gram of sugar provides 4.2 kJ of work.

(b) m = 54.4 kg

$$w = (9.79 \times 10^{-3})(54.4 \text{ kg})(4158 \text{ m}) = 2210 \text{ kJ}$$

$$2210 \text{ kJ} \times \frac{1.0 \text{ sugar}}{4.2 \text{ kJ}} = 530 \text{ g sugar}$$

81. We have

$$q_{reaction} = -C_{cal}/\Delta T$$
$$= -(22.51 \text{ kJ/K})(1.76) = -37.59 \text{ kJ}$$

$$-37.59 \text{ kJ} \times \frac{1 \text{ mol sucrose}}{-5.64 \times 10^3 \text{ kJ}} \times \frac{342.30 \text{ g sucrose}}{1 \text{ mol. sucrose}} = 2.28 \text{ g sucrose}$$

$$\text{mass \% sucrose} = \frac{2.28 \text{ g}}{3.0 \text{ g}} \times 100\% = 76\%$$

82. Here,

$$\Delta G° = \Delta H° - T\Delta S° \text{ and,}$$
$$\Delta G° = -RT \ln K, \text{ but } \ln(1.0) = 0, \text{ thus,}$$
$$\Delta G° = 0 = \Delta H° - T\Delta S°$$

Calculate $\Delta H°$ and $\Delta S°$, and then solve for T.

$$\Delta H_{(reaction)} = \Sigma\Delta H^\circ_{f\,(products)} - \Sigma\Delta H^\circ_{f\,(reactants)}$$
$$\Delta H_{(reaction)} = [(1 \text{ mol.})(0 \text{ kJ mol}^{-1}) + (1 \text{ mol.})(0 \text{ kJ mol}^{-1}) - [(1 \text{ mol.})(-520 \text{ kJ mol}^{-1})]$$
$$\Delta H_{(reaction)} = +520 \text{ kJ}$$
$$\Delta S_{(reaction)} = \Sigma S^\circ_{f\,(products)} - \Sigma S^\circ_{f\,(reactants)}$$
$$\Delta S_{(reaction)} = [(1 \text{ mol.})(0.032 \text{ kJ mol}^{-1} \cdot \text{K}^{-1}) + (1 \text{ mol.})(0.2050 \text{ kJ mol}^{-1} \cdot \text{K}^{-1})] - [(1 \text{ mol.})(0.053 \text{ kJ mol}^{-1} \cdot \text{K}^{-1})]$$
$$\Delta S_{(reaction)} = 0.184 \text{ kJ K}^{-1}$$
$$\Delta G° = 0 = 520 \text{ kJ} - (T)(0.184 \text{ kJ K}^{-1})$$

or $\quad T = 2.83 \times 10^3$ K

83. The heat lost, q, by the water(s) with a temperature higher than the final temperature must be equal to the heat gained, q, by the water(s) with a temperature lower than the final temperature.

q(lost by water at higher temp) = q(gained by water at lower temp)

$$\Sigma(s \times m \times \Delta t) = \Sigma(s \times m \times \Delta t)$$

Divide both sides of the equation by the specific heat, s, to eliminate this term, and substitute the other values. Since the mass of the water at 313 K is greater than the sum of the masses of the other two waters, assume that the final temperature will be greater than 310 K and greater than 288 K. Use this to set up the three Δt expressions, and write one equation in one unknown, letting t equal the final temperature. Simplifying by omitting the "grams" from 45 g, 25 g, and 15g.

$$\Sigma(m \times \Delta t) = \Sigma(m \times \Delta t)$$

$$45 \times (323\text{ K} - t) = 25 \times (t - 288\text{ K}) + 15 \times (t - 310\text{ K})$$

$$14535\text{ K} - 45\,t = 25\,t - 7200\text{ K} + 15\,t - 4650\text{ K} = 40\,t - 11850\text{ K}$$

or $$-85\,t = -11850 - 14535\text{ K} = -26385\text{ K} \qquad \therefore t = \frac{26385}{85}$$

$$t = 310.4\text{ K (the final temperature)}$$

84. First calculate the mole fraction of each gas in the product, assuming 100 g product
Mol CO = 33 g CO × 1 mol CO/28.01 g CO = 1.178 mol CO
Mol CO_2 = 67 g CO_2 × 1 mol CO_2/44.01 g CO_2 = 1.522 mol CO_2

$$\text{Mol frac. CO} = \frac{1.178\text{ mol CO}}{(1.178 + 1.522)\text{ mol}} = 0.4364$$

$$\text{Mol frac. CO}_2 = \frac{1.522\text{ mol CO}_2}{(1.178 + 1.522)\text{ mol}} = 0.5637$$

Now calculate the starting moles of C, which equals the total moles of CO and

Starting mol C(s) = 1.0 C × 1 mol C/12.01 g C
= 0.08236 mol C = mol CO + CO_2

Use the mol fractions to convert mol CO + CO_2 to mol CO and mol CO_2

$$\text{Mol CO} = \frac{0.4363\text{ mol CO}}{1\text{ mol total}} \times 0.8326\text{ mol total} = 0.03633\text{ mol}$$

$$\text{Mol CO}_2 = \frac{0.5637\text{ mol CO}_2}{1\text{ mol total}} \times 0.08326\text{ mol total} = 0.04693\text{ mol CO}_2$$

Now use enthalpies of formation to calculate the heat combustion for both:

C(s)	+ 1/2O_2(g)	→ CO(g)
0.03633	excess	0.03633 (mol)
0	0	–110.5 (kJ mol^{-1})
0	0	–4.014 (kJ/0.3633 mol)

C(s)	+ O_2(g)	→ CO_2(g)
0.04693	excess	0.04693 (mol)
0	0	–393.5 (kJ/mol)
0	0	–18.47 (kJ/0.4693 mol)

Total ΔH = –4.014 + (–18.47) = –22.48 .2 kJ
Heat released = 22 kJ

85. This problem is similar to the one above. Calculate the moles of trichloroethane and the pressure that would result if all the material vaporized.
Trichloroethane M.wt = 133.39 g

$$236.5\,\text{mL}\left(\frac{1.325\,\text{g}}{1\,\text{mL}}\times\frac{1\,\text{mol}}{133.39\,\text{g}}\right) = 2.35\ \text{mol. C2H3Cl}_3$$

Volume = 510 L; Temperature = 280 K

$$P = \frac{\pi RT}{V} = \frac{(2.35\,\text{mol})\,(0.082)\,\text{L}\cdot\text{atm/mol}^{-1}\text{K}^{-1})\,(280\,\text{K})}{510\,\text{L}} = 0.11\ \text{atm}$$

Now we must calculate the vapor pressure of trichloroehane at 277 K and compare the vapor pressure to the pressure calculated above. Calculate ΔH_{vap} and from that, the vapour pressure at 277 K.

$$\ln\left(\frac{P_2}{P_1}\right) = \frac{+\Delta H_{vap}}{R}\left(\frac{1}{T_1}-\frac{1}{T_2}\right)$$

$$\ln\left(\frac{760}{100}\right) = \frac{+\Delta H_{vap}}{8.31\,\text{J mol}^{-1}\ \text{K}^{-1}}\left(\frac{1}{293\,\text{K}}-\frac{1}{347.3\,\text{K}}\right)$$

$$2.03 = \frac{+\Delta H_{vap}}{8.31\,\text{J/mol}\cdot\text{K}}\ (5.34\times10^{-4}\ \text{K}^{-1})$$

ΔH_{vap} = 3.17 × 104 J/mol. (Now calculate vapour pressure at 277 K

$$\ln\left(\frac{760}{P_1}\right) = \frac{3.17\times10^4\ \text{J mol}^{-1}}{8.31\,\text{J mol}^{-1}\text{K}^{-1}}\left(\frac{1}{277\,\text{K}}\ \frac{1}{347.3\,\text{K}}\right) = 2.79$$

$$\frac{760\ \text{mm Hg}}{P_1} = e^{2.79} = 16.3$$

$$P_1 = 46.6\ \text{mm Hg}\times\frac{1\,\text{atm}}{760\ \text{mm Hg}} = 0.0613\ \text{atm}$$

Since the pressure that would result from complete vaporization is greater than the vapour pressure of triochloroethane, not all the $C_2H_3Cl_3$ would evaporate. To calculate the percentage that remains as liquid, we need to determine the amount that would evaporate to produce a pressure of 0.0613 atm. The rest would remain as liquid.

$$n = \frac{PV}{RT} = \frac{(0.0613\,\text{atm})\,(510\,\text{L})}{(0.0821\,\text{L}\cdot\text{atm mol}^{-1}\text{K}^{-1})\,(277\,\text{K})}\ 1.27\ \text{mol. (would vaporize)}$$

2.35 mol. –1.37 mol. = 0.98 mol. (would remain)

$$\text{Percent remaining} = \frac{0.98}{2.35}\times100\% = 42\%$$

86. Write the net ionic equation for the reaction, then calculate $\Delta H°$ for the reaction ($\Delta H° = \Sigma\Delta H^\circ_{f\,(\text{products})} - \Sigma\Delta H^\circ_{f\,(\text{reactants})}$). Calculate ΔH for the 0.10 mol; given in the problem.
(a) $Na^+(aq) + OH^-(aq) + H^+(aq) + Cl^-(aq) \rightleftharpoons Na^+(a) + Cl^-(aq) + H_2O(l)$

$OH^-(aq) + H^+(aq) \rightleftharpoons H_2O(l)$

$\Delta H° = [(1 \text{ mol.})(-285.8 \text{ kJ mol}^{-1})] - [(1 \text{ mol.})(-230 \text{ kJ mol}^{-1}) + (1 \text{ mol.})(0.0 \text{ kJ mol}^{-1})]$

$\Delta H° = -55.8 \text{ J}$ (This is for 1 mole, calculate for 0.1 moles.)

$\Delta H°_{(0.10 \text{ mol})} = -55.8 \text{ kJ mol}^{-1} \times 0.10 \text{ mol.} = 5.58 \text{ kJ}$

(b) $HF(aq) + Na^+(aq) + OH^-(aq) \rightleftharpoons Na^+(aq) + F^-(aq) + H_2O(l)$

$HF(aq) + OH^-(aq) \rightleftharpoons F^-(aq) + H_2O(l)$

$\Delta H° = [(1 \text{ mol.})(-332.6 \text{ kJ mol}^{-1}) - (1 \text{ mol.})(285.5 \text{ kJ mol}^{-1})] - [(1 \text{ mol.})(-320.1 \text{ kJ mol}^{-1}) - (1 \text{ mol.})(230 \text{ kJ mol}^{-1})]$

$\Delta H° = -68.3 \text{ kJ}$ (This is for 1 mole, calculate for 0.10 moles.)

$\Delta H°_{(0.10 \text{ mol})} = -68.3 \text{ kJ mol}^{-1} \times 0.1 \text{ mol.} = -6.83 \text{ kJ}$

87. $CaH_2(s) \rightarrow Ca(s) + H_2(g)$

$\Sigma H° = —\Delta H°_f + 186.21 \text{ kJ}$

$\Delta S° = \Sigma S°_{\text{products}} - \Sigma S°_{\text{reactants}}$

$\Delta S° = [(1 \text{ mol.})(0.0414 \text{ kJ mol}^{-1} \cdot K^{-1}) + (1 \text{ mol.})(0.1306 \text{ kJ mol}^{-1} \cdot K^{-1})] - [(1 \text{ mol.})(0.0420 \text{ kJ mol}^{-1} \cdot K^{-1})]$

$\Delta S° = 0.13 \text{ kJ K}^{-1}$

$\Delta G° = -RT \ln K = -RT \ln (1 \text{ atm } H_2) = 0 \text{ kJ}$

$\Delta G° = \Delta H° - T \Delta S°$

$0 \text{ kJ} = 186.2 \text{ kJ} - (T)(0.130 \text{ kJ K}^{-1})$

$T = 1432 \text{ K}$

This is not a feasible source of H_2 fuel.

88. $Cu(s) + \frac{1}{2} O_2(g) \rightarrow CuO(s)$

$\Delta G° = \Sigma \Delta G°_{f \text{ products}} - \Sigma \Delta G°_{f \text{ reactants}}$

$\Delta G° = [(1 \text{ mol.})(-129.7 \text{ kJ mol}^{-1})] - [(1 \text{ mol.})(0.0\text{- kJ mol}^{-1}) + (\frac{1}{2} \text{ mol.})(0.0 \text{ kJ mol}^{-1})]$

$\Delta G° = -129.7 \text{ kJ}$

$\Delta G°$ is negative, so the black CuO forms spontaneously at room temperature

$2CuO(s) \rightarrow Cu_2(O) + O_2(g)$

$\Delta H° = \Sigma \Delta H°_{f \text{ products}} - \Sigma \Delta H°_{f \text{ reactants}}$

$\Delta H° = [(1 \text{ mol.})(-168.8 \text{ kJ mol}^{-1}) + (\frac{1}{2} \text{ mol.})(0.0 \text{ kJ mol}^{-1})] - [(2 \text{ mol.})(-157.3 \text{ kJ mol}^{-1})]$

$\Delta H° = 146 \text{ kJ}$

$\Delta S° = \Sigma S°_{\text{products}} - \Sigma S°_{\text{reactants}}$

$\Delta S° = [(1 \text{ mol.})(0.0931 \text{ kJ mol}^{-1} \cdot K^{-1}) + (\frac{1}{2} \text{ mol.})(0.205 \text{ kJ mol}^{-1} \cdot K^{-1})] - [(2 \text{ mol.})(0.0426 \text{ kJ mol}^{-1})]$

$\Delta S° = 0.1104 \text{ kJ K}^{-1}$

The reaction becomes spontaneous when $\Delta G° < 0$, giving

$$\Delta G° = \Delta H° - T\Delta S°$$
$$0 = 140.6 \text{ kJ} - (T)(0.1104 \text{ kJ K}^{-1})$$
$$T = 1322 \text{ K} \qquad \text{(Reaction becomes spontaneous above 1322 K)}$$

Thus the black CuO converts to red Cu_2O when the temperature exceeds 1322 K.

$$Cu_2O(s) \rightarrow 2Cu(s) + \frac{1}{2}O_2(g)$$
$$\Delta H° = \Delta H^\circ_{f\,\text{products}} - \Sigma\Delta H^\circ_{f\,\text{reactants}}$$
$$\Delta H° = [(2 \text{ mol.})(0.0 \text{ kJ mol}^{-1}) + (\frac{1}{2} \text{ mol.})(0.0 \text{ kJ mol}^{-1})] - [(1 \text{ mol.})(-168.6 \text{ kJ mol}^{-1})]$$
$$\Delta H° = 168.6 \text{ kJ}$$
$$\Delta S° = \Sigma S^\circ_{\text{products}} - \Sigma S^\circ_{\text{reactants}}$$
$$\Delta S° = [(2 \text{ mol.})(0.0322 \text{ kJ mol}^{-1}) + (\frac{1}{2} \text{ mol.})(0.2050 \text{ kJ mol}^{-1})] - [(1 \text{ mol.})(-0.0931 \text{ kJ mol}^{-1})]$$
$$\Delta S° = 0.0758 \text{ kJ}$$
$$\Delta G° = \Delta H° - \Delta S°$$

The reaction becomes spontaneous when $\Delta G° < 0$, giving

$$0 = 168.6 \text{ kJ} - (T)(0.0758 \text{ kJ K}^{-1})$$
$$T = 2220 \text{ K} \qquad \text{(Reaction becomes spontaneous above 2220 K)}$$

Thus the red Cu_2O convert to metallic Cu when the temperature exceeds 2220 K.

89.
$$Na^+(g) + Cl^-(g) \rightarrow NaCl(s) \qquad \Delta H = -787 \text{ kJ mol}^{-1}$$
$$NaCl(s) \rightarrow Na^+(aq) + Cl^-(aq) \qquad \Delta H = +4 \text{ kJ mol}^{-1}$$
$$Na^+(g) + Cl^-(g) \rightarrow Na^+(aq) + Cl^-(aq) \qquad \Delta H = -783 \text{ kJ mol}^{-1}$$

The heat of hydration of Na^+ is

$$Na^+(g) + Cl^-(g) \rightarrow Na^+(aq) + Cl^-(aq) \qquad \Delta H = -783 \text{ kJ mol}^{-1}$$
$$Cl^-(aq) \rightarrow Cl^-(g) \qquad \Delta H = +338 \text{ KJ mol}^{-1}$$
$$Na^+(g) \rightarrow Na^+(g) \qquad \Delta H = -445 \text{ KJ mol}^{-1}$$

90. Neither of these liquids is hydrogen bonded, so that we expect Trouton's Rule to apply to them. We must calculate ΔH_{vap} as $T_{bp}\Delta S_{vap}$. Assuming ΔS_{vap} is 88 J · mol^{-1} K^{-1}, we obtain for acetone

$$T_{bp} = 329.2 \quad \text{and} \quad \Delta H_{vap} \sim (329.2 \text{ K})(88 \text{ J} \cdot \text{mol}^{-1} \text{ K}^{-1}) = 29 \text{ kJ} \cdot \text{mol}^{-1}$$

Similarly, for nitrobenzene,

$$T_{bp} = 483.8 \quad \text{and} \quad \Delta H_{vap} \sim (483.8 \text{ K})(88 \text{ J} \cdot \text{mol}^{-1} \text{ K}^{-1}) = 43 \text{ kJ} \cdot \text{mol}^{-1}$$

91. (a) Before you begin the numerical calculation it is helpful to ask yourself "Will this $\Delta S°$ value per positive be negative?" The left-hand side of this equation contains a solid plus a gas, whereas there is only a solid on the right. There is considerably less molecular disorder on the right-hand side, and therefore the entropy decreases as this reaction proceeds, so that $\Delta S°$ will be negative. Using $\Delta S°$ values from Tables we have,

$$\Delta S^\circ = S^\circ_{ZnO} - S^\circ_{Zn} - \frac{1}{2}S^\circ_{O_2}$$

$$= 43.64 - 41.64 - (0.5)(205.03) = -100.5 \text{ J} \cdot \text{K}^{-1}$$

(b) The dimerization of NO_2 is also a reaction for which ΔS° is negative. There are 2 mol of gas on the left-hand side of the equation, and only 1 mol of gas on the right. In general, gases have so much more disorder than liquids and solids, that the side of an equation with the large number of moles of gas is the side with greater entropy. Using S° values from Tables.

$$\Delta S^\circ = S^\circ_{N_2O_4} \quad 2S^\circ_{NO_2} = 304.18 - 2(239.95) = -175.72 \text{ J} \cdot \text{K}^{-1}$$

92. (a) First we calculate ΔH°_{298} from Tables:

$$\Delta H^\circ = \Delta H^\circ_f(N_2O_4, g) - 2\Delta H^\circ_{298}(NO_2, g) = 9.16 - 2(33.18) = -57.20 \text{ kJ}$$

$$\Delta S^\circ = -175.72 \text{ J} \cdot \text{K}^{-1} \text{ (as calculated in problem 91).}$$

It is particularly important to note that ΔH° is given in kilojoules while ΔS° is in joules per kelvin both terms must be in the *same* units. The term $T\Delta S^\circ$ must be converted top kilojoules before it can be combined with ΔH°. Probably the most common student error is to have different units (kilojoules and joules) for the two terms. As it is customary to report ΔG° in kilojoules, multiply the ΔS° value by 10^{-3} so that $T\Delta S^\circ$ is also in kilojoules. For the dimerziation of N_2O_4 we obtain

$$\Delta G^\circ = -57.30 - (298.15)(-175.72 \times 10^{-3}) = -57.20 + 52.39 = -4.81 \text{ kJ}$$

The negative sign of ΔG° indicates that the dimerzation reaction $2NO_2(g) \rightleftharpoons N_2O_4(g)$ proceeds spontaneously to the right at 298 K when both gases are at 1-atm pressure.

(b) First calculate ΔH°_{298} using Hess' Law.

$$\Delta H^\circ_{298} = \Delta H^\circ_f(CaO) + \Delta H^\circ_f(CO_2) - \Delta H^\circ_f(CaO_3). \text{ Using values from Tables.}$$

$$= -635.1 - 393.51 - (-1206.9) = +178.3 \text{ kJ}$$

Note that is an endothermic reaction; heat must be supplied to decompose calcium carbonate to calcium oxide and carbon dioxide.

Next calculate ΔS° using Tables.

$$\Delta S^\circ = S^\circ_{CaO} + S^\circ_{CO_2} - S^\circ_{CaCO_3} = 39.75 + 213.63 - 92.9 = 160.5 \text{ J} \cdot \text{K}^{-1}$$

The entropy changes is positive for this reaction. There is a large increase in molecular disorder when a solid compound is decomposed into a gas and a simpler solid compound

$$\Delta G^\circ = \Delta H^\circ - T\Delta S^\circ = 178.3 - (298.15)(160.5 \times 10^{-3}) \text{ kJ}$$

$$= 178.3 - 47.85 = 130.4 \text{ kJ}$$

Since $\Delta G^\circ > 0$, this reaction does not proceed spontaneously to the right at 298 K and 1 atm.

93. The formation reaction is

$N_2(g) + 2O_2(g) \rightleftharpoons N_2O_4(g)$

$\Delta H^\circ_f(N_2O_4, g)$ from is tables is 9.16 kJ · mol^{-1}.

$\Delta S(N_2O_4)$ is also obtained using Tables as :

$\Delta S_f^\circ (N_2O_4) = S^\circ_{N_2O_4} - S^\circ_{N_2} - 2S^\circ_{O_2} = 304.18 - 191.50 - 2(205.03) = -297.38\ J \cdot mol^{-1}\ K^{-1}$

Therefore

$$\Delta G_f^\circ (N_2O_4) = 9.16 - (298.15)(-297.38 \times 10^{-3}) = 9.16 + 88.66 = +97.82\ kJ \cdot mol^{-1}$$

94. (a) Here,

$$\Delta G^\circ = \Delta G_f^\circ (N_2O_4) - 2\Delta G_f^\circ (NO_2) = 97.82 - 2(51.29) = -4.76\ kJ \cdot mol^{-1}$$

(b) $\Delta G^\circ = \Delta G_f^\circ (CaO) + \Delta G_f^\circ (CaCO_3)$. Using values from tables

$$= -604.05 - 394.36 - (-1128.84) = 130.43 \cdot mol^{-1}$$

95. (a) For the dimerziation of NO_2, ΔG° has been calculated to be $-4.76\ kJ \cdot mol^{-1}$ g at 298 K in Problem 94(a). Substituting this value into Eq. $\Delta G^\circ = -RT\ In\ LK$ yields

$$-4.76\ kJ \cdot mol^{-1} = -\left(8.3145 \times 10^{-3} \frac{kJ}{mol \cdot K}\right)(298.15\ K)\ \ln K_{eq}$$

so that; $\ln K_{eq} = +1.92$ and $K_{eq} = e^{1.92} = 6.8 = P_{N_2O_4} / P^2_{NO_2}$

The pressures of both gases must be given in atmospheres in the equilibrium constant, because the standard state is defined as the state in which each gas is 1-atm pressure.

(b) For the decomposition of $CaCO_2$, ΔG° was calculated to be $+130.4\ kJ \cdot mol^{-1}$ in problem 92 (b)

$$\therefore \quad +130.4\ kJ \cdot mol^{-1} = \left(8.3145 \times 10^{-3} \frac{kJ}{mol \cdot K}\right)(298.15\ K) \ln K_{eq}$$

$$\ln K_{eq} = -52.60 \quad \text{and} \quad K_{eq} = e^{-52.60} = 1.4 \times 10^{-23} = P_{CO_2}$$

96. The reaction quotient, Q, for these pressures is

$$Q = P_{N_2O_4} / P^2_{NO_2} = 0.50/(0.10)^2 = 50$$

We have already calculated ΔG° for this reaction in Problem 94 (a) above. We can now calculate ΔG at the pressures given,

$$\Delta G = \Delta G^\circ + RT \ln Q = -4.76\ kJ \cdot mol^{-1} + (8.3145 \times 10^{-3}) \times (298.15)(\ln 50)\ kJ \cdot mol^{-1}$$

$$= -4.76 + (2.479)(3.912) = -4.76 + 9.70 = +4.94\ kJ \cdot mol^{-1}$$

As ΔG is positive this reaction proceeds to the left, and some of the N_2O_4 decomposes to yield more NO_2. The reaction will continue until $Q = K_{eq} = 6.8$.

97. According to Hess's Law

$$\Delta G^\circ = \frac{1}{2} \Delta G_f^\circ (Br_2) + \Delta G_f^\circ (NO) - \Delta G_f^\circ (NOBr)$$

$$= \frac{1}{2}(3.142) + 86.57 - 82.42 = +5.72\ kJ \cdot mol^{-1}$$

The reaction quotient, Q, at the instant of mixing is

$$Q = \frac{P^{1/2}_{Br_2} P_{NO}}{P_{NOBr}} = \frac{(0.16)^{1/2}(0.10)}{(0.80)} = 0.050\ atm^{1/2}$$

Thus, $\Delta G = 5.72 + RT \ln(0.050) = 5.72 + (8.3145 \times 10^{-3})\,(298.15)\,(-2.996)$
$= 5.72 - 7.43 = -1.71 \text{ kJ} \cdot \text{mol}^{-1}$

Since $\Delta G < 0$, the reaction will proceed spontaneously to the right when the three gases are mixed at these pressures. Note that $\Delta G°$ is positive, but ΔG is negative. If the three gases are each at 1 atm when mixed, the reaction proceeds spontaneously to the left, but at the pressures stipulated at the instant of mixing, the reaction proceeds to the right.

98. We have already calculated ΔH_f° in Problem 92 (b) above.

$$\Delta G^\circ_{298} = 178.3 - (298.15)\,(0.1605) = 178.3 - 47.85 = 130.4 \text{ kJ}$$

Similarly, $\Delta G^\circ_{600} = 176 - (600.0)\,(0.1563) = 176 - 93.8 = 82 \text{ kJ}$

Note that the entropy factor ($T\Delta S°$) is much larger at the higher temperature.

99. (a) $\Delta G^\circ_{1000} = -R(1000) \ln K_{1000} = -(8.314 \times 10^{-3})\,(1000) \ln 0.513$
$= -(8.314)\,(-0.667) = +5.55 \text{ kJ}$

Note that to obtain ΔG°_{1000} in kilojoules we must use R in kilojoules per mole per kelvin. We obtain an average value of $\Delta H°$ between 900 and 1000 K

$$\ln\left(\frac{K_{1000}}{K_{900}}\right) = \ln\left(\frac{0.513}{0.517}\right) = \frac{\Delta H^\circ}{8.3145 \times 10^{-3}}\left[\frac{9\times10^2 - 1\times10^3}{(9\times10^2)\,(1\times10^3)}\right]$$

$$= \ln(3.27) = +\Delta H°(100)/(8.3145)\,(9 \times 10^2)$$

so that

$$\Delta H° = (8.3145)\,(9)\,(1.184) = 88.6 \text{ kJ}$$

(b) Again, we use the van't Hoff equation,

$$\ln\left(\frac{K_{1000}}{K_{1000}}\right) = \frac{88.6}{8.314 \times 10^{-3}}\left[\frac{1000 - 1200}{(1000)\,(1200)}\right] = \frac{(88.6)\,(200)}{(8.314\times10^{-3})\,(10^3)(1.2\times10^3)}$$

$$= 1.776 = 1.78$$

Therefore

$$K_{1200}/K_{1000} = e^{1.776} = 5.906 = 5.91$$

$$K_{1200} = (5.906)\,(K_{1000}) = (5.906)\,(0.513) = 3.03$$

From K_{1200} we determine ΔG°_{1200} as:

$$\Delta G^\circ_{1200} = -(8.314 \times 10^{-3})\,(1.2 \times 10^3)\,(\ln 3.03) = -11.1 \text{ kJ}$$

(c) $\Delta H^\circ_{298} = \Delta H_f^\circ(SO_2) - \Delta H_f^\circ(SO_3)$
$= -296.8 - (-395.7) = +98.9 \text{ kJ}$

$$\Delta G^\circ_{298} = \Delta G_f^\circ(SO_2) - \Delta G_f^\circ(SO_3) = -300.2 - (-371.1) = +70.9 \text{ kJ}$$

It is useful to tabulate $\Delta H°$ and $\Delta G°$ at two widely separated temperature and compare the percentage change in each.

	298 K	*100 K*	*Change*
$\Delta H°$ (kJ)	98.9	88.6	–9.9%
$\Delta G°$ (kJ)	70.9	5.5	–92.2%

We observe that over a temperature rage of 700°, $\Delta H°$ has changed relatively little, $\Delta G°$ has changed by a large amount.

100. We must determine the final volume of the gas in order to calculate the work of expansion. Since the gas id ideal, and the temperature and number of moles of gas are both constant, we can use Boyle's law to calculate the final volume, V_2.

$$P_1V_1 = P_2V_2 \quad \text{so that} \quad V_2 = P_1V_1/P_2$$

Hence,

$$V_2 = (6.0 \text{ atm})\ (0.4)\ \text{L})/(1.5 \text{ atm}) = 1.6 \text{ L}$$

$$\text{Work of expansion} = -P_{ext}\Delta V = -(1.5)\ (1.6 - 0.4)\ \text{L} \cdot \text{atm} = -(1.5)\ (1.2) = -1.8 \text{ L} \cdot \text{atm}$$

We convert to joules

$$\text{Work of expansion} = -1.8 \text{ L} \cdot \text{atm})\ (101.325 \text{ J} \cdot \text{L}^{-1} \cdot \text{atm}^{-1}) = -182 \text{ J}$$

The work done by the gas is therefore 182 J.

101. When the stopcock is opened, the gas expands to fill both bulbs. The final volume is the combined volumes of both bulbs, or 1.6 L. Because the gas expanded into a vaccum, it did not have to push back any external pressure, and so no work was done. When $P_{ext} = 0$, the work of expansion is 0.

102. (a) The value of Δ_{ngas} is $4 - \frac{13}{2} - 1 = \frac{-7}{2}$ for this reaction which is the combustion of butane.

$$\therefore\ \Delta H - \Delta E = \frac{-7}{2} RT = -(3.5)\ (8.3145)\ (298.15) = -8.68 \times 10^3 \text{ J} = -8.68 \text{ kJ}$$

(b) The values of Δ_{ngas} for this reaction $6 - 6 = 0$. Hence $\Delta H \simeq \Delta E$.

(c) For the vaporization of water (on any liquid) $\Delta_{ngas} = 1$, as mol of gas is produced from 1 mol of liquid. Thus,

$$\Delta H - \Delta E = RT = (8.3145)\ (298.15) = 2.48 \times 10^3 \text{ J} = 2.48 \text{ kJ}$$

103. We have

$$\Sigma H° = 3\Delta H_f^\circ\ (\text{HCl}) + \Delta H_f^\circ\ (\text{CHCl}_3) - \Delta H_f^\circ\ (\text{CH}_4)$$

Note that since $Cl_2(g)$ is an element in its standard state, ΔH_f° (Cl_2) is 0. Using the ΔH_f° values, From Tables:

$$\Delta H° = 3(-92.31) + (-134.5) - (-74.81)$$
$$= -276.93 - 134.5 + 74.81 = -336.6 \text{ kJ}$$

104. To calculate $\Delta H°$ for this reaction at 298 K we use data from tables. Since both aluminium and iron are elements are therefore have zero heats of formation,

$$\therefore \quad \Delta H° = \Delta H_f^\circ\ (\text{Al}_2\text{O}_3) - \Delta H_f^\circ\ (\text{Fe}_2\text{O}_3)$$
$$= -1675.7 - (-824.2) = -851.5 \text{ kJ}$$

When 2 mol of Al react in the thermite process 851.5 kJ of heat are released. The atomic weight of Al is 26.98 g · mol^{-1}. Hence 10.00 g is 0.3076 mol of Al. The amount of heat released when 10.00 g of powered Al react in the thermite process is therefore

$$\left(\frac{851.5 \text{ kJ}}{2 \text{ mol Al}}\right) (0.03076 \text{ mol Al}) = 157.8 \text{ kJ}$$

105. Using heats of combustion, we write the combustion equations corresponding to the values from Tables for C(gr) and $CH_3COOH(\ell)$, as follows:

$$CH_3COOH(\ell) = 2O_2(g) \rightarrow 2CO_2(g) + 2H_2O(\ell) \quad \text{(a)}$$

$$\Delta H_a = \Delta H^\circ_{comb}(CH_3COOH) = -874.2 \text{ kJ} \cdot \text{mol}^{-1}$$

$$C(gr) + O_2(g) \rightarrow CO_2(g) \qquad \Delta H_b = \Delta H^\circ_{comb}(C, gr) = -393.51 \text{ kJ} \cdot \text{mol}^{-1} \quad \text{(b)}$$

The desired equation is 2 (b) – (a). We have to multiply (b) by 2 in order to have 2 mole of C(gr) on the left-hand side, and we must subtract (a) in order or get CH_3COOH on the right-hand side. Thus, using Hess' law

$$\Delta H^\circ = 2\Delta H_b - \Delta H_a = 2(-393.51) - (-874.2) = +87.2 \text{ kJ}$$

Alternatively, we can use heats of formation.

$$\Delta H^\circ = \Delta H^\circ_f(CH_3COOH) - 2\Delta H^\circ_f(H_2O, \ell) = -484.5 - 2(-285.83) = +87.2 \text{ kJ}$$

Clearly, it makes no difference which path is followed to obtain the desired.

106. First write a correctly balanced equation for the combustion of ethanol,

$$CH_3CH_2OH(\ell) + 3O_2(g) \rightarrow 2CO_2(g) + 3_2H_2O(\ell)$$

Applying combustion reaction, we obtain

$$\Delta H^\circ_{comb} = 2\Delta H^\circ_f(CO_2) + 3\Delta H^\circ_f(H_2O, \ell) - \Delta H^\circ_f(CH_3CH_2OH, \ell)$$

The only unknown in this equation is $\Delta H_f(CH_3CH_2OH, \ell)$. Substitution yields

$$-1366.8 = 2(-393.51) + 3(-285.83) - \Delta H^\circ_f(CH_3CH_2OH, \ell)$$

By rearranging the equation we obtain

$$\Delta H^\circ_f(CH_3CH_2OH, \ell) = 2(-393.51) + 3(-285.83) + 1366.8 = -277.7 \text{ kJ} \cdot \text{mol}^{-1}$$

107. Using Hess' Law, ΔH for this reaction can be obtained by combining the three reactions

$$H_2(g) \rightarrow 2H(g) \qquad \Delta H(H_2) = 435.9 \text{ kJ} \cdot \text{mol}^{-1}$$

$$O_2(g) \rightarrow 2O(g) \qquad \Delta H(O_2) = 495.0 \text{ kJ} \cdot \text{mol}^{-1}$$

$$H_2(g) + \frac{1}{2}O_2(g) \rightarrow H_2O(g) \qquad \Delta H^\circ_f(H_2O, g) = -241.82 \text{ kJ} \cdot \text{mol}^{-1}$$

$$\Delta H = \Delta H(H_2) + \frac{1}{2}\Delta H(O_2) - \Delta H^\circ_f(H_2O, g)$$

$$= 435.9 + \frac{1}{2}(495.0) + 241.82 = 925.2 \text{ kJ} \cdot \text{mol}^{-1}$$

The O—H bond enthalpy is therefore $\frac{1}{2}(952.2) = 463 \text{ kJ} \cdot \text{mol}^{-1}$

108. From Hess' law we obtain ΔH for $CH_3CH_2OH(g) \rightarrow 2C(g) + 6H(g) + O(g)$

$$\Delta H = 2\Delta H_{subl}(C) + 3\Delta H(H_2) + \frac{1}{2}\Delta H(O_2) - \Delta H^\circ_f(C_2H_5OH, g)$$

$$= 2(716.7) + 3(435.9) + \frac{1}{2}(495.0) - (-235.1) = 3224 \text{ kJ}$$

Now,

The ethanol molecule, CH_3CH_2OH, has the structure

```
    H  H
    |  |
H—C—C—O—H
    |  |
    H  H
```

so that if we dissociate ethanol completely into individual atoms we break five C—H bonds, one C—C bond, one O—H bond, and one C—O bond. Using 413 kJ for the C—H bond enthalpy, 348 kJ for the C—C bond enthalpy, and 463 kJ for the O—H bond enthaly, we obtain

$$\varepsilon_{C—O} = 3224 - 5(413) - 348 - 463 = 348 \text{ kJ}$$

The value usually listed as the C—O single bond enthalpy, obtained by averaging calculated values for various molecules, is 351 kJ.

109. (a) Dimethyl ether, $CH_3—O—CH_3$, has a total of six C—H bonds and two C—O bonds. The amount of energy required to break all bonds in dimethyl ether is therefore 6(413) + 2(351) = 3180 kJ per mole of CH3—O—CH_3. Ethanol, CH_3CH_2OH, has five C—H bonds, are C—C bond, O—H bond, and one C—O bond. The amount of energy required to break all bonds is 5(413) + 348 + 463 + 351 = 3227 kJ · mol^{-1}.
The isomerization reaction is the sum of the two equations:

$$CH_3—O—CH_3 \rightarrow 2C(g) + 6H(g) + O(g) \qquad \Delta H_1 = 3180 \text{ kJ} \qquad (1)$$

$$2C(g) + 6H(g) + O(g) \rightarrow CH_3CH_2OH(g) \qquad \Delta H_2 = -3227 \text{ kJ} \qquad (2)$$

Therefore the desired ΔH = 3180 – 3227 = –47 kJ.

(b) The desired ΔH can be written as

$$\Delta H = \Delta H_f^\circ (CH_3CH_2OH, g) - \Delta H_f^\circ (CH_3OCH_3, g)$$

The ΔH_f° values can be found in tables in text book.

$$\Delta H = -235.1 - (-184.0) = -51.1 \text{ kJ}$$

The estimated value is in error by a little more than 4 kJ. This is certainly as good as can be expected in view of the limits of reliability on bond enthalpy values. The percentage error of the estimate is given by $\left(\frac{-47 + 51.1}{51.1}\right) \times 100 = 8.0\%$.

110. For solids many liquids $C_p \cong C_v$, so that for $H_2O(l)$ and Pb(s), C_p is the same as C_p. For the gases, $C_v = C_p - R$, and R is 8.3145 J · mol^{-1} K^{-1}. Values of C_p and C_v for these four substances are, therefore,

Substance	C_p(*J · mol^{-1} K^{-1}*)	C_v(*J · mol^{-1} K^{-1}*)
Ar(g)	20.8	12.5
N_2(g)	29.3	21.0
$H_2O(l)$	75.4	75.4
Pb(s)	26.4	26.4

111. **(a)** Sucrose is $C_{12}H_{22}O_{11}$. The combustion reaction is

$$C_{12}H_{22}O_{11}(s) + 12O_2(g) \rightarrow 12CO_2(g) + 11H_2O(l)$$

In balancing the oxygen atoms, do not forget the oxygen that is contained in the sucrose.

(b) The molecular weight of sucrose is 342.30. Thus the sample consists of (2.006 g)/(342.30 g · mol^{-1}) = 5.8504 × 10^{-3} ml of sucrose. The amount of heat released by the combustion of this sample of sucrose is +(5.8504 × 10^{-3}) (5647) kJ = 33.04 kJ. The release of this much heat causes the temperature of the calorimeter and its contents to rise 2.966°,

$$\therefore \qquad C_p(\text{calorimeter}) = \frac{(5.8504 \times 10^{-3})(5.647)}{2.966} = 11.14 \text{ kJ} \cdot \text{K}^{-1}$$

(c) $C_6H_6(l) + \frac{15}{2}O_2(g) \rightarrow 6CO_2(g) + 3H_2O(l)$

The molecular weight of benazene is 78.11, so that the sample of benzene contains (0.8021 g)/(78.11 g · mol–1) = 1.027×10^{-2} mol of benzene. Since the same calorimeter is used, $\Delta E(C_6H_6)$ can be calculated

$$-(1.027 \times 10^{-2} \text{ mol})\Delta E(C_6H_6) = (11.14)\ (3.009) \text{ kJ}$$

So that $\Delta E(C_6H_6) = -3264 \text{ kJ} \cdot \text{mol}^{-1}$.

For the combustion of C_6H_6, $\Delta n_{gas} = 6 - \frac{15}{2} = -\frac{3}{2}$,

$$\therefore \qquad \Delta H = \Delta E + \Delta n_{gas}(RT) = \Delta E - (1.5)RT$$

$$(1.5)RT = (1.5 \text{ mol})\ (8.3145 \text{ J} \cdot \text{mol}^{-1}\ \text{K}^{-1})\ (298.15 \text{ K}) = 3.72 \times 10^3 \text{ J}$$

As $\Delta E(C_6H_6)$ is given kilojoules, we obtain

$$\Delta H(C_6H_6) = \Delta E(C_6H_6) - 3.72 = -3264 - 3.72 = -3268 \text{ kJ}$$

112. Energy from 100 g glucose = 1560 kJ

$$\text{Energy not used} = \frac{1560}{2} = 780 \text{ kJ}$$

$$\text{Amount of water to be converted into vapours from this energy} = \frac{780 \text{ kJ}}{33 \text{ kJ mol}^{-1}} = 17.73 \text{ mol}$$

∴ Mass of water perspired = 17.36 × 18 = 312.48 g

113. ΔH = (Bond breaking) – (Bond making)

$$= \left(\frac{1}{2}H_2 + \frac{1}{2}Cl_2\right) - (HCl)$$

$$= \left(\frac{1}{2} \times 434.7 + \frac{1}{2} \times 242.5\right) - 430.5 = 338.6 - 430.5 = 91.96 = 430.5 \text{ J mole}^{-1}$$

114. Given that

(i) $H_2(g) + O_2(g) \rightarrow H_2(g)$ $\Delta H = -285.8$ kJ

(ii) $CH_4(g) + 2O_2(g) \rightarrow 2O_2(g) + H_2O(g)$ $\Delta H = -890.0$ kJ

(iii) $C_2H_6(g) + \frac{7}{2}O_2(g) \rightarrow 2CO_2(g) + 3H_2O(g)$ $\Delta H = -1560.0$ kJ

(iv) $C_{graphite} + O_2 \rightarrow CO_2(g)$ $\Delta H = -393.5$ kJ

(v) $3C_{graphite} + 4H_2(g) \rightarrow C_3H_8(g)$ $\Delta H = -103.8$ kJ

To get the equation

$$C_3H_8 + H_2(g) \rightarrow C_2H_6(g) + CH_4(g)$$

following operations are carried out

–eq.(v) + 5.eq.(i) – eq.(iii) – eq(ii) + 3.eq.(iv)

$\therefore$ $$\Delta H = (103.8 - 5 \times 285.8 + 1560.0 + 90.0 - 3 \times 393.5) = -55.7 \text{ kJ}.$$

115. The combustion reactions are

$$C_2H_2(g) + O_2(g) \rightarrow 2CO_2 + 2H_2O$$
1 vol 2 vol

$$CH_4(g) + 2O_2(g) \rightarrow O_2 + 2H_2O$$
1 vol 1 vol

and the heats of combustion are

$$\Delta H_{C_2H_4} = -1423 \text{ kJ mol}^{-1}$$

$$\Delta H_{CH_4} = -891 \text{ kJ mol}^{-1}$$

Now the total volume of ethylene and methane is 6.11.1. Suppose the volume of ethylene in mixture is V1, Then the volume of methane in the mix becomes (3.67 – V)l. Accordingly the volume of CO_2 generated in equation (i) = 2 × V and that in equation (ii) = (3.67 – V). thus the total vol of CO_2 generated in both equations = 2V + (3.67 – V), which means that

$$2V + (3.67 - V) = 6.11 \text{ l (as given)}$$

Simplifying we get; V = 2.44 liter

Thus the volume of C_2H_4 = V = 2.44 liter

and the total of CH_4 = 3.67 – V = 3.67 – 2.44 = 1.23 liter

Accordingly at 298 K 25°C the volume of 1 mole of gas is 22.414 L × $\frac{298}{273}$ = 24.467 liter

Thus the number of moles of ethylene and methane in gaseous mix are $\frac{2.44\,l}{24.467\,l}$ and $\frac{1.23}{24.467}$ i.e., 0.1 mol and 0.05 mol respectively.

Thus the no of moles of V_2H_4 in 1 lit gaseous mix = $\frac{0.1}{3.67}$ mol = 0.027 mol and no. of mole of H4 in 1 lit gaseous mix = $\frac{0.05}{3.67}$ = 0.013 mol.

Thus the amount of heat energy.

$$= [-1423 \times 0.027 + (-891 \times 0.013) = -38.7 - 12.1 = 59.8 \text{ kJ}.$$

116. In the polymerization process, Bond double bond breaks to form two single bonds. But the molecule is a polymeric form of n units.

Breaking of 'n' > $c = c$<bond = 590 × n kJ = ΔH_1

Formation of 2C—C bond = 2 × 331 KJ = ΔH_2

The complete chain is formed by n molecule thus

$$\Delta H_2 = 662 \text{ KJ mole}^{-1}$$

Hence net energy charge = $\Delta H_1 + \Delta H_2$ = 590n + (–662)n = –72 n.

The per mole energy charge = $\frac{-72n}{n}$ = –72 KJ mole^{-1}

117. The equations of combustion of CH_4 and C_4H_{10} are as follows

$$CH_4 + 2O_2 \rightarrow CO_2 + 2H_2O, \Delta H = -804 \text{ kJ mol}^{-1}$$

$$C_4H_{10} + 13/2O_2 \rightarrow 4CO_2 + 5H_2O, \Delta H = -2878 \text{ kJ mol}^{-1}$$

In order to get the same calorific output due to C_4H_{10}, the rate of supply

$$= \frac{x \times 804}{2876} = 0.2811\ xl/h^2$$

And the rate of supply of oxygen $= 0.2811x \times 6.5 \times 3 = 5.4814\frac{xl}{hr}$

118. Target equation

$$B_2H_6 + 3O_2 \rightarrow B_2O_3 + 3H_2O \qquad \Delta H \text{ (KJ mol}^{-1})$$

1. $2B(s) + \left(\frac{3}{2}\right)O_2 \rightarrow B_2O_{(3)}(s) = -1273$

2. $H_{(2)}(g) + \left(\frac{1}{2}\right)O_2 + H_2O(l) = -286$

3. $H_2O(l) \rightarrow H_2O(g) = +44$
4. $2B(s) + 3H_2(s) \rightarrow B_2H_6(g) = +36$

Add 1, 2 and 3 & subtract from 4 before that multiply 2 & 3 by factor three.
Hence $\Delta H = -2035$ kJ mol^{-1}

119. Target equation

$$\text{cyclopropane} + \frac{9}{2}O_2 \rightarrow 3\ CO_2 + 3H_2O$$

$$\Delta H_{com} = [3 \times \Delta H_f\ (CO_2) + 3 \times \Delta H_f\ \ (H_2O) - \Delta\ H_f \text{ cyclopropane}]$$

Since cyclopropane first isomerises into propane

$$\Delta H_f \text{ propane} = \Delta H_f \text{ cyclopropane} - 33$$

$$\Delta H_f \text{ cyclopropane} = 20.42 + 33.3 = +53.42 \text{ KJ mol}^{-1}$$

$$\Delta H \text{ comb.} = 3 \times (-393.5) + 3 \times (-285.8) - 53.42 = -2091.31 \text{ KJ mol}^{-1}$$

120. $C_6H_{12}O_6 + 6O_2 \rightarrow 5CO_2 + 6H_2O$

$M = 72 + 12 + 96\Delta = -2880$ KJ mol^{-1} $= 180$ gm

1 mole of glucose or 180 gms produce 2880 KJ energy

$\Rightarrow$ 120 gms produce $\frac{2800}{180} \times 120 = -1920$ kJ enrgy (ii)

Only 25% of this is used for muscular work as per question

$\Rightarrow$ 25% of -1920 kJ $= 480$ kJ

As 100 kJ energy reqruied to walk 1 km

480 kJ energy required to walk $\frac{480}{100} = 4.8$ km

121. $S(g) + 6F(g) \rightarrow SF_6\ (g)$

$\Delta H_f = -1100$ KJ mol^{-1}

$S(g) \rightarrow S(g)\ \Delta H = +275$ KJ mol^{-1}

$\frac{1}{2}F_2\ (g) \rightarrow F(g)\ \Delta H = 80$ KJ mol^{-1}

$\Delta H_r = \Sigma B_R - \Sigma B_P$

B_R = Bond energy of reactant

B_P = Bond energy of products

$-1100 = [275 + 6 \times 80] - [6 \times S - F]$

Thus bond energy of S – F = 309 KJ

122. Given that, ΔH_f Cyclohexane (1) = –156 kJ mol^{-1}

ΔH_f Benzene (l) = +49 kJ mol^{-1}

Cyclohexane (l) + $H_2 \rightarrow$ Cyclohexane $\Delta H = -119$

Thus ΔH_f cyclohexane = –156 – (–119) = –37 KJ mol^{-1}

From the given data it is clear that in the process of insertion of one double bond the decrease in energy is +119 KJ/mol.

So in Benzene this will be 3 × = 119 = 357 KJ mol^{-1}

$\therefore$ ΔH_f Benzene (Theoretical) = 357 – 56 = 201 KJ mol^{-1}

Thus Resonance energy ΔH_f (Th.) – ΔH_f (obs) = 152 KJ mol^{-1}

123. $CO + \frac{1}{2}O_2 \rightarrow CO_2$

$$\Delta G^\circ = \Delta G^\circ_{\text{products}} - \Sigma G^\circ_{\text{reactants}}$$

$$= -394.4 - (-137 + 0) = -257.2 \text{ KJ mol}^{-1}$$

Since ΔG is negative, reaction is spontaneous

From Gibbs helmoltz equation

$$\Delta G = \Delta H - T \cdot \Delta S$$

$$\Delta H = -285.4 \text{ KJ mole}^{-1}$$

Since ΔH is negative, reaction is Exothermic

124. (i) $N_2O_4(g) \rightleftharpoons 2NO_2(g)$

$$\text{Reaction quotient} = \frac{P^2_{NO_2}}{P_{N_2O_4}} = \frac{100}{10} = 10 \text{ atm}$$

$$\Delta G^\circ \text{ reaction} = 2\ \Delta G_f(NO_2) - \Delta G_f(N_2O_4)$$

$$0 = 100 - 100$$

$$\Delta G = \Delta G^\circ + RT \ln K$$

$$\therefore \quad \Delta G = RT \ln Q$$

$$= 2.303 \times .082 \times 298 \times \log 9.9$$

$$= 56.0304 \text{ Lit atm.} = \text{Positive}$$

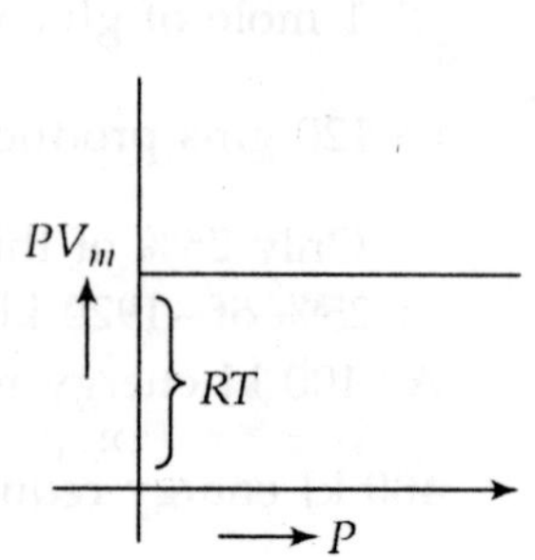

(ii) Since initial Gibbs free energy change of the reaction is positive, so the reverse reaction will take place.

$$PV = nRT$$

$$PV_m - RT$$

So, y-intercept of the graph = RT

125. (i) In reversible Isobaric (constant pressure) system expansion pressure is constant (i.e., one atmosphere) and volume increased from 20.0 liter 40.0 Litres (from A to B).

$$\text{Work done } W_1 = P(V_2 - V_2)$$
$$= -1 \times (40 - 20) = -1 \times 20 = -20\text{atm–litre.}$$

(ii) In reversible isochoric (constant volume) change of state (from B to C) volume is constant ({i.e., 40 Litre) while pressure is decreased from 1 to 0.5 atm. In this change $\Delta V = 0$ so work done $W_2 = 0$.

(iii) In reversible isothermal process (constant temperature) volume and pressure both are changing. Thus system is reached at state A (change takes place from C to A) Hence in this change work done.

$$W_3 = -2.303\ nRT\ \text{Log}_{10}\frac{V_2}{V_1} = -2.303 \times 2 \times 0.0821 \times 298\ \text{Log}_{10}\frac{20}{40}$$

$$= -2.303 \times 2 \times 0.0821 \times 298 \times -0.3010 \text{ Litre} - \text{atm} = +33.92 \text{ Litre} - \text{atm}$$

(II) Total work done $W = W_1 + W_2 + W_3$

$$= -20 + 0\ 33.92 = 13.92 \text{ Liter-atm} = 13.92 \times \frac{8.314}{0.0821}\text{J} = \mathbf{1.409 \times 10^3\ J}$$

(III) ΔU = O, ΔS = O, and ΔH, = O (Because it is a cyclic process>)

126. (a) If e is the efficiency, T_1 is the temperature of hot reservoir and T_2 that of cold reservoir, then

$$e = \frac{T_1 - T_2}{T_1} = \frac{320°\text{K} - 260°\text{K}}{320°\text{K}} = \frac{3}{16}$$

Work done $W = eq_1$, where q_1 is the heat absorbed at the hot reservoir.

$$W = \left(\frac{3}{16}\right) \text{(500 joules)} = 93.75 \text{ joules.}$$

(b) $e = 1 - \dfrac{q_2}{q_1}$, where $e = \dfrac{3}{16}$ and q_2 = 1000 joules.

$$\frac{3}{16} = 1 - \frac{1000 \text{ joules}}{q_1}$$

whence, $$q_1 = (1000)\left(\frac{16}{13}\right) \text{joules} = 231 \text{ J}$$

Work that must be supplied to remove 1000 joules of heat from the cold reservoir is

$$W = q_1 - q_2 = 1231\text{–}1000 = 231 \text{ J.}$$

127. (a) During isothermal expansion, since temperature remains constant, the internal energy is unchanged. Therefore, according to first law of thermodynamics, work done, $W_1 = q_1 = 2090$ Joules.

(b) During isothermal compression, heat rejected from the gas is q_2.

But $$\frac{q_1}{q_2} = \frac{T_1}{T_2}$$

$\therefore$ $$q_2 = q_1 \frac{T_2}{T_1} = 2090\left(\frac{300\text{ K}}{400\text{ K}}\right) = 1568 \text{ J}$$

(c) Work done on the gas during isothermal compression, $W_2 = q_2 = 1568$ joules.

128. For the conduction q calories of heat through the brass rod, from the hot reservoir at T_1 = 400K to a cold reservoir at T_2 = 300K, the changes in entropy are

$$\text{For hot reservoir, } \Delta S_1 = \frac{1}{T_1} = -\frac{5000\text{ Joules}}{400\text{K}} = -12.5 \text{ JK}^{-1}$$

$$\text{For cold reservoir, } \Delta S_2 = \frac{q}{T_2} = \frac{5000\text{ Joules}}{300\text{K}} = +16.67 \text{ JK}^{-1}$$

Total change in entropy arising from the process of conduction.

$$\Delta S = \Delta S_1 + \Delta S_2 = -12.5 + 16.67 = 4.17 \text{ Joules K}^{-1}$$

The entropy of the rod remains constant since for any portion of the rod, as much heat is passed out at absorbed at a given temperature.

129. Here, q_1 = 2.5 KJ = 2500 Joules, T_1 = 400 K, T_2 = 200 K.

Now, the heat engine operating according to the Carnot cycle performs work equal to $W = q_1 - q_2 = nq_1$, where q_1 is the amount of heat received by the engine form the hot source, q_2 the amount of heat rejected to the cold sink and η the efficiency of the engine. In our case $\eta = \frac{T_1 - T_2}{T_1} = \frac{400-300}{400} = \frac{1}{4}$ = 0.25. Hence $W = nq_1$ = 6.52 J, and $q_2 = q_1 - W$ = 2500 − 625 = 1825 Joules.

130. Here, let initial volume of Hydrogen be V and so final volume is 2V. Also we have,

$m = 6.6$, $M = 1.0$, $\gamma = 1.33$

and $$\frac{V_2}{V_1} = \frac{2V}{V} = 2$$

$\therefore$ $$\Delta S = \frac{m}{M}C_v \log_e \frac{P_2}{P_1} + \frac{m}{M}C_p \log_e \frac{V_2}{V_1}.$$ In the isobaric process $p_1 = p_2$.

$\therefore$ $$\Delta S = \frac{m}{M}C_p \log_e \frac{V_2}{V_1}.$$

or $$\Delta S = \frac{m}{M}\cdot\frac{\gamma R}{\gamma - 1}\log\frac{V_2}{V_1}$$

Upon inserting numerical data, we get

$$\Delta S = \frac{6.6 \times 1.33 \times 8.314 \times 0.3010}{0.33} = 66.5 \text{ JK}^{-1}$$

131. For adiabatic expn we have $TV^{\gamma-1} = T_0 V_0^{\gamma-1} \Rightarrow T = T_0 \cdot n^{1-\gamma} = T_1$

where $n = \frac{V_1}{V_2}$, V_1 is volume end of a addition process. So there is no entropy change.

But $$\frac{V_1}{T_1} = \frac{V_2}{T_f} \text{ or } T_f = T_1 \frac{V_2}{V_1} = T_o n^{-\gamma}$$

Now, For isobarically process,

$$\Delta S = \left(\frac{m}{M} C_p\right) \ln \frac{T}{T_1} = \left(\frac{m}{M} C_p\right) \ln \frac{1}{n} = -\frac{m}{M} C_p \ln n$$

or $$\Delta S = -\frac{m}{M} \frac{\gamma R}{\gamma - 1} \ln n = \frac{-1.7}{4.003} \cdot \frac{1.66 \times 8.314}{(1.66 - 1)}, 2.30 \log_3 = -9.77 \text{ J/K.} \quad \left[n = \frac{V_1}{V_2} = 3\right]$$

$r = 1.66$
For He

132. Here, $P_1 = 10^5$ N/m^2, $P_2 = 0.5 \times 10^5$ N/m^2, $m = 6$ gm, $M = 2$ for H_2,
We have

$$\Delta S = \frac{m}{M} C_p \log_e \frac{T_2}{T_1} - \frac{m}{M} R \log_e \frac{P_2}{P_1}.$$ In the isothermal process $T_1 = T_2$.

$$\therefore \quad \Delta S = -\frac{m}{M} R \log_e \frac{P_2}{P_1} = \frac{m}{M} R \ln \frac{P_1}{P_2}.$$

Upon inserting the numerical date, $\Delta S = \frac{6}{2} \times 8.314 \times 0.3010 \times 2.303 = 17.27 \text{ JK}^{-1}$

133. We have $\Delta S = S_2 - S_1 = \int_1^2 \frac{dq}{T}$. (But $dq = \frac{m}{M} C_v \, dT + pdV$ and, $pV = \frac{m}{M} RT$)

$$= \int_1^2 \frac{m}{M} \frac{C_v dt}{\not{T}} + \int_1^2 \frac{\frac{m}{M} RdV \not{T}}{V \not{T}}$$

or $$\Delta S = \frac{m}{M} C_v \log_e \frac{T_2}{T_1} + \frac{m}{M} R \log_e \frac{V_2}{V_1}$$

$$= \frac{m}{M}\left(\frac{R}{\gamma - 1}\right) \log \frac{T_2}{T_1} + \frac{m}{M} R \log_e \frac{V_2}{V_1} = \frac{mR}{M}\left[\frac{1}{\gamma - 1} \log \frac{T_2}{T_1} + \log_e \frac{V_2}{V_1}\right]$$

Now, $m = 8$, $M = 32$

$T_1 = 353$, $T_2 = 573$, $V_1 = 10$ L, $V_2 = 40$ L

γ for $O_2 = 1.66$.

Substituting the values

$$\Delta S = \frac{8 \times 8.314 \times 2.303}{0.66}\left[\log\frac{573}{373} + \log\frac{40}{10}\right] = 5.4\ \text{JK}^{-1}$$

134. (a) 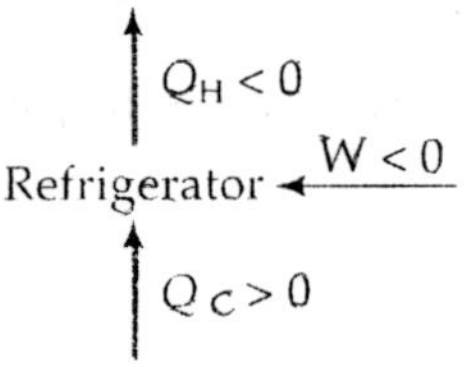 Refrigerator ← $W < 0$; $Q_H < 0$ (out, upward) ; $Q_C > 0$ (in, from below)

$T_H = 295$ K; $T_C = 273$ K

The amount of heat taken out of the water to make the liquid → solid phase change of $Q = -mL_f = -(40\ \text{kg})\ (334 \times 103\ \text{J/kg}) = -1.336 \times 107$ J.

This amount of heat must go into the working substance of the refrigerator ⇒ $Q_C = +1.336$

For carnot cycle $|Q_H| = |Q_C|\left(\frac{T_H}{T_C}\right) = 1.336 \times 107\ \text{J}\left(\frac{295\ \text{K}}{273\ \text{K}}\right) = 1.44 \times 107$ J

(b) $W = Q_C + Q_H = +1.336 \times 107\ \text{J} - 1.444 \times 107\ \text{J} = -1.08 \times 106$ J
(W is negative because this much energy must be supplied to the refrigerator rather than obtained from it.)

135. From Gibb's Helmoltz equation we have

$$\Delta G = \Delta H + T\frac{d(\Delta G)}{dT}$$

$$G_1 = -33.089\ \text{kJ};\ T_1 = 298\ \text{K}$$

$$G_2 = -28.018\ \text{kJ};\ T_2 = 308\ \text{K}$$

$$\therefore \quad \frac{d\Delta G}{dT} = \frac{-28.018 - (-33.089)}{308 - 298}\ 0.507\ \text{KJ}$$

$$\Delta G = -33.089\ \text{kJ};\ T = 308\ \text{K}$$

Making the required substitutions

$$-33.089 = \Delta H + 308 + 0.507 = -33.089 - 156.156$$

$$\Delta H = -189.24\ \text{kJ}$$

136. Here we are given that

$$T_1 = 373\ \text{K}$$

$$\Delta H_V = 2260\ \text{Jg}^{-1} = 2260 \times 18\ \text{J mol}^{-1} = 2260 \times 18\ \text{J mol}^{-1}$$

$$\Delta S = \frac{\Delta H_V}{T_b} = \frac{40680}{373} = 109\ \text{Jk}^{-1}\ \text{mol}^{-1}$$

137. $\Delta H_m = 2510 \text{ J mol}^{-1}$, $T_m = 300$ K.

$$\Delta S = \frac{\Delta H_m}{T_m} = \frac{2510}{300} = 8.37 \text{ JK–1 mol}^{-1}$$

138. From equation in equilibrium $\Delta S = \dfrac{\Delta H}{T}$

Here we are given $\Delta H = 30.58 \text{ kJ} = 30580$ J

$\Delta S = 66.11 \text{ JK}^{-1}$

Substituting the values

$$66.11 = \frac{30580}{T}$$

$\Rightarrow$ $T = 462.6$ K

139. $\Delta S = C_V \ln \dfrac{T_2}{T_1} + R \ln \dfrac{V_2}{V_1}$ for 1 mole of an ideal gas.

For isothermal process,

$T_1 = T_2$ so that

$$\Delta S = R \ln \frac{V_2}{V_1} \text{ (for 1 mol)}$$

or $\Delta S = \text{nR} \ln \dfrac{V_2}{V_1}$ (for n moles) $= 2.303 \text{ nR} \log \dfrac{V_2}{V_1}$

Here $n = 5$; $V_1 = 10$ litres

$V_2 = 100$ litres, $R = 8.324 \text{ JK}^{-1} \text{ mol}^{-1}$

$\therefore$ $\Delta S = 2.303 \times 5 \times 8.134 \times \log \dfrac{100}{10} = 9575 \text{ JK}^{-1} \text{ mol}^{-1}$

140. Here it is given that

$n = 1$ mol; $T = 330$ K; $P_1 = 5$ atm; $P_2 = 50$ atm

Taking $R = 8.314 \text{ J K}^{-1} \text{ mol}^{-1}$ we get

$(\Delta G)_T = 2.303 \text{ RT} \log P_2$

$= 2.303 \times 1 \times 8.314\ P_1 \times 303 \times \log \dfrac{50}{1} = 6312$ Joules.

141. Here it is given that

$n = 1$ mole; $T = 300$ K; $V_1 = 5$ litres; $V_2 = 50$ litres

$R = 8.314 \text{ JK}^{-1} \text{ mol}^{-1}$

$(\Delta G)_T = 2.303 \text{ nRT} \log \dfrac{V_1}{V_2} = 2.303 \times 1 \times 8.314 \times 314 \times 300 \log \dfrac{5}{50} = -5744$ Joules.

142. $\Delta G° = -2.303 \text{ RT} \log \dfrac{[\text{Product}]}{[\text{Reactant}]}$

For equilibrium $B \rightleftarrows A$

$$\Delta G^\circ = -2.303\ \text{RT}\log\frac{[A]}{[B]}$$

Given, $[A] = 13\%$ $[B] = 95.2\%$
and $R = 8.314\%$ $T = 448$ k
Putting these values, we have,

$$\Delta G_1^\circ = (-2.303 \times 8.314 \times 448)\log\frac{1.3}{95.2} = 15.998\ \text{mole}^{-1}$$

Now for the equilibrium $B \rightleftarrows C$

$$\Delta G^\circ = -\ 2.303\ \text{RT}\log\frac{[C]}{[B]}$$

where $[C] = 3.5\%$ & $[B] = 95.2\%$
Putting these values we get,

$$\Delta G_2^\circ = (-2.303 \times 8.314 \times 448)\log\frac{3.5}{95.2} = 12.305\ \text{kJ mole}^{-1}$$

Stability order : $B > C > A$

143. Given, $\text{HCN(g)} \rightarrow \text{H(g)} + \text{C(g)} + \text{N(g)}$
(135 218 715.0 473) ΔH_f (kJ mol^{-1})

$\Delta H_{nm} = 1271\ \text{kJ mol}^{-1}$
$\text{BE(C}\equiv\text{N)} = \Delta H - \text{BE(C—H)} = [1.271 - 411]\ \text{kJ mol}^{-1}$
$= 860\ \text{kJ mol}^{-1}$

144. Use the O—H bond and its bond energy of 459 kJ/mol to calculate X_0.

$$\text{BE(O—H)} = ½(\text{BE(H—H)} + \text{BE(O—O)}] + k(X_O - X_H)^2$$

$$459\ \text{kJ mol}^{-1} = ½(432\ \text{kJ mol}^{-1} + 142\ \text{kJ mol}^{-1}) + 98.6\ \text{kJ}\ (X_O - X_H)$$

Collecting the terms gives

$$\frac{459-287}{98.6} = [X_O - X_H)^2$$

Taking the square root of both sides gives

$$1.3207 = 1.32 = (X_O - X_H)$$

Because $X_H = 2.1,\ X_O = 2.1 + 1.32 = 3.42 = 3.4$

145. Calculate the moles of HCN in 10 mL of the solution (density = 0.687 g HCN/mL HCN):

$$10\ \text{mL HCN} \times \frac{0.687\ \text{g HCN}}{\text{mL HCN}} \times \frac{1\text{mol HCN}}{27.03\ \text{g HCN}} = 0.254\ \text{mol HCN}$$

$0.254\ \text{mol HCN(l)} \rightarrow 0.254\ \text{mol HCN(g)}$
$(\Delta H_f^\circ = 105\ \text{kJ mol}^{-1})(\Delta H_f^\circ = 135\ \text{kJ mol}^{-1})$
$\Delta H^\circ = 0.254\ \text{mol} \times (135\ \text{kJ mol}^{-1} - 105\ \text{kJ mol}^{-1}) = 7.62 = 7.6\ \text{kJ}$

146. First convert mass to moles; then multiply by the standard heat of formation to obtain the heat absorbed in vaporizing this mass:

$$25\text{ g P}_4 \times \frac{1\text{ mol P}_4}{123.92\text{ g P}_4} = 0.2017\text{ mol P}_4$$

$$0.2017\text{ mol P}_4 \times \frac{95.4\text{ J}}{\text{K} \bullet \text{mol P}_4} \times (317.01\text{–}298)\text{K} = 367.6\text{ J} = 0.3676\text{ kJ}$$

$$2.63\text{ kJ mol}^{-1}\text{ P}_4 \times 0.2017\text{ mol P}_4 = 0.5306\text{ kJ}$$

$$\text{Total heat} = 0.5306\text{ kJ} + 0.3676\text{ kJ} = 0.89818 = 898\text{ J}$$

Chapter 7 : CHEMICAL KINETICS

1. rate = $k \cdot (\text{acetone})^x \cdot ((I_2)_0)^y \cdot (H^+)^2$

$$\text{rate I} = 4.2 \times 10^{-6} = k \cdot (0.8)^x \cdot (0.001)^y \cdot (0.2)^z$$
$$\text{rate II} = 8.3 \times 10^{-6} = k \cdot (1.6)^x \cdot (0.001)^y \cdot (0.2)^z$$

Ratio of Rate II to Rate I:

$$\frac{\text{rate II}}{\text{rate I}} = \frac{8.3 \times 10^{-6}}{4.2 \times 10^{-6}} = 2.0 = \frac{\cancel{k} \cdot (1.6)^x \cdot \cancel{(0.001)^y \cdot (0.2)^z}}{\cancel{k} \cdot (0.8)^x \cdot \cancel{(0.001)^y \cdot (0.2)^z}}$$

$$\text{rate III} = 8.3 \times 10^{-6} = k \cdot (0.8)^x \cdot (0.001)^y \cdot (0.4)^z$$
$$\text{rate IV} = 4.2 \times 10^{-6} = k \cdot (0.8)^x \cdot (0.0005)^y \cdot (0.2)^z$$

Ratio of rate III to rate IV:

$$\frac{\text{rate III}}{\text{rate IV}} = \frac{8.3 \times 10^{-6}}{4.2 \times 10^{-6}} = 2.0 = \frac{\cancel{k \cdot (0.8)^x \cdot (0.0005)^y} \cdot (0.4)^z}{\cancel{k \cdot (0.8)^x \cdot (0.0005)^y} \cdot (0.2)^z} \quad (2)^z$$

Ratio of rate IV to rate I:

$$\frac{\text{rate IV}}{\text{rate I}} = \frac{4.2\ x\ 10^6}{4.2 \times 10^{-6}} = 1.0 = \frac{\cancel{k \cdot (0.8)^x} \cdot (0.0005)^y \cdot \cancel{(0.2)^z}}{\cancel{k \cdot (0.8)^x} \cdot (.001)^y \cdot \cancel{(0.2)^z}} = (0.5)^y$$

So,
As the concentration of the acetone doubles, the rate doubles.
As the concentration of the H^+ doubles, the rate doubles;
The concentration of the I_2 has no affect on the rate;

2. We have,

$$k = \frac{\text{rate}}{(\text{acetone})(H^+)}$$

Mixture 1: $k = \dfrac{4.2 \times 10^{-6}}{(0.8)(0.2)} = 2.6 \times 10^{-5}$; Mixture 2: $k = \dfrac{8.3 \times 10^{-6}}{(1.6)(0.2)} = 2.6 \times 10^{-5}$

Mixture 3: $k = \dfrac{8.3 \times 10^{-6}}{(0.8)(0.2)} = 2.6 \times 10^{-5}$; Mixture 4: $k = \dfrac{4.2 \times 10^{-6}}{(0.8)(0.2)} = 2.6 \times 10^{-5}$

Average: $k = 2.6 \times 10^{-5}$

3. Total volume of mixture = 25 mL + 25 mL + 25 mL + 25 mL = 1.0×10^2 mL = 0.1L
Concentration of each after mixing

$$\text{Acetone}: \frac{0.025\ \cancel{L}}{1} \times \frac{3.2\ \text{moles}}{\cancel{L}} \times \frac{1}{0.1\ \text{L sol'n}} = 0.8\ M$$

$$\text{HCl}: \frac{0.025\ \cancel{L}}{1} \times \frac{4\ \text{moles}}{\cancel{L}} \times \frac{1}{0.1\ \text{L sol'n}} = 1.0\ \text{M}$$

$$I_2: \quad \frac{0.025 \cancel{L}}{1} \times \frac{0.02 \text{ moles}}{\cancel{L}} \times \frac{1}{0.1 \text{ L sol'n}} \quad 0.005 \text{ M}$$

$$\text{rate} = k \cdot (\text{acetone})^x \cdot (I_2)^y \cdot (H^+)^z$$

$$= 2.6 \times 10^{-5} \cdot (0.8)^1 \cdot (0.005)^0 \cdot (1.0)^1 = \mathbf{2.1 \times 10^{-5} \text{ moles} \cdot L^{-1} \cdot sec^{-1}}$$

4. Here

298 K; 240 seconds
313 K; 45 seconds

Being by creating a table summarizing the known information and leaving room for calculated results:

Acetone	H^+	$(I_2)_o$	*Temp*	*Time (sec)*	*rate*	*k*	*log k*	*1/T (K)*
0.8	0.2	0.001	298K	240	4.2×10^{-6}	2.6×10^{-6}	–4.58	3.36×10^{-3}
0.8	0.2	0.001	313K	45	2.2×10^{-5}	1.4×10^{-5}	–3.86	3.19×10^{-3}

$$\text{rate} = k \cdot (\text{acetone})^1 \cdot ((I_2)_o)^0 \cdot (H^+)^1$$
$$\text{rate} = (I_2)_0 \cdot \text{time}^{-1}$$

At 298 K

$$\text{rate} = \frac{0.001 \text{ M}}{240 \text{ sec}} = 4.2 \times 10^{-6}; \Rightarrow k = \frac{4.2 \times 10^{-6}}{(0.8)(0.2)} = 2.6 \times 10^{-5}$$

At 313 K

$$\text{rate} = \frac{0.001 \text{ M}}{45 \text{ sec}} = 2.2 \times 10^{-5}; \Rightarrow k = \frac{2.2 \times 10^{-6}}{(0.8)(0.2)} = 1.4 \times 10^{-4}$$

$$\text{Slope log } k \text{ vs. } 1/T = \frac{y_2 - y_1}{x_2 - x_2} = \frac{-3.86 + 4.58}{3.19 \times 10^{-3} - 3.36 \times 10^{-3}} = 4.24 \times 10^3$$

$$E_o = 2.3 \log k \cdot R \cdot T = -2.3 \cdot \frac{\log k}{T^{-1}} \cdot R$$

$$E_o = -2.3 \times 8.31 \times (-4240) = 8.1 \times 10^4 \text{ Joules}$$

$$\% \text{ error} = \frac{\text{obs} - \text{exp}}{\text{exp}} \times 100\% = \frac{8.1 \times 10^4 - 8.6 \times 10^4}{8.6 \times 10^4} \times 100\% = -5.8\%$$

5. 20% of 1 mole = 0.2 mole = the number of moles of HI that dissociate. So, 0.8 mole of HI is not dissociated. The equation, 2 HI = H_2 + I_2, tells us that 2 moles of HI yield 1 mole of H_2 and 1 mole of I_2. Therefore 0.2 mole of HI yields 0.1 mole of H_2 and 0.1 mole of I_2.
6. First write the equilibrium formula

$$K = \frac{[COCl_2]}{[CO] \times [Cl_2]}$$

The notation $[COCl_2]$, by definition, means concentration of $COCl_2$ in moles of $COCl_2$ per liter. Likewise, [CO] and $[Cl_2]$ mean, respectively, moles of CO per liter and moles of Cl_2 per liter. Substituting the values in the above equation, we get,

$$K = \frac{0.8}{0.3 \times 0.2} = 13 \text{ liters mole}^{-1}$$

7. First, write the equilibrium formula.

$$K = \frac{[SO_3]^2}{[SO_2]^2 \times [O_2]}$$

The total concentrations of O_2, SO_3, and SO_2 are substituted in this formula to give

$$K = \frac{(0.6)^2}{(0.2)^2 \times (0.3)} = \frac{0.36}{0.012} = 3 \times 10^1 \text{ liters mole}^{-1}$$

8. First calculate the equilibrium constant.

$$k = \frac{[SO_3] \times [NO]}{[SO_2] \times [NO_2]} = \frac{0.6 \times 0.4}{0.8 \times 0.1} = 3.0$$

Let X = moles of NO that must be added. We can see from the equation, $SO_2 + NO_2 \rightleftharpoons SO_3 + NO$, that in order to produce 0.2 more mole of NO_2 we must also produce 0.2 more mole of SO_2 and we must use up 0.2 mole of SO_2 and 0.2 mole of NO. therefore, when we have 0.3 mole of NO_2 we will have 1 mole (0.8 + 0.2) of SO_2, 0.4 mole (0.6 – 0.2) of SO_3 and X + 0.2 mole (0.4 + X – 0.2) of NO. Substituting the values in the equilibrium formula, we have

$$\frac{0.4 \times (X + 0.2)}{1 \times 0.3} = 3.0 \Rightarrow X = 2.05 \text{ moles}$$

9. To calculate V when T is constant we must know P and n. Before any reaction has occurred we know the number of moles, 3, but we do not know the pressure. At equilibrium, we know P, 0.92 min, but we do not know the number o moles. Therefore, to calculate V we must calculate either the number of moles of equilibrium or the pressure at the start.

To calculate moles of equilibrium, let X = moles of CS_2. Then X = moles of H_2, 2 – X = moles of CH_4, and 1 – 2X = moles of H_2S. Total moles of equilibrium = the sum of the above quantities = 3 + 2X. In a system at constant V and T the number of moles is directly proportional to the pressure. Therefore,

$$\frac{\text{total moles}}{\text{moles of } H_2} = \frac{\text{total pressure}}{\text{partial pressure of } H_2}$$

$$\frac{3 + 2X}{4X} = \frac{0.92}{0.2}$$

Solving $X = 0.183$ mole

total moles = 3.366, and V = 300 liters.

To calculate the total pressure of the 3 moles at the start let us assume that we take the system at equilibrium and, somehow, force the reaction back to the left so that all of the CS_2 and H_2 interact to form CH_4 and H_2S; we would then be back at the start. Since P_{H_4} = 0.2 atm, P_{CS_2}, must be $\frac{1}{4}$ of that amount or 0.05 atm; the sum of the two partial pressures will be 0.25 atm. Since when the reaction goes from right to left, the number of moles *decreases* from 5 to 3, that is by a factor of $\frac{2}{5}$, the pressure will decreases by a factor of $\frac{2}{5} \cdot \frac{2}{5} \times 0.25$ atm = 0.1 atm. Since the equilibrium pressure was 0.92 atm and the calculated decreases is 0.1 atm, the original pressure must have been 0.82 atm. Substituting this value in $PV = nRT$ when n = 3 moles and T= 1000 K gives a volume of 300 liters.

10. Let X equal the moles per liter of O_2 derived from the N_2O_5 and Y equal the moles per liter of O_2 derived from the N_2O_3. Then:

$$[O_2] = X + Y \qquad [N_2O_5] = 4.0 - x$$
$$[N_2O_3] = X - Y \qquad [N_2O] = Y$$

Since three equations involving X and Y are available, the value of X and Y, and hence the concentrations of all species, can be calculated.

11. (a)

$$\begin{array}{ccccccc} CO_2 & + & H_2 & \rightleftharpoons & CO & + & H_2O \\ 63.1 \text{ atm} & & 21.1 \text{ atm} & & 84.2 \text{ atm} & & 31.6 \text{ atm} \end{array}$$

$$K_p = \frac{P_{CO} \times P_{H_2O}}{P_{CO_2} \times P_{H_2}} = \frac{84.2 \text{ atm} \times 31.6 \text{ atm}}{63.1 \text{ atm} \times 21.1 \text{ atm}} = 2.0$$

Let X = the partial pressure of the CO_2 in the system after removal of CO_2. Since the partial pressure of the CO is reduced to 63.0 atm, a quantity of CO with a partial pressure of 21.2 atm must have reacted with H_2O to form CO_2 and H_2. Therefore, 21.2 atm worth of H_2O must have reacted and 21.2 atm worth of both CO_2 and H_2 must have been produced. The partial pressure of each reactant in the new equilibrium system will then be

$$\begin{array}{ccccccc} CO_2 & + & H_2 & \rightleftharpoons & CO & + & H_2O \\ X \text{ atm} & & 42.3 \text{ atm} & & 63.0 \text{ atm} & & 10.4 \text{ atm} \end{array}$$

$$K_p = \frac{63.0 \text{ atm} \times 10.4 \text{ atm}}{X \text{ atm} \times 42.3 \text{ atm}} = 2.0$$

Solving, X = 7.8 atm

(b) Since there is no change in the number of moles, $K_c = K_p$.

(c) Since, in the gaseous equilibrium represented by the reaction, $CO_2 + H_2 \rightleftharpoons CO = H_2O$, there is no change in the number of moles, increase in pressure by reducing the volume to one half its original value will not shift the equilibrium. All that will happen will be that the partial pressure of each reactant will be doubled. therefore, the partial pressure of the CO_2 will be 2 × 7.8 atm or 15.6 atm.

12. $K = \frac{[HCl]^2}{[H_2O]}$

Assume that, in the first equilibrium [HCl] = a and $[H_2O] = b$. Then $K = a^2/b$. Since T is constant K will have the value a^2/b in the second equilibrium.

Let X = the factor by which the concentration of HCl is multiplied in the second equilibrium; since, at constant volume and temperature the concentration of a gas is directly proportional to its partial pressure, this will be the factor by the which is partial pressure is multiplied. Then, at the second equilibrium, [HCl] = Xa and [H_2O] = $2b$

$$K = \frac{[Xa]^2}{2b} = \frac{X^2 a^2}{2b} \quad \text{But } K = a^2/b$$

Therefore, $\frac{X^2 a^2}{2b} = \frac{a^2}{b}$. Solving, $X = \sqrt{2} = 1.41$.

13. First calculate the equilibrium constant. Its value is 3.0. We can see from the equation, $SO_2 + NO_2 \rightleftharpoons SO_3 + NO$, that if the concentration of NO is increased the equilibrium will be shifted to the left. Let X be the number of additional moles of SO_2 formed as a result of this shift. There will then be 0.8 + X moles of SO_2. But when X new moles of SO_2 are formed, X new moles of NO_2 will also be formed; X moles of SO_3 and X moles of NO will be used up. There will , therefore, be present in the new mixture 0.8 + X moles of SO_2, 0.1 + X moles of NO_2, 0.6 – X moles of SO_3, and 1 + 0.4 – X or 1.4 – X moles of NO. If we insert these values in the equilibrium formula, for which the constant 3 has been calculated, we have

$$\frac{[SO_3] \times [NO]}{[SO_2] \times [NO_2]} = \frac{(0.6 - X) \times (1.4 - X)}{(0.8 + X) \times (0.1 + X)} = 3.0$$

Solving, X = 0.12 mole. Substituting this value of X we find the concentrations of the 4 reactants to be SO_3 = 0.48 mole, NO = 1.3 moles SO_2 = 0.92 mole, and NO_2 = 0.2 mole.

14. Since this reaction is the reaction for the formation of 2 moles of $NO_{2(g)}$ from its elements in their standard states, $\Delta G° = 2(\Delta G_f^o)$ = 103.66 KJ = 103660 J. Substitution into $\Delta G° = -2.3\ RT \log K$ gives

$$\log K = \frac{103660}{(8.314)(298)} = -41.9 = 1.26 \times 10^{-42}$$

15. The average rate of appearance of CO_2 is $\Delta P_{CO_2}/\Delta t$. Initially P_{CO_2} = 0, after 5 minutes P_{CO_2} = 100 atm. So

$$\frac{\Delta P_{CO_2}}{\Delta t} = \frac{100 - 0}{5 - 0} = 20 \text{ min}^{-1}$$

During the second 5-minues period P_{CO_2}, increases from 100 mm to 200 m and the time increases form 5 minutes to 10 minutes

$$\frac{\Delta P_{CO_2}}{\Delta t} = \frac{200 - 100}{10 - 5} = 20 \text{ min}^{-1}$$

For the 10-minues period

$$\frac{\Delta P_{CO_2}}{\Delta t} = \frac{200 - 0}{10 - 0} = 20 \text{ min}^{-1}$$

16. The rate of disappearance of $N_2O_3(g)$ is given by $\frac{\Delta P_{N_2O_3}}{\Delta t}$

Thus we must calculate $P_{N_2O_3}$ after 75 seconds ad after 150 seconds from the data given on the total pressure. It should be noted that the pressure increases as the reaction proceeds because each mole of $N_2O_{3(g)}$ which decompose produces 2 moles of gaseous products. If we let X = mm of $N_2O_{3(g)}$ which decomposes after 75 seconds, then at this time $P_{N_2O_3}$ = $100 - X$ $P_{NO} = P_{NO_2}$ = X. Since $P = P_{N_2O_3} + P_{NO} + P_{NO_2}$ then

$$150 = (100 - X) + X + X$$

$$X = 50 \text{ mm and } P_{N_2O_3} = 100 - 50 = 50 \text{ mm}$$

$$\frac{-\Delta P_{N_2O_3}}{\Delta t} = \frac{50-100}{75} = 0.67 \text{ mm sec}^{-1}$$

After 150 seconds, let Y = the mm of N_2O_3 which has decomposed. The 175 = (100 – Y) + Y + Y, Y = 75 mm, and $P_{N_2O_3}$ = 100 – 75 = 25 mm

$$\frac{-\Delta P_{N_2O_3}}{\Delta t} = \frac{25-50}{150-75} = 0.33 \text{ mm sec}^{-1}$$

for the interval from 75 to 150 seconds. For the entire time interval

$$\frac{-\Delta P_{N_2O_3}}{\Delta t} = -\frac{25-100}{150-0} = 0.50 \text{ mm sec}^{-1}$$

17. Since the reaction is first order in C_4H_9I we can write

$$\frac{-\Delta[C_6H_6I]}{\Delta t} = k[C_4H_9I]_{ave} = \frac{\Delta[H^+]}{\Delta t}$$

For each 30-minute time interval we can calculate the $[C_4H_9I]_{ave}$ since $[C_4H_9I] + [H^+]$ = 0.2 M. We can also calculate the average rate. At t = 0, $[C_4H_9I]$ = 0.2 M. At t = 30 min, $[C_4H_9I]$ = 0.2 – $[H^+]$ = 0.141 M; $C_4H_9I]_{ave}$ = (0.2 + 0.141)/2 = 0.17 M. At t= 60 min; $[C_4H_9I]$ = 0.2 – 0.1 = 0.1 M and $[C_4H_9I]_{ave}$ from 30 to 60 min = (0.1 + 0.141)/2 = 0.121 M. by the same method $[C_4H_9I]_{ave}$ = 0.85 M from 60 to 90 min and = 060 M from 90 to 120 min.

The average rates for the successive time intervals can be calculated directly from $[H^+]$:

$$\frac{\Delta[H^+]}{\Delta t} = \frac{0.0587-0}{30} = 1.96 \times 10^{-3} \text{ moles liter}^{-1} \text{ min}^{-1}$$

for the first 30 minutes: $= \frac{0.1 - 0.0567}{30} = 1.38 \times 10^{-3}$ moles liter^{-1} min^{-1}

for the second 30 minutes; and so on

To calculate k we substitute each pair of values of $[C_4H_9I]_{ave}$ and $\frac{\Delta[H^+]}{\Delta t}$.

Thus $1.96 \times 10^{-3} = k(0.171)$, $k = 0.0115$ min^{-1}. (Note that the units moles/liter appear on both sides of the equation and therefore cancel each other.) For the second 30-minutte interval $1.38 \times 10^{-3} = k(0.121)$, $k = 0.0114$ min^{-1}. This process can be repeated to obtain two more values of k and the best value of k can be obtained by averaging all the values so obtained. It should be noted that in order for the values of k obtained from each interval to be close to the correct one it is necessary to know the order of the reaction. Often it is possible to deduce the order of a reaction by testing the data in the way done here for different orders and seeing which order gives the best agreement between the calculated values of k for each time interval.

18. In the form of the integrated rate equation

$$2.3 \log \frac{[A]}{[A_0]} = -kt,$$

$[A]$ and $[A_0]$ refer to C_4H_9I. Since the $[C_4H_9I]$ is given to be 0.2 M initially, that $[C_4H_9I]$ at any time is $0.2 - [H^+]$. The calculation of k simply involves substitution of the appropriate pairs of values into the equation. For example at 30 minutes $[C_4H_9I] \doteq 0.2 - 0.0587 = 0.141$. Substituting

$$2.3 \log \frac{(0.1413)}{0.2} = -k(30)$$

$$k = 1.16 \times 10^{-2} \text{ min}^{-1}$$

At 60 minutes $[C_4H_9I] = 0.2 - 0.1 = 0.1$. Substituting

$$2.3 \log \frac{(0.1)}{(0.2)} = -k60$$

$$k = 1.16 \times 10^{-2} \text{ min}^{-1}$$

This process can b repeated for each measurement and the values of k so obtained averaged to give the best value.

19. Since this reaction is known to be first order in $[C_4H_9I]$ the integrated first order rate equation may be used.

$$t = \left(2.3 \log \frac{A}{A_0}\right)\left(-\frac{1}{k}\right) = \left(2.3 \log \frac{0.125}{0.2}\right)\left(\frac{1}{1.16 \times 10^{-2}}\right) = 40.5 \text{ min}$$

20. Since the initial concentration and the concentration at different times are given, k can be calculated by substitution into the integrated second order rate equation.

$$\frac{1}{A} - \frac{1}{A_0} = kt \text{ for } t = 250 \text{ sec}$$

$$\frac{1}{200} - \frac{1}{300} = k(250) \text{ mol}^{-1} \text{ sec}^{-1} \; k = 6.67 \times 10^{-6} \text{ liters mol}^{-1} \text{ sec}^{-1}$$

Repeating for another time: $\frac{1}{151} - \frac{1}{300} = k(500)$

$\Rightarrow$ $k = 6.58 \times 10^{-6}$ liters mol^{-1} sec^{-1}. The best value of k is then obtained by averaging the value of k obtained from each measurement.

21. The balanced equation is: $2N_2O_5 \rightarrow 4NO_2 + O_2$
Note that the moles of O_2 formed will be one half that of the moles of N_2O_5 decomposed. Now use the integrated form of the firs order rate law to calculate the fraction of the 1.0 mole N_2O_5 decomposing in 20 hr or 1,200 min.

$$\log \frac{[N_2O_5]_t}{[N_2O_5]_O} = \frac{-kt}{2.303} = \frac{-(6.2\times10^{-4})(1.200)}{2.303} = -0.32305$$

$$\therefore \quad \frac{[N_2O_5]_t}{[N_2O_5]_O} = 0.47527$$

Fraction of N_2O_5 decomposed = 1.0 – 0.47527 = 0.52473

Since there is 1.0 mol of N_2O_5 present at the start, the moles of N_2O_5 decomposed is 0.52473 mol. Thus,

$$\text{Mol } O_2 \text{ formed} = (1\ O_2/2N_2O_5) \times 0.52473 \text{ mol } N_2O_5 = 0.26237 \text{ mol } O_2$$

Now use ideal gas law to calculate the volume of this number of moles of O_2 gas at 318 K and 780 mmHg.

$$V = \frac{nRT}{P} = \frac{(0.26237\text{ mol})(0.08206\text{ L}\cdot\text{atm}/(\text{K}\cdot\text{mol})(318\text{ K})}{(780/760)\text{ atm}} = 6.67$$

22. Using the first-order rate law, the initial rate of decomposition is given by

$$\text{Rate} = k[H_2O_2] = (7.4 \times 10^{-4}/\text{s}) \times (1.5\text{ M } H_2O_2) = 1.11 \times 10^{-3}\text{ M } H_2O_2/\text{s}$$

That heat liberated per second per mol of H_2O_2 can be found by first calculating the standard enthalpy of the decomposition of 1 mol of H_2O_2.

	$H_2O_2(aq)$	$\rightarrow$	$H_2O(l)$	+	$1/2O_2(g)$
$\Delta H^0 f$	$= -191.2\text{ kJ mol}^{-1}$		$-285.84\text{ kJ mol}^{-1}$		0 kJ mol^{-1}

For the reaction, the standard enthalpy change is

$$\Delta H^0 = -285.84\text{ kJ mol}^{-1} - (-191.2\text{ kJ mol}^{-1}) = -94.64\text{ kJ mol}^{-1}\ H_2O_2$$

The heat liberated per second is:

$$\frac{94.64\text{ kJ}}{\text{mol } H_2O_2} \times \frac{1.11\times10^{-3}\text{ mol } H_2O_2}{\text{L}\cdot\text{s}} \times 2.0\text{ L} = 0.2101 = 0.21\text{ kJ s}^{-1}$$

23. Use the ideal gas law (P/RT = n/V) to calculate the mol L^{-1} of each gas

$$[O_2] = \frac{(345/760)\text{ atm}}{0.0820\text{ L}\cdot\text{atm}/(\text{K}^{-1}\text{mol}^{-1})\times 612\text{ K}} = 0.009039\text{ mol } O_2\text{ L}^{-1}$$

$$[NO] = \frac{(155/760)\text{ atm}}{0.082\text{ L}\cdot\text{atm}\text{K}^{-1}\text{ mol}^{-1}\times 612\text{ K}} = 0.004061\text{ mol NO L}^{-1}$$

The rate of decreases of NO is

$$\frac{1.16\times10^{-5}\text{ L}^2}{\text{mol}^2\cdot\text{s}} \times \left(\frac{4.061\times10^{-3}\text{ mol}}{1\text{L}}\right) \times \frac{9.039\times10^{-3}\text{ mol}}{1\text{L}}$$

$$= 1.729 \times 10^{-12}\text{ mol L}^{-1}\text{ s}^{-1})$$

The rate of decrease in atm/s is found by multiplying by RT

$$\frac{1.729 \times 10^{-12}\ \text{mol}}{\text{L} \cdot \text{s}} \times \frac{0.082 \cdot \text{atm}}{\text{K} \cdot \text{mol}} \times 612\ \text{K} = 8.682 \times 10^{-11}\ \text{atm/s}$$

The rate of decrease in mmHg/s is

$$8.682 \times 10^{-11}\ \text{atm/s} \times (760\ \text{mmHg/atm}) = 6.598 \times 10^{-8} = 6.6 \times 10^{-8}\ \text{mmHg/s}$$

24. In both the trails depicted, the amount of reactant is reduced to half, and thus the time is the half-time. Calculate k for each trial assuming a 0 order reaction, a 1st order reaction and 2nd order reaction. The equation that gives a constant k for both trials indicates order the reaction.

0 order: half-life = $[A]_0/2k$

trial 1: 1 min = $8/2\ k \Rightarrow k = 4$

trial 2: 2 min = $4/2\ k \Rightarrow k = 1$

The reaction is not 0 order

1st order: half-life = $0.693/\text{k}$

trial 1: 1 min = $893/k \Rightarrow k = 0.693$

trial 2: 2 min = $0.693/k \Rightarrow k = 0.347$

the reaction is not 1st order

2nd order: half-life = $1/k[A]_0$

trial 1: 1 min = $1/k(8) \Rightarrow k = 0.125$

trial 2: 2 min = $1/\text{k}(4) \Rightarrow k = 0.125$

The reaction is 2nd order

25. We have

(a) rate = $k[A]^2[B]$

(b) rate = $k(3\ M)^2\ (4M)$

rate = $k\ 36\ M^3$

(c) Combine the equations from (a) and (b) above, substitute for A (squares) and solve the B (circles).

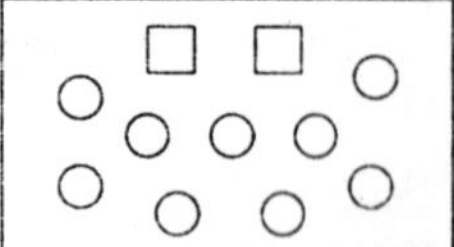

$$k[A]^2[B] = \text{rate} = k(36\ M^3]$$

$$k[2]^2[B] = k(36\ M^3)$$

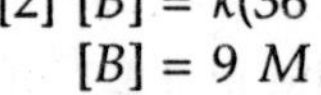

$$[B] = 9\ M$$

26. (a) (1) rate = $k(0.1\ M)\ (0.1\ M) = 0.01\ k$

(2) rate = $k(0.15\ M)(0.15\ M) = 0.023\ k$

(3) rate = $k(0.06\ M)\ (1.0\ M)\ (1.0\ M) = 0.06\ k$

Test tube (1) will have the smallest rate.

(b) When the temperature increases, rate increases

Since rate and k are directly proportional, k also increase.

E_a is constant for a given reaction, thus E_a remains the same. The increase in rate is due to an increase in the kinetic motion of the molecules.

27. (a) $\Delta H° = \Sigma \Delta H°_{f(\text{prod})} - \Sigma \Delta H°_{f(\text{react.})}$

$\Delta H° = (2\ \text{mol})\ (26.48\ \text{kJ mol}^{-1}) - [(1\ \text{mol})\ (62.44\ \text{kJ mol}^{-1}) + (1\ \text{mol})\ (0.0\ \text{kJ mol}^{-1})]$

$\Delta H° = -9.48\ \text{kJ}$

$$\Delta H^\circ = E_{a(\text{forward})} - E_{a(\text{reverse})}$$

$$E_{a(\text{reverse})} = 165 \text{ kJ mol}^{-1} - (-9.48 \text{ kJ mol}^{-1}) = 174 \text{ kJ mol}^{-1}.$$

(b) $k_{\text{forward}} = Ae^{-Ea/RT}$

$$138 \text{ L mol}^{-1} \cdot \text{s}^{-1} = Ae^{(-165000 \text{ J})(8.31 \text{ J mol}^{-1} \cdot \text{K}^{-1}) (973 \text{ K})}$$

$$138 \text{ L mol}^{-1} \cdot \text{s}^{-1} = Ae^{-20.4}$$

$$138 \text{ L mol}^{-1} \cdot \text{s}^{-1} = A(1.38\times 10^{-9})$$

$$A = 1.0 \times 10^{11}$$

$$k_{\text{reserve}} = Ae^{-Ea/RT}$$

$$k_{\text{reserve}} = (1.0 \times 10^{11})\ e^{(-174000 \text{ J}) (8.31 \text{ J mol} \cdot \text{K}) (973 \text{ K})}$$

$$k_{\text{reserve}} = 46 \text{ L mol}^{-1} \cdot \text{s}$$

(c) $\text{rate}_{\text{reserve}} = k_{\text{reserve}} [\text{Hi}]^2$

$$\text{rate}_{\text{reserve}} = (46 \text{ L mol}^{-1} \cdot \text{s})\ 0.2\ M)^2$$

$$\text{rate}_{\text{reserve}} = 1.8 \text{ mol L}^{-1} \cdot \text{s}$$

28. For first order reactions rate $= \dfrac{-d[A]}{a \cdot dt} = k[A]$

$$-d[A] = ak[A]dt \Rightarrow \int \frac{-d[A]}{[A]} = \int adkt$$

$$\ln \frac{[A]_0}{[A]} = akt$$

Since both a and k are constants, they can be replaced with a new constant k.

$$\ln \frac{[A]_0}{[A]} = kt$$

29. (a) rate $= \dfrac{-d[A]}{a \cdot dt} = k[A]^2$

$$\int \frac{-d[A]}{[A]^2} = \int akdt$$

$$\frac{1}{[A]} - \frac{1}{[A]_o} = a\ kt$$

Since both a and k are constants, they can be replaced with a new constant k.

$$\frac{1}{[A]} - \frac{1}{[A]_o} = kt$$

(b) rate $= \dfrac{-d[A]}{a \cdot dt} = k[A]^3$

$$\int \frac{-d[A]}{[A]^3} = \int akdt$$

or $$\frac{1}{2[A]^2} - \frac{1}{2[A]_o^2} = a\,kt$$

Since both a and k are constants, they can be replaced with a new constant k.

$$\frac{1}{[A]^2} - \frac{1}{[A]_o^2} = 2kt$$

30. To determine the order with respect to A, use two trials in which $[A]$ changes and $[B]$ and $[C]$ are constant (such as experiments 1 and 4, us used below), and solve for m.

$$\frac{\text{rate}_1}{\text{rate}_2} = \frac{k[A]_1^m\ [B]_1^n\ [C]_1^p}{k[A]_2^m\ [B]_2^n\ [C]_2^p}$$

$$\frac{4x}{x} = \left(\frac{0.4}{0.2}\right)^m \left(\frac{0.4}{0.4}\right)^n \left(\frac{0.1}{0.1}\right)^p$$

$$4 = (2)^m$$

$m = 2$ Thus the reaction is 2nd order in A

Determining the order with respect to B involves a small twist of the usual procedure. Use two trials in which $[B]$ changes and $[C]$ is constant such as experiments 2 and 3, as used below), and substitute $m = 2$ into the equation

$$\frac{8x}{x} = \left(\frac{0.4}{0.2}\right)^2 \left(\frac{0.4}{0.2}\right)^n \left(\frac{0.2\text{ mol L}^{-1}}{0.2\text{ mol L}^{-1}}\right)^p$$

$$2 = (2)^n$$

$n = 1$ Thus the reaction is 1st order in B

To determine the order with respect to CC, use to trials in which $[C]$ changes (such as experiments 1 and 2, as used below) and substitute $m = 2$ and $n = 1$ into the equation.

$$\frac{8x}{x} = \left(\frac{0.4}{0.2}\right)^2 \left(\frac{0.4}{0.4}\right)^1 \cdot \left(\frac{0.2}{0.1}\right)^p$$

$$2 = (2)^p$$

$p = 1$ Thus the reaction is 1st order in C

$$\text{rate} = k[A]^2[B][C]$$

31. As 3 times per day, $\Delta t = \dfrac{24\text{ hr}}{3} = 8\text{ hr}$

$$t_{1/2} = 2\text{ days} = 48\text{ hr}$$

$$\text{saturation value} = \frac{X}{1-10^{-0.30\frac{\Delta t}{t_{\frac{1}{2}}}}} = \frac{0.1\text{ g}}{1-10^{-0.30\frac{8}{40}}}$$

$$\text{saturation value} = \frac{0.1\,\text{g}}{1-10^{-0.05}} = \frac{0.1\,\text{g}}{1-0.89} = 0.91\ \text{g}$$

In the second part of the problem, the goal is to find a mass (*X*) of drug that gives a saturation value of 0.5 g.

$$\text{saturation value} = 0.5\ \text{g} = \frac{X}{1-0.89} = \frac{X}{0.11}$$

$$X = 0.055\ \text{g} = 55\ \text{mg}$$

Thus a does of 55 mg, three times a day would produce an accumulation of 0.5 g drug in the body, and thus the patent would experience side effects. To prevent the side effects pharmacist would assign a dosage of 54 mg, three times a day.

32. $K_p = \dfrac{P_{NO}\,P_{O_2}^{1/2}}{P_{NO_2}}\ \text{atm}^{1/2}$ $\quad K_c = \dfrac{[NO][O_2]^{1/2}}{[NO_2]}\left(\dfrac{\text{mol}}{\text{L}}\right)^{1/2}$

For this reaction $\Delta n_{gas} = 1 + \frac{1}{2} - 1 = \frac{1}{2}$. Therefore,

$$K_p = K_c(RT)^{1/2}$$

33. (a) At the instant of measuring these partial pressures, the reaction quotient, *Q*, for the reaction $3H_2(g) + N_2(g) \rightleftharpoons 2NH_3(g)$ has the value

$$Q = P_{NH_3}^2 / P_{N_2} P_{H_2}^3 = [(60)^2\ \text{atm}^2]/[(80)\ (150)^3\ \text{atm}^4]$$
$$= 1.3 \times 10^{-5}\ \text{atm}^{-2}$$

Since $Q < K_p = 6.5 \times 10^{-3}\ \text{atm}^{-2}$, a reaction will occur to make *Q* bigger. The number of moles of NH_3 present and therefore the partial pressure of NH_3, will increase, and some of the H_2 and N_2 present will be used up, so that the partial pressure of H_2 and N_2 will both decrease. Reaction will occur spontaneously to the right, and will continue until $Q = K_p$ and the system is at equilibrium.

(b) Here, $Q = (60)^2/(2)^3(5)\ \text{atm}^{-2} = 90\ \text{atm}^{-2}$

Since $Q > K_p$, a reaction will occur to make *Q* smaller. Some of the NH_3 present will be used up, and more N_2 and H_2 will be formed. Reaction will occur spontaneously to the left, and will continue until $Q = K_p$ and the system is at equilibrium.

34. (a) The reaction quotient expression for this equilibrium is

$$Q = P_{NH_3}\,P_{HCl}$$

Initially $Q = 0$, as there is neither NH_3 nor HCl in the container. The equilibrium constant is $1.04 \times 10^{-2}\ \text{atm}^2$. Since $Q < K_p$, the reaction will occur spontaneously to the right. Some NH_3 and HCl will be formed and there will be some decomposition of the NH_4Cl.

(b) In this reaction NH_3 and HCl are formed in a 1 : 1 molar ratio. Since the partial pressure of each gas is directly proportional to the number of moles of that gas, the partial pressures of NH_3 and HCl must be equal. Let

$$x = P_{NH_3} = P_{HCl}\ \text{at equilibrium}$$

The equilibrium constant expression is

$$K_p = P_{NH_3} P_{HCl} = 1.04 \times 10^{-2} \text{ atm}^2$$

Hence,
$$x^2 = 1.04 \times 10^{-2} \text{ atm}^3$$
$$x = 1.02 \times 10^{-1} \text{ atm}$$

When the system has come to equilibrium at 548 K, the partial pressure of NH_3 and of HCl is 1.02×10^{-1} atm, or 77.5 mmHg.

(c) We can calculate the number of moles of HCl or. NH_3 in the container assuming the gases are ideal.

$$n_{NH_3} = n_{HCl} = \frac{(1.02 \times 10^{-1} \text{ atm})(1.0 \text{ L})}{(0.08206 \text{ L} \cdot \text{atm} \cdot \text{mol}^{-1} \text{ K}^{-1})\ 548.0 \text{ K})} = 2.268 \times 10^{-3} \text{ mol}$$

Since

$$\frac{\text{No. mol } NH_4Cl \text{ decomposed}}{\text{No. mol } NH_3 \text{ formed}} = \frac{1}{1}$$

2.27×10^{-3} mol of solid NH_4Cl must have decomposed. The formula weigh of NH_4Cl is 53.49 g $\cdot$ mol^{-1}, so that the mass of NH_4Cl that has decomposed is

$$(53.49 \text{ g} \cdot \text{mol}^{-1})(2.268 \times 10^{-3} \text{ mol}) = 0.121 \text{ g}$$

As the mass of the original sample was 0.98 g, the mass of solid NH_4Cl remaining in the container at equilibrium is 0.98 – 0.121 = 0.859 g.

35. (a) At the instant of mixing, the reaction quotient is

$$Q = P_{NH_3} P_{HCl} = (0.5)\ (0.8) \text{ atm}^2 = 0.4 \text{ atm}^2$$

Since $K_p = 1.04 \times 10^{-2}$ atm^2, $Q > K_p$. Thus the backward reaction, the reaction to the left, will occur spontaneously, and NH_3 and HCl will combine to form solid NH_4Cl.

(b) At equilibrium, the partial pressure of NH_3 will be lss than 0.500 atm, and the partial pressure of HCl will be less than 0.8 atm, because some NH_3 and some HCl will be used up to form $NH_4Cl(s)$. For every mole of NH_3 that is used up, a mole of HCl is also used up, since the two gases combine in a 1: 1 molar ratio. Thus the decrease in the partial pressure of HCl is exactly the same as the decrease in the partial pressure of NH_3. Let

x = decrease in the partial pressure of each gas when the system has come to equilibrium

Then at equilibrium

$$P_{NH_3} = (0.5 - x) \text{ atm}$$
$$P_{HCl} = (0.8 - x) \text{ atm}$$

The equilibrium constant-expression is therefore

$$(0.5 - x)\ (0.8 - x) = 1.04 \times 10^{-2} \text{ atm}^2$$

Multiplying this out and collecting terms, we obtain a quadratic equation in x:

$$x^2 - 1.3\ x + 0.3896 = 0$$

To solve this, use the quadratic formula refer to Appendix B4)

$$x = \frac{1.300 \pm [1.69 - (4)(0.3896)^{1/2}}{2}$$

which yields $x = 0.469$ or 0.831. The value 0.831 must be rejected, as ix must be smaller than 0.500, since both P_{NH_3} and P_{HCl} must be positive. With $x = 0.469$, we obtain

$$P_{NH_3} = 0.500 - 0.469 = 0.031 \text{ atm}$$

$$P_{HCl} = 0.800 - 0.469 = 0.331 \text{ atm}$$

As the partial pressure of H_3 has dropped from its initial value of 0.500 to an equilibrium value of 0.031 atm, we see that most of the NH_3 introduced into the flask has been used u to form NH_4Cl.

36. (a) $K_p = P^2_{SO_3} / P^2_{SO_2} P_{O_2}$

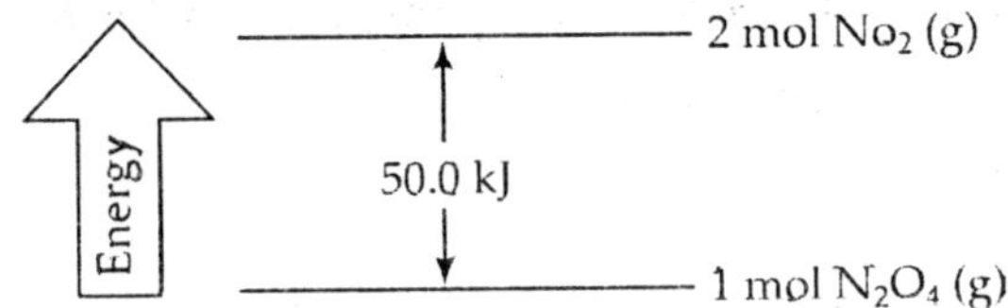

Difference in energy between one mole of N_2O_4 and two moles of NO_2.

(b) The equilibrium constant at 300 K will be greater than the equilibrium constant at 900 K. This is an exothermic reaction. If we decrease the temperature from 900 to 300 K, the equilibrium will be displaced to the right, releasing heat, to offset the stress of decreased temperature. If more SO_3 is produced and SO_2 and O_2 are used up, the equilibrium constant increases. In fact, there is a very large change in the numerical value of the equilibrium constant. At 298 K.

$$K_p = 3.37 \times 10^{24} \text{ atm}^{-1}$$

37. (a) $K_{eq} = K_{sp} = [Mn^{2+}][S^{2-}] \text{ mol}^2 \cdot L^{-2}$ or M^2

(b) Adding excess $Mn(NO_3)_2$ solution will decrease the solubility of MnS. A solution of $Mn(NO_3)_2$ is 100% Mn^{2+} ions and NO_3^- ions. When Mn^{2+} ins are added, the equilibrium shifts to the left to use up some of the added Mn^{2+} ions, and inimze the disturbance. Some Mn^{2+} ions and some S^{2-} ions react to precipitate more MnS. Since the amount of solid MnS precipitated has increased, the solubility of MnS has decreased. This is an example of the common ion effect.

(c) Adding excess HCl will greatly increase the solubility of MnS. Indeed, it is possible to completely dissolve MnS in excess HCl. Hydronium ions combine with S^{2-} ions to form the weak acids HS^- and H_2S. If excess HCl is used H_2S will be the principal product. Adding $H^+(aq)$ ions decrease the $[S^{2-}]$ in solution, and the equilibrium shifts to the right, producing more Mn^{2+} and S^{2-} ions, to offset the stress. The net ionic equation for dissolving MnS in HCl is

$$MnS(s) + 2H^+(aq) \rightleftharpoons Mn^{2+}(aq) + H_2S(aq)$$

38. For this reaction

$$K_p = P^2_{SO_2} P_{O_2} / P^2_{SO_3}$$

This is just the inverse of the expression used in Q. 36(a). Hence, as written here,

$$K_p \text{ at } 900\text{ K} = \frac{1}{40.5\text{ atm}^{-1}} = 2.47 \times 10^{-2}\text{ atm}$$

$$K_p \text{ at } 298\text{ K} = \frac{1}{3.37 \times 10^{24}\text{ atm}^{-1}} = 2.97 \times 10^{-25}\text{ atm}$$

39. $K_{pI} = P_{NOCl}/P_{NO}\,P_{Cl_2}^{1/2}$

$$K_{pII} = P_{Cl_2}\,P_{NO}^2 / P_{NOCl}^2$$

Equation (II) is obtained from Eq. (1) (see problem) by multiplying the equation by 2 and reversing its direction. Inspection of the expression for K_{pI} and K_{pII} shows that

$$K_{pII} = \frac{1}{K_{pI}^2}$$

Hence,

$$K_{pII} = \frac{1}{(3.7 \times 10^3\text{ atm}^{-1/2})^2} = 7.3 \times 10^{-8}\text{ atm}$$

40. Equation (III) is simply the algebraic sum of Eqs. (I) and (II) (see problem). The equilibrium constants for these three reactions are

$$K_I = \frac{P_{H_2}\,P_{Cl_2}}{P_{HCl}^2} \quad K_{II} = \frac{P_{ICl}^2}{P_{I_2}\,P_{Cl_2}} \quad \text{and} \quad K_{III} = \frac{P_{H_2}\,P_{ICl}^2}{P_{HCl}^2\,P_{I_2}}$$

We can see that

$$K_I K_{II} = \left(\frac{P_{H_2}\,P_{Cl_2}}{P_{HCl}^2}\right)\left(\frac{P_{ICl}^2}{P_{I_2}\,P_{Cl_2}}\right) = K_{III} = (4.17 \times 10^{-34})\ (2.1 \times 10^5)$$

$$K_{III} = 8.8 \times 10^{-29} \text{ at } 298\text{ K}$$

41. To obtain Eq. (3) we must reverse Eq. (1) and then add it to Eq. (2). Let us call the reverse of Eq. (1), i.e. (see problem)

$$N_2O(g) \rightleftharpoons N_2(g) + \frac{1}{2}O_2(g) \qquad \ldots\ldots (4)$$

Then $$K_4 = \frac{P_{N_2}\,P_{O_2}^{1/2}}{P_{N_2O}} = \frac{1}{7.1 \times 10^{-19}} = 1.4 \times 10^{18}$$

Adding Eqs. (2) and (4) yields Eq. (3). Thus K_3 is the product of K_2 and K_4:

$$K_3 = \frac{P_{NO}^2}{P_{N_2O}\,P_{O_2}^{1/2}} = \left(\frac{P_{NO}^2}{P_{N_2}\,P_{O_2}}\right)\left(\frac{P_{N_2}\,P_{O_2}^{1/2}}{P_{N_2O}}\right) = K_2K_4 = K_2K_1^{-1}$$

$$= (4.23 \times 10^{-31})\ (1.4 \times 10^{18}) = 6.0 \times 10^{-13} \text{ at } 298\text{ K}$$

42. Since; $\frac{\text{No. mol } SO_2 \text{ used up}}{\text{No. mol } SO_3 \text{ formed}} = \frac{1}{1}$

the *decreases* in the partial pressure of SO_2 from its initial value will be equal to the partial pressure of So_3 in the flask at equilibrium, because the initial pressure of SO_3 was zero. Remember that the partial pressure of a gas is directly proportional to the number of moles of that gas, at fixed volume and temperature. Hence, let at equilibrium

$$x = P\, P_{SO_3}. \text{ Then } (1.0 - x) = P_{SO_2}$$

The decreases in the partial pressure of O_2 is only half as large a the decrease in the partial pressure of SO_2 because

$$\frac{\text{No. mol } O_2 \text{ used up}}{\text{No. mol } SO_2 \text{ used up}} = \frac{1}{2}$$

$\therefore \quad (0.5 - x/2) = P_{O_2}$

By definition, $P_{total} = P_{SO_3} + P_{SO_2} + P_{O_2}$. Substituting the values into this equation yields

$1.35 = x + (1.0 - x) + (0.5 - x/2)$

or $\quad 1.35 = 1.5 - x/2$

$\therefore \quad x = 3.0 - 2.70 = 0.3$ atm

Thus at equilibrium,

$P_{SO_3} = x = 0.3$ atm

$P_{SO_2} = (1.0 - x) = 0.70$ atm

and, $\quad P_{O_2} = (0.5 - x/2) = 0.5 - 0.15 = 0.35$ atm

The value of the equilibrium constant at 1100 K is therefore

$$K_p = P^2_{SO_3} / P^2_{SO_2}\, P_{O_2} = (0.3)^2/(0.7)^2(0.35) = 0.52 \text{ atm}^{-1}$$

43. The equilibrium constant for this reaction is dimensionless, so that both K_c and K_p are 6.

$$K_e = \frac{[HI]^2}{[H_2][I_2]} = 6 = K_p = \frac{P^2_{HI}}{P_{H_2}\, P_{I_2}}$$

Let x equal the number of moles of I_2 used up when the system has come to equilibrium. Then, since H_2 and I_2 react in a 1 : 1 molar ratio, the number of moles of H_2 used up when the system comes to equilibrium is also x. The number of moles of each of these gases present at equilibrium is the initial number of moles *minus* the amount used up, that is, $1.0 - x$. Since

$$\frac{\text{No. mol HI formed}}{\text{No. mol } I_2 \text{ used up}} = \frac{2}{1}$$

the number of oles of HI formed when the system is at equilibrium is $2x$. The volume of the flask is 1.0 L. The concentration, in moles per liter, of each of these three gases is, therefore,

$$[H_2] - [I_2] = (1.0 - x) \text{ and } [HI] = 2x \text{ at equilibrium}$$

Substituting into the expression for K_c, we obtain

$$K_c = 6 = \frac{(2x)^2}{(1.0 - x)^2}$$

To solve this equation, take the square root of both sides. We obtain

$$2.45 = \frac{2x}{(1.0 - x)};\ \text{or } 2.45 - 2.45\,x = 2x$$

$$\Rightarrow 2.45 = 4.45\,x$$

so that $x = 0.55 \text{ mol} \cdot L^{-1}$

At equilibrium therefore,

$$[HI] = 2x = 1.1 \text{ mol} \cdot L^{-1}$$

$$[H_2] = [I_2] = 1.0 - x = 1.0 - 0.55 = 0.45 \text{ mol} \cdot L^{-1}$$

44. At the instant the partial pressure of HI is increased to 1.0 atm, the reaction quotient Q, becomes

$$Q = \frac{(1.0 \text{ atm})^2}{(0.08592 \text{ atm})(0.08592 \text{ atm})} = 1.35 > 48.7 = K_{eq}$$

Since $Q > K_{eq}$ the reaction must shift to the left to restore equilibrium. Let x equal the increase in the partial pressure of H_2 and I_2 that restores the system to equilibrium. Then the partial pressure of HI must decrease by an amount $2x$, because, for the revese reaction,

$$\frac{\text{mol HI used up}}{\text{mol } H_2 \text{ formed}} = \frac{2}{1}$$

We can make a table to organize these calculations

	P_{I_2} *(atm)*	P_{H_2} *(atm)*	P_{III} *(atm)*
Instant of disturbance	0.08592	0.08592	1.0
Change	x	x	$-2x$
New equilibrium	$(0.08592 + x)$	$(0.08592 + x)$	$(1.0 - 2x)$

At the new position of equilibrium,

$$\frac{(1.0\ \ 2x)^2}{(0.08592 + x)^2} = 48.7$$

Taking the square root of both sides of this equation yields

$$\frac{1.0 - 2x}{0.08592 + x} = 6.979$$

so that $1.0 - 2x = 0.5996 + 6.979x$. Solving this equation for x we obtain $0.4004 = 8.979x$, and $x = 0.04459$ atm. when equilibrium is restored,

$$P_{H_2} = P_{I_2} = 0.08592 + 0.04459 = 0.13051 \text{ atm}$$

$$P_{HI} = 1.0 - 2(0.04459) = 0.9108 \text{ atm}$$

You can check your calculations by verifying that $\dfrac{(0.9108)^2}{(0.1305)^2} = 48.7$

45.

	H_2	+	I_2	$\rightleftharpoons$	2HI
Initial	a		b		o
	$a - x$		$b - x$		$2x$
equib. conc.	$(a - x)/V$		$(b - x)/V$		$2x/V$

We have $$k_c = \frac{[HI]^2}{[H_2][I_2]} = \frac{(2x/V)^2}{[(a-x)/V][(b-x)/V]} = \frac{4x^2}{(a-x)(b-x)}$$

Here a = 2.94 moles; $b = 8.1$ moles; $2x = 5.64$ moles; $x = 2.82$ moles

$$\therefore \quad k_c = \frac{4 \times (2.82)^2}{(2.94 - 2.82)(8.1 - 2.82)}; \; k_c = \frac{4 \times 2.82 \times 2.82}{0.12 \times 5.28} = 50.12$$

46. Let x moles/litre of H_2 react with x moles litre^{-1} of I_2 to give $2x$ moles per litre of HI.

The reaction is	H_2 +	I_2	$\rightleftharpoons$ 2HI
Initial conc.	8	5	0
eqm. conc	$8 - x$	$5 - x$	$2x$

According to the Law of Mas Action

$$K_c = \frac{[HI]^2}{[H_2][I_2]} = \frac{2x^2}{(8-x)(5-x)} \quad \text{i.e, } 60 = \frac{4x^2}{40 - 13x + x^2}$$

Simplifying $x^2 - 13.93x + 42.86 = 0$

The quadratic equation is solved as follows:

$$x = \frac{13.39 \pm \sqrt{(13.93)^2 - 4 \times 42.86}}{2} \doteq \frac{13.39 \pm \sqrt{194.1715}}{2} = \frac{13.39 + 4.743}{2}$$

$\Rightarrow x = 4.595$ or 9.335

The value of x can not be 9.335 because it is greater than the initial concentration.

$\therefore \quad x = 4.6$ moles litre^{-1}

The amount of HI present at equilibrium is $2x = 2 \times 4.6 = 9.2$ moles litre^{-1}

47. Let the initial concentration of PCl_5 be 1 mole. The degree of dissociation of PCl_5 is 70% i.e. 0.7. At equilibrium the concentrations of PCl_3, Cl_2 and PCl_5 are 0.7, 0.7 and 0.3 respectively. Therefore the total number of moles in the reaction mixture are 1.7.

	at start PCl_5	$\rightarrow$ PCl_3	+ Cl_2
Initial mole	1	0	0
moles reacted	0.7	0.7	0.7
moles at equilibrium	1.07 = 0.3	0.7	0.7

The equilibrium constant in terms of partial pressure is given by

$$K_p = \frac{P_{PCl_3} \times P_{Cl_2}}{P_{PCl_5}}$$

Partial pressure of any component = mole fraction × total pressure

Thus $$P_{PCl_3} = \frac{0.7}{1.7} \times 2 = 0.82 \text{ atm},$$

$$P_{Cl_2} = \frac{0.7}{1.7} \times 2 = 0.82 \text{ atm}$$

$$P_{PCl_5} = \frac{0.3}{1.7} \times \ = 0.35 \text{ atm}$$

$$K_p = \frac{0.82 \text{ min} \times 0.82 \text{ atm}}{0.35 \text{ atm}} = 1.9 \text{ atm.}$$

The equilibrium constant is 1.9 atm.

48. Applying the law of mass action to the system.

H_2	+	I_2	$\rightleftharpoons$	2HIs	
2		3		0	(initial)
$2 - x$		$3 - x$		$2x$	(at equilibrium)

(where x is the degree of dissociation)

Applying the law of mass action to the system, we get

$$K = \frac{[HI]^2}{[H_2][I_2]}$$

or $$0.02 = \frac{2x \times 2x}{(2-x)(3-x)}$$

$$\frac{1}{50} = \frac{4x^2}{6 - 5x + x^2}$$

or $$199x^2 + 5x - 6 = 0$$

or $$x = \frac{-5 \pm \sqrt{5^2 - (4)(199)(-6)}}{2 \times 199} = \frac{-5 \pm 69.3}{398} = 0.1615$$

% of iodine which is converted into HI

$$= \frac{0.1615 \times 100}{3} = 5.38\%$$

49.

H_2	+	I_2	$\rightleftharpoons$	2HI	
0.2		0.2		0	(initially)
$0.2 - x$		$0.2 - x$		$2x$	(at equil.)

$$K_C = \frac{[HI]^2}{[H_2][I_2]} = \frac{(2x)^2}{(0.2-x)(0.2-x)}$$

$$49.5 = \frac{(2x)^2}{(0.2-x)^2}$$

or $x = 0.16$

$[H_2] = [I_2] = 0.2 - 0.16 =$ **0.04 mole, [HI] = 0.32 mole**

50. We have $t_{1/2} = \dfrac{0.693}{K}$ or $K = \dfrac{0.693}{t_{1/2}}$

Here, $t_{1/2} = 60$ minutes or 60×60 sec.

$$K = \frac{0.693}{60 \times 60} = 1.925 \times 10^{-4} \text{ sec}^{-1}$$

$\therefore$ Rate constant 1.925×10^{-4} sec^{-1}

51. Here, $t_{1/2} = \dfrac{0.693}{K}$; $K = 6.2 \times 10^{-4}$ sec^{-1}

$$\therefore\ t_{1/2} = \frac{0.693}{6.2 \times 10^{-4}} \text{ sec} = 0.1118 \times 10^{4} \text{ sec.} = 1118 \text{ sec.}$$

Thus, Half life period is 1118 seconds.

52. As, 50% of the reactant is consumed in 20 min. therefore at the end of 20 minutes 50% of the original amount remains.

Half of this remaining amount of the reactant is consumed in the next 20 min i.e. 50% of the reactant gets reduced to 25%.

Therefore totally 75% of the reaction is over in 40 minutes

53. Rate $= \dfrac{d[A]}{dt} = \dfrac{0.032 - 0.044}{25} \dfrac{\text{moles lit.}^{-1}}{\text{min}} = 0.00048$ mole lit^{-1} min^{-1}

Rate of the reaction $= 4.8 \times 10^{-4}$ mole lit^{-1} min.$^{-1}$

54. The expressions for the velocity constant of the first order reactions is

$$K = \frac{2.303}{t} \log \frac{a}{a - x}$$

For 50% complition, $t_{50\%} = \dfrac{2.303}{K} \log \dfrac{100}{100 - 50} = \dfrac{2.303}{K} \log 2$ (1)

and 80% complition, $t_{80\%} = \dfrac{2.303}{K} \log \dfrac{100}{100 - 80} = \dfrac{2.303}{K} \log 5$ (2)

Dividing (2) by (1) we have

$$\frac{t_{80\%}}{t_{50\%}} = \frac{2.303 \log 5}{K} \times \frac{K}{2.303 \log 2} = \frac{\log 5}{\log 2}$$

$$\therefore \text{ Time for 80\% change} = 69.3 \times \frac{\log 5}{\log 2} = \frac{69.3 \times 0.6990}{0.3010} = 160.9 \text{ min.}$$

55.
$$\begin{array}{ccccc} H_2 & + & I_2 & \rightleftharpoons & 2HI \\ a - x & & b - x & & 2x \end{array}$$

Molar concentration of HI at equilibrium= $2x = 5.64$

$$x = \frac{5.64}{2} = 2.82$$

Molar concentration of H_2 at equilibrium $= 8.1 - 2.82 = 5.28$
Molar concentration of I_2 at equilibrium $= 2.94 - 2.82 = 0.12$

$$\text{Equilibrium constant} = \frac{4 \times 2.82 \times 2.82}{5.28 \times 0.12} = 50.12$$

56. The expression for the velocity constant of the first order reaction is

$$K = \frac{2.303}{t} \log \frac{a}{a - x}$$

The initial concentration $a = 100$

t in min	$a - x$	$K = \frac{2.303}{t} \log \frac{a}{a-x}$ min^{-1}
30	100 − 32 = 68	$= \frac{2.303}{30} \log \frac{100}{68} = 0.01286$
53	100 − 50 = 50	$= \frac{2.303}{53} \log \frac{100}{50}$ 0.01308
65	100 − 57 = 43	$= \frac{2.303}{65} \log \frac{100}{43} = 0.01299$
80	100 − 65 = 35	$= \frac{2.303}{80} \log \frac{100}{35} = 0.01312$
100	100 − 73 = 27	$= \frac{2.303}{100} \log \frac{100}{27} = 0.01309$
120	1000 − 78	$= \frac{2.303}{120} \log \frac{100}{22} = 0.01262$

The values of K are fairly constant showing that the reaction is of the first order.

57. We have, $CH_3COOH + C_2H_5OH \rightleftharpoons CH_3COOC_2H_5 + H_2O$

	CH_3COOH	C_2H_5OH	$CH_3COOC_2H_5$	H_2O
(Initial moles)	240/60	46/46	0	0
(equilibrium conc.)	$(4 - x/V)$	$(1 - x/V)$	x/V	x/V

$$K_C = 4 = \frac{x^2/V^2}{(4-x)(1-x)/V^2} = \frac{x^2}{(4-x)(1-x)} \text{ or } \frac{x^2}{x^2 - 5x + 4} = 4 \text{ or } x = 0.93$$

At equilibrium
Ester $= x = 0.93$ moles; water $= x = 0.93$ mole; acid $= 4 - 0.93 = 3.07$ moles; alcohol $= 1 - 0.93 = 0.07$ mole

58. Total number of moles of equilibrium in the gaseous phase $= 2 + 1 = 3$

Mole fraction of NH_3 is $\frac{2}{3}$ and that of CO_3 is $\frac{1}{3}$

partial pressure of $NH_3 = \frac{2}{3} \times 0.225 = 0.15$ atm.

Partial pressure of $CO_2 = 0.0225 \times \frac{1}{3} = 0.075$ atm.

$$K_p = (p_{NH_3})^2 (p_{CO_2}) = 0.15 \times 0.15 \times 0.075 = 1.69 \text{ x. } 10^{-2} \text{ atm.}$$

The introduction of insert gas increases the pressure to 1 atm. due to the partial pressure of this introduced gas which does not affect the partial pressure of NH_3 and CO_2 and hence the value of K_p is not affected.

59. Equilibrium concentration per litre

$$[N_2] = \frac{3.4}{2.8 \times 5} = 0.024 \; [H_2] = \frac{0.16}{2 \times 5} = 0.016 \; [NH_3] = \frac{19}{17 \times 5} = 0.224$$

$$K_C = \frac{0.224 \times 0.224}{0.016 \times 0.016 \times 0.016 \times 0.024} = 5.1 \times 10^5 \text{ mole}^{-2} \text{ litre}^2$$

$$K_p = K_C \times [RT]^{\Delta n} = 5.1 \times 10^5 \times (0.0821 \times 620)^{-2} = 1.968 \times 10^2 \text{ atm, mole}^{-3}\text{, litre}^3$$

60. Since volume of the vessel is constant during the process, hence the pressure at 620 K 347°C can be calculated as,

$$\frac{P_1}{T_1} = \frac{P_2}{T_2}$$

or $$\frac{15 \text{ atm}}{300 \text{ K}} = \frac{P_2}{620 \text{ K}}$$

or $$P_2 = \frac{15 \text{ atm} \times 620 \text{ K}}{300 \text{ K}} = 31 \text{ atm}$$

It means initial pressure is 31 atm. at 620 K

Now,

$$2NH_3 \rightleftharpoons N_2 + 3H_2$$

$2NH_3$	N_2	$3H_2$	
1	0	0	(Initially)
$1 - x$	$\frac{x}{2}$	$\frac{3x}{2}$	(A equilibrium)

No. of moles initially = 1

NO. of moles at equilibrium

$$= 1 - x + \frac{x}{2} + \frac{3x}{2} = 1 + x$$

The ratio of no. of moles is equal to the ratio of the pressure. thus

$$\frac{1}{1+x} = \frac{31 \text{ atm}}{50 \text{ atm}}$$

or $$x = \frac{19}{31} = 0.6129 = \mathbf{61.29\%}$$

61. (i) $H_2 + I_2 \rightleftharpoons 2HI$

1 mole 1 mole 2 mole

Since $K_p = K_c(RT)^{\Delta n} = 50(RT)^0$ ($\because \Delta n = 0$)

or $K_p = 50$

(ii) $H_2 + I_2 \rightleftharpoons 2HI$

	H_2	I_2	$2HI$	
	0.5	0.5	0	(Initially)
	$(0.5 - x)$	$(0.5 - x)$	$2x$	(At equilibrium)

Now $$K_c = \frac{[HI]^2}{[H_2][I_2]} = \frac{(2x)^2}{(0.5-x)(0.5-x)}$$

or $$50/4 = \frac{(x)^2}{(0.5-x)^2}$$

or $$\frac{x}{0.5-x} = 3.53; \Rightarrow x = 0.39$$

Excess of iodine = 0.5 – 0.39 = **0.11 moles**

62. Pressure of NH_3 at 300 K = 15 atm.

By Charles' law, pressure of ammonia at 620 K = $\frac{15 \times 620}{300}$ = 31 atm.

It there were 1 mole of ammonia in the beginning and xmole decomposed with the rise of temperature the total no. of moles after decomposition at equilibrium is given by the following reaction.

$$NH_3 \rightleftharpoons \frac{1}{2}N_2 + \frac{3}{2}H_2$$

$$1 - x \quad \frac{x}{2} \quad \frac{3x}{2}$$

Total no. of moles after decomposition = $1 - x\ \frac{x}{2} + \frac{3x}{2}$ = 1 + x and this molar concentration corresponds to 50 atm. at the same volume and temperature. thus

$$\frac{\text{no. of moles without decomposition}}{\text{no. of moles after decomposition}} = \frac{1}{1+x} = \frac{31}{50}$$

Hence x = 0.613; % decomposition of NH_3 = 0.613 × 100 = 61.3%

63.

	PCl_5	$\rightleftharpoons$ PCl_3	+ Cl_2	
Initial conc.	2.0	0	0	
At equilibrium	$\left(2 - \frac{2 \times 40}{100}\right)$	$\frac{2 \times 40}{100}$	$\frac{2 \times 40}{100}$	(equil. conc.)

$$[PCl_5] = \frac{2-0.8}{2}; [PCl_3] = \frac{0.8}{2}; [Cl_2] = \frac{0.8}{2}$$

(Since the vessel is of two litre capcatiy)

$$K = \frac{[PCl_3][Cl_2]}{[PCl_5]} = \frac{0.4 \times 0.4}{0.6} = 0.267 \text{ moles litre}$$

64.

	H_2	+	I_2	$\rightleftharpoons$	$2HI$
(initial conc.)	60 ml		42 ml		0
(equl. conc)	46		28		28 ml

$$K = \frac{[HI]^2}{[h_2][I_2]} = \frac{28 \times 28}{46 \times 28} = \frac{14}{23}$$

For dissociation of HI equilibrium constant will be $\frac{1}{K} = \frac{23}{14} = 1.2817$. The dissociation of HI is

$2HI$	$\rightleftharpoons$	H_2	+	I_2
1		0		0
$(1 - \alpha)$		$(\alpha/2)$		$(\alpha/2)$

$$k = \frac{(\alpha/2)(\alpha/2)}{(1-\alpha)^2} = \frac{\alpha^2}{4(1-\alpha)^2} = 1.2817$$

$$\alpha^2 = 4(1.2817)(1 + \alpha^2 - 2\alpha) = 5.126 + 5.126\,\alpha^2 - 10.253\,\alpha$$

$\therefore$ $4.125\,\alpha^2 = 10.253\,\alpha + 5.126 = 0$

After evaluating for $\alpha = 0.69$ $\quad\therefore$ The degree of dissociation is 0.69

65.

	$2HI$	$\rightleftharpoons$	H_2	+	I_2
(initial conc)	5		0		0
(equl. conc)	$5(1 - x)$		$5x/2$		$5x/2$

$x = 22\% = 0.22$ per mole

At equilibrium the no. of moles of HI $= 5(1 - 0.22) = 3.90$
the no. of moles of $H_2 = 5 \times 0.11 = 0.55$
the no. of moles of $I_2 = 5 \times 0.11 = 0.55$

66. (a) False As greater K_f corresponds to more product at equilibrium. Thus larger K_f's, also corresponds to more stable products.

(b) True. $[Ag(NH_3)_2]^+$ undergoes the following reaction:

$$[Ag(NH_3)_2]^+ \rightleftharpoons Ag^+ + 2NH_3 \qquad \text{eq 1}$$

The NH_3 can then react with HNO_3 to form NH_4^+.

$$NH_3 + HNO_3 \rightleftharpoons NH_4^+ + NO_3^- \qquad \text{eq 2}$$

As this point, the NH_3 is effectively removed from solution. With no NH_3, equation 1 cannot reverse and reverse and reform $[Ag(NH_3)_2]^+$

(c) False $[AgBr_2]^-$ undergoes the following reaction

$$[AgBr_2]^- \rightleftharpoons Ag^+ + 2Br^-$$

Br^- is a spectator (neutral) anion and does not react with HNO_3. The nitric acid, therefore, does not cause $[AgBr_2]^-$ to dissociate.

(d) True AgI undergoes the following reactions in the pressure of CN^-:

$$AgI \rightleftharpoons Ag^+ + I^-$$
$$Ag^+ + 2CN^- \rightleftharpoons [Ag(CN)_2]^-$$

AgI is insoluble, thus the equilibrium lies far to the left (reactant). The reaction with cyanide forms the soluble $[ag(CN)_2]^-$ complex NaCN is salt, which dissociates completely to Na^+ and CN^+. HCN is a weak acid, thus only a small amount dissociates to form CN^-. Consequently, less NaCN would be needed

(e) True The reaction of interest here are

$$Ag^+ + 2Br^- \rightleftharpoons [AgBr_2]^-$$
$$Ag^+ + 2CN^- \rightleftharpoons [Ag(CN)_2]^-$$

The ratio of products reactants is given by the formation constant:

$$K_f = \frac{[AgL_2]^-}{[Ag^+][L^-]^2}, \text{ where } L^- = B^- \text{ or } CN^-$$

Since $[Br^-] = [CN^-]$, the species with the greater K_f will have the greater ratio of $[AgL_2]^-/Ag^+$

67. (a) The observed rate of reaction, depends only on the concentrations of acetone and hydronium ion, Thus

$$\text{rate in solution I} = k_2' (0.1)(0.02) = (k_2')(2.0 \times 10^{-3})$$
$$\text{rate in solution II} = k_2' (0.3)(0.1) = (k_2')(3.0 \times 10^{-3})$$
$$\text{rate in Solution III} = k_2' (0.2)(0.02) = (k_2')(4.0 \times 10^{-3})$$

We see that the rate of reaction between acetone and I_2 is fastest in Solution III and slowest in Solution I.

(b) Adding water to each of these solution increases the volume of solution and thereby decreases the $[CH_3COCH_3]$, $[H_3O^+]$ and $[I_2]$. Thus the rate of reaction will decrease. The value of the specific rate constant, k_2' will remain the same because rate constants change only if the temperature or the solvent is changed.

(c) Decreasing he temperature will cause both the specific rate constant k_2', and the rate pf reaction, to decrease. The rate of most chemical reactions increases by a factor of 2 or 3 for every 10° rise in temperature. A decrease in temperature of 20° therefore causes a significant decrease in the reaction rate.

68. A comparison of Experiments I and II shows us that doubling the initial concentration of B wile leaving the initial concentration of A constant has no effect at all on the initial rate of reaction. We can only conclude that the rate does not depend on $[B]$. a comparison of Experiments I and III reveals that doubling the initial concentration of A while keeping the initial concentration of B constant multiplies the initial rate by a factor of 4, for $4(3.20 \times 10^{-3}) = 1.28 \times 10^{-2}$. This indicates that the rate expression is of the form $k[A]^2$. We can check this by comparing, Experiments II and IV. Tripling the initial concentration of A while holding the initial concentration of B constant increases the rate by a factor of 9 times. We conclude therefore, that the rate is second order in A and zero order in B, and the rate expression is

$$\text{rate} = k[A]^2$$

The units of the specific rate constant, k, are liters per mole per minute.

69. A comparison of Experiments 1 and 2 shows us that doubling the $[I]_o$ while keeping the $[Ar]_o$ constant multiplies the initial rate by 3, since $4(8.70 \times 10^{-4}) = 3.48 \times 10^{-3}$. Comparing Experiments 2 and 3 confirms this result, since the $[I]_o$ is doubled again and the rate is increased by a factor of 4, $4(3.48 \times 10^{-3}) = 1.39 \times 10^{-2}$. A comparison of Experiments 1 and 4 shows that multiplying the $[Ar]_o$ by 5 while keeping the $[I]_o$ constant multiplies the initial rate by a factor of 5. This is confirmed by a comparison of Experiments 2 and 5. All the data are constant with the following rate law: rate = $k_3[I]^2[Ar]$.

This is a third-order reaction, second order in I and first order in Ar. The fact that the rate depends on [Ar] indicates that a three body collision, involving two I atoms and an Ar atoms, is necessary for the reaction to occur. The Ar atoms absorbs some of the energy of the colliding I atoms and prevents the I_2 molecule from dissociating as soon as it is formed.

The rate constant can be determined by substituting any set of data into the rate expression. For Experiment I we obtain.

$$\text{rate} = 8.70 \times 10^{-4} = k_3(1.0 \times 10^{-3})^2\ (1.0 \times 10^{-3}) = k_3(1.0 \times 10^{-13})$$

The value of k_3 is therefore 8.7×10^9 mol^{-2} L^2 s^{-1}.

70. For a gas-phase reaction, the pressure of a gas is used in the rate law instead of the concentration, as they are directly proportional: $P_B = [B]RT$, where $[B] = n_B/V$.

$$2B(g) \rightarrow B_2(g);\ \text{where B(g) represents Butadiene}$$

To determine whether the reaction is first or second order, we must plot both in P_1 versus time and $1/P_B$ versus time using the following tabulated data

t(min)	0	17.3	42.5	55.08	68.05	90.05	119	176.67
ln P_B	–0.184	–0.414	–0.676	–0.7	–0.874	–1.02	–1.18	–1.45
$1/P_B$	1.2	1.51	1.97	2.18	2.4	2.77	3.25	4.25

It is observed that the plot of ln P_B versus time is not linear, but the plot of $1/P_B$ versus time is a straight line. The dimerization of butadiene is therefore second order in butadiene. The rate law is $-dP_B/dt = kP_B^2$. The second-order rate constant is the slope of the plot of $1/P_B$ versus time, and is 1.73×10^{-2} atm^{-1} min^{-1}.

71. Here, $k_1 = 2.45 \times 10^{-4}$ and $T_1 = 285.2$ $k_2 = 20.8 \times 10^{-4}$ and $T_2 = 303.2$. Substitution into the equation yields

$$\text{In}\left(\frac{k_1}{k_2}\right) = -\frac{E_a}{R}\left(\frac{t}{T_1} - \frac{t}{T_2}\right)$$

$$\ln\left(\frac{2.45 \times 10^{-4}}{2.08 \times 10^{-3}}\right) = \frac{E_{act}}{R}\left[\frac{18}{(285.2)(303.2)}\right]$$

Since ln(0.245/2.08) = ln(0.1178) = –2.139, the equation becomes

$$-2.139 = \frac{E_{act}}{R}\left[\frac{18.0}{(285.2)(303.2)}\right]$$

or

$$E_{act}/R = \frac{(2.139)(285.2)(303.2)}{18} = 1.028 \times 10^4 = 1.03 \times 10^4$$

$$E_{act} = (1.028 \times 10^4)(8.3145) = 8.55 \times 10^4 \text{ J} \cdot \text{mol}^{-1} = 85.5 \text{ kJ} \cdot \text{mol}^{-1}$$

72. In addition to the activation energy, we need the specific rate constant (at some other temperature. We can use any value for K_2. Let us choose value at 297 K as 10×10^{-4} (from tables in text book).

$$\therefore \qquad \ln\left(\frac{k_{293}}{k_{297}}\right) = \frac{82.6 \times 10^3}{8.314}\left(\frac{1}{293.2} - \frac{1}{297.2}\right) = -0.456$$

Hence $\qquad \dfrac{k_{293}}{k_{297}} = e^{-0.456} = 0.0634$

$$\therefore \qquad k_{293} = (0.634)(10.0 \times 10^{-4}) = 6.34 \times 10^{-4}$$

Thus the specific rate constant at 293.2 K, is 6.34×10^{-4} mol^{-1} $\text{L} \cdot \text{s}^{-1}$.

73. (a) Let the order with respect to A and B are x and y respectively

$$r = K[A]^x [B]^y$$

$$0.05 = K[0.1]^x [0.1]^y$$

$$0.1 = K[0.2]^x [0.1]^y$$

or $\qquad 2 = [2]^x \Rightarrow x = 1$

$$0.05 = K[0.1]^x [0.2]^y$$

$$1 = [2]^y \Rightarrow y = 0$$

(b) rate equation $= r = K[A] [B]^0$

$$0.1 = K[0.2]$$

$$K = \frac{0.05}{0.1} = 5 \times 10^{-1} \text{ s}^{-1}$$

74. The reaction is $A + B \rightarrow$ products

$$\therefore \qquad r = K [A]^n [B]^m$$

It is given that the initial rate becomes double when the concentration of A is doubled. Therefore, the order or reaction with respect to A is 1. Thus $n = 1$. The reaction is a first order reaction with respect to B.

The rate does not change on doubling the concentration of B. Therefore $n = 0$. That is the reaction is a zero order reaction with respect to B.

Thus $\qquad r = [A]^1 [B]^0$

That is $0.005 = K[0.01]^1 [0.01]^0$

or $\qquad K = \dfrac{0.005}{0.01} = 0.5 \text{ min}^{-1}$

Half-life of A in the reaction $= \dfrac{0.69}{0.5} = 1.38$ min.

75. Let C_o be the initial concentration C_t the concentration after time t. The first order reaction is given by the equation

$$\ln \frac{C_t}{C_0} = -kt$$

(where k is specific rate constant of the reaction).

$$\therefore \qquad k = \frac{1}{t} \ln \frac{C_t}{C_0} \qquad (1)$$

(i) $t = 10$ minutes

$$\therefore k = \frac{1}{10} \ln\left(\frac{100}{100-20}\right) = \frac{1}{10} \ln \frac{100}{80} = \frac{1}{10} \times 2.3026 \times 0.0969 \text{ minute}^{-1} = 0.0223 \text{ minute}^{-1}$$

(ii) Let the time taken for the reaction to go to 75% completion be t minutes

$$\therefore t = \frac{1}{k} \ln\left(\frac{C_0}{C_t}\right) = \frac{1}{0.0223} \times \ln\left(\frac{100}{100-75}\right) = \frac{1}{0.0223} \times \ln \frac{100}{25}$$

$$= \frac{1}{0.0223} \times 2.3026 \times 0.6021 \text{ minutes} = 62.17 \text{ minutes.}$$

76. Representing $T_1 = 300$ K and $T_2 = 310$ K, we have

$$k_1 = A \exp(-\Delta E_A/RT_1)$$

$$k_2 = A \exp(-\Delta E_A/RT_2)$$

Thus, $$\frac{k_2}{k_2} = \exp\left[\frac{\Delta E_A}{R}\left(\frac{1}{T_1} - \frac{1}{T_2}\right)\right]$$

or $$\Delta E_A = \frac{R \ln(k_2/k_1) T_1 T_2}{T_2 - T_1} = \frac{R(\ln 2) T_1 T_2}{T_2 - T_1}$$

For $B \rightarrow$ products, we have

$$k_2 = 2\left(\frac{0.693}{30}\right) \text{min}^{-1} = 0.0462 \text{ min}^{-1}$$

Now, $$\Delta E_B = \frac{1}{2} \Delta E_A$$

$$\therefore \qquad \frac{k_1}{k_2} = \exp\left[\frac{\Delta E_B}{R}\left(\frac{1}{T_2} - \frac{1}{T_1}\right)\right]$$

Substituting the expression of ΔE_B, we get

$$\frac{k_1}{k_2} = \exp\left[-\frac{\ln 2}{2}\right] = 1\frac{1}{\sqrt{2}}$$

Hence, $$k_1 = \left(\frac{1}{\sqrt{2}}\right) (0.0462 \text{ min}^{-1}) = 0.0327 \text{ min}^{-1}$$

77. For first order reaction, we have; $k = \frac{0.693}{t_{0.5}}$

$\therefore$ Reaction rate constant at 300 K; $k_1 = \frac{0.693}{30} = 0.0231 \text{ min}^{-1}$

Reaction rte constant at 320 K; $k_2 = \frac{0.693}{10} = 0.0693 \text{ min}^{-1}$

E (energy of activation) is given by the expression

$$-\frac{E}{2.303 \times R}\left(\frac{T_2 - T_1}{T_1 T_2}\right) = \log \frac{k_1}{k_2}$$

or,

$$-\frac{E}{2.303 \times 8.314}\left(\frac{320 - 300}{320 \times 300}\right) = \log \frac{0.0231}{0.0961}$$

$\Rightarrow$ $E = 44.0 \text{ kJ mol}^{-1}$

78. Here $K_1 = \frac{2.303}{t}\left(\frac{a}{a-x}\right)$

At 298 K : $a = 100$, $a - x = 100 - 25$, $t = 20$ mts.

Thus $K_1 = \frac{2.303}{20} \log \frac{100}{75} = 0.014386 \text{ min}^{-1}$

Calculation of K at 313 K, i.e. K_2 can be done from Arthenius equation

$$\log \frac{K_2}{K_1} = \frac{E_a}{2.303\, R}\left(\frac{1}{T_1} - \frac{1}{T_2}\right)$$

$$\therefore \log \frac{K_2}{0.014386} = \frac{70 \times 10^3}{2.303 \times 8.314} \times \left(\frac{1}{298} - \frac{1}{313}\right) = \frac{70 \times 10^3 \times 15}{2.303 \times 8.314 \times 298 \times 313} = 0.587$$

$$\log K_2 = 0.014386 \times .587 \Rightarrow K_2 = 0.05570 \text{ min}^{-1}$$

And for % decomposition at 313 K
Let the % decomposition in this case = x
Here $a = 100$, $a - x = 100 - x$

$$t = 0 \text{ mts}, K_2 = 0.05570$$

Substituting the values to the first order reaction equation

$$K_2 = 0.05570 = \frac{2.303}{20} \log \frac{100}{100 - x}$$

On usual calculations $x = 67.169 = 67.17\%$

79. Given that rate constant for $A \rightarrow B$

$$K = 4.5 \times 10^{-3} \text{ min}^{-1}$$

$$a_o = \text{initial conc. of } A = 1\text{M}$$

time (t) = 1 h

On applying first order kinetics; $k = \frac{2.303}{t} \log \frac{(a)}{(a-x)}$

$(a - x)$ = conc. of reactant left after limit = 0.76
Rate of reaction = $K[A] = 4.5 \times 10^{-3} \times 0.76 = 3.44 \times 10^{-3}$

80. From question,

$$NH_4^+ \Leftrightarrow NH_3 + H^+ ; \quad K_1 = 5.6 \times 10^{-10}$$

$$NH_4^+ + OH^- \Leftrightarrow NH_3 + H_2O; \quad K_2 = 5.6 \times \frac{10^{-10}}{10^{-14}} = 5.6 \times 10^4$$

Thus, Rate constant of protonation = $\frac{3.4 \times 10^{10}}{5.6 \times 10^4} = 6.07 \times 10^5 \text{ sec}^{-5}$

81. Since $t_{0.5}$ = 10 min, we have

$$k = \frac{0.693}{t_{0.5}} = \frac{0.693}{10 \times 60\,s} = 1.155 \times 10^{-3} \text{ s}^{-1}$$

Substituting the above value of k alongwith the other data in the Arrhenius equation

$K = A\left(\frac{-E_a}{RT}\right)$,

we get $\quad 1.155 \times 10^{-2} = 4 \times 10^{13} \exp\left\{\frac{-98600}{8.314 \times T}\right\}$

or $\quad \ln \frac{1.155 \times 10^{-3}}{4 \times 10^{13}} = \frac{98600}{8.314 \times T}$

$$T = \frac{98600}{8.314 \ln(2.8875 \times 10^{-17})} = \frac{98600}{8.314 \times 2.303 \times 16.54} = 311.36 \text{ K}$$

82. We know that rate of reaction is given as $r = kc$

Let $\quad r_1 = k_1c_1$ and $r_2 = k_2c_2$

$\therefore \quad r_1/r_2 = c_1/c_2 = \frac{0.04}{0.03}$

According to first orer reaction; $k = \frac{2.303}{t} \log \frac{c_1}{c_2}$

On putting the values, we get $k = \frac{2.303}{20 \times 60} \log \frac{0.04}{0.63}$

$$k = 0.0287 \text{ min}^{-1}$$

Now, half life period

$$t_{1/2} = \frac{0.693}{k} = \frac{0.693}{0.0287} = 24.14 \text{ min}$$

83. According to Arrhenius

$$K = Ae^{-E_a/RT} \qquad (1)$$

$$\log \frac{K_2}{K_1} = \frac{E_a}{2.303\,R}\left[\frac{T_2 - T_1}{T_2 T_1}\right] \qquad (2)$$

$$K_1 = 1.5 \times 10^7,\ T_1 = 323\text{ K},\ K_2 = 4.5 \times 10^7,\ T = 373\text{ K}$$

According equation (2) on substituting the given values,

$$E_a = 22\text{ kJ mol}^{-1}$$

and from equation (1) at any tempt.

$$A = 5.4 \times 10^{10}\text{ sec}^{-1}$$

84. Let the rate be V_R = Rate = $k[A]^x\,[B]^y$

Where k_2 = rate constant at 300 K

y = order with respect to B

x = order with respect to A.

We shall have

$$V_1 = 5.0 \times 10^{-4} = [2.5 \times 10^{-4}]^x\ [3.0 \times 10^{-5}]^y$$
$$V_2 = 4.0 \times 10^{-3} = k[5.0 \times 10^{-4}]^x\ [6.0 <> 10^{-5}]^y$$
$$V_3 = 1.6 \times 10^{-2} = k[1.0 \times 10^{-3}]^x\ [6.0 \times 10^{-5}]^y$$

$$\therefore \quad \frac{V_1}{V_2} = \frac{5.0\times10^{-4}}{4.0\times10^{-3}} = \frac{k(2.5\times10^{-4})(3.0\times10^{-5})^y}{k(5.0\times10^{-4})^x\,(6.0\times10^{-3})^y}$$

$$\therefore \quad 1/8 = \left(\frac{1}{2}\right)^{x-y} \text{ i.e. } \left(\frac{1}{2}\right)^3 = \left(\frac{1}{2}\right)^{x+y}$$

$$\therefore \quad x + y = 3 \qquad \text{(i)}$$

$$\text{and} \quad \frac{V_2}{V_3} = \frac{4.0\times10^{-3}}{1.6\times10^{-2}} = \frac{(5.0\times10^{-4})\times(6.0\times10^{-5})^y}{(1.0\times10^{-3})^x\,(6.0\times10^{-5})^y}$$

$$\text{i.e.} \quad \left(\frac{1}{2}\right)^2 = \left(\frac{1}{2}\right)^x$$

(i) $\therefore$ x= order with respect to A = 2

(ii) $\therefore$ Rate constant

$$k = \frac{5.0\times10^{-4}\text{ mol lit}^{-1}\text{ s}^{-1}}{(2.5\times10^{-4}\text{ mol lit}^{-1})^2\,(3.0\times10^{-5}\text{ mol lit}^{-1})} = 2.66 \times 10^8\text{ lit}^2\text{ mol}^2\text{ s}^{-1}$$

(iii) Let K_1, be the rate constant at 320 K

$$k_1 = \frac{2.0\times10^{-3}\text{ mol lit}^{-1}\text{ s}^{-1}}{(2.5\times10^{-4}\text{ mol lit}^{-1})^2\,(3.0\times10^{-5}\text{ mol lit}^{-1})} = 10.66 \times 10^8\text{ ln}^2\text{ mol}^2\text{ s}^{-1}$$

$$\text{Now } \log\frac{k_1}{k} = \frac{E}{2.303\,R}\,\frac{(T_2 - T_1)}{T_1 T_2}$$

i.e. $$\log \frac{10.66}{2.66} = \frac{E}{2.303 \times 8.314} \times \frac{320 - 300}{320 \times 300}$$

i.e. $$\log 4.007 = \frac{E}{2.303 \times 8.314} \times \frac{20}{320 \times 300}$$

$\therefore$ $$E = \frac{0.6029 \times 2.303 \times 8.314 \times 320 \times 300}{20} = \mathbf{55.41\ KJ\ mol^{-1}}$$

(iv) $$\log K = \log A - \frac{E}{2.303\ RT}$$

$\therefore$ $$\log 2.66 \times 10^8 = \log A - \frac{55.41 \times 10^3}{2.303 \times 8.314 \times 300}$$

or $8.4249 + 9.6463 = \log A$

$\therefore$ $\log A = 18.0712$

or $A = 1.179 \times 10^{18}\ s^{-1}$

85. (i) Arrhenius equation is

$$k = A \exp\ (-E_\alpha/RT)$$

where E_α is the energy of activation

$\therefore$ $$2.303 \log k = A + \left(\frac{-E_\alpha}{RT}\right)$$

or $$\log k = \frac{A}{2.303} - \frac{E_\alpha}{2.303\ RT} \qquad \text{(i)}$$

comparing this with the equation given in the question i.e.

$$\log k(s^{-1}) = 14.34 - \frac{1.25 \times 10^4\ K}{T} \qquad \text{(ii)}$$

$$\frac{E_\alpha}{2.303\ RT} = \frac{1.25 \times 10^4\ K}{T}$$

$\therefore$ $E_\alpha = 2.303\ R \times 1.25 \times 10^4\ K$

$= 2.303 \times 8.314\ JK^{-1}\ mol^{-1} \times 1.25 \times 10^4\ K = 2.39 \times 10^2\ kJ \cdot mol^{-1}$.

(ii) Half-life period of a first order reaction $t_{0.5}$ is given by $t_{0.5} = \frac{0.693}{k}$ where k is the rate constant or $$k = \frac{0.693}{256\ min} = \frac{0.693}{256 \times 60 \times 60\ sec.}$$

$= 7.5 \times 10^{-7}\ sec^{-1}$

Putting the value of k in equation (ii)

$$\log\ (7.5 \times 10^{-7}\ sec^{-1}) = 14.34 - \frac{1.25 \times 10^4\ K}{T}$$

or
$$-6.12 - 14.34 = \frac{1.25 \times 10^4 \text{ K}}{T}$$

or
$$T = \frac{1.25 \times 10^4}{20.46} k = 6.1 \times 10^2 k \simeq 600 \text{ K}$$

86. For 1st order reaction

$$\log \frac{A}{A_o} = -\frac{kt}{2.303}$$

$$t_{298} \text{ for 10\% completion} = \frac{2.303}{k_1} \log \frac{1}{10} \quad \text{(i)}$$

$$t_{318} \text{ for 25\% completion} = \frac{2.303}{k_2} \log \frac{1}{4} \quad \text{(ii)}$$

According to question $t_{298} = t_{318}$
Equating (i) and (ii) gives

$$\frac{k_1}{k_2} = \log\left(\frac{10}{4}\right) \quad \text{(iii)}$$

From Arhenius equation

$$\ln k = \ln A - \frac{E_a}{RT} \quad \text{(iv)}$$

$$\log\left(\frac{k_1}{k_2}\right) = \frac{E_a}{2.303\, R}\left[\frac{1}{T_2} - \frac{1}{T_2}\right] \quad \text{(v)}$$

$$\Rightarrow \quad E_a = 2.303 \times R \frac{T_1 \times T_1}{(T_2 - T_1)} \log \frac{k_1}{k_2}$$

Putting value of $T_1 = 298\ k$ and $T_2 = 308$ k
$R = 8.314 \text{ JK}^{-1} \text{ mol}^{-1}$

$$E_a = 2.303 \times 8.314 \times \frac{298 \times 308}{10} = 4818.94 \text{ J} \approx 48.19 \text{ kJ}$$

Part B From equation (iv)

$$\log k_{318} = \log A - \frac{E_a}{2.303\, R \times 318}$$

Putting $A = 3.56 \times 10^9 \text{ s}^{-1}$; $T = 318$ K; Ea = 48187.94 J

$$\log k_{318} \approx -9.346$$

$$\therefore \quad k_{318} = \text{antilog } (-9.346)$$

$$k_{318} = 4.508 \times 10^{-10} \text{ s}^{-1}$$

87.

	$CH_3 \cdot O \cdot CH_3$	$\rightarrow CH_4$	+ CO	+ H_2	Total conc.
Initial state	a	0	0	0	a
After 12 mts.	$a - x$	x	x	x	$a + 2x$

We know that

$$K = \frac{2.303}{t} \log \frac{a}{a-x}$$

Here we know, t= 12 mts. For calculating 't' (initial concentration)

$$PV = nRT \text{ or } \frac{n}{V} = \frac{P}{RT}$$

$$n = \frac{P}{RT} = \frac{0.4}{0.082 \times 773} = 6.31 \times 10^{-3} \text{ moles } 1^{-1}$$

Now for K

$$K = \frac{0.693}{t_{1/2}} = \frac{0.693}{14.5} = 4.78 \,' \, 10^{-2} \text{ min}^{-1}$$

Substituting the value in the first order reaction equation

$$K = \frac{2.303}{t} \log \frac{a}{a-x}$$

$$4.78 \times 10^{-2} = \frac{2.303}{12} \log \frac{a}{a-x}$$

$$\frac{a}{a-x} = 1.77446$$

$$a - x = \frac{6.31 \times 10^{-3}}{1.77446} \text{ moles } l^{-1}$$

$$= 3.556 \times 10^{-3} \text{ moles } l^{-1}$$

$\therefore$ $x = (6.310 - 3.556) \times 10^{-3}$ moles l^{-1} = 2.754×10^{-3} moles l^{-1}

$\therefore$ After 12 min.

Total no. of moles l^{-1} = $a + 2x$

$$= 6.31 \times 10^{-3} + 2 \times 2.754 \times 10^{-3} = 11.818 \times 10^{-3}$$

$\therefore$ $p = \frac{n}{V} RT$

$$p = 11.818 \times 10^{-3} \times 0.082 \times 773 = 0.749 \text{ atm.}$$

Chapter 8 : CHEMICAL EQUILIBRIUM

1. $N_2O_4 \rightleftharpoons 2NO_2$

Initial conc. 1 0

equil. conc. 1-x 2x

Suppose the initial concentration of N_2O_4 is one mole and its degree of dissociation is x. At equilibrium x moles of N_2O_4 will have dissociated to $2x$ moles of N_2O_2 will have dissociated to $2x$ moles of NO_2 leaving 1-x moles of N_2O_4.

Total no. of moles at equilibrium = $1 - x + 2x = 1 + x$

Let one mole of gas occupy a volume of V lit.

$1 + x$ moles of gas will occupy a volume of $V(1 + x)$ litres.

Density being inversely proportional to volume, density of undissociated N_2O_4 at the beginning

$$D = \frac{1}{V} \text{ and density d, at equilibrium } = \frac{1}{V(V+x)}$$

$$\therefore \quad \frac{D}{d} = \frac{V(1+x)}{V} = 1 + x$$

Hence $x = \dfrac{D-d}{d}$; density (D) of N_2O_4 = Mol. Wt./2 = $\dfrac{28+64}{2} = 46$

$$x = \frac{46 - 24.8}{24.8} = 0.854$$

Percentage dissociation = 85.4

2. Wt. of alcohol = 92 gm; no. of moles of alcohol = $\dfrac{92}{46} = 2$

wt. of acetic acid = 120 g.; no. of moles of acid = 120/60 = 2

CH_3COOH +	C_2H_5OH $\rightleftharpoons$	$CH_3COOC_2H_5$ +	H_2O	
2	2	0	0	(initial conc.)
(2-x)	(2-x)	x	x	(equl. moles)

$$K_c = \frac{[CH_3COOC_2H_5][H_2O]}{[CH_3COOH][C_2H_5OH]} = \frac{x^2}{(2-x)^2};$$

$$4 = \frac{x^2}{(2-x)^2} \text{ or } 2 = \frac{x}{(2-x)}$$

$4 - 2x = x$ or $x = 4/3 = 1.33$ g. moles wt. of ethyl acetate formed
$= 1.33 \times 88 = 117.0$ g. (mol wt. ethyl acetate = 88)

3. $CH_3COOH + C_2H_5OH \rightleftharpoons CH_3COOC_2H_5 + H_2O$

CH_3COOH	C_2H_5OH	$CH_3COOC_2H_5$	H_2O	
1	1	0	0	(initial conc.)
1/3	1/3	2/3	2/3	(equil. conc.)

Let, volume of reaction flask be 'V'; $K_c = \dfrac{2/3V \times 2/3V}{1/3V \times 1/3V} = 4$

Now, $CH_3COOC_2H_5 + H_2O \rightleftharpoons CH_3COOH + C_2H_5OH$

1	3	0	0	(initial conc.)
1-x	3-x	x	x	(equl. Conc)

$$K_C^1 = \frac{[CH_3COOH][C_2H_5OH]}{[CH_3COOC_2H_5][H_2O]} = \frac{1}{K_c} = \frac{1}{4}$$

$$4 = \frac{[\text{ester}][H_2O]}{[\text{acid}][\text{alcohol}]} = \frac{[(1-x)/V][(3-x)/V]}{(x/V)(x/V)}$$

$$4 = \frac{(1-x)(3-x)}{x^2} \text{ or } 4x^2 = (1-x)(3-x) \text{ or } 3x^2 + 4x - 3 = 0$$

$x = 0.533$ mole $\therefore$ Amount of ester at equilibrium $= 1 - x = 1 - 0.53 = 0.47$ mole

4. $p_{NH_3} = \frac{30 \times 6}{100} = 1.8$ atm

Since gases are ideal, hence the rest of pressure will be due to unconverted N_2 and H_2 in the mixture. Thus

$$p_{N_2} + p_{H_2} = 30 - 1.8 = 28.2 \text{ atm.}$$

But $N_2 : H_2$ is 1 : 3 therefore the ratio of their pressure will be

$$p_{N_2} = \frac{1}{4} \times 28.2 = 7.05 \text{ atm}$$

$$p_{H_2} = \frac{3}{4} \times 28.2 = 21.15 \text{ atm}$$

Now, $N_2 + 3H_2 \rightleftharpoons 2NH_3$

$$\therefore K_p = \frac{(p_{NH_3})^2}{(p_{N_2})(p_{H_2})^3} = \frac{(1.8 \text{ atm})^2}{(7.05 \text{ atm})(21.15 \text{ atm})^3} = \mathbf{4.9 \times 10^{-5} \text{ atm}^{-2}}$$

5. $CO(g) + H_2O(g) \rightleftharpoons CO_2(g) + H_2(g)$

3	1	0	0	initial moles
$(3-x)$	$(1-x)$	x	x	x at equil.

Now $$K_p = \frac{P_{CO_2} \times P_{H_2}}{P_{CO} \times P_{H_2O}} = \frac{(x)(x)}{(3-x)(1-x)}$$

or $$0.63 = \frac{x^2}{x^2 - 4x + 3}$$

or $$x^2 = 0.63 = (x^2 - 4x + 3)$$
$$x = 0.68$$

Hence moles H_2 and O_2 at equilibrium = **0.68**

Since, partial pressure $= \frac{\text{Moles of substance}}{\text{Total moles}} \times$ Total pressure

Total moles = 3 + 1 = 4

$$P_{CO_2} = \frac{0.68 \times 2}{4} = 0.34 \text{ atm.}; \qquad P_{CO} = \frac{(3-0.68)\times 2}{4} = 1.16 \text{ atm.}$$

$$P_{H_2} = \frac{0.68 \times 2}{4} = 0.34 \text{ atm.}; \qquad P_{H_2O} = \frac{(1-0.68)\times 2}{4} = 0.16 \text{ atm.}$$

6. Total moles at equilibrium = moles of NH_3 + moles of H_2S at 298 K
   Partial pressure = mole fraction × *P*

$$P_{NH_3} = \frac{1}{2} \times P = 0.5\ P; \qquad P_{H_2S} = \frac{1}{2} \times P = 0.5\ P$$

$K_p = P_{NH_3} \times P_{H_2S}$

(NH_4 HS is a solid so its pressure is negligible)

$$K_{p_1} = 0.5P \times 0.5P = 0.25\ P^2 = 0.25 \times .66 \times 0.66 = 0.109; \quad \therefore\ P = 0.66$$

7. The decomposition of PCl_5 is

	PCl_5	$\rightleftharpoons$ PCl_3	+ Cl_2
initial conc.	1	0	0
equal. conc.	0.2	0.8	0.8

Total no. of moles at equilibrium = 0.2 + 0.8 + 0.8 = 1.8

$$K_p = \frac{P_{PCl_3} \times P_{Cl_2}}{P_{PCl_5}}; \qquad \text{Here, } P_{PCl_5} = \frac{02}{1.8} \times 1; \ P_{PCl_3} = P_{Cl_2} = \frac{0.8}{1.8} \times 1$$

$$\therefore \quad K_p = \frac{(0.8/1.8)(0.8/1.8)}{(0.2/1.8)} = \frac{0.8 \times 0.8}{1.8 \times 0.2} = 1.777$$

8. (a) For the calculation of rate constant; we have,

$$K = \frac{2.303}{1} \log \frac{a}{(a-x)}$$

Here, $t = 50$ min.; $a = 100$; $x = 40\%$

$\therefore \quad (a - x) = (100\text{-}40) = 60$

Hence, $$K = \frac{2.303}{K} \log \frac{100}{60} = 0.01021 \text{ min}^{-1}$$

(b) For the time for 80% completion of the reaction

$a = 100$; $x = 80$; $a\text{-}x = 20$

$K = 0.01021 \text{ min}^{-1}$

$$\therefore \quad t = \frac{2.303}{K} \log \frac{100}{20} = \frac{2.303}{0.01021} \times 0.699 = 157.6 \text{ min.}$$

9. Here we have : $t = 72$ min; $x = 75\%$; $a = 100$; $a - x = 25$

$$\therefore \quad K = \frac{2,303}{t} \log \frac{100}{25};\ K = \frac{2.303}{72} \times 0.6012 \text{ min}^{-1} = 0.01925 \text{ min}^{-1}/n = 0.01925$$

or, $K = \frac{2.303}{t}\log\frac{100}{50} = \frac{2.303}{t} \times 0.3010;\ t = \frac{2.30 \times 0.301}{0.01925} = 36$ min.

10. Here, $a = 100$; $a\text{-}x = 75$; $t = 30$ min.

Substituting the values we get,

(a) $K = \frac{2.303}{30}\log\frac{100}{75} = 9.6 \times 10^{-3}\ \text{min}^{-1}$

(b) $a = 100$; $(a\text{-}x) = 25$

$$t = \frac{2.303}{9.6 \times 10^{-3}}\log\frac{100}{25} = 144.4 \text{ min}$$

11. Using the formula, $K = \frac{2.303}{t}\log\frac{a}{(a-x)}$ and substituting the values we get,

$$4.29 \times 10^{-2}\ \text{min}^{-1} = \frac{2.303}{t_{1/2}}\log\frac{100}{50}$$

or, $$t_{1/2} = \frac{2.303}{4.29 \times 10^{-2}} \times 0.3010 = 16.15 \text{ min}$$

12. $N_2O_4 \rightleftharpoons 2NO_2$

N_2O_4	$2NO_2$	
1	0	(Initially)
$1 - \alpha$	2α	(At equil.)

Total no. of moles at equil. = $1 - \alpha + 2\alpha = 1 + \alpha$

But here, α is 20% i.e., 0.2

$\therefore$ Total moles at equil. = $1 + 0.2 = 1.2$

According to gas equation, $PV = nRT$

$$\therefore\ V = \frac{nRT}{P} = \frac{(1.2\ \text{mole})(0.82\ \text{L atm K}^{-1}\ \text{mol}^{-1})(300\ \text{K})}{1\ \text{atm}} = 29.52\ \text{L}$$

Density of the equilibrium mixture = $\frac{\text{Mass}}{\text{Vol.}}$

$$= \frac{92\ \text{g}}{29.52\ \text{L}}\ (\text{Mol. mass of } N_{24} = 92) = 3.12\ \text{g L}^{-1}$$

13. $N_2O_4(g) \rightleftharpoons 2NO_2(g)$

$N_2O_4(g)$	$2NO_2(g)$	
a	0	(Initially)
$a - x$	$2x$	(At equilibrium)

Total no. of moles at equilibrium $a - x + 2x = a + x$

Given that original volume = $\frac{75}{100}$ existing volume at equilibrium

But moles $\propto$ volume at constant P & T

$\therefore$ Original moles = $\dfrac{75}{100}$ moles at equilibrium

$a = 0.75(a + x)$

or $x = 0.33\ a$

$\therefore$% Decomposition = $\dfrac{x}{a} \times 100 = \dfrac{0.33\ a}{a} = 33\%$

14. Using the formula 2.303 $\dfrac{K_2}{K_1} = \dfrac{E_a}{R}\left[\dfrac{T_2 - T_1}{T_1 T_2}\right]$ or, $\dfrac{R}{E_a}$ 2.303 log $\dfrac{K_2}{K_1} = \left[\dfrac{T_2 - T_1}{T_1 T_2}\right]$

Substituting the values we get,

$$\frac{2 \times 2.303}{68400}\log\left[\frac{1.16 \times 10^4}{3.76 \times 10^3}\right] = \frac{T_2 - 1085}{1085 \times T_2}$$

or, $3.3 \times 10^{-5} = \dfrac{T_2 - 1085}{1085 \times T_2} = 3.3 \times 10^{-5} \times 1085\ T_2 = T_2 - 1085$

$\therefore$ $0.0358\ T_2 = -1085$; or, $0.9642\ T_2 = 1085$

$\therefore$ $T_2 = 1125.2\ K$

15. We have

$$CO(g) + 2H_2(g) \rightleftharpoons CH_3OH(g)$$

at $t = 0.02$ mol

at $t_{eq} = 0.1$ mol ? 0.1 mol

Total amount of gases at equilibrium $\dfrac{PV}{RT} = \dfrac{(4.92\ \text{atm})(5\ \text{L})}{(0.082\ \text{atm L mol}^{-1}\ \text{K}^{-1})(600\ \text{K})} = 0.5$ mol

Hence, amount of H_2 at equilibrium, n_{H_2} = 0.5 mol – 0.2 mol = 0.3 mol
Now

$$K_c = \frac{[CH_3OH]}{[CO][H_2]^2} = \frac{(0.1\ \text{mol}/5\text{L})}{(0.1\ \text{mol}/5\text{L})(0.3\ \text{mol}/5\text{L})^2} = 2.77.78\ (\text{mol L}^{-1})^{-2}$$

$$K_p = \frac{P_{CH_3OH}}{P_{CO} \cdot P_{H_2}^2} = \frac{(0.1/0.5)\ 4.92}{(0.1/0.5)\ 4.92 \times [(0.3/0.5)\ 4.92]^2} = 0.115\ \text{atm}^{-2}$$

16. The Arthenius equation is $k = A \exp(-E_a/RT)$

Since $t_{0.5} = 10$ min, we have $k = \dfrac{0.693}{t_{0.5}} = \dfrac{0.693}{10 \times 60\ \text{s}} = 1.155 \times 10^{-3}\ \text{s}^{-1}$

Substituting the above value along with other data in the above Arrhenius equation, we get $1.155 \times 10^{-3}\ \text{s}^{-1} = (4 \times 10^{13}\ \text{s}^{-1}) \exp\{(-98600)\ \text{mol}^{-1})/(8.314]\ \text{K}^{-1}\ \text{mol}^{-1})\ (T)\}$

or $\ln \dfrac{1.155 \times 10^{-3}}{4 \times 10^{13}} = -\dfrac{98600}{8.314\ (T)}$;

$$\log 2.8875 \times 10^{-17} = -\frac{98600}{2.303 \times 8.314 (T)} = 16.54$$

$$\therefore \qquad T = \frac{98600}{2.303 \times 8.314 \times 16.54} = 311.36$$

or, $\qquad T = 311.36$ K

17. For first order reaction, $K = 0.693/t_{1/2}$
Reaction rate constant at 300 K, $K_1 = 0.693/30 = 0.0231$ min^{-1}
Reaction rate constant at 320 K, $K_2 = 0.693/10 = 0.0693$ min^{-1}
Energy of activation, E can be evaluated using the relationship.

$$\log \frac{K_1}{K_2} = \frac{-E}{2.303 \times R}\left(\frac{T_2 - T_1}{T_1 T_2}\right)$$

$$\log \frac{0.0231}{0.0693} = \frac{-E}{2.303 \times 8.314}\left(\frac{320 - 300}{320 \times 300}\right)$$

$$-0.477 = \frac{-E}{2.303 \times 8.314}\left(\frac{320 - 300}{320 \times 300}\right)$$

This gives E = 43.839 KJ mole^{-1}

18. For simplicity, suppose the initial amounts of SO_3 and O_2 are 1 mol and 2 mol, respectively. Suppose the volume of the vessel is such that the above mixture produces 3 atm pressure so that the partial pressure of SO_3 and O_2 are equal to the given partial pressure.

	$2SO_2(g)$ +	$O_2(g)$ $\rightleftharpoons$	$2SO_3(g)$
At t =	0	2 mol	1 mol
At t_{eq}	$2x$	(2 mol-x)	(1 mol-$2x$)

Total amount of gases = 3 mol – x

Total pressure of gasses will be (3 mol – x) $\left(\frac{3 \text{ atm}}{3 \text{ mol}}\right)$

Partial pressure at equilibrium will be

$$P_{SO_2} = \left(\frac{2x}{3 \text{ mol} - x}\right)\left(\frac{3 \text{ mol} - x}{1 \text{ mol}}\right) \text{atm} = \frac{2x}{\text{mol}} \text{atm}$$

$$P_{O_2} = \left(\frac{2 \text{ mol} - x}{\text{mol}}\right) \text{atm} = \left(2 - \frac{x}{\text{mol}}\right) \text{atm}$$

$$P_{SO_3} = \frac{1 \text{ mol} - 2x}{\text{mol}} \text{atm} = \left(1 - \frac{2x}{\text{mol}}\right) \text{atm}$$

we have, $\qquad K_p = \dfrac{P^2_{SO_3}}{P^2_{SO_3} \times P_{O_2}}$; Substituting the values,

$$900\ \text{atm}^{-1} = \frac{\left(1-\frac{2x}{\text{mol}}\right)^2 \text{atm}^2}{\left(\frac{2x}{\text{mol}}\right)^2\left(2-\frac{x}{\text{mol}}\right)\text{atm}^3}$$

or $$900 = \frac{\left(\frac{\text{mol}}{2x}-1\right)^2}{(2\,\text{mol}-x)}$$

Assuming $x < 2$ mol the above expression reduces to

$$\left(\frac{\text{mol}}{2x}-1\right)^2 = 900 \text{ or } \frac{\text{mol}}{2x}-1=30 \Rightarrow x\frac{\text{mol}}{2\times 31}$$

Solving for x, we get $x = 0.0115$ mol

Thus, $$P_{SO_2} = \frac{2x}{\text{mol}}\text{atm} = 0.023 \text{ atm}$$

$$P_{O_2} = \left(2-\frac{x}{\text{mol}}\right) = 1.9885 \text{ atm}$$

$$P_{SO_3} = \left(1-\frac{2x}{\text{mol}}\right) = 0.977 \text{ atm}$$

19. Since, $NH_4OH \rightleftharpoons NH_4^+ + OH^-$

$$\Rightarrow \quad K_b = \frac{[NH_4^+][OH^-]}{[NH_4OH]}$$

$$\therefore \quad [OH^-] = \frac{K_b\,[NH_4OH]}{[NH_4^+]} = \frac{1.8\times10^{-5}\times0.05}{0.25} \text{ mol litre}^{-1} = 0.36\times10^{-5} \text{ mol litre}^{-1}$$

$$[Al^{3+}] = \frac{K_{b)}\,[NH_4OH]}{[NH_4^+]} = 6\times10^{-32}/(0.36\times10^{-5})^3 \text{ mol litre}^{-1} = 1.29\times10^{-19} \text{ mol litre}^{-1}$$

$$[Mg^{2+}] = \frac{K_{sp}\,[Mg(OH)_2]}{[OH^-]^2} = \frac{6\times10^{-10}}{(0.36\times10^{-5})^2} \text{ mol litre}^{-1} = 46.3 \text{ mol litre}^{-1}$$

20. Determine the number of moles of the initial substances:

$$n_{H_2} = \frac{0.0212\times10^{-3}}{2}\ 10^3 = 0.0166 \text{ mole } H_2$$

$$n_{I_2} = \frac{0.846\times10^{-3}}{254}\ 10^3 = 0.0333 \text{ mole } I_2$$

H_2	+	I_2	$= 2HI$	
10.6×10^{-3}		3.33×10^{-3}	0	numbers of moles in the initial state of the system
$10.6 \times 10^{-3} - x$		$3.33 \times 10^{-3} - x$	x	numbers of moles in the equilibrium state of the system

where, $x = n_i^0 - n_i$

$$K_C = \frac{(n_{HI}/V)^2}{(n_{H_2}/V)(n_{I_2}/V)} = \frac{4x^2/(10^{-3})^2}{[(10.8\times10^{-3} - x)/10^{-3}](3.33\times10^{-3} - x)/10^{-3}]} = 50.25$$

Solution of this equation gives

$$46.25x^2 - 0.7015x + 1.7737 \times 10^{-3} = 0$$

Hence $x_{1.2} = \dfrac{0.7015 \pm \sqrt{0.7015^2 - 4\times1.7737\times10^{-3}\times46.25}}{2\times46.25}$

$= \dfrac{0.7015 \pm 0.4049}{92.50}$, $x_1 = 11.961 \times 10^{-3}$ and $x_2 = 3.206 \times 10^{-3}$

The first root, x_1, makes no physical sense because x cannot exceed $n^0_{H_2}$ in magnitude; $x_2 = 3.206 \times 10^{-3}$ mole. The amount of HI resulting from establishment of the state of equilibrium will be $m_{HI} = 2xM_{HI} = 2 \times 3.206 \times 10^{-3} \times 128 \times 10^{-3} = 0.821 \times 10^{-3}$ kg.

21. 25% dissociated. The degree of dissociation is 0.25 at 1 atmosphere.

	N_2O_4	$\rightleftharpoons$	$2NO_2$
Initial conc.	1.0		0
Conc. at equilibrium	1 – 0.25 = 0.75		0.25 × 2 = 0.5

Total number of mole = 1.25

Partial pressure of $N_2O_4 = \dfrac{0.75}{1.25} \times 1$; Partial pressure of $NO_2 = \dfrac{0.5}{1.25} \times 1$

Equilibrium constant $K_p = \dfrac{P^2_{NO_2}}{P_{N_2O_4}} = \dfrac{(0.5/1.25)^2}{(0.75/1.25)^2} = 0.2666.$

At the same temperature and at 0.1 atmosphere pressure degree dissociation is a

	N_2O_4	$\rightleftharpoons$	$2NO_2$
Initial Conc.	1		0
Conc. at equilibrium	$1 - a$		$2a$

Total number of moles = $1 - a + 2a = 1 + a$

Partial pressure of $NO_2 = \left(\dfrac{2a}{1+a}\right) 0.1$; partial pressure of $N_2O_4 = \left(\dfrac{1-a}{1+a}\right) 0.1$

Equilibrium constant $k_p = 0.2666 = \dfrac{4a^2\,(0.1)^2}{(1+a)^2} \times \dfrac{(1+a)}{(1-a)\times0.1} = \dfrac{4a^2\;0.1}{1-a^2}$

$0.2666\,(1-a)^2 = 0.4\,a^2$; $0.2666 = 0.6666\,a^2 \Rightarrow$ a = 0.63.

22. $$A_2(g) + B_2(g) \rightleftharpoons 2AB(g)$$

	$A_2(g)$	$B_2(g)$	$2AB(g)$
Initial moles	1	2	0
moles unreacted	$1-x$	$2-x$	$2x$

$$K_{AB} = \frac{[AB]^2}{[A_2][B_2]};\ 50 = \frac{4x^2}{(1-x)(2-x)}$$

$46x^2 - 150x + 100 = 0 \Rightarrow x = 0.69$

No. of moles of AB formed = 1.38 oles

23. 1. Partial pressure is directly proportional to mole fraction. Therefore by knowing the mole fraction at different instants one can find the kinetics of the reaction as

$$K = \frac{2.303}{t}\log\frac{a}{a-x}$$

It ill come out to be straight line. Therefore it is first order reaction

2. The specific rate constant 'K' can be determined.
3. The time for half change, $t_{1/2}$ can be evaluated.

24. For the calculation of the specific rate constant:

Use $$K = \frac{2.303}{t}\log\frac{a}{(a-x)}$$

Where K is rate constant of the reaction,

t = time = 10 min; a = initial concentration = 100 moles litre^{-1}; $(a\text{-}x)$ = conc. after 10 min = (100 – 20) = 80 moles litre^{-1}

$$\therefore \quad K = \frac{2.303}{10} \times 0.09691 = 0.0223 \text{ min}^{-1}$$

Calculation of time for 75% completion

Given $K = 0.0223 \text{ min}^{-1}$

a = 100 moles litre^{-1}; $(a\text{-}x)$ = (100 – 75) = 25 mols litre^{-1}; t = ?

$$t = \frac{2.303}{k}\log\left(\frac{100}{25}\right)\text{min} = \frac{2.303}{0.223} \times 0.6020 = 62.18 \text{ min.}$$

25. $PCl_5 \rightleftharpoons PCl_3 + Cl_2$

PCl_5	PCl_3	Cl_2	
1 mole	0	0	initial moles
$1-x$	x	x	Moles at equilibrium
$3(1-x)$	$3x$	$3x$	Since 3 moles of PCl_5 present, moles present at equilibrium.

Total moles at equilibrium = $3 - 3x + 3x + 3x + 1 = 4 + 3x$ (includes one mole of N_2)

Now, $$PV = nRT;\ n = \frac{PV}{RT} = \frac{2.05 \times 100}{0.0821 \times 500} = 5$$

Thus $4 + 3x = 5$; $3x = 1$; $x = 1/3$ = 33% degree of dissociation is = 33%

$$K = \frac{P_{PCl_3}\,P_{Cl_3}}{P_{PCl_5}}$$ and Partial pressure = Mole fraction X Total pressure

$$P_{Cl_2} = P_{PCl_3} = \frac{3x}{4+3x} \times 2.05 \text{ atms.};\ P_{PCl_5} = \frac{3-3x}{4+3x} \times 2.05 \text{ atms.}$$

$$K_p = \frac{\left(\frac{3x}{4+x}\right)^2 (2.05)^2}{\left(\frac{3-3x}{4+3x}\right)\times 2.05}\text{; since } x = 1/3 \Rightarrow K_p = 0.205 \text{ atm.}$$

Note : Nitrogen has no effect on the position of equilibrium through it causes an increases in the total pressure.

26. As seen form the data, when we double the concentration of A, the rate is also doubled. But when we double the concentration of B, the rate remains constant.
Hence the rate is independent of 'B' Therefore rate $\propto [A]\,[B]^{\circ}$
i.e. the above reaction is 1st order with respect to A and zero order w.r.t. B.

The specific rate constant, $K = \frac{\text{rate}}{[A]} = \frac{0.005}{0.01} = 0.5$

For a 1st order reaction, the half life period

$$t_{1/2} = \frac{0.693}{K} = \frac{0.693}{0.5} = 1.386 \text{ min.}$$

27. Consider the following equilibrium

	N_2	+	$3H_2$	$\rightleftharpoons$	$2NH_3$
Initial moles	1		3		0
Equil. moles	(1-0.0025)		(3-0.0075)		(0.0025)
Equil. conc.	$\frac{1-0.025}{4}$		$\frac{3.0.0075}{4}$		$\frac{0.005}{4}$

$$K_c = \frac{[NH_3]^2}{[N_2][H_2]^3} = \frac{\left(\frac{0.005}{4}\right)^2}{\frac{1}{4}\times\left(\frac{3}{4}\right)^3}$$

(small fraction are neglected)

$$K = \frac{0.005}{4}\times\frac{0.005}{4}\times\frac{4}{1}\times\frac{4}{3}\times\frac{4}{3}\times\frac{4}{3} = \frac{0.000025\times 16}{27} = 1.48 \times 10^{-5} \text{ (lit. mol)}^2$$

For the equilibrium, $\frac{1}{2}N_2 + \frac{3}{2}H_2 = NH_3$

$$K_C^1 = \frac{[NH_3]}{[N_2]^{1/2}\,[H_2]^{3/2}} = \sqrt{K_C} = \sqrt{1.48\times 10^{-5} \text{ (lit. mol)}^2} = 3.82 \times 10^{-3} \text{ lit. mol.}$$

28. According to gas equation,

$$PV = \frac{w}{m}RT \Rightarrow m = \frac{w}{V}\times\frac{RT}{P} = d\times\frac{RT}{P} \quad (d = \text{density of the mixture})$$

Substituting the given values

$$m = \frac{(0.925 \text{ g L}^{-1})(0.0821 \text{ L atm K}^{-1} \text{ mol}^{-1})(900 \text{ K})}{1 \text{ atm}} = 68.34 \text{ g mol}^{-1}$$

$\therefore$ Experimental mol. mass of SO_3 = 68.34 g mol^{-1}
Normal mol. mass of SO_3 80 g mol^{-1}

$$\text{Now, } SO_3 \rightleftharpoons SO_2 + \frac{1}{2}O_2$$

1	0	0	(initially)
$1 - \alpha$	α	$\alpha/2$	(At equilibrium)

Total no. of moles at equilibrium

$$= 1 - \alpha + \alpha + \frac{\alpha}{2} = 1 + \frac{\alpha}{2}$$

According to Vant Hoff's factor

$$1 + \frac{\alpha}{2} = \frac{\text{Normal mol. mass}}{\text{Observed mol. mass}} = \frac{80}{68.34} = 1.1706$$

or $\qquad \alpha = 0.3412 = 34.12\%$

29. $N_2O_4 \rightleftharpoons 2NO_2$
$1 - \alpha \qquad 2\alpha$

moles of equilibrium = $1 - \alpha + 2\alpha = 1 + \alpha$

$$P_{NO_2} = \frac{2\alpha}{1+\alpha}P, \qquad P_{N_2O_4} = \frac{1-\alpha}{1+\alpha}P$$

$$\therefore \qquad K_p = \frac{(P_{NO_2})^2}{P_{N_2O_4}} = \frac{\left(\frac{2\alpha}{1+\alpha}P\right)^2}{\left(\frac{1-\alpha}{1+\alpha}\right)P} = \frac{4\alpha^2 P}{1-\alpha^2} = 640$$

or, $\qquad \dfrac{4\alpha^2 \times 160}{1-\alpha^2} = 640$

$\therefore \qquad 4\alpha^2 = (1 - \alpha^2) \times 4 = 4 - 4\alpha^2$

or, $\qquad 8\alpha^2 = 4 \Rightarrow \alpha^2 = 0.5$

$\therefore \qquad \alpha = 0.707$

Hence % dissociation = 70.7%

If $\qquad \alpha = 0.5$ (Disociation is 50%)

$\therefore \qquad \dfrac{4 \times (0.5)^2 P}{1-(0.5)^2} = 640$

or, $\qquad P = \dfrac{640 \times (1 - 0.25)}{4 \times 0.5 \times 0.5} = 480$ mm

30. K = 0.1 lit mole^{-1} sec^{-1} K should be in terms of ml-mole^{-1} – min^{-1}. We know,
1 litre = 100 ml and

$$1 \text{ sec} = \frac{1}{60} \text{ minute}$$

$$K = \frac{0.1 \text{ lit} \times 1000 \text{ ml lit}^{-1}}{1/60 \text{ min sec}^{-1}} = 0.1 \times 1000 \times 60 = 6000 \text{ ml mol}^{-1} \text{ min}^{-1}$$

31. (a) According to the equation of the reaction, write the numbers of moles in the initial and equilibrium states:

A	+	$4B$	=	D	
1		12		0	numbers of moles in the initial state
$1 - x$		$12 - 4x$		x	numbers of moles in the equilibrium state

According to the Clapeyron-Mendeleev equation $\left(PV = \sum_i n_i RT\right)$,

Here $\Sigma n_i = 1 - x + 12 - 4x + x = 13 - 4x$

or, $$\sum_i n_i = 13 - 4x = \frac{4.51 \times 10^5 \times 0.05}{8.3143 \times 298} = 0.1 \text{ moles}$$

$\Rightarrow$ $$x = \frac{13 - 9.1}{4} = 0.975 \text{ mole}$$

To determine the equilibrium concentration, use the equation $C = n/V$:

$$c_A = \frac{1 - 0.975}{0.05} = 0.5 \text{ moles m}^{-3}$$

$$c_B = \frac{12 - 4 \times 0.975}{9.05} \quad 162 \text{ moles m}^{-3}$$

$$c_D = \frac{0.975}{0.05} = 19.5 \text{ mols m}^{-3}$$

31. (b) According to the question we have,

$$K_c = \frac{c_D}{c_A c_B^4} = \frac{n_D/V}{n_A/V(n_B/V)^4} = \frac{0.975 \times (50)^4}{0.025 \times 8.(100)^4} = 5.66 \times 10^4$$

32. To determine the equilibrium yield one must know the equilibrium constant of the reaction. We have Gibb's energy equation

$$\Delta G^\circ_{298} = \Delta H^\circ_{298} - 298 \Delta S^\circ_{298}$$

$$\Delta G^\circ_{298} = 58.41 \times 10^3 - 298 \times 176.6 = 5.78 \times 10^3 \text{ J}$$

$\therefore$ $$\log K_p = \frac{5780}{2.3026\, x\, 8.3143 \times 298} = -1.0131 = 2.9869,$$

or, $$K_p = 0.097$$

Also, $$K_p = \frac{P x_{NO_2}^2}{1 - x_{NO_2}} = 0.097, \; x_{NO_2}^2 = 0.097 - 0.097 x_{NO_2}$$

$$x_{NO_2}^2 + 0.097 x_{NO_2} - 0.097 = 0$$

Solution of the second-degree equation give two roots one of which makes no physical sense. Then $x_{NO_2} = 0.2667$, and the equilibrium yield is $x_{NO_2} \times 100 = 26.67\%$.

33. If one proceeds from one mole of $H_2O(g)$ ($n^0_{H_2O} = 1$), then at the degree of dissociation α, $(1 - \alpha)$ mole H_2O (n_{H_2O}) will remain in the equilibrium mixture and $\alpha/2$ mole O_2 (n_{O_2}) will be formed. The total number of moles in the equilibrium mixture of the gases will be

$$= \alpha + \alpha/2 + 1 - \alpha = (\alpha + 2)/2$$

Then, the equilibrium partial pressure will be

$$P_{H_2} = P\frac{\alpha}{(\alpha+2)/2} = P\frac{2\alpha}{\alpha+2}$$

$$P_{O_2} = P\frac{\alpha/2}{(\alpha+2)/2} = P\frac{\alpha}{\alpha+2}$$

$$P_{H_2O} = P\frac{1-\alpha}{\frac{\alpha+2}{2}} = P\frac{2(1-\alpha)}{\alpha+2}$$

and the equilibrium constant of the reaction will be

$$K_P = \frac{P_{H_2O}}{P_{H_2}P_{O_2}^{1/2}} = \frac{P2(1-\alpha)(\alpha+2)(\alpha+2)^{1/2}}{(\alpha+2)P2\alpha P^{1/2}\alpha^{1/2}} = \frac{(1-\alpha)(\alpha+2)^{1/2}}{P^{1/2}\alpha^{3/2}}$$

34. Derive an equation in which K_p to be found is related to the known values of K_{P_1} and K_{P_2}:

$$\Delta G^\circ = \frac{3}{2}\Delta G^\circ_{f,H_2O} - \Delta G^\circ_{f,NH_3}$$

where ΔG°_f, is the standard change in Gibbs' energy during formation of a given compound from simple substances, and ΔG° is the standard change in Gibbs' energy of the ammonia oxidation reaction.

$$-RT \ln K_p = \frac{3}{2}RT \ln K_{P_2} + RT \ln K_{P_1}$$

or

$$K_p = \frac{K_{P_2}}{K_{P_1}} = \frac{(1.16\times10^{10})^{3/2}}{5.69\times10^{-4}} = 2.196 \times 10^{18}$$

35.

	$SO_3(g)$	$\rightleftharpoons$	$SO_2(g)$	+	$½O_2(g)$
P_o	1.0 atm		0 atm		0 atm
ΔP	$-2x$ atm		$2x$ atm		x atm
P_{eq}	$1.0 - 2x$ atm		$2x$ atm		x atm

$$K = 0.45 = \frac{(P_{SO_2})(P_{O_2})^{\frac{1}{2}}}{(P_{SO_3})} = \frac{(2x)(x)^{\frac{1}{2}}}{1.0-2x} = \frac{2x^{\frac{1}{2}}}{1.0-2x}$$

$$0.45 - 09x = 2x^{\frac{3}{2}}$$

This is best solved by successive approximation, which gives,

$$x = 0.24$$

$$\therefore \quad P_{SO_3} = 1.0 - 2(0.24) = 0.52$$

36. A 50% yield indicates that half the reactions were converted to product, and the partial pressure damages according

$$K = \frac{(P_{XeF_4})}{(P_{Xe})(P_{F_2})^2} = \frac{(0.1)}{(0.1)(0.2)^2} = 25$$

A 75% yield would mean that the pressure of Xe would decrease by 75% (to 0.025 atm) and the amount of XeF_4 would increase by that amount (to 0.15 atm). The F_2 would decrease by twice the amount of Xe (to $x - 2(0.15)$).

$$K = 25 = \frac{(P_{X_eF_4})}{(P_{Xe})(P_{F_2})} = \frac{(0.15)}{(0.05)(x - 0.3)^2} \Rightarrow 0 = x^2 - 0.6x - 0.03$$

$$x = \frac{0.6 \pm \sqrt{(0.6)^2 - 4(1)(-0.03)}}{2(1)} = 0.65$$

37.

	$I_2(g)$	$\rightleftharpoons$	$2I(g)$
P_o	1.0 atm		0 atm
ΔP	$-x$ atm		$2x$ atm
P_{eq}	$1.0 - x$ atm		$2x$ atm

Total pressure at equilibrium is 40% greater than initial pressure, so total pressure equals 1.4 and $1 - x + 2x = 1.4$ and $x = 0.4$

$$K = \frac{(P_I)^2}{P_{I_2}} = \frac{(0.8)^2}{0.6} = 1.1$$

38. (a) Use the ideal gas law to determine the pressure of the benzyl alcohol, then use an equilibrium table and the equilibrium expression to solve for x.

$$1.5 \text{ g} \times \frac{1 \text{ mol}}{108.1 \text{ g}} = 0.0139 \text{ mol.}$$

$$P = \frac{nRT}{V} = \frac{(0.0139 \text{ mol})(0.0821 \text{ L} \cdot \text{atm mol}^{-1} \text{ K}^{-1})(523 \text{ K})}{2.0 \text{L}} = 0.3 \text{ atm}$$

	$C_6H_5CH_2OH(g)$	$\rightleftharpoons$	$C_6H_5CHO(g)$	+	$H_2(g)$
P_o	0.3 atm		0 atm		0 atm
ΔP	$-x$ atm		x atm		x atm
P_{eq}	$0.3 - x$ atm		x atm		x atm

$$K = 0.56 = \frac{x^2}{0.3 - x} \quad \Rightarrow \quad x^2 + 0.56\,x - 0.168 = 0$$

$$\therefore \qquad X = \frac{-0.56 \pm \sqrt{(0.56)^2 - 4(1\,(-0.168)}}{2(1)} = 0.22 \text{ atm}$$

(b) $$n = \frac{PV}{RT} = \frac{(0.3 - 0.22 \text{ E atm})(2.0 \text{ L})}{(0.0821 \text{ L}\cdot\text{atm mol}^{-1}\text{K}^{-1})(523 \text{ K})} = 4\times10^{-3}\text{ mol} = 4\times10^{-3}\text{ mol}\times\frac{108.1\text{ g}}{1\text{ mol}} = 0.4\text{g}$$

39.

	$2NH_3(g)$	$\rightleftharpoons$	$N_2(g)$	+	$3H_2(g)$
P_o	1.0 atm		0 atm		0 atm
ΔP	$-2x$ atm		x atm		$3x$ atm
P_{eq}	$1.0 - 2x$ atm		x atm		$3x$ atm

$$K = 2.5 = \frac{(P_{N_2})(P_{H_2})^2}{(P_{NH_3})^2} = \frac{(x)(3\,x)^2}{(1-2\,x)^2} = \frac{27\,x^2}{1.0 - 4\,x + 4\,x^2}$$

$$2.5 - 10x + 10x^2 = 27x^4 \quad \Rightarrow \quad 2.5 = 10x - 10x^2 + 27x^4$$

This is best solved by successive approximations. As a first approximation $x - 2.5/10$. Successive approximation gives: $x = 0.33$.

$$P_{NH_3} = 1 - 2(0.33) = 0.34 \text{ atm}$$

$$P_{H_2} = 3(0.33) = 0.99 \text{ atm}$$

$$P_{H_2} = 0.33 \text{ atm}$$

40. Use the titration data to calculate the amount of I_2 that was present at equilibrium. Then calculate the moles of each species and calculate K using moles instead of pressure (this is acceptable since the R, T and V terms would cancel in the equilibrium constant expression).

$$0.037\ L \times \frac{0.2 \text{ mol } S_2O_3^{2-}}{1\text{ L}} \times \frac{1 \text{ mol } I_2}{2 \text{ mol } S_2O_3^{2-}} = 3.7 \times 10^{-3} \text{ mol } I_2$$

$$3.2 \text{ g HI} \times \frac{1\text{ mol}}{127.9\text{ g}} = 2.5 \times 10^{-2} \text{ mol. HI}$$

At equilibrium: $HI = 2.5 \times 10^{-2} - 2\,(3.7 \times 10^{-3}) = 0.0176$ mol

$I_2 = 3.7 \times 10^{-3}$ mol.

$H_2 = 3.7 \times 10^{-3}$ mol.

$$K = \frac{(P_{I_2})(P_{H_2})}{(P_{HI})^2} = \frac{(3.7\times10^{-3})\,(3.7\times10^{-3})}{(0.0176)^2} = 0.0442$$

41. For $Sb_2S_3(s) + 3H_2(g) \rightleftharpoons 2Sb(s) + 3H_2S(g)$ ($+3Pb^{2+} \rightarrow 3PbS(s) + 6H^+$]

Starting M of $H_2(g)$ = 0.01 mol ÷ 2.5 L = 0.004 M H_2

1.029 g PbS ÷ 239.26 g PbS/mol H_2S = 43007 × 10^{-3} mol H_2S

4.3007 × 10^{-3} mol H_2S ÷ 2.5 L = [1.7203 × 10^{-3}[= M of H_2S

Conc. (M)	$3H_2(g)$	+ $Sb_2S_3(s)$	$\rightleftharpoons$	$3H_2S(g)$	+ $2Sb(s)$
Starting	0.004			0	
Change	–0.0017203			+ 0.0017203	
Equilibrium	0.0022797			0.0017203	

Substituting into the equilibrium expression for K_c gives

$$K_c = \frac{[H_2S]^3}{[H_2]^3} = \frac{(0.0017203)^3}{(0.0022797)^3} = 0.4297 = 0.430$$

42. For $PCl_5(f) \rightleftharpoons PCl_3(g) + Cl_2(g)$

Starting M of PCl_5 = 0.01 mol ÷ 2.0 L = 0.005 M

Conc. (M)	$PCl_5(g)$	$\rightleftharpoons$	$PCl_3(g)$	+ $Cl_2(g)$
Starting	0.005		0	0
Change	$-x$		$+x$	$+x$
Equilibrium	0.005 – x		x	x

Substituting into the equilibrium expression for K_c gives

$$K_c = 4.15 \times 10^{-2} = \frac{[PCl_3][Cl_2]}{[PCl_3]} = \frac{(x)(x)}{(0.005 - x)}$$

$$x^2 + (4.15 \times 10^{-2})\,x - 2.075 \times 10^{-4} = 0 \text{ (quadratic)}$$

Solving the quadratic gives $x = 4.6 \times 10^{-2}$ (impossible) and $x = 4.51 \times 10^{-3}$ M (use)

Total M of gas = 0.00451 + 0.00049 = 0.00951 M

$P = (n/V)RT$ = (0.00951 M) (0.082057 L · atm K^{-1} mol^{-1}) (523 K) = 0.408 atm

43. (a) **True** Strong acids dissociates completely, thus $[Y^-] = [H^+]$, which equals the concentrations of the acid.

(b) **False**Strong acids dissociate completely, thus the concentration of the undissociated acid is approximately zero ([HY]) ≈ 0)

(c) **True** See (a) above

(d) **True**pH = $-\log[H^+] = -\log(0.1) = 1$.

(e) **True.** See (a) above

44. The graph from the date is as:

(a) $P_A = 0.78$, $P_B = 0.47$

(b) 0.62 atm, equilibrium was established after 80 seconds

(c) 0.61 atm, see the graph above

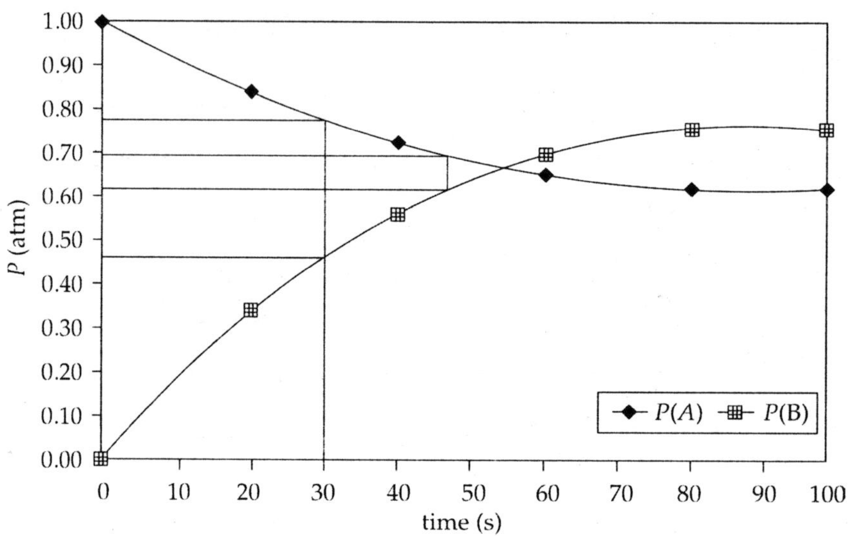

45. Given that K is quite high, we can estimate that at equilibrium the mixture is mostly product (CO). Thus can be verified by calculation.

	C(s)	+	$CO_2(g)$	$\rightleftharpoons$	$2CO(g)$
P_o	—		0.3 atm		0 atm
ΔP	—		$-x$ atm		$+2x$ atm
P_{eq}	—		$0.3 - x$ atm		$2x$ atm

$$K = \frac{(P_{CO})^2}{P_{CO_2}} \Rightarrow 168 = \frac{(2x)^2}{0.3 - x}$$

$$50.4 - 168x = 4x^2 \Rightarrow 4x^2 + 168x - 50.4 = 0$$

$$x = \frac{168 \pm \sqrt{(168)^2 - 4(4)(-50.4)}}{2(4)} = 0.25$$

There is actually a second value for x, –189, but a negative pressure is impossible, so this value is rejected.

$P_{eq}(CO_2) = 0.30 - 0.25 = 0.05$ atm

P_{eq} (CO) $= 0 + 2(0.25) = 0.5$ atm

As predicted, the mixture is mostly product $(CO)_2$ with 10 times as much CO as CO_2. Thus option (b) is correct.

46. At equilibrium there is much more NO_2 (product) than N_2O_4 (reactant), thus $K > 1$ and equilibrium equation is:

$$N_2O_4 \rightleftharpoons 2NO_2$$

If the pressure of N_2O_4 is suddenly increased to 1.0 atm, the system will no longer be at equilibrium. Since reactant had been added, the reaction will proceed to the right, increasing the amount of products (and reducing the reactants) until equilibrium is reestablished.

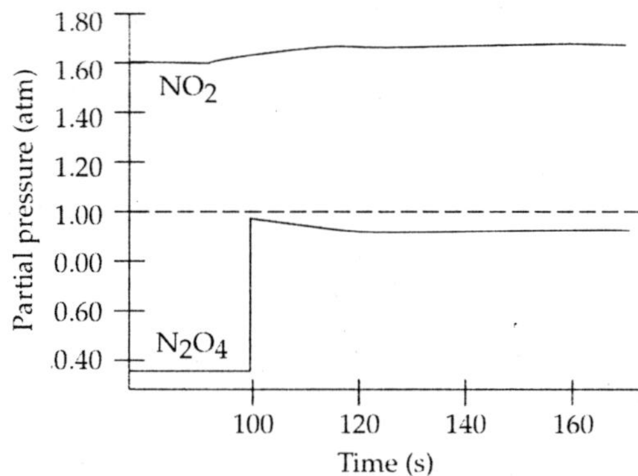

Thus the graph will show a sudden spike at 100 s (where the partial pressure was increased) and then, over time, a decrease, followed by an eventual leveling off as equilibrium is reestablished. The NO_2, on the other hand will initially increase, then level off.

47. $PCl_5 \rightleftarrows PCl_3 + Cl_2$

$0 \quad 0 \quad 0$

$1 - \alpha \quad \alpha \quad \alpha$ at equilibrium

At equilibrium; $[PCl_5] = 0.6$

and $[Cl_2] = [PCl_3] = 0.4$

Total conc. = 1.4

$$d = \frac{m}{V} \text{ and } V = \frac{nRT}{P}$$

So $$d = \frac{m}{nRT} \times P = \frac{208.5\,c \times 1}{14\,C \times 0.082 \times 400} = 4.53 \text{ g L}^{-1}.$$

48. (i) $NH_4HS_{(g)} \rightleftharpoons NH_{3(g)} + H_2S_{(g)}$

m.wt. of $NH_4HS = 17 + 5 + 32 = 54$

$$\text{Mole of } NH_4HS = \frac{3.06}{54}$$

$$\text{Active mass of } NH_4HS = \frac{3.06}{54 \times 2} \text{ moles lit}^{-1}$$

Degree of dissociation of NH_4HS = 30%

$$K_c = \frac{[NH_3(g)][H_2S(g)]}{[NH_4HS(g)]}$$

Activity of solid = 1

Thus $$K_c = 8.1 \times 10^{-5}$$

$$K_p = K_C[RT]^{\Delta n}$$

Δn= No. of moles of product – No. of moles of reactant = 2 – 1 = 2

$\therefore$ $K_p = 8.1 \times 10^{-5}\ [0.082 \times 0300]^2 = 0.049$

(ii) Since NH_4HS is solid, so it causes no change in equilibrium

49. Inverse sucrose is of 1st order reaction. Hence $K = \frac{2.303}{t}\log\frac{a}{(a-x)}$ sec^{-1}

T_o the rotation of zero time represents the conc. of sucrose and rotation due to sucrose.

T_∞ (infinite reading) is due to glucose and fructose which are obtained by hydrolysis with acid.

Initial conc. of sucrose (a) = $T_o - T_\infty$ = 32.4 – (–14.1) = 46.5

conc. at different time 't' (a-x) = $T_t - T_\infty$

$$K = \frac{2.303}{t}\log\frac{T_o - T_\infty}{T_t - T_\infty}$$

Time	T_t	$T_t - T_\infty$	$K = \frac{2.303}{t}\log\frac{a}{(a-x)}$
600	28.8	42.9	$K = \frac{2.303}{600}\log\frac{48.5}{42.9} = 1.343 \times 10^{-4}$
1200	25.5	29.6	$K = \frac{2.303}{1200}\log\frac{46.5}{39.6} = 1.339 \times 10^{-4}$
2400	19.6	33.7	$K = \frac{2.303}{2400}\log\frac{46.5}{33.7} = 1.341 \times 10^{-4}$
6000	6.7	20.8	$K = \frac{2.303}{6000}\log\frac{46.5}{20.8} = 1.341 \times 10^{-4}$

Average $K = 1.341 \times 10^{-4}$ sec

Calculation of time

The solution is optically inactive implies that its angles of rotation is zero i.e., $T = 0°$

$$t = \frac{2.303}{K}\log\frac{T_o - T_\infty}{T_t - T_\infty}\ \text{Sec..} = \frac{2.303}{1.341\times 10^{-4}}\log\frac{46.5}{0-(-14.1)} = 8.9 \times 10^3\ \text{Sec.}$$

50. The reaction is $CH_3COOC_2H_5 + H_2O \rightleftharpoons CH_3COOH + C_2H_5OH$

5 ml. of reaction mixture at every instant of time will contain acetic acid formed at that instant and litre reading corresponds to acetic acid formed and HCl in 5 ml. mixture which is proportional to ethyl acetate reacted.

Therefore the reading T_t represents 'x' + HCl is 5 ml. while reading T_o represents initial conc. and HCl in 5 ml represent HCl in 5 ml

$$T - T_o = a;\ T_t - T_o = x\ ;\ T - T_t = (a\text{-}x);\ \text{and } K = \frac{2.303}{t}\log\frac{T - T_o}{T - T_t}$$

Substituting the values from the given problem, we get,

1. $K = \frac{2.303}{75}\log\frac{21.05 - 9.62}{21.05 - 12.1} = 3.26 \times 10^{-3}\ min^{-1}$

2. $K = \frac{2.303}{119} \log \frac{21.05 - 9.62}{21.05 - 13.1} = 3.05 \times 10^{-3}\ \text{min}^{-1}$

3. $K = \frac{2.303}{183} \log \frac{21.05 - 9.62}{21.05 - 14.75} = 3.25 \times 10^{-3}\ \text{min}^{-1}$.

As K is constant – this reaction is of 1st order

51.

Initial Conc.	x		x		K_pO
	$N_{2(g)}$	+	$O_{(g)}$	$\rightleftarrows$	$2NO_{(g)}$
Equilibrium Conc	$x - \alpha$		$x - \alpha$		$2\alpha = 0.018$ mole

$$K_p = \frac{[NO]^2}{[N_2][O_2]} \text{ (from law of mass action)} \qquad \ldots\ldots \text{(i)}$$

since $\Delta n = 0$ for this rection

$$K_p = K_c \ (\because K_p = K_c (RT)^{\Delta n}) \qquad \ldots\ldots \text{(ii)}$$

Also if α = degree of dissociation and 'x' = Initial amount of N_2 and O_2 at the start of reaction.

Expression (i) can be expressed as

$$K_c = \frac{(2\alpha)^2}{(x - \alpha)^2}$$

$$\Rightarrow \quad (x - \alpha)^2 = \frac{(2\alpha)^2}{K_c}$$

$$\Rightarrow \quad (x - \alpha) = \frac{2\alpha}{\sqrt{K_c}} = \frac{0.018}{\sqrt{2.1 \times 10^{-3}}}$$

$$x - \alpha = 154.28 \times 10^{-3} \approx 0.154 \text{ moles}$$

or $\quad x = 0.154 + \alpha$

From given condition $2\alpha = 0.018$ mole

$$\alpha = 0.009 \text{ mole}$$

$\Rightarrow \quad x = 0.154 + 0.009 = 0.163$ moles

$\therefore \quad$ Initial amount of N_2 = 0.163 moles

Initial amount of O_2 = 0.163 moles

$\therefore \quad$ Mole fraction of $N_2 = \frac{n_1}{n_1 + n_2} = \frac{1}{2}$

$[n_1 = n_2 = 0.163)$

Mole fraction of $O_2 = \frac{1}{2}$

52. Given reaction is

$$[Ag(CN)_2]^- \rightarrow Ag^+ + 2CN^-$$

$$\therefore \qquad K = \frac{[Ag^+][CN^-]^2}{[Ag(CN)_2]^-} = 4.0 \times 10^{-19} \qquad \text{(i)}$$

KCN dissociates in the solution and produces 0.1 mol K^+ and produces 0.1 mol K^+ and 0.1 mol CN^-. Similarly $AgNO_3$ gives 0.3 mol Ag^+ and 0.03 mol NO_3^-

Since 1 mol of Ag^+ and 2 mol of CN^- gives 1 mol of $[Ag(CN_2)]^-$

∴ Initial concentration of $[Ag(CN)_2]^-$ = 0.03 mol Concentration of

$$CN^- = 0.1 - 0.03 \times 2 = 0.04.$$

By the equation (i)

$$4.0 \times 10^{-19} = \frac{[Ag^+][0.04]^2}{0.03}$$

$$\therefore \qquad [Ag^+] = \frac{4.0 \times 10^{-19} \times 0.03}{(0.04)^2} = 7.5 \times 10^{-18} \text{ M}$$

53. Reaction is $A \Leftrightarrow nB$

It is clear from the graph

Time (hours)	*Concentration [mol lit⁻¹]* A	B
1	0.5	0.2
5	0.3	0.6

(At equilibrium)

(i) Rate of consumption of A $= \dfrac{0.5 - 0.3}{4} \quad 0.5\ Mh^{-1}$

Rate of formation of $B = \dfrac{0.6 - 0.2}{4} = 0.1 \text{ Mh}^{-1}$

∴ Rate of consumption of A

$=> \dfrac{1}{n} \times$ Rate of formation of B

or $0.05 = \dfrac{0.1}{n} \quad \therefore n = 2$

(ii) $K = \dfrac{[B]^n}{[A]}$

At equilibrium $[A] = 0.3\ M$ and $[B] = 0.6$ M.

$$\therefore \qquad K = \frac{(0.6)^2}{0.3} = 1.2 \text{ mol litre}^{-1}$$

(iii) Initial rate of conversion of $A = \dfrac{0.5 - 0.3}{4} = 0.05 \text{ mol litre}^{-} \text{ hr}^{-1}$

54. The relevant equation is

$$CO(g) + 2H_2(g) \rightleftharpoons CH_3OH(g)$$

Initially the reaction vessel has 0.2 mole of CO(g) to which is added hydrogen at 600 K The total volume = 5 litre and the total pressure is 4.92 atm.
At this point, the total number of moles of gases in the flask are

$$n = \frac{PV}{RT} = \frac{5 \times 4.92}{0.082 \times 600} = 0.5 \text{ mole}$$

of which CO(g) is initially 0.2 mole.

After the equilibrium is attained, 0.1 mole of $CH_3OH(g)$ is formed. The number of moles of various gases are:

$CO(g) = 0.2 - 0.1 = 0.1$

$CH_3OH(g) = 0.1$

Total number of moles= 0.5

$\therefore$ Number of moles of $H_2 = 0.5 - [0.1 + 0.1] = 0.3$ mole
Concentration of gases in grams mole^{-1} are:

$$[CO] = \frac{0.1}{5}; \ [CH_2OH] = \frac{0.1}{5} : [H_2] = \frac{0.3}{5}$$

$$\therefore K_c = \frac{[CH_3OH]}{[CO][H_2]^2} = \frac{0.1/5}{\frac{0.1}{5} \times \frac{0.3}{5} \times \frac{0.3}{5}} = 277.8$$

Pressure of gases,

$$P_{CO} = \frac{0.1}{0.5} \times 4.92 = 0.984$$

$$p_{CH_2OH} = \frac{0.1}{0.5} \times 4.92 = 0.984$$

$$p_{H_2} = \frac{0.3}{0.5} \times 4.92 = 2.952$$

$$\therefore \quad K_p = \frac{p_{CH_3OH}}{p_{CO}\, p_{H_2}^2} = \frac{0.984}{0.984 \times (2.952)^2} = \frac{1}{(2.952)^2} = 0.113 \text{ atm}^{-1}$$

55. Suppose α is the degree of dissociation for the reaction $N_2O_4 \rightleftharpoons 2NO_2$

Then, $K_p = \dfrac{(2\alpha)^2}{1-x}$

Substituting the values as 310 K and 1 atm pressure, we have

(i) $K_p = \dfrac{(0.5/1.25)^2}{(0.75/1.25)}$ 0.267

(ii) At 310 K and 0.1 atm, α can be calculate from the expression

$$K_p = 0.267 = \frac{[2a/(1+a)\times 0.1]^2}{[(1-a)/(1+a)]\times 0.1}$$

$\Rightarrow$ $\alpha = \pm 0.65$ or 65%

56. Let 'a' moles of NO_2 react with SO_2 at equilibrium in the given reaction:
$SO_2(g) + NO_2(g) \rightleftharpoons SO_3(g) + NO(g)$
We have,
Initial concentration of SO_2 = 1 mole
Equilibrium concentration of $SO_2 = (1 - a)$
Initial concentration of NO_2 = 1 mole
Equilibrium concentration of $NO_2 = (1 - a)$
Initial concentration of SO_3 = 1 mole
Equilibrium concentration of $SO_3 = (1 + a)$
Initial concentration of NO = 1 mole
Equilibrium concentration of O = $(1 + a)$
Thus,

$$\frac{[SO_3][NO]}{[SO_2][NO_2]} = K_e = 16$$

or $$\frac{(1+a)(1+a)}{(1-a)(1-a)} = 16 \Rightarrow \frac{1+a}{1-a} = 4$$

or $$a = \frac{3}{5} = 0.6$$

$\therefore$ Equilibrium concentration of $NO_2 = (1 + a) = (1 + 0.6) = 1.6$ moles
Equilibrium concentration of $NO_2 = (1 - a) = (1 - 0.6) = 0.4$ mole.

57. $A_2(g) + B_2(g) \rightleftharpoons 2AB(g)$

Initial moles:	1 mole	2 mole	—
Equilibrium moles:	$1 - x$	$2 - x$	$2x$

Volume = 3 litres

$\therefore$ $$[Ag] = \frac{1-x}{3};\ [B_2] = \frac{2-x}{3};\ [AB] = \frac{2x}{3}$$

Now, $$K_{eq} = \frac{[AB]^2}{[A_2][B_2]}$$

$$\therefore\ 50 = \frac{\left[\frac{2x}{3}\right]^2}{\left[\frac{1-x}{3}\right]\left[\frac{2-x}{3}\right]} = \frac{4x^2}{(1-x)(2-x)} = \frac{4x^2}{x^2 - 3x - 2}$$

or $50x^2 - 150x + 100 = 4x^2$

or $46x^2 - 150x + 100 = 0$

Solving quadratic equation, $x = 0.93$

$\therefore$ Moles of AB formed $= 2x = 2 \times 0.93 = 1.86$ moles.

58. Given,

	N_2	+	$3H_2$	$\rightleftharpoons$	$2NH_3$
Initial concentration	a		$3a$	$\rightleftharpoons$	0
Equilibrium concentration	$(a - x)$		$3(a - x)$	$\rightleftharpoons$	$2x$

Now, $$K_c = \frac{[NH_3]^2}{[N_2][H_2]^3} = \frac{(2x)^2}{(a-x)(a-x)^3}$$

Initial concentration of $N_2(=) = \frac{1}{4} = 0.25$ mole litre^{-1}

Number of moles of N_2 converted to

$$NH_3(x) = 0.25 \times \frac{0.25}{100} = 6.25 \times 10^{-4} \text{ mole litre}^{-1}$$

Substituting the value of a and x in the expression for equilibrium constant we get

$$K_c = \frac{4 \times (6.25 \times 10^{-4})^2}{27(0.25 - 6.25 \times 10^{-4})^4} = \frac{4 \times (6.25 \times 10^{-4})^2}{27(0.2494)^4} = 1.48 \times 10^{-5} \text{ litre}^2 \text{ mol}^{-2}$$

The equilibrium constant for the reaction

$$\frac{1}{2}[N_2(g) + \frac{3}{2}H_2(g) \rightleftharpoons NH_3(g)]$$

$$K_c' = \frac{[NH_3]}{[N_2]^{1/2}[H_2]^{3/2}} = \sqrt{K_c} = \sqrt{1.5 \times 10^{-5}} = 3.82 \times 10^{-3} \text{ litre mol}^{-1}$$

59.

	$CO_{(g)}$	+	$2H_{2(g)}$	$\Leftrightarrow$	$CH_3OH_{(g)}$
Initial state	$\frac{0.15}{2.5} = 0.06$		0		0

Now consider 'y' mole of H_2 are introduced, so that at a pressure of 8.5 atm

	$CO_{(g)}$	+	$2H_{2(g)}$	$\Leftrightarrow$	$CH_3OH_{(g)}$
At equilibrium	$(0.06 - x)$		$(y - 2x)$		$(x = 0.032)$ ml^{-1}
	0.028		$y = 0.064$		0.032

Total pressure = 8.5 atm.

For ideal gas $PV = nRT$

or $$P = \frac{n}{V}RT$$

$$\therefore \quad P_{CO} = 0.028 \times 0.082 \times 750 = 1.722 \text{ atm}$$

$$P_{CH_3OH} = 0.032 \times 0.082 \times 750 = 1.968 \text{ atm}$$

$$\therefore \quad P_{H_2} = 8.5 - 1.722 - 1.968 = 8.5 - 3.690 = 4.810 \text{ atm}$$

$$\therefore \quad K_p = \frac{p_{CH_3OH}}{(p_{CO})(p_{H_2})^2} = \frac{1.968}{1.722 \times (4.81)^2} \quad 4.94 \times 10^{-2} \text{ atm}^{-2}$$

$$K_p = K_c(RT)^{\Delta n} \Rightarrow K_c = K_p\ (RT)^{-\Delta n}$$

Here $\Delta n = 1 - 3 = -2$

$\therefore$ $K_c = 4.94 \times 10^{-2} \times (0.082 \times 750)^2$

or, $K_c = \mathbf{1.8684 \times 10^2\ (m/l)^{-2}}$

(ii) For initial amount of H_2

$$p_{H_2} = \frac{n}{V} = RT = (a - 0.064) \times 0.082 \times 750$$

$$\therefore a - 0.064 = \frac{4.81}{0.082 \times 750} = 0.0782$$

$\therefore$ $a = 0.0782 + 0.064 = 0.1422\ \text{m l}^{-1}$

Total number of moles of H_2 in 2.5 litres

$$0.1422 \times 2.5 = 0.3555$$

No of moles of CO in 2.5 litres = 0.15

$\therefore$ Total number of moles of CO and H_2

$$0.15 + 0.335 = 0.5055$$

Substituting the value in the equation

$$PV = nRT$$

$$P \times 2.5 = 0.5055 \times 0.082 \times 750$$

$$\therefore \quad P = \frac{0.5055 \times 0.082 \times 750}{2.5} = 12.435 \text{ atm.}$$

60. Let the initial amount of SO_3 and O_2 be 1 mol and 2 mol respectively. Let thevolume of the vessel be such that the above mixture produces 3 atm pressure so that the partial pressures of SO_3 and O_2 are equal to the given partial pressures.

$2SO_2$	+ O_2	$\rightleftharpoons$ $2SO_3$	
	2 mol	1 mol	(at $t = 0$)
$2x$	(2 mol − x)	(2 mol − $2x$)	(at t_{equil})

Total amount of gases = 3 mol − x

$$\text{Total pressure of gases} = (3\text{ mol} - x)\left(\frac{3\text{ atm}}{3\text{ mol}}\right)$$

Partial pressure at equilibrium

$$p(SO_2) = \left(\frac{2x}{3\text{ mol} - x}\right)\left(\frac{3\text{ mol} - x}{1\ Emol}\right)\text{atm} = \frac{2x}{\text{mol}}\text{ atm}$$

$$p(O_2) = \frac{(2\text{ mol} - x)}{\text{mol}} = \left(2 - \frac{x}{\text{mol}}\right)\text{atm}$$

$$p(SO_2) = \frac{(1\text{ mol} - 2x)}{\text{mol}} = \left(1 - \frac{x}{\text{mol}}\right)\text{atm}$$

Substituting the values in the equation,

$$K_p = \frac{p^2_{SO_3}}{p^2_{SO_2} \times p_{O_2}}$$

$$900 \text{ atm}^{-1} = \frac{\left(1 - \frac{2x}{\text{mol}}\right)^2 \text{atm}^2}{\left[\left(\frac{2x}{\text{mol}}\right)^2 \text{atm}^2\right]\left[\left(2 - \frac{x}{\text{mol}}\right) \text{atm}\right]} = \frac{\left(1 - \frac{2x}{\text{mol}}\right)^2 \text{atm}^2}{\left(\frac{2x}{\text{mol}}\right)^2 \left(2 - \frac{x}{\text{mol}}\right)} = \frac{\left(\frac{\text{mol}}{2x} - 1\right)^2}{\left(2 - \frac{x}{\text{mol}}\right)}$$

Neglecting $\frac{x}{\text{mol}}$ (being very small),

$$900 = \frac{1}{2}\left(\frac{\text{mol}}{2x} - 1\right)^2 ; \; x = 0.0115 \text{ mol.}$$

Thus,

$$p(SO_2) = \frac{2x}{\text{mol}} \text{ atm} = 0.023 \text{ atm}$$

$$p(O_2) = \left(2 - \frac{x}{\text{mol}}\right) \text{ atm} = 1.9885 \text{ atm}$$

$$p(SO_2) = \left(1 - \frac{2x}{\text{mol}}\right). \text{ atm} = 0.917 \text{ atm.}$$

61. (a) According to Henderson's equation

$$p(OH) = pk_B + \frac{\log [\text{salt}]}{[\text{base}]}$$

At pH = 10.04; pOH = 14 – 10.04 = 3.96

At pH = 8.14; pOH = 14 – 9.14 = 4.86

Taking x ml to be the amount of HCl required for complete neutralization. Put the values Henderson equation we get

$$3.96 = pk_B + \log \frac{5}{x - 5} \quad (1)$$

and $$4.86 = pk_B + \log \frac{20}{x - 20} \quad (2)$$

Subtract (1) from (2) we get,

$$\log \frac{5}{x - 20} - \log \frac{20}{x - 20} = 0.90$$

$$\Rightarrow \quad \log \frac{20}{x - 20} \times \frac{x - 5}{5} = 0.90$$

$$\Rightarrow \quad \log \frac{4(x - 5)}{(x - 20)} = 0.90$$

Taking antilog we get

$$\frac{4(x-5)}{(x-20)} = 7.943 \text{ or } x = 35.2$$

Now $x = 40$ N where N is the normality of base BOH. x is the amount of HCl required for complete neutralization for 40 ml of the base BOH. Putting the values of x in above we get

$$35.2 = 40 \text{ N}$$

$$\Rightarrow \quad N = \frac{35.2}{40} = 0.08$$

$$3.96 = pk_b + \log \frac{5}{35.2-5} = pk_b + \frac{5}{30.2}$$

$$\Rightarrow \quad 4.741 = pk_b$$

$$pk_b = -\log pk_b = 4.741$$

$$\log k_b = -4.741 = 5.26$$

$$pk_b = 1.82 \times 10^{-5}$$

62. Consider the equation $AB_2(g) \leftrightarrow AB(g) + \frac{1}{2} B_2(g)$

Let x be the degree of dissociation

$\therefore$	$AB_2(g)$	$\leftrightarrow AB(g)$	$+ \frac{1}{2} B_2(g)$
Initial mole	1	0	0
Final mole	$1-x$	x	$x/2$

Total number of moles $= 1 - x + x + \frac{x}{2} = \frac{2+x}{2}$

$$\therefore \quad P_{AB_2} = \frac{2(1-x)}{2+x} P,\; P_{AB} = \frac{2x}{2+x} P,\; P_{B_2} = \frac{x}{2+x} P.$$

$$\therefore \quad K_p^1 = \frac{P_{AB} + P_{B_2}^{1/2}}{P_{AB_2}} = \frac{\frac{2x}{2+x} P \times \left(\frac{x}{2+x}\right)^{1/2}}{\frac{2(1-x)}{2+x} P} = \frac{x}{1-x}\left(\frac{x}{2+x} P\right)^{1/2} \qquad \text{(i)}$$

Now the given equation is

$2AB_2(g) \leftrightarrow 2AB + B_2$

It is clear that $K_p = (K_p^1)^2$.

By equation (i) $K_p = \frac{x^2}{(1-x)^2} \times \frac{x}{(2+x)} \times P = \frac{x^3 P}{(1-2x+x^2)(2+x)}$

Since x is small, therefore we can neglech higher powers of x

$$\therefore \quad K_p = \frac{x^3 P}{(1-2x)(2+x)} = \frac{x^3 P}{2-3x}$$

63. At boiling point randomness increases hence entropy increases (from liquid to vapour). The difference in disorderness between vapour and liquid is very high, this extent of disorderness depends on the liquid. The disorderness may be considered to be almost equal.

$$K_b = \frac{RT_b^2}{1000\, I_v} = \frac{RT_b^2 M}{1000\, \Delta H_v} = \frac{RT_B M}{1000\, \Delta S} \quad (\text{since } I_v = \frac{\Delta H_v}{M})$$

ΔS considered to be constant M of all liquids are same and R is constant

So $K_b = T_b$

Thus K_b of $(x) = 0.68$

K_b of $(y) = 0.53$

64. Suppose [HCN] = a gram mol litre^{-1}

$[CN^-] = (0.01 - a)$

Now, $pH = -\log [H^+]$

or $[H^+] = 10^{-pH} = 10^{-8.8} = 3.2 \times 10^{-9}$ mol litre^{-1}

Also, $HCN \rightleftharpoons H^+ + CN^-$

$$\therefore \quad k_{\text{dissociation}} = \frac{[H^+][CN^-]}{[HCN]} = 4.1 \times 10^{-19}$$

$$\text{or} \quad \frac{[3.2 \times 10^{-9}][CN^-]}{[HCN]} = 4.1 \times 10^{-10}$$

$$\text{or} \quad \frac{[CN^-]}{[HCN]} = \frac{4.1 \times 10^{-10}}{3.2 \times 10^{-9}} = 1.28$$

$$\text{or} \quad \frac{(0.01 - a)}{a} = 0.128$$

or $a = 0.009$.

∴ Gram mole of HCl needed = 0.009

65. $(H^+) = 6.4 \times 10^{-3}$ moles lit^{-1}

Concentration of acid = 0.2 molar

$K_a = 2.4 \times 10^{-4}$

[acid] = 0.2 + 0.0064 = 0.2064

[Salt] = 1 − 0.75 = 0.25

$$\begin{array}{ccccc} HCOONa & \rightleftharpoons & HCOO^- & + & Na^+ \\ C(1-\alpha) & & C\alpha & & C\alpha \\ C = 1 & & \alpha = 0.75 & & \end{array}$$

∴ (Salt) = 0.25

$$pH = pK_a + \log \frac{(\text{salt})}{(\text{acid})} = -\log (2.4 \times 10^{-4}) + \log \frac{0.25}{0.2064} = 3.7029$$

66. $[H^+] = \sqrt{\dfrac{K_w K_a}{K_b}}$; $p^H = \dfrac{1}{2}[pK_w + pK_a + pK_b] = 6.5$

67. We have

$CH_3COOH \rightleftharpoons CH_3COO^- + H^+$

$$K_a = \frac{[CH_3COO^-][H^+]}{[CH_3COOH]} = \frac{[H^+]^2}{[CH_3COOH]}$$

Hence, $[H^+] = (K_a(CH_3COOH))^{1/2}$

$= (1.8 \times 10^{-8})^{1/2} = 4.30 \times 10^{-3}$

∴ $p^H = -\log(4.30 \times 10^{-3}) = 2.37 = 4.74$

which gives $[H^+] = 1.897 \times 10^{-5}$

Substituting this in the expression $K_a = \dfrac{[H^+]^2}{[CH_2COOH]-[H^+]}$

$$1.8 \times 10^{-5} = \frac{(1.897 \times 10^{-5})^2}{[CH_3COOH] - 1.897 \times 10^{-5}}$$

Solving, we get $[CH_3COOH] = 3.896 \times 10^{-5}$

Volume of dilute acetic acid containing 1 mol of acid = $\dfrac{1}{3.896 \times 10^{-5}}$ L = 2.567×10^4 L

Thus 1 L of solution should be diluted to 2.567×10^4 L.

68. (i) Hydrochloric acid added = 0.2 mole

∴ $[H^+] = 0.2$ g lit^{-1} (∵ HCl is a strong acid)

When H^+ ions are added, concentration of CH_3COO^- ions will be decreased

$[CH_3COO^-] = (1.0 - 0.2) = 0.8$ mole

Hence $[CH_3COOH] = (1.0 + 0.2) = 1.2$ mole

Now, $p^H = -\log K_a + \log \dfrac{[\text{salt}]}{[\text{acid}]} = -\log 1.8 \times 10^{-5} + \log \dfrac{0.8}{1.2} = 4.57$

(ii) Hydrochloric acid added = 0.2 mole. From 0.2 mole of HCl added, 0.01 mole will combine with 0.1 mole of CH_3COO^- ions already present in 1 litre of the solution.

∴ $[CH_3COO] = (0.1 + 0.1) = 0.2$ mole

also, $[H^+] = (0.2 - 0.1) = 0.1$ mole

Because of the presence of $[H^+]$, acetic acid will remain unionized. the pH of the solution is, therefore, due to the presence of 0.1 mole of $[H^+]$. It means

$$p^H = \frac{1}{\log(0.1)} = 1$$

69. According to the Henderson equation

$$p^H = -\log K_a + \log \frac{[\text{salt}]}{[\text{Acid}]}$$

and we have $p^H = 7.4;\ K_a = 7.8;$

concentration of acid $= \frac{2 \times 10}{1000} = 0.02$ (no. of moles of H_2CO_3 in 10 ml of 2 MH_2CO_3)

Let the concentration of salt be x

$$\therefore \quad 7.4 = -\log 7.8 \times 10^{-7} + \log \frac{0.005\,x}{0.02}$$

$$\Rightarrow \quad x = 78.36 \text{ m}$$

70. $pH = pK_a + \log \frac{C_{salt}}{C_{acid}}$

On substituting the appropriate given values

$$4.75 = -\log 1.34 \times 10^{-5} + \log \frac{C_{salt}}{0.02}$$

$$C_{suit} = 0.015 \text{ moles}$$

Addition of 0.01 moles of HCl will decrease the concentration of propionate ion by 0.01 moles and in turn the concentration of propionic acid will increase by the same amount.

$$\underset{0.01\text{ M}}{HCl} \rightleftharpoons \underset{0.01\text{ M}}{H^+} \quad \underset{0.01\text{ M}}{Cl^-}$$

Thus at equilibrium

$$pH = -\log 1.34 \times 10^{-5} + \log \frac{0.015 - 0.01}{0.02 + 0.01}$$

$$= -\log 1.34 \times 10^{-5} + \log \frac{0.005}{0.03} = 4..$$

In constant to the pH of above solution, the pH of 0.01 molar HCl solution is given by

$$pH = -\log 0.01 = 2.0$$

71. (i) 500 mL of 0.2 M CH_3COOh solution is mixed with 500 mL of 0.2 M HCl at 298 K, the concentration of CH_3COOH is reduced to half i.e. 0.1 M. The concentration of HCl which is already ionized in H^+ & Cl^- also reduced by half i.e. 0.1 M H^+ ion.

	$CH_3COOH \rightleftharpoons$	CH_3COO^-	+	H^+
Initially	0.1 M	0		0.1 M
At equilibrium	$(0.1 - x)$M	xM		$(0.1 + N)$M

$$K_a \text{ of } CH_3COOH = \frac{[CH_3COO^-][H^+]}{[CH_3COOH]}$$

$$K_a \text{ of } CH_3COOH = 1.75 \times 10^{-5} \text{ mol litre}^{-1}$$

$$1.75 \times 10^{-5} = \frac{x(0.1 + x)}{(0.1 - x)}$$

Acetic acid is weak acid so the value of a x is negligible, so $0.1 + x = 0.1$ M and $0.1 = x = 0.1$ M

Hence 1.75×10^{-5}

$$\text{Degree of dissociated aicd} = \frac{1.75 \times 10^{-5} \times 1}{0.1} = 1.75 \times 10^{-4}$$

$$P_H = \log_{10}[H^+] = -\log_{10} 0.1 = \log_{10}10 = \mathbf{1\ Ans.}$$

(ii) 6 g or 0.15 M NaOH is added to the above solution will neutralize the solution of 0.1 M aicd and rest 0.05 M NaOH will react will CH_3COOH to Give CH_3COONa

	CH_3COOH +	NaOH ⇄	CH_3COONa +	H_2O
Initially	0.1 m	0.5 M	0	0
After raction	(0.1 – 0.05)		0.05 M	

In this reaction **acetate buffer** is formed

Now, H of buffer solution is given by Henderson Haselbatch equation

$$\text{For acidic buffer} \rightarrow p_H = pK_a + \log_1 \frac{[\text{Salt}]}{[\text{Acid}]}$$

$$= -\text{Log}_{10}K_a + \log_{10} \frac{[CH_3COONa]}{[CH_3COOH]}$$

$$= -\log_{10} 1.75 \times 10^{-5} + \text{Log}_{10} \frac{0.05}{0.05} = 0.2430 + 5 + 0 = 4.75\%$$

72.

	HF ⇌	H^+ +	F
Initial	0.6	0 M	0 M
Equilibrium	$(0.6 - x)$ M	xM	xM

$$K_a = \frac{[H^+][F]}{[HF]} = \frac{x^2}{(0.6 - x)}$$

Since $x \ll 0.6$, $K_a = \frac{x^2}{0.6}$ or $7.1 \times 10^{-4} = \frac{x^2}{0.6}$

$x^2 = 0.6 \times 7.1 \times 10^{-4}$ $\therefore x = 2.06 \times 10^{-2}$

$$\text{Percentage of ionistion} = \frac{0.0206}{0.6} \times 100 = 3.43\%$$

Chapter 9 : IONIC EQUILIBRIA

1. (a) We know that $P^H = -\log_{10}[H^+]$

0.1 N HCl contain 0.1 gm ion of H^+ per litre, $[H^+] = 0.1$ or 10^{-1}

$\therefore \quad p^H = -\log_{10}[0.1] = -(-1.0) = 1$

(b) $p^H = -\log[H^+]$

0.0032 M sulphuric acid is same as 0.0064 N H_2SO_4(dibasic acid)

$\therefore \quad p^H = -\log(0.0064) = -\log(6.4 \times 10^{-3})) = -(-3) - \log 6.4 = +3 - 0.8062 = 2.1938$

(c) $p^H = -\log[H^+]$

Here, $\quad [H^+] = 0.05$ gm. ion

$\therefore \quad p^H = -\log[0.05] = -\log[5 \times 10^{-2}] = +2 - 0.699 = 1.301$

2. (a) $p^H = -\log[H^+]; \Rightarrow [H^+] = 10^{P^H}$

$= 10^{-5.81} = 10^{-6.0} \times 10^{0.19} = 1.549 \times 10^{-6}$

(b) $[H^+] = 10^{P^H} = 10^{-9.63} = 10^{-10} \times 10^{0.37} = 2.344\times 10^{-10}$

(c) $[H^+] = 10^{P^H} = 10^{-11.9} = 10^{-12} \times 10^{0.2} = 1.585 \times 10^{-2}$

3. (a) $[H^+] = \dfrac{K_w}{[OH^-]} = \dfrac{1\times10^{-14}}{1\times10^{-3}} = 1 \times 10^{-11}$

$\therefore \quad p^H = -\log[H^+] = -(-11) = 11.$

(b) $[OH^-] = 0.02 = 2 \times 10^{-2}$

$\therefore \; [H^+] = \dfrac{K_w}{[OH^-]} = \dfrac{1\times10^{-14}}{2\times10^{-2}} = 0.5 \times 10^{-12} = 5 \times 10^{-13}$

$p^H = -\log[H^+] = -\log(5 \times 10^{-13}) = +13 - 0.6990 = 12.301$

4. 0.98 solution means 100 ml contains 0.98 gm of H_2SO_4. Hence 1000 ml contains 9.8 gm No. of gm equivalents present per litre = 98/4 = 0.2. The normality of H_2SO_4 = 0.2. As the H_2SO_4 is only 96% ionized, the number of gm. ions of hydrogen in the solution would be equal to $0.2 \times 0.96 = 0.192$

$[H^+] = 0.192 = 1.92 \times 10^{-1}$

$p^H = -\log[H^+] = -(\bar{1}.2833) = -(-0.7167) = 0.7167$

5. Ammonia solution is ammonium hydroxide. The ionisation of ammonium hydroxide gives ammonium and hydroxyl ion.

$$NH_4OH \rightleftharpoons NH_4^+ + OH^-$$

as the degree of ionization is 0.014, the concentration of hydroxyl ion would be

$$[OH^-] = 0.12 \times 0.01 = 0.00168$$

$p^{OH} = -\log_{10}[OH^-] = -\log_{10}(0.00168) = -\log 1.68 \times 10^{-3} = 3 - 0.2253 = 2.7747$

We know that for aqueous solutions, $P^H + P^{OH} = 14$

$\therefore \; P^H = 14 - P^{OH} = 14 - 2.7747 = 11.2253$

6. $P^H = -\log[H^+]$
0.1 M H_2SO_4 has a normality of 0.2
$[H^+]= 0.2= 2 \times 10^{-1}$; $P^H = -\log(2 \times 10^{-1}) = 1 - 0.301 = 0.699$
On dilution from 100 ml to 250 ml viz. 2.5 times, the H^+ ion concentration decreases by 2.5 times per litre. The molarity therefore decreases to 0.04. The normally now becomes 0.08 N

$\therefore$ $[H^+] = 0.08 = 8 \times 10^{-2}$. Thus, new $P^H = -\log(8 \times 10^{-2}) = (2-0.9031)$

7. $K_a = \dfrac{[H^+][NO_2^-]}{[HNO_2]} = \dfrac{x^2}{0.05 - x}$; Since $x \ll 0.05$

$\therefore\ 4.5 \times 10^{-4} = \dfrac{x^2}{0.05}$ or $x^2 = 4.5 \times 10^{-4} \times 5 \times 10^{-2} = 22.5 \times 10^{-6}$
$\therefore\ x = 4.74 \times 10^{-3}$ Moles.
$\therefore\ P^H = -\log H^+ = -\log(4.74 \times 10^{-3}) = (3 - 0.675) = 2.325$

8. $P^H = 2.85$ or $-\log[H^+] = 2.85$
$\therefore\ H^+= 10^{-2.85} = 10^{-3} \times 10^{+0.15} = 1.41 \times 10^{-3}$ M.
HA] = (0.1)
$\therefore\ [H^+]= [A^-] = 1.41 \times 10^{-3}$

$$K_a = \frac{(1.41 \times 10^{-3})^2}{0.1} = 1.995 \times 10^{-5}$$

9. (a) When the bottle is first opened $[H^+] = 3.2 \times 10^{-4}$ M
$\therefore\ P^H = -\log[H^+] = -\log 3.2 \times 10^{-4} = 3.49$
On the second occasion $[H^+] = 1.0 \times 10^{-3}$ M
$\therefore\ P^H = -\log(1.0 \times 10^{-3}) = 3$

(b) The neutralisation reaction is
$CH_3COOH(aq) + NaOH(aq) \rightleftharpoons CH_3COONa(aq) + H_2O(l)$
The number of moles of NaOH in ml is

$$\frac{10 \times 0.1}{1000} = 1 \times 10^{-3} \text{ moles lit.}^{-1}$$

The number of moles of acetic acid is $\dfrac{25 \times 0.1}{1000} = 2.5 \times 10^{-3}$ mole lit.$^{-1}$

No of moles of acetic acid left neutralization = $2.5 \times 10^{-3} - 1.0 \times 10^{-3} = 1.5 \times 10^{-3}$
At this stage a buffer system made up of CH_3 COONa and CH_3COOH is present.

$$\therefore\ P^H = P^{K_a} + \frac{\text{salt}}{\text{acid}} = 4.74 + \log\frac{1 \times 10^{-3}}{1.5 \times 10^{-3}} = 4.74 - 0.176 = 4.56$$

10. $P^H = 7.40$ buffer is prepared by $H_2PO_4^-$ and HPO_4^- (conjugate base is most suitable because $P^{K_2} = 7.21$ is close to the desired P^H

$$p^{K_2} = p^{K_a} + \log \frac{\text{conjugate base}}{\text{acid}}; \text{ or } 7.4 = 7.21 + \log \frac{[HPO_4^{-2}]}{[H_2PO_4^-]}$$

$$\text{or} \quad 0.19 = \log \frac{[HPO_4^{-2}]}{[H_2PO_4]} \quad \therefore \text{ talking antilog} \frac{[HPO_4^{-2}]}{[H_2PO_4^-]} = 1.5$$

Thus one way to prepare to phosphate buffer with P^H of 7.40 is dissolve disodium hydrogen phosphate (Na_2HPO_4) and sodium dihydrogen phosphate (NaH_2PO_4) in molar ratio of 1.5 : 1.0 in water.

11. $7.40 = 6.38 + \log \frac{[HCO_3^-]}{[H_2CO_3]}; \Rightarrow 7.40 - 6.38 = \log \frac{[HCO_3^-]}{[H_2CO_3]} = 0.02$

or $10.47 = \frac{[HCO_3^-]}{[H_2CO_3]}$ (anti log of 0.02)

Hence, HCO_3^- and H_2CO_3 ratio is 10.47 : 1.0
since HCO_3^- ion concentration is high it is more effective against added acid.

12. $[H^+] = 2 \times 10^{-3}$ moles lit^{-1}

$\therefore p^H = -\log[H^+] = -\log(2 \times 10^{-3}) = 3 - \log 2 = 3 - 0.301 = 2.699$

13. $[OH^-] = 5 \times 10^{-3}$ M; Now, $[H^+][OH^-] = K_w$

$$\therefore [H^+] = \frac{K_w}{[OH^{-3}]} = \frac{1 \times 10^{14}}{5 \times 10^{-3}} = 2 \times 10^{-12}$$

$\therefore \quad P^H = [OH^-] - \log[H^+] = -\log(2 \times 10^{-12}) = -\log 2 + 12 = -0.3010 + 12 = 11.6990$

14. $[H^+]$ = ? We know that, $[H^+] = 10^{P^H}$

$[H^+] = 10^{-4.62} = 10^{-5} \cdot 10^{0.38} = 2.399 \times 10^{-5}$ moles/lit.

15. $[H^+] = 10^{-P^H} = 10^{-8.62} = 10^{-9} \times 10^{.38} = 2.399 \times 10^{-9}$

Now, $\quad [H^+][OH^-] = K_w = 1 \times 10^{-14}$

$$\therefore \quad [OH^-] = \frac{1 \times 10^{-14}}{2.399 \times 10^{-9}} = 0.4169 \times 10^{-5} = 4.169 \times 10^{-6} \text{ moles lit}^{-1}.$$

Thus, concentration of NaOH = 4.169×10^{-6} moles lit^{-1}

16. Here, $\quad HCN \rightleftharpoons H^+ + CN^-$

at equilibrium $\quad 0.1 - x \quad x \quad x$

$$\therefore K_a = \frac{x^2}{0.1 - x} = 7.2 \times 10^{-10}$$

since $x \ll 0.1 \ \therefore \frac{x^2}{0.1} = 7.2 \times 10^{-10}$; or, $x^2 = 7.2 \times 10^{-11} = 72 \times 10^{-12} \Rightarrow x = 8.4 \times 10^{-6}$

$\therefore [H^+] = 8.4 \times 10^{-6}$; or, $P^H = -\log H^+ = -\log 8.4 \times 10^{-6} = 6 - 0.921 = 5.08$

17. $P^H = 2.5$ or, $-\log H^+ = 2.5$

∴ $[H^+] = 10^{-2.5} = 10^{-3} \cdot 10^{0.5} = 3.16 \times 10^{-3}$

Now, $50 \times 0.2 = x \times 3.16 \times 10^{-3}$ ∴ there are 2 H^+ in H_2O

$\Rightarrow$ $\dfrac{50 \times 0.2}{3.16 \times 10^{-2}} = x$; Now, or $3.164 \times 10^3 = 3164 = x$

Amount of water added = 3164 – 50 = 3114 ml = 3.114 lit.

18. (i) $P^H = 5.34$; or $-\log[H^+] = 5.34$

∴ $[H^+] = 10^{-5.34} = 10^{-6} \cdot 10^{0.66} = 4.57 \times 10^{-6}M$

(ii) $P^H = 2.3$; or $-\log[H^+] = 2.3$

∴ $[H^+] = 10^{-2.3} = 10^{-3} \cdot 10^{+0.7} = 0.5 \times 10^{-3}M$

(iii) $P^H = 8.9$; or $-\log[H^+] = 8.9$

∴ $[H^+] = 10^{-8.9} = 10^{-9} \cdot 10^{0.1} = 1.25 \times 10^{-9}M$

19. Here, $P^{K_a} = -\log K_a = -\log 1.75 \times 10^{-5} = 4.756$

∴ $P^H = P^{K_a} + \log \dfrac{\text{Salt}}{\text{Acid}} = 4.756 + \log \dfrac{0.1}{0.01}$ or

$= 4.576 + \log 10 = 4.756 + 1 = 5.756$

20. $P^{K_a} = -\log 1.75 \times 10^{-5} = 4.756$

But, $P^H = P^{K_a} + \log \dfrac{\text{Salt}}{\text{Acid}}$

or, $4.0 = 4.756 + \log \dfrac{\text{Salt}}{\text{Acid}}$; or, $-0.756 = \log \dfrac{\text{Salt}}{\text{Acid}}$

$5.70 = \dfrac{\text{Acid}}{\text{Salt}}$ = the required ratio.

21. $CH_3COOH \rightleftharpoons CH_3COO^- + H^+$

∴ $K_a = \dfrac{[CH_3COO^-][H^+]}{[CH_3COOH]} = \dfrac{[H^+]^2}{[CH_3COOH]}$;

∴ $[H^+] = \sqrt{K_a\,(CH_3COOH]} = \sqrt{1.8 \times 10^{-5} \times 1} = 4.3 \times 10^{-3}$

$\Rightarrow$ $P^H = 2.37$

New P^H, i.e., after dilution = 2 × 2.37 = 4.74

∴ $[H^+] = 1.897 \times 10^{-5}$

∴ $K_a = \dfrac{[H^+]^2}{[CH_3COOH]}$; or $1.8 \times 10^{-5} = \dfrac{[1.897 \times 10^{-5}]^2}{[CH_3COOH]}$

$\Rightarrow$ $[CH_3COOH] = \dfrac{[1.897 \times 10^{-5}]}{1.8 \times 10^{-5}} = 1.9 \times 10^{-5}$ mole lit^{-1}

Volume of acetic acid containing 1 mol acid is $V = \frac{1}{3.896} \times 10^{-5}$ L $= 2.56 \times 10^{4}$ L

22. Given, $P^{H} = 8.5$

$\therefore \quad [H^+] = 10^{-8.5} = 3.2 \times 10^{-9}$ mole. lit^{-1}

Also, $HCN \rightleftharpoons H^+ + CN^-$

$$\therefore K_{\text{dissociation}} = \frac{[H^+][CN^-]}{[HCN]} = 4.1 \times 10^{-10}; \therefore \frac{[CN^{-1}]}{[HCN]} = \frac{4.1 \times 10^{-10}}{3.2 \times 10^{-9}} = 0.128$$

Let [HCN] be x $\therefore$ $[CN^-] = 0.01 - x$ or $\frac{[CN^-]}{[HCN]} = \frac{0.01 - x}{x} = 0.128$

Gram mole of HCl required $x = 0.0089$

23. (i) Amount of HCl added = 0.2 mole or $[H^+] = 0.2$ g litre^{-1}

Addition of H^+s ions will reduce the concentration of CH_3COO^- ions by the following any

$$\underset{0.2}{CH_3COO^-}\ \underset{0.2}{H^+} \rightarrow \underset{0.2}{CH_3COOH}$$

$$[CH_3COO^-] = 1.0 - 0.2 = 0.8 \text{ mole}$$
$$[CH_3COOH] = 1.0 + 0.2 = 1.2 \text{ mole}$$

Now $P^{H} = -\log K_a + \log \frac{[\text{salt}]}{[\text{acid}]} = -\log 1.8 \times 10^{-5} + \log \frac{0.8}{1.2}$

$= 4.7447 + 0.3010 - 0.4771 = 4.5686$

(ii) Amount of HCl added = 0.2 mole

Out of 0.2 mole of H^+ added, 0.1 mole will continue with 0.1 mole of CH_3COO^- to form 0.1 mole CH_3COOH. Therefore the total $[CH_3COOH] = 0.1 + 0.1 = 0.2$

Due to the H^+ ions presence, CH_3 COOH does not ionize to give H^+ ions

The P^{H} ions presence, CH_3COOH does not ionize to give H^+ ions.

The P^{H} of the solution $= -\log[H^+] = -\log[0.1] = +1$

24. Consider the following equilibrium

	HA (aq) $\rightleftharpoons$	H^+(aq) +	A^-(aq)
Initial	1	0	0
equil. conc.	$c(1 - \alpha)$	$c\alpha$	$c\alpha$

The dissociation constant $K_a = \frac{[H^+][A^-]}{[HA]} = \frac{c\alpha \cdot c\alpha}{c(1-\alpha)} = \frac{c\alpha^2}{(1-\alpha)}$

Calculation of percentage ionization

$c\alpha^2 = 4.9 \times 10^{-8}$, Also c = 0.1—decimolar solution

$$\therefore \quad \alpha = \sqrt{\frac{4.9 \times 10^{-8}}{c}} = \sqrt{\frac{4.9 \times 10^{-8}}{0.1}} = \sqrt{4.8 \times 10^{-8}}$$

Thus, degree of dissociation, $\alpha = 7 \times 10^{-4}$

$\therefore$ % ionization = $7 \times 10^{-4} \times 100 = 7 \times 10^{-2}$

Calculation of P^H

$$[H^+] = c.\alpha = 0.1 \times 7 \times 10^{-4} = 7 \times 10^{-5}$$

$$\therefore \quad \log [H^+] = \log 7 + \log 10^{-5} = -5 + 0.8451 = -4.1459$$

or $\quad -\log[H^+] = P^H = 4.1459$

Calculation of $[OH^-]$

$$[OH^-] = \frac{1 \times 10^{-14}}{[H^+]} = \frac{1 \times 10^{-14}}{7 \times 10^{-5}} = 1.428 \times 10^{-10} \text{ moles lit}^{-1}$$

25. Use $P^{K_a} = -\log K_a = -\log[1.8 \times 10^{-5}] = 5 - \log 1.8 = 5 - 0.2553 = 4.7447$

Calculation of P^H

Volume = 20 ml of 0.2 M

excess of acetic acid = (50 – 20) ml 0.2 M = 30 ml of 0.2 M

$$\text{conc. of sod. acetate } [CH_3COONa] = \frac{20 \times 0.2}{70} = \frac{4}{70}$$

$$\text{conc. of acetic acid } [CH_3COOH] = \frac{30 \times 0.2}{70} = \frac{6}{70}$$

$$\therefore \ P^H = P^{K_a} + \log \frac{[\text{salt}]}{[\text{acid}]} = 4.7447 + \log \frac{4}{70} \times \frac{70}{6} = 4.5686$$

Calculation of additional volume of 0.2 M NaOH to make P^H 4.74

$$P^H = P^{K_a} + \log\frac{[\text{salt}]}{[\text{acid}]}; \quad \therefore 4.74 = 4.7447 + \log\frac{[\text{salt}]}{[\text{acid}]}$$

$$\text{or } \log \frac{[\text{salt}]}{[\text{acid}]} = 4.7447 - 4.74 \equiv 0.0047; \quad \therefore \frac{[\text{salt}]}{[\text{acid}]} = \text{antilog } (0.0047) = 0.9892$$

Let the additional volume of 0.2 M NaOH to be added to make the P^H of the solution 4.74 be x ml The volume will neutralize further 'x' ml of 0.2 M acetic acid and produce x ml of 0.2 M sodium acetate. Therefore the solution will contain.

(30 – x) ml of 0.2 M CH_3COOH in (70 + x) ml solution
(20 + x) ml of 0.2 M sodium acetate in (70 – x) ml solution
(20 + x) ml of 0.2 M sodium acetate in (70 + x) ml solution

$$\therefore \quad 0.9892 = \frac{(30 - x) \times 0.2 \times (70 + x)}{(70 + x)(20 + x) \times 0.2};$$

$$\text{or } 0.9892 (20 + x) = 30 - x; \Rightarrow x = \frac{10.216}{1.9892} = 5.13 \text{ ml.}$$

26. $K_a = 1.34 \times 10^{-5}$

$$\therefore \quad P^{K_a} = -\log K_a = -\log (1.34 \times 10^{-5}) = 4.8729$$

Now, $$p^H = p^{K_a} + \log\frac{[\text{salt}]}{[\text{acid}]}$$

Let the number of moles of sodium propionate [salt] be x; [acid] = 0.02

$$\therefore \quad 4.75 = 4.8729 + \log\frac{x}{0.02}$$

or $$\log\frac{0.02}{x} = 4.8729 - 4.75 = 0.1229; \therefore \frac{0.02}{x} = \text{antilog } 0.12229 = 1.327$$

$$\Rightarrow \quad 1.327\,x = 0.02$$

$$\therefore \quad x = \frac{0.02}{1.327} = 1.507 \times 10^{-3} \text{ M}$$

Consider the following dissociation

$$\underset{0.01 \text{ mole}}{HCl} \rightleftharpoons \underset{0.01 \text{ mole}}{H^+} + \underset{0.01 \text{ mole}}{Cl^-}$$

When 0.01 mole of HCl is added [acid] becomes as (0.01 + 0.02) M; [salt] becomes as (0.015– 0.01) M

$$p^H = K_a + \log\frac{[\text{salt}]}{[\text{acid}]} = 4.8729 + \log\frac{0.005}{0.03}$$

$$4.8729 - \log 6 = 4.8729 - 0.7782 = 4.0947$$

The p^H of a 0.01 molar HCl = $-\log 10^{-2} = 2$

27. Concentration of hydrogen ion in 0.4833% HCl = $\frac{0.4833 \times 1000}{100}$ = 4.833 gm lit^{-1}

$$\text{Normality of HCl} = \frac{4.833}{36.5} = 0.15$$

$$[H^+] = 0.15 = 1.5 \times 10^{-1}$$

$$\therefore \quad p^H = -\log[H^+] = -(1.1761) = (-0.8239) = 0.8239$$

28. The reaction between acetic acid and caustic soda (NaOH) is

$$CH_3COOH + NaOH = CH_3COONa + H_2O$$

When 20 ml of 0.01 M acetic acid is treated with 12.2 ml of 0.01 M NaOH the resulting mixture contains (20 – 12.2) = 7.8 ml of 0.01 M unreacted acetic acid. The total volume of the mixture becomes 32.2 ml (20 ml of CH_3COOH + 12.2 ml of NaOH)

The normality of the resulting mixture would be = $\frac{7.8 \times 0.01}{32.2}$ = 0.002422 w.r.t. to acetic acid

$$[H^+] = 0.002422 = 2.422 \times 10^{-3}$$

$$p^H = -\log[H^+] = (\bar{3}.38342) = (-2.6158) = 2.6158$$

The p^H of the resulting mixture = 2.6158.

29. $p^H = -\log[H^+]$

$$[H^+] = 10^{-p^H} = 10^{-4.54} = 10^{-5} \times 10^{0.46} = 2.886 \times 10^{-5}$$

Now, $[H^+]$ = molar concentration × degree of dissociation.

$2.886 \times 10^{-5} = 0.002 \times \alpha$

$$\alpha = \frac{2.886 \times 10^{-5}}{0.002} = \frac{2.886 \times 10^{-2}}{2} = 0.01443$$

Percent dissociation = 0.01443 × 100 = 1.443

30. Total volume of the mixture = 50 ml of acetic acid + 100 ml of sodium Acetate = 150 ml
The effective normality of acetic acid in the mixture

$$N = \frac{50 \times 0.12}{150} = 0.04$$

Similarly, effective normally of Potassium acetate in the mixture:

$$N = \frac{100 \times 0.12}{150} = 0.08$$

Hence [salt] = 0.08 gm. mole lit^{-2}; [acid] 0.04 gm mole lit^{-1}

$$P^H = PK_a + \log\frac{[\text{salt}]}{[\text{acid}]}$$

$$PK_a = -\log K_a = -\log(1.8 \times 10^{-5}) = 4.7447$$

$$P^H = 4.447 + \log\frac{0.08}{0.04} = 4.7447 + \log 2 = 4.7447 + 0.3010 = 5.0457$$

31. $P^H = P^{K_a} + \log\frac{[\text{salt}]}{[\text{acid}]}$

$\Rightarrow$ $P^{K_a} = -\log K_a = -\log(1.8 \times 10^{-5}) = 4.7447$

$$[\text{Salt}] = \frac{\text{wt. per litre}}{\text{mol. wt}} = \frac{16}{82} = 0.195 \text{ gm. mole}$$

$$[\text{acid}] = \frac{\text{wt. per litre}}{\text{mol. wt}} = \frac{8.3}{60} = 0.1383 \text{ gm mole}$$

When 0.7 ml. of 1 N HCl is added, the increases in volume is negligible, moles of HCl added = 0.7 ml. of 1 molar = 0.0007 gm mole

When HCl is added the concentration of H^+ increases by 0.0007 gm. ion.

Whereas the salt decreases by the same amount.

$$CH_3COO^- + H^+ \rightarrow CH_3COOH$$

Hence, after adding 0.7 ml of HCl, the new concentrations of acid; and salt are

[Acid] = 0.1383 + 0.0007 = 0.1390

{Salt] = 0.1960 – 0.0007 = 0.1943

Now, $P^H = P^{K_a} + \log\frac{[\text{salt}]}{[\text{acid}]} = 4.7447 + \log\frac{0.1943}{0.1390} = 4.7447 + 0.1455 = 4.8902$

Hence, P^H of the solution = 4.8902

32. $p^H = p^{K_a} + \log \frac{[\text{salt}]}{[\text{acid}]}$; $p^H = -\log[H]$; $p^{K_a} = -\log K_a$

$$-\log[H^+] = -\log K_a \text{ when } \log \frac{[\text{salt}]}{[\text{acid}]} = 0$$

$$[H^+] = K_a \text{ or } [\text{salt}] = [\text{acid}]$$

To get $[H^+] = K_a$ the concentration of salt must be equal to that of the acid. The buffer contains 2.0 gm of acetic acid per 250 ml or in other words 8.0 g. per litre. From this it follows.

$$[\text{acid}] = \frac{\text{wt per litre}}{\text{Mol. wt}} = \frac{8.0}{60} = 0.133 \text{ gm. mole}$$

[salt] = acid = 0.133 gm. mole

Weight of the salt per 1000 ml = 0.1.333 × 82 = 10.91

Weight of the salt per 250 ml = $10.9 \times \frac{250}{1000} = 2.73$

Amount of the sodium acetate required = 2.73 gm

33. Potassium acetate when dissolved in water under goes anion hydrolysis and produces free OH ion as $CH_3COO^- + H_2O \rightarrow CH_3COOH + OH^-$

$$\text{Degree of hydrolysis} = \frac{0.014}{100} = 0.00014 = 1.4 \times 10^{-4}$$

the concentration of $[OH^-]$ ion = molar concentration of salt × degree of hydrolysis

Now, $p^{OH} = -\log[OH^-] = -\log [1.082 \times 10^{-6}] = -(\bar{6}.2601) = -(-5.7399) = 5.7399$

But, $p^H + p^{OH} = 14$

$\therefore$ $p^H = 14 - p^{OH} = 14 - 5.7329 = 8.2601$

34. We know from the Ostwald's dilution law, $\frac{\alpha_1}{\alpha_2} = \sqrt{\frac{V_1}{V_2}}$

Where α_1 = Degree of dissociation when dilution is 0.1 = 0.011

α_2 = Degree of dissociation when dilution is 0.01 = ?

V_1 = 10 litres; V_2 = 100 litres $\therefore$ $\frac{0.011}{\alpha_2} = \sqrt{\frac{10}{100}} = \sqrt{0.1}$

$$\text{or } \alpha_2 = \frac{0.011}{\sqrt{0.1}} = 0.03478$$

Now, $[H^+] = \alpha \times C = 0.3478 \times 0.01 = 3.478 \times 10^4$

Again, $p^H = -\log[H^+] = -\log[3.478 \times 10^{-4}] = -(-4 + 0.5413) = 3.4587$

$\therefore$ p^H of the solution = 3.4587

35. We know that for a solution of monobasic acid.

$$[H^+] = \sqrt{K_a \times [\text{acid}]} = \text{Dissociation constant of the acid} = 1.85 \times 10^{-5}$$

$$[\text{Acid}] = \frac{[H^+]^2}{K_a} \doteq \frac{(3.981 \times 10^3)}{1.85 \times 10^{-5}} \doteq \frac{1.585 \times 10^{-7}}{1.85 \times 10^{-5}} \doteq 8.567 \text{ x. } 10^{-1} \text{ g mole lit}^{-1}$$

Concentration of the acid in gram per litre = $8.567 \times 10^{-1} \times 60 = 51.4$ g per litre

36. We know that for a mixture of monobasic acid

$$[H^+] = \sqrt{(K_{HA_1} \times [HA_1]) + (K_{HA_2} \times [HA]_2)}$$

Where, K_{HA_1} = Dissociation constant of $HA_1 = 1.6 \times 10^{-5}$

K_{HA_2} = Dissociation constant of $HA_2 = 1.8 \times 10^{-5}$

$\therefore$ $[HA_1]$ = Concentration the acid HA = $\frac{N}{5} \times \frac{1}{2} = \frac{N}{10} = 0.1$

$$[H^+] = \sqrt{(1.6 \times 10^{-5} \times 0.1) + (1.8 \times 10^{-5} \times 0.1)} = \sqrt{3.4 \times 10^{-6}}$$

$= 1.844 \times 10^{-6}$ gm. ion per lit^{-1}

Now, $p^H = -\log[H^+] = -\log\ (1.844 \times 10^{-3}) = -(-2.7343) = 2.7343$

p^H = of the solution = 2.7343

37. Since the first dissociation is 100 per cent the concentration of H^+ from 0.15 M, H_2SO_4 would be 0.15 gm ion per litre.

or $[H^+] = 0.15$ gm ion lit^{-1}

First dissociation

$$HSO_4 \rightleftharpoons H^+ + SO_4^{2-}; \text{ and } K_1 = \frac{[H^+][SO_4^{-2}]}{[HSO_4^-]}$$

Let a be the mole dissociated then $[H^+] = (0.15 + a)$ gm. ion lit^{-1} and $[SO_4^{2-}]$ = a gm. ion lit^{-1}

$\therefore$ $(HSO_4^-) = (0.15–a)$ gm. ion lit^{-1}

As, $K_2 = 1.26 \times 10^{-12}$; $\therefore$ $1.26 \times 10^{-12} = \frac{(0.15 + a)(a)}{(0.15 - a)} \Rightarrow a = 0.0109$

Concentration of H^+ from second dissociation = 0.0109

Hence, $H^+ = (0.15 + 0.0109) = 0.1609$ gm. ion lit^{-1}

So total concentration H^+ ion = 0.1609 gm ion lit^{-1}

38. The reaction between NH_3 and HCl is: $NH_3 + HCl = NH_4Cl$

This salt would be completely dissociated. Now with ammonium chloride the solution will be a buffer solution containing unreacted ammonia.

25 ml of 0.1 M NH_3 + 25 ml of 0.1 M HCl $\rightarrow$ 50 ml of 0.05 M NH_4Cl

Unreacted base (NH_3) = 5 ml of 0.1 M

Now, $p^{OH} = p^{K_b} + \log \frac{[\text{Salt}]}{[\text{Base}]}$

$p^{K_b} = -\log K_b = -\log 1.81 \times 10^{-5} = 4.7423$

$$[\text{Salt}] = \frac{50 \times 0.05}{1000} = 0.0025; \quad [\text{Base}] = \frac{5 \times 0.1}{1000} = 0.0005$$

$$P^{OH} = 4.7423 + \log \frac{0.0025}{0.0005} = 4.7423 + \log 5 = 4.7423 + 0.699 = 5.4413$$

Again, $P^H + P^{OH} = 14$

$\Rightarrow$ $P^H = 14 - P^{OH} = (14 - 5.4413) = 8.5587$

Hence, P^H of the solution = 8.5587

39. Given, $K_{\ln} = 10^{-9}$

$$P^{K_{\ln}} = -\log K_{\ln} = -\log 10^{-9} = 9$$

$$P^H \text{ range} = (P^{K_{\ln}} - 1) \text{ to } (P^{K_{\ln}} + 1) = 8 \text{ to } 10$$

i.e. P^H range would be between 8 and 10.

Therefore at P^H = 8 and less colour of acid from the indicator is observed i.e. solution is colour less while at P^H = 10 and more the colour of alkaline from red is seen. In the Range from P^H = 8 to 10 the colourless solution gradually turns bright red.

40. Limus; P^H – range = 5 to 8. It means with in this range only colour change wil take place. Beyound P^H = 8 the colour of indicator will remain the same. Therefore the solution of P^{-H} = 8.5 and 11 will give same colour with litmus.

Phenolphthalen; P^H Range = 8 – 10

Since P^H = 8.5 falls within the range and P^H = 11 out of this range, the two solution will produces different colour with phenolphthalein.

41. CH_3COONa is the salt of weak acid [CH_3COOH] and strong base (NaOH). It will hydrolysis and P^H of the resuling solution is given by

$$P^H = \frac{1}{2}(pK_w + pK_a + \log C)$$

$$\text{or } P^H = \frac{1}{2}[-\log 10^{-14} - \log(1.8 \times 10^{-5}) + \log 0.1] = \frac{1}{2}[15 - \log 1.8 + 5 - 1]$$

$$\text{or } P^H = \frac{1}{2}(14 - 0.2553 + 5 - 1) = \frac{17.7447}{2} = 8.8723$$

$$\text{Now, } P^H = P^{K_{In}} - \log \frac{[\text{acid form}]}{[\text{alk. form}]} \Rightarrow \log \frac{[\text{acid form}]}{[\text{alk. form}]} = P^{K_{In}} - P^H$$

$$= -\log(3.16 \times 10^{-10}) - 8.8726 = -\log 3.16 + 10 - 8.8726 = -0.4997 + 10 - 8.8726 = 0.6277$$

$$\text{Taking anti log, } \frac{[\text{acid form}]}{[\text{alk. form}]} = \text{Anti log } 0.6277 = 4.243$$

Therefore, fraction of indicator in coloured (alkaline) from $= \frac{1}{4.243 + 1} = \frac{1}{5.243} = 0.19$

42. $P^{K_{In}} = -\log K_{\ln} = -\log 10^{-5} = 5$

$$\log \frac{[\text{acid form}]}{[\text{alk. form}]} = P^{K_{In}} - P^H$$

(i) When $P^H = 4$, $\log \frac{[\text{acid form}]}{[\text{alk. form}]} = 5 - 4 = 1$

$$\frac{[\text{acid form}]}{[\text{alk. form}]} = \text{Anti log } 1 = 10$$

Acidic form is red and alkaline from is yellow

Hence $\frac{[\text{red}]}{[\text{yellow}]} = 10$; $s[\text{red}] = \frac{10}{1+1} = \frac{1}{2} = 0.909$

i.e. fraction of indicator in red form

(ii) When $P^H = 5$, $\log \frac{[\text{red}]}{[\text{yellow}]} = 5 - 5 = 0$

$$\frac{[\text{red}]}{[\text{yellow}]} = \text{Anti log } 0 = 1 \text{ or } [\text{red}] = \frac{1}{1+1} = \frac{1}{2} = 0.5$$

i.e. fraction of indicator in red form = 0.5

(iii) When $P^H = 6$, $\log \frac{[\text{red}]}{[\text{yellow}]} = 5 - 6 = -1$

$$\frac{[\text{red}]}{[\text{yellow}]} = \text{Anti log } 1.0 = 0.1 = \frac{1}{10} \text{ or } [\text{yellow}] = \frac{1}{1+10} = \frac{1}{11} = 0.0909$$

i.e. fraction of indicator in yellow form = 0.09

43. Suppose indicator is Hln, then

$Hln = H^+ + ln^-$; $K = 5.0 \times 10^{-8}$

yellow blue

On adding acid, $[H^+]$ increased and the equilibrium is shifted to the left (i.e. Chatelier principle)

Hence in acid solution the colour of the indicator is yellow (acid colour, colour of Hln). Similarly in presence of a base (OH^-), H^+ are removed the equilibrium shifts to the right with the formation of more ln^-. Hence in alkali solution the colour becomes blue (base colour, colour of ln^-)

$$K = \frac{[H^+][ln^-]}{[Hln]}$$

Yellow colour is visible when, $\frac{[\text{yellow form}]}{[\text{blue form}]} = \frac{[Hln]}{[ln^-]} = \frac{20}{1}$ $\therefore$ $\frac{[ln^-]}{[Hln]} = \frac{1}{20}$

Hence in this case $K = \frac{[H^+][ln^-]}{[Hln]}$ or $5.0 \times 10^{-8} = [H^+] \times \frac{1}{20}$

$\therefore$ $[H^+] = 20 \times 5.0 \times 10^{-8}$ or 1×10^{-6} mol l^{-1}

pH $= -\log [H^+] = -\log (1 \times 10^{-6}) = -(-6)$ or 6.

Blue colour is predominant when $\frac{[\text{blue form}]}{[\text{yellow form}]} = \frac{[\text{In}^-]}{[\text{HIn}]} = \frac{2}{1}$

$\therefore \quad K = \frac{[H^+][\text{In}^-]}{[\text{HIn}]}$ or $5.0 \times 10^{-8} = [H^+] \times \frac{2}{1}$

$[H^+] = 2.5 \times 10^{-8}$ mol l^{-1}

pH = $-\log [H^+] = -\log (2.5 \times 10^{-8}) = (8.398) = -(-7.6)$ or 7.6

$\therefore$ pH range of colour change for indicator is 6.0 to 7.6

44. (a) Henderson's equation;

$$p^H = p^{Ka} + \log \frac{[\text{salt}]}{[\text{acid}]}; \text{ Here, } PK_a = -\log K_a = -\log(2.4 \times 10^{-4}) = 3.6198$$

[salt] = 0.75 moles lit^{-1}; [acid] = 0.2 moles lit^{-1}

$$\therefore \quad p^H = 3.62 + \log \frac{0.75}{0.20} = 4.194$$

(b) Here we have;

$$K_b = \frac{K_w}{K_a} = \frac{1.0 \times 10^{-14}}{1.75 \times 10^{-5}} = 5.71 \times 10^{-10}.$$

$\therefore$ $pK_b = \log K_b = -\log 5.71 \times 10^{-10} = -(-10 + \log 5.71) = 9.244.$

45. The saturated solution contained 7.0 mg $CaCO_3$, dissolved in it.

This is 7×10^{-3} gm $CaCO_3 \times \frac{1 \text{ mol } CaCO_3}{100 \text{ gm } CaCO_3} = 7.0 \times 10^{-5}$ mole $CaCO_3$

Saturated solution now contains 7.0×10^{-5} mole Ca^{2+} ions litre^{-1} and 7.0×10^{-5} mole CO_3^{2-} ions litre^{-1}

According to definition, solubility product, K_{sp} ($CaCO_3$) = $[Ca^{2+}] \times [CO_3^{2-}]$

on substitution we have

$K_{sp}(CaCO_3) = (7.0 \times 10^{-5})(7.0 \times 10^{-5}) = 49 \times 10^{-10}$ (OR) $= 4.9 \times 10^{-9}$

46. 4.5 gm ($PbCl_2$) litre^{-1} $= \frac{4.5 \text{ gm } PbCl_2}{\text{litre}} \times \frac{1 \text{ mol } PbCl_2}{278 \text{ g } PbCl_2} = 1.62 \times 10^{-2}$ mole $PbCl_2$ litre^{-1}

But 1.62×10^{-2} mole $PbCl_2$ litre^{-1} in solution withproduce 1.62×10^{-2} mole Pb^{2+} ions litre^{-1} and $2 \times 1.62 \times 10^{-2}$ mole Cl^- ions litre^{-1} = 3.24×10^{-2} moles of Cl^- ions litre^{-1}

$K_{sp} = [Pb^{2+}][Cl^-]^2$. But, on substitution

$K_{sp} = (1.62 \times 10^{-2}) \times (3.24 \times 10^{-2}) = 1.62 \times 10^{-2} \times 10.5 \times 10^{-4} = 1.7 \times 10^{-6}$

47. $K_{sp} = [Ag^+] \times [Cl^-]$

Let us call the solubility of AgCl as x moles litre^{-1}

It produces x moles litre^{-1} of Ag^+ and $\times$ moles litre^{-1} of Cl^- ions.

$K_{sp} = (x)(x) = 1.6 \times 10^{-10}$; $x^2 = 1.6 \times 10^{-10}$

$$x = \sqrt{1.6 \times 10^{-10}} = 1.26 \times 10^{-5} \text{ moles litre}^{-1}$$

48. Mixture solution contains 0.1 M Ag^+ and 0.1 M Hg_2^+.

K_{sp} of Hg_2I_2 = 2.5×10^{-26} in much smaller than K_{sp} of AgI which is 8.5×10^{-17}.

$[I^-]$ concentration needed to precipitate $H_{g2}I_2$ is calculated as:

$$Hg_2I_2 \rightleftharpoons Hg_2^+ + 2I^-$$

$$K_{sp} = [Hg_2^+][I^-]^2$$

$$2.5 \times 10^{-26} = [1 \times 10^{-2}][I^-]^2$$

$$[I^-] = \sqrt{\frac{2.5 \times 10^{-25}}{1 \times 10^{-2}}} = 5.0 \times 10^{-13} \text{ M}$$

Similarly $[I^-]$ concentrated needed to ppt AgI is:

$$[Ag^+][I^-] = K_{sp}$$

$$[1 \times 10^{-1}][I^-] = 8.5 \times 10^{-17}$$

$$[I^-] = 8.5 \times 10^{-16} \text{ M}$$

Since $[I^-]$ concentration needed to ppt. AgI is smaller than that needed to ppt. Hg_2I_2, AgI is completely precipitated first.

Hg_2I_2 is precipitated only when molar concentration of I^- reaches 5.0×10^{-13} M.

$[Ag^+]$ left when $H_{g2}I_2$ begins to ppt. is given by

$$\frac{K_{sp} \text{ of AgI}}{[I^-]_{HgI_2}} = \frac{8.5 \times 10^{-17}}{5.0 \times 10^{-12}} = 1.7 \times 10^{-4} \text{ M}$$

$$\% [Ag^+] \text{ left} = \frac{1.7 \times 10^{-4}}{0.1} \times 100 = 0.17\%$$

$$\% Ag^+ \text{ ppted} = 99.83\%$$

49. $2AgCl + Na_2CO_3 \rightarrow Ag_2CO_3 + 2NaCl$

The reaction vessel has 0.0026 gm lit^{-1}, i.e., $\frac{0.0026}{35}$ mol lit^{-1} or 7.4×10^{-5} mol · l^{-1} of Cl^- ion.

Now, $$Ag_2CO_3 \rightleftharpoons 2Ag^+ + CO_3^{2-}$$

∴ $$K_{sp}\ Ag_2CO_3 = [Ag^+]^2 \times [CO_3^{2-}] \qquad \text{......(i)}$$

Na_2CO_3 is supposed to be completely dissociated in solution Now $[CO_3^{2-}]$ is 1.5 M Substituting this in equation (i)

$$\frac{8.2 \times 10^{-12}}{1.5} \text{ mol} \cdot l^{-1} = [Ag^+]^2 = 2.34 \times 10^{-6} \text{ mol} \cdot l^{-1}$$

Now K_{Sp} AgCl= $[Ag^+][Cl^-] = (2.34 \times 10^{-6}) \times (7.4 \times 10^{-5}) = 1.71 \times 10^{-10}$

50. Solubility of $Mg(OH)_2$.

$$s = 9.57 \times 10^{-3} \text{ g litre}^{-1} = \frac{9.57 \times 10^{-3}}{58} = 1.65 \times 10^{-4} \text{ mole litre}^{-1}$$

$$Mg(OH)_2 \rightleftharpoons Mg^{2+} + 2OH^-$$

$$K_{sp} = (S)(2S)^2 = 4S^3 = 4(1.65 \times 10^{-4})^3 = 1.8 \times 10^{-11} \text{ approx.}$$

$$K_{sp} = [Mg^{2+}]\,[OH^-]^2$$

$$\therefore\ [OH^-]^2 = \frac{1.8 \times 10^{-11}}{0.02} = 9 \times 10^{-10} \qquad \text{(In 0.02 M } Mg(NO_3)_2]$$

or $[OH^-] = 3 \times 10^{-5}$

$$[Mg^{2+}] = \frac{1}{2}[OH^-] = \frac{3 \times 10^{-5}}{2} = 1.5 \times 10^{-5} \text{ mole litre}^{-1}$$

$$= 1.5 \times 10^{-5} \times 58 = 8.7 \times 10^{-4} \text{ g litre}^{-1}$$

51. $AgCl \rightleftharpoons Ag^+ + Cl^-$ and $Ag(NH_3)_2^+ \rightleftharpoons Ag^+ + 2NH_3$

At equilibrium $[NH_3] = 1$ M

$$[Cl^-] = [Ag^+] + [Ag(NH_3)_+^2]$$

$$\cong [Ag(NH_3)_2^+]$$

Since $[Ag(NH_3)_+^2] + Cl^- \rightarrow 2NH_3 + AgCl$

$$K = \frac{6.2 \times 10^{-8}}{1.8 \times 10^{-10}}$$

$$\text{i.e.} = \frac{[NH_3]^2}{[Ag(NH_3)_2^+\,[Cl^-]} = \frac{[NH_3]^2}{[Ag(NH_3)_2^+]}$$

$$[NH_3] = 1 \text{ M}$$

So, $[Ag(NH_3)_2^+] = 0.054$ M

52. $Ca(OH)_2 \leftrightarrow Ca^{2+} + 2OH^-$

If x is solubility of $Ca(OH)_2$

K_{sp} of $Ca(OH)_2 = [Ca^{2+}]\,[OH^-]^2 = (x)\,(2x)^2 = 4x^3$

$$\text{Hence, } x = \left(\frac{K_{sp}}{4}\right)^{1/3} = \left(\frac{4.42 \times 10^{-5}}{4}\right)^{1/3} = 0.0223 \text{ M}$$

Number of moles of Ca^{2+} ions in 500 ml of the solution = 0.01115.

Now when 500 ml of saturated solution is mixed with 500 ml of 0.4 M NaOH, the resultant volume is 100 ml. the molarity of OH^- ions in the resultant solution would be 0.2 M. Now

$$Ca^{2+} = \frac{K_{sp}}{[OH^-]^2} = \frac{4.42 \times 10^{-5}}{(0.2)^2} = 0.001105 \text{ M.}$$

The number of moles of Ca^{2+} ions or $Ca(OH)_2$ precipitated
= 0.01115 – 0.001105 = 0.010045

Mass of $Ca(OH)_2$ precipitated = (0.010045) (74) = 0.7433 g = 743.3 mg.

53. $NH_4OH \rightleftharpoons NH_4^+ + OH^-$

$$K_b = \frac{[NH_4^+][OH^-]}{[NH_4OH]}$$

$$\therefore\ [OH^-] = \frac{K_b[NH_4OH]}{[NH_4^+]} = \frac{1.8\times10^{-5}\times0.05}{0.25}\ \text{mol L}^{-1} = 0.36\times10^{-5}\ \text{mol L}^{-1}$$

From solubility product equation,

$$[Al^{3+}] = \frac{K_{sp}\,[Al(OH_3)]}{[OH^-]^3} = \frac{6\times10^{-32}}{(0.30\times10^{-5})^3}\ \text{mol L}^{-1} = 1.29\times10^{-19}\ \text{mol L}^{-1}$$

$$[Mg^{2+}] = \frac{K_{sp}\,[Mg(OH)_2]}{[OH^-]^2} = \frac{6\times10^{-10}}{(0.36\times10^{-5})^2}\ \text{mol L}^{-1} = 46.3\ \text{mol L}^{-1}$$

54. Ionization equation of $Ag_2C_2O_4$ is $Ag_2C_2O_4 \leftrightarrow 2Ag^+ + C_2O_4^{2-}$

$$K_{sp}(Ag_2C_2O_4) = [Ag^+][C_2O_4^{2-}]$$

$$\Rightarrow\quad [Ag^+]^2 = \frac{k_{sp}[Ag_2C_2O_4]}{[C_2O_4^{2-}]} \qquad \ldots\ldots\ (1)$$

and $AgCO_3$ ionizes as $Ag_2CO_3 \leftrightarrow 2Ag^+ + CO_3^{2-}$
and we have $K_{sp}(Ag_2CO_3)= [Ag^+][CO_3^{2-}]$

$$\Rightarrow\ [Ag^+]^2 = \frac{K_{sp}\,(Ag_2CO_3)}{[O_3^{2-}]} \qquad \ldots\ldots\ (2)$$

Now since the concentration of Ag^{2+} is same in both (1) and (2) we have

$$\frac{K_{sp}[Ag_2C_2O_4]}{[C_2O_4^{2-}]} = \frac{k_{sp}\,[Ag_2CO_3]}{[CO_3^{2-}]} \qquad \ldots\ldots\ (3)$$

The main equation is

$$Ag_2CO_3 + K_2C_2O_4 \leftrightarrow Ag_2C_2O_4 + K_2CO_3$$

Number of moles

	$K_2C_2O_4$	K_2CO_3
Initial	0.152	0
At equilibrium	0.152 – 0.0358 = 0.1162	0.0358
	0.1162 × 2	0.0358 × 2

Given that $K_{sp}(Ag_2C_2O_4) = 1.29\times10^{-11}\ \text{mol}^3\ \text{lit}^{-3}$

Putting these value in (3) we get

$$K_{sp}[Ag_2CO_3] = \frac{K_{sp}[Ag_2C_2O_4][C_2O_4^{2-}]}{[CO_3^{2-}]}$$

$$= 1.29 \times 10^{-11}\ \text{mol}^3 \times \frac{2\times 0.0358}{2\times 0.1162} = 3.97 \times 10^{-12}\ \text{mol}^3\ \text{lit}^{-3}$$

55. k_{sp} for MS = 6.0×10^{-21}, concentration of saturated H_2S = 0.1 M;

$\therefore$ $K_1 = 10^{-7}$ and $K_2 = 1.3 \times 10^{-13}$,

for H_2S $H_2S \xleftrightarrow{K_1} H^+ + HS^-$

$$\therefore \quad K_1 = \frac{[H^+][HS^-]}{[H_2S]}$$

Further, HS^- K_2 $H^+ + S^{2-}$

$$\therefore \quad K_2 = \frac{[H^+][S^{2-}]}{[HS^-]}$$

Dissociation constant of H_2S, $K = K_1 \times K_2 = 1 \times 10^{-7} \times 1.3 \times 10^{-13} = 1.3 \times 10^{-20}$

Now, we know that, $K_{sp} = [M^{2+}]\ [S^{2-}]$

or $6 \times 10^{-21} = 0.05 \times [S^{2-}]$

$$\therefore \quad [S_{2-}] = \frac{6\times 10^{-21}}{0.05} = 1.2 \times 10^{-19}$$

substituting the various values in the following relation

$$K = \frac{[H^+]^2\ [S^{2-}]}{[H_2S]}$$

$$1.3 \times 10^{-20} = \frac{[H^+]^2\ [1.2\times 10^{-19}]}{0.1} \quad \text{(since } [H_2S] = 0.1 \text{ M)}$$

$$\text{or} \quad [H^+]^2 = \frac{1.3\times 10^{-20}\times 0.1}{1.2\times 10^{-19}}$$

$$\Rightarrow \quad [H^+] = \sqrt{\left[\frac{1.3\times 10^{-20}\times 0.1}{1.2\times 10^{-19}}\right]} = 1.04 \times 10^{-1}$$

$\therefore$ $p^H = -\log[H^+]$, $p^H = -\log(1.04 \times 10^{-1}) = 1.0 - \log 1.04 = 1.0 - 0.017 = 0.983.$

56. Here, solubility of $Pb(OH)_2S$, in water is 6.7×10^{-6} m

Now, $Pb(OH)_2 \rightleftarrows Pb^{+2} + 2OH^-$

$$K_{sp} = [Pb^{2+}]\ [OH^-]^2$$

$$K_{sp} = 4S^3$$

$K_{sp} = 4 \times (6.7 \times 10^{-6})^3$

$K_{sp} = 1.2 \times 10^{-15}$ mol^3 lit^{-3}

In the Buffer sol. of $P^H = 8$

$[OH^-] = 10^{-6}$ mol^3 lit^{-3}

Assume S_1 solubility in Buffer of Pb $(OH)_2$

$K_{sp} = S_1 \times [10^{-6}]^2$

$$S_1 = \frac{1.2 \times 10^{-15}}{[10^{-6}]^2} = 1.2 \times 10^{-3}$$

Solubility $(S_1) = 1.2 \times 10^{-3}$ m

57. m. eq of HCl $= \dfrac{M \times V}{1000}$

$$= \frac{10^{-2} \times 200}{1000} = 2 \times 10^{-3} \text{ M}$$

$$\text{m.eq of NaOH} = \frac{10^{-2} \times 300}{1000} = 3 \times 10^{-3} \text{ M}$$

Thus 200 ml of HCl (pH = 2) containing 2×10^{-3} M of H^+ and 300 ml of NaOH (pH = 12) containing 3×10^{-2} M of OH^- when mixed resulting solution has 2×10^{-3} M OH^-

Thus

$$\text{pOH} = -\log[\text{OH}] = 2.7$$

and, $$\text{pH} = 14 - \text{pOH} = 11.3$$

58. The isonisation of the indicator HIn may be shown an HIn $\rightleftarrows H^+ + In^-$
The equation for the above equilibrium may be written as

$$K_a = \frac{[H^+][In^-]}{[HIn]} \quad \ldots\ldots \text{(i)}$$

Color difference is visible only when $\dfrac{[In^-]}{[HIn]}$ is greater than 10 or smaller than 0.1

From (i), when $\dfrac{[In^-]}{[HIn]}$ is greater than 10,

$$1.0 \times 10^{-5} = [H^+] \times 10$$

or $$[H^+] = 1 \times 10^{-6}$$

$\therefore$ $$\text{pH} = -\log[H^+] = 6$$

Similarly from (i) when $\frac{[In^-]}{[HIn]}$ is less than 0.1

$$[H^+] = \frac{1.0 \times 10^{-5}}{0.1} = 1 \times 10^{-4}$$

$\therefore$ $$pH = -\log [H^+] = 4$$

Thus the minimum change in the pH of the solution will be between 4 and 6, a unit of 2, to observe a complete colour change.

59. On adding sodium hydroxide to acetic acid sodium acetate will be formed. The quantity of acetic acid is more than that of sodium hydroxide, therefore, the resulting solution will contain sodium acetate and acetic acid. Such a mixture will form a buffer solution
The pH of a buffer is calculated by

$$pH = pKa + \log \frac{[salt]}{[acid]}$$

Volume of 0.2 M NaOH = 20 ml
Volume of 0.2 M acetic acid = 50 ml
Total volume of buffer = (20 + 50) ml = 70 ml
Volume of 0.2 M sodium acetate = 20 ml
Volume of acetic acid left after adding sodium hydroxide = (50 – 20) ml = 30 ml

$$[acid] = \frac{30 \times 0.2}{70}$$

$$[salt] = \frac{20 \times 0.2}{70}$$

$$K_a = 1.8 \times 10^{-5}$$
$$pK_a = -\log Ka = 4.74$$

Therefore,

$$pH = 4.74 + \log \left[\frac{20 \times 0.2}{70} \times \frac{30}{30 \times 0.2}\right] = 4.74 + \log \frac{2}{3} = 4.74 - 0.18 = 4.36$$

The pH of the solution will be 4.74 when [salt] = [acid]. It will be possible when 25 ml 0.2 M sodium hydroxide is added to 50 ml of 0.2 M acetic acid.
Volume of sodium hydroxide already added = 20 ml
Volume of sodium hydroxide needed to make pH 4.74 = 25 – ml = 5 ml.

60. Dissociation constant (K_a) of a weak acid HA = 4.9×10^{-8}

$$HA \rightleftharpoons H^+ + A^-$$

$\therefore$ $$c(1-x) \quad cx \quad cx$$

(where x stands for degree of ionization and $c = 0.1$)
By the Law of Mas Action at equilibrium conditions.

$$K_a = \frac{x^2 x^2}{x(1-z)} \cong cx^2$$

or $$x^2 = \frac{K_a}{c}; \Rightarrow x = \left(\frac{K_a}{c}\right)^{\frac{1}{2}}$$

or $$x = \left(\frac{4.9 \times 10^{-8}}{0.1}\right)^{\frac{1}{2}}$$ [∵ For decimolar sol. $x = 0.1$]

or $$x = (49 \times 10^{-8})^{\frac{1}{2}} = 7 \times 10^{-4}$$

(i) ∴ Percentage ionization = $7 \times 10^{-4} \times 100 = 7 \times 10^{-2} = \frac{7}{100} = 0.07\%$

(ii) Concentration of

$H^+ = cx = 0.1 \times 7 \times 10^{-4}$ mole litre^{-1} = 7×10^{-5} mole litre^{-1}

∴ $p^H = -\log C_{H^+} = -(\log 7 - 5) = 5 - \log 7 = 5 - 0.8451 = 4.1549 = 4.15$

(iii) Ionic product of water i.e. $K_w = 10^{-14}$

$C_{H^+} \times C_{OH^-} = K_w$

$$\therefore C_{OH^-} = \frac{K_w}{C_{H^+}} = \frac{10^{-14}}{7 \times 10^{-5}} \times 10^{-14+5} = 0.143 \times 10^{-9} \text{ mole litre}^{-1}$$

$= 1.43 \times 10^{-10}$ mole litre^{-1}

∴ OH^- concentration in a decimolar solution of acid = 1.43×10^{-10} mole litre^{-1}

61. Sodium cyanide is a salt of strong base weak acid its hydrolysis can be given as

$$CN^- + H_2O \rightarrow OH^- + HCN$$

$$K = \frac{[OH^-][HCN]}{[CN^-]}$$

Or, $[OH^-] = \sqrt{PK_b \times C}$ C = Concen. of CN^{-1}

$$[OH^-] = \sqrt{10^{-4.7} \times 0.5} = 3.15 \times 10^{-3} \text{ M}$$

Thus, pH = 11.5

62. (a) In first case

1. $\Delta T_b = K_b \times m$; Substituting the values;

$$0.17 = 1.7 \times \frac{1.22}{M \times 100 \times 10^{-3}} \Rightarrow M = 122$$

(2) In second case

$$\Delta T_b = K_b \times m$$

$$\therefore \quad 0.13 = 2.6 \times \frac{1.22}{M' \times 100 \times 10^{-3}}$$

$$\Rightarrow \quad M' = 244$$

Hence, Benzoic acid dimerises in benzene.

(b) It means that benzoic acid remains as it is acetone while it dimensises in benzene as

$$C_6H_5-C(=O\cdots H-O)(-O-H\cdots O=)C-C_6H_5$$

(ii) Since at end point molarity of salt = $\frac{0.1}{2}$ M

since pH of salt of weak acid and strong base.

$$\text{pH} = \frac{(pK_w + pK_a + \log_c)}{2} = \frac{1}{2}(14 + 5.3010 + [-1.3010]$$

or pH = 9

63. By definition: $K_{sp}[Pb^{2+}][Br^-]^2$
Let x = solubility of Pb Br_2 mole litre^{-1}
But x mole of Pb Br_2/litre^{-1} produce x mole of Pb^{2+} litre^{-1} and $2x$ mole of Br^- ions litre^{-1}

$$K_{sp}(PbBr_2) = (x)(2x)^2 = 6.3 \times 10^{-6};\ 4x^3 = 6.3 \times 10^{-6}$$

$$x^3 = \frac{6.3 \times 10^{-6}}{4} = 1.58 \times 10^{-6}$$

$$x = 3\sqrt{1.58 \times 10^{-6}} = 1.17 \times 10^{-2} \text{ moles litre}^{-1}$$

64. By definition $K_{sp}(AgCl) = [Ag^+][Cl^-]$;
K_{sp} 1.6×10^{-10}; $[Cl^-] = 1 \times 10^{-2}$
Then on solving for $[Ag^+]$ ions and substitution

$$[Ag^+] = \frac{K_{sp}[AgCl]}{[Cl^-]} = \frac{1.6 \times 10^{-10}}{1.0 \times 10^{-2}} = 1.6 \times 10^{-8} \text{ mole litre}^{-1}$$

65. By definition, $K_{sp}(Ag_2CrO_4) = [Ag^+]^2[CrO_4^{2-}]$
We are given, $[CrO_4^{2-}] = 0.40$; $K_{sp} = 1.1 \times 10^{-11}$

$$[Ag^+]^2 = \frac{K_{sp}}{(CrO_4^{2-}]} = \frac{1.1 \times 10^{-11}}{0.40}$$

$$[Ag^+]^2 = 27.5 \times 10^{-12};\ [Ag^+] = \sqrt{27.5 \times 10^{-12}} = 5.2 \times 10^{-6} \text{ mole litre}^{-1}$$

66. By definition, $K_{sp}(PbCl_2) = [Pb^{2+}][Cl^{-2}]^2$
This relation holds true only in a solution saturated with $PbCl_2$. If the solution is not saturated (It more $PbCl_2$ can be dissolved in it) then the product of the $[Pb^{2+}][Cl^-]^2$ will be less than $K_{sp}(PbCl_2)$ in the problem given:
$[Pb^{2+}][Cl^-]^2 = (2 \times 10^{-3})(3 \times 10^{-2})^2 = 2 \times 10^{-3} \times 9$ x. $10^{-4} = 18 \times 10^{-7}$ or = $.8 \times 10^{-6}$
But 1.8×10^{-6} is clearly smaller than 1.7×10^{-5}, which is $K_{sp}(PbCl_2)$, and so the solution is not saturated and no ppt. of $PbCl_2$ will form.

67. $K_{sp}(PbSO_4) = [Pb^{2+}]\,[SO_4]^{2-}$

$$\therefore\ [SO_4]^{2-} = \frac{K_{sp}}{[Pb^{2+}]} = \frac{1.8\times10^{-8}}{2\times10^{-6}} \text{ or } 9.0 \times 10^{-3} \text{ mole litre}^{-1}$$

68. (a) The equilibrium in a saturated solution of $BaSO_4$ can be represented as:

$$BaSO_4 \rightleftharpoons Ba^{2+} + SO_4^{2-}$$

if the solubility of $BaSO_4$ is assumed to be 's' then $[Ba^{2+}] = [SO_4^{2-}] = s$

$$\therefore\ K_{sp} = [Ba^{2+}]\,[SO_4^{2-}] = s^2 = 1.5 \times 10^{-9};\ s = \sqrt{1.5\times10^{-9}} = 4.87 \times 10^{-5}$$

solubility of $BaSO_4$ is 3.87×10^{-5} mole litre^{-1}

(b) The equilibrium in a saturated solution of CaF_2 can be represented as

$CaF_2 \rightleftharpoons Ca^{2+} + 2F^-;\ \therefore\ K_{sp} = [Ca^{2+}]\,[F^-]^2 = 1.08 \times 10^{-10}$

If the solubility of CF_2 is assumed to be 's' then $[Ca^{2+}] = s$ and $[F]^2 = 2s$

$K_{sp} = [Ca^{2+}]\,[F^-]^2 = s(2s)^2 = 4s^3 = 1.08 \times 10^{-10}$

$$s^3 = 0.27\times 10^{-10} = 27 \times 10^{-12} \text{ or } s = \sqrt[3]{27\times10^{-12}} = 3 \times 10^{-4}$$

solubility of CaF_2 is 3×10^{-4} mole litre^{-1}

(c) The equilibrium in a saturated solution of aluminum hydroxide is

$$Al(OH)_3 \rightleftharpoons Al^{3+} + 3OH^-;\ K_{sp} = [Al^{3+}]\,[OH^-]^3 = 1 \times 10^{-33}$$

if the solubility of $Al(OH)_3$ is assumed as 's'

$$K_{sp} = s \times (3s)^3 = 27s^4 = 1 \times 10^{-33}$$

$$s^4 = \frac{10^{-33}}{27} \text{ or } s = \sqrt[4]{\frac{10^{-33}}{27}} = 2.47 \times 10^{-9}$$

solubility of aluminum hydroxide is 2.47×10^{-9} moles litre^{-1}

69. (a) $AgI \rightleftharpoons Ag^+ + I^-;\ [Ag^+] = [I^-] = 1.23 \times 10^{-8}$

$$K_{sp} = [Ag^+]\,[I^-] = [1.23 \times 10^{-8}]\,[1.23 \times 10^{-8}] = 1.51 \times 10^{-16}$$

The solubility product of AgI is 1.51×10^{-16}

(b) $Ag_2CO_3 \rightleftharpoons 2Ag^+ + CO_3^{2-}$

If the solubility of Ag_2CO_3 is 1.27×10^{-4} mole litre^{-1}

$$[Ag^+] = 2.54 \times 10^{-4} \text{ and } [CO_3^{2-}] = 1.27 \times 10^{-4}$$

$$K_{sp} = [Ag^+]^2\,[CO_3^{2-}] = [2.54 \times 10^{-4}]^2\,[1.27 \times 10^{-4}] = 8.19 \times 10^{-12}$$

The solubility product of $Ag_2\,CO_3$ is 8.19×10^{-12}

(c) $Sr\,F_2 \rightleftharpoons Sr^{2+} + 2F^-$

Solubility of SrF_2 is 1.22×10^{-2} gm litre^{-1}; and molecular weight of SrF_2 = 1.25.6

$$\text{Molar solubility of } SrF_2 = \frac{1.22\times10^{-2}}{125.6} = 9.712 \times 10^{-5} \text{ moles litre}^{-1}$$

$$SrF_2 \rightleftharpoons Sr^{2+} + 2F^-;\ [Sr^{2+}] = 9.712 \times 10^{-5};\ [F^-] = 2 \times 9.712 \times 10^{-5}$$

$$K_{sp} = [Sr^{2+}]\,[F^-]^2 = [9.712 \times 10^{-5}]\,[\,2 \times 9.712 \times 10^{-5}]^2 = 3.66 \times 10^{-12}$$

The solubility product of SrF_2 is = 3.66×10^{-12}

70. Let the solubility of $Mg(OH)_2$ be '*s*';

$$Mg(OH)_2 \rightleftharpoons Mg^{2+} + 2OH^-$$

$\therefore$ $K_{sp}= [Mg^{2+}] [OH^-]^2 = s(2s)^2 = 4 s^3$

$4s^3 = 8.9 \times 10^{-12}$

$$s^3 = \frac{8.9 \times 10^{-12}}{4} = 2.22 \times 10^{-4}$$

the solubility of $Mg(OH)_2$ is therefore 1.3×10^{-4} moles $litre^{-1}$

$$[OH^-] = 2 \times 1.3 \times 10^{-4} \text{ moles litre}^{-1}; [H^+] = \frac{1 \times 190^{-14}}{2.6 \times 10^{-4}} = \frac{10^{-10}}{2.6}$$

$$p^{H} = -\log[H^+] = -\log \frac{10^{-10}}{2.6} = -[\log 10^{-10} - \log 2.6) = 10 + 0.415 = 10.415$$

p^H of the saturated solution of $Mg(OH)_2$ is 10.415

71. There are two equivalent ways to solve this problem (see below) and we will use both, to ensure that both methods are clear to you. For either method we begin by writing the proton-transfer reaction between acetic acid and water, and the expression for K_a:

$$CH_3COOH(aq) + H_2O \rightleftharpoons CH_3COO^-(aq) + H_2O^-(aq)$$

$$K_a = \frac{[H_3O^+][CH_3COO^-]}{[CH_3COOH]} = 1.8 \times 10^{-5}$$

This numerical value of K_a for acetic acid at 298 K is obtained from Tables. The stoichiometric concentration of acetic acid, is 0.1 *M* in this problem.

We can calculate both the pH and α if we know the $[H_3O^+]$ in this solution, so let us begin by setting $x= [H_3O^+]$ at equilibrium in 0.1 *F* Ch_3COOH

Now, any solution prepared by mixing only acetic acid and water, the molar ratio of acetate ions to hydronium ions is 1: 1, that is, the $[CH_3COO^-]$ must be equal to the $[H_3O^+]$. therefore $[CH_3COO^-]= [H_3O^+] = x$

The concentration of acetic acid molecules in this solution is slightly smaller that the stoichiometric concentration, 0.1 *M*, because although 0.1 mol of CH_3COOH molecules were dissolved per liter of solution, a fraction of those molecules exists as the ions, CH_3COO^- and H_3O^+, at equilibrium the equation for the proton-transfer reaction, tells us.

$$\frac{\text{No. mol } CH_3COO^- \text{ formed}}{\text{No. mol } CH_3COOH \text{ used up}} = \frac{1}{1}$$

and hence the concentration of CH_3COOH that is used up is also equal to *x*.

In solving equilibrium problems, it is helpful to write down the initial concentration of all species, the change in concentration that occurs as the system comes to equilibrium and then the final concentrations at equilibrium. The following format may be used:

Concentration (M)	CH_3COOH	CH_3COO^-	H_3O^+
Initial value	0.100	0	0
Change	$-x$	x	x
Equilibrium value	$0.1 - x$	x	x

Note that at equilibrium the [CH_3COOH] is *less* than its initial value, as some CH_3COOH molecules are used up during the proton-transfer reaction with H_2O.

Since at equilibrium [CH_3COOH] = $(0.1 - x)$, the expression for the acidity constant of acetic acid is

$$K_a(\text{HOAc}) = \frac{[H_3O^+][CH_3COO^-]}{[CH_3COOH]} = \frac{x^2}{0.1-x} = 1.8 \times 10^{-5}$$

This is a quadratic equation in one unknown and can be solved for x in a straight for ward manner. Since we know that acetic acid, a weak acid, is only slightly ionized, we expect x, the [H_3O^+], to be significantly smaller that 0.1 M. Assume x is small compared to 0.1, so that $0.1 - x \sim 0.1$. Or equation then become

$$K_a(\text{HOAc}) = \frac{x^2}{0.1-x} = \frac{x^2}{0.1-x} = 1.8 \times 10^{-5}$$

Thus, $x^2 = 1.8 \times 10^{-6}$

and $x = 1.3 \times 10^{-3}\ M$

i.e. $[H_3O^+] = [CH_3COO^-] = 1.3 \times 10^{-3}\ M$

The concentration of acetic acid molecules is given by

$$[CH_3COOH] = 0.1 - 1.3 \times 10^{-3} = 0.1 - 0.001 = 0.99\ M$$

just slightly less than the stoichiometric concentration.

72. (a) The conjugate acid of S^{2-} ion is HS^- ion, the bisulfide ion or hydrogen sulfide ion. The acidity constant at 298 K, K_a (HS^-) = 1.3×10^{-13} (from tables). Thus we obtain

$$K_b(S^{2-}) = \frac{K_w}{K_a(HS^-)} = \frac{1.0 \times 10^{-14}}{1.3 \times 10^{-13}} = 7.7 \times 10^{-2}$$

(b) The conjugate acid of HPO_4^{2-} is $H_2PO_4^-$, the dihydrogen phosphate ion. In Table we find Now, $K_a(H_2PO_4^-) = 6.2 \times 10^{-8}$ (from tables). Thus,

$$K_a(HPO_4^{2-}) = \frac{K_w}{K_a(H_2PO_4^-)} = \frac{1.0 \times 10^{-14}}{6.2 \times 10^{-8}} = 1.6 \times 10^{-2}$$

73. The proton-transfer reaction between NH_3 and H_2O is

$$NH_3(aq) + H_2O \rightleftharpoons NH_4^+(aq) + OH^-(aq)$$

$$K_a(NH_3) = \frac{[NH_4^+][OH^-]}{[NH_3]} = 1.8 \times 10^{-3}$$

Let $x = [NH_4^+]$ in 0.25 $F NH_3$ at equilibrium at 298 K

Therefore we have $[NH_4^+] = [OH^-] = x$

The concentration of NH_3 *used up* in the proton-transfer reaction is also x, so that

$[NH_3] = 0.250 - x$

The equilibrium constant expression therefore becomes

$$K_a(NH_3) = \frac{[NH_4^+][OH^-]}{[NH_3]} = \frac{x^2}{0.25 - x} = 1.8 \times 10^{-3}$$

Assume x is small compared to 0.25, that is assume x is less than 10% of 0.25. Then $0.25 - x \simeq 0.25$, and the expression for K_b becomes

$$\frac{x^2}{0.25 - x} = \frac{x^2}{0.25} = 1.8 \times 10^{-3}$$

or $x^2 = 0.45 \times 10^{-3} = 4.5 \times 10^{-6}$ and $x = 2.1 \times 10^{-3}\ M$

Since, $2.1 \times 10^{-3} < 2.5 \times 10^{-2}$. Hence the assumption made above is valid, and

$[NH_4^+] = [OH^-] = 2.1 \times 10^{-3}\ M$

$[NH_3] = 0.25 - 0.002 = 0.248\ M$

$$[H_3O^+] = \frac{K_w}{[OH^-]} = \frac{1.0 \times 10^{-14}}{2.1 \times 10^{-3}} = 4.8 \times 10^{-12}\ M$$

$pH = -\log[H_3O^+] = 12 - \log 4.8 = 12 - 0.68 = 11.32$

74. Here we are asked to find is the degree of dissociation, α,. We begin by writing the Ostwald Dilution Law, which relates the degree of dissociation to the basicity constant,

In utilizing the Ostward Dilution Law for calculation, the first step in computation is usually to assume that α is small compared to 1. It is justified only if $\alpha < 0.1$, or $\alpha^2 < 0.01$. Since

$$K_b(CH_3NH_2) = \frac{[CH_3NH_3^+][OH^-]}{[CH_3NH_3]} = \frac{\alpha^2 C}{1-\alpha} = 4.20 \times 10^{-4}$$

In this problem C, is 0.02

$$\therefore \frac{\alpha^2}{1-\alpha} = \frac{4.2 \times 10^{-4}}{2.0 \times 10^{-2}} = 2.1 \times 10^{-2}$$

First approximation: to solve this equation for α, assume that α is less than 10% of 1, so that $(1 - \alpha) \sim 1$. The equation then becomes

$\alpha^2 = 2.10 \times 10^{-2}$ and $\alpha = 1.45 \times 10^{-1}$

Since 1.45×10^{-1} (0.145) is *not* less than 10% of 1 (0.1), our first approximation not valid. We therefore make a second approximation that is better than first.

Second approximation: Assume $1 - \alpha = 1 - 0.145 = 0.855$. Note that we use the answer obtained from the first approximation to get a better value for $(1 - \alpha)$. We are no longer assuming that α is small compared to 1: we are assuming that 0.855 is a better approximation to the quantity $(1 - \alpha)$ than 1 is. We therefore obtain

$$\frac{\alpha^2}{1-\alpha} = \frac{\alpha^2}{0.855} = 2.1 \times 10^{-2}$$

Then α^2 = (0.855) (2.1 × 10^{-2}) = 1.8 × 10^{-2}, and α = 1.34 × 10^{-1}

Third approximation: Assume 1 – α = 1 – 0.134 = 0.866. We use the result obtained from the second approximation to obtain a still better value for (1 – α). Hence,

$$\frac{\alpha^2}{1-\alpha} = \frac{\alpha^2}{0.866} = 2.1 \times 10^{-2} \text{ and } \alpha^2 = (0.866)\,(2.1 \times 10^{-2})$$

so that $\alpha^2 = 1.82 \times 10^{-2}$, and $\alpha = 1.35 \times 10^{-1} = 0.135.$

Fourth approximation: 1 – α = 1 – 0.135 = 0.865. then

$$\alpha^2 = (0.865)\,(2.1 \times 10^{-2}) = 1.82 \times 10^{-2} \text{ and } \alpha = 0.135 \text{ or } 13.5$$

The results of the third and fourth approximations for α are exactly the same to three significant figures. When two successive approximations give identical results to the correct number of significant figures, the answer obtained is the correct answer. The same answer would be obtained if the quadratic formula were employed, which you may want to verify for yourself.

Note that the first approximation gave α = 0.145, whereas the correct answer is 0.135 Thus the first answer was not so terribly far off; the two values differ only by 1 in the second significant figure. One way that you can check that you have obtained the correct answer is to substitute back into the originals equation:

$$K_b = \alpha^2 C/(1 - \alpha) = (0.135)^2(2.00 \times 10^{-2})/0.865 = 4.2 \times 10^{-4}$$

75. The following equilibrium is established in this solution:

$$CH_3COOH(aq) + H_2O \rightleftharpoons CH_3COO^-(aq) + H_3O^+(aq)$$

Let us consider the initial concentration of these species, and the changes that occur on reaction.

Concentration (M)	CH_3COOH	CH_3COO^-	H_3O^+
Initial value	0.12	0	0.1
Change	$-x$	x	x
Equilibrium value	$0.12 - x$	x	$0.1 + x$

There are two sources of H_3O^+ in this solution. The strong acid, HCl, is 100% ionized. In addition to the 0.1 H_3O^+ from the strong acid, there is a contribution to the [H_3O^+] from the dissociation of the acetic acid. The [CH_3COOH] at equilibrium is slightly less than the initial 0.12 *M* because some of it has ionized.

The equilibrium constant expression for the dissociation of acetic acid is

$$K_a(\text{HOAc}) = 1.8 \times 10^{-5} = \frac{[\text{H}_3\text{O}^+][\text{OAc}^-]}{[\text{HOAc}]} = \frac{(0.1 + x)(x)}{0.12 - x}$$

Because we know that acetic acid ionized only slightly, and that its ionization is repressed by the extra H_3O^+ ions from the HCl, we expect x to be considerably smaller than 0.1 *M*. Assume x is less than 10% of 0.100. It is then necessarily also less than 10% of 0.120. Thus (0.1 + x) ~ 0.1, and (0.12 – x) ~ 0.12. The equilibrium constant expression becomes

$$K_a(\text{HOAc}) = 1.8 \times 10^{-3} = \frac{(0.1)(x)}{0.12}$$

and $x = (1.2)\,(1.8 \times 10^{-5}) = 2.2 \times 10^{-5}\ M$

Is 2.2×10^{-5} less than 10% of 0.1? Yes, $2.2 \times 10^{-5} < 1.0 \times 10^{-2}$. Hence the approximation made is valid.

$$[CH_3COO^-] = 2.2 \times 10^{-5}\ M$$

$$[H_3O^+] = 0.1 + 2.2 \times 10^{-5} = 0.1\ M \quad pH = 1.0$$

The contribution to the total $[H_3O^+]$ from the ionization of acetic acid ís negligible compared to the contribution from the strong acid. The pH of this solution in the same as that of 0.1 *F* HCl.

76. It is important to remember that the hydronium ion concentration produced by the two stages of ionization of H_2S is very small; both H_2S and HS^- are extremely weak acids, with acidity constants very much less than 1. Thus the $[H_3O^+]$ in this solution is essentially identical to that provided by the strong acid, HCl. All the terms in equation.

$$K_{overall} = \frac{[H_3O^+][S^{2-}]}{[H_2S]} = 1 \times 10^{-20}$$ are therefore known except for the $[S^{2-}]$,

$$\therefore \quad K_{overall} = 1 \times 10^{-20} = \frac{[H_3O^+]^2\,[S^{2-}]}{[H_2S]} = \frac{(0.3)^2\,[S^{2-}]}{0.1}$$

$$[S^{2-}] = \frac{1 \times 10^{-21}}{9 \times 10^{-2}} = 1 \times 10^{-20}\ M$$

77. Ammonium bromide, a strong electrolyte, is 100% ions in aqueous solution. Bromide ions do not react with water. The Br^- ion has virtually no tendency to combine with H^+, because HBr is a strong acid. Thus $[Br^-] = 0.3\ M$ in this solution. Ammonium ion is an acid, and domates a proton to water

Let, $\quad x = [NH_3] = [H_3O^+]$ at equilibrium in 0.3 *F* NH_4Br

It is useful to choose as the unknown the concentration of a species present in very small concentration, as this makes it easier to make approximations that greatly simplify the mathematical solution of the equilibrium constant equation. The concentrations of ammonia and hydronium ions are equal because these two substances are formed in a 1 : 1 molar ratio by reaction. Let us tabulate the initial concentrations and the changes in concentration of the three species involved in the equilibrium

Concentration (M)	NH_4^+	NH_3	H_3O^-
Initial value	0.3	0	0
Change	$-x$	x	x
Equilibrium value	$0.3 - x$	x	x

Remember that the concentration of NH_4^+ ions present at equilibrium is the initial concentration *minus* the concentration that is used up is reaction

The equilibrium constant is the acidity constant of NH_4^+. The numerical value of $K_a(NH_4^+)$ is 5.7×10^{-10} (from tables) and we therefore write

$$\therefore\ K_a(NH_4^+) = 5.7 \times 10^{-10} = \frac{[NH_3][H_3O^+]}{[NH_4^+]} = \frac{x^2}{0.3 - x}$$

Assume that x is small compared to 0.3, that is, that x is less than 10% of 0.3. Then,

$$\frac{x^2}{0.3} = 5.7 \times 10^{-10} \text{ or, } x^2 = 1.7 \times 10^{-10} \text{ and hence, } x = 1.3 \times 10^{-5} M$$

78. 1. We have from the data given in the problem:

$p^{OH} = 14.0 - p^{H} = 14.0 - 1.0 = 13.0$

$[H^+] = 10^{-pH} = 10^{-1.0} = 0.100$ M

2. $p^{OH} = -\log[OH^-] = -\log(2.0 \times 10^{-12}) = 11.7$

pH $= 14.0 - p^{OH} = 14.0 - 1.7 = 2.3$

3. We have, $[H_2O^+] = 10^{-2.87} = 1.35 \times 10^{-3}$ M at equilibrium

$$HAc_{(aq)} + H_2O(l) = Ac_{(aq)} + H_2O_{(aq)}$$

	HAc	*Ac*⁻	*H_3O^-*
1	0.00 M	0	0
C	–0.00135	+0.00135	+0.00135
E	–0.099	0.00135	0.00135

$$K_a = \frac{[Ac^-][H_3O^+]}{[HAc]} = \frac{(0.00135)(0.00135)}{0.099} = 1.8 \times 10^{-5}$$

4. From collected data, $p^H = 5.13$, so that we get, $[H_3O^+] = 7.4 \times 10^{6}$ M

$$NH_4^+ + H_2O \rightleftharpoons NH_3 + H_3O^+$$

	NH_4^+	*NH_3*	*H_3O^+*
I	0.1 M	0	0
C	-7.4×10^{-6}	7.4×10^{-6}	7.4×10^{-6}
E	–0.1	7.4×10^{-6}	7.4×10^{-6}

$$K_h = \frac{[NH_3][H_3O^+]}{[NH_4^+]} = \frac{(7.4 \times 10^{-6})^2}{0.1} = 5.5 \times 10^{-10}$$

5. From the given data

$$0.2 \cancel{L} \times \frac{0.25 \text{ mol HAc}}{\cancel{L}} = 0.05 \text{ mol HAc}$$

$$\frac{4 \text{ g }\cancel{\text{NaAc}}}{1} \times \frac{1 \cancel{\text{mol NaAc}}}{82.034 \text{ g }\cancel{\text{NaAc}}} \times \frac{1 \text{ mol Ac}}{1 \cancel{\text{mol NaAc}}} = 0.0488 \text{ mol Ac}$$

Since, $$\frac{[HAc]}{[Ac^-]} = \frac{n_{HAc}/V_{sol'n}}{nAc^-/Vsol'n}$$

Then, $$[H^+] = K_a \frac{n_{HAc}}{n_{Ac^-}} = 1.8 \times 10^{-5} \cdot \frac{0.05 - x \text{ mol HAc}}{0.0488 - x \text{ mol Ac}} = 1.84 \times 10^{-5} \text{ M}$$

where x is negligible

$\therefore$ pH = $-\log[H^+] = -\log(1.84 \times 10^{-5}) = 4.73$

6. From the data, we have:

$$0.05\,\text{L} \times \frac{0.25\ \text{HAc}}{\text{L}} = 0.0125\ \text{mol HAc}$$

$$0.05\,\text{L} \times \frac{0.488\ \text{mol Ac}^-}{0.2\ \text{L}} = 0.0122\ \text{mol Ac}^-$$

$$0.025\,\text{L} \times \frac{0.25\ \text{mol H}^+}{\text{L}} = 0.00625\ \text{mol H}^+$$

$$Ac^- + H^+ \rightleftharpoons HAc$$

	Ac⁻	*H⁺*	*HAc*
I	0.0122	0.00625	0.0125
C	–0.00625	–0.00625	+0.00625
E	0.00595	–0	0.01875

$$[H^+] = K_a \times \frac{n_{HAc}}{n_{Ac^-}} = 1.8 \times 10^{-5} \times \frac{0.01875\ \text{mol HAc}}{0.00595\ \text{mol Ac}} = 5.67 \times 10^{-5}\ \text{M}$$

$$pH = -\log[H^+] = -\log(5.67 \times 10^{-5}) = 4.25$$

7. Here we have: $0.05\,\text{L} \times \dfrac{0.25\ \text{mol HAc}}{\text{L}} = 0.0122\ \text{mol HAc}$

$$0.05\,\text{L} \times \frac{0.488\ \text{mol Ac}^-}{0.200\ \text{L}} = 0.0122\ \text{mol Ac}^-$$

$$0.025\,\text{L} \times \frac{0.25\ \text{mol OH}^-}{\text{L}} = 0.00625\ \text{mol OH}^-$$

$$HAc + OH^- \rightleftharpoons Ac^- + H_2O$$

	HA_c	*OH⁻*	*Ac⁻*
1	0.0125	0.00625	0.0122
C	–0.00625	–0.00625	+0.00625
E	0.00625	–0	0.0815

$$[H^+] = K_a \times \frac{n_{HAc}}{n_{Ac^-}} = 1.8 \times 10^{-3} \times \frac{0.00625\ \text{mol HAc}}{0.01845\ \text{mol Ac}^-} = 6.10 \times 10^{-6}\ \text{M}$$

$$pH = -\log[H^+] = -\log(6.10 \times 10^{-6}) = 5.21$$

8. From data given $[H^+] = 0.025\ \dfrac{0.25\ \text{mol H}^+}{\text{L}} \times \dfrac{1}{(0.25 + 0.05)\ \text{L}} = 8.33 \times 10^{-3}\ \text{M}$

$$pH = -\log[H^+] = -\log(8.33 \times 10^{-3}) = 1.08$$

9. $[OH^-] = 0.025\ \cancel{L} \times \dfrac{0.25 \text{ mol } OH^-}{\cancel{L}} \times \dfrac{1}{(0.025\ \ 0.05)\ L} = 8.33 \times 10^{-2}$ M

pOH = $-\log[OH^-] = -\log(8.33 \times 10^{-2}) = 1.08$
pH = 14 – pH = 14 – 1.08 = 12.92

79. Hydrochloric acid is a strong acid, completely ionized in dilute aqueous solution. The self-ionization of water occurs to only a very slight extent in pure H_2O, and in the presence of the H_3O^+ from the HCl, the self-ionization of water is repressed. This means that the equilibrium

$$2H_2O \rightleftharpoons H_3O^+(aq) + OH^-(aq) \qquad \ldots\ldots (i)$$

is driven to the left, because the reaction between HCl and H_2O produces such a high $[H_3O^+]$ compared to that in pure water. this is another example of Le Chateller Principle. In pure water, the $[H_3O^+] = 1.0 \times 10^{-7}$ M. In 0.400 *F* HCl, the total $[H_3O^+]$ comes from two sources: the HCl and the self-ionization of H_2O. The concentration of H_3O^+ due to the self-ionization reaction is very much smaller that 1×10^{-7} *M*, as reaction (i) is driven to the left by the H_3O^+ form the HCl. Thus

$[H_3O^+]$ = contribution from HCl + contribution from self-ionization of H_2O
= 0.4 + ≪ 0.0000001 = 0.04

Using the expression for K_w, we can calculate the $[OH^-]$

$$[OH^-] = \frac{K_w}{[H_3O^+]} = \frac{1.0 \times 10^{-14}\ M^2}{0.4\ M} = 2.5 \times 10^{-14}\ M$$

Note that in pure water $[OH^-] = 1.0 \times 10^{-7}$ *M* at 298 K, but it has been reduced to 0.25× 10^{-14} *M* in 0.4 *F*Cl. Thus illustrates quantitatively the operation of Le Chatelier's Principle,

80. These three acids are all strong acid; 100% ions in dilute aqueous solution.

(a) In 2.00 *F* HCl, $[H_3O^+] = 2.00$ *M*
Hence, pH= –log 2.0 = – 0.3
Note that the pH is negative in this solution. The pH is negative in any solution for which the $[H_3O^+]$ is greater than 1.0 *M*.

(b) Since perchloric acid is a strong acid, $[H_3O^+] = 0.2$ *M* in this solution. Thus,

$$pH = -\log(2.0 \times 10^{-1}) = +0.7$$

(c) In 2.0×10^{-4} *F* HNO_3, $[H_3O^+] = 2.0 \times 10^{-4}$ *M*
Thus, pH= $-\log(2.0 \times 10^{-4}) = 3.7$
We see that the greater the $[H_3O^+$, the lower the pH.

81. Skim milk is dilute aqueous solution. Since the pH = $-\log[H_3O^+] = -6.6$, ∴ $[H_3O^+]$ = antilog (–6.6) = $10^{-6.6}$.

or $[H_3O^+] = 10^{-6.6} = 2.5 \times 10^{-7}$ $\left(\because 10^{-6.6} = \dfrac{1}{10^{6.6}} = \dfrac{1}{4.0 \times 10^6} = 0.25 \times 10^{-6} = 2.5 \times 10^{-7}\right)$

Hence, We can calculate the $[OH^-]$ in this skim milk as follows:

$$[OH^-] = \frac{K_w}{[H_3O^+]} = \frac{1.0 \times 10^{-14}}{2.5 \times 10^{-7}} = 4.0 \times 10^{-8}\ M$$

82. (a) For a neutral solution pH = $\frac{1}{2}pK_w$. From Table, we find $K_w = 1.2 \times 10^{-15}$.

Thus, pK_w =14.92 and pH = 7.66 for a neutral solution at 273 K

(b) At 333K, $K_w = 9.6 \times 10^{-14}$, so that pK_w = 13.02 an pH = 6.15 for a neutral solution.

83. Since $Ba(OH)_2$ is a strong base, $[OH^-]$ = 0.12 *M* in 0.060 *F* $Ba(OH)_2$ (from tables).

$\therefore$ therefore, $$[H_3O^+] = \frac{1.0 \times 10^{-14}\ M^2}{0.12\ M} = 8.3 \times 10^{-14}\ M$$

$$pH = -\log(8.3 \times 10^{-14}) = 13.08$$
$$pOH = 14 - pH = 0.92$$

we could have obtained pOH directly from the $[OH^-]$ as

$$pOH = -\log(0.12) = +0.92$$

84. Sodium ions and acetic ions are the principal species in an aqueous solution of sodium acetate. Sodium ions do not react with water, so that $[Na^+]$ = 0.045 *M* in 0.045 *F* NaOAc. Acetate ion is a weak base, and accepts a proton form water:

$$OAc^-(aq) + H_2O \rightleftharpoons HOAc(aq) + OH^-(aq)$$

Since $K_b(OAc^-) = 5.7 \times 10^{-10}$ (from table), the proton-transfer reaction between OAc^- and H_2O does not proceed very far to the right, and we expect the concentration of HOAc molecules and OH^- ions in this solution to be low. Let

x = [HOAc] = $[OH^-]$ at equilibrium in 0.045 *F* NaOAc at 298 K.

Then $[OAc^-] = 0.045 - x$

because one acetate ion is used up to form one HOAc molecule and one OH^- ion. substitution of these expression into the basicity constant of acetate ion yields.

$$K_b(OAc^-) = 5.7 \times 10^{-10} = \frac{[HOAc][OH^-]}{[OAc]} = \frac{x^2}{0.045 - x}$$

Assume x is small compared to 0.045 *M*, that is $(0.045 - x) - 0.045$. The equilibrium constant equation them becomes.

$$5.7 \times 10^{-10} = \frac{x^2}{0.045} \text{ so that } x^2 = (0.045)(5.7 \times 10^{-10}) = 25.7 \times 10^{-12}$$

and $$x = 5.1 \times 10^{-6}\ M$$

We must now check the approximation made. Is x less than 10 of 0.045? Yes $5.1 \times 10^{-6} < 4.5 \times 10^{-3}$. Hence

$$[HOAc = [OH^-] = 5.1 \times 10^{-6}\ M$$
$$[OAc] = 0.045 - 5.1 \times 10^{-6} = -0.045\ M$$

We obtain the $[H_2O^+]$ from the relation

$$[H_2O^+] = \frac{K_w}{[OH^-]} = \frac{1.0 \times 10^{-14}}{5.1 \times 10^{-6}} = 2.0 \times 10^{-9}\ M$$

Hence,

$$pH = -\log(2.0 \times 10^{-9}) = 9 - \text{lo } 2.0 = 8.70$$

85. It is helpful first to classify each of these solution as either acidic, basic or neutral. There are three acidic solutions:

(a) 0.2 *F* NH_4NO_3 — NH_4^+ ion is a weak acid
(d) 0.2 *F* $HClO_4$ — perchloric acid is a strong acid
(f) 0.2 *F* HNO_2 — nitrous acid is a weak acid

There are two basic solutions:

(b) 0.2 *F* KNO_2 — NO_2^- ion is a weak base
(e) 0.2 *F* Na_2CO_3 — CO_3^{2-} ion is a weak base

The remaining solution, (c) 0.2 *F* KNO_2, is neutral. Neither K^+ nor NO_3^- ions either accept or donate protons to water. Another way of recognizing that KNO_3 is a neutral salt is to note that it is produced by the reaction of a strong base (KOH) and a strong acid (HNO_3).

To decide on the relative acidities of the two weak acids, we must look up their k_a values (in Tables). Nitrous acid is a stronger acid than ammonium ion, since $K_a(HNO_2)$ $= 4.5 \times 10^{-4}$ is significantly larger than $K_a(NH_4^+) = 5.7 \times 10^{-10}$.

To decide on the relative basicities of the two weak bases, we must calculate their K_b values from the K_a values of their conjugate weak acids. The conjugate acid of NO_2^- ions is nitrous acid, HNO_2. Thus

$$K_b(NO^-) = \frac{K_w}{K_a(HNO_2)} = \frac{1.0 \times 10^{-14}}{4.5 \times 10^{-4}} = 2.2. \times 10^{-11}$$

The conjugate acid of CO_3^{2-} is HCO_3^-, bicarbonate ion.

$$K_b = (CO_3^{2-}) = \frac{K_w}{K_a(HCO_3^-)} = \frac{1.0 \times 10^{-14}}{4.7 \times 10^{-11}} = 2.1 \times 10^{-4}$$

Thus carbonate ion is a considerable stronger base than nitrite ion.

We can now arrange these six solutions in order of increasing pH, beginning with the most acidic, the strong acid, $HClO_4$.

lowestpH (most acidic)	(d) 0.2 *F* $HClO_4$	strong acid
	(f) 0.2 *F* HNO_2	$K_a(HNO_2) = 4.5 \times 10^{-4}$
	(a) 0.2 *F* NH_4NO_3	$K_a(NH_4^+) = 5.7 \times 10^{-10}$
	(c) 0.2 *F* KNO_3	neutral
	(b) 0.2 *F* KNO_2	$K_b(NO_2^-) = 2.2 \times 10^{-11}$
highest pH (most basic)	(e) 0.2 *F* $Na_2CO_3^{2-}$	$K_b(CO_3^{2-}) = 2.1 \times 10^{-4}$

86. (a) Ammonium carbonate is a basic salt. Its cation, NH_4^+, is a weak acid, $K_a(NH_4^+) = 5.7 \times 10^{-10}$. Its anion, CO_3^{2-}, is a Weak base.

$$K_b(CO_3^{2-}) = \frac{K_w}{K_a(HCO_3^-)} = \frac{1.0 \times 10^{-14}}{4.7 \times 10^{-11}} = 2.1 \times 10^{-4}$$

Since $K_b(CO_3^{2-}) > K_a(NH_4^+)$, solutions of ammonium carbonate are basic.

(b) Potassium perchlorate, $KClO_4$, is a neutral salt, Neither K^+ nor ClO_4^- has any tendency of donate or accept a proton in dilute aqueous solution. The reaction between the strong base KOH and the strong acid $HClO_4$ produces $KClO_4$.

(c) Potassium hydrogen sulfate (also called potassium bisulfate)is an acidic salt. The anion, HSO_4^-, is a moderately strong acid with $K_a = 1.2 \times 10^{-2}$. The cation K^+ is neither an acid nor a base.

(d) Barium oxide, BaO, is a basic salt. Oxides of the alkaline earth metals are basic. The reaction that occurs when BaO dissolves in water is

$$BaO(s) + H_2O \rightleftharpoons Ba^{2+}(aq) + 2OH^-(aq)$$

(e) The anion HSO_3^- is an ampholyte.

$$K_a(HSO_3^-) = 6.2 \times 10^{-8}$$

$$K_b(HSO_3^-) = K_w/K_a(H_2SO_3) = (1.0 \times 10^{-14})/(1.2 \times 10^{-2}) = 8.3 \times 10^{-13}$$

Since $K_a(HSO_3^-) > K_b(HSO_3^-)$, solutions of $NaHSO_3$ (sodium bisulfite or sodium hydrogen sulfite) are acidic.

(f) Methylammonium nitrate, $CH_3NH_3NO_3$, is an acidic salt. Its cation, $CH_3NH_3^+$, is a weak acid. Its anion, NO_3^-, has virtually no tendency to accept a proton in dilute aqueous solution.

87. Acid OH^- will react with one HX to give one X^- an a water molecule (not shown).

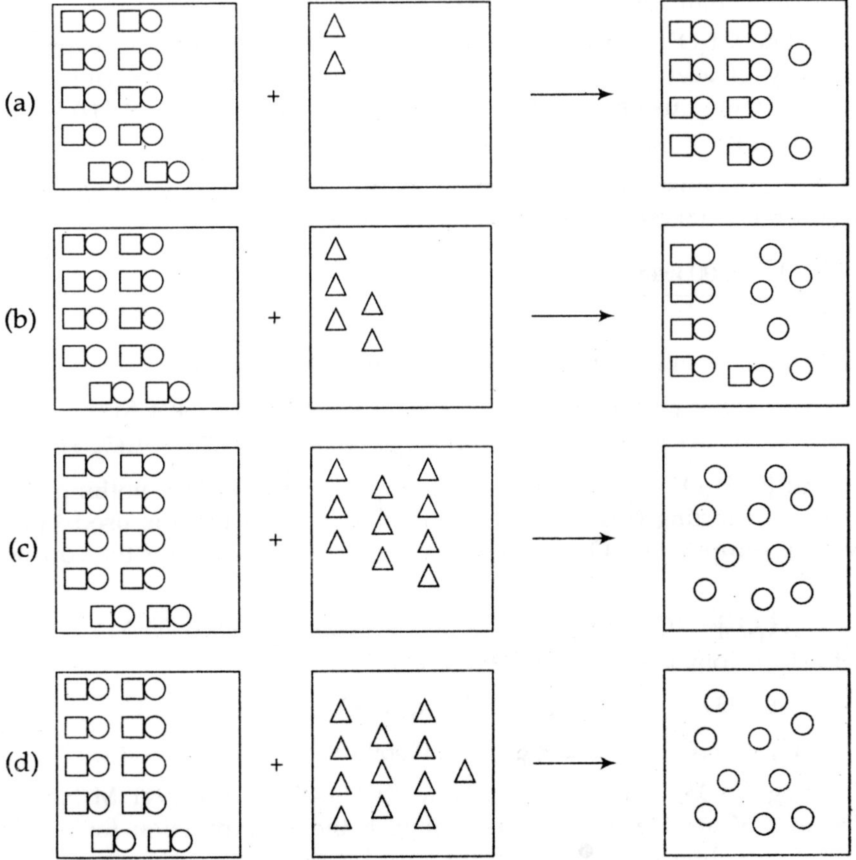

(b) above represent neutralization halfway to the equivalence point. Half the HX molecule have been converted to X^-, and half remain as HX.

88. Acid in the blood reacts with the bicarbonate buffer, establishing the equilibrium

$$H^+(aq) + HCO_3^-(aq) \rightleftharpoons H_2CO_3(aq) \quad \ldots\ldots (1)$$

Carbonic acid also establishes an equilibrium decomposing to carbon dioxide and water

$$H_2CO_3(aq) \rightleftharpoons H_2O(l) + CO_2(g) \quad \ldots\ldots (2)$$

According to Le Chateller's principle. If the concentration of H^+ (a reactant) increases then the equilibrium will shift to produce more products, that partially offsetting that increase in reactant concentration. Such an increase would result in an increase in the concentration of carbonic acid. The increase in $[H_2CO_3]$ would result in the equilibrium of (2) also shifting to more products, thus producing more CO_2.

89. (a) **False**: The formate ion concentration in 0.1 *M* $HCHO_2$ is much less than the formate ion concentration in 0.1 *M* $NaCHO_2$

$HCHO_2$ is a weak acid, dissociating very little, value $NaCHO_2$ is a salt and dissociates completely.

(b) **False** : Since K_a for HCO_3^- is 4.7×10^{-11}, K_b for CO_3^{2-} is 2.1×10^{-4}.

$K_b = K_w/K_a$ for the conjugate base (B) of the weak and HB.

90. Use the pH to calculate $[H_3O^+]$ and then use the K_a of 1.7×10^{-5} and the K_a expression to calculate the molarity of acetic acid (HAc), assuming ionization is negligible. Convert moalrity to mass percentage using the formula weight of 60.05 g mol^{-1} of HAc.

$$[H_3O^+] = \text{antilog}\ (-2.45) = 3548 \times 10^{-3}\ M$$

Write the equilibrium-constant expression in terms of chemical symbols and then substitute the x and the (0.003548 M) terms into the expression

$$K_a = \frac{[H_3O^+][Ac^-]}{[HAc]} = \frac{(0.003548)^2}{(x)}$$

Solve the equation for x, assuming that 0.003548 is much smaller than x

$$x = (0.003548)^2 \div (1.7 \times 10^{-5}) = [HAc] = 0.7404\ M$$

$(0.7404\ \text{mol Hac}\ L^{-1}) \times (60.05\ g\ mol^{-1}) \times (1\ L/1090\ g)\ (100\%) = 4.078 = 4.1\%$ HAc

91. Find the $[H_3O^+]$ and $[C_2H_3O_2^-]$ by solving for the approximate $[H_3O^+]$, nothing that x is much smaller than the starting 0.92 M of acetic acid. The usual table is used but is not shown, only the final setup for $[H_3O^+]$ is shown. Use $K_f = 1.858°C/m$ for the constant for water

$$[H_3O^+] = [C_2H_2O_2^-] = (1.7 \times 10^{-5} \times 0.92)^{1/2} = 0.003954\ M$$

Total molarity of acid + ions = 0.92 + 0.003954 = 0.9239

$$\text{Moalrity} = m = 0.9239\ \text{mol} \div (1.0\ L \times 0.953\ kg\ H_2O\ L^{-1}) = 0.9695\ m$$

Freezing point $= -\Delta T_f = -K_f C_m = (-1.858°C/m) \times 0.9695\ m = -1.801 = -1.8°C$

92. The $[H_3O^+]$ = -antilog $(-4.35) = 4.46 \times 10^{-5}$. Note that 0.465 L of 0.0941 M NaOH will produce 0.043756 mol of acetate Ac^-, ion. rearranging the K_a expression for acetic acid (HAc) and Ac^-, and canceling the volume in the mol L^{-1} of each.

$$\frac{[\text{Hac}]}{[\text{Ac}^-]} = \frac{[\text{H}^+]}{K_a} = \frac{4.46 \times 10^{-5}}{1.7 \times 10^{-5}} = \frac{2.63}{1.0} = \frac{x \text{ mol HAc}}{0.043756 \text{ mol Ac}^-}$$

x = 0.1149 mol HAc

Total mol HAc added = 0.1149 + 0.043756 = 0.1587 mol HAc

Mol L^{-1} of pure HAc = 1049 g HAc L^{-1} x (1 mol HAc/60.05 g) = 17.467 mol L^{-1}

L of pure HAc needed = 0.587 mol HAc × (L/17.467 mol) = 0.00908 L (9.1 mL)

93. Write the equation for the reaction. Calculate $[H^+]$ (from pH) and moles of acetate (from the reaction stoichiometry) and from that, moles of HB still in solution. Then calculate the moles of acetic acid added and finally, the volume of acetic acid

$HC_2H_3O_2$(aq_ + OH^-(aq) $\rightleftharpoons$ $C_2H_3O_2^-$(aq) + $H_2O(l)$

$$[H^+] = 10^{-pH} = 10^{-4.20} = 6.31 \times 10^{-5}\ M$$

$$0.1 \text{ L} \times \frac{1.25 \text{ mol NaOH}}{1 \text{ L}} \times \frac{1 \text{ mol } C_2H_3O_2}{1 \text{ mol NaOH}} = 0.125 \text{ mol } C_2H_3O_2$$

$$[H^+] = K_a \times \frac{n_{HC_2H_3O_2}}{n_{C_2H_3O_2^-}} \Rightarrow 6.31 \times 10^{-3} = 1.8 \times 10^{-5} \times \frac{n_{HC_2H_3O_2}}{0.125}$$

$$n_{HC_2H_3O_2} = 0.44 \text{ mol}$$

This is the moles of acid present *after* reaction has occurred. The total moles of acid that must have been present *before* reaction is

$$n_{total} = n_{HC_2H_3O_2} + n_{C_2H_3O_2} = 0.44 \text{ mol} + 0.125 \text{ mol.} = 0.56 \text{ mol}$$

Not that this solution contains roughly equal amounts of HB and B^-, acid is therefore a buffer

$$0.56 \text{ mol} \times \frac{60.05 \text{ g } HC_2H_3O_2}{1 \text{ mol}} \times \frac{100 \text{ g solution}}{98 \text{ g } HC_2H_3O_2} = 33 \text{ mL solution}$$

94. Calculate the mole ratio of NH_4^+ to NH_3 needed to make a solution with the requistic pH. Consider the mole ratio and the definition of buffer

$$[H^+] = 10^{-pH} = 10^{-6.50} = 3.16 \times 10^{-7}\ M$$

$$[H^+] = K_a \times \frac{n_{NH_4^+}}{n_{NH_3}} \Rightarrow 3.16 \times 10^{27} = 5.6 \times 10^{-10} \times \frac{n_{NH_4^+}}{n_{NH_3}}$$

$$\Rightarrow \frac{n_{NH_4^+}}{n_{NH_3}} = 560$$

A buffer requires roughly equal amounts of a weak acid and its conjugate base. A 600:1 ratio does not meet this requirement.

95. After solution *A* is titrated to its equivalence point, all the acid will have been converted to the conjugate base. also, when solution *A* ad *B* are combined, [HB] = $[B^-]$, thus

$[HB]/[B^-] = 1$. Calculate the $[H^+]$ of this solution, and the K_a for the acid. From K_a and pH, of the original solution, one can calculate [HB] and thus the moalr mass.

from the final solution

$$[H^+] = 10^{-pH} = 10^{-4.26} = 5.5 \times 10^{-5}\ M$$

$$\therefore \quad [H^+] = K_b \times \frac{[HB]}{[B^-]} \Rightarrow 5.5 \times 10^{-5} = K_a \times 1$$

$$\therefore \quad K_a = 5.5 \times 10^{-5}$$

From the initial solution,

$$[H^+] = 10^{-pH} = 10^{-2.56} = 2.75 \times 10^{-5}\ M$$

$$\therefore \quad K_a = \frac{[B^-][H^+]}{[HB]} \Rightarrow 5.5 \times 10^{-5} = \frac{(2.75 \times 10^{-3})^2}{[HB]}$$

and $\quad [HB] = 0.138\ M$

$$\therefore \quad 0.25\ L \times \frac{0.138 \text{ mol HB}}{1\ L} = 0.0345 \text{ mol. HB}$$

$$\frac{4.0\ g}{0.0345 \text{ mol HB}} = 116 \text{ g mol}^{-1} \text{ HB}$$

96. (a) $K_a = \dfrac{[NO_2^-][H^+]}{[HNO_2]} \Rightarrow 6.0 \times 10^{-4} = \dfrac{x^2}{1.0\ M}$

$$x^2 = 6.0 \times 10^{-4}$$

$$x = [H^+] = 0.024\ M$$

$$\therefore \quad p^H = -\log (0.0024) = 1.62$$

(b) $0.05\ L \times \dfrac{1.0 \text{ mol}}{1\ L} = 0.05 \text{ mol. } HNO_2$

As half neutralization half of the HNO_2 will be converted to NO_2^-, thus

$$[NO_2^-] = [HNO_2] = 0.025\ M$$

$$[H^+] = K_a \times \frac{n_{HNO_2}}{n_{NO_2^-}} \ 6.0 \times 10^{-4} \times \frac{0.025}{0.025} = 6.0 \times 10^{-4}\ M$$

$$\therefore \ p^H = -\log (6.0 \times 10^{-4}) = 3.22$$

(c) At equivalence point, all the nitrous acid has been converted to nitrate ion and the solution is weakly base.

$$NO_2^-(aq) \ H_2O(l) \rightleftharpoons HNO_2(aq) + OH^-(aq)$$

Calculate the volume of base add and the total volume of the final solution. Then calculate $[NO_2]$, $[OH^-]$, pOH and finally pH

$$0.05 \text{ mol } HNO_2 \times \frac{1 \text{ mol NaOH}}{1 \text{ mol } HNO_2} \times \frac{1\ L}{0.85 \text{ mol NaOH}} = 0.0588\ L$$

$$V_{total} = 0.05 \text{ L } (HNO_2)+0.0588 \text{ L } (NaOH) = 0.1088 \text{ L}$$

$$[NO_2^-] = \frac{0.05 \text{ mol}}{0.1088 \text{ L}} = 0.4596\ M$$

$$K_b = \frac{[HNO_2][OH^-]}{[NO_2^-]} \Rightarrow 1.7 \times 10^{-11} = \frac{x^2}{0.4596\ M}$$

$$x^2 = 7.8 \times 10^{-12}$$

$$x = [OH^-]= 2.8 \times 10^{-6}\ M$$

$$p^{OH} = -\log (2.8 \times 10^{-4}6) = 5.55$$

$$\therefore \quad p^H = 14.0 - p^{OH} = 14.0 - 5.55 = 8.45$$

(d) as calculated in part (c) above, 0.0588 L (58.8 mL) of NaOH solution were needed to rach equivalence 0.1 mL les that volume is 0.0587 L (58.7 mL).

$$0.0587 \times \frac{0.85 \text{ mol NaOH}}{1 \text{ L}} \times \frac{1 \text{ mol } HNO_2}{1 \text{ mol NaOH}} = 0.0499 \text{ mol. } HNO_2$$

$$0.04 \text{ mol. } HNO_2 - 0.0499 \text{ mol. } HNO_2 - 0.0001 \text{ mol. } HNO_2$$

$$[H^+] = K_a \times \frac{n_{HNO_2}}{n_{NO_2^-}} \; 6.0 \times 10^{-4} \times \frac{0.0001}{0.0499} = 1.2 \times 10^{-6}\ M$$

$$p^H = -\log (1.2 \times 10^{-6}) = 5.92$$

(e) At this point, any NaOH added in addition to the 58.8 mL (from part (c) above) is excess. Calculate the moles of excess NaOH [OH⁻] in the resulting solution.

$$n_{NaOH} = 0.1 \text{ mL} \times \frac{1 \text{ L}}{1000 \text{ mL}} \times \frac{0.85 \text{ mol.}}{1 \text{ L}} = 8.5 \times 10^{-5} \text{ mol. (in excess)}$$

$$[OH^-] = \frac{8.50 \times 10^{-5} \text{ mol}}{0.1089 \text{ L}} = 7.81 \times 10^{-4}\ M$$

Note that we can neglect the OH^- formed by te nitrate (see part (c) above) since its contribution is no small, considering that K_b has only two significant figures.

$$p^{OH} = -\log (7.81 \times 10^{-4}) = 3.11$$

$$p^H = 14 - p^{OH} = 14 - 3.11 = 10.89$$

97. The indicator will be in the midst of changing color when $[HIn^-] = [In]$. Calculate the $[H^+]$ at this point. Determine the amount of $C_2H_3O_2^-$ that is present, and thus the amount of NaOH used to produce this amount of $C_2H_3O_2^-$.

$$[H^+] = K_a \times \frac{[HIn]}{[In^-]} \Rightarrow [H^+] = 1.0 \times 10^{-5} \times 1 = 1.0 \times 10^{-5}\ M$$

$$[H^+] = K_a \times \frac{n_{HC_2H_3O_2}}{n_{C_2H_3O_2^-}} \Rightarrow 1.0 \times 10^{-5} = 1.8 \times 10^{-5} \times \frac{n_{HC_2H_3O_2}}{n_{C_2H_3O_2^-}}$$

$$\frac{n_{HC_2H_3O_2}}{n_{C_2H_3O_2}} = 0.56 \Rightarrow n_{HC_2H_3O_2} = 0.56\ n_{C_2H_3O_2^-} \quad \text{eq. 1}$$

Before titration $0.05\ L \times \dfrac{1.0\ mol}{1\ L} = 0.05$ mol. $HC_2H_3O_2$

After titration $n_{HC_2H_3O_2} = x$, $\therefore\ n_{C_2H_3O_2} = 0.05 - x$.

Substituting into eq. (1) we get,

$$x = 0.56\ (0.05 - x)$$

$$\Rightarrow \quad x = [HC_2H_3O_2] = 0.018\ M$$

$$n_{C_2H_3O_2^-} = 0.05 - x = 0.032\ \text{mol.}$$

$$0.032\ \text{mol. } C_2\,H_3O_2^- \times \frac{1\ \text{mol. NaOH}}{1\ \text{mol } C_2H_3O_2^-} \times \frac{1\ L}{1.0\ \text{mol NaOH}} = 0.032\ \text{NaOH}$$

Thus, a better indicator would be phenolphithalen

98. Here we have,

(a) $H_2SO_4(aq) \rightleftharpoons H^+(aq) + HSO_4^-(aq)$

This is a strong acid and ionizes completely.

$[H^+] = -\log\ (0.15) = 0.8239$

99. Using the K_{sp} of 1.1×10^{-21} for ZnS (from tables) we calculate the $[S^{2-}]$ needed to maintain a saturated solution (without precipitation)

$$[S^{2-}] = \frac{1.1 \times 10^{-21}}{1.5 \times 10^{-4}\ M\ Zn^{2+}} = 7.33 \times 10^{-18}\ M$$

Next use the overall H_2S ionization to calculate the $[H_3O^+]$ needed to achieve this $[S^{2-}]$ level.

$$[H_3O^+] = \sqrt{\frac{1.1 \times 10^{-20}\ (0.10\,M)}{7.33 \times 10^{-8}\ M}} = 1.225 \times 10^{-2}\ M$$

Finally, calculate the buffer ratio of $[SO_4^{2-}]/[HSO_4^-]$ from the H_2SO_4 K_{a2} expression where K_{a2} has the value 1.1×10^{-2}

$$\frac{[SO_4^{2-}]}{[HSO_4^-]} = \frac{K_{a_2}}{[H_3O^+]} = \frac{1.1 \times 10^{-2}}{1.225 \times 10^{-2}} = \frac{0.8979}{1.0}$$

If $[HSO_4^-] = 0.2$ M then

$$[SO_4^{2-}] = 0.8979 \times 0.2\ M = 0.1795 = 0.18\ M$$

100. Being by solving for $[H_3O^+]$ in the buffer. Ignoring charges in $[HCHO_2]$ as a result of ionization in the buffer, we obtain

$$[H_3O^+] \cong 1.7 \times 10^{-4} \times \frac{0.45\ M}{0.2\ M} = 3.825 \times 10^{-4}\ M$$

Verify that this approximation is valid. This equilibrium for this dissolution of CaF_2 in acidic solution is obtained by substracting twice the acid ionization of HF from the solubility equilibrium of CaF_2:

$$CaF_2(s) \rightleftharpoons Ca^{2+}(aq) + 2F^-(aq) \qquad K_{sp}$$

$$2H_3O^+(aq) + 2F^-(aq) \rightleftharpoons 2HF(aq) + 2H_2O(l) \qquad 1/(K_a)^2$$

$$2H_3O^+(aq) + CaF_2(s) \rightleftharpoons Ca^{2+}(aq) + 2HF(aq) + 2H_2O(l) \qquad K_c = K_{sp}/(K_a)^2$$

Therefore $K_c = (3.4 \times 10^{-11})/(6.6 \times 10^{-4})^2 = 7.35 \times 10^{-5}$ in order to solve the equilibrium constant equation, we require the concentration of HF, which is obtained from the acid-ionization constant of HF.

$$K_a = \frac{[H_3O^+][F^-]}{[HF]}$$

or, $$6.8 \times 10^{-4} = 3.825 \times 10^{-4} \times \frac{[F^-]}{[HF]}$$

$$\frac{[F^-]}{[HF]} = 1.778 \text{ or } [F^-] = 1.778[HF]$$

Let x be line solubility of CaF_2 in the buffer. Then $[Ca^{2+}] = x$, and $[F^-] + [HF] = 2x$. Therefore, from the previous equation,

$$2x = 1.778[HF] + [HF] = 2.778[HF], \text{ or } [HF] = 2x/2.778$$

Now substitute for $[H_3O^+]$ and [HF] into the equation for K_c

$$K_c = \frac{[Ca^{2+}][HF]^2}{[H_3O^+]^2} = \frac{x(2x/2.778)^2}{(3.825 \times 10^{-4})^2} = 7.35 \times 10^{-5}$$

or, $7.35 \times 10^{-5} = (3.543 \times 10^6)\, x^3$

$\Rightarrow \quad x^3 = 2.075 \times 10^{-11}$

$\therefore \quad x = 2.748 \times 10^{-4} = 2.7 \times 10^{-4}$ M

101. The net ionic equation is

$$Ba^{2+}(aq) + 2OH^-(aq) + Mg^{2+}(aq) + SO_4^{2-}(aq) \rightleftharpoons BaSO_4(s) + Mg(OH)_2(s)$$

Start by calculating the mol L^{-1} of each ion after mixing and before precipitation. Use a total volume of 0.45 + 0.67 L = 0.112 L

M of SO_4^{2-} and Mg^{2+}= (0.35 mol L^{-1} × 0.67 L) ÷ 0.112 L = 0.2094

M of Ba^{2+}= (0.25 mol L^{-1} × 0.045 L) ÷ 0.112 L = 0.1004 M

M of OH^-= (2 × 0.25 mol L^{-1} × 0.045 L) ÷ 0.112 L = 0.20089 M

Assemble a table showing the precipitation of $BaSO_4$ and $Mg(OH)_2$

Conc. (M)	Ba^{2+} +	SO_4^{2-} +	Mg^{2+} +	OH^-	$\rightleftharpoons$ $BaSO_4(s)$ + $Mg(OH)_2(s)$
Starting	0.1004	0.2094	0.2094	0.20089	
Change	–0.1004	–0.1004	–0.1004	–0.20089	
Equilibrium	0.0	0.109	0.109	0.0	

To calculate $[Ba^{2+}]$ use the K_{sp} expression for $BaSO_4$

$$[BA^{2+}] = \frac{1.1 \times 10^{-10}}{0.109\text{ M } SO_4^{2-}} = 1.0 \times 10^{-9}\text{ M}$$

$$[SO_4^{2-}] = 0.109\text{ M}$$

Calculate the $[OH^-]$ using the K_{sp} expression for $Mg(OH)_2$

$$[OH^-] = \sqrt{\frac{1.8 \times 10^{-11}}{0.109\text{ M } Mg^{2+}}} = 1.28 \times 10^{-5}\text{ M}$$

$$[Mg^{2+}] = 0.109\text{ M}$$

102. Since the top box in the problem represents a saturated solution, we know that 4 moles of MX_2 will dissolve in one liter. For the subsequent boxes, dissolve up to 4 MX_2's and leave any remainder undissolved.

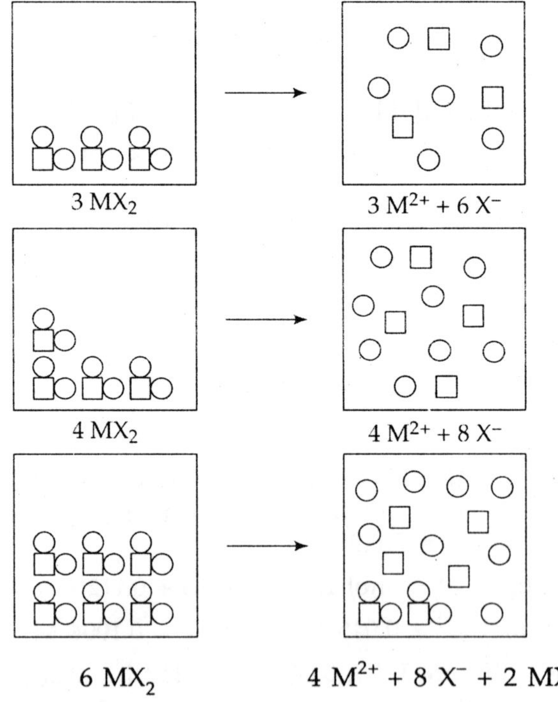

103. The lower the pH, the higher$[H^+]$. Acid rain has a lower pH and therefore a higher $[H^+]$ than normal rain. By LeChatelier's principle, increasing $[H^+]$, one of the reactants, will cause the equilibrium to shift to more products, the dissolving more $CaCO_3$.

104. Calculate the solubility of $Mg(OH)_2$ in water using K_{sp}. Then calculate K for the reaction given. Using K, calculate the solubility of $Mg(OH)_2$ in 0.2 M NH_4Cl and compare the two solubilities.

solubility in water: $K_{sp} = 6 \times 10^{-12} = [Mg^{2+}][OH^-]^2 = (x)(2x)^2 = 4x^3$

$\therefore\ x^3 = 1.5 \times 10^{-12} \Rightarrow x = 1 \times 10^{-4}\ M$

Solubility in 0.2 M NH_4Cl.

$Mg(OH)_2(s) \rightleftharpoons Mg^{2+}(aq) + 2OH^-(aq)$	$K_{sp} = 6 \times 10^{-12}$
$2[NH_4^+(aq) \rightleftharpoons H^+(aq) + NH_3(aq)]$	$K_1 = (K_a)^2 = 3.1 \times 10^{-19}$
$2[H^+(aq) + OH^-(aq) \rightleftharpoons H_2O(l)]$	$K_2(1/K_w)^2 = 1 \times 10^{28}$
$Mg(OH)_2(s) + NH_4^+(aq) \rightleftharpoons Mg^{2+}(aq) + 2NH_3(aq) + 2H_2O(l)$	$K = K_{sp} \times K_1 \times K_2 = 0.02$

$$K = \frac{[Mg^2][NH_3]^2}{[NH_4^+]^2}; \text{ Let } [Mg^{2+}] = x, \text{ then } [NH_3] = 2x \text{ and } [NH_4^+] = 0.2 - x$$

$$\therefore \quad K = 0.02 = \frac{(x)(2x)^2}{(0.2 - x)^2} \Rightarrow .0008 - 0.008\ x + x^2 = 4x^3$$

Using successive approximations; $x = 0.06\ M$

$Mg(OH)_2$ is approximately 600 times more soluble in 0.2M NH_4Cl than in "pure" water.

105. Calculate K for the given reaction. Determine $[H^+]$ using the equatin for buffers, and with that, calculate $[Ca^{2+}]$

$CaF_2(s) \rightleftharpoons Ca^{2+}(aq) + 2F^-(aq)$	$K_{sp} = 1.5 \times 10^{-10}$
$2[H^+(aq) + F^-(aq) \rightleftharpoons HF(aq)]$	$K_1 = (1/K_a)^2 = 2.1 \times 10^6$
$CaF_2(s) + 2H^+(aq) \rightleftharpoons Ca^{2+}(aq) + 2HF(aq)$	$K = K_{sp} \times K_1 = 3.2 \times 10^{-4}$

$$[H^+] = K_a \times \frac{[HCHO_2]}{[CHO_2^-]} = 1.9 \times 10^{-4} \times 10\ \frac{0.3}{0.2} = 2.9 \times 10^{-4}\ M$$

$$K = \frac{[Ca^{2+}][HF]^2}{[H^+]^2} \Rightarrow 3.2 \times 10^{-4} = \frac{(x)(2x)^2}{(2.9 \times 10^{-4})^2} = \frac{4x^3}{8.4 \times 10^{-8}}$$

$x^3 = 6.8 \times 10^{-12}; \Rightarrow x = 1.9 \times 10^{-4}\ M$

106. Since AgI has the smaller K_{sp} it will precipitate first. Thereafter, this is a common ion problem. Calculate the $[Ag^+]$ in solution at the point at which AgCl will just begin to precipitate. Using the concentration and K_{sp} (AgI), calculate $[I^-]$.

$$K_{sp}(AgCl) = 1.8 \times 10^{-10} = [Ag^+][Cl^-] = (x)(0.02\ M)$$

$$x = 9.0 \times 10^{-9}\ M\ Ag^+$$

$$K_{sp}(AgI) = 1 \times 10^{-16} = [Ag^+][I^-] = (9.0 \times 10^{-9}\ M)\ (x)$$

$$x = 1 \times 10^{-8}\ M\ I^-$$

107. (a) $K_{sp} = 6 \times 10^{-12} = [Mg^{2+}][OH^-]^2 = (0.056)(x)^2$

$x^2 = 1 \times 10^{-10}$

$x = 1 \times 10^{-5}\ M$

(b) Na^+:No. NaOH is extremely soluble

Ca^{2+}: $P = [Ca^{2+}][OH^-]^2 = (0.01)(1 \times 10^{-5})^2 = 1 \times 10^{-12}$

$P < K_{sp}$, therefore Ca^{2+} does not precipitate

Al^{3+}: $P = [Al^{3+}][OH^-]^3 = (4 \times 10^{-7})(1\times 10^{-5})^3 = 4 \times 10^{-22}$

$P > K_{sp}$, therefore Al^{3+} does not precipitate

Fe^{3+}: $P = [Fe^{3+}][OH^-]^3 = (2 \times 10^{-7})(1 \times 10^{-5})^3 = 2 \times 10^{-22}$

$P > K_{sp}$, therefore Fe^{3+} does precipitate

(c) Calculate $[OH^-]$ when half the Mg^{2+} is precipitated, then calculate $[Al^{3+}]$ and $[Fe^{3+}]$ in solution and the percent precipitated

$K_{sp} = 6 \times 10^{-12} = [Mg^2][OH^-]^2 = (0.028)(x)^2$

$x = 1.5 \times 10^{-5}\ M$

$K_{sp}\{Al(OH)_3\} = [Al^{3+}][OH^-]^3 \Rightarrow 2 \times 10^{-31} = (x)(1.5 \times 10^{-5})^3$

$\Rightarrow \quad x = 6 \times 10^{-17}$ M

$$\text{percent in solution} = \frac{6\times 10^{-17}}{4\times 10^{-7}} \times 100\% = 1 \times 10^{-8}\%$$

Essentially all the Al^{3+} will have precipitated

$K_{sp}\{Fe(OH)_3\} = [Fe^{3+}][OH^-]^3 \Rightarrow 3 \times 10^{-39} = (x)(1.5 \times 10^{-5})^3$

$\Rightarrow \quad x = 9 \times 10^{-25}$ M

$$\text{percent in solution} = \frac{9\times 10^{-25}}{2\times 10^{-7}} \times 100\% = 4 \times 10^{-16}\%$$

Essentially all the Fe^{3+}will have precipitated

(d) Calculate the acid up masses of the species that precipitate.

$$0.028 \text{ mol } Mg(OH)_2 \times \frac{58.32 \text{ g}}{1 \text{mol}} = 1.6 \text{ g}$$

$$4 \times 10^{-7} \text{ mol. } Al(OH)_3 \times \frac{70.0 \text{ g}}{1 \text{ mol.}} = 3 \times 10^{-5} \text{ g}$$

$$2 \times 10^{-7} \text{ mol. } Fe(OH)_3 \times \frac{106.9 \text{ g}}{1 \text{ mol.}} = 2 \times 10^{-5} \text{ g}$$

Total mass = 1.6 g

108. (a)

$$Zn(NH_3)_4^{2+}\,(aq) \rightleftharpoons Zn^{2+}(aq) = 4NH_3(aq) \qquad K_1 = 1/K_f = 2.8 \times 10^{-9}$$
$$Zn^{2+}(aq) + 4OH^-(aq) \rightleftharpoons Zn(OH)_4^{2-}(aq) \qquad K_f = 3 \times 10^{14}$$

$$Zn(NH_3)_4^{2-}(aq) + 4OH(aq) \rightleftharpoons Zn(OH)_4^{2-}(aq) + 4HN_3(aq) \qquad K = K_1 \times K_f = 8 \times 10^5$$

(b) Calculate the $[OH^-]$ in a 1.0 M NH_3 solution, then calculate the ratio of the two complexes using the equilibrium concentration expression.

$$NH_3(aq) + H_2O \rightleftharpoons NH_4^+(aq) + OH^-(aq)$$

$$K_b = \frac{[NH_4^+][OH^-]}{[NH_3]} \Rightarrow 1.8 \times 10^{-5} = \frac{x^2}{1.0} \quad (\text{let, } [OH^-] = [NH_4^+] = x)$$

$\therefore$ $x^2 = 1.8 \times 10^{-5}$

or $x = 0.042\ M\ OH^-$

$$K = \frac{[Zn(OH)_4^{2-}][NH_3]^4}{[Zn(NH_3)_4^{2+}][OH^-]^4} \Rightarrow 8 \times 10^5 = \frac{[Zn(OH)_4^{2-}]}{[Zn(NH_3)_4^{2+}]} \times \frac{(1.0)^2}{(0.0042)^4}$$

or, $$\frac{[Zn(OH)_4^{2-}]}{[Zn(NH_3)_4^{2+}]} = 2 \times 10^{-4}$$

or, $$\frac{[Zn(NH_3)_4^{2+}]}{[Zn(OH)_4^{2-}]} = 4 \times 10^3$$

109.
$$Ag(NH_3)_2^+(aq) \rightleftharpoons Ag^+(aq) + 2NH_3(aq) \qquad K_1 = 1/K_f = 5.9 \times 10^{-8}$$
$$2[H^+(aq) + NH_3(aq) \rightleftharpoons NH_4^+(aq)] \qquad K_2 = (1/K_a)^2 = 3.2 \times 10^{18}$$
$$Ag^+(aq) + Cl^-(aq) \rightleftharpoons AgCl(s) \qquad K_3 = 1/K_{sp} = 5.6 \times 10^9$$

$$Ag(NH_3)_2^+(aq) + 2H^+(aq) + Cl^-(aq) \rightleftharpoons AgCl(s) + NH_4^+(aq) \qquad K = K_1 \times K_2 \times K_3$$
$$\therefore K = 1.0 \times 10^{21}$$

110. Use the solubility rules to product which ions will precipitate. Since only sodium, potassium, and ammonium carbonates are soluble, it follows that both $CaCO_3$ and Ag_2CO_3 will precipitate according to the following equations. Either equation could be written first, the precipitation of $CaCO_3$ is arbitrarily written first, the same results could be obtained by writing the precipitation of Ag_2CO_3 first and that of $CaCO_3$ second. From the ratios of moles given, it can be seen that the Ca^{2+} ion and the Ag^+ ion will be precipitated completely, leaving excess CO_3^{2-} ion in solution

	$Ca(NO_3)_2(aq)$	+ $Na_2CO_3(aq)$	→ $2NaNO_3(aq)$	+ $CaCO_3(s)$
Before any pptn:	0.125 mol	0.375 mol	0 mol	0 mol
After $CaCO_3$ pptn:	0 mol	0.25 mol	0.25 mol	0.125 mol
	$2AgNO_3(aq)$	+ $Na_2CO_3(aq)$	→ $2\ NaNO_3(aq)$	+ $Ag_2CO_3(s)$
After $CaCO_3$ pptn:	0.2 mol	0.25 mol	0.25 mol	0 mol
After Ag_2CO_3 pptn:	0 mol	0.15 mol	0.45 mol	0.1 mol

The molarities of the various ions are calculated by dividing the moles from the second equation by the final volume of 2 L. Note that the moles of Na^+ ion comes from the moles of Na_2CO_3 left as well as the moles o $NaNO_3$ formed

M of CO_3^{2-} left = 0.15 mol ÷ 2 L = 0.075 M

M of NO_3^- left = 0.45 mol ÷ 2 L = 0.225 M

mol Na^+ left = 0.45 mol (from $NaNO_3$) = 2 × 0.15 (from Na_2CO_3) = 0.75 mol

M of Na^+ = 0.75 mol ÷ 2 L = 0.375 M

111. Recall that freezing point lowering is a colligative property and thus depends on the total moles of solute (ions and molecules) present in solution. Determine the moles of H^+, B^- and HB in solution, calculate the molarity of the solution and the freeing point depression. Assume one liter of solution.

$$1\ \text{L} \times \frac{1000\ \text{mL}}{1\ \text{L}} \times \frac{1.006\ \text{g}}{1\ \text{mL}} = 1006\ \text{g solution}$$

$$1006\ \text{g solution} \times \frac{5.0\ \text{g}\ HC_2H_3O_2}{100\ \text{g solution}} = 50.3\ \text{g}\ HC_2H_3O_4$$

$$50.3\ \text{g}\ HC_2H_3O_2 \times \frac{1\ \text{mol}}{65.05\ \text{g}} = 0.838\ \text{mol}\ HC_2H_3O_2$$

	$HC_2H_3O_2$	⇌	$C_2H_3O_2(aq)$	+	$H^+\ (aq)$
$[\ ]_0$	0.838		0		0
$\Delta[\ ]$	$-x$		$+x$		$+x$
$[\ ]_{eq}$	$0.838 - x$		x		x

$$K_a = 1.8 \times 10^{-5} = \frac{[H^+][C_2H_3O_2^-]}{[HC_2H_3O_2]} = \frac{x^2}{0.838 - x}$$

$$K_a/a = 1.8 \times 10^{-5}/0.838 = 2.1 \times 10^{-5}$$
$$K_a/a = 2.1 \times 10^{-5} < 2.5 \times 10^{-3}$$

$$1.8 \times 10^{-5} = \frac{x^2}{0.838} \Rightarrow 1.5 \times 10^{-5} = x^2$$

$$x = 3.9 \times 10^{-3}\ M$$

moles solute = $[HC_2H_3O_2] + C_2H_3O_2^-] + [H^+] = (0.838 - x) + x + x = 0.838 + x$

moles solute = 0.838 + 0.0039 = 0.842

$$m = \text{molarity} = \frac{\text{mol. solute}}{\text{kg solvent}} = \frac{0.842\ \text{mol.}}{1.006\ \text{kg solution} - 0.0503\ \text{kg}\ HC_2H_3O_2} = 0.877$$

$$\Delta T_f = k_f(m) = 1.86°\ \text{Kg} \cdot \text{k} \cdot \text{mol}^{-1}\ (0.877) = 1.64°$$

112. Write K_a expressions for the concentrated acid $[HA]_c$, and the diluted acid $[HA]_d$, assuming that $[HA]_0 = [HA]_{eq}$ (which is true for a weak acid). Recallin that K_a is constant for a given acid, equate the two expressions and solve for $[H^+]$ and substitute the results into the percent ionization expression. Note the we using the equation

$$HA(aq) \rightleftharpoons H^+(aq) + A^-(aq)$$

$$K_a = \frac{[H^+]_c [A^-]_c}{[HA]_c} = \frac{x^2}{[HA]_c}, \text{ where } [H^+]_c = [A^-]_c = x$$

$$K_a = \frac{[H^-]_c [A^-]_d}{[HA]_d} = \frac{y^2}{[HA]_d}, \text{ where } [H^+]_d = [A^-]_d = y$$

If $[HA]_d$ is one tenth the concentration of $[HA]_c$, then $10[HA]_d = [HA]_c$ and

$$K_a = \frac{x^2}{[HA]_c} = \frac{y^2}{[HA]_d} \Rightarrow \frac{x^2}{10[HA]_d} = \frac{y^2}{[HA]_d} \Rightarrow x^2 = 10\, y^2$$

$$x = y\sqrt{10}$$

Substituting into the percent ionization expression.

$$\%\text{ion}_{(\text{conc.})} = \frac{[H^+]_c}{[HA]_c} \times 100\% = \frac{100 - x}{[HA]_c} = \frac{100(y\sqrt{10})}{10[HA]_d} = \frac{y - 10\sqrt{10}}{[HA]_d}$$

$$\%\text{ion}_{(\text{dil.})} = \frac{[H^+]_d}{[HA]_d} \times 100\% = \frac{y}{[HA]_d} \times 100\% = \frac{100 - y}{[HA]_d}$$

Obviously the two terms differ by a factor of $\sqrt{10}$,

$$\frac{y - 100}{[HA]_d} = \frac{y - 10\sqrt{10}}{[HA]_d} \cdot \sqrt{10} \quad \text{Thus , } \%\text{ion}_{(\text{dil})} = (\%\text{ion}_{(\text{conc})}) \times \sqrt{10}$$

113. 1. $pH = \log_{10}(H_3O^+)$

Step 1: Enter pH

Step 2: Change sign by pressing +/– key

Step 3: Press "2nd" then "log" or "inv" then "log"

X: 3.16×10^{-7} MH_2O^+

Y: 2.0×10^{-7} M H_3O^+

Z: 7.94×10^{-8} M H_3O^+

2. The moles of acid was constant for each solution,

$$\frac{50 \cancel{mL}}{1} \times \frac{1 \cancel{L}}{100 \cancel{mL}} \times \frac{0.1 \text{ moles HA}}{1 \cancel{L}} = 5 \times 10^{-3} \text{ mol HA}$$

3. X: $\frac{12 \cancel{mL}}{1} \times \frac{1 \cancel{L}}{1000 \cancel{mL}} \times \frac{0.1 \text{ moles } OH^-}{1 \cancel{L}} = 9 \times 10^{-4} \text{ mol } OH^-$

Y: $\frac{18.5 \text{ mL}}{1} \times \frac{1 \text{ L}}{100 \text{ mL}} \times \frac{7.5 \times 10^{-2} \text{ mol OH}^-}{\text{L}} = 1.39 \times 10^{-3} \text{ mol OH}^-$

Z: $\frac{35 \text{ mL}}{1} \times \frac{1 \text{ L}}{1000 \text{ mL}} \times \frac{7.5 \times 10^{-2} \text{ mol OH}^-}{\text{L}} = 2.69 \times 10^{-3} \text{ mol OH}^-$

4. X: $[\text{HA}] = \frac{\text{initial moles HA} - \text{moles OH}^- \text{ added}}{\text{volume sol'n}}$

$$[\text{HA}] = \frac{5 \times 10^{-3} - 9 \times 10^{-4}}{0.1 \text{ L}} = 4.1 \times 10^{-2} \text{ M}$$

moles A^- formed = moles OH^- addd

$$[\text{A}^-] = \frac{9 \times 10^{-4} \text{ mo A}^-}{0.1 \text{ L}} = 9 \times 10^{-3} \text{ M}$$

Y: $[\text{HA}] = \frac{\text{initial moles HA} - \text{moles OH}^- \text{ added}}{\text{volume sol'n}}$

$$[\text{HA}] = \frac{5 \times 10^{-2} - 1.39 \times 10^{-3}}{0.1 \text{L}} = 3.61 \times 10^{-2} \text{ M}$$

moles A^- formed = moles OH^- added

$$[\text{A}^-] = \frac{1.39 \times 10^{-4} \text{ mol A}^-}{0.1 \text{ L}} = 1.39 \times 10^{-2} \text{ M}$$

Z: $[\text{HA}] = \frac{\text{initial moles HA} - \text{moles OH}^- \text{ added}}{\text{volume sol'n}}$

$$[\text{HA}] = \frac{5 \times 10^{-3} - 2.63 \times 10^{-3}}{0.1 \text{ L}} = 2.37 \times 10^{-2} \text{ M}$$

moles A^- formed= moles OH^- added

$$[\text{A}^-] = \frac{2.63 \times 10^{-3} \text{ mol A}^-}{0.1 \text{ L}} = 2.63 \times 10^{-2} \text{ M}$$

5. X: $\frac{1}{9 \times 10^{-3}} = 1.11 \times 10^{-2} \text{ M}^{-1}$

Y: $\frac{1}{1.39 \times 10^{-2}} = 7.19 \times 10^{1} \text{ M}^{-1}$

Z: $\frac{1}{2.63 \times 10^{-2}} = 3.8 \times 10^{1} \text{ M}^{-1}$

6.

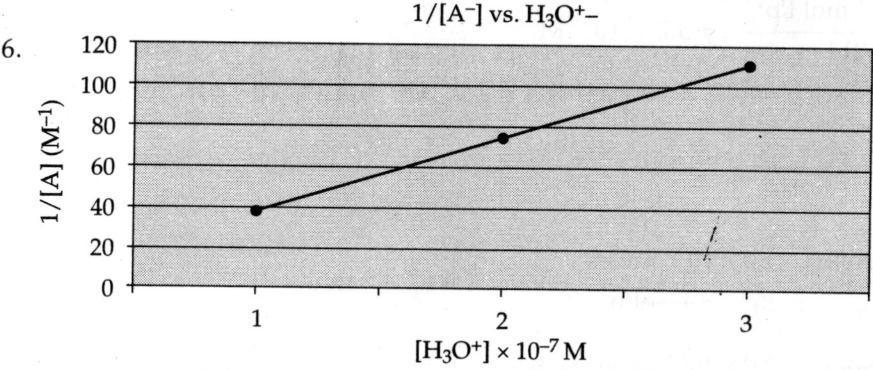

7. Slope $= \dfrac{\Delta y}{\Delta x} = \dfrac{\Delta[1/A^-]}{\Delta[H_3O^+]} = \dfrac{1.11 \times 10^2 - 3.8 \times 10^1}{(3.16 \times 10^{-7}) - (7.94 \times 10^{-8})} = 3.08 \times 10^{-8}$

8. $= \dfrac{5 \times 10^{-3} \text{ mol HA}}{0.1 \text{ L}} = 5 \times 10^{-2}$ M

9. $\dfrac{1}{\text{slope}} = \dfrac{\Delta[H_3O^+]}{\Delta[1/A^-]} = (H_3O^+) \times (A^-)$

$$K_a = \frac{1}{\text{HA slope}} = \frac{1}{5 \times 10^{-2} \cdot (3.08 \times 10^8)} = 6.49 \times 10^{-4} \text{ M}$$

114. 1. $\dfrac{5.0 \text{ mL sol'n}}{1} \times \dfrac{1 \text{ L sol'n}}{1000 \text{ mL sol'n}} \times \dfrac{0.05 \text{ mol Pb(NO}_3)_2}{1 \text{ L sol'n}}$

$= 7.5 \times 10^{-5}$ mol $Pb(NO_3)_2$ yielding 7.5×10^{-5} mol Pb^{2+}

2. $\dfrac{3.0 \text{ mL sol'n}}{1} \times \dfrac{1 \text{ L sol'n}}{1000 \text{ mL sol'n}} \times \dfrac{0.03 \text{ mol KI}}{1 \text{ L sol'n}}$

$= 9.0 \times 10^{-5}$ mol KI yielding 9.0×10^{-5} mol I^-

3. $\dfrac{1.4 \times 10^{-3} \text{ moles}}{\text{L}} \times \dfrac{0.01 \text{ L}}{1} = 1.4 \times 10^{-4}$ mol I^- in sol'n at equil.

4. moles I^- originally present – moles of I^- in solution

$= 9.0 \times 10^{-5} - 1.4 \times 10^{-5} = 7.6 \times 10^{-5}$ moles I^- ppt.

5. There are twice is many moles of I^- as there are Pb^{2+}, therefore,

$$\frac{7.6 \times 10^{-5} \text{ mol I}^-}{1} \times \frac{1.0 \text{ mol Pb}^{2+}}{2.0 \text{ mol I}^-} = 3.8 \times 10^{-5} \text{ moles Pb}^{2+}$$

6. Total moles of Pb^{2+} – moles of Pb^{2+} in ppt

$= 7.5 \times 10^{-5} - 3. \times 10^{-5} = 3.7 \times 10^{-3}$ moles Pb^{2+} in sol'n

7. $\dfrac{3.7 \times 10^{-5} \text{ mol } Pb^{2+}}{0.01 \text{ L}} = 3.7 \times 10^{-5}$ M

8. $K_{sp} = [Pb^{2+}]\,[I^-]^2 = (3.7 \times 10^{-3} \text{ M})\,(1.4 \times 10^{-3} \text{ M})^2 = 7.3 \times 10^{-9} \text{ M}^3$

9. $\% \text{ error} = \dfrac{\text{obs} - \text{exp}}{\text{exp}} \times 100\% = \dfrac{(7.3 \times 10^{-9} - 7.1 \times 10^{-9})}{7.1 \times 10^{-9}} \times 100\% = 2.8\%$

115. 1. $\dfrac{0.005 \cancel{\text{ L sol'n}}}{1} \times \dfrac{2.0 \times 10^{-3} \text{ moles } Fe^{3+}}{\cancel{\text{L sol'n}}} = 1.0 \times 10^{-5} \text{ moles } Fe^{3+}$

$\dfrac{0.005 \cancel{\text{ L sol'n}}}{1} \times \dfrac{2.0 \times 10^{-5} \text{ moles } SCN^-}{\cancel{\text{L sol'n}}} = 1.0 \times 10^{-5} \text{ mols } SCN^-$

2. Total volume of sol'n = 5.0 mL + 5.0 mL = 10.0 mL

$$\dfrac{1.3 \times 10^{-4} \text{ moles } FeSCN^{2+}}{\cancel{\text{L sol'n}}} \times \dfrac{0.1 \cancel{\text{ L sol'n}}}{1} = 1.3 \times 10^{-6} \text{ moles } FeSCN^{2+}$$

3. The balanced equation of equilibrium is:

$$Fe^{3+}_{(aq)} + SCN^-_{(aq)} \rightleftharpoons FeSCN^{2+}_{(aq)}$$

The ratio of Fe^{3+} to SCN^- is: 1 : 1 and the ratio of Fe^{3+} and SCN^- to $FeSCN^{2+}$ is 1 : 1

Therefore, 1.3×10^{-6} moles of $Fe^{3+}$and 1.3×10^{-6} moles of SCN^- were used in making the $FeSCN^{2+}$

4. initial No. of moles – No. of moles used up in making $FeSCN^{2+}$

$= 1.0 \times 10^{-5} \text{ moles } Fe^{3+} - 1.3 \times 10^{-6} \text{ moles } Fe^{3+} = 8.7 \times 10^{-6} \text{ moels } Fe^{3+}$

$= 1.0 \times 10^{-5} \text{ moles } SCN^- - 1.3 \times 10^{-6} \text{ moles } SCN^- = 8.7 \times 10^{-6} \text{ moles } SCN^-$

5. $[Fe^{3+}] = \dfrac{8.7 \times 10^{-6} \text{ moles } Fe^{2+}}{0.01 \text{ L sol'n}} = 8.7 \times 10^{-4} \text{ M } Fe^{3+}$

$$[SCN^-] = \dfrac{8.7 \times 10^{-6} \text{ moles } SCN^-}{0.01 \text{ L sol'n}} = 8.7 \times 10^{-4} \text{ M } SCN^-$$

$$[FESCN^{2+}] = 1.3 \times 10^{-4} \text{ M } FeSCN^{2+}$$

6. $Fe^{3+}_{(aq)} + SCN^-_{(aq)} \rightleftharpoons FeSCN^{2+}_{(aq)}$

$$K_c = \dfrac{[FeSCN^{2+}]}{[Fe^{3+}][SCN^-]} = \dfrac{1.3 \times 10^{-4} \text{ M}}{(8.7 \times 10^{-4})^2 \text{ M}^2} = 1.7 \times 10^{-7} \text{ M}^{-1}$$

116. 1. $50 \text{ mL} \times \dfrac{1.0 \cancel{\text{ mol HCl}}}{1000 \text{ mL}} \times \dfrac{1 \text{ mol NaOH}}{1 \cancel{\text{ mol HCl}}} \times \dfrac{1}{0.0487 \text{ L}} = 0.103 \text{ M NaOH}$

2. $K = 10^{-pKa} = 10^{-5} = 1 \times 10^{-5}$

$$K = \frac{(HA)(OH^-)}{(A^-)} \cdot \frac{(H^+)}{(H^+)} = \frac{K_w}{K_a} = \frac{(OH^-)^2}{(A^-)}$$

$$K_A = \frac{K_w\,(A^-)}{(OH^-)^2}$$

3. The p^H at the equivalence pint appears to be around 9. Therefore $(OH^-) = 10^{-5}$ M

$$(A^-) = \frac{(43.2 \text{ mL} \times 0.013 \text{ M}) - (5.2 \text{ mL} \times 0.1 \text{ M})}{43.2 \text{ mL} + 5.2 \text{ mL}} = 0.0812 \text{ M at equiv. point}$$

$$K_a = K_w \frac{(A^-)}{(OH^-)} = \frac{(1 \times 10^{-14})(0.0812)}{(1 \times 10^{-5})^2} \; 8.12 \times 10^{-6}$$

4. $43.2 \cancel{\text{mL}} \times \dfrac{0.103 \text{ mol } OH^-}{1000 \cancel{\text{mL}}} = 0.00445$ mol OH^- dispensed from buret

$5.2 \cancel{\text{mL}} \times \dfrac{0.1 \text{ mol } H^+}{1000 \cancel{\text{mL}}} = 0.00052$ mol H^+ used in back-titratiion

(0.00445 – 0.00052) mol = 0.00393 mol OH^- actually used to neutralize

Since H^+ from the acid reacts in a 1 : 1 mole ratio with OH^-, the number of moles of H^+ furnished by the acid must also be 0.00393

$$\text{equivalent mass} = \frac{\text{grams of acid}}{\text{moles of } H^+} = \frac{0.5 \text{ g}}{0.0393 \text{ mol}} = 1.27 \text{ g mol}^{-1}$$

117. The equilibrium involved in this buffer solution is

$$HOA(Aq) + H_2O \rightleftharpoons OAc^-(aq) + H_3O^+(aq)s$$

Let $x = [H_3O^+]$

Then $[OAc^-] = 0.4 + x$ and $[HOAc] = 0.1 - x$

Note particularly that the $[H_3O^+$ and the $[OAc^-]$ are *not* equal in this solution. The great bulk of the acetate ions presents in the solution comes from the sodium acetate which is a strong electrolyte. The acetate ion from the sodium acetate represses the ionization of the acetic acid, so that the $[H_3O^+]$ in this solution is quite small relative to the $[OAc^-]$. Thus we expect x to be small relative to both 0.1 and 0.4. If we make this approximation the equilibrium constant expression for HOAc is

$$K_a = \frac{[H_3O^+][OAc^-]}{[HOAc]} = \frac{(x)(0.4)}{0.1} = 1.8 \times 10^{-5}$$

$\Rightarrow \quad x = 4.5 \times 10^{-6}$

The p^H is therefore 6 – log 4.5 = 5.35. This is an effective buffer solution because the $[OAc^-]/[HOAc]$ ratio is 0.4/0.1 = 4. The value of 4 lie between the acceptable limits of 10 and 0.10.

Note that we could have done this problem in a different, but equivalent way,

$$p^H = P^{K_a} + \log (0.4/0.1) = P^{K_a} + \log 4.0$$

Since K_a of HOAc is 1.8×10^{-3}

$$p^H = 4.745 + 0.602 \simeq 5.35$$

118. The p^H of this buffer is given as

$$p^H = P^{K_a}(NH_4^+) + \log \frac{[NH_3]}{[NH_4^+]}$$

Since $K_a(NH_4^+) = 5.7 \times 10^{-10}$ (from tables), $P^{K_a}(NH_4^+) = 9.24$, so that $p^H = 9.24 + \log (NH_3)/[NH_4^+])$.

Originally, both NH_4^+ and $[NH_3]$ are 0.2 *M*. The dilution with water decreases both these concentrations to 0.1 *M*, but the $[NH_3]/[NH_4^+]$ *ratio* remains 1 : 1. Thus the p^H remains constant on dilution.

119. Let, C_A= molarity of the acid solution
Then, No. mmol acid= (25 ml) (C_A mmol · mL^{-1})

At the equivalence point, the number of millimoles of added base is exactly equal to the number of millimoles of acid originally present.

∴ No. mmol OH^- added = (32.8 mL) (0.15 *M*) = (25 mL) (C_A)

$$\text{or, } C_A = \frac{(32.8)(0.15)}{(25)} = 0.1968\ M$$

120. The number of millimoles of OH^- added during the titration was (43.72 mL) (0.5 mmol · mL^{-1}) = 21.86 mmol. Therefore 21.86 mmol of H_3O^+ reacted with the added OH^- during the titration. As the tablet was dissolved in a solution containing 50 mmol of H_3O^+, the amount of acid neutralized by the tablet was 50 – 21.86 = 28.14 mmol. or 0.02814 mol acid.

121. When 24.85 mL of base have been added

No. mmol OH^- added = (24.85 mL) (0.01 mmol · mL^{-1})= 0.2485 mmol

No. mmol H_2O^+ untitrated = 0.25 – 0.28 = 0.0015 mmol

volume of solution = 24.85 + 25 = 49.85 mL

$[H_3O^+] = (1.5 \times 10^{-3}$ mmol)/(49.85 mL) $= 3.0 \times 10^{-5}$ *M* and H = 4.52

Since pH = 4.52 is in the middle of the pH range in which bromerosol green changes colour, when 24.85 mL of base have been added, the solution will be green and the titration will be ended. The true equivalence point, of course, occurs when 25 ml, of base have been added. Thus the titration error is

$$\left(\frac{24.85 - 25}{25}\right) \times 100 = -0.60\%$$

122. (a) The concentration of the methylamine is:
(25.57 mL) (0.1009 mmol · mL^{-1}) = (25.00 mL) (C_B)

$$C_B = \frac{(25.57)(0.1009)}{25} = 0.1032\ M$$

(b) The number of millimoles of methylamine titrated was

(25 mL) (0.1032 mmol · mL^{-1}) = 0.58 mmol

The titration reaction is $CH_3NH_2(aq) + H_3O^+(aq) \rightleftharpoons CH_3NH_3^+(aq) + H_2O$

A the equivalence point, the titration reaction has gone to completion and there are 2.58 mmol of methylammonium ion, $CH_3NH_3^+$, in solution. the total volume of the solution at the equivalence point is 50.57 mL. The solution is therefore identical with a solution of methylammonium chloride of concentration

$$\frac{2.58 \text{ mmol}}{50.57 \text{ mL}} = 0.05102\ M$$

The acidity constant of methylammonium ion From tables is 2.4×10^{-11}. A solution of methylammonium chloride is acidic due to the proton transfer reaction between $CH_3NH_3^+$ and water:

$CH_3NH_3^+(aq) + H_2O \rightleftharpoons CH_3NH_2(aq) + H_3O^+(aq)$

Note that above equation is the reverse of the titration reaction.

$$[CH_3NH_2] = [H_3O^+] \text{ at the equivalence point}$$

Let, $x = [H_3O^+]$ at the equivalence point

Then, $[CH_3NH_3^+] = 0.05102 - x$

The equilibrium constant equation is

$$K_a(CH_3NH_3^+) = \frac{[CH_3NH_3][H_3O^+]}{[CH_3NH_3^+]} \quad \frac{x^2}{0.05102 - x} \quad = \ = 2.4 \times 10^{-11}$$

Since methylammonium ion is such a weak acid, we expect the $[H_3O^+]$ to be small compared to 0.05102. If we assume $0.05102 - x$ is approximately 0.05102, the equilibrium constant becomes

$x^2 = (2.40 \times 10^{-11})\ (5.102 \times 10^{-2}) = 1.22 \times 10^{-12}$

and; $x \quad = 1.11 \times 10^{-6}$

Since 1.11×10^{-6} is small compared to 0.05102, the assumption we made valid, and therefore at the equivalence point in this titration the $[H_3O^+] = 1.11 \times 10^{-6}\ M$, and the pH = 5.96. The best indicator to use, is chlorophenol red. Note that the pH at the equivalence point is less than 7.

123. The concentration of CO_2 in water in equilibrium with air is obtained

$$[CO_2] = 0.0337;\ P_{CO_2} = (3.37 \times 10^{-2}\ M \cdot atm^{-1})\ (3.0 \times 10^{-4}\ atm) = 1.0 \times 10^{-5}\ M$$

Since K_1 is almost 10,000 times larger than K_2, we can ignore the second single of ionization of carbonic acid and write

$$[H_3O^+] = [HCO_3^-]$$

The equation for the first acidity constant of carbonic acid, therefore becomes simply

$$K_1 = 4.3 \times 10^{-7} = \frac{[HCO_3^-][H_3O^+]}{[CO_2]} = \frac{[H_3O^+]^2}{1.0 \times 10^{-5}}$$

so that $[H_3O^+]^2 = 4.3 \times 10^{-12}$ and $[H_3O^+] = 2.1 \times 10^{-6}\ M$

Note that in the example, $[CO_2]$ is constant, 1×10^{-5} *M*, because the CO_2 in solution is in equilibrium with the air which has a constant partial pressure of CO_2, 3.0×10^{-4} atm.

We can verify the assumptions we made when we neglected the second stage of ionization by calculating the $[CO_3^{2-}]$ in this solution, using K_2.

$$K_2 = \frac{[H_3O^+][CO_3^{2-}]}{[HCO_3^-]} = 4.7 \times 10^{-11}$$

If $[H_3O^+] = [HCO_3^-]$, we obtain $[CO_3^{2-}] = K_2 = 4.7 \times 10^{-11}$. Since 4.7×10^{-11} is very small compared to 2.1×10^{-6}, it is correct to say that the $[CO_3^{2-}]$ is negligible compared to the $[HCO_3^-]$

We can check to see whether our neglect of the self-ionization of water was justifiable by calculating the $[OH^-]$.

$$[OH^-] = \frac{K_w}{[H_3O^+]} = \frac{1.0 \times 10^{-14}}{2.1 \times 10^{-6}} = 4.8 \times 10^{-9}\ M$$

Therefore, $[OH^-] \ll [HCO_3^-]$

Thus, the concentration of all species in this solution are as follows:

$[CO_2] = 1.0 \times 10^{-5}\ M$

$[H_3O^+] = [HCO_3^-] = 2.1 \times 10^{-6}\ M$

$[CO_3^{2-}] = 4.7 \times 10^{-11}\ M$

$[OH^-] = 4.8 \times 10^{-9}\ M$

The pH is $6 - \log(2.1) = 5.7$. Thus pure water exposed to the air has a pH = 5.7.

124. The bicarbonate ion, HCO_3^-, is an ampholyte. Now, the acidity constant for HCO_3^-, which is the second ionization constant of carbonic acid, is 4.7×10^{-11}. When HCO_3^- functions as a base, its conjugate acid is carbonic acid, or aqueous CO_2,

$$\therefore K_b = \frac{K_w}{K_1(CO_2 + H_2O)} = \frac{1.0 \times 10^{-14}}{4.3 \times 10^{-7}} = 2.3 \times 10^{-8}$$

Since $K_b(HCO_3^-) > K_a(HCO_3^-)$, a solution of $NaHCO_3$ is basic.

The $[H_3O^+]$ in a solution of $NaHCO_3$ can be obtained as

$$[H_3O^+]^2 = K_1K_2 = (4.3 \times 10^{-7})(4.7 \times 10^{-11}) = 20 \times 10^{-18}$$

so that, $[H_3O^+] = 4.5 \times 10^{-9}$ *M* and pH = 8.35

The initial concentration of HCO_3^- is 0.1 *M*, but at equilibrium the $[HCO_3^-]$ is somewhat less than 0.1 *M*, because both CO_3^{2-} and CO_2 have been formed and HCO_3^- reacts with water, as an acid and as a base. Conservation of mass requires that

$$0.1 = [HCO_3^-] + [CO_3^{2-}] + [CO_2]$$

This equation is the material balance equation.

The sum of the acidity and basicity reaction for HCO_3^- is

$$2HCO_3^-(aq) \rightleftharpoons CO_3^{2-}(aq) + CO_2(aq) + H_2O$$

This is the principal reaction in a solution of $NaHCO_3$, Here, the species H_2CO_3 is written as $H_2O + CO_2$. According to this overall reaction, $[CO_2] = [CO_3^{2-}]$, and we have already used that approximation when we determined the pH. If we introduce that relation into the material balance equation, we obtain

$0.1 = [HCO_3^-] + 2[CO_3^{2-}] = [HCO_3^-]\ \ 2[CO_2]$

This can be rearranged to solve for the $[HCO_3^-]$

$[HCO_3^-] = 0.1 - 2[CO_3^{2-}] = 0.1 - 2[CO_2]$

The equilibrium constant for the principal (prin) reaction,

$$K_{prin} = \frac{[CO_2][CO_3^{2-}]}{[HCO_3^-]^2} = \frac{[CO_2]}{[H_3O^+][HCO_3^-]} \cdot \frac{[CO_3^{2-}][H_3O^+]}{[HCO_3^-]} = \frac{K_2}{K_1}$$

$$= (4.7 \times 10^{-11})/(4.3 \times 10^{-7}) = 1.1. \times 10^{-4}$$

Let $x = [CO_2] = [CO_3^{2-}]$ in 0.1 F $NaHCO_3$

Then $[HCO_3^-]\ = 0.1 - 2x$

and $K_{prin} = 1.1. \times 10^{-4} = \dfrac{x^2}{(0.1-2x)^2}$

Taking the square root of both sides of this equation we obtain

$$1.05 \times 10^{-2} = \frac{x}{0.1 - 2x}$$

which can be multiplied out and solved for x. We find

$x = [CO_3^{2-}] = [CO_2] = 1.0 \times 10^{-3}\ M$

and $[HCO_3^-] = 0.1 - 2(0.001) = -0.098\ M$

We had already found that $[H_3O^+] = 4.5 \times 10^{-9}\ M$, so that

$$[OH^-] = \frac{K_w}{[H_3O^+]} = \frac{1.0 \times 10^{-14}}{4.5 \times 10^{-9}} = 2.2 \times 10^{-6}\ M$$

125. The carbonate ion is a base, and the principal reaction in a solution of Na_2CO_3 is

$$CO_3^{2-}(aq) + H_2O \rightleftharpoons HCO_3^-(aq) + OH^-(aq)$$

The basicity constant of carbonate ion is

$$K_b(CO_3^{2-}) = \frac{[HCO_3^-][OH^-]}{[CO_3^{2-}]} = \frac{K_w}{K_a\ (HCO_3^-)} = \frac{1.0 \times 10^{-14}}{4.7 \times 10^{-11}} = 2.11 \times 10^{-4}$$

The basicity constant for HCO_3^- is found to be 2.3×10^{-8} in example above. This value is so much smaller than 2.1×10^{-4} that we can neglect the basicity of HCO_3^- compared to the basicity of CO_3^{2-}, and assume that $[HCO_3^-] = [OH^-]$ in a solution of Na_2CO_3.

As the solution is 0.05 F Na_2CO_3, the mass balance equation, which indicates the conversion of mass, is

$0.05 = [CO_3^{2-}] + [HCO_3^-] + [CO_2]$

The $[CO_2$, however, is very much smaller than either the $[CO_3^{2-}]$ of the $[HCO_3^-]$, because it is formed by the basicity reaction of HCO_3^-, and $K_b(HCO_3^-)$ is so much smaller than $K_b(CO_3^{2-})$. Hence can neglect the $[CO_2]$ in the mass balance equation and rearrange it to solve for the $[CO_3^{2-}]$, as follows:

$$[CO_3^{2-}] = 0.05 - [HCO_3^-]$$

Let, $y = [HCO_3^-] = [OH^-]$ in 0.05 F Na_2CO_3
Then, $[CO_3^{2-}] = 0.05 - y$
and the equilibrium constant equation is

$$K_b(CO_3^{2-}) = 2.1 \times 10^{-4} = \frac{y^2}{0.05 - y}$$

To solve this equation assume that $0.05 - y = 0.05$. Then
$y^2 = (5 \times 10^{-2})\ (2.1 \times 10^{-4}) = 10.5 \times 10^{-6} \Rightarrow y = 3.24 \times 10^{-3}$

Since 3.2×10^{-3} is not very much smaller than 5×10^{-2} (it is 6.4% of 0.05) we might make a second approximation and set

$0.05 = y = 0.05 - 0.0032 = 0.0468$

Te equilibrium constant equation then becomes
$2.1 \times 10^{-4} = y^2/(4.68 \times 10^{-2})$ or $y^2 = 9.83 \times 10^{-6}$ and therefore $y = 3.1 \times 10^{-3}$
If we make a third approximation and set

$$0.05 \sim y = 0.05 - 0.0031 = 0.0469$$

We still obtain $y = 3.1 \times 10^{-3}$ on solving the quilibrium constant expression. Since two successive approximations yield the same value, this is the correct answer.

Thus, $[OH^-] = [HCO_3^-] = 3.1 \times 10^{-3}\ M$

$[CO_3^{2-}] = 0.05 - 0.0031 = 0.0469\ M$

$[H_3O^+] = K_w/[OH^-] = 3.2 \times 10^{-12}\ M$ and pH $= 12 - \log(3.2) = 11.49$

Note that this solution is distinctly basic. Carbonate ion is a moderately strong base, with $K_b = 2.1 \times 10^{-4}$.

We can also calculate the $[CO_2]$, using K_1 for carbonic acid.

$$K_1 = 4.3 \times 10^{-7} = \frac{[H_3O^+][HCO_3^-]}{[CO_2]} = \frac{(3.2 \times 10^{-12})(3.1 \times 10^{-3})}{[CO_2]}$$

and therefore $[CO_2] = 2.3 \times 10^{-8}\ M$

Since 2.3×10^{-8} is negligible compared to the $[CO_3^{2-}]$ and the $[HcO_3^-]$, it is valid to neglect the $[CO_2]$ in the material balance equation.

126. To determine the change to the p^H we must first know the p^H of the solution before adding the 0.5 mL of 0.2 F NaOH. The equilibrium involved in this buffer is

$$HOAc(aq) + H_2O \rightleftharpoons H_3O^+(aq) + OAc^-(aq)$$

Let, $x = [H_3O^+]$ in this solution

Then $[OAc^-] = 1.0 + x$ and $[HOAc] = 1.0 - x$

If we assume that x is negligible compared to 1.0 because the acetate ions from the potassium acetate repress the ionization of acetic acid, the equilibrium constant expression becomes

$$K_a = \frac{[H_3O^+]}{[HOAc]} = \frac{[H_3O^+](1.0)}{1.0} = 1.8 \times 10^{-5}$$

Since 1.8×10^{-5} is negligible compared to 1.0, the $[H_3O^+] = 1.8 \times 10^{-5}$ in this buffer, and $p^H = pK_a = 4.74$.

In order to calculate how much the p^H changes when we add 0.5 mL of 0.2 NaOH, we must know how many millimoles of HOAc and OAc^- we had before the base is added.

No. mmol HOAc originally present = (10 mL) (1.0 *M*) = 10 mmol

No. mmol OAc^- originally present = (10 mL) (1.0 *M*) = 10 mmol

How many millimoles of OH^- are being added to this solution?

No. mmol OH^- added = (0.5 mL) (0.2 *M*) = 0.1 mmol

The OH^- added to the solution reacts with HOAc present in the buffer as follows

$$HOAc(aq) + OH^-(aq) \rightleftharpoons OAc^-(aq) + H_2O$$

This reaction proceeds essentially to completion so adding the OH^- serve to use up some of the HOAc and to increase the amount of OAc^- in the solution.

Since OH^- and HOAc combine in a 1 : 1 molar ratio, 0.10 mmol of OH^- combines with 0.1 mmol of HOAc and forms 0.1 mmol of OAc^-. Thus after the NaOH has been added, there are 10 – 0.1 = 9.9 mol of HOAc and 10 + 0.1 = 10.1 mmol of OAc^- present in the solution. These numbers should give you a feeling for what is meant by the requirement that the number of millimoles of weak acid and weak base in a buffer must be relatively large in order for the buffer to be effective at maintaining constant p^H. The 10 mmol of both HOAc and OAc^- originally present is large relative to the 0.1 mmol of OH^-added to the solution.

The volume of the solution in 10.5 mL after the NaOH has been adde, so that in the resultant solution

[HOAc]= (0.9 mmol)/(10.5 mL) = 0.943 *M*

and $[OAc^-]$= (10.1 mmol)/(10.5 mL) = 0.962 *M*

Again we can assume that the $[H_3O^+]$ is small compared to the [HOAc] and $[OAc^-]$, so that

$$[H_3O^+] = \frac{K_a[HOAc]}{[OAc^-]} = \frac{(1.8 \times 10^{-5})(0.943)}{0.962} = 1.76 \times 10{-}3$$

Therefore p^H = 5 – log 1.76 = 4.75. Note that the p^H has changed by only 0.01 unit after adding 0.5 mL of 0.2 *F* NaOH!

127. The weak acid in this buffer is NH_4^+, for which K_b is 5.7×10^{-10}. Its P^{K_a} is therefore 9.24,

$\therefore$ $p^H = P^{K_a} + \log(NH_3)/(NH_4^+) = 9.24 + \log([NH_3]/[NH_4^+])$

Setting the $p^H = 8.54$, we obtain

$$8.54 = 9.24 + \log([NH_3]/[NH_4^+])$$

or

$$\log([NH_3]/[NH_4^+,]) = 0.7$$

Thus the desired molar ratio is $[NH_3]/[NH_4^+,]$ = antilog(–0.7) = 0.2

(b) Propanoic acid is an acid closely related to acetic acid, with very similar properties. Note that K_a for propanoic acid is very close in value to K_a for acetic acid. The formula for propanoic acid is CH_3CH_2COOH, but we will denote it HOPr, using an abbreviation analogous to the HOAc used for acetic acid. In aqueous solution, propanoic acid is in equilibrium with hydronium ion and propanoate ion, the anion formed when the acid donates a proton to some base. The equilibrium reaction is:

$$HOPr(aq) + H_2O \rightleftharpoons H_3O^-(aq) + OPr^-(aq)$$

When OH^- ions are added to a solution containing HOPr, the reaction

$$OH^-(aq) + HOPr(aq) \rightleftharpoons H_2O + OPr^-(aq)$$

proceeds essentially to completion. Let us calculate how many millimoles of HOPr and how many millimoles of OH^- were mixed together to prepare this buffer.

$$\text{No. mmol HOPr} = (40 \text{ mL})\ (1.0 \text{ mmol} \cdot \text{mL}^{-1}) = 40 \text{ mmol}$$

$$\text{No. mmol OH}^- = (60 \text{ mL})\ (0.1 \text{ mmol} \cdot \text{mL}^{-1}) = 6 \text{ mmol}$$

Note that there is excess HOPr. The number of millimoles of HOPr is greater than the number of millimoles of OH^- so that when OH^- and HOPr react there will be HOPr remaining after the reaction.

Because the reaction between OH^- and HOPr proceeds virtually to completion, 6 mmol of OH^- react with 6 mmol of HOPr, forming 6 mmol of OPr^- (propanoate ion) and leaving 40.0 – 6 = 34 mmol of HOPr in excess. The total volume of solution is 100 mL. The solution is identical to a mixture of 34 mmol of propanoic acid and 6 mmol of sodium propanoate in enough water to make 100 mL of solution. Let

$$x = [H_3O^+] \text{ in this solution at equilibrium}$$

Then

$$[OPr^-] = \frac{(6 \text{ mmol})}{(100 \text{ mL})} + x = 0.06 + x - 0.06\ M$$

$$[HOPr] = \frac{(34 \text{ mmol})}{(100 \text{ mL})} - x = 0.34 - x = 0.34\ M$$

the propanoate ion concentration is slightly greater than 0.06 *M*because every time a hydronium ion is formed fro the proton-transfer reaction between HOPr and H a propanoate ion is also formed. We have assumed that the $[H_3O^+]$ in this solution is small relative to either the [HOPr] or the $[OPr^-]$ because the OPr^- ions from the sodium propanoate repress the ionization of HOPr.

The equilibrium constant equation for propanoic acid is

$$K_a = \frac{[H_3O^+][OPr^-]}{[HOPr]} = \frac{(x)(0.06)}{0.34} = 1.3 \times 10^{-5}$$

If we solve this equation for x we obtain

$$x = \left(\frac{34}{6}\right)\ (1.3 \times 10^{-5}) = 7.4 \times 10^{-5}$$

since 74×10^{-5} is small compared to 0.06, our assumption is justified. In this buffer the $[H_3O^+] = 7.4 \times 10^{-5}\ M$ and the $p^H = 4.13$. we can show that this is indeed an effective buffer solution by calculating the [conjugate base]/[weak acid] ratio.

$$\frac{[OPr^-]}{[HOPr]} = \frac{0.06}{0.34} = 0.176$$

which is within the limits of 10 and 0.1.

128. (a) The reaction that occurs when sodium acetate solution and HCl are mixed is described by Eq. (10.9b). We must therefore find out how many millimoles of acetate ion and how many millimoles of hydronium ion we are mixing together. The formula weight of sodium acetate, CH_3COONa, is 82.03 g · mol^{-1}. Thus we have added

$$\frac{2.461}{82.03\ \text{g}\cdot\text{mol}^{-1}} = 3 \times 10^{-2}\ \text{mol} = 30\ \text{mmol.}$$

of sodium acetate. The solution therefore contains 30.00 mmol of acetate ions.
The number of millimoles of H_3O^+ added is

$$(100\ \text{mL})\ (0.12\ \text{mmol}\cdot\text{mL}^{-1}) = 12\ \text{mmol}$$

Note that there is excess weak base.
Now, 12 mmol of H_3O^+ react with 12 mmol of CH_3COO^- to form 12 mmol of CH_3COOH. That reaction goes essentially to completion, and the number of millimoles of acetate ion in excess is 30 – 12 = 18 mmol.

The solution resulting after mixing the two reagents contains 12 mmol of acetic acid and 18 mmol of aetate ion in a total volume of 150 mL. There is no. significant contraction or expansion of volume on mixing two dilute aqueous solutions. The equilibrium

$$\text{HOAc(aq)} + H_2O \rightleftharpoons \text{OAc}^-\text{(Aq)} + H_2O^+\text{(aq)}$$

is of course established in this solution. Let, $x = [H_3O^+]$ at equilibrium

Then $$[\text{OAc}^-] = \frac{18\ \text{mmol}}{150\ \text{mL}} + x = 0.12 + x = 0.12\ M$$

and, $$[\text{HOAc}] = \frac{12\ \text{mmol}}{150\ \text{mL}} - x = 0.08\ M$$

We have made the assumption that x is small compared to 0.08 because we expect the $[H_3O^+]$ in this solution to be small relative to the [HOAc] and $[\text{OAc}^-]$. The equilibrium constant expression is therefore

$$K_a(\text{HOAc}) = \frac{[H_3O^+][\text{OAc}^-]}{[\text{HOAc}]} = 1.8 \times 10^{-5}$$

Solving this equation for x we obtain

$$x = \left(\frac{8}{12}\right)(1.8 \times 10^{-5}) = \frac{2}{3}(1.8 \times 10^{-5}) = 1.2 \times 10^{-5}$$

Since 1.2×10^{-5} is small compared to 0.080, our assumption is valid. That, $[H_3O^+] = 1.2 \times 10^{-5}\ M$, and $p^H = 4.92$

This is an effective buffer solution because the ratio

$$\frac{[\text{OAc}^-]}{[\text{HOAc}]} = \frac{0.12}{0.08} = 1.5$$

(b) If we add 5 mL of 0.120 *F* HCl to the buffer, we are adding (5.0 mL) (0.12 mmol · mL^{-1}) = 0.6 mmol of H_3O^+. This 0.6 mmol of H_3O^+ reacts with 0.6 mmol of OAc^-

forming 0.6 mmol of HOAc. Thus After the addition the solution will contain 12 + 0.6 = 12.6 mmol of HOAc and 18 – 0.6 = 17.4 mmol of OAc^- in a total volume of 155 mL.

We may assume that the $[H_3O^+]$ is small compared to the [HOAc] and $[OAc^-]$, so that

$$[OAc^-] = (17.4 \text{ mmol})\ (155.0 \text{ mL}) = 0.112\ M$$

and

$$[HOAc] = (12.6 \text{ mmol})/(155 \text{ mL}) = 0.0813\ M$$

The $[OAc^-]/[HOAc]$ ratio has changed from 0.12/0.08 = 150 before the addition of the 5 mL of HCl, to 0.112/0.813 = 1.38 after the HCl was added. The $[H_3O^+]$ is now

$$[H_3O^+] = \frac{(1.8 \times 10^{-5})(0.0813)}{0.112} = 1.3 \times 10^{-5}\ M$$

and the p^H is therefore 4.89

The p^H changed from 4.92 to 4.89 when 5 mL of 0.12 *F* HCl were added to the buffer. A p^H change of only 0.03 units upon the addition of a moderately large amount of strong acid is a very small p^H change, and shows how effective the buffer solution is at maintaining the p^H approximately constant.

129. We find the acidity constant of formic acid, HCOOH, is 1.8×10^{-4} (from tables). The P^{K_a} of formic acid is $-\log (1.8 \times 10^{-4}) = 3.74$, therefore,

$$p^H = 4 = 3.74 + \log \frac{[HCOO^-]}{[HCOOH]}$$

so $\log \frac{[HCOO^-]}{[HCOOH]} = 0.26$ and $\frac{[HCOO^-]}{[HCOOH]} = 10^{0.26} = 1.8$

We see, therefore, that in order to make a buffer with $p^H = 4$ we must make the $[HCOO^-]/[HCOOH]$ ratio equal to 1.8.

The net ionic equation for the reaction that occurs when HCOOH and NaOH solutions are mixed is

$$HCOOH(aq) + OH^-(aq) \rightleftharpoons HCOO^-(aq) + H_2O$$

Let *V* be the volume (in mL) of 2 *F* NaOh tht must be added to 100 mL of 2.5 *F* HCOOH to make the $[HCOO^-]/[HCOOH]$ ratio = 1.8. A sample of 2 *F* NaOH of volume *V* mL contains (2 *V*) mmol of OH^-. Since HCOOH and OH^- react in a 1 : 1 molar ratio, 2 *V* mmol of OH^- combine with 2 *V* mmol of HCOOH, and produce 2 *V* mmol of $HCOO^-$ ions. We began with 100 mL of 2.50 *F* HCOOH, or 25 mmol of HCOOH. Since 2 *V* mmol of HCOOH are used up in the reaction with OH^-, the amount of excess HCOOH is (250 – 2 *V*) mmol. Therefore,

$$\frac{\text{No. mmol } HCOO^- \text{ formed}}{\text{No. mmol HCOOH remainin}} = \frac{2\,V \text{ vmmol}}{(250 - 200\,V)\text{mmol}}$$

130. In 1 L of solution the number of grams of CaF_2 that can be dissolved is 10 times the number that can be dissolved in 100 mL ad is thus 1.7×10^{-2} g. The gram formula weight of CaF_2 is 78.08, so the number of moles of CaF_2 that can be dissolved per liter is

$$\frac{1.7 \times 10^{-2}\ g \cdot L^{-1}}{78.08\ g \cdot mol^{-1}} = 2.18 \times 10^{-4} = 2.2. \times 10^{-4} \text{ mol} \cdot L^{-1}$$

The equation for the dissolution of CaF_2 in water is $CaF_2(s) \rightleftharpoons Ca^{2+}(aq) + 2F^-(aq)$

Thus, $[Ca^{2+}] = 2.2. \times 10^{-4}\ M$ and $[F^-] = 2(2.18 \times 10^{-4}) = 4.36 \times 10^{-4}\ M$

$$K_{sp}(CaF_2) = [Ca^{2+}][F^-]^2 = (2.18 \times 10^{-4})\ (4.36 \times 10^{-4})^2 = 4.1 \times 10^{-11}$$

131. Begin by writing the equation for the reaction for which the equilibrium constant is the K_{sp} of $BaSO_4$.

$$BaSO_4(s) \rightleftharpoons Ba^{2+}(aq) + SO_4^{2-}(aq)$$

Remember that by convention the insoluble solid is written on the left-hand side of the equation. Barium sulfate is a strong electrolyte. Although its solubility is very low, 100% of the small amount that does dissolve is present in solution as the hydrated ions.

Let x be the molar solubility of $BaSO_4$ at 298 K. when x moles of $BaSO_4$ dissolve in enough water to make 1 L of solution, there are x moles of Ba^{2+} ions and x moles of SO_4^{2-} ions in that solution. Hence,

$$[Ba^{3+}] = [SO_4^{2-}] = x$$

$$K_{sp}[BaSO_4] = [Ba^{2+}][SO_4^{2-}] = x^2 = 1.1 \times 10^{-10}\ M^3$$

$$x = 1.0 \times 10^{-5}\ M$$

A saturated aqueous solution of $BaSO_4$ at 298 K is $1.0 \times 10^{-5}\ M$.

To convert the molar solubility to the gram solubility, we need to know the gram formula weight of $BaSO_4$, which is 233.4. Hence the number of grams of $BaSO_4$ that dissolve in 1 L of solution at 298 is

$$(233.4\ g \cdot mol^{-1})\ (1.0 \times 10^{-5}\ mol \cdot L^{-1}) = 2.3 \times 10^{-3}\ g\ L^{-1}$$

The solubility in grams per 100 mL of solution is therefore 2.3×10^{-4}, or 0.00023 g.

132. The molar ratios needed in order to express the concentrations of the ions in terms of the molar solubility are

$$\frac{\text{No. mol } Pb^{2+} \text{ in soluion}}{\text{No. mol } Pb(IO_3)_4 \text{ dissolved}} = \frac{1}{1}$$

and $$\frac{\text{No. mol } IO_3^- \text{ solution}}{\text{No. mol } Pb(IO_3)_2 \text{ dissolved}} = \frac{2}{1}$$

Let s = molar solubility of $Pb(IO_3)_2$ in water at 298 K

Then $[Pb^{2+}] = s$ and $[IO_3^-] = 2s$

$$K_{sp} = [Pb^{2+}]\ [IO_3^-] = (s)\ (2s)^2 = 2.6 \times 10^{-13}$$

We therefore obtain $4s^2 = 2.6 \times 10^{-13}$

To determine the power of 10 of the answer in your head, make the power of 10 divisible by 3 before taking the cube root. We then write

$$4s^3 = 2.5 \times 10^{-13} = 260 \times 10^{-15}$$

$$s^3 = 65 \times 10^{-15} \text{ and } s = 4.0 \times 10^{-5}\ M$$

We are only entitled to two significant figures in the answer since K_{sp} is only given to two figures. Thus 4×10^{-5} mol of solid $Pb(IO_3)_2$ dissolves per liter of solution in a saturated solution of $Pb(IO_3)_2$ in water, $[Pb^{2+}] = 4 \times 10^{-5}\ M$ and $[IO_3^-] = 8 \times 10^{-5}\ M$.

133. $La(OH)_3(s) \rightleftharpoons La^{3+}(aq) + 3OH^-(aq)$
Let, z = molar solubility of $La(OH)_3$ in water at 298 K
Then, $[La^{3+}] = z$ and $[OH^-] = 3z$

$\therefore \quad K_{sp} = [La^{3+}][OH^-]^3 = (z)\,(3z)^3 = 27z^4 = 1 \times 10^{-19}$

$\Rightarrow \quad z^4 = 3.7 \times 10^{-24}\ M^4$

or, $\quad z = 7.8 \times 10^{-6}\ M$

Extract the fourth root using your calculator by taking the square root twice. Since 7.8×10^{-6} mol of $La(OH)_3$ dissolve per liter of solution, in a saturated solution of $La(OH)_3$ in water:

$[La^{3+}] = 7.8 \times 10^{-6}\ M$ and $[OH^-] = 3(7.8 \times 10^{-6}\ M) = 2.3 \times 10^{-6}\ M$

134. The equation for dissolving AgCl is

$AgCl(s) \rightleftharpoons Ag^+(aq) + Cl^-(aq)$

In this solution there are two sources of chloride ion: The soluble NaCl is completely dissociated into Na^+ ions and Cl^- ions, and the small amount of AgCl that dissolves also provides Cl^- ions. The Cl^- ion is common to both the insoluble electrolyte, AgCl and the soluble salt, NaCl. There is, however, only one source of Ag^+ ions, namely the dissolution of solid AgCl. Let,

s = molar solubility of AgCl in 0.04 *F* NaCl at 298 K

Then $\quad [Ag^+] = s$ and $[Cl^-] = 0.04 + s$

$$K_{sp}\ (AgCl) = [Ag^+][Cl^-] = (s)\ (0.04 + is) = 1.8 \times 10^{-10}$$

Assume *s* is small compared to 0.04 an that 0.04 + *s* = 0.04.
The solubility product equation therefore simplities to

$$[Ag^+][Cl^-] = (s)\ (0.04) = 1.8 \times 10^{-10}$$

$$s = \left(\frac{1.8}{4}\right) \times 10^{-4} = 4.5 \times 10^{-9}\ M$$

Now; is 0.04 + *s* = 0.040? Yes, 0.04 + *s* = 0.040 + 0.0000000045 = 0.04. The value of *s* is too small to be significantly added to 0.04, so that our assumption is certainly justified. Only 4.5×10^{-9} mol of AgCl dissolve per liter of 0.04 *F* NaCl.

135. The equilibrium we have, $Hg_2Br(s) \rightleftharpoons Hg_2^{2+}(aq) + 2Br^-(aq)$

Note that mercurous ion is a diatomic cation

There are two sources of bromide ion in this solution, the soluble KBr and the insoluble Hg_2Br_2. Tere is only one source of mercurous ion, namely, the very small amount of Hg_2Br_2 that dissolves. Let, *y* = molar solubility of Hg_2Br_2 in 0.034 *F* KBr at 298 K

Then $[Hg_2^{2+}] = y$ and $[Br^-] = 0.034 + 2y$

The equilibrium constant expression is, therefore,

$$K_{sp}(Hg_2Br_2) = [Hg_2^{2+}][Br^-]^2 = (y)\ (0.034 + 2y)^2 = 1.3 \times 10^{-21}$$

This is a cubic equation in *y*. Let us assume that 2*y* is small compared to 0.034, so that 0.034 + 2*y* = 0.034. Note that the sum involves 2*y* and therefore out assumption must be about the relative magnitudes of the two terms in the sum namely. 2*y* and 0.034. With this assumption, the solubility product expression simplifies to:

$$y(0.034)^2 = 1.3 \times 10^{-21}$$

$$\Rightarrow \qquad y = \frac{1.3 \times 10^{-21}}{1.16 \times 10^{-3}} = 1.1 \times 10^{-18}$$

Now, is $2y = 2.2 \times 10^{-18}$ less than 10% of 0.034? Yes 2.2×10^{-18} is very much smaller than 3.4×10^{-3}, and therefore our assumption above is justified. The molar solubility of mercurous bromide in 0.034 *F* KBr is exceedingly small, only 1.1×10^{-18} *M* You should be able to prove that the solubility of Hg_2Br_2 in 0.034 *F* KBr is about 30 million times smaller than it is in pure water.

136. $CaSO_4(s) \rightleftharpoons Ca^{2+}(aq) + SO_4^{2-}(aq)$
Let, s = molar solubility of $CaSO_4$ in 8.0×10^{-3} $FCaCl_2$ at 298 K
Then, $[Ca^{2+}] = 8.0 \times 10^{-3} + s$ and $[SO_4^{2-}] = s$

$$K_{sp}(CaSO_4) = [Ca^{2+}]\,[SO_4^{2-}] = (8.0 \times 10^{-3} + s)\,(s) = 2.4 \times 10^{-5}$$

First approximation.
Assume, $8.0 \times 10^{-3} + s = 8.0 \times 10^{-3}$
The solubility product equation then simplifies to:

$$(8.0 \times 10^{-3}) \text{ is} = 2.4 \times 10^{-5} \text{ and hence } s = 3.0 \times 10^{-3}$$

Now, Is 3.0×10^{-3} less than 10% of 8.0×10^{-3}? No, 10% of 8.0×10^{-3} is 8.0×10^{-4} and $3 \times 10^{-3} > 8.0 \times 10^{-4}$. Therefore our assumption is not valid. We can, however, make a second approximation, which is better than the first one. We use the result of the first approximation to estimate the value of the sum $8.0 \times 10^{-3} + s$.

Second approximation.
Assume $8.0 \times 10^{-3} + s = 8.0 \times 10^{-3} + 3.0 \times 10^{-3} = 1.1 \times 10^{-2}$
The solubility product equation now becomes

$$[Ca^{2+}][SO_4^{2-}] = (1.1. \times 10^{-2})\,(s) = 2.4 \times 10^{-5}$$

and hence $\quad s = 2.2. \times 10^{-3}$

In the method of successive approximation we must continue making approximation until two approximation in succession give the same answer. This we now utilize the result of the second approximation to make a still better approximation to the sum $8.0 \times 10^{-3} + s$.

Third approximation.
Assume, $8.0 \times 10^{-3} + s = 8.0 \times 10^{-3} + 2.2. \times 10^{-3} = 1.02 \times 10^{-2}$
The solubility product equation is

$$(1.02 \times 10^{-2})s = 2.4 \times 10^{-5} \quad \Rightarrow s = 2.35 \times 10^{-3}$$

Fourth approximation.
Assume, $8.0 \times 10^{-3} + s = 8.0 \times 10^{-3} + 2.35 \times 10^{-3} = 1.04 \times 10^{-2}$
Then, $\quad (1.04 \times 10^{-2})s = 2.4 \times 10^{-5} \Rightarrow s = 2.31 \times 10^{-3}$

. The successive approximations are converting on the correct answer, which is therefore between 2.35×10^{-3} and 2.31×10^{-3} and consequently must be 2.3×10^{-3} to two figures. Note that the first approximation gave too large a result, the second too small, and so on. The first approximation yielded 3.0×10^{-3} *M*, which is at least the correct order of magnitude, although it is in error by more than 10%. The molar solubility of $CaSO_4$ in 0.08 *F* $CaCl_2$ is 2.3×10^{-3} mol · L^{-1}.

137. To answer a question like this, we must imagine the "instant of mixing", an instant before any reaction that may happen has had a chance to take place. We then calculate Q at the instant of mixing, and compare it with the equilibrium constant. Bear in mind that when two solutions are mixed, dilution always occurs. Because $NaNO_3$ is a soluble, strong electrolyte the only reaction than can possibly occur when these two solutions are mixed is the formation of insoluble BaF_2. Let us calculate both calculate both the $[Ba^{2+}]$ and $[F^-]$ at the instant of mixing.

No. mmol Ba^{2+} added = (70 mL) (0.05 mmol mL^{-1}) = 3.5 mmol

total volume after mixing = 100 mL

Hence at the instant of mixing

$$[Ba^{2+}] = \frac{\text{mmol } Ba^{2+} \text{ added}}{\text{total volume}} = \frac{(70)(0.05)}{100} = 0.035\ M$$

Note that we can consider the ratio of the initial volume of the solution containing Ba^{2+} ions to the final volume of the mixed solution a *dilution facto* and obtain the concentration at the instant of mixing by multiplying the initial concentration by the dilution factor. Similarly, the $[F^-]$ at the instant of mixing is given by

$$[F^-] = \frac{\text{mmol } I^- \text{ added}}{\text{total volume}} = \frac{(30)(0.02)}{100} = 0.006\ M$$

We are considering the reaction

$$BaF_2(s) \rightleftharpoons Ba^{2+}(aq) + 2F^-(aq) \qquad \ldots\ldots\ (1)$$

This is the reaction whose equilibrium constant is the K_{sp} of BaF_2. The reaction quotient for reaction (1) is

$$Q = [Ba^{2+}][F^-]^2 = (3.5 \times 10^{-2})\ (6 \times 10^{-3})^2 = 1.3 \times 10^{-6}$$

Since $Q < K_{sp}$ the reaction does *not* proceed to the left. The solution is unsaturated and no BaF_2 will form.

138. First let us calculate the $[Zn^{2+}]$ and $[C_2O_4^{2-}]$ at the instant of mixing

NO. mmol Zn^{2+} added = (15 mL) (0.12 mmol · mL^-) = 1.8 mmol

No. mmol $C_2O_4^{2-}$ added = (10 mL) (0.1 mmol · mL^{-1}) = 1 mmol

total volume after mixing = 25 mL

At the instant of mixing, therefore,

$$[Zn^{2+}] = \frac{1.8 \text{ mmol}}{25 \text{ mL}} = 0.072\ M$$

$$[C_2O_4^{2-}] = \frac{1.0 \text{ mmol}}{25 \text{ mL}} = 0.04\ M$$

The value of the reaction quotient at the instant of mixing is

$$Q = [Zn^{2+}]\ [C_2O_4^{2-}] = (7.2 \times 10^{-2})\ (4.0 \times 10^{-2}) = 2.9 \times 10^{-3}$$

Since $Q > K_{sp}(ZnC_2O_4)$, ZnC_2O_4 will precipitate. The precipitation reaction is

$$Zn^{2+}(aq) + C_2O_4^{2-}(aq) \rightleftharpoons ZnC_2O_4\downarrow$$

Because Zn^{2+} and $C_2O_4^{2-}$ combine in a 1 : 1 molar ratio, the reaction mixture has excess Zn^{2+} that cannot be precipitated, as we have added 1.8 mmol of Zn^{2+} but only 1.0 mmol of $C_2O_4^{2-}$, Virtuall all of the $C_2O_4^{2-}$ will be in the precipitate. First let us calculate the concentration of Zn^{2+} that is in excess,

No. mmol excess Zn^{2+} = 1.8 – 1.0 = 0.8 mmol

$$\text{excess}[Zn^{2+}] = \frac{0.8 \text{ mmol}}{25.0 \text{ mL}} = 0.032\ M$$

Let x [$C_2O_4^{2-}$] after the precipitation is complete. We know this will be a very small value, although it will not be zero. For every $C_2O_4^{2-}$ ion that is not in the precipitate but remains in solution, there is also a Zn^{2+} ion in solution. S the total $[Zn^{2+}]$ consists of the excess plus an amount equal to the $[C_2O_4^{2-}]$. Thus, $[Zn^{2+}] = 0.032 + x$

When equilibrium is attained, $[Zn^{2+}]\ [C_2O_4^{2-}] = 2.5 \times 10^{-9} = K_{sp}$

Thus, $(0.032 \times x) = 2.5 \times 10^{-9} = K_{sp}$

Assume x is small compared to 0.032. Then $0.032 + x = 0.032$, and the equation becomes simply $0.032 = 2.5 \times 10^{-9}$. We obtain

$$x = \frac{2.5 \times 10^{-9}}{3.2 \times 10^{-2}} = 7.8 \times 10^{-8}\ M$$

We must check our assumption is 7.8×10^{-8} small compared to 0.032? Yes, 7.8 <x. 10^{-8} is certainly much less than 10% of 3.2×10^{-2}. Hence,

$$[C_2O_4^{2-}] = 7.8 \times 10^{-8}\ M$$

$$[Zn^{2+}] = 3.2 \times 10^{-2} + 7.8 \times 10^{-8} = 3.2 \times 10^{-2}\ M$$

139. The equation for dissolving Mns in HCl is

$$MnS(s) + 2H_3O^+(aq) \rightleftharpoons Mn^{2+}(aq) + H_2S(aq) + 2H_2O \quad \text{....... (i)}$$

Manganous sulfide is pale pinkish-beige insoluble solid. The solubility product of MnS is 5×10^{-15} at 298 K, so that is solubility in water is extremely small. To find out whether it is possible to dissolve much MnS in strɔng acid, we must evaluate the equilibrium constant of Eq. (i)

$$K_{eq} = \frac{[Mn^{2+}][H_2S]}{[H_3O^+]^3} \quad \text{....... (ii)}$$

The equilibrium constant for any reaction involving an insoluble electrolyte depends, in some way, on the K_{sp} of that insoluble solid. Therefore this equilibrium constant must depend on K_{sp}(MnS). We introduce the $[S^{2-}]$ into the expression for K_{eq} by multiplying both the numerator and denominator and denominator of Eq. (ii) by $[S^{2-}]$

$$K_{eq} = \frac{9Mn^{2+}][H_2S]}{[H_3O^+]^2} \cdot \frac{[S^{2-}]}{[S^{2-}]} = [Mn^{2+}][S^{2-}] \frac{[H_2S]}{[H_3O^+]^2[S^{2-}]} \quad \text{....... (iii)}$$

The equilibrium constant for the overall two stages of ionization of H_2S is

$$K_{overall} = K_a(H_2S) \cdot K_a(HS^-) = \frac{[H_3O^+]^2[S^{2-}]}{[H_2S]}$$

The reciprocal of this expression appears in the K_{eq} we are trying to evaluate, Eq. (iii) Thus we have

$$K_{eq} = K_{sp}(\text{MnS}) \cdot \frac{1}{K_a\,(H_2S) \cdot K_a\,(HS^-)} = \frac{5 \times 10^{-15}}{(1 \times 10^{-7})(1.3 \times 10^{-13})} = 4 \times 10^5 \qquad \text{(iv)}$$

Since the value of the equilibrium constant for Eq. (i) is much greater than 1, the position of equilibrium for reaction (i) is to the right. this tells us that it is indeed possible to dissolve manganous sulfide by adding a strong acid such as HCl.

140. The first thing to do in answering a question like this is to think about the reason why $BaSO_4$ is more soluble in HCl than in pure water. Use Le Chatelier's Principle (to dissolve insoluble solids). When $BaSO_4$ dissolves in water, the two ion present in solution are Ba^{2+} and SO_4^{2-} ions. If we decrease the $[SO_4^{2-}]$ by adding H_2O^+ ions to form HSO_4^- ions:

$$H_3O^+(aq) + SO_4^{2-}(aq) \rightleftharpoons HSO_4^- + H_2O \qquad \text{.......(i)}$$

we drive the dissolution of $BaSO_4$

$$BaSO_4(s) \rightleftharpoons Ba^{2+}(aq) + SO_4^{2-}(aq) \qquad \text{.......(ii)}$$

to the right. Thus the overall reaction for the dissolution of $BaSO_4$ in strong acid is the sum of reaction (i) and (ii), namely,

$$BaSO_4(s) + H_3O^+(aq) \rightleftharpoons Ba^{2+}(aq) + HSO_4^-(aq) + H_2O \qquad \text{.......(iii)}$$

To find out whether reaction (iii) proceeds very far to the right, we msut evaluate in equilibrium constant,

$$K_{eq} = \frac{[Ba^{2+}][HSO_4^-]}{[H_3O^+]} \qquad \text{.......(iv)}$$

No table of equilibrium constant contains the value of K_{eq} desired. We can, however, readily calculate it form the solubility product of $BaSO_4$ and the acidity constant of SO_4^-, (from Tables)

$$K_{sp}(BaSO_4 = [Ba^{2+}][SO_4^{2-}] = 1.1 \times 10^{-10}$$

$$K_a(HSO_4^-) = \frac{[H_3O^+][SO_4^{2-}]}{[HSO_4^-]} = 1.2 \times 10^{-2}$$

One way to approach this problem is to utilize the fact that Eq. (ii) is the sum of Eqs. (i) and (ii). The equilibrium constant for reaction (i) is

$$K = \frac{[HSO_4^-]}{[H_3O^+][SO_4^{2-}]} = \frac{1}{K_a\,(HSO_4^-)}$$

The equilibrium constant for reaction (ii) is the K_{sp} of $BaSO_4$. Now, if an overall reaction is the algebraic sum of two other equations, denoted (i) and (ii) then the equilibrium constant for the overall reaction is the product of the equilibrium constants of reaction (i) and (ii).

$$\therefore\ K_{eq} = \frac{[Ba^{2+}][HSO_4^-]}{[H_3O^+]} = \frac{[HSO_4^-]}{[H_3O^+\,[\cancel{SO_4^{2-}}]} \cdot [Ba^{2+}]\,[\cancel{SO_4^{2-}}] = \frac{1}{K_a(HSO_4^-)}\ K_{sp}(BaSO_4)$$

$$= \frac{1.1 \times 10^{-10}}{1.2 \times 10^{-2}} = 9.2 \times 10^{-9}$$

Thus the value of the equilibrium constant 9.2×10^{-9}.

We see that this value is very small compared to 1. this cells us that not much $BaSO_4$ is going to dissolve in HCl or any other strong acid

In order to calculate how many grams of $BaSO_4$ dissolve in 1 L of 2.0 *F* HCl, we must consider the molar ratios specified in Eq. (iii). If x moles of solid $BaSO_4$ dissovle per liter, then

$$[Ba^{2+}] = [HSO_4^-] = x$$

The $[H_3O^+]$ was originally 2.0 *M* but x moles per liter are used up in forming HSO_4^-. thus $[H_3O^+] = 2.0 - x$. The equilibrium constant expression then becomes

$$K_{eq} = 9.2 \times 10^{-9} = \frac{[Ba^{2+}][HSO_4^-]}{[H_3O^+]} = \frac{x^2}{2-x}$$

assume x is small compared to 2. Then $2 - x = 2$, and

$$x^2 = 2(9.2 \times 10^{-9}) = 18.4 \times 10^{-9} = 1.84 \times 10^{-8}$$

$$x = 1.36 \times 10^{-4} = 1.4 \times 10^{-4}\ M$$

Now, Is 1.4×10^{-4} less than 10% of 2.0? Yes, it is considerably less than 0.2, so our assumption is justified. The gram formula weight of $BaSO_4$ is 233.4, so the mass of $BaSO_4$ that dissolves in 1 L of 2.0 *F* HCl is

$$(1.36 \times 10^{-4}\ \text{mol L}^{-1})\ (233.4\ \text{g} \cdot \text{mol}^{-1}) = 3.2 \times 10^{-2}\ \text{g} \cdot \text{L}^{-1}$$

Thus 0.032 g of $BaSO_4$ dissolve per liter of 2.0 *F* HCl

141. There are two methods of approaching this problem and we will consider both of them.

Method. First let us calculate the $[S^{2-}]$ in the solution at the instant of mixing that is, before any MnS can precipitate. By equilibrium constant for the two stages of ionization of H_2S we have,

$$\frac{[H_3O^+][S^{2-}]}{[H_2S]} = K_a(H_2S) \cdot K_a(HS^-) = 1.3 \cdot 10^{-20}$$

Substituting $[H_3O^+] = 0.3$ and $[H_2S] = 0.1$ into this equation we obtain

$$[S^{2-}] = \frac{(1.3 \times 10^{-20})(0.10)}{(0.3)^2} = \frac{1.3 \times 10^{-21}}{9 \times 10^{-2}} = 1.4 \times 10^{-20}$$

Now, calculate the reaction quotient Q at the instant of mixing for the reaction

$$MnS(s) \rightleftharpoons Mn^{2+}(aq) + S^{2-}(aq)$$

and compare it with the K_{sp} of MnS, which is 5×10^{-15}

$$Q = [Mn^{2+}]\ [S^{2-}] = (0.2)\ (1.4 \times 10^{-20}) = 2.8 \times 10^{-21}$$

Since the value of Q is much smaller than K_{sp}(MnS), the solution is under saturated with respect to Mn^{2+} and S^{2-} ions, and o MnS will precipitate.

Method 2. Consider the overall reaction for the precipitation of MnS in this souion:

$$Mn^{2+}(aq) + H_2S(aq) \rightleftharpoons MnS(s) + 2H^+(aq)$$

To find out whether this reaction proceeds to the right with the concentrations specified, we must compare Q for this reaction with the equilibrium constant for this reaction. First we evaluate the equilibrium constant, K_{eq}.

$$K_{eq} = \frac{[H^+]^2}{[Mn^{2+}][H_2S]} = \frac{[H^+][S^{2-}]}{[H_2S]} \cdot \frac{1}{[Mn^{2+}][S^{2-}]} = \frac{K_a(H_2S) \cdot K_b(HS^-)}{K_{sp}(MnS)}$$

$$= \frac{1.3 \times 10^{-20}}{5 \times 10^{-15}} = 2.6 \times 10^{-6}$$

At the instant of mixing, $Q = \frac{[H^+]^2}{[Mn^{2+}][H_2S]} = \frac{(0.3)^2}{(0.2)(0.1)} = 4.5$

Since $Q > K_{eq}$, the only possible reaction is the one that makes Q smaller, that is the backward reaction. If there were any solid MnS present, it would dissolve. No reaction occurs because there is nothing that can happen to decrease Q. Certainly the reaction will not proceed to the right, and therefore no MnS will be formed.

Both methods of approach to this problem of course lead to the same conclusion : No MnS will precipitate from a solution that is saturated with H_2S and has $[H_3O^+] = 0.3\ M$ and $[Mn^{2+}] = 0.1\ M$.

142. The overall reaction that occurs on dissolving Mns in water is

$$MnS(s) + H_2O \rightleftharpoons Mn^{2+}(aq) + HS^-(aq) + OH^-(aq)$$

$$S^{2-}(aq) + H_2O \rightleftharpoons HS^-(aq) + OH^-(aq)$$

and of $\quad MnS(s) \rightleftharpoons Mn^{2+}(aq) + S^{2+}(aq) \quad \ldots\ldots \text{(i)}$

Now, $\quad K_b(S^{2-}) = \frac{[HS^-][OH^-]}{[S^{2-}]} = \frac{K_w}{K_a(HS^-)} = \frac{1.0 \times 10^{-14}}{1.3 \times 10^{-13}} = 7.7 \times 10^{-7}$

The equilibrium constant for Eq. (i) is the solubility product of MnS

$$K_{sp}(MnS) = [Mn^{2+}][S^{2-}] = 5 \times 10^{-15}$$

Also, $\quad K_{eq} = K_{sp}(MnS) \cdot K_b(S^{2-}) = (7.7 \times 10^{-2})(5 \times 10^{-15}) = 3.8 \times 10^{-16}$

(Product of two equilibrium constants)

Let $\quad x$ = molar solubility of MnS in water

Then $x = [Mn^{2+}] = [HS^-] = [OH^-]$ according to the overall reaction.

$\therefore\ x^3 = 3.8 \times 10^{-16} \Rightarrow x = 7.2 \times 10^{-6}\ M$

143. (a) We first calculate how large a concentration of carbonate ion is required to precipitate 99.9% of the Ca^2 ions present. *Remember*. If the ions are not part of the precipitate, they are still in solution! Since 99.9% of all the Ca^{2+} is in the precipitate 0.1% remains in solution. We began with 0..08 M Ca^{2+}. Hence the $[Ca^{2+}]$ remaining in solution after the precipitation of the $CaCO_3$ is 0.1% of the 0.08 M, or

$$[Ca^2 = (1 \times 10^{-3})(8.0 \times 10^{-2}) = 8.0 \times 10^{-5}\ M$$

What must the $[CO_3^{2-}]$ be, in order for the precipitate of $CaCO_3$ to be in equilibrium with a solution with $[Ca^{2+}] = 8 \times 10^{-5}\ M$? Since the precipitate and the solution are in equilibrium,

$$[Ca^{2+}][CO_3^{2-}] = K_{sp}(CaCO_3) = 4.8 \times 10^{-9}$$

and hence, $[CO_3^{2-}] = \frac{4.8 \times 10^{-9}}{8.0 \times 10^{-5}} = 0.6 \times 10^{-4} = 6 \times 10^{-5}$

Thus the $[CO_3^{2-}]$ must be maintained at 6×10^{-5} M in order to precipitate 99.9% of the Ca^{2+} originally present in solution.

To determine whether or not any Mg^{2+} will precipitate at this $[CO_3^{2-}]$, we evaluate Q for the equilibrium

$$MgCO_2(s) \rightleftharpoons Mg^{2+}(aq) + CO_3^{2-}(aq)$$

$$Q = [Mg^{2+}]\,[CO_3^{2-}] = (0.06)\,(6 \times 10^{-3}) = 3.6 \times 10^{-6}$$

Since Q is smaller that the K_{sp} of $MgCO_3$, which is 4×10^{-5}, no $MgCO_3$ will precipitate. It is possible, therefore to precipitate 99.9% of the Ca^{2+} in this solution at $CaCO_3$ and not precipitate any of the Mg^{2+}.

(b) No $MgCO_3$ will precipitate until Q for $MgCO_3$ is equal to the K_{sp} of $MgCO_3$. The $[CO_3^{2-}]$ at which the very first speck of $MgCO_3$ will precipitate is, therefore,

$$[CO_3^{2-}] = \frac{K_{sp}(MgCO_3)}{[Mg^{2+}]} = \frac{4 \times 10^{-5}}{6 \times 10^{-2}} = 6.7 \times 10^{-4}\ M = 7 \times 10^{-4}\ M$$

The upper limit on the $[CO_3^{2-}]$ we can allow in this solution is 6.7×10^{-4} I. as long as the $[CO_3^{2-}]$ does not exceed that value, no $MgCO_3$ will precipitate.

These calculation tell us that an analytical separation of Mg^{2+} and Ca^{2+} is possible if there is a way to control the $[CO_3^{2-}]$ in the solution so that it is at least 6.0×10^{-5} M but never exceed 6.7×10^{-4} M.

(c) If we maintain the $[CO_3^{2-}]$ at 6.7×10^{-4} M, the $[Ca^{2+}]$ in solution at equilibrium is

$$[Ca^{2+}] = \frac{K_{sp}(CaCO_3)}{[CO_3^{2-}]} = \frac{4.8 \times 10^{-9}}{6.7 \times 10^{-4}} = 7.2 \times 10^{-6}\ M$$

Thus the fraction of the Ca^{2+} ion originally present that is left unprecipitated is $(7.2 \times 10^{-6})/(8.0 \times 10^{-2})$ or 0.009%.

Chapter 10 : ELECTRO CHEMISTRY

1. m = mass of silver = 50 g.

M = mass of 1 mole silver = 107.9 g.

Z = valency = 1 ($Ag^+ + e^- \rightarrow Ag$)

F = 96500 coulombs

We know that $m/M = Q/ZF$ or

$$Q = mzF/M = \frac{50 \times 1 \times 96500}{107.9} = 4.5 \times 10^4 \text{ columbs.}$$

Therefore to deposit 50 g of Ag at cathode, the number of coulombs required is 4.5×10^4

2. The reaction taking place at electrode is $2Cl^- \rightarrow Cl_2(g) + 2e^-$

$Q = 1 \times t = 100$ (ampere) $\times 5 \times 60 \times 60$ (seconds) $= 1.8 \times 10^6$ coulombs.

M = molecular weight of chlorine = 70.9

$Z = 2$

From the equation $\frac{m}{M} = \frac{Q}{ZF}$; $m = \frac{M \times Q}{ZF} = \frac{70.9 \times 1.8 \times 10^6}{2 \times 96500}$ g.

We know that 1 mole of any gas at STP occupies 22.4 lit.

i.e. 70.9g of Cl_2 at S.T.P. occupies 22.4 litres.

∴ Volume of m g of chlorine at STP occupies $22.4 \times m/70.9$ litres.

$$= \frac{22.4}{70.9} \times \frac{70.9 \times 1.8 \times 10^6}{2 \times 96500} = 208.9 \text{ litres}$$

Volume of chlorine liberated at STP is 208.9 litres

3. The number of equivalents of different substances discharged by the same quantity of electricity are equal.

Number of equivalents of silver = Number of equivalents of hydrogen.

Mass of Ag/Mass of H_2 = Eq. wt. of Ag/Eq. wt. of H_2 :

0.54/Mass of H_2 = 108/1

∴ Mass of Hydrogen liberated = 0.54 × 1/108 = 0.005 gm.

The mass can be converted into volume by using the equation

$PV = (W/M)\,RT$ or $V = (WRT)/PM = (0.005 \times 0.0821 \times 300 \times 760)/(2 \times 750) = 0.06234$ lit or = 62.34 ml.

4. The electrode reaction for the discharge of Fe^{2+} ions from

$$Fe^{2+} + 2e^- \rightarrow Fe$$

Therefore, gram-equivalent weight of Fe in Fe^{2+} = 56/2 = 28 g.

1 faraday of electricity will deposit 28 g. of Fe

3 faradays of electricity will deposit = 3 × 28 = 84 g of Fe

5. Amount of Cu likely to be deposited when 1.307 g. of Ag are deposited, is given by the equation: Wt. of Cu deposited/Wt of Ag deposited = Eq wt of Cu/Eq wt of Ag

or Weight of Cu deposited or removed from solution = (1.307 × 63.5) (108 × 2) = 0.384 g.

∴ Weight of Cu originally present = 63.5 × 10/249.5 = 2.545 g/500 ml.

[Gram mol. wt of $CuSO_4 \cdot 5H_2O$, i.e. 249.5 g contains 63.5 g of Cu]

Wt of Cu left after electrolysis = 2.545 – 0.3842 = 2.1608 g/500 mol ml = 4.3216 g/litre.

6. According to Faraday' second law of electrolysis;

$$\frac{m_1}{m_2} = \frac{E_1}{E_2} \Rightarrow \frac{10\ g}{6.08\ g} = \frac{108}{E_2}$$

or $$E_2 = \frac{108 \times 6.08}{10} = 65.66$$

Since, charge on cation = $\frac{\text{At. wt.}}{\text{Eq. wt.}} = \frac{197}{65.66} = 3$

7. We know that, w = Z.i.t. = $\frac{108}{96500} \times 8.46 \times 8 \times 60 \times 60 = 272.684$ g

∴ Mass of Ag deposited = 272.684 g

∵ Mass= Volume × Density = Area × thickness × Density

or Area = $\frac{\text{Mass}}{\text{Thickness} \times \text{Density}} = \frac{272.684\ \text{g}}{0.00254\ \text{cm} \times 10.5\ \text{g cm}^{-3}} = 10224.37\ \text{cm}^2$

∴ Area of the tray = **10224.37 cm²**

8. (i) Aluminium is liberated by the reaction

$$Al^{3+} + 3e^- \rightarrow Al$$

Hence, gram eq. wt of Al = atomic wt./3 = 27/3 = 9.g

27.0 g of Al = 27/9 × 1 gm. eq. wt. = 3g equivalents of Al

(ii) Since at. wt. of Al is 27, 27 gm of Al = 1 gram atom of Al.

(iii) No. of atoms in 1 gram atom of Al = 6.02×10^{23}

(iv) An Al atom is produced when an Al^{3+} ion takes up 3 electrons. Total no. of electrons involved = $3 \times 6.02 \times 10^{23} = 18.06 \times 10^{23}$

(v) 1 g. eq. wt. of Al is liberated by one faraday of electricity.

∴ 3 g eq. wt. of Al is liberated by 3 faradays of electricity.

(vi) Total quantity of electricity used up = 0.5 × t sec = 0.5t coulombs.

Actual quantity of electricity used up = 3 faradays = 3 × 96500 coulombs

0.5 t coulombs= 3 × 96500 C

or, t = 3 × 96500/0.5 = 579000 sec = 160 hr 50 min

(vii) Since 3 g. eq. wt. of Al are produced, the quantity of chlorine produced will also be 3 g equivalents.

1 g. eq. wt. of chlorine = 11.2 litres of chlorine at STP

∴ 3 g eq. wt. of chlorine = 33.6 litres of chlorine at STP

9. $Cu^{2+} + 2e^- \rightarrow Cu$

For the deposit of gram atomic weight or 1 mole of Cu atoms, 2 faradays or 96500 × 2 = 19300 coulombs are required

Coulombs used = 0.5 × 30 × 60 = 900

Now, 900 coulombs deposit (w) = 0.2964 g

193000 coulombs will deposit = $\frac{0.2964 \times 193000}{900} \times 63.56$

Hence, the atomic wt. = 63.56

10. Faradays of electricity passed = 1 × 965/96500 = 0.01
Equivalents of bromine formed = 0.01.
∴ Amount of bromine formed = 0.01 × 80 = 0.80 g
Thus, Cathode half-reaction $Cu^{2+} + 2e^- \rightarrow Cu$
and, Anode half-reaction $2Br - 2e \rightarrow Br_2$

11. Coulombs of electricity passed = 3.7 × 6 × 60 × 60 = 7.99×10^4
Equivalents of nickel nitrate decomposed or Ni^{2+} discharged at the electrode
= $(7.99 \times 10^4)/(96500 \times 2)$ = 0.8282
No moles discharged at electrodes = 0.8282/2 = 0.414
[1 mole of nickel nitrate = 2 equivalents]
Moles of nickel nitrate present in solution before electroysis = 0.5 × 2 = 1.0 mole
Moles of nickel nitreate left after electrolysis = 1.0 – 0.414 = 0.685.
Moles of nickel nitrate left after electrolysis = 1.0 – 0.414 = 0.685.
0.586 moles being present in 0.5 litres of the solution, the molarity of the solution after electrolysis = 0.586 × 2 = 1.172 M.

12. (a) Cd/Cd^{2+} (1 M || Cl^- (1 M), AgCl(s)/Ag

left hand side electode is anode
Right hand side electrode is cathode

Oxidation occurs at anode $\qquad Cd \rightarrow Cd^{2+} + 2e$

reduction occurs at cathode $\qquad [AgCl + e^- \rightarrow Ag(s) + Cl^{-1}] \times 2$

Total reaction $\qquad Cd + 2AgCl \rightarrow Cd^{2+} + 2Ag(s) + 2Cl^-$

(b) Zn | Zn^{2+} (1 M) || Cl^- | Cl_2, Pt.

Left hand side electrode is anode
Right hand side electrode is cathode

Oxidation occurs at anode: $\qquad Zn \rightarrow Zn^{2+} + 2e^-$

Reduction occurs at cathode: $\qquad Cl_2 + 2e^- \rightarrow 2Cl^-$

Total reaction $\qquad Zn + Cl_2 \rightarrow Zn^{2+} + 2Cl^-$

(c) Pl, H_2 (1 atm) | HCl (0.1 M) | Cl_2, Pt

Oxidation occurs at anode: $\qquad H_2 \rightarrow 2H^+ + 2e^-$

reduction at occur cathode: $\qquad Cl_2 + 2e^- \rightarrow 2Cl^-$

Total reaction $\qquad H_2 + Cl_2 \rightarrow 2H^+ + 2Cl^-$

13. (a) $Fe + Cu^{2+} = Cu + Fe^{2+}$

Oxidation reaction $\qquad Fe \rightarrow Fe^{2+} + 2e^-$

Reduction reaction $\qquad Cu^{2+} + 2e^- \rightarrow Cu$

Anode half cell is $\qquad Fe \mid Fe^{2+}$

Cathode half cell is $\qquad Cu \mid Cu^{2+}$

∴ Cell notation is $\qquad Fe \mid Fe^{2+} \parallel Cu^{2+} \mid Cu$

(b) $Cu + 2Ag^+ = 2Ag + Cu^{++}$

Oxidation reaction $\qquad Cu \rightarrow Cu^{2+} + 2e^-$

Reduction reaction $\qquad Ag^+ + e^- \rightarrow Ag$

Anode hall cell is $Cu \mid Cu^{2+}$

Cathode half cells is $Ag \mid Ag^{+}$

∴ Cell notation is $Cu \mid Cu^{2+} \parallel Ag^{+} \mid Ag$.

(c) $6Fe^{2+} + Cr_2O_7^{2-} + 14H^{+} \rightarrow 6Fe^{3+} + 2Cr^{3+} + 7H_2O$

Oxidation Reaction $Fe^{2+} \rightarrow Fe^{3+} + 1e^{-}$

Reduction Reaction $Cr_2O_7^{2-} + 14H^{+} + 16e^{-} \rightarrow 2Cr^{3+} + 7H_2O$

Anode half cell $Pt \mid Fe^{2+}, Fe^{3+}$

Cathode half cell $Pt \mid Cr^{3+}, Cr_2O_7^{2-}, H^{+}$

Cell representation $Pt \mid Fe^{2+}, Fe^{3+} \parallel H^{+}, Cr_2O_7^{2-}, Cr^{3+} \mid Pt$.

14. (a) $Cu \mid Cu^{2+} \parallel Ag^{+} \mid Ag$ $E^{\circ}_{Cu^{2+}/Cu} = 0.34$ V; $E^{\circ}_{Ag^{-}/Ag} = 0.8$ V

anode cathode

$E_{cell} = E_R - E_L = 0.8 - 0.34 = 0.46$ V

(b) $Zn \mid Zn^{2+} \parallel Fe^{2+} \mid Fe$

Electrode reduction potentials $E^{2+}_{Fe/Fe} = -0.44$ V, $E^{\circ}_{Zn^{2+}/Zn} = -0.76$ V

$E_{cell} = E_R - E_L = -0.44 - (-0.76) = -0.44 + 0.76 = 0.32$ V

15. $E^{\circ}_{Ag^{+}/Ag} = 0.8$ V; $E^{\circ}_{Zn^{2+}/Zn} = -0.76$ V

$E_{cell} = E_R - E_L = E^{\circ}_{Zn^{2+}/Zn} - E^{\circ}_{Ag^{+}/Ag} = -0.76 - 0.8 = -1.56$ V

Since E is negative, it will not produce electricity.

16. Since in the reaction,

$$2Cl^{-}(aq) + 2H_2O(l) \rightarrow 2OH^{-}(aq) + H_2(g) + Cl_2(g)$$

Cl^{-} is oxidized to Cl_2, hence the reactions are as follow:

Anode: $2Cl^{-}(aq) \rightarrow Cl_2(g) + 2e^{-}$

Cathode: $2H_2O(l) + 2e^{-} \rightarrow 2OH^{-}(aq) + H_2(g)$

Number of moles of Cl_2 to be produced $= \dfrac{1000\text{ g}}{71\text{ g mol}^{-1}} = 14.08$ mole

Electricity required to produce 14.08 mole of $Cl_2 = 2 \times 14.08 \times 96500$ C

Since current efficiency is 62%, hence 25 amp. current is equivalent to $25 \times 0.62 = 15.5$ amp.

We know that $Q = i.t$

or $$t = \frac{Q}{i} = \frac{2 \times 14.08 \times 96500\text{ C}}{15.5\text{ A}} = 1.75300\text{ s} = \frac{175300\text{ s}}{60 \times 60\text{ s hr}^{-1}} = 48.69\text{ hours}$$

Since on the production of 1 mole $Cl_2(g)$ 2 moles $OH^{-}(aq)$ are produced hence the molarity of $OH^{-}(aq)$ ions

$$= \frac{2 \times 14.08\text{ mole}}{20\text{ L}} = 1.408\text{ M}$$

17. (i) $Ni_2O_3 + Fe \rightarrow 2NiO + FeO$
this is obtained by **subtracting** the second half cell reaction from the first.
(ii) **E = 1.27 V**
This is obtained by subtracting II half cell from the 1 half cell. Since the cell reaction is independent on OH^- ions therefore e.m.f. cell is certainly independent on KOH concentration.
(iii) Max. energy that can be released $= -\Delta G = nFE$
$= 2 \times 1.27 \times 96500 = 245.11$ kJ

18. $Cd + Cu^{2+} = Cd^{2+} + Cu$

$E_{Cd^{2+}/Cd} = -0.403$ V $\qquad E_{Cu^{2+}/Cu} = 0.134$ V

Oxidation reaction is $Cd \rightarrow Cd^{2+} + 2e$ (anode)
Reduction reaction is $Cu^{2+} + 2e^- \rightarrow Cu$ (cathode)
$E_R - E_L = E_{cathode} - E_{anode} = 0.34 - (0.403) = +0.743$

19. Wattage of the bulb = 200 W; Voltage of the bulb = 110 V
Current = Wattage/Voltage = 200/110 amp; Time = 30 min. = 1800 sec.
Quantity of electricity, Ct = (200 × 1800)/110 coulombs = 36000/11 coulombs
The relevant reactions are $ZnCl_2 \rightarrow Zn^{2+} + 2Cl^-$

$$Zn^2 + 2e^- \rightarrow Zn$$

Eq. wt. of Zn = 65.4/2 = 32.7

96500 coulombs deposit 32.7 g of Zn
36000/11 coulombs deposit (32.7/96500) × (36500/11) = 1.109 g. of Zn

20. (a) The reaction is $1/2\ H_2 \rightarrow H^+ + e^-$ or $_2 \rightarrow 2H^- + 2e^-$

$$E = E^\circ_{H_2} - \frac{0.059}{n}\log\frac{(a_{H^+})}{p_{H_2}} = 0 - \frac{0.059}{2}\log\frac{0.02\times0.02}{2} = 0.109 \text{ V}$$

(b) The reaction is $Fe \rightarrow Fe^{2+} + 2e^-$

$$E = E^\circ_{Fe^-} - \frac{0.059}{2}\log(a_{Fe^{2+}}) = 0.44 - \frac{0.059}{2}\log\frac{0.1}{1} = 0.4695 \text{ V}$$

(c) The reaction is $2Cl^- \rightarrow Cl_2 + 2e^-$

$$E = E^\circ_{Cl_1} - \frac{0.059}{n}\log\frac{(PCl_2)}{(a_{Cl^-})^2}$$

$$E = E^\circ - \frac{0.059}{2}\log\frac{10}{0.1\times0.1} = -1.36 - \frac{0.059}{2}\log\frac{10}{0.1\times0.1} = -1.4485 \text{ V}$$

21. (a) $E = E^\circ - RT/nF \ln a^{2+}_{Zn}/Zn$

$$= 0.763 - \frac{0.059}{2}\log a_{Zn^{2+}} = 0.763 - \frac{0.059}{2}\log(0.025)$$

$= 0.763 - (-0.0471) = 0.810$ V.

(b) E_{cell} = S.O.P. of Zn electrode – S.O.P. of Cu electrode = 0.763 + 0.34 = 1.103 v.

$$E = E^\circ - (2.303/nF)\log a_{Zn^{2+}} / a_{Cu^{2+}}$$

$= 1.103 - (0.059/2) \log 0.5/0.2 = 1.103 - 0.0117 = 1.088$ volts.

22. Total coulombs used = Ct = 5 × 60 × 60 × 2 = 3.6×10^4

Coulombs used for electroplating (70%) = $(3.6 \times 10^4 \times 70/100) = 2.52 \times 10^4$

Eq. wt. of Ni = 58.71/2 = 29.35

(a) 9.65×10^4 C deposits 29.35 g. Ni

2.52×10^4 C deposits $(29.35 \times 2.52 \times 10^4)/(9.654 \times 10^4) = 7.66$ g.

(b) Volume of Ni deposited = mass/density = 7.66/8.9 = 0.86 cm^3

Area of cathode plated = 2 × 4 × 4 = 32.Sq. cm.

(Since cathode plate is electroplated on both faces)

Thickness of plating = volume/Area = 0.86/32 = 0.2687 cm.

23. Current efficiency = Wt. of metal deposited in experiment/Theoretical wt. that can be deposited

Theoritical wt. CtE/F = 3 × 2 × 3600 × 31.75/96500 = 7.107 g

Current efficiency = 3/7.107 = 0.422 = 42.2%

24. According to second law of electrolysis

$$\frac{m_1}{E_1} = \frac{m_2}{E_2} = \frac{m_3}{E_3} = \frac{m_4}{E_4}$$

$$\Rightarrow \quad \frac{m_{Ag}}{E_{Ag}} = \frac{m_{Al}}{E_{Al}}; \text{ or } \frac{0.36}{108} = \frac{m_{Al}}{9}$$

or $\quad m_{Al} = 0.03$ g.

∴ 0.03 g Al will deposit in beaker A and B separately.

$$\text{Similarly,} \quad \frac{m_{Ag}}{E_{Ag}} = \frac{m_{OH}}{E_{OH}} \quad \text{or} \quad \frac{0.36}{108} = \frac{m_{OH}}{17}$$

$$\Rightarrow \quad m_{OH} = \frac{17}{300} \text{ g}$$

$$\text{or No. of moles of } OH^- = \frac{17}{300} \times \frac{1}{17} = \frac{1}{300} \text{ mole}$$

∵ $\quad 4OH^- \rightarrow 2H_2O + O_2$

∵ $4OH^-$ mole will liberate 22.4 L O_2 at N.T.P.

$$\therefore \frac{1}{300} OH^- \text{ mole liberate} = \frac{22.4}{4} \times \frac{1}{300} \text{ L} = 1.866 \times 10^{-2} \text{ L } O_2$$

25. Fe | Fe^{++} || Sn^{++} | Sn

$$E^\circ_{cell} = E^\circ_R - E^\circ_L = 0.14 - (-0.44) = 0.58 \text{ V}$$

The cell reaction is Fe + $Sn^{++} \rightarrow Fe^{++}$ + Sn

According to Nernst's equation.

$$E = E^\circ - \frac{0.059}{n} \log \frac{[Fe^{2+}]}{[Sn^{++}]} = 0.58 - \frac{0.059}{2} \log \frac{0.6}{0.2}$$

$$= 0.58 - 0.0295 \log 3 = 0.58 - 0.295 \times 0.4771 = 0.58 - 0.014 = 5.66 \text{ V}$$

26. Molecular weight of O_2 = 32
Equivalent wt. of O_2 = 8
Vol. occupied by 8 g. at NTP = 5600 cm^3
quantity of electricity required to liberate 5600 cm^3 at
NTP = 96,500 C.
1 cm^3 of O_2 at NTP = 96500/5600 = 17.23 coulombs
$Q = Ct$ or $C = Q/t$ = 17.23/1 = 17.23 amperes.

27. (a) No. of moles of substance = Molarity × V lit.
Initial no. of moles of Cu = (100/1000) 0.4 = 0.04 moles
No. of equivalents of current passed = Ct/F = (1.5 × 50 × 60)/96500 = 0.0466
No. of moles = No. of equivalents/Valency = 0.0466/2 = 0.233 mole
Mole of Cu^{2+} remained = 0.04 – 0.233 = 0.0167 mole
∴ molarity of Cu = no. of moles/V lit. = 0.0167/0.1 = 0.167 M
No. of mole of H^+ formed = 0.0466 mol.
Molarity of H^+ = 0.0466/0.1 = 0.466 M
SO_4^{2-} dos not take part in electrolysis so its conc. unchanged = 0.4 M
(b) 1 micro ampere = 1×10^{-6} amp.
C = 60×10^{-4} amp; t = 20 × 60 sec.
No. of equivalents of copper deposited
= C × t/F = $(60 \times 10^{-6} \times 20 \times 60)/96500 = 7.47 \times 10^{-17}$
Therefore no. of atoms of copper = $3.73 \times 10^{-7} \times 6.02 \times 10^{23} = 2.245 \times 10^{17}$
Area of cathode covered = $2.245 \times 10^{17} \times 8.7 \times 10^{-16}$ = 195.35 cm^2
Total area = 195.35/53 × 100 = 386.6 cm^2

28. 1 Ampere = 10^6 micro ampere
$1 = 40 \times 10^{-6}$ A, t = 32 × 60 = 1920 sec
Equivalent weight and atomic weight of Ag are equal.
No. of equivalents of Ag deposited = No. of gram atoms of Ag deposited.

$$= \frac{it}{F} = \frac{40 \times 10^{-16} \times 1920}{96500} = 7.96 \times 10^{-7} \text{ g. atom.}$$

No. of atoms of Ag = No. of g atoms × Avogadro's number
= $7.96 \times 10^{-7} \times 6.02 \times 10^{23} = 4.792 \times 10^{17}$
Area of cathode covered= $4.792 \times 10^{17} \times 5.4 \times 10^{-16}$ = 258.764 cm^2
Total area = 258.764 × 100/43 cm = 601.776 cm^2

29. 88 gm of Hg require 12 g. of Cd;
22 g. of Hg, require = (12 × 22)/88 = 3 g
Eq. wt of Cd = 112/2 = 56
For 56 g of Cd→ 96500 C are required
For 3 g of Cd→ 3 × 96500/58 = 5169.6 C
current = 5 amp.
∴ time= Q/C = 5169/5 = 1034 sec. = 17 min. 14 sec.

30. $AgI \rightleftharpoons Ag^+ + I^-$

Here $[Ag^+] = [I^-]$

$$K_{sp} = [Ag^+]\,[I^-] = [Ag^+]^2$$

$$\therefore \quad [Ag^+] = \sqrt{K_{sp}} = \sqrt{8.7 \times 10^{-17})} = 9.33 \times 10^{-9} \text{ mole L}^{-1}$$

Now electrode potential of the Ag^+/Ag electrode $= E° - 0.059 \log \dfrac{1}{[Ag^+]}$

$$= 0.799 - 0.059 \log \frac{1}{9.33 \times 10^{-9}} = 0.325 \text{ V}$$

Reduction half reaction for $I^-/AgI/Ag$ electrode is

$AgI(s) + e^- \rightarrow Ag(s) + I^-(aq)$ I

$Ag^+(aq) + e^- \rightarrow Ag(s),\ E° = 0.799$ V II

$\therefore$ (I)-(II)

$AgI(s) \rightarrow Ag^+(aq) + I^-$ (aq)

for the cell reaction given above

$E = E° - E° = -0.059 \log [Ag^+|I^-]\,[I^-]$

Now at equilibrium $\Delta G = 0$

$nFE = 0 \Rightarrow E = 0$

At equilibrium; $[Ag^+][I^-] = K_{sp}$

$\therefore \qquad E' - E° = 0.059 \log K_{sp} = 0.059 \log 8.7 \times 10^{-17} = 0.948$

We get $\qquad E' = E° + (-0.948) = 0.799 - 0.948 = -0.149$ V

31. (i) The reaction is: $Zn \rightarrow Zn^{++} + 2e^-$

$$\therefore\ E = E^\circ_x - \frac{0.059}{2} \log[Zn^{++}] = 0.76 - \frac{0.059}{2} \log (0.1) = 0.7895 \text{ V}$$

(ii) The reaction is: $2Cl^- \rightarrow Cl_2 + 2e^-$

$$\therefore\ = E^\circ_x - \frac{0.059}{n} \log \frac{P_{Cl_2}}{[Cl^-]^2} = -1.36 - \frac{0.059}{2} \log \frac{10}{0.1 \times 0.1} = -1.4485 \text{ V}$$

(iii) The reaction is: $H_2 \rightarrow 2H^+ + 2e^-$

$$\therefore\quad E = E_x - \frac{0.059}{n} \log \frac{[H^+]^2}{P_{H_2}} = 0.0 - \frac{0.059}{2} \log \frac{0.2 \times 0.2}{4} = 0.118 \text{ V}$$

32. (a) $1/2\ H_2 \rightarrow H^+ + e^-$

$E = E° - (0.059/1) \log [H^+] \quad \because a^\circ_H = 0.059\ p^H$

$[\therefore E° = 0; -\log [H^+] = p^H]$

$$\therefore\ p^H = \frac{E}{0.059} = \frac{0.25}{0.059} = 4.24$$

(b) $p^H = E/0.059 = 0.3/0.059 = 5.08$

33. The above reaction cell is, Fe | Fe^{2+} | Cu^{2+} | Cu

The equilibrium E = 0 and $E = (E^\circ_{Fe^-} - E^\circ_{Cu}) - (0.059/2) \log (a_{Fe^{2+}} / a_{Cu^{2+}})$

At equilibrium $E = 0$ and $a_{Fe^{2+}} / a_{Cu^{2+}} = K$

$$E - E° = \frac{0.059}{2} \log K;$$

$$\text{or} \quad 0.44 - [-0.337] = \frac{0.059}{2} \log K$$

$$\therefore \quad \frac{0.059}{2} \log K = 0.777.$$

$\Rightarrow \log K = 26.339$, Hence, $K= 2.183 \times 10^{26}$

34. $E°$ $(Cu^{2+}/Cu) = 0.337$ V. $E°$ $(Ag^+/Ag) = 0.799$
anode cathode
Cell notation $Cu|Cu^+ \| Ag^+ | Ag$
$E_{cell}= E_{Right} - E_{Left} = 0.799 - 0.337 = 0.462$ volts
At cathode $Cu \rightarrow Cu^{2+} + 2e^-$ $E = -0.337$
At anode $(Ag^+ + e^- \rightarrow Ag)\ 2$ $E = 0.799$
Total cell reaction $Cu + 2Ag^+ \rightarrow Cu^{2+} + 2Ag$ $E = 0.462$
By using Nernst equation $E = E° - (0.0591/n) \log K$
$E= E° - (0.0591/n) \log ([Cu^{2+}] [Ag]^2/[Cu][Ag^+]^2)$
$E= 0$; $E° = 0.462$; $n = 2$; $Cu^{2+} = 0.1$
$\therefore\ 0 = 0.462 - (0.0591/2) (\log 0.01/[Ag^+]^2)$
$0.462= (0.0591/2) (\log 0.01/[Ag^+]^2)$

$$\frac{2\times 0.462}{0.0591} = \log 0.01 + \log[Ag^+]^2 = -2-2 \log [Ag^+]$$

$$15.63 + 2 = -2 \log [Ag^+]$$

$-17.63/2 = \log [Ag^+]$; $-8.817 = \log [Ag^+]$
$10^{-8.817} = [Ag^+]$; $1.523 \times 10^{-9} = [Ag^+]$

35. $C_6H_5NO_2 + 6H^+ + 6e^- \rightarrow C_6H_5NH_2 + 2H_2O$
Equivalent weight of Nitrobenzene = 123/6 = 20.5
20.5 grams nitrobenzene is reduced by = 1 Faraday
12.3 grams nitrobenzene is reduced by = 12.3/20.5 = 0.6 F
Current efficiency is 50%
The amount current needed to reduce 12.3 grams = 0.6 × 2F, or 115800 coulombs.
Since the potential drop is 3 volts
Energy consumed is → 115800 × 3 = 347400 joules.

36. 0.4 grams is deposited at cathode
Current used for 0.4 grams Cu → 0.4/31.78 = 0.0125 Faraday
0.0125 Faradays liberate → 8 × 0.0125 gr. of O_2
Volume occupied by the gas → 8 × 0.0125 × 22.4/32 = 0.0705 litre
After deposition electrolysis is carried out for seven min. with 1.2 amp.
It is equivalent to → 7 × 60 × 1.2/96500 = 0.00522 Faradays.
1 Faradays release → 1g of 22.4/2 lit of H_2 at cathode
0.00522 Faradays → 0.05849 litre of H_2

1 Faraday release → 8 g of O_2 or 22.4/4 litre O_2 at anode
0.00522 Faradays → 000522 × 22.4/4 = 0.0292
Total volume of gasses evolved → 0.0705 + 0.05849 + 0.0292 → 0.1582 liter.

37.

$$2H_2O + 2e^- \rightarrow H_2 + 2OH^- \qquad E\ -0.8277$$
$$H_2 \rightarrow 2H^+ + 2e^- \qquad E = 0.00$$
$$2H_2O \rightarrow H^+ + 2OH^- \qquad E = -0.8277$$

$$E_0 = (0.0591/n) \log k$$
$$-0.8277 = (0.0591/2) \log ([H^+][OH^-]/[H_2O])$$
$$-0.8277/0.0591 = \log k; \text{ or } -14.00507 = \log K$$
$$K = 10^{-14.00507} = 10^{-15} \times 10^{0.9949} = 9.883 \times 10^{-15} = 1 \times 10^{-14}$$

38. $E_2 = E° + 0.0591 \log x$
$E_1 = E° = 0.0591$ lo 10^{-4}
$E_2 - E_1 = 0.118 = 0.0591 (\log x - \log 10^{-6})$
Solving, $x = 10^{-4}$
∴ concentration of H^+ ions = 10^{-4} mol litre^{-1}

39. No moles of H_2 reacted = 67.2/22.4 = 3.0
∴ 1 mole of H_2 = 2F
∴ 3 moles of H_2= 3 × 2 = 6F
∴ Current produced = 57900/(15 × 60)= 643.33 C
∵ 2 F = 63.5 g Cu
∴ Cu deposited = 63.5 × 6/2 = 190.5 g

40. Walt = volt × current
100 = 110 × Current
or Current I = 100/110 = 10/11 amp.
Quantity of electricity = I × t = (10/11) × 10 × 3600
No. of equivalents = (10/11) × 10 × 3600 × (1/96500) = 0.339
Wt. of Cd deposited = 0.339 × 112.4/2 = 19.06 g.

41. Volume of 1 mole of benzene (l) = $\dfrac{78\text{ g}}{0.877\text{ g mL}^{-1}}$ = 88.94 mL

Volume of 1 mole of benzene (g) = 2750 × 88.94 = 244585 mL = 244.58 L

Volume of 1 mole toluene (l) = $\dfrac{92\text{ g}}{0.867\text{ g mL}^{-1}}$ = 106.113 mL

Volume of 1 mole of toluene (g) = 7720 × 106.113 = 819192 mL = 819.19 L
Now according to gas equation, $PV = nRT$

$$\therefore \quad P^\circ_{\text{benzene}} = \frac{nRT}{V} = \frac{(1\text{ mol})(0.082\text{ L atm K}^{-1}\text{ mol}^{-1})(293\text{ K})}{244.58\text{ L}} = 0.098\text{ atm}$$

and

$$p^\circ_{\text{toluene}} = \frac{(1\text{ mol})(0.082\text{ L atm K}^{-1}\text{ mol}^{-1})(293\text{ K})}{819.19\text{ L}} = 0.029$$

For the solution of benzene and toluene

$$X_{benzene} \times P^{\circ}_{benzene} + 1 - X_{benzene} \times p^{\circ}_{toluene} = \text{Total Pressure}$$

or

or $$X_{benzene} \times P^{\circ}_{benzene} + X_{toluene} \times p^{\circ}_{toluene} = \text{Total pressure}$$

$$X_{benzene} \times 0.098 \text{ atm} + (1 - X_{benzene}) \times 0.029 \text{ atm} = \frac{46 \text{ atm}}{760} = 0.0605 \text{ atm}$$

or

$$0.098\ X_{benzene} - 0.029\ X_{benzene} = 0.0605 - 0.029$$

or $$0.069\ X_{benzene} = 0.0315$$

$\therefore$ $$X_{benzene} = \frac{0.315}{0.069} = 0.456$$

$\therefore$ Mole fraction of benzene in vapour phase $= \dfrac{0.456 \times 0.098}{0.0605} = 0.74$

42. Volume of bettery = 3.5 lit = 3500 ml
Density of 39% H_2SO_4 before discharge = 1.294 g ml^{-1}
Amount of H_2SO_4 in solution before discharge = (3500 × 1.294 × 39)/100 = 1767 g
Density of 20% H_2SO_4 after discharge = 1.139 g ml^{-1}
Amount of 20% H_2SO_4 in solution after discharge = (3500 × 1.139 × 20)/100 = 797.1 g
Amount of H_2SO_4 consumed = 1767– 797.1 = 969.9 g.
H_2SO_4 used during charging = 98 g. = 2 g. equivalent
Therefore 969.9 g of H_2SO_4 = 969.9/49 = 19.974 equivalent.
1 equivalent of H_2SO_4 ≡ 1 F
1 Faraday = 96500 coulombs
19.976 Faradayas (96500 × 19.974)
The number of ampere hour which battery worked = (96500 × 19974)/(60 × 60) = 535.40

43. Area of the metal surface = 80 cm^2
Thickness = 0.005 mm= 0.0005 cm
Volume of the metal= 80 × 0.0005 = 0.04 cm^3
1 cm^3 of Silver weighs 10.5 g
0.04 cm^3 Silver weighs = 0.04 × 10.5 = 0.42 g.
As, 96500 coulombs will deposit 108 g. of Ag
$\therefore$ Coulombs of electricity required to deposit 0.42 g. of Ag = (96500 × 0.42)/108 = 375.3
Current strength = 3 ampere.
Therefore time = 375.3/3 = 125.1 sec.

44. At. wt. of gold = 197 (valency = 3) Eq. wt. = 197/3 = 65.67
At wt. of copper = 63.54 (valency = 2) $\therefore$ Eq. wt. = 63.54/2 = 31.77
wt. of copper deposited $w_1/w_2 = E_1/E_2$
wt. of copper = 31.77 × 9.85/65.67 = 4.765 g.
For, Calculation of current in amperes we have:
m= I wt; Here, m = 4.765; w = 31.77/96500; t = 5 × 60 × 60
$\therefore$ $I = m/wt$ = (4.765 × 96500)/(31.77 × 60 × 60 × 5) = 0.8046 ampere

45. The relevant electro chemical reactions are

$Sn^{2+} \rightarrow Sn^{4+} + 2e^-$ (at anode) $Sn^{2+} + 2e^- \rightarrow Sn$ (at cathode)

$2\ SnCl_2 \rightarrow Sn + SnCl_4$

380 g 119 g 261 g

119 g of Sn deposition carries 380 g of $SnCl_2$ (m) to react and 261 g. of $SnCl_4$ (m)

$\therefore$ 0.119 g. of Sn deposition = 380 × 0.119/119 = 0.380 g of $SnCl_2$

and $SnCl_4$ = 261 × 0.119/119 = 0.261 g. $SnCl_4$

Wt of $SnCl_2$ left unused = 19–0.38 = 18.62 g.

Thus, ratio between $SnCl_2$ and $SnCl_2$ and $SnCl_4$ after electrolysis = 18.62 : 0.261 = 71.34 : 1

46. $H^+ + e^- \rightarrow 1/2\ H_2$ (at cathode)

$Cl^- \rightarrow 1/2\ Cl^2 + e^-$ (at anode)

Since, diaphragm is not used, the liberated chlorine will combine with NaOH in hot condition

$$3Cl_2^- + 6\ NaOH \rightarrow 5\ NaCl + NaClO + 3H_2O$$

47. In the N^{3-} ion, there are 3 units of electronic charge.

i.e. one ion of N^{3-} carries 3 electronic charge.

Therefore, 1 g ion of N^{3-} carries $3 \times 6.023 \times 10^{23}$ electronic charges

i.e., $3 \times 6.023 \times 10^{23} \times 1.6 \times 10^{-19} = 2.89 \times 10^5$ coulombs.

48. (a) Since the e.m.f. of the cell is the same as the standard $E°$ of the electrode, the other electrode must be the hydrogen electrode with zero reduction potential with its half cell reaction being $2H^+ + 2e^- \rightarrow H_2$

(b) Multiply equation by 2 and substract the same from equation (1)

Thus, we have

$$2Hg^- - 2Hg = Hg_2^{2+} - 2Hg^{2+} + 2e^- - 4e^- \text{ or } E = (-0.796 + 0.85) = 0.055\ V$$

$$\therefore \quad 2Hg^{2+} + 2e^- = Hg_2^{2+};\ E° - 0.055\ V$$

49. Volume of copper required for deposit = 10 × 10 × 10 = 1 cm^3

Weight of copper required = 8.94 g (Since density of copper is 8.94 g ml^{-1}

For the deposition of 1 g. atom or 63.5 g. of Cu from Cu^{2+} ion; amount of electricity required = 2 Faraday = 2 × 96500 coulombs.

Hence the quantity of electricity required to deposit of 8.94 g. of Cu

= (2 × 96500/63.5) 8.94 coulombs = 2.717×10^4 coulombs.

50. Coulombs of electricity passed = 0.2 × 3 × 60 × 60 = 2160

One Faraday or 96500 coulombs deposit one gm. eq. wt of silver or 108 g. of Ag.

Amount of Ag deposited by 2160 coulombs = 108 × 2160/96500 = 2.417 g.

Density of Ag = 10.47 g cm^3

Volume of Ag deposited = 2.417/10.47 = 0.2310 cm^3

Area of the vessel = 800 sq. cm

thickness of Ag deposited = 0.2310/800 = 2.89×10^{-4} cm.

51. For liberating 1 gm. eq. wt of 35 g. chlorine. i.e. amount of electricity required = 1 Faraday = 96,500 coulombs.

Therefore, For liberating 710 g. or 20 g. wt. of chlorine, amount of electricy required = 20 Faraday.

(i) Again 1 g. eq. wt. of chlorine = 1g. eq. wt of hydrogen = 1 g. eq. wt. NaOH = 1 Faraday

35 g chlorine = 40 g NaOH; or 710 g chlorine = 800 g NaOH;
1g hydrogen = 1 Faraday ∴ Also, 20 g hydrogen need 20 Faraday

(ii) Vol. of 20 g of H_2 at N.T.P. = 22.4 × 20/2 = 224 litres
Vol at 300 K and 1 atm. pressure = 224 × (300/273) × (760/760) = 246.2 litres.

52. Equivalents of NaOH present in 600 ml. of 1 N solution 600 × 1/1000 = 0.6
Equivalents of copper produced = 31.8/31.75 = 1 Eq. wt. Cu = Gram. mol. wt./2
Theoretically the amount of NaOH produced simultaneously should also have been 1 g. equivalent.
Hence, the percentage of theoretical yield of NaOH = 0.6 × 100/1 = 60

53. Faradays of electricity passed = 5 × 30 × 60/96500 = 18/193
moles of $CuSO_4$ electrolysed = (18/193) 0.5
Moles of $CuSO_4$ left = 0.08 – 9/193 = 15.44 – 9.00/193 = 6.44/193
Mol. wt. of $CuSO_4$ = 63.5 + 96 = 159.5
Grams of $CuSO_4$ left in solution = 6.44 × 159.5/193 = 5.32 g.

54. The reaction involved in the cell:

$$Ag|AgCl(s)\,|\,KCl\ (0.02\ M)\,||\,KBr\ (0.001\ M)\,|\,AgBr(s)\,|\,Ag$$

are:

Anode: $Ag + Cl^- \rightarrow AgCl + e^-$
Cathode: $AgBr + e^- \rightarrow Ag + Br^-$
Over all: $AgBr + Cl^- \rightarrow AgCl + Br^-$

Accroding to Nernst's equation, the EMF of the cell is

$$E_{cell} = E^\circ = \frac{RT}{nF}\ln\frac{[Br^-]}{[Cl^-]} = E^\circ - \frac{0.0591}{n}\log\frac{[Br^-]}{[Cl^-]}$$

where $E^\circ = E^\circ_{Br^-|AgBr|Ag} - E^\circ_{Cl^-|AgCl|Ag} = \left(E^\circ_{Ag^+|Ag} + \frac{RT}{F}\ln K_{sp}^{AgBr}\right) - \left(E^\circ_{Ag^+|Ag} + \frac{RT}{F}\ln K_{sp}^{AgCl}\right)$

$$= \frac{RT}{F}\ln\frac{K_{sp}^{AgBr}}{K_{sp}^{AgCl}} = 0.0591\log\left(\frac{3.3\times10^{-13}}{2.8\times10^{-8}}\right) = 0.1732$$

Hence

$$E_{cell} = (-0.7132) - (0.0591)\log\frac{0.01}{0.2} = -0.037\ V$$

Since E is negative, **the cell reaction is nonspontaneous**. The spontaneous direction is

$$AgCl + Br^- \rightarrow AgBr + Cl^-$$

with Ag/AgCl as cathode (positive electrode) an Ag/AgBr as anode (negative electrode)

55. Electrolytic equations:
∴ Cathode $[Cu^{2+} + 2e^- \rightarrow Cu] \times 2$
Anode $[H_2O \rightarrow 2H^+ + 1/2\ O_2 + 2e]$

According to the equations given above, four faradays of electricity are consumed to deposit 2 g. atoms or 2 × 63.5 = 127.0 g. of copper and liberate 32 g. of oxygen and simultaneously produce 4 equivalents of acid.

Hence will the passage of 0.01 faraday of electricity.
The amount of copper deposited at the cathode = 127.0 × 0.01/4 = 0.3175 g.
And the amount of oxygen liberated at the anode = 32 × 0.01/4 = 0.08 g.
Total loss of weight = 0.3175 + 0.08 = 0.3975 g.
Hence, the weight of resulting solution = 10 – 0.3975 = 9.6025 g.
Again, equivalents of acid produced = 4 × 0.01/4 = 0.01

56. (a) Oe faraday or 98.500 couloumb charge is possessed by avogadro number or 6.02×10^{23} electrons.

∴ Charge on 1 electron in coulombs: = $96500/6.02 \times 10^{23} = 1.602 \times 10^{-19}$

(b) (i) In the cell A at anode $2H_2O \rightarrow 4H^+ + O_2 + 4e^-$ at cathode $Ag^+ + e^- \rightarrow Ag$
In the cell *B* at anode $Cu \rightarrow Cu^{2+} + 2e^-$ at cathode $Cu^{++} + 2e^- \rightarrow Cu$

(ii) Amount of O_2 liberated at anode in the cell = 1 g = 1/8 g. eq. of O_2
∴ Wt of Cu/Wt of O_2 = Eq. wt. of Cu/Eq. wt. of O_2
Wt. of Cu = 31.75 × 1/8 = 3.97 g; no. of g eq. of Ag = 1/8;
Wt. of Ag = 1.8 × 108 = 13.5 g.

57. (i) From the equation, it is clear that for the production of 1 mole of ClO_4^-; ions or $NaClO_4$ 2 mole electrode and 2 Faradays of electricity would be required.

(ii) Mol. wt. of $NaClO_4$ = 23 + 35.5 + 64 = 122.5
Eq. wt of $NaClO_4$ = Mol wt/2 = 122.5/2 = 61.25
Equivalents of $NaClO_4$ in 12.25 = (12.25/61.25) = 0.2
Equivalents of H_2 produced during the same time = 0.2 [=0.1 mole]
Volume of H_2 at S.T.P. = 0.1 × 22.4 lit. = 22.4 litres.

58. (i) Cathode $Ni^{2+} + 2e^- = Ni$ Anode $2H_2O = 4H^+ + O_2 + 4e^-$

(ii) Coulombs = 0.05 × 40 × 60 = 120 coulombs.

(iii) From (i) it is clear that 2 Faradays i.e. 96,500 × 2 coulombs of electricity are required to deposit one gram atom or 59 g. Ni.
Hence, the amount of product deposited on by 120 coulombs of electricity
= 59 × 120/96500 × 2 = 0.037 g

(iv) Total number of moles of Ni^{2+} present in 40 ml. of 0.125 M $NiSO_4$
= 40 × 0.125/1000 = 0.005
Total quantity of electricity required to deposit = 2 × 96,500 × 0.005 coul.
Strength of current = 0.05 Ampere.
Total time required = 2 × 96500 × 0.005/0.05 = 19,300 seconds.
Extra time required for the deposit of the rest of the metal
= 19,300 – 40 × 60 = 16,900 seconds.

(v) For 0.005 moles of Ni to deposit according to equations in (1)
the amount of oxygen liberated at the anode = 0.005/2 mole = 0.005 × 32/2 = 0.08 g

59. Overall reaction when 2 faradays of electricity is passed through the battery

$$2PbSO_4 + 2H_2O = Pb + PbO_2 + 2H_2SO_4$$

Amount of H_2SO_4 present in 2 litres of battery liquid before charging
= (1.11 × 2000)/1 × (15.7/100) = 348.5 g.

Amount of H_2SO_4 present in 2 litres of battery liquid after charging
= (1.28 × 2000)/1 × (358/100) = 944.5 g.

Thus the amount of H_2SO_4 produced during charging = 944.5 – 348.5 = 596.0 gm.
Now 98 g H_2SO_4 require 2 faraday.
Therefore, 596 gm H_2SO_4 will require = 596/98 × 2 × 96500 = 1.1737×10^6 coulombs.
Time in seconds = 100 × 60 × 60 = 3.6×10^5 seconds
The strength of average current = coulombs/Seconds = $1.1737 \times 10^6/3.6 \times 10^5$ = 3.26 amp.

60. Amount of Cu likely to be deposited when 1.307 gm. of Ag are deposited is given by the equation.
Weight of Cu deposited/Weight of Ag deposited = Eq. wt. of Cu/Eq. wt. of Ag
or Weight of Cu deposited or removed from the solution = (1.307 × 63.5)/(108 × 2) = 0.3842 g
Weight of Cu originally present in 500 ml = (63.5 × 10)/249.5 = 2.545 g.
[G. mol wt. of $CuSO_4$ $5H_2O$ i.e. 249.5 g contains 63.5 g. Cu]
Weight of Cu left after electrolysis in 500 ml. = 2.545 – 0.3842 = 2.1608
Weight of Cu left after electrolysis = (2.1608/500) × 1000 = 4.3216 g. $litre^{-1}$

61. At equilibrium the potentials of Zn and Cu in against the solution are equal i.e.
$E_{Zn} = E_{Cu}$
Reactions $Zn \rightarrow Zn^{2+} + 2^{e-}$

$$Cu^{2+} + 2e^- \rightarrow Cu$$

$$Zn + Cu^{2+} \rightleftharpoons Zn^{2+} + Cu$$

The electrode oxidation potentials are given by

$E_{Zn} = E° - 0.059/2 \log a_{Zn^{2+}}$ $(n + 2)$;

$E_{Cu} = E^{\circ}_{Cu} - 0.059/2 \log (a_{Cu^{2+}})$
At equilibrium, since $E_{Zn} = E_{Cu}$

∴ $E_{Zn} - E_{Cu} = 0.0295 \log a_{Cu^{2+}}$

or $0.76 - (-0.34) = 0.0295 \log \Rightarrow a_{Zn^{2+}}/a_{Cu^{2+}} = 1.10$

Therefore $\log a_{Zn^{2+}}/a_{Cu^{2+}} = 1.10/0.0295 = 37.2881 \Rightarrow a_{Zn^{2+}}/a_{Cu^{2+}} = 1.941 \times 10^{27}$

Hence, Concentration of $Zn_{Zn^{2+}}$ ion : $Cu_{Cu^{2+}}$ ion is 1.941×10^{37} : 1

62. Length (l) = 2.1 cm; Area (a) = 4.2 sq. cm.; Resistance (R) = 25 ohms; and Concentrations (C)= 0.5 N
∴ Specific conductance; K = 1/R × 1/a mho. cm^{-1} = 1/25 × 2.1/4.2 = 0.02 mho. cm^{-1}
Equivalent conductance, = 1000 K/C mho cm^2/eq. = 1000 × 0.02/0.05 = 40 mho cm/equl.

63. Cell reaction is,
$Ag|Ag^+ (Ag_2CrO_4 \text{ saturated}) \| Ag^+(0.1 M) | Ag$
According to Nernst's equation

$$E_{cell} = \frac{0.059}{n} \log \frac{[Ag^+]_{anode}}{[Ag^+]_{cathode}} \Rightarrow 0.164 = \frac{0.059}{1} \log \frac{[Ag^+]_{anode}}{[Ag^+]_{cathode}}$$

or $\log \frac{[Ag^+]_{cathode}}{[Ag^+]_{anode}} = \frac{0.164}{0.059}$

i.e. $\log \frac{0.1}{[Ag^+]_{anode}} = 2.78$ or $\frac{0.1}{[Ag^+]_{anode}} = 6.025 \times 10^2$

$\Rightarrow [Ag^+]_{anode} = \frac{0.1}{602.5} = 1.66 \times 10^{-4}$

Since there are two Ag^+ ions in Ag_2CrO_4 hence

$$[Ag^+] = \frac{1.66 \times 10^{-4}}{2} = 0.83 \times 10^{-4}$$

For Ag_2CrO_4 or the compound of the type AB_2

$$K_{sp} = 4s^3$$
$$= 4(0.83 \times 10^{-4})^5 = 2.3 \times 10^{-12}$$

64. Let us first determine how much moles of Zn metal and of Cu^{2+} ions were in the cell initially. The atomic weight of Zn is 65.38 g · mol^{-1} so that we have (23.72 g)/(65.36 g · mol^{-1}) = 0.3628 mol of Zn to begin with. Also (0.6 L) (1.0 *M*) = 0.6 mol of Cu^{2+} ions are there initially. Thus the Zn will be used up before the Cu^{2+} ions, as the net cell reaction is $Zn(s) + Cu^{2+} \rightarrow Cu(s) + Zn^{2+}$.

So, the question can be : How long it takes to oxidize 0.3628 mol of Zn to Zn^{2+} ions, if the current being drawn is 0.1 A? Since $Zn(s) \rightarrow Zn^{2+}(aq) + 2e^-$. So $2f$ of charges are transferred per mole of Zn oxidized.

Therefore, the total amount of charge that must pass through the wire is

$$(0.3628 \text{ mol}) (2\mathscr{F} \cdot \text{mol}^{-1}) (96{,}485 \text{ C}/\mathscr{F}) = 7 \times 10^4 \text{ C}$$

To calculate the length of time it will take for this amount of charge to be transferred. Now, we have,

Charge in coulombs = (current in amps) (time in seconds)

Solving we obtain:

$$t = \frac{7 \times 10^4 \text{ C}}{0.1 \text{ C} \cdot \text{s}^{-1}} = 7 \times 10^5 \text{ s} = 194 \text{ h} = 8.1 \text{ days}$$

The cell will operate continuously for 8.1 days if a steady current of 0.1 A is drawn (an unlikely situation). If smaller currents are drawn the cell will last longer, and if larger currents are drawn the cell will be depleted more quickly.

65. Reduction occurs at the cathode, which is the right-hand electrode of the cell notation. Thus, Cathode reaction $Cu^{2+}(a) + 2e^- \rightarrow Cu(s)$

Oxidation occurs at the anode, which is left-hand electrode of the cell notation. Thus

Anode reaction $Zn\ (s) \rightarrow Zn^{2+}_{(aq)} + 2e^-$

The net cell reaction is $Zn(s) + Cu^{2+}(aq) \rightarrow Zn^{2+}(aq) + Cu(s)$. This is, of cource, a Daniell cell.

$$\Delta \mathscr{E}^\circ_{cell} = 1.103 \text{ V} = \mathscr{E}^\circ_{Cu^{2+}/Cu} - \mathscr{E}^\circ_{Zn^{2+}/Zn}$$

$$1.103\text{ V} = \mathscr{E}^\circ_{Cu^{2+}/Cu} - (-0.763\text{ V}) = \mathscr{E}^\circ_{Cu^{2+}/Cu} + 0.763\text{ V}$$

Thus $\mathscr{E}^\circ_{Cu^{2+}/Cu} = 1.103\text{ V} - 0.763\text{ V} = 0.340\text{ V}$

66. We calculate $\Delta\mathscr{E}^\circ_{cell}$ for the reaction in which MnO_4^- oxidizes Br^-, to see whether it is negative or positive.

Reduction $MnO_4^- + 8H^+(aq) + 5e^- \rightarrow Mn^{2+} + 4H_2O$

Oxidation $2Br^- \rightarrow Br_2(aq) + 2e^-$

Net cell reaction $2MnO_4^- + 16H^+(aq) + 10Br^- \rightarrow 5Br_2(aq) + 2Mn^{2+} + 8H_2O$

If a galvanic cell is constructed using this reaction with all substances in their standard states, the cell potential will be

$$\Delta\mathscr{E}^\circ_{cell} = \mathscr{E}^\circ_{Pt/MnO_4^-, Mn^{2+}} - \mathscr{E}^\circ_{Pt|Br_2, Br^-} = 1.49 - 1.09 = +0.40\text{ V}$$

Because the cell potential is positive, $\Delta\mathscr{E}^\circ_{cell} > 0$, the reaction proceeds spontaneously. Permanganate ion is a very strong oxidizing agent and oxidizes Br^- to Br_2.

67. Here we calculate the value of $\Delta\mathscr{E}^\circ_{cell}$ and use its sign to predict the direction of reaction. If Pb reduces Zn^{2+} the half-reactions are

Reduction $Zn^{2+}(aq) + 2e^- \rightarrow Zn(s)$

Oxidation $Pb(s) \rightarrow Pb^{2+}(aq) + 2e^-$

Net cell reaction $Pb(s) + Zn^{2+}(aq) \rightarrow Zn(s) + Pb^{2+}(aq)$

If a galvanic cell were constructed using this reaction with all substances in their standard states, the cell potential would be

$$\Delta\mathscr{E}^\circ_{cell} = \mathscr{E}^\circ_{Zn^{2+}|Zn} - \mathscr{E}^\circ_{Pb^{2+}|Pb} = -0.763\text{ V} - (-0.126\text{ V}) = -0.637\text{ V}$$

Since the calculated standard cell potential is negative, reaction will not occur as written; rather, the reverse reaction will occur. Zinc will reduce Pb^{2+} ions. The spontaneous reaction is $Zn(s) + Pb^{2+}(aq) \rightarrow Pb(s) + Zn^{2+}(aq)$

Lead will not reduce Zn^{2+} ions i.e. if a piece of Pb metal is immersed in a solution of a zinc salt, nothing at all will happen.

68. From the given table, both half-reactions are reductions. During reaction, one of them must be reversed to be an oxidation, and that one will have it $\mathscr{E}^\circ$ value subtracted in determining the cell potential. The only way we can obtain a positive cell potential subtracting one of these values is for Ni(s) to be oxidized to $Ni(OH)_2$, that is, for –0.72 to be subtracted.

Reduction : $CrO_4^{2-} + 4H_2O + 3e^- \rightarrow Cr(OH)_3(s) + 5\,OH^-$

Oxidation : $Ni(s) + 2OH^- \rightarrow Ni(OH)_2(s) + 2e^-$

To balance, multiply the oxidation half-reaction by 3 and the reduction half-reaction by 2. The spontaneous redox reaction is

$$3Ni(s) + 2CrO_4^{2-} + 8H_2O \rightarrow 4\,OH^- + 3\,Ni(OH)_2(s) + 2Cr(OH)_3(s)$$

and $\Delta\mathscr{E}^\circ_{cell} = -0.12\text{ V} - (-0.72\text{ V}) = +0.60\text{ V}$

69. (a) Every substance to the right of the double vertical bars is part of the cathode compartment, where reduction occurs. The Pt is an inert metal electrode. Therefore,

Cathode reaction $Fe^{3+} + e^- \rightarrow Fe^{2+}$

Also, every substance to the left to of the double vertical bars is part of the anode compartment, where oxidation occurs. Hence,

Anode reaction $2Cl^- \rightarrow Cl_2(g) + 2e^-$

Net cell reaction $2Fe^{3+} + 2Cl^- \rightarrow 2Fe^{2+} + Cl_2(g)$

$$\Delta\mathscr{E}^\circ_{cell} = \mathscr{E}^\circ_{Fe^{3+}|Fe^{2+}} - \mathscr{E}^\circ_{Cl_2|Cl^-} = +0.770 \text{ V} - 1.358 \text{ V} = -0.588 \text{ V}$$

The reaction as written does *not* proceed spontaneously to the right. The reverse reaction, $2Fe^{2+} + Cl_2(g) \rightarrow 2Fe^{3+} + 2Cl^-$, proceeds spontaneously. Chlorine will oxidize ferrous ion to ferric ion.

(b) The only element with two different oxidation states on the right-hand side of the double vertical bars (the cathode compartment) is manganese, which is in the +4 state in $MnO_2(s)$ and the +2 states in Mn^{2+}. Aqueous $MnCl_2$ consists of Mn^{2+} and Cl^- ions. Thus,

Cathode reaction $4H^+(aq) + MnO_2(s) + 2e^- \rightarrow Mn^{2+} + 2H_2O$

Anode reaction $Cu(s) \rightarrow Cu^{2+} + 2e^-$

Net cell reaction $4H^+(aq) + MnO_2(s) + Cu(s) \rightarrow Cu^{2+} + Mn^{2+} + 2H_2O$

$$\Delta\mathscr{E}^\circ_{cell} = \mathscr{E}^\circ_{MnO_2|Mn^{2+}} - \mathscr{E}^\circ_{Cu^{2+}|Cu} = 1.208 \text{ V} - 0.340 \text{ V} = 0.868 \text{ V}$$

Since the standard cell voltage is positive, the net cell reaction is spontaneous in the direction written when all substances are in their standard states.

70. First write the anode and cathode reactions.

Anode reaction $2\,I^- \rightarrow I_2 + 2e^-$

Cathode reaction $2Fe^{3+} + 2e^- \rightarrow 2Fe^{2+}$

Net cell reaction $2Fe^{3+} + 2\,I^- \rightarrow I_2 + 2Fe^{2+}$

The values of ΔG°_f needed are taken from Tables.

$$\Delta G^\circ = 2\Delta G^\circ_f\,(Fe^{2+}) + \Delta G^\circ_f\,(I_2) - 2\Delta G^\circ_f\,(I^-) - 2\Delta G^\circ_f\,(Fe^{3+})$$
$$= 2(-78.9) + 0 - 2(-51.57) - 2(-4.7) = -45.3 \text{ kJ}$$

To obtain $\Delta\mathscr{E}^\circ_{cell}$ in volts using Eq. $\Delta G^\circ = -n\mathscr{F}\Delta\mathscr{E}^\circ_{cell}$, ΔG° must be in joules, since volts × coulombs = joules

For the net cell reaction, as it is written above, $n = 2$. Substitution into above Eq. yields

$$\Delta G^\circ = -4.53 \times 10^4 \text{ J} = -2(29{,}485 \text{ C})\Delta\mathscr{E}^\circ_{cell} \quad \text{so that} \quad \Delta\mathscr{E}^\circ_{cell} = 0.235 \text{ V}$$

(*Note:* ΔG° is negative and $\Delta\mathscr{E}^\circ_{cell}$ is positive. Both these criteria predict the reaction written occurs spontaneously)

71. (a) When M_1 and M_2 are both 1.0 *M*, all substance are in their standard states and the cell voltage is equal to the standard cell potential, $\Delta\mathscr{E}^\circ_{cell}$, which has already been determined to be 1.562 V. The net cell reaction for the cell, is

$$Zn(s) + 2Ag^+(aq) \rightarrow 2Ag(s) + Zn^{2+}(aq)$$

and therefore Q, the reaction quotient, is given by $Q = \dfrac{[Zn^{2+}]}{[Ag^+]^2}$

If both the $[Zn^{2+}]$ and $[Ag^+]$ are 1.0 M, $Q = 1$ and $\ln Q = 0$, so that the Nernst equation, becomes simply

$\Delta\mathscr{E}_{cell} = \Delta\mathscr{E}^{\circ}_{cell} = 0.799\ V - (-0.763\ V) = 1.562\ V$

(b) When $[Zn^{2+}] = 0.1\ M$ and $[Ag^+] = 0.;01\ M$

$$Q = \frac{(0.1)}{(0.01)^2} = \frac{1.0 \times 10^{-1}}{1.0 \times 10^{-4}} = 1.0 \times 10^{+3} \quad \text{and} \quad \log Q = 3.0$$

For this reaction $n = 2$, because the two half-reactions that combine to give the net cell reaction are

Reduction $2Ag^+ + 2e^- \rightarrow 2Ag(s)$

Oxidation $Zn(s) \rightarrow Zn^{2+} + 2e^-$

Thus Nernst equation

$\Delta\mathscr{E}_{cell} = 1.562 - (0.0592/2)\log\ (1.0 \times 10^{+3}) = 1.562 - (0.0296)\ (3.0)$

$= 1.562 - 0.089 = 1.473\ V$

Note that when $Q > 1$ the cell potential is less than the standard cell potential.

(c) When $[Zn^{2+}] = 1.2 \times 10^{-2}\ M$ and $[Ag^+] = 2.0 \times 10^{-1}\ M$

$$Q = \frac{[Zn^{2+}]}{[Ag^+]^2} = \frac{1.2 \times 10^{-2}}{4.0 \times 10^{-2}} = 0.3 = 3.0 \times 10^{-1}$$

Since, $\ln(0.3) = -1.2$, the Nernst equation becomes

$\Delta\mathscr{E}_{cell} = 1.562 - (0.02569/2)\ (-1.2) = 1.562 + 0.015 = 1.577\ V$

When $Q < 1$, the cell potential is greater than the standard cell potential.

72. Use Tables to look up the standard reduction potential of the $Cl_2(g)|Cl^-$ electrode. Its value is 1.358 V. For this cell,

$$\Delta\mathscr{E}^{\circ}_{cell} = \mathscr{E}^{\circ}_{Cl_2|Cl^-} - \mathscr{E}^{\circ}_{Cu^{2+}|Cu} = 1.358\ V - 0.34\ V = 1.018\ V$$

The half-cell reactions are

Reduction	$Cl_2(g) + 2e^- \rightarrow 2Cl^-$
Oxidation	$Cu(s) \rightarrow Cu^{2+} + 2e^-$
Net cell reaction	$Cu(s) + Cl_2(g) \rightarrow Cu^{2+} + 2Cl^-$

So that the reaction quotient is $Q = \dfrac{[Cu^{2+}][Cl^-]^2}{P_{Cl_2}}$

The Nernst equation for this cell at 298 K is

$\Delta\mathscr{E}_{cell} = 1.018\ V - (RT/2\mathscr{F})\ln Q = 1.018 - (0.01285)\ln Q$

We can therefore calculate the value of Q required to make the cell potential 1.122 V. Substitution into the Nernst equation yields

$1.122 = 1.018 - (0.01285)\ln Q$

Therefore, $0.104 = (-0.01285)\ln Q$ and $\ln Q = -8.09$

Hence, $Q = e^{-8.09} = 3.1 \times 10^{-4}$

Since, $[Cl^-] = 0.1\ M$ in this cell, and $P_{Cl_2} = 1.0$ atm,

$Q = 3.1 \times 10^{-4} = [Cu^{2+}](1.0 \times 10^{-2})(/1.0$

and $[Cu^{2+}] = 3.1 \times 10^{-2}\ M$

Thus, if $[Cu^{2+}] = 3.1 \times 10^{-2}\ M$, the cell voltage will be 1.122 V.

73. (a) The cell reactions are

Reduction	$Zn^{2+}(0.48\ M) + 2e^- $	$\rightarrow Zn(s)$
Oxidation	$Zn(s)$	$\rightarrow Zn^{2+}(0.024\ M) + 2e^-$
Net cell reaction	$Zn^{2+}(0.48\ M)$	$\rightarrow Zn^{2+}(0.024\ M)$

The reaction quotient, Q, for this reaction is

$Q = (0.024)/(0.48) = 5.0 \times 10^{-2}$

and therefore $\log Q = \log(5.0 \times 10^{-2}) = -1.3$. Substituting into Eq.

$$\Delta\mathscr{E}_{cell} = \left(\frac{0.05916}{n}\right)\log\frac{M_1}{M_2} \text{ we get,}$$

$$\Delta\mathscr{E}_{cell} = -\left(\frac{0.0592}{2}\right)\log Q = -\left(\frac{0.0592}{2}\right)(-1.30) = +0.0358 \text{ V}$$

(b) If water is added to the more dilute solution, the disparity between the two concentrations will become greater, and the voltage will increase. Specifically the reaction quotient Q will become smaller, and $\log Q$ will become more negative making $\Delta\mathscr{E}_{cell}$ larger. If more dilute solution becomes 0.012 M, the reaction quotient will become $Q = (0.012)/(0.48) = 2.5 \times 10^{-2}$, and $\log Q = -1.60$. The cell potential after dilution is therefore

$\Delta\mathscr{E}_{cell} = -0.0592/2)\ (-1.60) = +0.0474$ V

Any change that causes the reaction quotient, Q, to be further unity (that causes the concentrations to be further apart) will increase the absolute value of the EMF of a concentration cell.

74. (a) Write the two half-cell reactions and calculate the standard cell potential. For this reaction we have

Reduction	$Br_2(aq) + 2e^- \rightarrow 2Br^-(aq)$
Oxidation	$Cu(s) \rightarrow Cu^{2+}(aq) + 2e^-$

Hence, $\Delta\mathscr{E}^\circ_{cell} = \mathscr{E}^\circ_{Pt|Br_2,\,Br^-} - \mathscr{E}^\circ_{Cu^{2+}|Cu} = 1.087 - 0.340 = 0.747$ V

The value of n for this reaction is 2. We have the equation $l_n K_{eq} = \dfrac{(n\Delta\mathscr{E}^\circ_{cell})}{0.02569}$.

$$\text{or, } \ln K_{eq} = \frac{(2)(0.747)}{0.02569} = 58.2$$

$\therefore\ K_{eq} = e^{58.2} = 2 \times 10^{25}$

Both the large value of K_{eq} and the positive sign of $\Delta\mathscr{E}^\circ_{cell}$ convey the same information about this reaction; Cu(s) and Br_2 in aqueous solution will spontaneously react to form a solution of cupric bromide. The express for the equilibrium constant is

$$K_{eq} = \frac{[Cu^{2+}][Br^-]^2}{[Br_2]} = 2 \times 10^{25}$$

(b) The two half-reactions are

Reduction	$2H^+(aq) + 2e^- \rightarrow H_2(g)$
Oxidation	$2Ag(s) \rightarrow 2Ag^+(aq) + 2e^-$

So that the value of n for this reaction is 2. The expression for the equilibrium constant is

$$K_{eq} = \frac{P_{H_2}[Ag^+]^2}{[H^+]^2}$$

$$\Delta\mathscr{E}^\circ_{cell} = \mathscr{E}^\circ_{H^+|H_2|Pt} - \mathscr{E}^\circ_{Ag^+|Ag} = 0 - 0.799 = -0.799 \text{ V}$$

Now, we have,

$$\ln K_{eq} = \frac{(2)(-0.799)}{0.02569} = -62.2$$

$$\therefore\ K_{eq} = e^{-62.2} = 1 \times 10^{-27}$$

Note that both the negative sign of $\Delta\mathscr{E}^\circ_{cell}$ and the very small value of K_{eq} tell you the same thing about this reaction: It proceeds to a negligible extent. There will be virtually no reaction if a piece of silver metal is inserted into an aqueous acidic solution.

(c) The two half-reactions are now

Reduction $\quad H^+(aq) + e^- \rightarrow \frac{1}{2}H_2(g)$

Oxidation $\quad Ag(s) \rightarrow Ag^+(aq) + e^-$

so that the value of n is 1. When the equation is written in this way the expression for the equilibrium constant is

$$K_{eq} = \frac{[Ag^+]P_{H_2}^{1/2}}{[H_3O^+]}$$

$\Delta\mathscr{E}^\circ_{cell} = -0.799$ V. The standard cell potential does not change because we have multiplied the net cell reaction of part (b) by $\frac{1}{2}$. Now, since $n = 1$,

$$\therefore\ \ln K_{eq} = (-0.799)/(0.02569) = -31.1$$

$$\text{or } K_{eq} = e^{-31.1} = 3 \times 10^{-14}$$

75. We must first write the two half-reactions.

Reduction $\quad 4H^+(aq) + O_2(g) + 4e^- \rightarrow 2H_2O$

Oxidation $\quad 4Fe^{2+} \rightarrow 4Fe^{3+} + 4e^-$

Net cell reaction $\quad 4Fe^{2+} + O_2(g) + 4H^+(aq) \rightarrow 4Fe^{3+} + 2H_2O$

With the half-cell potential values from Tables we get,

$$\Delta\mathscr{E}^\circ_{cell} = \mathscr{E}^\circ_{cathode} - \mathscr{E}^\circ_{anode} = 1.229 - 0.77 = 0.459 \text{ V}$$

Since $\Delta\mathscr{E}^\circ_{cell}$ is positive, the reaction will proceed to the right spontaneously if all substances are in their standard states. As the value of n for this reaction is 4.

$$\therefore\ \Delta G^\circ = -4(96{,}485 \text{ C})(0.459 \text{ V}) = -1.77 \times 10^5 \text{ J} = -177 \text{ kJ}$$

The large negative value of ΔG° also tells us that the reaction proceeds spontaneously to the right. We can obtain the equilibrium constant.

$$\Delta G^\circ = -177 \text{ kJ} = -RT \ln K_{eq}$$

$\therefore$ $\ln K_{eq} = \dfrac{177}{(8.31 \times 10^{-3})(298)} = 71.4$

So that $K_{eq} = e^{-71.4} = 1 \times 10^{31}$

This is a very large equilibrium constant. Ferrous ions in acidic solution are indeed readily air-oxidized to ferric ions.

76. (a) First write the half-reactions.

Cathode reaction $\quad Ni^{2+} + 2e^- \rightarrow Ni(s)$

Anode reaction $\quad Co(s) \rightarrow Co^{2+} + 2e^-$

Net cell reaction $\quad Ni^{2+} + Co(s) \rightarrow Ni(s) + Co^{2+}$

(b) $\Delta\mathscr{E}^\circ_{cell} = \mathscr{E}^\circ_{cathode} - E^\circ_{anode} = -0.23 - (-0.28) = +0.05$ V

$\Delta G^\circ = -n\mathscr{F}\Delta\mathscr{E}^\circ_{cell} = -(2)\ (96{,}485\ )\ (0.05\ V) = -9.6 \times 10^3\ J = -10\ kJ$

We now calculate the equilibrium constant

$\ln K_{eq} = (n\Delta\mathscr{E}^\circ_{cell})/0.02569 = (2)\ (0.05)/0.02569 = 3.8$

$\therefore \quad K_{eq} = e^{3.8} = 4 \times 10^1$

Note : We use one significant figure, as $\Delta\mathscr{E}^\circ_{cell}$ is only known to one figure.

(c) $Q = [Co^{2+}]/[Ni^{2+}] = c_1/c_2 = 1.0/0.001 = 1.0 \times 10^3$

$\Delta\mathscr{E}_{cell} = \Delta\mathscr{E}^\circ_{cell} - (0.0592/n)\log Q = +0.05 - (0.0592/2)\log 10^3$

$= 0.05 - (3)\ (0.0592)/2 = 0.05 - 0.089 = -0.04$ V

$\Delta G = -n\mathscr{F}\Delta\mathscr{E}_{cell} = -2(96{,}485\ C)\ (-0.04\ V) = +7.7 \times 10^3\ J = 8\ kJ$

(d) When all substance are in their standard states $c_1 = c_2 = 1.0\ M$, and $Q = 1$. Since $K_{eq} = 40$, $Q < K_{eq}$, and the reaction will proceed spontaneously to the right. Alternatively, one could say that since $\Delta\mathscr{E}^\circ_{cell} = +0.05 > 0$, the reaction will proceed spontaneously to the right when all substances are in their standard states. A third way of reaching the same conclusion is to note that when all substances are in their standard states $\Delta G = \Delta G^\circ = -10$ kJ. Since $\Delta G < 0$, the reaction will proceed spontaneously to the right.

(e) When $[CO^{2+}] = 1.0\ M$ and $[Ni^{2+}] = 1.0 \times 10^{-3}\ M$,

$\Delta\mathscr{E}_{cell} = -0.02$ V and $Q = 1.0 \times 10^3$ and $\Delta G + 8$ kJ

Since $\Delta\mathscr{E}_{cell} = < 0$, $Q > K_{eq}$, and $\Delta G > 0$, the net cell reaction given in part (a) will not proceed spontaneously. It is the reverse reaction that proceeds spontaneously: $Ni(s) + CO^{2+} \rightarrow Co(s) + Ni^{2+}$.

77. The half-reactions that must be combined are

$Cu(OH)_2(s) + 2e^- \rightleftharpoons Cu(s) + 2\ OH^- \qquad \mathscr{E}^\circ = -0.224$ V

$Cu^{2+} + 2e^- \rightleftharpoons Cu(s) \qquad \mathscr{E}^\circ = +0.34$ V

To yield the solubility product equation $Cu(OH)_2(s) \rightleftharpoons Cu^{2+} + 2\ OH^-$

We must make the second of these half-reactions an oxidation. Note that $n = 2$ for the half-reactions that are combined. For the galvanic cell

$Cu(s)|\ Cu^{2+}\ ||\ OH^-\ |\ Cu(OH)_2\ |\ Cu(s)$

$\Delta\mathscr{E}^\circ_{cell} = -0.224 - 0.34 = -0.564$ V

$$\ln K_{sp} = \frac{(2)(-0.564)}{0.02569} = -43.9$$

$$\therefore \qquad K_{sp} = e^{-43.9} = 9 \times 10^{-20}$$

78. (a) The atomic weight of aluminium is 26.98, so that 1000 g is 1000/26.98 = 37.06 mol of Al. Now, $Al^{3+} + 3^{e-} \rightarrow Al(l)$, thus 3$\mathscr{F}$ of charge are required for each mole of Al produced, so that 3(37.06) = 111.2 $\mathscr{F}$ are needed per kilogram of Al. As 1 $\mathscr{F}$ = 96,485 C, the number of coulombs required to produce a kilogram of Al is

(111.2 $\mathscr{F}$)(96,485 C/$\mathscr{F}$) = 1.073×10^7 C.

(b) Substituting into equation $Q = It$, yields

$$1.073 \times 10^7 \text{ C} = (100 \text{ A})\ (t)$$

where t is the time in seconds. Therefore the time it will take to produce 1 kg of Al with a current of 100 A is

$t = 1.073 \times 10^5$ s = 1788 min = 29.8 h

(c) Five minutes is 300 s. The current required to produce 1 kg of Al in 300 s is

$I = (1.073 \times 10^7 \text{ C})/(3 \times 10^2 \text{ s}) = 3.58 \times 10^4$ A or 35,800 A

79. (a) It is more difficult to reduce K^+ than to reduce H_2O or H^+ (aq) ions, and it is more difficult to oxidisze SO_4^{2-} ions than it is so oxidize H_2O or OH^- ions, so that the electrolysis of aqueous K_2SO_4 is just the electrolysis of water

Cathode $\qquad 4H_2O + 4e^- \rightarrow 2H_2(g) + 4OH^-(aq)$

Anode $\qquad 2H_2O \rightarrow O_2(g) + 4e^- + 4H^+(aq)$

Overall reaction $\qquad 2H_2O \rightarrow 2H_2(g) + O_2(g)$

On the right-hand side we have combined 4 OH^- + $4H^+$ to form $4H_2O$, so that there is a net of $2H_2O$ on the left.

(b) Calcium is below H_2 in the Table of Standard Reduction Potentials so that H_2 will be formed at the cathode, but I^- is readily oxidize to I_2.

Cathode $\qquad 2H_2O + 2e^- \rightarrow H_2(g) + 2OH^-$

Anode $\qquad 2I^- \rightarrow I_2 + 2e^-$

Overall reaction $\qquad 2H_2O + 2I^- \rightarrow H_2(g) + 2OH^- + I_2$

Thus the products of the electrolyis of an aqueous solution of CaI_2 are $H_2(g)$, I_2, and an aqueous solution of $Ca(OH)_2$.

(c) Mercury is above H_2 in the Table of Standard Reduction Potentials, so that Hg will be formed at the cathode. It is easier to oxidize H_2O than nitrate ions, and therefore O_2 will be formed at the anode.

Cathode $\qquad Hg^{2+} + 2e^- \rightarrow Hg(l)$

Anode $\qquad 2H_2O \rightarrow O_2(g) + 4e^- + 4H^+(aq)$

Overall reaction $\qquad 2Hg^{2+} + 2H_2O \rightarrow O_2(g) + 2Hg(l) + 4H^+(aq)$

(d) Cupric ions wil be reduced; bromide ions will be oxidize

Cathode $\qquad Cu^{2+} + 2e^- \rightarrow Cu(s)$

anode $\qquad 2Br^- \rightarrow Br_2 + 2e^-$

Overall reaction $\qquad Cu^{2+} + 2Br^- \rightarrow Cu(s) + Br_2$

80. (a) For Step 1.

$\Delta G^\circ = -nFE^\circ$ Substituting the values, we have

$\Delta G^\circ = -(2 \text{ mol.}) (9.648 \times 10^4 \text{ J/mol} \cdot \text{V}) (-0.581 \text{ V}) = 1.12 \times 10^5 \text{ J}$

For step 2:

$\Delta G^\circ = -nFE^\circ$

$\Delta G^\circ = -(2 \text{ mol}) (9.648 \times 10^4 \text{ J/mol} \cdot \text{V}) (-0.197 \text{ V}) = 3.80 \times 10^4 \text{ J}$

Overall:

$\Delta G^\circ = 1.12 \times 10^5 \text{ J} + 3.80 \times 10^4 \text{ J} = 1.50 \times 10^5 \text{ J}$

(b) In the overall process, 4 electrons are transferred

$\Delta G^\circ = -nFE^{\circ\prime}$

$\therefore 1.50 \times 10^5 \text{ J} = -(4 \text{ mol}) (9.648 \times 10^4 \text{ J/mol} \cdot \text{V})(E^\circ)$

or $E^\circ = -0.389$

81.

$$H_2(g) \rightarrow 2H^+(aq) + 2e^- \qquad E^\circ_{OX} = 0.0 \text{ V}$$

$$2H^+(aq) + 2e^- \rightarrow H_2(g) \qquad E^\circ_{red} = 0.0 \text{ V}$$

$$H_{2(anode)}(g) + 2H^+_{(cathode)}(aq) \rightarrow 2H^+_{(anode)}(aq) + H_{2\,(cathode)}(g) \qquad E^\circ = 0.00 \text{ V}$$

$$[H^+]_{(anode)} = 10^{-pH} = 10^{-7.0} = 1.0 \times 10^{-7}\ M$$

$$[H^+]_{(cathode)} = 10^{-pH} = 10^{0.0} = 1.0\ M$$

$$E = E^\circ - \frac{0.0257}{n} \cdot \ln \frac{(P_{H_2})_{cathode}[H^+]^2_{anode}}{(P_{H_2})_{anode}[H^+]^2_{cathode}} = 0.0 - \frac{0.0257}{2} \cdot \ln \frac{(1.0)(1.0 \times 10^{-7})^2}{(1.0)(1.0)^2}$$

$$E = 0.414 \text{ V}$$

82. 1.

Reduction Rxn	*Voltages Using Zn*	*E°_{red} voltages using H_2 (Chart)*	*Difference*
$Ag^+ + e^- \rightarrow Ag$	1.4	0.8	0.6
$Cu^{2+} + 2e^- \rightarrow Cu$	0.99	0.34	0.65
$Fe^{3+} + 3e^- \rightarrow Fe$	0.55	–0.04	0.59
$Mg^{2+} + 2e^- \rightarrow Mg$	0.6	–2.37	2.97
$Pb^{2+} + 2e^- \rightarrow Pb$	0.48	–0.13	0.61
$Zn^{2+} + 2e^- \rightarrow Zn$	0.0	–0.76	0.76

2. Compairing potentials of predicted values with measured values as obtained from tables.

Redox Rxn	*Predicted Voltages*	*Measured Voltage*
$Cu^{2+}_{(aq)} + Fe_{(s)} \rightarrow Cu_{(s)} + Fe^{2+}_{(aq)}$	0.99 + (–0.55) = 0.44	0.5
$Pb^{2+}_{(aq)} = Mg_{(s)} \rightarrow Pb_{(s)} + Mg^{2+}_{(aq)}$	0.48 + 0.6 = 1.08	1.0
$Cu^{2+}_{(aq)} + Pb_{(s)} \rightarrow Cu_{(s)} + Pb^{2+}_{(aq)}$	0.99 + (–0.48) = 0.51	0.51

3. We have from above

$$Zn_{(s)} + Cu^{2+}_{(aq)} \rightarrow Zn^{2+}_{(aq)} + Cu_{(s)} \qquad E^\circ = 1.0 \text{ V}$$

$$E = E° \frac{0.0592}{n} \log Q = 1.0 \text{ V} - \frac{0.0592}{2} \log \frac{[Zn^{2+}]}{[Cu^{2+}]}$$

$$= 1.0 \text{ V} \frac{0.0592}{2} \log \frac{1}{0.01} = 0.94 \text{ V}$$

This compares favorably to the 1.01 V that was actually measured.

4. We have from data above

$$Zn_{(s)} + 2Ag^{2+}_{(aq)} \rightarrow Zn^{2+}_{(aq)} + 2Ag_{(s)} \qquad E° = 1.56 \text{ V}$$

oxidation $Zn_{(s)} \rightarrow$ && + $2e^-$ $\qquad E^\circ_{ox} = 0.76$ V

reduction $Ag^{2+}_{(aq)} + e^- \rightarrow Ag_{(s)}$ $\qquad E^\circ_{red} = 0.8$ V

$$E° = E^\circ_{ox} + E^\circ_{red} = 0.76 \text{ V} + 0.80 \text{ V} + 1.56 \text{ V}$$

$$E = E° \frac{0.0592}{n} \log Q = E° - \frac{0.0592}{2} \log \frac{[Zn^{2+}]}{[Ag^+]^2}$$

$$0.91 = 1.56 = \frac{0.0592}{2} \log \frac{1}{[Ag^+]^2}$$

$$\log \frac{1}{[Ag^+]^2} = 21.959 \Rightarrow \frac{1}{[Ag^+]^2} = 9.1 \times 10^{21}$$

$\therefore$ $[Ag^+] = 1.0 \times 10^{-11}$

$\therefore$ $K_{sp} = [Ag^+]\,[Cl^-] = (1.0 \times 10^{-11}) \times (1) = 1.0 \times 10^{-11}$

The actual K_{sp} for AgCl is 1.56×10^{-10} at 298 K

5. From the data in problem we get:

$Zn_{(s)} + Cu^{2+}_{(aq)} \rightarrow Zn^{2+}_{(aq)} + Cu_{(s)}$; $E = 0.56$ V; $E° = 1.1$ V

In a 6.0 M NH_3 solution most of the $Cu^{2+}_{(aq)}$ will combine with NH_3 to form $Cu(NH_3)^{2+}_{4(aq)}$ as evidenced by the magnitude of K_f.

$$Cu^{2+}_{(aq)} + NH_{3(aq)} \rightleftharpoons Cu(NH_3)^{2+}_{4(aq)}$$

It is the uncomplexed $Cu^{2+}_{(aq)}$ ion concentration that needs to be determined through the Nernst equation

$$E = E° - \frac{0.0592}{n} \log Q = E° - \frac{0.0592}{2} \log \frac{[Zn^{2+}]}{[Cu^{2+}]}$$

or, $\qquad 0.56 \text{ V} = 1.10 \text{ V} - \frac{0.0592}{2} \log \frac{1}{[Cu^{2+}]}$

$\therefore \qquad \log \frac{1}{[Cu^{2+}]} = 18.2 \Rightarrow \frac{1}{[Cu^{2+}]} = 1.75 \times 10^{18} \text{ M}^{-1}$

Hence, $\qquad [Cu^{2+}] = 5.7 \times 10^{-19}$ M

Since 20 drops = 1 mL, therefore, 1 drop = 0.05 mL; and 2 drops were added,
$\therefore\ V_1 = 0.1$ mL

$$\therefore \qquad [Cu(NH_3)_4^{2+}] = \frac{V_1 \times M_1}{V_2} = \frac{0.1\,\text{mL} \times 1.0\,\text{M}}{10.1\,\text{mL}} = 0.01\ \text{M}$$

The formation constant, K_f, is derived from the following equation
$Cu^{2+} + 4NH_3 \rightleftharpoons Cu(NH_3)_4^{2+}$

$$K_f = \frac{[Cu(NH_3)_4^{2+}]}{[Cu^{2+}][NH_3]^4} = \frac{0.099}{(5.71 \times 10^{-19}) \times (6.0)^4} = 1.3 \times 10^{13}$$

$$\%\ \text{eror} = \frac{\text{obs} - \text{exp}}{\text{exp}} \times 100\% = \frac{1.3 \times 10^{13} - 2 \times 10^{13}}{2 \times 10^{13}} \times 100\% = -3\%$$

83. The standard emf of this call is calculated by summing the electrode potentials of the oxidizing and reducing agents. The use $E°$ cell to calculate the $\Delta G°$ from $-nFE°$ cel.

$3Zn(s) \rightarrow 3Zn^{2+}(aq) + 6e^-$ $\qquad -E^{\circ}_{Zn} = 0.76$ V
$2Cr^{3+}(aq) + 6e^- \rightarrow 2Cr(s)$ $\qquad E^{\circ}_{Cr} = -0.74$ V
$2Cr^{3+}(aq) + 3Zn(s) \rightarrow 3Zn^{2+}(aq) + 2Cr(s)$ $\qquad E^{\circ}_{cell} = 0.02$ V

$\Delta G° = -nFE^{\circ}_{cell} = -(6 \times 9.65 \times 10^4\ C \times 0.02\ V) = -1.1580 \times 10^4\ J = -11.58$ kJ

Use the data from tables, and calculate $\Delta H°$ for the overall cell reaction:

$$2Cr^{3+}(aq) + 3Zn(s) \rightarrow 3Zn^{2+}(aq) + 2Cr(s)$$
$$2(-1971) \qquad 3(0) \qquad 3(-152.4) \qquad 2(0)$$

$\Delta H° = (3\Delta H^{\circ}_f(Zn^{2+}) + 2\Delta H^{\circ}_f(Cr) - [2\Delta H^{\circ}_f(Cr^{3+}) + 3\Delta H^{\circ}_f(Zn)]$

$\Delta H° = [3(-152.4)\ \ 2(0)[\ -\ [2(-1971) + 3(0)]$

$\Delta H° = 3484.8 = 3485$ kJ

Now calculate $\Delta S°$ from the $\Delta G° = \Delta H° - T\Delta S°$ relation.

$T\Delta S° = \Delta H° - \Delta G° = 3484.8\ \text{kJ} - (-11.58\ \text{kJ}) = 3496.38$ kJ

$\Delta S° = 3496.38\ \text{kJ} \div 298\ \text{K} = 11.73 = 11.7\ \text{kJ K}^{-1}$

84. First calculate E°_{cell}

$E^{\circ}_{cell} = -(1.07\ V) + (1.23\ V) = 0.16$ V

The cell reaction is $O_2(g) + 4H^+(aq) + 4Br^-(aq) \rightarrow 2H_2O(l) + 2Br_2(l)$; $n = 4$

Convert pH to $[H^+]$: $[H^+]$ = antilog (–3.6) = 2.51×10^{-4}

Substitute $E^{\circ}_{cell} = 0.16$ V, 1 M Br^-, 1 M Br_2, and 2.51×10^{-4} M H^+ into the Nernst equation and solve for E_{cell}.

$$E_{cell} = E^{\circ}_{cell} - \frac{0.0592}{n}\log Q = 0.16\ V - \frac{0.0592}{4}\log\frac{1}{(2.51 \times 10^{-4})^4}$$

$$= 0.16\ V - \frac{0.0592}{4}\log(2.519 \times 10^{14}) = 0.16\ V - \frac{0.0592}{4}(14.401)$$

or, $E_{cell} = 0.16\ V - 0.2131\ V = -0.0531 = -0.05$ V

85. We have E°_{cell} was found to be 0.16 V in problem 86. Neglecting ionization of HOCN, the $[H^+]$ of the buffer is

$$[H^+] = 3.5 \times 10^{-4} \times \frac{(0.1\,M\,HOCN)}{(0.1\,M\,OCN^-)} = 3.5 \times 10^{-4}\ M$$

From the Nernst equation for the reaction of O_2 with $4Br^-$ and $4H^+$

$$E_{cell} = E^\circ_{cell} - \frac{0.0592}{n}\log Q = 0.16\ V - \frac{0.0592}{4}\log\frac{1}{(3.5\times 10^{-4})^4}$$

$$= 0.16\ V - \frac{0.0592}{4}\log(6.6639 \times 10^{13}) = 0.16\ V - \frac{0.0592}{4}(13.8237)$$

$\therefore E_{cell} = 0.16\ V - 0.20459\ V = -0.04459 = -0.04\ V$ (nonspontaneous reaction)

86. The cell reaction is $1/2H_2(g) + Ag^+(aq) \rightarrow Ag(s) + H^+(aq)$; $n = 1$. Now, calculate E°_{cell}:

$E^\circ_{cell} = -0.0\ V + 0.8\ V = 0.8\ V$

Substitute $E_{cell} = 0.45$ V, $E^\circ_{cell} = 0.8$ V, and 1.0 M H^+ into the Nernst equation and solve for log $[Ag^+]$

$$\therefore \quad E_{cell} = E^\circ_{cell} - \frac{0.0592}{n}\log Q;\ \text{or, } 0.46\ V = 0.80\ V = \frac{0.0592}{1}\log\frac{(1.0)}{[Ag^+]}$$

$$\therefore \quad \log [Ag^+] = \frac{(0.45 - 0.80)}{0.0592} = -5.912\ V$$

$[Ag^+] = 10^{-5.912} = 1.224 \times 10^{-6}$ m

$K_{sp} = [Ag^+][SCN^-] = (1.224 \times 10^{-6})(0.1) = 1.224 \times 10^{-7} = 1 \times 10^{-7}$

87. Write the equation for the cell reaction and evaluate Q. Use the Nernst equation to determine the effect on E.

$$Co(s) \rightarrow Co^+(aq) + 2e^-$$

$$2H^+(aq) + 2e^- \rightarrow H_2(g)$$

$$Co(s) + 2H^+(aq) \rightarrow H_2(g) + Co^{2+}(aq)$$

$$E = E^\circ - \frac{RT}{nF}\cdot \ln Q = E^\circ - \frac{0.0257\ V}{2}\cdot \ln\frac{[Co^{2+}](P_{H_2})}{[H^+]^2}$$

(a) The voltage is dependent on the concentration of Co^{2+}, not the volume. So changing the volume will have no effect on the voltage of the cell.

(b) The voltage of the cell is dependent on the concentration of H^+. Increasing $[H^+]$ will decrease Q, which will decrease in Q. Consequently, the value being subtracted from E° will be smaller, and E will be larger.

(c) The voltage of the cell is dependent on the pressure of H_2. Increasing $P(H_2)$ will increase Q, which will increase ln Q. Consequently, the value being substracted from E° will be larger, and E will be smaller.

(d) the voltage of the cell does not depend on the amount of Co(s) present. Changing this value will have no effect on the voltage of the cell.

(e) The voltage of the cell does not depend on the surface area of the electrode. Changing this value will have no effect on the voltage of the cell.

Thus, only (b) will increase the voltage of the cell (E)

88.

$$Ag(s) \rightarrow Ag^+(aq) + e^- \qquad E^\circ_{ox} = -0.799 \text{ V}$$
$$AgSCN(s) + e^- \rightarrow Ag(s) + SCN^-(aq) \qquad E^\circ_{red} = 0.0895 \text{ V}$$

$$AgSCN(s) \rightarrow Ag^+(aq) + SCN^-(aq) \qquad E^\circ = -0.170 \text{ V}$$

Note that the equation above is the equation for K_{sp}. Thus one can use the Nernst equation to calculate K_{sp} from the calculated E° for the cell.

$$E^\circ = \frac{RT}{nF} \ln K = -0.710 \text{ V} = \frac{0.0257}{1} \ln K$$

or, $\ln K = -27.6$

$\therefore$ $K = K_{sp} = e^{-27.6} = 1.0 \times 10^{-22}$

89. Calculate E°, $[H^+]$ in the acid solution, and from K_a, the $[SO_4^{2-}]$. Substitute these value into the Nernst equation and calculate E.

$$Pb(s) + SO_4^{2-}(aq) \rightarrow PbSO_4(s) + 2e^- \qquad E^\circ_{ox} = 0.356 \text{ V}$$
$$PbO_2(s) + SO_4^{2-}(aq) + 4H^-(aq) + 2e^- \rightarrow PbSO_4(s) + 2H_2O(l) \qquad E^\circ_{red} = 1.687 \text{ V}$$

$$Pb(s + PbO_2(s) + 2SO_4^{2-}(aq) + 4H^+(aq) + 2e^- \rightarrow 2PbSO_4(s) + 2H_2O(l) + 2e^-$$
$$\therefore\ Pb(s) + PbSO_2(s) + 2SO_4^{2-}(aq) + 4H^+(aq) \rightarrow 2PbSO_4(s) + 2H_2O(l) \qquad E^\circ = 2.043 \text{ V}$$

Now, $\dfrac{1.286 \text{ g soution}}{1 \text{ cm}^2} \times \dfrac{1 \text{ cm}^3}{1 \text{mL}} \times \dfrac{1000 \text{ mL}}{1 \text{ L}} \times \dfrac{38 \text{ g } H_2SO_4}{100 \text{ g sol'n}} \times \dfrac{1 \text{ mol}}{98.1 \text{ g}} = 4.98 \text{ M } H_2SO_4$

$$[H^+] = [HSO_4^-] = [H_2SO_4] = 4.98\ M$$

$$K_a = \frac{[H^+][SO_4^{2-}]}{[HSO_4^-]} \Rightarrow 1.0 \times 10^{-2} = \frac{(4.98 + x)(x)}{(4.98 - x)}$$

$$K_a/a = 1.0 \times 10^{-2}/4.98 = 2.0 \times 10^{-3} < 2.5 \times 10^{-3}$$

$$\therefore\ 1.0 \times 10^{-2} = \frac{(4.98)(x)}{4.98} \Rightarrow x = [SO_4^{2-}] = 0.01\ M$$

$$\therefore\ E = E^\circ = \frac{0.0257}{n} \cdot \ln \frac{1}{[H^+]^4\,[SO_4^{2-}]^2} = 2.043 - \frac{0.0257}{2} \ln \frac{1}{(4.98)^4 (0.01)^2} = 2.01 \text{ V}$$

90. (a)

$$Zn(s) \rightarrow Zn^{2+}(aq) + 2e^- \qquad E^\circ_{ox} = 0.762 \text{ V}$$
$$Sn^{2+}(aq) + 2e^- \rightarrow Sn(s) \qquad E^\circ_{red} = -0.141 \text{ V}$$

$$Sn^{2+}(aq) + Zn(s) \rightarrow Zn^{2+}(aq) + Sn(s) \qquad E^\circ = 0.621 \text{ V}$$

(b) E° is positive, so the reaction is spontaneous in the forward direction. As the cell operates, $[Zn^{2+}]$ increases and $[Sn^{2+}]$ decrease.

(c) $E = E^\circ - \dfrac{0.0257}{n} \cdot \ln \dfrac{[Zn^{2+}]}{[Sn^{2+}]} \Rightarrow 0 = 0.621 - \dfrac{0.0257}{2} \cdot \ln \dfrac{[Zn^{2+}]}{[Sn^{2+}]}$

$$\ln \frac{[Zn^{2+}]}{[Sn^{2+}]} = 48.3 \Rightarrow \frac{[Zn^{2+}]}{[Sn^{2+}]} = 1 \times 10^{21}$$

(d) $[Zn^{2+}] = [Zn^{2+}] + x$ and $[Sn^{2+}] = [Sn^{2+}] - x$

$[Zn^{2+}] = 1.0\ M + x$ and $[Sn^{2+}] = 1.0\ M - x$

$$\therefore\ 1 \times 10^{21} = \frac{[Zn^{2+}]}{[Sn^{2+}]} = \frac{1.0\ M + x}{1.0\ M - x}$$

$$1 \times 10^{21} - 1 \times 10^{21}\ x = 1 + x$$
$$1 \times 10^{21} = 1 \times 10^{21}\ x$$
$$x = 1.0$$
$$[Zn^{2+}] = 1.0\ M + x = 2.0\ M$$
$$[Sn^{2+}] = 1.0\ M - x = 0.0\ M \text{ or}$$
$$[Sn^{2+}] = [Zn^{2+}]/1 \times 10^{21} = 2.0\ M/1 \times 10^{21} = 2 \times 10^{-21}\ M\ (\approx 0.0\ M)$$

Since K is so large, the reaction lies almost completely to the right at equilibrium (when $E = 0$). Thus we can assume that essentially all the Sn^{2+} reacts to form Zn^{2+}. Consequently, $[Zn^{2+}] = 1.0\ M + 1.0\ M = 2.0\ M$. The residual $[Sn^{2+}]$ can be calculated from

$$K = 1 \times 10^{21} = \frac{[Zn^{2+}]}{[Sn^{2+}]}$$

reduction (+1 → 0)

91. $Mn^{2+}_{aq} - 2H_2O \rightarrow \overset{+4}{MnO_2}(s) + \overset{+1}{2H^+_{aq}} - \overset{0}{H_2}(g)$

oxidation

Reaction occurring

(i) At cathode : $Ma^{2+} + 2OH^- \rightarrow MnO_2 + 2H^+$

(ii) At Anode : $H_2O \rightarrow \frac{1}{2} H_2 + OH^-$

or, $2H_2O \rightarrow H_2 + 2OH^-$

Given, $i = 27$ A and $t = 24$ hrs $= 24 \times 3600$ sec.

Mass of MnO_2 = 1kg = 1000 gm

Molar mass of MnO_2 = 55 + 32 = 87 gm.

Total charge = 27 amp × 3600 × 24 sec.

In 24 hrs. = 23.328×10^5 coulomb (i)

no. of moles of $MnO_2 = \frac{1000}{87} = 1.50$ moles (ii)

Current efficiency means no. of moles deposited per ampere of electric current per second

⇒ From (i) and (ii) we see that

23.328×10^5C electricity deposits 11.5 moles of MnO_2

$$\Rightarrow 1C \text{ deposits} = \frac{11.5}{23.328 \times 10^5} = 0.492 \times 10^{-5} \text{ moles}$$

Thus current efficiency value is 0.492×10^{-5} moles/C

92. Number of electronic charges on one $N^{3-} = 3$
Number of ions in 14 g of $N^{3-} = 6.023 \times 10^{23}$

Number of ions in 1 g of $N^{3-} = \dfrac{6.028 \times 10^{23}}{14}$

Number of electronic charges on 1 g $N^{3-} = \dfrac{6.023 \times 10^{23}}{14} \times 3$

Charge on 1 electron = 1.6×10^{-19} coulomb

$\therefore$ Charge on 1 g of $N^{3-} = \dfrac{6.023 \times 10^{23} \times 3 \times 1.6 \times 10^{-9}}{14}$ = **1.29×10^4 coulombs.**

93. The reaction $2H_2O \rightleftharpoons H_3O^+ + OH^-$ can be obtained as follows:

Reduction at RHC $H_2O + e^- \rightarrow \frac{1}{2}H_2 + OH^-$

Oxidation at LHC $H_2O + \frac{1}{2}H_2 \rightarrow H_3O^+ + e^-$

$$2H_2O \rightarrow H_3O^+ + OH^-$$

Thus, the standard e.m.f. ($E°$ cell) of the cell producing the above reaction is given by

$E^\circ_{cell} = E^\circ_{RHC} - E^\circ_{LHC} = -0.8277$ V

Now, $\log K = \dfrac{nFE°}{2.303\,RT} = \dfrac{1 \times 96500\,C(-0.8277\,V)}{2.303 \times 8.314\,JK^{-1}\,(298\,K)} = -13.9985$

$\therefore \quad K = 1.00 \times 10^{-14}$

94. $2Fe^{+3} + 3I^- \rightleftharpoons 2Fe^{+2} + I_3^-$

$E^\circ_{cell} = E^\circ_{Fe^{+3}/Fe^{+2}}$

Since, $-\Delta G = nFE° = RT \ln K_f$

$\ln K = \dfrac{2 \times 96500 \times 0.23}{8.314 \times 298}$

or, $K = 6.04 \times 10^{-7}$

95. Amount of electricity passed = 1.70×230 c

Number of faraday of electricity passed = $\dfrac{1.70 \times 230}{96500} = 0.00405$

Given that efficiency = 90%

Therefore, the number of faraday actually used = $0.00405 \times \dfrac{90}{100} = 0.00365$

Now the equation of electronic exchange is $Zn^{2+} + 2e^- \rightarrow Zn$

Thus number of moles of Zn^{2+} ions reduced = $\frac{0.00365}{2}$

Thus the strength of Zn^{2+} after reduction will be $0.160 - \left(\frac{0.00365}{2} \times \frac{1000}{300}\right) = 0.154\ M$

96. $CuSO_4 \rightarrow Cu^{+2} + SO_4^{-2}$

We have, $\frac{W}{E} = \frac{Q \cdot t}{F}$

Hence, number of faraday of electricity passed = $\frac{2 \times 10^{-3} \times 16 \times 60}{96500} = 1.9 \times 10^{-5}$ M

No. of g-equivalent of Cu^{+2} discharged = 1.9×10^{-5}

Hence initial concentration = $\frac{1.989 \times 2 \times 10^{-5}}{0.25 \times 2} = 7.95 \times 10^{-5}$

97. (a) Cell is Ag | Ag^+ || Cu^{+2} | Cu
At anode $2Ag \rightarrow Ag^+ + 2e^-$
At cathode $Cu^{+2} + 2e^- \rightarrow Cu$
Cell reaction is represented as : $2Ag + Cu^{+2} \rightarrow 2Ag^+ + Cu$

According to Nernst equation, $E = E^\circ_{cell} - \frac{RT}{nF} \ln \frac{[Ag^+]^2}{[Cu^{+2}]}$

Since $[Ag^+] = [Cu^{+2}]$
So $E = E^\circ_{cell}$

Amount of electricity passed $Q = 1 \cdot t = \frac{9.65 \times 60 \times 60}{96500} = 0.36\ F$

Thus amount of ions after passing the electricity for on hour are
$[Ag^+] = 1.36$; and $[Cu^{+2}] = 0.82$

$$E = E^\circ - \frac{0.0591}{2} \log \frac{[1.36]^2}{[0.82]}$$

$$\Delta E = E^\circ - E = \frac{0.059}{2} \log \frac{(1.36)^2}{(0.82)^2}$$

Now, $\Delta E = 0.0104$ V

98. **In the given reaction** Ce^{4+} is reduced to Ce^{3+} (and hence constitutes RHE) and Fe^{2+} is oxidized to Fe^{3+} (constitutes LHE). Hence the cell producing the given reaction is
Pt | Fe^{2+}, Fe^{3+} || Ce^{4+}, Ce^{3+} | Pt
Its standard potential is

$$E^\circ_{cell} = E^\circ_{Ce^{4+}/Ce^{3+}} - E^\circ_{Fe^{3+}/Fe^{2+}} = 1.44\ V - 0.68\ V = 0.76\ V$$

$$\Delta G^\circ = -RT \ln K_{eq}$$

where K_{eq} is the equilibrium constant but $\Delta G^\circ = -nF E^\circ_{cell}$

$\therefore$ $RT \ln K_{eq} = nF E^\circ_{cell}$

$$\text{or } \log K_{eq} = \frac{nFE^\circ_{cell}}{2.303\, RT} = \frac{1 \times (96500 \text{ C} \cdot \text{mol}^{-1}) \times (0.76 \text{ V})}{2.303 \times (8.314 \text{ JK}^{-1} \text{ mol}^{-1}) \times (298 \text{ K})} = 12.85$$

Taking anti log $\therefore$ $K_{ef} = 7.1 \times 10^{12}$

99. The two half reaction are

Oxidation $\frac{1}{2} H_2 \rightarrow H^+ + e^-$

Reduction $AgCl(s) + e^- \rightarrow Ag(s) + Cl^-$

(i) The cell reaction is : $1/2H_2(g) + AgCl(s) \rightarrow HClaq) + Ag(s)$

(ii) E° at the average temperature of 298 K $= \dfrac{0.23 + 0.21}{2} = 0.22$ V

$$\text{Now, } E^\circ = \frac{\Delta H^\circ}{nF} + T\left(\frac{dE}{dT}\right)_P \Rightarrow 0.22 = \frac{\Delta H^\circ}{96500} + 298\left(\frac{0.21 - 0.23}{20}\right) = \frac{\Delta H^\circ}{96500} = 0.298$$

$\Rightarrow$ $\Delta H^\circ = -49987$ kJ mole^{-1}

Since $\Delta G^\circ = \Delta H^\circ - T\Delta S^\circ$

$$\Rightarrow \Delta S^\circ = \frac{\Delta H^\circ - \Delta G^\circ}{T} = \frac{-49987 - (-96500 \times 0.22)}{298} = -96.5 \text{ JK}^{-1} \text{ mol}^{-1}$$

(iii) Given, $E^\circ = 0.22 - 0.8 = -0.58$ V

But, $E^\circ = 0.059 \log K_{sp}$

$\Rightarrow$ $-0.58 = 0.059 \log K_{sp}$

$\Rightarrow$ $K_{sp} = 1.48 \times 10^{-10}$ mol^{2}L^{-2}

100.

$Cu^{+2} + e^- \rightarrow Cu^+$ $\quad \Delta G^\circ_1 = -0.15$ F

$In^{+2} + e^- \rightarrow In^+$ $\quad \Delta G^\circ_2 + 0.4$ F

$In^+ \rightarrow In^{+3} + 2e^-$ $\quad \Delta G^\circ_3 = -0.84$ F

$Cu^{+2} + In^{+2} \rightleftharpoons Cu^+ + In^{+3}$ $\quad \Delta G^\circ = -0.59$ F

$-nFE^\circ = -0.59$ F

$-E^\circ_{cell}F = -0.59$

$E^\circ_{cell} = 0.59$

$$E^\circ_{cell} = E^\circ - \frac{0.0591}{n} \log K_c$$

$$0.59 = \frac{0.0591}{1} \log K_c$$

$K_c = 10^{10}$

101. Let the emf of 1st cell be x volt.

Emf of 2nd cell= $(x + 0.03)$ V

$[Cu^{2+}]$ in 2nd cell = 0.5 M

$[Cu^{2+}]$ In 1st cell = ?

$$E_1 = E_1^\circ - \frac{2.303\,RT}{2F}\log\frac{[Zn^{2+}]}{[Cu^{2+}]} \text{ and } E_2 = E_1^\circ - \frac{2.303\,RT}{2F}\log\frac{[Zn^{2+}]}{[Cu^{2+}]_2}$$

$$\therefore\ E_2 = E_1 = E_1^\circ - \frac{2.303\,RT}{2F}\log\frac{[Zn^{2+}]}{[Cu^{2+}]_1} = \left(E_1^\circ - \frac{2.303\,RT}{2F}\log\frac{[Zn^{2+}]}{[Cu^{2+}]_1}\right)$$

$$\text{or, } x + 0.03 - x = \frac{2.303\ldots}{2F}\left[\log\frac{[Zn^{2+}]}{[Cu^{2+}]_1} - \log\frac{[Zn^{+}]}{[Cu^{2+}]_2}\right]$$

$$\text{or, } 0.03 = 0.03\left[\log\frac{[Zn^{2+}]}{[Cu^{2+}]_1} \times \log\frac{[Cu^{2+}]_2}{[Zn^{2+}]}\right]$$

$$\text{or, } 1 = \log\frac{[Cu^{2+}]}{[Cu^{2+}]_1} = \log\frac{0.5}{[Cu^{2+}]_1}$$

$\therefore\ [Cu^{2+}]_1 = 0.05$ M

102. NO_3^- (a.q.) + $2H^+$(aq.) + e^- $\rightarrow$ NO_2(g) + H_2O

(i) Conc. of H^+ = 8 M and of all other species is unity

$$E = E^\circ - \frac{0.059}{n}\log\frac{[NO_2][H_2O]}{[NO_3^-][H^+]^2}$$

$$E = 0.78 - \frac{0.059}{1}\log\left(\frac{1}{(8)^2}\right)$$

$E = 0.78 - 0.059 \times (-1.8)$

$E = 0.78 + 0.1062$ $E = 0.8862$ V

(ii) In neutral solution $[H^+] = 10^{-7}$ M

$$\therefore E = 0.78 - \frac{0.059}{1}\log\left(\frac{1}{10^{-14}}\right)$$

$E = 0.78 - 0.059 \times 14$

$E = 0\ 78 - 0.826 = -0.046$

103. Cell reaction:

$$Zn(s) + Ni^{2+}(aq) \rightleftharpoons Zn^{2+}(aq) + Ni(s)$$

So the cell is Zn | Zn^{2+} || Ni^{2+} | Ni

Now the standard cell potential is given by:

$E^\circ_{cell} = E^\circ_{cathode} - E^\circ_{anode} = -0.24$ V – (0.75 V) = 0.51 V

$$\text{So the cell potential is; } E = E^\circ_{cell} - \frac{0.0592}{n}\log\frac{[Zn^{2+}]}{[Ni^{2+}]}$$

And at equilibrium $E = 0$

$$\Rightarrow \quad E^\circ_{cell} = \frac{0.0592}{n} \log \frac{[Zn^{2+}]}{[Ni^{2+}]}$$

$$0.51 = \frac{0.0592}{n} \log \frac{[Zn^{2+}]}{[Ni^{2+}]}$$

$$\frac{\log[Zn^{2+}]}{[Ni^{2+}]} = \frac{0.51 \times 2}{0.592} \; 17.23$$

$$\therefore \quad \frac{[Zn^{2+}]}{[Ni^{2+}]} = 1.698 \times 10^{17}$$

Suppose x to be the concentration of Ni^{2+} at equilibrium. we have at equilibrium

$$Zn(s) + Ni^{2+}(aq) \leftrightarrow Zn^{2+}(aq) + Ni(s)$$

So, $[Zn^{2+}] = 1 - x$

$$\text{or} \quad K_{eq} = \frac{[Zn^{2+}]}{[Ni^{2+}]} = 1.698 \times 10^{17}$$

$$\Rightarrow \quad \frac{1-x}{x} = 1.698 \times 10^{17}$$

$$\Rightarrow \quad x = 5.86 \times 10^{-18} \text{ mol } l^{-1}$$

104. (i) Reaction at the copper electrode : $Cu^{2+} + 2e^- \rightarrow Cu$
Reaction at the zinc electrode : $Zn \rightarrow Zn^{2+} + 2e^-$
The cell reaction is $Zn + Cu^{2+} \rightarrow Zn^{2+} + Cu$
(ii)e.m.f. of the cell is = 0.35 – (–0.763) volts = 1.113 volts.

105. Reactions taking place during electrolysis of acidic Cu^{2+} solution are:

$H_2O \rightarrow 2H^+ + \frac{1}{2}O_2 + 2e^-$ (at anode)

$Cu^{2+} + 2e^- \rightarrow Cu$ (at cathode)
After complete deposition of copper, the reactions are:

$H_2O \rightarrow 2H^+ + \frac{1}{2}O_2 + 2e^-$ (at anode)

$2H_3 + 2e^- \rightarrow H_2 + 2OH^-$ (at cathode)

$$\text{Amount of copper deposited} = \frac{0.4 \text{ g}}{63.6 \text{ g mol}^{-1}} = 0.0629 \text{ mol}$$

Amount of oxygen liberated during electrolysis = $\frac{1}{2} \times 0.0629$ mol

Quantity of electricity passed after the electrolysis = (1.2 A) (7 × 60 s) = 504 C

Amount of electricity passed = $\dfrac{500\ C}{96500\ C\ mol^{-1}}$ = 0.00522 mol

After the electrolysis of Cu^{2+}, the amount of oxygen liberated

$= \dfrac{1}{4} \times 0.00522\ mol = 0.00261\ mol$

total amount of gases evolved = (0.03145 + 0.001305 + 0.00261) mol = 0.00706 mol.

∴ Volume of gases evolved at NTP = (22400 ml mol^{-1}) (0.00706 mol) = 158.2 ml.

106. Greater value of O.P. leads for the oxidation Greater value of R.P. lead for the reduction

Thus, $E^{\circ}_{cell} = E^{\circ}_{ce^{+4}|Ce^{+3}}\ T\ E_{Fe^{+3}|Fe^{+2}}$

$E^{\circ}_{cell} = 0.84\ V$

Since E°_{cell} is greater than zero it is spontaneous

As reaction goes on completion, current will decreases and it flows from cathode to anode.

107. Half-cell reactions are: $Zn^{2+} + 2e^- \rightleftharpoons Zn$

$$H^+ + e^- \rightleftharpoons \frac{1}{2}H_2$$

$$E_{Zn/Zn^{2+}} = E^{\circ}_{Zn/Zn^{2+}} - \frac{RT}{nF}\ln\frac{[Zn^{2+}]}{[Zn]} = 0.76 - \frac{0.05915}{2}\log_{10}\frac{0.1}{1} = 0.76 + 0.03 = 0.79 \text{ volt.}$$

$$E_{H^+/H_2} = 0 - \frac{0.05913}{2}\log_{10}\frac{1}{[H^+]^2} = \frac{0.05915}{2} \times (+\log_{10}[H^+]) = 0.05915 \times p^H$$

Thus, 0.28= 0.79 – 0.05915 p^H

or 0.05915 pH= 0.79 – 0.28 = 0.51

or $p^H = \dfrac{0.51}{0.05915} = 8.62.$

108. $2Hg\ \ 2Fe^{3+} \leftrightarrow Hg_2^{2+} + 2Fe^{2+}$

The system is in equilibrium ∴ $E = 0$

$$E = 0 = \left(E^{\circ}_{Fe^{3+}/Fe^{2+}} - E^{\circ}_{Hg_2^{2+}/Hg^{2+}}\right)$$

$$\ldots = \frac{0.059}{2}\log\frac{[Fe^{2+}]^2\ [Hg_2^{2+}]}{[Fe^{3+}]^2}$$

$[Fe^{2+}] = 0.95 \times 10^{-3}$; $[Fe^{3+}] = 0.05 \times 10^{-3}$

$$[Hg_2^{2+}] = \frac{0.95 \times 10^{-3}}{2}$$

$$\therefore\ E^{\circ}_{Hg_2^{2+}/Hg^{2+}} = \mathbf{0.792\ V}$$

Chapter 11 : NUCLEAR CHEMISTRY

1. Mass defect $\Delta M = (ZM_H + (A - Z)\, M_n) - M$

$= 14 \times 1.007825 + (28 - 14)\, 1.008665] - 27.9769 = 0.2431$ a.m.u.

Total binding energy $= \Delta M \times 931.5$ MeV $= 0.2431 \times 931.5$ MeV $= 226.2$ MeV

$$\text{Average binding energy} = \frac{\text{Total binding energy}}{\text{Total number of nucleous}} = \frac{226.2}{28} = 8.071 \text{ MeV}$$

2. Mass defect $\Delta M = (ZM_4 + (A - Z)\, M_n) - M$

$= [20 \times 1.007826 + (40 - 20)\, 1.008665] - 40.08 = 0.24982$ a.m.u.

Total binding energy $= (0.24982 \times 931.5)$ MeV $= 232.7$ MeV

$$\text{Average binding energy} = \frac{\text{Total binding energy}}{\text{Total number of nucleous}} = \frac{232.7}{40} = 5.8175$$

3. $_{11}Na^{23} + {_2}He^4 \rightarrow {_{13}}Al^{26} + {_0}n^1$

$_3Li^7 + {_2}He^4 \rightarrow {_5}B^{10} + {_0}n^1$

4. The change can be represented as

$$^{232}_{90}\text{Th} \rightarrow {^{208}_{82}}\text{Pb}$$

Atomic number of parent nuclide $(Z_p) = 90$
Atomic number of daughter nuclide $(Z_d) = 82$
Mass number of parent nuclide $(A_p) = 232$
Mass number of daughter nuclide $(A_d) = 208$

$$\text{No. of alpha particles emitted} = \frac{A_p - A_d}{4} = \frac{232 - 208}{4} = 6$$

$$\text{No. of beta particles emitted} = 2\left(\frac{A_p - A_d}{4}\right) - (Z_p - Z_d) = 2(6) - (90 - 82) = 12 - 8 = 4$$

5. We know that

$$\lambda = \frac{0.693}{t_{1/2}} = \frac{0.693}{28} = 0.02475 \text{ day}^{-1}$$

$$\text{But, } \lambda = \frac{2.303}{t}\log\frac{a}{a - x}$$

$$0.02475 = \frac{2.303}{t}\log\frac{1}{1/10} = \frac{2.303}{t}\log 10$$

$$\Rightarrow t = \frac{2.303}{0.02475} = 93.05 \text{ days.}$$

6. Mass defect, $\Delta M = [ZM_H + (A - Z)\, M_n] - M$

$= [26 \times 1.00785 + 30 \times 1.998665 - 55.9375$

$= 56.4634 - 55.9375 = 0.5259$ a.m.u

7. Given, $t_{\frac{1}{2}}$ = 22yrs. $\therefore$ t = 11 yrs.; Hence, $n = \frac{11}{22} = \frac{1}{2}$

$$\text{Amount left after } n \text{ half-lives} = \frac{a_0}{2^n}$$

$$\text{Amount left after } \frac{1}{2} \text{ half life.} = \frac{a_0}{2^{1/2}} = \frac{2}{2^{1/2}} = \sqrt{2} = 1.141 \text{ g.}$$

8. $R = R_0A^{1/3} = 1.4 \times 10^{-13} \times (27)^{1/3} = 1.4 \times 30^{-13} \times 3 = 4.2 \times 10^{-13}$ cm. = 64.

9. $R = R_0A^{1/3}$

$$\therefore A^{1/3} = \frac{R}{R_0} = \frac{5.6}{1.4} = 4$$

$A = 4^3 = 64.$

10. Mass number is $M = 4$
Atomic number is $Z = 2$

11. $t_{\frac{1}{2}}$ = 11 mi. 33sec = 693 sec.

$$\therefore \lambda = \frac{0.693}{t_{1/2}} = \frac{0.693}{693} = 10^{-3} \text{ sec}^{-1}$$

12. By applying the expression

$$A^{1/3} = R/R_0 = \frac{8.4 \times 10^{-13}}{1.4 \times 10^{-13}} = 6$$

$\therefore A = 6^3 = 216$
Mass number is 216.

13. By applying the formula

$$A^{1/3} = R/R_0 = \frac{2.52 \times 10^{-13}}{1.4 \times 10^{-13}} = 1.8$$

$\therefore A = (1.8)^3 = 6$ (approximately)
Mass number is 6.

14. By applying $R = R_0 \times \sqrt[3]{A} = 1.4 \times \sqrt[3]{64} = 5.6$
Nuclear radius = 5.6 Fm.

15. By applying the formula

$$t_{\frac{1}{2}} = \frac{0.693}{\lambda}; \text{ where, } t_{1/2} = \text{half life}$$

$$\Rightarrow \lambda = \text{disintegration constant} = \frac{0.693}{t_{1/2}} = \frac{0.693}{5.8} = 0.1195 \text{ disintegrations year}^{-1}$$

16. By applying the equation

$$\lambda = \frac{0.693}{t_{1/2}} = \frac{0.693}{50};\ \lambda = 0.0139 \text{ month}^{-1}$$

17. By applying the expression

$$\lambda = \frac{0.693}{t_{1/2}} = \frac{0.693}{15} = 0.0462 \text{ disintegrations hour}^{-1}$$

18. By applying the formula

$$t_{1/2} = \frac{0.693}{\lambda} = \frac{0.693}{0.0581} = 11.93 \text{ years}$$

Half life is 11.93 years.

19. $t_{1/2} = \frac{0.693}{\lambda} = \frac{0.693}{0.00781} = 88.73$ hours

Half life is 88.73 hours.

20. By applying Einstein's equation

$E = mc^2 = 0.05 \times (3.0 \times 10^{10})^2$ erg $= 0.45 \times 10^{20}$ erg $= 4.5 \times 10^{12}$ J

21. $E = mc^2 = 0.08 \times (3.0 \times 10^{10})^2 = 0.72 \times 10^{20}$ erg $= 7.2 \times 10^{12}$ J

22. By applying the formula

$A^{1/3} = R/R_0 = 3.5/1.4 = 2.5$

$A = (2.5)^3 = 16$ (approx.)

Mass number = 16

23. $E = mc^2$; or, $20.9 \times 10^{10} = m\ (3\times 10^{10})^2$

$$\therefore\ m = \frac{20.9\times10^{10}\times10^{7}}{9\times10^{20}} = 2.322\times10^{-3}\ g.$$

24. Mass defect of a nuclide is calculated by using the expression

$$\Delta M = ZM_H + (A - Z)M_n - M$$

$$\Delta M = (14 \times 1.007825) + (28 - 14) \times 1.00865 - 27.9769$$

Mass binding energy = 0.25375 a.m.u.

Total binding energy = 0.25875 × 931.5 = 236.368 MeV.

$$\text{Average binding energy} = \frac{\text{Total binding energy}}{\text{No. of nucleons (protons + neutrons)}} = \frac{236.368}{28} = 8.442 \text{ MeV.}$$

25. (a) Mass defect = ΔM

= Mass of the reactants – Mass of the products

= (26.9815 + 2.0141) – (24.9858 + 4.0026) = 0.0072 a.m.u.

Energy of the reaction = 0.0072 × 931.5 = 6.7068 MeV.

(b) ΔM = (10.01294 + 1.008665) – (7.016 + 4.0026) = 0.003005 a.m.u.

Energy = 0.003005 × 931.5 = 2.7992 MeV.

(c) ΔM = (12 + 20141) – (13.00574 + 1.008665)

= (–) 0.000305 a.m.u.

Energy= –0.000205 × 931.5 = –0.284 MeV

26. By applying the formula;
$A^{1/3} = R/R_0$ or $R = R_0 \times A^{1/3}$

$R = 1.4 \times \sqrt[3]{125}$

Nuclear radius= 6.995 Fm.

27. We know that 1 curie is the weight of a radio-active substance which gives 3.7×10^{10} disintegrations per second.
We have Rate of disintegration = $KN = 3.7 \times 10^{10}$ per sec.

Where, K = Decay constant = $\frac{0.693}{T_{\frac{1}{2}}} = \frac{0.693}{3.825 \times 24 \times 60 \times 60} = 2.096 \times 10^{-6}\ \text{sec}^{-1}$.

N = Number of atoms = $\frac{\text{Rate of disintegration}}{\text{Decay constant}} = \frac{3.7 \times 10^{10}}{2.096 \times 10^{23}} = 1.7653 \times 10^{16}$ atoms.

We know that 6.02×10^{23} atoms of radon weigh 222 gm. hence 1.7653 × 1015 atoms would weigh.

$$= \frac{1.7653 \times 10^{16} \times 222}{6.02 \times 10^{23}} = 6.51 \times 10^{-6}\ \text{gm.}$$

Weight of Radon = 6.51×10^{-6} or 6.51 microgram

28. We know that,
Rate of disintegration = $K \times N = 3.3 \times 10^{10}$ per day where,
K = Decay constant = $0.693/t_{1/2}$

$$= \frac{0.693}{1.1 \times 10^5 \times 365} = 1.726 \times 10^{-8}\ \text{per day}$$

N = Number of atoms of calcium = $\frac{\text{Rate of disintegration}}{\text{decay constant}}$

$$= \frac{3.3 \times 10^{10}}{1.726 \times 10^{-8}} = 1.912 \times 10^{18}\ \text{atoms}$$

Now, we know that, 6.02 × 1023 atoms would weight 40.97 gm. Hence, weight of 1.912×10^{18} atoms.

$$= \frac{1.912 \times 10^{18} \times 40.97}{6.02 \times 10^{23}} = 1.30 \times 10^{-4}\ \text{gm.}$$

Weight of calcium= 1.30×10^{-4} gm

29. We have in radioactive equilibrium, $\frac{N_A}{N_B} = \frac{\lambda_B}{\lambda_A}$

Where N_A = Number of atoms of Radium A = ?
N_B = Number of atoms of radium B

$$= \frac{6.02 \times 10^{23} \times 4}{214} = 0.1125 \times 10^{23} = 1.125 \times 10^{25}\ \text{atoms}$$

λ_A = Disintegration constant of Radius $B = \frac{0.693}{26.8} = 2.586 \times 10^{-2}$ minute^{-1}

l_B = Disintegration constant of Radium $A = \frac{0.693}{3.05} = 2.272 \times 10^{-1}$ minute^{-1}.

$$N_A = \frac{\lambda_B}{\lambda_A} \times N_B = \frac{2.596 \times 10^{-2}}{2.272 \times 10^{-1}} \times 1.125 \times 10^{22} = 1.2804 \times 10^{21} \text{ atoms}$$

The weight of 1.2804×10^{21} atoms would be $= \frac{1.2804 \times 10^{21} \times 218}{6.02 \times 10^{23}} = 4.637 \times 10^{-1}$

Weight of radium $A = 4.637 \times 10^{-1}$ gm

30. We know that radioactive disintegrations are first order reactions, hence.
we get,

$$\therefore \lambda = \frac{0.693}{t_{1/2}} = \frac{0.693}{54.5} = 0.01272 \text{ sec}^{-1}$$

Now, let initial concentration be 'a', then the concentration at the end of the one minute would be $(a - x)$

$$\therefore \lambda = \frac{2.303}{t} \log \frac{a}{a - x}$$

$$\text{or, } 0.01272 = \frac{2.303}{60} \log \frac{a}{a-x} = 0.03838 \log \frac{a}{a-x}$$

$$\therefore \log \frac{a}{a-x} = 0.3314 \Rightarrow \frac{a}{a-x} = 2.14$$

$$\text{or } a - x = \frac{a}{2.14} = 0.47 \text{ a}$$

When $a = 1$, fraction left behind = $(a - x) = (1 - x) = 0.47$

$\therefore$ Fraction left behind = 47%

31. We know that,
Rate of disintegration per sec. = Decay constant per sec. × Number of atoms.

$$\text{Decay constant per sec.} = \frac{0.693}{T_{1/2}} = \frac{0.693}{2.33 \times 24 \times 60 \times 60} = 3.442 \times 10^{-6} \text{ per sec.}$$

We know that 239 gm of Neptunium contain 6.023×10^{23} atoms.

Hence 0.1 milligram, i.e., 0.0001 gm would contain

$$= \frac{6.02 \times 10^{23} \times 0.0001}{239} = 2.51882 \times 10^{17} \text{ atoms}$$

To get 0.0001 gm of plutonium 2.51882×10^{17} atoms of Neptunium must disintegrate.
Rate of disintegration per second

$$= 3.442 = \times 10^{-6} \times 2.51882 \times 10^{17} = 8.6698 \times 10^{11}$$

We know that, 1 curie = 3.7×10^{10} disintegrations/sec

$$\text{Number of curies} = \frac{\text{Rate of disintegration}}{3.7 \times 10^{10}} = \frac{8.6698 \times 10^{11}}{3.7 \times 10^{10}} = 23.43$$

Number of curies = 23.43

32. We know that,
No. of counts in unit time = Rate of disintegration.
We have,

Rate of disintegration = decay constant × Number of atoms

20.4 = decay constant × No. of atoms

$$\text{Decay constant} = \frac{0.693}{5760 \times 365 \times 24 \times 60} = 2.289 \times 10^{-10} \text{ per minute}$$

$$\text{Number of } C^{14} \text{ atoms} = \frac{\text{Rate of disintegration}}{\text{decay constant}} = 8.912 \times 10^{10} \text{ atoms}$$

Initial Number of atoms = 8.912×10^{10}

Now, we know that,

$$N = N_o e^{-\lambda t}$$

$$\text{or, } \log \frac{N_0}{N} = \frac{\lambda t}{2.303}$$

$$\text{or } \log \frac{8.912 \times 10^{10}}{N} = \frac{2.289 \times 10^{-10} \times 100 \times 365 \times 24 \times 60}{2.303} = 5.224 \times 10^{-3}$$

$$N = 5.320 \times 10^{8} \text{ atoms}$$

Again, Rate of disintegration after 100 yrs.
= No. of counts after 100 yrs. = decay constant × number of atoms
= $2.289 \times 10^{-10} \times 5.320 \times 10^{8}$ = 12.18 counts/min^{-1}
Number of counter per minute = 12.18

33. We know that $N = N_0 e^{-\lambda t}$
where

N_0 = Initial number of atoms = 100

N = Number of atoms after time = 100 – 71.28 = 28.72

$$\lambda = \text{Decay constant} = \frac{0.693}{\text{half life period}}$$

t = Time = 10,000 yrs.

$$2.303 \log \frac{N_0}{N} = \lambda t$$

$$2.303 \log \frac{100}{28.72} = \lambda \times 10{,}000$$

$$2.303 \times 0.5418 = \lambda \times 10{,}000$$

$$\lambda = \frac{2.303 \times 0.5418}{10000} = 1.248 \times 10^{-4} \text{ per year}$$

Now, $\lambda = \dfrac{0.693}{T_{\frac{1}{2}}}$

or, $t_{\frac{1}{2}} = \dfrac{0.693}{1.241 \times 10^{-4}} = 5552$ yrs.

Half life of carbon = 5552 yrs.

34. We know that, $N = N_0 e^{-\lambda t}$
Where, N = Number of atoms at time t, N_0 = Initial number of atoms, λ = Decay constant

We have, $N_{235} = N_{\circ 235}\, e^{-\lambda t}$ and $N_{238} = N_{\circ 238}\, e^{-\frac{\lambda}{2}t}$
If $N_{\circ 235} = N_{\circ (238)}$

Then, $\dfrac{N_{238}}{N_{235}} = \dfrac{99.2}{0.80} = 124$

$\lambda_2 = \dfrac{0.693}{t_{\frac{1}{2}}} = \dfrac{0.693}{4.5 \times 10^9} = 1.54 \times 10^{-10}$ year

$\lambda_1 = \dfrac{0.693}{t_{\frac{1}{2}}} = \dfrac{0.693}{7.1 \times 10^8} = 9.75 \times 10^{-10}$ year

$\therefore$ 124 = 2.303 log 124= $-t(l_2 - l_1) - t(1./52 - 9.75) \times 10^{-10}$

$t = \dfrac{2.303 \log 124}{-(1.54 - 9.75) \times 10^{-10}} = \dfrac{4.82}{8.21 \times 10^{10}} = 5.871 \times 10^9$ yrs.

Age of the earth = 5.871×10^9 yrs.

35. We know that, Binding energy = [Mass defect (a.m.u) × 931.48] MeV
Mass defect = Expected mass – Actual mass
Expected Mass = Mass of electrons + Mass of protons + Mass of neutrons
In $^{18}_{8}O$, we have

No. of protons = 8
No. of neutrons = 10,
No. of electrons = 8

$\therefore$ Expected Mass = Mass of 8 protons + Mass of 10 neutrons + Mass of 8 electrons
= (1.00759 × 8) + (1.00898 × 10) + (0.00055 × 8)
= 8.06072 + 10.0898 + 0.0044 = 18.1592 a.m.u

Actual mass = 18.00369 a.m.u
Mass defect = ΔM = (18.15492 – 18.00369) = 0.15123 a.m.u
Binding energy = (0.15123 × 931.48) MeV = 140.87 MeV

Oxygen atom contains 8 protons and 10 neutron
Hence, number of nucleons = 8 + 10 = 18

Binding energy per nucleon $= \dfrac{140.87}{18} = 7.826$ Mev

Amount of energy equivalent to 1 a.m.u of mass = 931.5 MeV
Amount of energy equivalent to 0.5259 a.m.u of a mass = 931.5 × 0.5259 = 489.8 MeV
Binding energy of the nucleus = 489.8 MeV

Binding energy per nucleon = $\frac{489.8}{55}$ = 8.746 MeV

36. We know that an α-particle is a doubly charged Helium ion. Atomic weight of Helium is 4.0. According to the Avogadro's hypothesis 6.02×10^{23} atoms would be present in 4.0 gm of Helium.

The weight of one atom of Helium would be $\frac{4.0}{6.02 \times 10^{23}} = 6.644 \times 10^{-24}$

But, α-particle = 2 proton + 2 neutrons
Weight of one Helium atom = weight of 1 a particle.
Weight of α-particle = 6.664×10^{-24} gm
Now, 1α-particle = 2 positive charges = $2 \times 4.8 \times 10^{-10}$ esu = q

Hence, $\frac{q}{m} = \frac{2 \times 4.8 \times 10^{-10}}{6.644 \times 10^{24}} = 1.4449 \times 10^{-14}$ esu/gm.

Ratio of charge to mass of α-particle = 1.4449×10^{14} esu/gm

37. Change in atomic weight = 234 – 206 = 28
No. of alpha particles emitted can be calculated by dividing the change in atomic weight by 4 (If 1 particle is emitted At. wt. decrease by 4 units).

No. of alpha particles = $\frac{28}{4} = 7$

Expected decrease in atomic number due to the emission of 7 alpha particles = 7 × 2 = 14 (if 1 alpha particle is emitted At. No. decreases by two units).
Expected atomic number = 90 – 14 = 76
The atomic no. of the product is 82.
The number of particles emitted = 82 – 76 = 6
(If 1 Beta particle is emitted atomic number increases by one unit)
During the transformation of $^{234}_{90}$Th into $^{206}_{82}$Pb, the number of particles emitted are:
alpha (a) particles = 7
beta (b) particles = 6

38. Given, t = 5770 yrs.

$\therefore$ Rate constant (K)= $\frac{0.693}{t_{1/2}} = \frac{0.693}{5770} = 1.2 \times 10^{-4}$ yrs.

Fraction remaining after 11540 yrs.
Ist method : we have, $t_{1/2}$ = 5770 yrs.
Initial amount = 1

Amount left after 5770 yrs. = $\frac{1}{2}$

$\therefore$ Amount left after 11540 yrs. = $\frac{1}{2} \times \frac{1}{2} = \frac{1}{4}$ = 25%

II method:

$K = \frac{2.303}{t} \log \frac{a}{a-x}$ yrs

t = 11540 yrs.
K= 1.2 × 10^{-4} yrs. rate constant
a = 100

$$1.2 \times 10^{-4} = \frac{2.303}{11540} \log \frac{100}{100-x}$$

$\Rightarrow x = 75$
Fraction remaining after 11540 years.
$= a - x = 100 - 75 = 25\%$

39. Half life period of uranium-237 = 2.5 × 10^5 yrs.
Amount of uranium-237 remained after first 2.5 × 10^5 yrs. = 50%
Amount of uranium-237 remained after next 2.5 × 10^5 yrs. = 25%
$\therefore$ The time taken to remain 25% of the original mass = (2.5 + 2.5) 10^5 yrs. = 5.0 × 10^5 yrs.

40. After the emission of two β-particles,
Mass number of the atom = 232
Atomic number of the atom = 92
After the emission of two particles, the change can be represented as

$${}^{232}_{92}X \rightarrow {}^{212}_{82}Y$$

Decrease in mass = 232 – 212 = 20
Decrease in atomic number = 92 – 82 = 10
Each alpha particle has 4 units of mass and two units of nuclear charge.
$\therefore$ No. of alpha particles emitted = 5.

41. ${}_7N^{14} + {}_0n^1 + \rightarrow {}_6C^{14} + {}_1H^1$
Mass of the reactants = 13.9824 + 1.008665 = 14.991065
Mass of the products = 13.9817 + 1.007825 = 14.989525
Mass disappeared = 14.991065 – 14.989525 = 0.001540
Amount of energy equivalent to 1 a.m.u of mass = 931.5 MeV
Amount of energy equivalent to 0.00154 a.m.u mass = 931.5 × 0.001540 = 1.432 MeV.
Here, the mass of the reactants is greater than that of products.
$\therefore Q = -1.432$

42. Half life period, $t_{1/2} = \frac{0.693}{\lambda}$

or, $5600 = \frac{0.693}{\lambda}$

$\therefore$ Decay constant, $\lambda = \frac{0.693}{5600}$ year^{-1}

Let "a" be the initial amount of C-14 isotope and "t" be the time during which its activity fall to $\frac{a}{6}$

Then according to formula, $\lambda = \frac{2.303}{t}\log\frac{a}{a/6}$

or, $1.237 \times 10^{-4} = \frac{2.303}{t}\log 6$

$\therefore \quad t = \frac{2.303}{1.237\times10^{-4}}\log 6 = \frac{2.303}{1.237\times10^{-4}} \times 0.7782 = 1.448 \times 10^{4}$ yrs.

43. (a). $\lambda = \frac{0.693}{4.8} = 0.1444\ \text{min}^{-1}$

Let the amount of substance remained after 24 minutes be x.
Then according to formula

$\lambda = \frac{2.303}{t}\log\frac{1}{x}\ 0.144 = \frac{2.303}{24}\log\frac{1}{x}$

$\log\frac{1}{x} = \frac{1.444\times24}{2.303} = 1.504$

$\log 1 - \log x = 1.504$

$\log x = \log 1 - 1.504 = -1.504$

Since the mantissa has to be positive, the above expression assumes the value of $\bar{2}.4996$
i.e. $\log x = \bar{2}.4996$

The amount of substance remained after 24 minutes = 0.03159 g.

(b) No. of atoms in 232 g. of element = 6.023×10^{23}

No. of atoms in 1 g. of the element = $\frac{6.023\times10^{23}}{232}$

$\lambda = \frac{0.693}{t_{1/2}} = \frac{0.693}{1.39\times10^{9}}\ /\text{year}^{-1}$

Hence the number of atoms disintegrating per second = $\frac{0.693}{1.39\times10^{9}\times365\times24\times3600}$

No. of atoms disintegrating by one gram of the substance

$= \frac{6.023\times10^{23}}{232} \times \frac{0.693}{1.39\times10^{9}\times365\times24\times3600} = 4.104 \times 10^{4}$ disint sec.$^{-1}$

Hence no. of alpha particles emitter per second by one gram of the substance $= 4.104 \times 10^{4}$

44. We know that If = 10^{-13} cm.)
Nuclear radius $R = R_0\ A^{1/3}$
Where,
R_0 = Proportionality constant = 1.5×10^{-13} cm (given)
A = mass number

1. Thorium, mass number = 232;

$R_0 = 1.5 \times 10^{-13}$ cm

Radius of nucleus = $1.5 \times 10^{-13} \times (232)^{1/3} = (1.5 \times 10^{-13} \times 6.145)$
$= 9.22 \times 10^{-13}$ cm = 9.22 fermi

2. Astatine (mass number = 210)
$= 1.5 \times 10^{-13} \times (210)^{1/3} = (1.5 \times 10^{-13} \times 5.944) = 8.916 \times 10^{-13}$ cm = 8.916 fermi

45. Loss of 8 alpha particles decreases mass by 8 × 4 = 32 units and atomic number by 8 × 2 = 16 units
Loss of 6 beta particles causes no change in mass but increase the atomic number by 6 units.
Hence the atomic mass of the product = 238 – 32 = 206
The atomic number = 92 – 16 + 6 = 82

46. Change in atomic mass = 235 – 207 = 28 a.m.u.

Number of alpha particles corresponding to 28 a.m.u $= \frac{1}{4} \times 28 = 7$

Decrease in atomic number caused by 7 alpha particles = 7 × 2 = 14
Atomic number of the resulting element = 92 – 14 = 78
But the atomic number of the actual product, $^{207}_{82}PB$ = 82.
As beta emission increases atomic no. by one unit.
The no. of beta particles = 82 – 78 = 4

In the transformation of $^{207}_{82}U$ to $^{207}_{82}Pb$ 7 alpha particles and 4 beta particles are emitted.

47. **(a)** By applying the expression N_0 = 30 gm, N = ?
λ = 0.0737 disint. hour^{-1}, t = 5 hours and 0.4343 as constant
$\log N_0 - \log N = 0.4343 \times \lambda \times t$
$\log 30 - \log N = 0.4343 \times 0.0737 \times 5 = 0.1600395$
$\log N = \log 30 - 0.1600395 = 1.4771 - 0.1600395 = 1.31706$
N = antilog of (1.31706) = 20.75
wt. of the substance left after 5 hours = 20.75 g.

(b) By applying the expression
$\log N_0 - \log N = 0.4343 \times \lambda \times t$
$\log 15 - \log N = 0.4343 \times 0.00283 \times 200 = 0.2458$
$\log N = \log 15 - 0.2458$
$= 1.1761 - 0.2458 = 0.9303$
N = anti log of (0.9303) = 8.5
wt. of zinc left after 200 days is 8.5 g.
wt. of zinc dsintegrated 15.0 – 8.5 = 6.5 g.

48. By applying the expression
$\log N_0 - \log N = 0.4343 \times \lambda \times t$

$$\lambda = \frac{\log N_0 - \log N}{0.4343 \times t} = \frac{\log 6 - \log(4.226)}{0.4343 \times 20} = \frac{0.1522}{8.689} = 0.0175224$$

Disintegration constant = 0.0175224 disint. hour^{-1}.

49. By applying equation
$\log N_0 - \log N = 0.4343 \times \lambda \times t$

$$\lambda = \frac{\log N_0 - \log N}{0.4343 \times t} = \frac{\log 15 - \log 7}{0.4343 \times 5} = 0.1524\ y$$

50. The decay constant radon is

$$\lambda = \frac{0.693}{t_{\frac{1}{2}}} = \frac{0.693}{(3.8\,d)(86{,}400\ \text{s/d})} = 2.1 \times 10^{-6}\ \text{s}^{-1}$$

The mass of an atom is very close to its mass number expressed in atomic mass units, so the number N of atoms in 1 g of ^{222}Rn is

$$N = \frac{10^{-6}\ \text{kg}}{(222)(1.66 \times 10^{-27}\ \text{kg})} = 2.7 \times 10^{18}\ \text{atoms}$$

Hence, $R = \lambda N = (2.1 \times 10^{-6}\ \text{s}^{-1})\ (2.7 \times 10^{18}\ \text{nuclei}) = 5.7 \times 10^{12}$ events/s

$= 5.7 \times 10^{3}$ GBq $= 153$ Ci

In this case the "events" are alpha decays.

51. We have

$$\frac{N}{N_0} = e^{-\lambda t} \Rightarrow -\lambda t = \ln \frac{N}{N_0}$$

$$\text{or } \lambda t = \ln \frac{N_0}{N}$$

$$\therefore\ t = \frac{1}{\lambda} \ln \frac{N_0}{N}$$

Here, $\lambda = 0.693\ t_{\frac{1}{2}} = 0.693/3.8$ d and $N = (1 - 0.6)N_0 = 0.4\ N_0$, so that

$$t = \frac{3.8\,d}{0.693} \ln \frac{1}{0.4} = 5.02\ \text{d}$$

52. The decay constant λ, of radiocarbon, is $\lambda = 0.693\ t_{\frac{1}{2}} = 0.693/5760\ y$. here $R_0/R = 16/13$

$$\text{and so } t = \frac{1}{\lambda} \ln \frac{R_0}{R} = \frac{5760\ y}{0.693} \ln \frac{16}{13} = 1726\ y$$

53. The radioactive decays of ^{238}U leads, after 14 steps, to the stable lead isotope ^{206}Pb. The first of these steps, the α decay of ^{238}U with a $4.5 \times 10^{9}\ y$ half-life, in intrinsically more than 10^{4} times as slow as any of the subsequent steps. As a result, the time required for the step accounts for essentially all the time required for the entire 14-step sequence.

In a sample containing 1.0 g U, there is

$$\frac{0.214\ \text{g Pb}}{206\ \text{g mol}^{-1}} = 1.04 \times 10^{-3}\ \text{mol Pb} \quad \text{and} \quad \frac{1.0\ \text{g}}{238\ \text{g mol}^{-1}} = 4.20 \times 10^{-3}\ \text{mol}$$

if each atom of lead in the mineral today is the daughter of a uranium atom that existed at the time of the formation of the minerals then the original number of moles of uranium in the sample would have seen

$$(1.04 + 4.20) \times 10^{-3} = 5.24 \times 10^{-3}$$

Then the fraction remaining is $\dfrac{[R]}{[R]_0} = \dfrac{4.20 \times 10^{-3}}{5.24 \times 10^{-3}} = 0.802$

Letting t be the elapsed time from the formation of the mineral in the earth's crust to the present, we have

$$k = \frac{0.693}{t_{1/2}} = \frac{0.693}{4.5 \times 10^9\ y} = 1.54 \times 10^{-10}\ y^{-1}$$

$$t = \frac{(2.303)(\log [R]/[R]_0)}{k} = -\frac{(2.303)(\log 0.802)}{1.54 \times 10^{-10}\ y^{-1}} = 1.4 \times 10^9\ y$$

54. The activity at the accessation of bombardment is given by :

$$(2.30)\left(\log \frac{R}{R_0}\right) = (2.30)\left(\log \frac{10\ \text{min}^{-1}}{R_0}\right) = -\lambda t = -\left(\frac{0.693}{0.50\ h}\right)(3.0\ \text{h})$$

$$\log \frac{10\ \text{min}^{-1}}{R_0} = -1.8 \frac{10\ \text{min}^{-1}}{R_0} = 0.0156 \therefore R_0 = \frac{10\ \text{min}^{-1}}{0.0156} = 640\ \text{min}^{-1}$$

$$\therefore N = \frac{R_0}{\lambda} = \frac{(640\ \text{min}^{-1})(30\ \text{min})}{0.693} = 2.8 \times 10^4$$

55. We have

$$\left(\frac{1}{t}\right)\left(\log \frac{N}{N_0}\right) = \frac{\lambda}{2.30}\left(\frac{1}{4}\right)\left(\frac{\log 9.0 \times 10^5}{2.00 \times 10^6}\right) = \left(\frac{1}{40}\right)\left(\log \frac{N}{2.00 \times 10^6}\right)$$

$$\log \frac{N}{2.00 \times 10^6} = -3.47 \qquad \frac{N}{2.00 \times 10^6} = 3.41 \times 10^{-4} \quad N = 6.8 \times 10^2\ \text{dis min}^{-1}$$

56. We have

$$A = 2.8\ nCi = (2.8 \times 10^{-9}\ \text{Ci})\left(\frac{3.7 \times 10^{-10}\ \text{dis. s}^{-1}}{\text{Ci}}\right) = 104\ \text{dis. s}^{-1}$$

$$N = 30\ \mu\text{g} = (30 \times 10^{-6}\ \text{g})\left(\frac{1\ \text{mol}}{247\ \text{g}}\right)\left(\frac{6.0 \times 10^{23}\ \text{atoms}}{\text{mol}}\right)$$

$$= 7.3 \times 10^{16}\ \text{atoms}$$

$$A = \lambda N \lambda = \frac{A}{N} = \frac{104\ s^{-1}}{7.3 \times 10^{16}} = 1.4 \times 10^{-15}\ s^{-1}$$

$$t_{1/2} = \frac{0.693}{\lambda} = (4.9 \times 10^{14}\ \text{s})\left(\frac{1\ \text{h}}{3600\ \text{s}}\right)\left(\frac{1\ day}{24\ \text{h}}\right)\left(\frac{1\ y}{365\ \text{days}}\right) = 1.5 \times 10^7\ y$$

57. **(a)** 49 years is almost exactly four half-lives ($4 \times 12.3 = 49.2\ y$). It is simpler in this case to avoid the logarithmic formula. The fraction of the tritium atoms remaining after four half-lives would be $(1/2)^4 = 1/16$. The normal hydrogen atoms are not radioactive, and therefore their number does not change. Thus the final ratio of tritium to normal hydrogen will be $\left(\frac{1}{16}\right)(8 \times 10^{-18}) = 5 \times 10^{-19}$.

(b) 10 g of water contains (10 g)/(18 g mol^{-1}) or 10/8 mol of water or 10/9 mol of hydrogen atoms. The total number of hydrogen atoms is $\left(\frac{10}{9}\right)(6 \times 10^{23}) = 6.7 \times 10^{23}$. Of these, the original number of tritium atoms is $(8 \times 10^{-18}) \times (6.7 \times 10^{23}) = 5 \times 10^6$. Only a fraction of this number $[N]/[N_0]$, remains after 40 years.

$$\therefore \quad k = \frac{0.693}{t_{1/2}} = \frac{0.693}{12.3\ y} = 5.6 \times 10^{-2}\ y^{-1}$$

$$\log \frac{[N]}{[N]_0} = \frac{(5.6 \times 10^{-2}\ y^{-1})(40\ y)}{2.303} = -0.973$$

from which $[N]/[N_0] = 0.106$. The number of remaining tritium atoms is $(0.106)(5 \times 10^6) = 6 \times 10^5$.

58. Let the initial activity of A and B be a and b curie respectively.

$$\therefore \quad a + b = 1 \text{ curie} \qquad \ldots\ldots \text{(i)}$$

Since rate of disintegration $\propto$ No. of atoms

$$\therefore \text{ For '}A\text{'; } t = \frac{2.303}{k} \log \frac{N_0}{N} = \frac{2.303}{k} \log \frac{\gamma_0}{\gamma_1}$$

$$\text{or } 20 = \frac{2.303}{k} \log \frac{a}{\gamma_1} = \frac{2.303 \times 14}{0693} \log \frac{a}{\gamma_1}$$

$$\text{or } \log \frac{a}{\gamma_1} = \frac{20 \times 0.693}{2.303 \times 14} = 0.43$$

$$\text{or } \frac{a}{\gamma_1} = 2.69 \Rightarrow \gamma_1 = \frac{a}{2.69} = 0.3717\ a$$

$$\text{Similarly for '}B\text{', } t = \frac{2.303}{k} \log \frac{b}{\gamma_2}$$

$$\text{or } 20 = \frac{2.303 \times 25}{0.693} \log \frac{b}{\gamma_2}$$

$$\text{or } \log \frac{b}{\gamma_2} = \frac{20 \times 0.693}{2.303 \times 25} = 0.2407 \Rightarrow \frac{b}{\gamma_2} = 1.74$$

$$\text{or } \gamma_2 = \frac{b}{1.74} = 0.5747\ b$$

Given that after 20 days activity become = 1/2 curie

$\therefore \quad 0.3717\ a + 0.5747\ b = \frac{1}{2}$ ...... (ii)

Solving (i) and (ii) we get a = 0.38 and b = 0.632

59. ${}^2_1\text{H} + {}^2_1\text{He} \rightarrow {}^3_2\text{He} + {}^1_0n$

Mass of reactants = 2 × 2.0141 = 4.0282 amu

Mass of products = 3.0160 + 1.0087 = 4.0247

Mass defeat (Δm) = 4.0282 – 4.0247 = 0.0035

Energy released = Δm × 931 MeV = 0.0035 × 931 MeV

= 3.2584 MeV = 3.2585 × 1.6021 × 10^{-13} J ($\because$ 1 eV = 1.6021 × 10^{-19} J) = 5.22 × 10^{-13} J

60. One atomic mass unit is $\frac{1}{12}$ the mass of a ${}^{12}_6\text{C}$ atom. Since one ${}^{12}_6\text{C}$ atom has a mass of $12/N_A$ grams, where N_A is Avogadro's number, the mass in grams of 1 amu is

$$1/N_A = 1/(6.022137 \times 10^{23}) = 1.66054 \times 10^{-24}\ \text{g} = 1.66054 \times 10^{-27}\ \text{kg}$$

Using the Einstein relation, $E = mc^2$,

the energy equivalent of 1 amu = 1.6605 × 10^{-27} kg) (3.0 × 10^8 m · $s^{-1})^2$

= 1.49 × 10^{-10} J

The charge on an electron is 1.602×10^{-19} C, so that there are 1.602×10^{-19} J · eV^{-1}, and hence

$$\text{the energy equivalent of 1 amu} = \frac{1.492 \times 10^{-10}\ \text{J}}{1.602 \times 10^{-19}\ \text{J} \cdot \text{eV}^{-1}} = 9.31 \times 10^8\ \text{eV} = 9.31 \times 10^2\ \text{MeV}$$

61. The binding energy of ${}^4_2\text{He}$ is defined by the reaction

$$2n + 2p + 2e^- \rightarrow {}^4_2\text{He(atom)} + \text{binding energy}$$

The mass of the left-hand side of this equation is the mass of two neutrons and two hydrogen atoms.

Mass of $2n$ = 2(1.00866) = 2.0173

+ mass of (${}^1_1$H) = 2(1.0078) = 2.0156

---

total mass of left-hand side = 4.0329

Virtually the only isotope of helium found in nature is ${}^4_2$He (there is 0.00013% of ^{3}He). The mass of ${}^4_2$He is therefore equal to the atomic weight of He. For most elements, several isotopes occur naturally, and the atomic weight is a weighted average of the isotopic masses. The mass of the specific isotope if needed to calculate its binding energy.

The mass of ${}^4_2$He is 4.0026 amu. The loss of mass on forming the atom from its nucleons and electrons is

4.0329 – 4.0026 = 0.03038 amu

Since the energy equiavelnt of 1 amu is 931.5 MeV,

$$\text{binding energy of } {}^4_2\text{He} = (0.0303\ \text{amu})\left(931.5\frac{\text{MeV}}{\text{amu}}\right) = 28.2\ \text{MeV}$$

As there are 4 nucleons ($2p$, $2n$), the binding energy per nucleon is

E_B/A = 28.20/4 = 7.05 MeV

62. There are $7n$, $7p$, and $7e^-$ in 14_N.
mass nucleons = 7(1.0078) + 7(1.0086) = 14.115 amu
mass loss = 14.1115 – 14.003 = 0.112 amu

$$E_B/A = \frac{(0.112 \text{ amu})(931.5 \text{ MeV/amu})}{14 \text{ nucleons}} = \frac{104.33}{14} = 7.452 \frac{\text{MeV}}{\text{nucleon}}$$

There are $8n$, $7p$, and $7e^-$ in ^{15}N.
mass nucleons = 7(1.0078) + 8(1.0086) = 15.123 amu
mass loss = 15.123 – 15.00011 = 0.1239 amu

$$EB/A = \frac{(0.1232 \text{ amu})(931.5 \text{ MeV/amu})}{15 \text{ nucleons}} = \frac{115}{15} = 7.66 \frac{\text{MeV}}{\text{nucleon}}$$

As $E_B/A > 15_N$ than 14_N; 15_N is more stable.

63. The energy equivalent of the mass difference for this decay process is

$$(172.939 - 172.9383)\ (93.15) = 0.65 \text{ MeV}$$

Since this is less than 1.022 MeV; β^+ emission is not possible. Only electron capture is a possible decay process for $^{173}_{71}Lu$.

64. (a) For $^{13}_{7}N$ the n/p ratio is 6/7 or 0.86, which is too low for stability. Either electron capture or β^+ emission will produce $^{13}_{6}C$. The mass of $^{13}_{7}N$ is 13.00574 amu, and that of $^{13}_{6}C$ is 13.00335. The energy equivalent of the mass difference is
(13.00574 – 13.00335) (931.5) = 2.23 MeV

At this is greater than 1.022 MeV, positron emission is possible and is the most likely decay mode, as $Z < 30$. In fact ^{13}N is observed to be a β^+ emitter.

$$^{13}_{7}N \rightarrow {}^{13}_{6}C + {}^{0}_{1}\beta^+ + \nu$$

(b) For $^{25}_{11}Na$, the n/p ratio is 14/11 = 1.27, which is too large for Z = 11. The only possible mode of decay is β^- emission.

$$^{25}_{11}Na \rightarrow {}^{25}_{12}Mg^+ + {}^{0}_{-1}\beta^- + \nu$$

(c) For $^{132}_{57}La$, the n/p ratio is 75/57 = 1.316, which is too low for stability at Z = 57. Both electron capture and positron emission produce $^{132}_{56}Ba$, a stable isotope of barium. The mass of the two isotopes are 131.9103 for $^{132}_{57}La$ and 131.9051 for $^{132}_{56}Ba$. Since
(131.9103 – 131.9051) (931.5) = 4.8 meV > 1.022 MeV
both electron capture and β^+ emission are possible. In fact $^{132}_{57}La$ decays by both paths.

$$^{132}_{57}La \rightarrow {}^{132}_{56}Ba + \nu$$

$$^{132}_{56}Ba \rightarrow {}^{132}_{56}Ba + {}^{0}_{-1}\beta^- + \nu$$

(d) For $^{210}_{85}At$, the n/p ratio is 125/85 = 1.47, which is too low of Z = 85. At this value of Z both electron capture and α decay are possible, but β+ decay is not as likely. In fact, both electron capture and α decay are observed for $^{210}_{85}At$.

$$^{210}_{85}At \rightarrow {}^{210}_{84}Po + \nu$$

$$^{210}_{85}At \rightarrow {}^{206}_{83}Bi^{2-} + {}^{4}_{2}He^{2+}$$

65. The relation between the energy of the radiation and its wavelength is $E = h\nu = hc/\lambda$, so that $\lambda = hc/E$. We must express the energy in joules if we use Planck's constant in joule seconds. The energy of this transition is

$$(0.1427 \times 10^6 \text{ eV})\ (1.602177 \times 10^{-19} \cdot \text{eV}^{-1}) = 2.2863 \times 10^{-14} \text{ J}$$

The wavelength of the emitted radiation is

$$\lambda = \frac{(6.6206 \times 10^{-34} \text{ J} \cdot \text{s})\ (2.997925 \times 10^6 \text{ m} \cdot \text{s}^{-1})}{2.2863 \times 10^{-14} \text{ J}} = 8.688 \times 10^{-12} \text{ m}$$

or, $\lambda = 8.688$ pm $= 0.08688$ Å

66. The activity of a sample is given by λN. Since N is the number of radioactive nuclei, λ must be in reciprocal seconds if the activity is expressed in disintegrations per seconds. To obtain λ in reciprocal seconds we must express the half-life, $t_{1/2}$, in seconds.

$$t_{1/2} = (5730 \text{ yr}) \left(365 \frac{\text{day}}{\text{yr}}\right)\left(24 \frac{\text{h}}{\text{day}}\right)\left(60 \frac{\text{min}}{\text{h}}\right)\left(60 \frac{\text{s}}{\text{min}}\right) = 1.807 \times 10^{11} \text{ s}$$

Thus $\lambda = 0.6931/t_{1/2} = 0.6931/(1.807 \times 10^{11}) = 3.84 \times 10^{-12} \text{ s}^{-1}$
How many ^{14}C atoms are needed for the activity to be 3.7×10^{10} dps?
$\lambda N = (3.84 \times 10^{-12} \text{ s}^{-1})$ (N atoms) $= 3.7 \times 10^{10}$ dps

$$N = \frac{3.7 \times 10^{10}}{3.84 \times 10^{-12}} = 0.9645 \times 10^{22} = 10^{21} \text{ atoms}$$

Since 1 mol of carbon-14 contains 6.022×10^{23} atoms, the number of moles of carbon-14 needed to have an activity of 1 Ci is

$$\frac{9.65 \times 10^{21} \text{ atoms}}{6.022 \times 10^{23} \text{ atoms} \cdot \text{mol}^{-1}} = 1.6 \times 10^{-2} \text{ mol of } ^{14}\text{C}$$

A mole of ^{14}C weighs 14 g, and therefore we need
$(14 \text{ g} \cdot \text{mol}^{-1})\ (1.6 \times 10^{-2} \text{ mol}) = 0.224 \text{ g } ^{14}\text{C}$
to have a sample with an activity of 1 Ci.
Remember: The 0.224 g is the mass of the carbon along. If the ^{14}C were in $^{14}CO_2$, we would need 1.6×10^{-2} mol of $^{14}CO_2$, or
$(46 \text{ g} \cdot \text{mol}^{-1})\ (1.6 \times 10^{-2} \text{ mol}) = 0.736 \text{ g } ^{14}\text{CO}_2$

67. The activity now is 1×10^{-3} Ci. We want to calculate the length of time it takes for the activity to decrease to 1×10^{-8} Ci, as 0.01 μCi is $10^{-2} \times 10^{-6}$ Ci.

$$\text{We have, } \ln \frac{N}{N_0} = \ln \frac{10^{-8}}{10^{-3}} = \frac{-0.693\,t}{t_{1/2}} = \frac{-0.693\,t}{14.3}$$

Simplifying this expression we have
$\ln(10^{-5}) = -11.513 = -0.693t/14.3$

$$t = \frac{(11.513)\ (14.3 \text{ days})}{0.693} = 238 \text{ days}$$

Note that there is no need to convert the half-life from 14.3 days to any other unit of time. The units of t and $t_{1/2}$ must be identical. If $t_{1/2}$ is expressed in days, then t will automatically be in days.

68. Let, V = volume of blood in the animal
total activity before mixing = 1.8×10^6 cps

$$\text{Total activity after mixing} = (V\text{mL})\left(\frac{1.2\times10^4 \text{ CPS}}{2\text{ ml}}\right)$$

Hence, $1.8 \times 10^6 = V(6.0 \times 10^3)$

$$\therefore\ V = \frac{1.8\times10^6}{6.0\times10^3} = 3.0 \times 10^2 = 300 \text{ mL}$$

69. At the time of death, which is zero time, the specific activity of the wood was 15.3 dpm · g^{-1} of carbon, (the value for living wood all over the earth).

$$\therefore\ \ln\left(\frac{10.8}{15.3}\right) = -\frac{0.693\,t}{5730 \text{ yr}} \text{ where } t \text{ is in years}$$

or, $\ln(0.7059) = -0.3483 = -(0.693t)/5730$

$\therefore\ t = (5730 \text{ yr})\ (0.3483)/0.693 = 2.88 \times 10^3$ yr

The tree therefore died about 2880 years ago.

70. The first thing to calculate is the amount of ^{238}U that decayed to yield the 0.232 g of ^{206}Pb. Since 238 g of uranium will decay to yield only 206 g of lead, the mass of uranium that decayed to lead-206 was more than 0.232 g.

$$\text{mass of } ^{238}\text{U that decayed} = \left(\frac{238}{206}\right)(0.232 \text{ g}) = 0.268 \text{ g}$$

Thus the rock originally contained 1.605 + 0.268 = 1.872 g of uranium-238.
Over the years, 0.268 g of ^{238}U decayed, leaving 1.605 g of ^{238}U and producing 0.232 g of ^{206}Pb.

If zero time is the time when the rock was formed and there was no ^{206}Pb present, the time, t, Now, half-life of $^{238}U = 4.5 \times 10^9$ yr.

$\ln(1.605/1.873) = -(0.693t)/(4.5 \times 10^9)$

or, $\ln(0.8569) = -0.1544 = -(0.693t)/(4.5 \times 10^9)$

$\therefore\ t = (0.1544)\ (4.5 \times 10^9)/0.693 = 1.0 \times 10^9$

This rock was found 1 billion years ago.

71. (a) $t_{1/2} = \dfrac{\ln 2}{k} = \dfrac{0.693}{1.42\times10^{-3}\ \text{s}^{-1}} = 488 \text{ s} = 8.13 \text{ min}$

(b) The partial pressure of a gas is directly proportional to its concentration in moles per liter at constant temperature: $PA = [A]RT$.

If the partial pressure of N_2O_5 falls from 2.84 to 0.355 atm it has decreased to $\frac{1}{8}$ its original value. A decrease by a factor of $\frac{1}{8}$ takes three half-lives. At the end of the first 8.13 min, the partial pressure falls to $\frac{1}{2}$ its initial value (1.42 atm); after 16.26 min (two half-lives) the partial pressure is again cut in half, to 0.710 atm. At the end of the three half-lives, or after 24.4 min, the partial pressure is $\frac{1}{8}$ its original value, 0.355 atm.

72. A plot of the natural logarithm of the activity as a function of time should be drawn. A table of the values plotted should always appear along with the plot. These values and the plot are shown here.

Tim (h)	*ln (activity)*
0	9.08
2	8.92
4	8.77
6	8.61
8	8.46
12	8.15

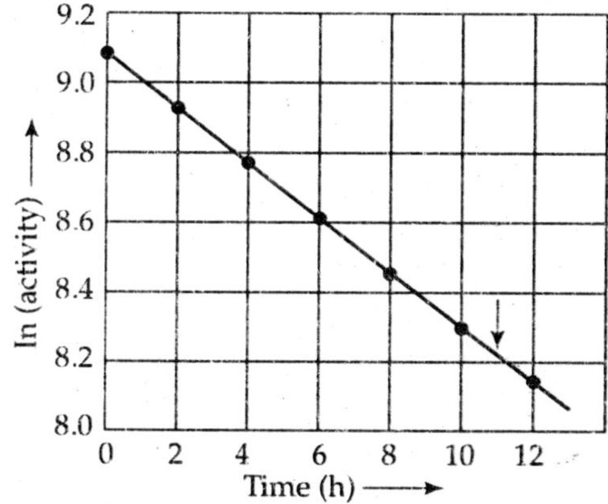

The slope of the plot is determined using two widely separated points (indicated by arrows on the plot above).

$$\text{Slope} = -\lambda = \frac{8.23 - 9.0}{11.0 - 1.0} = -0.077$$

The first-order decay constant is therefore 0.077 h^{-1}.

The half-life is: $t_{\frac{1}{2}} = (\ln 2)/\lambda = 0.693/0.077 = 9.0$ h

The isotope is $^{234}_{94}Pu$.

73. Because the rate = kN_{t}, first find the value of k in reciprocal seconds:

$$k = \frac{0.693}{t_{1/2}} = \frac{0.693}{14.3\,d} \times \frac{1d}{24\text{ h}} \times \frac{1\text{ h}}{3600\text{ s}} = 5.609 \times 10^{-7}/\text{s}$$

Now calculating the number of P-32 nuclei in the sample (N_t). For the formula weight of Na_3PO_4 containing 15.6% of P-32 and (100.0 – 15.6)% = 84.4% naturally occurring *P*. Usually, the formula weight of Na_3PO_4 containing naturally occurring P is 163.9 amu; the formula weight of Na_3PO_4 when the P is 100% P-32 is 165.0 amu (the atomic mass of P-32 is 32.0 amu). The formula of Na_3PO_4 containing 15.6% P-32 is :

(163.9 amu × 0.844) × (165.0 × 0.156) = 164.1 amu

The moles of *P* in the sample

$$= 0.0545 \text{ g } Na_3PO_4 \times \frac{1 \text{ mol } Na_3PO_4}{164.1 \text{ g } Na_3PO_4} \times \frac{1 \text{ mol P}}{1 \text{ mol } Na_3PO_4} = 3.321 \times 10^{-4} \text{ mol P}$$

and the moles of *P*-32

$= (3.321 \times 10^{-4} \text{ mol } P) \times 0.156 = 5.181 \times 10^{-5} \text{ mol } P\text{-}32$

Then, the number of P-32 nuclei is

$$(5.181 \times 10^{-5} \text{ mol P-32}) \times \frac{6.022 \times 10^{23} \text{ P - 32 nuclel}}{1 \text{ mol P} - 32} = 3.120 \times 10^{19} \text{ P-32 nuclei}$$

Finally, the rate of disintegrations:

Rate = $kN_t = (5.609 \times 10^{-7}/\text{s}) \times (3.120 \times 10^{19})$ P-32 nuclei
$= 1.750 \times 10^{13} = 1.75 \times 10^{13}$ P-32 nuclei/s

74. First calculate the value of the decay constant, *k*, for Po-210 using 2.303 log 2 = 0.6931:

$$k = \frac{0.6931}{t_{1/2}} = \frac{0.6931}{138.4 \text{ d}} = 5.00794 \times 10^{-3} \text{ d}^{-1}$$

Now, to find, g, of PoO_2 left after 2 days;
The first-order relationship between time and concentration is:

$$-\log \frac{[A]_t}{[A]_0} = -\log \frac{[g PoO_2]_t}{[g\ PoO_2]_0} = -\log \frac{[g\ PoO_2]_{2d}}{[1.0\ g]} = \frac{kt}{2.303}$$

$$\text{or } -\log \frac{[g\ PoO_2]_{2d}}{[1.0\ g]} = \frac{(5.00794 \times 10^{-3}\ d^{-1}) \times (2.0\ d)}{2.303} = 4.349 \times 10^{-3}$$

$-\log [g\ PoO_2)_{2d} = (4.349 \times 10^{-3}) - \log [1.0\ g] = 4.349 \times 10^{-3}$

$\log [g\ PoO_2]_{2d} = -(4.349 \times 10^{-3})$

or, $[g\ PoO_2]_{2d} = 0.9900358$ g left after 2 days

g PoO_2 decomposed after 2d = 1.0 g – 0.99 = 0.01 g

$$\text{Mol He} = \text{mol } PoO_2 \text{ decomp.} = 0.0099641\ 0.01 \text{ g Po} \times \frac{1 \text{ mol } PoO_2}{242\ g\ PoO_2} \times \frac{1 \text{ mol He}}{1 \text{ mol } PoO_2}$$

= 0.000041 mol He

Now calculate the volume of He at 298 K and 735 mmHg using R = 0.8206.

$$V = \frac{nRT}{P} = \frac{(0.000041 \text{ mol})(0.08206 \text{ L} \cdot \text{atm})/(\text{K} \cdot \text{mol})(298 \text{ K})}{(735/760) \text{ atm}} = 0.001 \text{ L (1 mL)}$$

75. From the problem we have:

$$2p + 2n \rightarrow \text{He-4}$$

On a mole basis, the mass difference, Δm, is

–2.01456 g/2 mol protons

–2.01733 g/2 mol neutrons

+4.00260 g mol^{-1} He

Δm = –0.02929 g mol^{-1} He = -2.929×10^{-5} kg mol^{-1} He

$\Delta E = (\Delta m)c^2 = -2.929 \times 10^{-5}$ kg mol^{-1} He $\times (2.998 \times 10^8\ m^{-1})^2$

$= -2.632 \times 10^{12}\ kg \cdot m^2\,s^{-2} = -2.632 \times 10^{12}\ J = -2.632 \times 10^9$ KJ

Next $\Delta H°$ for burning of ethane is:

	$C_2H_6(g)$ + 7/2 $O_2(g)$		→	$2CO_2(g)$	+	$3H_2O(g)$
ΔH_f° =	–84.66 KJ	0		2(–393.5 KJ)		3(–241.826 KJ)

ΔH_∞ = [2(–393.5) + 3(–241.826) – (–84.667) + 0] kJ = 1427.81

$= -1.43781 \times 10^3$ KJ mol^{-1} ethane

Now the mols of ethane needed to obtain 2.632×10^9 kJ heat is:

$$-2.632 \times 10^9\ \text{kJ} \times \frac{1\ \text{mol ethane}}{-1.42781 \times 10^3\ \text{kJ}} = 1.854. \times 10^6\ \text{mol ethane}$$

Finally, convert moles to liters at 298 K and 725 mmHg using R = 0.8206.

$$V = \frac{nRT}{P} = \frac{1.854 \times 10^6\ \text{mol})\,[0.08206\ \text{L} \bullet \text{am/K} \bullet \text{atm})](298\ \text{K})}{(725/760)\ \text{atm}} = 4.73 \times 10^7\ \text{L}$$

76. Let us

Calculate the percent Am-241 that remains after one year, setting X_0 to 100%.

$$\ln\left(\frac{X_0}{X}\right) = kt \Rightarrow \ln\left(\frac{100}{X}\right) = (1.51 \times 10^{-3}/\text{yr})(1\ \text{yr})$$

ln (100)– ln (X) = 1.51×10^{-3}

ln (X) = 4.6

X = $e^{4.6}$ = 99.8 (per cent Am-241 remaining)

Thus 99% of the Am-241 remains. At that rate, the Am–241 will last a life time (only 14% decays in 100 years).

77. Plots ln(A) versus time (in hours). The slope of the line is $-k$, which is used to calculate

Time (h)	0.0	0.5	1.0	1.5	2.0	2.5
A (disintegrations h^{-1})	14,472	13,095	11,731	10,615	9,605	8.504
ln (A)	9.58	9.48	9.37	9.27	9.17	9.05

$$-k = \text{slope} = \frac{\Delta Y}{\Delta X} = \frac{-0.32}{1.5} = -0.21 \Rightarrow k = 0.21$$

$$\text{and } -k = \frac{0.693}{t_{1/2}} \Rightarrow 0.21 = \frac{0.693}{t_{1/2}}$$

$$\therefore\ t_{\frac{1}{2}} = 3.3 \text{ hr}$$

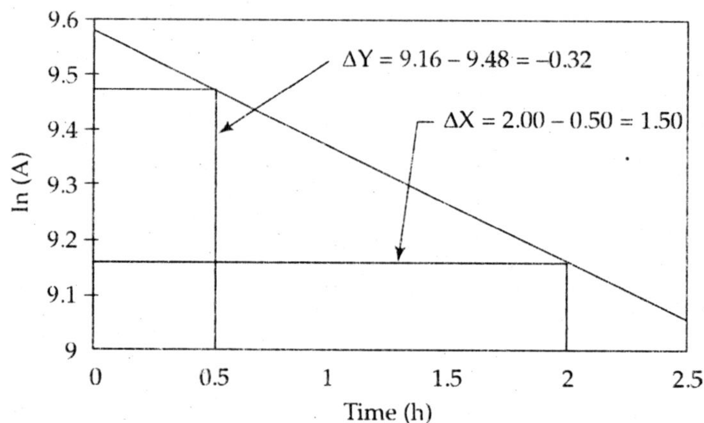

78. From $t_{1/2}$, calculate k and, from k and A, calculate N.

$$k = \frac{0.693}{3.82\,d} \times \frac{1d}{24\,\text{hr}} \times \frac{1\text{ hr}}{3600\text{ s}} = 2.10 \times 10^{-6}\text{ s}^{-1}$$

$$N = \frac{20 \times 10^{-12}\text{ Ci}}{2.10 \times 10^{-6}\text{ s}^{-1}} \times \frac{3.700 \times 10^{10}\text{ atoms s}^{-1}}{1\text{Ci}} = 3.5 \times 10^{5}\text{ atoms}$$

This is concentration of 3.54×10^{-5} atoms L.

$$= \frac{3.5 \times 10^{-5}\text{ atoms}}{1\text{ L}} \times \frac{1\text{ mol}}{6.02 \times 10^{23}\text{ atoms}} = 5.8 \times 10^{-19}\text{ mol. L}^{-1}$$

79. (a) $1.00\text{ g} \times 5.5 \times 10^{-11}\text{ min}^{-1}. \times 45\text{ min} = 2.5 \times 10^{-9}\text{ g}$

(b) $\Delta m = 234.9934 + 4.0015 - 239.0006 = -0.0057\text{ g}$

$$\Delta m_{(1\text{ g})} = 2.5 \times 10^{-9}\text{ g} \times \frac{1\text{ mol.}}{239.0\text{ g}} \times \frac{-0.0057\text{ g}}{1\text{ mol.}} = -6.0 \times 10^{-14}\text{ g}$$

$$\Delta E = 9.0 \times 10^{10}\text{ kJ g}^{-1} \times (-6.0 \times 10^{-14})\text{g} = -5.4 \times 10^{-3}\text{ kJ} = 5.4\text{ J}$$

Thus 5.4 J of energy is given off.

(c) 1 rad = 10^{-2} j kg^{-1}

rems = n(rads) $n = 10$ for α rays

$$\frac{5.4\text{ J}}{75\text{ kg}} \times \frac{1\text{ rad}}{10^{-2}\text{ J kg}^{-1}} = 7.2\text{ rads}$$

rems = 7.2 rads × 10 rems rad^{-1} = 72 rems

80. From the given data of the problem we have

(a) $E = 8.99 \times 10^9 q_1q_2/r = (8.99 \times 10^9)(1.60 \times 10^{-19}\ C)^2/(2 \times 10^{-15}\ m) = 1 \times 10^{-13}\ J$

(b) Convert the mass of the deuteron to kilograms.

$$2 \text{ deuterons} \times \frac{2.01355 \text{ g}}{1 \text{ mol.}} \times \frac{1 \text{ mol.}}{6.022 \times 10^{23}} \times \frac{1 \text{ kg}}{1000 \text{ g}} = 6.687 \times 10^{-27} \text{ kg}$$

$E = mv^2/2 \Rightarrow 2E/m = v^2$

$\therefore\ v = [(2)(1 \times 10^{-13}\ J)/(6.687 \times 10^{-27}\ kg)]^{1/2} = 5 \times 10^6\ m/s$

81. Number of C atoms in 14 g of $^{14}C = 6.02 \times 10^{23}$

$\therefore$ Number of C atoms in 1000 mg of $^{14}C = \dfrac{6.02 \times 10^{23}}{14}$

number of C atoms in 7 mg of $^{14}C = \dfrac{6.02 \times 10^{23} \times 7}{14 \times 1000} = 3.01 \times 10^{20}$

Number of neutrons in 1 carbon atom = 7

$\therefore$ Total number of neutrons in 7 mg of $^{14}C = 3.01 \times 10^{20} \times 7$

Assuming, weight of 1 neutron = weight of hydrogen

Weight of 1 hydrogen atom $= \dfrac{1}{6.02 \times 10^{23}}$ g

$\therefore$ Weight of $3.01 \times 10^{20} \times 7$ hydrogen atom $= \dfrac{3.01 \times 10^{20} \times 7}{6.02 \times 10^{23}}\ g = \dfrac{7}{2 \times 10^3}\ g = 3.5 \times 10^{-3}\ g$

82. $^{234}_{90}Th \rightarrow {}^{206}_{82}Pb$

$\therefore$ Decrease in atomic mass = 234 – 206 = 28.

and No. of α-particles emitted $= \dfrac{28}{4} = 7.$

The atomic number decreases to 90 – (7 × 2) = 76

Now, atomic number of Pb = 82

Therefore, β-particles emitted = 82–76 = 6

83. $t_{\frac{1}{2}} = \dfrac{0.693}{\lambda}$ where $t_{\frac{1}{2}}$ = half life

and λ = disintegration constant $= \dfrac{0.693}{28.1}\ 0.0246\ yr^{-1}$

Now, $M_t = M_0, e^{-\lambda t} = 1\ \mu\ g.\ e\ 0.0246 \times 20 =$ **0.061** μg

84. For first order gas reaction, $= \dfrac{2.303}{t} \times \log \dfrac{1}{(1-x)}$

Substituting the values,

$$1.5 \times 10^{-6} = \frac{2.303}{10 \times 3600} \log \frac{1}{(1-x)}$$

$$\text{or } \log \frac{1}{(1-x)} = \frac{1.5 \times 10^{-6} \times 10 \times 3600}{2.303} = 0.0234$$

$\Rightarrow x = 0.052$

Initial concentration charged into product is, therefore, 5.2%.

Also,

$$t_{\frac{1}{2}} = \frac{9.693}{K} = \frac{0.693}{1.5 \times 10^{-6}\ s^{-1}} = 462000 \text{ seconds} = 128.33 \text{ hours.}$$

85. According to question ^{227}Ac follow two paths for decay

$$^{223}Fe \xleftarrow{\lambda_1} {}^{227}Ac \xrightarrow{\lambda_2} {}^{227}Th \quad t_{\frac{1}{2}} = 22 \text{ Yrs.}$$

$$\text{Total} \quad \lambda = \frac{0.693}{t_{\frac{1}{2}}} = \frac{0.693}{22} = 0.0315 \text{ Yr}^{-1}$$

Now, $\lambda = \lambda_1 + \lambda_2$ (1)

While, $\frac{\lambda_1}{\lambda_2} = \frac{98}{2} = 49$

$\lambda_1 = 49\ \lambda_2$ (2)

$0.0315 = 49\ \lambda_2 + \lambda_2$

$\lambda = 6.3 \times 10^{-4}\ \text{Yr}^{-1}$ & $\lambda_1 = 0.03087\ \text{Yr}^{-1}$

86. $\lambda_{653\ k} = 0.693/360 = 1.925 \times 10^{-3}\ \text{min}^{-1}$

$$\frac{\lambda_{723k}}{\lambda_{653k}} = e^{\frac{-200 \times 10^3}{8.314}\left[\frac{1}{723k} - \frac{1}{653}\right]} = e^{\frac{-14000 \times 10^3}{8.314 \times 723 \times 653}} = e^{3.566}$$

Solving, $\lambda_{723\ k} = e^{3.566} \times 1.925 \times 10^{-3} = 6.81 \times 10^{-2}\ \text{min}^{-1}$

$\therefore$ t = 20.34 min.

87. Minimum number of beta particles needed = 346 min^{-1}

Number of beta particles needed to carry out the experiment for 6.909 hours

= 346 909 × 60 = 143431

Amount of beta particles needed

$$= \frac{143431}{6.023 \times 10^{22}\ \text{mol}^{-1}} = 2.3814 \times 10^{-19} \text{ mol}$$

$$\text{Now, } \lambda = \frac{0.693}{t_{1/2}} = \frac{0.693}{66.6 \text{ hr}} = 0.014 \text{ hr}^{-1}$$

Integrated rate expression for first order kinetics is given by

$$\log \frac{[A_0] - [A]_{consumed}}{[A_0]} = \frac{\lambda}{2.303}$$

$$= \frac{0.01404 \text{ hr}^{-1}}{2.303} \times 6.909 \text{ hr} = -0.04212$$

or $\dfrac{[A_0]-[A]_{consumed}}{[A_0]} = 0.9076$

Substituting the volume of
$[A]_{consumed} = 2.3814 \times 10^{-19}$, we obtain

$$[A_0] = \frac{[A]_{consumed}}{1-0.9074} = \frac{2.3814 \times 10^{-19} \text{ mol}}{0.0926} = 2.572 \times 10^{-18} \text{ mol.}$$

88.

$$_{29}Cu^{64} \xrightarrow[\lambda_1]{B^- \text{ emission}} {}_{30}Zn^{64} (38\%)$$

$$_{29}Cu^{64} \xrightarrow[\lambda_2]{B^+ \text{ emission}} {}_{29}Ni^{64} (19\%)$$

$$_{29}Cu^{64} \xrightarrow[\lambda_3]{\text{Electron Capture}} {}_{28}Ni^{64} (43\%)$$

Half life period of $_{29}Cu^{64}$ is 12.8 hrs.

$$\text{decay constant } \lambda = \frac{0.693}{12.8} = 0.0541 \text{ hr}^{-1}$$

$$\lambda = \lambda_1 + \lambda_2 + \lambda_3$$

Where, $\lambda_1 \rightarrow$ 1st decay constant of $_{29}Cu^{64}$
$\lambda_2 \rightarrow$ 2nd decay constant of $_{29}Cu^{64}$
$\lambda_3 \rightarrow$ 3rd decay constant of $_{29}Cu^{64}$

From given data $\dfrac{\lambda_1}{\lambda_2} = \dfrac{38}{19} = 2$ and $\dfrac{\lambda_3}{\lambda_1} = \dfrac{43}{38}$

$\lambda_3 = 1.1315\ \lambda_1;\ \lambda_2 = 0.5\ \lambda_1$
$0.0541 = \lambda_1 + 0.5\ \lambda_1 + 1.1315\ \lambda_1$
$0.0541 = 1.631\ \lambda_1$

$$\lambda_1 = \frac{0.0541}{1.631} = 0.03315 \text{ hr}^{-1}$$

$\lambda_2 = 0.5 \times 0.03315 = 0.01637 \text{ hr}^{-1}$
$\lambda_3 = 1.1315 \times .03315\ \ 0.03750 \text{ hr}^{-1}$

Now, half life of different emission are calculated as follows.

$$(t_{\frac{1}{2}})_1 = \frac{0.693}{\lambda_1} = \frac{0.693}{0.03315} = 20.90 \text{ hrs.}$$

$$(t_{\frac{1}{2}})_2 = \frac{0.693}{\lambda_2} = \frac{0.693}{0.01657} = 41.82 \text{ hrs.}$$

$$(t_{\frac{1}{2}})_3 = \frac{0.693}{\lambda_3} = \frac{0.693}{0.03750} = 18.4 \text{ hrs.}$$

89. Given, $t_{1/2} = 1$ hour

$$\therefore \qquad k = \frac{0.693}{t_{1/2}} = \frac{0.693}{1} = 0.593 \text{ hr}^{-1};$$

$$K = \frac{2.303}{1} \log \frac{a}{(a-x)} \quad (t = 8 \text{ hrs})$$

$$\text{or,} \qquad \log \frac{a}{(a-x)} = \frac{Kt}{2.303} = \frac{0.693 \times 8}{2.303} = 2.407;$$

$a/(a\text{-}x)$ = anti log 2.407 = 255.3
or, $(a\text{-}x)/a = 1/255.3 = 3.9 \times 10^{-3}$ g. or $a - x = 3.9 \times 10^{-3}$ g.

Chapter 12 : CRYSTALLOGRAPHY

1. Once mole of NaCl has a mass of 58.5×10^{-3} kg and therefore has a volume

$$(58.5 \times 10^{-3} \text{ kg})\left(\frac{1 \text{ m}^3}{2.17 \times 10^3 \text{ kg}}\right) = 2.7 \times 10^{-5} \text{ m}^3$$

A unit cell, containing 4 NaCl formula units, has sides of length of 0.564 mm. The volume of a unit cell is $(0.564 \times 10^{-9})^3 = 1.79 \times 10^{-28}$ m^3
The number of units cell per mole is then

$$\frac{2.7 \times 10^{-5} \text{ m}^3}{1.79 \times 10^{-28} \text{ m}^3} = 1.51 \times 10^{23}$$

Since there are 4 NaCl formula units per unit cell, there are $N_A = (4)(1.51 \times 10^{23})$ $= 6.04 \times 10^{23}$ formula units per mole.

2. All distance in angstrom units:

		I–	2.19
RbI	3.67	Rb^+	3.67 – 2.19 = 1.48
RbBr	3.44	Br^-	3.44 – 1.48 = 1.96
KBr	3.29	K^+	3.29 – 1.96 = 1.33
KCl	3.14	Cl^-	3.14 – 1.33 = 1.81
NaCl	2.79	Na^+	2.79 – 1.81 = 0.98
NaF	2.31	F^-	2.31 – 0.98 = 1.33

3. (a) The body diagonal is $\sqrt{3}(5.20 \text{ Å}) = 9.01$ Å. The nearest neighbours along that diagonal are half that distance apart, 4.50 Å.

(b) 5.20 Å, along the cell edge.

(c) 8 (For example, the body center is next to 8 corners.)

(d) 6 (The corner has neighbour along each cell edge: up, down, in, out, left, right.)

(e) Two K atoms per unit cell: one of the body center and one-eighth at each of eight corners. Their mass:

$$= 2\left(\frac{39.1 \text{ g}}{6.02 \times 10^{23}}\right) = 1.3 \times 10^{-22} \text{ g}$$

$$\text{The volume is} = \left[(5.20 \text{ Å})\left(\frac{10^{-8} \text{ cm}}{1 \text{ Å}}\right)\right]^3 = 1.41 \times 10^{-22} \text{ cm}^3$$

$$\text{The density} = \frac{1.3 \times 10^{-23} \text{ g}}{1.41 \times 10^{-22} \text{ cm}^3} = 0.925 \text{ g cm}^{-3}$$

4. (a) Four of each: $\frac{1}{4}$ Br along each of 12 edges plus one at the center; $\frac{1}{8}$ K at each of eight corners, plus $\frac{1}{2}$ K at each of six face centers.

(b) a = 2(K^+ radius + Br^- radius) = 2(1.33 Å + 1.95Å) = 6.56 Å

(c) $V = 6.56 \times 10^{-8}\ \text{cm})^3 = 2.82 \times 10^{-22}\ \text{cm}^3$

$$m = (4\ \text{formula units})[(39.1 + 79.9)\ \text{g mol}^{-1}]\left(\frac{1\ \text{mol}}{6.02 \times 10^{23}\ \text{formula units}}\right) = 7.91 \times 10^{-22}\text{g}$$

$$d = \frac{m}{V} = \frac{7.91 \times 10^{-22}\ \text{g}}{2.82 \times 10^{-22}\ \text{cm}^3} = 2.08\ \text{g cm}^{-3}$$

(d) Side = $a_{+}, + a_{-}$ Diagonal = $2a_{-} = (\sqrt{2})(a_{+} + a_{-})$

$\therefore (\sqrt{2})a = a_{+} + a_{-}$ or, $(\sqrt{2} - 1)a_{-} = a_{+}$

Hence, $a_{+}/a_{-} = 0.414$

5. (a) The volume is calculated form the density, per unit cell :

$$\text{Mass} = (2\ \text{atoms})\left(\frac{24.3\ \text{g}}{6.02 \times 10^{23}\ \text{atoms}}\right) = 8.07 \times 10^{-23}\ \text{g}$$

$$\therefore\ \text{Volume} = (8.07 \times 10^{-23}\ \text{g})\left(\frac{(1.0 \times 10^{8}\ \text{Å})^3}{1.74\ \text{g}}\right) = 46.4\ \text{Å}^3$$

(b) The base of the cell (Fig. b) has an area calculated as follows:

$$\text{Sin } 60^\circ = \frac{x}{a} \Rightarrow x = a\ (\sin 60^\circ) = a\ (0.866)$$

$\therefore$ Area = ax = $0.866\ a^2$

The volume is $(0.8660a^2)/c = (0.8860a^2)(1.633a) = 1.414a^3 = 46.4\ \text{Å}^3$.

$\therefore\ a = 3.20$ Å

(c) 3.20 Å. Nearest neighbours are along the base edge.

(d) 12. The 12 are most easily viewed as 6 in the same plane, 3 above, and 3 below.

6. Using Bragg's law, $n\lambda = 2d \sin \theta$, and assuming that the diffractions are first order (N = 1), the distances are calculated to be 0.204 nm, 0.176 nm, and 0.124 nm. The distance 0.176 nm is seen to be half the unit cell length, corresponding to be perpendicular distance between layers of atoms in one fact of the unit cell and the center (see Figure a) The distance 0.204 nm turns out to be $1/\sqrt{3}$ times the unit cell length, corresponding to the perpendicular distance between a corner of the unit cell and the plane of the three adjacent corners (see Figure c). The distance 0.124 nm corresponds to $\sqrt{2}/4$ times the unit cell dimension—one fourth of the distance form one edge of the cell to the opposite edge (see Figure b). The plane intersecte the fce-centered atoms of two sides. The plane including the edge atoms is equivalent to the plane across the center of the unit cell.

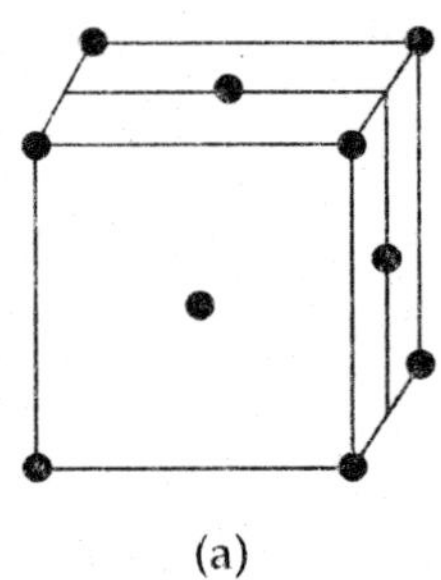

(a)

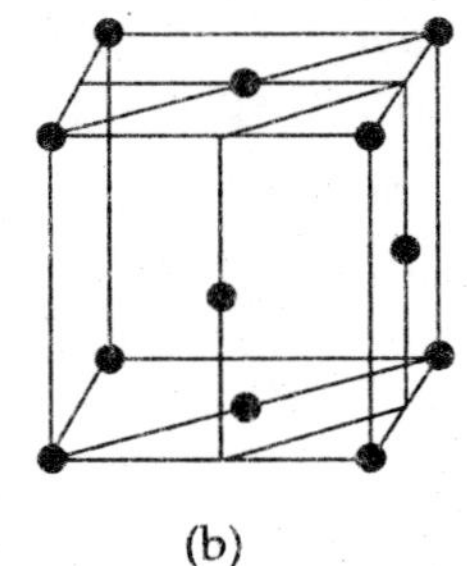

(b)

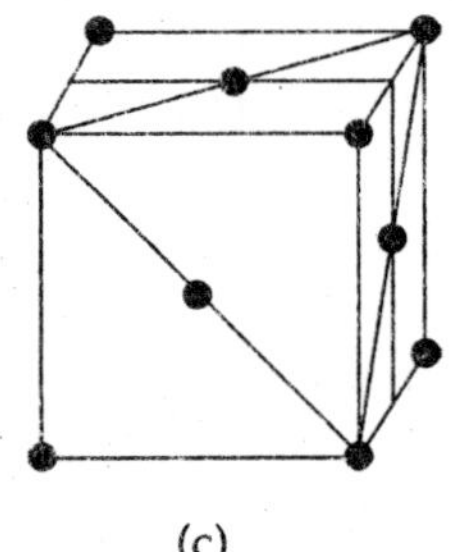

(c)

7. Length of the edge of the unit cell = 654 pm = 6.54×10^{-8} cm

$\therefore$ Volume (V) of the unit cell = $(6.54 \times 10^{-8} \text{ cm})^3$

Molecular mass of KBr = 39 × 80 = 119 g mol^{-1}

Density of KBr = 2.75 g cm^{-3}

$$\because \quad \rho = \frac{n \times M}{V \times N_0} \Rightarrow n = \frac{\rho \times V \times N_0}{M}$$

$$= \frac{(2.75 \text{ g cm}^{-3})(6.54 \times 10^{-8} \text{ cm})^3 \ (6.023 \times 10^{23} \text{ mol}^{-1})}{119 \text{ g mol}^{-1}} = \frac{2.76 \times (6.54)^3 \ (6.023)\ (10^{-1})}{119} = 3.9$$

Since number of atoms (or ions in a unit cell is approximately four hence the crystal must be face-centered cubic.

8. Length of the edge of the unit cell = 200 pm = 2×10^{-8} cm

$\therefore$ Volume (V) of the unit cell = $(2 \times 10^{-8} \text{ cm})^3 = 8 \times 10^{-24} \text{ cm}^3$

Mass of element (M) = 200 g

Number of the atoms (N_0) = 24×10^{23}

The number of atoms per unit cell cell (fcc) = 4

$$\rho = \frac{n \times M}{V \times N_0} = \frac{(4)\ (200 \text{ g})}{(8 \times 10^{-24} \text{ cm}^{-3})\ (24 \times 10^{23})} = 41.7 \text{ g cm}^{-3}$$

9 Since NaCl has octahedral structural arrangement hence for such arrangement maximum radius ratio (r^+/r^-) is 0.732.

$$\therefore \quad \frac{A^+}{B^-} = 0.732 \Rightarrow B^- = \frac{A^+}{0.732} = \frac{110 \text{ pm}}{0.732} = 150 \text{ pm}$$

10. (a) $\frac{r_{A^+}}{r_{B^-}}$ = ideal ratio = 0.225

$$\therefore \ r_{B^-} = \frac{r_{A^+}}{0.225} = \frac{22.5}{0.225} = 100 \text{ pm.}$$

(b) $\frac{r_{A^+}}{r_{B^-}} = \frac{50}{100} = 0.5$

It crystallizes in an NaCl types of structure. Co-ordination number changes from 4 to 6.

11. We know that if the sequence of layers is AB AB AB ... then it gives hexagonal close packing (hcp). The number of atoms in hcp is 6 per unit cell.

$\because$ Packing fraction $= \dfrac{\text{Volume of 6 spheres}}{\text{Area of the base} \times C}$

Now, volume of 6 spheres $= 6 \times \dfrac{4}{3} = 8\pi r^3$

Area of the base $= 6 \times \dfrac{\sqrt{3}}{4}(2r)^2 = 6\sqrt{3}r^2$

and $C = 4r\sqrt{(2/3)}$

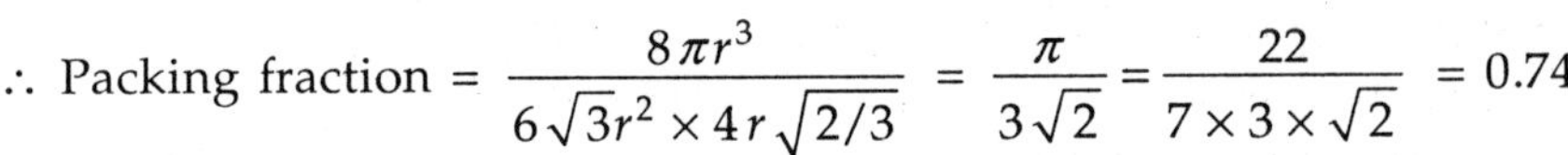

$$\therefore \text{ Packing fraction } = \frac{8\pi r^3}{6\sqrt{3}r^2 \times 4r\sqrt{2/3}} = \frac{\pi}{3\sqrt{2}} = \frac{22}{7 \times 3 \times \sqrt{2}} = 0.74$$

$\therefore$ Void $= 1 - 0.74 = 0.26$

or Void % $= 26\%$.

12. In a face-centered cell there are molecules at each of the eight corners and at the center of each of the six faces.

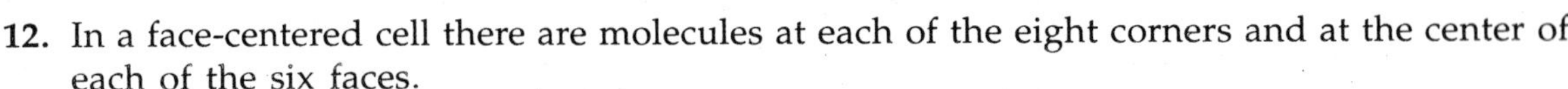

$\therefore$ (8 corners) $\left(\frac{1}{8} \text{ per corner}\right)$ + (6 face centers) $\left(\frac{1}{2} \text{ per center}\right)$ = 1 + 3 = 4 molecules per unit cell.

13. The radius of an atom in a *bcc* lattice is $a\sqrt{3}/4$, where a is the unit cell length, that is, the side of the cube has length a. As the volume of a sphere is $\frac{4}{3}\pi r^3$,

$$\text{atomic volume in a } bcc \text{ cell} = \frac{4}{3}\pi a^3 \frac{(3\sqrt{3})}{64} = \frac{\pi a^3\sqrt{3}}{16}$$

As there are two atoms per unit cell, the occupied volume is $(\pi a^3\sqrt{3})/8$. The volume of the unit cell itself is a^3, so that

$$\text{fraction of } bcc \text{ cell occupied} = \sqrt{3}\pi/8 = 0.68$$

14, If the side of the cubic unit cell is 3.923 Å or 3.923×10^{-8} cm, the volume of one unit cell is $(3.9231 \times 10^{-8})^3$ cm^3, or 6.0379×10^{-23} cm^3. This volume is occupied by four platinum atoms.

As the density is mass per unit volume, we must calculate the mass of four platinum atoms. The mass of 1 mol of Pt is 195.08 g, so that

$$\text{mass 4 Pt atoms} = \frac{4(195.08 \text{ g} \cdot \text{mol}^{-1})}{6.022 \times 10^{23} \text{ mol}^{-1}} = 1.296 \times 10^{-21} \text{ g}$$

$$\text{Thus, density} = \frac{1.296 \times 10^{-21} \text{ g}}{6.037 \times 10^{-23} \text{ cm}^3} = 21.46 \text{ f} \cdot \text{cm}^{-3}$$

The atomic radius is given by

$r = a\sqrt{2}/4 = (3.923 \text{ Å})(\sqrt{2})/4 = 1.387 \text{ Å} = 1.38.7 \text{ pm}$

15. A face of the LiCl unit cell, assuming anion–anion contact. Small colored circles are Li^+ ions, large open circle are Cl^- ions. (a) Structure if Li^+ ions are too small for anion–cation contact. This is the correct structure. (b) Structure if there were contact between Li^+ and Li^- ions.

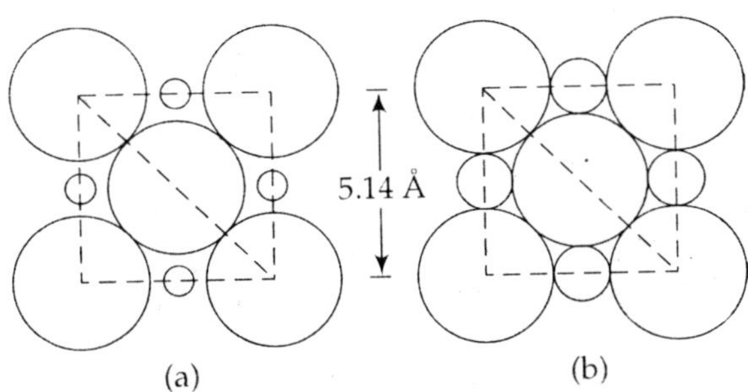

The Li^+ ions may be so small relative to the Cl^- ions in LiCl that they are not in actual contact with the anions or they may just fit exactly in the space between the Cl^- ions,. In either case, the assumption of anion-anion contact allows us to calculate a value for the radius of the chloride ion.

The length of the unit cell is 5.14 Å. Using the Pythagorean theorem, the length of a face diagonal is $(\sqrt{2})$ (5.14 Å) = 7.27 Å, and this must be four times the radius of the chloride ion. Thus,

$r_{Cl^-} = 7.27/4 = 1.82$ Å = 182 pm

If the assume the Li^+ ions are in contat with all the adjacent Cl^- ions, the length of one side of the unit cell is $2r_{Cl^-} + 2r_{Li^+}$. Thus

$5.14 \text{ Å} = 2r_{Cl^-} + 2r_{Li^+} = 2(1.82 \text{ Å}) + 2r_{Li^+}$

and $r_{Li^+} = \frac{1}{2}(5.14 - 3.64) = 0.75 \text{ Å} = 75 \text{ pm}$

In fact, by measuring internuclear distances in many lithium compouds, We find that the Li^+ ion is smaller than 75 pm.

The radius of the Li^+ ion, obtained from a number of measured internuclear distances in lithium compounds, is 0.6 Å or 60 pm. The value of the radius of the Cl^- ion that fits best for a large number of chlorides is 1.81 Å or 181 pm.

16. With ions that are roughly in size, anions are not in contact with one another. The following figure shows the face of the unit cell.

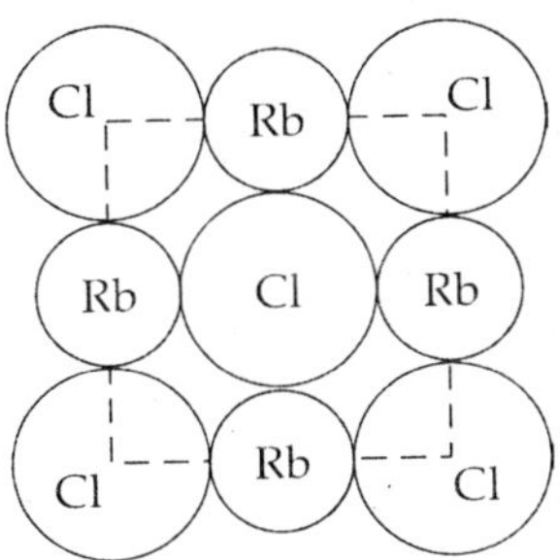

Since the length of the size of the cube is $2r_{Cl^-} + 2r_{Rb^+}$

$$6.54\ \text{Å} = 2r_{Cl-} + 2r_{Rb+} \quad \text{and} \quad r_{Rb+} = 3.27\ \text{Å} - 1.81\ \text{Å} = 1.46\ \text{Å}$$

The value of r_{Rb+} that fits best for many compounds is 1.48 Å.

17. The only factor in the equation $U = \dfrac{N_A Z_+ Z_-}{r_0}$ that varies for these four compounds is r_o, which can be considered the sun of the ionic radii, $r_c + r_a$. The cation is the same in all these crystals, but the antion radius increase regfularly as the atomic number of the halide increases. The lattice energy is inversely proportional to $r_c + r_a$, so as that sum increases the lattice energy decreases. As r_a increases regularly for the series F^-, Cl^-, Br^-, I^-, the lattice energy decreases regularly for KF, Kl, KBr, and KI.

18. (a) The value of $r_c + r_a$ is 321 pm for AgBr and 342 pm for AgI. If the only difference between AgBr and AgI were in the size of the anion, we would expect the lattice energies to be proportional to the inverse ratio of $r_c + r_a$. Hence we expect the lattice energy of AgI to be (895) (321/342) = 840 kJ · mol^{-1}.

(b) The Ag^+ ion does not have rare gas electronic configuration and is a good polarizer of anions. The I^- ion is larger and more polarizable than the Br^- ion. Therefore the cation–anion bond has more covalent character in AgI than in AgBr. The lattice energy of AgI is therefore larger than the value calculated assuming the only difference between U for AgBr for AgI is in the value of $r_c + r_a$ above.

19. LiCl is a face centered cube, thus $4r = s\sqrt{2}$

The length of one side $= s = 2(r_{cation} + r_{anion})$

The diagonal $= 4r_{anion} = s\sqrt{2}$

Substituting: $4r_{anion} = 2(r_{cation} + r_{anion})\sqrt{2}$

$$4r_{anion} = 2\sqrt{2}r_{caction} + 2\sqrt{2}r_{anion}$$

$$\frac{4r_{anion}}{2\sqrt{2}r_{anion}} = \frac{2\sqrt{2}r_{cation}}{2\sqrt{2}r_{anion}} + \frac{2\sqrt{2}r_{anion}}{2\sqrt{2}r_{anion}} = \frac{r_{cation}}{r_{anion}}\ 1$$

$$\frac{r_{cation}}{r_{anion}} = \frac{4}{2\sqrt{2}} - 1 = 0.414$$

20. True. If $V_{bc} = V_{fc}$, then $s_{bc} = s_{fc}$

$$4r_{bc} = s\sqrt{2} \quad \text{and} \quad 4r_{fc} = s\sqrt{3}$$

$$\frac{4r_{bc}}{\sqrt{2}} = s \text{ and } \frac{4r_{fc}}{\sqrt{3}} = s$$

$$\text{thus, } \frac{4r_{bc}}{\sqrt{2}} = \frac{4r_{fc}}{\sqrt{3}} \quad \text{and} \quad \frac{4\sqrt{3}}{4\sqrt{2}} = \frac{r_{fc}}{r_{bc}}$$

which gives: $\frac{r_{fc}}{r_{bc}} = \frac{\sqrt{3}}{\sqrt{2}} = \frac{\sqrt{3}}{\sqrt{2}} = \sqrt{1.5}$

21. First calculate the volume of one cell. Using density as a conversion factor, calculate the mass per cell and mass per atom of Fe. Using the mass of a single Fe atom and the mass of a mole of Fe (atomic mass), calculate the number of atoms in a mole.

For a body centered cube: $4r = s\sqrt{3}$. Thus

$$s = \frac{4(0.124 \text{ nm})}{\sqrt{3}} = 0.286 \text{ nm} = 0.286 \text{ nm} \times \frac{1 \text{ m}}{1 \times 10^9 \text{ nm}} \times \frac{100 \text{ cm}}{1 \text{ m}} = 2.86 \times 10^{-8} \text{ cm}$$

$$\therefore V = s^3 = (2.86 \times 10^{-8} \text{ cm})^3 = 2.34 \times 10^{-23} \text{ cm}^3 = \frac{2.34 \times 10^{-23} \text{ cm}^3}{1 \text{ cell}} \times \frac{7.86 \text{ g}}{1 \text{ cm}^3} \times \frac{1 \text{ cell}}{2 \text{ atoms}}$$

$$= 9.19 \times 10^{-23} \text{ g atom}^{-1} = \frac{55.847 \text{ g}}{1 \text{ mole}} \times \frac{1 \text{ atom}}{9.19 \times 10^{-23} \text{ g}} = 6.08 \times 10^{23} \text{ atoms mole}^{-1}$$

22. Edge length of FCC unit cell = 2 × shortest distance between A and B = $2\ Y^{1/3}$

$$\text{Density of the crystal of } AB = \frac{Z \times M}{N_0 \times a^3} = \frac{4 \times (6.023\ Y)}{6.023 \times 10^{23} \times (2Y^{1/3} \times 10^{-9})^3}$$

$$= 5 \times 10^3 \text{ gm}^{-3} = 5 \text{ kg m}^{-3}$$

23. In b.c.c arrangement, atoms are touching one another along the cross-diagonal of the cuble (dotted line) making distance equal to 4 r (r = atomic radius). Thus if the length of the unit cell is 'a', then

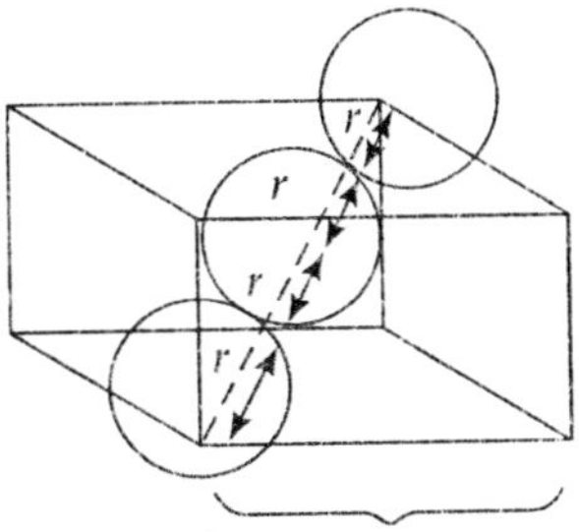

$$4r = \sqrt{3}\ a \text{ i.e, } r = \frac{\sqrt{3}}{4} = \frac{\sqrt{3}}{4} \times 287 \text{ pm.}$$

$$= \frac{\sqrt{3}}{4} \times 287 \times 10^{-12} \text{ m} = 124.2 \times 10^{-12} \text{ m}$$

$$\text{Now, density} = \frac{\text{mass}}{\text{volume}} = \frac{n \times M}{a^3 \times \text{Na}}$$

(where n = number of atoms per unit cell, which is 2 for b.c.c.; M = Molecular weight; and Na = Avogadro's Number)

$$= \frac{2 \times 52 \text{ gm} \cdot \text{mol}^{-1}}{(287 \text{ pm})^3 \times (6.023 \times 10^{23} \text{ mol}^{-1})} = \frac{2 \times 52 \text{ gm} \cdot \text{mol}^{-1}}{(287 \times 10^{-10} \text{ cm})^3\ (6.023 \times 10^{23} \text{ mol}^{-1})}$$

$$= \frac{2 \times 52 \text{ gm} \cdot \text{mol}^{-1}}{2.3 \times 10^{-23} \text{ cm}^3 \times 6.023 \times 10^{23} \text{ mol}^{-1}} = 7.5 \text{ gm} \cdot \text{cm}^{-3}$$

24. Density of Unit cell = $\frac{Z \times M}{N_0 \times (a)^3}$; where,

Z = atomic contribution per unit cell; M = Atomic weight.
N_0 = Avagadro's number; and
a = edge length

$$\text{density } d_1 \text{ for FCC} = \frac{4 \times M}{N_0 \times (3.5 \times 10^{-10})^3}$$

$$\text{density } d_2 \text{ for BCC} = \frac{2 \times M}{N_0 \times (3 \times 10^{-10})^3}$$

$$\therefore \frac{d_1 \text{ FCC}}{d_2 \text{ BCC}} = \frac{4M}{N_0 \ (3.5 \times 10^{-10})} \times \frac{N_0 \cdot (3 \times 10^{-10})^3}{2M} = \frac{2 \times 27}{42.87} = 1.259$$

25. (a) Face plane - 5 atoms are on the face

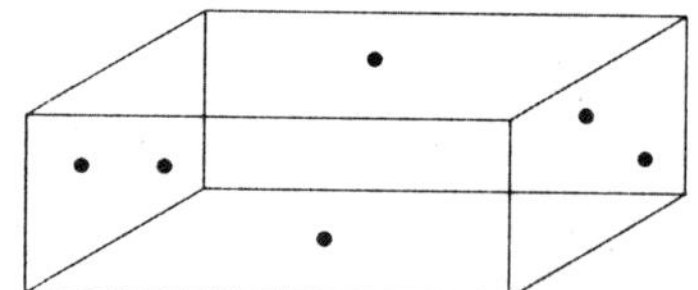

Face diagonal plane

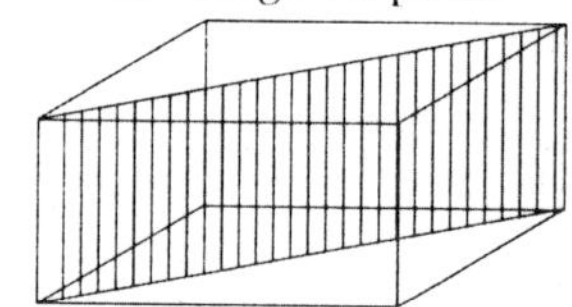

Diagonal plane

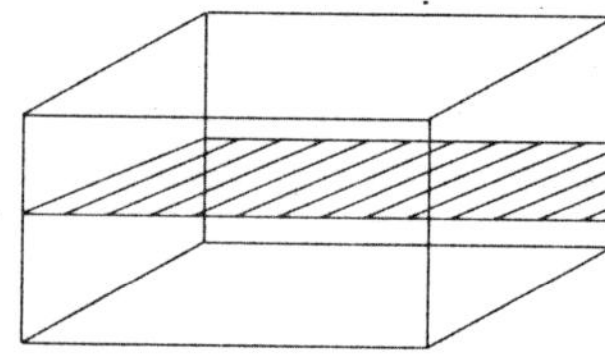

Combined representation

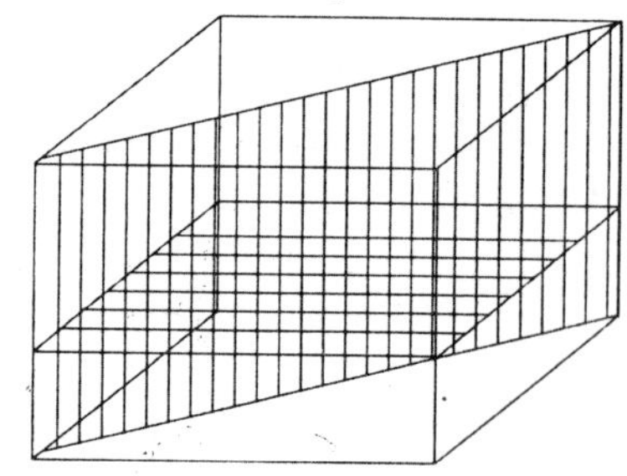

26. Given, diameter of marble = 10 mm
Side of square = 40 mm

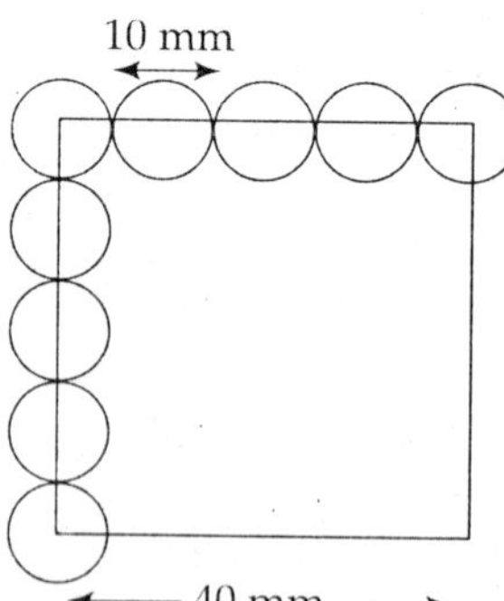

If the center of marbles is lying in square sides then, the number of marble in one side = 5.
So, maximum number of marble per unit area = 5 × 5 = 25.

27. No. of moles of acetic acid taken initially = $100 \times 0.5 \times 10^{-3} = 50 \times 10^{-3}$
After absorption concn. Reduces to 0.49 M.
So, final no. of moles of acetic acid = $100 \times 0.49 \times 10^{-3} = 49 \times 10^{-3}$
No. of moles of acetic acid absorbed = $(50 - 49) \times 10^{-3} = 10^{-3}$
So, number of moles of acetic acid absorbed = $6.023 \times 10^{23} \times 10^{-3} = 6.023 \times 10^{20}$
given, 1g charcoal has area = $3.01 \times 10^2 m^2$
Since molecules of acetic acid are absorbed in $3.01 \times 10^2 m^2$ area

$\therefore$ 1 molecules of acetic acid is absorbed in $\dfrac{3.01 \times 10^2}{6.023 \times 10^{20}} = \dfrac{1}{2} \times 10^{-18} = 5 \times 10^{-19}\ m^2$

28. Avogadro's number = 6.023×10^{13}
At. Wt. of mercury Hg = 200

$\because$ In 1g of Hg, the total number of atoms = $\dfrac{6.023 \times 10^{13}}{200} = 3.0115 \times 10^{21} = 3.012 \times 10^{21}$

Density of mercury (Hg) = $13.6\ g\ cm^{-3}$

Volume of 1 atom of mercury (Hg) = $\dfrac{1}{3.012 \times 10^{21} \times 13.6}$ c.c. = $\dfrac{0.17}{409632}$ c.c = 2.44×10^{-23} c.c.

Since each mercury atom occupies a cube of edge length equal to its diameter, therefore

Diameter of one Hg atoms = $(2.44 \times 10^{-23})^{\frac{1}{3}}$ cm. = 2.905×10^{-8} cm $\cong$ 2.91 Å cm.

29. In ABAB ...... structure (i.e. hcp) number of atoms is 6 per unit cell. The space occupied by the atoms can be calculated by

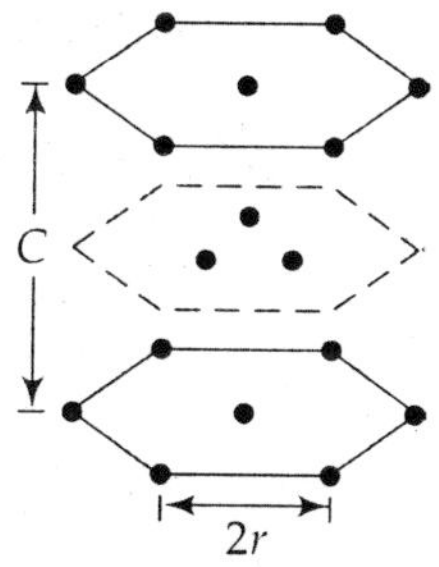

packing fraction $x = \dfrac{6 \times 4/3\, \pi r^3}{\text{Base area} \times C}$

Base area = $6 \times \sqrt{3}/4(4r^2)$ and $C = 4r\sqrt{2/3}$ $\therefore x = 0.76$

Thus % empty space or, void = 26%

C

2r

30. (i) From the data we have; $r^+ = 0.95$ Å; and $r^- = 1.81$ Å

$\therefore \quad \dfrac{r^+}{r^-} = \dfrac{0.95}{1.81} = 0.525$

The radius ratio lies between 0.414 and 0.732. Hence the coordination number of $Na^+ = 6$.

(ii) From the data we have; $r^+ = 1.48$ Å; and $r^- = 1.95$ Å

$\therefore \quad \dfrac{r^+}{r^-} = \dfrac{1.48}{1.95} = 0.759$

The radius ratio is more than 0.732. Hence, the coordination number $Rb^+ = 8$.

(iii) From the data we have; $r^+ = 1.48$ Å; and $r^- = 2.16$ Å

$\therefore \quad \dfrac{r^+}{r^-} = \dfrac{1.48}{2.16} = 0.685$

The radius ratio lies between 0.414 and 0.732. Hence the coordination number of Rb^+ = 6.

(iv) From the data we have; r^+ = 1.69 Å; and r^- = 1.81 Å

$$\therefore \frac{r^+}{r^-} = \frac{1.69}{1.81} = 0.934$$

The radius ratio is more than 0.732, Hence, the coordination number of Cs^+ = 8.

(v) From the data we have; r^+ = 0.68 Å; and r^- = 1.36 Å

$$\therefore \frac{r^+}{r^-} = \frac{0.68}{1.36} = 0.5$$

The radius ratio lies between 0.414 and 0.732. Hence the coordination number of Li^+ = 6.

31. For NaCl structure, the radius ratio lies between the range 0.414 and 0.732,i.e.,

$$0.414 < \frac{r^+}{r^-} < 0.732 \text{ or } 0.414 < \frac{100}{r^-} < 0.732$$

or $$\frac{1}{0.414} > \frac{r^-}{100} > \frac{1}{0.732} \text{ or } \frac{100}{0.414} > r^- > \frac{100}{0.732}$$

or $241.54 > r^- > 136.61$

Thus, the radius of the anion lies between the range 136.61 pm to 241.54 pm.

32. From the given data, r^+ = 88 pm; and r^- = 200 pm

$$\therefore \text{ The radius ratio } = \frac{r^+}{r^-} = \frac{88}{200} = 0.44$$

It lies within the range 0.441 to 0.732. **Hence, the coordination number of A^+ = 6.**

33.

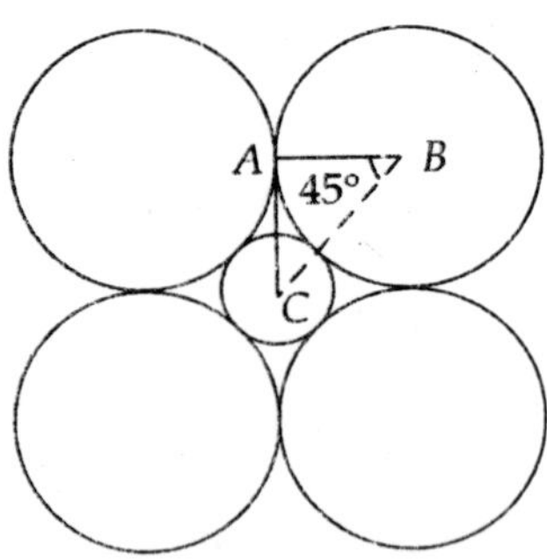

In the given figure, the big circles represent the atoms in close packed structure and the small circle represents the octahedral void. Though an octahedral void is surrounded by six spheres, only four are shown; one sphere above and one below the void are not shown.

Let, AB = Radius of an atom = R

and BC = Radius of an atom + Radius of the void = $R + r$

Angle ABC $= 45°$

Now, in right angled triangle ABC, $AB = BC \cos 45° = \frac{BC}{\sqrt{2}} = 0.7071\ BC$

$\therefore$ $R = 0.7071\ (R + r) = 0.7071\ R + 0.7071\ r$
or by solving $\quad r = 0.414\ R$

34. Here,
Number of tetrahedral holes per anion (O^{2-} ion) in the array = 2
Number of octahedral holes per anion (2^- ion) in the array = 1

$\therefore$ Number of Fe atoms present in tetrahedral holes $= \frac{1}{8} \times 2 = \frac{1}{4}$

Number of Fe atoms present in octahedral holes $= \frac{1}{2} \times 1 = \frac{1}{2}$

35. The distance between the centers of K^+ and Cl^- is half of the edge length, i.e., $\frac{1}{2} \times 6.28 =$ 3.14 Å. Assuming anion–cation contact. 3.14 Å is the sum of the ionic radii of K^+ and Cl^-, i.e.,

$$r_{K^+} + r_{Cl^-} = 3.14; \text{ or } r_{K^+} + 1.81 = 3.14$$

or $r_{K^+} = 3.14 - 1.81 = 1.33$ Å

(Here, chloride ions do not touch each other.)

36. The NaCl structure has a right angled triangle involving two chloride ions and one sodium ion. In the isomorphous LiCl structure, the lithium ion is so small that all ions in the structure touch.

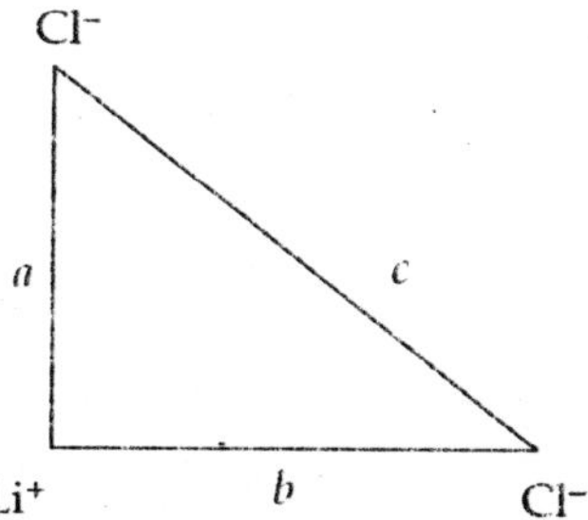

Here, $a = b$ = the distance between the centers of Cl^- ion and Li^+ ion, is half of the edge length of the cubic unit cell, i.e.,

$\therefore$ $a = b = \frac{514}{2} = 2.57$ Å

By the Pythagorus theorem, $c^2 = a^2 + b^2 = (2.57)^2 + (2.57)^2 = 13.21$

$\therefore$ $c = \sqrt{1321} = 3.63$ Å

Since the anions are assumed to touch each other, c is twice the radius of a chloride ion.

$\therefore$ $r_{Cl^-} = \frac{1}{2}\ c = \frac{1}{2} \times 3.63 = 1.81$ Å

37. NaCl exhibits FCC structure.

$\therefore$ A unit cell of NaCl contains $4Na^+$ and $4Cl^-$ ions.

$$\text{Mass of a unit cell of NaCl} = \frac{23 \times 4}{N_0} + \frac{35.5 \times 4}{N_0} = \frac{234}{N_0}$$

$$\text{Volume of the unit cell} = \frac{\text{Mass of the unit cell}}{\text{Density of the unit cell}} = \frac{234/N_0}{2.165}$$

Now, edge of the unit cell, $a = r_{Cl^-} + 2r_{Na^+} + r_{Cl^-} = 2\,(r_{Na^+} + r_{Cl^-})$

$= 2$ (distance between Na^+ and Cl^-) $= 2 \times 281 \times 10^{-10} = 5.62 \times 10^{-8}$ cm

Volume of the unit cell $= (5.62 \times 10^{-8})^3 = 1.77 \times 10^{-22}$ cm^3

$$\therefore \quad \frac{234/N_0}{2.165} = 1.77 \times 10^{-22}$$

Hence, $N_0 = 6.1 \times 10^{23}$

38. Given,

$a = 3.5239$ Å $= 3.5239 \times 10^{-8}$ cm

M $= 58.71$ g mol^{-1}; and z = 4 Ni has FCC structure) $\rho = 8.9$ g cm^{-3}

$$\text{We know that, } \rho = \frac{z \times M}{a^3 N_0} \Rightarrow N_0 = \frac{z \times M}{\rho \times a^3} = \frac{4 \times 58.71}{8.9 \times (3.5239 \times 10^{-8})^3} = 6.03 \times 10^{23}$$

39. Since KF has NaCl structure, z = 4

Molar mass of KF, M = 39 + 19 = 58 g mol^{-1}

$\rho = 2.48$ g cm^{-3} (given)

$$\text{We know that, } \rho = \frac{zM}{a^3 N_0} \Rightarrow a = \left(\frac{z \times M}{\rho N_0}\right) = \left(\frac{4 \times 58}{2.48 \times 6.023 \times 10^{23}}\right) = 5.375 \times 10^{-8} \text{ cm}$$

$$\text{Now, distance between } K^+ \text{ and } F^- = \frac{a}{2} = \frac{5.375 \times 10^{-8}}{2} = 2.69 \times 10^{-8} \text{ cm} = 269 \text{ pm}$$

40. $\because$ CuCl has ZnS (FCC) structure

$\therefore$ z = 4

Molar mass of CuCl, M = 63.5 + 35.5 = 99 g mol^{-1}

$\rho = 3.4$ g cm^{-3} (given)

$$\text{We know that } \rho = \frac{zM}{a^3 N_0}$$

$$\Rightarrow \quad \therefore a = \left(\frac{z \times M}{\rho N_0}\right)^{1/3} = \left(\frac{4 \times 99}{3.4 \times 6.023 \times 10^{23}}\right)^{1/3} = 578 \times 10^{-10} \text{ cm} = 578 \text{ pm}$$

41. For BCC structure.

In the unit cell of a BCC structure, atoms touch each other along the body diagonal.

Now, for a cube, body diagonal = $\sqrt{3}$(edge of the cube).

or $4r = \sqrt{3}\,a$

or $a = \frac{4}{\sqrt{3}}r = \frac{4}{\sqrt{3}} \times 1.24 = 2.86$ Å $= 2.86 \times 10^{-8}$ cm

$$\therefore \text{ Density, } \rho = \frac{zM}{a^3 N_0} = \frac{2 \times 56}{(2.86 \times 10^{-8})^3 \times 6.023 \times 10^{23}} = 7.95 \text{ g cm}^{-3} = 7950 \text{ kg m}^{-3}$$

For FCC structure.

In the unit cell of a FCC structure, atoms touch each other along the face diagonal.

$\therefore$ Edge length, $a = 2\sqrt{2}r = 2\sqrt{2} \times 1.24 = 3.51$ Å $= 3.51 \times 10^{-8}$ cm

$$\therefore \text{ Density, } \rho = \frac{z \times M}{a^3 N_0} = \frac{4 \times 56}{(3.51 \times 10^{-8})^3 \times 6.023 \times 10^{23}} = 8.6 \text{ g cm}^{-3} = 8600 \text{ kg m}^{-3}$$

42. (i) Since silver has FCC structure, in a unit cell, atoms touch each other along the face diagonal.

$\therefore$ Edge length, $a = 2\sqrt{2}r \Rightarrow r = \frac{1}{2\sqrt{2}}a = \frac{1}{2\sqrt{2}} \times 4.0862 = 1.445$ Å

(ii) We know that,

$$\rho = \frac{zM}{a^3 N_0} = \frac{4 \times 108}{(4.0862 \times 10^{-8})^3 \times 6.023 \times 10^{23}} = 10.51 \text{ g cm}^{-3}$$

Also, $M = 108$ g mol^{-1}

Now, we know that, $\rho = \frac{z\,M}{a^3\,N_0}$

$$\therefore z = \frac{\rho \times a^3 \times N_0}{M} = \frac{10.6 \times (408 \times 10^{-10}) \times 6.02 \times 10^{23}}{108} = 4$$

$\because$ There are 4 atoms per unit cell,

$\therefore$ Silver has face-centered cubic lattice.

43. Silver crystallizes in FCC form.

$\therefore$ A unit cell of silver contains 4 atoms.

Now, in a FCC crystal, the distance between the nearest neighbour atoms is half of the face diagonal.

If a is the cell edge, the face diagonal = $\sqrt{2}$

$$\therefore \frac{a\sqrt{2}}{2} = 2.88 \Rightarrow a = \frac{2.88 \times 2}{\sqrt{2}} = 4.07 \text{ Å} = 4.07 \times 10^{-8} \text{ cm}$$

Now, we know that, $\rho = \frac{zM}{a^3 N_0}$

Substituting the proper values in the above expression,

$$\rho = \frac{4 \times 108}{(4.07 \times 10^{-8})^3 \times 6.023 \times 10^{23}} = 10.64 \text{ g cm}^{-3}.$$

The density of silver is 10.64 g cm^{-3}.

44. From the given data

Volume, $\upsilon = 24 \times 10^{-24}$ cm^3

Density, $\rho = 7.2$ g cm^{-3}

Number of atoms in a unit cell,

$$z = \frac{1}{8} \times 8 + 2 = 1 + 2 = 3$$

Now, $\rho = \frac{zM}{\upsilon N_0}$

$$\therefore M = \frac{\rho \upsilon N_0}{z} = \frac{7.2 \times 24 \times 10^{-24} \times 6.023 \times 10^{23}}{2} = 34.69 \text{ g mol}^{-1}$$

$\because$ Number of atoms in 34.69 g = 6.023×10^{23}

$$\therefore \text{Number of atoms in 200 g} = \frac{6.023 \times 10^{23} \times 200}{34.69} = 3.47 \times 10^{24}$$

45. From the given data,

$M = 100$ g mol^{-1}

$a = 400$ pm $= 400 \times 10^{-10} = 4 \times 10^{-8}$ m

$z = 2$ (since the structure is B.C.C)

$$\therefore \rho = \frac{zM}{a^3 N_0} = \frac{2 \times 100}{(4 \times 10^{-8})^3 \times 6.023 \times 10^{23}} = 5.19 \text{ g cm}^{-3}$$

$\therefore$ The density of A is 5.19 g cm^{-3}

Now, mass of a unit cell = density of a unit cell × volume of a unit cell

$= 5.19 \times (4 \times 10^{-8})^3 = 3.32 \times 10^{-22}$ g

$$\therefore \text{Number of unit cells in 10 g of A} = \frac{10}{3.32 \times 10^{-22}} = 3.01 \times 10^{22}$$

46. Edge length of the unit cell, (a) = $2 \times 297 \times 10^{-12}$ m = 5.94×10^{-10} m

Volume of the unit cell (a^3) = $(5.94 \times 10^{-10})^3 = 2.9058 \times 10^{-28}$ m^3

47. For a cubic unit cell,

$$\text{Density} = \frac{z \times M}{a^3 \times N_0} = \frac{4 \times 60}{a^3 \times 6.02 \times 10^{23}} = 6.23$$

$$\therefore \quad a^3 = \frac{4 \times 60}{6.02 \times 10^{23} \times 6.23}$$

or $a^3 = 64 \times 10^{-24}$ cm^3

or $a = 4 \times 10^{-8}$ cm.

Chapter 13 : SURFACE CHEMISTRY

1. Number of molecules of H_2 in 1.36 cc = $\frac{1.36}{22400} \times N_A = 3.65 \times 10^{19}$

   Molar volume (H_2) = $\frac{M}{\rho}$

   $\Rightarrow$ Volume of one H_2 molecule = $\frac{M}{\rho N_A} = 4.743 \times 10^{-23}$ cm^3

   radius = $\left(\frac{3V}{4\pi}\right)^{1/3} = 2.246 \times 10^{-8}$ cm

   $A(H_2) = \pi r^2 = 1.583 \times 10^{-15}$ cm^2

   Surface area covered on Cu per gram = $3.65 \times 10^{19} \times 1.583 \times 10^{-15} = 5.78 \times 10^4$ cm^2

2. Let V(ml) be the volume of solution required

   $\therefore$ $N(\text{acid}) = \frac{0.016\,V}{1000} \times 6.023 \times 10^{23} = 9.637 \times 10^{18} V$

   Area covered = $9.637 \times 10^{18} V \times 0.2 \times 10^{-14}$ cm^2 = $19274V$ = 500 cm^2

   $\Rightarrow$ $V = 25.94 \times 10^{-3}$ mL

3. Total A = 25000 m^2 = 25×10^7 cm^2, $r = 1.5 \times 10^{-8}$ cm

   $A(NH_3) = \pi r^2 = 7.068 \times 10^{-16}$ m^2

   $\Rightarrow N = \frac{25 \times 10^7}{7.068 \times 10^{-16}} = 3.53 \times 10^{23}$

   moles of NH_3 = 0.587

   $\Rightarrow$ Vol. of NH_3 at STP required = 13153.5 mL

4. Here, $N = \frac{100}{22400} \times 6.023 \times 10^{23} = 2.69 \times 10^{21}$

   A/molecule = 0.16×10^{-14} cm^2

   Total area covered by N_2(g)/gram of catalyst = $1.6 \times 10^{-15} \times 2.69 \times 10^{21} = 43.04 \times 10^5$ cm^2

5. According to Lanamuir's equation y/y_∞ corresponds to the slope of the straight line representing P/y as a function P, whereas $1/ky_\infty$ is the Y-intercept of this line. Calculating the values of P/y:

P, Pa	31.9×10^3	130.5×10^3	290×10^3
P/y	6.4×10^3	8.7×10^3	12.1×10^3

   Plot P on the abscissa and P/y on the ordinate. As can be seen from the graph below:

   $1/y = (11 - 7)/(240 - 60) = 0.0222$

   Hence, $y = 45.1/ky_\infty = 5.7 \times 10^3$. Consequently, $k = 1/5.7 \times 45 = 0.004$.

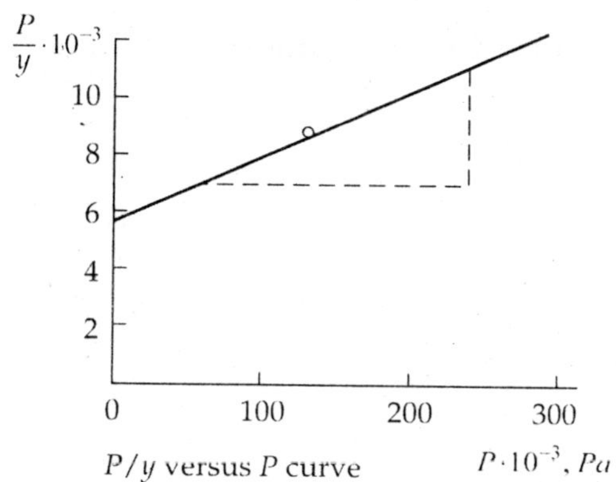

P/y versus P curve

6. To calculate the area use the equation (7)

$$S = \frac{(0.129 \text{ liter g}^{-1})(6.02 \times 10^{23} \text{ mole}^{-1})(16.2 \times 10^{-20} \text{ m}^2)}{22.4 \text{ litre mole}^{-1}} = 560 \text{ m}^2 \text{ g}^{-1}$$

7. Zeigler Natta catalyst is used for polymerisation of ethylene. It is a heterogeneous catalyst formed by the reaction of transition metal halide (e.g., $TiCl_4$, VCl_3) with organometallic compounds (e.g., R_3 Al, R_2 AlCl, R_2Zn) in hydrocarbon solvent.

Chapter 14 : Organic Chemisty (Section I)

1. Compound A, optically inactive, has plane of symmetry. Oxime optically active and resolved.

```
        CH3                      CH3
         |                        |
   H     N     H            H     N     H
C6H5▸           ◂C6H5    C6H5▸           ◂C6H5
         ||                       ||
         N                        N
          \                      /
           OH                  HO
```

(Nonseuperposable mirror images)

2. (R)-3-Methylcyclopentene.
3. CCl_4 is the least soluble, because it is a nonpolar molecule and very slightly soluble in a polar liquid like water. The three electronegative halogen atoms in $CHCl_3$ and CHBr increase the electrnegatively of the carbon, which gives the C—H bond sufficient polarity to produce some hydrogen bonding of the C—H bond with water molecules. This is not large but does produce an increase in the solubilities of these haloforms in water, which is a hundredfold in the case of $CHCl_3$ compared to CCl_4. Since Cl is more electronegative than Br, the effect is larger for $CHCl_3$ than for $CHBr_3$.
4. (a) Freon (b) Naphthalene or *p*-dichlorobenzene
   (c) Cyclopropane (d) Iodoform
   (e) CCl_4 (f) Butyl rubber
5. A test for water solubility would identify acetone. Test with 2,4-dinitrophenyl-hydrazine to identify the aldehyde or ceric ammonium nitrate to identify the alcohol; (b) Tollens' reagent to identify the aldehyde or iodoform test to detect the methyl ketone; (c) the iodoform test, will identify the 2-bromohexane.

```
                      CH2
                       |
                      CH2    Cl             CH3
                       |      |              |
       Br             CH2    CH—CH2—CH—CH2
        |              |      |
6. H—C—CH2CH2—C——C————CH2—CH—CH2—CH2
        |              |      |                |
       Br             CH    C(CH3)3          CH2Cl
                     /  \
                  CH3    CH3
```

7. 2,7-Dimethyl-3,3-dichloro-4-sec-butyl-4-isopropyl-5-*n*-butyl-5-isobutyl-6-bromo-6-chloromethyl nonane.
8. If isobutyl bromide were used in a Wuriz reaction, it would produce 2,5-dimethylhexane. see-Butyl bromide would yield 3,-4-dimethylhexane. Comparison of the properties of either product with that of the unknown would identify the unknown. One could compare their indexes of refraction, infrared absorption spectra, or some other physical property.

9. Relative electronegativities of Cl, H and CH_3 decrease in that order. Hence, the inductive effect in CCl_3COOH, CH_3COOH, and $(CH_3)_3C{-}COOH$ decrease the acid strength of these acids in the order given: On this basis, relative acid strengths would be (a) $Cl{-}CH_2COOH > Br{-}CH_2COOH$; (b) $CH_3{-}CHCl{-}COOH > Cl{-}CH_3{-}CH_2COOH$.

10. (a) $CH_3CH_2CH_2Br \xrightarrow[\Delta]{base} CH_3CH{=}CH_2 \xrightarrow{Cl_2} CH_3{-}CHCl{=}CH_2Cl$

(b) $CH_3{-}CHOH{-}CH_3 \xrightarrow{H_2SO_4} CH_3{-}CH{=}CH_2 \xrightarrow[(air)]{HBr} CH_3{-}CH_2{-}CH_2Br$

(c) $CH_3CH_2Br \xrightarrow[\Delta]{base} CH_2{=}CH_2 \xrightarrow{HOCl} CH_2Cl{-}CH_2OH$

(d) $CH_2{=}\underset{\displaystyle CH_3}{\underset{|}{C}}\,CH_2CH_2 \xrightarrow{O_2,\ then\ H_2O/Zn} O{=}\underset{\displaystyle CH_3}{\underset{|}{C}}\,CH_2CH_3$

$\downarrow$ Δ base

$BrCH_2{-}\underset{\displaystyle CH_3}{\underset{|}{C}}{-}CH_2CH_3$

(e) $CH_3CHBrCH_3 \xrightarrow[\Delta]{base} CH_3CHCH_2 \xrightarrow[(air)]{HBr} CH_3CH_2CH_2Br \xrightarrow[\Delta]{Na} CH_3(CH_2)_4CH_3$
(*n*-hexane)

(f) $CH_3\underset{\displaystyle CH_3}{\underset{|}{C}}CH{=}CH_3 \xrightarrow[OH^-]{Cr_2O_7^-} CH_3\underset{\displaystyle CH_3}{\underset{|}{C}}{-}CO_2^- \xrightarrow{electrolysis} CH_3\underset{\displaystyle CH_3}{\underset{|}{C}}H_3\,\overset{\displaystyle CH_3}{\overset{|}{C}}H\,CH_3$

11. $C{-}\underset{\displaystyle C}{\underset{|}{\overset{\displaystyle C}{\overset{|}{C}}}}{-}\underset{\displaystyle Br}{\underset{|}{C}}{-}C \xrightarrow[\Delta]{base} C{-}\underset{\displaystyle C}{\underset{|}{\overset{\displaystyle C}{\overset{|}{C}}}}{-}C{=}C \xrightarrow[\text{Anti-Markownikoff}]{HBr} C{-}\underset{\displaystyle C}{\underset{|}{\overset{\displaystyle C}{\overset{|}{C}}}}{-}C{-}C{-}Br$

$\downarrow$ Na

$C{-}\underset{\displaystyle C}{\underset{|}{\overset{\displaystyle C}{\overset{|}{C}}}}{-}C{-}C{-}C{-}C{-}\underset{\displaystyle C}{\underset{|}{\overset{\displaystyle C}{\overset{|}{C}}}}{-}C$

12. $\underset{\displaystyle Br}{\underset{|}{C}}H_2CH_2CH{=}CH{-}\underset{\displaystyle CH_3}{\underset{|}{C}}\,CH_2Br \xrightarrow{Zn/DMF}$ (ring, CH_3) $\xrightarrow{Br_2}$ (ring, CH_3, Br, Br) $\xrightarrow[alc]{2\ KOH}$ (ring, CH_3)

13. The reaction for preparation is:

$$C{-}COOH \xrightarrow{LiAlH_4} \underset{I}{C{-}C{-}OH} \xrightarrow[H_2O^+]{dil.\ Cr_2O_7^-} \underset{II}{C{-}CHO}$$

$$\downarrow Br_2, P$$

$$\underset{|}{\overset{}{C}}(Br){-}COOH \xrightarrow[H_2SO_4]{EtOH} C(Br){-}CO_2Et \xrightarrow[2.\ H_2O^+]{1.\ Zn,\ +II} C{-}C{-}C(OH){-}CO_2H \xrightarrow[ZnCl_2]{HBr} C{-}C(Br){-}C{-}COOH$$

14. By following reaction we have the required conversion.

$$C_6H_5NH_2 \xrightarrow{Ac_2O} C_6H_5NHAc \xrightarrow{H_2SO_4} p\text{-}HO_3SC_6H_5NHAc \xrightarrow[H_2SO_4]{HNO_3}$$

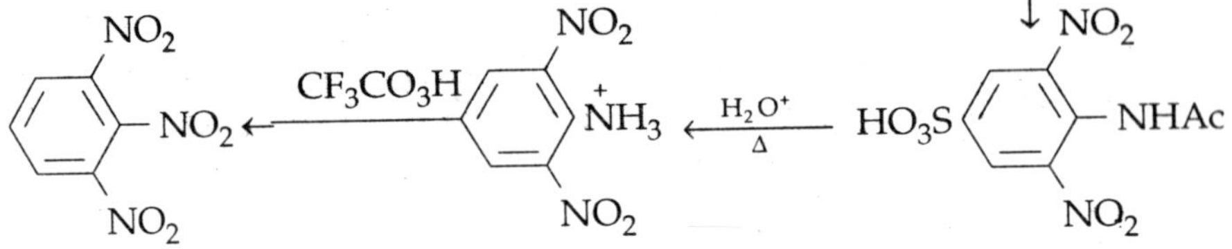

$$HO_3S\text{-}C_6H_2(NO_2)_2\text{-}NHAc \xrightarrow[\Delta]{H_2O^+} C_6H_3(NO_2)_2\overset{+}{N}H_3 \xrightarrow{CF_3CO_3H} C_6H_3(NO_2)_3$$

$$C{-}CN$$
$$|$$
$$OCOCH_3$$

15. (a) $(CH_3)_2CH{-}(CH_2)_2{-}CH_2OH \longrightarrow (CH_3)_2CH{-}(CH_2)_3{-}CH_2NO_2$

It can be seen that an alcohol is to be converted into a nitroalkane containing one more CH_2 group. So, plan backwards and recall that nitroalkanes are preparede from alkyl halides; that alkyl halides may be prepared from alcohols; that alcohols may be prepared by reduction of carboxylic acids or aldehydes; that of the chain-lengthening methods, either the aldehyde, acid, or alcohol may be obtained from the Grigard, and that the Grignard is made from the alkyl halide, which may be prepared from the given alcohol. This can be expressed by equations below, where $R=(CH_3)_2CH{-}(CH_2)_3{-}$

$$ROH \xrightarrow{PBr_3} RBr \xrightarrow{Mg} RMgBr \xrightarrow[2.\ H_2O^+]{1.\ H_2CO} RCH_2OH \xrightarrow{KBr,\ H_2SO_4} RCH_2Br \xrightarrow{KNO_2} RCH_2NO_2$$

Of course, several other routes are possible, for example:

$$ROH \xrightarrow{PBr_3} RBr \xrightarrow{KCN} RCN \xrightarrow{LiAlH_4} RCH_2NH_2 \xrightarrow[2.\ PX_3]{1.\ C_6H_5COCl} RCH_2X \xrightarrow{KNO_2} RCH_2NO_2$$

OR

$$ROH \xrightarrow{SOBr_2} RBr \xrightarrow{Mg} RMgBr \xrightarrow[2.\ H_2O^+]{1.\ CO_2} RCOH \xrightarrow{R'OH,\ H_2SO_4} RCOOR' \xrightarrow{NH_4} RCONH_2$$

$$RCOH \xrightarrow[2.\ H_2O^-]{1.\ HC(OE+)_3} RCHO \qquad RCONH_2 \xrightarrow{LiAlH_4} RCH_2NH_2$$

$$RCHO \xrightarrow{LiAlH_4} RCH_2OH \xleftarrow{HONO} RCH_2NH_2$$

$$\downarrow HX,\ ZnCl_2$$

$$RCH_2NO_2 \xleftarrow{KNO_2} RCH_2X$$

(b) $CH_2—(CH_3)_4COOC_2H_5 \longrightarrow CH_3—(CH_2)_3—CN$

Here, a CH_2 group is to be cut off. Of the ways to make a nitrile, one is from the alkyl halide and one from the amide. Among the degradation methods one changes an acid to an alkyl halide. Thus, a route is suggested below, where $R{=}CH_3—(CH_2)_3—$;

$$R—CH_2—COOC_2H_5 \xrightarrow[\Delta]{H_2O^+} R—CH_2COOH \xrightarrow[NH_4OH]{AgNO_2} RCH_2CO_2Ag$$

$$\downarrow Br_2, CCl_4, \Delta$$

$$RCO_2Ag \xleftarrow[NH_4OH]{AgNO_2} RCOOH \xleftarrow{KMnO_4} RCH_2OH \xleftarrow{KOH} RCH_2Br$$

$$\downarrow Br_2, CCl_4, \Delta$$

$$RBr \xrightarrow{KCN} RCN\ [CH_3—(CH_2)_3—CN]$$

Other routes can be, for example:

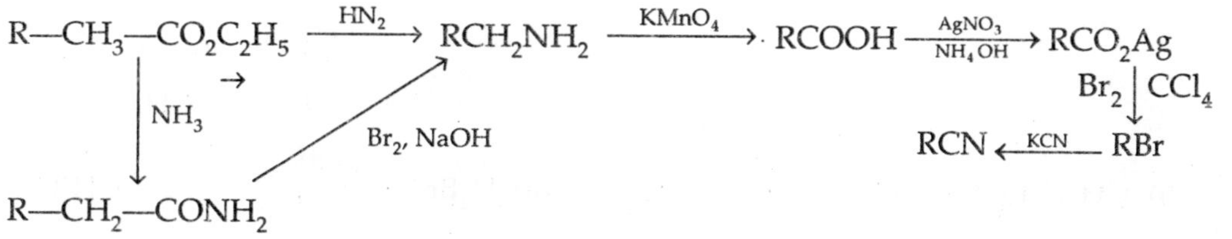

$$R—CH_3—CO_2C_2H_5 \xrightarrow{HN_2} RCH_2NH_2 \xrightarrow{KMnO_4} RCOOH \xrightarrow[NH_4OH]{AgNO_3} RCO_2Ag$$

$$R—CH_3—CO_2C_2H_5 \xrightarrow{NH_3} R—CH_2—CONH_2 \xrightarrow{Br_2, NaOH} RCH_2NH_2$$

$$RCO_2Ag \xrightarrow{Br_2, CCl_4} RBr \xrightarrow{KCN} RCN$$

16.

(I) $>C(H)—COOH$

(II) $>C(H)—COCl$

(III) $>C(H)—CON(CH_3)_2$

(IV) $>C(H)—CH_2N(CH_3)_2$

(V) $>C(H)—CH_2N^+(CH_3)_3I^-$

(VI) $>C{=}CH_2$

(VII) $>C(H)—CH_2OH$

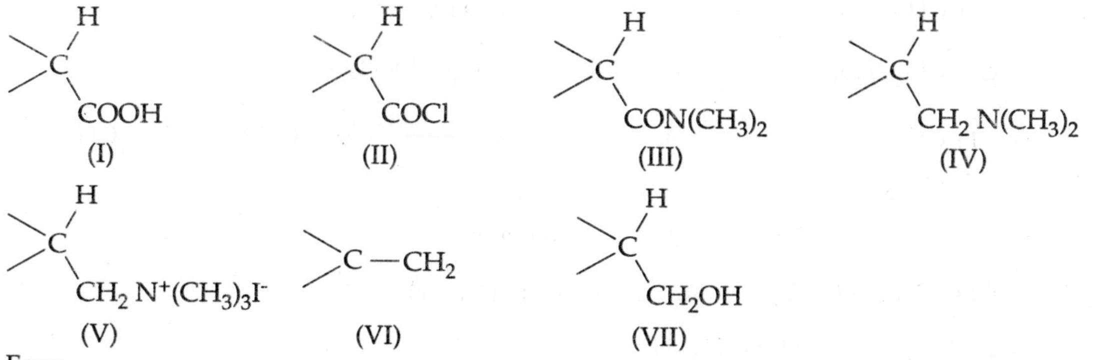

17. Four

18. (a) Benzene to *m*-fluorobenzoic acid:

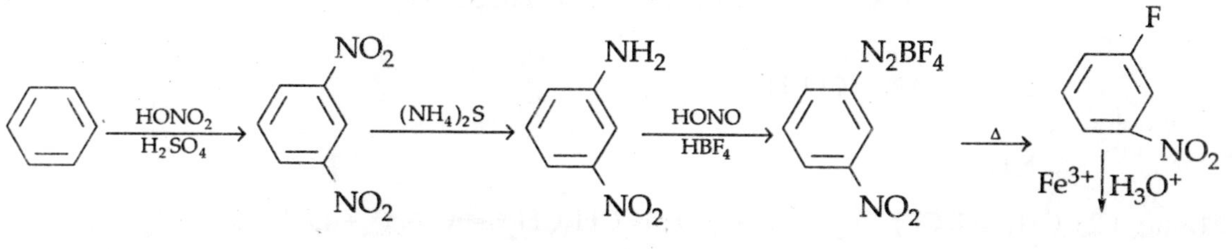

$$C_6H_6 \xrightarrow[H_2SO_4]{HONO_2} m\text{-}C_6H_4(NO_2)_2 \xrightarrow{(NH_4)_2S} m\text{-}NH_2C_6H_4NO_2 \xrightarrow[HBF_4]{HONO} m\text{-}N_2BF_4C_6H_4NO_2 \xrightarrow{\Delta} m\text{-}FC_6H_4NO_2$$

$$\downarrow Fe^{3+}, H_3O^+$$

(3-fluorobenzoic acid) $\xleftarrow[H_3O^+]{\Delta}$ (3-fluorobenzonitrile, F / CN) $\xleftarrow{CuCN}$ (3-fluorobenzenediazonium, F / N_2^+) $\xleftarrow{HONO,\ 273\ K}$ (3-fluoroaniline, F / NH_3)

F, CO_2H ; F, CN ; F, N_2^+ ; F, NH_3

(b) $EtO_2C{-}(CH_2)_5{-}CO_2Et \xrightarrow[2.\ H_3O^+,\ \Delta]{1.\ OR^-} (CH_2)_5\ C{=}O \xrightarrow[2.\ H_3O^+]{1.\ EtMgBr}$ $(CH_2)_5\ C(C_2H_5)(OH) \xrightarrow{PX_3} (CH_2)_5\ C(C_2H_5)(X) \xrightarrow[2.\ H_3O^+]{1.\ Mg.\ CH_2O} (CH_2)_5\ C(C_2H_5)(CH_2OH)$

19. (a) $CH_3CHOHCH_3 \xrightarrow{H_2SO_4} CH_3CH{=}CH_2 \xrightarrow{Br_2} CH_3CHBrCH_2Br \xrightarrow{KOH\ (alc.)} CH_3C{=}CH$

(b) $CH_3CH_2CH{=}CH_2 \xrightarrow{Br_2} CH_3CH_2CHBrCH_2Br \xrightarrow[2.\ HCl\ eq.]{1.\ NaNH_2\ excess} CH_3CH_2C{=}CH$

(c) $CH_3CH_2CH_2CH_2CH_2Br \xrightarrow[\Delta]{base} CH_3CH_2CH_2CH{=}CH_2 \xrightarrow{Br_2}$ $CH_3CH_2CH_2CHBr{-}CH_2Br \xrightarrow[HCl\ aq]{NaNH_2\ excess} CH_3CH_2CH_3{-}C{\equiv}CH$

(d) $CH_3CH_2C(CH_3)C\equiv C_2 \xrightarrow{MnO_4^-} CH_3CH_2C(CH_3)CO_2^- \xrightarrow{electrolysis} CH_3CH_2C(CH_3)CH_2CH_2$

20. (a) C_6H_5COOH, $H_3\overset{+}{N}CH_2COOH$, and $\overset{+}{N}H_4$ ion.

(b) $CH_3(CH_2)_3\overset{+}{N}H_3$, CO_2, $\overset{+}{N}H_4$, and $HO_2C(CH_2)_3CH_3$

(c) $(C_2H_5)_2CO$ and $H_3\overset{+}{N}NHC_6H_5$

(d) (benzene ring with CO_3H and OH), $HOCH_2CH_2\overset{+}{N}H(C_2H_5)_2$, and HO_2CCH_3

(e) (benzene ring with $CH_3-CH_2-CO_2H$ and $\overset{+}{N}H_3$)

21. (a) $CH_3CH_2NH_3^+Cl^-$; (b) $CH_3CH_2C{\equiv}C^-Na^+ + NH_3$

(c) $(CH_3CH_2CH_2)_2\overset{+}{N}HCH_2CH_3Br^-$ (d) $CH_3CH_2CH_2CHBrCH_2CH_3$

(e) $CH_3CH_2CH_2S^-Na^+ + H_2O$ (f) $CH_3CH_2CHBrCHBrCH_3$

(g) $CH_3CH_2CH_2CH_2CH_2CH_3$ (h) $CH_3CH_2Cl + CH_3CHCl_2 + ClCH_2CH_2Cl + HCl$

22. (a) $CH_3CH_2CH_2CH_2CH(OH)CH_3$ racemic mixture (b) $CH_3C(CH_3){=}CH_2$ (c) $CH_3CH{=}CH_2$

23. (a) $CH_3CH_2CH_2CH_2CH_2CH_3$ b) $CH_3CH_2CH_2CH(CH_3)CH_3$

(c) $CH_3CH_2CH_2CH_2CH_2CH_2CH_3$

24. Here, (a) will be analyzed in detal. Only the equations are shown for (b)

1. To what functional group class do the reactants belong?

   (One reactant is an alkene. The double bond of the alkene is a nucleophile. The other reagent is an halogen, an electrophile.)

2. Does either reactant have a good leaving group?

   (No)

3. Are any of the reactants good acids, bases, nucleophiles, or electrophiles or is there an ionizing solvent present?

   (Bromine, above the arrow, is an electrophile. The solvent, carbon tetrachloride, is not an ionizing solvent)

4. What is the most likely first step for the reaction? Most common reactions can be classified either as protonation-deprotonation, or as reactions of a nucleophilic with an electrophile.

   Here attack by the nucleophilic double bond on the electrophilic bromine leads to the formation of a bromonium ion.

:Br:–:Br: + $CH_3CH_2(H)C{=}C(H)CH_2CH_3$ ⟶ bromonium ion [CH_3CH_2, H, :Br:, H, CH_2CH_3 on C–C ring] + :Br:⁻

5. What are the properties of the species present in the reaction mixture after the first step? Is any further reaction likely?

   (The species present in the reaction mixture are the bromonium ion, an electrophile, and bromide ion, a nucleophile. The bromonium ion will react with bromide ion to give the dibromide product.)

$$\underset{\text{H}}{\overset{\text{CH}_3\text{CH}_2}{}}\!\!\!\overset{:\ddot{\text{Br}}:}{\text{C}-\text{C}}\!\!\!\underset{\text{CH}_2\text{CH}_3}{\overset{\text{H}}{}} \;\; :\ddot{\text{Br}}: \longrightarrow \underset{\text{H, Br}}{\overset{\text{CH}_3\text{CH}_2}{\text{C}}}-\underset{\text{H, CH}_2\text{CH}_3}{\overset{\text{Br}}{\text{C}}}$$

4. What is the stereochemistry of the reaction?
(The opening of the bromonium ion by bromide leads to an overall anti addition and the creation of two stereocentres. The product, however, has a plane of symmetry and is therefore a meso compound.)

25. (a) $CH_3\overset{O}{\overset{\|}{C}}CH_3 + CH_3\overset{O}{\overset{\|}{C}}H$ (b) $CH_3\overset{CH_3}{\overset{|}{C}}H{=}CHO + CH_3CH_2\overset{O}{\overset{\|}{C}}CH_2CH_3$

(c) $CH_3\overset{CH_3}{\overset{|}{C}}H{=}\overset{O}{\overset{\|}{C}}OH + CH_3CH_2\overset{O}{\overset{\|}{C}}CH_2CH_3$ (d) $CH_3CH_2\overset{O}{\overset{\|}{C}}CH_3 + H\overset{O}{\overset{\|}{C}}H$

(e) $CH_3(CH_2)_{13}\overset{O}{\overset{\|}{C}}OH + H\overset{O}{\overset{\|}{C}}OH$ (f) $CH_3CH_2\overset{CH_3}{\overset{|}{\underset{Br}{\underset{|}{C}}}}CH_2CH_3$

(g) [Br(H)(HOC=O)C–C(H)(Br)COOH] + [Br(H)(HOC=O)C–C(H)(OH)COOH]

(h) $\underset{\text{H}}{\overset{\text{CH}_3}{}}\text{C}{=}\text{C}\underset{\text{CH}_2\text{CH}_3}{\overset{\text{H}}{}} + \underset{\text{H}}{\overset{\text{CH}_3}{}}\text{C}{=}\text{C}\underset{\text{H}}{\overset{\text{CH}_2\text{CH}_3}{}}$ (i) $CH_3\overset{CH_3}{\overset{|}{C}}{=}CHCH_3 + CH_2{=}\overset{CH_3}{\overset{|}{C}}CH_2CH_3$

26. (a) $CH_3CH_2CH_2C{\equiv}C^-Na^+$ (b) $CH_3CH_2CH_2C{\equiv}CMgBr + CH_3CH_3\uparrow$

(c) $CH_3\overset{CH_3}{\overset{|}{C}}HCH_2CH_2\overset{OCH_3}{\overset{|}{C}}HOCH_3$ (d) $CH_3\underset{\underset{\underset{CH_3}{|}}{NCH_2CHCH_3}}{\underset{\|}{C}}CH_2\overset{CH_3}{\overset{|}{C}}HCH_3$

(e) $CH_3CH_2{\equiv}C\overset{OTHP}{\overset{|}{C}}HCH_2CH_2CH{=}CH\overset{NOCH_3}{\overset{\|}{C}}CH_3$ (f) $H\overset{O}{\overset{\|}{C}}CH_2\overset{CH_3}{\overset{|}{\underset{CH_3}{\underset{|}{C}}}}CH_2CH{=}CH\overset{O}{\overset{\|}{C}}OCH_3$

(g) $CH_3CH_2(CH_2)CH_3$

(h)
$$\begin{array}{c}\quad CH_3 \qquad\qquad\qquad\quad CH_3 \\ \quad | \qquad\qquad\qquad\qquad\quad | \\ CH_3CHCH_2CH_2CH_2CHCH_3\end{array}$$

27. (a) $\underset{\textbf{A}}{CH_3CH_2CH_2C{\equiv}C{:}^-Na^+}$ $\underset{\textbf{B}}{CH_3CH_2CH_2C{\equiv}CCH_2CH_2CH_3}$

(b) $\underset{\textbf{C}}{HC{\equiv}C{:}^-Na^+}$; $\underset{\textbf{D}}{CH_3\overset{\overset{\displaystyle CH_3}{|}}{C}HCH_2CH_2C{\equiv}CH}$

(c) $\underset{\textbf{E}}{CH_3CH_2CH_2C{\equiv}CH}$; $\underset{\textbf{F}}{CH_3CH_2CH_2C{\equiv}C{:}^-Na^+}$ $\underset{\textbf{G}}{CH_3CH_2CH_2C{\equiv}CCH_3}$

The leaving group in dimethyl sulfate, $(CH_3)_2SO_4$, is the methyl sulfate aion, $^-OSO_3CH_3$

(d) $\underset{\textbf{H}}{CH_3CH_2\underset{\underset{\displaystyle Br}{|}}{\overset{\overset{\displaystyle Br}{|}}{C}}CH_3}$ (e) $\underset{\textbf{I}}{CH_3CH_2CH_2\underset{\underset{\displaystyle Cl}{|}}{C}HCH_3}$

(f) $\underset{\textbf{J}}{CH_3CHBr_2}$ (g) $\underset{\textbf{K}}{CH_3\underset{\underset{\displaystyle HO}{|}}{\overset{\overset{\displaystyle CH_3}{|}}{C}}-\underset{\underset{\displaystyle OH}{|}}{\overset{\overset{\displaystyle CH_3}{|}}{C}}CH_3}$

(h) $\underset{\textbf{L}}{CH_3\overset{\overset{\displaystyle O}{||}}{C}CH_3}$ + $\underset{\textbf{M}}{H\overset{\overset{\displaystyle O}{||}}{C}CCH_2CH_2\overset{\overset{\displaystyle O}{||}}{C}OCH_3}$

28. (a) $CH_3CH_2C_2CH_2CH{=}CH_2$

(b)
```
CH3CH2CH2CH2       CH3
            \     /
             C = C
            /     \
           H       H
```

(c)
```
CH3CH2CH2CH2       H
            \     /
             C = C
            /     \
           H       CH3
```

(d) $CH_3CH_2CH_2CH_2CH_2CH_3$

(e) $CH_3CH_2CH_2CH_2CH_2CH_2CH_3$

(f) $CH_3\underset{\underset{\displaystyle CH_3}{|}}{C}H\overset{\overset{\displaystyle O}{||}}{C}OH + CH_3CH_2\overset{\overset{\displaystyle O}{||}}{C}OH$

(g) $CH_3\overset{\overset{\displaystyle O}{||}}{C}OH + CH_3CH_2\underset{\underset{\displaystyle CH_3}{|}}{C}H\overset{\overset{\displaystyle O}{||}}{C}OH$

29. (a) 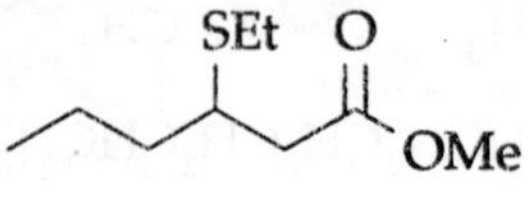 SEt O OMe (b) 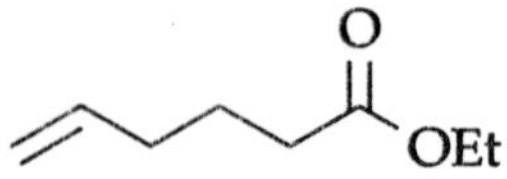 O OEt (c) 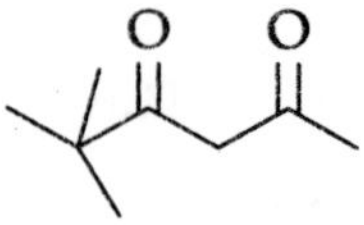 O O

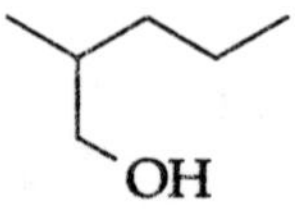

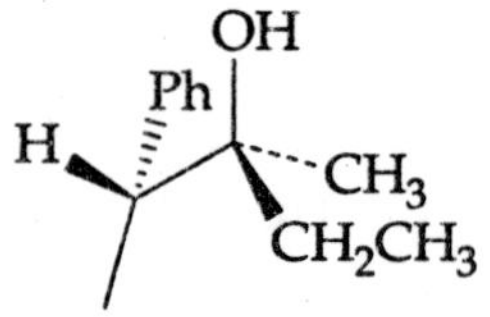

30. (a)  SCH_2CH_3 (b) OH (c) OH Ph H CH_3 CH_2CH_3

(d) $CH_3CH(OH)—COO^-$ (e) 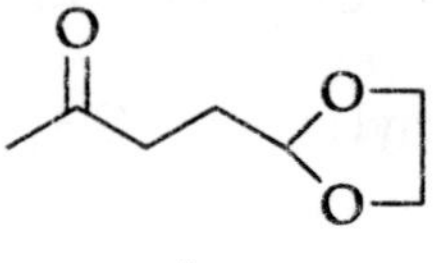 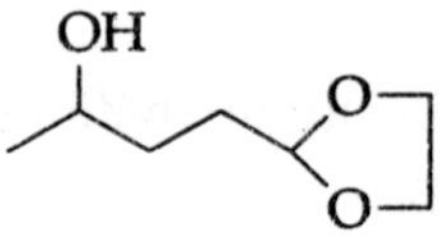 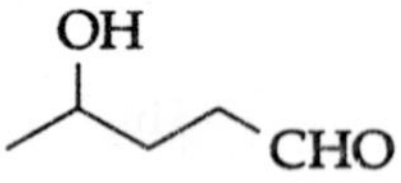 O O O (A); OH O O (B); OH CHO (C)

(f) 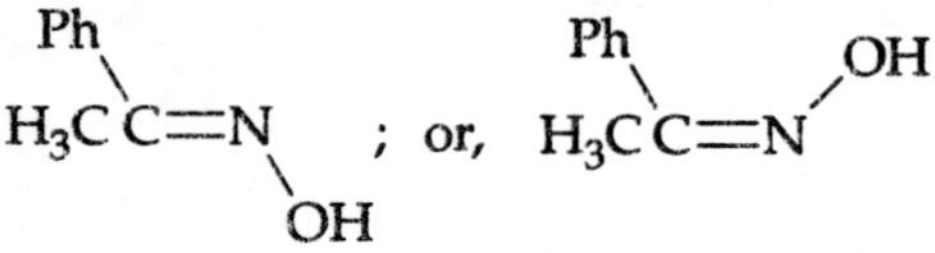 $H_3C(Ph)C=N—OH$; or, $H_3C(Ph)C=N—OH$ (D); $CH_3\overset{O}{\overset{||}{C}}NHPh$ (E); CH_3COO^- (F)

$Pb—\overset{O}{\overset{||}{C}}NHCH_3$ (G); $PhCOO^-$ (H); CH_3NH_2 (I)

31. (i) $PhCH_2NH—\overset{O}{\overset{||}{C}}—CH=CH_2$ (ii) $PhCH_2COCH_2Cl$
(iii) ClCOOEt (iv) $BrCH_2CH_2COOH$
(v) $(CH_3CH_2CH_2CH_2)_3$ (A) $CH_3CH_2CH_2CH_2NH_2$ (B) (vi) C_2H_5NHCN (C) $C_2H_5NHCH_2NH_2$ (D)

(vii) $CH_2=C(OH)—CN$ (E); $CH_2=C(OCOCH_3)—CN$ (F)

32. (a) $CH_3CH(NHPh)—C(CH_3)_2CH_3$ (b) $CH_3C(CH_3)(Br)—CH(CH_3)CH_2CH_3$

(c) $Ph\ CH=CH\overset{O}{\overset{||}{C}}CH_3 + CH_3\overset{O}{\overset{||}{C}}CH_2C(OH)(CH_3)—CH_3$

A B

(d) $C_6H_5CH{=}CH{-}\overset{O}{\overset{||}{C}}C_6H_5 + C_6H_5\overset{O}{\overset{||}{C}}CH{=}\underset{CH_3}{\underset{|}{C}}C_6H_5$

C D

(e) $C_6H_5CH{=}CHNO_2$ (f) 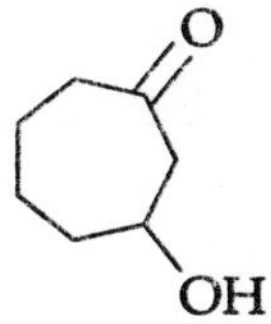 O, OH (g) 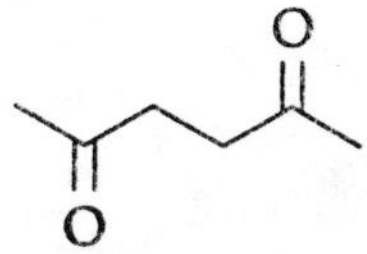 O, O

33. (a) 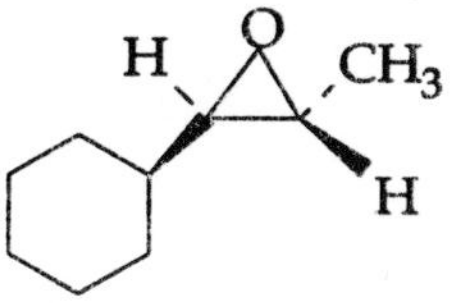 H, CH_3, H (b) 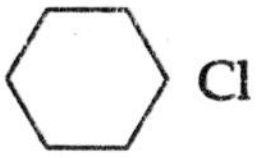 Cl (c) 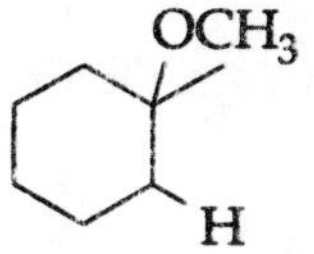 OCH_3, H

(d) 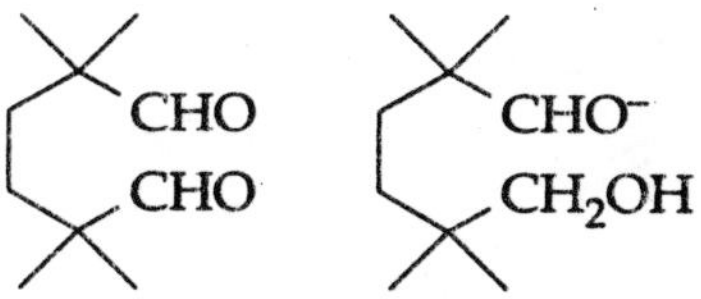 CHO, CHO (A); CHO^-, CH_2OH (B)

(e) 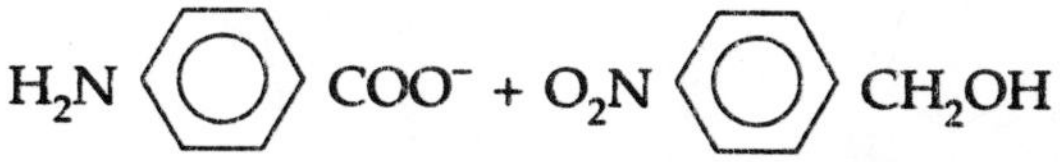 H_2N–C_6H_4–$COO^- + O_2N$–C_6H_4–CH_2OH

(f) 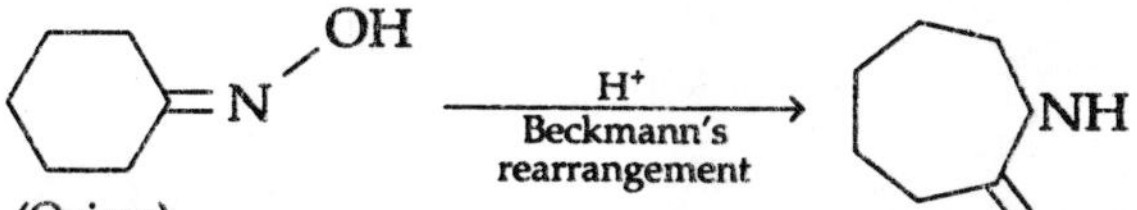 =N–OH (Oxime) C $\xrightarrow[\text{Beckmann's rearrangement}]{H^+}$ NH, O (Caprolactum) D

34. (a) 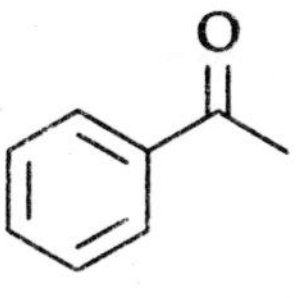 O (b) 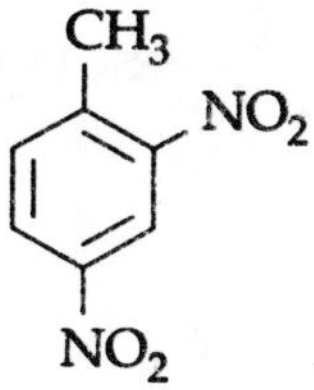 CH_3, NO_2, NO_2 (c) 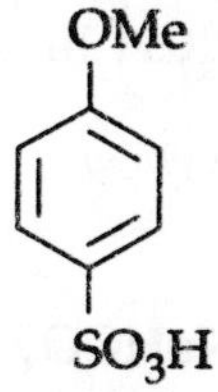 OMe, SO_3H (d) 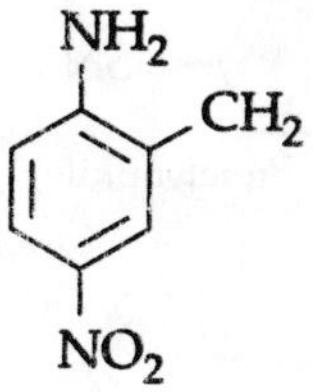 NH_2, CH_2, NO_2

35. (I) (A) CH_3ONa/CH_3OH (B) $(CH_3)_2NH$

(C) $CH_3CH{=}CH_2MGBr$; H_3O^+ (D) CH_3OH/H_2SO_4

(II) (a) $H(CH_3)C{=}CH(CH_2Cl)$ $\xrightarrow{NaCN}$ $H(CH_3)C{=}CH(CH_2CN)$ $\xrightarrow{H_3O^+}$ $H(CH_3)C{=}CH(CH_2COOH)$ $\xrightarrow{LiAlH_4}$ $H(CH_3)C{=}CH(OH)$ $\xrightarrow{TsCl}$ $\xrightarrow{NaBr}$ Product

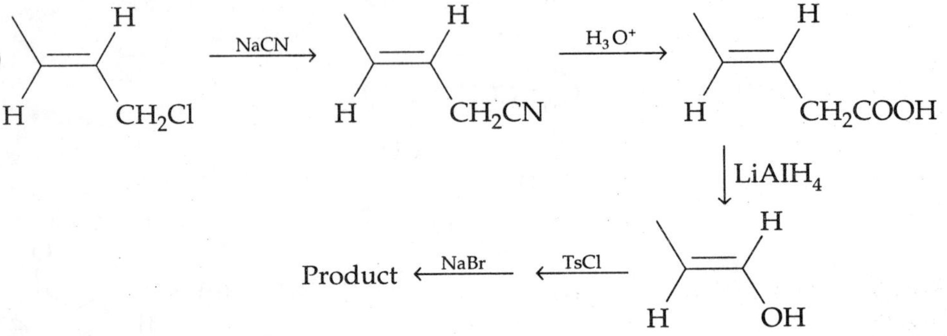

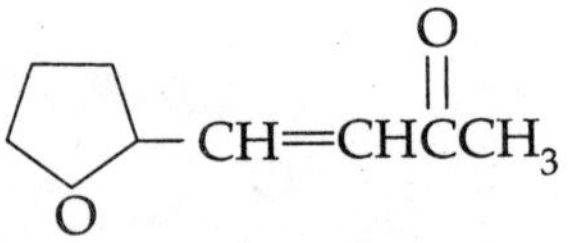

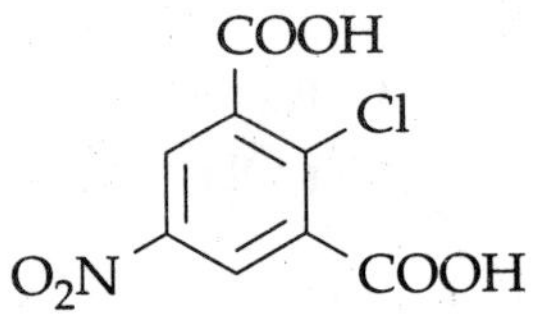

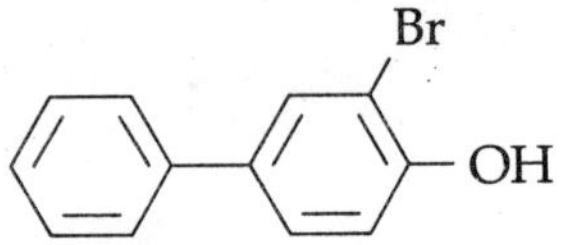

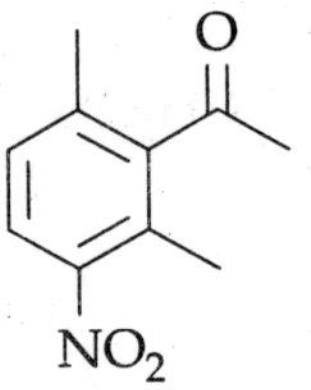

(b) $H(CH_3)C{=}CH(COOEt)$ $\xrightarrow{(CH_3\ \overset{CH_3}{CH_3}\ CH_2)_2AlH}$ $H(CH_3)C{=}CH(CHO)$

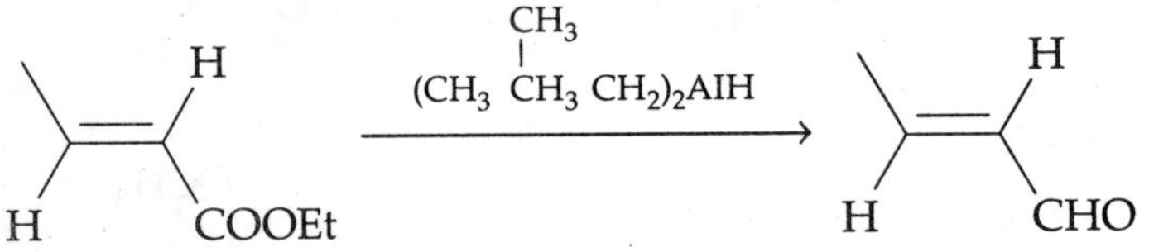

(c) $\{(CH_3)_2CO\}_3$ Al,

36. (a) (tetrahydrofuran-2-yl)–CH=CH$\overset{O}{\overset{\|}{C}}CH_3$ (b) benzene ring with COOH, Cl, COOH and O_2N

(c) phenyl–(benzene ring with Br)–OH (d) benzene ring with CH_3, $\overset{O}{\overset{\|}{C}}CH_3$, CH_3, NO_2 (e) benzene ring with OCH_3, $C(CH_3)_3$, CH_3

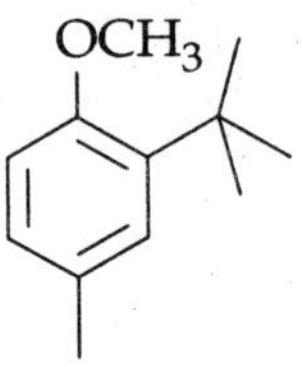

37. (a) $C_6H_5\overset{+}{N}H_2CH_3\ Cl^- + NaOH \longrightarrow C_6H_5NHCH_3 + NaCl + H_2O$

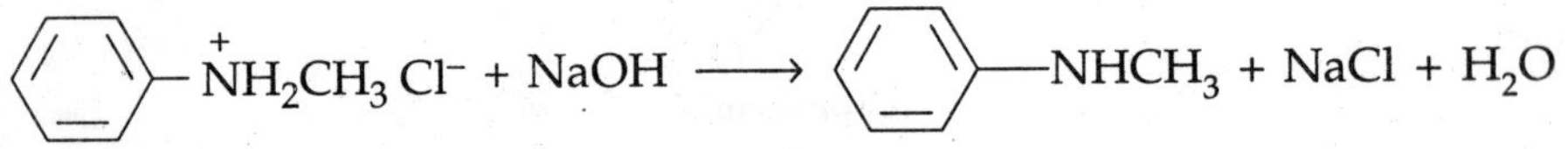

(acid) (base)

(b) $C_6H_5SNa + CH_3CH_2CH_2Br \longrightarrow C_6H_5SCH_2CH_2CH_3 + NaBr$

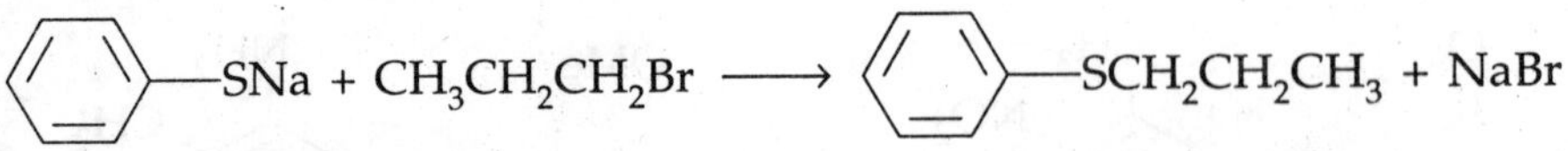

(nucleophile) (electrophile)

(c) $Br{-}C_6H_4{-}\overset{O}{\overset{\|}{C}}OH + NaHCO_3 \longrightarrow Br{-}C_6H_4{-}\overset{O}{\overset{\|}{C}}O^-Na^+ + H_2O + CO_2\uparrow$

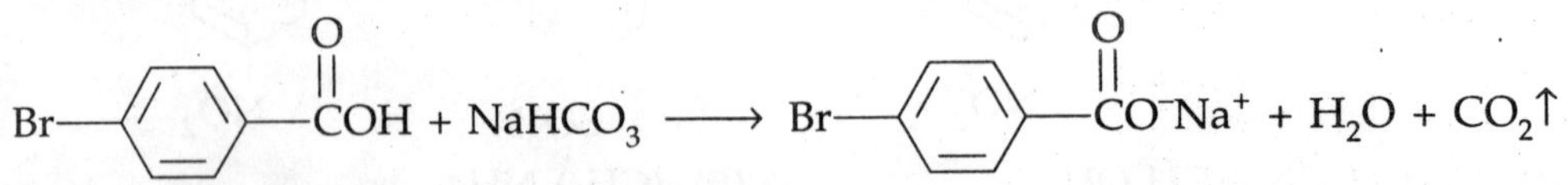

(acid) (base)

38. A concept triangle be used to answer part (a). Only the equations will be shown for the rest of the problem.

(a) 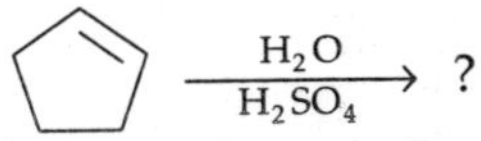 cyclopentene $\xrightarrow[H_2SO_4]{H_2O}$?

The starting material is an alkene. The regent is a mixture of water with a strong acid; therefore, the most abundant species in solution will be the hydronium ion, H_3O^+, an electrophile. The type of reaction is an electrophile addition to the double bond. Because the electrophile in H_2O is H_3O^+, the overall adition is that of H—OH to the double bond.

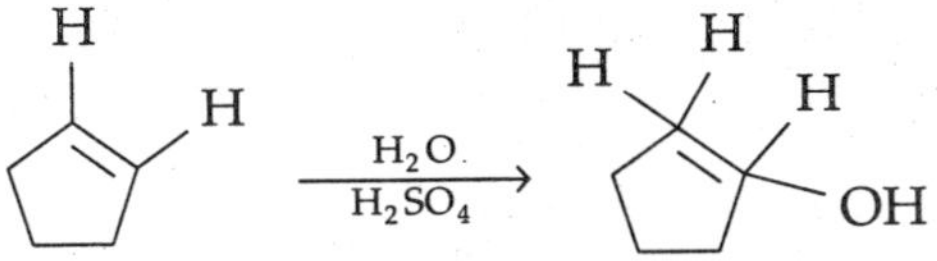

C=C — Markovnikov electropilic addition → —C—C— (H, OH)

$\xrightarrow[H_2SO_4]{H_2O}$ H, H, H, OH

The alkene is symmetric so we do not have to worry about regioelectivity in this case.

(b) 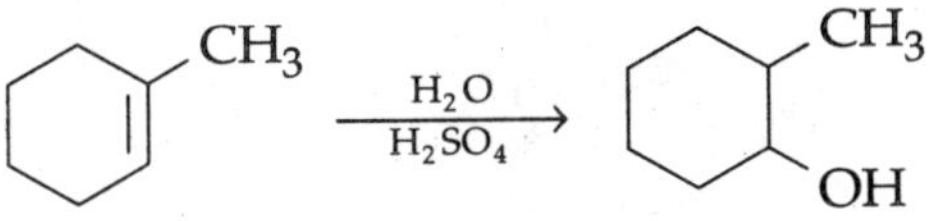 CH_3 $\xrightarrow[H_2SO_4]{H_2O}$ CH_3 OH

39. (a) 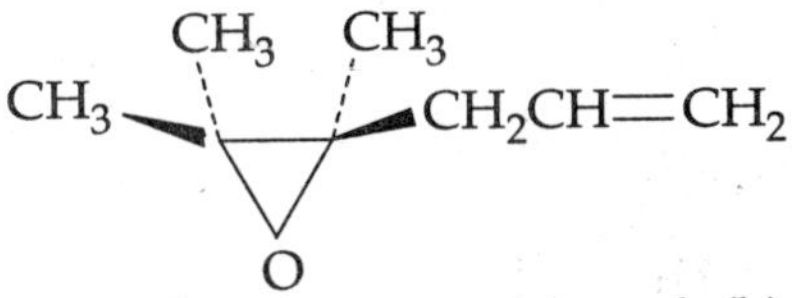 CH_3 CH_3 CH_3 $CH_2CH{=}CH_2$ O (b) 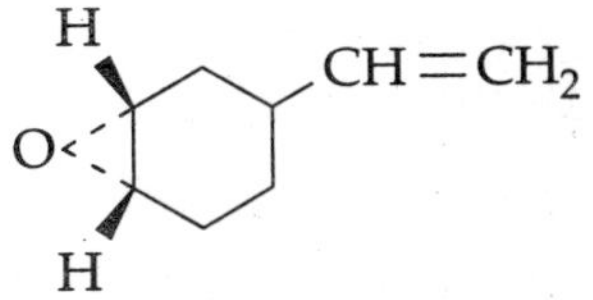 O H H $CH{=}CH_2$

Note that in parts (a) and (b), the more highly substituted double bond is more nucleophilic and therefore reacts faster with the peroxyacid than the less substituted double bond.

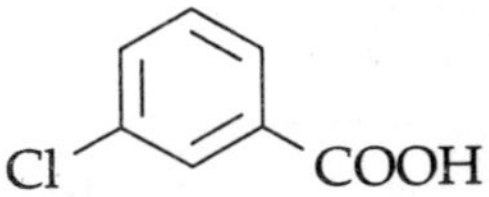

Cl COOH

(c) 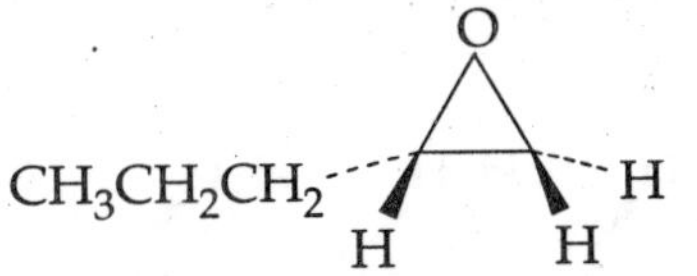 $CH_3CH_2CH_2$ O H H H

enantiomer

(d) 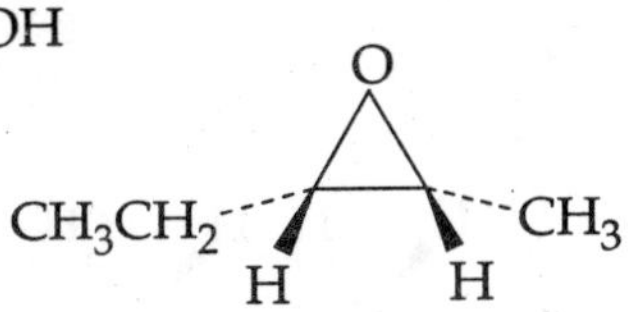 CH_3CH_2 O H H CH_3

enantiomer

(e) 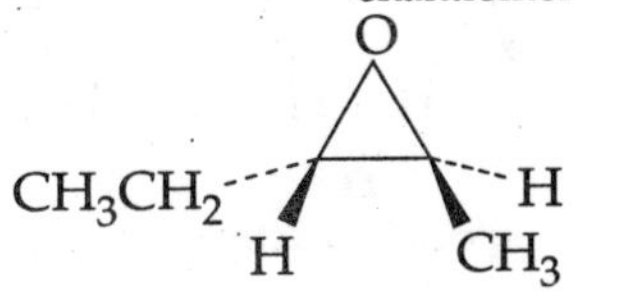 CH_3CH_2 O H H CH_3

enantiomer

40. (I) (a) 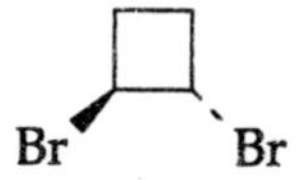 Br Br

(b) 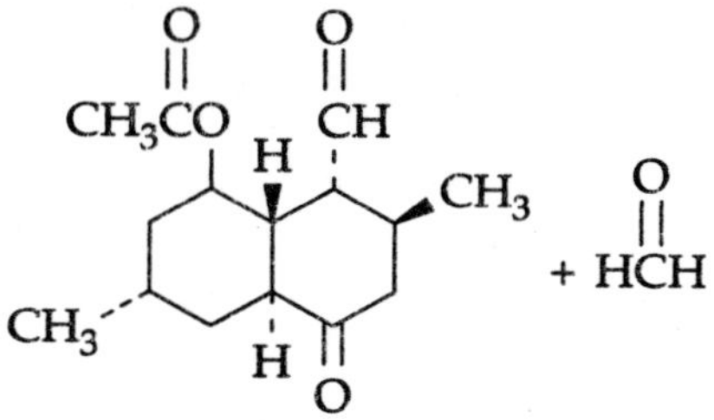 $CH_3\overset{O}{\overset{||}{C}}O$, $\overset{O}{\overset{||}{C}}H$, H, CH_3, CH_3, H, O + $H\overset{O}{\overset{||}{C}}H$

(c) Br, H, CH_3, C—C, H, CH_3, Br

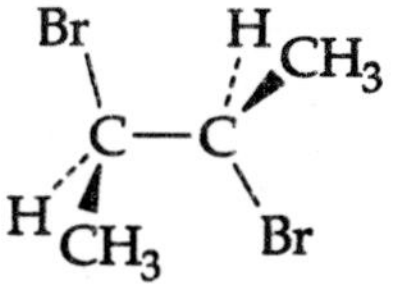

(d) 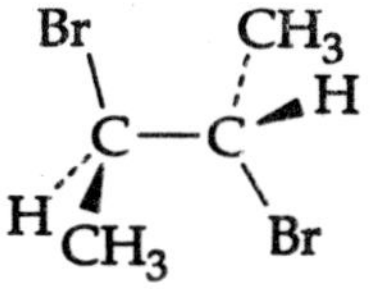 Br, CH_3, H, C—C, H, CH_3, Br

(e) 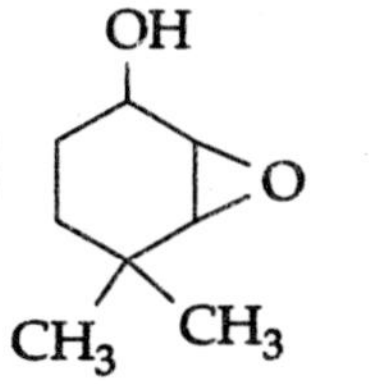 OH, O, CH_3 CH_3

mixture of diastereomers

(f) 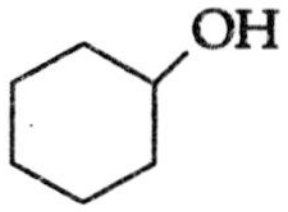 OH

(g) 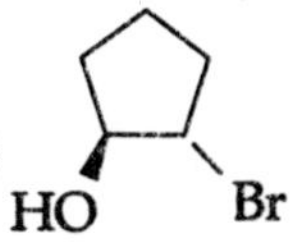 HO Br

(h) 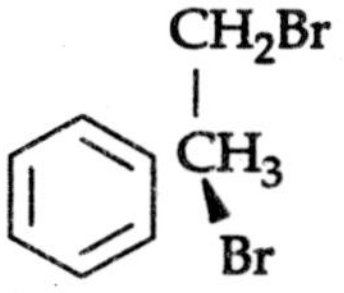 CH_2Br, CH_3, Br

(i) 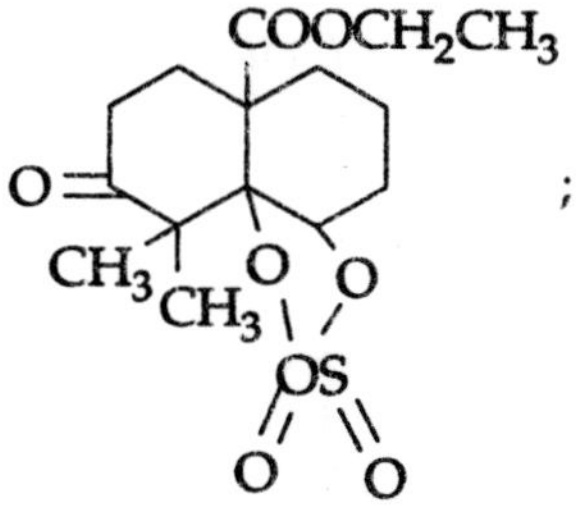 $COOCH_2CH_3$, O=, CH_3, CH_3, O, O, OS, O O ; $COOCH_2CH_3$, O=, CH_3, CH_3, OH, OH

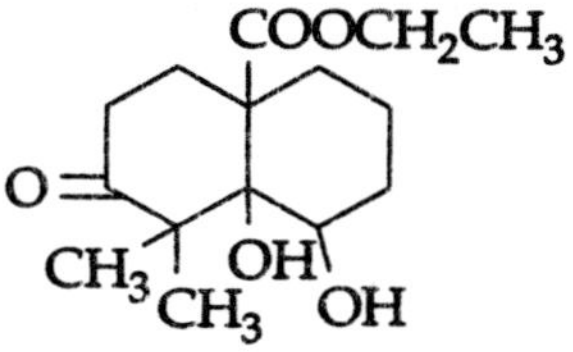

II. (a)  O

A

(b) $2CH_3\overset{O}{\overset{||}{C}}CH_2CH_2\overset{O}{\overset{||}{C}}CH_3$

B

(c) 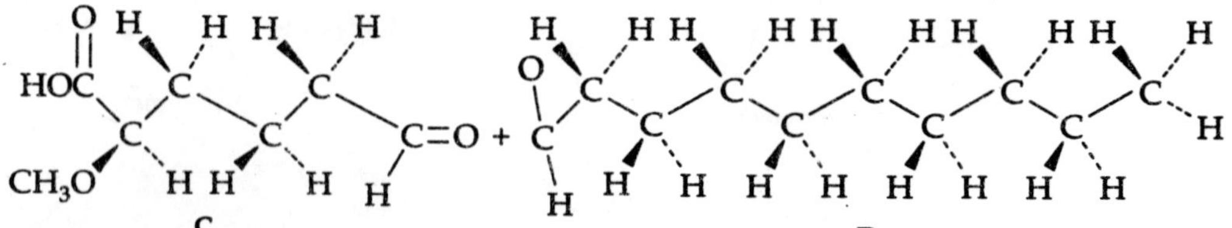 O H H H H, HOC, C, C, C, C=O, CH_3O, H H H H + H H H H H H H H H H, O, C, C, C, C, C, C, C, C, C, C, C, H, H, H H H H H H H H H

C

optically active

D

optically inactive

41. (a)  Cl, Br (b) 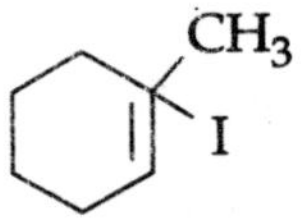 CH_3, I (c) 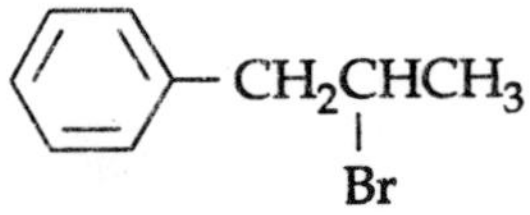 CH_2CHCH_3 Br

42. (a) 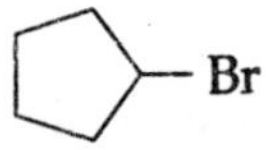 Br (b) 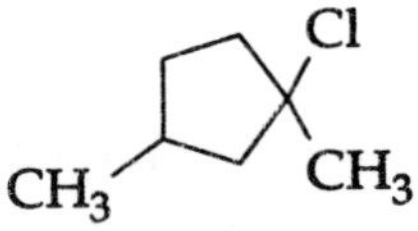 Cl, CH_3, CH_3 (c) $CH_3(CH_2)_9CH_2Cl$

(d) $CH_3(CH_2)_9CH_2Br$ (e) $CH_3SCH_2CH_2Cl$ (f) $CH_3CH_2OCH_2CH_2Br$

(g) 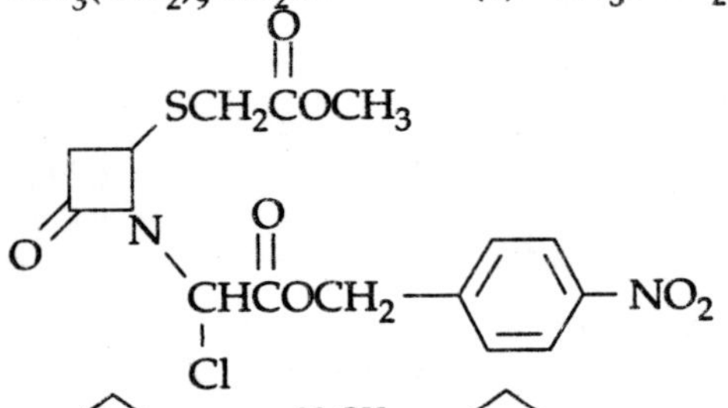 O, SCH_2COCH_3, O, N, O, $CHCOCH_2$ — NO_2, Cl

43. (a) 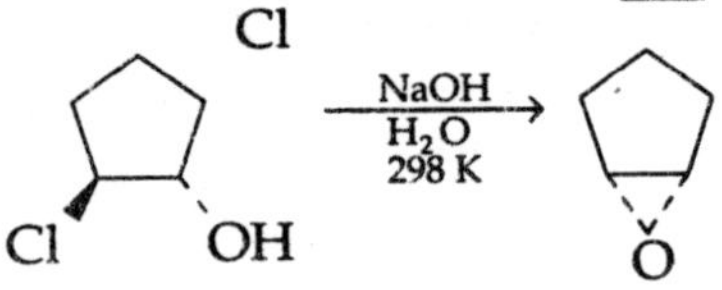 Cl, OH $\xrightarrow[\text{298 K}]{\text{NaOH, } H_2O}$ O

A B

(b) 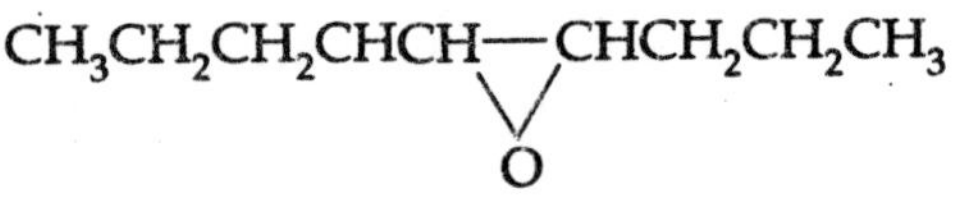 $CH_3CH_2CH_2CHCH—CHCH_2CH_2CH_3$ (O)

C

(c) 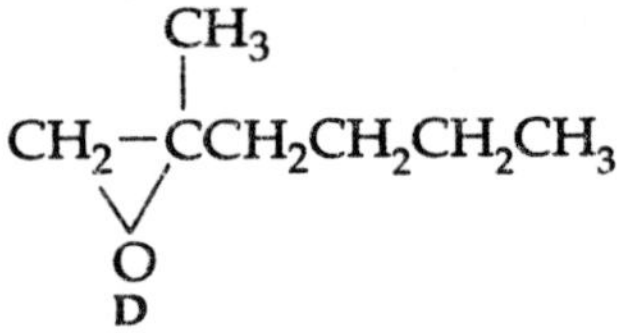 CH_3, $CH_2-CCH_2CH_2CH_2CH_3$ (O)

D

(d)  O

E

(e) 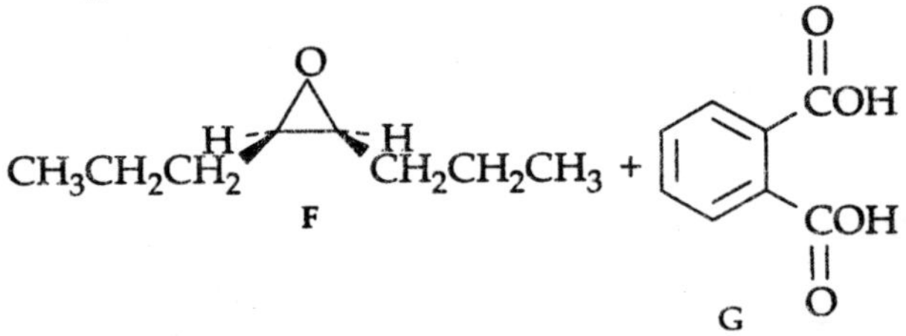 $CH_3CH_2CH_2$ H O H $CH_2CH_2CH_3$ + COOH, COOH

F G

44. (a) 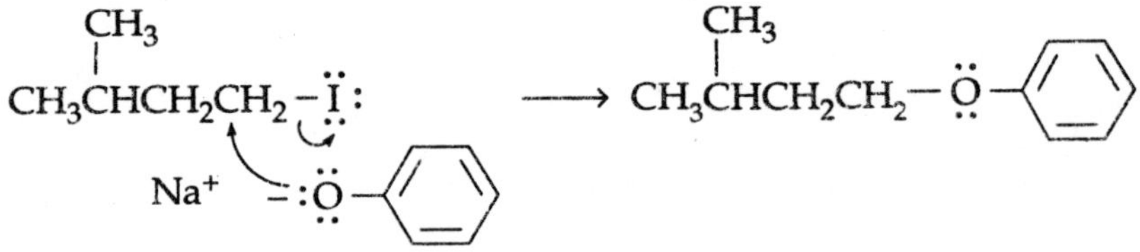 $CH_3CHCH_2CH_2-I:$ (CH_3) Na⁺ :Ö— ⟶ $CH_3CHCH_2CH_2-O-$ (CH_3)

(b) 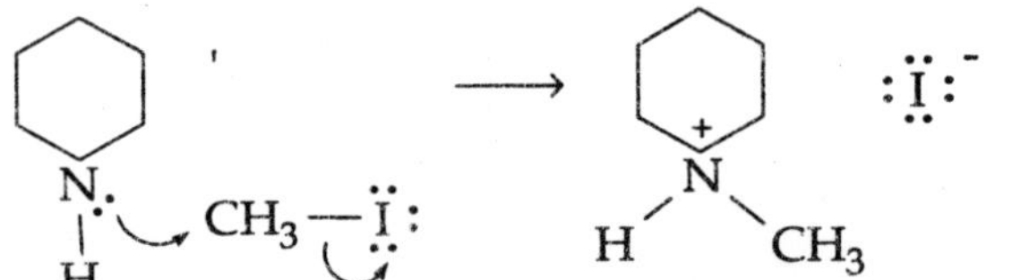 N—H $CH_3-I:$ ⟶ N⁺ H CH_3 :I:⁻

first product;
acid with p*ka* - 10

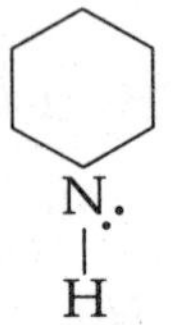

(cyclohexyl)–N̈–H is a base that has a conjugate acid with $pK_a \simeq 10$.

In the presence of excess base, the following equilibrium will be established.

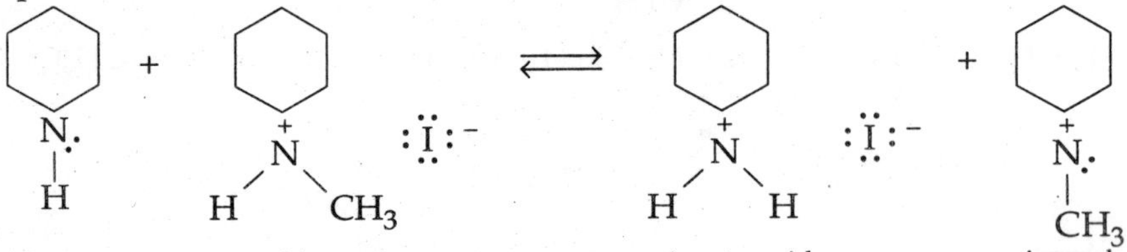

$$C_6H_{11}\ddot{N}H + C_6H_{11}\overset{+}{N}H(CH_3)\ :\ddot{I}:^- \rightleftharpoons C_6H_{11}\overset{+}{N}H_2\ :\ddot{I}:^- + C_6H_{11}\ddot{N}(CH_3)$$

base in large excess; acid $pK_a \simeq 10$; conjugate acid $pK_a \simeq 10$; conjugae base

Because the two acids have roughly the same acidity, the relative amounts of the different species at equilibrium will be determined by the relative concentrations of the different regents. Because there is a larege excess of the amine used as a starting material in the reaction, the portion will spend most of its time with that amine and the product amine will be mostly in the unprodonated from.

(c) 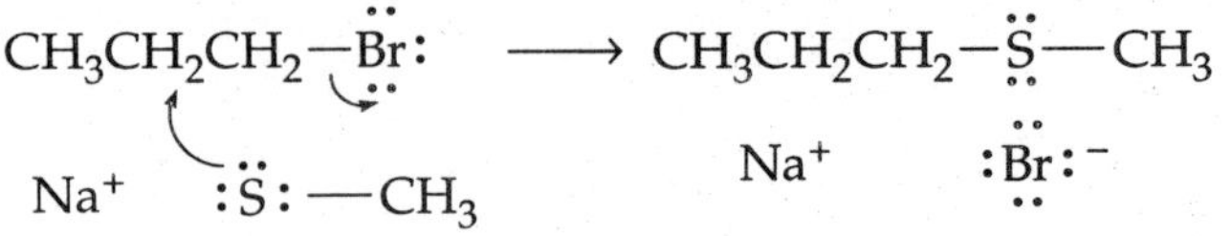 $CH_3CH_2CH_2-\ddot{Br}: \longrightarrow CH_3CH_2CH_2-\ddot{S}-CH_3$

$Na^+\ :\ddot{S}:^- -CH_3$ $\qquad Na^+\ :\ddot{Br}:^-$

45. (a) 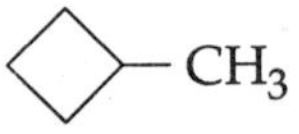 (cyclobutyl)–CH_3 (b) 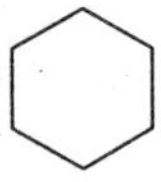 (cyclohexane) (c) $CH_3C(CH_3)_2CH_2CH_3$

(d) 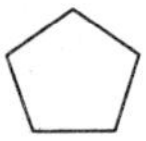 (cyclopentane) (e) $CH_3\overset{O}{\overset{\|}{C}}CH_2CHCH_2CH_2\overset{O}{\overset{\|}{C}}H$

(f) $2Na^{+-}O\overset{O}{\overset{\|}{C}}CH_2CH_2CH_2CH_2\overset{O}{\overset{\|}{C}}O^-Na^+$

46. (I) (a) 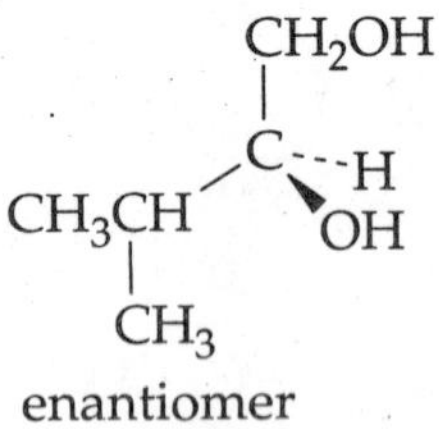 $CH_3CH(CH_3)-C(H)(OH)-CH_2OH$ (b) 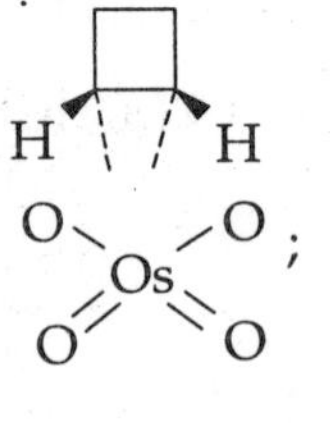 (cyclobutane osmate ester: H, H; O, O; Os; O, O) ; (cyclobutane-1,2-diol: H, H; HO OH)

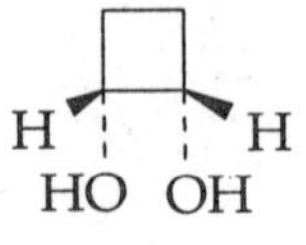

enantiomer A B C

(c) 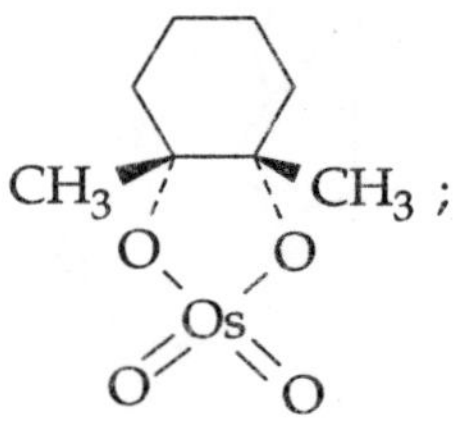 CH_3 ... CH_3 ; (cyclic osmate ester) **D**

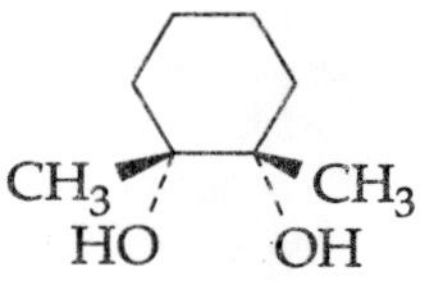

CH_3 CH_3, HO OH **E**

(d) 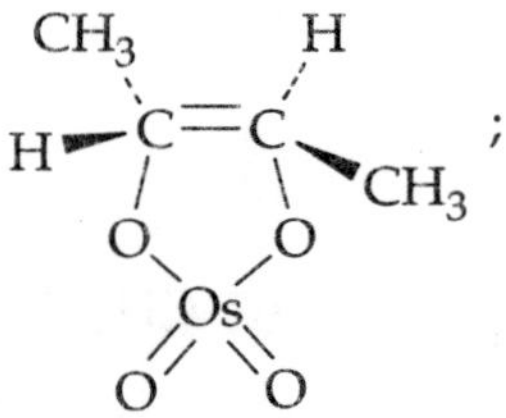 CH_3, H, H, CH_3 — C=C — O, O, Os, O, O **F** ;

CH_3 H, H–C=C–CH_3, HO OH

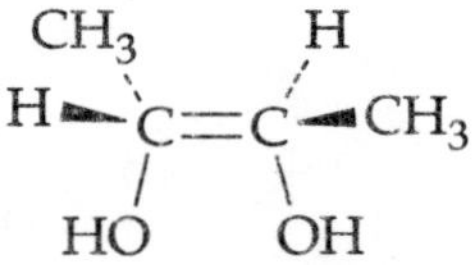

and enantiomer (higher energy conformation)

G

(e) 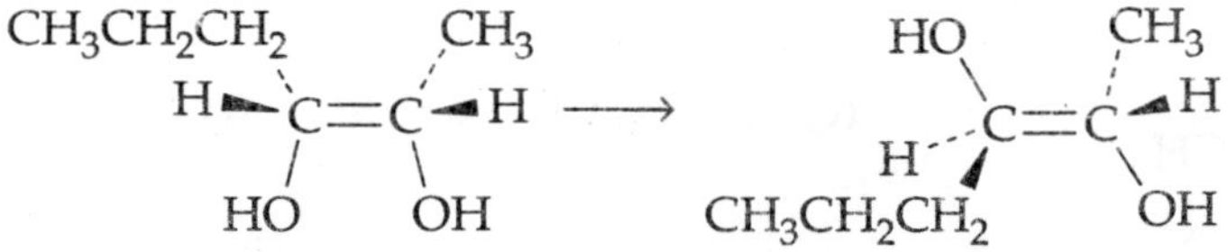 $CH_3CH_2CH_2$ CH_3, H–C=C–H, HO OH ⟶ HO CH_3, C=C–H, H, $CH_3CH_2CH_2$ OH

enantiomer (higher energy conformation) — enantiomer (lower energy conformation)

H

II. (a) 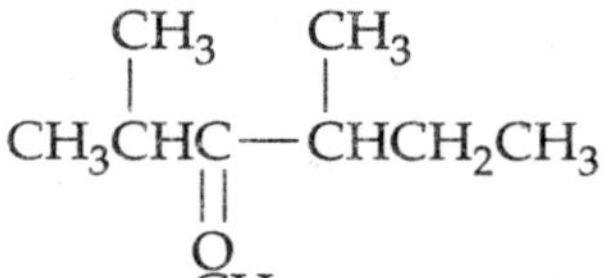 $CH_3CHC(=O)-CHCH_2CH_3$ with CH_3 groups

(b) 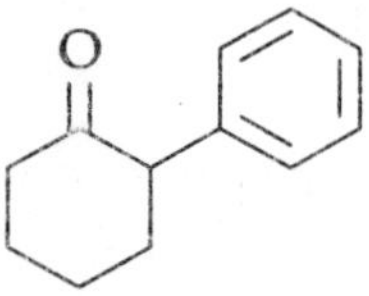 (2-phenylcyclohexanone)

(c) 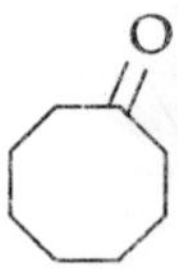 (cyclooctanone)

(d) 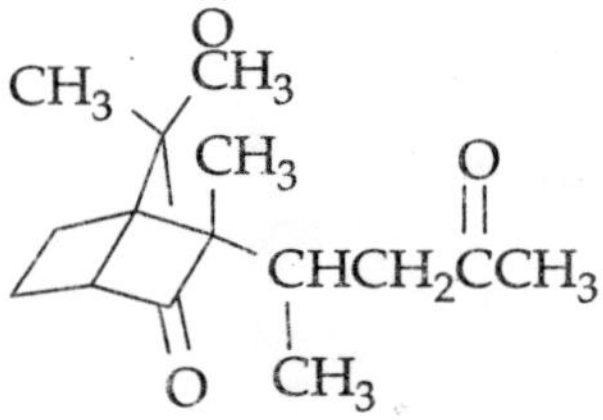 CH_3, CH_3, CH_3, O, $CHCH_2CCH_3$, O, CH_3

(e) 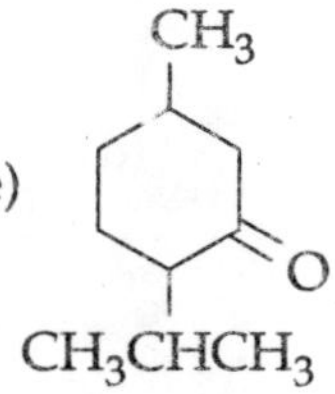 CH_3, O, CH_3CHCH_3

III. (a) 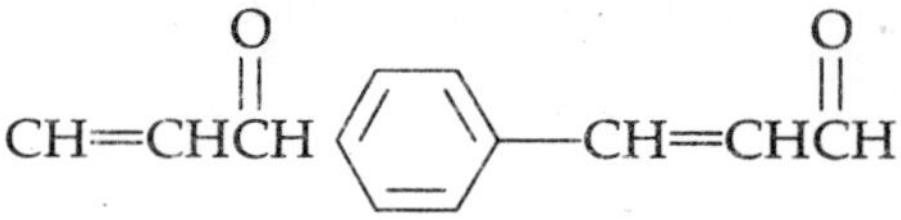 $CH{=}CHCH(=O)$–(benzene)–$CH{=}CHCH(=O)$

(b) 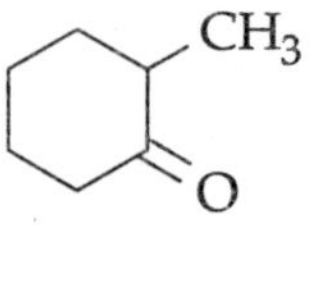 CH_3, O

(c) 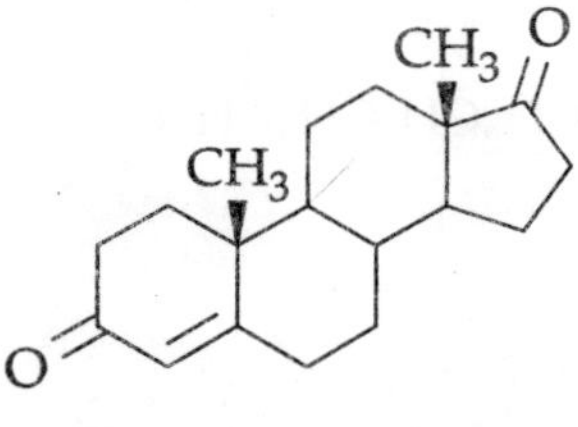 CH_3, O, CH_3, O

IV. (a) $2HO\overset{O}{\overset{\|}{C}}\underset{\underset{O}{\|}}{C}H$ (b) 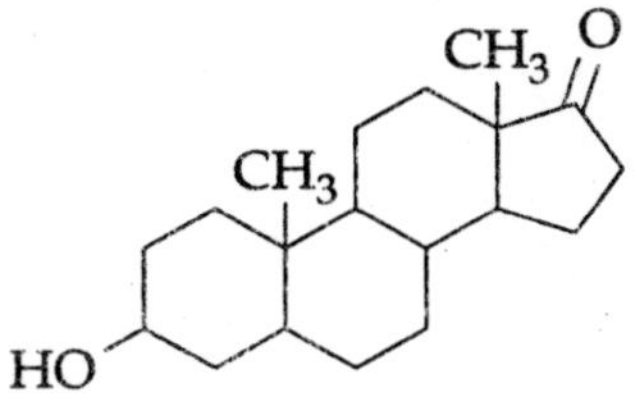 [steroid structure: HO-substituted tetracyclic ring with two CH_3 groups and ring C=O] $+ CH_3\overset{O}{\overset{\|}{C}}OH + H\overset{O}{\overset{\|}{C}}H$

(c) $2\ C_6H_5-\overset{O}{\overset{\|}{C}}H$

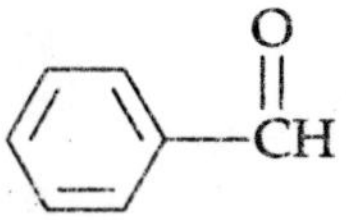

47. I (a) $CH_3\overset{CH_3}{\overset{|}{C}}HCH_2Cl$ (b) $CH_3\underset{Cl}{\underset{|}{\overset{CH_3}{\overset{|}{C}}}}CH_2CH_3$ (c) $CH_3CH_2\underset{Br}{\underset{|}{C}}HCH_3$

(d) 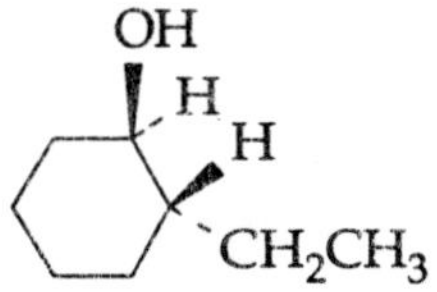 [cyclohexane ring with OH (wedge) and H on C-1, H and CH_2CH_3 (wedge) on C-2] (e) $BrCH_2(CH_2)_4CH_2Br$ (f) 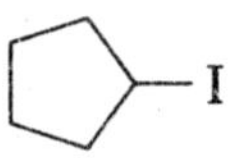 [cyclopentyl]–I

(g) $CH_3\overset{O}{\overset{\|}{C}}-\underset{CH_3}{\underset{|}{\overset{CH_3}{\overset{|}{C}}}}CH_2O\underset{CH_3}{\underset{|}{\overset{CH_3}{\overset{|}{C}}}}CH_3$ (h) 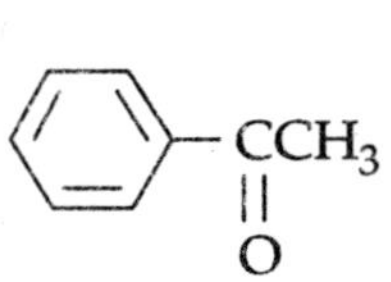 $C_6H_5-\underset{O}{\underset{\|}{C}}CH_3$

II. (a) 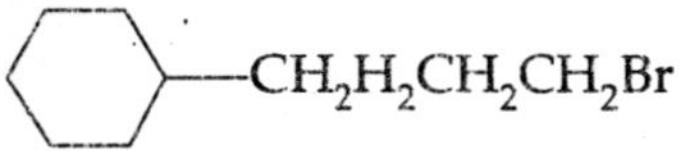 [cyclohexyl]$-CH_2H_2CH_2CH_2Br$ (b) 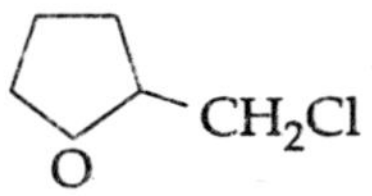 [tetrahydrofuran ring, O]$-CH_2Cl$

(c) $H\overset{O}{\overset{\|}{C}}CH_2CH_2CH_2CH_2\overset{O}{\overset{\|}{C}}H$ (d) 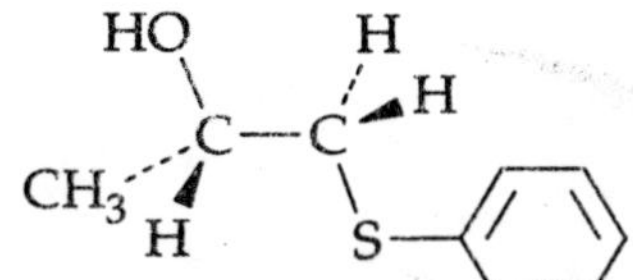 [HO, CH_3, H on C; C–C; H, H on C bearing S–C_6H_5]

48. I. (a) 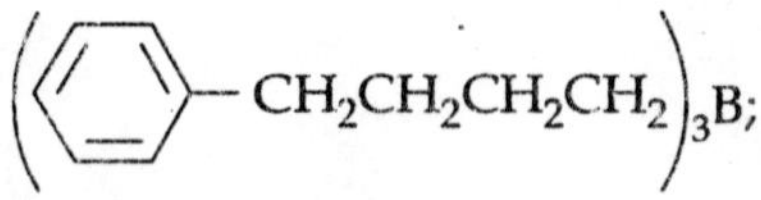 $(C_6H_5-CH_2CH_2CH_2CH_2)_3B$; $C_6H_5-CH_2CH_2CH_2CH_2OH$

A 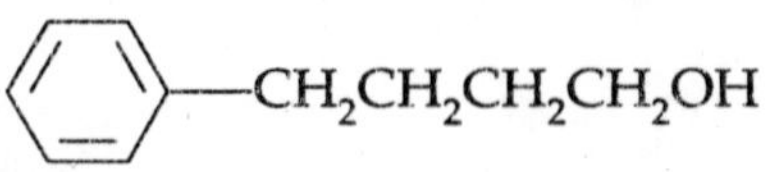 B

$C_6H_5-CH_2CH_2CH_2\overset{O}{\overset{\|}{C}}H$

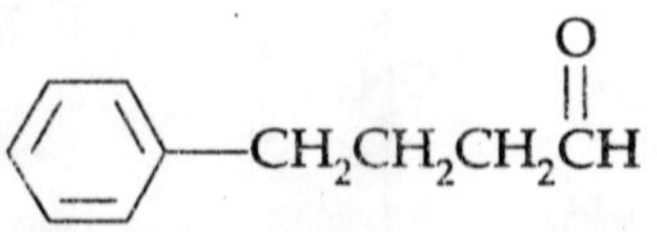

C

(b)

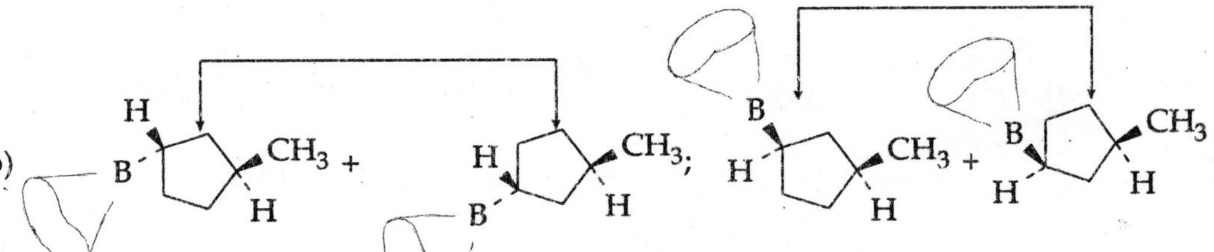

D
major isomers from less hindered transition state

minor isomers
E

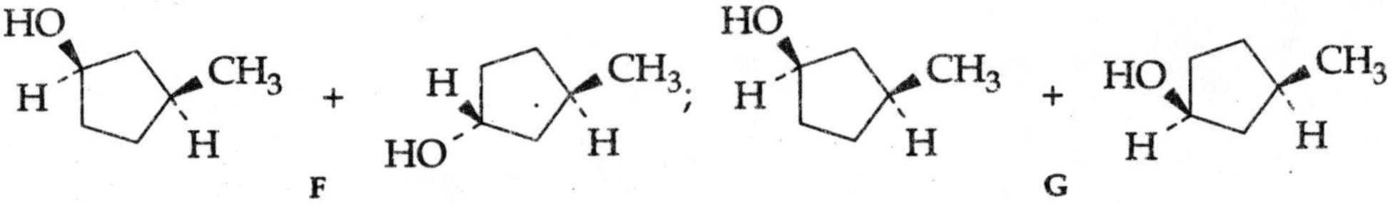

F
enatiomers

G
enatiomers

(c)

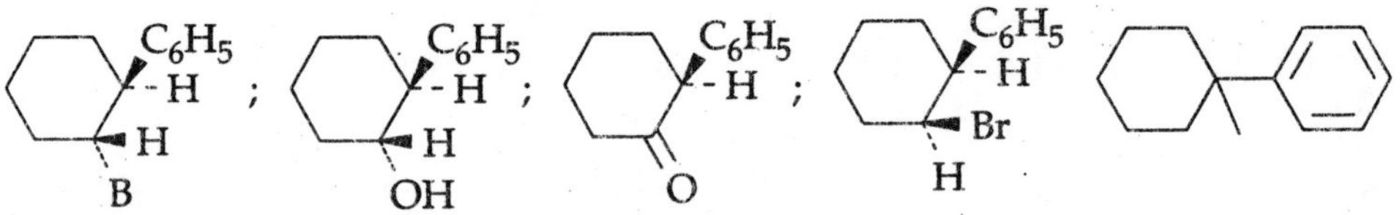

enantiomer
H

enantiomer
I

enantiomer
J

enantiomer
K

L

(d) 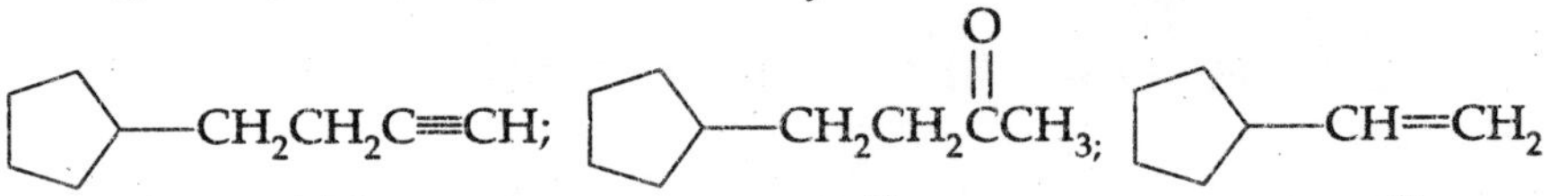 $\bigcirc$—$CH_2CH_2C{\equiv}CH$; $\bigcirc$—$CH_2CH_2CCH_3$ (C=O); $\bigcirc$—$CH{=}CH_2$

M N O

$\bigcirc$—$CHCH_2Br$ (Br); $\bigcirc$—$CHCH_3$ (Br); $\bigcirc$—$CHCH_3$ (CN)

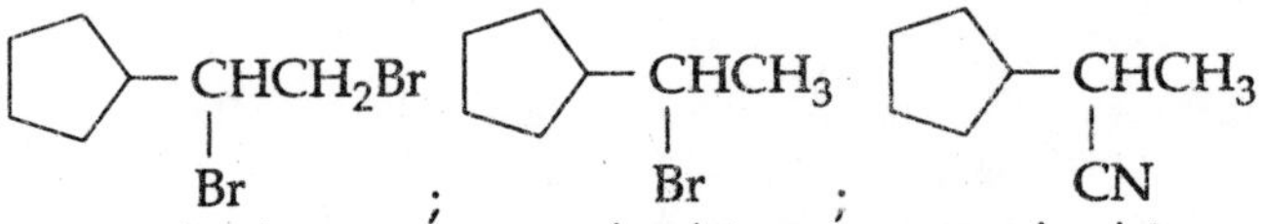

racemic mixture
P

racemic mixture
Q

racemic mixture
R

II. (a)

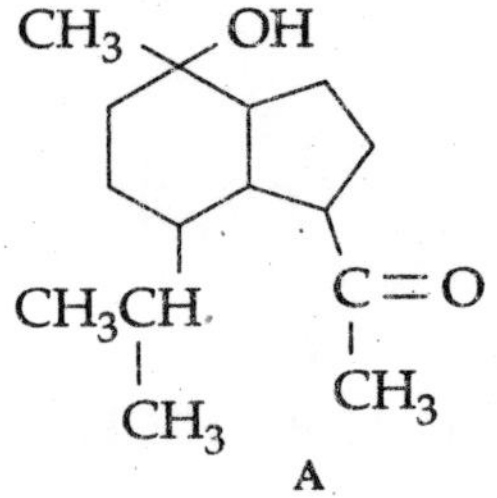

A

(b)

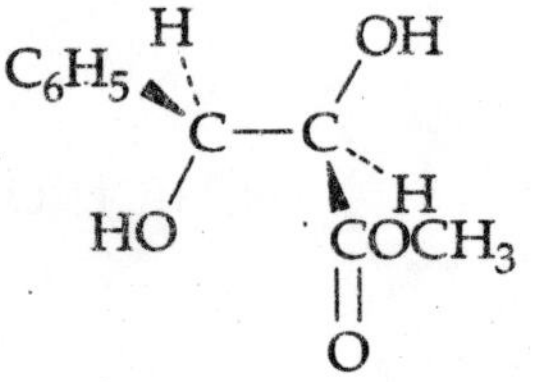

and enantiomer
B

(c)

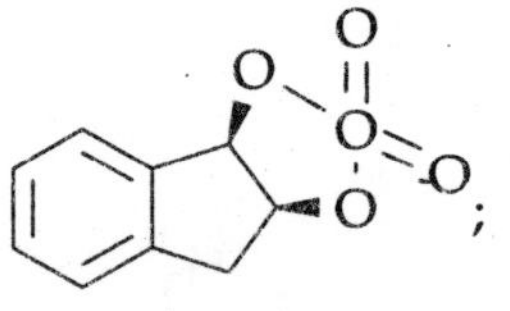

and enantiomer
C

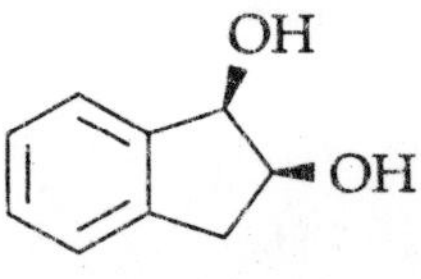

and enantiomer
D

(d) $\frac{HOCl}{E}$; $\frac{NaOH}{F}$ (e) 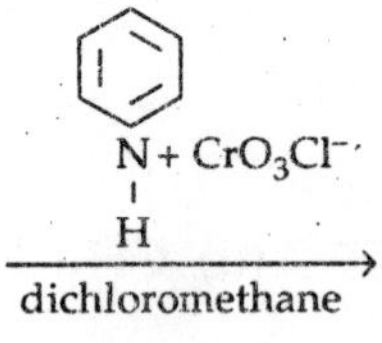 $\xrightarrow[\text{dichloromethane}]{\text{Pyridinium-H } CrO_3Cl^-}$

G

49. I. (a) 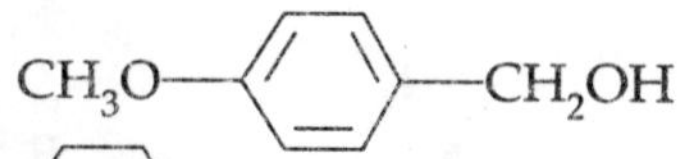 $CH_3O-C_6H_4-CH_2OH$ (para-substituted benzene ring)

(b) $CH_3(CH_2)_3CH(CH_3)CH_2CH_2OH$

(c) 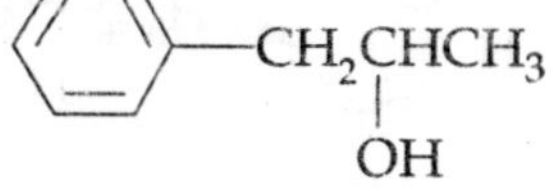 $C_6H_5-CH_2CH(OH)CH_3$

(d) 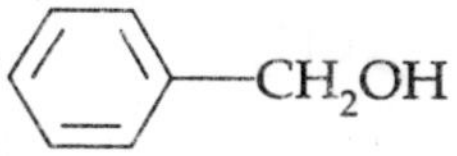 $C_6H_5-CH_2OH$

(e) $CH_3CH(OH)CH_2CH(CH_3)CH_3$

(f) $CH_3CH{=}CHCH(OH)CH_3$

(g) 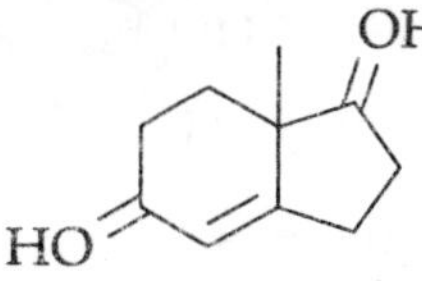 [fused bicyclic structure: six-membered ring bearing =C–HO and a ring double bond, fused to a five-membered ring bearing an angular methyl and OH]

(h) 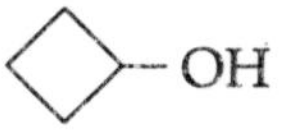 cyclobutyl–OH

II. (a) $CH_3CH_2CH_2CH_2Li + Li^+Br^-$

(b) 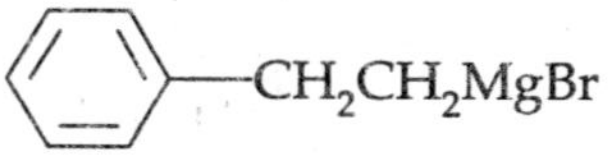 $C_6H_5-CH_2CH_2MgBr$

(c) $CH_2{=}CHMgCl$

(d)  $C_6H_5-C{\equiv}CMgBr + CH_3CH_3\uparrow$

(e) 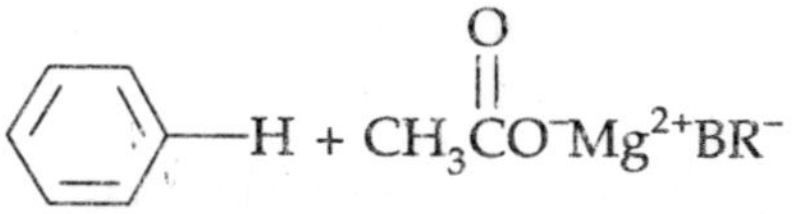 $C_6H_5-H + CH_3\overset{O}{\overset{\|}{C}}O^-Mg^{2+}BR^-$

(f) $CH_3CH_2C{\equiv}CH\uparrow + Na^+OH^-$

(g) 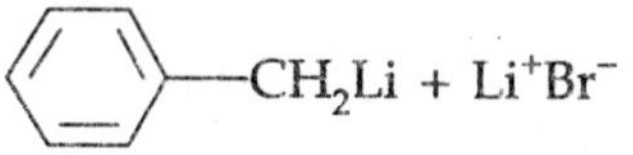 $C_6H_5-CH_2Li + Li^+Br^-$

(h) $CH_3CH_2CH_2\uparrow + Li^+NH_2^-$

(i) $CH_3CH_3\uparrow + CH_3O^-Mg^{2+}I^-$

(j) 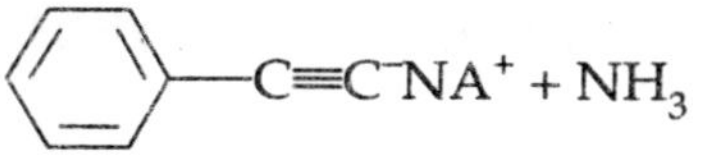 $C_6H_5-C{\equiv}C^-NA^+ + NH_3$

50. (a) 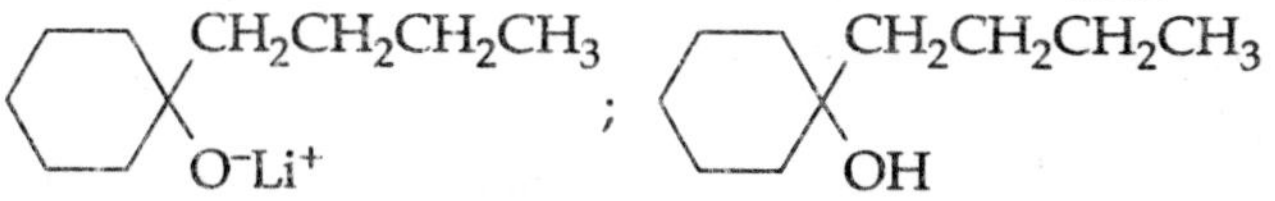 1-$CH_2CH_2CH_2CH_3$-cyclohexyl-O^-Li^+ (A) ; 1-$CH_2CH_2CH_2CH_3$-cyclohexan-1-OH (B)

(b) 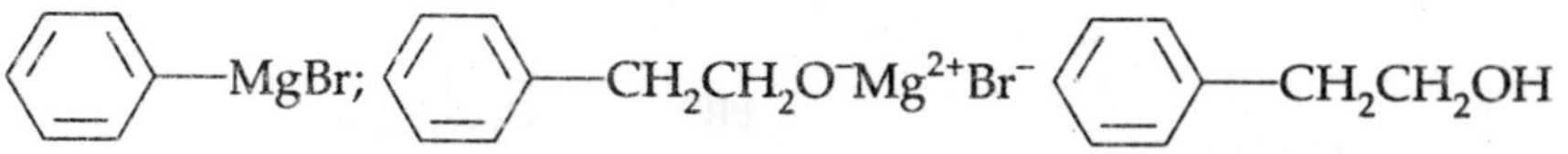 C_6H_5-MgBr; (C) $C_6H_5-CH_2CH_2O^-Mg^{2+}Br^-$ (D) $C_6H_5-CH_2CH_2OH$ (E)

(c) 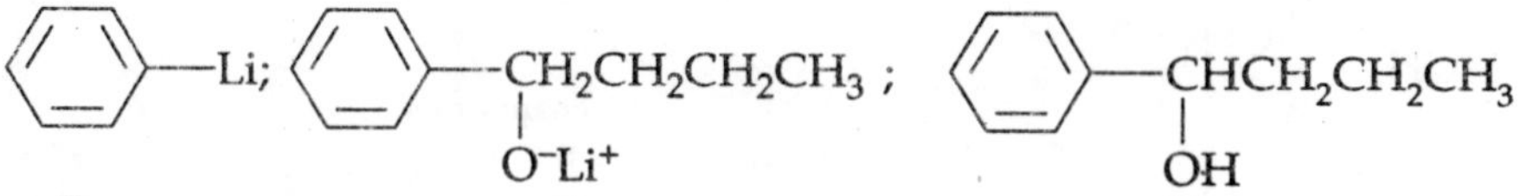 C_6H_5-Li; (F) $C_6H_5-CH(O^-Li^+)CH_2CH_2CH_2CH_3$; (G) $C_6H_5-CH(OH)CH_2CH_2CH_3$ (H)

(d) $CH_3CH_2C{\equiv}C^-Na^+$; (I) 1-O^-Na^+-1-($C{\equiv}CCH_2CH_3$)cyclopentane (J) ; 1-OH-1-($C{\equiv}CCH_2CH_3$)cyclopentane (K)

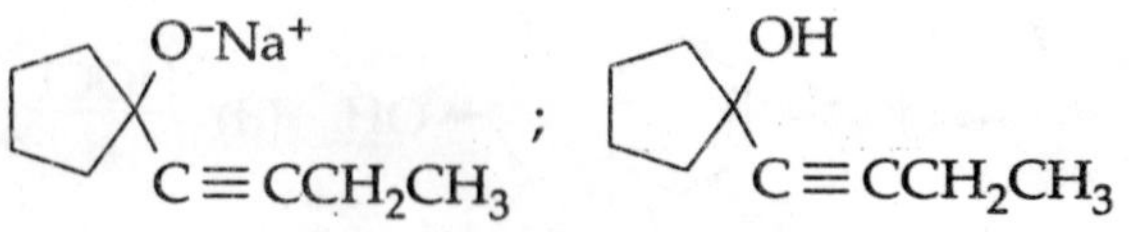

(e) 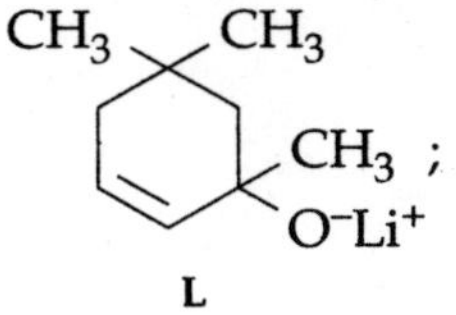 L 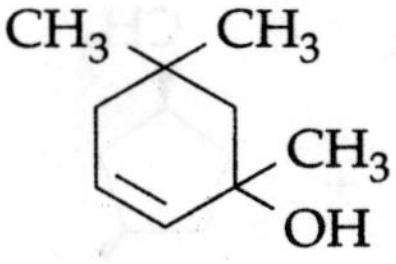 ; M

51. (a) 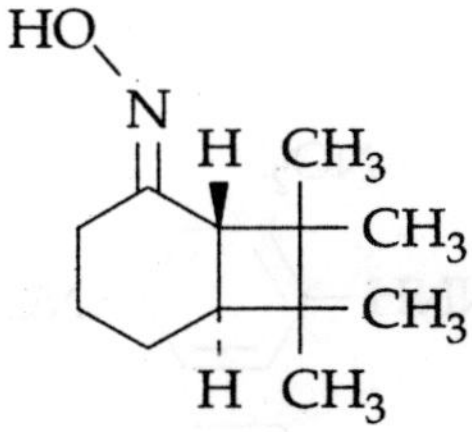 (b) 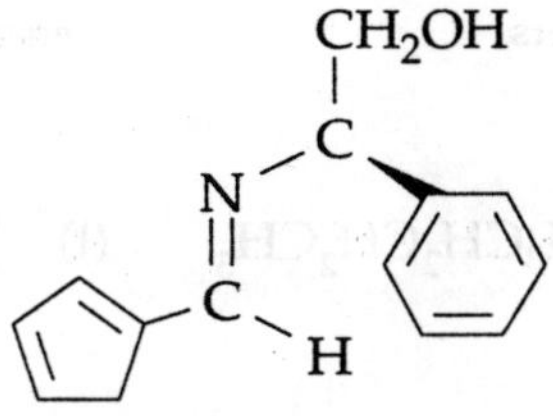 (c)

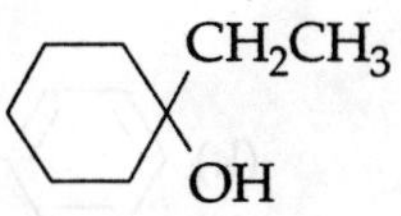

(d) 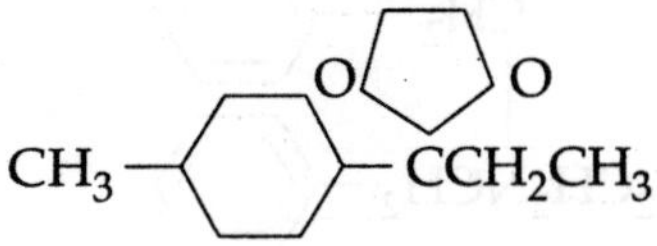 CH_3— —CCH_2CH_3 (e)

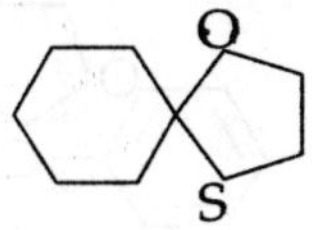

52. I. (a)  =NNH— (b)

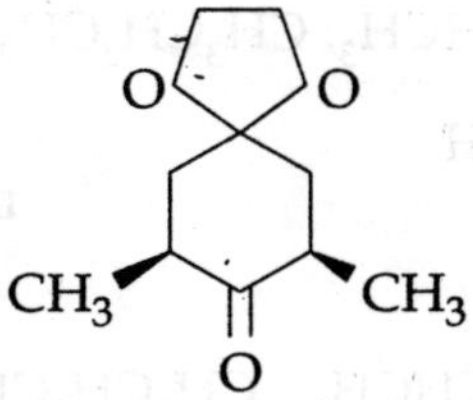

(c) 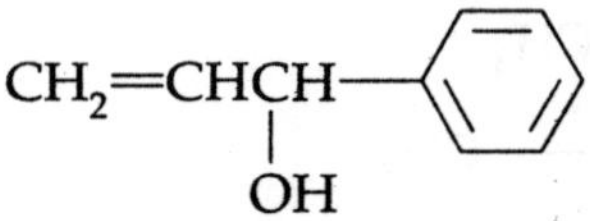 CH_2=CHCH— with OH (d)

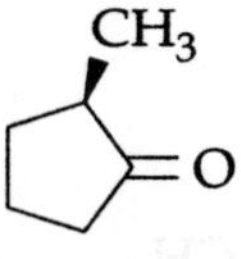

(e) 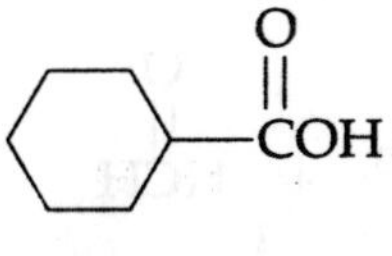 —COH (f) 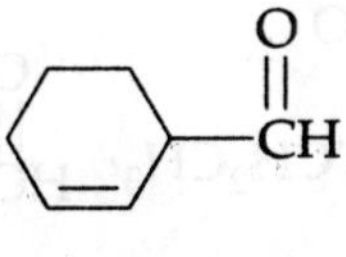 —CH

(g) 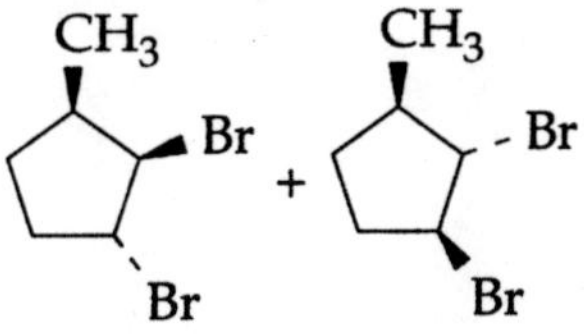 + (h) 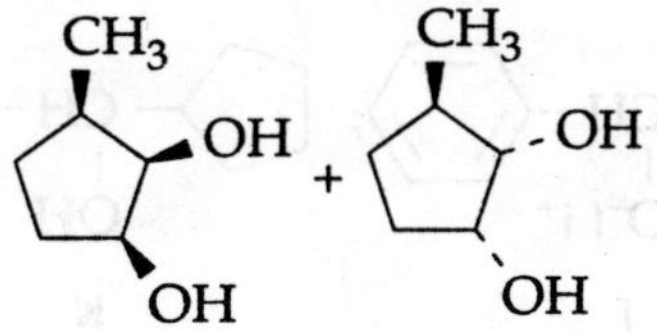 +

(i) 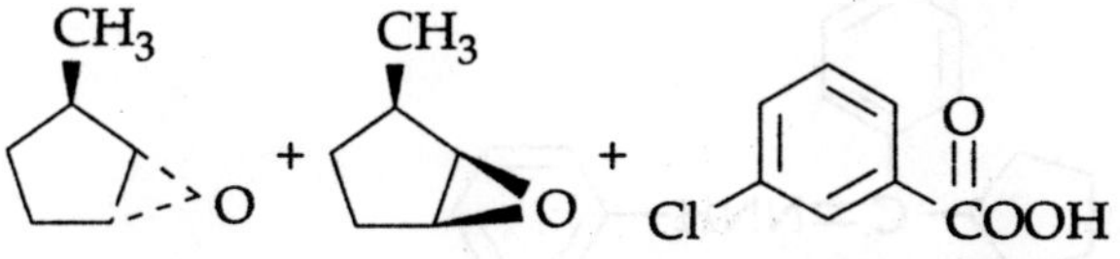 + + Cl— —COOH

(j) [cis/trans 3-methylcyclopentanol structures: CH_3 … HO + CH_3 … HO — enantiomers; CH_3 … HO + CH_3 … HO — enantiomers]

(k) C_6H_5–C(CH_3)(CH_3)–$CH_2CH_2CH_2CH_3$ (l) C_5H_9–CH=NNH–$C_6H_3(NO_2)_2$

II. (a) $C_6H_5C(OCH_2CH_2O)CH_2CH_2CH_2Cl$ (A); $C_6H_5C(OCH_2CH_2O)CH_2CH_2CH_2N(CH_2C_6H_5)CH_2C_6H_5$ (B)

(b) $CH_3CH_2CH_2CH(OH)CH_3$ (C); $CH_3CH_2CH_2CH(OH)CH_3$ (D)

$CH_3CH_2CH{=}CHCH_3$ (major product, E); $CH_3CH_2CH{-}CHCH_3$ (epoxide, bridged by O) (F)

(c) $CH_2{=}CHCH_2CH_2C(CH_2CH_3)$ (acetonide, $O{-}C(CH_3)_2{-}O$) (G); $HC(=O)CH_2CH_2C(CH_2CH_3)$ (acetonide) (H) + $HC(=O)H$ (I)

(d) C_5H_9–CH(O^-Li^+)–C_6H_5 (J); C_5H_9–CH(OH)–C_6H_5 (K)

C_5H_9–C(=O)–C_6H_5 (L); C_5H_9–C(C_6H_5)=NNH–C_6H_5 (M)

(e) 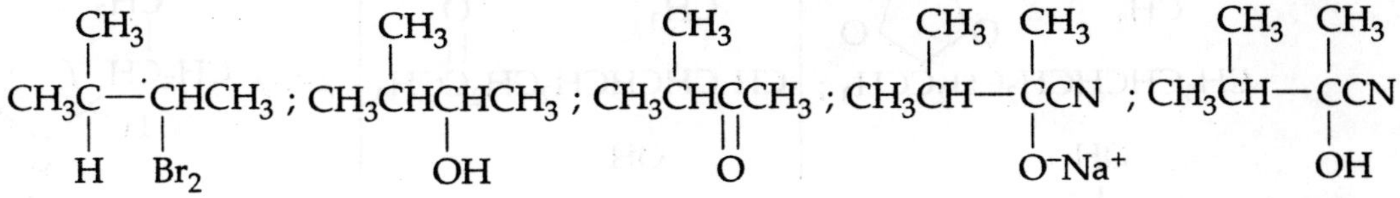 $CH_3\overset{CH_3}{\underset{H}{C}}—\underset{Br_2}{\dot{C}}HCH_3$; $CH_3\overset{CH_3}{C}H\underset{OH}{C}HCH_3$; $CH_3\overset{CH_3}{C}H\underset{O}{\overset{||}{C}}CH_3$; $CH_3\overset{CH_3}{C}H—\overset{CH_3}{\underset{O^-Na^+}{C}}CN$; $CH_3\overset{CH_3}{C}H—\overset{CH_3}{\underset{OH}{C}}CN$

N **O** **P** **Q** **R**

(f) $HC{\equiv}C^-Na^+$; $C_6H_5—CH_2C{\equiv}CH$; $C_6H_5—CH_2C{\equiv}CMgBr$

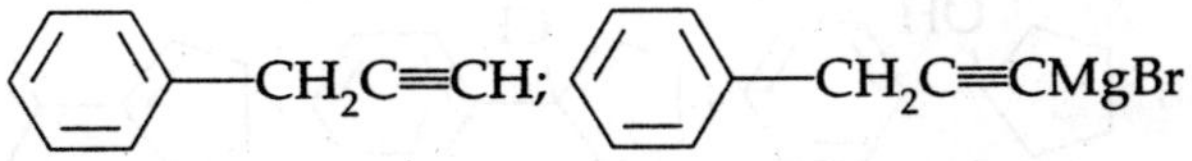

T **U**

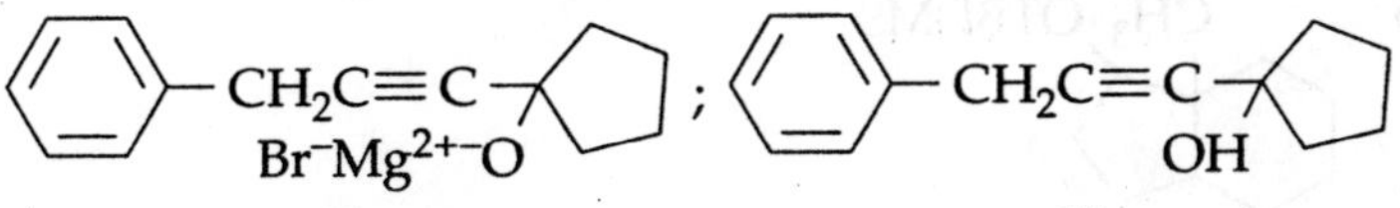

$C_6H_5—CH_2C{\equiv}C$–(cyclopentyl)–$O^-Mg^{2+}Br^-$; $C_6H_5—CH_2C{\equiv}C$–(cyclopentyl)–OH

V **W**

(g) 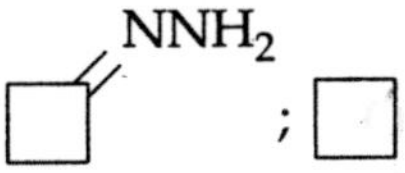 cyclobutylidene=NNH_2 ; cyclobutane (h) (Z)-dec-alkenal chain, $CH=O$ (with H)

X **Y** **Z**

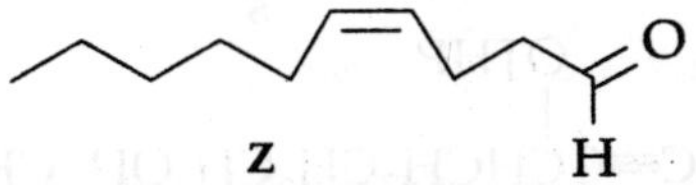

53. (a) $HC{\equiv}CCH_2CH_2OTHP$; $BrMgC{\equiv}CCH_2CH_2OTHP$; $CH_3(CH_2)_3\underset{O^-Mg^{2+}Br^-}{C}HC{\equiv}CCH_2CH_2OTHP$

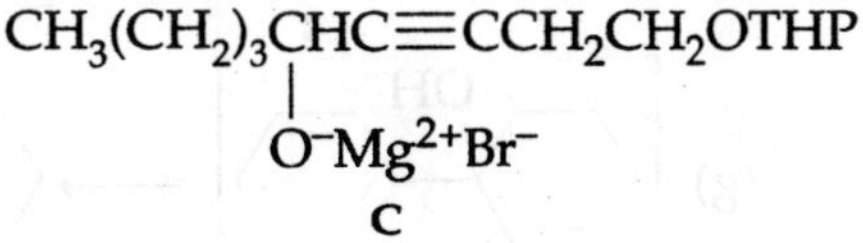

A **B** **C**

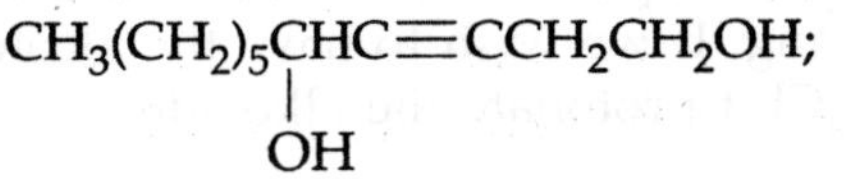

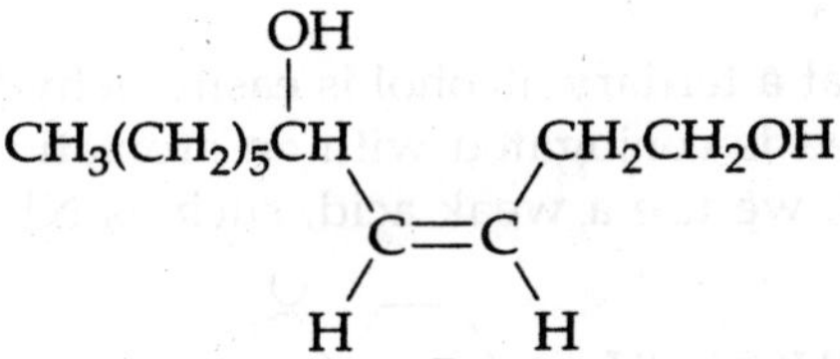

$CH_3(CH_2)_5\underset{OH}{C}HC{\equiv}CCH_2CH_2OH$; $CH_3(CH_2)_5\overset{OH}{C}H(H)C{=}C(H)CH_2CH_2OH$

D **E**

(b) 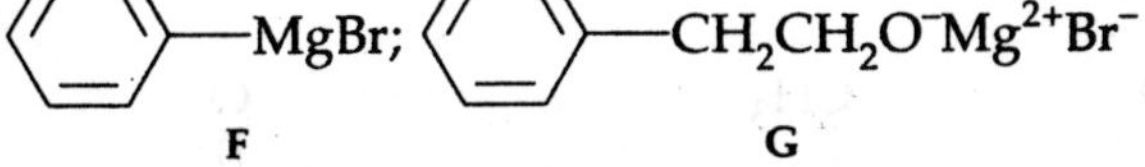 $C_6H_5—MgBr$; $C_6H_5—CH_2CH_2O^-Mg^{2+}Br^-$

F **G**

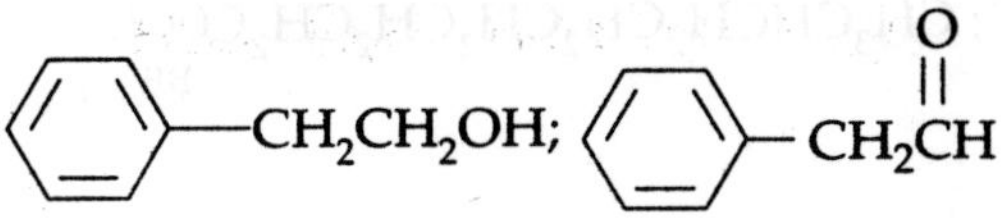

$C_6H_5—CH_2CH_2OH$; $C_6H_5—CH_2\overset{O}{\overset{||}{C}}H$

H **I**

(c) 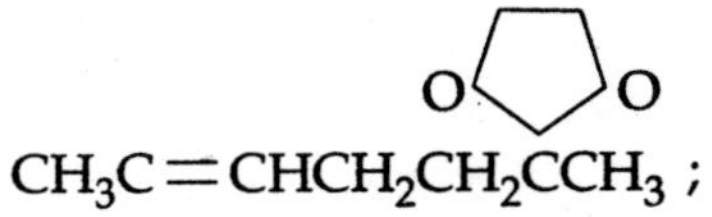 $CH_3C{=}CHCH_2CH_2\overset{O\ \ O}{C}CH_3$ (1,3-dioxolane ketal) ; $CH_3\overset{CH_3}{C}H\underset{Br_2}{C}HCH_2CH_2\overset{O\ \ O}{C}CH_3$ (1,3-dioxolane ketal)

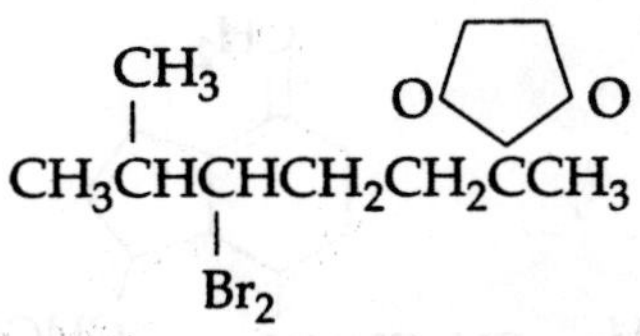

J **K**

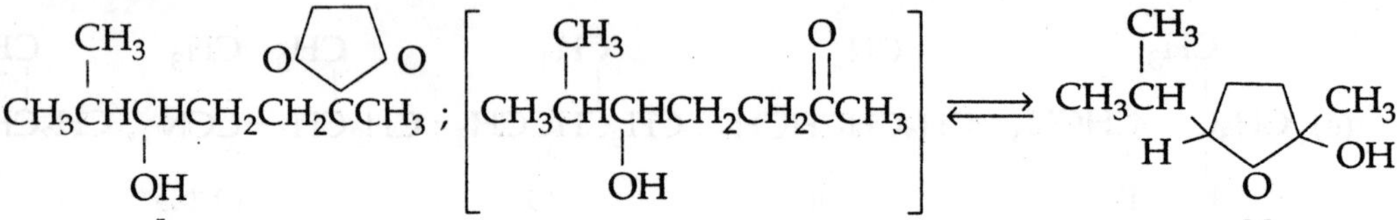

$$\underset{\textbf{L}}{CH_3CH(CH_3)CH(OH)CH_2CH_2C(OCH_2CH_2O)CH_3}\ ;\ \left[CH_3CH(CH_3)CH(OH)CH_2CH_2C(=O)CH_3\right] \rightleftharpoons \textbf{M}$$

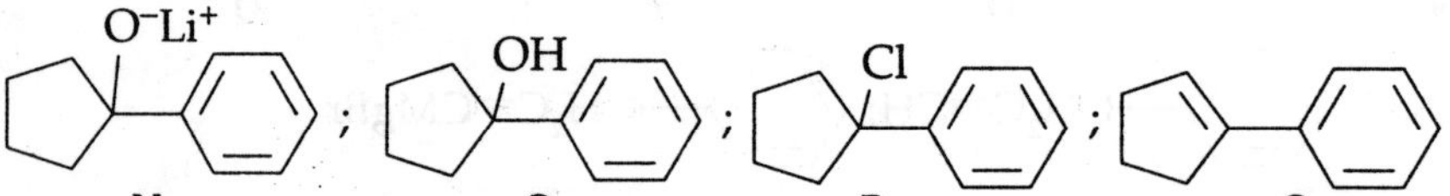

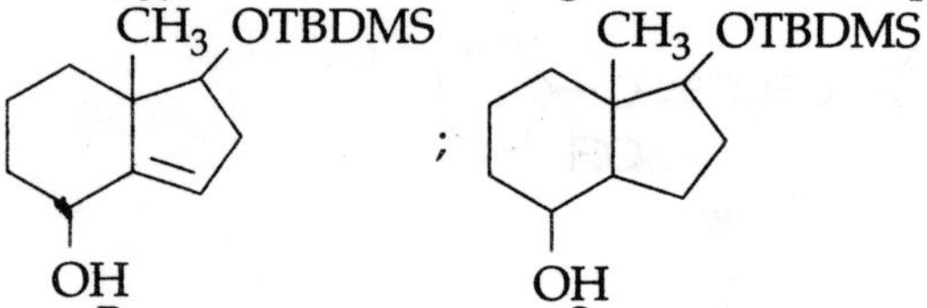

(M: five-membered ring hemiketal bearing $CH_3CH(CH_3)$, H on one ring carbon and CH_3, OH on the other, ring O between them)

(d) N (1-phenylcyclopentan-1-olate, O^-Li^+) ; O (1-phenylcyclopentanol, OH) ; P (1-chloro-1-phenylcyclopentane, Cl) ; Q (1-phenylcyclopentene)

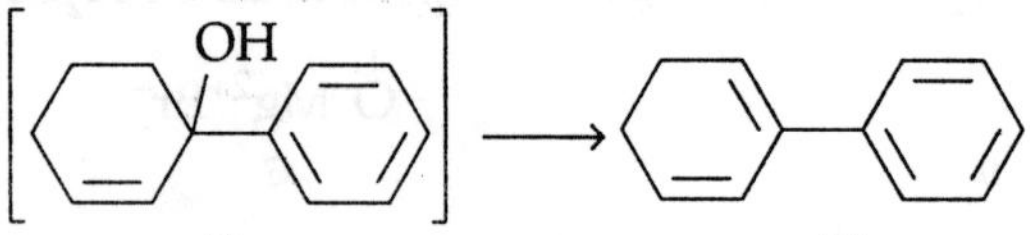

(e) R (CH_3, OTBDMS, OH; bicyclic with double bond) ; S (CH_3, OTBDMS, OH)

(f) $\underset{\textbf{T}}{CH_3CH_2C{\equiv}CCH(OTHP)CH_2CH_2CH_2OH}$; $\underset{\textbf{U}}{CH_3CH_2C{\equiv}CCH(OTHP)CH_2CH_2CH{=}O}$

(g) [V (1-phenylcyclohex-2-en-1-ol, OH)] ⟶ W (1-phenylcyclohexa-1,3-diene)

Note that a tertiary alcohol is easily dehydratel, especially when the double bond that is formed is conjugated with an aromatic ring. If we wish to have the alcohol as the product, we use a weak acid, such as NH_4Cl, t protonate the alkoxide.

(i) $\underset{\textbf{X}}{BrCH_2CH(Br)CH_2CH_2}$— (cyclohexane spiro-dioxolane)

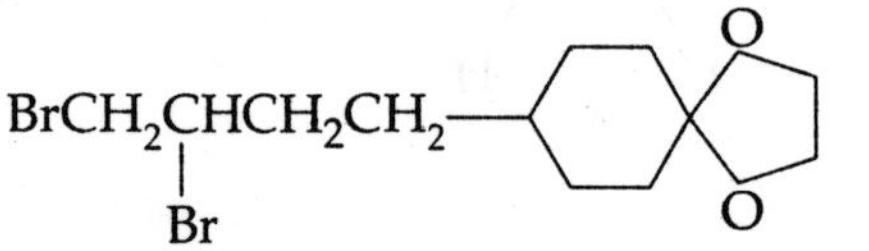

(j) $\underset{\textbf{AA}}{CH_3CH(CH_3)CH_2CH_2CH_2CH_2CH_2CO^-Na^+}$; $\underset{\textbf{BB}}{CH_3CH(CH_3)CH_2CH_2CH_2CH_2CH_2COH}$ (C=O)

(k) $CH_3CH_2CH_2CH_2CH_2$—(benzene ring)

(l) CC: (CH_2=, CH_3CHCH_3, $CHCH_3$ with O^-Li^+) ; (CH_2=, CH_3CHCH_3, $CHCH_3$ with OH)

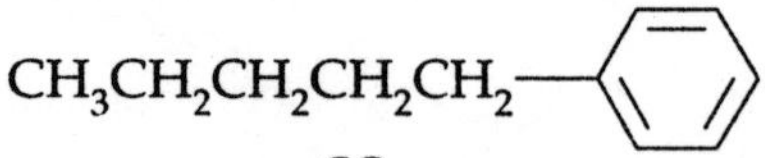

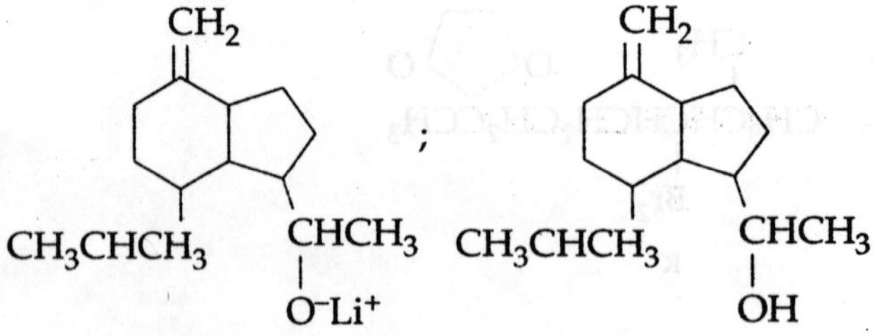

DD EE

54. (a) $CH_3C(=O)OC(CH_3)_2CH_3 + CH_3C(=O)OH$ (b) $CH_3CH_2CH_2C(=O)OCH_2CH_2CH_2CH_3$

(c) $C_6H_5C(=O)O{-}C_6H_{11}$ (d) $HOC(=O)CH{=}CHC(=O)OCH_2CH_3$ (cis; H, H)

55. (a) $C_6H_5{-}CH_2CH_2C(=O)Cl$ (b) $CH_3CH_2CH_2C(=O)Cl$

(c) $CH_3CH_2CH_2C(=O)OC(=O)CH_2CH_2CH_3$ (d) $CH_3(CH_2)_{10}C(=O)Cl$

(e) succinic anhydride (cyclic, O) (f) $CH_3CH_2C(=O)OC(=O)CH_2CH_3$ (g) $ClC(=O)(CH_2)_5C(=O)Cl$

56. (a) $CH_3CH(CH_3)CH_2CH_2CH{=}NOH$ (b) $CH_3CH_2CH_2C(CH_3){=}NNH{-}C_6H_5$

(c) $C_5H_8{=}N{-}C_6H_4{-}CH_3$ (d) $C_6H_5{-}CH{=}NNH{-}C_6H_3(NO_2)_2$

(e) $C_6H_5{-}C(NNH_2)CH_2CH_2CH_3$ (f) $CH_3C(CH_3){=}NCH_2CH(CH_3)CH_3$

(g) $C_6H_5{-}CH{=}NCH_2CH_2CH_3$

57. (a) cyclohexane (b) $C_6H_5{-}CH_2CH_3$

(c) $HO{-}C_6H_4{-}CH_3$ (d) $C_6H_5{-}CH_3$ (e) $C_5H_9{-}CH_2CH_2C(=O)OH$

58. (a) CH_3MgCl; PCC,

(b) $$C_6H_5NHCH_3 \xrightarrow{CH_3COCl} C_6H_5N\overset{\overset{O}{\|}}{C}CH_3 \xrightarrow{LiAlH_4} C_6H_5\underset{\underset{CH_3}{|}}{N}\ CH_2CH_3$$

(c) $$C_6H_5NH_2 \xrightarrow[HCl\ 273\ K]{NaNO_2} C_6H_5N_2Cl \xrightarrow{CuCN} \xrightarrow{LiAlH_4} C_6H_5CH_2NH_2$$

$$\xrightarrow{CH_3COCl} \xrightarrow{LiAlH_4} C_6H_5CH_2NHCH_2CH_3$$

(d) *X*: HNO_3/H_2SO_4 followed by reduction with NH_4/OH^-/Heat
Y: Reduction with N_2H_4/OH^-/Heat followed by nitration.

59. (a) $$(CH_3)_3CCH{=}CH_2 \xrightarrow[H_2O_2/NaOH]{B_2H_6} (CH_3)_3\ CCH_2CH_2OH \xrightarrow{H_2CrO_4} \xrightarrow{SOCl_2}$$

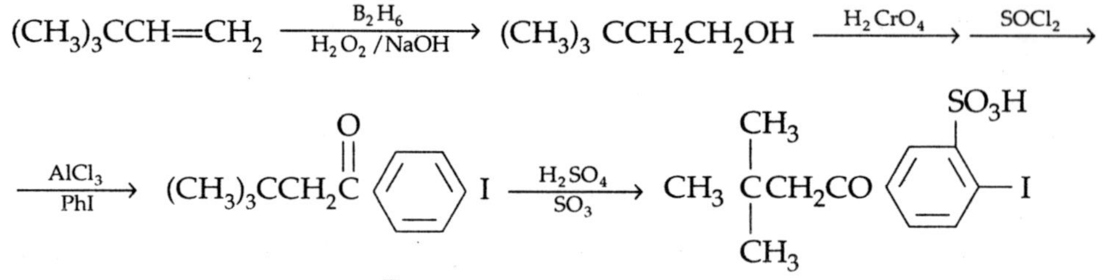

$$\xrightarrow[PhI]{AlCl_3} (CH_3)_3CCH_2\overset{\overset{O}{\|}}{C}-C_6H_4-I \xrightarrow[SO_3]{H_2SO_4} CH_3\underset{\underset{CH_3}{|}}{\overset{\overset{CH_3}{|}}{C}}CH_2CO-C_6H_3(SO_3H)-I$$

(b) $$PhCH_2CH_3 \xrightarrow[CCl_4]{NBS/hv} Ph\overset{\overset{Br}{|}}{C}HCH_3 \xrightarrow[C_2H_5OH]{C_2H_5ONa} PhCH{=}CH_2 \xrightarrow{mCPBA}$$

$$\text{Ph-epoxide} \xrightarrow{HBr} Ph\overset{\overset{Br}{|}}{C}HCH_2OH$$

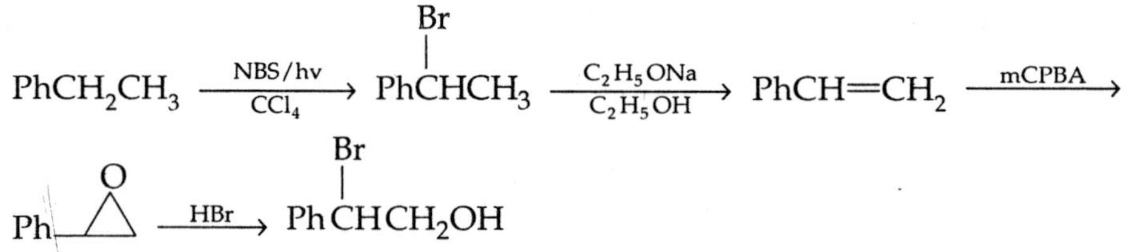

(c) $$CH_3CH_2CH_2CH_2Br \xrightarrow[C_2H_5OH]{C_2H_5ONa} CH_3CH_2CH{=}CH_2 \xrightarrow[H_2O]{Hg(OAc)_2} \xrightarrow{NaBH_4}$$

$$CH_3CH_2\underset{\underset{CH_2CH_3}{|}}{\overset{\overset{OH}{|}}{C}}CH_2 \xleftarrow[(ii)\ H_2O]{(i)\ C_2H_5MgBr} CH_3CH_2\overset{\overset{O}{\|}}{C}\ CH_3 \xleftarrow[CH_2Cl_2]{PCC} CH_3CH_2\overset{\overset{OH}{|}}{C}HCH_3$$

(d) Cl, CHO $\xrightarrow[C_2H_5OH]{C_2H_5ONa}$ CHO $\xrightarrow[CCl_4]{NBS/hv}$ Br, CHO, I $\xrightarrow[CCl_4]{NBS/hv}$

Br $\xrightarrow[Et_2O]{Mg}$ $\xrightarrow[H_3O^+]{I}$ Br, OH $\xrightarrow{KCN}$ NC, OH

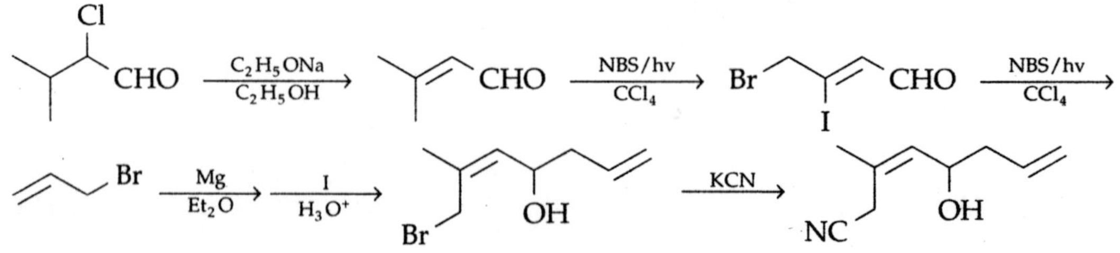

(e) 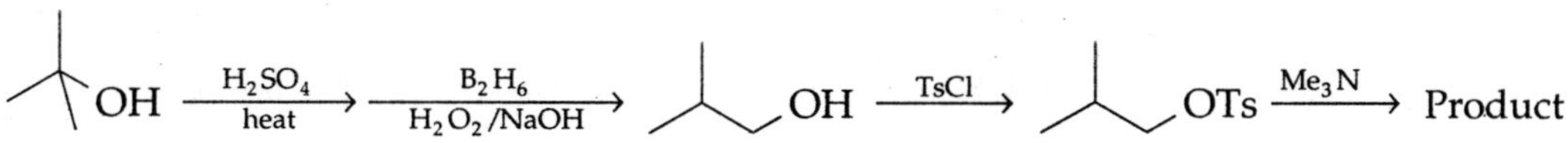 (CH$_3$)$_3$C–OH $\xrightarrow[\text{heat}]{H_2SO_4}$ $\xrightarrow[H_2O_2/NaOH]{B_2H_6}$ (CH$_3$)$_2$CHCH$_2$OH $\xrightarrow{TsCl}$ (CH$_3$)$_2$CHCH$_2$OTs $\xrightarrow{Me_3N}$ Product

60. (a) 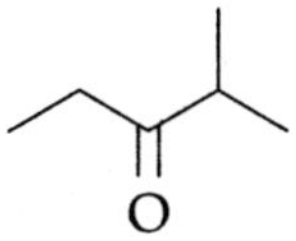 O (b) 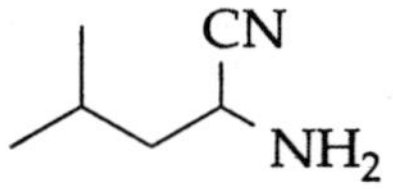 CN, NH$_2$

(c) 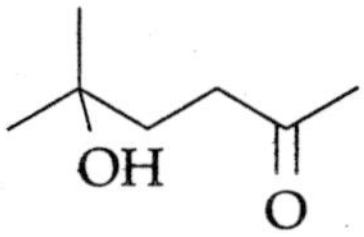 OH, O (d) $PhCH_2CH{=}C(CHO)Ph$

(e) 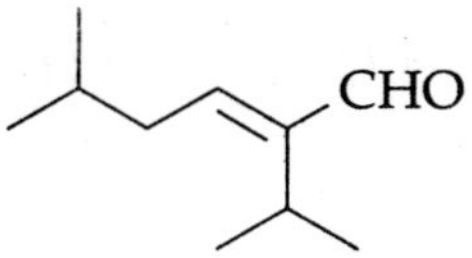 CHO (f) 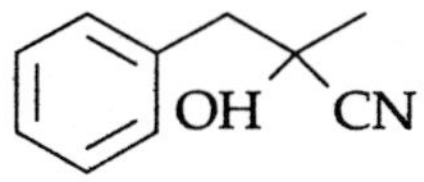 OH, CN (g) 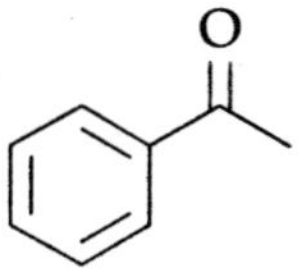 O

(h) $PhCONHCH_3$ (i) 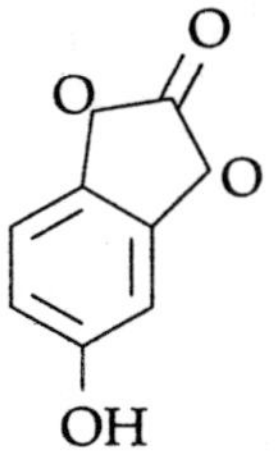 O, O, O, OH (j) 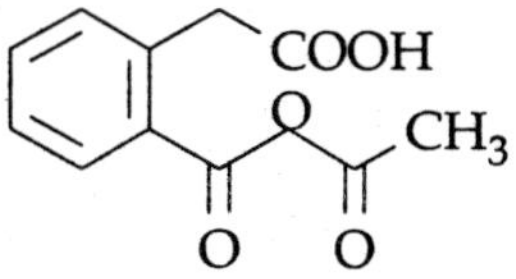 COOH, O, CH$_3$, O, O

(k) $CH_3OH + Na_2CO_3$ (l) 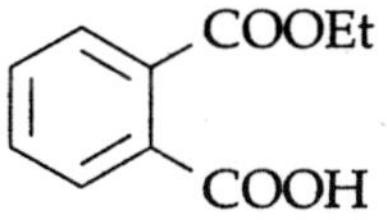 COOEt, COOH (m) $Ph{-}CH(COPh){-}COOEt$

(n) $CH_3CH_2CH(COOEt)_2$ (o) (i)  $H_3C{-}C_6H_4{-}COOH$

(ii) 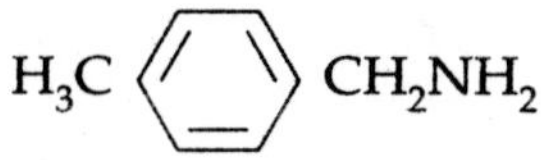 $H_3C{-}C_6H_4{-}CH_2NH_2$ (iii) 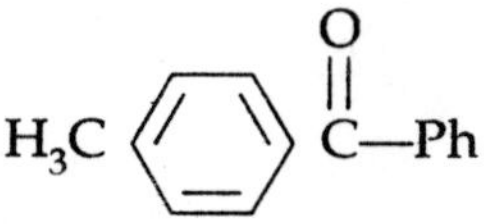 $H_3C{-}C_6H_4{-}C(=O){-}Ph$

61. (a) 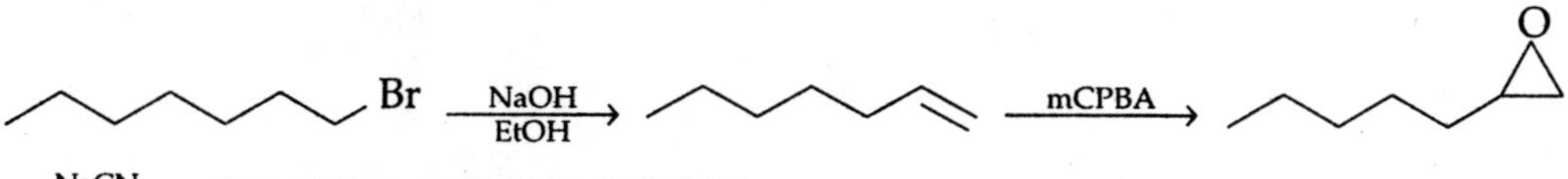 Br $\xrightarrow[EtOH]{NaOH}$ $\xrightarrow{mCPBA}$ O

$\xrightarrow[H_2O]{NaCN}$ $CH_3(CH_3)_4CH(OH)\ CH_2CN$

(b) $PhCH_2CH_2Br \xrightarrow[C_2H_5OH]{C_2H_5ONa} PhCH_2CH{=}CH_2 \xrightarrow[H_2O]{Hg(OAc)_2} \xrightarrow{NaBH_4}$

$PhCH_2CH(OH)CH_3 \xrightarrow[CH_2Cl_2]{PCC} PhCH_2C(=O)CH_3$

(I)

$$CH_3C{\equiv}CH \xrightarrow[\text{heat}]{Na} CH_3C{\equiv}C^-Na^+ \xrightarrow{(I)} \xrightarrow{H_3O^+} CH_3C{\equiv}C{-}\overset{OH}{\underset{CH_2Ph}{C}}{-}CH_3$$

$$\xrightarrow[CaCO_3/PbO]{Pd} \underset{H}{\overset{H_3C}{}}C{=}C\overset{C(OH)-CH_2Ph}{\underset{H}{}}\ CH_3$$

(c) $$CH_3CH_2CH_2CH_2OH \xrightarrow{SOCl_2} \xrightarrow[C_2H_5OH]{C_2H_5ONa} \xrightarrow{mCPBA} CH_2CH_2CH\overset{O}{—}CH_2$$

$$\xrightarrow{NaCN} \xrightarrow{H_2O} CH_2CH_2\overset{OH}{C}HCH_2CN \xrightarrow[CH_2Cl_2]{PCC} CH_3CH_2COCH_2CN$$

(d) $$\text{methylenecyclopentane} \xrightarrow[H_2O_2/NaOH]{B_2H_6} C_5H_9CH_2OH \xrightarrow[CH_2Cl_2]{PCC} C_5H_9\overset{O}{\overset{\|}{C}}H \ \text{(I)}$$

$$C_6H_6 \xrightarrow[AlCl_3]{Cl_2} C_6H_5Cl \xrightarrow[\text{ether}]{Mg} \xrightarrow{(I)} \xrightarrow{H_3O^+} C_5H_9\overset{OH}{C}H{-}C_6H_5$$

$$\xrightarrow[AlCl_3]{Cl_2} C_5H_9{-}\overset{OH}{C}H_2{-}C_6H_4{-}Cl \xrightarrow[CH_2CH_2]{PCC} C_5H_9{-}CO{-}C_6H_4{-}Cl$$

(e) $$\text{(CH}_3)_2CHCH_2CHClCH_3 \xrightarrow[(CH_3)_3COH]{(CH_3)_3COK} (CH_3)_2CHCH_2CH{=}CH_2 \xrightarrow[CCl_4]{NBS/hv} (CH_3)_2CHCHBrCH{=}CH_2 \xrightarrow[\Delta]{KOH}$$

$$(CH_3)_2C{=}CHCH{=}CH_2 \xrightarrow[CCl_4]{NBS/hv} Br{-}CH_2C(CH_3){=}CHCH{=}CH_2 \xrightarrow[H_2O]{NaOH} HO{-}CH_2C(CH_3){=}CHCH{=}CH_2$$

62. (a) $$(CH_3)_3C\overset{O}{\overset{\|}{C}}CH_3 \xrightarrow{OH^-} \xrightarrow[\text{Aldol}]{C_6H_5CHO} (CH_3)_3C\overset{O}{\overset{\|}{C}}CH{=}CHC_6H_5$$

(b) $$\text{cyclopentanone} \xrightarrow{OH^-} \xrightarrow[\text{Aldol}]{C_6H_5CHO} \text{2-}(C_6H_5CH{=})\text{cyclopentanone}$$

(c) $CH_3CH_2NO_2 \xrightarrow{OH^-} CH_3\overset{-}{C}HNO_2 \xrightarrow[H_2O]{C_6H_5CHO} C_6H_5CH{=}C(NO_2)CH_3$

$\xrightarrow{LiAlH_4} C_6H_5CH{=}C(NH_2)CH_3 \xrightarrow{BH_3} \xrightarrow{CH_3COOH} C_6H_5CH_2CH(NH_2)CH_3$

(d) 2,7-Octanedione $\xrightarrow[Aldol]{OH^-}$ 1-acetyl-2-methylcyclopentene $\xrightarrow{NaBH_4}$ Product

63. (a) 1-(bromomethyl)cyclohexanol (HO, Br) $\xrightarrow{NaOH}$ epoxide (O) $\xrightarrow[H_3O^-]{NaN_2}$ HO, N_3 $\xrightarrow{NaH} \xrightarrow{CH_3I}$ CH_3O, N_3

(b) Benzyl bromide (Br) $\xrightarrow{NaOH} \xrightarrow[H_2SO_4]{K_2Cr_2O_7}$ C_6H_5COOH $\xrightarrow[EtOH]{Na/NH_3} \xrightarrow{H_3O^+}$

$C_6H_5COOH \xrightarrow{CH_2N_2} C_6H_5COOCH_3$

(c) $C_6H_5C(CH_3)_2CH_2OH \xrightarrow[H_2SO_4]{K_2Cr_2O_7} \xrightarrow{SOCl_2} C_6H_5C(CH_3)_2COCl \xrightarrow{NH_3}$

$\xrightarrow{POCl_3} C_6H_5C(CH_3)_2CN$

(d) (i) 3-methylcyclohexanol (OH, CH_3) $\xrightarrow{TsCl}$ (OTs, CH_3) $\xrightarrow[S_N2]{NaCl}$ (Cl, CH_3)

(ii) $\xrightarrow[S_N2]{NaCl}$ (I, CH_3) $\xrightarrow[Et_2O,\ S_N2]{NaCl}$ (Cl, CH_3)

(e) [bicyclic structure with two O atoms]

64. (a) Nitrobenzene ($C_6H_5NO_2$) $\xrightarrow{Zn/HCl}$ $\xrightarrow{CH_3COCl}$ $C_6H_5NHCOCH_3$ $\xrightarrow[AlCl_3]{CH_3Cl}$ p-$CH_3C_6H_4NHCOCH_3$ $\xrightarrow{H_3O^+}$ $\xrightarrow[NaOH/Heat]{KMnO_4}$ p-$NO_2C_6H_4COOH$

(b) C_6H_5CHO $\xrightarrow[OH^-/Heat]{N_2H_4}$ $\xrightarrow[AlCl_3]{Cl_2}$ H_3C–C_6H_4–Cl $\xrightarrow[Ac_2O]{CrO_2Cl_2}$ Product

(c) m-$C_6H_4(NO_2)_2$ $\xrightarrow{Na_2S}$ $\xrightarrow[HCl/273\,K]{NaNO_2}$ $\xrightarrow{CuCl}$ Product

(d) o-Cl–C_6H_4–NO_2 $\xrightarrow[\Delta]{KCN}$ o-CN–C_6H_4–NO_2 $\xrightarrow{H_3O^+}$ $\xrightarrow{Zn\text{-}HCl}$ $\xrightarrow[HCl/273\,K]{NaNO_2}$ $\xrightarrow[aq]{KCN}$ Product

(e) H_2N–C_6H_4–CH_3 $\xrightarrow{CF_3CO_3H}$ O_2N–C_6H_4–CH_3 $\xrightarrow[FeBr_2]{Br_2}$

O_2N–$C_6H_3(Br)$–CH_3 $\xrightarrow{Zn\text{-}HCl}$ H_2N–$C_6H_3(Br)$–CH_3

(f) 4-amino-3-bromotoluene (CH_3, Br, NH_2) $\xrightarrow[HCl/273\,K]{NaNO_2}$ $\xrightarrow{H_3PO_2}$ 3-bromotoluene (CH_3, Br)

(g) o-$NO_2C_6H_4CHO$ $\xrightarrow[HOCH_2CH_2OH]{H^+}$ o-$NO_2C_6H_4$–$CH(OCH_2CH_2O)$ $\xrightarrow{Zn\text{-}HCl}$ $\xrightarrow[HCl/273\,K]{NaNO_2}$

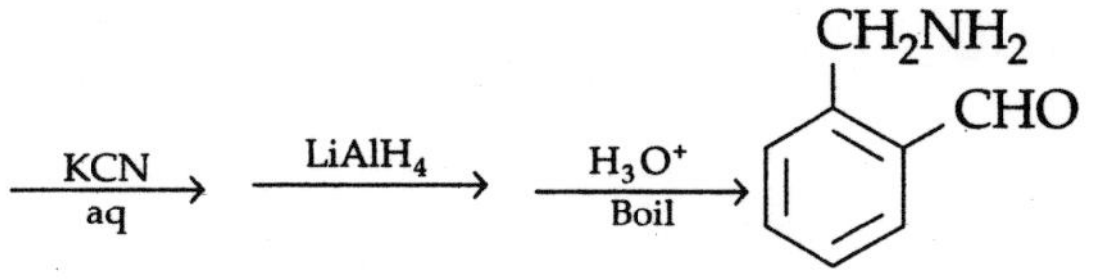

$\xrightarrow[aq]{KCN} \xrightarrow{LiAlH_4} \xrightarrow[Boil]{H_3O^+}$ (ring: CH_2NH_2, CHO)

(h) (ring: CH_3) $\xrightarrow[H_2SO_4]{HNO_3}$ (ring: CH_3, NO_2) $\xrightarrow{Zn\text{-}HCl} \xrightarrow{CH_3COCl}$ (ring: CH_3, $NHCOCH_3$) $\xrightarrow[H_2SO_4]{HNO_3}$

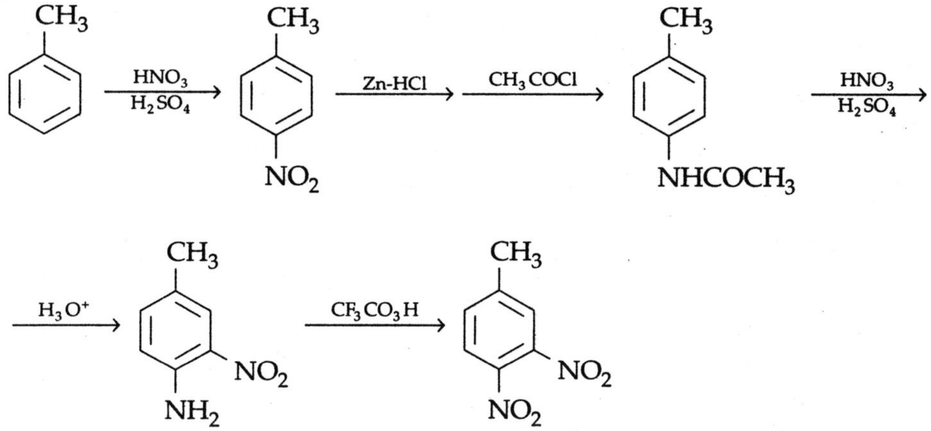

$\xrightarrow{H_3O^+}$ (ring: CH_3, NO_2, NH_2) $\xrightarrow{CF_3CO_3H}$ (ring: CH_3, NO_2, NO_2)

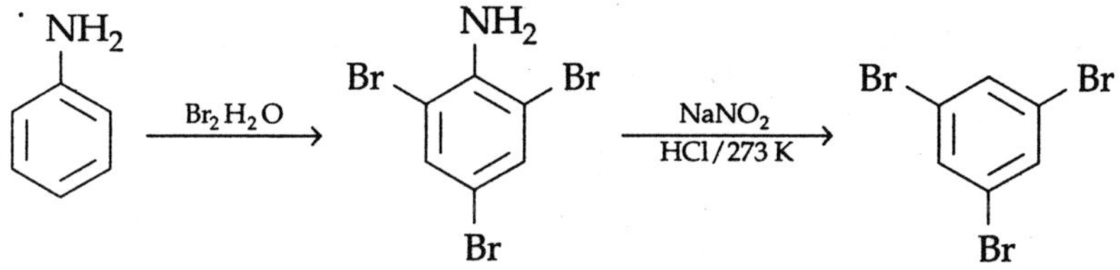

(i) (ring: NH_2) $\xrightarrow{Br_2\,H_2O}$ (ring: NH_2, Br, Br, Br) $\xrightarrow[HCl/273\,K]{NaNO_2}$ (ring: Br, Br, Br)

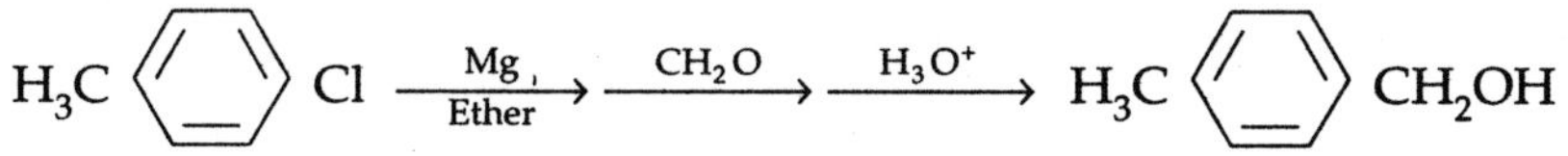

65. (a) H_3C (ring) Cl $\xrightarrow[Ether]{Mg,} \xrightarrow{CH_2O} \xrightarrow{H_3O^+}$ H_3C (ring) CH_2OH

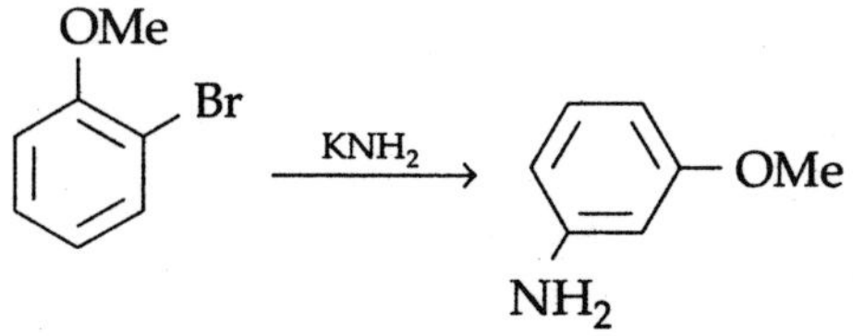

(b) (ring: OMe, Br) $\xrightarrow{KNH_2}$ (ring: OMe, NH_2)

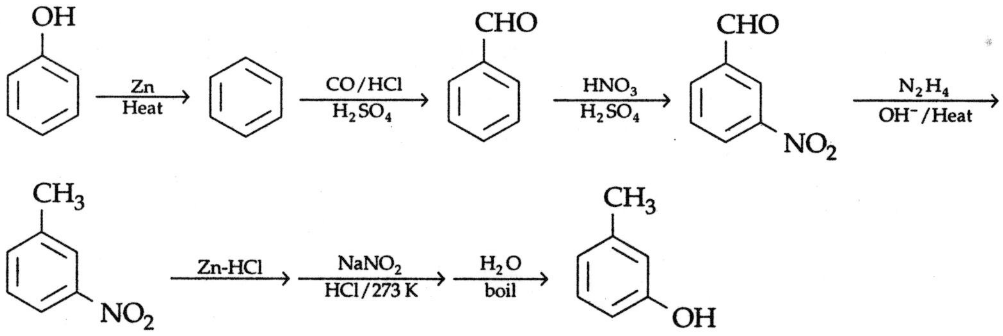

(c) (ring: OH) $\xrightarrow[Heat]{Zn}$ (ring) $\xrightarrow[H_2SO_4]{CO/HCl}$ (ring: CHO) $\xrightarrow[H_2SO_4]{HNO_3}$ (ring: CHO, NO_2) $\xrightarrow[OH^-/Heat]{N_2H_4}$

(ring: CH_3, NO_2) $\xrightarrow{Zn\text{-}HCl} \xrightarrow[HCl/273\,K]{NaNO_2} \xrightarrow[boil]{H_2O}$ (ring: CH_3, OH)

(d) Benzene + CO + HCl $\xrightarrow{AlCl_3}$ C₆H₅CHO (benzaldehyde) $\xrightarrow[AlCl_3]{Cl_2}$ m-chlorobenzaldehyde (CHO, Cl) $\xrightarrow[Heat]{N_2H_4/OH^-}$ m-chlorotoluene (CH_3, Cl)

(e) Benzene $\xrightarrow[H_2SO_4]{HNO_3}$ nitrobenzene (NO_2) $\xrightarrow{Zn\text{-}HCl}$ aniline (NH_2) $\xrightarrow{CH_3COCl}$ acetanilide ($NHCOCH_3$) $\xrightarrow{Conc\ H_2SO_4}$ $\xrightarrow[H_2SO_4]{HNO_3}$ $\xrightarrow[Warm]{H_3O^+}$ o-nitroaniline (NH_2, NO_2)

(f) H_3C–C₆H₄–NO_2 $\xrightarrow{Zn\text{-}HCl}$ $\xrightarrow{CH_3COCl}$ $\xrightarrow[H_2SO_4]{HNO_3}$ H_3C–C₆H₃(NO_2)–$NHCOCH_3$

$\xrightarrow{H_3O^+}$ $\xrightarrow[HCl/273\ K]{NaNO_2}$ $\xrightarrow{H_3PO_2}$ H_3C–C₆H₄–NO_2

66. (a) Benzene $\xrightarrow[H_2SO_4]{HNO_3}$ $\xrightarrow{Zn\text{-}HCl}$ $\xrightarrow{CH_3COCl}$ C₆H₅NHOAc $\xrightarrow[Blocking]{Conc.\ H_2SO_4}$ $\xrightarrow[H_2SO_4]{HNO_3}$ (NHOAc, NO_2, SO_3H)

$\xrightarrow[Reflux]{H_3O^+}$ o-nitroaniline (NH_2, NO_2) $\xrightarrow[HCl/273\ K]{NaNO_2}$ ($N_2^+Cl^-$, NO_2) $\xrightarrow{PhLi}$ LiCl + N_2 + (Ph, NO_2)

(b) PhBr $\xrightarrow[ether]{Mg}$ $\xrightarrow{CH_2O}$ $\xrightarrow{H_3O^+}$ $PhCH_2OH$ $\xrightarrow{PCC}$ Product

(c) $PhCH_3$ $\xrightarrow[\substack{Ac_2O \\ \text{Etard's reaction}}]{CrO_2Cl_2}$ Product.

(d) Phenol (OH) + $CHCl_3$ $\xrightarrow[\substack{\text{Reimer-Tieman} \\ \text{reaction}}]{NaOH}$ $\xrightarrow{H_3O^+}$ product

(e) Phenol (OH) + $CHCl_3$ $\xrightarrow{\text{Conc } H_2SO_4}$ p-hydroxybenzenesulphonic acid (OH, SO_3H) $\xrightarrow[CH_3COCl]{AlCl_3\text{(excess)}}$ $\xrightarrow[\text{Reflux}]{H_3O^+}$ product

(f) Nitrobenzene (NO_2) $\xrightarrow{\text{Zn-HCl}}$ $\xrightarrow[\text{aqueous}]{Br_2}$ (NO_2, Br, Br, Br) $\xrightarrow[HCl/273\ K]{NaNO_2}$ $\xrightarrow{H_3PO_2}$ (Br, Br, Br)

67. (a) Benzene $\xrightarrow[AlCl_3]{PhCH_2COCl}$ (C(=O), Ph) $\xrightarrow[AlCl_3]{Cl_2}$ (C(=O), Ph, Cl) $\xrightarrow[OH^-/\text{Heat}]{N_2H_4}$ (Ph, Cl)

(b) Benzaldehyde (CHO) $\xrightarrow{CH_3MgBr}$ $\xrightarrow{H_3O^+}$ (OH, CH, CH_3) $\xrightarrow[AlCl_3]{CH_3Cl}$ (OH, $CH-CH_3$) $\xrightarrow{PCC}$ Product

(c) Isopropylbenzene $\xrightarrow[\text{Heat}]{H_2SO_4}$ (SO_3H) $\xrightarrow[NaOH]{KMnO_4}$ $\xrightarrow{H_3O^+}$ (COOH, HO_3S)

(d) Benzene + $CH_3CH_2CH_2Cl$ $\xrightarrow{AlCl_3}$ Isopropylbenzene $\xrightarrow[H_2SO_4]{HNO_3}$ (NO_2)

(e) $H_3C-C_6H_5 \xrightarrow[H_2SO_4]{HNO_3} \xrightarrow{Zn\text{-}HCl} \xrightarrow[HCl/273\,K]{NaNO_2} \xrightarrow{CuCN} H_3C-C_6H_4-CN$

(f) $C_6H_6 + CH_3CH_2CH_2COCl \xrightarrow{AlCl_3} PhCOCH_2CH_2CH_3 \xrightarrow{NaBH_4} PhCH(OH)CH_2CH_2CH_3$

68. (a) $PhCOCH_2CH_3 \xrightarrow[heat]{Cl_2} PhCOCH(Cl)CH_3 \xrightarrow[C_2H_5OH]{C_2H_5ONa}$

$PhCOCH{=}H_2 \xrightarrow{HBr} PhCOCH_2CH_2Br \xrightarrow[\underset{OH\quad OH}{CH_2-CH_2}]{H^+}$

$\text{Product} \xleftarrow[H_2O]{Ag_2O} PhCOCH_2CH_2CH_2CH_2OH\,Ph-\underset{}{\overset{O-O}{C}}-CH_2CH_2Br$

(b) $PhCOOH \xrightarrow{SOCl_2} PhCOCl \xrightarrow{PhC\equiv CNa} PhCOC\equiv C-Ph \xrightarrow[OH^-/Heat]{N_2H_4}$

$\text{Product} \xleftarrow[BaSO_4/H_2]{Pd} PhCH_2C\equiv CPh$

(c) $\xrightarrow{C_2H_5ONa} \xrightarrow{PhCH_2CH_2CH_2Br} Ph(CH_2)_3CH(COOCH_3)_2$

69. (a) $CH_3CH_2C\equiv CH \xrightarrow[heat]{Na} CH_3CH_2C\equiv C^-Na^+ \xrightarrow[H_2O]{\text{ethylene oxide}} CH_3CH_2C\equiv CCH_2CH_2OH \xrightarrow{Na\,|\,liq.\ NH_3} \xrightarrow{H_2O} \text{Product}$

(b) $HCOCH_2CH_2COCH_3 \xrightarrow[H^+]{HOCH_2CH_2OH} (\text{dioxolane})CH\,CH_2CH_2COCH_3 \xrightarrow{CH_3CH_2CH_2CH{=}PPh_3}$

$(\text{dioxolane})CH\,CH_2CH_2\underset{CH_3}{C}{=}CH(CH_2)_2CH_3 \xrightarrow{H_3O^+} HCOCH_2CH_2C(CH_3){=}CHCH_2CH_2CH_3 \xrightarrow[OH^-/Heat]{N_2H_4} \xrightarrow{H_2/Ni} \text{Product}$

(c) $CH_3COOH \xrightarrow[I_2]{P_4} ICH_2COOH \xrightarrow[NaCN]{aqueous} \xrightarrow[H_2O]{H_3O^+} HOOC—CH_2—COOH$

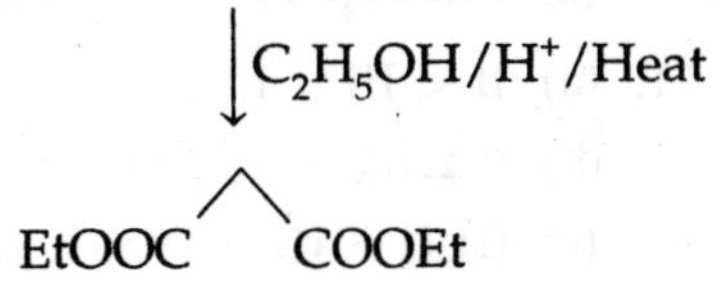

$\downarrow C_2H_5OH/H^+/Heat$

EtOOC‸COOEt

(d) 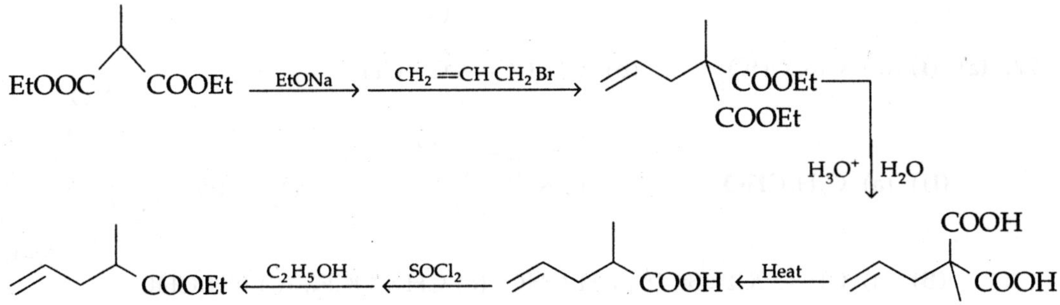 EtOOC‸COOEt (with methyl) $\xrightarrow{EtONa} \xrightarrow{CH_2{=}CH\,CH_2\,Br}$ $CH_2{=}CHCH_2C(CH_3)(COOEt)COOEt$

$\downarrow H_3O^+ \mid H_2O$

$CH_2{=}CHCH_2CH(CH_3)COOEt \xleftarrow{C_2H_5OH} \xleftarrow{SOCl_2} CH_2{=}CHCH_2CH(CH_3)COOH \xleftarrow{Heat} CH_2{=}CHCH_2C(CH_3)(COOH)COOH$

(e) $PhCOOH \xrightarrow{SOCl_2} PhCOCl \xrightarrow[Excess]{CH_3MgCl} \xrightarrow{H_3O^+} PhC(CH_3)_2CH_3$

(f) $PhCH_2OH \xrightarrow{SOCl_2} PhCH_2Cl \xrightarrow{NaCN} PhCH_2CN$

70. (a) 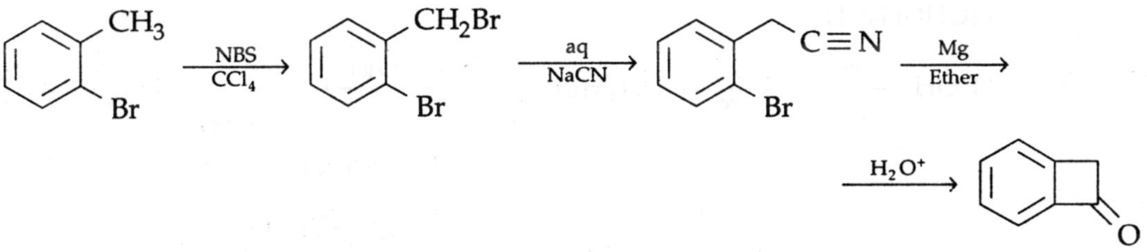 o-bromotoluene (CH_3, Br) $\xrightarrow[CCl_4]{NBS}$ o-bromobenzyl bromide (CH_2Br, Br) $\xrightarrow[NaCN]{aq}$ o-bromobenzyl cyanide ($CH_2C{\equiv}N$, Br) $\xrightarrow[Ether]{Mg}$

$\xrightarrow{H_2O^+}$ benzocyclobutenone (C=O)

(b) $PhC(CH_3)_2OMe \xrightarrow{H_3O^+} PhC(OH)_2CH_3 \xrightarrow{-H_2O} PhC(=O)CH_4$

(c) $Ph\overset{O}{\overset{||}{C}}OMe$ + MeMgBr(excess) $\xrightarrow{H_3O^+} PhC(OH)(CH_3)CH_3$

(d) $Ph\overset{O}{\overset{||}{C}}OMe \xrightarrow{H_3O^+} Ph\overset{O}{\overset{||}{C}}OH \xrightarrow{LiAlH_4} PhCH_2OH \xrightarrow{PCC} PhCHO$

(e) $PhCH_2OH \xrightarrow{SOCl_2} \xrightarrow[\text{Ether}]{Mg} PhCH_2MgCl \xrightarrow{\text{(epoxide)}} PhCH_2CH_2CH_2OmgCl \xrightarrow{H_2O}$ Product

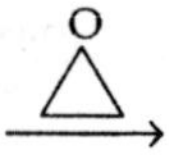

71. (a) II < III < I.

(b) $ClCH_2-$ >H > $(CH_3)_2C$ –> Ph

(c) (i) II > III > IV> 1, (ii) I > II > III > IV, (iii) III > II > I .

72. (a) (I) (a) Cl_3CCHO (b) $CH_3CCl_2—\overset{O}{\overset{||}{C}}—CH_3$

(II) (a) C_6H_5CHO (b) 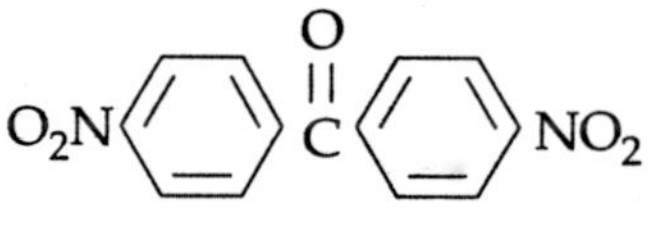 O_2N–C_6H_4–CO–C_6H_4–NO_2 (c) (NO_2)C_6H_4–CO–C_6H_4(NO_2)

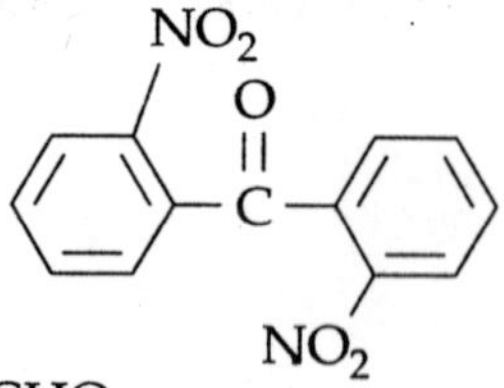

(B) (a) CH_3COCH_2Br, (b) CH_3CHO, (c) CH_3COCF_3, (d) CH_3CHO.

73. (a) $CH_2{=}CHCH_3 \xrightarrow[CCl_4]{NBS/hv} CH_2{=}CH\ CH_2Br \xrightarrow[THF]{Mg} CH_2{=}CHCH_2MgBr$

$\xrightarrow{\text{Proponal}} \xrightarrow{H_3O^+} CH_2{=}CHCH_2\overset{OH}{\overset{|}{C}}HCH_2CH_3 \xrightarrow{NaH} \xrightarrow{CH_3I} CH_2CH$

$\overset{OCH_3}{\overset{|}{CHC}}HCH_2CH_3$

(b) 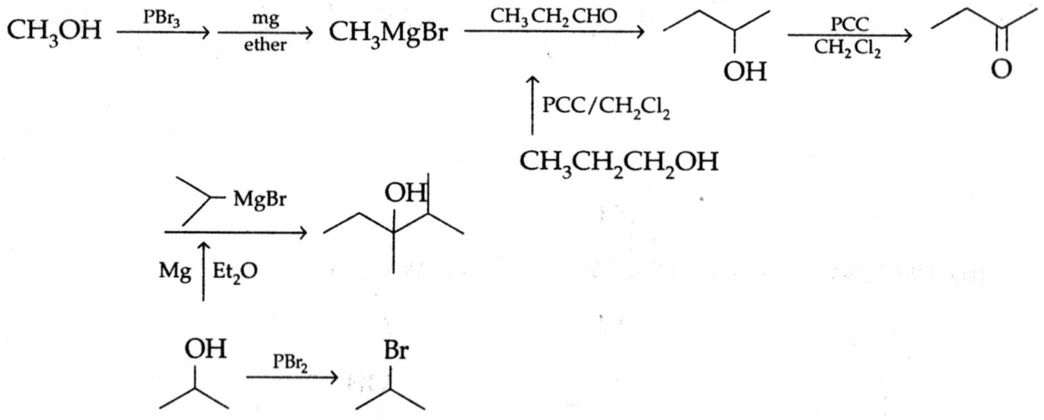 $CH_3OH \xrightarrow{PBr_3} \xrightarrow[\text{ether}]{mg} CH_3MgBr \xrightarrow{CH_3CH_2CHO}$ (butan-2-ol, OH) $\xrightarrow[CH_2Cl_2]{PCC}$ (butan-2-one, O)

↑ PCC/CH_2Cl_2

$CH_3CH_2CH_2OH$

$\xrightarrow{\text{(isopropyl)MgBr}}$ (OH product)

↑ Mg | Et_2O

(isopropanol, OH) $\xrightarrow{PBr_2}$ (isopropyl bromide, Br)

(c) (i) 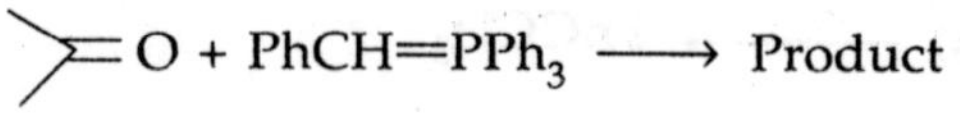 (acetone)=O + $PhCH{=}PPh_3 \longrightarrow$ Product

(ii) 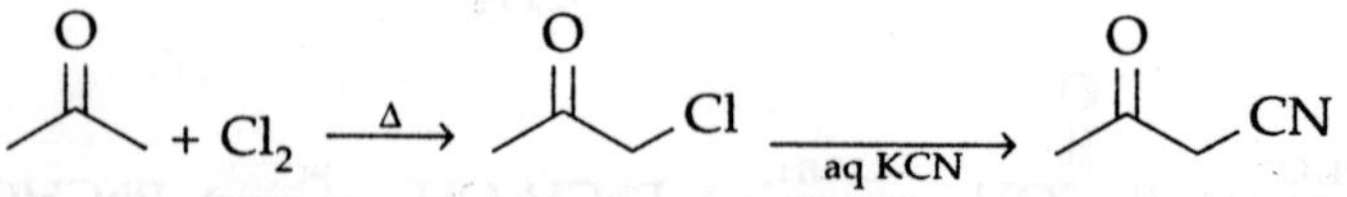 (acetone, O) + $Cl_2 \xrightarrow{\Delta}$ (chloroacetone, O, Cl) $\xrightarrow{\text{aq KCN}}$ (O, CN)

(iii) 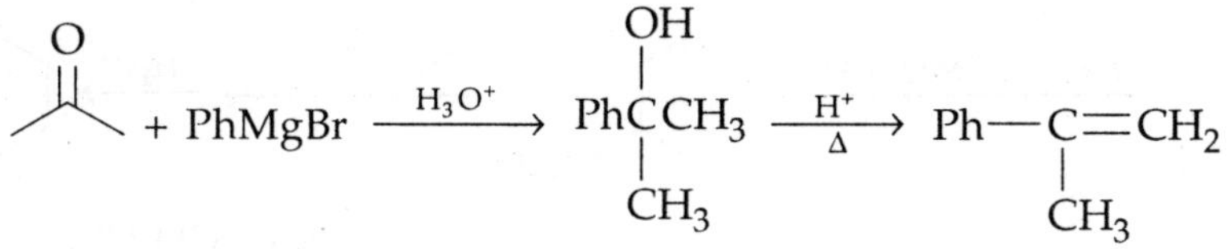 CH_3COCH_3 + PhMgBr $\xrightarrow{H_3O^+}$ $PhC(OH)(CH_3)CH_3$ $\xrightarrow[\Delta]{H^+}$ $Ph-C(CH_3)=CH_2$

(iv) 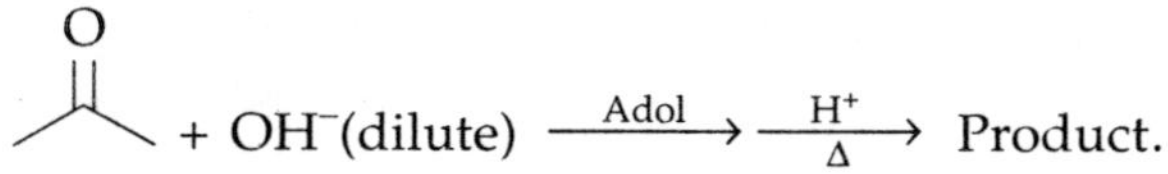 CH_3COCH_3 + OH^-(dilute) $\xrightarrow{\text{Adol}}$ $\xrightarrow[\Delta]{H^+}$ Product.

74. (a) $C_6H_5CH_2CH_3$; (b) $C_6H_5CH(Br)CH_3$; (c) $C_6H_5CH{=}CH_2$;

(d) C_6H_5CHO; (e) $C_6H_5CH(OH)CH_3$; (f) $C_6H_5COCH_3$;

75. (a) $CH_3CH_2C(Br){=}CH_2 \xrightarrow[NH_3]{NaNH_2} CH_3H_2C{\equiv}CNa \xrightarrow{CH_3-CO-CH_3} \xrightarrow[H_2O]{H_3O^+}$

$CH_3CH(OH)CH_3 \xrightarrow{KMnO_4}$ (to CH_3COCH_3)

$CH_3CH_2C{\equiv}C\,C(OH)(CH_3)CH_3 \xrightarrow{PBr_3} \xrightarrow[EtOH]{KOH}$ Product

(b) 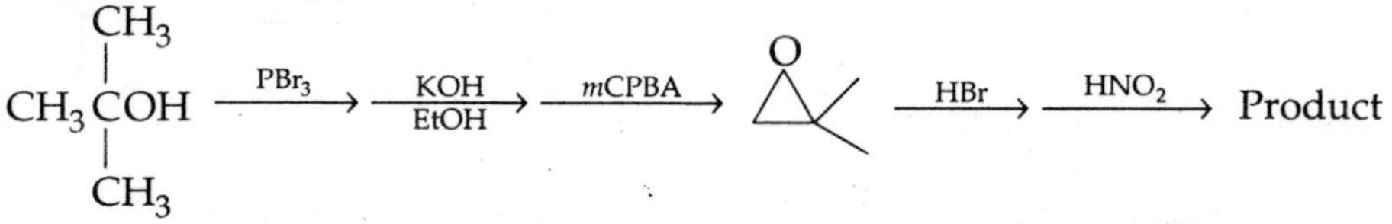 $CH_3C(OH)(CH_3)CH_3 \xrightarrow{PBr_3} \xrightarrow[EtOH]{KOH} \xrightarrow{mCPBA}$ (2,2-dimethyloxirane) $\xrightarrow{HBr} \xrightarrow{HNO_2}$ Product

(c) $CH_3CH_2CH(CH_3)CH{=}CH_2 \xrightarrow[H_2O]{Hg(OAc)_2} \xrightarrow{NaBH_4} \xrightarrow{H_2CrO_4} CH_3CH_2CH_2C(CH_3)HC(=O)CH_3$

$CH_2{=}CH_2 \xrightarrow{Br_2} CH_2(Br)-CH_2(Br) \xrightarrow[\text{(ii) } H_3O^+ \text{ (iii) } Mg/Et_2O]{\text{(i) } NaNH_2/NH_3}$

(d) 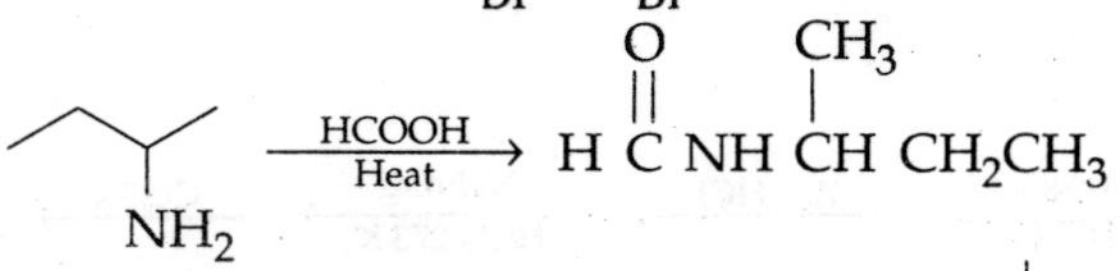 $CH_3CH(NH_2)CH_2CH_3 \xrightarrow[\text{Heat}]{HCOOH} H\overset{O}{\overset{\|}{C}}\,NH\,CH(CH_3)\,CH_2CH_3$

(e) 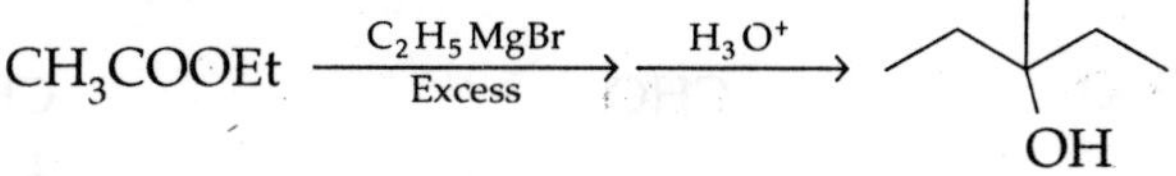 $CH_3COOEt \xrightarrow[\text{Excess}]{C_2H_5MgBr} \xrightarrow{H_3O^+}$ $CH_3C(OH)(C_2H_5)_2$

76. (a) 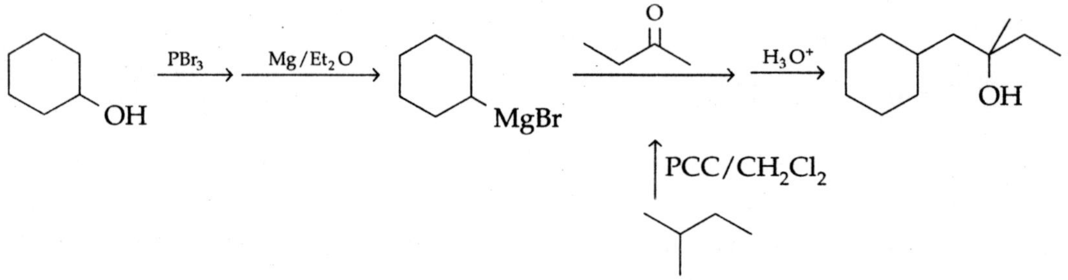 Cyclohexanol $\xrightarrow{PBr_3}$ $\xrightarrow{Mg/Et_2O}$ cyclohexyl-MgBr $\xrightarrow{\text{2-butanone}}$ $\xrightarrow{H_3O^+}$ product (OH)

2-butanone obtained from 2-butanol: $\uparrow$ PCC/CH_2Cl_2

(b) 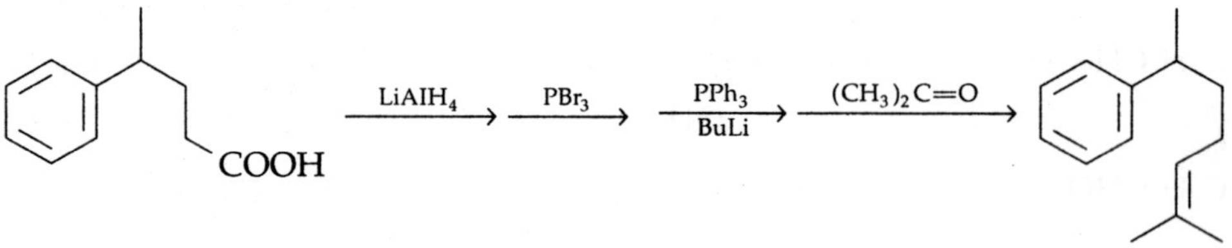 Ph–CH(CH_3)–CH_2CH_2COOH $\xrightarrow{LiAlH_4}$ $\xrightarrow{PBr_3}$ $\xrightarrow[BuLi]{PPh_3}$ $\xrightarrow{(CH_3)_2C{=}O}$ Ph–CH(CH_3)–$CH_2CH_2CH{=}C(CH_3)_2$

(c) $C_2H_5Br \xrightarrow[NaCN]{aq} C_2H_5CN$

$C_2H_5Br \xrightarrow[ether]{Mg} C_2H_5MgBr$

$\longrightarrow C_2H_5C(=NMgBr)C_2H_5 \xrightarrow{H_3O^+}$ 3-pentanone.

(d) 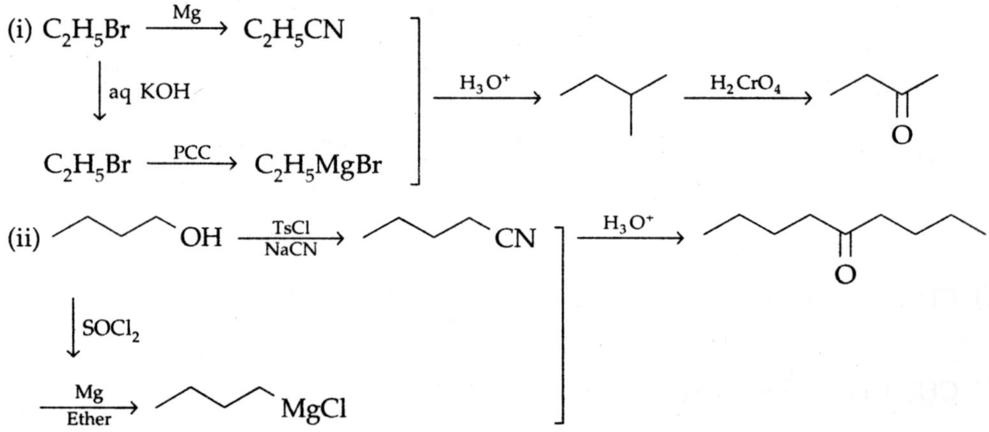 (i) $C_2H_5Br \xrightarrow{Mg} C_2H_5CN$

$\downarrow$ aq KOH

$C_2H_5Br \xrightarrow{PCC} C_2H_5MgBr$

$\xrightarrow{H_3O^+}$ 2-butanol $\xrightarrow{H_2CrO_4}$ 2-butanone

(ii) 1-Butanol (OH) $\xrightarrow[NaCN]{TsCl}$ butyl CN $\xrightarrow{H_3O^+}$ 5-nonanone (O)

$\downarrow SOCl_2$

$\xrightarrow[Ether]{Mg}$ butyl MgCl

77. (a) 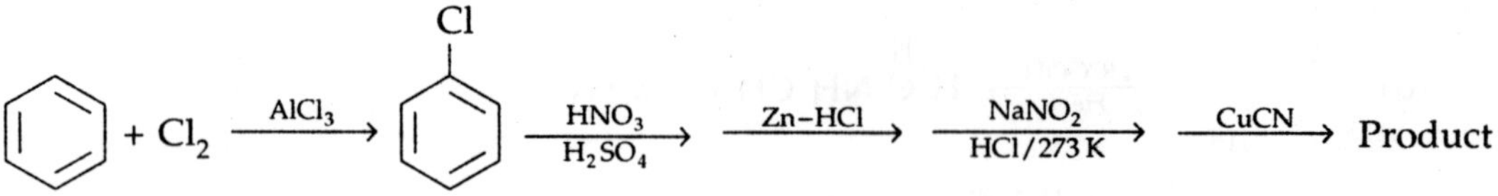 Benzene $+ Cl_2 \xrightarrow{AlCl_3}$ chlorobenzene (Cl) $\xrightarrow[H_2SO_4]{HNO_3}$ $\xrightarrow{Zn-HCl}$ $\xrightarrow[HCl/273\ K]{NaNO_2}$ $\xrightarrow{CuCN}$ Product

(b) 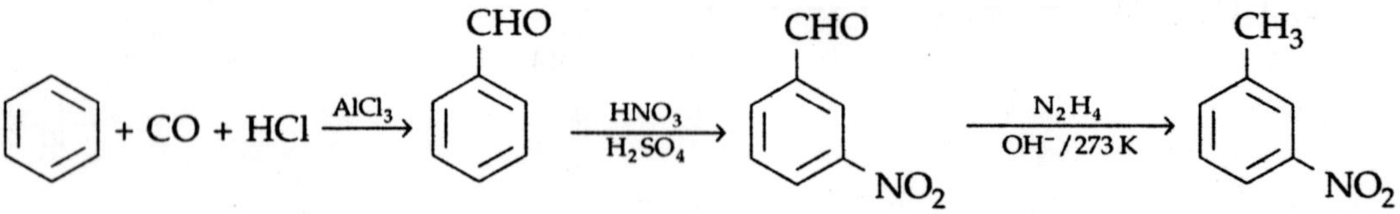 Benzene $+ CO + HCl \xrightarrow{AlCl_3}$ benzaldehyde (CHO) $\xrightarrow[H_2SO_4]{HNO_3}$ m-nitrobenzaldehyde (CHO, NO_2) $\xrightarrow[OH^-/273\ K]{N_2H_4}$ m-nitrotoluene (CH_3, NO_2)

(c) Benzene $\xrightarrow[H_2SO_4]{HNO_3}$ $\xrightarrow[AlCl_3]{Cl_2}$ (m-chloronitrobenzene: NO_2, Cl) $\xrightarrow{Zn-HCl}$ $\xrightarrow[HCl/273\ K]{NaNO_2}$ $\xrightarrow{CuBr}$ Product

(d) Benzene $\xrightarrow[AlCl_3]{CH_3Cl}$ (toluene: CH_3) $\xrightarrow[H_2SO_4]{HNO_3}$ (CH_3, NO_2) $\xrightarrow[NaOH/Heat]{KMnO_4}$ (COOH, NO_2) $\xrightarrow{Sn/HCl}$

$\xrightarrow[HCl/273\ K]{NaNO_2}$ $\xrightarrow[Boil]{H_2O}$ Product

(e) (CH_3, NO_2) $\xrightarrow{Sn-HCl}$ $\xrightarrow[HCl/273\ K]{NaNO_2}$ $\xrightarrow{CuCN}$ $\xrightarrow{H_3O^+}$ Product

(f) Benzene $\xrightarrow[H_2SO_4]{HNO_3}$ (NO_2) $\xrightarrow{Zn-HCl}$ $\xrightarrow{CH_3COCl}$ ($NHCOCH_3$) $\xrightarrow[H_2SO_4\ Heat]{HNO_3}$

($NHCOCH_3$, NO_2, NO_2) $\xrightarrow{H_3O^+}$ $\xrightarrow[HCl/273\ K]{NaNO_2}$ $\xrightarrow[Heat]{HBF_4}$ (F, NO_2, NO_2)

(g) Benzene $+ I_2$ $\xrightarrow{HNO_3}$ (I) $\xrightarrow[H_2SO_4]{HNO_3}$ (I, NO_2) $\xrightarrow{Sn-HCl}$ $\xrightarrow[HCl/273\ K]{NaNO_2}$ $\xrightarrow[Boil]{-H_2O}$ (I, OH)

78. (a) Benzene $+ HNO_3/H_2SO_4$ $\longrightarrow$ (NO_2) $\xrightarrow[FeCl_3]{Cl_2}$ (NO_2, Cl) $\xrightarrow{Zn-HCl}$ $\xrightarrow[HCl/273\ K]{NaNO_2}$

$\xrightarrow[KI]{aq}$ Product

(b) Benzene + $Cl_2/FeCl_3$ ⟶ Chlorobenzene $\xrightarrow[H_2SO_4]{HNO_3}$ p-chloronitrobenzene (Cl, NO_2) $\xrightarrow{Zn-HCl}$ $\xrightarrow[HCl/273\ K]{NaNO_2}$ $\xrightarrow[KI]{aq}$ Product

(c) Nitrobenzene (NO_2) $\xrightarrow[FeCl_3(excess)]{Cl_2}$ 3,5-dichloronitrobenzene (NO_2, Cl, Cl) $\xrightarrow{Zn-HCl}$ $\xrightarrow[HCl/273\ K]{NaNO_2}$ $\xrightarrow[KI]{aq}$ Product

(d) Benzene $\xrightarrow[H_2SO_4\ (Excess)]{HNO_3}$ m-dinitrobenzene (NO_2, NO_2) $\xrightarrow{Zn-HCl}$ $\xrightarrow[HCl/273\ K]{NaNO_2}$ $\xrightarrow{CuCl}$ Product

(e) m-dinitrobenzene (NO_2, NO_2) [from (d)] $\xrightarrow{Zn-HCl}$ $\xrightarrow[HCl/273\ K]{NaNO_2}$ $\xrightarrow{CuCl}$ Product

(f) m-dinitrobenzene (NO_2, NO_2) [from (d)] $\xrightarrow{Na_2S}$ m-nitroaniline (NO_2, NH_2) $\xrightarrow[HCl/273\ K]{NaNO_2}$ $\xrightarrow{KI}$ m-iodonitrobenzene (NO_2, I) $\xrightarrow{Zn-HCl}$ $\xrightarrow[HCl/273\ K]{NaNO_2}$ $\xrightarrow[Boil]{-H_2O}$ Product

(g) m-dinitrobenzene (NO_2, NO_2) [from (f)] Product $\xrightarrow[HCl/273\ K]{NaNO_2}$ $\xrightarrow{CuBr}$ m-bromonitrobenzene (NO_2, Br) $\xrightarrow{Zn-HCl}$ $\xrightarrow[HCl/273\ K]{NaNO_2}$ $\xrightarrow{KCN}$

(h) Toluene $\xrightarrow[H_2SO_4\ (excess)]{HNO_3}$ 2,4-dinitrotoluene (CH_3, NO_2, NO_2) $\xrightarrow{Zn-HCl}$ $\xrightarrow[HCl/273\ K]{NaNO_2}$ $\xrightarrow{CuBr}$ 2,4-dibromotoluene (CH_3, Br, Br) $\xrightarrow[AlCl_3]{CH_3Cl}$ Product

(i) Benzene $\xrightarrow[H_2SO_4]{HNO_3}$ Nitrobenzene $\xrightarrow[FeBr_3]{Br_2\,(excess)}$ 1,3-Dibromo-5-nitrobenzene $\xrightarrow{Zn-HCl}$ Product

79. (a) Benzene $\xrightarrow[H_2SO_4]{HNO_3}$ $C_6H_5NO_2$ $\xrightarrow{Zn-HCl}$ $C_6H_5NH_2$ $\xrightarrow[273\,K]{HNO_2}$ $\xrightarrow[\Delta]{HBF_4}$ C_6H_5F

$C_6H_5NH_2$ $\xrightarrow{CH_3COCl}$ $C_6H_5NHOC-CH_3$

$C_6H_5NHOC-CH_3$ $\xrightarrow[H_2SO_4]{HNO_3}$ p-$NO_2C_6H_4NHCOCH_3$ $\xrightarrow{H_3O^+}$ $\xrightarrow[273\,K]{HNO_2}$ $\xrightarrow[\Delta]{HBF_4}$ p-$FC_6H_4NO_2$

$C_6H_5NHOC-CH_3$ $\xrightarrow[AlCl_3]{CH_3Cl}$ $\xrightarrow{H_3O^+}$ p-$CH_3C_6H_4NH_2$ $\xrightarrow[\Delta]{HBF_4}$ $\xrightarrow[273\,K]{HNO_2}$ p-$FC_6H_4CH_3$ $\xrightarrow[NaOH]{KMnO_4}$ p-FC_6H_4COOH

(b) C_6H_5Cl $\xrightarrow[H_2SO_4]{HNO_3}$ p-$ClC_6H_4NO_2$ $\xrightarrow[ether]{Mg}$ $\xrightarrow{\text{oxirane}}$ $O_2N-C_6H_4-CH_2CH_2OH$ $\xrightarrow[CCl_4]{NBS}$ Product

(c) $C_6H_6 + Cl_2 \xrightarrow{AlCl_3} C_6H_5Cl \xrightarrow[ether]{Mg} C_6H_5MCl \xrightarrow[H_3O^+]{Oxirane} C_6H_5CH_2CH_2OH \xrightarrow{PCC}$

$\xrightarrow{C_6H_5MgCl} \xrightarrow{H_3O^+} C_6H_5CH_2CH(OH)C_6H_5 \xrightarrow{PCC} C_6H_5CH_2C(=O)C_6H_5$

(d) $PhCH_2CH_2CH_2Br \xrightarrow[C_2H_5OH]{C_2H_5ONa} PhCH_2CH{=}CH_2 \xrightarrow[H_2O]{(CH_3COO)_2Hg} \xrightarrow{NaBH_4}$

$\xleftarrow{H_3O^+} \xleftarrow{CH_3C{\equiv}CNa} PhCH_2\overset{O}{\overset{\|}{C}}CH_3 \xleftarrow{PCC} PhCH_2\overset{OH}{\overset{|}{C}}HCH_3 \xleftarrow{NaBH_4}$

$\rightarrow Ph{-}CH_3\overset{OH}{\overset{|}{C}}CH_3(C{\equiv}C{-}CH_3) \xrightarrow[H_2]{Pd/BaSO_4}$ Product

(e) Benzene $+ Cl_2 \xrightarrow{AlCl_3}$ Chlorobenzene $\xrightarrow[ether]{Mg}$ $\xrightarrow{\text{ethylene oxide}}$ $\xrightarrow{H_3O^+}$ $PhCH_2CH_2OH$

80. (a) $PhBr \xrightarrow[ether]{Mg} PhMgBr \xrightarrow{\text{methyloxirane}} \xrightarrow{H_2O} PhCH_2\overset{OH}{\overset{|}{C}}HCH_3 \xrightarrow{PCC} PhCH_2\overset{O}{\overset{\|}{C}}CH_3$

(b) $PhBr \xrightarrow[ether]{Mg} PhMgBr \xrightarrow{\text{ethylene oxide}} \xrightarrow{H_2O} PhCH_2CH_2OH \xrightarrow{PBr_3} PhCH_2CH_2Br$

(c) $PhBr \xrightarrow[ether]{Mg} PhMgBr \xrightarrow{CH_3CH_2CN} \xrightarrow{H_3O^+} Ph\overset{O}{\overset{\|}{C}}CH_2CH_3$

(d) Product from (d) $\xrightarrow[Heat]{N_2H_4/OH^-} PhCH_2CH_2CH_3$

81. I. (a) 6-hepten-1-ol $\xrightarrow{PCl_5} \xrightarrow{NH_3}$ $H_2N(CH_2)_5CH{=}CH_2$ $\xrightarrow[H_2O_2]{O_3}$ $H_2N(CH_2)_5COOH \xrightarrow{Heat}$ caprolactam (cyclic amide, NH, C=O)

(b) $PhCH_2Br \xrightarrow{NaCN} \xrightarrow{NaBH_4} PhCH_2CH_2NH_2$

(c) $PhCHO + CH_3MgBr \xrightarrow{H_3O^+} Ph\underset{OH}{\underset{|}{C}}HCH_3 \xrightarrow[Heat]{H_2SO_4} PhCH{=}CH_2$

$\xrightarrow[Peroxide]{HBr} \xrightarrow{NaCN} PhCH_2CH_2CN$

II. (a) MeO–C_6H_4–$NHCOCH_3$ (b) $Ph\overset{O}{\overset{\|}{C}}NHOH$

(c) 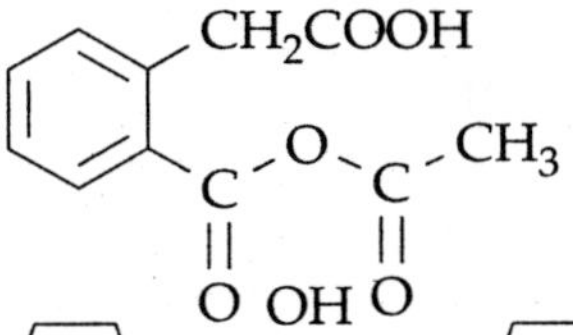 Benzene ring with CH_2COOH and $C(=O)-O-C(=O)-CH_3$ substituents (ortho)

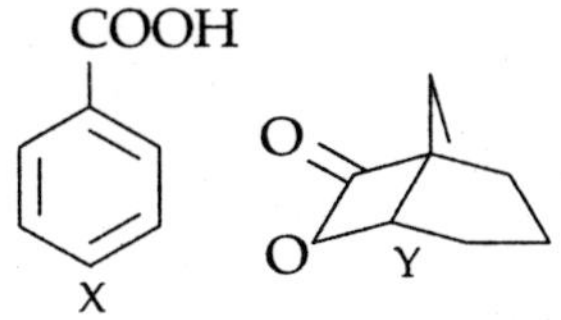

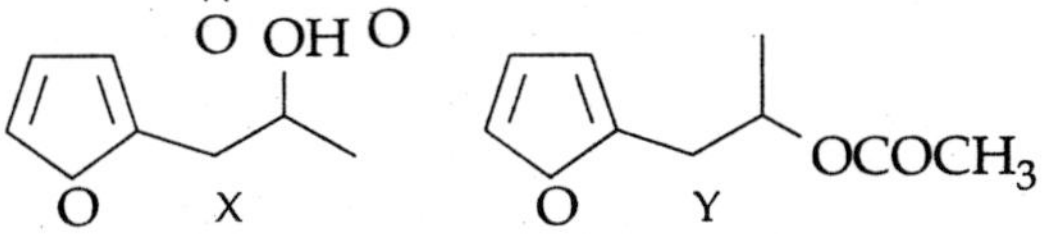

(d) X: C_6H_5COOH (benzoic acid); Y: bicyclic lactone

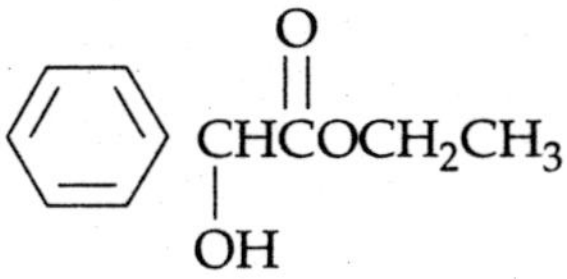

(e) X: furan–$CH_2CH(OH)CH_3$; Y: furan–$CH_2CH(OCOCH_3)CH_3$

82. (a) $CH_3OC(=O)(CH_2)_9C(=O)OCH_3$
A

(b) 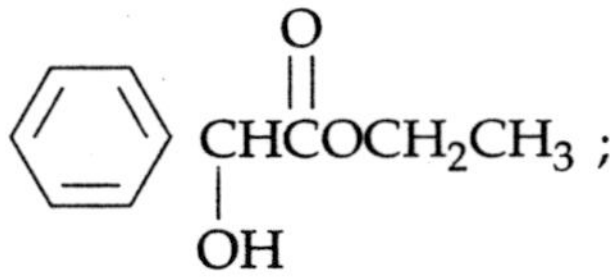 $C_6H_5CH(OH)C(=O)OCH_2CH_3$; $C_6H_5CH(OH)C(=O)OCH_2CH_3$
B ; C

(c) 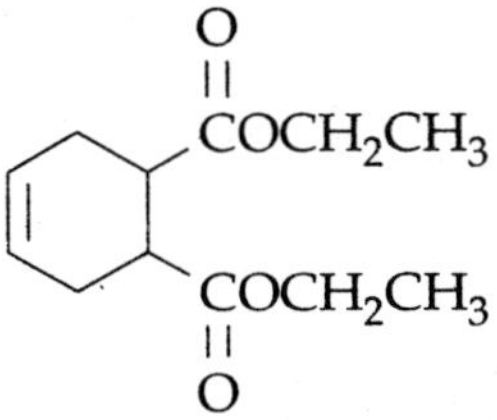 Cyclohexene with two $C(=O)OCH_2CH_3$ groups
D

(d) $CH_3OC(=O)C{\equiv}CC(=O)OCH_3$
E

(e) $BrCH_2C(=O)OCH_2CH_3$
F

83. (a) $H_2NC(=O)CN{=}CHC(=O)NH_2 + 2CH_3CH_2OH$
A

(b) $H_2NHHC(=O)CH_2CH_2CH_2CH_2C(=O)NHNH_2 + 2CH_3CH_2OH$
D

(c) $CH_3(CH_2)_4CH(CH_2CH_3)C(=O)Cl$; $CH_3(CH_2)_4CH(CH_2CH_3)C(=O)NH_2$
E ; F

(d) $N{\equiv}CCH_2C(=O)NH_2 + CH_3CH_2OH$
H

(e) 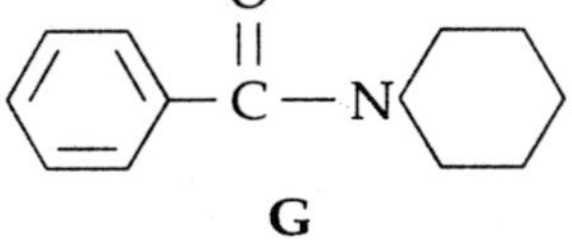 $C_6H_5-C(=O)-N$(piperidine ring)
G

(f) 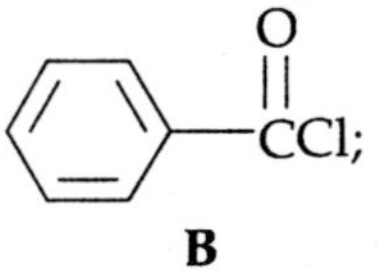 $C_6H_5C(=O)Cl$; $(CH_3)_2NH_2^+Cl^+$
B ; C

84. (a) $CH_3O-C_6H_4-NHCOCH_3 + CH_3COOH$ (A) (b) C_6H_5COBr (B)

(c) $C_6H_5CO^-Na^+$ (C) (d) $CH_3(CH_2)_5COOC(CH_2)_5CH_3$ (D)

(e) $C_6H_5CH(CH_2CH_3)COOH$ (E) (f) $o\text{-}CH_3C_6H_4COOH$ (F)

(g) CH_3CH_2COCl (G) (h) $C_6H_5CONHOH + CH_3CH_2OH$ (H)

85. (a) CH_3, $CH_2CH_2CO^-Li^+$ on 1,3-dioxolane (A); CH_3, $CH_2CH_2C(O^-Li^+)(CH_3)O^-Li^+$ on 1,3-dioxolane (B); $CH_3COCH_2CH_2COCH_3 + HOCH_2CH_2OH$ (C)

(b) $CH_3CH_2CH_2CH_2$–C(cyclohexyl)(:Ö:---$Mg^{2+}Br^-$)–N(CH_3)–:Ö:–CH_3 (D); $C_6H_5COCH_2CH_2CH_2CH_3$ (E) + $CH_3\overset{+}{N}H_2OCH_3$ (F)

(c) (cycloheptyl)(phenyl)C(:Ö:$^-$ - - Li^+)–N(CH_3)–:Ö:–CH_3 (G); $C_6H_{11}-CO-C_6H_5$ (H) + $H_3\overset{+}{N}H_2OCH_3$ (I)

(d) 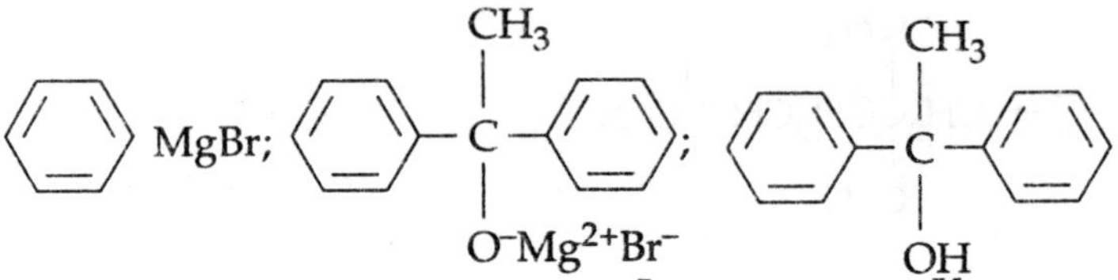 C_6H_5MgBr; $(C_6H_5)_2C(CH_3)O^-Mg^{2+}Br^-$; $(C_6H_5)_2C(CH_3)OH$

I J K

(e) 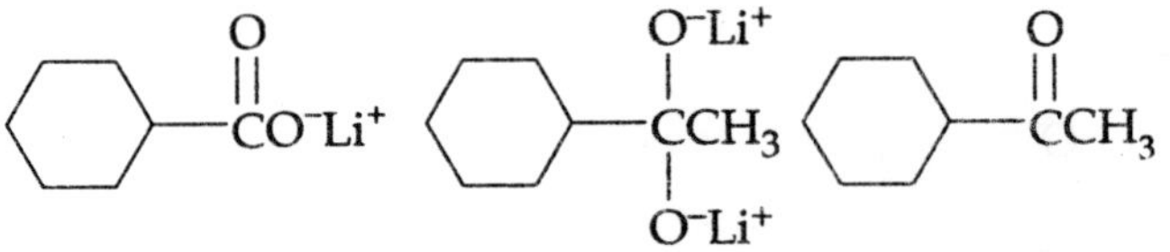 $C_6H_{11}CO^-Li^+$ (with C=O); $C_6H_{11}C(O^-Li^+)_2CH_3$; $C_6H_{11}CCH_3$ (with C=O)

L M N

86. (a) 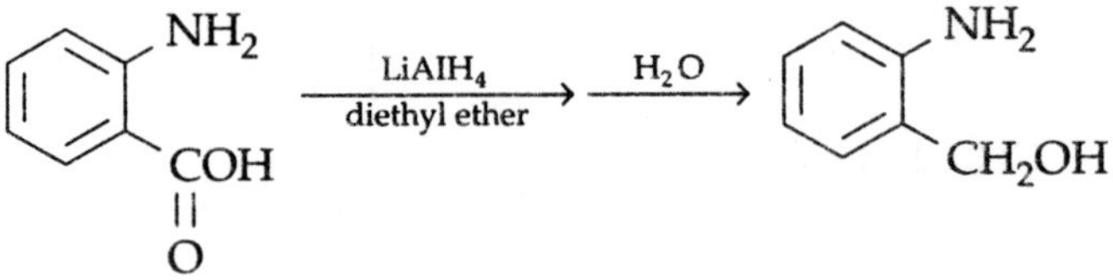 2-aminobenzoic acid (NH_2, COH with C=O) $\xrightarrow[\text{diethyl ether}]{LiAlH_4}$ $\xrightarrow{H_2O}$ 2-aminobenzyl alcohol (NH_2, CH_2OH)

(b) 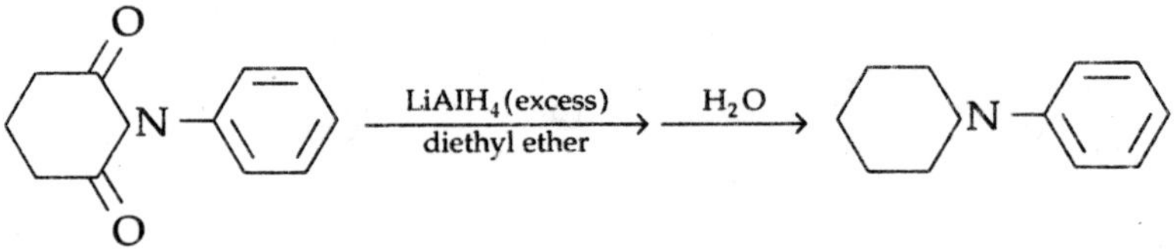 N-phenylglutarimide (piperidine-2,6-dione with N–C_6H_5) $\xrightarrow[\text{diethyl ether}]{LiAlH_4\text{(excess)}}$ $\xrightarrow{H_2O}$ N-phenylpiperidine

(c) $CH_3CH{=}CHCH_2CH_2\overset{O}{\overset{\|}{C}}OCH_3 \xrightarrow[\text{diethyl ether}]{LiAlH_4} \xrightarrow{H_3O^+} CH_3CH{=}CHCH_2CH_2CH_2OH$

(d) $CH_3CH_2O\overset{O}{\overset{\|}{C}}(CH_2)_4\overset{O}{\overset{\|}{C}}OCH_2CH_3 \xrightarrow[\text{diethyl ether}]{LiAlH_4} \xrightarrow{H_3O^+} HOCH_2(CH_2)_4CH_2OH$
$+ CH_3CH_2OH$

(e) $CH_3CH_2\overset{O}{\overset{\|}{C}}NHCH_2CH_3 \xrightarrow[\text{diethyl ether}]{LiAlH_4} \xrightarrow{H_3O^+} CH_3CH_2CH_2NHCH_2CH_3$

87. (a) 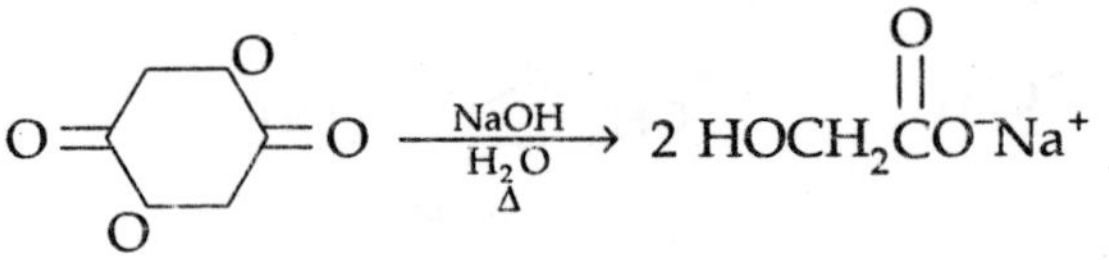 glycolide (1,4-dioxane-2,5-dione) $\xrightarrow[H_2O,\ \Delta]{NaOH}$ $2\ HOCH_2\overset{O}{\overset{\|}{C}}O^-Na^+$

(b) 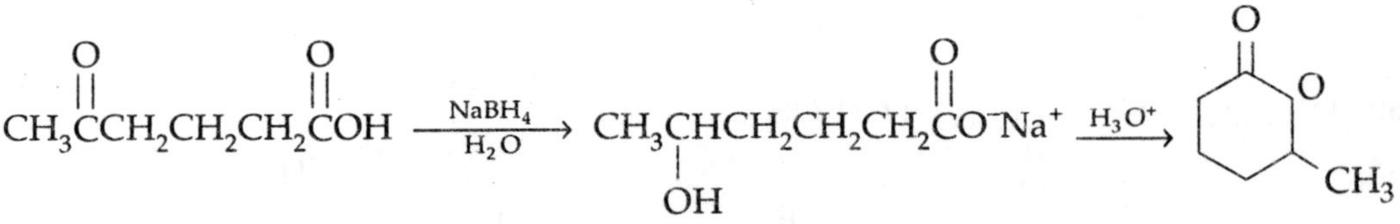 $CH_3\overset{O}{\overset{\|}{C}}CH_2CH_2CH_2\overset{O}{\overset{\|}{C}}OH \xrightarrow[H_2O]{NaBH_4} CH_3\underset{OH}{\underset{|}{C}}HCH_2CH_2CH_2\overset{O}{\overset{\|}{C}}O^-Na^+ \xrightarrow{H_3O^+}$ 6-methyltetrahydropyran-2-one (lactone with CH_3)

(c) 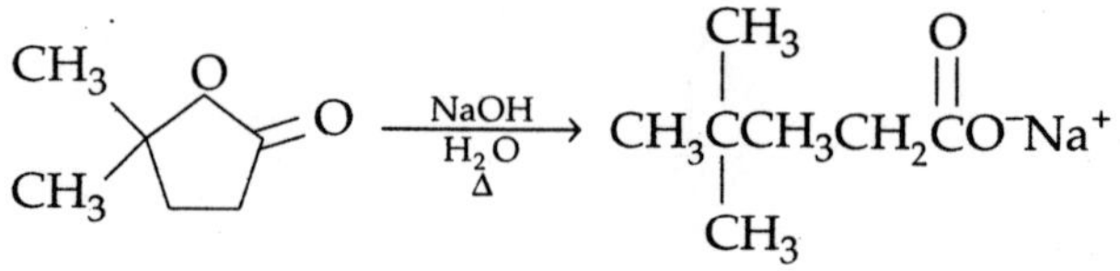 CH_3, CH_3 (lactone) $\xrightarrow[\Delta]{NaOH,\ H_2O}$ $CH_3\overset{CH_3}{\underset{CH_3}{C}}CH_3CH_2\overset{O}{\overset{\|}{C}}O^-Na^+$

(d) 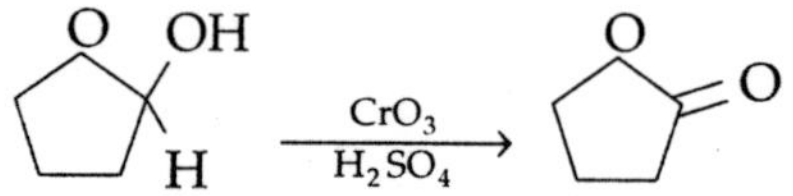 O, OH, H $\xrightarrow[H_2SO_4]{CrO_3}$ O, =O

(e) 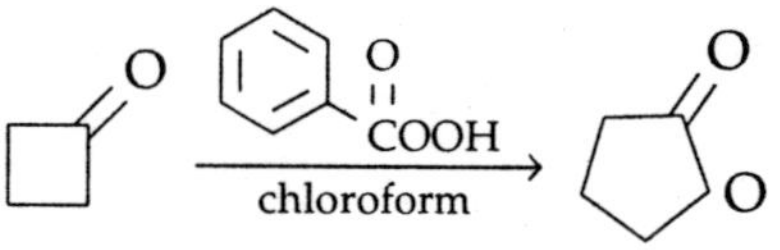 =O $\xrightarrow[\text{chloroform}]{C_6H_5\overset{O}{\overset{\|}{C}}OOH}$ O, O

88. (a) 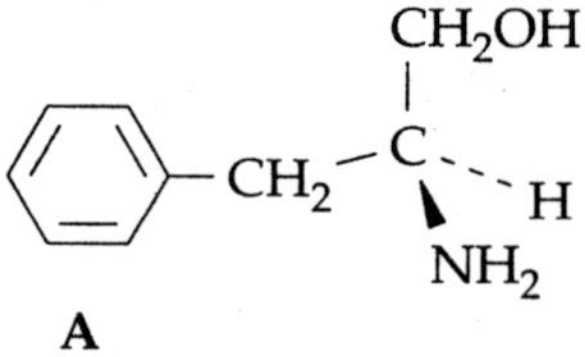 $C_6H_5-CH_2-\overset{CH_2OH}{\underset{NH_2}{C}}\cdots H$ + CH_3CH_2OH

A B

(b) 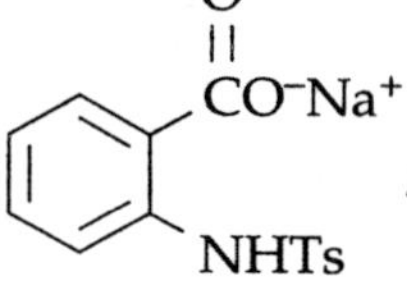 $\overset{O}{\overset{\|}{C}}O^-Na^+$, NHTs ; $\overset{O}{\overset{\|}{C}}OH$, NHTs

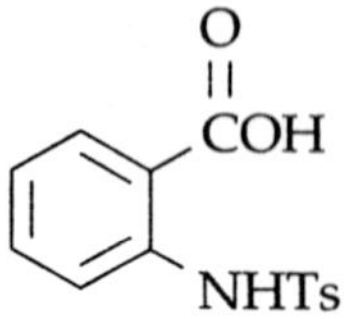

C D

(c) 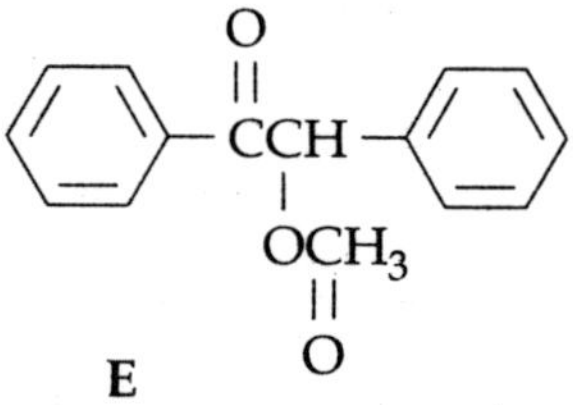 $C_6H_5-\overset{O}{\overset{\|}{C}}CH-C_6H_5$, OCH_3, =O

E

(d) 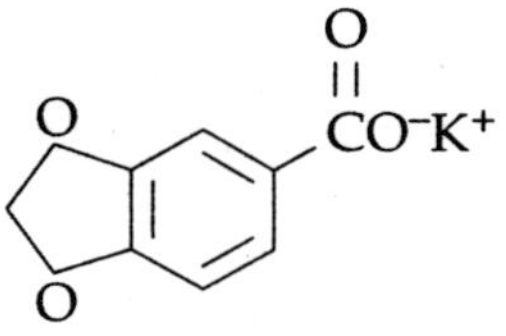 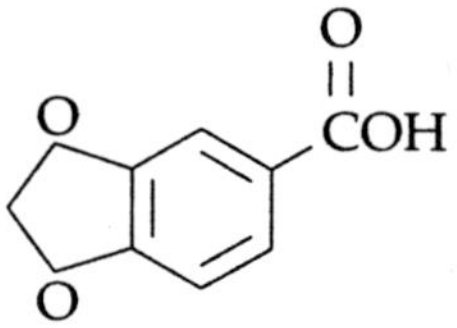 O, O, $\overset{O}{\overset{\|}{C}}O^-K^+$; O, O, $\overset{O}{\overset{\|}{C}}OH$

F G

Note: The cyclic acetal on the aromatic ring is stable and does not open during the protonation of the carboxylate anion.

(e) $CH_3(CH_2)_4CH_2OTs$

H

(f) 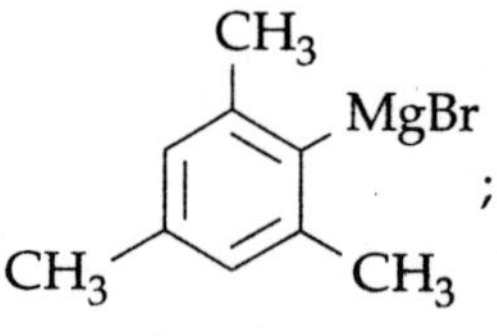 CH_3, MgBr, CH_3, CH_3 ; CH_3, $\overset{O}{\overset{\|}{C}}O^-Mg^{2+}Br^-$, CH_3, CH_3 ; CH_3, $\overset{O}{\overset{\|}{C}}OH$, CH_3, CH_3

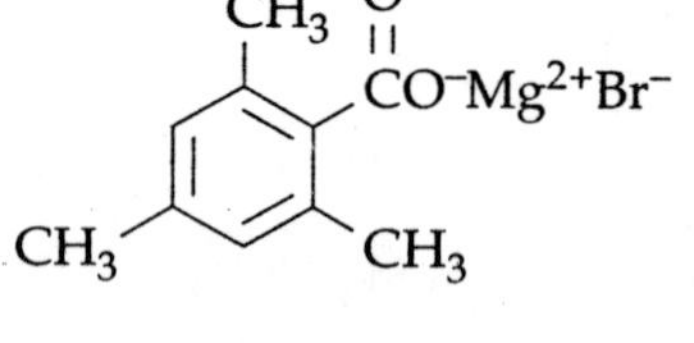

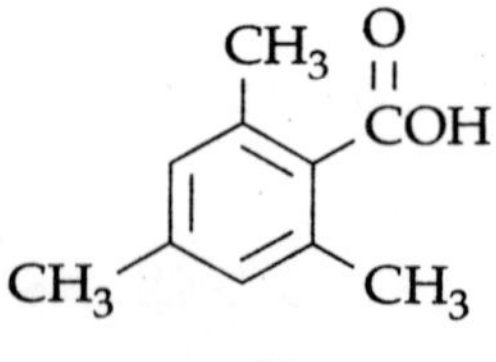

I J K

89. (a) 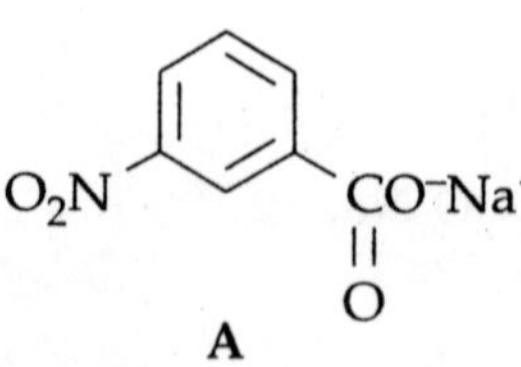 O_2N, $\underset{O}{\underset{\|}{C}}O^-Na^+$; CH_3OH

A B

(b) O, O, O, CH_3

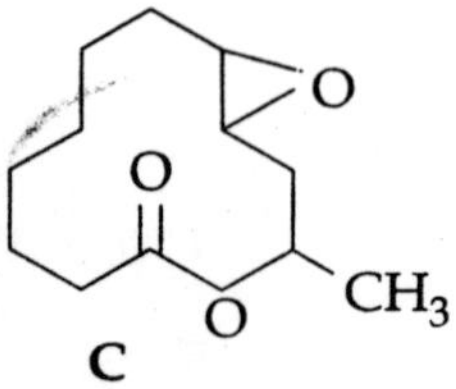

C

(c) 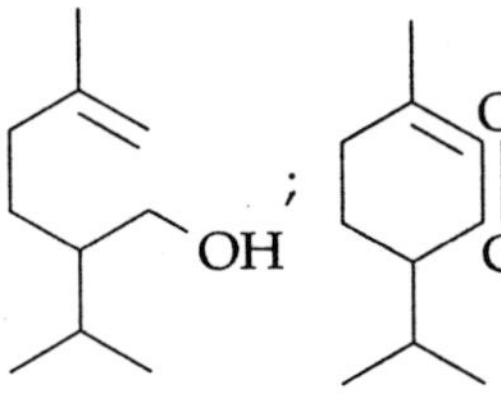 [structure D: OH] ; [structure E: CH=O] (d) 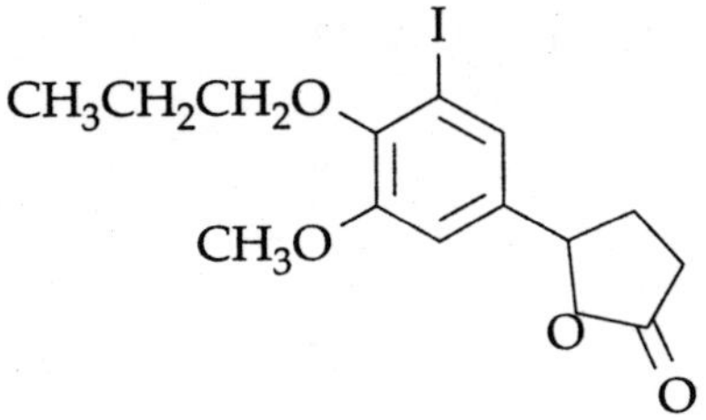 $CH_3CH_2CH_2O$, I, CH_3O [structure F] $+ H_2O\uparrow$

D E F

[pyridinium ion, N^+–H] is an acid with pK_a 5.2, close enough to the pK_a of the carboxylic acid (pK_a – 4.8) to maintain an equilibrium concentration of the acid. The carboxylic acid then forms the lactone under conditions in which water is removed from the reaction mixture.

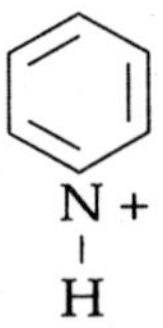

(e) 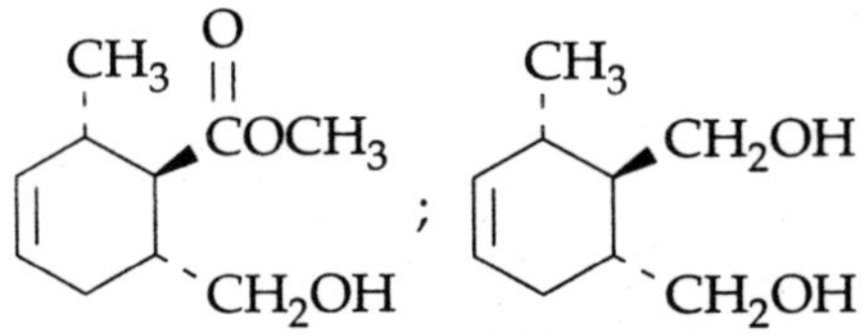 [structure G: CH_3, $COCH_3$, CH_2OH] ; [structure H: CH_3, CH_2OH, CH_2OH] $+ CH_3OH$ (f) [structure J: NC(=O), (O=)CN, O]

G H I J

(g) 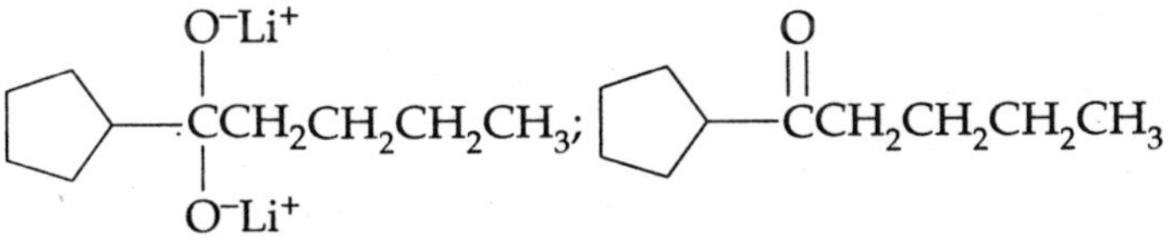 [cyclopentyl]–$C(O^-Li^+)_2CH_2CH_2CH_2CH_3$; [cyclopentyl]–$\overset{O}{\overset{\|}{C}}CH_2CH_2CH_2CH_3$ (h) [structure M: H, H, $OCCH_3$ (C=O), H]

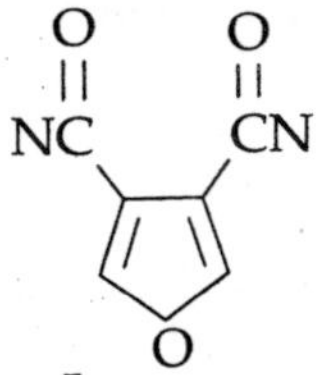

K L M

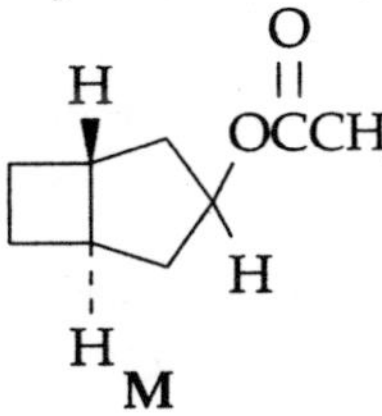

90. (a) $CH_2{=}CHCH{=}CHCH_2OH$; $CH_2{=}CHCH{=}CH\overset{O}{\overset{\|}{C}}H$

A B

(b) $CH_3CH_2CH_2CH_2C(CH_2CH_3)(O^-Mg^{2+}Br^-)CH_2CH_3$; $CH_3CH_2CH_2CH_2C(CH_2CH_3)(OH)CH_2CH_3$

C D

(c) $CH_3CH_2CH_2NHCH_2CH_3$ (d) 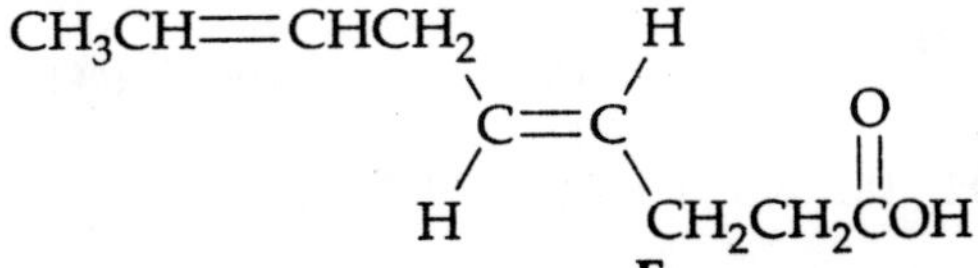 $CH_3CH{=}CHCH_2$(H)$C{=}C$(H)$CH_2CH_2\overset{O}{\overset{\|}{C}}OH$

E F

(e) 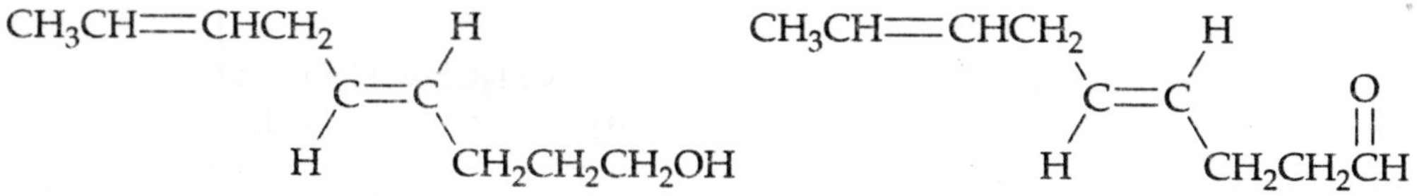 $CH_3CH{=}CHCH_2(H)C{=}C(H)CH_2CH_2CH_2OH$ **G** $CH_3CH{=}CHCH_2(H)C{=}C(H)CH_2CH_2CHO$ **H**

(f) 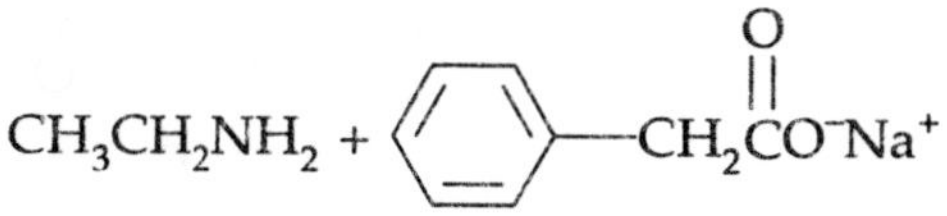 $CH_3CH_2NH_2$ + $C_6H_5{-}CH_2CO^-Na^+$

I J

(g) 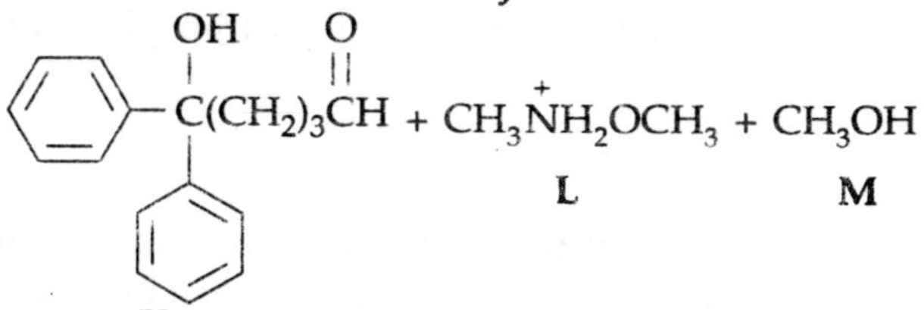 $(C_6H_5)_2C(OH)(CH_2)_3CHO$ + $CH_3\overset{+}{N}H_2OCH_3$ + CH_3OH

K L M

91. (a) 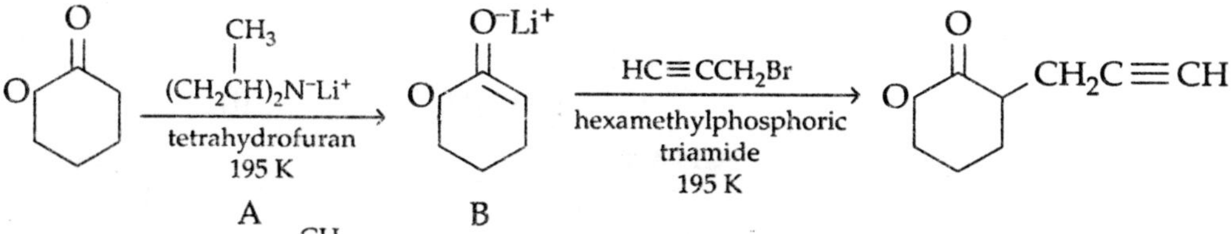 O-heterocyclic ketone (A) → [$(CH_3CH)_2N^-Li^+$ ... $(CH_2CH)_2N^-Li^+$ with CH_3; tetrahydrofuran; 195 K] → enolate O^-Li^+ (B) → [$HC{\equiv}CCH_2Br$; hexamethylphosphoric triamide; 195 K] → product with $CH_2C{\equiv}CH$

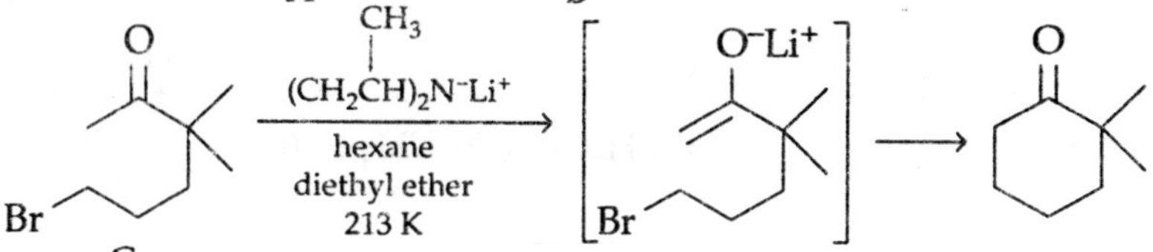

A B

(b) C (bromo ketone, Br) → [$(CH_2CH)_2N^-Li^+$ (CH_3); hexane; diethyl ether; 213 K] → [enolate O^-Li^+, Br] → 2,2-dimethylcyclohexanone

C

(c) $CH_3CH_2CH_2\overset{O}{\overset{||}{C}}OCH_3$ $\xrightarrow[\text{tetrahydrofuran, 195 K}]{(CH_2\overset{CH_3}{\overset{|}{C}}H)_2N^-Li^+}$ $CH_3CH_2CH{=}\overset{O^-Li^+}{\overset{|}{C}}OCH_3$

D E

$\xrightarrow[\text{hexamethyl-phosphoric triamide, 195 K}]{ClCH_2OCH_3}$ $CH_3CH_2\underset{CH_2OCH_3}{\underset{|}{C}}H\overset{O}{\overset{||}{C}}OCH_3$

F

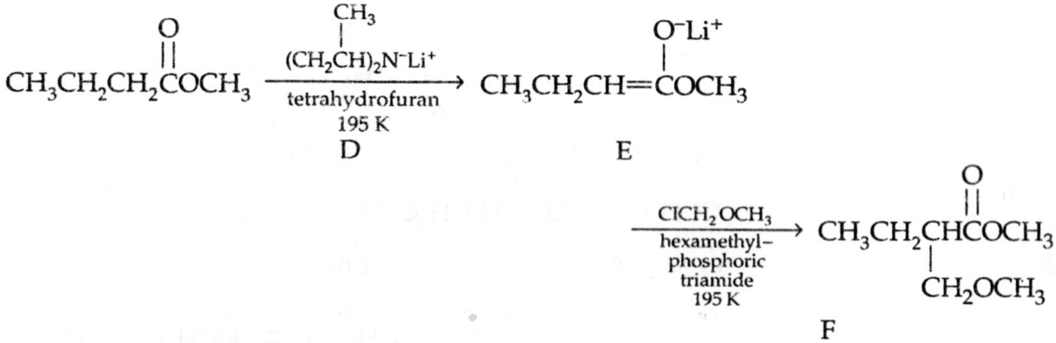

92. (a) $Na^{+-}CH(\overset{O}{\overset{||}{C}}OCH_2CH_3)_2$; $CH_3\overset{CH_3}{\overset{|}{C}}HCH(\overset{O}{\overset{||}{C}}OCH_2CH_3)_2$

A B

(b) $CH_3CH_2\overset{-}{C}(\overset{O}{\overset{||}{C}}OCH_2CH_3)_2$; $CH_2{=}CHCH_2C(\overset{O}{\overset{||}{C}}OCH_2CH_3)_2$ with CH_2CH_3 on C

C Na^+ **D**

(c) $Na^+CH(\overset{O}{\overset{||}{C}}OCH_2CH_3)_2$; C_6H_5—$CH_2CH(\overset{O}{\overset{||}{C}}OCH_2CH_3)_2$

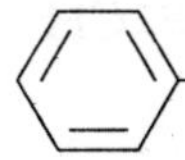

E **F**

(d) $CH_3\overset{O}{\overset{||}{C}}\overset{-}{C}H\overset{O}{\overset{||}{C}}OCH_2CH_3$ Na^+ ; $CH_3\overset{O}{\overset{||}{C}}CH\overset{O}{\overset{||}{C}}OCH_2CH_3$ with $CH_2CH_2\overset{O}{\overset{||}{C}}OCH_2CH_3$ on CH

G **H**

(e) $CH_3\overset{O}{\overset{||}{C}}\overset{-}{C}H\overset{O}{\overset{||}{C}}OCH_2CH$; $CH_3\overset{O}{\overset{||}{C}}CH\overset{O}{\overset{||}{C}}OCH_2CH_3$ with $CH_2CH_2CH_2Cl$ on CH

I **J**

(Br^- is a better leaving group than Cl^-)

(f) $CH_3\overset{O}{\overset{||}{C}}\overset{-}{C}H\overset{O}{\overset{||}{C}}OCH_2CH_3$; $CH_3\overset{O}{\overset{||}{C}}CH\overset{O}{\overset{||}{C}}OCH_2CH_3$ with $CH_2CH_2CH{=}CH_2$ on CH   Na/toulene;

K **L** **M**

$CH_3\overset{O}{\overset{||}{C}}—\overset{Na^+}{\overset{-}{C}}—\overset{O}{\overset{||}{C}}OCH_2CH_3$ with $CH_2CH_2CH{=}CH_2$ on C ;

N

C bearing $CH_3CH(CH_3)$, $\overset{O}{\overset{||}{C}}OCH_2CH_3$, $CH_3\underset{O}{\underset{||}{C}}$ and $CH_2CH_2CH{=}CH_2$

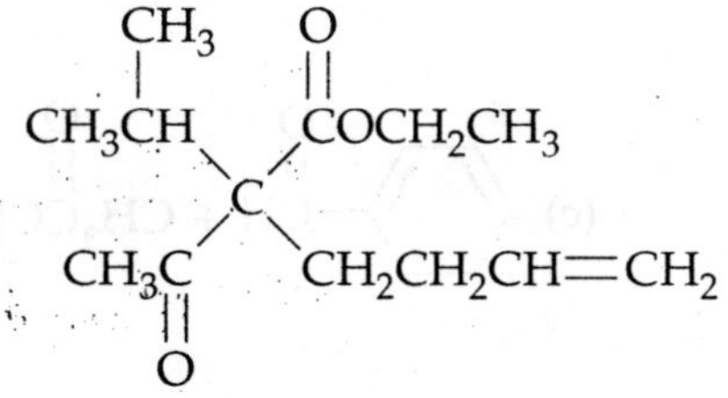

O

93. (a) $CH_3\overset{O}{\overset{||}{C}}\overset{-}{C}H\overset{O}{\overset{||}{C}}OCH_2CH_3$ Na^+; $CH_3\overset{O}{\overset{||}{C}}CH\overset{O}{\overset{||}{C}}OCH_2CH_3$ with $CH_2\underset{O}{\underset{||}{C}}OCH_2CH_3$ on CH

A **B**

$$CH_3\overset{O}{\overset{||}{C}}CH_2CH_2\overset{O}{\overset{||}{C}}OH + CO_2\uparrow + CH_3CH_2OH$$

C

(b) $C_6H_5-\overset{O}{\overset{||}{C}}\bar{C}H\overset{O}{\overset{||}{C}}OCH_2CH_3 \; Na^+$ **D**

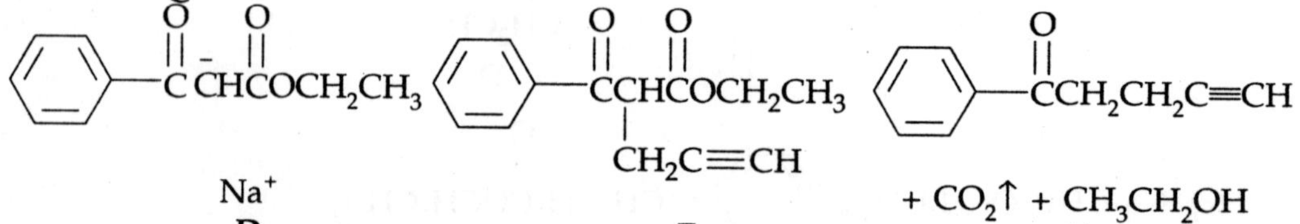

$C_6H_5-\overset{O}{\overset{||}{C}}\underset{CH_2C\equiv CH}{\underset{|}{C}H}\overset{O}{\overset{||}{C}}OCH_2CH_3$ **E**

$C_6H_5-\overset{O}{\overset{||}{C}}CH_2CH_2C\equiv CH + CO_2\uparrow + CH_3CH_2OH$ **F**

(c) $Na^+(\overset{O}{\overset{||}{C}}OCH_2CH_3)_2$ **G**

$ClCH_2CH_2CH_2CH(\overset{O}{\overset{||}{C}}OCH_2CH_3)_2$ **H**

(Br^- is a better leaving group than Cl^-)

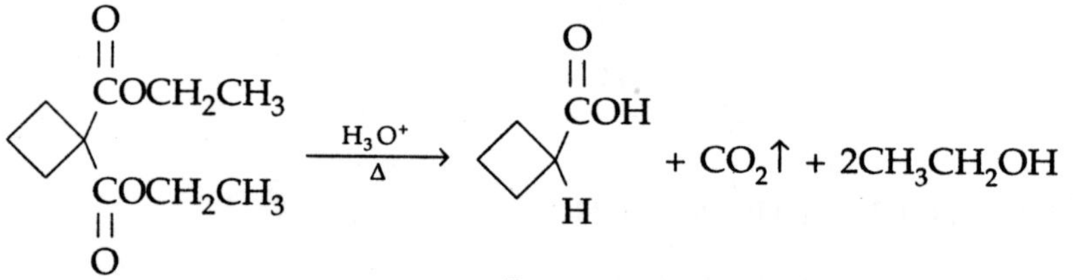

$$\text{cyclobutane-1,1-}(\overset{O}{\overset{||}{C}}OCH_2CH_3)_2 \xrightarrow[\Delta]{H_3O^+} \text{cyclobutane-}\overset{O}{\overset{||}{C}}OH\,(H) + CO_2\uparrow + 2CH_3CH_2OH$$

I **J**

94. (a) tetrahydrofuran-2-yl$-\underset{O}{\underset{||}{C}}H + CH_3\overset{O}{\overset{||}{C}}CH_3 \xrightarrow[H_2O]{NaOH} \xrightarrow{H_3O^+}$ tetrahydrofuran-2-yl$-CH=CH\overset{O}{\overset{||}{C}}CH_3$

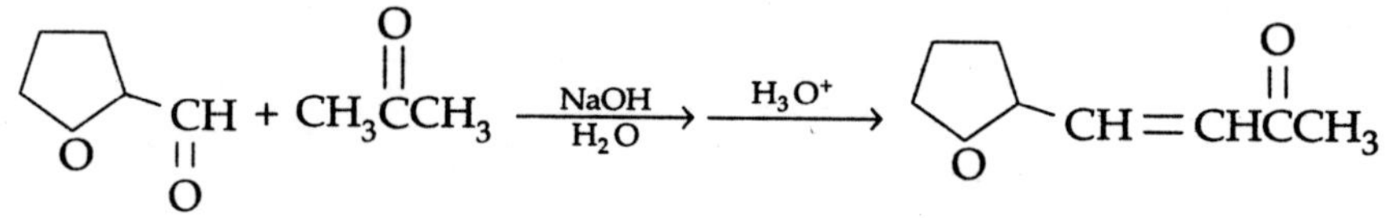

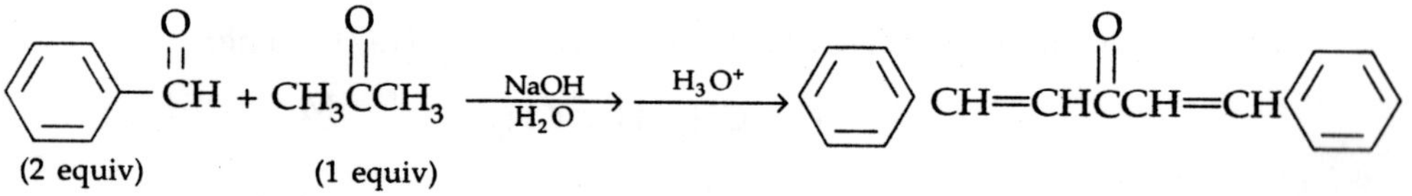

(b) $C_6H_5-\overset{O}{\overset{||}{C}}H$ (2 equiv) $+ CH_3\overset{O}{\overset{||}{C}}CH_3$ (1 equiv) $\xrightarrow[H_2O]{NaOH} \xrightarrow{H_3O^+} C_6H_5CH=CH\overset{O}{\overset{||}{C}}CH=CHC_6H_5$

(c) $C_6H_5-\overset{O}{\overset{||}{C}}H + CH_3\overset{O}{\overset{||}{C}}CH_2CH_3 \xrightarrow{HCl} C_6H_5-CH=\underset{CH_3}{\underset{|}{C}}\overset{O}{\overset{||}{C}}CH_3$

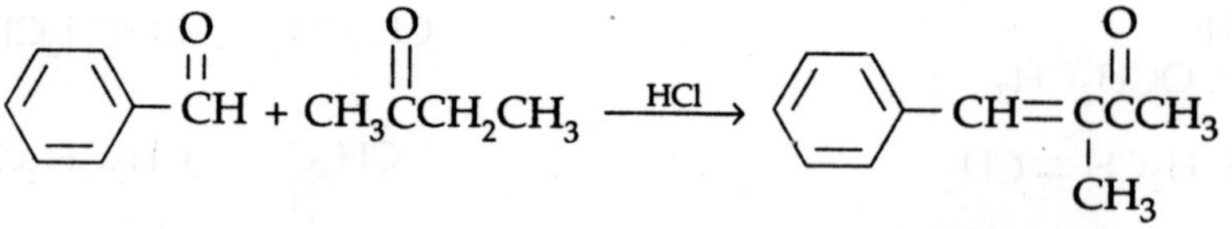

(The more stable enol from 2-butanone has the more highly substituted double bond)

(d) $C_6H_5-CH_2\overset{O}{\overset{||}{C}}H$ (2 equiv) (i) $\xrightarrow[H_2O]{NaOH} C_6H_5\overset{CHO}{\overset{|}{C}}=CHCH_2C_6H_5$ (ii)

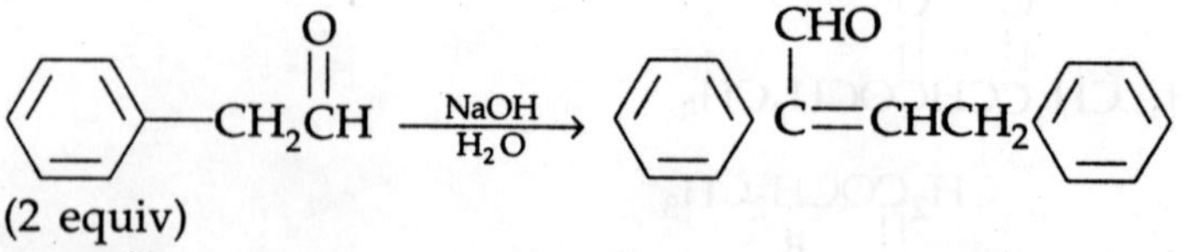

95. (a) $CH_3O-C_6H_4-\overset{O}{\overset{\|}{C}}H + CH_3CH_2\overset{O}{\overset{\|}{C}}H \longrightarrow (p\text{-}CH_3OC_6H_4)(H)C{=}C(H)(CHO)$

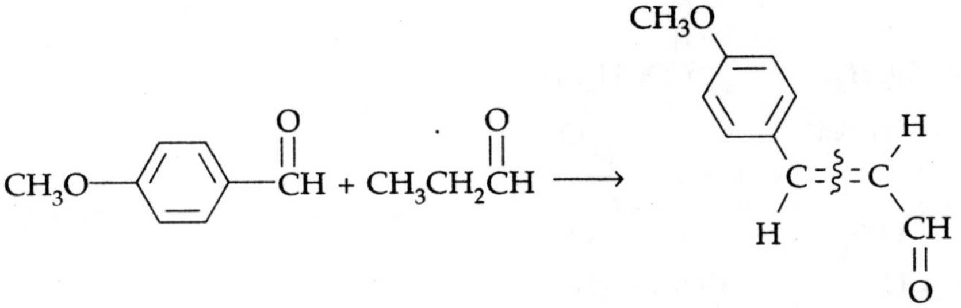

(b) $2\ CH_3CH_2CH_2CH_2\overset{O}{\overset{\|}{C}}H \longrightarrow CH_3CH_2CH_2CH_2\overset{OH}{\overset{|}{C}}H-\underset{CH_2CH_2CH_3}{\underset{|}{C}}H\overset{O}{\overset{\|}{C}}H$

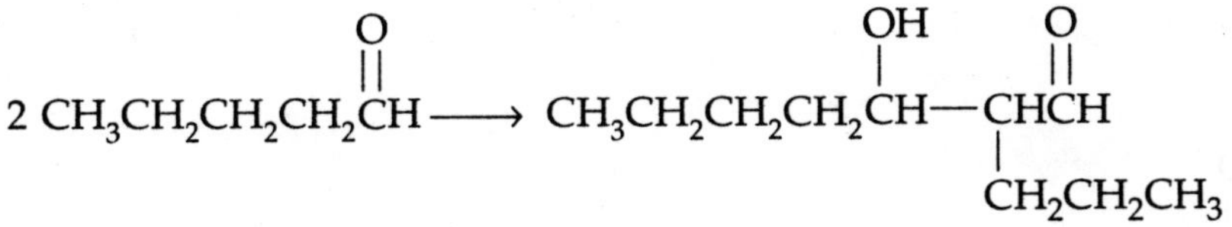

(c) $C_6H_5\overset{O}{\overset{\|}{C}}CH_2CH_2CH_2CH_2CH_2\overset{O}{\overset{\|}{C}}-C(CH_3)_3 \longrightarrow$ 1-$(\overset{O}{\overset{\|}{C}}-C(CH_3)_3)$-2-phenylcyclohexene

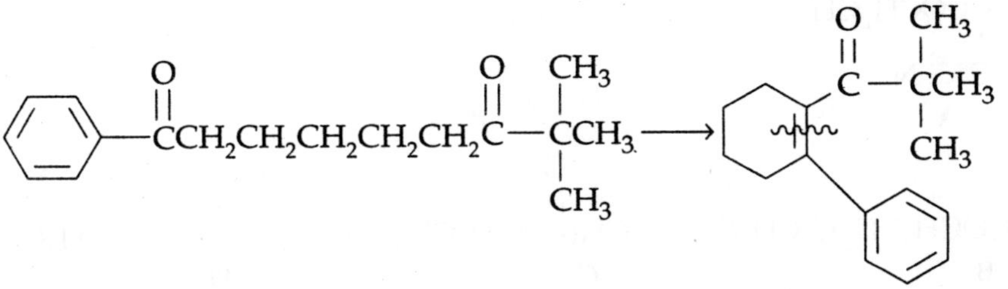

(d) $C_6H_5\overset{O}{\overset{\|}{C}}H + CH_3\overset{O}{\overset{\|}{C}}-C_6H_5 \longrightarrow (C_6H_5)(H)C{=}C(H)(\overset{O}{\overset{\|}{C}}-C_6H_5)$

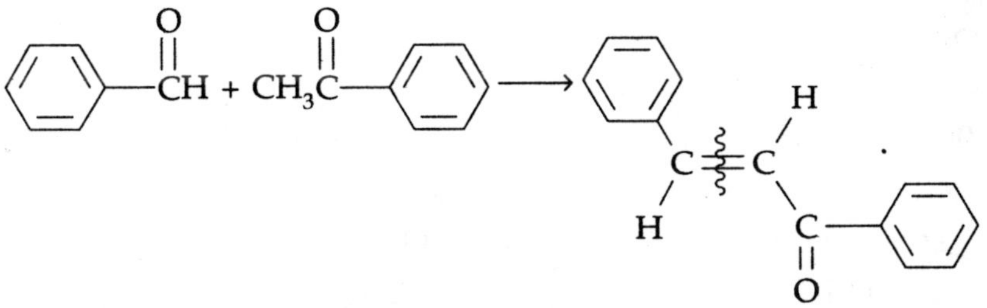

96. (a) $C_6H_5\overset{C_6H_5}{\overset{|}{C}}{=}\underset{\underset{O}{\|}}{\underset{COCH_2CH_3}{\underset{|}{C}}}CH_2\overset{O}{\overset{\|}{C}}OCH_2CH_3$

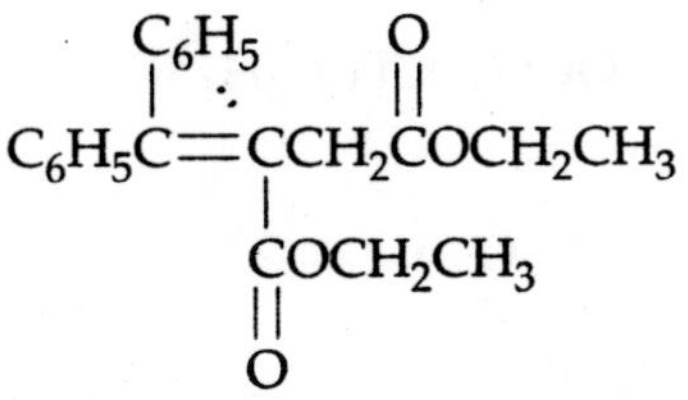

A

(b) **B**: cyclooct-2-en-1-one ; **C**: $Na^{+-}O$–cyclooctene–$\overset{O}{\overset{\|}{C}}OCH_2CH_3$; **D**: 2-oxocyclooctene–$\overset{O}{\overset{\|}{C}}OCH_2CH_3$

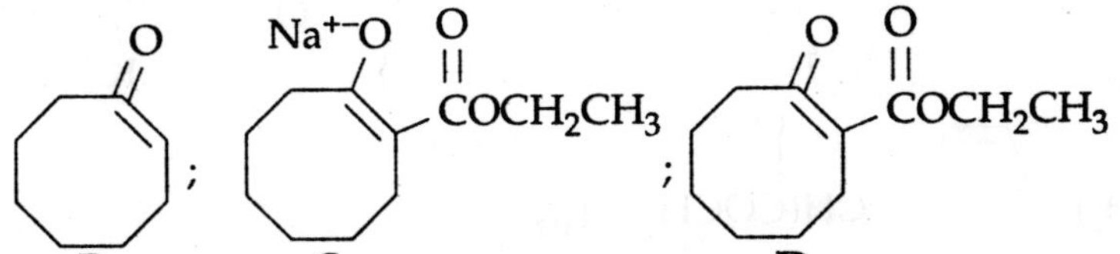

B C D

(c) 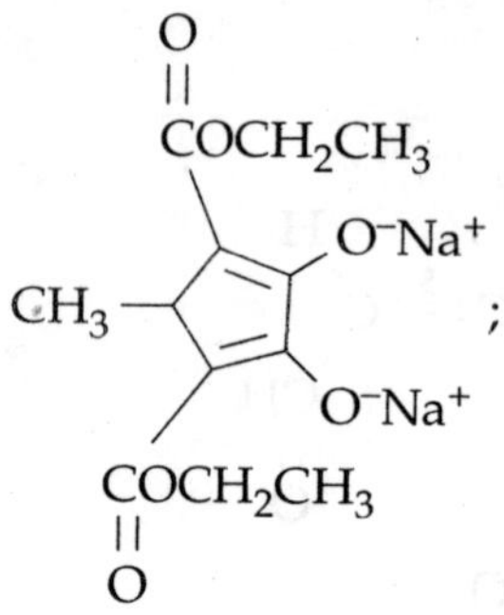 CH_3 — cyclopentene ring bearing two $COCH_2CH_3$ (C=O) groups, O^-Na^+, O^-Na^+ ; CH_3 — cyclopentane-dione ring bearing two $COCH_2CH_3$ (C=O) groups, =O, =O

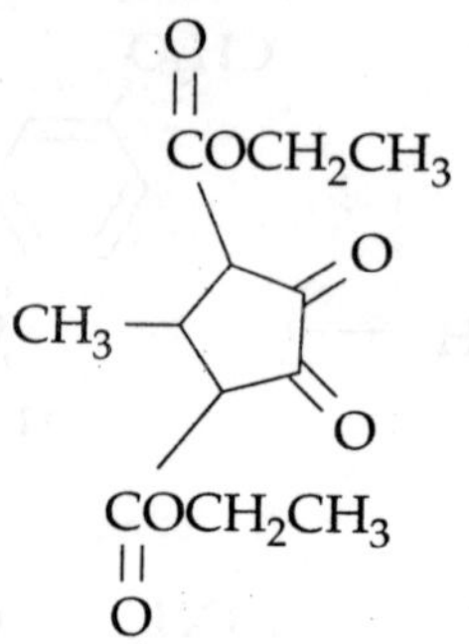

E F

97. (a) 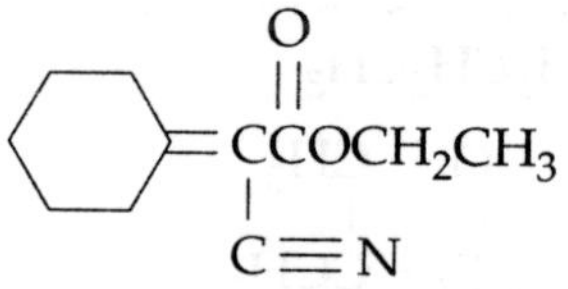 cyclohexylidene $=\overset{O}{\overset{\|}{C}}COCH_2CH_3$ with $C{\equiv}N$

A

(b) $Na^+CH(\overset{O}{\overset{\|}{C}}OCH_2CH_3)_2$; $CH_3\overset{CH_3}{\overset{|}{C}}HCH_2CH(\overset{O}{\overset{\|}{C}}OCH_2CH_3)_2$; $CH_3\overset{CH_3}{\overset{|}{C}}HCH_2(CH_2CH_2\overset{O}{\overset{\|}{C}}OH$

B C D

(c) 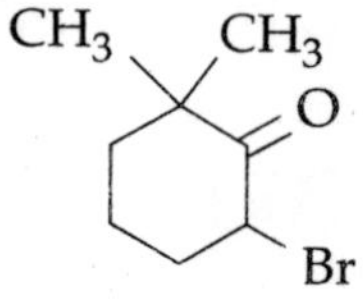 2-Bromo-6,6-dimethylcyclohexanone (CH_3, CH_3, O, Br)

E

(d) 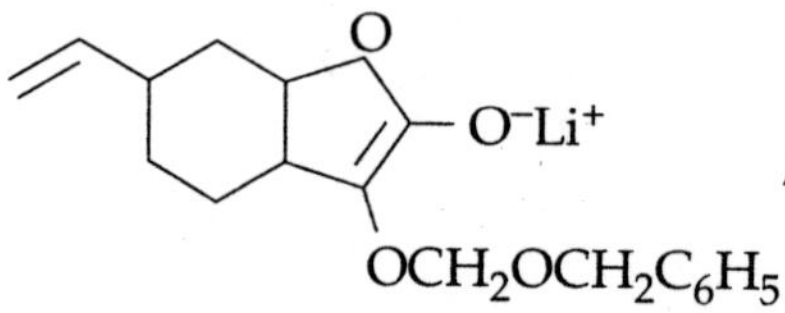 bicyclic enolate with O^-Li^+ and $OCH_2OCH_2C_6H_5$; bicyclic lactone =O with $C_6H_5CH_2OCH_2$ and $OCH_2OCH_2C_6H_5$

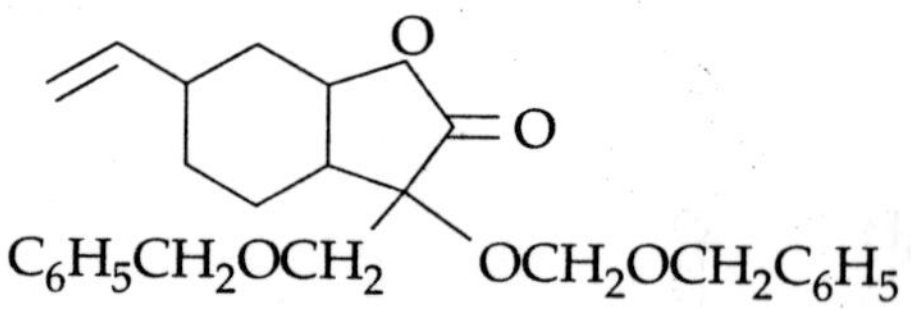

F G

(e) 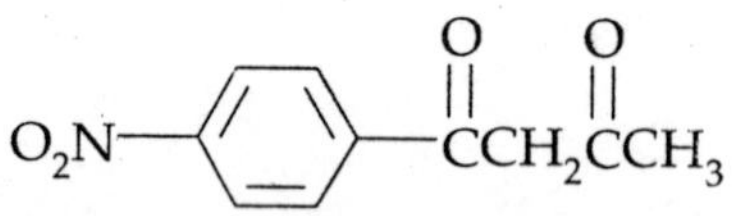 O_2N—C_6H_4—$\overset{O}{\overset{\|}{C}}CH_2\overset{O}{\overset{\|}{C}}CH_3 + CH_3OH$ (f) 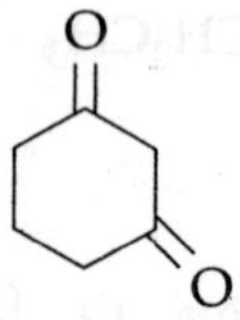 cyclohexane-1,3-dione (O, O) $+ CH_3CH_2OH$

H I J K

(g) $Na^{+-}CH(\overset{O}{\overset{\|}{C}}OCH_2CH_3)_2$ cyclopentenyl—$CH(\overset{O}{\overset{\|}{C}}OCH_2CH_3)_2$

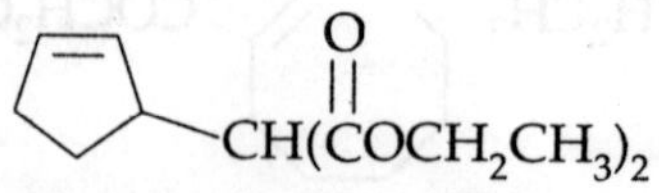

L M

98. (a) 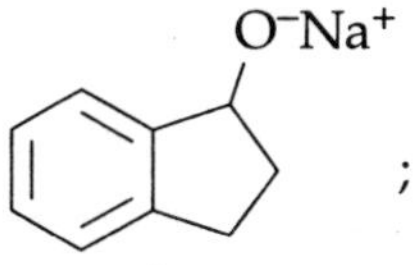 A ; B

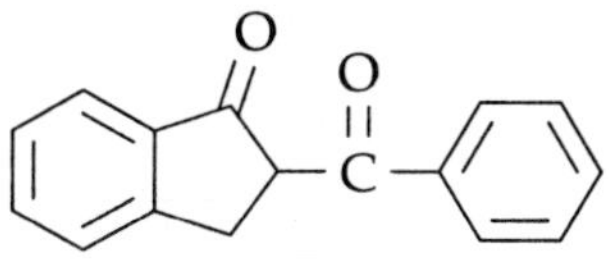

(b) $CH_3\overset{O}{\overset{\|}{C}}\overset{-}{C}H\overset{O}{\overset{\|}{C}}OCH_2CH_3$; Na^+ C

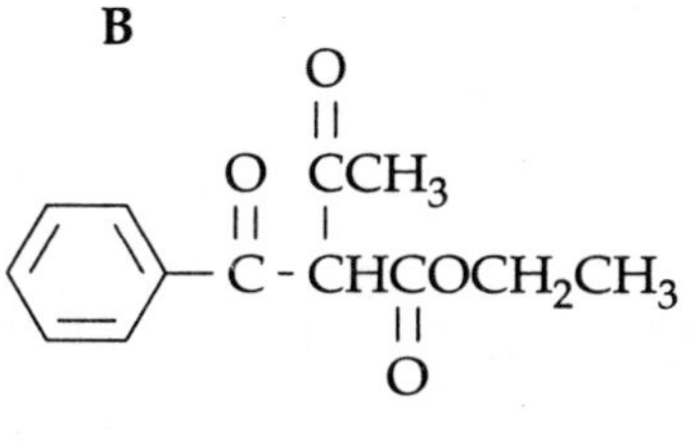

D: $C_6H_5-\overset{O}{\overset{\|}{C}}-CH(\overset{O}{\overset{\|}{C}}CH_3)\overset{O}{\overset{\|}{C}}OCH_2CH_3$

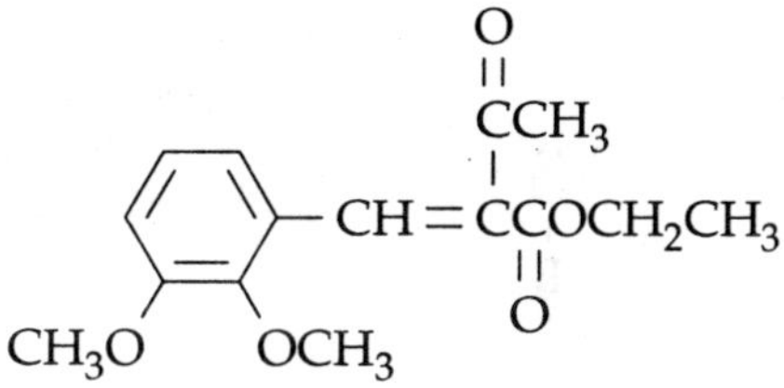

(c) E: (2,3-dimethoxyphenyl) $-CH=C(\overset{O}{\overset{\|}{C}}CH_3)\overset{O}{\overset{\|}{C}}OCH_2CH_3$ (CH_3O, OCH_3)

(d) $CH_2=\overset{O^-Li^+}{\overset{|}{C}}CH_2\overset{O}{\overset{\|}{C}}CH_2CH_2CH_2CH=CH_2$; $CH_3\overset{OH}{\overset{|}{C}}HCH_2\overset{O}{\overset{\|}{C}}CH_2\overset{O}{\overset{\|}{C}}CH_2CH_2CH_2OH=CH_2$

F G

(e) 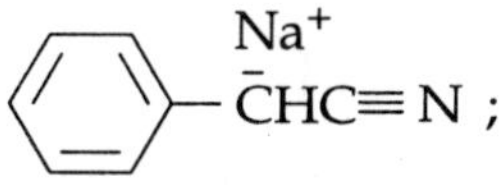 $C_6H_5-\overset{-}{C}HC\equiv N$; Na^+ $C_6H_5-\overset{\overset{O}{\|}{COCH_2CH_3}}{\overset{|}{C}}HC\equiv N$

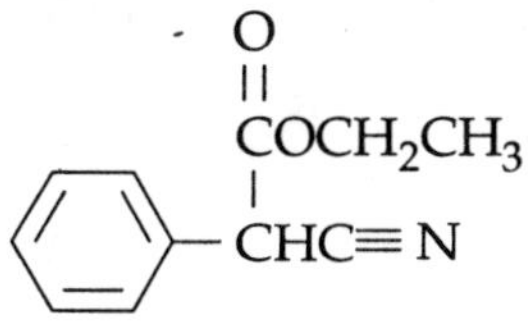

H I

(f) 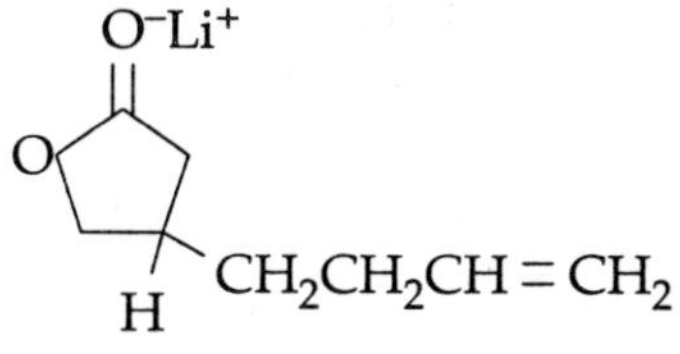 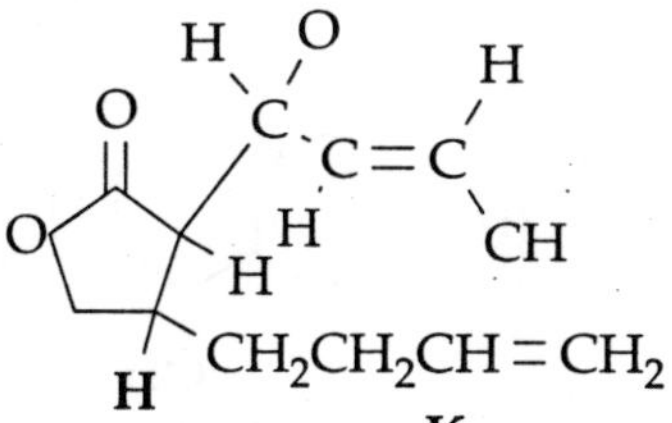 J: O^-Li^+ ... $CH_2CH_2CH=CH_2$, H ; K: ... $CH_2CH_2CH=CH_2$, H

J K

(g) 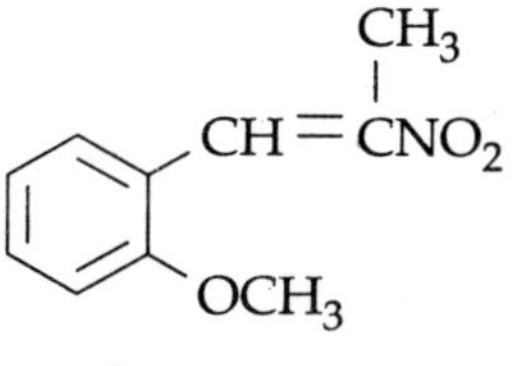 L: $CH=\overset{CH_3}{\overset{|}{C}}NO_2$, OCH_3

(h) $CH_3(CH_2)_4CH_2\overset{O}{\overset{\|}{C}}CH_2CH_2\overset{O}{\overset{\|}{C}}H$; $(CH_2)_4CH_3$

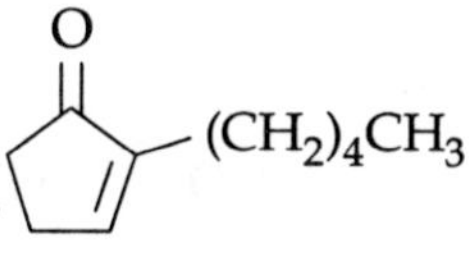

L M N

99. 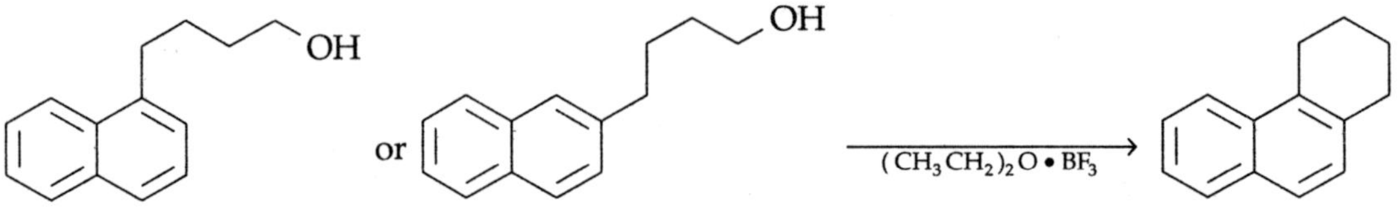 OH or OH $\xrightarrow{(CH_3CH_2)_2O \bullet BF_3}$

The detailed mechanism is as follows :

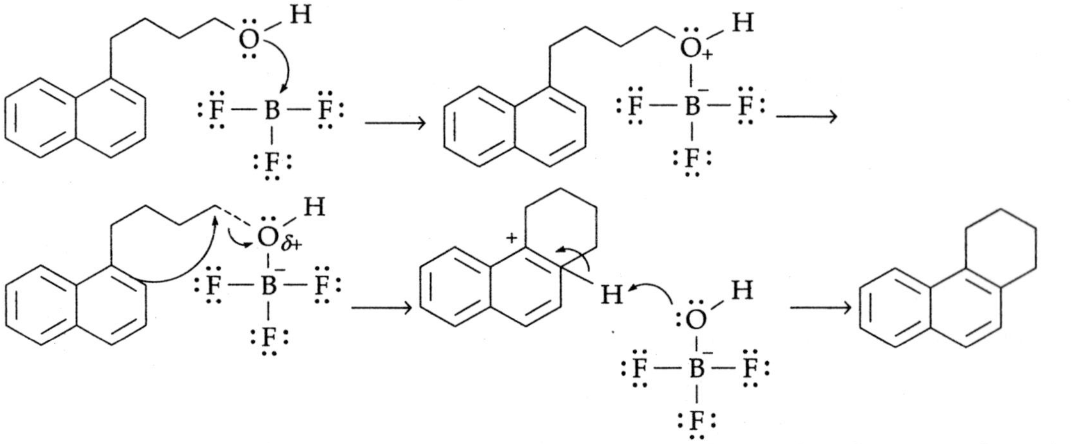

The same process would take place with the other alcohol except that the point of attack of the chain would be different. Attack would be at carbon 1 of the ring instead of carbon 2.

100. (a) 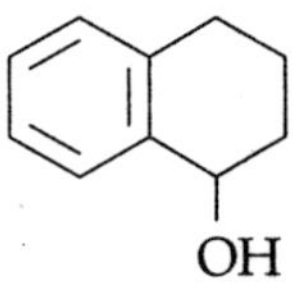 OH **A**

(b) 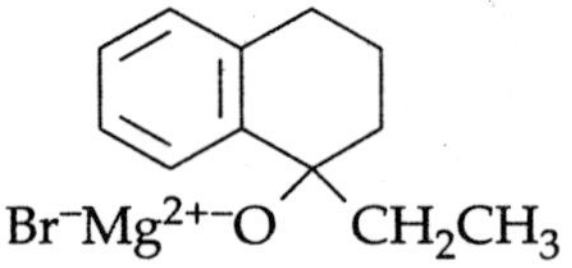 $Br^{-}Mg^{2+-}O$ CH_2CH_3 **B** ; HO CH_2CH_3 **C**

(c) 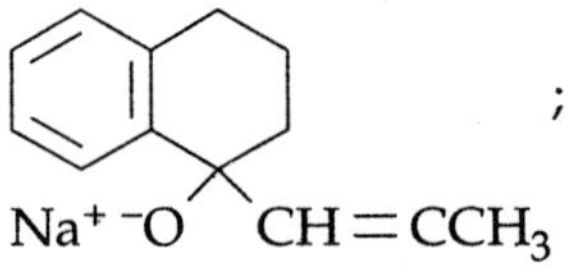 $Na^{+}\,{}^{-}O$ $CH{=}CCH_3$ **D** ; HO $CH{=}CCH_3$ **E**

(d) 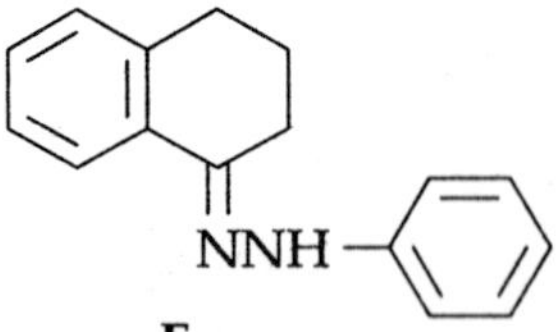 NNH— **F**

(e) 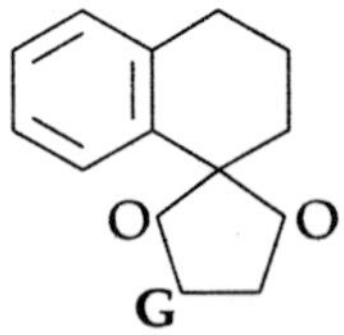 O O **G**

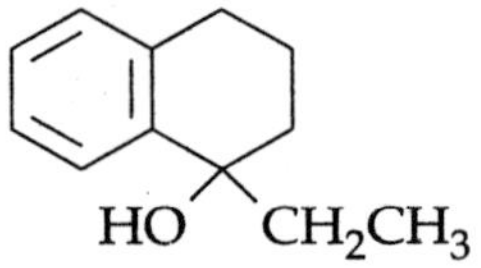

(f) 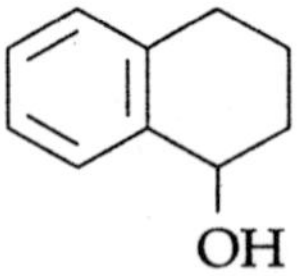 OH **H** ; **I** ; OH **J**

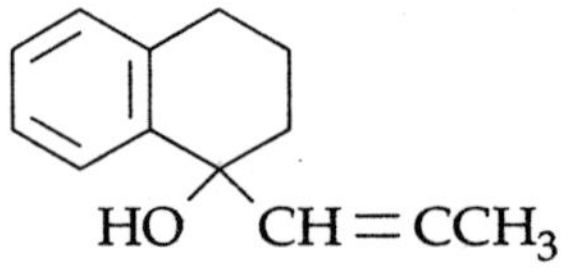

(g) 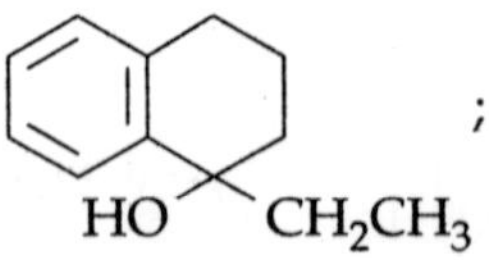 HO CH_2CH_3 **K** ; CH_2CH_3 **L** (major product)

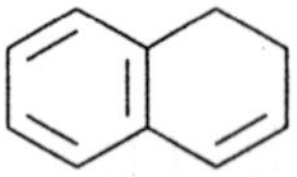

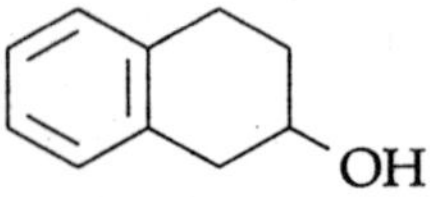

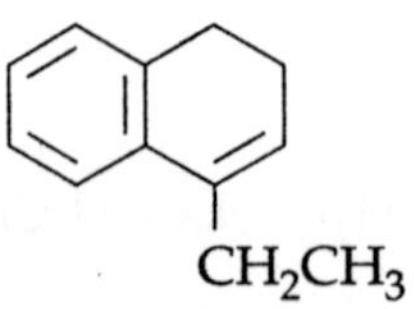

(h) HO $C\equiv CCH_3$; HO $C=C$ CH_3 H H

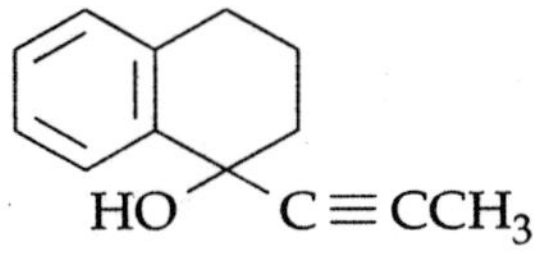

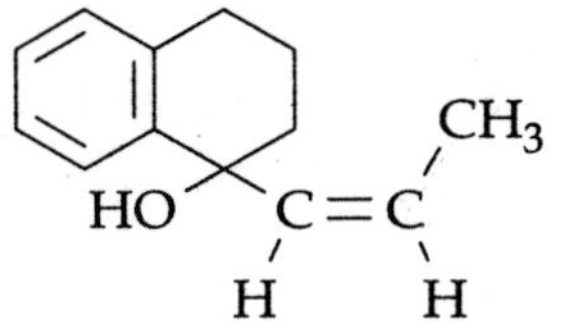

M **N**

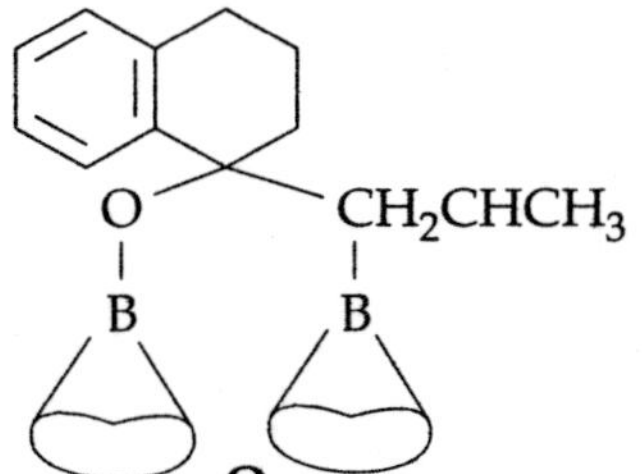

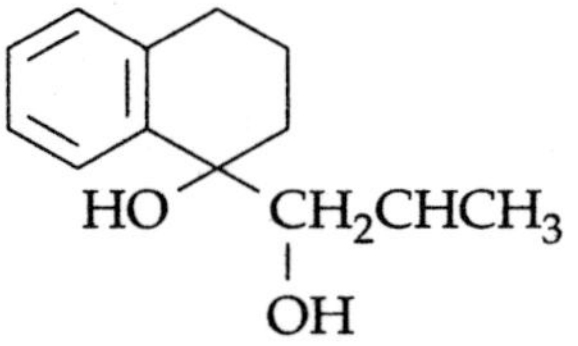

O CH_2CHCH_3 ; HO CH_2CHCH_3
B B OH

O **P**

101. (a) 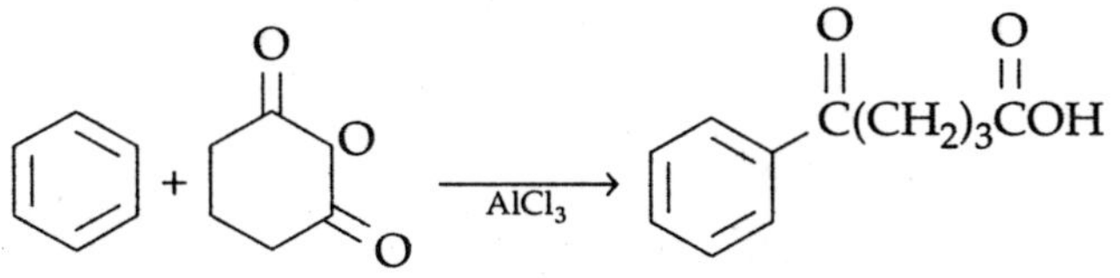 + O O O $\xrightarrow{AlCl_3}$ O O $C(CH_2)_3COH$

(b) 2 + O CCl CCl O $\xrightarrow[\text{(excess)}]{AlCl_3}$ O O $C(CH_2)_3COH$

(c) O CH_2CCl + $\xrightarrow{AlCl_3}$ O CH_2C

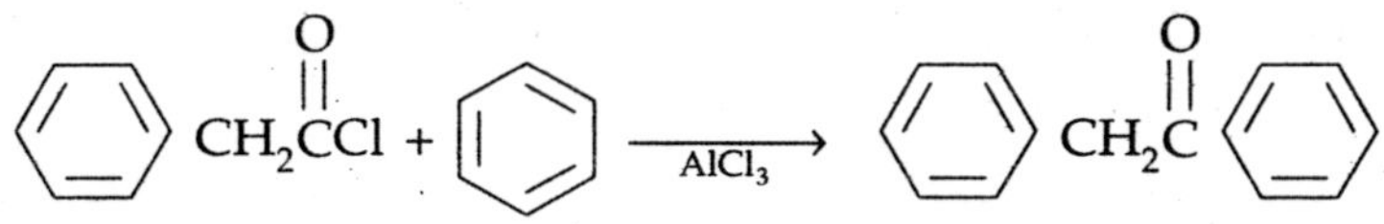

(d) 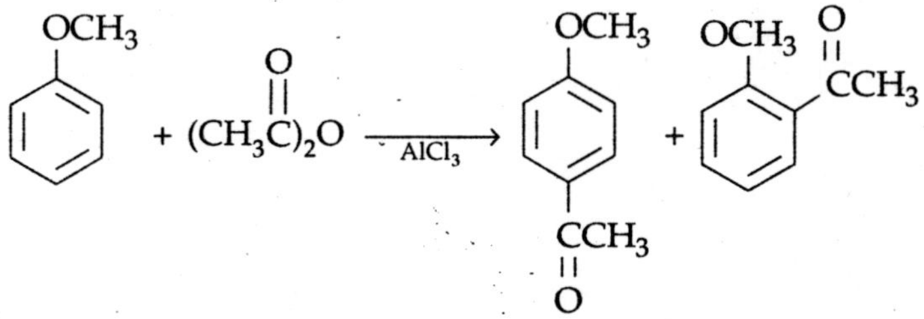 OCH_3 + $(CH_3\overset{O}{C})_2O$ $\xrightarrow{AlCl_3}$ OCH_3 CCH_3 O + OCH_3 O CCH_3

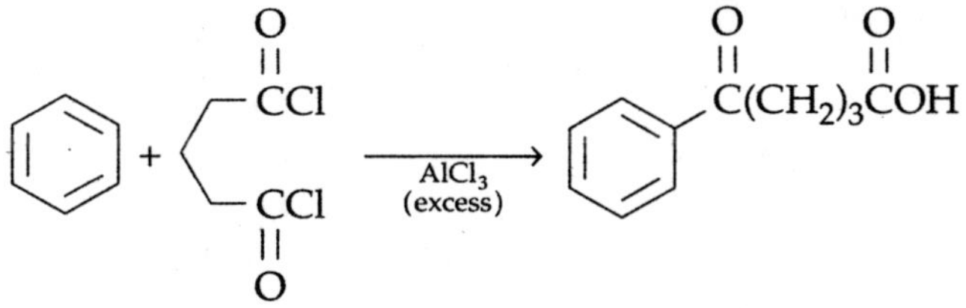

102. (a) 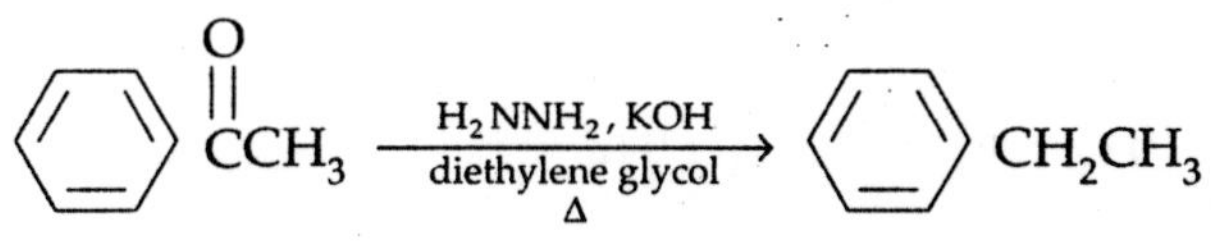 O CCH_3 $\xrightarrow[\text{diethylene glycol, }\Delta]{H_2NNH_2,\ KOH}$ CH_2CH_3

(b) $C_6H_5CH_2COCH_2CH_3 \xrightarrow[\Delta]{Zn(Hg),\ HCl} C_6H_5CH_2CH_2CH_2CH_3$

(c) $C_{10}H_7COCH_2CH_2COOH \xrightarrow[\Delta]{Zn(Hg),\ HCl} C_{10}H_7CH_2CH_2CH_2COOH$

(d) 3,4,5-$(CH_3O)_3C_6H_2CHO \xrightarrow[\text{diethyl glycol}]{H_2NNH_2,\ KOH}$ 3,4,5-$(CH_3O)_3C_6H_2CH_3$

(e) $C_6H_5COCH_2CH_2CH_3 \xrightarrow[BF_3,\ \Delta]{HSCH_2CH_2SH}$ (S,S-dithiolane) $C_6H_5C(SCH_2CH_2S)CH_2CH_2CH_3 \xrightarrow[\text{ethanol},\ \Delta]{\text{Raney Ni}} C_6H_5CH_2CH_2CH_2CH_3$

103. (a) o-$CH_3OC_6H_4NO_2 \xrightarrow[\text{HF, 283-293 K}]{CH_3CH(OH)CH_3}$ (2-OCH_3, 1-NO_2, 4-CH_3CHCH_3 product) + (2-OCH_3, 3-$CH_3CH(CH_3)$, 1-NO_2 product)

(b) p-$HOC_6H_4NO_2 \xrightarrow[\text{acetic acid, 298 K}]{Br_2}$ 2-Br-4-$NO_2C_6H_3OH$

(c) o-$C_6H_4(CH_3)_2 \xrightarrow[Fe]{Br_2}$ 3-bromo-1,2-dimethylbenzene (Br, CH_3, CH_3) + 4-bromo-1,2-dimethylbenzene (Br, CH_3, CH_3)

104. (a) 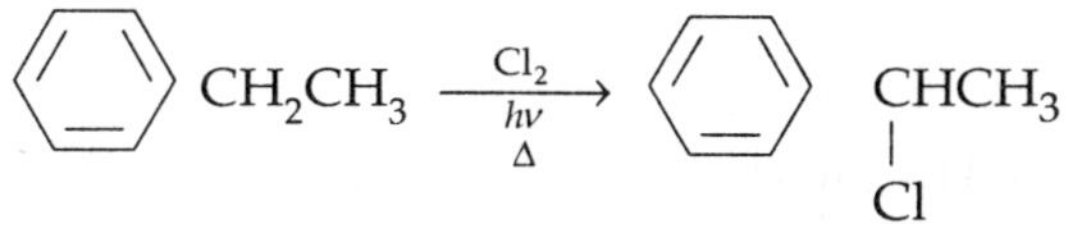 $C_6H_5CH_2CH_3 \xrightarrow[h\nu,\ \Delta]{Cl_2} C_6H_5CH(Cl)CH_3$

major product

(b) 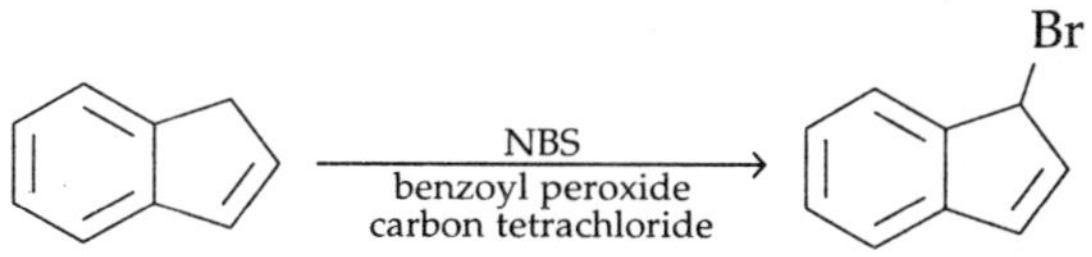 Indene $\xrightarrow[\text{benzoyl peroxide, carbon tetrachloride}]{NBS}$ 1-bromoindene (Br on the CH of the five-membered ring)

(c) 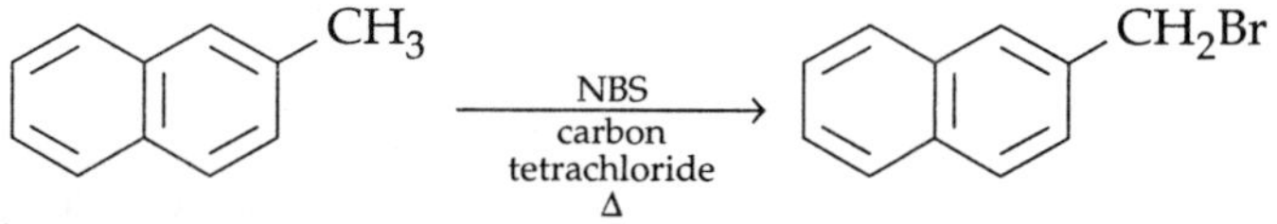 2-methylnaphthalene (CH_3) $\xrightarrow[\text{carbon tetrachloride, }\Delta]{NBS}$ 2-(bromomethyl)naphthalene (CH_2Br)

(d) 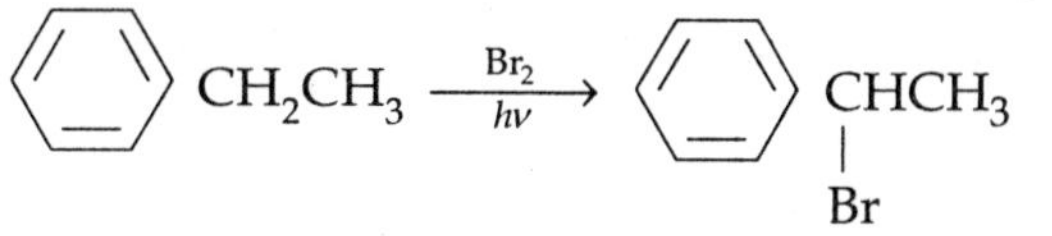 $C_6H_5CH_2CH_3 \xrightarrow[h\nu]{Br_2} C_6H_5CH(Br)CH_3$

major product

105. (a) 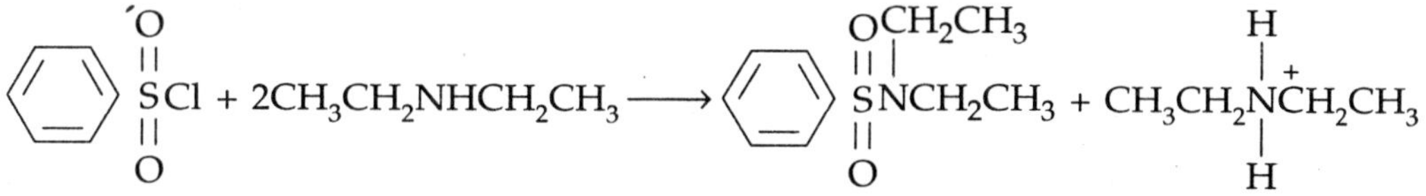 $C_6H_5SO_2Cl + 2CH_3CH_2NHCH_2CH_3 \longrightarrow C_6H_5SO_2N(CH_2CH_3)CH_2CH_3 + CH_3CH_2\overset{+}{N}H_2CH_2CH_3$

(b) $C_6H_5SO_2Cl + CH_3CH_2CH_2CH_2NH_2 \xrightarrow[H_2O]{NaOH} C_6H_5SO_2\bar{N}CH_2CH_2CH_2CH_3\ Na^+$

(c) $C_6H_5SO_2\bar{N}CH_2CH_2CH_2CH_3\ Na^+ \xrightarrow{H_3O^+} C_6H_5SO_2NHCH_2CH_2CH_2CH_3$

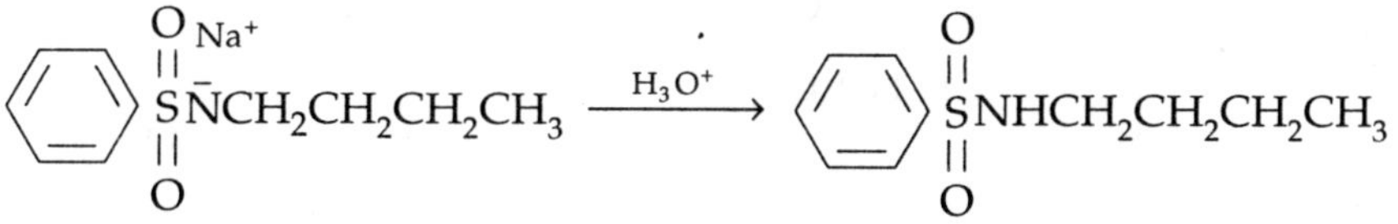

(d) 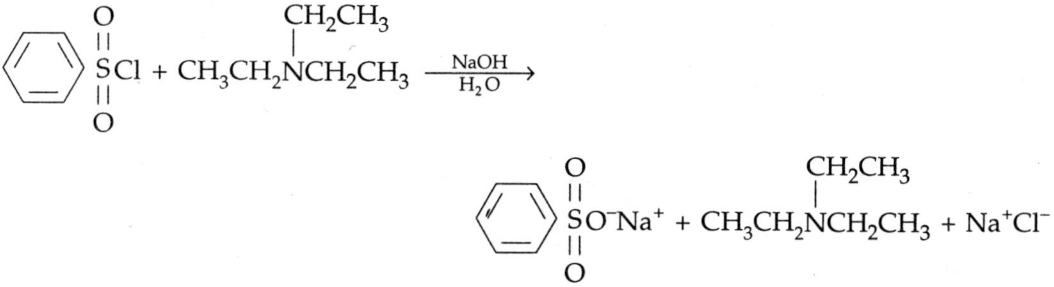 $C_6H_5SO_2Cl + CH_3CH_2N(CH_2CH_3)CH_2CH_3 \xrightarrow[H_2O]{NaOH}$

$C_6H_5SO_2O^-Na^+ + CH_3CH_2N(CH_2CH_3)CH_2CH_3 + Na^+Cl^-$

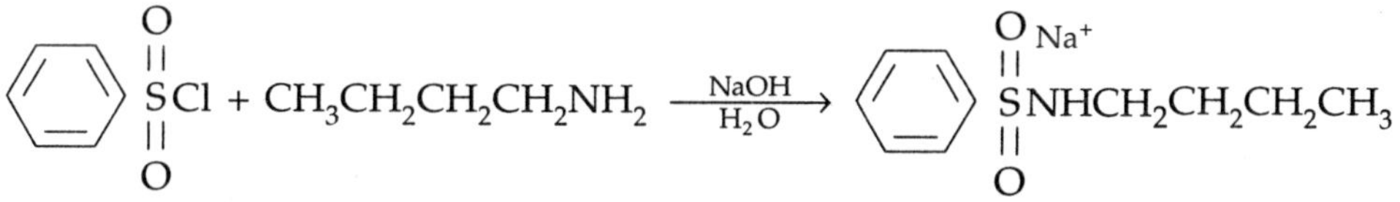

106. (a) 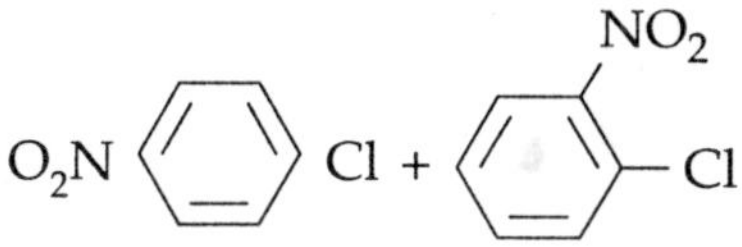 O_2N–C_6H_4–Cl (para) + 2-nitrochlorobenzene (NO_2 and Cl ortho)

A B

(b) 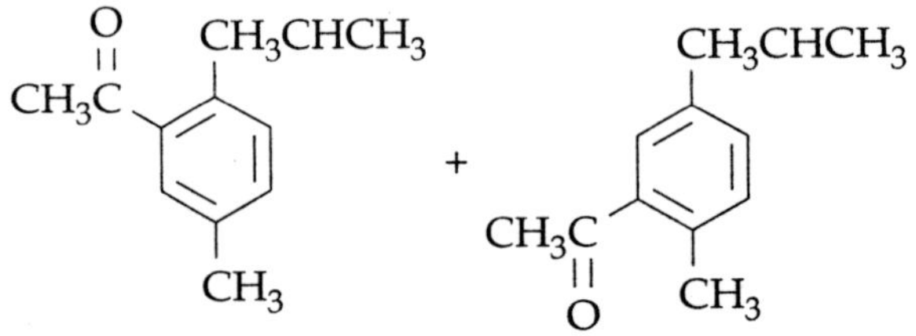 $CH_3\overset{O}{\overset{\|}{C}}$, CH_3CHCH_3, CH_3 + CH_3CHCH_3, $CH_3\underset{O}{\underset{\|}{C}}$, CH_3

C D

(c) 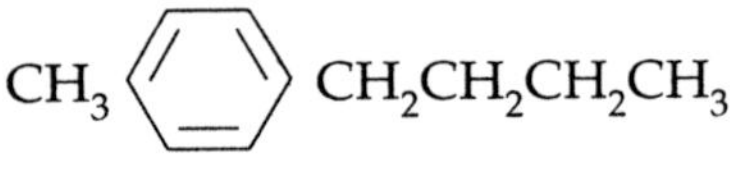 CH_3–C_6H_4–$CH_2CH_2CH_2CH_3$

E

(d) 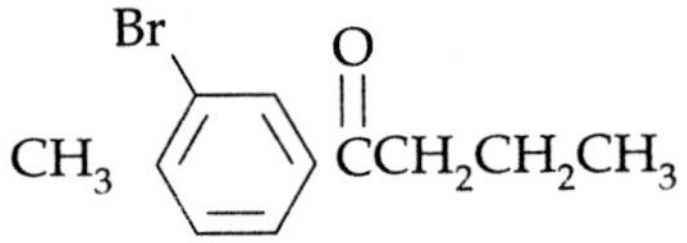 CH_3, Br, $\overset{O}{\overset{\|}{C}}CH_2CH_2CH_3$

F

(e) 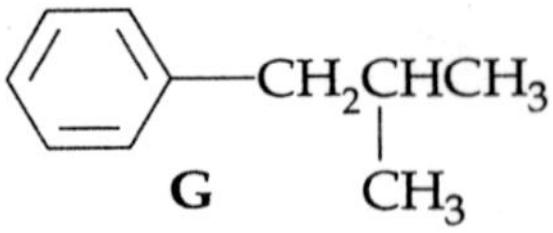 C_6H_5–$CH_2\underset{CH_3}{\underset{|}{C}}HCH_3$

G

(f) 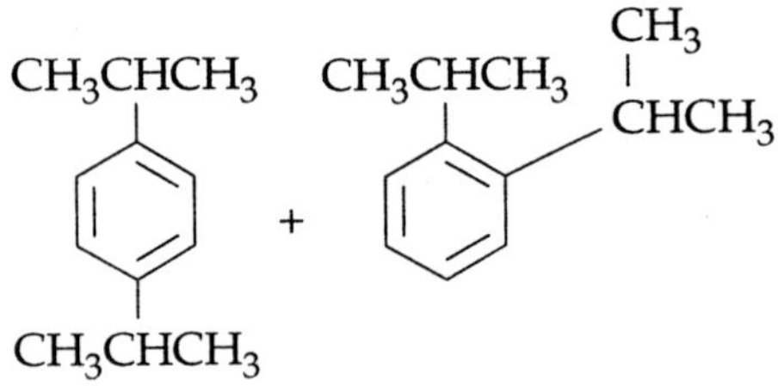 CH_3CHCH_3, CH_3CHCH_3 + CH_3CHCH_3, $\overset{CH_3}{\overset{|}{C}}HCH_3$

H I

(g) 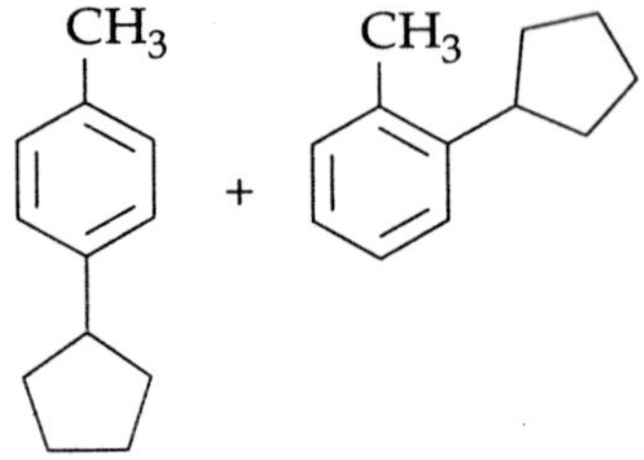 CH_3 + CH_3

J K

(h) 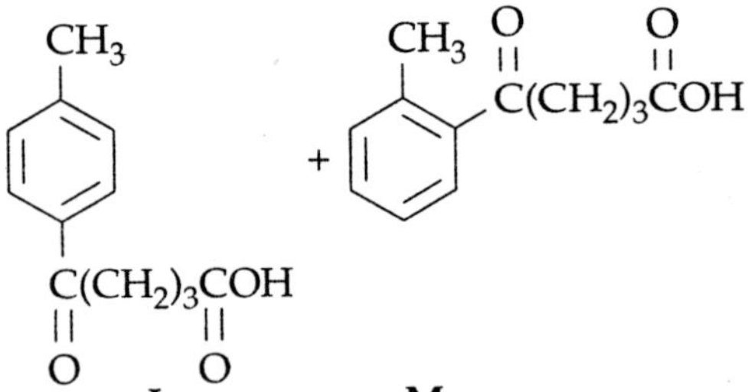 CH_3, $\underset{O}{\underset{\|}{C}}(CH_2)_3\underset{O}{\underset{\|}{C}}OH$ + CH_3, $\overset{O}{\overset{\|}{C}}(CH_2)_3\overset{O}{\overset{\|}{C}}OH$

L M

(i) 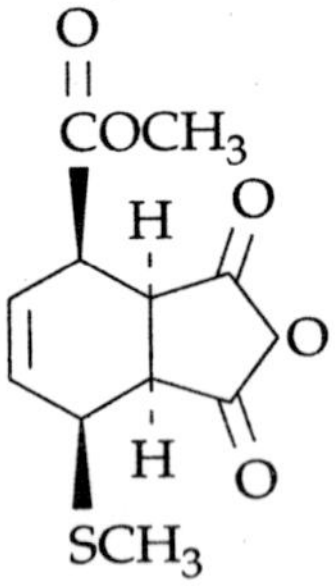 $\overset{O}{\overset{\|}{C}}OCH_3$, H, O, O, H, O, SCH_3

N

(j) 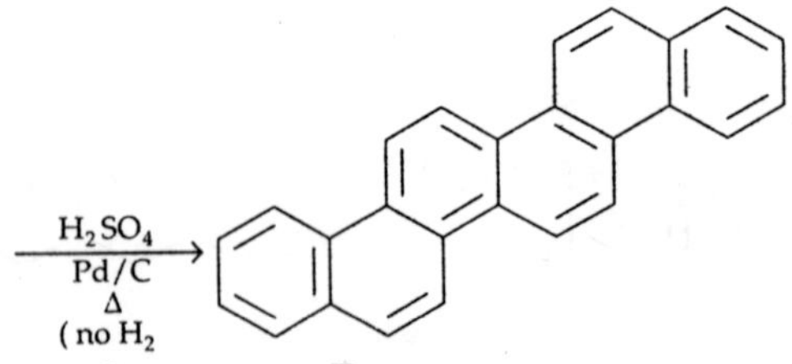 $\xrightarrow[\substack{\Delta \\ (\text{no } H_2}]{H_2SO_4 \\ Pd/C}$

O P

$C_{26}H_{16}$

107. (a) 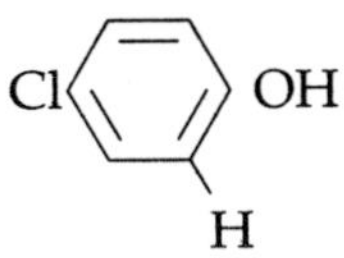 Cl OH H **A** (b) 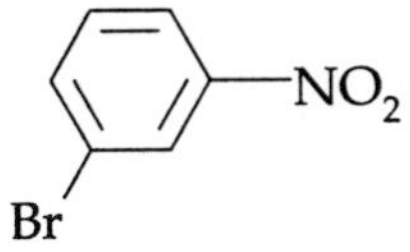 NO_2 Br **B**

(c) 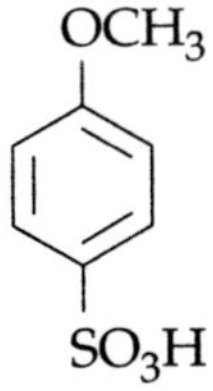 OCH_3 SO_3H + OCH_3 SO_3H (d) CH_3 O=C HOC O + CH_3 O C HOC O

C **D** 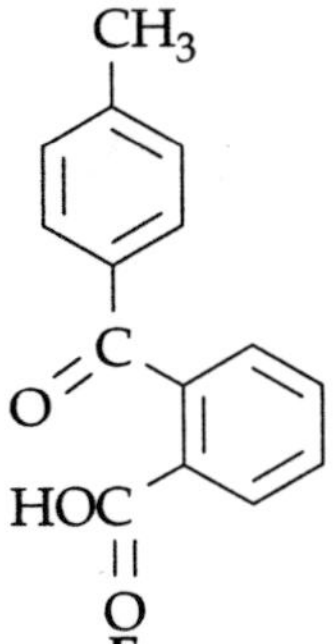 **E** **F**

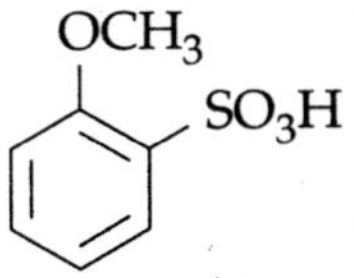

(e) 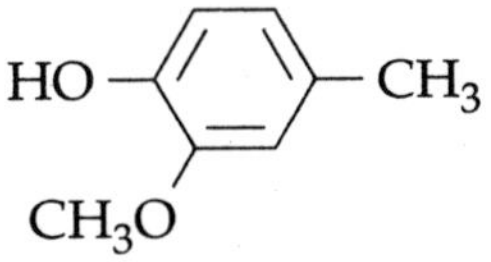 HO CH_3 CH_3O (f) 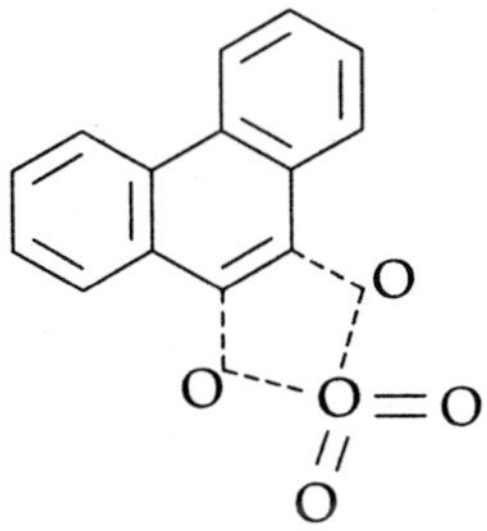 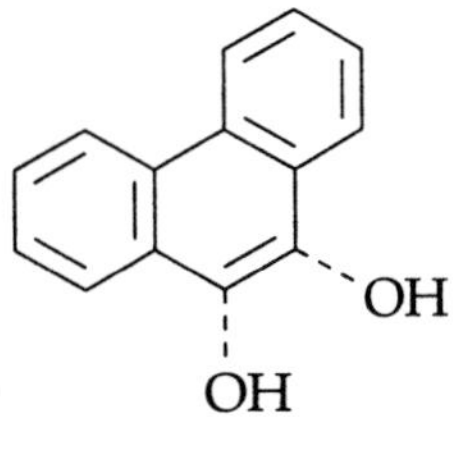 O O O=O O OH OH

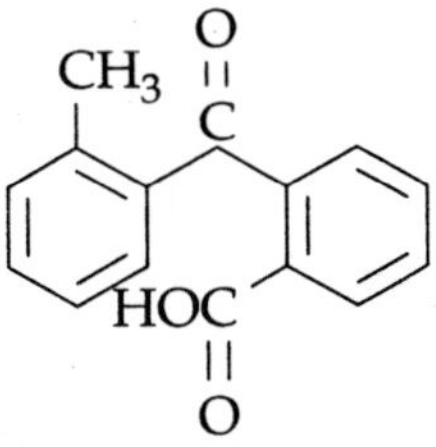

G **H** **I**

(g) 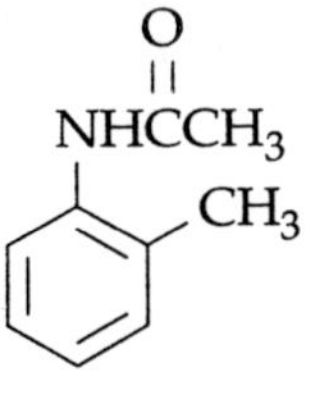 O NHCCH$_3$ CH_3 ; O NHCCH$_3$ CH_3 NO_2 ; O O_2N NHCCH$_3$ CH_3 ; NH_3 CH_3 NO_2 + O_2N $^+NH_3$ CH_3

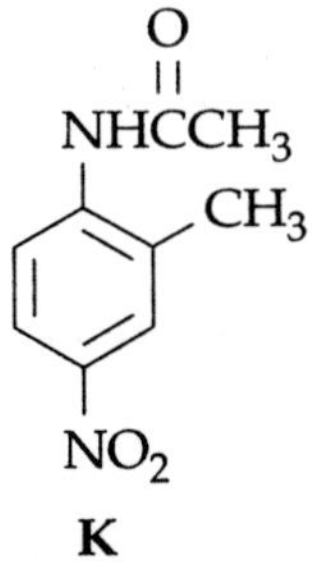

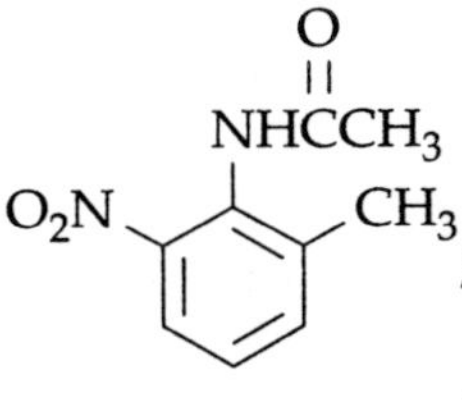

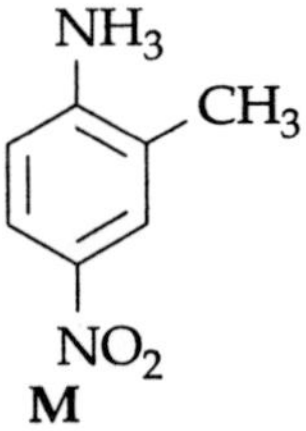

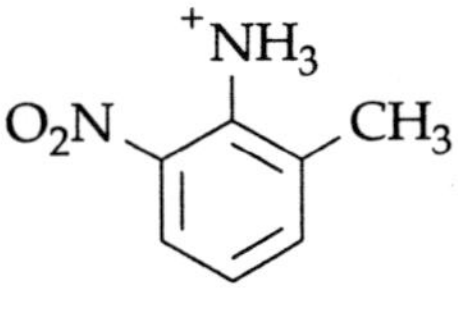

J **K** **L** **M** **N**

(h) 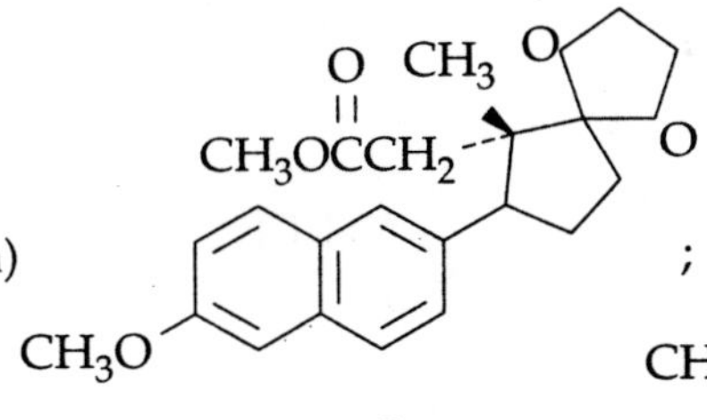 O CH_3 O $CH_3O\overset{\|}{C}CH_2$ O CH_3O ; O CH_3 O K^+ $^-O\overset{\|}{C}CH_2$ O CH_3O ; CH_3 O O CH_3O

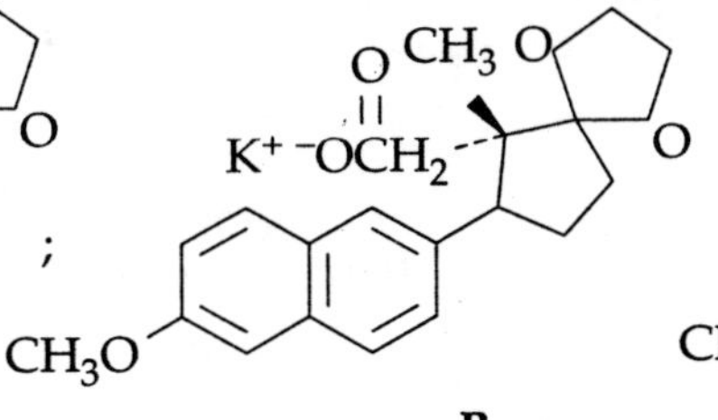

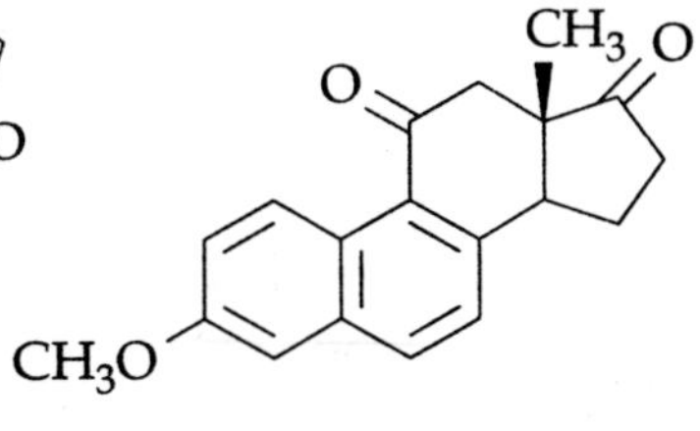

O **P** **Q**

(i) 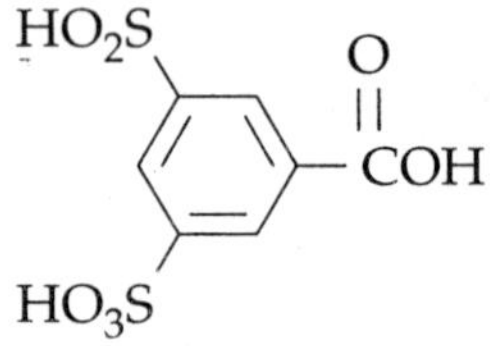 HO_2S, HO_3S ... $\overset{O}{\overset{||}{C}}OH$

R

(j) 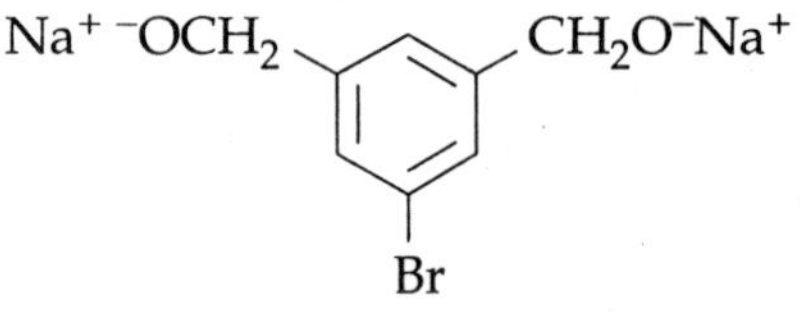 $Na^+\ ^-OCH_2$ ... $CH_2O^-Na^+$ ... Br ; CH_3OCH_2 ... CH_2OCS_3 ... Br

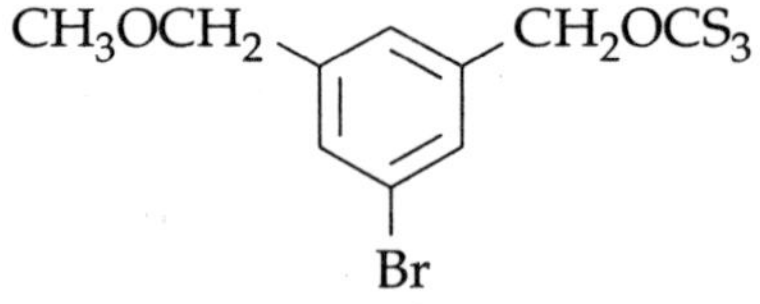

S **T**

108. (a) 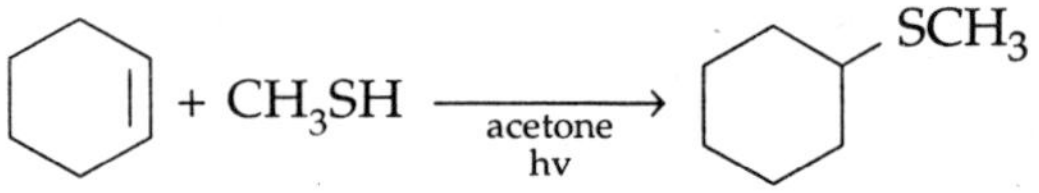 cyclohexene + CH_3SH $\xrightarrow[hv]{acetone}$ SCH_3-cyclohexane

(b) $CH_3(CH_2)_3CH{=}CH_2 \xrightarrow[peroxides]{HBr} CH_3(CH_2)_7Br$

(c) 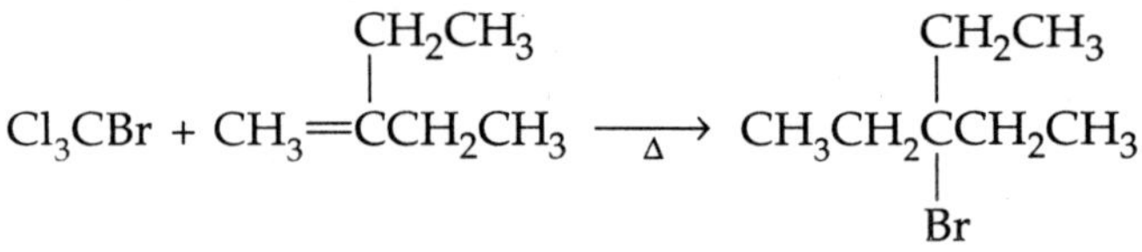 $Cl_3CBr + CH_3{=}C(CH_2CH_3)CH_2CH_3 \xrightarrow{\Delta} CH_3CH_2C(CH_2CH_3)(Br)CH_2CH_3$

(d) $CH_3(CH_2)_4SiH_3 + CH_3(CH_2)_3CH{=}CH_2 \xrightarrow{proxides} CH_3(CH_2)_4SiH_2(CH_2)_7CH_3$

109. (a) 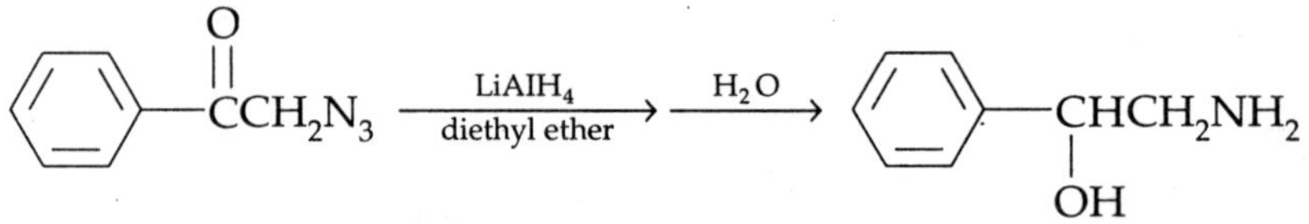 $C_6H_5\overset{O}{\overset{||}{C}}CH_2N_3 \xrightarrow[diethyl\ ether]{LiAlH_4} \xrightarrow{H_2O} C_6H_5CH(OH)CH_2NH_2$

(b) $CH_3(CH_2)_4C{\equiv}C(CH_2)_8N_3 \xrightarrow[Pd/CaCO_3,\ quinoline]{H_2}$ $CH_3(CH_2)_4$(H)C=C(H)$(CH_2)_8NH_2$

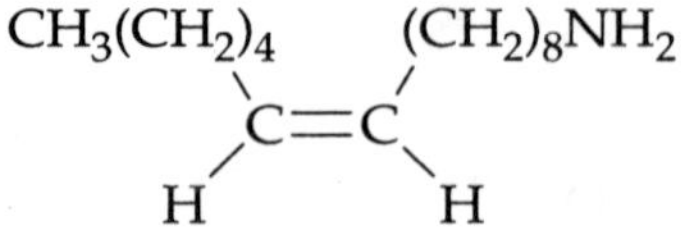

110. 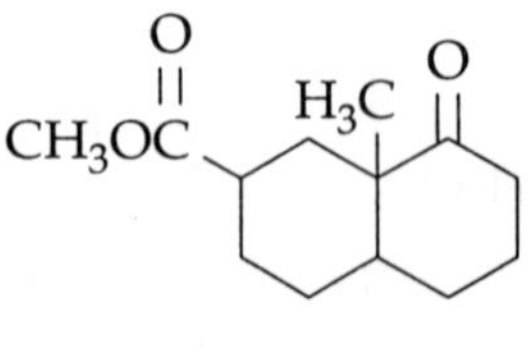 $CH_3O\overset{O}{\overset{||}{C}}$, H_3C, O ; $CH_3O\overset{O}{\overset{||}{C}}$, H_3C, NOH ; $CH_3O\overset{O}{\overset{||}{C}}$, H_3C, $^+NH_3Cl^-$

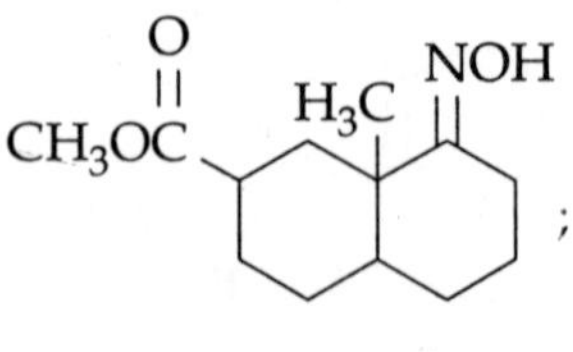

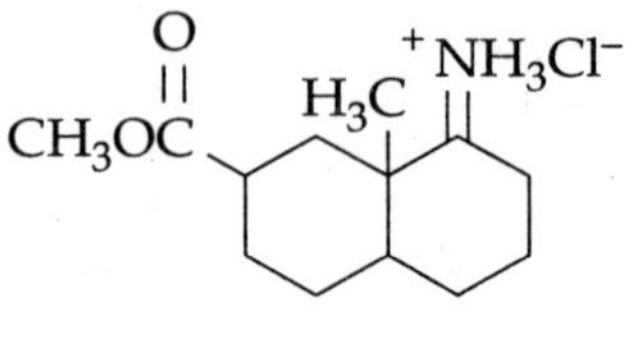

A **B** **C**

111. A: methyl (S)-2-amino-3-hydroxypropanoate, $HOCH_2CH(NH_2)COOCH_3$; B: [$HOCH_2CH(N{=}CH{-}C_6H_5)COOCH_3$] ; C: $HOCH_2CH(\overset{+}{N}H_2CH_2{-}C_6H_5)COOCH_3$ Cl^- ; D: NH_3

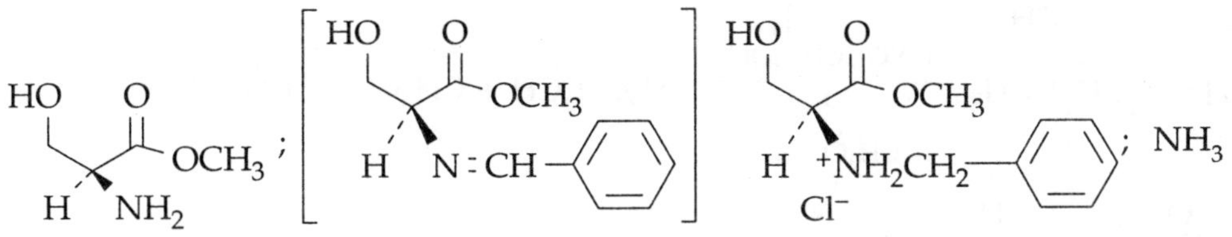

A **B** **C** **D**

112. (a) $C_6H_5CHO + NH_3 \xrightarrow[RaNi]{H_2} C_6H_5CH_2NH_2$

(b) $C_6H_5CH_2CH_2N_3 \xrightarrow[\text{diethyl ether}]{LiAlH_4} \xrightarrow{H_2O} C_6H_5CH_2CH_2NH_2$

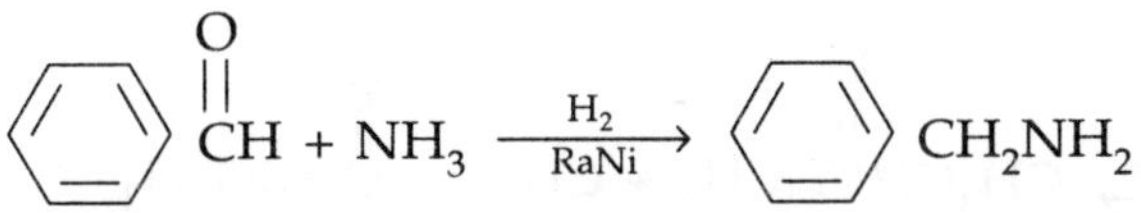

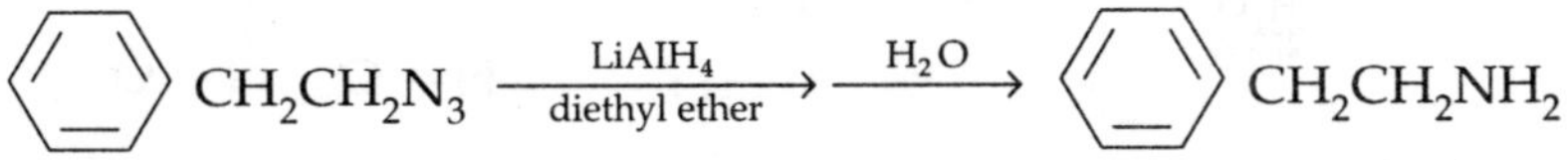

(c) $O_2N{-}C_6H_4{-}CH_2COOH \xrightarrow{(NH_4)_2S} H_2N{-}C_6H_4{-}CH_2COOH$

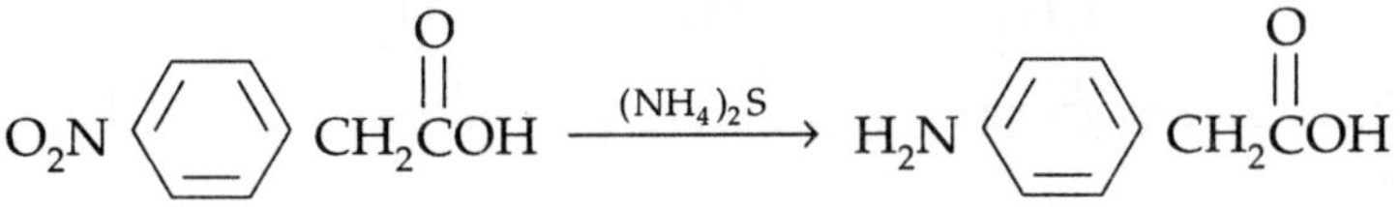

(d) cyclohexene oxide $\xrightarrow[\substack{H_2O \\ \text{dioxane}}]{NaN_3}$ trans-2-azidocyclohexanol (N₃, OH) $\xrightarrow[PtO_2]{H_2}$ trans-2-aminocyclohexanol (NH_2, OH)

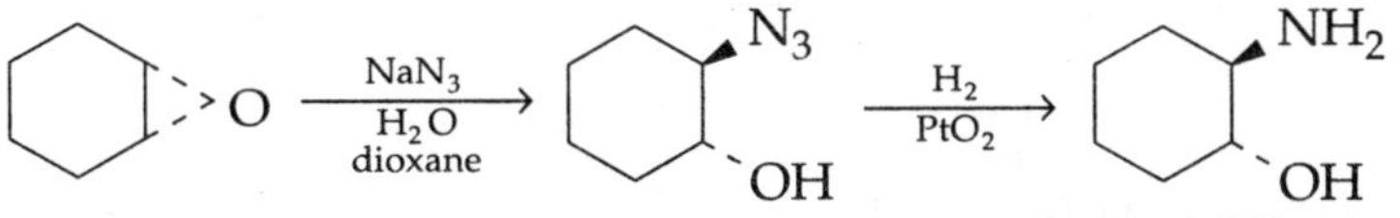

and enantiomer and enantiomer

(e) 1-cyclohexenyl-CH_2CN $\xrightarrow[\text{diethyl ether}]{LiAlH_4} \xrightarrow{H_2O}$ 1-cyclohexenyl-$CH_2CH_2NH_2$

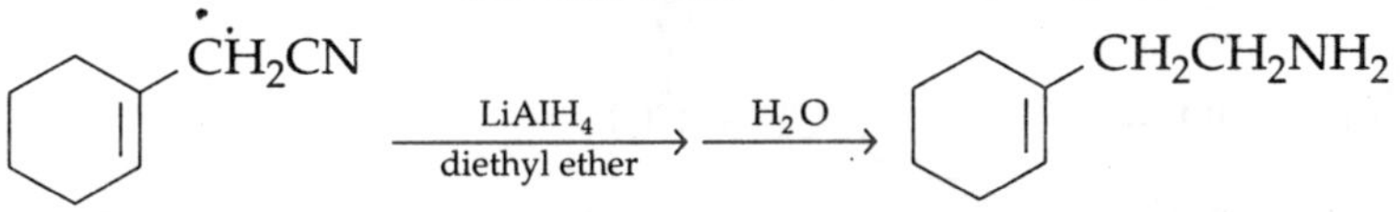

113. (a) $CH_3CH_2CH_2CH(NH_2)CH_3 \xrightarrow[\substack{H_2O \\ 273\text{ K}}]{NaNO_2,\ HCl} CH_3CH_2CH_2CH(Cl)CH_3 + CH_3CH_2CH_2CH(OH)CH_3$

$CH_3CH_2CH(Cl)CH_2CH_3 + CH_3CH_2CH(OH)CH_2CH_3 +$ (H)(CH₃)C=C(CH₂CH₃)(H) (trans)

$+$ (CH₃)(H)C=C(CH₂CH₃)(H) (cis) $+ CH_3CH_2H_2CH{=}CH_2 + N_2\uparrow$

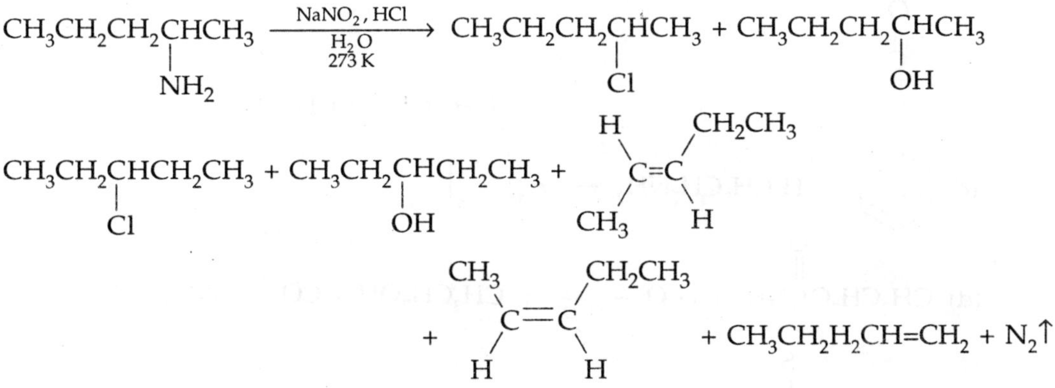

(b) piperidine (N–H) $\xrightarrow[\substack{H_2O \\ \Delta}]{NaNO_2,\ HCl}$ N-nitrosopiperidine (N–N=O)

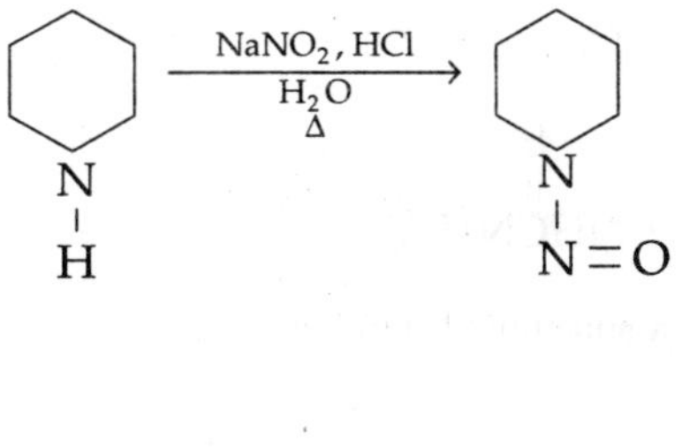

(c) $$CH_3CH_2CH_2\underset{}{\overset{CH_3}{\overset{|}{N}}}CH_3 \xrightarrow[\substack{\overset{O}{\overset{||}{}}\\ CH_3CO^-Na^+\\ \Delta}]{\substack{\overset{O}{\overset{||}{}}\\ NaNO_2,\ CH_3COH}} CH_3CH_2\overset{O}{\overset{||}{C}}H + CH_3CH_2CH_2\overset{CH_3}{\overset{|}{N}}{-}N{=}O$$

$$+ H\overset{O}{\overset{||}{C}}H + CH_3\overset{CH_3}{\overset{|}{N}}{-}N{=}O$$

(d) $$CH_3NH\overset{O}{\overset{||}{C}}NHCH_3 \xrightarrow[\substack{H_2O\\ 273\text{ K}}]{NaNO_2,\ HCl} CH_3NH\overset{O}{\overset{||}{C}}\underset{\underset{N=O}{|}}{N}CH_3 + CH_3\underset{\underset{N=O}{|}}{N}{-}\overset{O}{\overset{||}{C}}{-}\underset{\underset{N=O}{|}}{N}CH_3$$

(e) $$CH_3\underset{\underset{CH_3}{|}}{\overset{CH_3}{\overset{|}{C}}}NH_2 \xrightarrow[\substack{H_2O\\ 273\text{ K}}]{NaNO_2,\ HCl} CH_3\underset{\underset{Cl}{|}}{\overset{CH_3}{\overset{|}{C}}}CH_3 + CH_3\underset{\underset{OH}{|}}{\overset{CH_3}{\overset{|}{C}}}CH_3 + CH_3\overset{CH_3}{\overset{|}{C}}{=}CH_2 + N_2\uparrow$$

114. (a) $$Cl\overset{O}{\overset{||}{C}}Cl + NH_3(\text{excess}) \longrightarrow H_2N\overset{O}{\overset{||}{C}}NH_2$$

(b) $$Cl\overset{O}{\overset{||}{C}}Cl + \underset{(1\text{ molar equiv})}{C_6H_5CH_2OH} \longrightarrow C_6H_5CH_2O\overset{O}{\overset{||}{C}}Cl$$

(c) $$C_6H_5{-}N{=}C{=}O + CH_3CH_2CH_2NH_2 \longrightarrow C_6H_5{-}NH\overset{O}{\overset{||}{C}}NHCH_2CH_2CH_3$$

(d) $$CH_3CH_2O\overset{O}{\overset{||}{C}}NH_2 + H_2O \xrightarrow[\Delta]{H_3O^+} CH_3CH_2OH + CO_2\uparrow + NH_4^+$$

(e) $$\underset{\text{phenyl isothiocyanate}}{C_6H_5{-}N{=}C{=}S} + C_6H_5NH_2 \longrightarrow \underset{\text{a substituted thiourea}}{C_6H_5NH\overset{S}{\overset{||}{C}}NHC_6H_5}$$

115. (a) A: $CH_3\overset{+}{N}HCH_3$ / cyclohexadiene ring / $CH_3\overset{+}{N}HCH_3$; B: CH_3NCH_3 / ring / CH_3NCH_3 ; C: $CH_3\overset{+}{N}CH_3$ / ring / $CH_3\overset{+}{N}CH_3$ I^- with CH_3 ; D (benzene) + $2CH_3\overset{CH_3}{N}CH_3$ E

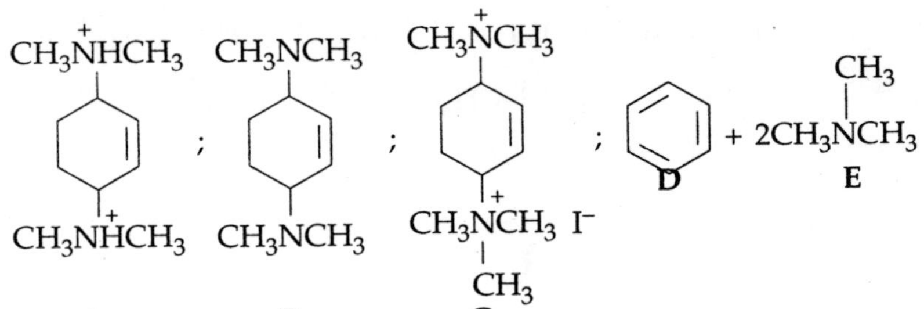

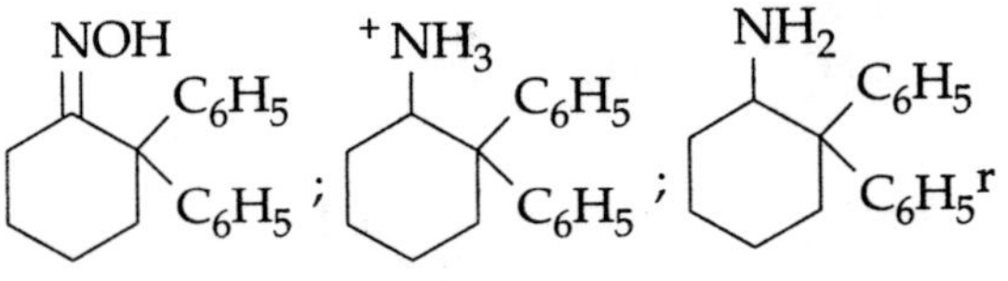

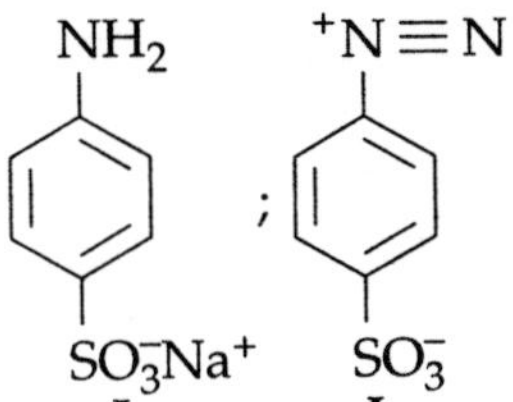

A B C

(b) F: NOH, C_6H_5, C_6H_5 ; G: $^+NH_3$, C_6H_5, C_6H_5 ; H: NH_2, C_6H_5, C_6H_5

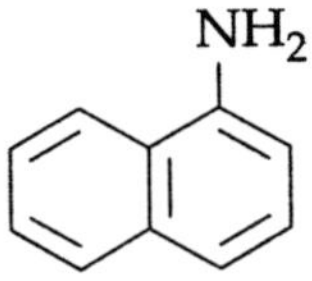

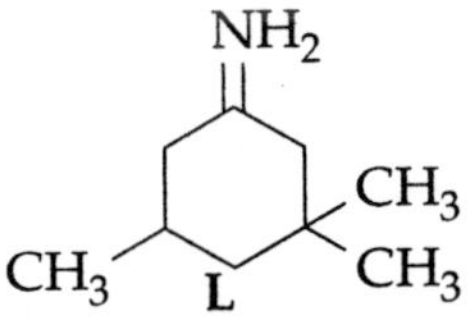

F G H

(c) I: NH_2 / benzene ring / $SO_3^-Na^+$; J: $^+N{\equiv}N$ / benzene ring / SO_3^-

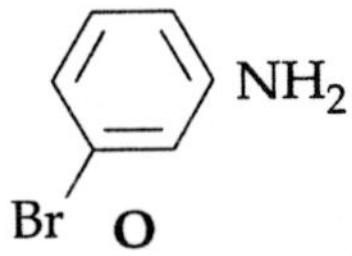

I J

(d) K: NH_2 on naphthalene

K

(e) L: NH_2 (=), CH_3, CH_3, CH_3

L

(f) M: $H\overset{+}{N}{=}\overset{CH_3}{C}CH_3$ on cyclohexane ; N: $H\overset{+}{N}{=}\overset{CH_3}{C}CH_3$ on cyclohexane

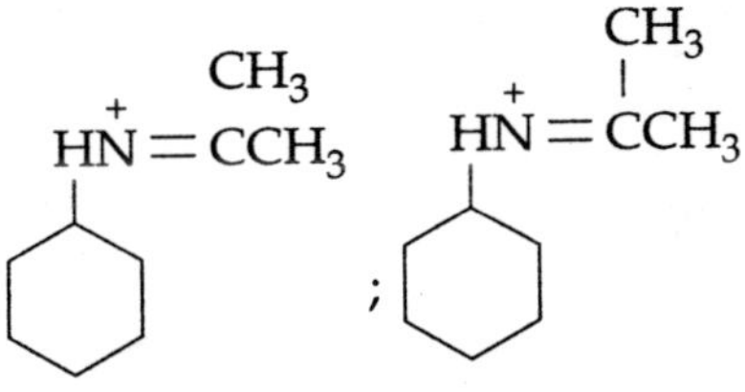

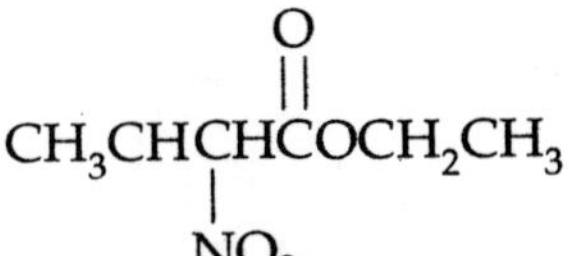

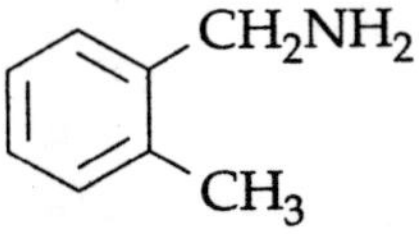

M N

(g) O: Br, NH_2

O

(h) $\xrightarrow[\text{pyridine}]{\text{TsCl}}$ P; N, H–N, Cl, Cl → Q

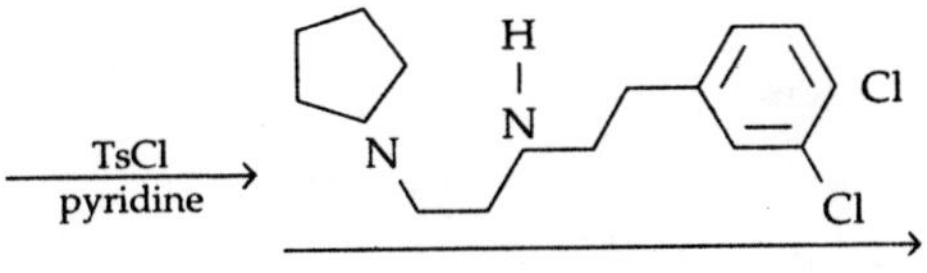

P Q

116. (a) $CH_3CH\underset{NO_2}{C}H\overset{O}{\overset{||}{C}}OCH_2CH_3$ A

(b) CH_2NH_2, CH_3 B

(c) $CH_3NH_2 + HCO_3^-$ C

(d) CH_3-ring-N=CH-ring D ; CH_3-ring-$NHCH_2$-ring E

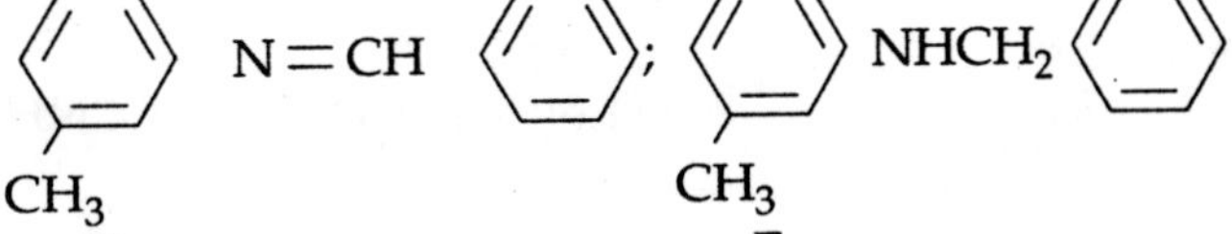

D E

(e) 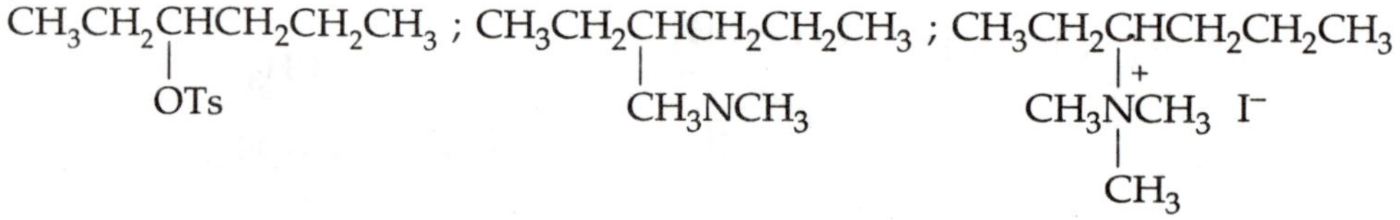 $CH_3CH_2CH(OTs)CH_2CH_2CH_3$; $CH_3CH_2CH(N(CH_3)CH_3)CH_2CH_2CH_3$; $CH_3CH_2CH(\overset{+}{N}(CH_3)_3)CH_2CH_2CH_3\ I^-$

F **G** **H**

(f) 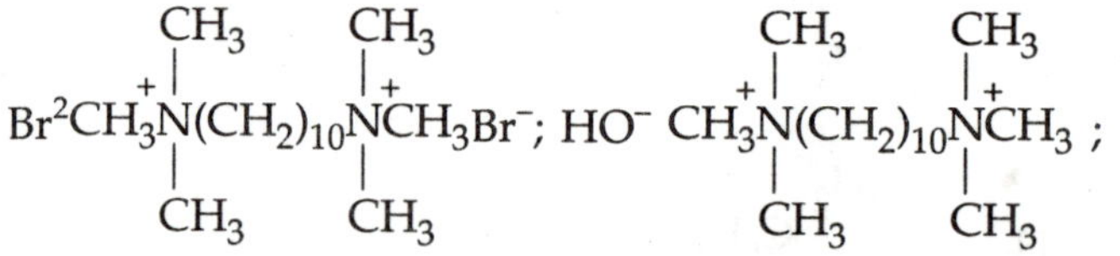 $Br^{2}CH_3\overset{+}{N}(CH_3)_2(CH_2)_{10}\overset{+}{N}(CH_3)_2CH_3Br^-$; $HO^-\ CH_3\overset{+}{N}(CH_3)_2(CH_2)_{10}\overset{+}{N}(CH_3)_2CH_3$;

I **J**

$CH_2{=}CH(CH_2)_6CH{=}CH_2 + 2CH_3N(CH_3)CH_3$

K

(g) 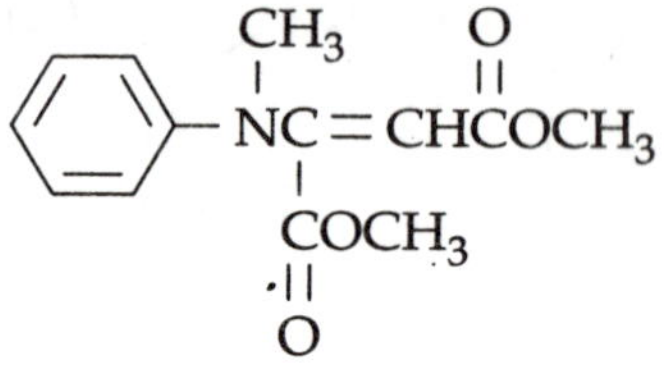 $C_6H_5{-}N(CH_3)C(COOCH_3){=}CHCOOCH_3$

L

(h) 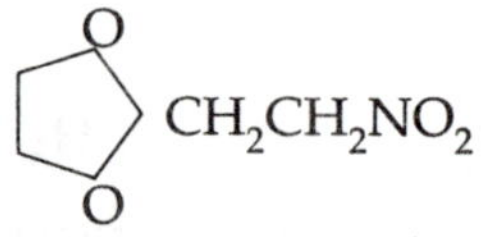 1,3-dioxolan-2-yl $CH_2CH_2NO_2$

M

(i) 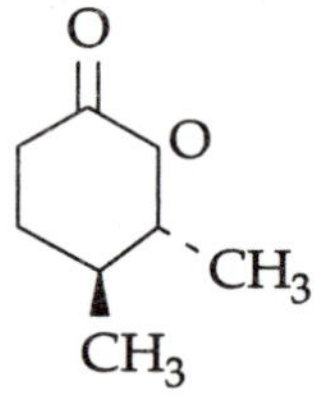 six-membered lactone (C=O, ring O) with two CH_3 substituents

N

(j) $\xrightarrow{LiAlH_4}$

O

117. (a) 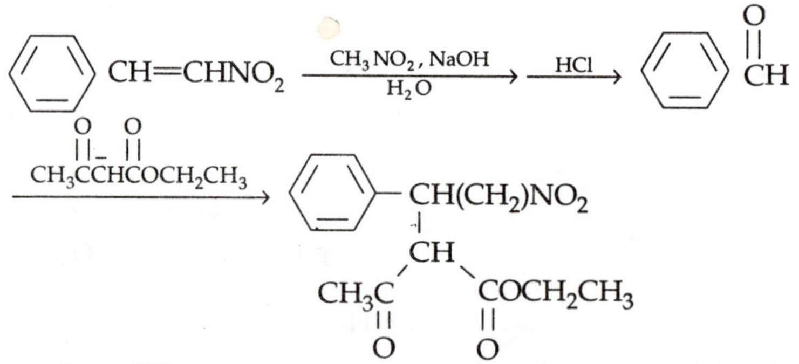 $C_6H_5CH{=}CHNO_2 \xrightarrow[H_2O]{CH_3NO_2,\ NaOH} \xrightarrow{HCl} C_6H_5CHO$

$\xrightarrow{CH_3CO\bar{C}HCOOCH_2CH_3} C_6H_5CH(CH_2)NO_2$ with CH bearing $CH_3C(=O)$ and $COCH_2CH_3(=O)$, i.e. $C_6H_5CH(CH_2NO_2)CH(COCH_3)COOCH_2CH_3$

(b) 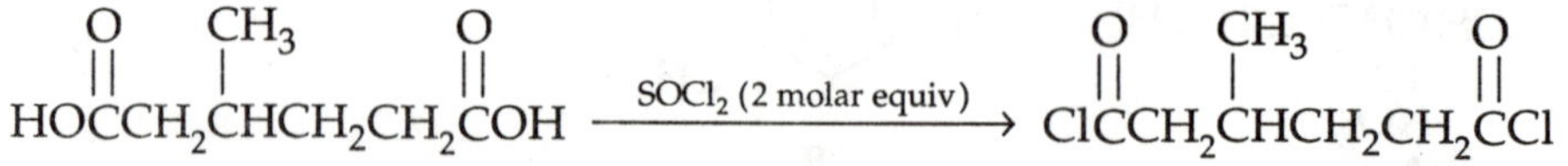 $HOOCCH_2CH(CH_3)CH_2CH_2COOH \xrightarrow{SOCl_2\ (2\ molar\ equiv)} ClCOCH_2CH(CH_3)CH_2CH_2COCl$

$\xrightarrow{NH_3(excess)} H_2NCOCH_2CH(CH_3)CH_2CH_2CONH_2 \xrightarrow{Br_2\ NaOH} H_2NCH_2CH(CH_3)CH_2CH_2CH_2NH_2$

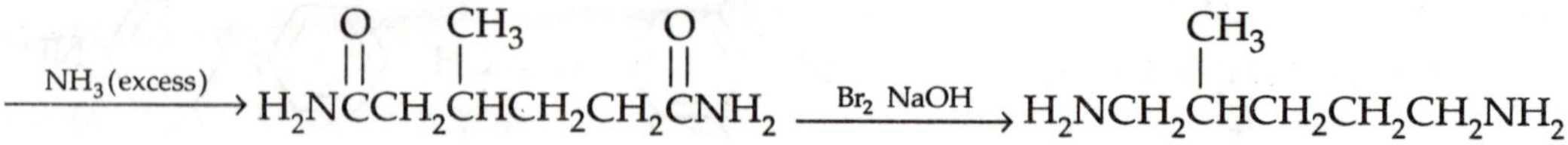

(c) $CH_3CH_2\overset{O}{\overset{\|}{C}}CH_2CH_3 \xrightarrow[Na_2CO_3]{HO\overset{+}{N}H_3\ Cl^-} CH_3CH_2\overset{NOH}{\overset{\|}{C}}CH_2CH_3 \xrightarrow[\Delta]{H_2SO_4} CH_3CH_2\overset{O}{\overset{\|}{C}}NHCH_2CH_3$

(d) 1. 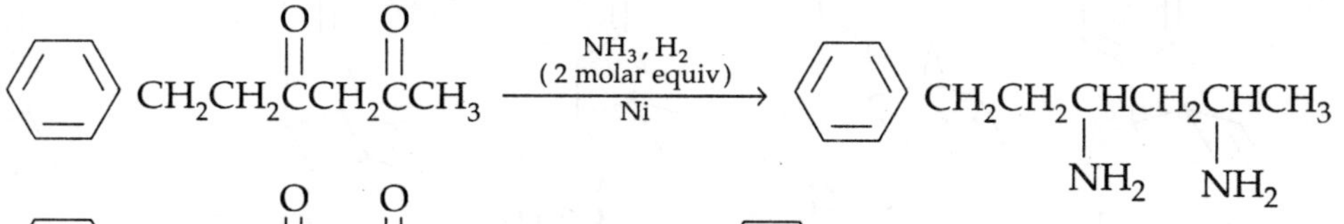 $C_6H_5CH_2CH_2\overset{O}{\overset{\|}{C}}CH_2\overset{O}{\overset{\|}{C}}CH_3 \xrightarrow[Ni]{NH_3, H_2\ (2\ molar\ equiv)} C_6H_5CH_2CH_2\underset{NH_2}{\underset{|}{C}}HCH_2\underset{NH_2}{\underset{|}{C}}HCH_3$

2. $C_6H_5CH_2CH_2\overset{O}{\overset{\|}{C}}CH_2\overset{O}{\overset{\|}{C}}CH_3 \xrightarrow[methanol]{NaBH_4} C_6H_5CH_2CH_2\underset{OH}{\underset{|}{C}}HCH_2\underset{OH}{\underset{|}{C}}HCH_3$

$\xrightarrow[pyridine]{TsCl} C_6H_5CH_2CH_2\underset{OTs}{\underset{|}{C}}HCH_2\underset{OTs}{\underset{|}{C}}HCH_3$

$\xrightarrow{NaN_3} C_6H_5CH_2CH_2\underset{N_3}{\underset{|}{C}}HCH_2\underset{N_3}{\underset{|}{C}}HCH_3$

$\xrightarrow[diethyl\ ether]{LiAlH_4} \xrightarrow{H_2O} C_6H_5CH_2CH_2\underset{NH_2}{\underset{|}{C}}HCH_2\underset{NH_2}{\underset{|}{C}}HCH_3$

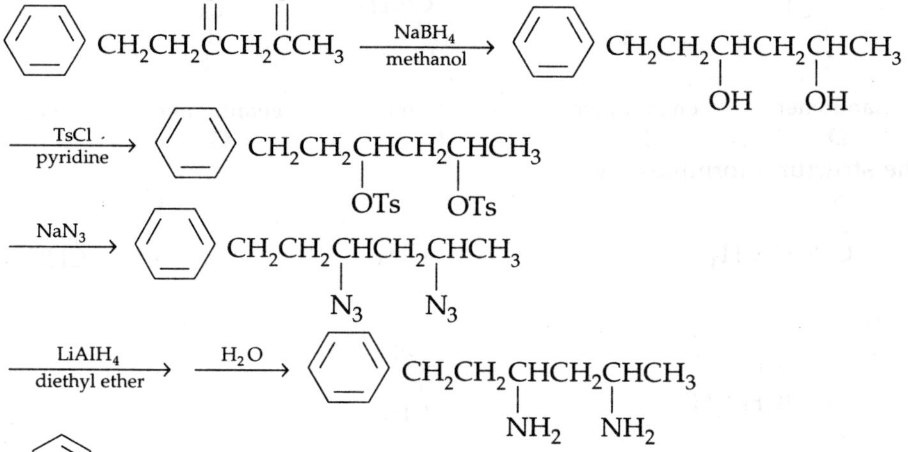

(e) 1. 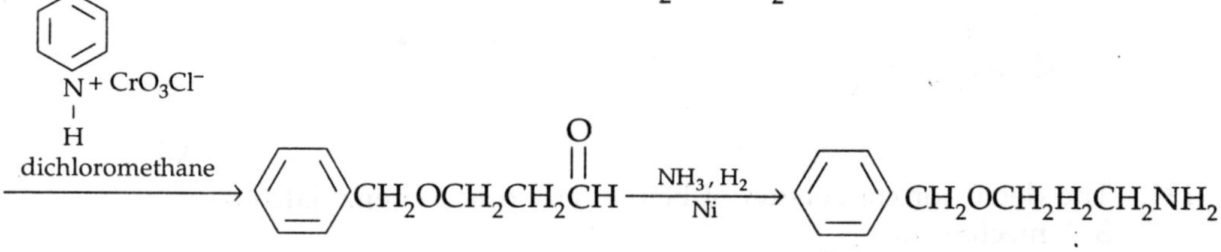 $\xrightarrow[dichloromethane]{C_5H_5\overset{+}{N}H\ CrO_3Cl^-} C_6H_5CH_2OCH_2CH_2\overset{O}{\overset{\|}{C}}H \xrightarrow[Ni]{NH_3, H_2} C_6H_5CH_2OCH_2H_2CH_2NH_2$

118. (a) 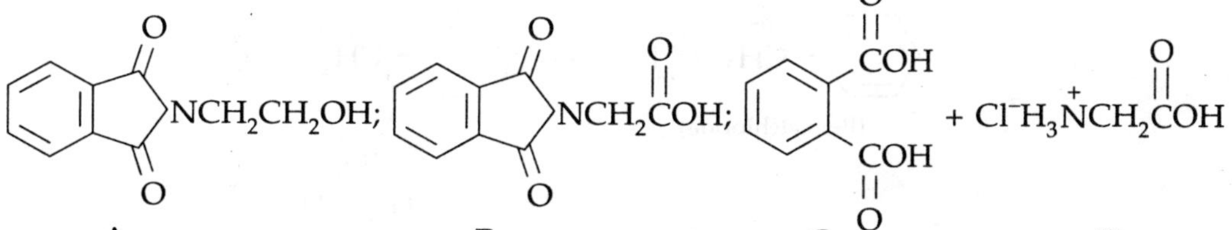 N-(2-hydroxyethyl)phthalimide, NCH_2CH_2OH; N-phthaloylglycine, $NCH_2\overset{O}{\overset{\|}{C}}OH$; phthalic acid (benzene with two $\overset{O}{\overset{\|}{C}}OH$); $+ Cl^-H_3\overset{+}{N}CH_2\overset{O}{\overset{\|}{C}}OH$

A B C D

119. (a) 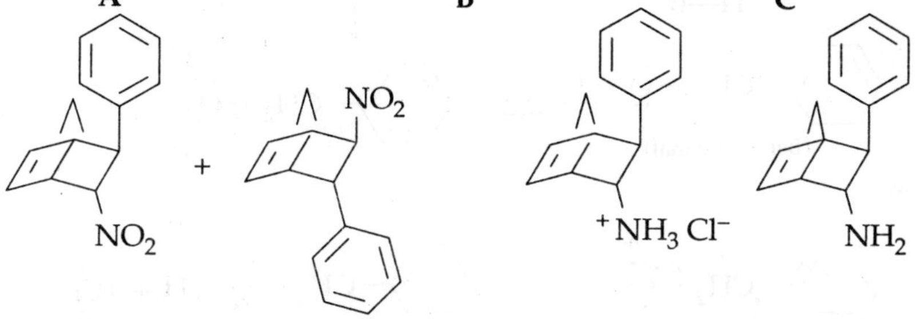 [bicyclic structure with phenyl and NO_2] + [bicyclic structure with NO_2 and phenyl]; [bicyclic structure with phenyl and $^+NH_3\ Cl^-$]; [bicyclic structure with phenyl and NH_2]

enantiomer enantiomer

A B C

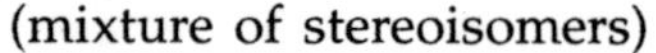

enantiomer enantiomer

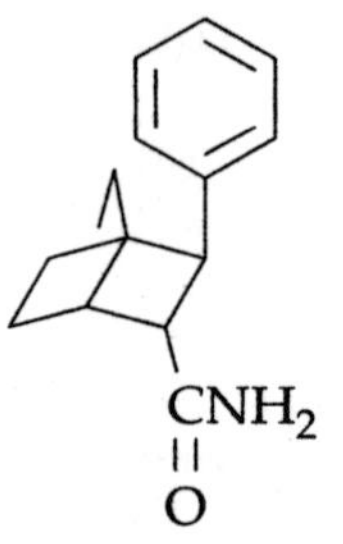

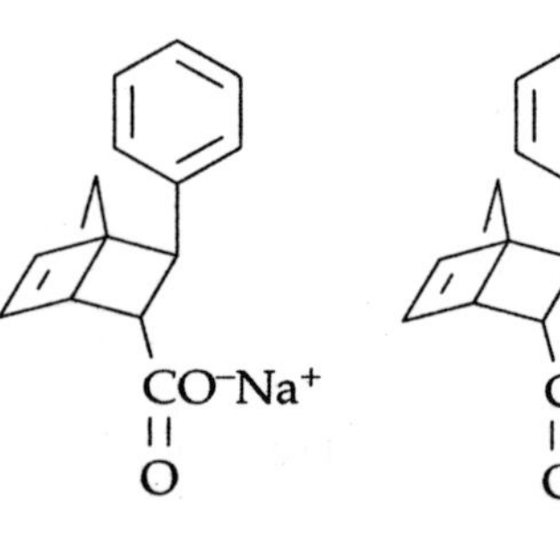

CCl, O — enantiomer **D**; CNH$_2$, O — enantiomer **E**; CNH$_2$, O — enantiomer **F**; CO$^-$Na$^+$, O — enantiomer **G**; COH, O — enantiomer **H**

120. The structural formulas are:

(a) 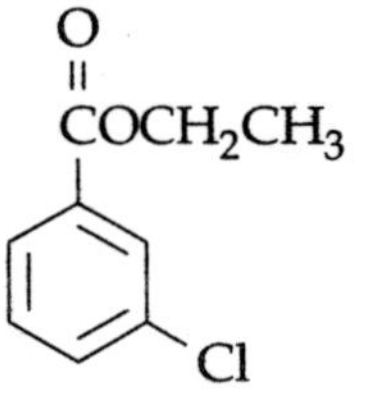 $C_6H_4(Cl)COCH_2CH_3$ (b) 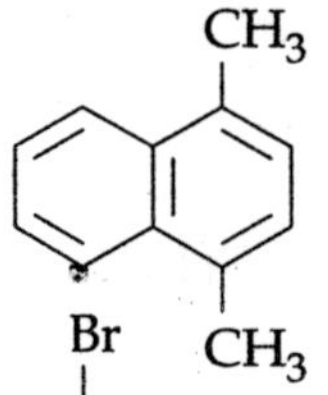 (c) 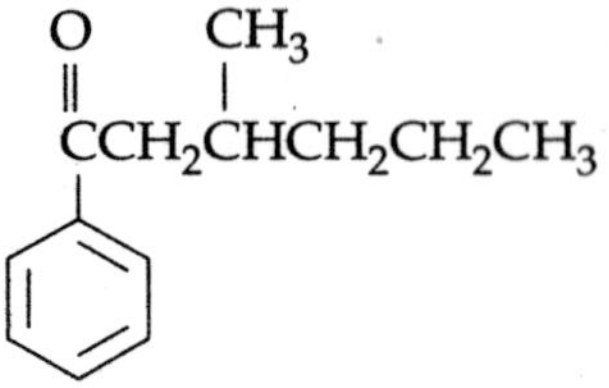 $C_6H_5COCH_2CH(CH_3)CH_2CH_2CH_3$

(d) 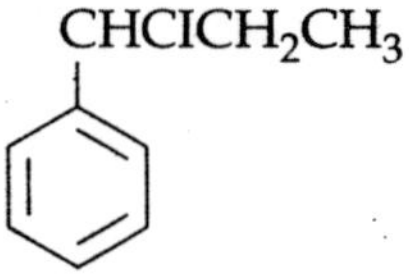 $C_6H_5CHClCH_2CH_3$ (e) 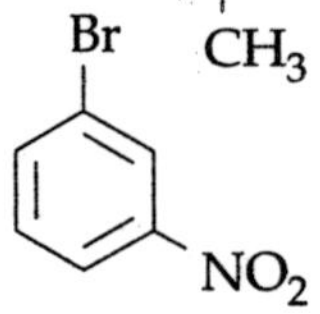 Br, NO_2 (f)  Br, Cl

121. The detail for conversion of benzyl chloride to benzylalcohal is :
S_N1 mechanisms

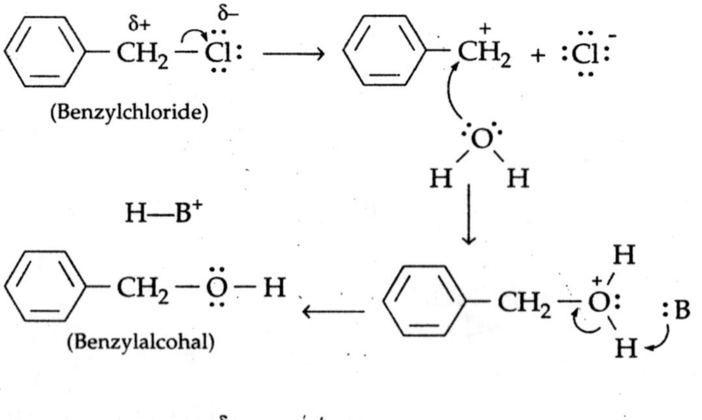

(Benzylchloride)

H—B$^+$

(Benzylalcohal)

S_N2 mechanism

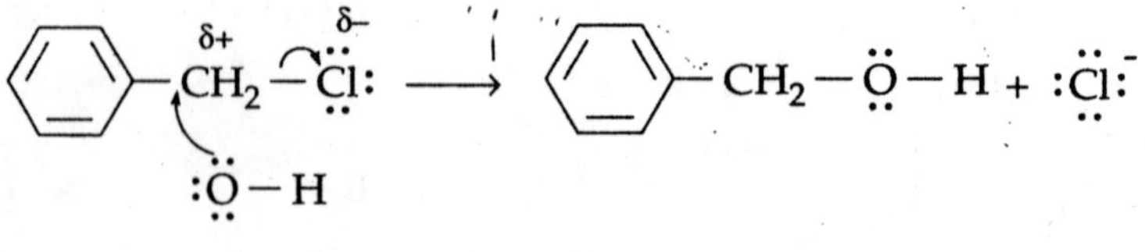

122. (a) (S)-2-bromopentane $\xrightarrow{NaBr}$ (R)-2-bromopentane

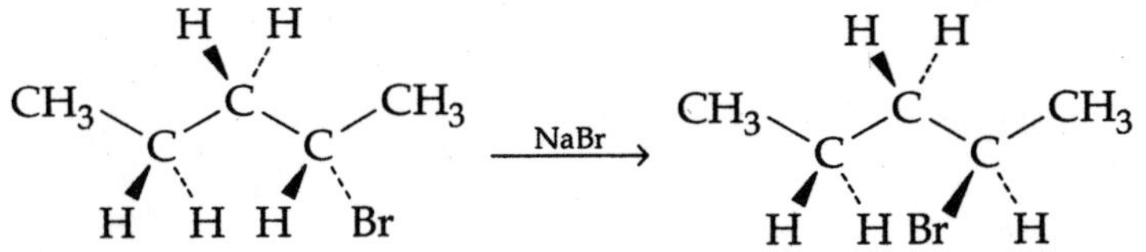

These compounds are enantimers.

(b) The results of B, in which the reaction rate increases with an increase in the concentration of bromide ion, indicates that the reaction proceeds by an S_N2 reaction. The rate of an S_N2 reaction depends on the concentration of the nucleophile as well as on the concentration of the alkyl halide.

123. The structure of alkene is :

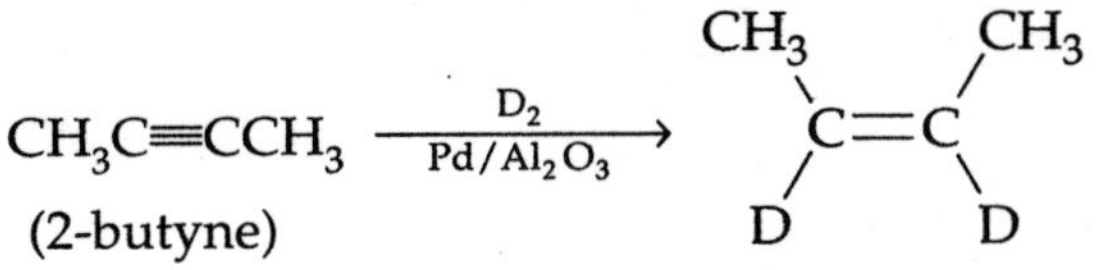

$$CH_3C{\equiv}CCH_3 \xrightarrow[Pd/Al_2O_3]{D_2} (CH_3)(D)C{=}C(CH_3)(D)$$

(2-butyne)

124. The reaction and structural formulas of the compounds are :

(a) $HOCH_2(CH_2)_7C{\equiv}CH \xrightarrow[NH_3\ (liq)]{Na^+NH_2^-\ (excess)} Na^{+-}OCH_2(CH_2)_7C{\equiv}C{:}^-\ Na^+$ (D)

$\xrightarrow{(1)\,CH_3CH_2CH_2CH_2Br} \xrightarrow{(2)\,H_3O^+} HOCH_2(CH_2)_7C{\equiv}CCH_2CH_2CH_2CH_3$ (E)

a carbon anion is more nucleophilic than an oxygen anion

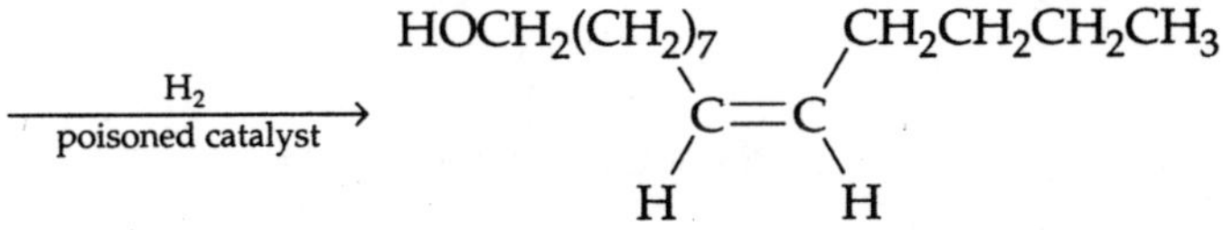

$\xrightarrow[\text{poisoned catalyst}]{H_2}$ $HOCH_2(CH_2)_7(H)C{=}C(H)CH_2CH_2CH_2CH_3$ (cis)

spodoptol
F

(b) E $\xrightarrow{red}$ $HOCH_2(CH_2)_7(H)C{=}C(H)CH_2CH_2CH_2CH_3$ (trans)

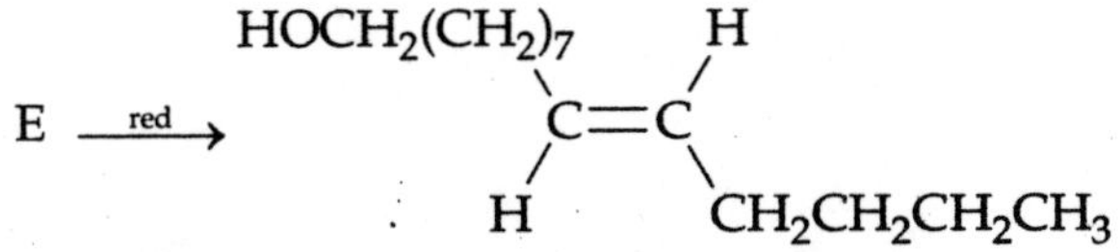

(c) The structures of aldehydes A and B are:

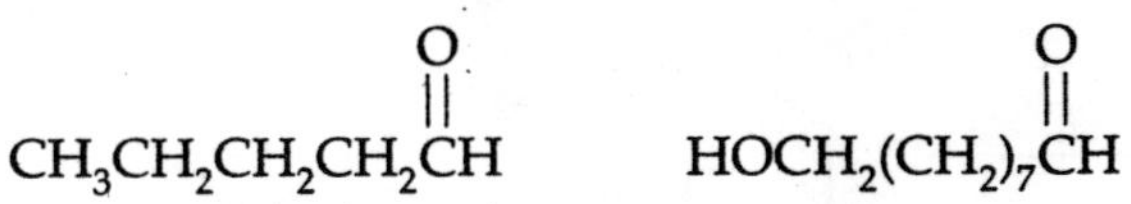

$CH_3CH_2CH_2CH_2CH{=}O$ — Aldehyde A

$HOCH_2(CH_2)_7CH{=}O$ — Aldehyde B

125. The structural formulas are :

(a) 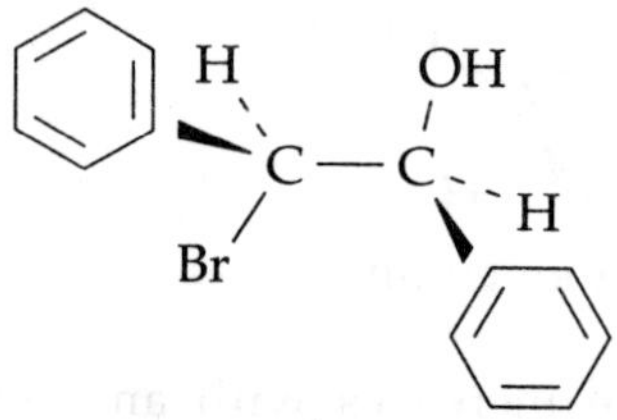 [structure: C_6H_5–CH(Br)–CH(OH)–C_6H_5] enantiomer

A

(b) 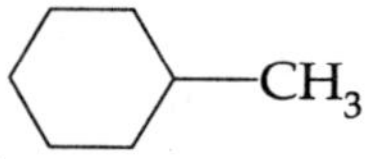 [structure: cyclohexane–CH_3]

B

(c) 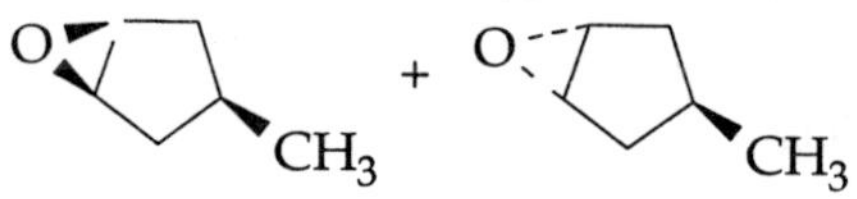 [structure: epoxide of methylcyclopentane with CH_3] + [structure: epoxide of methylcyclopentane with CH_3]

C **D**

(d) 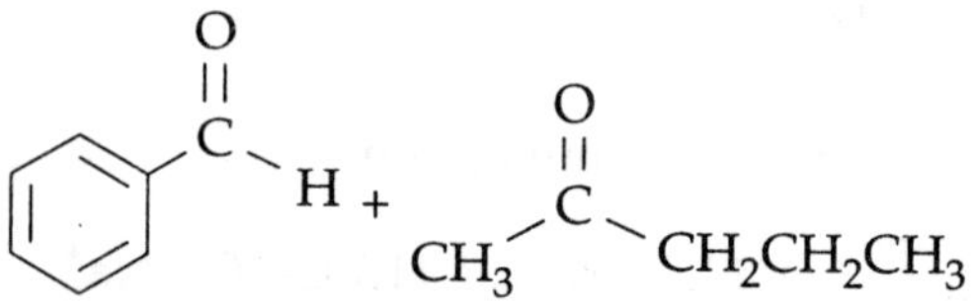 C_6H_5CHO + $CH_3COCH_2CH_2CH_3$

E **F**

(e) $(CH_3\overset{O}{\overset{\|}{C}}OCH_2CH_2CH_2CHD\underset{CH_3}{\underset{|}{C}}H—)_3B$ + $(CH_3\overset{O}{\overset{\|}{C}}OCH_2CH_2CH_2\underset{CHDCH_3}{\underset{|}{C}}H—)_3B$ $\xrightarrow{H_2O_2,\ NaOH}$

G **H**

$CH_3\overset{O}{\overset{\|}{C}}OCH_2CH_2CH_2CHD\underset{CH_3}{\underset{|}{C}}H—OH$ + $CH_3\overset{O}{\overset{\|}{C}}COCH_2CH_2CH_2\underset{CHDCH_3}{\underset{|}{C}}H—OH$

I **J**

(f) 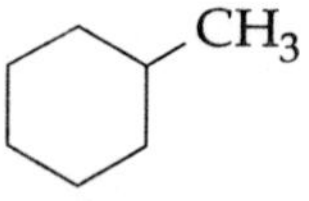 [structure: methylcyclohexane, CH_3]

K

(g) 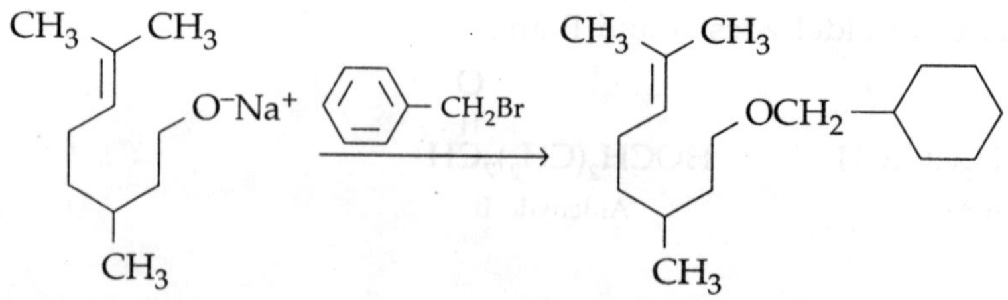 [structure: CH_3, CH_3, O^-Na^+, CH_3] $\xrightarrow{C_6H_5CH_2Br}$ [structure: CH_3, CH_3, OCH_2–cyclohexyl, CH_3]

L M

126. (a) 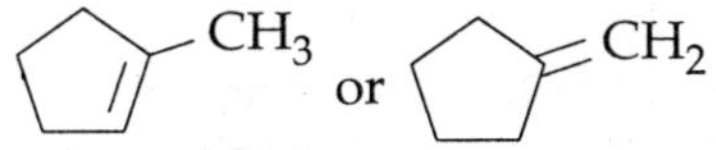 1-methylcyclopentene (CH_3) or methylenecyclopentane (CH_2)
A

(b) 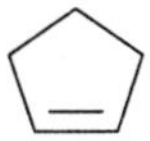 cyclopentene $\xrightarrow[H_2O]{Br_2}$
B C

(c) $CH_3\overset{CH_3}{\overset{|}{C}}HCH_2CH{=}CH_2$
D

(d) $CH_3\overset{CH_3}{\overset{|}{C}}{=}CHCH_3$
E

(e) $CH_3CH_2OCHCH{=}CH_2$ with CH_3CH_2O on CH
F

(f) 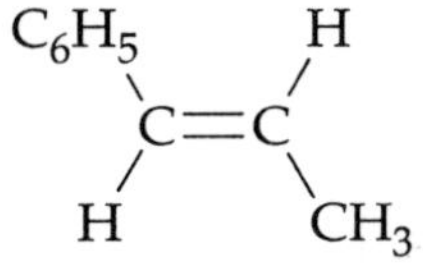 C_6H_5, H / C=C / H, CH_3
G

(g) $\xrightarrow{HBr}$
H

(h) 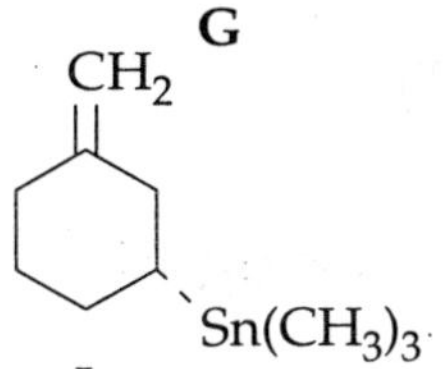 CH_2 = cyclohexane, $Sn(CH_3)_3$
I

(i) 
J

(j) $\xrightarrow[pyridine]{OsO_4}$
K

127. (a) 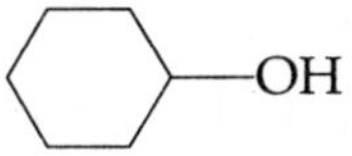 cyclohexane–OH

2. H_2SO_4, cold conc
4. CrO_3, H_2O, H_2SO_4

(b) 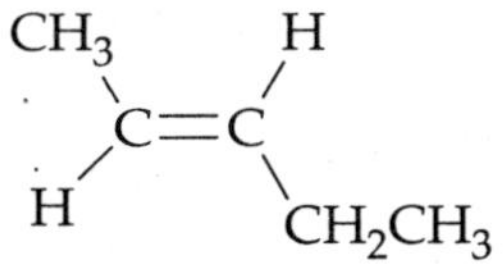 CH_3, H / C=C / H, CH_2CH_3

2. H_2SO_4, cold, conc
3. Br_2, carbon tetrachloride

(c) 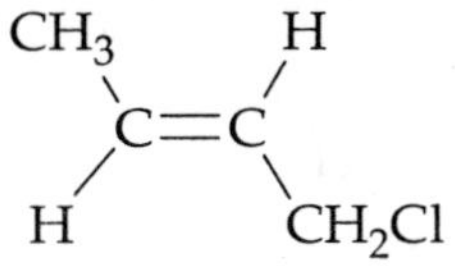 CH_3, H / C=C / H, CH_2Cl

2. H_2SO_4, cold, conc
3. Br_2, carbon tetrachloride
5. $AgNO_3$, C_3CH_2OH
6. NaI, acetone

(d)  cyclopentane with CH_3 and OH

2. H_2SO_4, coLd, conc

benzene–CH_2Cl

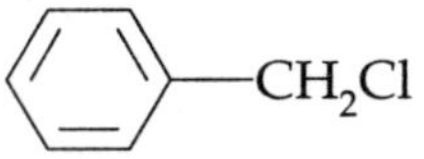

5. $AgNO_3$, CH_3CH_2OH
6. NaI, acetone

(f) $CH_3CH_2CH_2CH_2CH_2OH$ — 2. H_2SO_4, cold, conc; 4. CrO_3, H_2O, H_2SO_4

(g) $CH_3C{\equiv}CCH_2CH_3$ — 2. H_2SO_4, cold, conc; 3. Br_2, carbon tetrachloride

(h) 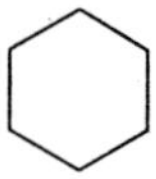 (cyclohexane) — none

(i) $HOCH_2CH_2OH$ — 1. H_2O, cold; 2. H_2SO_4, cold, conc; 4. CrO_3, H_2O, H_2SO_4

(j) 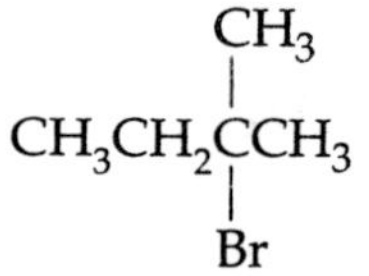 $CH_3CH_2C(CH_3)(Br)CH_3$ — 5. $AgNO_3$, CH_3CH_2OH

(k) 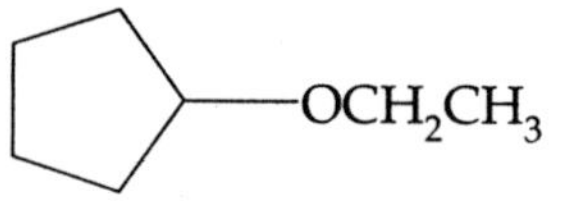 (cyclopentyl)$-OCH_2CH_3$ — 2. H_2SO_4, cold, conc

(l) $CH_3CH_2CH_2CH_2CH_2CH_2Cl$ — 6. NaI, acetone

128. The Refinal structures are:

(a) 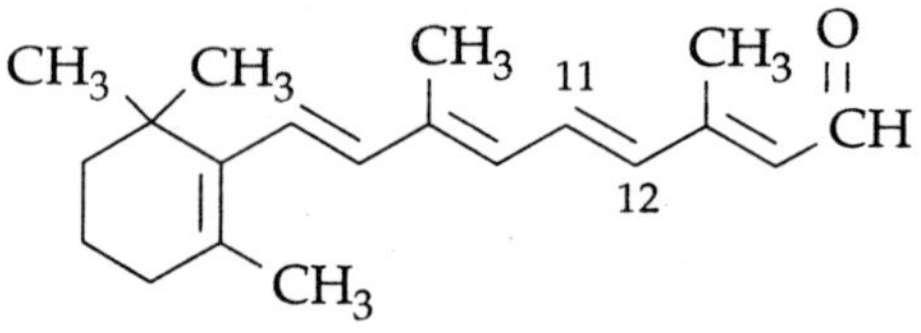 (11*E*)-retinal

(b) 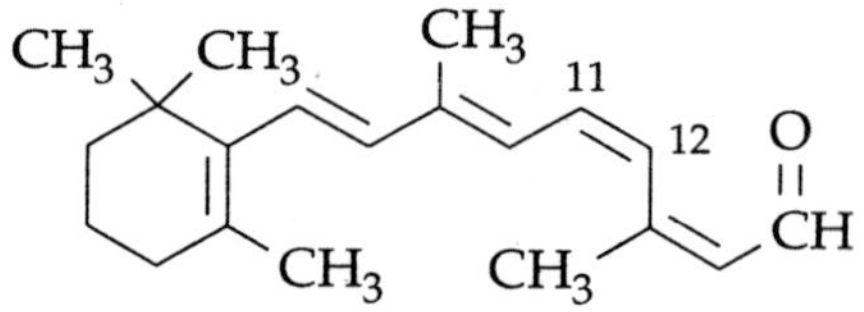 (11*Z*)-retinal

129. Following are the structural formulas for x and y

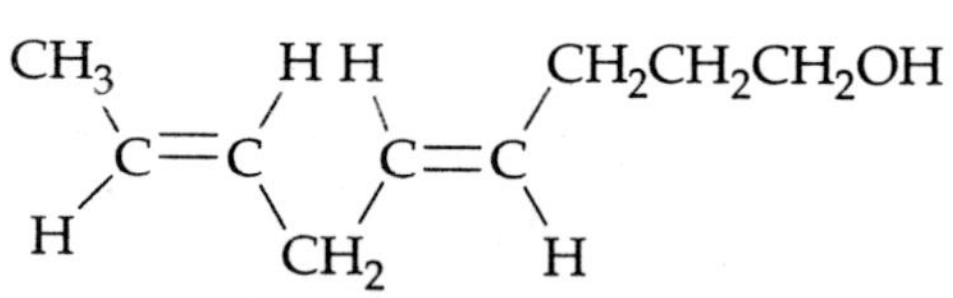

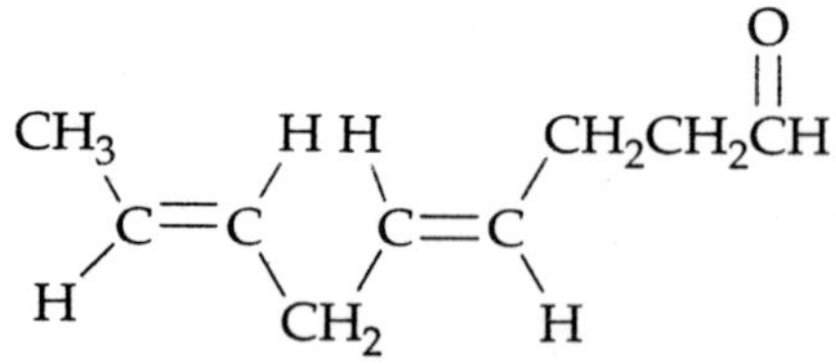

$CH_3(H)C{=}C(H)-CH_2-(H)C{=}C(H)CH_2CH_2CH_2OH$; $CH_3(H)C{=}C(H)-CH_2-(H)C{=}C(H)CH_2CH_2CHO$

Y

130. The required structure is a cyclic ketal. The reaction is:

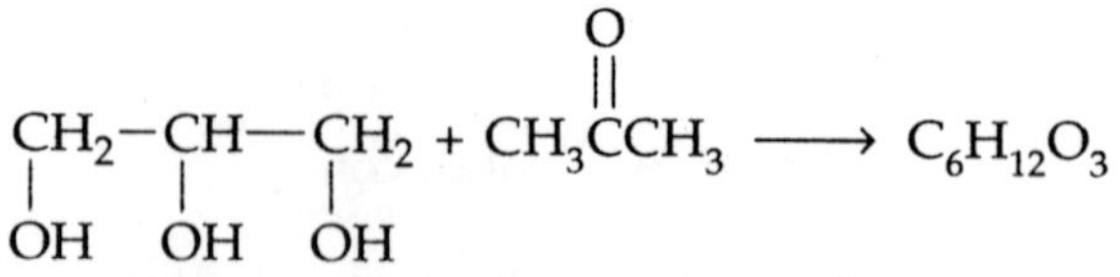

$$\underset{OH}{CH_2}-\underset{OH}{CH}-\underset{OH}{CH_2} + CH_3\overset{O}{\overset{\|}{C}}CH_3 \longrightarrow C_6H_{12}O_3$$

$$C_3H_6O_3 + C_3H_6O = C_6H_{14}O_4$$

$$C_6H_{14}O_4 - C_6H_{12}O_3 = H_2O$$

One molecule of water has been lost in the reaction.

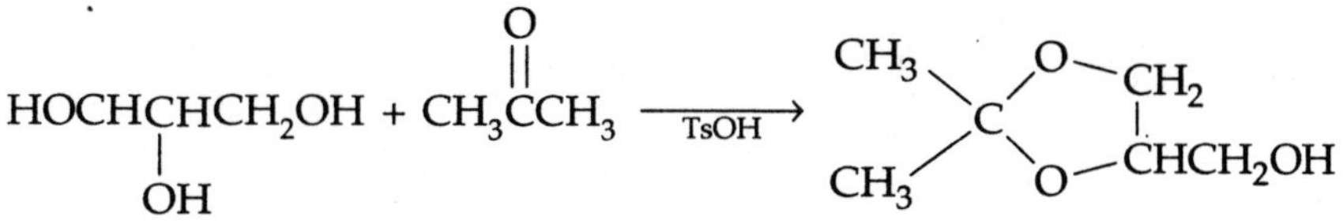

$$HOCH_2CH(OH)CH_2OH + CH_3\overset{O}{\overset{\|}{C}}CH_3 \xrightarrow{TsOH} \text{(CH}_3)_2C\text{ in ring } O{-}CH_2{-}CHCH_2OH{-}O$$

a cyclic ketal

131. The required conversions are as :

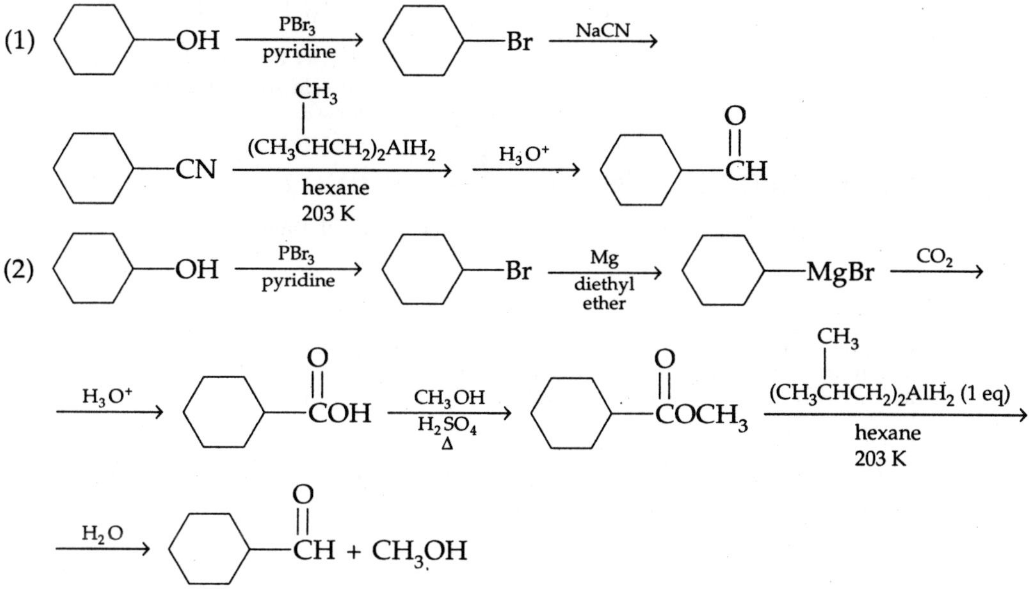

(1) Cyclohexyl–OH $\xrightarrow[\text{pyridine}]{PBr_3}$ Cyclohexyl–Br $\xrightarrow{NaCN}$

Cyclohexyl–CN $\xrightarrow[\text{hexane, 203 K}]{(CH_3CH(CH_3)CH_2)_2AlH_2}$ $\xrightarrow{H_3O^+}$ Cyclohexyl–$\overset{O}{\overset{\|}{C}}H$

(2) Cyclohexyl–OH $\xrightarrow[\text{pyridine}]{PBr_3}$ Cyclohexyl–Br $\xrightarrow[\text{diethyl ether}]{Mg}$ Cyclohexyl–MgBr $\xrightarrow{CO_2}$

$\xrightarrow{H_3O^+}$ Cyclohexyl–$\overset{O}{\overset{\|}{C}}OH$ $\xrightarrow[H_2SO_4, \Delta]{CH_3OH}$ Cyclohexyl–$\overset{O}{\overset{\|}{C}}OCH_3$ $\xrightarrow[\text{hexane, 203 K}]{(CH_3CH(CH_3)CH_2)_2AlH_2\ (1\ eq)}$

$\xrightarrow{H_2O}$ Cyclohexyl–$\overset{O}{\overset{\|}{C}}H + CH_3OH$

132. The structures of compounds A and B are:

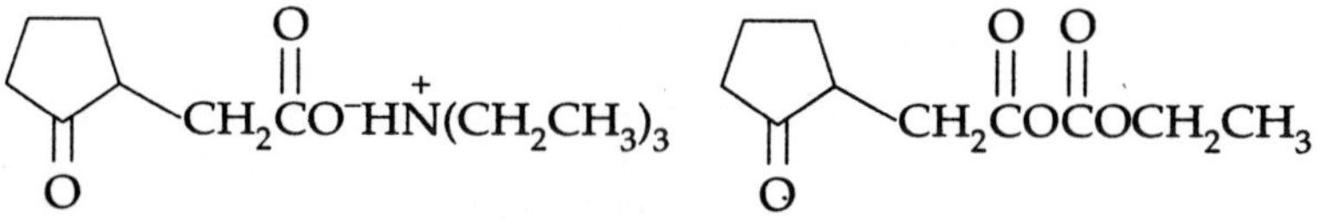

A: 2-oxocyclopentyl–$CH_2\overset{O}{\overset{\|}{C}}O^-\ H\overset{+}{N}(CH_2CH_3)_3$

B: 2-oxocyclopentyl–$CH_2\overset{O}{\overset{\|}{C}}\overset{O}{\overset{\|}{C}}OCH_2CH_3$

133. The synthesis for 3-ethyl-3-pentanol-3-^{14}c is carried out as following:

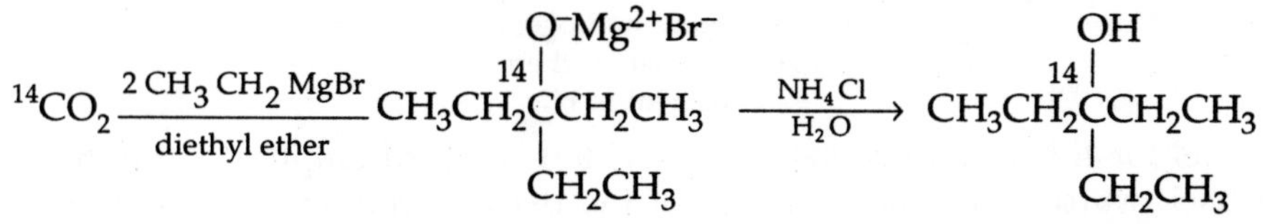

$$^{14}CO_2 \xrightarrow[\text{diethyl ether}]{2\,CH_3CH_2MgBr} CH_3CH_2\overset{O^-Mg^{2+}Br^-}{\overset{|}{^{14}C}}(CH_2CH_3)CH_2CH_3 \xrightarrow[H_2O]{NH_4Cl} CH_3CH_2\overset{OH}{\overset{|}{^{14}C}}(CH_2CH_3)CH_2CH_3$$

3 ethyl-3 pentanol-3-^{14}c

134. A good way to start each problem is to analyze the change in connectivity that occurs in going from the starting material or reagent to the product. Looking for possible nucleophilies and electophilies is also a good idea.

nucleopile

(a) $CH_3CH_2Br \xrightarrow{CH_3NH_2} CH_3CH_2\overset{+}{N}H_2CH_3\ H_2CH_3$

this portion of the product comes from the nucleophile; the rest must be from the electrophile it reacts with

(b) $CH_2{=}CH_2 \xrightarrow{HBr} CH_3CH_2Br \xrightarrow{CH_3O^-Na^+} CH_3CH_2OCH_3$

looks as thourh CH_3OH has added. CH_3OH is too weak an acid to acid to double bond: must use two consecutive reactions.

(c) $CH_3CH_2CH_2CH_2NH_2 \xrightarrow{HBr} CH_3CH_2CH_2CH_2\overset{+}{N}H_3Cl^-$

base — conjugate acid of base

(d) $CH_3CH_2C{\equiv}CH \xrightarrow[2\,Br_2]{electrophile} CH_3CH_2CBr_2CHBr_2$

nucleophile with carbon skeleton of product triple bond will react two times with Br_2

Br_2 added twice electrophile with carbon skeleton found in product

(e) $CH_3CH_2CH_2OH \xrightarrow{NH_2^-} CH_3CH_2CH_2HCH_2O^- \xrightarrow{CH_3Br} CH_3CH_2CH_2CH_2OCH_3$

proton lost in product therefore need base — nucleophile

(f) $CH_3CH{=}CHCH_3 \xrightarrow{HBr} CH_3\underset{\displaystyle Br}{\underset{|}{C}}HCH_2CH_3 \xrightarrow{CH_3Br} CH_3\underset{\displaystyle CN}{\underset{|}{C}}HCH_2CH_3$

nucleophile

looks as though CN has added; HCN is too weak an acid to add to double bond; must use to consecutive reactions

(g) $CH_3\underset{\displaystyle OH}{\underset{|}{C}}HCH_2CH{=}CH_2 \xrightarrow[PtO_2]{H_2} CH_3\underset{\displaystyle OH}{\underset{|}{C}}HCH_2CH_2CH_3$

has two more hydrogens than starting material, therefore reagent is H_2

135. The equations for the given problems are written so as to emphasize the necessity of reasoning back wards from the structure of the dsired product to the structure of the starting material.

(a) $CH_3CH{=}CH_2 \xrightarrow{HBr} CH_3\underset{\displaystyle Br}{\underset{|}{C}}HCH_3 \xrightarrow{CH_3CH_2S^-Na^+} CH_3\underset{\displaystyle SCH_2CH_3}{\underset{|}{C}}HCH_3$

(b) $CH_3CH_2CH_2CH_2CH_2Br \xrightarrow{Na^+ :^- CH(\overset{O}{\overset{\|}{C}}OCH_2CH_3)_2} CH_3(CH_2)_5CH(\underset{\underset{O}{\|}}{C}OCH_2CH_3)_2$

(c) 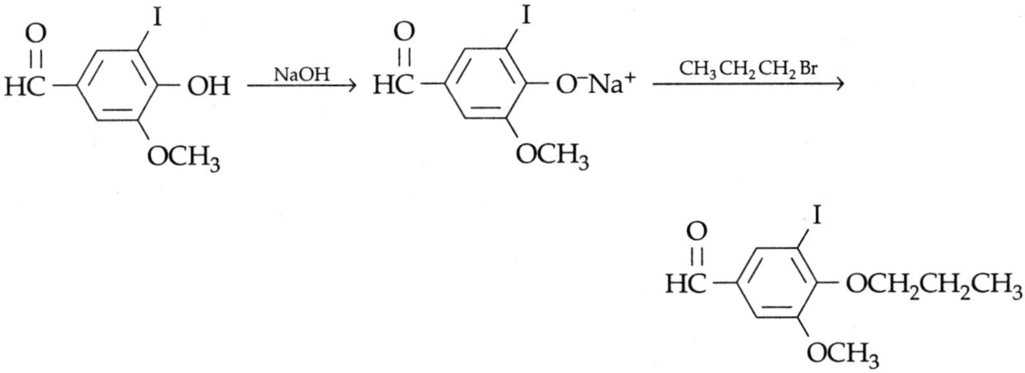 $HC(=O)$–(benzene ring with I and OCH_3)–OH $\xrightarrow{NaOH}$ $HC(=O)$–(benzene ring with I and OCH_3)–O^-Na^+ $\xrightarrow{CH_3CH_2CH_2Br}$

$HC(=O)$–(benzene ring with I and OCH_3)–$OCH_2CH_2CH_3$

136. 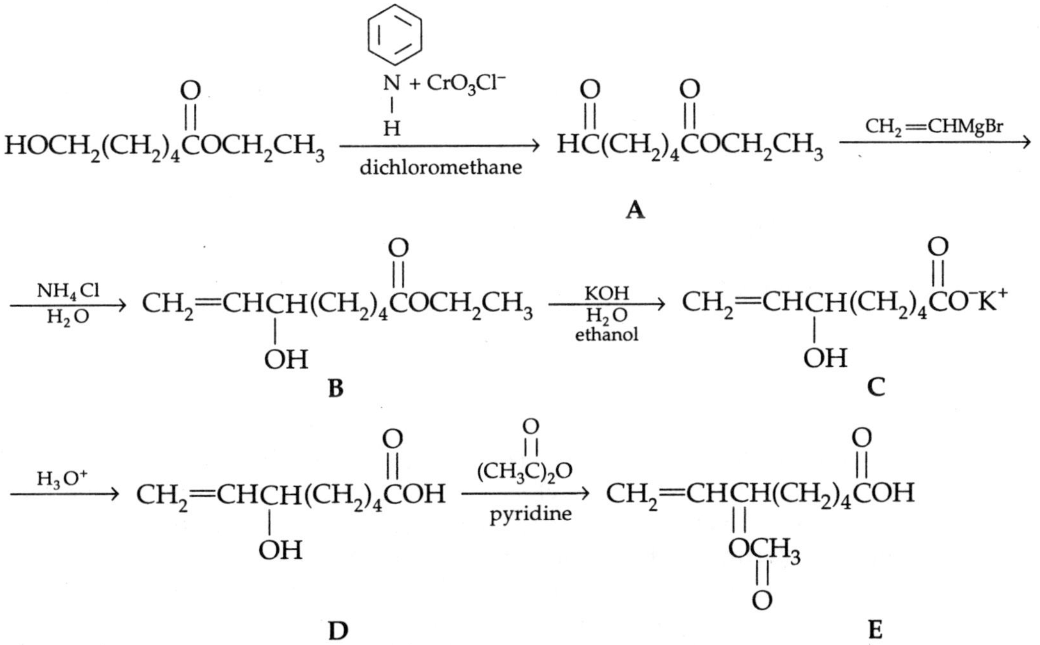 $HOCH_2(CH_2)_4\overset{O}{\overset{\|}{C}}OCH_2CH_3 \xrightarrow[\text{dichloromethane}]{C_5H_5\overset{+}{N}H + CrO_3Cl^-} H\overset{O}{\overset{\|}{C}}(CH_2)_4\overset{O}{\overset{\|}{C}}OCH_2CH_3 \xrightarrow{CH_2=CHMgBr}$

A

$\xrightarrow[H_2O]{NH_4Cl} CH_2{=}CH\underset{OH}{\underset{|}{C}}H(CH_2)_4\overset{O}{\overset{\|}{C}}OCH_2CH_3 \xrightarrow[H_2O,\ \text{ethanol}]{KOH} CH_2{=}CH\underset{OH}{\underset{|}{C}}H(CH_2)_4\overset{O}{\overset{\|}{C}}O^-K^+$

B **C**

$\xrightarrow{H_3O^+} CH_2{=}CH\underset{OH}{\underset{|}{C}}H(CH_2)_4\overset{O}{\overset{\|}{C}}OH \xrightarrow[\text{pyridine}]{(CH_3\overset{O}{\overset{\|}{C}})_2O} CH_2{=}CH\underset{O\overset{\|}{C}H_3}{\underset{\|}{C}}H(CH_2)_4\overset{O}{\overset{\|}{C}}OH$ (with O below)

D **E**

The carbonyl group of the aldehyde is less stabilized by resonance than the carbonyl group of the ester and, therefore, is more reactive.

137. The structural formulas are :

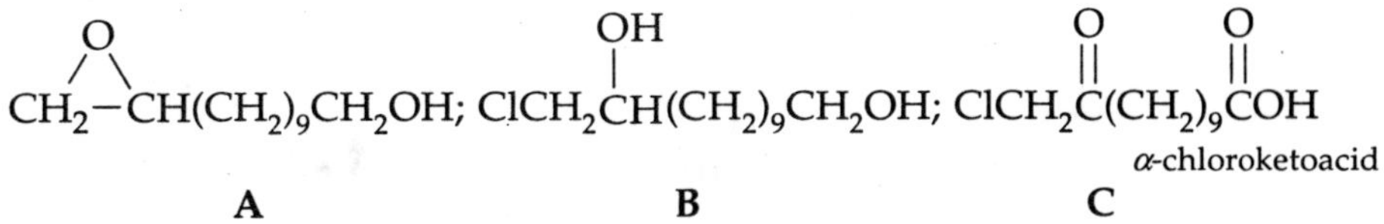

$\overset{O}{\overbrace{CH_2-CH}}(CH_2)_9CH_2OH$; $ClCH_2\overset{OH}{\overset{|}{C}}H(CH_2)_9CH_2OH$; $ClCH_2\overset{O}{\overset{\|}{C}}(CH_2)_9\overset{O}{\overset{\|}{C}}OH$

α-chloroketoacid

A **B** **C**

138. The structural formulas are:

A: $H\cdots C(CH_3)(O-CH(CH_3)OCH_2CH_3)-CH_2COCH_2CH_3$ (with C=O)

B: $H\cdots C(CH_3)(O-CH(CH_3)OCH_2CH_3)-CH_2CH_2OH$;

C: $H\cdots C(CH_3)(O-CH(CH_3)OCH_2CH_3)-CH_2CH_2OTs$

D: $H\cdots C(CH_3)(HO)-CH_2-CH_2-OTs$;

E: $H\cdots C(CH_3)$ in ring $C-CH_2-CH_2-O$

139. The answer to part (a) is analyzed below. The equation is written so as to emphasize the necessity of reasoning backward from the structure of the desired product to the structure of the starting material.

(a) $C_6H_5-CH_2OH \xrightarrow{?} C_6H_5-CH_2-SCH_2CH_3$

Look for following points:

1. What are the connectivities of the two compounds? How many carbon atoms does each contain? Are there any rings? What are the positions of branches and functional groups on the carbon skeletons?
The product has two more carbon atoms than the starting material. These carbon atoms are on the sulfur atom. The thioether group in the product is on the same carbon atom that the hydroxyl group was attached to in the starting material.
2. How do the functional groups change in going from starting material to product? Does the starting material have a good leaving group?
The hydroxyl group has been replaced by a thioether. The hydroxyl group is not a good leaving group but can be converted into one.
3. Is it possible to dissect the structure of the starting material and product to see which bonds must be broken and which formed?

$C_6H_5 \wr CH_2-OH$ (bond broken) $\qquad C_6H_5 \wr CH_2-SCH_2CH_3$ (bond formed)

4. New bonds are created when an electrophile reacts with a nucleophile. Do we recognize any part of the product molecule as coming from a good nucleophile or an electrophilic addition?
   The ethanethiolate atom, $CH_3 CH_2S^-$, is a good nucleophile.
5. What type of compound would be a good precursor to the product?
   The displacement of a leaving group (halide or tosylane ion) by the nucleophile will give us the product we want.

$$C_6H_5-CH_2-X \quad \text{or} \quad C_6H_5-CH_2-OTs$$

6. Now, how to get from starting material to product? If not, we need to analyze the structure obtained in step 5 by applying questions 4 and 5 to it.
   Complete this synthesis as:

$$C_6H_5-CH_2OH \xrightarrow[H_2O]{HBr} C_6H_5-CH_2Br \xrightarrow{CH_3CH_2S^-Na^+} C_6H_5-CH_2SCH_2CH_3$$

The compound could also have been synthesized using an S_N1 reaction with ethanethiol in an ionizing solvent such as water, but the product of the reaction with the solvent would also from. When possible, it is better to use the S_N2 reaction for synthesis, instead of an S_N1 reaction, because there will be fewer competing side reactions to lower the yield of the desired product.

(b) $CH_3CH_2CH_2CH_2CH_2OH \xrightarrow{HI} CH_3CH_2CH_2CH_2CH_2I$

(c) $CH_3CH_2CH_2CH_2Br \xrightarrow{HC\equiv\bar{C}Na^+} CH_3CH_2CH_2CH_2C\equiv CH$

(d) $C_6H_5SH \xrightarrow{NaOH} C_6H_5S^-Na^+ \xrightarrow{C_6H_5-CH_2CH_2Br} C_6H_5SCH_2CH_2C_6H_5$

(e) $HO-C(CH_3)(H)(CH_2CH_3) \xrightarrow[pyridine]{TsCl} TsO-C(CH_3)(H)(CH_2CH_3) \xrightarrow{NH_3 (excess)} H-C(CH_3)(NH_2)(CH_3CH_2)$

(f) $C_6H_5-CH_2CH_2CH=CH_2 \xrightarrow{HBr} C_6H_5-CH_2CH_2CH(Br)CH_3 \xrightarrow{NaN_3 \text{ ethanol}} C_6H_5-CH_2CH_2CH(N_3)CH_3$

(g $CH_3C(=O)OCH_2C(CH_3)=CHCH_2Cl \xrightarrow{H_2NCH(CH=CH_2)CH_2OH} CH_3C(=O)OCH_2C(CH_3)=CHCH_2NHCH(CH=CH_2)CH_2OH$

140. (a) This is a tertiary alcohol. It can be prepared in three ways corresponding to the three dissections shown below. Each will lead to an organometallic reagent and a ketone. The ketone will be the fragment containing the oxygen

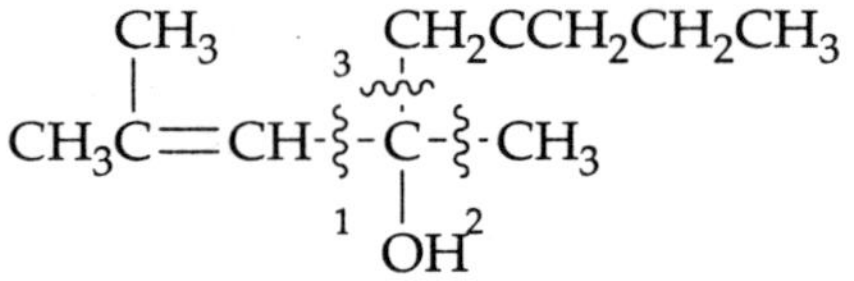

$$CH_3C(CH_3){=}CH\text{-}\S_1\text{-}C(OH)(\S_3\text{-}CH_2CCH_2CH_2CH_3)\text{-}\S_2\text{-}CH_3$$

Reagents for dissection 1:

$CH_3C(CH_3){=}CHMgBr$ or $CH_3C(CH_3){=}CHLi$ (prepared from $CH_3C(CH_3){=}CHBr$) and $CH_3C(=O)CH_2CH_2CH_2CH_3$

Regents for dissection 2:

CH_3MgBr or CH_3Li (from CH_3Br) and $CH_3C(CH_3){=}CHC(=O)CH_2CH_2CH_2CH_3$

Regents for dissection 3:

$CH_3CH_2CH_2CH_2MgBr$ or $CH_3CH_2CH_2CH_2Li$ (from $CH_3CH_2CH_2CH_2Br$) and $CH_3C(CH_3){=}CHC(=O)CH_3$

(b) This is a secondary alcohol. Two dissections are possible leading to an aldehyde and an organometallic reagent. A third dissection indicates that a reduction reaction, hydride addition, will also give this alcohol.

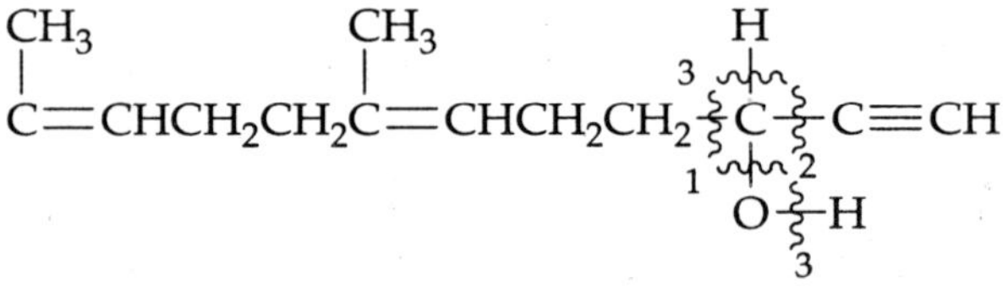

$$C(CH_3){=}CHCH_2CH_2C(CH_3){=}CHCH_2CH_2\text{-}\S_1\text{-}C(\S_3\text{-}H)(O\text{-}\S_3\text{-}H)\text{-}\S_2\text{-}C{\equiv}CH$$

Reagents for dissection 1:

$C(CH_3){=}CHCH_2CH_2C(CH_3){=}CHCH_2CH_2MgBr$ or $C(CH_3){=}CHCH_2CH_2C(CH_3){=}CHCH_2CH_2Li$ (prepared from $C(CH_3){=}CHCH_2CH_2C(CH_3){=}CHCH_2CH_2Br$) and $HC(=O)C{\equiv}CH$. This aldehyde may give problems depending on whether nucleophilic addition to the carbonyl group will compete with deprotonation of the alkyne.

Reagents for dissection 2:

$HC{\equiv}CMgBr$ or $HC{\equiv}CNa$ (from $HC{\equiv}CH$) and

$$\begin{array}{c} CH_3 \quad\quad\quad\quad CH_3 \quad\quad\quad\quad O \\ CH_3\overset{|}{C}{=}CHCH_2CH_2\overset{|}{C}{=}CHCH_2CH_2\overset{||}{C}H \end{array}$$

This is the better choice for an aldehyde and an organometallic reagent.

Reagents for dissection 3:

$$NaBH_4 \text{ and } CH_3\overset{CH_3}{\overset{|}{C}}{=}CHCH_2CH_2\overset{CH_3}{\overset{|}{C}}{=}CHCH_2CH_2\overset{O}{\overset{||}{C}}{-}C{\equiv}CH$$

(c) This is primary alcohol. a primary alcohol is prepared 1) by reduction of an aldehyde, 2) by addition of an organometalic reagent to formaldehyde or, sometimes, by addition of an organometallic reagent to oxirane. The dissections corresponding to these three routes are shown below.

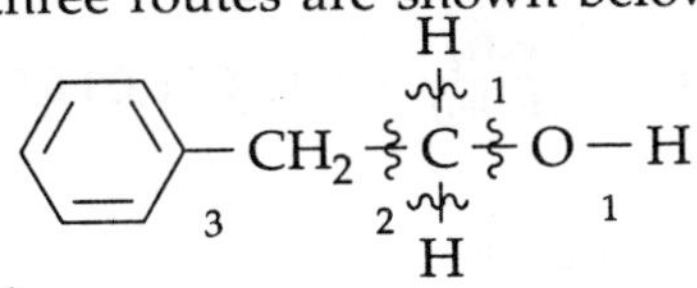

$$C_6H_5 \overset{3}{-} CH_2 \overset{2}{\wr} \underset{H}{\overset{H}{C}} \overset{1}{\wr} O-H$$

Reagents for dissection 1:

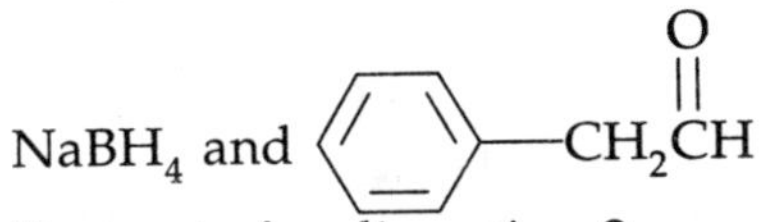

$$NaBH_4 \text{ and } C_6H_5{-}CH_2\overset{O}{\overset{||}{C}}H$$

Reagents for dissection 2:

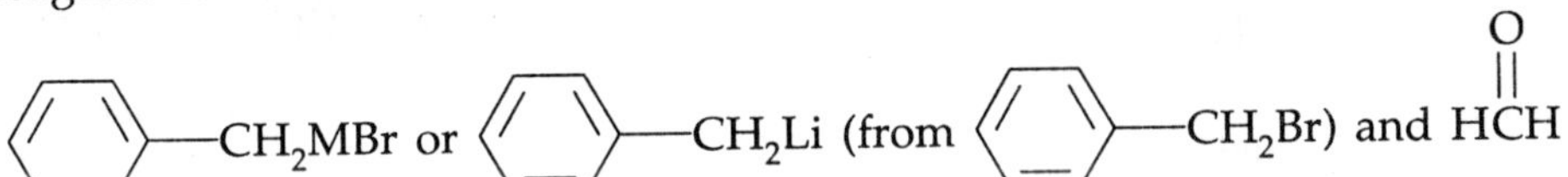

$$C_6H_5{-}CH_2MBr \text{ or } C_6H_5{-}CH_2Li \text{ (from } C_6H_5{-}CH_2Br) \text{ and } H\overset{O}{\overset{||}{C}}H$$

Regeants for dissection 3:

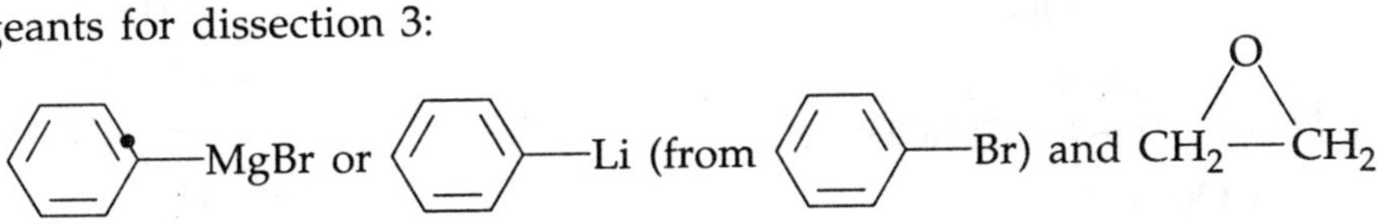

$$C_6H_5{-}MgBr \text{ or } C_6H_5{-}Li \text{ (from } C_6H_5{-}Br) \text{ and } CH_2{-}CH_2 \text{ (oxirane, } \overset{O}{\triangle})$$

141. (a) Rate = $K[CH_3CH_2CH_2CH_2Br][NaOH]$

(b) The rate will be doubled.

(c) The rate will be halved.

142. Rate = $k\left[\begin{array}{c} C_6H_5 \\ | \\ C_6H_5CCH_3 \\ | \\ Br \end{array}\right]$

143. The various steps and products for transformation are given below:

(a) $\underset{(i)}{HC{\equiv}CH} \xrightarrow[NH_3(liq)]{NaNH_2} \underset{(ii)}{Na^{+-}:C{\equiv}CH} \xrightarrow{CH_3CH_2CH_2Br} \underset{(iii)}{CH_3CH_2CH_2C{\equiv}CH}$

$$\xrightarrow[\text{quinoline}]{H_2\,Pd/BaSO_4} \underset{\text{(iv)}}{CH_3CH_2CH_2CH{=}CH_2} \xrightarrow{HBr} \underset{\text{(v)}}{CH_3CH_2CH_2\underset{\displaystyle Br}{\underset{|}{C}}HCH_3}$$

(b) $$\underset{\text{(i)}}{CH_3CH_2C{\equiv}CH} \xrightarrow[NH_3\,(liq)]{NaNH_2} \underset{\text{(ii)}}{CH_3CH_2{\equiv}C{:}^-Na^+} \xrightarrow{CH_3CH_2Br}$$

$$\underset{\text{(iii)}}{CH_3CH_2C{\equiv}CCH_2CH_3} \xrightarrow[H_2SO_4]{H_2O} \underset{\text{(iv)}}{CH_3CH_2CH_2\overset{\displaystyle O}{\overset{||}{C}}CH_2CH_3}$$

(c) $$\underset{\text{(i)}}{CH_3CH_2CH_2C{\equiv}CH} \xrightarrow[NH_3\,(liq)]{NaNH_2} \underset{\text{(ii)}}{CH_3CH_2CH_2C{\equiv}C{:}^-Na^+}$$

$$\xrightarrow{CH_3CH_2Br} \underset{\text{(iii)}}{CH_3CH_2CH_2CH{\equiv}CCH_2CH_3} \xrightarrow{H_2\,Pd/BaSO_4\ \text{quinoline}}$$

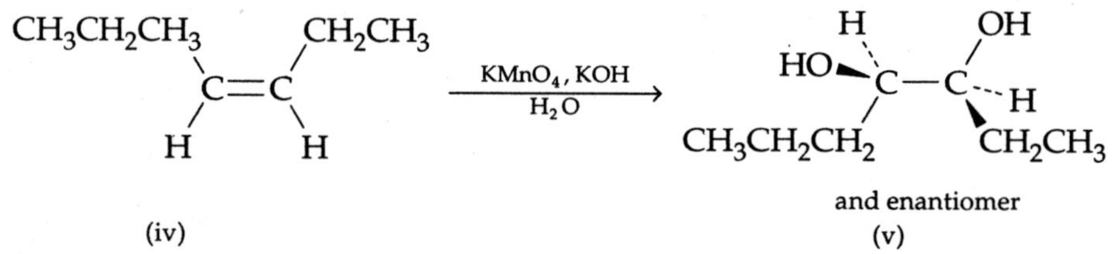

(iv): $CH_3CH_2CH_3$ and CH_2CH_3 on C=C, with H and H (cis) $\xrightarrow[H_2O]{KMnO_4,\ KOH}$ (v): HO, H on C — C with OH, H; $CH_3CH_2CH_2$ and CH_2CH_3

and enantiomer

(v)

144. (a) Cyclohexane: H, $\underset{\displaystyle O}{\overset{||}{C}}OCH_3$ (COCH₃ with =O), OH, H, $CH_2CH{=}CH_2$, H $\xrightarrow[\text{pyridine}]{TsCl}$ Cyclohexane: H, $COCH_3$ (=O), OTs, H, $CH_2CH{=}CH_2$, H $\xrightarrow{NaH_3}$

Cyclohexane: H, $COCH_3$ (=O), H, N_3, $CH_2CH{=}CH_2$, H $\xrightarrow{BH_3} \xrightarrow[H_2O]{H_2O_2,\ NaOH}$ Cyclohexane: H, $COCH_3$ (=O), H, N_3, $CH_2CH_2CH_2OH$, H

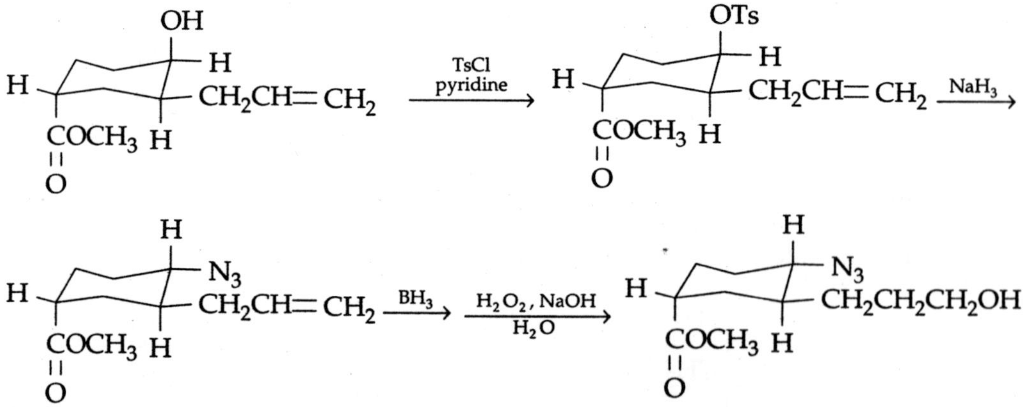

(b) $$CH_3CH_2C{\equiv}CCH_2CH_3 \xrightarrow[\text{quinoline}]{H_2\ Pd/CaCO_3}$$ CH_3CH_2 and CH_2CH_3 on C=C, with H and H (cis)

$\xrightarrow[\text{dichloromethane}]{\text{Cl–C}_6\text{H}_4\text{–COOH (m-chloroperbenzoic acid)}}$ epoxide (O bridging): H, CH_3CH_2 and H, CH_2CH_3

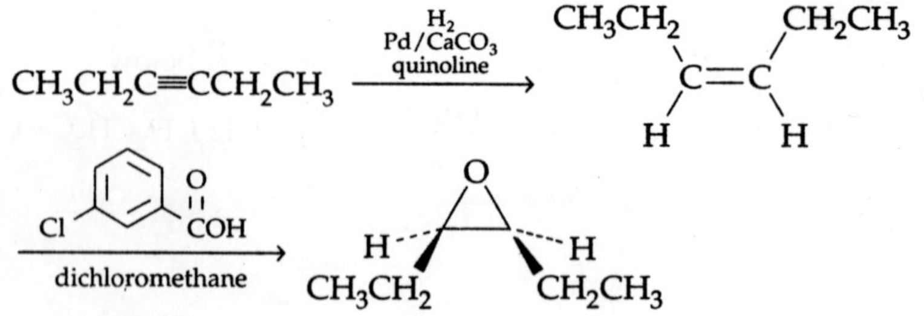

(c) 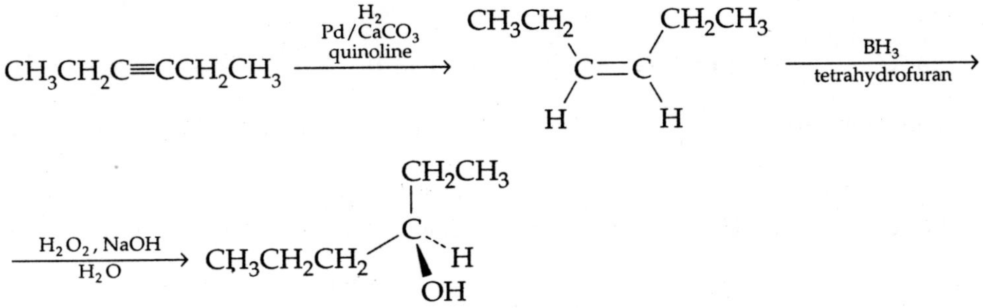 $CH_3CH_2C{\equiv}CCH_2CH_3 \xrightarrow[\text{quinoline}]{H_2,\ Pd/CaCO_3}$ (cis) $CH_3CH_2CH{=}CHCH_2CH_3$ $\xrightarrow[\text{tetrahydrofuran}]{BH_3}$ $\xrightarrow[H_2O]{H_2O_2,\ NaOH}$ $CH_3CH_2CH_2CH(OH)CH_2CH_3$

and enantiomer

Another Method

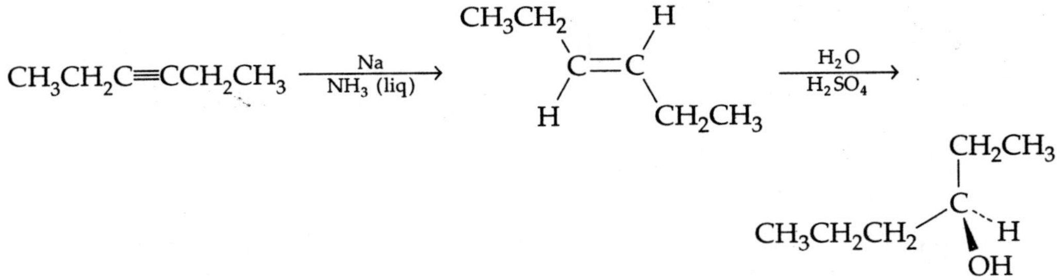

$CH_3CH_2C{\equiv}CCH_2CH_3 \xrightarrow[NH_3\ (liq)]{Na}$ (trans) $CH_3CH_2CH{=}CHCH_2CH_3$ $\xrightarrow[H_2SO_4]{H_2O}$ $CH_3CH_2CH_2CH(OH)CH_2CH_3$

and enantiomer

(d) 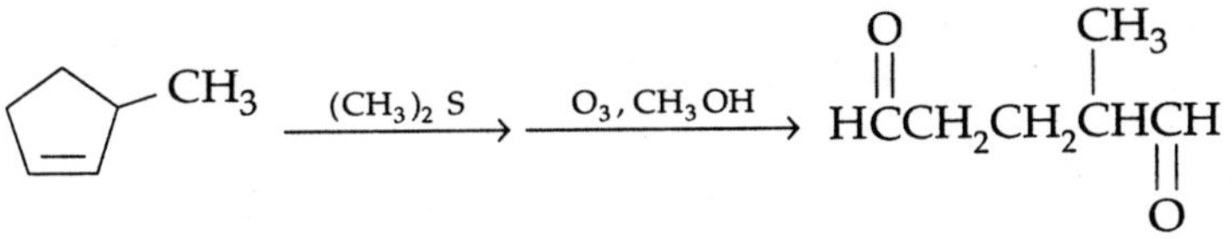 3-methylcyclopentene $\xrightarrow{O_3,\ CH_3OH}$ $\xrightarrow{(CH_3)_2S}$ $HCCH_2CH_2CH(CH_3)CH$ (with C=O at both ends: $OHCCH_2CH_2CH(CH_3)CHO$)

(e) 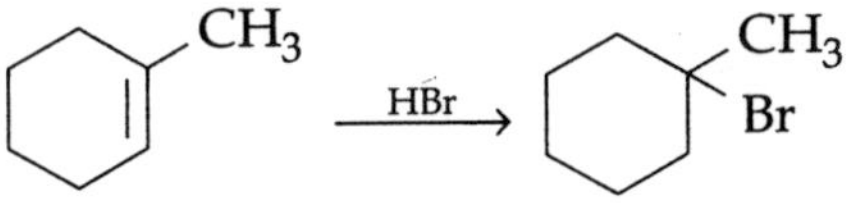 1-methylcyclohexene $\xrightarrow{HBr}$ 1-bromo-1-methylcyclohexane (CH_3, Br)

(f) 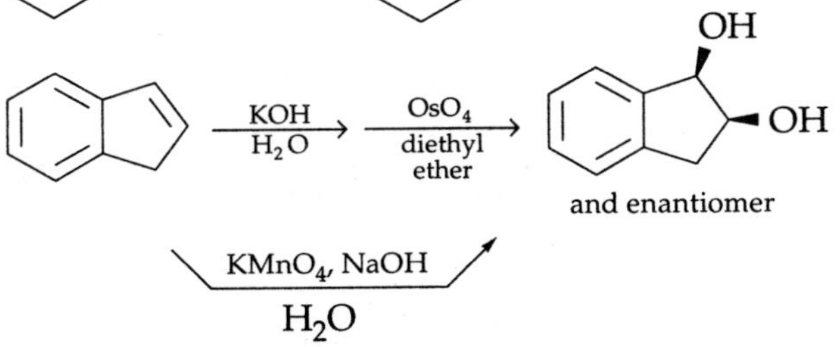 1H-indene (with KOH, H_2O) $\xrightarrow{KOH,\ H_2O}$ $\xrightarrow[\text{diethyl ether}]{OsO_4}$ cis-indane-1,2-diol (OH, OH)

and enantiomer

(alternatively: $\xrightarrow[H_2O]{KMnO_4,\ NaOH}$)

145. (a) 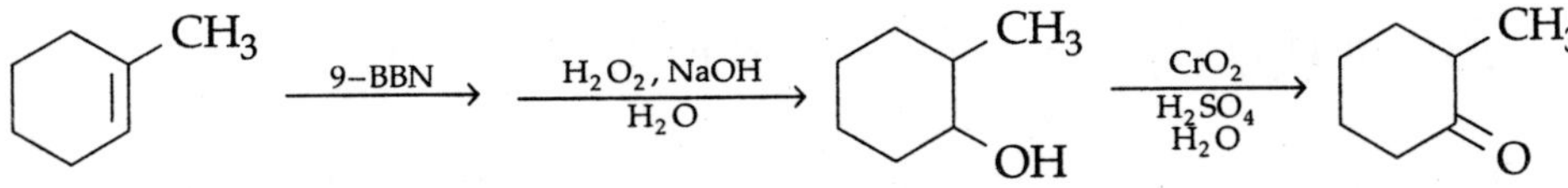 1-methylcyclohexene $\xrightarrow{\text{9-BBN}}$ $\xrightarrow[H_2O]{H_2O_2,\ NaOH}$ trans-2-methylcyclohexanol (CH_3, OH) $\xrightarrow[H_2SO_4,\ H_2O]{CrO_2}$ 2-methylcyclohexanone (CH_3, O)

(b)

$CH_3CH_2CH{=}CH_2 \xrightarrow{HBr} CH_3CH_2CH(Br)CH_3 \xrightarrow{NH_3\ (excess)} CH_3CH_2CH(NH_2)CH_3 + NH_4^+Br^-$

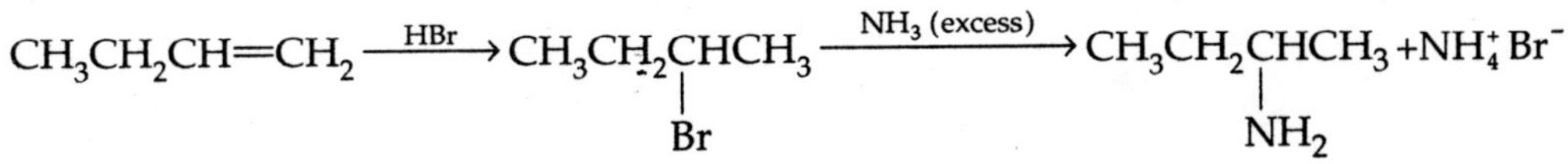

(c) $C_6H_5CH_2OH \xrightarrow[\text{(conc)}]{HBr} C_6H_5CH_2Br \xrightarrow{NH_3\ (excess)} C_6H_5CH_2C{\equiv}CCH_3$

146. The required reagents and intermediate compounds are given in the equations below:

(a) $$\underset{\displaystyle CH\overset{CH_3}{\overset{|}{C}}HCH_2CH{=}CH_2}{} \xrightarrow{H_3O^+} CH_3\overset{CH_3}{\overset{|}{C}}HCH_2\underset{OH}{\underset{|}{C}}HCH_3 \xrightarrow[H_2SO_4,\ H_2O]{CrO_3} CH_3\overset{CH_3}{\overset{|}{C}}HCH_2\overset{O}{\overset{||}{C}}CH_3CH_3$$

(b) $$CH_3(CH_2)_4CH{=}CH_2 \xrightarrow[\text{diglyme}]{BH_3} (CH_3(CH_2)_4CH_2CH_2)_3B \xrightarrow[H_2O]{H_2O_2,\ NaOH}$$

$$CH(CH_2)_4CH_2CH_2OH \xrightarrow[\text{dichloromethane}]{C_5H_5\overset{+}{N}H\ CrO_3Cl^-} CH_3(CH_2)CH_2\overset{O}{\overset{||}{C}}H$$

(c) $$C_6H_5CH_2CH_2CH{=}CH_2 \xrightarrow{BH_3} \xrightarrow[H_2O]{H_2O_2,\ NaOH} CH_2CH_2CH_2CH_2OH$$

$$\xrightarrow[\text{pyridine}]{PBr_3} C_6H_5\ CH_2(CH_2)_2CH_2Br \xrightarrow{C_6H_5O^-Na^+} C_6H_5\ CH_2(CH_2)_2CH_2OC_6H_5$$

(d) N-acetyl spiro compound (with ring C=C) $\xrightarrow[Pt]{H_2}$ N-acetyl saturated spiro compound

(e) $$CH_3CH_2C(H)(D)OH \xrightarrow[\text{pyridine}]{TsCl} CH_3CH_2C(H)(D)OTs \xrightarrow{NaN_3} CH_3CH_2C(D)(H)N_3$$

(f) $$CH_3(CH_2)_3CH{=}CH_2 \xrightarrow{BH_3} \xrightarrow[H_2O]{H_2O_2,\ NaOH} CH_3(CH_2)_4CH_2OH \xrightarrow{Na}$$

$$CH_3(CH_2)_4CH_2O^-Na^+ \xrightarrow{CH_3I} CH_3(CH_2)_4CH_2OCH_3$$

(g) $$C_6H_5CH{=}CH_2 \xrightarrow[\text{dichloromethane}]{m\text{-}ClC_6H_4COOOH} C_6H_5\overset{O}{\widehat{CH-CH_2}} \xrightarrow{(CH_3)_2NH} C_6H_5\underset{OH}{\underset{|}{C}}HCH_2N(CH_3)_2$$

147. The required reagents and structures are :

$$CH_3C(CH_3){=}CHCH_2CH{=}CH_2 \xrightarrow{\text{9-BBN}} \xrightarrow[H_2O]{H_2O_2,\ NaOH} \underset{\text{A}}{CH_3C(CH_3){=}CHCH_2CH_2CH_2OH}$$

$$\xrightarrow[\text{dichloromethane}]{C_5H_5\overset{+}{N}H\ CrO_3Cl^-} \underset{\text{B}}{CH_3C(CH_3){=}CHCH_2CH_2CH{=}O} \xrightarrow{CH_3MgI} \xrightarrow[H_2O]{NH_4Cl} CH_3C(CH_3){=}CHCH_2CH_2CH(OH)CH_3$$

148. These transformations are given below :

(a)
$$\underset{\text{(}CH_3\text{, H / H, }CH_2Cl\text{)}}{CH_3CH{=}CHCH_2Cl} \xrightarrow[\text{diethyl ether}]{Li} CH_3CH{=}CHCH_2Cl \xrightarrow{HCHO} \xrightarrow{H_3O^+}$$

$$CH_3CH{=}CHCH_2CH_2OH \xrightarrow[\text{pyridine}]{PBr_3} CH_3CH{=}CHCH_2CH_2Br$$

(b)
$$CH_2(CH_2)_3CH{=}CHCOCH_2CH_3\ (\text{ester, C=O}) \xrightarrow[\text{hexane, 203 K}]{(CH_3CH(CH_3)CH_2)_2AlH} \xrightarrow{H_2O} CH_2(CH_2)_3CH{=}CHCH{=}O$$

(c)
$$CH_3CH(OH)CH_2CH_2CH_2CH(OCH_3)OCH_3 \xrightarrow[\text{pyridine}]{(CH_3C(O))_2O} CH_3CH(OC(O)CH_3)CH_2CH_2CH_2CH(OCH_3)OCH_3$$

$$\xrightarrow[\text{cold, short time}]{H_3O^+} CH_3CH(OC(O)CH_3)CH_2CH_2CH_2CH{=}O$$

(f)
$$HC{\equiv}CC(O)OCH_2CH_3 \xrightarrow[NH_3\ (liq)]{NaNH_2} Na^{+-}C{\equiv}CC(O)OCH_2CH_3$$

$$\xrightarrow{CH_3(CH_2)_7\overset{O}{\overset{||}{C}}H} \xrightarrow[H_2O]{NH_4Cl} CH_3(CH_2)_7\underset{OH}{\underset{|}{C}}HC{\equiv}C\overset{O}{\overset{||}{C}}OCH_2CH_3$$

(g) $$CH_3(CH_2)_7\underset{OH}{\underset{|}{C}}HC{\equiv}C\overset{O}{\overset{||}{C}}OCH_2CH_3 \xrightarrow[\text{quinoline}]{H_2,\ Pd/BaSO_4} \underset{\text{cis}}{(H)C{=}C(H)}:\ CH_3(CH_2)_7\underset{OH}{\underset{|}{C}}H{-}CH{=}CH{-}\underset{O}{\underset{||}{C}}OCH_2CH_3$$

$$\xrightarrow[\Delta]{NaOH,\ H_2O} CH_3(CH_2)_7\underset{OH}{\underset{|}{C}}H{-}CH{=}CH{-}\underset{O}{\underset{||}{C}}O^-Na^+ \text{ (cis)} \xrightarrow{H_3O^+} CH_3(CH_2)_7\text{-(2-oxo-2,5-dihydrofuran-5-yl)}$$

149. These transformations are done as below :

(a) $$CH_3\overset{CH_3}{\overset{|}{C}}{=}\underset{\underset{O}{\underset{||}{C}}OCH_2CH_3}{\underset{|}{C}}CH_2\overset{O}{\overset{||}{C}}OCH_2CH_3 \xrightarrow[Pt]{H_2} CH_3\overset{CH_3}{\overset{|}{C}}H\underset{\underset{O}{\underset{||}{C}}OCH_2CH_3}{\underset{|}{C}}HCH_2\overset{O}{\overset{||}{C}}OCH_2CH_3 \xrightarrow[\text{diethyl ether}]{LiAlH_4}$$

$$\xrightarrow{H_3O^+} CH_3\overset{CH_3}{\overset{|}{C}}H\underset{CH_2OH}{\underset{|}{C}}HCH_2CH_2OH \xrightarrow[\text{pyridine}]{CH_3\overset{O}{\overset{||}{C}}Cl\ (\text{excess})} CH_3\overset{CH_3}{\overset{|}{C}}H\underset{CH_2O\underset{O}{\underset{||}{C}}CH_3}{\underset{|}{C}}HCH_2CH_2O\overset{O}{\overset{||}{C}}CH_3$$

(b) $$HOCH_2\overset{OH}{\overset{|}{C}}HCH_2\overset{O}{\overset{||}{C}}OCH_3 \xrightarrow[TsOH]{(CH_3)_2C{=}O} \text{(acetonide)}CH_2\overset{O}{\overset{||}{C}}OCH_3 \xrightarrow{LiAlH_4} \xrightarrow{H_2O}$$

$$\text{(acetonide)}CH_2CH_2OH \xrightarrow[\text{pyridine}]{TsCl} \text{(acetonide)}CH_2CH_2OTs \xrightarrow[\text{acetone}]{NaI} \text{(acetonide)}CH_2CH_2I$$

150. These transformations are done as below:

(a) $$\text{cyclohexanone} \xrightarrow[Ni]{NH_3,\ H_2} C_6H_{11}NH_2 \xrightarrow{2\,CH_3I} C_6H_{11}\overset{+}{N}HCH_3\ I^- \xrightarrow[H_2O]{NaOH} C_6H_{11}NCH_3$$

(b) $C_6H_5OCH_3 \xrightarrow[H_2SO_4]{HNO_3}$ $p\text{-}O_2NC_6H_4OCH_3 \xrightarrow{Fe\text{ , }HCl} \xrightarrow[H_2O]{NaOH}$ $p\text{-}H_2NC_6H_4OCH_3$

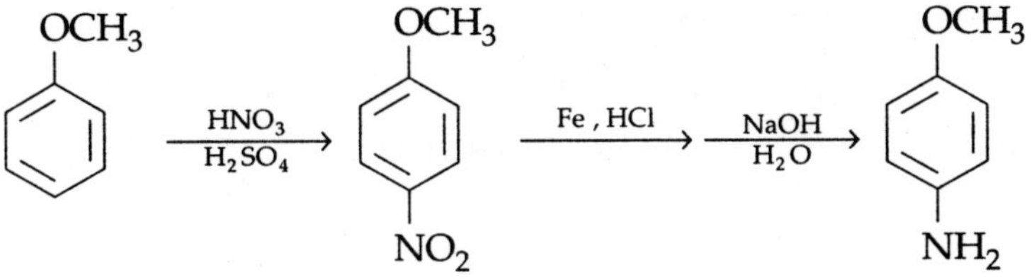

(c) $CH_3CH_2CH_2\overset{O}{\overset{||}{C}}H \xrightarrow{C_6H_5NH_2} C_6H_5{-}N{=}CHCH_2CH_2CH_3$

$\xrightarrow[H_2O]{NaBH_4} C_6H_5{-}NHCH_2CH_2CH_2CH_3 \xrightarrow{CH_3CH_2CH_2CH_2Br} \xrightarrow[H_2O]{NaOH}$

$C_6H_5{-}N(CH_2CH_2CH_2CH_3)CH_2CH_2CH_2CH_3$

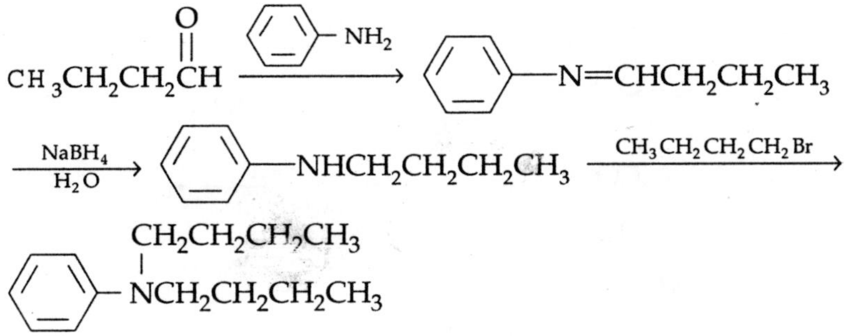

(d) 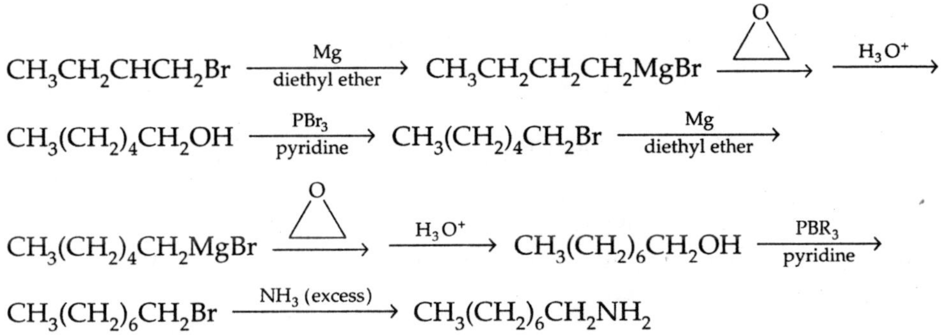 $CH_3CH_2CHCH_2Br \xrightarrow[\text{diethyl ether}]{Mg} CH_3CH_2CH_2CH_2MgBr \xrightarrow{\text{ethylene oxide}} \xrightarrow{H_3O^+}$

$CH_3(CH_2)_4CH_2OH \xrightarrow[\text{pyridine}]{PBr_3} CH_3(CH_2)_4CH_2Br \xrightarrow[\text{diethyl ether}]{Mg}$

$CH_3(CH_2)_4CH_2MgBr \xrightarrow{\text{ethylene oxide}} \xrightarrow{H_3O^+} CH_3(CH_2)_6CH_2OH \xrightarrow[\text{pyridine}]{PBR_3}$

$CH_3(CH_2)_6CH_2Br \xrightarrow{NH_3\text{ (excess)}} CH_3(CH_2)_6CH_2NH_2$

151. To convert,

$H{-}\underset{H}{\overset{H}{C}}{-}\overset{O}{\overset{||}{C}}OC(CH_3)_3 \xrightarrow{?}$ [skeletal structure: diene chain bearing $\overset{O}{\overset{||}{C}}OC(CH_3)_3$ and H on a carbon, with chain ending in $\overset{O}{\overset{||}{C}}H$]

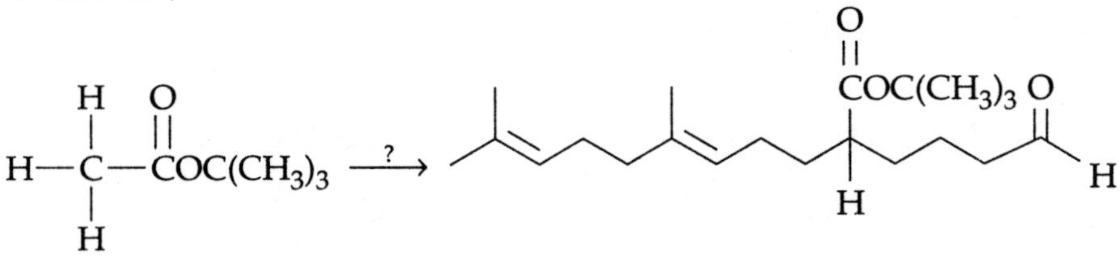

the conversion is done as given below:

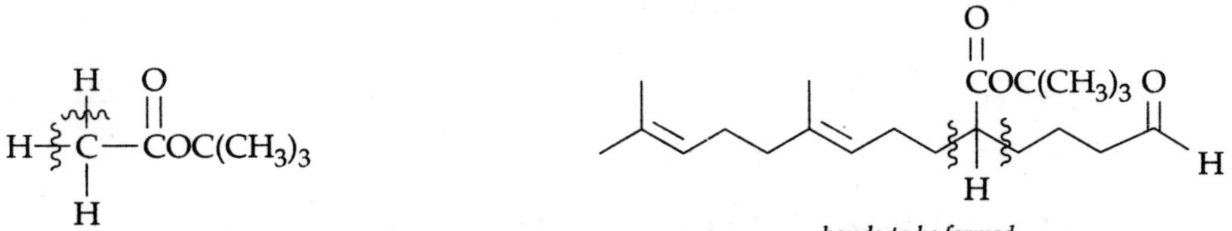

$H{-}\underset{H}{\overset{H}{C}}{-}\overset{O}{\overset{||}{C}}OC(CH_3)_3$

bonds to be broken

[skeletal structure with $\overset{O}{\overset{||}{C}}OC(CH_3)_3$, H and terminal $\overset{O}{\overset{||}{C}}H$]

bonds to be formed

We need two deprotonation and alkylation steps.

$$H_3C\text{—}\overset{O}{\overset{\|}{C}}OC(CH_3)_3 \xrightarrow{((CH_3)_2CH)_2N^-Li^+} H_2C{=}\overset{O^-Li^+}{\overset{|}{C}}OC(CH_3)_3 \xrightarrow[\text{LG = I, Br, OTs, etc.}]{\text{(geranyl)–LG}}$$

$$\text{(geranyl)}CH_2\overset{O}{\overset{\|}{C}}OC(CH_3)_3 \xrightarrow{((CH_3)_2CH)_2N^-Li^+}$$

$$\text{(geranyl)}CH{=}\overset{O^-Li^+}{\overset{|}{C}}OC(CH_3)_3 \xrightarrow[\text{LG = I, Br, OTs, etc.}]{LG\text{–}CH_2CH{=}CHCH_2CH(OCH_3)_2}$$

$$\text{(geranyl)}CH(COC(CH_3)_3)\text{–chain–}CH(OCH_3)OCH_3 \xrightarrow{H_3O^+}$$

$$\text{(geranyl)}CH(COC(CH_3)_3)\text{–chain–}CHO$$

152. The shortest routes for synthesis from bromobenzene and reagents having three or fewer carbon atoms are:

(a) $C_6H_5\text{—}Br \xrightarrow[\text{diethyl ether}]{Mg} C_6H_5\text{—}MgBr \xrightarrow{\overset{O}{CH_2\text{—}CHCH_3}}$

$C_6H_5\text{—}CH_2\underset{O^-Mg^{2+}Br^-}{\underset{|}{C}H}CH_3 \xrightarrow[H_2O]{NH_4Cl} C_6H_5\text{—}CH_2\underset{OH}{\underset{|}{C}H}CH_3$

$\xrightarrow[H_3O^+]{Na_2Cr_2O_7} C_6H_5\text{—}CH_2\overset{O}{\overset{\|}{C}}CH_3$

(b) $HC{\equiv}CH \xrightarrow[NH_3\text{(liq)}]{NaNH_2} Na^+C{\equiv}CH \xrightarrow{CH_3CH_2Br} CH_3CH_2C{\equiv}CH \xrightarrow{CH_3CH_2MgBr}$

$CH_3CH_2C{\equiv}CMgBr \xrightarrow{\text{ethylene oxide}} CH_3CH_2C{\equiv}CCH_2CH_2O^-Mg^{2+}Br^-$

$$\xrightarrow{H_3O^+} CH_3CH_2C{\equiv}CCH_2CH_2OH \xrightarrow[\text{quinoline}]{H_2,\ Pd/CaCO_3} \underset{H\qquad\qquad H}{\overset{CH_3CH_2\qquad CH_2CH_2OH}{C{=}C}}$$

(c)
$$CH_3CH_2CH_2Br \xrightarrow[\text{ether}]{Mg} CH_3CH_2CH_2MgBr \xrightarrow{CH_3CH_2\overset{O}{\overset{\|}{C}}H} CH_3CH_2CH_2\underset{O^-Mg^{2+}Br^-}{\underset{|}{C}}HCH_2CH_3$$

$$\xrightarrow{H_3O^+} CH_3CH_2CH_2\underset{OH}{\underset{|}{C}}HCH_2CH_3 \xrightarrow[H_3O^+]{Na_2Cr_2O_7} CH_3CH_2CH_2\overset{O}{\overset{\|}{C}}CH_2CH_3$$

(d)
$$CH_3CH_2Br \xrightarrow[\text{diethyl ether}]{Mg} CH_3CH_2MgBr \xrightarrow{\text{ethylene oxide}} CH_3CH_3CH_2CHO^-Mg^{2+}Br^- \xrightarrow{H_3O^+}$$

$$CH_3CH_2CH_2CH_2OH \xrightarrow[\text{pyridine}]{PBr_3} CH_3CH_2CH_2CH_2Br \xrightarrow[\text{diethyl ether}]{Mg}$$

$$CH_3CH_2CH_2CH_2MgBr \xrightarrow{CH_3\overset{O}{\overset{\|}{C}}CH_3} CH_3\underset{O^-Mg^{2+}Br^-}{\underset{|}{\overset{CH_3}{\overset{|}{C}}}}CH_2CH_2CH_2CH_3 \xrightarrow{H_3O^+}$$

$$CH_3\underset{OH}{\underset{|}{\overset{CH_3}{\overset{|}{C}}}}CH_2CH_2CH_2CH_3 \xrightarrow[\Delta]{H_3PO_4} CH_3\overset{CH_3}{\overset{|}{C}}{=}CHCH_2CH_2CH_3$$

(major product)

(e)
$$C_6H_5\text{—}Br \xrightarrow[\text{diethyl ether}]{Mg} C_6H_5\text{—}MgBr \xrightarrow{\text{ethylene oxide}} C_6H_5\text{—}CH_2CH_2O^-Mg^{2+}Br^-$$

$$\xrightarrow{H_3O^+} C_6H_5\text{—}CH_2CH_2OH \xrightarrow[\text{pyridine}]{PBr_3} C_6H_5\text{—}CH_2CH_2Br$$

(f)
$$HC{\equiv}CH \xrightarrow[NH_3(liq)]{NaNH_2} Na^{+-}C{\equiv}CH \xrightarrow{CH_3CH_2CH_2Br} CH_3CH_2CH_2C{\equiv}CH$$

$$\xrightarrow[NH_3\ (liq)]{NaNH_2} CH_3CH_2CH_2C{\equiv}C^-N^+ \xrightarrow{CH_3I} CH_3CH_2CH_2{\equiv}CCH_3 \xrightarrow[NH_2\ (liq)]{Na}$$

$$\underset{H\qquad\qquad CH_3}{\overset{CH_3CH_2CH_2\qquad H}{C{=}C}} \xrightarrow[\text{carbon tetrachloride}]{Br_2} \text{(H, Br, }CH_3CH_2CH_2\text{)C—C(Br, H, }CH_3\text{)}$$

and enantiomer

(g) $C_6H_5Br \xrightarrow[\text{diethyl ether}]{Mg} C_6H_5MgBr \xrightarrow[NH_4Cl/H_2O]{CH_3CH_2CHO} C_6H_5CH(OH)CH_2CH_3$

$\xrightarrow{CH_3SCH_3,\ ClC(=O)-C(=O)Cl} \xrightarrow{(CH_3CH_2)_3N} C_6H_5C(=O)CH_2CH_3$

(h) $HC{\equiv}CCH_3 \xrightarrow[NH_3\ (liq)]{NaNH_2} Na^{+-}C{\equiv}CCH_3 \xrightarrow{CH_3I} CH_3C{\equiv}CCH_3$

(a) $\xrightarrow[\text{quinoline}]{H_2,\ Pd/CaCO_3}$ cis-$CH_3CH{=}CHCH_3$ (both CH_3 on the same side, both H on the same side)

(b) $\xrightarrow{Na,\ NH_3\ (liq)}$ trans-$CH_3CH{=}CHCH_3$ $\xrightarrow[\text{dichloromethane}]{m\text{-}ClC_6H_4COOOH}$ trans-2,3-dimethyloxirane (H, CH_3 / H, CH_3) and enantiomer

$\xrightarrow[\text{cold, dilute}]{KMnO_4,\ H_2O}$ $CH_3CH(OH)-CH(OH)CH_3$ (HO, H, CH_3 / OH, H, CH_3) and enantiomer $\rightleftarrows$ (H, CH_3, HO / OH, H, CH_3) and enantiomer

(i) $CH_3CH(Br)CH_3 \xrightarrow[\text{diethyl ether}]{Mg} CH_3CH(MgBr)CH_3 \xrightarrow{\text{ethylene oxide}} \xrightarrow{H_3O^+} CH_3CH(CH_3)CH_2CH_2OH$

$\xrightarrow[\text{pyridine}]{PBr_3} CH_3CH(CH_3)CH_2CH_2Br \xrightarrow[\text{diethyl ether}]{Li} CH_3CH(CH_3)CH_2CH_2Li \xrightarrow{CH_3CHO}$

$\xrightarrow{H_3O^+} CH_3CH(CH_3)CH_2CH_2CH(OH)CH_3$

(j) $C_6H_5Br \xrightarrow[\text{diethyl ether}]{Mg} C_6H_5MgBr \xrightarrow{CH_3CH_2CHO} \xrightarrow[H_2O]{NH_4Cl}$

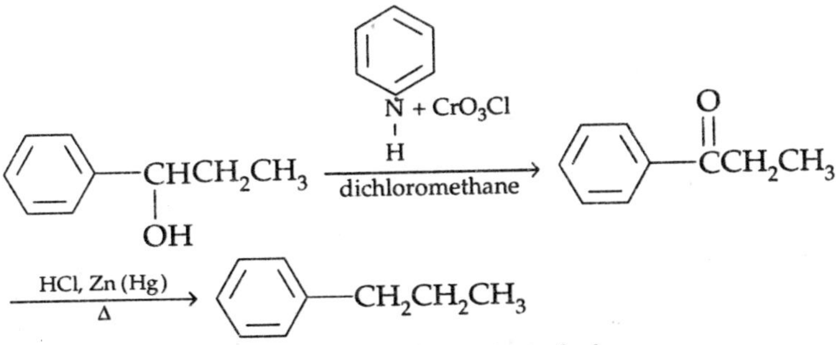

$$C_6H_5-\underset{OH}{CH}CH_2CH_3 \xrightarrow[\text{dichloromethane}]{C_5H_5\overset{+}{N}H + CrO_3Cl^-} C_6H_5-\overset{O}{\overset{\|}{C}}CH_2CH_3$$

$$\xrightarrow[\Delta]{\text{HCl, Zn (Hg)}} C_6H_5-CH_2CH_2CH_3$$

153. These compounds are synthesized as given below :

(a) Toluene $\xrightarrow[Fe]{Br_2}$ p-$BrC_6H_4CH_3$ $\xrightarrow[\text{diethyl ether}]{Mg}$ p-$BrMgC_6H_4CH_3$ $\xrightarrow{H_3O^+}\xrightarrow{CO_2}$ p-$HO\overset{O}{\overset{\|}{C}}C_6H_4CH_3$

separate from ortho isomer

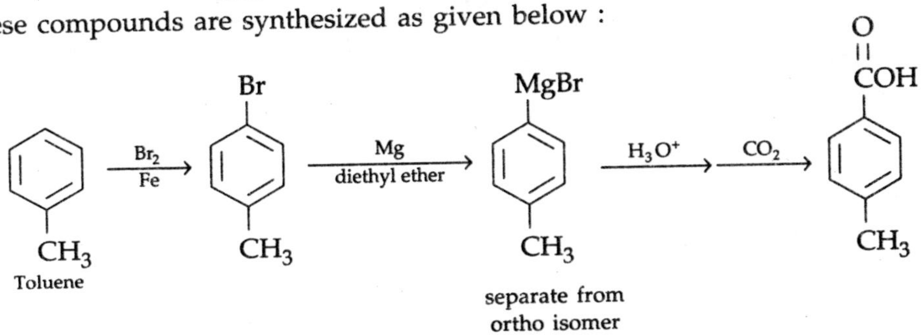

(b) Benzene $\xrightarrow[H_2SO_4]{HNO_3}$ $C_6H_5NO_2$ $\xrightarrow[FeCl_3]{Cl_2}$ m-$ClC_6H_4NO_2$

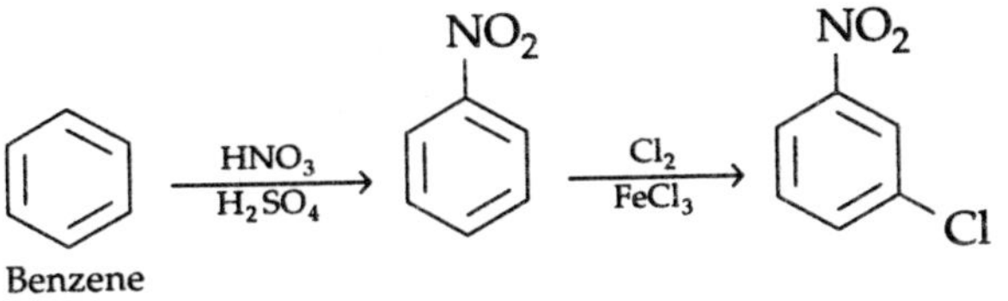

(Note: reversing the order in which these regents are used would give ortho- and para-substituted products.)

(c) Benzene $\xrightarrow[Fe]{Br_2}$ C_6H_5Br $\xrightarrow[FeCl_3]{Cl_2}$ p-BrC_6H_4Cl

separate from ortho isomer

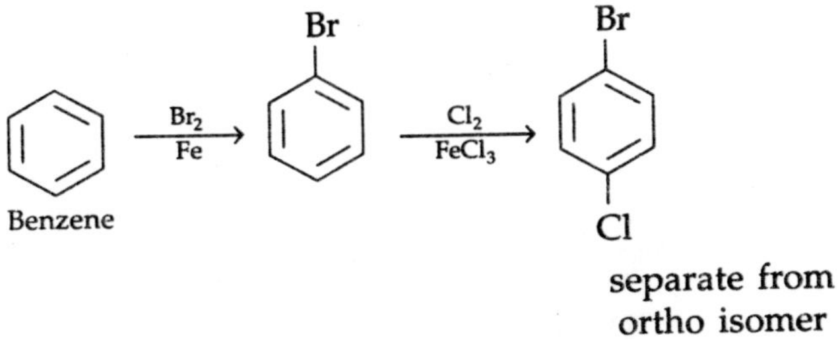

(d) Toluene $\xrightarrow[H_2SO_4]{HNO_3}$ o-$CH_3C_6H_4NO_2$ + p-$CH_3C_6H_4NO_2$ $\xrightarrow[H_2SO_4]{HNO_3}$ 2,4-dinitrotoluene (CH_3, NO_2, NO_2)

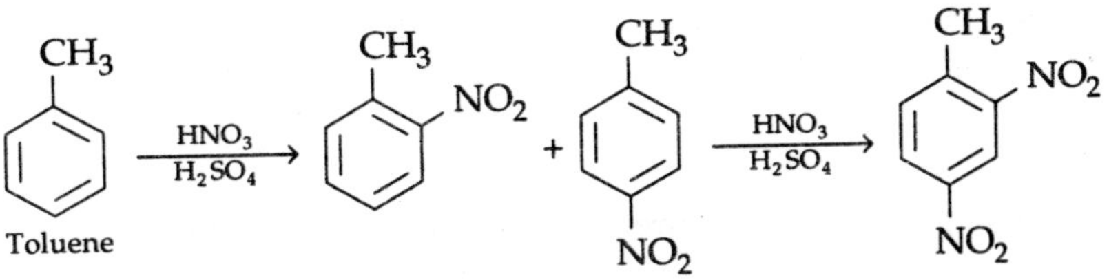

(e) 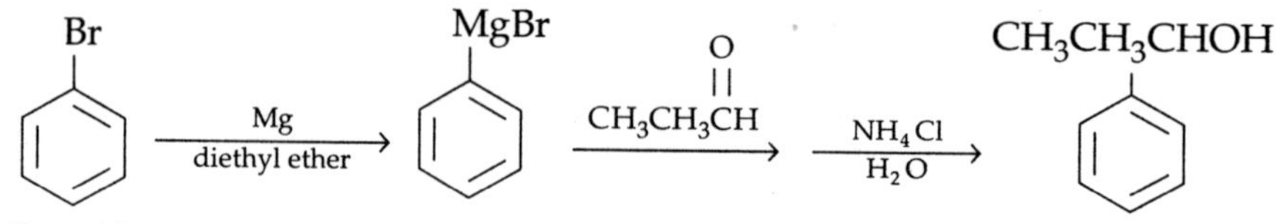 Bromobenzene (From (c)) $\xrightarrow[\text{diethyl ether}]{Mg}$ C_6H_5MgBr $\xrightarrow{CH_3CH_3\overset{O}{\overset{||}{C}}H}$ $\xrightarrow[H_2O]{NH_4Cl}$ C_6H_5–CH_3CH_3CHOH

(f) 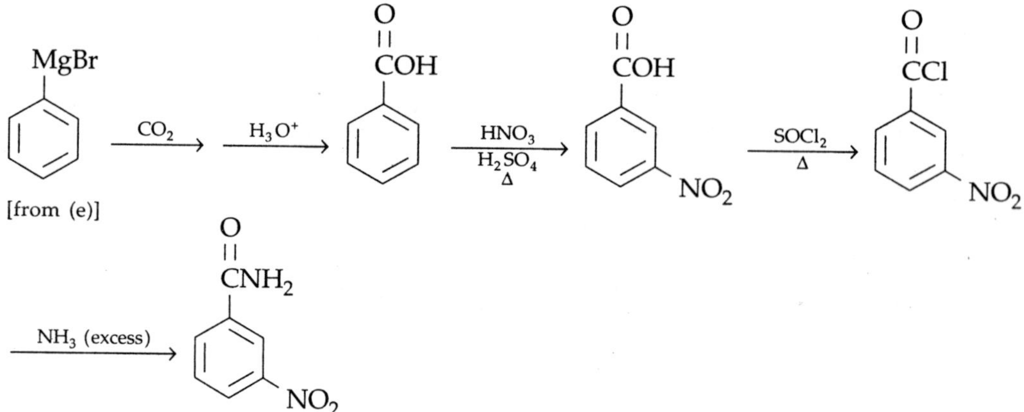 C_6H_5MgBr [from (e)] $\xrightarrow{CO_2}$ $\xrightarrow{H_3O^+}$ $C_6H_5\overset{O}{\overset{||}{C}}OH$ $\xrightarrow[H_2SO_4,\ \Delta]{HNO_3}$ m-$NO_2C_6H_4\overset{O}{\overset{||}{C}}OH$ $\xrightarrow[\Delta]{SOCl_2}$ m-$NO_2C_6H_4\overset{O}{\overset{||}{C}}Cl$

$\xrightarrow{NH_3\ (excess)}$ m-$NO_2C_6H_4\overset{O}{\overset{||}{C}}NH_2$

154. The required structures and steps are given below :

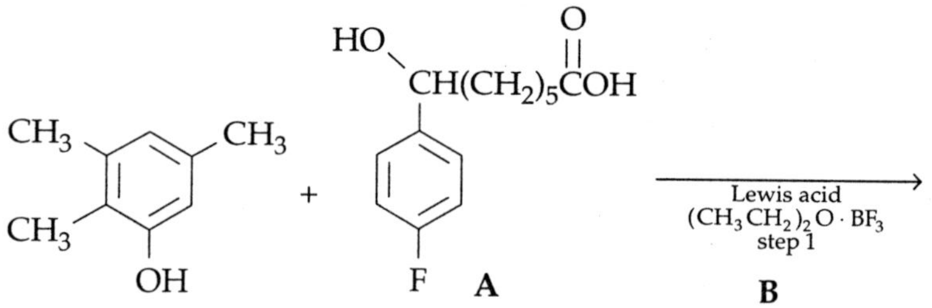

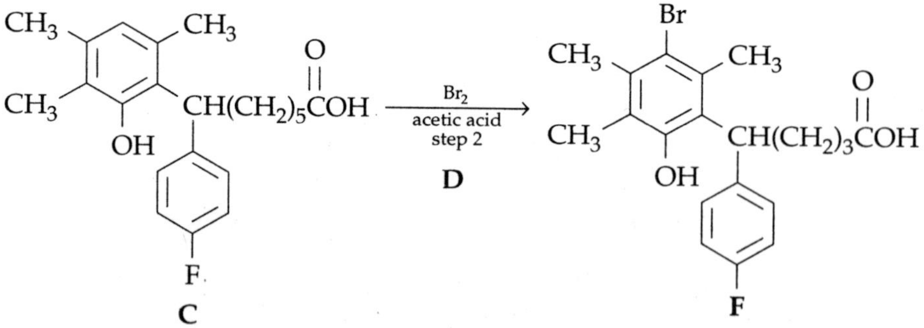

2,3,5-Trimethylphenol (CH_3, CH_3, CH_3, OH) + A: HO–$CH(CH_2)_5\overset{O}{\overset{||}{C}}OH$ attached to p-fluorophenyl (F) $\xrightarrow[\text{step 1}]{\text{Lewis acid } (CH_3CH_2)_2O \cdot BF_3}$ **B**

C: trimethylphenol ring (CH_3, CH_3, CH_3, OH) bearing $CH(CH_2)_5\overset{O}{\overset{||}{C}}OH$ with p-fluorophenyl (F) $\xrightarrow[\text{step 2}]{Br_2,\ \text{acetic acid}}$ **D**: Br-substituted ring (CH_3, CH_3, CH_3, OH, Br) bearing $CH(CH_2)_3\overset{O}{\overset{||}{C}}OH$ with p-fluorophenyl (**F**)

155. The following steps give the required synthesis for the compounds.

(a) $CH_3CH_2CH_2Br \xrightarrow[\text{ethnol}]{NaCN} CH_3CH_2CH_2CN \xrightarrow[H_2O,\ \Delta]{NaOH} \xrightarrow{H_3O^+} CH_3CH_2CH_2\overset{O}{\overset{||}{C}}OH$

$\xrightarrow{SOCl_2}$ $CH_3CH_2CH_2\overset{O}{\overset{\|}{C}}Cl$ $\xrightarrow[AlCl_3]{C_6H_6}$ $C_6H_5-\overset{O}{\overset{\|}{C}}CH_2CH_2CH_3$ $\xrightarrow[\Delta]{Zn(Hg),\ HCl}$

$C_6H_5-CH_2CH_2CH_2CH_3$

(b) 1. $C_6H_6 \xrightarrow[Fe]{Br_2} C_6H_5-Br \xrightarrow[\text{diethyl ether}]{Mg} C_6H_5-MgBr \xrightarrow{H\overset{O}{\overset{\|}{C}}H} \xrightarrow{H_3O^+}$

$C_6H_5-CH_2OH \xrightarrow[\text{pyridine}]{PBr_3} C_6H_5-CH_2Br$

2. $HC{\equiv}CH \xrightarrow[NH_3\ (liq)]{NaNH_2} HC{\equiv}CNa \xrightarrow{CH_3CH_2Br} HC{\equiv}CCH_2CH_3 \xrightarrow[NH_3\ (liq)]{NaNH_2}$

$NaC{\equiv}CCH_2CH_3 \xrightarrow{C_6H_5-CH_2Br} C_6H_5-CH_2C{\equiv}CCH_2CH_3 \xrightarrow[\text{Pd/BaSO}_4,\ \text{quinoline}]{H_2}$

$C_6H_5-CH_2(H)C{=}C(H)CH_2CH_3$ (cis)

(c) 1. $C_6H_6 \xrightarrow[AlCl_3]{CH_3I} C_6H_5-CH_3$

(*Note* : The reagents used increase the number of carbon atoms.)

2. $BrCH_2CH_2Br \xrightarrow[\text{ethanol}]{NaCN\ (excess)} NCCH_2CH_2CN \xrightarrow[H_2O,\ \Delta]{NaOH} \xrightarrow{H_3O^+}$

$HO\overset{O}{\overset{\|}{C}}CH_2CH_2\overset{O}{\overset{\|}{C}}OH \xrightarrow[\Delta]{DH\ \|O}$ (succinic anhydride) $\xrightarrow[AlCl_3]{C_6H_5-CH_3}$ (7-methyl-4-oxo-1,2,3,4-tetrahydronaphthalene bearing $HO\overset{O}{\overset{\|}{C}}$ and CH_3)

$\xrightarrow[\Delta]{Zn(Hg),\ HCl}$ (tetrahydronaphthalene bearing $HO\overset{O}{\overset{\|}{C}}$ and CH_3) $\xrightarrow[\Delta]{SOCl_2}$ (tetrahydronaphthalene bearing $Cl\overset{O}{\overset{\|}{C}}$ and CH_3)

$\xrightarrow{AlCl_3}$ (7-methyl-1-tetralone: CH_3 on aromatic ring, C=O)

(d) Benzene $\xrightarrow[AlCl_3]{CH_3CH_2\overset{O}{\overset{\|}{C}}Cl}$ $C_6H_5\overset{O}{\overset{\|}{C}}CH_2CH_3$ $\xrightarrow[\text{diethyl ether}]{CH_3MgI}$ $\xrightarrow[H_2O]{NH_4Cl}$ $C_6H_5\overset{CH_3}{\overset{|}{\underset{OH}{\underset{|}{C}}}}CH_2CH_3$

156. The shortest routes for synthesis of the compounds are given below:

(a) C_6H_5MgBr [from 153(e)] $\xrightarrow{CO_2}$ $\xrightarrow{H_3O^+}$ $C_6H_5\overset{O}{\overset{\|}{C}}OH$ $\xrightarrow[Fe]{Br_2}$ m-$BrC_6H_4\overset{O}{\overset{\|}{C}}OH$

$\xrightarrow[H_2SO_4,\ \Delta]{CH_3\overset{CH_3}{\overset{|}{C}}HOH}$ m-$BrC_6H_4\overset{O}{\overset{\|}{C}}O\overset{CH_3}{\overset{|}{C}}HCH_3$

(b) Succinic anhydride $\xrightarrow[AlCl_3]{\text{benzene}}$ $C_6H_5\overset{O}{\overset{\|}{C}}CH_2CH_2\overset{O}{\overset{\|}{C}}OH$ $\xrightarrow[\Delta]{SOCl_2}$ $C_6H_5\overset{O}{\overset{\|}{C}}CH_2CH_2\overset{O}{\overset{\|}{C}}Cl$

$\xrightarrow{AlCl_3}$ (1,4-naphthalenedione, tetralin-1,4-dione) $\xrightarrow[\text{diethylene glycol},\ \Delta]{H_2NNH_2,\ KOH}$ (tetralin)

(c) $C_6H_5CH_3$ $\xrightarrow[Fe]{Br_2}$ p-$BrC_6H_4CH_3$ $\xrightarrow[\text{diethyl ether}]{Mg}$ p-$BrMgC_6H_4CH_3$ $\xrightarrow{\text{ethylene oxide}}$ $\xrightarrow{H_3O^+}$ p-$HOCH_2CH_2C_6H_4CH_3$

$\xrightarrow[\text{pyridine}]{PBr_3}$ p-$CH_3C_6H_4CH_2CH_2Br$ $\xrightarrow[\text{diethyl ether}]{Mg}$ p-$CH_3C_6H_4CH_2CH_2MgBr$ $\xrightarrow{\text{ethylene oxide}}$ $\xrightarrow{H_3O^+}$

p-$CH_3C_6H_4CH_2CH_2CH_2CH_2OH$

(d) $CH_3CH_2CH_2Br \xrightarrow[\text{ethanol}]{NaCN} CH_3CH_2CH_2C{\equiv}N \xrightarrow[\Delta]{H_3O^+} CH_3CH_2CH_2\overset{O}{\overset{\|}{C}}OH$

$\xrightarrow[\Delta]{SOCl_2} CH_3CH_2CH_2\overset{O}{\overset{\|}{C}}Cl \xrightarrow[AlCl_3]{C_6H_6} C_6H_5\overset{O}{\overset{\|}{C}}CH_2CH_2CH_3 \xrightarrow[\Delta]{Zn(Hg),\ HCl}$

$C_6H_5CH_2CH_2CH_2CH_3$

(e) 1. p-$CH_3C_6H_5$ $\xrightarrow[FeBr_3]{Br_2}$ p-$BrC_6H_4CH_3$ $\xrightarrow[\text{diethyl ether}]{Mg}$ $\xrightarrow{H\overset{O}{\overset{\|}{C}}H}$ $\xrightarrow{H_3O^+}$ p-$CH_3C_6H_4CH_2OH$

$\xrightarrow[\text{dichloromethane}]{C_5H_5NH^+\ CrO_3Cl^-}$ p-$CH_3C_6H_4\overset{O}{\overset{\|}{C}}H$

2. $CH_3\underset{}{\overset{CH_3}{\overset{|}{C}}}HBr \xrightarrow[\text{diethyl ether}]{Mg} \xrightarrow{\text{ethylene oxide}} \xrightarrow{H_3O^+} CH_3\overset{CH_3}{\overset{|}{C}}HCH_2CH_2OH$

$\xrightarrow[\text{pyridine}]{PBr_3}$ $CH_3\underset{}{C}HCH_2CH_2Br$ (with CH_3 on C-2) $\xrightarrow[\text{(from 1 above)}]{CH_3-C_6H_4-\overset{O}{\overset{\|}{C}}H}$ $\xrightarrow[H_2O]{NH_4Cl}$

p-$CH_3C_6H_4-\overset{OH}{\overset{|}{C}}HCH_2CH_2\overset{CH_3}{\overset{|}{C}}HCH_3$ $\xrightarrow[\text{dichloromethane}]{C_5H_5\overset{+}{N}H + CrO_3Cl^-}$ p-$CH_3C_6H_4-\overset{O}{\overset{\|}{C}}CH_2CH_2\overset{CH_3}{\overset{|}{C}}HCH_3$

(f) Benzene $\xrightarrow[\Delta]{HNO_3,\ H_2SO_4}$ $C_6H_5\overset{O}{\overset{\|}{C}}CH_3$ $\xrightarrow[AlCl_3]{(CH_3\overset{O}{\overset{\|}{C}})_2O}$ m-$NO_2C_6H_4\overset{O}{\overset{\|}{C}}CH_3$

(g) Succinic anhydride [from 155(c)] + Toluene ($CH_3C_6H_5$) $\xrightarrow{AlCl_3}$ 6-methyl-1-oxo-tetralin-4-carboxylic acid (CH_3, $HO\overset{\|}{C}=O$) $\xrightarrow[\Delta]{Zn(Hg)HCl}$

(keto acid, CH_3, $HOC(=O)$) $\xrightarrow[\Delta]{SOCl_2}$ (keto acid chloride, CH_3, $ClC(=O)$) $\xrightarrow{AlCl_3}$ CH_3-substituted tetralone (C=O)

$\xrightarrow[\text{methanol}]{NaBH_4}$ CH_3-tetralin-ol (OH) $\xrightarrow[\Delta]{H_3PO_4}$ CH_3-tetralin

(h) Succinic anhydride [from 155(c)] + Benzene $\xrightarrow{AlCl_3}$ 4-oxo-tetralin carboxylic acid ($HO\overset{\|}{C}=O$) $\xrightarrow[\Delta]{Zn(Hg)}$ tetralin carboxylic acid ($HO\overset{\|}{C}=O$) $\xrightarrow[\Delta]{SOCl_2}$

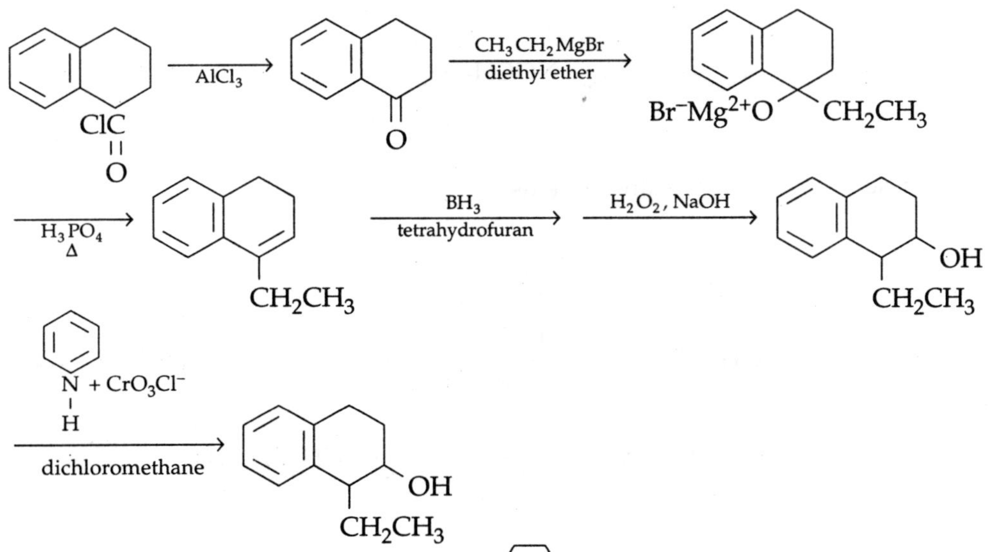

$\xrightarrow{AlCl_3}$ $\xrightarrow[\text{diethyl ether}]{CH_3CH_2MgBr}$ $Br^-Mg^{2+}O$ CH_2CH_3

ClC
O

$\xrightarrow[\Delta]{H_3PO_4}$ CH_2CH_3 $\xrightarrow[\text{tetrahydrofuran}]{BH_3}$ $\xrightarrow{H_2O_2,\ NaOH}$ OH CH_2CH_3

N + CrO_3Cl^-
H
$\xrightarrow{\text{dichloromethane}}$ OH CH_2CH_3

(i) 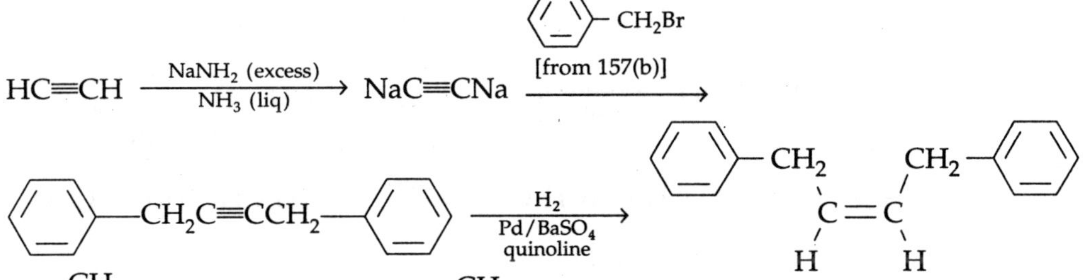 $HC{\equiv}CH \xrightarrow[NH_3\ (liq)]{NaNH_2\ (excess)} NaC{\equiv}CNa$ $\xrightarrow[\text{[from 157(b)]}]{C_6H_5CH_2Br}$

$C_6H_5CH_2C{\equiv}CCH_2C_6H_5 \xrightarrow[\text{Pd/BaSO}_4\text{ quinoline}]{H_2}$ $C_6H_5CH_2$(H)C=C(H)$CH_2C_6H_5$

157. CH_3, CH_3, CH_3 $\xrightarrow{I{-}Cl}$ CH_3, I, CH_3, CH_3 + HCl

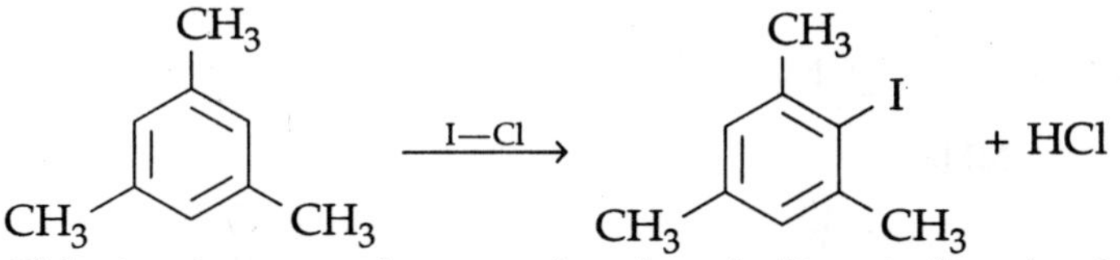

Chlorine is more electrogative than iodine; iodine is, therefore, the electrophile.

158. The structural formulas A, B, C and D are:

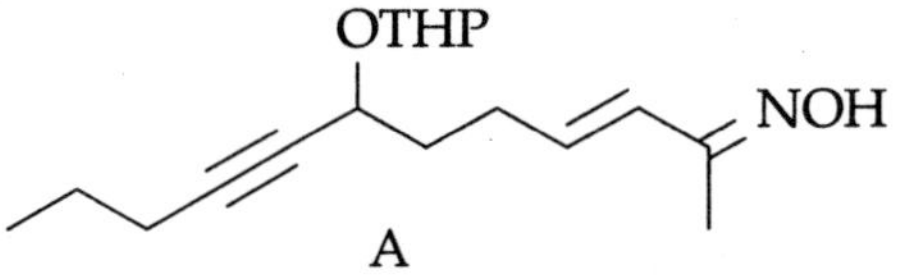

OTHP
NOH
A

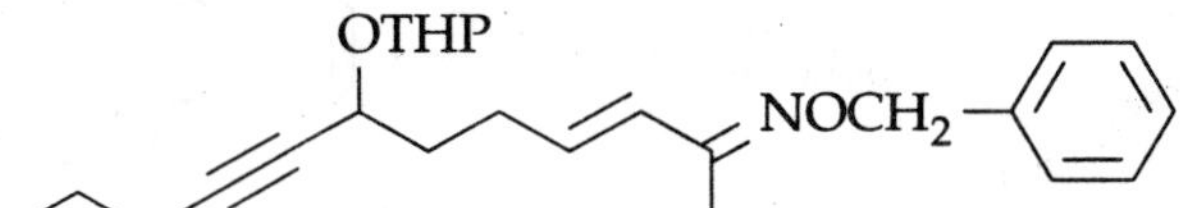

OTHP
$NOCH_2$—C_6H_5
B

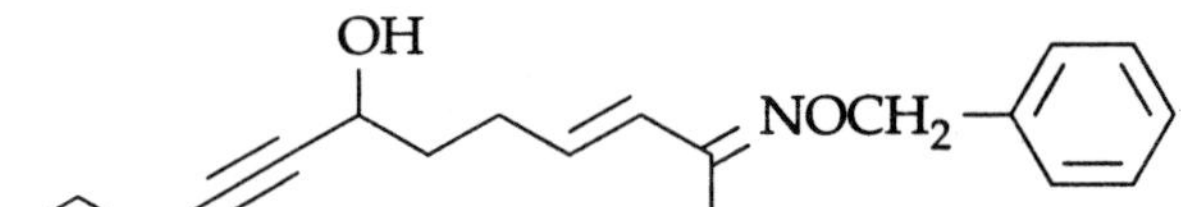

OH
$NOCH_2$—C_6H_5
C

D

159. These transformations are done as:

(a) $C_6H_5-CH_2OH \xrightarrow[\text{dichloromethane}]{C_5H_5\overset{+}{N}H\ CrO_3Cl^-} C_6H_5-\overset{O}{\overset{\|}{C}}H \xrightarrow{CH_3CH_2,\ NaOH}$

$C_6H_5-CH{=}CHNO_2 \xrightarrow[PrO_2]{H_2} C_6H_5-CH_2CH_2NH_2$

(b) $CH_3CH_2CH_2CH_2OH \xrightarrow[\text{dichloromethane}]{C_5H_5\overset{+}{N}H\ CrO_3Cl^-} CH_3CH_2CH_2\overset{O}{\overset{\|}{C}}H \xrightarrow[\text{diethyl ether}]{CH_3CH_2MgBr}$

$\xrightarrow[H_2O]{NH_4Cl} CH_3CH_2CH_2\underset{OH}{\underset{|}{C}}HCH_2CH_3 \xrightarrow[\text{acetone}]{CrO_3} CH_3CH_2CH_2\overset{O}{\overset{\|}{C}}CH_2CH_3$

$\xrightarrow[Ni]{NH_3,\ H_2} CH_3CH_2CH_2\underset{NH_2}{\underset{|}{C}}HCH_2CH_3$

(c) $C_5H_9-\overset{O}{\overset{\|}{C}}OH \xrightarrow[\Delta]{SOCl_2} C_5H_9-\overset{O}{\overset{\|}{C}}Cl \xrightarrow{NH_3\text{(excess)}} C_5H_9-\overset{O}{\overset{\|}{C}}NH_2$

$\xrightarrow[\text{diethyl ether}]{LiAlH_4} \xrightarrow{H_2O} C_5H_9-CH_2NH_2$

(d) $NC-CH{=}CH-CN$ + $CH_2{=}CH-CH{=}CH_2$ $\longrightarrow$ (cyclohexene with two $C{\equiv}N$ groups) $\xrightarrow[\text{diehyl ether}]{LiAlH_4} \xrightarrow{H_2O}$ (cyclohexene with two CH_2NH_2 groups)

$\xrightarrow[PrO_2]{H_2}$ (ring with two CH_2NH_2 groups)

160. (a) Benzene $\xrightarrow[H_2SO_4]{HNO_3}$ $C_6H_5NO_2$ $\xrightarrow{Sn,\ HCl}$ $\xrightarrow{NaOH}$ $C_6H_5NH_2$ $\xrightarrow{2\ CH_3I}$ $C_6H_5\overset{+}{N}HCH_3\ CH_3\ I^-$

$\xrightarrow[H_2O]{NaOH}$ $C_6H_5N(CH_3)CH_3$

(b) Benzene $\xrightarrow[H_2SO_4,\ \Delta]{HNO_3}$ $C_6H_5NO_2$ $\xrightarrow{Fe,\ HCl}$ $\xrightarrow[H_2O]{NaOH}$ $C_6H_5NH_2$ $\xrightarrow[\text{pyridine}]{(CH_3\overset{O}{\overset{\|}{C}})_2O}$ $C_6H_5NH\overset{O}{\overset{\|}{C}}CH_3$

$\xrightarrow[H_2SO_4]{HNO_3}$ $p\text{-}O_2NC_6H_4NH\overset{O}{\overset{\|}{C}}CH_3$ $\xrightarrow[H_2O,\ \Delta]{NaOH}$ $p\text{-}O_2NC_6H_4NH_2$

(c) $CH_3\overset{CH_3}{\overset{|}{C}}HBr$ $\xrightarrow[\text{diethyl ether}]{Mg}$ $CH_3\overset{CH_3}{\overset{|}{C}}HMgBr$ $\xrightarrow{H\overset{O}{\overset{\|}{C}}H}$ $\xrightarrow{H_3O^+}$ $CH_3\overset{CH_3}{\overset{|}{C}}HCH_2OH$

$\xrightarrow[\text{dichloromethane}]{C_5H_5\overset{}{N}H\ +\ CrO_3Cl^-}$ $CH_3\overset{CH_3}{\overset{|}{C}}H\underset{\underset{O}{\|}}{C}H$ $\xrightarrow{CH_3NO_2,\ NaOH}$ $CH_3\overset{CH_3}{\overset{|}{C}}HCH{=}CHNO_2$

$\xrightarrow[\text{Pd, acetic acid}]{H_2}$ $CH_3\overset{CH_3}{\overset{|}{C}}HCH_2CH_2NH_2$

161. These compounds are synthesized as:

(a) Benzene $\xrightarrow[H_2SO_4]{HNO_3}$ $C_6H_5NO_2$ $\xrightarrow[FeCl_3]{Cl_2}$ $m\text{-}ClC_6H_4NO_2$ $\xrightarrow[HCl]{Fe}$ $\xrightarrow{NaOH}$ $m\text{-}ClC_6H_4NH_2$

$\xrightarrow[H_2O,\ 273\ K]{NaNO_2,\ HBr}$ m-chlorobenzenediazonium bromide ($\overset{+}{N}\equiv N\ Br^-$, Cl) $\xrightarrow{CuBr}$ (Br, Cl)

(b) CH_3-benzene (Toluene) $\xrightarrow[\Delta]{HNO_3,\ H_2SO_4}$ (CH_3, NO_2) $\xrightarrow[HCl]{Fe}$ $\xrightarrow{NaOH}$ (CH_3, NH_2) $\xrightarrow[H_2O,\ 273\ K]{NaNO_2,\ H_2SO_4}$

(CH_3, $N\equiv N\ HSO_4^-$) $\xrightarrow[H_2SO_4,\ \Delta]{H_2O}$ (CH_3, OH)

(c) (CH_3, MgBr) $\xrightarrow{CO_2}$ $\xrightarrow{H_3O^+}$ (CH_3, $O=\overset{|}{C}OH$)

(d) (NO_2) [from (a)] $\xrightarrow[HCl]{Sn}$ $\xrightarrow{NaOH}$ (NH_2) $\xrightarrow[H_2O,\ 273\ K]{NaNO_2,\ H_2SO_4}$ ($^+N\equiv N\ HSO_4^-$)

$\xrightarrow{HBF_4}$ ($^+N\equiv N\ BF_4^-$) $\xrightarrow{\Delta}$ (F) $\xrightarrow[H_2SO_4,\ \Delta]{HNO_3\ (funing)}$ (F, NO_2, NO_2)

(e) 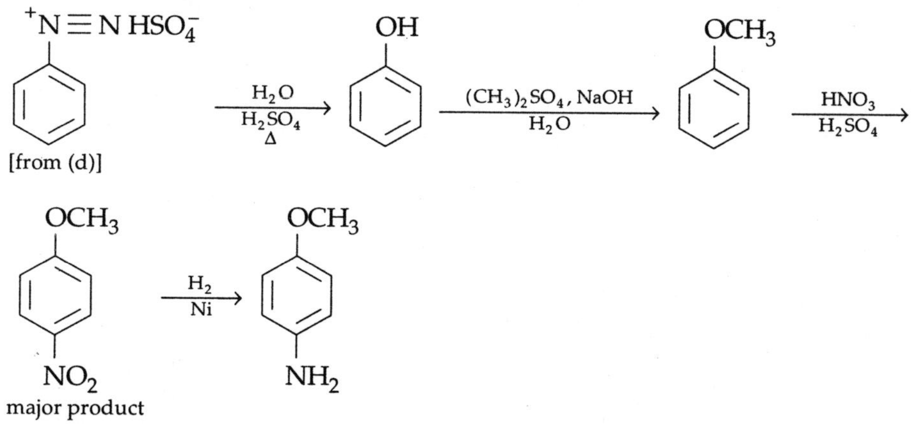 [from (d)] $\overset{+}{N}\equiv N\ HSO_4^-$ (benzenediazonium) $\xrightarrow[H_2SO_4,\ \Delta]{H_2O}$ phenol (OH) $\xrightarrow[H_2O]{(CH_3)_2SO_4,\ NaOH}$ anisole (OCH_3) $\xrightarrow[H_2SO_4]{HNO_3}$

p-nitroanisole (OCH_3, NO_2) major product $\xrightarrow[Ni]{H_2}$ p-anisidine (OCH_3, NH_2)

(f) 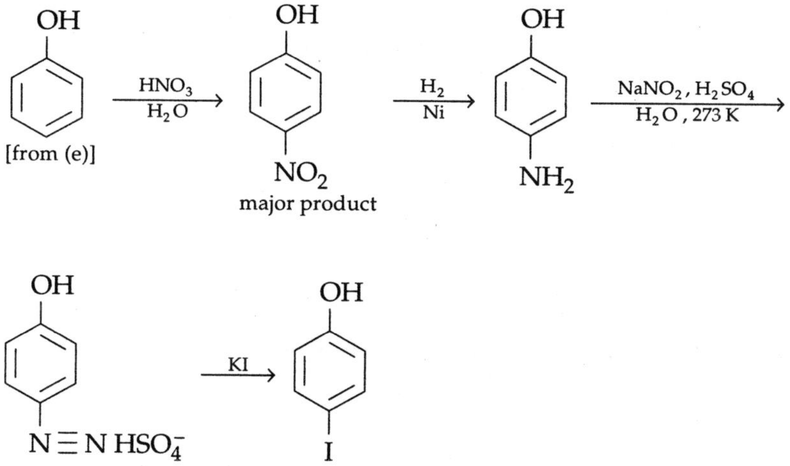 [from (e)] phenol (OH) $\xrightarrow[H_2O]{HNO_3}$ p-nitrophenol (OH, NO_2) major product $\xrightarrow[Ni]{H_2}$ p-aminophenol (OH, NH_2) $\xrightarrow[H_2O,\ 273\ K]{NaNO_2,\ H_2SO_4}$

p-hydroxybenzenediazonium (OH, $N\equiv N\ HSO_4^-$) $\xrightarrow{KI}$ p-iodophenol (OH, I)

162. The designated letters have the following structural formulas :

(a) 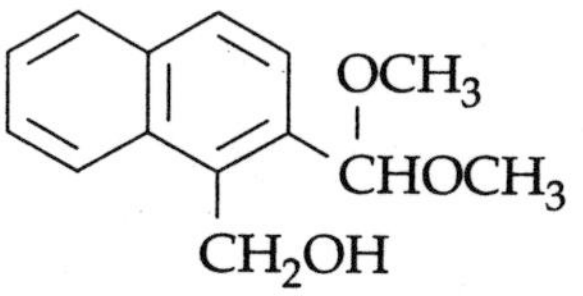 **A**: naphthalene bearing $CH(OCH_3)OCH_3$ ($\overset{OCH_3}{|}CHOCH_3$) and CH_2OH; **B**: naphthofuran hemiacetal with OH, H, O

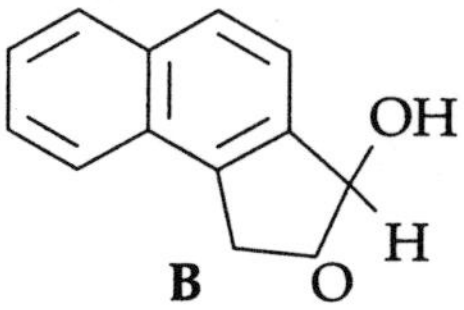

(b) **C**: OCH_3, isopropyl-substituted benzene with butenyl side chain; **D**: OCH_3, isopropyl-substituted benzene with side chain ending in OH

(c) **E**: OCH_3, isopropyl benzene with side chain $CH(CH_3)CH_2C(=O)OH$ (HOC=O); **F**: indanone with OCH_3, isopropyl, methyl (=O); indanol with OCH_3, isopropyl, methyl, OH

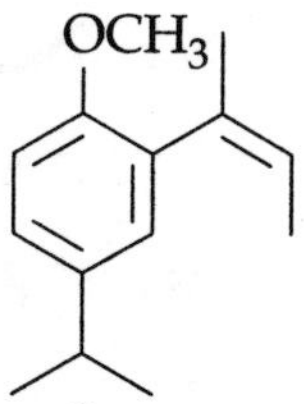

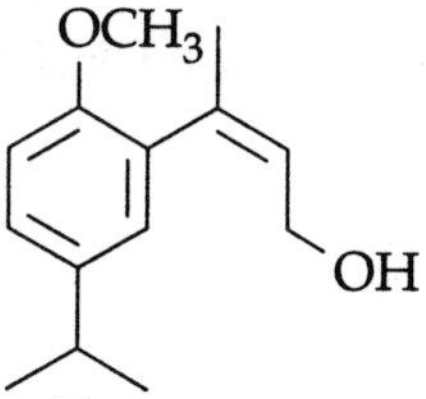

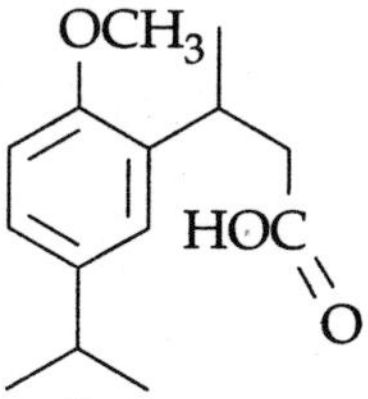

(d) **H**: tricyclic diketone (O, O); **I**: tricyclic diketone (O, O)

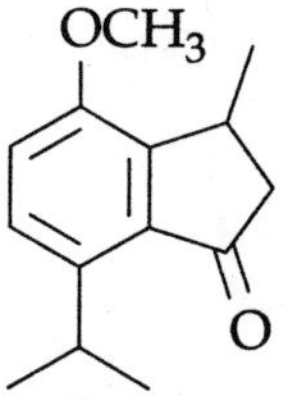

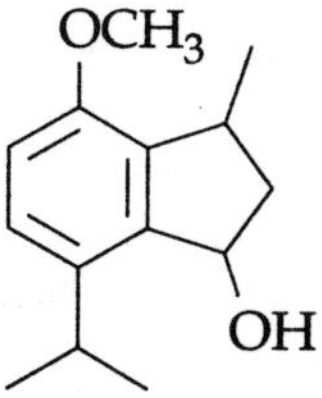

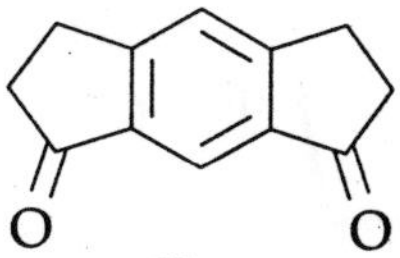

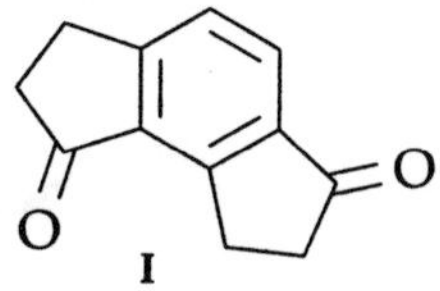

(e) J: 2,4-dinitroanisole — OCH_3, NO_2, NO_2

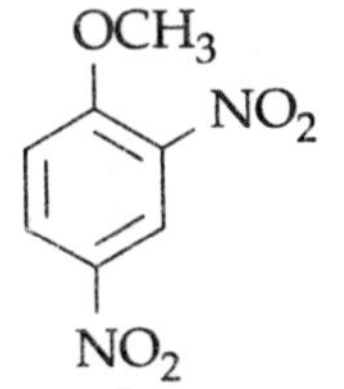

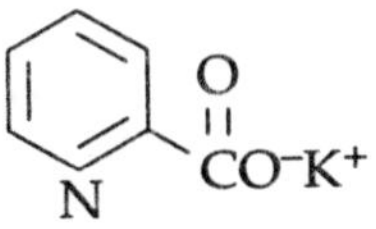

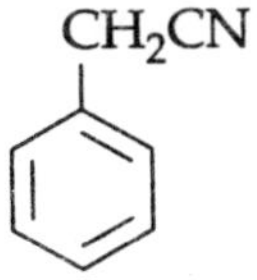

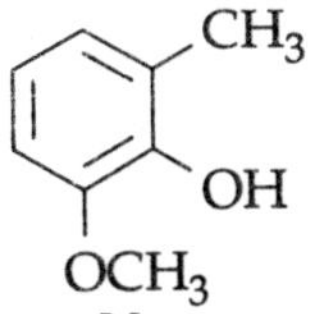

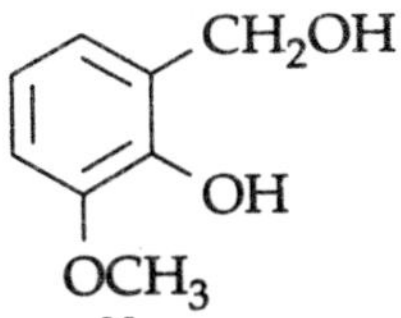

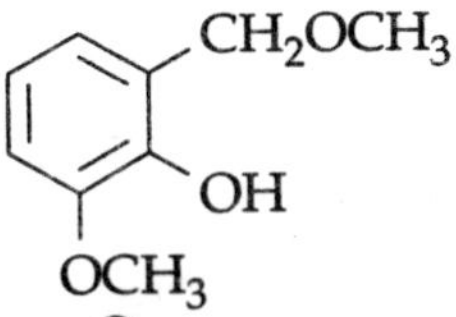

(f) K: 2-pyridinecarboxylate — CO^-K^+, N

(g) L: CH_2CN on benzene

(h) M: CH_3, OH, OCH_3

(i) N: CH_2OH, OH, OCH_3

O: CH_2OCH_3, OH, OCH_3

(an alkoxide ion is a beter nucleophile than a phenolate ion)

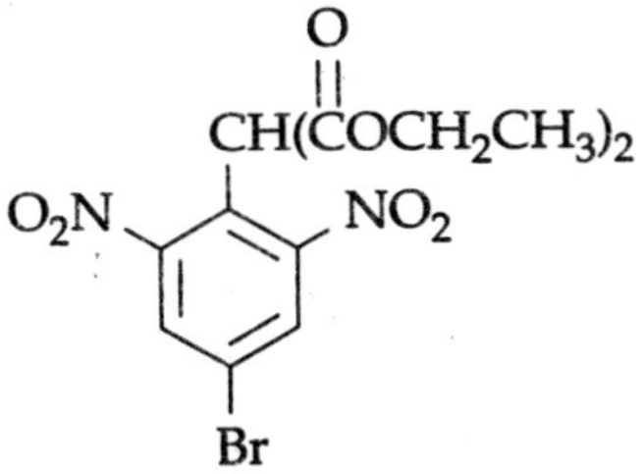

(j) 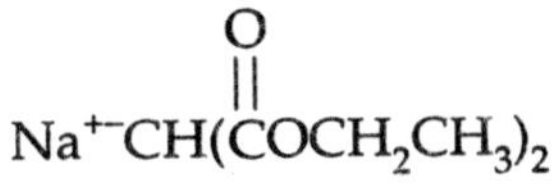 $Na^{+-}CH(\overset{O}{\overset{||}{C}}OCH_2CH_3)_2$

P

Q: $CH(\overset{O}{\overset{||}{C}}OCH_2CH_3)_2$, O_2N, NO_2, Br

(k) $R = \xrightarrow{NaOH, (CH_3)_2SO_4 \text{ or } CH_3I}$, $S = \xrightarrow{LiAlH_4}$, $T = \xrightarrow[\text{(NBS used)}]{Br_2}$

163. The designated letters have the following structural formulas.

(a) 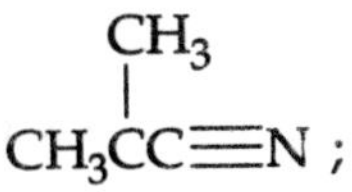 $CH_3\overset{CH_3}{\overset{|}{C}}C{\equiv}N$;

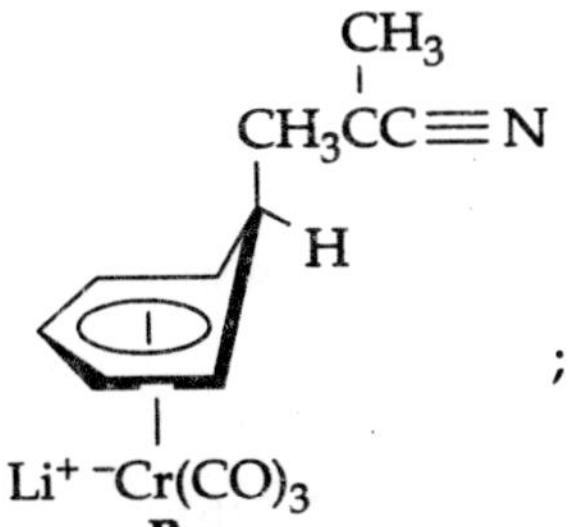

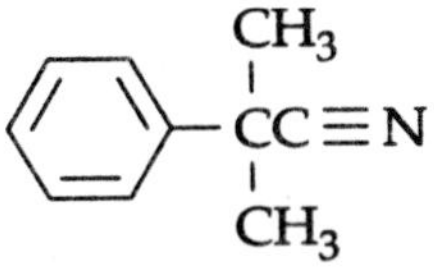

A

B: $CH_3\overset{CH_3}{\overset{|}{C}}C{\equiv}N$, H, $Li^+ \, ^-Cr(CO)_3$

; C: C_6H_5–$\overset{CH_3}{\underset{CH_3}{C}}C{\equiv}N$

(b) D: $CH_3C_6H_4$, H, $Li^+ \, ^-Cr(CO)_3$; E: $CH_3C_6H_4$–C_6H_5

(c) F: $CH_2\overset{O}{\overset{||}{C}}O\overset{CH_3}{\underset{CH_3}{C}}CH_3$, H, $Li^+ \, ^-Cr(CO)_3$

G: $CH_2\overset{O}{\overset{||}{C}}O\overset{CH_3}{\underset{CH_3}{C}}CH_3$ on C_6H_5

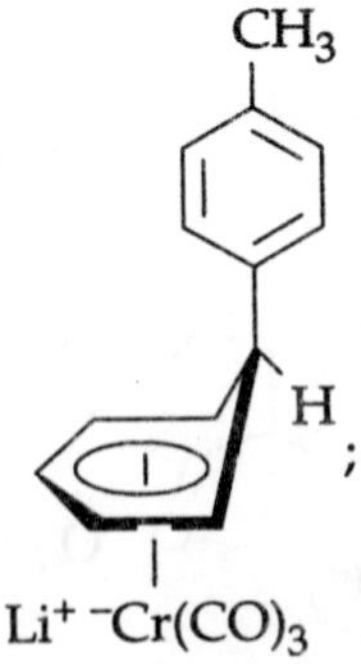

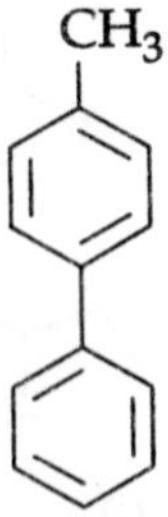

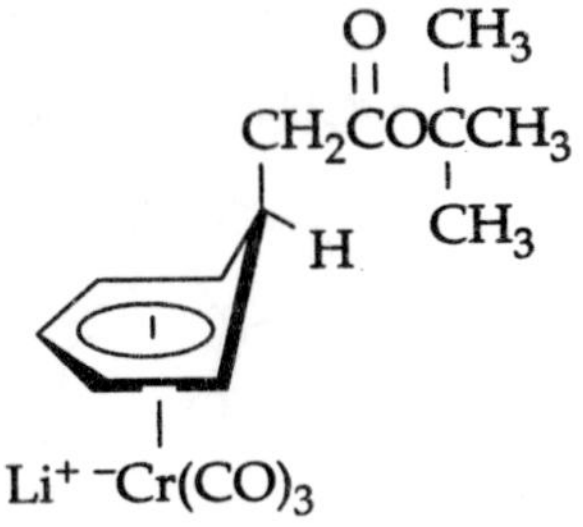

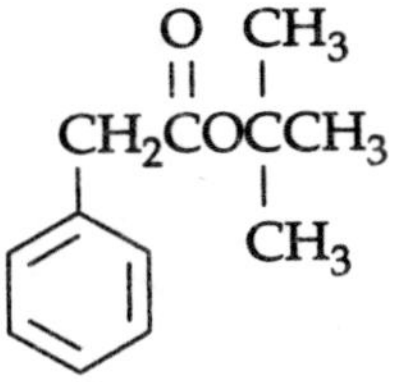

(d) $C_6H_5C(CH_3)_2CN$ (CN, CH_3CCH_3, benzene ring, H)

164. (a) O_2N–$C_6H_3(NO_2)$–N (piperidine)

(b) O_2N–$C_6H_3(NO_2)$–S–C_6H_5

(c) O_2N–$C_6H_2(NO_2)_2$–NH–C_6H_5

(d) O_2N–$C_6H_2(NO_2)_2$–NH—N$(C_6H_5)_2$

(e) O_2N–$C_6H_3(NO_2)$–NHCH(CH_3)COH (C=O)

165. The reagents are needed in more than one step.

(a) $C_6H_5CH_3$ $\xrightarrow[H_2SO_4]{HNO_3}$ p-$CH_3C_6H_4NO_2$ $\xrightarrow[Ni]{H_2}$ p-$CH_3C_6H_4NH_2$ $\xrightarrow[H_2O,\ 273\,K]{NaNO_2,\ HCl}$ p-$CH_3C_6H_4\overset{+}{N}\equiv N\ Cl^-$

$\xrightarrow{HBF_4}$ p-$CH_3C_6H_4\overset{+}{N}\equiv N\ BF_4^-$ $\xrightarrow{\Delta}$ p-$CH_3C_6H_4F$ $\xrightarrow[H_2O,\ \Delta]{KMnO_4,\ NaOH}$ $\xrightarrow{H_3O^+}$ p-FC_6H_4COH (C=O)

$\xrightarrow[H_2SO_4,\ \Delta]{CH_3CH_2OH}$ p-$F-C_6H_4-COCH_2CH_3$ (with C=O)

(b) p-$CH_3-C_6H_4-N^+\equiv N\ Cl^-$ [from (a)] $\xrightarrow[H_2SO_4]{H_2O}$ p-$CH_3-C_6H_4-OH$ $\xrightarrow[\text{acetic acid}]{Br_2}$ 2-bromo-4-methylphenol (CH_3, Br, OH)

(c) $C_6H_5CH_2Br$ $\xrightarrow[\text{ethanol}]{CH_2(COCH_2CH_3)_2,\ CH_3CH_2ONa}$ $C_6H_5CH_2CH(COCH_2CH_3)_2$ (C=O) $\xrightarrow[\text{ethanol}]{CH_3CH_2ONa}$

$\xrightarrow{CH_3CH_2Br}$ $C_6H_5CH_2C(CH_2CH_3)(COCH_2CH_3)_2$ (C=O) $\xrightarrow[\Delta]{H_3O^+}$ $C_6H_5CH_2CH(CH_2CH_3)COH$ (C=O) $\xrightarrow{SOCl_2}$

$C_6H_5CH_2CH(CH_2CH_3)CCl$ (C=O) $\xrightarrow{NH_3}$ $C_6H_5CH_2CH(CH_2CH_3)CNH_2$ (C=O) $\xrightarrow[\text{diethyl ether}]{LiAlH_4}$ $\xrightarrow{H_2O}$ $C_6H_5CH_2CH(CH_2CH_3)CH_2NH_2$

(d) C_6H_5Br $\xrightarrow[H_2SO_4]{HNO_3}$ p-$Br-C_6H_4-NO_2$ (major product) $\xrightarrow[Pd/C]{H_2}$ p-$Br-C_6H_4-NH_2$ $\xrightarrow[H_2O,\ 273\ K]{NaNO_2,\ H_2SO_4}$ p-$Br-C_6H_4-N^+\equiv N\ HSO_4^-$

$\xrightarrow{KI}$ p-$Br-C_6H_4-I$

(e) [Chlorobenzene] $\xrightarrow[H_2SO_4]{HNO_3}$ [p-chloronitrobenzene: Cl, NO_2] major product $\xrightarrow[HCl]{Sn}$ $\xrightarrow{NaOH}$ [p-chloroaniline: Cl, NH_2]

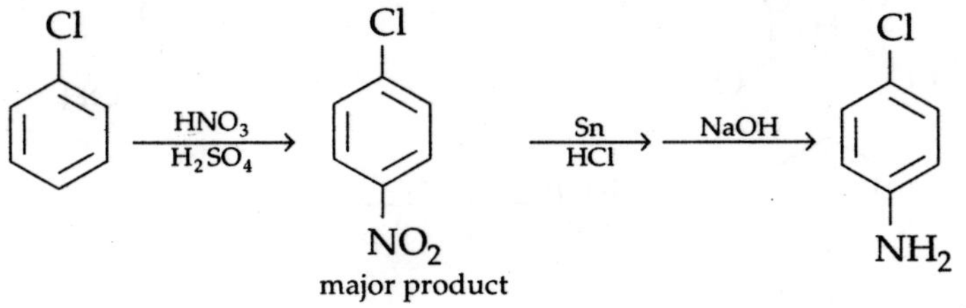

(f) [Toluene: CH_3] $\xrightarrow[H_2SO_4]{HNO_3}$ [CH_3, NO_2] $\xrightarrow[H_2SO_4, H_2O, \Delta]{Na_2Cr_2O_7}$ [COH (C=O), NO_2] $\xrightarrow[Ni]{H_2}$ [COH (C=O), NH_2]

$\xrightarrow[pyridine]{CH_3I\,(2\ molarequiv)}$ [COH (C=O), CH_3NCH_3]

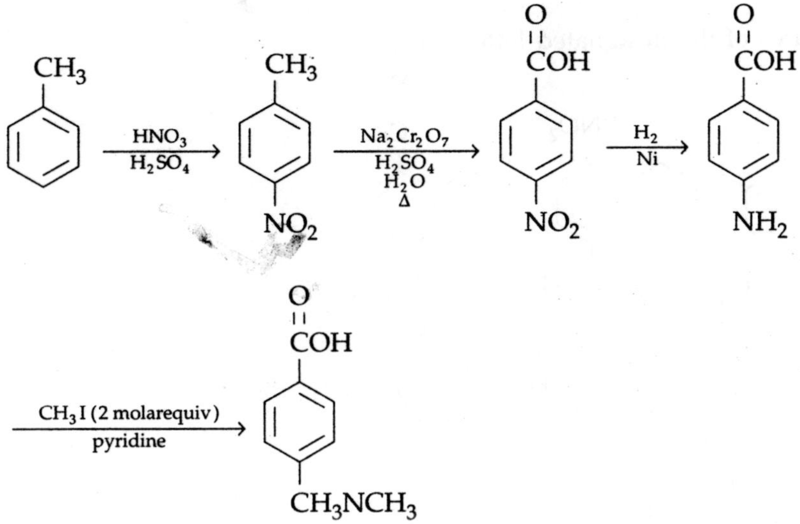

166. The necessary steps to synthesise ibuprofen from benzene are as

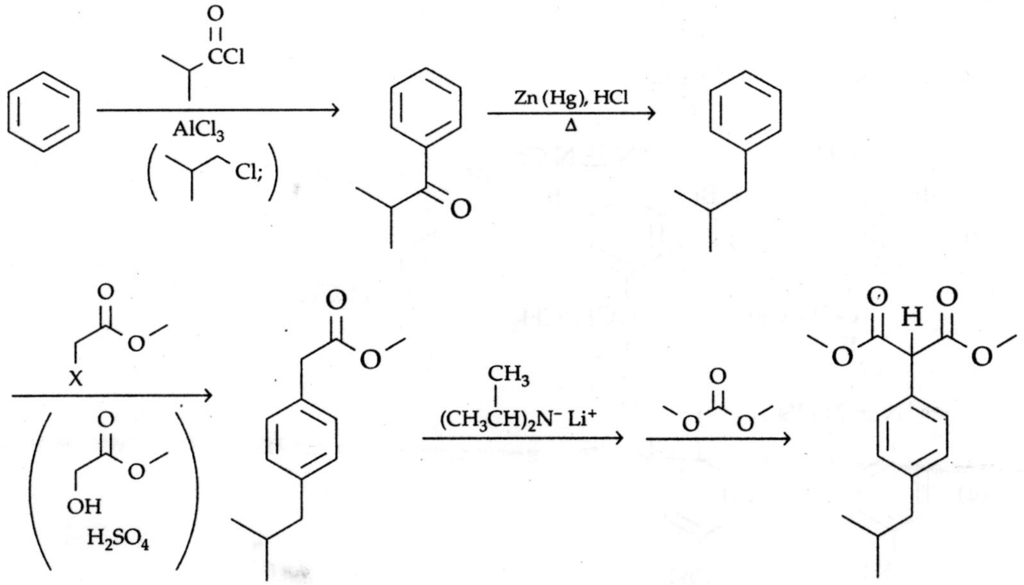

[Benzene] $\xrightarrow[AlCl_3]{(CH_3)_2CHCCl\,(C=O)}$ (Cl;) [isobutyrophenone, O] $\xrightarrow[\Delta]{Zn(Hg),\ HCl}$ [isobutylbenzene]

$\xrightarrow{X\text{–}CH_2C(=O)OCH_3}$ ($HOCH_2C(=O)OCH_3$, H_2SO_4) [methyl 4-isobutylphenylacetate] $\xrightarrow{(CH_3CH)_2N^- Li^+ \ (CH_3)}$ $\xrightarrow{CH_3OC(=O)OCH_3}$ [$CH_3OC(=O)$–CH(H)–$C(=O)OCH_3$ on 4-isobutylphenyl]

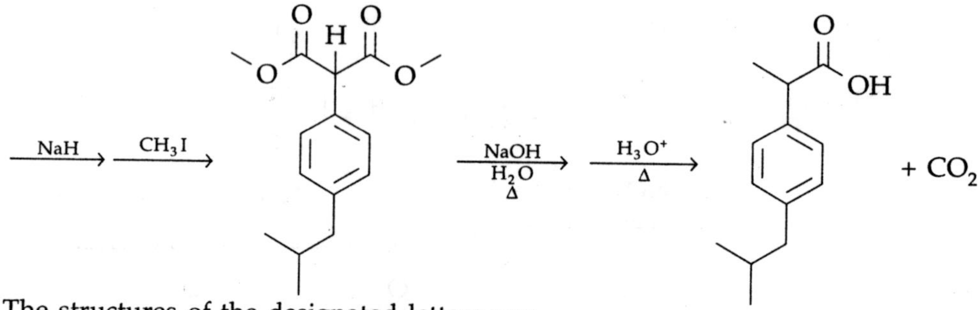

$\xrightarrow{NaH}$ $\xrightarrow{CH_3I}$ (dimethyl 2-(4-isobutylphenyl)malonate structure with O, H, O labels) $\xrightarrow[H_2O, \Delta]{NaOH}$ $\xrightarrow[\Delta]{H_3O^+}$ (2-(4-isobutylphenyl)propanoic acid: O, OH) $+ CO_2$

167. The structures of the designated letters are:

(a) A: 1-methoxynaphthalene (OCH_3); B: 4-nitro-1-methoxynaphthalene (NO_2, OCH_3); C: 4-methoxy-1-naphthonitrile ($N \equiv C$, OCH_3)

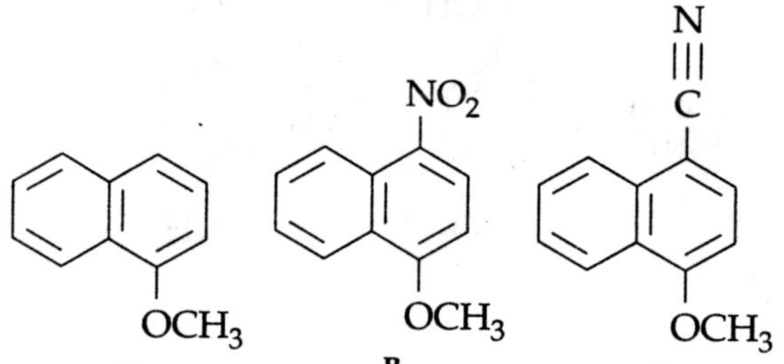

A B C

(b) O_2N–C₆H₃(NO_2)–N(piperidine) + HO–C₆H₄–NO_2

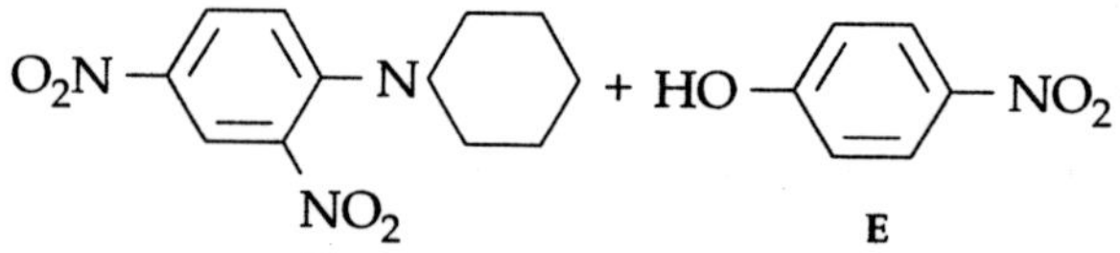

D E

(c) F: 2,6-diiodo-4-nitroaniline (NH_2, I, I, NO_2) ; G: 1,2,3-triiodo-5-nitrobenzene (I, I, I, NO_2)

F G

(d) H: 2,6-dibromo-4-ethoxyaniline (NH_2, Br, Br, OCH_2CH_3) ; I: $^+N \equiv N\ Cl^-$ (Br, Br, OCH_2CH_3) ; J: 3,5-dibromoethoxybenzene (Br, Br, OCH_2CH_3)

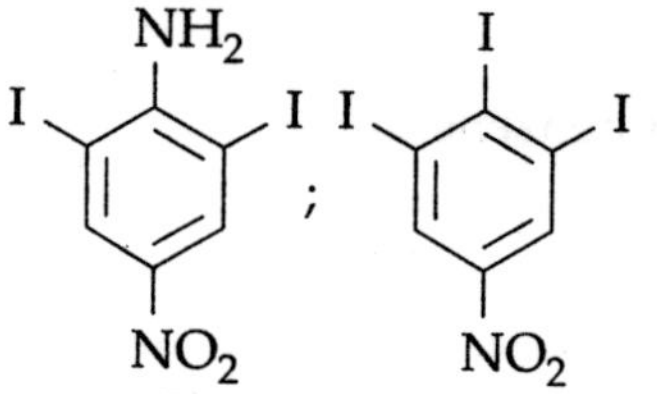

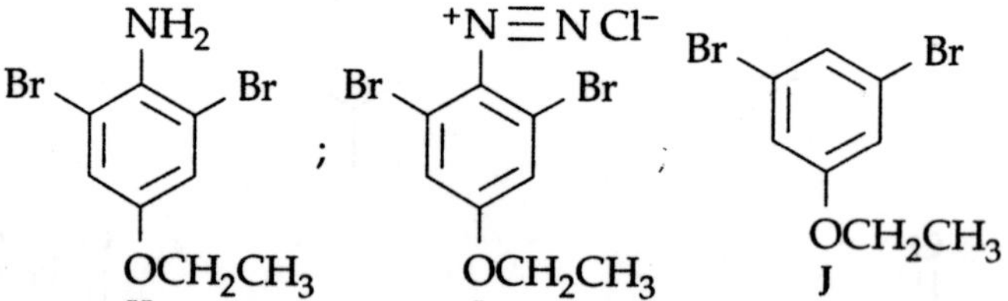

H I J

(e) K: $^+N \equiv N\ HSO_4^-$ (OH) ; L: 4-iodophenol (I, OH)

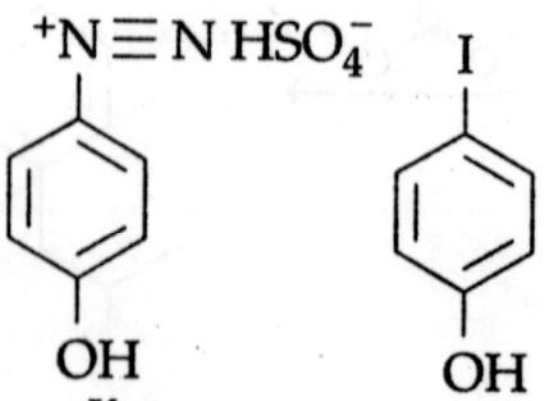

K L

(f) $^{+}N{\equiv}N\,Cl^-$ (M) $^{+}N{\equiv}N\,BF_4^-$ (N) F (O)

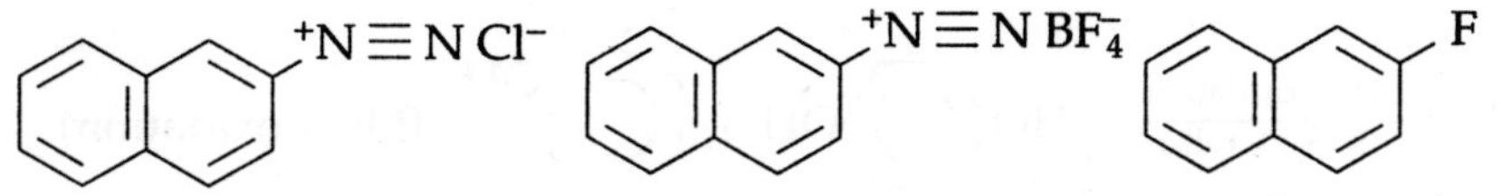

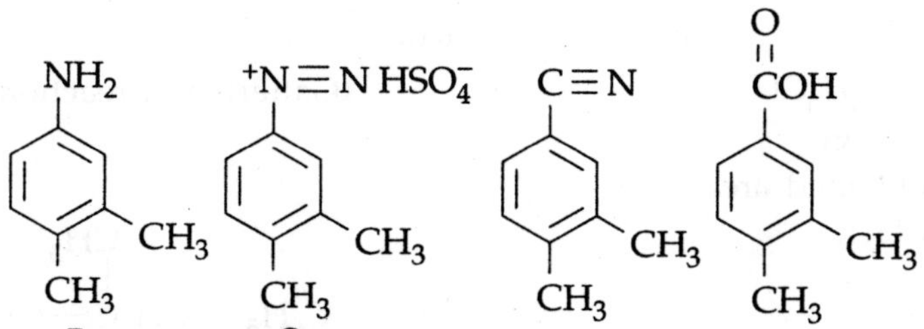

(g) NH_2, CH_3, CH_3 (P) $^{+}N{\equiv}N\,HSO_4^-$, CH_3, CH_3 (Q) $C{\equiv}N$, CH_3, CH_3 (R) COOH, CH_3, CH_3 (S)

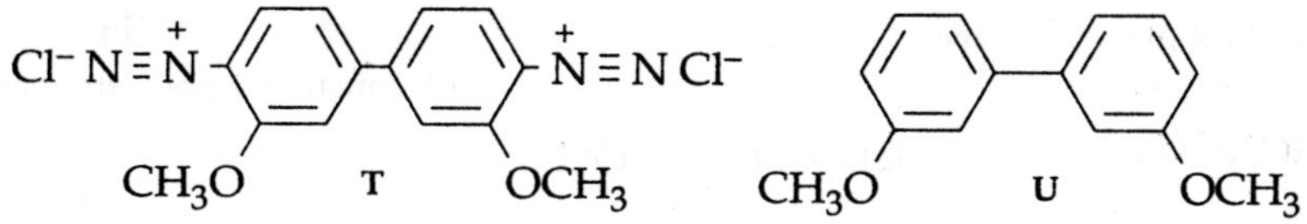

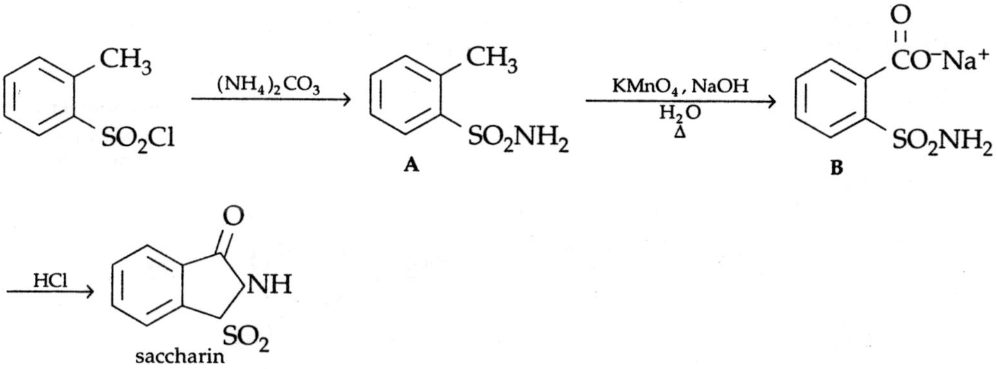

(h) $Cl^-\,N{\equiv}\overset{+}{N}$ ... $\overset{+}{N}{\equiv}N\,Cl^-$, CH_3O T OCH_3 CH_3O U OCH_3

168. CH_3, SO_2Cl $\xrightarrow{(NH_4)_2CO_3}$ CH_3, SO_2NH_2 (A) $\xrightarrow[\Delta]{KMnO_4,\ NaOH,\ H_2O}$ CO^-Na^+, SO_2NH_2 (B)

$\xrightarrow{HCl}$ O, NH, SO_2
saccharin

169. (i) $\underset{(A)}{(CH_3)_2C{=}O} \xrightarrow{\text{dry HCl}} \underset{(B)}{(CH_3)_2C{=}CHCOCH{=}C(CH_3)_2}$

(ii) $\underset{(C)}{CH_3COCl} + H_2 \xrightarrow[\Delta]{\text{conc. } H_2SO_4} \underset{(D)}{CH_3CHO} + HCl$ (Rosenmund's reaction)

(iii) $\underset{(E)}{(CH_3)_2CO} \xrightarrow[\text{quinolene}]{Pd/BaSO_4} \underset{(F)}{CH_3CH_2CH_3}$ (Wolf Kishner reaction)

(iv) $C_6H_5OH \xrightarrow[K_2S_2O_8\text{ (alk)}]{\text{Elb's oxidation}}$ HO–C_6H_4–OH + $C_6H_4(OH)_2$ (Elb's oxidation)

(quinol) (catechol)

(G) (H_1) (H_2)

(v) $C_6H_5N_2Cl \xrightarrow[HCl]{Cu} C_6H_5Cl$ (Gattermann reaction)

(J) (K)

170. The major products obtained are:

(i) $C_6H_5C(CH_3)_3$

(A) 2-methyl, 2-phenylpropane

(ii) $C_6H_5—CH_2C(OH)(CH_3)—CH_3$

(B) 2-methyl, 1-phenylpropan-2-ol

(iii) $C_6H_5CHCNCH_3$

(C) 2-phenylpropane nitrile

(iv) CH_3CH_2, OH, H, CH_2NH_3 on C

(D)

(v) H—C—Ph ; MeO—C—H ; Ph

(E) enantiomer of the orignal compound is formed following S_{N^2} mechanism

(vi) 1,2-dimethylcyclopentene (E) $\xrightarrow{\text{dil. KMnO}_4}$ diol **(F)** $\xrightarrow{HIO_4}$ diketone **(G)** $\xrightarrow{OH^-}$ 3-hydroxy-3-methylcyclohexanone **(H)**

171. (a) $CH_3CH_2C{\equiv}CH \xrightarrow{Na} CH_3CH_2C{\equiv}CNa \xrightarrow{CH_3X}$ (A)

(A) or $CH_3CH_2C{\equiv}CCH_3 \xrightarrow[H_2SO_4]{HgSO_4\ +} CH_3CH_2CH_2\overset{O}{\overset{\|}{C}}CH_3$

(b) Allene is $\overset{I}{CH_2}=\overset{II}{C}=\overset{III}{CH_2}$

sp^2 sp sp^2

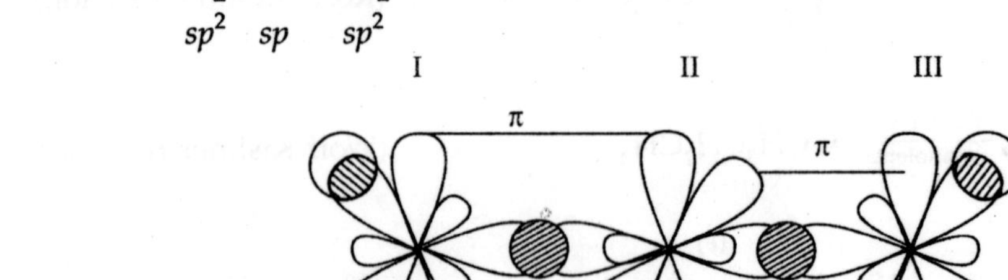

(c) *o*-hydroxy benzaldehyde shows intramolecular H-bonding, (chelation) a weaker H-bonding than intermolecular H-bonding noticed in p-hydroxy benzaldehyde.

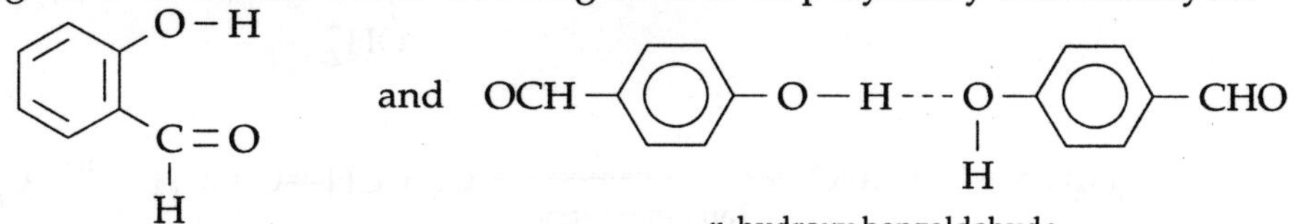

o-hydroxy benzaldehyde (benzene ring bearing O–H and CH=O on adjacent carbons) and $OCH\text{–}C_6H_4\text{–}O\text{–}H\cdots O(H)\text{–}C_6H_4\text{–}CHO$

o-hydroxy benzaldehyde

p-hydroxy benzaldehyde

172. (a)

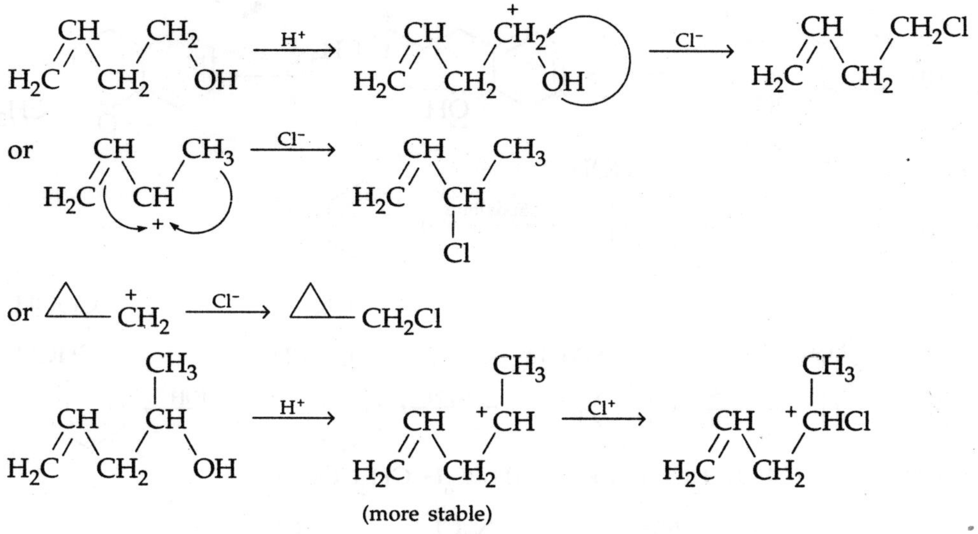

$$H_2C{=}CH\text{–}CH_2\text{–}CH_2\text{–}OH \xrightarrow{H^+} H_2C{=}CH\text{–}CH_2\text{–}\overset{+}{C}H_2 \xrightarrow{Cl^-} H_2C{=}CH\text{–}CH_2\text{–}CH_2Cl$$

or

$$H_2C{=}CH\text{–}\overset{+}{C}H\text{–}CH_3 \xrightarrow{Cl^-} H_2C{=}CH\text{–}CH(Cl)\text{–}CH_3$$

or

$$\triangle\text{–}\overset{+}{C}H_2 \xrightarrow{Cl^-} \triangle\text{–}CH_2Cl$$

$$H_2C{=}CH\text{–}CH_2\text{–}CH(CH_3)\text{–}OH \xrightarrow{H^+} H_2C{=}CH\text{–}CH_2\text{–}\overset{+}{C}H\text{–}CH_3 \xrightarrow{Cl^+} H_2C{=}CH\text{–}CH_2\text{–}\overset{+}{C}HCl\text{–}CH_3$$

(more stable)

(b)

$$C_6H_4(Br)(OCH_3) \xrightarrow[-Br^-]{NaNH_2 + NH_3} \text{2,3-aryne (with } OCH_3) \xrightarrow{NH_2^+} C_6H_4(OCH_3)(NH_2)$$

2,3-aryne

(c)

$$H_2C(COOC_2H_5)_2 \xrightarrow{NaOC_2H_5} NaHC(COOC_2H_5)_2 \xrightarrow{CH_2CHBrCOOC_2H_5} CH_3\text{–}CH(COOC_2H_5)\text{–}CH(COOC_2H_5)_2$$

$$\xrightarrow{HOH} CH_3\text{–}CH(COOC_2H_5)\text{–}CH(COOH)_2 \xrightarrow[-CO_2]{\Delta} HOOC\text{–}CH(CH_3)CH_2COOH$$

(COOH / CHCH₂COOH / CH₃)

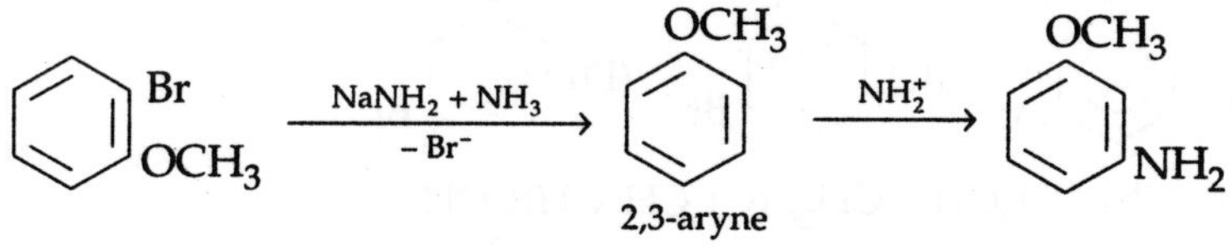

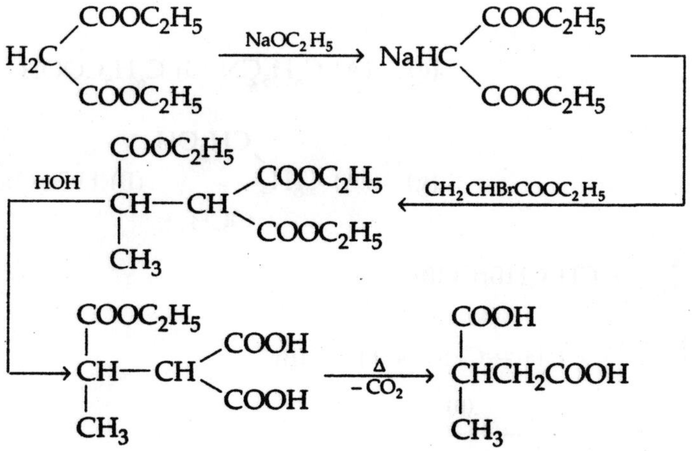

173. (a) (i) $C_6H_5CH(OH)C{\equiv}CH \xrightarrow{H_3\overset{+}{O}} C_6H_5CH(\overset{+}{O}H_2){-}C{\equiv}CH \xrightarrow{-H_2O} C_6H_5CH{=}C{=}\overset{+}{C}H$

$C_6H_5CH{=}C{=}\overset{+}{C}H \xrightarrow{OH^-} C_6H_5CH{=}C{=}CH(OH) \underset{\text{Tautomerisation}}{\rightleftharpoons} C_6H_5CH{=}CH{\cdot}CHO$

(ii) 2-(but-3-enyl)cyclohexadienol (OH) $\xrightarrow{H^+}$ secondary carbocation ($\overset{+}{C}H$–CH_3), attacked by $\ddot{O}H$ $\longrightarrow$ cyclic ether (ring O, CH_3)

(b) (i) $C_6H_5C_2H_5 \xrightarrow{[O]} C_6H_5COOH \xrightarrow{\text{Sodalime}} C_6H_6$

(ii) $C_6H_5CH_2CH_3 \xrightarrow[hv]{X_2} C_6H_5CHXCH_3 \xrightarrow{KCN} C_6H_5CH(CN)CH_3 \xrightarrow{HOH} C_6H_5CH(COOH)CH_3$

174. (a) CH_3Cl, (B) $C_6H_5CH_2Cl$, (C) KCN, (D) $C_6H_5CH_2COOH$

(b) (A) $C_6H_4(NO_2)Br$ (B) $C_6H_4(NH_2)Br$ (C) $C_6H_4(N_2Cl)Br$ (D) $C_6H_4(Cl)Br$ (Br meta in each)

(c) (A) $CH_3CH_2CH_2Br$, (B) $CH_3CH{=}CH_2$, (C) $CH_3CHBrCH_3$
(D) $CH_3CHNH_2CH_3$

(d) (A) $C_6H_4(COOC_6H_5)(NO_2)$ (e) (A) C_6H_5CN (B) C_6H_5COOH

(f) (A) $C_6H_5C(CH_3)_3$ (g) (A) $C_6H_4(CH_2OH)(OCH_3)$ (B) HCOOK

(h) $CH_3CH_2COOH \xrightarrow[\text{bromine}]{\text{P and}} \underset{(A)}{CH_3CHBrCOBr}$

$CH_3CHBrCOBr \xrightarrow[\text{(ii) } H^+]{\text{(i) KOH alc.}} \underset{(B)}{CH_2{=}CHCOOH} + HBr$

(i) 2,4-Dinitroaniline (ring with NH_2, NO_2, NO_2) $\xrightarrow[\text{2. Anisole}]{\text{1. NaNO}_2\text{ and HCl at 278 K}}$ O_2N–$C_6H_3(NO_2)$–N=N–C_6H_4–OCH_3

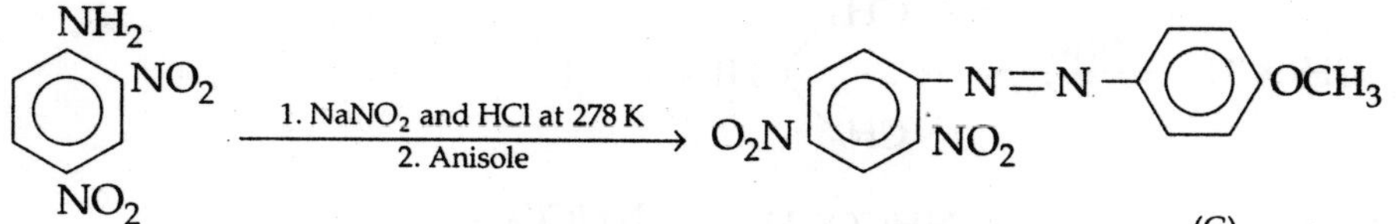

(C)

(j) $C_6H_5CHO + CH_3COOHC_2H_5 \xrightarrow[\text{absolute}C_2H_5OH,\ \Delta\ \text{(Claisencondensation)}]{NaOC_2H_5\ \text{in}} C_6H_5CH{=}CHCOOC_2H_5$

(D)

(k) o-HOOC—C_6H_4—CH_2—C_6H_5 $\xrightarrow{SOCl_2}$ $C_6H_4(COCl)$–CH_2–C_6H_5 (H)

$\downarrow$ AlCl$_3$ anhydrous, Intramolecular Friedel Craft's acetylation

(I) $\xleftarrow[\text{HCl}]{\text{Zn/Hg}}$ (H) [cyclic ketone, C=O]

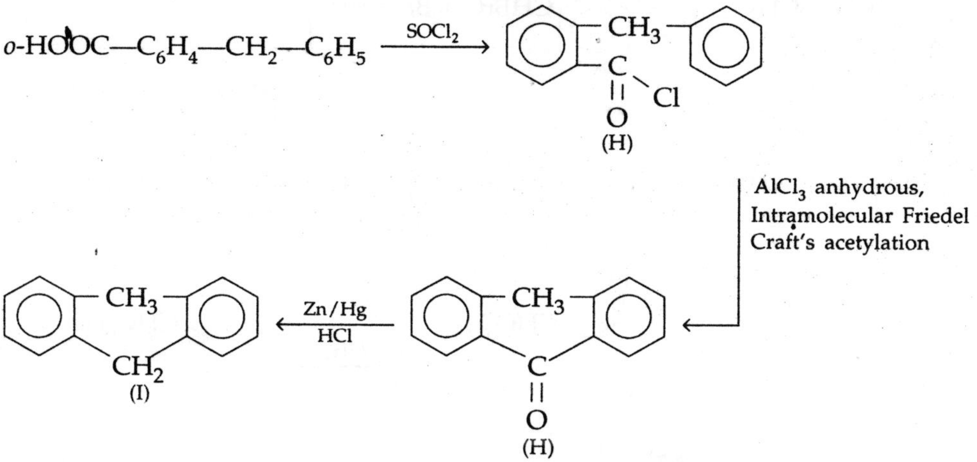

(l) (A) : CH_3COH, (B) CH_3COCl, (C) CH_3CH_2OH, (D) CH_3CH_2Br

(m) $R{-}\underset{\underset{NOH}{\|}}{C}{-}R' \xrightarrow[\text{(Beckmann's rearrangement)}]{\text{Acidic reagent}} R{-}NH{-}\underset{\underset{O}{\|}}{C}{-}R'$

(n) $R{-}COCl + H_2 \xrightarrow{Pd\,BaSO_4} RCHO$

(o) $BrCH_2CH_2Br \xrightarrow{KOH\ (alc.)} C_2H_2$

(p) $C_6H_{11}{-}C{\equiv}CH \xrightarrow[H_2SO_4]{Hg^{+2}} C_6H_{11}{-}COCH_3$

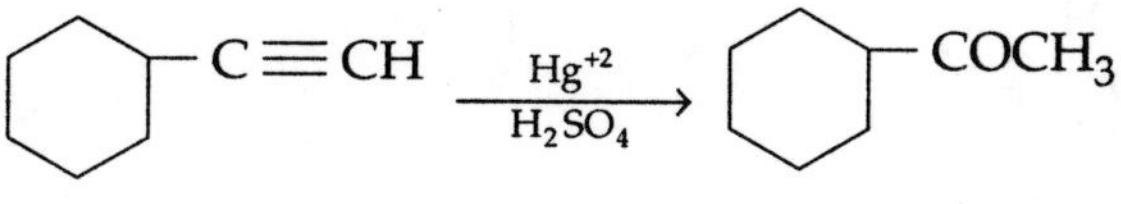

(q) Benzene + butadiene $\longrightarrow$ 1,4-dihydronaphthatene

(r) $C_6H_5CHH_2CHClC_6H_5 \xrightarrow[\Delta]{KOH_{(alc.)}} C_6H_5CH{=}CHC_6H_5$

(*Cis* and *Trains* forms)

(s) $CH_3CH_2NH_2 \xrightarrow[\Delta]{(CH_3CO_2)_2O} CH_3CH_2NHCOCH_3 + CH_3COOH$

(t) 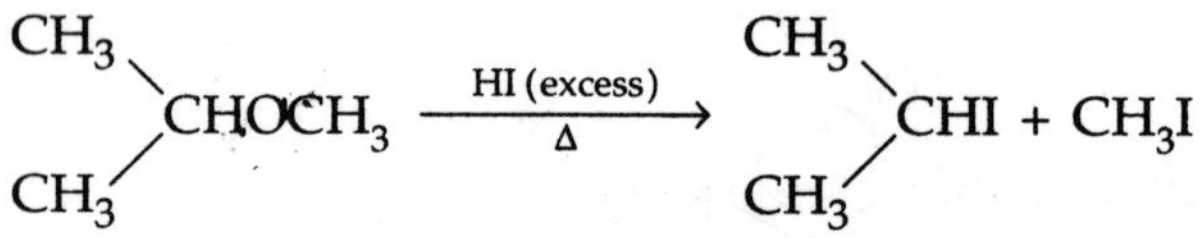 $(CH_3)_2CHOCH_3 \xrightarrow[\Delta]{\text{HI (excess)}} (CH_3)_2CHI + CH_3I$

(u) 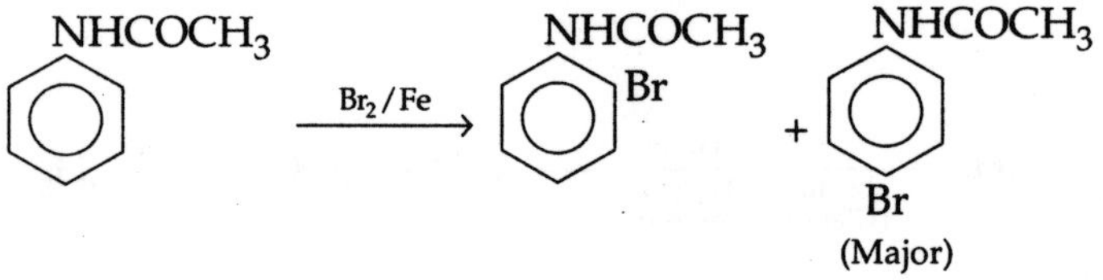 $C_6H_5NHCOCH_3 \xrightarrow{Br_2/Fe}$ o-$BrC_6H_4NHCOCH_3$ + p-$BrC_6H_4NHCOCH_3$ (Major)

(v) 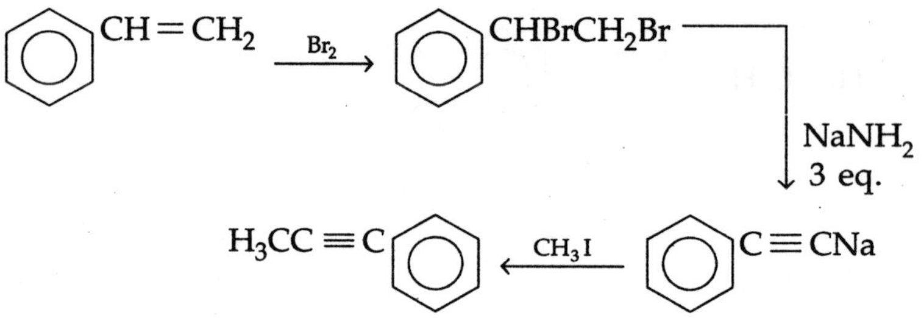 $C_6H_5CH{=}CH_2 \xrightarrow{Br_2} C_6H_5CHBrCH_2Br \xrightarrow{\text{NaNH}_2\text{, 3 eq.}} C_6H_5C{\equiv}CNa \xrightarrow{CH_3I} H_3CC{\equiv}CC_6H_5$

(w) 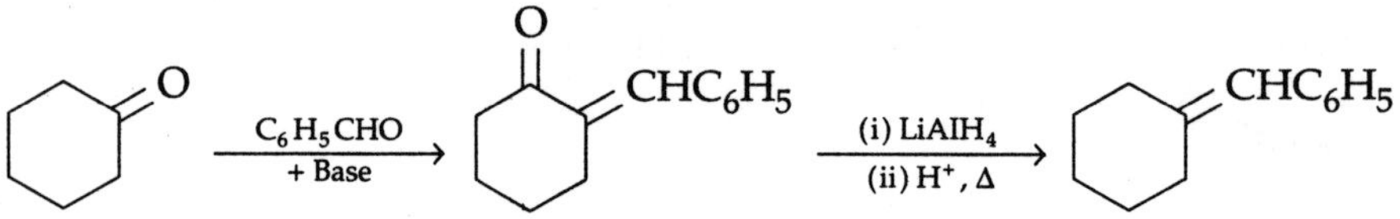 Cyclohexanone $\xrightarrow[\text{+ Base}]{C_6H_5CHO}$ 2-benzylidenecyclohexanone ($=CHC_6H_5$) $\xrightarrow[\text{(ii) }H^+,\ \Delta]{\text{(i) LiAlH}_4}$ benzylidenecyclohexane ($=CHC_6H_5$)

(x) 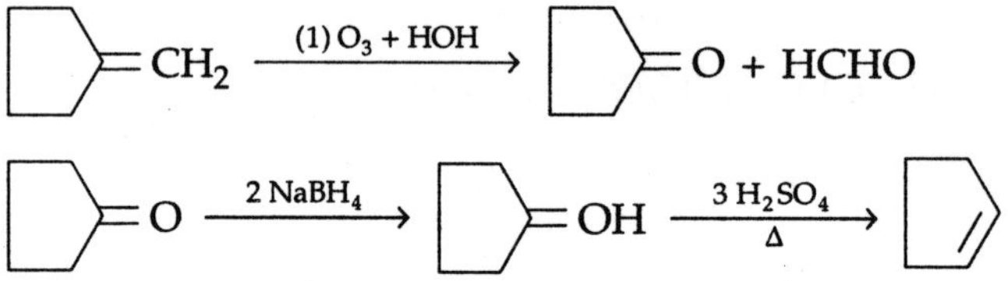 Methylenecyclopentane ($=CH_2$) $\xrightarrow{(1)\ O_3 + HOH}$ cyclopentanone ($=O$) + HCHO

Cyclopentanone ($=O$) $\xrightarrow{2\ NaBH_4}$ cyclopentanol ($=OH$) $\xrightarrow[\Delta]{3\ H_2SO_4}$ cyclopentene

(y) 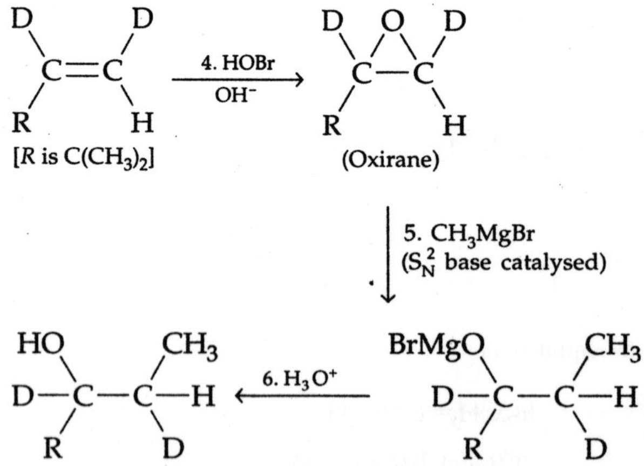 $(D)(R)C{=}C(D)(H)$ [R is $C(CH_3)_2$] $\xrightarrow[OH^-]{4.\ HOBr}$ $(D)(R)C\text{—}C(D)(H)$ with O bridge (Oxirane)

$\xrightarrow{\text{5. } CH_3MgBr\ (S_N^2 \text{ base catalysed})}$ $BrMgO\text{—}C(D)(R)\text{—}C(CH_3)(H)(D)$

$\xrightarrow{6.\ H_3O^+}$ $HO\text{—}C(D)(R)\text{—}C(CH_3)(H)(D)$

(z) Acetanilide ($C_6H_5NHCOCH_3$) $\xrightarrow{\text{7. } HNO_3 + H_2SO_4}$ $NHCOCH_3$ (benzene ring) $\xrightarrow{\text{8. HOH}}$ p-nitroaniline (NH_2, NO_2) $\xrightarrow{\text{9. ICl + } CH_3COOH}$ 2,6-diiodo-4-nitroaniline (NH_2, I, I, NO_2) $\xrightarrow[\text{and } H_3PO_2]{\text{10. } NaNO_2 + HCl}$ 1,3-diiodo-5-nitrobenzene (I, I, NO_2)

175. A: $CH_3CH_2-C(CH_3)_2-Br$ B: $(CH_3)_2CH-CH_2-CH_2-Br$

176. R=$CH_3-CBr_2-CH_2-CH(CH_3)_2$ or $CH_3-CHBr-CHBr-CH(CH_3)_2$
S=$CH_3-C\equiv C-CH(CH_3)_2$

177. Y = $(CH_3)_2-CH-C(CH_3)(Br)-CH=C(CH_2)-CH_2Br$

X = $(CH_3)_2-CH-C(CH_3)=C-C(CH_3)=CH_2$

W = $(CH_3)_2CH-CH(CH_3)-CH_2Br-C(CH_3)=CH_2$ or $(CH_3)_2CH-C(CH_3)=CH-C(CH_3)(Br)-CH_3$ or

$(CH_3)_2CH-C(CH_3)=CH-CH(CH_3)-CH_2Br$

178. $CH_3-CH_2-C(OH)(CH_3)-CH(OH)-CH(CH_3)_2$

179. $CH_3-CH_2-CH_2-C(CH_3)=O$ $O=C(CH_3)-CH_2-CH_3$

180. A = neopentane, B = cyclohexene, C = 1-hexyne

181. H_3C, H_3C, H, H, COOH, COOH $\xrightarrow{-CO_2}$ H_3C, H_3C, H, H, COOH, H + H_3C, H_3C, H, H, H, COOH

H, H_3C, CH_3, H, COOH, COOH $\xrightarrow{-CO_2}$ H, H_3C, CH_3, H, COOH, H + H, H_3C, CH_3, H, H, COOH

182. Its solubility properties and immediate reaction with Ag^+ indicates it is ionic. Hence, the $C_7H_7^+$ ion is stable, presumably the tropylium ion. The isomeric benzyl carbonium ion is not as stable.

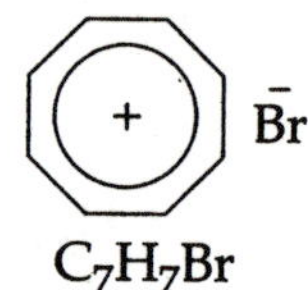

C_7H_7Br

183. C_9H_8 unsaturated with one olefinic bond. R gives C_9H_{16}, which must have two rings.

R yields phthalic acid; hence it must have in its structure (benzene ring bearing two C)

Adding the double bond and the second ring to fit C_9H_8, we get: (indene)

184. Reduces Tollens's regent; therefore CHO
Na → H_2; therefore active H, probably OH.
Iodoform; therefore CO—CH_2 or CHOH—CH_2. Cannot be the ketone, since it would have only two oxygen atoms and has already used two for CHO and OH groups.
Thus, possible structure is C—C(OH)—C—CHO

185. Compound A should have a single methylenic peak, whereas B should have two –one a doublet and one a singlet. Also B should have a >C—H triplet peak. Found A, T = 6.31 (s); B, r = 5.45 (t) (>C—H); r = 6.47 (s) and 6.94 (d). Since the COOH groups are fairly close together in B. K_2 is markedly reduced by the ionization of the first carboxyl group, whereas in B the two groups act more or less independently and have K's in the same order of magnitude. Found A, K_1 and K_2 = 1.23×10^{-4} and 1.95×10^{-2}, respectively, B, K_1 and K_2 = 7.25×10^{-5} and 1.23×10^{-5} respectively.

186. (a) A, Gives 2,4-DNP; therefore $>C=O$ is present.

Ce^{+4} red color; therefore, ROH (No ϕOH, since only four C's)

(b) $C_6H_5O_2 \xrightarrow{AcCl} C_5H_{10}O_3$; therefore one OH

(c) A $\xrightarrow{oxid.}$ RCOOH; therefore, oxidizing CH_2OH or CHO

(d) Therefore, B is C—C—CCOOH or 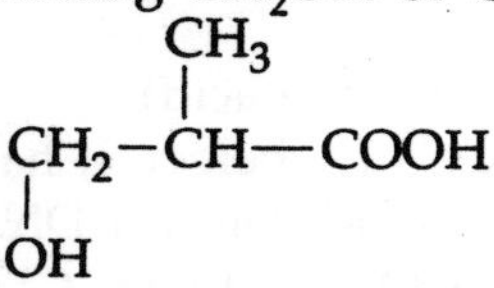

Therefore, A is 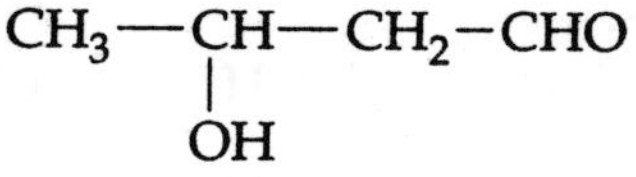 or 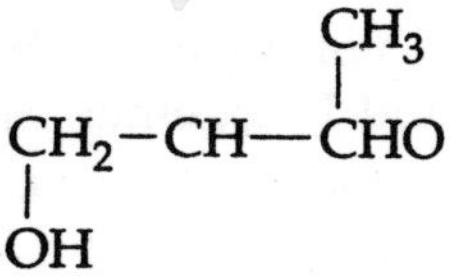

187. A $\xrightarrow[\Delta]{H_2O}$ B + C

B is H_2O insol, alkali sol; therefore , and RCOOH is the only structure of $C_5H_{10}O_2$ to give asymmetric carbon:

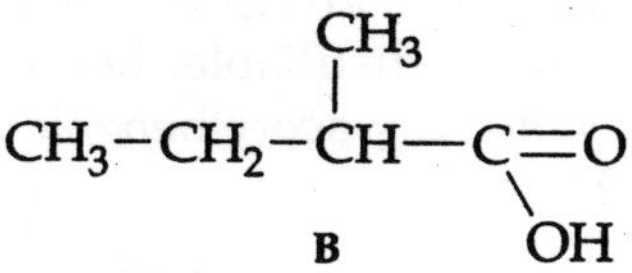

C is water soluble; therefore, CH_3CH_2OH

Therefore, A = CH_3—CH_2—$CH(CH_3)$—$CO_2C_2H_5$.

188. (a) Amide or imide. (b) Amine is secondary; therefore, at least two C's are attached to N. Therefore acid can have maximum of five C's, and only one structure with C, H, and O solely meets this condition:

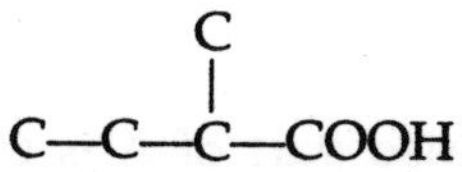

Therefore, $C_7H_{15}NO = CH_2CH_2CH(CH_3)$—C(=O)—$N(CH_3)_2$

189. 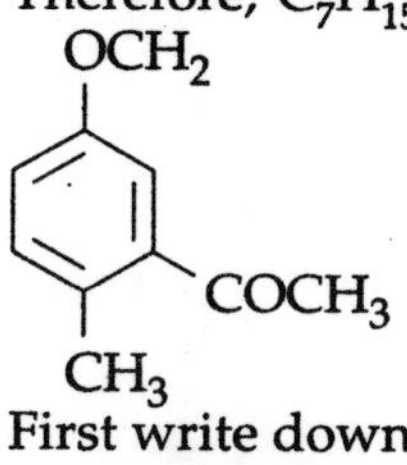

190. First write down the deductions which can be made from each experimental observation

A is = $C_5H_{10}O_4$ ($\therefore$ there are five C's to work with)

A + Na $\rightarrow$ H_2 ($\therefore$ at least one active H)

Neg, 1,2-glycol ($\therefore$ no two OH's on adjacent C's)

Pos. iodoform test ($\therefore$ CH_3—COOH— or CH_3—CO— structure present)

No 2,4-DPH ($\therefore$ no ketone or aldehyde; hence have CH_3COOH and not CH_3CO^-)
$C_5H_{10}O_4 + Ac_2O \rightarrow C_9H_{14}O_6$— since the gain is $2(C_2H_2O)$, have 2 OH's.
Hydrolysis $\rightarrow$ C + D ($\therefore$ A cannot be an amide because there is no nitrogen; nor can A be an ether, since ethers do not hydrolyze in dil. acid; so we suspect A is an ester. Hence, C and D are alcohol and acid)
C neutral ($\therefore$ suspect it to be the alcohol)
C gives pos. 1,2-glycol test, so must contain two adjacent OH's. One OH arose from the hydrolysis.
D acidic ($\therefore$ suspect D is the acid)
D changes color of Ce^{+4} ($\therefore$ beleve it has an OH on side chain)
D gives pos. iodoform and neg. 2,4 DNPH ($\therefore$ must have CH_3—CHOH— structure)
Since D must have COOH and CH_2 CHOH (structures, this accounts for 3 C's in A)
C has at least 2 C's of A (since it has, 1,2-glycol group) ($\therefore$ all carbons in A are accounted for). Putting C and D together, one gets.

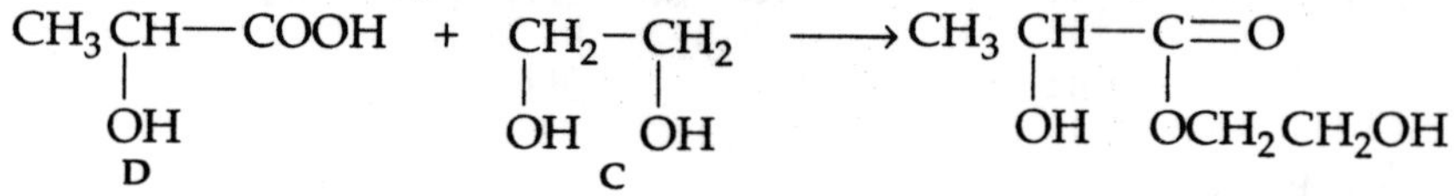

$$\underset{\text{D}}{CH_3\underset{|}{CH}(OH)—COOH} + \underset{\text{C}}{\underset{OH}{CH_2}—\underset{OH}{CH_2}} \longrightarrow CH_3\,\underset{OH}{CH}—\underset{OCH_2CH_2OH}{C}{=}O$$

Checking back, the structure for A would agree with the properties given in the problem.

191. Carbonyl; ketone; aromatic; acyl group A possibily an ester; could be an amide. C is phenol and NO_2 group, steam distillable, hence *o*-nitrophenol. B : still ketone; methyl ketone; acid. Therefore, *o*, *m*, or *p*-acetylbenzoic acid. Oxid, $\rightarrow$ phthalic acid; therefore, B is *o*-$CH_3COC_6H_4COOH$.

thus, A [structure: benzene ring bearing $COCH_3$ and —C(=O)—O— linked to a benzene ring bearing O_2N (ortho)]

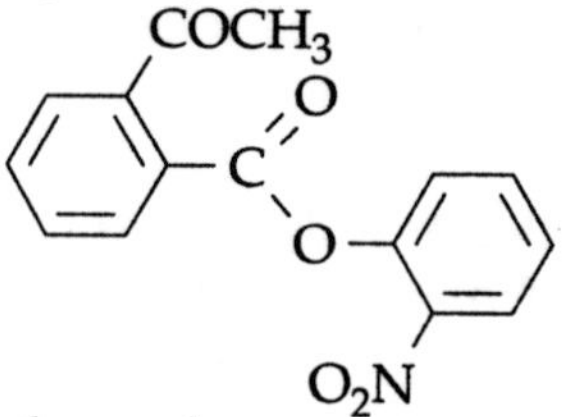

192. (a) No plane of symmetry.

(b) A is amide or imide. Also part of A is $CH_3\underset{OH}{CH}COOH$, only structure possible for asymmetric $C_2H_6O_2$ (optically active).

(c) Therefore, amine is secondary: N(C)(C)

Therefore A is $CH_3—\underset{OH}{CH}—C(=O)—N(CH_3)—CH_2$

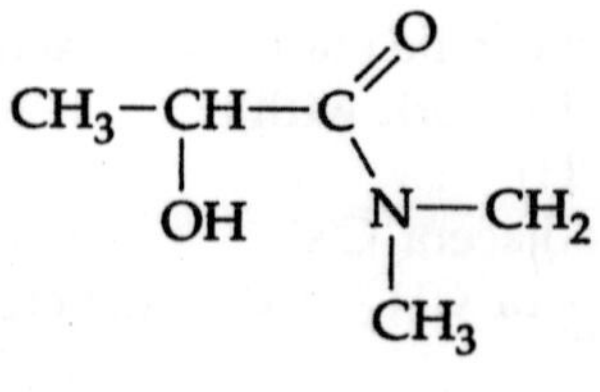

193. Since A is removed from alkali unchanged, it is an acid, not an acid halide; hence, A is $CH_3CH(Cl)CO_2H$. Therefore, Cl must be on α C for A to be optically active.

194. $R_2CO \xrightarrow{H_2NOH} R_2C{=}NOH \xrightarrow{red.} R_2C(H)(NH_2)$

$R_1 = CH_3$, $R_2 = CH_2CH_2CH_2$—
$R_1 = CH_2$, $R_2 = (CH_3)_2$ CH—
$R_1 \neq R_2$ or have no asymmetric carbon

195. (a) A is amide or cyclic imide

(b) Therefore, alcohol of four carbon atoms, has asymmetric carbon. Only one such alcohol can be C—C—C(OH)—C

(c) Iodoform test only confirmatory. Therefore, A is CH_3—CH_2—CH(CH_3)—$CONH_2$

196. A is aldehyde; phenol or enol; two OH groups:

(Structure: benzene ring with CHO, OH ortho to CHO, and OH para to CHO)

Need more information about positions of second OH group. One is probably ortho to CHO to give deep color in $FeCl_3$ test.

197. (Structure: benzene ring with Br and CH_2Cl para to each other, H_3C ortho to Br)

198. (Structure: benzene ring with $CH(OCH_3)_2$ and CH_3 ortho to each other)

199. (Structure: benzene ring with OH, H_3C and H_3CNH in 1,3,5-positions)

200. 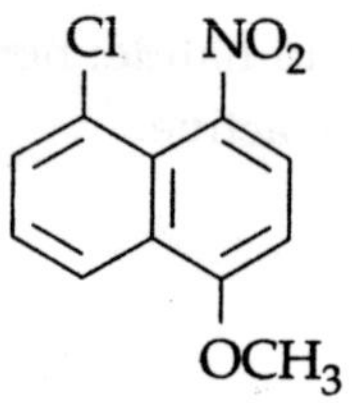 Cl NO$_2$ OCH$_3$ (1-chloro-8-nitro-5-methoxy naphthalene structure)

201. IR shows conjugated and isolated C=C bonds, all *cis*, UV shows only two C=C bonds to be conjugated, not three. The n.m.r. shows two allyl CH_2 methylenes, one bisallyl methylene, and six vinyl protons.

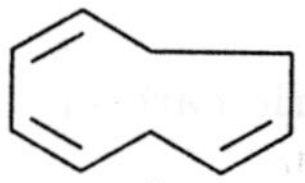

A

202. $CH_2{=}CH{-}CH{=}CH{-}CH(CH_3){-}CH{=}CH_2$

203. A is $CF_3{-}CO{-}CH_2{-}CH{=}CH_2$ + $CF_3{-}C(OH){=}CH{-}CH{=}CH_2$
B is $CF_2{-}CO{-}CH{=}CH{-}CH_2$
The infrared carbonyl frequencies are unusually high because of the large in ductive effect of the CF_3 group.

204. S is $CH_3{-}CH(CH_3){-}CH{=}CH{-}CH_2CH_3$

T is $CH_3{-}CH(CH_3){-}CH\langle O_2 \rangle CH{-}CH_2{-}CH_3$

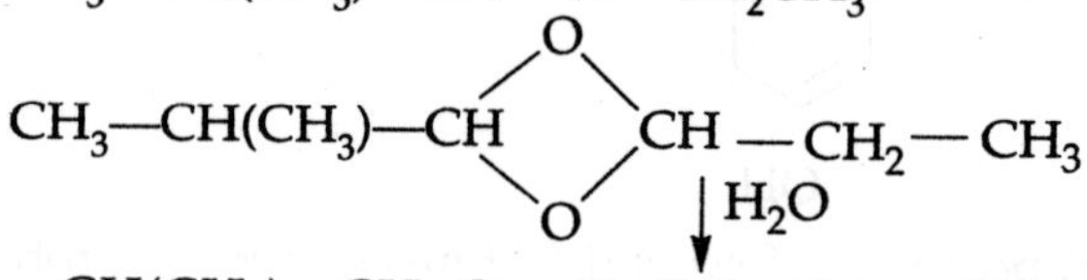

$\downarrow H_2O$

$CH_3{-}CH(CH_3){-}CH{=}O + O{=}CH{-}CH_2{-}CH_3$

205. The empirical formula and the lack of unsaturation indicate one ring in X. The IR bonds show the presence of an OH group (secondary OH region) and absence of a carbonyl group. Oxidation occurs without loss of carbon to produce two acid grops; hence oxidation must have opned a ring. Possible structure: cycloheptanol, a methylcyclohexanol (2-methyl, 3-methyl, etc). a dimethylcyclopentanol, etc.

206. First write down the deductions which can be made from each experimental observation:

A $\xrightarrow{Na}$ H_2 ∴ A has an active H, perhaps an alcohol, acid,; or 1-alkyne (≡C—H).
A insol, alkali, ∴ not an acid.
A decolorizes MnO_4^-, ∴ unsaturated and with active H.

A $\xrightarrow{Pb(OAc)_4}$ B + C Note that B and C together have the same number of O's and two has H's so A is probably a 1,2-glycol.

C $\xrightarrow{Tollens}$ Ag, ∴ C is an aldehyde.

C decolorizes MnO_4^-, ∴ unsaturated. This gives us C=C—CHO, and to comply with the molecular formula, merely the correct number of H's need be added. C then, is CH_2=CH—CHO

B does not reduce Tollens', ∴ not an aldehyde. B forms bisulfite addition compound; since it is not an aldehyde, B must be is methyl ketone.

B gives pos. iodoform test, ∴ has $COCH_3$.

B $\xrightarrow{Na}$ H_2, ∴ has active H. Not enough O's for an acid; hence may be OH. This gives for B: CH_3—CO—CH_2—CHOH—CH_3 or CH_2—CO—CHOH—C_2H_5.

B $\xrightarrow[\Delta]{Cu}$ D, losing two H's hence have dehydrogenation Thus, D is diketone. Has properties of an enol; thus is 1,3-diketone, CH_3—CO—CH_2—CO—CH_3,. ∴ B is CH_3—CO—CH_4—CHOH—CH_3 and A has structure

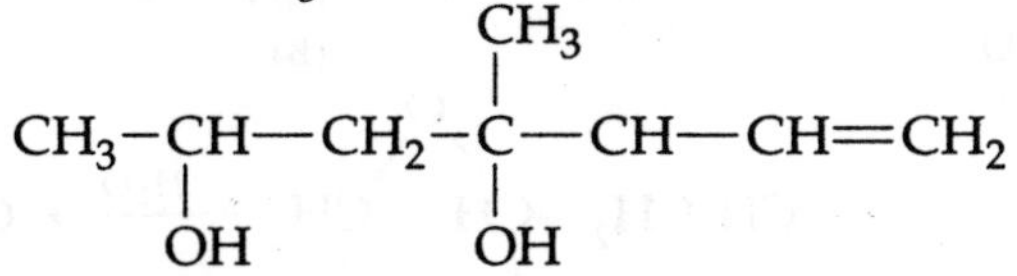

$$\begin{array}{c} \qquad\qquad\qquad CH_3 \\ \qquad\qquad\qquad | \\ CH_3-\underset{\underset{OH}{|}}{CH}-CH_2-\underset{\underset{OH}{|}}{C}-CH-CH=CH_2 \end{array}$$

207. A is oxidized to the 1,2-diketone, but a fraction of the product enolizes and the two tautomers are isolated. Thus, B is the yellow diketone, has the IR band at 1715 cm^{-1}, C is the enol, gives a $FeCl_2$ test, has IR bands for hydrogen bonded OH and C=O groups and a C=C band (1630 cm^{-1})

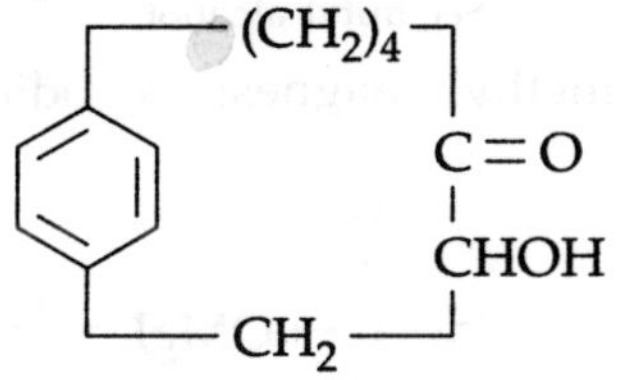

(para-phenylene ring bridged: ring—$(CH_2)_4$—C=O—CHOH—CH_2—ring)

$(CH_2)_4$
C=O
CHOH
CH_2

208. 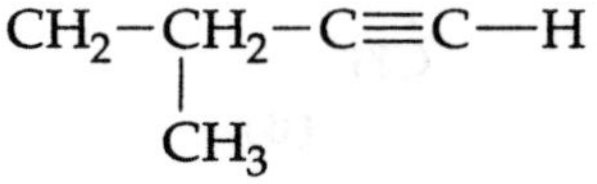 $CH_2-\underset{\underset{CH_3}{|}}{CH_2}-C\equiv C-H$

209. A is $CH_3-\underset{\underset{CH_3}{|}}{\overset{\overset{CH_3}{|}}{C}}-CH_3$; B is CH_3—CH=CH—CH=CH—CH_3

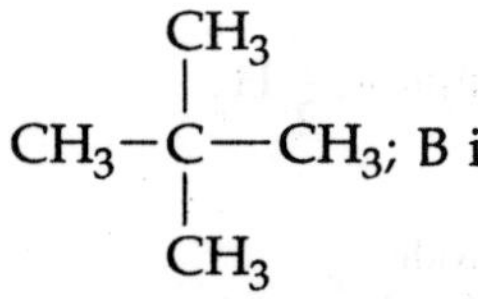

C is CH_3—C=C—$CH_3CH_2CH_3$ or CH_3CH_2—C≡C—CH_2CH_3

210. A is symmetric center; alcohol, phenol; primary alcohol; 2 OH's. Therefore,

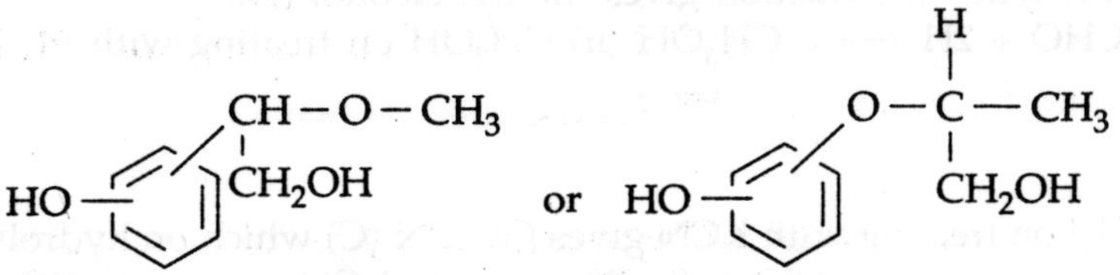

HO—C_6H_3(—CH—O—CH_3)(—CH_2OH) or HO—C_6H_4—O—$\underset{\underset{CH_2OH}{|}}{\overset{\overset{H}{|}}{C}}$—$CH_3$

$$\text{C is } CH_3O\text{—}C_6H_4\text{—}CO_2H$$

Therefore
$$\text{A is } HO\text{—}C_6H_4\text{—}\underset{\displaystyle CH_2OH}{\underset{|}{CH}}-OCH_3$$

211. To get formaldehyde and proponaidehyde on ozonolysis, the orginal compound B must have at least four carbon atoms. It must be butane-1. Then 'A' must be 1,2 dihalbutane. B on hydration gives secbutyl alcohol because ketone is obtained on oxidation.

$$\underset{\textbf{(A)}}{CH_3CH_2\cdot\underset{X}{\underset{|}{CH}}\underset{X}{\underset{|}{CH_2}}} + Zn \longrightarrow \underset{\textbf{(B)}}{CH_3CH_2CH{=}CH_2} + ZnX_2$$

$$\underset{\text{(1-butene)}}{CH_3CH_2CH{=}CH_2} + O_3 \longrightarrow CH_3CH_2\text{—}CH\overset{O}{\diagup\diagdown}CH_2 \text{ (O—O bridge) } \xrightarrow{H_2O} CH_3CH_2CHO + HCHO$$

$$\underset{\textbf{(B)}}{CH_3CH_2CH{=}CH_2} + H_2O \longrightarrow \underset{\text{Sec-butyl alcohol}}{CH_3CH_2CHOH\ CH_3} \xrightarrow{[O]} \underset{\textbf{(C)}\text{ 2-butanone}}{CH_3CH_2COCH_3}$$

212. Formaldehyde reacts with methyl magnesium iodide and on hydrolysis gives ethyl alcohol.

$$\underset{\textbf{(A)}}{H\text{—}\overset{\displaystyle H}{\overset{|}{C}}{=}O} + CH_3\ MgI \longrightarrow H\text{—}\underset{\displaystyle CH_3}{\underset{|}{\overset{\displaystyle H}{\overset{|}{C}}}}\text{—}OMgI \xrightarrow{HOH} \underset{\textbf{(B)}}{H\text{—}\underset{\displaystyle CH_3}{\underset{|}{\overset{\displaystyle H}{\overset{|}{C}}}}\text{—}OH} + Mg(OH)\ I$$

Ethyl alcohol reacts with Sodium to liberate hydrogen

$$\underset{\textbf{(B)}}{CH_3CH_2OH} + Na \longrightarrow \underset{\substack{\textbf{(C)}\\ \text{Sodium ethoxide}}}{CH_3CH_2ONa} + \frac{1}{2}H_2$$

Sodium ethoxide (c) reacts with methyl iodide to give methoxyethane.

$$CH_3CH_2ONa{+}I\text{—}CH_3 \longrightarrow \underset{\text{Methoxyethane}}{CH_3CH_2OCH_3} + NaI$$

213. Formaldehyde on reduction gives methyl alcohol (A).

(i) $HCHO + 2H \longrightarrow \underset{\textbf{(A)}}{CH_3OH}$ (ii) CH_3OH on treating with PI_3 gives CH_3I [$CH_3OH + PI_3 \longrightarrow \underset{\textbf{(B)}}{CH_3I}$]

(ii) CH_3I on treating with KCN gives CH_3CN (C) which on hydrolysis gives acetic acid.

$$CH_3\ I{+}K\ CN \longrightarrow \underset{\textbf{(C)}}{CH_3CN} \longrightarrow \underset{\textbf{(D)}}{CH_3COOH}$$

A – Methyl alcohol CH_3OH; B – Methyl iodide CH_3I; C-Methyl cyanide CH_3CN, D- Acetic acid CH_3COOH

214. Ethyl alcohol on oxidation gives acetaldehyde

$$\underset{}{C_2H_5OH} + (O) \longrightarrow \underset{(A)}{CH_3CHO}$$

Acetaldehyde on further oxidation gives acetic acid (B).

$$\underset{\textbf{(A)}}{CH_3CHO} \longrightarrow \underset{\textbf{(B)}}{CH_3COOH}$$

Acetic acid (B) on treating with ethyl alcohol in presence on mineral acids gives ethyl acetate (C).

$$CH_3COO\ ;\ H{+}OHC_2H_5 \longrightarrow\ ;\ CH_3COOC_2H_5$$

A- Acetaldehyde ; B- Acetic acid ; C-Ethylacetate

215. $C_6H_5COOH + NH_3 \longrightarrow \underset{\textbf{(A)}}{C_6H_5\ COONH_4} \xrightarrow[-H_2O]{\Delta} \underset{\textbf{(B)}}{C_6H_5CONH_2} \xrightarrow[-H_2O]{P_2O_5} \underset{\textbf{(C)}}{C_6H_5CN}$

A. Ammonium benzoate $C_6H_5COONH_4$; B-benzamide-$C_6H_5\ CONH_2$; C-Phenyl cyanide· C_6H_5CN

216. From its composition the compound A must be an aromatic compound. The chlorine atoms must be present in the side chain because it reacts with caustic soda solution. B on oxidation gives an acid so B must be an aromatic aldehyde and C an aromatic acid.

$$C_6H_5CHCl_2 \xrightarrow{NaOH} C_6H_{45}CH(CH)_2 \xrightarrow[-H_2O]{} C_6H_5CHO \xrightarrow{[O]} C_6H_5COOH$$

A is Benzal chloride ($C_7H_6Cl_2$) (Unstable); B is Benzaldehyde (C_7H_6O); C is Benzoic acid ($C_7H_6O_2$).

217. Pentane has three isomers – they are: A – n-pentane, gives three different monochloroderivatives.

A : $CH_3—CH_2—CH_2—CH_3$

n-pentane

$CH_3—CH_2—CH_2—CH_2—CH_2Cl$; 1-chloropentane

$CH_3—CH_2—CH_2—CHCl—CH_3$ 2-chloropentane

$CH_3—CH_2—CHCl—CH_2—CH_3$ 3-chloropentane

(B) Neopentane : It gives one monochloropentane

$$CH_3—\overset{\displaystyle CH_3}{\underset{\displaystyle CH_3}{\overset{|}{\underset{|}{C}}}}—CH_3 \qquad CH_3—\overset{\displaystyle CH_3}{\underset{\displaystyle CH_3}{\overset{|}{\underset{|}{C}}}}—CH_2Cl$$

Neopentyl choloride

Isopentane

$$CH_3—\underset{\displaystyle CH_3}{\underset{|}{CH}}—CH_2-CH_3$$

C : isopentane

1. $CH_3-\underset{\displaystyle CH_3}{\underset{|}{C}H}-CH_2-CH_2Cl$

Chloro-3-methylbutane

2. $CH_3\underset{\displaystyle CH_3}{\underset{|}{C}H}-CHCl-CH_3$

2-chloro-3-methylbutane

3. $CH_3-\underset{\displaystyle CH}{\underset{|}{C}Cl}-CH_2-CH_3$

2-Chloro-2-methylbutane

4. $ClCH_2-\underset{\displaystyle CH_3}{\underset{|}{C}H}-CH_2-CH_3$

1 chloro-2-methylbutane

218. Carbon monoxide on reduction with hydrogen gives methyl alcohol (A). This on treating with Pl_3. methyl iodide is obtained (B) CH_3O Na reacts with CH_3I gives dimethyl ether.

$$\underset{}{CO + 2H_2} \xrightarrow[Ni]{2H_2} \underset{\textbf{(A)}}{CH_3OH} \xrightarrow{Pl_3} \underset{\textbf{(B)}}{CH_3I}$$

$$CH_3\,I\text{+}NaOCH_3 \longrightarrow \underset{\textbf{(C)}}{CH_3OCH_3} + NaI$$

A – Methyl alcohol **B**- Methyl odide **C**-Dimethyl ether

219. (1) Molecular weight of bromo-derivative **(A)** = $\dfrac{0.369 \times 22400}{67.2} = 123$

(2) Let brom o-derivative be R Br

Then Molecular weight of $R = 123 - 80 = 43$

$\therefore \qquad C_nH_{2n+1} = 43$

$\therefore \qquad N = 3$

Thus, bromo-derivative is C_3H_7Br

(3) Given,

(a) $\underset{\textbf{(A)}}{C_3H_7Br} \xrightarrow{NaOH} \underset{\textbf{(B)}}{C_3H_7OH}$

(b) $\underset{\textbf{(B)}}{C_3H_7OH}$ — $\xrightarrow{Al_2O_3,\,723\,K} \underset{\textbf{(C)}}{C_3H_7OC_3H_7}$

— $\xrightarrow{Al_2O_3,\,823\,K} \underset{\textbf{(D)}}{C_3H_6} + H_2O$

(c) $\underset{\textbf{(D)}}{C_3H_6} \xrightarrow{HBr} \underset{\text{Isomer of (A)}}{C_3H_7Br}$

(4) (A) may be $\underset{\textbf{(I)}}{CH_3{\cdot}CHBr{\cdot}CH_3}$ or $\underset{\textbf{(II)}}{CH_3{\cdot}CH_2{\cdot}CH_3Br}$

(5) Reaction (c) suggests that isomer of (A) is 2-bromopropane.

$$\underset{\textbf{(D)}}{CH_3{\cdot}CH{=}CH_2} \xrightarrow{HBr} CH_3{\cdot}CHBrCH_3$$

Therefore, (A) is 1-bromopropane, i.e., $CH_3{\cdot}CH_2{\cdot}CH_2Br$
Thus

$$\underset{\textbf{(A)}}{CH_3{\cdot}CH_2{\cdot}CH_2Br} \xrightarrow{NaOH} \underset{\text{(B) Propane-I-ol}}{CH_3{\cdot}CH_2{\cdot}CH_2OH}$$

$$\underset{\textbf{(B)}}{CH_3CH_2{\cdot}CH_2OH} \xrightarrow{Al_2O_3,\,723\,K} \underset{\text{(C) Dipropyl ether}}{CH_3{\cdot}CH_2{\cdot}CH_2{\cdot}O{\cdot}CH_2{\cdot}CH_2{\cdot}CH_3}$$

$$\underset{\textbf{(B)}}{CH_3CH_2{\cdot}CH_2OH} \xrightarrow{Al_2O_3,\,823\,K} \underset{\text{(D) Prop-1-ene}}{CH_3{\cdot}CH{=}CH_2}$$

$$\underset{\textbf{(D)}}{CH_3{\cdot}CH{=}CH_2} \xrightarrow{HBr} \underset{\text{Isomer of (A)}}{CH_3CHBr{\cdot}CH_3}$$

$$\underset{\textbf{(D)}}{CH_3{\cdot}CH{=}CH_2} \xrightarrow[+H_2O]{H_2SO_4} \underset{\text{(E) Propane-2-ol}}{CH_3{\cdot}\underset{\displaystyle OH}{\underset{|}{C}}H{\cdot}CH_3}$$

220. (1) Molecular wt. of $A = \dfrac{1000 \times K\,' \times w}{\Delta T \times W} = \dfrac{1000 \times 40 \times 0.0088}{8 \times 0.5} = 88$

(2) For empirical formula

Element	%	*Relative no. f atoms*	*Simplest ratio*
C	68.18	5.68	4.8 ≈ 5
H	13.16	13.16	11.2 ≈ 11
O	18.66	1.17	1

∴ Empirical formula of (A) = $C_5H_{11}O$; Empirical formul weight = 87 but molecular weight = 88 and therefore, molecular, formula = $C_5H_{12}O$.

(3) (A) reacts with acety chloride and gives H_2 with Na so, (A) is alcohol

(4) (A) reacts with Lucas reagent ($HCl + ZnCl_2$), a dense oily layer separated immediately hence, (A) is a tertiary alcohol.

(5) Keeping in view of above facts, (A) is $CH_3{-}\overset{\displaystyle C_2H_5}{\overset{|}{\underset{\displaystyle CH_3}{\underset{|}{C}}}}{-}OH$, the only tertiary alcohol of 5 carbon atoms.

Thus,

$$\underset{\substack{\text{2-metylbutan-2-ol}\\ \textbf{(A)}}}{CH_3{-}\overset{\displaystyle CH_3}{\overset{|}{\underset{\displaystyle C_2H_5}{\underset{|}{C}}}}{-}OH} \xrightarrow{CH_3COCl} \underset{\substack{\text{methyl, 2,}\\ \text{2-dimethylbutanoate}}}{CH_3{-}\overset{\displaystyle CH_3}{\overset{|}{\underset{\displaystyle C_2H_5}{\underset{|}{C}}}}{-}OOC{\cdot}CH_3} \xrightarrow{Na} CH_3{-}\overset{\displaystyle CH_3}{\overset{|}{\underset{\displaystyle C_2H_5}{\underset{|}{C}}}}{-}ONa + 1/2H$$

$$\underset{\text{(A)}}{CH_3-\overset{\overset{CH_3}{|}}{\underset{\underset{C_2H_5}{|}}{C}}-OH} \xrightarrow[623\,K]{Al_2O_3} \underset{\substack{\text{2-methylbut-2-ene}\\ \text{(B)}}}{(CH_3)_2C{=}CH{\cdot}CH_3}$$

$$\underset{\text{(B)}}{(CH_3)_2C{=}CH{\cdot}CH_3} \xrightarrow{\text{Ozonolysis}} \underset{\text{(D) Acetone}}{(CH_3)_2C{=}CO} + \underset{\text{Ethanal (C)}}{CH_3{\cdot}CHO}$$

Both (C) and (D) give test with carbonyl reagents such as HCN, $NaHSO_3$ forming addition products but only (C) gives test with Fehlin solution. Thus, (C) is aldehyde and (D) is ketone, each having CH_3CO—unit.

221. (1) Ester is $R-\underset{\underset{O}{||}}{C}-OR'$; Molecular weight of ester = 102 and M.wt of $> CO_2 = 44$;

$\therefore$ Molecular weight of $R + R' = 102 - 44 = 58$

(2) Given $R\text{—}COOR' \xrightarrow{HOH} \underset{\text{Acid}}{RCOOH} + \underset{\text{Alcohol}}{R'OH}$

mM of acid = mM of NaOH

$$\frac{0.185}{M/1} \times 1000 = 25 \times 0.1$$

$\therefore$ Molecular weight of acid $R-\underset{\underset{O}{||}}{C}-OH = 74$

$\therefore$ Molecular weight of $R = 74 -$ (Molecualr weight of COOH) $= 74 - 45 = 29$

$\therefore$ Molecualr weight of $R' = 58 - 29 = 29$

Thus, $R = R'$, also R may be $C_nH_{2n+1} = 29$

$\therefore$ $n = 2$ i.e. R' and R is C_2H_5

(3) Thus, ester is $CH_3{\cdot}CH_2COOCH_2{\cdot}CH_3$

Thus:

$$\underset{\substack{\text{Ester}\\ \text{(Ethyl propanoate)}}}{CH_3CH_2COOC_2H_5} + H_2O \longrightarrow \underset{\substack{\text{Acid}\\ \text{(Propanoic acid)}}}{CH_3CH_2COOH} + \underset{\substack{\text{Alcohol}\\ \text{(Ethyl alcohol)}}}{C_2H_5OH}$$

222. (i) (A) undergoes acetylation and thus, contains OH groups

$$\underset{\text{(A)}}{C_4H_{10}O_4} \xrightarrow{CH_3COCl} \underset{\text{(B)}}{C_{12}H_{18}O_8}$$

(ii) Presence of one OH group enhances C_2H_2O structure in molecule. Since (B) shows an increase in $4(C_2H_2O)$ units and thus, (A) has four OH groups.

$$\text{—}OH \xrightarrow{CH_3COCl} \text{—}OOCCH_3 + HCl$$

(iii) Thus, (A) is $\underset{OH}{\underset{|}{CH_2}}\cdot\underset{OH}{\underset{|}{CH}}\cdot\underset{OH}{\underset{|}{CH}}\cdot\underset{OH}{\underset{|}{CH_2}}$ (Butan-1, 2, 3, 4-tetra-ol)

(B) $\underset{OOC\cdot CH_3}{\underset{|}{CH_2}}$——$\underset{OOC\cdot CH_3}{\underset{|}{CH}}$——$\underset{OOC\cdot CH_3}{\underset{|}{CH}}$——$\underset{OOC\cdot CH_3}{\underset{|}{CH_2}}$

223. (i) (A) is an acid as it shows reaction in presence of P + Br_2. It suggests that **(A)** is $R\cdot CH_2\cdot COOH$.

i.e. $R—CH_2\cdot COOH \xrightarrow{Br_2/P} RCHBr\cdot COOH$ where R is C_3H_7.

(ii) $CH_3CHBrCOOH$ **(B)** has asymmetric C atom represented by C, i.e. $CH_3CHBr\cdot COOH \xrightarrow{-HBr}$ (C);

(iii) Since C has no geometrical isomers it may be

$$(CH_3)_2C{=}CH\cdot COOH \quad \textbf{(C)}$$

(iv) $$(CH_3)_2C{=}CH\cdot COOH \xrightarrow{\text{Decarboxylate}} \underset{\textbf{(D)}}{(CH_3)_2C{=}CH_2}$$

(v) $$\underset{\textbf{(D)}}{(CH_3)_2C{=}CH_2} \xrightarrow{\text{Ozonolysis}} \underset{\textbf{(F)}}{(CH_3)_2CO} + \underset{\textbf{(E)}}{HCHO}$$

(F) Negative Schiff's reagent Test an thus ketone

(E) Positive Schiff's reagent test and thus aldehyde

Thus, structures A to F are obtained above.

224. (i) (C) is an amide which on heating with P_2O_5 gives propannitrile and so, (C) is propanamide.

$$\underset{\text{(C)}}{CH_3\cdot CH_2\cdot ONH_2} \xrightarrow[\Delta]{P_2O_5} CH_3\cdot CH_2CN$$

(ii) (C) is formed by the action of NH_3 on acid (B) so, acid (B) is propanoic acid.

$$\underset{\text{(B)}}{CH_3\cdot CH_2COOH} + NH_3 \xrightarrow{\Delta} CH_3CH_2COONH_4 \xrightarrow{\Delta} \underset{\text{(C)}}{CH_3CH_2CONH_2}$$

(iii) Acid (B) is formed from hydrolysis of (A) as well as (A) fumes in most air so (A) is acid halide. Thus (A) is

$$\underset{\text{(A)}}{CH_3\cdot CH_2COCl} + \underset{\text{Moist air}}{HOH} \longrightarrow \underset{\text{(B)}}{CH_3CH_2COOH} + \underset{\text{Fumes in air}}{HCl\uparrow}$$

Hence, **(A)** : $CH_3 \cdot CH_2COCl$ (Propanoyl chloride)
(B) : CH_3CH_2COOH (Propanoic acid)
(C) : $CH_3 \cdot CH_2CONH_2$ (Propanamide)

Isomers of (A) are
(i) $CH_3 \cdot CO \cdot CH_2Cl$ 1-chloropropanone
(ii) $Cl \cdot CH_2 \cdot CH_2 \cdot CHO$ 3-chloropanal
(iii) $CH_3 \cdot CHCl \cdot CHO$ 2-chloropropanal

225. (1) (D) gives 2,2-dimethylpropanal and 1-butanal on ozonolysis and thus, (D) should be

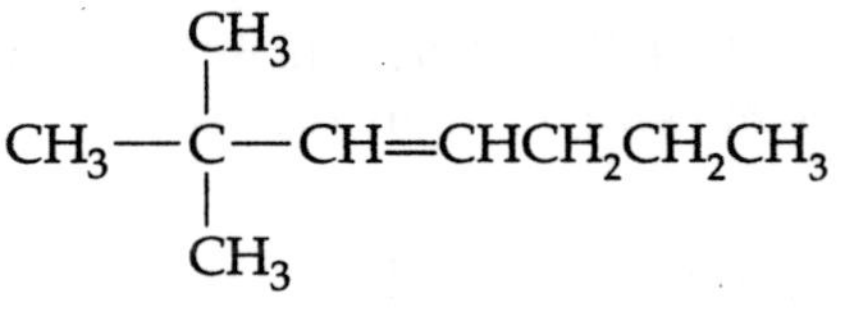

$$CH_3-\underset{\underset{CH_3}{|}}{\overset{\overset{CH_3}{|}}{C}}-CH{=}CHCH_2CH_2CH_3$$

(D)

$$CH_3-\underset{\underset{CH_3}{|}}{\overset{\overset{CH_3}{|}}{C}}-CH{=}CHCH_2CH_2CH_3 \xrightarrow{\text{Ozonolysis}} CH_3-\underset{\underset{CH_3}{|}}{\overset{\overset{CH_3}{|}}{C}}-CHO + CH_3CH_2CH_2CHO$$

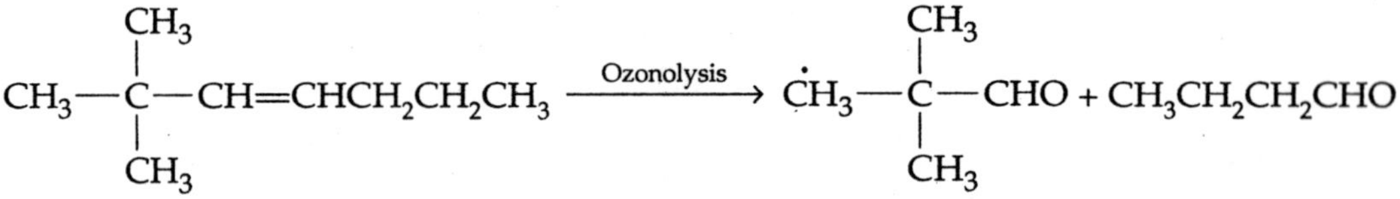

(2,2-dimethyllhept-3-ene)
(D)

(2) (D) is formed by partial hydrogenation (in presence of Lindlar's catalyst) of (C) and thus, (C) is

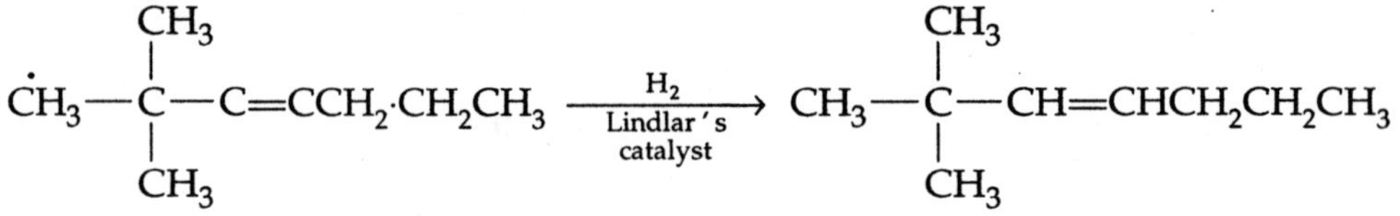

$$CH_3-\underset{\underset{CH_3}{|}}{\overset{\overset{CH_3}{|}}{C}}-C{\equiv}CCH_2 \cdot CH_2CH_3 \xrightarrow[\text{Lindlar's catalyst}]{H_2} CH_3-\underset{\underset{CH_3}{|}}{\overset{\overset{CH_3}{|}}{C}}-CH{=}CHCH_2CH_2CH_3$$

(2,2-dimethylhept-3-yne) (C)

(3) (A) reacts with $NaNH_2/NH_3$ to give (B) which indicates a terminal triple bond in (A)

$$\underset{\textbf{(A)}}{C_6H_{10}} \xrightarrow[NH_3]{NaNH_2} \underset{\textbf{(B)}}{C_4H_9C{\equiv}C \cdot Na}$$

(4) (B) reacts with I-chloropropane to give (C)

$$\underset{\textbf{(B)}}{C_4H_9C{\equiv}C \cdot Na} + Cl \cdot CH_2 \cdot CH_2 \cdot CH_3 \longrightarrow \underset{\textbf{(C)}}{C_4H_9C{\equiv}C \cdot CH_2 \cdot CH_2 \cdot CH_3}$$

But (C) is

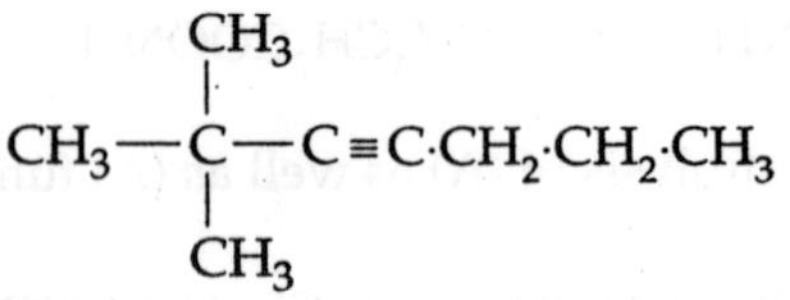

$$CH_3-\underset{\underset{CH_3}{|}}{\overset{\overset{CH_3}{|}}{C}}-C{\equiv}C \cdot CH_2 \cdot CH_2 \cdot CH_3$$

A comparison of two structure s of (C) suggest that (A) and (B) are

$$\underset{\textbf{(A)}}{CH_3-\overset{\displaystyle CH_3}{\underset{\displaystyle CH_3}{\overset{|}{\underset{|}{C}}}}-C\equiv CH} \qquad \underset{\textbf{(B)}}{CH_3-\overset{\displaystyle CH_3}{\underset{\displaystyle CH_3}{\overset{|}{\underset{|}{C}}}}-C\equiv CNa}$$

(3,3-dimethylbut-I-yne)

226. (A) $C_4H_8O_3$ loses its optical activity on strong heating and gives (B) $C_4H_6O_2$. Thus (A) may be $CH_3{\cdot}CH_2CHOH{\cdot}COOH$ or $CH_3CHOH{\cdot}CH_2COOH$
Since (B) is $C_4H_6O_2$ and thus (A) is β-hydroxy acid and not α-hydroxy acid, because only β-hydroxy acids lose H_2O on heating.

Thus (A) is $\underset{\textbf{(A)}}{CH_3{\cdot}CHOH{\cdot}CH_2COOH} \xrightarrow{\Delta} \underset{\textbf{(B)}}{CH_2{=}CH{\cdot}CH_2{\cdot}COOH} + H_2O$

$$\underset{\textbf{(B)}}{CH_2{=}CH{\cdot}CH_2COOH} \xrightarrow{SOCl_2} \underset{\textbf{(C)}}{CH_2{=}CH{\cdot}CH_2COCl}$$

$$\downarrow NH(CH_3)_2$$

$$\underset{\textbf{(D)}}{CH_2{=}CH{\cdot}CH_2{\cdot}CON\begin{matrix}\diagup CH_3\\ \diagdown CH_3\end{matrix}}$$

$$\underset{\textbf{(A)}}{CH_3{\cdot}CHOH{\cdot}CH_2COOH} \xrightarrow{[O]} \underset{\textbf{(E)}}{CH_3{\cdot}CO{\cdot}CH_2COOH}$$

$$\textbf{(E)} \xrightarrow{\text{Decarboxylation}} \underset{\textbf{(F)}}{CH_3COCH_3} \xrightarrow{Zn/HCl} \underset{\textbf{(G)}}{CH_3CH_2CH_3}$$

227. Molecular weight of (A)

$$m = \frac{1000\,K \times w}{\Delta T \times W}$$

where, $w = 0.816$ g, $W = 7.5$ g, $K = 4.9$

$$\Delta T = 5.51 - 1.59 = 3.92$$

$$m = \frac{1000 \times 4.9 \times 0.816}{3.92 \times 7.5} = 136$$

For (A)

Element	%	*Relative no. of atoms*	*Simplest ratio*
C	70.58	5.88	4
H	5.88	5.88	4
O	23.54	1.47	1

∴ Empirical formula of (A) C_4H_4O

Empirical formula weight = 68

Molecular formula of (A) is $(C_4H_4O)_2 = C_8H_8O_2$

Now,

$$\underset{(A)}{C_8H_8O_2} \xrightarrow[-CO_2]{\text{Sodalime}} \underset{(B)}{C_7H_8} \xrightarrow[H^+]{[O]} \underset{(C)}{C_6H_5\text{—COOH}} \xrightarrow{\text{Sodalime}} \underset{(D)}{\text{Benzene}}$$

Since (D) is benzene obtained by the action of sodalime on (C) and thus, (C) an acid with equivalent weight 122 should be benzoic acid; (B) should be toluene.

Thus, **(A)** should be $C_6H_4\begin{smallmatrix}CH_3\\COOH\end{smallmatrix}$ or $C_6H_5CH_2COOH$

o-*m*-or *p*-toluic acid   Phenylacetic acid

Thus:

$$\underset{(A)}{C_6H_4\begin{smallmatrix}CH_3\\COOH\end{smallmatrix}} \xrightarrow{\text{Sodalime}} \underset{(B)}{C_6H_4\begin{smallmatrix}CH_3\\H\end{smallmatrix}} \xrightarrow{[O]} \underset{\substack{(C)\\(\text{Benzoic acid})}}{C_6H_5COOH} \xrightarrow{\text{Sodalime}} \underset{\substack{(D)\\(\text{Benzene})}}{C_6H_6}$$

(Toluene) ↑ Sodalime

$\underset{(A)}{C_6H_5CH_2COOH}$

228. (C) forms monoozonide (D) which on hydrolysis gives only acetaldehyde (CH_3CHHO) and thus, (C)is symmetrical alkene of four carbon atom, i.e.,

$$\underset{(C)}{CH_3{\cdot}CH{=}CH{\cdot}CH_3}$$

(A), a ketone gives haloform reaction and thus, contains $CH_3\text{—}\underset{\underset{O}{\|}}{C}\text{—}$unit.

(A) on reduction gives (B) which must be an alcohol which on heating with H_2SO_4 gives (C) ($CH_3{\cdot}CH{=}CH{\cdot}CH_3$).

Thus, (A), (B), (C), have four C atoms each; since (A) has CH_3CO—unit

So, **(A)** is $\underset{\text{Butanone}}{CH_3{\cdot}\underset{\underset{O}{\|}}{C}{\cdot}CH_2{\cdot}CH_3}$;   **(B)** is $\underset{\text{Butan-2-ol}}{CH_3{\cdot}CHOH{\cdot}CH_2CH_2{\cdot}CH_3}$

and **(C)** is $\underset{\text{But-2-ene}}{CH_3{\cdot}CH{=}CH{\cdot}CH_3}$

Hence, $\underset{\textbf{(A)}}{CH_3{\cdot}CO{\cdot}CH_2{\cdot}CH_3} \xrightarrow{\text{Reduction}} \underset{\textbf{(B)}}{CH_3{\cdot}CHOH{\cdot}CH_2{\cdot}CH_3}$

$\downarrow H_2SO_4, \Delta$

$$\underset{\text{Ethanol}}{2CH_3CHO} \xleftarrow{HOH} \underset{\substack{\text{Ozonide}\\ \text{(D)}}}{CH_3-\underset{\underset{O\text{———}}{|}}{C}H_2-O-\underset{\underset{\text{———}O}{|}}{C}H-CH_3} \xleftarrow{O_3} \underset{\text{(C)}}{CH_3CH{=}CH{\cdot}CH_3}$$

229. Since (X(is hydrolysed to give an acid (Y) and an alcohol (Y) and thus, (X) is ester $R-\underset{\underset{O}{||}}{C}-OR'$. Therefore, $\underset{\text{(X)}}{R-\underset{\underset{O}{||}}{C}-OR'} \xrightarrow[H^+]{HOH} \underset{\text{(Y)}}{RCOOH} + \underset{\text{(Z)}}{R'OH}$

Oxidation of alcohol (Z) gives acid (Y), it shows that (Z) is primary alcohol $R—CH_2OH$

$$\underset{\text{(Z)}}{R'OH} \xrightarrow{\text{Oxidation}} \underset{\text{(Y)}}{RCOOH}$$

$\therefore$ $R{\cdot}CH_2{\cdot}OH \longrightarrow RCOOH$ $\quad$ ($\therefore$ R' is $R—CH_2$)

So (X) is $R-\underset{\underset{O}{||}}{C}-OCH_2R$

$\therefore$ **(X)** is $CH_3{\cdot}CH_2\underset{\underset{O}{||}}{C}-OCH_2CH_2CH_3$ (Propylpropanoate)

(Y) is CH_3CH_2COOH (Propaoic acid)

(Z) is $CH_3CH_2CH_2OH$ (Propan-1-ol)

230. (A) is sodium salt of a carboxylic acid produced by a gas (B) on reacting with NaOH at high *P* and *T*. The reaction is a characteristic reaction of CO_2 and NaOH at high *P* and *T*.

$$\underset{\textbf{(B)}}{CO_2} + NaOH \xrightarrow{P,T} HCOONa \xrightarrow{HOH} \underset{\textbf{(A)}}{HCOOH}$$

(A) on heating in presence of NaOH followed with H_2SO_4 gives a dibasic acid (C)

$$\underset{\textbf{(A)}}{HCOH} \xrightarrow[\Delta]{NaOH} \underset{\text{Sod. Formate}}{HCOONa} \xrightarrow{\Delta} \begin{matrix} COONa \\ | \\ COONa \end{matrix} \xrightarrow[\Delta]{H_2SO_4} \underset{\substack{\text{Oxalic acid}\\ \textbf{(C)}}}{\begin{matrix} COOH \\ | \\ COOH \end{matrix}}$$

The data given for (C) also confirms that it is oxalic acid.
Empirical formula of (C)

Element	%	Relative no. of atoms	Simplest ratio
C	$\frac{12}{44} \times \frac{0.39}{0.40} \times 100 = 26.59$	2.21	1
H	$\frac{2}{18} \times \frac{0.08}{0.40} \times 100 = 2.22$	2.22	1
O	100 – (26.59 + 2.22) = 71.19	4.44	2

Empirical formula of (C) is CHO_2; Empiricial formula weight = 45
Also molecular weight of (C)

$$\frac{\text{Wt. of Ag}}{\text{E. wt. of Ag}} = \frac{\text{Wt. of Ag salt}}{\text{E} + 108}$$, where E is Eq. weight of anion of acid

$$\therefore \quad \frac{0.71}{108} = \frac{1}{\text{E} + 108}$$

$\therefore \Rightarrow$ E = 44

$\therefore$ E. wt. of acid = 44 + 1 = 45

$\therefore$ M. wt. of acid = 90

$\therefore$ M. formula of (C) = $(CHO_2)_2$ = COOH·COOH

231. For empirical formula of 'Y'

Element	%	Relative no. of atoms	Simplest ratio
C	49.31	4.11	3
H	9.59	9.59	7
N	19.18	1.37	1
O	21.92	1.37	1

$\therefore$ Empirical formula of (Y) is C_3H_7NO.

Now, (Y) reacts with Br_2 and NaOH to give (Z) and (Z) reacts with HNO_2 to give ethanol and thus, empirical fornula of (Y) seems to be molecular formula with $—CONH_2$ group.

$$\underset{\substack{\text{(Y)}\\\text{(Propanamide)}}}{CH_3CH_2CONH_2} + Br_2 + 4NaOH \longrightarrow \underset{\text{(Z) (Ethanamine) basic}}{CH_3CH_2NH_2}$$

Hofmann's bromamide reaction

$$CH_3CH_2NH_2 \xrightarrow{HNO_2} \underset{\text{Ethanol}}{CH_3{\cdot}CH_2OH}$$

(Y) is formed from (X) having Cl on treatment with NH_3 and so (X) is CH_3CH_2COCl, i.e., Propanoyl chloride.

$$\underset{\text{(X)}}{CH_3{\cdot}CH_2COCl} + NH_3 \longrightarrow \underset{\text{(Y)}}{CH_3CH_2CONH_2}$$

232. (i) (A) on mild oxidation gives (C) which reacts with 50% KOH to generate (A) and (D). Thus, (C) is aldehyde having no α H atom (as it gives Cannizzaro reaction). Hence, (A) is an alcohol.

$$\underset{(A)}{—CH_2OH} \xrightarrow{O} \underset{(C)}{—CHO} \xrightarrow{KOH} —COOK + \underset{(A)}{—CH_2OH}$$

$$—COOK \xrightarrow{HOH} \underset{(D)}{—COOH}$$

(ii) (D) on reacting with PCl_5 followed with action of NH_3 gives (E) which on dehydration produces HCN.

$$\underset{(D)}{—COOH} \xrightarrow{PCl_5} —COCl \xrightarrow{NH_3} \underset{(E)}{—CONH_2} \xrightarrow{\text{Dehydration}} H—CN$$

(iii) Step II clearly suggests that
(E) is $HCONH_2$
(D) is $HCOOH$
Hence, (C) is $HCHO$
(A) is CH_3OH

(iv) (A) also reacts with acid to form ester (B)

$$\underset{(A)}{CH_3OH} + CH_3COOH \longrightarrow \underset{\text{Ester (B)}}{CH_3COOCH_3} + H_2O$$

Hence, **(A)** is CH_3OH (Methanol)
(B) is CH_3COOCH_3(Methyl ethanoate)
(C) is $HCHO$ (Methanal)
(D) is $HCOOH$ (Methanoic acid)
(E) is $HCONH_2$ (Formamide)

233. (E) C_3H_6O does not give Tollen's test and does not reduce Fehiling solution but forms hydrazone, so, (E) is a ketone, A ketone of 3 C atoms is only (acetone). $\underset{\textbf{(E)}}{CH_3 \cdot CO \cdot CH_3}$

(E) is formed by heating (B) with $Ca(OH)_2$ thus, (B) is CH_3COOH, (acetic acid).

$$\underset{(B)}{2CH_3COOH} + Ca(OH)_2 \xrightarrow{\Delta} (CH_3COO)_2Ca \xrightarrow{\Delta} \underset{(E)}{CH_3COCH_3}$$

(B) is formed by the oxidation of (D) and thus, (D) is C_2H_5OH, (ethyl alcohol).

$$\underset{(D)}{C_2H_5OH} \xrightarrow{\text{Oxidation}} \underset{(B)}{CH_3COOH}$$

(B) and (D) both are formed by (C) hydrolysis in acid medium and thus (C) should be an ester $CH_3COOC_2H_5$. (ethyl acetate)

$$\underset{\textbf{(C)}}{CH_3COOC_2H_5} \xrightarrow[H^+]{HOH} \underset{\textbf{(B)}}{CH_3COOH} + \underset{\textbf{(D)}}{C_2H_5OH}$$

Now (B) and (C) both are formed with action of (A) over C_2H_5OH and so, **(A)** can be acetic anhydride.

234. (A) adds HBr and its formula C_5H_{10} suggests it to be an alkene.
(A) on hydrogenation gives 2-methylbutane and thus, it has the chain of

$$C—C—\underset{\underset{CH_3}{|}}{C}—C$$

Addition of HBr on (A) gives (B) which on treating with AgOH gives an alcohol (C) which on oxidation gives a ketone (D). Thus, (C) is secondary alcohol and (B) is secondary bromide which can be obtained only when (A) is $\underset{(A)}{CH_2{=}CH{\cdot}\underset{\large|\atop CH_3}{C}H{-}CH_3}$

Thus, the reactions are :

(i)

$$\underset{(A)}{CH_2{=}CH{-}CH(CH_3)_2} \xrightarrow{HBr} \underset{(B)}{CH_3{-}\underset{\large|\atop Br}{C}H{-}HC(CH_3)_2}$$

$$CH_3{\cdot}CO{\cdot}CH(CH_3)_2 \xleftarrow{Oxidation} \underset{(C)}{CH_3{\cdot}\underset{\large|\atop OH}{C}H{\cdot}CH(CH_3)_2} \xleftarrow[AgOH]{}$$

(ii) $$\underset{\substack{\text{3-methylbut-1-ene}\\(A)}}{CH_2{=}CH{-}CH(CH_3)_2} \xrightarrow[Catalyst]{H_2} \underset{\text{2-methylbutane}}{CH_3{-}CH_2{\cdot}\underset{\large|\atop CH_3}{C}H{-}CH_3}$$

235. The compound (X) has two isomers (Y) and (Z).
The compound (X) has C, H and Cl atoms.

$$\%\text{ of Cl in (X)} = \frac{35.5 \times 2.9}{143.5} \times 100 = 71.72\%$$

For empirical formula of (X)

Element	%	*Relative no. of atoms*	*Simplest ratio*
C	24.24	2.02	1
H	4.04	4.04	2
Cl	71.72	2.02	1

Empirical formula of (X) is **CH_2Cl**.
Since (X) has two isomers (Y) an (Z); both reacts with $KOH_{(aq)}$, but the isomer (Y) of the compound gives a dihydroxy compound on treatment with aqueous KOH, it follows that the compound must contain two Cl atoms. Hence the molecular formula of **(X)** is $C_2H_4Cl_2$. Its two isomers are $\underset{\textbf{(Z)}}{CH_3CHCl_2}$ and $\underset{\textbf{(Y)}}{Cl{\cdot}CH_2{\cdot}CH_2Cl}$.

Also given Y $\xrightarrow{KOH_{(aq.)}}$ Dihydroxy compound

and Z $\xrightarrow{KOH_{(aq.)}}$ CH_3CHO i.e. Z should have 2 Cl atoms on one C atom
Thus, Z should be $\underset{\textbf{(1, 1-dichloroethane)}}{CH_3{\cdot}CHCl_2}$ and Y should be $\underset{\textbf{(1, 2-dichlorometahen)}}{CH_2Cl{\cdot}CH_2Cl}$

Hence,

$$\underset{\textbf{(Z)}}{CH_3CHCl_2} \xrightarrow{KOH_{(aq.)}} CH_3CH(OH)_2 \longrightarrow \underset{\textbf{(Ethanal)}}{CH_3CHO}$$

$$\underset{(Y)}{CH_2ClCH_2Cl} \xrightarrow{KOH_{(aq.)}} \underset{\text{ethan-1-2-dil or ethylene glycol}}{CH_2OHCH_2OH} \text{ dihydroxy compound}$$

236. For empirical formula :

Element Percentage	Relative next atoms	Simple Ratio
C 71.11%	5.92	8
H 6.67%	6.67	9
N 10.3%	0.74	1
O 11.85	0.74	1

Empirical formula of (A) C_8H_9NO; (A) contains only one N atom in its molecule and thus molecular formula of (A) is C_8H_9NO

(A) gives NH_3 with NaOH and thuis has $CONH_2$ gp.

Molecular formula of (A) suggests it to be $C_6H_4\begin{matrix} CH_3 \\ CONH_2 \end{matrix}$

$$\text{Now, } \underset{(A)}{C_6H_4\begin{matrix} CH_3 \\ CONH_2 \end{matrix}} \xrightarrow{NaOH} \underset{(B)}{C_6H_4\begin{matrix} CH_3 \\ COONa \end{matrix}} \xrightarrow{HOH} \underset{(C)}{C_6H_4\begin{matrix} CH_3 \\ COOH \end{matrix}}$$

Electrophilic substitution on (C) gives only one product.

$$\text{(v)} \quad \underset{(C)}{CH_3C_6H_4COOH} \xrightarrow{PCl_3} \underset{(D)}{CH_3C_6H_4COCl} \xrightarrow{CH_3C_6H_4COONa\ (B)} \underset{(E)}{H_3C{-}C_6H_4{-}\overset{O}{\overset{\|}{C}}{-}O{-}\overset{O}{\overset{\|}{C}}{-}C_6H_4{-}CH_3}$$

237. $\underset{\text{but-2-ene}}{CH_3CH{=}CHCH_3}$

$$\xrightarrow{HBr} CH_3CHBrCH_2CH_3 \quad \text{2-bromobutane}$$

$$\xrightarrow[\text{Peroxide}]{HBr} CH_3CHBrCH_2CH_3$$

$\underset{\text{2-methylpropene}}{CH_3{-}\underset{CH_3}{\underset{|}{C}}{=}CH_2}$

$$\xrightarrow{HBr} (CH_3)_2CBrCH_3 \text{ 2-bromo, 2-methylpropane}$$

$$\xrightarrow[\text{Peroxide}]{HBr} (CH_3)_2CHCH_2Br \text{ 1-bromo, 2-methylpropane}$$

238. Mol. Wt. of (A) suggest it to be C_4H_6 i.e., mol. wt. = 54

C_4H_6 has terminal triple bond as it reacts with CH_3CH_2Br in presence of $NaNH_2$

$$\underset{\substack{(A) \\ \text{Mol. wt.} = 54}}{CH_3CH_2C{\equiv}CH} \xrightarrow{Br_3} \underset{\substack{(B) \\ \text{Mol. wt.} = 374}}{CH_3CH_2CBr_2CHBr_2}$$

Thus, Mol, wt. of $B = \dfrac{593 \times 54}{100} + 54 = 320 + 54 = 374$

$$\underset{\substack{\textbf{(C)} \\ \text{Mol. wt.} = 58}}{CH_3{-}CH_2C{\equiv}CH} \xrightarrow{H_2} CH_3CH_2CH_2CH_2$$

Thus, Mol. wt of $C = \dfrac{7.4}{100} \times 54 = 58$

$$CH_3CH_2C{\equiv}CH \xrightarrow[CH_3 \cdot CH_2Br]{NaNH_2} \underset{\textbf{(D)}}{CH_3CH_2C{\equiv}CCH_2CH_3}$$

$$\text{(D)} \xrightarrow{\text{Ozonolysis}} \underset{\textbf{(E)}}{CH_3CH_2{-}\underset{\underset{O}{\|}}{C}{-}\underset{\underset{O}{\|}}{C}{-}CH_2CH_3} \xrightarrow{[O]} \underset{\textbf{(F)}}{CH_3CH_2COOH}$$

239. (A) reacts with Na to give H_2 and thus it contains —OH group.
M. wt. of (A) $(C_4H_{10}O_2) = 90$
if one OH group then,
90 g (A) with a gives 11200 mL H_2 (i.e. half mole H_2)

$\therefore$ 0.90 with Na gives $\dfrac{11200 \times 0.90}{90}$ mL H_2 at NTP

$= 12$ mL H_2 at NTP

Since 0.90 g (A) gives 224 mL H_2 at NTP and thus, it contains two —OH gp.

Keeping in view of above facts (A) is $C_4H_8\begin{smallmatrix} / OH \\ \backslash OH \end{smallmatrix}$

Now, (A) Shows optical isomerism of which (B) form is optically active having two isomers (D) and (E); (C) being inactive form, therefore, (A), (B), (C) and (D) are

(A)	(C) meso form	(D) 'l' form	(E) 'd' form
CH_3	CH_3	CH_3	CH_3
CHOH	HO—C—H	HO—C—H	H—C—OH
CHOH	HO—C—H	H—C—OH	HO—C—H
CH_3	CH_3	CH_3	CH_3

(D) and (E) together: **(B)** 'd, l' form

240. Alkane (A) C_5H_{12} on chlorination gives four mono chloro products and thus, (A) is isopentane

$$\underset{\textbf{(A)}}{CH_3\underset{|}{\overset{}{C}H}(CH_3)-CH_2\cdot CH_3} \xrightarrow{\text{Chlorination}} \underset{\textbf{(B)}}{CH_2Cl\cdot CH(CH_3)\cdot CH_2\cdot CH_3} \xrightarrow{\text{Dehydro-halogenation}} CH_2{=}C(CH_3)-CH_2CH_3$$

$$\underset{\textbf{(C)}}{CH_3CCl(CH_3)CH_2CH_3} \longrightarrow \underset{\textbf{(F)}}{(CH_3)_2C{=}CHCH_3}$$

$$\underset{\textbf{(D)}}{CH_3CH(CH_3)CHClCH_3} \longrightarrow \underset{\textbf{(F)}}{(CH_3)_2C{=}CHCH_3}$$

$$\underset{\textbf{(E)}}{CH_3CH(CH_3)CH_2CH_2Cl} \longrightarrow (CH_3)_2CH{=}CH_2$$

$$\underset{\textbf{(F)}}{(CH_3)_2C = CHCH_3} \xrightarrow{[O]} \underset{\textbf{(G)}}{(CH_3)_2CO} + \underset{\textbf{(H)}}{CH_3COOH}$$

241. The statements given from (i) to (iv) suggest A to be:

$$C_6H_5CH_2\overset{\bullet}{C}HOHCH_3$$

(i) It is optically active has C centre.

(ii) It reacts with Na to give H_2

$$C_6H_5CH_2CHOHCH_3 \xrightarrow{Na} C_6H_5CH_2CH(ONa)CH_3 + \frac{1}{2}H_2$$

(iii) It gives iodoform with I_2 + NaOH

$$C_6H_5CH_2CHOHCH_3 \xrightarrow{I_2 + NaOH} C_6H_5CH_2COONa + CHI_3$$

Yellow

(iv) It does not react with Br_2 + CCl_4

(v) $C_6H_5CH_2CHOHCH_3 \xrightarrow[KMnO_4]{[O]} C_6H_5COOH + CH_3COOH$

$C_6H_6 + COCl_2 \xrightarrow{AlCl_3} C_6H_5COCl \xrightarrow{HOH} C_6H_5COOH$

(vi) $CH_2CHOHCH_3 \xrightarrow{HI/P} C_6H_5CH_2CH_2CH_3$

Optically inactive (1-phenyl propane)

(vii) Being secondary alcohol, it gives test with Lucas reagent in about 5 minute.

242. From the statements of the problem; (A) can be acetic acid. The reactions are given below:

$\underset{(A)}{CH_3COOH} \xrightarrow[1173\,K]{AlPO_4} \underset{\textbf{(B) ketone}}{CH_2{=}C{=}O}$

$\underset{(B)}{CH_2{=}C{=}O} + \underset{(A)}{CH_3COOH} \longrightarrow \underset{(C)}{(CH_3CO)_2O}$

$\underset{(C)}{(CH_3CO)_2O} + \underset{\text{1, 3, 5-trimethyl benzene}}{C_6H_3(CH_3)_3} \xrightarrow{AlCl_3} \underset{(D)}{C_6H_2(CH_3)_3COCH_3} + CH_3COOH$

$\underset{(A)}{C_6H_2(CH_3)_3CO{\cdot}CH_3} \xrightarrow{Na/Hg\ HCl} \underset{(E)}{C_6H_2(CH_3)_3CH_2CH_3}$

243. Compound (A) ($C_6H_{12}O_2$) on reduction with $LiAlH_4$ gives two compounds (B) and (C) and thus it seems to be an ester *RCOOR′*

$\underset{(A)}{RCOOR'} \xrightarrow{\text{Reduction}} \underset{(B)}{RCH_2OH} + \underset{(C)}{R'OH}$

(F) is monobasic acid of molecular weight 60 therefore, **(F)** is CH_3COOH. ($\therefore$ RCOOH = 60. $\therefore$ *R* = 15, i.e., CH_3COOH)

(F) is obtained by oxidation of (D) and (D) is obtained by oxidation of (B) therefore **(D)** is CH_3CHO and **(B)** is CH_3CH_2OH

Hence, $\underset{\text{Acealdehyde (D)}}{CH_3CHO} \xrightarrow{O} \underset{(F)}{CH_3COOH}$ and $\underset{\text{Ethyl alcohol (B)}}{CH_3CH_2OH} \xrightarrow{O} \underset{(D)}{CH_3CHO}$

Thus, ester **(A)** is $C_3H_7COOC_2H_5$ or $CH_3COOC_4H_9$

The reactions are:

$$CH_3CH_2CH_2COOCH_2CH_3 \text{ or } CH_3COOCH_2CH_2CH_2CH_2$$

$$\downarrow \text{Reduction} \qquad \downarrow \text{Reduction}$$

$$\underset{\text{(C)}}{CH_3CH_2CH_2CH_2OH} + \underset{\text{(B)}}{CH_3CH_2OH} \qquad \underset{\text{(B)}}{CH_3CH_2OH} + \underset{\text{(C)}}{CH_3CH_2CH_2CH_2OH}$$

$$\underset{\text{(B)}}{CH_3CH_2OH} \xrightarrow{[O]} \underset{\text{(D)}}{CH_3CHO} \xrightarrow{[O]} \underset{\text{(F)}}{CH_3COOH}$$

(D) on treating with alkali and subsequent heating gives (E) which on catalytic hydrogenation gives (C)

$$\underset{\text{(D)}}{2CH_3CHO} \xrightarrow[\text{Aldol condensation}]{\text{NaOH (aq.)}} \underset{\text{Aidol}}{CH_3CH(OH)CH_2CHO} \xrightarrow{\Delta} \underset{\substack{\text{Crotanaldehyde} \\ \text{(E)}}}{CH_3CH{=}CHCHO}$$

$$\xrightarrow{\text{Reduction}} \underset{\substack{\text{Butanol-1} \\ \text{(C)}}}{CH_3CH_2CH_2CH_2OH} \xrightarrow{\text{HBr}} CH_3{\cdot}CHBr{\cdot}CH_2{\cdot}CH_3$$

244. The statements provided suggest (A) to be $C_6H_5CH_2CONH_2$ having mol. wt. = 135.

$$\underset{\text{(A)}}{C_6H_5CH_2CONH_2} \xrightarrow{\text{NaOH}} C_6H_5CH_2COONa \xrightarrow{\text{HOH}} \underset{\text{(B)}}{C_6H_5CH_2COOH}$$

(A) $\xrightarrow{\text{HNO}_2}$ (B)

$$\text{(A)} \xrightarrow{Br_2 + KOH} \underset{\text{(C)}}{C_6H_5CH_2NH_2} \xrightarrow{\text{HNO}_2} \underset{\text{(D)}}{C_6H_5CH_2OH}$$

gives red colour wih ceric amm. nitrate due to the presence of alcoholic group.

An isomer (R) of (A) is $C_6H_4(CONH_2)(CH_3)$ (ortho)

$$\underset{\text{E}}{C_6H_4(CONH_2)(CH_3)} \xrightarrow[\text{HOH}]{\text{HCl}} \underset{\text{F}}{C_6H_4(COOH)(CH_3)} \xrightarrow[\Delta]{[O]} \underset{\text{G}}{C_6H_4(CO)_2O}$$

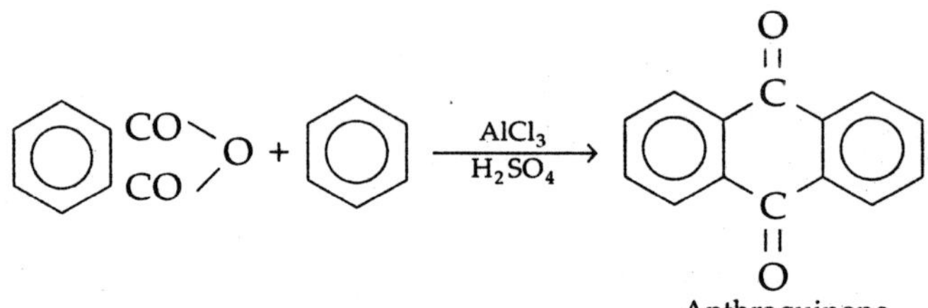

$$C_6H_4(CO)_2O + C_6H_6 \xrightarrow[H_2SO_4]{AlCl_3} C_6H_4(CO)_2C_6H_4$$

Anthraquinone

245. The given compound (A) is $CH_3CH_2CH_2C{\equiv}CCH_3$. This is confirmed by the following findings.

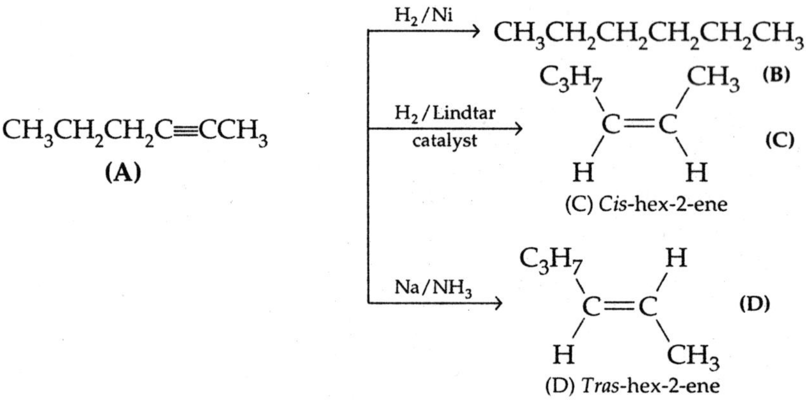

$$CH_3CH_2CH_2C{\equiv}CCH_3 \text{ (A)} \xrightarrow{H_2/Ni} CH_3CH_2CH_2CH_2CH_2CH_3 \text{ (B)}$$

$$CH_3CH_2CH_2C{\equiv}CCH_3 \text{ (A)} \xrightarrow[\text{catalyst}]{H_2/\text{Lindtar}} (C_3H_7)(H)C{=}C(CH_3)(H) \text{ (C)}$$

(C) *Cis*-hex-2-ene

$$CH_3CH_2CH_2C{\equiv}CCH_3 \text{ (A)} \xrightarrow{Na/NH_3} (C_3H_7)(H)C{=}C(H)(CH_3) \text{ (D)}$$

(D) *Tras*-hex-2-ene

(A) is non-terminl alkyne as it does not give white salt with ammonical $AgNO_3$.
(A) forms (E) with $NaNH_2$ which reacts with CH_3I to give (F).

$$\underset{\text{Hexyne-2}}{CH_3CH_2C{\equiv}CCH_3} \xrightarrow[\Delta]{NaNH_2} \underset{(E)}{CH_3CH_2CH_2CH_2C{\equiv}CNa}$$

(A)

$$\xrightarrow{CH_3I} \underset{(F)}{CH_3CH_2CH_2CH_2C{\equiv}CCH_3}$$

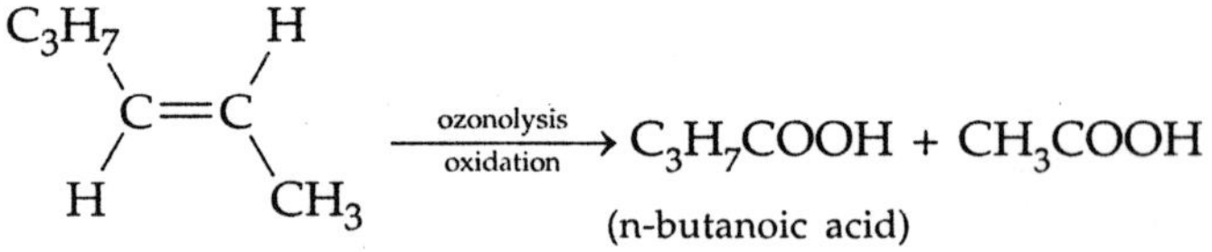

$$(C_3H_7)(H)C{=}C(H)(CH_3) \xrightarrow[\text{oxidation}]{\text{ozonolysis}} \underset{\text{(n-butanoic acid)}}{C_3H_7COOH} + CH_3COOH$$

246. The given compound (*A*) is

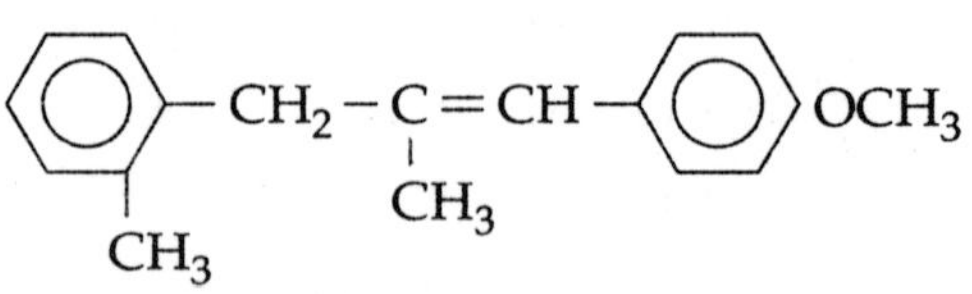

$$CH_3C_6H_4{-}CH_2{-}C(CH_3){=}CH{-}C_6H_4{-}OCH_3$$

This is proved by following steps.

$$\underset{\text{(A)}}{o\text{-}CH_3C_6H_4-CH_2-\underset{\underset{CH_3}{|}}{C}=CH-C_6H_4-OCH_3} \xrightarrow{\text{Ozonolysis}}$$

$$\underset{\text{(B)}}{o\text{-}CH_3C_6H_4-CH_2-\underset{\underset{CH_3}{|}}{C}=O} + \underset{\text{(C)}}{OHC-C_6H_4-OCH_3}$$

(*B*) give iodoform test as it has $CH_3-\underset{\underset{O}{\|}}{C}-$unit and produces oxime with NH_2OH

$$o\text{-}CH_3C_6H_4-CH_2-\underset{\underset{O}{\|}}{C}-CH_3 + I_2 + NaOH \longrightarrow CHI_3 + o\text{-}CH_3C_6H_4-CH_2COONa$$

$$o\text{-}CH_3C_6H_4-CH_2-\underset{\underset{CH_3}{|}}{C}=O + H_2NOH \longrightarrow \underset{\text{(D) (Oxime)}}{o\text{-}CH_3C_6H_4-CH_2-\underset{\underset{CH_3}{|}}{C}=NOH}$$

$$\underset{\text{(D)}}{o\text{-}CH_3C_6H_4-CH_2-\underset{\underset{CH_3}{|}}{C}=NOH} \xrightarrow[\text{ether}]{PCl_5} \underset{\text{(E)}}{o\text{-}CH_3C_6H_4-CH_2-NHCOCH_3} \xrightarrow{HOH} \underset{\text{(F)}}{o\text{-}CH_3C_6H_4-CH_2NH_2} + CH_3COOH$$

$$\underset{\text{(F)}}{o\text{-}CH_3C_6H_4-CH_2NH_2} \xrightarrow{HNO_2} o\text{-}CH_3C_6H_4-CH_2OH \xrightarrow{[O]} \underset{\text{(Phthalic acid)}}{C_6H_4(COOH)_2}$$

This confirms ortho position of CH_3 in ring.

$$\underset{\text{(C)}}{p\text{-}OHC-C_6H_4-OCH_3} \xrightarrow{[O]} \underset{\text{(G)}}{p\text{-}HOOC-C_6H_4-OCH_3} \xrightarrow{[H]} \underset{p\text{-hydroxy benzoic acid}}{p\text{-}HOOC-C_6H_4-OH}$$

Formation of p-hydroxy benzoic acid confirms that two groups are at para position in (C).

247. Let isobutane be a; g and isobutene be b; g. Thus, $a + b = 10$.

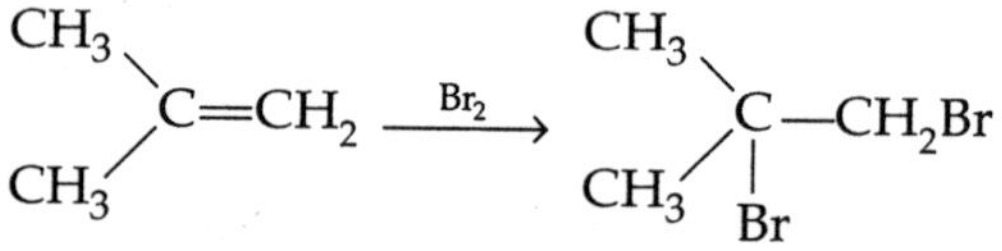

$$(CH_3)_2C{=}CH_2 \xrightarrow{Br_2} (CH_3)_2C(Br){-}CH_2Br$$

(isobutene)

$$(CH_3)_2CCHCH_3 \xrightarrow{Br_2} \text{No reaction of addition}$$

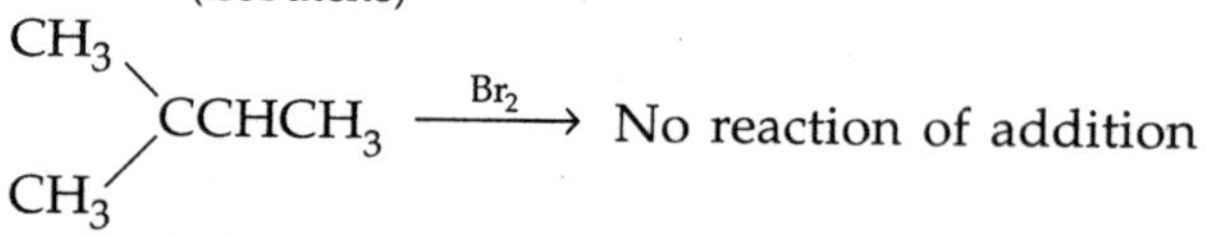

(isobutene)

Now $\because$ 160 g Br_2 is needed for 56 g isobutene

$\therefore$ 20 g Br_2 is needed for $\dfrac{56 \times 20}{160} = 7$ g

Thus, isobutane, i.e., $b = 7$ g
isobutene, i.e., $a = 3$ g

Now 7 g isobutene is converted to isobutene by reduction to give $\dfrac{58 \times 7}{56}$ g isobutene = 7.25 g.

Now isobutene reacts with Br_2 to give monobromo product in light.

$$C_4H_{10} + Br_2 \longrightarrow C_4H_9Br + HBr$$

58 g C_4H_{10} give product = 137 g C_4H_9Br

$\therefore$ (7.25 + 3) g C_4H_{10} gives product = $\dfrac{137 \times 10.25}{58}$ g C_4H_9Br = 24.21 g C_4H_9Br

248. A and B ($C_{11}H_{13}OCl$) both are isomers and unsaturated

$A \xrightarrow{H_2} C$ and $B \xrightarrow{H_2} C$

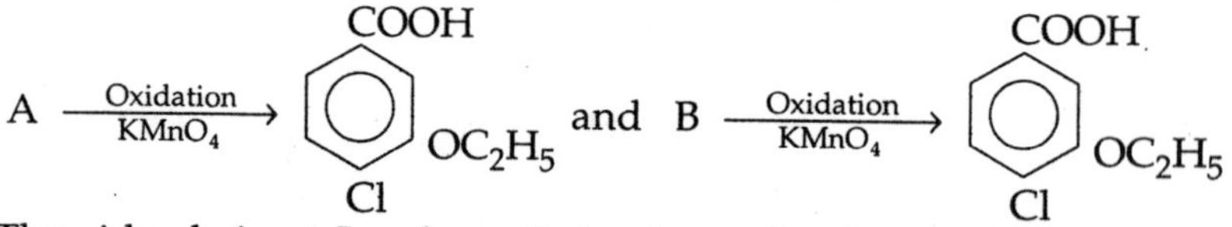

A $\xrightarrow[KMnO_4]{\text{Oxidation}}$ (benzene ring with COOH, OC_2H_5, Cl) and B $\xrightarrow[KMnO_4]{\text{Oxidation}}$ (benzene ring with COOH, OC_2H_5, Cl)

The side chain at I carbon of ring is oxidized to —COOH

$\therefore$ **A is** (benzene ring with $CH{=}CHCH_3$, OC_2H_5, Cl) with two geometrical isomers

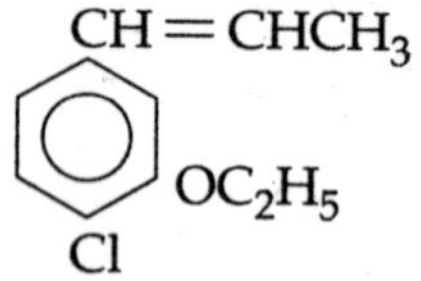

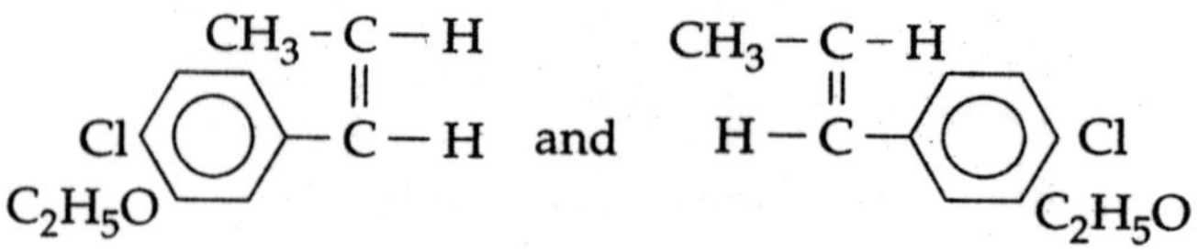

$$\begin{matrix} CH_3-C-H \\ \| \\ Cl(C_2H_5O)C_6H_3-C-H \end{matrix} \quad \text{and} \quad \begin{matrix} CH_3-C-H \\ \| \\ H-C-C_6H_3(Cl)(OC_2H_5) \end{matrix}$$

and **B** is

$$\text{(benzene ring with } CH_2CH{=}CH_2,\ OC_2H_5,\ Cl)$$

249. Compound (B) is obtained from pantanol-1 vapours of passing through Cu at 570 K and thus it is ạn aldehyde (1° alcohols give aldehyde).(iodoform)

$$CH_3CH_2CH_2CH_2CH_2OH \xrightarrow[570\ K]{Cu} CH_3CH_2CH_2CH_2CHO$$

Pentanal-1 (B)

Compound (B), i.e., pentanal-1 is also obtained along with CO_2 and H_2O from (A) on heating with H_2SO_4. Thus, (A) is $CH_3(CH_2)_3{\cdot}CHOH{\cdot}COOH$.

$$\underset{\textbf{(A)}}{CH_3(CH_2)_3CHOHCOOH} \xrightarrow{H_2SO_4} \underset{\textbf{(B)}}{CH_3CH_2CH_2CH_2CHO} + H_2O + CO_2$$

(A) on heating gives (C), i.e.,

$$CH_3(CH_2)_3CHOHCOOH \xrightarrow{\Delta} \begin{array}{c} CH_3(CH_2)_3CH-C{=}O \\ |\quad\ \ | \\ O\quad\ O \\ |\quad\ \ | \\ O{=}C\cdot CH-(CH_2)_3CH_3 \end{array}$$

(C)

250. $C_7H_9N \xrightarrow{\text{Diazotization}} \text{diazotized product} \xrightarrow[HCl]{Cu_2Cl_2} (B) \xrightarrow{[O]} (C)$

$$(C) \xrightarrow{\text{Sodalime, heat}} \text{Chlorobenzene}$$

Since chlorobenzene is obtained from (C) on sodalime treatment therefore, (C) is chlorobenzoic acid.

(C) is obtained from (B) on oxidation, taking molecular formula of (A) —CH_3 group on C_6H_6 ring which is oxidized to —COOH group.

(B) is obtained after carrying out diazotization and S and Meyer's rection over (A) so,

(A) is $C_6H_4\begin{smallmatrix} CH_3 \\ NH_2 \end{smallmatrix}$ i.e., toluidine.

Thus:

$$C_6H_4\begin{smallmatrix} CH_3 \\ NH_2 \end{smallmatrix} \xrightarrow{\text{Diazotization}} C_6H_4\begin{smallmatrix} CH_3 \\ N_2Cl \end{smallmatrix} \xrightarrow[HCl]{Cu_2Cl_3} C_6H_4\begin{smallmatrix} CH_3 \\ Cl \end{smallmatrix}$$

(A) (*o*-, *m*- or *p*-toluidine) **(B)** Chlorotoluene

$$C_6H_4\begin{smallmatrix} CH_3 \\ Cl \end{smallmatrix} \xrightarrow{\text{Oxidation}} C_6H_4\begin{smallmatrix} COOH \\ Cl \end{smallmatrix} \xrightarrow{\text{Sodalime, }\Delta} C_6H_5Cl$$

(B) **(C)** Chlorobenzene

Chlorobenzoic acid

251. Compound (A) has —OH gp. along with unsaturation as it reacts with Na and decolourises Br_2/CCl_4.

(A) gives iodoform test and thus it should have $CH_3—CH(OH)—R$ structure

Thus (A) may be $C_6H_5—CH{=}CH—CH(OH)CH_3$

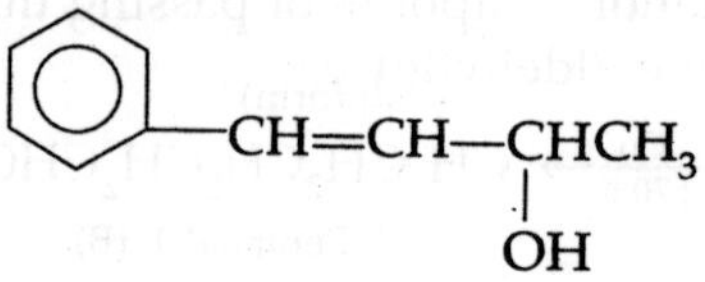

Thus, the reactions are:

$$C_6H_5—CH{=}CH—CH(OH)—CH_3 \xrightarrow{Na} C_6H_5—CH{=}CH—CH(ONa)—CH_3 + \frac{1}{2}H_2 \quad \text{(A)}$$

$$C_6H_5—CH{=}CH—CH(OH)—CH_3 \xrightarrow{Br_2,\ CCl_4} C_6H_5—CH(Br)—CH(Br)—CH(OH)—CH_3 \quad \text{(B)}$$

$$C_6H_5—CH{=}CH—CH(OH)—CH_3 \xrightarrow{I_2,\ NaOH} C_6H_5—CH{=}CHCOOH + CI_3 \quad \text{(C)}$$

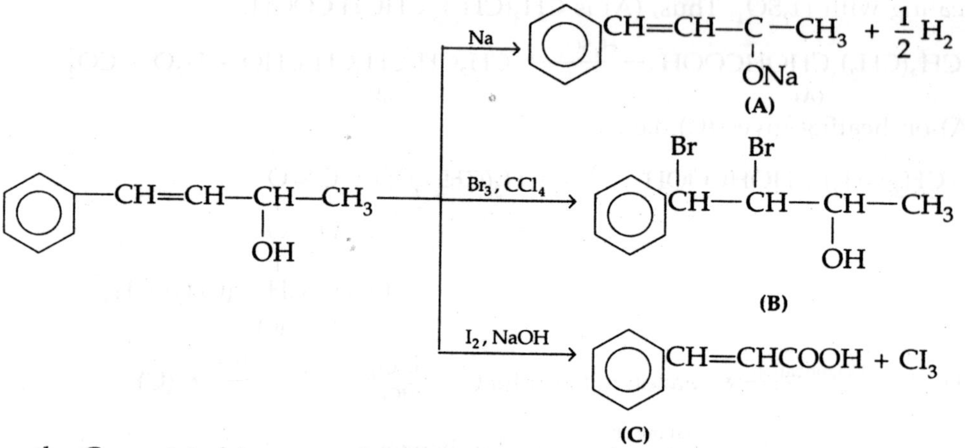

Now, the Geometrical isomers of A are:

$$\begin{matrix} C_6H_5—C—H \\ \| \\ CH_3CHOH—C—H \end{matrix} \quad \text{and} \quad \begin{matrix} H—C—C_6H_5 \\ \| \\ CH_3CHOH—C—H \end{matrix}$$

and Optical isomers of A are:

$$\begin{matrix} CH{=}CHC_6H_5 \\ | \\ H—C—OH \\ | \\ CH_3 \end{matrix} \quad \text{and} \quad \begin{matrix} CH{=}CHC_6H_5 \\ | \\ HO—C—H \\ | \\ CH_3 \end{matrix}$$

252. For Empirical formula

Element	%	*Relative no. of atoms*	*Simplest ratio*
C	76.6	6.38	6
H	6.38	6.38	6
O	17.02	1.05	1

∴ Empirical formula of (A) = $\mathbf{C_6H_5O}$

Empirical formula weight = 94

Molecular weight of (A) = 2 × 47 = 94

∴ Molecular formula (A) is C_6H_6O

Now, (A) gives characteristic colour with $FeCl_3$ and thus, it has phenolic group. Therefore,

(A) should be C_6H_5OH (Phenol).
Thus:

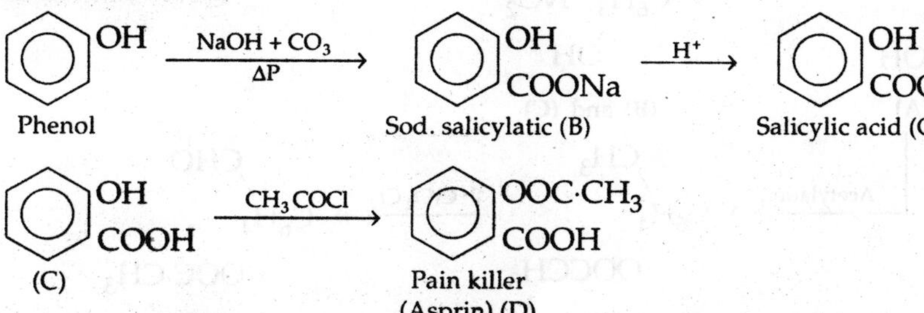

253. The problem **suggests** that isomers of **(A)** is

(a) $C_6H_3(OC_2H_5)(Cl)(CH{=}CH{\cdot}CH_3)$ $\xrightarrow{Br_2/H_2O}$ decolourise

(b) or $C_6H_3(OC_2H_5)(Cl)(CH_2{-}CH{=}CH_3)$ $\xrightarrow{Br_2/H_2O}$ decolourise

(a) Shows geometrical isomerism while (b) does not,

(D) *Cis*: H, H / C=C \ CH_3 (aryl ring with Cl, OC_2H_5)

(E) *Trans*: H, CH_3 / C=C \ H (aryl ring with Cl, OC_2H_5)

Now, $C_6H_5(OC_2H_5)(Cl)(CH_2CH{=}CH_3)$ or $C_6H_3(OC_2H_5)(Cl)(CH_2CH{=}CH_2)$ isomer (b) $\xrightarrow{H_2}$ $C_6H_3(OC_2H_5)(Cl)(CH_2CH_2{-}CH_3)$ **(B)**

and, $C_6H_3(OC_2H_5)(Cl)(CH_2CH_2CH_3)$ (B) $\xrightarrow{[O]}$ $C_6H_3(OC_2H_5)(Cl)(COOH)$ (C) $\xrightarrow[NaOH]{Ni/Al}$ $C_6H_4(OC_2H_5)(COOH)$

254. The reactions suggest that (A) is $C_6H_4(CH_3)(OH)$

$$\underset{(A)}{C_6H_4\begin{cases}CH_3\\OH\end{cases}} \xrightarrow{\text{Nitration}} \underset{(B)\text{ and }(C)}{C_6H_3-NO_2\begin{cases}CH_3\\OH\end{cases}}$$

$$(A) \xrightarrow{\text{Acetylation}} \underset{(D)}{C_6H_4\begin{cases}CH_3\\OOCCH_3\end{cases}} \xrightarrow{CrO_2Cl_2} \underset{(E)}{C_6H_4\begin{cases}CHO\\OOC\cdot CH_3\end{cases}}$$

$$(E) \xrightarrow[\text{[O], HOH}]{KMnO_4} \underset{(F)}{C_6H_4\begin{cases}COOH\\OH\end{cases}} \xrightarrow{\Delta} \underset{\text{Phenol}}{C_6H_5OH}$$

Since (F) on heating gives phenol and thus (F) is ortho derivative i.e., salicylic acid. Thus (A), (D) and (E) are also ortho derivatives.

Also, as per problem,

$$\underset{\substack{\text{ortho cresol}\\(A)}}{C_6H_4\begin{cases}CH_3\\OH\end{cases}} \xrightarrow{C_6H_5SO_2Cl} \underset{(G)}{C_6H_4\begin{cases}CH_3\\OSO_2C_6H_5\end{cases}} \xrightarrow{KMnO_4} \underset{(H)}{C_6H_4\begin{cases}COOH\\OSO_2C_6H_5\end{cases}}$$

$$\xrightarrow{HoH} \underset{(F)}{C_6H_4\begin{cases}COOH\\OH\end{cases}}$$

255. For empirical formula we have:

Element	*%*	*Relative no. of atoms*	*Simplest ratio*
C	68.32	5.59	8
H	6.4	6.40	9
Cl	25.26	0.71	1

Hence, Empirical formula of A = C_8H_9Cl
Empirical formula weight = 140.5
Molecular weight = 140.5
Molecular formula = C_8H_9Cl

Given,

$$\underset{(A)}{C_8H_9Cl} \xrightarrow{HOH} \underset{(B)}{C_8H_9OH} \xrightarrow{\text{Oxidation}} \underset{(C)}{C_8H_8O}$$

(C) reacts with phenyl hydrazine and thus, a carbonyl compound.
(C) gives positive iodoform test and thus, has CH_3CO—unit. Thus (C) is

(C) Acetophenone

Since (C) is formed by the oxidation of (B) and thus, (B) is

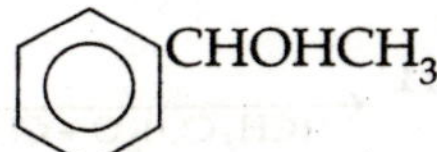

(1-phenylethanol)

Since (B) is obtained by hydrolysis of (A) so, (A) is

$C_6H_5CHClCH_3$

(1-chloro, 1-phenylethane)

Thus

$$\underset{\textbf{(A)}}{C_6H_5CHClCH_3} \xrightarrow[H^+]{HOH} \underset{\textbf{(B)}}{C_6H_5CHOHCH_3} \xrightarrow{\text{Oxidation}} \underset{\textbf{(C)}}{C_6H_5COCH_3}$$

$$C_6H_5COCH_3 + H_2NNHPh \longrightarrow \underset{\text{Phenyl hydrazone}}{C_6H_5—\underset{\underset{N\cdot NHPh}{||}}{C}—CH_3}$$

$$C_6H_5COCH_3 + 3I_2 + 4NaOH \longrightarrow C_6H_5COONa + \underset{\text{Yellow ppt.}}{CHI_3} + 3NaI + 3H_2O$$

256. The given statement about (A) and (B) suggest the following changes

$$\underset{\textbf{(A)}}{(CH_3)_2CHMgBr} + \underset{\textbf{(B)}}{(CH_3)_2CHCH_2CHO} \xrightarrow{\text{ether}} (CH_3)_2CHCH_2\underset{\underset{OMgBr}{|}}{CH}—CH(CH_3)_2$$

$$(CH_3)_2CHCH_2CHCH(OMgBr)CH(CH_3)_2 \xrightarrow[{[O]}]{HOH} \underset{\text{2,5-dimethyl hexanone-3}}{(CH_3)_2CH\cdot CH_2\underset{\underset{O}{||}}{C}CH(CH_3)_2}$$

257. The organic compound (A) is phenol as it undergoes Rimer-Telmenn reaction

$$\underset{\textbf{(A)}}{C_6H_5OH} \xrightarrow{CHCl_3/KOH} \underset{\textbf{(C)}}{o\text{-}HOC_6H_4CHO} + \underset{\textbf{(B)}}{p\text{-}HOC_6H_4CHO}$$

(C) does not reuce Schiff's reagent due to intramolecular H-bonding and is also steam volatile.

$$\underset{(C)}{C_6H_4(OH)CHO} \xrightarrow{LiAlH_4} \underset{(D)}{C_6H_4(OH)CH_2OH} \xrightarrow{[O]} \underset{(E)}{C_6H_4(OH)COOH}$$

$$\underset{(E)}{C_6H_4(OH)COOH} \xrightarrow{(CH_3CO)_2O + CH_3COOH} \underset{(F)}{C_6H_4(OCOCH_3)COOH}$$

Acetyl salicylic acid or aspirin, a pain killer

258. The given compound (A) is $C_6H_4(CH_2Br)(CH_3)$. This is confirmed by the following findings.

$$\underset{(A)}{C_6H_4(CH_2Br)CH_3} + CH_2(COOC_2H_5)_2 \xrightarrow{C_2H_5ONa} C_6H_4(CH_2 \cdot CH(COOC_2H_5)_2)CH_3$$

$$\underset{(B)}{C_6H_4(CH_2 \cdot CH(COOC_2H_5)_2)CH_3} + HOH \xrightarrow{H_2SO_4} \underset{(C)}{C_6H_4(CH_2CH(COOH)_2)CH_3}$$

$$\underset{(C)}{C_6H_4(CH_2CH(COOH)_2)CH_3} \xrightarrow{[O]} \underset{(D)}{C_6H_4(COOH)COOH} \xrightarrow{Nitration} \underset{(E)}{O_2N\text{-}C_6H_3(COOH)COOH}$$

The compound (E) a monomitroderivative in which each group is at same distance in benzene mucleus means that it is symmetrical trisubstituted (1, 3, 5) product and thus two groups in (A), (B), (C), (D) are at meta positions.

259. Bromobenzene is chloroinated, since bromine is *o*- and *p*-directing and thus, products (A) and (B) are

$$C_6H_5Br \xrightarrow{Chlorination} o\text{-}C_6H_4(Br)Cl + p\text{-}C_6H_4(Br)Cl$$

Monobromination of (A) gives several products whereas (B) give only two isomers and thus, (A) and (B) are

(A) $\xrightarrow{Br_2}$ (I) + (II) + (III) + (IV)

(B) $\xrightarrow{Br_2}$ (D) + (C)

260. Given,

$$C_7H_8 \xrightarrow[hv]{Cl_2} \underset{(B)}{\text{Chlorinated product}} \xrightarrow{H_2O} (C)$$

(C) gives positive test with Tollen's reagent and thus, (C) is aldehyde.
(C) is obtained by hydrolysis of (B) and thus, (B) is terminal gemdihalide, —$CHCl_2$ group.
(B) is obtained by chloination of (A) thus, (A) is $C_{65}CH_3$.
Hence the reactions are:

$C_6H_5CH_3$ (A) Toluene $\xrightarrow[hv]{Cl_2}$ $C_6H_5CHCl_2$ (B) Benzal dichloride

$C_6H_5CHCl_2$ (B) $\xrightarrow{HOH}$ C_6H_5CHO (C) Benzaldehyde

C_6H_5CHO + Ag_2O (Tollen's reagent) $\longrightarrow$ C_6H_5COOH + 2Ag (Silver mirror)

$$\underset{(C)}{C_6H_5CHO} + (CH_3CO)_2O \xrightarrow{CH_3COONa} \underset{(D)}{C_6H_5CH{=}CH{\cdot}COOH}$$

This is Perkins reaction, (D) has molecular weight 148.15

Monobasic acid ∴ E. wt. = 148.15

261. For empirical formula of (A) we have:

Element	%	*Relative no. of atoms*	*Simplest ratio*
C	90.56	7.55	1 or 4
H	9.44	9.44	1.25 or 5

∴ Empirical formula of (A) = C_4H_5
Empirical formula weight = 53
Given, Molecular weight = 53 × = 106
Molecular formula of (A) = $(C_4H_5) = C_8H_{10}$

$$(A) \xrightarrow[\text{oxidation}]{\text{Vigorous}} (B) \text{ Dibasic acid, } R\begin{matrix} COOH \\ COOH \end{matrix}$$

Meq. of dibasic acid = Meq. of NaOH

$$\frac{0.1}{E} \times 1000 = 24.1 \times 0.05$$

∴ Equivalent wt. of dibasic acid = 83
∴ Molecular wt. of dibasic acid = 166
∴ Mol. wt. of $R = 166 - 2 \times 45 = 76$
This also suggests R to be C_6H_4.
(B) on heating with sodalime gives benzene and thus, (B) is

$$C_6H_4\begin{matrix} COOH \\ COOH \end{matrix} \text{ and so (A) is } C_6H_4\begin{matrix} CH_3 \\ CH_3 \end{matrix}$$

Also (B) shows mono nitro derivative so (B) is

COOH–C_6H_4–COOH (benzene ring with COOH at top and bottom) a para derivative and so (A) is CH_3–C_6H_4–CH_3 (benzene ring with CH_3 at top and bottom)

Thus the reactions are:

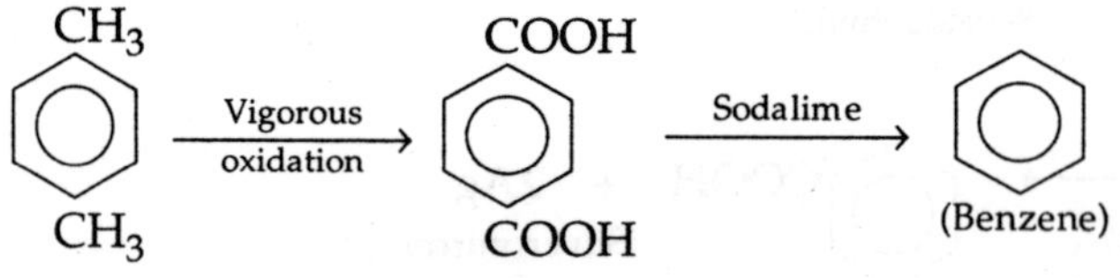

$$\underset{\text{(A) (}p\text{-xylene)}}{CH_3\text{–}C_6H_4\text{–}CH_3} \xrightarrow[\text{oxidation}]{\text{Vigorous}} \underset{\text{(B) (tere phthatic acid)}}{COOH\text{–}C_6H_4\text{–}COOH} \xrightarrow{\text{Sodalime}} \underset{\text{(Benzene)}}{C_6H_6}$$

$$\underset{\text{(B)}}{COOH\text{–}C_6H_4\text{–}COOH} \xrightarrow{\text{Nitration}} COOH\text{–}C_6H_3(NO_2)\text{–}COOH$$

262. The given reactions suggests that A is $C_6H_4(CH_2OH)(OH)$ (benzene ring with CH_2OH and OH on adjacent positions)

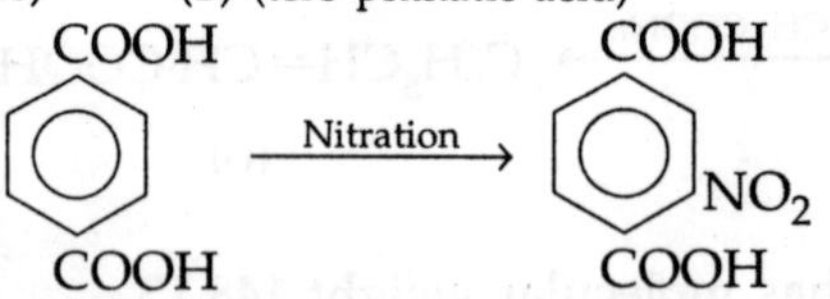

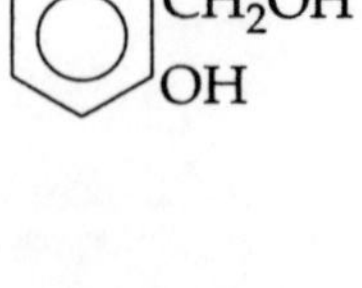

$C_6H_4(CH_2OH)(OH)$ **(A)** $\xrightarrow{[O]}$ $C_6H_4(CHO)(OH)$ oil **(B)**

Volatile due to Intramolecular H-bonding

(A) $\xrightarrow{(CH_3)_2SO_4 \, : \, NaOH}$ $C_6H_4(OH)(CH_2OCH_3)$ (C)

(C) $\longrightarrow$ $C_6H_4(OH)(COOH)$ **(D)** $\xrightarrow{Br_2}$ 2,4,6-tribromophenol **(E)** (72% bromine)

263. (a) The alkane having C_8H_{18} structure gives only single monochloride on substitution and thus, it should be 2,2,3,3-tetra methyl butane.

$$H_3C-C(CH_3)_2-C(CH_3)_2-CH_3 \xrightarrow{Cl_2} CH_3-C(CH_3)_2-C(CH_3)_2-CH_2Cl$$

(b) The following are reaction with Bromine.

$C_6H_5COOC_6H_5 \xrightarrow{Br^+}$ $COOC_6H_5$ ring with Br (meta)

OCH_3 / C_6H_5 (para-substituted benzene) $\xrightarrow{Br^+}$ OCH_3, Br (ortho to OCH_3), C_6H_5

C_6H_5 / C_6H_5 (para-substituted benzene) $\xrightarrow{Br^+}$ C_6H_5, Br, C_6H_5

$NHCOCH_3$ / SO_3H (para-substituted benzene) $\xrightarrow{Br^+}$ $NHCOCH_3$, Br (ortho to $NHCOCH_3$), SO_3H

264. The compound and reactions are:

$$\underset{\text{(A)}}{C_6H_3(CH_2NO_2)(Cl)-Cl} \xrightarrow{\text{Sn/HCl}} \underset{\text{(B)}}{C_6H_3(CH_2NH_2)(Cl)-Cl} \xrightarrow{\text{NaNO}_2\text{/HCl}} \underset{\text{(C)}}{C_6H_3(CH_2OH)(Cl)-Cl}$$

$$\text{(C)} \xrightarrow{[O]} \underset{\text{(D)}}{C_6H_3(COOH)(Cl)-Cl}$$

$$\underset{\text{(F)}}{C_6H_3(Cl)(Cl)-NO_2} \xleftarrow{\text{Nitration}} \underset{\text{(E)}}{C_6H_4(Cl)(Cl)} \xleftarrow{\text{Decarboxylatrion}} \underset{\text{(D)}}{C_6H_3(COOH)(Cl)-Cl}$$

265. Given,

$$A(C, H, N, O) \xrightarrow[\text{Br}_2]{\text{NaOH}} (B) \xrightarrow[\text{HCl}]{\text{NaNO}_2} CH_3CH_2OH$$

(B) on treatment with $NaNO_2$ and HCl gives ethanol and thus, (*B*) is ethanamine ($C_2H_5NH_2$).

Thus the reactions are:

(A) on reaction with NaOH and HCl gives ethanol and thus, (*B*) is ethanamine ($C_2H_5NH_2$).

$$\underset{\substack{\text{(A)}\\\text{(Propanamide)}}}{CH_3{\cdot}CH_2CONH_2} \xrightarrow[\substack{\text{hofmani's}\\\text{bromamide reaction}}]{\text{Br}_2 + 4\,\text{NaOH}} \underset{\text{(B)}}{CH_3CH_2{\cdot}NH_2} + 2NaBr + Na_2CO_3 + 2H_2O$$

$$\underset{\substack{\text{(B)}\\\text{(Ethanamine)}}}{CH_3{\cdot}CH_2NH_2} \xrightarrow[\text{or HNO}_2]{\text{NaNO}_2 + \text{HCl}} \underset{\text{Ethanol}}{CH_3CH_2OH} + N_2 + H_2O$$

266. The statements given in problem suggest (A) to be $CH_3\overset{\overset{\large O}{||}}{C}CH_2CH_2COOH$

$$\underset{\text{(A)}}{CH_3COCH_2CH_2COOH} \xrightarrow{\text{Sodalime}} \underset{\text{(B)}}{CH_3CH_2COCH_3} \xrightarrow{\text{HCN}}$$

Cimamic acid $\quad (CH_3CH_2)(CH_3)C(OH)(CN)$

$$\underset{\text{(A)}}{CH_3COCH_2CH_2COOH} \xrightarrow{[O]} \underset{\text{(Acetic acid)}}{CH_3COOH} + \underset{\text{(Malonic acid)}}{CH_2(COOH)_2}$$

$$\underset{\text{(C)}}{(CH_3CH_2)(CH_3)C(OH)(CN)} \xrightarrow{SOCl_2} \underset{\text{(D)}}{(CH_3CH_2)(CH_3)C(Cl)(CN)} \xrightarrow{KCN} \underset{\text{(E)}}{(CH_3CH_2)(CH_3)C(CN)(CN)}$$

$$\underset{\text{(E)}}{(CH_3CH_2)(CH_3)C(CN)(CN)} \xrightarrow[NaOH]{HOH} \underset{\text{Sod. salt (F)}}{(CH_3CH_2)(CH_3)C(COONa)(COONa)} \xrightarrow{\text{Sodalime}} \underset{n\text{-butane}}{CH_3CH_2CH_2CH_3}$$

267. (A) $C_5H_8O_2$ liberates CO_2 with $NaHCO_3$ so, (A) is acid, i.e., (A) is C_4H_7COOH.
(A) seems to be unsaturated and thus shows geometrical isomerism

$$\underset{\text{(A)}}{C_4H_7COOH} \xrightarrow{H_2} \underset{\text{(B)}}{C_5H_{10}O_2}\text{; Since, (B) is optically active}$$

and thus, acid (B) may be $CH_3CH_2C(H)(CH_3)—COOH$

Since (B) is formed by hydrogenation of acid (A) having geometrical isomers, thus (A) can only be

$CH_3{\cdot}CH{=}C(CH_3)—COOH$ (2-methylbut-2-enoic acid)

268. (A) on treating with alcoholic KOH gives a mixutre of two isomeric alkenes (B) and (C).

$$\underset{\text{(A)}}{C_6H_{13}Cl} \xrightarrow[-HCl]{\text{alc. KOH}} \underset{\text{(B)}}{C_6H_{12}} + \underset{\text{(C)}}{C_6H_{12}} \quad \text{(isomers)}$$

Mixture of (B) and (C) on ozonolysis gives four products (i) CH_3CHO (ii) C_2H_5CHO (iii) CH_3COCH_3 (iv) $(CH_3)_2CHCHO$
Since (B) and (C) are of six carbon atoms each and thus, product formed on ozonolysis for each alkene must contain in all six carbon atoms. That is why one alkene gives CH_3CHO + $(CH_3)_2CHCHO$ and other alkene gives C_2H_5CHO and CH_3COCH_3
Thus alkenes are:

$$(CH_3)_2CHCH{=}CHCH_3\ ;\ C_2H_5CH{=}C(CH_3)_2$$

Since (B) and (C) are formed from halide (A), thus, structure of halide should be

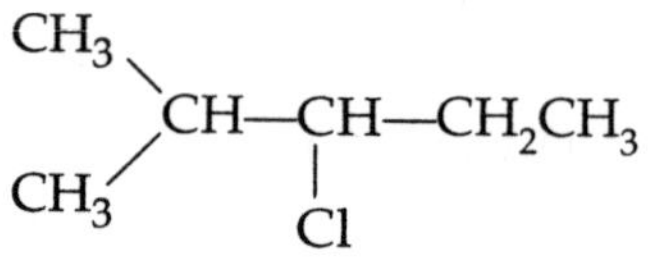

$$\begin{array}{l} CH_3\diagdown \\ \quad\quad CH-CH-CH_2CH_3 \\ CH_3\diagup \quad\quad | \\ \quad\quad\quad\quad Cl \end{array}$$

(A) 3-chloro, 2-methyl pentane

269. 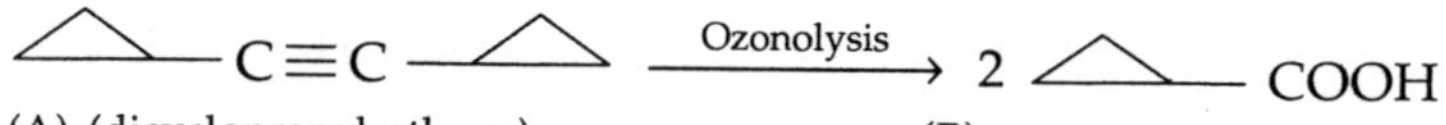 $\triangle - C \equiv C - \triangle \xrightarrow{\text{Ozonolysis}} 2\ \triangle - COOH$

(A) (dicyclopropyl ethyne) **(B)**

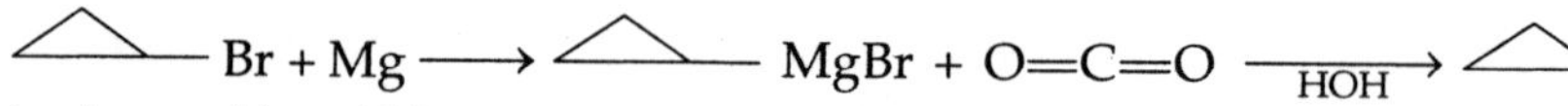

$\triangle - Br + Mg \longrightarrow \triangle - MgBr + O{=}C{=}O \xrightarrow[\text{HOH}]{} \triangle - COOH$

(cyclopropyl bromide)

270. 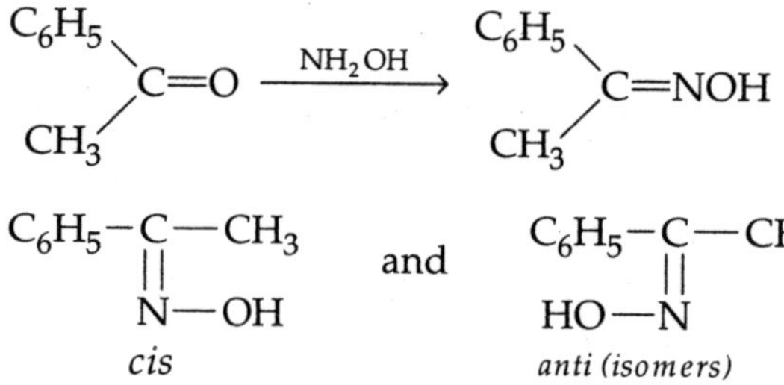 $\begin{array}{l} C_6H_5\diagdown \\ \quad\quad C{=}O \\ CH_3\diagup \end{array} \xrightarrow{NH_2OH} \begin{array}{l} C_6H_5\diagdown \\ \quad\quad C{=}NOH \\ CH_3\diagup \end{array}$

$\begin{array}{c} C_6H_5-C-CH_3 \\ \| \\ N-OH \end{array}$ and $\begin{array}{c} C_6H_5-C-CH_3 \\ \| \\ HO-N \end{array}$

cis *anti (isomers)*

271. Here, the problem says that,

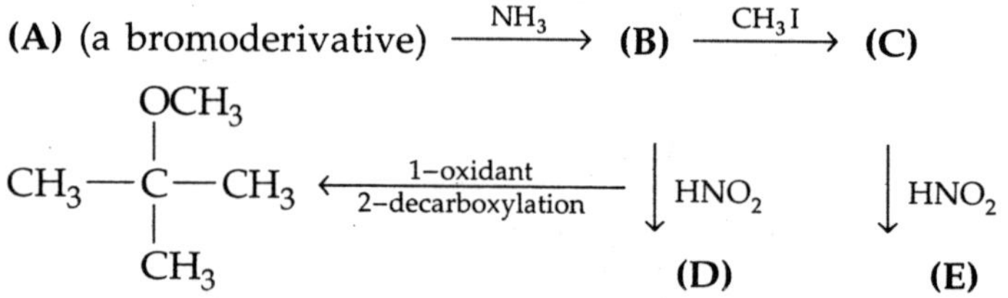

(A) (a bromoderivative) $\xrightarrow{NH_3}$ **(B)** $\xrightarrow{CH_3I}$ **(C)**

$$\begin{array}{c} OCH_3 \\ | \\ CH_3-C-CH_3 \\ | \\ CH_3 \end{array} \xleftarrow[\text{2-decarboxylation}]{\text{1-oxidant}} \quad \downarrow HNO_2 \text{ (on B)} \rightarrow \textbf{(D)} \qquad \downarrow HNO_2 \text{ (on C)} \rightarrow \textbf{(E)}$$

Now, 2-methoxy, 2-methylpropane is obtained by oxidation of (D) followed with decarboxylation and so, (D) is

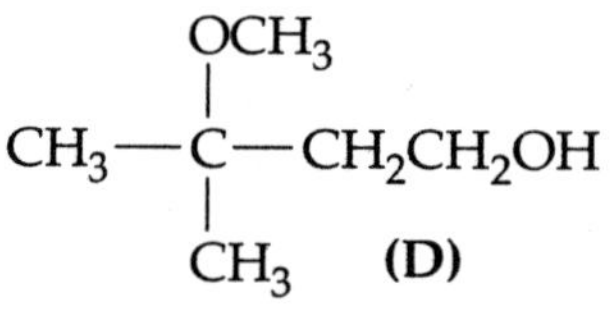

$$\begin{array}{c} OCH_3 \\ | \\ CH_3-C-CH_2CH_2OH \\ | \\ CH_3 \end{array} \quad \textbf{(D)}$$

(D) is formed by the action of HNO_2 on (B) and so, (B) is

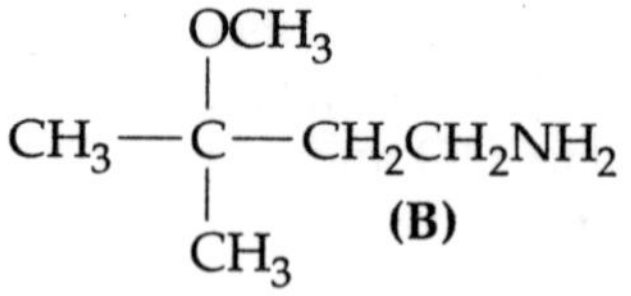

$$\begin{array}{c} OCH_3 \\ | \\ CH_3-C-CH_2CH_2NH_2 \\ | \\ CH_3 \end{array} \quad \textbf{(B)}$$

(B) is formed by the action of NH_3 over (A) bromoderivative

$$\therefore \textbf{(A)} \text{ is } CH_3-\underset{\displaystyle CH_3}{\overset{\displaystyle OCH_3}{\underset{|}{\overset{|}{C}}}}-CH_2CH_2Br$$

(C) is obtained by action of CH_3I over (B)

$$\therefore \textbf{(C)} \text{ is } CH_3-\underset{\displaystyle CH_3}{\overset{\displaystyle OCH_3}{\underset{|}{\overset{|}{C}}}}-CH_2CH_2\overset{\displaystyle CH_3}{\overset{|}{N}}H$$

$$\therefore \textbf{(E)} \text{ is } CH_3-\underset{\displaystyle CH_3}{\overset{\displaystyle OCH_3}{\underset{|}{\overset{|}{C}}}}-CH_2-CH_2-\overset{\displaystyle CH_3}{\overset{|}{N}}-NO$$

272. (A) reacts with sodium in ammonia and thus, it is terminal alkyne, *i.e.*, $C_3H_7C{\equiv}CH$

$$\text{Now, } \underset{\textbf{(A)}}{C_3H_7C{\equiv}CH} \xrightarrow{Na} C_3H_7-C{\equiv}CNa \xrightarrow{CH_3CH_2CH_2I} \underset{\textbf{(B)}}{C_3H_7-C{\equiv}C-CH_2-\underset{\displaystyle CH_3}{\underset{|}{CH_2}}}$$

$$\underset{\textbf{(A)}}{C_3H_7-C{\equiv}CH} \xrightarrow[H_2SO_4]{HgSO_4} \underset{\textbf{(C)}}{C_3H_7-CO-CH_3}$$

$$\underset{\textbf{(B)}}{C_3H_7-C{\equiv}C-CH_2-CH_2-CH_3} \xrightarrow[KMnO_4]{[O]} \underset{\textbf{(D)}}{CH_3CH_2CH_2COOH} + \underset{\textbf{(E)}}{C_3H_7COOH}$$

Since (D) and (E) are isomers and thus, (E) is $(CH_3)_2CHCOOH$

$$\therefore \text{(A) is } \begin{matrix} CH_3 \diagdown \\ \quad CH-C{\equiv}CH; \\ CH_3 \diagup \end{matrix} \text{(B)} \begin{matrix} CH_3 \diagdown \\ \quad CH-C{\equiv}C-CH_2CH_2CH_3 \\ CH_3 \diagup \end{matrix}$$

273. Since alkene (A) on oxidation gives a carbonyl compound (D) and acid (E) of four carbon atoms each and thus

$$\underset{\textbf{(A)}}{\begin{matrix} CH_3 \diagdown \\ \quad C{=}CHCH_2CH_2CH_3 \\ CH_3{\cdot}CH_3 \diagup \end{matrix}} \xrightarrow[KMnO_4]{[O]} \underset{\substack{\textbf{(D)} \\ \text{Butanone}}}{CH_3COCH_2CH_3} + \underset{\substack{\textbf{(E)} \\ \text{Butanoic acid}}}{CH_3CH_2CH_2COOH}$$

Alkene (A) and (B) give same alcohol on hydration and thus, (B) can only be (also no two groups on same carbon atom).

$$\underset{\textbf{(B)}}{\begin{matrix} CH_3CH_2CH_2CH_2 \diagdown \\ \quad C{=}CHCH_3 \\ CH_3 \diagup \end{matrix}} \xrightarrow[KMnO_4]{[O]} \underset{\substack{\textbf{(G)} \\ \text{Hexanone-2}}}{CH_3COCH_2CH_2CH_2CH_3} + \underset{\substack{\textbf{(F)} \\ \text{Acetic acid}}}{CH_3COOH}$$

The hydration of (A) and (B) give same alcohol

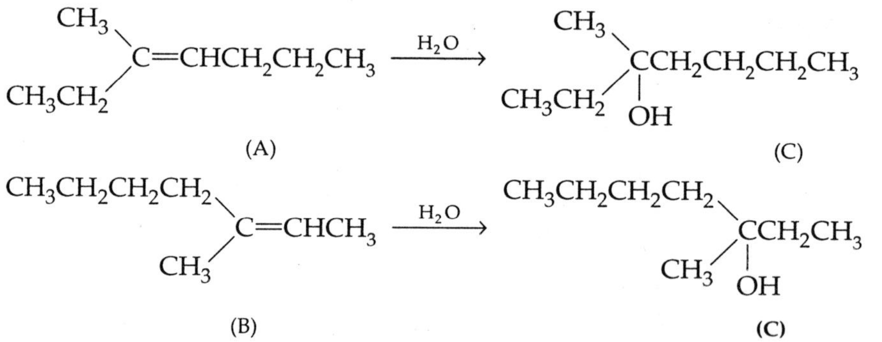

$$\underset{\text{(A)}}{\begin{matrix} CH_3 \diagdown \\ C{=}CHCH_2CH_2CH_3 \\ CH_3CH_2 \diagup \end{matrix}} \xrightarrow{H_2O} \underset{\text{(C)}}{\begin{matrix} CH_3 \diagdown \\ CCH_2CH_2CH_2CH_3 \\ CH_3CH_2 \diagup \ | \\ OH \end{matrix}}$$

$$\underset{\text{(B)}}{\begin{matrix} CH_3CH_2CH_2CH_2 \diagdown \\ C{=}CHCH_3 \\ CH_3 \diagup \end{matrix}} \xrightarrow{H_2O} \underset{\text{(C)}}{\begin{matrix} CH_3CH_2CH_2CH_2 \diagdown \\ CCH_2CH_3 \\ CH_3 \diagup \ | \\ OH \end{matrix}}$$

(3-methylheptan-3-ol)

274. (A) C_6H_{10} either has one triple bond or two double bonds. Ozonolysis of (A) suggests that it has two double bonds.

$$\underset{\text{(A)}}{C_6H_{10}} \xrightarrow{\text{Ozonolysis}} \underset{\text{(D)}}{C_2H_4O} + \underset{\text{(E)}}{C_2H_2O_2}$$

Since (D) and (E) are aldehydes i.e.,

$$\underset{\text{Ethanal (D)}}{C_6H_{10}} \quad \text{and} \quad \underset{\text{Ethane-1-2-dial (E)}}{\begin{matrix} CHO \\ | \\ CHO \end{matrix}}$$

∴ **(A)** is $\underset{\text{Hexa-2, 4-diene}}{CH_3CH{=}CH{-}CH{=}CHCH_3}$

$$CH_3CH{=}CH{-}CH{=}CHCH_3 \xrightarrow{\text{Ozonolysis}} \underset{\text{(D)}}{2CH_3CHO} + \underset{\text{(E)}}{\begin{matrix} CHO \\ | \\ CHO \end{matrix}}$$

(A) gives (B) on reduction and finally gives (C).

$$\underset{\text{(A)}}{CH_3CH{=}CH{-}CH{=}CH{-}CH_3} \xrightarrow{H_2} \underset{\text{(B) Hex-2-ene}}{CH_3CH_2CH_2CH{=}CHCH_3}$$

$$\downarrow H_2$$

$$\underset{\text{Hexane (C)}}{CH_3CH_2CH_2CH_2CH_2CH_3}$$

$$CH_3CH_2CH_2CH{=}CHCH_3 \xrightarrow{H_2} \underset{\text{(F)}}{CH_3CH_2CH_2COOH} + CH_3COOH$$

275. (A) contains carbon and hydrogen, decolourizes Br_2 in CCl_4 and so (A) is unsaturated hydrocarbon.

$$\text{(A)} \xrightarrow{HBr} \underset{\text{Bromide}}{\text{(B)}} \xrightarrow{NaOH} \underset{\text{Alcohol}}{\text{(C)}} \xrightarrow{\text{Oxidation}} \text{Hexanone-3}$$

Since oxidation of (C) gives ketone and so, (C) should be secondary alcohol
The position of OH in (C) should be at '3' because hexan-3-one is formed

∴ **(C)** is $CH_3CH_2\underset{\displaystyle OH}{\underset{|}{C}}HCH_2CH_2CH_3$

(C) is obtained by hydrolysis of (B) and thus, **(B)** is

$$CH_3CH_2\underset{\displaystyle Br}{\underset{|}{C}}HCH_2CH_2CH_3 \quad \text{3-bromohexane}$$

(B) is formed by the action of HBr on (A) and so, **(A)** can be

$$\underset{\text{(A) Hex-2-ene}}{CH_3CH{=}CHCH_2CH_2CH_3} \text{ or } \underset{\text{(A) Hex-3-ene}}{CH_3CH_2CH{=}CHCH_2CH_3}$$

276. (i) *Cis*-2-butene (H(H₃C)C=C(H)CH₃) $\xrightarrow{Br_2}$ $CH_3CH(Br)-CH(Br)CH_3$

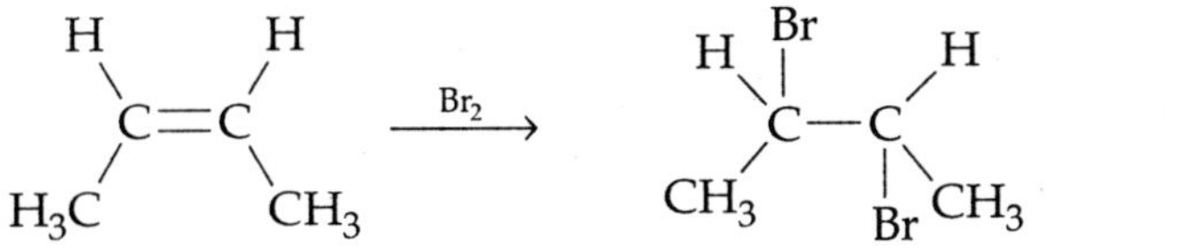

Cis-2-butene — recemic mixtuer of '*d*' and '*l*' forms

(ii) *Cis*-2-butene $\xrightarrow[H_2O+[O]]{KMnO_4}$ $CH_3CH(OH)-CH(OH)CH_3$ (meso form) or

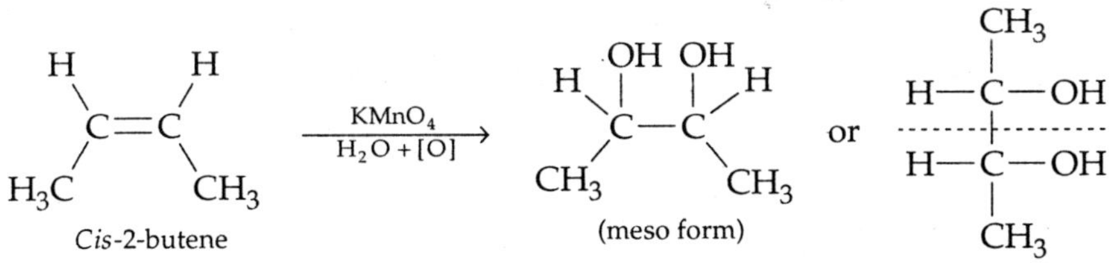

$$\begin{array}{c} CH_3 \\ | \\ H-C-OH \\ \text{-------}|\text{-------} \\ H-C-OH \\ | \\ CH_3 \end{array}$$

277. (a) The given statements suggest that following reactions are possible:

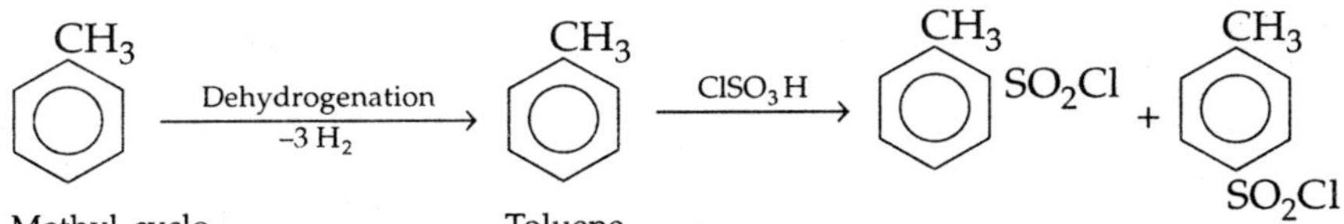

Methyl cyclo hexane $\xrightarrow[-3H_2]{\text{Dehydrogenation}}$ Toluene $\xrightarrow{ClSO_3H}$ o-$CH_3C_6H_4SO_2Cl$ **(A)** + p-$CH_3C_6H_4SO_2Cl$ **(B)**

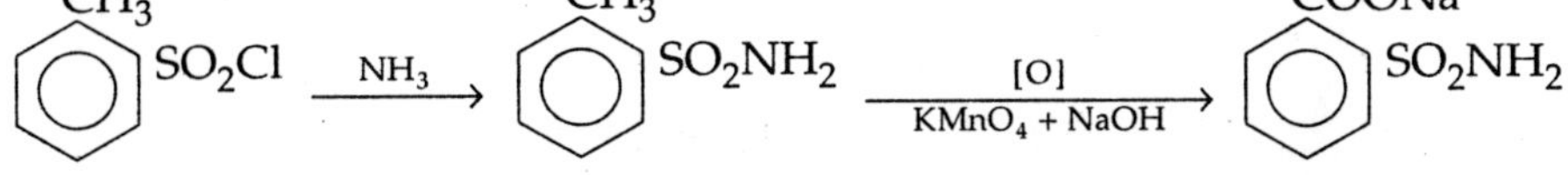

o-$CH_3C_6H_4SO_2Cl$ (A) $\xrightarrow{NH_3}$ o-$CH_3C_6H_4SO_2NH_2$ **(C)** $\xrightarrow[KMnO_4+NaOH]{[O]}$ o-$NaOOC\text{-}C_6H_4SO_2NH_2$ **(D)**

$\xrightarrow{H_3O^+ \Delta}$ $C_6H_4(CO)(SO_2)NH$

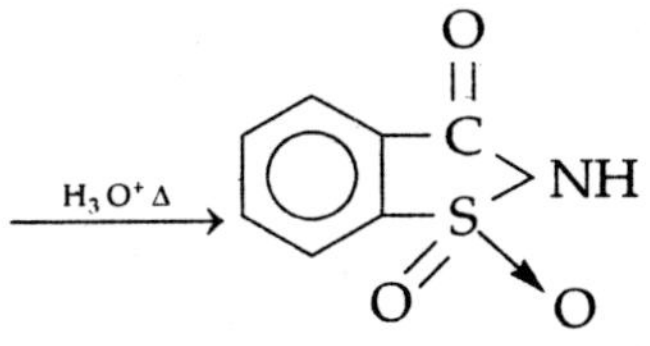

(saccharin) **(E)**

Also, $CH_3C_6H_4SO_2Cl$ (B) $\xrightarrow{NH_2}$ $CH_3C_6H_4SO_2NH_2$ (F) $\xrightarrow{NaClO/NaOH}$ $CH_3C_6H_4SO_2N\bar{C}l Na^+$ (G)

(Chloramine-T) antiseptic

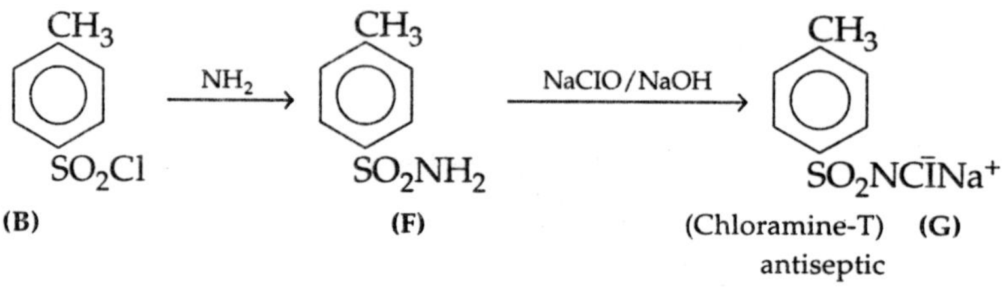

278. $R-\overset{O}{\overset{||}{C}}-O-\overset{O}{\overset{||}{C}}-R' \xrightarrow[Na_2CO_3]{\text{Hot aq.}} R-\overset{O}{\overset{||}{C}}-ONa + R'\overset{O}{\overset{||}{C}}-ONa$

(A) (B) (C)

$\xrightarrow{H_2SO_4}$ ROCOOH + R'COOH (Acids)

$\xrightarrow{\text{Distill.}}$ $R\overset{O}{\overset{||}{C}}-O-\overset{O}{\overset{||}{C}}-R'$ (A)

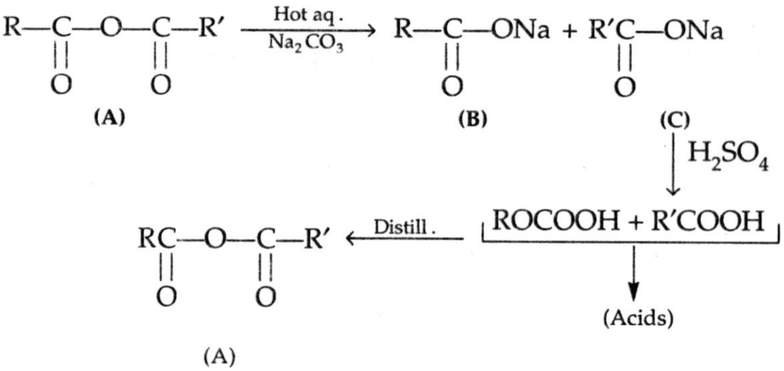

279. $CH_2=CHCH_2CH=CH_2$ (1, 4-Pentadiene) $\xrightarrow{HCl}$ $CH_3-\underset{Cl}{CH}-CH_2-\underset{Cl}{CH}CH_3$ (X)

$\xrightarrow[\text{dry ether}]{\text{excess Mg}}$ $CH_3-\underset{MgCl}{CH}-CH_2-\underset{MgCl}{CH}-CH_3$ (Y) $\longrightarrow$ $CH_3\overset{O}{\overset{||}{C}}-OC_2H_5$

$$CH_3-\overset{OMgCl}{C}(-OC_2H_5)-\underset{}{}CH(CH_3)-CH_2-CH(MgCl)-CH_3$$

(Intermediate)

$\longrightarrow CH_3-C(=O)-CH(CH_3)-CH_2-CH(MgCl)-CH_3$

$\longrightarrow$ cyclic alkoxide: $CH_3-\overset{OMgCl}{C}-CH-CH_3$ ring with $CH_3-\underset{H}{C}-CH_2$ $\xrightarrow{HOH}$ $CH_3-\overset{OMgCl}{C}-CH-CH_3$ ring with $CH_3-\underset{H}{C}-CH_2$

(Z)

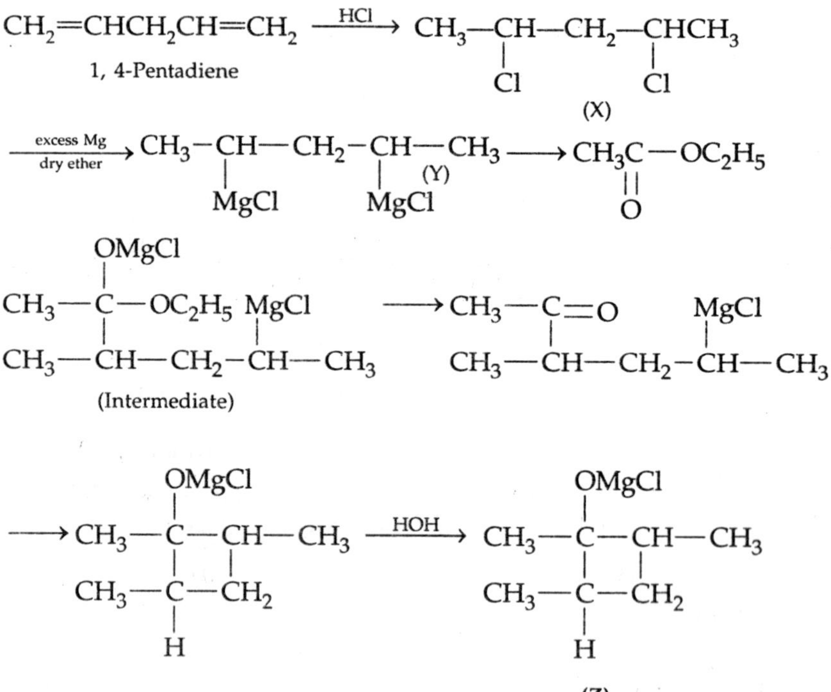

(Z) can also be $CH_3-\overset{O}{\overset{||}{C}}-\underset{CH_3}{CH}-CH_2-\underset{CH_3}{CH}-\underset{O}{\underset{||}{C}}CH_3$

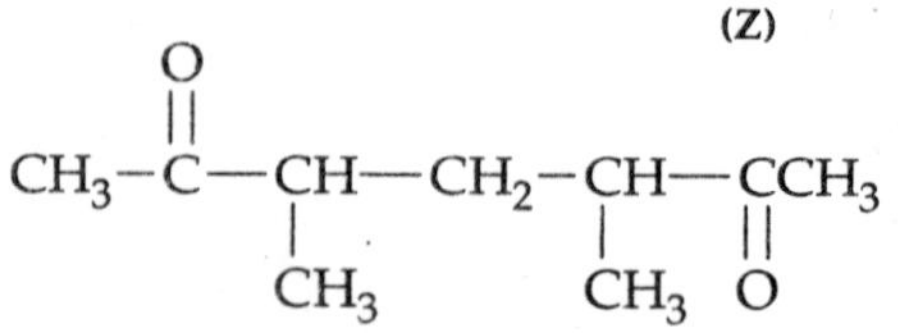

280. (D) is terminal alkyne since it forms precipitate with ammonical $AgNO_3$, i.e., R—C≡CH.

$$\underset{(A)}{C_4H_9Br} \xrightarrow[-\,HBr]{(alc.)\,KOH} \underset{\substack{(B)\\ \text{Alkene}}}{C_4H_8} \xrightarrow{Br_2} \underset{\substack{(C)\\ \text{Dibromide}}}{C_4H_8Br_2} \xrightarrow[-2\,HBr]{NaNH_2} \underset{(D)}{C_4H_6}$$

Thus, if (D) is R—C≡CH, then (C) should be $R-\underset{Br}{\underset{|}{CH}}-\underset{Br}{\underset{|}{CH_2}}$

because it is formed by the action of Br_2 on (B). Thus, (B) should be $R-CH=CH_2$. Since (B) is obtained by dehydrobromination of (A) and thus, (A) should be $R-CH_2CH_2Br$. (A) is of 4 carbon atoms and therefore, (A) is $CH_3CH_2CH_2CH_2Br$.
Thus the reactions involved are:

$$\underset{\substack{\text{1-bromobutane}\\ \textbf{(A)}}}{CH_3CH_2CH_2CH_2Br} \xrightarrow[-\,HBr]{(alc.)\,KOH} \underset{\textbf{(B)}}{CH_3CH_2CH{=}CH_2}$$

$$\underset{\substack{\text{But-1-ene}\\ (B)}}{CH_3CH_2CH{=}CH2_2} \xrightarrow{Br_2} \underset{(C)}{CH_3CH_2CHBrCH_2Br}$$

$$\underset{\substack{\text{1,2-dibnromobutane}\\ \textbf{(C)}}}{CH_3CH_2CHBrCH_2Br} \xrightarrow[-2\,HBr]{NaNH_2} \underset{\textbf{(D)}}{CH_3CH_2C{\equiv}CH}$$

$$\underset{\text{But-1-yne}}{CH_3CH_2C{\equiv}CH} + \underset{\text{Tollen's reagent}}{Ag_2O} \longrightarrow \underset{\text{White ppt.}}{CH_3CH_2C{\equiv}CAg}$$

281. The structure of A and B is given below:

(A) $\xrightarrow{\text{Ozonolysis}}$ (B): ring containing $\overset{O}{\overset{\|}{C}}-CH_2-CH_2-CH_2-\overset{O}{\overset{\|}{C}}$ joined through $CH_2-CH_2-CH_2$

282. (a) 2,3-dihydroxybutane
NOTE: This name is wrong. It should be butan-2, 3-diol.

$$CH_3\overset{*}{C}HOH\overset{*}{C}HOHCH_3$$

where * shows asymmetric carbon atoms, thus compound has optical isomers.

meso form inactive	laevo form	dextro form
CH_3 / HO—C—H / HO—C—H / CH_3	CH_3 / HO—C—H / H—C—OH / CH_3	CH_3 / HO—C—H / H—C—OH / CH_3

(laevo form and dextro form: active)

(b) 3-phenyl-2-propenoic acid:

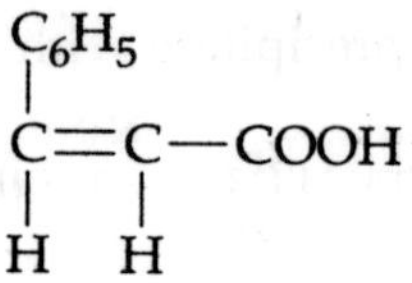

$$\begin{array}{c} C_6H_5 \\ | \\ C{=}C{-}COOH \\ |\quad | \\ H\quad H \end{array}$$

It has two geometrical isomers.

$$\begin{array}{c} C_6H_5{-}C{-}H \\ \| \\ HOOC{-}C{-}H \end{array} \qquad \begin{array}{c} C_6H_5{-}C{-}H \\ \| \\ H{-}C{-}COOH \end{array}$$

Cis-form *Trans*-form

283. (A) gives characteristic colour with ceric ammonium nitrate and thus it has OH group. Reactions according to the problem are:

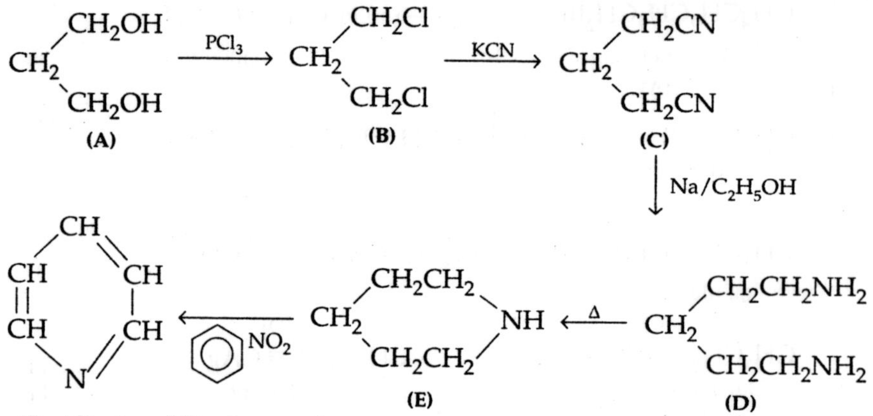

$$\underset{\textbf{(A)}}{CH_2\begin{array}{l} CH_2OH \\ CH_2OH \end{array}} \xrightarrow{PCl_3} \underset{\textbf{(B)}}{CH_2\begin{array}{l} CH_2Cl \\ CH_2Cl \end{array}} \xrightarrow{KCN} \underset{\textbf{(C)}}{CH_2\begin{array}{l} CH_2CN \\ CH_2CN \end{array}} \xrightarrow{Na/C_2H_5OH}$$

$$\underset{\text{(Pyridine)}}{C_5H_5N} \xleftarrow[\text{(Nitrobenzene)}]{C_6H_5NO_2} \underset{\textbf{(E)}}{CH_2\begin{array}{l} CH_2CH_2 \\ CH_2CH_2 \end{array}\!\!\!>NH} \xleftarrow{\Delta} \underset{\textbf{(D)}}{CH_2\begin{array}{l} CH_2CH_2NH_2 \\ CH_2CH_2NH_2 \end{array}}$$

284. The given reactions in Wurtz reazction i.e., reaction of alkyl halides with Na in presence of ether.

When a mixture of halide reacts with Na in presence of ether, the alkanes formed are:

(a) $\underset{\text{1-Bromopropane}}{CH_3CH_2CH_2Br} + 2Na + BrCH_2CH_2CH_3 \xrightarrow{Ether} \underset{\text{(Hexane)}}{CH_3CH_2CH_2CH_2CH_2CH_3}$

(b) 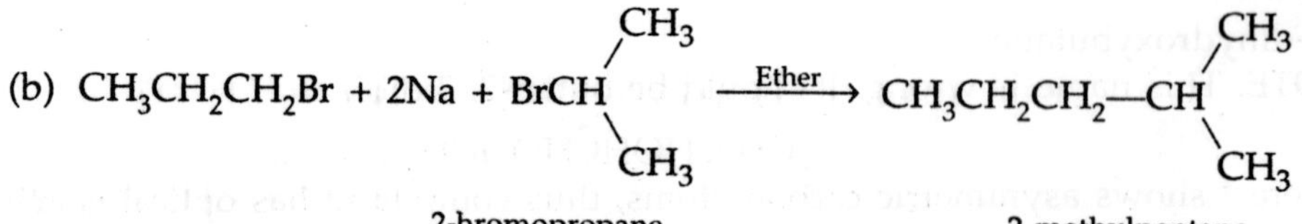 $CH_3CH_2CH_2Br + 2Na + \underset{\text{2-bromopropane}}{BrCH\begin{array}{l} CH_3 \\ CH_3 \end{array}} \xrightarrow{Ether} \underset{\text{2-methylpentane}}{CH_3CH_2CH_2{-}CH\begin{array}{l} CH_3 \\ CH_3 \end{array}}$

(c) 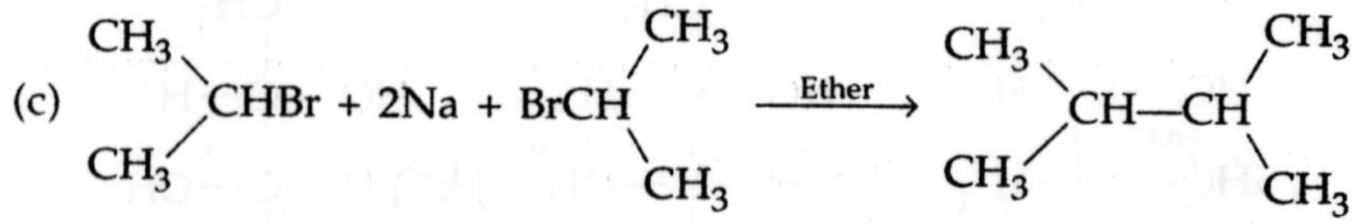 $\begin{array}{r} CH_3 \\ CH_3 \end{array}\!\!>CHBr + 2Na + BrCH\begin{array}{l} CH_3 \\ CH_3 \end{array} \xrightarrow{Ether} \begin{array}{r} CH_3 \\ CH_3 \end{array}\!\!>CH{-}CH\begin{array}{l} CH_3 \\ CH_3 \end{array}$

285. Alkene gives only pentan-2-one thus, it should have $\begin{array}{r} CH_3CH_2CH_3 \\ CH_3 \end{array}\!\!>C$—units round double

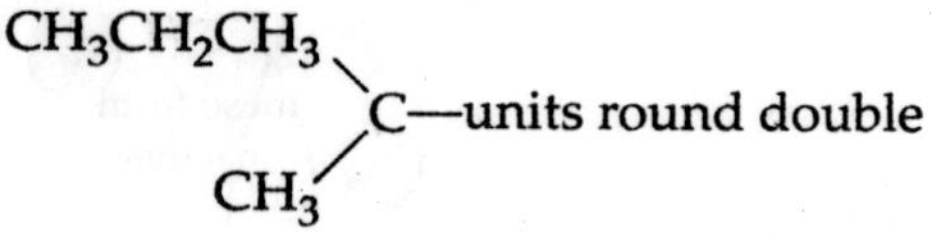

bond, and thus, two molecules of pentan-2-one are to be formed.
Thus, alkene is $CH_3CH_2CH_2—\underset{CH_3}{\underset{|}{C}}=\underset{CH_3}{\underset{|}{C}}CH_2CH_2CH_3$

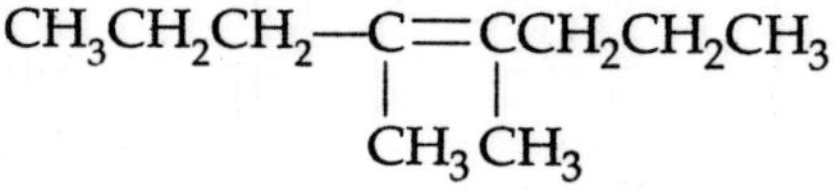

(4,5-dimethyloct-4-ene)

Thus:

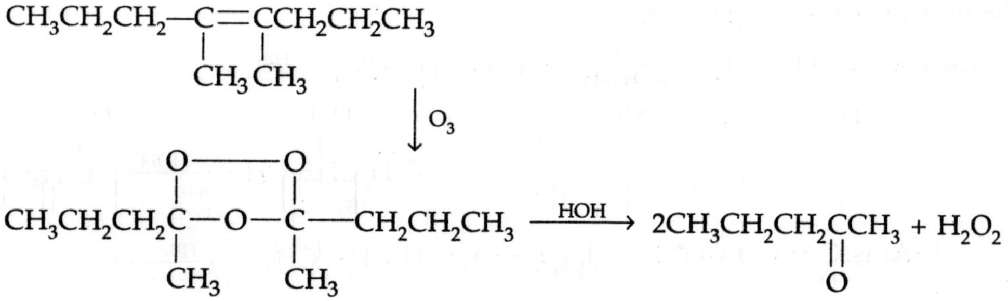

$$CH_3CH_2CH_2—\underset{CH_3}{\underset{|}{C}}=\underset{CH_3}{\underset{|}{C}}CH_2CH_2CH_3 \xrightarrow{O_3} CH_3CH_2CH_2\underset{CH_3}{\underset{|}{\overset{O———O}{C}}}—O—\underset{CH_3}{\underset{|}{C}}—CH_2CH_2CH_3 \xrightarrow{HOH} 2CH_3CH_2CH_2\underset{O}{\underset{||}{C}}CH_3 + H_2O_2$$

Pentan-2-one

286. (a)

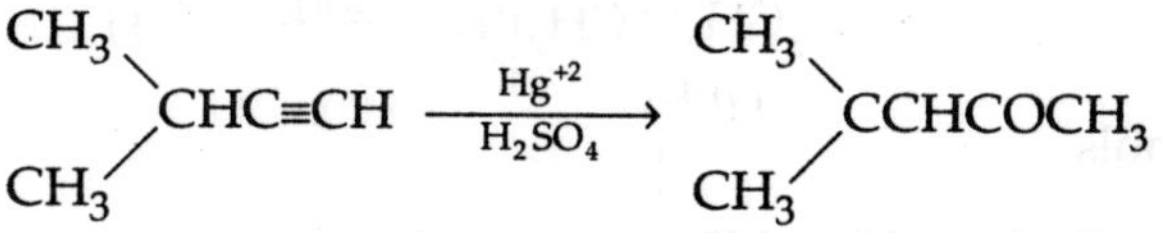

$$\underset{\text{(3-methyl butyne)}}{\underset{\textbf{(A)}}{(CH_3)_2CHC\equiv CH}} \xrightarrow[H_2SO_4]{Hg^{+2}} \underset{\text{(3-methyl butanone)}}{(CH_3)_2CCHCOCH_3}$$

$$(CH_3)_2CHCOCH_3 \xrightarrow{NH_2OH} \underset{\text{(Oxime)}}{(CH_3)_2CH\underset{NOH}{\underset{||}{C}}CH_3}$$

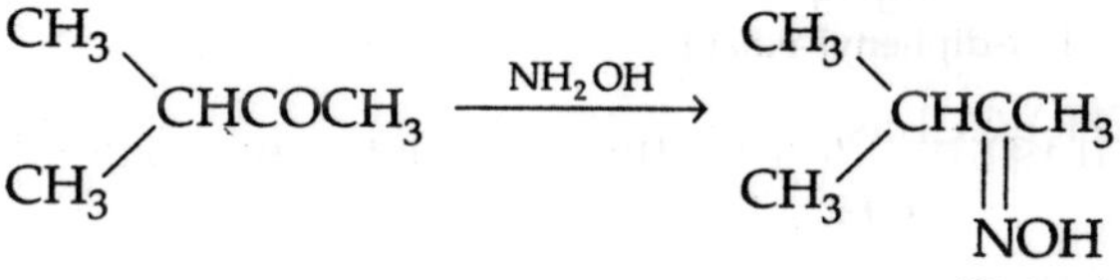

$$(CH_3)_2CHCOCH_3 \xrightarrow{[O]} (CH_3)_2CHCOOH + HCOOH$$

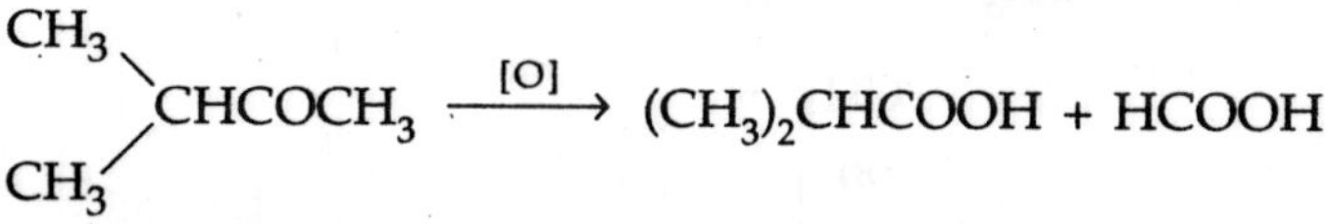

(b) (A) Decolourizes bromine water and $KMnO_4$ and thus it has unsaturation.

The reactions suggest (A) to be $CH_3—CH=\underset{Br}{\underset{|}{C}}—CH_2—CH_3$

$$\underset{\textbf{(A)}}{CH_3—CH=\underset{Br}{\underset{|}{C}}—CH_2—CH_3} \xrightarrow{Sn/HCl} \underset{\textbf{(B)}}{CH_3—CH_2—\underset{Br}{\underset{|}{C}}—CH_2CH_3}$$

$$\textbf{(A)} \xrightarrow{NaNH_2} \underset{\textbf{(C)}}{CH_3—C\equiv C—CH_2—CH_3}$$

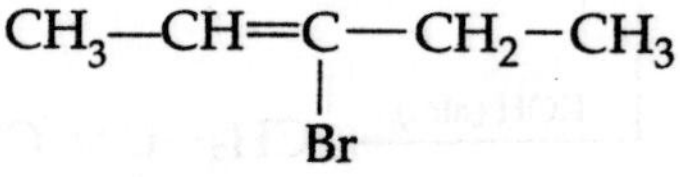

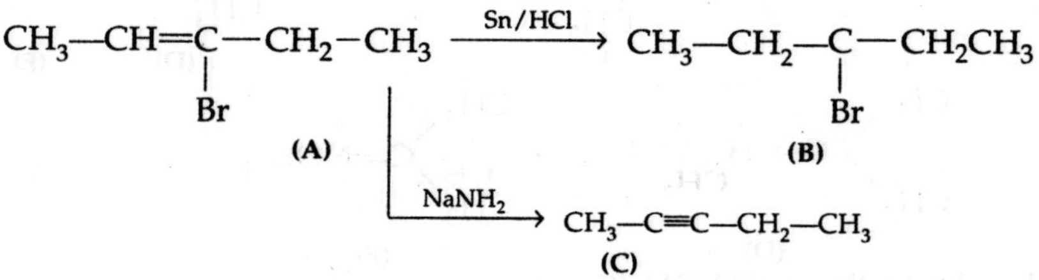

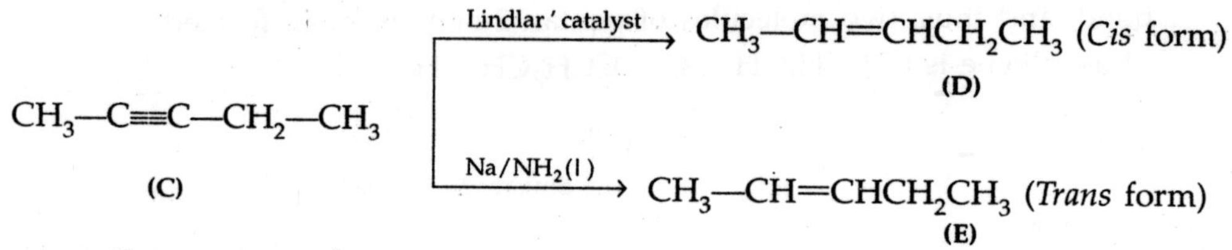

$$CH_3{-}C{\equiv}C{-}CH_2{-}CH_3 \ (C) \xrightarrow{\text{Lindlar' catalyst}} CH_3{-}CH{=}CHCH_2CH_3 \ (\textit{Cis form}) \ (D)$$

$$CH_3{-}C{\equiv}C{-}CH_2{-}CH_3 \ (C) \xrightarrow{Na/NH_2(l)} CH_3{-}CH{=}CHCH_2CH_3 \ (\textit{Trans form}) \ (E)$$

287. The given reactions suggest that:

(B) is $C_6H_5CHBrCH_3 \xrightarrow[\text{KOH (alc.)}]{-HBr} \underset{(C)}{C_6H_5CH{=}CH_2} \xrightarrow{HBr}$

$$\underset{\textbf{(B)}}{C_6H_5CHBrCH_3} \xrightarrow{HOH} \underset{\textbf{(E)}}{C_6H_5CHOHCH_3}$$

and **(A)** is $C_6H_5CH_2CH_2Br \xrightarrow[\text{KOH (alc.)}]{-HBr} \underset{\textbf{(C)}}{C_6H_5CH{=}CH_2} \xrightarrow[\text{Peroxide}]{HBr}$

$$\underset{\textbf{(A)}}{C_6H_5CH_2{-}CH_2Br} \xrightarrow{HOH} \underset{\textbf{(D)}}{C_6H_5CH_2CH_2OH}$$

(D) and (E) are isomeric alcohols

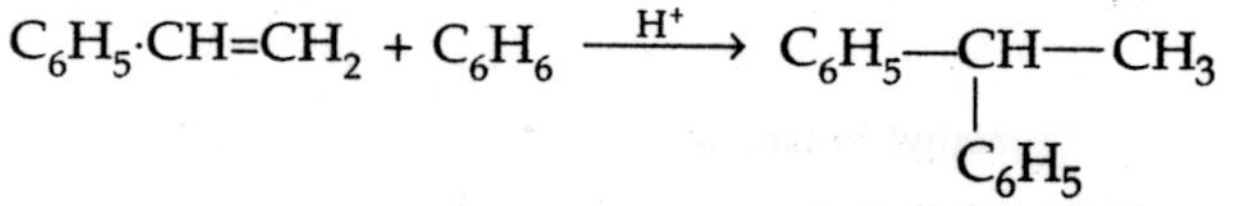

$$C_6H_5{\cdot}CH{=}CH_2 + C_6H_6 \xrightarrow{H^+} C_6H_5{-}\underset{\underset{C_6H_5}{|}}{CH}{-}CH_3$$

1, 1-diphenyl ethane

288. The reactions suggest that (A) is $(CH_3)_3CCl$ as:

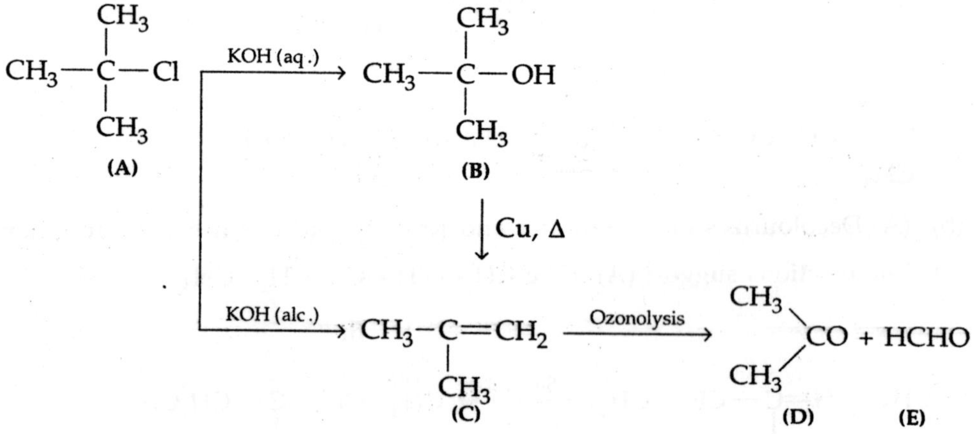

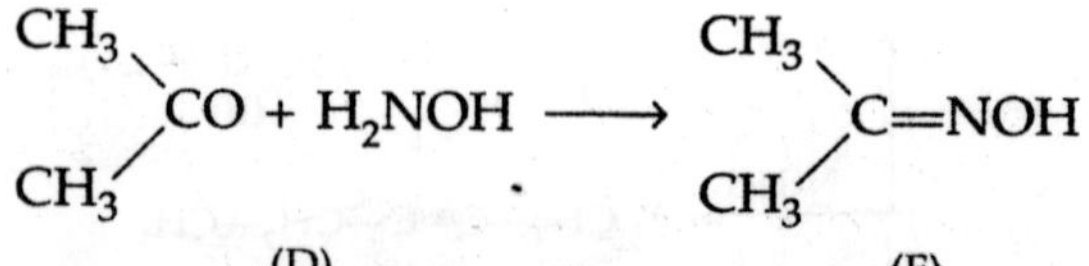

$$\underset{\textbf{(A)}}{CH_3{-}\overset{\overset{CH_3}{|}}{\underset{\underset{CH_3}{|}}{C}}{-}Cl} \xrightarrow{\text{KOH (aq.)}} \underset{\textbf{(B)}}{CH_3{-}\overset{\overset{CH_3}{|}}{\underset{\underset{CH_3}{|}}{C}}{-}OH} \xrightarrow{Cu,\ \Delta} \text{(C)}$$

$$\textbf{(A)} \xrightarrow{\text{KOH (alc.)}} \underset{\textbf{(C)}}{CH_3{-}\underset{\underset{CH_3}{|}}{C}{=}CH_2} \xrightarrow{\text{Ozonolysis}} \underset{\textbf{(D)}}{\begin{matrix}CH_3\\ CH_3\end{matrix}\!\!>\!CO} + \underset{\textbf{(E)}}{HCHO}$$

$$\underset{\text{(D)}}{\begin{matrix}CH_3\\ CH_3\end{matrix}\!\!>\!CO} + H_2NOH \longrightarrow \underset{\text{(F)}}{\begin{matrix}CH_3\\ CH_3\end{matrix}\!\!>\!C{=}NOH}$$

$$\underset{\text{(E)}}{2HCHO} + NaOH \rightarrow \underset{\textbf{(H)}}{HCOONa} + \underset{\textbf{(G)}}{CH_3OH} \text{ (Cannizzaro reaction)}$$

(D) is obtained with propyne as $CH_3—C\equiv CH \xrightarrow[Hg^{+2},\, H_2SO_4]{H_2O} CH_3COCH_3$ and thus (D) is acetone.
(E) undergoes Cannizzaro reaction and thus HCHO.
The alkene (C) formed by the action of KOH(alc.) on (A), on ozonolysis gives (*D*) and (*E*).
Thus (C) is $(CH_3)_2C{=}CH_2$ and (*A*) is $(CH_3)_3CCl$.

289.

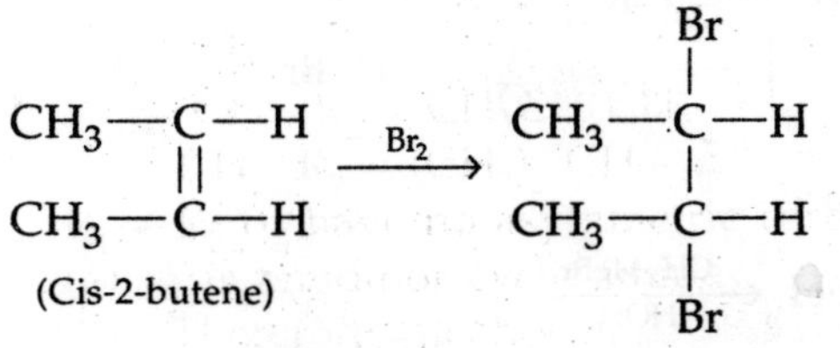

$$\begin{matrix} CH_3—C—H \\ \| \\ CH_3—C—H \end{matrix} \xrightarrow{Br_2} \begin{matrix} Br \\ | \\ CH_3—C—H \\ | \\ CH_3—C—H \\ | \\ Br \end{matrix}$$

(Cis-2-butene)

The stereoisomers are racemic products (d, l)

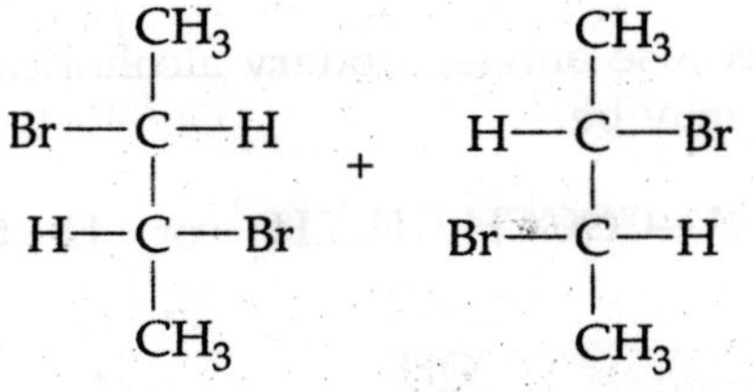

$$\begin{matrix} CH_3 \\ | \\ Br—C—H \\ | \\ H—C—Br \\ | \\ CH_3 \end{matrix} \quad + \quad \begin{matrix} CH_3 \\ | \\ H—C—Br \\ | \\ Br—C—H \\ | \\ CH_3 \end{matrix}$$

'd' or 'l' *'l' or 'd'*

290.

$$C_6H_5COCH_3 \xrightarrow{NH_2OH\cdot HCl} \begin{matrix} C_6H_5—C—CH_3 \\ \| \\ N—OH \end{matrix} + \begin{matrix} C_6H_5—C—CH_3 \\ \| \\ HO—N \end{matrix}$$

(A) (B) Syn (C) Anti

Oximes

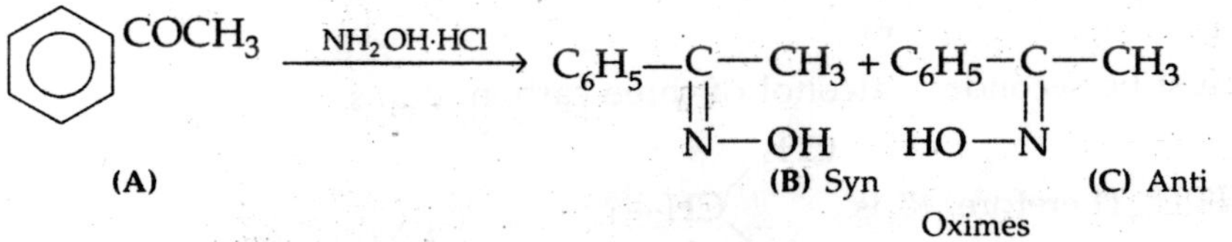

Oximes undergo Beckmann rearrangement as:?

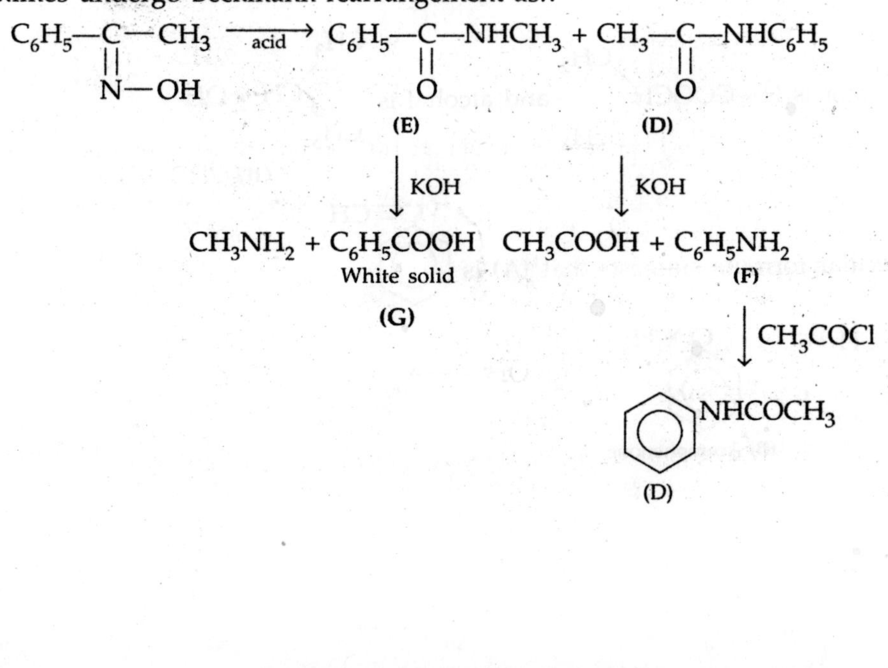

$$\begin{matrix} C_6H_5—C—CH_3 \\ \| \\ N—OH \end{matrix} \xrightarrow{acid} \begin{matrix} C_6H_5—C—NHCH_3 \\ \| \\ O \end{matrix} + \begin{matrix} CH_3—C—NHC_6H_5 \\ \| \\ O \end{matrix}$$

(E) (D)

$$(E) \xrightarrow{KOH} CH_3NH_2 + C_6H_5COOH$$

White solid

(G)

$$(D) \xrightarrow{KOH} CH_3COOH + C_6H_5NH_2 \; (F)$$

$$C_6H_5NH_2 \; (F) \xrightarrow{CH_3COCl} C_6H_5NHCOCH_3 \; (D)$$

291. Let the ester A be $R-\overset{\overset{O}{||}}{C}-OR'$, i.e., $C_4H_8O_2$

It reacts with CH_3MgBr; the product *(B)* on acidification gives an alcohol (B) which on oxidation with NaOCl followed with acidification gives acetic acid. Thus,

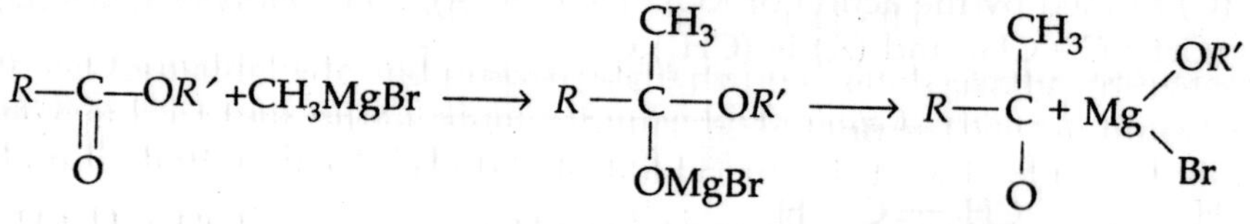

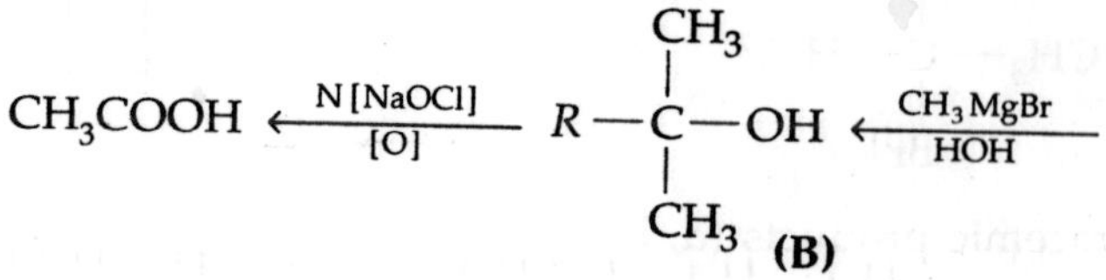

$$R-\underset{\underset{O}{||}}{C}-OR' + CH_3MgBr \longrightarrow R-\underset{\underset{OMgBr}{|}}{\overset{\overset{CH_3}{|}}{C}}-OR' \longrightarrow R-\underset{\underset{O}{|}}{\overset{\overset{CH_3}{|}}{C}} + Mg\begin{matrix} OR' \\ Br \end{matrix}$$

$$CH_3COOH \xleftarrow[\text{[O]}]{\text{N [NaOCl]}} R-\underset{\underset{CH_3}{|}}{\overset{\overset{CH_3}{|}}{C}}-OH \xleftarrow[\text{HOH}]{CH_3MgBr}$$

(B)

Thus, *R* should be H because only secondary alcohol can be oxidized by NaOCl and not tertiary. Thus, the ester may be

$$H-COOCH_2CH_2CH_3 \quad \text{or} \quad H-COOCH\begin{matrix} CH_3 \\ CH_3 \end{matrix}$$

$$\text{Also } Mg\begin{matrix} OR' \\ Br \end{matrix} \xrightarrow{HOH} Mg\begin{matrix} OH \\ Br \end{matrix} + R'OH \xrightarrow{[O]} CH_3COOH$$

Thus *R*′OH must be secondary alcohol of three carbon atoms,

i.e., $\begin{matrix} CH_3 \\ CH_3 \end{matrix}\!\!>CHOH$, Therefore, *R*′ is $\begin{matrix} CH_3 \\ CH_3 \end{matrix}\!\!>CH-$

Thus, ester is $H-COOCH\begin{matrix} CH_3 \\ CH_3 \end{matrix}$ and alcohal is $\begin{matrix} CH_3 \\ CH_3 \end{matrix}\!\!>CH-OH$

(B)

292. The molecular formula suggests that **(A)** is $C_6H_5-C\equiv CH$

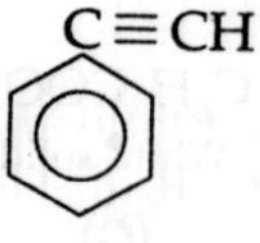

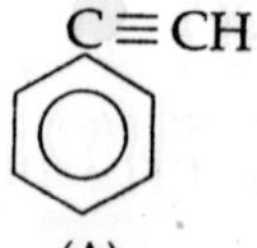

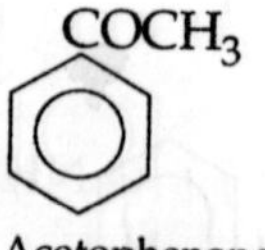

$C_6H_5-C\equiv CH$ (A) $\quad$ $C_6H_5-COCH_3$ **(B) Acetophenone**

(B) can also be obtained as

$$C_6H_6 + CH_3COCl \xrightarrow{AlCl_3\ (anhy.)} C_6H_5COCH_3\ (B)$$

$$C_6H_5COCH_3 + 3I_2 + 4KOH \longrightarrow \underset{(C)}{C_6H_5COOK} + \underset{\substack{(D)\\ yellow}}{CH_3} + 3KI + 3H_2O$$

$$C_6H_5OH + CHCl_3 + KOH \xrightarrow{H^+} \underset{(D)\ Ortho}{o\text{-}HOC_6H_4CHO} + \underset{(E)\ Para}{p\text{-}HOC_6H_4CHO}$$

Hydroxybenzaldehydes

293. For empirical formula of (*A*)

Element	%	*Relative of atoms*	*Simplest ratio*
C	$\frac{12}{44} \times 10 = 77.77$	6.46	7
H	$\frac{2}{18} \times 100 = 7.40$	7.40	8
O	$100 - (77.77 + 7.40) = 14.83$	0.93	1

∴ Empirical formula of (*A*) = $\mathbf{C_7H_8O}$

Since (*B*) is gives $C_7H_5OBr_3$ and (*B*) is isomer to (*A*). So the, molecular formula of (*A*) and (*B*) is C_7H_8O.

Now, (*A*) is insoluble in NaOH and $NaHCO_3$ so **(A)** is $C_6H_5OCH_3$

(*A*) is also confirmed by action with HI

$$\underset{(A)}{C_6H_5OCH_3} + HI \longrightarrow \underset{\text{(Soluble in NaOH) (D)}}{C_6H_5OH} + \underset{\substack{\text{(Soluble in ethanoic)}\\ AgNO_3\ (C)}}{CH_3I}$$

(*B*) is soluble in NaOH and gives (*E*) $C_7H_5OBr_3$ with Br_2 water and so **(B)** is $o\text{-}CH_3C_6H_4OH$ i.e.,

m-cresol since only meta position can provide tribromo derivative.

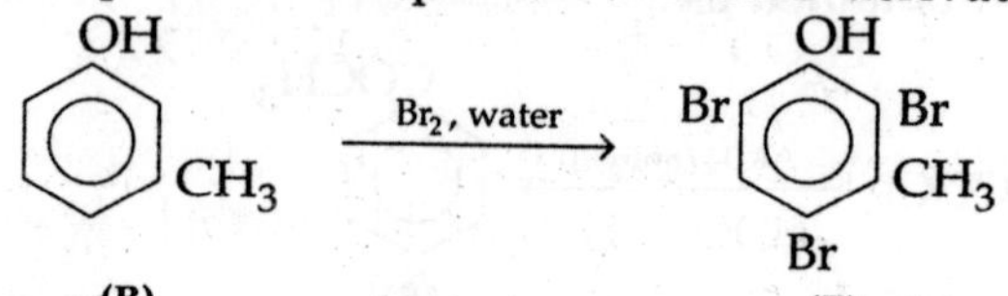

OH-C$_6$H$_4$-CH$_3$ (B) $\xrightarrow{Br_2,\ water}$ Br, Br, Br, OH, CH$_3$ substituted benzene (E)

294. The statement provides that the compound A, $C_{11}H_6O$ may be

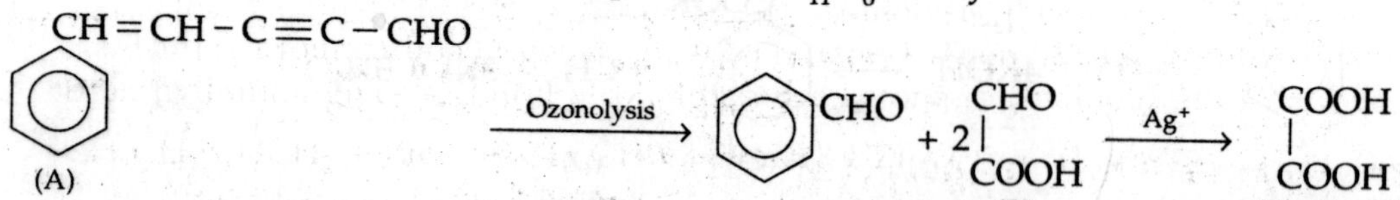

C_6H_5–CH=CH–C≡C–CHO (A) $\xrightarrow{\text{Ozonolysis}}$ C_6H_5CHO + 2 CHO–COOH (B) $\xrightarrow{Ag^+}$ COOH–COOH (Oxalic acid)

295. (A) gives characteristic colour with aqueous $FeCl_3$ is so it has phenolic group. (A) on treatment wih Br_2 gives $C_7H_5OBr_3$ and thus, taking a account of molecular ormula of (A); (A) seems to be a methyl substituted phenol, i.e, cresol.

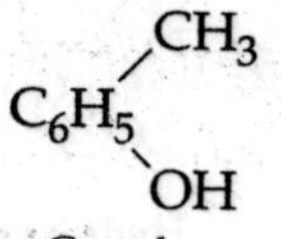

C_6H_5(CH$_3$)(OH)
Cresol

On the basis of bromination. (A) can be (OH, CH$_3$ meta-substituted benzene) because metaposition can form tribromo derivative. Thus the possible reactions are:

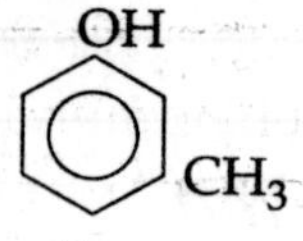

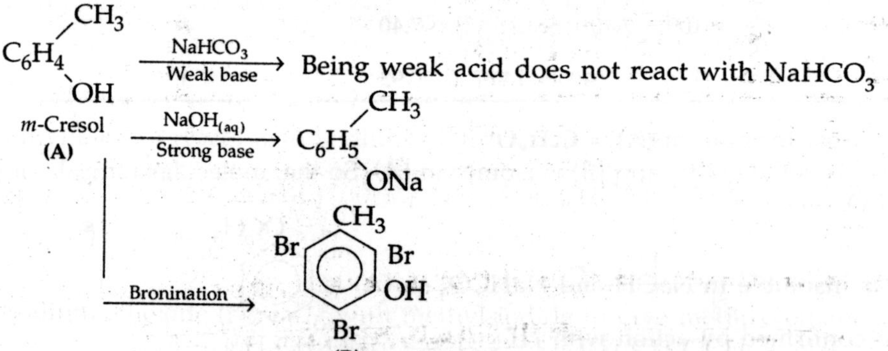

C_6H_4(CH$_3$)(OH) m-Cresol (A) $\xrightarrow[\text{Weak base}]{NaHCO_3}$ Being weak acid does not react with $NaHCO_3$

$\xrightarrow[\text{Strong base}]{NaOH_{(aq)}}$ C_6H_5(CH$_3$)(ONa)

$\xrightarrow{\text{Bronination}}$ CH$_3$, Br, Br, OH, Br substituted benzene (B)

(A) neither dissolves in aqueous NaOH, nor it gives characteristic colour with $FeCl_3$, so it is a neutral compound, ie, either

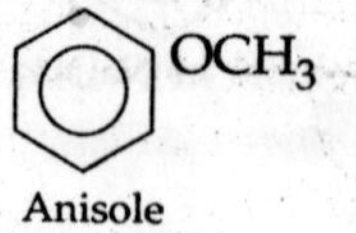

$C_6H_5OCH_3$ (Anisole) or $C_6H_5CH_2OH$ (Benzyl alcohol)

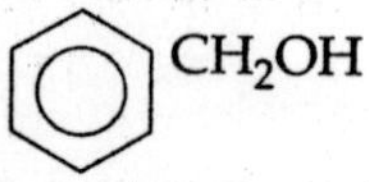

296. Given, Mixture of (*A*) and (*B*) $\xrightarrow[+\text{KOH}_{aq}]{\text{CHCl}_3}$ Organic layer + Alkaline aqueous layer

(A) (B)

Organic layer on treating with $KOH_{(alc.)}$ produce (C) of unpleasant odour (C_7H_5N) and thus, (C) is C_6H_5NC. This is carbylamine reaction. Therefore, (*A*) is $C_6H_5NH_2$.

$$\underset{\text{(Amiline (A)}}{C_6H_5NH_2} + CHCl_3 + 3KOH_{(alc.)} \longrightarrow \underset{\text{Phenyl isocyanide (C)}}{C_6H_5NC} + 3KCl + 3H_2O$$

Alkaline layer on treating with $CHCl_3$ followed with acidification gives two isomers **(D)** and **(E)** ($C_7H_6O_2$). This is Riemer-Teimenn reaction and thus, (*B*) is C_6H_5OH

297. For empirical formula of (*A*),

Element	%	*Relative no. of atoms*	*Simplest ratio*
C	49.32	4.11	3
H	9.59	9.59	7
N	19.18	1.37	1
O	21.91	1.37	1

∴ Empirical formula of (*A*) is C_3H_7NO and Empirical formula wt. of (*A*) = 73

Molecular weight of acid (*B*) can be derived as:

$$\frac{\text{Wt. of Ag}}{\text{E. wt. of Ag}} = \frac{\text{Wt. of salt}}{\text{E. wt. ofAg salt}}$$

$$\frac{59.67}{108} = \frac{100}{E + 108}$$

∴ $E = 73$ i.e., equivalent weight of anion of salt

∴ Equivalent weight of acid = $E + 1 = 73 + 1 = 74$

∴ Molecular weight of acid = 74×1 ∵ basicity = 1

Acid (B) is monobasic and thus (B) is $C_6H_{2n+1}COOH$

∴ $12n + 2n + 1 + 12 + 32 + 1 = 74$

$n = 2$

So, acid (*B*) is CH_3CH_2COOH

Since, (*B*) is obtained by the action of (*A*) with NaOH followed with hydrolysis and so (*A*) is $CH_3CH_2CONH_2$.

$$\underset{\text{Propanamide }(A)}{CH_3CH_2CONH_2} \xrightarrow{\text{NaOH}} CH_3CH_2COONa + NH_3$$

$$\xrightarrow{\text{HOH}} \underset{\textbf{(B)}}{CH_3CH_2COOH}$$

298. Compound gives foul odour with $CHCl_3$ and alcohotic KOH and thus, it is primary amine say $R—NH_2$. The reaction is carbylamine reaction.

$$RNH_2 \xrightarrow[273\text{ K}]{\text{HCl + NaNO}_2} ROH + N_2$$

∵ 112 mL N_2 is given by 0.295 g

∴ 22400 mL N_2 is given by $\frac{0.295 \times 22400}{112} = 59$

$\therefore$ Molecular weight of $RNMH_2 = 59$

$\therefore$ $R = 59 - (14 + 2) = 43$

i.e. $C_nH_{2n+1} = 43$

$\therefore$ $n = 3$

$$\text{Now, Given, } C_3H_7NH_2 \xrightarrow{HCl + NaNO_2} C_3H_7OH \xrightarrow{I_2 + NaOH} \underset{\text{precipitate}}{\text{yellow}}$$

Since alcohol gives iodoform test and thus, it is secondary alcohol, i.e., $CH_3CHOH{\cdot}CH_3$ (Propan 2-ol).

Thus, original compound should be $CH_3\underset{\underset{NH_2}{|}}{C}HCH_3$

Propan-2-amine

299. $C_5H_{10}O$ has two isomers (A) and (B). Molecular formula suggests that they may be carbonyl compounds.

(A) on treating with $NaOH_{(aq)}$, gives 2,2-dimethylpropanol an 2, 2-dimethylpropanoic acid salts. Thus, (*A*) is an aldehyde lacking α-C atom; since this is Cannizzaro reaction.

$$\underset{\substack{\text{2,2-dimethylpropanal} \\ \textbf{(A)}}}{(CH_3)CCHO} \xrightarrow{NaOH} (CH_3)_3CCH_2OH + (CH_3)CCOONa$$

Isomer (*B*) reacts with NaOH to show aldol condensation product 3-hydroxy-2-propylheptanal and thus, (*B*) is pentanal-1. as condensation occurs at α-carbon atom.

$$\underset{\text{(Pentan-I-al)}}{CH_3CH_2CH_2CH_2\underset{\underset{O}{||}}{C}H} + \underset{\text{(B)}}{\underset{\underset{C_3H_7}{|}}{CH_2}CHO} \xrightarrow{NaOH_{(aq)}} CH_3CH_2CH_2CH_2\underset{\underset{OH}{|}}{C}H{-}\underset{\underset{C_3H_7}{|}}{C}HCHO$$

300. $C_4H_8Cl_2$ on hydrolysis gives (*Y*) which is an aldehyde as it reduces Fehling solution so (*X*) is terminal gem dihalide $C_3H_7CHCl_2$. Given,

$$\underset{(X)}{C_3H_7CHCl_2} \xrightarrow{HOH} \underset{(Y)}{C_3H_7CHO} \xrightarrow{\text{Oxidation}} \underset{(Z)}{C_3H_7COOH} \xrightarrow[\text{LiAlH}_4]{\text{Reduction}} \underset{\text{Alcohol}}{C_3H_7CH_2OH}$$

The given alcohol is $CH_3{-}\underset{\underset{CH_3}{|}}{C}HCH_2OH$ and thus,

2-methylpropan-1-ol

So, the structural formulas are :

X is $CH_3{-}\underset{\underset{CH_3}{|}}{C}HCHCl_2$ 1,1-dichloro, 2-methylpropane

Y is $CH_3{-}\underset{\underset{CH_3}{|}}{C}HCHO$ 2-methylpropanal

Z is CH_3—CHCOOH 2-methylpropanoic acid
$$CH_3-\underset{\displaystyle CH_3}{\underset{|}{C}}HCOOH$$

301. For empirical formula,

Element	%	*Relative no. of atoms*	*Simplest ratio*
C	40	3.33	1
H	6.7	6.70	2
O	53.3	3.33	1

∴ Empirical formula of (A) = CH_2O;
Empirical formula weight = 30
∵ Molecular weight = 2 × 15 = 30
∴ Molecular formula of (*A*) = CH_2O

Since (A) reacts with $KOH_{(aq.)}$ to give (*B*) and (C) thus, formula of (*A*) is HCHO
The reactions of the problem are,

$$\underset{\substack{\text{Methanol}\\ \textbf{(A)}}}{HCHO} \xrightarrow{KOH} \underset{\textbf{(B)}}{CH_3OH} = \underset{\textbf{(C)}}{HCOOK}$$

$$\underset{\substack{\text{Methanol}\\ \text{(B)}}}{CH_3OH} \xrightarrow{\text{Oxidaized}} \underset{\textbf{(A)}}{HCHO}$$

$$\underset{\substack{\text{Pot. methanoate}\\ \text{(C)}}}{HCOOK + HOH} \xrightarrow{HCl} \underset{\textbf{(D)}}{HCOOH}$$

$$\underset{\substack{\text{Mathanoic acid}\\ \textbf{(D)}}}{HCOOH + Ag_2O} \longrightarrow H_2O + CO_2 + 2Ag \text{ (silver mirror)}$$

$$\underset{\text{(D)}}{HCOOH + 2CuO} \longrightarrow H_2O + CO_2 + Cu_2O \text{ (red ppt.)}$$

$$HCCOH + NaHCO_3 \longrightarrow HCOONa + H2O + CO_2\uparrow$$

302. (A) C_5H_8O adds one Br_2 molecule and thus, has one double bond
It forms phenyl hydrazone with phenyl hydrazine and thus, carbonyl group is present in (*A*)
(*A*) does not reduce Tollen's regent so carbony group is ketone and not aldehyde.
Ozonolysis of (*A*) gives CH_3CHO and $C_3H_4O_2$ and thus, in consideration of above facts (*A*) is

$$\underset{\textbf{(A)}}{CH_3CH{=}CHCOCH_3} \xrightarrow{\text{Ozonolysis}} CH_3CHO + CH_3COCHO$$
$$\downarrow$$
$$CH_3CH + CO$$

Note: α-keto aldehydes lose CO moderately to form aldehydes.

303. (i) $CH_3CH{=}CH_2 + BrCCl_3 \xrightarrow{\text{Peroxide}} CH_3\underset{\underset{Br}{|}}{C}HCH_2CCl_3$

(ii) $CH_3CH{=}\underset{\underset{CH_3}{|}}{C}{-}CH_3 + HCl \xrightarrow{\text{Peroxide}} CH_3{-}CH_2{-}\overset{\overset{CH_3}{|}}{\underset{\underset{Cl}{|}}{C}}{-}CH_3$

304. (C) is β-keto acid and thus, it may be

$$R{-}\underset{\underset{O}{\|}}{C}{-}\underset{\underset{R'}{|}}{C}H{-}COOH$$

(C) on decarboxylation gives 3-pentanone, i.e.,

$$R{-}\underset{\underset{O}{\|}}{C}{-}\underset{\underset{R'}{|}}{C}H{-}COOH \xrightarrow{\text{Sodalime}} R{-}\underset{\underset{O}{\|}}{C}{-}\underset{\underset{R'}{|}}{C}H_2 \text{ or } \underset{\text{3-Pentanone}}{CH_3{-}CH_2{-}\underset{\underset{O}{\|}}{C}{-}\underset{\underset{CH_3}{|}}{C}H_2}$$

Thus R is C_2H_5 and R' is CH_3

∴ (C) is $C_2H_5CO\underset{\underset{CH_3}{|}}{C}HCOOH$

(C) is formed along with ethanol from (*B*) on acid hydrolysis so (*B*) is β-keto ester (ethyl ester).

$$\underset{\text{(B) (Ethyl (2 methyl, 3- ketopentonate)}}{C_2H_5CO{\cdot}\underset{\underset{CH_3}{|}}{C}H{\cdot}COOC_2H_5} \xrightarrow{\text{HOH}} \underset{\text{(C) (2 methyl, 3 keto- pentonic acid)}}{C_2H_5CO\underset{\underset{CH_3}{|}}{C}H{\cdot}COOH} + C_2H_5OH$$

(*B*) is formed as a condensation of two mode of ester (*A*), since the condensation occurs at 'α' C atom i.e., Claisen condensation or Claiser reaction so, the probable structure of (*A*) is

$$\underset{\text{(A) (Ethyl propanonate)}}{CH_3CH_2COOC_2H_5}$$

Thus:

$$CH_3CH_2\underset{\underset{O}{\|}}{C}OC_2H_5 + \overset{\alpha\text{-carbon}\downarrow}{CH_3CH_2COOC_2H_5} \xrightarrow{C_2H_5ONa} \underset{\text{(B)}}{CH_3{-}CH_2{-}\underset{\underset{O}{\|}}{C}{-}\underset{\underset{CH_3}{|}}{C}HCOOC_2H_5}$$

$$+ C_2H_5OH$$

305. HO–(chain with CH$_3$)–CHO (A) $\xrightarrow[\text{Audification}]{\text{Dichronate}}$ HOOC–(chain, H, CH$_3$)–COOH (achiral) (B)

A $\xrightarrow[CH_2Cl_2]{PCC}$ OHC–(chain, CH$_3$, CH$_3$)–CHO (achiral) (C) $\xrightarrow{\text{dil Na, CO}}$ (cyclohexane with CH$_3$, CHO, OH) (D) $\xrightarrow[\Delta]{H_2SO_4}$

(cyclohexene with CH$_3$, CHO) E Chiral $\xrightarrow[\Delta]{H_2/Pt}$ 1,3-dimethyl cyclo hexane

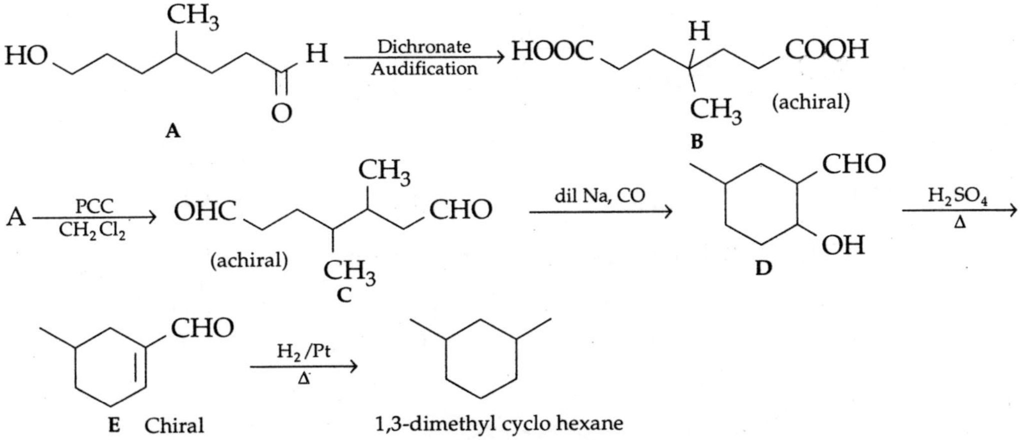

306. The reductive ozonolysis tells us there is a double bond in *A* so an aldehyde & a ketone is produced from the two stereomeric forms of *A*; and *D* and *E* are obtained by catalytic hydrogenation of *B* & *C* which are aldehyde & ketone.

∴ *D* is an alcohol. Here it is n-propyl alcohol, and *E* is isobutyl alcohol.

B & *C* on treatment with HCN followed by hydrolysis will give *F* & *G*.

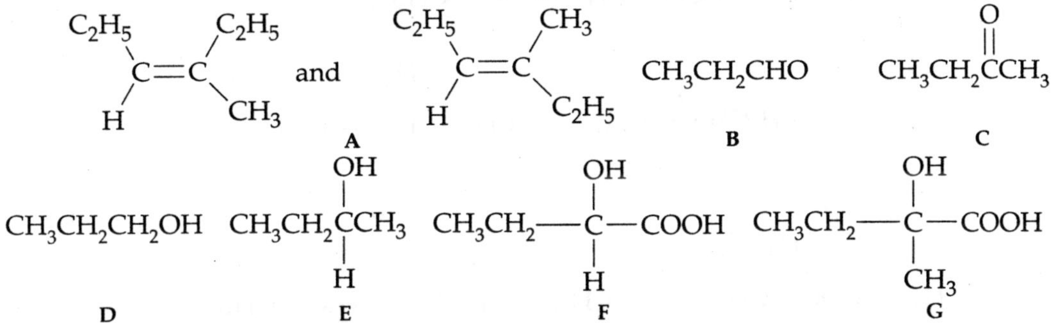

A: $(C_2H_5)(H)C{=}C(C_2H_5)(CH_3)$ and $(C_2H_5)(H)C{=}C(CH_3)(C_2H_5)$

B: CH_3CH_2CHO

C: $CH_3CH_2\overset{O}{\overset{\|}{C}}CH_3$

D: $CH_3CH_2CH_2OH$

E: $CH_3CH_2C(OH)(H)CH_3$

F: $CH_3CH_2{-}C(OH)(H){-}COOH$

G: $CH_3CH_2{-}C(OH)(CH_3){-}COOH$

For Solutions 304 to 334 only answers are given below. The student should do the reactions himself.

307. A: cyclohexene; B: OHC–(CH$_2$)$_4$–CHO; C: cyclopentene–CHO; D: cyclopentyl–CH$_2$OH; E: cyclopentyl–COOH; F: cyclopentane

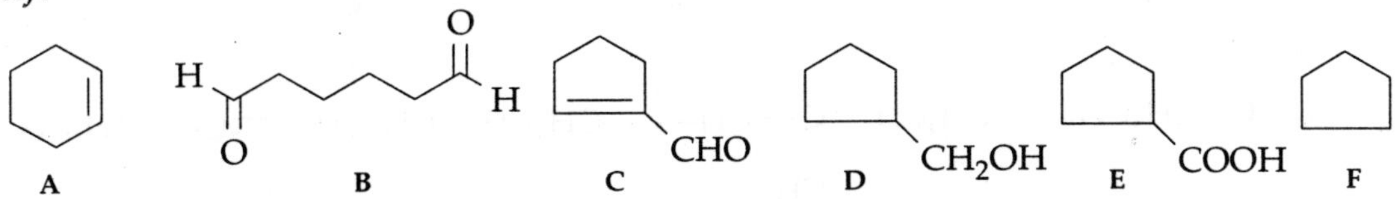

308. A: $Ph{-}\overset{O}{\overset{\|}{C}}{-}CH(CH_3)(C_2H_5)$

B: $(Ph)(C_2H_5{-}CH(CH_3))C{=}N{-}OH$

C: $(Ph)(C_2H_5{-}CH(CH_3))C{=}N{-}OH$

D: $Ph{-}\overset{O}{\overset{\|}{C}}{-}NH{-}CH(CH_3){-}C_2H_5$

E: $C_2H_5{-}CH(CH_3){-}\overset{O}{\overset{\|}{C}}{-}NHPh$

F: $CH_3{-}CH(C_2H_5){-}NH_2$

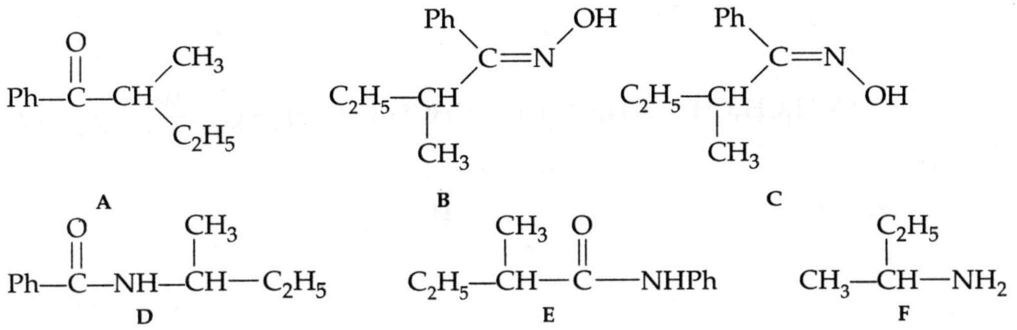

309. O, Cl

A B C D E

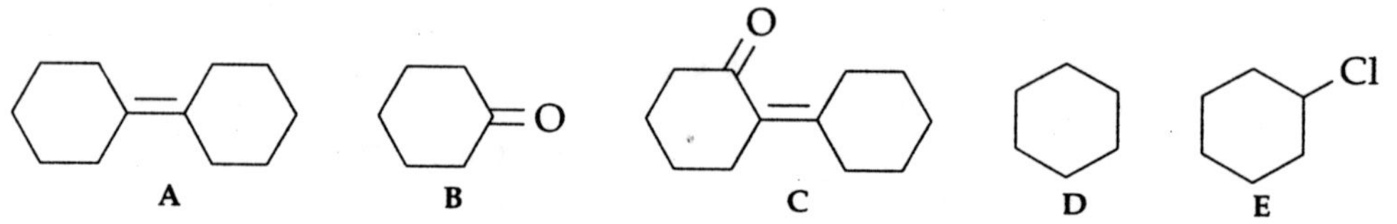

310. $CH_3C(CH_3)=CHCH(CH_3)CH_3$; $CH_3CH(CH_3)COOH$; $CH_3C(CH_3)(Br)COOH$ $CH_3C(CH_3)(OH)COOH$

A B C D

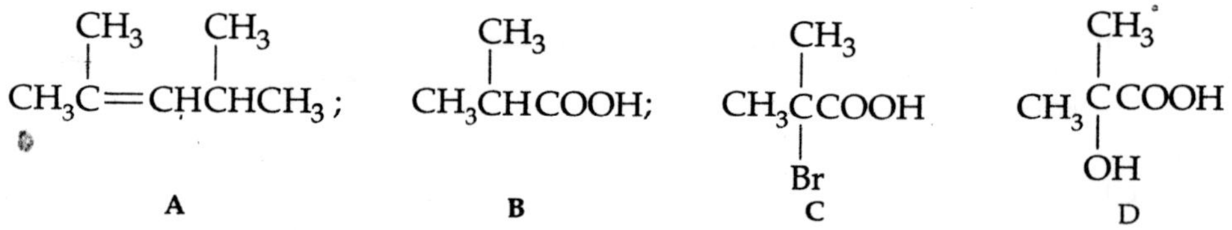

311. Cl, Cl, CHO

A B C D

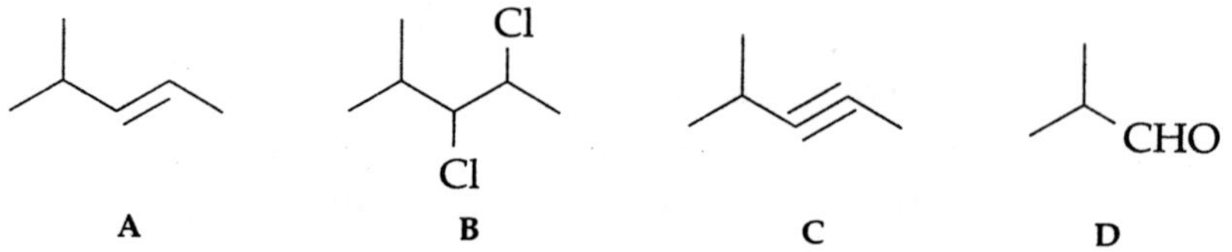

312. $CH_3-CH(CH_3)-C\equiv CH$ $CH_3-CH(CH_3)-C\equiv C-CH_2CH_2CH_3$

A B

$CH_3-CH(CH_3)-C(=O)-CH_2CH_2CH_2CH_3$ $CH_3-CH(CH_3)-COOH$ $CH_3CH_2CH_2COOH$

C D E

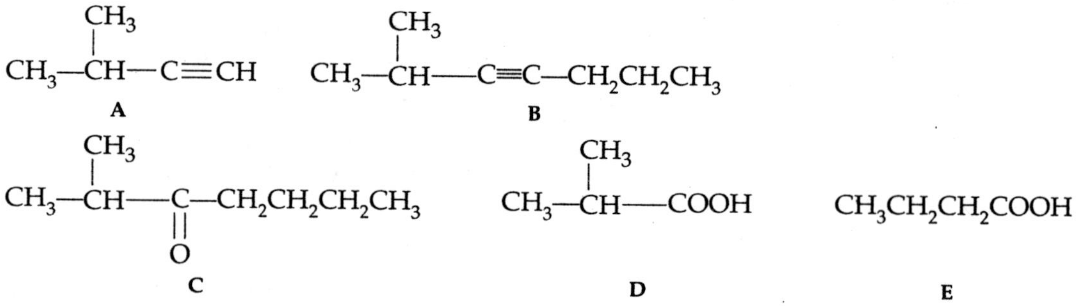

313. $CH_3CH_2CH_2COCH(CH_3)\ CH_2CH_3$ (A) $CH_3CH_2CHCH_3$ with OH (B) $CH_3CH_2CH_2CH_2OH$ (C)

A B C

$CH_3-C(=O)-CH_2CH_3$ $CH_3CH_2-C(CH_3)-CH-C(=O)-CH_2CH_3$ $CH_3CH_2C(CH_3)=CH-CH_2CH_2CH_3$

D E F

$CHCH_2CH(CH_3)=C(OH)HCH_2CH_2CH_3$ $CH_3CH_2CH(CH_3)\ C(=O)CHCH_2CH_2CH_3 \xrightarrow[\text{Bayervillgeroxidtion}]{\text{Mepba}} A$

G H

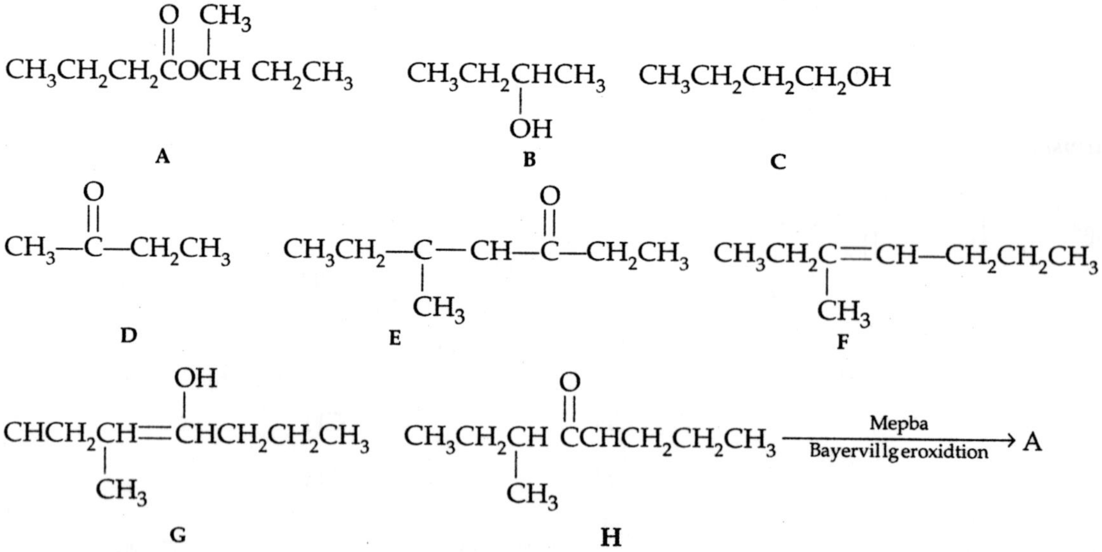

314.

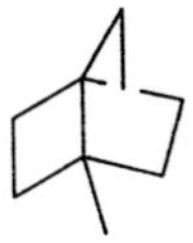

A

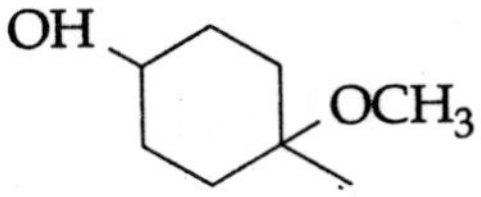

OH, OCH_3

B

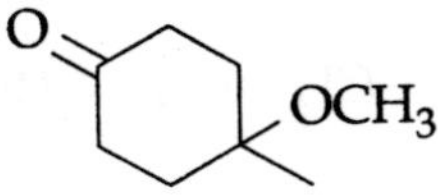

O, OCH_3

C

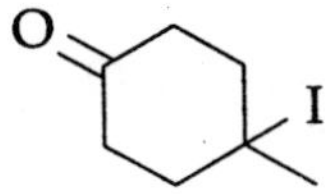

O, I

D

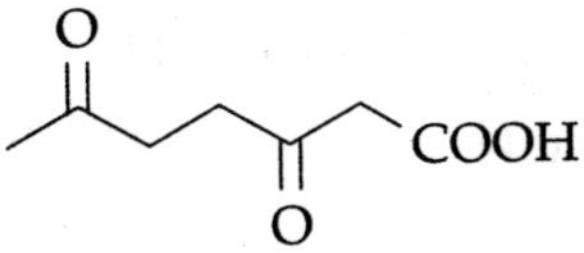

O, COOH, O

E

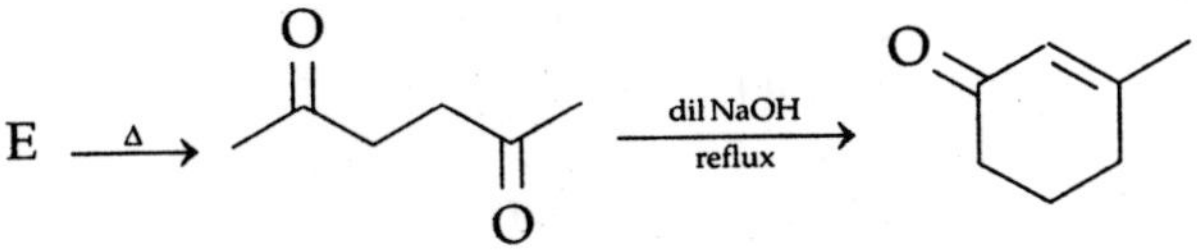

E $\xrightarrow{\Delta}$ O, O $\xrightarrow[\text{reflux}]{\text{dil NaOH}}$ O

315.

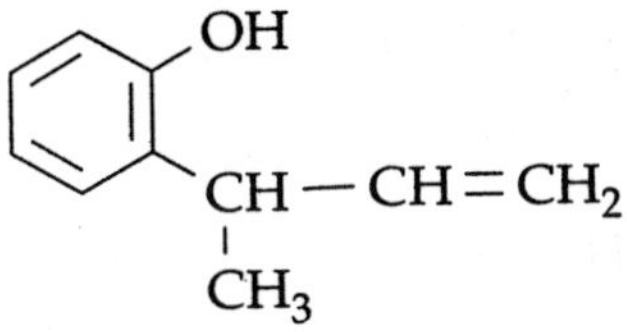

OH, $CH-CH=CH_2$, CH_3

A

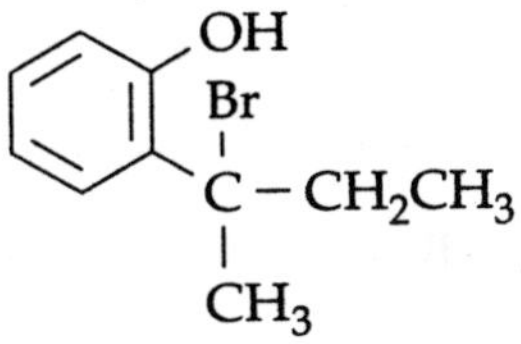

OH, Br, $C-CH_2CH_3$, CH_3

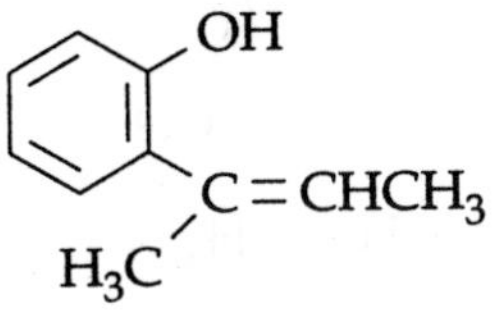

B

OH, $C=CHCH_3$, H_3C

C

HO, OH, CH_2, CH_2, CH, CH_3

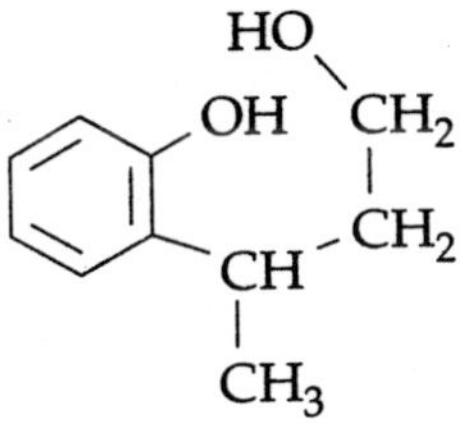

D

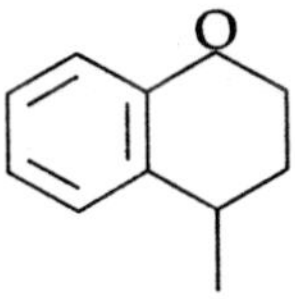

O

E

316.

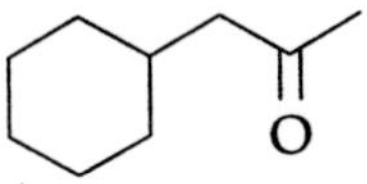

O

A

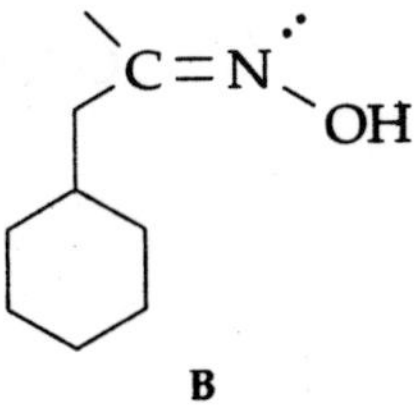

$C=N$, OH

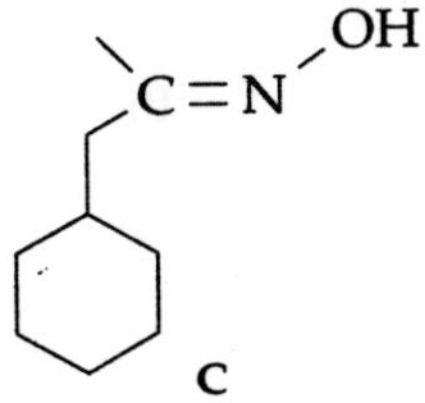

B

OH, $C=N$

C

$NHCH_3$, C, O

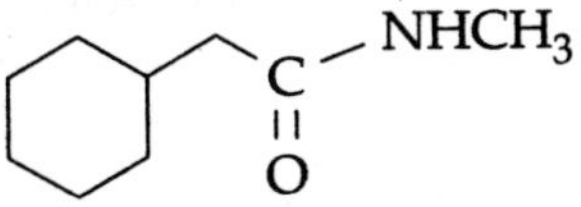

D

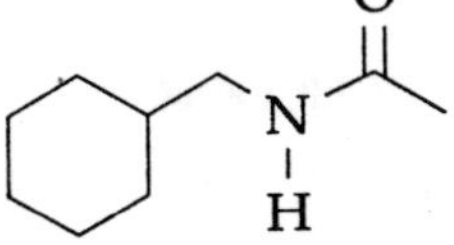

O, N, H

E

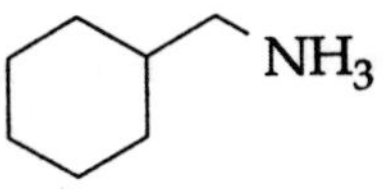

NH_3

F

COOH

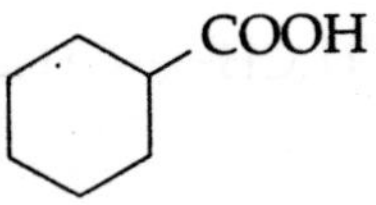

G

317.

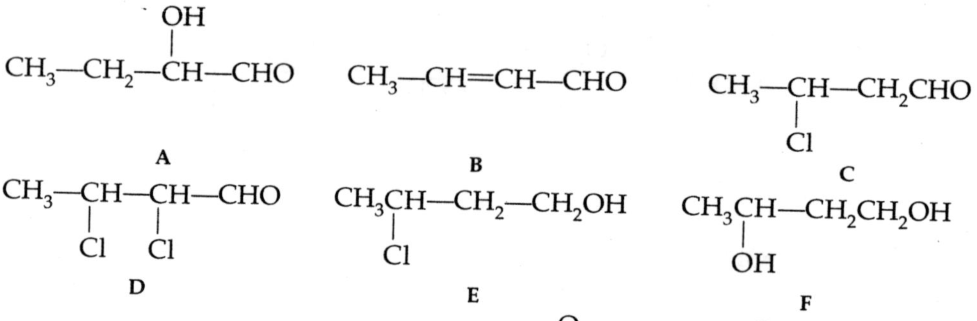

318.

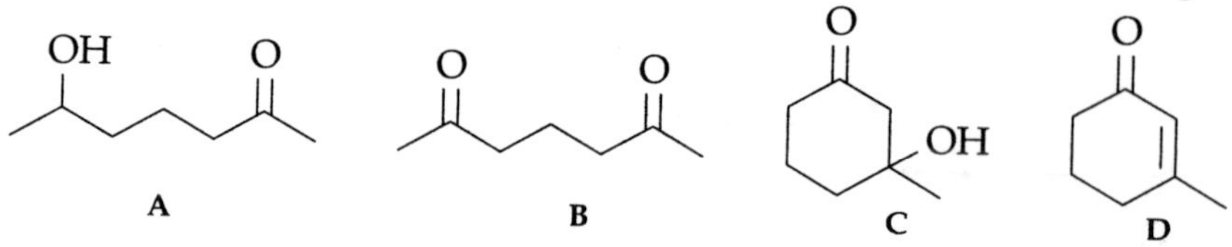

319.

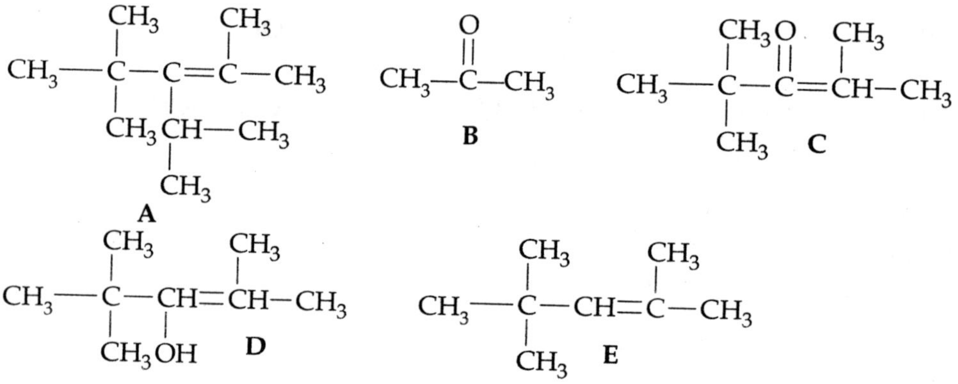

320. A: 2-methylmalonic acid; B: Propanoic acid

321.

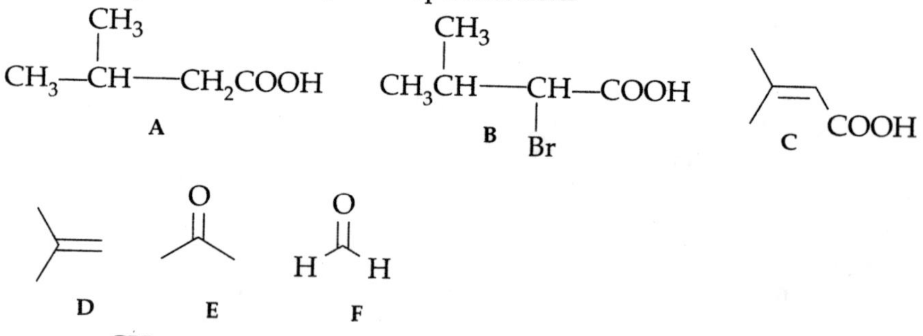

322.

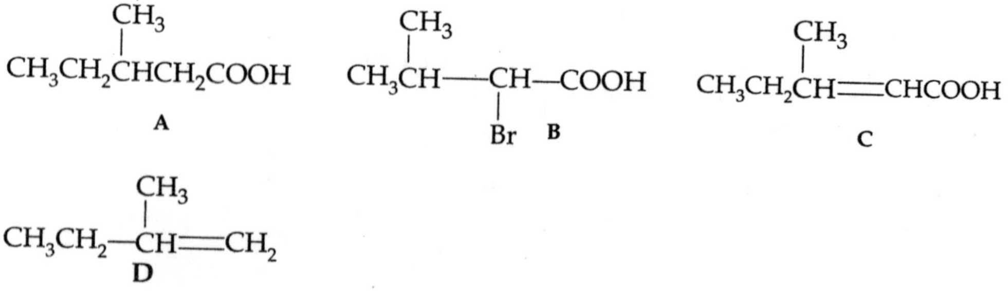

323.

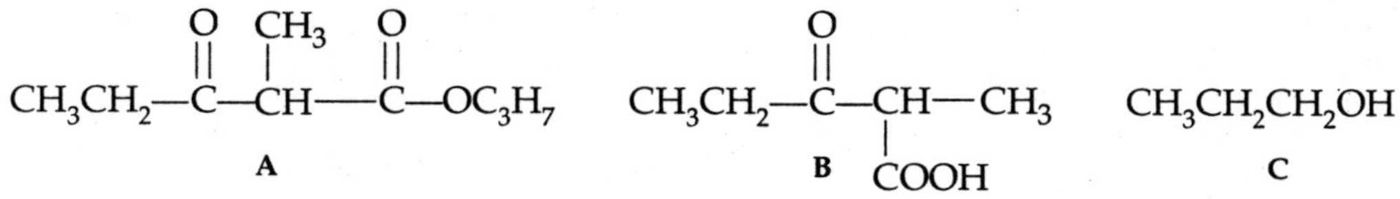

$$CH_3CH_2-\overset{\overset{O}{\|}}{C}-\overset{\overset{CH_3}{|}}{CH}-\overset{\overset{O}{\|}}{C}-OC_3H_7$$
A

$$CH_3CH_2-\overset{\overset{O}{\|}}{C}-\underset{\underset{COOH}{|}}{CH}-CH_3$$
B

$CH_3CH_2CH_2OH$
C

CH_3CH_2CHO
D

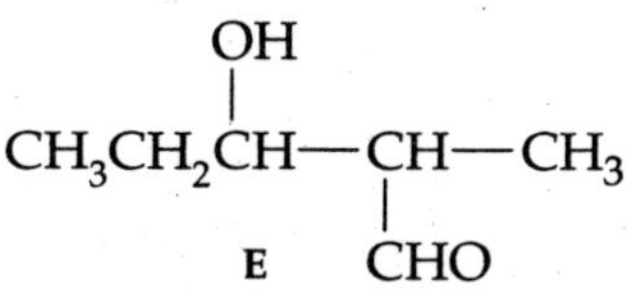

$$CH_3CH_2\overset{\overset{OH}{|}}{CH}-\underset{\underset{CHO}{|}}{CH}-CH_3$$
E

324.

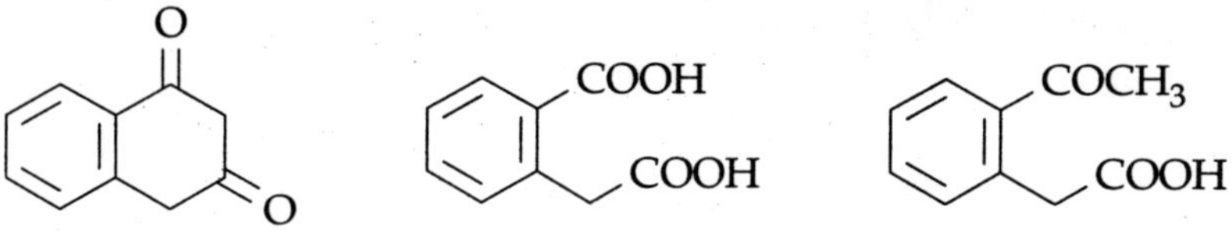

O, O
A

COOH, COOH
B

$COCH_3$, COOH
C

325.

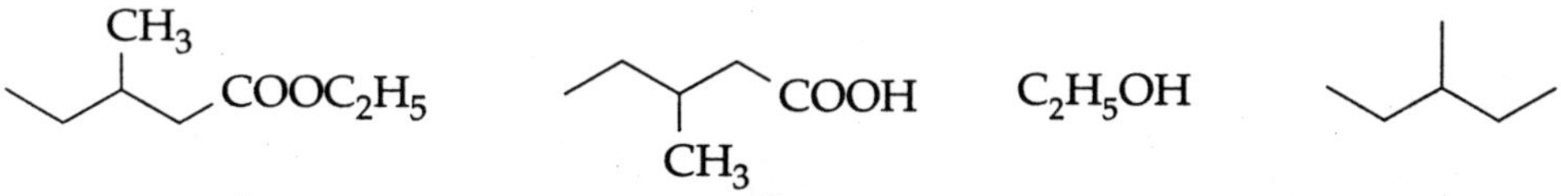

CH_3, $COOC_2H_5$
A

COOH, CH_3
B

C_2H_5OH
C

D

326.

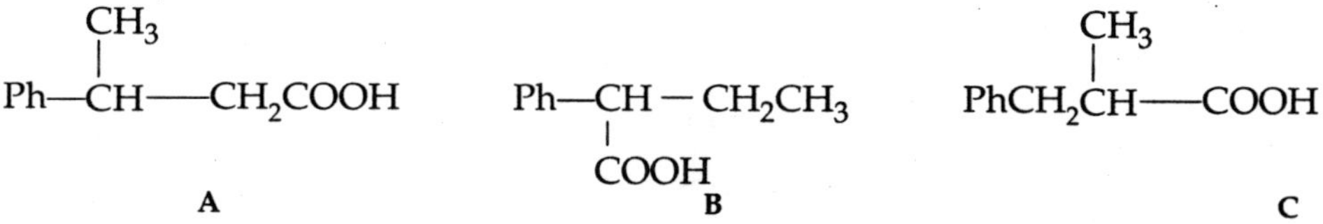

$$Ph-\overset{\overset{CH_3}{|}}{CH}-CH_2COOH$$
A

$$Ph-\underset{\underset{COOH}{|}}{CH}-CH_2CH_3$$
B

$$PhCH_2\overset{\overset{CH_3}{|}}{CH}-COOH$$
C

$PhCH_2CH_2CH_3$
D

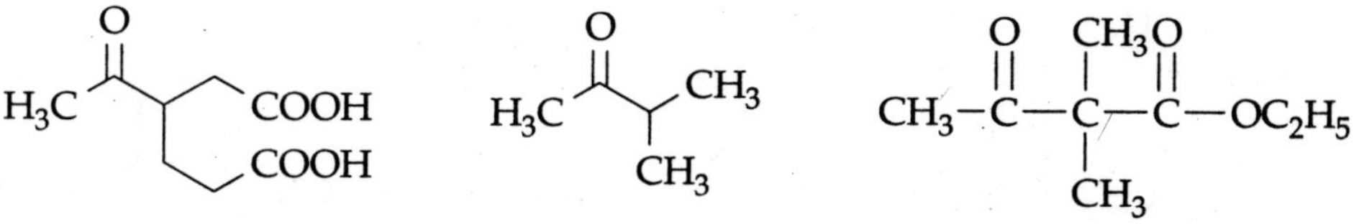

$PhCH(CH_3)_2$
E

327.

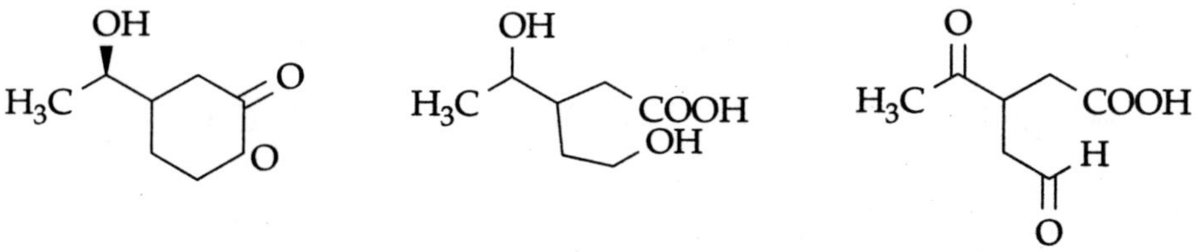

OH, H_3C, O, O
A

OH, H_3C, COOH, OH
B

O, H_3C, COOH, H, O
C

O, H_3C, COOH, COOH
D

O, H_3C, CH_3, CH_3
E

$$CH_3-\overset{\overset{O}{\|}}{C}-\underset{\underset{CH_3}{|}}{\overset{\overset{CH_3}{|}}{C}}-\overset{\overset{O}{\|}}{C}-OC_2H_5$$
F

328. **A:** $C_6H_5COOCH_2CH_3$; **B:** CH_3CH_2OH; **C:** CH_3CHO; **D:** C_6H_5COOH.

329. A: $CH(COOCH_3)=CH-COOCH(CH_3)-C_6H_5$

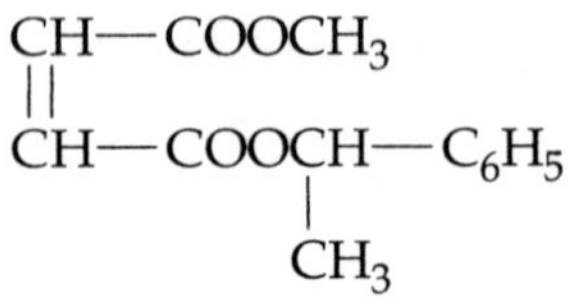

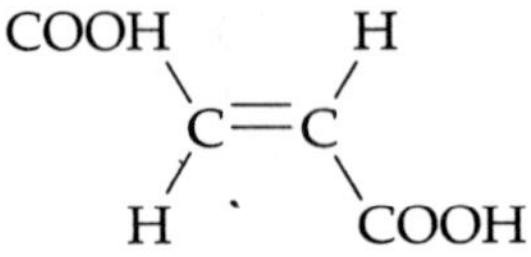

B: $HOOC(H)C=C(H)COOH$ (trans)

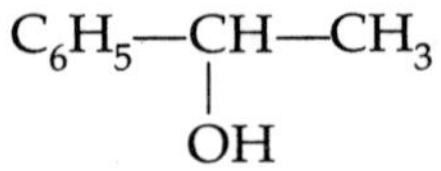

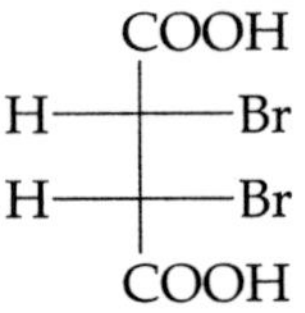

C: $C_6H_5-CH(OH)-CH_3$

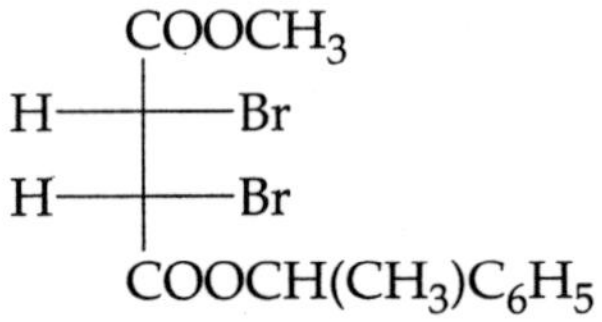

D: $HOOC-CH(Br)-CH(Br)-COOH$ (Fischer projection: H–Br, H–Br)

(E and F): $CH_3OOC-CH(Br)-CH(Br)-COOCH(CH_3)C_6H_5$ (Fischer projection: H–Br, H–Br)

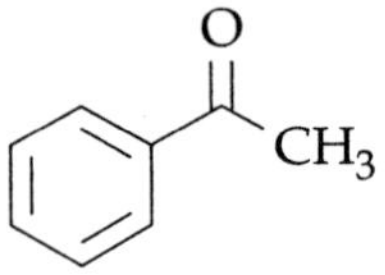

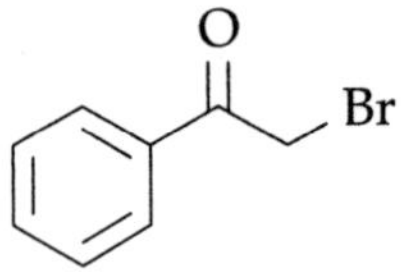

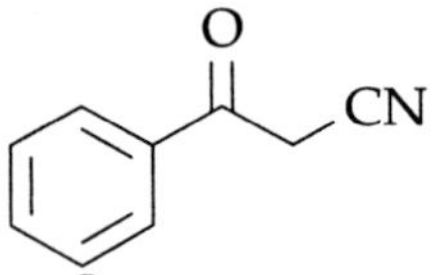

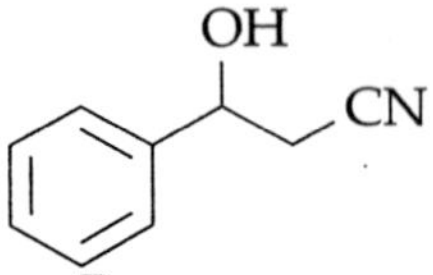

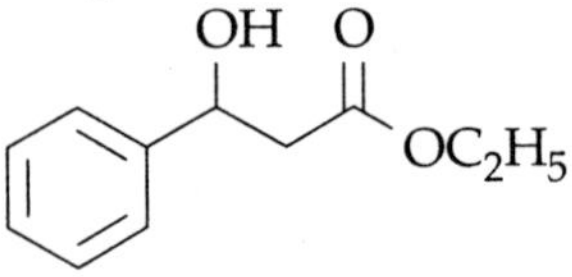

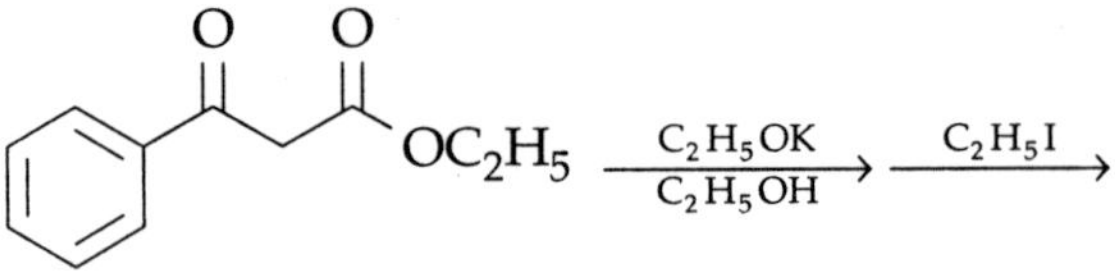

330. A: $C_6H_5COCH_3$ B: $C_6H_5COCH_2Br$ C: $C_6H_5COCH_2CN$ D: $C_6H_5CH(OH)CH_2CN$

E: $C_6H_5CH(OH)CH_2COOC_2H_5$

F: $C_6H_5COCH_2COOC_2H_5 \xrightarrow[C_2H_5OH]{C_2H_5OK} \xrightarrow{C_2H_5I}$

$C_6H_5COCH(C_2H_5)COOC_2H_5 \xrightarrow[\Delta]{H^+} C_6H_5COCH_2CH_2CH_3$

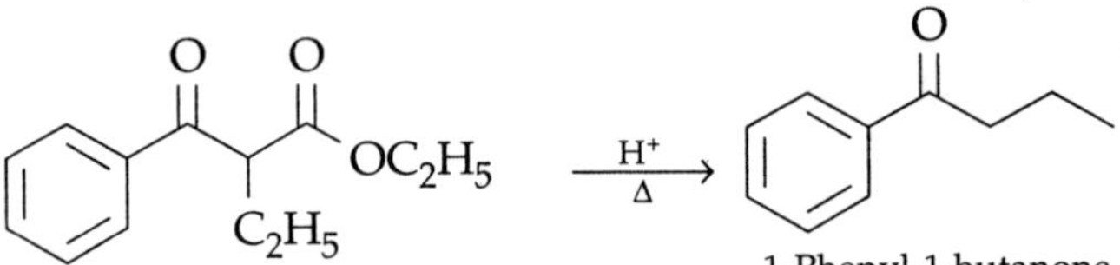

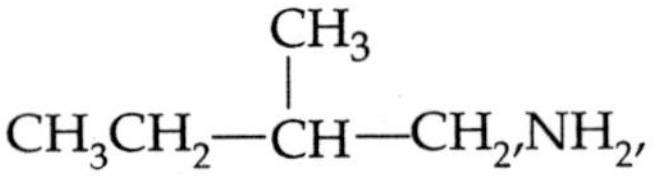

1-Phenyl-1-butanone

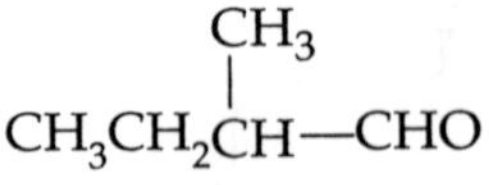

331. A: $CH_3CH_2-CH(CH_3)-CH_2NH_2$, B: $CH_3CH_2CH(CH_3)-CH_2OH$, C: $CH_3CH_2CH(CH_3)-CHO$

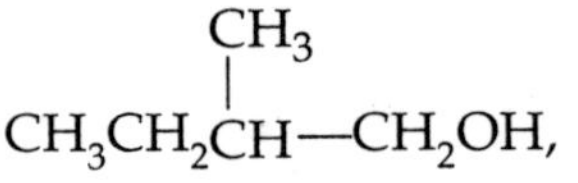

D: $CH_3CH_2CH(CH_3)_2$

332. $H_2N-C_6H_4-CH_3$ (para)

333. A: $H_3C-CH(OC_2H_5)-C_6H_4-CH_2OH$ (meta) B: $H_3C-CH(I)-C_6H_4-CH_2I$ (meta) C: $H_3C-CH(OH)-C_6H_4-CH_2OH$ (meta)

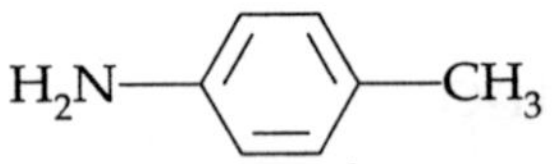

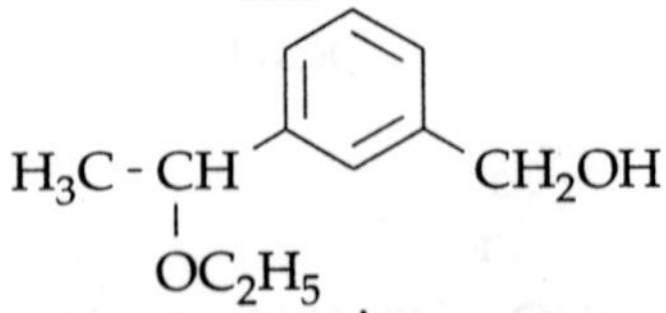

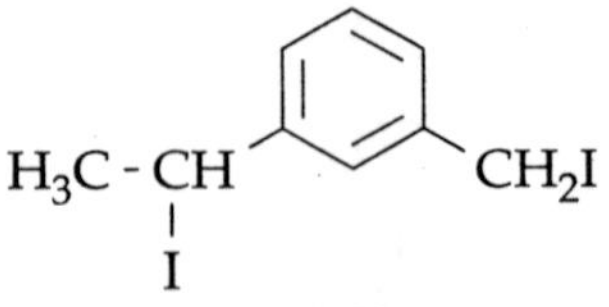

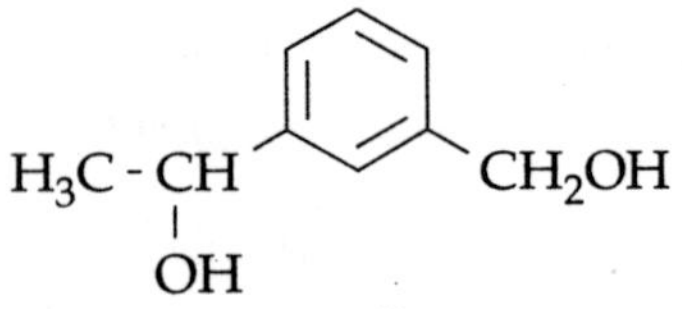

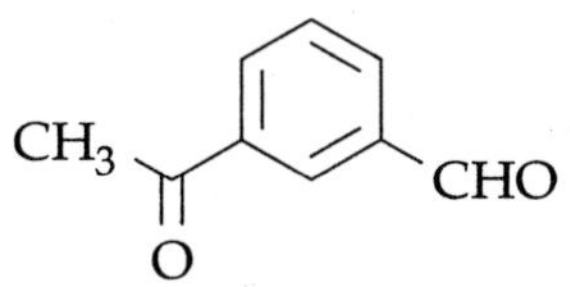

Cl

CH_3 CHO CH_3 CHO

O O

D 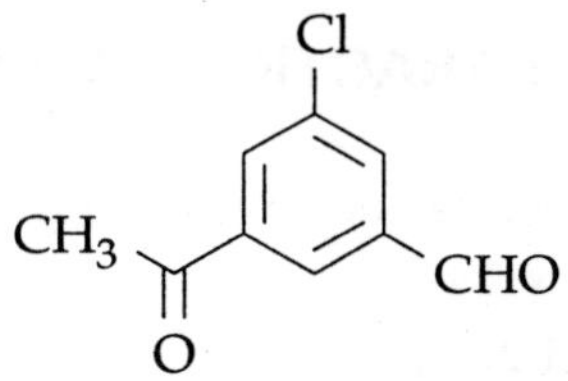 **E**

334. C_2H_5 C_2H_5 SO_3H C_2H_5 C_2H_5 CN C_2H_5 COOH

SO_3H

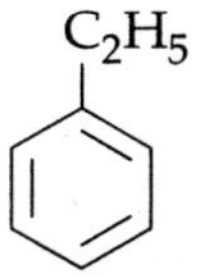

A 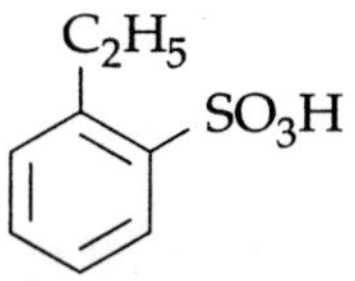 **B** 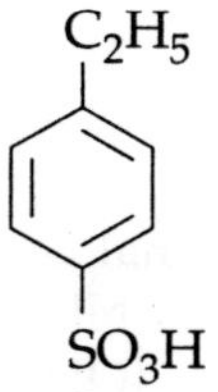 **C** 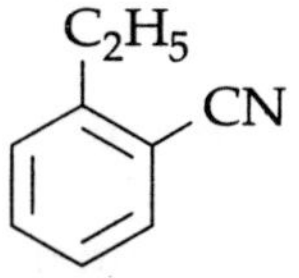 **D** 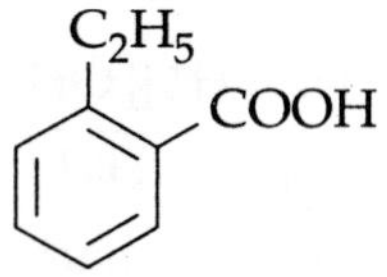 **E**

COOH O

COOH O

O

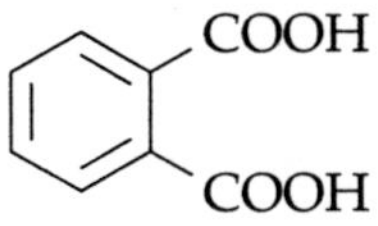

F 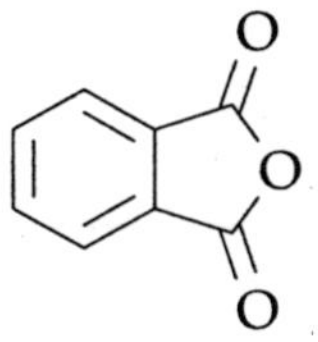 **G**

Chapter 15 : PRACTICE PAPER I

1. Given $C_2H_5NH_2(aq) + H_2O(l) = C_2H_5NH_3^+(aq) + OH^-(aq)$ ..... (i)

K_b for $C_2H_3NH_2 = 5.6 \times 10^{-4}$

(a) Given 65.987 mL of 0.25 M $C_2H_5NH_2$

To find $[OH^-]$

The base-dissociation constant for eq. (i) is :

$$K_b = \frac{[C_2H_5NH_3^+][OH^-]}{[C_2H_5NH_2]} = 5.6 \times 10^{-4}$$

Now create a chart showing initial and final concentrations (at equilibrium) of the involved species. Let x be the amount of $C_2H_5NH_3^+$ that forms from $C_2H_5NH_2$. Because $C_2H_5NH_3^+$ is an a 1:1 molar ratio with OH^-, $[OH^-]$ also equals x.

Species	*Initial Concentration*	*Final Concentration (at equilibrium)*
$C_2H_5NH_2$	0.25 M	$0.25 - x$
$C_2H_5NH_3^+$	0 M	x
OH^-	0 M	x

$$\therefore\ K_b = \frac{[C_2H_5NH_3^+][OH^-]}{C_2H_5NH_2} = 5.6 \times 10^{-4} = \frac{(x)(x)}{0.25 - x}$$

You have a choice in solving for x. The first method would require the quadratic equation—not a good idea because compared to the magnitude of 0.25, the value of x is negligible. If you used to quadratic, you would be wasting time. The second method would assume that $[C_2H_5NH_2]$ remains constant at 0.25 M; $5.6 \times 10^{-4} = x^2/0.25$.

$\Rightarrow x = [OH^-] = 0.012$ M

By the way, the 65.987 mL is not needed because concentration is independent of the amount of solution measured.

(b) To find P^{OH} of solution.

$P^{OH} = -\log [OH^-]$

$P^{OH} = -\log [0.012] = 1.92$

(c) To find % ionization of ethylamine

$$\% = \frac{\text{part}}{\text{whole}} \times 100\% = \frac{0.012}{0.25} \times 100\% = 4.8\%$$

(d) Given: 15 g $C_2H_5NH_3Br$ + 250 mL 0.1 M $C_2H_5NH_2$

To Find P^H of solution

Note that when $C_2H_5NH_2Br$ dissolves in water, it dissociates into $C_2H_5NH_3^+$ and Br^- Furthemore, $C_2H_5NH_3^+$ is a weak acid.

Rewrite the balanced equation at equilibrium for the reaction

$$C_2H_5NH_3^+ \rightleftharpoons C_2H_5NH_2 + H^+$$

$\therefore$ equilibrium expression is:

$$K_a = K_w/K_b = \frac{[C_2H_5NH_2][H^+]}{[C_2H_5NH_3^+]} = \frac{10^{-14}}{5.6\times10^{-4}} = 1.8 \times 10^{-11}$$

Calculate the initial concentrations of the species of interest.
$[C_2H_5NH_4]$= 0.1 M (given)

$$[C_2H_5NH_3^+] = \frac{15\text{ g } C_2H_5NH_3Br}{1} \times \frac{1\text{ mole } C_2H_5NH_3^+}{126.05\text{ g } H_5NH_3Br} \times \frac{1}{0.25\text{ liter}} = 0.476\text{ M}$$

$[H^+] = 0$

Species	*Initial Concentration*	*Final Concentration (at equilibrium)*
$C_2H_5NH_2$	0.1 M	$0.1 + x$
$C_2H_5NH_3^+$	0.476 M	$0.476 - x$
H^+	0 M	x

$$\therefore \frac{(0.1+x)}{(0.476-x)} = 1.8 \times 10^{-4}$$

$$\Rightarrow x = (H^+) = 8.57 \times 10^{-11}$$

$$\therefore P^H = \log(8.57 \times 10^{-11}) = 10.07$$

(e) Given: 0.125 g AgBr(s) with $AgNO_2(s)$

K_{sp} AgOH = 1.52×10^{-4}

To find Will AgOH precipitate?

The equation in equilibrium for the dissociation of AgOH is,.

$AgOH(s) \rightleftharpoons Ag^+(aq) + OH^-(aq)$

and, the concentration of the ion present is:

$$[Ag^+] = \frac{0.125\text{ g AgBr}}{0.065987\text{ liter sol'n}} \times \frac{1\text{ mole AgBr}}{187.772\text{ g AgBr}} \times \frac{1\text{ mole } Ag^+}{1\text{ mole AgBr}} = 0.0101\text{ M}$$

$[OH^-] = 0.012$ M

So, the ion product, Q will be

$Q = [Ag^+][OH^-] = (0.0101)(0.012) = 1.32 \times 10^{-4}$

$\therefore K_{sp}$= AgOH = 1.52×10^{-8}

Since $Q > K_{sp}$, AgOH will precipitate

2. Given: H_2O + 2.51 g $SbCl_3 \rightarrow$ 1.906 g $Sb_xO_yCl_z$
0.802 g gas, 97.20% Cl, and 2.75% H

(a) To find, Simplest formula for gas.

If you had 100, grams of the gas, 97.2 grams would be due to the weigh of chlorine atoms, and 2.75 grams would be due to the weight of hydrogen atoms.

97.2 grams Cl/(1 mole Cl/35.453 g mole^{-1}) = 2.742 moles Cl

2.75 grams H/(1 mole H/1.00794 g mole^{-1}) = 2.73 moles H

Because this is essentially a 1;1 molar ratio, the empirical formula of the gas is HCl.

(b) To find Fraction of chlorine in the solid product and in the gas phase.

1. The mass of chlorine in the original compound:

$$2.51 \text{ g of } SbCl_3 \times 106.36 \text{ g Cl}/228.10 \text{ g } SbCl_3 = 1.17 \text{ g Cl}$$

2. Fraction of chlorine in the gas.

According to the Law of Conservation of Mass, if you have 1.17 grams of chlorine in th original compound, you must account for 1.17 grams of chlorine on the product side.

Because the gas is 97.2% by mass chlorine, the fraction of chlorine in the gas can be found as follows:

$$\frac{\text{part}}{\text{whole}} \times 100\% = \frac{0802 \text{ g gas } (0.972)}{1.17 \text{ g total chlorine}} \times 100\% = 66.6\% \text{ gas}$$

fraction of chlorine in the solid product = 100% – 66.6% =33.4% solid

(c) To find: Formula of solid product.

You know form the question that the solid product contains Sb, Cl, and O atoms. the weight of Sb can be found by taking the weight of antimony chloride (which has all of the antimony atoms in it) and getting rid of the weight of chlorine atoms, which you have determined to be 1.17 g. Therefore, 2.51 g $SbCl_3$ – 1.17 grams chlorine atoms = 1.34 grams of antimony.

The weight of chlorine in the solid product can be determined by taking the weight of chlorine in the original compound (which has all of the chlorine atoms in it) and multiplying it by the percent of chlorine found in the solid product. This becomes 1.17 grams of chlorine × 0.334 = 0.391 g chlorine in the solid product.

The weight of oxygen in the solid product can be found by taking the total weight of the solid product and subtracting the amount of antimony and chlorine previously determined. The becomes 1.906 g solid product –1.34 g antimony –0.391 g chlorine = 0.175 g oxygen atoms in the solid product.

Expressing these weight as moles yields.

1.34 g Sb → 0.011 mole Sb

0.175 g O → 0.0109 mole O

0.391 g Cl → 0.011 mole Cl

Thus they are in essentially a 1:1:1 molar ratio, which indicates the molecular formula SbOCl.

(d) Given: Empirical formula is true formula.

Balanced equation is:

$$SbCl_3(s) + H_2O(l) \rightarrow SbOCl(s) + 2\ HCl(g)$$

3. Given: $CH_3OH(aq) + O_2(g) \rightarrow HCOOH(aq) + H_2O(l)$. From the data given in the table,

(a) To find: $\Delta H°$ for the oxidation of methyl alcohol.

$\Delta H_f^\circ HCOOH(aq)$= –409 kJ mole^{-1}

$\Delta H_f^\circ H_2O(l)$= 285.84 kJ mole^{-1}

$\Delta H_f^\circ CH_3OH(aq)$= –238.6 kJ mole^{-1}

$\Delta H° = \Sigma\Delta H_f^\circ - \Sigma\Delta H_f^\circ = (\Delta H_f^\circ HCOOH + \Delta H_f^\circ H_2O) - (\Delta H_f^\circ CH_3OH)$

= [(–409 kJ mole^{-1}) + (–285.84 kJ mole^{-1})] – (238.6 kJ mole^{-1}) = –456 kJ mole^{-1}

(b) To find: $\Delta S°$ for the oxidation of methyl alcohol.

$S°$ HCOOH(*aq*) = 129 J mole^{-1} · K

$S°\ H_2O\ (l)$= 69.94 J mole^{-1} · K

$S°\ CH_3OH(aq)$= 127.0 J mole^{-1} · K

$S°\ O_2(g)$= 205.0 J mole^{-1} · K

$\Delta S° = \Sigma S°_{products} - \Sigma S°_{reactants} = (S°\ HCOOH + S°\ H_2O) - (S°\ CH_3OH + S°O_2)$

$= (129\ J\ K^{-1} \cdot mole^{-1} + 69.94\ J\ K^{-1} \cdot mole^{-1}) - (127.0\ J\ K^{-1} \cdot mole^{-1} + 205.0\ J\ K^{-1} \cdot mole^{-1})$

$= -133\ J\ K^{-1} \cdot mole^{-1}$

(c) (1) To find: Is the reaction spontaneous at 298 K, we have,

$\Delta G° = \Delta H° - T\Delta S = -456$ kJ mole^{-1} – 298 K (–0.133 kJ mole · K) = –416 kJ mole^{-1}

Reaction is spontaneous because $\Delta G°$ is negative.

(2) Restatement: If the temperature isincreased to 373 K, will the reaction be spontaneous? We have,

$\Delta G = \Delta H - T\Delta S = -456$ kJ mole^{-1} – 373 K (–0.133 kJ mole^{-1} · K) = –406. kJ mole^{-1}

Reaction is spontaneous because ΔG is still negative.

(d) Given: ΔH_{fas} HCOOH = 12.71 kJ mole at 281.3 K

To find: Calculate ΔS for the reaction

$$HCOOH(l) \rightarrow HCOOH(s)$$

The temperature at which liquid HCOOH converts to solid HCOOH is known as the freezing point, it is also the melting point. Because at this particular temperature a state of equilibrium exists—that is, HCOOH(l) $\leftrightarrow$ HCOOH(s)—you can set ΔG = 0. Substituting into the ibbs. Helmholtz equation yields.

$\Delta G = \Delta H - T\Delta S$

$0 = 12.71$ kJ mole^{-1} – 281.3 K(ΔS)

(Did you remember to make 12.71 negative, because you want ΔH for freezing, which is on exothermic process?)

$$\Delta S = \frac{-12.71\ kJ\ mole^{-1}}{281.3\ K} = -0.04518\ kJ\ mole^{-1} \cdot K^{-1} = -45.18\ J\ mole^{-1}\ K^{-1}$$

(e) Given: $S°$ HCOOH(l) = 109.1 J mole · K^{-1}

To find: The standard molar entropy of HCOOH(s)?, we have,

HCOOH(l) → HCOOH(s)

$\Delta S° = \Sigma S°_{products} - \Sigma S°_{reactants} = S°\ HCOOH(s) - S°HCOOH(l)$

or, –45.18 J mole^{-1} · K^{-1} = $S°$ HCOOH(s) – 109.1 J mole^{-1} · K^{-1}

$\therefore\ S°$ HCOOH(s) = 63.9 J mole^{-1} · K^{-1}

(f) Given: ($K_b = 1.9 \times 10^{-4}$)

$\Delta G°$ for the ionization of methanoic (formic) acid at 298 K.

$\Delta G° = -2.303\ R \cdot T \log K_a$ Here, R = 8.314 J K^{-1}

$= -2.303\ (8.314\ J\ K^{-1}) \cdot 298\ K\ (-3.72) = 2.1 \times 10^4 \approx 21$ kJ

4. Here, Give a formula for each reaction, showing the reactions and products.

(a) I. $Sn + Cl_2 \rightarrow SnCl_4$

(Usually pick the higher oxidation state of the metal ion.)

(b) II. $C_2H_6 + O_2 \rightarrow CO_2 + H_2O$

All hydrocarbons burn in oxygen gas to produce CO_2 and H_2O ("Air" almost always means oxygen gas.) Note the use of the word "completely". Unless this word was in the problem, a mixture of CO and CO_2 gases would result.

(c) IXCu + H^+ + NO^{3-} → Cu^{2+} + H_2O + NO

This reaction is well known and is covered quite extensively in textbooks. Note how it departs from the rubric. Copper metal does not react directly with H^+ ions, since it has a negative standard oxidation voltage. However, it will react with 6M HNO_3 because the ion is a much stronger oxidizing agent than H^+. The fact that copper metal is difficult to oxidize indicates that s readily reduced. This fact allows one to qualitatively test for the presence of by reacting it with dithionite (hydrosulfite) ion, as the reducing agent to produce copper.

$$Cu^{2+}_{(aq)} + S_2O_4^{2-} + 2H_2O \rightarrow Cu + 2SO_3^{2-} + 4H^+$$

(d) $SO_4^{2-} + Hg^{2+} \rightarrow HgSO_4$

(e) $SO_3 + CaO \rightarrow CaSO_4$

(f) $CuSO_4 \cdot 5H_2O \rightarrow CuSO_4 + 5\ H_2O$

(g) $Zn(OH)_2 + NH_3 \rightarrow Zn\ (NH_3)_4^{2+} + OH^-$

(h) $C_2H_6 + Br_2 \rightarrow C_2H_2r + HBr$

5. Here: Explain each of the following:

(a) Water can act as either an acid or a base.

(i) Water can provide both H^+ and OH^-

$$H_2O \rightleftharpoons H^+ + OH^-$$

(ii) According to Brønsted-Lowry theory, a water molecule can accept a proton, thereby becoming hydronium ion. In this case, water is acting as a base (proton acceptor)

$$H_2O + H^+ \rightarrow H_3O^+$$

(ii) According to Lewis theory, water can act as a Lewis base (electron pair donor). Water contains an unshared pair of electrons that is utilizd in accepting a proton to form the hydronium ion.

$$\underset{\text{base}}{H_2O} + \underset{\text{acid}}{H_2O} \rightleftharpoons \underset{\text{conjugate base}}{OH^-} + \underset{\text{conjugate acid}}{H_3O^+}$$

(iii) When water acts as a Brøonsted-Lowry acid, it donates a proton to another species, thereby converting to the hydroxide ion.

$$\text{H}\ddot{\underset{\cdot\cdot}{\text{O}}}\text{:H} + H^+ \longrightarrow \left[\text{H:}\ddot{\underset{\cdot\cdot}{\text{O}}}\text{:H}\,\text{(H)}\right]$$

(b) HF is a weaker acid than HCl.

(i) Fluorine is more electronegative than Cl.

(ii) The bond between electronegative than Cl.

(iii) Acid strength is measured in terms of how easy it is for the H to ionize. The stronger the acid, the weaker the bond between the H atom and the rest of the acid molecule, measured as K_a or, if the acid is polyprotic, $K_{a'}$ $K_{a2'}$ $K_{a3'}$ ...

(c) For the tripotic acid H_3PO_4, K_{a1} is 7.5×10^{-3}, whereas K_{a2} is 6.2×10^{-8}.

(i) K_{a1} represents the first hydrogen to depart the H_2PO_4 molecule, leaving the conjugate base $H_2PO_4^-$.

(ii) The conjugate base, $H_2PO_4^-$, has an overall negative charge.

(iii) The overall negative charge of the $H_2PO_4^-$ species increases the attraction of its own conjugate base HPO_4^{2-} to the departing proton. This creates a *stronger* bond which indicates that it is a weaker acid.

(d) Pure HCl is not an acid.

(i) An acid is measured by its concentration of H^+ (its pH)

(ii) Pure HCl would not ionize, the sample would remain as molecular HCl (a gas).

(iii) In order to ionize, a water solution of HCl is required:

$$HCl(aq) \rightarrow H^+(aq) + Cl^-(aq)$$

(e) $HClO_4$ is a stronger acid than $HClO_3$, HSO_3^-, or H_2SO_3

(i) As the number of ione oxygen atoms (those not bonded to H) increases, the strength of the acid increases. Thus, $HClO_4$ is a stronger acid than $HClO_3$.

(ii) As electronegativity of central atom increases, acid strength increases,. Thus, Cl is more electronegative than S.

(iii) Loss of H^+ by a neutral acid molecule (H_2SO_3) reduces acid strength. Thus H_2SO_3 is a stronger acid than HSO_3^-

(iv) as effective nuclear charge (Z_{eff}) on the central atom increases, acid strength is likewise increases. Thus, a larger nuclear charge draws the electrons closer to the nuclus and binds them more tightly.

6. Interpret using bonding principles

(a) Compare carbon-to-carbon bod lengths in C_2H_2 and C_2H_6.

(i) Lewis structure of C_2H_2;

H—C≡C—H

(iii) Lewis structure of C_2H_6;

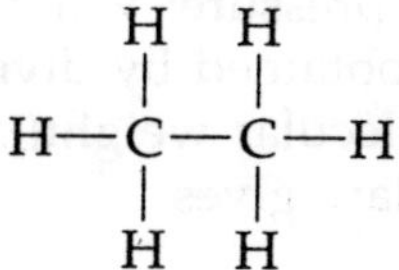

(iii) C_2H_2 has a triple bond, whereas C_2H_6 consists only of single bonds.

(iv) Triple bonds are shorter than single bonds since bond energy is larger for a multiple bond. Te extra electron pairs strengthen the bond, making it more difficult to separate the bonded atoms from each other.

(b) Here, The O—H bond angle of H_3O^+s is less than 109.5°.

(i) Lewis structure of H_2O:

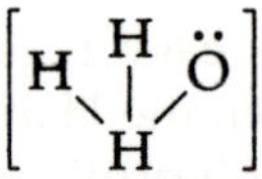

$$\left[\begin{array}{c} \text{H} \quad \text{H} \quad \ddot{\text{O}} \\ \diagdown \ | \ \diagup \\ \text{H} \end{array}\right]$$

(ii) H_2O^+ is pyramidal in geometry due to single pair of unshared electrons.

(iii) Angle of tetrahedron is 109.5°, this exists only if there are to unshared electrons.

(iv) Repulsion between shared pairs of electrons is less than repulsion between an unshared pair and a shared pair. This stronger repulsion found in the shared-unshared pair condition as seen in H_2O, decreases the bond angle of the pure tetrahedrom (109.5°)

(c) To compare C—O bond lengths as found in CO and CO_3^{2-}.

(i) Lewis structure of CO:

$$:C\equiv O:$$

(ii) Lewis structures of CO_3^{2-}

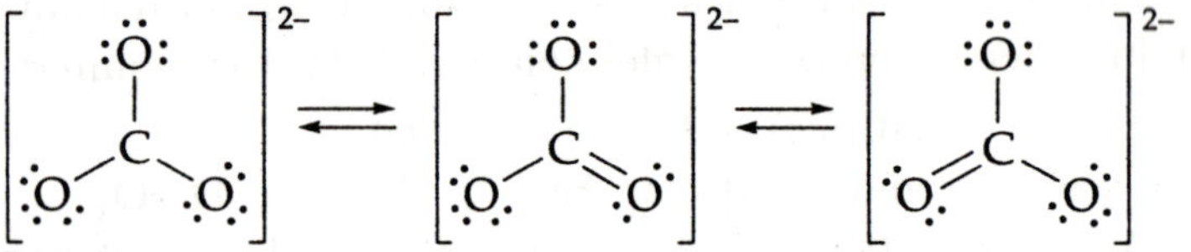

$$\left[\begin{array}{c} :\ddot{O}: \\ | \\ C \\ :\ddot{O}\diagup \quad \diagdown \ddot{O}: \end{array}\right]^{2-} \rightleftharpoons \left[\begin{array}{c} :\ddot{O}: \\ | \\ C \\ :\ddot{O}\diagup \quad \diagdown\!\!= \ddot{O} \end{array}\right]^{2-} \rightleftharpoons \left[\begin{array}{c} :\ddot{O}: \\ | \\ C \\ \ddot{O}=\!\!\diagup \quad \diagdown \ddot{O}: \end{array}\right]^{2-}$$

CNO^- exists in three "resonance" forms. CO bond length is considered to be the average of the lengths of all single and double bonds.

(d) Given: CNO^- ion is linear.

(i) Lewis structure of CNO^-:

$$[\ddot{C}=N=\ddot{O}]^-$$

(ii) There are no unshared pairs of electrons around the central atom N, resulting in a linear molecule.

(iii) The molecule is polar because O is more electronegative than C.

7. Let us explain how MW measured just above boiling point derivates from its ideal value in terms of the ideal gas law.

The ideal gs equation $PV = nRTs$, stems from three relationships known to be true for gases:

(i) The volume is directly proportional to the number of moles: $V \alpha n$

(ii) The volume is directly proportional to the absolute temperature: $V \alpha T$.

(iii) The volume is inversely proportional to the pressure: $V \alpha 1/P$

n the symbol used for the moles of gas, can be obtained by dividing the mass of the gas by the molecular weight in effect, n = mass/molecular weight($n = m/\text{MW}$).

Substituting this relationship into the ideal gas law gives

$$PV = \frac{m \cdot R \cdot T}{MW}$$

Solving this equation for the molecular weight yields.

$$MW = \frac{m \cdot R \cdot T}{PV}$$

Real gas behavior deviates from the values obtained using the ideal gas equation because the ideal equation assumes that (1) the molecules do not occupy space and 2) there is no attractive force between the individual molecules. However, at low temperature (just above the boiling point of the liquid) these two postulates are not true and one must use an alternative equation known as the van der Waals equation, which accounts for these factors.

Because the attraction between the molecules becomes more significant at the lower temperatures, the compressibility of the gas is increased. This causes the product $P.\ V$ to be smaller than predicted. PV is found in the denominator in the equatin listed above, so the molecular weight tends to be higher than its ideal value.

8. Given $F_2 < PH_3 < H_2O$

Discuss boiling point (BP) order.

General Trends

(i) BP is result of the strength of intermolecular forces—the forces between *molecules.*
(ii) A direct relationship exists between the strength of intermolecular forces and the BP. The stronger the intermolecular force, the higher the BP.
(iii) Relative strength of intermolecular forces: H bonds > dipole forces > dispersion forces.
(iv) BP is directly proportional to increasing MW-dispersion force (van der Waals force).
(v) Greater MW results in greater dispersion forces.
(vi) Strength of dispersion force depends on how readily electrons can be polarized.
(vii) Large molecules are easier to polarize than small, compact molecules. Hence, for comparable MW, compact molecules have lower BP.
(viii) Polar compounds have slightly higher BP than nonpolar compounds of comparable MW.
(ix) Hydrogen bonds are very strong intermolecular forces, causing very high BP.

Lowest BP: F_2

(i) F_2 is nonpolar; the only intermolecular attraction present is due to dispersion forces.
(ii) F_2 has a MW of 38 g $mole^{-1}$.
(iii) F_2 is covalently bonded.

Intermediate BP: PH_3

(i) PH_3 is polar; geometry is trigonal pyramidal; presence of ione pair of electrons.
(ii) PH_3 is primarily covalently bonded; two nometals.
(iii) There are dipole forces present between PH_3 molecules because PH_3 is polar.
(iv) PH_3 has a MW of 34 g $mole^{-1}$ (even though PH_3 has a lower MW than F_2 and might be expected to have a lower BP, the effect of the polarity out weight any effect of MW).

Highest BP; H_2O

(i) H_2O is covalently bonded.
(ii) H_2O is a beat molecule; hence, it is polar.
(iii) H_2O has a MW of 18 g $mole^{-1}$.
(iv) Between H_2O molecules then exist hydrogen bonds.
(v) Even though H_2O has the lowest MW of all three compounds, the hydrogen bonds out weight any effect of MW or polarity.

PRACTICE PAPER 2

1. (a) Electron configuration of P.
   $1s^2\ 2s^2\ 2p^6\ 3s^2\ 3p^3$

   (b) Quantum numbers for valence electrons in P.
   Here's a good place to use the chart format.

Electron Number	n	*l*	m_l	m_2
11	3	0	0	+1/2
12	3	0	0	−1/2
13	3	1	1	+1/2
14	3	1	0	+1/2
15	3	1	−1	+1/2

   (c) Phosphorus is paramagnetic, because a paramagnetic atom is defined as having magnetic properties caused by *unpaired* electrons. The unpaired electrons are found in this 3*p* orbitals, each of which is half-filled.

   (d) Consider the chart below:

	PCl_3	PCl_5	PCl_4^+	PCl_6^-	P_4
Type of bond	covalent	covalent	covalent	covalent	covalent
Lewis structure	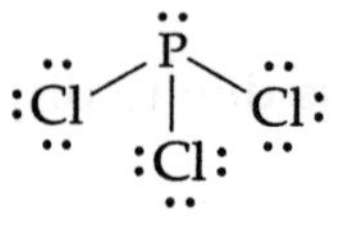	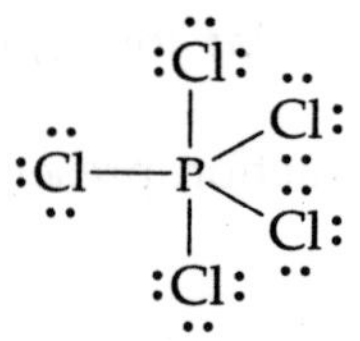	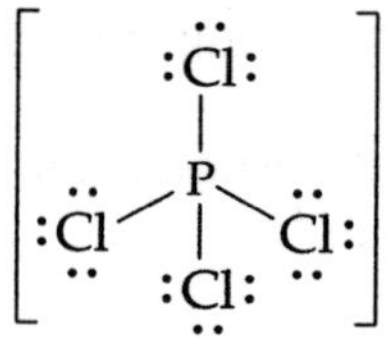	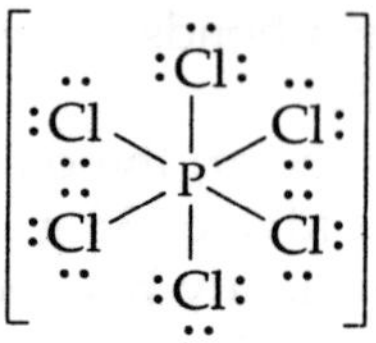	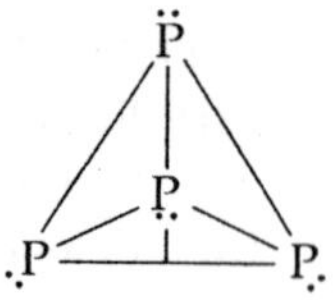
Geometry	pyramidal	triangular bipyramidal	tetrahedral	octahedral	tetrahedral
Hybrization	*sp*³	*sp*³*d*	*sp*³	*sp*³*d*²	*sp*³

2. This problem is best done in the chart format. Note how the column headings record what is given and how the first entries in the rows serve as a restatement of what is wanted.

	ClO_2^-	ClO_4^-	Cl_2O	ClO_3^-	ClO_2
(a) Lewis structure	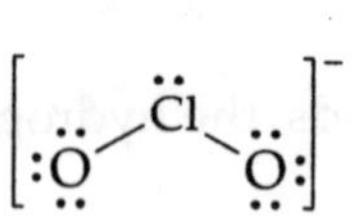	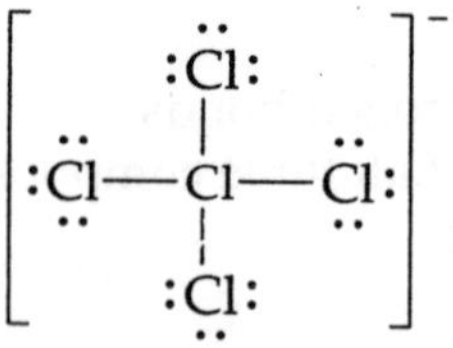	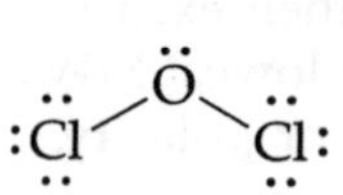	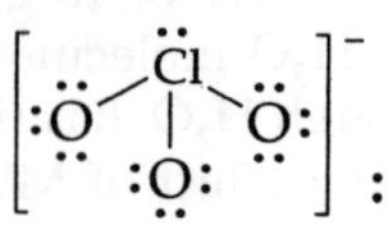	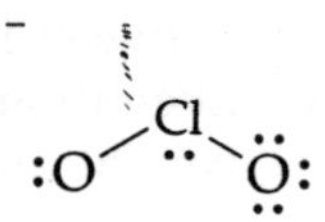
(b) Bond angles	less than 109.5°	109.5°	less than 109.5°	less than 109.5°	less than 109.5°
(c) Geometry	bent	tetrahedral	bent	pyramidal	bent

(*Contd.*)

(*Contd.*)

(d) Hybridization	sp^3	sp^3	sp^3	sp^3	sp^3
(e) Dimerization	no	no	no	no	yes-unpaired electron will give $O_2Cl{-}ClO_2$
(f) Polarity	polar	nonpolar	polar	polar	polar

3. Given: First three ionization energies I, I_2 and I_3 of Be and Ne.
   (a) Electron configuration of Be and Ne.

   Be: $1s^2\ 2s^2$

   Ne: $1s^2\ 2s^2\ 2p^6$

   (b) Significant trends/discrepancies in the first three ionization energies of Be and Ne are.

   I. Note that in the case of both beryllium and neon, ionization energies increase as one moves from I_1 to I_2 to I_3.

   II. The general trend is for ionization energy to increase as one moves from left to right across the periodic table and to decrease as one moves down; this is the inverse of the trend one finds in examining the atomic radius.

   III. Note further that beryllium and neon are in the second period.

   IV. Beryllium

   A. There is generally not enough energy available in chemical reactions to remove inner electrons, as noted by the significantly higher third ionization energy.

   B. The Be^{2+} ion is a very stable species with a noble-gas configuration, so removing the third outer most electron from beryllium requires significantly greater energy.

   V. Neon

   A. Neon is an inert element with a full complement of 8 electrons in its valence shell.

   B. It is significantly more difficult to remove neon's most loosely held electron (I1) than that of beryllium's I_1. This trend is also noted when examining I_2's and I_3's. Neon also has a greater nuclear charge than beryllium, which is, if all factors are held constant, would result in a smaller atomic radius.

   (c) The only compound formed would be $BeCl_2$. The Be atom readily loses 2 electrons to form the stable Be^{2+} ion. The third ionization energy is too high to form Be^{3+}. The electron affinity of neon is very low because it has a stable octet of electrons in its valence shell and the ionization energies of neon are too high.

   (d) The first ionization energy (I_1) of element X is relatively low when compared to I_2 and I_3. This means that X is probably a member of the Group alkali metals. Thus, the formation of X^{2+} and X^{3+} would be difficult to achieve. Therefore, the formula is most likely to be XCl.

4. Given: 185 gm. $F_2(g)$; 4.0 moles $Xe(g)$; T = 273 K P = 2.5 atm

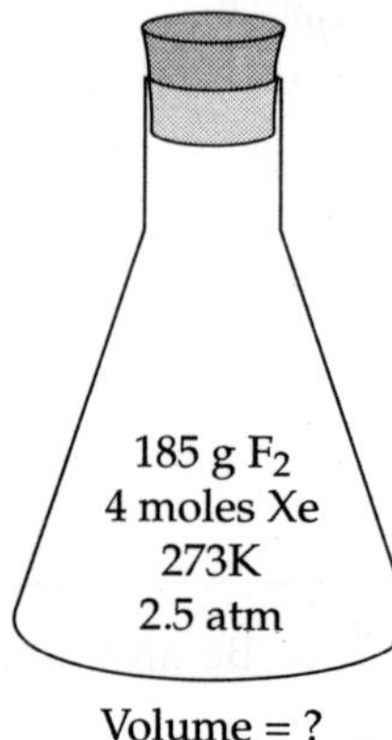

Volume = ?
P F$_2$ = ?
P Xe = ?

23.99 g Li
185 g F$_2$
4 moles Xe
(violent reaction)

g LiF = ?

(a) (1) To find: Volume of flask.

$PV = nRT$

n = total moles = moles F_2 + moles xenon

$$\text{moles } F_2 = \frac{185 \text{ g } F_2}{1} \times \frac{1 \text{ mole } F_2}{38 \text{ g } F_2} = 4.868 \text{ moles } F_2$$

total moles of gas = 4.868 + 4.0 = 8.9 moles

Solve for the volume.

$$V = \frac{nRT}{P} = \frac{8.9 \text{ moles } (0.821 \text{ liter} \cdot \text{atm}) \cdot 273 \text{ K}}{\text{mole} \cdot \text{K} \cdot 2.5 \text{ atm}} = 8.0 \times 10^2 \text{ litres}$$

(2) To find partial pressure of each gas.

$$P_{\text{total}} = P_{\text{fluorine}} + P_{\text{xenon}}$$

$$P_{\text{fluorine}} = \frac{n_{\text{flourine}} \cdot R \cdot T}{V} = \frac{4.868 \text{ moles} \cdot 0.0821 \text{ liter} \cdot \text{atm} \cdot 273 \text{ K}}{\text{mole} \cdot \text{K} \cdot 8.0 \times 10^1 \text{ liters}} = 1.4 \text{ atm}$$

$$P_{\text{xenon}} = \frac{r_{\text{xenon}} \cdot R \cdot T}{V} = \frac{4.0 \text{ moles} \cdot 0.0821 \text{ liter} \cdot \text{atm} \cdot 273 \text{ K}}{\text{mole} \cdot \text{K} \cdot 8.0 \cdot 10^1 \text{ liters}} = 1.1 \text{ atm}$$

(b) 23 g Li added to flask. Weight of LiF formed = ?

As, Xe is inert (no reaction with lithium).

The balanced equation for the reaction is thus $F_2(g) + 2Li(s) \rightarrow 2\,LiF(s)$.

$$\frac{23 \text{ g Li}}{1} \times \frac{1 \text{ mole Li}}{6.94 \text{ g Li}} = 3.31 \text{ mole Li}$$

Therefore, lithium is the limiting reagent. So 3.31 moles of Li is used up along with 3.31/2 or 1.66 moles of F_2. This leaves 3.21 moles (4.868 – 1.66) or F_2.

$$\frac{3.31\ \cancel{\text{moles Li}}}{1} \times \frac{2\ \cancel{\text{moles LiF}}}{2\ \cancel{\text{moles Li}}} \times \frac{25.94\ \text{g Lif}}{1\ \text{mole Lif}} = 86.0\ \text{g LiF}$$

(c) Partial pressures of gases present after reaction, 273 K
Xenon is inert (no reaction): 1.1 atm pressure

$$P_{\text{fluorine}} = \frac{n_{\text{florine}} \cdot R \cdot T}{V} - \frac{3.21\ \cancel{\text{moles}} \cdot 0.0821\ \cancel{\text{liter}} \cdot \text{atm} \cdot 273\ \cancel{K}}{\cancel{\text{mole}} \cdot \cancel{K}\ 3.0 \times 10^1\ \cancel{\text{liters}}} = 0.90\ \text{atm}$$

5. Do a restatement of the general; experiment. In this case, I would draw a sketch of the apparatus before and after the reaction, labeling everything. This will get rid of all the words and enable you to visualize the experiment.

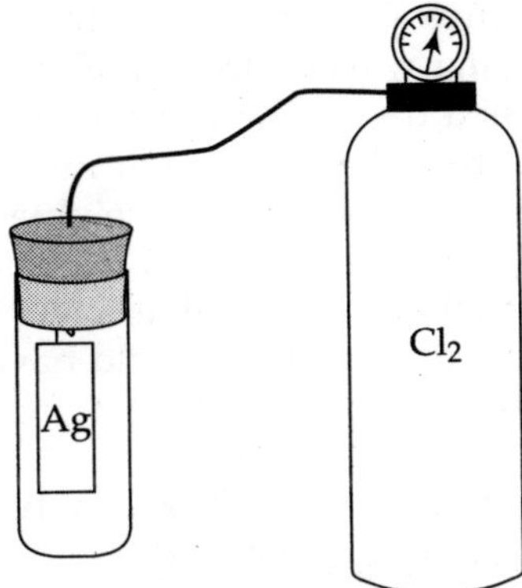

Now, Write a balanced chemical equation that describes the reaction.

$$2\ Ag(s) + Cl_2(g) \rightarrow 2AgCl(s)$$

Begin to answer the questions asked. Remember to given brief restatement for each question, to label each specific question so that the grader knows which question ou are answering, and to underline the conclusion(s) where necessary.

1. (a) (1) To find the number of moles of chlorine gas that reacted, we have,
mass of chlorine reacted = (mass of silver strip + compound) – mass of original silver strip

$$= 1.9342\ \text{g} - 1.7838\ \text{g} = 0.1504\ \text{g of chlorine atoms}$$

and moles of chlorine atoms that reacted = mass of chlorine atoms/atomic mass of chlorine

$$0.1504\ \text{g}/135.45\ \text{g} \cdot \text{mole}^{-1} = 0.004242\ \text{mole of chlorine atoms}$$

(2) Moles of silver that reacted

$$= \frac{\text{(mass of original silver strip – mass of dry strip after washig)}}{\text{atomic mass of silver}} = \frac{1.7838 - 1.3258}{107.87}$$

$$= 0.43580\text{g}/107.87\ \text{g} \cdot \text{mole}^{-1} = 0.004246\ \text{mole of silver atoms}$$

(b) Empirical formula for silver chloride.
= moles of silver atoms/moles of chlorine atoms

$$\frac{0.004246 \text{ mole of silver}}{0.004242 \text{ mole of chlorine}} = 1.001 \rightarrow AgCl$$

(c) (1) The effect on the empirical formula if some of the white compound had been washed down the sink before the coated strip was dried and reweighted is that the white product is silver chloride. Had she lost some before she weighed it, the mass of silver chloride will less than what it should have been. So, the number of grams of chlorine appear too low, which in turn will make the number of moles of chlorine will appear low. Thus, in the ratio of moles of silver to moles of chlorine, the denominator will be lower than expected and the ratio would be large. Because the mass of the compound does not enter into the calculations for the mole of silver that reacted, the moles of silver will not affected.

(2) The effect on empirical formula if the silver strip had not been dried thoroughly after being washed free of silver chloride will be that because the strip has been washed free of the compound (silver chloride), you assume that any silver missing from the strip; went into the making of the silver chloride. If the strip had been wet when you weighted it, you would have been led to think that the strip was heavier than expected, and therefore that less silver had gone into the making of the silver chloride. Thinking that less silver had been involved in the reaction, you would have calculated fewer moles of silver. The calculation ratio of moles of silver to moles of chlorine would have been less than expected.

6. (a) Unknown hydrocarbon is combusted in O_2 and subjected to colligative tests.
Given: 9.01 g H_2O + 11.2 liters CO_2 produced.

$$\frac{9.01 \text{ g } H_2O}{18.02 \text{ g mole}^{-1}} = 0.5 \text{ mole } H_2O$$

$$\therefore \quad \frac{0.5 \text{ mole } H_2O}{1} \times \frac{2 \text{ moles H}}{1 \text{ mole } H_2O} = 1.00 \text{ mole H}$$

$$\text{Also,} \quad \frac{11.2 \text{ liters } CO_2}{22.4 \text{ liter mole}^{-1}} = 0.5 \text{ mole } CO_2$$

$$\therefore \quad \frac{0.5 \text{ mole } CO_2}{1} \times \frac{1 \text{ mole C}}{1 \text{ mole } CO_2} = 0.5 \text{ mole C}$$

$$\text{empirical formula} = C_{0.5}H_1 \rightarrow CH_2$$

(b) To calculate the mass of O_2 required for complete combustion the reactionn is

$$2\,CH_2(g) + 3O_2(g) \rightarrow 2\,CO_2(g) + 2\,H_2O(l)$$

$$\therefore \quad \frac{0.5 \text{ mole } H_2O}{1} \times \frac{2 \text{ moles } O_2}{2 \text{ moles } H_2O} \times \frac{32 \text{ g } O_2}{1 \text{ mole } O_2} = 24 \text{ g } O_2$$

(c) Given:

- The unknown hydrocarbon dissolves in CCl_4.
- 135 grams of CCl_4 + 4.36 grams of hydrocarbon.
- B.P. of CCl_4 = 349.8 K
- New B.P. = 351.7 K
- K_b = 5.02°C/m

∴ To calculate the MW of hydrocarbon we have,

$$\Delta T = K_b \cdot m \Rightarrow \Delta T = \frac{(K_b)(\text{g/MW})}{\text{kg solvent}}$$

$$\Rightarrow \text{MW} = \frac{(K_b)(\text{grams solute})}{(\Delta T)(\text{kg solvent})}$$

$$\therefore \text{MW} = \frac{(5.02\ \cancel{C} \cdot \cancel{\text{kg solvent}}/\text{moles solute})(4.36\text{ g})}{(351.7 - 340.8)(0.135\ \cancel{\text{kg solvent}})} = 85.2\text{ g mol}^{-1}$$

(d) To determine the molecular formula of hydrocarbonwe use,

$$\frac{85\ \cancel{\text{g/mol}}\ (\text{molecular mass})}{14.03\text{ g}\ \cancel{\text{g/mol}}\ (\text{empirical mass})} = 6.1$$

Thus, the molecular mass is 6 times greater than the empirical mass.

∴ Molecular formula = $\mathbf{C_6H_{12}}$

7. (a) Balanced equation is:

$2Al(s) + Fe_2O_3(s) \rightarrow Al_2O_3(s) + 2Fe(s)$

(b) To Calculate ΔH we have,

$$\Delta H = \Sigma\Delta H_f^\circ - \Sigma\Delta H_{f\ \text{reactants}}^\circ$$

or, $\Delta H = [\Delta H_f Al_2O_3(s) + 2\ \Delta H_f Fe(s)] - [2\ \Delta H_f Al(s) + \Delta H_f Fe_2O_3(s)]$

$= [-1669.8 + 2(0)] - [2(0) + (-822.2)]$ Remember that ΔH_f for elements is 0.

Hence, $\Delta H = -847.6$ kJ

(c) Given: Specific heat of aluminium oxide = 0.79 J/g · °C

Given: Specific heat or iron = 0.48 J/g · °C

Going back to the balanced reaction (answer a), note that 1 mole Al_2O_3 is being produced for every 2 moles of iron—they are not being produced in a ratio of 1 gram to 1 gram. So for each of the two products, determine how much energy (in joules) is required for a 1°C rise in temperature.

$$\text{For } Al_2O_3\text{:}\ \frac{1\text{ mole }\cancel{ALO_3}}{1} \times \frac{101.96\ \cancel{\text{g }ALO_3}}{1\ \cancel{\text{mole }ALO_3}} \times \frac{0.79\text{ J}}{\text{g} \cdot \text{K}} = 81\text{ J K}^{-1}$$

$$\text{For Fe: } \frac{2\ \cancel{\text{mole Fe}}}{1} \times \frac{55.85\ \cancel{\text{g Fe}}}{\text{g} \cdot \text{C}} = 54\text{ J K}^{-1}$$

Together, the products are absorbing 81 + 54 = 135 joules for every 1°K rise in temperature. Because the reaction is producing 847.6 kJ of energy (you know this from you r answer to part b), the change in temperature, from the initial conditions of the reactants (presumably at room temperature) to those of the hot products, is found by dividing the heat of reaction by the energy absorbed per degree Kelvin (J divided by J/K = K).

Thus, $\frac{847,600 \text{ J}}{135 \text{ J/K}} = 6.28 \times 10^3$K change in temperature

(d) Given: H_{fus} Fe = 270 J g^{-1}

Because ΔH_{fus} represents the amount of energy absorbed in melting iron, you must subtract this component from the ΔH you obtained in part (b). And because ΔH reflects the amount of heat produced when 2 moles of iron are produced, ΔH_{fus} for the reaction is

$$\frac{2 \text{ mole of Fe}}{1} \times \frac{44.85 \text{ g Fe}}{1 \text{ mole of Fe}} \times \frac{270 \text{ J}}{1 \text{ g Fe}} = 30,200 \text{ J}$$

The 30,200 J of energy was absorbed in melting iron, so you subtract it from the ΔH value obtained in part (b) to find how much energy was actually in the reaction.

847,600 J – 30,200 J = 817400 J

Finally, from part (c), you know that the amount of energy released divided by the energy absorbed per degree Kelvin (J divided by J K = K) equal the change in temperature.

Thus, $\frac{81,400 \text{ J}}{135 \text{ J/K}} = 6.05 \times 10^3$K

which is above the melting point of iron, 1808 K

8. Given: ΔH_f° and ΔG_f° information for the equation

$$N_2O_4(g) \rightarrow 2\ NO_2(g)$$

(a) To calculate ΔG° we know that,

$\Delta G^\circ = \Sigma\Delta G^\circ$ products – $\Sigma\Delta G^\circ$ reactants = 2(51.30) – (97.82) = 4.78 kJ · mole^{-1}

(b) We have, $\Delta H^\circ = \Sigma\Delta H_f^\circ$ products = $-\Sigma\Delta H_f^\circ$ reactants = 2(33.2) – 9.16 = 57.2 kJ · mole^{-1}

(c) The equilibrium constant K_p at 298 K and 1 atm is;

$$K_p = \frac{P^2_{NO_2}}{P_{N_2O_4}}$$ (where P represents the partial pressure of a gas in a atmospheres)

$\Delta G^\circ = -2.303\ RT \log K$ (R = 8.314 J · K^{-1})

$$\log K_p = \frac{\Delta G^\circ}{-2.303\ RT} = \frac{4.78 \text{ kJ} \cdot \text{mole}^{-1}}{-2.303\ (0.008314 \text{ kJ} \cdot \text{K}^{-1})(298 \text{ K})} = -0.838$$

$\therefore$ K_p= 0.145 (at standard temperature of 298 K)

(d) Now, to find calculate K at 773 K and 1 atm we have,

$$\frac{\Delta H^\circ (T_2 - T_1)}{2.303\, RT_1 T_2} = \log \frac{K_{T_2}}{K_{T_1}}$$

or $$\frac{57,200 \text{ J } (773 \not{K} - 298 \not{K})}{(2.303)(8.314 \text{ J} \cdot \not{K}^{-1})(773 \not{K})(298 \not{K})} = \log \frac{K_{773}}{K_{298}}$$

$$6.16 = \log \frac{K_{773}}{K_{298}}$$

$\log K_{773} - K_{298} = 6.16$
$\log K_{298} = -0.838$ from part (c)
$\log K773 = 6.16 + (-0.838) = 5.32$
$\Rightarrow K = 2.09 \times 10^5$

(e) To find, ΔS° at 298 K and 1 atm we have,
$\Delta G^\circ = \Delta H^\circ - T\Delta S^\circ$

$$\therefore\ \Delta S^\circ = \frac{\Delta H^\circ - \Delta G^\circ}{T} = \frac{57,200 \text{ J} - 4,780 \text{ J}}{298 \text{ K}} = 176 \text{ J} \cdot \text{K}^{-1}$$

(f) To calculate the temperature at which ΔG° is equal to zero at 1 atom, assuming that ΔH° and ΔS° of not change significantly as the temperature increases.
$\Delta G^\circ = \Delta H^\circ - T\Delta S^\circ$
$0 = 57,200 \text{ J} - T(176 \text{ J} \cdot \text{K}^{-1})$

$$T = \frac{57,200 \not{J}}{176 \not{J} \cdot \text{K}} = 325 \text{ K}$$

9. Here, we have, $2\ NO_2(g) + CO_2(g) \rightarrow 2\ NO_2Cl(g)$

(a) Using rate law.

$\text{Rate} = k[NO_2]^n[Cl_2]^m$
Expt. 1: $\text{rate} = 1.35 \times 10^{-7}$ mole liter^{-1} · sec^{-1}
$= k(0.1\ M)^n(0.005\ M)^m$
Expt. 2: $\text{rate} = 2.7 \times 10^{-7}$ mole liter^{-1} · sec^{-1}
$= k(0.1\ M)^n(0.01\ M)^m$
Expt. 3: $\text{rate} = 5.4 \times 10^{-7}$ mole liter^{-1} · sec^{-1}
$= k(0.2\ M)^n(0.01\ M)^m$

$$\therefore\ \frac{\text{rate 2}}{\text{rate 1}} = \frac{2.7 \times 10^{-7} \text{ mole liter}^{-1} \cdot \text{sec}^{-1}}{1.35 \times 10^{-7} \text{ mole liter}^{-1} \cdot \text{sec}^{-1}}$$

$$= \frac{k\,\not{(0.10\,\text{M})^n}\ (0.01\,\text{M})^m}{k\,\not{(0.10\,\text{M})^n}\ (0.005\,\text{M})^m}$$

or $2.0 = (2.0)^m \Rightarrow m = 1$

$$\frac{\text{rate 3}}{\text{rate 2}} = \frac{5.4 \times 10^{-7} \text{ mole liter}^{-1} \cdot \text{sec}^{-1}}{2.7 \times 10^{-7} \text{ mole liter}^{-1} \cdot \text{sec}^{-1}}$$

$$= \frac{k(0.2\,\text{M})^n\,(0.01\,\text{M})^m}{k(0.1\,\text{M})^n\,(0.01\,\text{M})^m} \text{ or } 2.0 = (2.0)^n \Rightarrow n = 1$$

rate= $k[NO_2]^1[Cl_2]^1$.

(b) We have here,
The overall order = $m + n = 1 = 2$
The rate is proportional to the product of the concentrations of the two reactants:
$2\,NO_2(g) + Cl_2(g) \rightarrow 2\,NO_2Cl(g)$

$$\text{rate} = \frac{-\Delta[NO_2]}{\Delta t} = k[NO_2]\,[Cl_2]$$

$$\text{or rate} = \frac{-2\Delta[Cl_2]}{\Delta t} = k[NO_2]\,[Cl_2]$$

(c) Rate-specific constant k.
rate = $k[NO_2][Cl_2]$
k = rate/$[NO_2]Cl_2]$

$$= \frac{1.35 \times 10^{-7} \text{ mole} \cdot \text{liter}^{-1} \cdot \text{sec}^{-1}}{(0.1 \text{ mole} \cdot \text{liter}^{-1})(0.005 \text{ mole} \cdot \text{liter}^{-1})} = 2.7 \times 10^{-4} \text{ liter mole}^{-1} \cdot \text{sec}^{-1}$$

(d) In experiment.3, initial rate of decrease of $[Cl_2]$.
$2\,NO_2(g) + Cl_2(g) \rightarrow 2\,NO_2Cl(g)$

$$\frac{-d[Cl_2]}{dt} = \frac{1}{2}\left(\frac{d(NO_2Cl]}{dt}\right)$$

$= -(5.40 \times 10^{-7})/2 = -2.7 \times 10^{-7}$ mole · liter^{-1} · sec^{-1}

10. Given: 250.0 g $Cu(NO_3)_2$ in 4.0-liter flask
Heated to 523 K, reaches equilibrium
Total pressure at equilibrium = 5.50 atmospheres

(a) Balanced reaction.

$$2\,Cu(NO3)_2(s) \leftrightarrow 2\,CuO(s) + 4\,NO_2(g) + O_2(g)$$

(b) To find moles of $O_2(g)$ at equilibrium we have,

$$PV = nRT \Rightarrow N_{gas} = \frac{PV}{RT} = \frac{(5.50\,\cancel{\text{atm}})(4\,\cancel{\text{liters}})}{(0.0821\,\cancel{\text{liter}} \cdot \cancel{\text{atm}}/\text{mole} \cdot \cancel{K})(523\,\cancel{K})} = 0.51 \text{ mole gas}$$

$$= \frac{0.51\,\cancel{\text{mole gas}}}{1} \times \frac{2 \text{ moles Cu }(NO_3)}{1\,\cancel{\text{mole Cu }(NO_3)_2}} = 0.10 \text{ mole } O_2$$

(c) To find grams of solid $Cu(NO_3)_2$ in flask at equilibrium we have,
Moles of $Cu(NO_3)_2$ that decomposed:

$$= \frac{0.1 \text{ mole } O_2}{1} \times \frac{18.57 \text{ g } Cu(NO_3)_2}{1 \text{ mole } Cu(NO_3)_2} = 0.2 \text{ mole } Cu(NO_3)_2$$

∴ mass of $Cu(NO_3)_2$ that decomposed:

$$\frac{0.20 \text{ mole } Cu(NO_3)_2}{1} \times \frac{187.57 \text{ g } Cu(NO_3)_2}{1 \text{ mole } Cu(NO_3)_2} = 38 \text{ g } Cu(NO_3)_2$$

∴ mass of $Cu(NO_3)_2$ that remains in flask:
= 250 g $Cu(NO_3)_2$ originally – 38 g $Cu(NO_3)_2$ decomposed
= 212 g $Cu(NO_3)_2$ remain

(d) Equilibrium expression for K_p and value.

$$K_p = (\text{pressure } NO_2)^4 \times \text{pressure } O_2$$

Dalton's law of partial pressures:

$$\frac{4 \text{ moles } NO_2(g)}{5 \text{ total moles gas}} \times \frac{5.5 \text{ atm}}{1} = 4.4 \text{ atm } NO_2$$

$5.5 \text{ atm}_{tot} - 4.4 \text{ atm}_{NO_2} = 1.1 \; O_2$

$K_p = (4.40 \text{ atm})^4(1.1 \text{ atm}) = 412 \text{ atm}^5$

(e) Here 420 grams of $Cu(NO_3)_2$ are placed in flask,
To find : What is total pressure at equilibrium?

Because the temperature was kept constant, as was the size of the flask, and because some of the original 250 grams of $Cu(NO_3)_2$ was left as solid in the flask at equilibrium, any extra $Cu(NO_3)_2$ introduced into the flask would remain as solid—there would be *no changing in the pressure.*

11. Given: 0.25 g solid acid
45.77 mL f 0.15 M NaOH required for phenolphthalein end point
Back-titration: 1.5-mL of 0.01 M HCl

(a) To find: Moles of OH^- used in the titration, we have
moles of H^+= moles of NaOH

$$= \frac{0{,}04577 \text{ liter}}{1} \times \frac{0.15 \text{ mole NaOH}}{1 \text{ liter}} = 1.5 \times 10^{-3} \text{ mole } OH^+$$

(b) To find: Moles of H^+ used in back-titration, we have
moles of H^+ = moles of HCl

$$= \frac{0.0015 \text{ liter}}{1} \times \frac{0.01 \text{ mole HCl}}{1 \text{ liter}} = 1.5 \times 10^{-5} \text{ mole } H^+$$

(c) To find: Moles of H^+ in solid., we have
moles H^+ in solid acid= mole OH^- – moles H^+ = $(M_{NaOH} + V_{NaOH}) \times (M_{HCl} \times V_{HCl})$
= 6.87×10^{-3} mole $OH^- \times 1.5 \times 10^{-5}$ mole H^+ = 6.86×10^{-3} mole H^+

(d) To find Equivalent mass of unknown acid we have,

$$\text{GEM} = \frac{\text{grams of acid}}{\text{moles of H}^+ \text{ furnished}} = \frac{0.250 \text{ g acid}}{6.86 \times 10^{-3} \text{ mole H}^+} = 36.4 \text{ g/mole acid}$$

12. Here, we have, Cu electrodes in 1 M $CuSO_4$.
and Ag electrode in 1 M $AgNO_3$.
Voltage = 0.42 volt.

(a) Diagram the cell is:

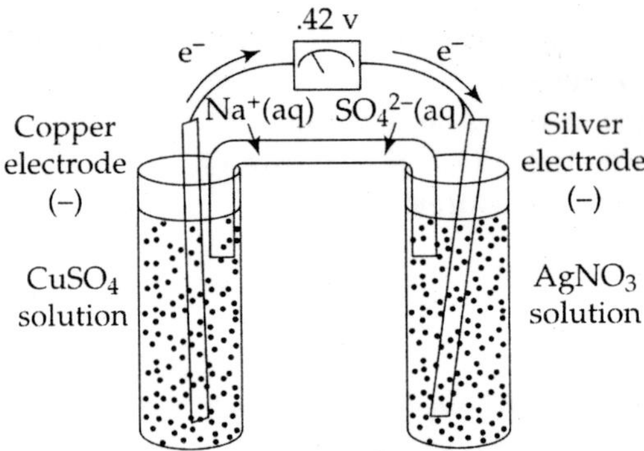

(b) Reduction always occurs at the cathode. Note that E°_{red} for silver is +0.7991 volt, according to the Table of Standard Reduction Potentials. E°_{red} for copper is +0.337. This means that the copper metal is higher in the activity series than the silver metal, so copper metal will reduce the silver ion. The equation that describes reduction (or the cathode reaction) is therefore

Red: $Ag^+(aq) + e^- \rightarrow Ag(s)$

(c) Oxidation occurs at the anode. Silver is lower in the activity series than copper. Therefore, the oxidation half-reaction is

ox: $Cu(s) \rightarrow Cu^{2+}(aq) + 2e^-$

(d) Overall cell equation.

Combing these two half-equations into one cell equation produces

ox:	$Cu(s)$	$\rightarrow Cu^{2+}(aq) + \cancel{2e^-}$	$E^\circ_{cell} = -0.337$
	red: $2Ag+(aq) + \cancel{2e^-}$	$\rightarrow 2Ag(s)$	$E^\circ_{cell} = +0.7991$
	$Cu(s) + 2Ag^+(aq)$	$\rightarrow Cu^{2+}(aq) + 2Ag(s)$	$E^\circ_{cell} = 0.462$

The overall theoretical E°_{cell} is 0.7991 – 0.337 = +0.462 volt. Therefore, this reaction is spontaneous, because when the cell was measured experimentally it produced 0.42 volt; this is in agreement with our cell designations.

(e) Cell notation.

$$Cu(s)\ |\ Cu^{2+}(aq)\ ||\ Ag^+(aq)\ |\ Ag(s)$$

(f) Given:

4 M $NH_3(aq)$ added to $CuSO_4$ solution $\rightarrow Cu(NH_3)_4^{2+}(aq)$

Voltage = 0.88 volt

By Nernst equation: $E_{cell} = E^{\circ}_{cell} - \frac{0.0592}{n} - \log Q$

The variable take on the following values: $E_{cell} = 0.88$ volt and $E^{\circ}_{cell} = 0.46$ volt

$$\therefore\ Q = \frac{[Cu^{2+}(aq)]}{[Ag^{+}(aq)]^2} = \frac{x}{1^2} = x$$

Substituting into the Nernst equation yields

$$0.88 = 0.46 - \frac{0.0592}{2}\log x$$

or, $0.0296 \log x = -0.42$

$$\log x = \frac{0.42}{0.0296} = -14$$

So $[Cu^{2+}(aq)] = 1.0 \times 10^{-14}$ M

13. Given: $^{208}_{81}Tl$ (β-decay) $\quad t_{1/2} = 3.1$ min

(a) The Nuclear equation is $^{208}_{81}Tl + {}^{208}_{82}Pb + {}^{0}_{-1}e$

(b) Time for 1.0 g to reduce to 0.2 g.

$$\ln\frac{[A]_0}{[A]} = \ln\frac{1.0}{0.2} = kt$$

or, $0.693 = kt_{1/2}$

$$\therefore\ k = \frac{0.693}{3.1 \text{ min}} = 0.22 \text{ min}^{-1}$$

$$t = \ln\frac{1.0}{0.2} \times \frac{1}{0.22 \text{ min}} = 7.3 \text{ min}$$

(c) Given: 11 g of $^{208}_{81}Tl$ initially.
To find how many grams remain after 23 min?

We have, $\log\frac{x_0}{x} = \frac{kt}{2.303}$ $\qquad$ Here, $k = \frac{0.693}{3.1 \text{ min}} = 0.22 \text{ min}^{-1}$

$$\therefore\ \log\frac{11}{x} = \frac{0.22 \text{ min}^{-1} \times 23 \text{ min}}{2.303} = 2.2$$

$$\log\frac{11}{x_0} = 2.2 \rightarrow \frac{11}{x_0} = 10^{22}$$

$x_0 = 0.07$ g

PRACTICE PAPER 3

1. (a) ICl has a significantly lower B.P. than NaCl is because:
    - ICl is a covalently bonded, molecular solid; NACl is an ionic solid.
    - There are dipole forces between ICl molecules but electroststic forces between Na^+ Cl^- ions.
    - Dipole forces ICl are much weaker then the ionic bonds in NaCl.
    - I and Cl are similar in electronegativity—generated only partial δ^+ and δ^- around molecules.
    - Na and Cl differ greatly in elctronegativity—greater electrostatic force.
    - When heated supply, ICl boils because energy supplied (heat) overcomes weak dipole forces.

   (b) KJ is water soluble; I_2 is not is because
    - KI is an ionic solid, held together by ionic bonds.
    - I_2 is a molecular solid, held together by covalent bonds.
    - KI dissociated into K^+ and I^- ions.
    - I_2 slightly dissolves in water, maintaining its covalent bond.
    - Solubility rule: Like dissolves like H_2O is polar; KI is polar; I_2 is not polar.

   (c) Ag conducts; $AgNO_3$ does not is because
    - Ag is a metal.
    - $AgNO_3$ is an ionic solid.
    - As structure consists of Ag^+ cations surrounded by mobile or "free" electrons.
    - $AgNO_3$ structure consists of Ag^+ cations electrostatically attracted to NO_3^- polyatomic anions—no free or mobile electrons.

   (d) PCl_3 has a dipole; PCl_5 does not is because
    - PCl_3 Lewis diagram:

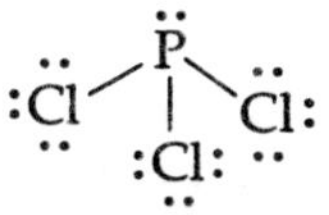

```
         ..
         P
   ..  /  |  \ ..
  :Cl     |    Cl:
   ..   :Cl:  ..
          ..
```

    - PCl_3—note the lone pair of unshared electrons.
    - PCl_3 is pyramidal, and all pyramidal structures are polar.
    - PCl_5—Lewis diagram is:

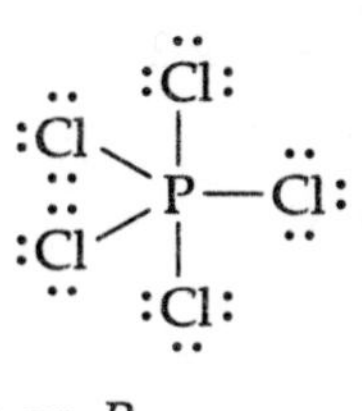

```
            ..
           :Cl:
   ..       |
  :Cl \     |    ..
   ..   >  P — Cl:
  :Cl /     |    ..
   ..       |
           :Cl:
            ..
```

    - PCl_5—no unshared electrons on *P*
    - PCl_5 is trigonal bipyramidal and thus perfectly symmetrical, so there is no polarity all diploes cancel.

2. Begin this problem by numbering the problem and listing the critical information, getting rid of superfluous words, and clarifying the situation in your mind.
   Given: flask + stopper = 62.371 g

volatile liquid = 4.4 mL
boiling water bath = 374.1 K
flask, stopper, condensed vapor = 63.088 g
volum,e of stoppered flask = 261.9 mL
barometric pressure = 733 mm Hg

Then, number each section; restate the question in as a few words as possible; list the necessary information; write down any generic formulas; substitute the data from the problem into the generic formulas: do the math; and underline or box your answer.

(a) Pressure inside the flask after vapor cooled when the flask was opened momentarily.

$$= \frac{733 \cancel{\text{mm Hg}}}{1} \times \frac{1 \text{ atm}}{760 \cancel{\text{mm Hg}}} = 0.964 \text{ atm}$$

The students opened the flask and thus made the pressure inside the flask equal to the pressure outside it. Thus,

(b) Mass of vapor inside the flask.

$$62.088 \text{ g} - 62.371 \text{ g} = 0.717 \text{ g}$$

(c) Mass of 1 mole of vapor.

$$PV = nRT \rightarrow \text{PV} = \frac{\text{g} \cdot \text{R} \cdot \text{T}}{\text{MW}}$$

$$\Rightarrow \text{MW} = \frac{\text{g} \cdot \text{R} \cdot \text{T}}{\text{P} \cdot \text{V}} = \frac{(0.717 \text{ g}) \cdot (0.0821 \cancel{\text{liter}} \cdot \cancel{\text{atm}}) \cdot (374.1 \cancel{\text{K}})}{\text{mole} \cdot \cancel{\text{K}} \; (0.964 \cancel{\text{atm}}) \cdot (0.2619 \cancel{\text{liter}})} = 87.2 \text{ g mole}^{-1}$$

3. Given: Compounds *D*, *E* and *F* with % (by weight of element *G*) and their respective MW's

(a) The mass of element G in 1.0 mole of compounds *D*, *E* and *F is:*

$0.539 \times 131.7 \text{ g mole}^{-1} = 71.0 \text{ g G mole}^{-1}$ *D*
$0.642 \times 165.9 \text{ g mole}^{-1} = 107 \text{ g G mole}^{-1}$ *E*
$0.477 \times 74.5 \text{ g mole}^{-1} = 35.5 \text{ g G mole}^{-1}$ *F*

(b) For, most likely atomic weight of G.

According to the law of multiple proportions, the ratios of the mass of element G to the masses compounds *D*, *E* and *F* must be small, whole numbers. The largest common denominator of 71, 106.5, and 35.5 is 35.5, so out best estimate is that the atomic weight of G is 35.5 (chlorine).

(c) Compound $F = C_xH_yG_z$ or $C_xH_yCl_z$?

$C_xH_yCl_z + O_2(g) \rightarrow CO_2(g) + H_2O(l) + Cl_2(g)$
$2.19 \text{ g} + ? \text{ g} \rightarrow 3.88 \text{ f} + 0.8 \text{ g} + ? \text{ g}Cl_2(g)$

$$\text{moles of carbon} = \text{moles of } CO_2 = \frac{3.88 \text{ g } CO_2}{44.01 \text{ g} \cdot \text{mole}} = 0.0882$$

$$\text{moles of hydrogen} = 2 \times \text{moles of } H_2O = 2 \times \frac{0.8 \text{ g } H_2O}{18.02 \text{ g} \cdot \text{mole}} = 0.088$$

(2.19 grams of F) (1 mole F/74.5 g F) = 0.0294 mole of compound F

This means that each mole of F contains 3 mole of C(0.0882/0.0294) and 3 mole of H (0.088), or 39 grams of CH. This leaves 74.5 – 39 = 36 grams, corresponding to 1 mole of element G(Cl). Therefore, the empirical formulas C_3H_3Cl.

4. Given: 6 aluminium cans (empty) at 50 g each
each can = 355 mL cola
density of cola = 1.23 g mL^{-1}
specific heat of aluminium = 0.902 J $g^{-1} \cdot K^{-1}$
specific heat of cola = 4.0 J $g^{-1} \cdot K^{-1}$
t_i = 303 K and t_f = 283 K

(a) Amount of heat absorbed by ice
= quantity of heat released by six pack = heat released by cane + cola

or $q_{\text{six-pack}} = q_{\text{cans}} + q_{\text{cola}}$

Now, q = mass × specific heat × Δt

$$\therefore\ q_{\text{cans}} = \frac{6\ \text{cans}}{1} \times \frac{50\ \text{g Al}}{1\ \text{can}} \times \frac{0.902\ \text{J}}{\text{g K}} \times \frac{(283 - 303\ \text{K})}{1} = -5412\ \text{J}$$

$$q_{\text{cola}} = \frac{6\ \text{cans}}{1} \times \frac{335\ \text{mL cola}}{1\ \text{can}} \times \frac{1.23\ \text{g coln}}{1\ \text{mL cola}} \times \frac{(283 - 303\ \text{K})}{1} = -2.1 \times 10^5\ \text{J}$$

$q_{\text{six-pack}} = -5412\ \text{J} + (-2.1 \times 10^5\ \text{J}) = -2.15 \times 10^5$ J, or –215 kJ

Therefore, the amount of heat absorbed by the ice is 215 kJ.

(b) Minimum amount of ice required to accomplish the necessary cooling
For the system to react 283 K the heat absorbed includes the warming of the water.
Let x= mass of ice

$$215\ \text{kJ} = \left[\left(\frac{6.00\ \text{kJ/mole}}{18.02\ \text{g/mole}}\right)\right] + (4.184\ \text{J/g} \cdot {}^\circ\text{C} \times 10^\circ\text{C})$$

$$x = \frac{215\ \text{kJ}}{0.375\ \text{kJ/g}} = 573\ \text{g ice}$$

5. In each set, choose which would have the lower entropy (greatest order).

1. SiC(s)
   - $H_2O(s)$ is a polar covalent molecule. Between the individual molecules would be hydrogen bonds.
   - SiC(s) exists as a structure and ordered covalent network.
   - Melting point of SiC(s) is much higher than that of $H_2O(s)$, so it would take more energy to vaporize the more ordered SiC(s) than to vaporize $H_2O(s)$.
2. $O_2(g)$ at 3.0 atm
   - At higher pressures, the oxygen molecules have less space to move within and are thus more ordered.
3. $NH_3(\ell)$
   - $NH_3(\ell)$ has hydrogen bonds (favors order).
   - $C_6H_6(\ell)$ has more atoms and so more vibrations—thus greater disorder.

4. $SiO_2(s)$
   - Na(*s*) has high entropy. It exhibits metallic bonding, forming soft crystals with high amplitudes of vibration.
   - $SiO_2(s)$ forms an ordered, structural covalent network.
   - $SiO_2(s)$ has a very high melting point, so much more energy is necessary to break the order system.

6. Given H_2O_2, first-order decomposition

k = 0.015 min at 873 K

$[H_2O_2]_0$ = 0.5 M

(a) To calculate $[H_2O_2]$ after 10 min.

For first-order reactions: rate = $kX \rightarrow -dX/dt = kX$

Integrating from X_0 to X given

$$\log\left(\frac{X_0}{X}\right) = \frac{kt}{2.3} \text{ or, } \log\left(\frac{0.5\text{ M}}{X}\right) = 0.065]$$

$$\text{or } \frac{0.5}{X} = 10^{0.065}\ 222 = 1.2$$

$$\therefore\ X = \frac{0.5}{4.2} = 0.42\text{ M}$$

(b) Time for H_2O_2 to go from 0.5 M to 0.15 M.

For first-order reactions:

$$\log\left(\frac{X_0}{X}\right) = \frac{Kt}{2.3}$$

$$\Rightarrow t = \frac{2.3}{k}\log\left(\frac{X_0}{X}\right) = \frac{2.3}{0.015\text{ min}^{-1}}\log\left(\frac{0.5\text{ M}}{0.15\text{ M}}\right) = \frac{2.3}{0.015\text{ min}} \times 2.523 = 8.0 \times 10^1\text{ min}$$

(c) To calculate the half-life we have

For the first-order reactions:

$$t_{1/2} = \frac{0.693}{k} = \frac{0.693}{0.015\text{ min}^{-1}} = 46\text{ min.}$$

7. Given: $Mg(OH)_2$ solubility = 9.24×10^{-4} g/100 mL H_2O at 298 K

Balanced equation in equilibrium is:

$$Mg(OH)_2\text{ (s)} \leftrightarrow Mg^{2+}\text{(aq)} + 2OH^-\text{(aq)}$$

Equilibrium expression is:

$$K_{sp} = [Mg^{2+}]\,[OH^-]^2$$

Value of K_{sp}.

MW $Mg(OH)_2$ = 58.33 g mole^{-1}

$$\frac{9.24 \times 10^{-4}\text{ gMg(OH)}_2}{100\text{ mL H}_2\text{O}} \times \frac{1\text{ mole Mg (OH)}}{58.33\text{ g Mg (OH)}_2} \times \frac{1000\text{ mL H}_2\text{O}}{1\text{ liter H}_2\text{O}}$$

$= 1.58 \times 10^{-4}$ M $Mg(OH)_2$ = 1.58×10^{-4} M Mg^{2+}

$= 2(1.58 \times 10^{-4}) = 3.16 \times 10^{-4}$ M OH^-

$K_{sp}= [Mg^{2+}]\ [OH^-]^2 = (1.58 \times 10^{-4})\ (3.16 \times 10^{-4})^2 = 1.568 \times 10^{-11}$

(d) To find P^H and P^{OH} we have,

$$P^{OH} = -\log[OH^-] = -\log\ (3.16 \text{ M} \times 10^{-4}) = 3.5$$

$$\therefore\ P^H = 14.0 - P^{OH} = 10.5$$

(e) Given: Add 75 mL of 4.00×10^{-4} M $MgCl^2$ to 75 mL of 4×10^{-4} M KOH.

To know that would a precipitate form we consider,

$MgCl_2 \rightarrow Mg^{2+}(aq) + 2\ Cl^-(aq)$

$KOH \rightarrow K^+(aq) + OH^-(aq)$

$\therefore$ Total volume of solution = 75 mL + 75 mL = 15 mL

$M_1V_1 \qquad = M_2V_2$

$(4 \times 10^{-4}$ mole liter$^{-1})$ (0.0075 liter) = (x)(0.1500 liter)

$x = [Mg^{2+}] = 2.0 \times 10^{-4}$ M

The same would be true for $[OH^-]$

$Q = [Mg^{2+}][OH^-]^2 = (2.0 \times 10^{-4})^3 = 8.0 \times 10^{-12}$

$K_{sp} = 1.58 \times 10^{-11}$

A precipitate would not form because $Q < K_{sp}$.

8. Given K_a for HA is 1.74×10^{-5}.

(a) To find $[H^+]$ in 0.50 M HA we have

The equation for the ionization of acetic acid,

$Ha(aq) \leftrightarrow H^+(aq) + A^-(aq)$

the equilibrium expression K_a is

$$K_a = \frac{[H^+][A^-]}{[HA]}$$

Now, let x equal the amount of H^+ that ionizes from HA. Because the molar ratio of $[H^+]$: $[A^-]$ is 1 : 1, $[A^-]$ also equals x, and we can approximate $0.5 - x$ as 0.5 (5% rule).

$$1.74 \times 10^{-5} = \frac{x^2}{0.5}$$

Solve for x.

$$x^2 = (1.74 \times 10^{-5})(0.5) = 8.7 \times 10^{-6}$$

$$\Rightarrow\ x = 2.9 \times 10^{-3} \text{ M} = [H^+]$$

(b) To find P^H and P^{OH} in 0.5 M HA we have

$$P^H = -\log[H^+] = -\log\ (2.9 \times 10^{-3}) = 2.47$$

Now, $P^H + P^{OH} = 14$

$$\therefore\ P^{OH} = 14.0 - 2.47 = 11.53$$

(c) $$\%\frac{\text{part}}{\text{Whole}} \times 100\% = \frac{[H^+]}{[HA]} \times 100\% = \frac{2.9 \times 10^{-3}}{0.5} \times 100\% = 0.58\ \%$$

However, 0.58% represents the percentage of the HA molecules that *do* ionize. Therefore, 100 – 0.58 = 99.42% of the HA molecules do not ionize.

(d) Given: Buffer $P^H = 6.5$

$[HA]/[A^-]$?

The Henderson-Hasselbalch equation is $P^H = pK_a + \log_{10}\frac{[\text{base}]}{[\text{acid}]}$

Substituting into the equation the known (and unknown) information we get

$p^{K_a} = -\log K_a = 4.76$

or $6.5 = 4.76 + \log\frac{[A^-]}{[HA]}$

$\log\frac{[A^-]}{[HA]} = 6.5 - 4.76 = 1.74$

$\therefore \frac{[A^-]}{[HA]} = 5.5 \times 10^1$

Because the question is asking for [HA]/[A–], you need to take the reciprocal: 0.018

(e) Given: volume = 0.5 liter

$0.35\ M = [NaC_2H_3O_2] \rightarrow Na^+ + C_2H_3O_2^-$

[HA] = 0.5 M

1.5 g LiOH added to solution

$$\frac{15\text{ g } \cancel{\text{LiOH}}}{1} \times \frac{1\text{ mole LiOH}}{23.95\text{ g } \cancel{\text{LiOH}}} = 0.063\text{ mole LiOH}$$

P^H = ?

Write a balanced equation expressing the reaction between acetic acid and lithium hydroxide.

$$HA + OH^- \rightarrow A^- + H_2O$$

Create a chart that express initial an final concentrations (at equilibrium) of the species.

Species	*Initial concentration*	*Final Concentration*
HA	0.5 M	$0.5 - \frac{0.063}{0.5} = 0.375$ M
A^-	0.35 M	$0.35 + \frac{0.063}{0.5} = 0.475$ M

Write an equilibrium expression for the ionization of acetic acid.

$$K_a + \frac{[H^+][A^-]}{[HA]} = 1.74 \times 10^{-5} = \frac{[H^+][0.48]}{[0.374]}$$

$H^+] = 1.37 \times 10^{-5}$ N

Solve for the P^H.

$P^H = -\log[H^+] = -\log(1.37 \times 10^{-5}) = 4.86$

9. Given: 25 grams of HA.
Dissolved fin H_2O to make 500 mL of solution.
Fifty-mL aliquots (samples) of acidic solution.
0.1 M NaOH used for titration.
See results of titration in given table.

(a) Moles of HA in fifty-mL aliquots.
At the end of titration, moles of HA = moles of NaOH.

$$\text{Average volume NaOH} = \frac{87.45 + 84.68 + 91.23}{3} = 87.79 \text{ mL}$$

$$\text{Moles HA} = \text{moles NaOH} = V_{\text{NaOH}} \times M_{\text{NaOH}}$$

$$= \frac{0.08779 \text{ liter}}{1} \times \frac{0.1 \text{ mole}}{1 \text{ liter}} = 8.78 \times 10^{-3} \text{ mole}$$

(b) Molecular weight of HA.

$$\text{MW} = \frac{48.99 \text{ mL HA sol'n}}{8.78 \times 10^{-3} \text{ mole HA}} \times \frac{25 \text{ g HA}}{500 \text{ mL HA sol'n}} = 279 \text{ g mole}^{-1}$$

(c) P^H of fifty-mL aliquots (assume 100% ionization).
Average volume of fifty-mL aliquots = 48.99 mL

$$[\text{H}^+] = \frac{\text{moles H}^+}{\text{liters solution}} = \frac{8.78 \times 10^{-3} \text{ mole H}^+}{0.04899 \text{ liter HA sol'n}} = 0.179 \text{ M}$$

$P^H = -\log[0.179] = 0.747$

(d) P^{OH} of fifty-mL aliquot.
$P^{OH} = 14 - P^H = 14 - 0.747 = 13.253$

(e) Effects of following errors.

1. Balance reading 0.01 gram too low.
   - Student would think she or he had 25 grams when there were actually 25.01 grams.
   - In the calculation of molecular weight, grams mole^{-1}, grams would be too low, so the effect would be a lower MW than expected.
2. An impurity in the sample of HA.
   - Student would have less HA than expected.
   - In the calculation of molecular weight, g/mole, there would be less HA available than expected. Therefore, in the titration against NaOH, it would take less NaOH than expected to reach the equivalence point. This error would cause a larger MW than expected, because the denominator (moles) would be smaller.
   - These results assume that the impurity does not have more H^+/mass of impurity than the HA.

3. NaOH was 0.15 M instead of 0.1 M.
   - It would take less NaOH to reach the equivalence point because the NaOH is stronger.
   - Because it would take less NaOH, the number of moles NaOH would be less than expected, causing the denominator (moles) to be smaller than expected, making the calculated MW larger then expected.

10. Given: $Fe^{2+}(aq) + MnO_4^-(aq) \rightarrow Fe^{3+}(aq)$

6.893 grams of ore.

All of the iron in the ore wasconverted to $Fe^{2+}(aq)$.

$Fe^{2+}(aq)$ treated with 0.001 M $KMnO_4$ solution $\rightarrow Fe^{3+}$ (aq).

13.889 mL of $KMnO_4$ required to reach end point.

(a) Oxidation half-reaction.

$$Fe^{2+}(aq) \rightarrow Fe^{3+}(aq) + e^- \text{ OIL (Oxidation IS Losing)}$$

(b) Reduction half-reaction.

$$MnO_4^-(aq) \rightarrow Mn^{2+}(aq) + 4H_2O(1) \text{ (balance O's)}$$
$$MnO_4^-(aq) + 8H^+ (aq) \rightarrow Mn^{2+} (aq) + 4H_2O(1) \text{ (balance H's)}$$
$$MnO_4^-(aq) + 8H^+ + 5e^- - \rightarrow Mn^{2+}(aq) + 4H_2O\ (1) \text{ (balance charge)}$$

(c) Balanced redox reaction.

ox: $5Fe^{2+} \rightarrow 5Fe^{3+} + 5e^-$

red: $MnO_4^- + 8H^+ + 5e^- \rightarrow Mn^{2+} + 4H_2O$

$$MnO_4^- + 8H^+ + 5Fe^{2+} \rightarrow Mn^{2+} + 4H_2O + 5Fe^{3+}$$

(d) To Identify:

oxidizing agent = $MnO_4^-(aq)$ ∴ a species that accepts electron from another
reducing agent = $Fe^{2+}(aq)$ ∴ a species that furnishes electrons to another
species oxidized = Fe^{2+}(aq) ∴ Oxidation is Losing (electrons)
species reduced = $MnO_4^-(aq)$ ∴ Reduction IS Gaining (electrons)

(e) To find moles of iron in the sample of ore we use the factor-label method.

$$= \frac{13.889 \text{ mL } KMnO_4}{1} \times \frac{1 \text{ liter } KMnO_4 \text{ sol'n}}{1000 \text{ mL } KMnO_4 \text{ sol'n}}$$

$$\times \frac{0.100 \text{ mole } KMnO_4}{1 \text{ liter } KMnO_4 \text{ sol'n}} \times \frac{5 \text{ moles } Fe^{2+}}{1 \text{ mole } KMnO_4} = 6.94 \times 10^{-3} \text{ mole } Fe^{2+} = 6.94 \times 10^{-3} \text{ mole Fe}$$

(because all of the Fe was converted to Fe^{2+})

(f) To find mass perfect of iron in the ore we have,

$$\% = \frac{\text{part}}{\text{whole}} \times 100\% = \frac{0.0694 \text{ mole Fe}}{6.893 \text{ g ore}} \times \frac{55.85 \text{ g Fe}}{1 \text{ mole Fe}} \times 100\% = 5.62\%$$

11. Given: ${}^{222}_{86}Rn$ produced by α-decay of ${}^{226}_{88}Ra$.

(a) The Nuclear reaction is

$${}^{226}_{88}Ra \rightarrow {}^{222}_{86}Rn + {}^{4}_{2}He$$

(b) Given: 7 g of ${}^{226}_{88}Ra$ decays.

Now, $\Delta E = \Delta mc^2 = 9.00 \times 10^{10} \frac{\text{kJ}}{\text{g}} \times \Delta m$

We have, Δm = mass products – mass reactants

$= 4.0015 \text{ g} + 221.9703 \text{ g} - 225.9771 \text{ g} = -0.0053 \text{ g}$

$$\therefore \quad \Delta E = \frac{7 \text{ g } ^{226}_{88}\text{Ra}}{1} \times \frac{1 \text{ mole } ^{226}_{88}\text{Ra}}{225.9771 \text{ g } ^{226}_{88}\text{Ra}} \cdot \frac{-0.0053 \text{ g}}{1 \text{ mole } ^{226}_{88}\text{Ra}} \times \frac{9 \times 10^{10} \text{ kJ}}{\text{g}} = -1.5 \times 10^7 \text{ kJ}$$

(c) For the Mass defect of $^{226}_{88}$Ra we have,

$\therefore$ 88 moles protons 88 × 1.00728 g = 88.6406 g

138 moles neutrons = 138 × 1.00867 g = +139.196 g 227.837 g

$\therefore$ mass defect = 227.837 g – 225.9771 g = 1.,86 g

(d) Binding energy (kJ mole)$^{-1}$ of $^{226}_{88}$Ra.

$$\Delta E = \frac{9 \times 10^{10} \text{ kJ}}{\text{g}} \times \frac{1.86 \text{ g}}{1 \text{ mole}} = 1.67 \times 10^{11} \text{ kJ mole}^{-1}$$

(e) Given $^{226}_{88}$Ra has $t_{1/2} = 1.62 \times 10^3$ yr.

Rate constant for $^{226}_{88}$Ra.

$$k = \frac{0.693}{1.62 \times 10^3 \text{ yr}} = 4.28 \times 10^{-4} \text{ yr}$$

(f) fraction of $^{226}_{88}$Ra that remains after 100 yr.

$$\ln \frac{x_0}{x} = kt = 4.28 \times 10^{-4} \text{ yr}^{-1} \times 100 \text{ yr} = 4.28 \times 10^{-2}$$

$$e^{0.0428} = 1.04 = \frac{x_0}{x}$$

fraction remaining: $\frac{x}{x_0} = \frac{1}{1.04} = 0.962 = 96.2\%$

UNITS AND CONVERSION

Base Units of the International System (SI Units)

Physical Quantity	*Name of Unit*	*Symbol*
Length	Meter	m
Mass	Kilogram	kg
Time	Second	s
Electric current	Ampere	A
Temperature	Kelvin	K
Luminous intensity	Candela	cd
Amount of substance	Mole	mol

Derived SI Units

Physical Quantity	*Name of Units*	*Symbol*	*Basic Units*
Area	Square meter	m^2	m^2
Volume	Cubic meter	m^3	m^3
Density	Kilogram per cubic meter	$kg \cdot m^{-3}$	$kg \cdot m^{-3}$
Velocity	Meter per second	$m \cdot s^{-1}$	$m \cdot s^{-1}$
Acceleration	Meter per second per second	$m \cdot s^{-2}$	$m \cdot s^{-2}$
Force	Newton	N	$kg \cdot m \cdot s^{-2}$
Pressure	Pascal	Pa	$kg \cdot m^{-1}s^{-2}$ or $N \cdot m^{-2}$
Energy	Joule	J	$kg \cdot m^2s^{-2}$ or $N \cdot m$
Electric charge	Coulomb	C	$A \cdot s$
Electric potential difference	Volt	V	$kg \cdot m^2s^{-3}A^{-1}$ or $J \cdot C^{-1}$

Conversion Factors for Units of Length, Volume

Length	*Volume*
1 kilometer (km) = 1×10^3 m	(L) = 1×10^{-3} m^3
1 centimeter (cm) = 1×10^{-2} m	= 1 dm^3
1 millimeter (mm) = 1×10^{-3} m	= 1×10^3 cm^3
1 nanometer (nm) = 1×10^{-9} m	1 millimeter (mL) = 1×10^{-6} m^3
1 picometer (pm) = 1×10^{-12} m	= 1 cm^3
1 angstrom (A) = 1×10^{-10} m	
= 1×10^{-8} cm	
= 1×10^2 pm	

Conversion Factors for Units of Mass and Weight

1 gram (g) = 1×10^{-3} kg
1 milligram (mg) = 1×10^{-3} g
= 1×10^{-6} kg
1 metric ton = 1×10^{3} kg
1 atomic mass unit (amu) = 1.66054×10^{-27} kg = 12.66054×10^{-24} g

Units of Force

1 dyne (dyn) = 1×10^{-5} N   1 Newton (N) = 1×10^{5} dyn

Definitions

One Newton (N) is the force that imparts an acceleration of one meter per second per second to a mass of one kilogram.

One dyne (dyn) is the force that imparts an acceleration of one centimeter per second per second to a mass of one gram.

Conversion Factors for Units of Pressure

1 atmosphere (atm) = 760 mmHg = 1.01325×10^{5} Pa

1 bar = 1×10^{5} Pa   1 torr = 1 mmHg

1 dyne per square centimeter ($dyn \cdot cm^{-2}$) = 1×10^{-1} Pa

Definition

Pressure is force per unit area. One pascal (Pa) is the pressure exerted when a force of one Newton is applied per square meter.

$$1 \text{ Pa} = 1 \text{ N} \cdot \text{m}^{-2}$$

Conversion Factors for Other Units of Energy

1 calorie (cal) = 4.184 J (exactly)   1 erg = 1×10^{-7} J   1 J = 1×10^{7} erg

1 liter × atmosphere (L · atom) = 101.325 J = 24.217 cal

1 electron volt (eV) = 1.60218×10^{-19} J = 1.60218×10^{-12} erg

1 eV per particle is equivalent to 96.485 $kJ \cdot mol^{-1}$ or 23.06 $kcal \cdot mol^{-1}$

1 million electron volts (MeV) = 1.602177 × 10–13 J

Definitions

One joule of energy is expended when a force of one Newton is applied through a distance of one meter.

$$1 \text{ J} = 1 \text{ N} \cdot \text{m}$$

One calorie was originally defined as the amount of heat required to raise the temperature of one gram of water from 283.5K to 284.5 K.

TABLE 1 : Some Physical Constants

S.No.	Name	Symbol	S.I. unit value
1.	Universal Gas constant	(R)	8.314 J K^{-1} mol^{-1}
2.	Avogadro's No.	N	6.022×10^{23} mol^{-1}
3.	Atomic mass unit	amu	1.66×10^{-23} kg
4.	Mass of electron in rest	m_e	9.109×10^{-31} g
5.	Mass of proton	m_p	1.6726×10^{-27} kg
6.	Mass of neutron	m_n	1.675×10^{-27} kg
7.	Speed of light	C	2.998×10^{8} m sec^{-1}
8.	Planck's constant	(h)	6.626×10^{-34} J sec
9.	Boltzmann constant	(K)	1.380×10^{-23} JK^{-1}
10.	Charge on electron	e	1.602×10^{19} C
11.	Molar volume at STP	$\overline{v}$	0.0224 m^3
12.	Acceleration due to gravity	(g)	9.8066 m sec^{-2}
13.	Atmospheric pressure	(P)	1.0132×105 N m^{-2}
14.	I. Bohr's radius (for H)	r_1 H	0.529×10^{-10} m
15.	Rydberg constant	R_H	1.097×10^{7} m^{-1} or 2.18×10^{-18} J
16.	Faraday	(F)	9.648×10^{7} C/kg
17.	Specific charge of electron	(e/m)	1.759×10^{11} Ckg^{-1}
18.	Triple point H_2O		273.16 K

TABLE 2 : Vapor Pressure of Water

Temperature (K)	Pressure (mmHg)	Temperature (K)	Pressure (mmHg)
273	4.6	294	18.5
274	4.9	295	19.8
275	5.3	296	20.9
276	6.6	297	22.2
277	6.1	298	23.6
278	6.5	299	25.1
279	7.0	300	26.5
280	7.5	301	28.1
281	8.0	302	29.8
282	8.6	303	31.5
283	9.2	304	33.4
284	9.8	305	35.4
285	10.5	306	37.4
286	11.2	307	39.6
287	11.9	308	41.9
288	12.7	309	44.2
289	13.5	310	46.7
290	14.4	311	49.4
291	15.4	312	52.1
292	16.3	313	55.0
293	17.4	373	760.0

TABLE 3

The Molal Freezing Point Depression Constants of Some Substances Commonly Used as the Solvent for Freezing Point Depression Measurements

Substance	K_f*(kg . K · mol*$^{-1}$*)*	*Normal Freezing point (K)*
Acetic acid, CH_3COOH	3.90	289.6
Benzene, C_6H_6	5.1	278.5
Camphor, $C_{10}H_{16}O$	40.	447.0
Cyclohexane, C_6H_{12}	19.2	279.5
Napthalene, $C_{10}H_8$	6.8	353.2
Water, H_2O	1.86	273.1

TABLE 4 : The Modal Boiling Point Elevation Constants of Some Common Solvents

Substance	K_f*(kg <215> mol*$^{-1}$*)*	*Normal Freezing point (K)*
Acetone, C_3H_6O	1.71	329.2
Benzene, C_6H_6	2.53	353.2
Camphor tetrachloride, CCl_4	5.03	349.5
Chloroform, $CHCl_3$	3.63	334.2
Ethanol, C_2H_5OH	1.22	351.5
Water, H_2O	0.51	373.1

TABLE 5 : Standard Chemical Thermodynamic Date at 298 K and 101.325 kPa (1 atm)*

Substance	ΔH_f°*(kJ · mol*$^{-1}$*)*	ΔH_f°*(kJ · mol*$^{-1}$*)*	S°_{298}*(J · mol*$^{-1}$ *K*$^{-1}$*)*
Aluminum			
Al(s)	0.00	0.00	28.33
Al(g)	326.4	285.7	164.4
Al^{3+}(aq)	–531.	–485.	–321.7
Al_2O_3(s)	–1675.7	–1582.3	50.92
AlF_3(s)	–1504.	–1425.	66.44
$AlCl_3$(s)	–704.2	–628.8	110.67
$AlBr_3$(s)	–526.3	–505.0	184.
$Al_2(SO_4)_3$(s)	–3440.8	–3099.94	239.3
Antimony			
Sb(s)	0.00	0.00	45.69
Sb_4O_6(s, cubic)	–1440.6	–1268.2	220.9
$SBCl_3$(g)	–314.	–301.	337.7
$SBCl_3$(s)	–382.2	–323.7	184.1
Arsenic			
As(s, gray)	0.00	0.00	35.1
As_4(g)	143.9	92.4	314.

(Contd.)

Table (*Contd.*)

Substance	$\Delta H_f^\circ (kJ \cdot mol^{-1})$	$\Delta H_f^\circ (kJ \cdot mol^{-1})$	$S_{298}^\circ (J \cdot mol^{-1}\ K^{-1})$
$AsCl_3$(g)	–261.5	–248.9	327.1
$AsCl_3$(l)	–305.0	–259.4	216.3
Barium			
Ba(s)	0.00	0.00	62.8
Ba^{2+}(aq)	–537.64	–560.77	9.6
BaO(s)	–553.3	–525.1	70.42
BaF_2(s)	–1207.1	–1156.8	96.36
$BaCl_2$(s)	–858.6	–810.4	123.68
$BaCl_2$ <215> $2H_2O$(s)	–1460.13	–1296.3	202.9
Beryllium			
Be(s)	0.00	0.00	9.50
Be(g)	324.3	286.6	136.16
BeO(s)	–609.6	–580.3	14.14
Boron			
B(s)	0.00	0.00	5.86
B(g)	562.7	518.8	153.34
B_2O_3(s)	–1272.8	–11983.7	53.97
B_2H_6(g)	35.6	86.6	232.0
B_5H_9(g)	73.2	174.9	275.81
H_3BO_3(s)	–1094.3	–969.0	88.83
BF_3(g)	–1137.0	–1120.35	254.01
Bromide			
Br_2((l))	0.00	0.00	152.23
Br_2(g)	30.907	3.144	245.35
Br(g)	111.88	82.43	174.91
HBr(g)	–36.40	–53.43	198.59
Br^-(aq)	–121.55	–103.96	82.4
Cadmium			
Cd(s)	0.00	0.00	51.67
Cd(g)	112.01	77.44	167.64
Cd^{2+}(aq)	–75.90	–77.61	–73.2
CdO(s)	–258.2	–228.4	54.8
$Cd(OH)_2$(s)	560.7	–473.6	96.
$CdCl_2$	–391.5	–343.93	96.
CdS(s)	–161.9	–156.5	64.9
$CdSO_4$(s)	–933.28	–822.72	123.04

(*Contd.*)

Table (*Contd.*)

Substance	$\Delta H_f^\circ(kJ \cdot mol^{-1})$	$\Delta H_f^\circ(kJ \cdot mol^{-1})$	$S^\circ_{298}(J \cdot mol^{-1}\ K^{-1})$
Calcium			
Ca(s)	0.00	0.00	41.42
Ca(g)	178.2	144.3	154.77
Ca^{2+}(aq)	–542.83	–553.58	–53.1
CaO(s)	–635.09	–604.05	39.75
$Ca(OH)_2$(s)	–986.09	–898.56	83.39
CaF_2(s)	–1219.6	–1167.3	68.87
$CaBr_2$(s)	–682.8	–663.6	130.
$CaCO_3$(calcite)	–1206.92	–1128.84	92.9
$CaSO_4$(s,anh)	–1434.11	–1321.86	106.7
$CaSO_4 \cdot 2H_2O$(s)	–2022.63	–1797.44	194.1
$Ca(NO_3)_2$(s)	–938.39	–743.20	193.3
Carbon			
C(graphite)	0.00	0.00	5.740
C(diamond)	1.895	2.900	2.377
C(g)	716.68	671.29	157.99
CO(g)	–110.525	–137.15	197.56
CO_2(g)	–393.51	–394.36	213.63
CH_4(g)	–74.81	–50.75	186.15
C_2H_2(g)	226.73	209.20	200.83
C_2H_4(g)	52.26	68.12	219.45
C_2H_6(g)	–84.68	–32.89	229.49
C_3H_8(g)	–103.85	–23.49	269.91
n-C_4H_{10}(g)	–126.1	–17.51	310.1
C_6H_6(g)	82.927	129.66	269.2
C_6H_6(l)	49.028	124.5	172.8
CH_3OH(g)	–200.66	–162.01	239.70
CH_3OH(l)	–238.66	–166.35	126.8
CH_3CH_2OH(g)	–235.10	–168.57	282.59
CH_3CH_2OH(l)	–277.69	–174.89	160.7
HCHO(g)	108.57	–102.55	218.66
CH_3CHO(g)	166.19	–128.91	250.5
HCOOH(l)	–424.72	–361.42	128.95
HCOOH(aq)	–425.43	–372.3	163.
CH_3COOH(l)	–484.5	–390.0	159.8
CH_3COOH(aq)	–485.76	–396.46	178.7
CH_3COO^-(aq)	–486.01	–369.31	86.6
$(CH_3)_2CO$(l)	–248.	–155.4	200.
CH_3OCH_3(g)	–184.05	–112.67	266.27
$H_2C_2O_4$(s)	–827.2	...	...

(*Contd.*)

Table (*Contd.*)

Substance	$\Delta H_f^\circ (kJ \cdot mol^{-1})$	$\Delta H_f^\circ (kJ \cdot mol^{-1})$	$S^\circ_{298} (J \cdot mol^{-1}\ K^{-1})$
$C_2O_4^-(aq)$	–825.1	–673.9	45.6
$CHCl_3(g)$	–103.14	–70.37	295.60
$CHCl_3(\ell)$	–0134.47	–73.73	201.7
$CCl_4(g)$	–102.9	–60.62	309.74
$CCl_4(\ell)$	–135.44	–65.28	216.4
HCN(g)	135.1	124.7	201.67
$COCl_2(g)$	–218.8	–204.6	283.53
$CO(NH_2)_2(s)$	–333.51	–197.44	104.60
$NH_2CH_2COOH(s)$	–528.1	–368.57	103.51
Chlorine			
$Cl_2(g)$	0.00	0.00	22.96
Cl(g)	121.68	105.71	165.09
$Cl^-(aq)$	–167.16	–131.23	56.5
$ClO_2(g)$	102.5	120.5	256.73
HCl(g)	–92.307	–95.299	186.80
Chromium			
Cr(s)	0.00	0.00	23.77
$Cr_2O_3(s)$	–1139.7	–1058.1	81.2
$CrCl_3(s)$	–556.5	–486.1	123.0
Cobalt			
Co(s)	0.00	0.00	30.04
$CoCl_2(S)$	–212.5	–269.8	109.16
$CoSO_4(s)$	–888.3	–782.4	118.0
Copper			
Cu(s)	0.00	0.00	33.15
Cu(g)	338.32	298.61	166.27
$Cu^{2+}(aq)$	64.77	65.49	–99.6
$CuCl_2(s)$	–220.1	–175.7	108.07
CuO(s)	–157.3	–129.7	42.63
CuS(s)	–53.1	–53.6	66.5
$CuSO_4$(s,anh)	–771.36	–661.9	109.
$CuSO_4 \cdot 5H_2O(s)$	–2279.65	–1180.0	300.4
Fluorine			
$F_2(g)$	0.00	0.00	202.67
F(g)	78.99	61.94	158.64
$F^-(aq)$	–322.6	–278.8	–13.8
HF(g)	–271.1	–273.2	173.67

(*Contd.*)

Table (*Contd.*)

Substance	$\Delta H_f^\circ (kJ \cdot mol^{-1})$	$\Delta H_f^\circ (kJ \cdot mol^{-1})$	$S_{298}^\circ (J \cdot mol^{-1} K^{-1})$
Hydrogen			
H_2(g)	0.00	0.00	130.57
H(G)	217.965	203.263	114.60
H^+(aq)	0.00	0.00	0.00
H_3O^+(aq)	–285.83	–237.18	69.91
Iodine			
I_2(s)	0.00	0.00	116.14
I_2(g)	62.438	19.36	260.58
I(g)	106.84	70.283	180.68
I^-(aq)	–55.19	–51.57	111.3
I_3^-(aq)	–51.5	–51.4	239.3
HI(g)	26.48	1.72	206.48
ICl(g)	17.78	–5.44	247.44
IBr(g)	40.84	3.71	258.66
Iron			
Fe(s)	0.00	0.00	27.78
Fe^{2+}(aq)	–89.1	–78.90	–137.7
Fe^{3+}(aq)	–48.5	–4.7	–315.9
Fe_2O_3(s)	–824.2	–742.2	87.40
Fe_3O_4(s)	–1118.4	–1015.5	146.4
$Fe(OH)_3$(s)	–823.0	–696.6	106.7
$FeCl_3$(s)	–399.49	–334.05	142.3
Lead			
Pb(s)	0.00	0.00	64.81
Pb^{2+}(aq)	–1.7	–24.43	10.5
PbO(s, red)	–218.99	–188.95	66.5
PbO(s, yellow)	–217.32	–187.91	68.70
PbO_2(s)	–277.4	–217.4	68.6
PbF_2(s)	–664.0	–617.1	110.5
$PbCl_2$(s)	–359.41	–314.13	136.0
$PbBr_2$(s)	–278.7	–261.92	161.5
PbI_2(s)	–175.48	–173.64	174.85
PbS_2(s)	–100.4	–98.7	91.2
$PbSO_4$(s)	–919.94	–813.21	148.57
Mercury			
Hg(l)	0.00	0.00	76.02
Hg(g)	61.317	31.853	174.85
HgO(s, red)	–90.83	–58.555	70.29

(*Contd.*)

Table (*Contd.*)

Substance	$\Delta H_f^\circ (kJ \quad mol^{-1})$	$\Delta H_f^\circ (kJ \cdot mol^{-1})$	$S^\circ_{298} (J \cdot mol^{-1} \; K^{-1})$
HgS(s, black)	–53.6	–47.7	88.3
HgCl(s)	–224.3	–178.6	146.0
Hg_2Cl_2(s)	–265.22	–210.78	192.5
Nitrogen			
N_2(g)	0.00	0.00	191.50
N(g)	472.704	455.579	153.19
NO(g)	90.25	86.55	210.652
NO_2(g)	33.18	51.29	239.95
N_2O(g)	82.05	104.18	219.74
N_2O_4(g)	9.16	97.82	304.18
N_2O_5(s)	–43.1	113.8	178.2
NH_3(g)	–46.11	–16.48	192.34
NH_3(aq)	–80.29	–26.50	111.3
NH_4^+(aq)	–132.51	–79.31	113.4
NOCl(g)	51.71	66.06	261.58
N_2H_4(*l*)	50.63	149.24	121.21
NH_4Cl(s)	–314.43	–202.97	94.6
NH_4Br(s)	–270.83	–175.3	113.
$(NH_4)_2SO_4$(s)	–1180.85	–901.90	220.1
NH_4NO_3(s)	–365.56	–184.02	151.08
Oxygen			
O_2(g)	0.00	0.00	205.03
O(g)	249.170	231.747	160.946
O_3(g)	142.7	163.2	238.82
OH^-(aq)	–229.99	–157.24	–10.75
H_2O(g)	–241.82	–228.59	188.72
H_2O(*l*)	–285.83	–237.18	69.91
H_2O_2(*l*)	–187.78	–120.42	109.6
Phosphorus			
P(s, white)	0.00	0.00	41.09
P_4(g)	58.91	24.47	279.87
PH_3(g)	5.4	13.4	210.12
PCl_3(g)	–287.0	–267.8	311.67
PCl_5(g)	–374.9	–305.0	364.47
P_3O_{10}(s)	–2984.0	–2697.9	228.86
Potassium			
K(s)	0.00	0.00	64.18
K(g)	89.24	60.62	160.227

(*Contd.*)

Table (*Contd.*)

Substance	$\Delta H_f^\circ (kJ \cdot mol^{-1})$	$\Delta H_f^\circ (kJ \cdot mol^{-1})$	$S^\circ_{298} (J \cdot mol^{-1}\ K^{-1})$
KF(s)	–567.27	–537.77	66.57
KCl(s)	–436.747	–409.16	82.59
KBr(s)	393.80	–380.66	95.90
KI(s)	–327.90	324.89	106.32
Silver			
Ag(s)	0.00	0.00	42.55
Ag^+(aq)	105.569	77.11	72.68
Ag_2O(s)	–31.05	–11.22	121.3
AgCl(s)	–127.068	–109.805	96.2
AgBr(s)	–100.37	–96.90	107.1
AgI(s)	–61.84	–66.19	115.5
Ag_2S(rhombic)	–32.59	–40.67	144.01
Ag_2SO_4(s)	–715.88	–618.48	200.4
Sodium			
Na(s)	0.00	0.00	51.21
Na(g)	107.32	76.79	153.60
Na_2O(s)	–414.22	–375.48	75.06
NaCl(s)	–411.15	–384.15	72.13
$NaNO_3$(s)	–467.85	–367.07	116.52
Sulfur			
S(rhombic)	0.00	0.00	31.80
S(monoclinic)	0.33	...	...
S(g)	278.805	238.282	167.712
SO_2(g)	–296.830	–300.194	248.11
SO_3(g)	–395.72	–371.08	256.65
SO_4^{2-}(aq)	–909.27	–744.52	20.1
H_2S(g)	–20.63	–33.56	205.68
H_2SO_4(*l*)	–813.99	690.10	156.90
SF_6(g)	–1209.	–1105.4	291.71
Zinc			
Zn(s)	0.00	0.00	41.63
Zn^{2+}(aq)	–153.89	–147.06	–112.1
ZnO(s)	–348.28	–218.32	43.64
ZnS(sphalerite)	–205.98	–201.29	57.7
$ZnCl_2$(s)	–415.05	–369.43	111.46
$ZnBr_2$(s)	–328.65	–312.13	138.5
ZnI_2(s)	–208.03	–208.95	161.1
$ZnSO_4$(s,anh)	–982.8	–871.6	110.5
$ZnSO_4 \cdot 7H_2O$(s)	–3077.75	–2563.08	388.7

Table 6 : Standard Electrode Potentials at 298K

Half-Reaction	*ε° (V)*
$F_2(g) + 2e^- \rightarrow 2F^-$	+2.87
$Co^{3+} + e^- \rightarrow Co^{2+}$	+1.82
$H_2O_2 + 2H^+ + 2e^- \rightarrow 2H_2O$	+1.776
$PbO_2(s) + SO_4^{2-} + 4H^+ + 2e^- \rightarrow PbSO_4(s) + H_2O$	+1.685
$MnO_4^- + 4H^+ + 3e^- \rightarrow MnO_2(s) + 2H_2O$	+1.63
$BrO_3^- + 6H^+ + 5e^- \rightarrow \frac{1}{2} Br_2(l) + 3H_2O$	+1.52
$MnO_4^- + 8H^+ + 5e^- \rightarrow Mn^{2+} + 4H_2O$	+1.491
$ClO_3^- + 6H^+ + 5e^- \rightarrow \frac{1}{2} Cl^{2+} + 3H_2O$	+1.47
$PbO_2(s) + 4H^+ + 2e^- \rightarrow Pb^{2+} + 2H_2O$	+1.46
$Ce^{4+} + e^- \rightarrow Ce^{3+}$	+1.443
$Au^{3+} + 3e^- \rightarrow Au(s)$	+1.42
$Cl_2(g) + 2e^- \rightarrow 2Cl^-$	+1.358
$Cr_2O_7^{2-} + 14H^+ + 6e^- \rightarrow 2Cr^{3+} + 7H_2O$	+1.33
$O_2(g) + 4H^+ + 4e^- \rightarrow 2H_2O$	+1.229
$MnO_2(s) + 4H^+ + 2e^- \rightarrow Mn^{2+} + 2H_2O$	+1.208
$2\ IO_3^- + 12H^+ + 10e^- \rightarrow I_2 + 6H_2O$	+1.195
$ClO_4^- + 2H^+ + 2e^- \rightarrow ClO_3^- + H_2O$	+1.19
$Br_2(aq) + 2e^- \rightarrow 2Br^-$	+1.087
$Br_2(l) + 2e^- \rightarrow 2Br^-$	+1.0652
$NO_3^- + 4H^+ + 3e^-\ NO(g) + 2H_2O$	+0.96
$NO_3^- + 3H^+ + 2e^- \rightarrow HONO + H_2O$	+0.94
$2Hg^{2+} + 2e^- \rightarrow Hg_2^{2+}$	+0.920
$Hg^{2+} + 2e^- \rightarrow Hg(l)$	+0.855
$Ag^+ + e^- \rightarrow Ag(s)$	+0.799
$Hg_2^{2+} + 2e^- \rightarrow 2Hg(l)$	+0.789
$Fe^{3+} + e^- \rightarrow Fe^{2+}$	+0.770
$O_2(g) + 2H^+ + 2e^- \rightarrow H_2O_2$	+0.682
$Ag_2SO_4(s) + 2e^- \rightarrow 2Ag(s) + SO_4^{2-}$	+0.653
$MnO_4^- + 2H_2O + 3e^- \rightarrow MnO_2(s) + 4\ OH^-$	+0.588
$MnO_4^- + e^- \rightarrow MnO_4^{2-}$	+0.564
$I_2(s) + 2e^- \rightarrow 2\ I^-$	+0.535
$Cu^+ + e^- \rightarrow Cu(s)$	+0.521
$Ag_2CrO_4(s) + 2e^- \rightarrow 2Ag(s) + CrO_4^{2-}$	+0.446
$O_2(g) + 2H_2O + 4e^- \rightarrow 4\ OH^-$	+0.401
$Cu^{2+} + 2e^- \rightarrow Cu(s)$	+0.340
$Hg_2Cl_2(s) + 2e^- \rightarrow 2Hg(l) + 2Cl^-$	+0.268
$AgCl(s) + e^- \rightarrow Ag(s) + Cl^-$	+0.223
$Co(OH)_3(s) + e^- \rightarrow Co(OH)_2(s) + OH^-$	+0.17
$Cu^{2+} + e^- \rightarrow Cu^+$	+0.153
$Sn^{4+} + 2e^- \rightarrow Sn^{2+}$	+0.15
$S(s) + 2H^+ + 2e^- \rightarrow H_2S(aq)$	+0.14

(*Contd.*)

Table (*Contd.*)

Half-Reaction	$\varepsilon°$ (*V*)
$S_4O_6^{2-} + 2e^- \rightarrow 2S_2O_3^{2-}$	+0.09
$AgBr(s) + e^- \rightarrow Ag(s) + Br^-$	+0.0713
$NO_3^- + H_2O + 2e^- \rightarrow NO_2^- + 2\ OH^-$	+0.01
$2H^+(aq) + 2e^- \rightarrow H_2(g)$	0.0000
$Fe^{3+} + 3e^- \rightarrow Fe(s)$	–0.036
$CrO_4^{2-} + 4H_2O + 3e^- \rightarrow Cr(OH)_3(s) + 5\ OH^-$	–0.12
$Pb^{2+} + 2e^- \rightarrow Pb(s)$	–0.126
$Sn^{2+} + 2e^- \rightarrow Sn(s)$	–0.136
$O_2(g) + 2H_2O + 2e^- \rightarrow H_2O_2 + 2\ OH^-$	–0.146
$AgI(s) + e^- \rightarrow Ag(s) + I^-$	–0.152
$Cu(OH)_2(s) + 2e^- \rightarrow Cu(s) + 2\ OH^-$	–0.224
$Ni^{2+} + 2e^- \rightarrow Ni(s)$	–0.23
$Co^{2+} + 2e^- \rightarrow Co(s)$	–0.28
$PbSO_4(s) + 2e^- \rightarrow Pb(s) + SO_4^{2-}$	–0.356
$Cd^{2+} + 2e^- \rightarrow Cd(s)$	–0.403
$Cr^{3+} + e^- \rightarrow Cr^{2+}$	–0.41
$Fe^{2+} + 2e^- \rightarrow Fe(s)$	–0.44
$2CO_2(g) + 2H^+ + 2e^- \rightarrow H_2C_2O_4$	–0.49
$S(s) + 2e^- \rightarrow S^{2-}$	–0.508
$Ag_2S(s) + 2e^- \rightarrow 2AG(s) + S^{2-}$	–0.705
$Ni(OH)_2(s) + 2e^- \rightarrow Ni(s) + 2\ OH^-$	–0.72
$Cr^{3+} + 3e^- \rightarrow Cr(s)$	–0.74
$Zn^{2+} + 2e^- \rightarrow Zn(s)$	–0.763
$2H_2O + 2e^- \rightarrow H_2(g) + 2\ OH^-$	–0.8277
$Cr^{3+} + 2e^- \rightarrow Cr(s)$	–0.91
$SO_4^{2-} + H_2O + 2e^- \rightarrow SO_3^{2-} + 2\ OH–$	–0.93
$Mn^{2+} + 2e^- \rightarrow Mn(s)$	–1.18
$Cr(OH)_3(s) + 3e^- \rightarrow Cr(s) + 3\ OH^-$	–1.3
$Ti^{2+} + 2e^- \rightarrow Ti(s)$	–1.63
$Al^{3+} + 3e^- \rightarrow Al(s)$	–1.662
$Be^{2+} + 2e^- \rightarrow Be(s)$	–1.85
$Ti^{3+} + e^- \rightarrow Ti^{2+}$	–2.0
$Sc^{3+} + 3e^- \rightarrow Sc(s)$	–2.08
$Mg^{2+} + 2e^- \rightarrow Mg(s)$	–2.363
$Na^{2+} + e^- \rightarrow Na(s)$	–2.714
$Ca^{2+} + 2e^- \rightarrow Ca(s)$	–2.866
$Sr^{2+} + 2e^- \rightarrow Sr(s)$	–2.89
$Ba^{2+} + 2e^- \rightarrow Ba(s)$	–2.90
$Cs^+ + e^- \rightarrow Cs(s)$	–2.92
$K^+ + e^- \rightarrow K(s)$	–2.924
$Li^+ + e^- \rightarrow Li(s)$	–3.045

TABLE 7 : Formation Constants of Complex Ions (K_f)

$AgBr_2^-$	2×10^{7}	$Cu(CN)_2^-$	1×10^{24}	$PbBr_4^{2-}$	6×10^{13}
$AgCl_2^-$	1.8×10^{5}	$Cu(NH_3)_4^{2+}$	2×10^{12}	$PtBr_4^{2-}$	6×10^{17}
$Ag(CN)_2^-$	2×10^{20}	$FeSCN^{2+}$	9.2×10^{2}	$PtCl_4^{2-}$	1×10^{16}
AgI_2^-	5×10^{10}	$Fe(CN)_6^{3-}$	4×10^{52}	$Pt(CN)_4^{2-}$	1×10^{41}
$Ag(NH_3)_2^+$	1.7×10^{7}	$Fe(CN)_6^{4-}$	4×10^{45}	$c\text{-}Pt(NH_3)_2Cl_2$	3×10^{29}
$Al(OH)_4^{2-}$	1×10^{33}	$HgBr_4^{2-}$	1×10^{21}	$t\text{-}Pt(NH_3)_2Cl_2$	3×10^{28}
$Cd(CN)_4^{2-}$	2×10^{18}	$HgCl_4^{2-}$	1×10^{15}	$c\text{-}Pt(NH_3)_2(H_2O)_2^{2+}$	4×10^{23}
CdI_4^{2-}	4.0×10^{5}	$Hg(CN)_4^{2-}$	2×10^{41}	$c\text{-}Pt(NH_3)_2I_2$	2×10^{33}
$Cd(NH_3)_4^{2-}$	2.8×10^{7}	HgI_4^{2-}	1×10^{30}	$t\text{-}Pt(NH_3)_2I_2$	5×10^{32}
$Cd(OH_3)_4^{2-}$	1.2×10^{9}	$Hg(NH_3)_4^{2-}$	2×10^{19}	$c\text{-}Pt(NH_3)_2(OH)_2$	1×10^{39}
$Co(NH_3)_6^{2+}$	1×10^{5}	$Ni(CH)_4^{2-}$	1×10^{30}	$Pt(NH_3)_4^{2+}$	2×10^{35}
$Co(NH_3)_6^{3+}$	2×10^{23}	$Ni(NH_3)_6^{2+}$	9×10^{8}	$Zn(CH)_4^{2-}$	6×10^{16}
$Co(NH_3)_5Cl^{2+}$	2×10^{28}	PbI_4^{2-}	1.7×10^{4}	$Zn(NH_3)_4^{2+}$	3.6×10^{8}
$Co(NH_3)_5NO_4^{2-}$	1×10^{24}	$Pb(OH)_3^-$	8×10^{13}	$Zn(OH)_4^{2-}$	3×10^{14}

TABLE 8 : Solubility Product Constants (K_{sp})

$AgBr$	5×10^{-13}	$Co(OH)_2$	2×10^{-16}	$NiCO_3$	1.4×10^{-7}
$AgC_2H_3O_2$	1.9×10^{-3}	$Co_3(PO_4)_2$	1×10^{-35}	$Ni_3(PO_4)_2$	1×10^{-32}
Ag_2CO_3	8×10^{-12}	$CuCl$	1.7×10^{-7}	NiS	1×10^{-21}
$AgCl$	1.8×10^{-10}	$CuBr$	6.3×10^{-9}	$PbBr_2$	6.6×10^{-6}
Ag_2CrO_4	1×10^{-12}	Cul	1.2×10^{-12}	$PbCO_3$	1×10^{-13}
AgI	1×10^{-16}	$Cu_3(PO_4)_2$	1×10^{-37}	$PbCl_2$	1.7×10^{-5}
Ag_3PO_4	1×10^{-16}	CuS	1×10^{-36}	$PbCrO_4$	2×10^{-14}
Ag_2S	1×10^{-49}	Cu_2S	1×10^{-48}	PbF_2	7.1×10^{-7}
$AgSCN$	1.0×10^{-12}	$FeCo_3$	3.1×10^{-11}	PbI_2	8.4×10^{-9}
AlF_3	1×10^{-18}	$Fe(OH)_2$	5×10^{-17}	$Pb(OH)_2$	1×10^{-20}
$Al(OH)_3$	2×10^{-31}	$Fe(OH)_3$	3×10^{-39}	PbS	1×10^{-28}
$AlPO_4$	1×10^{-20}	FeS	2×10^{-19}	$Pb(SCN)_2$	2.1×10^{-5}
$BaCO_3$	2.6×10^{-9}	GaF_3	2×10^{-16}	$PbSO_4$	1.8×10^{-8}
$BaCrO_4$	1.2×10^{-10}	$Ga(OH)_3$	1×10^{-35}	$Sc(OH)_3$	2×10^{-31}
BaF_2	1.8×10^{-7}	$GaPO_4$	1×10^{-21}	SnS	3×10^{-28}
$BaSO_4$	1.1×10^{-10}	Hg_2Br_2	6×10^{-23}	$SrCO_3$	5.6×10^{-10}
Bi_2S_3	1×10^{-99}	Hg_2Cl_2	1×10^{-18}	$SrCrO_4$	3.6×10^{-5}
$CaCO_3$	4.9×10^{-9}	Hg_2I_2	5×10^{-29}	SrF_2	4.3×10^{-9}
CaF_2	1.5×10^{-10}	HgS	1×10^{-52}	$SrSO_4$	3.4×10^{-7}
$Ca(OH)_2$	4.0×10^{-6}	Li_2CO_3	8.2×10^{-4}	$TlCl$	1.9×10^{-4}
$Ca_3(PO_4)_2$	1×10^{-33}	$MgCO_3$	6.8×10^{-6}	$TlBr$	3.7×10^{-6}
$CaSO_4$	7.1×10^{-5}	MgF_2	7×10^{-11}	TlI	5.6×10^{-8}
$CdCO_3$	6×10^{-12}	$Mg(OH)_2$	6×10^{-12}	$Tl(OH)_3$	2×10^{-44}
$Cd(OH)_2$	5×10^{-15}	$Mg_3(PO_4)_2$	1×10^{-24}	Tl_2s	1×10^{-20}
$Cd_3(PO_4)_2$	1×10^{-33}	$Mn(OH)_2$	2×10^{-13}	$ZnCO_3$	1.1×10^{-10}
CdS	1×10^{-29}	MnS	5×10^{-14}	$Zn(OH)_2$	4×10^{-17}

TABLE 9 : Equilibrium Constants of Weak Acids (K_a)

Acid	K_a	Acid	K_a	Acid	K_a
H_3AsO_4	5.7×10^{-3}	HNO_2	6.0×10^{-4}	$N_2H_5^+$	1.0×10^{-8}
$H_2AsO_4^-$	1.8×10^{-7}	H_3PO_4	7.1×10^{-3}	$Al(H_2O)_6^{3+}$	1.2×10^{-5}
$HAsO_4^{2-}$	2.5×10^{-12}	$H_2PO_4^-$	6.2×10^{-8}	$Ag(H_2O)_2^+$	1.2×10^{-12}
$HBrO$	2.6×10^{-9}	HPO_4^{2-}	4.5×10^{-13}	$Ca(H_2O)_6^{2+}$	2.2×10^{-13}
$HCHO_2$	1.9×10^{-4}	H_2S	1.0×10^{-7}	$Cd(H_2O)_4^{2+}$	4.0×10^{-10}
$HC_2H_3O_2$	1.8×10^{-5}	H_2SO_3	1.7×10^{-2}	$Fe(H_2O)_6^{3+}$	6.7×10^{-3}
HCN	5.8×10^{-10}	HSO_3^-	6.0×10^{-8}	$Fe(H_2O)_6^{2+}$	1.7×10^{-7}
H_2CO_3	4.4×10^{-7}	HSO_4^-	1.0×10^{-2}	$Mg(H_2O)_6^{2+}$	3.7×10^{-12}
HCO_3^-	4.7×10^{-11}	H_2Se	1.5×10^{-4}	$Mn(H_2O)_6^{2+}$	2.8×10^{-11}
$HClO_2$	1.0×10^{-2}	H_2SeO_3	2.7×10^{-3}	$Ni(H_2O)_6^{2+}$	2.2×10^{-10}
$HClO$	2.8×10^{-8}	$HseO_3^-$	5.0×10^{-8}	$Pb(H_2O)_6^{2+}$	6.7×10^{-7}
HF	6.9×10^{-4}	$CH_3NH_3^+$	2.4×10^{-11}	$Sc(H_2O)_6^{3+}$	1.1×10^{-4}
HIO	2.4×10^{-11}	NH_2^+	5.6×10^{-10}	$Zn(H_2O)_4^{2+}$	3.3×10^{-10}
HN_3	2.4×10^{-5}				

TABLE 10 : Equilibrium Constants of Weak Acids (K_a)

Species	K	Species	K	Species	K
AsO_4^{3-}	4.0×10^{-3}	N_3^-	4.2×10^{-10}	$HseO_3^-$	3.7×10^{-12}
$HAsO_4^{2-}$	5.6×10^{-8}	NH_3	1.8×10^{-5}	$AlOH^{2+}$	8.3×10^{-10}
$H2AsO_4^{2-}$	1.8×10^{-12}	N_2H_4	1.0×10^{-6}	$AgOH^+$	8.3×10^{-3}
BrO^-	3.8×10^{-6}	NO_2^-	1.7×10^{-11}	$CaOH^+$	4.5×10^{-2}
CH_3NH_2	4.2×10^{-4}	PO_4^{3-}	2.2×10^{-2}	$CdOH^+$	2.5×10^{-5}
CHO_2^-	5.3×10^{-11}	HPO_4^{2-}	1.6×10^{-7}	$FeOH^{2+}$	1.5×10^{-12}
$C_2H_3O_2^-$	5.6×10^{-10}	$H_2PO_4^-$	1.4×10^{-12}	$FeOH^+$	1.9×10^{-8}
CN^-	1.7×10^{-5}	HS^-	1.0×10^{-7}	$MgOH^+$	2.7×10^{-3}
CO_3^{2-}	2.1×10^{-4}	SO_3^{2-}	1.7×10^{-7}	$MnOH^+$	3.6×10^{-4}
HCO_3^-	2.3×10^{-8}	HSO_3^-	5.9×10^{-13}	$NiOH^+$	4.5×10^{-5}
ClO_2^-	2.3×10^{-12}	SO_4^{2-}	1.0×10^{-12}	$PbOH^+$	1.5×10^{-8}
ClO^-	3.6×10^{-7}	HSe^-	2.7×10^{-11}	$ScOH^2$	9.1×10^{-11}
F^-	1.4×10^{-11}	SeO_2^{2-}	2.0×10^{-7}	$ZnOH$	3.0×10^{-5}
IO^-	4.2×10^{-4}